ALABAMA

# CHEMISTRY

MATTER AND CHANGE

McGraw Hill Education

**COVER:** Science Photo Library/Getty Images

**MHEonline.com**



Send all inquiries to:
McGraw-Hill Education
STEM Science
8787 Orion Place
Columbus, OH 43240

ISBN: 978-0-07-667227-1
MHID: 0-07-667227-1

Printed in the United States of America.

2 3 4 5 6 7 8 9 10 DOW 20 19 18 17 16

# CONTENTS IN BRIEF


Chemistry Online ........ viii

1 Introduction to Chemistry ........ 2
2 Analyzing Data ........ 30
3 Matter—Properties and Changes ........ 68
4 The Structure of the Atom ........ 100
5 Electrons in Atoms ........ 134
6 The Periodic Table and Periodic Law ........ 172
7 Ionic Compounds and Metals ........ 204
8 Covalent Bonding ........ 238
9 Chemical Reactions ........ 280
10 The Mole ........ 318
11 Stoichiometry ........ 366
12 States of Matter ........ 400
13 Gases ........ 440
14 Mixtures and Solutions ........ 474
15 Energy and Chemical Change ........ 514
16 Reaction Rates ........ 558
17 Chemical Equilibrium ........ 592
18 Acids and Bases ........ 632
19 Redox Reactions ........ 678
20 Electrochemistry ........ 706
21 Hydrocarbons ........ 742
22 Substituted Hydrocarbons and Their Reactions ........ 784
23 The Chemistry of Life ........ 824
24 Nuclear Chemistry ........ 858

Student Resources ........ 900

Magnus Hjorleifsson/Getty Images

# A² DEVELOPMENT PROCESS

## Why A²?

**Accuracy Assurance** is central to McGraw-Hill's commitment to high-quality, learner-oriented, real-world, and error-free products. Also at the heart of our A² Development Process is a commitment to make the text **Accessible** and **Approachable** for both students and teachers. A collaboration among authors, content editors, academic advisors, and classroom teachers, the A² Development Process provides opportunities for continual improvement through customer feedback and thorough content review.

The A² Development Process begins with a review of the previous edition and a look forward to state and national standards. The authors for *Chemistry: Matter and Change* combine expertise in teacher training and education with a mastery of chemistry content knowledge. As manuscript is created and edited, consultants review the accuracy of the content while our Teacher Advisory Board members examine the program from the points of view of both teacher and student. Student labs and teacher demonstrations are reviewed for both accuracy of content and safety. As design elements are applied, chapter content is again reviewed, as are photos and diagrams.

Throughout the life of the program, Glencoe/McGraw-Hill continues to troubleshoot and incorporate improvements. Our goal is to deliver to you a program that has been created, refined, tested, and validated as a successful tool for your continued **Academic Achievement.**

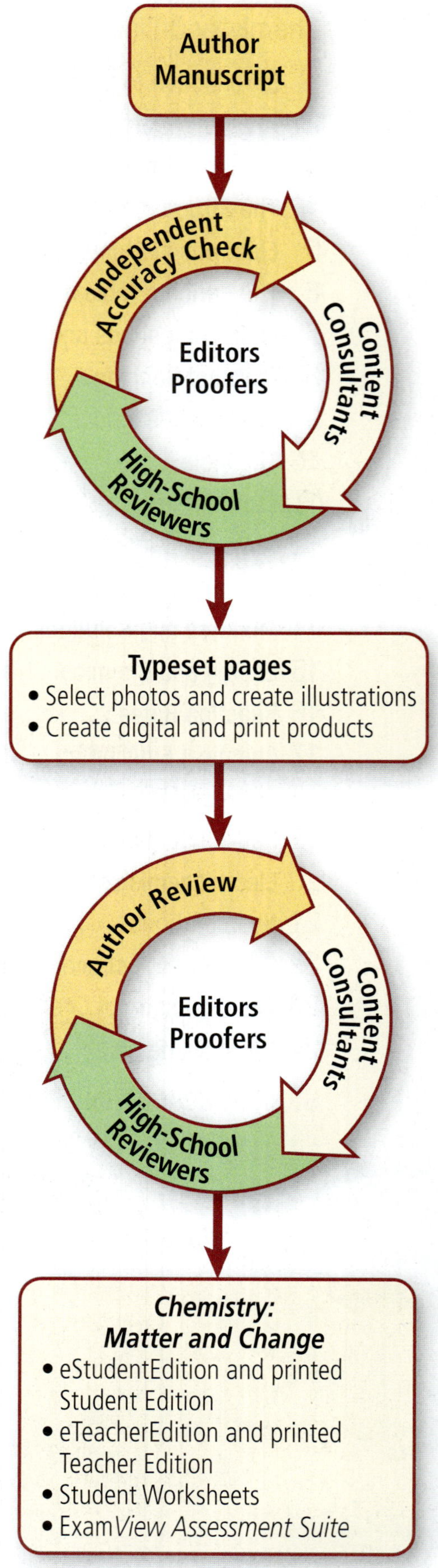

## Authors

**The authors of *Chemistry: Matter and Change* used their content knowledge and teaching expertise to craft manuscript that is accessible and accurate, geared toward student achievement.**

### Thandi Buthelezi

is Associate Professor of Chemistry at Western Kentucky University, Bowling Green, KY. She earned her BA in Chemistry from Williams College, Williamstown, MA, and PhD in Experimental Physical Chemistry from the University of Florida, Gainesville, FL. Dr. Buthelezi has taught Chemistry at the undergraduate and graduate (master's) level for seven years. She is the co-founder and co-director of the Girls in Science Outreach Program at WKU. She is a member of the American Chemical Society, the American Association for the Advancement of Science, and Sigma Xi. She has co-authored over two dozen research papers published in peer-reviewed journals.

### Laurel Dingrando

is currently serving as the Secondary Science Coordinator for the Garland Independent School District. Mrs. Dingrando has a BS in Microbiology with a minor in Chemistry from Texas Tech University and an MAT in Science from the University of Texas at Dallas. She taught Chemistry for 25 years in the Garland Independent School District. She is a member of the American Chemical Society, National Science Teachers Association, Science Teachers Association of Texas, Texas Science Educators Leadership Association, and T3 (Teachers Teaching with Technology).

### Nicholas Hainen

taught chemistry and physics in the Worthington City Schools, Worthington, Ohio, for 31 years. Mr. Hainen holds BS and MA degrees in Science Education from The Ohio State University, majoring in chemistry and physics. His honors and awards include: American Chemical Society Outstanding Educator in Chemical Sciences; The Ohio State University Honor Roll of Outstanding High School Teachers; Ashland Oil Company Golden Apple Award; and Who's Who Among America's Teachers. Mr. Hainen is a member of the American Chemical Society and the ACS Division of Chemical Education.

### Cheryl Wistrom

is an associate professor of chemistry at Saint Joseph's College in Rensselaer, Indiana, where she has been honored with both the Science Division and college faculty teaching awards. She has taught chemistry, biology, and science education courses at the college level since 1990 and is also a licensed pharmacist. She earned her BS degree in biochemistry at Northern Michigan University, a BS in pharmacy at Purdue University, and her MS and PhD in biological chemistry at the University of Michigan. Dr. Wistrom is a member of the Indiana Academy of Science, the National Science Teachers Association, and the American Society of Health-System Pharmacists.

### Dinah Zike

is an international curriculum consultant and inventor who has developed educational products and three-dimensional, interactive graphic organizers for over 30 years. As president and founder of Dinah-Might Adventures, L.P., Dinah is the author of more than 100 award-winning educational publications, including *The Big Book of Science.* Dinah has a BS and an MS in educational curriculum and instruction from Texas A&M University. Dinah Zike's *Foldables* are an exclusive feature of McGraw-Hill textbooks.

(t to b)courtesy of Thandi Buthelezi, courtesy of Laurel Dingrando, courtesy of Nicholas Hainen, Dinah Zike

## Teacher Advisory Board

**The Teacher Advisory Board gave the editorial staff and design team feedback on the content and design of both the Student Edition and Teacher Edition. We thank these teachers for their hard work and creative suggestions.**

**Ann Cooper**
Science Teacher
United Local Schools
Hanoverton, OH

**David L. French**
Chemistry Teacher
Milford High School
Milford, OH

**Richard Glink**
Chemistry/Physics Teacher
Indian Lake High School
Lewistown, OH

**Susan Godez**
Chemistry/Physics Teacher
Grandview Heights High School
Columbus, OH

**Judith Johnston**
Science Teacher, Department Chair
Wilmington High School
Wilmington, OH

**Christine Lewis**
Science Teacher
Martins Ferry High School
Martins Ferry, OH

**Jennifer L. Most**
Chemistry Teacher,
Science Department Chair
West Holmes High School
Millersburg, OH

**Sandra Petrie-Forgey**
National Board Certified
Science Teacher
Gallia Academy High School
Gallipolis, OH

**Jason J. Zaros**
Chemistry/Physics Teacher
Waterford High School
Waterford, OH

## Teacher Reviewers

**Each teacher reviewed selected chapters of *Chemistry: Matter and Change* and provided feedback and suggestions regarding the effectiveness of the instruction.**

**Bridget B. Adkins**
Ravenwood High School
Brentwood, TN

**Deborah Bennett**
Canoga Park High School
Canoga Park, CA

**James Breaux**
Stratford High School
Goose Creek, SC

**Bob Callender**
Warren Mott High School
Warren, MI

**Betsy Hamrick**
Crest High School
Shelby, NC

**Treva Jeffries**
Scott High School
Toledo, OH

**Dr. Aruna Kailasa**
Benjamin E. Mays High School
Atlanta, GA

**Phil Lampe**
Upper Arlington High School
Columbus, OH

**Les McSparrin**
Sharpsville Area High School
Sharpsville, PA

**Delores Miller**
Alden High School
Alden, NY

**Leon Olivier**
Union Grove High School
McDonough, GA.

**Dan Reid**
Central High School
Champaign, IL

**Jay Wilder**
Franklin County High School
Frankfort, KY

## Content Consultants

**Content consultants each reviewed selected chapters of *Chemistry: Matter and Change* for content accuracy and clarity.**

**Alton J. Banks, PhD**
Professor of Chemistry
North Carolina State University
Raleigh, NC

**Howard Drossman, PhD**
Professor of Chemistry and Environmental Science
Colorado College
Colorado Springs, CO

**Michael O. Hurst, Sr., PhD**
Associate Professor of Chemistry
Georgia Southern University
Statesboro, GA

**Kristen Kulinowski, PhD**
Faculty Fellow, Department of Chemistry
Rice University
Houston, TX

**Maria Pacheco, PhD**
Associate Professor of Chemistry
Buffalo State College
Buffalo, NY

## Safety Consultant

**The safety consultant reviewed labs and lab materials for safety and implementation.**

**Kenneth R. Roy, PhD**
Director of Environmental Health and Safety
Glastonbury Public Schools
Glastonbury, CT

## Contributing Writers

**Additional science writers added feature content, teacher materials, assessment, and laboratory investigations.**

**Peter Carpico**
Louisville, OH

**Jennifer Gonya**
Galena, OH

**Cindy Klevickis**
Elkton, VA

**Jack Minot**
Columbus, OH

**Richard G. Smith**
Ocean Isle Beach, NC

**Stephen Whitt**
Columbus, OH

**Jenipher Willoughby**
Forest, VA

**Margaret K. Zorn**
Yorktown, VA

# CHEMISTRY ONLINE

## Introducing

**iLab Station**

At MHEonline.com, you will find this one-stop online resource for laboratory investigations, procedures, and worksheets. Look for the iLab Station button throughout your eStudentEdition for links to these laboratory experiences:

- LaunchLABs
- MiniLABs
- ChemLABs

Online, editable lab worksheets from the iLab Station will let your teacher tailor the content to your specific needs.

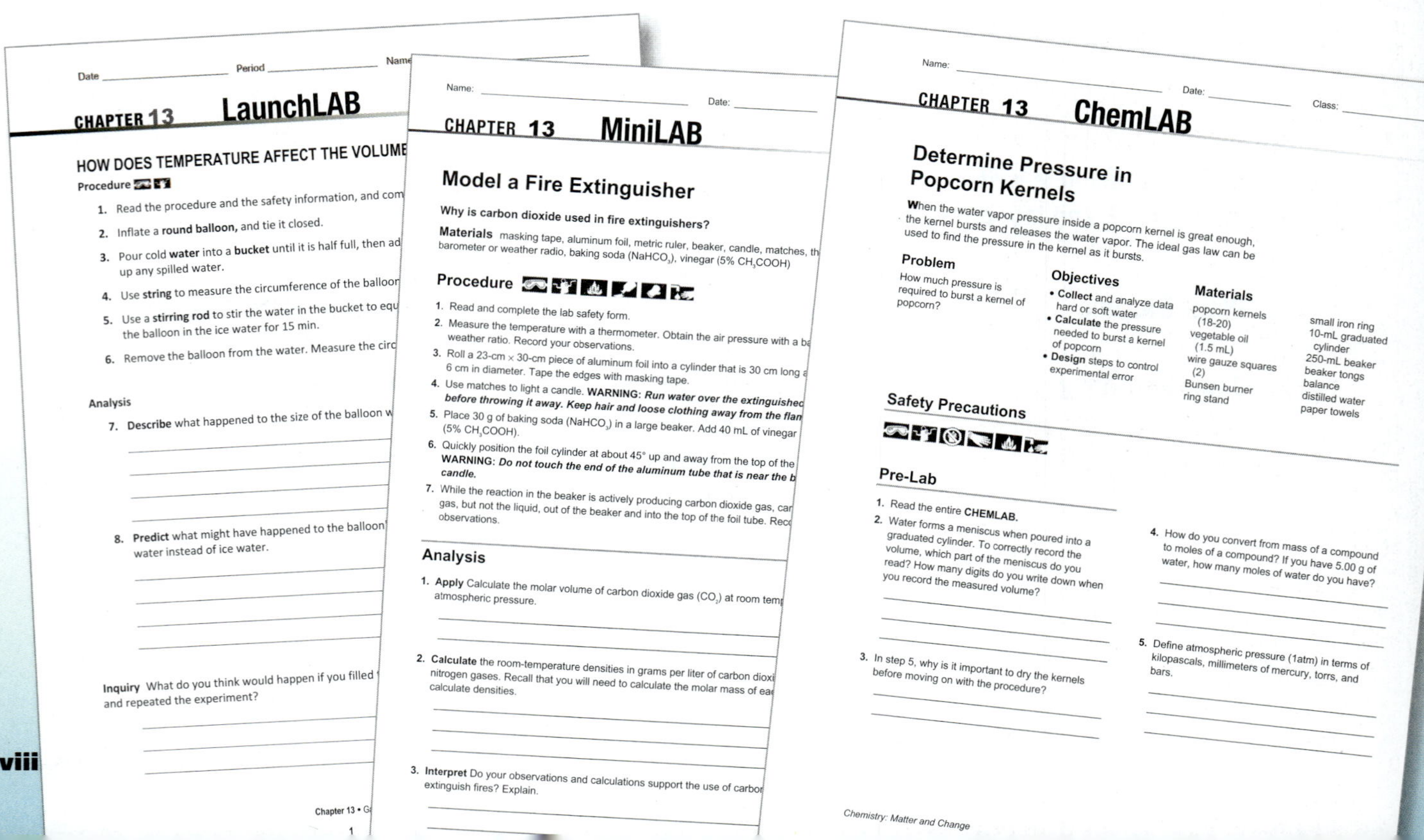

Go online!
MHEonline.com

# “It’s EASY to get my assignments online and QUICK to find everything I need.”

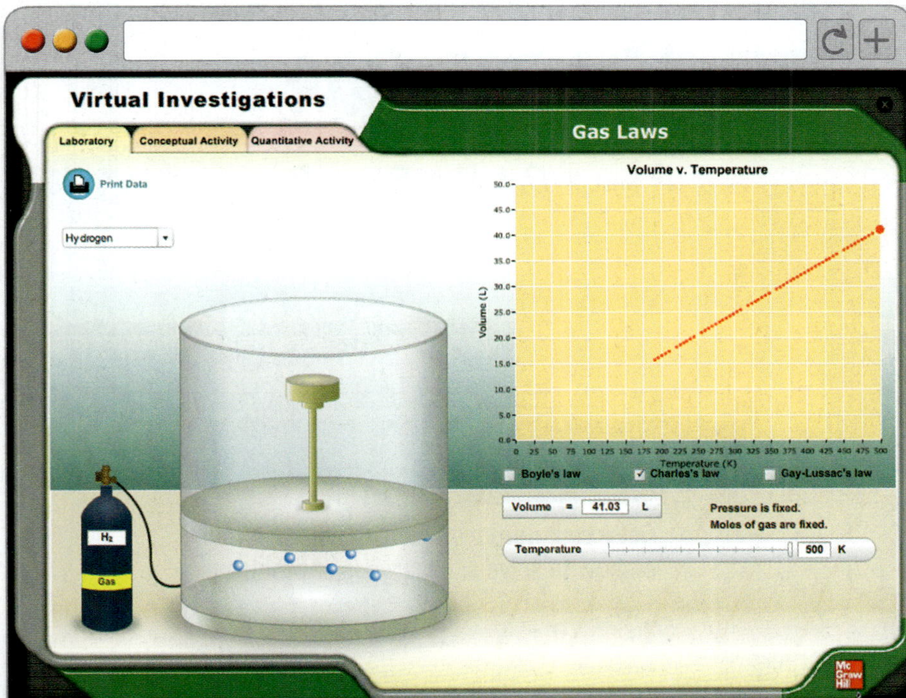

Figure 4 Tethers attached at the sides of a weather balloon hold it in place while it is being filled with helium or hydrogen gas. Weather balloons carry instruments that send data, such as air temperature, pressure, and humidity, to receivers on the ground. As the balloon rises, its volume responds to changes in temperature and pressure, expanding until the sides burst. A small parachute returns the instruments to Earth.

### The Combined Gas Law

In a number of applications involving gases, such as the weather balloon in **Figure 4,** pressure, temperature, and volume might all change. Boyle's, Charles's, and Gay-Lussac's laws can be combined into a single law. This **combined gas law** states the relationships between pressure, temperature, and volume of a fixed amount of gas. All three variables have the same relationship to each other as they have in the other gas laws: pressure is inversely proportional to volume and directly proportional to temperature, and volume is directly proportional to temperature. The combined gas law can be expressed mathematically as follows.

Explore the **gas laws.**
Virtual Investigations

**The Combined Gas Law**

$$\frac{P_1V_1}{T_1} = \frac{P_2V_2}{T_2}$$

*P* represents pressure. *V* represents volume. *T* represents temperature.

For a given amount of gas, the product of pressure and volume, divided by the Kelvin temperature, is a constant.

**Using the combined gas law** The combined gas law enables you to solve problems involving changes in more than one variable. It also provides a way for you to remember the other three laws without memorizing each equation. If you can write out the combined gas law equation, equations for the other laws can be derived from it by remembering which variable is held constant in each case.

For example, if temperature remains constant as pressure and volume vary, then $T_1 = T_2$. After simplifying the combined gas law under these conditions, you are left with $P_1V_1 = P_2V_2$, which you should recognize as the equation for Boyle's law.

Get help with the **combined gas law.**
Personal Tutor

READING CHECK **Derive** Charles's and Gay-Lussac's laws from the combined gas law.

Section 1 • The Gas Laws 449

## Virtual Investigations

Enter an **online laboratory** where you can perform investigations that would be difficult or impossible to complete in the classroom. Manipulate atoms, build molecules, experiment with radioactive elements, and more!

# ONLINE PROBLEMS & HELP

**Example Problems** throughout each chapter give you step-by-step instructions and hints on how to solve each type of problem. Each Example Problem is followed by a number of **Practice Problems** on the same topic, allowing you to practice the skill.

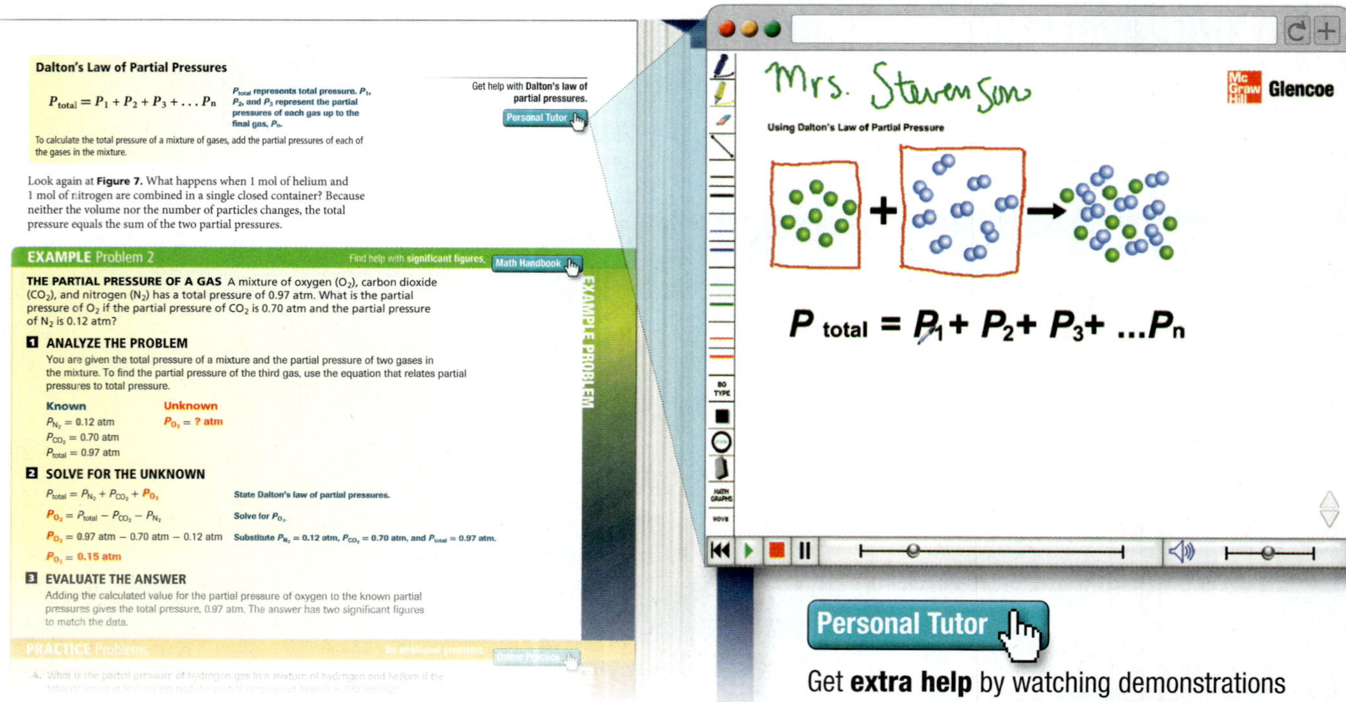


**Personal Tutor**

Get **extra help** by watching demonstrations on how to solve similar types of problems.

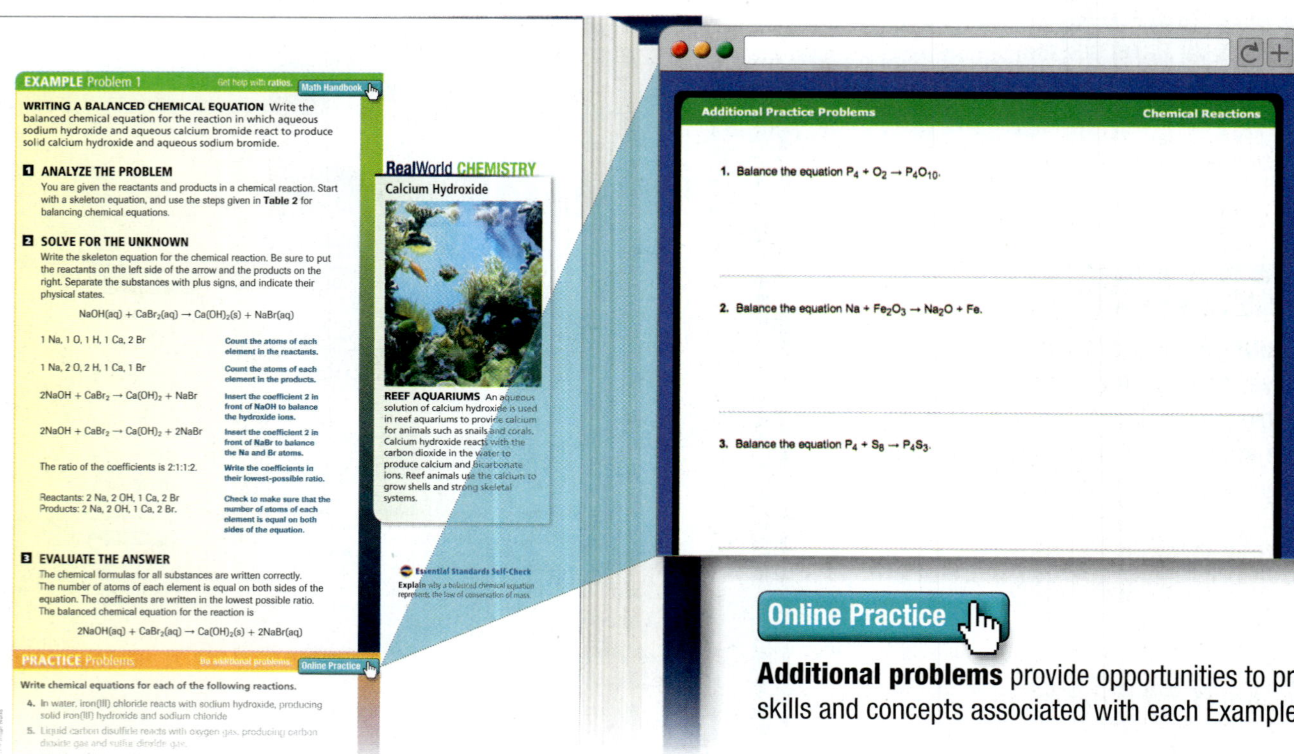


**Online Practice**

**Additional problems** provide opportunities to practice the skills and concepts associated with each Example Problem.

Go online!
MHEonline.com

# “The e-book includes VIDEOS, ANIMATIONS, and tools to help me LEARN.”

## Concepts In Motion

See chemistry content come to life in **animated figures** and moving diagrams.

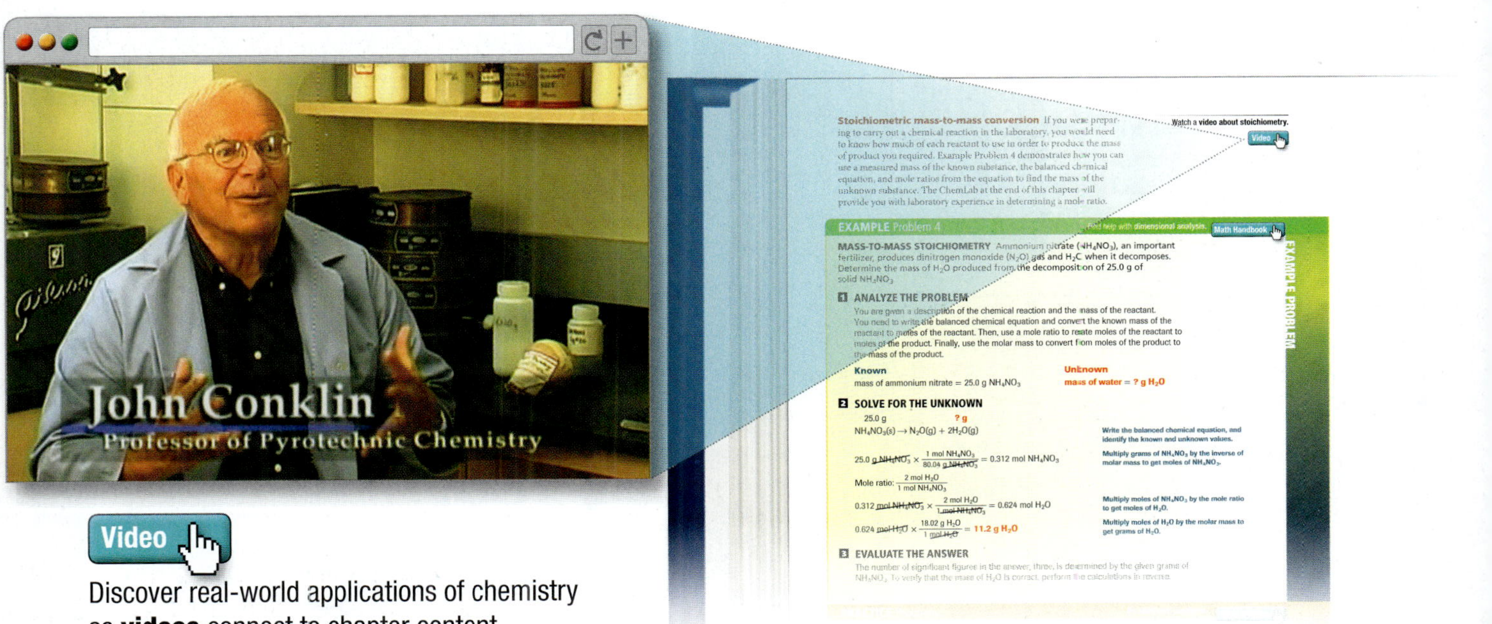


## Video

Discover real-world applications of chemistry as **videos** connect to chapter content.

# ONLINE STUDY TOOLS

**Quizzes, assessments, and study tools provide opportunities for self-assessment, review, and additional practice.**

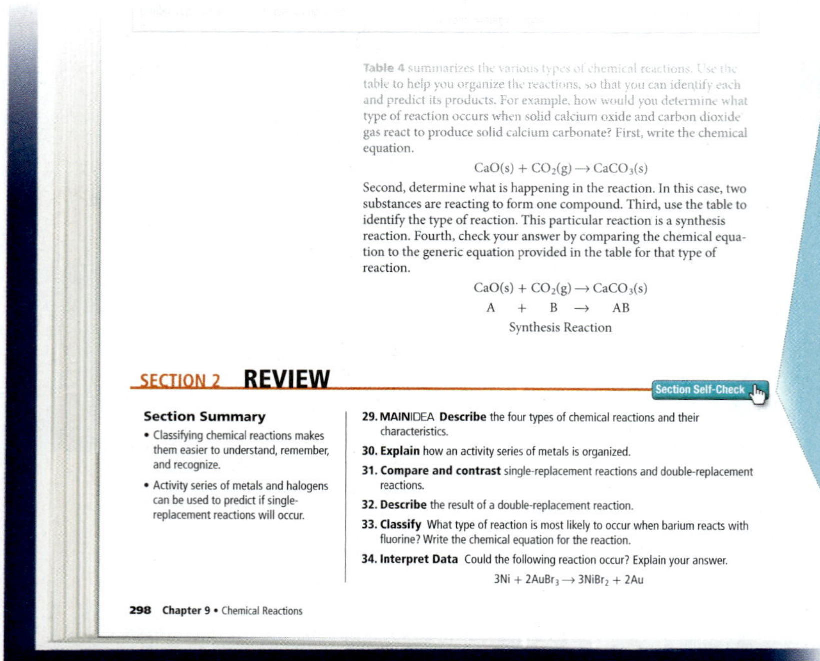

Table 4 summarizes the various types of chemical reactions. Use the table to help you organize the reactions, so that you can identify each and predict its products. For example, how would you determine what type of reaction occurs when solid calcium oxide and carbon dioxide gas react to produce solid calcium carbonate? First, write the chemical equation.

$$CaO(s) + CO_2(g) \rightarrow CaCO_3(s)$$

Second, determine what is happening in the reaction. In this case, two substances are reacting to form one compound. Third, use the table to identify the type of reaction. This particular reaction is a synthesis reaction. Fourth, check your answer by comparing the chemical equation to the generic equation provided in the table for that type of reaction.

$$CaO(s) + CO_2(g) \rightarrow CaCO_3(s)$$
$$A + B \rightarrow AB$$

Synthesis Reaction

SECTION 2 REVIEW

Section Self-Check

Section Summary

- Classifying chemical reactions makes them easier to understand, remember, and recognize.
- Activity series of metals and halogens can be used to predict if single-replacement reactions will occur.

29. MAINIDEA Describe the four types of chemical reactions and their characteristics.
30. Explain how an activity series of metals is organized.
31. Compare and contrast single-replacement reactions and double-replacement reactions.
32. Describe the result of a double-replacement reaction.
33. Classify What type of reaction is most likely to occur when barium reacts with fluorine? Write the chemical equation for the reaction.
34. Interpret Data Could the following reaction occur? Explain your answer.

$$3Ni + 2AuBr_3 \rightarrow 3NiBr_2 + 2Au$$

298 Chapter 9 • Chemical Reactions

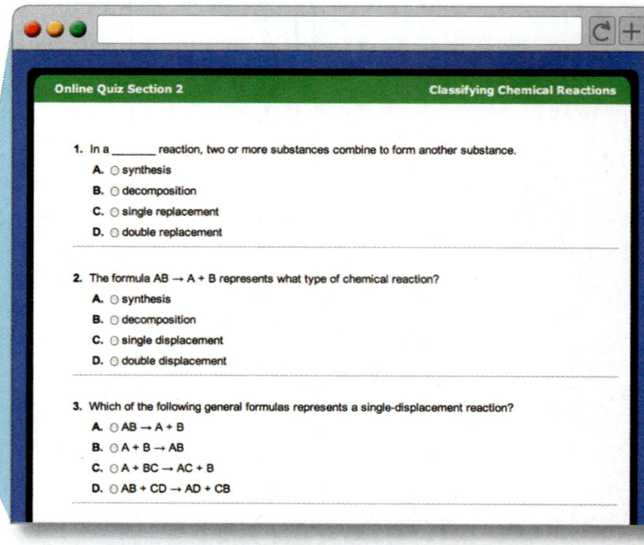

Online Quiz Section 2 — Classifying Chemical Reactions

1. In a ______ reaction, two or more substances combine to form another substance.
   A. synthesis
   B. decomposition
   C. single replacement
   D. double replacement
2. The formula AB → A + B represents what type of chemical reaction?
   A. synthesis
   B. decomposition
   C. single displacement
   D. double displacement
3. Which of the following general formulas represents a single-displacement reaction?
   A. AB → A + B
   B. A + B → AB
   C. A + BC → AC + B
   D. AB + CD → AD + CB

**Section Self-Check**

**Review questions and problems** for each section help you spot concepts that require additional study.

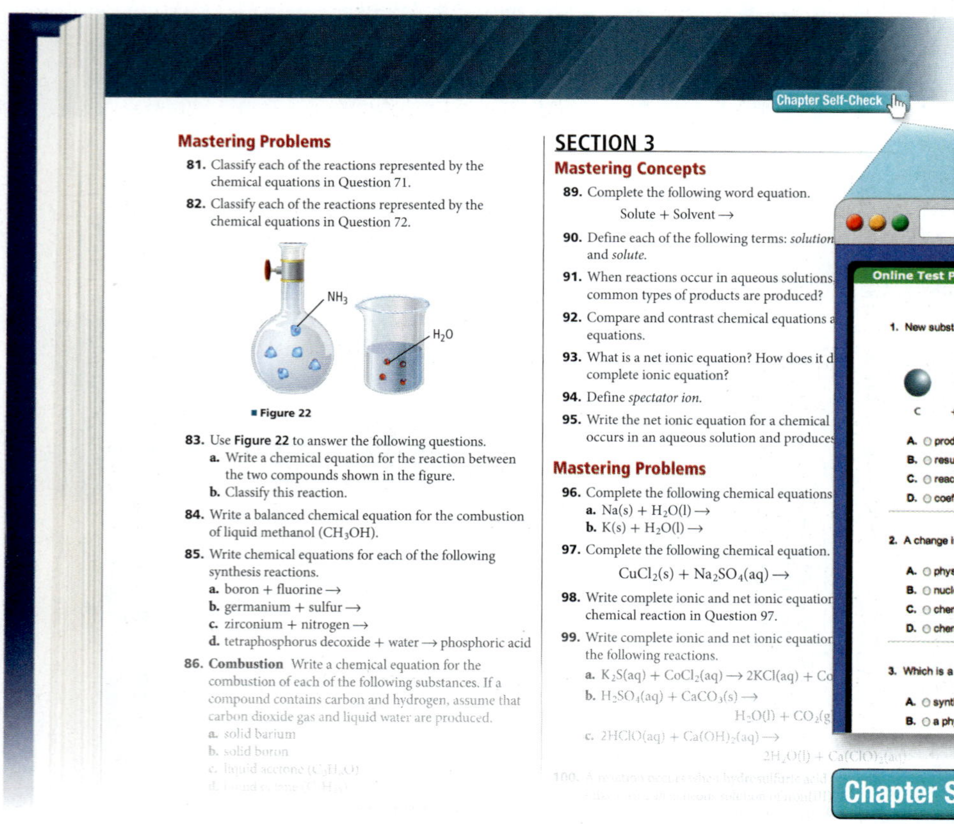

Chapter Self-Check

Mastering Problems

81. Classify each of the reactions represented by the chemical equations in Question 71.
82. Classify each of the reactions represented by the chemical equations in Question 72.



Figure 22

83. Use Figure 22 to answer the following questions.
   a. Write a chemical equation for the reaction between the two compounds shown in the figure.
   b. Classify this reaction.
84. Write a balanced chemical equation for the combustion of liquid methanol ($CH_3OH$).
85. Write chemical equations for each of the following synthesis reactions.
   a. boron + fluorine →
   b. germanium + sulfur →
   c. zirconium + nitrogen →
   d. tetraphosphorus decoxide + water → phosphoric acid
86. Combustion Write a chemical equation for the combustion of each of the following substances. If a compound contains carbon and hydrogen, assume that carbon dioxide gas and liquid water are produced.
   a. solid barium
   b. solid boron

SECTION 3

Mastering Concepts

89. Complete the following word equation.
   Solute + Solvent →
90. Define each of the following terms: *solution* and *solute*.
91. When reactions occur in aqueous solutions common types of products are produced?
92. Compare and contrast chemical equations a equations.
93. What is a net ionic equation? How does it d complete ionic equation?
94. Define *spectator ion*.
95. Write the net ionic equation for a chemical occurs in an aqueous solution and produce

Mastering Problems

96. Complete the following chemical equations
   a. $Na(s) + H_2O(l) \rightarrow$
   b. $K(s) + H_2O(l) \rightarrow$
97. Complete the following chemical equation.
   $CuCl_2(s) + Na_2SO_4(aq) \rightarrow$
98. Write complete ionic and net ionic equatior chemical reaction in Question 97.
99. Write complete ionic and net ionic equatior the following reactions.
   a. $K_2S(aq) + CoCl_2(aq) \rightarrow 2KCl(aq) + C$
   b. $H_2SO_4(aq) + CaCO_3(s) \rightarrow H_2O(l) + CO_2(g$
   c. $2HClO(aq) + Ca(OH)_2(aq) \rightarrow 2H_2O(l) + Ca(ClO)_2(aq)$

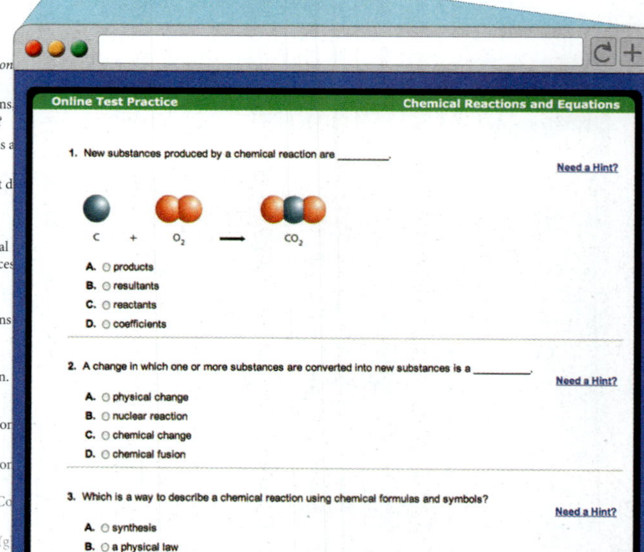

Online Test Practice — Chemical Reactions and Equations

1. New substances produced by a chemical reaction are ______. Need a Hint?



   A. products
   B. resultants
   C. reactants
   D. coefficients
2. A change in which one or more substances are converted into new substances is a ______. Need a Hint?
   A. physical change
   B. nuclear reaction
   C. chemical change
   D. chemical fusion
3. Which is a way to describe a chemical reaction using chemical formulas and symbols? Need a Hint?
   A. synthesis
   B. a physical law

**Chapter Self-Check**

Chapter-level **study tools** provide more opportunities for review and practice.

Go online!
MHEonline.com

# "I have the online STUDY TOOLS I need to help me SUCCEED."

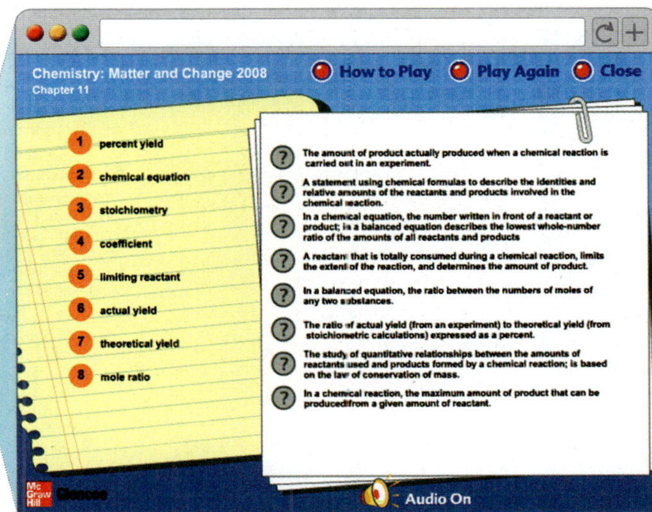


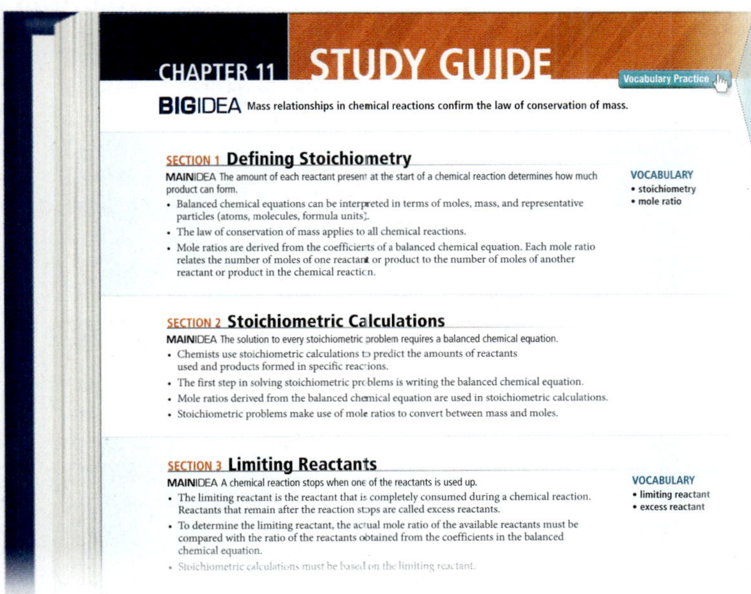


## Vocabulary Practice

The **multilingual e-Glossary** and vocabulary **study tools** drive home important concepts.

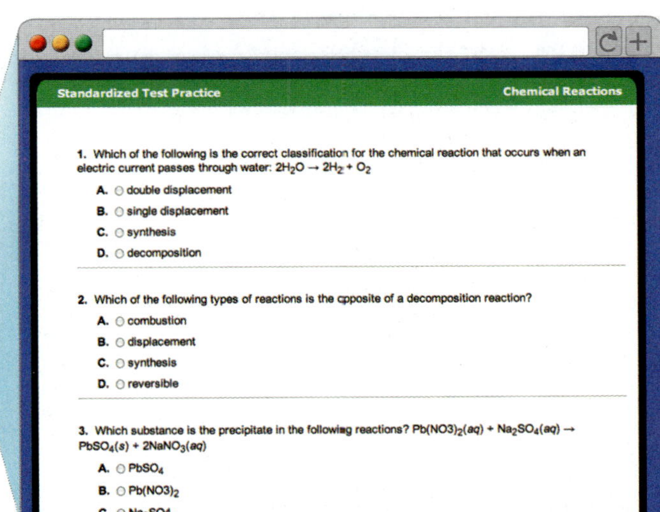


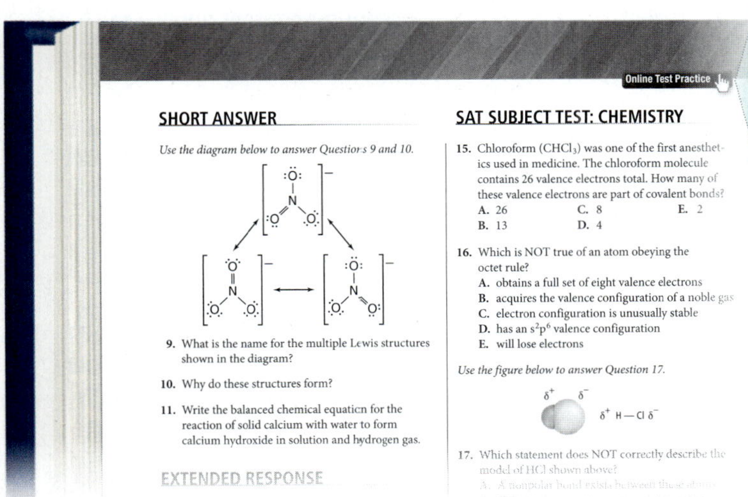


## Online Test Practice

Practice taking **standardized tests** as you also review chapter content.

PhotoDisc/SuperStock

# REAL-WORLD CHEMISTRY

***Chemistry: Matter and Change*** **connects chemistry to your world. Throughout the text, find personal science connections, surprising examples of chemistry in careers, and how chemists are engaged in cutting-edge science research.**

SECTION 4

## Metallic Bonds and the Properties of Metals

**MAINIDEA** Metals form crystal lattices and can be modeled as cations surrounded by a "sea" of freely moving valence electrons.

**Essential Questions**

- What are the characteristics of a metallic bond?
- How does the electron sea model account for the physical properties of metals?
- What are alloys, and how can they be categorized?

**Review Vocabulary**

**physical property:** a characteristic of matter that can be observed or measured without altering the sample's composition

**New Vocabulary**

electron sea model
delocalized electron
metallic bond
alloy

**CHEM 4 YOU** Imagine a buoy in the ocean, bobbing by itself surrounded by a vast expanse of open water. Though the buoy stays in the same area, the ocean water freely flows past. In some ways, this description also applies to metallic atoms and their electrons.

### Metallic Bonds

Although metals are not ionic, they share several properties with ionic compounds. The bonding in both metals and ionic compounds is based on the attraction of particles with unlike charges. Metals often form lattices in the solid state. These lattices are similar to the ionic crystal lattices discussed earlier. In such a lattice, 8 to 12 other metal atoms closely surround each metal atom.

**A sea of electrons** Although metal atoms always have at least one valence electron, they do not share these valence electrons with neighboring atoms, nor do they lose their valence electrons. Instead, within the crowded lattice, the outer energy levels of the metal atoms overlap. This unique arrangement is described by the electron sea model. The **electron sea model** proposes that all the metal atoms in a metallic solid contribute their valence electrons to form a "sea" of electrons. This sea of electrons surrounds the metal cations in the lattice.

The electrons present in the outer energy levels of the bonding metallic atoms are not held by any specific atom and can move ...

**CHEM 4 YOU** at the beginning of each section connects chemistry content to your life.

**Table 8 Monatomic Metal Ions**

| Group | Common Ions |
|---|---|
| 3 | $Sc^{3+}$, $Y^{3+}$, $La^{3+}$ |
| 4 | $Ti^{2+}$, $Ti^{3+}$ |
| 5 | $V^{2+}$, $V^{3+}$ |
| 6 | $Cr^{2+}$, $Cr^{3+}$ |
| 7 | $Mn^{2+}$, $Mn^{3+}$, $Tc^{2+}$ |
| 8 | $Fe^{2+}$, $Fe^{3+}$ |
| 9 | $Co^{2+}$, $Co^{3+}$ |
| 10 | $Ni^{2+}$, $Pd^{2+}$, $Pt^{2+}$, $Pt^{4+}$ |
| 11 | $Cu^{+}$, $Cu^{2+}$, $Ag^{+}$, $Au^{+}$, $Au^{3+}$ |
| 12 | $Zn^{2+}$, $Cd^{2+}$, $Hg_2^{2+}$, $Hg^{2+}$ |
| 13 | $Al^{3+}$, $Ga^{2+}$, $Ga^{3+}$, $In^{+}$, $In^{2+}$, $In^{3+}$, $Tl^{+}$, $Tl^{3+}$ |
| 14 | $Sn^{2+}$, $Sn^{4+}$, $Pb^{2+}$, $Pb^{4+}$ |

**Oxidation numbers** The charge of a monatomic ion is equal to its oxidation number, or oxidation state. As shown in **Table 8,** most transition metals and group 13 and 14 metals have more than one possible ionic charge. Note that the ionic charges given in **Table 8** are the most common ones, not the only ones possible.

The oxidation number of an element in an ionic compound equals the number of electrons transferred from the atom to form the ion. For example, a sodium atom transfers one electron to a chlorine atom to form sodium chloride. This results in $Na^+$ and $Cl^-$. Thus, the oxidation number of sodium in the compound is 1+ because one electron was transferred from the sodium atom. Because an electron is transferred to ...

**CAREERS IN CHEMISTRY**

**Food Scientist** Have you ever thought about the science behind the food you eat? Food scientists are concerned about the effects of processing on the appearance, aroma, taste, and the vitamin and mineral content of food. They also develop and improve foods and beverages. Food scientists often maintain "tasting notebooks" as they learn the characteristics of individual and blended flavors.

WebQuest

Throughout the book, **CAREERS IN CHEMISTRY** demonstrates how the chapter content applies to everyday careers.

**End-of-chapter features** highlight chemistry as it applies to careers, how it connects to the real world, and what chemistry means to your life.

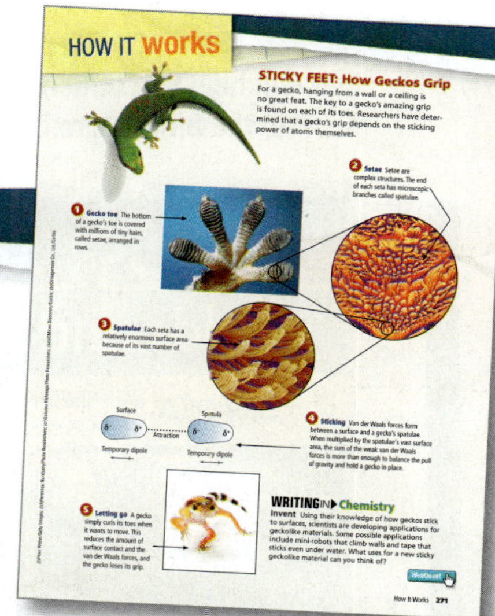

## HOW IT works

Explore the chemistry involved in everyday objects or natural phenomena by discovering how they work.

## everyday CHEMISTRY

Did you use chemistry today? You might be surprised to discover how chemistry can be found in everyday experiences.

## CHEMISTRY & health

Did you know that chemistry happens inside your body? Investigate how chemistry and health are interrelated.

## CHEMISTRY & careers

Put chemistry to work on the job. Explore jobs that unexpectedly rely on an understanding of chemistry.

# UNDERSTANDING CHEMISTRY

**At the start of each chapter,** you will see the **BIG**IDEA that will help you understand how what you are about to investigate fits into the big picture of science.

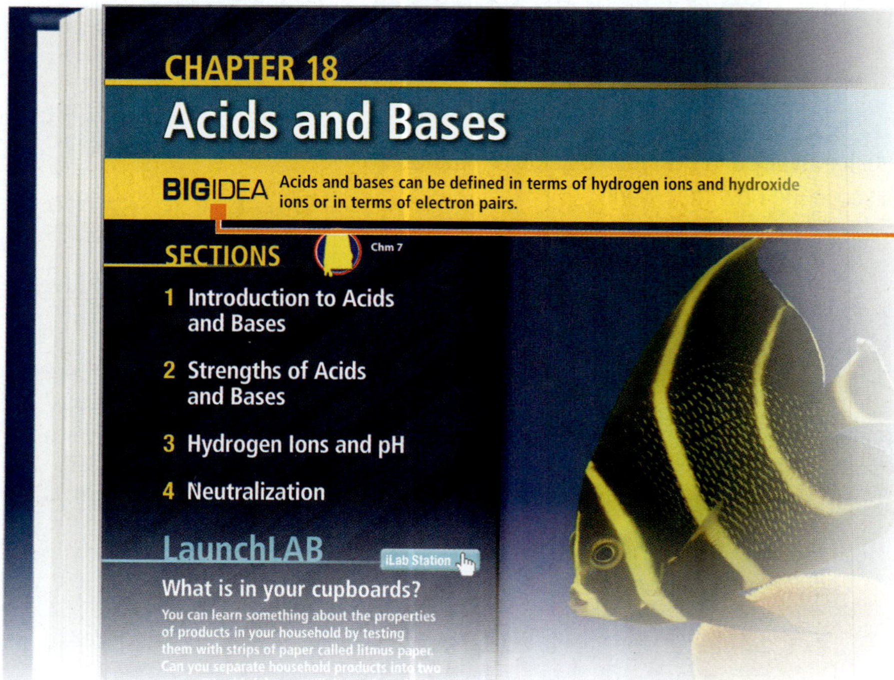
CHAPTER 18

Acids and Bases

BIGIDEA Acids and bases can be defined in terms of hydrogen ions and hydroxide ions or in terms of electron pairs.

SECTIONS

Chm 7

1 Introduction to Acids and Bases

2 Strengths of Acids and Bases

3 Hydrogen Ions and pH

4 Neutralization

LaunchLAB

iLab Station

What is in your cupboards?

You can learn something about the properties of products in your household by testing them with strips of paper called litmus paper. Can you separate household products into two

The **BIG**IDEA is the focus of the chapter. The labs, text, and other chapter content will build an in-depth understanding of these major concepts.

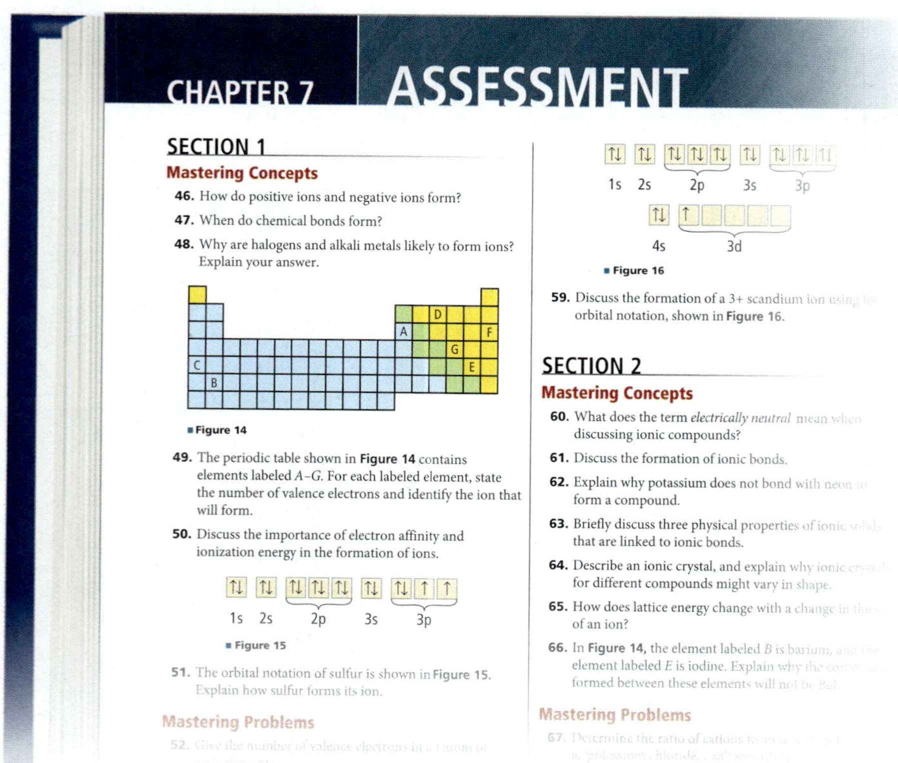
CHAPTER 7 ASSESSMENT

SECTION 1

Mastering Concepts

46. How do positive ions and negative ions form?

47. When do chemical bonds form?

48. Why are halogens and alkali metals likely to form ions? Explain your answer.



Figure 14

49. The periodic table shown in Figure 14 contains elements labeled *A–G*. For each labeled element, state the number of valence electrons and identify the ion that will form.

50. Discuss the importance of electron affinity and ionization energy in the formation of ions.



Figure 15

51. The orbital notation of sulfur is shown in Figure 15. Explain how sulfur forms its ion.

Mastering Problems



Figure 16

59. Discuss the formation of a 3+ scandium ion using its orbital notation, shown in Figure 16.

SECTION 2

Mastering Concepts

60. What does the term *electrically neutral* mean when discussing ionic compounds?

61. Discuss the formation of ionic bonds.

62. Explain why potassium does not bond with neon to form a compound.

63. Briefly discuss three physical properties of ionic solids that are linked to ionic bonds.

64. Describe an ionic crystal, and explain why ionic crystals for different compounds might vary in shape.

65. How does lattice energy change with a change in the size of an ion?

66. In Figure 14, the element labeled *B* is barium, and the element labeled *E* is iodine. Explain why the compound formed between these elements will not be BaI.

Mastering Problems

The Chapter Assessment will help you evaluate your understanding of the **BIG**IDEA.

**At the start of each section,** you will find a reading preview that summarizes what you will learn while exploring the section.

The **MAIN**IDEA is the core concept covered in the section. Together, the Main Ideas from all the sections in the chapter support the chapter's Big Idea.

**Essential Questions** reflect the important goals of the section. Together, an understanding of these questions will lead toward an understanding of the section's Main Idea.

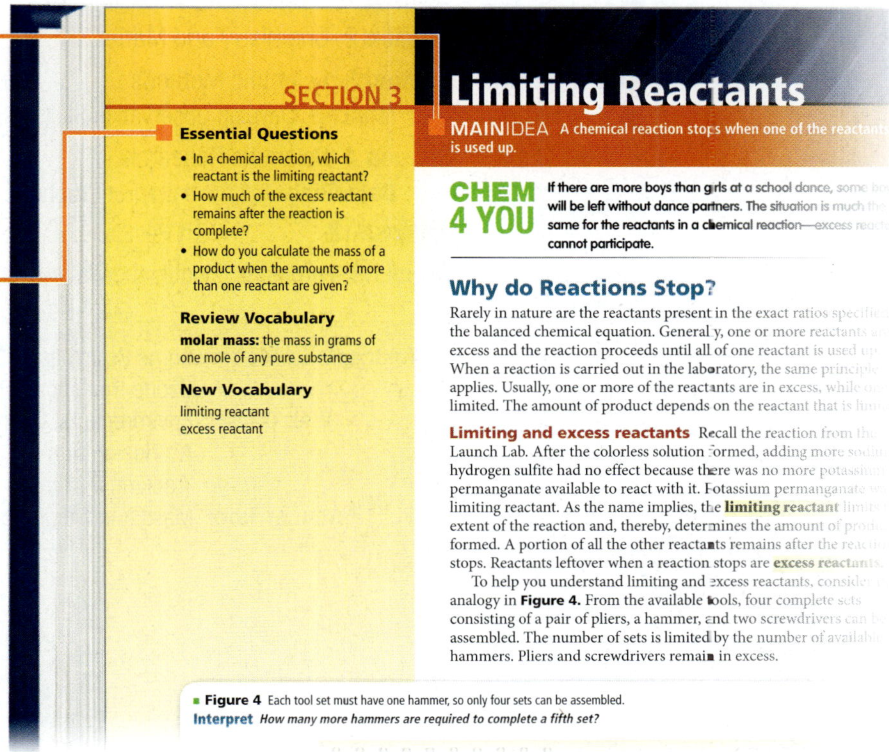

SECTION 3

**Essential Questions**

- In a chemical reaction, which reactant is the limiting reactant?
- How much of the excess reactant remains after the reaction is complete?
- How do you calculate the mass of a product when the amounts of more than one reactant are given?

**Review Vocabulary**

**molar mass:** the mass in grams of one mole of any pure substance

**New Vocabulary**

limiting reactant
excess reactant

# Limiting Reactants

MAINIDEA A chemical reaction stops when one of the reactants is used up.

CHEM 4 YOU If there are more boys than girls at a school dance, some bo... will be left without dance partners. The situation is much the same for the reactants in a chemical reaction—excess reac... cannot participate.

## Why do Reactions Stop?

Rarely in nature are the reactants present in the exact ratios specifi... the balanced chemical equation. Generally, one or more reactants a... excess and the reaction proceeds until all of one reactant is used up. When a reaction is carried out in the laboratory, the same principle applies. Usually, one or more of the reactants are in excess, while o... limited. The amount of product depends on the reactant that is lim...

**Limiting and excess reactants** Recall the reaction from the Launch Lab. After the colorless solution formed, adding more sodiu... hydrogen sulfite had no effect because there was no more potassium permanganate available to react with it. Potassium permanganate w... limiting reactant. As the name implies, the **limiting reactant** limits the extent of the reaction and, thereby, determines the amount of produ... formed. A portion of all the other reactants remains after the reaction stops. Reactants leftover when a reaction stops are **excess reactants**.

To help you understand limiting and excess reactants, consider the analogy in **Figure 4.** From the available tools, four complete sets consisting of a pair of pliers, a hammer, and two screwdrivers can be assembled. The number of sets is limited by the number of availabl... hammers. Pliers and screwdrivers remain in excess.

**Figure 4** Each tool set must have one hammer, so only four sets can be assembled.
**Interpret** *How many more hammers are required to complete a fifth set?*

In the Section Review, you will find a question that will help you to assess your understanding of the section's **MAIN**IDEA.

**Essential Questions** are assessed by the remaining review questions.

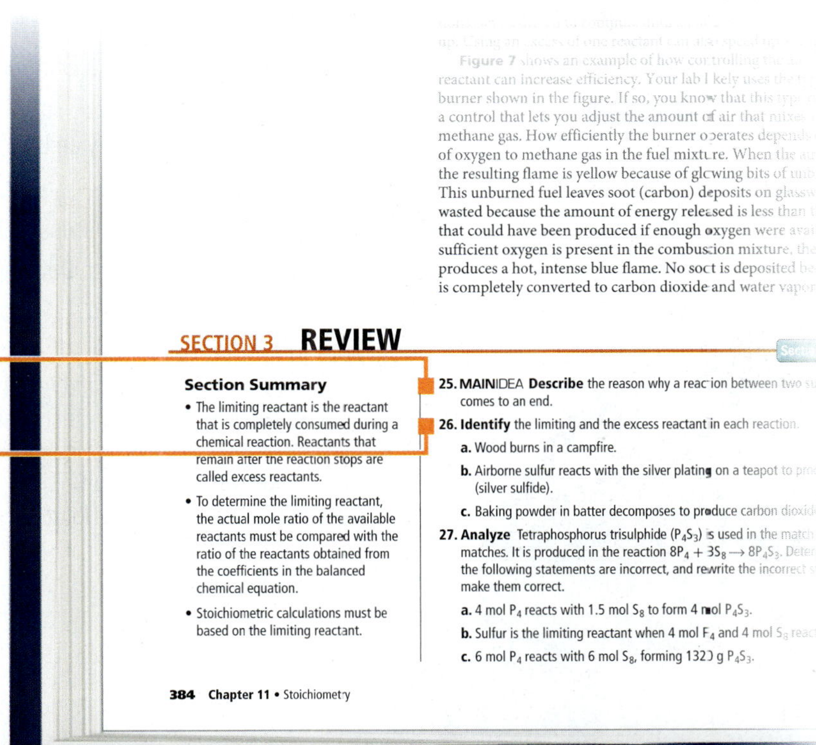

Figure 7 shows an example of how controlling the ... reactant can increase efficiency. Your lab likely uses the ... burner shown in the figure. If so, you know that this typ... a control that lets you adjust the amount of air that mixe... methane gas. How efficiently the burner operates depends ... of oxygen to methane gas in the fuel mixture. When the ai... the resulting flame is yellow because of glowing bits of unb... This unburned fuel leaves soot (carbon) deposits on glassw... wasted because the amount of energy released is less than th... that could have been produced if enough oxygen were avai... sufficient oxygen is present in the combustion mixture, the ... produces a hot, intense blue flame. No soot is deposited be... is completely converted to carbon dioxide and water vapor.

SECTION 3 **REVIEW**

**Section Summary**

- The limiting reactant is the reactant that is completely consumed during a chemical reaction. Reactants that remain after the reaction stops are called excess reactants.
- To determine the limiting reactant, the actual mole ratio of the available reactants must be compared with the ratio of the reactants obtained from the coefficients in the balanced chemical equation.
- Stoichiometric calculations must be based on the limiting reactant.

**25.** MAINIDEA **Describe** the reason why a reaction between two su... comes to an end.

**26. Identify** the limiting and the excess reactant in each reaction.
  a. Wood burns in a campfire.
  b. Airborne sulfur reacts with the silver plating on a teapot to pro... (silver sulfide).
  c. Baking powder in batter decomposes to produce carbon dioxid...

**27. Analyze** Tetraphosphorus trisulphide ($P_4S_3$) is used in the match ... matches. It is produced in the reaction $8P_4 + 3S_8 \rightarrow 8P_4S_3$. Determ... the following statements are incorrect, and rewrite the incorrect st... make them correct.
  a. 4 mol $P_4$ reacts with 1.5 mol $S_8$ to form 4 mol $P_4S_3$.
  b. Sulfur is the limiting reactant when 4 mol $P_4$ and 4 mol $S_8$ react.
  c. 6 mol $P_4$ reacts with 6 mol $S_8$, forming 1320 g $P_4S_3$.

384 Chapter 11 • Stoichiometry

# TABLE OF CONTENTS

## CHAPTER 1


**Introduction to Chemistry** .......... **2**

LaunchLAB Where did the mass go? .......... 2

**Section 1** A Story of Two Substances .......... 4

**Section 2** Chemistry and Matter .......... 9

**Section 3** Scientific Methods .......... 12

MiniLAB Develop Observation Skills .......... 13

**Section 4** Scientific Research .......... 17

Data Analysis LAB Interpret Graphs .......... 21

CHEMISTRY&careers Art Restorer .......... 23

ChemLAB **Forensics:** Identify the Water Source .......... 24


Go online!

| | |
|---|---|
| **Concepts in Motion** | Some Branches of Chemistry |
| | Ozone Depletion |
| **Web Quests** | Environmental Chemist |
| | Art Restoration |
| **Video** | Tracking a Crime |
| **Personal Tutor** | Mass and Weight Relationships |

## CHAPTER 2


**Analyzing Data** .......... **30**

LaunchLAB How can you form layers of liquids? .......... 30

**Section 1** Units and Measurement .......... 32

MiniLAB Determine Density .......... 39

**Section 2** Scientific Notation and Dimensional Analysis .......... 40

**Section 3** Uncertainty in Data .......... 47

Problem-Solving LAB Identify an Unknown .......... 50

**Section 4** Representing Data .......... 55

CHEMISTRY&health **Toxicology:** Assessing Health Risk .......... 59

ChemLAB **Forensics:** Use Density to Date a Coin .......... 60


Go online!

| | |
|---|---|
| **Concepts in Motion** | SI Prefixes |
| | Precision and Accuracy |
| **Web Quests** | Calibration Technician |
| | Assessing Health Risk |
| **Video** | Tracking a Crime |
| **Personal Tutor** | Scientific Notation |
| | Graphing |
| **Virtual Investigations** | Density |

Philippe Psaila/Science Photo Library/Photo Researchers

Go online! MHEonline.com

CHAPTER 3


## Matter—Properties and Changes ................ 68

LaunchLAB How can you observe chemical change? ................ 68

**Section 1** Properties of Matter ................ 70

Problem-Solving LAB Cause and Effect ................ 72

**Section 2** Changes in Matter ................ 76

**Section 3** Mixtures of Matter ................ 80

MiniLAB Observe Dye Separation ................ 82

**Section 4** Elements and Compounds ................ 84

CHEMISTRY&careers Arson Investigator ................ 91

ChemLAB Identify the Products of a Chemical Reaction ................ 92


Go online!

| | |
|---|---|
| **Concepts in Motion** | States of Matter |
| | Conservation of Mass |
| | Types of Solution Systems |
| **Web Quest** | Materials Scientist |
| | Arson Investigation |
| **Video** | Matter is Made Up of Atoms |
| | Open the Faucet |
| | Hair: A Chemical Storehouse |
| **Personal Tutor** | Writing an Equation |
| | Percentages |
| **Virtual Investigations** | Laws of Definite and Multiple Proportions |

CHAPTER 4


## The Structure of the Atom ................ 100

LaunchLAB How can the effects of electric charges be observed? ...... 100

**Section 1** Early Ideas About Matter ................ 102

**Section 2** Defining the Atom ................ 106

Data Analysis LAB Interpret Scientific Illustrations ................ 113

**Section 3** How Atoms Differ ................ 115

MiniLAB Model Isotopes ................ 120

**Section 4** Unstable Nuclei and Radioactive Decay ................ 122

HOW IT works Mass Spectrometer: Chemical Detective ................ 125

ChemLAB Model Atomic Mass ................ 126


Go online!

| | |
|---|---|
| **Concepts in Motion** | Gold Foil Experiment |
| | Properties of Subatomic Particles |
| | Structure of the Atom |
| | Radiation Deflection |
| **Web Quests** | Chemistry Teacher |
| | Mass Spectrometry |
| **Personal Tutor** | Finding an Average |
| **Virtual Investigations** | Decoding the Periodic Table |

Charles D. Winters/Photo Researchers

# TABLE OF CONTENTS

## CHAPTER 5


**Electrons in Atoms .................................. 134**

LaunchLAB How do you know what is inside an atom? ............ 134

**Section 1** Light and Quantized Energy ............ 136

MiniLAB Identify Compounds ............ 144

**Section 2** Quantum Theory and the Atom ............ 146

Problem-Solving LAB Interpret Scientific Illustrations ............ 150

**Section 3** Electron Configuration ............ 156

CHEMISTRY&health Micro Scissors and Tiny Tweezers ............ 163

ChemLAB Analyze Line Spectra ............ 164


Go online!

**Concepts in Motion** Balmer Series
Electron Transitions
Electron Configurations and Orbital Diagrams for Elements 1–10
Electron Configurations
Electron Configurations and Dot Structures

**Web Quests** Spectroscopist
Laser Medicine

**Video** Lighting Up the Night Sky

**Virtual Investigation** Electron Configuration

## CHAPTER 6


**The Periodic Table and Periodic Law .......... 172**

LaunchLAB How can you recognize trends? ............ 172

**Section 1** Development of the Modern Periodic Table ............ 174

Problem-Solving LAB Analyze Trends ............ 180

**Section 2** Classification of the Elements ............ 182

**Section 3** Periodic Trends ............ 187

MiniLAB Organize Elements ............ 193

CHEMISTRY&health Elements of the Body ............ 195

ChemLAB Investigate Descriptive Chemistry ............ 196


Go online!

**Concepts in Motion** The Periodic Table
Noble Gas Electron Configuration
Atomic Radii
Electronegativity

**Web Quests** Research Chemist
Elements of the Body

**Video** It's Elemental

**Personal Tutor** Identifying Trends

**Virtual Investigations** Periodic Properties of the Elements

Go online! MHEonline.com

CHAPTER 7


## Ionic Compounds and Metals .................... 204

LaunchLAB What compounds conduct electricity in solution? .......... 204

**Section 1** Ion Formation .......... 206

**Section 2** Ionic Bonds and Ionic Compounds .......... 210

Data Analysis LAB Interpret Data .......... 216

**Section 3** Names and Formulas for Ionic Compounds .......... 218

**Section 4** Metallic Bonds and the Properties of Metals .......... 225

MiniLAB Observe Properties .......... 227

everydayCHEMISTRY Killer Fashion .......... 229

ChemLAB Synthesize an Ionic Compound .......... 230


Go online!

**Concepts in Motion** Electron-Dot Structures
Formation of Sodium Chloride

**Web Quests** Food Scientist
Lead Poisoning

**Video** Formation of Compounds

**Personal Tutor** Naming Ionic Compounds

**Virtual Investigations** Ionic Compounds

CHAPTER 8


## Covalent Bonding .................... 238

LaunchLAB What type of compound is used to make a super ball? ...... 238

**Section 1** The Covalent Bond .......... 240

MiniLAB Compare Melting Points .......... 242

**Section 2** Naming Molecules .......... 248

**Section 3** Molecular Structures .......... 253

**Section 4** Molecular Shapes .......... 261

**Section 5** Electronegativity and Polarity .......... 265

Data Analysis LAB Interpret Data .......... 269

HOW IT works Sticky Feet: How Geckos Grip .......... 271

ChemLAB Model Molecular Shapes .......... 272


Go online!

**Concepts in Motion** Sigma and Pi Bonding
Prefixes in Covalent Compounds
Formulas and Names of Some Covalent Compounds
Molecular Shapes (Animation)
Molecular Shapes (Interactive Table)
Bond Types

**Web Quests** Flavor Chemist
How Geckos Grip

**Video** Chemical Bonding

**Personal Tutor** Multiple Covalent Bonds
Greatest Common Factors

**Virtual Investigations** Molecular Shapes
Covalent Compounds

# TABLE OF CONTENTS

## CHAPTER 9


**Chemical Reactions** .......... **280**

LaunchLAB How do you know when a chemical change has occurred? ... 280

**Section 1** Reactions and Equations .......... 282

**Section 2** Classifying Chemical Reactions .......... 289

Problem-Solving LAB Analyze Trends .......... 294

**Section 3** Reactions in Aqueous Solutions .......... 299

MiniLAB Observe a Precipitate-Forming Reaction .......... 301

HOW IT works Lighting up the Night: Bioluminescence .......... 309

ChemLAB **Small Scale:** Develop an Activity Series .......... 310

Go online!

| | |
|---|---|
| Concepts in Motion | Steps for Balancing Equations |
| | Precipitate Formation |
| | Predicting Products of Chemical Reactions |
| Web Quests | Biochemist |
| | Bioluminescence |
| Video | Chemical Reactions and Equations |
| Personal Tutor | Balancing Chemical Equations |
| Virtual Investigations | Balancing Chemical Reactions |


## CHAPTER 10


**The Mole** .......... **318**

LaunchLAB How much is a mole? .......... 318

**Section 1** Measuring Matter .......... 320

**Section 2** Mass and the Mole .......... 325

Problem-Solving LAB Formulate a Model .......... 326

**Section 3** Moles of Compounds .......... 333

**Section 4** Empirical and Molecular Formulas .......... 341

MiniLAB Analyze Chewing Gum .......... 342

**Section 5** Formulas of Hydrates .......... 351

everydayCHEMISTRY History in a Glass of Water .......... 355

ChemLAB Determine the Formula of a Hydrate .......... 356

Go online!

| | |
|---|---|
| Concepts in Motion | Molar Mass |
| | Formulas of Hydrates |
| Web Quests | Medicinal Chemist |
| | Back-of-the-Envelope Calculations |
| Personal Tutor | Using Conversion Factors |
| Virtual Investigations | Moles, Mass, and Molecules |

Mihaela Ninic/Alamy

Go online! MHEonline.com

CHAPTER 11


## Stoichiometry ........ 366

LaunchLAB What evidence can you observe that a reaction has stopped? .. 366

**Section 1** Defining Stoichiometry ........ 368

**Section 2** Stoichiometric Calculations ........ 373

MiniLAB Apply Stoichiometry ........ 378

**Section 3** Limiting Reactants ........ 379

**Section 4** Percent Yield ........ 385

Data Analysis LAB Analyze and Conclude ........ 387

CHEMISTRY&health Battling Resistant Strains ........ 389

ChemLAB Determine the Mole Ratio ........ 390


Go online!

**Concepts in Motion** Relationships Derived from a Balanced Chemical Equation; Limiting Reactants

**Web Quests** Pharmacist; Drug Resistant Pathogens

**Video** Stoichiometry

**Personal Tutor** Ratios

**Virtual Investigations** Limiting Reactants

CHAPTER 12


## States of Matter ........ 400

LaunchLAB How do different liquids affect the speed of a sinking ball bearing? ........ 400

**Section 1** Gases ........ 402

Data Analysis LAB Make and Use Graphs ........ 408

**Section 2** Forces of Attraction ........ 411

**Section 3** Liquids and Solids ........ 415

MiniLAB Model Crystal Unit Cells ........ 423

**Section 4** Phase Changes ........ 425

everydayCHEMISTRY Cocoa Chemistry ........ 431

ChemLAB **Small Scale:** Compare Rates of Evaporation ........ 432


Go online!

**Concepts in Motion** Unit Cells; Types of Crystalline Solids; Phase Diagram

**Web Quests** Metallurgist; Cocoa Chemistry

**Video** Enzymes and Surfactants

**Personal Tutor** Law of Partial Pressures

**Virtual Investigations** Kinetic Theory

# TABLE OF CONTENTS

## CHAPTER 13


**Gases** .......... **440**

LaunchLAB How does temperature affect the volume of a gas? .......... 440

**Section 1** The Gas Laws .......... 442

Problem-Solving LAB Apply Scientific Explanations .......... 444

**Section 2** The Ideal Gas Law .......... 452

MiniLAB Model a Fire Extinguisher .......... 457

**Section 3** Gas Stoichiometry .......... 460

CHEMISTRY&health Health Under Pressure .......... 465

ChemLAB Determine Pressure in Popcorn Kernels .......... 466


Go online!

**Concepts in Motion** The Gas Laws (Animation)
The Gas Laws (Interactive Table)
**Web Quests** Meteorologist
Hyperbaric Oxygen
**Video** Gases
Leagues Under the Sea
**Personal Tutor** Combined Gas Law
**Virtual Investigations** The Gas Laws

## CHAPTER 14


**Mixtures and Solutions** .......... **474**

LaunchLAB How does energy change when solutions form? .......... 474

**Section 1** Types of Mixtures .......... 476

Data Analysis LAB Design an Experiment .......... 478

**Section 2** Solution Concentration .......... 480

**Section 3** Factors Affecting Solvation .......... 489

**Section 4** Colligative Properties of Solutions .......... 498

MiniLAB Examine Freezing Point Depression .......... 502

CHEMISTRY&careers Environmental Chemist .......... 505

ChemLAB Investigate Factors Affecting Solubility .......... 506


Go online!

**Concepts in Motion** Types and Examples of Solutions
Dissolution of Compounds
Strong, Weak, and Nonelectrolytes
Osmosis
**Web Quests** Pharmacy Technician
Environmental Chemistry
**Video** Water and Its Solutions
**Personal Tutor** Calculating Molarity and Molality
**Virtual Investigations** Salts and Solubility
Colligative Properties

Steve Allen/Brand X Pictures

CHAPTER 15


## Energy and Chemical Change .................. 514

LaunchLAB How can you make a cold pack? .................. 514

**Section 1** Energy .................. 516

**Section 2** Heat .................. 523

MiniLAB Determine Specific Heat .................. 526

**Section 3** Thermochemical Equations .................. 529

Problem-Solving LAB Make and Use Graphs .................. 531

**Section 4** Calculating Enthalpy Change .................. 534

**Section 5** Reaction Spontaneity .................. 542

HOW IT works Driving the Future: Flexible Fuel Vehicles .................. 549

ChemLAB Measure Calories .................. 550


Go online!

**Concepts in Motions** Calorimetry
Heat Flow in Endothermic and Exothermic Reactions
Reaction Spontaneity
**Web Quests** Heating and Cooling Specialist
Alternative Fuel Vehicles
**Personal Tutor** Probability
**Virtual Investigations** Phase Changes
Calorimetry

CHAPTER 16


## Reaction Rates .................. 558

LaunchLAB How can you accelerate a reaction? .................. 558

**Section 1** A Model for Reaction Rates .................. 560

Problem-Solving LAB Interpret Data .................. 566

**Section 2** Factors Affecting Reaction Rates .................. 568

MiniLAB Examine Reaction Rate and Temperature .................. 571

**Section 3** Reaction Rate Laws .................. 574

**Section 4** Instantaneous Reaction Rates and Reaction Mechanisms .................. 578

CHEMISTRY&health Reaction Rates and Body Temperature .................. 583

ChemLAB Observe How Concentration Affects Reaction Rate .................. 584


Go online!

**Concepts in Motion** Molecular Orientation and Collision Effectiveness
**Web Quests** Chemical Engineer
Reaction Rate and Metabolism
**Personal Tutor** Solving Reaction Rates
**Virtual Investigations** Reaction Rates

Stephen Frisch/McGraw-Hill Education

CHAPTER 17


**Chemical Equilibrium** .......... **592**

LaunchLAB What is equal about equilibrium? .......... 592

**Section 1** A State of Dynamic Balance .......... 594

**Section 2** Factors Affecting Chemical Equilibrium .......... 606

MiniLAB Observe Shifts in Equilibrium .......... 611

**Section 3** Using Equilibrium Constants .......... 612

Problem-Solving LAB Apply Scientific Explanations .......... 622

CHEMISTRY&health Hemoglobin Rises to the Challenge .......... 623

ChemLAB **Small Scale:** Compare Two Solubility Product Constants .......... 624

Go online!

| | |
|---|---|
| Concepts in Motion | Equilibrium Shifts |
| | Precipitation Reactions |
| Web Quests | Science Writer |
| | High-Altitude Metabolism |
| Personal Tutor | Equilibrium Constant Expressions |
| Virtual Investigations | Salts and Solubility |


CHAPTER 18


**Acids and Bases** .......... **632**

LaunchLAB What is in your cupboards? .......... 632

**Section 1** Introduction to Acids and Bases .......... 634

**Section 2** Strengths of Acids and Bases .......... 644

MiniLAB Compare Acid Strengths .......... 648

**Section 3** Hydrogen Ions and pH .......... 650

**Section 4** Neutralization .......... 659

Problem-Solving LAB Apply Scientific Explanations .......... 668

everydayCHEMISTRY Acid-Base Reactions on the Rise .......... 669

ChemLAB Standardize a Base .......... 670

Go online!

| | |
|---|---|
| Concepts in Motion | Three Models for Acids and Bases |
| | Ionization Equations |
| | Neutralization Reaction |
| | Titration |
| | Neutralization Reactions |
| Web Quests | Nursery Worker |
| | Acids, Bases, and Cooking |
| Video | Acids and Bases |
| Personal Tutor | Logarithms |
| Virtual Investigations | Titrations |

Go online! MHEonline.com

CHAPTER 19


## Redox Reactions .................................. 678

LaunchLAB What happens when iron and copper(II) sulfate react? ......678

**Section 1** Oxidation and Reduction........................................680

MiniLAB Observe a Redox Reaction..........................................683

**Section 2** Balancing Redox Equations ......................................689

Data Analysis LAB Analyze and Conclude .......................................691

CHEMISTRY&careers Crime Scene Investigator.......................................697

ChemLAB **Small Scale Forensics:** Identify the Damaging Dumper ..........698


Go online!

Concepts in Motions Redox Reaction
Summary of Redox Reactions
Web Quests Potter
Crime Scene Investigation
Personal Tutor Balancing Redox Reactions
Virtual Investigations Redox Titration

CHAPTER 20


## Electrochemistry .................................. 706

LaunchLAB How can you make a battery from a lemon? .............706

**Section 1** Voltaic Cells ..........................................708

**Section 2** Batteries ..........................................718

Data Analysis LAB Interpret Graphs ..........................................724

MiniLAB Observe Corrosion ..........................................726

**Section 3** Electrolysis..........................................728

HOW IT works The Pacemaker: Helping a Broken Heart.................733

ChemLAB **Small Scale:** Measure Voltaic Cell Potentials.................734


Go online!

Concepts in Motion Voltaic Cell
Web Quests Alternative Energy Specialist
Batteries and Medical Devices
Video Hydrogen Fuel Cells
Personal Tutor Calculating Potentials
Virtual Investigations Electrochemical Cells

Terry Whittaker/Alamy

# TABLE OF CONTENTS


CHAPTER 21

**Hydrocarbons** .......... **742**

LaunchLAB How can you model simple hydrocarbons? .......... 742

**Section 1** Introduction to Hydrocarbons .......... 744

**Section 2** Alkanes .......... 750

**Section 3** Alkenes and Alkynes .......... 759

MiniLAB Synthesize and Observe Ethyne .......... 763

**Section 4** Hydrocarbon Isomers .......... 765

Data Analysis LAB Interpret Data .......... 768

**Section 5** Aromatic Hydrocarbons .......... 770

HOW IT works Pooch to Power: How a Methane Digester Works .......... 775

ChemLAB **Forensics:** Analyze Hydrocarbon Burner Gases .......... 776

Go online!

**Concepts in Motion** Examples of Alkenes
Examples of Alkynes
Isomers of Pentane
Sigma- and Pi-Bonding in Benzene

**Web Quests** Petroleum Technician
Methane Digesters

**Personal Tutor** Naming Hydrocarbons

CHAPTER 22

**Substituted Hydrocarbons and Their Reactions** .......... **784**

LaunchLAB How do you make slime? .......... 784

**Section 1** Alkyl Halides and Aryl Halides .......... 786

**Section 2** Alcohols, Ethers, and Amines .......... 792

**Section 3** Carbonyl Compounds .......... 796

MiniLAB Make an Ester .......... 800

**Section 4** Other Reactions of Organic Compounds .......... 802

Data Analysis LAB Interpret Data .......... 805

**Section 5** Polymers .......... 809

everydayCHEMISTRY Garlic: Pleasure and Pain .......... 815

ChemLAB **Small Scale:** Observe Properties of Alcohols .......... 816

Go online!

**Concepts in Motion** Organic Compounds and Their Functional Groups
Common Polymers

**Web Quests** Polymer Chemist
Chemistry and Sensation

**Video** Substituted Hydrocarbons and Their Reactions
Recycling

**Personal Tutor** Identifying Organic Reactions

**Virtual Investigations** Functional Groups

Cordelia Molloy/Photo Researchers

## CHAPTER 23


**The Chemistry of Life** .......... **824**

LaunchLAB How do you test for simple sugars? .......... 824

**Section 1** Proteins .......... 826

**Section 2** Carbohydrates .......... 832

**Section 3** Lipids .......... 835

MiniLAB Observe a Saponification Reaction .......... 837

**Section 4** Nucleic Acids .......... 840

Problem-Solving LAB Formulate a Model .......... 842

**Section 5** Metabolism .......... 844

CHEMISTRY&careers Molecular Paleontologist .......... 849

ChemLAB Observe Temperature and Enzyme Action .......... 850


Go online!

**Concepts in Motion** Amino Acid Examples
Structure of DNA
**Web Quests** Baker
Molecular Paleontology
**Video** Enzymes and Surfactants
**Personal Tutor** Photosynthesis and Respiration

## CHAPTER 24


**Nuclear Chemistry** .......... **858**

LaunchLAB How do chain reactions occur? .......... 858

**Section 1** Nuclear Radiation .......... 860

**Section 2** Radioactive Decay .......... 865

MiniLAB Model Radioactive Decay .......... 873

**Section 3** Nuclear Reactions .......... 875

**Section 4** Applications and Effects of Nuclear Reactions .......... 885

Problem-Solving LAB Interpret Graphs .......... 890

CHEMISTRY&careers Archaeologist .......... 891

ChemLAB Investigate Radiation Dosage .......... 892


Go online!

**Concepts in Motion** Summary of Radioactive Decay Processes
Nuclear Chain Reaction
Critical Mass
Nuclear Power
**Web Quests** Radiation Therapist
Neutron Activation Analysis
**Video** Radioisotopes
**Personal Tutor** Exponential Graphing
**Virtual Investigations** Half-Life

Chris Johnson/Alamy

# STUDENT RESOURCES

## Elements Handbook 901


Hydrogen . . . . . 904
Group 1: Alkali Metals . . . . . 906
Group 2: Alkaline Earth Metals . . . . . 910
Groups 3–12: Transition Elements . . . . . 916
Group 13: Boron Group . . . . . 922
Group 14: Carbon Group . . . . . 926
Group 15: Nitrogen Group . . . . . 932
Group 16: Oxygen Group . . . . . 936
Group 17: Halogen Group . . . . . 940
Group 18: Noble Gases . . . . . 944


## Math Handbook 946


Scientific Notation . . . . . 946
Operations with Scientific Notation . . . . . 948
Square and Cube Roots . . . . . 949
Significant Figures . . . . . 949
Solving Algebraic Equations . . . . . 954
Dimensional Analysis . . . . . 956
Unit Conversion . . . . . 957
Drawing Line Graphs . . . . . 959
Using Line Graphs . . . . . 961
Ratios, Fractions, and Percents . . . . . 964
Operations Involving Fractions . . . . . 965
Logarithms and Antilogarithms . . . . . 966


## Reference Tables 968


**R-1** Color Key . . . . . 968
**R-2** Symbols and Abbreviations . . . . . 968
**R-3** Solubility Product Constants . . . . . 969
**R-4** Physical Constants . . . . . 969
**R-5** Names and Charges of Polyatomic Ions . . . . . 970
**R-6** Ionization Constants . . . . . 970
**R-7** Properties of Elements . . . . . 971
**R-8** Solubility Guidelines . . . . . 974
**R-9** Specific Heat Values . . . . . 975
**R-10** Molal Freezing Point Depression and Boiling Point Elevation Constants . . . . . 975
**R-11** Heat of Formation Values . . . . . 975
**R-12** Safety Symbols . . . . . 976


## Supplemental Practice Problems 977

## Selected Solutions 993

## Glossary • Glosario 1006

## Index 1032

Rubberball/Getty Images

FOLDABLES® by Dinah Zike

# Folding Instructions

**The following pages offer step-by-step instructions to make the Foldables study guides.**

## Layered-Look Book

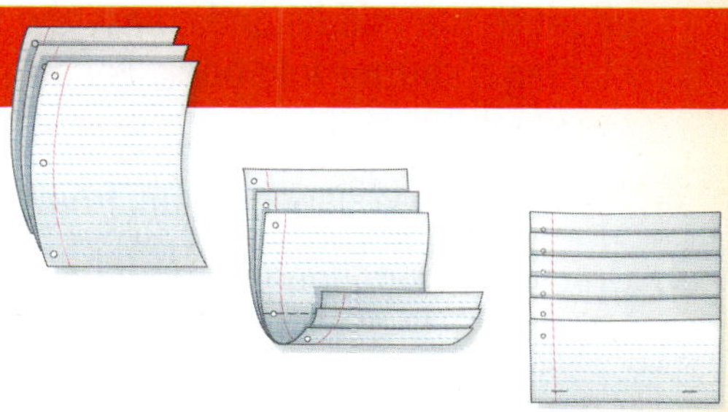

1. Collect three sheets of paper and layer them about 1 cm apart vertically. Keep the edges level.
2. Fold up the bottom edges of the paper to form 6 equal tabs.
3. Fold the papers and crease well to hold the tabs in place. Staple along the fold. Label each tab.

## Trifold Book

1. Fold a vertical sheet of paper into thirds.
2. Unfold and label each row.

## Three-Tab Book

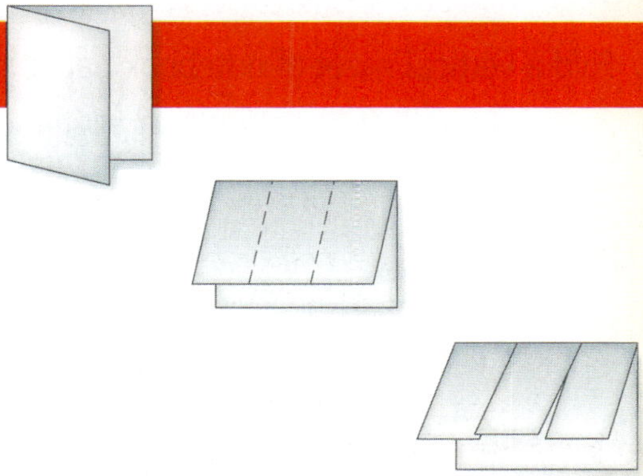

1. Fold a horizontal sheet of paper from side to side. Make the front edge about 2 cm shorter than the back edge.
2. Turn length-wise and fold into thirds.
3. Unfold and cut only the top layer along both folds to make three tabs. Label each tab.

## Two- and Four-Tab Books

1. Fold a sheet of paper in half.
2. Fold in half again. If making a four-tab book, then fold in half again to make three folds.
3. Unfold and cut only the top layer along the folds to make two or four tabs. Label each tab.

# FOLDABLES®

by Dinah Zike

## Shutter-Fold and Four-Door Books

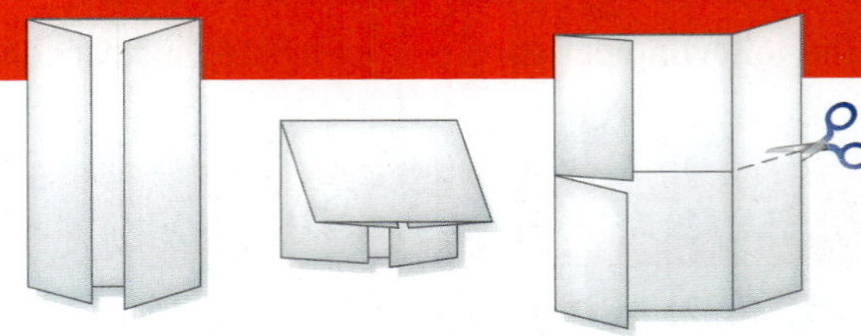

1. Find the middle of a horizontal sheet of paper. Fold both edges to the middle and crease the folds. Stop here if making a shutter-fold book. For a four-door book, complete the steps below.
2. Fold the folded paper in half, from top to bottom.
3. Unfold and cut along the fold lines to make four tabs. Label each tab.

## Concept-Map Book

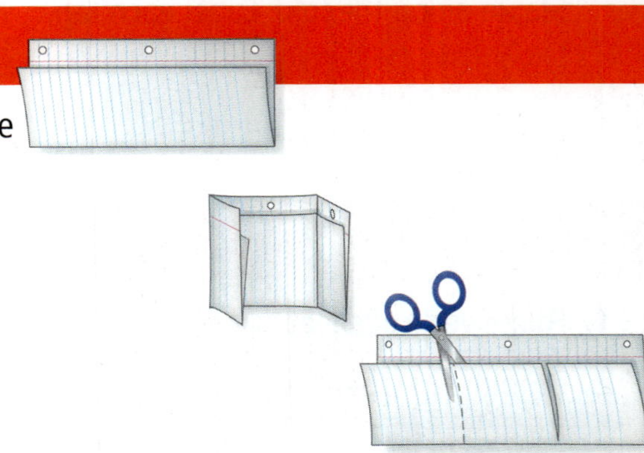

1. Fold a horizontal sheet of paper from top to bottom. Make the top edge about 2 cm shorter than the bottom edge.
2. Fold width-wise into thirds.
3. Unfold and cut only the top layer along both folds to make three tabs. Label the top and each tab.

## Vocabulary Book

1. Fold a vertical sheet of notebook paper in half.
2. Cut along every third line of only the top layer to form tabs. Label each tab.

## Folded Chart

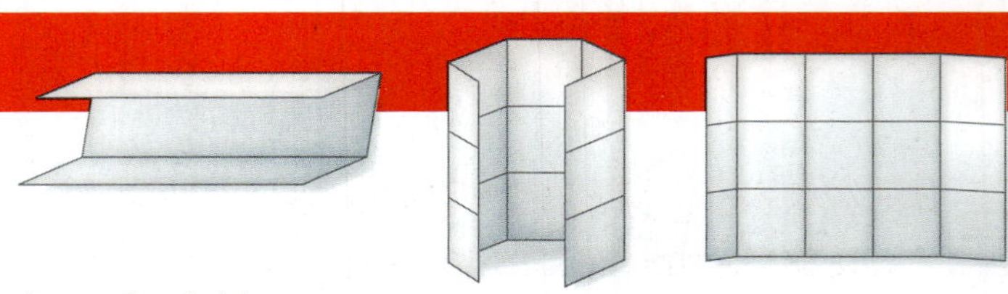

1. Fold a sheet of paper length-wise into thirds.
2. Fold the paper width-wise into fifths.
3. Unfold, lay the paper length-wise, and draw lines along the folds. Label the table.

## Pocket Book

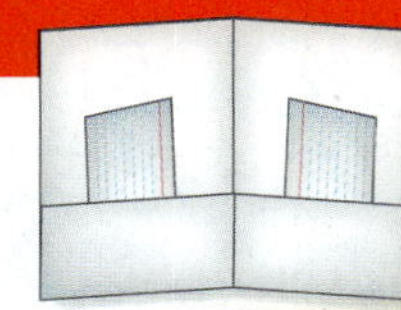

1. Fold the bottom of a horizontal sheet of paper up about 3 cm.
2. If making a two-pocket book, fold in half. If making a three-pocket book, fold in thirds.
3. Unfold once and dot with glue or staple to make pockets. Label each pocket.

## Bound Book

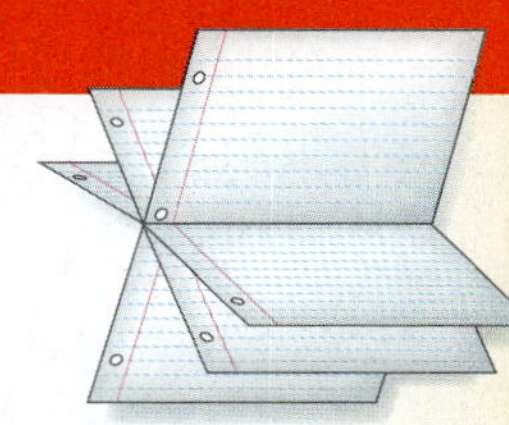

1. Fold several sheets of paper in half to find the middle. Hold all but one sheet together and make a 3-cm cut at the fold line on each side of the paper.
2. On the final page, cut along the fold line to within 3-cm of each edge.
3. Slip the first few sheets through the cut in the final sheet to make a multi-page book.

## Top-Tab Book

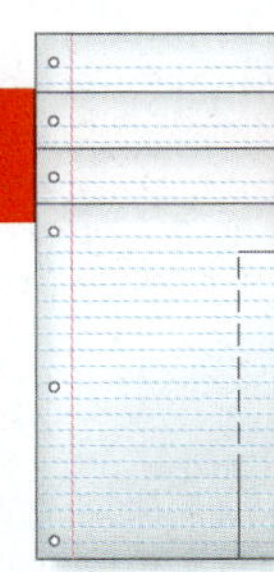

1. Layer multiple sheets of paper so that about 2–3 cm of each can be seen.
2. Make a 2–3-cm horizontal cut through all pages a short distance (3 cm) from the top edge of the top sheet.
3. Make a vertical cut up from the bottom to meet the horizontal cut.
4. Place the sheets on top of an uncut sheet and align the tops and sides of all sheets. Label each tab.

## Accordion Book

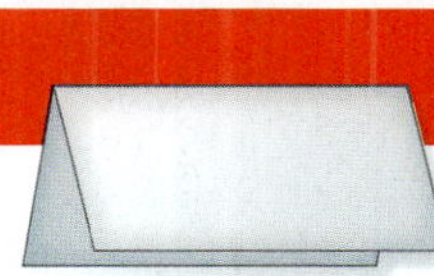

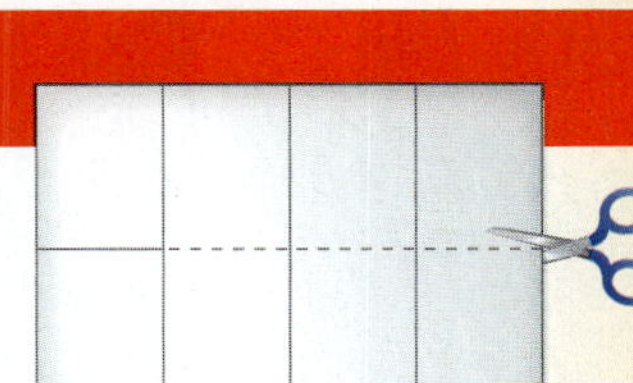

1. Fold a sheet of paper in half. Fold in half and in half again to form eight sections.
2. Cut along the long fold line, stopping before you reach the last two sections.
3. Refold the paper into an accordion book. You may want to glue the double pages together.

CHAPTER 1

# Introduction to Chemistry

**BIGIDEA** Chemistry is a science that is central to our lives.

## SECTIONS

1 A Story of Two Substances

2 Chemistry and Matter

3 Scientific Methods

4 Scientific Research

## LaunchLAB

iLab Station

### Where did the mass go?

When an object burns, the mass of what remains is less than the original object. What happens to the mass of the object?

**FOLDABLES® Study Organizer**

### Scientific Methods

Make a concept-map book. Label the tabs as follows: *Observation, Hypothesis, Experiments,* and *Conclusion.* Use it to summarize what you learn about scientific methods.

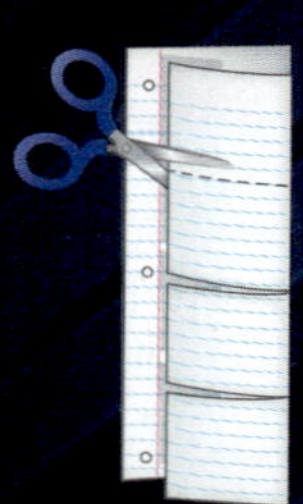

(t)TED KINSMAN/SCIENCE PHOTO LIBRARY/Photo Researchers; (tr)Beateworks Inc./Alamy; (br)Daniel Sambraus/Photo Researchers; (bkgd)ping han/Alamy

Many of the processes that occur around you—such as the freezing of water, the burning of a log, and the rusting of a nail— are examples of chemistry in action.

## SECTION 1

### Essential Questions

- What is a substance?
- How does ozone form and why is it important?
- What are chlorofluorocarbons and how do they get into the atmosphere?

### Review Vocabulary

**matter:** anything that has mass and takes up space

### New Vocabulary

chemistry
substance

# A Story of Two Substances

**MAINIDEA** Chemistry is the study of everything around us.

**CHEM 4 YOU** Have you ever moved a piece of furniture to a new location, only to discover that the new location won't work? Sometimes, moving furniture creates a new problem, such as a door will not open all the way or an electric cord will not reach an outlet. Solving a problem only to find that the solution creates a new problem also occurs in science.

## Why study chemistry?

Take a moment to observe your surroundings and **Figure 1.** Where did all the "stuff" come from? All the stuff in the universe, including everything in the photos, is made from building blocks formed in stars. Scientists call these building blocks and the "stuff" made from these building blocks *matter.*

As you begin your study of **chemistry**—the study of matter and the changes that it undergoes—you are probably asking yourself, "Why is chemistry important to me?" The answer to this question can be illustrated by real-life events that involve two discoveries. One discovery involves something that you probably use every day—refrigeration. If you go to school in an air-conditioned building or if you protect your food from spoilage by using a refrigerator, this discovery is important to you. The other discovery involves energy from the Sun. Because you eat food and spend time outdoors, this discovery is also important to you. These two seemingly unrelated discoveries became intertwined in an unexpected way—as you will soon learn.

**Figure 1** Everything in the universe, including particles in space and things around you, is composed of matter.

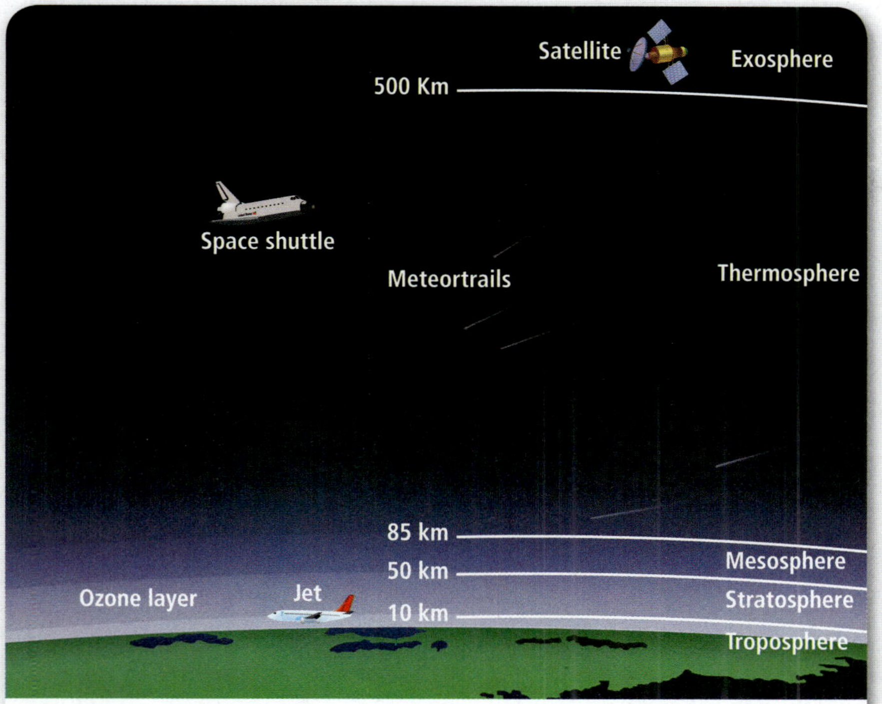


■ **Figure 2** Earth's atmosphere consists of several layers. The protective ozone layer is located in the stratosphere.

## The Ozone Layer

If you have ever had a sunburn, you have experienced the damaging effects of ultraviolet radiation from the Sun. Overexposure to ultraviolet radiation is harmful to both plants and animals. Increased levels of a type of ultraviolet radiation called UVB can cause cataracts and skin cancer in humans, lower crop yields in agriculture, and disrupted food chains in nature.

Living organisms have evolved in the presence of UVB, and cells have some ability to repair themselves when exposed to low levels of UVB. However, some scientists believe that when UVB levels reach a certain point, the cells of living organisms will no longer be able to cope, and many organisms will die.

**Earth's atmosphere** Living organisms on Earth exist because they are protected from high levels of UVB by ozone. Ozone, which is made up of oxygen, is a substance in the atmosphere that absorbs most harmful radiation before it reaches Earth's surface. A **substance,** which is also known as a chemical, is matter that has a definite and uniform composition.

About 90% of Earth's ozone is spread out in a layer that surrounds and protects our planet. As you can see in **Figure 2,** Earth's atmosphere consists of several layers. The lowest layer is called the troposphere and contains the air we breathe. The troposphere is where clouds occur and where airplanes fly. All of Earth's weather occurs in the troposphere. The stratosphere is the layer above the troposphere. It extends from about 10 to 50 kilometers (km) above Earth's surface. The ozone layer that protects Earth is located in the stratosphere.

☑ READING CHECK **Explain** the benefits of ozone in the atmosphere.

### RealWorld CHEMISTRY

### The Ozone Layer

**SUNSCREEN** To offer some protection from harmful UV radiation, sunscreen can be applied to the skin. Sunscreen helps prevent sunburn and skin cancer. Health professionals recommend the use of sunscreen anytime that you are outdoors and exposed to the Sun's ultraviolet radiation.

**VOCABULARY**

**WORD ORIGIN**

**Ozone**
comes from the Greek word *ozōn,* which means *to smell*

■ **Figure 3** Ultraviolet radiation from the Sun causes some oxygen gas ($O_2$) to break into individual particles of oxygen (O). These individual particles combine with oxygen gas ($O_2$) to form ozone ($O_3$).
**Explain** ***why there is a balance between oxygen gas and ozone levels in the stratosphere.***

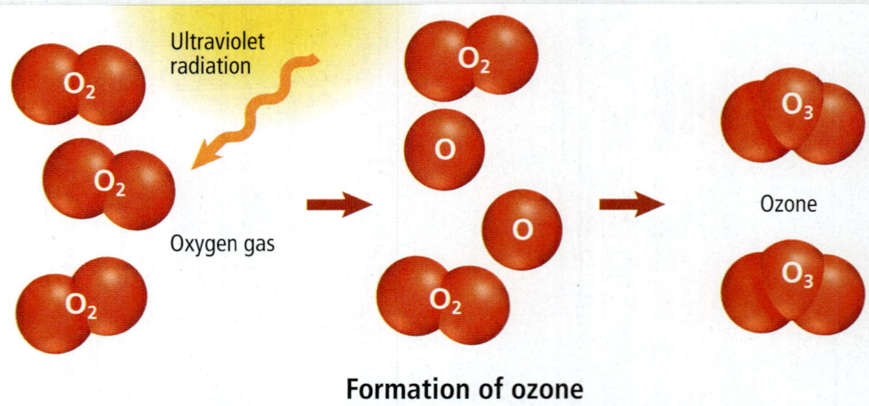


**Ozone formation** How does ozone enter the stratosphere? When oxygen gas ($O_2$) is exposed to ultraviolet radiation in the upper regions of the stratosphere, ozone ($O_3$) is formed. Molecules of oxygen gas are made of two smaller oxygen particles. The energy of the radiation breaks the oxygen gas into individual oxygen particles (O), which then interact with $O_2$ to form $O_3$. **Figure 3** illustrates this process. Ozone can also absorb radiation and break apart to reform oxygen gas. Thus, there tends to be a balance between oxygen gas and ozone levels in the stratosphere.

Ozone was first identified and measured in the late 1800s, so its presence has been studied for a long time. It was of interest to scientists because air currents in the stratosphere move ozone around Earth. Ozone forms over the equator, where the rays of sunlight are the strongest, and then flows toward the poles. Thus, ozone makes a convenient marker to follow the flow of air in the stratosphere.

In the 1920s, British scientist G.M.B. Dobson (1889–1976) began measuring the amount of ozone in the atmosphere. Although ozone is formed in the higher regions of the stratosphere, most of it is stored in the lower stratosphere. Ozone can be measured in the lower stratosphere by instruments on the ground or in balloons, satellites, and rockets. Dobson's measurements helped scientists determine the normal amount of ozone that should be in the stratosphere. Three hundred Dobson units (DU) is considered the normal amount of ozone in the stratosphere. Instruments, like those shown in **Figure 4,** monitor the amount of ozone present in the stratosphere today.

■ **Figure 4** Scientists use a variety of equipment, including this Brewer spectrometer, to take ozone measurements.

David Hay Jones/Science Photo Library/Photo Researchers

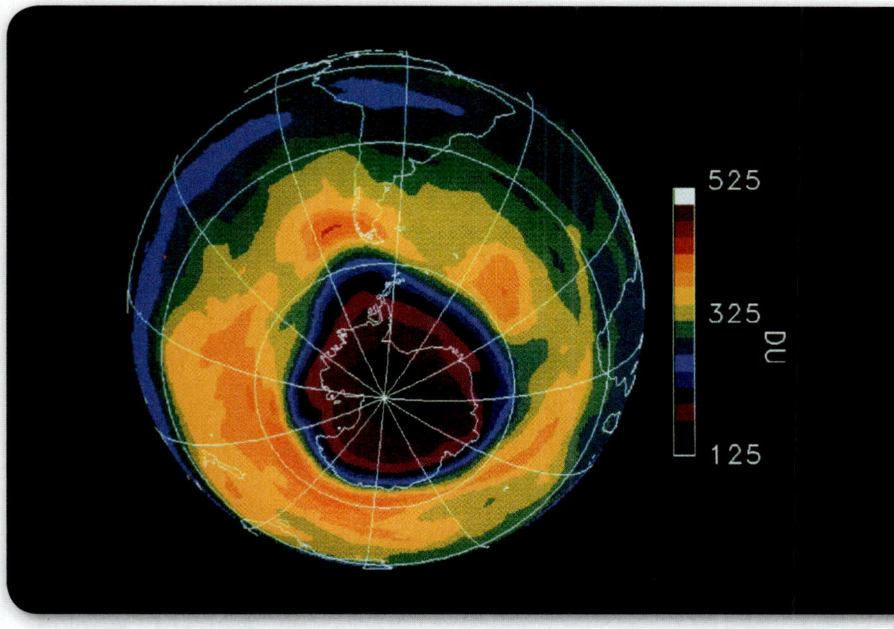


■ **Figure 5** Satellite photos confirmed the British Antarctic Survey team's measurements that the ozone layer was thinning over Antarctica. On this satellite map, the area over Antarctica appears pink, purple, and black. The color-key on the right indicates that the ozone level ranges from 125 to about 200 Dobson Units, which is well below the normal level of 300 Dobson units.

Between 1981 and 1983, a research group from the British Antarctic Survey was monitoring the atmosphere above Antarctica. They measured surprisingly low levels of ozone—readings as low as 160 DU—especially during the Antarctic spring in October. They checked their instruments and repeated their measurements. In October 1985, they reported a confirmed decrease in the amount of ozone in the stratosphere and concluded that the ozone layer was thinning. **Figure 5** shows how the thinning ozone layer looked in October 1990.

Although the thinning of the ozone layer is often called the ozone hole, it is not a hole. The ozone is still present in the atmosphere. However, the protective layer is much thinner than normal. This fact has alarmed scientists, who never expected to find such low levels. Measurements made from balloons, high-altitude planes, and satellites have supported the measurements made from the ground. What could be causing the ozone hole?

## Chlorofluorocarbons

The story of the second substance in this chapter begins in the 1920s. Large-scale production of refrigerators, which at first used toxic gases such as ammonia as coolants, was just beginning. Because ammonia fumes could escape from the refrigerator and harm the members of a household, chemists began to search for safer coolants. Thomas Midgley, Jr. synthesized the first chlorofluorocarbons in 1928. A chlorofluorocarbon (CFC) is a substance that consists of chlorine, fluorine, and carbon. Several different substances are classified as CFCs. They are all made in the laboratory and do not occur naturally. CFCs are nontoxic and stable—they do not readily react with other substances. At the time, they seemed to be ideal coolants for refrigerators. By 1935, the first self-contained home air-conditioning units and eight million new refrigerators in the United States used CFCs as coolants. In addition to their use as refrigerants, CFCs were also used in plastic foams, solvents, and as propellants in spray cans.

☑ READING CHECK **Explain** why scientists thought CFCs were safe for the environment.

### CAREERS IN CHEMISTRY

**Environmental Chemist** An environmental chemist uses tools from chemistry and other sciences to study how chemicals interact with the physical and biological environment. This includes identifying the sources of pollutants such as ozone and their effects on living organisms.

WebQuest

■ **Figure 6** Scientists collected data on the global use of CFCs and the accumulation of CFCs over Antarctica. CFC-11 is one particular type of CFC. In the graph, the concentration of CFC-11 in the atmosphere is shown in parts per trillion (ppt).

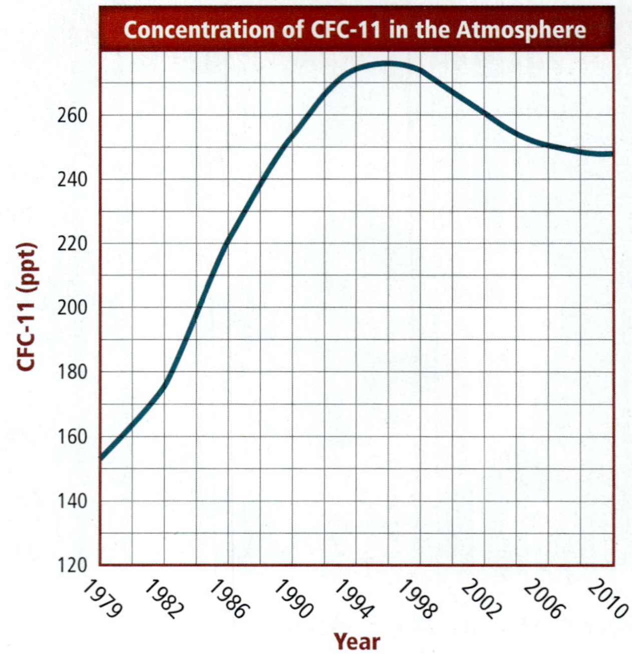


**GRAPH CHECK**

**Describe** the trend in the data from 1979 through 2010.

Scientists first began to detect the presence of CFCs in the atmosphere in the 1970s. They decided to measure the amount of CFCs in the stratosphere and found that quantities in the stratosphere increased year after year. By 1990, the concentration of CFCs had reached an all-time high, as shown in **Figure 6.** However, it was widely thought that CFCs did not pose a threat to the environment because they are so stable, and consequently many scientists were not alarmed.

Scientists had noticed and measured two separate phenomena: the protective ozone layer in the atmosphere was thinning, and increasingly large quantities of CFCs were drifting into the atmosphere. Could there be a connection between the two occurrences? Before you learn the answer to this question, you need to understand some of the basic ideas of chemistry and know how chemists—and most scientists—solve scientific problems.

## SECTION 1 REVIEW

Section Self-Check

### Section Summary

- Chemistry is the study of matter.
- Chemicals are also known as substances.
- Ozone is a substance that forms a protective layer in Earth's atmosphere.
- CFCs are synthetic substances made of chlorine, fluorine, and carbon that were originally thought to be the ideal coolants for refrigeration.

1. **MAINIDEA Explain** why the study of chemistry should be important to everyone.
2. **Define** *substance* and give two examples of things that are substances.
3. **Describe** how the ozone layer forms and why it is important.
4. **Explain** why chlorofluorocarbons were developed and how they were used.
5. **Explain** If cells have the ability to repair themselves after exposure to UVB, why do the increasing levels of UVB in the atmosphere concern scientists?
6. **Explain** why the concentration of CFCs increased in the atmosphere.
7. **Evaluate** why it was important for Dobson's data to be confirmed by satellite photos.

SECTION 2

# Chemistry and Matter

**MAIN IDEA** Branches of chemistry involve the study of different kinds of matter.

**CHEM 4 YOU** Chemistry is sometimes called the central science. Research and technology, such as greener energy and cures for disease, rely on chemistry. Even when you brush your teeth or digest your breakfast, important chemical processes are at work.

### Essential Questions

- How do mass and weight compare and contrast?
- Why are chemists interested in a submicroscopic description of matter?
- What defines the various branches of chemistry?

### Review Vocabulary

**technology:** a practical application of scientific information

### New Vocabulary

mass
weight
model

## Matter and its Characteristics

Matter, the stuff of the universe, has many different forms. Everything around you, like the things in **Figure 7,** is matter. Some matter occurs naturally, such as ozone, and other substances are not natural, such as CFCs, which you read about in Section 1.

You might realize that everyday objects are composed of matter, but how do you define matter? Recall that matter is anything that has mass and takes up space. Also recall that **mass** is a measurement that reflects the amount of matter. You know that your textbook has mass and takes up space, but is air matter? You cannot see it and you cannot always feel it. However, when you inflate a balloon, it expands to make room for the air. The balloon gets heavier. Thus, air must be matter. Is everything matter? The thoughts and ideas that fill your head are not matter; neither are heat, light, radio waves, nor magnetic fields. What else can you name that is not matter?

**Mass and weight** Have you ever used a bathroom scale to measure your weight? **Weight** is a measure not only of the amount of matter but also of the effect of Earth's gravitational pull on that matter. This force is not exactly the same everywhere on Earth and actually becomes less as you move away from Earth's surface at sea level. You might not notice a difference in your weight from one place to another, but subtle differences do exist.

**Figure 7** Everything in this photo is matter and has mass and weight.
**Compare and contrast** *mass and weight.*

Fuse/Getty Images

■ **Figure 8** Scientists use models to visualize complex ideas, such as the materials and structure used to build office buildings. They also use models to test a concept, such as a new airplane design, before it is mass produced.

**Infer** *why chemists use models to study atoms.*

It might seem more convenient for scientists to simply use weight instead of mass. Why is it so important to think of matter in terms of mass? Scientists need to be able to compare the measurements that they make in different parts of the world. They could identify the gravitational force every time they weigh something, but that would not be practical or convenient. They use mass as a way to measure matter independently of gravitational force.

**Structure and observable characteristics** What can you observe about the outside of your school building? You know that there is more to the building than what you can observe from the outside. Among other things, there are beams inside the walls that give the building structure, stability, and function. Consider another example. When you bend your arm at the elbow, you observe that your arm moves, but what you cannot see is that muscles under the skin contract and relax to move your arm.

**FOLDABLES®**
Incorporate information from this section into your Foldable.

Much of matter and its behavior is macroscopic; that is, you do not need a microscope to observe it. You will learn in Chapter 3 that the tremendous variety of stuff around you can be broken down into more than a hundred types of matter called elements, and that elements are made up of particles called atoms. Atoms are so tiny that they cannot be seen even with optical microscopes. Thus, atoms are submicroscopic. They are so small that over a trillion atoms could fit onto the period at the end of this sentence. The structure, composition, and behavior of all matter can be explained on a submicroscopic level—or the atomic level. All that we observe about matter depends on atoms and the changes they undergo.

**VOCABULARY**
**SCIENCE USAGE V. COMMON USAGE**

**Weight**

***Science usage:*** the measure of the amount of matter in and the gravitational force exerted on an object

*The weight of an object is the product of its mass and the local acceleration of gravity.*

***Common usage:*** the relative heaviness of an object

*The puppy grew so quickly it doubled its weight in a matter of weeks.*

Chemistry seeks to explain the submicroscopic events that lead to macroscopic observations. One way this can be done is by making a model. A **model** is a visual, verbal, or mathematical explanation of experimental data. Scientists use many types of models to represent things that are hard to visualize, such as the structure and materials used in the construction of a building and the computer model of the airplane shown in **Figure 8.** Chemists also use several different types of models to represent matter, as you will soon learn.

☑ READING CHECK **Identify** two additional types of models that are used by scientists.

(l)TANG CHHIN SOTHY/AFP/Getty Images, (r)NASA Ames Research Center/Photo Researchers

Explore the **branches of chemistry with an interactive table.** Concepts In Motion

**Table 1** Some Branches of Chemistry

| Branch | Area of Emphasis | Examples of Emphasis |
|---|---|---|
| **Organic chemistry** | most carbon-containing chemicals | pharmaceuticals, plastics |
| **Inorganic chemistry** | in general, matter that does not contain carbon | minerals, metals and nonmetals, semiconductors |
| **Physical chemistry** | the behavior and changes of matter and the related energy changes | reaction rates, reaction mechanisms |
| **Analytical chemistry** | components and composition of substances | food nutrients, quality control |
| **Biochemistry** | matter and processes of living organisms | metabolism, fermentation |
| **Environmental chemistry** | matter and the environment | pollution, biochemical cycles |
| **Industrial chemistry** | chemical processes in industry | paints, coatings |
| **Polymer chemistry** | polymers and plastics | textiles, coatings, plastics |
| **Theoretical chemistry** | chemical interactions | many areas of emphasis |
| **Thermochemistry** | heat involved in chemical processes | heat of reaction |

## Chemistry: The Central Science

Recall from Section 1 that chemistry is the study of matter and the changes that it undergoes. A basic understanding of chemistry is central to all sciences—biology, physics, Earth science, ecology, and others. Because there are so many types of matter, there are many areas of study in the field of chemistry. Chemistry is traditionally broken down into branches that focus on specific areas, such as those listed in **Table 1.** Although chemistry is divided into specific areas of study, many of the areas overlap. For example, as you can see from **Table 1,** an organic chemist might study plastics, but an industrial chemist or a polymer chemist could also focus on plastics.

Get help with **mass and weight relationships.**

Personal Tutor

## SECTION 2 REVIEW

Section Self-Check

### Section Summary

- Models are tools that scientists, including chemists, use.
- Macroscopic observations of matter reflect the actions of atoms on a submicroscopic scale.
- There are several branches of chemistry, including organic chemistry, inorganic chemistry, physical chemistry, analytical chemistry, and biochemistry

8. **MAIN**IDEA **Explain** why there are different branches of chemistry.
9. **Explain** why scientists use mass instead of weight for their measurements.
10. **Summarize** why it is important for chemists to study changes in the world at a submicroscopic level.
11. **Infer** why chemists use models to study submicroscopic matter.
12. **Identify** three models that scientists use, and explain why each model is useful.
13. **Evaluate** How would your mass and weight differ on the Moon? The gravitational force of the Moon is one-sixth the gravitational force on Earth.
14. **Evaluate** If you put a scale in an elevator and weigh yourself as you ascend and then descend, does the scale have the same reading in both instances? Explain your answer.

SECTION 3

# Scientific Methods

**MAINIDEA** Scientists use scientific methods to systematically pose and test solutions to questions and assess the results of the tests.

### Essential Questions

- What are the common steps of scientific methods?
- What are the similarities and differences between qualitative data and quantitative data?
- In an experiment, which variable is the independent variable, which is the dependent variable, and which are controls?
- What is the difference between a theory and a scientific law?

### Review Vocabulary

**systematic approach:** an organized method of solving a problem

### New Vocabulary

scientific method
qualitative data
quantitative data
hypothesis
experiment
independent variable
dependent variable
control
conclusion
theory
scientific law

**CHEM 4 YOU** **When packing for a long trip, how do you start? Do you throw all of your clothes into a suitcase, or do you plan what you are going to wear? Usually, it is most effective to make a plan. Similarly, scientists develop and follow a plan that helps them investigate the world.**

## A Systematic Approach

You might have worked with a group on an experiment in the laboratory in a previous science course. If so, you know that each person in the group probably has a different idea about how to do the lab. Having many different ideas about how to do the lab is one of the benefits of many people working together. However, communicating ideas effectively to one another and combining individual contributions to form a solution can be difficult in group work.

Scientists approach their work in a similar way. Each scientist tries to understand his or her world based on a personal point of view and individual creativity. Often, the work of many scientists is combined in order to gain new insight. It is helpful if all scientists use common procedures as they conduct their experiments.

A **scientific method** is a systematic approach used in scientific study, whether it is chemistry, biology, physics, or another science. It is an organized process used by scientists to do research, and it provides a method for scientists to verify the work of others. An overview of the typical steps of a scientific method is shown in **Figure 9.** The steps are not meant to be used as a checklist, or to be done in the same order each time. Therefore, scientists must describe their methods when they report their results. If other scientists cannot confirm the results after repeating the method, then doubt arises over the validity of the results.

**Figure 9** The steps in a scientific method are repeated until a hypothesis is supported or discarded.

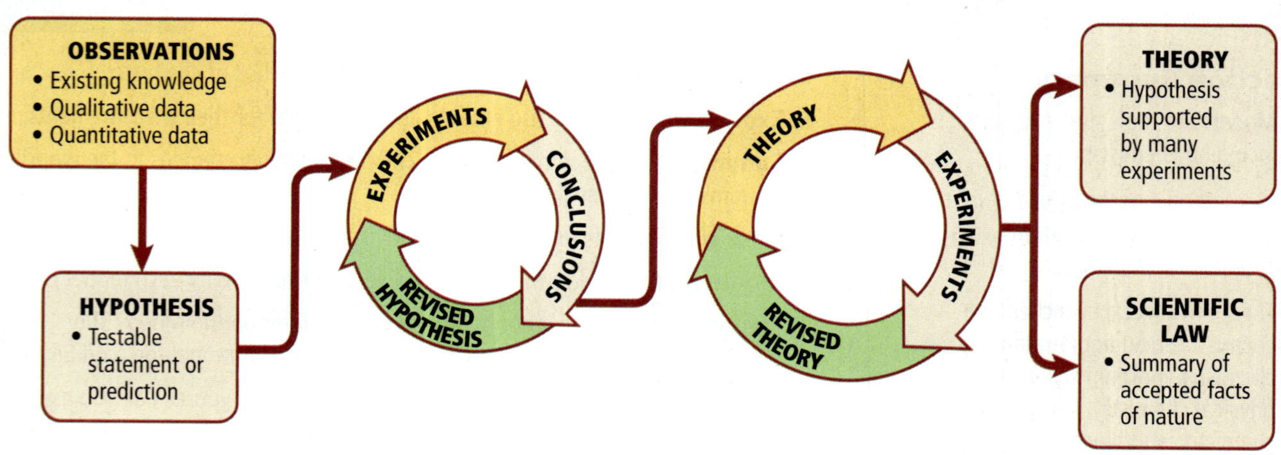

# MiniLAB

iLab Station

## Develop Observation Skills

**Why are observation skills important in chemistry?** Observations are often used to make inferences. An inference is an explanation or interpretation of observations.

### Procedure

1. Read and complete the lab safety form.
2. Add **water** to a **petri dish** to a height of 0.5 cm. Use a **graduated cylinder** to measure 1 mL of **vegetable oil,** then add it to the petri dish.
3. Dip the end of a **toothpick** into **liquid dishwashing detergent.**
4. Touch the tip of the toothpick to the water at the center of the petri dish. Record your detailed observations.
5. Add **whole milk** to a **second petri dish** to a height of 0.5 cm.
6. Place one drop each of **four different food colorings** in four different locations on the surface of the milk. Do not put a drop of food coloring in the center.
7. Repeat Steps 3 and 4.

### Analysis

1. **Describe** what you observed in Step 4.
2. **Describe** what you observed in Step 7.
3. **Infer** Oil, the fat in milk, and grease belong to a class of substances called lipids. What can you infer about the addition of detergent to dishwater?
4. **Explain** why observations skills were important in this chemistry lab.

**Observation** You make observations throughout your day in order to make decisions. Scientific study usually begins with simple observation. An observation is the act of gathering information. Often, the types of observations scientists make first are **qualitative data**—information that describes color, odor, shape, or some other physical characteristic. In general, anything that relates to the five senses is qualitative: how something looks, feels, sounds, tastes, or smells.

Chemists frequently gather another type of data. For example, they can measure temperature, pressure, volume, the quantity of a chemical formed, or how much of a chemical is used up in a reaction. This numerical information is called **quantitative data.** It tells how much, how little, how big, how tall, or how fast. What kind of qualitative and quantitative data can you gather from **Figure 10?**

**Hypothesis** Recall the stories of the two substances that you read about in Section 1. Even before quantitative data showed that ozone levels were decreasing in the stratosphere, scientists observed CFCs there. Chemists Mario Molina and F. Sherwood Rowland were curious about how long CFCs could exist in the atmosphere.

Molina and Rowland examined the interactions that can occur among various chemicals in the troposphere. They determined that CFCs were stable there for long periods of time, but they also knew that CFCs drift upward into the stratosphere. They formed a hypothesis that CFCs break down in the stratosphere due to interactions with ultraviolet light from the Sun. In addition, the calculations they made led them to hypothesize that chlorine produced by this interaction would break down ozone.

A **hypothesis** is a tentative, testable statement or prediction about what has been observed. Molina and Rowland's hypothesis stated what they believed to be happening, even though there was no formal evidence at that point to support the statement.

**READING CHECK** **Infer** why a hypothesis is tentative.

■ **Figure 10** Quantitative data are numerical information. Qualitative data are observations made by using the human senses.

**Identify** *the quantitative and qualitative data in the photo.*

Art Vandalay/Digital Vision/Getty Images

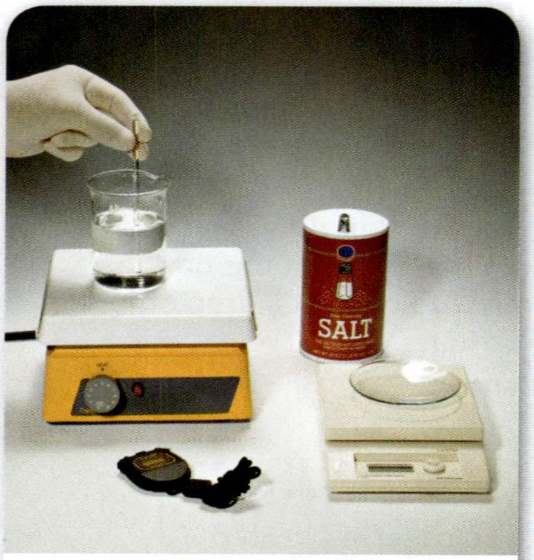


■ **Figure 11** These materials can be used to determine the effect of temperature on the rate at which table salt dissolves.

**Experiments** A hypothesis is meaningless unless there are data to support it. Thus, forming a hypothesis helps the scientist focus on the next step in a scientific method—the experiment. An **experiment** is a set of controlled observations that test the hypothesis. The scientist must carefully plan and set up one or more laboratory experiments in order to change and test one variable at a time. A variable is a quantity or condition that can have more than one value.

Suppose your chemistry teacher asks your class to use the materials shown in **Figure 11** to design an experiment to test the hypothesis that table salt dissolves faster in hot water than in water at room temperature (20°C). Because temperature is the variable that you plan to change, it is the **independent variable.** Your group determines that a given quantity of salt completely dissolves within 1 min at 40°C, but that the same quantity of salt dissolves after 3 min at 20°C. Thus, temperature affects the rate at which the salt dissolves. This rate is called the **dependent variable** because its value changes in response to a change in the independent variable. Although your group can determine the way in which the independent variable changes, it has no control over the way the dependent variable changes.

☑ READING CHECK **Explain** the difference between a dependent and an independent variable.

**Other factors** What other factors could you vary in your experiment? Would the amount of salt you try to dissolve make a difference? The amount of water you use? Would stirring the mixture affect your results? The answer to all of these questions might be yes. You must plan your experiment so that these variables are the same at each temperature, or you will not be able to tell clearly what caused your results. In a well-planned experiment, the independent variable should be the only condition that affects the experiment's outcome. A constant is a factor that is not allowed to change during the experiment. The amount of salt, water, and stirring must be constant at each temperature.

In many experiments, it is valuable to have a **control,** that is, a standard for comparison. In the above experiment, the room-temperature water is the control. **Figure 12** shows a different type of control. A chemical indicator has been added to each of three test tubes. An acidic solution is in the test tube on the left, and the indicator turns red. The test tube in the middle contains water, and the indicator is yellow. The test tube on the right contains a basic solution, and the indicator turns blue.

■ **Figure 12** Because the acidity of the solutions in these test tubes is known, these solutions can be used as controls in an experiment.

**Infer** ***If the same chemical indicator were added to a solution of unknown acidity, how could you determine if it was acidic, neutral, or basic?***

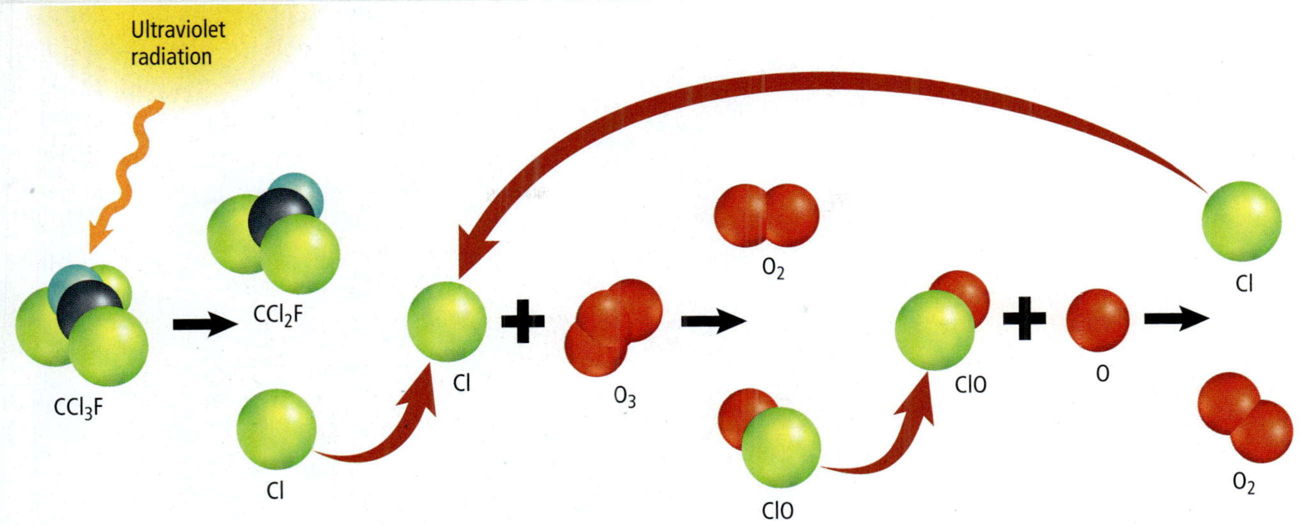


Molina and Rowland's model predicted that ultraviolet radiation causes chlorine (Cl) to split off from a CFC ($CCl_3F$).

The chlorine then destroys the ozone by combining with it to form oxygen gas ($O_2$) and chlorine monoxide (ClO).

A free oxygen (O) combines with the chlorine monoxide (ClO) to form oxygen gas ($O_2$) and a chlorine (Cl). Then, the chlorine is free to combine with another ozone. The process begins again.

**Figure 13** Molina and Rowland's model showed how CFCs could destroy ozone.

**Controlling variables** The interactions described between CFCs and ozone in Molina and Rowland's hypothesis take place high overhead. Many variables are involved. For example, there are several gases present in the stratosphere. Thus, it would be difficult to determine which gases, or if all gases, are causing decreasing ozone levels. Winds, variations in ultraviolet light, and other factors could change the outcome of any experiment on any given day, making comparisons difficult. Sometimes, it is easier to simulate conditions in a laboratory, where the variables can be more easily controlled.

View an **animation about ozone depletion.**

Concepts In Motion

**Conclusion** An experiment might generate a large amount of data. Scientists take the data, analyze it, and check it against the hypothesis to form a conclusion. A **conclusion** is a judgment based on the information obtained. A hypothesis can never be proven. Therefore, when the data support a hypothesis, this only indicates that the hypothesis might be true. If further evidence does not support it, then the hypothesis must be discarded or modified. The majority of hypotheses are not supported, but the data might still yield new and useful information.

Molina and Rowland formed a hypothesis about the stability of CFCs in the stratosphere. The data that they gathered supported their hypothesis. They developed a model in which the chlorine formed by the breakdown of CFCs would react over and over again with ozone.

A model can be tested and used to make predictions. Molina and Rowland's model predicted the formation of chlorine and the depletion of ozone, as shown in **Figure 13.** Another research group found evidence of interactions between ozone and chlorine when taking measurements in the stratosphere, but they did not know the source of the chlorine. Molina and Rowland's model predicted a source of the chlorine. They came to the conclusion that ozone in the stratosphere could be destroyed by CFCs, and they had enough support for their hypothesis to publish their discovery. They won the Nobel Prize in 1995.

FOLDABLES®
Incorporate information from this section into your Foldable.

**Figure 14** It does not matter how many times skydivers leap from a plane; Newton's law of universal gravitation applies every time.

## Theory and Scientific Law

A **theory** is an explanation of a natural phenomenon based on many observations and investigations over time. You might have heard of Einstein's theory of relativity or the atomic theory. A theory states a broad principle of nature that has been supported over time. All theories are still subject to new experimental data and can be modified. Also, theories often lead to new conclusions. A theory is considered valid if it can be used to make predictions that are proven true.

Sometimes, many scientists come to the same conclusion about certain relationships in nature and they find no exceptions to these relationships. For example, you know that no matter how many times skydivers, like those shown in **Figure 14,** leap from a plane, they always return to Earth's surface. Sir Isaac Newton was so certain that an attractive force exists between all objects that he proposed his law of universal gravitation. Newton's law is a **scientific law**—a relationship in nature that is supported by many experiments. It is up to scientists to develop further hypotheses and experiments to explain why these relationships exist.

## SECTION 3 REVIEW

Section Self-Check

### Section Summary

- Scientific methods are systematic approaches to problem solving.
- Qualitative data describe an observation; quantitative data use numbers.
- Independent variables are changed in an experiment. Dependent variables change in response to the independent variable.
- A theory is a hypothesis that is supported by many experiments.

15. MAINIDEA **Explain** why scientists do not use a standard set of steps for every investigation they conduct.
16. **Differentiate** Give an example of quantitative and qualitative data.
17. **Evaluate** You are asked to study the effect of temperature on the volume of a balloon. The balloon's size increases as it is warmed. What is the independent variable? The dependent variable? What factor is held constant? How would you construct a control?
18. **Distinguish** Jacques Charles described the direct relationship between temperature and volume of all gases at constant pressure. Should this be called Charles's law or Charles's theory? Explain.
19. **Explain** Good scientific models can be tested and used to make predictions. What did Molina and Rowland's model of the interactions of CFCs and ozone in the atmosphere predict would happen to the amount of ozone in the stratosphere as the level of CFCs increased?

moodboard/Getty Images

SECTION 4

# Scientific Research

**MAIN**IDEA Some scientific investigations result in the development of technology that can improve our lives and the world around us.

**Essential Questions**

- How do pure research, applied research, and technology compare and contrast?
- What are some of the important rules for laboratory safety?

**Review Vocabulary**

**synthetic:** something that is human-made and does not necessarily occur in nature

**New Vocabulary**

pure research
applied research

**CHEM 4 YOU** Much of the information that scientists obtain through basic research is used to meet a specific need. For example, X-rays were discovered by scientists conducting basic research on electrical discharge through gases. Later, it was discovered that X-rays could be used to diagnose medical problems.

## Types of Scientific Investigations

Every day in the media—through TV, newspapers, magazines, or the Internet—the public is bombarded with the results of scientific investigations. Many deal with the environment, medicine, or health. As a consumer, you are asked to evaluate the results of scientific research and development. How do scientists use qualitative and quantitative data to solve different types of scientific problems?

Scientists conduct **pure research** to gain knowledge for the sake of knowledge itself. Molina and Rowland were motivated by curiosity and, thus, conducted research on CFCs and their interactions with ozone as pure research. No environmental evidence at the time indicated that there was a correlation to their model in the stratosphere. Their research showed only that CFCs could speed the breakdown of ozone in a laboratory setting.

By the time the ozone hole was reported in 1985, scientists had made measurements of CFC levels in the stratosphere that supported the hypothesis that CFCs could be responsible for the depletion of ozone. The early pure research done only for the sake of knowledge became applied research. **Applied research** is research undertaken to solve a specific problem. Scientists continue to monitor the amount of CFCs in the atmosphere and the annual changes in the amount of ozone in the stratosphere, as shown in **Figure 15.** Applied research is also being conducted to find replacement chemicals for the CFCs that are now banned.

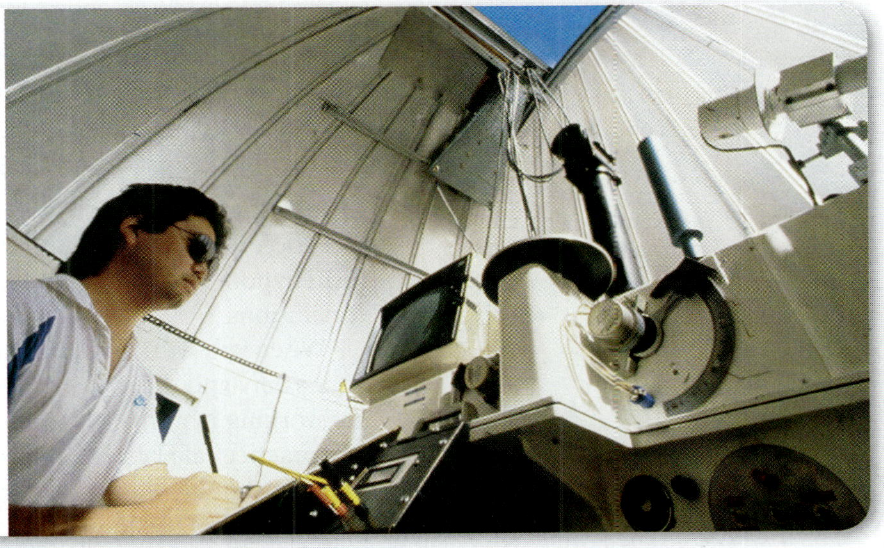

**Figure 15** This UV-visible spectrometer (UV-Vis) is used to measure ozone and other stratospheric gases during the dark winter months in Antarctica.

Hank Morgan/Photo Researchers

**Figure 16** After its discovery, nylon was used mainly for war materials and was not available for home use until after World War II. Today it is used in a variety of products.

Strands of nylon can be pulled from the top layer of solution.

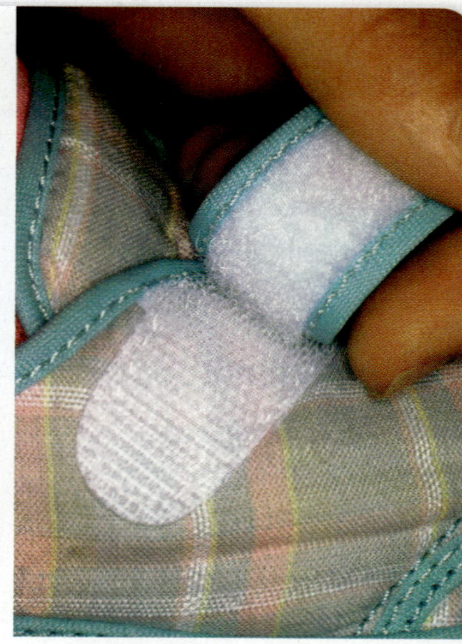

Nylon fibers are used to make hook-and-loop fastener tape.

**Chance discoveries** Often, a scientist conducts experiments and reaches a conclusion that is far different from what was predicted. Some truly wonderful discoveries in science have been made unexpectedly. You might be familiar with the two examples described below.

**Connection to Biology** Alexander Fleming is famous for making several accidental discoveries. In one accidential discovery, Fleming found that one of his plates of *Staphylococcus* bacteria had been contaminated by a greenish mold, later identified as *Penicillium.* He observed it carefully and saw a clear area around the mold where the bacteria had died. In this case, a chemical in the mold—penicillin—was responsible for killing the bacteria.

The discovery of nylon is another example of an accidental discovery. In 1930, Julian Hill, an employee of E.I. DuPont de Nemours and Company, dipped a hot glass rod in a mixture of solutions and unexpectedly pulled out long fibers similar to those shown in **Figure 16.** Hill and his colleagues pursued the development of these fibers as a synthetic silk that could withstand high temperatures. They eventually developed nylon in 1934. During World War II, nylon was used as a replacement for silk in parachutes. Today, nylon is used extensively in textiles and some kinds of plastics. It is also used to make hook-and-loop tape, as shown in **Figure 16.**

## Students in the Laboratory

Watch a video about **criminal science investigation.**

Video

In your study of chemistry, you will learn many facts about matter. You will also do investigations and experiments in which you will be able to form hypotheses, test hypotheses, gather data, analyze data, and draw conclusions.

When you work in the chemistry laboratory, you are responsible for your safety and the safety of people working nearby. Often, many people are working in a small space during a lab, so it is important that everyone practice safe laboratory procedures. **Table 2** lists some safety rules that you should follow each time you enter the lab. Chemists and all other scientists use these safety rules as well.

## Table 2 Safety in the Laboratory

1. Study your lab assignment before you come to the lab. If you have any questions, ask your teacher for help.
2. Do not perform experiments without your teacher's permission. Never work alone in the laboratory. Know how to contact help, if necessary.
3. Use the table on the inside front cover of this textbook to understand the safety symbols. Read and adhere to all **WARNING** statements.
4. Wear safety goggles and a laboratory apron whenever you are in the lab. Wear gloves whenever you use chemicals that cause irritations or can be absorbed through the skin. If you have long hair, you must tie it back.
5. Do not wear contact lenses in the lab, even under goggles. Lenses can absorb vapors and are difficult to remove during an emergency.
6. Avoid wearing loose, draping clothing and dangling jewelry. Wear only closed-toe shoes in the lab.
7. Keep food, beverages, and chewing gum out of the lab. Never eat in the lab.
8. Know where to find and how to use the fire extinguisher, safety shower, fire blanket, first-aid kit, and gas and electrical power shutoffs.
9. Immediately clean up spills on the floor and keep all walkways clear of objects, such as backpacks, to prevent accidental falls or tripping. Report any accident, injury, incorrect procedure, or damaged equipment to your teacher.
10. If chemicals come in contact with your eyes or skin, flush the area immediately with large quantities of water. Immediately inform your teacher of the nature of the spill.
11. Handle all chemicals carefully. Check the labels of all bottles before removing the contents. Read the label three times: before you pick up the container, when the container is in your hand, and when you put the bottle back.
12. Do not take reagent bottles to your work area unless instructed to do so. Use test tubes, paper, or beakers to obtain your chemicals. Take only small amounts. It is easier to get more than to dispose of excess.
13. Do not return unused chemicals to the stock bottle.
14. Do not insert droppers into reagent bottles. Pour a small amount of the chemical into a beaker.
15. Never taste any chemicals. Never draw any chemicals into a pipette with your mouth.
16. Keep combustible materials away from open flames.
17. Handle toxic and combustible gases only under the direction of your teacher. Use the fume hood when such materials are present.
18. When heating a substance in a test tube, be careful not to point the mouth of the test tube at another person or yourself. Never look down into the mouth of a test tube.
19. Do not heat graduated cylinders, burettes, or pipettes with a laboratory burner.
20. Use caution and proper equipment when handling a hot apparatus or glassware. Hot glass looks the same as cool glass.
21. Dispose of broken glass, unused chemicals, and products of reactions only as directed by your teacher.
22. Know the correct procedure for preparing acid solutions. Always add the acid to the water slowly.
23. Keep the balance area clean. Never place chemicals directly on the pan of a balance.
24. After completing an experiment, clean and put away your equipment. Clean your work area. Make sure the gas and water are turned off. Wash your hands with soap and water before you leave the lab.

Matt Meadows

■ **Figure 17** This graph shows the concentration of two common CFCs in the atmosphere over Antarctica and the global consumption of CFCs from 1980 to 2000. While CFC consumption began to decrease drastically a few years after the signing of the Montreal Protocol, the concentration of CFCs above Antarctica continued to increase for awhile before slowly leveling out.

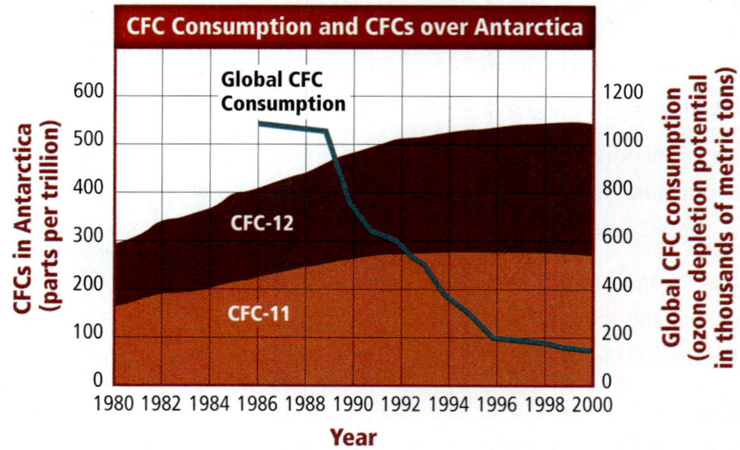


☑ GRAPH CHECK **Identify** when CFCs in Antarctica began to level off after national leaders signed the Montreal Protocol.

## The Story Continues

Now, back to the two substances that you have been reading about. A lot has happened since the 1970s, when Molina and Rowland hypothesized that CFCs broke down stratospheric ozone. The National Oceanic and Atmospheric Administration (NOAA) and many other groups are actively collecting historic and current data on CFCs in the atmosphere and ozone concentrations in the stratosphere. Through applied research, scientists determined that not only do CFCs react with ozone, but a few other substances react as well. Carbon tetrachloride and methyl chloroform are two additional substances that harm the ozone. Substances that contain bromine can also damage the ozone.

FOLDABLES® Incorporate information from this section into your Foldable.

**The Montreal Protocol** Because ozone depletion is an international concern, nations banded together to try to solve this problem. In 1987, leaders from many nations met in Montreal, Canada, and signed the Montreal Protocol. By signing this agreement, nations agreed to phase out the use of these compounds and place restrictions on how they should be used in the future. As you can see from **Figure 17,** the global use of CFCs began to decline after the Montreal Protocol was signed. However, the graph shows that the amount of CFCs measured over Antarctica did not decline immediately.

**The ozone hole today** Scientists have also learned that the ozone hole forms each year over Antarctica during the spring. Stratospheric ice clouds form over Antarctica when temperatures there drop below −78°C. These clouds produce changes that promote the production of chemically active chlorine and bromine. When temperatures begin to warm in the spring, this chemically active chlorine and bromine react with ozone, causing ozone depletion. This ozone depletion causes the ozone hole to form over Antarctica. Some ozone depletion also occurs over the Arctic, but temperatures do not remain low for as long, which means less ozone depletion in the Arctic. Upon further research, scientists have also determined that ozone thinning has occurred above every continent.

☑ READING CHECK **Explain** what triggers the formation of the ozone hole over Antarctica.

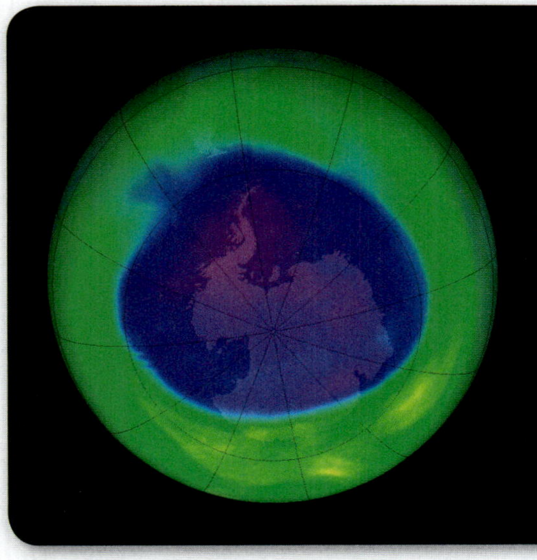

■ **Figure 18** The ozone hole over Antarctica reached its maximum level of thinning in September 2005. The color-key below shows what the colors represent in this colorized satellite image.

**Compare** ***How do these ozone levels compare with what is considered normal?***

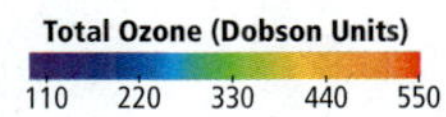


**Figure 18** shows the ozone hole over Antarctica in September 2005. The ozone thinning over Antarctica reached its maximum for the year during this month. If you compare the color-coded key to the satellite image, you can see that the ozone level is between 110 and 200 DU. Notice the area surrounding the ozone hole. Much of this area has ozone levels around 300 DU, which is considered normal.

Scientists are not sure when the ozone layer will begin to recover. Originally, scientists predicted that it would begin to recover in 2050. However, new computer models predict that it will not begin to recover until 2068. The exact date of its recovery is not as important as the fact that it will recover given time.

**VOCABULARY**

**WORD ORIGIN**

**Recover**
to bring back to normal
*It takes several days to recover from the flu.*

## Data Analysis LAB

Based on Real Data*

### Interpret Graphs

**How do ozone levels vary throughout the year above Antarctica?** Many agencies monitor the concentration of ozone in the stratosphere over Antarctica.

#### Think Critically

1. **Describe** the trend in the data for 1979–2008.
2. **Evaluate** how the 2009 data compare with the data from 1979–2008.
3. **Identify** the month during which the ozone levels were the lowest in 1979–2008. In 2009?
4. **Assess** Do these data points back up what you learned in this chapter about ozone depletion? Explain your answer.

#### Data and Observation

This graph displays data from NASA collected over Antarctica. Data values below 220 Dobson Units are defined as the region of the ozone hole area.

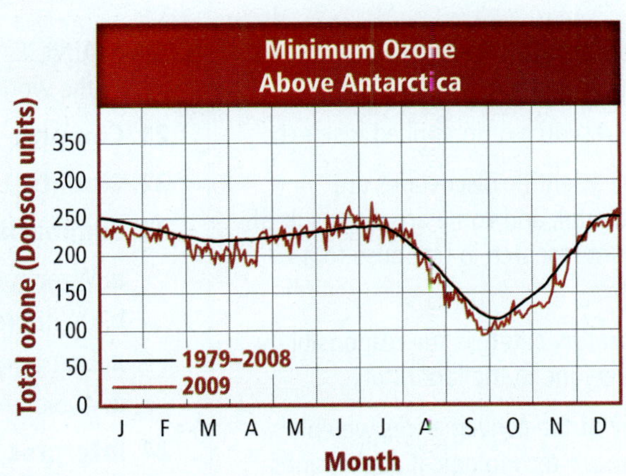


*Data obtained from Ozone Hole Watch. 2010. *National Aeronautics and Space Administration.*

NASA/Science Photo Library/Photo Researchers

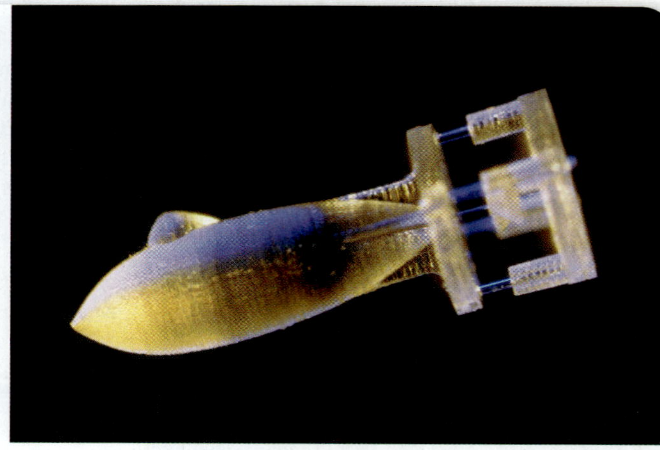

■ **Figure 19** This car, which is powered by compressed air, and this tiny submarine, which is only 4 mm long, are examples of technologies that are made possible by the study of matter.

## The Benefits of Chemistry

Chemists are an important part of the team of scientists that solve many of the problems or issues that we face today. Chemists are not only involved in resolving the ozone depletion problem. They are also involved in finding cures or vaccines for diseases, such as AIDS and influenza. Almost every situation that you can imagine involves a chemist, because everything in the universe is made of matter.

**Figure 19** shows some of the advances in technology that are possible because of the study of matter. The car on the left is powered by compressed air. When the compressed air is allowed to expand, it pushes the pistons that move the car. Because the car is powered by compressed air, no pollutants are released. The photo on the right shows a tiny submarine that is made by computer-aided lasers. This submarine, which is only 4 mm long, might be used for detecting and repairing defects in the human body.



## SECTION 4 REVIEW

Section Self-Check

### Section Summary

- Scientific methods can be used in pure research or in applied research.
- Some scientific discoveries are accidental, and some are the result of diligent research in response to a need.
- Laboratory safety is the responsibility of everyone in the laboratory.
- Many of the conveniences we enjoy today are technological applications of chemistry.

20. MAINIDEA **Name** three technological products that have improved our lives or the world around us.

21. **Compare and contrast** pure research and applied research.

22. **Classify** Is technology a product of pure research or applied research? Explain.

23. **Summarize** the reason behind each of the following.
    - **a.** Wear goggles and an apron in the lab even if you are only an observer.
    - **b.** Do not return unused chemicals to the stock bottle.
    - **c.** Do not wear contact lenses in the laboratory.
    - **d.** Avoid wearing loose, draping clothing and dangling jewelry.

24. **Interpret Scientific Diagrams** What safety precautions should you take when the following safety symbols are listed?

# CHEMISTRY & careers

## Career: Art Restorer

### Painting Restoration

Art does not last forever. It is damaged by events such as people sneezing on it, touching it, or by smoke during a fire. The repair of damage to artwork is the job of art restorers. Art repair is not always an easy task, because the materials used to correct the damage can also damage the artwork.

**Help from above** Oxygen makes up 21% of Earth's atmosphere. Near the ground, almost all the oxygen exists as oxygen gas ($O_2$). However, high in the atmosphere, ultraviolet light from the Sun splits oxygen gas into atomic oxygen (O). While oxygen gas is chemically reactive, atomic oxygen is even more reactive. It can damage spacecraft in orbit, which is why NASA actively studies the reactions between atomic oxygen and other substances.

**Oxygen and art** Atomic oxygen is especially reactive with the element carbon—the main substance found in soot from a fire. When NASA scientists treated the soot-damaged painting shown in **Figure 1** with atomic oxygen, the carbon in the soot reacted with oxygen to produce gases that floated away.

**Figure 1** The photo on the left shows soot damage to an oil painting. The photo on the right shows the painting after oxygen treatment. Removal of a small amount of glossy binder was the only damage to the painting during the treatment.

NASA

**Figure 2** The lipstick stain could not be removed using conventional techniques. However, atomic oxygen removed the stain without damage to the painting.

**On the surface** Because atomic oxygen acts only on what it touches, paint layers below the soot or other surface impurities are unaffected. If you compare the image on the left with the image on the right in **Figure 1,** you will notice that the soot was removed, but the painting was not harmed. This is in contrast to more conventional treatments, in which organic solvents are used to remove the soot. These solvents often react with the paint as well as the soot.

**The kiss** Another successful restoration was the Andy Warhol painting called *The Bathtub.* It was damaged when a lipstick-wearing viewer kissed the canvas. Most conventional restoration techniques would have driven the lipstick deeper into the painting, leaving a permanent pink stain. When atomic oxygen was applied to the stain using the equipment shown in **Figure 2,** the pink color vanished.

### WRITING IN ▸ Chemistry

**Research** a recent art restoration project. Prepare a news article explaining why the object needed restoration, the challenges presented, and the chemistry used to complete the project.

WebQuest

# ChemLAB

iLab Station

## Forensics: Identify the Water Source

**Background:** **The contents of tap water vary from community to community. Water is classified as hard or soft based on the amount of calcium or magnesium in the water, measured in milligrams per liter (mg/L). Imagine a forensics lab has two samples of water. One sample comes from Community A, which has soft water. The other sample comes from Community B, which has hard water.**

**Question:** *From which community did each water sample originate?*

### Materials

- test tubes with stoppers (3)
- test-tube rack
- grease pencil
- graduated cylinder (25-mL)
- distilled water
- dropper
- beaker (250-mL)
- Water sample 1
- Water sample 2
- dish detergent
- metric ruler

### Safety Precautions

### Procedure

1. Read and complete the lab safety form.
2. Prepare a data table like the one shown. Then, use a grease pencil to label three large test tubes: *D* (for distilled water), *1* (for Sample 1), and *2* (for Sample 2).
3. Use a graduated cylinder to measure out 20 mL of distilled water. Pour the water into Test Tube D.
4. Place Test Tubes 1 and 2 next to Test Tube D and make a mark on each test tube that corresponds to the height of the water in Test Tube D.
5. Obtain 50 mL of Water Sample 1 in a beaker from your teacher. Slowly pour the water sample into Test Tube 1 until you reach the marked height.
6. Obtain 50 mL of Water Sample 2 in a beaker from your teacher. Slowly pour Water Sample 2 into Test Tube 2 until you reach the marked height.
7. Add one drop of dish detergent to each test tube. Stopper the tubes tightly. Then, shake each sample for 30 s to produce suds. Use a metric ruler to measure the height of the suds.
8. Use some of the soapy solutions to remove the grease marks from the test tubes.

**Data Table**

| Sample | Height of Suds |
|---|---|
| D | |
| 1 | |
| 2 | |

9. **Cleanup and Disposal** Rinse all of the liquids down the drain with tap water. Return all lab equipment to its designated location.

### Analyze and Conclude

1. **Compare and Contrast** Which sample produced the most suds? Which sample produced the least amount of suds?
2. **Conclude** Soft water produces more suds than hard water. Determine from which community each water sample originated.
3. **Calculate** If the 50 mL of hard water that you obtained contained 6.3 mg of magnesium, how hard would the water be according to the table below? (50 mL = 0.05 L)

**Classification of Water Hardness**

| Classification | mg of Calcium or Magnesium /L |
|---|---|
| Soft | 0–60 |
| Moderate | 61–120 |
| Hard | 121–180 |
| Very hard | >180 |

4. **Apply Scientific Methods** Identify the independent and dependent variables in this lab. Was there a control in this lab? Explain. Did all your classmates have the same results as you? Why or why not?
5. **Error Analysis** Could the procedure be changed to make the results more quantitative? Explain.

### INQUIRY EXTENSION

**Investigate** There are a number of products that claim to soften water. Visit a grocery store or home-improvement store to find these products and design an experiment to test their claims.

# CHAPTER 1 STUDY GUIDE

Vocabulary Practice

**BIGIDEA** Chemistry is a science that is central to our lives.

## SECTION 1 A Story of Two Substances

**MAINIDEA** Chemistry is the study of everything around us.

- Chemistry is the study of matter.
- Chemicals are also known as substances.
- Ozone is a substance that forms a protective layer in Earth's atmosphere.
- CFCs are synthetic substances made of chlorine, fluorine, and carbon that were originally thought to be the ideal coolants for refrigeration.

**VOCABULARY**

- chemistry
- substance

## SECTION 2 Chemistry and Matter

**MAINIDEA** Branches of chemistry involve the study of different kinds of matter.

- Models are tools that scientists, including chemists, use.
- Macroscopic observations of matter reflect the actions of atoms on a submicroscopic scale.
- There are several branches of chemistry, including organic chemistry, inorganic chemistry, physical chemistry, analytical chemistry, and biochemistry.

**VOCABULARY**

- mass
- weight
- model

## SECTION 3 Scientific Methods

**MAINIDEA** Scientists use scientific methods to systematically pose and test solutions to questions and assess the results of the tests.

- Scientific methods are systematic approaches to problem solving.
- Qualitative data describe an observation; quantitative data use numbers.
- Independent variables are changed in an experiment. Dependent variables change in response to the independent variable.
- A theory is a hypothesis that is supported by many experiments.

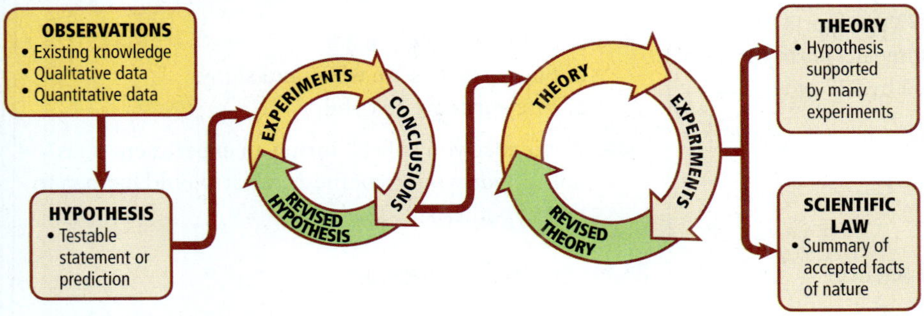


**VOCABULARY**

- scientific method
- qualitative data
- quantitative data
- hypothesis
- experiment
- independent variable
- dependent variable
- control
- conclusion
- theory
- scientific law

## SECTION 4 Scientific Research

**MAINIDEA** Some scientific investigations result in the development of technology that can improve our lives and the world around us.

- Scientific methods can be used in pure research or in applied research.
- Some scientific discoveries are accidental, and some are the result of diligent research in response to a need.
- Laboratory safety is the responsibility of everyone in the laboratory.
- Many of the conveniences we enjoy today are technological applications of chemistry.

**VOCABULARY**

- pure research
- applied research

## SECTION 1

### Mastering Concepts

**25.** Define *substance* and *chemistry*.

**26. Ozone** Where is ozone located in Earth's atmosphere?

**27.** What three elements are found in chlorofluorocarbons?

**28. CFCs** What were common uses of CFCs?

**29.** Scientists noticed that the ozone layer was thinning. What was occurring at the same time?

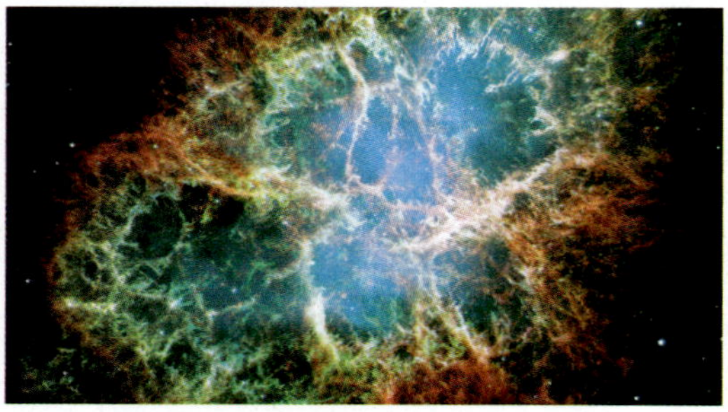

■ **Figure 20**

**30.** Why do chemists study regions of the universe, such as the one shown in **Figure 20?**

### Mastering Problems

**31.** If three oxygen particles are needed to form ozone, how many units of ozone could be formed from 6 oxygen particles? From 9? From 27?

**32. Measuring Concentration** **Figure 6** shows that the CFC level was measured at about 272 ppt (parts per thousand) in 1995. Because percent means *parts per hundred,* what percent is represented by 272 ppt?

## SECTION 2

### Mastering Concepts

**33.** Why is chemistry called the central science?

**34.** Which measurement depends on gravitational force—mass or weight? Explain.

**35.** Which branch of chemistry studies the composition of substances? The environmental impact of chemicals?

### Mastering Problems

**36.** Predict whether your weight in the city of Denver, which has an altitude of 1.7 km above sea level, will be the same as, more than, or less than your weight in New Orleans, a city located at sea level.

**37.** The text tells you that 1 trillion atoms could fit onto a period at the end of this sentence. Write out the number 1 trillion using the correct number of zeros.

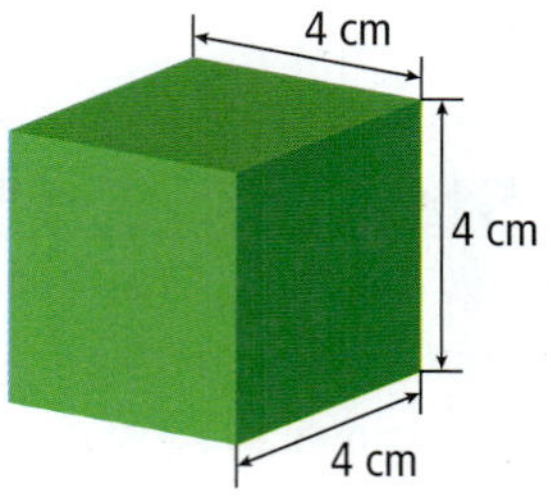


■ **Figure 21**

**38.** How much mass will the cube in **Figure 21** have if a 2-$cm^3$ cube of the same material has a mass of 4.0 g?

## SECTION 3

### Mastering Concepts

**39.** How does qualitative data differ from quantitative data? Give an example of each.

**40.** What is the function of a control in an experiment?

**41.** What is the difference between a hypothesis, a theory, and a law?

**42. Laboratory Experiments** You are asked to study how much table sugar can be mixed or dissolved in water at different temperatures. The amount of sugar that can dissolve in water goes up as the water's temperature goes up. What is the independent variable? Dependent variable? What factor is held constant?

**43.** Label each of the following pieces of data as qualitative or quantitative.
- **a.** A beaker weighs 6.6 g.
- **b.** Sugar crystals are white and shiny.
- **c.** Fireworks are colorful.

**44.** If evidence you collect during an experiment does not support your hypothesis, what should happen to that hypothesis?

### Mastering Problems

**45.** One carbon (C) and one ozone ($O_3$) react to form one carbon monoxide (CO) and one oxygen gas ($O_2$) particle. How many ozone particles are needed to form 24 particles of oxygen gas ($O_2$)?

## SECTION 4

### Mastering Concepts

**46. Laboratory Safety** Finish each statement about laboratory safety so that it correctly states a safety rule.
- **a.** Study your lab assignment
- **b.** Keep food, beverages, and
- **c.** Know where to find and how to use the

## Mastering Problems

**47.** If your lab procedure instructs you to add two parts acid to each one part of water and you start with 25 mL of water, how much acid will you add, and how will you add it?

## THINK CRITICALLY

**48. Compare and Contrast** Match each of the following research topics with the branch of chemistry that would study it: water pollution, the digestion of food in the human body, the composition of a new textile fiber, metals to make new coins, and a treatment for AIDS.

**49. Interpret Scientific Diagrams** Decide whether each of the diagrams shown below in **Figure 22a and b** is displaying qualitative or quantitative data.

**a** **Types of Apples Grown in Bioscience Greenhouse**

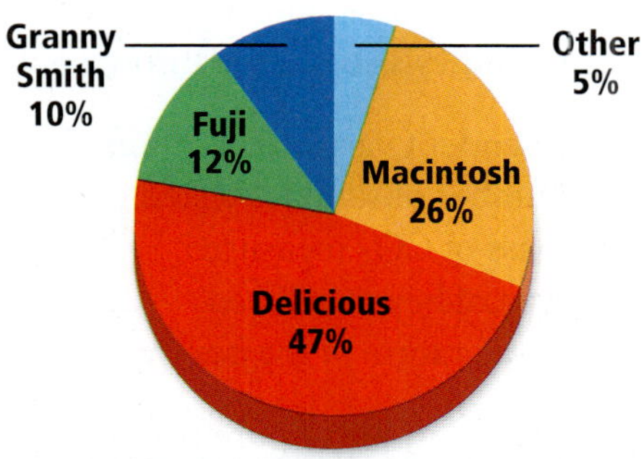


**b**

| Data: Characteristics of Product Formed | |
|---|---|
| Color | white |
| Crystal Form | needles |
| Odor | none |

■ **Figure 22**

**50. Classify** CFCs break down to form chemicals that react with ozone. Is this a macroscopic or a microscopic observation?

**51. Infer** A newscaster reports, "The air quality today is poor. Visibility is only 1.7 km. Pollutants in the air are expected to rise above 0.085 parts per million (ppm) in the next eight-hour average. Spend as little time outside today as possible if you suffer from asthma or other breathing problems." Which of these statements are qualitative and which are quantitative?

## CUMULATIVE REVIEW

In Chapters 2 through 24, this heading will be followed by questions that review your understanding of previous chapters.

## WRITING IN ▶ Chemistry

**52. Ozone Depletion** Based on your knowledge of chemistry, describe the research into depletion of the ozone layer by CFCs in a timeline.

**53. CFC Reduction** Research the most recent measures taken by countries around the world to reduce CFCs in the atmosphere since the Montreal Protocol. Write a short report describing the Montreal Protocol and more recent environmental measures to reduce CFCs.

**54. Technology** Name a technological application of chemistry that you use everyday. Prepare a booklet about its discovery and development.

## DBQ Document-Based Questions

**Ozone Depletion** *The area of low-ozone varies over the Antarctic. NOAA collects data and monitors the low-ozone area at the poles.*

**Figure 23** *shows the single-day maximum area of the ozone hole for each year from 1995 to 2009.*

Data obtained from: Southern Hemisphere Winter Summary. 2009. *National Oceanic and Atmospheric Administration.*

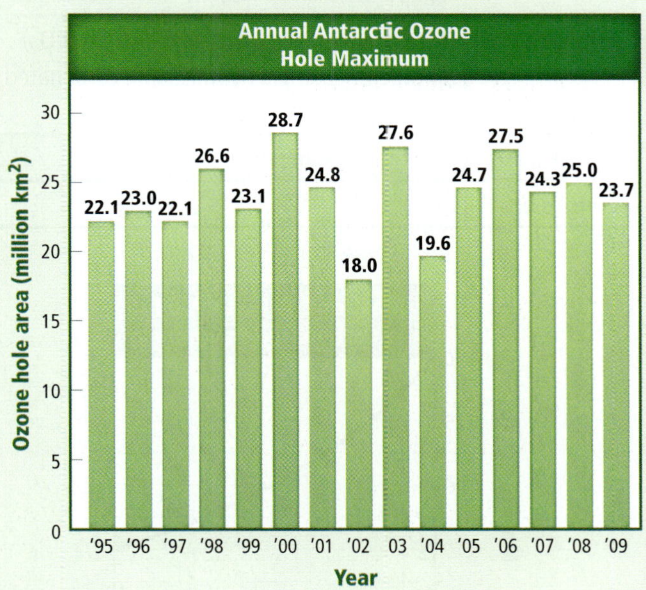


■ **Figure 23**

**55.** In what year did the largest maximum area of the ozone hole occur? The smallest?

**56.** What is the average maximum area of the ozone hole between the years 2005 and 2009? Between 2000 and 2004? Between 1995 and 1999?

## MULTIPLE CHOICE

1. When working with chemicals in the laboratory, which is something you should NOT do?
   - **A.** Read the label of chemical bottles before using their contents.
   - **B.** Pour any unused chemicals back into their original bottles.
   - **C.** Use a lot of water to wash skin that has been splashed with chemicals.
   - **D.** Take only as much as you need of shared chemicals.

*Use the table and graph below to answer Questions 2–5.*

| Page From a Student's Laboratory Notebook | |
|---|---|
| **Step** | **Notes** |
| Observation | Carbonated beverages taste fizzier when they are warm than when they are cold. (Carbonated beverages are fizzy because they contain dissolved carbon dioxide gas.) |
| Hypothesis | At higher temperatures, greater amounts of carbon dioxide gas will dissolve in a liquid. This is the same relationship between temperature and solubility seen with solids. |
| Experiment | Measure the mass of carbon dioxide ($CO_2$) in different samples of the same carbonated beverage at different temperatures. |
| Data analysis | See graph below. |
| Conclusion | |

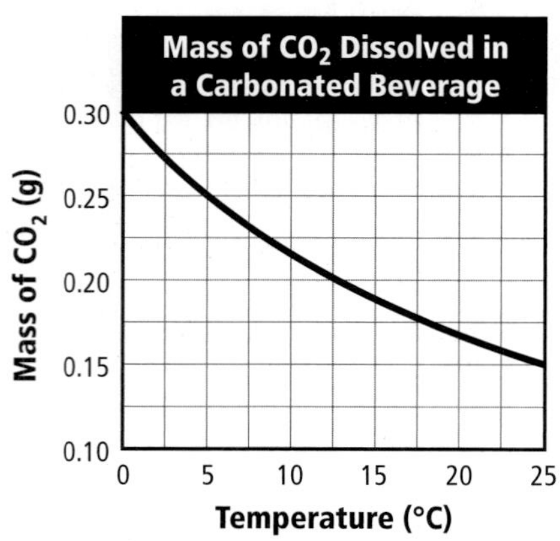


2. What must be a constant during the experiment?
   - **A.** temperature
   - **B.** mass of $CO_2$ dissolved in each sample
   - **C.** amount of beverage in each sample
   - **D.** independent variable

3. Assuming that all of the experimental data are correct, what is a reasonable conclusion for this experiment?
   - **A.** Greater amounts of $CO_2$ dissolve in a liquid at lower temperatures.
   - **B.** The different samples of beverage contained the same amount of $CO_2$ at each temperature.
   - **C.** The relationship between temperature and solubility seen with solids is the same as the one seen with $CO_2$.
   - **D.** $CO_2$ dissolves better at higher temperatures.

4. The scientific method used by this student showed that
   - **A.** the hypothesis is supported by the experimental data.
   - **B.** the observation accurately describes what occurs in nature.
   - **C.** the experiment is poorly planned.
   - **D.** the hypothesis should be thrown out.

5. The independent variable in this experiment is
   - **A.** the number of samples tested.
   - **B.** the mass of $CO_2$ measured.
   - **C.** the type of beverage used.
   - **D.** the temperature of the beverage.

6. Which is an example of pure research?
   - **A.** creating synthetic elements to study their properties
   - **B.** producing heat-resistant plastics for use in household ovens
   - **C.** finding ways to slow down the rusting of iron ships
   - **D.** searching for fuels other than gasoline to power cars

*Use the table below to answer Question 7.*

| What is the effect of drinking soda on heart rate? | | |
|---|---|---|
| **Student** | **Cans of Soda** | **Heart Rate (beats per minute)** |
| 1 | 0 | 73 |
| 2 | 1 | 84 |
| 3 | 2 | 89 |
| 4 | 4 | 96 |

7. In this experiment testing the effects of soda on students' heart rates, which student serves as the control?
   - **A.** Student 1
   - **B.** Student 2
   - **C.** Student 3
   - **D.** Student 4

## SHORT ANSWER

*Use the table below to answer Questions 8 and 9.*

**Physical Properties of Three Elements**

| Element | Symbol | Melting Point (°C) | Color | Density ($g/cm^3$) |
|---|---|---|---|---|
| Sodium | Na | 897.4 | Grey | 0.986 |
| Phosphorus | P | 44.2 | White | 1.83 |
| Copper | Cu | 1085 | Orange | 8.92 |

**8.** Give examples of qualitative data that are true for the element sodium.

**9.** Give examples of quantitative data that are true for the element copper.

**10.** A student in class announces that he has a theory to explain why he scored poorly on a quiz. Is this a proper use of the term *theory*? Explain your answer.

## EXTENDED RESPONSE

**11.** Explain why scientists use mass for measuring the amount of a substance instead of using weight.

*Consider the following experiment as you answer Questions 12 and 13.*

A chemistry student is investigating how particle size affects the rate of dissolving. In her experiment, she adds a sugar cube, sugar crystals, or crushed sugar to each of three beakers of water, stirs the mixtures for 10 seconds, and records how long it takes the sugar to dissolve in each beaker.

**12.** Identify the independent and dependent variables in this experiment. How can they be distinguished?

**13.** Identify a feature of this experiment that should be kept constant. Explain why it is important to keep this feature constant.

## SAT SUBJECT TEST: CHEMISTRY

**14.** A scientist from which field of chemistry investigates a new form of packaging material that breaks down rapidly in the environment?
**A.** biochemistry
**B.** theoretical chemistry
**C.** environmental chemistry
**D.** inorganic chemistry
**E.** physical chemistry

*Use the safety symbols below to answer Questions 15–18. Some choices may be used more than once; others will not be used at all.*

**A.** 
**B.** 
**C.** 
**D.** 
**E.** 

Select the symbol for the safety rule being described in each case.

**15.** Safety goggles should be worn whenever you are working in the lab.

**16.** Use chemicals in rooms with proper ventilation in case of strong fumes.

**17.** Wear proper protective clothing to prevent stains and burns.

**18.** Objects may be extremely hot or extremely cold; use hand protection.

**19.** Which statement is NOT true about mass?
**A.** It has the same value everywhere on Earth.
**B.** It is independent of gravitational forces.
**C.** It becomes less in outer space, farther from Earth.
**D.** It is a constant measure of the amount of matter.
**E.** It is found in all matter.

**NEED EXTRA HELP?**

| If You Missed Question . . . | 1 | 2 | 3 | 4 | 5 | 6 | 7 | 8 | 9 | 10 | 11 | 12 | 13 | 14 | 15 | 16 | 17 | 18 | 19 |
|---|---|---|---|---|---|---|---|---|---|---|---|---|---|---|---|---|---|---|---|
| Review Section . . . | 1.4 | 1.3 | 1.3 | 1.3 | 1.3 | 1.4 | 1.3 | 1.3 | 1.3 | 1.3 | 1.2 | 1.3 | 1.3 | 1.2 | 1.4 | 1.4 | 1.4 | 1.4 | 1.2 |

# CHAPTER 2
# Analyzing Data

**BIGIDEA** Chemists collect and analyze data to determine how matter interacts.

## SECTIONS

1 Units and Measurements

2 Scientific Notation and Dimensional Analysis

3 Uncertainty in Data

4 Representing Data

## LaunchLAB

iLab Station

### How can you form layers of liquids?

You know that this skydiver falls through the air. And you know that ice floats in water, whereas a rock sinks. In this lab you will investigate what can happen when you pour one liquid into another.

**FOLDABLES® Study Organizer**

### Types of Graphs

Make a layered-look book. Label it as shown. Use it to organize information about types of graphs.

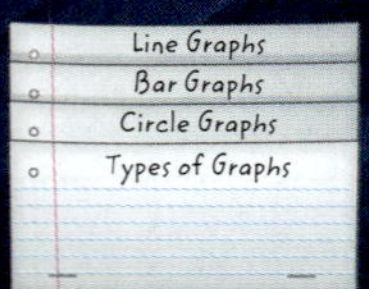

©Joggie Botma/Alamy

Go online!
MHEonline.com

Skydivers typically jump from an altitude of about 4000 m. They can reach speeds of 190 km/h and even greater. Collecting and analyzing data is important for a skydiver in order to maximize safety.

# Units and Measurements

**MAINIDEA** Chemists use an internationally recognized system of units to communicate their findings.

**CHEM 4 YOU** Have you ever noticed that a large drink varies in volume depending on where it is purchased? Wouldn't it be better if you always knew how much drink you would get when you ordered the large size? Chemists use standard units to ensure the consistent measurement of a given quantity.

### Essential Questions

- What are the SI base units for time, length, mass, and temperature?
- How does adding a prefix change a unit?
- How are the derived units different for volume and density?

### Review Vocabulary

**mass:** a measurement that reflects the amount of matter an object contains

### New Vocabulary

base unit
second
meter
kilogram
kelvin
derived unit
liter
density

## Units

You use measurements almost every day. For example, reading the bottled water label in **Figure 1** helps you decide what size bottle to buy. Notice that the label uses a number and a unit, such as 700 mL, to give the volume. The label also gives the volume as 23.7 fluid ounces. Fluid ounces, pints, and milliliters are units used to measure volume.

**Système Internationale d'Unités** For centuries, units of measurement were not exact. A person might measure distance by counting steps, or measure time using a sundial or an hourglass filled with sand. Such estimates worked for ordinary tasks. Because scientists need to report data that can be reproduced by other scientists, they need standard units of measurement. In 1960, an international committee of scientists met to update the existing metric system. The revised international unit system is called the Système Internationale d'Unités, which is abbreviated SI.

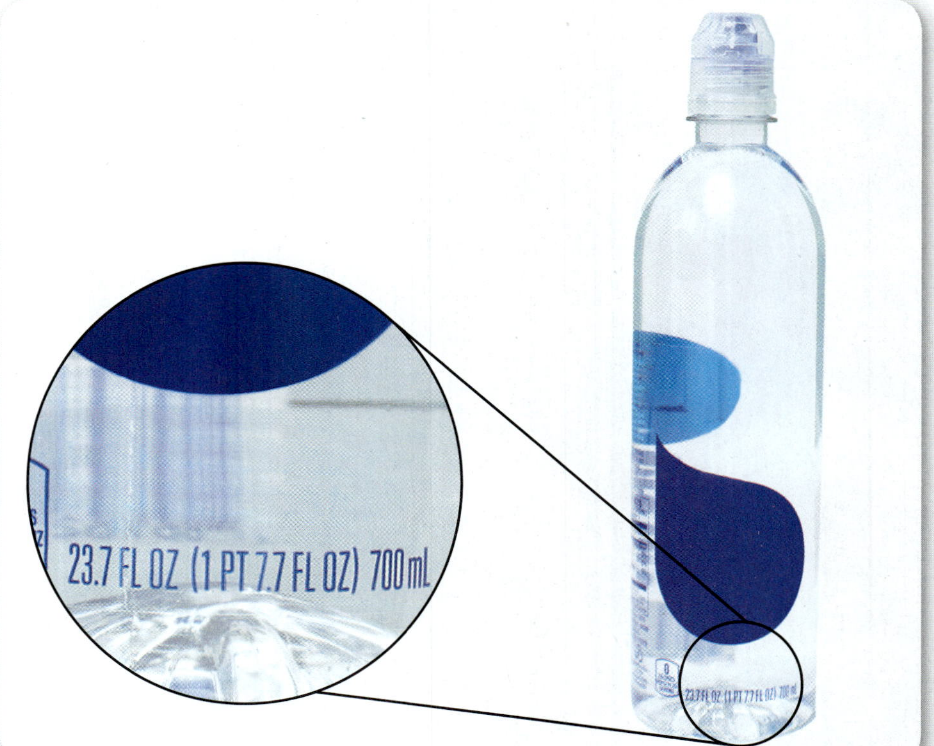


■ **Figure 1** The label gives the volume of water in the bottle in three different units: fluid ounces, pints, and milliliters. Notice that each volume includes a number and a unit.

**Infer** *Which is the larger unit of volume: a fluid ounce or a milliliter?*

## Base Units and SI Prefixes

There are seven base units in SI. A **base unit** is a defined unit in a system of measurement that is based on an object or event in the physical world. A base unit is independent of other units. **Table 1** lists the seven SI base units, the quantities they measure, and their abbreviations. Some familiar quantities that are expressed in base units are time, length, mass, and temperature.

To better describe the range of possible measurements, scientists add prefixes to the base units. This task is made easier because the metric system is a decimal system—a system based on units of 10. The prefixes in **Table 2** are based on factors of ten and can be used with all SI units. For example, the prefix *kilo-* means one thousand; therefore, 1 km equals 1000 m. Similarly, the prefix *milli-* means one-thousandth; therefore, 1 mm equals 0.001 m. Many mechanical pencils use lead that is 0.5 mm in diameter. How much of a meter is 0.5 mm?

**Time** The SI base unit for time is the **second** (s). The physical standard used to define the second is the frequency of the radiation given off by a cesium-133 atom. Cesium-based clocks are used when highly accurate timekeeping is required. For everyday tasks, a second seems like a short amount of time. In chemistry, however, many chemical reactions take place within a fraction of a second.

**Length** The SI base unit for length is the **meter** (m). A meter is the distance that light travels in a vacuum in 1/299,792,458 of a second. A vacuum exists where space contains no matter.

A meter is close in length to a yard and is useful for measuring the length and width of a small area, such as a room. For larger distances, such as between cities, you would use kilometers. Smaller lengths, such as the diameter of a pencil, are likely to be given in millimeters.

**Table 1 SI Base Units**

| Quantity | Base Unit |
|---|---|
| Time | second (s) |
| Length | meter (m) |
| Mass | kilogram (kg) |
| Temperature | kelvin (K) |
| Amount of a substance | mole (mol) |
| Electric current | ampere (A) |
| Luminous intensity | candela (cd) |

**Vocabulary**

**Science usage v. Common usage**

**Meter**

***Science usage:*** the SI base unit of length

*The metal rod was 1 m in length.*

***Common usage:*** a device used to measure

*The time ran out on the parking meter.*

Explore **SI prefixes with an interactive table.** Concepts In Motion

**Table 2 SI Prefixes**

| Prefix | Symbol | Numerical Value in Base Units | Power of 10 Equivalent |
|---|---|---|---|
| Giga | G | 1,000,000,000 | $10^{9}$ |
| Mega | M | 1,000,000 | $10^{6}$ |
| Kilo | k | 1000 | $10^{3}$ |
| – | – | 1 | $10^{0}$ |
| Deci | d | 0.1 | $10^{-1}$ |
| Centi | c | 0.01 | $10^{-2}$ |
| Milli | m | 0.001 | $10^{-3}$ |
| Micro | $\mu$ | 0.000001 | $10^{-6}$ |
| Nano | n | 0.000000001 | $10^{-9}$ |
| Pico | p | 0.000000000001 | $10^{-12}$ |

■ **Figure 2** Scientists at the National Institute of Standards and Technology are experimenting with redefining the kilogram using an apparatus known as a watt balance. The watt balance uses electric current and a magnetic field to measure the force required to balance a one-kilogram mass against the force of gravity. Other scientists are counting the number of atoms in a one-kilogram mass to redefine the kilogram.

**Mass** Recall that mass is a measure of the amount of matter an object contains. The SI base unit for mass is the **kilogram** (kg). Currently, a platinum and iridium cylinder kept in France defines the kilogram. The cylinder is stored in a vacuum under a triple bell jar to prevent the cylinder from oxidizing. As shown in **Figure 2,** scientists are working to redefine the kilogram using basic properties of nature.

A kilogram is equal to about 2.2 pounds. Because the masses measured in most laboratories are much smaller than a kilogram, scientists often measure quantities in grams (g) or milligrams (mg). For example, a laboratory experiment might ask you to add 35 mg of an unknown substance to 350 g of water. When working with mass values, it is helpful to remember that there are 1000 g in a kilogram. How many milligrams are in a gram?

**Temperature** People often use qualitative descriptions, such as hot and cold, when describing the weather or the water in a swimming pool. Temperature, however, is a quantitative measurement of the average kinetic energy of the particles that make up an object. As the particle motion in an object increases, so does the temperature of the object.

Measuring temperature requires a thermometer or a temperature probe. A thermometer consists of a narrow tube that contains a liquid. The height of the liquid indicates the temperature. A change in temperature causes a change in the volume of the liquid, which results in a change in the height of the liquid in the tube. Electronic temperature probes make use of thermocouples. A thermocouple produces an electric current that can be calibrated to indicate temperature.

Several different temperature scales have been developed. Three temperature scales—Kelvin, Celsius, and Fahrenheit—are commonly used to describe how hot or cold an object is.

**Fahrenheit** In the United States, the Fahrenheit scale is used to measure temperature. German scientist Gabriel Daniel Fahrenheit devised the scale in 1724. On the Fahrenheit scale, water freezes at 32°F and boils at 212°F.

Get help with **conversions.**

Personal Tutor

**Celsius** Another temperature scale, the Celsius scale, is used throughout much of the rest of the world. Anders Celsius, a Swedish astronomer, devised the Celsius scale. The scale is based on the freezing and boiling points of water. He defined the freezing point of water as 0 and the boiling point of water as 100. He then divided the distance between these two fixed points into 100 equal units, or degrees. To convert from degrees Celsius (°C) to degrees Fahrenheit (°F), you can use the following equation.

$$°F = 1.8(°C) + 32$$

Imagine a friend from Canada calls you and says that it is 35°C outside. What is the temperature in degrees Fahrenheit? To convert to degrees Fahrenheit, substitute 35°C into the above equation and solve.

$$1.8(35) + 32 = 95°F$$

If it is 35°F outside, what is the temperature in degrees Celsius?

$$\frac{35°F - 32}{1.8} = 1.7\ °C$$

☑ READING CHECK **Infer** Which is warmer, 25°F or 25°C?

Robert Rathe

**Kelvin** The SI base unit for temperature is the **kelvin** (K). The Kelvin scale was devised by a Scottish physicist and mathematician, William Thomson, who was known as Lord Kelvin. Zero kelvin is a point where all particles are at their lowest possible energy state. On the Kelvin scale, water freezes at 273.15 K and boils at 373.15 K. Later, you will learn why scientists use the Kelvin scale to describe properties of a gas.

**Figure 3** compares the Celsius and Kelvin scales. It is easy to convert between the Celsius scale and the Kelvin scale using the following equation.

**Kelvin-Celsius Conversion Equation**

$$K = °C + 273$$

K represents temperature in kelvins.
°C represents temperature in degrees Celsius.

Temperature in kelvins is equal to temperature in degrees Celsius plus 273.

As shown by the equation above, to convert temperatures reported in degrees Celsius to kelvins, you simply add 273. For example, consider the element mercury, which melts at −39°C. What is this temperature in kelvins?

$$-39°C + 273 = 234 \text{ K}$$

To convert from kelvins to degrees Celsius, just subtract 273. For example, consider the element bromine, which melts at 266 K. What is this temperature in degrees Celsius?

$$266 \text{ K} - 273 = -7°C$$

You will use these conversions frequently throughout chemistry, especially when you study how gases behave. The gas laws you will learn are based on kelvin temperatures.

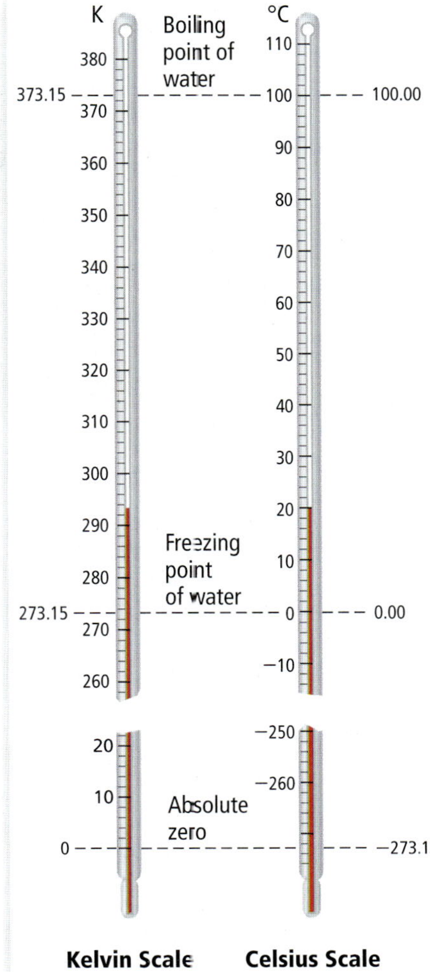


**Figure 3** A change of 1 K on the Kelvin scale is equal in size to a change of 1°C on the Celsius scale. Notice also that the degree sign ( ° ) is not used with the Kelvin scale.

## Derived Units

Not all quantities can be measured with SI base units. For example, the SI unit for speed is meters per second (m/s). Notice that meters per second includes two SI base units—the meter and the second. A unit that is defined by a combination of base units is called a **derived unit.** Two other quantities that are measured in derived units are volume ($cm^3$) and density ($g/cm^3$).

**Volume** Volume is the space occupied by an object. The volume of an object with a cubic or rectangular shape can be determined by multiplying its length, width, and height dimensions. When each dimension is given in meters, the calculated volume has units of cubic meters ($m^3$). In fact, the derived SI unit for volume is the cubic meter. It is easy to visualize a cubic meter; imagine a large cube whose sides are each 1 m in length. The volume of an irregularly shaped solid can be determined using the water displacement method, a method used in the MiniLab in this section.

The cubic meter is a large volume that is difficult to work with. For everyday use, a more useful unit of volume is the liter. A **liter** (L) is equal to one cubic decimeter ($dm^3$), that is, 1 L equals 1 $dm^3$. Liters are commonly used to measure the volume of water and beverage containers. One liter has about the same volume as one quart.

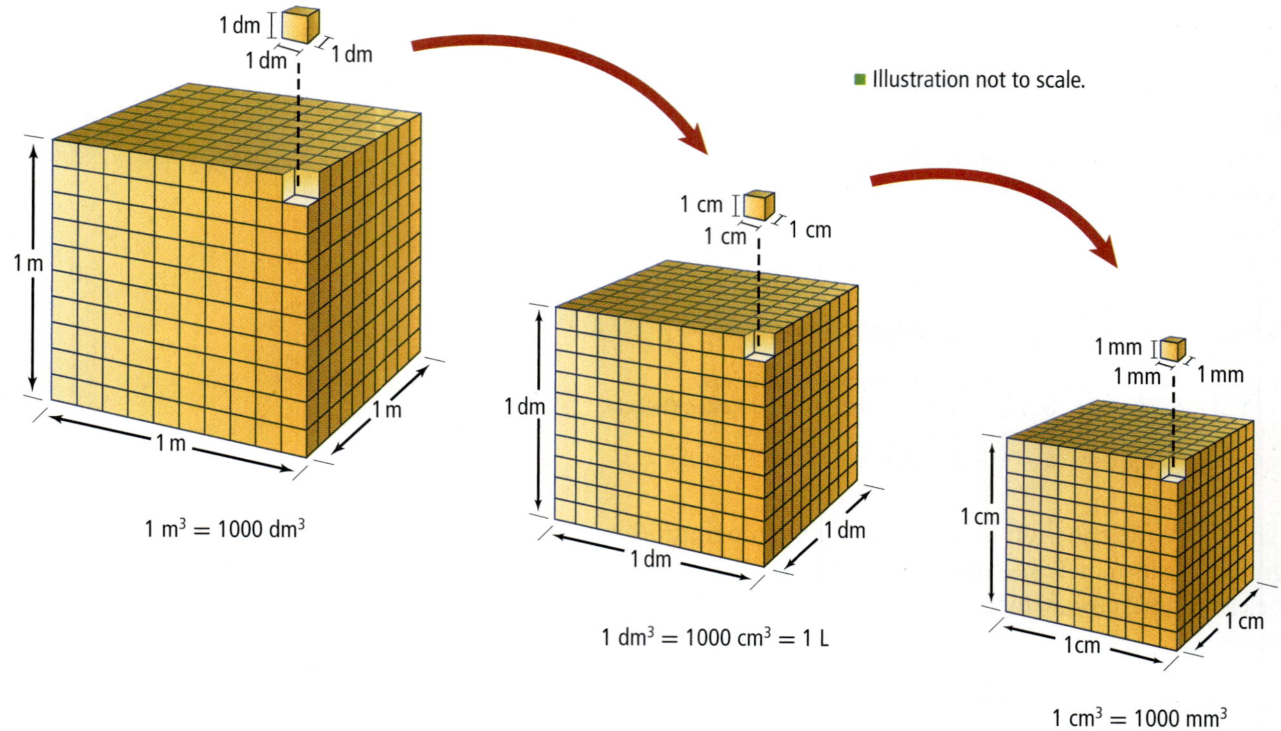


**Figure 4** The three cubes show volume relationships between cubic meters ($m^3$), cubic decimeters ($dm^3$), cubic centimeters ($cm^3$), and cubic millimeters ($mm^3$). As you move from left to right, the volume of each cube gets $10 \times 10 \times 10$, or 1000 times, smaller.
**Interpret** *How many cubic centimeters ($cm^3$) are in 1 L?*

For smaller quantities of liquids in the laboratory, volume is often measured in cubic centimeters ($cm^3$) or milliliters (mL). A milliliter and a cubic centimeter are equal in size.

$$1 \text{ mL} = 1 \text{ cm}^3$$

Recall that the prefix *milli-* means one-thousandth. Therefore, one milliliter is equal to one-thousandth of a liter. In other words, there are 1000 mL in 1 L.

$$1 \text{ L} = 1000 \text{ mL}$$

**Figure 4** shows the relationships among several different SI units of volume.

Explore **density**
Virtual Investigations

**Density** Why is it easier to lift a backpack filled with gym clothes than the same backpack filled with books? The answer can be thought of in terms of density—the book-filled backpack contains more mass in the same volume. **Density** is a physical property of matter and is defined as the amount of mass per unit volume. Common units of density are grams per cubic centimeter ($g/cm^3$) for solids and grams per milliliter (g/mL) for liquids and gases.

Consider the grape and the piece of foam in **Figure 5.** Although both have the same mass, they clearly occupy different amounts of space. Because the grape occupies less volume for the same amount of mass, its density must be greater than that of the foam.

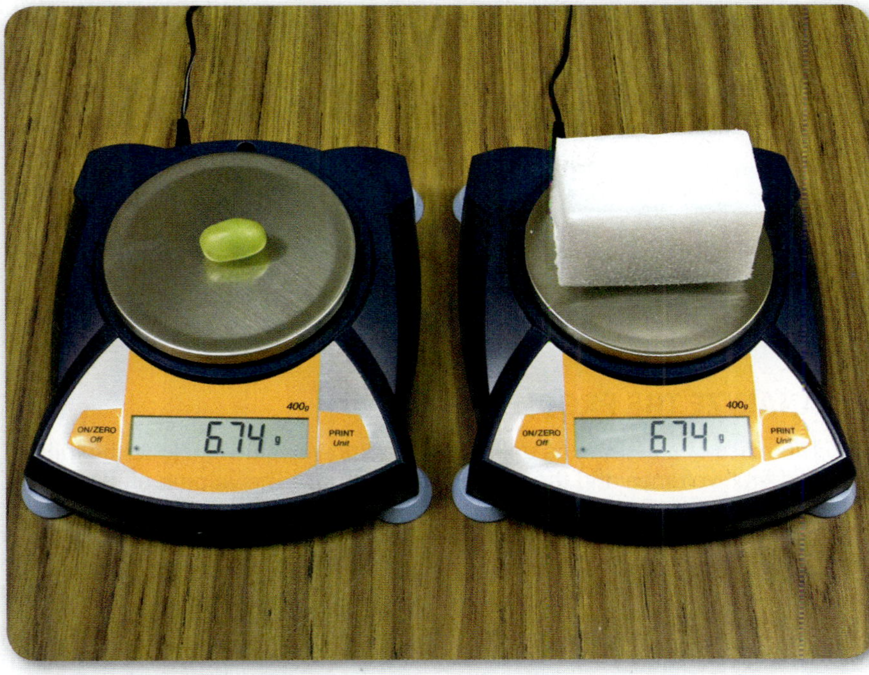


■ **Figure 5** The grape and the foam have the same mass but different volumes because the grape is more dense.

**Interpret** ***How would the masses compare if the volumes were equal?***

The density of a substance usually cannot be measured directly. Rather, it is calculated using mass and volume measurements. You can calculate density using the following equation.

**Density Equation**

$$\text{density} = \frac{\text{mass}}{\text{volume}}$$

The density of an object or a sample of matter is equal to its mass divided by its volume.

Because density is a physical property of matter, it can sometimes be used to identify an unknown element. For example, imagine you are given the following data for a piece of an unknown metallic element.

$$\text{volume} = 5.0\ \text{cm}^3$$
$$\text{mass} = 13.5\ \text{g}$$

Substituting these values into the equation for density yields:

$$\text{density} = \frac{13.5\ \text{g}}{5.0\ \text{cm}^3} = 2.7\ \text{g/cm}^3$$

Now go to **Table R–7** and scan through the given density values until you find one that closely matches the calculated value of 2.7 g/cm$^3$. What is the identity of the unknown element?

**Connection to Earth Science** As air at the equator is warmed, the particles in the air move farther apart and the air density decreases. At the poles, the air cools and its density increases as the particles move closer together. When a cooler, denser air mass sinks beneath a rising warm air mass, winds are produced. Weather patterns are created by moving air masses of different densities.

☑ READING CHECK **State** the quantities that must be known in order to calculate density.

**RealWorld CHEMISTRY**

**Liquid Density Measurement**

**HYDROMETERS** A hydrometer is a device that measures the specific gravity (the ratio of the fluid's density to that of water) of a fluid. Fluids of different densities result in different readings. Hydrometers are often used at service stations to diagnose problems with an automobile's battery.

Your textbook includes many Example Problems, each of which is solved using a three-step process. Read Example Problem 1 and follow the steps to calculate the mass of an object using density and volume.

**THE PROBLEM**

1. Read the problem carefully.
2. Be sure that you understand what is being asked.

**ANALYZE THE PROBLEM**

1. Read the problem again.
2. Identify what you are given, and list the known data. If needed, gather information from graphs, tables, or figures.
3. Identify and list the unknowns.
4. Plan the steps you will follow to find the answer.

**SOLVE FOR THE UNKNOWN**

1. Determine whether you need a sketch to solve the problem.
2. If the solution is mathematical, write the equation and isolate the unknown factor.
3. Substitute the known quantities into the equation.
4. Solve the equation.
5. Continue the solution process until you solve the problem.

**EVALUATE THE ANSWER**

1. Reread the problem. Is the answer reasonable?
2. Check your math. Are the units and the significant figures correct? (Refer to Section 3.)

## EXAMPLE Problem 1

**USING DENSITY AND VOLUME TO FIND MASS** When a piece of aluminum is placed in a 25-mL graduated cylinder that contains 10.5 mL of water, the water level rises to 13.5 mL. What is the mass of the aluminum?

### 1 ANALYZE THE PROBLEM

The mass of aluminum is unknown. The known values include the initial and final volumes and the density of aluminum. The volume of the sample equals the volume of water displaced in the graduated cylinder. According to **Table R-7,** the density of aluminum is 2.7 g/mL. Use the density equation to solve for the mass of the aluminum sample.

| **Known** | **Unknown** |
|---|---|
| density = 2.7 g/mL | **mass = ? g** |
| initial volume = 10.5 mL | |
| final volume = 13.5 mL | |

### 2 SOLVE FOR THE UNKNOWN

volume of sample = final volume − initial volume — **State the equation for volume.**

volume of sample = 13.5 mL − 10.5 mL — **Substitute final volume = 13.5 mL and initial volume = 10.5 mL.**

volume of sample = 3.0 mL

$$\text{density} = \frac{\text{mass}}{\text{volume}}$$ **State the equation for density.**

**mass** = volume × density — **Solve the density equation for mass.**

**mass** = 3.0 mL × 2.7 g/mL — **Substitute volume = 3.0 mL and density = 2.7 g/mL.**

**mass** = $3.0\,\cancel{\text{mL}} \times 2.7\,\text{g}/\cancel{\text{mL}}$ = **8.1 g** — **Multiply, and cancel units.**

### 3 EVALUATE THE ANSWER

Check your answer by using it to calculate the density of aluminum.

$$\text{density} = \frac{\text{mass}}{\text{volume}} = \frac{8.1\ \text{g}}{3.0\ \text{mL}} = 2.7\ \text{g/mL}$$

Because the calculated density for aluminum is correct, the mass value must also be correct.

## PRACTICE Problems

Do additional problems. Online Practice

1. Is the cube pictured at right made of pure aluminum? Explain your answer.
2. What is the volume of a sample that has a mass of 20 g and a density of 4 g/mL?
3. **Challenge** A 147-g piece of metal has a density of 7.00 g/mL. A 50-mL graduated cylinder contains 20.0 mL of water. What is the final volume after the metal is added to the graduated cylinder?

Mass = 20 g
Volume = 5 $cm^3$

# MiniLAB

iLab Station

## Determine Density

**What is the density of an unknown and irregularly shaped solid? To calculate the density of an object, you need to know its mass and volume. The volume of an irregularly shaped solid can be determined by measuring the amount of water it displaces.**

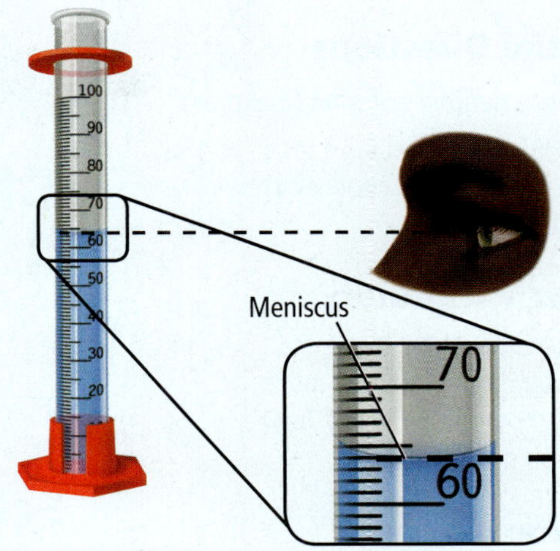


### Procedure

1. Read and complete the lab safety form.
2. Obtain several **unknown objects** from your teacher. *Note: Your teacher will identify each object as A, B, C, and so on.*
3. Create a data table to record your observations.
4. Measure the mass of the object using a **balance.** Record the mass and the letter of the object in your data table.
5. Add about 15 mL of **water** to a **graduated cylinder.** Measure and record the initial volume in your data table. Because the surface of the water in the cylinder is curved, make volume readings at eye level and at the lowest point on the curve, as shown in the figure. The curved surface is called a meniscus.
6. Tilt the graduated cylinder, and carefully slide the object down the inside of the cylinder. Be sure not to cause a splash. Measure and record the final volume in your data table.

### Analysis

1. **Calculate** Use the initial and final volume readings to calculate the volume of each mystery object.
2. **Calculate** Use the calculated volume and the measured mass to calculate the density of each unknown object.
3. **Explain** Why can't you use the water displacement method to find the volume of a sugar cube?
4. **Describe** how you can determine a washer's volume without using the water displacement method. Note that a washer is similar to a short cylinder with a hole through it.

# SECTION 1 REVIEW

Section Self-Check

### Section Summary

- SI measurement units allow scientists to report data to other scientists.
- Adding prefixes to SI units extends the range of possible measurements.
- To convert to kelvin temperature, add 273 to the Celsius temperature.
- Volume and density have derived units. Density, which is a ratio of mass to volume, can be used to identify an unknown sample of matter.

4. **MAIN**IDEA **Define** the SI units for length, mass, time, and temperature.
5. **Describe** how adding the prefix *mega-* to a unit affects the quantity being described.
6. **Compare** a base unit and a derived unit, and list the derived units used for density and volume.
7. **Define** the relationships among the mass, volume, and density of a material.
8. **Apply** Why does oil float on water?
9. **Calculate** Samples A, B, and C have masses of 80 g, 12 g, and 33 g, and volumes of 20 mL, 4 $cm^3$, and 11 mL, respectively. Which of the samples have the same density?
10. **Design** a concept map that shows the relationships among the following terms: *volume, derived unit, mass, base unit, time,* and *length.*

SECTION 2

# Scientific Notation and Dimensional Analysis

**MAINIDEA** Scientists often express numbers in scientific notation and solve problems using dimensional analysis.

### Essential Questions

- Why use scientific notation to express numbers?
- How is dimensional analysis used for unit conversion?

### Review Vocabulary

**quantitative data:** numerical information describing how much, how little, how big, how tall, how fast, and so on

### New Vocabulary

scientific notation
dimensional analysis
conversion factor

**CHEM 4 YOU** If you have ever had a job, one of the first things you probably did was figure out how much you would earn per week. If you make 10 dollars per hour and work 20 hours per week, how much money will you make? Performing this calculation is an example of dimensional analysis.

## Scientific Notation

The Hope Diamond, which is shown in **Figure 6,** contains approximately 460,000,000,000,000,000,000,000 atoms of carbon. Each of these carbon atoms has a mass of 0.00000000000000000000002 g. If you were to use these numbers to calculate the mass of the Hope Diamond, you would find that the zeros would get in your way. Using a calculator offers no help, as it won't let you enter numbers this large or this small. Numbers such as these are best expressed in scientific notation. Scientists use this method to conveniently restate a number without changing its value.

**Scientific notation** can be used to express any number as a number between 1 and 10 (known as the coefficient) multiplied by 10 raised to a power (known as the exponent). When written in scientific notation, the two numbers above appear as follows.

Coefficient — Exponent

$$\text{carbon atoms in the Hope Diamond} = 4.6 \times 10^{23}$$

$$\text{mass of one carbon atom} = 2 \times 10^{-23}\ \text{g}$$

■ **Figure 6** At more than 45 carats, the Hope Diamond is the world's largest deep-blue diamond. Originally mined in India, the diamond's brilliant blue color is due to trace amounts of boron within the diamond. Diamonds are formed from a unique structure of carbon atoms, creating one of nature's hardest known substances. Note that a carat is a unit of measure used for gemstones (1 carat = 200 mg).

Let's look at these two numbers more closely. In each case, the number 10 raised to an exponent replaced the zeros that preceded or followed the nonzero numbers. For numbers greater than 1, a positive exponent is used to indicate how many times the coefficient must be multiplied by 10 in order to obtain the original number. Similarly, for numbers less than 1, a negative exponent indicates how many times the coefficient must be divided by 10 in order to obtain the original number.

Determining the exponent to use when writing a number in scientific notation is easy: simply count the number of places the decimal point must be moved to give a coefficient between 1 and 10. The number of places moved equals the value of the exponent. The exponent is positive when the decimal moves to the left and the exponent is negative when the decimal moves to the right.

Get help with **scientific notation.**

Personal Tutor

$$460{,}000{,}000{,}000{,}000{,}000{,}000{,}000. \rightarrow 4.6 \times 10^{23}$$

**Because the decimal point moves 23 places to the left, the exponent is 23.**

$$0.00000000000000000000002 \rightarrow 2 \times 10^{-23}$$

**Because the decimal point moves 23 places to the right, the exponent is −23.**

## EXAMPLE Problem 2

Find help with **scientific notation.** Math Handbook

**SCIENTIFIC NOTATION** Write the following data in scientific notation.

**a.** The diameter of the Sun is 1,392,000 km.

**b.** The density of the Sun's lower atmosphere is 0.000000028 g/cm$^3$.

### 1 ANALYZE THE PROBLEM

You are given two values, one much larger than 1 and the other much smaller than 1. In both cases, the answers will have a coefficient between 1 and 10 multiplied by a power of 10.

### 2 SOLVE FOR THE UNKNOWN

Move the decimal point to give a coefficient between 1 and 10. Count the number of places the decimal point moves, and note the direction.

1,392,000. — **Move the decimal point six places to the left.**

0.000000028 — **Move the decimal point eight places to the right.**

**a.** $1.392 \times 10^6$ km

**b.** $2.8 \times 10^{-8}$ g/cm$^3$

**Write the coefficients, and multiply them by $10^n$ where $n$ equals the number of places moved. When the decimal point moves to the left, $n$ is positive; when the decimal point moves to the right, $n$ is negative. Add units to the answers.**

### 3 EVALUATE THE ANSWER

The answers are correctly written as a coefficient between 1 and 10 multiplied by a power of 10. Because the diameter of the Sun is a number greater than 1, its exponent is positive. Because the density of the Sun's lower atmosphere is a number less than 1, its exponent is negative.

EXAMPLE PROBLEM

## PRACTICE Problems

Do additional problems. Online Practice

**11.** Express each number in scientific notation.

**a.** 700
**b.** 38,000
**c.** 4,500,000
**d.** 685,000,000,000
**e.** 0.0054
**f.** 0.00000687
**g.** 0.000000076
**h.** 0.0000000008

**12. Challenge** Express each quantity in regular notation along with its appropriate unit.

**a.** $3.60 \times 10^5$ s
**b.** $5.4 \times 10^{-5}$ g/cm$^3$
**c.** $5.060 \times 10^3$ km
**d.** $8.9 \times 10^{10}$ Hz

■ **Figure 7** The uneven heating of Earth's surface causes wind, which powers these turbines and generates electricity.

**Vocabulary**

**Academic vocabulary**

**Sum**
the whole amount; the result of adding numbers
*At the checkout counter, all of the items came to a sizable sum.*

**Addition and subtraction** In order to add or subtract numbers written in scientific notation, the exponents must be the same. Suppose you need to add $7.35 \times 10^2$ m and $2.43 \times 10^2$ m. Because the exponents are the same, you can simply add the coefficients.

$$(7.35 \times 10^2 \text{ m}) + (2.43 \times 10^2 \text{ m}) = 9.78 \times 10^2 \text{ m}$$

How do you add numbers in scientific notation when the exponents are not the same? To answer this question, consider the amounts of energy produced by renewable energy sources in the United States. Wind-powered turbines, shown in **Figure 7,** are one of several forms of renewable energy used in the United States. Other sources of renewable energy include hydroelectric, biomass, geothermal, and solar power. In 2008, the energy production amounts from renewable sources were as follows.

| | |
|---|---|
| Hydroelectric | $2.643 \times 10^{18}$ J* |
| Biomass | $4.042 \times 10^{18}$ J |
| Geothermal | $3.89 \times 10^{17}$ J |
| Wind | $5.44 \times 10^{17}$ J |
| Solar | $7.8 \times 10^{16}$ J |

* J stands for joules, a unit of energy.

To determine the sum of these values, they must be rewritten with the same exponent. Because the two largest values have an exponent of $10^{18}$, it makes sense to convert the other numbers to values with this exponent. These other exponents must increase to become $10^{18}$. As you learned earlier, each place the decimal shifts to the left increases the exponent by 1. Rewriting the values with exponents of $10^{18}$ and adding yields the following.

| | |
|---|---|
| Hydroelectric | $2.643 \times 10^{18}$ J |
| Biomass | $4.042 \times 10^{18}$ J |
| Geothermal | $0.389 \times 10^{18}$ J |
| Wind | $0.544 \times 10^{18}$ J |
| Solar | $0.078 \times 10^{18}$ J |
| Total | $7.696 \times 10^{18}$ J |

☑ READING CHECK **Restate** the process used to add two numbers that are expressed in scientific notation.

## PRACTICE Problems

Do additional problems. Online Practice

**13.** Solve each problem, and express the answer in scientific notation.

**a.** $(5 \times 10^{-5}) + (2 \times 10^{-5})$
**b.** $(7 \times 10^{8}) - (4 \times 10^{8})$
**c.** $(9 \times 10^{2}) - (7 \times 10^{2})$
**d.** $(4 \times 10^{-12}) + (1 \times 10^{-12})$

**14. Challenge** Express each answer in scientific notation in the units indicated.

**a.** $(1.26 \times 10^{4}$ kg$) + (2.5 \times 10^{6}$ g$)$ in kg
**b.** $(7.06$ g$) + (1.2 \times 10^{-4}$ kg$)$ in kg
**c.** $(4.39 \times 10^{5}$ kg$) - (2.8 \times 10^{7}$ g$)$ in kg
**d.** $(5.36 \times 10^{-1}$ kg$) - (7.40 \times 10^{-2}$ kg$)$ in g

**Multiplication and division** Multiplying and dividing numbers in scientific notation is a two-step process, but it does not require the exponents to be the same. For multiplication, multiply the coefficients and then add the exponents. For division, divide the coefficients, then subtract the exponent of the divisor from the exponent of the dividend.

To calculate the mass of the Hope Diamond, multiply the number of carbon atoms by the mass of a single carbon atom.

$$(4.6 \times 10^{23} \text{ atoms})(2 \times 10^{-23} \text{ g/atom}) = 9.2 \times 10^{0} \text{ g} = 9.2 \text{ g}$$

Note that any number raised to a power of 0 is equal to 1; thus, $9.2 \times 10^{0}$ g is equal to 9.2 g.

## EXAMPLE Problem 3

Find help with **scientific notation.** Math Handbook

**MULTIPLYING AND DIVIDING NUMBERS IN SCIENTIFIC NOTATION** Solve the following problems.

**a.** $(2 \times 10^3) \times (3 \times 10^2)$

**b.** $(9 \times 10^8) \div (3 \times 10^{-4})$

### 1 ANALYZE THE PROBLEM

You are given numbers written in scientific notation to multiply and divide. For the multiplication problem, multiply the coefficients and add the exponents. For the division problem, divide the coefficients and subtract the exponent of the divisor from the exponent of the dividend.

$$\frac{9 \times 10^8}{3 \times 10^{-4}}$$

The exponent of the dividend is 8.

The exponent of the divisor is −4.

### 2 SOLVE FOR THE UNKNOWN

| | |
|---|---|
| **a.** $(2 \times 10^3) \times (3 \times 10^2)$ | State the problem. |
| $2 \times 3 = 6$ | Multiply the coefficients. |
| $3 + 2 = 5$ | Add the exponents. |
| $6 \times 10^5$ | Combine the parts. |
| **b.** $(9 \times 10^8) \div (3 \times 10^{-4})$ | State the problem. |
| $9 \div 3 = 3$ | Divide the coefficients. |
| $8 - (-4) = 8 + 4 = 12$ | Subtract the exponents. |
| $3 \times 10^{12}$ | Combine the parts. |

### 3 EVALUATE THE ANSWER

To test the answers, write out the original data and carry out the arithmetic. For example, Problem **a** becomes $2000 \times 300 = 600{,}000$, which is the same as $6 \times 10^5$.

EXAMPLE PROBLEM

## PRACTICE Problems

Do additional problems. Online Practice

**15.** Solve each problem, and express the answer in scientific notation.

**a.** $(4 \times 10^2) \times (1 \times 10^8)$

**b.** $(2 \times 10^{-4}) \times (3 \times 10^2)$

**c.** $(6 \times 10^2) \div (2 \times 10^1)$

**d.** $(8 \times 10^4) \div (4 \times 10^1)$

**16. Challenge** Calculate the areas and densities. Report the answers in the correct units.

**a.** the area of a rectangle with sides measuring $3 \times 10^1$ cm and $3 \times 10^{-2}$ cm

**b.** the area of a rectangle with sides measuring $1 \times 10^3$ cm and $5 \times 10^{-1}$ cm

**c.** the density of a substance having a mass of $9 \times 10^5$ g and a volume of $3 \times 10^{-1}$ $cm^3$

**d.** the density of a substance having a mass of $4 \times 10^{-3}$ g and a volume of $2 \times 10^{-2}$ $cm^3$

PRACTICE PROBLEMS

■ **Figure 8** Dimensional analysis can be used to calculate the number of pizzas that must be ordered for a party. How many pizzas will you need if 32 people eat 3 slices per person and there are 8 slices in each pizza?

$$(32\ \cancel{\text{people}})\left(\frac{3\ \cancel{\text{slices}}}{\cancel{\text{person}}}\right)\left(\frac{1\ \text{pizza}}{8\ \cancel{\text{slices}}}\right) = 12\ \text{pizzas}$$

## Dimensional Analysis

When planning a pizza party for a group of people, you might want to use dimensional analysis to figure out how many pizzas to order. **Dimensional analysis** is a systematic approach to problem solving that uses conversion factors to move, or convert, from one unit to another. A **conversion factor** is a ratio of equivalent values having different units.

How many pizzas do you need to order if 32 people will attend a party, each person eats 3 slices of pizza, and each pizza has 8 slices? **Figure 8** shows how conversion factors are used to calculate the number of pizzas needed for the party.

**Writing conversion factors** As you just read, conversion factors are ratios of equivalent values. Not surprisingly, these conversion factors are derived from equality relationships, such as 12 eggs = 1 dozen eggs, or 12 inches = 1 foot. Multiplying a quantity by a conversion factor changes the units of the quantity without changing its value.

Most conversion factors are written from relationships between units. For example, the prefixes in **Table 2** are the source of many conversion factors. From the relationship 1000 m = 1 km, the following conversion factors can be written.

$$\frac{1\ \text{km}}{1000\ \text{m}} \quad \text{and} \quad \frac{1000\ \text{m}}{1\ \text{km}}$$

A derived unit, such as a density of 2.5 g/mL, can also be used as a conversion factor. The value shows that 1 mL of the substance has a mass of 2.5 g. The following two conversion factors can be written.

$$\frac{2.5\ \text{g}}{1\ \text{mL}} \quad \text{and} \quad \frac{1\ \text{mL}}{2.5\ \text{g}}$$

Percentages can also be used as conversion factors. A percentage is a ratio; it relates the number of parts of one component to 100 total parts. For example, a fruit drink containing 10% sugar by mass contains 10 g of sugar in every 100 g of fruit drink. The conversion factors for the fruit drink are as follows.

$$\frac{10\ \text{g sugar}}{100\ \text{g fruit drink}} \quad \text{and} \quad \frac{100\ \text{g fruit drink}}{10\ \text{g sugar}}$$

## PRACTICE Problems

Do additional problems. Online Practice

**17.** Write two conversion factors for each of the following.

**a.** a 16% (by mass) salt solution

**b.** a density of 1.25 g/mL

**c.** a speed of 25 m/s

**18. Challenge** What conversion factors are needed to convert:

**a.** nanometers to meters?

**b.** density given in $g/cm^3$ to a value in $kg/m^3$?

**Using conversion factors** A conversion factor used in dimensional analysis must accomplish two things: it must cancel one unit and introduce a new one. While working through a solution, all of the units except the desired unit must cancel. Suppose you want to know how many meters there are in 48 km. The relationship between kilometers and meters is 1 km = 1000 m. The conversion factors are as follows.

$$\frac{1\text{ km}}{1000\text{ m}} \quad \text{and} \quad \frac{1000\text{ m}}{1\text{ km}}$$

Because you need to convert km to m, you should use the conversion factor that causes the km unit to cancel.

$$48\ \cancel{\text{km}} \times \frac{1000\text{ m}}{1\ \cancel{\text{km}}} = 48{,}000\text{ m}$$

When converting a value with a large unit, such as km, to a value with a smaller unit, such as m, the numerical value increases. For example, 48 km (a value with a large unit) converts to 48,000 m (a larger numerical value with a smaller unit). **Figure 9** illustrates the connection between the numerical value and the size of the unit for a conversion factor.

Now consider this question: How many eight-packs of water would you need if the 32 people attending your party each had two bottles of water? To solve the problem, identify the given quantities and the desired result. There are 32 people and each of them drinks two bottles of water. The desired result is the number of eight-packs. Using dimensional analysis yields the following.

$$32\ \cancel{\text{people}} \times \frac{2\ \cancel{\text{bottles}}}{\cancel{\text{person}}} \times \frac{1\text{ eight-pack}}{8\ \cancel{\text{bottles}}} = 8\text{ eight-packs}$$

$$\frac{1\text{ km}}{1000\text{ m}}$$

**Figure 9** The two quantities shown above are equivalent; that is, 1 km = 1000 m. Note that a smaller numerical value (1) accompanies the larger unit (km), and a larger numerical value (1000) accompanies the smaller unit (m).

## PRACTICE Problems

Do additional problems. Online Practice

**Use Table 2 to solve each of the following.**

**19.** **a.** Convert 360 s to ms.

**b.** Convert 4800 g to kg.

**c.** Convert 5600 dm to m.

**d.** Convert 72 g to mg.

**e.** Convert $2.45 \times 10^2$ ms to s.

**f.** Convert 5 $\mu$m to km.

**g.** Convert $6.800 \times 10^3$ cm to km.

**h.** Convert $2.5 \times 10^1$ kg to Mg.

**20. Challenge** Write the conversion factors needed to determine the number of seconds in one year.

## EXAMPLE Problem 4

Find help with **unit conversion.** Math Handbook

EXAMPLE PROBLEM

**USING CONVERSION FACTORS** In ancient Egypt, small distances were measured in Egyptian cubits. An Egyptian cubit was equal to 7 palms, and 1 palm was equal to 4 fingers. If 1 finger was equal to 18.75 mm, convert 6 Egyptian cubits to meters.

### 1 ANALYZE THE PROBLEM

A length of 6 Egyptian cubits needs to be converted to meters.

**Known**

length = 6 Egyptian cubits
7 palms = 1 cubit
1 palm = 4 fingers
1 finger = 18.75 mm
1 m = 1000 mm

**Unknown**

**length = ? m**

### 2 SOLVE FOR THE UNKNOWN

Use dimensional analysis to convert the units in the following order.

cubits → palms → fingers → millimeters → meters

$$6 \text{ cubits} \times \frac{7 \text{ palms}}{1 \text{ cubit}} \times \frac{4 \text{ fingers}}{1 \text{ palm}} \times \frac{18.75 \text{ mm}}{1 \text{ finger}} \times \frac{1 \text{ meter}}{1000 \text{ mm}} = ?\ \text{m}$$

Multiply by a series of conversion factors that cancels all the units except meter, the desired unit.

$$6 \cancel{\text{cubits}} \times \frac{7 \cancel{\text{palms}}}{1 \cancel{\text{cubit}}} \times \frac{4 \cancel{\text{fingers}}}{1 \cancel{\text{palm}}} \times \frac{18.75 \cancel{\text{mm}}}{1 \cancel{\text{finger}}} \times \frac{1 \text{ meter}}{1000 \cancel{\text{mm}}} = 3.150 \text{ m}$$

Multiply and divide the numbers as indicated, and cancel the units.

### 3 EVALUATE THE ANSWER

Each conversion factor is a correct restatement of the original relationship, and all units except for the desired unit, meters, cancel.

## PRACTICE Problems

Do additional problems. Online Practice

**21.** The speedometer at right displays a car's speed in miles per hour. What is the car's speed in km/h? (1 km = 0.62 mile)

**22.** How many seconds are in 24 h?

**23. Challenge** Vinegar is 5.00% acetic acid by mass and has a density of 1.02 g/mL. What mass of acetic acid, in grams, is present in 185 mL of vinegar?

## SECTION 2 REVIEW

Section Self-Check

### Section Summary

- A number expressed in scientific notation is written as a coefficient between 1 and 10 multiplied by 10 raised to a power.
- To add or subtract numbers in scientific notation, the numbers must have the same exponent.
- To multiply or divide numbers in scientific notation, multiply or divide the coefficients and then add or subtract the exponents, respectively.
- Dimensional analysis uses conversion factors to solve problems.

**24.** MAINIDEA **Describe** how scientific notation makes it easier to work with very large or very small numbers.

**25. Express** the numbers 0.00087 and 54,200,000 in scientific notation.

**26. Write** the measured distance quantities $3 \times 10^{-4}$ cm and $3 \times 10^{4}$ km in regular notation.

**27. Write** a conversion factor relating cubic centimeters and milliliters.

**28. Solve** How many millimeters are there in $2.5 \times 10^{2}$ km?

**29. Explain** how dimensional analysis is used to solve problems.

**30. Apply Concepts** A classmate converts 68 km to meters and gets 0.068 m as the answer. Explain why this answer is incorrect, and identify the likely source of the error.

**31. Organize** Create a flowchart that outlines when to use dimensional analysis and when to use scientific notation.

## SECTION 3 Uncertainty in Data

**MAINIDEA** Measurements contain uncertainties that affect how a calculated result is presented.

### Essential Questions

- How do accuracy and precision compare?
- How can the accuracy of experimental data be described using error and percent error?
- What are the rules for significant figures and how can they be used to express uncertainty in measured and calculated values?

### Review Vocabulary

**experiment:** a set of controlled observations that test a hypothesis

### New Vocabulary

accuracy
precision
error
percent error
significant figure

**CHEM 4 YOU** When making cookies from a recipe, amounts are measured in cups, tablespoons, and teaspoons. Would a batch of cookies turn out well if you measured all of the ingredients using only a teaspoon? Most likely not, because measurement errors would build up.

## Accuracy and Precision

Just as each teaspoon you measure in the kitchen contains some amount of error, so does every scientific measurement made in a laboratory. When scientists make measurements, they evaluate both the accuracy and the precision of the measurements. Although you might think that the terms accuracy and precision basically mean the same thing, to a scientist, they have very different meanings.

**Accuracy** refers to how close a measured value is to an accepted value. **Precision** refers to how close a series of measurements are to one another. The archery target in **Figure 10** illustrates the difference between accuracy and precision. For example, you measure in the lab the mass of an object three times. The arrows represent each measurement and the center of the target is the accepted value.

■ **Figure 10** An archery target illustrates the difference between accuracy and precision. An accurate shot is located near the bull's-eye; precise shots are grouped closely together.
**Apply** *Why is it important to measure the same data more than once?*

View an **animation about precision and accuracy.**

Concepts In Motion

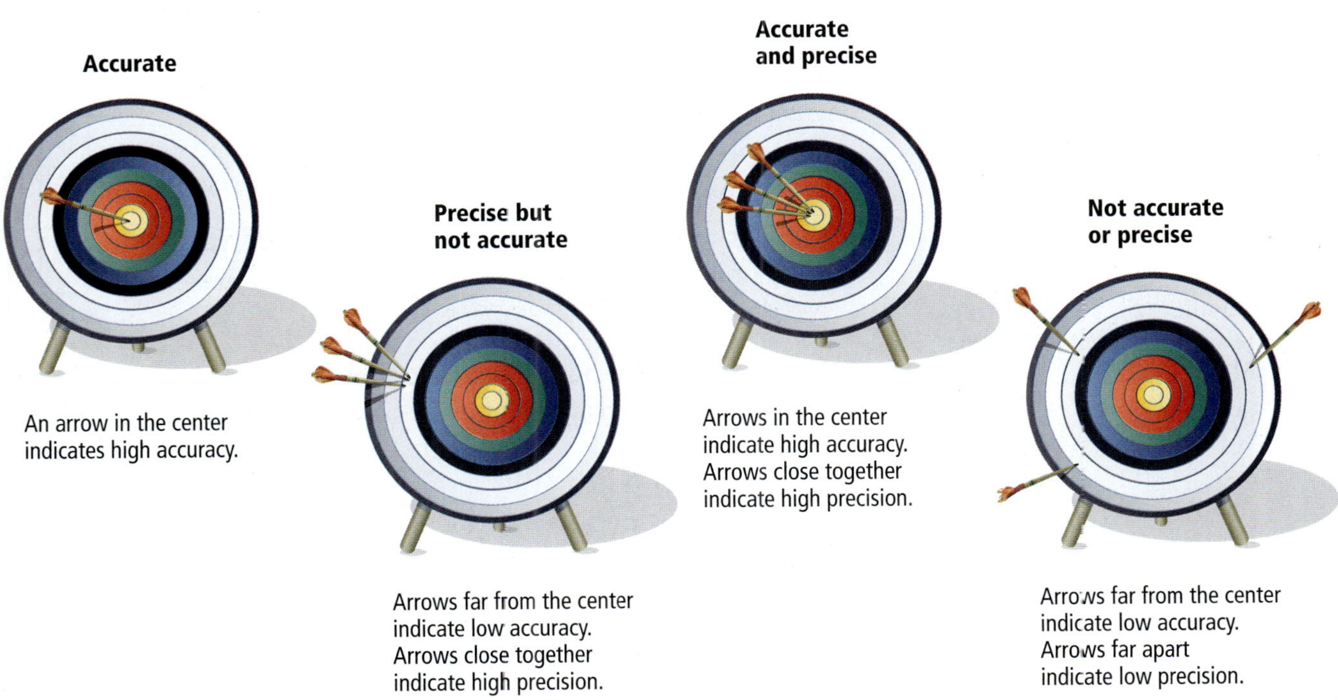

**Table 3 Student Density and Error Data (Unknown was sucrose; density = 1.59 g/cm³)**

| | Student A | | Student B | | Student C | |
|---|---|---|---|---|---|---|
| | Density | Error (g/cm³) | Density | Error (g/cm³) | Density | Error (g/cm³) |
| Trial 1 | 1.54 g/cm³ | −0.05 | 1.40 g/cm³ | −0.19 | (a) 1.70 g/cm³ | +0.11 |
| Trial 2 | 1.60 g/cm³ | +0.01 | 1.68 g/cm³ | +0.09 | 1.69 g/cm³ | +0.10 |
| Trial 3 | 1.57 g/cm³ | −0.02 | 1.45 g/cm³ | −0.14 | 1.71 g/cm³ | + 0.12 |
| Average | (b) 1.57 g/cm³ | | 1.51 g/cm³ | | 1.70 g/cm³ | |

**(a) These trial values are the most precise.**

**(b) This average is the most accurate.**

Consider the data in **Table 3.** Students were asked to find the density of an unknown white powder. Each student measured the volume and mass of three separate samples. They reported calculated densities for each trial and an average of the three calculations. The powder, sucrose (table sugar), has a density of 1.59 g/cm³. Which student collected the most accurate data? Who collected the most precise data? Student A's measurements are the most accurate because they are closest to the accepted value of 1.59 g/cm³. Student C's measurements are the most precise because they are the closest to one another.

Recall that precise measurements might not be accurate. Looking at just the average of the densities can be misleading. Based solely on the average, Student B appears to have collected fairly reliable data. However, on closer inspection, Student B's data are neither accurate nor precise. The data are not close to the accepted value, nor are they close to one another.

Get help with **precision measurement.**

Personal Tutor

**Error and percent error** The density values reported in **Table 3** are experimental values, which means they are values measured during an experiment. The known density of sucrose is an accepted value, which is a value that is considered true. To evaluate the accuracy of experimental data, you can compare how close the experimental value is to the accepted value. **Error** is defined as the difference between an experimental value and an accepted value. The errors for the experimental density values are also given in **Table 3.**

**Error Equation**

$$\text{error} = \text{experimental value} - \text{accepted value}$$

The error associated with an experimental value is the difference between the experimental value and the accepted value.

**VOCABULARY**

**WORD ORIGIN**

**Percent**

comes from the Latin words *per,* which means *by,* and *centum,* which means *100*

Scientists often want to know what percent of the accepted value an error represents. **Percent error** expresses error as a percentage of the accepted value.

**Percent Error Equation**

$$\text{percent error} = \frac{|\text{error}|}{\text{accepted value}} \times 100$$

The percent error of an experimental value equals the absolute value of its error divided by the accepted value, multiplied by 100.

Notice that the percent-error equation uses the absolute value of the error. This is because only the size of the error matters; it does not matter whether the experimental value is larger or smaller than the accepted value.

**READING CHECK** **Summarize** why error is important.

Percent error is an important concept for the machinist who made the nut shown in **Figure 11.** The machinist must check the tolerances of the nut. Tolerances are a narrow range of allowable dimensions based on acceptable amounts of error. If the dimensions of the nut do not fall within the acceptable range—that is, the nut exceeds its tolerances—it will be retooled or possibly discarded.

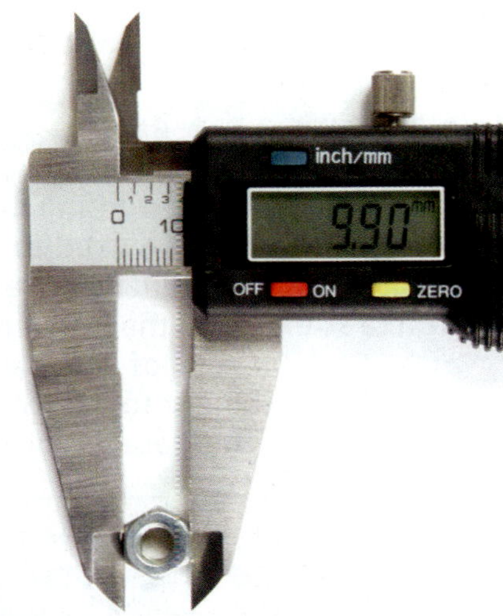


**Figure 11** This digital caliper is being used to check the size of a nut to one-hundredth of a millimeter (0.01 mm). Skill is required to correctly position the part in the caliper. Experienced machinists will obtain more precise and more accurate readings than inexperienced machinists.

## EXAMPLE Problem 5

Find help with **percents.** Math Handbook

**CALCULATING PERCENT ERROR** Use Student A's density data in **Table 3** to calculate the percent error in each trial. Report your answers to two places after the decimal point.

### 1 ANALYZE THE PROBLEM

You are given the errors for a set of density calculations. To calculate percent error, you need to know the accepted value for density, the errors, and the equation for percent error.

**Known**

accepted value for density = 1.59 g/cm$^3$

errors: −0.05 g/cm$^3$; 0.01 g/cm$^3$; −0.02 g/cm$^3$

**Unknown**

**percent errors = ?**

### 2 SOLVE FOR THE UNKNOWN

$$\text{percent error} = \frac{|\text{error}|}{\text{accepted value}} \times 100$$

**State the percent error equation.**

$$\text{percent error} = \frac{|-0.05\ \text{g/cm}^3|}{1.59\ \text{g/cm}^3} \times 100 = 3.14\%$$

**Substitute error = −0.05 g/cm$^3$, and solve.**

$$\text{percent error} = \frac{|0.01\ \text{g/cm}^3|}{1.59\ \text{g/cm}^3} \times 100 = 0.63\%$$

**Substitute error = 0.01 g/cm$^3$, and solve.**

$$\text{percent error} = \frac{|-0.02\ \text{g/cm}^3|}{1.59\ \text{g/cm}^3} \times 100 = 1.26\%$$

**Substitute error = −0.02 g/cm$^3$, and solve.**

### 3 EVALUATE THE ANSWER

The percent error is greatest for Trial 1, which had the largest error, and smallest for Trial 2, which was closest to the accepted value.

## PRACTICE Problems

Do additional problems. Online Practice

**Answer the following questions using data from Table 3.**

**32.** Calculate the percent errors for Student B's trials.

**33.** Calculate the percent errors for Student C's trials.

**34.** **Challenge** Based on your calculations in questions 32 and 33, which student's trial was the most accurate? The least accurate?

# Problem-Solving LAB

## Identify an Unknown

**How can mass and volume data for an unknown sample be used to identify the unknown?** A student collected several samples from a stream bed that looked like gold. She measured the mass of each sample and used water displacement to determine each sample's volume. Her data are given in the table.

**Mass and Volume Data for an Unknown Sample**

| Sample | Mass | Initial Volume (water only) | Final Volume (water + sample) |
|---|---|---|---|
| 1 | 50.25 g | 50.1 mL | 60.3 mL |
| 2 | 63.56 g | 49.8 mL | 62.5 mL |
| 3 | 57.65 g | 50.2 mL | 61.5 mL |
| 4 | 55.35 g | 45.6 mL | 56.7 mL |
| 5 | 74.92 g | 50.3 mL | 65.3 mL |
| 6 | 67.78 g | 47.5 mL | 60.8 mL |

### Analysis

For a given sample, the difference in the volume measurements made with the graduated cylinder yields the volume of the sample. Thus, for each sample, the mass and volume are known, and the density can be calculated. Note that density is a property of matter that can often be used to identify an unknown sample.

### Think Critically

1. **Calculate** the volume and density for each sample and the average density of the six samples. Be sure to use significant figure rules.
2. **Apply** The student hopes the samples are gold, which has a density of 19.3 g/cm$^3$. A local geologist suggested the samples might be pyrite, which is a mineral with a density of 5.01 g/cm$^3$. What is the identity of the unknown sample?
3. **Calculate** the error and percent error of each sample. Use the density value given in Question 2 as the accepted value.
4. **Conclude** Was the data collected by the student accurate? Explain your answer.

## Significant Figures

Often, precision is limited by the tools available. For example, a digital clock that displays the time as 12:47 or 12:48 can record the time only to the nearest minute. With a stopwatch, however, you might record time to the nearest hundredth second. As scientists have developed better measuring devices, they have been able to make more precise measurements. Of course, for measurements to be both accurate and precise, the measuring devices must be in good working order. Additionally, accurate and precise measurements rely on the skill of the person using the instrument; the user must be trained and use proper techniques.

The precision of a measurement is indicated by the number of digits reported. A value of 3.52 g is more precise than a value of 3.5 g. The reported digits are called significant figures. **Significant figures** include all known digits plus one estimated digit. Consider the rod in **Figure 12.** The end of the rod falls between 5.2 cm and 5.3 cm. The 5 and 2 are known digits corresponding to marks on the ruler. To these known digits, an estimated digit is added. This last digit estimates the rod's location between the second and third millimeter marks. Because it is an estimate, one person might report the measurement as 5.22 cm and another as 5.23 cm. Either way, the measurement has three significant figures—two known and one estimated.

Remember that measurements reported with a lot of significant figures might be precise but not accurate. For example, some chemistry labs have balances that report mass to the nearest hundreth of a gram. If you and each of your classmates measured the same copper cylinder on the same balance, you would probably have a group of very precise measurements. But what if the balance had been previously damaged by an object that was too large for it? Your precise measurements would not be very accurate.

**Figure 12** The markings on the ruler represent known digits. The reported measurement includes the known digits plus the estimated digit. The measurement is 5.23 cm.

**Infer** ***What is the estimated digit if the length of an object being measured falls exactly on the 5-cm mark?***

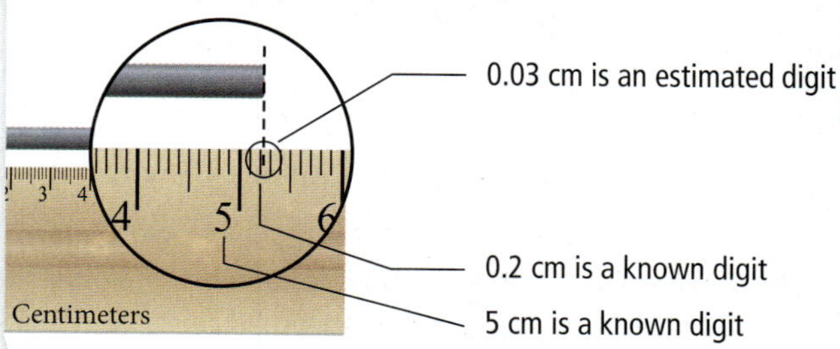

## PROBLEM-SOLVING STRATEGY

### Recognizing Significant Figures

Learning these five rules for recognizing significant figures will help you when solving problems. Examples of each rule are shown below. Note that each of the highlighted examples has three significant figures.

| | |
|---|---|
| **Rule 1.** Nonzero numbers are always significant. | 72.3 g has three. |
| **Rule 2.** All final zeros to the right of the decimal are significant. | 6.20 g has three. |
| **Rule 3.** Any zero between significant figures is significant. | 60.5 g has three. |
| **Rule 4.** Placeholder zeroes are not significant. To remove placeholder zeros, rewrite the number in scientific notation. | 0.0253 g and 4320 g (each has three) |
| **Rule 5.** Counting numbers and defined constants have an infinite number of significant figures. | 6 molecules<br>60 s = 1 min |

## EXAMPLE Problem 6

Find help with **significant figures.** Math Handbook

**SIGNIFICANT FIGURES** Determine the number of significant figures in the following masses.

**a.** 0.00040230 g

**b.** 405,000 kg

### 1 ANALYZE THE PROBLEM

You are given two measured mass values. Apply the appropriate rules to determine the number of significant figures in each value.

### 2 SOLVE FOR THE UNKNOWN

Count all nonzero numbers, zeros between nonzero numbers, and final zeros to the right of the decimal place. **(Rules 1, 2, and 3)**
Ignore zeros that act as placeholders. **(Rule 4)**

**a.** 0.00040230 g has five significant figures.

**b.** 405,000 kg has three significant figures.

### 3 EVALUATE THE ANSWER

One way to verify your answers is to write the values in scientific notation: $4.0230 \times 10^{-4}$ g and $4.05 \times 10^{5}$ kg. Without the placeholder zeros, it is clear that 0.00040230 g has five significant figures and that 405,000 kg has three significant figures.

## PRACTICE Problems

Do additional problems. Online Practice

**Determine the number of significant figures in each measurement.**

**35. a.** 508.0 L **c.** $1.0200 \times 10^{5}$ kg
**b.** 820,400.0 L **d.** 807,000 kg

**36. a.** 0.049450 s **c.** $3.1587 \times 10^{-4}$ g
**b.** 0.000482 mL **d.** 0.0084 mL

**37. Challenge** Write the numbers 10, 100, and 1000 in scientific notation with two, three, and four significant figures, respectively.

## Rounding Numbers

Calculators perform flawless arithmetic, but they are not aware of the number of significant figures that should be reported in the answer. For example, a density calculation should not have more significant figures than the original data with the fewest significant figures. To report a value correctly, you often need to round. Consider an object with a mass of 22.44 g and volume of 14.2 $cm^3$. When you calculate the object's density using a calculator, the displayed answer is 1.5802817 $g/cm^3$, as shown in **Figure 13.** Because the measured mass had four significant figures and the measured volume had three, it is not correct to report the calculated density value with eight significant figures. Instead, the density must be rounded to three significant figures, or 1.58 $g/cm^3$.

■ **Figure 13** You need to apply the rules of significant figures and rounding to report a calculated value correctly.

Consider the value 3.515014. How would you round this number to five significant figures? To three significant figures? In each case, you need to look at the digit that follows the desired last significant figure.

To round to five digits, first identify the fifth significant figure, in this case 0, and then look at the number to its right, in this case 1.

3.515014 — **Last significant figure**: 0; **Number to right of last significant figure**: 1

Do not change the last significant figure if the digit to its right is less than five. Because a 1 is to the right, the number rounds to 3.5150. If the number had been 5 or greater, you would have rounded up.

To round to three digits, identify the third significant figure, in this case 1, and then look at the number to its right, in this case 5.

3.515014 — **Last significant figure**: 1; **Number to right of last significant figure**: 5

If the digits to the right of the last significant figure are a 5 followed by 0, then look at the last significant figure. If it is odd, round it up; if it is even, do not round up. Because the last significant digit is odd (1), the number rounds up to 3.52.

### PROBLEM-SOLVING STRATEGY

**Rounding Numbers**

Learn these four rules for rounding, and use them when solving problems. Examples of each rule are shown below. Note that each example has three significant figures.

| Example | Rule |
|---|---|
| 2.532 → 2.53 | **Rule 1.** If the digit to the right of the last significant figure is less than 5, do not change the last significant figure. |
| 2.536 → 2.54 | **Rule 2.** If the digit to the right of the last significant figure is greater than 5, round up the last significant figure. |
| 2.5351 → 2.54 | **Rule 3.** If the digits to the right of the last significant figure are a 5 followed by a nonzero digit, round up the last significant figure. |
| 2.5350 → 2.54<br>2.5250 → 2.52 | **Rule 4.** If the digits to the right of the last significant figure are a 5 followed by 0 or no other number at all, look at the last significant figure. If it is odd, round it up; if it is even, do not round up. |

Matt Meadows

## PRACTICE Problems

Do additional problems. Online Practice

**38.** Round each number to four significant figures.

**a.** 84,791 kg  **c.** 256.75 cm

**b.** 38.5432 g  **d.** 4.9356 m

**39. Challenge** Round each number to four significant figures, and write the answer in scientific notation.

**a.** 0.00054818 g  **c.** 308,659,000 mm

**b.** 136,758 kg  **d.** 2.0145 mL

**Addition and subtraction** When you add or subtract measurements, the answer must have the same number of digits to the right of the decimal as the original value having the fewest number of digits to the right of the decimal. For example, the measurements 1.24 mL, 12.4 mL, and 124 mL have two, one, and zero digits to the right of the decimal, respectively. When adding or subtracting, arrange the values so that the decimal points align. Identify the value with the fewest places after the decimal point, and round the answer to that number of places.

**Multiplication and division** When you multiply or divide numbers, your answer must have the same number of significant figures as the measurement with the fewest significant figures.

## EXAMPLE Problem 7

Find help with **significant figures.** Math Handbook

EXAMPLE PROBLEM

**ROUNDING NUMBERS WHEN ADDING** A student measured the length of his lab partners' shoes. If the lengths are 28.0 cm, 23.538 cm, and 25.68 cm, what is the total length of the shoes?

### 1 ANALYZE THE PROBLEM

The three measurements need to be aligned on their decimal points and added. The measurement with the fewest digits after the decimal point is 28.0 cm, with one digit. Thus, the answer must be rounded to only one digit after the decimal point.

### 2 SOLVE FOR THE UNKNOWN

$$\begin{array}{r} 28.0 \text{ cm} \\ 23.538 \text{ cm} \\ +\ 25.68 \text{ cm} \\ \hline 77.218 \text{ cm} \end{array}$$

**Align the measurements and add the values.**

The answer is **77.2 cm.** **Round to one place after the decimal; Rule 1 applies.**

### 3 EVALUATE THE ANSWER

The answer, 77.2 cm, has the same precision as the least-precise measurement, 28.0 cm.

## PRACTICE Problems

Do additional problems. Online Practice

**40.** Add and subtract as indicated. Round off when necessary.

**a.** 43.2 cm + 51.0 cm + 48.7 cm

**b.** 258.3 kg + 257.11 kg + 253 kg

**41. Challenge** Add and subtract as indicated. Round off when necessary.

**a.** $(4.32 \times 10^3 \text{ cm}) - (1.6 \times 10^4 \text{ mm})$

**b.** $(2.12 \times 10^7 \text{ mm}) + (1.8 \times 10^3 \text{ cm})$

EXAMPLE PROBLEM

## EXAMPLE Problem 8

Find help with **rounding.** Math Handbook

**ROUNDING NUMBERS WHEN MULTIPLYING** Calculate the volume of a book with the following dimensions: length = 28.3 cm, width = 22.2 cm, height = 3.65 cm.

### 1 ANALYZE THE PROBLEM

Volume is calculated by multiplying length, width, and height. Because all of the measurements have three significant figures, the answer also will.

**Known**

length = 28.3 cm
width = 22.2 cm
height = 3.65 cm

**Unknown**

**Volume = ? $cm^3$**

### 2 SOLVE FOR THE UNKNOWN

Calculate the volume, and apply the rules of significant figures and rounding.

**Volume** = length × width × height — **State the formula for the volume of a rectangle.**

**Volume** = 28.3 cm × 22.2 cm × 3.65 cm = 2293.149 $cm^3$ — **Substitute values and solve.**

**Volume** = **2290 $cm^3$** — **Round the answer to three significant figures.**

### 3 EVALUATE THE ANSWER

To check if your answer is reasonable, round each measurement to one significant figure and recalculate the volume. Volume = 30 cm × 20 cm × 4 cm = 2400 $cm^3$. Because this value is close to your calculated value of 2290 $cm^3$, it is reasonable to conclude the answer is correct.

## PRACTICE Problems

Do additional problems. Online Practice

**Perform the following calculations. Round the answers.**

**42.** **a.** 24 m × 3.26 m **b.** 120 m × 0.10 m
**c.** 1.23 m × 2.0 m **d.** 53.0 m × 1.53 m

**43.** **a.** 4.84 m ÷ 2.4 s **b.** 60.2 m ÷ 20.1 s
**c.** 102.4 m ÷ 51.2 s **d.** 168 m ÷ 58 s

**44.** **Challenge** $(1.32 \times 10^3 \text{ g}) \div (2.5 \times 10^2 \text{ cm}^3)$

# SECTION 3 REVIEW

Section Self-Check

### Section Summary

- An accurate measurement is close to the accepted value. A set of precise measurements shows little variation.
- The measurement device determines the degree of precision possible.
- Error is the difference between the measured value and the accepted value. Percent error gives the percent deviation from the accepted value.
- The number of significant figures reflects the precision of reported data.
- Calculations are often rounded to the correct number of significant figures.

**45. MAINIDEA State** how a measured value is reported in terms of known and estimated digits.

**46. Define** *accuracy* and *precision.*

**47. Identify** the number of significant figures in each of these measurements of an object's length: 76.48 cm, 76.47 cm, and 76.59 cm.

**48. Apply** The object in Question 47 has an actual length of 76.49 cm. Are the measurements in Question 47 accurate? Are they precise?

**49. Calculate** the error and percent error for each measurement in Question 47.

**50. Apply** Write an expression for the quantity 506,000 cm in which it is clear that all the zeros are significant.

**51. Analyze Data** Students collected mass data for a group of coins. The mass of a single coin is 5.00 g. Determine the accuracy and precision of the measurements.

| Number of coins | 5 | 10 | 20 | 30 | 50 |
|---|---|---|---|---|---|
| Mass (g) | 23.2 | 54.5 | 105.9 | 154.5 | 246.2 |

SECTION 4

# Representing Data

**MAIN IDEA** Graphs visually depict data, making it easier to see patterns and trends.

### Essential Questions

- Why are graphs created?
- How can graphs be interpreted?

### Review Vocabulary

**independent variable:** the variable that is changed during an experiment

### New Vocabulary

graph

**CHEM 4 YOU** Have you ever heard the saying, "A picture is worth a thousand words"? A graph is a "picture" of data. Scientists use graphs to present data in a form that allows them to analyze their results and communicate information about their experiments.

## Graphing

When you analyze data, you might set up an equation and solve for an unknown, but this is not the only method scientists have for analyzing data. A goal of many experiments is to discover whether a pattern exists in a certain situation. Does raising the temperature change the rate of a reaction? Does a change in diet affect a rat's ability to navigate a maze? When data are listed as shown in **Table 4,** a pattern might not be obvious. However, using data to create a graph can help to reveal a pattern if one exists. A **graph** is a visual display of data.

**Circle graphs** Newspapers and magazines often feature circle graphs. A circle graph, like the one shown in **Figure 14,** is sometimes called a pie chart because it is divided into wedges that look like a pie. A circle graph is useful for showing parts of a fixed whole. The parts are usually labeled as percents with the whole circle representing 100%. The circle graph shown in **Figure 14** is based on the percentage data given in **Table 4.**

■ **Figure 14** Although the percentage data presented in the table and the circle graph are basically the same, the circle graph makes it much easier to analyze.

**Table 4 Sources of Chlorine in the Stratosphere**

| Source | Percent |
|---|---|
| Hydrogen chloride (HCl) | 3 |
| Methyl chloride ($CH_3Cl$) | 15 |
| Carbon tetrachloride ($CCl_4$) | 12 |
| Methyl chloroform ($C_2H_3Cl_3$) | 10 |
| CFC-11 | 23 |
| CFC-12 | 28 |
| CFC-13 | 6 |
| HCFC-22 | 3 |

**Chlorine in the Stratosphere**

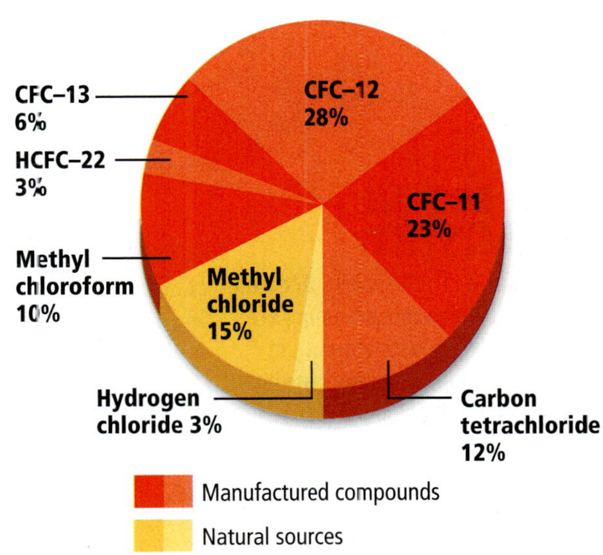


**GRAPH CHECK**

**Analyze** What percent of the chlorine sources are natural? What percent are manufactured compounds?

■ **Figure 15** A bar graph is an effective way to present and compare data. This graph shows various dietary sources of the element magnesium. Magnesium plays an important role in the health of your muscles, nerves, and bones.

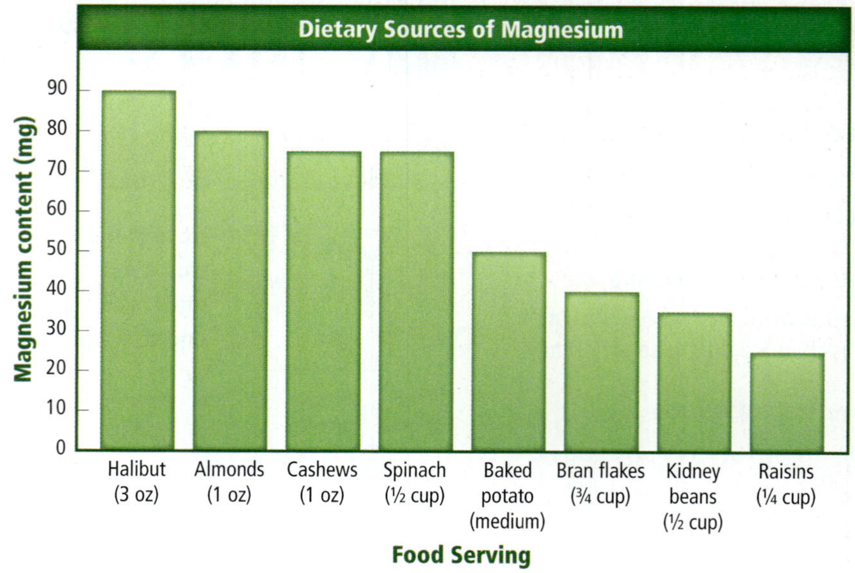


☑ GRAPH CHECK

**Interpret** Which two food servings provide equal amounts of magnesium?

Get help with **graphing.**

Personal Tutor

FOLDABLES®

Incorporate information from this section into your Foldable.

**Bar graphs** A bar graph is often used to show how a quantity varies across categories. Examples of categories include time, location, and temperature. The quantity being measured appears on the vertical axis ($y$-axis). The independent variable appears on the horizontal axis ($x$-axis). The relative heights of the bars show how the quantity varies. A bar graph can be used to compare population figures for a single country by decade or the populations of multiple countries at the same point in time. In **Figure 15,** the quantity being measured is magnesium, and the category being varied is food servings. When examining the graph, you can quickly see how the magnesium content varies for these food servings.

**Line Graphs** In chemistry, most graphs that you create and interpret will be line graphs. The points on a line graph represent the intersection of data for two variables.

**Independent and dependent variables** The independent variable is plotted on the $x$-axis. The dependent variable is plotted on the $y$-axis. Remember that the independent variable is the variable that a scientist deliberately changes during an experiment. In **Figure 16a,** the independent variable is volume and the dependent variable is mass. What are the values for the independent variable and the dependent variable at Point B? **Figure 16b** is a graph of temperature versus elevation. Because the data points do not fit perfectly, the line cannot pass exactly through all of the points. The line must be drawn so that about as many points fall above the line as fall below it. This line is called a best-fit line.

**Relationships between variables** If the best-fit line for a set of data is straight, there is a linear relationship between the variables and the variables are said to be directly related. The relationship between the variables can be described further by analyzing the steepness, or slope, of the line.

## CAREERS IN CHEMISTRY

**Calibration Technician** Accurate and repeatable measurements are essential to chemists working in research and industry. An instrument calibration technician adjusts, troubleshoots, maintains, and repairs the instruments used in laboratories and manufacturing plants. Their jobs require an understanding of the instrument's electronics and the use of computers and calibration software.

WebQuest

If the best-fit line rises to the right, then the slope of the line is positive. A positive slope indicates that the dependent variable increases as the independent variable increases. If the best-fit line sinks to the right, then the slope of the line is negative. A negative slope indicates that the dependent variable decreases as the independent variable increases. In either case, the slope of the line is constant.

You can use two pairs of data points to calculate the slope of the line. The slope is the rise, or change in $y$, denoted as $\Delta y$, divided by the run, or change in $x$, denoted as $\Delta x$.

**Slope Equation**

$$\text{slope} = \frac{\text{rise}}{\text{run}} = \frac{\Delta y}{\Delta x} = \frac{y_2 - y_1}{x_2 - x_1}$$

$y_2$, $y_1$, $x_2$, and $x_1$ are values from data points $(x_1, y_1)$ and $(x_2, y_2)$.

The slope of a line is equal to the change in $y$ divided by the change in $x$.

When the mass of a material is plotted against its volume, the slope of the line represents the material's density. An example of this is shown in **Figure 16a.** To calculate the slope of the line, substitute the $x$ and $y$ values for Points A and B in the slope equation and solve.

$$\text{slope} = \frac{54\text{ g} - 27\text{ g}}{20.0\text{ cm}^3 - 10.0\text{ cm}^3}$$
$$= \frac{27\text{ g}}{10.0\text{ cm}^3}$$
$$= 2.7\text{ g/cm}^3$$

Thus, the slope of the line, and the density, is 2.7 g/cm$^3$.

When the best-fit line is curved, the relationship between the variables is nonlinear. In chemistry, you will study nonlinear relationships called inverse relationships. Refer to the Math Handbook for more discussion of graphs.

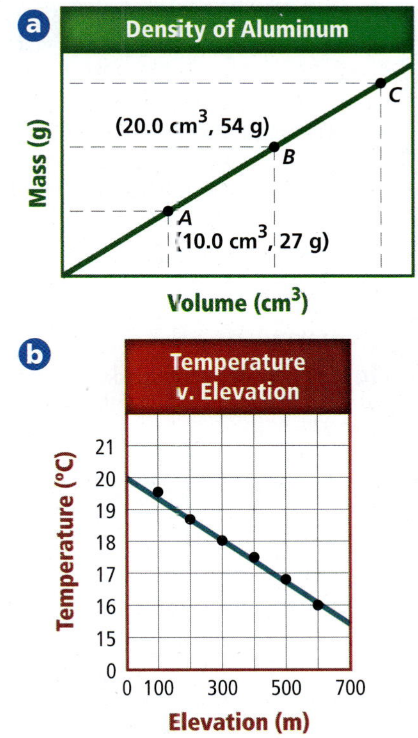


**Figure 16** Both of these line graphs show linear relationships. The slope of each line is defined as the ratio of rise over run.

GRAPH CHECK **Identify** the graph that shows a direct relationship.

## Interpreting Graphs

You should use an organized approach when analyzing graphs. First, note the independent and dependent variables. Next, decide if the relationship between the variables is linear or nonlinear. If the relationship is linear, is the slope positive or negative?

**Interpolation and extrapolation** When points on a line graph are connected, the data are considered to be continuous. Continuous data allow you to read the value from any point that falls between the recorded data points. This process is called interpolation. For example, from **Figure 16b,** what is the temperature at an elevation of 350 m? To interpolate this value, first locate 350 m on the $x$-axis; it is located halfway between 300 m and 400 m. Project upward until you hit the plotted line, and then project that point horizontally to the left until you reach the $y$-axis. The temperature at 350 m is approximately 17.8°C.

You can also extend a line beyond the plotted points in order to estimate values for the variables. This process is called extrapolation. It is important to be very careful with extrapolation, however, as it can easily lead to errors and result in very inaccurate predictions.

READING CHECK **Explain** why extrapolation might be less reliable than interpolation.

Watch a video about **tracking crime.**

Video

■ **Figure 17** The two lines in this graph represent average ozone levels for two time periods, 1957–1972 and 1979–2010. The graph shows clearly that ozone levels in recent years have been lower overall than in 1957–1972. The ozone hole is generally considered to be the area where total ozone is less than 220 Dobson Units (DU).

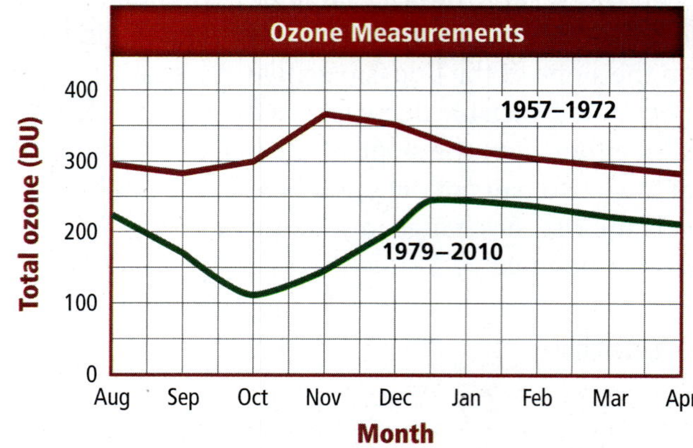


**GRAPH CHECK**

**Interpret** By how much did the total ozone vary during the 9-month period shown for 1979–2010?

**Interpreting ozone data** The value of using graphs to visualize data is illustrated by **Figure 17.** These important ozone measurements were taken at the Halley Research Station in Antarctica. The graph shows how ozone levels vary from August to April. The independent and dependent variables are the month and the total ozone, respectively.

Each line on the graph represents a different period of time. The red line represents average ozone levels from 1957 to 1972, during which time ozone levels varied from about 285 DU (Dobson units) to 360 DU. The green line shows the ozone levels from 1979 to 2010. At no point during this period were the ozone levels as high as they were at corresponding times during 1957–1972.

The graph makes the ozone hole clearly evident—the dip in the green line indicates the presence of the ozone hole. Having data from two time periods on the same graph allows scientists to compare recent data with data from a time before the ozone hole existed. Graphs similar to **Figure 17** helped scientists identify a significant trend in ozone levels and verify the depletion in ozone levels over time.

## SECTION 4 REVIEW

Section Self-Check

### Section Summary

- Circle graphs show parts of a whole. Bar graphs show how a factor varies with time, location, or temperature.
- Independent ($x$-axis) variables and dependent ($y$-axis) variables can be related in a linear or a nonlinear manner. The slope of a straight line is defined as rise/run, or $\Delta y/\Delta x$.
- Because line-graph data are considered continuous, you can interpolate between data points or extrapolate beyond them.

**52.** MAINIDEA **Explain** why graphing can be an important tool for analyzing data.

**53. Infer** What type of data must be plotted on a graph for the slope of the line to represent density?

**54. Relate** If a linear graph has a negative slope, what can you say about the dependent variable?

**55. Summarize** What data are best displayed on a circle graph? On a bar graph?

**56. Construct** a circle graph for the composition of air: 78.08% N, 20.95% $O_2$, 0.93% Ar, and 0.04% $CO_2$ and other gases.

**57. Infer** from **Figure 17** how long the ozone hole lasts.

**58. Apply** Graph mass versus volume for the data given in the table. What is the slope of the line?

| Volume (cm$^3$) | 7.5 | 12 | 15 | 22 |
|---|---|---|---|---|
| Mass (g) | 24.1 | 38.5 | 48.0 | 70.1 |

# CHEMISTRY&health

## Toxicology: Assessing Health Risk

It is likely that a closet or cupboard in your home or school contains products labeled with the symbol shown in **Figure 1.** Many cleaning, painting, and gardening products contain poisonous chemicals. Exposure to these chemicals can be dangerous. Possible effects are headaches, nausea, rashes, convulsions, coma, and even death. A toxicologist works to protect human health by studying the harmful effects of the chemicals and determining safe levels of exposure to them.

**Figure 1** A skull-and-crossbones is the symbol for poison.

**Keys to toxicity** Warfarin is a drug used to prevent blood clots in people who have had a stroke or heart attack. It is also an effective rat poison. How is this possible? One key to toxicity is the dose—the amount of the chemical taken in by an organism. Exposure time can also be a factor; even low-dose exposure to some chemicals over long periods of time can be hazardous. Toxicity is also affected by the presence of other chemicals in the body, the age and gender of the individual, and the chemical's ability to be absorbed and excreted.

A dose-response curve, such as the one shown in **Figure 2,** relates the toxicity of a substance to its physical effects. This dose-response curve shows the results of an experiment in which different doses of a possible carcinogen were given to mice. The mice were checked for tumors 90 days after exposure. The graph indicates a noticeable increase in the incidence of tumors.

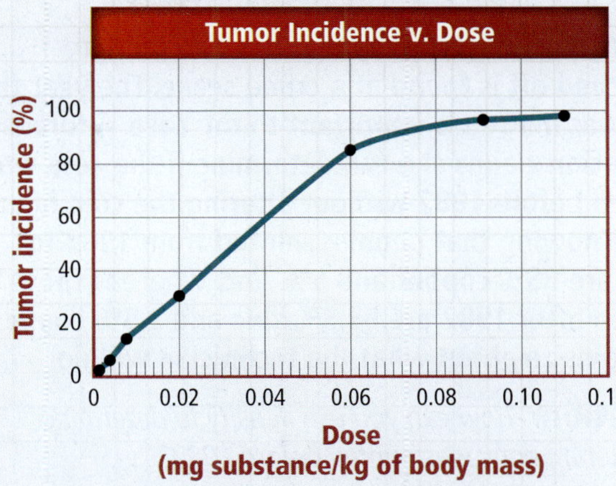


**Figure 2** The seven data points correspond to seven groups of mice that were given different doses of a possible carcinogen.

**Applying toxicity data** How do toxicologists predict health risks to people? Toxicity data might be available from studies of routine chemical exposure in the workplace, as well as from medical records of accidental chemical contact. Toxicity testing is often carried out using bacteria and cell cultures. Toxicologists observe the effect of chemical doses on bacteria. If mutations occur, the chemical is considered potentially harmful.

**MSDS** Toxicologists apply mathematical models and knowledge of similar substances to toxicity data to estimate safe human exposure levels. How can you obtain this information? Every employer is required to keep Material Safety Data Sheets (MSDS) of the potentially hazardous chemicals they use in their workplace. The MSDS describe possible health effects, clothing and eye protection that should be worn, and first-aid steps to follow after exposure. You can also consult the Household Products Database, which provides health and safety information on more than 5000 commonly used products.

### WRITING IN ▶ Chemistry

**Research** Access the MSDS for several products used at home. Compare the possible adverse health effects of exposure to the products and list the first aid requirements.

WebQuest

# ChemLAB

iLab Station

## Forensics: Use Density to Date a Coin

**Background:** A penny that has had its date scratched off is found at a crime scene. The year the coin was minted is important to the case. A forensics technician claims she can determine if the coin was minted before 1982 without altering the coin in any way. Knowing that pennies minted from 1962 to 1982 are 95% copper and 5% zinc, whereas those minted after 1982 are 97.5% zinc and 2.5% copper, hypothesize about what the technician will do.

**Question:** *How can you use density to determine whether a penny was minted before 1982?*

### Materials

water
small plastic cup
pre-1982 pennies (25)
metric ruler
graph paper
100-mL graduated cylinder
balance
post-1982 pennies (25)
pencil
graphing calculator (optional)

### Safety Precautions

### Procedure

1. Read and complete the lab safety form.
2. Record all measurements in your data table.
3. Measure the mass of the plastic cup.
4. Pour about 50 mL of water into the graduated cylinder. Record the actual volume.
5. Add 5 pre-1982 pennies to the cup, and measure the mass again.
6. Add the 5 pennies to the graduated cylinder, and read the volume.
7. Repeat Steps 5 and 6 four times. After five trials there will be 25 pennies in the graduated cylinder.
8. **Cleanup and Disposal** Pour the water from the graduated cylinder down a drain, being careful not to lose any of the pennies. Dry the pennies with a paper towel.
9. Repeat Steps 3 through 7, using post-1982 pennies.

**Data Table for the Density of a Penny**

| Trial | Mass of Pennies Added (g) | Total Number of Pennies | Total Mass of Pennies (g) | Total Volume of Water Displaced (mL) |
|---|---|---|---|---|
| 1 | | 5 | | |
| 2 | | 10 | | |
| 3 | | 15 | | |
| 4 | | 20 | | |
| 5 | | 25 | | |

### Analyze and Conclude

1. **Calculate** Complete the data table by calculating the total mass and the total volume of water displaced for each trial.
2. **Make and Use Graphs** Graph total mass versus total volume for the pre-1982 and post-1982 pennies. Plot and label two sets of points on the graph, one for pre-1982 pennies and one for post-1982 pennies.
3. **Make and Use Graphs** Draw a best-fit line through each set of points. Use two points on each line to calculate the slope.
4. **Infer** What do the slopes of the lines tell you about the two groups of pennies?
5. **Apply** Can you determine if a penny was minted before or after 1982 if you know only its mass? Explain how the relationships among volume, mass, and density support using a mass-only identification technique.
6. **Error Analysis** Determine the percent error in the density of each coin.

### INQUIRY EXTENSION

**Compare** your results with those from the rest of the class. Are they consistent? If not, explain how you could refine your investigation to ensure more accurate results. Calculate a class average density of the pre-1982 pennies and the density of the post-1982 pennies. Determine the percent error of each average.

# CHAPTER 2 STUDY GUIDE

Vocabulary Practice

**BIGIDEA** Chemists collect and analyze data to determine how matter interacts.

## SECTION 1 Units and Measurements

**MAINIDEA** Chemists use an internationally recognized system of units to communicate their findings.

- SI measurement units allow scientists to report data to other scientists.
- Adding prefixes to SI units extends the range of possible measurements.
- To convert to Kelvin temperature, add 273 to the Celsius temperature.

$$K = {}^{\circ}C + 273$$

- Volume and density have derived units. Density, which is a ratio of mass to volume, can be used to identify an unknown sample of matter.

$$\text{density} = \frac{\text{mass}}{\text{volume}}$$

**VOCABULARY**
- base unit
- second
- meter
- kilogram
- kelvin
- derived unit
- liter
- density

## SECTION 2 Scientific Notation and Dimensional Analysis

**MAINIDEA** Scientists often express numbers in scientific notation and solve problems using dimensional analysis.

- A number expressed in scientific notation is written as a coefficient between 1 and 10 multiplied by 10 raised to a power.
- To add or subtract numbers in scientific notation, the numbers must have the same exponent.
- To multiply or divide numbers in scientific notation, multiply or divide the coefficients and then add or subtract the exponents, respectively.
- Dimensional analysis uses conversion factors to solve problems.

**VOCABULARY**
- scientific notation
- dimensional analysis
- conversion factor

## SECTION 3 Uncertainty in Data

**MAINIDEA** Measurements contain uncertainties that affect how a calculated result is presented.

- An accurate measurement is close to the accepted value. A set of precise measurements shows little variation.
- The measurement device determines the degree of precision possible.
- Error is the difference between the measured value and the accepted value. Percent error gives the percent deviation from the accepted value.

$$\text{error} = \text{experimental value} - \text{accepted value}$$

$$\text{percent error} = \frac{|\text{error}|}{\text{accepted value}} \times 100$$

- The number of significant figures reflects the precision of reported data.
- Calculations are often rounded to the correct number of significant figures.

**VOCABULARY**
- accuracy
- precision
- error
- percent error
- significant figure

## SECTION 4 Representing Data

**MAINIDEA** Graphs visually depict data, making it easier to see patterns and trends.

- Circle graphs show parts of a whole. Bar graphs show how a factor varies with time, location, or temperature.
- Independent ($x$-axis) variables and dependent ($y$-axis) variables can be related in a linear or a nonlinear manner. The slope of a straight line is defined as rise/run, or $\Delta y/\Delta x$.

$$\text{slope} = \frac{y_2 - y_1}{x_2 - x_1} = \frac{\Delta y}{\Delta x}$$

- Because line graph data are considered continuous, you can interpolate between data points or extrapolate beyond them.

**VOCABULARY**
- graph

# CHAPTER 2 ASSESSMENT

## SECTION 1

### Mastering Concepts

**59.** Why must a measurement include both a number and a unit?

**60.** Explain why standard units of measurement are particularly important to scientists.

**61.** What role do prefixes play in the metric system?

**62.** How many meters are in one kilometer? In one decimeter?

**63. SI Units** What is the relationship between the SI unit for volume and the SI unit for length?

**64.** Explain how temperatures on the Celsius and Kelvin scales are related.

**65.** Examine the density values for several common liquids and solids given in **Table 5.** Sketch the results of an experiment that layered each of the liquids and solids in a 1000-mL graduated cylinder.

**Table 5 Density Values**

| Liquids (g/mL) | | Solids ($g/cm^3$) | |
|---|---|---|---|
| Ethyl alcohol | 0.789 | Bone | 1.85 |
| Glycerin | 1.26 | Cork | 0.24 |
| Isopropyl alcohol | 0.870 | Plastic | 0.91 |
| Corn syrup | 1.37 | Wood (oak) | 0.84 |
| Motor oil | 0.860 | | |
| Vegetable oil | 0.910 | | |
| Water at 4°C | 1.000 | | |

### Mastering Problems

**66.** A 5-mL sample of water has a mass of 5 g. What is the density of water?

**67.** The density of aluminum is 2.7 g/mL. What is the volume of 8.1 g?

**68.** An object with a mass of 7.5 g raises the level of water in a graduated cylinder from 25.1 mL to 30.1 mL. What is the density of the object?

**69. Candy Making** The directions in the candy recipe for pralines instruct the cook to remove the pot containing the candy mixture from the heat when the candy mixture reaches the soft-ball stage. The soft-ball stage corresponds to a temperature of 236°F. After the soft-ball stage is reached, the pecans and vanilla are added. Can a Celsius thermometer with a range of −10°C to 110°C be used to determine when the soft-ball stage is reached in the candy mixture?

## SECTION 2

### Mastering Concepts

**70.** How does scientific notation differ from ordinary notation?

**71.** If you move the decimal place to the left to convert a number to scientific notation, will the power of 10 be positive or negative?

**72.** Two undefined numbers expressed in regular notation are shown below, along with the number of places the decimal must move to express each in scientific notation. If each X represents a significant figure, write each number in scientific notation.

**a.** XXX.XX
**b.** 0.000000XXX

**73.** When dividing numbers in scientific notation, what must you do with the exponents?

**74.** When you convert from a small unit to a large unit, what happens to the number of units?

**75.** When converting from meters to centimeters, how do you decide which values to place in the numerator and denominator of the conversion factor?

### Mastering Problems

**76.** Write the following numbers in scientific notation.

**a.** 0.0045834 mm
**b.** 0.03054 g
**c.** 438,904 s
**d.** 7,004,300,000 g

**77.** Write the following numbers in ordinary notation.

**a.** $8.348 \times 10^6$ km
**b.** $3.402 \times 10^3$ g
**c.** $7.6352 \times 10^{-3}$ kg
**d.** $3.02 \times 10^{-5}$ s

**78.** Complete the following addition and subtraction problems in scientific notation.

**a.** $(6.23 \times 10^6\ \text{kL}) + (5.34 \times 10^6\ \text{kL})$
**b.** $(3.1 \times 10^4\ \text{mm}) + (4.87 \times 10^5\ \text{mm})$
**c.** $(7.21 \times 10^3\ \text{mg}) + (43.8 \times 10^2\ \text{mg})$
**d.** $(9.15 \times 10^{-4}\ \text{cm}) + (3.48 \times 10^{-4}\ \text{cm})$
**e.** $(4.68 \times 10^{-5}\ \text{cg}) + (3.5 \times 10^{-6}\ \text{cg})$
**f.** $(3.57 \times 10^2\ \text{mL}) - (1.43 \times 10^2\ \text{mL})$
**g.** $(9.87 \times 10^4\ \text{g}) - (6.2 \times 10^3\ \text{g})$
**h.** $(7.52 \times 10^5\ \text{kg}) - (5.43 \times 10^5\ \text{kg})$
**i.** $(6.48 \times 10^{-3}\ \text{mm}) - (2.81 \times 10^{-3}\ \text{mm})$
**j.** $(5.72 \times 10^{-4}\ \text{dg}) - (2.3 \times 10^{-5}\ \text{dg})$

**79.** Complete the following multiplication and division problems in scientific notation.

**a.** $(4.8 \times 10^5\ \text{km}) \times (2.0 \times 10^3\ \text{km})$
**b.** $(3.33 \times 10^{-4}\ \text{m}) \times (3.00 \times 10^{-5}\ \text{m})$
**c.** $(1.2 \times 10^6\ \text{m}) \times (1.5 \times 10^{-7}\ \text{m})$
**d.** $(8.42 \times 10^8\ \text{kL}) \div (4.21 \times 10^3\ \text{kL})$
**e.** $(8.4 \times 10^6\ \text{L}) \div (2.4 \times 10^{-3}\ \text{L})$
**f.** $(3.3 \times 10^{-4}\ \text{mL}) \div (1.1 \times 10^{-6}\ \text{mL})$

**80.** Convert the following measurements.

- **a.** 5.70 g to milligrams
- **b.** 4.37 cm to meters
- **c.** 783 kg to grams
- **d.** 45.3 mm to meters
- **e.** 10 m to centimeters
- **f.** 37.5 g/mL to kg/L

**81. Gold** A troy ounce is equal to 480 grains, and 1 grain is equal to 64.8 milligrams. If the price of gold is $560 per troy ounce, what is the cost of 1 g of gold?

**82. Popcorn** The average mass of a kernel of popcorn is 0.125 g. If 1 pound = 16 ounces, and 1 ounce = 28.3 g, then how many kernels of popcorn are there in 0.500 pounds of popcorn?

**83. Blood** You have 15 g of hemoglobin in every 100 mL of your blood. 10.0 mL of your blood can carry 2.01 mL of oxygen. How many milliliters of oxygen does each gram of hemoglobin carry?

**84. Nutrition** The recommended calcium intake for teenagers is 1300 mg per day. A glass of milk contains 305 mg of calcium. One glass contains a volume of 8 fluid ounces. How many liters of milk should a teenager drink per day to get the recommended amount of calcium? One fluid ounce equals 29.6 mL.

# SECTION 3

## Mastering Concepts

**85.** Which zero is significant in the number 50,540? What is the other zero called?

**86.** Why are percent error values never negative?

**87.** If you report two measurements of mass, 7.42 g and 7.56 g, are the measurements accurate? How might you evaluate the precision of these measurements? Explain your answers.

**88.** Which produce the same number when rounded to three significant figures: 3.456, 3.450, or 3.448?

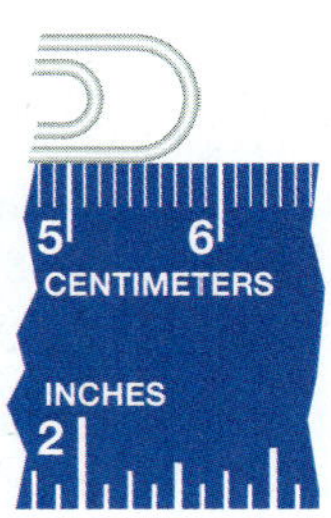


■ **Figure 18**

**89.** Record the measurement shown in **Figure 18** to the correct number of significant figures.

**90.** When subtracting 61.45 g from 242.6 g, which value determines the number of significant figures in the answer? Explain.

## Mastering Problems

**91.** Round each number to four significant figures.

- **a.** 431,801 kg
- **b.** 10,235.0 mg
- **c.** 1.0348 m
- **d.** 0.004384010 cm
- **e.** 0.00078100 mL
- **f.** 0.0098641 cg

**92.** Round the answer for each problem to the correct number of significant figures.

- **a.** $(7.31 \times 10^{4}) + (3.23 \times 10^{3})$
- **b.** $(8.54 \times 10^{-3}) - (3.41 \times 10^{-4})$
- **c.** 4.35 dm $\times$ 2.34 dm $\times$ 7.35 dm
- **d.** 4.78 cm + 3.218 cm + 5.82 cm
- **e.** 38,736 km $\div$ 4784 km

**93.** The accepted length of a steel pipe is 5.5 m. Calculate the percent error for each of these measurements.

**a.** 5.2 m **b.** 5.5 m **c.** 5.7 m **d.** 5.1 m

**94.** The accepted density for copper is 8.96 g/mL. Calculate the percent error for each of these measurements.

- **a.** 8.86 g/mL
- **b.** 8.92 g/mL
- **c.** 9.00 g/mL
- **d.** 8.98 g/mL

# SECTION 4

## Mastering Concepts

**95. Heating Fuels** Which type of graph would you use to depict how many households heat with gas, oil, or electricity? Explain.

**96. Gasoline Consumption** Which type of graph would you choose to depict gasoline consumption over a 10-year period? Explain.

**97.** How can you find the slope of a line graph?

## Mastering Problems

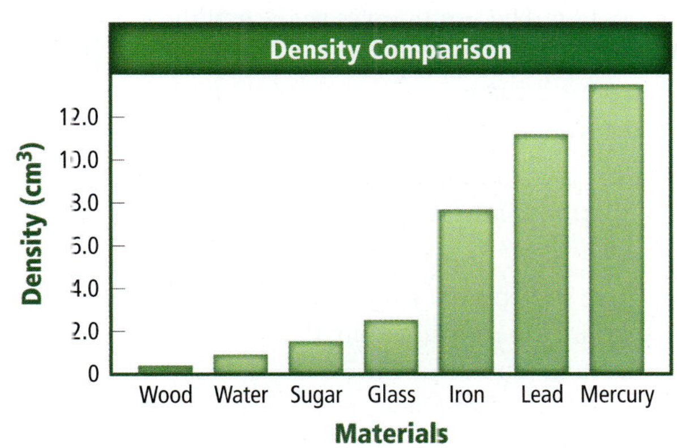


■ **Figure 19**

**98.** Use **Figure 19** to answer the following questions.

- **a.** Which substance has the greatest density?
- **b.** Which substance has the least density?
- **c.** Which substance has a density of 7.87 g/cm$^{3}$?
- **d.** Which substance has a density of 11.4 g/cm$^{3}$?

## MIXED REVIEW

**99.** Complete these problems in scientific notation. Round to the correct number of significant figures.
**a.** $(5.31 \times 10^{-2}\ \text{cm}) \times (2.46 \times 10^{5}\ \text{cm})$
**b.** $(3.78 \times 10^{3}\ \text{m}) \times (7.21 \times 10^{2}\ \text{m})$
**c.** $(8.12 \times 10^{-3}\ \text{m}) \times (1.14 \times 10^{-5}\ \text{m})$
**d.** $(9.33 \times 10^{4}\ \text{mm}) \div (3.0 \times 10^{2}\ \text{mm})$
**e.** $(4.42 \times 10^{-3}\ \text{kg}) \div (2.0 \times 10^{2}\ \text{kg})$
**f.** $(6.42 \times 10^{-2}\ \text{g}) \div (3.21 \times 10^{-3}\ \text{g})$

**100.** Convert each quantity to the indicated units.
**a.** 3.01 g → cg
**b.** 6200 m → km
**c.** $6.24 \times 10^{-7}$ g → μg
**d.** 0.2 L → $dm^3$
**e.** 0.13 cal/g → kcal/g
**f.** 3.21 mL → L

**101.** Students used a balance and a graduated cylinder to collect the data shown in **Table 6.** Calculate the density of the sample. If the accepted density of this sample is 6.95 g/mL, calculate the percent error.

**Table 6 Volume and Mass Data**

| | |
|---|---|
| Mass of sample | 20.46 g |
| Volume of water | 40.0 mL |
| Volume of water + sample | 43.0 mL |

**102.** Evaluate the following conversion. Will the answer be correct? Explain.

$$\text{rate} = \frac{75\ \text{m}}{1\ \text{s}} \times \frac{60\ \text{s}}{1\ \text{min}} \times \frac{1\ \text{h}}{60\ \text{min}}$$

**103.** You have a 23-g sample of ethanol with a density of 0.7893 g/mL. What volume of ethanol do you have?

**104. Zinc** Two separate masses of zinc were measured on a laboratory balance. The first zinc sample had a mass of 210.10 g, and the second zinc sample had a mass of 235.10 g. The two samples were combined. The volume of the combined sample was found to be 62.3 mL. Express the mass and density of the zinc sample in the correct number of significant figures.

**105.** What mass of lead (density 11.4 g/$cm^3$) would have a volume identical to 15.0 g of mercury (density 13.6 g/$cm^3$)?

**106.** Three students use a meterstick with millimeter markings to measure a length of wire. Their measurements are 3 cm, 3.3 cm, and 2.87 cm, respectively. Explain which answer was recorded correctly.

**107. Astronomy** The black hole in the M82 galaxy has a mass about 500 times the mass of the Sun. It has about the same volume as the Moon. What is the density of this black hole?

mass of the Sun = $1.9891 \times 10^{30}$ kg
volume of the Moon = $2.1968 \times 10^{10}$ $km^3$

**108.** The density of water is 1 g/$cm^3$. Use your answer from Question 107 to compare the densities of water and a black hole.

**109.** When multiplying 602.4 m by 3.72 m, which value determines the number of significant figures in the answer? Explain.

**110.** Round each figure to three significant figures.
**a.** 0.003210 g
**b.** 3.8754 kg
**c.** 219,034 m
**d.** 25.38 L
**e.** 0.08763 cm
**f.** 0.003109 mg

**111.** Graph the data in **Table 7,** with the volume on the $x$-axis and the mass on the $y$-axis. Then calculate the slope of the line.

**Table 7 Density Data**

| Volume (mL) | Mass (g) |
|---|---|
| 2.0 | 5.4 |
| 4.0 | 10.8 |
| 6.0 | 16.2 |
| 8.0 | 21.6 |
| 10.0 | 27.0 |

**112. Cough Syrup** A common brand of cough syrup comes in a 4-fluid ounce bottle. The active ingredient in the cough syrup is dextromethorphan. For an adult, the standard dose is 2 teaspoons, and a single dose contains 20.0 mg of dextromethorphan. Using the relationships, 1 fluid ounce = 29.6 mL and 1 teaspoon = 5.0 mL, determine how many grams of dextromethorphan are contained in the bottle.

## THINK CRITICALLY

**113. Interpret** Why does it make sense for the line in **Figure 16a** to extend to (0, 0) even though this point was not measured?

**114. Infer** Which of these measurements was made with the most precise measuring device: 8.1956 m, 8.20 m, or 8.196 m? Explain your answer.

**115. Apply** When subtracting or adding two numbers in scientific notation, why do the exponents need to be the same?

**116. Compare and Contrast** What advantages do SI units have over the units commonly used in the United States? Are there any disadvantages to using SI units?

**117. Hypothesize** Why do you think the SI standard for time was based on the distance light travels through a vacuum?

**118. Infer** Why does knowing the mass of an object not help you identify what material the object is made from?

**119. Conclude** Why might property owners hire a surveyor to determine property boundaries rather than measure the boundaries themselves?

**Nutrition Facts**

Serving Size ¾ cup (29 g)
Servings Per Container about 17

| Amount Per Serving | |
|---|---|
| **Calories** 120 | Calories from Fat 10 |
| | **% Daily Value *** |
| **Total Fat** 1g | 2% |
| Saturated Fat 1 g | 5% |
| **Cholesterol** 0 mg | 0% |
| **Sodium** 160 mg | 7% |
| **Potassium** 25 mg | 1% |
| **Total Carbohydrate** 25 g | 9% |
| Dietary Fiber less than 1 g | 2% |
| Sugars 13 g | |
| **Protein** 1 g | |
| Vitamin A | 4% |

■ **Figure 20**

**120. Apply Dimensional Analysis** Evaluate the breakfast cereal nutritional label shown in **Figure 20.** This product contains 160 mg of sodium in each serving. If you eat 2.0 cups of cereal a day, how many grams of sodium are you ingesting? What percent of your daily recommended sodium intake does this represent?

**121. Predict** Four graduated cylinders each contain a different liquid: A, B, C, and D.

Liquid A: mass = 18.5 g; volume = 15.0 mL
Liquid B: mass = 12.8 g; volume = 10.0 mL
Liquid C: mass = 20.5 g; volume = 12.0 mL
Liquid D: mass = 16.5 g; volume = 8.0 mL

Examine the information given for each liquid, and predict the layering of the liquids if they were carefully poured into a larger graduated cylinder.

## Challenge Problem

**122. Carboplatin** ($C_6H_{12}N_2O_4Pt$) is a platinum-containing compound that is used to treat certain forms of cancer. This compound contains 52.5% platinum. If the price for platinum is $1047/troy ounce, what is the cost of the platinum in 2.00 g of this compound? A troy ounce is equal to 480 grains, and one grain is equal to 64.8 mg.

## CUMULATIVE REVIEW

**123.** You record the following in your lab book: a liquid is thick and has a density of 4.58 g/mL. Which data is qualitative? Which is quantitative?

## WRITING IN ▶ Chemistry

**124. Kilogram Standard** Although the standard kilogram is stored at constant temperature and humidity, unwanted matter can build up on its surface. Scientists have been looking for a more reliable standard for mass. Research and describe alternative standards that have been proposed. Find out why no alternative standard has been chosen.

**125. Units** Research and report on unusual units of measurement such as bushels, pecks, firkins, and frails.

**126. Product Volume** Research the range of volumes used for packaging liquids sold in supermarkets.

**127. Dosing Error** In hospitals, medicines are given by dose. Find out what amount of error in the administered dose is acceptable for various medicines.

## DBQ Document-Based Questions

**Ocean Water** *The density of pure water is 1.00 g/cm³ at 4°C. Ocean water is denser because it contains salt and other dissolved substances. The graph in* **Figure 21** *shows the relationships among temperature, density, and salinity versus depth for ocean water.*

Data obtained from: *Windows to the Universe*, at the University Corporation for Atmospheric Research (UCAR).

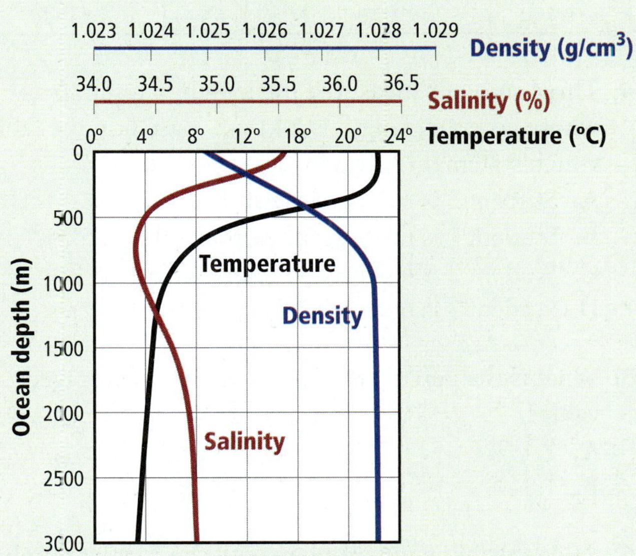


■ **Figure 21**

**128.** How is temperature related to the density of ocean water at depths less than 1000 m?

**129.** Describe the effect of depth on salinity.

**130.** Describe how salinity changes as the ocean water cools.

## MULTIPLE CHOICE

**1.** Which is NOT an SI base unit?
**A.** second
**B.** kilogram
**C.** degree Celsius
**D.** meter

**2.** Which value is NOT equivalent to the others?
**A.** 500 m
**B.** 0.5 km
**C.** 5000 cm
**D.** $5 \times 10^{11}$ nm

**3.** What is the correct representation of 702.0 g in scientific notation?
**A.** $7.02 \times 10^3$ g
**B.** $70.20 \times 10^1$ g
**C.** $7.020 \times 10^2$ g
**D.** $70.20 \times 10^2$ g

*Use the table below to answer Questions 4 and 5.*

| Measured Values for a Stamp's Length | | | |
|---|---|---|---|
| | Student 1 | Student 2 | Student 3 |
| Trial 1 | 2.60 cm | 2.70 cm | 2.75 cm |
| Trial 2 | 2.72 cm | 2.69 cm | 2.74 cm |
| Trial 3 | 2.65 cm | 2.71 cm | 2.64 cm |
| Average | 2.66 cm | 2.70 cm | 2.71 cm |

**4.** Three students measured the length of a stamp whose accepted length is 2.71 cm. Based on the table, which statement is true?
**A.** Student 2 is both precise and accurate.
**B.** Student 1 is more accurate than Student 3.
**C.** Student 2 is less precise than Student 1.
**D.** Student 3 is both precise and accurate.

**5.** What is the percent error for Student 1's averaged value?
**A.** 1.48%
**B.** 1.84%
**C.** 3.70%
**D.** 4.51%

**6.** Solve the following problem with the correct number of significant figures.

$$5.31 \text{ cm} + 8.4 \text{ cm} + 7.932 \text{ cm}$$

**A.** 22 cm
**B.** 21.64 cm
**C.** 21.642 cm
**D.** 21.6 cm

**7.** Chemists found that a complex reaction occurred in three steps. The first step takes $2.5731 \times 10^2$ s to complete, the second step takes $3.60 \times 10^{-1}$ s, and the third step takes $7.482 \times 10^1$ s. What is the total amount of time elapsed during the reaction?
**A.** $3.68 \times 10^1$ s
**B.** $7.78 \times 10^1$ s
**C.** $1.37 \times 10^1$ s
**D.** $3.3249 \times 10^2$ s

**8.** How many significant figures are there in a distance measurement of 20.070 km?
**A.** 2
**B.** 3
**C.** 4
**D.** 5

*Use the graph below to answer Questions 9 and 10.*

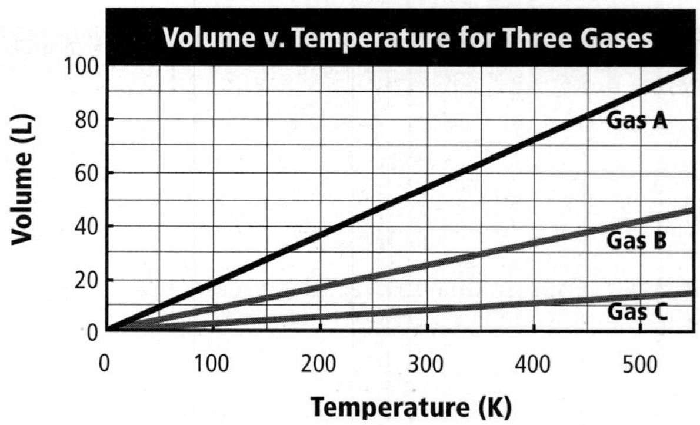


**9.** What volume will Gas A have at 450 K?
**A.** 23 L
**B.** 31 L
**C.** 38 L
**D.** 82 L

**10.** At what temperature will Gas B have a volume of 30 L?
**A.** 170 K
**B.** 350 K
**C.** 443 K
**D.** 623 K

**11.** Which is NOT a quantitative measurement of a pencil?
**A.** length
**B.** mass
**C.** color
**D.** diameter

## SHORT ANSWER

*Use the diagram below to answer Questions 12 and 13.*

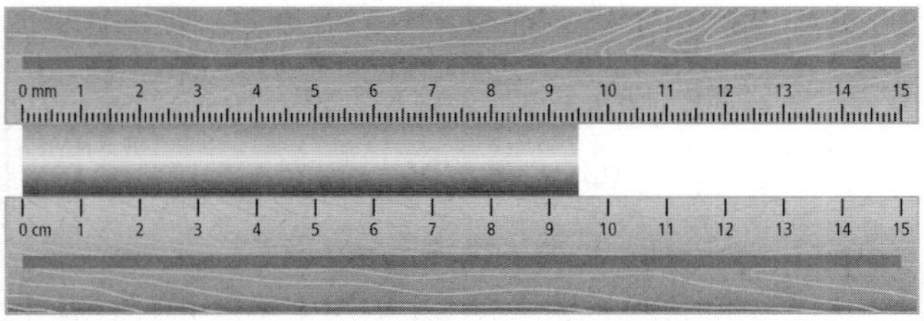


**12.** Explain which ruler you would use to make the more precise measurement.

**13.** What is the length of the rod using significant digits?

## EXTENDED RESPONSE

*Use the table below to answer Questions 14 to 16.*

| Temperature of a Solution While Heating | |
|---|---|
| Time (s) | Temperature (°C) |
| 0 | 22 |
| 30 | 35 |
| 60 | 48 |
| 90 | 61 |
| 120 | 74 |
| 150 | 87 |
| 180 | 100 |

**14.** A student recorded the temperature of a solution every 30 s for 3 min while the solution was heating on a Bunsen burner. Graph the data.

**15.** Show the setup to calculate the slope of the graph you created in Question 14.

**16.** Choose and explain two safety precautions the student should use with this experiment.

## SAT SUBJECT TEST: CHEMISTRY

*Use the graph below to answer Questions 17 to 20.*

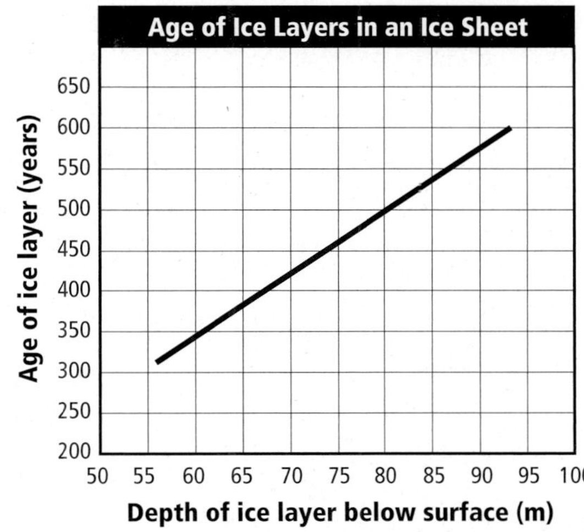


**17.** A student reported the age of an ice layer at 70 m as 425 years. The accepted value is 427 years. What is the percent error of the student's value?
**A.** 0.468%
**B.** 0.471%
**C.** 1.00%
**D.** 49.9%
**E.** 99.5%

**18.** What is the approximate slope of the line?
**A.** 0.00 m/y
**B.** 0.13 m/y
**C.** 0.13 y/m
**D.** 7.5 m/y
**E.** 7.5 y/m

**19.** What is the depth of an ice layer 450 years old?
**A.** 74 m
**B.** 75 m
**C.** 76 m
**D.** 77 m
**E.** 78 m

**20.** What is the relationship between ice depth and age?
**A.** linear, positive slope
**B.** linear, negative slope
**C.** linear, slope = 0
**D.** nonlinear, positive slope
**E.** nonlinear, negative slope

| NEED EXTRA HELP? | | | | | | | | | | | | | | | | | | | | |
|---|---|---|---|---|---|---|---|---|---|---|---|---|---|---|---|---|---|---|---|---|
| If You Missed Question . . . | 1 | 2 | 3 | 4 | 5 | 6 | 7 | 8 | 9 | 10 | 11 | 12 | 13 | 14 | 15 | 16 | 17 | 18 | 19 | 20 |
| Review Section . . . | 2.1 | 2.1 | 2.2 | 2.3 | 2.3 | 2.3 | 2.2 | 2.3 | 2.4 | 2.4 | 1.3 | 2.1 | 2.3 | 2.4 | 2.4 | 1.4 | 2.4 | 2.4 | 2.4 | 2.4 |

CHAPTER 3

# Matter—Properties and Changes

**BIGIDEA** Everything is made of matter.

## SECTIONS

Chm 4

1 Properties of Matter

2 Changes in Matter

3 Mixtures of Matter

4 Elements and Compounds

## LaunchLAB

iLab Station

### How can you observe chemical change?

Many objects in the everyday world do not change very much over time. However, when substances are mixed together, change is possible. In this lab, you will observe the changes that occur when two substances are combined.

**FOLDABLES® Study Organizer**

### Properties and Changes

Make a pocket book. Label it as shown. Use it to organize your study of chemical and physical changes and properties of matter.

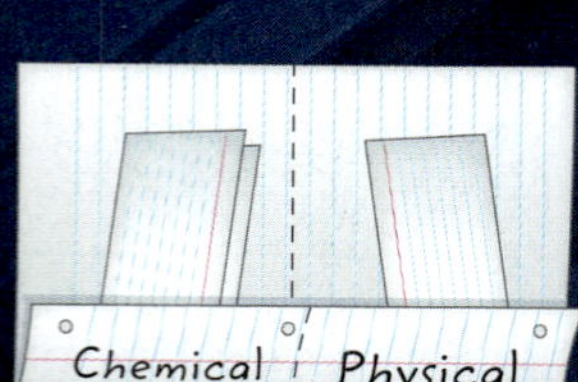

Magnus Hjorleifsson/Getty Images

Go online!
MHEonline.com

About 70% of Earth's surface is covered with water. Water is the only common substance on Earth that exists naturally as a solid, a liquid, and a gas.

SECTION 1

# Properties of Matter

**MAINIDEA** Most common substances exist as solids, liquids, and gases, which have diverse physical and chemical properties.

**CHEM 4 YOU** **Picture a glass of ice water. The ice floats, and you know the ice will eventually melt if left long enough at room temperature. When the water changes from solid to liquid, does the composition of the water also change?**

### Essential Questions

- What characteristics identify a substance?
- What distinguishes physical properties from chemical properties?
- How do the properties of the physical states of matter differ?

### Review Vocabulary

**density:** a ratio that compares the mass of an object to its volume

### New Vocabulary

states of matter
solid
liquid
gas
vapor
physical property
extensive property
intensive property
chemical property

## Substances

As you know, matter is anything that has mass and takes up space. Everything around us is matter, including things that we cannot see, such as air and microbes. For example, table salt is a simple type of matter that you are probably familiar with. Table salt has a unique and unchanging chemical composition. Its chemical name is sodium chloride. It is always 100% sodium chloride, and its composition does not change from one sample to another. Salt harvested from the sea or extracted from a mine, as shown in **Figure 1,** always has the same composition and properties.

Recall that matter with a uniform and unchanging composition is called a substance, also known as a pure substance. Table salt is a pure substance. Another example of a pure substance is pure water. Water is always composed of hydrogen and oxygen. Seawater and tap water, on the other hand, are not pure substances because samples taken from different locations will often have different compositions. That is, the samples will contain different amounts of water, minerals, and other dissolved substances. Substances are important; much of your chemistry course will be focused on the composition of substances and how they interact with one another.

**Figure 1** Whether harvested from the sea or extracted from a mine, salt always has the same composition.

Salt from the sea

Salt from a mine

(l)Luca Trovato/Getty Images, (r)Gabriela Insuratelu/iStock/Getty Images

■ **Figure 2** A solid has a definite shape and does not take the shape of its container. Particles in a solid are tightly packed.

## States of Matter

Imagine you are sitting on a bench, breathing heavily and drinking water after playing a game of soccer. You are in contact with three different forms of matter—the bench is a solid, the water is a liquid, and the air you breathe is a gas. In fact, all matter that exists naturally on Earth can be classified as one of these physical forms, which are called **states of matter.** Each of the three common states of matter can be distinguished by the way it fills a container. Scientists also recognize other states of matter. One of them is called plasma. It can occur in the form of lightning bolts and in stars.

☑ READING CHECK **Name** the common states of matter.

**Solids** A **solid** is a form of matter that has its own definite shape and volume. Wood, iron, paper, and sugar are all examples of solids. The particles of matter in a solid are tightly packed; when heated, a solid expands, but only slightly. Because its shape is definite, a solid might not conform to the shape of the container in which it is placed. If you place a rock into a container, the rock will not take the shape of the container, as shown in **Figure 2.** The tight packing of particles in a solid makes it incompressible; that is, it cannot be pressed into a smaller volume. It is important to understand that a solid is not defined by its rigidity or hardness. For instance, although concrete is rigid and wax is soft, they are both solids.

**Liquids** A **liquid** is a form of matter that flows, has constant volume, and takes the shape of its container. Common examples of liquids include water, blood, and mercury. The particles in a liquid are not rigidly held in place and are less closely packed than the particles in a solid. Liquid particles are able to move past each other. This property allows a liquid to flow and take the shape of its container, as shown in **Figure 3,** although it might not completely fill the container.

A liquid's volume is constant: regardless of the size and shape of the container in which the liquid is held, the volume of the liquid remains the same. Because of the way the particles of a liquid are packed, liquids are virtually incompressible. Like solids, however, liquids tend to expand when they are heated.

☑ READING CHECK **Compare** the properties of solids and liquids in terms of their particle arrangements.

■ **Figure 3** A liquid takes the shape of its container. Particles in a liquid are not held in place rigidly.

■ **Figure 4** Gases take the shape and volume of their containers. Particles in a gas are very far apart.

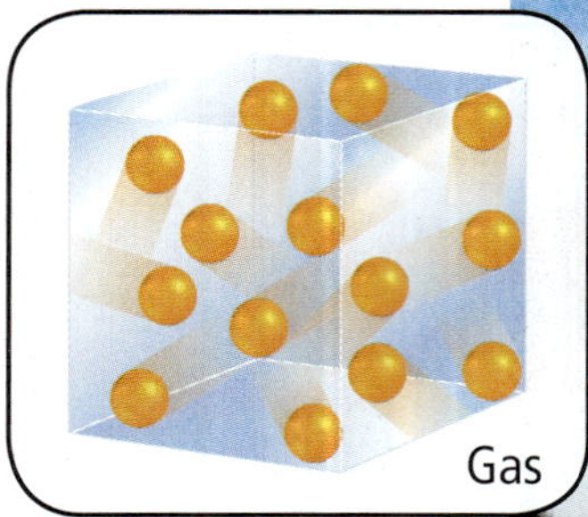

View an **animation about the three common states of matter.**

**Concepts In Motion**

**Gases** A **gas** is a form of matter that not only flows to conform to the shape of its container but also fills the entire volume of its container, as shown in **Figure 4.** If you flow gas into a container and close the container, the gas will expand to fill the container. Compared to solids and liquids, the particles of gases are far apart. Because of the significant amount of space between particles, gases are easily compressed.

You are probably familiar with the word *vapor* as it relates to the word *gas*. However, the words *gas* and *vapor,* while similar, do not mean the same thing, and should not be used interchangeably. The word *gas* refers to a substance that is naturally in the gaseous state at room temperature. The word **vapor** refers to the gaseous state of a substance that is a solid or a liquid at room temperature. For example, steam is a vapor because water exists as a liquid at room temperature.

☑ READING CHECK **Differentiate** between gas and vapor.

# Problem-Solving LAB

## Recognize Cause and Effect

**How is compressed gas released?** Tanks of compressed gases are common in chemistry laboratories. For example, nitrogen is often flowed over a reaction in progress to displace other gases that might interfere with the experiment. Given what you know about gases, explain how the release of compressed nitrogen is controlled.

### Analysis

The particles of gases are far apart, and gases tend to fill their containers—even if the container is a laboratory room. Tanks of compressed gas come from the supplier capped to prevent the gas from escaping. In the lab, a chemist or technician attaches a regulator to the tank and secures the tank to a stable fixture.

### Think Critically

1. **Explain** why the flow of a compressed gas must be controlled for practical and safe use.
2. **Predict** what would happen if the valve on a full tank of compressed gas were suddenly opened all the way or if the tank were accidentally punctured.

**Table 1 Physical Properties of Common Substances**

| Substance | Color | State at 25°C | Melting Point (°C) | Boiling Point (°C) | Density ($g/cm^3$) at 25°C |
|---|---|---|---|---|---|
| Oxygen | colorless | gas | −219 | −183 | 0.0013 |
| Mercury | silver | liquid | −39 | 357 | 13.5 |
| Water | colorless | liquid | 0 | 100 | 0.997 |
| Sucrose | white | solid | 186 | decomposes | 1.58 |
| Sodium chloride | white | solid | 801 | 1465 | 2.17 |

## Physical Properties of Matter

You are probably used to identifying objects by their properties—their characteristics and behavior. For example, you can easily identify a pencil in your backpack because you recognize its shape, color, weight, or some other property. These characteristics are all physical properties of the pencil. A **physical property** is a characteristic of matter that can be observed or measured without changing the sample's composition. Physical properties also describe pure substances. Because substances have uniform and unchanging compositions, they also have consistent and unchanging physical properties. Density, color, odor, hardness, melting point, and boiling point are common physical properties that scientists record as identifying characteristics of a substance. **Table 1** lists several common substances and their physical properties.

☑ READING CHECK **Define** *physical property* and provide examples.

**Extensive and intensive properties** Physical properties can be further described as being one of two types. **Extensive properties** are dependent on the amount of substance present. For example, mass is an extensive property. Length and volume are also extensive properties. Density, on the other hand, is an example of an intensive property of matter. **Intensive properties** are independent of the amount of substance present. For example, the density of a substance (at constant temperature and pressure) is the same no matter how much substance is present.

A substance can often be identified by its intensive properties. In some cases, a single intensive property is unique enough for identification. For instance, most of the spices shown in **Figure 5** can be identified by their scent.

**Figure 5** Many spices can be identified by their scent, which is an intensive property.

**Infer** *Name an extensive property of one of the spices pictured.*

**RealWorld CHEMISTRY**

**Physical Properties**

**MINERALS** Scientists use physical properties such as color and hardness to identify minerals. For instance, malachite is always green and relatively soft. Malachite was used as a pigment in paint and is now mainly used to make jewelry.

■ **Figure 6** One of the physical properties of copper is that it can be shaped into different forms, such as the wires on circuit boards. The fact that copper turns from reddish to green when reacting with substances in the air is a chemical property.

## Chemical Properties of Matter

**FOLDABLES®** Incorporate information from this section into your Foldable.

Some properties of a substance are not obvious unless the substance has changed composition as a result of its contact with other substances or the application of thermal or electric energy. The ability or inability of a substance to combine with or change into one or more other substances is called a **chemical property.**

Iron forming rust when combined with the oxygen in air is an example of a chemical property of iron. Similarly, the inability of a substance to change into another substance is also a chemical property. For example, when iron is placed in nitrogen gas at room temperature, no chemical change occurs.

☑ READING CHECK **Compare** physical and chemical properties.

## Observing Properties of Matter

Watch a video about the **composition of matter.** Video

Every substance has its own unique set of physical and chemical properties. **Figure 6** shows physical and chemical properties of copper. Copper can be shaped into different forms, which is a physical property. When copper is in contact with air for a long time, it reacts with the substances in the air and turns green. This is a chemical property. **Table 2** lists several physical and chemical properties of copper.

**Table 2** Properties of Copper

| Physical Properties | Chemical Properties |
|---|---|
| • reddish brown, shiny<br>• easily shaped into sheets (malleable) and drawn into wires (ductile)<br>• a good conductor of heat and electricity<br>• density = 8.96 g/cm$^3$<br>• melting point = 1085°C<br>• boiling point = 2562°C | • forms green copper carbonate compound when in contact with moist air<br>• reacts with nitric acid and sulfuric acid, forming new substances<br>• one type of compound forms a deep-blue solution when in contact with ammonia |

■ **Figure 7** Because the density of ice is lower than the density of water, icebergs float on the ocean.

**Properties and states of matter** The properties of copper listed in **Table 2** might vary depending on the conditions under which they are observed. Because the particular form, or state, of a substance is a physical property, changing the state introduces or adds another physical property to its characteristics. It is important to state the specific conditions, such as temperature and pressure, under which observations are made because both physical and chemical properties depend on these conditions. Resources that provide tables of physical and chemical properties of substances, such as the *CRC Handbook of Chemistry and Physics,* generally include the physical properties of substances in all of the states in which they can exist.

Consider the properties of water, for example. You might think of water as a liquid (physical property) which is not particularly chemically reactive (chemical property). You might also find that water has a density of 1.00 g/cm$^3$ (physical property). These properties, however, apply only to water at standard temperature and pressure. At temperatures greater than 100°C, water is a gas (physical property) with a density of about 0.0006 g/cm$^3$ (physical property) that reacts rapidly with many different substances (chemical property). Below 0°C, water is a solid (physical property) with a density of about 0.92 g/cm$^3$ (physical property). The lower density of ice accounts for the fact that icebergs float on the ocean, as shown in **Figure 7.** Clearly, the properties of water are dramatically different under different conditions.

**VOCABULARY**
**ACADEMIC VOCABULARY**
**Resource**
a source of information or expertise
*The library contains resources for many different subjects.*

**Alabama Applying Practices**
Go to the Resource tab in ConnectED to find the Applying Practices worksheet *Investigate Properties of Matter.*

## SECTION 1 REVIEW

Section Self-Check

### Section Summary

- The three common states of matter are solid, liquid, and gas.
- Physical properties can be observed without altering a substance's composition.
- Chemical properties describe a substance's ability to combine with or change into one or more new substances.
- External conditions can affect both physical and chemical properties.

**1.** MAINIDEA **Create** a table that describes the three common states of matter in terms of their shape, volume, and compressibility.

**2. Describe** the characteristics that identify a sample of matter as a substance.

**3. Classify** each of the following as a physical or a chemical property.

**a.** Iron and oxygen form rust.

**b.** Iron is more dense than aluminum.

**c.** Magnesium burns brightly when ignited.

**d.** Oil and water do not mix.

**e.** Mercury melts at −39°C.

**4. Organize** Create a chart that compares physical and chemical properties. Give two examples for each type of property.

SECTION 2

# Changes in Matter

**MAINIDEA** Matter can undergo physical and chemical changes.

### Essential Questions

- What is a physical change and what are several common examples?
- What defines a chemical change? How can you tell that a chemical change has taken place?
- How does the law of conservation of mass apply to chemical reactions?

### Review Vocabulary

**observation:** orderly, direct information gathering about a phenomenon

### New Vocabulary

physical change
phase change
chemical change
law of conservation of mass

**CHEM 4 YOU** Charcoal is a black solid. As it burns, it glows red. Eventually it ends up as ashes, carbon dioxide, and water. The dramatic changes to charcoal's appearance result from an important chemical property – the ability to burn.

## Physical Changes

A substance often undergoes changes that result in a dramatically different appearance yet leave the composition of the substance unchanged. An example is the crumpling of aluminum foil. While the foil goes from a smooth, flat, mirrorlike sheet to a round, compact ball, the actual composition of the foil is unchanged—it is still aluminum. A change such as this, which alters a substance without changing its composition, is a **physical change.** Cutting a sheet of paper and breaking a crystal are other examples of physical changes in matter.

**Phase change** As with other physical properties, the state of matter depends on the temperature and pressure of the surroundings. As temperature and pressure change, most substances undergo a change from one state (or phase) to another. A **phase change** is a transition of matter from one state to another.

**Connection to Earth Science** **The water cycle** This is the case with the water cycle, which allows life to exist on Earth. At atmospheric pressure and at temperatures of 0°C and below, water is in its solid state, which is known as ice. As heat is added to the ice, it melts and becomes liquid water. This change of state is a physical change because even though ice and water have different appearances, they have the same composition. If the temperature of the water increases to 100°C, the water begins to boil and liquid water is converted to steam. Melting and formation of a gas are both physical changes and phase changes. **Figure 8** shows condensation and solidification, two common phase changes. Terms such as *boil, freeze, condense, vaporize,* or *melt* in chemistry generally refer to a phase change in matter.

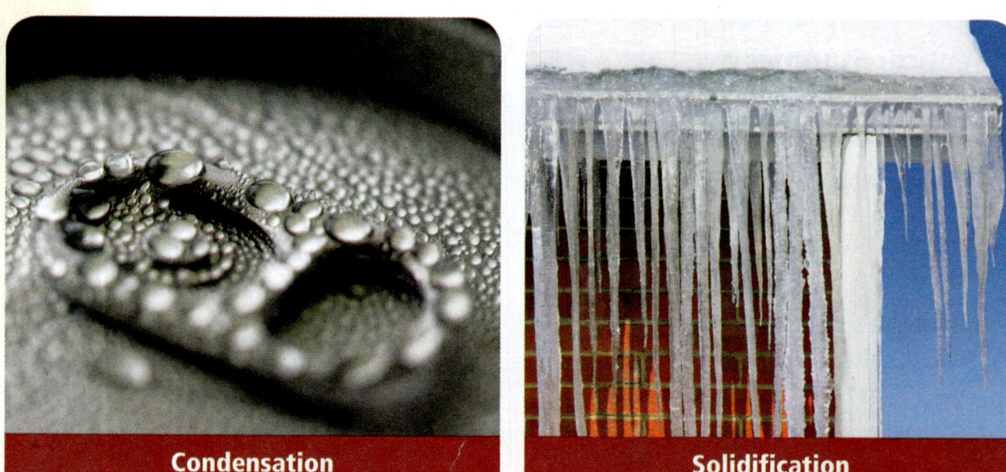


**Figure 8** Condensation can occur when a gas is in contact with a cool surface, causing droplets to form. Solidification occurs when a liquid cools. Water dripping from the roof forms icicles as it cools.

(l)i creative/Alamy; (r)Design Pics Inc./Alamy

The temperature and pressure at which a substance undergoes a phase change are important physical properties. These properties are called the melting and boiling points of the substance. Look again at **Table 1** to see this information for several common substances. Like density, the melting and boiling points are intensive physical properties that can be used to identify unknown substances. Tables of intensive properties, such as those given at the end of this textbook or in the *CRC Handbook of Chemistry and Physics,* are useful tools in identifying unknown substances from experimental data.

Watch a video about **wastewater treatment.**

Video

## Chemical Changes

A process that involves one or more substances changing into new substances is called a **chemical change,** commonly referred to as a chemical reaction. The new substances formed in the reaction have different compositions and different properties from the substances present before the reaction occurred. For example, the formation of rust when iron reacts with oxygen in moist air is a chemical change. Rust, shown in **Figure 9,** is a chemical combination of iron and oxygen.

**FOLDABLES®** Incorporate information from this section into your Foldable.

In chemical reactions, the starting substances are called reactants, and the new substances that are formed are called products. Terms such as *decompose, explode, rust, oxidize, corrode, tarnish, ferment, burn,* or *rot* generally refer to chemical reactions.

**READING CHECK** **Define** *chemical change.*

**Evidence of a chemical reaction** As **Figure 9** shows, rust is a brownish-orange powdery substance that looks very different from iron and oxygen. Rust is not attracted to a magnet, whereas iron is. The observation that the product (rust) has different properties than the reactants (iron and oxygen) is evidence that a chemical reaction has taken place. A chemical reaction always produces a change in properties. Spoiled food, such as rotten fruit and bread, is another example of chemical reactions. The properties of spoiled food, like its taste and its digestibility, differ from fresh food. Examples of food that have undergone chemical reactions are shown in **Figure 9.**

**Figure 9** When iron rusts and food rots, new substances are formed due to chemical change.
**Identify** *the reactants and the products in the formation of rust.*

## Conservation of Mass

It was only in the late eighteenth century that scientists began to use quantitative tools to study and monitor chemical changes. The analytical balance, which was capable of measuring small changes in mass, was developed at that time. By carefully measuring mass before and after many chemical reactions, it was observed that, although chemical changes occurred, the total mass involved in the reaction remained constant. Assuming this was true for all reactions, chemists summarized this observation in a scientific law. The **law of conservation of mass** states that mass is neither created nor destroyed during a chemical reaction—it is conserved. In other words, the mass of the reactants equals the mass of the products. The equation form of the law of conservation of mass is as follows.

**The Law of Conservation of Mass**

$$\text{mass}_{\text{reactants}} = \text{mass}_{\text{products}}$$

Mass is conserved in a chemical reaction; products have the same mass as reactants.

EXAMPLE PROBLEM

## EXAMPLE Problem 1

Find help with **algebraic equations.** Math Handbook

**CONSERVATION OF MASS** In an experiment, 10.00 g of red mercury(II) oxide powder is placed in an open flask and heated until it is converted to liquid mercury and oxygen gas. The liquid mercury has a mass of 9.26 g. What is the mass of oxygen formed in the reaction?

### 1 ANALYZE THE PROBLEM

You are given the mass of a reactant and the mass of one of the products in a chemical reaction. According to the law of mass conservation, the total mass of the products must equal the total mass of the reactants.

**Known**

$m_{\text{mercury(II) oxide}} = 10.00\text{ g}$

$m_{\text{mercury}} = 9.26\text{ g}$

**Unknown**

$m_{\text{oxygen}} = ?\text{ g}$

Get help with **writing equations.** Personal Tutor

### 2 SOLVE FOR THE UNKNOWN

$\text{Mass}_{\text{reactants}} = \text{Mass}_{\text{products}}$ — State the law of conservation of mass.

$m_{\text{mercury(II) oxide}} = m_{\text{mercury}} + m_{\text{oxygen}}$

$m_{\text{oxygen}} = m_{\text{mercury(II) oxide}} - m_{\text{mercury}}$ — Solve for $m_{\text{oxygen}}$.

$m_{\text{oxygen}} = 10.00\text{ g} - 9.26\text{ g}$ — Substitute $m_{\text{mercury(II) oxide}} = 10.00\text{ g}$ and $m_{\text{mercury}} = 9.26\text{ g}$.

$m_{\text{oxygen}} = \mathbf{0.74\text{ g}}$

### 3 EVALUATE THE ANSWER

The sum of the masses of the two products equals the mass of the reactant, verifying that mass has been conserved. The answer is correctly expressed to the hundredths place, making the number of significant digits correct.

PRACTICE PROBLEMS

## PRACTICE Problems

Do additional problems. Online Practice

**5.** Use the data in the table to answer the following questions.

| Aluminum and Liquid Bromine Reaction | | |
|---|---|---|
| **Substance** | **Before Reaction** | **After Reaction** |
| Aluminum | 10.3 g | 0.0 g |
| Liquid bromine | 100.0 g | 8.2 g |
| Compound | 0.0 g | |

How many grams of bromine reacted? How many grams of compound were formed?

**6.** From a laboratory process designed to separate water into hydrogen and oxygen gas, a student collected 10.0 g of hydrogen and 79.4 g of oxygen. How much water was originally involved in the process?

**7.** A student carefully placed 15.6 g of sodium in a reactor supplied with an excess quantity of chlorine gas. When the reaction was complete, the student obtained 39.7 g of sodium chloride. Calculate how many grams of chlorine gas reacted. How many grams of sodium reacted?

**8.** A 10.0-g sample of magnesium reacts with oxygen to form 16.6 g of magnesium oxide. How many grams of oxygen reacted?

**9. Challenge** 106.5 g of HCl(g) react with an unknown amount of $NH_3$(g) to produce 156.3 g of $NH_4Cl$(s). How many grams of $NH_3$(g) reacted? Is the law of conservation of mass observed in the reaction? Justify your answer.

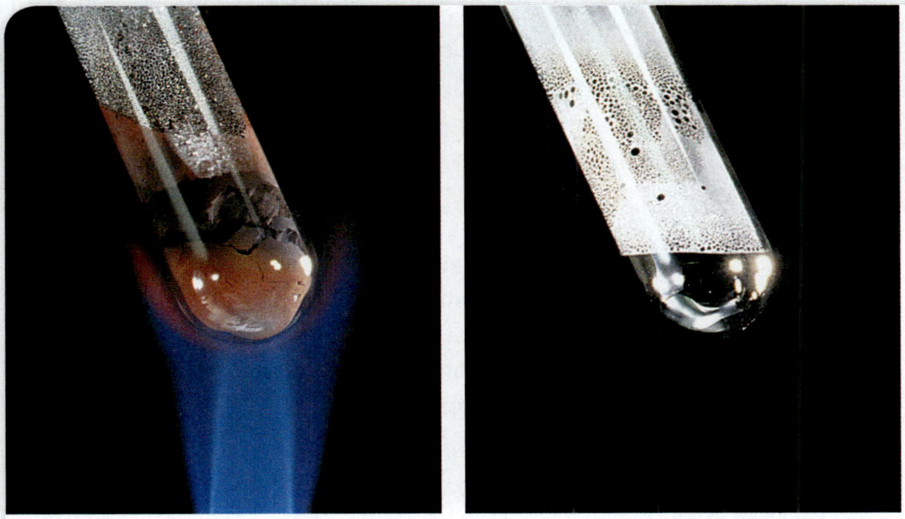

View an **animation about the conservation of mass.**

**Concepts In Motion**

■ **Figure 10** When mercury(II) oxide is heated, it reacts to form liquid mercury and oxygen gas. The sum of the masses of liquid mercury and oxygen gas produced during the reaction equals the mass of the mercury(II) oxide.

French scientist Antoine Lavoisier (1743–1794) was one of the first to use an analytical balance to monitor chemical reactions. He studied the thermal decomposition of mercury(II) oxide, known then as *calx of mercury*. Mercury(II) oxide, shown in **Figure 10,** is a powdery red solid. When it is heated, the red solid reacts to form silvery liquid mercury and colorless oxygen gas.

The color change and production of a gas are indicators of a chemical reaction. When the reaction occurs in a closed container, the oxygen gas cannot escape and the mass before and after the reaction can be measured. The masses will be the same. The law of conservation of mass is one of the most fundamental concepts of chemistry.

## SECTION 2 REVIEW

**Section Self-Check**

### Section Summary

- A physical change alters the physical properties of a substance without changing its composition.
- A chemical change, also known as a chemical reaction, involves a change in a substance's composition.
- In a chemical reaction, reactants form products.
- The law of conservation of mass states that mass is neither created nor destroyed during a chemical reaction; it is conserved.

**10. MAIN**IDEA **Classify** each example as a physical change or a chemical change.
   - **a.** crushing an aluminum can
   - **b.** recycling used aluminum cans to make new aluminum cans
   - **c.** aluminum combining with oxygen to form aluminum oxide

**11. Describe** the results of a physical change and list three examples of physical change.

**12. Describe** the results of a chemical change. List four indicators of chemical change.

**13. Calculate** Solve each of the following.
   - **a.** In the complete reaction of 22.99 g of sodium with 35.45 g of chlorine, what mass of sodium chloride is formed?
   - **b.** A 12.2-g sample of *X* reacts with a sample of *Y* to form 78.9 g of *XY*. What is the mass of *Y* that reacted?

**14. Evaluate** A friend tells you, "Because composition does not change during a physical change, the appearance of a substance does not change." Is your friend correct? Explain.

SECTION 3

# Mixtures of Matter

**MAINIDEA** Most everyday matter occurs as mixtures—combinations of two or more substances.

### Essential Questions

- How do mixtures and substances differ?
- Why are some mixtures classified as homogeneous, while others are classified as heterogeneous?
- What are several techniques used to separate mixtures?

### Review Vocabulary

**substance:** a form of matter that has a uniform and unchanging composition; also known as a pure substance

### New Vocabulary

mixture
heterogeneous mixture
homogeneous mixture
solution
filtration
distillation
crystallization
sublimation
chromatography

**CHEM 4 YOU** That familiar hiss when you open a soft-drink bottle is the sound of gas escaping. You might have noticed that when you leave the bottle open, eventually most of bubbles are gone and the drink is flat. But the soft drink remains sweet no matter how long you leave the bottle open.

## Mixtures

You have already read that a pure substance has a uniform and unchanging composition. What happens when two or more subtances are combined? A **mixture** is a combination of two or more pure substances in which each pure substance retains its individual chemical properties. The composition of mixtures is variable, and the number of mixtures that can be created by combining substances is infinite. Although much of the focus of chemistry is the behavior of substances, it is important to remember that most everyday matter occurs as mixtures. Substances tend to mix naturally; it is difficult to keep any substance pure.

Two mixtures are shown in **Figure 11.** Although you cannot distinguish between the components of the mercury-silver mixture in **Figure 11a,** you can separate them by heating the mixture. The mercury will evaporate before the silver does, and you will obtain two separate substances: mercury vapor and solid silver. The mercury and silver physically mixed to form the mixture but did not chemically react with each other. They could be separated by the physical method of boiling. When oil, seasonings, and vinegar are mixed, as shown in **Figure 11b,** the substances are in contact, but they do not react. In fact, you can still distinguish all of the substances. If the mixture remains undisturbed long enough, the oil will form a layer on top of the vinegar.

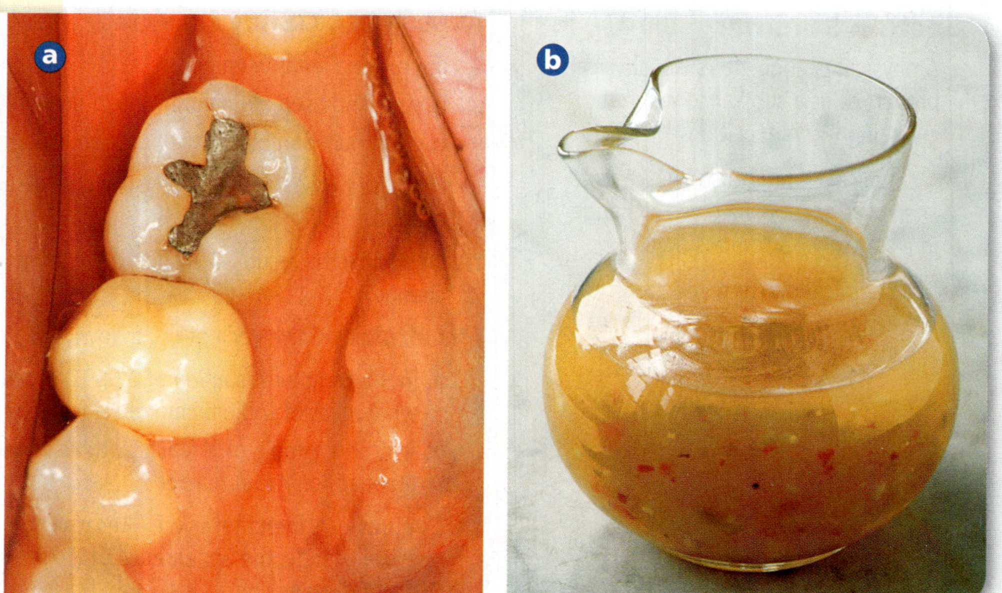


**Figure 11** There are different types of mixtures. **a.** It is not possible to see the different components of some mixtures, such as this mercury-silver filling. **b.** The components of other types of mixtures are visible, as in this salad dressing.

(l)Alexandru Kacso/iStock/Getty Images; (r)©Envision/Corbis

**Types of mixtures** The combinations of pure substances shown in **Figure 11** are both mixtures, despite their obvious visual differences. Mixtures can be defined in different ways and are classified as either heterogeneous or homogeneous.

A **heterogeneous mixture** is a mixture that does not blend smoothly throughout and in which the individual substances remain distinct. The salad dressing mixture is an example of a heterogeneous mixture. Its composition is not uniform—the substances have not blended smoothly and remain distinct. In another example, fresh-squeezed orange juice is a heterogeneous mixture of juice and pulp. The pulp component floats in the juice component. We can therefore say that the existence of two or more distinct areas indicates a heterogeneous mixture.

A **homogeneous mixture** is a mixture that has constant composition throughout; it always has a single phase. If you cut two pieces out of a silver mercury amalgam, their compositions will be the same. They will contain the same relative amounts of silver and mercury, no matter the size of each piece.

READING CHECK **Compare and contrast** heterogeneous and homogeneous mixtures. Give examples of each.

Homogeneous mixtures are also referred to as **solutions.** You are probably most familiar with solutions in a liquid form, such as tea and lemonade, but solutions can be solids, liquids, or gases. They can be a mixture of a solid and a gas, a solid and a liquid, a gas and a liquid, and so on. **Table 3** lists the various types of solution systems and examples. Each solution system in the table is also represented in **Figure 12**.

The solid-solid solution known as steel is called an alloy. An alloy is a homogeneous mixture of metals, or a mixture of a metal and a nonmetal in which the metal substance is the major component. For instance, steel is a mixture of iron and carbon. Adding carbon atoms increases the hardness of the metal.

Manufacturers combine the properties of various metals in an alloy to achieve greater strength and durability in their products. Jewelry is often made of alloys such as bronze, sterling silver, pewter, and 14-karat gold.

Explore **solution systems with an interactive table.** Concepts In Motion

**Table 3** Types of Solution Systems

| System | Example |
|---|---|
| Gas-gas | Air in a scuba tank is primarily a mixture of nitrogen, oxygen, and argon gases. |
| Gas-liquid | Oxygen and carbon dioxide are dissolved in seawater. |
| Liquid-gas | Moist air exhaled by the scuba diver contains water droplets. |
| Liquid-liquid | When it is raining, fresh water mixes with seawater. |
| Solid-liquid | Solid salts are dissolved in seawater. |
| Solid-solid | The air tank is made of an alloy—a mixture of two metals. |

**VOCABULARY**

**WORD ORIGIN**

**Mixture**
from the Latin word *misceo,* meaning *to mix* or *blend*

**CAREERS IN CHEMISTRY**

**Materials Scientist** Materials scientists synthesize new materials and analyze their properties. They work in national laboratories, in industry, and in academia. For example, scientists at NASA developed new aluminum-silicon alloys that can be employed to build lighter and stronger engines.

WebQuest

■ **Figure 12** All types of solution systems are represented in this photo.

Tane Casserley NOAA/NMS Maritime Heritage Program

■ **Figure 13** As the mixture passes through the filter, the solids remain in the filter, while the filtrate (the remaining liquid) is collected in the beaker.

## Separating Mixtures

Most matter exists naturally in the form of mixtures. To gain a thorough understanding of matter, it is important to be able to separate mixtures into their component substances. Because the substances in a mixture are physically combined, the processes used to separate a mixture are physical processes that are based on differences in the physical properties of the substances. For instance, a mixture of iron and sand can be separated into its components with a magnet because a magnet will attract iron but not sand. Numerous techniques have been developed that take advantage of different physical properties in order to separate various mixtures.

**Filtration** Heterogeneous mixtures composed of solids and liquids are easily separated by filtration. **Filtration** is a technique that uses a porous barrier to separate a solid from a liquid. As **Figure 13** shows, the mixture is poured through a piece of filter paper that has been folded into a cone shape. The liquid passes through, leaving the solids trapped in the filter paper.

**Distillation** Most homogeneous mixtures can be separated by distillation. **Distillation** is a physical separation technique that is based on differences in the boiling points of the substances involved. In distillation, a mixture is heated until the substance with the lowest boiling point boils to a vapor that can then be condensed into a liquid and collected. When precisely controlled, distillation can separate substances that have boiling points differing by only a few degrees.

# MiniLAB

iLab Station

### Observe Dye Separation

**How does paper chromatography allow you to separate substances?** Chromatography is an important diagnostic tool used by chemists and forensic technicians to separate and analyze substances.

#### Procedure

1. Read and complete the lab safety form.
2. Fill a **9-oz wide-mouth plastic cup** with **water** to about 2 cm from the top. Wipe off any water drops on the lip of the cup.
3. Place a piece of **round filter paper** on a clean, dry surface. Make a concentrated ink spot in the center of the paper by firmly pressing the tip of a **black water-soluble pen or marker** onto the paper.
4. Use **scissors** or another sharp object to create a small hole, about the diameter of a pen tip, in the center of the ink spot.
   **WARNING:** ***Sharp objects can puncture skin.***
5. Roll one quarter of an **11-cm round filter paper** into a tight cone. This will act as a wick to draw the ink. Work the pointed end of the wick into the hole in the center of the round filter paper.
6. Place the paper/wick apparatus on top of the cup of water, with the wick in the water. The water will move up the wick and outward through the round paper.
7. When the water has moved to within about 1 cm of the edge of the paper (about 20 min), carefully remove the paper from the water-filled cup and put it on a second empty **cup.**

#### Analysis

1. **Record** the number of distinct dyes you can identify on a drawing of the round filter paper. Label the color bands.
2. **Infer** why you see different colors at different locations on the filter paper.
3. **Compare** your chromatogram with those of your classmates. Explain any differences you might observe.

Martyn F. Chillmaid/Photo Researchers

**Crystallization** Making rock candy from a sugar solution is an example of separation by crystallization. **Crystallization** is a separation technique that results in the formation of pure solid particles of a substance from a solution containing the dissolved substance. When the solution contains as much dissolved substance as it can possibly hold, the addition of even a tiny amount more often causes the dissolved substance to come out of solution and collect as crystals on any available surface.

In the rock candy example, as water evaporates from the sugar-water solution, the solution becomes more concentrated. This is equivalent to adding more of the dissolved substance to the solution. As more water evaporates, the sugar forms a solid crystal on the string, as shown in **Figure 14.** Crystallization produces highly pure solids.

**Sublimation** Mixtures can also be separated by **sublimation,** which is the process during which a solid changes to vapor without melting, i.e. without going through the liquid phase. Sublimation can be used to separate two solids present in a mixture when one of the solids sublimates but not the other.

**Chromatography** **Chromatography** is a technique that separates the components of a mixture dissolved in either a gas or a liquid (called the mobile phase) based on the ability of each component to travel or to be drawn across the surface of a fixed substrate (called the stationary phase). For example, chromatography paper is a stationary phase with a solid substrate. During paper chromatography, the separation occurs because the various components of the mixture in the liquid mobile phase spread through the paper at different rates. Components with the strongest attraction for the paper travel slower.

■ **Figure 14** As the water evaporates from the water-sugar solution, the sugar crystals form on the string.

Watch a video about **analyzing chemicals in hair.**

Video

## SECTION 3 REVIEW

Section Self-Check

### Section Summary

- A mixture is a physical blend of two or more pure substances in any proportion.
- Solutions are homogeneous mixtures.
- Mixtures can be separated by physical means. Common separation techniques include filtration, distillation, crystallization, sublimation, and chromatography.

**15. MAINIDEA Classify** each of the following as either a heterogeneous or a homogeneous mixture.

**a.** tap water **b.** air **c.** raisin muffin

**16. Compare** mixtures and substances.

**17. Describe** the separation technique that could be used to separate each of the following mixtures.

**a.** two colorless liquids

**b.** a nondissolving solid mixed with a liquid

**c.** red and blue marbles of the same size and mass

**18. Design** a concept map that summarizes the relationships among matter, elements, mixtures, compounds, pure substances, and homogeneous and heterogeneous mixtures.

Tony Freeman/PhotoEdit

SECTION 4

# Elements and Compounds

**MAINIDEA** A compound is a combination of two or more elements.

## Essential Questions

- What distinguishes elements from compounds?
- How is the periodic table organized?
- What are the laws of definite and multiple proportions and why are they important?

## Review Vocabulary

**proportion:** the relation of one part to another or to the whole with respect to quantity

## New Vocabulary

element
periodic table
compound
law of definite proportions
percent by mass
law of multiple proportions

**CHEM 4 YOU** **When you eat fruit salad, you can separate your favorite fruit from the others. However, the jelly in your sandwich cannot be easily separated into its ingredients—fruit juice, sugar, and pectin. Similarly, compounds cannot be separated into their component atoms by physical processes.**

## Elements

Earlier in this chapter, you considered the diversity of your surroundings in terms of matter. Although matter can take many different forms, all matter can be broken down into a relatively small number of basic building blocks called elements. An **element** is a pure substance that cannot be separated into simpler substances by physical or chemical means. On Earth, over 90 elements occur naturally. Copper, oxygen, and gold are examples of naturally occurring elements. There are also several elements that do not exist naturally but have been developed by scientists.

Each element has a unique chemical name and symbol. The chemical symbol consists of one, two, or three letters; the first letter is always capitalized, and the remaining letter(s) are always lowercase. The names and symbols of the elements are universally accepted by scientists in order to make the communication of chemical information possible.

The naturally occurring elements are not equally abundant. For example, hydrogen is estimated to make up approximately 75% of the mass of the universe. Oxygen and silicon together comprise almost 75% of the mass of Earth's crust, while oxygen, carbon, and hydrogen account for more than 90% of the human body. Francium, on the other hand, is one of the least-abundant naturally-occurring elements. There is probably less than 20 g of francium dispersed throughout Earth's crust. Elements are found in different physical states in normal conditions, as shown in **Figure 15**.

■ **Figure 15** In normal conditions, elements exist in different states.

Copper pot—solid

Mercury switch—liquid

Helium balloon—gas

(l)Barry Mason/Alamy; (c)Tony Freeman/PhotoEdit; (r)Randy Brooke/AP Images

**Tabelle I.**

| Typische Elemente | | | | | | | |
|---|---|---|---|---|---|---|---|
| | | | K = 39 | Rb = 85 | Cs = 133 | — | — |
| | | | Ca = 40 | Sr = 87 | Ba = 137 | — | — |
| | | | — | ?Yt = 88? | ?Di = 138? | Er = 178? | — |
| | | | Ti = 48? | Zr = 90 | Ce = 140? | ?La = 180? | Th = 231 |
| | | | V = 51 | Nb = 94 | — | Ta = 182 | — |
| | | | Cr = 52 | Mo = 96 | — | W = 184 | U = 240 |
| | | | Mn = 55 | — | — | — | — |
| | | | Fe = 56 | Ru = 104 | — | Os = 195? | — |
| | | | Co = 59 | Rh = 104 | — | Ir = 197 | — |
| | | | Ni = 59 | Pd = 106 | — | Pt = 198? | — |
| H = 1 | Li = 7 | Na = 23 | Cu = 63 | Ag = 108 | — | Au = 199? | — |
| | Be = 9,4 | Mg = 24 | Zn = 65 | Cd = 112 | — | Hg = 200 | — |
| | B = 11 | Al = 27,3 | — | In = 113 | — | Tl = 204 | — |
| | C = 12 | Si = 28 | — | Sn = 118 | — | Pb = 207 | — |
| | N = 14 | P = 31 | As = 75 | Sb = 122 | — | Bi = 208 | — |
| | O = 16 | S = 32 | Se = 78 | Te = 125? | — | — | — |
| | F = 19 | Cl = 35,5 | Br = 80 | J = 127 | — | — | — |

■ **Figure 16** Mendeleev was one of the first scientists to organize elements in a periodic manner, as shown in this chart, and to observe periodic patterns in the properties of the elements.

**A first look at the periodic table** As many new elements were being discovered in the early nineteenth century, chemists began to observe and study patterns of similarities in the chemical and physical properties of particular sets of elements. In 1869, Russian chemist Dmitri Mendeleev (1834–1907) devised a chart, shown in **Figure 16,** which organized all of the elements that were known at the time. His classification was based on the similarities and masses of the elements. Mendeleev's table was the first version of what has been further developed into the periodic table of the elements. The **periodic table** organizes the elements into a grid of horizontal rows called periods and vertical columns called groups or families. Elements in the same group have similar chemical and physical properties. The table is called periodic because the pattern of similar properties repeats from period to period. The periodic table can be found at the end of this book and will be examined in great detail throughout your chemistry courses.

## Compounds

Many pure substances can be classified as compounds. A **compound** is made up of two or more different elements that are combined chemically. Most matter in the universe exists in the form of compounds.

Today, there are more than 50 million known compounds, and new compounds continue to be developed and discovered at the rate of about 100,000 per year. There appears to be no limit to the number of compounds that can be made or that will be discovered. Considering this virtually limitless potential, several organizations have assumed the task of collecting data and indexing the known chemical compounds. The information is stored in databases.

☑ **READING CHECK Define** *element* and *compound.*

The chemical symbols of the periodic table make it easy to write the formulas for chemical compounds. For example, table salt, which is called sodium chloride, is composed of one part sodium (Na) and one part chlorine (Cl), and its chemical formula is NaCl. Water is composed of two parts hydrogen (H) and one part oxygen (O), and its chemical formula is $H_2O$. The subscript 2 indicates that two hydrogen elements combine with one oxygen element to form water.

**VOCABULARY**

**SCIENCE USAGE V. COMMON USAGE**

**Element**

***Science usage:*** a pure substance that cannot be separated into simpler substances by ordinary chemical means

*Lead is one of the heaviest elements.*

***Common usage:*** the state or sphere that is natural or suited to any person or thing

*In snow, huskies are in their element.*

**Figure 17** An electric current breaks down water into its components, oxygen and hydrogen.

**Determine** ***What is the ratio between the amount of hydrogen and the amount of oxygen released during electrolysis?***

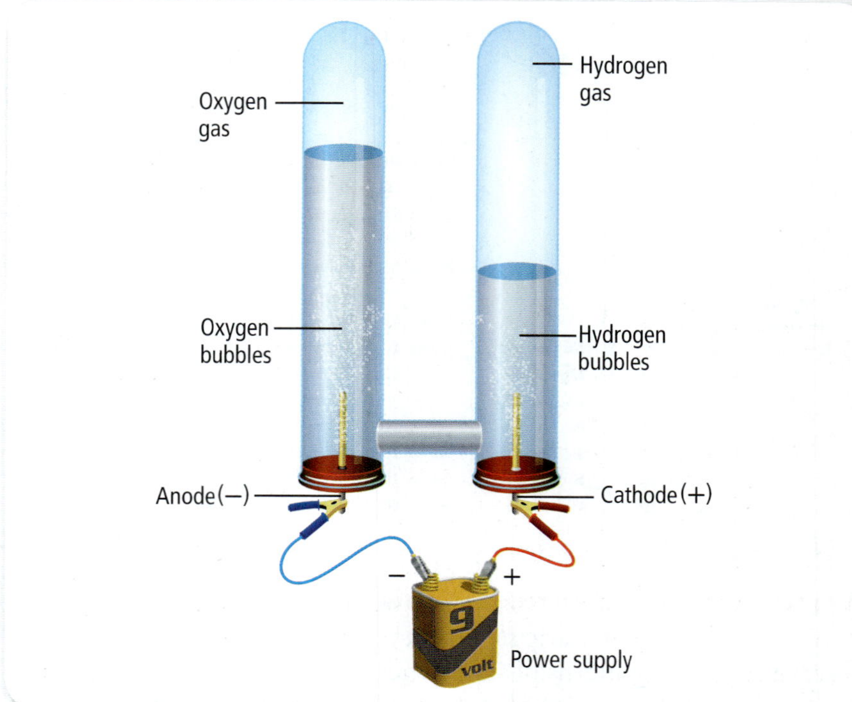


**Separating compounds into components** As you have read earlier in this chapter, elements can never be separated into simpler substances. However, compounds can be broken down into simpler substances by chemical means. In general, compounds that occur naturally are more stable than the individual component elements. Separating a compound into its elements often requires external energy, such as heat or electricity. **Figure 17** shows the setup used to produce the chemical change of water into its component elements—hydrogen and oxygen—through a process called electrolysis. During electrolysis, one end of a long platinum electrode is exposed to the water in a tube and the other end is attached to a power source. An electric current splits water into hydrogen gas in the compartment on the right and oxygen gas in the compartment on the left. Because water is composed of two parts hydrogen and one part oxygen, there is twice as much hydrogen gas as there is oxygen gas.

☑ READING CHECK **Explain** the process of electrolysis.

**Figure 18** When potassium and iodine react, they form potassium iodide, a compound with different properties.

**Properties of compounds** The properties of a compound are different from those of its component elements. The example of water in **Figure 17** illustrates this fact. Water is a stable compound that is liquid at room temperature. When water is broken down, its components, hydrogen and oxygen, are dramatically different than the liquid they form when combined. Oxygen and hydrogen are colorless, odorless gases that undergo vigorous chemical reactions with many elements. This difference in properties is a result of a chemical reaction between the elements. **Figure 18** shows the component elements—potassium and iodine—of the compound called potassium iodide. Note how different the properties of potassium iodide are from its component elements. Potassium is a light silver metal that reacts with water. Iodine is a black solid that changes into a purple gas at room temperature. Potassium iodide is a white salt.

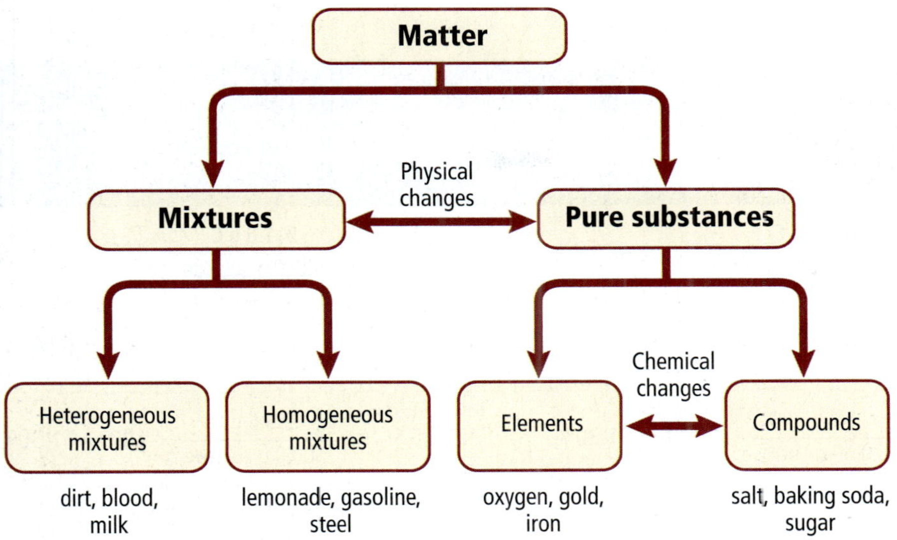


**Figure 19** Matter can be classified into different categories that have defined properties.

**Examine** ***How are mixtures and substances related? Elements and compounds?***

Recall what you have read about the organization of matter. You know that matter is classified as pure substances and mixtures. As you learned in the previous section, a mixture can be homogeneous or heterogeneous. You also know that an element is a pure substance that cannot be separated into simpler substances, whereas a compound is a chemical combination of two or more elements and can be separated into its components. Use **Figure 19** to review the classification of matter and how its components are related to each other.

☑ READING CHECK **Summarize** the different types of matter and how they are related to each other.

## Law of Definite Proportions

An important characteristic of compounds is that the elements comprising them always combine in definite proportions by mass. This observation is so fundamental that it is summarized as the law of definite proportions. The **law of definite proportions** states that a compound is always composed of the same elements in the same proportion by mass, no matter how large or small the sample. The mass of the compound is equal to the sum of the masses of the elements that make up the compound.

The relative amounts of the elements in a compound can be expressed as percent by mass. The **percent by mass** is the ratio of the mass of each element to the total mass of the compound expressed as a percentage.

**Percent by Mass**

$$\text{percent by mass (\%)} = \frac{\text{mass of element}}{\text{mass of compound}} \times 100$$

Percent by mass is obtained by dividing the mass of the element by the mass of the compound and then by multiplying this ratio by 100 to express it as a percentage.

☑ READING CHECK **State** the law of definite proportions.

**Table 4 Sucrose Analysis**

| Element | 20.00 g of Granulated Sugar | | 500.0 g of Sugarcane | |
|---|---|---|---|---|
| | Analysis by Mass (g) | Percent by Mass (%) | Analysis by Mass (g) | Percent by Mass (%) |
| Carbon | 8.44 | $\frac{8.44\text{ g C}}{20.00\text{ g sucrose}} \times 100 = 42.20\%$ | 211.0 | $\frac{211.0\text{ g C}}{500.0\text{ g sucrose}} \times 100 = 42.20\%$ |
| Hydrogen | 1.30 | $\frac{1.30\text{ g H}}{20.00\text{ g sucrose}} \times 100 = 6.50\%$ | 32.5 | $\frac{32.50\text{ g H}}{500.0\text{ g sucrose}} \times 100 = 6.500\%$ |
| Oxygen | 10.26 | $\frac{10.26\text{ g O}}{20.00\text{ g sucrose}} \times 100 = 51.30\%$ | 256.5 | $\frac{256.5\text{ g O}}{500.0\text{ g sucrose}} \times 100 = 51.30\%$ |
| Total | 20.00 | 100% | 500.0 | 100% |

Get help with **percentages.**

Personal Tutor

For example, consider the compound granulated sugar (sucrose). This compound is composed of three elements—carbon, hydrogen, and oxygen. The analysis of 20.00 g of sucrose from a bag of granulated sugar is given in **Table 4.** Note that the sum of the individual masses of the elements found in the sugar equals 20.00 g, which is the amount of the granulated sugar sample that was analyzed. This demonstrates the law of conservation of mass as applied to compounds: the mass of a compound is equal to the sum of the masses of the elements that make up the compound.

Suppose you analyzed 500.0 g of sucrose from a sample of sugarcane. The analysis is shown in **Table 4.** The percent-by-mass values for the sugarcane are equal to the values obtained for the granulated sugar. According to the law of definite proportions, samples of a compound from any source must have the same mass proportions. Conversely, compounds with different mass proportions must be different compounds. Thus, you can conclude that samples of sucrose will always be composed of 42.20% carbon, 6.50% hydrogen, and 51.30% oxygen, no matter their sources.

## PRACTICE Problems

Do additional problems. Online Practice

**19.** A 78.0-g sample of an unknown compound contains 12.4 g of hydrogen. What is the percent by mass of hydrogen in the compound?

**20.** 1.0 g of hydrogen reacts completely with 19.0 g of fluorine. What is the percent by mass of hydrogen in the compound that is formed?

**21.** If 3.5 g of element X reacts with 10.5 g of element Y to form the compound XY, what is the percent by mass of element X in the compound? The percent by mass of element Y?

**22.** Two unknown compounds are tested. Compound I contains 15.0 g of hydrogen and 120.0 g of oxygen. Compound II contains 2.0 g of hydrogen and 32.0 g of oxygen. Are the compounds the same? Explain your answer.

**23.** **Challenge** All you know about two unknown compounds is that they have the same percent by mass of carbon. With only this information, can you be sure the two compounds are the same? Explain.

## Law of Multiple Proportions

Compounds composed of different elements are obviously different compounds. However, different compounds can also be composed of the same elements. This happens when those different compounds have different mass compositions. The **law of multiple proportions** states that when different compounds are formed by a combination of the same elements, different masses of one element combine with the same fixed mass of the other element in a ratio of small whole numbers. Ratios compare the relative amounts of any items or substances. The comparison can be expressed using numbers separated by a colon or as a fraction. With regard to the law of multiple proportions, ratios express the relationship of elements in a compound.

☑ READING CHECK **State** the law of multiple proportions in your own words.

**Water and hydrogen peroxide** The two distinct compounds water ($H_2O$) and hydrogen peroxide ($H_2O_2$) illustrate the law of multiple proportions. Each compound contains the same elements (hydrogen and oxygen). Water is composed of two parts hydrogen and one part oxygen. Hydrogen peroxide is composed of two parts hydrogen and two parts oxygen. Hydrogen peroxide differs from water in that it has twice as much oxygen. When you compare the mass of oxygen in hydrogen peroxide to the mass of oxygen in water, you get the ratio 2:1.

**Compounds made of copper and chlorine** In another example, copper (Cu) reacts with chlorine (Cl) under different sets of conditions to form two different compounds. **Table 5** provides an analysis of their compositions. The two copper compounds must be different because they have different percents by mass. Compound I contains 64.20% copper; Compound II contains 47.27% copper. Compound I contains 35.80% chlorine; Compound II contains 52.73% chlorine.

Using **Figure 20** and **Table 5,** compare the ratio of the mass of copper to the mass of chlorine for each compound. Notice that the mass ratio of copper to chlorine in Compound I (1.793) is exactly 2 times the mass ratio of copper to chlorine in Compound II (0.8964).

$$\frac{\text{mass ratio of Compound I}}{\text{mass ratio of Compound II}} = \frac{1.793 \text{ g Cu/g Cl}}{0.8964 \text{ g Cu/g Cl}} = 2.000$$

☑ GRAPH CHECK **Explain** why the ratio of the relative masses of copper in both compounds is 2:1.

| **Table 5** Analysis Data of Two Copper Compounds | | | | | |
|---|---|---|---|---|---|
| Compound | % Cu | % Cl | Mass Cu (g) in 100.0 g of Compound | Mass Cl (g) in 100.0 g of Compound | Mass Ratio $\left(\frac{\text{mass Cu}}{\text{mass Cl}}\right)$ |
| I | 64.20 | 35.80 | 64.20 | 35.80 | $\frac{1.793 \text{ g Cu}}{1 \text{ g Cl}}$ |
| II | 47.27 | 52.73 | 47.27 | 52.73 | $\frac{0.8964 \text{ g Cu}}{1 \text{ g Cl}}$ |

■ **Figure 20** Copper and chlorine can form different compounds.

a
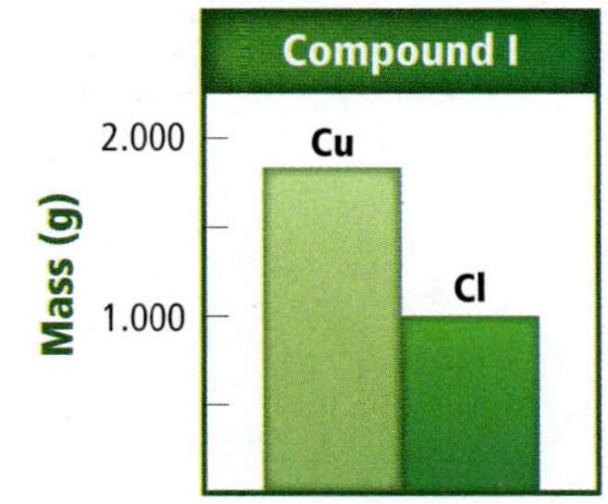


Bar graph **a** compares the relative masses of copper and chlorine in Compound I.

b
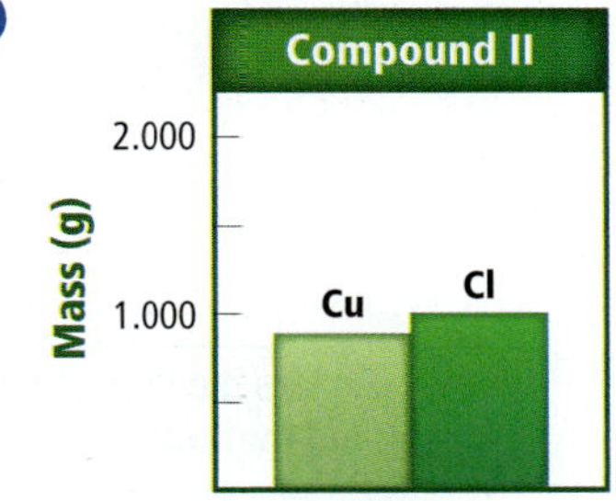


Bar graph **b** compares the relative masses of copper and chlorine in Compound II.

c
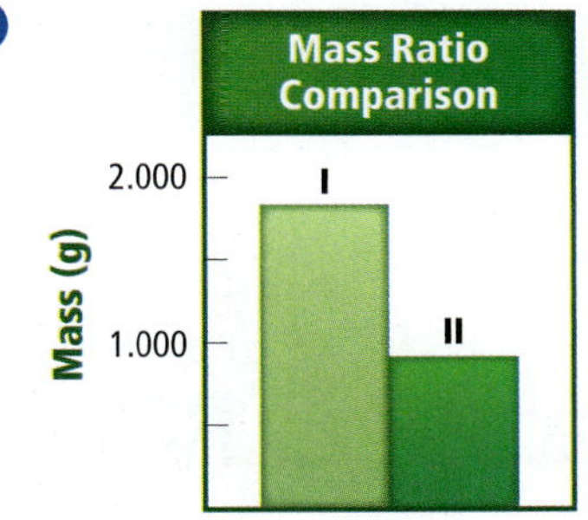


Bar graph **c** shows a comparison between the relative masses of copper in both compounds. The ratio is 2:1.

■ **Figure 21** Different compounds are formed when different relative masses of each element are combined. Although they are both made of copper and chlorine, Compound I has a greenish color, whereas Compound II has a bluish color.

Explore the laws of **definite and multiple proportions.**

**Virtual Investigations**

**Figure 21** shows the two compounds formed by the combination of copper and chlorine and presented in **Table 5** and **Figure 20.** These compounds are called copper(I) chloride and copper(II) chloride. As the law of multiple proportions states, the different masses of copper that combine with a fixed mass of chlorine in the two different copper compounds can be expressed as a small whole-number ratio. In this case, the ratio is 2:1.

Considering that there is a finite number of elements that exist today and an exponentially greater number of compounds that are composed of these elements under various conditions, it becomes clear how important the law of multiple proportions is in chemistry.

## SECTION 4 REVIEW

**Section Self-Check**

### Section Summary

- Elements cannot be broken down into simpler substances.
- Elements are organized in the periodic table of the elements.
- Compounds are chemical combinations of two or more elements, and their properties differ from the properties of their component elements.
- The law of definite proportions states that a compound is always composed of the same elements in the same proportions.
- The law of multiple proportions states that if elements form more than one compound, those compounds will have compositions that are whole-number multiples of each other.

**24.** MAINIDEA **Compare and contrast** elements and compounds.

**25. Describe** the basic organizational feature of the periodic table of the elements.

**26. Explain** how the law of definite proportions applies to compounds.

**27. State** the type of compounds that are compared in the law of multiple proportions.

**28. Complete** the table, and then analyze the data to determine if Compounds I and II are the same compound. If the compounds are different, use the law of multiple proportions to show the relationship between them.

**Analysis Data of Two Iron Compounds**

| Compound | Total Mass (g) | Mass Fe (g) | Mass O (g) | Mass Percent Fe | Mass Percent O |
|---|---|---|---|---|---|
| I | 75.00 | 52.46 | 22.54 | | |
| II | 56.00 | 43.53 | 12.47 | | |

**29. Calculate** the mass percent of hydrogen in water and the mass percent of oxygen in water.

**30. Graph** Create a graph that illustrates the law of multiple proportions.

McGraw-Hill Education

# CHEMISTRY & careers

## Career: Arson Investigator

### Forensic Ignitable Liquid Detection

Inside a burning warehouse, havoc and destruction reign. Intense heat and smoke fill closed spaces. Leaping flames spread; walls and ceilings collapse. Was the fire accidental or the work of an arsonist?

**Ignitable liquids** Fire investigators analyze evidence to determine how a fire began and spread. If arson is suspected, it is likely that ignitable liquids—chemicals that speed the spread of a fire—were involved.

**The properties of an ignitable liquid** The properties that make ignitable liquids useful as fuels also make them dangerous in fire situations. Ignitable liquids are readily absorbed and are powerful solvents. They do not mix well with water, often floating on top and at room temperature, and they form vapors that can ignite and burn.

**Evidence of an ignitable liquid** What evidence indicates the presence of an ignitable liquid? One indicator is an unusual burn pattern, like that present on the floor joists in **Figure 1.** In this case, called a "rundown" burn pattern, an ignitable liquid was likely poured in this area, running down between the floorboards to the joists below.

**Figure 1** Ignitable liquids can cause a rundown burn pattern.

Robert Corry

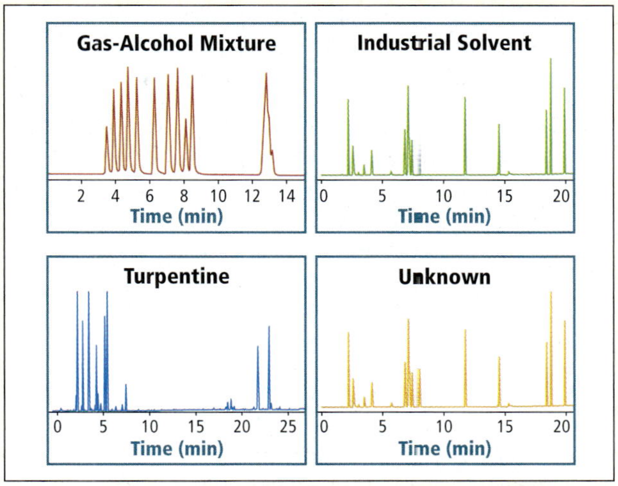


**Figure 2** Chromatograms display individual components of mixtures.

Another indicator is a small slick on top of any wet material, similar to the automobile-oil slick floating on a puddle on a wet street. If investigators see such clues, they can take samples of the affected materials for testing.

**Chemical analysis** Investigators take any samples they collect to the lab for chemical analysis. In the lab, volatiles are extracted from a sample. The extract is separated and analyzed using a process called gas chromatography-mass spectrometry. The components of the mixture are displayed in a chromatogram, like the ones shown in **Figure 2.** By comparing the chromatogram of the unknown with those of known compounds, the identity of the ignitable liquid can be determined. Mass spectra can be obtained for each sample component for confirmation.

### WRITING IN ▶ Chemistry

**Think Critically** Look at the chromatogram of the unknown sample and compare it to the three known samples. Can you determine which accelerant was used? Could that knowledge give you any insight into who might have committed the crime? Explain your answer.

WebQuest

# ChemLAB

iLab Station

## Identify the Products of a Chemical Reaction

**Background:** **Chemical changes can be studied by observing chemical reactions. Products of the reaction can be identified using a flame test.**

**Question:** *Is there a chemical reaction between copper and silver nitrate? Which elements react, and what is the compound they form?*

### Materials

$AgNO_3$ solution
sandpaper
stirring rod
funnel
filter paper
50-mL beaker
50-mL graduated cylinder
250-mL Erlenmeyer flask
small iron ring
ring stand
plastic petri dish
Bunsen burner
tongs
paper clip
copper wire

### Safety Precautions

WARNING: ***Silver nitrate is highly toxic. Avoid contact with eyes and skin.***

### Procedure

1. Read and complete the lab safety form.
2. Rub 8 cm of copper wire with sandpaper until it is shiny. Observe and record its physical properties.
3. Measure 25 mL $AgNO_3$ (silver nitrate) solution into a 50-mL beaker. Record its physical properties.
4. Coil the copper wire so that it fits into the beaker. Make a hook and suspend it from the stirring rod.
5. Place the stirring rod across the top of the beaker, immersing part of the coil in the $AgNO_3$ solution.
6. Make and record observations of the wire and the solution every 5 min for 20 min.
7. Set up a filtration apparatus: attach the iron ring to the ring stand, and adjust its height so the end of the funnel is inside the neck of the Erlenmeyer flask.
8. Fold the circle filter paper in half twice to form a quarter of a circle. Tear off the lower-right corner of the flap facing you. Open the folded paper into a cone, and place it into the funnel.
9. Remove the coil from the beaker, and dispose of it as directed by your teacher.

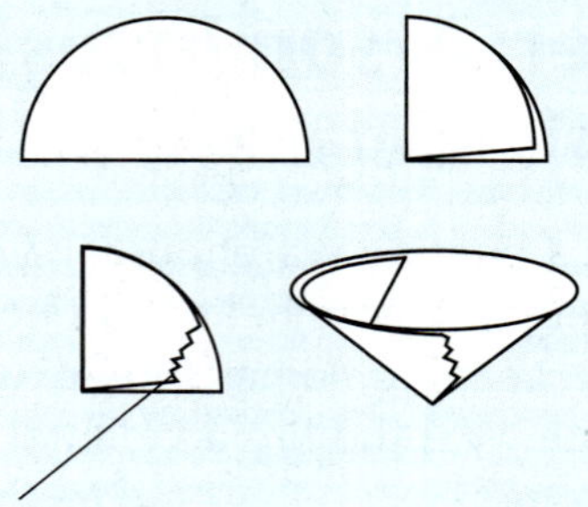


10. Slowly pour the liquid down the stirring rod into the funnel to catch the solid products in the filter paper.
11. Collect the filtrate in the Erlenmeyer flask, and transfer it to a petri dish.
12. Adjust a Bunsen burner flame until it is blue. Hold the paper clip in the flame with tongs until no additional color is observed.
13. Using tongs, dip the hot paper clip into the filtrate. Then, hold the paper clip in the flame. Record the color you observe. After removing the clip from the burner, let it cool before handling.
14. **Cleanup and Disposal** Dispose of materials as directed by your teacher. Clean and return all lab equipment to its proper place.

### Analyze and Conclude

1. **Observe and Infer** Describe the changes you observed in Step 6. Is there evidence that a chemical change occurred? Predict the products formed.
2. **Compare** Use resources such as the *CRC Handbook of Chemistry and Physics* to determine the colors of silver metal and copper nitrate in water. Compare this information with your observations of the reactants and products in Step 6.
3. **Identify** Copper emits a blue-green light in flame tests. Do your observations confirm the presence of copper in the filtrate collected in Step 11?
4. **Classify** Which type of mixture is silver nitrate in water? Which type of mixture is formed in Step 6? In Step 10?

### INQUIRY EXTENSION

**Compare** your recorded observations with those of several other lab teams. Form a hypothesis to explain any differences; design an experiment to test it.

# CHAPTER 3 STUDY GUIDE

Vocabulary Practice

**BIGIDEA** Everything is made of matter.

## SECTION 1 Properties of Matter

**MAINIDEA** Most common substances exist as solids, liquids, and gases, which have diverse physical and chemical properties.

- The three common states of matter are solid, liquid, and gas.
- Physical properties can be observed without altering a substance's composition.
- Chemical properties describe a substance's ability to combine with or change into one or more new substances.
- External conditions can affect both physical and chemical properties.

**VOCABULARY**

- states of matter
- solid
- liquid
- gas
- vapor
- physical property
- extensive property
- intensive property
- chemical property

## SECTION 2 Changes in Matter

**MAINIDEA** Matter can undergo physical and chemical changes.

- A physical change alters the physical properties of a substance without changing its composition.
- A chemical change, also known as a chemical reaction, involves a change in a substance's composition.
- In a chemical reaction, reactants form products.
- The law of conservation of mass states that mass is neither created nor destroyed during a chemical reaction; it is conserved.

$$\text{mass}_{\text{reactants}} = \text{mass}_{\text{products}}$$

**VOCABULARY**

- physical change
- phase change
- chemical change
- law of conservation of mass

## SECTION 3 Mixtures of Matter

**MAINIDEA** Most everyday matter occurs as mixtures—combinations of two or more substances.

- A mixture is a physical blend of two or more pure substances in any proportion.
- Solutions are homogeneous mixtures.
- Mixtures can be separated by physical means. Common separation techniques include filtration, distillation, crystallization, sublimation, and chromatography.

**VOCABULARY**

- mixture
- heterogeneous mixture
- homogeneous mixture
- solution
- filtration
- distillation
- crystallization
- sublimation
- chromatography

## SECTION 4 Elements and Compounds

**MAINIDEA** A compound is a combination of two or more elements.

- Elements cannot be broken down into simpler substances.
- Elements are organized in the periodic table of the elements.
- Compounds are chemical combinations of two or more elements and their properties differ from the properties of their component elements.
- The law of definite proportions states that a compound is always composed of the same elements in the same proportions.

$$\text{percent by mass} = \frac{\text{mass of the element}}{\text{mass of the compound}} \times 100$$

- The law of multiple proportions states that if elements form more than one compound, those compounds will have compositions that are whole-number multiples of each other.

**VOCABULARY**

- element
- periodic table
- compound
- law of definite proportions
- percent by mass
- law of multiple proportions

## SECTION 1

### Mastering Concepts

**31.** List three examples of substances. Explain why each is a substance.

**32.** Is carbon dioxide gas a pure substance? Explain.

**33.** List at least three physical properties of water.

**34.** Identify each physical property as extensive or intensive.
**a.** melting point
**b.** mass
**c.** density
**d.** length

**35.** "Properties are not affected by changes in temperature and pressure." Is this statement true or false? Explain.

**36.** List the three states of matter, and give an example for each state. Differentiate between a gas and a vapor.

**37.** Classify each as either a solid, a liquid, or a gas at room temperature.
**a.** milk
**b.** air
**c.** copper
**d.** helium
**e.** diamond
**f.** candle wax

**38.** Classify each as a physical property or a chemical property.
**a.** Aluminum has a silvery color.
**b.** Gold has a density of 19 g/cm$^3$.
**c.** Sodium ignites when dropped in water.
**d.** Water boils at 100°C.
**e.** Silver tarnishes.
**f.** Mercury is a liquid at room temperature.

**39.** A carton of milk is poured into a bowl. Describe the changes that occur in the milk's shape and volume.

**40. Boiling Water** At what temperature would 250 mL of water boil? 1000 mL? Is the boiling point an intensive or extensive property? Explain.

### Mastering Problems

**41. Chemical Analysis** A scientist wants to identify an unknown compound on the basis of its physical properties. The substance is a white solid at room temperature. Attempts to determine its boiling point were unsuccessful. Using **Table 6,** name the unknown compound.

**Table 6** Physical Properties of Common Substances

| Substance | Color | State at 25°C | Boiling Point (°C) |
|---|---|---|---|
| Oxygen | colorless | gas | −183 |
| Water | colorless | liquid | 100 |
| Sucrose | white | solid | decomposes |
| Sodium chloride | white | solid | 1465 |

## SECTION 2

### Mastering Concepts

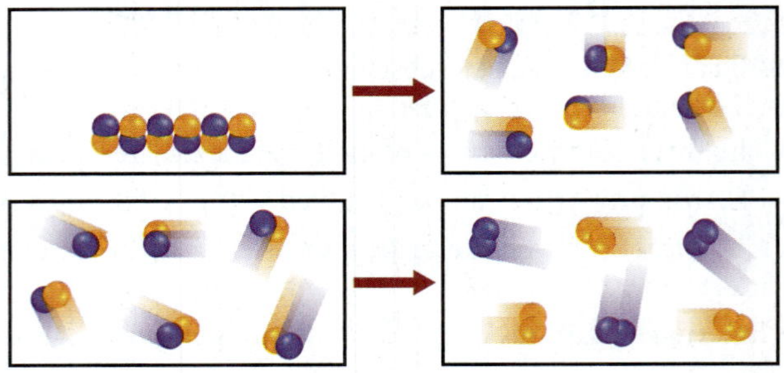

■ **Figure 22**

**42.** Label each set of diagrams in **Figure 22** as a physical or a chemical change.

**43.** Classify each as a physical change or a chemical change.
**a.** breaking a pencil in two
**b.** water freezing and forming ice
**c.** frying an egg
**d.** burning wood
**e.** leaves changing colors in the fall

**44. Ripening** Is the process of bananas ripening a chemical change or a physical change? Explain.

**45.** Is a change in phase a physical change or a chemical change? Explain.

**46.** List four indicators that a chemical change has probably occurred.

**47. Salt** Sodium and chlorine combine to form sodium chloride, or table salt. List the reactants and products of this reaction.

**48. Burning Candle** After burning for three hours, a candle has lost half of its mass. Explain why this example does not violate the law of conservation of mass.

**49.** Describe the difference between a physical change and a chemical change.

### Mastering Problems

**50. Ammonia Production** A 28.0-g sample of nitrogen gas combines completely with 6.0 g of hydrogen gas to form ammonia. What is the mass of ammonia formed?

**51.** A 13.0-g sample of $X$ combines with a 34.0-g sample of $Y$ to form the compound $XY_2$. What is the mass of the reactants?

**52.** If 45.98 g of sodium combines with an excess of chlorine gas to form 116.89 g of sodium chloride, what mass of chlorine gas is used in the reaction?

**53.** A substance breaks down into its component elements when it is heated. If 68.0 g of the substance is present before it is heated, what is the combined mass of the component elements after heating?

Chapter Self-Check

**54.** Copper(I) sulfide is formed when copper and sulfur are heated together. In this reaction, 127 g of copper reacts with 41 g of sulfur. After the reaction is complete, 9 g of sulfur remains unreacted. What is the mass of copper sulfide formed?

**55.** When burning 180 g of glucose in the presence of 192 g of oxygen, water and carbon dioxide are produced. If 108 g of water is produced, how many grams of carbon dioxide are produced?

## SECTION 3

### Mastering Concepts

**56.** Describe the characteristics of a mixture.

■ **Figure 23**

**57.** Name the separation method illustrated in **Figure 23.**

**58.** Describe a method that could be used to separate each mixture.
- **a.** iron filings and sand
- **b.** sand and salt
- **c.** the components of ink
- **d.** helium and oxygen gases

**59.** "A mixture is the chemical bonding of two or more substances in any proportion." Is this statement true or false? Explain.

**60.** Which of the following are the same and which are different?
- **a.** a substance and a pure substance
- **b.** a heterogeneous mixture and a solution
- **c.** a substance and a mixture
- **d.** a homogeneous mixture and a solution

**61.** Describe how a homogeneous mixture differs from a heterogeneous mixture.

**62. Seawater** is composed of salt, sand, and water. Is seawater a heterogeneous or homogeneous mixture? Explain.

**63. Iced Tea** Use iced tea with and without ice cubes as examples to explain homogeneous and heterogeneous mixtures. If you allow all of the ice cubes to melt, what type of mixture remains?

**64. Chromatography** What is chromatography, and how does it work?

## SECTION 4

### Mastering Concepts

**65.** State the definition of element.

**66.** Correct the following statements.
- **a.** An element is a combination of two or more compounds.
- **b.** When a small amount of sugar is completely dissolved in water, a heterogeneous solution is formed.

**67.** Name the elements contained in the following compounds.
- **a.** sodium chloride ($NaCl$)
- **b.** ammonia ($NH_3$)
- **c.** ethanol ($C_2H_6O$)
- **d.** bromine ($Br_2$)

**68.** What was Dmitri Mendeleev's major contribution to the field of chemistry?

**69.** Is it possible to distinguish between an element and a compound? Explain.

**70.** How are the properties of a compound related to those of the elements that comprise it?

**71.** Which law states that a compound always contains the same elements in the same proportion by mass?

**72.** **a.** What is the percent by mass of carbon in 44 g of carbon dioxide ($CO_2$)?

**b.** What is the percent by mass of oxygen in 44 g of carbon dioxide ($CO_2$)?

**73.** Complete **Table 7** by classifying the compounds as 1:1 or 2:2, 1:2 or 2:1, and 1:3 or 3:1.

**Table 7 Ratios of Elements in Compounds**

| Compound | Simple Whole-Number Ratios of Elements |
|---|---|
| NaCl | |
| CuO | |
| $H_2O$ | |
| $H_2O_2$ | |

### Mastering Problems

**74.** A 25.3-g sample of an unknown compound contains 0.8 g of oxygen. What is the percent by mass of oxygen in the compound?

**75.** Magnesium combines with oxygen to form magnesium oxide. If 10.57 g of magnesium reacts completely with 6.96 g of oxygen, what is the percent by mass of oxygen in magnesium oxide?

**76.** When mercury oxide is heated, it decomposes into mercury and oxygen. If 28.4 g of mercury oxide decomposes, producing 2.0 g of oxygen, what is the percent by mass of mercury in mercury oxide?

**77.** Carbon reacts with oxygen to form two different compounds. Compound I contains 4.82 g of carbon for every 6.44 g of oxygen. Compound II contains 20.13 g of carbon for every 53.7 g of oxygen. What is the ratio of carbon to a fixed mass of oxygen for the two compounds?

**78.** A 100-g sample of an unknown salt contains 64 g of chlorine. What is the percent by mass of chlorine in the compound?

**79.** Which law would you use to compare CO and $CO_2$? Explain. Without doing any calculations, determine which of the two compounds has the highest percent by mass of oxygen in the compound.

**80.** Complete **Table 8.**

**Table 8 Elements in Compounds**

| Compound | Mass of Compound (g) | Mass of Oxygen (g) | Mass % of Oxygen | Mass of Second Element in the Compound (g) |
|---|---|---|---|---|
| CuO | 80.0 | 16 | | |
| $H_2O$ | 18.0 | 16 | | |
| $H_2O_2$ | 34.0 | 32 | | |
| CO | 28.0 | 16 | | |
| $CO_2$ | 44.0 | 32 | | |

## MIXED REVIEW

**81.** Which state(s) of matter are compressible? Which state(s) of matter are not compressible? Explain.

**82.** Classify each mixture as homogeneous or heterogeneous.
**a.** brass (an alloy of zinc and copper)
**b.** a salad
**c.** blood
**d.** powdered drink mix dissolved in water

**83.** Phosphorus combines with hydrogen to form phosphine. In this reaction, 108.3 g of phosphorus combines with excess hydrogen to produce 129.9 g of phosphine. After the reaction, 11.0 g of hydrogen remains unreacted. What was the initial mass of hydrogen before the reaction? What mass of hydrogen is used in the reaction?

**84.** If you have 100 particles of hydrogen and 100 particles of oxygen, how many units of water can you form? Will you use all the particles of both elements? If not, what will remain?

**85.** Classify each substance as a pure substance, a homogeneous mixture, or a heterogeneous mixture.
**a.** air **c.** soil **e.** sediment
**b.** aerosol **d.** water **f.** muddy water

**86.** Identify each as a homogenous mixture, a heterogeneous mixture, a compound, or an element.
**a.** pure drinking water **d.** seawater
**b.** salty water **e.** air
**c.** helium

**87. Cooking** List physical properties of eggs before and after they are cooked. Based on your observations, does a physical change or chemical change occur when eggs are cooked? Justify your answer.

**88. Ice Cream** You might have noticed that while eating ice cream on a hot day, some of the ice cream begins to melt. Is the observed change in the state of the ice cream a physical or a chemical change? Justify your answer.

**89. Pizza** Is a pizza a homogeneous or heterogeneous mixture? Explain.

**90.** Sodium reacts chemically with chlorine to form sodium chloride. Is sodium chloride a mixture or a compound?

**91.** Is air a solution or a heterogeneous mixture? What technique can be used to separate air into its components?

**92.** Indicate whether combining the following elements yields a compound or a mixture.
**a.** $H_2(g) + O_2(g) \rightarrow$ water
**b.** $N_2(g) + O_2(g) \rightarrow$ air

## THINK CRITICALLY

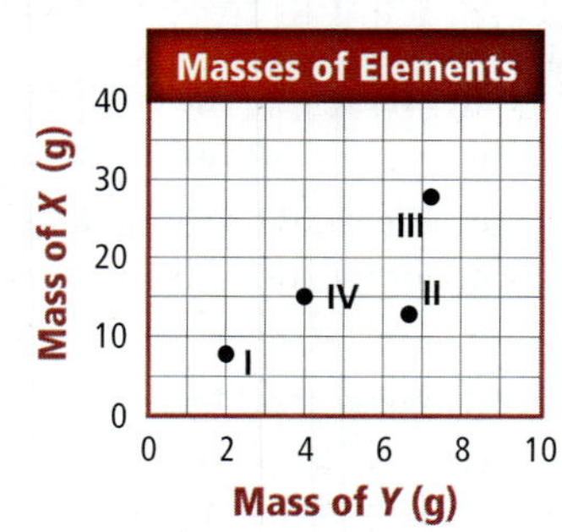


■ **Figure 24**

**93. Interpret Data** A compound contains the elements *X* and *Y*. Four samples with different masses were analyzed, and the masses of *X* and *Y* in each sample were plotted on a graph shown in **Figure 24.** The samples were labeled *I, II, III,* and *IV.*
**a.** Which samples are from the same compound? How do you know?
**b.** What is the approximate ratio of the mass of *X* to the mass of *Y* in the samples that are from the same compound?
**c.** What is the approximate ratio of the mass of *X* to the mass of *Y* in the sample(s) that are not from the same compound?

Chapter Self-Check

**94. Apply** Air is a mixture of many gases, primarily nitrogen, oxygen, and argon. Could distillation be used to separate air into its component gases? Explain.

**95. Analyze** Is gas escaping from an opened soft drink an example of a chemical or a physical change? Explain.

**96. Apply** Give examples of heterogeneous mixtures for the systems listed in **Table 9.**

**Table 9 Heterogeneous Mixtures**

| System | Example |
|---|---|
| Liquid-liquid | |
| Solid-liquid | |
| Solid-solid | |

## Challenge Problem

**97. Identify Lead Compounds** A sample of a certain lead compound contains 6.46 g of lead for each gram of oxygen. A second sample has a mass of 68.54 g and contains 28.76 g of oxygen. Are the two samples the same? Explain.

## CUMULATIVE REVIEW

**98.** What is chemistry?

**99.** What is mass? Weight?

**100.** Express the following numbers in scientific notation.

**a.** 34,500 **d.** 789
**b.** 2665 **e.** 75,600
**c.** 0.9640 **f.** 0.002189

**101.** Perform the following operations.

**a.** $10^7 \times 10^3$
**b.** $(1.4 \times 10^{-3}) \times (5.1 \times 10^{-5})$
**c.** $(2 \times 10^{-3}) \times (4 \times 10^{5})$

**102.** Convert 65°C to kelvins.

**103.** Graph the data in **Table 10.** What is the slope of the line?

**Table 10 Energy Released by Carbon**

| Mass (g) | Energy Released (kJ) |
|---|---|
| 1.00 | 33 |
| 2.00 | 66 |
| 3.00 | 99 |
| 4.00 | 132 |

## WRITING IN Chemistry

**104. Synthetic Elements** Select a synthetic element, and prepare a short written report on its development. Be sure to discuss recent discoveries, list major research centers that conduct this type of research, and describe the properties of the synthesized element.

## DBQ Document-Based Questions

**Pigments** *Long before scientists understood the properties of elements and compounds, artists used chemistry to create pigments from natural materials.* **Table 11** *gives some examples of such pigments used in ancient times.*

Data obtained from: Orna, Mary Virginia. 2001. Chemistry, color, and art. *Journal of Chemical Education* 78 (10): 1305

**Table 11 Common Artists' Pigments Used in Early Times**

| Common Name | Chemical Identity | Comments |
|---|---|---|
| Charcoal | elemental carbon (carbon black) | produced by dry distillation of wood in a closed vessel |
| Egyptian blue | calcium copper tetrasilicate, $CaCuSi_4O_{10}$ | crystalline compound containing some glass impurity |
| Indigo | indigotin, $C_{16}H_{10}N_2O_2$ | derived from different plants of the genus *Indigofera* |
| Iron oxide red | $Fe_2O_3$ | in continuous use in all geographic regions and time periods |
| Verdigris | dibasic acetate of copper, $Cu(C_2H_3O_2)_2{\cdot}2Cu(OH)_2$ | other copper compounds, including carbonate, are also called verdigris |

**105. a.** Compare the mass percent of carbon in charcoal, indigo, and verdigris.

**b.** Compare the mass percent of oxygen in iron oxide and Egyptian blue.

**106.** List an example of an element and a compound from **Table 11.**

**107.** Is the production of charcoal from the dry distillation of wood a chemical or a physical change? Explain.

# CHAPTER 3 STANDARDIZED TEST PRACTICE

## MULTIPLE CHOICE

*Use the table below to answer Questions 1 and 2.*

| Mass Analysis of Two Chlorine-Fluorine Samples | | | | |
|---|---|---|---|---|
| Sample | Mass of Chlorine (g) | Mass of Fluorine (g) | % Cl | % F |
| I | 13.022 | 6.978 | 65.11 | 34.89 |
| II | 5.753 | 9.248 | ? | ? |

**1.** What are the values for % Cl and % F, respectively, for Sample II?
**A.** 0.6220 and 61.65
**B.** 61.65 and 38.35
**C.** 38.35 and 0.6220
**D.** 38.35 and 61.65

**2.** Which statement best describes the relationship between the two samples?
**A.** The compound in Sample I is the same as in Sample II. Therefore, the mass ratio of Cl to F in both samples will obey the law of definite proportions.
**B.** The compound in Sample I is the same as in Sample II. Therefore, the mass ratio of Cl to F in both samples will obey the law of multiple proportions.
**C.** The compound in Sample I is not the same as in Sample II. Therefore, the mass ratio of Cl to F in both samples will obey the law of definite proportions.
**D.** The compound in Sample I is not the same as in Sample II. Therefore, the mass ratio of Cl to F in both samples will obey the law of multiple proportions.

**3.** After two elements react to completion in a closed container, the ratio of their masses in the container will be the same as before the reaction. Which law describes this principle?
**A.** law of definite proportions
**B.** law of multiple proportions
**C.** law of conservation of mass
**D.** law of conservation of energy

**4.** Which is NOT a physical property of table sugar?
**A.** forms solid crystals at room temperature
**B.** appears as white crystals
**C.** breaks down into carbon and water vapor when heated
**D.** tastes sweet

**5.** Which describes a substance that is in the solid state?
**A.** Its particles can flow past one another.
**B.** It can be compressed into a smaller volume.
**C.** It takes the shape of its container.
**D.** Its particles of matter are close together.

*Use the diagram below to answer Questions 6 and 7.*

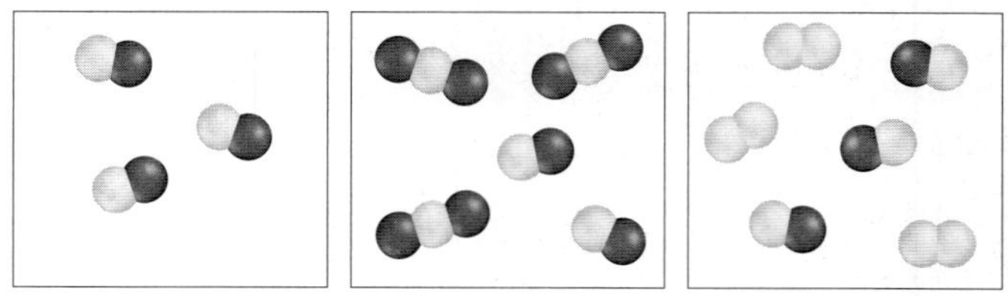

A B C

**6.** Which best describes Figure A?
**A.** element
**B.** mixture
**C.** solution
**D.** compound

**7.** Which statement is false?
**A.** Figure B is composed of two different compounds.
**B.** Figure C is composed of two different compounds.
**C.** Figure B represents 13 total atoms.
**D.** Three different types of elements are represented in Figure C.

**8.** Na, K, Li, and Cs all share similar chemical properties. In the periodic table of elements, they most likely belong to the same
**A.** row.
**B.** period.
**C.** group.
**D.** element.

**9.** Magnesium reacts explosively with oxygen to form magnesium oxide. Which is NOT true of this reaction?
**A.** The mass of magnesium oxide produced equals the mass of magnesium consumed plus the mass of oxygen consumed.
**B.** The reaction describes the formation of a new substance.
**C.** The product of the reaction, magnesium oxide, is a chemical compound.
**D.** Magnesium oxide has physical and chemical properties similar to both oxygen and magnesium.

## SHORT ANSWER

10. Compare and contrast the independent variable in an experiment with the dependent variable.

11. A student reports the melting point of a gas as −295°C. Explain why his claim is unlikely to be correct.

12. Place the following metric prefixes in order from the smallest value to the largest value: deci, kilo, centi, micro, mega, milli, giga, nano.

## EXTENDED RESPONSE

*Use the table below to answer Questions 13 to 15.*

**Selected Properties of Substances in a Mixture**

| Item | Soluble in Water? | Soluble in Alcohol? | Density ($g/cm^3$) | Particle Size (mm) |
|---|---|---|---|---|
| Sawdust | no | no | 0.21 | 1 |
| Mothball flakes | no | yes | 1.15 | 3 |
| Table salt | yes | no | 2.17 | 2 |

13. Is the mixture described in the table homogeneous or heterogeneous? Explain how you can tell.

14. Do the data describe chemical or physical properties? Explain your answer.

15. Propose a method to separate the three substances based on the properties described above.

16. Explain the difference between a chemical change and a physical change. Is the combustion of gasoline a chemical change or a physical change? Explain your answer.

## SAT SUBJECT TEST: CHEMISTRY

17. Which is a correct statement about methods for separating mixtures?
    A. Distillation results in the formation of solid particles of a dissolved substance.
    B. Filtration depends on differences in sizes of particles.
    C. Separations depend on the chemical properties of the substances involved.
    D. Chromatography depends on the different boiling points of substances.
    E. Sublimation can be used to separate two gases present in a mixture.

*Use the table below to answer Questions 18 and 19.*

**Percent by Mass of Carbon, Hydrogen, and Oxygen in Selected Compounds**

| Compound | % H | % C | % O |
|---|---|---|---|
| Carbonic acid ($H_2CO_3$) | 3.2 | 19.4 | 77.4 |
| Acetic acid ($CH_3COOH$) | 6.7 | 40.0 | 53.3 |
| Methanol ($CH_3OH$) | 12.5 | 37.5 | 40.0 |
| Methanal ($H_2CO$) | 6.7 | 40.0 | 53.3 |
| Isopropanol ($C_3H_8O$) | 13.3 | 60.0 | 26.6 |

18. You have a 125-g sample of one of these substances. You determine that it is made of 16.7 g H, 75.0 g C, and 33.3 g O. Which compound is it?
    A. acetic acid
    B. carbonic acid
    C. methanal
    D. methanol
    E. isopropanol

19. In another experiment, you determine that a sample of acetic acid consists of 56.8% oxygen. What is your percent error?
    A. 3.50%
    B. 6.57%
    C. 1.07%
    D. 12.6%
    E. 2.06%

| NEED EXTRA HELP? | | | | | | | | | | | | | | | | | | | |
|---|---|---|---|---|---|---|---|---|---|---|---|---|---|---|---|---|---|---|---|
| If You Missed Question . . . | 1 | 2 | 3 | 4 | 5 | 6 | 7 | 8 | 9 | 10 | 11 | 12 | 13 | 14 | 15 | 16 | 17 | 18 | 19 |
| Review Section . . . | 3.4 | 3.4 | 3.2 | 3.1 | 3.1 | 3.4 | 3.4 | 3.4 | 3.1 | 1.3 | 2.1 | 2.1 | 3.3 | 3.1 | 3.3 | 3.2 | 3.3 | 3.4 | 2.3 |

CHAPTER 4

# The Structure of the Atom

**BIGIDEA** Atoms are the fundamental building blocks of matter.

## SECTIONS

Chm 1, 2

1 Early Ideas About Matter

2 Defining the Atom

3 How Atoms Differ

4 Unstable Nuclei and Radioactive Decay

## LaunchLAB

iLab Station

### How can the effects of electric charges be observed?

Electric charge plays an important role in atomic structure. In this lab, you will study the effects of like and opposite charges.

## FOLDABLES® Study Organizer

### The Atom

Make a concept-map book. Label it as shown. Use it to help you organize your study of the structure of the atom.

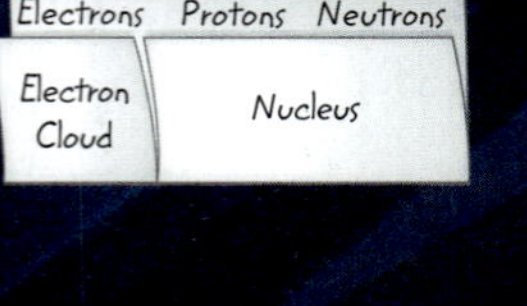

The graphite in this pencil and a diamond are made from atoms of the same element—carbon.

SECTION 1

# Early Ideas About Matter

**MAINIDEA** The ancient Greeks tried to explain matter, but the scientific study of the atom began with John Dalton in the early 1800s.

**Essential Questions**

- What are the similarities and differences of the atomic models of Democritus, Aristotle, and Dalton?
- How was Dalton's theory used to explain the conservation of mass?

**Review Vocabulary**

**theory:** an explanation supported by many experiments; is still subject to new experimental data, can be modified, and is considered successful if it can be used to make predictions that are true

**New Vocabulary**

Dalton's atomic theory

**CHEM 4 YOU** A football team might practice and try different plays in order to develop the best-possible game plan. As they see the results of their plans, coaches can make adjustments to refine the team's play. Similarly, scientists over the last 200 years have hypothesized different models of the atom, refining their models as they collected new data.

## The Roots of Atomic Theory

Science as we know it today did not exist several thousand years ago. No one knew what a controlled experiment was, and there were few tools for scientific exploration. In this setting, the power of the mind and intellectual thought were considered the primary avenues to the truth. Curiosity sparked the interest of scholarly thinkers known as philosophers who considered the many mysteries of life. As they speculated about the nature of matter, many of the philosophers formulated explanations based on their own life experiences.

Many of them concluded that matter was composed of things such as earth, water, air, and fire, as shown in **Figure 1.** It was also commonly accepted that matter could be endlessly divided into smaller and smaller pieces. While these early ideas were creative, there was no method available to test their validity.

**Figure 1** Many Greek philosophers thought that matter was composed of four elements: earth, air, water, and fire. They also associated properties with each element. The pairing of opposite properties, such as hot and cold, and wet and dry, mirrored the symmetry they observed in nature. These early ideas were non-scientific.

**Democritus** The Greek philosopher Democritus (460–370 B.C.) was the first person to propose the idea that matter was not infinitely divisible. He believed matter was made up of tiny individual particles called *atomos,* from which the English word *atom* is derived. Democritus believed that atoms could not be created, destroyed, or further divided. Democritus and a summary of his ideas are shown in **Table 1.**

While a number of Democritus's ideas do not agree with modern atomic theory, his belief in the existence of atoms was amazingly ahead of his time. However, his ideas were met with criticism from other philosophers who asked, "What holds the atoms together?" Democritus could not answer the question.

**Aristotle** Other criticisms came from Aristotle (384–322 B.C.), one of the most influential Greek philosophers. He rejected the notion of atoms because it did not agree with his own ideas about nature. One of Aristotle's major criticisms concerned the idea that atoms moved through empty space. He did not believe that empty space could exist. His ideas are also presented in **Table 1.** Because Aristotle was one of the most influential philosophers of his time, Democritus's atomic theory was eventually rejected.

In fairness to Democritus, it was impossible for him or anyone else of his time to determine what held the atoms together. More than two thousand years would pass before scientists would know the answer. However, it is important to realize that Democritus's ideas were just that—ideas, not science. Without the ability to conduct controlled experiments, Democritus could not test the validity of his ideas.

Unfortunately for the advancement of science, Aristotle was able to gain wide acceptance for his ideas on nature—ideas that denied the existence of atoms. Incredibly, the influence of Aristotle was so great and the development of science so primitive that his denial of the existence of atoms went largely unchallenged for two thousand years!

**VOCABULARY**
**WORD ORIGIN**
**Atom**
comes from the Greek word *atomos,* meaning *indivisible*

☑ READING CHECK **Infer** why it was hard for Democritus to defend his ideas.

**Table 1 Ancient Greek Ideas About Matter**

| Philosopher | | Ideas |
|---|---|---|
| **Democritus (460–370 B.C.)** |   | • Matter is composed of atoms, which move through empty space. • Atoms are solid, homogeneous, indestructible, and indivisible. • Different kinds of atoms have different sizes and shapes. • Size, shape, and movement of atoms determine the properties of matter. |
| **Aristotle (384–322 B.C.)** |  | • Empty space cannot exist. • Matter is made of earth, fire, air, and water. |

## Table 2 Dalton's Atomic Theory

| Scientist | | Ideas |
|---|---|---|
| **Dalton (1766–1844)** |  | • Matter is composed of extremely small particles called atoms.<br>• Atoms are indivisible and indestructible.<br>• Atoms of a given element are identical in size, mass, and chemical properties.<br>• Atoms of a specific element are different from those of another element.<br>• Different atoms combine in simple whole-number ratios to form compounds.<br>• In a chemical reaction, atoms are separated, combined or rearranged. |

**John Dalton** Although the concept of the atom was revived in the eighteenth century, it took another hundred years before significant progress was made. The work done in the nineteenth century by John Dalton (1766–1844), a schoolteacher in England, marks the beginning of the development of modern atomic theory. Dalton revived and revised Democritus's ideas based on the results of scientific research he conducted. In many ways, Democritus's and Dalton's ideas are similar.

Thanks to advancements in science since Democritus's day, Dalton was able to perform experiments that allowed him to refine and support his hypotheses. He studied numerous chemical reactions, making careful observations and measurements along the way. He was able to determine the mass ratios of the elements involved in those reactions. The results of his research are known as **Dalton's atomic theory,** which he proposed in 1803. The main points of his theory are summarized in **Table 2.** Dalton published his ideas in a book, an extract of which is shown in **Figure 2.**

☑ READING CHECK **Compare and contrast** Democritus' and Dalton's ideas.

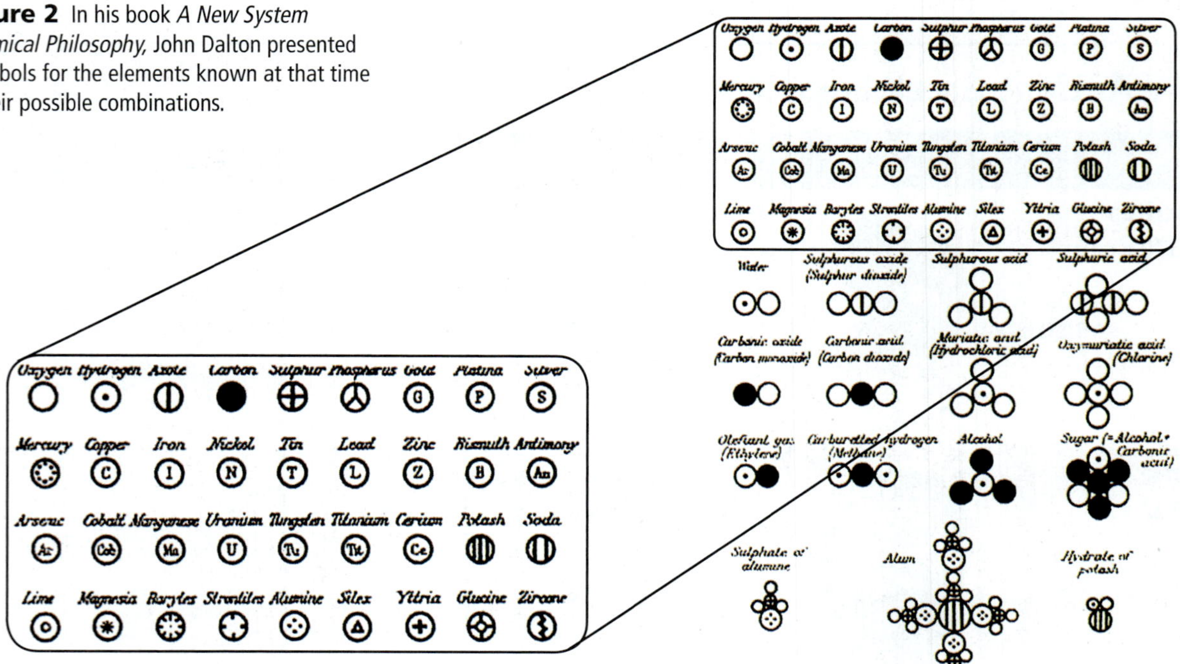


■ **Figure 2** In his book *A New System of Chemical Philosophy,* John Dalton presented his symbols for the elements known at that time and their possible combinations.

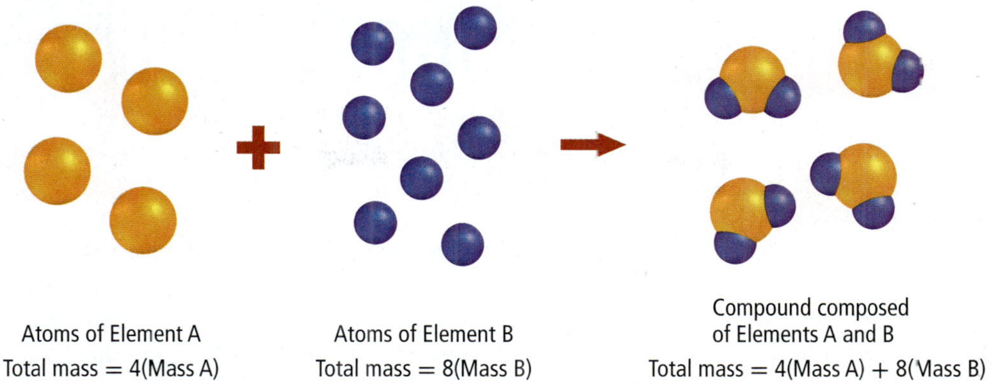


**Figure 3** When atoms of two or more elements combine to form a compound, the number of atoms of each element is conserved. Thus, the mass is conserved as well.

**Conservation of mass** Recall that the law of conservation of mass states that mass is conserved in any process, such as a chemical reaction. Dalton's atomic theory easily explains that the conservation of mass in chemical reactions is the result of the separation, combination, or rearrangement of atoms—atoms that are not created, destroyed, or divided in the process. The formation of a compound from the combining of elements and the conservation of mass during the process are shown in **Figure 3.** The number of atoms of each type is the same before and after the reaction. Dalton's convincing experimental evidence and clear explanation of the composition of compounds, and conservation of mass led to the general acceptance of his atomic theory.

Dalton's atomic theory was a huge step toward the current atomic model of matter. However, not all of Dalton's theory was accurate. As is often the case in science, Dalton's theory had to be revised as additional information was learned that could not be explained by the theory. As you will learn in this chapter, Dalton was wrong about atoms being indivisible. Atoms are divisible into several subatomic particles. Dalton was also wrong about all atoms of a given element having identical properties. Atoms of the same element can have slightly different masses.

## SECTION 1 REVIEW

Section Self-Check

### Section Summary

- Democritus was the first person to propose the existence of atoms.
- According to Democritus, atoms are solid, homogeneous, and indivisible.
- Aristotle did not believe in the existence of atoms.
- John Dalton's atomic theory is based on numerous scientific experiments.

1. **MAIN IDEA Contrast** the methods used by the Greek philosophers and Dalton to study the atom.
2. **Define** *atom* using your own words.
3. **Summarize** Dalton's atomic theory.
4. **Explain** how Dalton's theory of the atom and the conservation of mass are related.
5. **Apply** Six atoms of Element A combine with eight atoms of Element B to produce six compound particles. How many atoms of Elements A and B does each particle contain? Are all of the atoms used to form compounds?
6. **Design** a concept map that compares and contrasts the atomic ideas proposed by Democritus and John Dalton.

SECTION 2

# Defining the Atom

**MAINIDEA** An atom is made of a nucleus containing protons and neutrons; electrons move around the nucleus.

### Essential Questions

- What is an atom?
- How can the subatomic particles be distinguished in terms of relative charge and mass?
- Where are the locations of the subatomic particles within the structure of the atom?

### Review Vocabulary

**model:** a visual, verbal, and/or mathematical explanation of data collected from many experiments

### New Vocabulary

atom
cathode ray
electron
nucleus
proton
neutron

**CHEM 4 YOU** If you have ever accidentally bitten into a peach pit, you know that your teeth pass easily through the fruit, but cannot dent the hard pit. Similarly, many particles that pass through the outer parts of an atom are deflected by the dense center of the atom.

## The Atom

Many experiments since Dalton's time have proven that atoms do exist. So what exactly is the definition of an atom? To answer this question, consider a gold ring. Suppose you decide to grind the ring down into a pile of gold dust. Each fragment of gold dust still retains all of the properties of gold. If it were possible—which it is not without special equipment—you could continue to divide the gold dust particles into still smaller particles. Eventually, you would encounter a particle that could not be divided any further and still retain the properties of gold. This smallest particle of an element that retains the properties of the element is called an **atom.**

To get an idea of its size, consider the population of the world, which was about $6.5 \times 10^9$ in 2006. By comparison, a typical solid-copper penny contains $2.9 \times 10^{22}$ atoms, almost five trillion times the world population! The diameter of a single copper atom is $1.28 \times 10^{-10}$ m. Placing $6.5 \times 10^9$ copper atoms side by side would result in a line of copper atoms less than 1 m long. **Figure 4** illustrates another way to visualize the size of an atom. Imagine that you increase the size of an atom to be as big as an orange. To keep the proportions between the real sizes of the atom and of the orange, you would have to increase the size of the orange and make it as big as Earth. This illustrates how small atoms are.

**Figure 4** Imagine that you could increase the size of an atom to make it as big as an orange. At this new scale, an orange would be as big as Earth.

(l)©Stockdisc/PunchStock; (r)European Space Agency/Photo Researchers

**Connection to Biology** **Looking at atoms** You might think that because atoms are so small, there would be no way to see them. However, an instrument called the scanning tunneling microscope (STM) allows individual atoms to be seen. Just as you need a microscope to study cells in biology, the STM allows you to study atoms. An STM works as follows: a fine point is moved above a sample and the interaction of the point with the superficial atoms is recorded electronically. **Figure 5** illustrates how individual atoms look when observed with an STM. Scientists are now able to move individual atoms around to form shapes, patterns, and even simple machines. This capability has led to the exciting new field of nanotechnology. The promise of nanotechnology is molecular manufacturing—the atom-by-atom building of machines the size of molecules. As you will later read, a molecule is a group of atoms that are bonded together and act as a unit.

**Figure 5** This image, recorded with an STM, shows the individual atoms of a fatty acid on a graphite surface. The false colors were added later on to improve the contrast between each atom.

## The Electron

Once scientists were convinced of the existence of atoms, a new set of questions emerged. What is an atom like? Is the composition of an atom uniform throughout, or is it composed of still-smaller particles? Although many scientists researched the atom in the 1800s, it was not until almost 1900 that some of these questions were answered.

**The cathode-ray tube** As scientists tried to unravel the atom, they began to make connections between matter and electric charge. For instance, has your hair ever clung to your comb? To explore the connection, some scientists wondered how electricity might behave in the absence of matter. With the help of the newly invented vacuum pump, they passed electricity through glass tubes from which most of the air had been removed. Such tubes are called cathode-ray tubes.

A typical cathode-ray tube used by researchers for studying the relationship between mass and charge is illustrated in **Figure 6.** Note that metal electrodes are located at opposite ends of the tube. The electrode connected to the negative terminal of the battery is called the cathode, and the electrode connected to the positive terminal is called the anode.

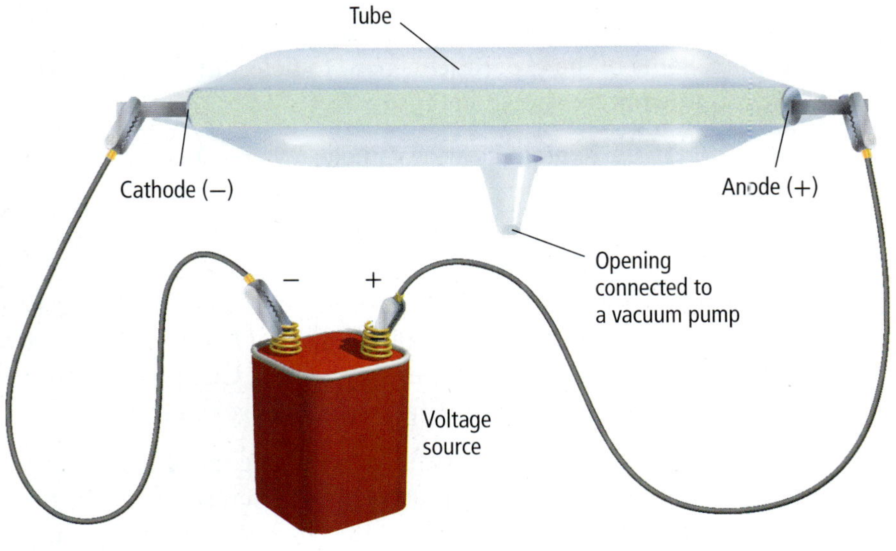


**Figure 6** A cathode-ray tube is a tube with an anode at one end and a cathode at the other end. When a voltage is applied, electricity travels from the cathode to the anode.

**Figure 7** A tiny hole located in the center of the anode produces a thin beam of electrons. A phosphor coating allows the position of the beam to be determined as it strikes the end of the tube.

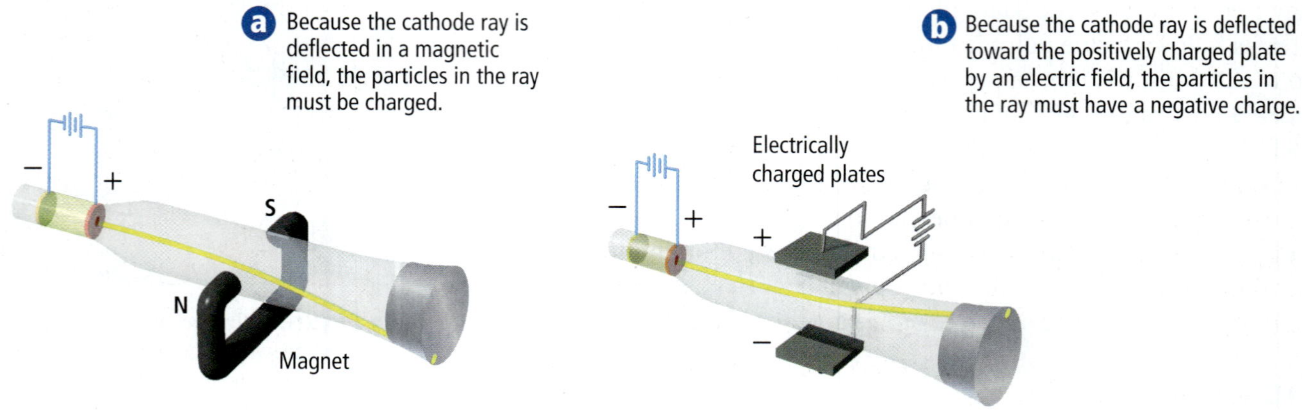


## RealWorld CHEMISTRY

### Cathode Ray

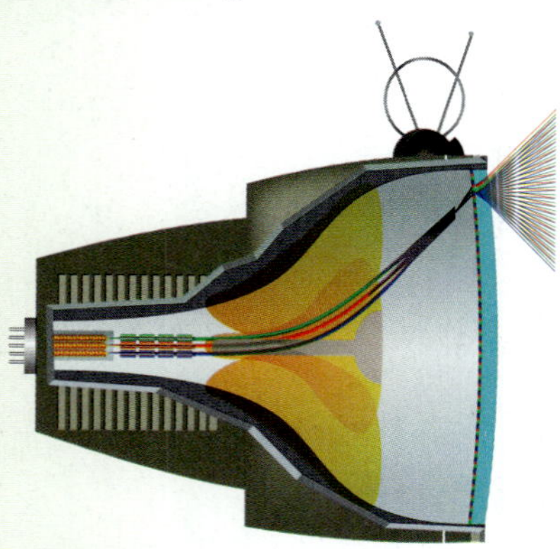

**TELEVISION** Television was invented in the 1920's. Conventional television images are formed as cathode rays strike light-producing chemicals that coat the back of the screen.

**Sir William Crookes** While working in a darkened laboratory, English physicist Sir William Crookes noticed a flash of light within one of the cathode-ray tubes. A green flash was produced by some form of radiation striking a zinc-sulfide coating that had been applied to the end of the tube. Further work showed that there was a ray (radiation) going through the tube. This ray, originating from the cathode and traveling to the anode, was called a **cathode ray.** The accidental discovery of the cathode ray led to the invention of television. A conventional television is nothing more than a cathode-ray tube.

Scientists continued their research using cathode-ray tubes, and they were fairly convinced by the end of the 1800s of the following:

- Cathode rays were a stream of charged particles.
- The particles carried a negative charge. (The exact value of the negative charge was not known.)

Because changing the metal that makes up the electrodes or varying the gas (at very low pressure) in the cathode-ray tube did not affect the cathode ray produced, researchers concluded that the ray's negative particles were found in all forms of matter. These negatively charged particles that are part of all forms of matter are now known as **electrons.** Some of the experiments used to determine the properties of the cathode ray are shown in **Figure 7.**

READING CHECK **Explain** how the cathode ray was discovered.

FOLDABLES®
Incorporate information from this section into your Foldable.

**Mass and charge of the electron** In spite of the progress made from all of the cathode-ray tube experiments, no one succeeded in determining the mass of a single cathode-ray particle. Unable to measure the particle's mass directly, English physicist J. J. Thomson (1856–1940) began a series of cathode-ray tube experiments at Cambridge University in the late 1890s to determine the ratio of its charge to its mass.

**Charge-to-mass ratio** By carefully measuring the effects of both magnetic and electric fields on a cathode ray, Thomson was able to determine the charge-to-mass ratio of the charged particle. He then compared that ratio to other known ratios.

Thomson concluded that the mass of the charged particle was much less than that of a hydrogen atom, the lightest known atom. The conclusion was shocking because it meant there were particles smaller than the atom. In other words, Dalton had been incorrect—atoms were divisible into smaller subatomic particles. Because Dalton's atomic theory had become so widely accepted and Thomson's conclusion was so revolutionary, many other scientists found it hard to accept this new discovery. But Thomson was correct. He had identified the first subatomic particle—the electron. He received a Nobel Prize in 1906 for this discovery.

**READING CHECK** **Summarize** how Thomson discovered the electron.

**The oil-drop experiment and the charge of an electron** The next significant development came in the early 1910s, when the American physicist Robert Millikan (1868–1953) determined the charge of an electron using the oil-drop apparatus shown in **Figure 8.** In this apparatus, oil is sprayed into the chamber above the two parallel charged plates. The top plate has a small hole through which the oil drops. X-rays knock out electrons from the air particles between the plates and the electrons stick to the droplets, giving them a negative charge. By varying the intensity of the electric field, Millikan could control the rate of a droplet's fall. He determined that the magnitude of the charge on each drop increased in discrete amounts and determined that the smallest common denominator was $1.602 \times 10^{-19}$ coulombs. He identified this number as the charge of the electron. This charge was later equated to a single unit of negative charge noted 1−; in other words, a single electron carries a charge of 1−.

So good was Millikan's experimental setup and technique that the charge he measured almost one hundred years ago is within 1% of the currently accepted value.

**Mass of an electron** Knowing the electron's charge and using the known charge-to-mass ratio, Millikan calculated the mass of an electron. The equation below shows how small the mass of an electron is.

$$\text{Mass of an electron} = 9.1 \times 10^{-28}\ \text{g} = \frac{1}{1840}\ \text{the mass of a hydrogen atom}$$

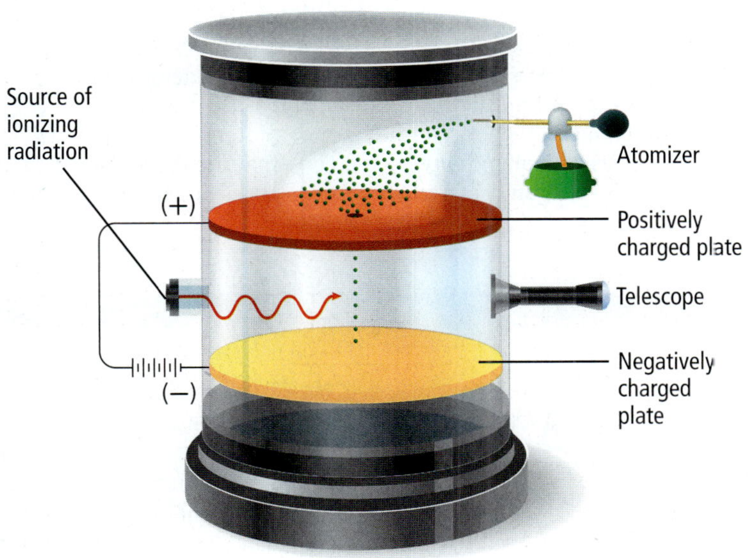


■ **Figure 8** The motion of the oil droplets within Millikan's apparatus depends on the charge of droplets and on the electric field. Millikan observed the droplets with the telescope. He could make the droplets fall more slowly, rise, or pause as he varied the strength of the electric field. From his observations, he calculated the charge on each droplet.

■ **Figure 9** J. J. Thomson's plum pudding model of the atom states that the atom is a uniform, positively charged sphere containing electrons.

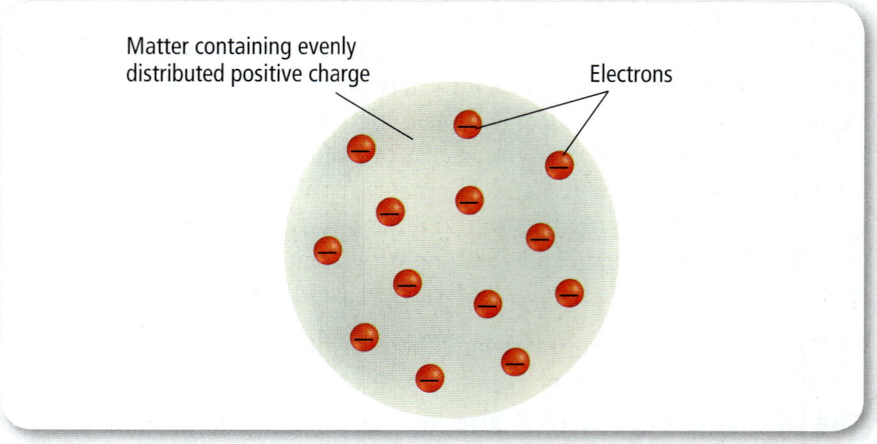


**The plum pudding model** The existence of the electron and the knowledge of some of its properties raised some interesting new questions about the nature of atoms. It was known that matter is neutral—it has no electric charge. You know that matter is neutral from everyday experience: you do not receive an electric shock (except under certain conditions) when you touch an object. If electrons are part of all matter and they possess a negative charge, how can all matter be neutral? Also, if the mass of an electron is so small, what accounts for the rest of the mass in a typical atom?

In an attempt to answer these questions, J. J. Thomson proposed a model of the atom that became known as the plum pudding model. As you can see in **Figure 9,** Thomson's model consisted of a spherically shaped atom composed of a uniformly distributed positive charge in which the individual negatively charged electrons resided. As you are about to read, the plum pudding model of the atom did not last for long. **Figure 10** summarizes the numerous steps in understanding the structure of the atom.

☑ READING CHECK **Explain** why Thomson's model was called the plum pudding model.

SSPL/The Image Works

■ **Figure 10**

## Development of Modern Atomic Theory

Current understanding of the properties and behavior of atoms and subatomic particles is based on the work of scientists worldwide during the past two centuries.

1860 — 1885 — 1910

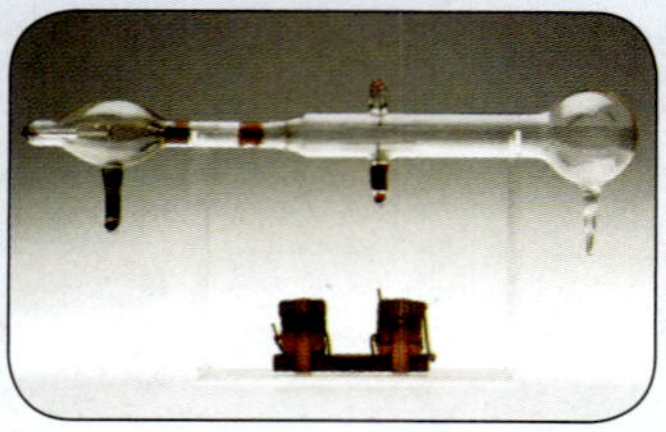

**1897** Using cathode-ray tubes, J. J. Thomson identifies the electron and determines the ratio of the mass of an electron to its electric charge.

**1911** With the gold foil experiment, Ernest Rutherford determines properties of the nucleus, including charge, relative size, and density.

**1913** Niels Bohr publishes a theory of atomic structure relating the electron arrangement in atoms and atomic chemical properties.

**1932** Scientists develop a particle accelerator to fire protons at lithium nuclei, splitting them into helium nuclei and releasing energy.

**1932** James Chadwick proves the existence of neutrons.

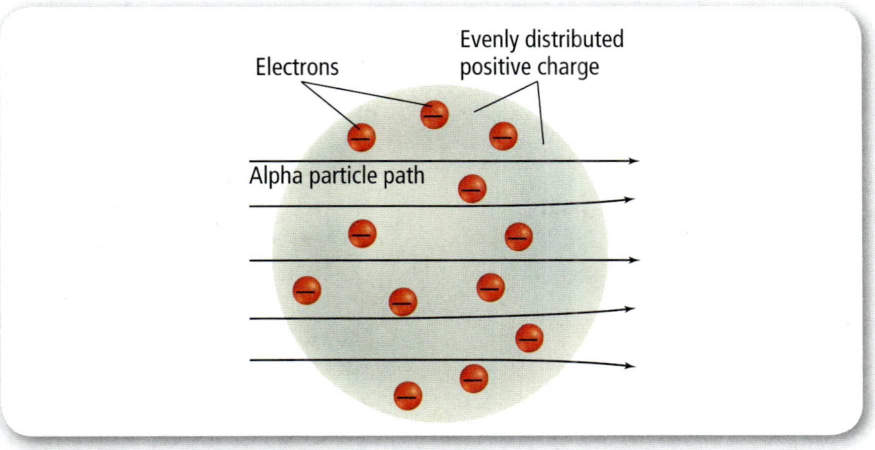


**Figure 11** Based on Thomson's model, Rutherford expected the light alpha particles to pass through gold atoms. He expected only a few of them to be slightly deflected.

## The Nucleus

In 1911, Ernest Rutherford (1871–1937) began to study how positively charged alpha particles (radioactive particles you will read more about later in this chapter) interacted with solid matter. With a small group of scientists, Rutherford conducted an experiment to see if alpha particles would be deflected as they passed through a thin gold foil.

**Rutherford's experiment** In the experiment, a narrow beam of alpha particles was aimed at a thin sheet of gold foil. A zinc-sulfide-coated screen surrounding the gold foil produced a flash of light when struck by an alpha particle. By noting where the flashes occurred, the scientists could determine if the atoms in the gold foil deflected the alpha particles.

Rutherford was aware of Thomson's plum pudding model of the atom. He expected the paths of the massive and fast-moving alpha particles to be only slightly altered by a collision with an electron. And because the positive charge within the gold atoms was thought to be uniformly distributed, he thought it would not alter the paths of the alpha particles, either. **Figure 11** shows the results Rutherford expected from the experiment.

**1938** Lise Meitner, Otto Hahn, and Fritz Strassman split uranium atoms in a process they called fission.

**1954** CERN, the world's largest nuclear physics research center, located in Switzerland, is founded to study particle physics.

**2010** The Large Hadron Collider at CERN enters a new era when it smashes beams of protons at the highest energy ever.

1960 | 1985 | 2010

**1939–1945** Scientists in the United States and Germany each work on projects to develop the first atomic weapon.

**1968** Scientists provide the first experimental evidence for subatomic particles known as quarks.

**1995** Scientists discover the top quark, the last of the six quarks predicted to exist. The top quark mass has helped scientists to theorize about the mass of the yet discovered Higgs boson particle.

View an **animation about the gold foil experiment.**

**Concepts In Motion**

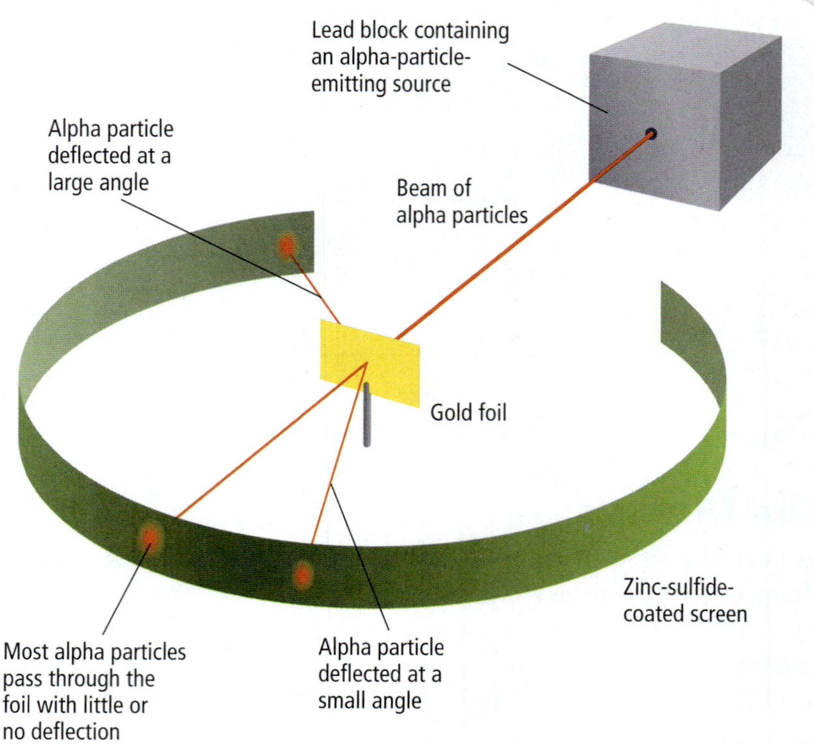


■ **Figure 12** During Rutherford's experiment, a beam of alpha particles bombarded a thin gold foil. Most of the alpha particles went through the gold foil. However, a few of them bounced back, some at large angles.

The actual results observed by Rutherford and his colleagues are shown in **Figure 12.** A few of the alpha particles were deflected at large angles. Several particles were deflected straight back toward the source. Rutherford likened the results to firing a large artillery shell at a sheet of paper and the shell coming back at the cannon.

**Rutherford's model of the atom** Rutherford concluded that the plum pudding model was incorrect because it could not explain the results of the gold foil experiment. Considering the properties of the alpha particles and the electrons, and the frequency of the deflections, he calculated that an atom consisted mostly of empty space through which the electrons move. He also concluded that almost all of the atom's positive charge and almost all of its mass were contained in a tiny, dense region in the center of the atom, which he called the **nucleus.** The negatively charged electrons are held within the atom by their attraction to the positively charged nucleus. Rutherford's nuclear atomic model is shown in **Figure 13.**

■ **Figure 13** In Rutherford's nuclear model, the atom is composed of a dense, positively charged nucleus that is surrounded by negative electrons. Alpha particles passing far from the nucleus are only slightly deflected. Alpha particles directly approaching the nucleus are deflected at large angles.

**Infer** ***what force causes the deflection of alpha particles.***

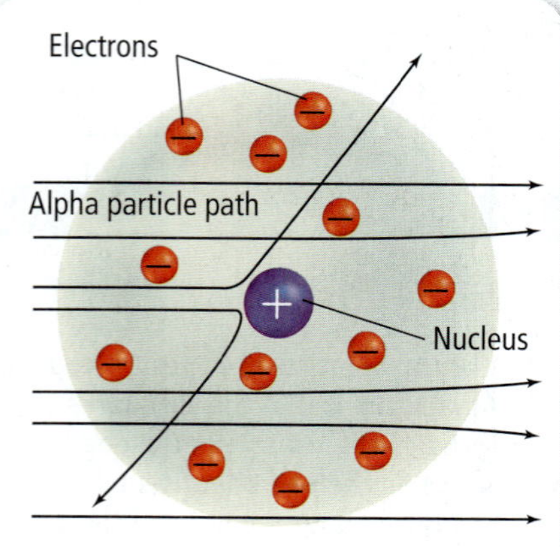


Because the nucleus occupies such a small space and contains most of an atom's mass, it is incredibly dense. If a nucleus were the size of the dot in the exclamation point at the end of this sentence, its mass would be approximately as much as that of 70 automobiles! The volume of space through which the electrons move is huge compared to the volume of the nucleus. A typical atom's diameter is approximately 10,000 times the diameter of the nucleus. If an atom had a diameter of two football fields, the nucleus would be the size of a nickel.

☑ READING CHECK **Describe** Rutherford's model of the atom.

The repulsive force produced between the positive nucleus and the positive alpha particles causes the deflections. **Figure 13** illustrates how Rutherford's nuclear atomic model explained the results of the gold foil experiment. The nuclear model also explains the neutral nature of matter: the positive charge of the nucleus balances the negative charge of the electrons. However, the model still could not account for all of the atom's mass.

**The proton and the neutron** By 1920, Rutherford had refined the concept of the nucleus and concluded that the nucleus contained positively charged particles called protons. A **proton** is a subatomic particle carrying a charge equal to but opposite that of an electron; that is, a proton has a charge of 1+. In 1932, Rutherford's coworker, English physicist James Chadwick (1891–1974), showed that the nucleus also contained another subatomic neutral particle, called the neutron. A **neutron** is a subatomic particle that has a mass nearly equal to that of a proton, but it carries no electric charge. In 1935, Chadwick received the Nobel Prize in Physics for proving the existence of neutrons.

**VOCABULARY**

**SCIENCE USAGE V. COMMON USAGE**

**Neutral**

***Science usage:*** to have no electric charge

*Neutrons have a charge of zero. They are neutral particles.*

***Common usage:*** not engaged in either side

*Switzerland remained neutral during World War II.*

## Data Analysis LAB

Based on Real Data*

### Interpret Scientific Illustrations

**What are the apparent atomic distances of carbon atoms in a well-defined crystalline material?** To visualize individual atoms, a group of scientists used a scanning tunneling microscope (STM) to test a crystalline material called highly ordered pyrolytic graphite (HOPG). An STM is an instrument used to perform surface atomic-scale imaging.

**Data and Observations**

The image shows all of the carbon atoms in the surface layer of the graphite material. Each hexagonal ring, indicated by the drawing in the figure, consists of three brighter spots separated by three fainter spots. These bright spots are from alternate carbon atoms in the surface layer of the graphite structure. The cross-sectional view below the photo corresponds to the line drawn in the image. It indicates the atomic periodicity and apparent atomic distances.

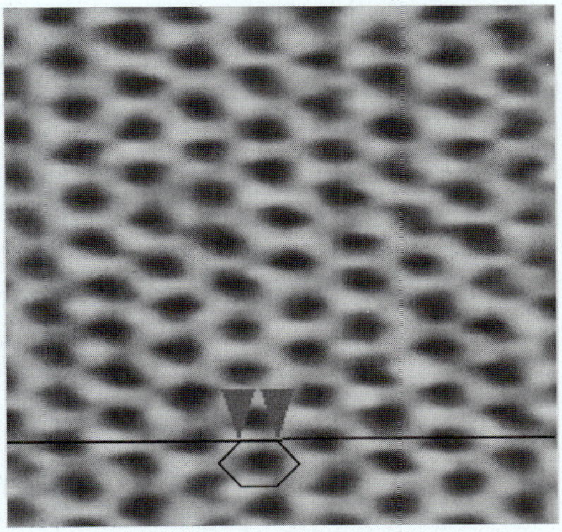

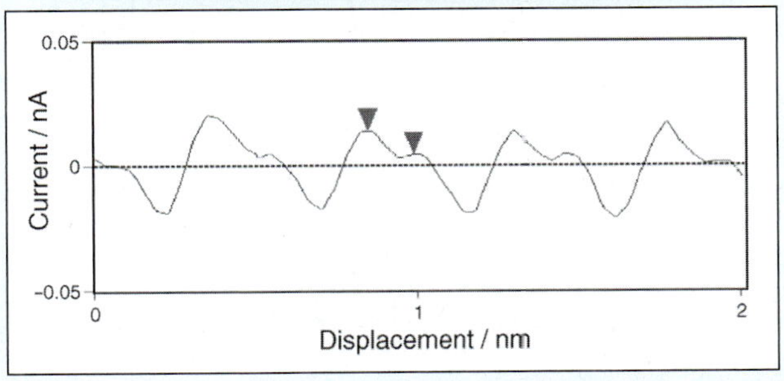


*Data obtained from: Chaun-Jian Zhong et al. 2003. Atomic scale imaging: a hands-on scanning probe microscopy laboratory for undergraduates. *Journal of Chemical Education* 80: 194–197.

**Think Critically**

1. **Estimate** the distance between two nearest bright spots.
2. **Estimate** the distance between two nearest neighbor spots (brighter–fainter, marked with triangles in the figure).
3. **State** What do the black spots in the image represent?
4. **Explain** How many carbon atoms are across the line drawn in the image?

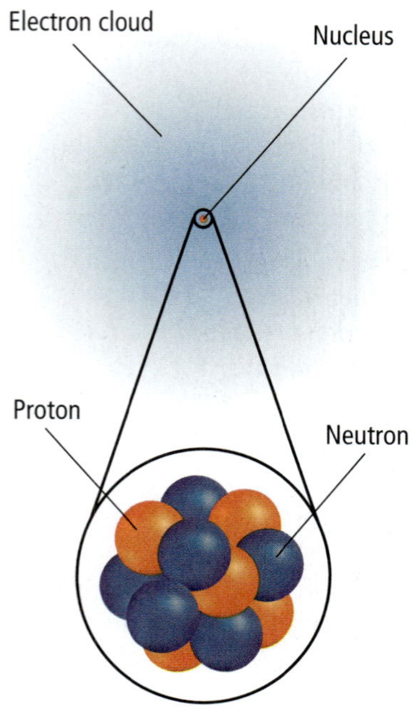


■ **Figure 14** Atoms are composed of a nucleus containing protons and neutrons, and surrounded by a cloud of electrons.

View an **animation about the structure of the atom.**

Concepts In Motion

Explore the **properties of subatomic particles with an interactive table.**

Concepts In Motion

**Table 3 Properties of Subatomic Particles**

| Particle | Symbol | Location | Relative Electric Charge | Relative Mass | Actual Mass (g) |
|---|---|---|---|---|---|
| Electron | $e^-$ | In the space surrounding the nucleus | 1− | $\frac{1}{1840}$ | $9.11 \times 10^{-28}$ |
| Proton | p | In the nucleus | 1+ | 1 | $1.673 \times 10^{-24}$ |
| Neutron | n | In the nucleus | 0 | 1 | $1.675 \times 10^{-24}$ |

**Completing the model of the atom** All atoms are made up of the three fundamental subatomic particles—the electron, the proton, and the neutron. Atoms are spherically shaped, with a small, dense nucleus of positive charge surrounded by one or more negatively charged electrons. Most of an atom consists of fast-moving electrons traveling through the empty space surrounding the nucleus. The electrons are held within the atom by their attraction to the positively charged nucleus. The nucleus, which is composed of neutral neutrons (hydrogen's single-proton nucleus is an exception) and positively charged protons, contains all of an atom's positive charge and more than 99.97% of its mass. It occupies only about one ten-thousandth of the volume of the atom. Because an atom is electrically neutral, the number of protons in the nucleus equals the number of electrons surrounding the nucleus. The features of a typical atom are shown in **Figure 14,** and the properties of the fundamental subatomic particles are summarized in **Table 3.**

Subatomic particle research is still a major interest to modern scientists. In fact, scientists have determined that protons and neutrons have their own structures. They are composed of subatomic particles called quarks. These particles will not be covered in this textbook because scientists do not yet understand if or how they affect chemical behavior. As you will learn in later chapters, chemical behavior can be explained by considering only an atom's electrons.

## SECTION 2 REVIEW

Section Self-Check

### Section Summary

- An atom is the smallest particle of an element that maintains the properties of that element.
- Electrons have a 1− charge, protons have a 1+ charge, and neutrons have no charge.
- An atom consists mostly of empty space surrounding the nucleus.

7. **MAIN**IDEA **Describe** the structure of a typical atom. Identify where each subatomic particle is located.
8. **Compare and contrast** Thomson's plum pudding atomic model with Rutherford's nuclear atomic model.
9. **Evaluate** the experiments that led to the conclusion that electrons are negatively charged particles found in all matter.
10. **Compare** the relative charge and mass of each of the subatomic particles.
11. **Calculate** What is the difference expressed in kilograms between the mass of a proton and the mass of an electron?

SECTION 3

# How Atoms Differ

**MAINIDEA** The number of protons and the mass number define the type of atom.

### Essential Questions

- How is the atomic number used to determine the identity of an atom?
- What is an isotope?
- Why are atomic masses not whole numbers?
- Given the mass number and atomic number, how are the number of electrons, protons, and neutrons in an atom calculated?

### Review Vocabulary

**periodic table:** a chart that organizes all known elements into a grid of horizontal rows (periods) and vertical columns (groups or families) arranged by increasing atomic number

### New Vocabulary

atomic number
isotope
mass number
atomic mass unit (amu)
atomic mass

**CHEM 4 YOU** **You are probably aware that numbers are used every day to identify people and objects. For example, people can be identified by their Social Security numbers and computers by their IP addresses. Atoms and nuclei are also identified by numbers.**

## Atomic Number

As shown in the periodic table of the elements inside the back cover of this textbook, there are more than 110 different elements. What makes an atom of one element different from an atom of another element?

Not long after Rutherford's gold foil experiment, the English scientist Henry Moseley (1887–1915) discovered that atoms of each element contain a unique positive charge in their nuclei. Thus, the number of protons in an atom identifies it as an atom of a particular element. The number of protons in an atom is referred to as the **atomic number.** The information provided by the periodic table for hydrogen is shown in **Figure 15.** The number 1 above the symbol for hydrogen (H) is the number of protons, or the atomic number. Moving across the periodic table to the right, you will next come to helium (He). It has two protons in its nucleus, and thus it has an atomic number of 2. The next row begins with lithium (Li), atomic number 3, followed by beryllium (Be), atomic number 4, and so on. The periodic table is organized left-to-right and top-to-bottom by increasing atomic number.

Because all atoms are neutral, the number of protons and electrons in an atom must be equal. Thus, once you know the atomic number of an element, you know the number of protons and the number of electrons an atom of that element contains. For example, an atom of lithium, atomic number 3, contains three protons and three electrons.

### Atomic number

$$\text{atomic number} = \text{number of protons}$$
$$= \text{number of electrons}$$

The atomic number of an atom equals its number of protons and its number of electrons.

**Figure 15** In the periodic table, each element is represented by its chemical name, atomic number, chemical symbol, and average atomic mass.
**Determine** *the number of protons and the number of electrons in an atom of gold.*

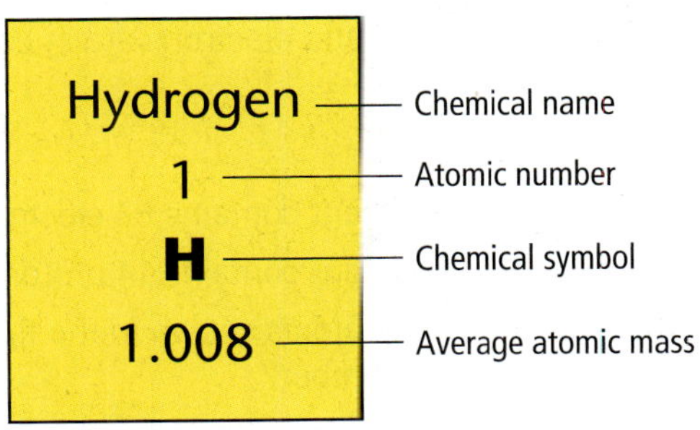

## EXAMPLE Problem 1

Find help with **solving algebraic equations.** Math Handbook

**ATOMIC NUMBER** Complete the following table.

**Composition of Several Elements**

| | Element | Atomic Number | Protons | Electrons |
|---|---|---|---|---|
| a. | Pb | 82 | | |
| b. | | | 8 | |
| c. | | | | 30 |

### 1 ANALYZE THE PROBLEM

Apply the relationship among atomic number, number of protons, and number of electrons to complete most of the table. Then, use the periodic table to identify the element.

**Known**

**a.** element = Pb, atomic number = 82

**b.** number of protons = 8

**c.** number of electrons = 30

**Unknown**

**a.** number of protons **($N_p$)**, number of electrons **($N_e$) = ?**

**b.** **element,** atomic number **($Z$)**, **$N_e$ = ?**

**c.** **element, $Z$, $N_p$ = ?**

### 2 SOLVE FOR THE UNKNOWN

**a.** number of protons = atomic number — **Apply the atomic-number relationship.**

**$N_p = 82$** — **Substitute atomic number = 82.**

number of electrons = number of protons

**$N_e = 82$**

The number of protons and the number of electrons is 82.

**b.** atomic number = number of protons — **Apply the atomic-number relationship.**

**$Z = 8$** — **Substitute number of protons = 8.**

number of electrons = number of protons

**$N_e = 8$**

The atomic number and the number of electrons is 8.

The **element** is **oxygen (O).** — **Consult the periodic table to identify the element.**

**c.** number of protons = number of electrons — **Apply the atomic-number relationship.**

**$N_p = 30$** — **Substitute number of electrons = 30.**

atomic number = number of protons

**$Z = 30$**

The atomic number and the number of protons is 30.

The **element** is **zinc (Zn).** — **Consult the periodic table to identify the element.**

### 3 EVALUATE THE ANSWER

The answers agree with atomic numbers and element symbols given in the periodic table.

## PRACTICE Problems

Do additional problems. Online Practice

**12.** How many protons and electrons are in each atom?

**a.** radon

**b.** magnesium

**13.** An atom of an element contains 66 electrons. Which element is it?

**14.** An atom of an element contains 14 protons. Which element is it?

**15. Challenge** Do the atoms shown in the figure to the right have the same atomic number?

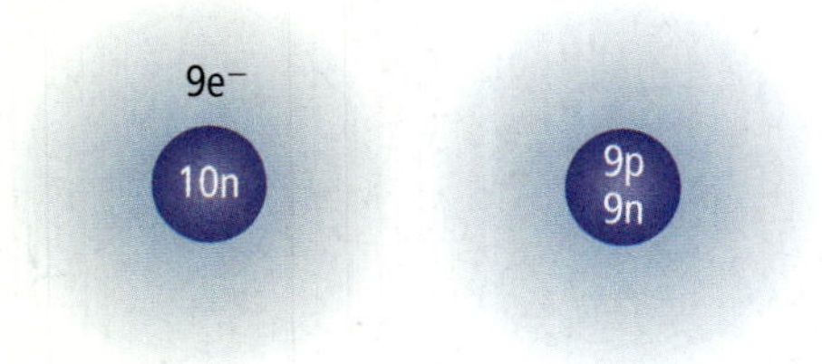

## Isotopes and Mass Number

Dalton was incorrect about atoms being indivisible and in stating that all atoms of an element are identical. All atoms of an element have the same number of protons and electrons, but the number of neutrons might differ. For example, there are three types of potassium atoms that occur naturally. All three types contain 19 protons and 19 electrons. However, one type of potassium atom contains 20 neutrons, another 21 neutrons, and still another 22 neutrons. Atoms with the same number of protons but different numbers of neutrons are called **isotopes.**

**Mass of isotopes** Isotopes containing more neutrons have a greater mass. In spite of these differences, isotopes of an atom have the same chemical behavior. As you will read later in this textbook, chemical behavior is determined only by the number of electrons an atom has.

**Isotope notation** Each isotope of an element is identified with a number called the mass number. The **mass number** is the sum of the atomic number (or number of protons) and neutrons in the nucleus.

**Mass number**

$$\text{mass number} = \text{atomic number} + \text{number of neutrons}$$

The mass number of an atom is the sum of its atomic number and its number of neutrons.

For example, copper has two isotopes. The isotope with 29 protons and 34 neutrons has a mass number of 63 (29 + 34 = 63), and is called copper-63 (also written $^{63}$Cu or Cu-63). The isotope with 29 protons and 36 neutrons is called copper-65. Chemists often write out isotopes using a notation involving the chemical symbol, atomic number, and mass number, as shown in **Figure 16.**

**Natural abundance of isotopes** In nature, most elements are found as mixtures of isotopes. Usually, no matter where a sample of an element is obtained, the relative abundance of each isotope is constant. For example, in a banana, 93.26% of the potassium atoms have 20 neutrons, 6.73% have 22 neutrons, and 0.01% have 21 neutrons. In another banana, or in a different source of potassium, the percentage composition of the potassium isotopes will still be the same. The three potassium isotopes are summarized in **Figure 17.**

Mass number

$^{63}_{29}\text{Cu}$ $^{65}_{29}\text{Cu}$

Atomic number

**Figure 16** Cu is the chemical symbol for copper. Copper, which was used to make this Chinese gong, is composed of 69.2% copper-63 and 30.8% copper-65.

**Alabama Applying Practices**

Go to the Resource tab in ConnectED to find the Applying Practices worksheet *Model Isotopes.*

| | Potassium-39 | Potassium-40 | Potassium-41 |
|---|---|---|---|
| Protons | 19 | 19 | 19 |
| Neutrons | 20 | 21 | 22 |
| Electrons | 19 | 19 | 19 |

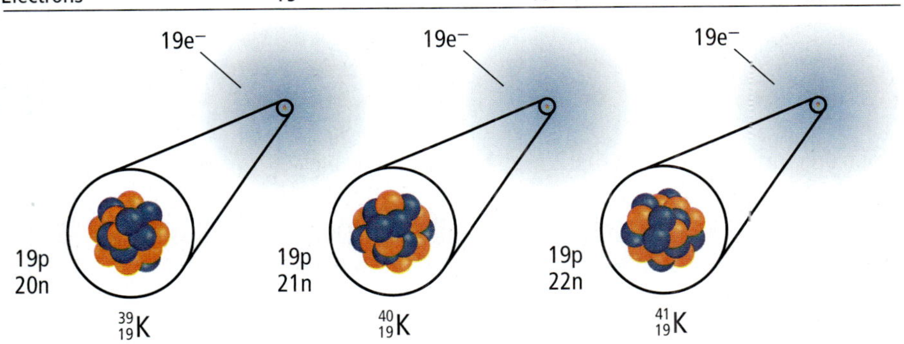


**Figure 17** Potassium has three naturally occurring isotopes: potassium-39, potassium-40, and potassium-41.

***List** the number of protons, neutrons, and electrons in each potassium isotope.*

## EXAMPLE Problem 2

EXAMPLE PROBLEM

**USE ATOMIC NUMBER AND MASS NUMBER** A chemistry laboratory has analyzed the composition of isotopes of several elements. The composition data is given in the table below. Determine the number of protons, electrons, and neutrons in the isotope of neon. Name the isotope and give its symbol.

**Isotope Composition Data**

| | Element | Atomic Number | Mass Number |
|---|---|---|---|
| a. | Neon | 10 | 22 |
| b. | Calcium | 20 | 46 |
| c. | Oxygen | 8 | 17 |
| d. | Iron | 26 | 57 |
| e. | Zinc | 30 | 64 |
| f. | Mercury | 80 | 204 |

### 1 ANALYZE THE PROBLEM

You are given some data for neon in the table. The symbol for neon can be found on the periodic table. From the atomic number, the number of protons and electrons in the isotope are known. The number of neutrons in the isotope can be found by subtracting the atomic number from the mass number.

**Known**

element: neon

atomic number = 10

mass number = 22

**Unknown**

number of protons ($N_p$), electrons ($N_e$), and neutrons ($N_n$) = ?

**name** of isotope = ?

**symbol** for isotope = ?

### 2 SOLVE FOR THE UNKNOWN

number of protons = atomic number = **10**

number of electrons = atomic number = **10**

*Apply the atomic number relationship.*

number of neutrons = mass number − atomic number

*Use the atomic number and the mass number to calculate the number of neutrons.*

$N_n = 22 - 10 = 12$

*Substitute mass number = 22 and atomic number = 10*

The **name** of the isotope is **neon-22.**

*Use the element name and mass number to write the isotope's name.*

The **symbol** for the isotope is $^{22}_{10}Ne$.

*Use the chemical symbol, mass number, and atomic number to write out the isotope in symbolic notation form.*

### 3 EVALUATE THE ANSWER

The relationships among number of electrons, protons, and neutrons have been applied correctly. The isotope's name and symbol are in the correct format. Refer to the Elements Handbook to learn more about neon.

## PRACTICE Problems

Do additional problems. Online Practice

**16.** Determine the number of protons, electrons, and neutrons for isotopes **b.–f.** in the table above. Name each isotope, and write its symbol.

**17. Challenge** An atom has a mass number of 55. Its number of neutrons is the sum of its atomic number and five. How many protons, neutrons, and electrons does this atom have? What is the identity of this atom?

**Table 4 Masses of Subatomic Particles**

| Particle | Mass (amu) |
|---|---|
| Electron | 0.000549 |
| Proton | 1.007276 |
| Neutron | 1.008665 |

## Mass of Atoms

Recall from **Table 3** that the masses of both protons and neutrons are approximately $1.67 \times 10^{-24}$ g. While this is a small mass, the mass of an electron is even smaller—only about 1/1840 that of a proton or a neutron.

**Atomic mass unit** Because these extremely small masses expressed in scientific notation are difficult to work with, chemists have developed a method of measuring the mass of an atom relative to the mass of a specific atomic standard. That standard is the carbon-12 atom. Scientists assigned the carbon-12 atom a mass of exactly 12 atomic mass units. Thus, one **atomic mass unit (amu)** is defined as one-twelfth the mass of a carbon-12 atom. Although a mass of 1 amu is nearly equal to the mass of a single proton or a single neutron, it is important to realize that the values are slightly different. **Table 4** gives the masses of the subatomic particles in terms of amu.

**Vocabulary**

**Academic vocabulary**

**Specific**

characterized by precise formulation or accurate restriction

*Some diseases have specific symptoms.*

**Atomic mass** Because an atom's mass depends mainly on the number of protons and neutrons it contains, and because protons and neutrons have masses close to 1 amu, you might expect the atomic mass of an element to always be nearly a whole number. However, this is often not the case. The explanation involves how atomic mass is defined. The **atomic mass** of an element is the weighted average mass of the isotopes of that element. Because isotopes have different masses, the weighted average is not a whole number. The calculation of the atomic mass of chlorine is illustrated in **Figure 18.**

Explore **decoding the periodic table.**

Virtual Investigations

■ **Figure 18** To calculate the weighted average atomic mass of chlorine, you first need to calculate the mass contribution of each isotope.

Get help with **finding an average.**

Personal Tutor

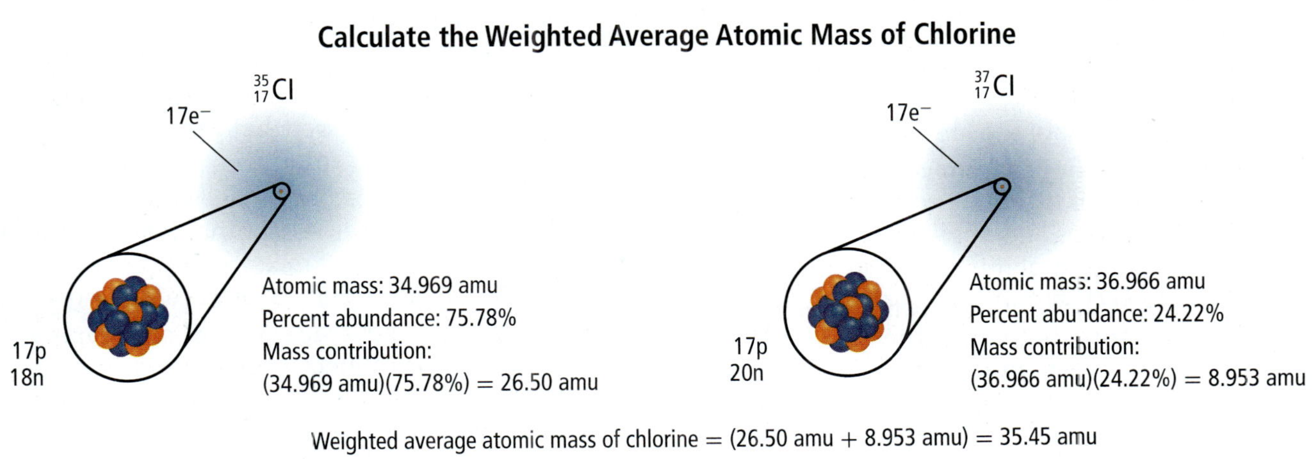

■ **Figure 19** Bromine is extracted from sea water and salt lakes. The Dead Sea area in Israel is one of the major bromine production sites in the world. Applications of bromine include microbe and algae control in swimming pools. It is also used in medicines, oils, paints, pesticides, and flame-retardants.

Eitan Simanor/Alamy

Chlorine exists naturally as a mixture of about 76% chlorine-35 and 24% chlorine-37. It has an atomic mass of 35.453 amu. Because atomic mass is a weighted average, the chlorine-35 atoms, which exist in greater abundance than the chlorine-37 atoms, have a greater effect in determining the atomic mass. The atomic mass of chlorine is calculated by multiplying each isotope's percent abundance by its atomic mass and then adding the products. The process is similar to calculating an average grade. You can calculate the atomic mass of any element if you know the number of naturally occurring isotopes, their masses, and their percent abundances.

☑ READING CHECK **Explain** how to calculate atomic mass.

**Isotope abundances** Analyzing an element's mass can indicate the most abundant isotope for that element. For example, fluorine (F) has an atomic mass that is extremely close to 19 amu. If fluorine had several fairly abundant isotopes, its atomic mass would not likely be so close to a whole number. Thus, you might conclude that all naturally occurring fluorine is probably in the form of fluorine-19 ($^{19}_{9}F$). Indeed, 100% of naturally occurring fluorine is in the form of fluorine-19. While this type of reasoning generally works well, it is not foolproof. Consider bromine (Br). It has an atomic mass of 79.904 amu. With a mass so close to 80 amu, it seems likely that the most common bromine isotope would be bromine-80. However, bromine's two isotopes are bromine-79 (78.918 amu, 50.69%) and bromine-81 (80.917 amu, 49.31%). There is no bromine-80 isotope. **Figure 19** shows one of the major production sites of bromine, located in the Dead Sea area. Refer to the Elements Handbook to learn more about chlorine, fluorine, and bromine.

## MiniLAB

iLab Station

### Model Isotopes

**How can you calculate the atomic mass of an element using the percentage abundance of its isotopes?** Because they have different compositions, pre- and post-1982 pennies can be used to model an element with two naturally occurring isotopes. From the penny 'isotope' data, you can determine the mass of each penny isotope and the average mass of a penny.

**Procedure**

1. Read and complete the lab safety form.
2. Get a bag of **pennies** from your teacher, and sort the pennies by date into two groups: pre-1982 pennies and post-1982 pennies. Count and record the total number of pennies and the number in each group.
3. Using a **balance,** determine the mass of 10 pennies from each group. Record each mass to the nearest 0.01 g. Divide the total mass of each group by 10 to get the average mass of a pre- and post-1982 penny isotope.

**Analysis**

1. **Calculate** the percentage abundance of each group using data from Step 2. To do this, divide the number of pennies in each group by the total number of pennies.
2. **Determine** the atomic mass of a penny using the percentage abundance of each "isotope" and data from Step 3. To do this, use the following equation:

   mass contribution = (% abundance)(mass)

   Total the mass contributions to determine the atomic mass. Remember that the percent abundance is a percentage.
3. **Infer** whether the atomic mass would be different if you received another bag of pennies containing a different mixture of pre- and post-1982 pennies. Explain your reasoning.
4. **Explain** why the average mass of each type of penny was determined by measuring 10 pennies instead of by measuring and using the mass of a single penny from each group.

## EXAMPLE Problem 3

**CALCULATE ATOMIC MASS** Given the data in the table, calculate the atomic mass of unknown Element X. Then, identify the unknown element, which is used medically to treat some mental disorders.

### 1 ANALYZE THE PROBLEM

Calculate the atomic mass and use the periodic table to confirm.

**Known**

$^6$X: mass = 6.015 amu

abundance = 7.59% = 0.0759

$^7$X: mass = 7.016 amu

abundance = 92.41% = 0.9241

**Unknown**

**atomic mass of X = ? amu**

**element X = ?**

**Isotope Abundance for Element X**

| Isotope | Mass (amu) | Percent Abundance |
|---|---|---|
| $^6$X | 6.015 | 7.59% |
| $^7$X | 7.016 | 92.41% |

### 2 SOLVE FOR THE UNKNOWN

$^6$X: mass contribution = (mass)(percent abundance) — **Calculate $^6$X's contribution.**

mass contribution = (6.015 amu)(0.0759) = 0.456 amu — **Substitute mass = 6.015 amu and abundance = 0.0759.**

$^7$X: mass contribution = (mass)(percent abundance) — **Calculate $^7$X's contribution.**

mass contribution = (7.016 amu)(0.9241) = 6.483 amu — **Substitute mass = 7.016 amu and abundance = 0.9241.**

**atomic mass of X** = (0.4565 amu + 6.483 amu) = **6.939 amu** — **Total the mass contributions to find the atomic mass.**

The **element** with a mass nearest 6.939 amu is **lithium (Li)**. — **Identify the element using the periodic table.**

### 3 EVALUATE THE ANSWER

The result of the calculation agrees with the atomic mass given in the periodic table. The masses of the isotopes have four significant figures, so the atomic mass is also expressed with four significant figures. Refer to the Elements Handbook to learn more about lithium.

EXAMPLE PROBLEM

## PRACTICE Problems

Do additional problems. Online Practice

**18.** Boron (B) has two naturally occurring isotopes: boron-10 (abundance = 19.8%, mass = 10.013 amu) and boron-11 (abundance = 80.2%, mass = 11.009 amu). Calculate the atomic mass of boron.

**19. Challenge** Nitrogen has two naturally occurring isotopes, N-14 and N-15. Its atomic mass is 14.007. Which isotope is more abundant? Explain your answer.

## SECTION 3 REVIEW

Section Self-Check

### Section Summary

- The atomic number of an atom is given by its number of protons. The mass number of an atom is the sum of its neutrons and protons.
- Atoms of the same element with different numbers of neutrons are called isotopes.
- The atomic mass of an element is a weighted average of the masses of all of its naturally occurring isotopes.

**20. MAIN**IDEA **Explain** how the type of an atom is defined.

**21. Recall** Which subatomic particle identifies an atom as that of a particular element?

**22. Explain** how the existence of isotopes is related to the fact that atomic masses are not whole numbers.

**23. Calculate** Copper has two isotopes: Cu-63 (abundance = 69.2%, mass = 62.930 amu) and Cu-65 (abundance = 30.8%, mass = 64.928 amu). Calculate the atomic mass of copper.

**24. Calculate** Three magnesium isotopes have atomic masses and relative abundances of 23.985 amu (78.99%), 24.986 amu (10.00%), and 25.982 amu (11.01%). Calculate the atomic mass of magnesium.

SECTION 4

# Unstable Nuclei and Radioactive Decay

**MAINIDEA** Unstable atoms emit radiation to gain stability.

### Essential Questions

- What is the relationship between unstable nuclei and radioactive decay?
- How are alpha, beta, and gamma radiation characterized in terms of mass and charge?

### Review Vocabulary

**element:** a pure substance that cannot be broken down into simpler substances by physical or chemical means

### New Vocabulary

radioactivity
radiation
nuclear reaction
radioactive decay
alpha radiation
alpha particle
nuclear equation
beta radiation
beta particle
gamma ray

**CHEM 4 YOU** Try dropping a rock from the height of your waist. The rock goes from a higher energy state at your waist to a lower energy state on the floor. A similar process happens with nuclei in an unstable state.

## Radioactivity

Recall that a chemical reaction is the change of one or more substances into new substances and involves only an atom's electrons. Although atoms might be rearranged, their identity remains the same. Another type of reaction, called a nuclear reaction, can change an element into a new element.

**Nuclear reactions** In the late 1890s, scientists noticed that some substances spontaneously emitted radiation in a process they named **radioactivity.** The rays and particles emitted by the radioactive material were called **radiation.** Scientists discovered that radioactive atoms undergo changes that can alter their identities. A reaction that involves a change in an atom's nucleus is called a **nuclear reaction.** The discovery of these nuclear reactions was a major breakthrough, as no chemical reaction had ever resulted in the formation of new kinds of atoms.

Radioactive atoms emit radiation because their nuclei are unstable. Unstable systems, whether they are atoms or people doing handstands, as shown in **Figure 20,** gain stability by losing energy.

**Radioactive decay** Unstable nuclei lose energy by emitting radiation in a spontaneous process called **radioactive decay.** Unstable atoms undergo radioactive decay until they form stable atoms, often of a different element. Just as a rock loses gravitational potential energy and reaches a stable state when falling to the ground, an atom can lose energy and reach a stable state when emitting radiation.

Image Source/Getty Images

**Figure 20** Being in a handstand position is an unstable state. Like unstable atoms, people doing handstands eventually return to a more stable state—standing on their feet—by losing potential energy.

## Types of Radiation

Scientists began researching radioactivity in the late 1800s. They investigated the effect of electric fields on radiation. By directing radiation from a radioactive source between two electrically charged plates, scientists were able to identify three different types of radiation based on their electric charge. As shown in **Figure 21,** radiation was deflected toward the negative plate, the positive plate, or not at all.

**Alpha radiation** The radiation that was deflected toward the negatively charged plate was named **alpha radiation.** It is made up of alpha particles. An **alpha particle** contains two protons and two neutrons, and thus has a 2+ charge, which explains why alpha particles are attracted to the negatively charged plate as shown in **Figure 21.** An alpha particle is equivalent to a helium-4 nucleus and is represented by ${}^{4}_{2}\text{He}$ or $\alpha$. The alpha decay of radioactive radium-226 into radon-222 is shown below.

$$\underset{\text{radium-226}}{{}^{226}_{88}\text{Ra}} \rightarrow \underset{\text{radon-222}}{{}^{222}_{86}\text{Rn}} + \underset{\text{alpha particle}}{\alpha}$$

Note that a new element, radon (Rn), is created as a result of the alpha decay of the unstable radium-226 nucleus. The type of equation shown above is known as a **nuclear equation.** It shows the atomic numbers and mass numbers of the particles involved. The mass number is conserved in nuclear equations.

**Beta radiation** The radiation that was deflected toward the positively charged plate was named **beta radiation.** This radiation consists of fast-moving beta particles. Each **beta particle** is an electron with a 1− charge. The negative charge of the beta particle explains why it is attracted to the positively charged plate shown in **Figure 21.** Beta particles are represented by the symbol $\beta$ or $e^-$. The beta decay of carbon-14 into nitrogen-14 is shown below. The beta decay of unstable carbon-14 results in the creation of the new atom, nitrogen (N).

$$\underset{\text{carbon-14}}{{}^{14}_{6}\text{C}} \rightarrow \underset{\text{nitrogen-14}}{{}^{14}_{7}\text{N}} + \underset{\text{beta particle}}{\beta}$$

CAREERS IN CHEMISTRY

**Chemistry Teacher** Chemistry teachers work in high schools and colleges. They lecture, guide discussions, conduct experiments, supervise lab work, and lead field trips. High school teachers might also be asked to monitor study halls and serve on committees. College instructors might be required to do research and publish their findings.

WebQuest

**Figure 21** An electric field will deflect radiation in different directions, depending on the electric charge of the radiation.

**Explain** *why beta particles are deflected toward the positive plate, alpha particles are deflected toward the negative plate, and gamma rays are not deflected.*

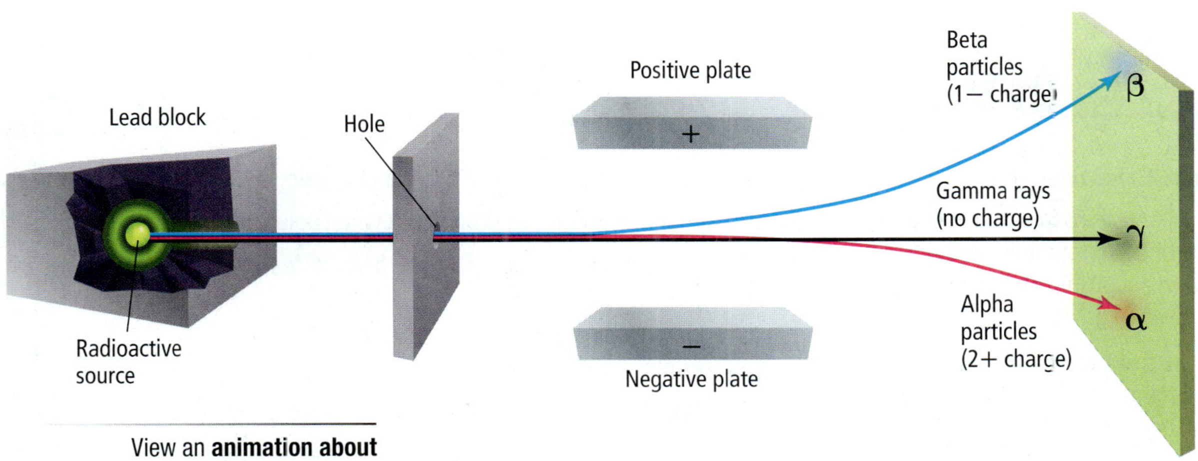


View an **animation about radiation deflection.**

Concepts In Motion

**Table 5 Characteristics of Radiation**

| | Alpha | Beta | Gamma |
|---|---|---|---|
| Symbol | ${}^{4}_{2}He$ or $\alpha$ | $e^-$ or $\beta$ | $\gamma$ |
| Mass (amu) | 4 | $\frac{1}{1840}$ | 0 |
| Mass (kg) | $6.65 \times 10^{-27}$ | $9.11 \times 10^{-31}$ | 0 |
| Charge | 2+ | 1− | 0 |

**Gamma radiation** The third common type of radiation is called gamma radiation, or gamma rays. A **gamma ray** is a high-energy radiation that possesses no mass and is denoted by the symbol $\gamma$. Because they are neutral, gamma rays are not deflected by electric or magnetic fields. They usually accompany alpha and beta radiation, and they account for most of the energy lost during radioactive decays. For example, gamma rays accompany the alpha decay of uranium-238.

$$\underset{\text{uranium-238}}{{}^{238}_{92}U} \rightarrow \underset{\text{thorium-234}}{{}^{234}_{90}Th} + \underset{\text{alpha particle}}{\alpha} + \underset{\text{gamma rays}}{2\gamma}$$

Because gamma rays are massless, the emission of gamma rays by themselves cannot result in the formation of a new atom. **Table 5** summarizes the basic characteristics of alpha, beta, and gamma radiation.

**Nuclear stability** The primary factor in determining an atom's stability is its ratio of neutrons to protons. Atoms that contain either too many or too few neutrons are unstable and lose energy through radioactive decay to form a stable nucleus. They emit alpha and beta particles and these emissions affect the neutron-to-proton ratio of the newly created nucleus. Eventually, radioactive atoms undergo enough radioactive decay to form stable, nonradioactive atoms. This topic will be covered later in the book.

## SECTION 4 REVIEW

Section Self-Check

### Section Summary

- Chemical reactions involve changes in the electrons surrounding an atom. Nuclear reactions involve changes in the nucleus of an atom.
- There are three types of radiation: alpha (charge of 2+), beta (charge of 1−), and gamma (no charge).
- The neutron-to-proton ratio of an atom's nucleus determines its stability.

25. **MAIN**IDEA **Explain** how unstable atoms gain stability.
26. **State** what quantities are conserved when balancing a nuclear reaction.
27. **Classify** each of the following as a chemical reaction, a nuclear reaction, or neither.
    **a.** Thorium emits a beta particle.
    **b.** Two atoms share electrons to form a bond.
    **c.** A sample of pure sulfur emits heat energy as it slowly cools.
    **d.** A piece of iron rusts.
28. **Calculate** How much heavier is an alpha particle than an electron?
29. **Create** a table showing how each type of radiation affects the atomic number and the mass number of an atom.

HOW IT **works**

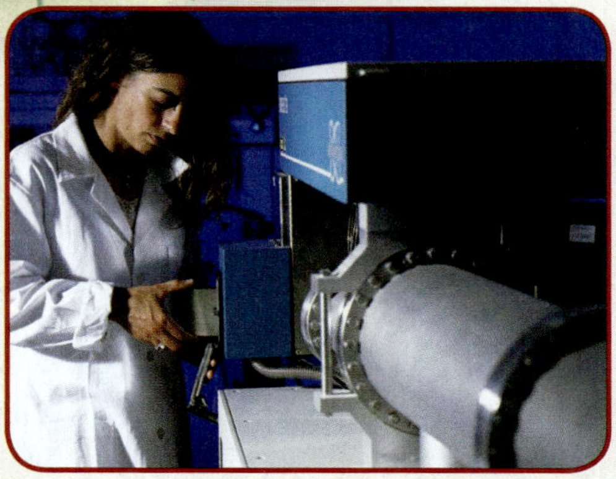

## Mass Spectrometer: Chemical Detective

Imagine a forensic scientist needs to identify the inks used on a document to test for possible counterfeiting. The scientist can analyze the inks using a mass spectrometer, such as the one shown at left. A mass spectrometer breaks the compounds in a sample of an unknown substance into smaller fragments. The fragments are then separated according to their masses, and the exact composition of the sample can be determined. Mass spectrometry is one of the most important techniques for studying unknown substances.

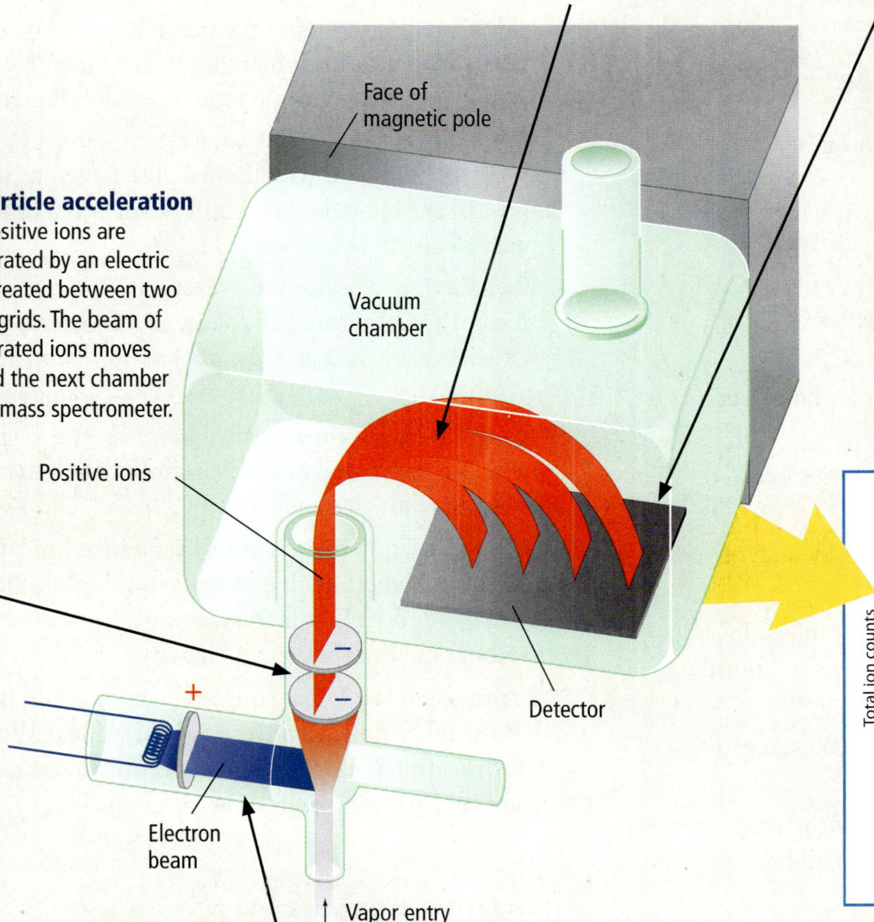


**1 Electron bombardment** A beam of high-energy electrons bombards the vaporized sample, knocking electrons from its atoms and forming positive ions.

**2 Particle acceleration** The positive ions are accelerated by an electric field created between two metal grids. The beam of accelerated ions moves toward the next chamber of the mass spectrometer.

**3 Ion deflection** The ions in the vacuum chamber are deflected by a magnetic field. The amount of deflection depends on the mass-to-charge ratio of the ions. The greater the mass-to-charge ratio, the less the ions are deflected.

**4 Ion detection** A detector measures the deflection and the amount of the ions.

**5 Data analysis** A data system generates a graphic display of results. The lines are located at the mass-to-charge ratio corresponding to the components found in one of the ink samples. A similar analysis can be performed with a different ink sample. The samples can then be compared to determine whether they originated from the same pen.

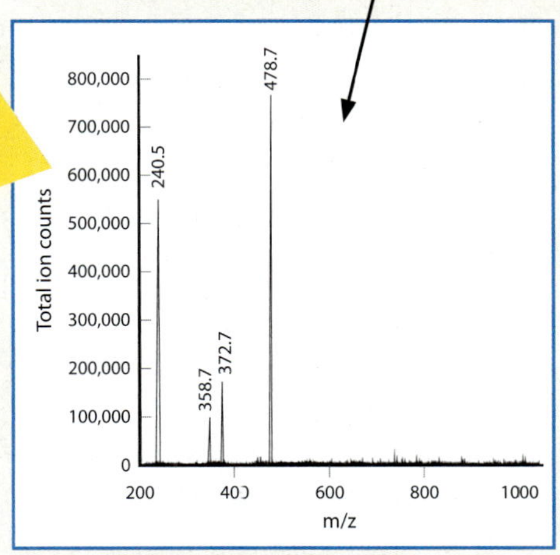


### WRITING IN ▶ Chemistry

**Summarize** Research a case in which a mass spectrometer was used to distinguish between different types of ink, and write a summary of the procedure and results.

WebQuest

# ChemLAB

iLab Station

## Model Atomic Mass

**Background:** Most elements in nature occur as a mixture of isotopes. The weighted average atomic mass of an element can be determined from the atomic mass and the relative abundance of each isotope. In this activity, you will model the isotopes of the imaginary element "snackium." The measurements you make will be used to calculate a weighted average mass that represents the average atomic mass of snackium.

**Question:** *How are the atomic masses of the natural isotopic mixtures calculated?*

### Materials

balance
calculator
bag of snack mix

### Safety Precautions

WARNING: ***Do not eat food used in lab work.***

### Procedure

1. Read and complete the lab safety form.
2. Create a table to record your data. The table will contain the mass and the abundance of each type of snack present in the mixture.
3. Open your snack-mix bag. Handle the pieces with care.
4. Organize the snack pieces into groups based on their types.
5. Count the number of snack pieces in each of your groups.
6. Record the number of snack pieces in each group and the total number of snack pieces in your data table.
7. Measure the mass of one piece from each group and record the mass in your data table.
8. **Cleanup and Disposal** Dispose of the snack pieces as directed by your teacher. Return all lab equipment to its designated location.

### Analyze and Conclude

1. **Calculate** Find the percent abundance of the pieces by dividing the individual-piece quantity by the total number of snack pieces.
2. **Calculate** Use the isotopic percent abundance of the snack pieces and the mass to calculate the weighted average atomic mass for your element snackium.
3. **Interpret** Explain why the weighted average atomic mass of the element snackium is not equal to the mass of any of the pieces.
4. **Peer Review** Gather the average atomic mass data from other lab groups. Explain any differences between your data and the data obtained by other groups.
5. **Apply** Why are the atomic masses on the periodic table not expressed as whole numbers like the mass number of an element?
6. **Research** Look in a chemical reference book to determine whether all elements in the periodic table have isotopes. What is the range of the number of isotopes chemical elements have?
7. **Error Analysis** What sources of error could have led the lab groups to different final values? What modifications could you make in this investigation to reduce the incidence of error?

### INQUIRY EXTENSION

**Predict** Based on your experience in this lab, look up the atomic masses of several elements on the periodic table and predict the most abundant isotope for each element.

# CHAPTER 4 STUDY GUIDE

Vocabulary Practice

**BIGIDEA** Atoms are the fundamental building blocks of matter.

## SECTION 1 Early Ideas About Matter

**MAINIDEA** The ancient Greeks tried to explain matter, but the scientific study of the atom began with John Dalton in the early 1800s.

- Democritus was the first person to propose the existence of atoms.
- According to Democritus, atoms are solid, homogeneous, and indivisible.
- Aristotle did not believe in the existence of atoms.
- John Dalton's atomic theory is based on numerous scientific experiments.

**VOCABULARY**
- Dalton's atomic theory

## SECTION 2 Defining the Atom

**MAINIDEA** An atom is made of a nucleus containing protons and neutrons; electrons move around the nucleus.

- An atom is the smallest particle of an element that maintains the properties of that element.
- Electrons have a 1− charge, protons have a 1+ charge, and neutrons have no charge.
- An atom consists mostly of empty space surrounding the nucleus.

**VOCABULARY**
- atom
- cathode ray
- electron
- nucleus
- proton
- neutron

## SECTION 3 How Atoms Differ

**MAINIDEA** The number of protons and the mass number define the type of atom.

- The atomic number of an atom is given by its number of protons. The mass number of an atom is the sum of its neutrons and protons.

$$\textbf{atomic number} = \textbf{number of protons} = \textbf{number of electrons}$$

$$\textbf{mass number} = \textbf{atomic number} + \textbf{number of neutrons}$$

- Atoms of the same element with different numbers of neutrons are called isotopes.
- The atomic mass of an element is a weighted average of the masses of all of its naturally occurring isotopes.

**VOCABULARY**
- atomic number
- isotope
- mass number
- atomic mass unit (amu)
- atomic mass

## SECTION 4 Unstable Nuclei and Radioactive Decay

**MAINIDEA** Unstable atoms emit radiation to gain stability.

- Chemical reactions involve changes in the electrons surrounding an atom. Nuclear reactions involve changes in the nucleus of an atom.
- There are three types of radiation: alpha (charge of 2+), beta (charge of 1−), and gamma (no charge).
- The neutron-to-proton ratio of an atom's nucleus determines its stability.

**VOCABULARY**
- radioactivity
- radiation
- nuclear reaction
- radioactive decay
- alpha radiation
- alpha particle
- nuclear equation
- beta radiation
- beta particle
- gamma ray

## SECTION 1

### Mastering Concepts

**30.** Who originally proposed the concept that matter is composed of tiny, indivisible particles?

**31.** Whose work is credited with being the beginning of modern atomic theory?

**32.** Distinguish between Democritus's ideas and Dalton's atomic theory.

**33. Ideas and Scientific Methods** Was Democritus's proposal of the existence of atoms based on scientific methods or ideas? Explain.

**34.** Explain why Democritus was unable to experimentally verify his ideas.

**35.** What was Aristotle's objection to the atomic theory?

**36.** State the main points of Dalton's atomic theory using your own words. Which parts of Dalton's theory were later found to be erroneous? Explain why.

**37. Conservation of Mass** Explain how Dalton's atomic theory offered a convincing explanation of the observation that mass is conserved in chemical reactions.

**38.** Define *matter* and give two everyday examples.

## SECTION 2

### Mastering Concepts

**39.** What particles are found in the nucleus of an atom? What is the charge of the nucleus?

**40.** How was the overall charge distributed in the plum pudding model?

**41.** How did the charge distribution in the plum pudding model affect alpha particles passing through an atom?

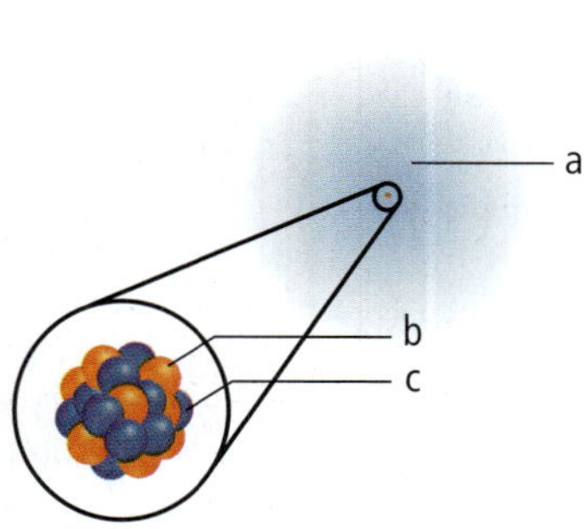


■ **Figure 22**

**42.** Label the subatomic particles shown in **Figure 22.**

**43.** Arrange the following subatomic particles in order of increasing mass: neutron, electron, and proton.

**44.** Explain why atoms are electrically neutral.

**45.** What is the charge of the nucleus of element 89?

**46.** Which particles account for most of an atom's mass?

**47.** If you had a balance that could determine the mass of a proton, how many electrons would you need to weigh on the same balance to measure the same mass as that of a single proton?

**48. Cathode-Ray Tubes** Which subatomic particle was discovered by researchers working with cathode-ray tubes?

**49.** What experimental results led to the conclusion that electrons were part of all forms of matter?

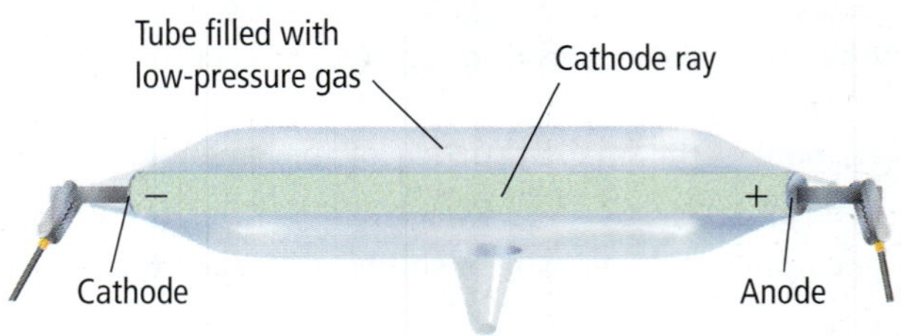


■ **Figure 23**

**50. Cathode Ray** Use the elements labeled in **Figure 23** to explain the direction of a cathode ray inside a cathode-ray tube.

**51.** Briefly explain how Rutherford discovered the nucleus.

**52. Particle Deflection** What caused the deflection of the alpha particles in Rutherford's gold foil experiment?

**53. Charge of Cathode Rays** How was an electric field used to determine the charge of a cathode ray?

**54.** Explain what keeps the electrons confined in the space surrounding the nucleus.

**55.** What is the approximate size of an atom?

**56. Visualizing Atoms** What technique can be used to visualize individual atoms?

**57.** What are the strengths and weaknesses of Rutherford's nuclear model of the atom?

## SECTION 3

### Mastering Concepts

**58.** How do isotopes of a given element differ? How are they similar?

**59.** How is an atom's atomic number related to its number of protons? To its number of electrons?

**60.** How is the mass number related to the number of protons and neutrons an atom has?

**61.** How can you determine the number of neutrons in an atom if its mass number and its atomic number are known?

**62.** What do the superscript and subscript in the notation $^{40}_{19}K$ represent?

**63. Standard Units** Define the atomic mass unit. What were the benefits of developing the atomic mass unit as a standard unit of mass?

**64. Isotopes** Are $^{24}_{12}Mg$, $^{25}_{12}Mg$, $^{26}_{12}Mg$ isotopes of each other? Explain.

**65.** Does the existence of isotopes contradict part of Dalton's original atomic theory? Explain.

## Mastering Problems

**66.** How many protons and electrons are contained in an atom of element 44?

**67. Carbon** A carbon atom has a mass number of 12 and an atomic number of 6. How many neutrons does it have?

**68. Mercury** An isotope of mercury has 80 protons and 120 neutrons. What is the mass number of this isotope?

**69. Xenon** An isotope of xenon has an atomic number of 54 and contains 77 neutrons. What is the xenon isotope's mass number?

**70.** If an atom has 18 electrons, how many protons does it have?

**71. Sulfur** Determine the atomic mass of sulfur given the following naturally occurring isotopes: $^{32}_{16}S$ (31.972 amu, 94.99%), $^{33}_{16}S$ (32.971 amu, 0.75%), $^{34}_{16}S$ (33.968 amu, 4.25%), and $^{36}_{16}S$ (35.967 amu, 0.01%).

**72.** Fill in the blanks in **Table 6.**

**Table 6 Chlorine and Zirconium**

| Element | Cl | Cl | Zr | Zr |
|---|---|---|---|---|
| Atomic number | 17 | | 40 | |
| Mass number | 35 | 37 | | 92 |
| Protons | | | | 40 |
| Neutrons | | | 50 | |
| Electrons | | 17 | | |

**73.** How many electrons, protons, and neutrons are contained in each atom?

**a.** $^{132}_{55}Cs$ **c.** $^{163}_{69}Tm$
**b.** $^{59}_{27}Co$ **d.** $^{70}_{30}Zn$

**74.** How many electrons, protons, and neutrons are contained in each atom?

**a.** gallium-69 **c.** titanium-48
**b.** fluorine-23 **d.** tantalum-181

**75.** For each chemical symbol, determine the number of protons and electrons an atom of the element contains.

**a.** V **c.** Ir
**b.** Mn **d.** S

**76.** Gallium, which has an atomic mass of 69.723 amu, has two naturally occurring isotopes, Ga-69 and Ga-71. Which isotope occurs in greater abundance? Explain.

**77. Atomic Mass of Silver** Silver has two isotopes: $^{107}_{47}Ag$, which has a mass of 106.905 amu and a percent abundance of 52.00%, and $^{109}_{47}Ag$, which has a mass of 108.905 amu and a percent abundance of 48.00%. What is the atomic mass of silver?

**78.** Data for chromium's four naturally occurring isotopes are provided in **Table 7.** Calculate chromium's atomic mass.

**Table 7 Chromium Isotope Data**

| Isotope | Percent Abundance | Mass (amu) |
|---|---|---|
| Cr-50 | 4.35 | 49.946 |
| Cr-52 | 83.79 | 51.941 |
| Cr-53 | 9.50 | 52.941 |
| Cr-54 | 2.36 | 53.939 |

# SECTION 4

## Mastering Concepts

**79.** What is radioactive decay?

**80.** Why are some atoms radioactive?

**81.** Discuss how radioactive atoms gain stability.

**82.** Define *alpha particle, beta particle,* and *gamma ray.*

**83.** Write the symbols used to denote alpha, beta, and gamma radiation and give their mass and charge.

**84.** What type of reaction involves changes in the nucleus of an atom?

**85. Radioactive Emissions** What change in mass number occurs when a radioactive atom emits an alpha particle? A beta particle? A gamma particle?

**86.** What is the primary factor that determines whether a nucleus is stable or unstable?

**87.** Explain how energy loss and nuclear stability are related to radioactive decay.

**88.** Explain what must occur before a radioactive atom stops undergoing further radioactive decay.

**89. Boron-10** emits alpha particles and cesium-137 emits beta particles. Write balanced nuclear reactions for each radioactive decay.

## MIXED REVIEW

**90.** Determine what was wrong with Dalton's theory and provide the most recent version of the atomic structure.

**91. Cathode-Ray Tube** Describe a cathode-ray tube and how it operates.

**92. Subatomic Particles** Explain how J. J. Thomson's determination of the charge-to-mass ratio of the electron led to the conclusion that atoms were composed of subatomic particles.

**93. Gold Foil Experiment** How did the actual results of Rutherford's gold foil experiment differ from the results he expected?

**94.** If a nucleus contains 12 protons, how many electrons are in the neutral atom? Explain.

**95.** An atom's nucleus has 92 protons and its mass number is 235. How many neutrons are in the nucleus? What is the name of the atom?

**96.** Complete **Table 8.**

**Table 8** Composition of Various Isotopes

| Isotope | | | Zn-64 | | |
|---|---|---|---|---|---|
| Atomic number | | | | 9 | 11 |
| Mass number | 32 | | | | 23 |
| Number of protons | 16 | | | | |
| Number of neutrons | | 24 | | 10 | |
| Number of electrons | | 20 | | | |

**97.** Approximately how many times greater is the diameter of an atom than the diameter of its nucleus? Knowing that most of an atom's mass is contained in the nucleus, what can you conclude about the density of the nucleus?

**98.** Is the charge of a nucleus positive, negative, or zero? The charge of an atom?

**99.** Why are electrons in a cathode-ray tube deflected by electric fields?

**100.** What was Henry Moseley's contribution to the modern understanding of the atom?

**101.** What is the mass number of potassium-39? What is the isotope's charge?

**102.** Boron-10 and boron-11 are the naturally occurring isotopes of elemental boron. If boron has an atomic mass of 10.81 amu, which isotope occurs in greater abundance?

**103. Semiconductors** Silicon is important to the semiconductor manufacturing industry. The three naturally occurring isotopes of silicon are silicon-28, silicon-29, and silicon-30. Write the symbol for each.

**104. Titanium** Use **Table 9** to calculate the atomic mass of titanium.

**Table 9** Titanium Isotopes

| Isotope | Atomic Mass (amu) | Relative Abundance (%) |
|---|---|---|
| Ti-46 | 45.953 | 8.00 |
| Ti-47 | 46.952 | 7.30 |
| Ti-48 | 47.948 | 73.80 |
| Ti-49 | 48.948 | 5.50 |
| Ti-50 | 49.945 | 5.40 |

**105.** Describe how each type of radiation affects an atom's atomic number and mass number.

**106. Relative Abundances** Magnesium constitutes about 2% of Earth's crust and has three naturally occurring isotopes. Suppose you analyze a mineral and determine that it contains the three isotopes in the following proportions: Mg-24 (abundance = 79%), Mg-25 (abundance = 10%), and Mg-26 (abundance = 11%). If your friend analyzes a different mineral containing magnesium, do you expect her to obtain the same relative abundances for each magnesium isotope? Explain your reasoning.

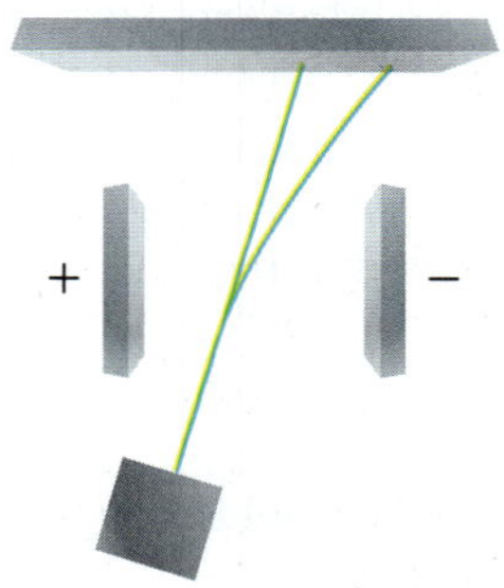


■ **Figure 24**

**107. Radiation** Identify the two types of radiation shown in **Figure 24.** Explain your reasoning.

## THINK CRITICALLY

**108. Formulate** How were scientific methods used to determine the model of the atom? Why is the model considered a theory?

**109. Discuss** What experiment led to the dispute of J. J. Thomson's plum pudding atomic model? Explain your answer.

**110. Apply** Which is greater, the number of compounds or the number of elements? The number of elements or the number of isotopes? Explain.

**111. Analyze** An element has three naturally occurring isotopes. What other information must you know in order to calculate the element's atomic mass?

**112. Apply** If atoms are primarily composed of empty space, explain why you cannot pass your hand through a solid object.

**113. Formulate** Sketch a modern atomic model of a typical atom and identify where each type of subatomic particle would be located.

**114. Apply** Indium has two naturally occurring isotopes and an atomic mass of 114.818 amu. In-113 has a mass of 112.904 amu and an abundance of 4.3%. What is the identity and percent abundance of indium's other isotope?

**115. Infer** Sulfur's average atomic mass is close to the whole number 32. Chlorine's average atomic mass is 35.453, which is not a whole number. Suggest a possible reason for this difference.

## Challenge Problem

**116. Magnesium Isotopes** Compute the mass number, *X*, of the third isotope of magnesium given that the respective abundances of the naturally occurring isotopes are: 79.0%, 10%, and 11% for $^{24}_{12}Mg$, $^{25}_{12}Mg$, and $^{X}_{12}Mg$. The atomic mass of magnesium is 24.305 amu.

## CUMULATIVE REVIEW

**117.** How is a qualitative observation different from a quantitative observation? Give an example of each.

**118.** A 1.0-$cm^3$ block of gold can be flattened to a thin sheet that averages $3.0 \times 10^{-8}$ cm thick. What is the area (in $cm^2$) of the flattened gold sheet?

**119.** A piece of paper has an area of 603 $cm^2$. How many sheets of paper would the sheet of gold mentioned in problem 118 cover?

**120.** Classify each mixture as heterogeneous or homogeneous.
**a.** salt water
**b.** vegetable soup
**c.** 14-K gold
**d.** concrete

**121.** Determine whether each change is physical or chemical.
**a.** Water boils.
**b.** A match burns.
**c.** Sugar dissolves in water.
**d.** Sodium reacts with water.
**e.** Ice cream melts.

**122. Television and Computer Screens** Describe how cathode rays are used to generate television and computer monitor images.

**123. The Standard Model** The standard model of particle physics describes all of the known building blocks of matter. Research the particles included in the standard model. Write a short report describing the known particles and those thought to exist but not yet detected experimentally.

**124. STM** Individual atoms can be seen using a sophisticated device known as a scanning tunneling microscope. Write a short report on how the scanning tunneling microscope works and create a gallery of this microscope's images from sources such as books, magazines, and the Internet.

## DBQ Document-Based Questions

**Zirconium** *is a lustrous, gray-white metal. Because of its high resistance to corrosion and its low cross section for neutron absorption, it is often used in nuclear reactors. It can also be processed to produce gems that look like diamonds that is used in jewelry.*

**Table 10** *shows the relative abundances of zirconium isotopes.*

**Table 10 Relative Abundances of Zirconium Isotopes**

| Element | Mass (amu) | Abundance (%) |
|---|---|---|
| Zirconium-90 | 89.905 | 51.45 |
| Zirconium-91 | 90.906 | 11.22 |
| Zirconium-92 | 91.905 | 17.15 |
| Zirconium-94 | 93.906 | 17.38 |
| Zirconium-96 | 95.908 | 2.80 |

Data obtained from: Lide, David R., ed. 2005. *CRC Handbook of Chemistry and Physics*. Boca Raton: CRC Press.

**125.** What is the mass number of each zirconium isotope?

**126.** Compute the number of protons and neutrons for each zirconium isotope.

**127.** Does the number of protons or neutrons remain the same for all isotopes? Explain.

**128.** Based on the relative abundances of each isotope, predict to which isotope's mass the average atomic mass of zirconium is going to be closest.

**129.** Calculate the weighted average atomic mass of zirconium.

## MULTIPLE CHOICE

1. Which describes an atom of plutonium?
   - **A.** It can be divided into smaller particles that retain all the properties of plutonium.
   - **B.** It cannot be divided into smaller particles that retain all the properties of plutonium.
   - **C.** It does not possess all the properties of a larger quantity of plutonium.
   - **D.** It has an atomic number of 244.

2. Neptunium's only naturally occurring isotope, $^{237}_{93}Np$, decays by emitting one alpha particle, one beta particle, and one gamma ray. What is the new atom formed from this decay?
   - **A.** $^{233}_{92}U$
   - **B.** $^{241}_{93}Np$
   - **C.** $^{233}_{90}Th$
   - **D.** $^{241}_{92}U$

3. Which type of matter has a definite composition throughout and is made of more than one type of element?
   - **A.** heterogeneous mixture
   - **B.** homogeneous mixture
   - **C.** element
   - **D.** compound

*Use the diagram below to answer Question 4.*

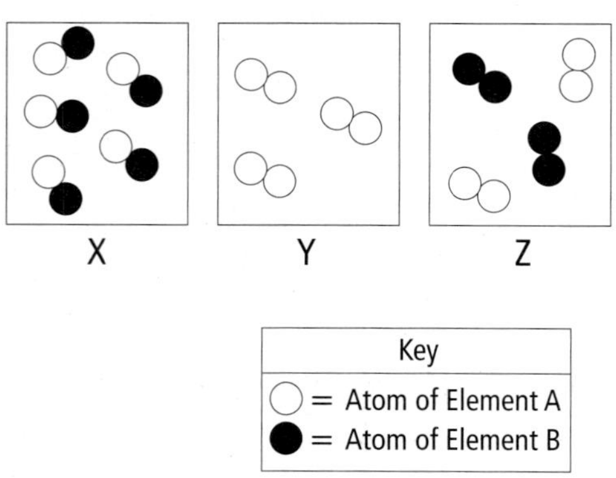


4. Which diagram shows a mixture?
   - **A.** X
   - **B.** Y
   - **C.** Z
   - **D.** both X and Z

5. The Moon is approximately 384,400 km from Earth. What is this value in scientific notation?
   - **A.** $384.4 \times 10^3$ km
   - **B.** $3.844 \times 10^5$ km
   - **C.** $3.844 \times 10^{-5}$ km
   - **D.** $3844 \times 10^{-2}$ km

6. Why does an atom have no net electric charge?
   - **A.** Its subatomic particles carry no electric charges.
   - **B.** The positively charged protons cancel out the negatively charged neutrons.
   - **C.** The positively charged neutrons cancel out the negatively charged electrons.
   - **D.** The positively charged protons cancel out the negatively charged electrons.

7. How many neutrons, protons, and electrons does $^{126}_{52}Te$ have?
   - **A.** 126 neutrons, 52 protons, and 52 electrons
   - **B.** 74 neutrons, 52 protons, and 52 electrons
   - **C.** 52 neutrons, 74 protons, and 74 electrons
   - **D.** 52 neutrons, 126 protons, and 126 electrons

*Use the figure below to answer Question 8.*

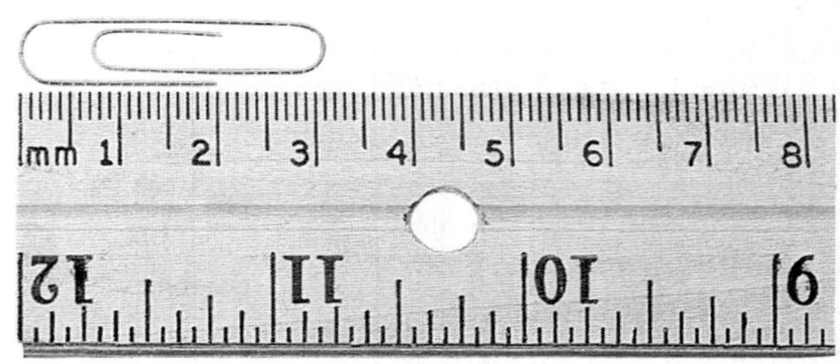


8. Record the length of this paper clip to the appropriate number of significant digits.
   - **A.** 31 mm
   - **B.** 31.1 mm
   - **C.** 30.1 mm
   - **D.** 31.15 mm

9. Element X has an unstable nucleus due to an over-abundance of neutrons. All are likely to occur EXCEPT
   - **A.** element X will undergo radioactive decay.
   - **B.** element X will remain an unstable radioactive element.
   - **C.** element X will gain more protons to balance the neutrons it possesses.
   - **D.** element X will spontaneously lose energy.

10. What makes up most of the volume of an atom?
    - **A.** protons
    - **B.** neutrons
    - **C.** electrons
    - **D.** empty space

## SHORT ANSWER

11. A 36.41-g sample of calcium carbonate ($CaCO_3$) contains 14.58 g of calcium and 4.36 g of carbon. What is the mass of oxygen contained in the sample? What is the percent by mass of each element in this compound?

*Use the table below to answer Questions 12 and 13.*

**Characteristics of Naturally Occurring Neon Isotopes**

| Isotope | Atomic Number | Mass (amu) | Percent Abundance |
|---|---|---|---|
| $^{20}Ne$ | 10 | 19.992 | 90.48 |
| $^{21}Ne$ | 10 | 20.994 | 0.27 |
| $^{22}Ne$ | 10 | 21.991 | 9.25 |

12. For each isotope listed above, write the number of protons, electrons, and neutrons it contains.

13. Using the data in the table above, calculate the average atomic mass of neon.

## EXTENDED RESPONSE

14. Assume that Element Q has the following three isotopes: $^{248}Q$, $^{252}Q$, and $^{259}Q$. If the atomic mass of Q is 258.63, which of its isotopes is most abundant? Explain your answer.

15. Iodine-131 undergoes radioactive decay to form an isotope with 54 protons and 77 neutrons. What type of decay occurs in this isotope? Explain how you can tell.

16. You are given an aluminum cube. Your measurements show that its sides are 2.14 cm and its mass is 25.1 g. Explain how you would find its density. If the density of aluminum is known to be 2.70 g/cm$^3$, what is your percent error?

## SAT SUBJECT TEST: CHEMISTRY

For each question below, indicate whether Statement I is true or false and indicate whether Statement II is true or false. If Statement II is a *correct explanation* of Statement I, write CE on your paper.

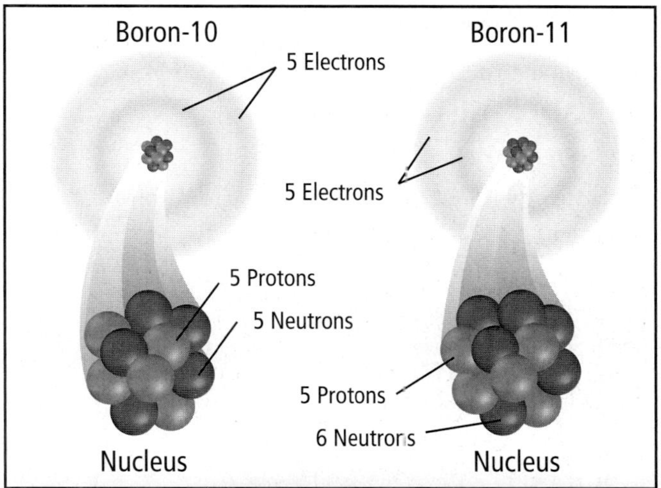


| Statement I | | Statement II |
|---|---|---|
| 17. The two atoms of boron pictured above are isotopes | BECAUSE | they have the same number of protons but a different number of neutrons. |
| 18. Most alpha particles shot at a piece of gold foil travel through it | BECAUSE | an atom has a large nucleus compared to its overall size. |
| 19. A beam of neutrons is attracted to the charged plates surrounding it | BECAUSE | neutrons have no charge. |
| 20. Carbon and oxygen can form either CO or $CO_2$ | BECAUSE | carbon and oxygen obey the law of definite composition. |
| 21. A mixture of sand and water is heterogeneous | BECAUSE | water is a compound formed from hydrogen and oxygen. |

**NEED EXTRA HELP?**

| If You Missed Question . . . | 1 | 2 | 3 | 4 | 5 | 6 | 7 | 8 | 9 | 10 | 11 | 12 | 13 | 14 | 15 | 16 | 17 | 18 | 19 | 20 | 21 |
|---|---|---|---|---|---|---|---|---|---|---|---|---|---|---|---|---|---|---|---|---|---|
| Review Section . . . | 4.2 | 4.4 | 3.4 | 4.2 | 4.3 | 3.1 | 2.2 | 2.3 | 2.2 | 4.2 | 3.4 | 4.3 | 4.3 | 4.3 | 4.4 | 2.3 | 4.3 | 4.2 | 4.2 | 4.1 | 3.3 |

CHAPTER 5

# Electrons in Atoms

**BIGIDEA** The atoms of each element have a unique arrangement of electrons.

## SECTIONS

Chm 1, 3

1 Light and Quantized Energy

2 Quantum Theory and the Atom

3 Electron Configuration

## LaunchLAB

iLab Station

### How do you know what is inside an atom?

Imagine that it is your birthday, and there is one wrapped present that is different from all the rest. Unlike the other gifts that you can open, you can only guess what is inside this package. In trying to determine the structure of the atom, early chemists had a similar experience. How good are your skills of observation and deduction? In this lab, you will try to determine what is inside a box through observation methods.

**FOLDABLES® Study Organizer**

### Electron Configuration

Make a concept-map book. Label it as shown. Use it to help you summarize the three rules that define how electrons are arranged in an atom.

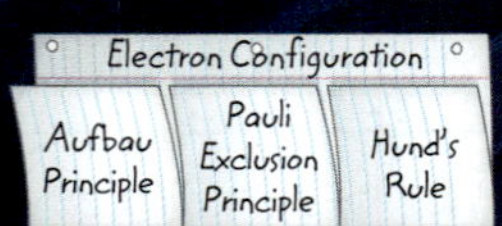


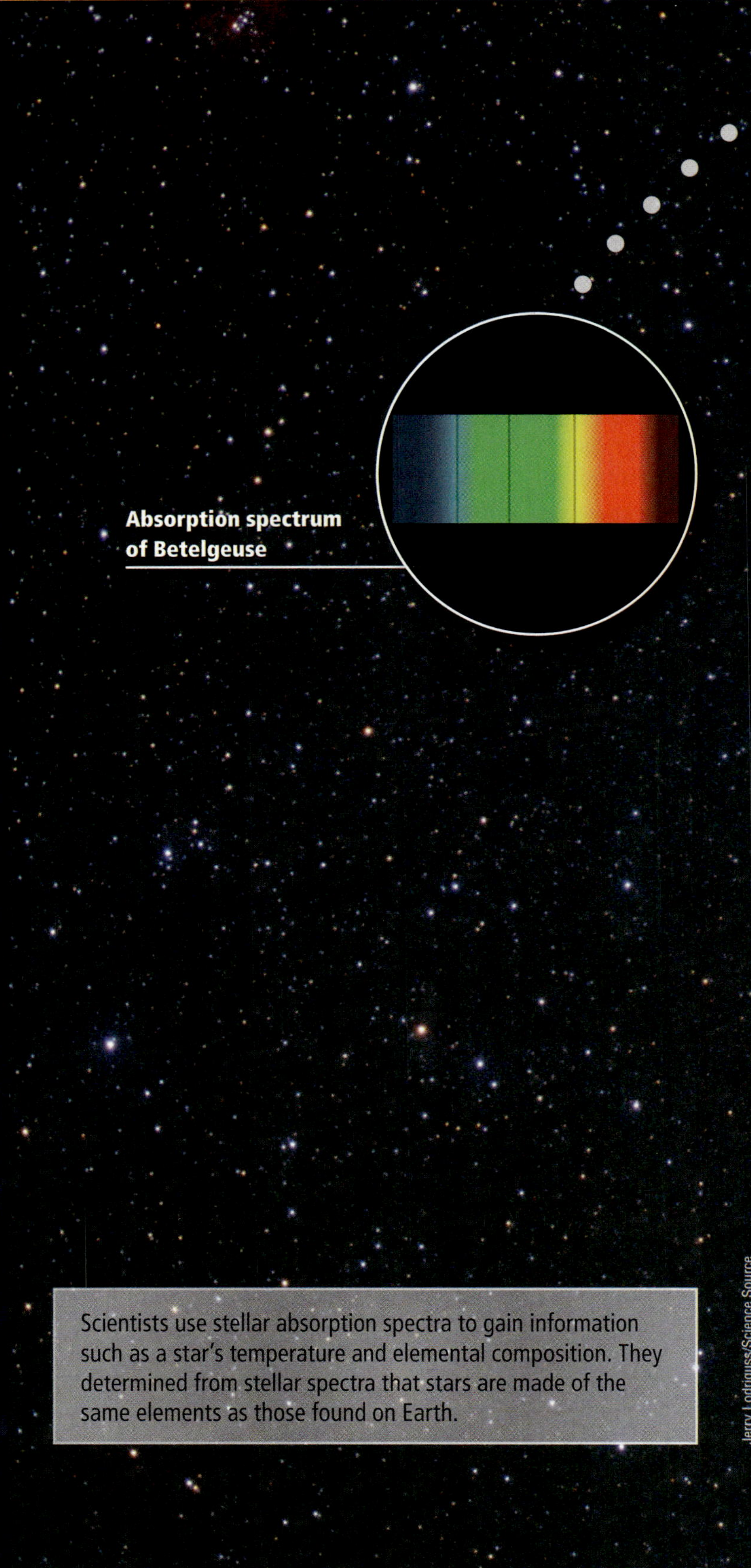


Scientists use stellar absorption spectra to gain information such as a star's temperature and elemental composition. They determined from stellar spectra that stars are made of the same elements as those found on Earth.

Jerry Lodriguss/Science Source

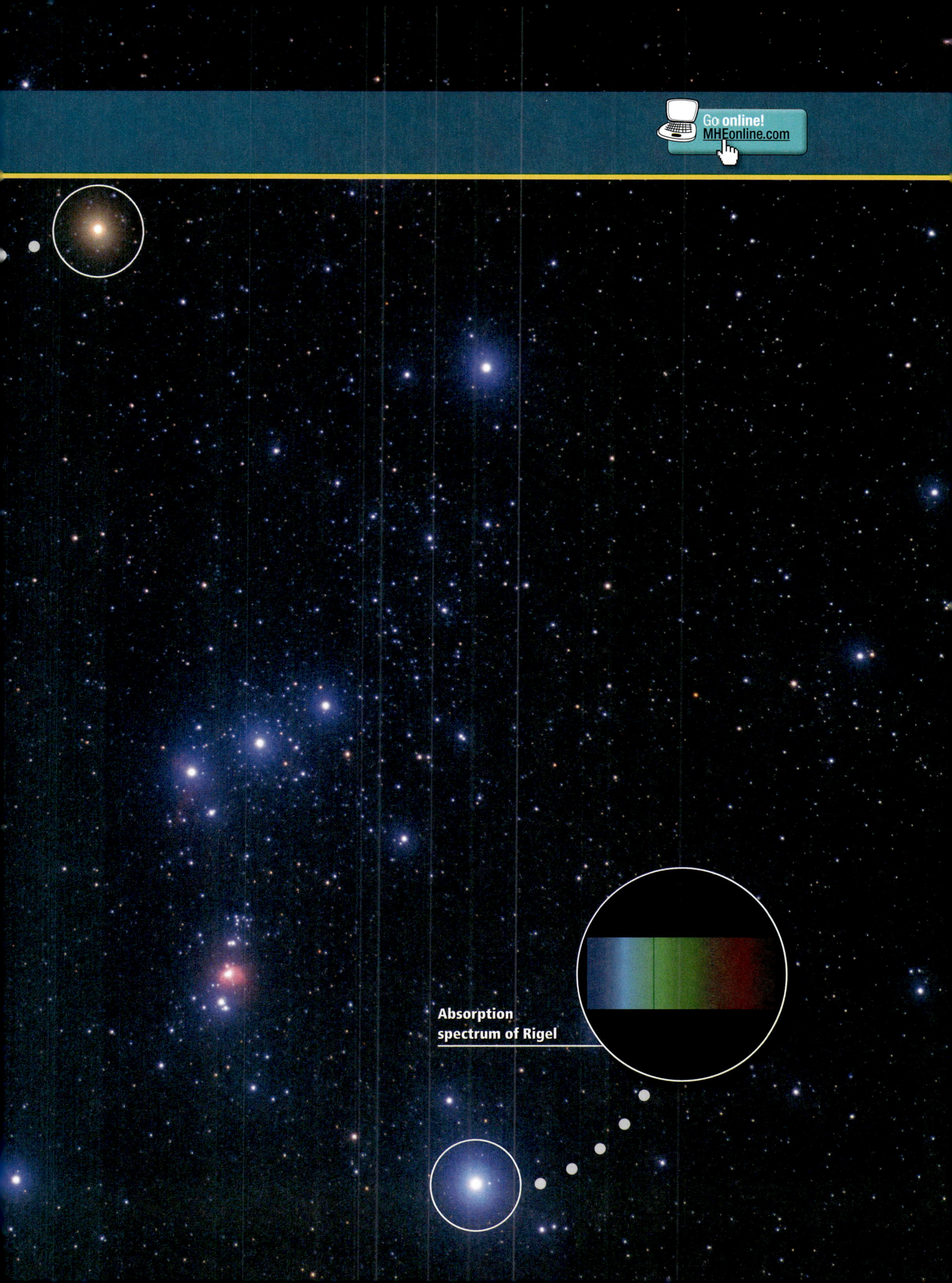
Go online!
MHEonline.com
Absorption
spectrum of Rigel

SECTION 1

# Light and Quantized Energy

**MAIN**IDEA Light, a form of electromagnetic radiation, has characteristics of both a wave and a particle.

### Essential Questions

- How do the wave and particle natures of light compare?
- What is a quantum of energy and how is it related to an energy change of matter?
- How do continuous electromagnetic spectra and atomic emission spectra compare and contrast?

### Review Vocabulary

**radiation:** the rays and particles—alpha particles, beta particles, and gamma rays—that are emitted by radioactive material

### New Vocabulary

electromagnetic radiation
wavelength
frequency
amplitude
electromagnetic spectrum
quantum
Planck's constant
photoelectric effect
photon
atomic emission spectrum

**CHEM 4 YOU** **Have you ever come inside on a cold day, headed for the kitchen, and popped a cold snack into the microwave oven? When the microwaves reached your snack, small packets of energy warmed it in practically no time at all.**

## The Atom and Unanswered Questions

After discovering three subatomic particles in the early 1900s, scientists continued their quest to understand atomic structure and the arrangement of electrons within atoms.

Rutherford proposed that all of an atom's positive charge and virtually all of its mass are concentrated in a nucleus that is surrounded by fast-moving electrons. The model did not explain how the atom's electrons are arranged in the space around the nucleus. Nor did it address the question of why the negatively charged electrons are not pulled into the atom's positively charged nucleus. Rutherford's nuclear model did not begin to account for the differences and similarities in chemical behavior among the various elements.

For example, consider the elements lithium, sodium, and potassium, which are found in different periods on the periodic table but have similar chemical behaviors. All three elements appear metallic in nature, and their atoms react vigorously with water to liberate hydrogen gas. In fact, as shown in **Figure 1,** both sodium and potassium react so violently that the hydrogen gas can ignite and even explode.

In the early 1900s, scientists began to unravel the puzzle of chemical behavior. They observed that certain elements emitted visible light when heated in a flame. Analysis of the emitted light revealed that an element's chemical behavior is related to the arrangement of the electrons in its atoms. To understand this relationship and the nature of atomic structure, it will be helpful to first understand the nature of light.

**Figure 1** Different elements can have similar reactions with water.

## The Wave Nature of Light

Visible light is a type of **electromagnetic radiation**—a form of energy that exhibits wavelike behavior as it travels through space. Other examples of electromagnetic radiation include microwaves that cook your food, X rays that doctors and dentists use to examine bones and teeth, and waves that carry radio and television programs into homes.

**Characteristics of waves** All waves can be described by several characteristics, a few of which might be familiar to you from everyday experience. You might have seen concentric waves when dropping an object into water, as shown in **Figure 2a.**

The **wavelength** (represented by $\lambda$, the Greek letter lambda) is the shortest distance between equivalent points on a continuous wave. For example, in **Figure 2b,** the wavelength is measured from crest to crest or from trough to trough. Wavelength is usually expressed in meters, centimeters, or nanometers ($1\ \text{nm} = 1 \times 10^{-9}\ \text{m}$).

The **frequency** (represented by $\nu$, the Greek letter nu) is the number of waves that pass a given point per second. One hertz (Hz), the SI unit of frequency, equals one wave per second. In calculations, frequency is expressed with units of waves per second, (1/s) or ($s^{-1}$); the term *waves* is understood. A particular frequency can be expressed in the following ways: $652\ \text{Hz} = 652\ \text{waves/second} = 652/\text{s} = 652\ \text{s}^{-1}$.

The **amplitude** of a wave is the wave's height from the origin to a crest, or from the origin to a trough, as illustrated in **Figure 2b.** Wavelength and frequency do not affect the amplitude of a wave.

All electromagnetic waves, including visible light, travel at a speed of $3.00 \times 10^{8}$ m/s in a vacuum. Because the speed of light is such an important and universal value, it is given its own symbol, *c*. The speed of light is the product of its wavelength ($\lambda$) and its frequency ($\nu$).

**Electromagnetic Wave Relationship**

$$c = \lambda\nu$$

*c* is the speed of light in a vacuum.
$\lambda$ is the wavelength.
$\nu$ is the frequency.

The speed of light in a vacuum is equal to the product of the wavelength and the frequency.

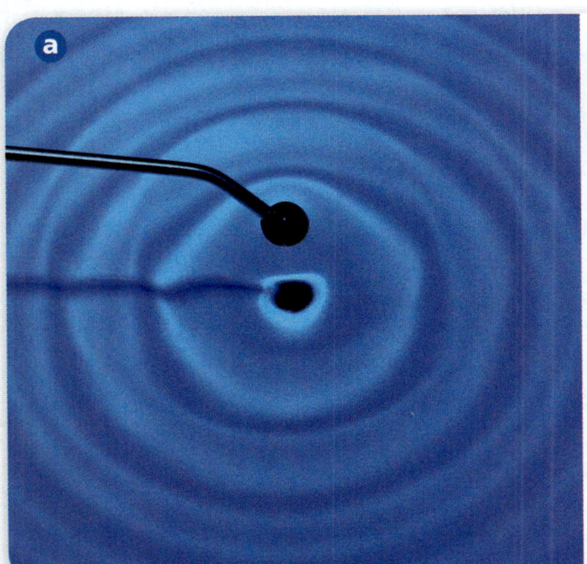


■ **Figure 2** **a.** The concentric waves in the water show the characteristic properties of all waves. **b.** Amplitude, wavelength, anc frequency are the main characteristics of waves.

**Identify** *a crest, a trough, and one wavelength in the photo.*

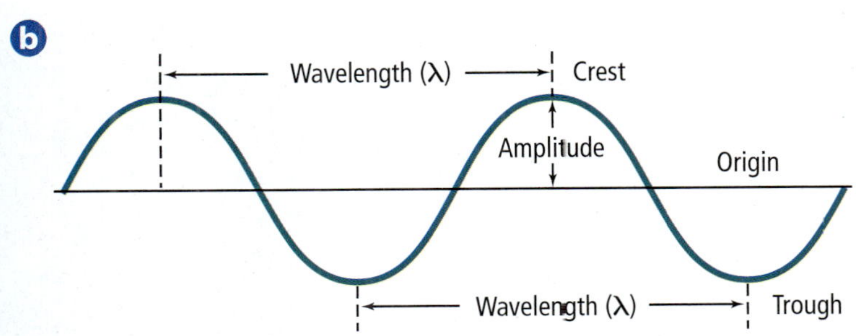

■ **Figure 3** These waves illustrate the relationship between wavelength and frequency. As wavelength increases, frequency decreases.

**Infer** ***Does frequency or wavelength affect amplitude?***

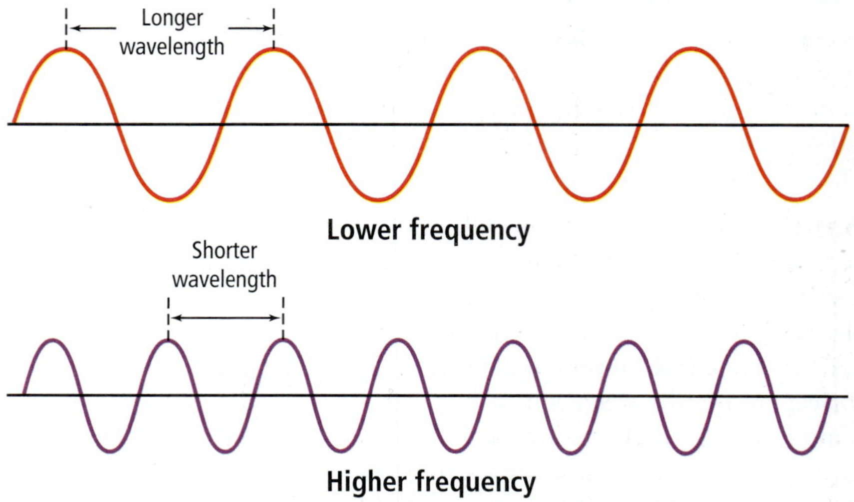


Although the speed of all electromagnetic waves in a vacuum is the same, waves can have different wavelengths and frequencies. As you can see from the equation on the previous page, wavelength and frequency are inversely related; in other words, as one quantity increases, the other decreases. To better understand this relationship, examine the two waves illustrated in **Figure 3.** Although both waves travel at the speed of light, you can see that the red wave has a longer wavelength and lower frequency than the violet wave.

**Electromagnetic spectrum** Sunlight, which is one example of white light, contains a nearly continuous range of wavelengths and frequencies. White light passing through a prism separates into a continuous spectrum of colors similar to the spectrum in **Figure 4.** These are the colors of the visible spectrum. The spectrum is called continuous because each point of it corresponds to a unique wavelength and frequency. You might be familiar with the colors of the visible spectrum. If you have ever seen a rainbow, you have seen all of the visible colors at once. A rainbow is formed when tiny drops of water in the air disperse the white light from the Sun into its component colors, producing a spectrum that arches across the sky.

■ **Figure 4** When white light passes through a prism, it is separated into a continuous spectrum of its different components—red, orange, yellow, green, blue, indigo, and violet light.

DAVID PARKER/SCIENCE PHOTO LIBRARY/Photo Researchers

The visible spectrum of light shown in **Figure 4,** however, comprises only a small portion of the complete electromagnetic spectrum, which is illustrated in **Figure 5.** The **electromagnetic spectrum,** also called the EM spectrum, includes all forms of electromagnetic radiation, with the only differences in the types of radiation being their frequencies and wavelengths. Note in **Figure 4** that the bend varies with the wavelengths as they pass through the prism, resulting in the sequence of the colors red, orange, yellow, green, blue, indigo, and violet. In examining the energy of the radiation shown in **Figure 5,** note that energy increases with increasing frequency. Thus, looking back at **Figure 3,** the violet light, with its greater frequency, has more energy than the red light. This relationship between frequency and energy will be explained in the next section.

Because all electromagnetic waves travel at the same speed in a given medium, you can use the formula $c = \lambda\nu$ to calculate the wavelength or frequency of any wave.

**READING CHECK** **State** the relationship between the energy and the frequency of electromagnetic radiation.

**Connection to Physics** Electromagnetic radiation from diverse origins constantly bombards us. In addition to the radiation from the Sun, human activities also produce radiation which include radio and TV signals, phone relay stations, lightbulbs, medical X-ray equipment, and particle accelerators. Natural sources on Earth, such as lightning, natural radioactivity, and even the glow of fireflies, also contribute. Our knowledge of the universe is based on electromagnetic radiation emitted by distant objects and detected with instruments on Earth.

## CAREERS IN CHEMISTRY

**Spectroscopist** Spectroscopy is the study of the spectra absorbed or emitted by matter. Because each element's spectrum is unique, it is like a fingerprint. Astrophysicists use spectroscopy to investigate what a star, such as the Sun, is made of. A star's absorption spectrum shows many dark lines, which allow spectroscopists to identify the elements present in the star.

WebQuest

■ **Figure 5** The electromagnetic spectrum covers a wide range of frequencies. The visible-light section of the spectrum is very narrow. As frequency and energy increase, wavelength decreases.

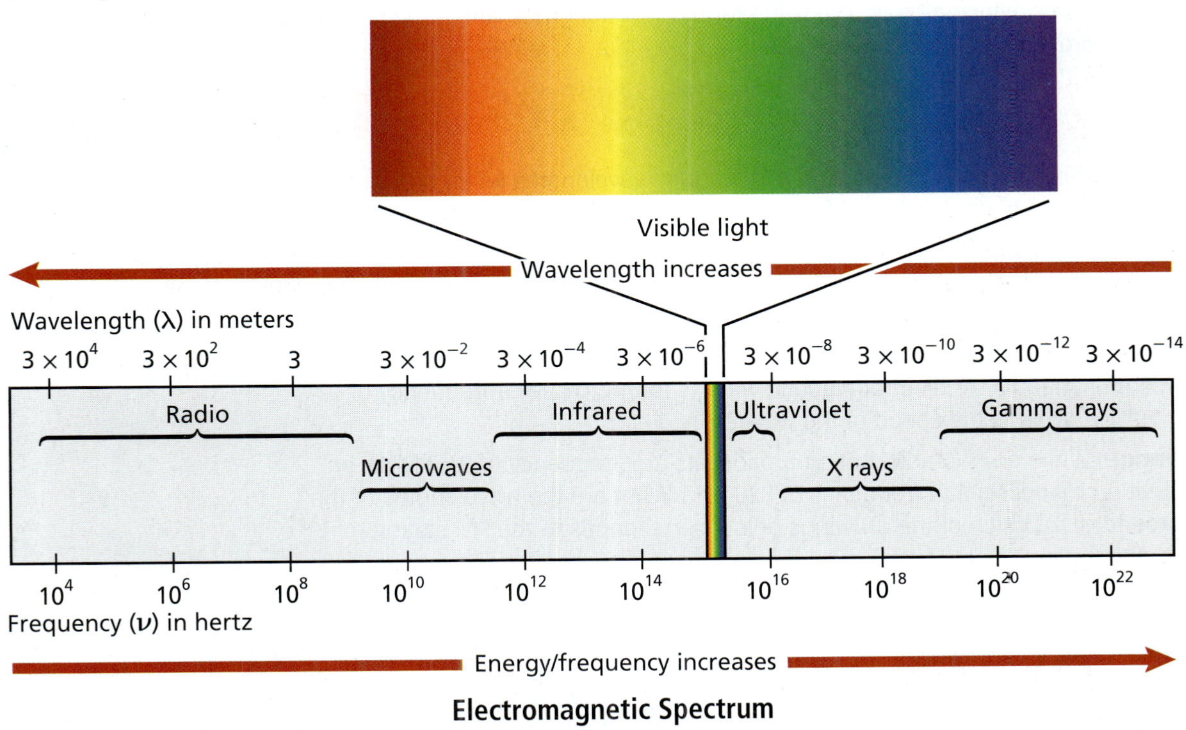

## EXAMPLE Problem 1

Find help with **Algebraic Equations.** Math Handbook

**CALCULATING WAVELENGTH OF AN EM WAVE** Microwaves are used to cook food and transmit information. What is the wavelength of a microwave that has a frequency of $3.44 \times 10^9$ Hz?

### 1 ANALYZE THE PROBLEM

You are given the frequency of a microwave. You also know that because microwaves are part of the electromagnetic spectrum, their speeds, frequencies, and wavelengths are related by the formula $c = \lambda\nu$. The value of $c$ is a known constant. First, solve the equation for wavelength, then substitute the known values and solve.

| **Known** | **Unknown** |
|---|---|
| $\nu = 3.44 \times 10^9$ Hz | $\lambda = ?$ m |
| $c = 3.00 \times 10^8$ m/s | |

### 2 SOLVE FOR THE UNKNOWN

Solve the equation relating the speed, frequency, and wavelength of an electromagnetic wave for wavelength ($\lambda$).

$c = \lambda\nu$ — **State the electromagnetic wave relationship.**

$\lambda = c/\nu$ — **Solve for $\lambda$.**

$\lambda = \dfrac{3.00 \times 10^8 \text{ m/s}}{3.44 \times 10^9 \text{ Hz}}$ — **Substitute $c = 3.00 \times 10^8$ m/s and $\nu = 3.44 \times 10^9$ Hz.**

Note that hertz is equivalent to 1/s or $s^{-1}$.

$\lambda = \dfrac{3.00 \times 10^8 \text{ m/}\cancel{s}}{3.44 \times 10^9 \cancel{s^{-1}}}$ — **Divide numbers and units.**

$\lambda = \mathbf{8.72 \times 10^{-2}}$ **m**

### 3 EVALUATE THE ANSWER

The answer is correctly expressed in a unit of wavelength (m). Both of the known values in the problem are expressed with three significant figures, so the answer should have three significant figures, which it does. The value for the wavelength is within the wavelength range for microwaves shown in **Figure 5.**

## PRACTICE Problems

**Do additional problems.** Online Practice

1. Objects get their colors from reflecting only certain wavelengths when hit with white light. Light reflected from a green leaf is found to have a wavelength of $4.90 \times 10^{-7}$ m. What is the frequency of the light?
2. X-rays can penetrate body tissues and are widely used to diagnose and treat disorders of internal body structures. What is the frequency of an X-ray with a wavelength of $1.15 \times 10^{-10}$ m?
3. After careful analysis, an electromagnetic wave is found to have a frequency of $7.8 \times 10^6$ Hz. What is the speed of the wave?
4. **Challenge** While an FM radio station broadcasts at a frequency of 94.7 MHz, an AM station broadcasts at a frequency of 820 kHz. What are the wavelengths of the two broadcasts? Which of the drawings below corresponds to the FM station? To the AM station?

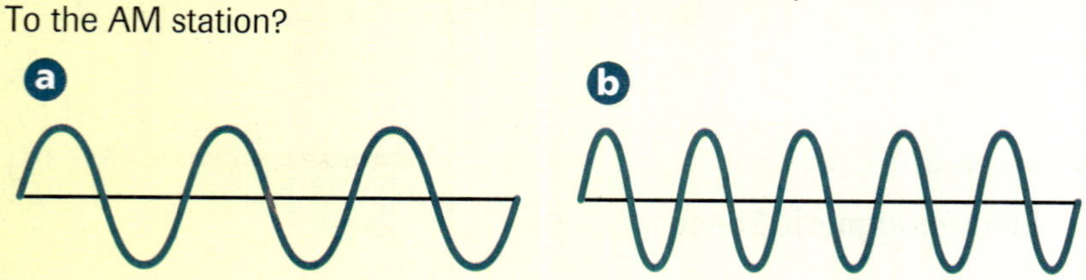

## The Particle Nature of Light

While considering light as a wave explains much of its everyday behavior, it fails to adequately describe important aspects of light's interactions with matter. The wave model of light cannot explain why heated objects emit only certain frequencies of light at a given temperature, or why some metals emit electrons when light of a specific frequency shines on them. Scientists realized that a new model or a revision of the wave model of light was needed to address these phenomena.

**The quantum concept** When objects are heated, they emit glowing light. **Figure 6** illustrates this phenomenon with iron. A piece of iron appears dark gray at room temperature, glows red when heated sufficiently, and turns orange, then bluish in color at even higher temperatures. As you will learn in later chapters, the temperature of an object is a measure of the average kinetic energy of its particles. As the iron gets hotter, it possesses a greater amount of energy and emits different colors of light. These different colors correspond to different frequencies and wavelengths.

The wave model could not explain the emission of these different wavelengths. In 1900, German physicist Max Planck (1858–1947) began searching for an explanation of this phenomenon as he studied the light emitted by heated objects. His study led him to a startling conclusion: matter can gain or lose energy only in small, specific amounts called quanta. A **quantum** is the minimum amount of energy that can be gained or lost by an atom.

**VOCABULARY**
**ACADEMIC VOCABULARY**
**Phenomenon**
an observable fact or event
*During rainstorms, electric currents often pass from the sky to Earth—a phenomenon we call lightning.*

☑ READING CHECK **Explain** why the color of heated objects changes with temperature.

Planck and other physicists of the time thought the concept of quantized energy was revolutionary, and some found it disturbing. Prior experience had led scientists to think that energy could be absorbed and emitted in continually varying quantities, with no minimum limit to the amount. For example, think about heating a cup of water in a microwave oven. It seems that you can add any amount of thermal energy to the water by regulating the power and duration of the microwaves. Instead, the water's temperature increases in infinitesimal steps as its molecules absorb quanta of energy. Because these steps are so small, the temperature seems to rise in a continuous, rather than a stepwise, manner.

■ **Figure 6** The wavelength of the light emitted by heated metal, such as the iron at left, depends on the temperature. At room temperature, iron is gray. When heated, it first turns red, then glowing orange.
**Identify** *the color of the piece of iron with the greatest kinetic energy.*

Planck proposed that the energy emitted by hot objects was quantized. He then went further and demonstrated mathematically that a relationship exists between the energy of a quantum and the frequency of the emitted radiation.

**Energy of a Quantum**

$$E_{quantum} = h\nu$$

$E_{quantum}$ represents energy.
$h$ is Planck's constant.
$\nu$ represents frequency.

The energy of a quantum is given by the product of Planck's constant and the frequency.

**Planck's constant** has a value of $6.626 \times 10^{-34}$ J•s, where J is the symbol for joule, the SI unit of energy. The equation shows that the energy of radiation increases as the radiation's frequency, $\nu$, increases.

According to Planck's theory, for a given frequency, $\nu$, matter can emit or absorb energy only in whole-number multiples of $h\nu$; that is, $1h\nu$, $2h\nu$, $3h\nu$, and so on. A useful analogy for this concept is that of a child building a wall of wooden blocks. The child can add to or take away height from the wall only in increments of whole numbers of blocks. Similarly, matter can have only certain amounts of energy—quantities of energy between these values do not exist.

**The photoelectric effect** Scientists also knew that the wave model of light could not explain a phenomenon called the photoelectric effect. In the **photoelectric effect,** electrons, called photoelectrons, are emitted from a metal's surface when light of a certain frequency, or higher than a certain frequency, shines on the surface, as shown in **Figure 7.**

The wave model predicts that given enough time, even low-energy, low-frequency light would accumulate and supply enough energy to eject photoelectrons from a metal. In reality, a metal will not eject photoelectrons below a specific frequency of incident light. For example, no matter how intensely or how long it shines, light with a frequency less than $1.14 \times 10^{15}$ Hz does not eject photoelectrons from silver. But even dim light with a frequency equal to or greater than $1.14 \times 10^{15}$ Hz ejects photoelectrons from silver.

**READING CHECK** **Describe** the photoelectric effect.

## RealWorld CHEMISTRY

### The Photoelectric Effect

**SOLAR ENERGY** is sometimes used to power road signs. Photovoltaic cells use the photoelectric effect to convert the energy of light into electric energy.

©Andrew Fox/Corbis

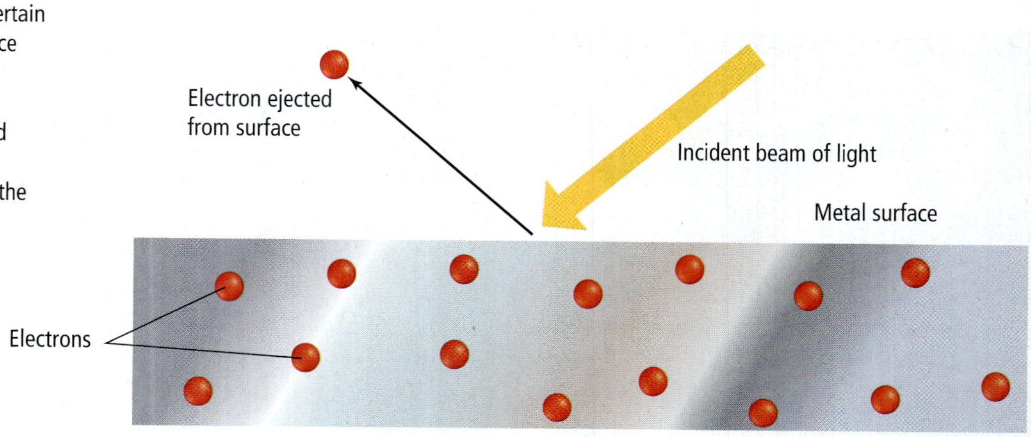


**Figure 7** The photoelectric effect occurs when light of a certain frequency strikes a metal surface and ejects electrons. When the intensity of the light increases, the number of electrons ejected increases. When the frequency (energy) of the light increases, the energy of the ejected electrons increases.

**Light's dual nature** To explain the photoelectric effect, Albert Einstein proposed in 1905 that light has a dual nature. A beam of light has wavelike and particlelike properties. It can be thought of as a beam of bundles of energy called photons. A **photon** is a massless particle that carries a quantum of energy. Extending Planck's idea of quantized energy, Einstein calculated that a photon's energy depends on its frequency.

**Energy of a Photon**

$$E_{\text{photon}} = h\nu$$

$E_{photon}$ represents energy.
$h$ is Planck's constant.
$\nu$ represents frequency.

The energy of a photon is given by the product of Planck's constant and the frequency.

Einstein also proposed that the energy of a photon must have a certain threshold value to cause the ejection of a photoelectron from the surface of the metal. Thus, even small numbers of photons with energy above the threshold value will cause the photoelectric effect. Einstein won the Nobel Prize in Physics in 1921 for this work.

## EXAMPLE Problem 2

Find help with **scientific notation.** Math Handbook

EXAMPLE PROBLEM

**CALCULATE THE ENERGY OF A PHOTON** Every object gets its color by reflecting a certain portion of incident light. The color is determined by the wavelength of the reflected photons, thus by their energy. What is the energy of a photon from the violet portion of the Sun's light if it has a frequency of $7.230 \times 10^{14}\ s^{-1}$?

### 1 ANALYZE THE PROBLEM

| Known | Unknown |
|---|---|
| $\nu = 7.230 \times 10^{14}\ s^{-1}$ | $E_{photon} = ?\ J$ |
| $h = 6.626 \times 10^{-34}\ J{\cdot}s$ | |

### 2 SOLVE FOR THE UNKNOWN

$E_{photon} = h\nu$ — State the equation for the energy of a photon.

$E_{photon} = (6.626 \times 10^{-34}\ J{\cdot}s)(7.230 \times 10^{14}\ s^{-1})$ — Substitute $h = 6.626 \times 10^{-34}\ J{\cdot}s$ and $\nu = 7.230 \times 10^{14}\ s^{-1}$.

$E_{photon} = \mathbf{4.791 \times 10^{-19}\ J}$ — Multiply and divide numbers and units.

### 3 EVALUATE THE ANSWER

As expected, the energy of a single photon of light is extremely small. The unit is joules, an energy unit, and there are four significant figures.

## PRACTICE Problems

Do additional problems. Online Practice

PRACTICE PROBLEMS

5. Calculate the energy possessed by a single photon of each of the following types of electromagnetic radiation.
   **a.** $6.32 \times 10^{20}\ s^{-1}$ **b.** $9.50 \times 10^{13}$ Hz **c.** $1.05 \times 10^{16}\ s^{-1}$
6. The blue color in some fireworks occurs when copper(I) chloride is heated to approximately 1500 K and emits blue light of wavelength $4.50 \times 10^2$ nm. How much energy does one photon of this light carry?
7. **Challenge** The microwaves used to heat food have a wavelength of 0.125 m. What is the energy of one photon of the microwave radiation?

# MiniLAB

iLab Station

## Identify Compounds

**How do flame colors vary for different elements?**

### Procedure

1. Read and complete the lab safety form.
2. Dip one of six **cotton swabs** into the **lithium chloride** solution. Put the swab into the flame of a **Bunsen burner.** Observe the color of the flame, and record it in your data table.
3. Repeat Step 2 for each of the metallic chloride solutions **(sodium chloride, potassium chloride, calcium chloride, and strontium chloride).** Record the color of each flame in your data table.
4. Compare your results to the flame tests shown in the Elements Handbook.
5. Repeat Step 2 using a sample of **unknown solution** obtained from your teacher. Record the color of the flame produced.
6. Dispose of the used cotton swabs as directed by your teacher.

### Analysis

1. **Suggest** a reason why each compound produced a flame of a different color, even though they each contain chlorine.
2. **Explain** how an element's flame test might be related to its atomic emission spectrum.
3. **Infer** the identity of the unknown crystals. Explain your reasoning.

## Atomic Emission Spectra

Have you ever wondered how light is produced in the glowing tubes of neon signs? This process is another phenomenon that cannot be explained by the wave model of light. The light of the neon sign is produced by passing electricity through a tube filled with neon gas. Neon atoms in the tube absorb energy and become excited. These excited atoms return to their stable state by emitting light to release that energy. If the light emitted by the neon is passed through a glass prism, neon's atomic emission spectrum is produced. The **atomic emission spectrum** of an element is the set of frequencies of the electromagnetic waves emitted by atoms of the element. Neon's atomic emission spectrum consists of several individual lines of color corresponding to the frequencies of the radiation emitted by the atoms of neon. It is not a continuous range of colors, as in the visible spectrum of white light.

☑ READING CHECK **Explain** how an emission spectrum is produced.

Each element's atomic emission spectrum is unique and can be used to identify an element or determine whether that element is part of an unknown compound. For example, when a platinum wire is dipped into a strontium nitrate solution and then inserted into a burner flame, the strontium atoms emit a characteristic red color. You can perform a series of flame tests by doing the MiniLab.

**Figure 8** shows an illustration of the characteristic purple-pink glow produced by excited hydrogen atoms and the visible portion of hydrogen's emission spectrum responsible for producing the glow. Note how the line nature of hydrogen's atomic emission spectrum differs from that of a continuous spectrum.

■ **Figure 8** The purple light emitted by hydrogen can be separated into its different components using a prism. Hydrogen has an atomic emission spectrum that comprises four lines of different wavelengths.
**Determine** *Which line has the highest energy?*

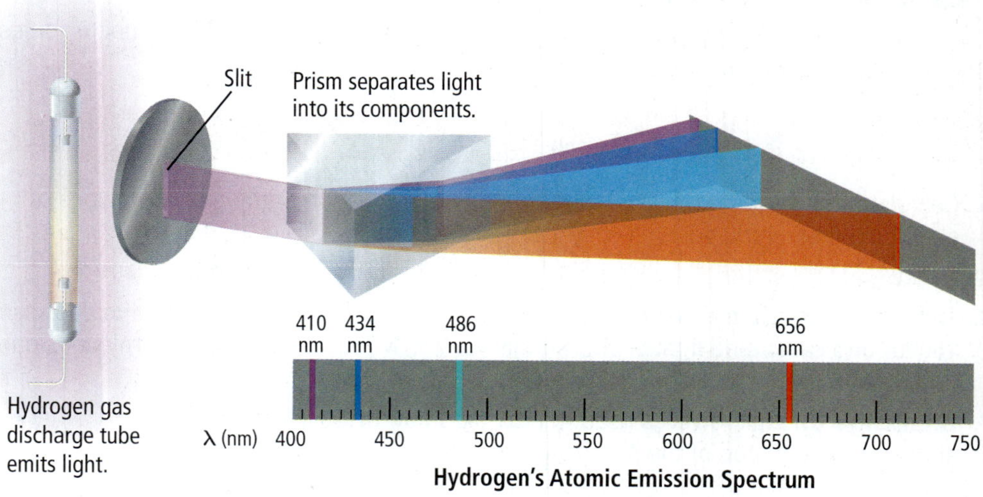

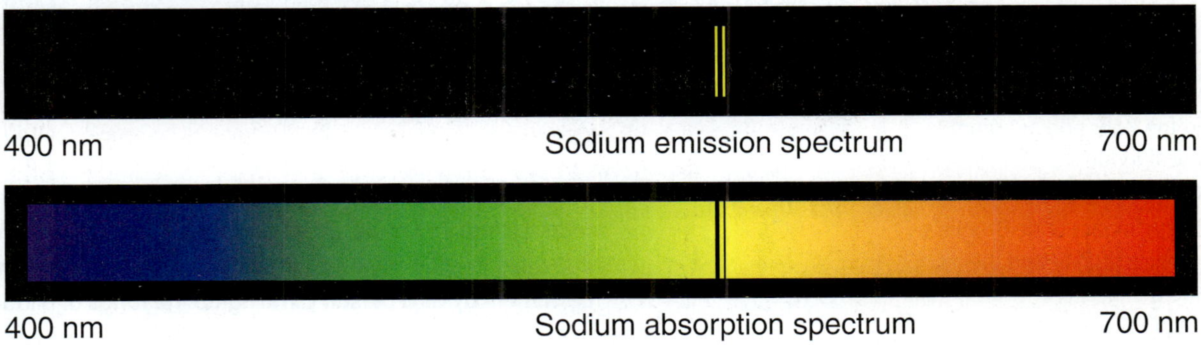


■ **Figure 9** The bottom spectrum is an absorption spectrum. It is composed of black lines on a continuous spectrum. The black lines correspond to certain frequencies absorbed by a given element, sodium in this case. They can be matched to the colored lines present in sodium's emission spectrum, shown above the absorption spectrum.

**Connection to Astronomy** An atomic emission spectrum is characteristic of the element being examined and can be used to identify that element. The fact that only certain colors appear in an element's atomic emission spectrum means that only specific frequencies of light are emitted. Because those emitted frequencies are related to energy by the formula $E_{photon} = h\nu$, only photons with specific energies are emitted. This was not predicted by the laws of classical physics. Scientists had expected to observe the emission of a continuous series of colors as excited electrons lost energy. Elements absorb the same specific frequencies of light as the frequencies they emit, thus creating an absorption spectrum. In an absorption spectrum, the absorbed frequencies appear as black lines, as shown in **Figure 9.** By comparing the black lines to the emission spectrum of elements, scientists are able to determine the composition of the outer layers of stars.

## SECTION 1 REVIEW

Section Self-Check

### Section Summary

- All waves are defined by their wavelengths, frequencies, amplitudes, and speeds.
- In a vacuum, all electromagnetic waves travel at the speed of light.
- All electromagnetic waves have both wave and particle properties.
- Matter emits and absorbs energy in quanta.
- White light produces a continuous spectrum. An element's emission spectrum consists of a series of lines of individual colors.

8. **MAIN**IDEA **Compare** the dual nature of light.
9. **Describe** the phenomena that can be explained only by the particle model of light.
10. **Compare and contrast** continuous spectrum and emission spectrum.
11. **Assess** Employ quantum theory to assess the amount of energy that matter gains and loses.
12. **Discuss** the way in which Einstein utilized Planck's quantum concept to explain the photoelectric effect.
13. **Calculate** Heating 235 g of water from 22.6°C to 94.4°C in a microwave oven requires $7.06 \times 10^4$ J of energy. If the microwave frequency is $2.88 \times 10^{10}\ s^{-1}$, how many quanta are required to supply the $7.06 \times 10^4$ J?
14. **Interpret Scientific Illustrations** Use **Figure 5** and your knowledge of electromagnetic radiation to match the numbered items with the lettered items. The numbered items may be used more than once or not at all.

| | |
|---|---|
| **a.** longest wavelength | **1.** gamma ray |
| **b.** highest frequency | **2.** infrared wave |
| **c.** greatest energy | **3.** radio waves |

SECTION 2

# Quantum Theory and the Atom

**MAINIDEA** Wavelike properties of electrons help relate atomic emission spectra, energy states of atoms, and atomic orbitals.

### Essential Questions

- How do the Bohr and quantum mechanical models of the atom compare?
- What is the impact of de Broglie's wave-particle duality and the Heisenberg uncertainty principle on the current view of electrons in atoms?
- What are the relationships among a hydrogen atom's energy levels, sublevels, and atomic orbitals?

### Review Vocabulary

**atom:** the smallest particle of an element that retains all the properties of that element; is composed of electrons, protons, and neutrons

### New Vocabulary

ground state
quantum number
de Broglie equation
Heisenberg uncertainty principle
quantum mechanical model of the atom
atomic orbital
principal quantum number
principal energy level
energy sublevel

**CHEM 4 YOU** Imagine climbing a ladder and trying to stand between the rungs. Unless you could stand on air, it would not work. When atoms are in various energy states, electrons behave in much the same way as a person climbing up the rungs of a ladder.

## Bohr's Model of the Atom

The dual wave-particle model of light accounted for several previously unexplainable phenomena, but scientists still did not understand the relationships among atomic structure, electrons, and atomic emission spectra. Recall that hydrogen's atomic emission spectrum is discontinuous; that is, it is made up of only certain frequencies of light. Why are the atomic emission spectra of elements discontinuous rather than continuous? Niels Bohr, a Danish physicist working in Rutherford's laboratory in 1913, proposed a quantum model for the hydrogen atom that seemed to answer this question. Bohr's model also correctly predicted the frequencies of the lines in hydrogen's atomic emission spectrum.

**Energy states of hydrogen** Building on Planck's and Einstein's concepts of quantized energy, Bohr proposed that the hydrogen atom has only certain allowable energy states. The lowest allowable energy state of an atom is called its **ground state.** When an atom gains energy, it is said to be in an excited state.

Bohr also related the hydrogen atom's energy states to the electron within the atom. He suggested that the electron in a hydrogen atom moves around the nucleus in only certain allowed circular orbits. The smaller the electron's orbit, the lower the atom's energy state, or energy level. Conversely, the larger the electron's orbit, the higher the atom's energy state, or energy level. Thus, a hydrogen atom can have many different excited states, although it contains only one electron. Bohr's idea is illustrated in **Figure 10.**

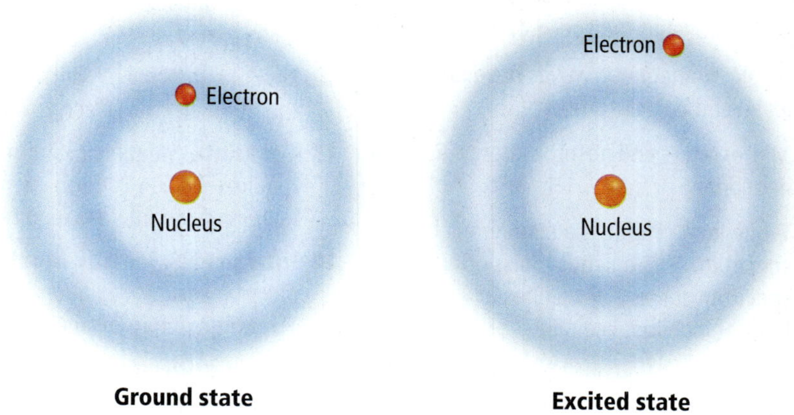


**Figure 10** The figure shows an atom that has one electron. Note that the illustration is not to scale. In its ground state, the electron is associated with the lowest energy level. When the atom is in an excited state, the electron is associated with a higher energy level.

**Table 1** Bohr's Description of the Hydrogen Atom

| Bohr's Atomic Orbit | Quantum Number | Orbit Radius (nm) | Corresponding Atomic Energy Level | Relative Energy |
|---|---|---|---|---|
| First | $n = 1$ | 0.0529 | 1 | $E_1$ |
| Second | $n = 2$ | 0.212 | 2 | $E_2 = 4E_1$ |
| Third | $n = 3$ | 0.476 | 3 | $E_3 = 9E_1$ |
| Fourth | $n = 4$ | 0.846 | 4 | $E_4 = 16E_1$ |
| Fifth | $n = 5$ | 1.32 | 5 | $E_5 = 25E_1$ |
| Sixth | $n = 6$ | 1.90 | 6 | $E_6 = 36E_1$ |
| Seventh | $n = 7$ | 2.59 | 7 | $E_7 = 49E_1$ |

In order to complete his calculations, Bohr assigned a number, $n$, called a **quantum number,** to each orbit. He also calculated the radius of each orbit. For the first orbit, the one closest to the nucleus, $n = 1$ and the orbit radius is 0.0529 nm; for the second orbit, $n = 2$ and the orbit radius is 0.212 nm; and so on. Additional information about Bohr's description of hydrogen's allowed orbits and energy levels is given in **Table 1.**

**The hydrogen line spectrum** Bohr suggested that the hydrogen atom is in the ground state, also called the first energy level, when its single electron is in the $n = 1$ orbit. In the ground state, the atom does not radiate energy. When energy is added from an outside source, the electron moves to a higher-energy orbit, such as the $n = 2$ orbit shown in **Figure 11.** Such an electron transition raises the atom to an excited state. When the atom is in an excited state, the electron can drop from the higher-energy orbit to a lower-energy orbit. As a result of this transition, the atom emits a photon corresponding to the energy difference between the two levels.

$$\Delta E = E_{\text{higher-energy orbit}} - E_{\text{lower-energy orbit}} = E_{\text{photon}} = h\nu$$

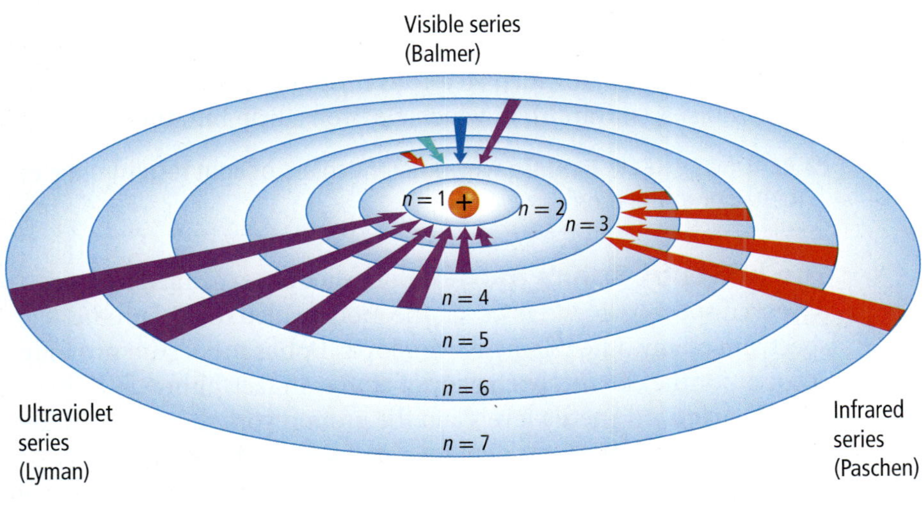


**Figure 11** When an electron drops from a higher-energy orbit to a lower-energy orbit, a photon is emitted. The ultraviolet (Lyman), visible (Balmer), and infrared (Paschen) series correspond to electrons dropping to $n = 1$, $n = 2$, and $n = 3$, respectively.

View an **animation about the Balmer Series.**

Concepts In Motion

■ **Figure 12** Only certain energy levels are allowed. The energy levels are similar to the rungs of a ladder. The four visible lines correspond to electrons dropping from a higher $n$ to the orbit $n = 2$. As $n$ increases, the hydrogen atom's energy levels are closer to each other.

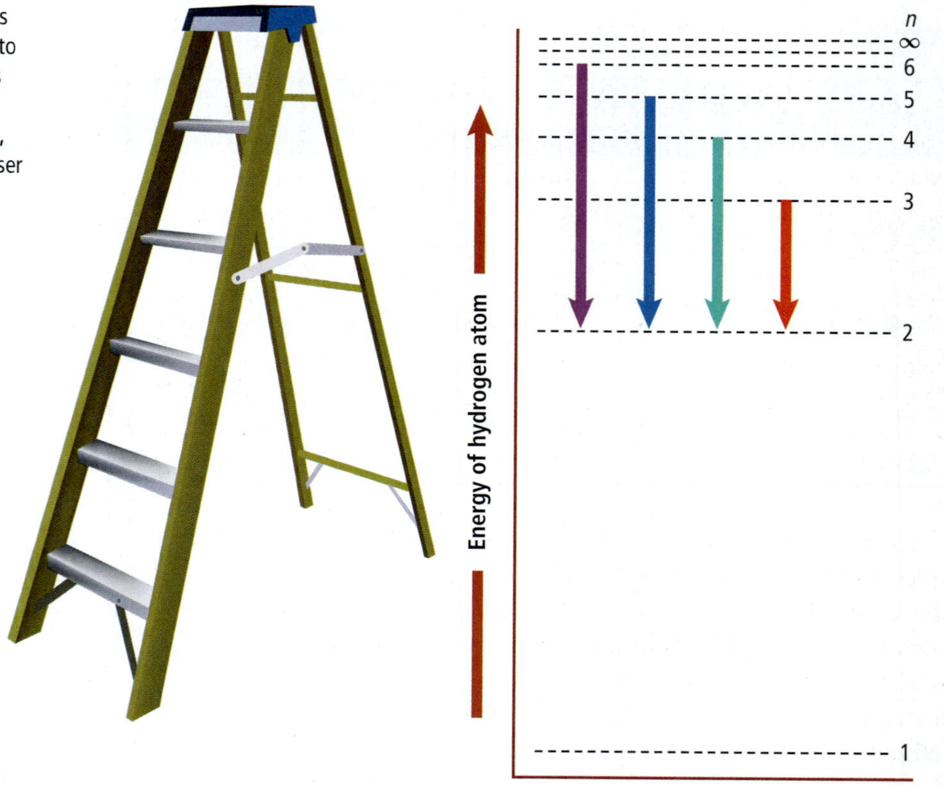


View an **animation about electron transitions.**

**Concepts In Motion**

Because only certain atomic energies are possible, only certain frequencies of electromagnetic radiation can be emitted. You might compare hydrogen's atomic energy states to rungs on a ladder. A person can climb up or down the ladder only from rung to rung. Similarly, the hydrogen atom's electron can move only from one allowable orbit to another, and therefore, can emit or absorb only certain amounts of energy, corresponding to the energy difference between the two orbits.

**Figure 12** shows that, unlike rungs on a ladder, however, the hydrogen atom's energy levels are not evenly spaced. **Figure 12** also illustrates the four electron transitions that account for visible lines in hydrogen's atomic emission spectrum, shown in **Figure 8.** Electron transitions from higher-energy orbits to the second orbit account for all of hydrogen's visible lines, which form the Balmer series. Other electron transitions have been measured that are not visible, such as the Lyman series (ultraviolet), in which electrons drop into the $n = 1$ orbit, and the Paschen series (infrared), in which electrons drop into the $n = 3$ orbit.

☑ READING CHECK **Explain** why different colors of light result from electron behavior in the atom.

**The limits of Bohr's model** Bohr's model explained hydrogen's observed spectral lines. However, the model failed to explain the spectrum of any other element. Moreover, Bohr's model did not fully account for the chemical behavior of atoms. In fact, although Bohr's idea of quantized energy levels laid the groundwork for atomic models to come, later experiments demonstrated that the Bohr model was fundamentally incorrect. The movements of electrons in atoms are not completely understood even now; however, substantial evidence indicates that electrons do not move around the nucleus in circular orbits.

# The Quantum Mechanical Model of the Atom

Scientists in the mid-1920s, by then convinced that the Bohr atomic model was incorrect, formulated new and innovative explanations of how electrons are arranged in atoms. In 1924, a French graduate student in physics named Louis de Broglie (1892–1987) proposed an idea that eventually accounted for the fixed energy levels of Bohr's model.

**Electrons as waves** De Broglie had been thinking that Bohr's quantized electron orbits had characteristics similar to those of waves. For example, as **Figures 13a** and **13b** show, only multiples of half-wavelengths are possible on a plucked harp string because the string is fixed at both ends. Similarly, de Broglie saw that only odd numbers of wavelengths are allowed in a circular orbit of fixed radius, as shown in **Figure 13c.** He also reflected on the fact that light—at one time thought to be strictly a wave phenomenon—has both wave and particle characteristics. These thoughts led de Broglie to pose a new question: If waves can have particlelike behavior, could the opposite also be true? That is, can particles of matter, including electrons, behave like waves?

■ **Figure 13** **a.** The string on the harp vibrates between two fixed endpoints. **b.** The vibrations of a string between the two fixed endpoints labeled *A* and *B* are limited to multiples of half-wavelengths. **c.** Electrons on circular orbits can only have odd numbers of wavelengths.

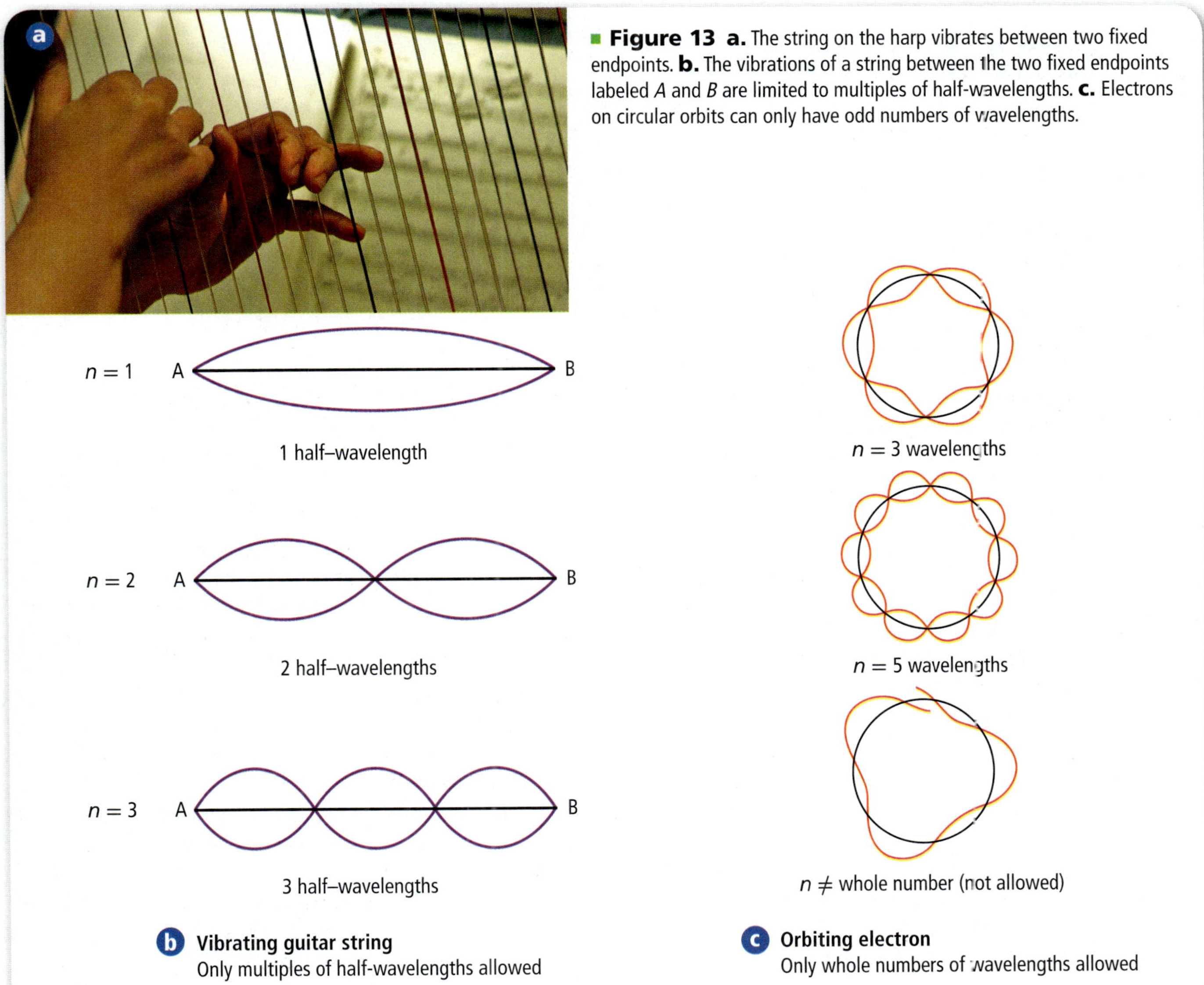

The **de Broglie equation** predicts that all moving particles have wave characteristics. It also explains why it is impossible to notice the wavelength of a fast-moving car. An automobile moving at 25 m/s and having a mass of 910 kg has a wavelength of $2.9 \times 10^{-38}$ m, a wavelength far too small to be seen or detected. By comparison, an electron moving at the same speed has the easily measured wavelength of $2.9 \times 10^{-5}$ m. Subsequent experiments have proven that electrons and other moving particles do indeed have wave characteristics. De Broglie knew that if an electron has wavelike motion and is restricted to circular orbits of fixed radius, only certain wavelengths, frequencies, and energies are possible. Developing his idea, de Broglie derived the following equation.

**Particle Electromagnetic–Wave Relationship**

$$\lambda = \frac{h}{m\nu}$$

$\lambda$ represents wavelength.
$h$ is Planck's constant.
$m$ represents mass of the particle.
$\nu$ represents velocity.

The wavelength of a particle is the ratio of Planck's constant and the product of the particle's mass and its velocity.

# Problem-Solving LAB

## Interpret Scientific Illustrations

**What electron transitions account for the Balmer series?** Hydrogen's emission spectrum comprises three series of lines. Some wavelengths are ultraviolet (Lyman series) and infrared (Paschen series). Visible wavelengths comprise the Balmer series. The Bohr atomic model attributes these spectral lines to transitions from higher-energy states with electron orbits in which $n = n_i$ to lower-energy states with smaller electron orbits in which $n = n_f$.

### Analysis

The image at right illustrates some of the transitions in hydrogen's Balmer series. These Balmer lines are designated $H_\alpha$ (6562 Å), $H_\beta$ (4861 Å), $H_\gamma$ (4340 Å), and $H_\delta$ (4101 Å). Each wavelength ($\lambda$) is related to an electron transition within a hydrogen atom by the following equation, in which $1.09678 \times 10^7$ m$^{-1}$ is known as the Rydberg constant.

$$\frac{1}{\lambda} = 1.09678 \times 10^7\left(\frac{1}{n_f^2} - \frac{1}{n_i^2}\right)\text{m}^{-1}$$

For hydrogen's Balmer series, electron orbit transitions occur from larger orbits to the $n = 2$ orbit; that is, $n_f = 2$.

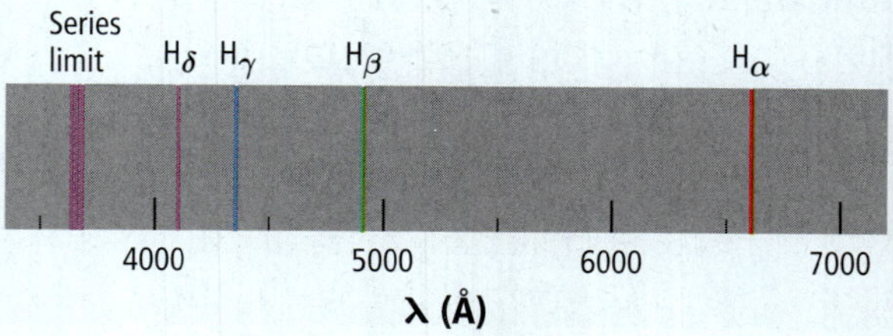


### Think Critically

1. **Calculate** the wavelengths for the following electron orbit transitions.
   a. $n_i = 3$; $n_f = 2$
   b. $n_i = 4$; $n_f = 2$
   c. $n_i = 5$; $n_f = 2$
   d. $n_i = 6$; $n_f = 2$
2. **Relate** the Balmer-series wavelengths you calculated in Question 1 to those determined experimentally. Allowing for experimental error and calculation uncertainty, do the wavelengths match? Explain your answer. One angstrom (Å) equals $10^{-10}$ m.
3. **Apply** the formula $E = hc/\lambda$ to determine the energy per quantum for each of the orbit transitions in Question 1.
4. **Extend** the Bohr model by calculating the wavelength and energy per quantum for the electron orbit transition for which $n_f = 3$ and $n_i = 5$. This transition accounts for a spectral line in hydrogen's Paschen series.

**The Heisenberg uncertainty principle** Step by step, scientists such as Rutherford, Bohr, and de Broglie had been unraveling the mysteries of the atom. However, a conclusion reached by the German theoretical physicist Werner Heisenberg (1901–1976) proved to have profound implications for atomic models.

Heisenberg showed that it is impossible to take any measurement of an object without disturbing the object. Imagine trying to locate a hovering, helium-filled balloon in a darkened room. If you wave your hand about, you can locate the balloon's position when you touch it. However, when you touch the balloon, you transfer energy to it and change its position. You could also detect the balloon's position by turning on a flashlight. Using this method, photons of light reflected from the balloon would reach your eyes and reveal the balloon's location. Because the balloon is a macroscopic object, the effect of the rebounding photons on its position is very small and not observable.

Imagine trying to determine an electron's location by "bumping" it with a high-energy photon. Because such a photon has about the same energy as an electron, the interaction between the two particles changes both the wavelength of the photon and the position and velocity of the electron, as shown in **Figure 14.** In other words, the act of observing the electron produces a significant, unavoidable uncertainty in the position and motion of the electron. Heisenberg's analysis of interactions, such as those between photons and electrons, led him to his historic conclusion. The **Heisenberg uncertainty principle** states that it is fundamentally impossible to know precisely both the velocity and position of a particle at the same time.

**READING CHECK** **Explain** the Heisenberg uncertainty principle.

Although scientists of the time found Heisenberg's principle difficult to accept, it has been proven to describe the fundamental limitations of what can be observed. The interaction of a photon with a macroscopic object such as a helium-filled balloon has so little effect on the balloon that the uncertainty in its position is too small to measure. But that is not the case with an electron moving at $6 \times 10^6$ m/s near an atomic nucleus. The uncertainty of the electron's position is at least $10^{-9}$ m, about 10 times greater than the diameter of the entire atom.

The Heisenberg uncertainty principle also means that it is impossible to assign fixed paths for electrons like the circular orbits in Bohr's model. The only quantity that can be known is the probability for an electron to occupy a certain region around the nucleus.

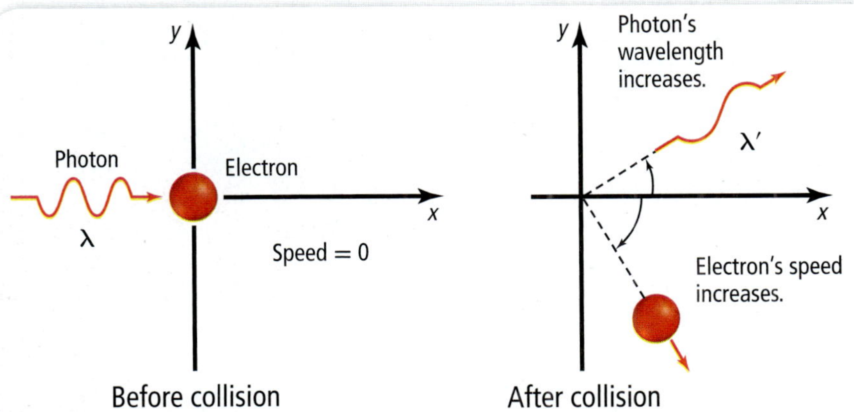


**Figure 14** When a photon interacts with an electron at rest, both the velocity and the position of the electron are modified. Th s illustrates the Heisenberg uncertainty principle. It is impossible to krow at the same time the position and the velocity of a particle.
**Explain** ***Why has the photon's energy changed?***

**The Schrödinger wave equation** In 1926, Austrian physicist Erwin Schrödinger (1887–1961) furthered the wave-particle theory proposed by de Broglie. Schrödinger derived an equation that treated the hydrogen atom's electron as a wave. Schrödinger's new model for the hydrogen atom seemed to apply equally well to atoms of other elements—an area in which Bohr's model failed. The atomic model in which electrons are treated as waves is called the wave mechanical model of the atom or, the **quantum mechanical model of the atom.** Like Bohr's model, the quantum mechanical model limits an electron's energy to certain values. However, unlike Bohr's model, the quantum mechanical model makes no attempt to describe the electron's path around the nucleus.

☑ READING CHECK **Compare and contrast** Bohr's model and the quantum mechanical model.

The Schrödinger wave equation is too complex to be considered here. However, each solution to the equation is known as a wave function, which is related to the probability of finding the electron within a particular volume of space around the nucleus. Recall from your study of mathematics that an event with a high probability is more likely to occur than one with a low probability.

**Electron's probable location** The wave function predicts a three-dimensional region around the nucleus, called an **atomic orbital,** which describes the electron's probable location. An atomic orbital is like a fuzzy cloud in which the density at a given point is proportional to the probability of finding the electron at that point. **Figure 15a** illustrates the probability map that describes the electron in the atom's lowest energy state. The probability map can be thought of as a time-exposure photograph of the electron moving around the nucleus, in which each dot represents the electron's location at an instant in time. The high density of dots near the nucleus indicates the electron's most probable location. However, because the cloud has no definite boundary, it is also possible that the electron might be found at a considerable distance from the nucleus.

☑ READING CHECK **Describe** where electrons are located in an atom.

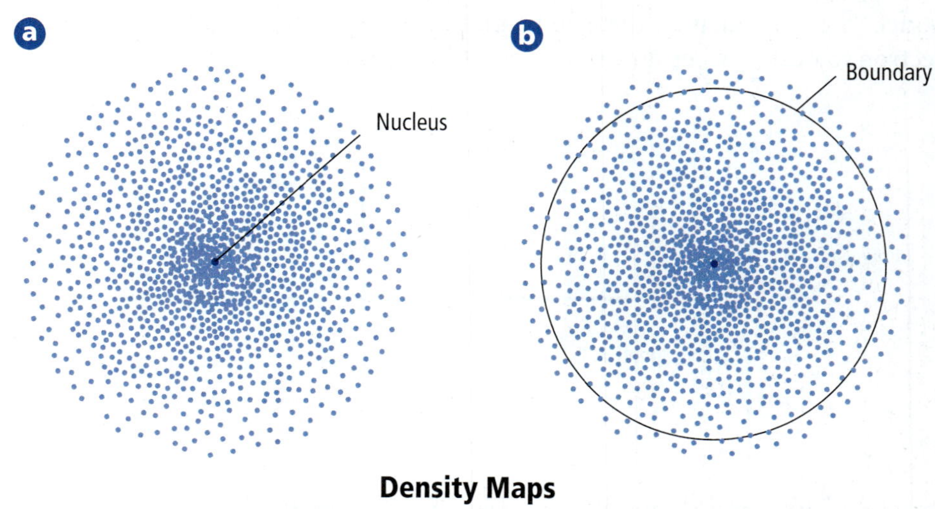


■ **Figure 15** The density map represents the probability of finding an electron at a given position around the nucleus. **a.** The higher density of points near the nucleus shows that the electron is more likely to be found close to the nucleus. **b.** At any given time, there is a 90% probability of finding the electron within the circular region shown. This surface is sometimes chosen to represent the boundary of the atom. In this illustration, the circle corresponds to a projection of the 3-dimensional sphere that contains the electrons.

## Hydrogen's Atomic Orbitals

Because the boundary of an atomic orbital is fuzzy, the orbital does not have an exact defined size. To overcome the inherent uncertainty about the electron's location, chemists arbitrarily draw an orbital's surface to contain 90% of the electron's total probability distribution. This means that the probability of finding the electron within the boundary is 0.9 and the probability of finding it outside the boundary is 0.1. In other words, it is more likely to find the electron close to the nucleus and within the volume defined by the boundary, than to find it outside the volume. The circle shown in **Figure 15b** encloses 90% of the lowest-energy orbital of hydrogen.

**Principal quantum number** Recall that the Bohr atomic model assigns quantum numbers to electron orbits. Similarly, the quantum mechanical model assigns four quantum numbers to atomic orbitals. The first one is the **principal quantum number** ($n$) and indicates the relative size and energy of atomic orbitals. As $n$ increases, the orbital becomes larger, the electron spends more time farther from the nucleus, and the atom's energy increases. Therefore, $n$ specifies the atom's major energy levels. Each major energy level is called a **principal energy level.** An atom's lowest principal energy level is assigned a principal quantum number of 1. When the hydrogen atom's single electron occupies an orbital with $n = 1$, the atom is in its ground state. Up to 7 energy levels have been detected for the hydrogen atom, giving $n$ values ranging from 1 to 7.

Watch a **video about fireworks.**

Video

**Energy sublevels** Principal energy levels contain **energy sublevels.** Principal energy level 1 consists of a single sublevel, principal energy level 2 consists of two sublevels, principal energy level 3 consists of three sublevels, and so on. To better understand the relationship between the atom's energy levels and sublevels, picture the seats in a wedge-shaped section of a theater, as shown in **Figure 16.** As you move away from the stage, the rows become higher and contain more seats. Similarly, the number of energy sublevels in a principal energy level increases as $n$ increases.

READING CHECK **Explain** the relationship between energy levels and sublevels.

**Figure 16** Energy levels can be thought of as rows of seats in a theater. The rows that are higher up and farther from the stage contain more seats. Similarly, energy levels related to orbitals farther from the nucleus contain more sublevels.

**Shapes of orbitals** Sublevels are labeled *s, p, d,* or *f* according to the shapes of the atom's orbitals. All s orbitals are spherical, and all p orbitals are dumbbell-shaped; however, not all d or f orbitals have the same shape. Each orbital can contain, at most, two electrons. The single sublevel in principal energy level 1 corresponds to a spherical orbital called the 1s orbital. The two sublevels in principal energy level 2 are designated 2s and 2p. The 2s sublevel corresponds to the 2s orbital, which is spherical like the 1s orbital but larger in size, as shown in **Figure 17a.** The 2p sublevel corresponds to three dumbbell-shaped p orbitals designated $2p_x$, $2p_y$, and $2p_z$. The subscripts *x, y,* and *z* merely designate the orientations of p orbitals along the *x, y,* and *z* coordinate axes, as shown in **Figure 17b.** Each of the p orbitals related to an energy sublevel has the same energy.

**READING CHECK** **Describe** the shapes of s and p orbitals.

Principal energy level 3 consists of three sublevels designated 3s, 3p, and 3d. Each d sublevel relates to five orbitals of equal energy. Four of the d orbitals have identical shapes but different orientations along the *x, y,* and *z* coordinate axes. However, the fifth orbital, $d_{z^2}$, is shaped and oriented differently than the other four. The shapes and orientations of the five d orbitals are illustrated in **Figure 17c.** The fourth principal energy level ($n = 4$) contains a fourth sublevel, called the 4f sublevel, which relates to seven f orbitals of equal energy. The f orbitals have complex, multilobed shapes.

**Figure 17** The shapes of atomic orbitals describe the probable distribution of electrons in energy sublevels.

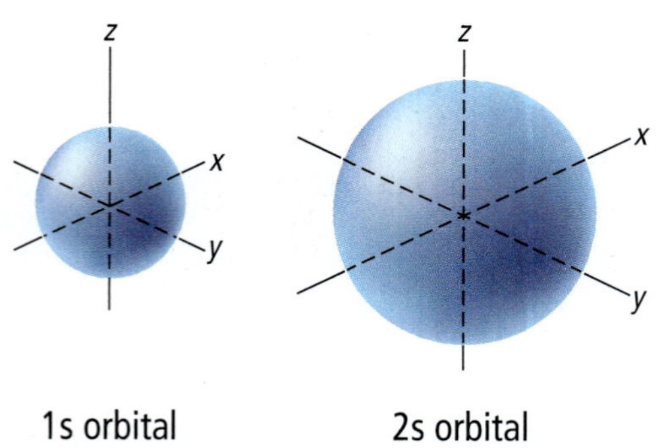


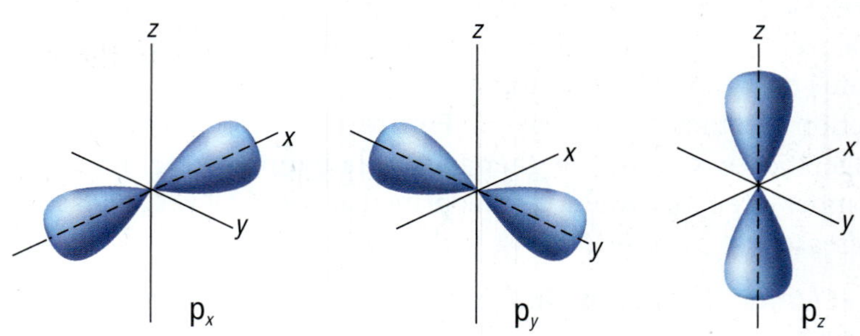


**a.** All s orbitals are spherical, and their size increases with increasing principal quantum number.

**b.** The three p orbitals are dumbbell-shaped and are oriented along the three perpendicular *x, y,* and *z* axes.

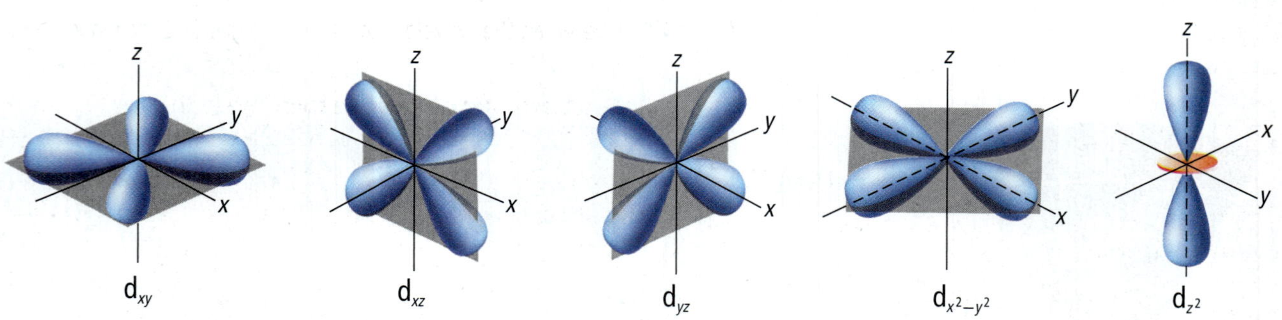


**c.** Four of the five d orbitals have the same shape but lie in different planes. The $d_{z^2}$ orbital has its own unique shape.

**Table 2** **Hydrogen's First Four Principal Energy Levels**

| Principal Quantum Number ($n$) | Sublevels (Types of Orbitals) Present | Number of Orbitals Related to Sublevel | Total Number of Orbitals Related to Principal Energy Level ($n^2$) |
|---|---|---|---|
| 1 | s | 1 | 1 |
| 2 | s<br>p | 1<br>3 | 4 |
| 3 | s<br>p<br>d | 1<br>3<br>5 | 9 |
| 4 | s<br>p<br>d<br>f | 1<br>3<br>5<br>7 | 16 |

Hydrogen's first four principal energy levels, sublevels, and related atomic orbitals are summarized in **Table 2.** Note that the number of orbitals related to each sublevel is always an odd number, and that the maximum number of orbitals related to each principal energy level equals $n^2$.

At any given time, the electron in a hydrogen atom can occupy just one orbital. You can think of the other orbitals as unoccupied spaces —spaces available should the atom's energy increase or decrease. For example, when the hydrogen atom is in the ground state, the electron occupies the 1s orbital. However, when the atom gains a quantum of energy, the electron is excited to one of the unoccupied orbitals. Depending on the amount of energy available, the electron can move to the 2s orbital, to one of the three 2p orbitals, or to any other orbital that is vacant.

## SECTION 2 REVIEW

Section Self-Check

### Section Summary

- Bohr's atomic model attributes hydrogen's emission spectrum to electrons dropping from higher-energy to lower-energy orbits.
- The de Broglie equation relates a particle's wavelength to its mass, its velocity, and Planck's constant.
- The quantum mechanical model assumes that electrons have wave properties.
- Electrons occupy three-dimensional regions of space called atomic orbitals.

**15. MAINIDEA Explain** the reason, according to Bohr's atomic model, why atomic emission spectra contain only certain frequencies of light.

**16. Differentiate** between the wavelength of visible light and the wavelength of a moving soccer ball.

**17. Enumerate** the sublevels contained in the hydrogen atom's first four energy levels. What orbitals are related to each s sublevel and each p sublevel?

**18. Explain** why the location of an electron in an atom is uncertain using the Heisenberg uncertainty principle and de Broglie's wave-particle duality. How is the location of electrons in atoms defined?

**19. Calculate** Use the information in **Table 1** to calculate how many times larger the hydrogen atom's seventh Bohr radius is than its first Bohr radius.

**20. Compare and contrast** Bohr's model and the quantum mechanical model of the atom.

SECTION 3

# Electron Configuration

**MAINIDEA** A set of three rules can be used to determine electron arrangement in an atom.

### Essential Questions

- How are the Pauli exclusion principle, the aufbau principle, and Hund's rule used to write electron configurations using orbital diagrams and electron configuration notation?
- What are valence electrons, and how do electron-dot structures represent an atom's valence electrons?

### Review Vocabulary

**electron:** a negatively charged, fast-moving particle with an extremely small mass that is found in all forms of matter and moves through the empty space surrounding an atom's nucleus

### New Vocabulary

electron configuration
aufbau principle
Pauli exclusion principle
Hund's rule
valence electron
electron-dot structure

**CHEM 4 YOU** As students board a bus, they each sit in a separate bench seat until they are all full. Then, they begin sharing seats. Electrons fill atomic orbitals in a similar way.

## Ground-State Electron Configuration

When you consider that atoms of the heaviest elements contain more than 100 electrons, the idea of determining electron arrangements in atoms with many electrons seems daunting. Fortunately, all atoms can be described with orbitals similar to hydrogen's. This allows us to describe arrangements of electrons in atoms using a few specific rules.

The arrangement of electrons in an atom is called the atom's **electron configuration.** Because low-energy systems are more stable than high-energy systems, electrons in an atom tend to assume the arrangement that gives the atom the lowest energy possible. The most stable, lowest-energy arrangement of the electrons is called the element's ground-state electron configuration. Three rules, or principles—the aufbau principle, the Pauli exclusion principle, and Hund's rule—define how electrons can be arranged in an atom's orbitals.

**The aufbau principle** The **aufbau principle** states that each electron occupies the lowest energy orbital available. Therefore, your first step in determining an element's ground-state electron configuration is learning the sequence of atomic orbitals from lowest energy to highest energy. This sequence, known as an aufbau diagram, is shown in **Figure 18.** In the diagram, each box represents an atomic orbital.

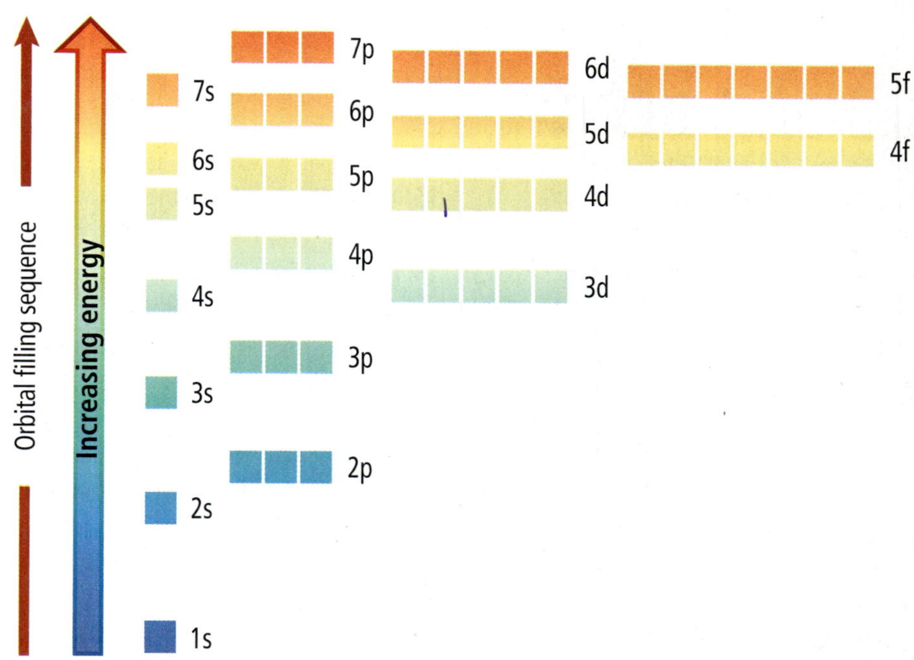


**Figure 18** The aufbau diagram shows the energy of each sublevel relative to the energy of other sublevels. Each box on the diagram represents an atomic orbital.

**Determine** *Which sublevel has the greater energy, 4d or 5p?*

**Table 3** Features of the Aufbau Diagram

| Feature | Example |
|---|---|
| All orbitals related to an energy sublevel are of equal energy. | All three 2p orbitals are of equal energy. |
| In a multi-electron atom, the energy sublevels within a principal energy level have different energies. | The three 2p orbitals are of higher energy than the 2s orbital. |
| In order of increasing energy, the sequence of energy sublevels within a principal energy level is s, p, d, and f. | If $n = 4$, then the sequence of energy sublevels is 4s, 4p, 4d, and 4f. |
| Orbitals related to energy sublevels within one principal energy level can overlap orbitals related to energy sublevels within another principal level. | The orbital related to the atom's 4s sublevel has a lower energy than the five orbitals related to the 3d sublevel. |

**Table 3** summarizes several features of the aufbau diagram. Although the aufbau principle describes the sequence in which orbitals are filled with electrons, it is important to know that atoms are not built up electron by electron.

**The Pauli exclusion principle** Electrons in orbitals can be represented by arrows in boxes. Each electron has an associated spin, similar to the way a top spins on its point. Like the top, the electron is able to spin in only one of two directions. An arrow pointing up [↑] represents the electron spinning in one direction, and an arrow pointing down [↓] represents the electron spinning in the opposite direction. An empty box [ ] represents an unoccupied orbital, a box containing a single up arrow [↑] represents an orbital with one electron, and a box containing both up and down arrows [↑↓] represents a filled orbital.

The **Pauli exclusion principle** states that a maximum of two electrons can occupy a single atomic orbital, but only if the electrons have opposite spins. Austrian physicist Wolfgang Pauli (1900–1958) proposed this principle after observing atoms in excited states. An atomic orbital containing paired electrons with opposite spins is written as [↑↓]. Because each orbital can contain, at most, two electrons, the maximum number of electrons related to each principal energy level equals $2n^2$.

**Hund's rule** The fact that negatively charged electrons repel each other has an important impact on the distribution of electrons in equal-energy orbitals. **Hund's rule** states that single electrons with the same spin must occupy each equal-energy orbital before additional electrons with opposite spins can occupy the same orbitals. For example, let the boxes below represent the 2p orbitals. One electron enters each of the three 2p orbitals before a second electron enters any of the orbitals. The sequence in which six electrons occupy three p orbitals is shown below.

1. [↑][ ][ ]
2. [↑][↑][ ]
3. [↑][↑][↑]
4. [↑↓][↑][↑]
5. [↑↓][↑↓][↑]
6. [↑↓][↑↓][↑↓]

**READING CHECK** **State** the three rules that define how electrons are arranged in atoms.

**VOCABULARY**

**WORD ORIGIN**

**Aufbau**
comes from the German word *aufbauen,* which means *to build up* or *arrange.*

**FOLDABLES®**
Incorporate information from this section into your Foldable.

## Electron Arrangement

You can represent an atom's electron configuration using one of two convenient methods: orbital diagrams or electron configuration notation.

**Orbital diagrams** As mentioned earlier, electrons in orbitals can be represented by arrows in boxes. Each box is labeled with the principal quantum number and sublevel associated with the orbital. For example, the orbital diagram for a ground-state carbon atom, shown below, contains two electrons in the 1s orbital, two electrons in the 2s orbital, and one electron in two of three separate 2p orbitals. The third 2p orbital remains unoccupied.

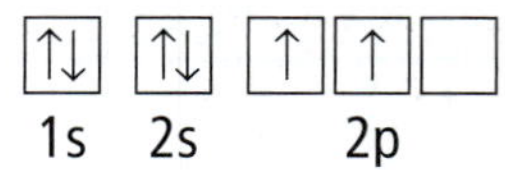


**Electron configuration notation** The electron configuration notation designates the principal energy level and energy sublevel associated with each of the atom's orbitals and includes a superscript representing the number of electrons in the orbital. For example, the electron configuration notation of a ground-state carbon atom is written $1s^22s^22p^2$. Orbital diagrams and electron configuration notations for the elements in periods one and two of the periodic table are shown in **Table 4. Figure 19** illustrates how the 1s, 2s, $2p_x$, $2p_y$, and $2p_z$ orbitals illustrated previously in **Figure 17** overlap in the neon atom.

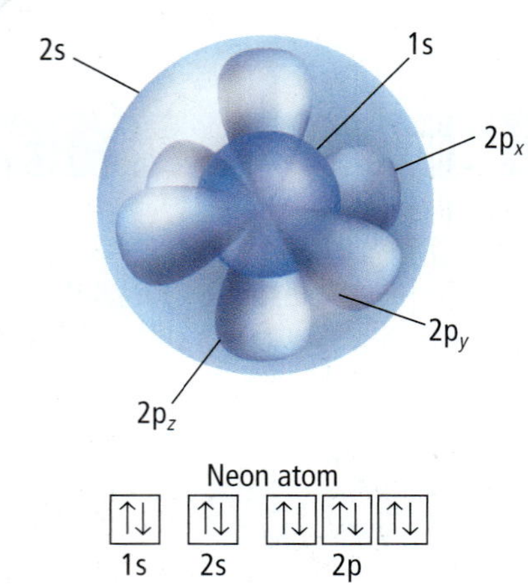


**Figure 19** The 1s, 2s, and 2p orbitals of a neon atom overlap.
***Determine how many electrons a neon atom has.***

Explore **electron configurations and orbital diagrams with an interactive table.** Concepts In Motion

**Table 4** Electron Configurations and Orbital Diagrams for Elements 1–10

| Element | Atomic Number | Orbital Diagram 1s | 2s | $2p_x$ | $2p_y$ | $2p_z$ | Electron Configuration Notation |
|---|---|---|---|---|---|---|---|
| Hydrogen | 1 | ↑ | | | | | $1s^1$ |
| Helium | 2 | ↑↓ | | | | | $1s^2$ |
| Lithium | 3 | ↑↓ | ↑ | | | | $1s^2 2s^1$ |
| Beryllium | 4 | ↑↓ | ↑↓ | | | | $1s^2 2s^2$ |
| Boron | 5 | ↑↓ | ↑↓ | ↑ | | | $1s^2 2s^2 2p^1$ |
| Carbon | 6 | ↑↓ | ↑↓ | ↑ | ↑ | | $1s^2 2s^2 2p^2$ |
| Nitrogen | 7 | ↑↓ | ↑↓ | ↑ | ↑ | ↑ | $1s^2 2s^2 2p^3$ |
| Oxygen | 8 | ↑↓ | ↑↓ | ↑↓ | ↑ | ↑ | $1s^2 2s^2 2p^4$ |
| Fluorine | 9 | ↑↓ | ↑↓ | ↑↓ | ↑↓ | ↑ | $1s^2 2s^2 2p^5$ |
| Neon | 10 | ↑↓ | ↑↓ | ↑↓ | ↑↓ | ↑↓ | $1s^2 2s^2 2p^6$ |

Get help with **electron configuration notation.**

Personal Tutor

Note that the electron configuration notation does not usually show the orbital distributions of electrons related to a sublevel. It is understood that a designation such as nitrogen's $2p^3$ represents the orbital occupancy $2p_x^1 2p_y^1 2p_z^1$.

For sodium, the first ten electrons occupy 1s, 2s, and 2p orbitals. Then, according to the aufbau sequence, the eleventh electron occupies the 3s orbital. The electron configuration notation and orbital diagram for sodium are written as follows.

$$1s^2 2s^2 2p^6 3s^1$$

| ↑↓ | ↑↓ | ↑↓ ↑↓ ↑↓ | ↑ |
|---|---|---|---|
| 1s | 2s | 2p | 3s |

**Noble-gas notation** is a method of representing electron configurations of noble gases. Noble gases are the elements in the last column of the periodic table. They have eight electrons in their outermost orbital and they are unusually stable. You will learn more about noble gases in a later chapter. The noble-gas notation uses bracketed symbols. For example, [He] represents the electron configuration for helium, $1s^2$, and [Ne] represents the electron configuration for neon, $1s^2 2s^2 2p^6$. Compare the electron configuration for neon with sodium's configuration above. Note that the inner-level configuration for sodium is identical to the electron configuration for neon. Using noble-gas notation, sodium's electron configuration can be shortened to the form $[Ne]3s^1$. The electron configuration for an element can be represented using the noble-gas notation for the noble gas in the previous period and the electron configuration for the additional orbitals being filled. The complete and abbreviated (using noble-gas notation) electron configurations of the period 3 elements are shown in **Table 5.**

**VOCABULARY**

**SCIENCE USAGE V. COMMON USAGE**

**Period**

***Science usage:*** a horizontal row of elements in the current periodic table

*There are seven periods in the current periodic table.*

***Common usage:*** an interval of time determined by some recurring phenomenon

*The period of Earth's orbit is one year.*

☑ READING CHECK **Explain** how to write the noble-gas notation for an element. What is the noble-gas notation for calcium?

Explore **electron configurations with an interactive table.** Concepts In Motion

**Table 5 Electron Configurations for Elements 11–18**

| Element | Atomic Number | Complete Electron Configuration | Electron Configuration Using Noble Gas |
|---|---|---|---|
| Sodium | 11 | $1s^2 2s^2 2p^6 3s^1$ | $[Ne]3s^1$ |
| Magnesium | 12 | $1s^2 2s^2 2p^6 3s^2$ | $[Ne]3s^2$ |
| Aluminum | 13 | $1s^2 2s^2 2p^6 3s^2 3p^1$ | $[Ne]3s^2 3p^1$ |
| Silicon | 14 | $1s^2 2s^2 2p^6 3s^2 3p^2$ | $[Ne]3s^2 3p^2$ |
| Phosphorus | 15 | $1s^2 2s^2 2p^6 3s^2 3p^3$ | $[Ne]3s^2 3p^3$ |
| Sulfur | 16 | $1s^2 2s^2 2p^6 3s^2 3p^4$ | $[Ne]3s^2 3p^4$ |
| Chlorine | 17 | $1s^2 2s^2 2p^6 3s^2 3p^5$ | $[Ne]3s^2 3p^5$ |
| Argon | 18 | $1s^2 2s^2 2p^6 3s^2 3p^6$ | $[Ne]3s^2 3p^6$ or [Ar] |

Explore **electron configurations.**

Virtual Investigations

**Exceptions to predicted configurations** You can use the aufbau diagram to write correct ground-state electron configurations for all elements up to and including vanadium, atomic number 23. However, if you were to proceed in this manner, your configurations for chromium, $[Ar]4s^2 3d^4$, and copper, $[Ar]4s^2 3d^9$, would be incorrect. The correct configurations for these two elements are $[Ar]4s^1 3d^5$ for chromium and $[Ar]4s^1 3d^{10}$ for copper. The electron configurations for these two elements, as well as those of several other elements, illustrate the increased stability of half-filled and filled sets of s and d orbitals.

## PROBLEM-SOLVING STRATEGY

### Filling Atomic Orbitals

By drawing a sublevel diagram and following the arrows, you can write the ground-state electron configuration for any chemical element.

1. Sketch the sublevel diagram on a blank piece of paper.
2. Determine the number of electrons in one atom of the element for which you are writing the electron configuration. The number of electrons in a neutral atom equals the element's atomic number.
3. Starting with 1s, write the aufbau sequence of atomic orbitals by following the diagonal arrows from the top of the sublevel diagram to the bottom. When you complete one line of arrows, move to the right, to the beginning of the next line of arrows. As you proceed, add superscripts indicating the numbers of electrons in each set of atomic orbitals. Continue only until you have sufficient atomic orbitals to accommodate the total number of electrons in one atom of the element.
4. Apply noble-gas notation.

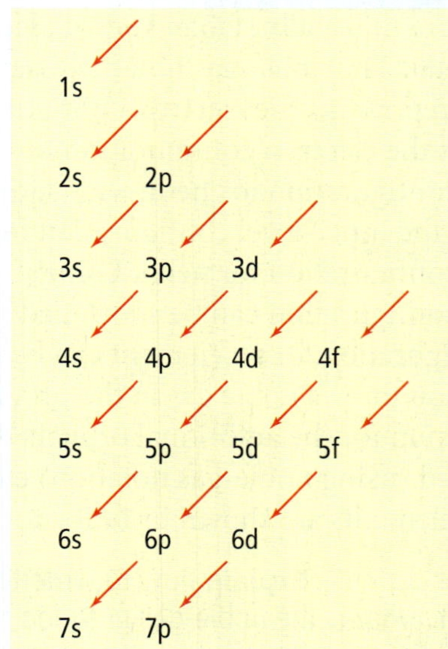

The sublevel diagram shows the order in which the orbitals are usually filled.

### Apply the Strategy

**Write** the ground-state electron configuration for zirconium.

## PRACTICE Problems

Do additional problems. Online Practice

**21.** Write ground-state electron configurations for the following elements.

**a.** bromine (Br) **c.** antimony (Sb) **e.** terbium (Tb)

**b.** strontium (Sr) **d.** rhenium (Re) **f.** titanium (Ti)

**22.** A chlorine atom in its ground state has a total of seven electrons in orbitals related to the atom's third energy level. How many of the seven electrons occupy p orbitals? How many of the 17 electrons in a chlorine atom occupy p orbitals?

**23.** When a sulfur atom reacts with other atoms, electrons in orbitals related to the atom's third energy level are involved. How many such electrons does a sulfur atom have?

**24.** An element has the ground-state electron configuration $[Kr]5s^2 4d^{10} 5p^1$. It is part of some semiconductors and used in various alloys. What element is it?

**25.** **Challenge** In its ground state, an atom of an element has two electrons in all orbitals related to the atom's highest energy level for which $n = 6$. Using noble-gas notation, write the electron configuration for this element, and identify the element.

## Valence Electrons

Only certain electrons, called valence electrons, determine the chemical properties of an element. **Valence electrons** are defined as electrons in the atom's outermost orbitals—generally those orbitals associated with the atom's highest principal energy level. For example, a sulfur atom contains 16 electrons, only six of which occupy the outermost 3s and 3p orbitals, as shown by sulfur's electron configuration. Sulfur has six valence electrons.

$$\text{S} \quad [\text{Ne}]3s^2 3p^4$$

Similarly, although a cesium atom contains 55 electrons, it has just one valence electron, the 6s electron shown in cesium's electron configuration.

$$\text{Cs} \quad [\text{Xe}]6s^1$$

**Electron-dot structures** Because valence electrons are involved in forming chemical bonds, chemists often represent them visually using a simple shorthand method, called electron-dot structure. An atom's **electron-dot structure** consists of the element's symbol, which represents the atomic nucleus and inner-level electrons, surrounded by dots representing all of the atom's valence electrons. American chemist G. N. Lewis (1875–1946) devised the method while teaching a college chemistry class in 1902.

In writing an atom's electron-dot structure, dots representing valence electrons are placed one at a time on the four sides of the symbol (they may be placed in any sequence) and then paired up until all are shown. The ground-state electron configurations and electron-dot structures for the elements in the second period are shown in **Table 6.**

Explore **electron-dot structures with an interactive table.** Concepts In Motion

**Table 6 Electron Configurations and Dot Structures**

| Element | Atomic Number | Electron Configuration | Electron-Dot Structure |
|---|---|---|---|
| Lithium | 3 | $1s^2 2s^1$ | Li· |
| Beryllium | 4 | $1s^2 2s^2$ | ·Be· |
| Boron | 5 | $1s^2 2s^2 2p^1$ | ·Ḃ· |
| Carbon | 6 | $1s^2 2s^2 2p^2$ | ·Ċ̣· |
| Nitrogen | 7 | $1s^2 2s^2 2p^3$ | ·Ṇ̈· |
| Oxygen | 8 | $1s^2 2s^2 2p^4$ | :Ọ̈· |
| Fluorine | 9 | $1s^2 2s^2 2p^5$ | :F̤̈· |
| Neon | 10 | $1s^2 2s^2 2p^6$ | :N̤̈e: |

## EXAMPLE Problem 3

EXAMPLE PROBLEM

**ELECTRON-DOT STRUCTURES** Some toothpastes contain stannous fluoride, a compound of tin and fluorine. What is tin's electron-dot structure?

### 1 ANALYZE THE PROBLEM

Consult the periodic table to determine the total number of electrons in a tin atom. Write out tin's electron configuration, and determine its number of valence electrons. Then use the rules for electron-dot structures to draw the electron-dot structure for tin.

### 2 SOLVE FOR THE UNKNOWN

Tin has an atomic number of 50. Thus, a tin atom has 50 electrons.

$$[Kr]5s^24d^{10}5p^2$$

**Write out tin's electron configuration using noble-gas notation. The closest noble gas is Kr.**

The two 5s and the two 5p electrons (the electrons in the orbitals related to the atom's highest principal energy level) represent tin's four valence electrons. Draw the four valence electrons around tin's chemical symbol (Sn) to show tin's electron-dot structure. 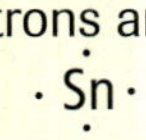

### 3 EVALUATE THE ANSWER

The correct symbol for tin (Sn) has been used, and the rules for drawing electron-dot structures have been correctly applied.

## PRACTICE Problems

PRACTICE PROBLEMS

**Do additional problems.** Online Practice

**26.** Draw electron-dot structures for atoms of the following elements.

**a.** magnesium **b.** thallium **c.** xenon

**27.** An atom of an element has a total of 13 electrons. What is the element, and how many electrons are shown in its electron-dot structure?

**28. Challenge** This element exists in the solid state at room temperature and at normal atmospheric pressure and is found in emerald gemstones. It is known to be one of the following elements: carbon, germanium, sulfur, cesium, beryllium, or argon. Identify the element based on the electron-dot structure at right.

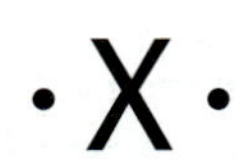

## SECTION 3 REVIEW

Section Self-Check

### Section Summary

- The arrangement of electrons in an atom is called the atom's electron configuration.
- Electron configurations are defined by the aufbau principle, the Pauli exclusion principle, and Hund's rule.
- An element's valence electrons determine the chemical properties of the element.
- Electron configurations can be represented using orbital diagrams, electron configuration notation, and electron-dot structures.

**29. MAINIDEA Apply** the Pauli exclusion principle, the aufbau principle, and Hund's rule to write out the electron configuration and draw the orbital diagram for each of the following elements.

**a.** silicon **b.** fluorine **c.** calcium **d.** krypton

**30. Define** *valence electron.*

**31. Illustrate** and describe the sequence in which ten electrons occupy the five orbitals related to an atom's d sublevel.

**32. Extend** the aufbau sequence through an element that has not yet been identified, but whose atoms would completely fill 7p orbitals. How many electrons would such an atom have? Write its electron configuration using noble-gas notation for the previous noble gas, radon.

**33. Interpret Scientific Illustrations** Which is the correct electron-dot structure for an atom of selenium? Explain.

**a.** 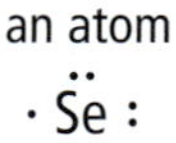 **b.** 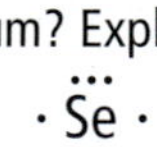 **c.** 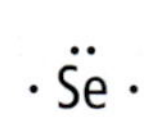 **d.** 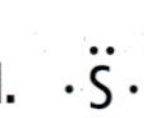

# CHEMISTRY&health

## Tiny Tweezers

Peering through a microscope, a cell biologist can grasp a single cell with a pair of "tweezers." But these are not the kind of tweezers you might find in a medicine cabinet. These tweezers are made from two laser beams and can hold very tiny things such as cells and even individual atoms.

You might have heard that lasers can be used to cut things. Laser "scissors" are used in some surgeries. But surprisingly, lasers can also trap living cells and other microscopic objects in their beams without damaging them. How can beams of light hold things in place?

**Gripping with light** When light rays pass through a cell, they change direction slightly. This is similar to how light rays bend when passing through water in an aquarium. When light rays are bent, they exert a force. Large objects, such as aquariums, are too massive to be affected by this miniscule force, but tiny cells respond to the force. If the light rays are positioned in just the right way, they can hold a small object in place, as shown in **Figure 1.**

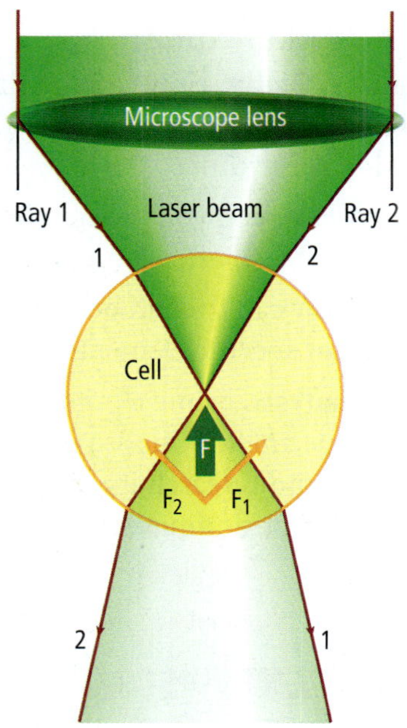


**Figure 1** The light rays in the laser beam bend as they pass through the cell membranes. The bending of the light rays creates forces on the cell. The combined forces hold the cell in place.

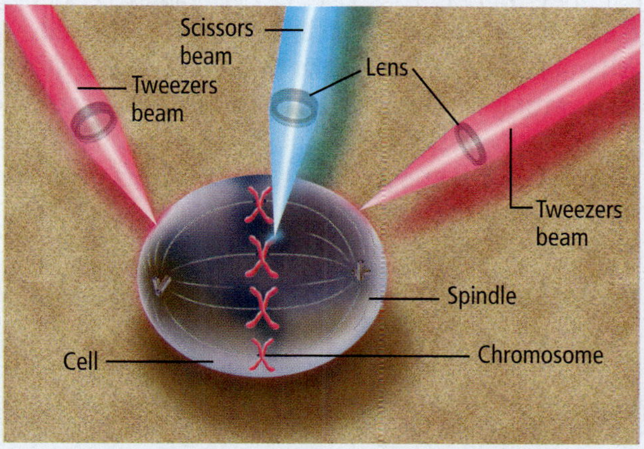


**Figure 2** Organelles found within living cells are accessible to the smallest lasers.

**Lasers and cancer** So what use do scientists have for these tiny tweezers? One group of scientists is using them to study cell organelles. They are studying the forces exerted by mitotic spindles—the grouping of microtubules that coordinates cell division. The spindles guide replicated chromosomes to opposite sides of the cell—a key role in cell division. However, scientists do not know exactly how the spindles perform this function.

Tiny laser scissors have been used to cut off pieces of chromosomes during cell division. Laser tweezers were then used to move the pieces around the cell and the spindles, as shown in **Figure 2.** Knowing the force with which the tweezers grasp the chromosomes, scientists can measure the opposing force exerted by the spindles. Scientists hope that learning how spindles function during cell division will help them learn more about diseases related to cell division, such as cancer—a disease in which cells divide uncontrollably.

### WRITING IN ▶ Chemistry

**Laser Light** Lasers can be found in a wide variety of everyday settings. Research the different types of lasers you might encounter daily, and find out what kind of light each laser uses. Summarize the results of your research in a presentation.

WebQuest

# ChemLAB

iLab Station

## Analyze Line Spectra

**Background:** **Emission spectra are produced when excited atoms return to a more stable state by emitting radiation of specific wavelengths. When white light passes through a sample, atoms in the sample absorb specific wavelengths. This produces dark lines in the continuous spectrum of white light and is called an absorption spectrum.**

**Question:** *What absorption and emission spectra do various substances produce?*

### Materials

ring stand with clamp
40-W tubular lightbulb
light socket with grounded power cord
275-mL polystyrene culture flask
Flinn C-Spectra® or similar diffraction grating
red, green, blue, and yellow food coloring
set of colored pencils
spectrum tubes (hydrogen, neon, and sodium)
spectrum-tube power supply (3)

### Safety Precautions

WARNING: ***Use care around the spectrum-tube power supplies. Spectrum tubes will become hot when used.***

### Procedure

1. Read and complete the lab safety form.
2. Use a Flinn C-Spectra® or similar diffraction grating to view an incandescent lightbulb. Draw the observed spectrum using colored pencils.
3. Use the Flinn C-Spectra® to view the emission spectra from tubes of gaseous hydrogen, neon, and sodium. Use colored pencils to draw the observed spectra.
4. Fill a 275-mL culture flask with about 100 mL of water. Add two or three drops of red food coloring to the water. Shake the solution.
5. Repeat Step 4 for the green, blue, and yellow food coloring.
6. Set up the 40-W lightbulb so that it is near eye level. Place the flask with red food coloring about 8 cm from the lightbulb so that you are able to see light from the bulb above the solution and light from the bulb projecting through the solution.

7. With the room lights darkened, view the light using the Flinn C-Spectra®. The top spectrum viewed will be a continuous spectrum from the white lightbulb. The bottom spectrum will be the absorption spectrum of the red solution. Use colored pencils to make a drawing of the absorption spectra you observe.
8. Repeat Steps 6 and 7 using the green, blue, and yellow solutions.
9. **Cleanup and Disposal** Turn off the light and spectrum-tube power supplies. Wait several minutes for the lightbulb and spectrum tubes to cool. Dispose of the liquids and store the lightbulb and spectrum tubes as directed by your teacher.

### Analyze and Conclude

1. **Think Critically** How can the single electron in a hydrogen atom produce all of the lines found in its emission spectrum?
2. **Predict** How can you predict the absorption spectrum of a solution by looking at its color?
3. **Apply** How can spectra be used to identify the presence of specific elements in a substance?
4. **Error Analysis** Name a potential source of error in this experiment. Choose one of the elements you observed, and research its absorption spectrum. Compare your findings with the results of your experiment.

### INQUIRY EXTENSION

**Hypothesize** What would happen if you mixed more than one color of food coloring with water and repeated the experiment? Design an experiment to test your hypothesis.

# CHAPTER 5 STUDY GUIDE

Vocabulary Practice

**BIGIDEA** **The atoms of each element have a unique arrangement of electrons.**

## SECTION 1 Light and Quantized Energy

**MAINIDEA** Light, a form of electromagnetic radiation, has characteristics of both a wave and a particle.

- All waves are defined by their wavelengths, frequencies, amplitudes, and speeds.

$$c = \lambda\nu$$

- In a vacuum, all electromagnetic waves travel at the speed of light.
- All electromagnetic waves have both wave and particle properties.
- Matter emits and absorbs energy in quanta.

$$E_{\text{quantum}} = h\nu$$

- White light produces a continuous spectrum. An element's emission spectrum consists of a series of lines of individual colors.

**VOCABULARY**
- electromagnetic radiation
- wavelength
- frequency
- amplitude
- electromagnetic spectrum
- quantum
- Planck's constant
- photoelectric effect
- photon
- atomic emission spectrum

## SECTION 2 Quantum Theory and the Atom

**MAINIDEA** Wavelike properties of electrons help relate atomic emission spectra, energy states of atoms, and atomic orbitals.

- Bohr's atomic model attributes hydrogen's emission spectrum to electrons dropping from higher-energy to lower-energy orbits.

$$\Delta E = E_{\text{higher-energy orbit}} - E_{\text{lower-energy orbit}} = E_{\text{photon}} = h\nu$$

- The de Broglie equation relates a particle's wavelength to its mass, its velocity, and Planck's constant.

$$\lambda = h \,/\, m\nu$$

- The quantum mechanical model assumes that electrons have wave properties.
- Electrons occupy three-dimensional regions of space called atomic orbitals.

**VOCABULARY**
- ground state
- quantum number
- de Broglie equation
- Heisenberg uncertainty principle
- quantum mechanical model of the atom
- atomic orbital
- principal quantum number
- principal energy level
- energy sublevel

## SECTION 3 Electron Configuration

**MAINIDEA** A set of three rules can be used to determine electron arrangement in an atom.

- The arrangement of electrons in an atom is called the atom's electron configuration.
- Electron configurations are defined by the aufbau principle, the Pauli exclusion principle, and Hund's rule.
- An element's valence electrons determine the chemical properties of the element.
- Electron configurations can be represented using orbital diagrams, electron configuration notation, and electron-dot structures.

**VOCABULARY**
- electron configuration
- aufbau principle
- Pauli exclusion principle
- Hund's rule
- valence electron
- electron-dot structure

# CHAPTER 5 ASSESSMENT

## SECTION 1

### Mastering Concepts

**34.** Define the following terms.
- **a.** frequency
- **b.** wavelength
- **c.** quantum
- **d.** ground state

**35.** Arrange the following types of electromagnetic radiation in order of increasing wavelength.
- **a.** ultraviolet light
- **b.** microwaves
- **c.** radio waves
- **d.** X-rays

**36.** A gamma ray has a frequency of $2.88 \times 10^{21}$ Hz. What does this mean?

**37.** What is the photoelectric effect?

**38. Neon Sign** How does light emitted from a neon sign differ from sunlight?

**39.** Explain Planck's quantum concept as it relates to energy lost or gained by matter.

**40.** How did Einstein explain the photoelectric effect?

**41. Rainbow** What are two differences between the red and green electromagnetic waves in a rainbow?

**42. Temperature** What happens to the light emitted by a heated, glowing object as its temperature increases?

**43.** What are three deficiencies of the wave model of light related to light's interaction with matter?

**44.** How are radio waves and ultraviolet waves similar? How are they different?

### Mastering Problems

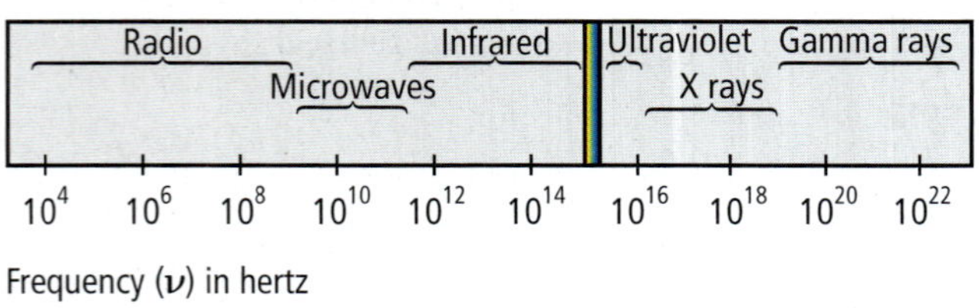


**Electromagnetic Spectrum**

**Figure 20**

**45. Radiation** Use **Figure 20** to determine the following types of radiation.
- **a.** radiation with a frequency of $8.6 \times 10^{11}\ s^{-1}$
- **b.** radiation with a wavelength of 4.2 nm
- **c.** radiation with a frequency of 5.6 MHz
- **d.** radiation that travels at a speed of $3.00 \times 10^8$ m/s

**46.** What is the wavelength of electromagnetic radiation with a frequency of $5.00 \times 10^{12}$ Hz? What kind of electromagnetic radiation is this?

**47.** What is the frequency of electromagnetic radiation with a wavelength of $3.33 \times 10^{-8}$ m? What type of electromagnetic radiation is this?

**48.** What is the speed of an electromagnetic wave with a frequency of $1.33 \times 10^{17}$ Hz and a wavelength of 2.25 nm?

**49.** What is the energy of a photon of red light that has a frequency of $4.48 \times 10^{14}$ Hz?

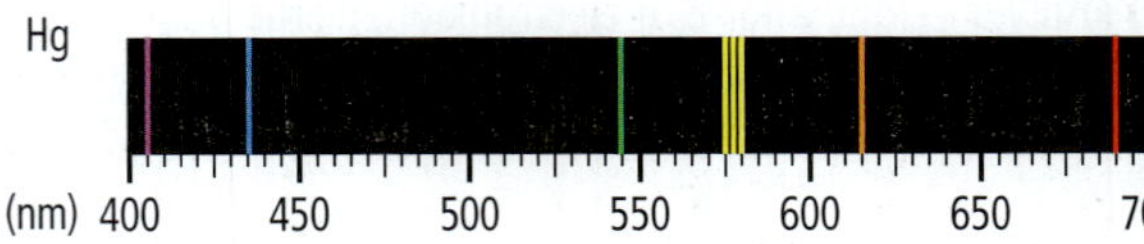


**Figure 21**

**50. Mercury** Mercury's atomic emission spectrum is shown in **Figure 21.** Estimate the wavelength of the orange line. What is its frequency? What is the energy of a photon corresponding to the orange line emitted by the mercury atom?

**51.** What is the energy of an ultraviolet photon that has a wavelength of $1.18 \times 10^{-8}$ m?

**52.** A photon has an energy of $2.93 \times 10^{-25}$ J. What is its frequency? What type of electromagnetic radiation is the photon?

**53.** A photon has an energy of $1.10 \times 10^{-13}$ J. What is the photon's wavelength? What type of electromagnetic radiation is it?

**54. Spacecraft** How long does it take a radio signal from the *Voyager* spacecraft to reach Earth if the distance between *Voyager* and Earth is $2.72 \times 10^9$ km?

**55. Radio Waves** If your favorite FM radio station broadcasts at a frequency of 104.5 MHz, what is the wavelength of the station's signal in meters? What is the energy of a photon of the station's electromagnetic signal?

**56. Platinum** What minimum frequency of light is needed to eject a photoelectron from atoms of platinum, which require at least $9.08 \times 10^{-19}$ J/photon?

**57. Eye Surgery** The argon fluoride (ArF) laser used in some refractive eye surgeries emits electromagnetic radiation of 193.3 nm wavelength. What is the frequency of the ArF laser's radiation? What is the energy of a single quantum of the radiation?

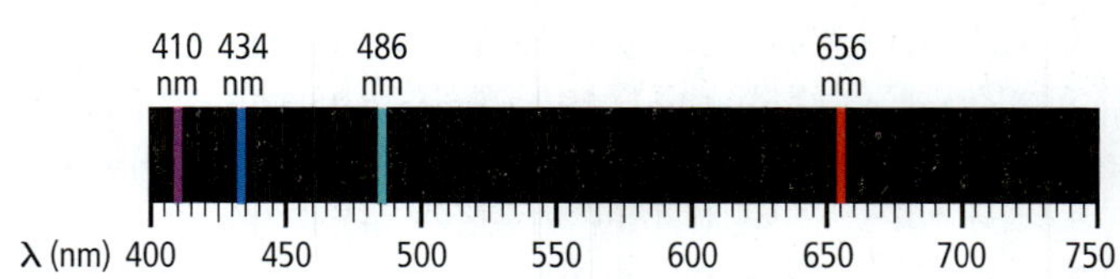


**Hydrogen's Atomic Emission Spectrum**

**Figure 22**

**58. Hydrogen** One line in hydrogen's emission spectrum has a wavelength of 486 nm. Examine **Figure 22** to determine the line's color. What is the line's frequency?

## SECTION 2

### Mastering Concepts

**59.** According to the Bohr model, how do electrons move in atoms?

**60.** What does *n* designate in Bohr's atomic model?

**61.** What is the difference between an atom's ground state and an excited state?

**62.** What is the name of the atomic model in which electrons are treated as waves? Who first wrote the electron wave equations that led to this model?

**63.** What is an atomic orbital?

**64.** What does *n* represent in the quantum mechanical model of the atom?

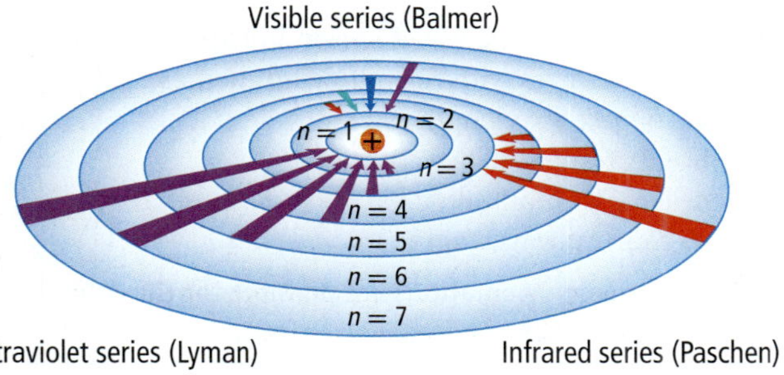


**Figure 23**

**65. Electron Transition** According to the Bohr model shown in **Figure 23,** what type of electron-orbit transitions produce the ultraviolet lines in hydrogen's Lyman series?

**66.** How many energy sublevels are contained in each of the hydrogen atom's first three energy levels?

**67.** What atomic orbitals are related to a p sublevel?

**68.** What do the sublevel designations s, p, d, and f specify with respect to the atom's orbitals?

**69.** How are the five orbitals related to an atom's d sublevel designated?

**70.** What is the maximum number of electrons an orbital can contain?

**71.** Describe the relative orientations of the orbitals related to an atom's 2p sublevel.

**72.** How many electrons can be contained in all the orbitals related to an argon atom's third energy level?

**73.** How does the quantum mechanical model of the atom describe the paths of an atom's electrons?

**74. Macroscopic Objects** Why do we not notice the wavelengths of moving objects such as automobiles?

**75.** Why is it impossible to know precisely the velocity and position of an electron at the same time?

## SECTION 3

### Mastering Concepts

**76.** In what sequence do electrons fill the atomic orbitals related to a sublevel?

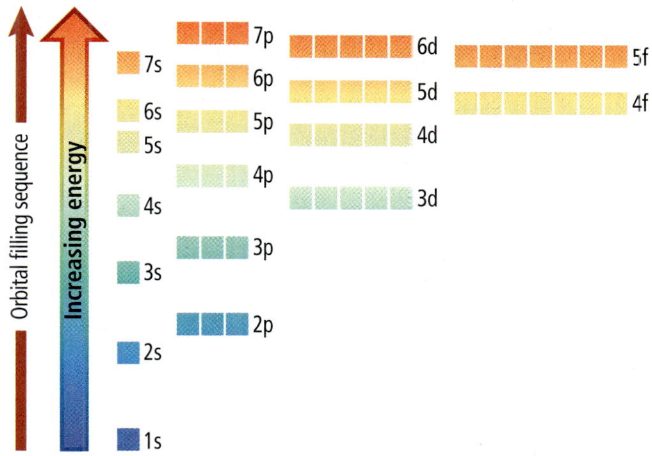


**Figure 24**

**77. Rubidium** Using **Figure 24,** explain why one electron in a rubidium atom occupies a 5s orbital rather than a 4d or 4f orbital.

**78.** What are valence electrons? How many of a magnesium atom's 12 electrons are valence electrons?

**79.** Light is said to have a dual wave-particle nature. What does this statement mean?

**80.** Describe the difference between a quantum and a photon.

**81.** How many electrons are shown in each element's electron-dot structure?

**a.** carbon
**b.** iodine
**c.** calcium
**d.** gallium

**82.** When writing the electron configuration notation for an atom, what three principles or rules should you follow?

**83.** Write the electron configuration and draw the orbital notation for atoms of oxygen and sulfur.

### Mastering Problems

**84.** List the aufbau sequence of orbitals from 1s to 7p.

**85.** Write each element's orbital notation and complete electron configuration.

**a.** beryllium
**b.** aluminum
**c.** nitrogen
**d.** sodium

**86.** Use noble-gas notation to describe the electron configurations of the elements represented by the following symbols.

**a.** Kr
**b.** P
**c.** Zr
**d.** Pb

87. What element is represented by each electron configuration?
   a. $1s^22s^22p^5$
   b. $[Ar]4s^2$
   c. $[Xe]6s^24f^4$
   d. $[Kr]5s^24d^{10}5p^4$
   e. $[Rn]7s^25f^{13}$
   f. $1s^22s^22p^63s^23p^64s^23d^{10}4p^5$

88. Which electron configuration notation describes an atom in an excited state?
   a. $[Ar]4s^23d^{10}4p^2$
   b. $[Ne]3s^23p^5$
   c. $[Kr]5s^24d^1$
   d. $[Ar]4s^23d^84p^1$

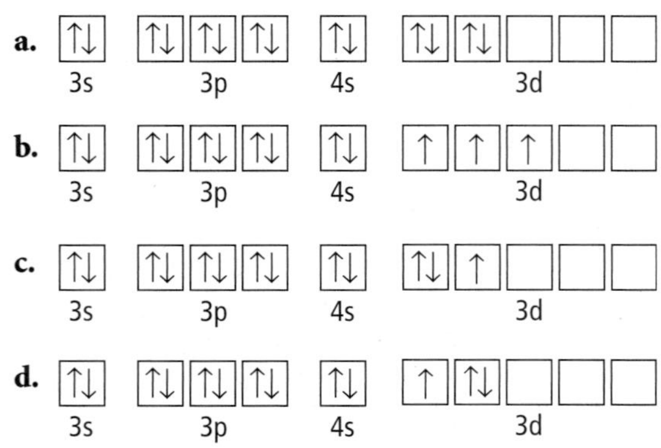


■ **Figure 25**

89. Which orbital diagram in **Figure 25** is correct for an atom in its ground state?

90. Draw an electron-dot structure for an atom of each element.
   a. carbon
   b. arsenic
   c. polonium
   d. potassium
   e. barium

91. **Arsenic** An atom of arsenic has how many electron-containing orbitals? How many of the orbitals are completely filled? How many of the orbitals are associated with the atom's $n = 4$ principal energy level?

■ **Figure 26**

92. Which element could have the ground-state electron-dot notation shown in **Figure 26**?
   a. manganese
   b. antimony
   c. calcium
   d. samarium

93. For an atom of tin in the ground state, write the electron configuration using noble-gas notation, and draw its electron-dot structure.

## MIXED REVIEW

94. What is the maximum number of electrons that can be contained in an atom's orbitals having the following principal quantum numbers?
   a. 3
   b. 4
   c. 6
   d. 7

95. What is the wavelength of light with a frequency of $5.77 \times 10^{14}$ Hz?

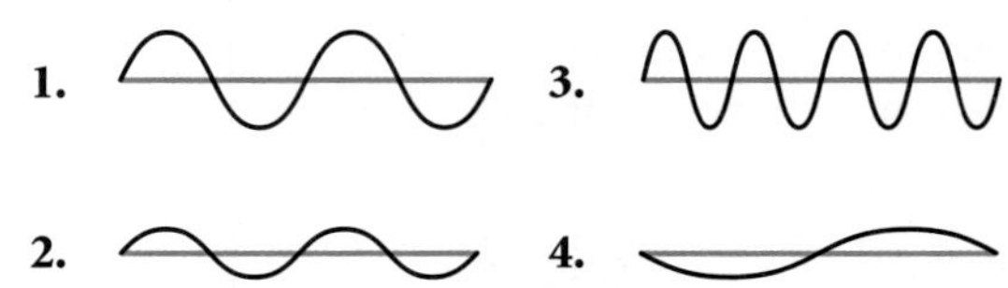


■ **Figure 27**

96. **Waves** Using the waves shown in **Figure 27**, identify the wave or waves with the following characteristics.
   a. longest wavelength
   b. greatest frequency
   c. largest amplitude
   d. shortest wavelength

97. How many orientations are possible for the orbitals related to each sublevels?
   a. s
   b. p
   c. d
   d. f

98. Which elements have only two electrons in their electron-dot structures: hydrogen, helium, lithium, aluminum, calcium, cobalt, bromine, krypton, or barium?

99. In Bohr's atomic model, what electron-orbit transition produces the blue-green line in hydrogen's atomic emission spectrum?

100. **Zinc** A zinc atom contains a total of 18 electrons in its 3s, 3p, and 3d orbitals. Why does its electron-dot structure show only two dots?

101. **X-Ray** An X-ray photon has an energy of $3.01 \times 10^{-18}$ J. What is its frequency and wavelength?

102. Which element has the ground-state electron configuration represented by the noble-gas notation $[Rn]7s^1$?

103. How did Bohr explain atomic emission spectra?

104. **Infrared Radiation** How many photons of infrared radiation with a frequency of $4.88 \times 10^{13}$ Hz are required to provide an energy of 1.00 J?

105. Light travels slower in water than it does in air; however, its frequency remains the same. How does the wavelength of light change as it travels from air to water?

106. According to the quantum mechanical model of the atom, what happens when an atom absorbs a quantum of energy?

Chapter Self-Check

## THINK CRITICALLY

**107. Compare and Contrast** Briefly discuss the difference between an orbit in Bohr's model of the atom and an orbital in the quantum mechanical view of the atom.

**108. Calculate** It takes $8.17 \times 10^{-19}$ J of energy to remove one electron from a gold surface. What is the maximum wavelength of light capable of causing this effect?

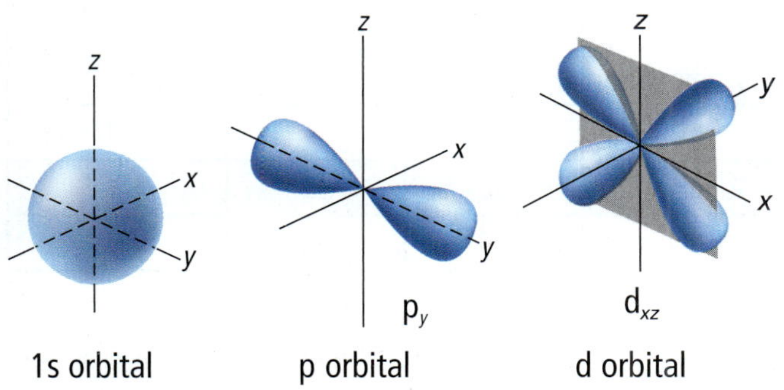


■ **Figure 28**

**109. Describe** the shapes of the atomic orbitals shown in **Figure 28.** Specify their orientations and relate each orbital to a particular type of energy sublevel.

**110. Infer** Suppose that you live in a universe in which the Pauli exclusion principle states that a maximum of three, rather than two, electrons can occupy a single atomic orbital. Evaluate and explain the new chemical properties of the elements lithium and phosphorus.

## CHALLENGE PROBLEM

**111. Hydrogen Atom** The hydrogen atom's energy is $-6.05 \times 10^{-20}$ J when the electron is in the $n = 6$ orbit and $-2.18 \times 10^{-18}$ J when the electron is in the $n = 1$. Calculate the wavelength of the photon emitted when the electron drops from the $n = 6$ orbit to the $n = 1$ orbit. Use the following values: $h = 6.626 \times 10^{-34}$ J•s and $c = 3.00 \times 10^{8}$ m/s.

## CUMULATIVE REVIEW

**112.** Round 20.56120 g to three significant figures.

**113.** Identify whether each statement describes a chemical property or a physical property.
- **a.** Mercury is a liquid at room temperature.
- **b.** Sucrose is a white, crystalline solid.
- **c.** Iron rusts when exposed to moist air.
- **d.** Paper burns when ignited.

**114.** An atom of gadolinium has an atomic number of 64 and a mass number of 153. How many electrons, protons, and neutrons does it contain?

## WRITING IN ▶ Chemistry

**115. Neon Signs** To make neon signs emit different colors, manufacturers often fill the signs with gases other than neon. Write an essay about the use of gases in neon signs and the colors produced by the gases.

**116. Rutherford's Model** Imagine that you are a scientist in the early twentieth century, and you have just learned the details of a new, nuclear model of the atom proposed by the prominent English physicist Ernest Rutherford. After analyzing the model, you discern what you believe to be important limitations. Write a letter to Rutherford in which you express your concerns regarding his model. Use diagrams and examples of specific elements to help you make your point.

## DBQ Document-Based Questions

**Sodium Vapor** *When sodium metal is vaporized in a gas-discharge lamp, two closely spaced, bright yellow-orange lines are produced. Because sodium vapor lamps are electrically efficient, they are used widely for outdoor lighting, such as streetlights and security lighting.*

**Figure 29** *shows the emission spectrum of sodium metal. The entire visible spectrum is shown for comparison.*

Data obtained from: Volland, W. March 2005. *Spectroscopy: Element Identification and Emission Spectra.*

■ **Figure 29**

**117.** Differentiate between the two spectra shown above.

**118.** Sodium's two bright lines have wavelengths of 588.9590 nm and 589.9524 nm. What is the ground-state electron configuration notation for sodium, and how does sodium's electron configuration relate to the lines?

**119.** Calculate the energies of photons related to the two lines using the relationships expressed in the following equations.

$$E_{photon} = h\nu;\ c = \lambda\nu;\ E = hc/\lambda$$

## MULTIPLE CHOICE

1. Cosmic rays are high-energy radiation from outer space. What is the frequency of a cosmic ray that has a wavelength of $2.67 \times 10^{-13}$ m when it reaches Earth? (The speed of light is $3.00 \times 10^{8}$ m/s.)
   **A.** $8.90 \times 10^{-22}\ s^{-1}$
   **B.** $3.75 \times 10^{12}\ s^{-1}$
   **C.** $8.01 \times 10^{-5}\ s^{-1}$
   **D.** $1.12 \times 10^{21}\ s^{-1}$

2. Which is the electron-dot structure for indium?
   **A.** $\cdot\text{In}$
   **B.** $\cdot\text{In}\cdot$
   **C.** $\cdot\overset{\cdot}{\text{In}}\cdot$
   **D.** $\cdot\underset{\cdot}{\overset{\cdot}{\text{In}}}\cdot$

*Use the figure below to answer Questions 3 and 4.*

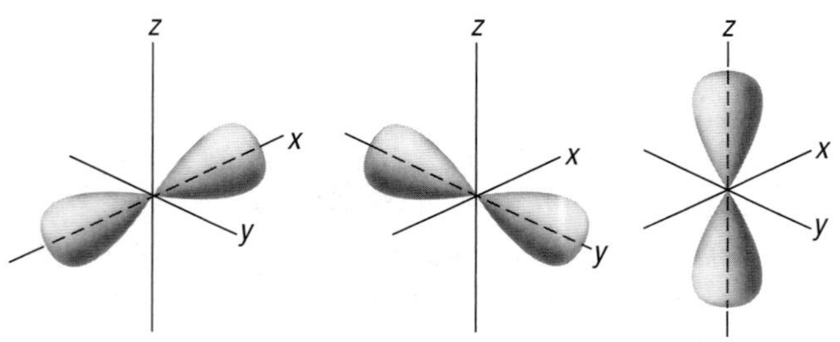


3. To which sublevel do all of these orbitals belong?
   **A.** s
   **B.** p
   **C.** d
   **D.** f

4. How many electrons total can reside in this sublevel?
   **A.** 2
   **B.** 3
   **C.** 6
   **D.** 8

5. What is the maximum theoretical number of electrons related to the fifth principal energy level of an atom?
   **A.** 2
   **B.** 8
   **C.** 18
   **D.** 32

*Use the periodic table and the table below to answer Questions 6 to 8.*

**Electron Configurations for Selected Transition Metals**

| Element | Symbol | Atomic Number | Electron Configuration |
|---|---|---|---|
| Vanadium | V | 23 | $[Ar]4s^2 3d^3$ |
| Yttrium | Y | 39 | $[Kr]5s^2 4d^1$ |
| | | | $[Xe]6s^2 4f^{14} 5d^6$ |
| Scandium | Sc | 21 | $[Ar]4s^2 3d^1$ |
| Cadmium | Cd | 48 | |

6. Using noble-gas notation, what is the ground-state electron configuration of Cd?
   **A.** $[Kr]4d^{10}4f^2$
   **B.** $[Ar]4s^2 3d^{10}$
   **C.** $[Kr]5s^2 4d^{10}$
   **D.** $[Xe]5s^2 4d^{10}$

7. What is the element that has the ground-state electron configuration $[Xe]6s^2 4f^{14} 5d^6$?
   **A.** La
   **B.** Ti
   **C.** W
   **D.** Os

8. What is the complete electron configuration of a scandium atom?
   **A.** $1s^2 2s^2 2p^6 3s^2 3p^6 4s^2 3d^1$
   **B.** $1s^2 2s^2 2p^7 3s^2 3p^7 4s^2 3d^1$
   **C.** $1s^2 2s^2 2p^5 3s^2 3p^5 4s^2 3d^1$
   **D.** $1s^2 2s^1 2p^7 3s^1 3p^7 4s^2 3d^1$

9. Which is NOT evidence that a chemical change has occurred?
   **A.** The properties of the substances involved in the reaction have changed.
   **B.** An odor is produced.
   **C.** The composition of the substances involved in the reaction have changed.
   **D.** The total mass of all substances involved has changed.

## SHORT ANSWER

*Use the data below to answer Questions 10 to 13.*

| Temperature of Water with Heating | |
|---|---|
| Time (s) | Temperature (°C) |
| 0 | 16.3 |
| 30 | 19.7 |
| 60 | 24.2 |
| 90 | 27.8 |
| 120 | 32.0 |
| 150 | 35.3 |
| 180 | 39.6 |
| 210 | 43.3 |
| 240 | 48.1 |

**10.** Make a graph showing temperature versus time.

**11.** Is the heating of this sample of water a linear process? Explain how you can tell.

**12.** Use your graph to find the approximate rate of heating in degrees per second. What is this value in degrees per minute?

**13.** Show the equation to convert the temperature at 180 s from degrees Celsius to Kelvin and to degrees Fahrenheit.

## EXTENDED RESPONSE

**14.** Compare the information provided in an electron-dot structure with the information in an electron configuration.

**15.** Explain why $1s^22s^22p^63s^23p^64s^24d^{10}4p^2$ is not the correct electron configuration for germanium (Ge). Write the correct electron configuration for germanium.

## SAT SUBJECT TEST: CHEMISTRY

*Use the diagram below to answer Questions 16 and 17.*

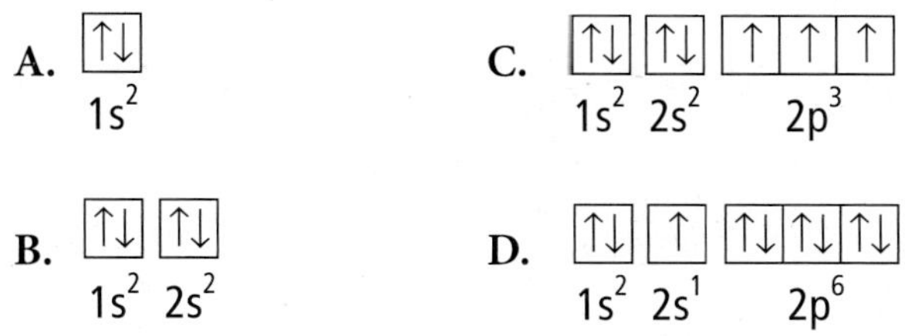


**16.** Which shows an orbital diagram that violates the aufbau principle?
**A.** A
**B.** B
**C.** C
**D.** D
**E.** none

**17.** Which shows the orbital diagram for the element beryllium?
**A.** A
**B.** B
**C.** C
**D.** D
**E.** none

**18.** A student performs an experiment to measure the boiling point of pentane and measures it at 37.2°C. The literature reports this value as 36.1°C. What is the student's percent error?
**A.** 97.0%
**B.** 2.95%
**C.** 1.1%
**D.** 15.5%
**E.** 3.05%

**19.** Which method of separating components of a mixture depends on the different boiling points of the components of the mixture?
**A.** chromatography
**B.** filtration
**C.** crystallization
**D.** distillation
**E.** sublimation

| NEED EXTRA HELP? | | | | | | | | | | | | | | | | | | | |
|---|---|---|---|---|---|---|---|---|---|---|---|---|---|---|---|---|---|---|---|
| If You Missed Question . . . | 1 | 2 | 3 | 4 | 5 | 6 | 7 | 8 | 9 | 10 | 11 | 12 | 13 | 14 | 15 | 16 | 17 | 18 | 19 |
| Review Section . . . | 5.1 | 5.3 | 5.2 | 5.2 | 5.3 | 5.3 | 5.3 | 5.3 | 3.2 | 2.4 | 2.4 | 2.4 | 2.1 | 5.3 | 5.3 | 5.3 | 5.3 | 2.3 | 3.3 |

CHAPTER 6

# The Periodic Table and Periodic Law

**BIGIDEA** Periodic trends in the properties of atoms allow us to predict physical and chemical properties.

## SECTIONS

Chm 1, 3, 3a

1 Development of the Modern Periodic Table

2 Classification of the Elements

3 Periodic Trends

## LaunchLAB

iLab Station

### How can you recognize trends?

The periodic table of the elements is arranged so that the properties of the elements repeat in a regular way. Such an arrangement can also be used for common items. In this lab, you will study ways to organize data according to trends.

## FOLDABLES Study Organizer

### Periodic Trends

Make a folded chart. Label it as shown. Use it to organize information about periodic trends.

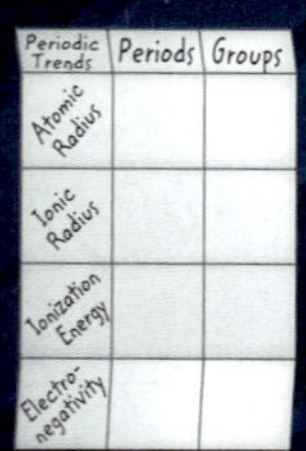


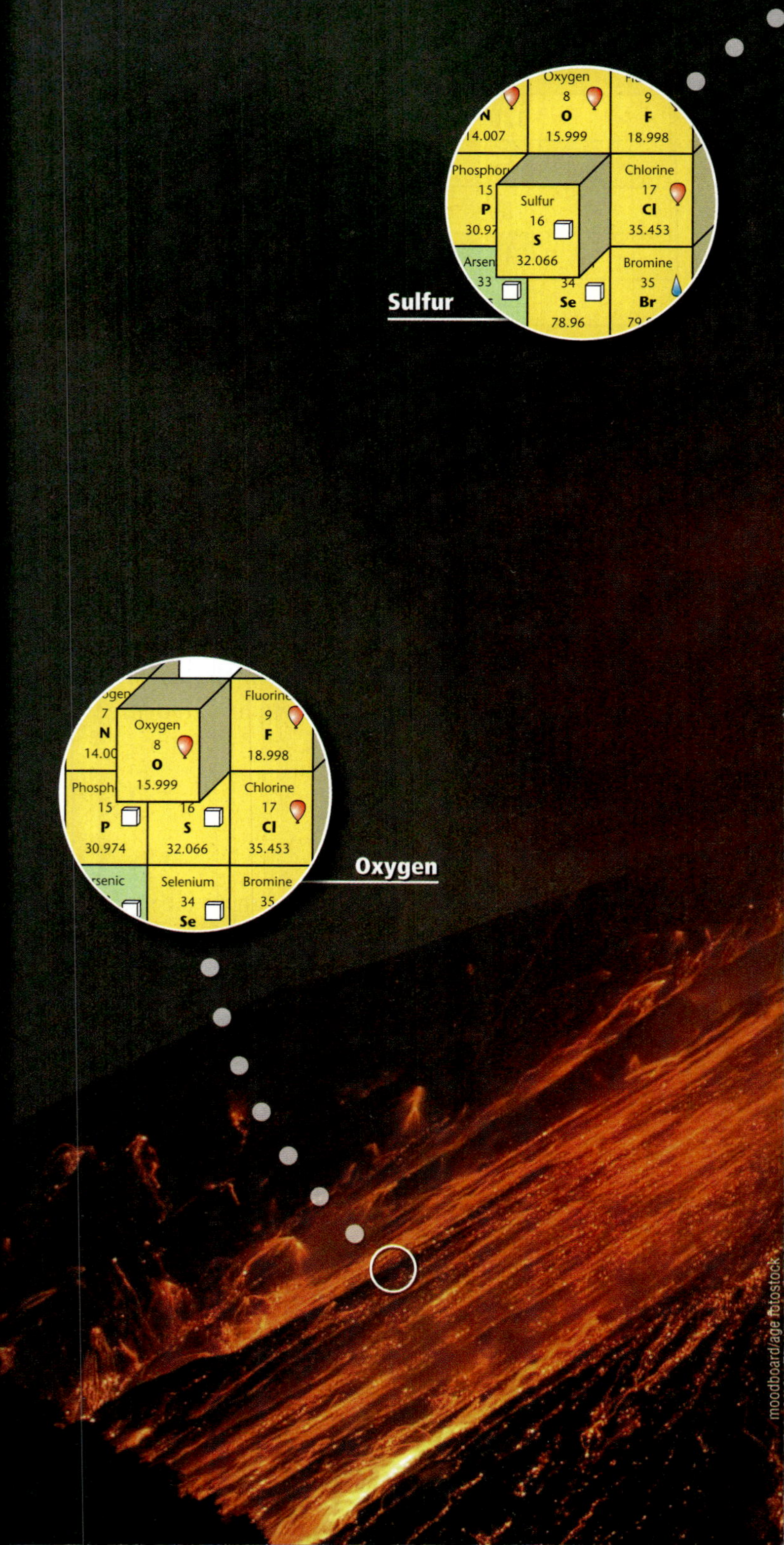


moodboard/age fotostock

Go **online!**
MHEonline.com

There are currently 117 elements in the periodic table. Approximately 90 of them occur naturally. By mass, oxygen is the most abundant element in Earth's crust (just under 50%).

SECTION 1

# Development of the Modern Periodic Table

**MAINIDEA** The periodic table evolved over time as scientists discovered more useful ways to compare and organize the elements.

### Essential Questions

- How was the periodic table developed?
- What are the key features of the periodic table?

### Review Vocabulary

**atomic number:** the number of protons in an atom

### New Vocabulary

periodic law
group
period
representative element
transition element
metal
alkali metal
alkaline earth metal
transition metal
inner transition metal
lanthanide series
actinide series
nonmetal
halogen
noble gas
metalloid

**CHEM 4 YOU** Imagine grocery shopping if all the apples, pears, oranges, and peaches were mixed into one bin at the grocery store. Organizing things according to their properties is often useful. Scientists organize the many different types of chemical elements in the periodic table.

## Development of the Periodic Table

In the late 1700s, French scientist Antoine Lavoisier (1743–1794) compiled a list of all elements that were known at the time. The list, shown in **Table 1,** contained 33 elements organized in four categories. Many of these elements, such as silver, gold, carbon, and oxygen, have been known since prehistoric times. The 1800s brought a large increase in the number of known elements. The advent of electricity, which was used to break down compounds into their components, and the development of the spectrometer, which was used to identify the newly isolated elements, played major roles in the advancement of chemistry. The industrial revolution of the mid-1800s also played a major role, which led to the development of many new chemistry-based industries, such as the manufacture of petrochemicals, soaps, dyes, and fertilizers. By 1870, there were over 60 known elements.

Along with the discovery of new elements came volumes of new scientific data related to the elements and their compounds. Chemists of the time were overwhelmed with learning the properties of so many new elements and compounds. What chemists needed was a tool for organizing the many facts associated with the elements. A significant step toward this goal came in 1860, when chemists agreed upon a method for accurately determining the atomic masses of the elements. Until this time, different chemists used different mass values in their work, making the results of one chemist's work hard to reproduce by another. With newly agreed-upon atomic masses for the elements, the search for relationships between atomic mass and elemental properties, and a way to organize the elements began in earnest.

**Table 1** Lavoisier's Table of Simple Substances (Old English Names)

| | |
|---|---|
| Gases | light, heat, dephlogisticated air, phlogisticated gas, inflammable air |
| Metals | antimony, silver, arsenic, bismuth, cobalt, copper, tin, iron, manganese, mercury, molybdena, nickel, gold, platina, lead, tungsten, zinc |
| Nonmetals | sulphur, phosphorus, pure charcoal, radical muriatique*, radical fluorique*, radical boracique* |
| Earths | chalk, magnesia, barote, clay, siliceous earth |

*no English name

**John Newlands** In 1864, English chemist John Newlands (1837–1898) proposed an organizational scheme for the elements. He noticed that when the elements were arranged by increasing atomic mass, their properties repeated every eighth element. A pattern such as this is called periodic because it repeats in a specific manner. Newlands named the periodic relationship that he observed in chemical properties the *law of octaves,* after the musical octave in which notes repeat every eighth tone. **Figure 1** shows how Newlands organized 14 of the elements known in the mid-1860s. Acceptance of the law of octaves was hampered because the law did not work for all of the known elements. Also, the use of the word *octave* was harshly criticized by fellow scientists, who thought that the musical analogy was unscientific. While his law was not generally accepted, the passage of a few years would show that Newlands was basically correct; the properties of elements do repeat in a periodic way.

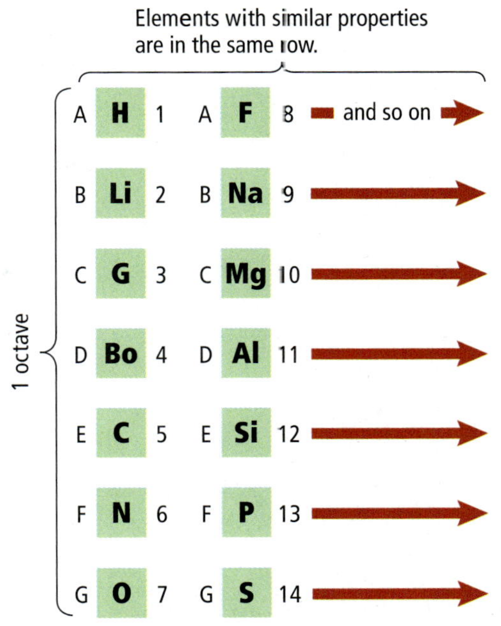


**Figure 1** John Newlands noticed that the properties of elements repeated every eighth element, in the same way musical notes repeat every eighth note and form octaves.

**Meyer and Mendeleev** In 1869, German chemist Lothar Meyer (1830–1895) and Russian chemist Dmitri Mendeleev (1834–1907) each demonstrated a connection between atomic mass and the properties of elements. Mendeleev, however, is generally given more credit than Meyer because he published his organizational scheme first. Like Newlands several years earlier, Mendeleev noticed that when the elements were ordered by increasing atomic mass, there was a periodic pattern in their properties. By arranging the elements in order of increasing atomic mass into columns with similar properties, Mendeleev organized the elements into a periodic table. Mendeleev's table, shown in **Figure 2,** became widely accepted because he predicted the existence and properties of undiscovered elements that were later found. Mendeleev left blank spaces in the table where he thought the undiscovered elements should go. By noting trends in the properties of known elements, he was able to predict the properties of the yet-to-be-discovered elements scandium, gallium, and germanium.

| Typische Elemente | | | | | | | |
|---|---|---|---|---|---|---|---|
| | | | K = 39 | Rb = 85 | Cs = 133 | — | — |
| | | | Ca = 40 | Sr = 87 | Ba = 137 | — | — |
| | | | — | ?Yt = 88? | ?Di = 138? | Er = 178? | — |
| | | | Ti = 48? | Zr = 90 | Ce = 140? | ?La = 180? | Th = 231 |
| | | | V = 51 | Nb = 94 | — | Ta = 182 | — |
| | | | Cr = 52 | Mo = 96 | — | W = 184 | U = 240 |
| | | | Mn = 55 | — | — | — | — |
| | | | Fe = 56 | Ru = 104 | — | Os = 195? | — |
| | | | Co = 59 | Rh = 104 | — | Ir = 197 | — |
| | | | Ni = 59 | Pd = 106 | — | Pt = 198? | — |
| H = 1 | Li = 7 | Na = 23 | Cu = 63 | Ag = 108 | — | Au = 199? | — |
| | Be = 9,4 | Mg = 24 | Zn = 65 | Cd = 112 | — | Hg = 200 | — |
| | B = 11 | Al = 27,3 | — | In = 113 | — | Tl = 204 | — |
| | C = 12 | Si = 28 | — | Sn = 118 | — | Pb = 207 | — |
| | N = 14 | P = 31 | As = 75 | Sb = 122 | — | Bi = 208 | — |
| | O = 16 | S = 32 | Se = 78 | Te = 125? | — | — | — |
| | F = 19 | Cl = 35,5 | Br = 80 | J = 127 | — | — | — |

**Figure 2** In the first version of his table, published in 1869, Mendeleev arranged elements with similar chemical properties horizontally. He left empty spaces for elements that were not yet discovered.

**Moseley** Mendeleev's table, however, was not completely correct. After several new elements were discovered and the atomic masses of the known elements were more accurately determined, it became apparent that several elements in his table were not in the correct order. Arranging the elements by mass resulted in several elements being placed in groups of elements with differing properties.

The reason for this problem was determined in 1913 by English chemist Henry Moseley (1887–1915). As you might recall, Moseley discovered that atoms of each element contain a unique number of protons in their nuclei—the number of protons being equal to the atom's atomic number. By arranging the elements in order of increasing atomic number, the problems with the order of the elements in the periodic table were solved. Moseley's arrangement of elements by atomic number resulted in a clear periodic pattern of properties. The statement that there is a periodic repetition of chemical and physical properties of the elements when they are arranged by increasing atomic number is called the **periodic law.**

**VOCABULARY**

**WORD ORIGIN**

**Periodic**
comes from the Greek word *periodos,* meaning *way around, circuit*

READING CHECK **Compare and contrast** the ways in which Mendeleev and Moseley organized the elements.

**Table 2** summarizes the contributions of Newlands, Meyer, Mendeleev, and Moseley to the development of the periodic table. The periodic table brought order to seemingly unrelated facts and became a significant tool for chemists. It is a useful reference for understanding and predicting the properties of elements and for organizing knowledge of atomic structure. Do the Problem-Solving Lab later in this chapter to see how the periodic law can be used to predict unknown elemental properties.

| Table 2 Contributions to the Classification of Elements |
|---|
| **John Newlands (1837–1898)**<br>• arranged elements by increasing atomic mass<br>• noticed the repetition of properties every eighth element<br>• created the law of octaves |
| **Lothar Meyer (1830–1895)**<br>• demonstrated a connection between atomic mass and elements' properties<br>• arranged the elements in order of increasing atomic mass |
| **Dmitri Mendeleev (1834–1907)**<br>• demonstrated a connection between atomic mass and elements' properties<br>• arranged the elements in order of increasing atomic mass<br>• predicted the existence and properties of undiscovered elements |
| **Henry Moseley (1887–1915)**<br>• discovered that atoms contain a unique number of protons called the atomic number<br>• arranged elements in order of increasing atomic number, which resulted in a periodic pattern of properties |

## The Modern Periodic Table

The modern periodic table consists of boxes, each containing an element name, symbol, atomic number, and atomic mass. A typical box from the table is shown in **Figure 3.** The boxes are arranged in order of increasing atomic number into a series of columns, called **groups** or families, and rows, called **periods.** The table is shown in **Figure 5** on the next page and on the inside back cover of your textbook.

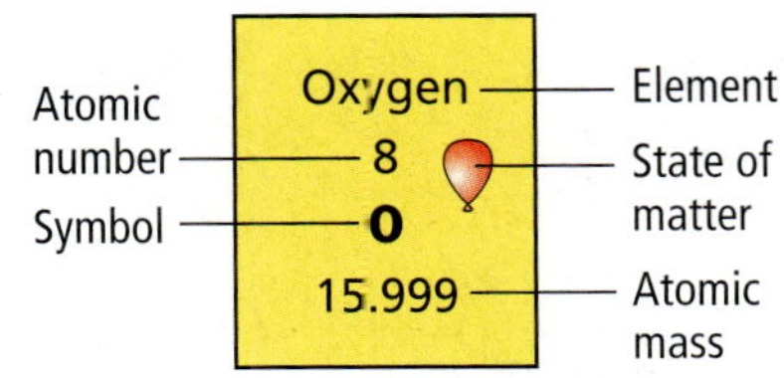


**Figure 3** A typical box from the periodic table contains the element's name, its chemical symbol, its atomic number, its atomic mass, and its state.

☑ READING CHECK **Define** *groups* and *periods.*

Beginning with hydrogen in period 1, there are a total of seven periods. Each group is numbered 1 through 18. For example, period 4 contains potassium and calcium. Scandium (Sc) is in the third column from the left, which is group 3. Oxygen is in group 16. The elements in groups 1, 2, and 13 to 18 possess a wide range of chemical and physical properties. For this reason, they are often referred to as the main group, or **representative elements.** The elements in groups 3 to 12 are referred to as the **transition elements.** Elements are classified as metals, nonmetals, and metalloids.

**Metals** Elements that are generally shiny when smooth and clean, solid at room temperature, and good conductors of heat and electricity are called **metals.** Most metals are also malleable and ductile, meaning that they can be pounded into thin sheets and drawn into wires, respectively. Most representative elements and all transition elements are metals. If you look at boron (B) in column 13, you will see a heavy stairstep line that zigzags down to astatine (At) at the bottom of group 17. This stairstep line is a visual divider between the metals and the nonmetals on the table. In **Figure 5,** metals are represented by the blue boxes.

**Alkali metals** Except for hydrogen, all of the elements on the left side of the table are metals. The group 1 elements (except for hydrogen) are known as the **alkali metals.** Because they are so reactive, alkali metals usually exist as compounds with other elements. Two familiar alkali metals are sodium (Na), one of the components of salt, and lithium (Li), often used in batteries.

**Alkaline earth metals** The **alkaline earth metals** are in group 2. They are also highly reactive. Calcium (Ca) and magnesium (Mg), two minerals important for your health, are examples of alkaline earth metals. Because magnesium is solid and relatively light, it is used in the fabrication of electronic devices, such as the laptop shown in **Figure 4.**

**Figure 4** Because magnesium is light and strong, it is often used in the production of electronic devices. For instance, this laptop case is made of magnesium.

**Figure 5**

# PERIODIC TABLE OF THE ELEMENTS

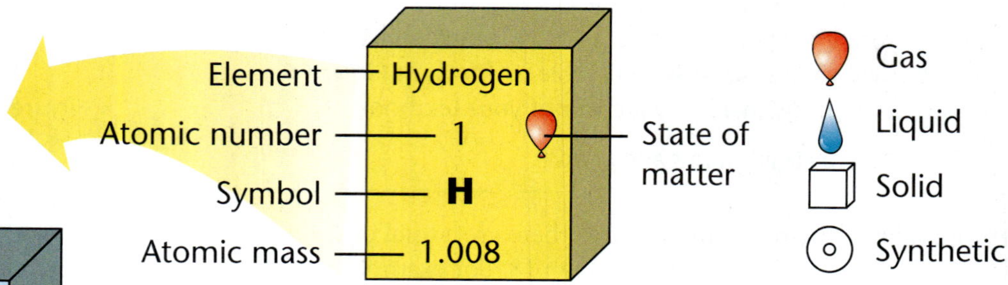


| | 1 | 2 | 3 | 4 | 5 | 6 | 7 | 8 | 9 |
|---|---|---|---|---|---|---|---|---|---|
| 1 | Hydrogen 1 **H** 1.008 | | | | | | | | |
| 2 | Lithium 3 **Li** 6.941 | Beryllium 4 **Be** 9.012 | | | | | | | |
| 3 | Sodium 11 **Na** 22.990 | Magnesium 12 **Mg** 24.305 | | | | | | | |
| 4 | Potassium 19 **K** 39.098 | Calcium 20 **Ca** 40.078 | Scandium 21 **Sc** 44.956 | Titanium 22 **Ti** 47.867 | Vanadium 23 **V** 50.942 | Chromium 24 **Cr** 51.996 | Manganese 25 **Mn** 54.938 | Iron 26 **Fe** 55.847 | Cobalt 27 **Co** 58.933 |
| 5 | Rubidium 37 **Rb** 85.468 | Strontium 38 **Sr** 87.62 | Yttrium 39 **Y** 88.906 | Zirconium 40 **Zr** 91.224 | Niobium 41 **Nb** 92.906 | Molybdenum 42 **Mo** 95.94 | Technetium 43 **Tc** (98) | Ruthenium 44 **Ru** 101.07 | Rhodium 45 **Rh** 102.906 |
| 6 | Cesium 55 **Cs** 132.905 | Barium 56 **Ba** 137.327 | Lanthanum 57 **La** 138.905 | Hafnium 72 **Hf** 178.49 | Tantalum 73 **Ta** 180.948 | Tungsten 74 **W** 183.84 | Rhenium 75 **Re** 186.207 | Osmium 76 **Os** 190.23 | Iridium 77 **Ir** 192.217 |
| 7 | Francium 87 **Fr** (223) | Radium 88 **Ra** (226) | Actinium 89 **Ac** (227) | Rutherfordium 104 **Rf** (261) | Dubnium 105 **Db** (262) | Seaborgium 106 **Sg** (266) | Bohrium 107 **Bh** (264) | Hassium 108 **Hs** (277) | Meitnerium 109 **Mt** (268) |

The number in parentheses is the mass number of the longest lived isotope for that element.

| | | | | | | |
|---|---|---|---|---|---|---|
| **Lanthanide series** | Cerium 58 **Ce** 140.115 | Praseodymium 59 **Pr** 140.908 | Neodymium 60 **Nd** 144.242 | Promethium 61 **Pm** (145) | Samarium 62 **Sm** 150.36 | Europium 63 **Eu** 151.965 |
| **Actinide series** | Thorium 90 **Th** 232.038 | Protactinium 91 **Pa** 231.036 | Uranium 92 **U** 238.029 | Neptunium 93 **Np** (237) | Plutonium 94 **Pu** (244) | Americium 95 **Am** (243) |

View an **animation about the periodic table.**

Concepts In Motion

Explore **updates to the periodic table.**

Periodic Table

Metal | Metalloid | Nonmetal | Recently observed

| 10 | 11 | 12 | 13 | 14 | 15 | 16 | 17 | 18 |
|---|---|---|---|---|---|---|---|---|
| | | | | | | | | Helium<br>2<br>**He**<br>4.003 |
| | | | Boron<br>5<br>**B**<br>10.811 | Carbon<br>6<br>**C**<br>12.011 | Nitrogen<br>7<br>**N**<br>14.007 | Oxygen<br>8<br>**O**<br>15.999 | Fluorine<br>9<br>**F**<br>18.998 | Neon<br>10<br>**Ne**<br>20.180 |
| | | | Aluminum<br>13<br>**Al**<br>26.982 | Silicon<br>14<br>**Si**<br>28.086 | Phosphorus<br>15<br>**P**<br>30.974 | Sulfur<br>16<br>**S**<br>32.066 | Chlorine<br>17<br>**Cl**<br>35.453 | Argon<br>18<br>**Ar**<br>39.948 |
| Nickel<br>28<br>**Ni**<br>58.693 | Copper<br>29<br>**Cu**<br>63.546 | Zinc<br>30<br>**Zn**<br>65.39 | Gallium<br>31<br>**Ga**<br>69.723 | Germanium<br>32<br>**Ge**<br>72.61 | Arsenic<br>33<br>**As**<br>74.922 | Selenium<br>34<br>**Se**<br>78.96 | Bromine<br>35<br>**Br**<br>79.904 | Krypton<br>36<br>**Kr**<br>83.80 |
| Palladium<br>46<br>**Pd**<br>106.42 | Silver<br>47<br>**Ag**<br>107.868 | Cadmium<br>48<br>**Cd**<br>112.411 | Indium<br>49<br>**In**<br>114.82 | Tin<br>50<br>**Sn**<br>118.710 | Antimony<br>51<br>**Sb**<br>121.757 | Tellurium<br>52<br>**Te**<br>127.60 | Iodine<br>53<br>**I**<br>126.904 | Xenon<br>54<br>**Xe**<br>131.290 |
| Platinum<br>78<br>**Pt**<br>195.08 | Gold<br>79<br>**Au**<br>196.967 | Mercury<br>80<br>**Hg**<br>200.59 | Thallium<br>81<br>**Tl**<br>204.383 | Lead<br>82<br>**Pb**<br>207.2 | Bismuth<br>83<br>**Bi**<br>208.980 | Polonium<br>84<br>**Po**<br>208.982 | Astatine<br>85<br>**At**<br>209.987 | Radon<br>86<br>**Rn**<br>222.018 |
| Darmstadtium<br>110<br>**Ds**<br>(281) | Roentgenium<br>111<br>**Rg**<br>(272) | Copernicium<br>112<br>**Cn**<br>(285) | Ununtrium<br>* 113<br>**Uut**<br>(284) | Flerovium<br>114<br>**Fl**<br>(289) | Ununpentium<br>* 115<br>**Uup**<br>(288) | Livermorium<br>116<br>**Lv**<br>(291) | Ununseptium<br>* 117<br>**Uus**<br>(294) | Ununoctium<br>* 118<br>**Uuo**<br>(294) |

* The names and symbols for elements 113, 115, 117, and 118 are temporary. Final names will be approved by IUPAC (International Union of Pure and Applied Chemistry).

| | | | | | | | |
|---|---|---|---|---|---|---|---|
| Gadolinium<br>64<br>**Gd**<br>157.25 | Terbium<br>65<br>**Tb**<br>158.925 | Dysprosium<br>66<br>**Dy**<br>162.50 | Holmium<br>67<br>**Ho**<br>164.930 | Erbium<br>68<br>**Er**<br>167.259 | Thulium<br>69<br>**Tm**<br>168.934 | Ytterbium<br>70<br>**Yb**<br>173.04 | Lutetium<br>71<br>**Lu**<br>174.967 |
| Curium<br>96<br>**Cm**<br>(247) | Berkelium<br>97<br>**Bk**<br>(247) | Californium<br>98<br>**Cf**<br>(251) | Einsteinium<br>99<br>**Es**<br>(252) | Fermium<br>100<br>**Fm**<br>(257) | Mendelevium<br>101<br>**Md**<br>(258) | Nobelium<br>102<br>**No**<br>(259) | Lawrencium<br>103<br>**Lr**<br>(262) |

# Problem-Solving LAB

## Analyze Trends

**Francium—solid, liquid, or gas?**
Francium was discovered in 1939, but its existence was predicted by Mendeleev in the 1870s. It is the least stable of the first 101 elements: Its most stable isotope has a half-life of just 22 minutes! Use your knowledge about the properties of other alkali metals to predict some of francium's properties.

### Analysis

In the spirit of Dmitri Mendeleev's prediction of the properties of then-undiscovered elements, use the given information about the known properties of the alkali metals to devise a method for determining the corresponding property of francium.

### Think Critically

1. **Devise** an approach that clearly displays the trends for each of the properties given in the table and allows you to extrapolate a value for francium. Use the periodic law as a guide.

**Alkali Metals Data**

| Element | Melting Point (°C) | Boiling Point (°C) | Radius (pm) |
|---|---|---|---|
| Lithium | 180.5 | 1342 | 152 |
| Sodium | 97.8 | 883 | 186 |
| Potassium | 63.4 | 759 | 227 |
| Rubidium | 39.30 | 688 | 248 |
| Cesium | 28.4 | 671 | 265 |
| Francium | ? | ? | ? |

2. **Predict** whether francium is a solid, a liquid, or a gas. How can you support your prediction?
3. **Infer** which column of data presents the greatest possible error in making a prediction. Explain.
4. **Determine** why producing 1 million francium atoms per second is not enough to make measurements, such as density or melting point.

**Transition and inner transition metals** The transition elements are divided into **transition metals** and **inner transition metals.** The two sets of inner transition metals, known as the **lanthanide series** and **actinide series,** are located along the bottom of the periodic table. The rest of the elements in groups 3 to 12 make up the transition metals. Elements from the lanthanide series are used extensively as phosphors, substances that emit light when struck by electrons. Because it is strong and light, the transition metal titanium is used to make frames for bicycles and eyeglasses.

**Connection to Biology** **Nonmetals** Nonmetals occupy the upper-right side of the periodic table. They are represented by the yellow boxes in **Figure 5.** **Nonmetals** are elements that are generally gases or brittle, dull-looking solids. They are poor conductors of heat and electricity. The only nonmetal that is a liquid at room temperature is bromine (Br). The most abundant element in the human body is the nonmetal oxygen, which constitutes 65% of the body mass.

Group 17 is comprised of highly reactive elements that are known as **halogens.** Like the group 1 and group 2 elements, the halogens are often part of compounds. Compounds made with the halogen fluorine (F) are commonly added to toothpaste and drinking water to prevent tooth decay. The extremely unreactive group 18 elements are commonly called the **noble gases** and are used in lasers, a variety of light bulbs, and neon signs.

**VOCABULARY**
**SCIENCE USAGE VS. COMMON USAGE**
**Conductor**
*Science usage:* a substance or body capable of transmitting electricity, heat, or sound
*Copper is a good conductor of heat.*
*Common usage:* a person who conducts an orchestra, chorus, or other group of musical performers
*The new conductor helped the orchestra perform at its best.*

■ **Figure 6** Scientists developing submarine technology created this robot that looks and swims like a real fish. Its body is made of a silicon resin that softens in water.

**Metalloids** The elements in the green boxes bordering the stairstep line in **Figure 5** are called metalloids, or semimetals. **Metalloids** have physical and chemical properties of both metals and nonmetals. Silicon (Si) and germanium (Ge) are two important metalloids, used extensively in computer chips and solar cells. Silicon is also used to make prosthetics or in lifelike applications, as shown in **Figure 6.**

This introduction to the periodic table touches only the surface of its usefulness. You can refer to the Elements Handbook at the end of your textbook to learn more about the elements and their various groups.

## SECTION 1 REVIEW

Section Self-Check

### Section Summary

- The elements were first organized by increasing atomic mass, which led to inconsistencies. Later, they were organized by increasing atomic number.
- The periodic law states that when the elements are arranged by increasing atomic number, there is a periodic repetition of their chemical and physical properties.
- The periodic table organizes the elements into periods (rows) and groups or families (columns); elements with similar properties are in the same group.
- Elements are classified as either metals, nonmetals, or metalloids.

1. **MAIN IDEA Describe** the development of the modern periodic table. Include contributions made by Lavoisier, Newlands, Mendeleev, and Moseley.
2. **Sketch** a simplified version of the periodic table, and indicate the location of metals, nonmetals, and metalloids.
3. **Describe** the general characteristics of metals, nonmetals, and metalloids.
4. **Identify** each of the following as a representative element or a transition element.

   **a.** lithium (Li) **b.** platinum (Pt) **c.** promethium (Pm) **d.** carbon (C)
5. **Compare** For each of the given elements, list two other elements with similar chemical properties.

   **a.** iodine (I) **b.** barium (Ba) **c.** iron (Fe)
6. **Compare** According to the periodic table, which two elements have an atomic mass less than twice their atomic number?
7. **Interpret Data** A company plans to make an electronic device. They need to use an element that has chemical behavior similar to that of silicon (Si) and lead (Pb). The element must have an atomic mass greater than that of sulfur (S), but less than that of cadmium (Cd). Use the periodic table to determine which element the company could use.

SECTION 2

# Classification of the Elements

**MAINIDEA** **Elements are organized into different blocks in the periodic table according to their electron configurations.**

### Essential Questions

- Why do elements in the same group have similar properties?
- Based on their electron configurations, what are the four blocks of the periodic table?

### Review Vocabulary

**valence electron:** electron in an atom's outermost orbital; determines the chemical properties of an atom

**CHEM 4 YOU** **A house number is not enough to deliver a letter to the correct address. More information, such as street name, city, and state, is necessary to deliver the letter. Similarly, chemical elements are identified according to details about the arrangement of their electrons.**

## Organizing the Elements by Electron Configuration

As you learned previously, electron configuration determines the chemical properties of an element. Writing out electron configurations using the aufbau diagram can be tedious. Fortunately, you can determine an atom's electron configuration and its number of valence electrons from its position on the periodic table. The electron configurations for some of the group 1 elements are listed in **Table 3.** All four configurations have a single electron in their outermost orbitals.

**Valence electrons** Recall that electrons in the highest principal energy level of an atom are called valence electrons. Each of the group 1 elements has one electron in its highest energy level; thus, each element has one valence electron. The group 1 elements have similar chemical properties because they all have the same number of valence electrons. This is one of the most important relationships in chemistry; atoms in the same group have similar chemical properties because they have the same number of valence electrons. Each group 1 element has a valence electron configuration of $s^1$. Each group 2 element has a valence electron configuration of $s^2$. Each column in groups 1, 2, and 13 to 18 on the periodic table has its own valence electron configuration.

**Valence electrons and period** The energy level of an element's valence electrons indicates the period on the periodic table in which it is found. For example, lithium's valence electron is in the second energy level and lithium is found in period 2. Now look at gallium, with its electron configuration of $[Ar]4s^23d^{10}4p^1$. Gallium's valence electrons are in the fourth energy level, and gallium is found in the fourth period.

**Table 3** Electron Configuration for the Group 1 Elements

| | | | |
|---|---|---|---|
| Period 1 | hydrogen | $1s^1$ | $1s^1$ |
| Period 2 | lithium | $1s^22s^1$ | $[He]2s^1$ |
| Period 3 | sodium | $1s^22s^22p^63s^1$ | $[Ne]3s^1$ |
| Period 4 | potassium | $1s^22s^22p^63s^23p^64s^1$ | $[Ar]4s^1$ |

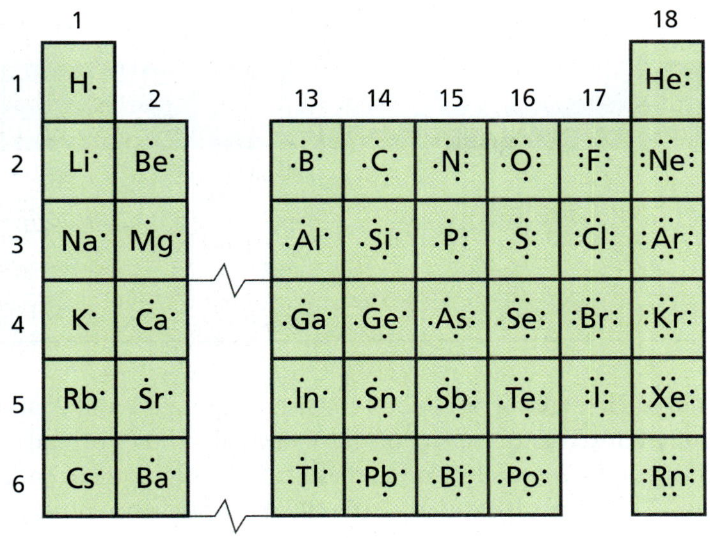


■ **Figure 7** The figure shows the electron-dot structure of most representative elements.

**Observe** ***How does the number of valence electrons vary within a group?***

### Valence electrons of the representative elements

Elements in group 1 have one valence electron; group 2 elements have two valence electrons. Group 13 elements have three valence electrons, group 14 elements have four, and so on. The noble gases in group 18 each have eight valence electrons, with the exception of helium, which has only two valence electrons. **Figure 7** shows how the electron-dot structures you studied previously illustrate the connection between group number and number of valence electrons. Notice that the number of valence electrons for the elements in groups 13 to 18 is ten less than their group number.

## The s-, p-, d-, and f-Block Elements

The periodic table has columns and rows of varying sizes. The reason behind the table's odd shape becomes clear if it is divided into sections, or blocks, representing the atom's energy sublevel being filled with valence electrons. Because there are four different energy sublevels (s, p, d, and f), the periodic table is divided into four distinct blocks, as shown in **Figure 8.**

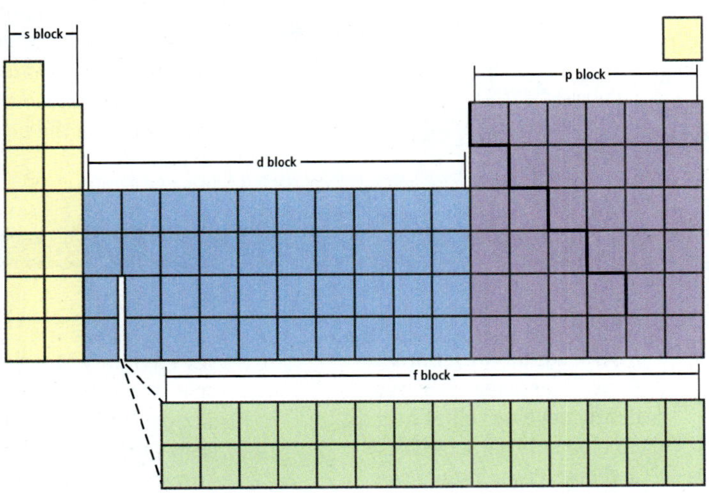


■ **Figure 8** The periodic table is divided into four blocks—s, p, d, and f.

**Analyze** ***What is the relationship between the maximum number of electrons an energy sublevel can hold and the number of columns in that block on the diagram?***

Explore **noble gas configuration with an interactive table.** Concepts In Motion

**Table 4** Noble Gas Electron Configuration

| Period | Principal Energy Level | Element | Electron Configuration |
|---|---|---|---|
| 1 | $n = 1$ | helium | $1s^2$ |
| 2 | $n = 2$ | neon | $[He]2s^2 2p^6$ |
| 3 | $n = 3$ | argon | $[Ne]3s^2 3p^6$ |
| 4 | $n = 4$ | krypton | $[Ar]4s^2 3d^{10} 4p^6$ |

Watch a **video about elements and the periodic table.**

Video

**s-Block elements** The s-block consists of groups 1 and 2, and the element helium. Group 1 elements have partially filled s orbitals containing one valence electron and electron configurations ending in $s^1$. Group 2 elements have completely filled s orbitals containing two valence electrons and electron configurations ending in $s^2$. Because s orbitals hold two electrons at most, the s-block spans two groups.

**p-Block elements** After the s sublevel is filled, the valence electrons next occupy the p sublevel. The p-block, comprised of groups 13 through 18, contains elements with filled or partially filled p orbitals. There are no p-block elements in period 1 because the p sublevel does not exist for the first principal energy level ($n = 1$). The first p-block element is boron (B), in the second period. The p-block spans six groups because the three p orbitals can hold a maximum of six electrons. The group 18 elements (noble gases) are unique members of the p-block. Their atoms are so stable that they undergo virtually no chemical reactions. The electron configurations of the first four noble gas elements are shown in **Table 4.** Here, both the s and p orbitals corresponding to the period's principal energy level are completely filled. This arrangement of electrons results in an unusually stable atomic structure. Together, the s- and p-blocks comprise the representative elements.

**VOCABULARY**

**ACADEMIC VOCABULARY**

**Structure**

something made up of more-or-less interdependent elements or parts

*Many scientists were involved in the discovery of the structure of the atom.*

■ **Figure 9**

**History of the Periodic Table**

The modern periodic table is the result of the work of many scientists over the centuries who studied elements and discovered periodic patterns in their properties.

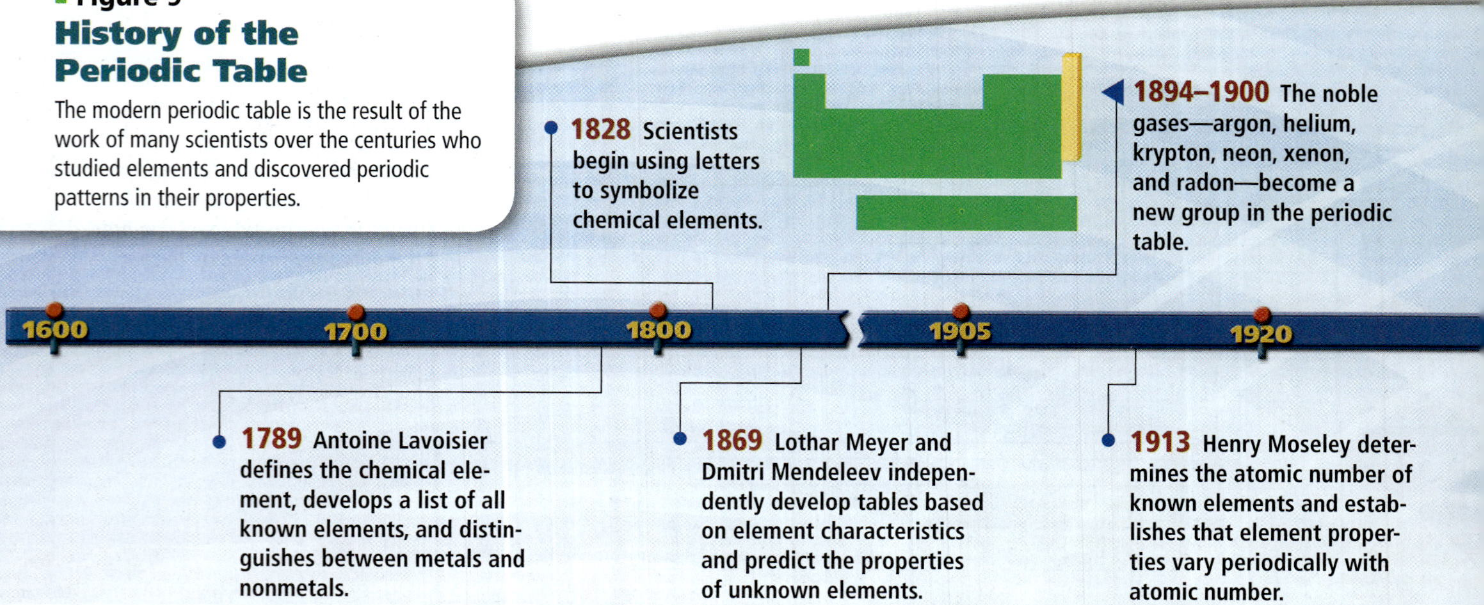

**d-Block elements** The d-block contains the transition metals and is the largest of the blocks. Although there are a number of exceptions, d-block elements are usually characterized by a filled outermost s orbital of energy level $n$, and filled or partially filled d orbitals of energy level $n$-1. As you move across a period, electrons fill the d orbitals. For example, scandium (Sc), the first d-block element, has an electron configuration of $[Ar]4s^2 3d^1$. Titanium, the next element on the table, has an electron configuration of $[Ar]4s^2 3d^2$. Note that titanium's filled outermost s orbital has an energy level of $n = 4$, while the d orbital, which is partially filled, has an energy level of $n = 3$. As you learned previously, the aufbau principle states that the 4s orbital has a lower energy level than the 3d orbital. Therefore, the 4s orbital is filled before the 3d orbital. The five d orbitals can hold a total of ten electrons; thus, the d-block spans ten groups on the periodic table.

**f-Block elements** The f-block contains the inner transition metals. Its elements are characterized by a filled, or partially filled outermost s orbital, and filled or partially filled 4f and 5f orbitals. The electrons of the f sublevel do not fill their orbitals in a predictable manner. Because there are seven f orbitals holding up to a maximum of 14 electrons, the f-block spans 14 columns of the periodic table.

Therefore, the s-, p-, d-, and f-blocks determine the shape of the periodic table. As you proceed down through the periods, the principal energy level increases, as does the number of orbitals containing electrons. Note that period 1 contains only s-block elements, periods 2 and 3 contain both s- and p-block elements, periods 4 and 5 contain s-, p-, and d-block elements, and periods 6 and 7 contain s-, p-, d-, and f-block elements.

The development of the periodic table took many years and is still an ongoing project as new elements are synthetized. Refer to **Figure 9** to learn more about the history of the periodic table and the work of the many scientists who contributed to its development.

**READING CHECK Summarize** how each block of the periodic table is defined.

## CAREERS IN CHEMISTRY

**Research Chemist** Some nuclear chemists specialize in studying the newest and heaviest elements. To produce heavy elements, a nuclear chemist works with a large team, including physicists, engineers, and technicians. Heavy elements are produced by collisions in a particle accelerator. The nuclear chemist analyzes the data from these collisions to identify the elements and understand their properties.

WebQuest

Lawrence Berkeley National Laboratory

**1940** Synthesized elements with an atomic number larger than 92 become part of a new block of the periodic table called the actinides.

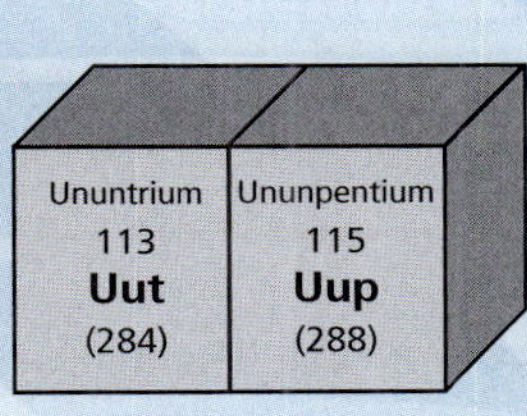


**2004** Scientists in Russia report the discovery of elements 113 and 115.

**2010** Scientists of the Joint Institute for Nuclear Research in Dubna, Russia, report synthesis of a new element with an atomic number of 117.

1950 — 1965 — 1980 — 1995 — 2010

**1969** Researchers at the University of California, Berkeley synthesize the first element heavier than the actinides. It has a half-life of 4.7 seconds and is named rutherfordium.

**1999** Researchers report the discovery of element 114, which they temporarily name ununquadium. Scientists believe this element might be the first of a series of relatively stable synthetic elements.

**2012** The International Union of Pure and Applied Chemistry (IUPAC) officially approves the names *Flerovium* for element 114 and *Livermorium* for element 116.

## EXAMPLE Problem 1

EXAMPLE PROBLEM

**ELECTRON CONFIGURATION AND THE PERIODIC TABLE** Strontium, which is used to produce red fireworks, has an electron configuration of $[Kr]5s^2$. Without using the periodic table, determine the group, period, and block of strontium.

### 1 ANALYZE THE PROBLEM

You are given the electron configuration of strontium.

| **Known** | **Unknown** |
|---|---|
| Electron configuration = $[Kr]5s^2$ | **Group = ?** |
| | **Period = ?** |
| | **Block = ?** |

### 2 SOLVE FOR THE UNKNOWN

The $s^2$ indicates that strontium's valence electrons fill the s sublevel. Thus, strontium is in group 2 of the **s-block.**

**For representative elements, the number of valence electrons can indicate the group number.**

The 5 in $5s^2$ indicates that strontium is in **period 5.**

**The number of the highest energy level indicates the period number.**

### 3 EVALUATE THE ANSWER

The relationships between electron configuration and position on the periodic table have been correctly applied.

## PRACTICE Problems

Do additional problems. Online Practice

PRACTICE PROBLEMS

**8.** Without using the periodic table, determine the group, period, and block of an atom with the following electron configurations.

**a.** $[Ne]3s^2$ **b.** $[He]2s^2$ **c.** $[Kr]5s^24d^{10}5p^5$

**9.** What are the symbols for the elements with the following valence electron configurations?

**a.** $s^2d^1$ **b.** $s^2p^3$ **c.** $s^2p^6$

**10. Challenge** Write the electron configuration of the following elements.

**a.** the group 2 element in the fourth period

**b.** the group 12 element in the fourth period

**c.** the noble gas in the fifth period

**d.** the group 16 element in the second period

## SECTION 2 REVIEW

Section Self-Check

### Section Summary

- The periodic table has four blocks (s, p, d, f).
- Elements within a group have similar chemical properties.
- The group number for elements in groups 1 and 2 equals the element's number of valence electrons.
- The energy level of an atom's valence electrons equals its period number.

**11. MAINIDEA Explain** what determines the blocks in the periodic table.

**12. Determine** in which block of the periodic table are the elements having the following valence electron configurations.

**a.** $s^2p^4$ **b.** $s^1$ **c.** $s^2d^1$ **d.** $s^2p^1$

**13. Infer** Xenon, a nonreactive gas used in strobe lights, is a poor conductor of heat and electricity. Would you expect xenon to be a metal, a nonmetal, or a metalloid? Where would you expect it to be on the periodic table? Explain.

**14. Explain** why elements within a group have similar chemical properties.

**15. Model** Make a simplified sketch of the periodic table, and label the s-, p-, d-, and f-blocks.

SECTION 3

# Periodic Trends

**MAINIDEA** Trends among elements in the periodic table include their sizes and their abilities to lose or attract electrons.

### Essential Questions

- What are the period and group trends of different properties?
- How are period and group trends in atomic radii related to electron configuration?

### Review Vocabulary

**principal energy level:** the major energy level of an atom

### New Vocabulary

ion
ionization energy
octet rule
electronegativity

**CHEM 4 YOU** A calendar is a useful tool for keeping track of activities. The pattern of days, from Sunday to Saturday, is repeated week after week. If you list an activity many weeks ahead, you can tell from the day of the week what else might happen on that day. In much the same way, the organization of the periodic table tells us about the behavior of many of the elements.

## Atomic Radius

Many properties of the elements tend to change in a predictable way, known as a trend, as you move across a period or down a group. Atomic size is one such periodic trend. The sizes of atoms are influenced by electron configuration.

Recall that the electron cloud surrounding a nucleus does not have a clearly defined edge. The outer limit of an electron cloud is defined as the spherical surface within which there is a 90% probability of finding an electron. However, this surface does not exist in a physical way, as the outer surface of a golf ball does. Atomic size is defined by how closely an atom lies to a neighboring atom. Because the nature of the neighboring atom can vary from one substance to another, the size of the atom itself also tends to vary somewhat from substance to substance.

For metals such as sodium, the atomic radius is defined as half the distance between adjacent nuclei in a crystal of the element, as shown in **Figure 10.** For elements that commonly occur as molecules, such as many nonmetals, the atomic radius is defined as half the distance between nuclei of identical atoms that are chemically bonded together. The atomic radius of a nonmetal diatomic hydrogen molecule ($H_2$) is shown in **Figure 10.**

**Figure 10** Atomic radii depend on the type of bonds that atoms form.

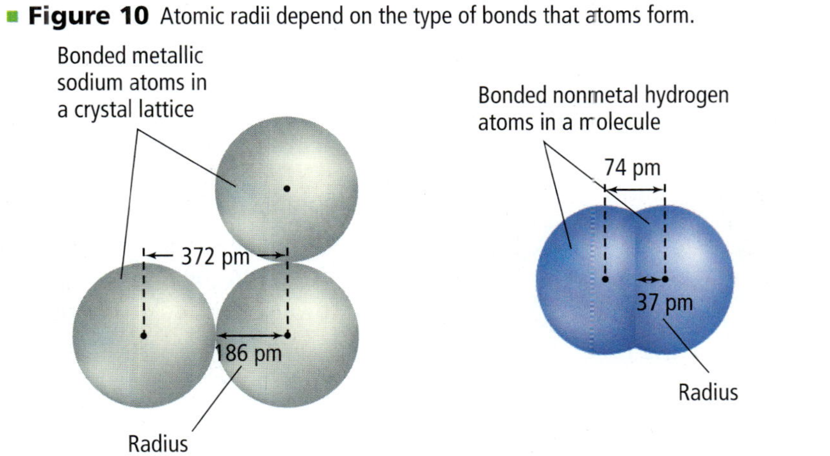


The radius of a metal atom is one-half the distance between two adjacent atoms in the crystal.

The radius of a nonmetal atom is often determined from a molecule of two identical atoms.

■ **Figure 11** The atomic radii of the representative elements, given in picometers ($10^{-12}$ m), vary as you move from left to right within a period and down a group.
**Infer** ***why the atomic radii increase as you move down a group.***

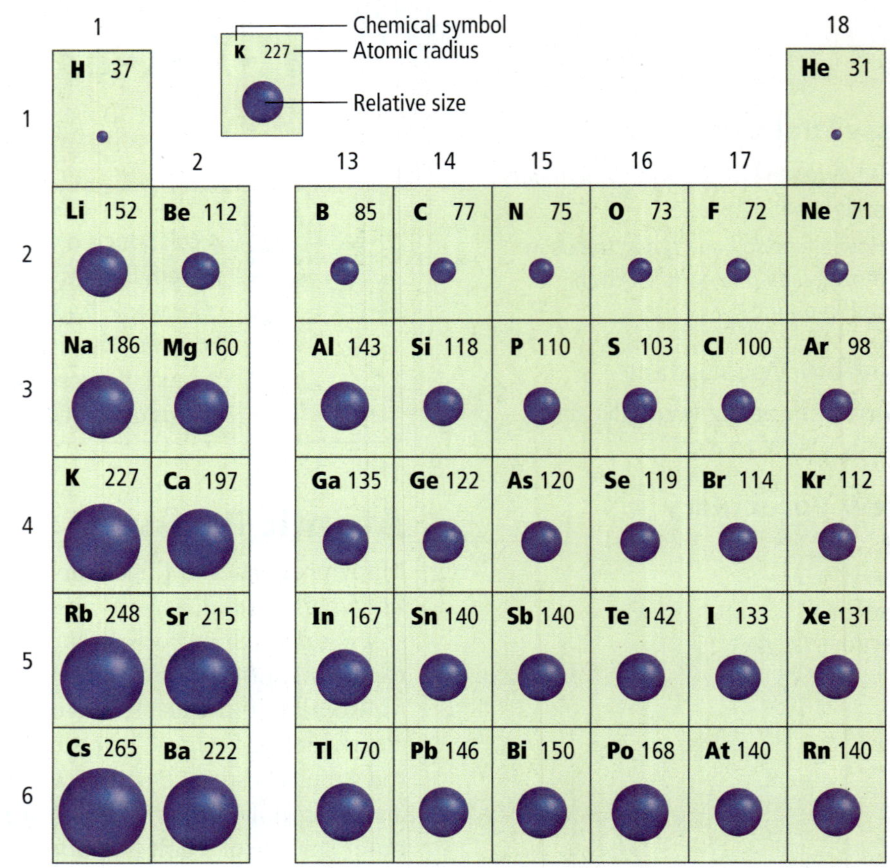


View an **animation about trends in atomic radii.**

Concepts In Motion

FOLDABLES®
Incorporate information from this section into your Foldable.

**Trends within periods** In general, there is a decrease in atomic radii as you move from left to right across a period. This trend, shown in **Figure 11,** is caused by the increasing positive charge in the nucleus and the fact that the principal energy level within a period remains the same. Each successive element has one additional proton and electron, and each additional electron is added to orbitals corresponding to the same principal energy level. Moving across a period, no additional electrons come between the valence electrons and the nucleus. Thus, the valence electrons are not shielded from the increased nuclear charge, which pulls the outermost electrons closer to the nucleus.

☑ READING CHECK **Discuss** how the fact that the principal energy level remains the same within a period explains the decrease in the atomic radii across a period.

■ **Figure 12** Atomic radii generally decrease from left to right in a period and generally increase as you move down a group.

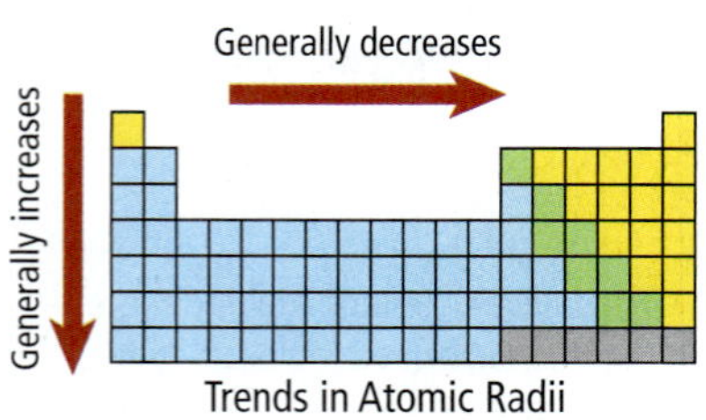


**Trends within groups** Atomic radii generally increase as you move down a group. The nuclear charge increases, and electrons are added to orbitals corresponding to successively higher principal energy levels. However, the increased nuclear charge does not pull the outer electrons toward the nucleus to make the atom smaller.

Moving down a group, the outermost orbital increases in size along with the increasing principal energy level; thus, the atom becomes larger. The larger orbital means that the outer electrons are farther from the nucleus. This increased distance offsets the pull of the increased nuclear charge. Also, as additional orbitals between the nucleus and the outer electrons are occupied, these electrons shield the outer electrons from the nucleus. **Figure 12** summarizes the group and period trends.

## EXAMPLE Problem 2

**INTERPRET TRENDS IN ATOMIC RADII** Which has the largest atomic radius: carbon (C), fluorine (F), beryllium (Be), or lithium (Li)? Answer without referring to **Figure 11.** Explain your answer in terms of trends in atomic radii.

### 1 ANALYZE THE PROBLEM

You are given four elements. First, determine the groups and periods the elements occupy. Then apply the general trends in atomic radii to determine which has the largest atomic radius.

### 2 SOLVE FOR THE UNKNOWN

From the periodic table, all the elements are found to be in period 2. **Determine the periods.**

Ordering the elements from left-to-right across the period yields: Li, Be, C, and F.

The first element in period 2, lithium, has the largest radius. **Apply the trend of decreasing radii across a period.**

### 3 EVALUATE THE ANSWER

The period trend in atomic radii has been correctly applied. Checking radii values in **Figure 11** verifies the answer.

## PRACTICE Problems

Do additional problems. Online Practice

**Answer the following questions using your knowledge of group and period trends in atomic radii. Do not use the atomic radii values in Figure 11 to answer the questions.**

**16.** Which has the largest atomic radius: magnesium (Mg), silicon (Si), sulfur (S), or sodium (Na)? The smallest?

**17.** The figure on the right shows helium, krypton, and radon. Which one is krypton? How can you tell?

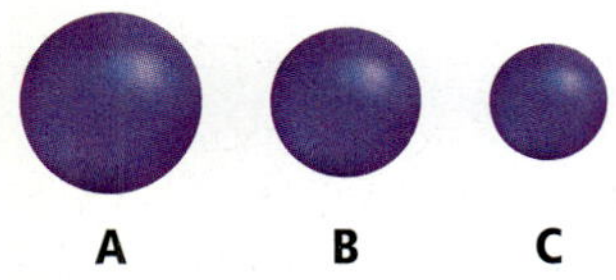

**18.** Can you determine which of two unknown elements has the larger radius if the only known information is that the atomic number of one of the elements is 20 greater than the other? Explain.

**19. Challenge** Determine which element in each pair has the largest atomic radius:

**a.** the element in period 2, group 1; or the element in period 3, group 18

**b.** the element in period 5, group 2; or the element in period 3, group 16

**c.** the element in period 3, group 14; or the element in period 6, group 15

**d.** the element in period 4, group 18; or the element in period 2, group 16

EXAMPLE PROBLEM

PRACTICE PROBLEMS

## Ionic Radius

Atoms can gain or lose one or more electrons to form ions. Because electrons are negatively charged, atoms that gain or lose electrons acquire a net charge. Thus, an **ion** is an atom or a bonded group of atoms that has a positive or negative charge. You will learn about ions later, but for now, consider how the formation of an ion affects the size of an atom.

When atoms lose electrons and form positively charged ions, they always become smaller. The reason is twofold. The electron lost from the atom will almost always be a valence electron. The loss of a valence electron can leave a completely empty outer orbital, which results in a smaller radius. Furthermore, the electrostatic repulsion between the now-fewer number of remaining electrons decreases. As a result, they experience a greater nuclear charge allowing these remaining electrons to be pulled closer to the positively charged nucleus.

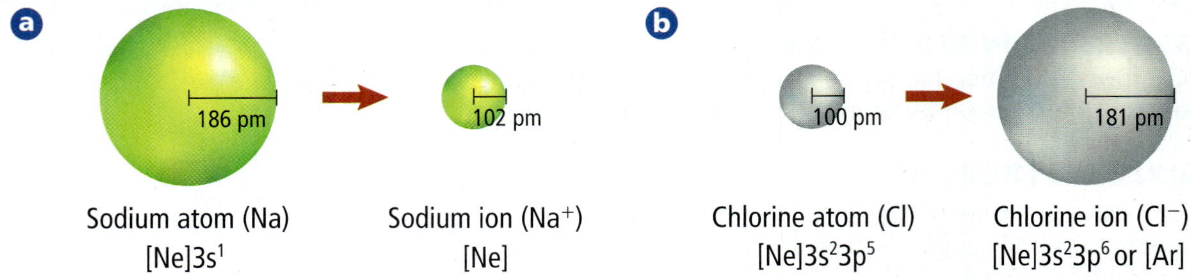


■ **Figure 13** The size of atoms varies greatly when they form ions.
**a.** Positive ions are smaller than the neutral atoms from which they form.
**b.** Negative ions are larger than the neutral atoms from which they form.

**FOLDABLES®** Incorporate information from this section into your Foldable.

When atoms gain electrons and form negatively charged ions, they become larger. The addition of an electron to an atom increases the electrostatic repulsion between the atom's outer electrons, forcing them to move farther apart. The increased distance between the outer electrons results in a larger radius.

**Figure 13a** illustrates how the radius of sodium decreases when sodium atoms form positive ions, and **Figure 13b** shows how the radius of chlorine increases when chlorine atoms form negative ions.

**Trends within periods** The ionic radii of most of the representative elements are shown in **Figure 14.** Note that elements on the left side of the table form smaller positive ions, and elements on the right side of the table form larger negative ions. In general, as you move from left to right across a period, the size of the positive ions gradually decreases. Then, beginning in group 15 or 16, the size of the much-larger negative ions also gradually decreases.

■ **Figure 14** The ionic radii of most of the representative elements are shown in picometers ($10^{-12}$ m).

**Explain** *why the ionic radii increase for both positive and negative ions as you move down a group.*

| Period | 1 | 2 | 13 | 14 | 15 | 16 | 17 |
|---|---|---|---|---|---|---|---|
| 2 | **Li** 76, 1+ | **Be** 31, 2+ | **B** 20, 3+ | **C** 15, 4+ | **N** 146, 3- | **O** 140, 2- | **F** 133, 1- |
| 3 | **Na** 102, 1+ | **Mg** 72, 2+ | **Al** 54, 3+ | **Si** 41, 4+ | **P** 212, 3- | **S** 184, 2- | **Cl** 181, 1- |
| 4 | **K** 138, 1+ | **Ca** 100, 2+ | **Ga** 62, 3+ | **Ge** 53, 4+ | **As** 222, 3- | **Se** 198, 2- | **Br** 196, 1- |
| 5 | **Rb** 152, 1+ | **Sr** 118, 2+ | **In** 81, 3+ | **Sn** 71, 4+ | **Sb** 62, 5+ | **Te** 221, 2- | **I** 220, 1- |
| 6 | **Cs** 167, 1+ | **Ba** 135, 2+ | **Tl** 95, 3+ | **Pb** 84, 4+ | **Bi** 74, 5+ | | |

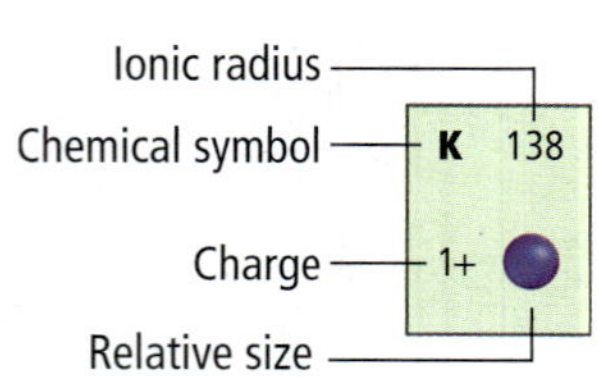

**Trends within groups** As you move down a group, an ion's outer electrons are in orbitals corresponding to higher principal energy levels, resulting in a gradual increase in ionic size. Thus, the ionic radii of both positive and negative ions increase as you move down a group. The group and period trends in ionic radii are summarized in **Figure 15.**

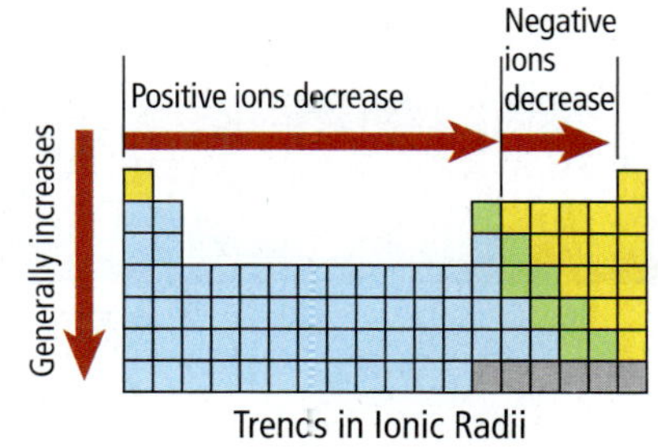


**Figure 15** The diagram summarizes the general trends in ionic radii.

## Ionization Energy

To form a positive ion, an electron must be removed from a neutral atom. This requires energy. The energy is needed to overcome the attraction between the positive charge of the nucleus and the negative charge of the electron. **Ionization energy** is defined as the energy required to remove an electron from a gaseous atom. For example, $8.64 \times 10^{-19}$ J is required to remove an electron from a gaseous lithium atom. The energy required to remove the first outermost electron from an atom is called the first ionization energy. The first ionization energy of lithium equals $8.64 \times 10^{-19}$ J. The loss of the electron results in the formation of a $Li^+$ ion. The first ionization energies of the elements in periods 1 through 5 are plotted on the graph in **Figure 16.**

**READING CHECK Define** *ionization energy.*

Think of ionization energy as an indication of how strongly an atom's nucleus holds onto its valence electrons. A high ionization energy value indicates the atom has a strong hold on its electrons. Atoms with large ionization energy values are less likely to form positive ions. Likewise, a low ionization energy value indicates an atom loses an outer electron easily. Such atoms are likely to form positive ions. Lithium's low ionization energy, for example, is important for its use in lithium-ion computer backup batteries, where the ability to lose electrons easily makes a battery that can quickly provide a large amount of electrical power.

Get help with **identifying trends.**

Personal Tutor

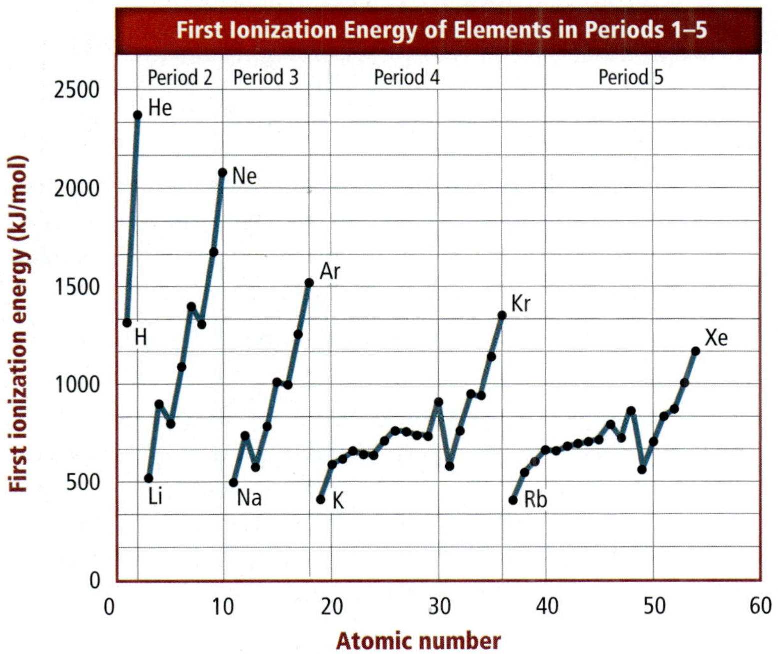


**Figure 16** The first ionization energies for elements in periods 1 through 5 are shown as a function of the atomic number.

**GRAPH CHECK**

**Describe** the trend in first ionization energies within a group.

**Table 5 Successive Ionization Energies for the Period 2 Elements**

| Element | Valence Electrons | Ionization Energy (kJ/mol)* | | | | | | | | |
|---|---|---|---|---|---|---|---|---|---|---|
| | | 1st | 2nd | 3rd | 4th | 5th | 6th | 7th | 8th | 9th |
| Li | 1 | 520 | 7300 | 11,810 | | | | | | |
| Be | 2 | 900 | 1760 | 14,850 | 21,010 | | | | | |
| B | 3 | 800 | 2430 | 3660 | 25,020 | 32,820 | | | | |
| C | 4 | 1090 | 2350 | 4620 | 6220 | 37,830 | 47,280 | | | |
| N | 5 | 1400 | 2860 | 4580 | 7480 | 9440 | 53,270 | 64,360 | | |
| O | 6 | 1310 | 3390 | 5300 | 7470 | 10,980 | 13,330 | 71,870 | 84,080 | |
| F | 7 | 1680 | 3370 | 6050 | 8410 | 11,020 | 15,160 | 17,870 | 92,040 | 106,430 |
| Ne | 8 | 2080 | 3950 | 6120 | 9370 | 12,180 | 15,240 | 20,000 | 23,070 | 115,380 |

*mol is an abbreviation for mole, a quantity of matter.

Each set of connected points on the graph in **Figure 16** represents the elements in a period. The group 1 metals have low ionization energies. Thus, group 1 metals (Li, Na, K, Rb) are likely to form positive ions. The group 18 elements (He, Ne, Ar, Kr, Xe) have high ionization energies and are unlikely to form ions. The stable electron configuration of gases of group 18 greatly limits their reactivity.

**Removing more than one electron** After removing the first electron from an atom, it is possible to remove additional electrons. The amount of energy required to remove a second electron from a 1+ ion is called the second ionization energy, the amount of energy required to remove a third electron from a 2+ ion is called the third ionization energy, and so on. **Table 5** lists the first through ninth ionization energies for elements in period 2.

Reading across **Table 5** from left to right, you will see that the energy required for each successive ionization always increases. However, the increase in energy does not occur smoothly. Note that for each element there is an ionization for which the required energy increases dramatically. For example, the second ionization energy of lithium (7300 kJ/mol) is much greater than its first ionization energy (520 kJ/mol). This means that a lithium atom is likely to lose its first valence electron but extremely unlikely to lose its second.

☑ READING CHECK **Infer** how many electrons carbon is likely to lose.

If you examine the table, you will notice that the ionization at which the large increase in energy occurs is related to the atom's number of valence electrons. Lithium has one valence electron and the increase occurs after the first ionization energy. Lithium easily forms the common lithium 1+ ion but is unlikely to form a lithium 2+ ion. The increase in ionization energy shows that atoms hold onto their inner core electrons much more strongly than they hold onto their valence electrons.

## RealWorld CHEMISTRY

### Ionization Energy

**SCUBA DIVING** The increased pressure that scuba divers experience far below the water's surface can cause too much oxygen to enter their blood, which would result in confusion and nausea. To avoid this, divers sometimes use a gas mixture called *heliox*—oxygen diluted with helium. Helium's high ionization energy ensures that it will not react chemically in the bloodstream.

**Trends within periods** As shown in **Figure 16** and by the values in **Table 5,** first ionization energies generally increase as you move from left to right across a period. The increased nuclear charge of each successive element produces an increased hold on the valence electrons.

**Trends within groups** First ionization energies generally decrease as you move down a group. This decrease in energy occurs because atomic size increases as you move down the group. Less energy is required to remove the valence electrons farther from the nucleus. **Figure 17** summarizes the group and period trends in first ionization energies.

**Octet rule** When a sodium atom loses its single valence electron to form a 1+ sodium ion, its electron configuration changes as shown below.

Sodium atom $1s^22s^22p^63s^1$ Sodium ion $1s^22s^22p^6$

Note that the sodium ion has the same electron configuration as neon ($1s^22s^22p^6$), a noble gas. This observation leads to one of the most important principles in chemistry, the octet rule. The **octet rule** states that atoms tend to gain, lose, or share electrons in order to acquire a full set of eight valence electrons. This reinforces what you learned earlier, that the electron configuration of filled s and p orbitals of the same energy level (consisting of eight valence electrons) is unusually stable. Note that the first-period elements are an exception to the rule, as they are complete with only two valence electrons.

The octet rule is useful for determining the type of ions likely to form. Elements on the right side of the periodic table tend to gain electrons in order to acquire the noble gas configuration; therefore, these elements tend to form negative ions. In a similar manner, elements on the left side of the table tend to lose electrons and form positive ions.

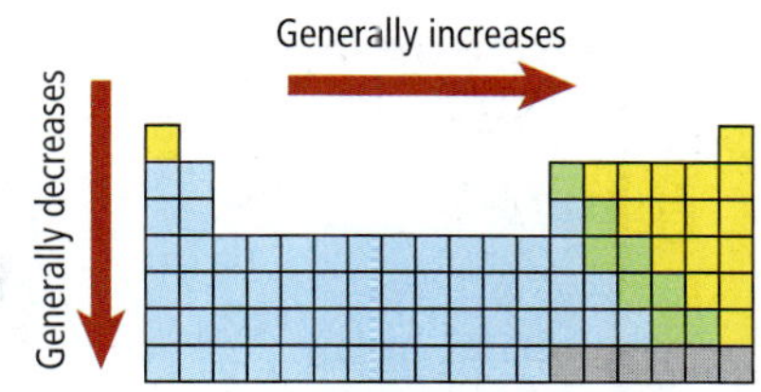


**Figure 17** Ionization energies generally increase from left to right in a period and generally decrease as you move down a group.

**FOLDABLES®** Incorporate information from this section into your Foldable.

Explore the **periodic properties of the elements.**

Virtual Investigations

# MiniLAB

iLab Station

## Organize Elements

**Can you find the pattern?**

### Procedure

1. Read and complete the lab safety form.
2. Make a set of element cards based on the information in the chart at right.
3. Organize the cards by increasing mass, and start placing them into a 4 column × 3 row grid.
4. Place each card based on its properties, and leave gaps when necessary.

| Symbol | Mass (g) | State | Color |
|---|---|---|---|
| Ad | 52.9 | solid/liquid | orange |
| Ax | 108.7 | ductile solid | light blue |
| Bp | 69.3 | gas | red |
| Cx | 112.0 | brittle solid | light green |
| Lq | 98.7 | ductile solid | blue |
| Pk | 83.4 | brittle solid | green |
| Qa | 68.2 | ductile solid | dark blue |
| Rx | 106.2 | liquid | yellow |
| Tu | 64.1 | brittle solid | hunter |
| Xn | 45.0 | gas | crimson |

### Analysis

1. **Make a table** listing the placement of each element.
2. **Describe** the period (across) and group (down) trends for the color in your new table.
3. **Describe** the period and group trends for the mass in your new table. Explain your placement of any elements that do not fit the trends.
4. **Predict** the placement of a newly found element, Ph, that is a fuchsia gas. What would be an expected range for the mass of Ph?
5. **Predict** the properties for the element that would fill the last remaining gap in the table.

**Increasing electronegativity** →

↓ **Decreasing electronegativity**

- electronegativity < 1.0
- 1.0 ≤ electronegativity < 2.0
- 2.0 ≤ electronegativity < 3.0
- 3.0 ≤ electronegativity < 4.0

| 1 **H** 2.20 | | | | | | | | | | | | | | | | | 2 **He** |
|---|---|---|---|---|---|---|---|---|---|---|---|---|---|---|---|---|---|
| 3 **Li** 0.98 | 4 **Be** 1.57 | | | | | | | | | | | 5 **B** 2.04 | 6 **C** 2.55 | 7 **N** 3.04 | 8 **O** 3.44 | 9 **F** 3.98 | 10 **Ne** |
| 11 **Na** 0.93 | 12 **Mg** 1.31 | | | | | | | | | | | 13 **Al** 1.61 | 14 **Si** 1.90 | 15 **P** 2.19 | 16 **S** 2.58 | 17 **Cl** 3.16 | 18 **Ar** |
| 19 **K** 0.82 | 20 **Ca** 1.00 | 21 **Sc** 1.36 | 22 **Ti** 1.54 | 23 **V** 1.63 | 24 **Cr** 1.66 | 25 **Mn** 1.55 | 26 **Fe** 1.83 | 27 **Co** 1.88 | 28 **Ni** 1.91 | 29 **Cu** 1.90 | 30 **Zn** 1.65 | 31 **Ga** 1.81 | 32 **Ge** 2.01 | 33 **As** 2.18 | 34 **Se** 2.55 | 35 **Br** 2.96 | 36 **Kr** |
| 37 **Rb** 0.82 | 38 **Sr** 0.95 | 39 **Y** 1.22 | 40 **Zr** 1.33 | 41 **Nb** 1.6 | 42 **Mo** 2.16 | 43 **Tc** 2.10 | 44 **Ru** 2.2 | 45 **Rh** 2.28 | 46 **Pd** 2.20 | 47 **Ag** 1.93 | 48 **Cd** 1.69 | 49 **In** 1.78 | 50 **Sn** 1.96 | 51 **Sb** 2.05 | 52 **Te** 2.1 | 53 **I** 2.66 | 54 **Xe** |
| 55 **Cs** 0.79 | 56 **Ba** 0.89 | 57 **La** 1.1 | 72 **Hf** 1.3 | 73 **Ta** 1.5 | 74 **W** 1.7 | 75 **Re** 1.9 | 76 **Os** 2.2 | 77 **Ir** 2.2 | 78 **Pt** 2.2 | 79 **Au** 2.4 | 80 **Hg** 1.9 | 81 **Tl** 1.8 | 82 **Pb** 1.8 | 83 **Bi** 1.9 | 84 **Po** 2.0 | 85 **At** 2.2 | 86 **Rn** |
| 87 **Fr** 0.70 | 88 **Ra** 0.90 | 89 **Ac** 1.1 | 104 **Rf** | 105 **Db** | 106 **Sg** | 107 **Bh** | 108 **Hs** | 109 **Mt** | 110 **Ds** | 111 **Rg** | 112 **Cn** | 113 **Uut** | 114 **Fl** | 115 **Uup** | 116 **Lv** | 117 **Uus** | 118 **Uuo** |

**Electronegativity Values in Paulings**

■ **Figure 18** The electronegativity values for most of the elements are shown. The values are given in Paulings.

**Infer** *why electronegativity values are not listed for the noble gases.*

View an **animation about trends in electronegativity.**

**Concepts In Motion**

## Electronegativity

The **electronegativity** of an element indicates the relative ability of its atoms to attract electrons in a chemical bond. As shown in **Figure 18,** electronegativity generally decreases as you move down a group. **Figure 18** also indicates that electronegativity generally increases as you move from left to right across a period.

Electronegativity values are expressed in terms of a numerical value of 3.98 or less. The units of electronegativity are arbitrary units called Paulings, named after American scientist Linus Pauling (1901–1994). Fluorine is the most electronegative element, with a value of 3.98, and cesium and francium are the least electronegative elements, with values of 0.79 and 0.70, respectively. In a chemical bond, the atom with the greater electronegativity more strongly attracts the bond's electrons. Note that because the noble gases form very few compounds, they do not have electronegativity values.

## SECTION 3 REVIEW

**Section Self-Check**

### Section Summary

- Atomic and ionic radii decrease from left to right across a period, and increase as you move down a group.
- Ionization energies generally increase from left to right across a period, and decrease as you move down a group.
- The octet rule states that atoms gain, lose, or share electrons to acquire a full set of eight valence electrons.
- Electronegativity generally increases from left to right across a period, and decreases as you move down a group.

**20.** **MAIN**IDEA **Explain** how the period and group trends in atomic radii are related to electron configuration.

**21.** **Indicate** whether fluorine or bromine has a larger value for each of the following properties.

**a.** electronegativity
**b.** ionic radius
**c.** atomic radius
**d.** ionization energy

**22.** **Explain** why it takes more energy to remove the second electron from a lithium atom than it does to remove the fourth electron from a carbon atom.

**23.** **Calculate** Determine the differences in electronegativity, ionic radius, atomic radius, and first ionization energy for oxygen and beryllium.

**24.** **Make and Use Graphs** Graph the atomic radii of the representative elements in periods 2, 3, and 4 versus their atomic numbers. Connect the points of elements in each period, so that there are three separate curves on the graph. Summarize the trends in atomic radii shown on your graph. Explain.

# CHEMISTRY & health

## Elements of the Body

Every time you eat a sandwich or take a breath, you are taking in elements your body needs to function normally. These elements have specific properties, depending on their location on the periodic table. **Figure 1** shows the percent by mass composition of cells in the human body.

**Oxygen** In an adult body, there are more than 14 billion billion billion oxygen atoms! Without a constant input of oxygen into the blood, the human body could die in just a few minutes.

**Carbon** Carbon can form strong bonds with itself and other elements. Carbon forms the long-chained carbon backbones that are an essential part of organic molecules such as carbohydrates, proteins, and lipids. The DNA molecule that determines your physical features relies on the versatility of carbon and its ability to bond with many different elements.

**Hydrogen** There are more hydrogen atoms in the body than atoms of all the other elements combined, although hydrogen represents only 10% of the composition by mass because of their significantly lower mass. The human body, requires hydrogen not in its elemental form, but in a variety of essential compounds, like water. With oxygen and carbon, hydrogen is also a crucial part of carbohydrates and other organic molecules that your body needs for energy.

**Percent by Mass of the Elements in the Human Body**

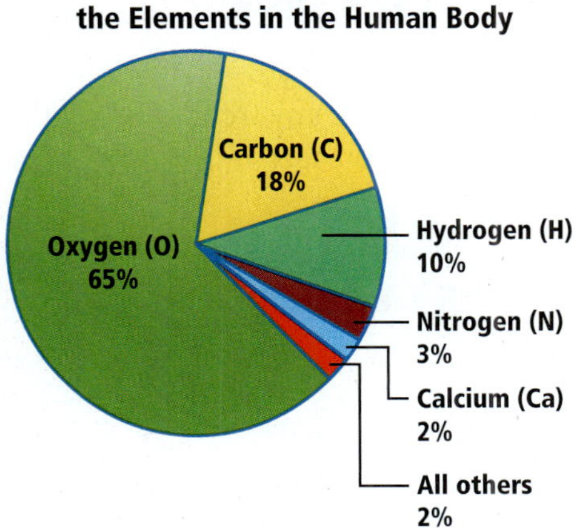


**Figure 1** The human body is composed of many different elements.

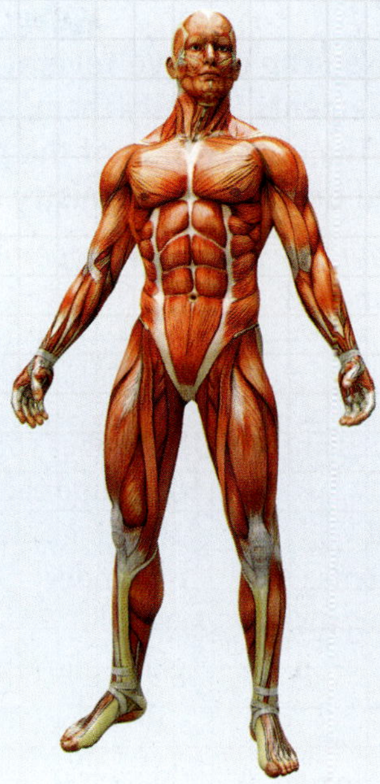

**Figure 2** The entire human body is covered with muscles.

**Nitrogen** As shown in **Figure 2,** the human body is entirely covered with muscle. Nitrogen atoms are found in compounds that make up the proteins your body needs to build muscle.

**Other elements in the body** Oxygen, carbon, hydrogen, and nitrogen are the most abundant elements in your body but only a few of the elements that your body needs to live and grow. Trace elements, which together make up less than 2% of the body's mass, are a critical part of your body. Your bones and teeth could not grow without the constant intake of calcium. Although sulfur comprises less than 1 percent of the human body by mass, it is an essential component and is found in the proteins in your fingernails for instance. Sodium and potassium are crucial for the transmission of electrical signals in your brain.

### WRITING IN ▶ Chemistry

Can you get all of the trace elements you need by eating only pre-packaged food? Why are trace elements necessary? Research trace elements and design a graphic novel that teaches elementary students about these nutrients.

WebQuest

# ChemLAB

iLab Station

## Investigate Descriptive Chemistry

**Background:** You can observe several of the representative elements, classify them, and compare their properties. The observation of the properties of elements is called descriptive chemistry.

**Question:** *What is the pattern of properties of the representative elements?*

### Materials

stoppered test tubes and plastic dishes containing small samples of elements
conductivity apparatus
1.0*M* HCl
small hammer
test tubes (6)
test-tube rack
10-mL graduated cylinder
spatula
glass-marking pencil

### Safety Precautions

***WARNING:*** *Never test chemicals by tasting. 1.0M HCl is harmful to eyes and clothing. Brittle samples might shatter into sharp pieces.*

### Procedure

1. Read and complete the lab safety form.
2. Observe and record the appearance (physical state, color, luster, texture, and so on) of the element sample in each test tube without removing the stoppers.
3. Remove a small sample of each of the elements contained in a plastic dish and place it on a hard surface. Gently tap each element sample with a small hammer. If the element is malleable, it will flatten. If it is brittle, it will shatter. Record your observations.
4. Use the conductivity tester to determine which elements conduct electricity. Clean the electrodes with water, and dry them before testing each element.
5. Label each test tube with the symbol for one of the elements in the plastic dishes. Using a graduated cylinder, add 5 mL of water to each test tube.
6. Use a spatula to put a small amount of each element into the corresponding test tubes. Using a graduated cylinder, add 5 mL of 1.0*M* HCl to each test tube. Observe each tube for at least 1 minute. The formation of bubbles is evidence of a reaction between the acid and the element. Record your observations.

**Observation of Elements**

| Classification | Properties |
|---|---|
| Metals | • malleable<br>• good conductor of electricity<br>• lustrous<br>• silver or white in color<br>• many react with acids |
| Nonmetals | • solids, liquids, or gases<br>• do not conduct electricity<br>• do not react with acids<br>• likely brittle if solid |
| Metalloids | • combine properties of metals and nonmetals |

7. **Cleanup and Disposal** Dispose of all materials as instructed by your teacher.

### Analyze and Conclude

1. **Interpret Data** Using the table above and your observations, list the element samples that display the general characteristics of metals.
2. **Interpret Data** Using the table above and your observations, list the element samples that display the general characteristics of nonmetals.
3. **Interpret Data** Using the table above and your observations, list the element samples that display the general characteristics of metalloids.
4. **Model** Construct a periodic table, and label the representative elements by group (1 through 17). Using your results and the periodic table presented in this chapter, record the identities of elements observed during the lab in the periodic table you have constructed.
5. **Infer** Describe any trends among the elements you observed in the lab.

### INQUIRY EXTENSION

**Investigate** Were there any element samples that did not fit into one of the three categories? What additional investigations could you conduct to learn even more about these elements' characteristics?

# CHAPTER 6 STUDY GUIDE

Vocabulary Practice

**BIGIDEA** **Periodic trends in the properties of atoms allow us to predict physical and chemical properties.**

## SECTION 1 Development of the Modern Periodic Table

**MAINIDEA** The periodic table evolved over time as scientists discovered more useful ways to compare and organize the elements.

- The elements were first organized by increasing atomic mass, which led to inconsistencies. Later, they were organized by increasing atomic number.
- The periodic law states that when the elements are arranged by increasing atomic number, there is a periodic repetition of their chemical and physical properties.
- The periodic table organizes the elements into periods (rows) and groups or families (columns); elements with similar properties are in the same group.
- Elements are classified as either metals, nonmetals, or metalloids.

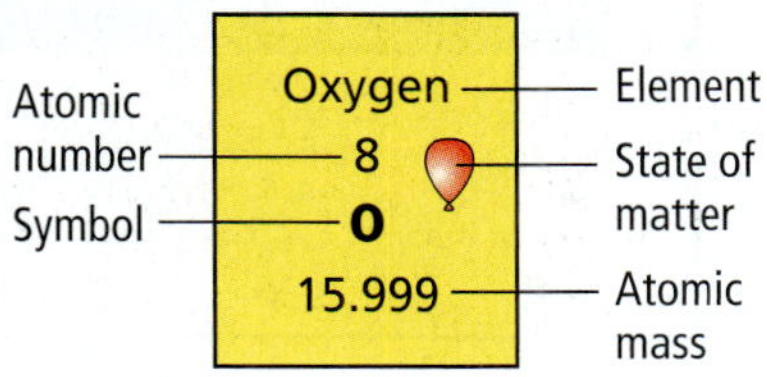


**VOCABULARY**
- periodic law
- group
- period
- representative element
- transition element
- metal
- alkali metal
- alkaline earth metal
- transition metal
- inner transition metal
- lanthanide series
- actinide series
- nonmetal
- halogen
- noble gas
- metalloid

## SECTION 2 Classification of the Elements

**MAINIDEA** Elements are organized into different blocks in the periodic table according to their electron configurations.

- The periodic table has four blocks (s, p, d, f).
- Elements within a group have similar chemical properties.
- The group number for elements in groups 1 and 2 equals the element's number of valence electrons.
- The energy level of an atom's valence electrons equals its period number.

## SECTION 3 Periodic Trends

**MAINIDEA** Trends among elements in the periodic table include their sizes and their abilities to lose or attract electrons.

- Atomic and ionic radii decrease from left to right across a period, and increase as you move down a group.
- Ionization energies generally increase from left to right across a period, and decrease as you move down a group.
- The octet rule states that atoms gain, lose, or share electrons to acquire a full set of eight valence electrons.
- Electronegativity generally increases from left to right across a period, and decreases as you move down a group.

**VOCABULARY**
- ion
- ionization energy
- octet rule
- electronegativity

# CHAPTER 6 ASSESSMENT

## SECTION 1

### Mastering Concepts

**25.** Explain how Mendeleev's periodic table was in error.

**26.** Explain the contribution of Newlands's law of octaves to the development of the modern periodic table.

**27.** Lothar Meyer and Dmitri Mendeleev both proposed similar periodic tables in 1869. Why is Mendeleev generally given credit for the periodic table?

**28.** What is the periodic law?

**29.** Describe the general characteristics of metals.

**30.** What are the general properties of a metalloid?

**31.** Identify each of the following as a metal, a nonmetal, or a metalloid.

- **a.** oxygen
- **b.** barium
- **c.** germanium
- **d.** iron

**32.** Match each item on the left with its corresponding group on the right.

- **a.** alkali metals — **1.** group 18
- **b.** halogens — **2.** group 1
- **c.** alkaline earth metals — **3.** group 2
- **d.** noble gases — **4.** group 17

**33.** Sketch a simplified periodic table, and use labels to identify the alkali metals, alkaline earth metals, transition metals, inner transition metals, noble gases, and halogens.

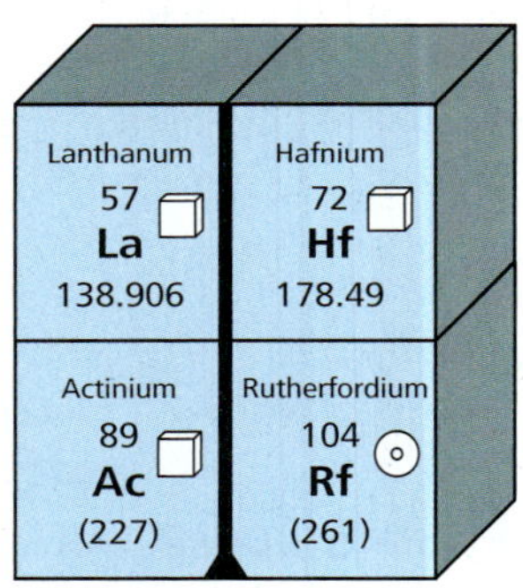


■ **Figure 19**

**34.** Explain what the dark line running down the middle of **Figure 19** indicates.

**35.** Give the chemical symbol of each of the following elements.

- **a.** a metal used in thermometers
- **b.** a radioactive gas used to predict earthquakes; the noble gas with the greatest atomic mass
- **c.** a coating for food cans; it is the metal in group 14 with the lowest atomic mass
- **d.** transition metal that is used to make burglar-proof vaults; also the name of a coin

**36.** If a new halogen and a new noble gas were discovered, what would be their atomic numbers?

### Mastering Problems

**37.** If the periodic table were arranged by atomic mass, which of the first 55 elements would be ordered differently than they are in the existing table?

**38. New Heavy Element** Scientists recently reported an element with 117 protons. What is its group and period? Would it be a metal, a metalloid, or a nonmetal?

**39. Naming New Elements** Recently discovered elements that have not been fully verified are given temporary names using the prefix words in **Table 6.** Based on this system, write names for elements 117 to 120.

**Table 6 Prefixes**

| 0 | 1 | 2 | 3 | 4 |
|---|---|---|---|---|
| nil | un | b(i) | tr(i) | quad |
| 5 | 6 | 7 | 8 | 9 |
| pent | hex | sept | oct | en(n) |

**40.** Give the chemical symbol for each element.

- **a.** the element in period 3 that can be used in making computer chips because it is a metalloid
- **b.** the group 13, period 5 metal used in making flat screens for televisions
- **c.** an element used as a filament in lightbulbs; has the highest atomic mass of the natural elements in group 6

## SECTION 2

### Mastering Concepts

**41. Household Products** Why do the elements chlorine, used in laundry bleach, and iodine, a nutrient added to table salt, have similar chemical properties?

**42.** How is the energy level of an atom's valence electrons related to its period in the periodic table?

**43.** How many valence electrons does each noble gas have?

**44.** What are the four blocks of the periodic table?

**45.** What electron configuration has the greatest stability?

**46.** Explain how an atom's valence electron configuration determines its place in the periodic table.

**47.** Write the electron configuration for the element fitting each of the following descriptions.

- **a.** the metal in group 15 that is part of compounds often found in cosmetics
- **b.** the halogen in period 3 that is part of a bleaching compound used in paper production
- **c.** the transition metal that is a liquid at room temperature; is sometimes used in outdoor security lights

48. Determine the group, period, and block in which each of the following elements is located in the periodic table.
   a. $[Kr]5s^24d^1$
   b. $[Ar]4s^23d^{10}4p^3$
   c. $[He]2s^22p^6$
   d. $[Ne]3s^23p^1$

49. Given any two elements within a group, is the element with the larger atomic number likely to have a larger or smaller atomic radius than the other element?

50. **Table 7** shows the number of elements in the first five periods of the periodic table. Explain why some of the periods have different numbers of elements.

**Table 7 Number of Elements in Periods 1–5**

| Period | 1 | 2 | 3 | 4 | 5 |
|---|---|---|---|---|---|
| Number of elements | 2 | 8 | 8 | 18 | 18 |

51. **Coins** One of the transition groups is often called the coinage group because at one time many coins were made of these metals. Which group is this? What element in this group is still used in many U.S. coins today?

52. Do any of the halogens have their valence electrons in orbitals of the same energy level? Explain.

53. The transition elements have their valence electrons in orbitals of more than one energy level, but the representative elements have their valence electrons in orbitals of only one energy level. Show this by using the electron configurations of a transition element and a representative element as examples.

## Mastering Problems

54. **Fireworks** Barium is a metal that gives a green color to fireworks. Write the electron configuration for barium. Classify it according to group, period, and block in the periodic table.

55. **Headphones** Neodymium magnets can be used in stereo headphones because they are powerful and lightweight. Write the electron configuration for neodymium. In which block of the periodic table is it?

56. **Soda Cans** The metal used to make soda cans has the electron configuration $[Ne]3s^23p^1$. Identify the metal and give its group, period, and block.

57. Identify each missing part of **Table 8.**

**Table 8 Electron Configuration**

| Period | Group | Element | Electron Configuration |
|---|---|---|---|
| 3 | | Mg | $[Ne]3s^2$ |
| 4 | 14 | Ge | |
| | 12 | Cd | $[Kr]5s^24d^{10}$ |
| 2 | 1 | | $[He]2s^1$ |

# SECTION 3

## Mastering Concepts

58. What is ionization energy?

59. An element forms a negative ion when ionized. On what side of the periodic table is the element located? Explain.

60. Of the elements magnesium, calcium, and barium, which forms the ion with the largest radius? The smallest? What periodic trend explains this?

61. Explain why each successive ionization of an electron requires a greater amount of energy.

62. How does the ionic radius of a nonmetal compare with its atomic radius? Explain the change in radius.

63. Explain why atomic radii decrease as you move from left to right across a period.

64. Which element has the larger ionization energy?
   a. Li, N
   b. Kr, Ne
   c. Cs, Li

65. Explain the octet rule. Why are hydrogen and helium exceptions to the octet rule?

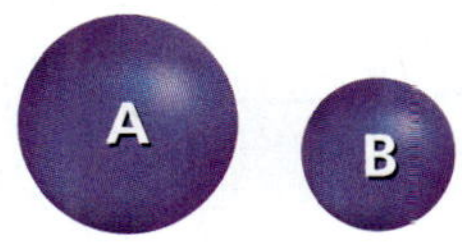


■ **Figure 20**

66. Use **Figure 20** to answer each of the following questions. Explain your reasoning for each answer.
   a. If A is an ion and B is an atom of the same element, is the ion a positive or negative ion?
   b. If A and B represent the atomic radii of two elements in the same period, what is their order?
   c. If A and B represent the ionic radii of two elements in the same group, what is their order?

67. How many valence electrons do elements in group 1 have? In group 18?

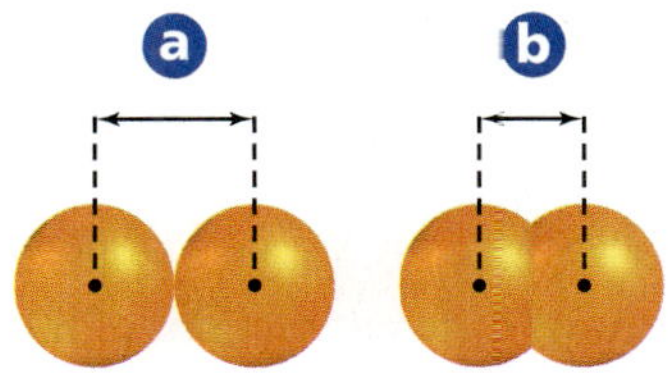


■ **Figure 21**

68. **Figure 21** shows two ways to define an atomic radius. Describe each method. When is each method used?

69. **Chlorine** The electron configuration of a chlorine atom is $[Ne]3s^23p^5$. When it gains an electron and becomes an ion, its electron configuration changes to $[Ne]3s^23p^6$, or [Ar], the electron configuration for argon. Has the chlorine atom changed to an argon atom? Explain.

### Mastering Problems

**70. Sport Bottles** Some sports bottles are made of Lexan, a plastic containing a compound of the elements chlorine, carbon, and oxygen. Order these elements from greatest to least according to atomic radius and ionic radius.

**71. Contact Lenses** Soft contact lenses are made of silicon and oxygen atoms bonded together. Create a table listing the atomic and ionic electron configurations, and the atomic and ionic radii for silicon and oxygen. When silicon bonds with oxygen, which atoms become larger and which become smaller? Why?

**72. Artificial Sweetener** Some diet sodas contain the artificial sweetener aspartame, a compound containing carbon, nitrogen, oxygen, and other atoms. Create a table showing the atomic and ionic radii of carbon, nitrogen, and oxygen. (Assume the ionization states shown in **Figure 14.**) Use the table to predict whether the sizes of carbon, nitrogen, and oxygen atoms increase or decrease in size when they form bonds in aspartame.

## MIXED REVIEW

**73.** Define an ion.

**74.** Explain why the radius of an atom cannot be measured directly.

**75.** What is the metalloid in period 2 of the periodic table that is part of compounds used as water softeners?

**76.** Do you expect cesium, a group 1 element used in infrared lamps, or bromine, a halogen used in firefighting compounds, to have the greatest electronegativity? Why?

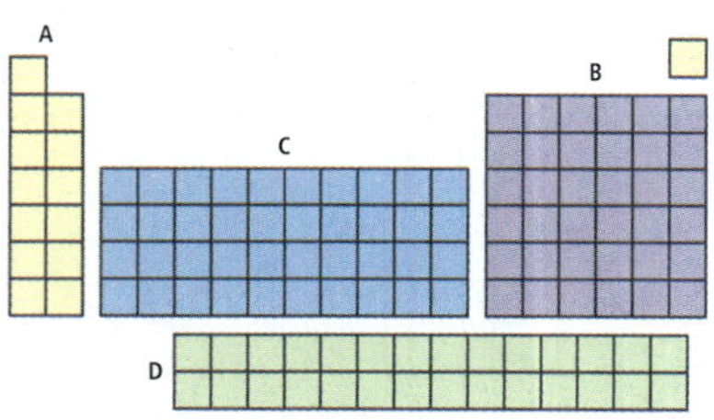


**Figure 22**

**77. Figure 22** shows different sections of the periodic table. Give the name of each section, and explain what the elements in each section have in common.

**78.** Which element in each pair is more electronegative?
**a.** K, As **b.** N, Sb **c.** Sr, Be

**79.** Explain why the s-block of the periodic table is two-groups wide, the p-block is six-groups wide, and the d-block is ten-groups wide.

**80.** Most of the atomic masses in Mendeleev's table are different from today's values. Explain why.

**81.** Arrange the elements oxygen, sulfur, tellurium, and selenium in order of increasing atomic radii. Is your order an example of a group trend or a period trend?

**82. Milk** The element with the electron configuration $[Ar]4s^2$ is an important mineral in milk. Identify this element's group, period, and block in the periodic table.

**83.** Why are there no p-block elements in the first period?

**84. Jewelry** What are the two transition metals that are used in making jewelry and are the group 11 elements with the lowest atomic masses?

**85.** Which has the largest ionization energy, platinum, an element sometimes used in dental crowns, or cobalt, an element that provides a bright blue color to pottery?

## THINK CRITICALLY

**86. Apply** Sodium forms a 1+ ion, while fluorine forms a 1− ion. Write the electron configuration for each ion. Why don't these two elements form 2+ and 2− ions, respectively?

**87. Make and Use Graphs** The densities of the group 15 elements are given in **Table 9.** Plot density versus atomic number, and state any trends you observe.

**Table 9 Group 15 Density Data**

| Element | Atomic Number | Density ($g/cm^3$) |
|---|---|---|
| Nitrogen | 7 | $1.14 \times 10^{-3}$ |
| Phosphorus | 15 | 1.82 |
| Arsenic | 33 | 5.22 |
| Antimony | 51 | 6.53 |
| Bismuth | 83 | 10.05 |

**88. Generalize** The outer-electron configurations of elements in group 1 can be written as $ns^1$, where $n$ refers to the element's period and its principal energy level. Develop a similar notation for all the other groups of the representative elements.

**89. Identify** A period 3 representative element is part of the rough material on the side of a match box used for lighting matches. **Table 10** shows the ionization energies for this element. Use the information in the table to infer the identity of the element. Explain.

**Table 10 Ionization Energies in kJ/mol**

| Number | 1st | 2nd | 3rd | 4th | 5th | 6th |
|---|---|---|---|---|---|---|
| Ionization energy | 1010 | 1905 | 2910 | 4957 | 6265 | 21,238 |

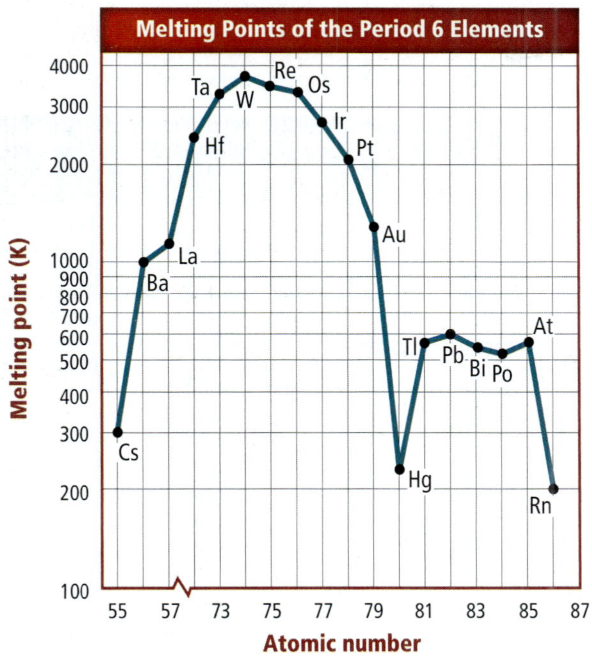


■ **Figure 23**

**90. Interpret Data** The melting points of the period 6 elements are plotted versus atomic number in **Figure 23.** Determine the trends in melting point and the orbital configurations of the elements. Form a hypothesis that explains the trends.

## CHALLENGE PROBLEM

**91.** Ionization energies are expressed in kilojoules per mole (one mole contains $6.02 \times 10^{23}$ atoms), but the energy to remove an electron from a gaseous atom is expressed in joules. Use the values in **Table 5** to calculate the energy, in joules, required to remove the first electron from an atom of Li, Be, B, and C. Then, use the relationship 1 eV $= 1.60 \times 10^{-19}$ J to convert the values to electron volts.

## CUMULATIVE REVIEW

**92.** Define *matter.* Identify whether or not each of the following is a form of matter.

**a.** microwaves
**b.** helium inside a balloon
**c.** heat from the Sun
**d.** velocity
**e.** a speck of dust
**f.** the color blue

**93.** Convert the following mass measurements as indicated.

**a.** 1.1 cm to meters
**b.** 76.2 pm to millimeters
**c.** 11 mg to kilograms
**d.** 7.23 µg to kilograms

**94.** How is the energy of a quantum of emitted radiation related to the frequency of the radiation?

**95.** What element has the ground-state electron configuration of $[Ar]4s^23d^6$?

## WRITING IN ▶ Chemistry

**96. Triads** In the early 1800s, German chemist J. W. Döbereiner proposed that some elements could be classified into sets of three, called triads. Research and write a report on Döbereiner's triads. What elements comprised the triads? How were the properties of elements within a triad similar?

**97. Affinity** Electron affinity is another periodic property of the elements. Write a summary on what electron affinity is, and describe its group and period trends.

## DBQ Document-Based Questions

*Mendeleev's original periodic table is remarkable given the knowledge of elements at that time, and yet it is different from the modern version. Compare Mendeleev's table, shown in* **Table 11,** *with the modern periodic table shown in* **Figure 5.**

Data obtained from: Dmitrii Mendeleev, *The Principles of Chemistry,* 1891.

**Table 11 Groups of Elements**

| Series | 0 | I | II | III | IV | V | VI | VII | VIII |
|---|---|---|---|---|---|---|---|---|---|
| 1 | — | H | — | — | — | | — | — | |
| 2 | He | Li | Be | B | C | N | O | F | |
| 3 | Ne | Na | Mg | Al | Si | P | S | Cl | |
| 4 | Ar | K | Ca | So | Ti | V | Cr | Mn | Fe |
| 5 | | Cu | Zn | Ga | Ge | As | Se | Br | Co |
| | | | | | | | | | Ni (Cu) |
| 6 | Kr | Rb | Sr | Y | Zr | Nb | Mo | — | Ru |
| 7 | | Ag | Cad | In | Sn | Sb | Te | I | Rh |
| | | | | | | | | | Pd (Ag) |
| 8 | Xe | Cs | Ba | La | — | — | — | — | — |
| 9 | | — | — | — | — | — | — | — | — |
| 10 | — | — | — | Yb | — | Ta | W | — | Os |
| 11 | | Au | Hg | Tl | | Bi | — | — | Ir |
| | | | | | | | | | Pt (Au) |
| 12 | — | — | Rd | — | Th | — | U | | |

**98.** Mendeleev placed the noble gases on the left of his table. Why does placement on the right of the modern table make more sense?

**99.** Which block on Mendeleev's table was most like today's placement? Which block was least like today's placement? Why?

**100.** Most of the atomic masses in Mendeleev's table differ from today's values. Why do you think this is so?

## MULTIPLE CHOICE

1. Elements in the same group of the periodic table have the same
   A. number of valence electrons.
   B. physical properties.
   C. number of electrons.
   D. electron configuration.

2. Which statement is NOT true?
   A. The atomic radius of Na is less than the atomic radius of Mg.
   B. The electronegativity of C is greater than the electronegativity of B.
   C. The ionic radius of $Br^-$ is greater than the atomic radius of Br.
   D. The first ionization energy of K is greater than the first ionization energy of Rb.

3. What is the group, period, and block of an atom with the electron configuration $[Ar]4s^2 3d^{10} 4p^4$?
   A. group 14, period 4, d-block
   B. group 16, period 3, p-block
   C. group 14, period 4, p-block
   D. group 16, period 4, p-block

*Use the table below to answer Questions 4 and 5.*

| Characteristics of Elements | | |
|---|---|---|
| Element | Block | Characteristic |
| X | s | soft solid; reacts readily with oxygen |
| Y | p | gas at room temperature; forms salts |
| Z | — | inert gas |

4. In which group does Element X most likely belong?
   A. 1
   B. 17
   C. 18
   D. 4

5. In which block is Element Z most likely found?
   A. s-block
   B. p-block
   C. d-block
   D. f-block

*Use the table below to answer Questions 6 and 7.*

| Percent Composition By Mass of Selected Nitrogen Oxides | | |
|---|---|---|
| Compound | Percent Nitrogen | Percent Oxygen |
| $N_2O_4$ | 30.4% | 69.6% |
| $N_2O_3$ | ? | ? |
| $N_2O$ | 63.6% | 36.4% |
| $N_2O_5$ | 25.9% | 74.1% |

6. What is the percent nitrogen in the compound $N_2O_3$?
   A. 44.7%
   B. 46.7%
   C. 28.1%
   D. 36.8%

7. A sample of a nitrogen oxide contains 1.29 g of nitrogen and 3.71 g of oxygen. Which compound is this most likely to be?
   A. $N_2O_4$
   B. $N_2O_3$
   C. $N_2O$
   D. $N_2O_5$

8. On the modern periodic table, metalloids are found only in
   A. the d-block.
   B. groups 13 through 17.
   C. the f-block.
   D. groups 1 and 2.

9. Which group is composed entirely of nonmetals?
   A. 1
   B. 13
   C. 15
   D. 18

10. It can be predicted that element 118 would have properties similar to a(n)
    A. alkali earth metal.
    B. halogen.
    C. metalloid.
    D. noble gas.

## SHORT ANSWER

11. Write the electron configuration for the element arsenic (As).

12. Write the nuclear decay equation for the beta decay of iodine-131.

13. Two students are identifying a sample of tap water. Student A says that tap water is a mixture, while Student B says that it is a compound. Which student is correct? Justify your answer.

## EXTENDED RESPONSE

*Use the table below to answer Questions 14 and 15.*

**Successive Ionization Energies for Selected Period 2 Elements, in kJ/mol**

| Element | Li | Be | B | C |
|---|---|---|---|---|
| Valence e– | 1 | 2 | 3 | 4 |
| First ionization energy | 520 | 900 | 800 | 1090 |
| Second ionization energy | 7300 | 1760 | 2430 | 2350 |
| Third ionization energy | | 14,850 | 3660 | 4620 |
| Fourth ionization energy | | | 25,020 | 6220 |
| Fifth ionization energy | | | | 37,830 |

14. Correlate the biggest jump in ionization energy to the number of valence electrons in each atom.

15. Predict which ionization energy will show the largest jump for magnesium. Explain your answer.

## SAT SUBJECT TEST: CHEMISTRY

*For Questions 16 to 19, answer true or false for the first statement, and true or false for the second statement. If the second statement is a correct explanation of the first statement, write* CE.

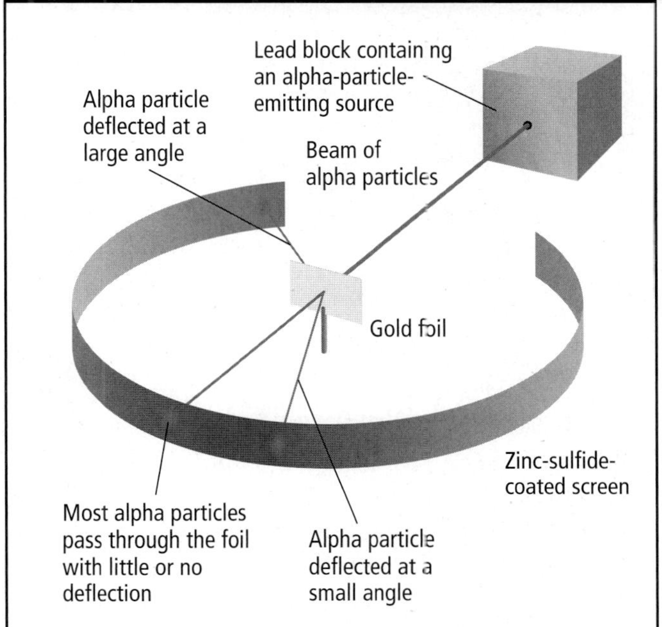


| Statement I | | Statement II |
|---|---|---|
| 16. Some particles bounce off the gold foil | BECAUSE | the nucleus is negatively charged. |
| 17. Some particles bounce off the gold foil | BECAUSE | they hit protons in the nucleus. |
| 18. Many particles pass through the gold foil | BECAUSE | atoms are made of protons, neutrons, and electrons. |
| 19. The symbol for an alpha particle is ${}^{4}_{2}He$ | BECAUSE | protons and neutrons have about the same mass. |

**NEED EXTRA HELP?**

| If You Missed Question . . . | 1 | 2 | 3 | 4 | 5 | 6 | 7 | 8 | 9 | 10 | 11 | 12 | 13 | 14 | 15 | 16 | 17 | 18 | 19 |
|---|---|---|---|---|---|---|---|---|---|---|---|---|---|---|---|---|---|---|---|
| Review Section . . . | 6.2 | 6.3 | 6.2 | 6.2 | 6.2 | 3.4 | 3.4 | 6.2 | 6.2 | 6.3 | 5.3 | 4.4 | 3.3 | 6.3 | 6.3 | 4.2 | 4.2 | 4.2 | 4.4 |

# CHAPTER 7

# Ionic Compounds and Metals

**BIGIDEA** Ionic compounds are held together by chemical bonds formed by the attraction of oppositely charged ions.

## SECTIONS

Chm.3b, 3c

1 Ion Formation

2 Ionic Bonds and Ionic Compounds

3 Names and Formulas for Ionic Compounds

4 Metallic Bonds and the Properties of Metals

## LaunchLAB

iLab Station

### What compounds conduct electricity in solution?

For a material to conduct an electric current, it must contain charged particles that can move throughout the substance. Electrical conductivity is a property of matter that tells you something about bonding.

**FOLDABLES® Study Organizer**

### Ionic Compounds

Make a trifold book. Label it as shown. Use it to help you organize information about ionic compounds.

Go online!
MHEonline.com

Metals and ionic compounds surround this scuba diver. For example, the metal tank holds the oxygen gas that keeps the diver alive under water. And the ionic compound calcium carbonate makes up much of the reef the diver is exploring. Even the water itself contains ionic compounds—there are about 35 g of ionic salts dissolved in every liter of sea water.

SECTION 1

# Ion Formation

**MAINIDEA** **Ions form when atoms gain or lose valence electrons to achieve a stable octet electron configuration.**

### Essential Questions

- What holds atoms together in a chemical bond?
- How do positive and negative ions form?
- How does ion formation relate to electron configuration?

### Review Vocabulary

**octet rule:** atoms tend to gain, lose, or share electrons in order to acquire eight valence electrons

### New Vocabulary

chemical bond
cation
anion

**CHEM 4 YOU** **Imagine that you and a group of friends go to a park to play soccer. There, you meet a larger group that also wants to play. To form even teams, one group loses members and the other group gains members. Atoms sometimes behave in a similar manner to form compounds.**

## Valence Electrons and Chemical Bonds

Imagine going on a scuba dive, diving below the ocean's surface and observing the awe-inspiring world below. You might explore the colorful and exotic organisms teeming around a coral reef, such as the one shown in **Figure 1.** The coral is formed from a compound called calcium carbonate, which is just one of thousands of compounds found on Earth. How do so many compounds form from the relatively few elements known to exist? The answer to this question involves the electron structure of atoms and the nature of the forces between atoms.

In previous chapters, you learned that elements within a group on the periodic table have similar properties. Many of these properties depend on the number of valence electrons the atom has. These valence electrons are involved in the formation of chemical bonds between two atoms. A **chemical bond** is the force that holds two atoms together. Chemical bonds can form by the attraction between the positive nucleus of one atom and the negative electrons of another atom, or by the attraction between positive ions and negative ions. This chapter discusses chemical bonds formed by ions, atoms that have acquired a positive or negative charge. You will learn about bonds that form from the sharing of electrons in a later chapter.

**Figure 1** As carbon dioxide dissolves in ocean water, carbonate ions are produced. Coral polyps capture these carbonate ions, producing crystals of calcium carbonate, which they secrete as an exoskeleton. Over time, the coral reef forms. A coral reef is a complex habitat that supports coral, algae, mollusks, echinoderms, and a variety of fishes.

David Nardini/Getty Images

Explore **electron-dot structures with an interactive table.** Concepts In Motion

**Table 1** Electron-Dot Structures

| Group | 1 | 2 | 13 | 14 | 15 | 16 | 17 | 18 |
|---|---|---|---|---|---|---|---|---|
| Diagram | Li· | ·Be· | ·B· | ·C· | ·N· | ·O: | :F: | :Ne: |

**Valence electrons** Recall that an electron-dot structure is a type of diagram used to keep track of valence electrons. Electron-dot structures are especially helpful when used to illustrate the formation of chemical bonds. **Table 1** shows several examples of electron-dot structures. For example, carbon, with an electron configuration of $1s^2 2s^2 2p^2$, has four valence electrons in the second energy level. These valence electrons are represented by the four dots around the symbol C in the table.

Also, recall that ionization energy refers to how easily an atom loses an electron and that electron affinity indicates how much attraction an atom has for electrons. Noble gases, which have high ionization energies and low electron affinities, show a general lack of chemical reactivity. Other elements on the periodic table react with each other, forming numerous compounds. The difference in reactivity is directly related to the valence electrons.

The difference in reactivity involves the octet—the stable arrangement of eight valence electrons in the outer energy level. Unreactive noble gases have electron configurations that have a full outermost energy level. This level is filled with two electrons for helium ($1s^2$) and eight electrons for the other noble gases ($ns^2np^6$). Elements tend to react to acquire the stable electron structure of a noble gas.

FOLDABLES® Incorporate information from this section into your Foldable.

## Positive Ion Formation

A positive ion forms when an atom loses one or more valence electrons in order to attain a noble gas configuration. A positively charged ion is called a **cation.** To understand the formation of a positive ion, compare the electron configurations of the noble gas neon (atomic number 10) and the alkali metal sodium (atomic number 11).

| | |
|---|---|
| Neon atom (Ne) | $1s^2 2s^2 2p^6$ |
| Sodium atom (Na) | $1s^2 2s^2 2p^6 3s^1$ |

Note that the sodium atom has one 3s valence electron; it differs from the noble gas neon by that single valence electron. When sodium loses this outer valence electron, the resulting electron configuration is identical to that of neon. **Figure 2** shows how a sodium atom loses its valence electron to become a sodium cation.

By losing an electron, the sodium atom acquires the stable outer-electron configuration of neon. It is important to understand that although sodium now has the electron configuration of neon, it is not neon. It is a sodium ion with a single positive charge. The 11 protons that establish the character of sodium still remain within its nucleus.

**READING CHECK** **Identify** the number of electrons in the outermost energy level that are associated with maximum stability.

**Figure 2** In the formation of a positive ion, a neutral atom loses one or more valence electrons. The atom is neutral because it contains equal numbers of protons and electrons; the ion, however, contains more protons than electrons and has a positive charge.

**Analyze** ***Does the removal of an electron from a neutral atom require energy or release energy?***

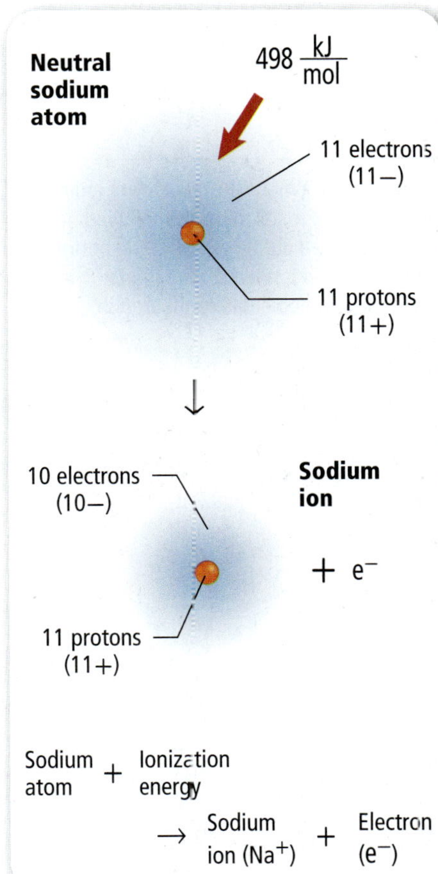

| Table 2 Group 1, 2, and 13 Ions | | |
|---|---|---|
| **Group** | **Configuration** | **Charge of Ion Formed** |
| 1 | [noble gas] $ns^1$ | 1+ when the $s^1$ electron is lost |
| 2 | [noble gas] $ns^2$ | 2+ when the $s^2$ electrons are lost |
| 13 | [noble gas] $ns^2np^1$ | 3+ when the $s^2p^1$ electrons are lost |

**Metal ions** Metals atoms are reactive because they lose valence electrons easily. The group 1 and 2 metals are the most reactive metals on the periodic table. For example, potassium and magnesium, group 1 and 2 elements, respectively, form $K^+$ and $Mg^{2+}$ ions. Some group 13 atoms also form ions. The ions formed by metal atoms in groups 1, 2, and 13 are summarized in **Table 2.**

**Transition metal ions** Recall that, in general, transition metals have an outer energy level of $ns^2$. Going from left to right across a period, atoms of each element fill an inner d sublevel. When forming positive ions, transition metals commonly lose their two valence electrons, forming 2+ ions. However, it is also possible for d electrons to be lost. Thus, transition metals also commonly form ions of 3+ or greater, depending on the number of d electrons in the electron structure. It is difficult to predict the number of electrons that will be lost. For example, iron (Fe) forms both $Fe^{2+}$ and $Fe^{3+}$ ions. A useful rule of thumb for these metals is that they form ions with a 2+ or a 3+ charge.

**Pseudo-noble gas configurations** Although the formation of an octet is the most stable electron configuration, other electron configurations can also provide some stability. For example, elements in groups 11–14 lose electrons to form an outer energy level containing full s, p, and d sublevels. These relatively stable electron arrangements are referred to as pseudo-noble gas configurations. In **Figure 3,** the zinc atom has the electron configuration of $1s^22s^22p^63s^23p^64s^23d^{10}$. When forming an ion, the zinc atom loses the two 4s electrons in the outer energy level, and the stable configuration of $1s^22s^22p^63s^23p^63d^{10}$ results in a pseudo-noble gas configuration.

**Figure 3** When zinc reacts with iodine, the heat of the reaction causes solid iodine to sublimate into a purple vapor. At the bottom of the tube, $ZnI_2$ is formed containing $Zn^{2+}$ ions with a pseudo-noble gas configuration.

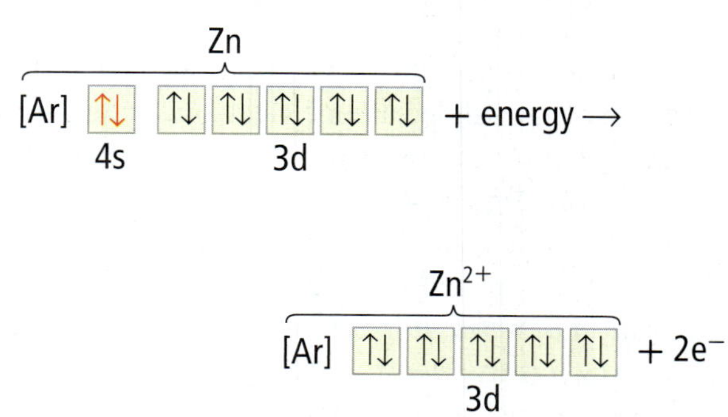


When the two 4s valence electrons are lost, a stable pseudo-noble gas configuration consisting of filled s, p, and d sublevels is achieved. Note that the filled 3s and 3p orbitals exist as part of the [Ar] configuration.

**Table 3 Group 15–17 Ions**

| Group | Configuration | Charge of Ion Formed |
|---|---|---|
| 15 | [noble gas] $ns^2np^3$ | 3– when three electrons are gained |
| 16 | [noble gas] $ns^2np^4$ | 2– when two electrons are gained |
| 17 | [noble gas] $ns^2np^5$ | 1– when one electron is gained |

## Negative Ion Formation

Nonmetals, which are located on the right side of the periodic table, easily gain electrons to attain a stable outer electron configuration. Examine **Figure 4.** To attain a noble-gas configuration, chlorine gains one electron, forming an ion with a 1– charge. After gaining the electron, the chloride ion has the electron configuration of an argon atom.

| | |
|---|---|
| Chlorine atom (Cl) | $1s^22s^22p^63s^23p^5$ |
| Argon atom (Ar) | $1s^22s^22p^63s^23p^6$ |
| Chloride ion ($Cl^-$) | $1s^22s^22p^63s^23p^6$ |

An **anion** is a negatively charged ion. To designate an anion, the ending *-ide* is added to the root name of the element. Thus, a chlorine atom becomes a chloride anion. What is the name of the nitrogen anion?

**Nonmetal ions** As shown in **Table 3,** nonmetals gain the number of electrons that, when added to their valence electrons, equals 8. For example, consider phosphorus, with five valence electrons. To form a stable octet, the atom gains three electrons and forms a phosphide ion with a 3– charge. Likewise, oxygen, with six valence electrons, gains two electrons and forms an oxide ion with a 2– charge.

Some nonmetals can lose or gain other numbers of electrons to form an octet. For example, in addition to gaining three electrons, phosphorus can lose five. However, in general, group 15 elements gain three electrons, group 16 elements gain two, and group 17 elements gain one to achieve an octet.

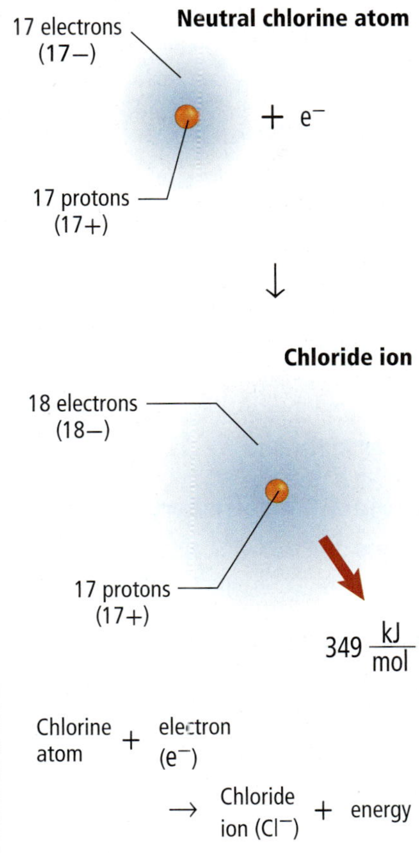


**Figure 4** During the formation of the negative chloride ion, a neutral atom gains one electron. The process releases 349 kJ/mol of energy.

**Compare** ***How do the energy changes accompanying positive ion and negative ion formation compare?***

## SECTION 1 REVIEW

Section Self-Check

### Section Summary

- A chemical bond is the force that holds two atoms together.
- Some atoms form ions to gain stability. A stable configuration involves a complete outer energy level, usually consisting of eight valence electrons.
- Ions are formed by the loss or gain of valence electrons.
- The number of protons remains unchanged during ion formation.

1. **MAIN**IDEA **Compare** the stability of a lithium atom with that of its ion, $Li^+$.
2. **Describe** two different causes of the force of attraction in a chemical bond.
3. **Apply** Why are all of the elements in group 18 relatively unreactive, whereas those in group 17 are very reactive?
4. **Summarize** ionic bond formation by correctly pairing these terms: *cation, anion, electron gain,* and *electron loss.*
5. **Apply** Write out the electron configuration for each atom. Then, predict the change that must occur in each to achieve a noble-gas configuration.
   **a.** nitrogen **b.** sulfur **c.** barium **d.** lithium
6. **Model** Draw models to represent the formation of the positive calcium ion and the negative bromide ion.

SECTION 2

# Ionic Bonds and Ionic Compounds

**MAINIDEA** Oppositely charged ions attract each other, forming electrically neutral ionic compounds.

### Essential Questions

- How do ionic bonds form and how are the ions arranged in an ionic compound?
- What can you conclude about the strength of ionic bonds based on the physical properties of ionic compounds?
- Is ionic bond formation exothermic or endothermic?

### Review Vocabulary

**compound:** a chemical combination of two or more different elements

### New Vocabulary

ionic bond
ionic compound
crystal lattice
electrolyte
lattice energy

**CHEM 4 YOU** Have you ever tried to separate sheets of plastic wrap that are stuck together? The hard-to-separate layers attract each other due to their oppositely charged surfaces.

## Formation of an Ionic Bond

What do the reactions shown in **Figure 5** have in common? In both cases, elements react with each other to form a compound. **Figure 5a** shows the reaction between the elements sodium and chlorine. During this reaction, a sodium atom transfers its valence electron to a chlorine atom and becomes a positive ion. The chlorine atom accepts the electron into its outer energy level and becomes a negative ion. The oppositely charged ions attract each other, forming the compound sodium chloride. The electrostatic force that holds oppositely charged particles together in an ionic compound is referred to as an **ionic bond.** Compounds that contain ionic bonds are **ionic compounds.** If ionic bonds occur between metals and the nonmetal oxygen, oxides form. Most other ionic compounds are called salts.

**Binary ionic compounds** Thousands of compounds contain ionic bonds. Many ionic compounds are binary, which means that they contain only two different elements. Binary ionic compounds contain a metallic cation and a nonmetallic anion. Sodium chloride (NaCl) is a binary compound because it contains two different elements, sodium and chlorine. Magnesium oxide (MgO), the reaction product shown in **Figure 5b,** is also a binary ionic compound.

■ **Figure 5** Each of these chemical reactions produces an ionic compound while releasing a large amount of energy. **a.** The reaction that occurs between elemental sodium and chlorine gas produces a white crystalline solid. **b.** When a ribbon of magnesium metal burns in air, it forms the ionic compound magnesium oxide.

(l)Andrew Lambert Photography/Photo Researchers, (r)Charles D. Winters/Photo Researchers

**Compound formation and charge** What role does ionic charge play in the formation of ionic compounds? To answer this question, examine how calcium fluoride forms. Calcium has the electron configuration $[Ar]4s^2$, and needs to lose two electrons to attain the stable configuration of argon. Fluorine has the configuration $[He]2s^2 2p^5$, and must gain one electron to attain the stable configuration of neon. Because the number of electrons lost and gained must be equal, two fluorine atoms are needed to accept the two electrons lost from the calcium atom.

$$1\ \text{Ca}\ \cancel{\text{ion}}\left(\frac{2+}{\text{Ca}\ \cancel{\text{ion}}}\right) + 2\ \text{F}\ \cancel{\text{ions}}\left(\frac{1-}{\text{F}\ \cancel{\text{ion}}}\right) = (1)(2+) + (2)(1-) = 0$$

As you can see, the overall charge of one unit of calcium fluoride ($CaF_2$) is zero. **Table 4** summarizes several ways in which the formation of an ionic compound such as sodium chloride can be represented.

**FOLDABLES®** Incorporate information from this section into your Foldable.

View an **animation of ionic bond formation in sodium chloride.** Concepts In Motion

**Table 4** Formation of Sodium Chloride

### Chemical Equation

$Na + Cl \rightarrow Na^+ + Cl^- + \text{energy}$

### Electron Configurations

One electron is transferred.

$\underbrace{[Ne]3s^1}_{Na} + \underbrace{[Ne]3s^2 3p^5}_{Cl} \rightarrow \underbrace{[Ne]}_{Na^+} + \underbrace{[Ar]}_{Cl^-} + \text{energy}$

### Orbital Notation

One electron is transferred.

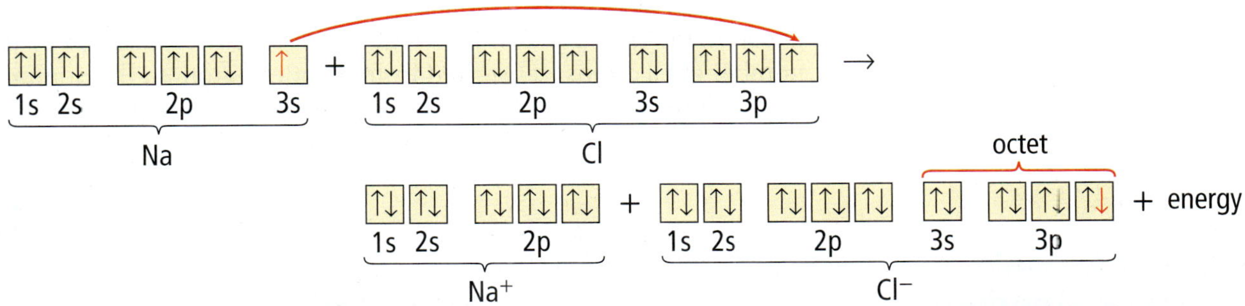


### Electron-Dot Structures

One electron is transferred.

$Na\cdot + \cdot\ddot{\underset{\cdot\cdot}{Cl}}: \rightarrow [Na]^+ + [:\ddot{\underset{\cdot\cdot}{Cl}}:]^- + \text{energy}$

### Atomic Models

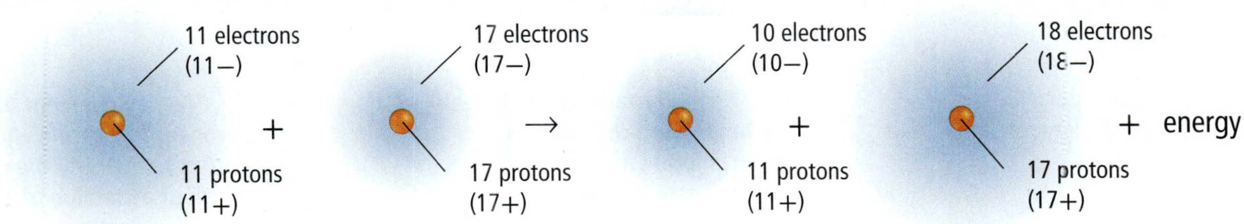


Sodium atom Chlorine atom Sodium ion Chloride ion

Next, consider aluminum oxide, the whitish coating that forms on aluminum chairs. To acquire a noble-gas configuration, each aluminum atom loses three electrons and each oxygen atom gains two electrons. Thus, three oxygen atoms are needed to accept the six electrons lost by two aluminum atoms. The neutral compound formed is aluminum oxide ($Al_2O_3$).

$$2\ \text{Al}\ \cancel{\text{ions}}\left(\frac{3+}{\text{Al}\ \cancel{\text{ion}}}\right) + 3\ \text{O}\ \cancel{\text{ions}}\left(\frac{2-}{\text{O}\ \cancel{\text{ion}}}\right) = 2(3+) + 3(2-) = 0$$

## PRACTICE Problems

Do additional problems. Online Practice

**Explain how an ionic compound forms from these elements.**

**7.** sodium and nitrogen

**8.** lithium and oxygen

**9.** strontium and fluorine

**10.** aluminum and sulfur

**11. Challenge** Explain how elements in the two groups shown on the periodic table at the right combine to form an ionic compound.

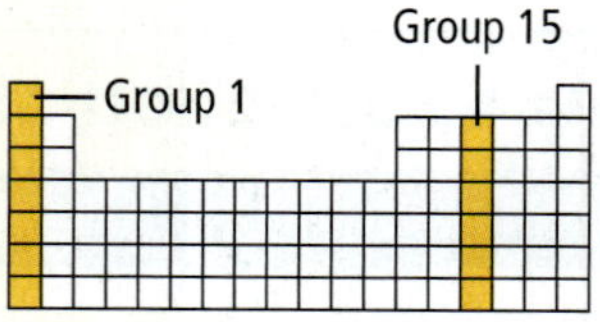


## Properties of Ionic Compounds

The chemical bonds in a compound determine many of its properties. For ionic compounds, the ionic bonds produce unique physical structures, unlike those of other compounds. The physical structures of ionic compounds also contribute to their physical properties. These properties have been used in many applications, as discussed in **Figure 6.**

**Physical structure** In an ionic compound, large numbers of positive ions and negative ions exist together in a ratio determined by the number of electrons transferred from the metal atom to the nonmetal atom. These ions are packed into a regular repeating pattern that balances the forces of attraction and repulsion between the ions.



**Figure 6**

### Milestones in Ionic and Metallic Bonding

A series of discoveries helped scientists understand the properties of ionic and metallic substances—leading to the creation of new tools and materials.

1900 1910 1930

**1897** J. J. Thomson speculates that electrons play a key role in chemical bonding.

**1913** X-ray crystallography reveals sodium ions and chloride ions in sodium chloride are arranged in regular geometric patterns.

**1916** Gilbert Lewis proposes a bonding theory based on the interaction of electrons among atoms.

**1932** The development of an electronegativity scale allows scientists to quantify the relative strength of attraction of each element for electrons.

**CIRCA 1940** Metallurgists develop alloys that perform under extreme temperature, pressure, and centrifugal force. Such alloys are later used in jet engines and spacecraft.

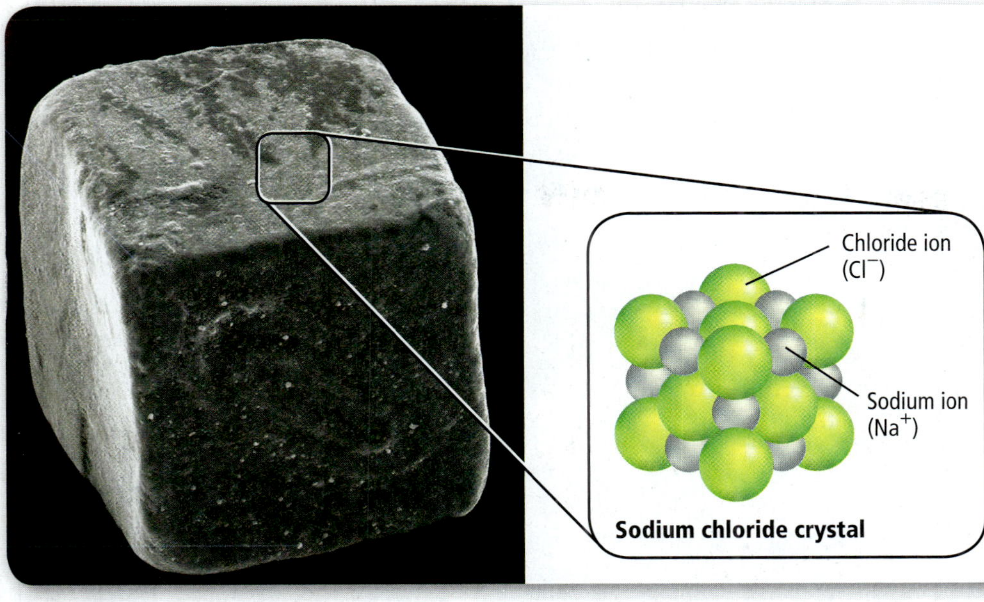


**Figure 7** The structure of a sodium chloride crystal is highly ordered. When viewed with a scanning electron microscope, the cubic shape of sodium chloride crystals is visible.

**Interpret** ***What is the ratio of sodium ions to chloride ions in the crystal?***

Examine the pattern of the ions in the sodium chloride crystal shown in **Figure 7.** Note the highly organized nature of an ionic crystal—the consistent spacing of the ions and the uniform pattern formed by them. Although the ion sizes are not the same, each sodium ion in the crystal is surrounded by six chloride ions, and each chloride ion is surrounded by six sodium ions. What shape would you expect a large crystal of this compound to be? As shown in **Figure 7,** the one-to-one ratio of sodium and chloride ions produces a highly ordered cubic crystal. As in all ionic compounds, in NaCl, no single unit consisting of only one sodium ion and one chloride ion is formed. Instead, large numbers of sodium ions and chloride ions exist together. If you can, obtain a magnifying lens and use it to examine some crystals of table salt (NaCl). What is the shape of these small salt crystals?

**READING CHECK** **Explain** what determines the ratio of positive ions to negative ions in an ionic crystal.

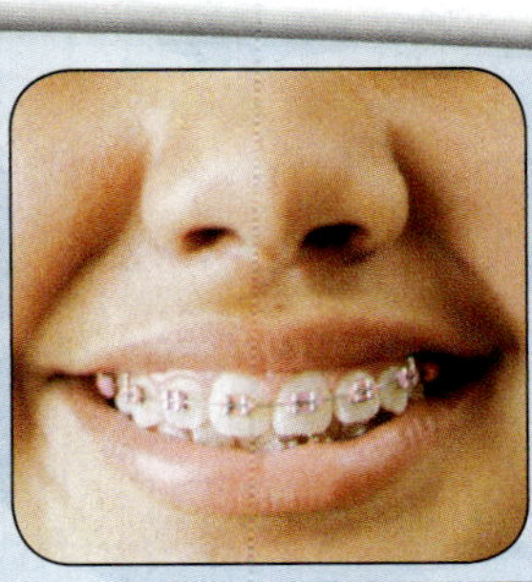

**1962** A nickel-titanium alloy with "shape memory" is discovered. The alloy reverts to its original shape after being deformed. Dental braces are one of many applications.

**2004** Scientists develop a nickel-gadolinium alloy that absorbs radioactive neutrons emitted by nuclear waste. Applications include transport and storage of highly radioactive fuel.

1970 1990 2000 2010

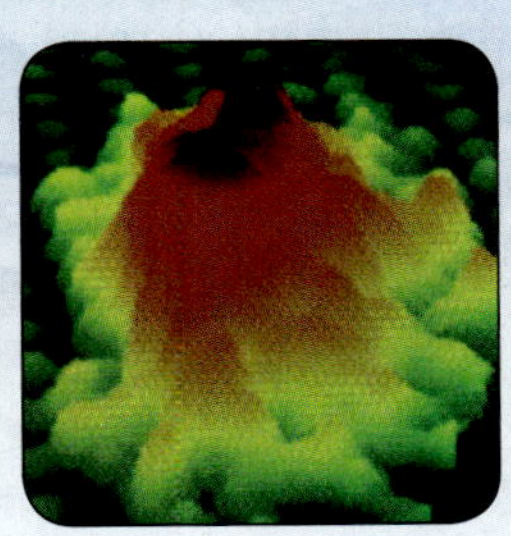

**1981** Invention of the scanning tunneling microscope allows researchers to study atomic-scale images in three dimensions.

**2009** U.S. aluminum can recycling rate reaches an all-time high of 57%. Recycling these cans saves roughly 95% of the energy required to extract aluminum from its ore.

**2014** NASA scientists develop a 3-D printing process that can build a single metal part of different alloys, rather than building the part from different pieces welded together. For example, a part could be printed of an unreactive alloy on one side and a magnetic alloy on the other.

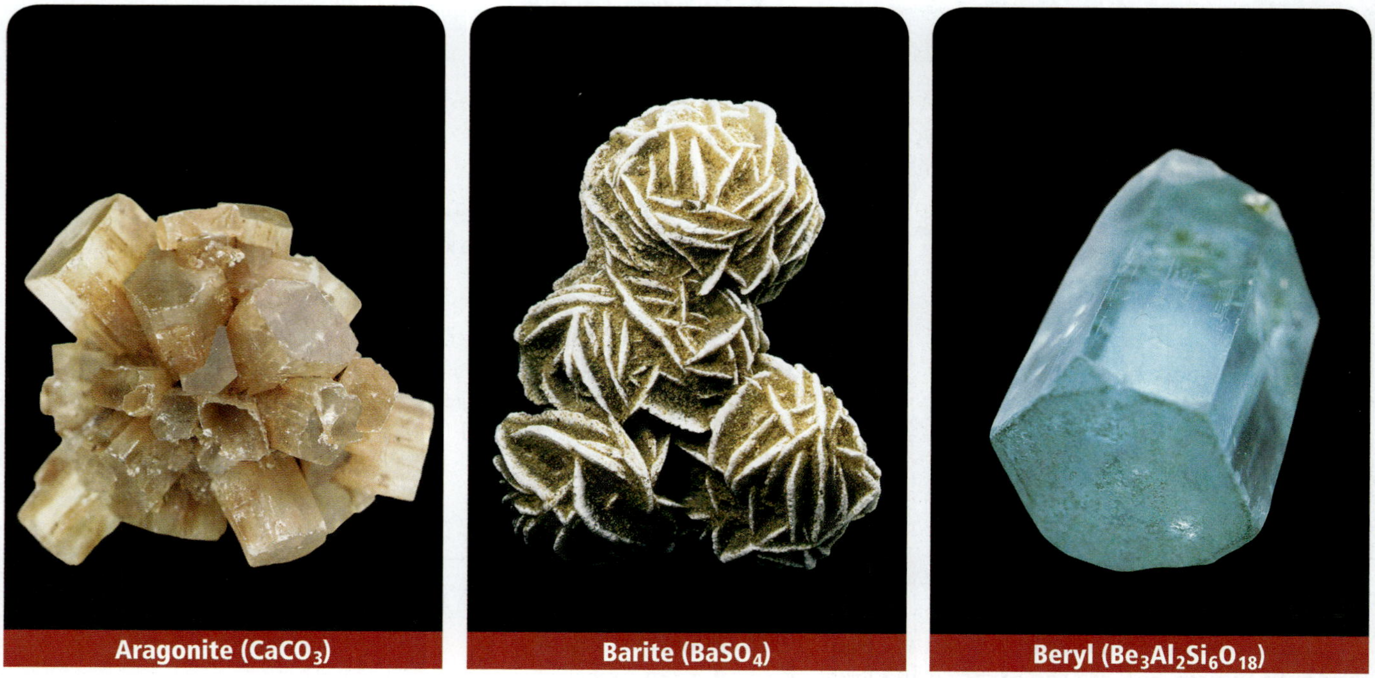


■ **Figure 8** Aragonite ($CaCO_3$), barite ($BaSO_4$), and beryl ($Be_3Al_2Si_6O_{18}$) are examples of minerals that are ionic compounds. The ions that form them are bonded together in a crystal lattice. Differences in ion size and charge result in different ionic crystal shapes, a topic that will be discussed later.

The strong attractions among the positive ions and the negative ions in an ionic compound result in the formation of a crystal lattice. A **crystal lattice** is a three-dimensional geometric arrangement of particles. In a crystal lattice, each positive ion is surrounded by negative ions, and each negative ion is surrounded by positive ions. Ionic crystals vary in shape due to the sizes and relative numbers of the ions bonded, as shown by the minerals in **Figure 8.**

**Connection to Earth Science** The minerals shown in **Figure 8** are just a few of the types studied by mineralogists, scientists who study minerals. They make use of several classification schemes to organize the thousands of known minerals. Color, crystal structure, hardness, chemical, magnetic, and electric properties, and numerous other characteristics are used to classify minerals. The types of anions minerals contain can also be used to identify them. For example, more than one-third of all known minerals are silicates, which are minerals that contain an anion that is a combination of silicon and oxygen. Halides contain fluoride, chloride, bromide, or iodide ions. Other mineral classes include boron-containing anions known as borates and carbon-oxygen containing anions known as carbonates.

☑ READING CHECK **Identify** the mineral shown in **Figure 8** that is a silicate. Identify the mineral that is a carbonate.

**Physical properties** Melting point, boiling point, and hardness are physical properties of matter that depend on how strongly the particles that make up the matter are attracted to one another. Another property—the ability of a material to conduct electricity—depends on the availability of freely moving charged particles. Ions are charged particles, so whether they are free to move determines whether an ionic compound conducts electricity. In the solid state, the ions in an ionic compound are locked into fixed positions by strong attractive forces. As a result, ionic solids do not conduct electricity.

**Table 5 Melting and Boiling Points of Some Ionic Compounds**

| Compound | Melting Point (°C) | Boiling Point (°C) |
|---|---|---|
| NaI | 660 | 1304 |
| KBr | 734 | 1435 |
| NaBr | 747 | 1390 |
| $CaCl_2$ | 782 | >1600 |
| NaCl | 801 | 1413 |
| MgO | 2852 | 3600 |

The situation changes dramatically, however, when an ionic solid melts to become a liquid or is dissolved in solution. The ions—previously locked in position—are now free to move and conduct an electric current. Both ionic compounds in solution and in the liquid state are excellent conductors of electricity. An ionic compound whose aqueous solution conducts an electric current is called an **electrolyte.** You will learn more about solutions of electrolytes later.

Because ionic bonds are relatively strong, ionic crystals require a large amount of energy to be broken apart. Thus, ionic crystals have high melting points and high boiling points, as shown in **Table 5.** Many crystals, including gemstones, have brilliant colors. These colors are due to the presence of transition metals in the crystal lattices.

Ionic crystals are also hard, rigid, brittle solids due to the strong attractive forces that hold the ions in place. When an external force is applied to the crystal—a force strong enough to overcome the attractive forces holding the ions in position within the crystal—the crystal cracks or breaks apart, as shown in **Figure 9.** The crystal breaks apart because the applied force repositions the like-charged ions next to each other; the resulting repulsive force breaks apart the crystal.

**VOCABULARY**
**SCIENCE USAGE V. COMMON USAGE**

**Conduct**

***Science usage:*** the ability to transmit light, heat, sound, or electricity
*The material did not conduct electricity well.*

***Common usage:*** to guide or lead
*It was the manager's job to conduct the training session.*

■ **Figure 9** Strong attractive forces hold the ions in place until a force strong enough to overcome the attraction is applied.

Undisturbed ionic crystal
Before the force is applied, the crystal has a uniform pattern of ions.

Applied force realigns particles.
If the applied force is strong enough, it pushes the ions out of alignment.

Forces of repulsion break crystal apart.
A repulsive force created by nearby like-charged ions breaks apart the crystal.

## Energy and the Ionic Bond

During every chemical reaction, energy is either absorbed or released. If energy is absorbed during a chemical reaction, the reaction is endothermic. If energy is released, it is exothermic.

The formation of ionic compounds from positive ions and negative ions is always exothermic. The attraction of the positive ion for the negative ions close to it forms a more stable system that is lower in energy than the individual ions. If the amount of energy released during bond formation is reabsorbed, the bonds holding the positive ions and negative ions together will break apart.

**Lattice energy** Because the ions in an ionic compound are arranged in a crystal lattice, the energy required to separate 1 mol of the ions of an ionic compound is referred to as the **lattice energy.** The strength of the forces holding ions in place is reflected by the lattice energy. The greater the lattice energy, the stronger the force of attraction.

Lattice energy is directly related to the size of the ions bonded. Smaller ions form compounds with more closely spaced ionic charges. Because the electrostatic force of attraction between opposite charges increases as the distance between the charges decreases, smaller ions produce stronger interionic attractions and greater lattice energies. For example, the lattice energy of a lithium compound is greater than that of a potassium compound containing the same anion because the lithium ion is smaller than the potassium ion.

# Data Analysis LAB

Based on Real Data*

### Interpret Data

**Can embedding nanoparticles of silver into a polymer give the polymer antimicrobial properties?** Researchers tested the antimicrobial properties of a new composite material—the polymer poly(4-vinyl-N-hexylpyridinium bromide), known as NPVP, which attracts cations. It is known that silver ions from silver bromide and silver nitrate exhibit antimicrobial activity. Silver bromide was embedded into the NPVP polymer. Scientists tested the antimicrobial properties of the composite material. Their results, illustrated in the graph, show the growth of *E. coli* bacteria over a period of approximately four hours. Each line represents the *E. coli* population in response to the introduction of a particular substance.

#### Data and Observations

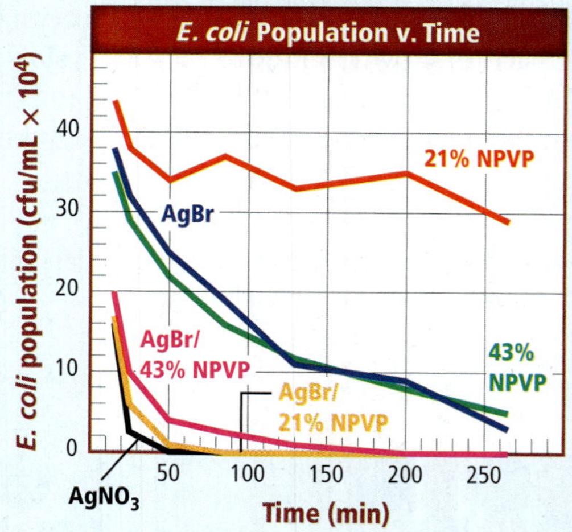


#### Think Critically

1. **Interpret** Does the addition of silver bromide (AgBr) ions to NPVP improve the antimicrobial properties of the composite?
2. **Interpret** Which composite reduced the *E. coli* population to zero? How long does it take for each substance to reduce the bacteria population to zero?
3. **Conclude** Does a composite polymer containing NPVP and silver bromide show antimicrobial properties? Explain your answer.

*Data obtained from: Sambhy, V., et al. Published on the Web 7/7/2006. Silver Bromide Nanoparticle/Polymer Composites. *Journal of the American Chemical Society.*

**Table 6** Lattice Energies of Some Ionic Compounds

| Compound | Lattice Energy (kJ/mol) | Compound | Lattice Energy (kJ/mol) |
|---|---|---|---|
| KI | 632 | KF | 808 |
| KBr | 671 | AgCl | 910 |
| RbF | 774 | NaF | 910 |
| NaI | 682 | LiF | 1030 |
| NaBr | 732 | $SrCl_2$ | 2142 |
| NaCl | 769 | MgO | 3795 |

The value of lattice energy is also affected by the charge of the ion. The ionic bond formed from the attraction of ions with larger positive or negative charges generally has a greater lattice energy. The lattice energy of MgO is almost four times greater than that of NaF because the charge of the ions in MgO is greater than the charge of the ions in NaF. The lattice energy of $SrCl_2$ is between the lattice energies of MgO and NaF because $SrCl_2$ contains ions with both higher and lower charges.

☑ READING CHECK **Summarize** the charges on each ion in the following ionic compounds: MgO, NaF, $SrCl_2$.

**Table 6** shows the lattice energies of some ionic compounds. Examine the lattice energies of RbF and KF. Because $K^+$ has a smaller ionic radius than $Rb^+$, KF has a greater lattice energy than RbF. This confirms that lattice energy is related to ion size. Notice the lattice energies of $SrCl_2$ and AgCl. How do they show the relationship between lattice energy and the charge of the ions involved?

## SECTION 2 REVIEW

Section Self-Check

### Section Summary

- Ionic compounds contain ionic bonds formed by the attraction of oppositely charged ions.
- Ions in an ionic compound are arranged in a repeating pattern known as a crystal lattice.
- Ionic compound properties are related to ionic bond strength.
- Ionic compounds are electrolytes; they conduct an electric current in the liquid phase and in aqueous solution.
- Lattice energy is the energy needed to remove 1 mol of ions from its lattice.

12. MAINIDEA **Explain** how an ionic compound made up of charged particles can be electrically neutral.
13. **Describe** the energy change associated with ionic bond formation, and relate it to stability.
14. **Identify** three physical properties of ionic compounds that are associated with ionic bonds, and relate them to bond strength.
15. **Explain** how ions form bonds, and describe the structure of the resulting compound.
16. **Relate** lattice energy to ionic-bond strength.
17. **Apply** Use electron configurations, orbital notation, and electron-dot structures to represent the formation of an ionic compound from the metal strontium and the nonmetal chlorine.
18. **Design** a concept map that shows the relationships among ionic bond strength, physical properties of ionic compounds, lattice energy, and stability.

SECTION 3

# Names and Formulas for Ionic Compounds

**MAINIDEA** In written names and formulas for ionic compounds, the cation appears first, followed by the anion.

### Essential Questions

- What is a formula unit and how does it relate to an ionic compound's composition?
- How do you write the formulas for compounds formed from different ions and oxyanions?
- What are the naming conventions for ionic compounds and oxyanions?

### Review Vocabulary

**nonmetal:** an element that is generally a gas or a dull, brittle solid and is a poor conductor of heat and electricity

### New Vocabulary

formula unit
monatomic ion
polyatomic ion
oxyanion

**CHEM 4 YOU** Although people have a wide range of names, most have both a first name and a last name. Ionic compound names are similar, in that they also consist of two parts.

## Formulas for Ionic Compounds

Because chemists around the world need to be able to communicate with one another, they have developed a set of rules for naming compounds. Using this standardized naming system, you can write a chemical formula from a compound's name and name a compound given its chemical formula.

Recall that an ionic compound is made up of ions arranged in a repeating pattern. The chemical formula for an ionic compound, called a **formula unit,** represents the simplest ratio of the ions involved. For example, the formula unit of magnesium chloride is $MgCl_2$ because the magnesium and chloride ions exist in a 1:2 ratio.

The overall charge of a formula unit is zero because the formula unit represents the entire crystal, which is electrically neutral. The formula unit for $MgCl_2$ contains one $Mg^{2+}$ ion and two $Cl^-$ ions, for a total charge of zero.

**Monatomic ions** Binary ionic compounds are composed of positively charged monatomic ions of a metal and negatively charged monatomic ions of a nonmetal. A **monatomic ion** is a one-atom ion, such as $Mg^{2+}$ or $Br^-$. **Table 7** indicates the charges of common monatomic ions according to their location on the periodic table. What is the formula for the beryllium ion? The iodide ion? The nitride ion?

Transition metals, which are in groups 3 through 12, and metals in groups 13 and 14 are not included in **Table 7** because of the variance in ionic charges of atoms in the groups. Most transition metals and metals in groups 13 and 14 can form several different positive ions.

**Table 7 Common Monatomic Ions**

| Group | Atoms that Commonly Form Ions | Charge of Ions |
|---|---|---|
| 1 | H, Li, Na, K, Rb, Cs | 1+ |
| 2 | Be, Mg, Ca, Sr, Ba | 2+ |
| 15 | N, P, As | 3− |
| 16 | O, S, Se, Te | 2− |
| 17 | F, Cl, Br, I | 1− |

**Table 8 Monatomic Metal Ions**

| Group | Common Ions |
|---|---|
| 3 | $Sc^{3+}$, $Y^{3+}$, $La^{3+}$ |
| 4 | $Ti^{2+}$, $Ti^{3+}$ |
| 5 | $V^{2+}$, $V^{3+}$ |
| 6 | $Cr^{2+}$, $Cr^{3+}$ |
| 7 | $Mn^{2+}$, $Mn^{3+}$, $Tc^{2+}$ |
| 8 | $Fe^{2+}$, $Fe^{3+}$ |
| 9 | $Co^{2+}$, $Co^{3+}$ |
| 10 | $Ni^{2+}$, $Pd^{2+}$, $Pt^{2+}$, $Pt^{4+}$ |
| 11 | $Cu^{+}$, $Cu^{2+}$, $Ag^{+}$, $Au^{+}$, $Au^{3+}$ |
| 12 | $Zn^{2+}$, $Cd^{2+}$, $Hg_2{}^{2+}$, $Hg^{2+}$ |
| 13 | $Al^{3+}$, $Ga^{2+}$, $Ga^{3+}$, $In^{+}$, $In^{2+}$, $In^{3+}$, $Tl^{+}$, $Tl^{3+}$ |
| 14 | $Sn^{2+}$, $Sn^{4+}$, $Pb^{2+}$, $Pb^{4+}$ |

**Oxidation numbers** The charge of a monatomic ion is equal to its oxidation number, or oxidation state. As shown in **Table 8,** most transition metals and group 13 and 14 metals have more than one possible ionic charge. Note that the ionic charges given in **Table 8** are the most common ones, not the only ones possible.

The oxidation number of an element in an ionic compound equals the number of electrons transferred from the atom to form the ion. For example, a sodium atom transfers one electron to a chlorine atom to form sodium chloride. This results in $Na^{+}$ and $Cl^{-}$. Thus, the oxidation number of sodium in the compound is 1+ because one electron was transferred from the sodium atom. Because an electron is transferred to the chlorine atom, its oxidation number is 1−.

**Formulas for binary ionic compounds** In the chemical formula for any ionic compound, the symbol of the cation is always written first, followed by the symbol of the anion. Subscripts, which are small numbers to the lower right of a symbol, represent the number of ions of each element in an ionic compound. If no subscript is written, it is assumed to be one. You can use oxidation numbers to write formulas for ionic compounds. Recall that ionic compounds have no charge. If you add the oxidation number of each ion multiplied by the number of these ions in a formula unit, the total must be zero.

Suppose you need to determine the formula for one formula unit of the compound that contains sodium and fluoride ions. Start by writing the symbol and charge for each ion: $Na^{+}$ and $F^{-}$. The ratio of ions in a formula unit of the compound must show that the number of electrons lost by the metal equals the number of electrons gained by the nonmetal. This occurs when one sodium atom transfers one electron to the fluorine atom; the formula unit is NaF.

☑ READING CHECK **Relate** the charge of an ion to its oxidation number.

CAREERS IN CHEMISTRY

**Food Scientist** Have you ever thought about the science behind the food you eat? Food scientists are concerned about the effects of processing on the appearance, aroma, taste, and the vitamin and mineral content of food. They also develop and improve foods and beverages. Food scientists often maintain "tasting notebooks" as they learn the characteristics of individual and blended flavors.

WebQuest

**VOCABULARY**

**ACADEMIC VOCABULARY**

**Transfer**
to cause to pass from one to another
*Carlos had to transfer to a new school when his parents moved to a new neighborhood.*

## EXAMPLE Problem 1

**FORMULA FOR AN IONIC COMPOUND** Determine the formula for the ionic compound formed from potassium and oxygen.

### 1 ANALYZE THE PROBLEM

You are given that potassium and oxygen ions form an ionic compound; the formula for the compound is the unknown. First, write out the symbol and oxidation number for each ion involved in the compound. Potassium, from group 1, forms 1+ ions, and oxygen, from group 16, forms 2− ions.

$$K^{+} \qquad O^{2-}$$

Because the charges are not the same, you need to determine the subscripts to use to indicate the ratio of positive ions to negative ions.

### 2 SOLVE FOR THE UNKNOWN

A potassium atom loses one electron, while an oxygen atom gains two electrons. If combined in a one-to-one ratio, the number of electrons lost by potassium will not balance the number of electrons gained by oxygen. Thus, two potassium ions are needed for each oxide ion. The formula is **$K_2O$.**

### 3 EVALUATE THE ANSWER

The overall charge of the compound is zero.

$$2\ \cancel{\text{K ion}}\text{s}\left(\frac{1+}{\cancel{\text{K ion}}}\right) + 1\ \cancel{\text{O ion}}\left(\frac{2-}{\cancel{\text{O ion}}}\right) = 2(1+) + 1(2-) = 0$$

## RealWorld CHEMISTRY

### Ionic Compounds

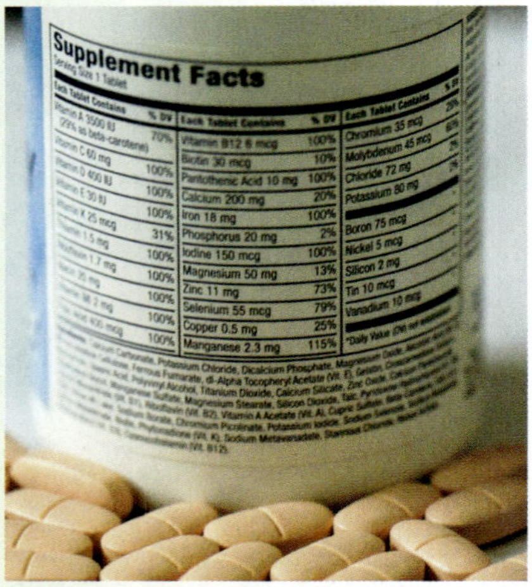


**MINERAL SUPPLEMENTS** To function properly, your body requires a daily intake of many different minerals. To ensure they are getting what they need, many people take a daily multivitamin and a mineral supplement. The minerals in these supplements come from a variety of ionic compounds. In fact, the majority of minerals found in mineral supplements come from ground-up rocks.

## EXAMPLE Problem 2

**FORMULA FOR AN IONIC COMPOUND** Determine the formula for the compound formed from aluminum ions and sulfide ions.

### 1 ANALYZE THE PROBLEM

You are given that aluminum and sulfur form an ionic compound; the formula for the ionic compound is the unknown. First, determine the charges of each ion. Aluminum, from group 13, forms 3+ ions, and sulfur, from group 16, forms 2− ions.

$$Al^{3+} \qquad S^{2-}$$

Each aluminum atom loses three electrons, while each sulfur atom gains two electrons. The number of electrons lost must equal the number of electrons gained.

### 2 SOLVE FOR THE UNKNOWN

The smallest number that can be divided evenly by both 2 and 3 is 6. Therefore, six electrons are transferred. Three sulfur atoms accept the six electrons lost by two aluminum atoms. The correct formula, **$Al_2S_3$,** shows two aluminum ions bonded to three sulfide ions.

### 3 EVALUATE THE ANSWER

The overall charge of one formula unit of this compound is zero.

$$2\ \cancel{\text{Al ion}}\text{s}\left(\frac{3+}{\cancel{\text{Al ion}}}\right) + 3\ \text{S}\ \cancel{\text{ions}}\left(\frac{2-}{\cancel{\text{S ion}}}\right) = 2(3+) + 3(2-) = 0$$

## PRACTICE Problems

Do additional problems. **Online Practice**

**Write formulas for the ionic compounds formed by the following ions.**

**19.** potassium and iodide

**20.** magnesium and chloride

**21.** aluminum and bromide

**22.** cesium and nitride

**23. Challenge** Write the general formula for the ionic compound formed by elements from the two groups shown on the periodic table at the right.

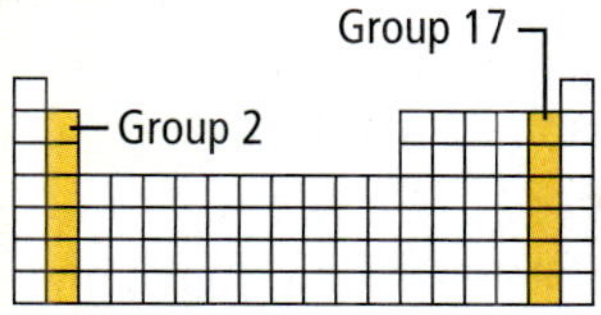


**Formulas for polyatomic ionic compounds** Many ionic compounds contain **polyatomic ions,** which are ions made up of more than one atom. **Table 9** and **Figure 10** list some common polyatomic ions. Also, refer to **Table R-5** in the Student Resources. A polyatomic ion acts as an individual ion in a compound and its charge applies to the entire group of atoms. Thus, the formula for a polyatomic compound follows the same rules used for a binary compound.

Because a polyatomic ion exists as a unit, never change subscripts of the atoms within the ion. If more than one polyatomic ion is needed, place parentheses around the ion and write the appropriate subscript outside the parentheses. For example, consider the compound formed from the ammonium ion ($NH_4^+$) and the oxide ion ($O^{2-}$). To balance the charges, the compound must have two ammonium ions for each oxide ion. To add a subscript to ammonium, enclose it in parentheses, then add the subscript. The correct formula is $(NH_4)_2O$.

**Table 9** Common Polyatomic Ions

| Ion | Name | Ion | Name |
|---|---|---|---|
| $NH_4^+$ | ammonium | $IO_4^-$ | periodate |
| $NO_2^-$ | nitrite | $C_2H_3O_2^-$ | acetate |
| $NO_3^-$ | nitrate | $H_2PO_4^-$ | dihydrogen phosphate |
| $OH^-$ | hydroxide | $CO_3^{2-}$ | carbonate |
| $CN^-$ | cyanide | $SO_3^{2-}$ | sulfite |
| $MnO_4^-$ | permanganate | $SO_4^{2-}$ | sulfate |
| $HCO_3^-$ | hydrogen carbonate | $S_2O_3^{2-}$ | thiosulfate |
| $ClO^-$ | hypochlorite | $O_2^{2-}$ | peroxide |
| $ClO_2^-$ | chlorite | $CrO_4^{2-}$ | chromate |
| $ClO_3^-$ | chlorate | $Cr_2O_7^{2-}$ | dichromate |
| $ClO_4^-$ | perchlorate | $HPO_4^{2-}$ | hydrogen phosphate |
| $BrO_3^-$ | bromate | $PO_4^{3-}$ | phosphate |
| $IO_3^-$ | iodate | $AsO_4^{3-}$ | arsenate |

**Figure 10** Ammonium and phosphate ions are polyatomic; that is, they are made up of more than one atom. Each polyatomic ion, however, acts as a single unit and has one charge.

**Identify** ***What are the charges of the ammonium ion and phosphate ion, respectively?***

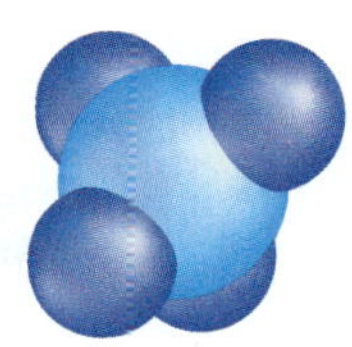

Ammonium ion ($NH_4^+$)

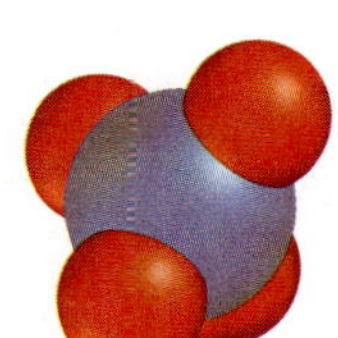

Phosphate ion ($PO_4^{3-}$)

EXAMPLE PROBLEM

## EXAMPLE Problem 3

**FORMULA FOR A POLYATOMIC IONIC COMPOUND** A compound formed by calcium ions and phosphate ions is often used in fertilizers. Write the compound's formula.

### 1 ANALYZE THE PROBLEM

You know that calcium and phosphate ions form an ionic compound; the formula for the compound is the unknown. First, write each ion along with its charge. Calcium, from group 2, forms 2+ ions, and the polyatomic phosphate acts as a single unit with a 3− charge.

$$Ca^{2+} \qquad PO_4^{3-}$$

Each calcium atom loses two electrons, while each polyatomic phosphate group gains three electrons. The number of electrons lost must equal the number of electrons gained.

### 2 SOLVE FOR THE UNKNOWN

The smallest number evenly divisible by both charges is 6. Thus, a total of six electrons are transferred. The negative charge from two phosphate ions equals the positive charge from three calcium ions. In the formula, place the polyatomic ion in parentheses and add a subscript to the outside. The correct formula for the compound is **$Ca_3(PO_4)_2$.**

### 3 EVALUATE THE ANSWER

The overall charge of one formula unit of calcium phosphate is zero.

$$3\ \text{Ca}\ \cancel{\text{ions}}\left(\frac{2+}{\text{Ca}\ \cancel{\text{ion}}}\right) + 2\ \text{PO}_4\ \cancel{\text{ions}}\left(\frac{3-}{\text{PO}_4\ \cancel{\text{ion}}}\right) = 3(2+) + 2(3-) = 0$$

## PRACTICE Problems

Do additional problems. Online Practice

**Write formulas for ionic compounds composed of the following ions.**

**24.** sodium and nitrate

**25.** calcium and chlorate

**26.** aluminum and carbonate

**27. Challenge** Write the formula for an ionic compound formed by ions from a group 2 element and polyatomic ions composed of only carbon and oxygen.

Get help **naming ionic compounds.**

Personal Tutor

Watch a **video about compound formation.**

Video

## Names for Ions and Ionic Compounds

Scientists use a systematic approach when naming ionic compounds. Because ionic compounds have both cations and anions, the naming system accounts for both of these ions.

**Naming an oxyanion** An **oxyanion** is a polyatomic ion composed of an element, usually a nonmetal, bonded to one or more oxygen atoms. More than one oxyanion exists for some nonmetals, such as nitrogen and sulfur. These ions are easily named using the rules in **Table 10.**

**Table 10 Oxyanion Naming Conventions for Sulfur and Nitrogen**

| |
|---|
| • Identify the ion with the greatest number of oxygen atoms. This ion is named using the root of the nonmetal and the suffix *-ate*. |
| • Identify the ion with fewer oxygen atoms. This ion is named using the root of the nonmetal and the suffix *-ite*.<br>Examples: $NO_3^-$ nitr*ate*, $NO_2^-$ nitr*ite*, $SO_4^{2-}$ sulf*ate*, $SO_3^{2-}$ sulf*ite* |

As shown in **Table 11,** chlorine forms four oxyanions that are named according to the number of oxygen atoms present. Names of similar oxyanions formed by other halogens follow the rules used for chlorine. For example, bromine forms the bromate ion ($BrO_3^-$), and iodine forms the periodate ion ($IO_4^-$) and the iodate ion ($IO_3^-$).

**Naming ionic compounds** Chemical nomenclature is a systematic way of naming compounds. Now that you are familiar with chemical formulas, you can use the following five rules to name ionic compounds.

1. Name the cation followed by the anion. Remember that the cation is always written first in the formula.
2. For monatomic cations, use the element name.
3. For monatomic anions, use the root of the element name plus the suffix *-ide.*
   Example:

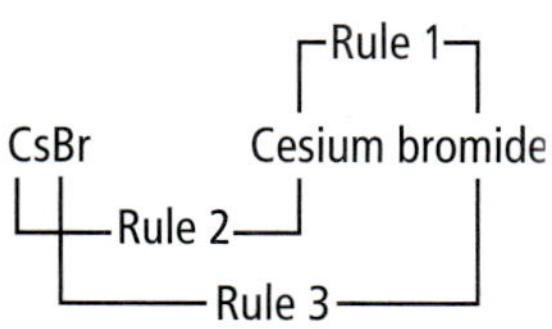


4. To distinguish between multiple oxidation numbers of the same element, the name of the chemical formula must indicate the oxidation number of the cation. The oxidation number is written as a Roman numeral in parentheses after the name of the cation.
   Note: This rule applies to the transition metals and metals on the right side of the periodic table, which often have more than one oxidation number. See **Table 8.** It does not apply to group 1 and group 2 cations, as they have only one oxidation number.
   Examples:
   $Fe^{2+}$ and $O^{2-}$ ions form FeO, known as iron(II) oxide.
   $Fe^{3+}$ and $O^{2-}$ ions form $Fe_2O_3$, known as iron(III) oxide.
5. When the compound contains a polyatomic ion, simply use the name of the polyatomic ion in place of the anion or cation.
   Examples:
   The name for NaOH is sodium hydroxide.
   The name for $(NH_4)_2S$ is ammonium sulfide.

**Table 11 Oxyanion Naming Conventions for Chlorine**

- The oxyanion with the greatest number of oxygen atoms is named using the prefix *per-*, the root of the nonmetal, and the suffix *-ate.*
- The oxyanion with one fewer oxygen atom is named using the root of the nonmetal and the suffix *-ate.*
- The oxyanion with two fewer oxygen atoms is named using the root of the nonmetal and the suffix *-ite.*
- The oxyanion with three fewer oxygen atoms is named using the prefix *hypo-*, the root of the nonmetal, and the suffix *-ite.*

Examples:

| | |
|---|---|
| $ClO_4^-$ perchlorate | $ClO_3^-$ chlorate |
| $ClO_2^-$ chlorite | $ClO^-$ hypochlorite |

Explore **ionic compounds.**

Virtual Investigations

## PRACTICE Problems

Do additional problems. Online Practice

**Name the following compounds.**

**28.** NaBr

**29.** $CaCl_2$

**30.** KOH

**31.** $Cu(NO_3)_2$

**32.** $Ag_2CrO_4$

**33. Challenge** The ionic compound $NH_4ClO_4$ is a key reactant used in solid rocket boosters, such as those that power the Space Shuttle into orbit. Name this compound.

## PROBLEM-SOLVING STRATEGY

### Naming Ionic Compounds

Naming ionic compounds is easy if you follow this naming-convention flowchart.

### Apply the Strategy

**Name** the compounds KOH and $Ag_2CrO_4$ using this flowchart.

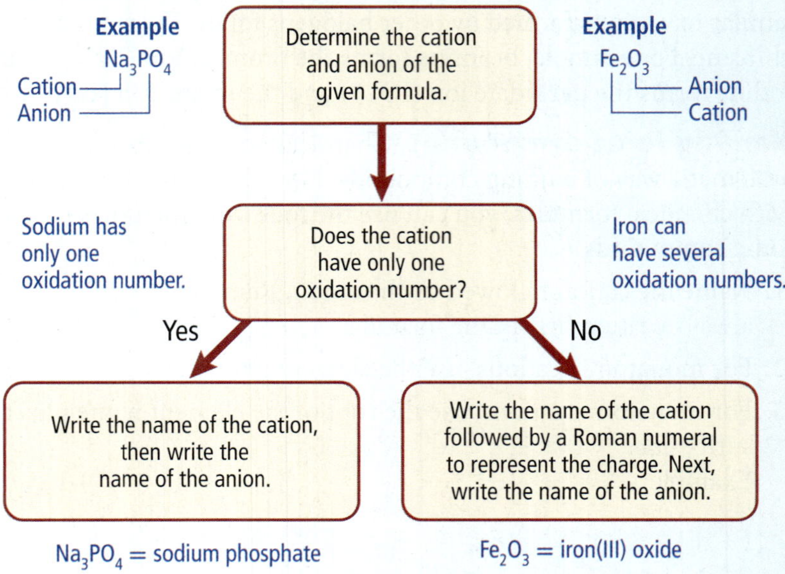


The Problem-Solving Strategy above reviews the steps used in naming ionic compounds if the formula is known. Naming ionic compounds is important in communicating the cation and anion present in a crystalline solid or aqueous solution. How might you change the diagram to help you write the formulas for ionic compounds if you know their names?

The ion-containing substances you have investigated so far have been ionic compounds. In the next section, you will learn how ions relate to the structure and properties of metals.

## SECTION 3 REVIEW

Section Self-Check

### Section Summary

- A formula unit gives the ratio of cations to anions in the ionic compound.
- A monatomic ion is formed from one atom. The charge of a monatomic ion is equal to its oxidation number.
- Roman numerals indicate the oxidation number of cations having multiple possible oxidation states.
- Polyatomic ions consist of more than one atom and act as a single unit.
- To indicate more than one polyatomic ion in a chemical formula, place parentheses around the polyatomic ion and use a subscript.

**34.** **MAIN**IDEA **State** the order in which the ions associated with a compound composed of potassium and bromine would be written in the chemical formula and the compound name.

**35. Describe** the difference between a monatomic ion and a polyatomic ion, and give an example of each.

**36. Apply** Ion *X* has a charge of 2+, and ion *Y* has a charge of 1−. Write the formula unit of the compound formed from the ions.

**37. State** the name and formula for the compound formed from Mg and Cl.

**38. Write** the name and formula for the compound formed from sodium ions and nitrite ions.

**39. Analyze** What subscripts would you most likely use if the following substances formed an ionic compound?

**a.** an alkali metal and a halogen

**b.** an alkali metal and a nonmetal from group 16

**c.** an alkaline earth metal and a halogen

**d.** an alkaline earth metal and a nonmetal from group 16

SECTION 4

# Metallic Bonds and the Properties of Metals

**MAIN**IDEA **Metals form crystal lattices and can be modeled as cations surrounded by a "sea" of freely moving valence electrons.**

### Essential Questions

- What are the characteristics of a metallic bond?
- How does the electron sea model account for the physical properties of metals?
- What are alloys, and how can they be categorized?

### Review Vocabulary

**physical property:** a characteristic of matter that can be observed or measured without altering the sample's composition

### New Vocabulary

electron sea model
delocalized electron
metallic bond
alloy

**CHEM 4 YOU** **Imagine a buoy in the ocean, bobbing by itself surrounded by a vast expanse of open water. Though the buoy stays in the same area, the ocean water freely flows past. In some ways, this description also applies to metallic atoms and their electrons.**

## Metallic Bonds

Although metals are not ionic, they share several properties with ionic compounds. The bonding in both metals and ionic compounds is based on the attraction of particles with unlike charges. Metals often form lattices in the solid state. These lattices are similar to the ionic crystal lattices discussed earlier. In such a lattice, 8 to 12 other metal atoms closely surround each metal atom.

**A sea of electrons** Although metal atoms always have at least one valence electron, they do not share these valence electrons with neighboring atoms, nor do they lose their valence electrons. Instead, within the crowded lattice, the outer energy levels of the metal atoms overlap. This unique arrangement is described by the electron sea model. The **electron sea model** proposes that all the metal atoms in a metallic solid contribute their valence electrons to form a "sea" of electrons. This sea of electrons surrounds the metal cations in the lattice.

The electrons present in the outer energy levels of the bonding metallic atoms are not held by any specific atom and can move easily from one atom to the next. Because they are free to move, they are often referred to as **delocalized electrons.** When the atom's outer electrons move freely throughout the solid, a metallic cation is formed. Each such ion is bonded to all neighboring metal cations by the sea of valence electrons, as shown in **Figure 11.** A **metallic bond** is the attraction of a metallic cation for delocalized electrons.

**Figure 11** The valence electrons in metals (shown as a blue cloud of minus signs) are evenly distributed among the metallic cations (shown in red). Attractions between positive cations and the negative "sea" hold the metal atoms together in a lattice.
**Explain** *Why are electrons in metals known as delocalized electrons?*

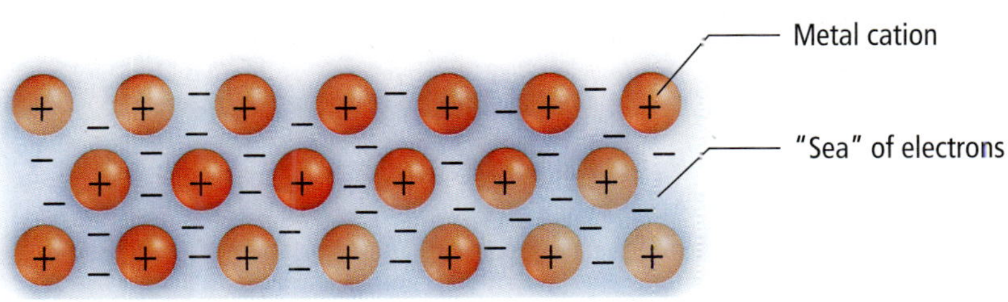

**Table 12 Melting and Boiling Points**

| Element | Melting Point (°C) | Boiling Point (°C) |
|---|---|---|
| Lithium | 180 | 1342 |
| Tin | 232 | 2602 |
| Aluminum | 660 | 2519 |
| Barium | 727 | 1897 |
| Silver | 962 | 2162 |
| Copper | 1085 | 2562 |

**Properties of metals** The physical properties of metals can be explained by metallic bonding. These properties provide evidence of the strength of metallic bonds.

**Melting and boiling points** The melting points of metals vary greatly. Mercury is a liquid at room temperature, which makes it useful in scientific instruments such as thermometers and barometers. On the other hand, tungsten has a melting point of 3422°C. Lightbulb filaments are usually made from tungsten, as are certain spacecraft parts.

In general, metals have moderately high melting points and high boiling points, as shown in **Table 12.** The melting points are not as extreme as the boiling points because the cations and electrons are mobile in a metal. It does not take an extreme amount of energy for them to be able to move past each other. However, during boiling, atoms must be separated from the group of cations and electrons, which requires much more energy.

**Malleability, ductility, and durability** Metals are malleable, which means they can be hammered into sheets, and they are ductile, which means they can be drawn into wire. **Figure 12** shows how the mobile particles involved in metallic bonding can be pushed or pulled past each other. Metals are generally durable. Although metallic cations are mobile in a metal, they are strongly attracted to the electrons surrounding them and are not easily removed from the metal.

**Thermal conductivity and electrical conductivity** The movement of mobile electrons around positive metallic cations makes metals good conductors. The delocalized electrons move heat from one place to another much more quickly than the electrons in a material that does not contain mobile electrons. Mobile electrons easily move as part of an electric current when an electric potential is applied to a metal. These same delocalized electrons interact with light, absorbing and releasing photons, thereby creating the property of luster in metals.

**Hardness and strength** The mobile electrons in transition metals consist not only of the two outer s electrons but also of the inner d electrons. As the number of delocalized electrons increases, so do the properties of hardness and strength. For example, strong metallic bonds are found in transition metals such as chromium, iron, and nickel, whereas alkali metals are considered soft because they have only one delocalized electron, $ns^1$.

☑ READING CHECK **Contrast** the behavior of metals and ionic compounds when each is struck by a hammer.

■ **Figure 12** An applied force causes metal ions to move through delocalized electrons, making metals malleable and ductile.

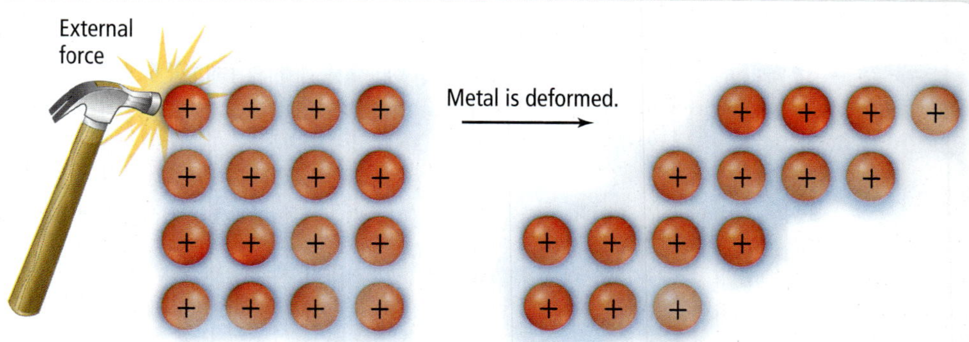

## Metal Alloys

Due to the nature of metallic bonds, it is relatively easy to introduce other elements into the metallic crystal, forming an alloy. An **alloy** is a mixture of elements that has metallic properties. Because of their unique blend of properties, alloys have a wide range of commercial applications. Stainless steel, brass, and cast iron are a few of the many useful alloys.

**VOCABULARY**

**WORD ORIGIN**

**Alloy**
comes from the Latin word *alligare,* which means *to bind.*

**Properties of alloys** The properties of alloys differ somewhat from the properties of the elements they contain. For example, steel is iron mixed with at least one other element. Some properties of iron are present, but steel has additional properties, such as increased strength. Some alloys vary in properties, depending on how they are manufactured. In the case of some metals, different properties can result based on heating and cooling.

## MiniLAB

iLab Station

### Observe Properties

**How do the properties of steel change when it is subjected to different types of heat treatment?** For centuries, people have treated metals with heat to change their properties. The final properties of the metal depend on the temperature to which the metal is heated and the rate at which it cools.

#### Procedure

1. Read and complete the lab safety form.
2. Examine a property of spring steel by trying to bend open one of three **hairpins.** Record your observations.
3. Next, hold each end of the hairpin with a pair of **forceps.** Place the curved central loop portion of the hairpin in the top of the blue flame from a **laboratory burner.** When the metal turns red, pull the hairpin open to form a straight piece of metal. Allow it to cool as you record your observations. Repeat Step 3 for the remaining two hairpins.
**WARNING:** ***Do not touch the hot metal. Do not hold your hand above the flame of the laboratory burner.***
4. To make softened steel, use a pair of forceps to hold all three hairpins vertically in the flame from the laboratory burner until the hairpins are glowing red all over. Slowly raise the three hairpins straight up and out of the flame so they cool slowly. Slow cooling results in the formation of large crystals.
5. After cooling, bend each of the three hairpins into the shape of the letter J. Record how the metal feels as you bend it.
6. To harden the steel, use the forceps to hold two of the bent hairpins in the flame until they are glowing red all over. Quickly plunge the hot metals into a **250-mL beaker** containing approximately 200 mL of **cold water.** Quick cooling causes the crystal size to be small.
7. Attempt to straighten one of the bends. Record your observations.
8. To temper the steel, use the forceps to hold the remaining hardened metal bend above the flame for a brief period of time. Slowly move the metal back and forth just above the flame until the gray metal turns to an iridescent blue-gray color. Do not allow the metal to become hot enough to glow red. Slowly cool the metal, and then try to unbend it using the end of your finger. Record your observations.

#### Analysis

1. **Analyze** your results, and identify the two types of steel that appear to have their properties combined in tempered steel.
2. **Hypothesize** how the different observed properties relate to crystal size.
3. **State** a use for spring steel that takes advantage of its unique properties.
4. **Infer** the advantages and disadvantages of using softened steel for body panels on automobiles.
5. **Apply** What is a major disadvantage of hardened steel? Do you think hardened steel would be wear-resistant and retain a sharpened edge? Explain your reasoning.

■ **Figure 13** Bicycle frames are sometimes made of 3/2.5 titanium alloy, an alloy of titanium containing 3% aluminum and 2.5% vanadium.



**Table 13 Commercial Alloys**

| Common Name | Composition | Uses |
|---|---|---|
| Alnico | Fe 50%, Al 20%, Ni 20%, Co 10% | magnets |
| Brass | Cu 67–90%, Zn 10–33% | plumbing, hardware, lighting |
| Bronze | Cu 70–95%, Zn 1–25%, Sn 1–18% | bearings, bells, medals |
| Cast iron | Fe 96–97%, C 3–4% | casting |
| Gold, 10-carat | Au 42%, Ag 12–20%, Cu 37.46% | jewelry |
| Lead shot | Pb 99.8%, As 0.2% | shotgun shells |
| Pewter | Sn 70–95%, Sb 5–15%, Pb 0–15% | tableware |
| Stainless steel | Fe 73–79%, Cr 14–18%, Ni 7–9% | instruments, sinks |
| Sterling silver | Ag 92.5%, Cu 7.5% | tableware, jewelry |

**Table 13** lists some commercially important alloys and their uses. An alloy of titanium and vanadium is used for the bicycle frame shown in **Figure 13.** Alloys such as this are classified into one of two basic types, substitutional alloys and interstitial alloys.

**Substitutional alloys** In a substitutional alloy, some of the atoms in the original metallic solid are replaced by other metals of similar atomic size. Sterling silver is an example of a substitutional alloy. In sterling silver, copper atoms replace some of the silver atoms in the metallic crystal. The resulting solid has properties of both silver and copper.

**Interstitial alloys** An interstitial alloy is formed when the small holes (interstices) in a metallic crystal are filled with smaller atoms. The best-known interstitial alloy is carbon steel. Holes in the iron crystal are filled with carbon atoms, and the physical properties of iron are changed. Iron is relatively soft and malleable. However, the presence of carbon makes the solid harder, stronger, and less ductile than pure iron.

## SECTION 4 REVIEW

Section Self-Check

### Section Summary

- A metallic bond forms when metal cations attract freely moving, delocalized valence electrons.
- In the electron sea model, electrons move through the metallic crystal and are not held by any particular atom.
- The electron sea model explains the physical properties of metallic solids.
- Metal alloys are formed when a metal is mixed with one or more other elements.

**40.** MAINIDEA **Contrast** the structures of ionic compounds and metals.

**41. Explain** how the conductivity of electricity and the high boiling points of metals are explained by metallic bonding.

**42. Contrast** the cause of the attraction in ionic bonds and metallic bonds.

**43. Summarize** alloy types by correctly pairing these terms and phrases: *substitutional, interstitial, replaced,* and *filled in.*

**44. Design an experiment** that could be used to distinguish between a metallic solid and an ionic solid. Include at least two different methods for comparing the solids. Explain your reasoning.

**45. Model** Draw a model to represent the physical property of metals known as ductility, or the ability to be drawn into a wire. Base your drawing on the electron sea model shown in **Figure 11.**

# everydayCHEMISTRY

## Killer Fashion

Shiny and colorful, costume jewelry can be inexpensive and fun. But is it safe? Usually the answer is yes. But some costume jewelry, particularly pieces made in developing countries, such as China and India, might pose a danger due to high levels of the toxic element lead (Pb).

**Poisoned plumbing** When lead gets wet, a certain amount of it dissolves, becoming lead ($Pb^{2+}$) ions. Inside the body, these ions can replace calcium ($Ca^{2+}$) ions. Other than their similar electric charges, lead and calcium are different (for one thing, lead ions are much heavier than calcium ions), and the presence of lead can cause learning disabilities, coma, or even death.

It might be surprising, then, to learn that lead was used by the Romans in, of all things, their water pipes! In fact, the symbol for lead—Pb—comes from the Latin word *plumbum,* which still appears in English as the root of the word *plumber,* one who works with pipes.

**Toxic pottery** While lead is not found in modern plumbing, it can still be found in other things. The pot shown in **Figure 1** was created with lead glaze and fired using traditional Mexican techniques to give it its distinctive black color. Glazes containing lead compounds can also create vibrant colors when fired under different conditions.

**Figure 1** Lead compounds in pottery glaze give this pot its distinctive look.



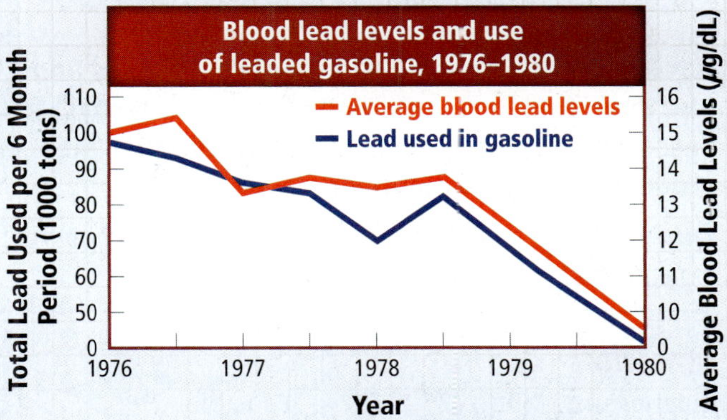


**Figure 2** Lead levels in Americans' blood dropped as leaded gasoline was phased out.

**A useful poison** Before it was known to be highly toxic, lead had a number of applications beyond pottery and plumbing. Lead has been used in paint and even gasoline, where its presence reduced "knock"—the tendency of gasoline to explode at the wrong time within the engine block. In the 1970s, when leaded gasoline was phased out in the United States, blood lead levels dropped immediately (see **Figure 2**).

But other avenues, such as jewelry or toys manufactured in other countries, can still contain lead. A lead-rich piece of costume jewelry might rest harmlessly against the skin until the metal finds its way into the mouth of a curious child or a daydreaming teenager.

**Chelation** Children are particularly susceptible to lead poisoning, due to their smaller body sizes and rapid rates of development. In serious cases, a process called chelation therapy might be the only way to save the child's life. Chelation therapy reverses one important effect of lead poisioning, replacing toxic lead with beneficial calcium in the body.

### WRITING IN ▶ Chemistry

**Sense of Danger** Our sense of taste can detect certain toxins found naturally in plants. Research other modern toxins, such as lead and antifreeze, to find out why they don't elicit a negative response from our taste buds.

WebQuest

# ChemLAB

iLab Station

## Synthesize an Ionic Compound

**Background:** You will form two compounds and test them to determine some of their properties. Based on your tests, you will decide whether the products are ionic compounds.

**Question:** *Can the physical properties of a compound indicate that they have ionic bonds?*

### Materials

magnesium ribbon (25 cm)
ring stand and ring
Bunsen burner
crucible tongs
100-mL beaker
conductivity tester
crucible
clay triangle
stirring rod
centigram balance
distilled water

### Safety Precautions

WARNING: ***Do not look directly at the burning magnesium; the intensity of the light can damage your eyes. Avoid handling heated materials until they have cooled.***

### Procedure

1. Read and complete the lab safety form.
2. Record all measurements in your data table.
3. Position the ring on the ring stand about 7 cm above the top of the Bunsen burner. Place the clay triangle on the ring.
4. Measure the mass of the clean, dry crucible.
5. Roll 25 cm of magnesium ribbon into a loose ball. Place it in the crucible. Measure the mass of the magnesium and crucible together.
6. Place the crucible on the triangle, and heat it with a hot flame (flame tip should be near the crucible).
7. Turn off the burner as soon as the magnesium ignites and begins to burn with a bright white light. Allow it to cool, and measure the mass of the magnesium product and the crucible.
8. Place the dry, solid product in the beaker.
9. Add 10 mL of distilled water to the beaker, and stir. Check the mixture with a conductivity tester.
10. **Cleanup and Disposal** Dispose of the product as directed by your teacher. Wash out the crucible with water. Return all lab equipment to its proper place.

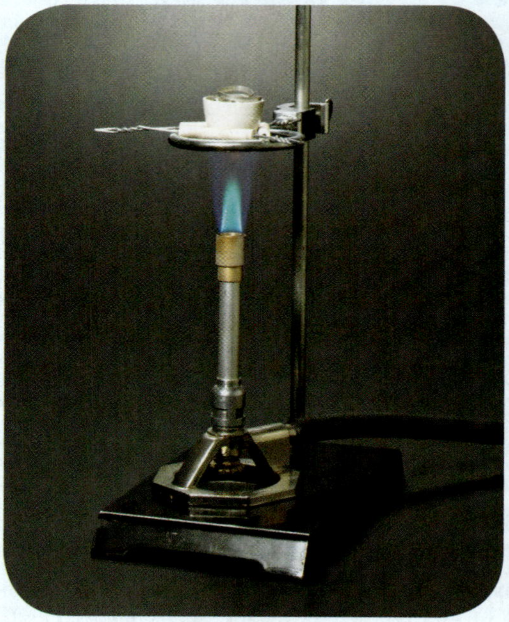

### Analyze and Conclude

1. **Analyze Data** Calculate the mass of the ribbon and the product. Record these masses in your table.
2. **Classify** the forms of energy released. What can you conclude about the stability of products?
3. **Infer** Does the magnesium react with the air?
4. **Predict** the ionic formulas for the two binary products formed, and write their names.
5. **Analyze and Conclude** The product of the magnesium-oxygen reaction is white, whereas the product of the magnesium-nitrogen reaction is yellow. Which compound makes up most of the product?
6. **Analyze and Conclude** Did the magnesium compounds conduct a current when in solution? Do these results verify that the compounds are ionic?
7. **Error Analysis** If the results show that the magnesium lost mass instead of gaining mass, cite possible sources of the error.

### INQUIRY EXTENSION

**Design an Experiment** If the magnesium compounds conduct a current in solution, can you affect how well they conduct electricity? If they did not conduct a current, could they? Design an experiment to find out.

# CHAPTER 7 STUDY GUIDE

Vocabulary Practice

**BIGIDEA** **Ionic compounds are held together by chemical bonds formed by the attraction of oppositely charged ions.**

## SECTION 1 Ion Formation

**MAINIDEA** Ions form when atoms gain or lose valence electrons to achieve a stable octet electron configuration.

- A chemical bond is the force that holds two atoms together.
- Some atoms form ions to gain stability. This stable configuration involves a complete outer energy level, usually consisting of eight valence electrons.
- Ions are formed by the loss or gain of valence electrons.
- The number of protons remains unchanged during ion formation.

**VOCABULARY**
- chemical bond
- cation
- anion

## SECTION 2 Ionic Bonds and Ionic Compounds

**MAINIDEA** Oppositely charged ions attract each other, forming electrically neutral ionic compounds.

- Ionic compounds contain ionic bonds formed by the attraction of oppositely charged ions.
- Ions in an ionic compound are arranged in a repeating pattern known as a crystal lattice.
- Ionic compound properties are related to ionic bond strength.
- Ionic compounds are electrolytes; they conduct an electric current in the liquid phase and in aqueous solution.
- Lattice energy is the energy needed to remove 1 mol of ions from its lattice.

**VOCABULARY**
- ionic bond
- ionic compound
- crystal lattice
- electrolyte
- lattice energy

## SECTION 3 Names and Formulas for Ionic Compounds

**MAINIDEA** In written names and formulas for ionic compounds, the cation appears first, followed by the anion.

- A formula unit gives the ratio of cations to anions in the ionic compound.
- A monatomic ion is formed from one atom. The charge of a monatomic ion is equal to its oxidation number.
- Roman numerals indicate the oxidation number of cations having multiple possible oxidation states.
- Polyatomic ions consist of more than one atom and act as a single unit.
- To indicate more than one polyatomic ion in a chemical formula, place parentheses around the polyatomic ion and use a subscript.

**VOCABULARY**
- formula unit
- monatomic ion
- polyatomic ion
- oxyanion

## SECTION 4 Metallic Bonds and the Properties of Metals

**MAINIDEA** Metals form crystal lattices and can be modeled as cations surrounded by a "sea" of freely moving valence electrons.

- A metallic bond forms when metal cations attract freely moving, delocalized valence electrons.
- In the electron sea model, electrons move through the metallic crystal and are not held by any particular atom.
- The electron sea model explains the physical properties of metallic solids.
- Metal alloys are formed when a metal is mixed with one or more other elements.

**VOCABULARY**
- electron sea model
- delocalized electron
- metallic bond
- alloy

## SECTION 1

### Mastering Concepts

**46.** How do positive ions and negative ions form?

**47.** When do chemical bonds form?

**48.** Why are halogens and alkali metals likely to form ions? Explain your answer.

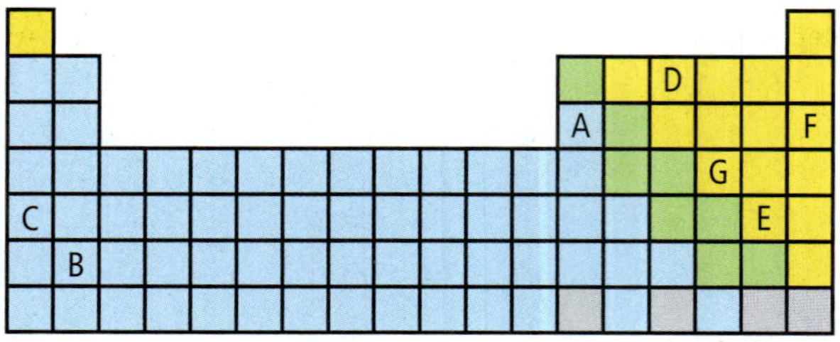


**Figure 14**

**49.** The periodic table shown in **Figure 14** contains elements labeled *A–G*. For each labeled element, state the number of valence electrons and identify the ion that will form.

**50.** Discuss the importance of electron affinity and ionization energy in the formation of ions.

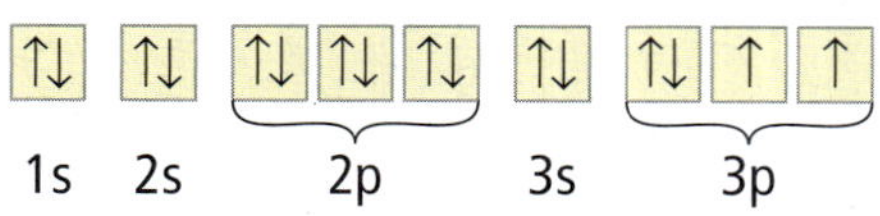


**Figure 15**

**51.** The orbital notation of sulfur is shown in **Figure 15.** Explain how sulfur forms its ion.

### Mastering Problems

**52.** Give the number of valence electrons in an atom of each element.

**a.** cesium
**b.** rubidium
**c.** gallium
**d.** zinc
**e.** strontium

**53.** Explain why noble gases are not likely to form chemical bonds.

**54.** Discuss the formation of the barium ion.

**55.** Explain how an anion of nitrogen forms.

**56.** The more reactive an atom, the higher its potential energy. Which atom has higher potential energy, neon or fluorine? Explain.

**57.** Explain how the iron atom can form both an iron 2+ ion and an iron 3+ ion.

**58.** Predict the reactivity of each atom based on its electron configuration.

**a.** potassium **b.** fluorine **c.** neon

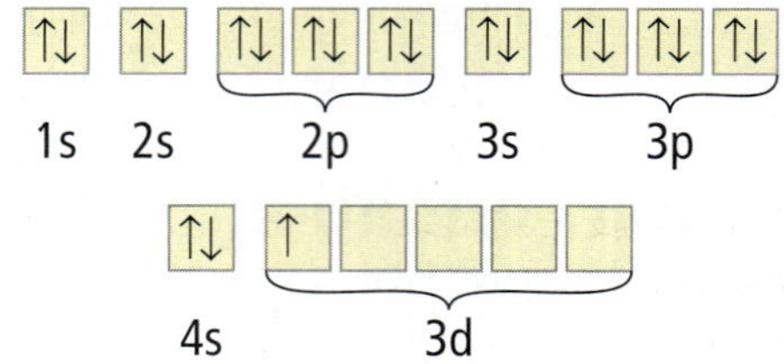


**Figure 16**

**59.** Discuss the formation of a 3+ scandium ion using its orbital notation, shown in **Figure 16.**

## SECTION 2

### Mastering Concepts

**60.** What does the term *electrically neutral* mean when discussing ionic compounds?

**61.** Discuss the formation of ionic bonds.

**62.** Explain why potassium does not bond with neon to form a compound.

**63.** Briefly discuss three physical properties of ionic solids that are linked to ionic bonds.

**64.** Describe an ionic crystal, and explain why ionic crystals for different compounds might vary in shape.

**65.** How does lattice energy change with a change in the size of an ion?

**66.** In **Figure 14,** the element labeled *B* is barium, and the element labeled *E* is iodine. Explain why the compound formed between these elements will not be BaI.

### Mastering Problems

**67.** Determine the ratio of cations to anions in each.

**a.** potassium chloride, a salt substitute
**b.** calcium fluoride, used in the steel industry
**c.** calcium oxide, used to remove sulfur dioxide from power-plant exhaust
**d.** strontium chloride, used in fireworks

**68.** Look at **Figure 14;** describe the ionic compound that forms from the elements represented by C and D.

**69.** Discuss the formation of an ionic bond between zinc and oxygen.

**70.** Using orbital notation, diagram the formation of an ionic bond between aluminum and fluorine.

**71.** Using electron configurations, diagram the formation of an ionic bond between barium and nitrogen.

**72. Conductors** Under certain conditions, ionic compounds conduct an electric current. Describe these conditions, and explain why ionic compounds are not always used as conductors.

**73.** Which of these compounds are not likely to occur: CaKr, $Na_2S$, $BaCl_3$, MgF? Explain your choices.

**74.** Use **Table 6** to determine which ionic compound has the highest melting point: MgO, KI, or AgCl. Explain your answer.

**75.** Which has the greater lattice energy, CsCl or KCl? $K_2O$ or CaO? Explain your choices.

## SECTION 3

### Mastering Concepts

**76.** What information do you need to write a correct chemical formula to represent an ionic compound?

**77.** When are subscripts used in formulas for ionic compounds?

**78.** Discuss how an ionic compound is named.

**79.** Using oxidation numbers, explain why the formula $NaF_2$ is incorrect.

**80.** Explain what the name chromium(III) oxide means in terms of electrons lost and gained, and identify the correct formula.

### Mastering Problems

**81.** Give the formula for each ionic compound.
- **a.** calcium iodide
- **b.** silver bromide
- **c.** copper(II) chloride
- **d.** potassium periodate
- **e.** silver acetate

**82.** Name each of the following ionic compounds.
- **a.** $K_2O$
- **b.** $CaCl_2$
- **c.** $Mg_3N_2$
- **d.** NaClO
- **e.** $KNO_3$

**83.** Complete **Table 14** by placing the symbols, formulas, and names in the blanks.

**Table 14 Identifying Ionic Compounds**

| Cation | Anion | Name | Formula |
|---|---|---|---|
| | | ammonium sulfate | |
| | | | $PbF_2$ |
| | | lithium bromide | |
| | | | $Na_2CO_3$ |
| $Mg^{2+}$ | $PO_4^{3-}$ | | |

**84. Chrome** Chromium, a transition metal used in chrome plating, forms both the $Cr^{2+}$ and $Cr^{3+}$ ions. Write the formulas for the ionic compounds formed when these ions react with fluorine ions and with oxygen ions.

**85.** Which are correct formulas for ionic compounds? For those that are not correct, give the correct formula and justify your answer.
- **a.** AlCl
- **b.** $Na_3SO_4$
- **c.** $BaOH_2$
- **d.** $Fe_2O$

**86.** Write the formulas for all of the ionic compounds that can be formed by combining each of the cations with each of the anions listed in **Table 15.** Name each compound formed.

**Table 15 List of Cations and Anions**

| Cations | Anions |
|---|---|
| $K^+$ | $SO_3^{2-}$ |
| $NH_4^+$ | $I^-$ |
| $Fe^{3+}$ | $NO_3^-$ |

## SECTION 4

### Mastering Concepts

**87.** Describe a metallic bond.

**88.** Briefly explain why metallic alloys are made.

**89.** Briefly describe how malleability and ductility of metals are explained by metallic bonding.

**90.** Compare and contrast the two types of metal alloys.

**91.** Explain how a metallic bond is similar to an ionic bond.

**92. Brass** Copper and zinc are used to form brass, an alloy. Briefly explain why these two metals form a substitutional alloy and not an interstitial alloy.

### Mastering Problems

**93.** How is a metallic bond different from an ionic bond?

**94. Silver** Briefly explain why silver is a good conductor of electricity.

**95. Steel** Briefly explain why steel, an alloy of iron, is used to build the supporting structure of many buildings.

**96.** The melting point of beryllium is 1287°C, while that of lithium is 180°C. Explain the large difference in values.

**97.** Titanium has a boiling point of 3287°C, and copper has a boiling point of 2562°C. Explain why there is a difference in the boiling points of these two metals.

**98. Alloys** Describe the difference between the metal alloy sterling silver and carbon steel in terms of the types of alloys involved.

## MIXED REVIEW

**99.** Give the number of valence electrons for atoms of oxygen, sulfur, arsenic, phosphorus, and bromine.

**100.** Explain why calcium can form a $Ca^{2+}$ ion but not a $Ca^{3+}$ ion.

**101.** Which ionic compound would have the greatest lattice energy: NaCl, KCl, or $MgCl_2$? Explain your answer.

**102.** Give the formula for each ionic compound.
- **a.** sodium sulfide
- **b.** iron(III) chloride
- **c.** sodium sulfate
- **d.** calcium phosphate
- **e.** zinc nitrate

**103.** Cobalt, a transition metal, forms both the $Co^{2+}$ and $Co^{3+}$ ions. Write the correct formulas, and give the name for the oxides formed by the two different ions.

**104.** Complete **Table 16**.

**Table 16 Element, Electron, and Ion Data**

| Element | Valence Electrons | Ion Formed |
|---|---|---|
| Selenium | | |
| Tin | | |
| Iodine | | |
| Argon | | |

**105. Gold** Briefly explain why gold can be used both in jewelry and as a conductor in electronic devices.

**106.** Discuss the formation of the nickel ion with a 2+ oxidation number.

**107.** Compare the oxyanions sulfate and sulfite.

**108.** Using electron-dot structures, diagram the formation of an ionic bond between potassium and iodine.

**109.** Magnesium forms both an oxide and a nitride when burned in air. Discuss the formation of magnesium oxide and magnesium nitride when magnesium atoms react with oxygen and nitrogen atoms.

**110.** An external force easily deforms sodium metal, while sodium chloride shatters when the same amount of force is applied. Why do these two solids behave so differently?

**111.** Name each ionic compound.
- **a.** CaO
- **b.** BaS
- **c.** $AlPO_4$
- **d.** $Ba(OH)_2$
- **e.** $Sr(NO_3)_2$

## THINK CRITICALLY

**112. Design** a concept map to explain the physical properties of both ionic compounds and metallic solids.

**113. Predict** which solid in each pair will have the higher melting point. Explain your answers.
- **a.** NaCl or CsCl
- **b.** Ag or Cu
- **c.** $Na_2O$ or MgO

**114. Compare and contrast** cations and anions.

**115. Observe and Infer** Identify the mistakes in the incorrect formulas and formula names, and design a flowchart to prevent the mistakes.
- **a.** copper acetate
- **b.** $Mg_2O_2$
- **c.** $Pb_2O_5$
- **d.** disodium oxide
- **e.** $Mg_3(PO_4)_2$

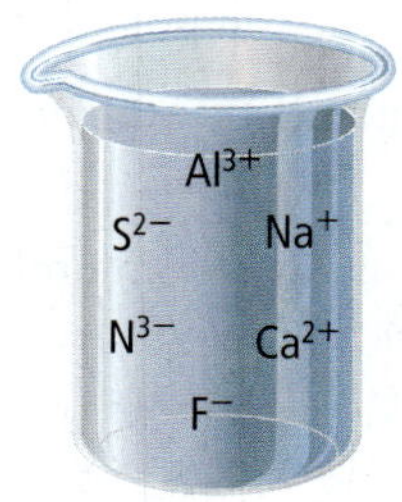


■ **Figure 17**

**116. Apply** Examine the ions in the beaker shown in **Figure 17.** Identify two compounds that could form using the available ions, and explain why this is possible.

**117. Apply** Praseodymium is a lanthanide element that reacts with hydrochloric acid, forming praseodymium(III) chloride. It also reacts with nitric acid, forming praseodymium(III) nitrate. Praseodymium has the electron configuration $[Xe]4f^36s^2$.
- **a.** Examine the electron configuration, and explain how praseodymium forms a 3+ ion.
- **b.** Write the correct formulas for both compounds formed by praseodymium.

**118. Hypothesize** Look at the locations of potassium and calcium on the periodic table. Form a hypothesis to explain why the melting point of calcium is considerably higher than the melting point of potassium.

**119. Assess** Explain why the term *delocalized* is an appropriate term for the electrons involved in metallic bonding.

**120. Apply** All uncharged atoms have valence electrons. Explain why elements such as iodine and sulfur do not have metallic bonds.

**121. Analyze** Explain why lattice energy is released upon formation of an ionic compound.

Chapter Self-Check

## CHALLENGE PROBLEM

**122.** **Ionic Compounds** Chrysoberyl is a transparent or translucent mineral that is sometimes opalescent. It is composed of beryllium aluminum oxide, $BeAl_2O_4$. Identify the oxidation numbers of each of the ions found in this compound. Explain the formation of this ionic compound.

## CUMULATIVE REVIEW

**123.** You are given a liquid of unknown density. The mass of a graduated cylinder containing 2.00 mL of the liquid is 34.68 g. The mass of the empty graduated cylinder is 30.00 g. Given this information, determine the density of the liquid.

**124.** In the laboratory, students used a balance and a graduated cylinder to collect the data shown in **Table 17.** Calculate the density of the sample. If the accepted value of this sample is 7.01 g/mL, calculate the percent error.

**Table 17** Volume and Mass Data

| | |
|---|---|
| Mass of sample | 19.21 g |
| Volume of water alone | 39.0 mL |
| Volume of water + sample | 43.1 mL |

**125.** A mercury atom drops in energy from $1.413 \times 10^{-18}$ J to $1.069 \times 10^{-18}$ J.

**a.** What is the energy of the photon emitted by the mercury atom?

**b.** What is the frequency of the photon emitted by the mercury atom?

**c.** What is the wavelength of the photon emitted by the mercury atom?

**126.** Which element has the greater ionization energy, chlorine or carbon?

**127.** Compare and contrast the ways in which metals and nonmetals form ions, and explain why they are different.

**128.** What are transition elements?

**129.** Write the symbol and name of the element that fits each description.

**a.** the second-lightest of the halogens

**b.** the metalloid with the lowest period number

**c.** the only group 16 element that is a gas at room temperature

**d.** the heaviest of the noble gases

**e.** the group 15 nonmetal that is a solid at room temperature

## WRITING IN Chemistry

**130.** **Free Radicals** Many researchers believe that free radicals are responsible for the effects of aging and cancer. Research free radicals, and write about the cause and what can be done to prevent free radicals.

**131.** **Growing Crystals** Crystals of ionic compounds can be easily grown in the laboratory setting. Research the growth of crystals, and design an experiment to grow a crystal in the laboratory.

## DBQ Document-Based Questions

**Oceans** *As part of an analysis of the world's oceans, scientists summarized the ion-related data shown in* **Table 18.**

Data from: Royal Society of Chemistry, *All at sea? The chemistry of the oceans.*

**Table 18** The Twelve Most-Common Ions in the Sea

| Ion | Concentration ($mg/dm^3$) | % by mass (of total dissolved solids) |
|---|---|---|
| $Cl^-$ | 19,000 | 55.04 |
| $Na^+$ | 10,500 | 30.42 |
| $SO_4^{2-}$ | 2655 | 7.69 |
| $Mg^{2+}$ | 1350 | 3.91 |
| $Ca^{2+}$ | 400 | 1.16 |
| $K^+$ | 380 | 1.10 |
| $CO_3^{2-}$ | 140 | 0.41 |
| $Br^-$ | 65 | 0.19 |
| $BO_3^{3-}$ | 20 | 0.06 |
| $SiO_3^{2-}$ | 8 | 0.02 |
| $Sr^{2+}$ | 8 | 0.02 |
| $F^-$ | 1 | 0.003 |

**132.** Identify the anions and cations listed in **Table 18.**

**133.** Create a bar graph of each ion's concentration. Explain why this is a difficult graph to draw.

**134.** Sodium chloride is not the only ionic compound that forms from sea water. Identify four other compounds that could be formed that contain the sodium ion. Write both the formula and the name for each compound.

CHAPTER 7 **STANDARDIZED TEST PRACTICE**

## MULTIPLE CHOICE

*Use the figure below to answer Question 1.*

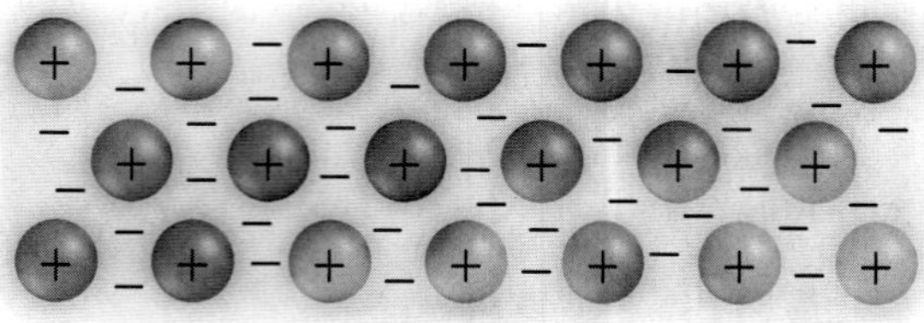

**1.** Which description is supported by the model shown?
  **A.** Metals are shiny, reflective substances.
  **B.** Metals are excellent conductors of heat and electricity.
  **C.** Ionic compounds are malleable compounds.
  **D.** Ionic compounds are good conductors of electricity.

**2.** Which is NOT true of the $Sc^{3+}$ ion?
  **A.** It has the same electron configuration as Ar.
  **B.** It is a scandium ion with three positive charges.
  **C.** It is considered to be a different element than a neutral Sc atom.
  **D.** It was formed by the loss of 4s and 3d electrons.

**3.** Of the salts below, which would require the most energy to break the ionic bonds?
  **A.** $BaCl_2$
  **B.** LiF
  **C.** NaBr
  **D.** KI

**4.** The high strength of its ionic bonds results in all of the following properties of NaCl EXCEPT
  **A.** hard crystals.
  **B.** high boiling point.
  **C.** high melting point.
  **D.** low solubility.

**5.** Which is the correct formula for the compound chromium(III) sulfate?
  **A.** $Cr_3SO_4$
  **B.** $Cr_2(SO_4)_3$
  **C.** $Cr_3(SO_4)_2$
  **D.** $Cr(SO_4)_3$

*Use the table below to answer Questions 6–8.*

**Physical Properties of Selected Compounds**

| Compound | Bond Type | Melting Point (°C) | Boiling Point (°C) |
|---|---|---|---|
| $F_2$ | Nonpolar covalent | −220 | −188 |
| $CH_4$ | Nonpolar covalent | −183 | −162 |
| $NH_3$ | Polar covalent | −78 | −33 |
| $CH_3Cl$ | Polar covalent | −64 | 61 |
| KBr | Ionic | 730 | 1435 |
| $Cr_2O_3$ | Ionic | ? | 4000 |

**6.** A compound is discovered to have a melting point of −100°C. Which could be true of this compound?
  **A.** It definitely has an ionic bond.
  **B.** It definitely has a polar covalent bond.
  **C.** It has either a polar covalent bond or a nonpolar covalent bond.
  **D.** It has either a polar covalent bond or an ionic bond.

**7.** Which could NOT be the melting point of $Cr_2O_3$?
  **A.** 2375°C
  **B.** 950°C
  **C.** 148°C
  **D.** 3342°C

**8.** Which is supported by the data in the table?
  **A.** Nonpolar covalent bonds have high boiling points.
  **B.** Polar covalent bonds have high melting points.
  **C.** Ionic bonds have low melting points.
  **D.** Ionic bonds have high boiling points.

**9.** Which is the correct orbital diagram for the third and fourth principal energy levels of vanadium?

  **A.** 3s [↑↓] 3p [↑↓][↑↓][↑↓] 4s [↑↓] 3d [↑↓][↑↓][ ][ ][ ]
  **B.** 3s [↑↓] 3p [↑↓][↑↓][↑↓] 4s [↑↓] 3d [↑][↑][↑][ ][ ]
  **C.** 3s [↑↓] 3p [↑↓][↑↓][↑↓] 4s [↑↓] 3d [↑↓][↑][ ][ ][ ]
  **D.** 3s [↑↓] 3p [↑↓][↑↓][↑↓] 4s [↑↓] 3d [↑][↑↓][ ][ ][ ]

## SHORT ANSWER

*Use the table below to answer Questions 10–12.*

Lutetium is a rare-earth element that can be used to speed up the chemical reactions involved in petroleum processing. It has two naturally occurring isotopes.

| Isotope | Form of Decay | Percent Abundance |
|---|---|---|
| $^{175}_{71}Lu$ | none | 97.41 |
| $^{176}_{71}Lu$ | beta | 2.59 |

**10.** Show the setup and calculate the average atomic mass of lutetium.

**11.** Identify the product when lutetium-176 goes through nuclear decay.

**12.** Compare the number of protons and neutrons in each of these isotopes.

## EXTENDED RESPONSE

**13.** Relate the change in atomic radius to the changes in atomic structure that occur across the periodic table.

*Use the diagram below to answer Question 14.*

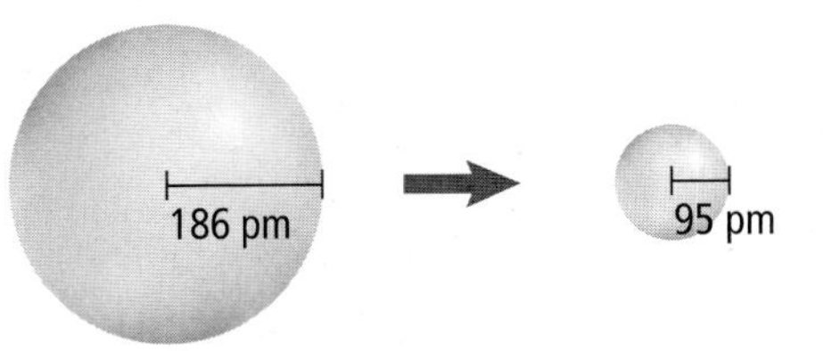


Sodium atom (Na)
[Ne]$3s^1$

Sodium ion ($Na^+$)
[Ne]

**14.** Relate the change in ionic radius to the changes in ion formation that occur across the periodic table.

## SAT SUBJECT TEST: CHEMISTRY

*Use the diagram below to answer Question 15.*

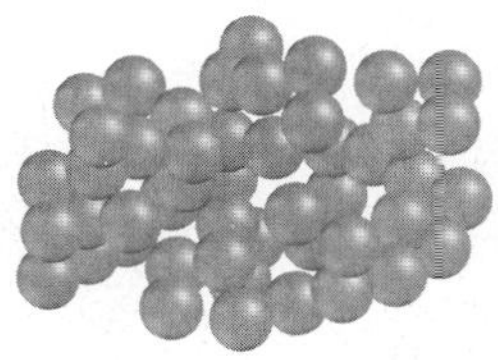

**15.** Which describes the state of matter shown?
- **A.** solid, because the particles are tightly packed against one another
- **B.** gas, because the particles are flowing past one another
- **C.** liquid, because the particles are able to move freely
- **D.** solid, because there is a regular pattern to the particles
- **E.** liquid, because the particles are in constant contact as they flow past each other

*Use the list of elements below to answer Questions 16–20.*

- **A.** sodium
- **B.** chromium
- **C.** boron
- **D.** argon
- **E.** chlorine

**16.** Which has its outermost electrons in an s-sublevel?

**17.** Which has seven valence electrons?

**18.** Which is a transition metal?

**19.** Which has an electron configuration of $1s^22s^22p^63s^23p^5$?

**20.** Which is a noble gas?

| NEED EXTRA HELP? | | | | | | | | | | | | | | | | | | | | |
|---|---|---|---|---|---|---|---|---|---|---|---|---|---|---|---|---|---|---|---|---|
| **If You Missed Question . . .** | 1 | 2 | 3 | 4 | 5 | 6 | 7 | 8 | 9 | 10 | 11 | 12 | 13 | 14 | 15 | 16 | 17 | 18 | 19 | 20 |
| **Review Section . . .** | 7.4 | 7.1 | 7.2 | 7.2 | 7.3 | 7.3 | 7.3 | 7.3 | 5.3 | 4.3 | 4.4 | 4.3 | 6.3 | 6.3 | 3.1 | 5.3 | 5.3 | 6.2 | 5.3 | 6.2 |

# CHAPTER 8
# Covalent Bonding

**BIGIDEA** Covalent bonds form when atoms share electrons.

## SECTIONS

Chm 3b, 3c, 12a

1 The Covalent Bond

2 Naming Molecules

3 Molecular Structures

4 Molecular Shapes

5 Electronegativity and Polarity

## LaunchLAB

iLab Station

### What type of compound is used to make a super ball?

Super balls are often made of a silicon compound called organosilicon oxide ($Si(OCH_2CH_3)_2O$). In this lab, you will compare the properties of organosilicon oxide to the properties of ionic compounds, which you studied previously.

**FOLDABLES® Study Organizer**

### Bond Character

Make a layered-look book. Label it as shown. Use it to help you organize your study of the three major types of bonding.

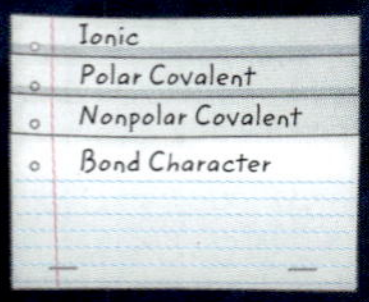


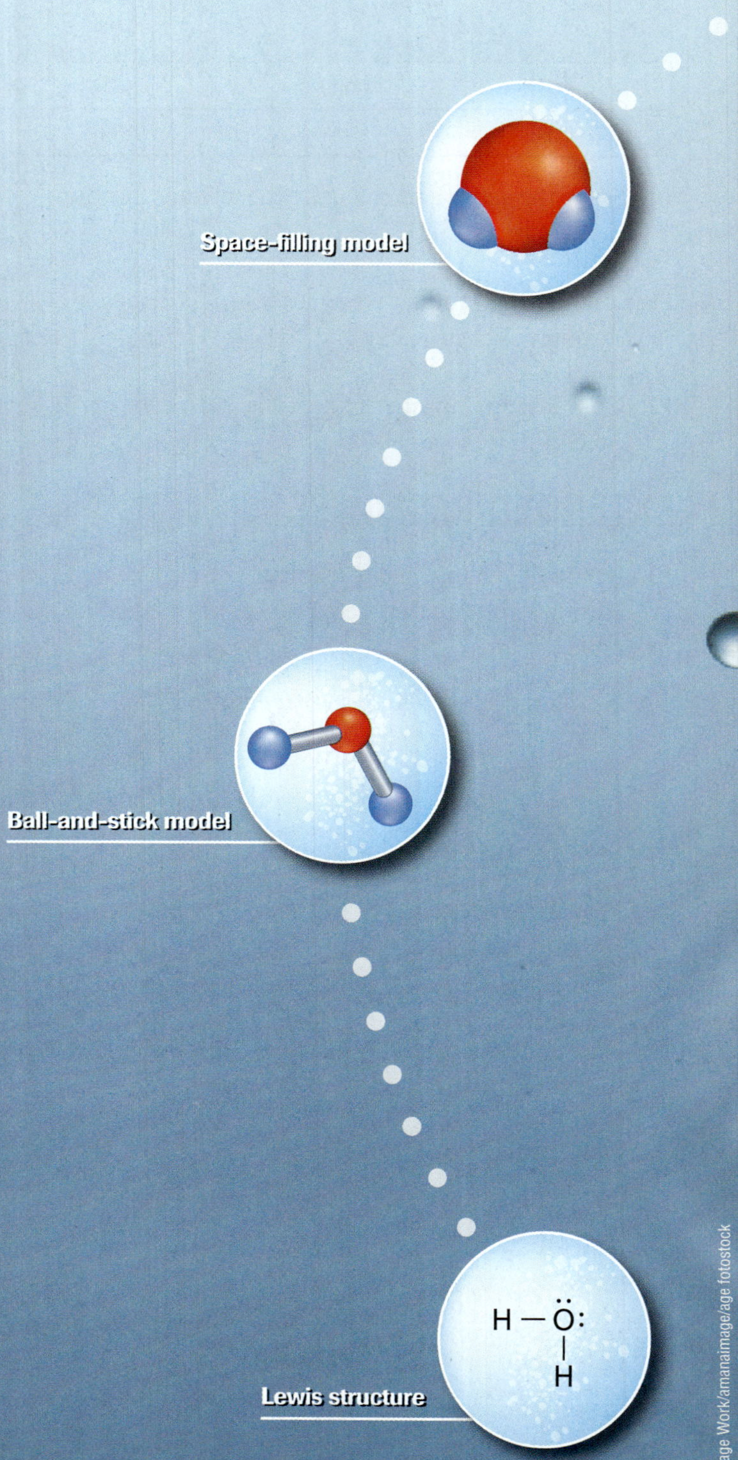

Go online!
MHEonlin.com

Water molecules are held together by different kinds of bonds than those that hold ionic compounds together. The forces between water molecules that result from these bonds help explain water's properties—for example, the spherical droplets it forms as it falls.

SECTION 1

# The Covalent Bond

**MAIN IDEA** Atoms gain stability when they share electrons and form covalent bonds.

### Essential Questions

- How does the octet rule apply to atoms that form covalent bonds?
- Why do atoms form single, double, and triple covalent bonds?
- What are sigma and pi bonds and how do they contrast?
- How are the strength of a covalent bond, its bond length, and its bond dissociation energy related?

### Review Vocabulary

**chemical bond:** the force that holds two atoms together

### New Vocabulary

covalent bond
molecule
Lewis structure
sigma bond
pi bond
endothermic reaction
exothermic reaction

**CHEM 4 YOU** Have you ever run in a three-legged race? Each person in the race shares one of their legs with a teammate to form a single three-legged team. In some ways, a three-legged race mirrors how atoms share electrons and join together as a unit.

## Why Do Atoms Bond?

Understanding the bonding in compounds is essential to developing new chemicals and technologies. To understand why new compounds form, recall what you know about elements that do not tend to form new compounds—the noble gases. You learned that all noble gases have stable electron arrangements. This stable arrangement consists of a full outer energy level and has lower potential energy than other electron arrangements. Because of their stable configurations, noble gases seldom form compounds.

**Gaining stability** The stability of an atom, ion, or compound is related to its energy; that is, lower energy states are more stable. From your study of ionic bonds, you know that metals and nonmetals can gain stability by transferring (gaining or losing) electrons to form ions. The resulting ions have stable noble-gas electron configurations. From the octet rule you know that atoms with a complete octet, a configuration of eight valence electrons, are stable. In this chapter, you will learn that the sharing of valence electrons is another way atoms can acquire the stable electron configuration of noble gases. The water droplets shown in **Figure 1** consist of water molecules formed when hydrogen and oxygen atoms share electrons.

■ **Figure 1** Each water droplet is made up of water molecules. Each water molecule is made up of two hydrogen atoms and one oxygen atom that have bonded by sharing electrons. The shapes of the drops are due to intermolecular forces acting on the water molecules.

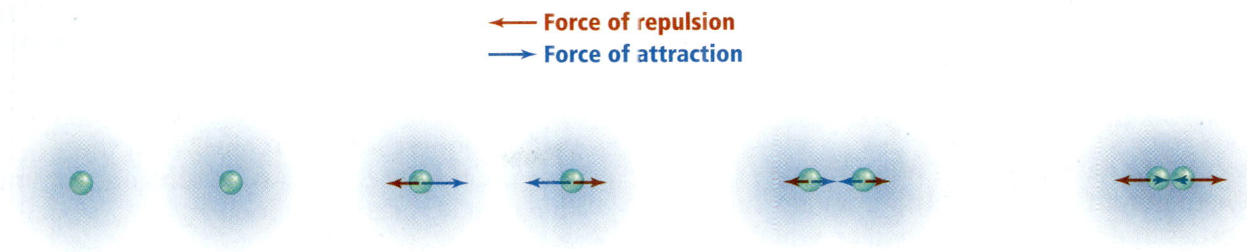


**Figure 2** The arrows in this diagram show the net forces of attraction and repulsion acting on two fluorine atoms as they move toward each other. The overall force between two atoms is the result of electron-electron repulsion, nucleus-nucleus repulsion, and nucleus-electron attraction. At the position of maximum net attraction, a covalent bond forms.

**Relate** ***How is the stability of the bond related to the forces acting on the atoms?***

## What is a covalent bond?

You just read that atoms can share electrons to form stable electron configurations. How does this occur? Are there different ways in which electrons can be shared? How are the properties of these compounds different from those formed by ions? Read on to answer these questions.

**Shared electrons** Atoms in nonionic compounds share electrons. The chemical bond that results from sharing valence electrons is a **covalent bond.** A **molecule** is formed when two or more atoms bond covalently. In a covalent bond, the shared electrons are considered to be part of the outer energy levels of both atoms involved. Covalent bonding generally can occur between elements that are near each other on the periodic table. The majority of covalent bonds form between atoms of nonmetallic elements.

**Covalent bond formation** Diatomic molecules, such as hydrogen ($H_2$), nitrogen ($N_2$), oxygen ($O_2$), fluorine ($F_2$), chlorine ($Cl_2$), bromine ($Br_2$), and iodine ($I_2$), form when two atoms of each element share electrons. They exist this way because the two-atom molecules are more stable than the individual atoms.

Consider fluorine, which has an electron configuration of $1s^22s^22p^5$. Each fluorine atom has seven valence electrons and needs another electron to form an octet. As two fluorine atoms approach each other, several forces act, as shown in **Figure 2.** Two repulsive forces act on the atoms, one from each atom's like-charged electrons and one from each atom's like-charged protons. A force of attraction also acts, as one atom's protons attract the other atom's electrons. As the fluorine atoms move closer, the attraction of the protons in each nucleus for the other atom's electrons increases until a point of maximum net attraction is achieved. At that point, the two atoms bond covalently and a molecule forms. If the two nuclei move closer, the repulsion forces increase and exceed the attractive forces.

The most stable arrangement of atoms in a covalent bond exists at some optimal distance between nuclei. At this point, the net attraction is greater than the net repulsion. Fluorine exists as a diatomic molecule because the sharing of one pair of electrons gives each fluorine atom a stable noble-gas configuration. As shown in **Figure 3**, each fluorine atom in the fluorine molecule has one pair of electrons that are covalently bonded (shared) and three pairs of electrons that are unbonded (not shared). Unbonded pairs are also known as lone pairs.

**Figure 3** Two fluorine atoms share a pair of electrons to form a covalent bond. Note that the shared electron pair gives each atom a complete octet.

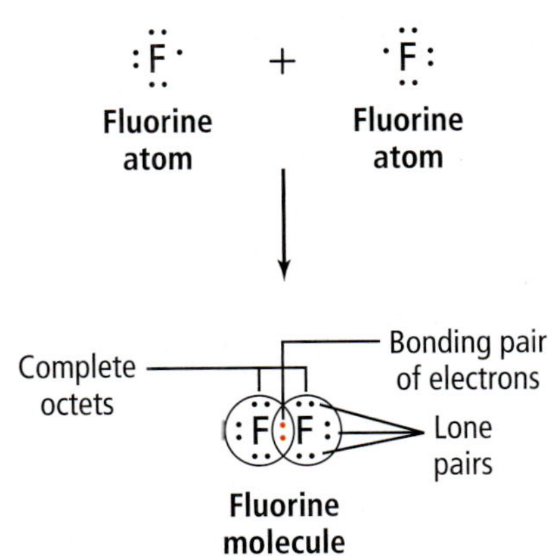

# MiniLAB

iLab Station

## Compare Melting Points

**How can you determine the relationship between bond type and melting point?** The properties of a compound depend on whether the bonds in the compound are ionic or covalent.

### Procedure

1. Read and complete the lab safety form.
2. Create a data table for the experiment.
3. Using a **permanent marker,** draw three lines on the inside bottom of a **disposable, 9-inch aluminum pie pan** to create three, equal wedges. Label the wedges, *A, B,* and *C.*
4. Set the pie pan on a **hot plate.** **WARNING:** ***Hot plate and metal pie pan will burn skin—handle with care.***
5. Obtain samples of the following from your teacher and deposit them onto the labeled wedges as follows: **sugar crystals** ($C_{12}H_{22}O_{11}$), *A;* **salt crystals** (NaCl) *B;* **paraffin** ($C_{23}H_{48}$), *C.*
6. Predict the order in which the compounds will melt.
7. Turn the temperature knob on the hot plate to the highest setting. You will heat the compounds for 5 min. Assign someone to time the heating of the compounds.
8. Observe the compounds during the 5-min period. Record which compounds melt and the order in which they melt.
9. After 5 min, turn off the hot plate and remove the pie pan using a **hot mitt** or **tongs.**
10. Allow the pie pan to cool, and then place it in the proper waste container.

### Analysis

1. **State** Which solid melted first? Which solid did not melt?
2. **Apply** Based on your observations and data, describe the melting point of each solid as low, medium, high, or very high.
3. **Infer** Which compounds are bonded with ionic bonds? Which are bonded with covalent bonds?
4. **Summarize** how the type of bonding affects the melting points of compounds.

## Single Covalent Bonds

When only one pair of electrons is shared, such as in a hydrogen molecule, it is a single covalent bond. The shared electron pair is often referred to as the bonding pair. For a hydrogen molecule, shown in **Figure 4,** each covalently bonded atom equally attracts the pair of shared electrons. Thus, the two shared electrons belong to each atom simultaneously, which gives each hydrogen atom the noble-gas configuration of helium ($1s^2$) and lower energy. The hydrogen molecule is more stable than either hydrogen atom is by itself.

Recall that electron-dot diagrams can be used to show valence electrons of atoms. In a **Lewis structure,** they can represent the arrangement of electrons in a molecule. A line or a pair of vertical dots between the symbols of elements represents a single covalent bond in a Lewis structure. For example, a hydrogen molecule is written as H—H or H:H.

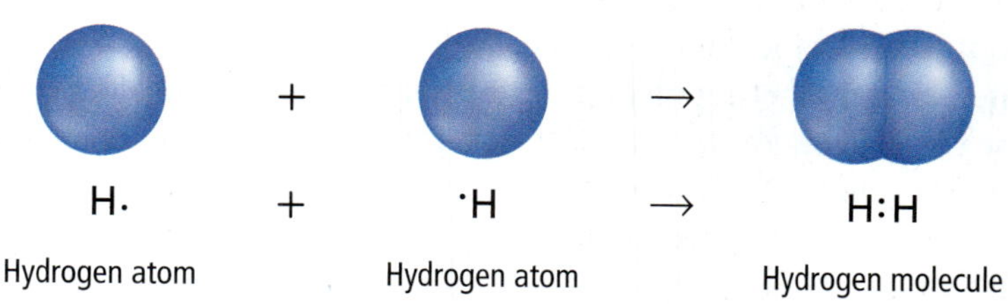


■ **Figure 4** When two hydrogen atoms share a pair of electrons, each hydrogen atom is stable because it has a full outer energy level.

**Group 17 and single bonds** The halogens—the group 17 elements, such as fluorine—have seven valence electrons. To form an octet, one more electron is needed. Therefore, atoms of group 17 elements form single covalent bonds with atoms of other nonmetals, such as carbon. You have already read that the atoms of some group 17 elements form covalent bonds with identical atoms. For example, fluorine exists as $F_2$ and chlorine exists as $Cl_2$.

**Group 16 and single bonds** An atom of a group 16 element can share two electrons and can form two covalent bonds. Oxygen is a group 16 element with an electron configuration of $1s^22s^22p^4$. Water is composed of two hydrogen atoms and one oxygen atom. Each hydrogen atom has the noble-gas configuration of helium when it shares one electron with oxygen. Oxygen, in turn, has the noble-gas configuration of neon when it shares one electron with each hydrogen atom. **Figure 5a** shows the Lewis structure for a molecule of water. Notice that the oxygen atom has two single covalent bonds and two unshared pairs of electrons.

**Group 15 and single bonds** Group 15 elements form three covalent bonds with atoms of nonmetals. Nitrogen is a group 15 element with the electron configuration of $1s^22s^22p^3$. Ammonia ($NH_3$) has three single covalent bonds. Three nitrogen electrons bond with the three hydrogen atoms leaving one pair of unshared electrons on the nitrogen atom. **Figure 5b** shows the Lewis structure for an ammonia molecule. Nitrogen also forms similar compounds with atoms of group 17 elements, such as nitrogen trifluoride ($NF_3$), nitrogen trichloride ($NCl_3$), and nitrogen tribromide ($NBr_3$). Each atom of these group 17 elements and the nitrogen atom share an electron pair.

**Group 14 and single bonds** Atoms of group 14 elements form four covalent bonds. A methane molecule ($CH_4$) forms when one carbon atom bonds with four hydrogen atoms. Carbon, a group 14 element, has an electron configuration of $1s^22s^22p^2$. With four valence electrons, carbon needs four more electrons for a noble gas configuration. Therefore, when carbon bonds with other atoms, it forms four bonds. Because a hydrogen atom, a group 1 element, has one valence electron, it takes four hydrogen atoms to provide the four electrons needed by a carbon atom. The Lewis structure for methane is shown in **Figure 5c.** Carbon also forms single covalent bonds with other nonmetal atoms, including those in group 17.

☑ READING CHECK **Describe** how a Lewis structure shows a covalent bond.

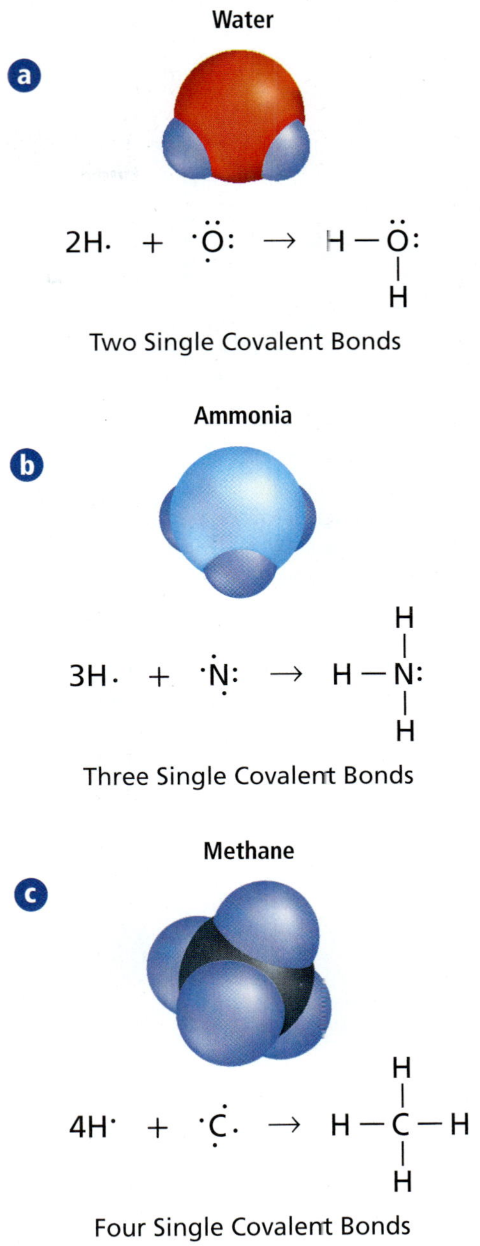


**Figure 5** These chemical equations show how atoms share electrons and become stable. As shown by the Lewis structure for each molecule, all atoms in each molecule achieve a full outer energy level.

**Describe** ***For the central atom in each molecule, describe how the octet rule is met.***

■ **Figure 6** The frosted-looking portions of this glass were chemically etched using hydrogen fluoride (HF), a weak acid. Hydrogen fluoride reacts with silica, the major component of glass, and forms gaseous silicon tetrafluoride ($SiF_4$) and water.

## EXAMPLE Problem 1

**LEWIS STRUCTURE OF A MOLECULE** The pattern on the glass shown in **Figure 6** was made by chemically etching its surface with hydrogen fluoride (HF). Draw the Lewis structure for a molecule of hydrogen fluoride.

### 1 ANALYZE THE PROBLEM

You are given the information that hydrogen and fluorine form the molecule hydrogen fluoride. An atom of hydrogen, a group 1 element, has only one valence electron. It can bond with any nonmetal atom when they share one pair of electrons. An atom of fluorine, a group 17 element, needs one electron to complete its octet. Therefore, a single covalent bond forms when atoms of hydrogen and fluorine bond.

### 2 SOLVE FOR THE UNKNOWN

To draw a Lewis structure, first draw the electron-dot diagram for each of the atoms. Then, rewrite the chemical symbols and draw a line between them to show the shared pair of electrons. Finally, add dots to show the unshared electron pairs.

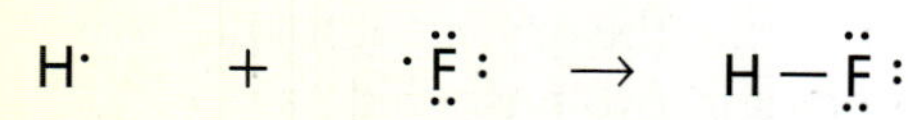

Hydrogen atom + Fluorine atom → Hydrogen fluoride molecule

### 3 EVALUATE THE ANSWER

Each atom in the new molecule now has a noble-gas configuration and is stable.

## PRACTICE Problems

Do additional problems. Online Practice

**Draw the Lewis structure for each molecule.**

1. $PH_3$
2. $H_2S$
3. HCl
4. $CCl_4$
5. $SiH_4$
6. **Challenge** Draw a generic Lewis structure for a molecule formed between atoms of group 1 and group 16 elements.

**The sigma bond** Single covalent bonds are also called **sigma bonds,** represented by the Greek letter sigma (σ). A sigma bond occurs when the pair of shared electrons is in an area centered between the two atoms. When two atoms share electrons, their valence atomic orbitals overlap end-to-end, concentrating the electrons in a bonding orbital between the two atoms. A bonding orbital is a localized region where bonding electrons will most likely be found. Sigma bonds can form when an s orbital overlaps with another s orbital or a p orbital, or two p orbitals overlap end-to-end. Water ($H_2O$), ammonia ($NH_3$), and methane ($CH_4$) have sigma bonds, as shown in **Figure 7.**

**VOCABULARY**
**ACADEMIC VOCABULARY**
**Overlap**
to occupy the same area in part
*The two driveways overlap at the street forming a common entrance.*

☑ READING CHECK **List** the orbitals that can form sigma bonds in a covalent compound.

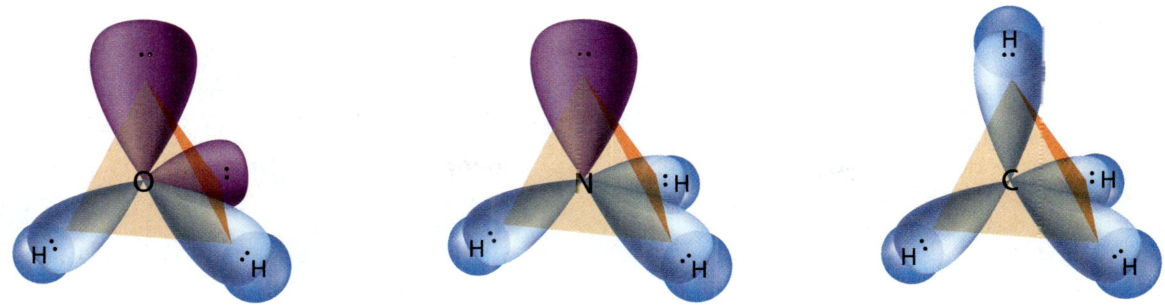


■ **Figure 7** Sigma bonds formed in each of these molecules when the atomic orbital of each hydrogen atom overlapped end-to-end with the orbital of the central atom.

**Interpret** ***Identify the number of sigma bonds in each molecule.***

## Multiple Covalent Bonds

In some molecules, atoms have noble-gas configurations when they share more than one pair of electrons with one or more atoms. Sharing multiple pairs of electrons forms multiple covalent bonds. A double covalent bond and a triple covalent bond are examples of multiple bonds. Carbon, nitrogen, oxygen, and sulfur atoms often form multiple bonds with other nonmetals. How do you know if two atoms will form a multiple bond? In general, the number of valence electrons needed to form an octet equals the number of covalent bonds that can form.

**Double bonds** A double covalent bond forms when two pairs of electrons are shared between two atoms. For example, atoms of the element oxygen only exist as diatomic molecules. Each oxygen atom has six valence electrons and must obtain two additional electrons for a noble-gas configuration, as shown in **Figure 8a.** A double covalent bond forms when each oxygen atom shares two electrons; a total of two pairs of electrons are shared between the two atoms.

**FOLDABLES®**
Incorporate information from this section into your Foldable.

**Triple bonds** A triple covalent bond forms when three pairs of electrons are shared between two atoms. Diatomic nitrogen ($N_2$) molecules contain a triple covalent bond. Each nitrogen atom shares three electron pairs, forming a triple bond with the other nitrogen atom as shown in **Figure 8b.**

**The pi bond** A multiple covalent bond consists of one sigma bond and at least one pi bond. A **pi bond,** represented by the Greek letter pi ($\pi$), forms when parallel orbitals overlap and share electrons. The shared electron pair of a pi bond occupies the space above and below the line that represents where the two atoms are joined together.

**a** :Ö· + ·Ö: → :O=O:

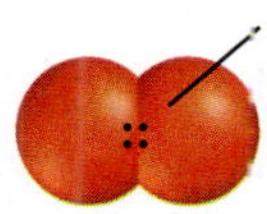


**b** :Ṅ· + ·Ṅ: → :N≡N:

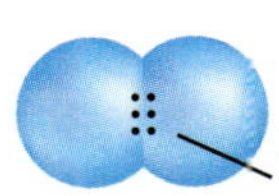


■ **Figure 8** Multiple covalent bonds form when two atoms share more than one pair of electrons. **a.** Two oxygen atoms form a double bond. **b.** A triple bond forms between two nitrogen atoms.

Get help with **multiple covalent bonds.**

Personal Tutor

■ **Figure 9** Notice how the multiple bond between the two carbon atoms in ethene ($C_2H_4$) consists of a sigma bond and a pi bond. The sigma bond is formed by the end-to-end overlap of orbitals directly between the two carbon atoms. The carbon atoms are close enough that the side-by-side p orbitals overlap and form the pi bond. This results in a doughnut-shaped cloud around the sigma bond.

View an **animation about sigma and pi bonding.**

**Concepts In Motion**

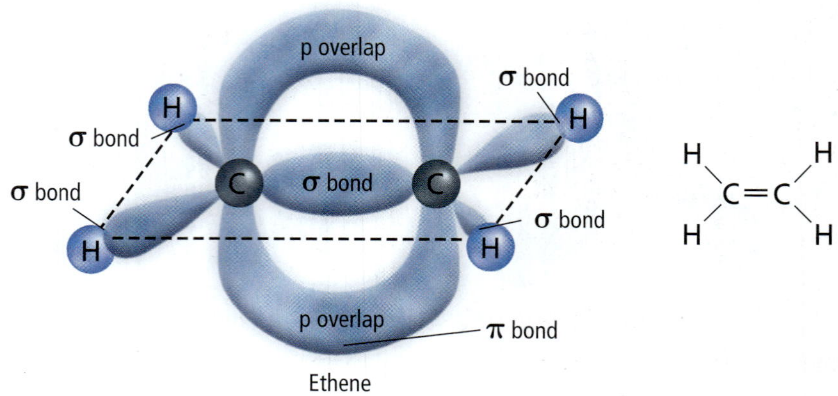


It is important to note that molecules having multiple covalent bonds contain both sigma and pi bonds. A double covalent bond, as shown in **Figure 9,** consists of one pi bond and one sigma bond. A triple covalent bond consists of two pi bonds and one sigma bond.

## The Strength of Covalent Bonds

Recall that a covalent bond involves attractive and repulsive forces. In a molecule, nuclei and electrons attract each other, but nuclei repel other nuclei, and electrons repel other electrons. When this balance of forces is upset, a covalent bond can be broken. Because covalent bonds differ in strength, some bonds break more easily than others. Several factors influence the strength of covalent bonds.

**Bond length** The strength of a covalent bond depends on the distance between the bonded nuclei. The distance between the two bonded nuclei at the position of maximum attraction is called bond length, as shown in **Figure 10.** It is determined by the sizes of the two bonding atoms and how many electron pairs they share. Bond lengths for molecules of fluorine ($F_2$), oxygen ($O_2$), and nitrogen ($N_2$) are listed in **Table 1.** Notice that as the number of shared electron pairs increases, the bond length decreases.

Bond length and bond strength are also related: the shorter the bond length, the stronger the bond. Therefore, a single bond, such as that in $F_2$, is weaker than a double bond, such as that in $O_2$. Likewise, the double bond in $O_2$ is weaker than the triple bond in $N_2$.

☑ READING CHECK **Relate** covalent bond type to bond length.

■ **Figure 10** Bond length is the distance from the center of one nucleus to the center of the other nucleus of two bonded atoms.

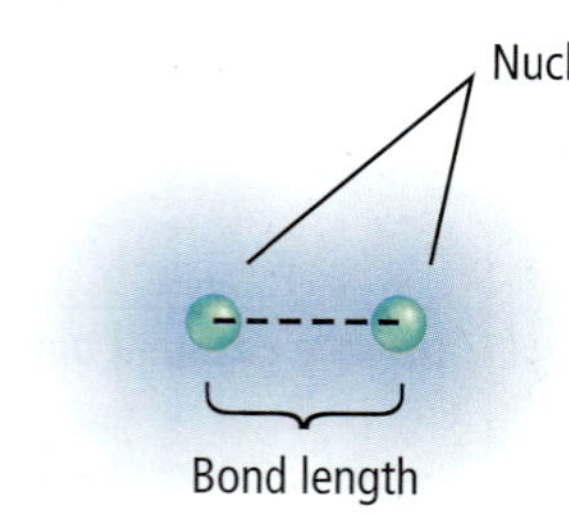


**Table 1** Covalent Bond Type and Bond Length

| Molecule | Bond Type | Bond Length |
|---|---|---|
| $F_2$ | single covalent | $1.43 \times 10^{-10}$ m |
| $O_2$ | double covalent | $1.21 \times 10^{-10}$ m |
| $N_2$ | triple covalent | $1.10 \times 10^{-10}$ m |

**Table 2 Bond-Dissociation Energy**

| Molecule | Bond-Dissociation Energy |
|---|---|
| $F_2$ | 159 kJ/mol |
| $O_2$ | 498 kJ/mol |
| $N_2$ | 945 kJ/mol |

**Bonds and energy** An energy change occurs when a bond between atoms in a molecule forms or breaks. Energy is released when a bond forms, but energy must be added to break a bond. The amount of energy required to break a specific covalent bond is called bond-dissociation energy and is always a positive value. The bond-dissociation energies for the covalent bonds in molecules of fluorine, oxygen, and nitrogen are listed in **Table 2.**

Bond-dissociation energy also indicates the strength of a chemical bond because of the inverse relationship between bond energy and bond length. As indicated in **Table 1** and **Table 2,** the smaller the bond length is, the greater the bond-dissociation energy. The sum of the bond-dissociation energy values for all of the bonds in a molecule is the amount of chemical potential energy in a molecule of that compound.

The total energy change of a chemical reaction is determined from the energy of the bonds broken and formed. An **endothermic reaction** occurs when a greater amount of energy is required to break the existing bonds in the reactants than is released when the new bonds form in the products. An **exothermic reaction** occurs when more energy is released during product bond formation than is required to break bonds in the reactants. **Figure 11** illustrates a common exothermic reaction. You will study exothermic and endothermic reactions in much greater detail when you study the energy changes in chemical reactions.

**Figure 11** Breaking the C–C bonds in charcoal and the O–O bonds in the oxygen in air requires an input of energy. Energy is released as heat and light when bonds form, producing $CO_2$. Thus, the burning of charcoal is an exothermic reaction.

## SECTION 1 REVIEW

Section Self-Check

### Section Summary

- Covalent bonds form when atoms share one or more pairs of electrons.
- Sharing one pair, two pairs, and three pairs of electrons forms single, double, and triple covalent bonds, respectively.
- Orbitals overlap directly in sigma bonds. Parallel orbitals overlap in pi bonds. A single covalent bond is a sigma bond but multiple covalent bonds are made of both sigma and pi bonds.
- Bond length is measured nucleus-to-nucleus. Bond-dissociation energy is needed to break a covalent bond.

7. **MAIN**IDEA **Identify** the type of atom that generally forms covalent bonds.
8. **Describe** how the octet rule applies to covalent bonds.
9. **Illustrate** the formation of single, double, and triple covalent bonds using Lewis structures.
10. **Compare and contrast** ionic bonds and covalent bonds.
11. **Contrast** sigma bonds and pi bonds.
12. **Apply** Create a graph using the bond-dissociation energy data in **Table 2** and the bond-length data in **Table 1.** Describe the relationship between bond length and bond-dissociation energy.
13. **Predict** the relative bond-dissociation energies needed to break the bonds in the structures below.

    **a.** H—C≡C—H

    **b.** 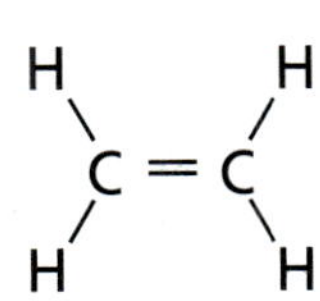

SECTION 2

# Naming Molecules

**MAINIDEA** Specific rules are used when naming binary molecular compounds, binary acids, and oxyacids.

### Essential Questions

- What rules do you follow to name a binary molecular compound from its molecular formula?
- How are acidic solutions named?

### Review Vocabulary

**oxyanion:** a polyatomic ion in which an element (usually a nonmetal) is bonded to one or more oxygen atoms

### New Vocabulary

oxyacid

**CHEM 4 YOU** You probably know that your mother's mother is your grandmother, and that your grandmother's sister is your great-aunt. But what do you call your grandmother's brother's daughter? Naming molecules requires a set of rules, just as naming family relationships requires rules.

## Naming Binary Molecular Compounds

Many molecular compounds have common names, but they also have scientific names that reveal their composition. To write the formulas and names of molecules, you will use processes similar to those described for ionic compounds.

Start with a binary molecular compound. Note that a binary molecular compound is composed only of two nonmetal atoms—not metal atoms or ions. An example is dinitrogen monoxide ($N_2O$), a gaseous anesthetic that is more commonly known as nitrous oxide or laughing gas. The naming of $N_2O$ is explained in the following rules.

1. The first element in the formula is always named first, using the entire element name. **N is the symbol for *nitrogen*.**
2. The second element in the formula is named using its root and adding the suffix *-ide*. **O is the symbol for oxygen so the second word is *oxide*.**
3. Prefixes are used to indicate the number of atoms of each element that are present in the compound. **Table 3** lists the most common prefixes used. **There are two atoms of nitrogen and one atom of oxygen, so the first word is *dinitrogen* and second word is *monoxide*.**

There are exceptions to using the prefixes shown in **Table 3.** The first element in the compound name never uses the *mono-* prefix. For example, CO is carbon monoxide, not monocarbon monoxide. Also, if using a prefix results in two consecutive vowels, one of the vowels is usually dropped to avoid an awkward pronunciation. For example, notice that the oxygen atom in CO is called monoxide, not monooxide.

Explore **naming covalent compounds with an interactive table.** Concepts In Motion

**Table 3** Prefixes in Covalent Compounds

| Number of Atoms | Prefix | Number of Atoms | Prefix |
|---|---|---|---|
| 1 | mono- | 6 | hexa- |
| 2 | di- | 7 | hepta- |
| 3 | tri- | 8 | octa- |
| 4 | tetra- | 9 | nona- |
| 5 | penta- | 10 | deca- |

## EXAMPLE Problem 2

EXAMPLE PROBLEM

**NAMING BINARY MOLECULAR COMPOUNDS** Name the compound $P_2O_5$, which is used as a drying and dehydrating agent.

### 1 ANALYZE THE PROBLEM

You are given the formula for a compound. The formula contains the elements and the number of atoms of each element in one molecule of the compound. Because only two different elements are present and both are nonmetals, the compound can be named using the rules for naming binary molecular compounds.

### 2 SOLVE FOR THE UNKNOWN

First, name the elements involved in the compound.

| | |
|---|---|
| phosphorus | **The first element, represented by P, is phosphorus.** |
| oxide | **The second element, represented by O, is oxygen. Add the suffix *–ide* to the root of oxygen, *ox-*.** |
| phosphorus oxide | **Combine the names.** |

Now modify the names to indicate the number of atoms present in a molecule.

| | |
|---|---|
| **diphosphorus pentoxide** | **From the formula $P_2O_5$, you know that two phosphorus atoms and five oxygen atoms make up a molecule of the compound. From Table 3, you know that *di-* is the prefix for two and *penta-* is the prefix for five. The *a* in *penta-* is not used because *oxide* begins with a vowel.** |

### 3 EVALUATE THE ANSWER

The name diphosphorus pentoxide shows that a molecule of the compound contains two phosphorus atoms and five oxygen atoms, which agrees with the compound's chemical formula, $P_2O_5$.

## PRACTICE Problems

Do additional problems. Online Practice

**Name each of the binary covalent compounds listed below.**

**14.** $CO_2$

**15.** $SO_2$

**16.** $NF_3$

**17.** $CCl_4$

**18. Challenge** What is the formula for diarsenic trioxide?

**Common names for some molecular compounds** Have you ever enjoyed an icy, cold glass of dihydrogen monoxide on a hot day? You probably have but you most likely called it by its common name, water. Recall that many ionic compounds have common names in addition to their scientific ones. For example, baking soda is sodium hydrogen carbonate and common table salt is sodium chloride.

Many binary molecular compounds, such as nitrous oxide and water, were discovered and given common names long before the present-day naming system was developed. Other binary covalent compounds that are generally known by their common names rather than their scientific names are ammonia ($NH_3$), hydrazine ($N_2H_4$), and nitric oxide (NO).

**READING CHECK** **Apply** What are the scientific names for ammonia, hydrazine, and nitric oxide?

## Naming Acids

Water solutions of some molecules are acidic and are named as acids. Acids are important compounds with specific properties. If a compound produces hydrogen ions ($H^+$) in solution, it is an acid. For example, HCl produces $H^+$ in solution and is an acid. Two common types of acids exist—binary acids and oxyacids.

**Naming binary acids** A binary acid contains hydrogen and one other element. The naming of the common binary acid known as hydrochloric acid is explained in the following rules.

1. The first word has the prefix *hydro-* to name the hydrogen part of the compound. The rest of the first word consists of a form of the root of the second element plus the suffix *-ic.* **HCl (hydrogen and chlorine) becomes *hydrochloric.***
2. The second word is always *acid.* **Thus, HCl in a water solution is called *hydrochloric acid.***

Although the term *binary* indicates exactly two elements, a few acids that contain more than two elements are named according to the rules for naming binary acids. If no oxygen is present in the formula for the acidic compound, the acid is named in the same way as a binary acid, except that the root of the second part of the name is the root of the polyatomic ion that the acid contains. For example, HCN, which is composed of hydrogen and the cyanide ion, is called *hydrocyanic acid* in solution.

**Naming oxyacids** An acid that contains both a hydrogen atom and an oxyanion is referred to as an **oxyacid.** Recall that an oxyanion is a polyatomic ion containing one or more oxygen atoms. The following rules explain the naming of nitric acid ($HNO_3$), an oxyacid.

1. First, identify the oxyanion present. The first word of an oxyacid's name consists of the root of the oxyanion and the prefix *per-* or *hypo-* if it is part of the oxyanion's name. The first word of the oxyacid's name also has a suffix that depends on the oxyanion's suffix. If the oxyanion's name ends with the suffix *-ate,* replace it with the suffix *-ic.* If the name of the oxyanion ends with the suffix *-ite,* replace it with the suffix *-ous.* **$NO_3^-$, the nitrate ion, becomes *nitric.***
2. The second word of the name is always *acid.* **$HNO_3$ (hydrogen and the nitrate ion) becomes *nitric acid.***

**Table 4** shows how the names of several oxyacids follow these rules. Notice that the hydrogen in an oxyacid is not part of the name.

**Table 4 Naming Oxyacids**

| Compound | Oxyanion | Acid Suffix | Acid Name |
|---|---|---|---|
| $HClO_3$ | chlorate | -ic | chloric acid |
| $HClO_2$ | chlorite | -ous | chlorous acid |
| $HNO_3$ | nitrate | -ic | nitric acid |
| $HNO_2$ | nitrite | -ous | nitrous acid |

Explore **naming covalent compounds with an interactive table.** Concepts In Motion

**Table 5 Formulas and Names of Some Covalent Compounds**

| Formula | Common Name | Molecular Compound Name |
|---|---|---|
| $H_2O$ | water | dihydrogen monoxide |
| $NH_3$ | ammonia | nitrogen trihydride |
| $N_2H_4$ | hydrazine | dinitrogen tetrahydride |
| HCl | muriatic acid | hydrochloric acid |
| $C_9H_8O_4$ | aspirin | 2-(acetyloxy)benzoic acid |

You have learned that naming covalent compounds follows different sets of rules depending on the composition of the compound. **Table 5** summarizes the formulas and names of several covalent compounds. Note that an acid, whether a binary acid or an oxyacid, can have a common name in addition to its compound name.

**PRACTICE Problems** Do additional problems. Online Practice

**Name the following acids. Assume each compound is dissolved in water.**

**19.** HI **20.** $HClO_3$ **21.** $HClO_2$ **22.** $H_2SO_4$ **23.** $H_2S$

**24. Challenge** What is the formula for periodic acid?

## Writing Formulas from Names

The name of a molecular compound reveals its composition and is important in communicating the nature of the compound. Given the name of any binary molecule, you should be able to write the correct chemical formula. The prefixes used in a name indicate the exact number of each atom present in the molecule and determine the subscripts used in the formula. If you are having trouble writing formulas from the names for binary compounds, you might want to review the naming rules listed on pages at the beginning of this section.

The formula for an acid can also be derived from the name. It is helpful to remember that all binary acids contain hydrogen and one other element. For oxyacids—acids containing oxyanions—you will need to know the names of the common oxyanions. If you need to review oxyanion names, see **Table 9** in the previous chapter.

**PRACTICE Problems** Do additional problems. Online Practice

**Give the formula for each compound.**

**25.** silver chloride

**26.** dihydrogen monoxide

**27.** chlorine trifluoride

**28.** diphosphorus trioxide

**29.** disulfur decafluoride

**30. Challenge** What is the formula for carbonic acid?

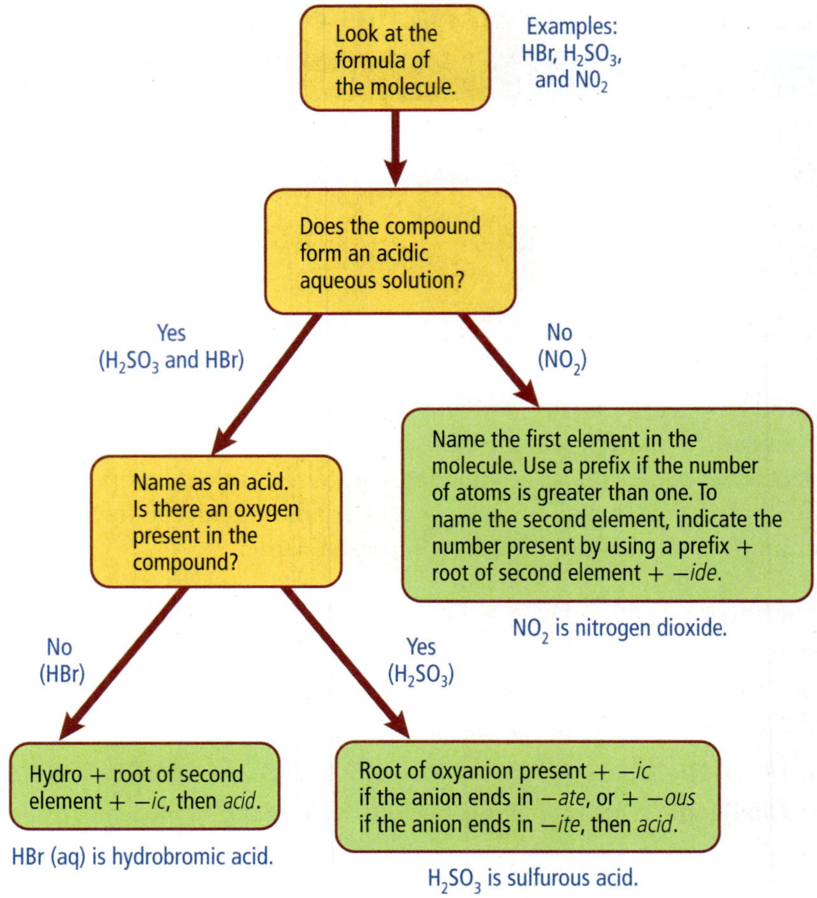


■ **Figure 12** Use this flowchart to name molecular compounds when their formulas are known.

**Apply** ***Which compound above is an oxyacid? Which is a binary acid?***

The flowchart in **Figure 12** can help you determine the name of a molecular covalent compound. To use the chart, start at the top and work downward by reading the text contained in the colored boxes. Apply the text at each step of the flowchart to the formula of the compound that you wish to name.

## SECTION 2 REVIEW

Section Self-Check

### Section Summary

- Names of covalent molecular compounds include prefixes for the number of each atom present. The final letter of the prefix is dropped if the element name begins with a vowel.
- Molecules that produce $H^+$ in solution are acids. Binary acids contain hydrogen and one other element. Oxyacids contain hydrogen and an oxyanion.

**31.** MAINIDEA **Summarize** the rules for naming binary molecular compounds.

**32. Define** a binary molecular compound.

**33. Describe** the difference between a binary acid and an oxyacid.

**34. Apply** Using the system of rules for naming binary molecular compounds, describe how you would name the molecule $N_2O_4$.

**35. Apply** Write the molecular formula for each of these compounds: *iodic acid, disulfur trioxide, dinitrogen monoxide,* and *hydrofluoric acid.*

**36. State** the molecular formula for each compound listed below.

**a.** dinitrogen trioxide
**b.** nitrogen monoxide
**c.** hydrochloric acid
**d.** chloric acid
**e.** sulfuric acid
**f.** sulfurous acid

SECTION 3

# Molecular Structures

**MAINIDEA** **Structural formulas show the relative positions of atoms within a molecule.**

**CHEM 4 YOU** **As a child, you might have played with plastic building blocks that connected only in certain ways. If so, you probably noticed that the shape of the object you built depended on the limited ways the blocks interconnected. Building molecules out of atoms works in a similar way.**

### Essential Questions

- What are the basic steps used to draw Lewis structures?
- Why does resonance occur, and what are some resonance structures?
- Which molecules are exceptions to the octet rule, and why do these exceptions occur?

### Review Vocabulary

**ionic bond:** the electrostatic force that holds oppositely charged particles together in an ionic compound

### New Vocabulary

structural formula
resonance
coordinate covalent bond

## Structural Formulas

You have already studied the structure of ionic compounds—substances formed from ionic bonds. The covalent molecules you have read about in this chapter have structures that are different from those of ionic compounds. In studying the molecular structures of covalent compounds, models are used as representations of the molecule.

The molecular formula, which shows the element symbols and numerical subscripts, tells you the type and number of each atom in a molecule. As shown in **Figure 13,** there are several different models that can be used to represent a molecule. Note that in the ball-and-stick and space-filling molecular models, atoms of each specific element are represented by spheres of a representative color, as shown in **Table R-1** in the Student Resources. These colors are used for identifying the atoms if the chemical symbol of the element is not present.

One of the most useful molecular models is the **structural formula,** which uses letter symbols and bonds to show relative positions of atoms. You can predict the structural formula for many molecules by drawing the Lewis structure. You have already seen some simple examples of Lewis structures, but more involved structures are needed to help you determine the shapes of molecules.

**Figure 13** All of these models can be used to show the relative locations of atoms and electrons in the phosphorus trihydride (phosphine) molecule.
**Compare and contrast** *the types of information contained in each model.*

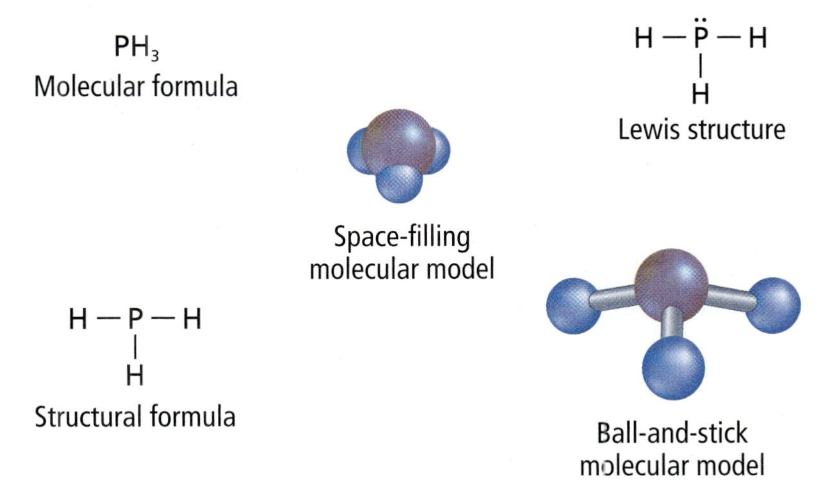

**Lewis structures** Although it is fairly easy to draw Lewis structures for most compounds formed by nonmetals, it is a good idea to follow a regular procedure. Whenever you need to draw a Lewis structure, follow the steps outlined in this Problem-Solving Strategy.

## PROBLEM-SOLVING STRATEGY

### *Drawing Lewis Structures*

1. Predict the location of certain atoms.

   The atom that has the least attraction for shared electrons will be the central atom in the molecule. This element is usually the one closer to the left side of the periodic table. The central atom is located in the center of the molecule; all other atoms become terminal atoms.

   Hydrogen is always a terminal, or end, atom. Because it can share only one pair of electrons, hydrogen can be connected to only one other atom.

2. Determine the number of electrons available for bonding.

   This number is equal to the total number of valence electrons in the atoms that make up the molecule.

3. Determine the number of bonding pairs.

   To do this, divide the number of electrons available for bonding by two.

4. Place the bonding pairs.

   Place one bonding pair (single bond) between the central atom and each of the terminal atoms.

5. Determine the number of electron pairs remaining.

   To do this, subtract the number of pairs used in Step 4 from the total number of bonding pairs determined in Step 3. These remaining pairs include lone pairs as well as pairs used in double and triple bonds. Place lone pairs around each terminal atom (except H atoms) bonded to the central atom to satisfy the octet rule. Any remaining pairs will be assigned to the central atom.

6. Determine whether the central atom satisfies the octet rule.

   Is the central atom surrounded by four electron pairs? If not, it does not satisfy the octet rule. To satisfy the octet rule, convert one or two of the lone pairs on the terminal atoms into a double bond or a triple bond between the terminal atom and the central atom. These pairs are still associated with the terminal atom as well as with the central atom. Remember that carbon, nitrogen, oxygen, and sulfur often form double and triple bonds.

### Apply the Strategy

**Study** Example Problems 3 through 5 to see how the steps in the Problem-Solving Strategy are applied.

## EXAMPLE Problem 3

Find help with **dimensional analysis.** Math Handbook

**LEWIS STRUCTURE FOR A COVALENT COMPOUND WITH SINGLE BONDS** Ammonia is a raw material used in the manufacture of many products, including fertilizers, cleaning products, and explosives. Draw the Lewis structure for ammonia ($NH_3$).

### 1 ANALYZE THE PROBLEM

Ammonia molecules consist of one nitrogen atom and three hydrogen atoms. Because hydrogen must be a terminal atom, nitrogen is the central atom.

### 2 SOLVE FOR THE UNKNOWN

Find the total number of valence electrons available for bonding.

$$1 \text{ N atom} \times \frac{5 \text{ valence electrons}}{1 \text{ N atom}} + 3 \text{ H atoms} \times \frac{1 \text{ valence electron}}{1 \text{ H atom}}$$
$$= 8 \text{ valence electrons}$$

There are 8 valence electrons available for bonding.

$$\frac{8 \text{ electrons}}{2 \text{ electrons/pair}} = 4 \text{ pairs}$$

**Determine the total number of bonding pairs. To do this, divide the number of available electrons by two.**

Four pairs of electrons are available for bonding.

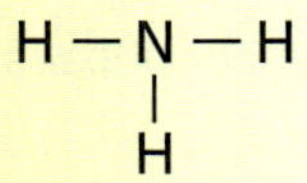

**Place a bonding pair (a single bond) between the central nitrogen atom and each terminal hydrogen atom.**

Determine the number of bonding pairs remaining.

4 pairs total − 3 pairs used
= 1 pair available

**Subtract the number of pairs used in these bonds from the total number of pairs of electrons available.**

The remaining pair—a lone pair—must be added to either the terminal atoms or the central atom. Because hydrogen atoms can have only one bond, they have no lone pairs.

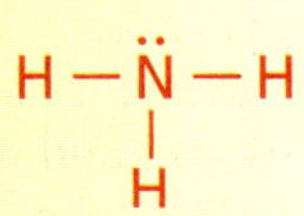

**Place the remaining lone pair on the central nitrogen atom.**

### 3 EVALUATE THE ANSWER

Each hydrogen atom shares one pair of electrons, as required, and the central nitrogen atom shares three pairs of electrons and has one lone pair, providing a stable octet.

## PRACTICE Problems

Do additional problems. Online Practice

**37.** Draw the Lewis structure for $BH_3$.

**38. Challenge** A nitrogen trifluoride molecule contains numerous lone pairs. Draw its Lewis structure.

## EXAMPLE Problem 4

**LEWIS STRUCTURE FOR A COVALENT COMPOUND WITH MULTIPLE BONDS** Carbon dioxide is a product of all cellular respiration. Draw the Lewis structure for carbon dioxide ($CO_2$).

### 1 ANALYZE THE PROBLEM

The carbon dioxide molecule consists of one carbon atom and two oxygen atoms. Because carbon has less attraction for shared electrons, carbon is the central atom, and the two oxygen atoms are terminal.

### 2 SOLVE FOR THE UNKNOWN

Find the total number of valence electrons available for bonding.

$$1 \text{ C atom} \times \frac{4 \text{ valence electrons}}{1 \text{ C atom}} + 2 \text{ O atoms} \times \frac{6 \text{ valence electrons}}{1 \text{ O atom}}$$
$$= 16 \text{ valence electrons}$$

There are 16 valence electrons available for bonding.

$$\frac{16 \text{ electrons}}{2 \text{ electrons/pair}} = 8 \text{ pairs}$$

**Determine the total number of bonding pairs by dividing the number of available electrons by two.**

Eight pairs of electrons are available for bonding.

O — C — O

**Place a bonding pair (a single bond) between the central carbon atom and each terminal oxygen atom.**

Determine the number of electron pairs remaining. Subtract the number of pairs used in these bonds from the total number of pairs of electrons available.

8 pairs total − 2 pairs used
= 6 pairs available

**Subtract the number of pairs used in these bonds from the total number of pairs of electrons available.**

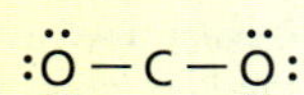

**Add three lone pairs to each terminal oxygen atom.**

Determine the number of electron pairs remaining.

6 pairs available − 6 pairs used
= 0 pairs available

**Subtract the lone pairs from the pairs available.**

Examine the incomplete structure above (showing the placement of the lone pairs). Note that the carbon atom does not have an octet and that there are no more electron pairs available. To give the carbon atom an octet, the molecule must form double bonds.

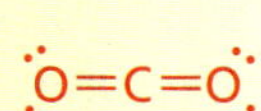

**Use a lone pair from each O atom to form a double bond with the C atom.**

### 3 EVALUATE THE ANSWER

Both carbon and oxygen now have an octet, which satisfies the octet rule.

## PRACTICE Problems

Do additional problems. Online Practice

**39.** Draw the Lewis structure for ethylene, $C_2H_4$.

**40.** **Challenge** A molecule of carbon disulfide contains both lone pairs and multiple-covalent bonds. Draw its Lewis structure.

Get help with **greatest common factors.**

Personal Tutor

**Lewis structures for polyatomic ions** Although the unit acts as anion, the atoms within a polyatomic ion are covalently bonded. The procedure for drawing Lewis structures for polyatomic ions is similar to drawing them for covalent compounds. The main difference is in finding the total number of electrons available for bonding. Compared to the number of valence electrons present in the atoms that make up the ion, more electrons are present if the ion is negatively charged and fewer are present if the ion is positive. To find the total number of electrons available for bonding, first find the number available in the atoms present in the ion. Then, subtract the ion charge if the ion is positive, and add the ion charge if the ion is negative.

## EXAMPLE Problem 5

**LEWIS STRUCTURE FOR A POLYATOMIC ION** Draw the correct Lewis structure for the polyatomic ion phosphate ($PO_4^{3-}$).

### 1 ANALYZE THE PROBLEM

You are given that the phosphate ion consists of one phosphorus atom and four oxygen atoms and has a charge of 3–. Because phosphorus has less attraction for shared electrons than oxygen, phosphorus is the central atom and the four oxygen atoms are terminal atoms.

### 2 SOLVE FOR THE UNKNOWN

Find the total number of valence electrons available for bonding.

$$1 \text{ P atom} \times \frac{5 \text{ valence electrons}}{\text{P atom}} + 4 \text{ O atoms} \times \frac{6 \text{ valence electrons}}{\text{O atom}}$$

$$+ 3 \text{ electrons from the negative charge} = 32 \text{ valence electrons}$$

$$\frac{32 \text{ electrons}}{2 \text{ electrons/pair}} = 16 \text{ pair}$$

**Determine the total number of bonding pairs.**

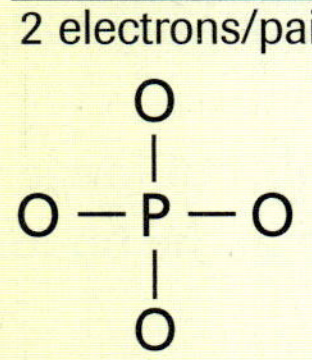

**Draw single bonds from each terminal oxygen atom to the central phosphorus atom.**

16 pairs total – 4 pairs used
= 12 pairs available

**Subtract the number of pairs used from the total number of pairs of electrons available.**

Add three lone pairs to each terminal oxygen atom.
12 pairs available – 12 lone pairs used = 0

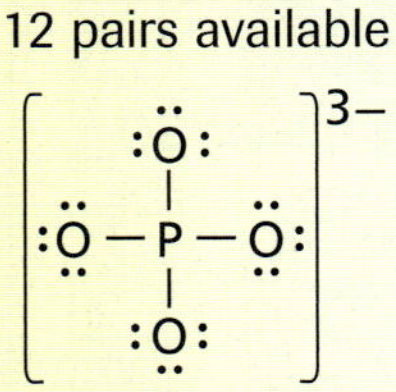

**Subtracting the lone pairs used from the pairs available verifies that there are no electron pairs available for the phosphorus atom. The Lewis structure for the phosphate ion is shown.**

### 3 EVALUATE THE ANSWER

All of the atoms have an octet, and the group has a net charge of 3–.

## PRACTICE Problems

Do additional problems. Online Practice

**41.** Draw the Lewis structure for the $NH_4^+$ ion.

**42. Challenge** The $ClO_4^-$ ion contains numerous lone pairs. Draw its Lewis structure.

## RealWorld CHEMISTRY

### Phosphorus and Nitrogen

**ALGAL BLOOMS** Phosphorus and nitrogen are nutrients required for algae growth. Both can enter lakes and streams from discharges of sewage and industrial waste, and in fertilizer runoff. If these substances build up in a body of water, a rapid growth of algae, known as an algal bloom, can occur, forming a thick layer of green slime over the water's surface. When the algae use up the supply of nutrients, they die and decompose. This process reduces the amount of dissolved oxygen in the water that is available to other aquatic organisms.

■ **Figure 14** The nitrate ion ($NO_3^-$) exhibits resonance. **a.** These resonance structures differ only in the location of the double bond. The locations of the nitrogen and oxygen atoms stay the same. **b.** The actual nitrate ion is like an average of the three resonance structures in **a.** The dotted lines indicate possible locations of the double bond.

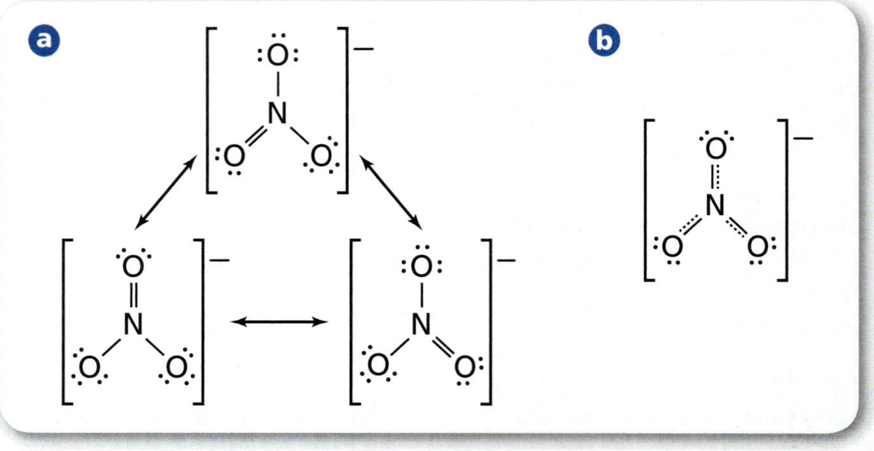

## Resonance Structures

Using the same sequence of atoms, it is possible to have more than one correct Lewis structure when a molecule or polyatomic ion has both a double bond and a single bond. Consider the polyatomic ion nitrate ($NO_3^-$), shown in **Figure 14a.** Three equivalent structures can be used to represent the nitrate ion.

**Resonance** is a condition that occurs when more than one valid Lewis structure can be written for a molecule or ion. The two or more correct Lewis structures that represent a single molecule or ion are referred to as resonance structures. Resonance structures differ only in the position of the electron pairs, never the atom positions. The location of the lone pairs and bonding pairs differs in resonance structures. The molecule $O_3$ and the polyatomic ions $NO_3^-$, $NO_2^-$, $SO_3^{2-}$, and $CO_3^{2-}$ all exhibit resonance.

It is important to note that each molecule or ion that exhibits resonance behaves as if it has only one structure. Refer to **Figure 14b.** Experimentally measured bond lengths show that the bonds are identical to each other. They are shorter than single bonds but longer than double bonds. The actual bond length is an average of the bonds in the resonance structures.

**VOCABULARY**
**SCIENCE USAGE V. COMMON USAGE**

**Resonance**

***Science usage:*** a phenomenon related to the stability of a molecule; a large vibration in a mechanical system caused by a small periodic stimulus
*The new molecule had several resonance structures.*

***Common usage:*** a quality of richness or variety
*The sound of the orchestra had resonance.*

**PRACTICE Problems** Do additional problems. Online Practice

**Draw the Lewis resonance structures for the following molecules.**

**43.** $NO_2^-$ **44.** $SO_2$ **45.** $O_3$

**46. Challenge** Draw the Lewis resonance structure for the ion $SO_3^{2-}$.

## Exceptions to the Octet Rule

Generally, atoms attain an octet when they bond with other atoms. Some molecules and ions, however, do not obey the octet rule. There are several reasons for these exceptions.

**Odd number of valence electrons** First, a small group of molecules might have an odd number of valence electrons and be unable to form an octet around each atom. For example, $NO_2$ has five valence electrons from nitrogen and 12 from oxygen, totaling 17, which cannot form an exact number of electron pairs. See **Figure 15.** $ClO_2$ and NO are other examples of molecules with odd numbers of valence electrons.

■ **Figure 15** The central nitrogen atom in this $NO_2$ molecule does not satisfy the octet rule; the nitrogen atom has only seven electrons in its outer energy level.

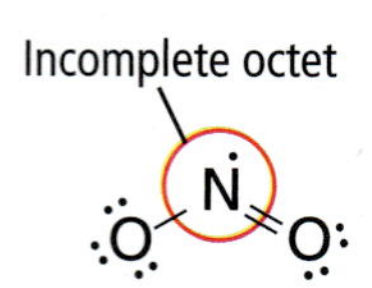

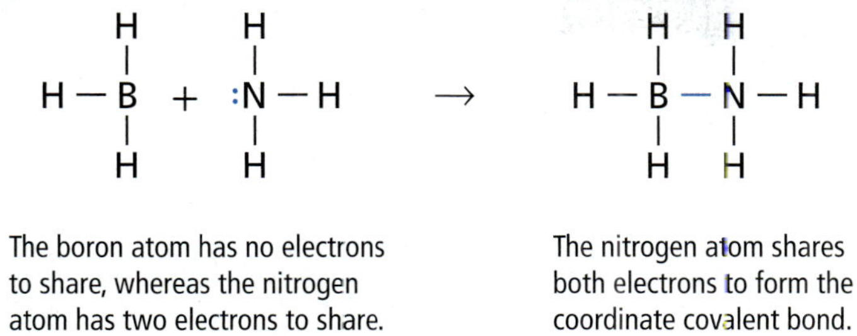


■ **Figure 16** In this reaction between boron trihydride ($BH_3$) and nitrogen ($NH_3$), the nitrogen atom donates both electrons that are shared by boron and nitrogen, forming a coordinate covalent bond.

**Interpret** ***Does the coordinate covalent bond in the product molecule satisfy the octet rule?***

**Suboctets and coordinate covalent bonds** Another exception to the octet rule is due to a few compounds that form suboctets—stable configurations with fewer than eight electrons present around an atom. This group is relatively rare, and $BH_3$ is an example. Boron, a group 13 metalloid, forms three covalent bonds with other nonmetallic atoms.

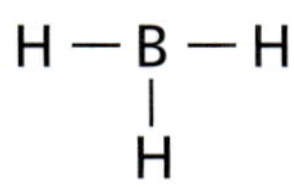

The boron atom shares only six electrons—too few to form an octet. Such compounds tend to be reactive and can share an entire pair of electrons donated by another atom.

A **coordinate covalent bond** forms when one atom donates both of the electrons to be shared with an atom or ion that needs two electrons to form a stable electron arrangement with lower potential energy. Refer to **Figure 16.** Atoms or ions with lone pairs often form coordinate covalent bonds with atoms or ions that need two more electrons.

**Expanded octets** The third group of compounds that does not follow the octet rule has central atoms that contain more than eight valence electrons. This electron arrangement is referred to as an expanded octet. An expanded octet can be explained by considering the d orbitals that occur in the energy levels of elements in period three or higher. An example of an expanded octet, shown in **Figure 17,** is the bond formation in the molecule $PCl_5$. Five bonds are formed with ten electrons shared in one s orbital, three p orbitals, and one d orbital. Another example is the molecule $SF_6$, which has six bonds sharing 12 electrons in an s orbital, three p orbitals, and two d orbitals. When you draw the Lewis structures for these compounds, either extra lone pairs are added to the central atom or more than four bonding atoms are present in the molecule.

☑ READING CHECK **Summarize** three reasons why some molecules do not conform to the octet rule.

■ **Figure 17** Prior to the reaction of $PCl_3$ and $Cl_2$, every reactant atom follows the octet rule. After the reaction, the product, $PCl_5$, has an expanded octet containing ten electrons.

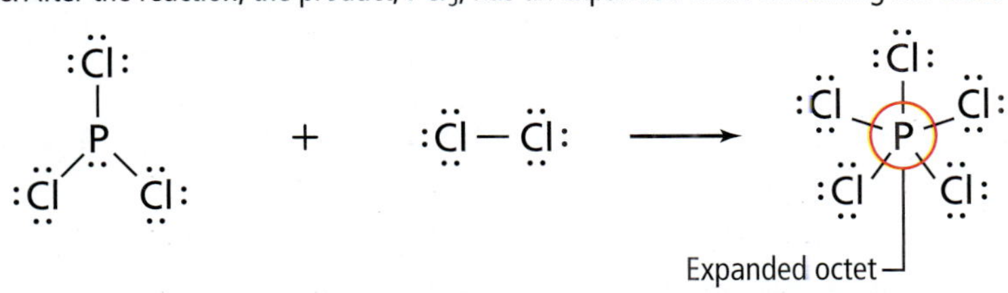

## EXAMPLE Problem 6

**LEWIS STRUCTURE: EXCEPTION TO THE OCTET RULE** Xenon is a noble gas that will form a few compounds with nonmetals that strongly attract electrons. Draw the correct Lewis structure for xenon tetrafluoride ($XeF_4$).

### 1 ANALYZE THE PROBLEM

You are given that a molecule of xenon tetrafluoride consists of one xenon atom and four fluorine atoms. Xenon has less attraction for electrons, so it is the central atom.

### 2 SOLVE FOR THE UNKNOWN

First, find the total number of valence electrons.

$$1 \text{ Xe atom} \times \frac{8 \text{ valence electrons}}{1 \text{ Xe atom}} + 4 \text{ F atoms} \times \frac{7 \text{ valence electrons}}{1 \text{ F atom}} = 36 \text{ valence electrons}$$

$$\frac{36 \text{ electrons}}{2 \text{ electrons/pair}} = 18 \text{ pairs}$$

**Determine the total number of bonding pairs.**

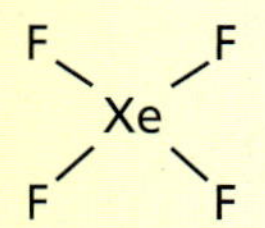

**Use four bonding pairs to bond the four F atoms to the central Xe atom.**

18 pairs available − 4 pairs used = 14 pairs available

**Determine the number of remaining pairs.**

$$14 \text{ pairs} - 4 \text{ F atoms} \times \frac{3 \text{ pairs}}{1 \text{ F atom}} = 2 \text{ pairs unused}$$

**Add three pairs to each F atom to obtain an octet. Determine how many pairs remain.**

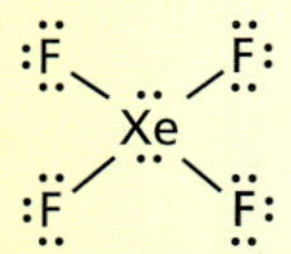

**Place the two remaining pairs on the central Xe atom.**

### 3 EVALUATE THE ANSWER

This structure gives xenon 12 total electrons, an expanded octet. Xenon compounds, such as the $XeF_4$ shown here, are toxic because they are highly reactive.

## PRACTICE Problems

Do additional problems. Online Practice

**Draw the expanded octet Lewis structure for each molecule.**

**47.** $ClF_3$

**48.** $PCl_5$

**49. Challenge** Draw the Lewis structure for the molecule formed when six fluorine atoms and one sulfur atom bond covalently.

## SECTION 3 REVIEW

Section Self-Check

### Section Summary

- Different models can be used to represent molecules.
- Resonance occurs when more than one valid Lewis structure exists for the same molecule.
- Exceptions to the octet rule occur in some molecules.

**50. MAIN**IDEA **Describe** the information contained in a structural formula.

**51. State** the steps used to draw Lewis structures.

**52. Summarize** exceptions to the octet rule by correctly pairing these molecules and phrases: odd number of valence electrons, $PCl_5$, $ClO_2$, $BH_3$, expanded octet, less than an octet.

**53. Evaluate** A classmate states that a binary compound having only sigma bonds displays resonance. Could the classmate's statement be true?

**54. Draw** the resonance structures for the dinitrogen oxide ($N_2O$) molecule.

**55. Draw** the Lewis structures for $CN^-$, $SiF_4$, $HCO_3^-$, and, $AsF_6^-$.

SECTION 4

# Molecular Shapes

**MAINIDEA** The VSEPR model is used to determine molecular shape.

## Essential Questions

- What is the VSEPR bonding theory?
- How can you use the VSEPR model to predict the shape of, and the bond angles in, a molecule?
- What is hybridization?

## Review Vocabulary

**atomic orbital:** the region around an atom's nucleus that defines an electron's probable location

## New Vocabulary

VSEPR model
hybridization

**CHEM 4 YOU** Have you ever rubbed two balloons in your hair to create a static electric charge on them? If you brought the balloons together, their like charges would cause them to repel each other. Molecular shapes are also affected by the forces of electric repulsion.

## VSEPR Model

The shape of a molecule determines many of its physical and chemical properties. Often, shapes of reactant molecules determine whether or not they can get close enough to react. Electron densities created by the overlap of the orbitals of shared electrons determine molecular shape. Theories have been developed to explain the overlap of bonding orbitals and can be used to predict the shape of the molecule.

The molecular geometry, or shape, of a molecule can be determined once a Lewis structure is drawn. The model used to determine the molecular shape is referred to as the **V**alence **S**hell **E**lectron **P**air **R**epulsion model, or **VSEPR model.** This model is based on an arrangement that minimizes the repulsion of shared and unshared electron pairs around the central atom.

**Bond angle** To understand the VSEPR model better, imagine balloons that are inflated to similar sizes and tied together, as shown in **Figure 18.** Each balloon represents an electron-dense region. The repulsive force of this electron-dense region keeps other electrons from entering this space. When a set of balloons is connected at a central point, which represents a central atom, the balloons naturally form a shape that minimizes interactions between the balloons.

The electron pairs in a molecule repel one another in a similar way. These forces cause the atoms in a molecule to be positioned at fixed angles relative to one another. The angle formed by two terminal atoms and the central atom is a bond angle. Bond angles predicted by VSEPR are supported by experimental evidence.

Unshared pairs of electrons are also important in determining the shape of the molecule. These electrons occupy a slightly larger orbital than shared electrons. Therefore, shared bonding orbitals are pushed together by unshared pairs.

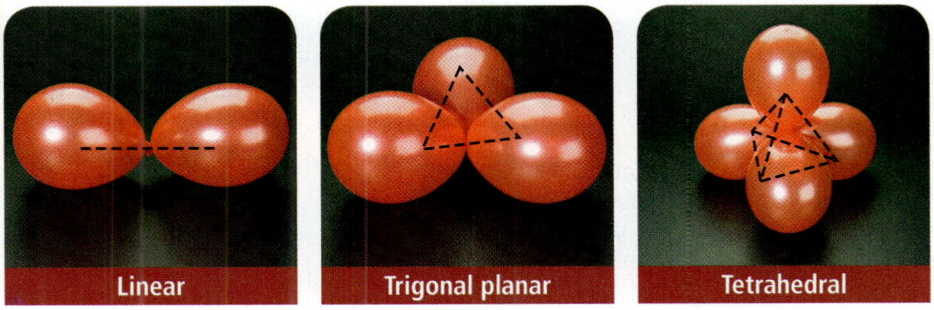


■ **Figure 18** Electron pairs in a molecule are located as far apart as they can be, just as these balloons are arranged. Two pairs form a linear shape. Three pairs form a trigonal planar shape. Four pairs form a tetrahedral shape.

**VOCABULARY**

**WORD ORIGIN**

**Trigonal planar**
comes from the Latin words *trigonum,* which means *triangular,* and *plan-,* which means *flat*

Explore **molecular shapes.**

**Virtual Investigations**

■ **Figure 19** A carbon atom's 2s and 2p electrons occupy the hybrid $sp^3$ orbitals. Notice that the hybrid orbitals have an intermediate amount of potential energy when compared with the energy of the original s and p orbitals. According to VSEPR theory, a tetrahedral shape minimizes repulsion between the hybrid orbitals in a $CH_4$ molecule.

**Identify** ***How many faces does the tetrahedral shape formed by the $sp^3$ orbitals have?***

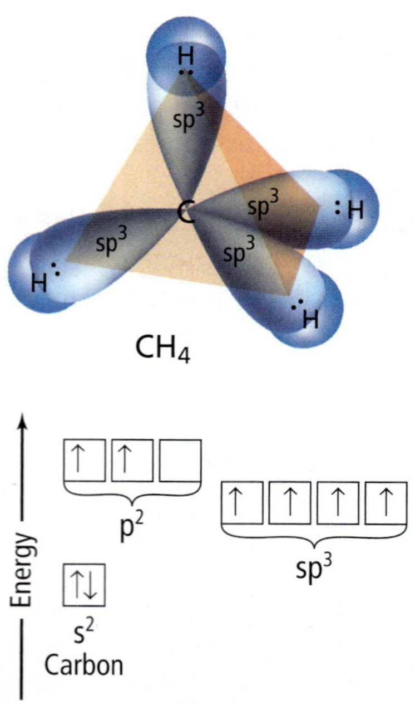


View an **animation about molecular shapes.**

**Concepts In Motion**

**Connection to Biology** The shape of food molecules is important to our sense of taste. The surface of your tongue is covered with taste buds, each of which contains from 50 to 100 taste receptor cells. Taste receptor cells can detect five distinct tastes—sweet, bitter, salty, sour, and umami (the taste of MSG, monosodium glutamate)—but each receptor cell responds best to only one taste.

The shapes of food molecules are determined by their chemical structures. When a molecule enters a taste bud, it must have the correct shape for the nerve in each receptor cell to respond and send a message to the brain. The brain then interprets the message as a certain taste. When such molecules bind to sweet receptors, they are sensed as sweet. The greater the number of food molecules that fit a sweet receptor cell, the sweeter the food tastes. Sugars and artificial sweeteners are not the only sweet molecules. Some proteins found in fruits are also sweet molecules. Some common molecular shapes are illustrated in **Table 6.**

## Hybridization

A hybrid occurs when two things are combined and the result has characteristics of both. For example, a hybrid automobile uses both gas and electricity as energy sources. During chemical bonding, different atomic orbitals undergo hybridization. To understand this, consider the bonding involved in the methane molecule ($CH_4$). The carbon atom has four valence electrons with the electron configuration[He]$2s^22p^2$. You might expect the two unpaired p electrons to bond with other atoms and the 2s electrons to remain an unshared pair. However, carbon atoms undergo **hybridization,** a process in which atomic orbitals mix and form new, identical hybrid orbitals.

The hybrid orbitals in a carbon atom are shown in **Figure 19.** Note that each hybrid orbital contains one electron that it can share with another atom. The hydrid orbital is called an $sp^3$ orbital because the four hybrid orbitals form from one s orbital and three p orbitals. Carbon is the most common element that undergoes hybridization.

The number of atomic orbitals that mix and form the hybrid orbital equals the total number of pairs of electrons, as shown in **Table 6.** In addition, the number of hybrid orbitals formed equals the number of atomic orbitals mixed. For example, $AlCl^3$ has a total of three pairs of electrons and VSEPR predicts a trigonal planar molecular shape. This shape results when one s and two p orbitals on the central atom, Al, mix and form three identical $sp^2$ hybrid orbitals.

Lone pairs also occupy hybrid orbitals. Compare the hybrid orbitals of $BeCl_2$ and $H_2O$ in **Table 6.** Both compounds contain three atoms. Why does an $H_2O$ molecule contain $sp^3$ orbitals? There are two lone pairs on the central oxygen atom in $H_2O$. Therefore, there must be four hybrid orbitals—two for bonding and two for the lone pairs.

Recall that multiple covalent bonds consist of one sigma bond and one or more pi bonds. Only the two electrons in the sigma bond occupy hybrid orbitals such as sp and $sp^2$. The remaining unhybridized p orbitals overlap to form pi bonds. It is important to note that single, double, and triple covalent bonds contain only one hybrid orbital. Thus, $CO_2$, with two double bonds, forms sp hybrid orbitals.

☑ READING CHECK **State** the number of electrons that are available for bonding in a hybrid $sp^3$ orbital.

Explore **molecular shapes with an interactive table.** Concepts In Motion

## Table 6 Molecular Shapes

| Molecule | Total Pairs | Shared Pairs | Lone Pairs | Hybrid Orbitals | Molecular Shape* |
|---|---|---|---|---|---|
| $BeCl_2$ | 2 | 2 | 0 | sp | 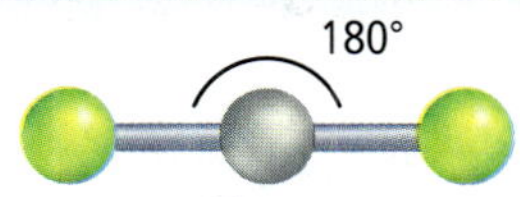 **Linear** |
| $AlCl_3$ | 3 | 3 | 0 | $sp^2$ | 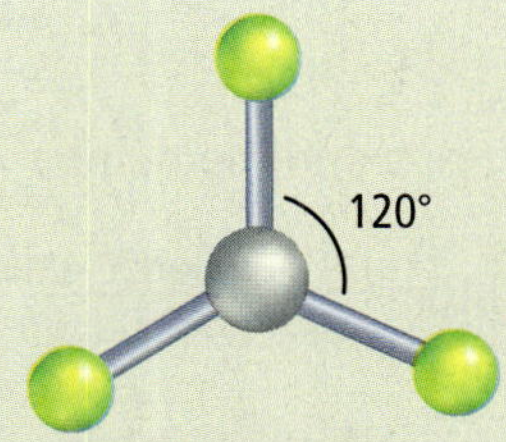 **Trigonal planar** |
| $CH_4$ | 4 | 4 | 0 | $sp^3$ | 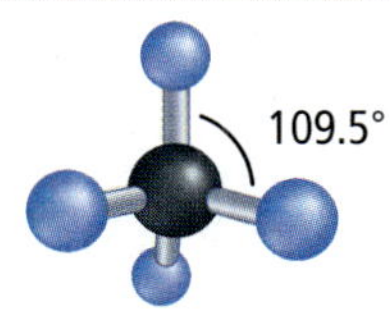 **Tetrahedral** |
| $NH_3$ | 4 | 3 | 1 | $sp^3$ | 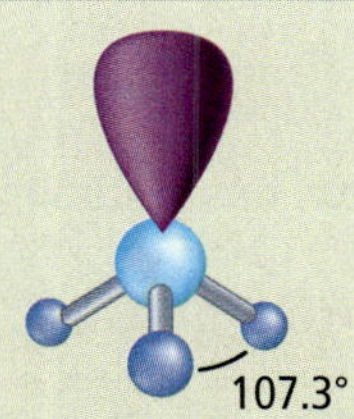 **Trigonal pyramidal** |
| $H_2O$ | 4 | 2 | 2 | $sp^3$ | 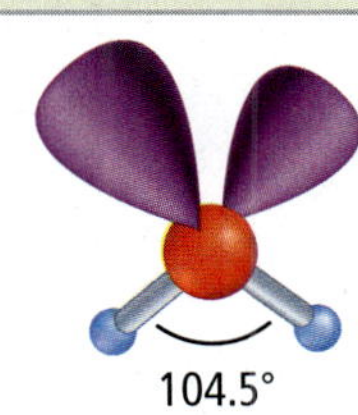 **Bent** |
| $NbBr_5$ | 5 | 5 | 0 | $sp^3d$ | 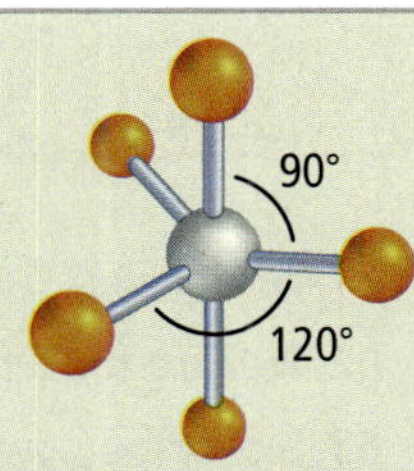 **Trigonal bipyramidal** |
| $SF_6$ | 6 | 6 | 0 | $sp^3d^2$ | 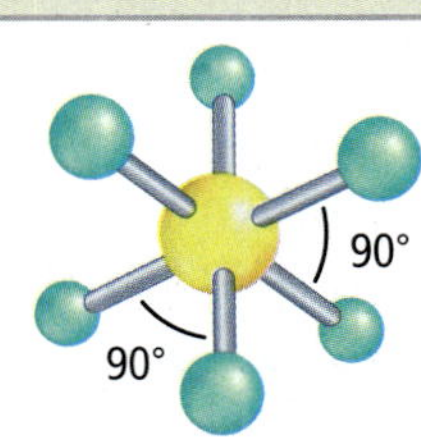 **Octahedral** |

*Balls represent atoms, sticks represent bonds, and lobes represent lone pairs of electrons.

**Alabama Applying Practices**
Go to the Resource tab in ConnectED to find the Activity *Modeling Molecular Shapes.*

The $BeCl_2$ molecule contains only two pairs of electrons shared with the central Be atom. These bonding electrons have the maximum separation, a bond angle of 180°, and the molecular shape is linear.

The three bonding electron pairs in $AlCl_3$ have maximum separation in a trigonal planar shape with 120° bond angles.

When the central atom in a molecule has four pairs of bonding electrons, as $CH_4$ does, the shape is tetrahedral. The bond angles are 109.5°.

$NH_3$ has three single covalent bonds and one lone pair. The lone pair takes up a greater amount of space than the shared pairs. There is stronger repulsion between the lone pair and the bonding pairs than between two bonding pairs. The resulting geometry is trigonal pyramidal, with 107.3° bond angles.

Water has two covalent bonds and two lone pairs. Repulsion between the lone pairs causes the angle to be 104.5°, less than both tetrahedral and trigonal pyramid. As a result, water molecules have a bent shape.

The $NbBr_5$ molecule has five pairs of bonding electrons. The trigonal bipyramidal shape minimizes the repulsion of these shared electron pairs.

As with $NbBr_5$, $SF_6$ has no unshared electron pairs on the central atom. However, six shared pairs arranged about the central atom result in an octahedral shape.

## EXAMPLE Problem 7

EXAMPLE PROBLEM

**FIND THE SHAPE OF A MOLECULE** Phosphorus trihydride, a colorless gas, is produced when organic materials, such as fish flesh, rot. What is the shape of a phosphorus trihydride molecule? Predict the bond angle and identify hybrid orbitals.

### 1 ANALYZE THE PROBLEM

A phosphorus trihydride molecule has three hydrogen atoms bonded to a central phosphorus atom.

### 2 SOLVE FOR THE UNKNOWN

Find the total number of valence electrons and the number of electron pairs.

$$1 \text{ P atom} \times \frac{5 \text{ valence electrons}}{1 \text{ P atom}} + 3 \text{ H atoms} \times \frac{1 \text{ valence electron}}{1 \text{ F atom}} = 8 \text{ valence electrons}$$

$$\frac{8 \text{ electrons}}{2 \text{ electrons/pair}} = 4 \text{ pairs}$$

**Determine the total number of bonding pairs.**

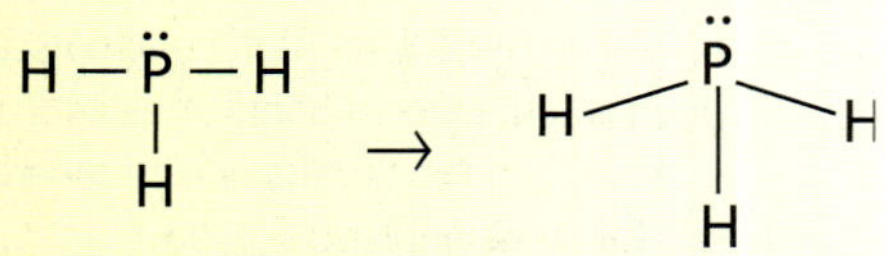

Lewis structure — Molecular shape

**Draw the Lewis structure, using one pair of electrons to bond each H atom to the central P atom and assigning the lone pair to the P atom.**

The molecular shape is trigonal pyramidal with a predicted 107° bond angle and $sp^3$ hybrid orbitals.

### 3 EVALUATE THE ANSWER

All electron pairs are used and each atom has a stable electron configuration.

## PRACTICE Problems

Do additional problems. Online Practice

**Determine the molecular shape, bond angle, and hybrid orbitals for each molecule.**

**56.** $BF_3$

**57.** $OCl_2$

**58.** $BeF_2$

**59.** $CF_4$

**60. Challenge** For a $NH_4^+$ ion, identify its molecular shape, bond angle, and hybrid orbitals.

## SECTION 4 REVIEW

Section Self-Check

### Section Summary

- VSEPR model theory states that electron pairs repel each other and determine both the shape of and bond angles in a molecule.
- Hybridization explains the observed shapes of molecules by the presence of equivalent hybrid orbitals.

**61.** MAINIDEA **Summarize** the VSEPR bonding theory.

**62. Define** the term *bond angle*.

**63. Describe** how the presence of a lone electron pair affects the spacing of shared bonding orbitals.

**64. Compare** the size of an orbital that has a shared electron pair with one that has a lone pair.

**65. Identify** the type of hybrid orbitals present and bond angles for a molecule with a tetrahedral shape.

**66. Compare** the molecular shapes and hybrid orbitals of $PF_3$ and $PF_5$ molecules. Explain why their shapes differ.

**67. List** in a table, the Lewis structure, molecular shape, bond angle, and hybrid orbitals for molecules of $CS_2$, $CH_2O$, $H_2Se$, $CCl_2F_2$, and $NCl_3$.

# Electronegativity and Polarity

**MAINIDEA** A chemical bond's character is related to each atom's attraction for the electrons in the bond.

### Essential Questions

- How is electronegativity used to determine bond type?
- How do polar and nonpolar covalent bonds and polar and nonpolar molecules compare? How do they contrast?
- What are the characteristics of covalently bonded compounds?

### Review Vocabulary

**electronegativity:** the relative ability of an atom to attract electrons in a chemical bond

### New Vocabulary

polar covalent bond

**CHEM 4 YOU** The stronger you are, the more easily you can do pull-ups. Just as people have different abilities for doing pull-ups, atoms in chemical bonds have different abilities to attract (pull) electrons.

## Electronegativity and Bond Character

The type of bond formed during a reaction is related to each atom's attraction for electrons. Electron affinity is a measure of the tendency of an atom to accept an electron. Excluding noble gases, electron affinity increases with increasing atomic number within a period and decreases with increasing atomic number within a group. The scale of electronegativities—shown in **Figure 20**—allows chemists to evaluate the electron affinity of specific atoms in a compound. Recall that electronegativity indicates the relative ability of an atom to attract electrons in a chemical bond. Note that electronegativity values were assigned, whereas electron affinity values were measured.

**Electronegativity** The version of the periodic table of the elements shown in **Figure 20** lists electronegativity values. Note that fluorine has the greatest electronegativity value (3.98), while francium has the least (0.7). Because noble gases do not generally form compounds, individual electronegativity values for helium, neon, and argon are not listed. However, larger noble gases, such as xenon, sometimes bond with highly electronegative atoms, such as fluorine.

**Figure 20** Electronegativity values are derived by comparing an atom's attraction for shared electrons to that of a fluorine atom's attraction for shared electrons. Note that the electronegativity values for the lanthanide and actinide series, which are not shown, range from 1.12 to 1.7.

**Electronegativity Values for Selected Elements**

Metal | Metalloid | Nonmetal

| 1 H 2.20 | | | | | | | | | | | | | | | | |
|---|---|---|---|---|---|---|---|---|---|---|---|---|---|---|---|---|
| 3 Li 0.98 | 4 Be 1.57 | | | | | | | | | | | 5 B 2.04 | 6 C 2.55 | 7 N 3.04 | 8 O 3.44 | 9 F 3.98 |
| 11 Na 0.93 | 12 Mg 1.31 | | | | | | | | | | | 13 Al 1.61 | 14 Si 1.90 | 15 P 2.19 | 16 S 2.58 | 17 Cl 3.16 |
| 19 K 0.82 | 20 Ca 1.00 | 21 Sc 1.36 | 22 Ti 1.54 | 23 V 1.63 | 24 Cr 1.66 | 25 Mn 1.55 | 26 Fe 1.83 | 27 Co 1.88 | 28 Ni 1.91 | 29 Cu 1.90 | 30 Zn 1.65 | 31 Ga 1.81 | 32 Ge 2.01 | 33 As 2.18 | 34 Se 2.55 | 35 Br 2.96 |
| 37 Rb 0.82 | 38 Sr 0.95 | 39 Y 1.22 | 40 Zr 1.33 | 41 Nb 1.6 | 42 Mo 2.16 | 43 Tc 2.10 | 44 Ru 2.2 | 45 Rh 2.28 | 46 Pd 2.20 | 47 Ag 1.93 | 48 Cd 1.69 | 49 In 1.78 | 50 Sn 1.96 | 51 Sb 2.05 | 52 Te 2.1 | 53 I 2.66 |
| 55 Cs 0.79 | 56 Ba 0.89 | 57 La 1.10 | 72 Hf 1.3 | 73 Ta 1.5 | 74 W 1.7 | 75 Re 1.9 | 76 Os 2.2 | 77 Ir 2.2 | 78 Pt 2.2 | 79 Au 2.4 | 80 Hg 1.9 | 81 Tl 1.8 | 82 Pb 1.8 | 83 Bi 1.9 | 84 Po 2.0 | 85 At 2.2 |
| 87 Fr 0.7 | 88 Ra 0.9 | 89 Ac 1.1 | | | | | | | | | | | | | | |

| Table 7 EN Difference and Bond Character | |
|---|---|
| **Electronegativity Difference** | **Bond Character** |
| $> 1.7$ | mostly ionic |
| $0.4 - 1.7$ | polar covalent |
| $< 0.4$ | mostly covalent |
| 0 | nonpolar covalent |

**Bond character** A chemical bond between atoms of different elements is never completely ionic or covalent. The character of a bond depends on how strongly each of the bonded atoms attracts electrons. As shown in **Table 7,** the character and type of a chemical bond can be predicted using the electronegativity difference of the elements that bond. Electrons in bonds between identical atoms have an electronegativity difference of zero—meaning that the electrons are equally shared between the two atoms. This type of bond is considered nonpolar covalent, or a pure covalent bond. On the other hand, because different elements have different electronegativities, the electron pairs in a covalent bond between different atoms are not shared equally. Unequal sharing results in a **polar covalent bond.** When there is a large difference in the electronegativity between bonded atoms, an electron is transferred from one atom to the other, which results in bonding that is primarily ionic.

Watch a **video about chemical bonding.**

Video

Bonding is not often clearly ionic or covalent. An electronegativity difference of 1.70 is considered 50 percent covalent and 50 percent ionic. As the difference in electronegativity increases, the bond becomes more ionic in character. Generally, ionic bonds form when the electronegativity difference is greater than 1.70. However, this cutoff is sometimes inconsistent with experimental observations of two nonmetals bonding together. **Figure 21** summarizes the range of chemical bonding between two atoms. What percent ionic character is a bond between two atoms that have an electronegativity difference of 2.00? Where would LiBr be plotted on the graph?

READING CHECK **Analyze** What is the percent ionic character of a pure covalent bond?

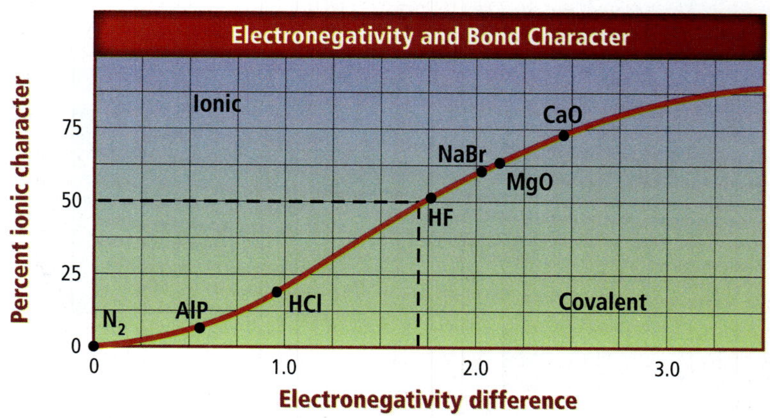


■ **Figure 21** This graph shows that the difference in electronegativity between bonding atoms determines the percent ionic character of the bond. Above 50% ionic character, bonds are mostly ionic.

GRAPH CHECK

**Determine** the percent ionic character of calcium oxide.

| Electronegativity | Cl = 3.16 |
|---|---|
| Electronegativity | H = 2.20 |
| Difference | = 0.96 |

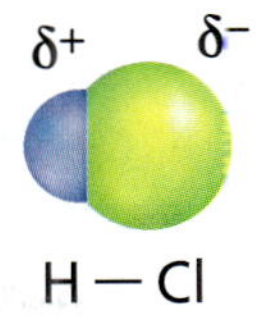


View an **animation about bond types.**

Concepts In Motion

■ **Figure 22** Chlorine's electronegativity is higher than that of hydrogen. Therefore, in a molecule containing hydrogen and chlorine, the shared pair of electrons is with the chlorine atom more often than it is with the hydrogen atom. Symbols are used to indicate the partial charge at each end of the molecule from this unequal sharing of electrons.

## Polar Covalent Bonds

As you just learned, polar covalent bonds form because not all atoms that share electrons attract them equally. A polar covalent bond is similar to a tug-of-war in which the two teams are not of equal strength. Although both sides share the rope, the stronger team pulls more of the rope toward its side. When a polar bond forms, the shared electron pair or pairs are pulled toward one of the atoms. Thus, the electrons spend more time around that atom than the other atom. This results in partial charges at the ends of the bond.

The Greek letter delta (δ) is used to represent a partial charge. In a polar covalent bond, $\delta^-$ represents a partial negative charge and $\delta^+$ represents a partial positive charge. As shown in **Figure 22,** $\delta^-$ and $\delta^+$ can be added to a molecular model to indicate the polarity of the covalent bond. The more-electronegative atom is at the partially negative end, while the less-electronegative atom is at the partially positive end. The resulting polar bond often is referred to as a dipole (two poles).

**Molecular polarity** Covalently bonded molecules are either polar or nonpolar; which type depends on the location and nature of the covalent bonds in the molecule. A distinguishing feature of nonpolar molecules is that they are not attracted by an electric field. Polar molecules, however, are attracted by an electric field. Because polar molecules are dipoles with partially charged ends, they have an uneven electron density. This results in the tendency of polar molecules to align with an electric field.

**Polarity and molecular shape** You can learn why some molecules are polar and some are not by comparing water ($H_2O$) and carbon tetrachloride ($CCl_4$) molecules. Both molecules have polar covalent bonds. According to the data in **Figure 20,** the electronegativity difference between a hydrogen atom and an oxygen atom is 1.24. The electronegativity difference between a chlorine atom and a carbon atom is 0.61. Although these electronegativity differences vary, a H—O bond and a C—Cl bond are considered to be polar covalent.

$$\overset{\delta^+}{H}\text{—}\overset{\delta^-}{O} \qquad \overset{\delta^+}{C}\text{—}\overset{\delta^-}{Cl}$$

According to their molecular formulas, both molecules have more than one polar covalent bond. However, only the water molecule is polar. Why might one molecule with polar covalent bonds be polar, while a second molecule with polar covalent bonds is nonpolar? The answer lies in the shapes of the molecules.

### CAREERS IN CHEMISTRY

**Flavor Chemist** A flavor chemist, or flavorist, must know how chemicals react and change in different conditions. A degree in chemistry is an asset, but is not required. Most flavorists work for companies that supply flavors to the food and beverage industries. A certified flavorist trains for five years in a flavor laboratory, passes an oral examination, and then works under supervision for another two years.

WebQuest

■ **Figure 23** A molecule's shape determines its polarity.

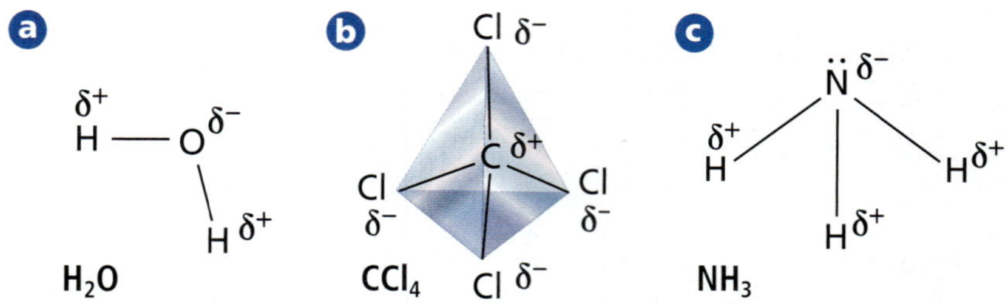


The bent shape of a water molecule makes it polar.

The symmetry of a $CCl_4$ molecule results in an equal distribution of charge, and the molecule is nonpolar.

The asymmetric shape of an ammonia molecule results in an unequal charge distribution and the molecule is polar.

The shape of an $H_2O$ molecule, as determined by VSEPR, is bent because the central oxygen atom has lone pairs of electrons, as shown in **Figure 23a.** Because the polar H—O bonds are asymmetric in a water molecule, the molecule has a definite positive end and a definite negative end. Thus, it is polar.

A $CCl_4$ molecule is tetrahedral, and therefore, symmetrical, as shown in **Figure 23b.** The electric charge measured at any distance from its center is identical to the charge measured at the same distance to the opposite side. The average center of the negative charge is located on each chlorine atom. The positive center is also located on the carbon atom. Because the partial charges are balanced, $CCl_4$ is a nonpolar molecule. Note that symmetric molecules are usually nonpolar, and molecules that are asymmetric are polar as long as the bond type is polar.

Is the molecule of ammonia ($NH_3$), shown in **Figure 23c,** polar? It has a central nitrogen atom and three terminal hydrogen atoms. Its shape is trigonal pyramidal because of the lone pair of electrons present on the nitrogen atom. Using **Figure 20,** you can find that the electronegativity difference of hydrogen and nitrogen is 0.84, making each N—H bond polar covalent. The charge distribution is unequal because the molecule is asymmetric. Thus, the molecule is polar.

**Solubility of polar molecules** The physical property known as solubility is the ability of a substance to dissolve in another substance. The bond type and the shape of the molecules present determine solubility. Polar molecules and ionic compounds are usually soluble in polar substances, but nonpolar molecules dissolve only in nonpolar substances, as shown in **Figure 24.**

■ **Figure 24** Symmetric covalent molecules, such as oil and most petroleum products, are nonpolar. Asymmetric molecules, such as water, are usually polar. As shown in this photo, polar and nonpolar substances usually do not mix.

**Infer** ***Will water alone clean oil from a fabric?***

## Properties of Covalent Compounds

Table salt, an ionic solid, and table sugar, a covalent solid, are similar in appearance. However, these compounds behave differently when heated. Salt does not melt, but sugar melts at a relatively low temperature. Does the type of bonding in a compound affect its properties?

Explore **covalent compounds.**

Virtual Investigations

**Intermolecular forces** Differences in properties are a result of differences in attractive forces. In a covalent compound, the covalent bonds between atoms in molecules are strong, but the attraction forces between molecules are relatively weak. These weak attraction forces are known as intermolecular forces, or van der Waals forces. Intermolecular forces vary in strength but are weaker than the bonds that join atoms in a molecule or ions in an ionic compound.

There are different types of intermolecular forces. Between nonpolar molecules, the force is weak and is called a dispersion force, or induced dipole. The force between oppositely charged ends of two polar molecules is called a dipole-dipole force. The more polar the molecule, the stronger the dipole-dipole force. The third force, a hydrogen bond, is especially strong. It forms between the hydrogen end of one dipole and a fluorine, oxygen, or nitrogen atom on another dipole. You will study intermolecular forces in more detail when you study states of matter.

# Data Analysis LAB

Based on Real Data*

## Interpret Data

**How does the polarity of the mobile phase affect chromatograms?** Chromatography is a technique in which a moving phase transports and separates the components of a mixture. A chromatograph is created by recording the intensity of each component carried in the moving phase versus time. The peak areas on the chromatograph indicate the amount of each component present in the mixture.

High-performance liquid chromatography, or HPLC, is used by analytical chemists to separate mixtures of solutes. During HPLC, components that are strongly attracted to the extracting solvent are retained longer by the moving phase and tend to appear early on a chromatograph. Several scientists performed HPLC using a methanol-water mixture as the extracting solvent to separate a phenol-benzoic acid mixture. Their results are shown in the graph to the right.

### Data and Observations

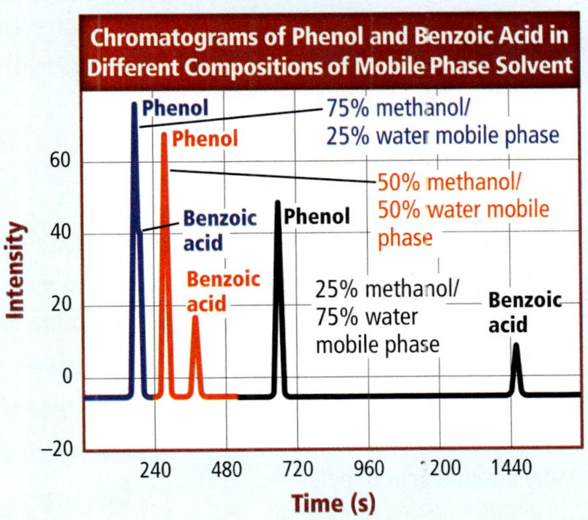


*Data obtained from: Joseph, Seema M. and Palasota, John A. 2001. The combined effects of pH and percent methanol on the HPLC separation of benzoic acid and phenol. *Journal of Chemical Education* 78:1381.

### Think Critically

1. **Explain** the different retention times shown on the chromatograms.
2. **Infer** from the graph the component, phenol or benzoic acid, that is in excess. Explain your answer.
3. **Infer** which component of the mixture has more polar molecules.
4. **Determine** the most effective composition of the mobile phase (of those tested) for separating phenol from benzoic acid. Explain.

■ **Figure 25** Network solids are often used in cutting tools because of their extreme hardness. Here, a diamond-tipped saw blade cuts through stone.

**Forces and properties** The properties of covalent molecular compounds are related to the relatively weak intermolecular forces holding the molecules together. These weak forces result in the relatively low melting and boiling points of molecular substances compared with those of ionic substances. That is why, when heated moderately, sugar melts but salt does not.

Weak intermolecular forces also explain why many molecular substances exist as gases or vaporize readily at room temperature. Oxygen ($O_2$), carbon dioxide ($CO_2$), and hydrogen sulfide ($H_2S$) are examples of covalent compounds that are gases at room temperature. Because the hardness of a substance depends on the intermolecular forces between individual molecules, many covalent molecules are relatively soft solids. Paraffin, found in candles and other products, is a common example of a covalent solid.

In the solid phase, molecules align to form a crystal lattice. This molecular lattice is similar to that of an ionic solid, but with less attraction between particles. The structure of the lattice is affected by molecular shape and the type of intermolecular force. Most molecular information has been determined by studying molecular solids.

## Covalent Network Solids

There are some solids, often called covalent network solids, that are composed only of atoms interconnected by a network of covalent bonds. Quartz and diamond are two common examples of network solids. In contrast to molecular solids, network solids are typically brittle, nonconductors of heat or electricity, and extremely hard. Analyzing the structure of a diamond explains some of its properties. In a diamond, each carbon atom is bonded to four other carbon atoms. This tetrahedral arrangement, which is shown in **Figure 25,** forms a strongly bonded crystal system that is extremely hard and has a very high melting point.

## SECTION 5 REVIEW

Section Self-Check

### Section Summary

- The electronegativity difference determines the character of a bond between atoms.
- Polar bonds occur when electrons are not shared equally forming a dipole.
- The spatial arrangement of polar bonds in a molecule determines the overall polarity of a molecule.
- Molecules attract each other by weak intermolecular forces. In a covalent network solid, each atom is covalently bonded to many other atoms.

**68.** MAINIDEA **Summarize** how electronegativity difference is related to bond character.

**69. Describe** a polar covalent bond.

**70. Describe** a polar molecule.

**71. List** three properties of a covalent compound in the solid phase.

**72. Categorize** bond types using electronegativity difference.

**73. Generalize** Describe the general characteristics of covalent network solids.

**74. Predict** the type of bond that will form between the following pair of atoms:

**a.** H and S
**b.** C and H
**c.** Na and S

**75. Identify** each molecule as polar or nonpolar: $SCl_2$, $CS_2$, and $CF_4$.

**76. Determine** whether a compound made of hydrogen and sulfur atoms is polar or nonpolar.

**77. Draw** the Lewis structures for the molecules $SF_4$ and $SF_6$. Analyze each structure to determine whether the molecule is polar or nonpolar.

# HOW IT works

## STICKY FEET: How Geckos Grip

For a gecko, hanging from a wall or a ceiling is no great feat. The key to a gecko's amazing grip is found on each of its toes. Researchers have determined that a gecko's grip depends on the sticking power of atoms themselves.

**1 Gecko toe** The bottom of a gecko's toe is covered with millions of tiny hairs, called setae, arranged in rows.

**2 Setae** Setae are complex structures. The end of each seta has microscopic branches called spatulae.

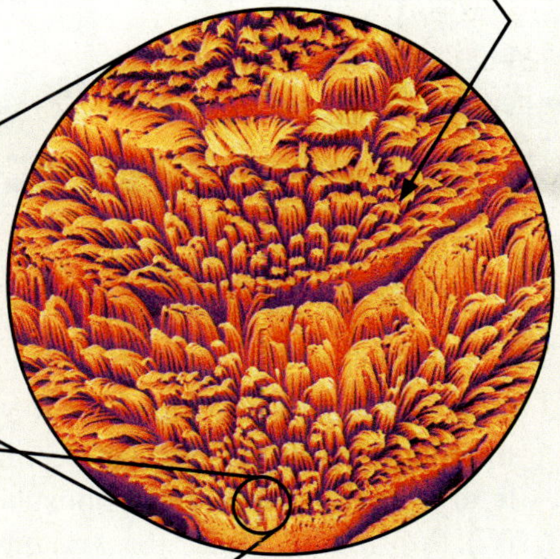

**3 Spatulae** Each seta has a relatively enormous surface area because of its vast number of spatulae.

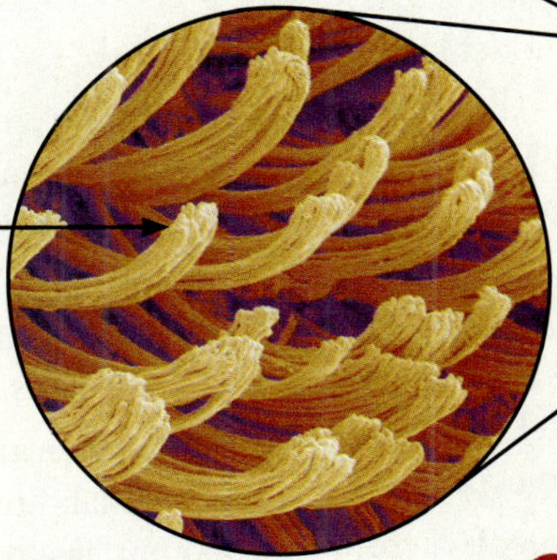

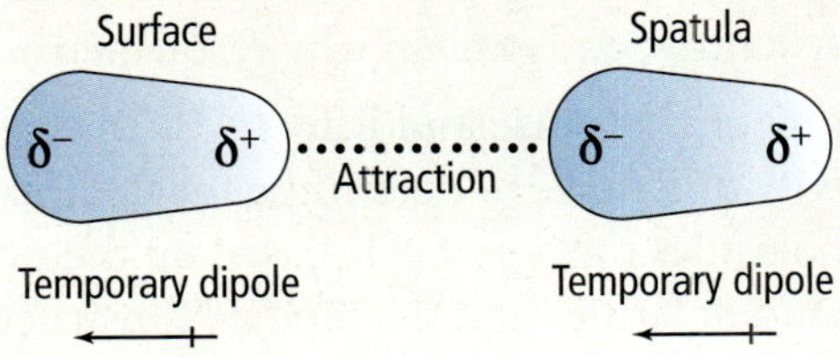


**4 Sticking** Van der Waals forces form between a surface and a gecko's spatulae. When multiplied by the spatulae's vast surface area, the sum of the weak van der Waals forces is more than enough to balance the pull of gravity and hold a gecko in place.

**5 Letting go** A gecko simply curls its toes when it wants to move. This reduces the amount of surface contact and the van der Waals forces, and the gecko loses its grip.

### WRITING IN ▶ Chemistry

**Invent** Using their knowledge of how geckos stick to surfaces, scientists are developing applications for geckolike materials. Some possible applications include mini-robots that climb walls and tape that sticks even under water. What uses for a new sticky geckolike material can you think of?

WebQuest

# ChemLAB

iLab Station

## Model Molecular Shapes

**Background:** Covalent bonding occurs when atoms share valence electrons. In the Valence Shell Electron Pair Repulsion (VSEPR) theory, the way in which valence electrons of bonding atoms are positioned is the basis for predicting a molecule's shape. This method of visualizing shape is also based on the molecule's Lewis structure.

**Question:** *How do the Lewis structure and the positions of valence electrons affect the shape of the covalent compound?*

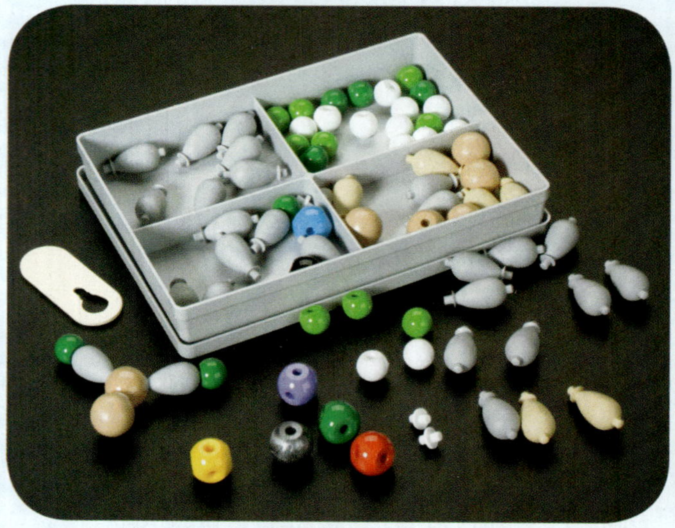

### Materials

molecular model kit

### Safety Precautions

### Procedure

1. Read and complete the lab safety form.
2. Create a table to record your data.
3. Note and record the color used to represent each of the following atoms in the molecular model kit: hydrogen (H), oxygen (O), phosphorus (P), carbon (C), fluorine (F), sulfur (S), and nitrogen (N).
4. Draw the Lewis structures of the $H_2$, $O_2$, and $N_2$ molecules.
5. Obtain two hydrogen atoms and one connector from the molecular model kit, and assemble a hydrogen ($H_2$) molecule. Observe that your model represents a single-bonded diatomic hydrogen molecule.
6. Obtain two oxygen atoms and two connectors from the molecular model kit, and assemble an oxygen ($O_2$) molecule. Observe that your model represents a double-bonded diatomic oxygen molecule.
7. Obtain two nitrogen atoms and three connectors from the molecular model kit, and assemble one nitrogen ($N_2$) molecule. Observe that your model represents a triple-bonded diatomic nitrogen molecule.
8. Recognize that diatomic molecules such as those formed in this lab are always linear. Diatomic molecules are made up of only two atoms and two points (atoms) can only be connected by a straight line.
9. Draw the Lewis structure of water ($H_2O$), and construct its molecule.
10. Classify the shape of the $H_2O$ molecule using information in **Table 6.**
11. Repeat Steps 9 and 10 for the $PH_3$, $CF_4$, $CO_2$, $SO_3$, HCN, and CO molecules.

### Analyze and Conclude

1. **Think Critically** Based on the molecular models you built and observed in this lab, rank single, double, and triple bonds in order of increasing flexibility and increasing strength.
2. **Observe and Infer** Explain why $H_2O$ and $CO_2$ molecules have different shapes.
3. **Analyze and Conclude** One of the molecules from this lab undergoes resonance. Identify the molecule that has three resonance structures, draw the structures, and explain why resonance occurs.
4. **Recognize Cause and Effect** Use the electronegativity difference to determine the polarity of the molecules in Steps 9–11. Based on their calculated bond polarities and the models constructed in this lab, determine the molecular polarity of each structure.

### INQUIRY EXTENSION

**Model** Use a molecular model kit to build the two resonance structures of ozone ($O_3$). Then, use Lewis structures to explain how you can convert between the two resonance structures by interchanging a lone pair for a covalent bond.

Matt Meadows

# CHAPTER 8 STUDY GUIDE

Vocabulary Practice

**BIG**IDEA **Covalent bonds form when atoms share electrons.**

## SECTION 1 The Covalent Bond

**MAIN**IDEA Atoms gain stability when they share electrons and form covalent bonds.

- Covalent bonds form when atoms share one or more pairs of electrons.
- Sharing one pair, two pairs, and three pairs of electrons forms single, double, and triple covalent bonds, respectively.
- Orbitals overlap directly in sigma bonds. Parallel orbitals overlap in pi bonds. A single covalent bond is a sigma bond but multiple covalent bonds are made of both sigma and pi bonds.
- Bond length is measured nucleus-to-nucleus. Bond dissociation energy is needed to break a covalent bond.

**VOCABULARY**
- covalent bond
- molecule
- Lewis structure
- sigma bond
- pi bond
- endothermic reaction
- exothermic reaction

## SECTION 2 Naming Molecules

**MAIN**IDEA Specific rules are used when naming binary molecular compounds, binary acids, and oxyacids.

- Names of covalent molecular compounds include prefixes for the number of each atom present. The final letter of the prefix is dropped if the element name begins with a vowel.
- Molecules that produce $H^+$ in solution are acids. Binary acids contain hydrogen and one other element. Oxyacids contain hydrogen and an oxyanion.

**VOCABULARY**
- oxyacid

## SECTION 3 Molecular Structures

**MAIN**IDEA Structural formulas show the relative positions of atoms within a molecule.

- Different models can be used to represent molecules.
- Resonance occurs when more than one valid Lewis structure exists for the same molecule.
- Exceptions to the octet rule occur in some molecules.

**VOCABULARY**
- structural formula
- resonance
- coordinate covalent bond

## SECTION 4 Molecular Shapes

**MAIN**IDEA The VSEPR model is used to determine molecular shape.

- VSEPR model theory states that electron pairs repel each other and determine both the shape of and bond angles in a molecule.
- Hybridization explains the observed shapes of molecules by the presence of equivalent hybrid orbitals.

**VOCABULARY**
- VSEPR model
- hybridization

## SECTION 5 Electronegativity and Polarity

**MAIN**IDEA A chemical bond's character is related to each atom's attraction for the electrons in the bond.

- The electronegativity difference determines the character of a bond between atoms.
- Polar bonds occur when electrons are not shared equally forming a dipole.
- The spatial arrangement of polar bonds in a molecule determines the overall polarity of a molecule.
- Molecules attract each other by weak intermolecular forces. In a covalent network solid, each atom is covalently bonded to many other atoms.

**VOCABULARY**
- polar covalent bond

# CHAPTER 8 ASSESSMENT

## SECTION 1

### Mastering Concepts

**78.** What is the octet rule, and how is it used in covalent bonding?

**79.** Describe the formation of a covalent bond.

**80.** Describe the bonding in molecules.

**81.** Describe the forces, both attractive and repulsive, that occur as two atoms move closer together.

**82.** How could you predict the presence of a sigma or pi bond in a molecule?

### Mastering Problems

**83.** Give the number of valence electrons in N, As, Br, and Se. Predict the number of covalent bonds needed for each of these elements to satisfy the octet rule.

**84.** Locate the sigma and pi bonds in each of the molecules shown below.

**a.**

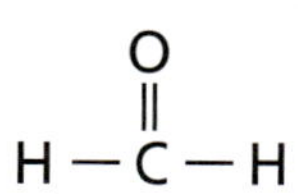

**b.** H—C≡C—H

**85.** In the molecules CO, $CO_2$, and $CH_2O$, which C—O bond is the shortest? Which C—O bond is the strongest?

**86.** Consider the carbon-nitrogen bonds shown below:

$C\equiv N^-$ and

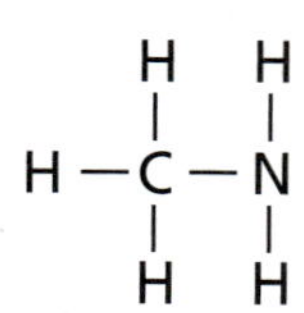

Which bond is shorter? Which is stronger?

**87.** Rank each of the molecules below in order of the shortest to the longest sulfur-oxygen bond length.

**a.** $SO_2$ **b.** $SO_3^{2-}$ **c.** $SO_4^{2-}$

## SECTION 2

### Mastering Concepts

**88.** Explain how molecular compounds are named.

**89.** When is a molecular compound named as an acid?

**90.** Explain the difference between sulfur hexafluoride and disulfur tetrafluoride.

**91. Watches** The quartz crystals used in watches are made of silicon dioxide. Explain how you use the name to determine the formula for silicon dioxide.

### Mastering Problems

**92.** Complete **Table 8.**

**Table 8 Acid Names**

| Formula | Name |
|---|---|
| $HClO_2$ | |
| $H_3PO_4$ | |
| $H_2Se$ | |
| $HClO_3$ | |

**93.** Name each molecule.

**a.** $NF_3$ **c.** $SO_3$
**b.** NO **d.** $SiF_4$

**94.** Name each molecule.

**a.** $SeO_2$ **c.** $N_2F_4$
**b.** $SeO_3$ **d.** $S_4N_4$

**95.** Write the formula for each molecule.

**a.** sulfur difluoride **c.** carbon tetrafluoride
**b.** silicon tetrachloride **d.** sulfurous acid

**96.** Write the formula for each molecule.

**a.** silicon dioxide **c.** chlorine trifluoride
**b.** bromous acid **d.** hydrobromic acid

## SECTION 3

### Mastering Concepts

**97.** What must you know in order to draw the Lewis structure for a molecule?

**98. Doping Agent** Material scientists are studying the properties of polymer plastics doped with $AsF_5$. Explain why the compound $AsF_5$ is an exception to the octet rule.

**99. Reducing Agent** Boron trihydride ($BH_3$) is used as reducing agent in organic chemistry. Explain why $BH_3$ often forms coordinate covalent bonds with other molecules.

**100.** Antimony and chlorine can form antimony trichloride or antimony pentachloride. Explain how these two elements can form two different compounds.

### Mastering Problems

**101.** Draw three resonance structures for the polyatomic ion $CO_3^{2-}$.

**102.** Draw the Lewis structures for these molecules, each of which has a central atom that does not obey the octet rule.

**a.** $PCl_5$ **c.** $ClF_5$
**b.** $BF_3$ **d.** $BeH_2$

**103.** Draw two resonance structures for the polyatomic ion $HCO_2^-$.

**104.** Draw the Lewis structure for a molecule of each of these compounds and ions.
**a.** $H_2S$ **c.** $SO_2$
**b.** $BF_4^-$ **d.** $SeCl_2$

**105.** Which elements in the list below are capable of forming molecules in which one of its atoms has an expanded octet? Explain your answer.
**a.** B **d.** O
**b.** C **e.** Se
**c.** P

# SECTION 4

## Mastering Concepts

**106.** What is the basis of the VSEPR model?

**107.** What is the maximum number of hybrid orbitals a carbon atom can form?

**108.** What is the molecular shape of each molecule? Estimate the bond angle for each molecule, assuming that there is not a lone pair.

**a.** A—B

**b.** A—B—A

**c.**
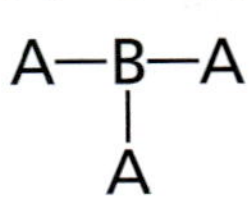

**d.**
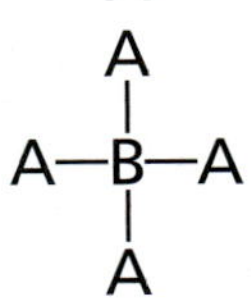

**109. Parent Compound** $PCl_5$ is used as a parent compound to form many other compounds. Explain the theory of hybridization and determine the number of hybrid orbitals present in a molecule of $PCl_5$.

## Mastering Problems

**110.** Complete **Table 9** by identifying the expected hybrid on the central atom. You might find drawing the molecule's Lewis structure helpful.

**Table 9 Structures**

| Formula | Hybrid Orbital | Lewis Structure |
|---|---|---|
| $XeF_4$ | | |
| $TeF_4$ | | |
| $KrF_2$ | | |
| $OF_2$ | | |

**111.** Predict the molecular shape of each molecule.
**a.** COS **b.** $CF_2Cl_2$

**112.** For each molecule listed below, predict its molecular shape and bond angle, and identify the hybrid orbitals. Drawing the Lewis structure might help you.
**a.** $SCl_2$ **c.** HOF
**b.** $NH_2Cl$ **d.** $BF_3$

# SECTION 5

## Mastering Concepts

**113.** Describe electronegativity trends in the periodic table.

**114.** Explain the difference between nonpolar molecules and polar molecules.

**115.** Compare the location of bonding electrons in a polar covalent bond with those in a nonpolar covalent bond. Explain your answer.

**116.** What is the difference between a covalent molecular solid and a covalent network solid? Do their physical properties differ? Explain your answer.

## Mastering Problems

**117.** For each pair, indicate the more polar bond by circling the negative end of its dipole.
**a.** C—S, C—O
**b.** C—F, C—N
**c.** P—H, P—Cl

**118.** For each of the bonds listed, tell which atom is more negatively charged.
**a.** C—H **c.** C—S
**b.** C—N **d.** C—O

**119.** Predict which bond is the most polar.
**a.** C—O **c.** C—Cl
**b.** Si—O **d.** C—Br

**120.** Rank the bonds according to increasing polarity.
**a.** C—H **d.** O—H
**b.** N—H **e.** Cl—H
**c.** Si—H

**121. Refrigerant** The refrigerant known as freon-14 is an ozone-damaging compound with the formula $CF_4$. Why is the $CF_4$ molecule nonpolar even though it contains polar bonds?

**122.** Determine if these molecules and ion are polar. Explain your answers.
**a.** $H_3O^+$ **c.** $H_2S$
**b.** $PCl_5$ **d.** $CF_4$

**123.** Use Lewis structures to predict the molecular polarities for sulfur difluoride, sulfur tetrafluoride, and sulfur hexafluoride.

## MIXED REVIEW

**124.** Write the formula for each molecule.
- **a.** chlorine monoxide
- **b.** arsenic acid
- **c.** phosphorus pentachloride
- **d.** hydrosulfuric acid

**125.** Name each molecule.
- **a.** $PCl_3$
- **b.** $Cl_2O_7$
- **c.** $P_4O_6$
- **d.** NO

**126.** Draw the Lewis structure for each molecule or ion.
- **a.** $SeF_2$
- **b.** $ClO_2^-$
- **c.** $PO_3^{3-}$
- **d.** $POCl_3$
- **e.** $GeF_4$

**127.** Determine which of the molecules are polar. Explain your answers.
- **a.** $CH_3Cl$
- **b.** ClF
- **c.** $NCl_3$
- **d.** $BF_3$
- **e.** $CS_2$

**128.** Arrange the bonds in order of least to greatest polar character.
- **a.** C—O
- **b.** Si—O
- **c.** Ge—O
- **d.** C—Cl
- **e.** C—Br

**129. Rocket Fuel** In the 1950s, the reaction of hydrazine with chlorine trifluoride ($ClF_3$) was used as a rocket fuel. Draw the Lewis structure for $ClF_3$ and identify the hybrid orbitals.

**130.** Complete **Table 10,** which shows the number of electrons shared in a single covalent bond, a double covalent bond, and a triple covalent bond. Identify the group of atoms that will form each of these bonds.

**Table 10** Shared Pairs

| Bond Type | Number of Shared Electrons | Atoms that Form the Bond |
|---|---|---|
| Single covalent | | |
| Double covalent | | |
| Triple covalent | | |

## THINK CRITICALLY

**131. Organize** Design a concept map that explains how VSEPR model theory, hybridization theory, and molecular shape are related.

**132. Compare and contrast** the two covalent compounds identified by the names arsenic(III) oxide and diarsenic trioxide.

**133. Make and Use Tables** Use your knowledge of ionic, metallic, and covalent bonds to complete **Table 11.**

**Table 11** Properties and Bonding

| Solid | Bond Description | Characteristic of Solid | Example |
|---|---|---|---|
| Ionic | | | |
| Covalent molecular | | | |
| Metallic | | | |
| Covalent network | | | |

**134. Apply** Urea, whose structure is shown below, is a compound used in manufacturing plastics and fertilizers. Identify the sigma bond, pi bonds, and lone pairs present in a molecule of urea.

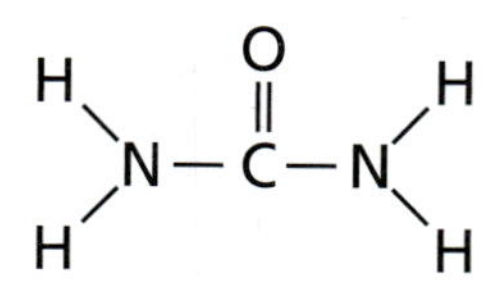

**135. Analyze** For each of the characteristics listed below, identify the polarity of a molecule with that characteristic.
- **a.** solid at room temperature
- **b.** gas at room temperature
- **c.** attracted to an electric current

**136. Apply** The structural formula for acetonitrile, $CH_3CN$, is shown below.

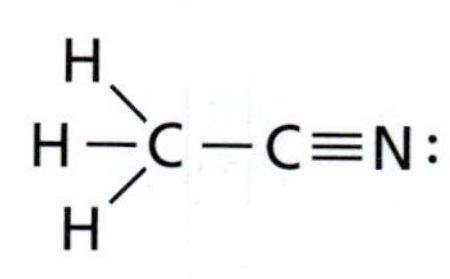

Examine the structure of the acetonitrile molecule. Determine the number of carbon atoms in the molecule, identify the hybrid present in each carbon atom, and explain your reasoning.

## CHALLENGE PROBLEM

**137.** Examine the bond-dissociation energies for the various bonds listed in **Table 12.**

**Table 12 Bond-Dissociation Energies**

| Bond | Bond-Dissociation Energy (kJ/mol) | Bond | Bond-Dissociation Energy (kJ/mol) |
|---|---|---|---|
| C—C | 348 | O—H | 467 |
| C=C | 614 | C—N | 305 |
| C≡C | 839 | O=O | 498 |
| N—N | 163 | C—H | 416 |
| N=N | 418 | C—O | 358 |
| N≡N | 945 | C=O | 745 |

**a.** Draw the correct Lewis structures for $C_2H_2$ and HCOOH.

**b.** Determine the amount of energy needed to break apart each of these molecules.

## CUMULATIVE REVIEW

**138.** **Table 13** lists a liquid's mass and volume data. Create a line graph of this data with the volume on the *x*-axis and the mass on the *y*-axis. Calculate the slope of the line. What information does the slope give you?

**Table 13 Mass v. Volume**

| Volume | Mass |
|---|---|
| 4.1 mL | 9.36 g |
| 6.0 mL | 14.04 g |
| 8.0 mL | 18.72 g |
| 10.0 mL | 23.40 g |

**139.** Write the correct chemical formula for each compound.

**a.** calcium carbonate
**b.** potassium chlorate
**c.** silver acetate
**d.** copper(II) sulfate
**e.** ammonium phosphate

**140.** Write the correct chemical name for each compound.

**a.** NaI
**b.** $Fe(NO_3)_3$
**c.** $Sr(OH)_2$
**d.** $CoCl_2$
**e.** $Mg(BrO_3)_2$

## WRITING IN ▶ Chemistry

**141. Antifreeze** Research ethylene glycol to learn its chemical formula. Explain how its structure makes it a useful antifreeze and coolant.

**142. Detergents** Choose a laundry detergent to research and write an essay about its chemical composition. Explain how it removes oil and grease from fabrics.

## DBQ Document-Based Questions

**Luminol** *Crime-scene investigators often use the covalent compound luminol to find blood evidence. The reaction between luminol, certain chemicals, and hemoglobin, a protein in blood, produces light.* **Figure 26** *shows a ball-and-stick model of luminol.*

Data obtained from: Fleming, Declan., 2002. The Chemiluminescence of Luminol, Exemplarchem, *Royal Society of Chemistry.*

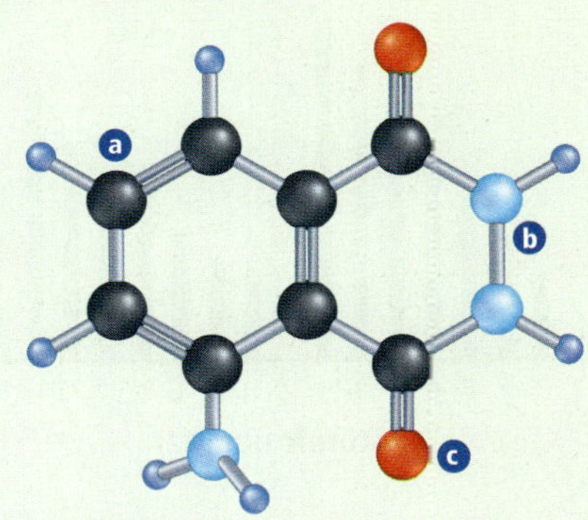


■ **Figure 26**

**143.** Determine the molecular formula for luminol and draw its Lewis structure.

**144.** Indicate the hybrid present on the atoms labeled *a, b,* and *c* in **Figure 26.**

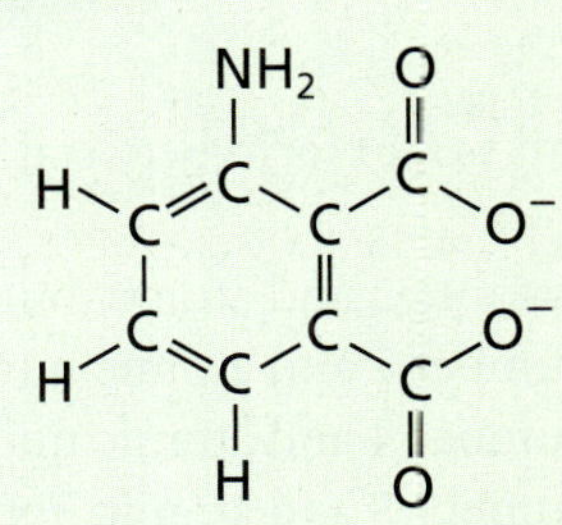


APA ion

■ **Figure 27**

**145.** When luminol comes in contact with the iron ion in hemoglobin, it reacts to produce $Na_2APA$, water, nitrogen, and light energy. Given the structural formula of the APA ion in **Figure 27,** write the chemical formula for the polyatomic APA ion.

## MULTIPLE CHOICE

1. The common name of $SiI_4$ is tetraiodosilane. What is its molecular compound name?
   A. silane tetraiodide
   B. silane tetraiodine
   C. silicon iodide
   D. silicon tetraiodide

2. Which compound contains at least one pi bond?
   A. $CO_2$
   B. $CHCl_3$
   C. $AsI_3$
   D. $BeF_2$

*Use the graph below to answer Questions 3 and 4.*

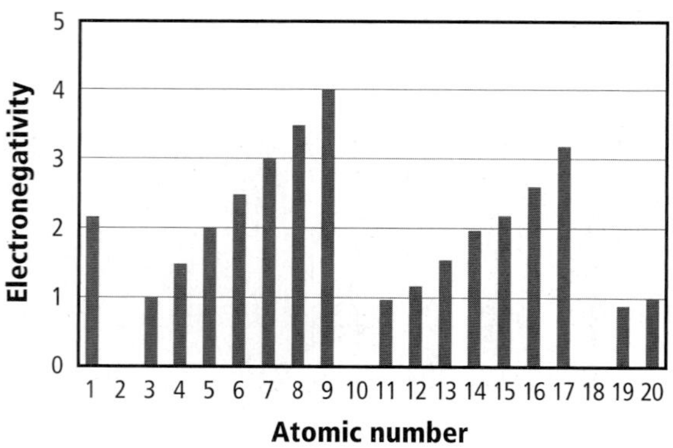


3. What is the electronegativity of the element with atomic number 14?
   A. 1.5
   B. 1.8
   C. 2.0
   D. 2.2

4. An ionic bond would form between which pairs of elements?
   A. atomic number 3 and atomic number 4
   B. atomic number 7 and atomic number 8
   C. atomic number 4 and atomic number 18
   D. atomic number 8 and atomic number 12

5. Which is the Lewis structure for silicon disulfide?
   A. :S::Si::S:
   B. 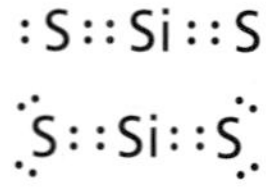
   C. 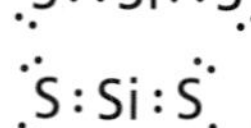
   D. 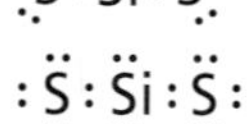

6. The central selenium atom in selenium hexafluoride forms an expanded octet. How many electron pairs surround the central Se atom?
   A. 4
   B. 5
   C. 6
   D. 7

*Use the table below to answer Questions 7 and 8.*

**Bond Dissociation Energies at 298 K**

| Bond | kJ/mol | Bond | kJ/mol |
|---|---|---|---|
| Cl–Cl | 242 | N≡N | 945 |
| C–C | 345 | O–H | 467 |
| C–H | 416 | C–O | 358 |
| C–N | 305 | C=O | 745 |
| H–I | 299 | O=O | 498 |
| H–N | 391 | | |

7. Which diatomic gas has the shortest bond between its two atoms?
   A. HI
   B. $O_2$
   C. $Cl_2$
   D. $N_2$

8. Approximately how much energy will it take to break all the bonds present in the molecule below?

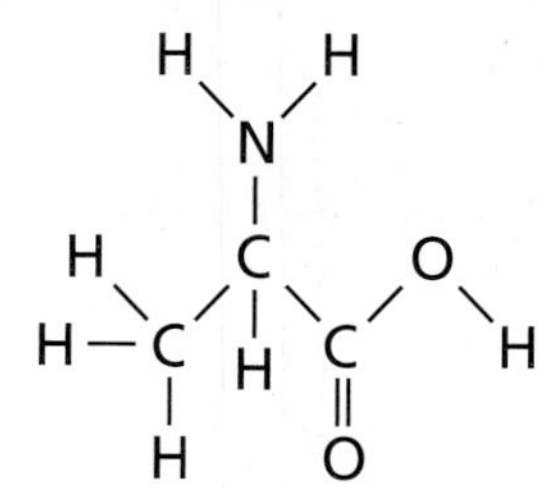


   A. 3024 kJ/mol
   B. 4318 kJ/mol
   C. 4621 kJ/mol
   D. 5011 kJ/mol

9. Which compound does NOT have a bent molecular shape?
   A. $BeH_2$
   B. $H_2S$
   C. $H_2O$
   D. $SeH_2$

10. Which compound is nonpolar?
   A. $H_2S$
   B. $CCl_4$
   C. $SiH_3Cl$
   D. $AsH_3$

## SHORT ANSWER

**11.** Oxyacids contain hydrogen and an oxyanion. There are two different oxyacids that contain hydrogen, nitrogen, and oxygen. Identify these two oxyacids. How can they be distinguished on the basis of their names and formulas?

*Use the atomic emission spectrum below to answer Questions 12 and 13.*

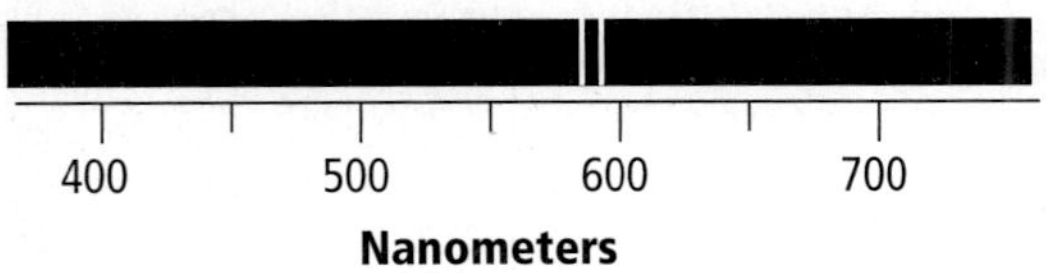


**12.** Estimate the wavelength of the photons being emitted by this element.

**13.** Find the frequency of the photons being emitted by this element.

## EXTENDED RESPONSE

*Use the table below to answer Question 14.*

**Percent Abundance of Silicon Isotopes**

| Isotope | Mass | Percent Abundance |
|---|---|---|
| $^{28}Si$ | 27.98 amu | 92.21 % |
| $^{29}Si$ | 28.98 amu | 4.70 % |
| $^{30}Si$ | 29.97 amu | 3.09 % |

**14.** Your lab partner calculates the average atomic mass of these three silicon isotopes. His average atomic mass value is 28.98 amu. Explain why your lab partner is incorrect, and show how to calculate the correct average atomic mass.

## SAT SUBJECT TEST: CHEMISTRY

*Use the list of separation techniques below to answer Questions 15 to 17.*

**A.** filtration
**B.** distillation
**C.** crystallization
**D.** chromatography
**E.** sublimation

**15.** Which technique separates components of a mixture with different boiling points?

**16.** Which technique separates components of a mixture based on the size of its particles?

**17.** Which technique is based on the stronger attraction some components have for the stationary phase compared to the mobile phase?

*Use the table below to answer Questions 18 to 19.*

**Electron-Dot Structures**

| Group | 1 | 2 | 13 | 14 | 15 | 16 | 17 | 18 |
|---|---|---|---|---|---|---|---|---|
| Diagram | Li· | ·Be· | ·Ḃ· | ·Ċ· | ·N̈· | ·Ö: | :F̈: | :N̈e: |

**18.** Based on the Lewis structures shown, which elements will combine in a 2:3 ratio?
**A.** lithium and carbon
**B.** beryllium and fluorine
**C.** beryllium and nitrogen
**D.** boron and oxygen
**E.** boron and carbon

**19.** How many electrons will beryllium have in its outer energy level after it forms an ion to become chemically stable?
**A.** 0
**B.** 2
**C.** 4
**D.** 6
**E.** 8

**NEED EXTRA HELP?**

| If You Missed Question . . . | 1 | 2 | 3 | 4 | 5 | 6 | 7 | 8 | 9 | 10 | 11 | 12 | 13 | 14 | 15 | 16 | 17 | 18 | 19 |
|---|---|---|---|---|---|---|---|---|---|---|---|---|---|---|---|---|---|---|---|
| Review Section . . . | 8.2 | 8.1 | 8.5 | 8.5 | 8.3 | 8.3 | 8.1 | 8.1 | 8.4 | 8.5 | 8.2 | 5.2 | 5.2 | 4.3 | 3.3 | 3.3 | 3.3 | 7.2 | 7.2 |

CHAPTER 9

# Chemical Reactions

**BIGIDEA** Millions of chemical reactions in and around you transform reactants into products, resulting in the absorption or release of energy.

## SECTIONS

Chm 5, 5b

1 Reactions and Equations

2 Classifying Chemical Reactions

3 Reactions in Aqueous Solutions

## LaunchLAB

iLab Station

### How do you know when a chemical change has occurred?

There are different ways to know when a chemical change has occurred, including temperature change, color change, and formation of a gas or solid. In this lab, you will observe evidence of a chemical change.

## FOLDABLES® Study Organizer

### Chemical Reactions

Make a concept-map book. Label it as shown. Use it to help you organize information about how chemical reactions are classified.

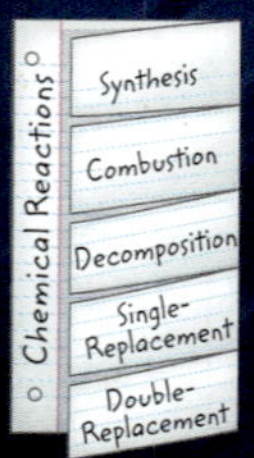

(t)Ingram Publishing; (b)©Kevin Schafer/Alamy; (bkgd)Photoshot Holdings Ltd/David Woodfall/Alamy

Before wood can burn, the water inside it boils off. Wood bursts into flames once it reaches 260°C. The smoke released during the reaction contains over 100 chemical substances.

SECTION 1

# Reactions and Equations

**MAINIDEA** Chemical reactions are represented by balanced chemical equations.

**Essential Questions**

- What is evidence of chemical change?
- How are chemical reactions represented?
- Why do chemical equations need to be balanced and how is this accomplished?

**Review Vocabulary**

**chemical change:** a process involving one or more substances changing into a new substance

**New Vocabulary**

chemical reaction
reactant
product
chemical equation
coefficient

**CHEM 4 YOU** When you purchase bananas from a grocery store, they might be green. Within a few days, the bananas turn yellow. This color change is one of the ways you can tell a chemical reaction has occurred.

## Chemical Reactions

Do you know that the foods you eat, the fibers in your clothes, and the plastic in your CDs have something in common? Foods, fibers, and plastics are produced when the atoms in substances are rearranged to form different substances. Atoms are rearranged during the forest fire shown in the photo at the beginning of the chapter. They were also rearranged when you dropped the effervescent tablet into the beaker of water and indicator during the Launch Lab.

The process by which the atoms of one or more substances are rearranged to form different substances is called a **chemical reaction.** A chemical reaction is another name for a chemical change, which you read about previously. Chemical reactions affect every part of your life. They break down your food, producing the energy you need to live. Chemical reactions in the engines of cars and buses provide the energy to power the vehicles. They produce natural fibers, such as cotton and wool, in plants and animals. In factories, they produce synthetic fibers such as nylon, shown in **Figure 1.**

**Evidence of a chemical reaction** How can you tell when a chemical reaction has taken place? Although some chemical reactions are hard to detect, many reactions provide physical evidence that they have occurred. A temperature change can indicate a chemical reaction. Many reactions, such as those that occur during the burning of wood, release energy in the form of heat and light. Other chemical reactions absorb heat.

**Figure 1** When adipoyl chloride in dichloromethane reacts with hexanediamine, nylon is formed. Nylon is used in many products, including carpeting, clothing, sports equipment, and tires.

■ **Figure 2** Each of these photos illustrates evidence of a chemical reaction.
**Describe** *the evidence in each photo that tells you a chemical reaction has occurred.*

In addition to a temperature change, other types of evidence might indicate that a chemical reaction has occurred. One indication of a chemical reaction is a color change. For example, you might have noticed that the color of some nails that are left outside changes from silver to orange-brown in a short time. The color change is evidence that a chemical reaction occurred between the iron in the nail and the oxygen in air. A banana changing from green to yellow is another example of a color change indicating that a chemical reaction has occurred. Odor, gas bubbles, and the formation of a solid are other indications of chemical change. Each of the photographs in **Figure 2** shows evidence of a chemical reaction.

## Representing Chemical Reactions

Chemists use statements called equations to represent chemical reactions. Equations show a reaction's **reactants,** which are the starting substances, and **products,** which are the substances formed during the reaction. Chemical equations do not express numerical equalities as mathematical equations do because during chemical reactions the reactants are used up as the products form. Instead, the equations used by chemists show the direction in which the reaction progresses. Therefore, an arrow rather than an equal sign is used to separate the reactants from the products. You read the arrow as *react to produce* or *yield.* The reactants are written to the left of the arrow, and the products are written to the right of the arrow. When there are two or more reactants, or when there are two or more products, a plus sign separates each reactant or each product. These elements of equation notation are shown below.

$$\text{Reactant 1} + \text{Reactant 2} \rightarrow \text{Product 1} + \text{Product 2}$$

In equations, symbols are used to show the physical states of the reactants and products. Reactants and products can exist as solids, liquids, and gases. When they are dissolved in water, they are said to be aqueous. It is important to show the physical states of a reaction's reactants and products in an equation because the physical states provide clues about how the reaction occurs. Some basic symbols used in equations are shown in **Table 1.**

**Table 1 Symbols Used in Equations**

| Symbol | Purpose |
|---|---|
| $+$ | separates two or more reactants or products |
| $\rightarrow$ | separates reactants from products |
| $\rightleftharpoons$ | separates reactants from products and indicates a reversible reaction |
| (s) | identifies a solid state |
| (l) | identifies a liquid state |
| (g) | identifies a gaseous state |
| (aq) | identifies a water solution |

■ **Figure 3** Science, like all other disciplines, has a specialized language that allows specific information to be communicated in a uniform manner. This reaction between aluminum and bromine can be described by a word equation, a skeleton equation, or a balanced chemical equation.

**Word equations** You can use statements called word equations to indicate the reactants and products of chemical reactions. The word equation below describes the reaction between aluminum (Al) and bromine (Br), which is shown in **Figure 3.** Aluminum is a solid, and bromine is a liquid. The brownish-red cloud in the photograph is excess bromine. The reaction's product, which is solid particles of aluminum bromide ($AlBr_3$), settles on the bottom of the beaker.

$$\text{Reactant 1} + \text{Reactant 2} \longrightarrow \text{Product 1}$$

$$\text{aluminum(s)} + \text{bromine(l)} \longrightarrow \text{aluminum bromide(s)}$$

This word equation reads, "Aluminum and bromine react to produce aluminum bromide."

**Skeleton equations** Although word equations help to describe chemical reactions, they lack important information. A skeleton equation uses chemical formulas rather than words to identify the reactants and the products. For example, the skeleton equation for the reaction between aluminum and bromine uses the formulas for aluminum, bromine, and aluminum bromide in place of words.

$$Al(s) + Br_2(l) \longrightarrow AlBr_3(s)$$

How would you write the skeleton equation that describes the reaction between carbon and sulfur to form carbon disulfide? Carbon and sulfur are solids. First, write the chemical formulas for the reactants to the left of the arrow. Then, separate the reactants with a plus sign and indicate their physical states.

$$C(s) + S(s) \longrightarrow$$

Finally, write the chemical formula for the product, liquid carbon disulfide, to the right of the arrow and indicate its physical state. The result is the skeleton equation for the reaction.

$$C(s) + S(s) \longrightarrow CS_2(l)$$

This skeleton equation tells us that carbon in the solid state reacts with sulfur in the solid state to produce carbon disulfide in the liquid state.

**VOCABULARY**

**ACADEMIC VOCABULARY**

**Formula**
an expression using chemical symbols to represent a chemical reaction *The chemical formula for water is $H_2O$.*

**PRACTICE Problems** Do additional problems. Online Practice

**Write skeleton equations for the following word equations.**

1. Hydrogen and bromine gases react to yield hydrogen bromide.

   hydrogen(g) + bromine(g) → hydrogen bromide(g)

2. When carbon monoxide and oxygen react, carbon dioxide forms.

   carbon monoxide(g) + oxygen(g) → carbon dioxide(g)

3. **Challenge** Write the word equation and the skeleton equation for the following reaction: when heated, solid potassium chlorate yields solid potassium chloride and oxygen gas.

$Al(s) + Br_2(g) \rightarrow AlBr_3(s)$

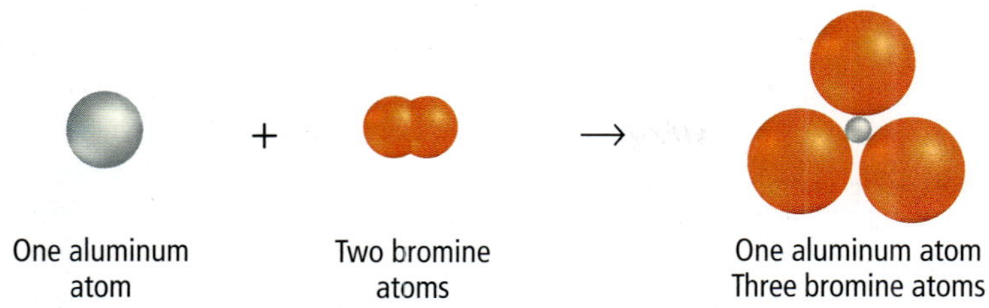


■ **Figure 4** The information conveyed by skeleton equations is limited. In this case, the skeleton equation is correct, but it does not show the exact number of atoms that interact. Refer to **Table R-1** in the Student Resources for a key to atom color conventions.

**Chemical equations** Like word equations, skeleton equations lack some information about reactions. Recall that the law of conservation of mass states that in a chemical change, matter is neither created nor destroyed. Chemical equations must show that matter is conserved during a reaction. Skeleton equations lack that information.

Look at **Figure 4.** The skeleton equation for the reaction between aluminum and bromine shows that one aluminum atom and two bromine atoms react to produce a substance containing one aluminum atom and three bromine atoms. Was a bromine atom created in the reaction? Atoms are not created in chemical reactions, and to accurately show what happened, more information is needed.

To accurately represent a chemical reaction by an equation, the equation must show equal numbers of atoms of each reactant and each product on both sides of the arrow. Such an equation is called a balanced chemical equation. A **chemical equation** is a statement that uses chemical formulas to show the identities and relative amounts of the substances involved in a chemical reaction.

## Balancing Chemical Equations

The balanced equation for the reaction between aluminum and bromine, shown in **Figure 5,** reflects the law of conservation of mass. To balance an equation, you must find the correct coefficients for the chemical formulas in the skeleton equation. A **coefficient** in a chemical equation is the number written in front of a reactant or product. Coefficients are usually whole numbers and are not usually written if the value is one. The coefficients in a balanced equation describe the lowest whole number ratio of the amounts of all of the reactants and products.

Get help with **balancing chemical equations.**

Personal Tutor

$2Al(s) + 3Br_2(g) \rightarrow 2AlBr_3(s)$

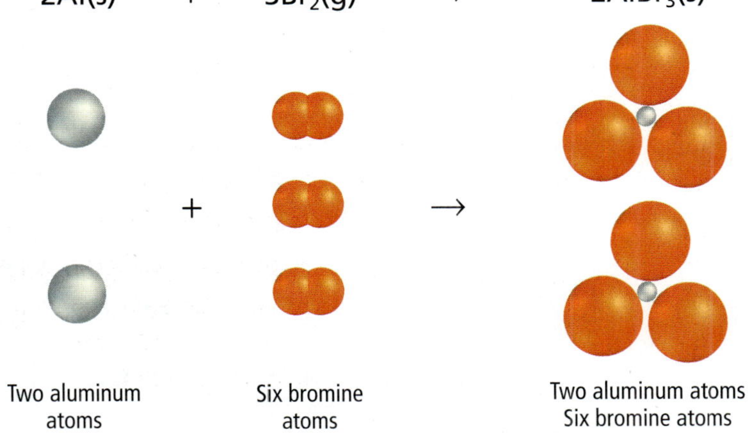


■ **Figure 5** In a balanced chemical equation, the number of particles on the reactant side of the equation equals the number of particles on the product side of the equation. In this case, two aluminum atoms and six bromine atoms are needed on both sides of the equation.

**Steps for balancing equations** Most chemical equations can be balanced by following the steps given in **Table 2.** For example, you can use these steps to write the chemical equation for the reaction between hydrogen ($H_2$) and chlorine ($Cl_2$) that produces hydrogen chloride (HCl).

Explore **balancing chemical equations with an interactive table.** Concepts In Motion

**Table 2** Steps for Balancing Equations

| Step | Process | Example |
|---|---|---|
| 1 | *Write the skeleton equation for the reaction.* Make sure that the chemical formulas correctly represent the substances. An arrow separates the reactants from the products, and a plus sign separates multiple reactants and products. Show the physical states of all reactants and products. | $H_2(g)$ + $Cl_2(g)$ → HCl(g)<br><br>Two hydrogen atoms + Two chlorine atoms → One hydrogen atom One chlorine atom |
| 2 | *Count the atoms of the elements in the reactants.* If a reaction involves identical polyatomic ions in the reactants and products, count each polyatomic ion as a single element. This reaction does not involve any poly-atomic ions. Two atoms of hydrogen and two atoms of chlorine are reacting. | $H_2$ + $Cl_2$ →<br>2 atoms H  2 atoms Cl |
| 3 | *Count the atoms of the elements in the products.* One atom of hydrogen and one atom of chlorine are produced. | HCl<br>1 atom H + 1 atom Cl |
| 4 | *Change the coefficients to make the number of atoms of each element equal on both sides of the equation.* Never change a subscript in a chemical formula to balance an equation because doing so changes the identity of the substance. | $H_2$ + $Cl_2$ → 2HCl<br>2 atoms H  2 atoms Cl  2 atoms H + 2 atoms Cl<br><br>Two hydrogen atoms + Two chlorine atoms → Two hydrogen atoms Two chlorine atoms |
| 5 | *Write the coefficients in their lowest possible ratio.* The coefficients should be the smallest possible whole numbers. The ratio 1 hydrogen to 1 chlorine to 2 hydrogen chloride (1:1:2) is the lowest-possible ratio because the coefficients cannot be reduced further and still remain whole numbers. | $H_2(g) + Cl_2(g) \longrightarrow 2HCl(g)$<br>1:1:2<br>1 $H_2$ to 1 $Cl_2$ to 2 HCl |
| 6 | *Check your work.* Make sure that the chemical formulas are written correctly. Then, check that the number of atoms of each element is equal on both sides of the equation. | $H_2$ + $Cl_2$ → 2HCl<br>2 atoms H  2 atoms Cl  2 atoms H + 2 atoms Cl<br>There are two hydrogen atoms and two chlorine atoms on both sides of the equation. |

## EXAMPLE Problem 1

Get help with **ratios.** Math Handbook

**WRITING A BALANCED CHEMICAL EQUATION** Write the balanced chemical equation for the reaction in which aqueous sodium hydroxide and aqueous calcium bromide react to produce solid calcium hydroxide and aqueous sodium bromide.

### 1 ANALYZE THE PROBLEM

You are given the reactants and products in a chemical reaction. Start with a skeleton equation, and use the steps given in **Table 2** for balancing chemical equations.

### 2 SOLVE FOR THE UNKNOWN

Write the skeleton equation for the chemical reaction. Be sure to put the reactants on the left side of the arrow and the products on the right. Separate the substances with plus signs, and indicate their physical states.

$$NaOH(aq) + CaBr_2(aq) \rightarrow Ca(OH)_2(s) + NaBr(aq)$$

1 Na, 1 O, 1 H, 1 Ca, 2 Br — **Count the atoms of each element in the reactants.**

1 Na, 2 O, 2 H, 1 Ca, 1 Br — **Count the atoms of each element in the products.**

$2NaOH + CaBr_2 \rightarrow Ca(OH)_2 + NaBr$ — **Insert the coefficient 2 in front of NaOH to balance the hydroxide ions.**

$2NaOH + CaBr_2 \rightarrow Ca(OH)_2 + 2NaBr$ — **Insert the coefficient 2 in front of NaBr to balance the Na and Br atoms.**

The ratio of the coefficients is 2:1:1:2. — **Write the coefficients in their lowest-possible ratio.**

Reactants: 2 Na, 2 OH, 1 Ca, 2 Br
Products: 2 Na, 2 OH, 1 Ca, 2 Br. — **Check to make sure that the number of atoms of each element is equal on both sides of the equation.**

### 3 EVALUATE THE ANSWER

The chemical formulas for all substances are written correctly. The number of atoms of each element is equal on both sides of the equation. The coefficients are written in the lowest possible ratio. The balanced chemical equation for the reaction is

$$2NaOH(aq) + CaBr_2(aq) \rightarrow Ca(OH)_2(s) + 2NaBr(aq)$$

## PRACTICE Problems

Do additional problems. Online Practice

**Write chemical equations for each of the following reactions.**

**4.** In water, iron(III) chloride reacts with sodium hydroxide, producing solid iron(III) hydroxide and sodium chloride.

**5.** Liquid carbon disulfide reacts with oxygen gas, producing carbon dioxide gas and sulfur dioxide gas.

**6. Challenge** A piece of zinc metal is added to a solution of dihydrogen sulfate. This reaction produces a gas and a solution of zinc sulfate.

## RealWorld CHEMISTRY

### Calcium Hydroxide

**REEF AQUARIUMS** An aqueous solution of calcium hydroxide is used in reef aquariums to provide calcium for animals such as snails and corals. Calcium hydroxide reacts with the carbon dioxide in the water to produce calcium and bicarbonate ions. Reef animals use the calcium to grow shells and strong skeletal systems.

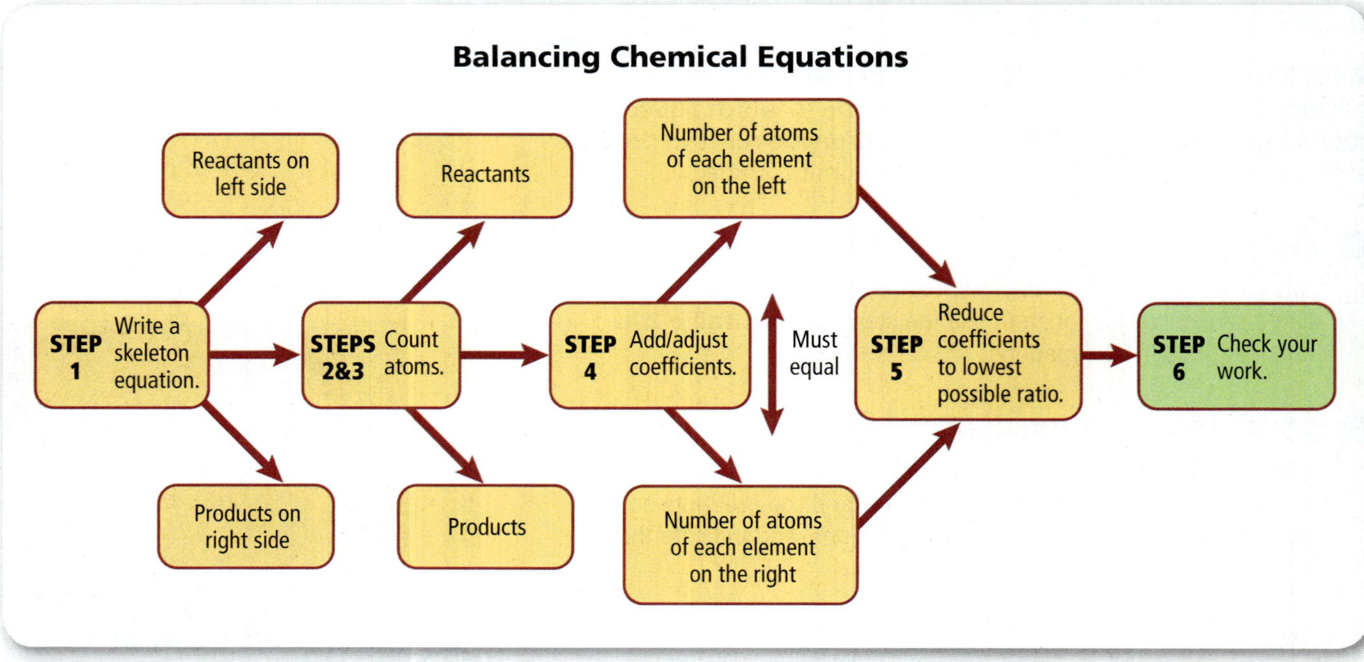


**Figure 6** It is imperative to your study of chemistry to be able to balance chemical equations. Use this flowchart to help you master the skill. Notice that the numbered steps correspond to the steps in **Table 2.**

Explore **balancing chemical reactions.**

Virtual Investigations

**Obeying the law of conservation of mass** Probably the most fundamental concept of chemistry is the law of conservation of mass that you encountered previously. All chemical reactions obey the law that matter is neither created nor destroyed. Therefore, it is also fundamental that the equations that represent chemical reactions include sufficient information to show that the reaction obeys the law of conservation of mass.

You have learned how to show the conservation of mass with balanced chemical equations. The flowchart shown in **Figure 6** summarizes the steps for balancing equations. You will probably find that some chemical equations can be balanced easily, whereas others are more difficult to balance. Most chemical equations can be balanced by the process you learned in this section.

## SECTION 1 REVIEW

Section Self-Check

### Section Summary

- Some physical changes are evidence that indicate a chemical reaction has occurred.
- Word equations and skeleton equations provide important information about a chemical reaction.
- A chemical equation gives the identities and relative amounts of the reactants and products that are involved in a chemical reaction.
- Balancing an equation involves adjusting the coefficients until the number of atoms of each element is equal on both sides of the equation.

7. **MAIN**IDEA **Explain** why it is important that a chemical equation be balanced.
8. **List** three types of physical evidence that indicate a chemical reaction has occurred.
9. **Compare and contrast** a skeleton equation and a chemical equation.
10. **Explain** why it is important to reduce coefficients in a balanced equation to the lowest-possible whole-number ratio.
11. **Analyze** When balancing a chemical equation, can you adjust the subscript in a formula? Explain.
12. **Assess** Is the following equation balanced? If not, correct the coefficients to balance the equation.

$$2K_2CrO_4(aq) + Pb(NO_3)_2(aq) \longrightarrow 2KNO_3(aq) + PbCrO_4(s)$$

13. **Evaluate** Aqueous phosphoric acid and aqueous calcium hydroxide react to form solid calcium phosphate and water. Write a balanced chemical equation for this reaction.

SECTION 2

# Classifying Chemical Reactions

**MAINIDEA** There are four types of chemical reactions: synthesis, combustion, decomposition, and replacement reactions.

**CHEM 4 YOU** It could take you a long time to find a specific novel in an unorganized bookstore. Bookstores classify and organize books into different categories to make your search easier. Chemical reactions are also classified and organized into different categories.

### Essential Questions

- How are chemical reactions classified?
- What are the characteristics of different classes of chemical reactions?

### Review Vocabulary

**metal:** an element that is a solid at room temperature, a good conductor of heat and electricity, and is generally shiny

### New Vocabulary

synthesis reaction
combustion reaction
decomposition reaction
single-replacement reaction
double-replacement reaction
precipitate

## Types of Chemical Reactions

Chemists classify chemical reactions in order to organize the many reactions that occur daily. Knowing the categories of chemical reactions can help you remember and understand them. It can also help you recognize patterns and predict the products of many chemical reactions. One way chemists classify reactions is to distinguish among the four types: synthesis, combustion, decomposition, and replacement reactions. Some reactions fit into more than one of these types.

## Synthesis Reactions

In **Figure 7,** sodium and chlorine react to produce sodium chloride. This reaction is a **synthesis reaction**—a chemical reaction in which two or more substances (A and B) react to produce a single product (AB).

$$A + B \rightarrow AB$$

When two elements react, the reaction is always a synthesis reaction.

Two compounds can also combine to form one compound. For example, the reaction between calcium oxide (CaO) and water ($H_2O$) to form calcium hydroxide ($Ca(OH)_2$) is a synthesis reaction.

$$CaO(s) + H_2O(l) \rightarrow Ca(OH)_2(s)$$

Another type of synthesis reaction involves a reaction between a compound and an element, as happens when sulfur dioxide gas ($SO_2$) reacts with oxygen gas ($O_2$) to form sulfur trioxide ($SO_3$).

$$2SO_2(g) + O_2(g) \rightarrow 2SO_3(g)$$

$$2Na(s) + Cl_2(g) \rightarrow 2NaCl(s)$$

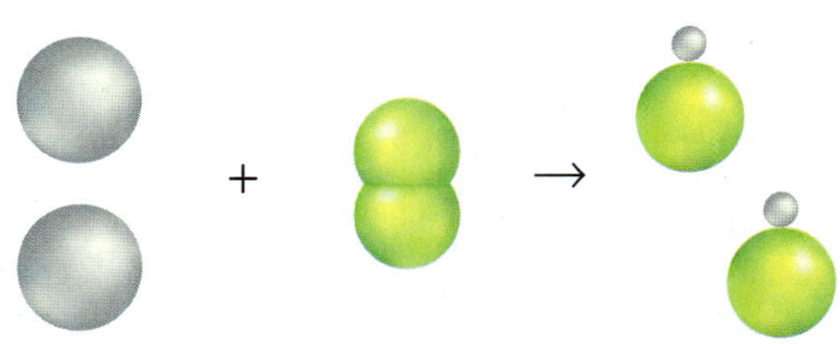

■ **Figure 7** In this synthesis reaction, two elements, sodium and chlorine, react to produce one compound, sodium chloride.

■ **Figure 8** The light produced by a sparkler is the result of a combustion reaction between oxygen and different metals.

**VOCABULARY**

**WORD ORIGIN**

**Combustion**
comes from the Latin word *comburere,* meaning *to burn.*

## Combustion Reactions

The synthesis reaction between sulfur dioxide and oxygen can also be classified as a combustion reaction. In a **combustion reaction,** such as the one shown in **Figure 8,** oxygen combines with a substance and releases energy in the form of heat and light. Oxygen can combine in this way with many different substances, making combustion reactions common. To learn more about the discovery of the chemical reaction for combustion and other reactions, review **Figure 9.**

A combustion reaction occurs between hydrogen and oxygen when hydrogen is heated, as illustrated in **Figure 10.** Water is formed during the reaction, and a large amount of energy is released. Another important combustion reaction occurs when coal is burned to produce energy. Coal is called a fossil fuel because it contains the remains of plants that lived long ago. It is composed primarily of the element carbon. Coal-burning power plants generate electric power in many parts of the United States. The primary reaction that occurs in these plants is between carbon and oxygen.

$$C(s) + O_2(g) \rightarrow CO_2(g)$$

■ **Figure 9**

### Real-World Chemical Reactions

Throughout history, people have worked to understand and apply the power of chemical reactions to solve problems.

1600 — 1700 — 1800 — 1905 — 1920

**1635** America's first chemical plant opens in Boston. Products include saltpeter, a component of gunpowder, and alum, a chemical used in tanning animal skins.

**1775** Antoine Lavoisier demonstrates that combustion is an exothermic chemical reaction involving oxygen.

**CIRCA 1800** Experiments with plants result in the discovery of the balanced chemical equation for photosynthesis.

**1885** The internal combustion engine is invented. It later becomes the prototype for the modern gas engine.

**1909–1910** German chemists Fritz Haber and Carl Bosch develop the Haber-Bosch process for synthesizing ammonia.

$$2H_2(g) + O_2(g) \rightarrow 2H_2O(g)$$

**Figure 10** During a combustion reaction between oxygen and hydrogen, water is formed.

**Analyze** ***Why is this chemical reaction both a synthesis reaction and a combustion reaction?***

Note that the combustion reactions just mentioned are also synthesis reactions. However, not all combustion reactions are synthesis reactions. For example, the reaction involving methane gas ($CH_4$) and oxygen illustrates a combustion reaction in which one substance replaces another in the formation of products.

$$CH_4(g) + 2O_2(g) \rightarrow CO_2(g) + 2H_2O(g)$$

Methane, which belongs to a group of substances called hydrocarbons, is the major component of natural gas. All hydrocarbons contain carbon and hydrogen and burn in oxygen to yield carbon dioxide and water. You will learn more about hydrocarbons in a later chapter.

## PRACTICE Problems

Do additional problems. Online Practice

**Write chemical equations for the following reactions. Classify each reaction into as many categories as possible.**

**14.** The solids aluminum and sulfur react to produce aluminum sulfide.

**15.** Water and dinitrogen pentoxide gas react to produce aqueous hydrogen nitrate.

**16.** The gases nitrogen dioxide and oxygen react to produce dinitrogen pentoxide gas.

**17.** **Challenge** Sulfuric acid ($H_2SO_4$) and sodium hydroxide solutions react to produce aqueous sodium sulfate and water.

**1974–1978** Researchers demonstrate that chlorofluorocarbons (CFCs) can deplete the ozone layer. The use of CFCs as spray propellants is banned in the United States.

**2004** Scientists discover that migrating birds are guided by chemical reactions in their bodies that are influenced by Earth's magnetic field.

1950 — 1965 — 1980 — 1995 — 2010

**1952** A heavy smog—sulfur dioxide and other coal-burning products—settles over London for five days in December, causing 4000 deaths.

**1995** Researchers use the atomic force microscope to create and observe chemical reactions as they occur molecule by molecule, paving the way for nanoscale engineering.

**2010** Deborah Jin and colleagues discover that chemical reactions can still occur at temperatures near absolute zero when they observed the formation of KRb from a reaction of diatomic potassium and rubidium molecules.

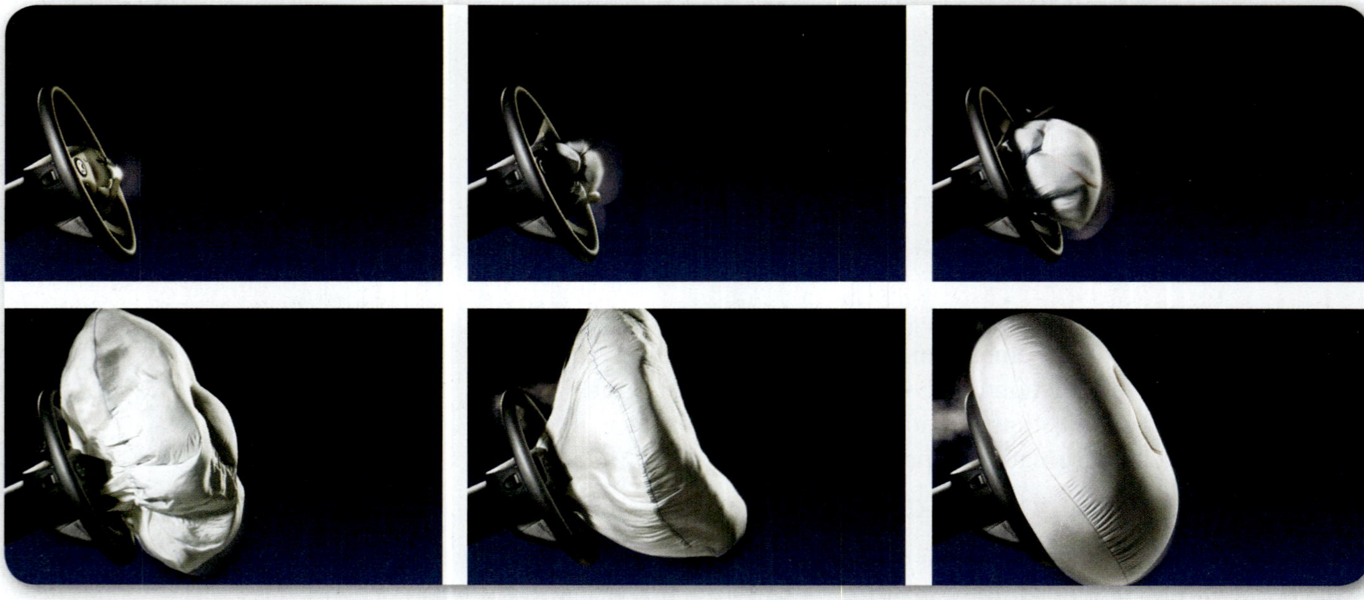

■ **Figure 11** The decomposition of sodium azide, which produces a gas, is the chemical reaction that inflates air bags.

## Decomposition Reactions

Some chemical reactions are essentially the opposite of synthesis reactions. These reactions are classified as decomposition reactions. A **decomposition reaction** is one in which a single compound breaks down into two or more elements or new compounds. In generic terms, decomposition reactions can be represented as follows.

$$AB \rightarrow A + B$$

**FOLDABLES®** Incorporate information from this section into your Foldable.

Decomposition reactions often require an energy source, such as heat, light, or electricity, to occur. For example, ammonium nitrate breaks down into dinitrogen monoxide and water when the reactant is heated to a high temperature.

$$NH_4NO_3(s) \rightarrow N_2O(g) + 2H_2O(g)$$

Notice that this decomposition reaction involves one reactant breaking down into more than one product.

The outcome of another decomposition reaction is shown in **Figure 11.** Automobile safety air bags inflate rapidly as sodium azide pellets decompose. A device that can provide an electric signal to start the reaction is packaged inside air bags along with the sodium azide pellets. When the device is activated, sodium azide decomposes, producing nitrogen gas that quickly inflates the air bag.

$$2NaN_3(s) \rightarrow 2Na(s) + 3N_2(g)$$

### PRACTICE Problems

Do additional problems. Online Practice

**Write chemical equations for the following decomposition reactions.**

**18.** Aluminum oxide(s) decomposes when electricity passes through it.

**19.** Nickel(II) hydroxide(s) decomposes to produce nickel(II) oxide(s) and water.

**20. Challenge** Heating sodium hydrogen carbonate(s) produces sodium carbonate(aq) and water. Carbon dioxide gas is also produced.

Courtesy of Mercedes-Benz Canada

**Figure 12** In a single-replacement reaction, the atoms of one element replace the atoms of another element in a compound.

## Replacement Reactions

In contrast to synthesis, combustion, and decomposition reactions, many chemical reactions are replacement reactions and involve the replacement of an element in a compound. These reactions are also called displacement reactions. There are two types of replacement reactions: single-replacement reactions and double-replacement reactions.

**Single-replacement reactions** The reaction between lithium and water is shown in **Figure 12.** The following chemical equation shows that a lithium atom replaces one of the hydrogen atoms in a water molecule.

$$2Li(s) + 2H_2O(l) \rightarrow 2LiOH(aq) + H_2(g)$$

A reaction in which the atoms of one element replace the atoms of another element in a compound is called a **single-replacement reaction.**

$$A + BX \rightarrow AX + B$$

**Metal replaces hydrogen or another metal** The reaction between lithium and water is one type of single-replacement reaction, in which a metal replaces a hydrogen atom in a water molecule. Another type of single-replacement reaction occurs when one metal replaces another metal in a compound dissolved in water. **Figure 12** shows a single-replacement reaction occurring when a bar of pure copper is placed in aqueous silver nitrate. The crystals that are accumulating on the copper bar are the silver atoms that the copper atoms replaced.

$$Cu(s) + 2AgNO_3(aq) \rightarrow 2Ag(s) + Cu(NO_3)_2(aq)$$

A metal will not always replace another metal in a compound dissolved in water because metals differ in their reactivities. Reactivity is the ability to react with another substance. An activity series of some metals is shown in **Figure 13.** This series orders metals by reactivity with other metals. Single-replacement reactions are used to determine a metal's position on the list. The most active metals are at the top of the list. The least active metals are at the bottom. Similarly, the reactivity of each halogen has been determined and listed, as shown in **Figure 13.**

**Figure 13** An activity series, similar to the series shown here for various metals and halogens, is a useful tool for determining whether a chemical reaction will occur and for determining the result of a single-replacement reaction.

Most active ↓ Least active

**METALS**
Lithium
Rubidium
Potassium
Calcium
Sodium
Magnesium
Aluminum
Manganese
Zinc
Iron
Nickel
Tin
Lead
Copper
Silver
Platinum
Gold

Most active ↓ Least active

**HALOGENS**
Fluorine
Chlorine
Bromine
Iodine

You can use the activity series to predict whether or not certain reactions will occur. A specific metal can replace any metal listed below it that is in a compound. It cannot replace any metal listed above it. For example, copper atoms replace silver atoms in a solution of silver nitrate. However, if you place a silver wire in aqueous copper(II) nitrate, the silver atoms will not replace the copper. Silver is listed below copper in the activity series, so no reaction occurs. The letters NR (no reaction) are commonly used to indicate that a reaction will not occur.

$$Ag(s) + Cu(NO_3)_2(aq) \rightarrow NR$$

**Nonmetal replaces nonmetal** A third type of single-replacement reaction involves the replacement of a nonmetal in a compound by another nonmetal. Halogens are frequently involved in these types of reactions. Like metals, halogens exhibit different activity levels in single-replacement reactions. The reactivities of halogens, determined by single-replacement reactions, are also shown in **Figure 13.** The most active halogen is fluorine, and the least active is iodine. A more reactive halogen replaces a less reactive halogen that is part of a compound dissolved in water. For example, fluorine replaces bromine in water containing dissolved sodium bromide. However, bromine does not replace fluorine in water containing dissolved sodium fluoride.

$$F_2(g) + 2NaBr(aq) \rightarrow 2NaF(aq) + Br_2(l)$$
$$Br_2(g) + 2NaF(aq) \rightarrow NR$$

☑ READING CHECK **Explain** how a single-replacement reaction works.

# Problem-Solving LAB

## Analyze Trends

**How can you explain the reactivities of halogens?** The location of all the halogens in group 17 in the periodic table tells you that halogens have common characteristics. Indeed, halogens are all nonmetals and have seven electrons in their outermost orbitals. However, each halogen also has its own characteristics, such as the ability to react with other substances.

### Analysis

Examine the accompanying data table. It includes data about the atomic radii, ionization energies, and electronegativities of the halogens.

**Properties of Halogens**

| Halogen | Atomic Radius (pm) | Ionization Energy (kJ/mol) | Electronegativity |
|---|---|---|---|
| Fluorine | 72 | 1681 | 3.98 |
| Chlorine | 100 | 1251 | 3.16 |
| Bromine | 114 | 1140 | 2.96 |
| Iodine | 133 | 1008 | 2.66 |
| Astatine | 140 | 920 | 2.2 |

### Think Critically

1. **Make graphs** Use the information in the data table to make three line graphs.
2. **Describe** any periodic trends that you identify in the data.
3. **Relate** any periodic trends that you identify among the halogens to the activity series of halogens shown in **Figure 13.**
4. **Predict** the location of the element astatine in the activity series of halogens. Explain.

## EXAMPLE Problem 2

**SINGLE-REPLACEMENT REACTIONS** Predict the products that will result when these reactants combine, and write a balanced chemical equation for each reaction.

**a.** $Fe(s) + CuSO_4(aq) \rightarrow$

**b.** $Br_2(l) + MgCl_2(aq) \rightarrow$

**c.** $Mg(s) + AlCl_3(aq) \rightarrow$

### 1 ANALYZE THE PROBLEM

You are given three sets of reactants. Using **Figure 13,** you must first determine if each reaction occurs. Then, if a reaction is predicted, you can determine the product(s) of the reaction. With this information you can write a skeleton equation for the reaction. Finally, you can use the steps for balancing chemical equations to write the complete balanced chemical equation.

### 2 SOLVE FOR THE UNKNOWN

**a.** Iron is listed above copper in the activity series. Therefore, the first reaction will occur because iron is more reactive than copper. In this case, iron will replace copper. The skeleton equation for this reaction is

$$Fe(s) + CuSO_4(aq) \rightarrow FeSO_4(aq) + Cu(s)$$

This equation is balanced.

**b.** In the second reaction, chlorine is more reactive than bromine because bromine is listed below chlorine in the activity series. Therefore, the reaction will not occur. The skeleton equation for this situation is

$$Br(l) + MgCl_2(aq) \rightarrow NR$$

No balancing is required.

**c.** Magnesium is listed above aluminum in the activity series. Therefore, the third reaction will occur because magnesium is more reactive than aluminum. In this case, magnesium will replace aluminum. The skeleton equation for this reaction is

$$Mg(s) + AlCl_3(aq) \rightarrow Al(s) + MgCl_2(aq)$$

This equation is not balanced. The balanced equation is

$$3Mg(s) + 2AlCl_3(aq) \rightarrow 2Al(s) + 3MgCl_2(aq)$$

### 3 EVALUATE THE ANSWER

The activity series shown in **Figure 13** supports the reaction predictions. The chemical equations are balanced because the number of atoms of each substance is equal on both sides of the equation.

## PRACTICE Problems

Do additional problems. Online Practice

**Predict whether the following single-replacement reactions will occur. If a reaction occurs, write a balanced equation for the reaction.**

**21.** $K(s) + ZnCl_2(aq) \rightarrow$

**22.** $Cl_2(g) + HF(aq) \rightarrow$

**23.** $Fe(s) + Na_3PO_4(aq) \rightarrow$

**24. Challenge** $Al(s) + Pb(NO_3)_2(aq) \rightarrow$

## RealWorld CHEMISTRY

### Single-Replacement Reactions

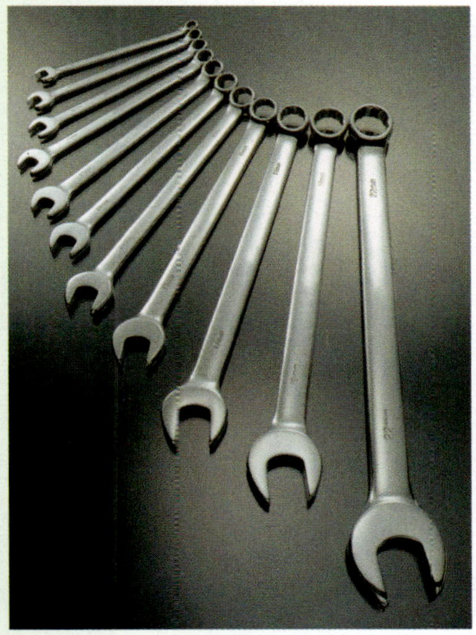

**ZINC PLATING** Tools made of steel are often covered with a layer of zinc to prevent corrosion. Zinc is more reactive than the lead in steel. During zinc plating, the zinc replaces some of the surface lead, coating the steel.

**Figure 14** The color-coding in the generic equation for a double-replacement reaction and in the equation for the reaction between calcium hydroxide and hydrochloric acid shows the anions changing places.

$$AX + BY \rightarrow AY + BX$$

$$Ca(OH)_2(aq) + 2HCl(aq) \rightarrow CaCl_2(aq) + 2H_2O(l)$$

**Double-replacement reactions** The final type of replacement reaction, which involves an exchange of ions between two compounds, is called a **double-replacement reaction.**

In the generic equation in **Figure 14,** A and B represent positively charged ions (cations), and X and Y represent negatively charged ions (anions). Notice that the anions have switched places and are now bonded to the other cations in the reaction. In other words, X replaces Y and Y replaces X—a double replacement. More simply, the positive and negative ions of two compounds switch places.

The reaction between calcium hydroxide and hydrochloric acid is a double-replacement reaction.

$$Ca(OH)_2(aq) + 2HCl(aq) \rightarrow CaCl_2(aq) + 2H_2O(l)$$

The ionic components of the reaction are $Ca^{2+}$, $OH^-$, $H^+$, and $Cl^-$. Knowing this, you can now see the two replacements of the reaction. The anions ($OH^-$ and $Cl^-$) have changed places and are now bonded to the other cations ($Ca^{2+}$ and $H^+$), as shown in **Figure 14.**

The reaction between sodium hydroxide and copper(II) chloride in solution is also a double-replacement reaction.

$$2NaOH(aq) + CuCl_2(aq) \rightarrow 2NaCl(aq) + Cu(OH)_2(s)$$

In this case, the anions ($OH^-$ and $Cl^-$) changed places and bonded to the other cations ($Na^+$ and $Cu^{2+}$). **Figure 15** shows that the result of this reaction is a solid product, copper(II) hydroxide. A solid produced during a chemical reaction in a solution is called a **precipitate.**

**Figure 15** When aqueous sodium hydroxide is added to a solution of copper(II) chloride, the anions ($OH^-$ and $Cl^-$) change places. The resulting products are sodium chloride, which remains in solution, and copper(II) hydroxide, the blue solid in the test tube.

View an **animation about precipitate formation.**

**Concepts In Motion**

## Table 3 Guidelines for Writing Double-Replacement Reactions

| Step | Example |
|---|---|
| **1.** Write the components of the reactants in a skeleton equation. | $Al(NO_3)_3 + H_2SO_4$ |
| **2.** Identify the cations and the anions in each compound. | $Al(NO_3)_3$ has $Al^{3+}$ and $NO_3^-$<br>$H_2SO_4$ has $H^+$ and $SO_4^{2-}$ |
| **3.** Pair up each cation with the anion from the other compound. | $Al^{3+}$ pairs with $SO_4^{2-}$<br>$H^+$ pairs with $NO_3^-$ |
| **4.** Write the formulas for the products using the pairs from Step 3. | $Al_2(SO_4)_3$<br>$HNO_3$ |
| **5.** Write the complete equation for the double-replacement reaction. | $Al(NO_3)_3 + H_2SO_4 \rightarrow Al_2(SO_4)_3 + HNO_3$ |
| **6.** Balance the equation. | $2Al(NO_3)_3 + 3H_2SO_4 \rightarrow Al_2(SO_4)_3 + 6HNO_3$ |

**Products of double-replacement reactions** One of the key characteristics of double-replacement reactions is the type of product that is formed when the reaction takes place. All double-replacement reactions produce either water, a precipitate, or a gas. Refer back to the two double-replacement reactions previously discussed in this section. The reaction between calcium hydroxide and hydrochloric acid produces water. A precipitate is produced in the reaction between sodium hydroxide and copper(II) chloride. An example of a double-replacement reaction that forms a gas is that of potassium cyanide and hydrobromic acid.

$$KCN(aq) + HBr(aq) \rightarrow KBr(aq) + HCN(g)$$

It is important to be able to evaluate the chemistry of double-replacement reactions and predict the products of these reactions. The basic steps to write double-replacement reactions are given in **Table 3.**

☑ READING CHECK **Describe** what happens to the anions in a double-replacement reaction.

## PRACTICE Problems

Do additional problems. Online Practice

**Write the balanced chemical equations for the following double-replacement reactions.**

**25.** The two substances at right react to produce solid silver iodide and aqueous lithium nitrate.

**26.** Aqueous barium chloride and aqueous potassium carbonate react to produce solid barium carbonate and aqueous potassium chloride.

**27.** Aqueous sodium oxalate and aqueous lead(II) nitrate react to produce solid lead(II) oxalate and aqueous sodium nitrate.

**28. Challenge** Acetic acid ($CH_3COOH$) and potassium hydroxide react to produce potassium acetate and water.

LiI(ac) $AgNO_3(aq)$

PRACTICE PROBLEMS

Explore **types of chemical reactions with an interactive table.** Concepts In Motion

**Table 4** **Predicting Products of Chemical Reactions**

| Type of Reaction | Reactants | Probable Products | Generic Equation |
|---|---|---|---|
| Synthesis | • two or more substances | • one compound | $A + B \rightarrow AB$ |
| Combustion | • a metal and oxygen<br>• a nonmetal and oxygen<br>• a compound and oxygen | • the oxide of the metal<br>• the oxide of the nonmetal<br>• two or more oxides | $A + O_2 \rightarrow AO$ |
| Decomposition | • one compound | • two or more elements and/or compounds | $AB \rightarrow A + B$ |
| Single-replacement | • a metal and a compound<br>• a nonmetal and a compound | • a new compound and the replaced metal<br>• a new compound and the replaced non-metal | $A + BX \rightarrow AX + B$ |
| Double-replacement | • two compounds | • two different compounds, one of which is a solid, water, or a gas | $AX + BY \rightarrow AY + BX$ |

**Alabama Applying Practices**

Go to the Resource tab in ConnectED to find the Applying Practices worksheet *Investigate Chemical Reactions.*

**Table 4** summarizes the various types of chemical reactions. Use the table to help you organize the reactions, so that you can identify each and predict its products. For example, how would you determine what type of reaction occurs when solid calcium oxide and carbon dioxide gas react to produce solid calcium carbonate? First, write the chemical equation.

$$CaO(s) + CO_2(g) \rightarrow CaCO_3(s)$$

Second, determine what is happening in the reaction. In this case, two substances are reacting to form one compound. Third, use the table to identify the type of reaction. This particular reaction is a synthesis reaction. Fourth, check your answer by comparing the chemical equation to the generic equation provided in the table for that type of reaction.

$$\begin{array}{ccccc} CaO(s) & + & CO_2(g) & \rightarrow & CaCO_3(s) \\ A & + & B & \rightarrow & AB \end{array}$$

Synthesis Reaction

## SECTION 2 REVIEW

Section Self-Check

### Section Summary

- Classifying chemical reactions makes them easier to understand, remember, and recognize.
- Activity series of metals and halogens can be used to predict if single-replacement reactions will occur.

29. MAINIDEA **Describe** the four types of chemical reactions and their characteristics.
30. **Explain** how an activity series of metals is organized.
31. **Compare and contrast** single-replacement reactions and double-replacement reactions.
32. **Describe** the result of a double-replacement reaction.
33. **Classify** What type of reaction is most likely to occur when barium reacts with fluorine? Write the chemical equation for the reaction.
34. **Interpret Data** Could the following reaction occur? Explain your answer.

$$3Ni + 2AuBr_3 \rightarrow 3NiBr_2 + 2Au$$

SECTION 3

# Reactions in Aqueous Solutions

**MAINIDEA** Double-replacement reactions occur between substances in aqueous solutions and produce precipitates, water, or gases.

### Essential Questions

- What are aqueous solutions?
- How are complete ionic and net ionic equations written for chemical reactions in aqueous solutions?
- How can you predict whether reactions in aqueous solutions will produce a precipitate, water, or a gas?

### Review Vocabulary

**solution:** a uniform mixture that might contain solids, liquids, or gases

### New Vocabulary

aqueous solution
solute
solvent
complete ionic equation
spectator ion
net ionic equation

**CHEM 4 YOU** One way to make lemonade involves using a powdered drink mix and water. When the powdered drink mix is added to the water, the flavor crystals dissolve in the water, forming a solution. The solution is lemonade-flavored drink.

## Aqueous Solutions

You read previously that a solution is a homogeneous mixture. Many of the reactions discussed in the previous section involve substances dissolved in water. When a substance dissolves in water, a solution forms. An **aqueous solution** contains one or more substances called **solutes** dissolved in the water. In this case, water is the **solvent**—the most plentiful substance in the solution.

**Molecular compounds in solution** Although water is always the solvent in aqueous solutions, there are many possible solutes. Some solutes, such as sucrose (table sugar) and ethanol (grain alcohol), are molecular compounds that exist as molecules in aqueous solutions. Other solutes are molecular compounds that form ions when they dissolve in water. For example, the molecular compound hydrogen chloride forms hydrogen ions and chloride ions when it dissolves in water, as shown in **Figure 16.** An equation can be used to show this ionization process.

$$HCl(aq) \rightarrow H^+(aq) + Cl^-(aq)$$

Compounds such as hydrogen chloride that produce hydrogen ions in aqueous solution are acids. In fact, an aqueous solution of hydrogen chloride is often referred to as hydrochloric acid.

■ **Figure 16** In water, hydrogen chloride (HCl) breaks apart into hydrogen ions ($H^+$) and chloride ions ($Cl^-$).

**VOCABULARY**
**SCIENCE USAGE V. COMMON USAGE**

**Compound**

***Science usage:*** a chemical combination of two or more different elements

*Salt is a compound comprised of the elements sodium and chlorine.*

***Common usage:*** a word that consists of two or more words

*Two compound words are* basketball *and* textbook.

**Ionic compounds in solution** In addition to molecular compounds, ionic compounds might be solutes in aqueous solutions. Recall that ionic compounds consist of positive ions and negative ions held together by ionic bonds. When ionic compounds dissolve in water, their ions can separate—a process called dissociation. For example, an aqueous solution of the ionic compound sodium chloride contains $Na^+$ and $Cl^-$ ions.

## Types of Reactions in Aqueous Solutions

When two aqueous solutions that contain ions as solutes are combined, the ions might react with one another. These reactions are always double-replacement reactions. The solvent molecules, which are all water molecules, do not usually react. Three types of products can form from the double-replacement reaction: a precipitate, water, or a gas.

**Reactions that form precipitates** Some reactions that occur in aqueous solutions produce precipitates. For example, recall from Section 2 that when aqueous solutions of sodium hydroxide and copper(II) chloride are mixed, a double-replacement reaction occurs in which the precipitate copper(II) hydroxide forms.

$$2NaOH(aq) + CuCl_2(aq) \rightarrow 2NaCl(aq) + Cu(OH)_2(s)$$

Note that the chemical equation does not show some details of this reaction. Sodium hydroxide and copper(II) chloride are ionic compounds. Therefore, in aqueous solutions they exist as $Na^+$, $OH^-$, $Cu^{2+}$, and $Cl^-$ ions, as shown in **Figure 17.** When their solutions are combined, $Cu^{2+}$ ions in one solution and $OH^-$ ions in the other solution react to form the precipitate copper(II) hydroxide, $Cu(OH)_2(s)$. The $Na^+$ and $Cl^-$ ions remain dissolved in the new solution.

■ **Figure 17** Like the aqueous solution of HCl in **Figure 16,** sodium hydroxide (NaOH) in an aqueous solution dissociates into sodium ($Na^+$) and hydroxide ($OH^-$) ions. Copper(II) chloride ($CuCl_2$) also dissociates into $Cu^{2+}$ and $Cl^-$ ions.

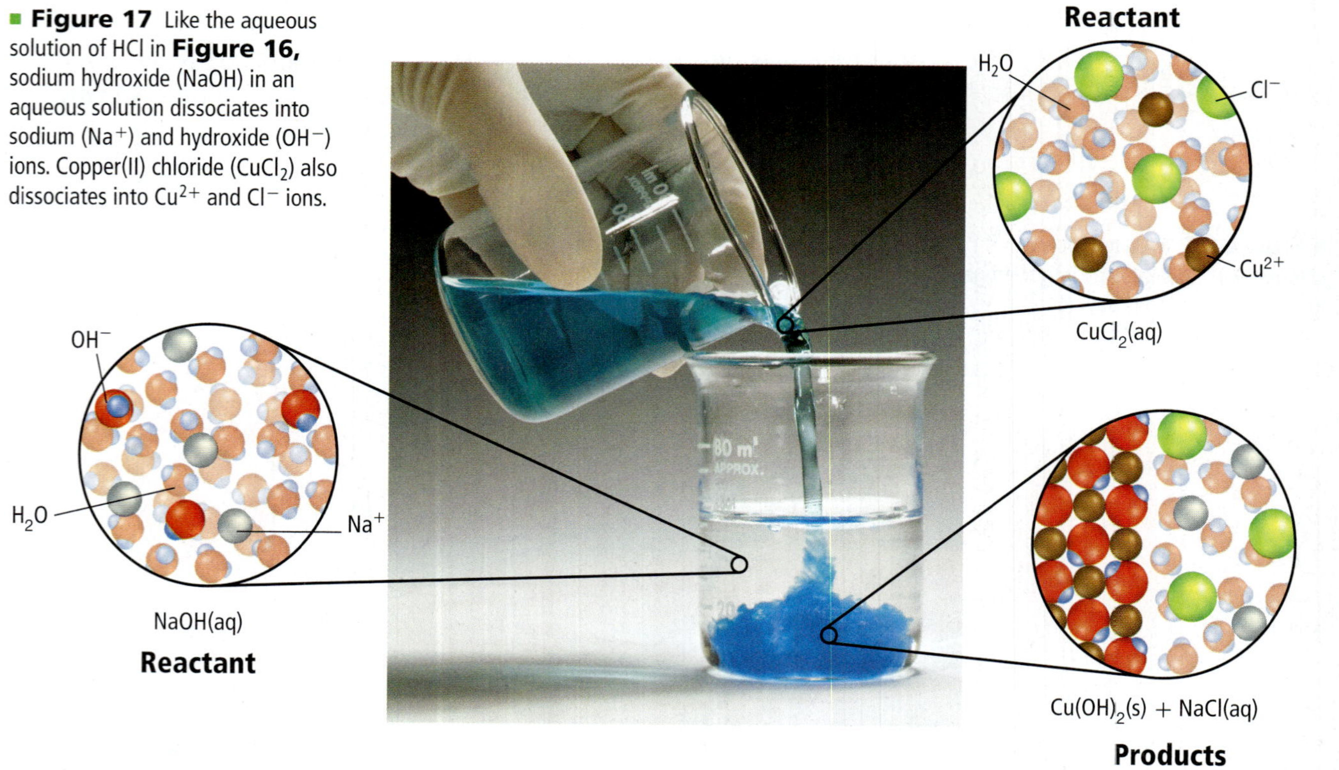

# MiniLAB

iLab Station

## Observe a Precipitate-Forming Reaction

**How do two liquids form a solid?**

### Procedure

1. Read and complete the lab safety form.
2. Place 50 mL **distilled water** in a **150-mL beaker.**
3. Measure about 4 g **NaOH pellets** on a **balance.** Add the NaOH pellets to the beaker one at a time. Mix with a **stirring rod** until each NaOH pellet dissolves before adding the next pellet.
4. Measure about 6 g **Epsom salts ($MgSO_4$)** and place it in another 150-mL beaker. Add 50 mL distilled water to the Epsom salts. Mix with another stirring rod until the Epsom salts dissolve.
5. Slowly pour the Epsom salts solution into the NaOH solution. Record your observations.
6. Stir the new solution. Record your observations.
7. Allow the precipitate to settle, then decant the liquid from the solid into a **100-mL graduated cylinder.**
8. Dispose of the solid as instructed by your teacher.

### Analysis

1. **Write** a balanced chemical equation for the reaction between the NaOH and $MgSO_4$. Note that most sulfate compounds exist as ions in aqueous solutions.
2. **Write** the complete ionic equation for this reaction.
3. **Determine** which ions are spectator ions, then write the net ionic equation for this reaction.

**Ionic equations** To show the details of reactions that involve ions in aqueous solutions, chemists use ionic equations. Ionic equations differ from chemical equations in that substances that are ions in solution are written as ions in the equation. Look again at the reaction between aqueous solutions of sodium hydroxide and copper(II) chloride. To write the ionic equation for this reaction, you must show the reactants, NaOH(aq) and $CuCl_2$(aq), and the product, NaCl(aq), as ions.

$$2Na^+(aq) + 2OH^-(aq) + Cu^{2+}(aq) + 2Cl^-(aq) \rightarrow 2Na^+(aq) + 2Cl^-(aq) + Cu(OH)_2(s)$$

An ionic equation that shows all of the particles in a solution as they exist is called a **complete ionic equation.** Note that the sodium ions and the chloride ions are both reactants and products. Because they are both reactants and products, they do not participate in the reaction. Ions that do not participate in a reaction are called **spectator ions** and are not usually shown in ionic equations. **Net ionic equations** are ionic equations that include only the particles that participate in the reaction. Net ionic equations are written from complete ionic equations by removing all spectator ions. For example, a net ionic equation is what remains after the sodium and chloride ions are crossed out of this complete ionic equation.

$$\cancel{2Na^+(aq)} + 2OH^-(aq) + Cu^{2+}(aq) + \cancel{2Cl^-(aq)} \rightarrow \cancel{2Na^+(aq)} + \cancel{2Cl^-(aq)} + Cu(OH)_2(s)$$

Only the hydroxide and copper ions are left in the net ionic equation shown below.

$$2OH^-(aq) + Cu^{2+}(aq) \rightarrow Cu(OH)_2(s)$$

**READING CHECK Compare** How are ionic equations different from chemical equations?

## EXAMPLE Problem 3

**REACTIONS THAT FORM A PRECIPITATE** Write the chemical, complete ionic, and net ionic equations for the reaction between aqueous solutions of barium nitrate and sodium carbonate that forms the precipitate barium carbonate.

### 1 ANALYZE THE PROBLEM

You are given the word equation for the reaction between barium nitrate and sodium carbonate. You must determine the chemical formulas and relative amounts of all reactants and products to write the balanced chemical equation. To write the complete ionic equation, you need to show the ionic states of the reactants and products. By crossing out the spectator ions from the complete ionic equation, you can write the net ionic equation. The net ionic equation will include fewer substances than the other equations.

### 2 SOLVE FOR THE UNKNOWN

Write the correct chemical formulas and physical states for all substances involved in the reaction.

$$Ba(NO_3)_2(aq) + Na_2CO_3(aq) \rightarrow BaCO_3(s) + NaNO_3(aq)$$

$$Ba(NO_3)_2(aq) + Na_2CO_3(aq) \rightarrow BaCO_3(s) + 2NaNO_3(aq)$$

**Balance the skeleton equation.**

$$Ba^{2+}(aq) + 2NO_3^-(aq) + 2Na^+(aq) + CO_3^{2-}(aq) \rightarrow BaCO_3(s) + 2Na^+(aq) + 2NO_3^-(aq)$$

**Show the ions of the reactants and the products.**

$$Ba^{2+}(aq) + \cancel{2NO_3^-(aq)} + \cancel{2Na^+(aq)} + CO_3^{2-}(aq) \rightarrow BaCO_3(s) + \cancel{2Na^+(aq)} + \cancel{2NO_3^-(aq)}$$

**Cross out the spectator ions from the complete ionic equation.**

$$Ba^{2+}(aq) + CO_3^{2-}(aq) \rightarrow BaCO_3(s)$$

**Write the net ionic equation.**

### 3 EVALUATE THE ANSWER

The net ionic equation includes fewer substances than the other equations because it shows only the reacting particles. The particles composing the solid precipitate that is the result of the reaction are no longer ions.

## PRACTICE Problems

Do additional problems. Online Practice

**Write chemical, complete ionic, and net ionic equations for each of the following reactions that might produce a precipitate. Use *NR* to indicate that no reaction occurs.**

**35.** Aqueous solutions of potassium iodide and silver nitrate are mixed, forming the precipitate silver iodide.

**36.** Aqueous solutions of ammonium phosphate and sodium sulfate are mixed. No precipitate forms and no gas is produced.

**37.** Aqueous solutions of aluminum chloride and sodium hydroxide are mixed, forming the precipitate aluminum hydroxide.

**38.** Aqueous solutions of lithium sulfate and calcium nitrate are mixed, forming the precipitate calcium sulfate.

**39. Challenge** When aqueous solutions of sodium carbonate and manganese(V) chloride are mixed, a precipitate forms. The precipitate is a compound containing manganese.

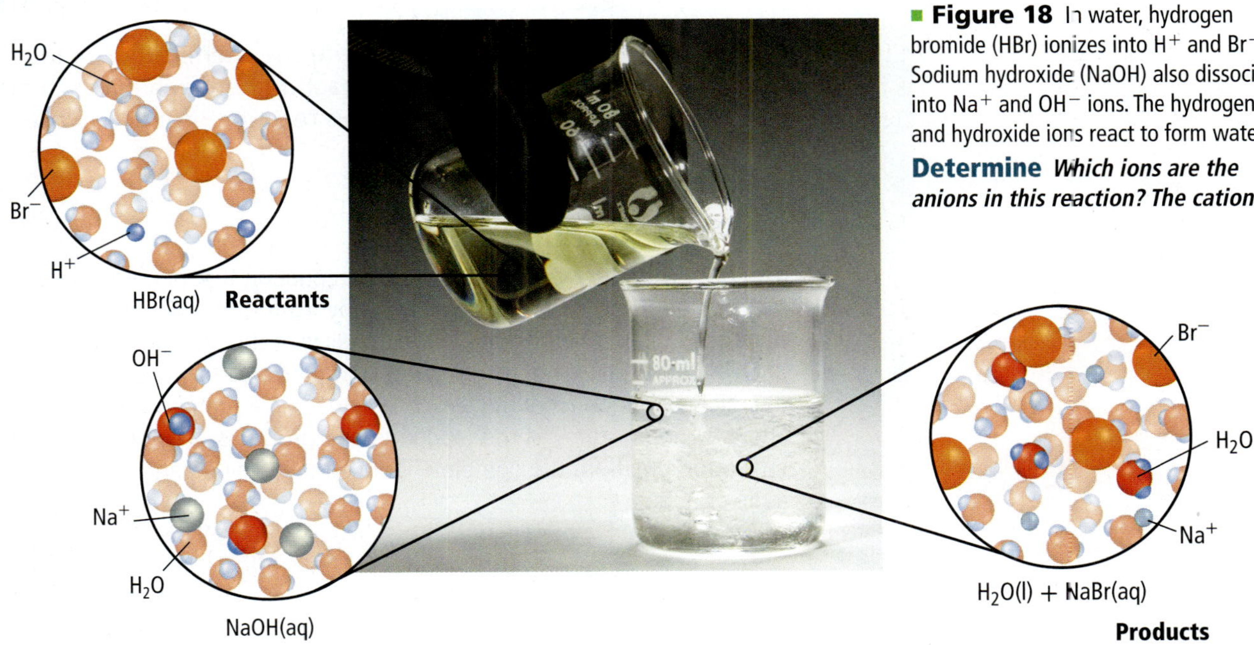


■ **Figure 18** In water, hydrogen bromide (HBr) ionizes into $H^+$ and $Br^-$ ions. Sodium hydroxide (NaOH) also dissociates into $Na^+$ and $OH^-$ ions. The hydrogen ions and hydroxide ions react to form water.
**Determine** ***Which ions are the anions in this reaction? The cations?***

**Reactions that form water** Another type of double-replacement reaction that occurs in an aqueous solution produces water molecules. The water molecules produced in the reaction increase the number of solvent particles. Unlike reactions in which a precipitate forms, no evidence of a chemical reaction is observable because water is colorless, odorless, and already makes up most of the solution. For example, when you mix hydrobromic acid (HBr) with a sodium hydroxide solution (NaOH), as shown in **Figure 18,** a double-replacement reaction occurs and water is formed. The chemical equation for this reaction is shown below.

$$HBr(aq) + NaOH(aq) \rightarrow H_2O(l) + NaBr(aq)$$

In this case, the reactants and the product sodium bromide exist as ions in an aqueous solution. The complete ionic equation for this reaction shows these ions.

$$H^+(aq) + Br^-(aq) + Na^+(aq) + OH^-(aq) \rightarrow H_2O(l) + Na^+(aq) + Br^-(aq)$$

Look carefully at the complete ionic equation. The reacting solute ions are the hydrogen ions and hydroxide ions because the sodium ions and bromine ions are both spectator ions. If you cross out the spectator ions, you are left with the ions that take part in the reaction.

$$H^+(aq) + \cancel{Br^-(aq)} + \cancel{Na^+(aq)} + OH^-(aq) \rightarrow H_2O(l) + \cancel{Na^+(aq)} + \cancel{Br^-(aq)}$$

This equation is the net ionic equation for the reaction.

$$H^+(aq) + OH^-(aq) \rightarrow H_2O(l)$$

☑ READING CHECK **Analyze** In the reaction between hydrobromic acid and sodium hydroxide, why are the sodium ions and bromine ions called spectator ions?

## EXAMPLE Problem 4

**REACTIONS THAT FORM WATER** Write the chemical, complete ionic, and net ionic equations for the reaction between hydrochloric acid and aqueous lithium hydroxide. This reaction produces water and aqueous lithium chloride.

### 1 ANALYZE THE PROBLEM

You are given the word equation for the reaction that occurs between hydrochloric acid and aqueous lithium hydroxide to produce water and aqueous lithium chloride. You must determine the chemical formulas for and relative amounts of all reactants and products to write the balanced chemical equation. To write the complete ionic equation, you need to show the ionic states of the reactants and products. By crossing out the spectator ions from the complete ionic equation, you can write the net ionic equation.

### 2 SOLVE FOR THE UNKNOWN

Write the skeleton equation for the reaction and balance it.

$$HCl(aq) + LiOH(aq) \rightarrow H_2O(l) + LiCl(aq)$$

$$H^+(aq) + Cl^-(aq) + Li^+(aq) + OH^-(aq) \rightarrow H_2O(l) + Li^+(aq) + Cl^-(aq)$$

**Show the ions of the reactants and the products.**

$$H^+(aq) + \cancel{Cl^-(aq)} + \cancel{Li^+(aq)} + OH^-(aq) \rightarrow H_2O(l) + \cancel{Li^+(aq)} + \cancel{Cl^-(aq)}$$

**Cross out the spectator ions from the complete ionic equation.**

$$H^+(aq) + OH^-(aq) \rightarrow H_2O(l)$$

**Write the net ionic equation.**

### 3 EVALUATE THE ANSWER

The net ionic equation includes fewer substances than the other equations because it shows only those particles involved in the reaction that produces water. The particles that compose the product water are no longer ions.

## PRACTICE Problems

**Do additional problems.** Online Practice

**Write chemical, complete ionic, and net ionic equations for the reactions between the following substances, which produce water.**

**40.** Mixing sulfuric acid ($H_2SO_4$) and aqueous potassium hydroxide produces water and aqueous potassium sulfate.

**41.** Mixing hydrochloric acid (HCl) and aqueous calcium hydroxide produces water and aqueous calcium chloride.

**42.** Mixing nitric acid ($HNO_3$) and aqueous ammonium hydroxide produces water and aqueous ammonium nitrate.

**43.** Mixing hydrosulfuric acid ($H_2S$) and aqueous calcium hydroxide produces water and aqueous calcium sulfide.

**44.** **Challenge** When benzoic acid ($C_6H_5COOH$) and magnesium hydroxide are mixed, water and magnesium benzoate are produced.

**Reactions that form gases** A third type of double-replacement reaction that occurs in aqueous solutions results in the formation of a gas. Some gases commonly produced in these reactions are carbon dioxide, hydrogen cyanide, and hydrogen sulfide.

A gas-producing reaction occurs when you mix hydroiodic acid (HI) with an aqueous solution of lithium sulfide. Bubbles of hydrogen sulfide gas form in the container during the reaction. Lithium iodide is also produced in this reaction and remains dissolved in the solution.

$$2HI(aq) + Li_2S(aq) \rightarrow H_2S(g) + 2LiI(aq)$$

The reactants hydroiodic acid and lithium sulfide exist as ions in aqueous solution. Therefore, you can write an ionic equation for this reaction. The complete ionic equation includes all of the substances in the solution.

$$2H^+(aq) + 2I^-(aq) + 2Li^+(aq) + S^{2-}(aq) \rightarrow H_2S(g) + 2Li^+(aq) + 2I^-(aq)$$

Note that there are many spectator ions in the equation. When the spectator ions are crossed out, only the substances involved in the reaction remain in the equation.

$$2H^+(aq) + \cancel{2I^-(aq)} + \cancel{2Li^+(aq)} + S^{2-}(aq) \rightarrow H_2S(g) + \cancel{2Li^+(aq)} + \cancel{2I^-(aq)}$$

This is the net ionic equation.

$$2H^+(aq) + S^{2-}(aq) \rightarrow H_2S(g)$$

If you completed the Launch Lab at the beginning of this chapter, you observed another gas-producing reaction. In that reaction, carbon dioxide gas was produced and bubbled out of the solution. Another reaction that produces carbon dioxide gas occurs in your kitchen when you mix vinegar and baking soda. Vinegar is an aqueous solution of acetic acid and water. Baking soda essentially consists of sodium hydrogen carbonate. Rapid bubbling occurs when vinegar and baking soda are combined. The bubbles are carbon dioxide gas escaping from the solution. You can see this reaction occurring in **Figure 19.**

A reaction similar to the one between vinegar and baking soda occurs when you combine any acidic solution and sodium hydrogen carbonate. In all cases, two reactions must occur almost simultaneously in the solution to produce the carbon dioxide gas. One of these is a double-replacement reaction and the other is a decomposition reaction.

For example, when you dissolve sodium hydrogen carbonate in hydrochloric acid, a gas-producing double-replacement reaction occurs. The hydrogen in the hydrochloric acid and the sodium in the sodium hydrogen carbonate replace each other.

$$HCl(aq) + NaHCO_3(aq) \rightarrow H_2CO_3(aq) + NaCl(aq)$$

Sodium chloride is an ionic compound, and its ions remain separate in the aqueous solution. However, as the carbonic acid ($H_2CO_3$) forms, it decomposes immediately into water and carbon dioxide.

$$H_2CO_3(aq) \rightarrow H_2O(l) + CO_2(g)$$

**Figure 19** When vinegar and baking soda (sodium hydrogen carbonate, $NaHCO_3$) combine, the result is a vigorous bubbling that releases carbon dioxide ($CO_2$).

## EXAMPLE Problem 5

**REACTIONS THAT FORM GASES** Write the chemical, complete ionic, and net ionic equations for the reaction between hydrochloric acid and aqueous sodium sulfide, which produces hydrogen sulfide gas.

### 1 ANALYZE THE PROBLEM

You are given the word equation for the reaction between hydrochloric acid (HCl) and sodium sulfide ($Na_2S$). You must write the skeleton equation and balance it. To write the complete ionic equation, you need to show the ionic states of the reactants and products. By crossing out the spectator ions in the complete ionic equation, you can write the net ionic equation.

### 2 SOLVE FOR THE UNKNOWN

Write the correct skeleton equation for the reaction.

$HCl(aq) + Na_2S(aq) \rightarrow H_2S(g) + NaCl(aq)$

$2HCl(aq) + Na_2S(aq) \rightarrow H_2S(g) + 2NaCl(aq)$ — **Balance the skeleton equation.**

$2H^+(aq) + 2Cl^-(aq) + 2Na^+(aq) + S^{2-}(aq) \rightarrow H_2S(g) + 2Na^+(aq) + 2Cl^-(aq)$ — **Show the ions of the reactants and the products.**

$2H^+(aq) + \cancel{2Cl^-(aq)} + \cancel{2Na^+(aq)} + S^{2-}(aq) \rightarrow H_2S(g) + \cancel{2Na^+(aq)} + \cancel{2Cl^-(aq)}$ — **Cross out the spectator ions from the complete ionic equation.**

$2H^+(aq) + S^{2-}(aq) \rightarrow H_2S(g)$ — **Write the net ionic equation in its smallest whole-number ratio.**

### 3 EVALUATE THE ANSWER

The net ionic equation includes fewer substances than the other equations because it shows only those particles involved in the reaction that produce hydrogen sulfide. The particles that compose the product are no longer ions.

## PRACTICE Problems

Do additional problems. Online Practice

**Write chemical, complete ionic, and net ionic equations for these reactions.**

**45.** Perchloric acid ($HClO_4$) reacts with aqueous potassium carbonate, forming carbon dioxide gas and water.

**46.** Sulfuric acid ($H_2SO_4$) reacts with aqueous sodium cyanide, forming hydrogen cyanide gas and aqueous sodium sulfate.

**47.** Hydrobromic acid (HBr) reacts with aqueous ammonium carbonate, forming carbon dioxide gas and water.

**48.** Nitric acid ($HNO_3$) reacts with aqueous potassium rubidium sulfide, forming hydrogen sulfide gas.

**49. Challenge** Aqueous potassium iodide reacts with lead nitrate in solution, forming solid lead iodide.

**Double-replacement reaction**

$$AX + BY \longrightarrow AY + BX$$

$$HCl(aq) + NaHCO_3(aq) \longrightarrow H_2CO_3(aq) + NaCl(aq)$$

$$AB \longrightarrow A + B$$

$$H_2CO_3(aq) \longrightarrow H_2O(l) + CO_2(g)$$

**Decomposition reaction**

**Figure 20** When HCl is combined with $NaHCO_3$, a double-replacement reaction takes place, followed immediately by a decomposition reaction.

**Overall equations** Recall that when you combine an acidic solution, such as hydrochloric acid, and sodium hydrogen carbonate, two reactions occur—a double-replacement reaction and a decomposition reaction. These reactions are shown in **Figure 20.** The two reactions can be combined and represented by one chemical equation in a process similar to adding mathematical equations. An equation that combines two reactions is called an overall equation. To write an overall equation, the reactants in the two reactions are written on the reactant side of the combined equation, and the products of the two reactions are written on the product side. Then, any substances that are on both sides of the equation are crossed out.

Watch a **video about chemical reactions and equations.**

Video

**Reaction 1** $HCl(aq) + NaHCO_3(aq) \rightarrow H_2CO_3(aq) + NaCl(aq)$

**Reaction 2** $H_2CO_3(aq) \rightarrow H_2O(l) + CO_2(g)$

**Combined equation** $HCl(aq) + NaHCO_3(aq) + H_2\cancel{CO_3}(aq) \rightarrow H_2\cancel{CO_3}(aq) + NaCl(aq) + H_2O(l) + CO_2(g)$

**Overall equation** $HCl(aq) + NaHCO_3(aq) \rightarrow H_2O(l) + CO_2(g) + NaCl(aq)$

In this case, the reactants in the overall equation exist as ions in aqueous solutions. Therefore, a complete ionic equation can be written for the reaction.

$$H^+(aq) + Cl^-(aq) + Na^+(aq) + HCO_3^-(aq) \rightarrow H_2O(l) + CO_2(g) + Na^+(aq) + Cl^-(aq)$$

Note that the sodium and chloride ions are the spectator ions. When you cross them out, only the substances that take part in the reaction remain.

$$H^+(aq) + \cancel{Cl^-(aq)} + \cancel{Na^+(aq)} + HCO_3^-(aq) \rightarrow H_2O(l) + CO_2(g) + \cancel{Na^+(aq)} + \cancel{Cl^-(aq)}$$

The net ionic equation shows that both water and carbon dioxide gas are produced in this reaction.

$$H^+(aq) + HCO_3^-(aq) \rightarrow H_2O(l) + CO_2(g)$$

READING CHECK **Describe** What is an overall equation?

■ **Figure 21** After a bicarbonate ion ($HCO_3^-$) enters a red blood cell, it reacts with a hydrogen ion ($H^+$) to form water and carbon dioxide ($CO_2$). The $CO_2$ is exhaled from the lungs during respiration.

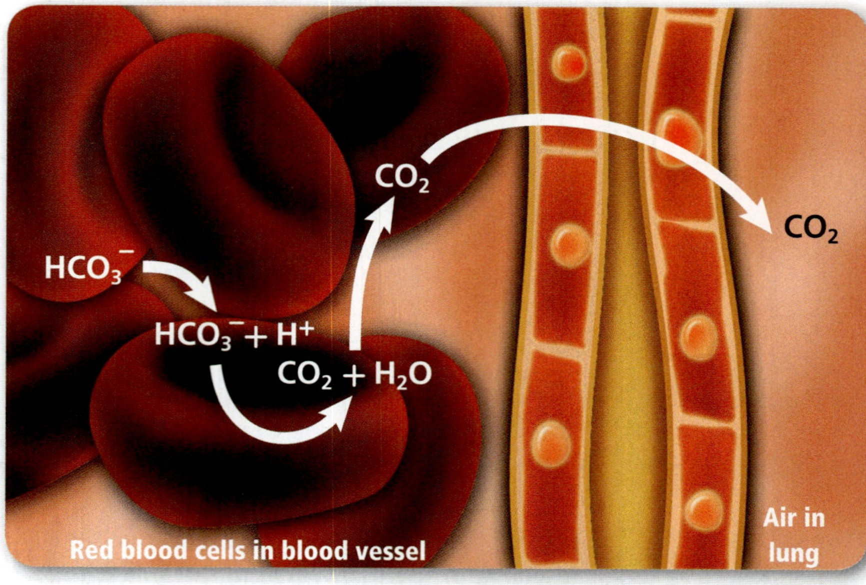


**Connection to Biology** The reaction between hydrogen ions and bicarbonate ions to produce water and carbon dioxide is an important one in your body. This reaction is occurring in the blood vessels of your lungs as you read these words. As shown in **Figure 21,** the carbon dioxide gas produced in your cells is transported in your blood in the form of bicarbonate ions ($HCO_3^-$). In the blood vessels of your lungs, the $HCO_3^-$ ions combine with $H^+$ ions to produce $CO_2$, which you exhale.

This reaction also occurs in products that are made with baking soda, which contains sodium bicarbonate. Sodium bicarbonate makes baked goods rise. It is used as an antacid and in deodorants to absorb moisture and odors. Baking soda can be added to toothpaste to whiten teeth and freshen breath. As a paste, sodium bicarbonate can be used in cleaning and scrubbing. It is also used as a fire-suppression agent in some fire extinguishers.

### CAREERS IN CHEMISTRY

**Biochemist** A biochemist is a scientist who studies the chemical processes of living organisms. A biochemist might study functions of the human body or research how food, drugs, and other substances affect living organisms.

WebQuest

## SECTION 3 REVIEW

Section Self-Check

### Section Summary

- In aqueous solutions, the solvent is always water. There are many possible solutes.
- Many molecular compounds form ions when they dissolve in water. When some ionic compounds dissolve in water, their ions separate.
- When two aqueous solutions that contain ions as solutes are combined, the ions might react with one another. The solvent molecules do not usually react.
- Reactions that occur in aqueous solutions are double-replacement reactions.

**50.** MAINIDEA **List** three common types of products produced by reactions that occur in aqueous solutions.

**51. Describe** solvents and solutes in an aqueous solution.

**52. Distinguish** between a complete ionic equation and a net ionic equation.

**53. Write** complete ionic and net ionic equations for the reaction between sulfuric acid ($H_2SO_4$) and calcium carbonate ($CaCO_3$).

$$H_2SO_4(aq) + CaCO_3(s) \rightarrow H_2O(l) + CO_2(g) + CaSO_4(aq)$$

**54. Analyze** Complete and balance the following equation.

$$CO_2(g) + HCl(aq) \rightarrow$$

**55. Predict** What type of product would the following reaction be most likely to produce? Explain your reasoning.

$$Ba(OH)_2(aq) + 2HCl(aq) \rightarrow$$

**56. Formulate Equations** A reaction occurs when nitric acid ($HNO_3$) is mixed with an aqueous solution of potassium hydrogen carbonate. Aqueous potassium nitrate is produced. Write the chemical and net ionic equations for the reaction.

# HOW IT works

## LIGHTING UP THE NIGHT: Bioluminescence

In the gathering darkness, a male firefly announces his presence by sending a signal in yellow-green light. A female near the ground answers his call, and he descends. The result might be a successful mating, or, if the female of another firefly species has fooled the male, he might be greedily devoured. The production of light by the firefly is the result of a chemical process called bioluminescence. This process is a strategy used by a wide variety of living things in many different environments. How does it work?

**1 Flashy Beetles** Fireflies (or lightning bugs) are not flies at all, but a group of beetles that flash their mating signals. They also use their light to lure their prey. The yellow-green light comes from cells in their lower abdomen. The wavelength for this light is between 510 and 670 nm.

**2 Bioluminescence** The glow of the firefly is the result of a chemical reaction. The reactants are oxygen and luciferin, a light-emitting substance found in some organisms. An enzyme, luciferase, speeds up the reaction. The products of this reaction are oxyluciferin and energy, in the form of light.

**3 Glowing Discoveries** Research into bioluminescence led to the discovery of green fluorescent protein (GFP), which is found in some species of jellyfish. GFP emits a green light when exposed to UV light. Researchers have inserted GFP into various organisms, such as mice, for research purposes. Examples of what scientists are using GFP to study include cancer, malaria, and cellular processes.

### WRITING IN ▶ Chemistry

**Research** Identify different life forms that use bioluminescence and create a pamphlet showing how bioluminescence is effective in each of these organisms.

WebQuest

# SMALL SCALE ChemLAB

iLab Station

## Develop an Activity Series

**Background:** **Some metals are more reactive than others. By comparing how different metals react with the known ions in aqueous solutions, an activity series for the tested materials can be developed. The activity series will reflect the relative reactivity of the tested metals.**

**Question:** *How is an activity series developed?*

### Materials

1.0*M* $Zn(NO_3)_2$
1.0*M* $Al(NO_3)_3$
1.0*M* $Cu(NO_3)_2$
1.0*M* $Mg(NO_3)_2$
pipettes (4)
wire cutters
Cu wire
Al wire
Mg ribbon
Zn metal strips (4)
Emery cloth or sandpaper
24-well microscale reaction plate

### Safety Precautions

### Procedure

1. Read and complete the lab safety form.
2. Create a table to record your data.
3. Use a pipette to fill the four wells in column 1 of the reaction plate with 2 mL of 1.0*M* $Al(NO_3)_3$ solution.
4. Repeat the procedure in Step 3 to fill the four wells in column 2 with 2 mL of 1.0*M* $Mg(NO_3)_2$.
5. Repeat the procedure in Step 3 to fill the four wells in column 3 with 2 mL of 1.0*M* $Zn(NO_3)_2$.
6. Repeat the procedure in Step 3 to fill the four wells in column 4 with 2 mL of 1.0*M* $Cu(NO_3)_2$.
7. With the emery cloth or sandpaper, polish 10 cm of aluminum wire until it is shiny. Use wire cutters to carefully cut the aluminum wire into four 2.5-cm pieces. Place a piece of the aluminum wire in each well of row A containing solution.
8. Repeat the procedure in Step 7 using 10 cm of magnesium ribbon. Place a piece of Mg ribbon in each well of row B containing solution.
9. Use the emery cloth or sandpaper to polish each small strip of zinc metal. Place a piece of Zn metal in each well of row C containing solution.
10. Observe what happens in each well. After 5 minutes, record your observations in the data table you made.

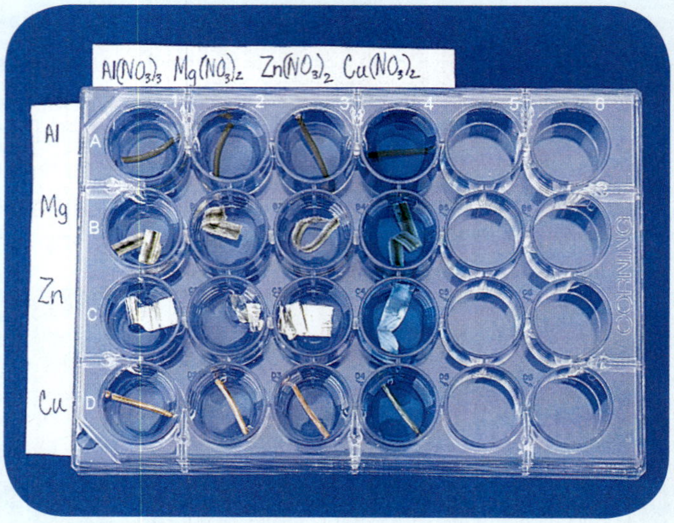


11. **Cleanup and Disposal** Dispose of the chemicals, solutions, and pipettes as directed by your teacher. Wash and return all lab equipment to the designated location. Wash your hands thoroughly.

### Analyze and Conclude

1. **Observe and Infer** In which wells of the reaction plate did chemical reactions occur? Which metal reacted with the most solutions? Which metal reacted with the fewest solutions? Which metal is the most reactive?
2. **Sequence** The most-active metal reacted with the most solutions. The least-active metal reacted with the fewest solutions. Order the four metals from most active to least active.
3. **Apply** Write a chemical equation for each single-replacement reaction that occurred on your reaction plate.
4. **Real-World Chemistry** Under what circumstances might it be important to know the activity tendencies of a series of elements?
5. **Error Analysis** How does your answer from Question 2 above compare with the activity series in **Figure 13?** What could account for the differences?

### INQUIRY EXTENSION

**Design an Experiment** Think of three "what if" questions about this investigation that might affect your results. Design an experiment to test one of them.

Matt Meadows

# CHAPTER 9 STUDY GUIDE

Vocabulary Practice

**BIGIDEA** **Millions of chemical reactions in and around you transform reactants into products, resulting in the absorption or release of energy.**

## SECTION 1 Reactions and Equations

**MAINIDEA** Chemical reactions are represented by balanced chemical equations.

- Some physical changes are evidence that indicate a chemical reaction has occurred.
- Word equations and skeleton equations provide important information about a chemical reaction.
- A chemical equation gives the identities and relative amounts of the reactants and products that are involved in a chemical reaction.
- Balancing an equation involves adjusting the coefficients until the number of atoms of each element is equal on both sides of the equation.

| $H_2$ | + | $Cl_2$ | → | 2HCl |
|---|---|---|---|---|
| 2 atoms H | | 2 atoms Cl | | 2 atoms H + 2 atoms Cl |

| Two hydrogen atoms | | Two chlorine atoms | | Two hydrogen atoms Two chlorine atoms |
|---|---|---|---|---|

**VOCABULARY**
- **chemical reaction**
- **reactant**
- **product**
- **chemical equation**
- **coefficient**

## SECTION 2 Classifying Chemical Reactions

**MAINIDEA** There are four types of chemical reactions: synthesis, combustion, decomposition, and replacement reactions.

- Classifying chemical reactions makes them easier to understand, remember, and recognize.
- Activity series of metals and halogens can be used to predict if single-replacement reactions will occur.

**VOCABULARY**
- **synthesis reaction**
- **combustion reaction**
- **decomposition reaction**
- **single-replacement reaction**
- **double-replacement reaction**
- **precipitate**

## SECTION 3 Reactions in Aqueous Solutions

**MAINIDEA** Double-replacement reactions occur between substances in aqueous solutions and produce precipitates, water, or gases.

- In aqueous solutions, the solvent is always water. There are many possible solutes.
- Many molecular compounds form ions when they dissolve in water. When some ionic compounds dissolve in water, their ions separate.
- When two aqueous solutions that contain ions as solutes are combined, the ions might react with one another. The solvent molecules do not usually react.
- Reactions that occur in aqueous solutions are double-replacement reactions.

**VOCABULARY**
- **aqueous solution**
- **solute**
- **solvent**
- **complete ionic equation**
- **spectator ion**
- **net ionic equation**

## SECTION 1

### Mastering Concepts

**57.** Define *chemical equation.*

**58.** Distinguish between a chemical reaction and a chemical equation.

**59.** Explain the difference between reactants and products.

**60.** What do the arrows and coefficients in equations communicate?

**61.** Does a conversion of a substance into a new substance always indicate that a chemical reaction has occurred? Explain.

**62.** Write formulas for the following substances and designate their physical states.
- **a.** nitrogen dioxide gas
- **b.** liquid gallium
- **c.** barium chloride dissolved in water
- **d.** solid ammonium carbonate

**63.** Identify the reactants in the following reaction: When potassium is dropped into aqueous zinc nitrate, zinc and aqueous potassium nitrate form.

**64.** Balance the reaction of hydrogen sulfide with atmospheric oxygen gas.

$$H_2S(g) + O_2(g) \rightarrow SO_2(g) + H_2O(g)$$

**65.** Write word equations for the following skeleton equations.
- **a.** $Cu(s) + O_2(g) \rightarrow CuO(s)$
- **b.** $K(s) + H_2O(l) \rightarrow KOH(aq) + H_2(g)$
- **c.** $CaCl_2(aq) + Na_2SO_4(aq) \rightarrow CaSO_4(s) + NaCl(aq)$

**66.** Balance the following reactions.
- **a.** $(NH_4)_2Cr_2O_7(s) \rightarrow Cr_2O_3(s) + N_2(g) + H_2O(g)$
- **b.** $CO_2(g) + H_2O(l) \rightarrow C_6H_{12}O_6(s) + O_2(g)$

### Mastering Problems

**67.** Hydrogen iodide gas breaks down into hydrogen gas and iodine gas during a decomposition reaction. Write a skeleton equation for this reaction.

**68.** Write skeleton equations for these reactions.
- **a.** sodium carbonate(s) → sodium oxide(s) + carbon dioxide(g)
- **b.** aluminum(s) + iodine(s) → aluminum iodide(s)
- **c.** iron(II) oxide(s) + oxygen(g) → iron(III) oxide(s)

**69.** Write skeleton equations for these reactions.
- **a.** butane ($C_4H_{10}$)(l) + oxygen(g) → carbon dioxide(g) + water(l)
- **b.** aluminum carbonate(s) → aluminum oxide(s) + carbon dioxide(g)
- **c.** silver nitrate(aq) + sodium sulfide(aq) → silver sulfide(s) + sodium nitrate(aq)

**70.** Write a skeleton equation for the reaction between lithium(s) and chlorine gas to produce lithium chloride(s).

**71.** Write skeleton equations for these reactions.
- **a.** iron(s) + fluorine(g) → iron(III) fluoride(s)
- **b.** sulfur trioxide(g) + water(l) → sulfuric acid(aq)
- **c.** sodium(s) + magnesium iodide(aq) → sodium iodide(aq) + magnesium(s)
- **d.** vanadium(s) + oxygen(g) → vanadium(V) oxide(s)

**72.** Write skeleton equations for these reactions.
- **a.** lithium(s) + gold(III) chloride(aq) → lithium chloride(aq) + gold(s)
- **b.** iron(s) + tin(IV) nitrate(aq) → iron(III) nitrate(aq) + tin(s)
- **c.** nickel(II) chloride(s) + oxygen(g) → nickel(II) oxide(s) + dichlorine pentoxide(g)
- **d.** lithium chromate(aq) + barium chloride(aq) → lithium chloride(aq) + barium chromate(s)

**73.** Balance the skeleton equations for the reactions described in Question 71.

**74.** Balance the skeleton equations for the reactions described in Question 72.

**75.** Write chemical equations for these reactions.
- **a.** When solid naphthalene ($C_{10}H_8$) burns in air, the reaction yields gaseous carbon dioxide and liquid water.
- **b.** Bubbling hydrogen sulfide gas through manganese(II) chloride dissolved in water results in the formation of the precipitate manganese(II) sulfide and hydrochloric acid.
- **c.** Solid magnesium reacts with nitrogen gas to produce solid magnesium nitride.
- **d.** Heating oxygen difluoride gas yields oxygen gas and fluorine gas.

## SECTION 2

### Mastering Concepts

**76.** List each of the four types of chemical reactions and give an example for each type.

**77.** How would you classify a chemical reaction between two reactants that produces one product?

**78.** Under what conditions does a precipitate form in a chemical reaction?

**79.** Will a metal always replace another metal in a compound dissolved in water? Explain.

**80.** In each of the following pairs, which element will replace the other in a reaction?
- **a.** tin and sodium
- **b.** fluorine and iodine
- **c.** lead and silver
- **d.** copper and nickel

## Mastering Problems

**81.** Classify each of the reactions represented by the chemical equations in Question 71.

**82.** Classify each of the reactions represented by the chemical equations in Question 72.

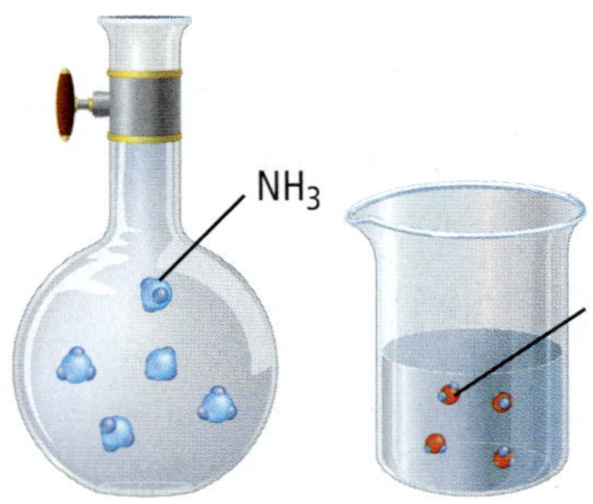


■ **Figure 22**

**83.** Use **Figure 22** to answer the following questions.
**a.** Write a chemical equation for the reaction between the two compounds shown in the figure.
**b.** Classify this reaction.

**84.** Write a balanced chemical equation for the combustion of liquid methanol ($CH_3OH$).

**85.** Write chemical equations for each of the following synthesis reactions.
**a.** boron + fluorine ⟶
**b.** germanium + sulfur ⟶
**c.** zirconium + nitrogen ⟶
**d.** tetraphosphorus decoxide + water ⟶ phosphoric acid

**86. Combustion** Write a chemical equation for the combustion of each of the following substances. If a compound contains carbon and hydrogen, assume that carbon dioxide gas and liquid water are produced.
**a.** solid barium
**b.** solid boron
**c.** liquid acetone ($C_3H_6O$)
**d.** liquid octane ($C_8H_{18}$)

**87.** Write chemical equations for each of the following decomposition reactions. One or more products might be identified.
**a.** magnesium bromide ⟶
**b.** cobalt(II) oxide ⟶
**c.** titanium(IV) hydroxide ⟶ titanium(IV) oxide + water
**d.** barium carbonate ⟶ barium oxide + carbon dioxide

**88.** Write chemical equations for the following single-replacement reactions that might occur in water. If no reaction occurs, write *NR* in place of the products.
**a.** nickel + magnesium chloride ⟶
**b.** calcium + copper(II) bromide ⟶
**c.** potassium + aluminum nitrate ⟶
**d.** magnesium + silver nitrate ⟶

# SECTION 3

## Mastering Concepts

**89.** Complete the following word equation.

Solute + Solvent ⟶

**90.** Define each of the following terms: *solution, solvent,* and *solute.*

**91.** When reactions occur in aqueous solutions, what common types of products are produced?

**92.** Compare and contrast chemical equations and ionic equations.

**93.** What is a net ionic equation? How does it differ from a complete ionic equation?

**94.** Define *spectator ion.*

**95.** Write the net ionic equation for a chemical reaction that occurs in an aqueous solution and produces water.

## Mastering Problems

**96.** Complete the following chemical equations.
**a.** $Na(s) + H_2O(l) \longrightarrow$
**b.** $K(s) + H_2O(l) \longrightarrow$

**97.** Complete the following chemical equation.

$$CuCl_2(s) + Na_2SO_4(aq) \longrightarrow$$

**98.** Write complete ionic and net ionic equations for the chemical reaction in Question 97.

**99.** Write complete ionic and net ionic equations for each of the following reactions.
**a.** $K_2S(aq) + CoCl_2(aq) \longrightarrow 2KCl(aq) + CoS(s)$
**b.** $H_2SO_4(aq) + CaCO_3(s) \longrightarrow H_2O(l) + CO_2(g) + CaSO_4(s)$
**c.** $2HClO(aq) + Ca(OH)_2(aq) \longrightarrow 2H_2O(l) + Ca(ClO)_2(aq)$

**100.** A reaction occurs when hydrosulfuric acid ($H_2S$) is mixed with an aqueous solution of iron(III) bromide. The reaction produces solid iron(III) sulfide and aqueous hydrogen bromide. Write the chemical and net ionic equations for the reaction.

**101.** Write complete ionic and net ionic equations for each of the following reactions.
**a.** $H_3PO_4(aq) + 3RbOH(aq) \longrightarrow 3H_2O(l) + Rb_3PO_4(aq)$
**b.** $HCl(aq) + NH_4OH(aq) \longrightarrow H_2O(l) + NH_4Cl(aq)$
**c.** $2HI + (NH_4)_2S(aq) \longrightarrow H_2S(g) + 2NH_4I(aq)$
**d.** $HNO_3(aq) + KCN(aq) + HCN(g) + KNO_3(aq)$

**102. Paper** A reaction occurs when sulfurous acid ($H_2SO_3$) is mixed with an aqueous solution of sodium hydroxide. The reaction produces aqueous sodium sulfite, a chemical used in manufacturing paper. Write the chemical and net ionic equations for the reaction.

## MIXED REVIEW

**103. Photosynthesis** Identify the products in the following reaction that occurs in plants: Carbon dioxide and water react to produce glucose and oxygen.

**104.** How will aqueous solutions of sucrose and hydrogen chloride differ?

**105.** Write the word equation for each of these skeleton equations. $C_6H_6$ is the formula for benzene.
- **a.** $C_6H_6(l) + O_2(g) \rightarrow CO_2(g) + H_2O(l)$
- **b.** $CO(g) + O_2(g) \rightarrow CO_2(g)$
- **c.** $Cl_2(g) + NaBr(s) \rightarrow NaCl(s) + Br_2(g)$
- **d.** $CaCO_3(s) \rightarrow CaO(s) + CO_2(g)$

**106.** Classify each of the reactions represented by the chemical equations in Question 105.

**107.** Write skeleton equations for the following reactions.
- **a.** ammonium phosphate(aq) + chromium(III) bromide(aq) → ammonium bromide(aq) + chromium(III) phosphate(s)
- **b.** chromium(VI) hydroxide(s) → chromium(VI) oxide(s) + water(l)
- **c.** aluminum(s) + copper(I) chloride(aq) → aluminum chloride(aq) + copper(s)
- **d.** potassium iodide(aq) + mercury(I) nitrate(aq) → potassium nitrate(aq) + mercury(I) iodide(s)

**108.** Balance the skeleton equations for the reactions described in Question 107.

**109.** Classify each of the reactions represented by the chemical equations in Question 108.

**110.** Predict whether each of the following reactions will occur in aqueous solutions. If you predict that a reaction will not occur, explain your reasoning. *Note: Barium sulfate and silver bromide precipitate in aqueous solutions.*
- **a.** sodium hydroxide + ammonium sulfate →
- **b.** niobium(V) sulfate + barium nitrate →
- **c.** strontium bromide + silver nitrate →

**111.** Complete the missing information in the following skeleton equation and balance the chemical equation:

$$NaOH(aq) + \_\_\_\_ \rightarrow 3NaCl(aq) + Al(OH)_3(aq)$$

**112. Precipitate Formation** The addition of hydrochloric acid to beakers containing solutions of either sodium chloride (NaCl) or silver nitrate ($AgNO_3$) causes a white precipitate in one of the beakers.
- **a.** Which beaker contains a precipitate?
- **b.** What is the precipitate?
- **c.** Write a chemical equation showing the reaction.
- **d.** Classify the reaction.

**113.** Write the skeleton equation and the balanced chemical equation for the reaction between iron and chlorine.

**114.** Write a chemical equation representing the decomposition of water into two gaseous products. What are the products?

**115.** Distinguish between an ionic compound and a molecular compound dissolved in water. Do all molecular compounds ionize when dissolved in water? Explain.

**116.** Classify the type of reactions that occur in aqueous solutions, and give an example to support your answer.

## THINK CRITICALLY

**117. Explain** how an equation can be balanced even if the number of reactant particles differs from the number of product particles.

**118. Apply** Describe the reaction of aqueous solutions of sodium sulfide and copper(II) sulfate, producing the precipitate copper(II) sulfide.

**119. Predict** A piece of aluminum metal is placed in aqueous KCl. Another piece of aluminum is placed in an aqueous $AgNO_3$ solution. Explain why a chemical reaction does or does not occur in each instance.

**120. Design an Experiment** You suspect that the water in a lake close to your school might contain lead in the form of $Pb^{2+}(aq)$ ions. Formulate your suspicion as a hypothesis and design an experiment to test your theory. Write the net ionic equations for the reactions of your experiment. *(Hint: In aqueous solution, $Pb^{2+}$ forms solid compounds with $Cl^-$, $Br^-$, $I^-$, and $SO_4^{2-}$ ions.)*

**121. Predict** When sodium metal reacts with water, it produces sodium hydroxide, hydrogen gas, and heat. Write balanced chemical equations for Li, Na, and K reacting with water. Use **Figure 13** to predict the order of the amount of heat released from least to most amount of heat released.

**122. Apply** Write the chemical equations and net ionic equations for each of the following reactions that might occur in aqueous solutions. If a reaction does not occur, write *NR* in place of the products. Magnesium phosphate precipitates in an aqueous solution.
- **a.** $KNO_3 + CsCl \rightarrow$
- **b.** $Ca(OH)_2 + KCN \rightarrow$
- **c.** $Li_3PO_4 + MgSO_4 \rightarrow$
- **d.** $HBrO + NaOH \rightarrow$

**123. Analyze** Explain why a nail exposed to air forms rust, whereas the same nail exposed to a pure nitrogen environment does not form rust.

**124. Evaluate** Write a balanced chemical equation for the reaction of aluminum with oxygen to produce aluminum oxide.

Chapter Self-Check

## CHALLENGE PROBLEM

**125.** A single-replacement reaction occurs between copper and silver nitrate. When 63.5 g of copper reacts with 339.8 g of silver nitrate, 215.8 g of silver is produced. Write a balanced chemical equation for this reaction. What other product formed? What is the mass of the second product?

## CUMULATIVE REVIEW

**126.** Complete the following problems in scientific notation. Round off to the correct number of significant figures.
**a.** $(5.31 \times 10^{-2}\ \text{cm}) \times (2.46 \times 10^{5}\ \text{cm})$
**b.** $(6.42 \times 10^{-2}\ \text{g}) \div (3.21 \times 10^{-3}\ \text{g})$
**c.** $(9.87 \times 10^{4}\ \text{g}) - (6.2 \times 10^{3}\ \text{g})$

**127.** Distinguish between a mixture, a solution, and a compound.

**128.** Data from chromium's four naturally occurring isotopes is provided in **Table 5.** Calculate chromium's atomic mass.

**Table 5** Chromium Isotope Data

| Isotope | Percent Abundance | Mass (amu) |
|---|---|---|
| Cr-50 | 4.35% | 49.946 |
| Cr-52 | 83.79% | 51.941 |
| Cr-53 | 9.50% | 52.941 |
| Cr-54 | 2.36% | 53.939 |

**129.** Differentiate between electron configuration and electron-dot structure.

**130.** Identify the elements by their electron configuration.
**a.** $1s^22s^22p^63s^23p^64s^23d^{10}4p^5$
**b.** $[Ne]3s^23p^4$
**c.** $[Xe]6s^2$

**131.** Write the electron configuration for the element fitting each description.
**a.** a metalloid in group 13
**b.** a nonmetal in group 15, period 3

**132.** Describe the formation of positive and negative ions.

**133.** Write the formula for the compounds made from each of the following pairs of ions.
**a.** copper(I) and sulfite
**b.** tin(IV) and fluoride
**c.** gold(III) and cyanide
**d.** lead(II) and sulfide

## WRITING IN ▶ Chemistry

**134. Kitchen Chemistry** Make a poster describing chemical reactions that occur in the kitchen.

**135. Mathematical Equations** Write a report that compares and contrasts chemical equations and mathematical equations.

**136. Balance Equations** Create a flowchart describing how to balance a chemical equation.

## DBQ Document-Based Questions

**Solubility** *Scientists, in determining whether a precipitate will occur in a chemical reaction, use a solubility rules chart.* **Table 6** *lists the solubility rules for ionic compounds in water.*

Data obtained from: Van Der Sluys, W.G. 2001, *J. Chem. Ed.* 78:111–115

**Table 6** Solubility Rules for Ionic Compound in Water

| Ionic Compound | Rule |
|---|---|
| Soluble salts | Group 1 cations and $NH_4^+$ ions form soluble salts. |
| | All nitrates are soluble. |
| | Most halides are soluble, except those of $Pb^{2+}$, $Hg_2^{2+}$, $Ag^+$, and $Cu^+$. |
| | Most sulfates are soluble, with the exception of those of $Ba^{2+}$; $Sr^{2+}$, $Pb^{2+}$, $Ag^+$, $Ca^{2+}$, and $Hg_2^{2+}$ form slightly soluble sulfates. |
| Insoluble salts | Hydroxides, oxides, and sulfides are usually insoluble, except that those of group 1 ions and $NH_4^+$ are soluble and those of group 2 ions are slightly soluble. |
| | Chromates, phosphates, and carbonates are usually insoluble, except that those of group 1 ions and $NH_4^+$ are soluble. |

Using the solubility rules provided in the table above, complete the following chemical equations. Indicate whether a precipitate forms or not. Identify the precipitate. If no reaction occurs, write *NR*.

**137.** $Ca(NO_3)_2(aq) + Na_2CO_3(aq) \rightarrow$

**138.** $Mg(s) + NaOH(aq) \rightarrow$

**139.** $PbS(s) + LiNO_3(aq) \rightarrow$

## MULTIPLE CHOICE

**1.** What type of reaction is described by the following equation?

$Cs(s) + H_2O(l) \rightarrow CsOH(aq) + H_2(g)$

**A.** synthesis
**B.** combustion
**C.** decomposition
**D.** single-replacement

*Use the figure below to answer Question 2.*

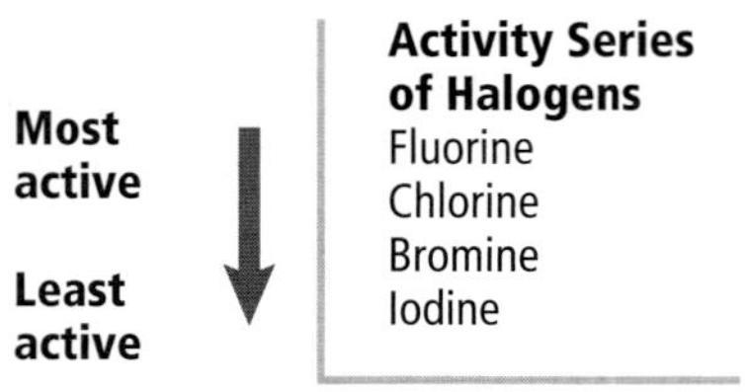


**2.** Which reaction between halogens and halide salts will occur?
**A.** $F_2(g) + FeI_2(aq) \rightarrow FeF_2(aq) + I_2(l)$
**B.** $I_2(s) + MnBr_2(aq) \rightarrow MnI_2(aq) + Br_2(g)$
**C.** $Cl_2(s) + SrF_2(aq) \rightarrow SrCl_2(aq) + F_2(g)$
**D.** $Br_2(l) + CoCl_2(aq) \rightarrow CoBr_2(aq) + Cl_2(g)$

**3.** Which is the electron configuration for iron?
**A.** $1s^22s^22p^63s^23p^64s^23d^6$
**B.** $[Ar]3d^6$
**C.** $1s^22p^63p^63d^6$
**D.** $[Ar]4s^24d^6$

**4.** Which is a description of a pattern displayed by elements in the periodic table?
**A.** repetition of their physical properties when arranged by increasing atomic radius
**B.** repetition of their chemical properties when arranged by increasing atomic mass
**C.** periodic repetition of their properties when arranged by increasing atomic number
**D.** periodic repetition of their properties when arranged by increasing atomic mass

**5.** When moving down a group on the periodic table, which two atomic properties follow the same trend?
**A.** atomic radius and ionization energy
**B.** ionic radius and atomic radius
**C.** ionization energy and ionic radius
**D.** ionic radius and electronegativity

*Use the table below to answer Questions 6 to 8.*

**Physical Properties of Select Ionic Compounds**

| Compound | Name | State at 25°C | Soluble in Water? | Melting Point (°C) |
|---|---|---|---|---|
| $NaClO_3$ | sodium chlorate | solid | yes | 248 |
| $Na_2SO_4$ | sodium sulfate | solid | yes | 884 |
| $NiCl_2$ | nickel(II) chloride | solid | yes | 1031 |
| $Ni(OH)_2$ | nickel(II) hydroxide | solid | no | 230 |
| $AgNO_3$ | silver nitrate | solid | yes | 210 |

**6.** An aqueous solution of nickel(II) sulfate is mixed with aqueous sodium hydroxide. Will a visible reaction occur?
**A.** No, solid nickel(II) hydroxide is soluble in water.
**B.** No, solid sodium sulfate is soluble in water.
**C.** Yes, solid sodium sulfate will precipitate out of the solution.
**D.** Yes, solid nickel(II) hydroxide will precipitate out of the solution.

**7.** What happens when $AgClO_3(aq)$ and $NaNO_3(aq)$ are mixed?
**A.** No visible reaction occurs.
**B.** Solid $NaClO_3$ precipitates out of the solution.
**C.** $NO_2$ gas is released during the reaction.
**D.** Solid Ag metal is produced.

**8.** Finely ground nickel(II) hydroxide is placed in a beaker of water. It sinks to the bottom of the beaker and remains unchanged. An aqueous solution of hydrochloric acid (HCl) is then added to the beaker, and the $Ni(OH)_2$ disappears. Which equation best describes what occurred in the beaker?
**A.** $Ni(OH)_2(s) + HCl(aq) \rightarrow NiO(aq) + H_2(g) + HCl(aq)$
**B.** $Ni(OH)_2(s) + 2HCl(aq) \rightarrow NiCl_2(aq) + 2H_2O(l)$
**C.** $Ni(OH)_2(s) + 2H_2O(l) \rightarrow NiCl_2(aq) + 2H_2O(l)$
**D.** $Ni(OH)_2(s) + 2H_2O(l) \rightarrow NiCl_2(aq) + 3H_2O(l) + O_2(g)$

## SHORT ANSWER

*Use the diagram below to answer Questions 9 and 10.*

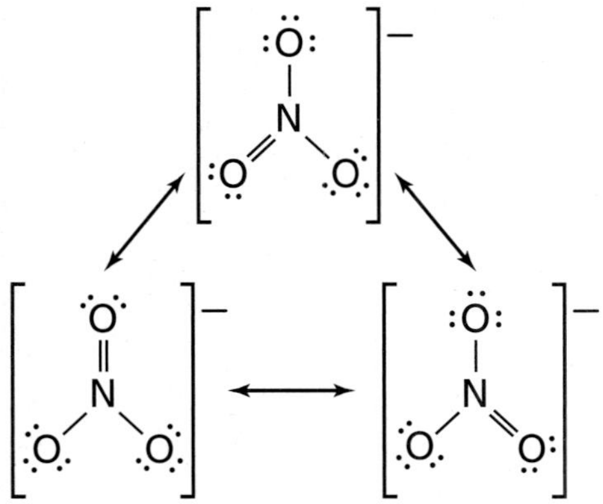

**9.** What is the name for the multiple Lewis structures shown in the diagram?

**10.** Why do these structures form?

**11.** Write the balanced chemical equation for the reaction of solid calcium with water to form calcium hydroxide in solution and hydrogen gas.

## EXTENDED RESPONSE

*Use the partial chemical equation below to answer Questions 12 and 13.*

$$AlCl_3(aq) + Fe_2O_3(aq) \rightarrow$$

**12.** What type of reaction will this be? Explain how you can tell from the reactants.

**13.** Predict what the products of this reaction will be. Use evidence from the reaction to support your answer.

**14.** What is the electron configuration for the ion $P^{3-}$? Explain how this configuration is different from the configuration for the neutral atom of phosphorus.

## SAT SUBJECT TEST: CHEMISTRY

**15.** Chloroform ($CHCl_3$) was one of the first anesthetics used in medicine. The chloroform molecule contains 26 valence electrons total. How many of these valence electrons are part of covalent bonds?
**A.** 26
**B.** 13
**C.** 8
**D.** 4
**E.** 2

**16.** Which is NOT true of an atom obeying the octet rule?
**A.** obtains a full set of eight valence electrons
**B.** acquires the valence configuration of a noble gas
**C.** electron configuration is unusually stable
**D.** has an $s^2p^6$ valence configuration
**E.** will lose electrons

*Use the figure below to answer Question 17.*

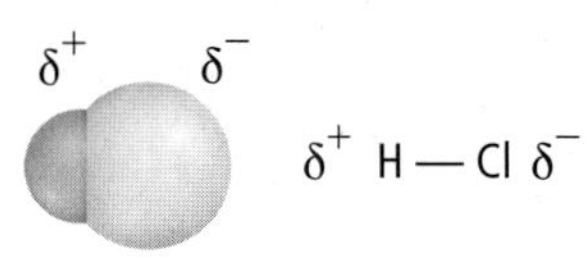


**17.** Which statement does NOT correctly describe the model of HCl shown above?
**A.** A nonpolar bond exists between these atoms.
**B.** Chlorine has a stronger attraction for electrons than does hydrogen.
**C.** The electrons in the bond are shared unequally.
**D.** This compound dissolves in a polar substance.
**E.** Chlorine is the more electronegative atom.

**18.** The combustion of ethanol ($C_2H_6O$) produces carbon dioxide and water vapor. What equation best describes this process?
**A.** $C_2H_6O(l) + O_2(g) \rightarrow CO_2(g) + H_2O(l)$
**B.** $C_2H_6O(l) \rightarrow 2CO_2(g) + 3H_2O(l)$
**C.** $C_2H_6O(l) + 3O_2(g) \rightarrow 2CO_2(g) + 3H_2O(g)$
**D.** $C_2H_6O(l) \rightarrow 3O_2(l) + 2CO_2(g) + 3H_2O(l)$
**E.** $C_2H_6O(l) \rightarrow 2CO_2(g) + 3H_2O(g)$

**NEED EXTRA HELP?**

| If You Missed Question . . . | 1 | 2 | 3 | 4 | 5 | 6 | 7 | 8 | 9 | 10 | 11 | 12 | 13 | 14 | 15 | 16 | 17 | 18 |
|---|---|---|---|---|---|---|---|---|---|---|---|---|---|---|---|---|---|---|
| Review Section . . . | 9.2 | 9.2 | 5.3 | 6.3 | 6.3 | 9.3 | 9.3 | 9.3 | 8.3 | 8.3 | 9.1 | 9.2 | 9.2 | 5.3 | 8.3 | 8.3 | 8.5 | 9.2 |

CHAPTER 10

# The Mole

**BIGIDEA** The mole represents a large number of extremely small particles.

## SECTIONS

Chm 5a

1 Measuring Matter

2 Mass and the Mole

3 Moles of Compounds

4 Empirical and Molecular Formulas

5 Formulas of Hydrates

## LaunchLAB

iLab Station

### How much is a mole?

Counting large numbers of items is easier when you use counting units, such as decades or dozens. Chemists use a counting unit called the mole. In this lab, you will use a common item to help you understand how much a mole really is.

**FOLDABLES® Study Organizer**

### Conversion Factors

Make a two-tab Book. Label it as shown. Use it to help you organize information about conversion factors.

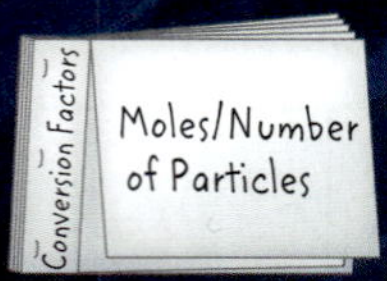

Go **online!**
MHEonline.com

Many people keep a jar full of pennies at home. Counting all those pennies individually can be pretty tedious, but grouping them in 50-cent rolls makes it a lot easier to tally how much money is in the jar. Similarly, scientists use the mole to count atoms, molecules, and ions.

SECTION 1

# Measuring Matter

**MAINIDEA** Chemists use the mole to count atoms, molecules, ions, and formula units.

### Essential Questions

- How is a mole used to indirectly count the number of particles of matter?
- What is a common everyday counting unit to which the mole can be related?
- How can moles be converted to number of representative particles and vice versa?

### Review Vocabulary

**molecule:** two or more atoms that covalently bond together to form a unit

### New Vocabulary

mole
Avogadro's number

**CHEM 4 YOU** Has your class ever had a contest to guess how many pennies or jelly beans were in a jar? You might have noticed that the smaller the object is, the harder it is to count.

## Counting Particles

If you were buying a bouquet of roses for a special occasion, you probably would not ask for 12 or 24; you would ask for one or two dozen. Similarly, you might buy a pair of gloves, a ream of paper for your printer, or a gross of pencils. Each of the units shown in **Figure 1**—a pair, a dozen, a gross, and a ream—represents a specific number of items. These units make counting objects easier. It is easier to buy and sell paper by the ream—500 sheets—than by the individual sheet.

Each of the counting units shown in **Figure 1** is appropriate for certain kinds of objects, depending primarily on their size and function. But regardless of the object—gloves, eggs, pencils, or paper—the number that the unit represents is always constant. Chemists also need a convenient method for accurately counting the number of atoms, molecules, or formula units in a sample of a substance. However, atoms are so small and there are so many of them in even the smallest sample that it is impossible to count them directly. Because of this, chemists created a counting unit called the mole. In the Launch Lab, you probably found that a mole of any object is an enormous number of items.

■ **Figure 1** Different units are used to count different types of objects. A pair is two objects, a dozen is 12, a gross is 144, and a ream is 500.
**List** *What other counting units are you familiar with?*

**The mole** The **mole,** abbreviated mol, is the SI base unit used to measure the amount of a substance. A mole is defined as the number of carbon atoms in exactly 12 g of pure carbon-12. Through years of experimentation, it has been established that a mole of anything contains $6.0221367 \times 10^{23}$ representative particles. A representative particle is any kind of particle, such as an atom, a molecule, a formula unit, an electron, or an ion. If you write out Avogadro's number, it looks like this.

$$602{,}213{,}670{,}000{,}000{,}000{,}000{,}000$$

The number $6.0221367 \times 10^{23}$ is called **Avogadro's number,** in honor of the Italian physicist and lawyer Amedeo Avogadro, who, in 1811, determined the volume of 1 mol of a gas. In this book, Avogadro's number is rounded to three significant figures, $6.02 \times 10^{23}$.

To count extremely small particles, such as atoms, Avogadro's number must be an enormous quantity. As you might imagine, Avogadro's number would not be convenient for measuring a quantity of marbles. Avogadro's number of marbles with a five-eighth inch diameter would cover the surface of Earth to a depth of more than fourteen kilometers! **Figure 2,** however, shows that it is convenient to use the mole to measure amounts of substances. One-mole quantities of water, copper, and salt are shown. Each one has a different representative particle. The representative particle in a mole of water is the water molecule, the representative particle in a mole of copper is the copper atom, and the representative particle in a mole of sodium chloride is the NaCl formula unit.

**VOCABULARY**

**SCIENCE USAGE v. COMMON USAGE**

**Mole**

***Science usage:*** an SI base unit used to measure the quantity of matter

*The chemist measured out a mole of the compound.*

***Common usage:*** a small burrowing animal

*The damage to the lawn was caused by a mole.*

■ **Figure 2** The amount of each substance shown is $6.02 \times 10^{23}$, or 1 mol, of representative particles. The representative particle for each substance is shown in a box. Refer to **Table R-1** in the Student Resources for a key to atom color conventions.

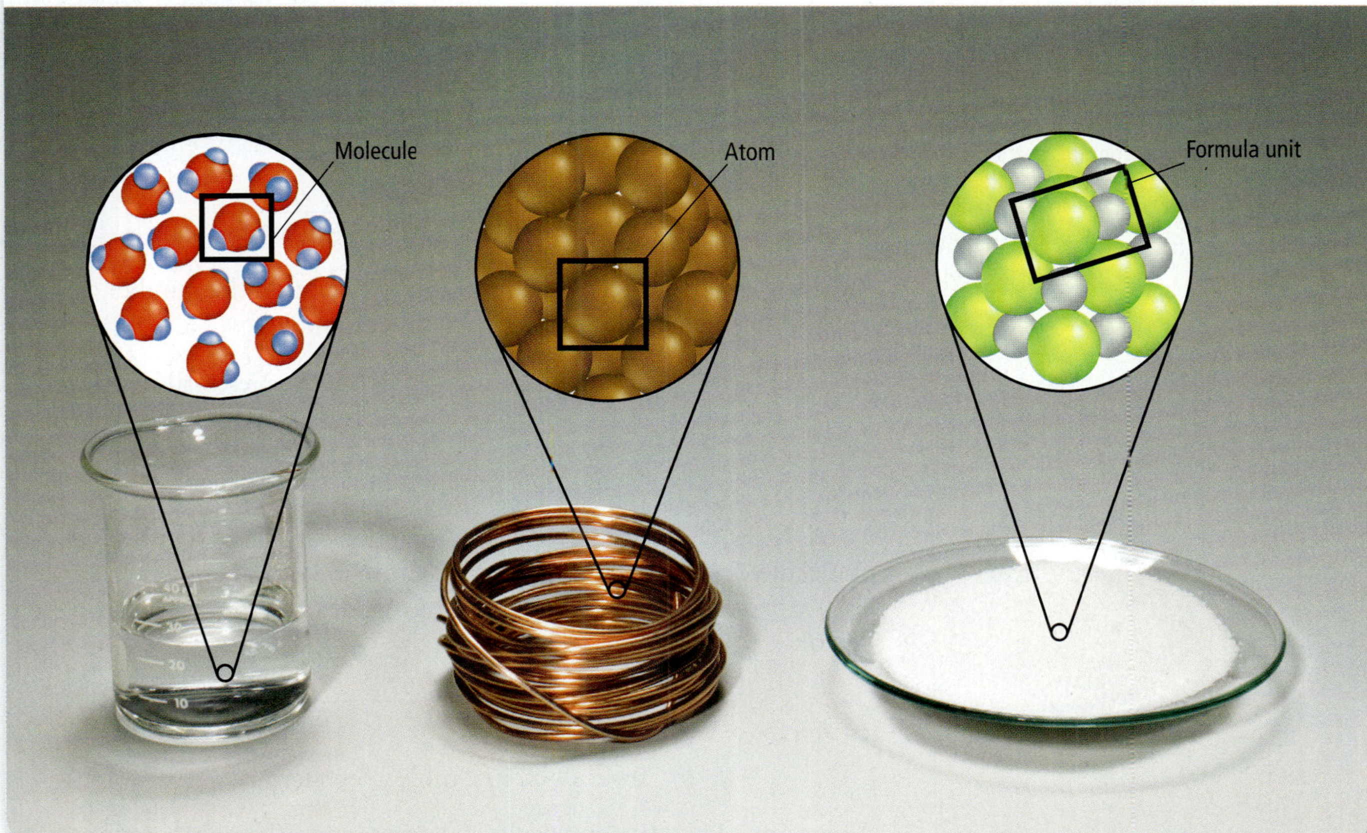

12 roses = 1 dozen roses

■ **Figure 3** A key to using dimensional analysis is correctly identifying the mathematical relationship between the units you are converting. The relationship shown here, 12 roses = 1 dozen roses, can be used to write two conversion factors.

## Converting Between Moles and Particles

Suppose you buy three-and-one-half dozen roses and want to know how many roses you have. Recall what you have learned about conversion factors. You can multiply the known quantity (3.5 dozen roses) by a conversion factor to express the quantity in the units you want (number of roses). First, identify the mathematical relationship that relates the given unit with the desired unit. **Figure 3** shows the relationship.

$$\text{Relationship: } 1 \text{ dozen roses} = 12 \text{ roses}$$

By dividing each side of the equality by the other side, you can write two conversion factors from the relationship.

$$\text{Conversion factors: } \frac{12 \text{ roses}}{1 \text{ dozen roses}} \text{ and } \frac{1 \text{ dozen roses}}{12 \text{ roses}}$$

Then choose the conversion factor that, when multiplied by the known quantity, results in the desired unit. When set up correctly, all units cancel except those required for the answer.

$$\text{Conversion: } 3.5 \cancel{\text{dozen roses}} \times \frac{12 \text{ roses}}{1 \cancel{\text{dozen roses}}} = 42 \text{ roses}$$

Here, *dozens of roses* cancels, leaving *roses* as the desired unit.

READING CHECK **Describe** how you can tell if the wrong conversion factor has been used.

**Moles to particles** Now suppose you want to determine how many particles of sucrose are in 3.50 mol of sucrose. The relationship between moles and representative particles is given by Avogadro's number.

$$1 \text{ mol of representative particles} = 6.02 \times 10^{23} \text{ representative particles}$$

Using this relationship, you can write two different conversion factors that relate representative particles and moles.

$$\frac{6.02 \times 10^{23} \text{ representative particles}}{1 \text{ mol}}$$

$$\frac{1 \text{ mol}}{6.02 \times 10^{23} \text{ representative particles}}$$

**FOLDABLES®** Incorporate information from this section into your Foldable.

By using the correct conversion factor, you can find the number of representative particles in a given number of moles.

$$\text{number of moles} \times \frac{6.02 \times 10^{23} \text{ representative particles}}{1 \text{ mol}}$$

$$= \text{number of representative particles}$$

As shown in **Figure 4,** the representative particle of sucrose is a molecule. To obtain the number of sucrose molecules contained in 3.50 mol of sucrose, you need to use Avogadro's number as a conversion factor.

$$3.50 \cancel{\text{mol sucrose}} \times \frac{6.02 \times 10^{23} \text{ molecules sucrose}}{1 \cancel{\text{mol sucrose}}}$$

$$= 2.11 \times 10^{24} \text{ molecules sucrose}$$

There are $2.11 \times 10^{24}$ molecules of sucrose in 3.50 mol of sucrose.

## PRACTICE Problems

Do additional problems. Online Practice

1. Zinc (Zn) is used to form a corrosion-inhibiting surface on galvanized steel. Determine the number of Zn atoms in 2.50 mol of Zn.
2. Calculate the number of molecules in 11.5 mol of water ($H_2O$).
3. Silver nitrate ($AgNO_3$) is used to make several different silver halides used in photographic films. How many formula units of $AgNO_3$ are there in 3.25 mol of $AgNO_3$?
4. **Challenge** Calculate the number of oxygen atoms in 5.00 mol of oxygen molecules. Oxygen is a diatomic molecule, $O_2$.

**Particles to moles** Now suppose you want to find out how many moles are represented by a certain number of representative particles. To do this, you can use the inverse of Avogadro's number as a conversion factor.

$$\text{number of representative particles} \times \frac{1 \text{ mol}}{6.02 \times 10^{23} \text{ representative particles}} = \text{number of moles}$$

For example, if instead of knowing how many moles of sucrose you have, suppose you knew that a sample contained $2.11 \times 10^{24}$ molecules of sucrose. To convert this number of molecules of sucrose to moles of sucrose, you need a conversion factor that has moles in the numerator and molecules in the denominator.

$$2.11 \times 10^{24} \cancel{\text{molecules sucrose}} \times \frac{1 \text{ mol}}{6.02 \times 10^{23} \cancel{\text{molecules sucrose}}} = 3.50 \text{ mol sucrose}$$

Thus, $2.11 \times 10^{24}$ molecules of sucrose is 3.50 mol of sucrose.

You can convert between moles and number of representative particles by multiplying the known quantity by the proper conversion factor. Example Problem 1 further illustrates the conversion process.

☑ READING CHECK **List** the two conversion factors that can be written from Avogadro's number.

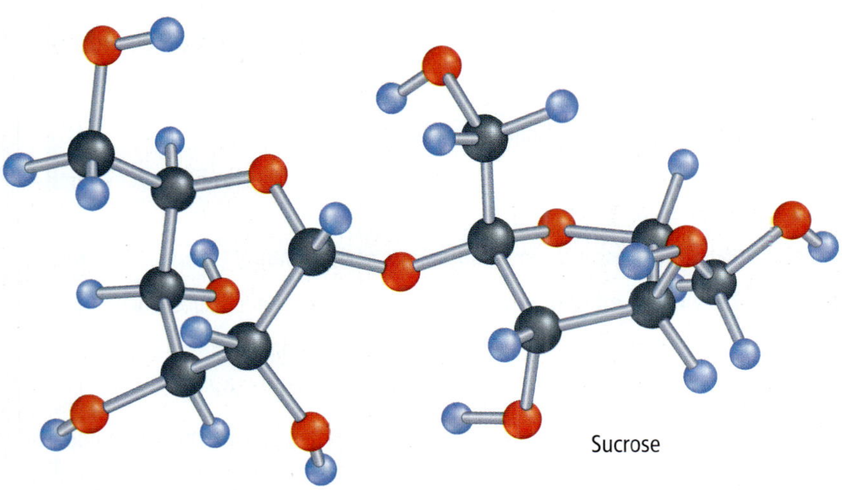


■ **Figure 4** The representative particle of sucrose is a molecule. The ball-and-stick model shows that a molecule of sucrose is a single unit made up of carbon, hydrogen, and oxygen.

**Analyze** *Use the ball-and-stick model of sucrose to write the chemical formula for sucrose.*

## EXAMPLE Problem 1

Find help with **dimensional analysis.** Math Handbook

EXAMPLE PROBLEM

**PARTICLES-TO-MOLES CONVERSION** Zinc (Zn) is used as a corrosion-resistant coating on iron and steel. It is also an essential trace element in your diet. Calculate the number of moles of zinc that contain $4.50 \times 10^{24}$ atoms.

### 1 ANALYZE THE PROBLEM

You are given the number of atoms of zinc and must find the equivalent number of moles. If you compare $4.50 \times 10^{24}$ atoms Zn with $6.02 \times 10^{23}$, the number of atoms in 1 mol, you can predict that the answer should be less than 10 mol.

**Known**

number of atoms = $4.50 \times 10^{24}$ atoms Zn

1 mol Zn = $6.02 \times 10^{23}$ atoms Zn

**Unknown**

**moles Zn = ? mol**

### 2 SOLVE FOR THE UNKNOWN

Use a conversion factor—the inverse of Avogadro's number—that relates moles to atoms.

$$\text{number of atoms} \times \frac{1 \text{ mol}}{6.02 \times 10^{23} \text{ atoms}} = \text{number of moles}$$

**Apply the conversion factor.**

$$4.50 \times 10^{24} \cancel{\text{atoms Zn}} \times \frac{1 \text{ mol Zn}}{6.02 \times 10^{23} \cancel{\text{atoms Zn}}} = 7.48 \text{ mol Zn}$$

**Substitute number of Zn atoms = $4.50 \times 10^{24}$. Multiply and divide numbers and units.**

### 3 EVALUATE THE ANSWER

Both the number of Zn atoms and Avogadro's number have three significant figures. Therefore, the answer is expressed correctly with three digits. The answer is less than 10 mol, as predicted, and has the correct unit, moles.

## PRACTICE Problems

**Do additional problems.** Online Practice

**5.** How many moles contain each of the following?

**a.** $5.75 \times 10^{24}$ atoms Al

**b.** $2.50 \times 10^{20}$ atoms Fe

**6. Challenge** Identify the representative particle for each formula, and convert the given number of representative particles to moles.

**a.** $3.75 \times 10^{24}$ $CO_2$

**b.** $3.58 \times 10^{23}$ $ZnCl_2$

## SECTION 1 REVIEW

Section Self-Check

### Section Summary

- The mole is a unit used to count particles of matter indirectly. One mole of a pure substance contains Avogadro's number of representative particles.
- Representative particles include atoms, ions, molecules, formula units, electrons, and other similar particles.
- One mole of carbon-12 atoms has a mass of exactly 12 g.
- Conversion factors written from Avogadro's relationship can be used to convert between moles and number of representative particles.

**7. MAIN**IDEA **Explain** why chemists use the mole.

**8. State** the mathematical relationship between Avogadro's number and 1 mol.

**9. List** the conversion factors used to convert between particles and moles.

**10. Explain** how a mole is similar to a dozen.

**11. Apply** How does a chemist count the number of particles in a given number of moles of a substance?

**12. Calculate** the mass in atomic mass units of 0.25 mol of carbon-12 atoms.

**13. Calculate** the number of representative particles of each substance.

**a.** 11.5 mol Ag

**b.** 18.0 mol $H_2O$

**c.** 0.150 mol NaCl

**d.** $1.35 \times 10^{-2}$ mol $CH_4$

**14. Arrange** these three samples from smallest to largest in terms of number of representative particles: $1.25 \times 10^{25}$ atoms of zinc (Zn), 3.56 mol of iron (Fe), and $6.78 \times 10^{22}$ molecules of glucose ($C_6H_{12}O_6$).

SECTION 2

# Mass and the Mole

**MAINIDEA** A mole always contains the same number of particles; however, moles of different substances have different masses.

**CHEM 4 YOU** When purchasing a dozen eggs, you can pick from several sizes—medium, large, and extra-large. The size of the egg does not affect how many come in the carton. A similar situation exists with the size of the atoms that make up a mole.

### Essential Questions

- Why can the mass of an atom be related to the mass of a mole of atoms?
- How can the number of moles be converted to the mass of an element and vice versa?
- How can the number of moles be converted to the number of atoms of an element and vice versa?

### Review Vocabulary

**conversion factor:** a ratio of equivalent values used to express the same quantity in different units

### New Vocabulary

molar mass

## The Mass of a Mole

You would not expect a dozen limes to have the same mass as a dozen eggs. Because eggs and limes differ in size and composition, it is not surprising that they have different masses, as shown in **Figure 5.** One-mole quantities of two different substances have different masses for the same reason—the substances have different compositions. For example, if you put one mole of carbon and one mole of copper on separate balances, you would see a difference in mass, just as you do for the eggs and the limes. This occurs because carbon atoms differ from copper atoms. Thus, the mass of $6.02 \times 10^{23}$ carbon atoms does not equal the mass of $6.02 \times 10^{23}$ copper atoms.

Recall that each atom of carbon-12 has a mass of 12 amu. The atomic masses of all other elements are established relative to carbon-12. For example, an atom of hydrogen-1 has a mass of approximately 1 amu, one-twelfth the mass of a carbon-12 atom. The mass of an atom of helium-4 is approximately 4 amu, one-third the mass of one atom of carbon-12.

You might have noticed, however, that the atomic-mass values given on the periodic table are not exact integers. For example, you will find 12.011 amu for carbon, 1.008 amu for hydrogen, and 4.003 amu for helium. These noninteger values occur because the values are weighted averages of the masses of all the naturally occurring isotopes of each element.

■ **Figure 5** A dozen limes has approximately twice the mass of a dozen eggs. The difference in mass is reasonable because limes are different from eggs in composition and size.

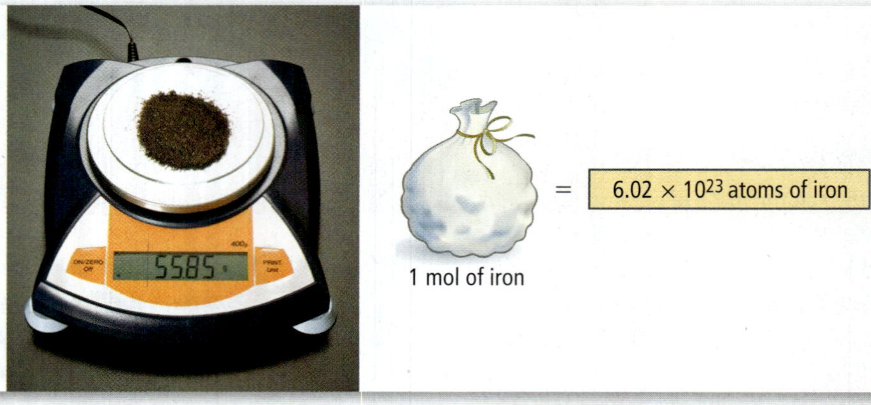


**Figure 6** One mole of iron, represented by a bag of particles, contains Avogadro's number of atoms and has a mass equal to its atomic mass in grams.
**Apply** *What is the mass of one mole of copper?*

View an **animation about molar mass.**

**Concepts In Motion**

**Molar Mass** How does the mass of one atom relate to the mass of one mole of that atom? Recall that the mole is defined as the number of carbon-12 atoms in exactly 12 g of pure carbon-12. Thus, the mass of one mole of carbon-12 atoms is 12 g. Whether you are considering a single atom or Avogadro's number of atoms (a mole), the masses of all atoms are established relative to the mass of carbon-12. The mass in grams of one mole of any pure substance is called its **molar mass.**

The molar mass of any element is numerically equal to its atomic mass and has the units g/mol. As given on the periodic table, an atom of iron has an atomic mass of 55.845 amu. Thus, the molar mass of iron is 55.845 g/mol, and 1 mol (or $6.02 \times 10^{23}$ atoms of iron) has a mass of 55.845 g. Note that by measuring 55.845 g of iron, you indirectly count out $6.02 \times 10^{23}$ atoms of iron. **Figure 6** shows the relationship between molar mass and one mole of an element.

# Problem-Solving LAB

## Formulate a Model

**How are molar mass, Avogadro's number, and the atomic nucleus related?**
A nuclear model of mass can provide a simple picture of the connections among the mole, molar mass, and the number of representative particles in a mole.

### Analysis

The diagram to the right shows the space-filling models of hydrogen-1 and helium-4 nuclei. The hydrogen-1 nucleus contains one proton with a mass of 1.007 amu. The mass of a proton, in grams, has been determined experimentally to be $1.672 \times 10^{-24}$ g. The helium-4 nucleus contains two protons and two neutrons and has a mass of approximately 4 amu.

### Think Critically

1. **Apply** What is the mass in grams of one helium atom? (The mass of a neutron is approximately the same as the mass of a proton.)
2. **Draw** Carbon-12 contains six protons and six neutrons. Draw the carbon-12 nucleus and calculate the mass of one atom in amu and g.
3. **Apply** How many atoms of hydrogen-1 are in a 1.007-g sample? Recall that 1.007 amu is the mass of one atom of hydrogen-1. Round your answer to two significant digits.
4. **Apply** If you had samples of helium and carbon that contained the same number of atoms as you calculated in Question 3, what would be the mass in grams of each sample?
5. **Conclude** What can you conclude about the relationship between the number of atoms and the mass of each sample?

## Using Molar Mass

Imagine that your class bought jelly beans in bulk to sell by the dozen at a candy sale. You soon realize that it is too much work to count out each dozen, so you instead decide to measure the jelly beans by mass. You find that the mass of 1 dozen jelly beans is 35 g. This relationship and the conversion factors that stem from it are as follows:

$$1 \text{ dozen jelly beans} = 35 \text{ g jelly beans}$$

$$\frac{35 \text{ g jelly beans}}{1 \text{ dozen jelly beans}} \text{ and } \frac{1 \text{ dozen jelly beans}}{35 \text{ g jelly beans}}$$

What mass of jelly beans should you measure if a customer wants 5 dozen jelly beans? To determine this mass, you would multiply the number of dozens of jelly beans to be sold by the correct conversion factor. Select the conversion factor with the units you are converting to in the numerator (g) and the units you are converting from in the denominator (dozen).

$$5 \cancel{\text{ dozen jelly beans}} \times \frac{35 \text{ g jelly beans}}{1 \cancel{\text{ dozen jelly beans}}} = 175 \text{ g jelly beans.}$$

A quantity of 5 dozen jelly beans has a mass of 175 g.

☑ READING CHECK **Compare** How are the jelly bean conversion factors used above similar to the molar mass of a compound?

Get help with **conversion factors.**

Personal Tutor

**Moles to mass** Now suppose that while working in a chemistry lab, you need 3.00 mol of copper (Cu) for a chemical reaction. How would you measure that amount? Like the 5 dozen jelly beans, the number of moles of copper can be converted to an equivalent mass and measured on a balance.

To calculate the mass of a given number of moles, simply multiply the number of moles by the molar mass.

$$\text{number of moles} \times \frac{\text{mass in grams}}{1 \text{ mole}} = \text{mass}$$

If you check the periodic table, you will find that copper, element 29, has an atomic mass of 63.546 amu. You know that the molar mass of an element (in g/mol) is equal to its atomic mass (given in amu). Thus, copper has a molar mass of 63.546 g/mol. By using the molar mass, you can convert 3.00 mol of copper to grams of copper.

$$3.00 \cancel{\text{ mol Cu}} \times \frac{63.546 \text{ g Cu}}{1 \cancel{\text{ mol Cu}}} = 191 \text{ g Cu}$$

So, as shown in **Figure 7,** you can measure the 3.00 mol of copper needed for the reaction by using a balance to measure out 191 g of copper. The reverse conversion—from mass to moles—also involves the molar mass as a conversion factor, but it is the inverse of the molar mass that is used. Can you explain why?

**FOLDABLES®** Incorporate information from this section into your Foldable.

**Connection to Biology** Cellular biologists continually discover new biologic proteins. After a new biomolecule is discovered, biologists determine the molar mass of the compound using a technique known as mass spectrometry. In addition to the molar mass, mass spectrometry also provides additional information that helps the biologist reveal the compound's composition.

**Figure 7** To measure 3.00 mol of copper, place a weighing paper on a balance, tare the balance, and then add the 191 g of copper.

## EXAMPLE Problem 2

**MOLE-TO-MASS CONVERSION** Chromium (Cr), a transition element, is a component of chrome plating. Chrome plating is used on metals and in steel alloys to control corrosion. Calculate the mass in grams of 0.0450 mol Cr.

### 1 ANALYZE THE PROBLEM

You are given the number of moles of chromium and must convert it to an equivalent mass using the molar mass of chromium from the periodic table. Because the sample is less than one-tenth of a mole, the answer should be less than one-tenth of the molar mass.

| Known | Unknown |
|---|---|
| number of moles = 0.0450 mol Cr | **mass Cr = ? g** |
| molar mass Cr = 52.00 g/mol Cr | |

### 2 SOLVE FOR THE UNKNOWN

Use a conversion factor—the molar mass—that relates grams of chromium to moles of chromium. Write the conversion factor with moles of chromium in the denominator and grams of chromium in the numerator. Substitute the known values into the equation and solve.

$$\text{moles Cr} \times \frac{\text{grams Cr}}{1\ \text{mol Cr}} = \textbf{grams Cr}$$

**Apply the conversion factor.**

$$0.0450\ \cancel{\text{mol Cr}} \times \frac{52.00\ \text{g Cr}}{1\ \cancel{\text{mol Cr}}} = \textbf{2.34 g Cr}$$

**Substitute 0.450 mol for moles Cr and 52.00 g/mol for molar mass of Cr. Multiply and divide numbers and units.**

### 3 EVALUATE THE ANSWER

The known number of moles of chromium has the smallest number of significant figures, three, so the answer is correctly stated with three digits. The answer is less than one-tenth the mass of 1 mol, as predicted, and is in grams, a mass unit.

## PRACTICE Problems

Do additional problems. **Online Practice**

**15.** Determine the mass in grams of each of the following.

**a.** 3.57 mol Al

**b.** 42.6 mol Si

**16. Challenge** Convert each given quantity in scientific notation to mass in grams expressed in scientific notation.

**a.** $3.45 \times 10^{2}$ mol Co

**b.** $2.45 \times 10^{-2}$ mol Zn

### RealWorld CHEMISTRY

**The Importance of Chromium**

**CHROMIUM** What gives these rims their mirrorlike finish? The metal alloy rim has been plated, or coated, with a thin layer of chromium. Chrome plating has been used in the automobile industry for decades because of its beauty and its corrosion resistance.

Explore **moles, mass, and molecules.**

**Virtual Investigations**

Many of the values for atomic mass given in the periodic table have five significant figures. However, in Example Problem 2, the periodic table value of 51.996 g/mol Cr was rounded to 52.00 g/mol. It is generally okay to round reference values as long as you keep one more significant figure than the answer will have. In the case of Example Problem 2, 52.00 has one more significant figure than 0.0450 mol Cr, which limits the answer to three significant figures.

## EXAMPLE Problem 3

**MASS-TO-MOLE CONVERSION** Calcium (Ca), the fifth most-abundant element on Earth, is always found combined with other elements because of its high reactivity. How many moles of calcium are in 525 g Ca?

### 1 ANALYZE THE PROBLEM

You must convert the mass of calcium to moles of calcium. The mass of calcium is more than ten times larger than the molar mass. Therefore, the answer should be greater than 10 mol.

**Known**
mass = 525 g Ca
molar mass Ca = 40.08 g/mol Ca

**Unknown**
**number of moles Ca = ? mol**

### 2 SOLVE FOR THE UNKNOWN

Use a conversion factor—the inverse of molar mass—that relates moles of calcium to grams of calcium. Substitute the known values and solve.

$$\text{mass Ca} \times \frac{1 \text{ mol Ca}}{\text{grams Ca}} = \textbf{moles Ca}$$

**Apply the conversion factor.**

$$525 \cancel{\text{g Ca}} \times \frac{1 \text{ mol Ca}}{40.08 \cancel{\text{g Ca}}} = \textbf{13.1 mol Ca}$$

**Substitute mass Ca = 525 g, and inverse molar mass of Ca = 1 mol/40.08 g. Multiply and divide numbers and units.**

### 3 EVALUATE THE ANSWER

The mass of calcium has the fewest significant figures, three, so the answer is expressed correctly with three digits. As predicted, the answer is greater than 10 mol and has the expected unit.

## PRACTICE Problems

Do additional problems. Online Practice

**17.** Determine the number of moles in each of the following.

**a.** 25.5 g Ag **b.** 300.0 g S

**18. Challenge** Convert each mass to moles. Express the answer in scientific notation.

**a.** $1.25 \times 10^3$ g Zn **b.** 1.00 kg Fe

**Converting between mass and atoms** So far, you have learned how to convert mass to moles and moles to mass. You can go one step further and convert mass to the number of atoms. Recall the jelly beans you were selling at the candy sale. At the end of the day, you find that 550 g of jelly beans is left unsold. Without counting, can you determine how many jelly beans that is? You know that 1 dozen jelly beans has a mass of 35 g and that 1 dozen is 12 jelly beans. Thus, you can first convert the 550 g to dozens of jelly beans by using the conversion factor that relates dozens and mass.

$$550 \cancel{\text{g jelly beans}} \times \frac{1 \text{ dozen jelly beans}}{35 \cancel{\text{g jelly beans}}} = 16 \text{ dozen jelly beans}$$

Next, you can determine how many jelly beans are in 16 dozen by multiplying by the conversion factor that relates number of particles (jelly beans) and dozens.

The conversion factor relating number of jelly beans and dozens is 12 jelly beans/dozen. Applying it yields the answer in jelly beans.

$$16 \text{ dozen} \times \frac{12 \text{ jelly beans}}{1 \text{ dozen}} = 192 \text{ jelly beans}$$

The 550 g of leftover jelly beans is equal to 192 jelly beans.

Just as you could not make a direct conversion from the mass of jelly beans to the number of jelly beans, you cannot make a direct conversion from the mass of a substance to the number of representative particles of that substance. You must first convert mass to moles by multiplying by a conversion factor that relates moles and mass. That conversion factor is the molar mass. The number of moles must then be multiplied by a conversion factor that relates the number of representative particles to moles. For this conversion, you will use Avogadro's number. This two-step process is shown in Example Problem 4.

## EXAMPLE Problem 4

Find help with **dimensional analysis.** Math Handbook

EXAMPLE PROBLEM

**MASS-TO-ATOMS CONVERSION** Gold (Au) is one of a group of metals called the coinage metals (copper, silver, and gold). How many atoms of gold are in a U.S. Eagle, a gold alloy bullion coin with a mass of 31.1 g Au?

### 1 ANALYZE THE PROBLEM

You must determine the number of atoms in a given mass of gold. Because you cannot convert directly from mass to the number of atoms, you must first convert the mass to moles using the molar mass. Then, convert moles to the number of atoms using Avogadro's number. The given mass of the gold coin is about one-sixth the molar mass of gold (196.97 g/mol), so the number of gold atoms should be approximately one-sixth Avogadro's number.

| **Known** | **Unknown** |
|---|---|
| mass = 31.1 g Au | **number of atoms Au = ?** |
| molar mass Au = 196.97 g/mol Au | |

### 2 SOLVE FOR THE UNKNOWN

Use a conversion factor—the inverse of the molar mass—that relates moles of gold to grams of gold.

$$\text{mass Au} \times \frac{1 \text{ mol Au}}{\text{grams Au}} = \text{moles Au}$$ **Apply the conversion factor.**

$$31.1 \text{ g Au} \times \frac{1 \text{ mol Au}}{196.97 \text{ g Au}} = 0.158 \text{ mol Au}$$ **Substitute mass Au = 31.1 g and the inverse molar mass of Au = 1 mol/196.97 g. Multiply and divide numbers and units.**

To convert the calculated moles of gold to atoms, multiply by Avogadro's number.

$$\text{moles Au} \times \frac{6.02 \times 10^{23} \text{ atoms Au}}{1 \text{ mol Au}} = \textbf{atoms Au}$$ **Apply the conversion factor.**

$$0.158 \text{ mol Au} \times \frac{6.02 \times 10^{23} \text{ atoms Au}}{1 \text{ mol Au}} = \mathbf{9.51 \times 10^{22} \text{ atoms Au}}$$ **Substitute moles Au = 0.158 mol, and solve.**

### 3 EVALUATE THE ANSWER

The mass of gold has the smallest number of significant figures, three, so the answer is expressed correctly with three digits. The answer is approximately one-sixth Avogadro's number, as predicted, and the correct unit, atoms, is obtained.

## EXAMPLE Problem 5

Find help with **significant figures.** Math Handbook

**ATOMS-TO-MASS CONVERSION** Helium (He) is an unreactive noble gas often found in underground deposits mixed with methane. The mixture is separated by cooling the gaseous mixture until all but the helium has liquefied. A party balloon contains $5.50 \times 10^{22}$ atoms of helium gas. What is the mass, in grams, of the helium?

### 1 ANALYZE THE PROBLEM

You are given the number of atoms of helium and must find the mass of the gas. First, convert the number of atoms to moles, then convert moles to grams.

**Known**

number of atoms He = $5.50 \times 10^{22}$ atoms He

molar mass He = 4.00 g/mol He

**Unknown**

**mass = ? g He**

### 2 SOLVE FOR THE UNKNOWN

Use a conversion factor—the inverse of Avogadro's number—that relates moles to number of atoms.

$$\text{atoms He} \times \frac{1 \text{ mol He}}{6.02 \times 10^{23} \text{ atoms He}} = \text{moles He}$$

**Apply the conversion factor.**

$$5.50 \times 10^{22} \cancel{\text{atoms He}} \times \frac{1 \text{ mol He}}{6.02 \times 10^{23} \cancel{\text{atoms He}}} = 0.0914 \text{ mol He}$$

**Substitute atoms He = $5.50 \times 10^{22}$ atoms. Multiply and divide numbers and units.**

Next, apply a conversion factor—the molar mass of helium—that relates mass of helium to moles of helium.

$$\text{moles He} \times \frac{\text{grams He}}{1 \text{ mol He}} = \textbf{grams He}$$

**Apply the conversion factor.**

$$0.0914 \cancel{\text{mol He}} \times \frac{4.00 \text{ g He}}{1 \cancel{\text{mol He}}} = \mathbf{0.366 \text{ g He}}$$

**Substitute moles He = 0.0914 mol, molar mass He = 4.00 g/mol, and solve.**

### 3 EVALUATE THE ANSWER

The answer is expressed correctly with three significant figures and is in grams, a mass unit.

EXAMPLE PROBLEM

## PRACTICE Problems

Do additional problems. Online Practice

**19.** How many atoms are in each of the following samples?

**a.** 55.2 g Li

**b.** 0.230 g Pb

**c.** 11.5 g Hg

**20.** What is the mass in grams of each of the following?

**a.** $6.02 \times 10^{24}$ atoms Bi

**b.** $1.00 \times 10^{24}$ atoms Mn

**c.** $3.40 \times 10^{22}$ atoms He

**d.** $1.50 \times 10^{15}$ atoms N

**e.** $1.50 \times 10^{15}$ atoms U

**21. Challenge** Convert each given mass to number of representative particles. Identify the type of representative particle, and express the number in scientific notation.

**a.** $4.56 \times 10^{3}$ g Si

**b.** 0.120 kg Ti

PRACTICE PROBLEMS

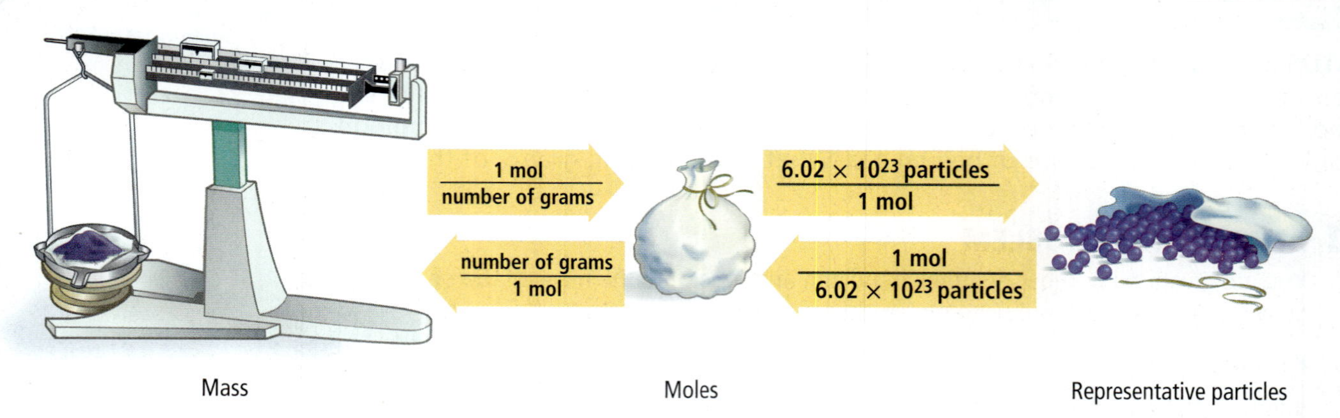


■ **Figure 8** The mole is at the center of conversions between mass and particles (atoms, ions, or molecules). In the figure, mass is represented by a balance, moles by a bag of particles, and representative particles by the contents that are spilling out of the bag. Two steps are needed to convert from mass to representative particles or the reverse.

Now that you have practiced conversions between mass, moles, and representative particles, you probably realize that the mole is at the center of these calculations. Mass must always be converted to moles before being converted to atoms, and atoms must similarly be converted to moles before calculating their mass. **Figure 8** shows the steps to follow as you complete these conversions. In the Example Problems, two steps were used to convert either mass to moles to atoms, or atoms to moles to mass. Instead of two separate steps, these conversions can be made in one step. Suppose you want to find out how many atoms of oxygen (O) are in 1.00 g of oxygen. This calculation involves two conversions—mass to moles and then moles to atoms. You could set up one equation like this.

$$1.00\ \cancel{g\ O} \times \frac{1\ \cancel{mol\ O}}{15.999\ \cancel{g\ O}} \times \frac{6.02 \times 10^{23}\ \text{atoms O}}{1\ \cancel{mol\ O}}$$

$$= 3.76 \times 10^{22}\ \text{atoms O}$$

## SECTION 2 REVIEW

Section Self-Check

### Section Summary

- The mass in grams of one mole of any pure substance is called its molar mass.
- The molar mass of an element is numerically equal to its atomic mass.
- The molar mass of any substance is the mass in grams of Avogadro's number of representative particles of the substance.
- Molar mass is used to convert from moles to mass. The inverse of molar mass is used to convert from mass to moles.

**22.** MAINIDEA **Summarize** in terms of particles and mass, one-mole quantities of two different monatomic elements.

**23. State** the conversion factor needed to convert between mass and moles of the atom fluorine.

**24. Explain** how molar mass relates the mass of an atom to the mass of a mole of atoms.

**25. Describe** the steps used to convert the mass of an element to the number of atoms of the element.

**26. Arrange** these quantities from smallest to largest in terms of mass: 1.0 mol of Ar, $3.0 \times 10^{24}$ atoms of Ne, and 20 g of Kr.

**27. Identify** the quantity that is calculated by dividing the molar mass of an element by Avogadro's number.

**28. Design** a concept map that shows the conversion factors needed to convert between mass, moles, and number of particles.

SECTION 3

# Moles of Compounds

**MAINIDEA** The molar mass of a compound can be calculated from its chemical formula and can be used to convert from mass to moles of that compound.

**CHEM 4 YOU** Imagine checking two pieces of luggage at the airport, only to find out that one of them is over the weight limit. Because the weight of each suitcase depends on the combination of the items packed inside, changing the combination of the items in the two suitcases changes the weight of each.

### Essential Questions

- What are the mole relationships shown by a chemical formula?
- How is the molar mass of a compound calculated?
- How can the number of moles be coverted to the mass of a compound and vice versa?
- What conversion factors are applied to determine the number of atoms or ions in a known mass of a compound?

### Review Vocabulary

**representative particle:** an atom, molecule, formula unit, or ion

## Chemical Formulas and the Mole

You have learned that different kinds of representative particles are counted using the mole. In the last section, you read how to use molar mass to convert among moles, mass, and number of particles of an element. Can you make similar conversions for compounds and ions? Yes, you can, but to do so you will need to know the molar mass of the compounds and ions involved.

Recall that a chemical formula indicates the numbers and types of atoms contained in one unit of the compound. Consider the compound dichlorodifluoromethane with the chemical formula $CCl_2F_2$. The subscripts in the formula indicate that one molecule of $CCl_2F_2$ consists of one carbon (C) atom, two chlorine (Cl) atoms, and two fluorine (F) atoms. These atom are chemically bonded together. The C-Cl-F ratio in $CCl_2F_2$ is 1:2:2.

Now suppose you had a mole of $CCl_2F_2$. The representative particles of the compound are molecules, and a mole of $CCl_2F_2$ contains Avogadro's number of molecules. The C-Cl-F ratio in one mole of $CCl_2F_2$ would still be 1:2:2, as it is in one molecule of the compound. **Figure 9** illustrates this principle for a dozen $CCl_2F_2$ molecules. Check for yourself that a dozen $CCl_2F_2$ molecules contains one dozen carbon atoms, two dozen chlorine atoms, and two dozen fluorine atoms. The chemical formula $CCl_2F_2$ not only represents an individual molecule of $CCl_2F_2$, it also represents a mole of the compound.

■ **Figure 9** A dozen molecules of $CCl_2F_2$ contains one dozen carbon atoms, two dozen chlorine atoms, and two dozen fluorine atoms.

**Interpret** *How many of each kind of atom—carbon, chlorine, and fluorine—are contained in 1 mol of $CCl_2F_2$?*

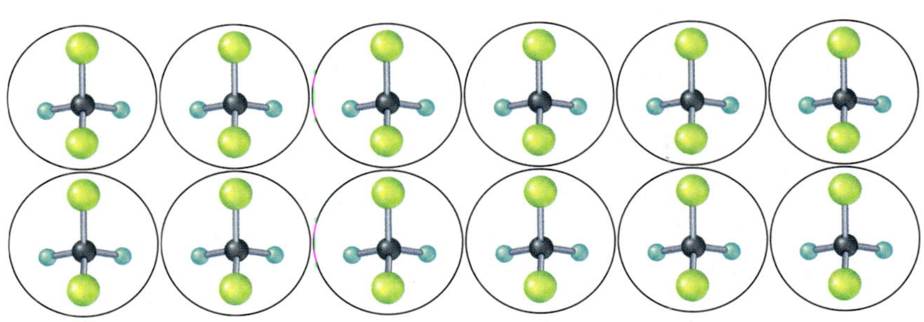

**VOCABULARY**

**ACADEMIC VOCABULARY**

**Ratio**
the relationship in size or quantity of two or more things; proportion
*The test results showed his LDL-to-HDL cholesterol ratio was too high.*

In some chemical calculations, you might need to convert between moles of a compound and moles of individual atoms in the compound. The following ratios, or conversion factors, can be written for use in these calculations for the molecule $CCl_2F_2$.

$$\frac{1 \text{ mol C atoms}}{1 \text{ mol } CCl_2F_2} \qquad \frac{2 \text{ mol Cl atoms}}{1 \text{ mol } CCl_2F_2} \qquad \frac{2 \text{ mol F atoms}}{1 \text{ mol } CCl_2F_2}$$

To find out how many moles of fluorine atoms are in 5.50 moles of freon, you multiply the moles of freon by the conversion factor relating moles of fluorine atoms to moles of freon.

$$\text{moles } CCl_2F_2 \times \frac{\text{moles F atoms}}{1 \text{ mol } CCl_2F_2} = \text{moles F atoms}$$

$$5.50 \cancel{\text{mol } CCl_2F_2} \times \frac{2 \text{ mol F atoms}}{1 \cancel{\text{mol } CCl_2F_2}} = 11.0 \text{ mol F atoms}$$

Conversion factors such as the one just used for fluorine can be written for any element in a compound. The number of moles of the element that goes in the numerator of the conversion factor is the subscript for that element in the chemical formula.

## EXAMPLE Problem 6

EXAMPLE PROBLEM

**MOLE RELATIONSHIPS FROM A CHEMICAL FORMULA** Aluminum oxide ($Al_2O_3$), often called alumina, is the principal raw material for the production of aluminum (Al). Alumina occurs in the minerals corundum and bauxite. Determine the moles of aluminum ions ($Al^{3+}$) in 1.25 mol of $Al_2O_3$.

### 1 ANALYZE THE PROBLEM

You are given the number of moles of $Al_2O_3$ and must determine the number of moles of $Al^{3+}$ ions. Use a conversion factor based on the chemical formula that relates moles of $Al^{3+}$ ions to moles of $Al_2O_3$. Every mole of $Al_2O_3$ contains 2 mol of $Al^{3+}$ ions. Thus, the answer should be two times the number of moles of $Al_2O_3$.

| **Known** | **Unknown** |
|---|---|
| number of moles = 1.25 mol $Al_2O_3$ | **number of moles = ? mol $Al^{3+}$ ions** |

### 2 SOLVE FOR THE UNKNOWN

Use the relationship that 1 mol of $Al_2O_3$ contains 2 mol of $Al^{3+}$ ions to write a conversion factor.

$$\frac{2 \text{ mol } Al^{3+} \text{ ions}}{1 \text{ mol } Al_2O_3}$$

**Create a conversion factor relating moles of $Al^{3+}$ ions to moles of $Al_2O_3$.**

To convert the known number of moles of $Al_2O_3$ to moles of $Al^{3+}$ ions, multiply by the ions-to-moles conversion factor.

$$\text{moles } Al_2O_3 \times \frac{2 \text{ mol } Al^{3+} \text{ ions}}{1 \text{ mol } Al_2O_3} = \textbf{moles } \mathbf{Al^{3+}} \textbf{ ions}$$

**Apply the conversion factor.**

$$1.25 \cancel{\text{mol } Al_2O_3} \times \frac{2 \text{ mol } Al^{3+} \text{ ions}}{1 \cancel{\text{mol } Al_2O_3}} = \mathbf{2.50 \text{ mol } Al^{3+} \text{ ions}}$$

**Substitute moles $Al_2O_3$ = 1.25 mol $Al_2O_3$ and solve.**

### 3 EVALUATE THE ANSWER

Because the conversion factor is a ratio of whole numbers, the number of significant digits is based on the moles of $Al_2O_3$. Therefore, the answer is expressed correctly with three significant figures. As predicted, the answer is twice the number of moles of $Al_2O_3$.

## PRACTICE Problems

Do additional problems. Online Practice

**29.** Zinc chloride ($ZnCl_2$) is used in soldering flux, an alloy used to join two metals together. Determine the moles of $Cl^-$ ions in 2.50 mol $ZnCl_2$.

**30.** Plants and animals depend on glucose ($C_6H_{12}O_6$) as an energy source. Calculate the number of moles of each element in 1.25 mol $C_6H_{12}O_6$.

**31.** Iron(III) sulfate [$Fe_2(SO_4)_3$] is sometimes used in the water purification process. Determine the number of moles of sulfate ions present in 3.00 mol of $Fe_2(SO_4)_3$.

**32.** How many moles of oxygen atoms are present in 5.00 mol of diphosphorus pentoxide ($P_2O_5$)?

**33. Challenge** Calculate the number of moles of hydrogen atoms in $1.15 \times 10^1$ mol of water. Express the answer in scientific notation.

## The Molar Mass of Compounds

The mass of your backpack is the sum of the mass of the pack and the masses of the books, notebooks, pencils, lunch, and miscellaneous items you put into it. You could find its mass by determining the mass of each item separately and adding them together. Similarly, the mass of a mole of a compound equals the sum of the masses of all the particles that make up the compound.

Suppose you want to determine the molar mass of the compound potassium chromate ($K_2CrO_4$). Start by looking up the molar mass of each element present in $K_2CrO_4$. Then, multiply each molar mass by the number of moles of that element in the chemical formula. Adding the masses of each element yields the molar mass of $K_2CrO_4$.

$$2\ \cancel{\text{mol K}} \times \frac{39.10\ \text{g K}}{1\ \cancel{\text{mol K}}} = 78.20\ \text{g}$$

$$1\ \cancel{\text{mol Cr}} \times \frac{52.00\ \text{g Cr}}{1\ \cancel{\text{mol Cr}}} = 52.00\ \text{g}$$

$$4\ \cancel{\text{mol O}} \times \frac{16.00\ \text{g O}}{1\ \cancel{\text{mol O}}} = 64.00\ \text{g}$$

$$\text{molar mass } K_2CrO_4 = 194.20\ \text{g}$$

The molar mass of a compound demonstrates the law of conservation of mass; the total mass of the reactants that reacted equals the mass of the compound formed. **Figure 10** shows equivalent masses of one mole of potassium chromate, sodium chloride, and sucrose.

## PRACTICE Problems

Do additional problems. Online Practice

**34.** Determine the molar mass of each ionic compound.

**a.** NaOH **b.** $CaCl_2$ **c.** $KC_2H_3O_2$

**35.** Calculate the molar mass of each molecular compound.

**a.** $C_2H_5OH$ **b.** HCN **c.** $CCl_4$

**36. Challenge** Identify each substance as a molecular compound or an ionic compound, and then calculate its molar mass.

**a.** $Sr(NO_3)_2$ **b.** $(NH_4)_3PO_4$ **c.** $C_{12}H_{22}O_{11}$

■ **Figure 10** Because each substance contains different numbers and kinds of atoms, their molar masses are different. The molar mass of each compound is the sum of the masses of all the elements contained in the compound.

Potassium chromate ($K_2CrO_4$)

Sodium chloride (NaCl)

Sucrose ($C_{12}H_{22}O_{11}$)

## Converting Moles of a Compound to Mass

Suppose you need to measure a certain number of moles of a compound for an experiment. First, you must calculate the mass in grams that corresponds to the necessary number of moles. Then, you can measure that mass on a balance. In Example Problem 2, you learned how to convert the number of moles of elements to mass using molar mass as the conversion factor. The procedure is the same for compounds, except that you must first calculate the molar mass of the compound.

### EXAMPLE Problem 7

Find help with dimensional analysis. **Math Handbook**

EXAMPLE PROBLEM

**MOLE-TO-MASS CONVERSION FOR COMPOUNDS** The characteristic odor of garlic is due to allyl sulfide [$(C_3H_5)_2S$]. What is the mass of 2.50 mol of $(C_3H_5)_2S$?

#### 1 ANALYZE THE PROBLEM

You are given 2.50 mol of $(C_3H_5)_2S$ and must convert the moles to mass using the molar mass as a conversion factor. The molar mass is the sum of the molar masses of all the elements in $(C_3H_5)_2S$.

**Known**

number of moles = 2.50 mol $(C_3H_5)_2S$

**Unknown**

**molar mass = ? g/mol $(C_3H_5)_2S$**

**mass = ? g $(C_3H_5)_2S$**

#### 2 SOLVE FOR THE UNKNOWN

Calculate the molar mass of $(C_3H_5)_2S$.

$1 \cancel{\text{mol S}} \times \frac{32.07 \text{ g S}}{1 \cancel{\text{mol S}}} = 32.07 \text{ g S}$ — **Multiply the moles of S in the compound by the molar mass of S.**

$6 \cancel{\text{mol C}} \times \frac{12.01 \text{ g C}}{1 \cancel{\text{mol C}}} = 72.06 \text{ g C}$ — **Multiply the moles of C in the compound by the molar mass of C.**

$10 \cancel{\text{mol H}} \times \frac{1.008 \text{ g H}}{1 \cancel{\text{mol H}}} = 10.08 \text{ g H}$ — **Multiply the moles of H in the compound by the molar mass of H.**

**molar mass** = (32.07 g + 72.06 g + 10.08 g) = **114.21 g/mol $(C_3H_5)_2S$** — **Total the mass values.**

Use a conversion factor—the molar mass—that relates grams to moles.

$\text{moles } (C_3H_5)_2S \times \frac{\text{grams } (C_3H_5)_2S}{1 \text{ mol } (C_3H_5)_2S} = \textbf{mass } (C_3H_5)_2S$ — **Apply the conversion factor.**

$2.50 \cancel{\text{mol } (C_3H_5)_2S} \times \frac{114.21 \text{ g } (C_3H_5)_2S}{1 \cancel{\text{mol } (C_3H_5)_2S}} = \textbf{286 g } (C_3H_5)_2S$ — **Substitute moles $(C_3H_5)_2S$ = 2.5 mol, molar mass $(C_3H_5)_2S$ = 114.21 g/mol, and solve.**

### PRACTICE Problems

Do additional problems. **Online Practice**

**37.** The United States chemical industry produces more sulfuric acid ($H_2SO_4$), in terms of mass, than any other chemical. What is the mass of 3.25 mol of $H_2SO_4$?

**38.** What is the mass of $4.35 \times 10^{-2}$ mol of zinc chloride ($ZnCl_2$)?

**39.** **Challenge** Write the chemical formula for potassium permanganate, and then calculate the mass in grams of 2.55 mol of the compound.

## Converting the Mass of a Compound to Moles

Imagine that an experiment you are doing in the laboratory produces 5.55 g of a compound. How many moles is this? To find out, you calculate the molar mass of the compound and determine it to be 185.0 g/mol. The molar mass relates grams and moles, but this time you need the inverse of the molar mass as the conversion factor.

$$5.50 \text{ g compound} \times \frac{1 \text{ mol compound}}{185.0 \text{ g compound}} = 0.0297 \text{ mol compound}$$

**EXAMPLE** Problem 8 — Find help **with significant figures.** Math Handbook

EXAMPLE PROBLEM

**MASS-TO-MOLE CONVERSION FOR COMPOUNDS** Calcium hydroxide [$Ca(OH)_2$] is used to remove sulfur dioxide from the exhaust gases emitted by power plants and for softening water by the elimination of $Ca^{2+}$ and $Mg^{2+}$ ions. Calculate the number of moles of calcium hydroxide in 325 g of the compound.

### 1 ANALYZE THE PROBLEM

You are given 325 g of $Ca(OH)_2$ and must solve for the number of moles of $Ca(OH)_2$. You must first calculate the molar mass of $Ca(OH)_2$.

**Known**

mass = 325 g $Ca(OH)_2$

**Unknown**

**molar mass = ? g/mol $Ca(OH)_2$**

**number of moles = ? mol $Ca(OH)_2$**

### 2 SOLVE FOR THE UNKNOWN

Determine the molar mass of $Ca(OH)_2$.

$$1 \text{ mol Ca} \times \frac{40.08 \text{ g Ca}}{1 \text{ mol Ca}} = 40.08 \text{ g}$$

**Multiply the moles of Ca in the compound by the molar mass of Ca.**

$$2 \text{ mol O} \times \frac{16.00 \text{ g O}}{1 \text{ mol O}} = 32.00 \text{ g}$$

**Multiply the moles of O in the compound by the molar mass of O.**

$$2 \text{ mol H} \times \frac{1.008 \text{ g H}}{1 \text{ mol H}} = 2.016 \text{ g}$$

**Multiply the moles of H in the compound by the molar mass of H.**

**molar mass** = (40.08 g + 32.00 g + 2.016 g) = **74.10 g/mol $Ca(OH)_2$**

**Total the mass values.**

Use a conversion factor—the inverse of the molar mass—that relates moles to grams.

$$325 \text{ g } Ca(OH)_2 \times \frac{1 \text{ mol } Ca(OH)_2}{74.10 \text{ g } Ca(OH)_2} = \mathbf{4.39 \text{ mol } Ca(OH)_2}$$

**Apply the conversion factor. Substitute mass Ca = 325 g, inverse molar mass $Ca(OH)_2$ = 1 mol/74.10 g, and solve.**

### 3 EVALUATE THE ANSWER

To check the reasonableness of the answer, round the molar mass of $Ca(OH)_2$ to 75 g/mol and the given mass of $Ca(OH)_2$ to 300 g. Seventy-five is contained in 300 four times. Thus, the answer is reasonable. The unit, moles, is correct, and there are three significant figures.

**PRACTICE** Problems — **Do additional problems.** Online Practice

**40.** Determine the number of moles present in each compound.

**a.** 22.6 g $AgNO_3$ **b.** 6.50 g $ZnSO_4$ **c.** 35.0 g HCl

**41. Challenge** Identify each as an ionic or molecular compound and convert the given mass to moles. Express your answers in scientific notation.

**a.** 2.50 kg $Fe_2O_3$ **b.** 25.4 mg $PbCl_4$

## Converting the Mass of a Compound to Number of Particles

**FOLDABLES®** Incorporate information from this section into your Foldable.

Example Problem 8 illustrated how to find the number of moles of a compound contained in a given mass. Now, you will learn how to calculate the number of representative particles—molecules or formula units—contained in a given mass and, in addition, the number of atoms or ions.

Recall that no direct conversion is possible between mass and number of particles. You must first convert the given mass to moles by multiplying by the inverse of the molar mass. Then, you can convert moles to the number of representative particles by multiplying by Avogadro's number. To determine numbers of atoms or ions in a compound, you will need conversion factors that are ratios of the number of atoms or ions in the compound to 1 mol of compound. These are based on the chemical formula. Example Problem 9 provides practice in solving this type of problem.

### EXAMPLE Problem 9

Find help with **significant fitures.** Math Handbook

EXAMPLE PROBLEM

**CONVERSION FROM MASS TO MOLES TO PARTICLES** Aluminum chloride ($AlCl_3$) is used in refining petroleum and manufacturing rubber and lubricants. A sample of aluminum chloride has a mass of 35.6 g.

**a.** How many aluminum ions are present?

**b.** How many chloride ions are present?

**c.** What is the mass, in grams, of one formula unit of aluminum chloride?

#### 1 ANALYZE THE PROBLEM

You are given 35.6 g of $AlCl_3$ and must calculate the number of $Al^{3+}$ ions, the number of $Cl^-$ ions, and the mass in grams of one formula unit of $AlCl_3$. Molar mass, Avogadro's number, and ratios from the chemical formula are the necessary conversion factors. The ratio of $Al^{3+}$ ions to $Cl^-$ ions in the chemical formula is 1:3. Therefore, the calculated numbers of ions should be in that same ratio. The mass of one formula unit in grams will be an extremely small number.

| **Known** | **Unknown** |
|---|---|
| mass = 35.6 g $AlCl_3$ | **number of ions = ? $Al^{3+}$ ions** |
| | **number of ions = ? $Cl^-$ ions** |
| | **mass = ? g/formula unit $AlCl_3$** |

#### 2 SOLVE FOR THE UNKNOWN

Determine the molar mass of $AlCl_3$.

$$1\ \cancel{\text{mol Al}} \times \frac{26.98\text{ g Al}}{1\ \cancel{\text{mol Al}}} = 26.98\text{ g Al}$$

**Multiply the moles of Al in the compound by the molar mass of Al.**

$$3\ \cancel{\text{mol Cl}} \times \frac{35.45\text{ g Cl}}{1\ \cancel{\text{mol Cl}}} = 106.35\text{ g Cl}$$

**Multiply the moles of Cl in the compound by the molar mass of Cl.**

$$\text{molar mass} = (26.98\text{ g} + 106.35\text{ g}) = 133.33\text{ g/mol AlCl}_3$$

**Total the molar mass values.**

Use a conversion factor–the inverse of the molar mass–that relates moles to grams.

$$\text{mass AlCl}_3 \times \frac{1\text{ mol AlCl}_3}{\text{grams AlCl}_3} = \text{moles AlCl}_3$$

**Apply the conversion factor.**

$$35.6\ \cancel{\text{g AlCl}_3} \times \frac{1\text{ mol AlCl}_3}{133.33\ \cancel{\text{g AlCl}_3}} = 0.267\text{ mol AlCl}_3$$

**Substitute mass $AlCl_3$ = 35.6 g and inverse molar mass $AlCl_3$ = 1 mol/133.33 g, and solve.**

Use Avogadro's number.

$0.267 \cancel{\text{mol AlCl}_3} \times \frac{6.02 \times 10^{23} \text{ formula units}}{1 \cancel{\text{mol AlCl}_3}}$ — Multiply and divide numbers and units.

$= 1.61 \times 10^{23}$ formula units $AlCl_3$

To calculate the number of $Al^{3+}$ and $Cl^-$ ions, use the ratios from the chemical formula as conversion factors.

$1.61 \times 10^{23} \cancel{\text{AlCl}_3 \text{ formula units}} \times \frac{1 \text{ Al}^{3+} \text{ ion}}{1 \cancel{\text{AlCl}_3 \text{ formula unit}}}$ — Multiply and divide numbers and units.

**$= 1.61 \times 10^{23}$ $Al^{3+}$ ions**

$1.61 \times 10^{23} \cancel{\text{AlCl}_3 \text{ formula units}} \times \frac{3 \text{ Cl}^- \text{ ions}}{1 \cancel{\text{AlCl}_3 \text{ formula unit}}}$ — Multiply and divide numbers and units.

**$= 4.83 \times 10^{23}$ $Cl^-$ ions**

Calculate the mass in grams of one formula unit of $AlCl_3$. Use the inverse of Avogadro's number as a conversion factor.

$\frac{133.33 \text{ g AlCl}_3}{1 \cancel{\text{mol}}} \times \frac{1 \cancel{\text{mol}}}{6.02 \times 10^{23} \text{ formula units}}$ — Substitute mass $AlCl_3$ = 133.33 g, and solve.

**$= 2.21 \times 10^{-22}$ g $AlCl_3$/formula unit**

## 3 EVALUATE THE ANSWER

A minimum of three significant figures is used in each value in the calculations. Therefore, the answers have the correct number of digits. The number of $Cl^-$ ions is three times the number of $Al^{3+}$ ions, as predicted. The mass of a formula unit of $AlCl_3$ can be checked by calculating it in a different way. Divide the sample mass of $AlCl_3$ (35.6 g) by the number of formula units contained in the mass ($1.61 \times 10^{23}$ formula units) to obtain the mass of one formula unit. The two answers are the same.

## PRACTICE Problems

Do additional problems. Online Practice

**42.** Ethanol ($C_2H_5OH$), a domestically produced fuel source, is often blended with gasoline. A sample of ethanol has a mass of 45.6 g.

**a.** How many carbon atoms does the sample contain?

**b.** How many hydrogen atoms are present?

**c.** How many oxygen atoms are present?

**43.** A sample of sodium sulfite ($Na_2SO_3$) has a mass of 2.25 g.

**a.** How many $Na^+$ ions are present?

**b.** How many $SO_3^{2-}$ ions are present?

**c.** What is the mass in grams of one formula unit of $Na_2SO_3$?

**44.** A sample of carbon dioxide ($CO_2$) has a mass of 52.0 g.

**a.** How many carbon atoms are present?

**b.** How many oxygen atoms are present?

**c.** What is the mass in grams of one molecule of $CO_2$?

**45.** What mass of sodium chloride (NaCl) contains $4.59 \times 10^{24}$ formula units?

**46. Challenge** A sample of silver chromate has a mass of 25.8 g.

**a.** Write the formula for silver chromate.

**b.** How many cations are present in the sample?

**c.** How many anions are present in the sample?

**d.** What is the mass in grams of one formula unit of silver chromate?

PRACTICE PROBLEMS

■ **Figure 11** Note the central position of the mole. To go from the left, right, or top of the diagram to any other place, you must go through the mole. The conversion factors on the arrows provide the means for making the conversions.

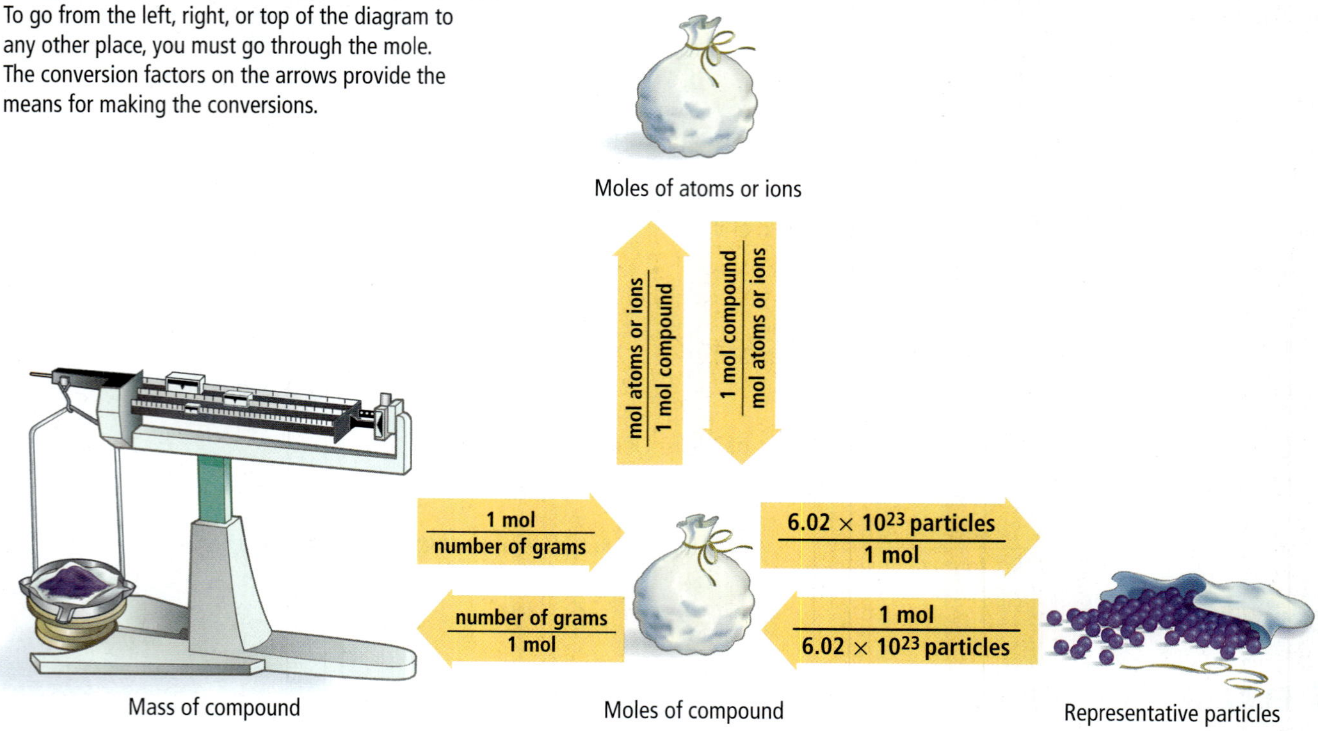


Conversions between mass, moles, and the number of particles are summarized in **Figure 11.** Note that molar mass and the inverse of molar mass are conversion factors between mass and number of moles. Avogadro's number and its inverse are the conversion factors between moles and the number of representative particles. To convert between moles and the number of moles of atoms or ions contained in the compound, use the ratio of moles of atoms or ions to 1 mole of compound or its inverse, which are shown on the upward and downward arrows in **Figure 11.** These ratios are derived from the subscripts in the chemical formula.

## SECTION 3 REVIEW

Section Self-Check

### Section Summary

- Subscripts in a chemical formula indicate how many moles of each element are present in 1 mol of the compound.
- The molar mass of a compound is calculated from the molar masses of all the elements in the compound.
- Conversion factors based on a compound's molar mass are used to convert between moles and mass of a compound.

**47. MAIN**IDEA **Describe** how to determine the molar mass of a compound.

**48. Identify** the conversion factors needed to convert between the number of moles and the mass of a compound.

**49. Explain** how you can determine the number of atoms or ions in a given mass of a compound.

**50. Apply** How many moles of K, C, and O atoms are there in 1 mol of $K_2C_2O_4$?

**51. Calculate** the molar mass of $MgBr_2$.

**52. Calculate** Calcium carbonate is the calcium source for many vitamin tablets. The recommended daily allowance of calcium is 1000 mg of $Ca^{2+}$ ions. How many moles of $Ca^{2+}$ does 1000 mg represent?

**53. Design** a bar graph that will show the number of moles of each element present in 500 g of a particular form of dioxin ($C_{12}H_4Cl_4O_2$), a powerful poison.

SECTION 4

# Empirical and Molecular Formulas

**MAINIDEA** A molecular formula of a compound is a whole-number multiple of its empirical formula.

### Essential Questions

- What is meant by the percent composition of a compound?
- How can the empirical and molecular formulas for a compound be determined from mass percent and actual mass data?

### Review Vocabulary

**percent by mass:** the ratio of the mass of each element to the total mass of the compound expressed as a percent

### New Vocabulary

percent composition
empirical formula
molecular formula

**CHEM 4 YOU** You might have noticed that some beverage bottles and food packages contain two or more servings instead of the single serving you expect. How would you determine the total number of calories contained in the package?

## Percent Composition

Chemists, such as those shown in **Figure 12,** are often involved in developing new compounds for industrial, pharmaceutical, and home uses. After a synthetic chemist (one who makes new compounds) has produced a new compound, an analytical chemist analyzes the compound to provide experimental proof of its composition and its chemical formula.

It is the analytical chemist's job to identify the elements a compound contains and determine their percents by mass. Gravimetric and volumetric analyses are experimental procedures based on the measurement of mass for solids and liquids, respectively.

**Percent composition from experimental data** For example, consider a 100-g sample of a compound that contains 55 g of Element X and 45 g of Element Y. The percent by mass of any element in a compound can be found by dividing the mass of the element by the mass of the compound and multiplying by 100.

$$\text{percent by mass (element)} = \frac{\text{mass of element}}{\text{mass of compound}} \times 100$$

**Figure 12** New compounds are first made on a small scale by a synthetic chemist like the one shown on the left. Then, an analytical chemist, like the one shown on the right, analyzes the compound to verify its structure and percent composition.

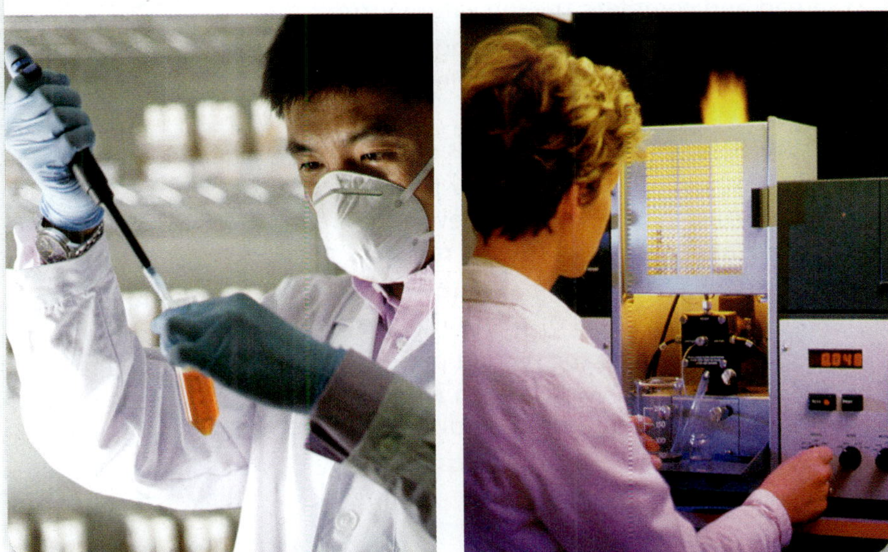

## CAREERS IN CHEMISTRY

**Medicinal Chemist** Much like a chef trying to perfect a recipe, the medicinal chemist works to perfect the best combination of reactants. They use their knowledge of the effects of toxins and medicines on the human body to synthesize new molecules that target disease.

WebQuest

Because percent means parts per 100, the percents by mass of all the elements of a compound must always add up to 100.

$$\frac{55\,\cancel{g}\text{ element X}}{100\,\cancel{g}\text{ compound}} \times 100 = 55\%\text{ element X}$$

$$\frac{45\,\cancel{g}\text{ element Y}}{100\,\cancel{g}\text{ compound}} \times 100 = 45\%\text{ element Y}$$

Thus, the compound is 55% X and 45% Y. The percent by mass of each element in a compound is the **percent composition** of a compound.

**Percent composition from the chemical formula** The percent composition of a compound can also be determined from its chemical formula. To do this, assume you have exactly 1 mol of the compound and use the chemical formula to calculate the compound's molar mass. Then, determine the mass of each element in a mole of the compound by multiplying the element's molar mass by its subscript in the chemical formula. Finally, use the equation below to find the percent by mass of each element.

**Percent by Mass from the Chemical Formula**

$$\text{percent by mass} = \frac{\text{mass of element in 1 mol of compound}}{\text{molar mass of compound}} \times 100$$

The percent by mass of an element in a compound is the mass of the element in 1 mol of the compound divided by the molar mass of the compound, multiplied by 100.

Example Problem 10 covers calculating percent composition.

# MiniLAB

iLab Station

## Analyze Chewing Gum

**Are sweetening and flavoring added as a coating or mixed throughout chewing gum?**

### Procedure

1. Read and complete the lab safety form.
2. Unwrap two pieces of **chewing gum.** Place each piece on a **weighing paper.** Measure and record each mass using a **balance.**
   **WARNING:** *Do not eat any items used in the lab.*
3. Add 150 mL of cold **tap water** to a **250-mL beaker.** Place one piece of chewing gum in the water, and stir with a **stirring rod** for 2 min.
4. Pat the gum dry using **paper towels.** Measure and record the mass of the dried gum.
5. Use **scissors** to cut the second piece of gum into small pieces. Repeat Step 3 using fresh water. Keep the pieces from clumping together.
   **WARNING:** *Use caution with scissors.*
6. Use a 10-cm × 10-cm piece of **window screen** to strain the water from the gum. Pat the gum dry using paper towels. Measure and record the mass of the dried gum.

### Analysis

1. **Calculate** For the uncut piece of gum, calculate the mass of sweeteners and flavorings that dissolved in the water. The mass of sweeteners and flavorings is the difference between the original mass of the gum and the mass of the dried gum.
2. **Calculate** For the gum cut into small pieces, calculate the mass of dissolved sweeteners and flavorings.
3. **Apply** For each piece of gum, determine the percent of the original mass from the soluble sweeteners and flavorings.
4. **Infer** What can you infer from the two percentages? Is the gum sugar-coated or are the sweeteners and flavorings mixed throughout?

## EXAMPLE Problem 10

Find help with **percents.** Math Handbook

EXAMPLE PROBLEM

**CALCULATING PERCENT COMPOSITION** Sodium hydrogen carbonate ($NaHCO_3$), also called baking soda, is an active ingredient in some antacids used for the relief of indigestion. Determine the percent composition of $NaHCO_3$.

### 1 ANALYZE THE PROBLEM

You are given only the chemical formula. Assume you have 1 mol of $NaHCO_3$. Calculate the molar mass and the mass of each element in 1 mol to determine the percent by mass of each element in the compound. The sum of all percents should be 100, although your answer might vary slightly due to rounding.

| **Known** | **Unknown** |
|---|---|
| formula = $NaHCO_3$ | **percent Na = ?** |
| | **percent H = ?** |
| | **percent C = ?** |
| | **percent O = ?** |

### 2 SOLVE FOR THE UNKNOWN

Determine the molar mass of $NaHCO_3$ and each element's contribution.

$$1\ \cancel{\text{mol Na}} \times \frac{22.99\text{ g Na}}{1\ \cancel{\text{mol Na}}} = 22.99\text{ g Na}$$

**Multiply the molar mass of Na by the number of Na atoms in the compound.**

$$1\ \cancel{\text{mol H}} \times \frac{1.008\text{ g H}}{1\ \cancel{\text{mol H}}} = 1.008\text{ g H}$$

**Multiply the molar mass of H by the number of H atoms in the compound.**

$$1\ \cancel{\text{mol C}} \times \frac{12.01\text{ g C}}{1\ \cancel{\text{mol C}}} = 12.01\text{ g C}$$

**Multiply the molar mass of C by the number of C atoms in the compound.**

$$3\ \cancel{\text{mol O}} \times \frac{16.00\text{ g O}}{1\ \cancel{\text{mol O}}} = 48.00\text{ g O}$$

**Multiply the molar mass of O by the number of O atoms in the compound.**

$$\text{molar mass} = (22.99\text{ g} + 1.008\text{ g} + 12.01\text{ g} + 48.00\text{ g})$$
$$= 84.01\text{ g/mol NaHCO}_3$$

**Total the mass values.**

Use the percent by mass equation.

$$\%\text{ mass element} = \frac{\text{mass of element in 1 mol of compound}}{\text{molar mass of compound}} \times 100$$

**State the equation.**

$$\text{percent Na} = \frac{22.99\text{ g/mol}}{84.01\text{ g/mol}} \times 100 = \mathbf{27.37\%\ Na}$$

**Substitute mass of Na in 1 mol compound = 22.99 g/mol and molar mass $NaHCO_3$ = 84.01 g/mol. Calculate % Na.**

$$\text{percent H} = \frac{1.008\text{ g/mol}}{84.01\text{ g/mol}} \times 100 = \mathbf{1.200\%\ H}$$

**Substitute mass of H in 1 mol compound = 1.008 g/mol and molar mass $NaHCO_3$ = 84.01 g/mol. Calculate % H.**

$$\text{percent C} = \frac{12.01\text{ g/mol}}{84.01\text{ g/mol}} \times 100 = \mathbf{14.30\%\ C}$$

**Substitute mass of C in 1 mol compound = 12.01 g/mol and molar mass $NaHCO_3$ = 84.01 g/mol. Calculate % C.**

$$\text{percent O} = \frac{48.00\text{ g/mol}}{84.01\text{ g/mol}} \times 100 = \mathbf{57.14\%\ O}$$

**Substitute mass of O in 1 mol compound = 48.00 g/mol and molar mass $NaHCO_3$ = 84.01 g/mol. Calculate % O.**

$NaHCO_3$ is 27.37% Na, 1.200% H, 14.30% C, and 57.14% O.

### 3 EVALUATE THE ANSWER

All masses and molar masses contain four significant figures. Therefore, the percents are correctly stated with four significant figures. When rounding error is accounted for, the sum of the mass percents is 100%, as required.

## PRACTICE Problems

Do additional problems. Online Practice

54. What is the percent composition of phosphoric acid ($H_3PO_4$)?
55. Which has the larger percent by mass of sulfur, $H_2SO_3$ or $H_2S_2O_8$?
56. Calcium chloride ($CaCl_2$) is sometimes used as a de-icer. Calculate the percent by mass of each element in $CaCl_2$.
57. **Challenge** Sodium sulfate is used in the manufacture of detergents.
    - **a.** Identify each of the component elements of sodium sulfate, and write the compound's chemical formula.
    - **b.** Identify the compound as ionic or covalent.
    - **c.** Calculate the percent by mass of each element in sodium sulfate.

## Empirical Formula

When a compound's percent composition is known, its formula can be calculated. First, determine the smallest whole-number ratio of the moles of the elements in the compound. This ratio gives the subscripts in the empirical formula. The **empirical formula** for a compound is the formula with the smallest whole-number mole ratio of the elements. The empirical formula might or might not be the same as the actual molecular formula. If the two formulas are different, the molecular formula will always be a simple multiple of the empirical formula. The empirical formula for hydrogen peroxide is HO; the molecular formula is $H_2O_2$. In both formulas, the ratio of oxygen to hydrogen is 1:1.

Percent composition or masses of the elements in a given mass of compound can be used to determine the formula for the compound. If percent composition is given, assume the total mass of the compound is 100.00 g and that the percent by mass of each element is equal to the mass of that element in grams. This can be seen in **Figure 13,** where 100.00 g of the 40.05% S and 59.95% O compound contains 40.05 g of S and 59.95 g of O. The mass of each element is then converted to moles.

$$40.05\ \cancel{g\ S} \times \frac{1\text{ mol S}}{32.07\ \cancel{g\ S}} = 1.249\text{ mol S}$$

$$59.95\ \cancel{g\ O} \times \frac{1\text{ mol O}}{16.00\ \cancel{g\ O}} = 3.747\text{ mol O}$$

Thus, the mole ratio of S atoms to O atoms in the oxide is 1.249:3.747.

When the values in a mole ratio are not whole numbers, they cannot be used as subscripts in a chemical formula. You can convert the ratio to whole numbers by recognizing that the element with the smallest number of moles might have the smallest subscript possible, 1. To make the mole value of sulfur equal to 1, divide both mole values by the moles of sulfur (1.249). This does not change the ratio between the two elements because both are divided by the same number.

$$\frac{1.249\text{ mol S}}{1.249} = 1\text{ mol S} \qquad \frac{3.747\text{ mol O}}{1.249} = 3\text{ mol O}$$

The simplest whole-number mole ratio of S to O is 1:3. Thus, the empirical formula is $SO_3$. Sometimes, dividing by the smallest mole value does not yield whole numbers. In such cases, each mole value must then be multiplied by the smallest factor that will make it a whole number. This is shown in Example Problem 11.

**READING CHECK** **List** the steps needed to calculate the empirical formula from percent composition data.

**Figure 13** Keep this figure in mind when doing problems using percent composition. You can always assume that you have a 100-g sample of the compound and use the percents of the elements as masses of the elements.

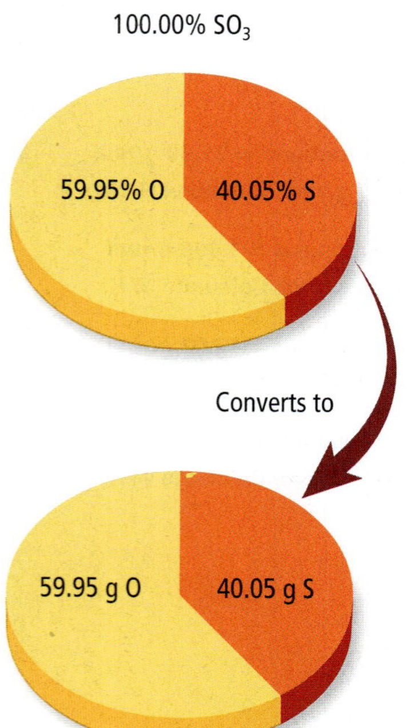

## EXAMPLE Problem 11

Find help with **ratios.** Math Handbook

EXAMPLE PROBLEM

**EMPIRICAL FORMULA FROM PERCENT COMPOSITION** Methyl acetate is a solvent commonly used in some paints, inks, and adhesives. Determine the empirical formula for methyl acetate, which has the following chemical analysis: 48.64% carbon, 8.16% hydrogen, and 43.20% oxygen.

### 1 ANALYZE THE PROBLEM

You are given the percent composition of methyl acetate and must find the empirical formula. Because you can assume that each percent by mass represents the mass of the element in a 100.00-g sample, the percent sign can be replaced with the unit grams. Then, convert from grams to moles and find the smallest whole-number ratio of moles of the elements.

**Known**

percent by mass C = 48.64% C

percent by mass H = 8.16% H

percent by mass O = 43.20% O

**Unknown**

**empirical formula** = ?

### 2 SOLVE FOR THE UNKNOWN

Convert each mass to moles using a conversion factor—the inverse of the molar mass—that relates moles to grams.

$$48.64\ \cancel{g\,C} \times \frac{1\ \text{mol C}}{12.01\ \cancel{g\,C}} = 4.050\ \text{mol C}$$

**Substitute mass C = 48.64 g, inverse molar mass C = 1 mol/12.01 g, and calculate moles of C.**

$$8.16\ \cancel{g\,H} \times \frac{1\ \text{mol H}}{1.008\ \cancel{g\,H}} = 8.10\ \text{mol H}$$

**Substitute mass H = 8.16 g, inverse molar mass H = 1 mol/1.008 g, and calculate moles of H.**

$$43.20\ \cancel{g\,O} \times \frac{1\ \text{mol O}}{16.00\ \cancel{g\,O}} = 2.700\ \text{mol O}$$

**Substitute mass O = 43.20 g, inverse molar mass O = 1 mol/16.00 g, and calculate moles of O.**

Methyl acetate has a mole ratio of (4.050 mol C):(8.10 mol H):(2.700 mol O).
Next, calculate the simplest ratio of moles of elements by dividing the moles of each element by the smallest value in the calculated mole ratio.

$$\frac{4.050\ \text{mol C}}{2.700} = 1.500\ \text{mol C} = 1.5\ \text{mol C}$$

**Divide moles of C by 2.700.**

$$\frac{8.10\ \text{mol H}}{2.700} = 3.00\ \text{mol H} = 3\ \text{mol H}$$

**Divide moles of H by 2.700.**

$$\frac{2.700\ \text{mol O}}{2.700} = 1.000\ \text{mol O} = 1\ \text{mol O}$$

**Divide moles of O by 2.700.**

The simplest mole ratio is (1.5 mol C):(3 mol H):(1 mol O). Multiply each number in the ratio by the smallest number—in this case 2—that yields a ratio of whole numbers.

$2 \times 1.5$ mol C = 3 mol C — **Multiply moles of C by 2 to obtain a whole number.**

$2 \times 3$ mol H = 6 mol H — **Multiply moles of H by 2 to obtain a whole number.**

$2 \times 1$ mol O = 2 mol O — **Multiply moles of O by 2 to obtain a whole number.**

The simplest whole-number ratio of atoms is (3 atoms C):(6 atoms H):(2 atoms O). Thus, the empirical formula of methyl acetate is **$C_3H_6O_2$**.

### 3 EVALUATE THE ANSWER

The calculations are correct, and significant figures have been observed. To check that the formula is correct, calculate the percent composition represented by the formula. The percent composition checks exactly with the data given in the problem.

## PRACTICE Problems

Do additional problems. Online Practice

**58.** The circle graph at the right gives the percent composition for a blue solid. What is the empirical formula for this solid?

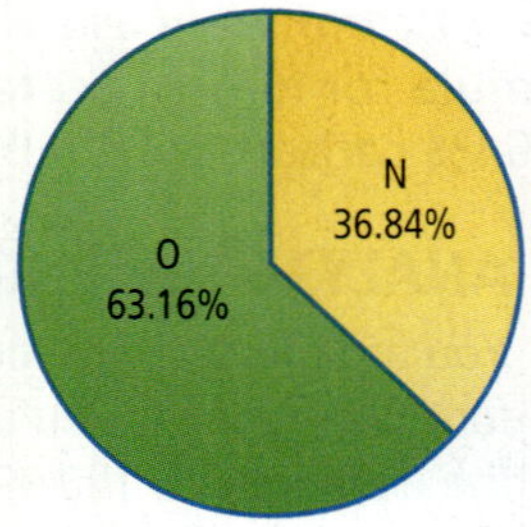


**59.** Determine the empirical formula for a compound that contains 35.98% aluminum and 64.02% sulfur.

**60.** Propane is a hydrocarbon, a compound composed only of carbon and hydrogen. It is 81.82% carbon and 18.18% hydrogen. What is the empirical formula?

**61. Challenge** Aspirin is the world's most-often used medication. The chemical analysis of aspirin indicates that the molecule is 60.00% carbon, 4.44% hydrogen, and 35.56% oxygen. Determine the empirical formula for aspirin.

## Molecular Formula

Would it surprise you to learn that substances with distinctly different properties can have the same percent composition and the same empirical formula? How is this possible? Remember that the subscripts in an empirical formula indicate the simplest whole-number ratio of moles of the elements in the compound. However, the simplest ratio does not always indicate the actual ratio in the compound. To identify a new compound, a chemist determines the **molecular formula,** which specifies the actual number of atoms of each element in one molecule or formula unit of the substance. **Figure 14** shows an important use of the gas acetylene. It has the same percent composition and the same empirical formula (CH) as benzene, which is a liquid. Yet chemically and structurally, acetylene and benzene are very different.

To determine the molecular formula for a compound, the molar mass of the compound must be determined through experimentation and compared with the mass represented by the empirical formula. For example, the molar mass of acetylene is 26.04 g/mol, and the mass of the empirical formula (CH) is 13.02 g/mol. Dividing the actual molar mass by the mass of the empirical formula indicates that the molar mass of acetylene is two times the mass of the empirical formula.

$$\frac{\text{experimentally determined molar mass of acetylene}}{\text{mass of empirical formula}} = \frac{26.04\ \cancel{\text{g/mol}}}{13.02\ \cancel{\text{g/mol}}} = 2.000$$

Because the molar mass of acetylene is two times the mass represented by the empirical formula, the molecular formula of acetylene must contain twice the number of carbon and hydrogen atoms as represented by the empirical formula.

■ **Figure 14** Acetylene is a gas used for welding because of the high-temperature flame produced when it is burned with oxygen.

Similarly, when the experimentally determined molar mass of benzene, 78.12 g/mol, is compared with the mass of the empirical formula, the molar mass of benzene is found to be six times the mass of the empirical formula.

$$\frac{\text{experimentally determined molar mass of benzene}}{\text{mass of the empirical formula CH}} = \frac{78.12 \text{ g mol}}{13.02 \text{ g mol}} = 6.000$$

The molar mass of benzene is six times the mass represented by the empirical formula, so the molecular formula for benzene must represent six times the number of carbon atoms and hydrogen atoms shown in the empirical formula. You can conclude that the molecular formula for acetylene is $2 \times$ CH, or $C_2H_2$, and the molecular formula for benzene is $6 \times$ CH, or $C_6H_6$.

A molecular formula can be represented as the empirical formula multiplied by an integer *n*.

$$\text{molecular formula} = (\text{empirical formula})n$$

The integer *n* is the factor (6 in the example of benzene above) by which the subscripts in the empirical formula must be multiplied to obtain the molecular formula.

The steps in determining empirical and molecular formulas from percent composition or mass data are outlined in **Figure 15.** As in other calculations, the route leads from mass through moles because formulas are based on the relative numbers of moles of elements in each mole of compound.

**Figure 15** Use this flowchart to guide you through the steps in determining the empirical and molecular formulas for compounds.
**Describe** *How is the integer* n *related to the empirical and molecular formulas?*

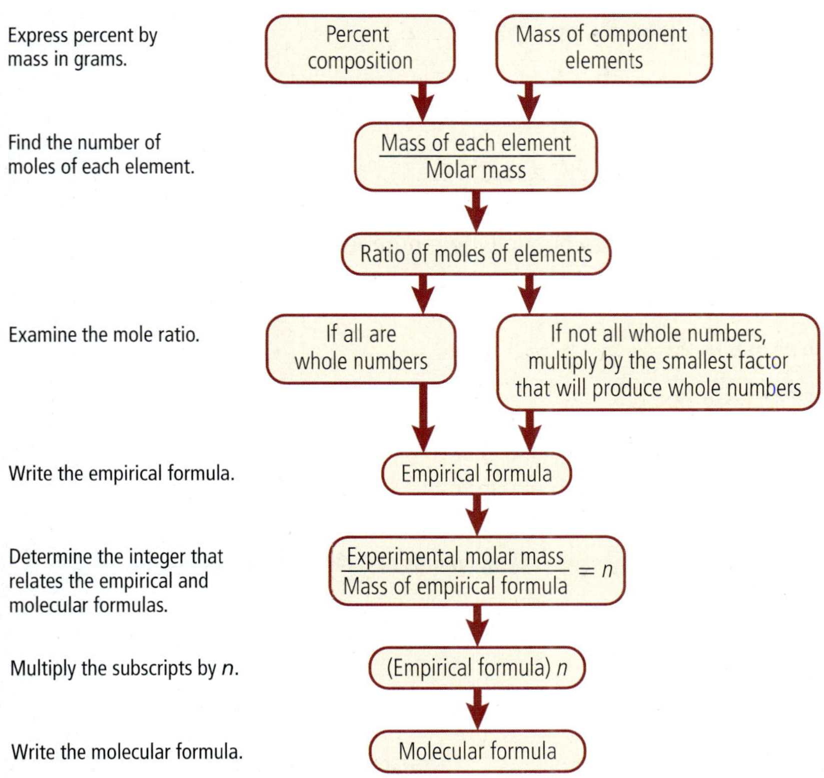

## EXAMPLE Problem 12

Find help with **ratios.** Math Handbook

EXAMPLE PROBLEM

**DETERMINING A MOLECULAR FORMULA** Succinic acid is a substance produced by lichens. Chemical analysis indicates it is composed of 40.68% carbon, 5.08% hydrogen, and 54.24% oxygen and has a molar mass of 118.1 g/mol. Determine the empirical and molecular formulas for succinic acid.

### 1 ANALYZE THE PROBLEM

You are given the percent composition. Assume that each percent by mass represents the mass of the element in a 100.00-g sample. You can compare the given molar mass with the mass represented by the empirical formula to find *n*.

| **Known** | **Unknown** |
|---|---|
| percent by mass C = 40.68% C | **empirical formula = ?** |
| percent by mass H = 5.08% H | **molecular formula = ?** |
| percent by mass O = 54.24% O | |
| molar mass = 118.1 g/mol succinic acid | |

### 2 SOLVE FOR THE UNKNOWN

Use the percents by mass as masses in grams, and convert grams to moles by using a conversion factor—the inverse of molar mass—that relates moles to mass.

$$40.68\ \cancel{g\ C} \times \frac{1\ \text{mol C}}{12.01\ \cancel{g\ C}} = 3.387\ \text{mol C}$$

**Substitute mass C = 40.68 g, inverse molar mass C = 1 mol/12.01 g, and solve for moles of C.**

$$5.08\ \cancel{g\ H} \times \frac{1\ \text{mol H}}{1.008\ \cancel{g\ H}} = 5.04\ \text{mol H}$$

**Substitute mass H = 5.08 g, inverse molar mass H = 1 mol/1.008 g, and solve for moles of H.**

$$54.24\ \cancel{g\ O} \times \frac{1\ \text{mol O}}{16.00\ \cancel{g\ O}} = 3.390\ \text{mol O}$$

**Substitute mass O = 54.24 g, inverse molar mass O = 1 mol/16.00 g, and solve for moles of O.**

The mole ratio in succinic acid is (3.387 mol C):(5.04 mol H):(3.390 mol O).

Next, calculate the simplest ratio of moles of elements by dividing the moles of each element by the smallest value in the calculated mole ratio.

$$\frac{3.387\ \text{mol C}}{3.387} = 1\ \text{mol C}$$

**Divide moles of C by 3.387.**

$$\frac{5.04\ \text{mol H}}{3.387} = 1.49\ \text{mol H} \approx 1.5\ \text{mol H}$$

**Divide moles of H by 3.387.**

$$\frac{3.390\ \text{mol O}}{3.387} = 1.001\ \text{mol O} \approx 1\ \text{mol O}$$

**Divide moles of O by 3.387.**

The simplest mole ratio is 1:1.5:1. Multiply all mole values by 2 to obtain whole numbers.

$2 \times 1\ \text{mol C} = 2\ \text{mol C}$ **Multiply moles of C by 2.**

$2 \times 1.5\ \text{mol H} = 3\ \text{mol H}$ **Multiply moles of H by 2.**

$2 \times 1\ \text{mol O} = 2\ \text{mol O}$ **Multiply moles of O by 2.**

The simplest whole-number mole ratio is 2:3:2. The empirical formula is **$C_2H_3O_2$.**

Calculate the empirical formula mass using the molar mass of each element.

$$2\ \cancel{\text{mol C}} \times \frac{12.01\ \text{g C}}{1\ \cancel{\text{mol C}}} = 24.02\ \text{g C}$$

**Multiply the molar mass of C by the moles of C atoms in the compound.**

$$3\ \cancel{\text{mol H}} \times \frac{1.008\ \text{g H}}{1\ \cancel{\text{mol H}}} = 3.024\ \text{g H}$$

**Multiply the molar mass of H by the moles of H atoms in the compound.**

$$2\ \cancel{\text{mol O}} \times \frac{16.00\ \text{g O}}{1\ \cancel{\text{mol O}}} = 32.00\ \text{g O}$$

**Multiply the molar mass of O by the moles of O atoms in the compound.**

molar mass $C_2H_3O_2$ = (24.02 g + 3.024 g + 32.00 g) = 59.04 g/mol **Total the mass values.**

Divide the experimentally determined molar mass of succinic acid by the mass of the empirical formula to determine *n*.

$$n = \frac{\text{molar mass of succinic acid}}{\text{molar mass of } C_2H_3O_2} = \frac{118.1 \text{ g/mol}}{59.04 \text{ g/mol}} = 2.000$$

Multiply the subscripts in the empirical formula by 2 to determine the actual subscripts in the molecular formula.

$2 \times (C_2H_3O_2) = C_4H_6O_4$

The molecular formula for succinic acid is **$C_4H_6O_4$.**

### 3 EVALUATE THE ANSWER

The calculation of the molar mass from the molecular formula gives the same result as the given, experimentally-determined molar mass.

## EXAMPLE Problem 13

Find help with **ratios.** Math Handbook

**CALCULATING AN EMPIRICAL FORMULA FROM MASS DATA** The mineral ilmenite is usually mined and processed for titanium, a strong, light, and flexible metal. A sample of ilmenite contains 5.41 g of iron, 4.64 g of titanium, and 4.65 g of oxygen. Determine the empirical formula for ilmenite.

### 1 ANALYZE THE PROBLEM

You are given the masses of the elements found in a known mass of ilmenite and must determine the empirical formula of the mineral. Convert the known masses of each element to moles, then find the smallest whole-number ratio of the moles of the elements.

| **Known** | **Unknown** |
|---|---|
| mass of iron = 5.41 g Fe | **empirical formula = ?** |
| mass of titanium = 4.64 g Ti | |
| mass of oxygen = 4.65 g O | |

### 2 SOLVE FOR THE UNKNOWN

Convert each known mass to moles by using a conversion factor—the inverse of molar mass—that relates moles to grams.

$$5.41 \text{ g Fe} \times \frac{1 \text{ mol Fe}}{55.85 \text{ g Fe}} = 0.0969 \text{ mol Fe}$$

**Multiply mass Fe = 5.41 g by the inverse molar mass Fe = 1 mol/55.85 g, and calculate moles of Fe.**

$$4.64 \text{ g Ti} \times \frac{1 \text{ mol Ti}}{47.87 \text{ g Ti}} = 0.0969 \text{ mol Ti}$$

**Multiply mass Ti = 4.64 g by the inverse molar mass Ti = 1 mol/47.87 g, and calculate moles of Ti.**

$$4.65 \text{ g O} \times \frac{1 \text{ mol O}}{16.00 \text{ g O}} = 0.291 \text{ mol O}$$

**Multiply mass O = 4.65 g by the inverse molar mass O = 1 mol/16.00 g, and calculate moles of O.**

The mineral ilmenite has a mole ratio of (0.0969 mol Fe):(0.0969 mol Ti):(0.291 mol O). Calculate the simplest ratio by dividing each mole value by the smallest value in the ratio.

$$\frac{0.0969 \text{ mol Fe}}{0.0969} = 1 \text{ mol Fe}$$

**Divide the moles of Fe by 0.0969.**

$$\frac{0.0969 \text{ mol Ti}}{0.0969} = 1 \text{ mol Ti}$$

**Divide the moles of Ti by 0.0969.**

$$\frac{0.291 \text{ mol O}}{0.0969} = 3 \text{ mol O}$$

**Divide the moles of O by 0.0969.**

Because all the mole values are whole numbers, the simplest whole-number mole ratio is (1 mol Fe):(1 mol Ti):(3 mol O). The empirical formula for ilmenite is **$FeTiO_3$.**

EXAMPLE PROBLEM

## 3 EVALUATE THE ANSWER

The mass of iron is slightly greater than the mass of titanium, but the molar mass of iron is also slightly greater than that of titanium. Thus, it is reasonable that the numbers of moles of iron and titanium are equal. The mass of titanium is approximately the same as the mass of oxygen, but the molar mass of oxygen is about one-third that of titanium. Thus, a 3:1 ratio of oxygen to titanium is reasonable.

## PRACTICE Problems

PRACTICE PROBLEMS

Do additional problems. Online Practice

**62.** A compound was found to contain 49.98 g of carbon and 10.47 g of hydrogen. The molar mass of the compound is 58.12 g/mol. Determine the molecular formula.

**63.** A colorless liquid composed of 46.68% nitrogen and 53.32% oxygen has a molar mass of 60.01 g/mol. What is the molecular formula?

**64.** When an oxide of potassium is decomposed, 19.55 g of K and 4.00 g of O are obtained. What is the empirical formula for the compound?

**65. Challenge** Analysis of a chemical used in photographic developing fluid yielded the percent composition data shown in the circle graph to the right. If the chemical's molar mass is 110.0 g/mol, what is its molecular formula?

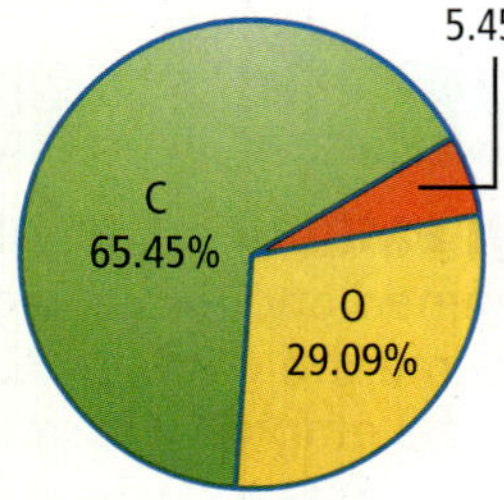


**66. Challenge** Analysis of the pain reliever morphine yielded the data shown in the table. Determine the empirical formula of morphine.

| Element | Mass (g) |
|---|---|
| carbon | 17.900 |
| hydrogen | 1.680 |
| oxygen | 4.225 |
| nitrogen | 1.228 |

## SECTION 4 REVIEW

Section Self-Check

### Section Summary

- The percent by mass of an element in a compound gives the percentage of the compound's total mass due to that element.
- The subscripts in an empirical formula give the smallest whole-number ratio of moles of elements in the compound.
- The molecular formula gives the actual number of atoms of each element in a molecule or formula unit of a substance.
- The molecular formula is a whole-number multiple of the empirical formula.

**67. MAINIDEA Assess** A classmate tells you that experimental data shows a compound's molecular formula to be 2.5 times its empirical formula. Is he correct? Explain.

**68. Calculate** Analysis of a compound composed of iron and oxygen yields 174.86 g of Fe and 75.14 g of O. What is the empirical formula for this compound?

**69. Calculate** An oxide of aluminum contains 0.545 g of Al and 0.485 g of O. Find the empirical formula for the oxide.

**70. Explain** how percent composition data for a compound are related to the masses of the elements in the compound.

**71. Explain** how you can find the mole ratio in a chemical compound.

**72. Apply** The molar mass of a compound is twice that of its empirical formula. How are the compound's molecular and empirical formulas related?

**73. Analyze** Hematite ($Fe_2O_3$) and magnetite ($Fe_3O_4$) are two ores used as sources of iron. Which ore provides the greater percent of iron per kilogram?

## SECTION 5

# Formulas of Hydrates

**MAINIDEA** Hydrates are solid ionic compounds in which water molecules are trapped.

**CHEM 4 YOU** Some products, such as electronic equipment, are boxed with small packets labeled *dessicant.* These packets control moisture by absorbing water. Some contain ionic compounds called hydrates.

### Essential Questions

- What is a hydrate and how does its name relate to its composition?
- How is the formula of a hydrate determined from laboratory data?

### Review Vocabulary

**crystal lattice:** a three-dimensional geometric arrangement of particles

### New Vocabulary

hydrate

## Naming Hydrates

Have you ever watched crystals slowly form from a water solution? Sometimes, water molecules adhere to the ions as the solid forms. The water molecules that become part of the crystal are called waters of hydration. Solid ionic compounds in which water molecules are trapped are called hydrates. A **hydrate** is a compound that has a specific number of water molecules bound to its atoms. **Figure 16** shows the beautiful gemstone known as opal, which is hydrated silicon dioxide ($SiO_2$). The unusual coloring is the result of water in the mineral.

In the formula of a hydrate, the number of water molecules associated with each formula unit of the compound is written following a dot—for example, $Na_2CO_3{\cdot}10H_2O$. This compound is called sodium carbonate decahydrate. In the word *decahydrate,* the prefix *deca-* means *ten* and the root word *hydrate* refers to *water.* A decahydrate has ten water molecules associated with one formula unit of compound. The mass of water associated with a formula unit is included in molar mass calculations. The number of water molecules associated with hydrates varies widely. Some common hydrates are listed in **Table 1.**

■ **Figure 16** The presence of water and various mineral impurities accounts for the variety of different-colored opals. Further changes in color occur when opals are allowed to dry out.

Explore **naming hydrates with an interactive table.** Concepts In Motion

**Table 1 Formulas of Hydrates**

| Prefix | Molecules $H_2O$ | Formula | Name |
|---|---|---|---|
| Mono- | 1 | $(NH_4)_2C_2O_4{\cdot}H_2O$ | ammonium oxalate monohydrate |
| Di- | 2 | $CaCl_2{\cdot}2H_2O$ | calcium chloride dihydrate |
| Tri- | 3 | $NaC_2H_3O_2{\cdot}3H_2O$ | sodium acetate trihydrate |
| Tetra- | 4 | $FePO_4{\cdot}4H_2O$ | iron(III) phosphate tetrahydrate |
| Penta- | 5 | $CuSO_4{\cdot}5H_2O$ | copper(II) sulfate pentahydrate |
| Hexa- | 6 | $CoCl_2{\cdot}6H_2O$ | cobalt(II) chloride hexahydrate |
| Hepta- | 7 | $MgSO_4{\cdot}7H_2O$ | magnesium sulfate heptahydrate |
| Octa- | 8 | $Ba(OH)_2{\cdot}8H_2O$ | barium hydroxide octahydrate |
| Deca- | 10 | $Na_2CO_3{\cdot}10H_2O$ | sodium carbonate decahydrate |

The hydrate cobalt(II) chloride hexahydrate is pink.

The hydrate can be heated to drive off the water of hydration.

Anhydrous cobalt(II) chloride is blue.

■ **Figure 17** Water of hydration can be removed by heating a hydrate, producing an anhydrous compound that can look very different from its hydrated form.

## Analyzing a Hydrate

When a hydrate is heated, water molecules are driven off leaving an anhydrous compound, or one "without water." See **Figure 17.** The series of photos show that when pink cobalt(II) chloride hexahydrate is heated, blue anhydrous cobalt(II) chloride is produced.

How can you determine the formula of a hydrate? You must find the number of moles of water associated with 1 mol of the hydrate. Suppose you have a 5.00-g sample of a hydrate of barium chloride. You know that the formula is $BaCl_2 \cdot xH_2O$. You must determine *x*, the coefficient of $H_2O$ in the hydrate formula that indicates the number of moles of water associated with 1 mol of $BaCl_2$. To find *x*, you would heat the sample of the hydrate to drive off the water of hydration. After heating, the dried substance, which is anhydrous $BaCl_2$, has a mass of 4.26 g. The mass of the water of hydration is the difference between the mass of the hydrate (5.00 g) and the mass of the anhydrous compound (4.26 g).

$$5.00 \text{ g } BaCl_2 \text{ hydrate} - 4.26 \text{ g anhydrous } BaCl_2 = 0.74 \text{ g } H_2O$$

You now know the masses of $BaCl_2$ and $H_2O$ in the sample. You can convert these masses to moles using the molar masses. The molar mass of $BaCl_2$ is 208.23 g/mol, and the molar mass of $H_2O$ is 18.02 g/mol.

$$4.26 \cancel{\text{g } BaCl_2} \times \frac{1 \text{ mol } BaCl_2}{208.23 \cancel{\text{g } BaCl_2}} = 0.0205 \text{ mol } BaCl_2$$

$$0.74 \cancel{\text{g } H_2O} \times \frac{1 \text{ mol } H_2O}{18.02 \cancel{\text{g } H_2O}} = 0.041 \text{ mol } H_2O$$

Now that the moles of $BaCl_2$ and $H_2O$ have been determined, you can calculate the ratio of moles of $H_2O$ to moles of $BaCl_2$ which is *x*, the coefficient that precedes $H_2O$ in the formula for the hydrate.

$$x = \frac{\text{moles } H_2O}{\text{moles } BaCl_2} = \frac{0.041 \text{ mol } H_2O}{0.0205 \text{ mol } BaCl_2} = \frac{2.0 \text{ mol } H_2O}{1.00 \text{ mol } BaCl_2} = \frac{2}{1}$$

The ratio of moles of $H_2O$ to moles of $BaCl_2$ is 2:1, so 2 mol of water is associated with 1 mol of barium chloride. The value of the coefficient *x* is 2 and the formula of the hydrate is $BaCl_2 \cdot 2H_2O$. What is the name of the hydrate? The ChemLab at the end of this chapter will give you practice in experimentally determining the formula of a hydrate.

☑ READING CHECK **Explain** why a dot is used in writing the formula of a hydrate.

**VOCABULARY**

**WORD ORIGIN**

**Anhydrous**
comes from the Greek prefix *an–*, meaning *not* or *without*, and *–hydrous*, from the Greek root *hydro* meaning *water*

## EXAMPLE Problem 14

Find help with **ratios.** Math Handbook

EXAMPLE PROBLEM

**Determining the Formula of a Hydrate** A mass of 2.50 g of blue, hydrated copper sulfate ($CuSO_4 \cdot xH_2O$) is placed in a crucible and heated. After heating, 1.59 g of white anhydrous copper sulfate ($CuSO_4$) remains. What is the formula for the hydrate? Name the hydrate.

### 1 ANALYZE THE PROBLEM

You are given a mass of hydrated copper sulfate. The mass after heating is the mass of the anhydrous compound. You know the formula for the compound, except for *x*, the number of moles of water of hydration.

**Known**

mass of hydrated compound = 2.50 g $CuSO_4 \cdot xH_2O$

mass of anhydrous compound = 1.59 g $CuSO_4$

molar mass $H_2O$ = 18.02 g/mol $H_2O$

molar mass $CuSO_4$ = 159.6 g/mol $CuSO_4$

**Unknown**

**formula of hydrate = ?**

**name of hydrate = ?**

### 2 SOLVE FOR THE UNKNOWN

Determine the mass of water lost.

| | |
|---|---|
| mass of hydrated copper sulfate | 2.50 g |
| mass of anhydrous copper sulfate | −1.59 g |
| mass of water lost | 0.91 g |

**Subtract the mass of anhydrous $CuSO_4$ from the mass of $CuSO_4 \cdot xH_2O$.**

Convert the known masses of $H_2O$ and anhydrous $CuSO_4$ to moles using a conversion factor—the inverse of molar mass—that relates moles and mass.

$$1.59\ \cancel{g\ CuSO_4} \times \frac{1\ mol\ CuSO_4}{159.6\ \cancel{g\ CuSO_4}} = 0.00996\ mol\ CuSO_4$$

**Substitute mass $CuSO_4$ = 1.59 g, inverse molar mass $CuSO_4$ = 1 mol/159.6 g, and solve.**

$$0.91\ \cancel{g\ H_2O} \times \frac{1\ mol\ H_2O}{18.02\ \cancel{g\ H_2O}} = 0.050\ mol\ H_2O$$

**Substitute mass $H_2O$ = 0.91 g, inverse molar mass $H_2O$ = 1 mol/18.02 g, and solve.**

$$x = \frac{moles\ H_2O}{moles\ CuSO_4}$$

**State the ratio of moles of $H_2O$ to moles of $CuSO_4$.**

$$x = \frac{0.050\ mol\ H_2O}{0.00996\ mol\ CuSO_4} \approx \frac{5.0\ mol\ H_2O}{1\ mol\ CuSO_4} = 5$$

**Substitute moles of $H_2O$ = 0.050 mol, moles of $CuSO_4$ = 0.00996 mol. Divide numbers, and cancel units to determine the simplest whole-number ratio.**

The ratio of $H_2O$ to $CuSO_4$ is 5:1, so the formula for the hydrate is **$CuSO_4 \cdot 5H_2O$.**
The name of the hydrate is **copper(II) sulfate pentahydrate.**

### 3 EVALUATE THE ANSWER

Copper(II) sulfate pentahydrate is a common hydrate listed in **Table 1.**

## PRACTICE Problems

Do additional problems. Online Practice

**74.** The composition of a hydrate is given in the circle graph shown at the right. What is the formula and name of this hydrate?

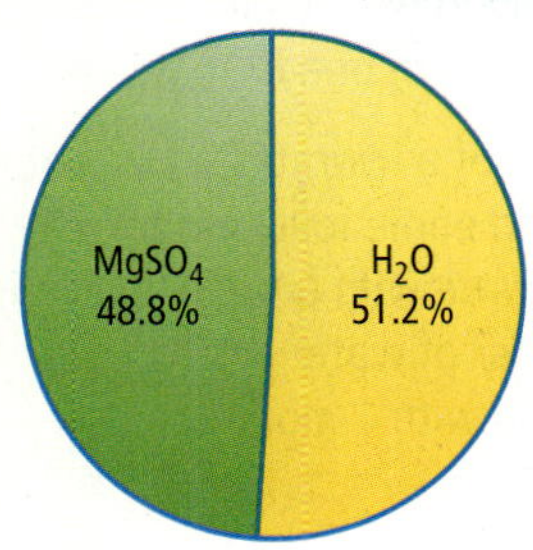


**75. Challenge** An 11.75-g sample of a common hydrate of cobalt(II) chloride is heated. After heating, 0.0712 mol of anhydrous cobalt chloride remains. What is the formula and the name of this hydrate?

■ **Figure 18** Calcium chloride, in the bottom of the desiccator, keeps the air inside the desiccator dry. In the chemistry lab, calcium chloride can also be packed into glass tubes called drying tubes. Drying tubes protect reactions from atmospheric moisture, but allow gases produced by reactions to escape.

## Uses of Hydrates

Anhydrous compounds have important applications in the chemistry laboratory. Calcium chloride forms three hydrates—a monohydrate, a dihydrate, and a hexahydrate. As shown in **Figure 18,** anhydrous calcium chloride is placed in the bottom of tightly sealed containers called desiccators. The calcium chloride absorbs moisture from the air inside the desiccator, creating a dry atmosphere in which other substances can be kept dry. Calcium sulfate is often added to solvents such as ethanol and ethyl ether to keep them free of water.

The ability of the anhydrous form of a hydrate to absorb water also has some important commercial applications. Electronic and optical equipment, particularly equipment that is transported overseas by ship, is often packaged with packets of desiccant. Desiccants prevent moisture from interfering with the sensitive electronic circuitry. While some types of desiccant simply absorb moisture, other types bond with moisture from the air and form hydrates.

Some hydrates, sodium sulfate decahydrate ($Na_2SO_4{\cdot}10H_2O$) for example, are used to store solar energy. When the Sun's energy heats the hydrate to a temperature greater than 32°C, the single formula unit of $Na_2SO_4$ in the hydrate dissolves in the 10 mol of water of hydration. In the process, energy is absorbed by the hydrate. This energy is released when the temperature decreases and the hydrate crystallizes again.

## SECTION 5 REVIEW

Section Self-Check

### Section Summary

- The formula of a hydrate consists of the formula of the ionic compound and the number of water molecules associated with one formula unit.
- The name of a hydrate consists of the compound name followed by the word *hydrate* with a prefix indicating the number of water molecules associated with 1 mol of the compound.
- Anhydrous compounds are formed when hydrates are heated.

**76.** MAINIDEA **Summarize** the composition of a hydrate.

**77.** **Name** the compound that has the formula $SrCl_2{\cdot}6H_2O$.

**78.** **Describe** the experimental procedure for determining the formula of a hydrate. Explain the reason for each step.

**79.** **Apply** A hydrate contains 0.050 mol of $H_2O$ to every 0.00998 mol of ionic compound. Write a generalized formula of the hydrate.

**80.** **Calculate** the mass of the water of hydration if a hydrate loses 0.025 mol of $H_2O$ when heated.

**81.** **Arrange** these hydrates in order of increasing percent water content: $MgSO_4{\cdot}7H_2O$, $Ba(OH)_2{\cdot}8H_2O$, and $CoCl_2{\cdot}6H_2O$.

**82.** **Apply** Explain how the hydrate in **Figure 17** might be used as a means of roughly determining the probability of rain.

# everydayCHEMISTRY

## History In a Glass of Water

Recall the last glass of water you drank. Although it seems unbelievable, that glass of water almost certainly contained water molecules that were also consumed by Albert Einstein, Joan of Arc, or Confucius! Just how can two glasses of water poured at different times in history contain some of the same molecules? Avogadro's number and molar calculations tell the story.

**Oceans and moles** The total mass of the water in Earth's oceans and from a variety of other sources is approximately $1.4 \times 10^{24}$ g. In contrast, an 8-fluid ounce glass of water contains about $2.3 \times 10^{2}$ g, or 230 g, of water. Using this data, you can calculate the total number of glasses of water available on Earth to drink, and the total number of water molecules contained in those glasses.

You know that one mol of water has a mass of about 18 g. Using dimensional analysis you can convert the grams of water in a glass to moles.

$$\frac{230\ \cancel{\text{g water}}}{\text{glass}} \times \frac{1\ \text{mol water}}{18\ \cancel{\text{g water}}} \approx$$
$$13\ \text{mol water/glass}$$

Thus, one glass of water contains around 13 moles of water. Now convert moles of water to molecules of water by using Avogadro's number.

$$\frac{13\ \cancel{\text{mol water}}}{\text{glass}} \times \frac{6 \times 10^{23}\ \text{molecules water}}{1\ \cancel{\text{mol water}}} \approx$$
$$8 \times 10^{24}\ \text{molecules water/glass}$$

Because you know the total mass of water and the mass of water per glass, you can calculate the total number of glasses of water available for drinking.

$$1.4 \times 10^{24}\ \cancel{\text{g water}} \times \frac{1\ \text{glass}}{230\ \cancel{\text{g water}}} \approx$$
$$6 \times 10^{21}\ \text{glasses}$$

So, there are $8 \times 10^{24}$ molecules in a single glass of water and there are $6 \times 10^{21}$ glasses of water on Earth. Comparing these numbers, you can see that there are about 1000 times more molecules in a single glass of water than there are glasses of water on Earth!

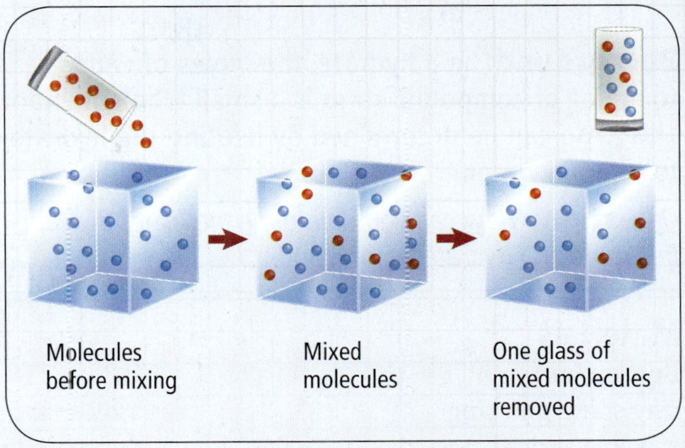


**Figure 1** Molecules from the first glass of water (red) are poured back into a container that holds all of Earth's water molecules (blue). A second glass of water taken from the container contains a small number of water molecules that were also in the first glass.

**Giant container** Suppose all the water on Earth was stored in a single, cube-shaped container. It would be enormous, with sides about 1100 km long! Imagine filling your glass with water from the container. Pour the water back into the container and wait for the water to mix completely. Then refill your glass. Would any of the molecules from the first glass be found in the second glass?

As shown in **Figure 1,** it is likely that the two glasses will share some number of water molecules. Why? Because there are 1000 times more molecules in a glass than there are glasses in the container, on average, the second glass will contain about 1000 molecules that were also in the first glass. This is true for any two glasses.

**The power of big numbers** Now, consider the amount of water—much more than a single glass—that passed through Einstein, Joan of Arc, or Confucius in their lifetimes. Assuming the molecules of water mixed evenly throughout the entire volume of Earth's water, you can understand how every glass of water must contain some of those same molecules.

### WRITING IN ▶ Chemistry

**Estimate** The estimating process used in this article is sometimes called a "back-of-the-envelope" calculation. Use this method to estimate the total mass of all of the students in your school.

WebQuest

# ChemLAB

iLab Station

## Determine the Formula of a Hydrate

**Background: In a hydrate, the moles of water to moles of compound ratio is a small whole number. This ratio can be determined by heating the hydrate to remove water.**

**Question:** *How can you determine the moles of water in a mole of a hydrated compound?*

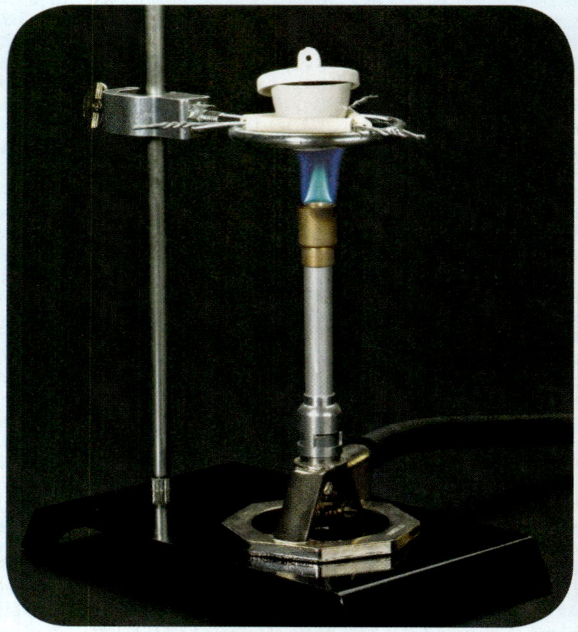

### Materials

Bunsen burner
ring stand and ring
crucible and lid
clay triangle
crucible tongs
balance
Epsom salts (hydrated $MgSO_4$)
spatula
spark lighter or matches

### Safety Precautions

WARNING: *Turn off the Bunsen burner when not in use. Crucible, lid, and triangle will be hot and can burn skin. Do not inhale fumes—they are respiratory irritants.*

### Procedure

1. Read and complete the lab safety form.
2. Prepare a data table.
3. Measure the mass of the crucible and its lid to the nearest 0.01 g.
4. Add about 3 g hydrated $MgSO_4$ to the crucible. Measure the mass of the crucible, lid, and hydrate to the nearest 0.01 g.
5. Record your observations of the hydrate.
6. Place the triangle on the ring of the ring stand. Adjust the ring stand so the triangle will be positioned near the tip of the Bunsen burner's flame. Do not light the Bunsen burner yet.
7. Carefully place the crucible in the triangle with its lid slightly ajar.
8. Begin heating with a low flame, then gradually progress to a stronger flame. Heat for about 10 min, then turn off the burner.
9. Use tongs to carefully remove the crucible from the triangle. Use tongs to place the lid on the crucible. Allow everything to cool.
10. Measure the mass of the crucible, lid, and $MgSO_4$.
11. Record your observations of the anhydrous $MgSO_4$.
12. **Cleanup and Disposal** Discard the anhydrous $MgSO_4$ as directed by your teacher. Return all lab equipment to its proper place and clean your station.

### Analyze and Conclude

1. **Calculate** Use your experimental data to calculate the formula for hydrated $MgSO_4$.
2. **Observe and Infer** How do the appearances of the hydrated and anhydrous $MgSO_4$ crystals compare? How are they different?
3. **Conclude** Why might the method used not be suitable for determining the water of hydration for all hydrates?
4. **Error Analysis** If the hydrate's formula is $MgSO_4 \cdot 7H_2O$, what is the percent error in your formula for hydrated $MgSO_4$? What are the possible sources for the error? What procedural changes could you make to reduce the error?
5. **Predict** the result of leaving the anhydrous crystals uncovered overnight.

### INQUIRY EXTENSION

**Design an experiment** to test whether a compound is hydrated or anhydrous.

Matt Meadows

# CHAPTER 10 STUDY GUIDE

Vocabulary Practice

**BIGIDEA** The mole represents a large number of extremely small particles.

## SECTION 1 Measuring Matter

**MAINIDEA** Chemists use the mole to count atoms, molecules, ions, and formula units.

- The mole is a unit used to count particles of matter indirectly. One mole of a pure substance contains Avogadro's number of representative particles.
- Representative particles include atoms, ions, molecules, formula units, electrons, and other similar particles.
- One mole of carbon-12 atoms has a mass of exactly 12 g.
- Conversion factors written from Avogadro's relationship can be used to convert between moles and number of representative particles.

**VOCABULARY**
- mole
- Avogadro's number

## SECTION 2 Mass and the Mole

**MAINIDEA** A mole always contains the same number of particles; however, moles of different substances have different masses.

- The mass in grams of 1 mol of any pure substance is called its molar mass.
- The molar mass of an element is numerically equal to its atomic mass.
- The molar mass of any substance is the mass in grams of Avogadro's number of representative particles of the substance.
- Molar mass is used to convert from moles to mass. The inverse of molar mass is used to convert from mass to moles.

**VOCABULARY**
- molar mass

## SECTION 3 Moles of Compounds

**MAINIDEA** The molar mass of a compound can be calculated from its chemical formula and can be used to convert from mass to moles of that compound.

- Subscripts in a chemical formula indicate how many moles of each element are present in 1 mol of the compound.
- The molar mass of a compound is calculated from the molar masses of all of the elements in the compound.
- Conversion factors based on a compound's molar mass are used to convert between moles and mass of a compound.

## SECTION 4 Empirical and Molecular Formulas

**MAINIDEA** A molecular formula of a compound is a whole-number multiple of its empirical formula.

- The percent by mass of an element in a compound gives the percentage of the compound's total mass due to that element.
- The subscripts in an empirical formula give the smallest whole-number ratio of moles of elements in the compound.
- The molecular formula gives the actual number of atoms of each element in a molecule or formula unit of a substance.
- The molecular formula is a whole-number multiple of the empirical formula.

**VOCABULARY**
- percent composition
- empirical formula
- molecular formula

## SECTION 5 Formulas of Hydrates

**MAINIDEA** Hydrates are solid ionic compounds in which water molecules are trapped.

- The formula of a hydrate consists of the formula of the ionic compound and the number of water molecules associated with one formula unit.
- The name of a hydrate consists of the compound name and the word *hydrate* with a prefix indicating the number of water molecules in 1 mol of the compound.
- Anhydrous compounds are formed when hydrates are heated.

**VOCABULARY**
- hydrate

## SECTION 1

### Mastering Concepts

**83.** What is the numerical value of Avogadro's number?

**84.** How many atoms of potassium does 1 mol of potassium contain?

**85.** Compare a mole of Ag-108 and a mole of Pt-195 using atoms, protons, electrons, and neutrons.

**86.** Why is the mole an important unit to chemists?

**87.** **Currency** Examine the information in **Table 2** and explain how rolls used to count pennies and dimes are similar to moles.

**Table 2 Rolled-Coin Values**

| Coin | Value of a Roll of Coins |
|---|---|
| Penny | \$0.50 |
| Dime | \$5.00 |

**88.** Explain how Avogadro's number is used as a conversion factor.

**89.** **Conversion** Design a flowchart that could be used to help convert particles to moles or moles to particles.

### Mastering Problems

**90.** Determine the number of representative particles in each substance.
**a.** 0.250 mol of silver
**b.** $8.56 \times 10^{-3}$ mol of sodium chloride
**c.** 35.3 mol of carbon dioxide
**d.** 0.425 mol of nitrogen ($N_2$)

**91.** Determine the number of representative particles in each substance.
**a.** 4.45 mol of $C_6H_{12}O_6$ **c.** 2.24 mol of $H_2$
**b.** 0.250 mol of $KNO_3$ **d.** 9.56 mol of Zn

**92.** How many molecules are contained in each compound?
**a.** 1.35 mol of carbon disulfide ($CS_2$)
**b.** 0.254 mol of diarsenic trioxide ($As_2O_3$)
**c.** 1.25 mol of water
**d.** 150.0 mol of HCl

**93.** Determine the number of moles in each substance.
**a.** $3.25 \times 10^{20}$ atoms of lead
**b.** $4.96 \times 10^{24}$ molecules of glucose
**c.** $1.56 \times 10^{23}$ formula units of sodium hydroxide
**d.** $1.25 \times 10^{25}$ copper(II) ions

**94.** Perform the following conversions.
**a.** $1.51 \times 10^{15}$ atoms of Si to mol of Si
**b.** $4.25 \times 10^{-2}$ mol of $H_2SO_4$ to molecules of $H_2SO_4$
**c.** $8.95 \times 10^{25}$ molecules of $CCl_4$ to mol of $CCl_4$
**d.** 5.90 mol of Ca to atoms of Ca

**95.** How many moles contain the given quantity?
**a.** $1.25 \times 10^{15}$ molecules of carbon dioxide
**b.** $3.59 \times 10^{21}$ formula units of sodium nitrate
**c.** $2.89 \times 10^{27}$ formula units of calcium carbonate

**96.** **RDA of Selenium** The recommended daily allowance (RDA) of selenium in your diet is $6.97 \times 10^{-7}$ mol. How many atoms of selenium is this?

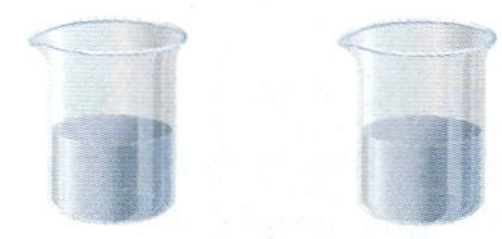

**Solution A** 0.250 mol $Cu^{2+}$ ions

**Solution B** 0.130 mol $Ca^{2+}$ ions

**Figure 19**

**97.** The two solutions shown in **Figure 19** are mixed. What is the total number of metal ions in the mixture?

**98.** **Jewelry** A bracelet containing 0.200 mol metal atoms is 75% gold. How many particles of gold atoms are in the bracelet?

**99.** **Snowflakes** A snowflake contains $1.9 \times 10^{18}$ molecules of water. How many moles of water does it contain?

**100.** If you could count two atoms every second, how long would it take you to count a mole of atoms? Assume that you counted continually for 24 hours every day. How does the time you calculated compare with the age of Earth, which is estimated to be $4.5 \times 10^9$ years old?

**101.** **Chlorophyll** The green color of leaves is due to the presence of chlorophyll, $C_{55}H_{72}O_5N_4Mg$. A fresh leaf was found to have $1.5 \times 10^{-5}$ mol of chlorophyll per $cm^2$. How many chlorophyll molecules are in 1 $cm^2$?

## SECTION 2

### Mastering Concepts

**102.** Explain the difference between atomic mass (amu) and molar mass (g).

**103.** Which contains more atoms, a mole of silver atoms or a mole of gold atoms? Explain your answer.

**104.** Which has more mass, a mole of potassium or a mole of sodium? Explain your answer.

**105.** Explain how you would convert from number of atoms of a specific element to its mass.

**106.** Discuss the relationships that exist between the mole, molar mass, and Avogadro's number.

**107.** **Barbed Wire** Barbed wire is often made of steel, which is primarily iron, and coated with zinc. Compare the number of particles and the mass of 1 mol of each.

Chapter Self-Check

## Mastering Problems

**108.** Calculate the mass of each element.
**a.** 5.22 mol of He
**b.** 0.0455 mol of Ni
**c.** 2.22 mol of Ti
**d.** 0.00566 mol of Ge

**109.** Perform the following conversions.
**a.** 3.50 mol of Li to g of Li
**b.** 7.65 g of Co to mol of Co
**c.** 5.62 g of Kr to mol of Kr
**d.** 0.0550 mol of As to g of As

**110.** Determine the mass in grams of each element.
**a.** $1.33 \times 10^{22}$ mol of Sb
**b.** $4.75 \times 10^{14}$ mol of Pt
**c.** $1.22 \times 10^{23}$ mol of Ag
**d.** $9.85 \times 10^{24}$ mol of Cr

**111.** Complete **Table 3.**

**Table 3 Mass, Mole, and Particle Data**

| Mass | Moles | Particles |
|---|---|---|
| | 3.65 mol Mg | |
| 29.54 g Cr | | |
| | | $3.54 \times 10^{25}$ atoms P |
| | 0.568 mol As | |

**112.** Convert each to mass in grams.
**a.** $4.22 \times 10^{15}$ atoms U
**b.** $8.65 \times 10^{25}$ atoms H
**c.** $1.25 \times 10^{22}$ atoms O
**d.** $4.44 \times 10^{23}$ atoms Pb

**113.** Calculate the number of atoms in each element.
**a.** 25.8 g of Hg
**b.** 0.0340 g of Zn
**c.** 150 g of Ar
**d.** 0.124 g of Mg

**114.** Arrange from least to most in moles: $3.00 \times 10^{24}$ atoms Ne, 4.25 mole Ar, $2.69 \times 10^{24}$ atoms Xe, 65.96 g Kr.

**115. Balance Precision** A sensitive electronic balance can detect masses of $1 \times 10^{-8}$ g. How many atoms of silver would be in a sample having this mass?

**116.** A sample of a compound contains 3.86 g of sulfur and 4.08 g of vanadium. How many atoms of sulfur and vanadium does the compound contain?

**117.** Which has more atoms, 10.0 g of C or 10.0 g of Ca? How many atoms does each have?

**118.** Which has more atoms, 10.0 mol of C or 10.0 mol of Ca? How many atoms does each have?

**119.** A mixture contains 0.250 mol of Fe and 1.20 g of C. What is the total number of atoms in the mixture?

**120. Respiration** Air contains several gases. Argon makes up 0.934% of the air. If a person takes a breath that contains 0.600 g of air, calculate the number of argon atoms inhaled.

# SECTION 3

## Mastering Concepts

**121.** What information is provided by the formula for potassium chromate ($K_2CrO_4$)?

**122.** In the formula for sodium phosphate ($Na_3PO_4$), how many moles of sodium are represented? How many moles of phosphorus? How many moles of oxygen?

**123.** Explain how you determine the molar mass of a compound.

**124. Insect Repellent** Many insect repellents use DEET as the active ingredient. DEET was patented in 1946 and is effective against many biting insects. What must you know to determine the molar mass of DEET?

**125.** Why can molar mass be used as a conversion factor?

**126.** List three conversion factors used in molar conversions.

**127.** Which of these contains the most moles of carbon atoms per mole of the compound: ascorbic acid ($C_6H_8O_6$), glycerin ($C_3H_8O_3$), or vanillin ($C_8H_8O_3$)? Explain.

## Mastering Problems

**128.** How many moles of oxygen atoms are contained in each compound?
**a.** 2.50 mol of $KMnO_4$
**b.** 45.9 mol of $CO_2$
**c.** $1.25 \times 10^{-2}$ mol of $CuSO_4 \cdot 5H_2O$

**129.** How many carbon tetrachloride ($CCl_4$) molecules are in 3.00 mol of $CCl_4$? How many carbon atoms? How many chlorine atoms? How many total atoms?

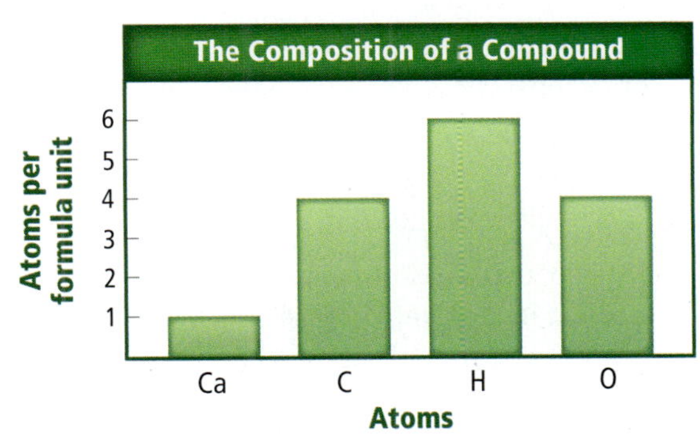


■ **Figure 20**

**130.** The graph in **Figure 20** shows the numbers of atoms of each element in a compound. What is the compound's formula? What is its molar mass?

**131.** Determine the molar mass of each compound.
**a.** nitric acid ($HNO_3$)
**b.** ammonium nitrate ($NH_4NO_3$)
**c.** zinc oxide (ZnO)
**d.** cobalt chloride ($CoCl_2$)

**132. Garlic** Determine the molar mass of allyl sulfide, the compound responsible for the smell of garlic. The chemical formula of allyl sulfide is $(C_3H_5)_2S$.

**133.** How many moles are in 100.0 g of each compound?
**a.** dinitrogen oxide ($N_2O$)
**b.** methanol ($CH_3OH$)

**134.** What is the mass of each compound?
**a.** $4.50 \times 10^{-2}$ mol of $CuCl_2$
**b.** $1.25 \times 10^2$ mol of $Ca(OH)_2$

**135. Acne** Benzoyl peroxide ($C_{14}H_{10}O_4$) is a substance used as an acne medicine. What is the mass in grams of $3.50 \times 10^{-2}$ mol $C_{14}H_{10}O_4$?

**136. Glass Etching** Hydrofluoric acid is a substance used to etch glass. Determine the mass of $4.95 \times 10^{25}$ HF molecules.

**137.** What is the mass of a mole of electrons if one electron has a mass of $9.11 \times 10^{-28}$ g?

**138.** How many moles of ions are in each compound?
**a.** 0.0200 g of $AgNO_3$
**b.** 0.100 mol of $K_2CrO_4$
**c.** 0.500 g of $Ba(OH)_2$
**d.** $1.00 \times 10^{-9}$ mol of $Na_2CO_3$

**139.** How many formula units are present in 500.0 g of lead(II) chloride?

**140.** Determine the number of atoms in 3.50 g of gold.

**141.** Calculate the mass of $3.62 \times 10^{24}$ molecules of glucose ($C_6H_{12}O_6$).

**142.** Determine the number of molecules of ethanol ($C_2H_5OH$) in 47.0 g.

**143.** What mass of iron(III) chloride contains $2.35 \times 10^{23}$ chloride ions?

**144.** How many moles of iron can be recovered from 100.0 kg of $Fe_3O_4$?

**145. Cooking** A common cooking vinegar is 5.0% acetic acid ($CH_3COOH$). How many molecules of acetic acid are present in 25.0 g of vinegar?

**146.** Calculate the moles of aluminum ions present in 250.0 g of aluminum oxide ($Al_2O_3$).

**147.** Determine the number of chloride ions in 10.75 g of magnesium chloride.

**148. Pain Relief** Acetaminophen, a common aspirin substitute, has the formula $C_8H_9NO_2$. Determine the number of molecules of acetaminophen in a 500-mg tablet.

**149.** Calculate the number of sodium ions present in 25.0 g of sodium chloride.

**150.** Determine the number of oxygen atoms present in 25.0 g of carbon dioxide.

**151. Espresso** There is $1.00 \times 10^2$ mg of caffeine in a shot of espresso. The chemical formula of caffeine is $C_8H_{10}N_4O_2$. Determine the moles of each element present in the caffeine in one shot of espresso.

**152.** The density of lead (Pb) is 11.3 g/cm$^3$. Calculate the volume of 1 mol of Pb.

## SECTION 4

### Mastering Concepts

**153.** Explain what is meant by percent composition.

**154.** What information must a chemist obtain in order to determine the empirical formula of an unknown compound?

**155.** What information must a chemist have to determine the molecular formula for a compound?

**156.** What is the difference between an empirical formula and a molecular formula? Provide an example.

**157.** When can the empirical formula be the same as the molecular formula?

**158. Antibacterial Soap** Triclosan is an antibacterial agent included in detergents, dish soaps, laundry soaps, deodorants, cosmetics, lotions, creams, toothpastes, and mouthwashes. The chemical formula for triclosan is $C_{12}H_7Cl_3O_2$. What information did the chemist need to determine this formula?

**159.** Which of the following formulas—NO, $N_2O$, $NO_2$, $N_2O_4$, and $N_2O_5$—represent the empirical and molecular formulas of the same compound? Explain your answer.

**160.** Do all pure samples of a given compound have the same percent composition? Explain.

### Mastering Problems

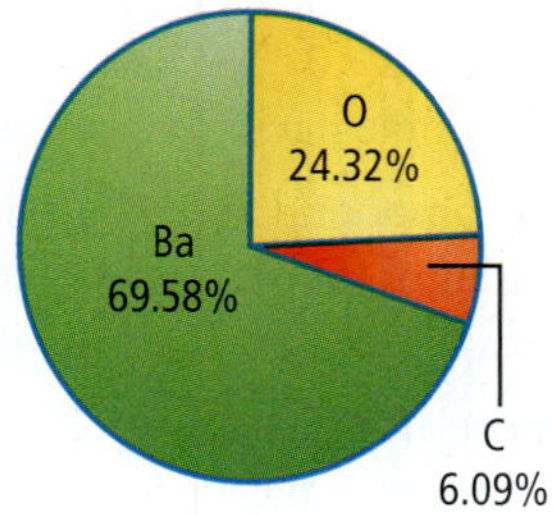


■ **Figure 21**

**161.** The circle graph in **Figure 21** shows the percent composition of a compound containing barium, carbon, and oxygen. What is the empirical formula of this compound?

**162. Iron** Three naturally occurring iron compounds are pyrite ($FeS_2$), hematite ($Fe_2O_3$), and siderite ($FeCO_3$). Which contains the greatest percentage of iron?

Chapter Self-Check

**163.** Express the composition of each compound as the mass percent of its elements (percent composition).
**a.** sucrose ($C_{12}H_{22}O_{11}$) **c.** magnetite ($Fe_3O_4$)
**b.** aluminum sulfate ($Al_2(SO_4)_3$)

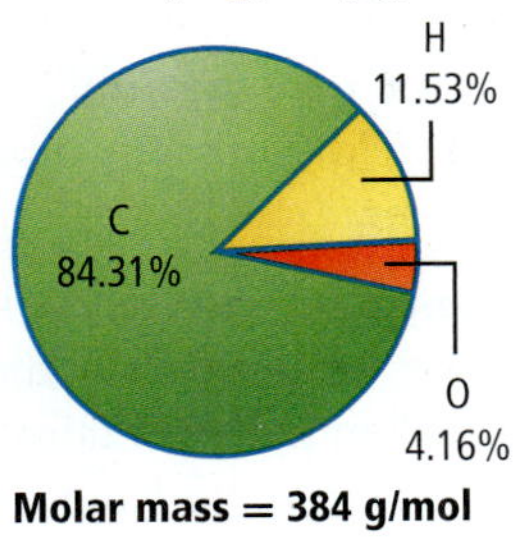


■ **Figure 22**

**164. Vitamin $D_3$** Your body's ability to absorb calcium is aided by vitamin $D_3$. Chemical analysis of vitamin $D_3$ yields the data shown in **Figure 22.** What are the empirical and molecular formulas for vitamin $D_3$?

**165.** When a 30.98-g sample of phosphorus reacts with oxygen, a 71.00-g sample of phosphorus oxide is formed. What is the percent composition of the compound? What is the empirical formula for this compound?

**166. Cholesterol** Heart disease is linked to high blood cholesterol levels. What is the percent composition of the elements in a molecule of cholesterol ($C_{27}H_{45}OH$)?

**167.** Determine the empirical formula for each compound.
**a.** ethylene ($C_2H_4$)
**b.** ascorbic acid ($C_6H_8O_6$)
**c.** naphthalene ($C_{10}H_8$)

**168. Caffeine** The stimulant effect of coffee is due to caffeine, $C_8H_{10}N_4O_2$. Calculate the molar mass of caffeine. Determine its percent composition.

**169.** Which titanium-containing mineral, rutile ($TiO_2$) or ilmenite ($FeTiO_3$), has the larger percentage of titanium?

**170. Vitamin E** Many plants contain vitamin E ($C_{29}H_{50}O_2$), a substance that some think slows the aging process in humans. What is the percent composition of vitamin E?

**171. Artificial Sweetener** Determine the percent composition of aspartame ($C_{14}H_{18}N_2O_5$), an artificial sweetener.

**172. MSG** Monosodium glutamate, known as MSG, is sometimes added to food to enhance flavor. Analysis determined this compound to be 35.5% C, 4.77% H, 8.29% N, 13.6% Na, and 37.9% O. What is its empirical formula?

**173.** What is the empirical formula of a compound that contains 10.52 g Ni, 4.38 g C, and 5.10 g N?

**174. Patina** The Statue of Liberty has turned green because of the formation of a patina. Two copper compounds, $Cu_3(OH)_4SO_4$ and $Cu_4(OH)_6SO_4$, form this patina. Find the mass percentage of copper in each compound.

# SECTION 5

## Mastering Concepts

**175.** What is a hydrated compound? Use an example to illustrate your answer.

**176.** Explain how hydrates are named.

**177. Desiccants** Why are certain electronic devices transported with desiccants?

**178.** In a laboratory setting, how would you determine if a compound was a hydrate?

**179.** Write the formula for the following hydrates.
**a.** nickel(II) chloride hexahydrate
**b.** cobalt(II) chloride hexahydrate
**c.** magnesium carbonate pentahydrate
**d.** sodium sulfate decahydrate

## Mastering Problems

**180.** Determine the mass percent of anhydrous sodium carbonate ($Na_2CO_3$) and water in sodium carbonate decahydrate ($Na_2CO_3{\cdot}10H_2O$).

**181.** **Table 4** shows data from an experiment to determine the formulas of hydrated barium chloride. Determine the formula for the hydrate and its name.

**Table 4** Data for $BaCl_2{\cdot}xH_2O$

| | |
|---|---|
| Mass of empty crucible | 21.30 g |
| Mass of hydrate + crucible | 31.35 g |
| Initial mass of hydrate | |
| Mass after heating 5 min | 29.87 g |
| Mass of anhydrous solid | |

**182.** Chromium(III) nitrate forms a hydrate that is 40.50% water by mass. What is its chemical formula?

**183.** Determine the percent composition of $MgCO_3{\cdot}5H_2O$ and draw a pie graph to represent the hydrate.

**184.** What is the formula and name of a hydrate that is 85.3% barium chloride and 14.7% water?

**185.** Gypsum is hydrated calcium sulfate. A 4.89-g sample of this hydrate was heated. After the water was removed, 3.87 g anhydrous calcium sulfate remained. Determine the formula for this hydrate and name the compound.

**186.** A 1.628-g sample of a hydrate of magnesium iodide is heated until its mass is reduced to 1.072 g and all water has been removed. What is the formula of the hydrate?

**187. Borax** Hydrated sodium tetraborate ($Na_2B_4O_7{\cdot}xH_2O$) is commonly called borax. Chemical analysis indicates that this hydrate is 52.8% sodium tetraborate and 47.2% water. Determine the formula and name the hydrate.

## MIXED REVIEW

**188.** Rank samples A–D from least number of atoms to greatest number of atoms. A: 1.0 mol of $H_2$; B: 0.75 mol of $H_2O$; C: 1.5 mol of NaCl; D: 0.50 mol of $Ag_2S$

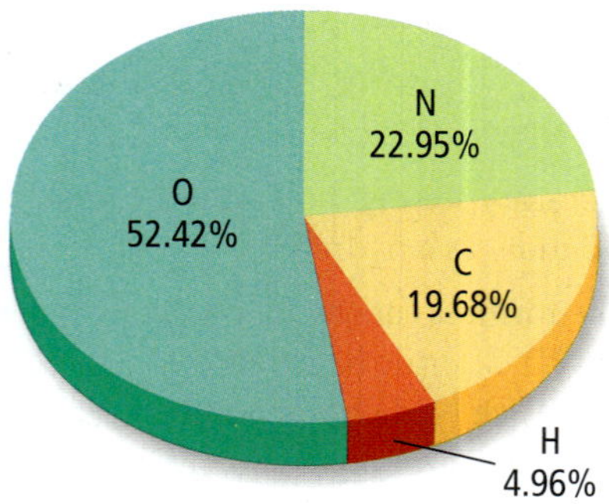


■ **Figure 23**

**189.** The graph in **Figure 23** shows the percent composition of a compound containing carbon, hydrogen, oxygen, and nitrogen. How many grams of each element are present in 100.0 g of the compound?

**190.** How many grams of $CoCl_2 \cdot 6H_2O$ must you measure out in a container to have exactly Avogadro's number of particles?

**191.** One atom of an unknown element has a mass of $6.66 \times 10^{-23}$ g. What is the identity of this element?

**192. Skunks** Analysis of skunk spray yields a molecule with 44.77% C, 7.46% H and 47.76% S. What is the chemical formula for this molecule found in the spray from skunks that scientists think is partly responsible for the strong odor?

**193.** How many moles are present in 1.00 g of each compound?

**a.** l-tryptophan ($C_{11}H_{12}N_2O_2$), an essential amino acid

**b.** magnesium sulfate heptahydrate, also known as Epsom salts

**c.** propane ($C_3H_8$), a fuel

**194.** A compound contains 6.0 g of carbon and 1.0 g of hydrogen, and has a molar mass of 42.0 g/mol. What are the compound's percent composition, empirical formula, and molecular formula?

**195.** Which of these compounds has the greatest percent of oxygen by mass: $TiO_2$, $Fe_2O_3$, or $Al_2O_3$?

**196. Mothballs** Naphthalene, commonly found in mothballs, is composed of 93.7% carbon and 6.3% hydrogen. The molar mass of naphthalene is 128 g/mol. Determine the empirical and molecular formulas for naphthalene.

**197.** Which of these molecular formulas are also empirical formulas: ethyl ether ($C_4H_{10}O$), aspirin ($C_9H_8O_4$), butyl dichloride ($C_4H_8O_2$), glucose ($C_6H_{12}O_6$)?

## THINK CRITICALLY

**198. Apply Concepts** A mining company has two possible sources of copper: chalcopyrite ($CuFeS_2$) and chalcocite ($Cu_2S$). If the mining conditions and the extraction of copper from the ore were identical for each of the ores, which ore would yield the greater quantity of copper? Explain your answer.

**199. Analyze and Conclude** On a field trip, students collected rock samples. Analysis of the rocks revealed that two of the rock samples contained lead and sulfur. **Table 5** shows the percent lead and sulfur in each of the rocks. Determine the empirical formula of each rock. What can the students conclude about the rock samples?

**Table 5 Lead and Sulfur Content**

| Rock Sample | % Lead | % Sulfur |
|---|---|---|
| 1 | 86.6% | 13.4% |
| 2 | 76.4% | 23.6% |

**200. Graph** A YAG, or yttrium aluminum garnet ($Y_3Al_5O_{12}$), is a synthetic gemstone which has no counterpart in nature. Design a bar graph to indicate the moles of each element present in a 5.67 carat yttrium aluminum garnet. (1 carat = 0.20 g)

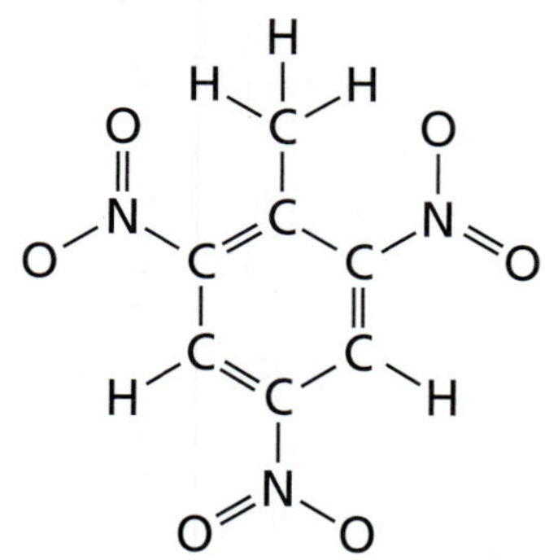

■ **Figure 24**

**201. Assess** The structure of the TNT molecule is shown in **Figure 24.** Critique the statement "Trinitrotoluene, TNT, contains 21 atoms per mole." What is correct about the statement and what is incorrect? Rewrite the statement.

**202. Design an Experiment** Design an experiment that can be used to determine the amount of water in alum ($KAl(SO_4)_2 \cdot xH_2O$).

**203.** Design a concept map that illustrates the mole concept. Include the terms *moles, Avogadro's number, molar mass, number of particles, percent composition, empirical formula, and molecular formula.*

Chapter Self-Check

## CHALLENGE PROBLEM

**204.** Two different compounds are composed of Elements X and Y. The formulas of the compounds are $X_2Y_3$ and XY. A 0.25 mol sample of XY has a mass of 17.96 g, and a 0.25 mol sample of $X_2Y_3$ has a mass of 39.92 g.
**a.** What are the atomic masses of elements X and Y?
**b.** What are the formulas for the compounds?

## CUMULATIVE REVIEW

**205.** Express each answer with the correct number of significant figures.
**a.** 18.23 − 456.7
**b.** 4.233 ÷ 0.0131
**c.** (82.44 × 4.92) + 0.125

**206. Making Candy** A recipe for pralines calls for the candy mixture to be heated until it reaches the "soft ball" stage, at about 236°F. Can a Celsius thermometer with a range of −10 to 110°C be used to determine when the "soft ball" stage is reached?

**207.** Contrast atomic number and mass number. Compare these numbers for isotopes of an element.

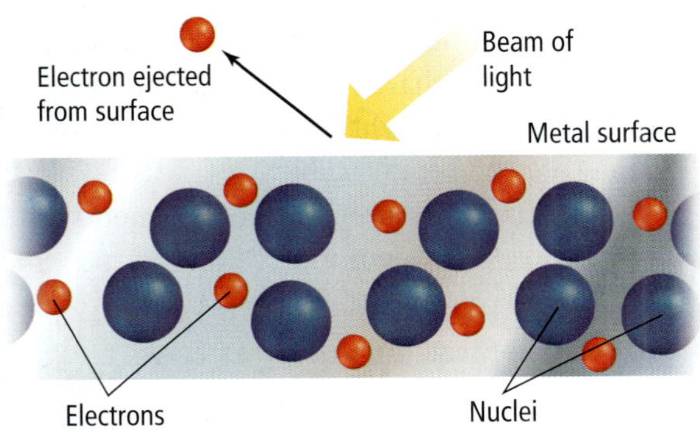


■ **Figure 25**

**208.** Describe the phenomenon in **Figure 25.** Why are the electrons not bound to the nuclei?

**209.** Given the elements Ar, Cs, Br, and Ra, identify those that form positive ions. Explain your answer.

**210.** Write the formula and name the compound formed when each pair of elements combine.
**a.** barium and chlorine
**b.** aluminum and selenium
**c.** calcium and phosphorus

**211.** Write balanced equations for each reaction.
**a.** Magnesium metal and water combine to form solid magnesium hydroxide and hydrogen gas.
**b.** Dinitrogen tetroxide gas decomposes into nitrogen dioxide gas.
**c.** Aqueous solutions of sulfuric acid and potassium hydroxide undergo a double-replacement reaction.

## WRITING IN ▸ Chemistry

**212. Natural Gas** Natural gas hydrates are chemical compounds known as clathrate hydrates. Research natural gas hydrates and prepare an educational pamphlet for consumers. The pamphlet should discuss the composition and structure of the compounds, the location of the hydrates, their importance to consumers, and the environmental impact of using the hydrates.

**213. Avogadro** Research and report on the life of Italian chemist Amedeo Avogadro (1776–1856) and how his work led scientists to determine the number of particles in a mole.

**214. Luminol** Crime-scene investigators use luminol to visualize blood residue. Research luminol and determine its chemical formula and percent composition.

## DBQ Document-Based Questions

**Space Shuttle Propellants** *At liftoff, the orbiter and an external fuel tank carry 3,164,445 L of the liquid propellants hydrogen, oxygen, hydrazine, monomethylhydrazine, and dinitrogen tetroxide. Their total mass is 727,233 kg. Data for the propellants carried at liftoff are given in* **Table 6.**

Data obtained from: "Space Shuttle Use of Propellants and Fluids." September 2001. *NASA Fact Sheet.*

**Table 6 Space Shuttle Liquid Propellants**

| Propellants | Molecular Formula | Mass (kg) | Moles | Molecules |
|---|---|---|---|---|
| Hydrogen | $H_2$ | | $5.14 \times 10^7$ | |
| Oxygen | $O_2$ | | | $1.16 \times 10^{31}$ |
| Hydrazine | | 493 | | |
| Monomethyl-hydrazine | $CH_3NHNH_2$ | 4909 | | |
| Dinitrogen tetroxide | $N_2O_4$ | | $8.64 \times 10^4$ | |

**215.** Hydrazine contains 87.45% nitrogen and 12.55% hydrogen, and has a molar mass of 32.04 g/mol. Determine hydrazine's molecular formula. Record the molecular formula in **Table 6.**

**216.** Complete **Table 6** by calculating the number of moles, mass in kilograms, or molecules for each propellant. Give all answers to three significant figures.

## MULTIPLE CHOICE

*Use the graph below to answer Questions 1 to 4.*

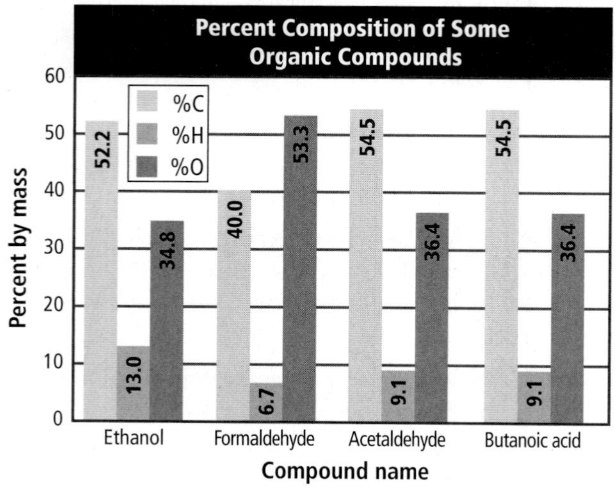


1. Acetaldehyde and butanoic acid must have the same
   A. molecular formula.
   B. empirical formula.
   C. molar mass.
   D. chemical properties.

2. If the molar mass of butanoic acid is 88.1 g/mol, what is its molecular formula?
   A. $C_3H_4O_3$
   B. $C_2H_4O$
   C. $C_5H_{12}O_1$
   D. $C_4H_8O_2$

3. What is the empirical formula of ethanol?
   A. $C_4HO_3$
   B. $C_2H_6O_2$
   C. $C_2H_6O$
   D. $C_4H_{13}O_2$

4. The empirical formula of formaldehyde is the same as its molecular formula. How many grams are in 2.000 mol of formaldehyde?
   A. 30.00 g
   B. 60.06 g
   C. 182.0 g
   D. 200.0 g

5. Which does NOT describe a mole?
   A. a unit used to count particles directly
   B. Avogadro's number of molecules of a compound
   C. the number of atoms in exactly 12 g of pure C-12
   D. the SI unit for the amount of a substance

*Use the graph below to answer Question 6.*

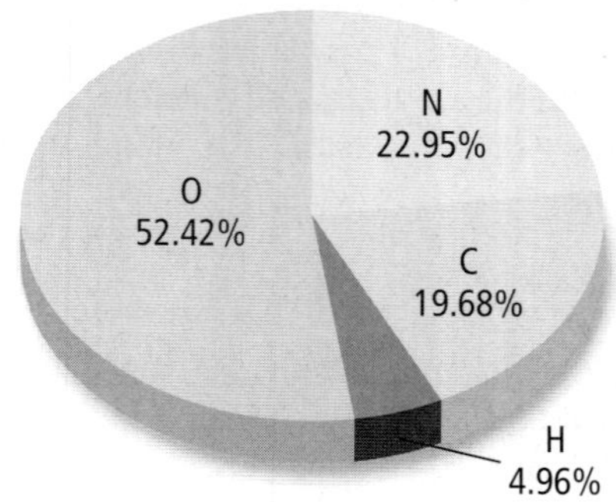


6. What is the empirical formula for this compound?
   A. $C_6H_2N_6O_3$
   B. $C_4HN_5O_{10}$
   C. $CH_3NO_2$
   D. $CH_5NO_3$

7. Which is NOT true of molecular compounds?
   A. Triple bonds are stronger than single bonds.
   B. Electrons are shared in covalent bonds.
   C. All atoms have eight valence electrons when they are chemically stable.
   D. Lewis structures show the arrangements of electrons in covalent molecules.

8. Which type of reaction is shown below?

   $$2HI + (NH_4)_2S \rightarrow H_2S + 2NH_4I$$

   A. synthesis
   B. decomposition
   C. single replacement
   D. double replacement

9. How many atoms are in 0.625 moles of Ge (atomic mass = 72.59 amu)?
   A. $2.73 \times 10^{25}$
   B. $6.99 \times 10^{25}$
   C. $3.76 \times 10^{23}$
   D. $9.63 \times 10^{23}$

10. What is the mass of one molecule of barium hexafluorosilicate ($BaSiF_6$)?
    A. $1.68 \times 10^{26}$ g
    B. $2.16 \times 10^{21}$ g
    C. $4.64 \times 10^{-22}$ g
    D. $6.02 \times 10^{-23}$ g

## SHORT ANSWER

*Use the table below to answer Question 11.*

| Charges of Some Ions | |
|---|---|
| Ion | Formula |
| Sulfide | $S^{2-}$ |
| Sulfite | $SO_3^{2-}$ |
| Sulfate | $SO_4^{2-}$ |
| Thiosulfate | $S_2O_3^{2-}$ |
| Copper(I) | $Cu^+$ |
| Copper(II) | $Cu^{2+}$ |

**11.** How many possible compounds can be made from the ions of copper, sulfur, and oxygen in the table above? Write their names and formulas.

## EXTENDED RESPONSE

*Use the figure below to answer Question 12.*

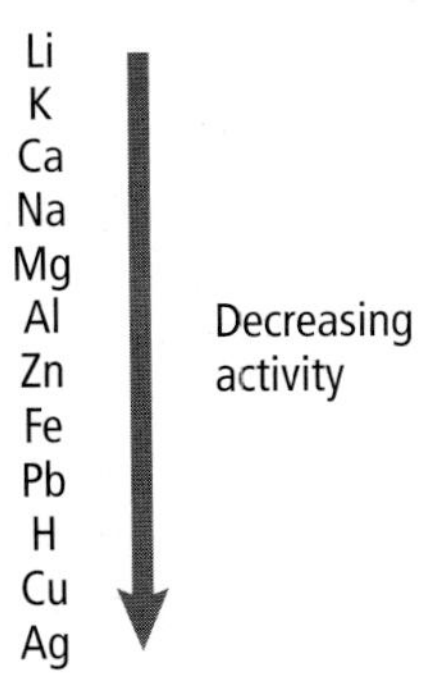


You have been asked to identify a sample of a metal. You have aqueous solutions of KCl, $AlCl_3$, $FeCl_3$, and $CuCl_2$ available. Could the metal be either lithium, zinc, or lead?

**12.** Explain how you would use these solutions to identify what metal your sample is made of.

## SAT SUBJECT TEST: CHEMISTRY

**13.** It takes 2 iron atoms and 6 chlorine atoms to make 2 iron(III) chloride particles. How many chlorine atoms are required to make 18 iron(III) chloride particles?

**A.** 9
**B.** 18
**C.** 27
**D.** 54
**E.** 72

**14.** What is the molar mass of fluorapatite ($Ca_5(PO_4)_3F$)?

**A.** 314 g/mol
**B.** 344 g/mol
**C.** 442 g/mol
**D.** 504 g/mol
**E.** 524 g/mol

**15.** Which is not a correct formula for an ionic compound?

**A.** $CaCl_2$
**B.** $Na_2SO_4$
**C.** $Al_3S_2$
**D.** $Mg(NO_3)_2$
**E.** NaCl

*Use the table below to answer question 16.*

| Percent composition of selected hydrocarbons | | | |
|---|---|---|---|
| Compound | %C | %H | %O |
| $C_4H_{10}O$ | 64.81 | 13.60 | 21.59 |
| $C_6H_{12}O_4$ | 48.64 | 8.108 | 43.24 |
| $C_7H_{16}O_3$ | 56.76 | 10.81 | 32.43 |
| $C_5H_8O_5$ | 40.54 | 5.405 | 54.05 |

**16.** A 25.0-g sample of an unknown hydrocarbon is composed of 12.16 g carbon, 2.027 g hydrogen, and 10.81 g oxygen. If its molar mass is 148 g/mol, what is the molecular formula for this compound?

**A.** $C_4H_{10}O$
**B.** $C_6H_{12}O_4$
**C.** $C_7H_{16}O_3$
**D.** $C_5H_8O_5$
**E.** $C_8H_5O_5$

| NEED EXTRA HELP? | | | | | | | | | | | | | | | | |
|---|---|---|---|---|---|---|---|---|---|---|---|---|---|---|---|---|
| If You Missed Question . . . | 1 | 2 | 3 | 4 | 5 | 6 | 7 | 8 | 9 | 10 | 11 | 12 | 13 | 14 | 15 | 16 |
| Review Section . . . | 10.4 | 10.4 | 10.4 | 10.4 | 10.1 | 10.4 | 8.3 | 9.2 | 10.1 | 10.2 | 7.3 | 9.2 | 10.3 | 10.3 | 7.3 | 10.4 |

CHAPTER 11

# Stoichiometry

**BIGIDEA** Mass relationships in chemical reactions confirm the law of conservation of mass.

## SECTIONS

Chm 5, 5a, 5b

1 Defining Stoichiometry

2 Stoichiometric Calculations

3 Limiting Reactants

4 Percent Yield

## LaunchLAB

iLab Station

### What evidence can you observe that a reaction has stopped?

During a chemical reaction, reactants are consumed as new products form. In this lab, you will look for signs a chemical reaction has stopped.

**FOLDABLES® Study Organizer**

### Steps in Stoichiometric Calculations

Make a four-tab book. Label the tabs with the steps in stoichiometric calculations. Use it to help you summarize the steps in solving a stoichiometric problem.

ASO FUJITA/age fotostock

Green plants make their own food through photosynthesis, a biological process that involves a series of chemical reactions. These reactions begin with carbon dioxide and water. On a summer day, one acre of corn produces enough oxygen through photosynthesis to meet the needs of 130 people.

SECTION 1

# Defining Stoichiometry

**MAINIDEA** The amount of each reactant present at the start of a chemical reaction determines how much product can form.

### Essential Questions

- Which relationships can be derived from a balanced chemical equation?
- How are mole ratios written from a balanced chemical equation?

### Review Vocabulary

**reactant:** the starting substance in a chemical reaction

### New Vocabulary

stoichiometry
mole ratio

**CHEM 4 YOU** Have you ever watched a candle burning? You might have watched the candle burn out as the last of the wick was used up. Or, maybe you used a candle snuffer to put out the flame. Either way, the combustion reaction ended when one of the reactants was used up.

## Particle and Mole Relationships

In doing the Launch Lab, were you surprised when the purple color of potassium permanganate disappeared as you added sodium hydrogen sulfite? If you concluded that the potassium permanganate had been used up and the reaction had stopped, you are right. Chemical reactions stop when one of the reactants is used up. When planning the reaction of potassium permanganate and sodium hydrogen sulfite, a chemist might ask, "How many grams of potassium permanganate are needed to react completely with a known mass of sodium hydrogen sulfite?" Or, when analyzing a photosynthesis reaction, you might ask, "How much oxygen and carbon dioxide are needed to form a known mass of sugar?" Stoichiometry is the tool for answering these questions.

**Stoichiometry** The study of quantitative relationships between the amounts of reactants used and amounts of products formed by a chemical reaction is called **stoichiometry.** Stoichiometry is based on the law of conservation of mass. Recall that the law states that matter is neither created nor destroyed in a chemical reaction. In any chemical reaction, the amount of matter present at the end of the reaction is the same as the amount of matter present at the beginning. Therefore, the mass of the reactants equals the mass of the products. Note the reaction of powdered iron (Fe) with oxygen ($O_2$) shown in **Figure 1.** Although iron reacts with oxygen to form a new compound, iron(III) oxide ($Fe_2O_3$), the total mass is unchanged.

**Figure 1** The balanced chemical equation for this reaction between iron and oxygen provides the relationships between amounts of reactants and products.

Explore **balanced chemical equations** with an interactive table. **Concepts In Motion**

**Table 1** **Relationships Derived from a Balanced Chemical Equation**

| $4Fe(s)$ | + | $3O_2(g)$ | → | $2Fe_2O_3(s)$ |
|---|---|---|---|---|
| iron | + | oxygen | → | iron(III) oxide |
| 4 atoms Fe | + | 3 molecules $O_2$ | → | 2 formula units $Fe_2O_3$ |
| 4 mol Fe | + | 3 mol $O_2$ | → | 2 mol $Fe_2O_3$ |
| 223.4 g Fe | + | 96.00 g $O_2$ | → | 319.4 g $Fe_2O_3$ |
| 319.4 g reactants | | | → | 319.4 g products |

The balanced chemical equation for the chemical reaction shown in **Figure 1** is as follows.

$$4Fe(s) + 3O_2(g) \rightarrow 2Fe_2O_3(s)$$

You can interpret this equation in terms of representative particles by saying that four atoms of iron react with three molecules of oxygen to produce two formula units of iron(III) oxide. Remember that coefficients in an equation represent not only numbers of individual particles but also numbers of moles of particles. Therefore, you can also say that four moles of iron react with three moles of oxygen to produce two moles of iron(III) oxide.

The chemical equation does not directly tell you anything about the masses of the reactants and products. However, by converting the known mole quantities to mass, the mass relationships become obvious. Recall that moles are converted to mass by multiplying by the molar mass. The masses of the reactants are as follows.

$$4\ \cancel{\text{mol Fe}} \times \frac{55.85\text{ g Fe}}{1\ \cancel{\text{mol Fe}}} = 223.4\text{ g Fe}$$

$$3\ \cancel{\text{mol O}_2} \times \frac{32.00\text{ g O}_2}{1\ \cancel{\text{mol O}_2}} = 96.00\text{ g O}_2$$

The total mass of the reactants is: (223.4 g + 96.00 g) = 319.4 g

Similarly, the mass of the product is calculated as follows:

$$2\ \cancel{\text{mol Fe}_2\text{O}_3} \times \frac{159.7\text{ g Fe}_2\text{O}_3}{1\ \cancel{\text{mol Fe}_2\text{O}_3}} = 319.4\text{ g}$$

Note that the mass of the reactants equals the mass of the product.

$$\text{mass of reactants} = \text{mass of products}$$
$$319.4\text{ g} = 319.4\text{ g}$$

As predicted by the law of conservation of mass, the total mass of the reactants equals the mass of the product. The relationships that can be determined from a balanced chemical equation are summarized in **Table 1.**

**VOCABULARY**

**WORD ORIGIN**

**Stoichiometry**
comes from the Greek words *stoikheion,* which means element, and *metron,* which means to measure

☑ **READING CHECK** **List** the types of relationships that can be derived from the coefficients in a balanced chemical equation.

## EXAMPLE Problem 1

Find help with **rounding.** Math Handbook

EXAMPLE PROBLEM

**INTERPRETING CHEMICAL EQUATIONS** The combustion of propane ($C_3H_8$) provides energy for heating homes, cooking food, and soldering metal parts. Interpret the equation for the combustion of propane in terms of representative particles, moles, and mass. Show that the law of conservation of mass is observed.

### 1 ANALYZE THE PROBLEM

The coefficients in the balanced chemical equation shown below represent both moles and representative particles, in this case molecules. Therefore, the equation can be interpreted in terms of molecules and moles. The law of conservation of mass will be verified if the masses of the reactants and products are equal.

**Known**

$C_3H_8(g) + 5O_2(g) \rightarrow 3CO_2(g) + 4H_2O(g)$

**Unknown**

**Equation interpreted in terms of molecules = ?**

**Equation interpreted in terms of moles = ?**

**Equation interpreted in terms of mass = ?**

### 2 SOLVE FOR THE UNKNOWN

The coefficients in the chemical equation indicate the number of molecules.

**1 molecule $C_3H_8$ + 5 molecules $O_2$ → 3 molecules $CO_2$ + 4 molecules $H_2O$**

The coefficients in the chemical equation also indicate the number of moles.

**1 mol $C_3H_8$ + 5 mol $O_2$ → 3 mol $CO_2$ + 4 mol $H_2O$**

To verify that mass is conserved, first convert moles of reactant and product to mass by multiplying by a conversion factor—the molar mass—that relates grams to moles.

$$\text{moles of reactant or product} \times \frac{\text{grams reactant or product}}{\text{1 mol reactant or product}} = \text{grams of reactant or product}$$

$$1\ \cancel{\text{mol } C_3H_8} \times \frac{44.09\text{ g } C_3H_8}{1\ \cancel{\text{mol } C_3H_8}} = 44.09\text{ g } C_3H_8$$

**Calculate the mass of the reactant $C_3H_8$.**

$$5\ \cancel{\text{mol } O_2} \times \frac{32.00\text{ g } O_2}{1\ \cancel{\text{mol } O_2}} = 160.0\text{ g } O_2$$

**Calculate the mass of the reactant $O_2$.**

$$3\ \cancel{\text{mol } CO_2} \times \frac{44.01\text{ g } CO_2}{1\ \cancel{\text{mol } CO_2}} = 132.0\text{ g } CO_2$$

**Calculate the mass of the product $CO_2$.**

$$4\ \cancel{\text{mol } H_2O} \times \frac{18.02\text{ g } H_2O}{1\ \cancel{\text{mol } H_2O}} = 72.08\text{ g } H_2O$$

**Calculate the mass of the product $H_2O$**

44.09 g $C_3H_8$ + 160.0 g $O_2$ = **204.1 g reactants** — **Add the masses of the reactants.**

132.0 g $CO_2$ + 72.08 g $H_2O$ = **204.1 g products** — **Add the masses of the products.**

**204.1 g reactants = 204.1 g products** — **The law of conservation of mass is observed.**

### 3 EVALUATE THE ANSWER

The sums of the reactants and the products are correctly stated to the first decimal place because each mass is accurate to the first decimal place. The mass of reactants equals the mass of products, as predicted by the law of conservation of mass.

## PRACTICE Problems

Do additional problems. Online Practice

1. Interpret the following balanced chemical equations in terms of particles, moles, and mass. Show that the law of conservation of mass is observed.
   **a.** $N_2(g) + 3H_2(g) \rightarrow 2NH_3(g)$
   **b.** $HCl(aq) + KOH(aq) \rightarrow KCl(aq) + H_2O(l)$
   **c.** $2Mg(s) + O_2(g) \rightarrow 2MgO(s)$
2. **Challenge** For each of the following, balance the chemical equation; interpret the equation in terms of particles, moles, and mass; and show that the law of conservation of mass is observed.
   **a.** ___Na(s) + ___$H_2O(l)$ → ___NaOH(aq) + ___$H_2(g)$
   **b.** ___Zn(s) + ___$HNO_3(aq)$ → ___$Zn(NO_3)_2(aq)$ + ___$N_2O(g)$ + ___$H_2O(l)$

PRACTICE PROBLEMS

**Mole ratios** You have read that the coefficients in a chemical equation indicate the relationships between moles of reactants and products. You can use the relationships between coefficients to derive conversion factors called mole ratios. A **mole ratio** is a ratio between the numbers of moles of any two of the substances in a balanced chemical equation. Consider the reaction between potassium (K) and bromine ($Br_2$) to form potassium bromide (KBr). The product of the reaction, the ionic salt potassium bromide, is prescribed by veterinarians, like the one in **Figure 2,** as an antiepileptic medication for dogs.

Get help with **ratios.** Personal Tutor

$$2K(s) + Br_2(l) \rightarrow 2KBr(s)$$

What mole ratios can be written for this reaction? Starting with the reactant potassium, you can write a mole ratio that relates the moles of potassium to each of the other two substances in the equation. Thus, one mole ratio relates the moles of potassium used to the moles of bromine used. The other mole ratio relates the moles of potassium used to the moles of potassium bromide formed.

$$\frac{2 \text{ mol K}}{1 \text{ mol Br}_2} \text{ and } \frac{2 \text{ mol K}}{2 \text{ mol KBr}}$$

Two other mole ratios show how the moles of bromine relate to the moles of the other two substances in the equation—potassium and potassium bromide.

$$\frac{1 \text{ mol Br}_2}{2 \text{ mol K}} \text{ and } \frac{1 \text{ mol Br}_2}{2 \text{ mol KBr}}$$

Similarly, two ratios relate the moles of potassium bromide to the moles of potassium and bromine.

$$\frac{2 \text{ mol KBr}}{2 \text{ mol K}} \text{ and } \frac{2 \text{ mol KBr}}{1 \text{ mol Br}_2}$$

These six ratios define all the mole relationships in this equation. Each of the three substances in the equation forms a ratio with the two other substances.

**READING CHECK Identify** the source from which a chemical reaction's mole ratios are derived.

**Figure 2** Potassium metal and liquid bromine react vigorously to form the ionic compound potassium bromide. Bromine is one of the two elements that are liquids at room temperature (mercury is the other). Potassium is a highly reactive metal. Potassium bromide is an ionic salt that is used to treat epilepsy in dogs.

## PRACTICE Problems

Do additional problems. Online Practice

3. Determine all possible mole ratios for the following balanced chemical equations.
   **a.** $4Al(s) + 3O_2(g) \rightarrow 2Al_2O_3(s)$
   **b.** $3Fe(s) + 4H_2O(l) \rightarrow Fe_3O_4(s) + 4H_2(g)$
   **c.** $2HgO(s) \rightarrow 2Hg(l) + O_2(g)$
4. **Challenge** Balance the following equations, and determine the possible mole ratios.
   **a.** $ZnO(s) + HCl(aq) \rightarrow ZnCl_2(aq) + H_2O(l)$
   **b.** butane ($C_4H_{10}$) + oxygen → carbon dioxide + water

The decomposition of potassium chlorate ($KClO_3$) is sometimes used to obtain small amounts of oxygen in the laboratory.

$$2KClO_3(s) \rightarrow 2KCl(s) + 3O_2(g)$$

The mole ratios that can be written for this reaction are as follows.

$$\frac{2 \text{ mol } KClO_3}{2 \text{ mol } KCl} \text{ and } \frac{2 \text{ mol } KClO_3}{3 \text{ mol } O_2}$$

$$\frac{2 \text{ mol } KCl}{2 \text{ mol } KClO_3} \text{ and } \frac{2 \text{ mol } KCl}{3 \text{ mol } O_2}$$

$$\frac{3 \text{ mol } O_2}{2 \text{ mol } KClO_3} \text{ and } \frac{3 \text{ mol } O_2}{2 \text{ mol } KCl}$$

Note that the number of mole ratios you can write for a chemical reaction involving a total of $n$ substances is $(n)(n-1)$. Thus, for reactions involving four and five substances, you can write 12 and 20 moles ratios, respectively.

$$\text{Four substances: } (4)(3) = 12 \text{ mole ratios}$$

$$\text{Five substances: } (5)(4) = 20 \text{ mole ratios}$$

**VOCABULARY**

**ACADEMIC VOCABULARY**

**Derive**
to obtain from a specified source
*The researcher was able to derive the meaning of the illustration from ancient texts.*

**Alabama Applying Practices**
Go to the Resource tab in ConnectED to find the Applying Practices worksheet *Conservation of Mass.*

## SECTION 1 REVIEW

Section Self-Check

### Section Summary

- Balanced chemical equations can be interpreted in terms of moles, mass, and representative particles (atoms, molecules, formula units).
- The law of conservation of mass applies to all chemical reactions.
- Mole ratios are derived from the coefficients of a balanced chemical equation. Each mole ratio relates the number of moles of one reactant or product to the number of moles of another reactant or product in the chemical reaction.

5. **MAIN**IDEA **Compare** the mass of the reactants and the mass of the products in a chemical reaction, and explain how these masses are related.
6. **State** how many mole ratios can be written for a chemical reaction involving three substances.
7. **Categorize** the ways in which a balanced chemical equation can be interpreted.
8. **Apply** The general form of a chemical reaction is $xA + yB \rightarrow zAB$. In the equation, A and B are elements, and $x$, $y$, and $z$ are coefficients. State the mole ratios for this reaction.
9. **Apply** Hydrogen peroxide ($H_2O_2$) decomposes to produce water and oxygen. Write a balanced chemical equation for this reaction, and determine the possible mole ratios.
10. **Model** Write the mole ratios for the reaction of hydrogen gas and oxygen gas, $2H_2(g) + O_2(g) \rightarrow 2H_2O$. Make a sketch of six hydrogen molecules reacting with the correct number of oxygen molecules. Show the water molecules produced.

SECTION 2

# Stoichiometric Calculations

**MAINIDEA** The solution to every stoichiometric problem requires a balanced chemical equation.

### Essential Questions

- What is the sequence of steps used in solving stoichiometric problems?
- How are these steps applied to solve stoichiometric problems?

### Review Vocabulary

**chemical reaction:** a process in which the atoms of one or more substances are rearranged to form different substances

**FOLDABLES®** Incorporate information from this section into your Foldable.

**CHEM 4 YOU** Baking requires accurate measurements. That is why it is necessary to follow a recipe when baking cookies from scratch. If you need to make more cookies than a recipe yields, what must you do?

## Using Stoichiometry

What tools are needed to perform stoichiometric calculations? All stoichiometric calculations begin with a balanced chemical equation. Mole ratios based on the balanced chemical equation are needed, as well as mass-to-mole conversions.

**Stoichiometric mole-to-mole conversion** The vigorous reaction between potassium and water is shown in **Figure 3.** The balanced chemical equation is as follows.

$$2K(s) + 2H_2O(l) \rightarrow 2KOH(aq) + H_2(g)$$

From the balanced equation, you know that two moles of potassium yield one mole of hydrogen. But how much hydrogen is produced if only 0.0400 mol of potassium is used? To answer this question, identify the given, or known, substance and the substance that you need to determine. The given substance is 0.0400 mol of potassium. The unknown is the number of moles of hydrogen. Because the given substance is in moles and the unknown substance to be determined is also in moles, this problem involves a mole-to-mole conversion.

To solve the problem, you need to know how the unknown moles of hydrogen are related to the known moles of potassium. Previously, you learned to derive mole ratios from the balanced chemical equation. Mole ratios are used as conversion factors to convert the known number of moles of one substance to the unknown number of moles of another substance in the same reaction. Several mole ratios can be written from the equation, but how do you choose the correct one?

**Figure 3** Potassium metal reacts vigorously with water, releasing so much heat that the hydrogen gas formed in the reaction catches fire.

As shown below, the correct mole ratio, 1 mol $H_2$ to 2 mol K, has moles of unknown in the numerator and moles of known in the denominator. Using this mole ratio converts the moles of potassium to the unknown number of moles of hydrogen.

$$\text{moles of known} \times \frac{\text{moles of unknown}}{\text{moles of known}} = \text{moles of unknown}$$

$$0.0400\ \cancel{\text{mol K}} \times \frac{1\ \text{mol H}_2}{2\ \cancel{\text{mol K}}} = 0.0200\ \text{mol H}_2$$

The following Example Problems show mole-to-mole, mole-to-mass, and mass-to-mass stoichiometry problems. The process used to solve these problems is outlined in the Problem-Solving Strategy below.

## PROBLEM-SOLVING STRATEGY

### Mastering Stoichiometry

The flowchart below outlines the steps used to solve mole-to-mole, mole-to-mass, and mass-to-mass stoichiometric problems.

1. Complete Step 1 by writing the balanced chemical equation for the reaction.
2. To determine where to start your calculations, note the unit of the given substance.
   - If mass (in grams) of the given substance is the starting unit, begin your calculations with Step 2.
   - If amount (in moles) of the given substance is the starting unit, skip Step 2 and begin your calculations with Step 3.
3. The end point of the calculation depends on the desired unit of the unknown substance.
   - If the answer must be in moles, stop after completing Step 3.
   - If the answer must be in grams, stop after completing Step 4.

### Apply the Strategy

**Apply** the Problem-Solving Strategy to Example Problems 2, 3, and 4.

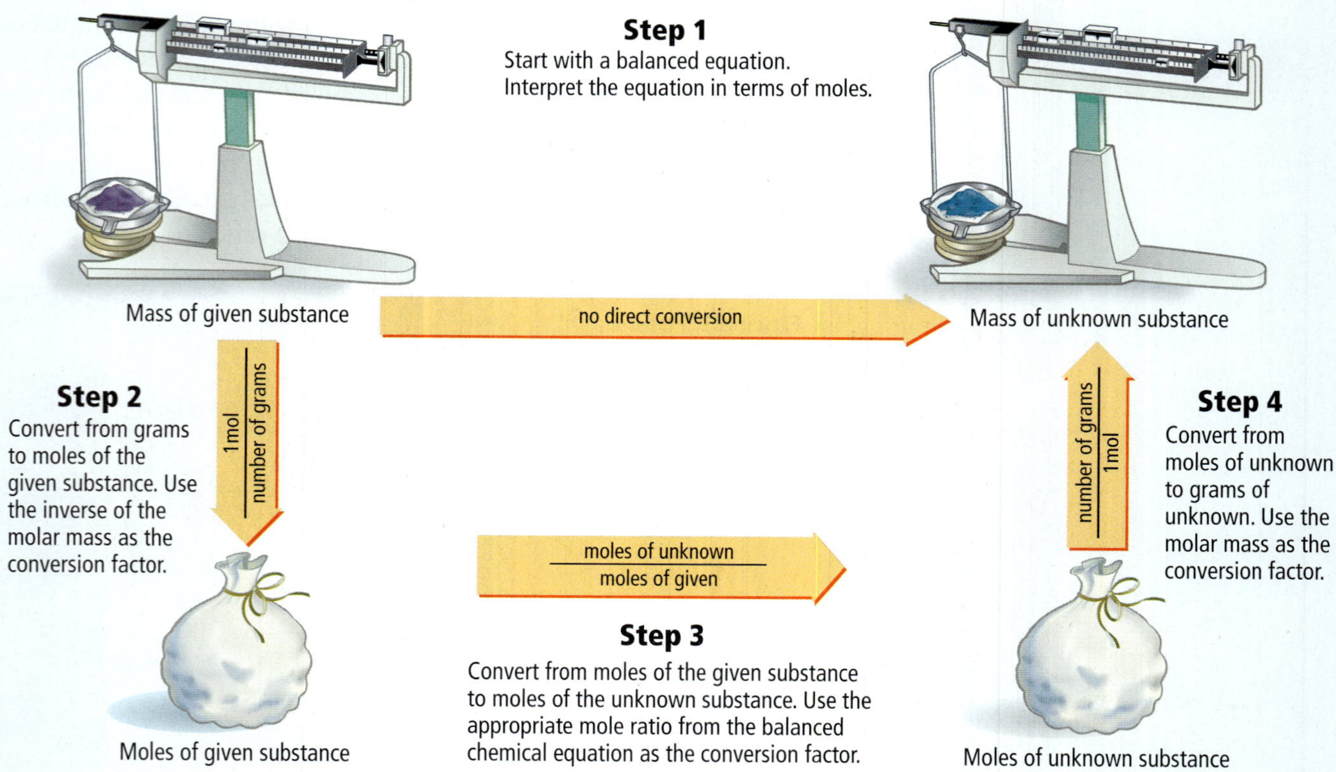

## EXAMPLE Problem 2

Find help with **ratios.** Math Handbook

**MOLE-TO-MOLE STOICHIOMETRY** One disadvantage of burning propane ($C_3H_8$) is that carbon dioxide ($CO_2$) is one of the products. The released carbon dioxide increases the concentration of $CO_2$ in the atmosphere. How many moles of $CO_2$ are produced when 10.0 mol of $C_3H_8$ are burned in excess oxygen in a gas grill?

### 1 ANALYZE THE PROBLEM

You are given moles of the reactant, $C_3H_8$ and must find the moles of the product, $CO_2$. First write the balanced chemical equation, then convert from moles of $C_3H_8$ to moles of $CO_2$. The correct mole ratio has moles of unknown substance in the numerator and moles of known substance in the denominator.

| **Known** | **Unknown** |
|---|---|
| moles $C_3H_8$ = 10.0 mol $C_3H_8$ | **moles $CO_2$ = ? mol $CO_2$** |

### 2 SOLVE FOR THE UNKNOWN

Write the balanced chemical equation for the combustion of $C_3H_8$. Use the correct mole ratio to convert moles of known ($C_3H_8$) to moles of unknown ($CO_2$).

10.0 mol **? mol**

$$C_3H_8(g) + 5O_2(g) \rightarrow 3CO_2(g) + 4H_2O(g)$$

$$\text{Mole ratio: } \frac{3 \text{ mol } CO_2}{1 \text{ mol } C_3H_8}$$

$$10.0 \cancel{\text{ mol } C_3H_8} \times \frac{3 \text{ mol } CO_2}{1 \cancel{\text{ mol } C_3H_8}} = \mathbf{30.0 \text{ mol } CO_2}$$

Burning 10.0 moles of $C_3H_8$ produces 30.0 moles $CO_2$.

### 3 EVALUATE THE ANSWER

Because the given number of moles has three significant figures, the answer also has three figures. The balanced chemical equation indicates that 1 mol of $C_3H_8$ produces 3 mol of $CO_2$. Thus, 10.0 mol of $C_3H_8$ produces three times as many moles of $CO_2$, or 30.0 mol.

## PRACTICE Problems

Do additional problems. Online Practice

**11.** Methane and sulfur react to produce carbon disulfide ($CS_2$), a liquid often used in the production of cellophane.

$$\_\_CH_4(g) + \_\_S_8(s) \rightarrow \_\_CS_2(l) + \_\_H_2S(g)$$

**a.** Balance the equation.

**b.** Calculate the moles of $CS_2$ produced when 1.50 mol $S_8$ is used.

**c.** How many moles of $H_2S$ are produced?

**12. Challenge** Sulfuric acid ($H_2SO_4$) is formed when sulfur dioxide ($SO_2$) reacts with oxygen and water.

**a.** Write the balanced chemical equation for the reaction.

**b.** How many moles of $H_2SO_4$ are produced from 12.5 moles of $SO_2$?

**c.** How many moles of $O_2$ are needed?

## RealWorld CHEMISTRY

### Outdoor Cooking

**GAS GRILLS** Using outdoor grills is a popular way to cook. Gas grills burn either natural gas or propane that is mixed with air. The initial spark is provided by a grill starter. Propane is more commonly used for fuel because it can be supplied in liquid form in a portable tank. Combustion of liquid propane also releases more energy than natural gas.

**Stoichiometric mole-to-mass conversion** Now, suppose you know the number of moles of a reactant or product in a reaction and you want to calculate the mass of another product or reactant. This is an example of a mole-to-mass conversion.

## EXAMPLE Problem 3

Find help with **significant figures.** Math Handbook

**MOLE-TO-MASS STOICHIOMETRY** Determine the mass of sodium chloride (NaCl), commonly called table salt, produced when 1.25 mol of chlorine gas ($Cl_2$) reacts vigorously with excess sodium.

### 1 ANALYZE THE PROBLEM

You are given the moles of the reactant, $Cl_2$, and must determine the mass of the product, NaCl. You must convert from moles of $Cl_2$ to moles of NaCl using the mole ratio from the equation. Then, you need to convert moles of NaCl to grams of NaCl using the molar mass as the conversion factor.

**Known**
moles of chlorine = 1.25 mol $Cl_2$

**Unknown**
**mass of sodium chloride = ? g NaCl**

### 2 SOLVE FOR THE UNKNOWN

$$\underset{}{2Na(s)} + \overset{1.25\ \text{mol}}{Cl_2(g)} \rightarrow \overset{?\ \text{g}}{2NaCl(s)}$$

**Write the balanced chemical equation, and identify the known and the unknown values.**

Mole ratio: $\frac{2\ \text{mol NaCl}}{1\ \text{mol } Cl_2}$

$$1.25\ \cancel{\text{mol } Cl_2} \times \frac{2\ \text{mol NaCl}}{1\ \cancel{\text{mol } Cl_2}} = 2.50\ \text{mol NaCl}$$

**Multiply moles of $Cl_2$ by the mole ratio to get moles of NaCl.**

$$2.50\ \cancel{\text{mol NaCl}} \times \frac{58.44\ \text{g NaCl}}{1\ \cancel{\text{mol NaCl}}} = \mathbf{146\ g\ NaCl}$$

**Multiply moles of NaCl by the molar mass to get grams of NaCl.**

### 3 EVALUATE THE ANSWER

Because the given number of moles has three significant figures, the mass of NaCl also has three. To quickly assess whether the calculated mass value for NaCl is correct, perform the calculations in reverse: divide the mass of NaCl by the molar mass of NaCl, and then divide the result by 2. You will obtain the given number of moles of $Cl_2$.

## PRACTICE Problems

Do additional problems. Online Practice

**13.** Sodium chloride is decomposed into the elements sodium and chlorine by means of electrical energy. How much chlorine gas, in grams, is obtained from the process diagrammed at right?

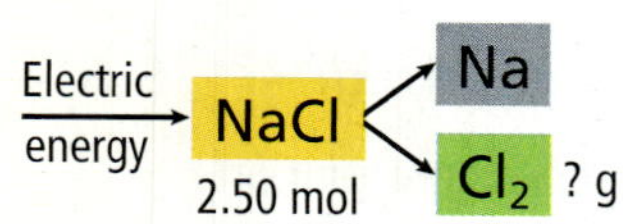


**14. Challenge** Titanium is a transition metal used in many alloys because it is extremely strong and lightweight. Titanium tetrachloride ($TiCl_4$) is extracted from titanium oxide ($TiO_2$) using chlorine and coke (carbon).

$$TiO_2(s) + C(s) + 2Cl_2(g) \rightarrow TiCl_4(s) + CO_2(g)$$

**a.** What mass of $Cl_2$ gas is needed to react with 1.25 mol of $TiO_2$?

**b.** What mass of C is needed to react with 1.25 mol of $TiO_2$?

**c.** What is the mass of all of the products formed by reaction with 1.25 mol of $TiO_2$?

**Stoichiometric mass-to-mass conversion** If you were preparing to carry out a chemical reaction in the laboratory, you would need to know how much of each reactant to use in order to produce the mass of product you required. Example Problem 4 demonstrates how you can use a measured mass of the known substance, the balanced chemical equation, and mole ratios from the equation to find the mass of the unknown substance. The ChemLab at the end of this chapter will provide you with laboratory experience in determining a mole ratio.

Watch a **video about stoichiometry.**

Video

## EXAMPLE Problem 4

Find help with **dimensional analysis.** Math Handbook

**MASS-TO-MASS STOICHIOMETRY** Ammonium nitrate ($NH_4NO_3$), an important fertilizer, produces dinitrogen monoxide ($N_2O$) gas and $H_2O$ when it decomposes. Determine the mass of $H_2O$ produced from the decomposition of 25.0 g of solid $NH_4NO_3$.

### 1 ANALYZE THE PROBLEM

You are given a description of the chemical reaction and the mass of the reactant. You need to write the balanced chemical equation and convert the known mass of the reactant to moles of the reactant. Then, use a mole ratio to relate moles of the reactant to moles of the product. Finally, use the molar mass to convert from moles of the product to the mass of the product.

**Known**
mass of ammonium nitrate = 25.0 g $NH_4NO_3$

**Unknown**
mass of water = ? g $H_2O$

### 2 SOLVE FOR THE UNKNOWN

$$\underset{25.0\ g}{NH_4NO_3(s)} \rightarrow N_2O(g) + \underset{?\ g}{2H_2O(g)}$$

Write the balanced chemical equation, and identify the known and unknown values.

$$25.0\ \cancel{g\ NH_4NO_3} \times \frac{1\ mol\ NH_4NO_3}{80.04\ \cancel{g\ NH_4NO_3}} = 0.312\ mol\ NH_4NO_3$$

Multiply grams of $NH_4NO_3$ by the inverse of molar mass to get moles of $NH_4NO_3$.

Mole ratio: $\dfrac{2\ mol\ H_2O}{1\ mol\ NH_4NO_3}$

$$0.312\ \cancel{mol\ NH_4NO_3} \times \frac{2\ mol\ H_2O}{1\ \cancel{mol\ NH_4NO_3}} = 0.624\ mol\ H_2O$$

Multiply moles of $NH_4NO_3$ by the mole ratio to get moles of $H_2O$.

$$0.624\ \cancel{mol\ H_2O} \times \frac{18.02\ g\ H_2O}{1\ \cancel{mol\ H_2O}} = 11.2\ g\ H_2O$$

Multiply moles of $H_2O$ by the molar mass to get grams of $H_2O$.

### 3 EVALUATE THE ANSWER

The number of significant figures in the answer, three, is determined by the given grams of $NH_4NO_3$. To verify that the mass of $H_2O$ is correct, perform the calculations in reverse.

EXAMPLE PROBLEM

## PRACTICE Problems

Do additional problems. Online Practice

**15.** One of the reactions used to inflate automobile air bags involves sodium azide ($NaN_3$): $2NaN_3(s) \rightarrow 2Na(s) + 3N_2(g)$. Determine the mass of $N_2$ produced from the decomposition of $NaN_3$ shown at right.

**16. Challenge** In the formation of acid rain, sulfur dioxide ($SO_2$) reacts with oxygen and water in the air to form sulfuric acid ($H_2SO_4$). Write the balanced chemical equation for the reaction. If 2.50 g of $SO_2$ reacts with excess oxygen and water, how much $H_2SO_4$, in grams, is produced?

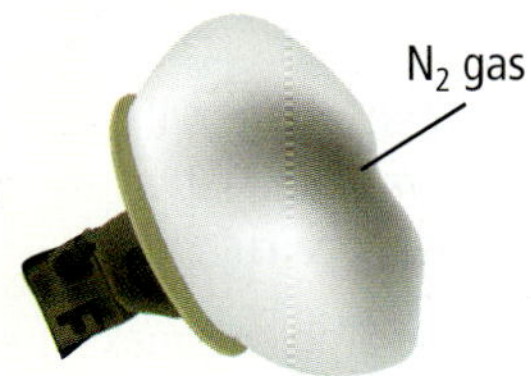


100.0 g $NaN_3 \rightarrow$ ? g $N_2(g)$

# MiniLAB

iLab Station

## Apply Stoichiometry

**How much sodium carbonate ($Na_2CO_3$) is produced when baking soda decomposes?** Baking soda is used in many baking recipes because it makes batter rise, which results in a light and fluffy texture. This occurs because baking soda, sodium hydrogen carbonate ($NaHCO_3$), decomposes upon heating to form carbon dioxide gas according to the following equation.

$$2NaHCO_3 \rightarrow Na_2CO_3 + CO_2 + H_2O$$

### Procedure

1. Read and complete the lab safety form.
2. Create a data table to record your experimental data and observations.
3. Use a **balance** to measure the mass of a clean, dry **crucible.** Add about 3.0 g of **sodium hydrogen carbonate ($NaHCO_3$),** and measure the combined mass of the crucible and $NaHCO_3$. Record both masses in your data table, and calculate the mass of the $NaHCO_3$.
4. Use this starting mass of $NaHCO_3$ and the balanced chemical equation to calculate the mass of $Na_2CO_3$ that will be produced.
5. Set up a **ring stand** with a **ring** and **clay triangle** for heating the crucible.
6. Heat the crucible with a **Bunsen burner,** slowly at first and then with a stronger flame, for 7–8 min. Record your observations during the heating.
7. Turn off the burner, and use **crucible tongs** to remove the hot crucible.

   **WARNING:** ***Do not touch the hot crucible with your hands.***
8. Allow the crucible to cool, and then measure the mass of the crucible and $Na_2CO_3$.

### Analysis

1. **Describe** what you observed during the heating of the baking soda.
2. **Compare** your calculated mass of $Na_2CO_3$ with the actual mass you obtained from the experiment.
3. **Calculate** Assume that the mass of $Na_2CO_3$ that you calculated in Step 4 is the accepted value for the mass of product that will form. Calculate the error and percent error associated with the experimentally measured mass.
4. **Identify** sources of error in the procedure that led to errors calculated in Question 3.

# SECTION 2 REVIEW

Section Self-Check

### Section Summary

- Chemists use stoichiometric calculations to predict the amounts of reactants used and products formed in specific reactions.
- The first step in solving stoichiometric problems is writing the balanced chemical equation.
- Mole ratios derived from the balanced chemical equation are used in stoichiometric calculations.
- Stoichiometric problems make use of mole ratios to convert between mass and moles.

17. MAINIDEA **Explain** why a balanced chemical equation is needed to solve a stoichiometric problem.
18. **List** the four steps used in solving stoichiometric problems.
19. **Describe** how a mole ratio is correctly expressed when it is used to solve a stoichiometric problem.
20. **Apply** How can you determine the mass of liquid bromine ($Br_2$) needed to react completely with a given mass of magnesium?
21. **Calculate** Hydrogen reacts with excess nitrogen as follows:

    $$N_2(g) + 3H_2(g) \rightarrow 2NH_3(g)$$

    If 2.70 g of $H_2$ reacts, how many grams of $NH_3$ is formed?
22. **Design** a concept map for the following reaction.

    $$CaCO_3(s) + 2HCl(aq) \rightarrow CaCl_2(aq) + H_2O(l) + CO_2(g)$$

    The concept map should explain how to determine the mass of $CaCl_2$ produced from a given mass of HCl.

SECTION 3

# Limiting Reactants

**MAINIDEA** **A chemical reaction stops when one of the reactants is used up.**

### Essential Questions

- In a chemical reaction, which reactant is the limiting reactant?
- How much of the excess reactant remains after the reaction is complete?
- How do you calculate the mass of a product when the amounts of more than one reactant are given?

### Review Vocabulary

**molar mass:** the mass in grams of one mole of any pure substance

### New Vocabulary

limiting reactant
excess reactant

**CHEM 4 YOU** **If there are more boys than girls at a school dance, some boys will be left without dance partners. The situation is much the same for the reactants in a chemical reaction—excess reactants cannot participate.**

## Why do Reactions Stop?

Rarely in nature are the reactants present in the exact ratios specified by the balanced chemical equation. Generally, one or more reactants are in excess and the reaction proceeds until all of one reactant is used up. When a reaction is carried out in the laboratory, the same principle applies. Usually, one or more of the reactants are in excess, while one is limited. The amount of product depends on the reactant that is limited.

**Limiting and excess reactants** Recall the reaction from the Launch Lab. After the colorless solution formed, adding more sodium hydrogen sulfite had no effect because there was no more potassium permanganate available to react with it. Potassium permanganate was a limiting reactant. As the name implies, the **limiting reactant** limits the extent of the reaction and, thereby, determines the amount of product formed. A portion of all the other reactants remains after the reaction stops. Reactants leftover when a reaction stops are **excess reactants.**

To help you understand limiting and excess reactants, consider the analogy in **Figure 4.** From the available tools, four complete sets consisting of a pair of pliers, a hammer, and two screwdrivers can be assembled. The number of sets is limited by the number of available hammers. Pliers and screwdrivers remain in excess.

**Figure 4** Each tool set must have one hammer, so only four sets can be assembled.
**Interpret** *How many more hammers are required to complete a fifth set?*

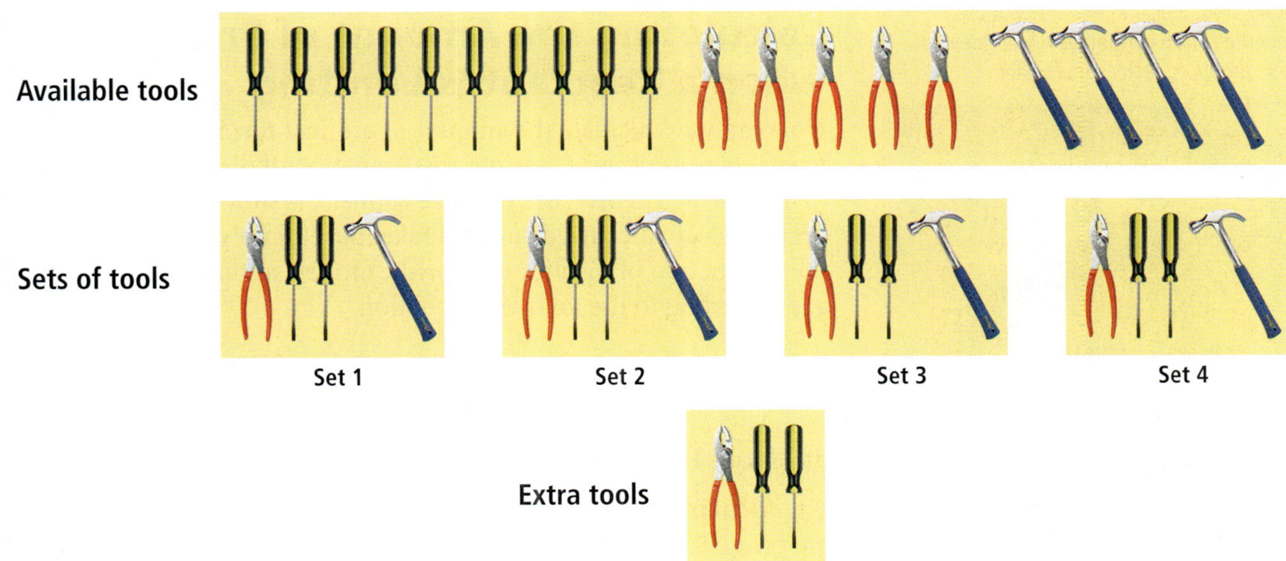

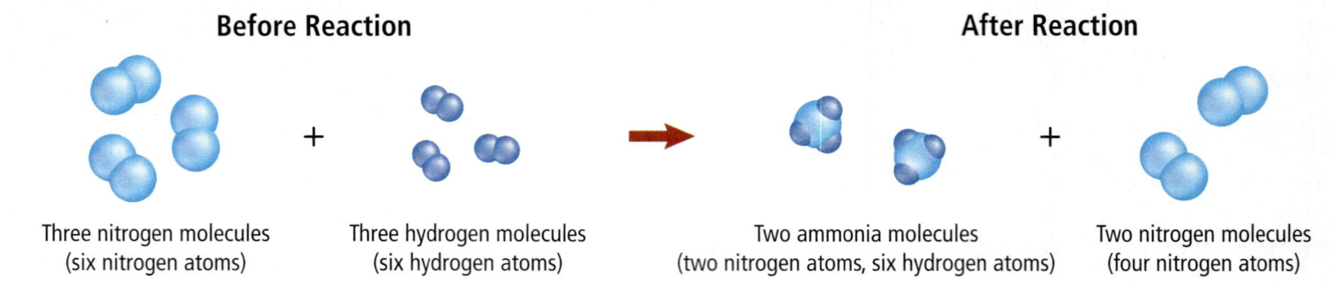


■ **Figure 5** If you check all the atoms present before and after the reaction, you will find that some of the nitrogen molecules are unchanged. These nitrogen molecules are the excess reactant.

View an **animation about limiting reactants.**

**Concepts In Motion**

**Determining the limiting reactant** The calculations you did in the previous section were based on having the reactants present in the ratio described by the balanced chemical equation. When this is not the case, the first thing you must do is determine which reactant is limiting.

Consider the reaction shown in **Figure 5,** in which three molecules of nitrogen ($N_2$) and three molecules of hydrogen ($H_2$) react to form ammonia ($NH_3$). In the first step of the reaction, all the nitrogen molecules and hydrogen molecules are separated into individual atoms. These atoms are available for reassembling into ammonia molecules, just as the tools in **Figure 4** are available to be assembled into tool kits. How many molecules of ammonia can be produced from the available atoms? Two ammonia molecules can be assembled from the hydrogen atoms and nitrogen atoms because only six hydrogen atoms are available—three for each ammonia molecule. When the hydrogen is gone, two unreacted molecules of nitrogen remain. Thus, hydrogen is the limiting reactant and nitrogen is the excess reactant. It is important to know which reactant is the limiting reactant because, as you have just read, the amount of product formed depends on this reactant.

**READING CHECK Extend** How many more hydrogen molecules would be needed to completely react with the excess nitrogen molecules shown in **Figure 5?**

■ **Figure 6** Natural rubber, which is soft and very sticky, is hardened in a chemical process called vulcanization. During vulcanization, molecules become linked together, forming a durable material that is harder, smoother, and less sticky. These properties make vulcanized rubber ideal for many products, such as this caster.

## Calculating the Amount of Product when a Reactant Is Limiting

How can you calculate the amount of product formed when one of the reactants is limiting? Consider the formation of disulfur dichloride ($S_2Cl_2$), which is used to vulcanize rubber. As shown in **Figure 6,** the properties of vulcanized rubber make it useful for many products. In the production of disulfur dichloride, molten sulfur reacts with chlorine gas according to the following equation.

$$S_8(l) + 4Cl_2(g) \rightarrow 4S_2Cl_2(l)$$

If 200.0 g of sulfur reacts with 100.0 g of chlorine, what mass of disulfur dichloride is produced?

**Calculating the limiting reactant** The masses of both reactants are given. First, determine which one is the limiting reactant, because the reaction stops producing product when the limiting reactant is used up.

joey333/iStock/Getty Images

**Moles of reactants** Identifying the limiting reactant involves finding the number of moles of each reactant. You can do this by converting the masses of chlorine and sulfur to moles. Multiply each mass by a conversion factor that relates moles and mass—the inverse of molar mass.

$$100.0\ \cancel{g\ Cl_2} \times \frac{1\ mol\ Cl_2}{70.91\ \cancel{g\ Cl_2}} = 1.410\ mol\ Cl_2$$

$$200.0\ \cancel{g\ S_8} \times \frac{1\ mol\ S_8}{256.5\ \cancel{g\ S_8}} = 0.7797\ mol\ S_8$$

**Using mole ratios** The next step involves determining whether the two reactants are in the correct mole ratio, as given in the balanced chemical equation. The coefficients in the balanced chemical equation indicate that 4 mol of chlorine is needed to react with 1 mol of sulfur. This 4:1 ratio from the equation must be compared with the actual ratio of the moles of available reactants just calculated above. To determine the actual ratio of moles, divide the number of available moles of chlorine by the number of available moles of sulfur.

$$\frac{1.410\ mol\ Cl_2\ available}{0.7797\ mol\ S_8\ available} = \frac{1.808\ mol\ Cl_2\ available}{1\ mol\ S_8\ available}$$

Only 1.808 mol of chlorine is available for every 1 mol of sulfur, instead of the 4 mol of chlorine required by the balanced chemical equation. Therefore, chlorine is the limiting reactant.

**Calculating the amount of product formed** After determining the limiting reactant, the amount of product in moles can be calculated by multiplying the given number of moles of the limiting reactant (1.410 mol $Cl_2$) by the mole ratio relating disulfur dichloride and chlorine. Then, moles of $S_2Cl_2$ are converted to grams of $S_2Cl_2$ by multiplying by the molar mass. These calculations can be combined as shown.

$$1.410\ \cancel{mol\ Cl_2} \times \frac{4\ \cancel{mol\ S_2Cl_2}}{4\ \cancel{mol\ Cl_2}} \times \frac{135.0\ g\ S_2Cl_2}{1\ \cancel{mol\ S_2Cl_2}} = 190.4\ g\ S_2Cl_2$$

Thus, 190.4 g $S_2Cl_2$ forms when 1.410 mol $Cl_2$ reacts with excess $S_8$.

**Analyzing the excess reactant** Now that you have determined the limiting reactant and the amount of product formed, what about the excess reactant, sulfur? How much of it reacted?

**Moles reacted** You need to make a mole-to-mass calculation to determine the mass of sulfur needed to react completely with 1.410 mol of chlorine. First, obtain the number of moles of sulfur by multiplying the moles of chlorine by the $S_8$-to-$Cl_2$ mole ratio.

$$1.410\ \cancel{mol\ Cl_2} \times \frac{1\ mol\ S_8}{4\ \cancel{mol\ Cl_2}} = 0.3525\ mol\ S_8$$

**Mass reacted** Next, to obtain the mass of sulfur needed, multiply 0.3525 mol of $S_8$ by its molar mass.

$$0.3525\ \cancel{mol\ S_8} \times \frac{265.5\ g\ S_8}{1\ \cancel{mol\ S_8}} = 90.42\ g\ S_8\ needed$$

**Excess remaining** Knowing that 200.0 g of sulfur is available and that only 90.42 g of sulfur is needed, you can calculate the amount of sulfur left unreacted when the reaction ends.

$$200.0\ g\ S_8\ available - 90.42\ g\ S_8\ needed = 109.6\ g\ S_8\ in\ excess$$

### CAREERS IN CHEMISTRY

**Pharmacist** Knowledge of drug composition, modes of action, and possible harmful interactions with other substances allows a pharmacist to counsel patients on their care. Pharmacists also mix chemicals to form powders, tablets, ointments, and solutions.

WebQuest

### VOCABULARY
**SCIENCE USAGE V. COMMON USAGE**

**Product**

***Science usage:*** a new substance formed during a chemical reaction
*The sole reaction product was a colorless gas.*

***Common usage:*** something produced
*The cosmetics counter in the department store had hundreds of products from which to choose.*

Explore **limiting reactants.**

Virtual Investigations

## EXAMPLE Problem 5

Find help with **dimensional analysis.** Math Handbook

EXAMPLE PROBLEM

**DETERMINING THE LIMITING REACTANT** The reaction between solid white phosphorus ($P_4$) and oxygen produces solid tetraphosphorus decoxide ($P_4O_{10}$). This compound is often called diphosphorus pentoxide because its empirical formula is $P_2O_5$.

**a.** Determine the mass of $P_4O_{10}$ formed if 25.0 g of $P_4$ and 50.0 g of oxygen are combined.

**b.** How much of the excess reactant remains after the reaction stops?

### 1 ANALYZE THE PROBLEM

You are given the masses of both reactants, so you must identify the limiting reactant and use it to find the mass of the product. From moles of the limiting reactant, the moles of the excess reactant used in the reaction can be determined. The number of moles of the excess reactant that reacted can be converted to mass and subtracted from the given mass to find the amount in excess.

| **Known** | **Unknown** |
|---|---|
| mass of phosphorus = 25.0 g $P_4$ | **mass of tetraphosphorus decoxide = ? g $P_4O_{10}$** |
| mass of oxygen = 50.0 g $O_2$ | **mass of excess reactant = ? g excess reactant** |

### 2 SOLVE FOR THE UNKNOWN

Determine the limiting reactant.

$$\underset{25.0\text{ g}}{P_4(s)} + \underset{50.0\text{ g}}{5O_2(g)} \rightarrow \underset{?\text{ g}}{P_4O_{10}(s)}$$

**Write the balanced chemical equation, and identify the known and the unknown.**

Determine the number of moles of the reactants by multiplying each mass by the conversion factor that relates moles and mass—the inverse of molar mass.

$$25.0\ \cancel{g\ P_4} \times \frac{1\text{ mol }P_4}{123.9\ \cancel{g\ P_4}} = 0.202\text{ mol }P_4$$

**Calculate the moles of $P_4$.**

$$50.0\ \cancel{g\ O_2} \times \frac{1\text{ mol }O_2}{32.00\ \cancel{g\ O_2}} = 1.56\text{ mol }O_2$$

**Calculate the moles of $O_2$.**

Calculate the actual ratio of available moles of $O_2$ and available moles of $P_4$.

$$\frac{1.56\text{ mol }O_2}{0.202\text{ mol }P_4} = \frac{7.72\text{ mol }O_2}{1\text{ mol }P_4}$$

**Calculate the ratio of moles of $O_2$ to moles of $P_4$.**

Determine the mole ratio of the two reactants from the balanced chemical equation.

Mole ratio: $\frac{5\text{ mol }O_2}{\text{mol }P_4}$

Because 7.72 mol of $O_2$ is available but only 5 mol is needed to react with 1 mol of $P_4$, $O_2$ is in excess and $P_4$ is the limiting reactant. Use the moles of $P_4$ to determine the moles of $P_4O_{10}$ that will be produced. Multiply the number of moles of $P_4$ by the mole ratio of $P_4O_{10}$ (the unknown) to $P_4$ (the known).

$$0.202\ \cancel{mol\ P_4} \times \frac{1\text{ mol }P_4O_{10}}{1\ \cancel{mol\ P_4}} = 0.202\text{ mol }P_4O_{10}$$

**Calculate the moles of product ($P_4O_{10}$) formed.**

To calculate the mass of $P_4O_{10}$, multiply moles of $P_4O_{10}$ by the conversion factor that relates mass and moles—molar mass.

$$0.202\ \cancel{mol\ P_4O_{10}} \times \frac{283.9\text{ g }P_4O_{10}}{1\ \cancel{mol\ P_4O_{10}}} = \mathbf{57.3\text{ g }P_4O_{10}}$$

**Calculate the mass of the product $P_4O_{10}$.**

Because $O_2$ is in excess, only part of the available $O_2$ is consumed. Use the limiting reactant, $P_4$, to determine the moles and mass of $O_2$ used.

$$0.202 \cancel{\text{mol } P_4} \times \frac{5 \text{ mol } O_2}{1 \cancel{\text{mol } P_4}} = 1.01 \text{ mol } O_2$$

**Multiply the moles of limiting reactant by the mole ratio to determine moles of excess reactant needed.**

Convert moles of $O_2$ consumed to mass of $O_2$ consumed.

$$1.01 \cancel{\text{mol } O_2} \times \frac{32.00 \text{ g } O_2}{1 \cancel{\text{mol } O_2}} = 32.3 \text{ g } O_2$$

**Multiply the moles of $O_2$ by the molar mass.**

Calculate the amount of excess $O_2$.

50.0 g $O_2$ available − 32.3 g $O_2$ consumed = **17.7 g $O_2$ in excess**

**Subtract the mass of $O_2$ used from the mass available.**

## 3 EVALUATE THE ANSWER

All values have a minimum of three significant figures, so the mass of $P_4O_{10}$ is correctly stated with three digits. The mass of excess $O_2$ (17.7 g) is found by subtracting two numbers that are accurate to the first decimal place. Therefore, the mass of excess $O_2$ correctly shows one decimal place. The sum of the $O_2$ that was consumed (32.3 g) and the given mass of $P_4$ (25.0 g) is 57.3 g, the calculated mass of the product $P_4O_{10}$.

## PRACTICE Problems

Do additional problems. Online Practice

23. The reaction between solid sodium and iron(III) oxide is one in a series of reactions that inflates an automobile airbag: $6Na(s) + Fe_2O_3(s) \rightarrow 3Na_2O(s) + 2Fe(s)$. If 100.0 g of Na and 100.0 g of $Fe_2O_3$ are used in this reaction, determine the following.
    - **a.** limiting reactant
    - **b.** reactant in excess
    - **c.** mass of solid iron produced
    - **d.** mass of excess reactant that remains after the reaction is complete
24. **Challenge** Photosynthesis reactions in green plants use carbon dioxide and water to produce glucose ($C_6H_{12}O_6$) and oxygen. A plant has 88.0 g of carbon dioxide and 64.0 g of water available for photosynthesis.
    - **a.** Write the balanced chemical equation for the reaction.
    - **b.** Determine the limiting reactant.
    - **c.** Determine the excess reactant.
    - **d.** Determine the mass in excess.
    - **e.** Determine the mass of glucose produced.

PRACTICE PROBLEMS

**Connection to Biology** Your body needs vitamins, minerals, and elements in small amounts to facilitate normal metabolic reactions. A lack of these substances can lead to abnormalities in growth, development, and the functioning of your body's cells. Phosphorus, for example, is an essential element in living systems; phosphate groups occur regularly in strands of DNA. Potassium is needed for proper nerve function, muscle control, and blood pressure. A diet low in potassium and high in sodium might be a factor in high blood pressure. Another example is vitamin B-12. Without adequate vitamin B-12, the body is unable to synthesize DNA properly, affecting the production of red blood cells.

**Alabama Applying Practices**

Go to the Resource tab in ConnectED to find the Applying Practices worksheet *Conservation of Mass.*

■ **Figure 7** With insufficient oxygen, the burner on the left burns with a yellow, sooty flame. The burner on the right burns hot and clean because an excess of oxygen is available to react completely with the methane gas.

**Why use an excess of a reactant?** Many reactions stop while portions of the reactants are still present in the reaction mixture. Because this is inefficient and wasteful, chemists have found that by using an excess of one reactant—often the least expensive one—reactions can be driven to continue until all of the limiting reactant is used up. Using an excess of one reactant can also speed up a reaction.

**Figure 7** shows an example of how controlling the amount of a reactant can increase efficiency. Your lab likely uses the type of Bunsen burner shown in the figure. If so, you know that this type of burner has a control that lets you adjust the amount of air that mixes with the methane gas. How efficiently the burner operates depends on the ratio of oxygen to methane gas in the fuel mixture. When the air is limited, the resulting flame is yellow because of glowing bits of unburned fuel. This unburned fuel leaves soot (carbon) deposits on glassware. Fuel is wasted because the amount of energy released is less than the amount that could have been produced if enough oxygen were available. When sufficient oxygen is present in the combustion mixture, the burner produces a hot, intense blue flame. No soot is deposited because the fuel is completely converted to carbon dioxide and water vapor.

## SECTION 3 REVIEW

Section Self-Check

### Section Summary

- The limiting reactant is the reactant that is completely consumed during a chemical reaction. Reactants that remain after the reaction stops are called excess reactants.
- To determine the limiting reactant, the actual mole ratio of the available reactants must be compared with the ratio of the reactants obtained from the coefficients in the balanced chemical equation.
- Stoichiometric calculations must be based on the limiting reactant.

25. **MAIN**IDEA **Describe** the reason why a reaction between two substances comes to an end.

26. **Identify** the limiting and the excess reactant in each reaction.
    - **a.** Wood burns in a campfire.
    - **b.** Airborne sulfur reacts with the silver plating on a teapot to produce tarnish (silver sulfide).
    - **c.** Baking powder in batter decomposes to produce carbon dioxide.

27. **Analyze** Tetraphosphorus trisulphide ($P_4S_3$) is used in the match heads of some matches. It is produced in the reaction $8P_4 + 3S_8 \rightarrow 8P_4S_3$. Determine which of the following statements are incorrect, and rewrite the incorrect statements to make them correct.
    - **a.** 4 mol $P_4$ reacts with 1.5 mol $S_8$ to form 4 mol $P_4S_3$.
    - **b.** Sulfur is the limiting reactant when 4 mol $P_4$ and 4 mol $S_8$ react.
    - **c.** 6 mol $P_4$ reacts with 6 mol $S_8$, forming 1320 g $P_4S_3$.

SECTION 4

# Percent Yield

**MAINIDEA** Percent yield is a measure of the efficiency of a chemical reaction.

**Essential Questions**

- What is the theoretical yield of a chemical reaction?
- How do you calculate the percent yield for a chemical reaction?

**Review Vocabulary**

**process:** a series of actions or operations

**New Vocabulary**

theoretical yield
actual yield
percent yield

**CHEM 4 YOU** Imagine that you are practicing free throws and you take 100 practice shots. Theoretically, you could make all 100 shots. In actuality, however, you know you will not make all of the shots. Chemical reactions also have theoretical and actual outcomes.

## How Much Product?

While solving stoichiometric problems in this chapter, you might have concluded that chemical reactions always proceed in the laboratory according to the balanced equation and produce the calculated amount of product. This, however, is not the case. Just as you are unlikely to make 100 out of 100 free throws during basketball practice, most reactions never succeed in producing the predicted amount of product. Reactions do not go to completion or yield as expected for a variety of reasons. Liquid reactants and products might adhere to the surfaces of their containers or evaporate. In some instances, products other than the intended ones might be formed by competing reactions, thus reducing the yield of the desired product. Or, as shown in **Figure 8,** some amount of any solid product is usually left behind on filter paper or lost in the purification process. Because of these problems, chemists need to know how to gauge the yield of a chemical reaction.

**Theoretical and Actual Yields** In many of the stoichiometric calculations you have performed, you have calculated the amount of product produced from a given amount of reactant. The answer you obtained is the theoretical yield of the reaction. The **theoretical yield** is the maximum amount of product that can be produced from a given amount of reactant.

A chemical reaction rarely produces the theoretical yield of product. A chemist determines the actual yield of a reaction through a careful experiment in which the mass of the product is measured. The **actual yield** is the amount of product produced when the chemical reaction is carried out in an experiment.

**Figure 8** Silver chromate is formed when potassium chromate is added to silver nitrate. Note that not all of the precipitate can be removed from the filter paper. Still more of the precipitate is lost because it adheres to the sides of the beaker.

**Percent yield** Chemists need to know how efficient a reaction is in producing the desired product. One way of measuring efficiency is by means of percent yield. **Percent yield** of product is the ratio of the actual yield to the theoretical yield expressed as a percent.

**Percent Yield**

$$\text{percent yield} = \frac{\text{actual yield}}{\text{theoretical yield}} \times 100$$

The actual yield divided by the theoretical yield multiplied by 100 is the percent yield.

## EXAMPLE Problem 6

Find help with **percents.** Math Handbook

EXAMPLE PROBLEM

**PERCENT YIELD** Solid silver chromate ($Ag_2CrO_4$) forms when excess potassium chromate ($K_2CrO_4$) is added to a solution containing 0.500 g of silver nitrate ($AgNO_3$). Determine the theoretical yield of $Ag_2CrO_4$. Calculate the percent yield if the reaction yields 0.455 g of $Ag_2CrO_4$.

### 1 ANALYZE THE PROBLEM

You know the mass of a reactant and the actual yield of the product. Write the balanced chemical equation, and calculate theoretical yield by converting grams of $AgNO_3$ to moles of $AgNO_3$, moles of $AgNO_3$ to moles of $Ag_2CrO_4$, and moles of $Ag_2CrO_4$ to grams of $Ag_2CrO_4$. Calculate the percent yield from the actual yield and the theoretical yield.

**Known**

mass of silver nitrate = 0.500 g $AgNO_3$

actual yield = 0.455 g $Ag_2CrO_4$

**Unknown**

**theoretical yield = ? g $Ag_2CrO_4$**

**percent yield = ? % $Ag_2CrO_4$**

### 2 SOLVE FOR THE UNKNOWN

$$\underset{0.500\text{ g}}{2AgNO_3(aq)} + K_2CrO_4(aq) \rightarrow \underset{?\text{ g}}{Ag_2CrO_4(s)} + 2KNO_3(aq)$$

**Write the balanced chemical equation, and identify the known and the unknown.**

$$0.500\ \cancel{\text{g AgNO}_3} \times \frac{1\text{ mol AgNO}_3}{169.9\ \cancel{\text{g AgNO}_3}} = 2.94 \times 10^{-3}\text{ mol AgNO}_3$$

**Use molar mass to convert grams of $AgNO_3$ to moles of $AgNO_3$.**

$$2.94 \times 10^{-3}\ \cancel{\text{mol AgNO}_3} \times \frac{1\text{ mol Ag}_2\text{CrO}_4}{2\ \cancel{\text{mol AgNO}_3}} = 1.47 \times 10^{-3}\text{ mol Ag}_2\text{CrO}_4$$

**Use the mole ratio to convert moles of $AgNO_3$ to moles of $Ag_2CrO_4$.**

$$1.47 \times 10^{-3}\ \cancel{\text{mol Ag}_2\text{CrO}_4} \times \frac{331.7\text{ g Ag}_2\text{CrO}_4}{1\ \cancel{\text{mol Ag}_2\text{CrO}_4}} = \mathbf{0.488\text{ g Ag}_2\text{CrO}_4}$$

**Calculate the theoretical yield.**

$$\frac{0.455\ \cancel{\text{g Ag}_2\text{CrO}_4}}{0.488\ \cancel{\text{g Ag}_2\text{CrO}_4}} \times 100 = \mathbf{93.2\%\text{ Ag}_2\text{CrO}_4}$$

**Calculate the percent yield.**

### 3 EVALUATE THE ANSWER

The quantity with the fewest significant figures has three, so the percent is correctly stated with three digits. The molar mass of $Ag_2CrO_4$ is about twice the molar mass of $AgNO_3$, and the ratio of moles of $AgNO_3$ to moles of $Ag_2CrO_4$ in the equation is 2:1. Therefore, 0.500 g of $AgNO_3$ should produce about the same mass of $Ag_2CrO_4$. The actual yield of $Ag_2CrO_4$ is close to 0.500 g, so a percent yield of 93.2% is reasonable.

## PRACTICE Problems

Do additional problems. Online Practice

**28.** Aluminum hydroxide ($Al(OH)_3$) is often present in antacids to neutralize stomach acid (HCl). The reaction occurs as follows: $Al(OH)_3(s) + 3HCl(aq) \rightarrow AlCl_3(aq) + 3H_2O(l)$. If 14.0 g of $Al(OH)_3$ is present in an antacid tablet, determine the theoretical yield of $AlCl_3$ produced when the tablet reacts with HCl.

**29.** Zinc reacts with iodine in a synthesis reaction: $Zn + I_2 \rightarrow ZnI_2$.

**a.** Determine the theoretical yield if 1.912 mol of zinc is used.

**b.** Determine the percent yield if 515.6 g of product is recovered.

**30. Challenge** When copper wire is placed into a silver nitrate solution ($AgNO_3$), silver crystals and copper(II) nitrate ($Cu(NO_3)_2$) solution form.

**a.** Write the balanced chemical equation for the reaction.

**b.** If a 20.0-g sample of copper is used, determine the theoretical yield of silver.

**c.** If 60.0 g of silver is recovered from the reaction, determine the percent yield of the reaction.

**Alabama Applying Practices**

Go to the Resource tab in ConnectED to find the Applying Practices worksheet *Determine the Quantities of Reactants and Products.*

# Data Analysis LAB

Based on Real Data[1, 2]

## Analyze and Conclude

**Can rocks on the Moon provide an effective oxygen source for future lunar missions?** Although the Moon has no atmosphere and thus no oxygen, its surface is covered with rocks and soil made from oxides. Scientists, looking for an oxygen source for future long-duration lunar missions, are researching ways to extract oxygen from lunar soil and rock. Analysis of samples collected during previous lunar missions provided scientists with the data shown in the table. The table identifies the oxides in lunar soil as well as each oxide's percent-by-weight of the soil.

### Think Critically

**1. Calculate** For each of the oxides listed in the table, determine the mass (in grams) that would exist in 1.00 kg of lunar soil.

**2. Apply** Scientists want to release the oxygen from its metal oxide using a decomposition reaction: metal oxide → metal + oxygen. To assess the viability of this idea, determine the amount of oxygen per kilogram contained in each of the oxides found in lunar soil.

**3. Identify** What oxide would yield the most oxygen per kilogram? The least?

**4. Determine** the theoretical yield of oxygen from the oxides present in a 1.00-kg sample of lunar soil.

### Data and Observation

Moon-Rock Data[1]

| Oxide | % Mass of Soil |
|---|---|
| $SiO_2$ | 47.3% |
| $Al_2O_3$ | 17.8% |
| CaO | 11.4% |
| FeO | 10.5% |
| MgO | 9.6% |
| $TiO_2$ | 1.6% |
| $Na_2O$ | 0.7% |
| $K_2O$ | 0.6% |
| $Cr_2O_3$ | 0.2% |
| MnO | 0.1% |

[1]Data obtained from: McKay, et al. 1994. JSC-1: A new lunar soil stimulant. *Engineering, Construction, and Operations in Space* IV: 857–866, American Society of Civil Engineers.

[2]Data obtained from: Berggren, et al. 2005. Carbon monoxide silicate reduction system. *Space Resources Roundtable* VII.

**5. Calculate** Using methods currently available, scientists can produce 15 kg of oxygen from 100 kg of lunar soil. What is the percent yield of the process?

■ **Figure 9** Sulfur, such as these piles at Vancouver Harbor, can be extracted from petroleum products by a chemical process. Sulfur is also mined by forcing hot water into underground deposits and pumping the liquid sulfur to the surface.

## Percent Yield in the Marketplace

Percent yield is important in the cost effectiveness of many industrial manufacturing processes. For example, the sulfur shown in **Figure 9** is used to make sulfuric acid ($H_2SO_4$). Sulfuric acid is an important chemical because it is a raw material used to make products such as fertilizers, detergents, pigments, and textiles.

The cost of sulfuric acid affects the cost of many of the consumer items you use every day. The first two steps in the manufacturing process are shown below.

Step 1 $$S_8(s) + 8O_2(g) \rightarrow 8SO_2(g)$$

Step 2 $$2SO_2(g) + O_2(g) \rightarrow 2SO_3(g)$$

In the final step, $SO_3$ combines with water to produce $H_2SO_4$.

Step 3 $$SO_3(g) + H_2O(l) \rightarrow H_2SO_4(aq)$$

The first step, the combustion of sulfur, produces an almost 100% yield. The second step also produces a high yield if a catalyst is used at the relatively low temperature of 400°C. A catalyst is a substance that speeds a reaction but does not appear in the chemical equation. Under these conditions, the reaction is slow. Raising the temperature increases the reaction rate but decreases the yield.

To maximize yield and minimize time in the second step, engineers have devised a system in which the reactants, $O_2$ and $SO_2$, are passed over a catalyst at 400°C. Because the reaction releases a great deal of heat, the temperature gradually increases with an accompanying decrease in yield. Thus, when the temperature reaches approximately 600°C, the mixture is cooled and then passed over the catalyst again. A total of four passes over the catalyst with cooling between passes results in a yield greater than 98%.

## SECTION 4 REVIEW

Section Self-Check

### Section Summary

- The theoretical yield of a chemical reaction is the maximum amount of product that can be produced from a given amount of reactant. Theoretical yield is calculated from the balanced chemical equation.
- The actual yield is the amount of product produced. Actual yield must be obtained through experimentation.
- Percent yield is the ratio of actual yield to theoretical yield expressed as a percent. High percent yield is important in reducing the cost of every product produced through chemical processes.

**31. MAIN IDEA Identify** which type of yield—theoretical yield, actual yield, or percent yield—is a measure of the efficiency of a chemical reaction.

**32. List** several reasons why the actual yield from a chemical reaction is not usually equal to the theoretical yield.

**33. Explain** how percent yield is calculated.

**34. Apply** In an experiment, you combine 83.77 g of iron with an excess of sulfur and then heat the mixture to obtain iron(III) sulfide.

$$2Fe(s) + 3S(s) \rightarrow Fe_2S_3(s)$$

What is the theoretical yield, in grams, of iron(III) sulfide?

**35. Calculate** the percent yield of the reaction of magnesium with excess oxygen: $2Mg(s) + O_2(g) \rightarrow 2MgO(s)$

| Reaction Data | |
|---|---|
| Mass of empty crucible | 35.67 g |
| Mass of crucible and Mg | 38.06 g |
| Mass of crucible and MgO (after heating) | 39.15 g |

# CHEMISTRY & health

## Battling Resistant Strains

The human immunodeficiency virus (HIV), the virus that causes AIDS, has proven to be among the most incurable foes ever faced by modern medical science. One reason for this is the virus's remarkable ability to adapt. Resistant strains of the virus appear quickly, rendering obsolete the newest and most powerful AIDS drugs. Now some researchers are using the virus's adaptability as a way to fight it.

**Selecting resistance** PA-457 is a promising new anti-HIV drug synthesized from betulinic acid, an organic compound derived from some plants, including the bark of birch trees. To find out just what PA-457 does to HIV, known as the drug's mechanism of action, researchers took what might seem a strange step: they encouraged samples of HIV to develop resistance to PA-457.

Researchers subjected HIV samples to small doses of PA-457. Using a low dose made it more likely that some of the virus would survive the treatment and possibly develop resistance. Those viruses that survived exposure were collected, and their genetic sequences were examined. The surviving viruses were found to have a mutation in the genes that control how the virus builds a structure called a capsid, shown in **Figure 1.**

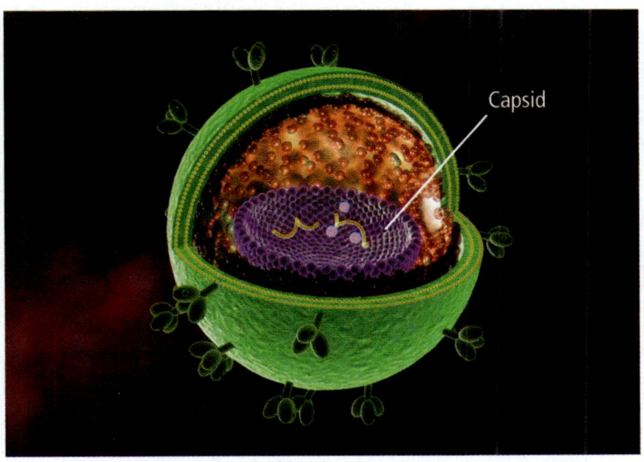


**Figure 1** In a normal HIV virus, the capsid forms a protective coating around the genetic material.

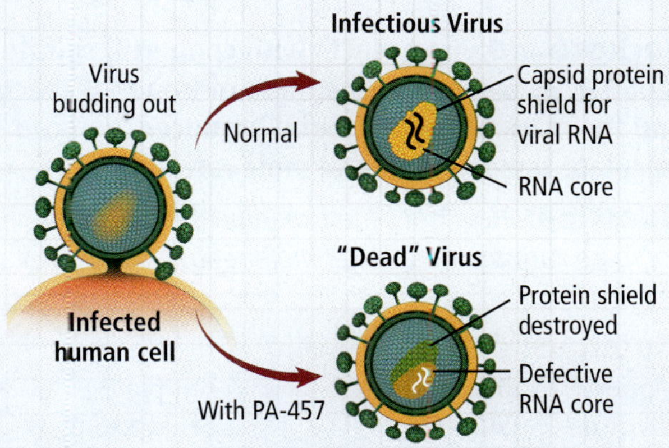


**Figure 2** When treated with PA-457, the HIV capsid becomes misshapen and collapses, resulting in the death of the virus.

**Surprise attack** This finding was surprising, because it showed that, unlike most drugs, PA-457 attacks the HIV structure, rather than the enzymes that help HIV reproduce, as illustrated in **Figure 2.** This makes PA-457 among the first of a new class of HIV drugs known as maturation inhibitors—drugs that can prevent the virus from maturing during the late stages in its development.

**Slowing evolution** The hope is that because PA-457 and other maturation inhibitors attack the HIV structure, resistance will be slower to develop. Even so, maturation inhibitors will likely be prescribed in combination with other AIDS drugs that attack HIV at different stages of its life cycle.

This practice, called multidrug therapy, makes it harder for HIV to develop resistance because any surviving virus would need to have multiple mutations—at least one for each anti-HIV drug. These mutations are less likely to occur at the same time.

### WRITING IN ▶ Chemistry

**Research** how scientists determine the safe dosing level for an experimental drug. Debate how a drug's effectiveness must be balanced with its potential toxicity and side effects.

WebQuest

# ChemLAB

iLab Station

## Determine the Mole Ratio

**Background:** **Iron reacts with copper(II) sulfate ($CuSO_4$). By measuring the mass of iron that reacts and the mass of copper metal produced, you can calculate the experimental mole ratio.**

**Question:** *How does the experimental mole ratio compare with the theoretical mole ratio?*

### Materials

copper(II) sulfate pentahydrate ($CuSO_4 \cdot 5H_2O$)
iron metal filings (20 mesh)
distilled water
150-mL beaker
100-mL graduated cylinder
hot plate
beaker tongs
balance
stirring rod
400-mL beaker
weighing paper

### Safety Precautions

WARNING: *Hot plates can cause burns. Turn off hot plates when not in use. Use only GFCI-protected circuits.*

### Procedure

1. Read and complete the lab safety form.
2. Measure the mass of a clean, dry 150-mL beaker. Record all measurements in a data table.
3. Place approximately 12 g $CuSO_4 \cdot 5H_2O$ into the 150-mL beaker, and measure the combined mass.
4. Add 50 mL of distilled water to the $CuSO_4 \cdot 5H_2O$. Place the mixture on a hot plate set at medium, and stir until all of the solid dissolves (do not boil). Using tongs, remove the beaker from the hot plate.
5. Measure about 2 g of iron filings onto a piece of weighing paper. Measure the mass of the filings.
6. While stirring, slowly add the iron filings to the hot copper(II) sulfate solution. Be careful not to splash the hot solution.
7. Allow the reaction mixture to sit for 5 min.
8. Use the stirring rod to decant (pour off) the liquid into a 400-mL beaker. Be careful to decant only the liquid—leave the solid copper metal behind.
9. Add 15 mL of distilled water to the copper solid, and carefully swirl the beaker to wash the copper. Decant the liquid into the 400-mL beaker.
10. Repeat Step 9 two more times.
11. Place the beaker containing the wet copper on the hot plate. Use low heat to dry the copper.
12. After the copper is dry, use tongs to remove the beaker from the hot plate and allow it to cool.
13. Measure the mass of the beaker and the copper.
14. **Cleanup and Disposal** The dry copper can be placed in a waste container. Moisten any residue that sticks to the beaker, and wipe it out using a paper towel. Pour the unreacted copper(II) sulfate and iron(II) sulfate solutions into a large beaker. Return all lab equipment to its proper place.

### Analyze and Conclude

1. **Apply** Write a balanced chemical equation for the reaction and calculate the mass of copper (Cu) that should have formed from the sample of iron (Fe) used. This mass is the theoretical yield.
2. **Interpret Data** Using your data, determine the mass and the moles of copper produced. Calculate the moles of iron used, and determine the whole-number iron-to-copper mole ratio and percent yield.
3. **Compare and Contrast** Compare the theoretical iron-to-copper mole ratio to the mole ratio you calculated using the experimental data.
4. **Error Analysis** Identify sources of the error that resulted in deviation from the mole ratio given in the balanced chemical equation.

### INQUIRY EXTENSION

**Compare** your results with those of other lab teams. Form a hypothesis to explain any differences. How would you test your hypothesis?

# CHAPTER 11 STUDY GUIDE

Vocabulary Practice

**BIGIDEA** **Mass relationships in chemical reactions confirm the law of conservation of mass.**

## SECTION 1 Defining Stoichiometry

**MAINIDEA** The amount of each reactant present at the start of a chemical reaction determines how much product can form.

- Balanced chemical equations can be interpreted in terms of moles, mass, and representative particles (atoms, molecules, formula units).
- The law of conservation of mass applies to all chemical reactions.
- Mole ratios are derived from the coefficients of a balanced chemical equation. Each mole ratio relates the number of moles of one reactant or product to the number of moles of another reactant or product in the chemical reaction.

**VOCABULARY**
- **stoichiometry**
- **mole ratio**

## SECTION 2 Stoichiometric Calculations

**MAINIDEA** The solution to every stoichiometric problem requires a balanced chemical equation.

- Chemists use stoichiometric calculations to predict the amounts of reactants used and products formed in specific reactions.
- The first step in solving stoichiometric problems is writing the balanced chemical equation.
- Mole ratios derived from the balanced chemical equation are used in stoichiometric calculations.
- Stoichiometric problems make use of mole ratios to convert between mass and moles.

## SECTION 3 Limiting Reactants

**MAINIDEA** A chemical reaction stops when one of the reactants is used up.

- The limiting reactant is the reactant that is completely consumed during a chemical reaction. Reactants that remain after the reaction stops are called excess reactants.
- To determine the limiting reactant, the actual mole ratio of the available reactants must be compared with the ratio of the reactants obtained from the coefficients in the balanced chemical equation.
- Stoichiometric calculations must be based on the limiting reactant.

**VOCABULARY**
- **limiting reactant**
- **excess reactant**

## SECTION 4 Percent Yield

**MAINIDEA** Percent yield is a measure of the efficiency of a chemical reaction.

- The theoretical yield of a chemical reaction is the maximum amount of product that can be produced from a given amount of reactant. Theoretical yield is calculated from the balanced chemical equation.
- The actual yield is the amount of product produced. Actual yield must be obtained through experimentation.
- Percent yield is the ratio of actual yield to theoretical yield expressed as a percent. High percent yield is important in reducing the cost of every product produced through chemical processes.

$$\text{Percent yield} = \frac{\text{actual yield}}{\text{theoretical yield}} \times 100$$

**VOCABULARY**
- **theoretical yield**
- **actual yield**
- **percent yield**

# CHAPTER 11 ASSESSMENT

## SECTION 1

### Mastering Concepts

**36.** Why must a chemical equation be balanced before you can determine mole ratios?

**37.** What relationships can be determined from a balanced chemical equation?

**38.** Explain why mole ratios are central to stoichiometric calculations.

**39.** What is the mole ratio that can convert from moles of A to moles of B?

**40.** Why are coefficients used in mole ratios instead of subscripts?

**41.** Explain how the conservation of mass allows you to interpret a balanced chemical equation in terms of mass.

**42.** When heated by a flame, ammonium dichromate decomposes, producing nitrogen gas, solid chromium(III) oxide, and water vapor.

$$(NH_4)_2Cr_2O_7 \rightarrow N_2 + Cr_2O_3 + 4H_2O$$

Write the mole ratios for this reaction that relate ammonium dichromate to the products.

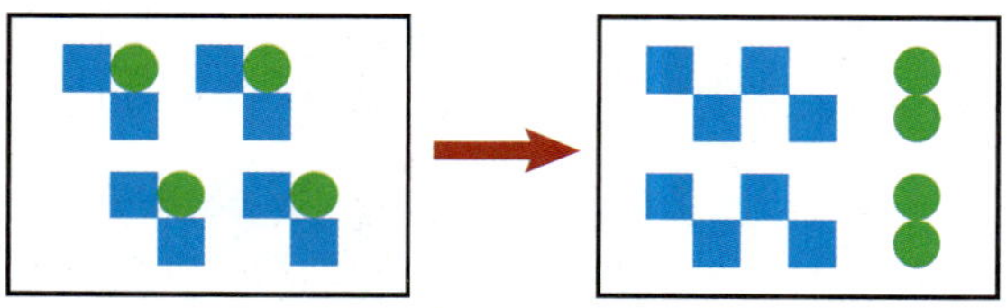

■ **Figure 10**

**43.** **Figure 10** depicts an equation with squares representing Element M and circles representing Element N. Write a balanced equation to represent the picture shown, using smallest whole-number ratios. Write mole ratios for this equation.

### Mastering Problems

**44.** Interpret the following equation in terms of particles, moles, and mass.

$$4Al(s) + 3O_2(g) \rightarrow 2Al_2O_3(s)$$

**45.** **Smelting** When tin(IV) oxide is heated with carbon in a process called smelting, the element tin can be extracted.

$$SnO_2(s) + 2C(s) \rightarrow Sn(l) + 2CO(g)$$

Interpret the chemical equation in terms of particles, moles, and mass.

**46.** When solid copper is added to nitric acid, copper(II) nitrate, nitrogen dioxide, and water are produced. Write the balanced chemical equation for the reaction. List six mole ratios for the reaction.

**47.** When hydrochloric acid solution reacts with lead(II) nitrate solution, lead(II) chloride precipitates and a solution of nitric acid is produced.

**a.** Write the balanced chemical equation for the reaction.

**b.** Interpret the equation in terms of molecules and formula units, moles, and mass.

**48.** When aluminum is mixed with iron(III) oxide, iron metal and aluminum oxide are produced, along with a large quantity of heat. What mole ratio would you use to determine moles of Fe if moles of $Fe_2O_3$ is known?

$$Fe_2O_3(s) + 2Al(s) \rightarrow 2Fe(s) + Al_2O_3(s) + \text{heat}$$

**49.** Solid silicon dioxide, often called silica, reacts with hydrofluoric acid (HF) solution to produce the gas silicon tetrafluoride and water.

**a.** Write the balanced chemical equation for the reaction.

**b.** List three mole ratios, and explain how you would use them in stoichiometric calculations.

**50.** **Chrome** The most important commercial ore of chromium is chromite ($FeCr_2O_4$). One of the steps in the process used to extract chromium from the ore is the reaction of chromite with coke (carbon) to produce ferrochrome ($FeCr_2$).

$$2C(s) + FeCr_2O_4(s) \rightarrow FeCr_2(s) + 2CO_2(g)$$

What mole ratio would you use to convert from moles of chromite to moles of ferrochrome?

**51.** **Air Pollution** The pollutant $SO_2$ is removed from the air in a reaction that also involves calcium carbonate and oxygen. The products of this reaction are calcium sulfate and carbon dioxide. Determine the mole ratio you would use to convert moles of $SO_2$ to moles of $CaSO_4$.

**52.** Two substances, W and X, react to form the products Y and Z. **Table 2** shows the moles of the reactants and products involved when the reaction was carried out. Use the data to determine the coefficients that will balance the equation W + X → Y + Z.

**Table 2 Reaction Data**

| Moles of Reactants | | Moles of Products | |
|---|---|---|---|
| W | X | Y | Z |
| 0.90 | 0.30 | 0.60 | 1.20 |

**53.** **Antacids** Magnesium hydroxide is an ingredient in some antacids. Antacids react with excess hydrochloric acid in the stomach to relieve indigestion.

$$\_\_Mg(OH)_2 + \_\_HCl \rightarrow \_\_MgCl_2 + \_\_H_2O$$

**a.** Balance the reaction of $Mg(OH)_2$ with HCl.

**b.** Write the mole ratio that would be used to determine the number of moles of $MgCl_2$ produced when HCl reacts with $Mg(OH)_2$.

# SECTION 2

## Mastering Concepts

**54.** What is the first step in all stoichiometric calculations?

**55.** What information does a balanced equation provide?

**56.** On what law is stoichometry based, and how do the calculations support this law?

**57.** How is molar mass used in some stoichiometric calculations?

**58.** What information must you have in order to calculate the mass of product formed in a chemical reaction?

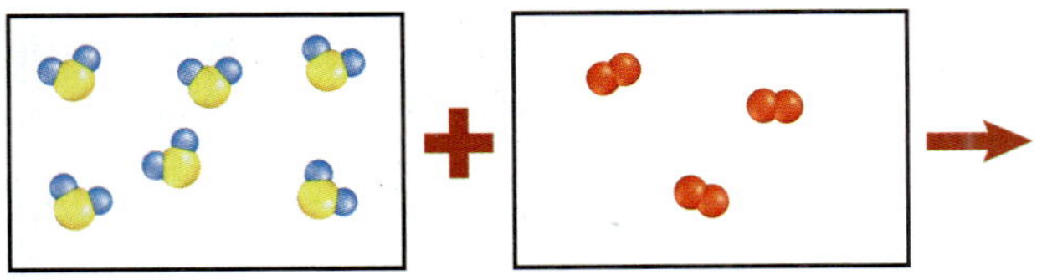

■ **Figure 11**

**59.** Each box in **Figure 11** represents the contents of a flask. One flask contains hydrogen sulfide, and the other contains oxygen. When the contents of the flasks are mixed, a reaction occurs and water vapor and sulfur are produced. In the figure, the red circles represent oxygen, the yellow circles represent sulfur, and blue circles represent hydrogen.

**a.** Write the balanced chemical equation for the reaction.

**b.** Using the same color code, sketch a representation of the flask after the reaction occurs.

## Mastering Problems

**60. Ethanol** ($C_2H_5OH$), also known as grain alcohol, can be made from the fermentation of sugar ($C_6H_{12}O_6$). The unbalanced chemical equation for the reaction is shown below.

$$\_\_C_6H_{12}O_6 \rightarrow \_\_C_2H_5OH + \_\_CO_2$$

Balance the chemical equation and determine the mass of $C_2H_5OH$ produced from 750 g of $C_6H_{12}O_6$.

**61. Welding** If 5.50 mol of calcium carbide ($CaC_2$) reacts with an excess of water, how many moles of acetylene ($C_2H_2$), a gas used in welding, will be produced?

$$CaC_2(s) + 2H_2O(l) \rightarrow Ca(OH)_2(aq) + C_2H_2(g)$$

**62. Antacid Fizz** When an antacid tablet dissolves in water, the fizz is due to a reaction between sodium hydrogen carbonate ($NaHCO_3$), also called sodium bicarbonate, and citric acid ($H_3C_6H_5O_7$).

$$3NaHCO_3(aq) + H_3C_6H_5O_7(aq) \rightarrow 3CO_2(g) + 3H_2O(l) + Na_3C_6H_5O_7(aq)$$

How many moles of $Na_3C_6H_5O_7$ can be produced if one tablet containing 0.0119 mol of $NaHCO_3$ is dissolved?

**63. Esterification** The process in which an organic acid and an alcohol react to form an ester and water is known as esterification. Ethyl butanoate ($C_3H_7COOC_2H_5$), an ester, is formed when the alcohol ethanol ($C_2H_5OH$) and butanoic acid ($C_3H_7COOH$) are heated in the presence of sulfuric acid.

$$C_2H_5OH(l) + C_3H_7COOH(l) \rightarrow C_3H_7COOC_2H_5(l) + H_2O(l)$$

Determine the mass of ethyl butanoate produced if 4.50 mol of ethanol is used.

**64. Greenhouse Gas** Carbon dioxide is a greenhouse gas that is linked to global warming. It is released into the atmosphere through the combustion of octane ($C_8H_{18}$) in gasoline. Write the balanced chemical equation for the combustion of octane and calculate the mass of octane needed to release 5.00 mol of $CO_2$.

**65.** A solution of potassium chromate reacts with a solution of lead(II) nitrate to produce a yellow precipitate of lead(II) chromate and a solution of potassium nitrate.

**a.** Write the balanced chemical equation.

**b.** Starting with 0.250 mol of potassium chromate, determine the mass of lead chromate formed.

**66. Rocket Fuel** The exothermic reaction between liquid hydrazine ($N_2H_4$) and liquid hydrogen peroxide ($H_2O_2$) is used to fuel rockets. The products of this reaction are nitrogen gas and water.

**a.** Write the balanced chemical equation.

**b.** How much hydrazine, in grams, is needed to produce 10.0 mol of nitrogen gas?

**67.** Chloroform ($CHCl_3$), an important solvent, is produced by a reaction between methane and chlorine.

$$CH_4(g) + 3Cl_2(g) \rightarrow CHCl_3(g) + 3HCl(g)$$

How much $CH_4$, in grams, is needed to produce 50.0 grams of $CHCl_3$?

**68. Oxygen Production** The Russian Space Agency uses potassium superoxide ($KO_2$) for the chemical oxygen generators in their space suits.

$$4KO_2 + 2H_2O + 4CO_2 \rightarrow 4KHCO_3 + 3O_2$$

Complete **Table 3.**

**Table 3 Oxygen Generation Reaction Data**

| Mass $KO_2$ | Mass $H_2O$ | Mass $CO_2$ | Mass $KHCO_3$ | Mass $O_2$ |
|---|---|---|---|---|
| | | | | 380 g |

**69. Gasohol** is a mixture of ethanol and gasoline. Balance the equation, and determine the mass of $CO_2$ produced from the combustion of 100.0 g of ethanol.

$$C_2H_5OH(l) + O_2(g) \rightarrow CO_2(g) + H_2O(g)$$

**70. Car Battery** Car batteries use lead, lead(IV) oxide, and a sulfuric acid solution to produce an electric current. The products of the reaction are lead(II) sulfate in solution and water.

**a.** Write the balanced equation for the reaction.

**b.** Determine the mass of lead(II) sulfate produced when 25.0 g of lead reacts with an excess of lead(IV) oxide and sulfuric acid.

**71.** To extract gold from its ore, the ore is treated with sodium cyanide solution in the presence of oxygen and water.

$$4Au(s) + 8NaCN(aq) + O_2(g) + 2H_2O(l) \rightarrow 4NaAu(CN)_2(aq) + 4NaOH(aq)$$

**a.** Determine the mass of gold that can be extracted if 25.0 g of sodium cyanide is used.

**b.** If the mass of the ore from which the gold was extracted is 150.0 g, what percentage of the ore is gold?

**72. Film** Photographic film contains silver bromide in gelatin. Once exposed, some of the silver bromide decomposes, producing fine grains of silver. The unexposed silver bromide is removed by treating the film with sodium thiosulfate. Soluble sodium silver thiosulfate ($Na_3Ag(S_2O_3)_2$) is produced.

$$AgBr(s) + 2Na_2S_2O_3(aq) \rightarrow Na_3Ag(S_2O_3)_2(aq) + NaBr(aq)$$

Determine the mass of $Na_3Ag(S_2O_3)_2$ produced if 0.275 g of AgBr is removed.

## SECTION 3

### Mastering Concepts

**73.** How is a mole ratio used to find the limiting reactant?

**74.** Explain why the statement, "The limiting reactant is the reactant with the lowest mass" is incorrect.

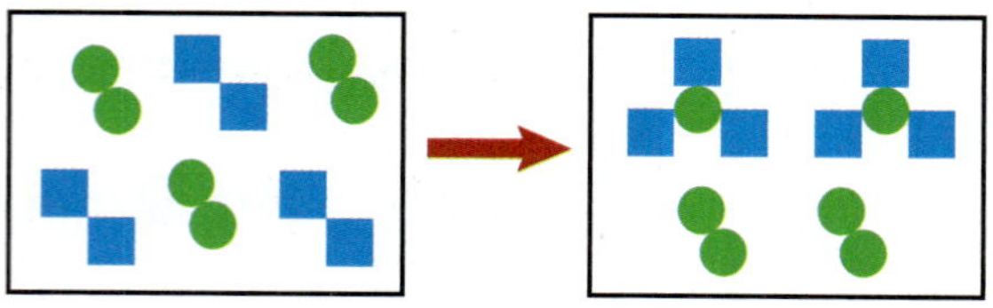

■ **Figure 12**

**75. Figure 12** uses squares to represent Element M and circles to represent Element N.

**a.** Write the balanced equation for the reaction.

**b.** If each square represents 1 mol of M and each circle represents 1 mol of N, how many moles of M and N were present at the start of the reaction?

**c.** How many moles of product form? How many moles of Element M and Element N are unreacted?

**d.** Identify the limiting reactant and the excess reactant.

### Mastering Problems

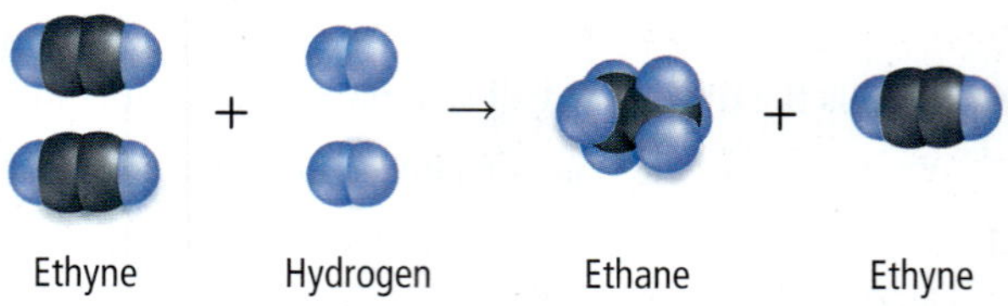


■ **Figure 13**

**76.** The reaction between ethyne ($C_2H_2$) and hydrogen ($H_2$) is illustrated in **Figure 13.** The product is ethane ($C_2H_6$). Which is the limiting reactant? Which is the excess reactant? Explain.

**77. Nickel-Iron Battery** In 1901, Thomas Edison invented the nickel-iron battery. The following reaction takes place in the battery.

$$Fe(s) + 2NiO(OH)(s) + 2H_2O(l) \rightarrow Fe(OH)_2(s) + 2Ni(OH)_2(aq)$$

How many mol of $Fe(OH)_2$ is produced when 5.00 mol of Fe and 8.00 mol of NiO(OH) react?

**78.** One of the few xenon compounds that form is cesium xenon heptafluoride ($CsXeF_7$). How many moles of $CsXeF_7$ can be produced from the reaction of 12.5 mol of cesium fluoride with 10.0 mol of xenon hexafluoride?

$$CsF(s) + XeF_6(s) \rightarrow CsXeF_7(s)$$

**79. Iron Production** Iron is obtained commercially by the reaction of hematite ($Fe_2O_3$) with carbon monoxide. How many grams of iron is produced when 25.0 mol of hematite reacts with 30.0 mol of carbon monoxide?

$$Fe_2O_3(s) + 3CO(g) \rightarrow 2Fe(s) + 3CO_2(g)$$

**80.** The reaction of chlorine gas with solid phosphorus ($P_4$) produces solid phosphorus pentachloride. When 16.0 g of chlorine reacts with 23.0 g of $P_4$, which reactant is limiting? Which reactant is in excess?

**81. Alkaline Battery** An alkaline battery produces electrical energy according to this equation.

$$Zn(s) + 2MnO_2(s) + H_2O(l) \rightarrow Zn(OH)_2(s) + Mn_2O_3(s)$$

**a.** Determine the limiting reactant if 25.0 g of Zn and 30.0 g of $MnO_2$ are used.

**b.** Determine the mass of $Zn(OH)_2$ produced.

**82.** Lithium reacts spontaneously with bromine to produce lithium bromide. Write the balanced chemical equation for the reaction. If 25.0 g of lithium and 25.0 g of bromine are present at the beginning of the reaction, determine

**a.** the limiting reactant.

**b.** the mass of lithium bromide produced.

**c.** the excess reactant and the excess mass.

## SECTION 4

### Mastering Concepts

**83.** What is the difference between actual yield and theoretical yield?

**84.** How are actual yield and theoretical yield determined?

**85.** Can the percent yield of a chemical reaction be more than 100%? Explain your answer.

**86.** What relationship is used to determine the percent yield of a chemical reaction?

**87.** What experimental information do you need in order to calculate both the theoretical and the percent yield of any chemical reaction?

**88.** A metal oxide reacts with water to produce a metal hydroxide. What additional information would you need to determine the percent yield of metal hydroxide from this reaction?

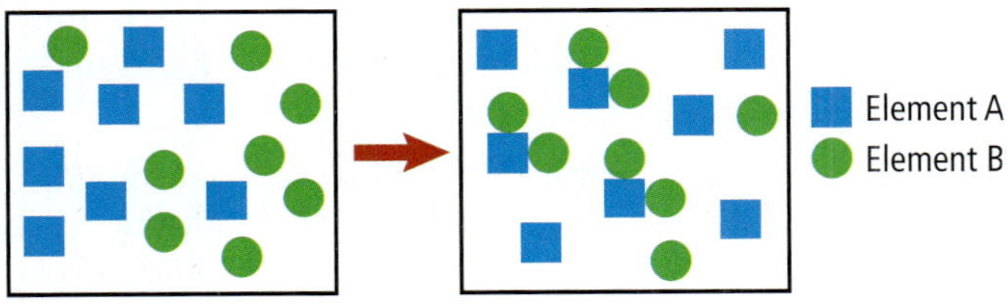


■ **Figure 14**

**89.** Examine the reaction represented in **Figure 14.** Determine if the reaction went to completion. Explain your answer, and calculate the percent yield of the reaction.

### Mastering Problems

**90.** **Ethanol** ($C_2H_5OH$) is produced from the fermentation of sucrose ($C_{12}H_{22}O_{11}$) in the presence of enzymes.

$$C_{12}H_{22}O_{11}(aq) + H_2O(g) \rightarrow 4C_2H_5OH(l) + 4CO_2(g)$$

Determine the theoretical yield and the percent yield of ethanol if 684 g of sucrose undergoes fermentation and 349 g of ethanol is obtained.

**91.** Lead(II) oxide is obtained by roasting galena, lead(II) sulfide, in air. The unbalanced equation is:

$$PbS(s) + O_2(g) \rightarrow PbO(s) + SO_2(g)$$

**a.** Balance the equation, and determine the theoretical yield of PbO if 200.0 g of PbS is heated.
**b.** What is the percent yield if 170.0 g of PbO is obtained?

**92.** Upon heating, calcium carbonate ($CaCO_3$) decomposes to calcium oxide (CaO) and carbon dioxide ($CO_2$).
**a.** Determine the theoretical yield of $CO_2$ if 235.0 g of $CaCO_3$ is heated.
**b.** What is the percent yield of $CO_2$ if 97.5 g of $CO_2$ is collected?

**93.** Hydrofluoric acid solutions cannot be stored in glass containers because HF reacts readily with silicon dioxide in glass to produce hexafluorosilicic acid ($H_2SiF_6$).

$$SiO_2(s) + 6HF(aq) \rightarrow H_2SiF_6(aq) + 2H_2O(l)$$

40.0 g $SiO_2$ and 40.0 g HF react to yield 45.8 g $H_2SiF_6$.
**a.** What is the limiting reactant?
**b.** What is the mass of the excess reactant?
**c.** What is the theoretical yield of $H_2SiF_6$?
**d.** What is the percent yield?

**94.** **Van Arkel Process** Pure zirconium is obtained using the two-step Van Arkel process. In the first step, impure zirconium and iodine are heated to produce zirconium iodide ($ZrI_4$). In the second step, $ZrI_4$ is decomposed to produce pure zirconium.

$$ZrI_4(s) \rightarrow Zr(s) + 2I_2(g)$$

Determine the percent yield of zirconium if 45.0 g of $ZrI_4$ is decomposed and 5.00 g of pure Zr is obtained.

**95.** Methanol, wood alcohol, is produced when carbon monoxide reacts with hydrogen gas.

$$CO + 2H_2 \rightarrow CH_3OH$$

When 8.50 g of carbon monoxide reacts with an excess of hydrogen, 8.52 g of methanol is collected. Complete **Table 4,** and calculate the percent yield.

**Table 4 Methanol Reaction Data**

| | CO(g) | $CH_3OH(l)$ |
|---|---|---|
| Mass | 8.50 g | |
| Molar mass | 28.01 g/mol | 32.05 g/mol |
| Moles | | |

**96.** **Phosphorus** ($P_4$) is commercially prepared by heating a mixture of calcium phosphate ($Ca_3(PO_4)_2$), sand ($SiO_2$), and coke (C) in an electric furnace. The process involves two reactions.

$$2Ca_3(PO_4)_2(s) + 6SiO_2(s) \rightarrow 6CaSiO_3(l) + P_4O_{10}(g)$$
$$P_4O_{10}(g) + 10C(s) \rightarrow P_4(g) + 10CO(g)$$

The $P_4O_{10}$ produced in the first reaction reacts with an excess of coke (C) in the second reaction. Determine the theoretical yield of $P_4$ if 250.0 g of $Ca_3(PO_4)_2$ and 400.0 g of $SiO_2$ are heated. If the actual yield of $P_4$ is 45.0 g, determine the percent yield of $P_4$.

**97.** Chlorine forms from the reaction of hydrochloric acid with manganese(IV) oxide. The balanced equation is:

$$MnO_2 + 4HCl \rightarrow MnCl_2 + Cl_2 + 2H_2O$$

Calculate the theoretical yield and the percent yield of chlorine if 86.0 g of $MnO_2$ and 50.0 g of HCl react. The actual yield of $Cl_2$ is 20.0 g.

## MIXED REVIEW

**98.** Ammonium sulfide reacts with copper(II) nitrate in a double replacement reaction. What mole ratio would you use to determine the moles of $NH_4NO_3$ produced if the moles of CuS are known?

**99. Fertilizer** The compound calcium cyanamide (CaNCN) is used as a nitrogen source for crops. To obtain this compound, calcium carbide is reacted with nitrogen at high temperatures.

$$CaC_2(s) + N_2(g) \rightarrow CaNCN(s) + C(s)$$

What mass of CaNCN can be produced if 7.50 mol of $CaC_2$ reacts with 5.00 mol of $N_2$?

**100.** When copper(II) oxide is heated in the presence of hydrogen gas, elemental copper and water are produced. What mass of copper can be obtained if 32.0 g of copper(II) oxide is used?

**101. Air Pollution** Nitrogen monoxide, which is present in urban air pollution, immediately converts to nitrogen dioxide as it reacts with oxygen.

**a.** Write the balanced chemical equation for the formation of nitrogen dioxide from nitrogen monoxide.

**b.** What mole ratio would you use to convert from moles of nitrogen monoxide to moles of nitrogen dioxide?

**102. Electrolysis** Determine the theoretical and percent yield of hydrogen gas if 36.0 g of water undergoes electrolysis to produce hydrogen and oxygen and 3.80 g of hydrogen is collected.

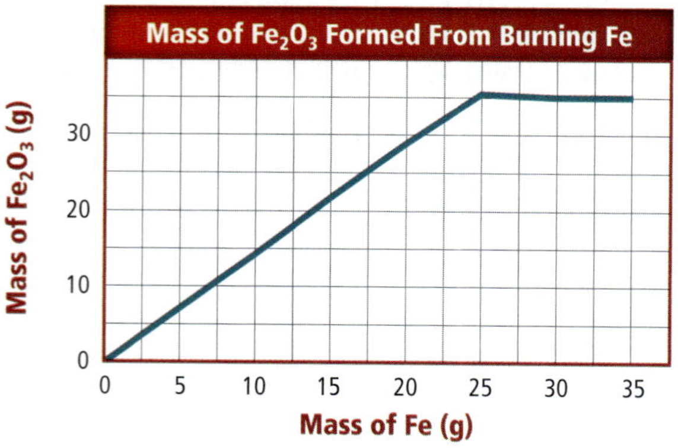


■ **Figure 15**

**103.** Iron reacts with oxygen as shown.

$$4Fe(s) + 3O_2(g) \rightarrow 2Fe_2O_3(s)$$

Different amounts of iron were burned in a fixed amount of oxygen. For each mass of iron burned, the mass of iron(III) oxide formed was plotted on the graph shown in **Figure 15.** Why does the graph level off after 25.0 g of iron is burned? How many moles of oxygen are present in the fixed amount?

## THINK CRITICALLY

**104. Analyze and Conclude** In an experiment, you obtain a percent yield of product of 108%. Is such a percent yield possible? Explain. Assuming that your calculation is correct, what reasons might explain such a result?

**105. Observe and Infer** Determine whether each reaction depends on a limiting reactant. Explain why or why not, and identify the limiting reactant.

**a.** Potassium chlorate decomposes to form potassium chloride and oxygen.

**b.** Silver nitrate and hydrochloric acid react to produce silver chloride and nitric acid.

**106. Design an Experiment** Design an experiment that can be used to determine the percent yield of anhydrous copper(II) sulfate when copper(II) sulfate pentahydrate is heated to remove water.

**107. Apply** When a campfire begins to die down and smolder, you can rekindle the flame by fanning the fire. Explain, in terms of stoichiometry, why the fire again begins to flare up when fanned.

**108. Apply** Students conducted a lab to investigate limiting and excess reactants. The students added different volumes of sodium phosphate solution ($Na_3PO_4$) to a beaker. They then added a constant volume of cobalt(II) nitrate solution ($Co(NO_3)_2$), stirred the contents, and allowed the beakers to sit overnight. The next day, each beaker had a purple precipitate at the bottom. The students decanted the supernatant from each beaker, divided it into two samples, and added one drop of sodium phosphate solution to one sample and one drop of cobalt(II) nitrate solution to the second sample. Their results are shown in **Table 5.**

**a.** Write a balanced chemical equation for the reaction.

**b.** Based on the results, identify the limiting reactant and the excess reactant for each trial.

**Table 5** Reaction Data for $Co(NO_3)_2$ and $Na_3PO_4$

| Trial | Volume $Na_3PO_4$ | Volume $Co(NO_3)_2$ | Reaction with Drop of $Na_3PO_4$ | Reaction with Drop of $Co(NO_3)_2$ |
|---|---|---|---|---|
| 1 | 5.0 mL | 10.0 mL | purple precipitate | no reaction |
| 2 | 10.0 mL | 10.0 mL | no reaction | purple precipitate |
| 3 | 15.0 mL | 10.0 mL | no reaction | purple precipitate |
| 4 | 20.0 mL | 10.0 mL | no reaction | purple precipitate |

Chapter Self-Check

## CHALLENGE PROBLEM

**109.** When 9.59 g of a certain vanadium oxide is heated in the presence of hydrogen, water and a new oxide of vanadium are formed. This new vanadium oxide has a mass of 8.76 g. When the second vanadium oxide undergoes additional heating in the presence of hydrogen, 5.38 g of vanadium metal forms.

- **a.** Determine the empirical formulas for the two vanadium oxides.
- **b.** Write balanced equations for the steps of the reaction.
- **c.** Determine the mass of hydrogen needed to complete the steps of this reaction.

## CUMULATIVE REVIEW

**110.** You observe that sugar dissolves more quickly in hot tea than in iced tea. You state that higher temperatures increase the rate at which sugar dissolves in water. Is this statement a hypothesis or a theory? Why?

**111.** Write the electron configuration for each of the following atoms.

- **a.** fluorine
- **b.** aluminum
- **c.** titanium
- **d.** radon

**112.** Explain why the gaseous nonmetals exist as diatomic molecules, but other gaseous elements exist as single atoms.

**113.** Write a balanced equation for the reaction of potassium with oxygen.

**114.** What is the molecular mass of $UF_6$? What is the molar mass of $UF_6$?

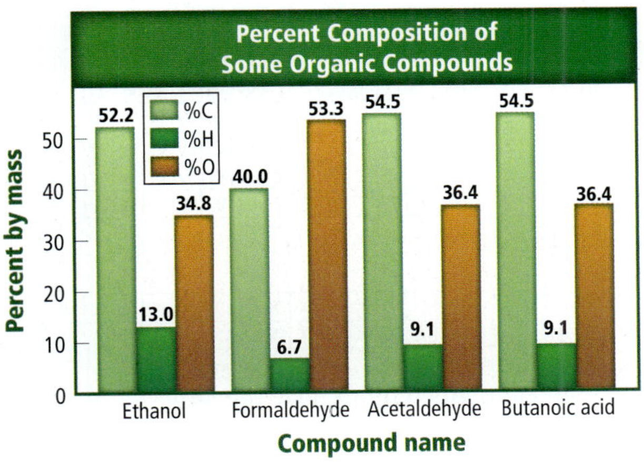


■ **Figure 16**

**115.** **Figure 16** gives percent composition data for several organic compounds.

- **a.** How are the molecular and empirical formulas of acetaldehyde and butanoic acid related?
- **b.** What is the empirical formula of butanoic acid?

## WRITING IN ▸ Chemistry

**116. Air Pollution** Research the air pollutants produced by combustion of gasoline in internal combustion engines. Discuss the common pollutants and the reaction that produces them. Show, through the use of stoichiometry, how each pollutant could be reduced if more people used mass transit.

**117. Haber Process** The percent yield of ammonia produced when hydrogen and nitrogen are combined under ordinary conditions is extremely small. However, the Haber Process combines the two gases under a set of conditions designed to maximize yield. Research the conditions used in the Haber Process, and find out why the development of the process was of great importance.

## DBQ Document-Based Question

**Chemical Defense** *Many insects secrete hydrogen peroxide ($H_2O_2$) and hydroquinone $C_6H_4(OH)_2$. Bombardier beetles take this a step further by mixing these chemicals with a catalyst. The result is an exothermic chemical reaction and a spray of hot, irritating chemicals for any would-be predator. Researchers hope to use a similar method to reignite aircraft turbine engines.*

**Figure 17** *below shows the unbalanced chemical reaction that results in the bombardier beetle's defensive spray.*

Data obtained from: Becker, Bob. April 2006. *ChemMatters.* 24: no. 2.

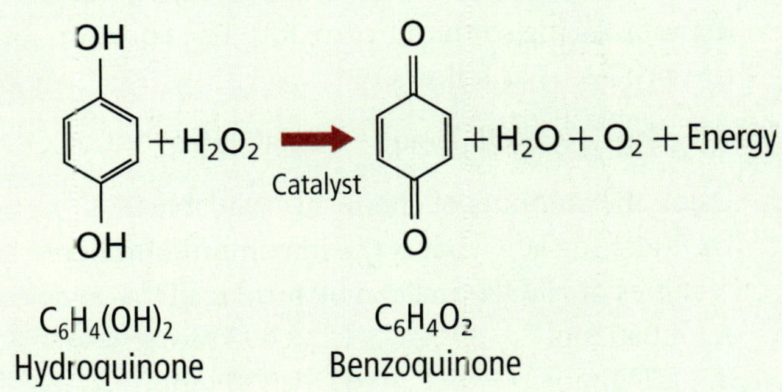


■ **Figure 17**

**118.** Balance the equation in **Figure 17.** If the bombardier beetle stores 100.0 mg of hydroquinone ($C_6H_4(OH)_2$) along with 50.0 mg of hydrogen peroxide ($H_2O_2$), what is the limiting reactant?

**119.** What is the excess reactant and how many milligrams are in excess?

**120.** How many milligrams of benzoquinone will be produced?

## MULTIPLE CHOICE

**1.** Stoichiometry is based on the law of
**A.** constant mole ratios.
**B.** Avogadro's constant.
**C.** conservation of energy.
**D.** conservation of mass.

*Use the graph below to answer Questions 2 to 5.*

**Supply of Various Chemicals in Dr. Raitano's Laboratory**

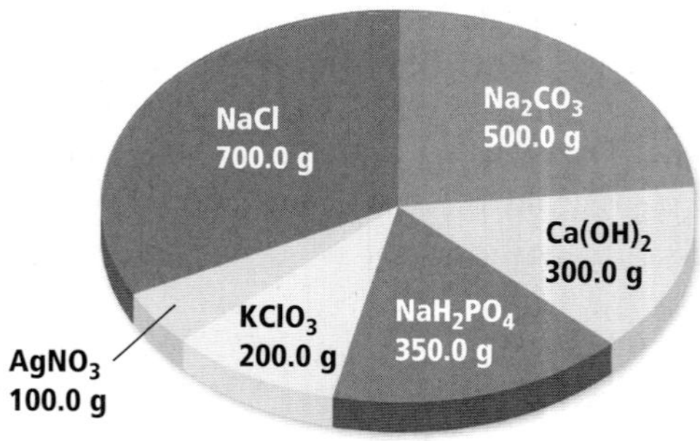


**2.** Pure silver metal can be made using the reaction shown below.

$$Cu(s) + 2AgNO_3(aq) \rightarrow 2Ag(s) + Cu(NO_3)_2(aq)$$

How many grams of copper metal will be needed to use up all of the $AgNO_3$ in Dr. Raitano's laboratory?
**A.** 18.70 g
**B.** 37.30 g
**C.** 74.70 g
**D.** 100.0 g

**3.** The LeBlanc process is the traditional method of manufacturing sodium hydroxide. The equation for this process is as follows.

$$Na_2CO_3(aq) + Ca(OH)_2(aq) \rightarrow 2NaOH(aq) + CaCO_3(s)$$

Using the amounts of chemicals available in Dr. Raitano's lab, what is the maximum number of moles of NaOH that can be produced?
**A.** 4.050 mol
**B.** 4.720 mol
**C.** 8.097 mol
**D.** 9.430 mol

**4.** Pure $O_2$ gas can be generated from the decomposition of potassium chlorate ($KClO_3$):

$$2KClO_3(s) \rightarrow 2KCl(s) + 3O_2(g)$$

If half of the $KClO_3$ in the lab is used and 12.8 g of oxygen gas is produced, what is the percent yield of this reaction?
**A.** 12.8%
**B.** 32.7%
**C.** 65.6%
**D.** 98.0%

**5.** Sodium dihydrogen pyrophosphate ($Na_2H_2P_2O_7$), more commonly known as baking powder, is manufactured by heating $NaH_2PO_4$ to a high temperature.

$$2NaH_2PO_4(s) \rightarrow Na_2H_2P_2O_7(s) + H_2O(g)$$

If 444.0 g of $Na_2H_2P_2O_7$ is needed, how much more $NaH_2PO_4$ will Dr. Raitano have to buy to make enough $Na_2H_2P_2O_7$?
**A.** 0.000 g
**B.** 94.00 g
**C.** 130.0 g
**D.** 480.0 g

**6.** Red mercury(II) oxide decomposes at high temperatures to form mercury metal and oxygen gas.

$$2HgO(s) \rightarrow 2Hg(l) + O_2(g)$$

If 3.55 mol of HgO decomposes to form 1.54 mol of $O_2$ and 618 g of Hg, what is the percent yield of this reaction?
**A.** 13.2%
**B.** 42.5%
**C.** 56.6%
**D.** 86.8%

*Use the diagram below to answer Questions 7 and 8.*

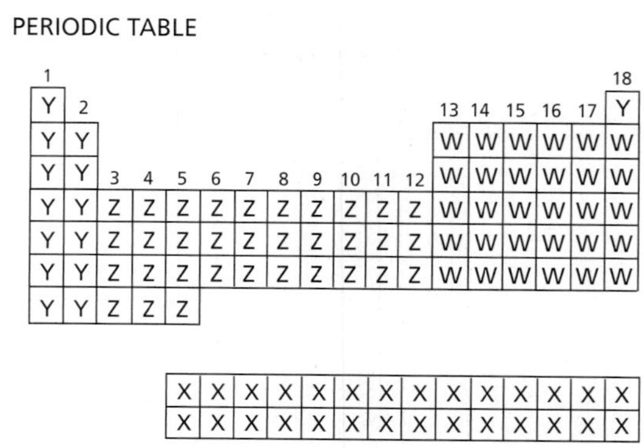


**7.** Which elements tend to have the largest atomic radius in their periods?
**A.** W
**B.** X
**C.** Y
**D.** Z

**8.** Elements labeled *W* have their valence electrons in which sublevel?
**A.** s
**B.** p
**C.** d
**D.** f

## SHORT ANSWER

9. Dimethyl hydrazine $(CH_3)_2N_2H_2$ ignites on contact with dinitrogen tetroxide $(N_2O_4)$.

$$(CH_3)_2N_2H_2(l) + 2N_2O_4(l) \rightarrow 3N_2(g) + 4H_2O(g) + 2CO_2(g)$$

Because this reaction produces an enormous amount of energy from a small amount of reactants, it was used to drive the rockets on the Lunar Excursion Modules (LEMs) of the Apollo space program. If 18.0 mol of dinitrogen tetroxide is consumed in this reaction, how many moles of nitrogen gas will be released?

## EXTENDED RESPONSE

*Use the table below to answer Questions 10 and 11.*

**First Ionization Energy of Period 3 Elements**

| Element | Atomic Number | 1st Ionization Energy, kJ/mol |
|---|---|---|
| Sodium | 11 | 496 |
| Magnesium | 12 | 736 |
| Aluminum | 13 | 578 |
| Silicon | 14 | 787 |
| Phosphorus | 15 | 1012 |
| Selenium | 16 | 1000 |
| Chlorine | 17 | 1251 |
| Argon | 18 | 1521 |

10. Plot the data from this data table. Place atomic numbers on the $x$-axis.

11. Summarize the general trend in ionization energy. How does ionization energy relate to the number of valence electrons in an element?

## SAT SUBJECT TEST: CHEMISTRY

12. How much cobalt(III) titanate $(Co_2TiO_4)$, in moles, is in 7.13 g of the compound?
    A. $2.39 \times 10^1$ mol
    B. $3.10 \times 10^{-2}$ mol
    C. $3.22 \times 10^1$ mol
    D. $4.17 \times 10^{-2}$ mol
    E. $2.28 \times 10^{-2}$ mol

*Use the pictures below to answer Questions 13 to 17.*

A.
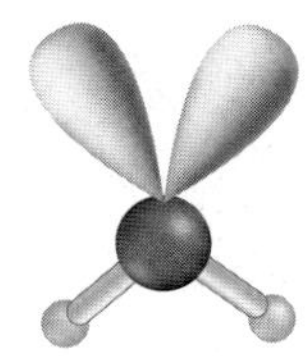

D.
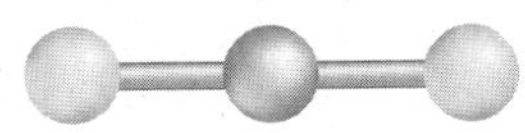

B.
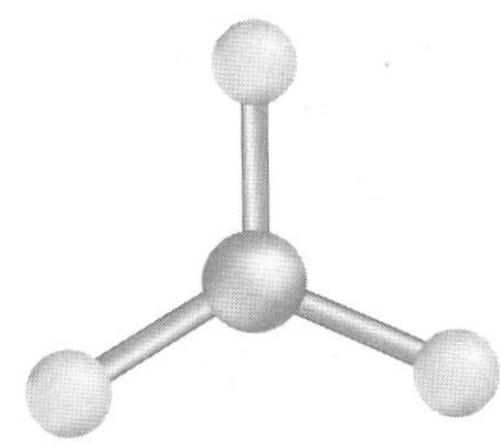

E.
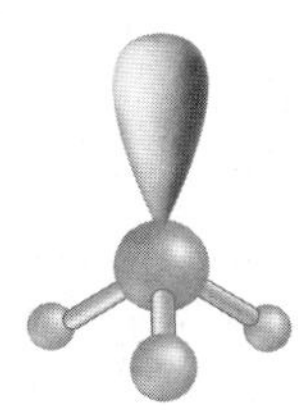

C.
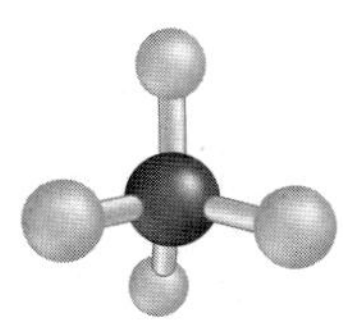

13. Hydrogen sulfide displays this molecular shape.

14. Molecules with this shape have four shared pairs of electrons and no lone pairs of electrons.

15. This molecular shape is known as *trigonal planar.*

16. Carbon dioxide displays this molecular shape.

17. This molecular shape displays $sp^2$ hybridization.

| NEED EXTRA HELP? | | | | | | | | | | | | | | | | | |
|---|---|---|---|---|---|---|---|---|---|---|---|---|---|---|---|---|---|
| **If You Missed Question . . .** | 1 | 2 | 3 | 4 | 5 | 6 | 7 | 8 | 9 | 10 | 11 | 12 | 13 | 14 | 15 | 16 | 17 |
| **Review Section . . .** | 11.1 | 11.2 | 11.2 | 11.4 | 11.3 | 11.4 | 6.3 | 5.3 | 11.2 | 6.3 | 6.3 | 10.3 | 8.4 | 8.4 | 8.4 | 8.4 | 8.4 |

CHAPTER 12

# States of Matter

**BIGIDEA** Kinetic-molecular theory explains the different properties of solids, liquids, and gases.

## SECTIONS

Chm 10, 11, 11a

1 Gases

2 Forces of Attraction

3 Liquids and Solids

4 Phase Changes

The iodine thermometer contains a few grams of iodine inside a sealed sphere. As the outdoor temperature increases, the iodine changes from a solid directly to a gas. The deeper the violet color, the higher the temperature.

## LaunchLAB

iLab Station

### How do different liquids affect the speed of a sinking ball bearing?

You've probably noticed that different liquids might have vastly different properties. For example, liquids such as maple syrup, corn oil, and vegetable oil are much thicker than liquids such as water. In this lab, you will study how the speed of an object is affected by the viscosity of two different liquids.

**FOLDABLES® Study Organizer**

### States of Matter

Make a three-tab book. Label it as shown. Use it to help you summarize information about three common states of matter.

Go online!
MHEonline.com

SECTION 1

# Gases

**MAINIDEA** Gases expand, diffuse, exert pressure, and can be compressed because they are in a low-density state consisting of tiny, constantly moving particles.

### Essential Questions

- How is the kinetic-molecular theory used to explain the behavior of gases?
- Why does mass affect the rates of diffusion and effusion?
- How is gas pressure measured and how is the partial pressure of a gas calculated?

### Review Vocabulary

**kinetic energy:** energy due to motion

### New Vocabulary

kinetic-molecular theory
elastic collision
temperature
diffusion
Graham's law of effusion
pressure
barometer
pascal
atmosphere
Dalton's law of partial pressures

**CHEM 4 YOU** **If you have gone camping, you might have slept on an air-filled mattress. How did the mattress compare to lying on the ground? It was probably warmer and more comfortable. The properties of the air mattress are due to the particles that make up the air inside it.**

## The Kinetic-Molecular Theory

You have learned that composition (the types of atoms present) and structure (their arrangement) determine the chemical properties of matter. Composition and structure also affect the physical properties of matter. Based solely on physical appearance, you can distinguish between the solids and liquids, as shown in **Figure 1.** By contrast, substances that are gases at room temperature usually display similar physical properties despite their different compositions. Why is there so little variation in behavior among gases? Why are the physical properties of gases different from those of liquids and solids?

By the eighteenth century, scientists knew how to collect gaseous products by displacing water. Now, they could observe and measure properties of individual gases. About 1860, chemists Ludwig Boltzmann and James Maxwell, who were working in different countries, each proposed a model to explain the properties of gases. That model is the kinetic-molecular theory. Because all of the gases known to Boltzmann and Maxwell contained molecules, the name of the model refers to molecules. The word *kinetic* comes from a Greek word meaning *to move.* Objects in motion have energy called kinetic energy. The **kinetic-molecular theory** describes the behavior of matter in terms of particles in motion. The model makes several assumptions about the size, motion, and energy of gas particles.

**Figure 1** You can distinguish some materials by looking at them, but this is not true for many gases.

(l)©Don Mason/Corbis; (c)©Lester V. Bergman/Corbis; (r)Dirk Wiersma/Photo Researchers

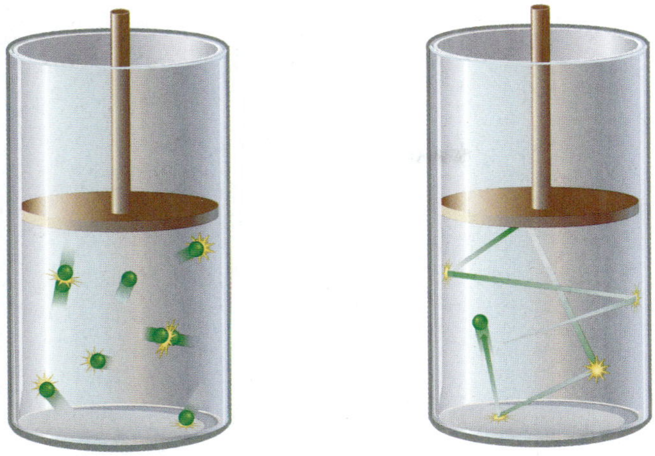

■ **Figure 2** Kinetic energy can be transferred between gas particles during an elastic collision. Between collisions, the particles move in straight lines.

**Explain** ***the influence that gas particles have on each other, both in terms of collisions and what happens to particles between collisions.***

**Particle size** Gases consist of small particles that are separated from one another by empty space. The volume of the particles is small compared with the volume of the empty space. Because gas particles are far apart, they experience no significant attractive or repulsive forces.

**Particle motion** Gas particles are in constant, random motion. Particles move in a straight line until they collide with other particles or with the walls of their container, as shown in **Figure 2.** Collisions between gas particles are elastic. An **elastic collision** is one in which no kinetic energy is lost. Kinetic energy can be transferred between colliding particles, but the total kinetic energy of the two particles does not change.

**Particle energy** Two factors determine the kinetic energy of a particle: mass and velocity. The kinetic energy of a particle can be represented by the following equation.

$$KE = \frac{1}{2}mv^2$$

$KE$ is kinetic energy, $m$ is the mass of the particle, and $v$ is its velocity. Velocity reflects both the speed and the direction of motion. In a sample of a single gas, all particles have the same mass, but all particles do not have the same velocity. Therefore, all particles do not have the same kinetic energy. **Temperature** is a measure of the average kinetic energy of the particles in a sample of matter.

## Explaining the Behavior of Gases

The kinetic-molecular theory helps explain the behavior of gases. For example, the constant motion of gas particles allows a gas to expand until it fills its container, such as when you inflate a beach ball. As you blow air into the ball, the air particles spread out and fill the inside of the container—the beach ball.

**Low density** Remember that density is mass per unit volume. The density of chlorine gas is $2.898 \times 10^{-3}$ g/mL at 20°C; the density of solid gold is 19.3 g/mL. Gold is more than 6700 times as dense as chlorine. This large difference cannot be due only to the difference in mass between gold atoms and chlorine molecules (about 3:1). As the kinetic-molecular theory states, a great deal of space exists between gas particles. Thus, there are fewer chlorine molecules than gold atoms in the same volume.

**VOCABULARY**

**WORD ORIGIN**

**Gas**
comes from the Latin word *chaos,* which means *space*

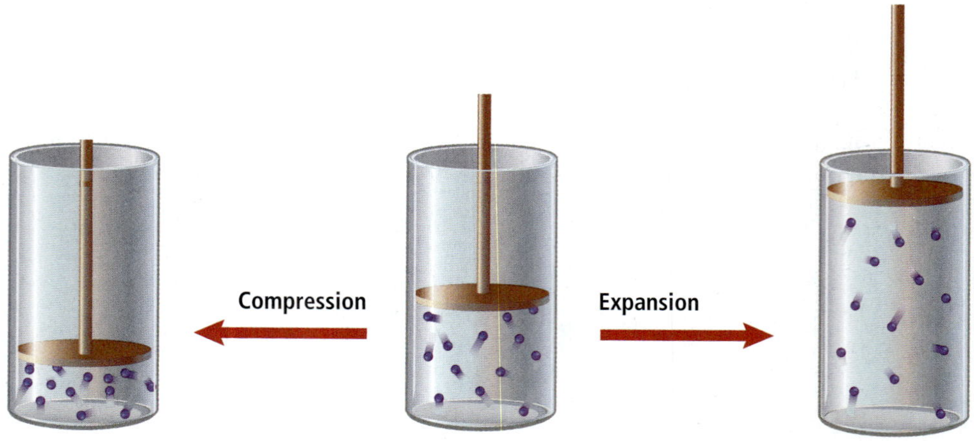


**Figure 3** In a closed container, compression and expansion change the volume occupied by a constant mass of particles.

**Relate** ***the change in volume to the density of the gas particles in each cylinder.***

**Compression and expansion** If you squeeze a pillow made of foam, you can compress it; that is, you can reduce its volume. Air, which is a mixture of gases, is also compressible. The large amount of empty space between the particles in the air allows air to be squeezed into a smaller volume. When the volume of a container is made larger, the random motion of the particles fills the available space. **Figure 3** illustrates what happens to the density of a gas in a container as it is compressed and as it is allowed to expand.

**Diffusion and effusion** According to the kinetic-molecular theory, there are no significant forces of attraction between gas particles. Thus, gas particles can flow easily past each other. Often, the space into which a gas flows is already occupied by another gas. The random motion of the gas particles causes the gases to mix until they are evenly distributed. **Diffusion** is the term used to describe the movement of one material through another. The term might be new, but you are probably familiar with the process. If food is cooking in the kitchen, you can smell it throughout the house because the gas particles diffuse. Particles diffuse from an area of high concentration (the kitchen) to one of low concentration (the other rooms in the house).

Effusion is a process related to diffusion. During effusion, a gas escapes through a tiny opening. What happens when you puncture a container, such as a balloon or a tire? In 1846, Thomas Graham conducted experiments to measure the rates of effusion for different gases at the same temperature. Graham designed his experiments so that the gases effused into a vacuum—space containing no matter. He discovered an inverse relationship between effusion rates and molar mass. **Graham's law of effusion** states that the rate of effusion for a gas is inversely proportional to the square root of its molar mass.

**Graham's Law**

$$\text{Rate of effusion} \propto \frac{1}{\sqrt{\text{molar mass}}}$$

The rate of diffusion or effusion of a gas is inversely proportional to the square root of its molar mass.

The rate of diffusion depends mainly on the mass of the particles involved. Lighter particles diffuse more rapidly than heavier particles. Recall that different gases at the same temperature have the same average kinetic energy as described by the equation KE = ½ $mv^2$. However, the mass of gas particles varies from gas to gas. For lighter particles to have the same average kinetic energy as heavier particles, they must have, on average, a greater velocity.

Graham's law also applies to rates of diffusion, which is logical because heavier particles diffuse more slowly than lighter particles at the same temperature. Using Graham's law, you can set up a proportion to compare the diffusion rates for two gases.

$$\frac{\text{Rate}_A}{\text{Rate}_B} = \sqrt{\frac{\text{molar mass}_B}{\text{molar mass}_A}}$$

Explore **kinetic theory.**

Virtual Investigations

**READING CHECK** **Explain** why the rate of diffusion depends on the mass of the particles.

## EXAMPLE Problem 1

Find help with **square and cube roots.** Math Handbook

**GRAHAM'S LAW** Ammonia has a molar mass of 17.0 g/mol; hydrogen chloride has a molar mass of 36.5 g/mol. What is the ratio of their diffusion rates?

### 1 ANALYZE THE PROBLEM

You are given the molar masses for ammonia and hydrogen chloride. To find the ratio of the diffusion rates for ammonia and hydrogen chloride, use the equation for Graham's law of effusion.

| **Known** | **Unknown** |
|---|---|
| molar $\text{mass}_{HCl}$ = 36.5 g/mol | ratio of diffusion rates = ? |
| molar $\text{mass}_{NH_3}$ = 17.0 g/mol | |

### 2 SOLVE FOR THE UNKNOWN

$$\frac{\text{Rate}_{NH_3}}{\text{Rate}_{HCl}} = \sqrt{\frac{\text{molar mass}_{HCl}}{\text{molar mass}_{NH_3}}}$$

State the ratio derived from Graham's law.

$$= \sqrt{\frac{36.5 \text{ g/mol}}{17.0 \text{ g/mol}}} = \mathbf{1.47}$$

Substitute molar $\text{mass}_{HCl}$ = 36.5 g/mol and molar $\text{mass}_{NH_3}$ = 17.0 g/mol.

The ratio of diffusion rates is 1.47.

### 3 EVALUATE THE ANSWER

A ratio of roughly 1.5 is logical because molecules of ammonia are about half as massive as molecules of hydrogen chloride. Because the molar masses have three significant figures, the answer also does. Note that the units cancel, and the answer is stated correctly without any units.

EXAMPLE PROBLEM

## PRACTICE Problems

Do additional problems. Online Practice

1. Calculate the ratio of effusion rates for nitrogen ($N_2$) and neon (Ne).
2. Calculate the ratio of diffusion rates for carbon monoxide and carbon dioxide.
3. **Challenge** What is the rate of effusion for a gas that has a molar mass twice that of a gas that effuses at a rate of 3.6 mol/min?

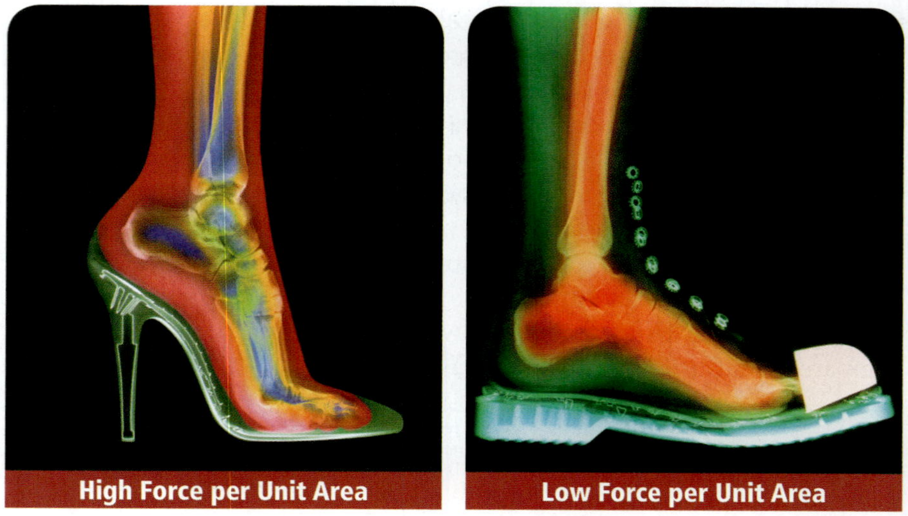


■ **Figure 4** High-heeled shoes increase the pressure on a surface because the area touching the floor is reduced. In flatter-heeled shoes, such as boots, the force is applied over a larger area.

**Infer** *where the highest pressure is located between the floor and high-heel shoe.*

## Gas Pressure

Have you watched someone try to walk across snow, mud, or hot asphalt in high heels? If so, you might have noticed that the heels sank into the soft surface. **Figure 4** shows why a person sinks when wearing high heels but does not sink when wearing boots. In each case, the force pressing down on the soft surface is related to the person's mass. With boots, the force is spread out over a larger area. **Pressure** is defined as force per unit area. The area of the bottom of a boot is much larger than the area of the bottom of a high-heeled shoe. So, the pressure on the soft surface is less with a boot than it is with high heels.

Gas particles also exert pressure when they collide with the walls of their container. Because an individual gas particle has little mass, it can exert little pressure. However, a liter-sized container could hold $10^{22}$ gas particles. With this many particles colliding, the pressure can be high.

**Air pressure** Earth is surrounded by an atmosphere that extends into space for hundreds of kilometers. Because the particles in air move in every direction, they exert pressure in all directions. This pressure is called atmospheric pressure, or air pressure. Air pressure varies at different points on Earth. Because gravity is greater at the surface of Earth, there are more particles than at higher altitudes where the force of gravity is less. Fewer particles at higher elevations exert less force than the greater concentration of particles at lower altitudes. Therefore, air pressure is less at higher altitudes than it is at sea level. At sea level, atmospheric pressure is about one-kilogram per square centimeter.

**Measuring air pressure** Italian physicist Evangelista Torricelli (1608–1647) was the first to demonstrate that air exerted pressure. He noticed that water pumps were unable to pump water higher than about 10 m. He hypothesized that the height of a column of liquid would vary with the density of the liquid. To test this idea, Torricelli designed the equipment shown in **Figure 5.** He filled a thin glass tube that was closed at one end with mercury. While covering the open end so that air could not enter, he inverted the tube and placed it (open end down) in a dish of mercury. The height of the mercury column fell to about one-fourteenth of a similar water column. This validated Terricelli's hypothesis because mercury is approximately fourteen times more dense than water.

■ **Figure 5** Torricelli was the first to design equipment to show that the atmosphere exerted pressure.

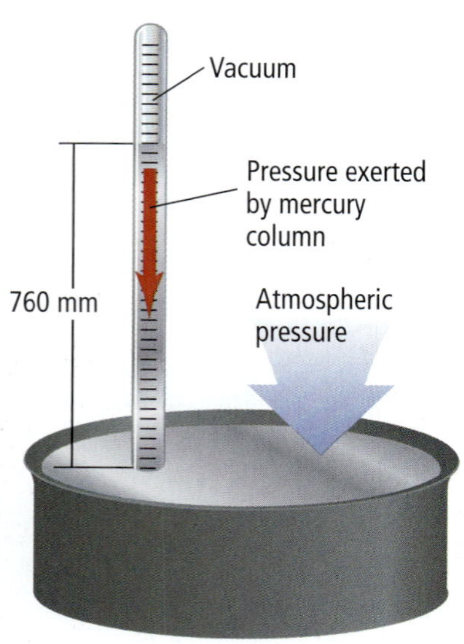


H. Turvey/Photo Researchers

**Barometers** The device that Torricelli invented is called a barometer. A **barometer** is an instrument used to measure atmospheric pressure. As Torricelli demonstrated, at sea level the height of the mercury in a barometer is usually about 760 mm. The exact height of the mercury is determined by two forces. Gravity exerts a constant downward force on the mercury. This force is opposed by an upward force exerted by air pressing down on the surface of the mercury. Changes in air temperature or humidity cause air pressure to vary.

**Manometers** A manometer is an instrument used to measure gas pressure in a closed container. In a manometer, a flask is connected to a U-tube that contains mercury, as shown in **Figure 6.** When the valve between the flask and the U-tube is opened, gas particles diffuse out of the flask into the U-tube. The released gas particles push down on the mercury in the tube. The difference in the height of the mercury in the two arms is used to calculate the pressure of the gas in the flask.

**Units of pressure** The SI unit of pressure is the pascal (Pa). It is named for Blaise Pascal (1623–1662), a French mathematician and philosopher. The pascal is derived from the SI unit of force, the newton (N). One **pascal** is equal to a force of one newton per square meter: 1 Pa equals 1 $N/m^2$. Many fields of science still use more traditional units of pressure. For example, engineers often report pressure as pounds per square inch (psi). The pressures measured by barometers and manometers can be reported in millimeters of mercury (mmHg).
There is also a unit called the torr and another unit called a bar.

At sea level, the average air pressure is 101.3 kPa when the temperature is 0°C. Air pressure is often reported in a unit called an atmosphere (atm). One **atmosphere** is equal to 760 mmHg or 760 torr or 101.3 kilopascals (kPa). **Table 1** compares different units of pressure. Because the units 1 atm, 760 mmHg, and 760 torr are defined units, they should have as many significant figures as needed when used in calculations.

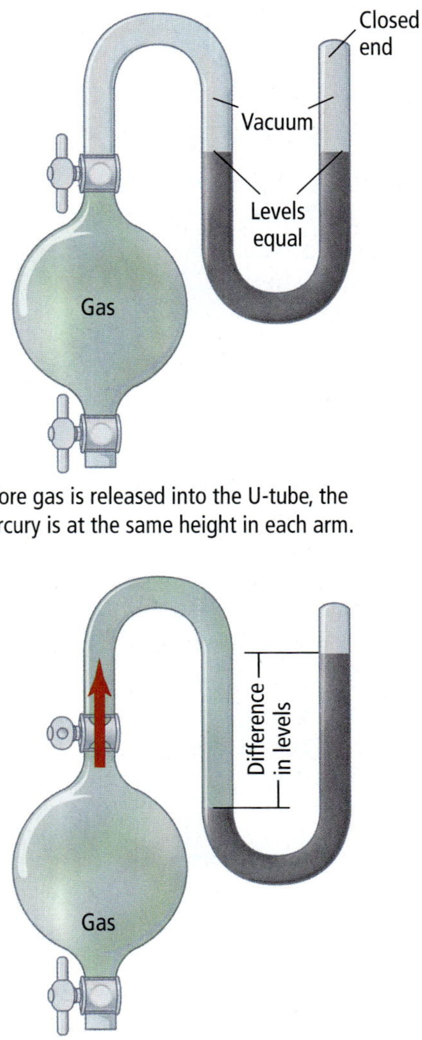

Before gas is released into the U-tube, the mercury is at the same height in each arm.

After gas is released into the U-tube, the heights in the two arms are no longer equal.

■ **Figure 6** A manometer measures the pressure of an enclosed gas.

**Table 1 Comparison of Pressure Units**

| Unit | Number Equivalent to 1 atm | Number Equivalent to 1 kPa |
|---|---|---|
| Kilopascal (kPa) | 101.3 kPa | — |
| Atmosphere (atm) | — | 0.009869 atm |
| Millimeters of mercury (mmHg) | 760 mmHg | 7.501 mmHg |
| Torr | 760 torr | 7.501 torr |
| Pounds per square inch (psi or $lb/in^2$) | 14.7 psi | 0.145 psi |
| Bar | 1.01 bar | 0.01 bar |

# Data Analysis LAB

Based on Real Data*

## Make and Use Graphs

**How are the depth of a dive and altitude related?** Most divers dive at locations that are at or near sea level in altitude. However, divers in Saskatchewan, Alberta, and British Columbia, Canada, as well as much of the northwestern United States, dive at higher altitudes.

### Data and Observations

The table shows the pressure gauge correction factor for high altitude underwater diving.

**Altitude Diving Correction Factors**

| Altitude (m) | Atmospheric Pressure (atm) | Pressure Gauge Correction Factor (m) |
|---|---|---|
| 0 | 1.000 | 0.0 |
| 600 | 0.930 | 0.7 |
| 1200 | 0.864 | 1.4 |
| 1800 | 0.801 | 2.0 |
| 2400 | 0.743 | 2.7 |
| 3000 | 0.688 | 3.2 |

*Data obtained from: Sawatzky, D. 2000. Diving at Altitude Part I. *Diver Magazine*. June 2000.

### Think Critically

1. **Compare** Use the data in the table to make a graph of atmospheric pressure versus altitude.
2. **Calculate** What is your actual diving depth if your depth gauge reads 18 m, but you are at an altitude of 1800 m and your gauge does not compensate for altitude?
3. **Infer** Dive tables are used to determine how long it is safe for a diver to stay under water at a specific depth. Why is it important to know the correct depth of the dive?

**Dalton's law of partial pressures** When Dalton studied the properties of gases, he found that each gas in a mixture exerts pressure independently of the other gases present. Illustrated in **Figure 7**, **Dalton's law of partial pressures** states that the total pressure of a mixture of gases is equal to the sum of the pressures of all the gases in the mixture. The portion of the total pressure contributed by a single gas is called its partial pressure. The partial pressure of a gas depends on the number of moles of gas, the size of the container, and the temperature of the mixture. It does not depend on the identity of the gas. At a given temperature and pressure, the partial pressure of 1 mol of any gas is the same. Dalton's law of partial pressures can be summarized by the equation at the top of the next page.

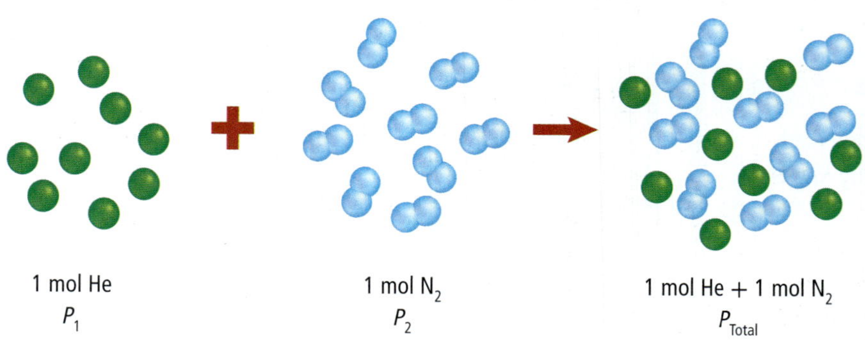


■ **Figure 7** When gases mix, the total pressure of the mixture is equal to the sum of the partial pressures of the individual gases.
**Determine** *How do the partial pressures of nitrogen gas and helium gas compare when a mole of nitrogen gas and a mole of helium gas are in the same closed container?*

## Dalton's Law of Partial Pressures

$$P_{total} = P_1 + P_2 + P_3 + \ldots P_n$$

$P_{total}$ represents total pressure. $P_1$, $P_2$, and $P_3$ represent the partial pressures of each gas up to the final gas, $P_n$.

To calculate the total pressure of a mixture of gases, add the partial pressures of each of the gases in the mixture.

Get help with **Dalton's law of partial pressures.**

Personal Tutor

Look again at **Figure 7.** What happens when 1 mol of helium and 1 mol of nitrogen are combined in a single closed container? Because neither the volume nor the number of particles changes, the total pressure equals the sum of the two partial pressures.

## EXAMPLE Problem 2

Find help with **significant figures.** Math Handbook

**THE PARTIAL PRESSURE OF A GAS** A mixture of oxygen ($O_2$), carbon dioxide ($CO_2$), and nitrogen ($N_2$) has a total pressure of 0.97 atm. What is the partial pressure of $O_2$ if the partial pressure of $CO_2$ is 0.70 atm and the partial pressure of $N_2$ is 0.12 atm?

### 1 ANALYZE THE PROBLEM

You are given the total pressure of a mixture and the partial pressure of two gases in the mixture. To find the partial pressure of the third gas, use the equation that relates partial pressures to total pressure.

**Known**

$P_{N_2} = 0.12$ atm

$P_{CO_2} = 0.70$ atm

$P_{total} = 0.97$ atm

**Unknown**

$P_{O_2} = ?$ atm

### 2 SOLVE FOR THE UNKNOWN

$P_{total} = P_{N_2} + P_{CO_2} + P_{O_2}$ — State Dalton's law of partial pressures.

$P_{O_2} = P_{total} - P_{CO_2} - P_{N_2}$ — Solve for $P_{O_2}$.

$P_{O_2} = 0.97 \text{ atm} - 0.70 \text{ atm} - 0.12 \text{ atm}$ — Substitute $P_{N_2} = 0.12$ atm, $P_{CO_2} = 0.70$ atm, and $P_{total} = 0.97$ atm.

$P_{O_2} = 0.15$ **atm**

### 3 EVALUATE THE ANSWER

Adding the calculated value for the partial pressure of oxygen to the known partial pressures gives the total pressure, 0.97 atm. The answer has two significant figures to match the data.

EXAMPLE PROBLEM

## PRACTICE Problems

Do additional problems. Online Practice

**4.** What is the partial pressure of hydrogen gas in a mixture of hydrogen and helium if the total pressure is 600 mmHg and the partial pressure of helium is 439 mmHg?

**5.** Find the total pressure for a mixture that contains four gases with partial pressures of 5.00 kPa, 4.56 kPa, 3.02 kPa, and 1.20 kPa.

**6.** Find the partial pressure of carbon dioxide in a gas mixture with a total pressure of 30.4 kPa if the partial pressures of the other two gases in the mixture are 16.5 kPa and 3.7 kPa.

**7. Challenge** Air is a mixture of gases. By percentage, it is roughly 78 percent nitrogen, 21 percent oxygen, and 1 percent argon. (There are trace amounts of many other gases in air.) If the atmospheric pressure is 760 mmHg, what are the partial pressures of nitrogen, oxygen, and argon in the atmosphere?

PRACTICE PROBLEMS

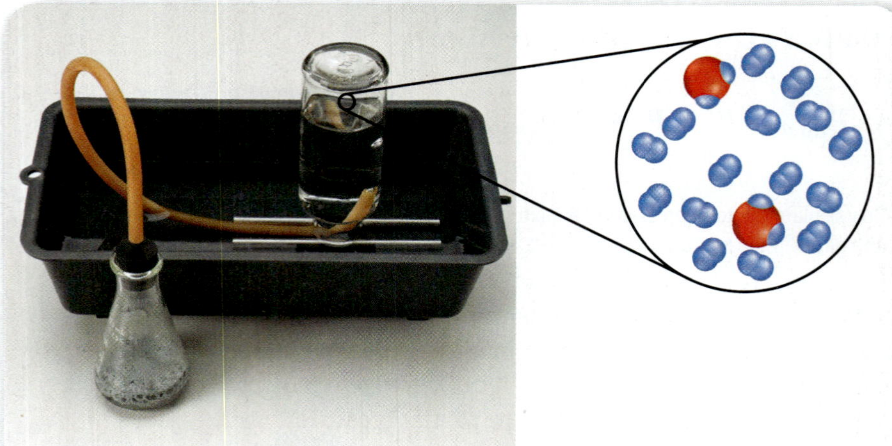

■ **Figure 8** In the flask, sulfuric acid ($H_2SO_4$) reacts with zinc to produce hydrogen gas. The hydrogen is collected at 20°C.

**Calculate** ***the partial pressure of hydrogen at 20°C if the total pressure of the hydrogen and water vapor mixture is 100.0 kPa.***

**FOLDABLES®** Incorporate information from this section into your Foldable.

**Using Dalton's law** Partial pressures can be used to determine the amount of gas produced by a reaction. The gas produced is bubbled into an inverted container of water, as shown in **Figure 8.** As the gas collects, it displaces the water. The gas collected in the container will be a mixture of hydrogen and water vapor. Therefore, the total pressure inside the container will be the sum of the partial pressures of hydrogen and water vapor.

The partial pressures of gases at the same temperature are related to their concentration. The partial pressure of water vapor has a fixed value at a given temperature. You can look up the value in a reference table. At 20°C, the partial pressure of water vapor is 2.3 kPa. You can calculate the partial pressure of hydrogen by subtracting the partial pressure of water vapor from the total pressure.

As you will read later, knowing the pressure, volume, and temperature of a gas allows you to calculate the number of moles of the gas. Temperature and volume can be measured during an experiment. Once the temperature is known, the partial pressure of water vapor is used to calculate the pressure of the gas. The known values for volume, temperature, and pressure are then used to find the number of moles.

## SECTION 1 REVIEW

Section Self-Check

### Section Summary

- The kinetic-molecular theory explains the properties of gases in terms of the size, motion, and energy of their particles.
- Dalton's law of partial pressures is used to determine the pressures of individual gases in gas mixtures.
- Graham's law is used to compare the diffusion rates of two gases.

8. **MAIN**IDEA **Explain** Use the kinetic theory to explain the behavior of gases.
9. **Describe** how the mass of a gas particle affects its rate of effusion and diffusion.
10. **Explain** how gas pressure is measured.
11. **Explain** why the container of water must be inverted when a gas is collected by displacement of water.
12. **Calculate** Suppose two gases in a container have a total pressure of 1.20 atm. What is the pressure of Gas B if the partial pressure of Gas A is 0.75 atm?
13. **Infer** whether or not temperature has any effect on the diffusion rate of a gas. Explain your answer.

# SECTION 2 Forces of Attraction

**MAINIDEA** **Intermolecular forces—including dispersion forces, dipole-dipole forces, and hydrogen bonds—determine a substance's state at a given temperature.**

### Essential Questions

- What are intramolecular forces?
- How do intermolecular forces compare? How do they contrast?

### Review Vocabulary

**polar covalent:** a type of bond that forms when electrons are not shared equally

### New Vocabulary

dispersion force
dipole-dipole force
hydrogen bond

**CHEM 4 YOU** **You might be aware that water is a substance that is found as a solid, a liquid, and a gas at temperatures and pressures common on Earth. This unique property, along with others that enable life as we understand it to exist, stems from the forces that exist between water molecules.**

## Intermolecular Forces

If all particles of matter at room temperature have the same average kinetic energy, why are some materials gases while others are liquids or solids? The answer lies with the attractive forces within and between particles. The attractive forces that hold particles together in ionic, covalent, and metallic bonds are called intramolecular forces. The prefix *intra-* means *within*. For example, intramural sports are competitions among teams from within a single school or district. The term *molecular* can refer to atoms, ions, or molecules. **Table 2** summarizes what you read previously about intramolecular forces.

Intramolecular forces do not account for all attractions between particles. There are forces of attraction called intermolecular forces. The prefix *inter-* means *between* or *among*. For example, an interview is a conversation between two people. These forces can hold together identical particles, such as water molecules in a drop of water, or two different types of particles, such as carbon atoms in graphite and the cellulose particles in paper. The three intermolecular forces that will be discussed in this section are dispersion forces, dipole-dipole forces, and hydrogen bonds. Although some intermolecular forces are stronger than others, all intermolecular forces are weaker than the intramolecular forces involved in bonding.

**Table 2** Comparison of Intramolecular Forces

| Force | Model | Basis of Attraction | Example |
|---|---|---|---|
| Ionic | 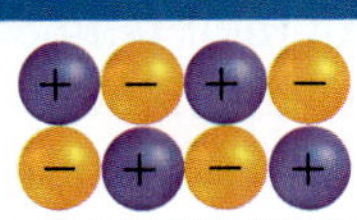 | cations and anions | NaCl |
| Covalent | 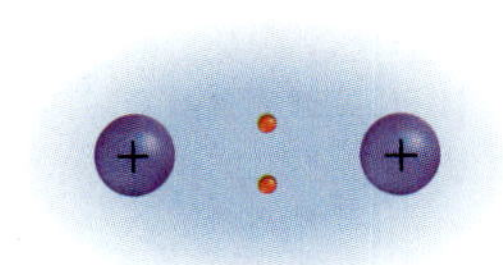 | positive nuclei and shared electrons | $H_2$ |
| Metallic | 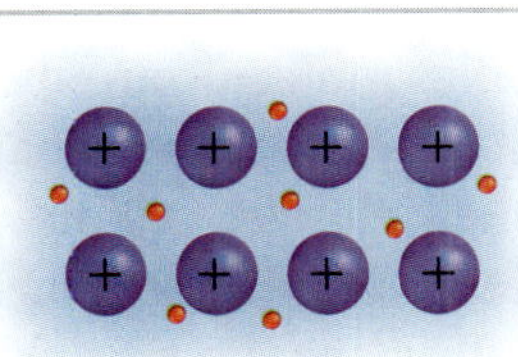 | metal cations and mobile electrons | Fe |

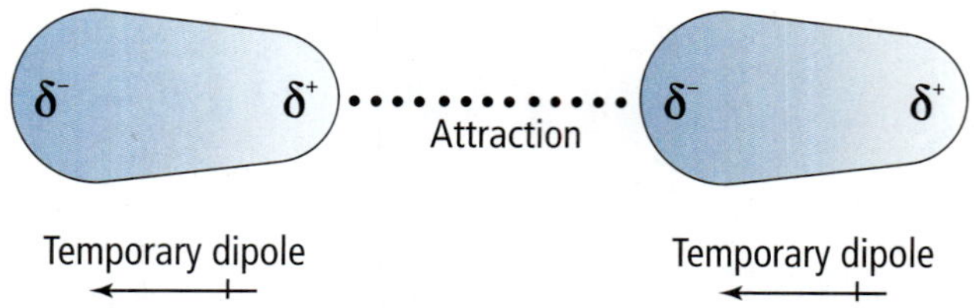


■ **Figure 9** When two molecules are close together, the electron clouds repel each other, creating temporary dipoles. The δ sign represents an area of partial charge on the molecule.
**Explain** ***what the δ+ and δ– signs on a temporary dipole represent.***

**Dispersion forces** Recall that oxygen molecules are nonpolar because electrons are evenly distributed between the equally electronegative oxygen atoms. Under the right conditions, however, oxygen molecules can be compressed into a liquid. For oxygen to condense, there must be some force of attraction between its molecules.

The force of attraction between oxygen molecules is called a dispersion force. **Dispersion forces** are weak forces that result from temporary shifts in the density of electrons in electron clouds. Dispersion forces are sometimes called London forces after the German-American physicist who first described them, Fritz London.

Remember that the electrons in an electron cloud are in constant motion. When two molecules are in close contact, especially when they collide, the electron cloud of one molecule repels the electron cloud of the other molecule. The electron density around each nucleus is, for a moment, greater in one region of each cloud. Each molecule forms a temporary dipole. When temporary dipoles are close together, a weak dispersion force exists between oppositely charged regions of the dipoles, as shown in **Figure 9.**

☑ READING CHECK **Explain** why dispersion forces form.

Dispersion forces exist between all particles. Dispersion forces are weak for small particles, and these forces have an increasing effect as the number of electrons involved increases. Thus, dispersion forces tend to become stronger as the size of the particles increase. For example, fluorine, chlorine, bromine, and iodine exist as diatomic molecules. Recall that the number of nonvalence electrons increases from fluorine to chlorine to bromine to iodine. Because the larger halogen molecules have more electrons, there can be a greater difference between the positive and negative regions of their temporary dipoles and, thus, stronger dispersion forces. This difference in dispersion forces explains why fluorine and chlorine are gases, bromine is a liquid, and iodine is a solid at room temperature.

☑ READING CHECK **Infer** the physical state of the element astatine at room temperature and explain your reasoning.

**VOCABULARY**
**ACADEMIC VOCABULARY**
**Orient**
to arrange in a specific position; to align in the same direction
*The blooms of the flowers were all oriented toward the setting Sun.*

**Dipole-dipole forces** Polar molecules contain permanent dipoles; that is, some regions of a polar molecule are always partially negative and some regions of the molecule are always partially positive. These attractions between oppositely charged regions of polar molecules are called **dipole-dipole forces.** Neighboring polar molecules orient themselves so that oppositely charged regions align.

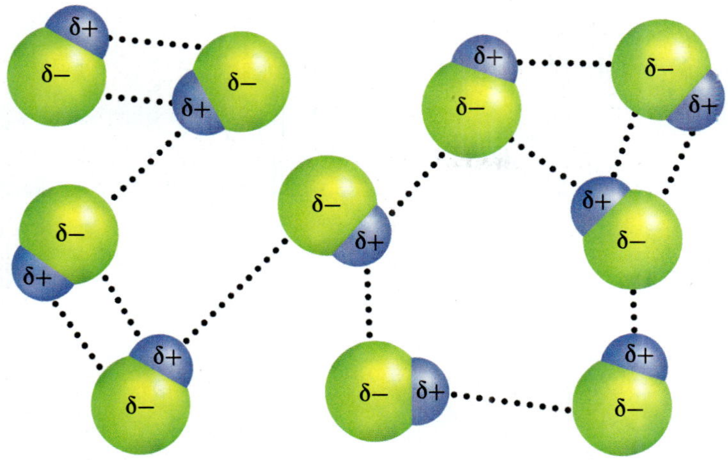


■ **Figure 10** Neighboring polar molecules orient themselves so that oppositely charged regions align.
**Identify** ***the types of forces that are represented in this figure.***

When hydrogen-chloride gas molecules approach, the partially positive hydrogen atom in one molecule is attracted to the partially negative chlorine atom in another molecule. **Figure 10** shows multiple attractions among hydrogen-chloride molecules. Because the dipoles are permanent, you might expect dipole-dipole forces to be stronger than dispersion forces. This prediction holds true for small polar molecules with large dipoles. However, for many polar molecules, including the HCl molecules in **Figure 10,** dispersion forces dominate dipole-dipole forces.

☑ READING CHECK **Compare** dipole-dipole forces and dispersion forces.

**Hydrogen bonds** One special type of dipole-dipole attraction is called a hydrogen bond. A **hydrogen bond** is a dipole-dipole attraction that occurs between molecules containing a hydrogen atom bonded to a small, highly electronegative atom with at least one lone electron pair. Hydrogen bonds typically dominate both dispersion forces and dipole-dipole forces. For a hydrogen bond to form, hydrogen must be bonded to either a fluorine, oxygen, or nitrogen atom. These atoms are electronegative enough to cause a large partial positive charge on the hydrogen atom, yet small enough that their lone pairs of electrons can come close to hydrogen atoms. For example, in a water molecule, the hydrogen atoms have a large partial positive charge and the oxygen atom has a large partial negative charge. When water molecules approach, a hydrogen atom on one molecule is attracted to the lone pair of electrons on the oxygen atom on the other molecule, as shown in **Figure 11.**

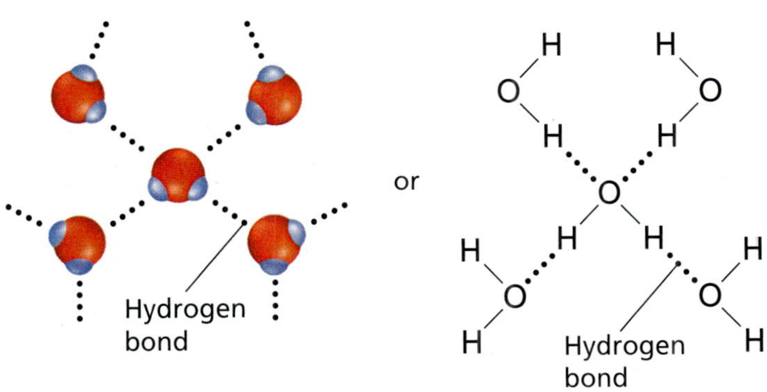


■ **Figure 11** The hydrogen bonds between water molecules are stronger than typical dipole-dipole attractions because the bond between hydrogen and oxygen is highly polar.

**Table 3** **Properties of Three Molecular Compounds**

| Compound | Molecular Structure | Molar Mass (g) | Boiling Point (°C) |
|---|---|---|---|
| Water ($H_2O$) | 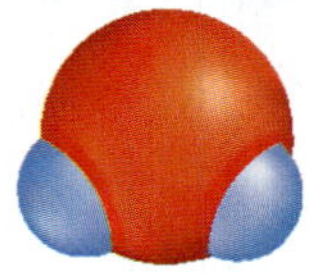 | 18.0 | 100 |
| Methane ($CH_4$) |  | 16.0 | −161.5 |
| Ammonia ($NH_3$) |  | 17.0 | −33.3 |

Hydrogen bonds explain why water is a liquid at room temperature, while compounds of comparable mass are gases. Look at the data in **Table 3.** The difference between methane and water is easy to explain. Because methane molecules are nonpolar, the only forces holding the molecules together are relatively weak dispersion forces. The difference between ammonia and water is not as obvious. Molecules of both compounds can form hydrogen bonds. Yet, ammonia is a gas at room temperature, which indicates that the attractive forces between ammonia molecules are not as strong. Because oxygen atoms are more electronegative than nitrogen atoms, the O–H bonds in water are more polar than the N–H bonds in ammonia. As a result, the hydrogen bonds between water molecules are stronger than the hydrogen bonds between ammonia molecules.

## SECTION 2 REVIEW

Section Self-Check

### Section Summary

- Intramolecular forces are stronger than intermolecular forces.
- Dispersion forces are intermolecular forces between temporary dipoles.
- Dipole-dipole forces occur between polar molecules.

**14. MAIN**IDEA **Explain** what determines a substance's state at a given temperature.

**15. Compare and contrast** intermolecular forces and describe intramolecular forces.

**16. Evaluate** Which of the molecules listed below can form hydrogen bonds? For which of the molecules would dispersion forces be the only intermolecular force? Give reasons for your answers.

**a.** $H_2$ **b.** $H_2S$ **c.** HCl **d.** HF

**17. Intepret Data** In a methane molecule ($CH_4$), there are four single covalent bonds. In an octane molecule ($C_8H_{18}$), there are 25 single covalent bonds. How does the number of bonds affect the dispersion forces in samples of methane and octane? Which compound is a gas at room temperature? Which is a liquid?

SECTION 3

# Liquids and Solids

**MAINIDEA** The particles in solids and liquids have a limited range of motion and are not easily compressed.

### Essential Questions

- How do the arrangements of particles in liquids and solids differ?
- What are the factors that affect viscosity?
- How are the unit cell and crystal lattice related?

### Review Vocabulary

**meniscus:** the curved surface of a column of liquid

### New Vocabulary

viscosity
surface tension
surfactant
crystalline solid
unit cell
allotrope
amorphous solid

**CHEM 4 YOU** Did you ever wonder why syrup that is stored in the refrigerator is harder to pour than syrup stored in the pantry? You probably know that warming syrup makes it pour more easily. But why does an increase in temperature help?

## Liquids

Although the kinetic-molecular theory was developed to explain the behavior of gases, the model also applies to liquids and solids. When applying the kinetic-molecular theory to the solid and liquid states of matter, you must consider the forces of attraction between particles as well as their energy of motion.

Previously, you read that a liquid can take the shape of its container but its volume is fixed. In other words, the particles can flow to adjust to the shape of a container, but the liquid cannot expand to fill its container, as shown in **Figure 12.** According to the kinetic-molecular theory, individual particles do not have fixed positions in the liquid. Forces of attraction between particles in the liquid limit their range of motion so that the particles remain closely packed in a fixed volume.

**Density and compression** At 25°C and 1 atm of air pressure, liquids are much denser than gases. The density of a liquid is much greater than that of its vapor at the same conditions. For example, liquid water is about 1250 times denser than water vapor at 25°C and 1 atm of pressure. Because they are at the same temperature, both gas and liquid particles have the same average kinetic energy. Thus, the higher density of liquids is due to the intermolecular forces that hold particles together.

Unlike gases, liquids are considered incompressible in many applications. The change in volume for liquids is much smaller because liquid particles are already tightly packed. An enormous amount of pressure must be applied to reduce the volume of a liquid by a very small amount.

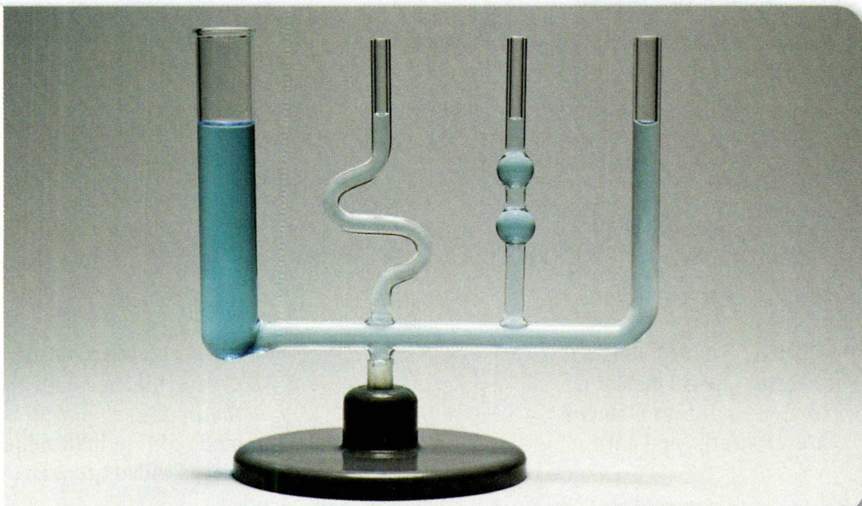

**Figure 12** Liquids flow and take the shape of their container, but they do not expand to fill their container like gases.

**Infer** *the reason that the liquid is at the same level in each of the interconnected tubes.*

■ **Figure 13** Gases and liquids have the ability to flow and diffuse. These photos show one liquid diffusing through another liquid.

**Fluidity** Gases and liquids are classified as fluids because they can flow and diffuse. **Figure 13** shows one liquid diffusing through another liquid. Liquids usually diffuse more slowly than gases at the same temperature, because intermolecular attractions interfere with the flow. Thus, liquids are less fluid than gases. A comparison between water and natural gas can illustrate this difference. When there is a leak in a basement water pipe, the water remains in the basement unless the amount of water released exceeds the volume of the basement.

A gas will not stay in the basement. For example, natural gas, or methane, is a fuel burned in gas furnaces, hot-water heaters, and stoves. Gas that leaks from a gas pipe diffuses throughout the house. Because natural gas is odorless, companies that supply the fuel include a compound with a distinct odor. Adding odor to natural gas warns the homeowner of the leak. The customer has time to shut off the gas supply, open windows to allow the gas to diffuse, and call the gas company to report the leak.

■ **Figure 14**

## Studying States of Matter

Scientific discoveries have led to a greater understanding of the states of matter.

**460 B.C.** Philosophers theorize that all matter is composed of four elements: fire, air, water, and earth.

**400 B.C.**

**400 B.C.** The philosopher Democritus develops the theory that all matter is made of tiny, indivisible pieces called *atomos.*

**360 B.C.** Aristotle, a philosopher with great influence, rejects Democritus's theory and supports the belief that all matter is composed of fire, air, water, and earth.

**1643 A.D.** The invention of the barometer proves that air has weight.

**A.D. 1700**

**1734** Daniel Bernoulli proposes that gas pressure results from gas particles colliding with the walls of the container.

**Viscosity** You are already familiar with viscosity if you have ever tried to get honey out of a bottle. **Viscosity** is a measure of the resistance of a liquid to flow. The particles in a liquid are close enough for attractive forces to slow their movement as they flow past one another. The viscosity of a liquid is determined by the type of intermolecular forces in the liquid, the size and shape of the particles, and the temperature.

You should note that not all liquids have viscosity. Scientists discovered superfluids in 1937. Scientists cooled liquid helium below −270.998°C and discovered that the properties of the liquid changed. The superfluid helium lost viscosity—the resistance to flow. The discovery of superfluidity and other milestones in our understanding of states of matter are shown in **Figure 14.**

**Attractive forces** In typical liquids, the stronger the intermolecular attractive forces, the higher the viscosity. If you have used glycerol in the laboratory to help insert a glass tube into a rubber stopper, you know that glycerol is a viscous liquid. **Figure 15** uses structural formulas to show the hydrogen bonding that makes glycerol so viscous. The hydrogen atoms attached to the oxygen atoms in each glycerol molecule are able to form hydrogen bonds with other glycerol molecules. The red dots in **Figure 15** show where the hydrogen bonds form between molecules.

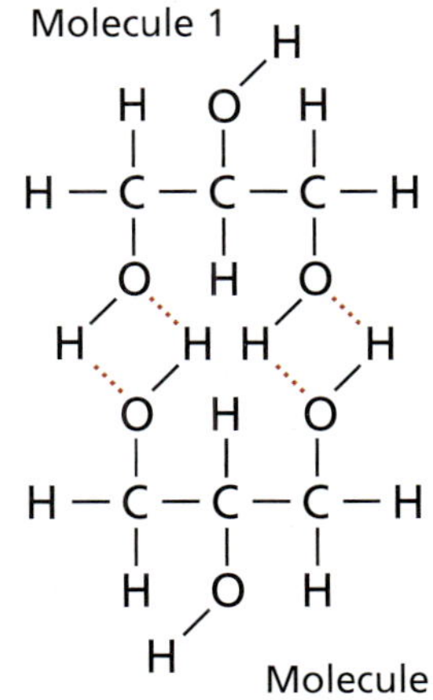


**Figure 15** This diagram shows two glycerol molecules and the hydrogen bonds between them.

**Determine** *the possible number of hydrogen bonds a glycerol molecule can form with a second molecule.*

**Particle size and shape** The size and shape of particles also affect viscosity. Recall that the overall kinetic energy of a particle is determined by its mass and velocity. Suppose the attractive forces between molecules in Liquid A and Liquid B are similar. If the molecules in Liquid A are more massive than the molecules in Liquid B, Liquid A will have a greater viscosity. Liquid A's molecules will, on average, move more slowly than the molecules in Liquid B. Molecules with long chains, such as cooking oils and motor oil, have a higher viscosity than shorter, more-compact molecules, assuming the molecules exert the same type of attractive forces. Within the long chains, there is less distance between atoms on neighboring molecules and, thus, a greater chance for attractions between atoms.

**1808** John Dalton proposes that all matter is composed of tiny particles.

**1937** Scientists discover superfluids—unusual fluids with properties not observed in ordinary matter.

**2003** Deborah S. Jin creates the first fermionic condensate—a superfluid considered to be a sixth state of matter.

1800 1900 2000

**1927** The term *plasma* is first used to describe a fourth state of matter, which is found in lightning.

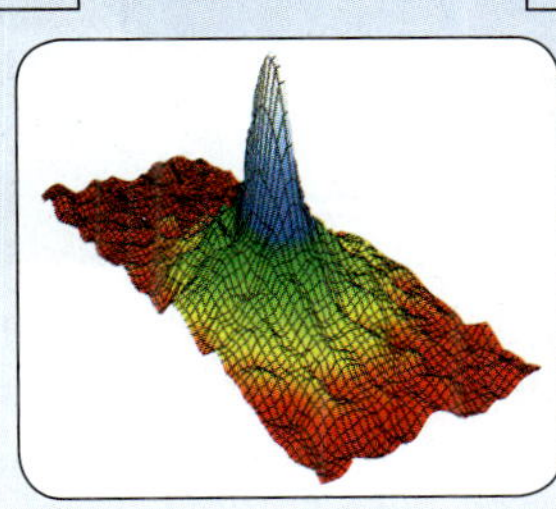

**1995** A fifth state of matter, a gaseous superfluid called a Bose-Einstein condensate, is created and named after Satyendra Nath Bose and Albert Einstein.

**2016** Scientists observe a new state of matter, called a quantum spin liquid. First predicted in the early 1970s, quantum spin liquids have very chaotic structures, even at extremely low temperatures.

**Temperature** Viscosity decreases with temperature. When you pour a small amount of cooking oil into a frying pan, the oil tends not to spread across the bottom of the pan until you heat it. With the increase in temperature, there is an increase in the average kinetic energy of the oil molecules. The added energy makes it easier for the molecules to overcome the intermolecular forces that keep the molecules from flowing.

Another example of the effects of temperature on viscosity is motor oil. Motor oil keeps the moving parts of an internal combustion engine lubricated. Because temperature changes affect the viscosity of motor oil, people once used different motor-oil blends in winter and summer. The motor oil used in winter was designed to flow at low temperatures. The motor oil used in summer was more viscous so that it could maintain sufficient viscosity on extremely hot days or during long trips. Today, additives in motor oil help adjust the viscosity so that the same oil blend can be used all year. Molecules in the additives are compact spheres with relatively low viscosity at cool temperatures. At high temperatures, the shape of the additive molecules changes to long strands. These strands get tangled with the oil molecules, which increases the viscosity of the oil.

☑ READING CHECK **Infer** why it is important for motor oil to remain viscous.

**Surface tension** Intermolecular forces do not have an equal effect on all particles in a liquid, as shown in **Figure 16.** Particles in the middle of the liquid can be attracted to particles above them, below them, and to either side. For particles at the surface of the liquid, there are no attractions from above to balance the attractions from below. Thus, there is a net attractive force pulling down on particles at the surface. The surface tends to have the smallest possible area and to act as though it is stretched tight like the head of a drum. For the surface area to increase, particles from the interior must move to the surface. It takes energy to overcome the attractions holding these particles in the interior. The energy required to increase the surface area of a liquid by a given amount is called **surface tension.** Surface tension is a measure of the inward pull by particles in the interior.

**Figure 16** At the surface of water, the particles are drawn toward the interior until attractive and repulsive forces are balanced.

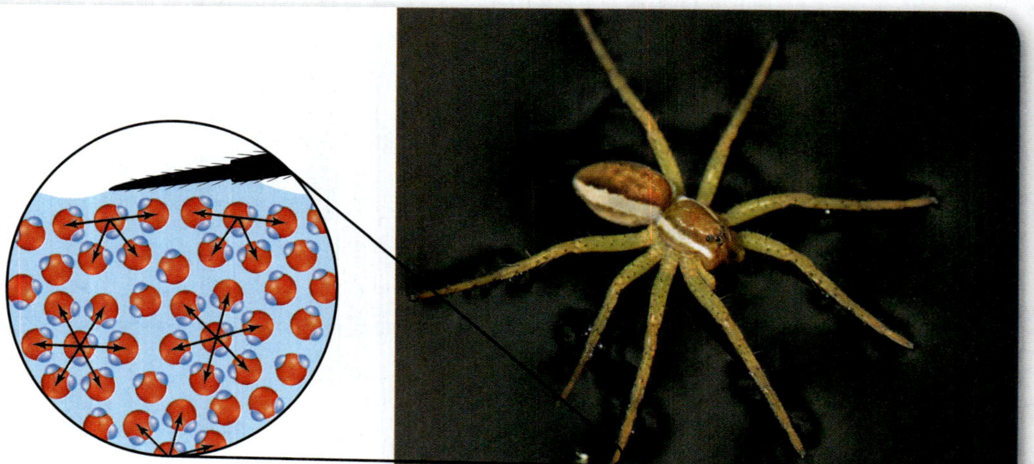

Side view
Intermolecular forces just below the surface of the water create surface tension.

The surface tension of the water allows this spider to walk on the surface of the water.

Alastair Thomson/Getty Images

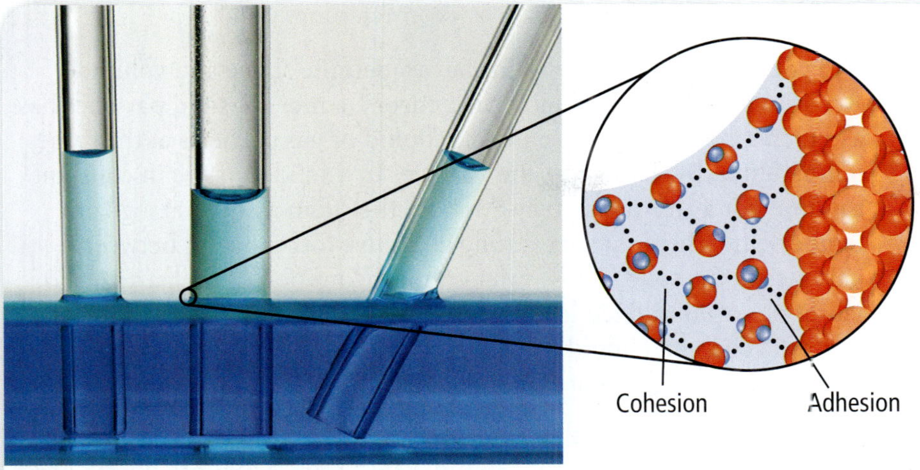


The force of attraction between the water molecules and the silicon dioxide in the glass causes the water molecules to creep up the glass.

Water molecules are attracted to each other—cohesion—and to the silicon dioxide molecules in the glass—adhesion.

■ **Figure 17** Water molecules have cohesive and adhesive properties.
**Infer** ***why the water level is higher in the smaller diameter tube.***

In general, the stronger the attractions between particles, the greater the surface tension. Water has a high surface tension because its molecules can form multiple hydrogen bonds. Drops of water are shaped like spheres because the surface area of a sphere is smaller than the surface area of any other shape of similar volume. Water's high surface tension is what allows the spider in **Figure 16** to walk on the surface of the pond.

The same forces that allow the spider to stay dry on the surface of a pond also makes it difficult to use water alone to remove dirt from skin and clothing. Because dirt particles cannot penetrate the surface of the waterdrops, water alone cannot remove the dirt. Soaps and detergents decrease the surface tension of water by disrupting the hydrogen bonds between water molecules. When the hydrogen bonds are broken, the water spreads out allowing the dirt to be carried away by the water. Compounds that lower the surface tension of water are called surface-active agents or **surfactants.**

Watch a **video about enzymes and surfactants.**

Video

**Cohesion and adhesion** When water is placed into a narrow container, such as the glass tubes in **Figure 17,** you can see that the surface of the water is not straight. The surface forms a concave meniscus; that is, the surface dips in the center. **Figure 17** models what is happening to the water at the molecular level. There are two types of forces at work: cohesion and adhesion. Cohesion describes the force of attraction between identical molecules. Adhesion describes the force of attraction between molecules that are different. Because the adhesive forces between water molecules and the silicon dioxide in glass are greater than the cohesive forces between water molecules, the water rises along the inner walls of the cylinder.

**Capillary action** If the cylinder is extremely narrow, a thin film of water will be drawn upward. Narrow tubes are called capillary tubes. This movement of a liquid such as water is called capillary action, or capillarity. Capillary action helps explain how paper towels can absorb large amounts of water. The water is drawn into the narrow spaces between the cellulose fibers in paper towels by capillary action. In addition, the water molecules form hydrogen bonds with cellulose molecules.

**Vocabulary**
**Science usage v. Common usage**
**Force**
***Science usage:*** a push or a pull, having both magnitude and direction, that is exerted on an object
*The gravitational force exists between any two objects with mass and is directly proportional to their masses.*

***Common usage:*** a group of people who have the power to work toward a desired outcome
*The U.S. labor force increased its productivity last year.*

■ **Figure 18** An iceberg can float because the rigid, three-dimensional structure of ice keeps water molecules farther apart than they are in liquid water. This open, symmetrical structure of ice results from hydrogen bonding.

## Solids

Did you ever wonder why solids have a definite shape and volume? According to the kinetic-molecular theory, a mole of solid particles has as much kinetic energy as a mole of liquid or gas particles at the same temperature. By definition, the particles in a solid must be in constant motion. For a substance to be a solid rather than a liquid at a given temperature, there must be strong attractive forces acting between particles in the solid. These forces limit the motion of the particles to vibrations around fixed locations in the solid. Thus, there is more order in a solid than in a liquid. Because of this order, solids are not fluid. Only gases and liquids are classified as fluids.

**Density of solids** In general, the particles in a solid are more closely packed than those in a liquid. Thus, most solids are more dense than most liquids. When the liquid and solid states of a substance coexist, the solid almost always sinks in the liquid. Solid cubes of benzene sink in liquid benzene because solid benzene is more dense than liquid benzene. There is about a 10% difference in density between the solid and liquid states of most substances. Because the particles in a solid are closely packed, ordinary amounts of pressure will not change the volume of a solid.

You cannot predict the relative densities of ice and liquid water based on benzene. Ice cubes and icebergs float because water is less dense as a solid than it is as a liquid. **Figure 18** shows the reason for the exception. As water freezes, each $H_2O$ molecule can form hydrogen bonds with up to four neighboring molecules. As a result, the water molecules in ice are less-closely packed together than in liquid water.

☑ READING CHECK **Describe** in your own words why ice floats in water.

**Crystalline solids** Although ice is unusual in its density, ice is typical of most solids in that its molecules are packed together in a predictable way. A **crystalline solid** is a solid whose atoms, ions, or molecules are arranged in an orderly, geometric structure. The locations of particles in a crystalline solid can be represented as points on a framework called a crystal lattice. **Figure 19** shows three ways that particles in a crystal lattice can be arranged to form a cube.

### CAREERS IN CHEMISTRY

**Metallurgist** Metallurgists are engineers who are involved in all stages of processing metals, from extracting and refining to casting the final product. At each stage, metallurgists must understand the physical and chemical properties of metals. A college degree is necessary to become a metallurgist, and many go on to earn advanced degrees.

WebQuest

■ **Figure 19** These drawings show three of the ways particles are arranged in crystal lattices. Each sphere represents a particle. **a.** Particles are arranged only at the corners of the cube. **b.** There is a particle in the center of the cube. **c.** There are particles in the center of each of the six cubic faces but no particle in the center of the cube itself.

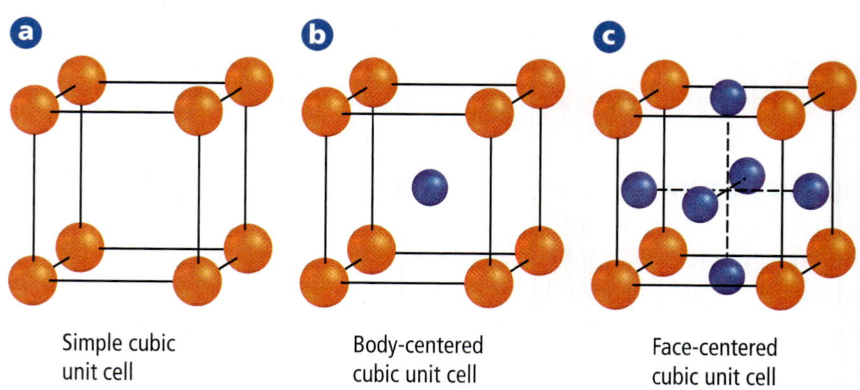

A **unit cell** is the smallest arrangement of atoms in a crystal lattice that has the same symmetry as the whole crystal. Like the formula unit that you read about previously, a unit cell is a small, representative part of a larger whole. The unit cell can be thought of as a building block whose shape determines the shape of the crystal.

**Table 4** shows seven categories of crystals based on shape. Crystal shapes differ because the surfaces, or faces, of unit cells do not always meet at right angles, and the edges of the faces vary in length. In **Table 4,** the edges are labeled *a, b,* and *c;* the angles at which the faces meet are labeled α, β, and γ.

Explore **unit cells with an interactive table.** **Concepts In Motion**

**Table 4** **Unit Cells**

| | | |
|---|---|---|
|  **Halite (rock salt)** |  **Vesuvianite** |  **Aragonite** |
| 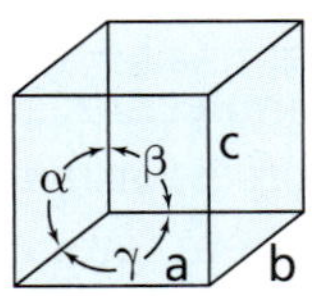 | 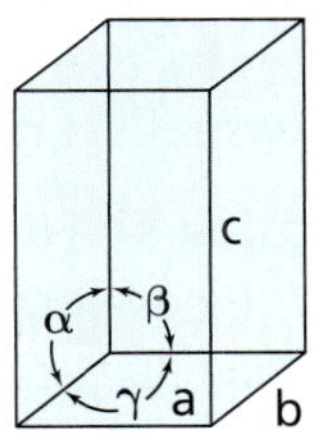 | 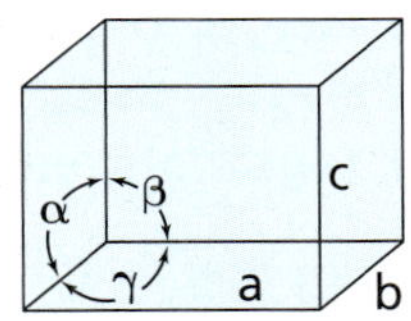 |
| a = b = c<br>α = β = γ = 90°<br>Cubic | a = b ≠ c<br>α = β = γ = 90°<br>Tetragonal | a ≠ b ≠ c<br>α = β = γ = 90°<br>Orthorhombic |

| | | | |
|---|---|---|---|
|  **Microcline** |  **Beryl (emerald)** |  **Tourmaline** |  **Crocoite** |
| 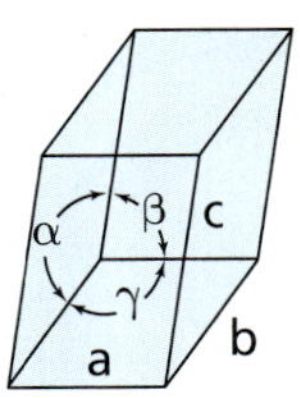 | 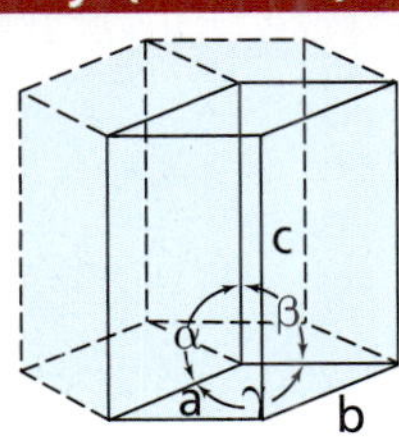 | 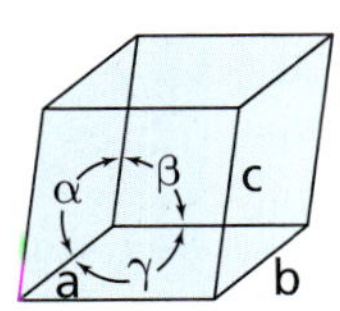 | 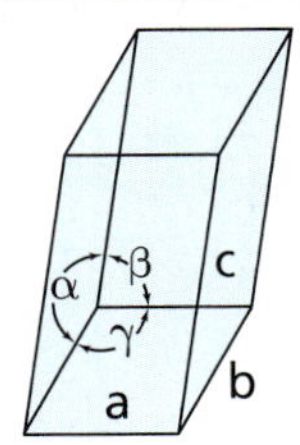 |
| a ≠ b ≠ c<br>α ≠ β ≠ γ ≠ 90°<br>Triclinic | a = b ≠ c<br>α = β = 90°, γ = 120°<br>Hexagonal | a = b = c<br>α = β = γ ≠ 90°<br>Rhombohedral | a ≠ b ≠ c<br>α = γ = 90° ≠ β<br>Monoclinic |

Explore types of crystalline solids with an interactive table. Concepts In Motion

**Table 5 Types of Crystalline Solids**

| Type | Unit Particles | Characteristics of Solid Phase | Examples |
|---|---|---|---|
| Atomic | atoms | soft to very soft; very low melting points; poor conductivity | group 18 elements |
| Molecular | molecules | fairly soft; low to moderately high melting points; poor conductivity | $I_2$, $H_2O$, $NH_3$, $CO_2$, $C_{12}H_{22}O_{11}$ (table sugar) |
| Covalent network | atoms connected by covalent bonds | very hard; very high melting points; often poor conductivity | diamond (C) and quartz ($SiO_2$) |
| Ionic | ions | hard; brittle; high melting points; poor conductivity | NaCl, KBr, $CaCO_3$ |
| Metallic | atoms surrounded by mobile valence electrons | soft to hard; low to very high melting points; malleable and ductile; excellent conductivity | all metallic elements |

**Categories of crystalline solids** Crystalline solids can be classified into five categories based on the types of particles that they contain and how thoses particles are bonded together: atomic solids, molecular solids, covalent network solids, ionic solids, and metallic solids. **Table 5** summarizes the general characteristics of each category and provides examples. The only atomic solids are noble gases. Their properties reflect the weak dispersion forces between the atoms.

**Molecular solids** In molecular solids, the molecules are held together by dispersion forces, dipole-dipole forces, or hydrogen bonds. Most molecular compounds are not solids at room temperature. Even water, which can form strong hydrogen bonds, is a liquid at room temperature. Molecular compounds such as sugar are solids at room temperature because of their large molar masses. With larger molecules, many weak attractions can combine to hold the molecules together. Because they contain no ions, molecular solids are poor conductors of heat and electricity.

**Figure 20** The most common kind of quartz has a hexagonal crystal structure.

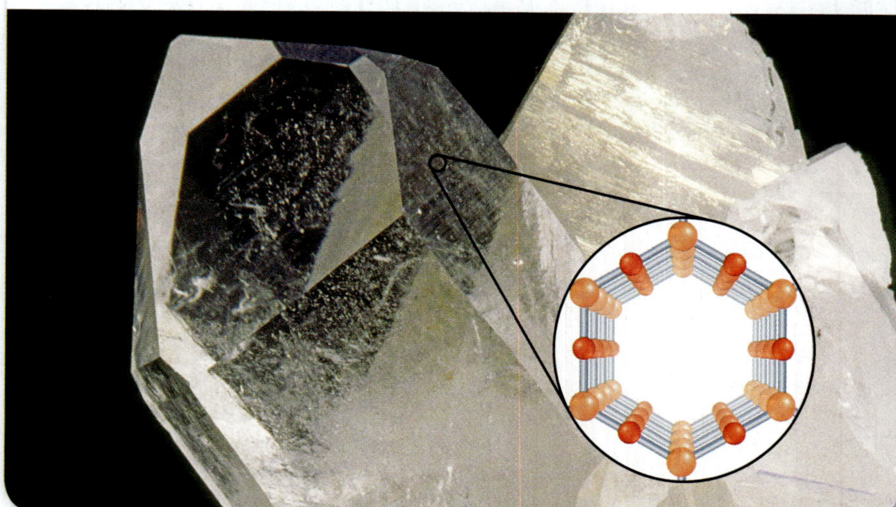

# MiniLAB

iLab Station

## Model Crystal Unit Cells

**How can you make physical models that illustrate the structures of crystals?**

### Procedure

1. Read and complete the lab safety form.
2. Cut **four soda straws** into thirds. Wire the straw pieces together to make a cube using **22- or 26-gauge wire.** Use **scissors** to cut the wire. Refer to **Table 4** for a guide to crystal shapes.
3. To model a rhombohedral crystal, deform the cube from Step 2 until no angles are 90°.
4. To model a hexagonal crystal, make the base angle $\gamma = 120°$ and the other two angles equal 90°.
5. To model a tetragonal crystal, cut **4 straws** in half. Cut 4 of the pieces in half again. Wire the 8 shorter pieces to make 4 square ends. Use the longer pieces to connect the square ends.
6. To model the orthorhombic crystal, cut **4 straws** in half. Cut one-third off 4 of the halves, creating 4 each of three different lengths. Connect the 4 long, 4 medium, and 4 short pieces so that each side is a rectangle.
7. To model the monoclinic crystal, deform the model from Step 6 along one axis. To model the triclinic crystal, deform the model from Step 6 until it has no 90° angles.

### Analysis

1. **Evaluate** Which two models have three axes of equal length? How do these models differ?
2. **Determine** which model includes a square and a rectangle.
3. **Determine** which models have three unequal axes.
4. **Infer** Do you think crystals are perfect, or do they have defects? Explain your answer.

**Covalent network solids** Atoms such as carbon and silicon, which can form multiple covalent bonds, are able to form covalent network solids. The covalent network structure of quartz, which contains silicon, is shown in **Figure 20.** Carbon forms three types of covalent network solids—diamond, graphite, and buckminsterfullerene. An element, such as carbon, that exists in different forms at the same state—solid, liquid, or gas—is called an **allotrope.** For more information about carbon allotropes see the Elements Handbook.

**Ionic solids** Remember that each ion in an ionic solid is surrounded by ions of opposite charge. The type of ions and the ratio of ions determine the structure of the lattice and the shape of the crystal. The network of attractions that extends throughout an ionic crystal gives these compounds their high melting points and hardness. Ionic crystals are strong, but brittle. When ionic crystals are struck, the cations and anions are shifted from their fixed positions. Repulsions between ions of like charge cause the crystal to shatter.

**Metallic solids** Recall that metallic solids consist of positive metal ions surrounded by a sea of mobile electrons. The strength of the metallic bonds between cations and electrons varies among metals and accounts for their wide range of physical properties. For example, tin melts at 232°C, but nickel melts at 1455°C. The mobile electrons make metals malleable—easily hammered into shapes—and ductile—easily drawn into wires. When force is applied to a metal, the electrons shift and thereby keep the metal ions bonded in their new positions. Mobile electrons make metals good conductors of heat and electricity. As shown in **Figure 21,** metal wiring is used to carry electricity to businesses and homes.

**READING CHECK** **Describe** the properties of metals that make them useful for making jewelry.

**Figure 21** Homes, business, and equipment of all types use metal wiring to carry electricity. The metal is usually copper, but other metals are used in special applications.

■ **Figure 22** Native Americans used the glass-like amorphous rock obsidian to make arrowheads and knives, because it can form sharp edges when broken. Obsidian rock forms when lava cools too quickly to form crystals.

**FOLDABLES®** Incorporate information from this section into your Foldable.

**Amorphous solids** An **amorphous solid** is one in which the particles are not arranged in a regular, repeating pattern. It does not contain crystals. The term *amorphous* is derived from a Greek word that means *without shape*. An amorphous solid often forms when a molten material cools too quickly to allow enough time for crystals to form. **Figure 22** shows an example of an amorphous solid.

Glass, rubber, and many plastics are amorphous solids. Recent studies have shown that glass might have some structure. When X-ray diffraction is used to study glass, there appears to be no pattern to the distribution of atoms. When neutrons are used instead, an orderly pattern of silicate units can be detected in some regions. Researchers hope to use this new information to control the structure of glass for optical applications and to produce glass that can conduct electricity.

## SECTION 3 REVIEW

Section Self-Check

### Section Summary

- The kinetic-molecular theory explains the behavior of solids and liquids.
- Intermolecular forces in liquids affect viscosity, surface tension, cohesion, and adhesion.
- Crystalline solids can be classified by their shape and composition.

18. MAINIDEA **Contrast** the arrangement of particles in solids and liquids.
19. **Describe** the factors that affect viscosity.
20. **Explain** why soap and water are used to clean clothing instead of water alone.
21. **Compare** a unit cell and a crystal lattice.
22. **Describe** the difference between a molecular solid and a covalent network solid.
23. **Explain** why water forms a meniscus when it is in a graduated cylinder.
24. **Infer** why the surface of mercury in a thermometer is convex; that is, the surface is higher at the center.
25. **Predict** which solid is more likely to be amorphous—one formed by allowing a molten material to cool slowly to room temperature or one formed by quickly cooling the same material in an ice bath.
26. **Design** an experiment to compare the relative abilities of water and isopropyl alcohol to support skipping stones. Include a prediction about which liquid will be better, along with a brief explanation of your prediction.

SECTION 4

# Phase Changes

**MAINIDEA** Matter changes phase when energy is added or removed.

## Essential Questions

- How can the addition and removal of energy cause a phase change?
- What is a phase diagram?

## Review Vocabulary

**phase change:** a change from one state of matter to another

## New Vocabulary

melting point
vaporization
evaporation
vapor pressure
boiling point
freezing point
condensation
deposition
phase diagram
triple point

**CHEM 4 YOU** **Have you ever wondered where the matter in a solid air freshener goes? The day it is opened and put in a room, it is a solid, fragrant mass. Day-by-day, the solid gets smaller and smaller. Finally, almost nothing is left and it is time to put a new one out. You never observe a puddle of liquid like you would see if it had melted.**

## Phase Changes That Require Energy

Most substances can exist in three states depending on the temperature and pressure. A few substances, such as water, exist in all three states under ordinary conditions. States of a substance are referred to as phases when they coexist as physically distinct parts of a mixture. Ice water is a heterogeneous mixture with two phases, solid ice and liquid water. When energy is added or removed from a system, one phase can change into another, as shown in **Figure 23.** Because you are familiar with the phases of water—ice, liquid water, and water vapor—and have observed changes between those phases, we can use water as the primary example in the discussion of phase changes.

**Melting** What does happen to ice cubes in a glass of ice water? When ice cubes are placed in water, the water is at a higher temperature than the ice. Heat flows from the water to the ice. Heat is the transfer of energy from an object at a higher temperature to an object at a lower temperature. At ice's melting point, the energy absorbed by the ice is not used to raise the temperature of the ice. Instead, it disrupts the hydrogen bonds holding the water molecules together in the ice crystal. When molecules on the surface of the ice absorb enough energy to break the hydrogen bonds, they move apart and enter the liquid phase. As molecules are removed, the ice cube shrinks. The process continues until all of the ice melts. If a tray of ice cubes is left on a counter, where does the energy to melt the cubes come from?

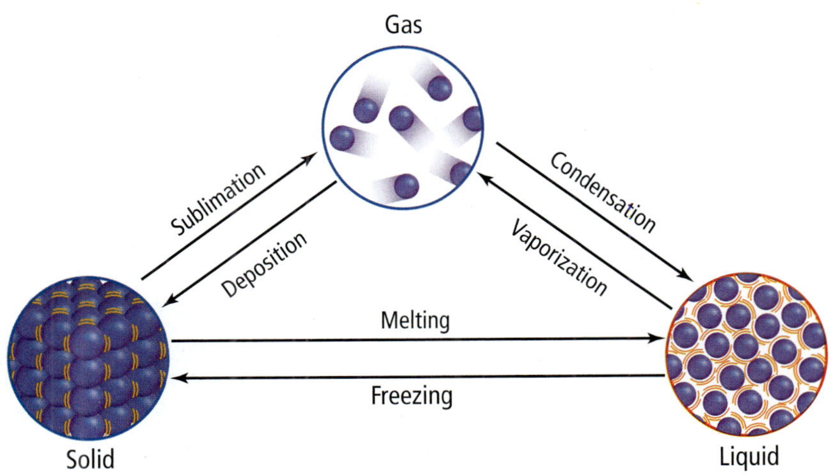


■ **Figure 23** The diagram shows the six possible transitions between phases.
**Determine** *what phase changes occur between solids and liquids.*

■ **Figure 24** This graph shows a typical distribution of kinetic energy of molecules in a liquid at 25°C. The most probable amount of kinetic energy for a molecule lies at the peak of the curve.

**Describe** *how the curve would look for the same liquid at 30°C.*

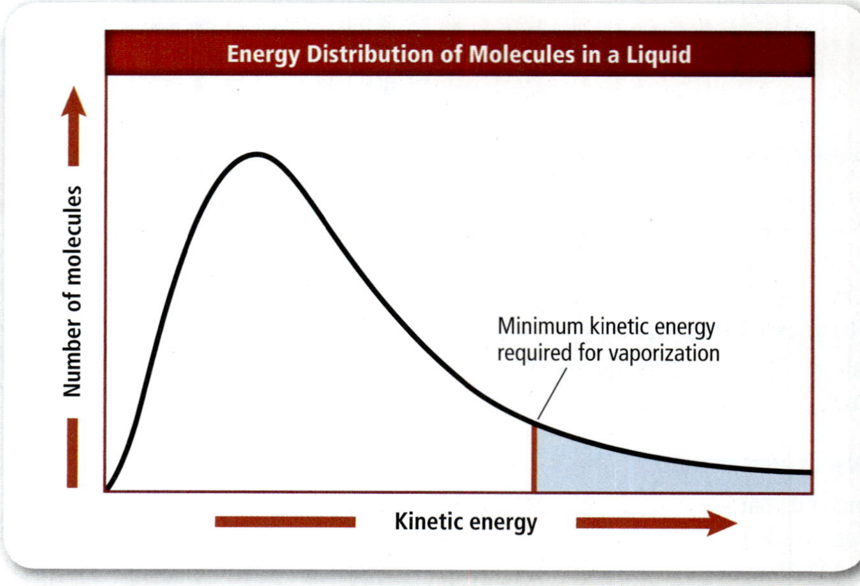


The amount of energy required to melt 1 mol of a solid depends on the strength of the forces keeping the particles together in the solid. Because hydrogen bonds between water molecules are strong, a relatively large amount of energy is required. However, the energy required to melt ice is much less than the energy required to melt table salt because the ionic bonds in sodium chloride are much stronger than the hydrogen bonds in ice.

The temperature at which the liquid phase and the solid phase of a given substance can coexist is a characteristic physical property of many solids. The **melting point** of a crystalline solid is the temperature at which the forces holding its crystal lattice together are broken and it becomes a liquid. It is difficult to specify an exact melting point for an amorphous solid because they tend to melt over a temperature range.

## RealWorld CHEMISTRY

### Evaporation

**PERSPIRATION** Evaporation is one way your body controls its temperature. When you become hot, your body releases sweat from glands in your skin. Water molecules in sweat can absorb heat energy from your skin and evaporate. Excess heat is carried from all parts of your body to your skin by your blood.

**Vaporization** When an ice cube melts, the temperature of the ice and the water produced remains constant. Once all of the ice has melted, additional energy added to the system increases the kinetic energy of the liquid molecules. The temperature of the system begins to rise. In liquid water, some molecules will have more kinetic energy than other molecules. **Figure 24** shows how energy is distributed among the molecules in a liquid at 25°C. The shaded portion indicates those molecules that have the energy required to overcome the forces of attraction holding the molecules together in the liquid.

☑ GRAPH CHECK **Describe** what happens to the particles in the shaded portion on the graph.

Particles that escape from the liquid enter the gas phase. For a substance that is ordinarily a liquid at room temperature, the gas phase is called a vapor. **Vaporization** is the process by which a liquid changes to a gas or vapor. If the input of energy is gradual, the molecules tend to escape from the surface of the liquid. Remember that molecules at the surface are attracted to fewer other molecules than are molecules in the interior. When vaporization occurs only at the surface of a liquid, the process is called **evaporation.** Even at cold temperatures, some water molecules have enough energy to evaporate. As the temperature rises, more and more molecules enter the gas phase.

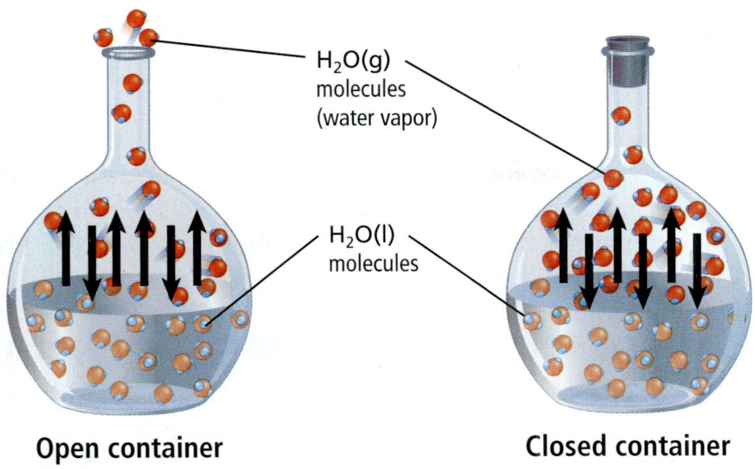


■ **Figure 25** Evaporation occurs in both open and closed containers. In an open container, water molecules that evaporate can escape from the container. Water vapor collects above the liquid in a closed container.

**Figure 25** compares evaporation in an open container with evaporation in a closed container. If water is in an open container, all the molecules will eventually evaporate. The time it takes for them to evaporate depends on the amount of water and the available energy. In a partially filled, closed container, the situation is different. Water vapor collects above the liquid and exerts pressure on the surface of the liquid. The pressure exerted by a vapor over a liquid is called **vapor pressure.**

**Boiling** The temperature at which the vapor pressure of a liquid equals the external or atmospheric pressure is called the **boiling point.** Use **Figure 26** to compare what happens to a liquid at temperatures below its boiling point with what happens to a liquid at its boiling point. At the boiling point, molecules throughout the liquid have enough energy to vaporize. Bubbles of vapor collect below the surface of the liquid and rise to the surface.

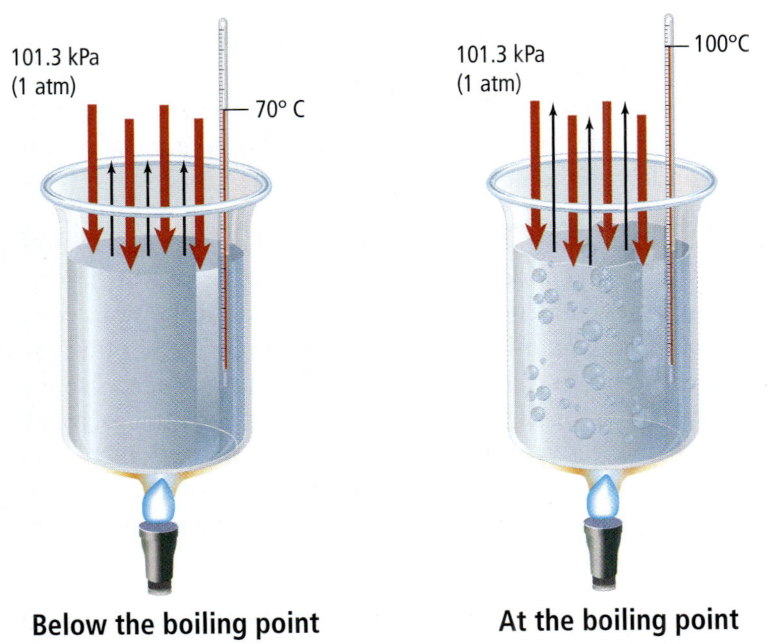


■ **Figure 26** As temperature increases, water molecules gain kinetic energy. Vapor pressure increases (black arrows) but is less than atmospheric pressure (red arrows). A liquid has reached its boiling point when its vapor pressure is equal to atmospheric pressure. At sea level, the boiling point of water is 100°C.

■ **Figure 27** Steaks, seafood, and other highly perishable food products often are shipped in a container with dry ice to keep the food cold during shipping.

**Explain** *why dry ice is preferred over regular ice for shipping steaks and other food products.*

**Sublimation** Many substances have the ability to change directly from the solid phase to the gas phase. Recall that sublimation is the process by which a solid changes directly to a gas without first becoming a liquid. Solid iodine and solid carbon dioxide (dry ice) sublime at room temperature. Dry ice, shown in **Figure 27,** keeps objects that could be damaged by melting water cold during shipping. Mothballs, which contain the compounds naphthalene or *p*-dichlorobenzene, also sublime, as do solid air fresheners.

## Phase Changes That Release Energy

Have you ever awakened on a chilly morning to see frost on your windows or the grass covered with water droplets? When you set a glass of ice water on a picnic table, do you notice beads of water on the outside of the glass? These events are examples of phase changes that release energy into the surroundings.

**Freezing** Suppose you place liquid water in an ice tray into a freezer. As heat is removed from the water, the molecules lose kinetic energy and their velocity decreases. The molecules are less likely to flow past one another. When enough energy has been removed, the hydrogen bonds between water molecules keep the molecules fixed, or frozen, into set positions. Freezing is the reverse of melting. The **freezing point** is the temperature at which a liquid is converted into a crystalline solid.

**Condensation** When a water vapor molecule loses energy, its velocity decreases. The water vapor molecule is more likely to form a hydrogen bond with another water molecule. The formation of a hydrogen bond releases thermal energy and indicates a change from the vapor phase to the liquid phase. The process by which a gas or a vapor becomes a liquid is called **condensation.** Condensation is the reverse of vaporization.

Different factors contribute to condensation. However, condensation always involves the transfer of thermal energy. For example, water vapor molecules can come in contact with a cold surface, such as the side of a glass of ice water. Thermal energy transfers from the water vapor molecules to the cool glass, causing condensation on the outside of the glass. A similar process can occur during the night when water vapor in the air condenses and dew forms on blades of grass.

■ **Figure 28** Normally, air becomes cooler as elevation increases. A temperature inversion occurs when the situation is reversed and the air becomes warmer at higher elevations. Inversions can trap smog over cities and fog in mountain valleys.

**Connection to Earth Science** Precipitation, clouds, and fog all result from condensation. They form as air cools when it rises or passes over cooler land or water. Their formations require a second factor, microscopic particles suspended in the air called condensation nuclei. These can be particles, such as soot and dust, or aerosols, such as sulfur dioxide and nitrogen oxide, on which water vapor condenses. In some circumstances, warm air can settle on top of cooler air, which is called a temperature inversion. **Figure 28** shows fog trapped in a mountain valley by such an inversion.

☑ READING CHECK **Describe** the condensation of water vapor in the atmosphere.

**Deposition** When water vapor comes in contact with a cold window in winter, it forms a solid deposit on the window called frost. **Deposition** is the process by which a substance changes from a gas or vapor to a solid without first becoming a liquid. Deposition is the reverse of sublimation. Snowflakes form when water vapor high up in the atmosphere changes directly into solid ice crystals. Energy is released as the crystals form.

## Phase Diagrams

There are two variables that combine to control the phase of a substance: temperature and pressure. These variables can have opposite effects on a substance. For example, a temperature increase causes more liquid to vaporize, but an increase in pressure causes more vapor to condense. A **phase diagram** is a graph of pressure versus temperature that shows in which phase a substance exists under different conditions of temperature and pressure.

**Figure 29** shows the phase diagram for water. You can use this graph to predict what phase water will be in for any combination of temperature and pressure. Note that there are three regions representing the solid, liquid, and vapor phases of water and three curves that separate the regions from one another. At points that fall along the curves, two phases of water can coexist. The short, yellow curve shows the temperature and pressure conditions under which solid water and water vapor can coexist. The long, blue curve shows the temperature and pressure conditions under which liquid water and water vapor can coexist. The red curve shows the temperature and pressure conditions under which solid water and liquid water can coexist.

Point A on the phase diagram of water—the point where the yellow, blue, and red curves meet—is the triple point for water. The **triple point** is the point on a phase diagram that represents the temperature and pressure at which three phases of a substance can coexist. All six phase changes can occur at the triple point: freezing and melting; evaporation and condensation; sublimation and deposition. Point B is called the critical point. This point indicates the critical pressure and critical temperature above which water cannot exist as a liquid. If water vapor is at the critical temperature, an increase in pressure will not change the vapor into a liquid.

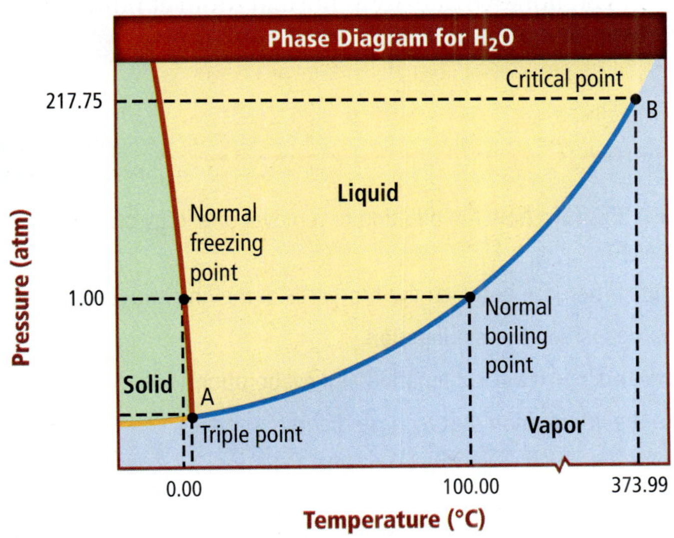


■ **Figure 29** This phase diagram shows the phase of water at different temperatures and pressures.

GRAPH CHECK

**Determine** the phase of water at 2.00 atm and 100.00°C.

View an **animation about a phase diagram.**

Concepts In Motion

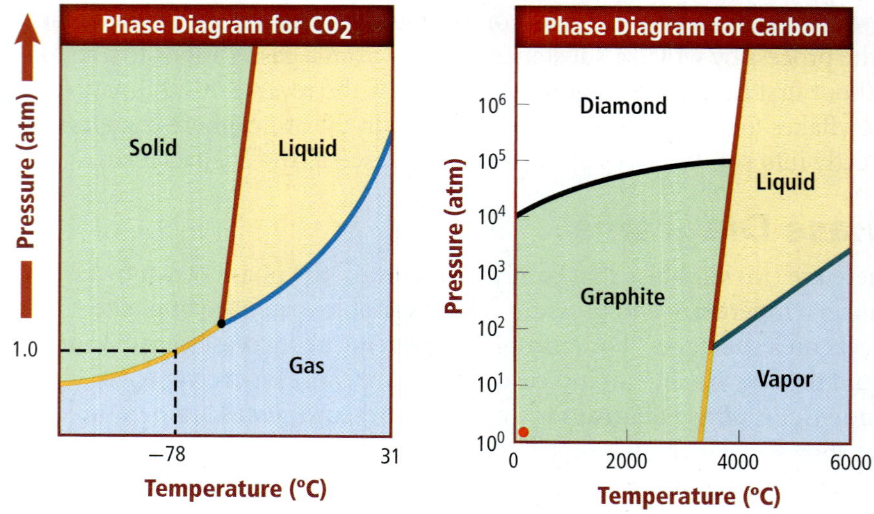


**Figure 30** Phase diagrams show useful information, such as why carbon dioxide sublimes at normal conditions and the existence of two forms of solid carbon.

**GRAPH CHECK**

**Contrast** the slope of the red line in water's phase diagram with that of the red line in carbon dioxide's phase diagram. How do water and carbon dioxide differ in their reaction to increased pressure at the solid/liquid boundary?

The phase diagram for each substance is different because the normal boiling and freezing points of substances are different. However, each diagram will supply the same type of data for the phases, including a triple point. Of course, the range of temperatures chosen will vary to reflect the physical properties of the substance.

Phase diagrams can provide important information for substances. For example, the phase diagram for carbon dioxide in **Figure 30** shows why carbon dioxide sublimes at normal conditions. Find 1.0 atm on the carbon dioxide graph and follow the dashed line to the yellow line. The graph shows that carbon dioxide changes from a solid to a gas at 1 atm. If you extend the dashed line past the yellow line, the graph shows that carbon dioxide does not liquefy as temperature increases. It remains a gas.

The diagram on the right is a phase diagram for carbon. Notice that the graph contains two allotropes of carbon in the solid region. Graphite is the standard state of carbon at normal temperatures and pressures, designated by a red dot. Diamond is more stable at higher temperatures and pressures. Diamonds that exist at normal room conditions originally formed at high temperature and pressure.

**Alabama Applying Practices**

Go to the Resource tab in ConnectED to find the Applying Practices worksheet *Investigate Interparticle Forces.*

## SECTION 4 REVIEW

Section Self-Check

### Section Summary

- States of a substance are referred to as phases when they coexist as physically distinct parts of a mixture.
- Energy changes occur during phase changes.
- Phase diagrams show how different temperatures and pressures affect the phase of a substance.

27. MAINIDEA **Explain** how the addition or removal of energy can cause a phase change.
28. **Explain** the difference between the processes of melting and freezing.
29. **Compare** deposition and sublimation.
30. **Compare and contrast** sublimation and evaporation.
31. **Describe** the information that a phase diagram supplies.
32. **Explain** what the triple point and the critical point on a phase diagram represent.
33. **Determine** the phase of water at 75.00°C and 3.00 atm using **Figure 29.**

# everyday CHEMISTRY

## Cocoa Chemistry

Chocolate is a food product that is native to Central America and Mexico. The Aztec ruler Montezuma served the bitter cocoa-bean drink to Hernan Cortéz in 1519. Cortéz took the cocoa beans and the recipe for the chocolate beverage to Spain where it became a very popular, but expensive beverage. Chocolate remained a food product for the wealthy until the mid-nineteenth century, when the price of chocolate became affordable and processing techniques improved. The chocolate served today bears little resemblance to the chocolate served in Montezuma's court. Processing techniques as well as additives create the smooth, sweet, delightful treat that you enjoy today.

**Melts in your mouth** Chocolate is a mixture of cocoa, cocoa butter, and other ingredients. This mixture is a solid at room temperature but melts in your mouth. Why? Because one of the main ingredients in chocolate—cocoa butter—is a fat that melts at or near body temperature.

**Particle size** Chocolate is a liquid during the mixing process. The cocoa butter in the melted chocolate coats the solid particles of cocoa, sugar, and milk solids. The solid particles in the mixture must not be too large, or the chocolate will have a gritty texture. Generally, the particles are ground to a maximum diameter of $2.0 \times 10^{-5}$ to $3.0 \times 10^{-5}$ m.

**Controlling flow** As you can see in **Figure 1,** a large number of small particles has a larger surface area than a single particle of the same mass.

**Figure 1** Although the mass of each particle or group of particles is the same, increasing the surface area allows more cocoa butter to coat the particles, which improves the flow of the chocolate.

Peter Scholey/Photographer's Choice RF/Getty Images

**Figure 2** Chocolate is carefully processed so that the proper crystal structure forms in the chocolate. These crystals give chocolate the characteristics found in popular chocolate bars.

Smaller particles in the chocolate requires more cocoa butter to coat the solid surfaces. It is the excess cocoa butter *between* the solid particles that allows chocolate to flow.

**Smooth texture** If the chocolate contains too little cocoa butter between the particles, the chocolate will be too thick to flow into a mold. To improve the flow of the chocolate without increasing particle size, manufacturers can either add more fat to the mixture or add an emulsifier, such as lecithin. Lecithin is a fat often obtained from soybeans that helps keep the fat molecules evenly suspended, or emulsified, in the chocolate.

**Crystallization** Another important process in chocolate manufacturing is tempering. During the tempering of the chocolate, the temperature of the chocolate is carefully controlled to ensure that the desired crystals form. When chocolate is not properly tempered, crystals form that create poor-quality chocolate. The desired crystals make the chocolate in **Figure 2** glossy and firm, and allow it to snap well and melt near body temperature.

### WRITING IN ▶ Chemistry

**Research** to find out more about chocolate and design a timeline that summarizes its changes over time.

WebQuest

# SMALL SCALE ChemLAB

iLab Station

## Compare Rates of Evaporation

**Background:** Several factors determine how fast a sample of liquid will evaporate. The volume of the sample is a key factor. A drop of water takes less time to evaporate than a liter of water. The amount of energy supplied to the sample is another factor.

**Question:** *How do intermolecular forces affect the evaporation rates of liquids?*

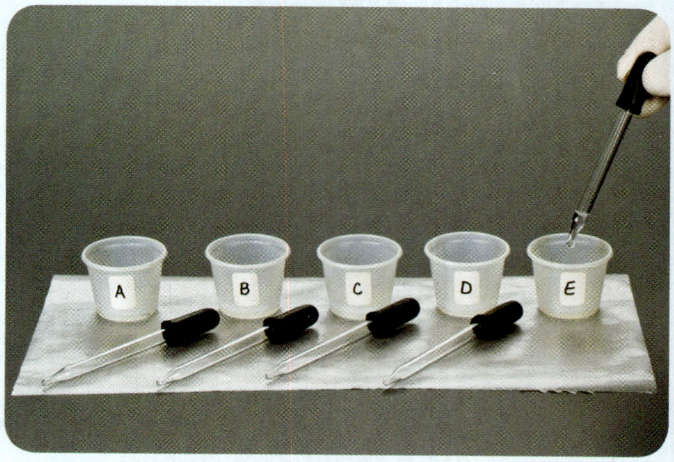


### Materials

distilled water
ethanol
isopropyl alcohol
acetone
household ammonia
droppers (5)
small plastic cups (5)
grease pencil or masking tape and a marking pen
paper towel
square of waxed paper
stopwatch

### Safety Precautions

### Procedure

1. Read and complete the lab safety form.
2. Make a data table to record data.
3. Use a grease pencil or masking tape to label each of 5 small plastic cups. Use *A* for distilled water, *B* for ethanol, *C* for isopropyl alcohol, *D* for acetone, and *E* for household ammonia. Place the plastic cups on a paper towel.
4. Use a dropper to collect about 1 mL of distilled water and place the water in the cup labeled *A*. Place the dropper on the paper towel directly in front of the cup. Repeat with the other liquids.
5. Place a square of waxed paper on your lab surface. Plan where on the waxed paper you will place each of the five drops that you will test to avoid mixing.
6. Have your stopwatch ready. Collect some water in your water dropper and place a single drop on the waxed paper. Begin timing. Time how long it takes for the drop to completely evaporate. While you wait, make a top-view and side-view drawing of the drop. If the drop takes longer than 5 min to evaporate, record $> 300$ s in your data table.
7. Repeat Step 6 with the four other liquids.
8. Use the above procedure to design an experiment in which you can observe the effect of temperature on the rate of evaporation of ethanol. Your teacher will provide a sample of warm ethanol.
9. **Cleanup and Disposal** Clean up lab materials as instructed by your teacher.

### Analyze and Conclude

1. **Classify** Which liquids evaporated quickly? Which liquids were slow to evaporate?
2. **Evaluate** Based on your data, in which liquid(s) are the attractive forces between molecules most likely dispersion forces?
3. **Consider** What is the relationship between surface tension and the shape of a liquid drop? What are the attractive forces that increase surface tension?
4. **Assess** The isopropyl alcohol you used was a mixture of isopropyl alcohol and water. Would pure isopropyl alcohol evaporate more quickly or more slowly compared to the alcohol and water mixture? Give a reason for your answer.
5. **Evaluate** Household ammonia is a mixture of ammonia and water. Based on the data you collected, is there more ammonia or more water in the mixture? Explain.
6. **Evaluate** How does the rate of evaporation of warm ethanol compare to ethanol at room temperature?
7. **Error Analysis** How could you change the procedure to make it more precise?

### INQUIRY EXTENSION

**Design an Experiment** How would different surfaces affect your results? Design an experiment to test your hypothesis.

Matt Meadows

# CHAPTER 12 STUDY GUIDE

Vocabulary Practice

**BIGIDEA** **Kinetic-molecular theory explains the different properties of solids, liquids, and gases.**

## SECTION 1 Gases

**MAINIDEA** Gases expand, diffuse, exert pressure, and can be compressed because they are in a low-density state consisting of tiny, constantly-moving particles.

- The kinetic-molecular theory explains the properties of gases in terms of the size, motion, and energy of their particles.
- Dalton's law of partial pressures is used to determine the pressures of individual gases in gas mixtures.
- Graham's law is used to compare the diffusion rates of two gases.

$$\frac{\text{Rate}_A}{\text{Rate}_B} = \sqrt{\frac{\text{molar mass}_B}{\text{molar mass}_A}}$$

**VOCABULARY**

- kinetic-molecular theory
- elastic collision
- temperature
- diffusion
- Graham's law of effusion
- pressure
- barometer
- pascal
- atmosphere
- Dalton's law of partial pressures

## SECTION 2 Forces of Attraction

**MAINIDEA** Intermolecular forces—including dispersion forces, dipole-dipole forces, and hydrogen bonds—determine a substance's state at a given temperature.

- Intramolecular forces are stronger than intermolecular forces.
- Dispersion forces are intermolecular forces between temporary dipoles.
- Dipole-dipole forces occur between polar molecules.

**VOCABULARY**

- dispersion force
- dipole-dipole force
- hydrogen bond

## SECTION 3 Liquids and Solids

**MAINIDEA** The particles in solids and liquids have a limited range of motion and are not easily compressed.

- The kinetic-molecular theory explains the behavior of solids and liquids.
- Intermolecular forces in liquids affect viscosity, surface tension, cohesion, and adhesion.
- Crystalline solids can be classified by their shape and composition.

**VOCABULARY**

- viscosity
- surface tension
- surfactant
- crystalline solid
- unit cell
- allotrope
- amorphous solid

## SECTION 4 Phase Changes

**MAINIDEA** Matter changes phase when energy is added or removed.

- States of a substance are referred to as phases when they coexist as physically distinct parts of a mixture.
- Energy changes occur during phase changes.
- Phase diagrams show how different temperatures and pressures affect the phase of a substance.

**VOCABULARY**

- melting point
- vaporization
- evaporation
- vapor pressure
- boiling point
- freezing point
- condensation
- deposition
- phase diagram
- triple point

# CHAPTER 12 ASSESSMENT

## SECTION 1

### Mastering Concepts

**34.** What is an elastic collision?

**35.** How does the kinetic energy of particles vary as a function of temperature?

**36.** Use the kinetic-molecular theory to explain the compression and expansion of gases.

**37.** List the three basic assumptions of the kinetic-molecular theory.

**38.** Describe the common properties of gases.

**39.** Compare diffusion and effusion. Explain the relationship between the rates of these processes and the molar mass of a gas.

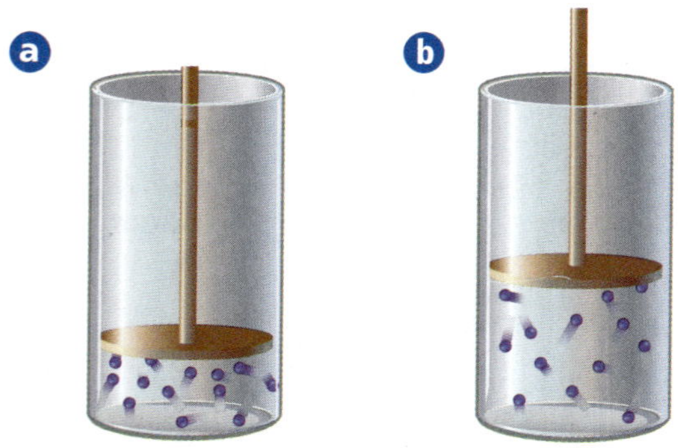


■ **Figure 31**

**40.** In **Figure 31,** what happens to the density of gas particles in the cylinder as the piston moves from Position A to Position B?

**41. Baking** Explain why the baking instructions on a box of cake mix are different for high and low elevations. Would you expect to have a longer or a shorter cooking time at a high elevation?

### Mastering Problems

**42.** What is the molar mass of a gas that takes three times longer to effuse than helium?

**43.** What is the ratio of effusion rates of krypton and neon at the same temperature and pressure?

**44.** Calculate the molar mass of a gas that diffuses three times faster than oxygen under similar conditions.

**45.** What is the partial pressure of water vapor in an air sample when the total pressure is 1.00 atm, the partial pressure of nitrogen is 0.79 atm, the partial pressure of oxygen is 0.20 atm, and the partial pressure of all other gases in air is 0.0044 atm?

**46.** What is the total gas pressure in a sealed flask that contains oxygen at a partial pressure of 0.41 atm and water vapor at a partial pressure of 0.58 atm?

**47. Mountain Climbing** The pressure atop the world's highest mountain, Mount Everest, is usually about 33.6 kPa. Convert the pressure to atmospheres. How does the pressure compare with the pressure at sea level?

**48. High Altitude** The atmospheric pressure in Denver, Colorado, is usually about 84.0 kPa. What is this pressure in atm and torr units?

**49.** At an ocean depth of 76.2 m, the pressure is about 8.4 atm. Convert the pressure to mmHg and kPa units.

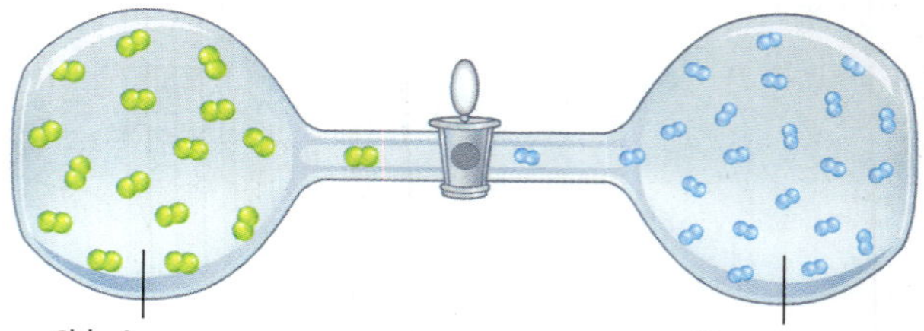


■ **Figure 32**

**50. Figure 32** represents an experimental set-up in which the left bulb is filled with chlorine gas and the right bulb is filled with nitrogen gas. Describe what happens when the stopcock is opened. Assume that the temperature of the system is held constant during the experiment.

## SECTION 2

### Mastering Concepts

**51.** Explain the difference between a temporary dipole and a permanent dipole.

**52.** Why are dispersion forces weaker than dipole-dipole forces?

**53.** Explain why hydrogen bonds are stronger than most dipole-dipole forces.

**54.** Compare intramolecular and intermolecular forces.

**55.** Hypothesize why long, nonpolar molecules would interact more strongly with one another than spherical nonpolar molecules of similar composition.

### Mastering Problems

**56. Polar Molecules** Use relative differences in electronegativity to label the ends of the polar molecules listed as partially positive or partially negative.

**a.** HF **b.** HBr **c.** NO **d.** CO

**57.** Draw the structure of the dipole-dipole interaction between two molecules of carbon monoxide.

**58.** Decide which of the substances listed can form hydrogen bonds.

**a.** $H_2O$ **b.** $H_2O_2$ **c.** HF **d.** $NH_3$

59. Decide which one of the molecules listed below can form intermolecular hydrogen bonds, and then draw it, showing several molecules attached together by hydrogen bonds.
    **a.** NaCl  **b.** $MgCl_2$  **c.** $H_2O_2$  **d.** $CO_2$

# SECTION 3

## Mastering Concepts

60. What is surface tension, and what conditions must exist for it to occur?
61. Explain why the surface of water in a graduated cylinder is curved.
62. Which liquid is more viscous at room temperature, water or molasses? Explain.
63. Explain how two different forces play a role in capillary action.

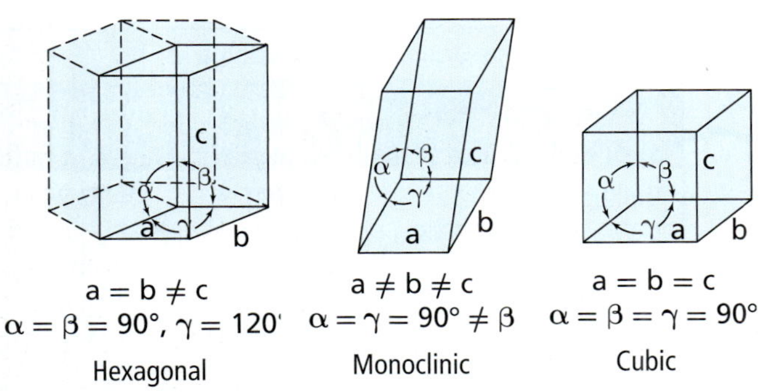


■ **Figure 33**

64. Use the drawings in **Figure 33** to compare the cubic, monoclinic, and hexagonal crystal systems.
65. What is the difference between a network solid and an ionic solid?
66. Explain why most metals bend when struck but most ionic solids shatter.
67. List the types of crystalline solids that are usually good conductors of heat and electricity.
68. How does the strength of a liquid's intermolecular forces affect its viscosity?
69. Explain why water has a higher surface tension than benzene, whose molecules are nonpolar.
70. Compare the numbers of particles in **Figure 19** that form the basic shapes of one unit cell for each of the following.
    **a.** simple cubic
    **b.** body-centered cubic
71. Predict which solid is more likely to be amorphous—one formed by cooling a molten material over 4 h at room temperature or one formed by cooling a molten material quickly in an ice bath.
72. **Conductivity** Predict which solid will conduct electricity better—sugar or salt.
73. Explain why ice floats in water but solid benzene sinks in liquid benzene. Which behavior is more "normal"?

## Mastering Problems

74. Given edge lengths and face angles, predict the shape of each of the following crystals.
    **a.** $a = 3$ nm, $b = 3$ nm, $c = 3$ nm; $\alpha = 90°$, $\beta° = 90$, $\gamma = 90°$
    **b.** $a = 4$ nm, $b = 3$ nm, $c = 5$ nm; $\alpha = 90°$, $\beta° = 100$, $\gamma = 90°$
    **c.** $a = 3$ nm, $b = 3$ nm, $c =$5nm; $\alpha = 90°$, $\beta° = 90$, $\gamma = 90°$
    **d.** $a = 3$ nm, $b = 3$ nm, $c = 5$ nm; $\alpha = 90°$, $\beta° = 90$, $\gamma = 120°$

# SECTION 4

## Mastering Concepts

75. How does sublimation differ from deposition?
76. Compare boiling and evaporation.
77. Define the term *melting point.*
78. Explain the relationships among vapor pressure, atmospheric pressure, and boiling point.
79. Explain why dew forms on cool mornings.
80. **Snow** Why does a pile of snow slowly shrink even on days when the temperature never rises above the freezing point of water?

## Mastering Problems

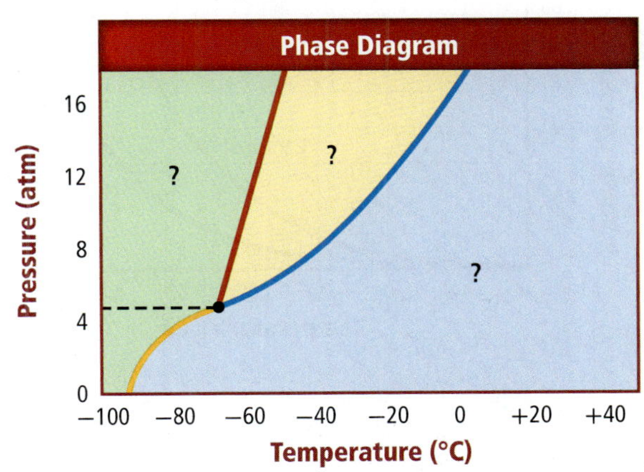


■ **Figure 34**

81. Copy and label the solid, liquid, and gas phases, triple point, and critical point on **Figure 34.**
82. Why does it take more energy to boil 10 g of liquid water than to melt an equivalent mass of ice?

## MIXED REVIEW

**83.** Use the kinetic-molecular theory to explain why both gases and liquids are fluids.

**84.** Use intermolecular forces to explain why oxygen is a gas at room temperature and water is a liquid.

**85.** Use the kinetic-molecular theory to explain why gases are easier to compress than liquids or solids.

**86.** At 25°C and a pressure of 760 mmHg, the density of mercury is 13.5 g/mL; water at the same temperature and pressure has a density of 1.00 g/mL. Explain this difference in terms of intermolecular forces and the kinetic-molecular theory.

**87.** If two identical containers each hold the same gas at the same temperature but the pressure inside one container is exactly twice that of the other container, what must be true about the amount of gas inside each container?

**88.** List three types of intermolecular forces.

**89.** When solid sugar crystals are dissolved in a glass of water, they form a clear homogeneous solution in which the crystals are not visible. If the beaker is left out at room temperature for a few days, the crystals reappear in the bottom and on the sides of the glass. Is this an example of freezing?

## THINK CRITICALLY

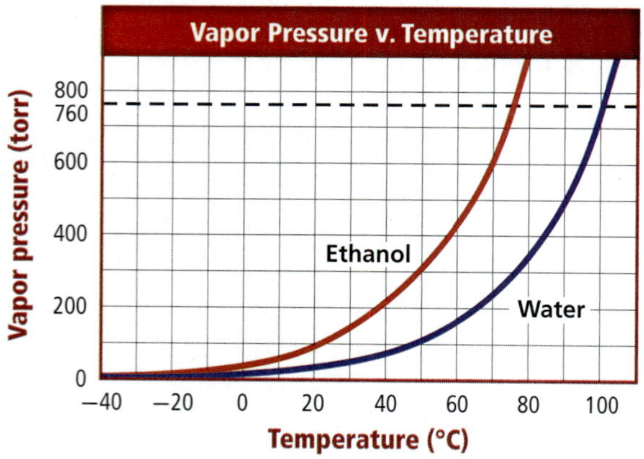


■ **Figure 35**

**90. Interpret Graphs** Examine **Figure 35,** which plots vapor pressure versus temperature for water and ethanol.

**a.** What is the boiling point of water at 1 atm?
**b.** What is the boiling point of ethanol at 1 atm?
**c.** Estimate the temperature at which water will boil when the atmospheric pressure is 0.80 atm.

**91. Hypothesize** What type of crystalline solid do you predict would best suit the following needs?

**a.** a material that can be melted and reformed at a low temperature
**b.** a material that can be drawn into long, thin wires
**c.** a material that conducts electricity when molten
**d.** an extremely hard material that is nonconductive

**92. Compare and Contrast** An air compressor uses energy to squeeze air particles together. When the air is released, it expands, allowing the energy to be used for purposes such as gently cleaning surfaces without using a more abrasive liquid or solid. Hydraulic systems use pressurized fluids to transmit and multiply force. What do you think are some advantages and disadvantages of these two types of technology?

**93. Graph** Use **Table 6** to construct a phase diagram for ammonia.

**Table 6 Phase Diagram for Ammonia**

| Selected Points | Pressure (atm) | Temperature (°C) |
|---|---|---|
| Triple point | 0.060 | −77.7 |
| Critical point | 112 | 132.2 |
| Normal boiling point | 1.0 | −33.5 |
| Normal freezing point | 1.0 | −77.7 |

**94. Apply** A solid being heated stays at a constant temperature until it is completely melted. What happens to the heat energy put into the system during that time?

**95. Communicate** Which process—effusion or diffusion—is responsible for your being able to smell perfume from an open bottle that is located across the room from you? Explain.

**96. Infer** A laboratory demonstration involves pouring bromine vapors, which are a deep red color, into a flask of air and then tightly sealing the top of the flask. The bromine is observed to first sink to the bottom of the beaker. After several hours have passed, the red color is distributed equally throughout the flask.

**a.** Is bromine gas more or less dense than air?
**b.** Would liquid bromine diffuse more or less quickly than gaseous bromine after you pour it into another liquid?

**97. Analyze** Use your knowledge of intermolecular forces to predict whether ammonia ($NH_3$) or methane ($CH_4$) will be more soluble in water.

**98. Evaluate** List three changes that require energy and three that release energy.

Chapter Self-Check

99. **Evaluate** Supercritical carbon dioxide is a liquid form of $CO_2$ used in the food industry to decaffeinate tea, coffee, and colas, as well as in the pharmaceutical industry to form polymer microparticles used in drug delivery systems. Use **Figure 36** to determine what conditions must be used to form supercritical carbon dioxide.

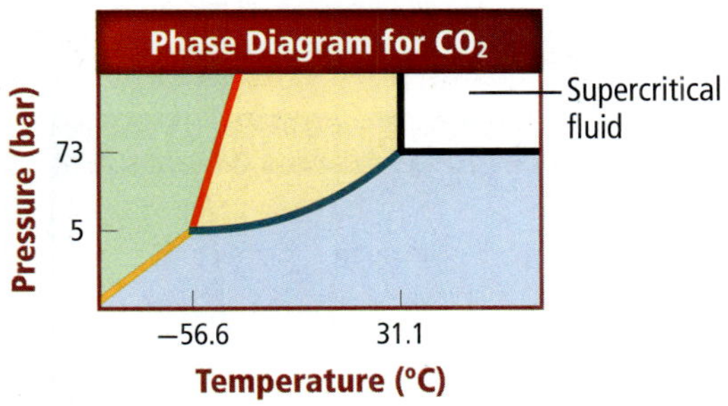


**Figure 36**

## Challenge Problem

100. You have a solution containing 135.2 g of dissolved KBr in 2.3 L of water. What volume of this solution, in mL, would you use to make 1.5 L of a 0.1 mol/L KBr solution? What is the boiling point of this new solution?

## CUMULATIVE REVIEW

101. Identify each of the following as an element, a compound, a homogeneous mixture, or a heterogeneous mixture.
   **a.** air
   **b.** blood
   **c.** antimony
   **d.** ammonia
   **e.** mustard
   **f.** water

102. You are given two clear, colorless aqueous solutions. You are told that one solution contains an ionic compound, and one contains a covalent compound. How could you determine which is an ionic solution and which is a covalent solution?

103. Which branch of chemistry would most likely study matter and phase changes?
   **a.** biochemistry
   **b.** organic chemistry
   **c.** physical chemistry
   **d.** polymer chemistry

104. What type of reaction is the following?

$$K_2CO_3(aq) + BaCl_2(aq) \rightarrow 2KCl(aq) + BaCO_3(s)$$

   **a.** combustion
   **b.** double-replacement
   **c.** single-replacement
   **d.** synthesis

105. Which chemist produced the first widely used and accepted periodic table?
   **a.** Dmitri Mendeleev
   **b.** Henry Moseley
   **c.** John Newlands
   **d.** Lothar Meyer

## WRITING IN ▸ Chemistry

106. **Musk** is the basic ingredient of many perfumes, soaps, shampoos, and even foods such as chocolates, licorice, and hard candies. Both synthetic and natural musk molecules have high molecular weights compared to other perfume ingredients, and as a result, have a slower rate of diffusion, assuring a slow, sustained release of fragrance. Write a report on the chemistry of perfume ingredients, emphasizing the importance of diffusion rate as a property of perfume.

107. **Birthstones** Find out what your birthstone is and write a brief report about the chemistry of that gem. Find out its chemical composition, which category its unit cell is in, how hard and durable it is, and what its approximate cost is at present.

108. **Propane gas** is a commonly used heating fuel for gas grills and homes. However, it is not packaged as a gas. It is liquefied and referred to as liquid propane or "LP gas." Make a poster explaining the advantages and disadvantages of storing and transporting propane as a liquid rather than a gas.

109. **Other States of Matter** Research and prepare an oral report about one of the following topics: plasma, superfluids, fermionic condensate, or Bose-Einstein condensate. Share your report with your classmates and prepare a visual aid that can be used to explain your topic.

## DBQ Document-Based Questions

**Iodine** *Solid iodine that is left at room temperature sublimates from a solid to a gas. But when heated quickly, a different process takes place, as described here.*

*"About 1 g of iodine crystals is placed in a sealed glass ampoule and gently heated on a hot plate. A layer of purple gas is formed at the bottom, and the iodine liquefies. If one tilts the tube, this liquid flows along the wall as a narrow stream and solidifies very quickly."*

Data obtained from: Leenson, 2005. Sublimination of Iodine at Various Pressures: Multipurpose Experiments in Inorganic and Physical Chemistry. *Journal of Chemical Education* 82(2):241–245.

110. Why does solid iodine sublime readily? Use your knowledge of intermolecular forces to explain.

111. Why is liquid iodine not usually visible if crystals are heated in the open air?

112. Why is it necessary to use a sealed ampoule in this investigation?

113. Infer why the iodine solidifies when the tube is tilted.

## MULTIPLE CHOICE

1. What is the ratio of diffusion rates for nitric oxide (NO) and nitrogen tetroxide ($N_2O_4$)?
   A. 0.326
   B. 0.571
   C. 1.751
   D. 3.066

2. Which is NOT an assumption of the kinetic-molecular theory?
   A. Collisions between gas particles are elastic.
   B. All the gas particles in a sample have the same velocity.
   C. A gas particle is not significantly attracted or repelled by other gas particles.
   D. All gases at a given temperature have the same average kinetic energy.

3. A sealed flask contains neon, argon, and krypton gas. If the total pressure in the flask is 3.782 atm, the partial pressure of Ne is 0.435 atm, and the partial pressure of Kr is 1.613 atm, what is the partial pressure of Ar?
   A. 2.048 atm
   B. 1.734 atm
   C. 1556 atm
   D. 1318 atm

*Use the figure below to answer Question 4.*

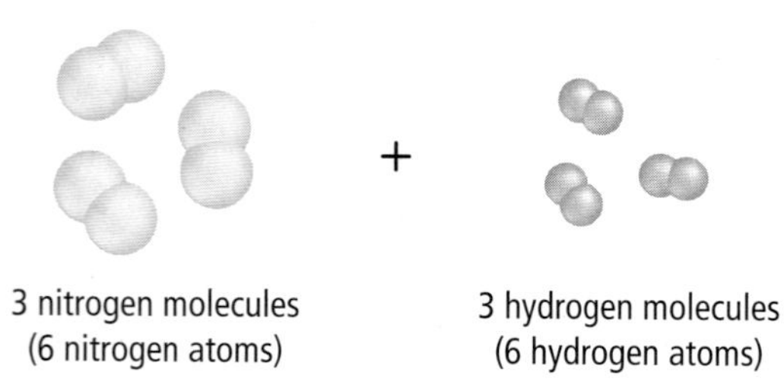


4. Hydrogen and nitrogen react as shown to form ammonia ($NH_3$). What is true of this reaction?
   A. Three ammonia molecules are formed, with zero molecules remaining.
   B. Two ammonia molecules are formed, with two hydrogen molecules remaining.
   C. Six ammonia molecules are formed, with zero molecules remaining.
   D. Two ammonia molecules are formed, with two nitrogen molecules remaining.

5. Which does not affect the viscosity of a liquid?
   A. intermolecular attractive forces
   B. size and shape of molecules
   C. temperature of the liquid
   D. capillary action

*Use the graph below to answer Questions 6 to 8.*

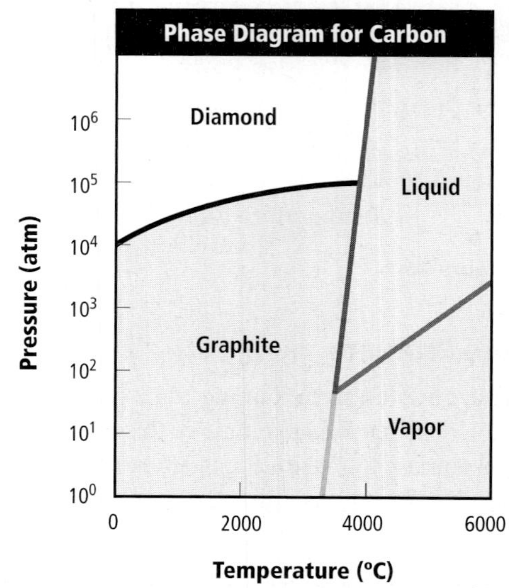


6. Under what conditions is diamond most likely to form?
   A. temperatures > 5000°C and pressures < 100 atm
   B. temperatures > 6000°C and pressures < 25 atm
   C. temperatures < 3500°C and pressures > $10^5$ atm
   D. temperatures < 4500°C and pressures < 10 atm

7. Find the point on the graph at which carbon exists in three phases: solid graphite, solid diamond, and liquid carbon. What are the approximate temperature and pressure at that point?
   A. 4700°C and $10^6$ atm
   B. 3000°C and $10^3$ atm
   C. 4000°C and $10^5$ atm
   D. 3500°C and 80 atm

8. In what form or forms does carbon exist at 6000°C and $10^5$ atm?
   A. diamond only
   B. liquid carbon only
   C. diamond and liquid carbon
   D. liquid carbon and graphite

## SHORT ANSWER

*Use the table below to answer Questions 9 and 10.*

| Properties of Single Bonds | | |
|---|---|---|
| Bond | Strength (kJ/mol) | Length (pm) |
| H – H | 435 | 74 |
| Br – Br | 192 | 228 |
| C – C | 347 | 154 |
| C – H | 393 | 104 |
| C – N | 305 | 147 |
| C – O | 356 | 143 |
| Cl – Cl | 243 | 199 |
| I – I | 151 | 267 |
| S – S | 259 | 208 |

**9.** Create a graph to show how bond length varies with bond strength. Place bond strength on the *x*-axis.

**10.** Summarize the relationship between bond strength and bond length.

## EXTENDED RESPONSE

*Use the table below to answer Question 11.*

| Geometry of $AlCl_3$ and $PCl_3$ | | |
|---|---|---|
| Compound | $AlCl_3$ | $PCl_3$ |
| Molecular Shape | 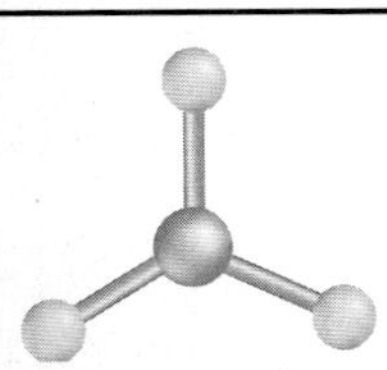 | 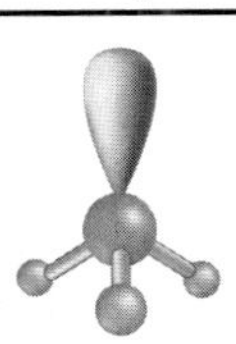 |

**11.** What are the names of the shapes of the molecules for each compound? Explain how the atomic arrangements in each compound result in their different shapes despite their similar formulas.

## SAT SUBJECT TEST: CHEMISTRY

**12.** Potassium chromate and lead(II) acetate are both dissolved in a beaker of water, where they react to form solid lead(II) chromate. What is the balanced net ionic equation describing this reaction?

**A.** $Pb^{2+}(aq) + C_2H_3O_2^-(aq) \rightarrow Pb(C_2H_3O_2)_2(s)$
**B.** $Pb^{2+}(aq) + 2CrO_4^-(aq) \rightarrow Pb(CrO_4)_2(s)$
**C.** $Pb^{2+}(aq) + CrO_4^{2-}(aq) \rightarrow PbCrO_4(s)$
**D.** $Pb^+(aq) + C_2H_3O_2^-(aq) \rightarrow PbC_2H_3O_2(s)$
**E.** $Pb^{2+}(aq) + CrO_4^-(aq) \rightarrow PbCrO_5(s)$

**13.** The solid phase of a compound has a definite shape and volume because its particles

**A.** are not in constant motion.
**B.** are always more tightly packed in the liquid phase.
**C.** can vibrate only around fixed points.
**D.** are held together by strong intramolecular forces.
**E.** have no intermolecular forces.

*Use the table below to answer Questions 14 and 15.*

| Properties of Sulfuric Acid | |
|---|---|
| Formula | $H_2SO_4$ |
| Molar mass | 98.08 g/mol |
| Density | 1.834 g/mL |

**14.** What is the mass of 75.0 mL of sulfuric acid?

**A.** 40.9 g
**B.** 138 g
**C.** 98.08 g
**D.** 180 g
**E.** 198.4 g

**15.** How many atoms of oxygen are present in 2.4 mol of sulfuric acid?

**A.** 940 atoms
**B.** 230 atoms
**C.** $1.5 \times 10^{24}$ atoms
**D.** $5.8 \times 10^{24}$ atoms
**E.** $6.02 \times 10^{23}$ atoms

| NEED EXTRA HELP? | | | | | | | | | | | | | | | |
|---|---|---|---|---|---|---|---|---|---|---|---|---|---|---|---|
| If You Missed Question . . . | 1 | 2 | 3 | 4 | 5 | 6 | 7 | 8 | 9 | 10 | 11 | 12 | 13 | 14 | 15 |
| Review Section . . . | 12.1 | 12.1 | 12.1 | 11.1 | 12.3 | 12.4 | 12.4 | 12.4 | 8.1 | 8.1 | 8.4 | 9.3 | 12.3 | 2.1 | 10.3 |

# CHAPTER 13

# Gases

**BIGIDEA** Gases respond in predictable ways to changes in pressure, temperature, volume, and number of particles.

## SECTIONS

Chm 8, 8a, 8b, 11, 11a

1 The Gas Laws

2 The Ideal Gas Law

3 Gas Stoichiometry

## LaunchLAB

iLab Station

### How does temperature affect the volume of a gas?

In a hot-air balloon, the burners raise the temperature of the air inside the balloon to keep it aloft. In this lab, you will examine how temperature changes affect the volume of a balloon.

Balloon basket

Propane burner

**FOLDABLES®**
**Study Organizer**

### The Gas Laws

Make a layered-look book. Label it with the gas laws. Summarize your notes on each tab.

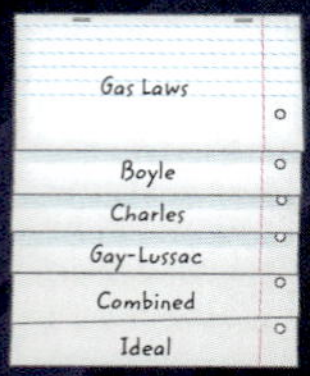

In the nineteenth century, Joseph Gay-Lussac used hot air balloon flights for research and experimentation. His work, along with that of other scientists, contributed to the development of the gas laws.

## SECTION 1

# The Gas Laws

**MAINIDEA** **For a fixed amount of gas, a change in one variable—pressure, temperature, or volume—affects the other two.**

### Essential Questions

- What are the relationships between pressure, temperature, and volume of a constant amount of gas?
- How can you use the gas laws to solve problems involving the pressure, temperature, and volume of a constant amount of gas?

### Review Vocabulary

**scientific law:** describes a relationship in nature that is supported by many experiments

### New Vocabulary

Boyle's law
absolute zero
Charles's law
Gay-Lussac's law
combined gas law

**CHEM 4 YOU** **What might happen to the gas in a balloon if you decreased its volume by squeezing it? You would feel increasing resistance as you squeeze and might see part of the balloon bulge.**

## Boyle's Law

As the balloon example illustrates, the pressure of a gas and its volume are related. Robert Boyle (1627–1691), an Irish chemist, described this relationship between the pressure and the volume of a gas.

**How are pressure and volume related?** Boyle designed experiments like the one shown in **Figure 1.** He showed that if the temperature and the amount of gas are constant, doubling the pressure decreases the volume by one-half. On the other hand, reducing the pressure by one-half doubles the volume. A relationship in which one variable increases proportionally as the other variable decreases is known as an inversely proportional relationship.

**Boyle's law** states that the volume of a fixed amount of gas held at a constant temperature varies inversely with the pressure. Look at the graph in **Figure 1,** in which volume versus pressure is plotted for a gas. The plot of an inversely proportional relationship results in a downward curve.

**Figure 1** As the external pressure on the cylinder's piston increases, the volume inside the cylinder decreases. The graph shows the inverse relationship between pressure and volume.

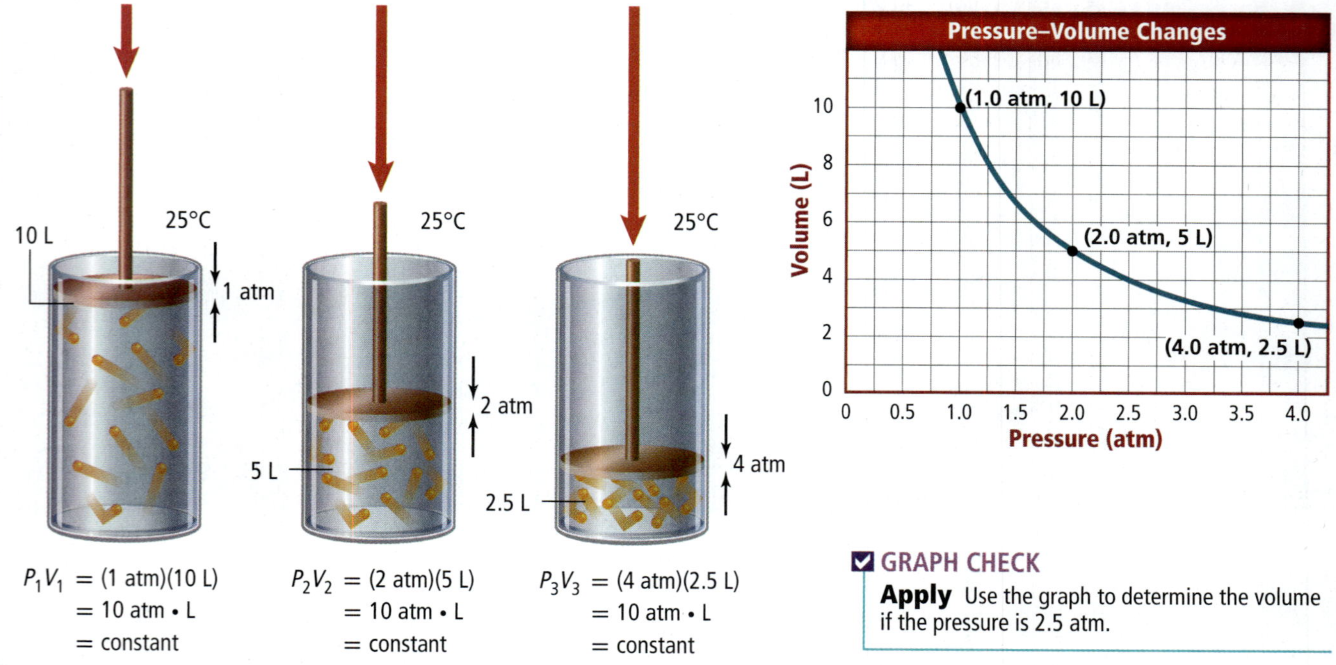


**GRAPH CHECK**

**Apply** Use the graph to determine the volume if the pressure is 2.5 atm.

Note that the product of the pressure and the volume for each point in **Figure 1** is 10 atm·L. Boyle's law can be expressed mathematically as follows.

**Boyle's Law**

$$P_1V_1 = P_2V_2$$ ***P* represents pressure. *V* represents volume.**

For a given amount of gas held at constant temperature, the product of pressure and volume is a constant.

$P_1$ and $V_1$ represent the initial conditions, and $P_2$ and $V_2$ represent new conditions. If you know any three of these values, you can solve for the fourth by rearranging the equation.

## EXAMPLE Problem 1

Find help with **inverse relationships.** Math Handbook

EXAMPLE PROBLEM

**BOYLE'S LAW** A diver blows a 0.75-L air bubble 10 m under water. As it rises to the surface, the pressure goes from 2.25 atm to 1.03 atm. What will be the volume of air in the bubble at the surface?

### 1 ANALYZE THE PROBLEM

According to Boyle's law, the decrease in pressure on the bubble will result in an increase in volume, so the initial volume should be multiplied by a pressure ratio greater than 1.

| Known | Unknown |
|---|---|
| $V_1 = 0.75$ L | $V_2 = ?$ L |
| $P_1 = 2.25$ atm | |
| $P_2 = 1.03$ atm | |

### 2 SOLVE FOR THE UNKNOWN

Use Boyle's law. Solve for $V_2$, and calculate the new volume.

$P_1V_1 = P_2V_2$ — **State Boyle's law.**

$V_2 = V_1\left(\frac{P_1}{P_2}\right)$ — **Solve for $V_2$.**

$V_2 = 0.75 \text{ L}\left(\frac{2.25 \text{ atm}}{1.03 \text{ atm}}\right)$ — **Substitute $V_1 = 0.75$ L, $P_1 = 2.25$ atm, and $P_2 = 1.03$ atm.**

$V_2 = 0.75 \text{ L}\left(\frac{2.25 \cancel{\text{atm}}}{1.03 \cancel{\text{atm}}}\right) = 1.6 \text{ L}$ — **Multiply and divide numbers and units.**

### 3 EVALUATE THE ANSWER

The pressure decreases by roughly half, so the volume should roughly double. The answer is expressed in liters, a unit of volume, and correctly contains two significant figures.

## PRACTICE Problems

Do additional problems. Online Practice

PRACTICE PROBLEMS

**Assume that the temperature and the amount of gas are constant in the following problems.**

1. The volume of a gas at 99.0 kPa is 300.0 mL. If the pressure is increased to 188 kPa, what will be the new volume?
2. The pressure of a sample of helium in a 1.00-L container is 0.988 atm. What is the new pressure if the sample is placed in a 2.00-L container?
3. **Challenge** Air trapped in a cylinder fitted with a piston occupies 145.7 mL at 1.08 atm pressure. What is the new volume when the piston is depressed, increasing the pressure by 25%?

## Problem-Solving LAB

### Apply Scientific Explanations

**What does Boyle's law have to do with breathing?** You take a breath about 20 times per minute, exchanging carbon dioxide gas for life-sustaining oxygen. How do pressure and volume change in your lungs as you breathe?

#### Analysis

The spongy, elastic tissue that makes up your lungs allows them to expand and contract in response to movement of the diaphragm, a strong muscle beneath the lungs. As your diaphragm moves downward, increasing lung volume, you inhale. As your diaphragm moves upward, decreasing lung volume, you exhale.

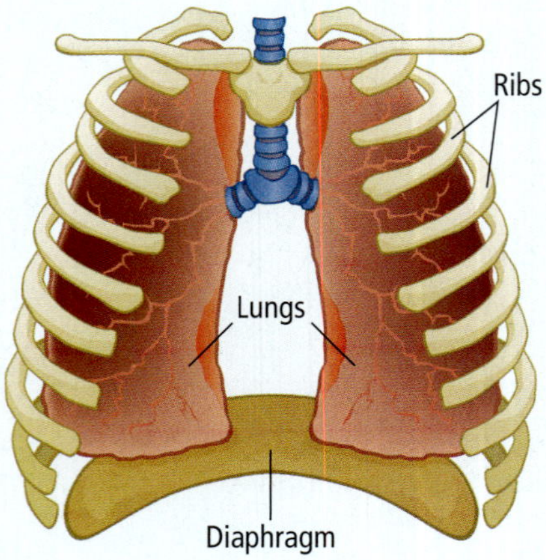


#### Think Critically

1. **Apply** Boyle's law to explain why air enters your lungs when you inhale and leaves when you exhale.
2. **Explain** what happens inside the lungs when a blow to the abdomen knocks the wind out of a person. Use Boyle's law to determine your answer.
3. **Infer** Parts of the lungs lose elasticity and become enlarged when a person has emphysema. From what you know about Boyle's law, why does this condition affect breathing?
4. **Explain** why beginning scuba divers are taught never to hold their breath while ascending from deep water.

## Charles's Law

In the Launch Lab, you observed that a balloon's circumference decreased after the balloon was submerged in ice water. Why did this happen? After a cool evening, a rubber pool raft can appear partially inflated. During a sunny afternoon, the same raft can appear fully inflated. Why did the appearance of the raft change? These questions can be answered by applying a second gas law—Charles's law.

**How are temperature and volume related?** Jacques Charles (1746–1823), a French physicist, studied the relationship between volume and temperature. He observed that as temperature increases, so does the volume of a gas sample when the amount of gas and the pressure remain constant. This property is explained by the kinetic-molecular theory: as temperature increases, gas particles move faster, striking the walls of their container more frequently and with greater force. Because pressure depends on the frequency and force with which gas particles strike the walls of their container, this would increase the pressure. For the pressure to stay constant, volume must increase so that the particles have farther to travel before striking the walls. Having to travel farther decreases the frequency with which the particles strike the walls of the container.

The cylinders in **Figure 2** show how the volume of a fixed amount of gas changes as the gas is heated. Unlike **Figure 1,** where pressure in addition to that of the atmosphere was applied to the piston, the piston in **Figure 2** is free to float. This means that the piston will be supported by the gas inside the cylinder at a level where the pressure of the gas exactly matches that of the atmosphere. As you can see, the volume occupied by a gas at 1 atm increases as the temperature in the cylinder increases. The distance the piston moves is a measure of the increase in volume of the gas as it is heated.

**Graphing the relationship of temperature and volume** **Figure 2** also shows graphs of the relationship between the temperature and the volume of a fixed amount of gas at constant pressure. The plot of volume versus temperature is a straight line. Note that you can predict the temperature at which the volume will reach 0 L by extrapolating the line to temperatures below the values that were measured.

In the first graph, the temperature that corresponds to 0 L is −273.15°C. This relationship is linear, but it is not a direct proportion. For example, you can see that the graph of the line does not pass through the origin and that doubling the temperature from 25°C to 50°C does not double the volume.

**Figure 2** When the cylinder is heated, the kinetic energy of the gas particles increases, causing them to push the piston outward. The graphs show the relationship of volume to Celsius and Kelvin temperature.

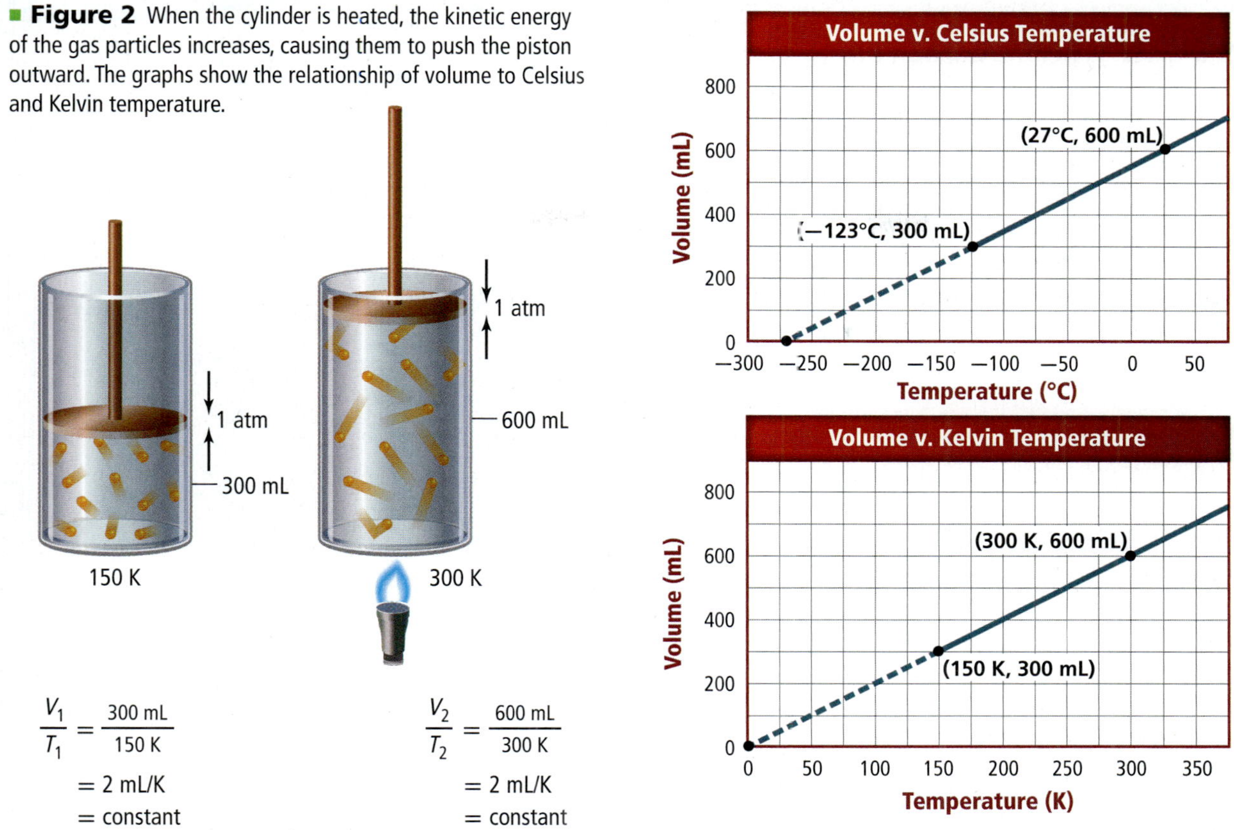


The second graph in **Figure 2,** which plots the Kelvin (K) temperature against volume, does show a direct proportion. A temperature of 0 K corresponds to 0 mL, and doubling the temperature doubles the volume. Zero on the Kelvin scale is also known as **absolute zero.** Absolute zero represents the lowest possible theoretical temperature. At absolute zero, the atoms are all in the lowest possible energy state.

**GRAPH CHECK** **Explain** why the second graph in **Figure 2** shows a direct proportion, but the first graph does not.

**Alabama Applying Practices**

Go to the Resource tab in ConnectED to find the Applying Practices worksheet *Model Average Kinetic Energy and Temperature.*

**FOLDABLES®**

Incorporate information from this section into your Foldable.

**Using Charles's law** **Charles's law** states that the volume of a given amount of gas is directly proportional to its Kelvin temperature at constant pressure. Charles's law can be expressed as follows.

### Charles's Law

$$\frac{V_1}{T_1} = \frac{V_2}{T_2}$$

*V* represents volume.
*T* represents temperature.

For a given amount of gas at constant pressure, the quotient of the volume and Kelvin temperature is a constant.

In the equation above, $V_1$ and $T_1$ represent initial conditions, while $V_2$ and $T_2$ are new conditions. As with Boyle's law, if you know three of the values, you can calculate the fourth.

The temperature must be expressed in kelvins when using the equation for Charles's law. To convert a temperature from Celsius degrees to kelvins, add 273 to the Celsius temperature:

$$T_K = 273 + T_C$$

## EXAMPLE Problem 2

Find help with **significant digits.** Math Handbook

**CHARLES'S LAW** A helium balloon in a closed car occupies a volume of 2.32 L at 40.0°C. If the car is parked on a hot day and the temperature inside rises to 75.0°C, what is the new volume of the balloon, assuming the pressure remains constant?

### 1 ANALYZE THE PROBLEM

Charles's law states that as the temperature of a fixed amount of gas increases, so does its volume, assuming constant pressure. Therefore, the volume of the balloon will increase. The initial volume should be multiplied by a temperature ratio greater than 1.

| **Known** | **Unknown** |
|---|---|
| $T_2 = 40.0°C$ | $V_2 = ?\ L$ |
| $V_1 = 2.32\ L$ | |
| $T_2 = 75.0°C$ | |

### 2 SOLVE FOR THE UNKNOWN

Convert degrees Celsius to kelvins.

$T_K = 273 + T_C$ — **Apply the conversion factor.**

$T_1 = 273 + 40.0°C = 313.0\ K$ — **Substitute $T_1 = 40.0°C$.**

$T_2 = 273 + 75.0°C = 348.0\ K$ — **Substitute $T_2 = 75.0°C$.**

Use Charles's law. Solve for $V_2$, and substitute the known values into the rearranged equation.

$\frac{V_1}{T_1} = \frac{V_2}{T_2}$ — **State Charles's law.**

$V_2 = V_1\left(\frac{T_2}{T_1}\right)$ — **Solve for $V_2$.**

$V_2 = 2.32\ L\left(\frac{348.0\ K}{313.0\ K}\right)$ — **Substitute $V_1 = 2.32\ L$, $T_1 = 313.0\ K$, and $T_2 = 348.0\ K$.**

$V_2 = 2.32\ L\left(\frac{348.0\ \cancel{K}}{313.0\ \cancel{K}}\right) = 2.58\ L$ — **Multiply and divide numbers and units.**

### 3 EVALUATE THE ANSWER

The increase in kelvins is relatively small, so the volume should show a small increase. The unit of the answer is liters, a volume unit, and there are three significant figures.

## PRACTICE Problems

Do additional problems. Online Practice

**Assume that the pressure and the amount of gas remain constant in the following problems.**

**4.** What volume will the gas in the balloon at right occupy at 250 K?

**5.** A gas at 89°C occupies a volume of 0.67 L. At what Celsius temperature will the volume increase to 1.12 L?

**6.** The Celsius temperature of a 3.00-L sample of gas is lowered from 80.0°C to 30.0°C. What will be the resulting volume of this gas?

**7. Challenge** A gas occupies 0.67 L at 350 K. What temperature is required to reduce the volume by 45%?

## Gay-Lussac's Law

In the Launch Lab, you saw Charles's law in action as the balloon's volume changed in response to temperature. What would have happened if the balloon's shape were rigid? If volume is constant, is there a relationship between temperature and pressure? The answer to that question is found in Gay-Lussac's law.

### How are temperature and pressure of a gas related?

Pressure is a direct result of collisions between gas particles and the walls of their container. An increase in temperature increases collision frequency and energy, so raising the temperature should also raise the pressure if the volume is not changed. Joseph Gay-Lussac (1778–1850) found that a direct proportion exists between Kelvin temperature and pressure, as illustrated in **Figure 3.** **Gay-Lussac's law** states that the pressure of a fixed amount of gas varies directly with the Kelvin temperature when the volume remains constant. It can be expressed mathematically as follows.

**Gay-Lussac's Law**

$$\frac{P_1}{T_1} = \frac{P_2}{T_2}$$

***P* represents pressure.**
***T* represents temperature.**

For a given amount of gas held at constant volume, the quotient of the pressure and the Kelvin temperature is a constant.

As with Boyle's and Charles's laws, if you know any three of the four variables, you can calculate the fourth using this equation. Remember that temperature must be in kelvins whenever it is used in a gas law equation.

**CAREERS IN CHEMISTRY**

**Meteorologist** Relationships among pressure, temperature, and volume of air help meteorologists understand and predict the weather. For example, winds and fronts result from pressure changes caused by the uneven heating of Earth's atmosphere by the Sun.

WebQuest

■ **Figure 3** When the cylinder is heated, the kinetic energy of the particles increases, increasing both the frequency and energy of the collisions with the container wall. The volume of the cylinder is fixed, so the pressure exerted by the gas increases.

View an **animation of the gas laws.**

Concepts In Motion

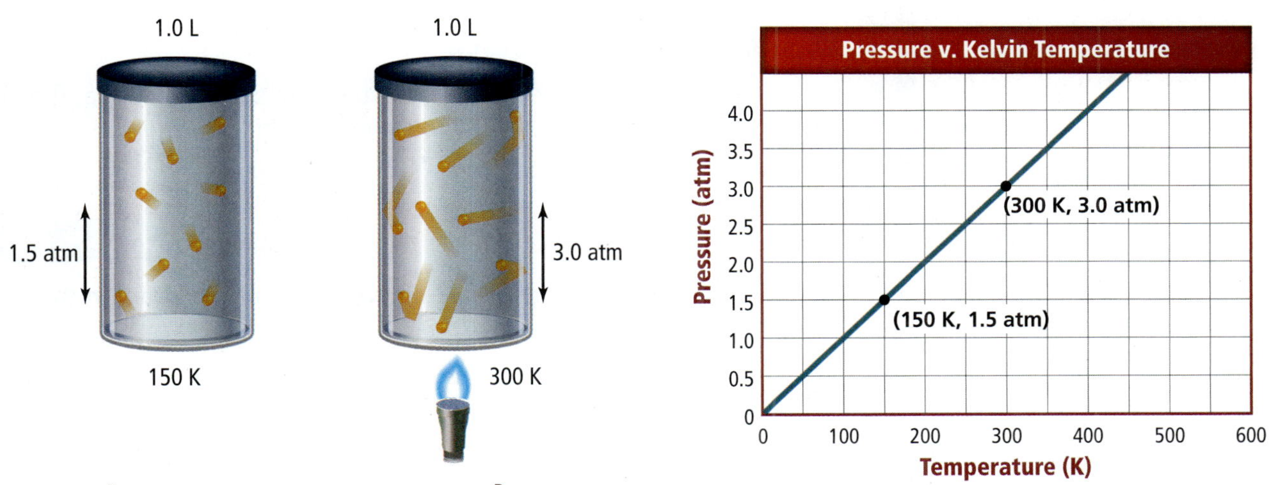


$$\frac{P_1}{T_1} = \frac{1.5 \text{ atm}}{150 \text{ K}} = 0.01 \text{ atm/K} = \text{constant}$$

$$\frac{P_2}{T_2} = \frac{3.0 \text{ atm}}{300 \text{ K}} = 0.01 \text{ atm/K} = \text{constant}$$

**GRAPH CHECK**

**Compare and contrast** the graphs in **Figures 2** and **3.**

## EXAMPLE Problem 3

**GAY-LUSSAC'S LAW** The pressure of the oxygen gas inside a canister is 5.00 atm at 25.0°C. The canister is located at a camp high on Mount Everest. If the temperature there falls to −10.0°C, what is the new pressure inside the canister?

### 1 ANALYZE THE PROBLEM

Gay-Lussac's law states that if the temperature of a gas decreases, so does its pressure when volume is constant. Therefore, the pressure in the oxygen canister will decrease. The initial pressure should be multiplied by a temperature ratio less than 1.

| Known | Unknown |
|---|---|
| $P_1 = 5.00$ atm | $P_2 = ?$ atm |
| $T_1 = 25.0°C$ | |
| $T_2 = -10.0°C$ | |

### 2 SOLVE FOR THE UNKNOWN

Convert degrees Celsius to kelvins.

$T_K = 273 + T_C$ — Apply the conversion factor.

$T_1 = 273 + 25.0°C = 298.0$ K — Substitute $T_1 = 25.0°C$.

$T_2 = 273 + (-10.0°C) = 263.0$ K — Substitute $T_2 = -10.0°C$.

Use Gay-Lussac's law. Solve for $P_2$, and substitute the known values into the rearranged equation.

$$\frac{P_1}{T_1} = \frac{P_2}{T_2}$$ State Gay-Lussac's law.

$$P_2 = P_1\left(\frac{T_2}{T_1}\right)$$ Solve for $P_2$.

$$P_2 = 5.00 \text{ atm}\left(\frac{263.0 \text{ K}}{298.0 \text{ K}}\right)$$ Substitute $P_1 = 5.00$ atm, $T_1 = 298.0$ K, and $T_2 = 263.0$ K.

$$P_2 = 5.00 \text{ atm}\left(\frac{263.0 \cancel{\text{K}}}{298.0 \cancel{\text{K}}}\right) = \mathbf{4.41 \text{ atm}}$$ Multiply and divide numbers and units.

### 3 EVALUATE THE ANSWER

Kelvin temperature decreases, so the pressure should decrease. The unit is atm, a pressure unit, and there are three significant figures.

## PRACTICE Problems

Do additional problems. Online Practice

**Assume that the volume and the amount of gas are constant in the following problems.**

**8.** The pressure in an automobile tire is 1.88 atm at 25.0°C. What will be the pressure if the temperature increases to 37.0°C?

**9.** Helium gas in a 2.00-L cylinder is under 1.12 atm pressure. At 36.5°C, that same gas sample has a pressure of 2.56 atm. What was the initial temperature in degrees Celsius of the gas in the cylinder?

**10. Challenge** If a gas sample has a pressure of 30.7 kPa at 0.00°C, by how many degrees Celsius does the temperature have to increase to cause the pressure to double?

### RealWorld CHEMISTRY
#### Gay-Lussac's Law

**PRESSURE COOKERS** A pressure cooker is a pot with a lid that locks into place. This seals the container, which keeps its volume constant. Heating the pot increases the pressure in the cooker. As pressure increases, the temperature continues to increase and foods cook faster.

■ **Figure 4** Tethers attached at the sides of a weather balloon hold it in place while it is being filled with helium or hydrogen gas. Weather balloons carry instruments that send data, such as air temperature, pressure, and humidity, to receivers on the ground. As the balloon rises, its volume responds to changes in temperature and pressure, expanding until the sides burst. A small parachute returns the instruments to Earth.

## The Combined Gas Law

In a number of applications involving gases, such as the weather balloon in **Figure 4,** pressure, temperature, and volume might all change. Boyle's, Charles's, and Gay-Lussac's laws can be combined into a single law. This **combined gas law** states the relationships between pressure, temperature, and volume of a fixed amount of gas. All three variables have the same relationship to each other as they have in the other gas laws: pressure is inversely proportional to volume and directly proportional to temperature, and volume is directly proportional to temperature. The combined gas law can be expressed mathematically as follows.

Explore the **gas laws.**

Virtual Investigations

### The Combined Gas Law

$$\frac{P_1V_1}{T_1} = \frac{P_2V_2}{T_2}$$

***P* represents pressure. *V* represents volume. *T* represents temperature.**

For a given amount of gas, the product of pressure and volume, divided by the Kelvin temperature, is a constant.

**Using the combined gas law** The combined gas law enables you to solve problems involving changes in more than one variable. It also provides a way for you to remember the other three laws without memorizing each equation. If you can write out the combined gas law equation, equations for the other laws can be derived from it by remembering which variable is held constant in each case.

For example, if temperature remains constant as pressure and volume vary, then $T_1 = T_2$. After simplifying the combined gas law under these conditions, you are left with $P_1V_1 = P_2V_2$, which you should recognize as the equation for Boyle's law.

Get help with the **combined gas law.**

Personal Tutor

☑ READING CHECK **Derive** Charles's and Gay-Lussac's laws from the combined gas law.

## EXAMPLE Problem 4

EXAMPLE PROBLEM

**THE COMBINED GAS LAW** A gas at 110 kPa and 30.0°C fills a flexible container with an initial volume of 2.00 L. If the temperature is raised to 80.0°C and the pressure increases to 440 kPa, what is the new volume?

### 1 ANALYZE THE PROBLEM

Both pressure and temperature change, so you will need to use the combined gas law. The pressure quadruples, but the temperature does not increase by such a large factor. Therefore, the new volume will be smaller than the starting volume.

**Known**

$P_1 = 110$ kPa $\quad P_2 = 440$ kPa

$T_1 = 30.0°C \quad T_2 = 80.0°C$

$V_1 = 2.00$ L

**Unknown**

$V_2 = ?$ L

### 2 SOLVE FOR THE UNKNOWN

Convert degrees Celsius to kelvins.

$T_K = 273 + T_C$ — **Apply the conversion factor.**

$T_1 = 273 + 30.0°C = 303.0$ K — **Substitute $T_1 = 30.0°C$.**

$T_2 = 273 + 80.0°C = 353.0$ K — **Substitute $T_2 = 80.0°C$.**

Use the combined gas law. Solve for $V_2$, and substitute the known values into the rearranged equation.

$$\frac{P_1V_1}{T_1} = \frac{P_2V_2}{T_2}$$ **State the combined gas law.**

$$V_2 = V_1\left(\frac{P_1}{P_2}\right)\left(\frac{T_2}{T_1}\right)$$ **Solve for $V_2$.**

$$V_2 = 2.00\text{ L}\left(\frac{110\text{ kPa}}{440\text{ kPa}}\right)\left(\frac{353.0\text{ K}}{303.0\text{ K}}\right)$$ **Substitute $V_1 = 2.00$ L, $P_1 = 110$ kPa, $P_2 = 440$ kPa, $T_2 = 353.0$ K, and $T_1 = 303.0$ K.**

$$V_2 = 2.00\text{ L}\left(\frac{110\text{ kPa}}{440\text{ kPa}}\right)\left(\frac{353.0\text{ K}}{303.0\text{ K}}\right) = 0.58\text{ L}$$ **Multiply and divide numbers and units.**

### 3 EVALUATE THE ANSWER

Because the pressure change is much greater than the temperature change, the volume undergoes a net decrease. The unit is liters, a volume unit, and there are two significant figures.

## PRACTICE Problems

PRACTICE PROBLEMS

Do additional problems. Online Practice

**Assume that the amount of gas is constant in the following problems.**

**11.** A sample of air in a syringe exerts a pressure of 1.02 atm at 22.0°C. The syringe is placed in a boiling-water bath at 100.0°C. The pressure is increased to 1.23 atm by pushing the plunger in, which reduces the volume to 0.224 mL. What was the initial volume?

**12.** A balloon contains 146.0 mL of gas confined at a pressure of 1.30 atm and a temperature of 5.0°C. If the pressure doubles and the temperature decreases to 2.0°C, what will be the volume of gas in the balloon?

**13. Challenge** If the temperature in the gas cylinder at right increases to 30.0°C and the pressure increases to 1.20 atm, will the cylinder's piston move up or down?

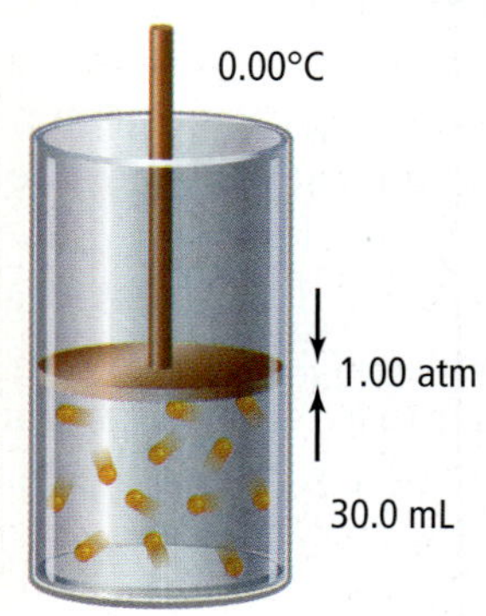

Explore the **gas laws with an interactive table.** Concepts In Motion

**Table 1** The Gas Laws

| Law | Boyle's | Charles's | Gay-Lussac's | Combined |
|---|---|---|---|---|
| Formula | $P_1V_1 = P_2V_2$ | $\frac{V_1}{T_1} = \frac{V_2}{T_2}$ | $\frac{P_1}{T_1} = \frac{P_2}{T_2}$ | $\frac{P_1V_1}{T_1} = \frac{P_2V_2}{T_2}$ |
| What is constant? | amount of gas, temperature | amount of gas, pressure | amount of gas, volume | amount of gas |
| Graphic organizer | 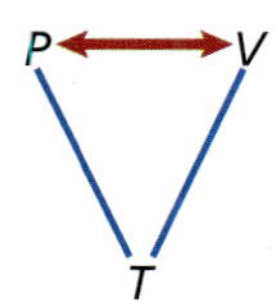 | 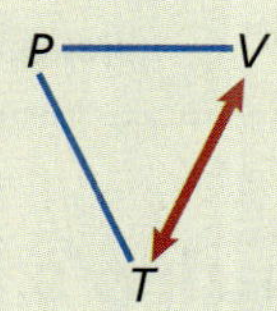 | 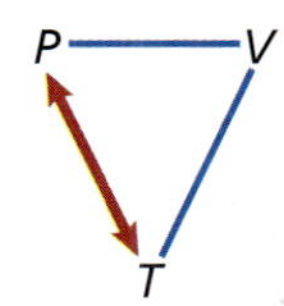 | 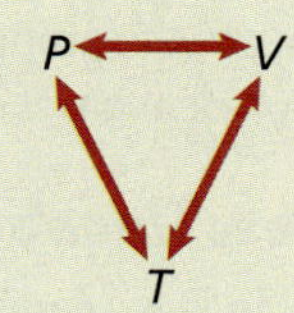 |

**Temperature scales and the gas laws** You might have noticed that the work done by Charles and Gay-Lussac preceded the development of the Kelvin scale, yet their laws require the use of temperature in kelvins. In the 1700s and early 1800s, scientists worked with several different scales. For example, a scale called the Réaumur scale was often used in France around Charles's time. On this scale—or any scale not based on absolute zero—the expression for Charles's law is more complex, requiring two constants in addition to $V$ and $T$. The Kelvin scale simplified matters, resulting in the familiar gas laws presented here.

You have now seen how pressure, temperature, and volume affect a gas sample. You can use the gas laws, summarized in **Table 1,** as long as the amount of gas remains constant. But what happens if the amount of gas changes? In the next section, you will add the fourth variable, amount of gas present, to the gas laws.

**Alabama Applying Practices**

Go to the Resource tab in ConnectED to find the Applying Practices worksheet *Investigate Kinetic Molecular Theory.*

## SECTION 1 REVIEW

Section Self-Check

### Section Summary

- Boyle's law states that the volume of a fixed amount of gas is inversely proportional to its pressure at constant temperature.
- Charles's law states that the volume of a fixed amount of gas is directly proportional to its Kelvin temperature at constant pressure.
- Gay-Lussac's law states that the pressure of a fixed amount of gas is directly proportional to its Kelvin temperature at constant volume.
- The combined gas law relates pressure, temperature, and volume in a single statement.

**14.** MAINIDEA **State** the relationships between pressure, temperature, and volume of a fixed amount of gas.

**15. Explain** Which of the three variables that apply to equal amounts of gases are directly proportional? Which are inversely proportional?

**16. Analyze** A weather balloon is released into the atmosphere. You know the initial volume, temperature, and air pressure. What information will you need to predict its volume when it reaches its final altitude? Which law would you use to calculate this volume?

**17. Infer** why gases such as the oxygen used at hospitals are compressed. Why must compressed gases be shielded from high temperatures? What must happen to compressed oxygen before it can be inhaled?

**18. Calculate** A rigid plastic container holds 1.00 L of methane gas at 660 torr pressure when the temperature is 22.0°C. How much pressure will the gas exert if the temperature is raised to 44.6°C?

**19. Design a concept map** that shows the relationships between pressure, volume, and temperature in Boyle's, Charles's, and Gay-Lussac's laws.

SECTION 2

# The Ideal Gas Law

**MAINIDEA** The ideal gas law relates the number of particles to pressure, temperature, and volume.

### Essential Questions

- How does Avogadro's principle relate the number of particles of gas to the gas's volume?
- How is the amount of gas present related to its pressure, temperature, and volume by the ideal gas law?
- What are the properties of real gases and of ideal gases?

### Review Vocabulary

**mole:** an SI base unit used to measure the amount of a substance; the amount of a pure substance that contains $6.02 \times 10^{23}$ representative particles

### New Vocabulary

Avogadro's principle
molar volume
standard temperature and pressure (STP)
ideal gas constant (R)
ideal gas law

**CHEM 4 YOU** You know that adding air to a tire causes the pressure in the tire to increase. But did you know that the recommended pressure for car tires is specified for cold tires? As tires roll over the road, friction causes their temperatures to increase. This also causes the pressure to increase.

## Avogadro's Principle

The particles that make up different gases can vary greatly in size. However, kinetic-molecular theory assumes that the particles in a gas sample are far enough apart that size has very little influence on the volume occupied by a gas. For example, 1000 relatively large krypton gas particles occupy the same volume as 1000 smaller helium gas particles at the same temperature and pressure. It was Avogadro who first proposed this idea in 1811. **Avogadro's principle** states that equal volumes of gases at the same temperature and pressure contain equal numbers of particles. **Figure 5** shows equal volumes of carbon dioxide, helium, and oxygen.

**Volume and moles** Recall that one mole of a substance contains $6.02 \times 10^{23}$ particles. The **molar volume** of a gas is the volume that 1 mol occupies at 0.00°C and 1.00 atm pressure. The conditions of 0.00°C and 1.00 atm are known as **standard temperature and pressure (STP).** Avogadro showed experimentally that 1 mol of any gas occupies a volume of 22.4 L at STP. Because the volume of 1 mol of a gas at STP is 22.4 L, you can use 22.4 L/mol as a conversion factor whenever a gas is at STP.

For example, suppose you want to find the number of moles in a sample of gas that has a volume of 3.72 L at STP. Use the molar volume to convert from volume to moles.

$$3.72 \text{ L} \times \frac{1 \text{ mol}}{22.4 \text{ L}} = 0.166 \text{ mol}$$

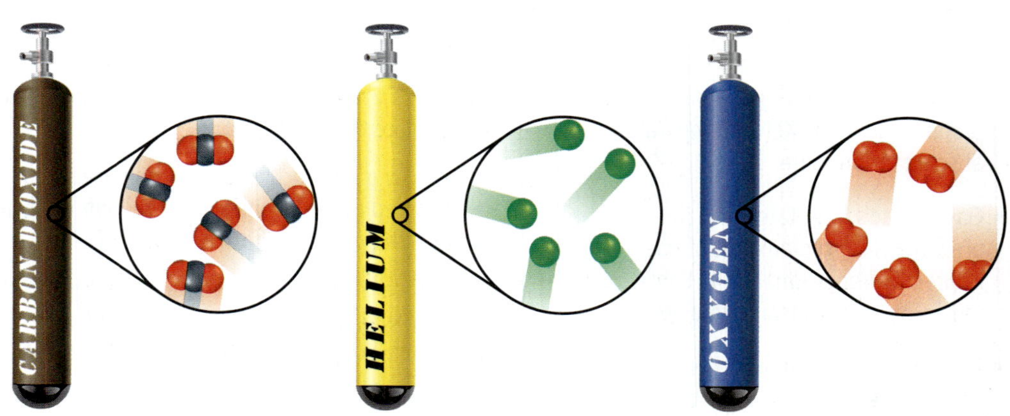


**Figure 5** Gas tanks of equal volume that are at the same pressure and temperature contain equal numbers of gas particles, regardless of which gas they contain.
**Infer** *Why doesn't Avogadro's principle apply to liquids and solids?*

## EXAMPLE Problem 5

Find help with **unit conversion.** Math Handbook

**MOLAR VOLUME** The main component of natural gas used for home heating and cooking is methane ($CH_4$). Calculate the volume that 2.00 kg of methane gas will occupy at STP.

### 1 ANALYZE THE PROBLEM

The number of moles can be calculated by dividing the mass of the sample, *m*, by its molar mass, M. The gas is at STP (0.00°C and 1.00 atm pressure), so you can use the molar volume to convert from the number of moles to the volume.

**Known**

$m = 2.00$ kg

$T = 0.00°C$

$P = 1.00$ atm

**Unknown**

$V = ?$ L

### 2 SOLVE FOR THE UNKNOWN

Determine the molar mass for methane.

$$M = 1\text{ C }\cancel{\text{atom}}\left(\frac{12.01\text{ amu}}{1\text{ C }\cancel{\text{atom}}}\right) + 4\text{ H }\cancel{\text{atoms}}\left(\frac{1.01\text{ amu}}{1\text{ H }\cancel{\text{atom}}}\right)$$

**Determine the molecular mass.**

$$= 12.01\text{ amu} + 4.04\text{ amu} = 16.05\text{ amu}$$
$$= 16.05\text{ g/mol}$$

**Express the molecular mass as g/mol to arrive at the molar mass.**

Determine the number of moles of methane.

$$2.00\ \cancel{\text{kg}}\left(\frac{1000\text{ g}}{1\ \cancel{\text{kg}}}\right) = 2.00 \times 10^3\text{ g}$$

**Convert the mass from kg to g.**

$$\frac{m}{M} = \frac{2.00 \times 10^3\ \cancel{\text{g}}}{16.05\ \cancel{\text{g}}/\text{mol}} = 125\text{ mol}$$

**Divide mass by molar mass to determine the number of moles.**

Use the molar volume to determine the volume of methane at STP.

$$V = 125\ \cancel{\text{mol}} \times \frac{22.4\text{ L}}{1\ \cancel{\text{mol}}} = \mathbf{2.80 \times 10^3\ L}$$

**Use the molar volume, 22.4 L/mol, to convert from moles to the volume.**

### 3 EVALUATE THE ANSWER

The amount of methane present is much more than 1 mol, so you should expect a large volume, which is in agreement with the answer. The unit is liters, a volume unit, and there are three significant figures.

## PRACTICE Problems

Do additional problems. Online Practice

**20.** What size container do you need to hold 0.0459 mol of $N_2$ gas at STP?

**21.** How much carbon dioxide gas, in grams, is in a 1.0-L balloon at STP?

**22.** What volume in milliliters will 0.00922 g of $H_2$ gas occupy at STP?

**23.** What volume will 0.416 g of krypton gas occupy at STP?

**24.** Calculate the volume that 4.5 kg of ethylene gas ($C_2H_4$) will occupy at STP.

**25. Challenge** A flexible plastic container contains 0.860 g of helium gas in a volume of 19.2 L. If 0.205 g of helium is removed at constant pressure and temperature, what will be the new volume?

EXAMPLE PROBLEM

PRACTICE PROBLEMS

■ **Figure 6** The volume and temperature of this tire stay the same as air is added. However, the pressure in the tire increases as the amount of air present increases.

## The Ideal Gas Law

Avogadro's principle and the laws of Boyle, Charles, and Gay-Lussac can be combined into a single mathematical statement that describes the relationships between pressure, volume, temperature, and number of moles of a gas. This formula works best for gases that obey the assumptions of the kinetic-molecular theory. Known as ideal gases, their particles occupy a negligible volume and are far enough apart that they exert minimal attractive or repulsive forces on one another.

**From the combined gas law to the ideal gas law** The combined gas law relates the variables of pressure, volume, and temperature for a given amount of gas.

$$\frac{P_1V_1}{T_1} = \frac{P_2V_2}{T_2}$$

For a specific sample of gas, this relationship of pressure, volume, and temperature is always the same. You could rewrite the relationship represented in the combined gas law as follows.

$$\frac{PV}{T} = \text{constant}$$

As **Figure 6** illustrates, increasing the amount of gas present in a sample will raise the pressure if temperature and volume are constant. Likewise, if pressure and temperature remain constant, the volume will increase as more particles of a gas are added. In fact, we know that both volume and pressure are directly proportional to the number of moles, $n$, so $n$ can be incorporated into the combined gas law as follows.

$$\frac{PV}{nT} = \text{constant}$$

**FOLDABLES®** Incorporate information from this section into your Foldable.

Experiments using known values of $P$, $T$, $V$, and $n$ have determined the value of this constant. It is called the **ideal gas constant,** and it is represented by the symbol R. If pressure is in atmospheres, the value of R is 0.0821 L·atm/mol·K. Note that the units for R are simply the combined units for each of the four variables. **Table 2** shows the numerical values for R in different units of pressure.

**Alabama Applying Practices**

Go to the Resource tab in ConnectED to find the Applying Practices worksheet *Investigate Gas Behavior.*

**READING CHECK** **Explain** why the number of moles, $n$, was added to the denominator of the equation above.

Substituting R for the constant in the equation above and rearranging the variables gives the most familiar form of the ideal gas law. The **ideal gas law** describes the physical behavior of an ideal gas in terms of the pressure, volume, temperature, and number of moles of gas present.

**Table 2 Values of R**

| Value of R | Units of R |
|---|---|
| 0.0821 | $\frac{\text{L·atm}}{\text{mol·K}}$ |
| 8.314 | $\frac{\text{L·kPa}}{\text{mol·K}}$ |
| 62.4 | $\frac{\text{L·mm Hg}}{\text{mol·K}}$ |

### The Ideal Gas Law

$$PV = n\text{R}T$$

***P*** **represents pressure. *V* represents volume. *n* represents number of moles. R is the ideal gas constant. *T* represents temperature.**

For a given amount of gas held at constant temperature, the product of pressure and volume is a constant.

If you know any three of the four variables, you can rearrange the equation to solve for the unknown.

## EXAMPLE Problem 6

Find help with **significant figures.** Math Handbook

**THE IDEAL GAS LAW** Calculate the number of moles of ammonia gas ($NH_3$) contained in a 3.0-L vessel at $3.00 \times 10^2$ K with a pressure of 1.50 atm.

### 1 ANALYZE THE PROBLEM

You are given the volume, temperature, and pressure of a gas sample. Use the ideal gas law, and select the value of R that contains the pressure units given in the problem. Because the pressure and temperature are close to STP, but the volume is much smaller than 22.4 L, it would make sense if the calculated answer were much smaller than 1 mol.

**Known**

$V = 3.0$ L

$T = 3.00 \times 10^2$ K

$P = 1.50$ atm

$R = 0.0821 \frac{\text{L·atm}}{\text{mol·K}}$

**Unknown**

$n = ?$ mol

### 2 SOLVE FOR THE UNKNOWN

Use the ideal gas law. Solve for ***n***, and substitute the known values.

$$PV = nRT$$
State the ideal gas law.

$$n = \frac{PV}{RT}$$
Solve for *n*.

$$n = \frac{(1.50 \text{ atm})(3.0 \text{ L})}{\left(0.0821 \frac{\text{L·atm}}{\text{mol·K}}\right)(3.00 \times 10^2 \text{ K})}$$
Substitute $V = 3.0$ L, $T = 3.00 \times 10^2$ K, $P = 1.50$ atm, and $R = 0.0821$ L·atm/mol·K.

$$n = \frac{(1.50 \cancel{\text{atm}})(3.0 \cancel{\text{L}})}{\left(0.0821 \frac{\cancel{\text{L}}\cdot\cancel{\text{atm}}}{\text{mol}\cdot\cancel{\text{K}}}\right)(3.00 \times 10^2 \cancel{\text{K}})} = 0.18 \text{ mol}$$
Multiply and divide numbers and units.

### 3 EVALUATE THE ANSWER

The answer agrees with the prediction that the number of moles present will be significantly less than 1 mol. The unit of the answer is the mole, and there are two significant figures.

## PRACTICE Problems

Do additional problems. Online Practice

**26.** Determine the Celsius temperature of 2.49 mol of a gas contained in a 1.00-L vessel at a pressure of 143 kPa.

**27.** Calculate the volume of a 0.323-mol sample of a gas at 265 K and 0.900 atm.

**28.** What is the pressure, in atmospheres, of a 0.108-mol sample of helium gas at a temperature of 20.0°C if its volume is 0.505 L?

**29.** If the pressure exerted by a gas at 25°C in a volume of 0.044 L is 3.81 atm, how many moles of gas are present?

**30. Challenge** An ideal gas has a volume of 3.0 L. If the number of moles of gas and the temperature are doubled, while the pressure remains constant, what is the new volume?

**VOCABULARY**

**WORD ORIGIN**

**Mole**

comes from the German word *Mol,* which is short for *Molekulargewicht,* meaning *molecular weight*

# The Ideal Gas Law—Molar Mass and Density

The ideal gas law can be used to solve for the value of any one of the four variables $P$, $V$, $T$, or $n$ if the values of the other three are known. However, you can also rearrange the $PV = nRT$ equation to calculate the molar mass and density of a gas sample.

**Molar mass and the ideal gas law** To find the molar mass of a gas sample, the mass, temperature, pressure, and volume of the gas must be known. Recall that the number of moles of a gas ($n$) is equal to the mass ($m$) divided by the molar mass (M). Therefore, the $n$ in the equation can be replaced by $m$/M.

$$PV = nRT \quad \xrightarrow{\text{substitute } n = \frac{m}{\text{M}}} \quad PV = \frac{mRT}{\text{M}}$$

You can rearrange the new equation to solve for the molar mass.

$$\text{M} = \frac{mRT}{PV}$$

Watch a **video about gases.**

Video

**Density and the ideal gas law** Recall that the density ($D$) of a substance is defined as mass ($m$) per unit volume ($V$). After rearranging the ideal gas equation to solve for molar mass, you can substitute $D$ for $m/V$.

$$\text{M} = \frac{mRT}{PV} \quad \xrightarrow{\text{substitute } \frac{m}{V} = D} \quad \text{M} = \frac{DRT}{P}$$

You can rearrange the new equation to solve for density.

$$D = \frac{\text{M}P}{RT}$$

Why might you need to know the density of a gas? Consider the requirements to fight a fire. One way to put out a fire is to remove its oxygen source by covering it with another gas that will neither burn nor support combustion, as shown in **Figure 7.** This gas must have a greater density than oxygen so that it will displace the oxygen at the source of the fire. You can observe a similar application of density by doing the MiniLab on the next page.

■ **Figure 7** To extinguish a fire, you need to take away fuel, oxygen, or heat. The fire extinguisher at right contains carbon dioxide, which displaces oxygen but does not burn. It also has a cooling effect due to the rapid expansion of the carbon dioxide as it is released from the nozzle.

**Explain** *Why does carbon dioxide displace oxygen?*

# MiniLAB

iLab Station

## Model a Fire Extinguisher

**Why is carbon dioxide used in fire extinguishers?**

### Procedure

1. Read and complete the lab safety form.
2. Measure the temperature with a **thermometer.** Obtain the air pressure with a **barometer** or **weather radio.** Record your data.
3. Roll a 23-cm × 30-cm piece of **aluminum foil** into a cylinder that is 30 cm long and roughly 6 cm in diameter. Tape the edges with **masking tape.**
4. Use **matches** to light a **candle.**
**WARNING:** ***Run water over the extinguished match before throwing it away. Keep hair and clothing away from the flame.***
5. Place 30 g of **baking soda ($NaHCO_3$)** in a large **beaker.** Add 40 mL of **vinegar (5% $CH_3COOH$).**
6. Quickly position the foil cylinder at about 45° up and away from the top of the candle flame.
**WARNING:** ***Do not touch the end of the aluminum tube that is near the burning candle.***
7. While the reaction in the beaker is actively producing carbon dioxide gas, carefully pour the gas, but not the liquid, out of the beaker and into the top of the foil tube. Record your observations.

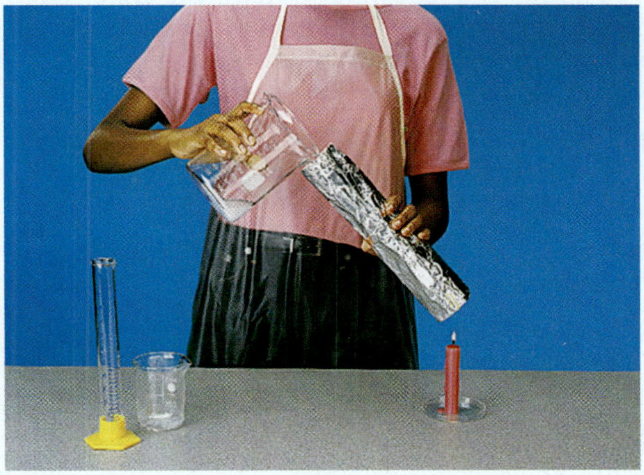

### Analysis

1. **Apply** Calculate the molar volume of carbon dioxide gas ($CO_2$) at room temperature and atmospheric pressure.
2. **Calculate** the room-temperature densities in grams per liter of carbon dioxide, oxygen, and nitrogen gases. Recall that you will need to calculate the molar mass of each gas in order to calculate densities.
3. **Interpret** Do your observations and calculations support the use of carbon dioxide gas to extinguish fires? Explain.

## Real Versus Ideal Gases

What does the term *ideal gas* mean? Ideal gases follow the assumptions of the kinetic-molecular theory. According to this theory, an ideal gas is one whose particles take up no space. Ideal gases experience no intermolecular attractive forces, nor are they attracted or repelled by the walls of their containers. The particles of an ideal gas are in constant, random motion, moving in straight lines until they collide with each other or with the walls of the container. Additionally, these collisions are perfectly elastic, which means that the kinetic energy of the system does not change. An ideal gas follows the gas laws under all conditions of temperature and pressure.

In reality, no gas is truly ideal. All gas particles have some volume, however small, and are subject to intermolecular interactions. Also, the collisions that particles make with each other and with the container are not perfectly elastic. Despite that, most gases will behave like ideal gases at a wide range of temperatures and pressures. Under the right conditions, calculations made using the ideal gas law closely approximate experimental measurements.

☑ READING CHECK **Explain** the relationship between the kinetic-molecular theory and an ideal gas.

Matt Meadows

## PROBLEM-SOLVING STRATEGIES

### Deriving Gas Laws

If you master the following strategy, you will need to remember only one gas law—the ideal gas law. Consider the example of a fixed amount of gas held at constant pressure. You need Charles's law to solve problems involving volume and temperature.

1. Use the ideal gas law to write two equations that describe the gas sample at two different volumes and temperatures. (Quantities that do not change are shown in **red**.)
2. Isolate volume and temperature–the two conditions that vary–on the same side of each equation.
3. Because n, R, and P are constant under these conditions, you can set the volume and temperature conditions equal, deriving Charles's law.

$$PV_1 = n\mathbf{R}T_1 \qquad PV_2 = n\mathbf{R}T_2$$

$$\frac{V_1}{T_1} = \frac{n\mathbf{R}}{P} \qquad \frac{V_2}{T_2} = \frac{n\mathbf{R}}{P}$$

$$\frac{V_1}{T_1} = \frac{V_2}{T_2}$$

### Apply the Strategy

**Derive** Boyle's law, Gay-Lussac's law, and the combined gas law based on the example above.

**Extreme pressure and temperature** When is the ideal gas law not likely to work for a real gas? Real gases deviate most from ideal gas behavior at high pressures and low temperatures. The nitrogen gas in the tanks shown in **Figure 8** behaves as a real gas. Lowering the temperature of nitrogen gas results in less kinetic energy of the gas particles, which means their intermolecular attractive forces are strong enough to affect their behavior. When the temperature is low enough, this real gas condenses to form a liquid. The propane gas in the tanks shown in **Figure 8** also behaves as a real gas. Increasing the pressure on a gas forces the gas particles closer together until the volume occupied by the gas particles themselves is no longer negligible. Real gases such as propane will liquefy if enough pressure is applied.

Watch a **video about gases, pressure, and scuba diving.**

Video

■ **Figure 8** Real gases do not follow the ideal gas law at all pressures and temperatures.

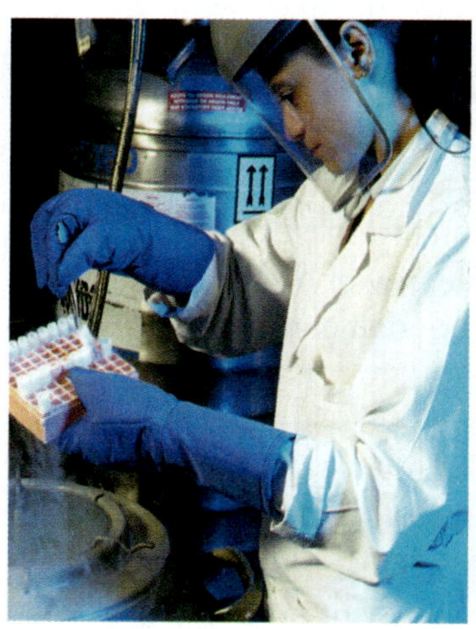

Nitrogen gas turns to liquid at −196°C. At this temperature, scientists can preserve biological specimens, such as body tissues, for future research or medical procedures.

About 270 times more propane can be stored as a liquid than as a gas in the same amount of space. Your family might use small tanks of liquid propane as fuel for your barbecue grill or larger tanks for heating and cooking.

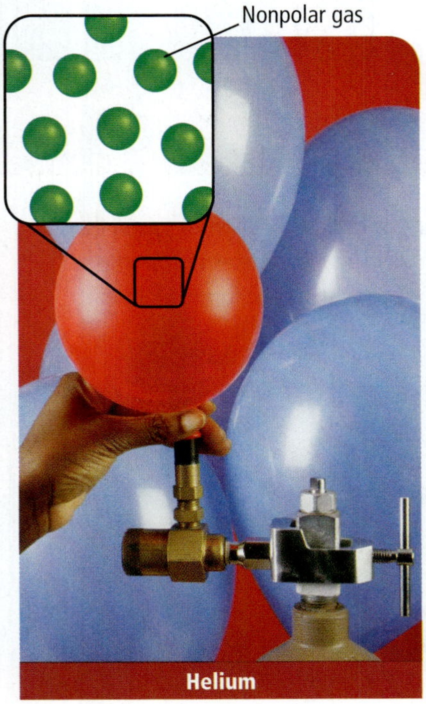


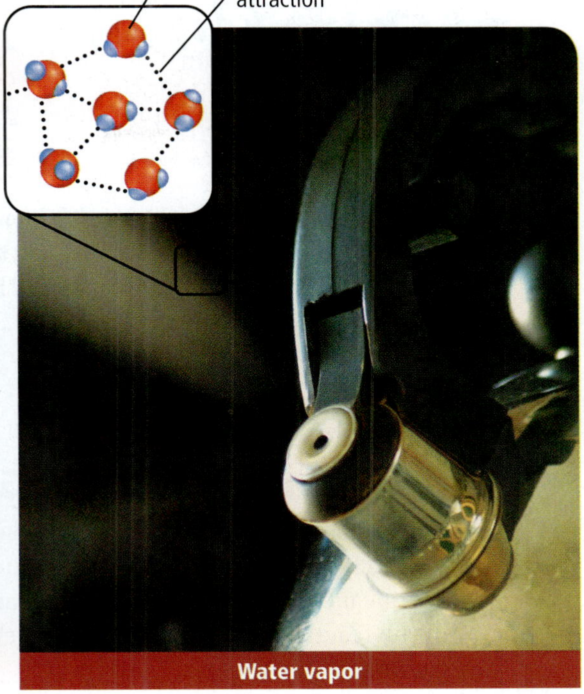


■ **Figure 9** In a nonpolar gas, there is minimal attraction between particles. However, polar gases, such as water vapor, experience forces of attraction between particles.

**Infer** ***Assuming the volume of the particles is negligible, how will the measured pressure for a sample of gas that experiences significant intermolecular attractive forces compare to the pressure predicted by the ideal gas law?***

**Polarity and size of particles** The nature of the particles making up a gas also affects how ideally the gas behaves. For example, polar gas molecules, such as water vapor, generally have larger attractive forces between their particles than nonpolar gases, such as helium. The oppositely charged ends of polar molecules are pulled together through electrostatic forces, as shown in **Figure 9.** Therefore, polar gases do not behave as ideal gases. Also, the particles of gases composed of larger nonpolar molecules, such as butane ($C_4H_{10}$), occupy more actual volume than an equal number of smaller gas particles in gases such as helium (He). Therefore, larger gas particles tend to exhibit a greater departure from ideal behavior than do smaller gas particles.

## SECTION 2 REVIEW

Section Self-Check

### Section Summary

- Avogadro's principle states that equal volumes of gases at the same pressure and temperature contain equal numbers of particles.
- The ideal gas law relates the amount of a gas present to its pressure, temperature, and volume.
- The ideal gas law can be used to find molar mass if the mass of the gas is known or the density of the gas if its molar mass is known.
- At very high pressures and very low temperatures, real gases behave differently than ideal gases.

**31.** MAINIDEA **Explain** why Avogadro's principle holds true for ideal gases that have small particles and for ideal gases that have large particles.

**32. State** the equation for the ideal gas law.

**33. Analyze** how the ideal gas law applies to real gases using the kinetic-molecular theory.

**34. Predict** the conditions under which a real gas might deviate from ideal behavior.

**35. List** common units for each variable in the ideal gas law.

**36. Calculate** A 2.00-L flask is filled with propane gas ($C_3H_8$) at a pressure of 1.00 atm and a temperature of −15.0°C. What is the mass of the propane in the flask?

**37. Make and Use Graphs** For every 6°C drop in temperature, the air pressure in a car's tires goes down by about 1 psi (14.7 psi = 1.00 atm). Make a graph illustrating the change in tire pressure from 20°C to −20°C (assume 30.0 psi at 20°C).

SECTION 3

# Gas Stoichiometry

**MAINIDEA** When gases react, the coefficients in the balanced chemical equation represent both molar amounts and relative volumes.

### Essential Questions

- What stoichiometric ratios can be determined for gaseous reactants and products from balanced chemical equations?
- How are the amounts of gaseous reactants and products in a chemical reaction calculated?

### Review Vocabulary

**coefficient:** the number written in front of a reactant or product in a chemical equation, which tells the smallest number of particles of the substance involved in the reaction

**CHEM 4 YOU** To make a cake, it is important to add the ingredients in the correct proportions. In a similar way, the correct proportions of reactants are needed in a chemical reaction to yield the desired products.

## Stoichiometry of Reactions Involving Gases

The gas laws can be applied to calculate the stoichiometry of reactions in which gases are reactants or products. Recall that the coefficients in chemical equations represent molar amounts of substances taking part in the reaction. For example, hydrogen gas can react with oxygen gas to produce water vapor.

$$2H_2(g) + O_2(g) \rightarrow 2H_2O(g)$$

From the balanced chemical equation, you know that 2 mol of hydrogen gas reacts with 1 mol of oxygen gas, producing 2 mol of water vapor. This tells you the molar ratios of substances in this reaction. Avogadro's principle states that equal volumes of gases at the same temperature and pressure contain equal numbers of particles. Thus, for gases, the coefficients in a balanced chemical equation represent not only molar amounts but also relative volumes. Therefore, 2 L of hydrogen gas would react with 1 L of oxygen gas to produce 2 L of water vapor.

## Stoichiometry and Volume–Volume Problems

To find the volume of a gaseous reactant or product in a reaction, you must know the balanced chemical equation for the reaction and the volume of at least one other gas involved in the reaction. Examine the reaction in **Figure 10,** which shows the combustion of methane. This reaction takes place every time you light a Bunsen burner.

Because the coefficients represent volume ratios for gases taking part in the reaction, you can determine that it takes 2 L of oxygen to react completely with 1 L of methane. The complete combustion of 1 L of methane will produce 1 L of carbon dioxide and 2 L of water vapor.

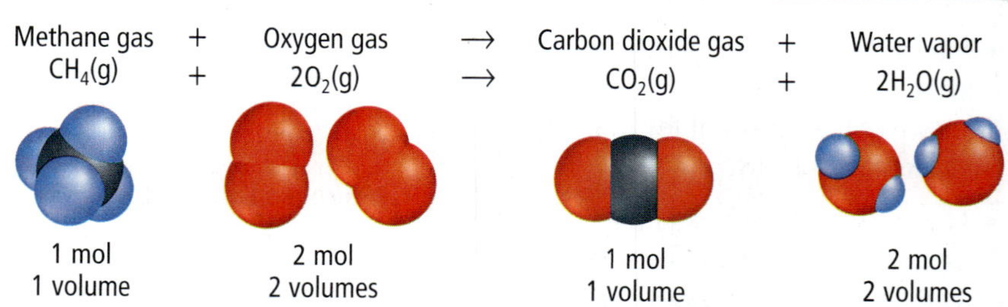


■ **Figure 10** The coefficients in a balanced equation show the relationships between numbers of moles of all reactants and products, and the relationships between volumes of any gaseous reactants or products. From these coefficients, volume ratios can be set up for any pair of gases in the reaction.

Note that no conditions of temperature and pressure are listed. They are not needed as part of the calculation because after mixing, both gases are at the same temperature and pressure. The temperature of the entire reaction might change during the reaction, but a change in temperature would affect all gases in the reaction the same way. Therefore, you do not need to consider pressure and temperature conditions.

## EXAMPLE Problem 7

Find help with **ratios.** Math Handbook

**VOLUME–VOLUME PROBLEMS** What volume of oxygen gas is needed for the complete combustion of 4.00 L of propane gas ($C_3H_8$)? Assume that pressure and temperature remain constant.

### 1 ANALYZE THE PROBLEM

You are given the volume of a gaseous reactant in a chemical reaction. Remember that the coefficients in a balanced chemical equation provide the volume relationships of gaseous reactants and products.

| Known | Unknown |
|---|---|
| $V_{C_3H_8} = 4.00$ L | $V_{O_2} = ?$ L |

### 2 SOLVE FOR THE UNKNOWN

Use the balanced equation for the combustion of $C_3H_8$. Find the volume ratio for $O_2$ and $C_3H_8$, then solve for $V_{O_2}$.

$C_3H_8(g) + 5O_2(g) \rightarrow 3CO_2(g) + 4H_2O(g)$ — **Write the balanced equation.**

$\frac{5 \text{ volumes } O_2}{1 \text{ volume } C_3H_8}$ — **Find the volume ratio for $O_2$ and $C_3H_8$.**

$V_{O_2} = (4.00 \text{ L } C_3H_8) \times \frac{5 \text{ volumes } O_2}{1 \text{ volume } C_3H_8}$ — **Multiply the known volume of $C_3H_8$ by the volume ratio to find the volume of $O_2$.**

$= 20.0 \text{ L } O_2$

### 3 EVALUATE THE ANSWER

The coefficients in the combustion equation show that a much larger volume of $O_2$ than $C_3H_8$ is used up in the reaction, which is in agreement with the calculated answer. The unit of the answer is liters, a unit of volume, and there are three significant figures.

## PRACTICE Problems

Do additional problems. Online Practice

38. How many liters of propane gas ($C_3H_8$) will undergo complete combustion with 34.0 L of oxygen gas?
39. Determine the volume of hydrogen gas needed to react completely with 5.00 L of oxygen gas to form water.
40. What volume of oxygen is needed to completely combust 2.36 L of methane gas ($CH_4$)?
41. **Challenge** Nitrogen and oxygen gases react to form dinitrogen monoxide gas ($N_2O$). What volume of $O_2$ is needed to produce 34 L of $N_2O$?

## RealWorld CHEMISTRY

### Using Stoichiometry

**KILNS** Correct proportions of gases are needed for many chemical reactions. Although many pottery kilns are fueled by methane, a precise mixture of propane and air can be used to fuel a kiln if methane is unavailable.

■ **Figure 11** Ammonia is essential in the production of fertilizers containing nitrogen. Proper levels of soil nitrogen lead to increased crop yields.

## Stoichiometry and Volume–Mass Problems

**Connection to Biology** What you have learned about stoichiometry can be applied to the production of ammonia ($NH_3$) from nitrogen gas ($N_2$). Fertilizer manufacturers use ammonia to make nitrogen-based fertilizers. Nitrogen is an essential element for plant growth. Natural sources of nitrogen in soil, such as nitrogen fixation by plants, the decomposition of organic matter, and animal wastes, do not always supply enough nitrogen for optimum crop yields. **Figure 11** shows a farmer applying fertilizer rich in nitrogen to the soil. This enables the farmer to produce a crop with a higher yield.

Example Problem 8 shows how to use a volume of nitrogen gas to produce a certain amount of ammonia. In doing this type of problem, remember that the balanced chemical equation allows you to find ratios for only moles and gas volumes, not for masses. All masses given must be converted to moles or volumes before being used as part of a ratio. Also, remember that the temperature units used must be kelvins.

**VOCABULARY**
**ACADEMIC VOCABULARY**
**Ratio**
the relationship in quantity between two things
*In a water molecule, the ratio of hydrogen to oxygen is 2:1.*

### EXAMPLE Problem 8

EXAMPLE PROBLEM

**VOLUME–MASS PROBLEMS** Ammonia is synthesized from hydrogen and nitrogen.

$$N_2(g) + 3H_2(g) \rightarrow 2NH_3(g)$$

If 5.00 L of nitrogen reacts completely with hydrogen at a pressure of 3.00 atm and a temperature of 298 K, how much ammonia, in grams, is produced?

#### 1 ANALYZE THE PROBLEM

You are given the volume, pressure, and temperature of a gas sample. The mole and volume ratios of gaseous reactants and products are given by the coefficients in the balanced chemical equation. Volume can be converted to moles and thus related to mass by using molar mass and the ideal gas law.

| Known | Unknown |
|---|---|
| $V_{N_2} = 5.00$ L | $m_{NH_3} = ?$ g |
| $P = 3.00$ atm | |
| $T = 298$ K | |

## 2 SOLVE FOR THE UNKNOWN

Determine how many liters of gaseous ammonia will be made from 5.00 L of nitrogen gas.

$$\frac{1 \text{ volume } N_2}{2 \text{ volumes } NH_3}$$

Find the volume ratio for $N_2$ and $NH_3$ using the balanced equation.

$$5.00 \text{ L } \cancel{N_2}\left(\frac{2 \text{ }\cancel{\text{volumes}}\text{ } NH_3}{1 \text{ }\cancel{\text{volume } N_2}}\right) = 10.0 \text{ L } NH_3$$

Multiply the known volume of $N_2$ by the volume ratio to find the volume of $NH_3$.

Use the ideal gas law. Solve for *n*, and calculate the number of moles of $NH_3$.

$$PV = nRT$$

State the ideal gas law.

$$n = \frac{PV}{RT}$$

Solve for *n*.

$$n = \frac{(3.00 \text{ atm})(10.0 \text{ L})}{\left(0.0821 \frac{\text{L·atm}}{\text{mol·K}}\right)(298 \text{ K})}$$

Substitute $P = 3.00$ atm, $V_{NH_3} = 10.0$ L, and $T = 298$ K.

$$n = \frac{(3.00 \cancel{\text{atm}})(10.0 \cancel{\text{L}})}{\left(0.0821 \frac{\cancel{\text{L}}·\cancel{\text{atm}}}{\text{mol·}\cancel{\text{K}}}\right)(298 \cancel{\text{K}})} = 1.23 \text{ mol } NH_3$$

Multiply and divide numbers and units.

$$M = \left(\frac{1 \cancel{\text{N atom}} \times 14.01 \text{ amu}}{1 \cancel{\text{N atom}}}\right) + \left(\frac{3 \cancel{\text{H atoms}} \times 1.01 \text{ amu}}{1 \cancel{\text{H atom}}}\right)$$

$$= 17.04 \text{ amu}$$

Find the molecular mass of $NH_3$.

$$M = 17.04 \text{ g/mol}$$

Express molar mass in units of g/mol.

Convert moles of ammonia to grams of ammonia.

$$1.23 \cancel{\text{mol } NH_3} \times \frac{17.04 \text{ g } NH_3}{1 \cancel{\text{mol } NH_3}} = \mathbf{21.0 \text{ g } NH_3}$$

Use the molar mass of ammonia as a conversion factor.

## 3 EVALUATE THE ANSWER

To check your answer, calculate the volume of reactant nitrogen at STP. Then, use molar volume and the mole ratio between $N_2$ and $NH_3$ to determine how many moles of $NH_3$ were produced. The unit of the answer is grams, a unit of mass. There are three significant figures.

## PRACTICE Problems

Do additional problems. Online Practice

**42.** Ammonium nitrate is a common ingredient in chemical fertilizers. Use the reaction shown to calculate the mass of solid ammonium nitrate that must be used to obtain 0.100 L of dinitrogen monoxide gas at STP.

$$NH_4NO_3(s) \rightarrow N_2O(g) + 2H_2O(g)$$

**43.** When solid calcium carbonate ($CaCO_3$) is heated, it decomposes to form solid calcium oxide (CaO) and carbon dioxide gas ($CO_2$). How many liters of carbon dioxide will be produced at STP if 2.38 kg of calcium carbonate reacts completely?

**44.** When iron rusts, it undergoes a reaction with oxygen to form iron(III) oxide.

$$4Fe(s) + 3O_2(g) \rightarrow 2Fe_2O_3(s)$$

Calculate the volume of oxygen gas at STP that is required to completely react with 52.0 g of iron.

**45. Challenge** An excess of acetic acid is added to 28 g of sodium bicarbonate at 25°C and 1 atm pressure. During the reaction, the gas cools to 20°C. What volume of carbon dioxide will be produced? The balanced equation for the reaction is shown below.

$$NaHCO_3(aq) + CH_3COOH(aq) \rightarrow NaCH_3COO(aq) + CO_2(g) + H_2O(l)$$

PRACTICE PROBLEMS

■ **Figure 12** To effectively manufacture a product, such as many of these plastics, it is essential to answer the following questions. How much of a reactant should be purchased? How much of a product will be produced?

Comstock/Getty Images

Stoichiometric problems, such as the ones in this section, are considered in industrial processes. For example, ethene gas ($C_2H_4$), also called ethylene, is the raw material for making polyethylene polymers. Polyethylene is produced when numerous ethene molecules join together in chains of repeating —$CH_2$–$CH_2$— units. These polymers are used to make many everyday items, such as the ones shown in **Figure 12.** The general formula for this polymerization reaction is shown below. In this formula, $n$ is the number of units used.

$$n(C_2H_4)(g) \rightarrow —(CH_2 - CH_2)_n—(s)$$

If you were a process engineer for a polyethylene manufacturing plant, you would need to know about the properties of ethene gas and the polymerization reaction. Knowledge of the gas laws would help you calculate both the mass and volume of raw material needed under different temperature and pressure conditions to make different types of polyethylene.

## SECTION 3 REVIEW

Section Self-Check

### Section Summary

- The coefficients in a balanced chemical equation specify volume ratios for gaseous reactants and products.
- The gas laws can be used along with balanced chemical equations to calculate the amount of a gaseous reactant or product in a reaction.

**46. MAIN**IDEA **Explain** When fluorine gas combines with water vapor, the following reaction occurs.

$$2F_2(g) + 2H_2O(g) \rightarrow O_2(g) + 4HF(g)$$

If the reaction starts with 2 L of fluorine gas, how many liters of water vapor react with the fluorine, and how many liters of oxygen and hydrogen fluoride are produced?

**47. Analyze** Is the volume of a gas directly or inversely proportional to the number of moles of a gas at constant temperature and pressure? Explain.

**48. Calculate** One mole of a gas occupies a volume of 22.4 L at STP. Calculate the temperature and pressure conditions needed to fit 2 mol of a gas into a volume of 22.4 L.

**49. Interpret Data** Ethene gas ($C_2H_4$) reacts with oxygen to form carbon dioxide and water. Write a balanced equation for this reaction, then find the mole ratios of substances on each side of the equation.

# CHEMISTRY & health

## Health Under Pressure

You live, work, and play in air that is generally about 1 atm in pressure and 21% oxygen. Have you ever wondered what might happen if the pressure and the oxygen content of the air were greater? Would you recover from illness or injury more quickly? These questions are at the heart of hyperbaric medicine.

**Hyperbaric medicine** The prefix *hyper-* means *above* or *excessive,* and a bar is a unit of pressure equal to 100 kPa, roughly normal atmospheric pressure. Thus, the term *hyperbaric* refers to pressure that is greater than normal. Patients receiving hyperbaric therapy are exposed to pressures greater than the pressure of the atmosphere at sea level.

**The oxygen connection** Greater pressure is most often combined with an increase in the concentration of oxygen a patient receives. The phrase *hyperbaric oxygen therapy* (HBOT) refers to treatment with 100% oxygen. **Figure 1** shows a chamber that might be used for HBOT. Inside the hyperbaric chamber, pressures can reach five to six times normal atmospheric pressure. At hyperbaric therapy centers across the country, HBOT is used to treat a wide range of conditions, including burns, decompression sickness, slow-healing wounds, anemia, and some infections.

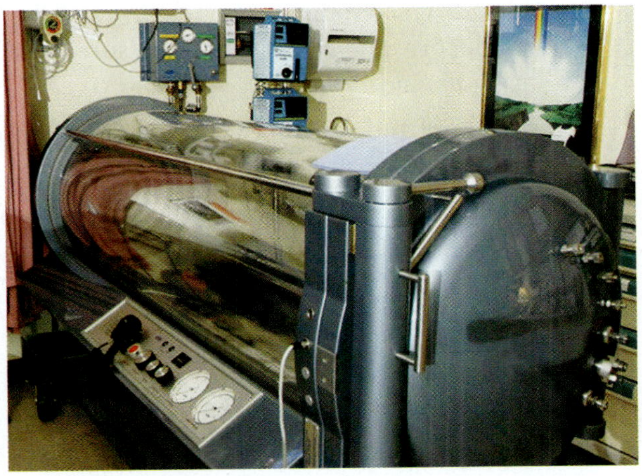



**Figure 1** During HBOT, the patient lies in a hyperbaric chamber. A technician controls the pressure and oxygen levels.

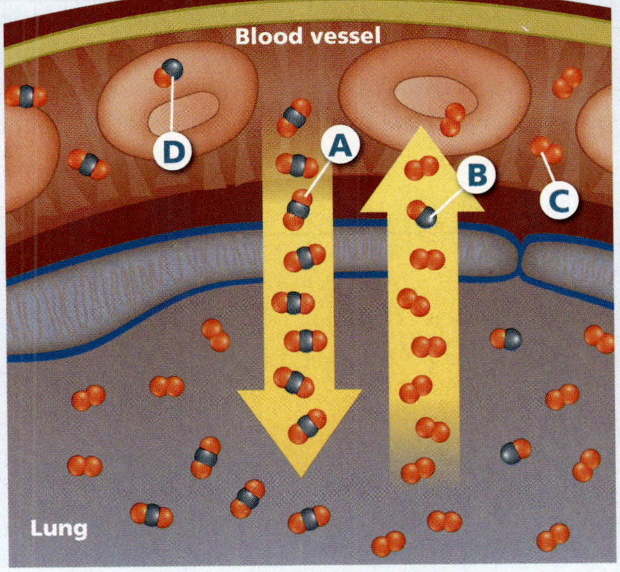


**Figure 2** Gases are exchanged between the lungs and the circulatory system.

**Carbon-monoxide poisoning** Use **Figure 2** to help you understand how HBOT aids in the treatment of carbon-monoxide poisoning.

**Normal gas exchange** Oxygen ($O_2$) moves from the lungs to the blood and binds to the hemoglobin in red blood cells. Carbon dioxide ($CO_2$) is released, as shown by **A**.

**Abnormal gas exchange** If carbon monoxide (CO) enters the blood, as shown by **B**, it, instead of oxygen, binds to the hemoglobin. Cells in the body begin to die from oxygen deprivation.

**Oxygen in blood plasma** In addition to the oxygen carried by hemoglobin, oxygen is dissolved in the blood plasma, as shown by **C**. HBOT increases the concentration of dissolved oxygen to an amount that can sustain the body.

**Eliminating carbon monoxide** Pressurized oxygen also helps remove any carbon monoxide bound to hemoglobin, as shown by **D**.

### WRITING IN ▸ Chemistry

**Research** and prepare an informational pamphlet about the use of HBOT to treat slow-healing wounds.

WebQuest

# ChemLAB

iLab Station

## Determine Pressure in Popcorn Kernels

**Background:** **When the water vapor pressure inside a popcorn kernel is great enough, the kernel bursts and releases the water vapor. The ideal gas law can be used to find the pressure in the kernel as it bursts.**

**Question:** *How much pressure is required to burst a kernel of popcorn?*

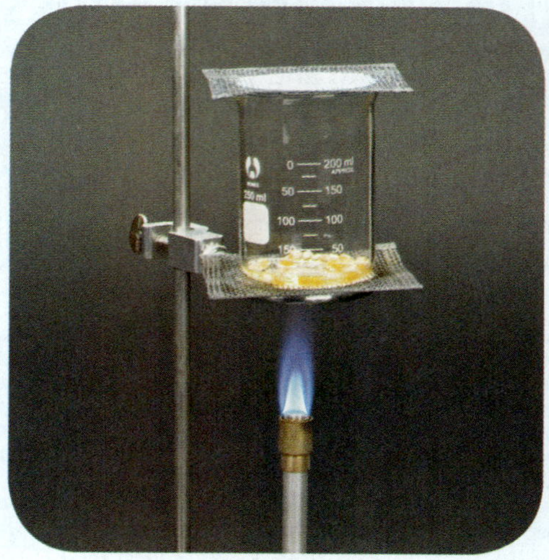

### Materials

| | |
|---|---|
| popcorn kernels (18–20) | 10-mL graduated cylinder |
| vegetable oil (1.5 mL) | 250-mL beaker |
| wire gauze squares (2) | beaker tongs |
| Bunsen burner | balance |
| ring stand | distilled water |
| small iron ring | paper towels |

### Safety Precautions

### Procedure

1. Read and complete the lab safety form.
2. Create a table to record your data.
3. Place approximately 5 mL of distilled water in the graduated cylinder, and record the volume.
4. Place 18–20 popcorn kernels in the graduated cylinder with the water. Tap the cylinder to force any air bubbles off the kernels. Record the new volume.
5. Remove the kernels from the graduated cylinder, and dry them.
6. Place the dry kernels and 1.0–1.5 mL of vegetable oil into the beaker.
7. Measure the total mass of the beaker, oil, and kernels.
8. Set up a Bunsen burner with a ring stand, ring, and wire gauze.
9. Place the beaker on the wire gauze and ring. Place another piece of wire gauze on top of the beaker.
10. Gently heat the beaker with the burner. Move the burner back and forth to heat the oil evenly.
11. Observe the changes in the kernels and oil while heating, then turn off the burner when the popcorn has popped and before any burning occurs.
12. Using the beaker tongs, remove the beaker from the ring and allow it to cool completely.
13. Measure the final mass of the beaker, oil, and popcorn once cooling is complete.
14. **Cleanup and Disposal** Dispose of the popcorn and oil as directed by your teacher. Wash and return all lab equipment to its designated location.

### Analyze and Conclude

1. **Calculate** the volume of the popcorn kernels, in liters, by the difference in the volumes of distilled water before and after adding popcorn.
2. **Calculate** the total mass of water vapor released using the mass measurements of the beaker, oil, and popcorn before and after popping.
3. **Convert** Use the molar mass of water to find the number of moles of water released.
4. **Use Formulas** Use the temperature of the boiling oil (225°C) as your gas temperature and the volume of popcorn, and calculate the pressure of the gas using the ideal gas law.
5. **Compare and contrast** atmospheric pressure to the pressure of the water vapor in the kernel.
6. **Infer** why all the popcorn kernels did not pop.
7. **Error Analysis** Identify a potential source of error for this lab, and suggest a method to correct it.

### INQUIRY EXTENSION

**Design an experiment** that tests the amount of pressure necessary to burst different types of popcorn kernels.

Matt Meadows

# CHAPTER 13 STUDY GUIDE

Vocabulary Practice

**BIGIDEA** **Gases respond in predictable ways to changes in pressure, temperature, volume, and number of particles.**

## SECTION 1 The Gas Laws

**MAINIDEA** For a fixed amount of gas, a change in one variable—pressure, temperature, or volume—affects the other two.

- Boyle's law states that the volume of a fixed amount of gas is inversely proportional to its pressure at constant temperature.

$$P_1V_1 = P_2V_2$$

- Charles's law states that the volume of a fixed amount of gas is directly proportional to its Kelvin temperature at constant pressure.

$$\frac{V_1}{T_1} = \frac{V_2}{T_2}$$

- Gay-Lussac's law states that the pressure of a fixed amount of gas is directly proportional to its Kelvin temperature at constant volume.

$$\frac{P_1}{T_1} = \frac{P_2}{T_2}$$

- The combined gas law relates pressure, temperature, and volume in a single statement.

$$\frac{P_1V_1}{T_1} = \frac{P_2V_2}{T_2}$$

**VOCABULARY**
- Boyle's law
- absolute zero
- Charles's law
- Gay-Lussac's law
- combined gas law

## SECTION 2 The Ideal Gas Law

**MAINIDEA** The ideal gas law relates the number of particles to pressure, temperature, and volume.

- Avogadro's principle states that equal volumes of gases at the same pressure and temperature contain equal numbers of particles.
- The ideal gas law relates the amount of a gas present to its pressure, temperature, and volume.

$$PV = n\mathrm{R}T$$

- The ideal gas law can be used to find molar mass if the mass of the gas is known or the density of the gas if its molar mass is known.

$$\mathrm{M} = \frac{m\mathrm{R}T}{PV} \qquad D = \frac{\mathrm{M}P}{\mathrm{R}T}$$

- At very high pressures and very low temperatures, real gases behave differently than ideal gases.

**VOCABULARY**
- Avogadro's principle
- molar volume
- standard temperature and pressure (STP)
- ideal gas constant (R)
- ideal gas law

## SECTION 3 Gas Stoichiometry

**MAINIDEA** When gases react, the coefficients in the balanced chemical equation represent both molar amounts and relative volumes.

- The coefficients in a balanced chemical equation specify volume ratios for gaseous reactants and products.
- The gas laws can be used along with balanced chemical equations to calculate the amount of a gaseous reactant or product in a reaction.

# CHAPTER 13 ASSESSMENT

## SECTION 1

### Mastering Concepts

**50.** State Boyle's law, Charles's law, Gay-Lussac's law, and the combined gas law in words and equations.

**51.** If two variables are inversely proportional, what happens to the value of one as the value of the other increases?

**52.** If two variables are directly proportional, what happens to the value of one as the value of the other increases?

**53.** List the standard conditions for gas measurements.

**54.** Identify the units most commonly used for *P*, *V*, and *T*.

### Mastering Problems

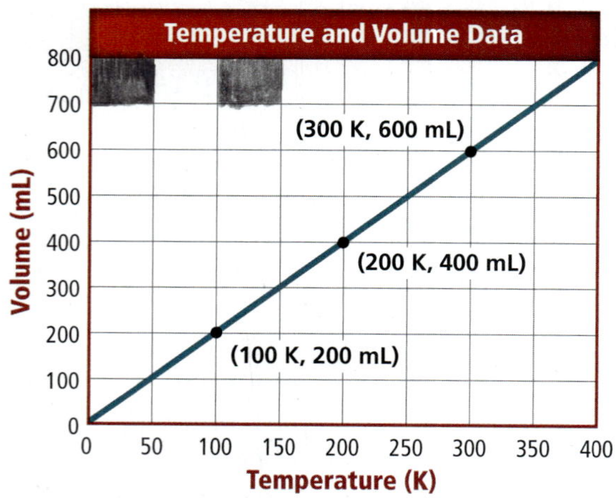


■ **Figure 13**

**55.** Use Charles's law to determine the accuracy of the data plotted in **Figure 13.**

**56. Weather Balloons** A weather balloon is filled with helium that occupies a volume of $5.00 \times 10^4$ L at 0.995 atm and 32.0°C. After it is released, it rises to a location where the pressure is 0.720 atm and the temperature is −12.0°C. What is the volume of the balloon at the new location?

**57.** Use Boyle's, Charles's, or Gay-Lussac's law to calculate the missing value in each of the following.
**a.** $V_1 = 2.0$ L, $P_1 = 0.82$ atm, $V_2 = 1.0$ L, $P_2 = ?$
**b.** $V_1 = 250$ mL, $T_1 = ?$, $V_2 = 400$ mL, $T_2 = 298$ K
**c.** $V_1 = 0.55$ L, $P_1 = 740$ mm Hg, $V_2 = 0.80$ L, $P_2 = ?$

**58. Hot-Air Balloons** A sample of air occupies 2.50 L at a temperature of 22.0°C. What volume will this sample occupy inside a hot-air balloon at a temperature of 43.0°C? Assume that the pressure inside the balloon remains constant.

**59.** What is the pressure of a fixed volume of hydrogen gas at 30.0°C if it has a pressure of 1.11 atm at 15.0°C?

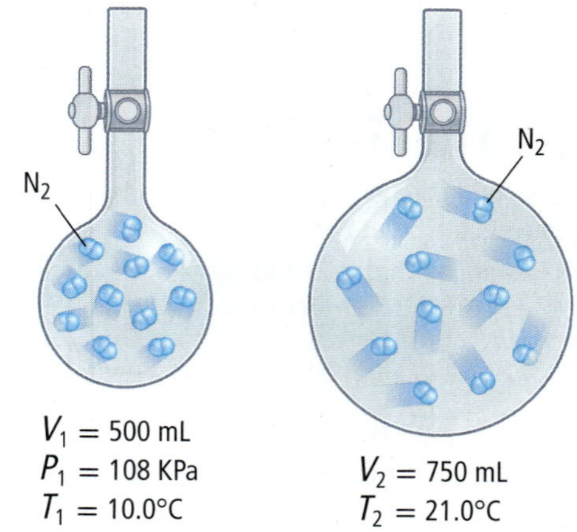


■ **Figure 14**

**60.** A sample of nitrogen gas is transferred to a larger flask, as shown in **Figure 14.** What is the pressure of nitrogen in the second flask?

## SECTION 2

**61.** State Avogadro's principle.

**62.** State the ideal gas law.

**63.** What volume is occupied by 1 mol of a gas at STP? What volume does 2 mol occupy at STP?

**64.** Define the term *ideal gas,* and explain why there are no true ideal gases in nature.

**65.** List two conditions under which a gas is least likely to behave ideally.

**66.** What units must be used to express the temperature in the equation for the ideal gas law? Explain.

### Mastering Problems

**67. Home Fuel** Propane ($C_3H_8$) is a gas commonly used as a home fuel for cooking and heating.
**a.** Calculate the volume that 0.540 mol of propane occupies at STP.
**b.** Think about the size of this volume and the amount of propane that it contains. Why do you think propane is usually liquefied before it is transported?

**68. Careers in Chemistry** A physical chemist measured the lowest pressure achieved in a laboratory—about $1.0 \times 10^{-15}$ mm Hg. How many molecules of gas are present in a 1.00-L sample at that pressure if the sample's temperature is 22.0°C?

**69.** Calculate the number of moles of $O_2$ gas held in a sealed, 2.00-L tank at 3.50 atm and 25.0°C. How many moles would be in the tank if the temperature was raised to 49.0°C and the pressure remained constant?

70. **Perfumes** Geraniol is a compound found in rose oil that is used in perfumes. What is the molar mass of geraniol if its vapor has a density of 0.480 g/L at a temperature of 260.0°C and a pressure of 0.140 atm?

71. Find the volume that 42 g of carbon monoxide gas occupies at STP.

72. Determine the density of chlorine gas at 22.0°C and 1.00 atm.

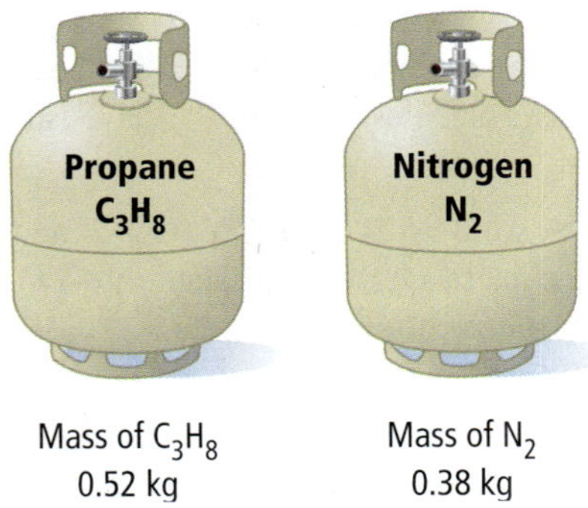


■ **Figure 15**

73. Which of the gases in **Figure 15** occupies the greatest volume at STP? Explain your answer.

74. If the containers in **Figure 15** each hold 4.00 L, what is the pressure inside each? Assume ideal behavior.

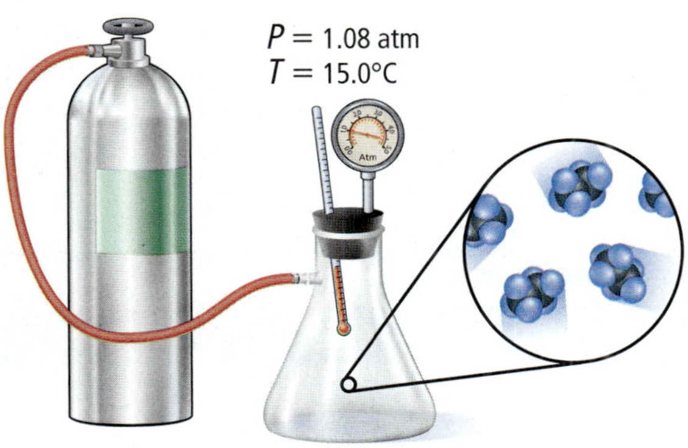


■ **Figure 16**

75. A 2.00-L flask is filled with ethane gas ($C_2H_6$) from a small cylinder, as shown in **Figure 16.** What is the mass of the ethane in the flask?

76. What is the density of a sample of nitrogen gas ($N_2$) that exerts a pressure of 5.30 atm in a 3.50-L container at 125°C?

77. How many moles of helium gas (He) would be required to fill a 22-L container at a temperature of 35°C and a pressure of 3.1 atm?

78. Before a reaction, two gases share a container at a temperature of 200 K. After the reaction, the product is in the same container at a temperature of 400 K. If both $V$ and $P$ are constant, what must be true of $n$?

## SECTION 3

### Mastering Concepts

79. Why must an equation be balanced before using it to determine the volumes of gases involved in a reaction?

80. It is not necessary to consider temperature and pressure when using a balanced equation to determine relative gas volume. Why?

81. What information do you need to solve a volume-mass problem that involves gases?

82. Explain why the coefficients in a balanced chemical equation represent not only molar amounts but also relative volumes for gases.

83. Do the coefficients in a balanced chemical equation represent volume ratios for solids and liquids? Explain.

### Mastering Problems

84. **Ammonia Production** Ammonia is often formed by reacting nitrogen and hydrogen gases. How many liters of ammonia gas can be formed from 13.7 L of hydrogen gas at 93.0°C and a pressure of 40.0 kPa?

85. A 6.5-L sample of hydrogen sulfide is treated with a catalyst to promote the reaction shown below.

$$2H_2S(g) + O_2(g) \rightarrow 2H_2O(g) + 2S(s)$$

If the $H_2S$ reacts completely at 2.0 atm and 290 K, how much water vapor, in grams, is produced?

86. To produce 15.4 L of nitrogen dioxide at 310 K and 2.0 atm, how many liters of nitrogen gas and oxygen gas are required?

87. Use the reaction shown below to answer these questions.

$$2CO(g) + 2NO(g) \rightarrow N_2(g) + 2CO_2(g)$$

**a.** What is the volume ratio of carbon monoxide to carbon dioxide in the balanced equation?
**b.** If 42.7 g of CO is reacted completely at STP, what volume of $N_2$ gas will be produced?

88. When 3.00 L of propane gas is completely combusted to form water vapor and carbon dioxide at 350°C and 0.990 atm, what mass of water vapor results?

89. When heated, solid potassium chlorate ($KClO_3$) decomposes to form solid potassium chloride and oxygen gas. If 20.8 g of potassium chlorate decomposes, how many liters of oxygen gas will form at STP?

90. **Welding** The gas acetylene, often used for welding, burns according to the following equation.

$$2C_2H_2(g) + 5O_2(g) \rightarrow 2H_2O(g) + 4CO_2(g)$$

If you have a 10.0-L tank of acetylene at 25.0°C and 1.00 atm pressure, how many moles of $CO_2$ will be produced if you burn all the acetylene in the tank?

## MIXED REVIEW

**91.** Gaseous methane ($CH_4$) undergoes complete combustion by reacting with oxygen gas to form carbon dioxide and water vapor.
**a.** Write a balanced equation for this reaction.
**b.** What is the volume ratio of methane to water in this reaction?

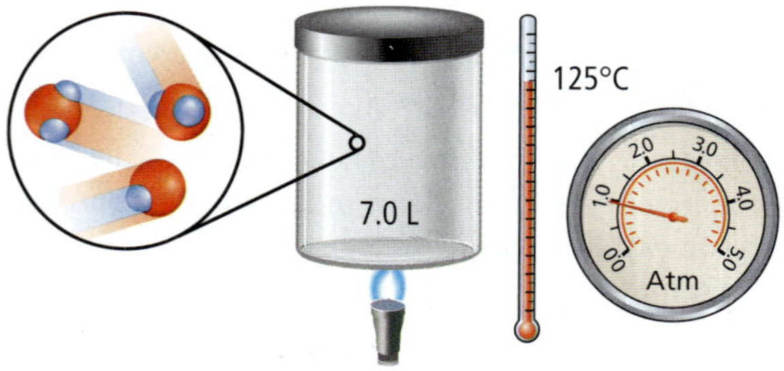


■ **Figure 17**

**92.** Calculate the amount of water vapor, in grams, contained in the vessel shown in **Figure 17.**

**93. Television** Determine the pressure inside a television picture tube with a volume of 3.50 L that contains $2.00 \times 10^{-5}$ g of nitrogen gas at 22.0°C.

**94.** Determine how many liters 8.80 g of carbon dioxide gas would occupy at:
**a.** STP
**b.** 160°C and 3.00 atm
**c.** 288 K and 118 kPa

**95. Oxygen Consumption** If 5.00 L of hydrogen gas, measured at a temperature of 20.0°C and a pressure of 80.1 kPa, is burned in excess oxygen to form water, what mass of oxygen will be consumed? Assume temperature and pressure remain constant.

**96.** A fixed amount of oxygen gas is held in a 1.00-L tank at a pressure of 3.50 atm. The tank is connected to an empty 2.00-L tank by a tube with a valve. After this valve has been opened and the oxygen is allowed to flow freely between the two tanks at a constant temperature, what is the final pressure in the system?

**97.** If 2.33 L of propane at 24°C and 67.2 kPa is completely burned in excess oxygen, how many moles of carbon dioxide will be produced?

**98. Respiration** A human breathes about 0.50 L of air during a normal breath. Assume the conditions are at STP.
**a.** What is the volume of one breath on a cold day atop Mt. Everest? Assume −60°C and 253 mm Hg pressure.
**b.** Air normally contains about 21% oxygen. If the $O_2$ content is about 14% atop Mt. Everest, what volume of air does a person need to breathe to supply the body with the same amount of oxygen?

## THINK CRITICALLY

**99. Apply** An oversized helium balloon in a floral shop must have a volume of at least 3.8 L to rise. When 0.1 mol is added to the empty balloon, its volume is 2.8 L. How many grams of He must be added to make it rise? Assume constant *T* and *P*.

**100. Calculate** A toy manufacturer uses tetrafluoroethane ($C_2H_2F_4$) at high temperatures to fill plastic molds for toys.
**a.** What is the density (in g/L) of $C_2H_2F_4$ at STP?
**b.** Find the molecules per liter of $C_2H_2F_4$ at 220°C and 1.0 atm.

**101. Analyze** A solid brick of dry ice ($CO_2$) weighs 0.75 kg. Once the brick has fully sublimated into $CO_2$ gas, what would its volume be at STP?

**102. Apply** Calculate the pressure of $4.67 \times 10^{22}$ molecules of CO gas mixed with $2.87 \times 10^{24}$ molecules of $N_2$ gas in a 6.00-L container at 34.8°C.

**103. Analyze** When nitroglycerin ($C_3H_5N_3O_9$) explodes, it decomposes into the following gases: $CO_2$, $N_2$, $O_2$, and $H_2O$. If 239 g of nitroglycerin explodes, what volume will the mixture of gaseous products occupy at 1.00 atm pressure and 2678°C?

**104. Make and Use Graphs** The data in **Table 3** show the volume of hydrogen gas collected at several different temperatures. Illustrate these data with a graph. Use the graph to complete the table. Determine the temperature at which the volume will reach a value of 0 mL. What is this temperature called?

**Table 3 Volume of $H_2$ Collected**

| Trial | *T* (°C) | *V* (mL) |
|---|---|---|
| 1 | 300 | 48 |
| 2 | 175 | 37 |
| 3 | 110 | |
| 4 | 0 | 22 |
| 5 | | 15 |
| 6 | −150 | 11 |

**105. Apply** What is the numerical value of the ideal gas constant (R) in $\frac{cm^3 \cdot Pa}{mol \cdot K}$?

**106. Infer** At very high pressures, will the ideal gas law calculate a pressure that is higher or lower than the actual pressure exerted by a sample of gas? How will the calculated pressure compare to the actual pressure at low temperatures? Explain your answers.

Chapter Self-Check

## CHALLENGE PROBLEM

**107. Baking** A baker uses baking soda as the leavening agent for his pumpkin-bread recipe. The baking soda decomposes according to two possible reactions.

$$2NaHCO_3(s) \rightarrow Na_2CO_3(s) + H_2O(l) + CO_2(g)$$

$$NaHCO_3(s) + H^+(aq) \rightarrow H_2O(l) + CO_2(g) + Na^+(aq)$$

Calculate the volume of $CO_2$ that forms per gram of $NaHCO_3$ by each reaction process. Assume the reactions take place at 210°C and 0.985 atm.

## CUMULATIVE REVIEW

**108.** Convert each mass measurement to its equivalent in kilograms.
- **a.** 247 g
- **b.** 53 mg
- **c.** 7.23 mg
- **d.** 975 mg

**109.** Write the electron configuration for each atom.
- **a.** iodine
- **b.** boron
- **c.** chromium
- **d.** krypton
- **e.** calcium
- **f.** cadmium

**110.** For each element, tell how many electrons are in each energy level and write the electron dot structure.
- **a.** Kr
- **b.** Sr
- **c.** P
- **d.** B
- **e.** Br
- **f.** Se

**111.** How many atoms of each element are present in five formula units of calcium permanganate?

**112.** You are given two clear, colorless aqueous solutions. One solution contains an ionic compound, and one contains a covalent compound. How could you determine which solution is the ionic solution and which solution is the covalent solution?

**113.** Write a balanced equation for the following reactions.
- **a.** Zinc displaces silver in silver chloride.
- **b.** Sodium hydroxide and sulfuric acid react to form sodium sulfate and water.

**114.** Terephthalic acid is an organic compound used in the formation of polyesters. It contains 57.8% C, 3.64% H, and 38.5% O. The molar mass is approximately 166 g/mol. What is the molecular formula of terephthalic acid?

**115.** The particles of which gas have the highest average speed? The lowest average speed?
- **a.** carbon monoxide at 90°C
- **b.** nitrogen trifluoride at 30°C
- **c.** methane at 90°C
- **d.** carbon monoxide at 30°C

## WRITING IN ▶ Chemistry

**116. Hot-Air Balloons** Many early balloonists dreamed of completing a trip around the world in a hot-air balloon, a goal not achieved until 1999. Write about what you imagine a trip in a balloon would be like, including a description of how manipulating air temperature would allow you to control altitude.

**117. Scuba** Investigate and explain the function of the regulators on the air tanks used by scuba divers.

## DBQ Document-Based Questions

**The Haber Process** *Ammonia ($NH_3$) is used in the production of fertilizer, refrigerants, dyes, and plastics. The Haber process is a method of producing ammonia through a reaction of molecular nitrogen and hydrogen. The equation for the reversible reaction is:*

$$N_2(g) + 3H_2(g) \rightleftharpoons 2NH_3(g) + 92 \text{ kJ}$$

**Figure 18** *shows the effect of temperature and pressure on the amount of ammonia produced by the Haber process.*

Data obtained from: Smith, M. 2004. *Science* 39:1021–1034.

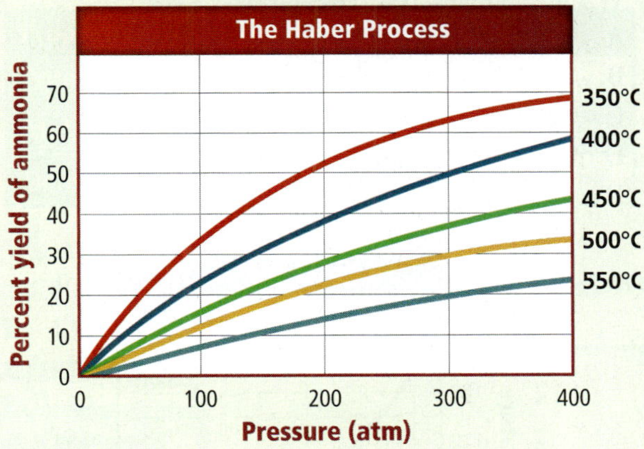


**Figure 18**

**118.** Explain how the percent yield of ammonia is affected by pressure and temperature.

**119.** The Haber process is typically run at 200 atm and 450°C, a combination proven to yield a substantial amount of ammonia in a short time.
- **a.** If the temperature goes from 450°C to 500°C, what pressure is necessary to maintain the typical percent yield?
- **b.** How do you think lowering the temperature of this reaction below 450°C would affect the amount of time required to produce ammonia?

## MULTIPLE CHOICE

*Use the graph below to answer Questions 1 and 2.*

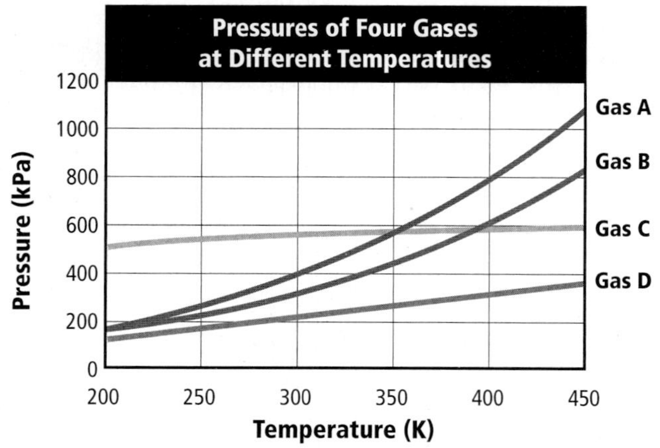


1. Which is evident in the graph above?
   - A. As temperature increases, pressure decreases.
   - B. As pressure increases, volume decreases.
   - C. As temperature increases, the number of moles decreases.
   - D. As pressure decreases, temperature decreases.

2. Which behaves as an ideal gas?
   - A. Gas A
   - B. Gas B
   - C. Gas C
   - D. Gas D

*Use the graph below to answer Question 3.*

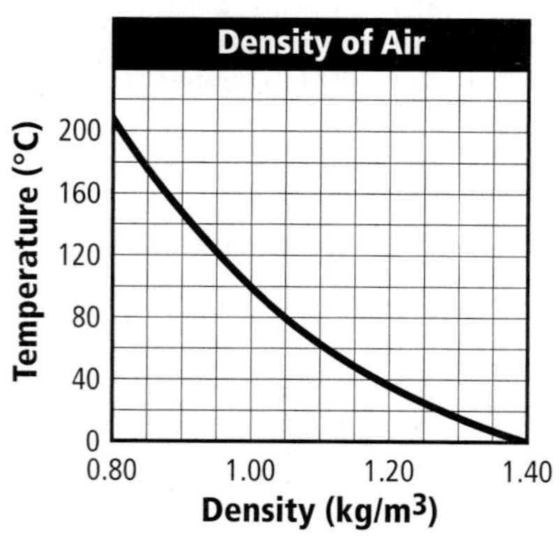


3. The graph shows data from an experiment that analyzed the relationship between temperature and air density. What is the independent variable in the experiment?
   - A. density
   - B. mass
   - C. temperature
   - D. time

4. Hydrofluoric acid (HF) is used in the manufacture of electronics equipment. It reacts with calcium silicate ($CaSiO_3$), a component of glass. What type of property prevents hydrofluoric acid from being transported or stored in glass containers?
   - A. chemical property
   - B. extensive physical property
   - C. intensive physical property
   - D. quantitative property

5. Sodium hydroxide (NaOH) is a strong base found in products used to clear clogged plumbing. What is the percent composition of sodium hydroxide?
   - A. 57.48% Na, 60.00% O, 2.52% H
   - B. 2.52% Na, 40.00% O, 57.48% H
   - C. 57.48% Na, 40.00% O, 2.52% H
   - D. 40.00% Na, 2.52% O, 57.48% H

*Use the circle graph below to answer Question 6.*

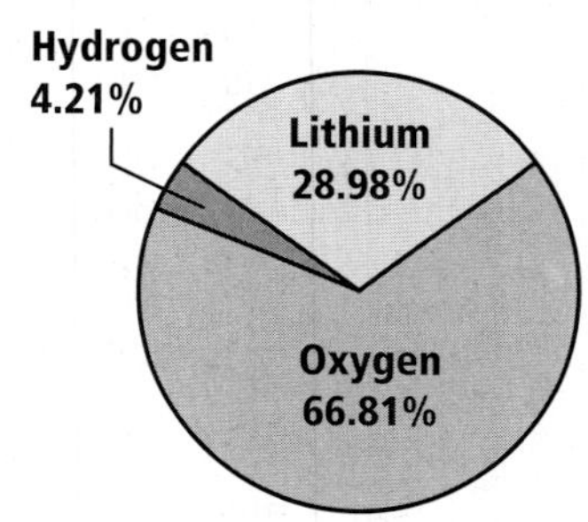


6. What is the empirical formula for this compound?
   - A. LiOH
   - B. $Li_2OH$
   - C. $Li_3OH$
   - D. $LiOH_2$

7. While it is on the ground, a blimp is filled with $5.66 \times 10^6$ L of He gas. The pressure inside the grounded blimp, where the temperature is 25°C, is 1.10 atm. Modern blimps are nonrigid, which means that their volumes can change. If the pressure inside the blimp remains the same, what will be the volume of the blimp at a height of 2300 m, where the temperature is 12°C?
   - A. $2.72 \times 10^6$ L
   - B. $5.40 \times 10^6$ L
   - C. $5.66 \times 10^6$ L
   - D. $5.92 \times 10^6$ L

## SHORT ANSWER

8. Describe several observations that provide evidence that a chemical change has occurred.

9. Identify five diatomic molecules that occur naturally, and explain why the atoms in these molecules share a pair, or pairs, of electrons.

10. The diagram below shows the Lewis structure for the polyatomic ion nitrate ($NO_3^-$). Define the term *polyatomic ion,* and give examples of other ions of this type.

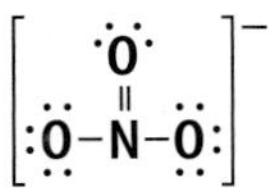

## EXTENDED RESPONSE

*Use the table below to answer Question 11.*

**Radon Levels August 2004 through July 2005**

| Date | Radon Level (pCi/L) | Date | Radon Level (pCi/L) |
|---|---|---|---|
| 8/04 | 0.26 | 2/05 | 0.087 |
| 9/04 | 0.052 | 3/05 | 0.087 |
| 10/04 | 0.087 | 4/05 | 0.10 |
| 11/04 | 0.052 | 5/05 | 0.22 |
| 12/04 | 0.069 | 6/05 | 0.087 |
| 1/05 | 0.035 | 7/05 | 0.16 |

11. Radon is a radioactive gas produced when radium in soil and rock decays. It is a known carcinogen. The data above show radon levels measured in a fictitious community. Select a method for graphing these data. Explain the reasons for your choice, and graph the data.

## SAT SUBJECT TEST: CHEMISTRY

12. Which diagram shows the relationship between volume and pressure for a gas at constant temperature?

A.
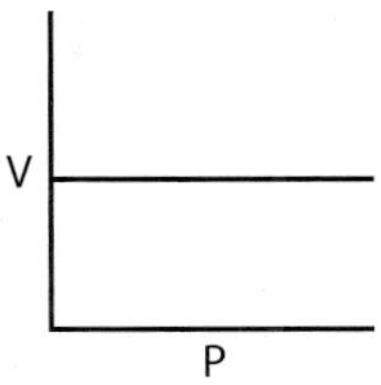


D.
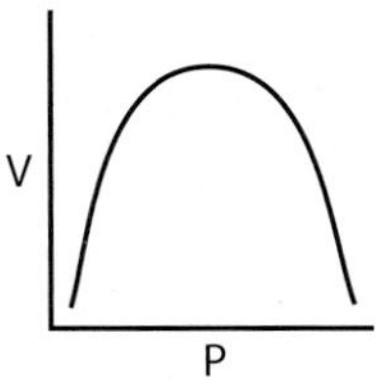


B.
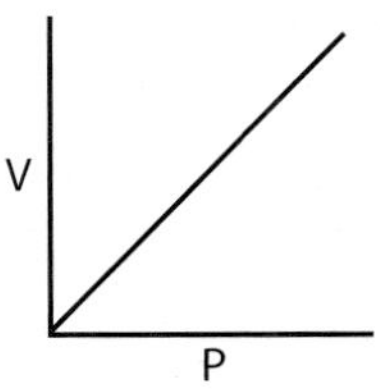


E.
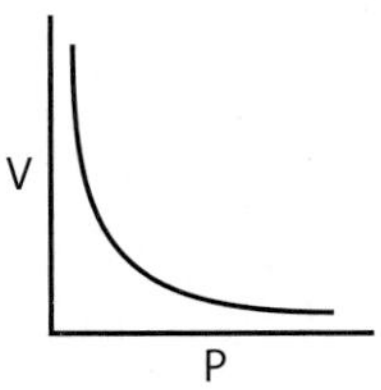


C.
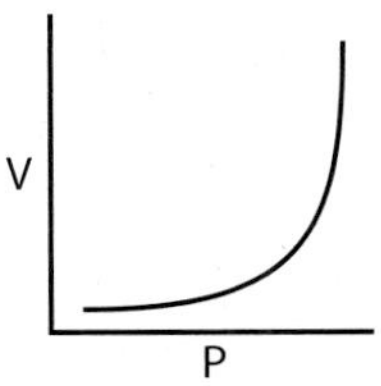


13. The reaction that provides blowtorches with their intense flame is the combustion of acetylene ($C_2H_2$) with oxygen to form carbon dioxide and water vapor. Assuming that the pressure and temperature of the reactants are the same, what volume of oxygen gas is required to completely burn 5.60 L of acetylene?
   A. 2.24 L
   B. 5.60 L
   C. 8.20 L
   D. 11.2 L
   E. 14.0 L

14. Assuming ideal behavior, how much pressure will 0.0468 g of ammonia ($NH_3$) gas exert on the walls of a 4.00-L container at 35.0°C?
   A. 0.0174 atm
   B. 0.296 atm
   C. 0.0126 atm
   D. 0.00198 atm
   E. 0.278 atm

**NEED EXTRA HELP?**

| If You Missed Question . . . | 1 | 2 | 3 | 4 | 5 | 6 | 7 | 8 | 9 | 10 | 11 | 12 | 13 | 14 |
|---|---|---|---|---|---|---|---|---|---|---|---|---|---|---|
| Review Section . . . | 13.1 | 13.2 | 1.3 | 3.1 | 10.4 | 10.4 | 13.1 | 3.2 | 8.1 | 7.3 | 2.4 | 13.1 | 13.1 | 13.1 |

CHAPTER 14

# Mixtures and Solutions

**BIGIDEA** Nearly all of the gases, liquids, and solids that make up our world are mixtures.

## SECTIONS

Chm 6, 6a, 6b, 6c, 10

1 Types of Mixtures

2 Solution Concentration

3 Factors Affecting Solvation

4 Colligative Properties of Solutions

## LaunchLAB

iLab Station

### How does energy change when solutions form?

When a solution forms, the forces of attraction between the solvent particles and the solute particles disrupt the intermolecular forces in the solute. In this lab, you will discover whether this interaction affects the energy of the solution.

**FOLDABLES® Study Organizer**

### Concentration

Make a bound book. Use it to help you organize information about the concentration of solutions.

Go **online!**
MHEonline.com

Two mixtures used in the construction of this building are steel and concrete. Iron is the major component of steel, but other elements, such as nickel, manganese, chromium, vanadium, and tungsten, can be added depending on the desired properties. Cement is used to make concrete and mortar to form building materials that are strong and can withstand everyday environmental stresses.

SECTION 1

# Types of Mixtures

**MAINIDEA** Mixtures can be either heterogeneous or homogeneous.

### Essential Questions

- How do the properties of suspensions, colloids, and solutions compare?
- What are the types of colloids and types of solutions?
- How are the electrostatic forces in colloids described?

### Review Vocabulary

**solute:** a substance dissolved in a solution

### New Vocabulary

suspension
colloid
Brownian motion
Tyndall effect
soluble
miscible
insoluble
immiscible

**CHEM 4 YOU** If you have ever filled a pail with ocean water, you might have noticed that some of the sediment settles to the bottom of the pail. However, the water will be salty no matter how long you let the pail sit. Why do some substances settle out but others do not?

## Heterogeneous Mixtures

Recall that a mixture is a combination of two or more pure substances in which each pure substance retains its individual chemical properties. Heterogeneous mixtures do not blend smoothly throughout, and the individual substances remain distinct. Two types of heterogeneous mixtures are suspensions and colloids.

**Suspensions** A **suspension** is a mixture containing particles that settle out if left undisturbed. The muddy water shown in **Figure 1** is a suspension. Pouring a liquid suspension through a filter will also separate out the suspended particles.

**Thixotropic mixtures** Some suspensions separate into a solidlike mixture on the bottom and water on the top. When the solidlike mixture is stirred or agitated, it flows like a liquid. Substances that behave in this way are said to be thixotropic. For example, toothpaste is thixotropic—it acts as a liquid when it is squeezed from the tube but as a solid when it sits on your brush. Some paints are thixotropic—you can stir them in the paint can yet they don't flow down the stirring stick or brush when you hold them up. Builders in earthquake zones must be aware that some clays are thixotropic. These clays form liquids in response to the agitation of an earthquake, which causes structures built on them to collapse.

■ **Figure 1** A suspension can be separated by allowing it to sit for a period of time. A liquid suspension can also be separated by pouring it through a filter.

**Table 1 Types of Colloids**

| Category | Example | Dispersed Particles | Dispersing Medium |
|---|---|---|---|
| Solid sol | colored gems | solid | solid |
| Sol | blood, gelatin | solid | liquid |
| Solid emulsion | butter, cheese | liquid | solid |
| Emulsion | milk, mayonnaise | liquid | liquid |
| Solid foam | marshmallow, soaps that float | gas | solid |
| Foam | whipped cream, beaten egg white | gas | liquid |
| Solid aerosol | smoke, dust in air | solid | gas |
| Liquid aerosol | spray deodorant, fog, clouds | liquid | gas |

**Colloids** Particles in a suspension are much larger than atoms and can settle out of solution. A heterogeneous mixture of intermediate-sized particles (between atomic-scale size of solution particles and the size of suspension particles) is a **colloid.** Colloid particles are between 1 nm and 1000 nm in diameter and do not settle out. Milk is a colloid. The components of homogenized milk cannot be separated by settling or by filtration.

The most abundant substance in the mixture is the dispersion medium. Colloids are categorized according to the phases of their dispersed particles and dispersing mediums. Milk is a colloidal emulsion because liquid particles are dispersed in a liquid medium. Other types of colloids are described in **Table 1.**

The dispersed particles in a colloid are prevented from settling out because they often have polar or charged atomic groups on their surfaces. These areas on their surfaces attract the positively or negatively charged areas of the dispersing-medium particles. This results in the formation of electrostatic layers around the particles, as shown in **Figure 2.** The layers repel each other when the dispersed particles collide; thus, the particles remain in the colloid.

If you interfere with the electrostatic layering, colloid particles will settle out of the mixture. For example, if you stir an electrolyte into a colloid, the dispersed particles clump together, destroying the colloid. Heating also destroys a colloid because it gives colliding particles enough kinetic energy to overcome the electrostatic forces and settle out.

**Brownian motion** The dispersed particles of liquid colloids make jerky, random movements. This erratic movement of colloid particles is called **Brownian motion.** It was first observed by, and later named for, the Scottish botanist Robert Brown (1773–1858), who noticed the random movements of pollen grains dispersed in water. Brownian motion results from collisions of particles of the dispersion medium with the dispersed particles. These collisions help to prevent the colloid particles from settling out of the mixture.

**READING CHECK Describe** two reasons why particles in a colloid do not settle out.

**Figure 2** The dispersing medium particles form charged layers around the colloid particles. These charged layers repel each other and keep the particles from settling out.

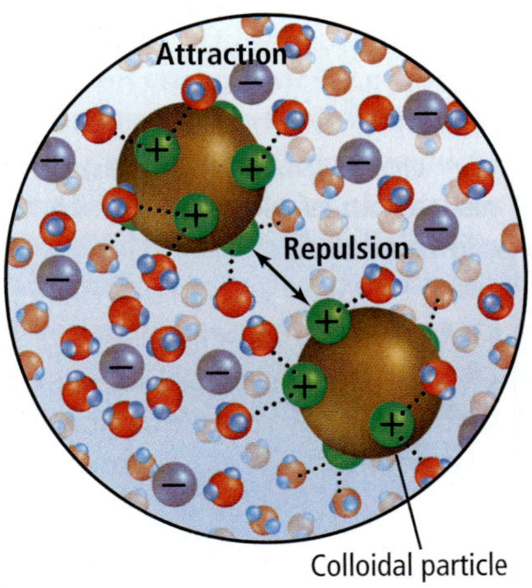

■ **Figure 3** Particles in a colloid scatter light, unlike particles in a solution. Called the Tyndall effect, the beam of light is visible in the colloid because of light scattering.

**Determine** *which mixture is the colloid.*

**Tyndall effect** Concentrated colloids are often cloudy or opaque. Dilute colloids sometimes appear as clear as solutions. Dilute colloids appear to be homogeneous solutions because their dispersed particles are so small. However, dispersed colloid particles scatter light, a phenomenon known as the **Tyndall effect.** In **Figure 3,** a beam of light is shone through two unknown mixtures. You can observe that dispersed colloid particles scatter the light, unlike particles in the solution. Suspensions also exhibit the Tyndall effect, but solutions never exhibit the Tyndall effect. You have observed the Tyndall effect if you have observed rays of sunlight passing through smoke-filled air, or viewed lights through fog. The Tyndall effect can be used to determine the amount of colloid particles in suspension.

## Homogeneous Mixtures

Cell solutions, ocean water, and steel might appear dissimilar, but they share certain characteristics. You learned earlier that solutions are homogeneous mixtures that contain two or more substances called the solute and the solvent. The solute is the substance that dissolves. The solvent is the dissolving medium. When you look at a solution, it is not possible to distinguish the solute from the solvent.

**Types of solutions** A solution might exist as a gas, a liquid, or a solid, depending on the state of its solvent, as shown in **Table 2.** Air is a gaseous solution, and its solvent is nitrogen gas. Braces that you wear on your teeth might be made of nitinol, a solid solution of titanium in nickel. Most solutions, however, are liquids. You read previously that reactions can take place in aqueous solutions, or solutions in which the solvent is water. Water is the most common solvent among liquid solutions.

## Data Analysis LAB

Based on Real Data*

### Design an Experiment

**How can you measure turbidity?** The National Primary Drinking Water Regulations set the standards for public water systems. Turbidity—a measure of the cloudiness of water that results from the suspension of solids in the water—is often associated with contamination from viruses, parasites, and bacteria. Most of these colloid particles come from erosion, industrial and human waste, algae blooms from fertilizers, and decaying organic matter.

#### Data and Observation

The Tyndall effect can be used to measure the turbidity of water. Your goal is to plan a procedure and develop a scale to interpret data.

#### Think Critically

1. **Identify** the variables that can be used to relate the ability of light to pass through the liquid and the number of the colloid particles present. What will you use as a control?
2. **Relate** the variables used in the experiment to the actual number of colloid particles that are present.
3. **Analyze** What safety precautions must be considered?
4. **Determine** the materials you need to measure the Tyndall effect. Select technology to collect or interpret data.

*Data obtained from U.S. Environmental Protection Agency. 2006. *The Office of Groundwater and Drinking Water.*

Explore **types of solutions with an interactive table.** Concepts In Motion

**Table 2** Types and Examples of Solutions

| Type of Solution | Example | Solvent | Solute |
|---|---|---|---|
| **Gas** | air | nitrogen (gas) | oxygen (gas) |
| **Liquid** | carbonated water | water (liquid) | carbon dioxide (gas) |
| | ocean water | water (liquid) | oxygen gas (gas) |
| | antifreeze | water (liquid) | ethylene glycol (liquid) |
| | vinegar | water (liquid) | acetic acid (liquid) |
| | ocean water | water (liquid) | sodium chloride (solid) |
| **Solid** | dental amalgam | silver (solid) | mercury (liquid) |
| | steel | iron (solid) | carbon (solid) |

Just as solutions can exist in different forms, the solutes in the solutions can be gases, liquids, or solids, also shown in **Table 2.** Solutions, such as ocean water, can contain more than one solute.

**Forming solutions** Some combinations of substances readily form solutions, and others do not. A substance that dissolves in a solvent is said to be **soluble** in that solvent. For example, sugar is soluble in water—a fact you might have learned by dissolving sugar in flavored water to make a sweetened beverage, such as tea or lemonade. Two liquids that are soluble in each other in any proportion, such as those that form the antifreeze listed in **Table 2,** are said to be **miscible.** A substance that does not dissolve in a solvent is said to be **insoluble** in that solvent. Sand is insoluble in water. The liquids in a bottle of oil and vinegar separate shortly after they are mixed. Oil is insoluble in vinegar. Two liquids that can be mixed together but separate shortly after are said to be **immiscible.**

## SECTION 1 REVIEW

Section Self-Check

### Section Summary

- The individual substances in a heterogeneous mixture remain distinct.
- Two types of heterogeneous mixtures are suspensions and colloids.
- Brownian motion is the erratic movement of colloid particles.
- Colloids exhibit the Tyndall effect.
- A solution can exist as a gas, a liquid, or a solid, depending on the solvent.
- Solutes in a solution can be gases, liquids, or solids.

1. MAINIDEA **Explain** Use the properties of seawater to describe the characteristics of mixtures.
2. **Distinguish** between suspensions and colloids.
3. **Identify** the various types of solutions. Describe the characteristics of each type of solution.
4. **Explain** Use the Tyndall effect to explain why it is more difficult to drive through fog using high beams than using low beams.
5. **Describe** different types of colloids.
6. **Explain** Why do dispersed colloid particles stay dispersed?
7. **Summarize** What causes Brownian motion?
8. **Compare and Contrast** Make a table that compares the properties of suspensions, colloids, and solutions.

SECTION 2

# Solution Concentration

**MAINIDEA** Concentration can be expressed in terms of percent or in terms of moles.

### Essential Questions

- How can the concentration be described using different units?
- How are the concentrations of solutions determined?
- What is the molarity of a solution and how can it be calculated?

### Review Vocabulary

**solvent:** the substance that dissolves a solute to form a solution

### New Vocabulary

concentration
molarity
molality
mole fraction

**CHEM 4 YOU** Have you ever tasted a glass of iced tea and found it too strong or too bitter? To adjust the taste, you could add water to dilute the tea, or you could add sugar to sweeten it. Either way, you are changing the concentration of the particles dissolved in the water.

## Expressing Concentration

The **concentration** of a solution is a measure of how much solute is dissolved in a specific amount of solvent or solution. Concentration can be described qualitatively using the words *concentrated* or *dilute.* Notice the containers of tea in **Figure 4.** One of the tea solutions is more concentrated than the other. In general, a concentrated solution contains a large amount of solute. The darker tea has more tea particles than the lighter tea. Conversely, a dilute solution contains a small amount of solute. The lighter tea in **Figure 4** is dilute and contains less tea particles than the darker tea.

Although qualitative descriptions of concentration can be useful, solutions are more often described quantitatively. Some commonly used quantitative descriptions are percent by mass, percent by volume, molarity, and molality. These descriptions express concentration as a ratio of measured amounts of solute and solvent or solution. **Table 3** lists each ratio's description.

Which quantitative description should be used? The description used depends on the type of solution analyzed and the reason for describing it. For example, a chemist working with a reaction in an aqueous solution most likely refers to the molarity of the solution because he or she needs to know the number of particles involved in the reaction.

**Figure 4** The strength of the tea corresponds to its concentration. The darker container of tea is more concentrated than the lighter container.

**Table 3 Concentration Ratios**

| Concentration Description | Ratio |
|---|---|
| Percent by mass | $\frac{\text{mass of solute}}{\text{mass of solution}} \times 100$ |
| Percent by volume | $\frac{\text{volume of solute}}{\text{volume of solution}} \times 100$ |
| Molarity | $\frac{\text{moles of solute}}{\text{liter of solution}}$ |
| Molality | $\frac{\text{moles of solute}}{\text{kilogram of solvent}}$ |
| Mole fraction | $\frac{\text{moles of solute}}{\text{moles of solute} + \text{moles of solvent}}$ |

**Percent by mass** The percent by mass is the ratio of the solute's mass to the solution's mass expressed as a percent. The mass of the solution equals the sum of the masses of the solute and the solvent.

**Percent by Mass**

$$\text{percent by mass} = \frac{\text{mass of solute}}{\text{mass of solution}} \times 100$$

Percent by mass equals the mass of the solute divided by the mass of the whole solution, multiplied by 100.

**FOLDABLES®**
Incorporate information from this section into your Foldable.

## EXAMPLE Problem 1

Find help with **percents.** Math Handbook

**CALCULATE PERCENT BY MASS** In order to maintain a sodium chloride (NaCl) concentration similar to ocean water, an aquarium must contain 3.6 g NaCl per 100.0 g of water. What is the percent by mass of NaCl in the solution?

### 1 ANALYZE THE PROBLEM

You are given the amount of sodium chloride dissolved in 100.0 g of water. The percent by mass of a solute is the ratio of the solute's mass to the solution's mass, which is the sum of the masses of the solute and the solvent.

**Known**

mass of solute = 3.6 g NaCl

mass of solvent = 100.0 g $H_2O$

**Unknown**

**percent by mass = ?**

### 2 SOLVE FOR THE UNKNOWN

Find the mass of the solution.

mass of solution = grams of solute + grams of solvent

mass of solution = 3.6 g + 100.0 g = 103.6 g — **Substitute mass of solute = 3.6 g, and mass of solvent = 100.0 g.**

Calculate the percent by mass.

$$\text{percent by mass} = \frac{\text{mass of solute}}{\text{mass of solution}} \times 100$$ **State the equation for percent by mass.**

$$\text{percent by mass} = \frac{3.6\ \text{g}}{103.6\ \text{g}} \times 100 = 3.5\%$$ **Substitute mass of solute = 3.6 g, and mass of solution = 103.6 g.**

### 3 EVALUATE THE ANSWER

Because only a small mass of sodium chloride is dissolved per 100.0 g of water, the percent by mass should be a small value, which it is. The mass of sodium chloride was given with two significant figures; therefore, the answer is also expressed with two significant figures.

EXAMPLE PROBLEM

## PRACTICE Problems

Do additional practice. Online Practice

**9.** What is the percent by mass of $NaHCO_3$ in a solution containing 20.0 g of $NaHCO_3$ dissolved in 600.0 mL of $H_2O$?

**10.** You have 1500.0 g of a bleach solution. The percent by mass of the solute sodium hypochlorite (NaOCl) is 3.62%. How many grams of NaOCl are in the solution?

**11.** In Question 10, how many grams of solvent are in the solution?

**12. Challenge** The percent by mass of calcium chloride in a solution is found to be 2.65%. If 50.0 g of calcium chloride is used, what is the mass of the solution?

■ **Figure 5** B20 is 20% by volume biodiesel and 80% by volume petroleum diesel. Biodiesel is an alternative fuel that can be produced from renewable resources, such as vegetable oil.

**Percent by volume** Percent by volume usually describes solutions in which both solute and solvent are liquids. The percent by volume is the ratio of the volume of the solute to the volume of the solution, expressed as a percent. The volume of the solution is the sum of the volumes of the solute and the solvent. Calculations of percent by volume are similar to those involving percent by mass.

**Percent by Volume**

$$\text{percent by volume} = \frac{\text{volume of solute}}{\text{volume of solution}} \times 100$$

Percent by volume equals the volume of solute divided by the volume of the solution, multiplied by 100.

Biodiesel, shown in **Figure 5,** is a clean-burning alternative fuel that is produced from renewable resources. Biodiesel can be used in diesel engines with little or no modifications. Biodiesel is simple to use, biodegradable, nontoxic, and it does not contain some of the pollutants found in regular gasoline. It does not contain petroleum, but it can be blended with petroleum diesel to create a biodiesel blend. B20 is 20% by volume biodiesel, 80% by volume petroleum diesel.

☑ READING CHECK **Compare** percent mass and percent volume.

**PRACTICE Problems** Do additional problems. Online Practice

13. What is the percent by volume of ethanol in a solution that contains 35 mL of ethanol dissolved in 155 mL of water?
14. What is the percent by volume of isopropyl alcohol in a solution that contains 24 mL of isopropyl alcohol in 1.1 L of water?
15. **Challenge** If 18 mL of methanol is used to make an aqueous solution that is 15% methanol by volume, how many milliliters of solution is produced?

**Molarity** Percent by volume and percent by mass are only two of the commonly used ways to quantitatively describe the concentrations of liquid solutions. One of the most common units of solution concentration is molarity. **Molarity** *(M)* is the number of moles of solute dissolved per liter of solution. Molarity is also known as molar concentration, and the unit *M* is read as molar. A liter of solution containing 1 mol of solute is a 1*M* solution, which is read as a one-molar solution. A liter of solution containing 0.1 mol of solute is a 0.1*M* solution. To calculate a solution's molarity, you must know the volume of the solution in liters and the amount of dissolved solute in moles.

**Molarity**

$$\text{molarity } (M) = \frac{\text{moles of solute}}{\text{liters of solution}}$$

The molarity of a solution equals the moles of solute divided by the liters of solution.

☑ READING CHECK **Determine** What is the molar concentration of a liter solution with 0.5 mol of solute?

## EXAMPLE Problem 2

**CALCULATING MOLARITY** A 100.5-mL intravenous (IV) solution contains 5.10 g of glucose ($C_6H_{12}O_6$). What is the molarity of this solution? The molar mass of glucose is 180.16 g/mol.

### 1 ANALYZE THE PROBLEM

You are given the mass of glucose dissolved in a volume of water. The molarity of the solution is the ratio of moles of solute per liter of solution.

**Known**

mass of solute = 5.10 g $C_6H_{12}O_6$

molar mass of $C_6H_{12}O_6$ = 180.16 g/mol

volume of solution = 100.5 mL

**Unknown**

**solution concentration = ? *M***

### 2 SOLVE FOR THE UNKNOWN

Calculate the number of moles of $C_6H_{12}O_6$.

$$(5.10\ \cancel{g\ C_6H_{12}O_6})\left(\frac{1\ mol\ C_6H_{12}O_6}{180.16\ \cancel{g\ C_6H_{12}O_6}}\right)$$

$$= 0.0283\ mol\ C_6H_{12}O_6$$

**Multiply grams of $C_6H_{12}O_6$ by the molar mass of $C_6H_{12}O_6$.**

Convert the volume of $H_2O$ to liters.

$$(100.5\ \cancel{mL}\ solution)\left(\frac{1\ L}{1000\ \cancel{mL}}\right) = 0.1005\ L\ solution$$

**Use the conversion factor 1 L/1000 mL.**

Solve for the molarity.

$$M = \frac{\text{moles of solute}}{\text{liters of solutions}}$$

**State the molarity equation.**

$$M = \left(\frac{0.0283\ mol\ C_6H_{12}O_6}{0.1005\ L\ solution}\right)$$

**Substitute moles of $C_6H_{12}O_6$ = 0.0283 and volume of solution = liters of solution = 0.1005 L.**

$$M = \left(\frac{0.0282\ mol\ C_6H_{12}O_6}{1\ L\ solution}\right) = 0.282M$$

**Divide numbers and units.**

### 3 EVALUATE THE ANSWER

The molarity value will be small because only a small mass of glucose was dissolved in the solution. The mass of glucose used in the problem has three significant figures; therefore, the value of the molarity also has three significant figures.

## PRACTICE Problems

Do additional problems. Online Practice

**16.** What is the molarity of an aqueous solution containing 40.0 g of glucose ($C_6H_{12}O_6$) in 1.5 L of solution?

**17.** Calculate the molarity of 1.60 L of a solution containing 1.55 g of dissolved KBr.

**18.** What is the molarity of a bleach solution containing 9.5 g of NaOCl per liter of bleach?

**19. Challenge** How much calcium hydroxide ($Ca(OH)_2$), in grams, is needed to produce 1.5 L of a 0.25*M* solution?

### CAREERS IN CHEMISTRY

**Pharmacy Technician** Most pharmacists rely on pharmacy technicians to prepare the proper medications to fill prescriptions. These technicians read patient charts and prescriptions in order to prepare the proper concentration, or dose, of medication that is to be administered to patients.

WebQuest

**Figure 6** Accurately preparing a solution of copper(II) sulfate involves several steps.

**Explain** ***why you cannot add 375 g of copper(II) sulfate directly to 1 L of water to make a 1.5M solution.***

Step 1: The mass of the solute is measured.

Step 2: The solute is placed in a volumetric flask of the correct volume.

Step 3: Distilled water is added to the flask to bring the solution level up to the calibration mark.

**Preparing molar solutions** Now that you know how to calculate the molarity of a solution, how do you think you would prepare 1 L of a 1.50*M* aqueous solution of copper(II) sulfate pentahydrate ($CuSO_4 \cdot 5H_2O$)? A 1.50*M* aqueous solution of $CuSO_4 \cdot 5H_2O$ contains 1.50 mol of $CuSO_4 \cdot 5H_2O$ dissolved in 1 L of solution. The molar mass of $CuSO_4 \cdot 5H_2O$ is about 249.70 g. Thus, 1.50 mol of $CuSO_4 \cdot 5H_2O$ has a mass of 375 g, an amount that you can measure on a balance.

$$\frac{1.50 \text{ mol } CuSO_4 \cdot 5H_2O}{1 \text{ L solution}} \times \frac{249.7 \text{ g } CuSO_4 \cdot 5H_2O}{1 \text{ mol } CuSO_4 \cdot 5H_2O} = \frac{375 \text{ g } CuSO_4 \cdot 5H_2O}{1 \text{ L solution}}$$

You cannot simply add 375 g of $CuSO_4 \cdot 5H_2O$ to 1 L of water to make the 1.50*M* solution. Like all substances, $CuSO_4 \cdot 5H_2O$ takes up space and will add volume to the solution. Therefore, you must use slightly less than 1 L of water to make 1 L of solution, as shown in **Figure 6.**

You will often do experiments that call for small quantities of solution. For example, you might need only 100 mL of a 1.50*M* $CuSO_4 \cdot 5H_2O$ solution for an experiment. Look again at the definition of molarity. As calculated above, a 1.50*M* solution of $CuSO_4 \cdot 5H_2O$ contains 1.50 mol of $CuSO_4 \cdot 5H_2O$ per 1 L of solution. Therefore, 1 L of solution contains 375 g of $CuSO_4 \cdot 5H_2O$.

This relationship can be used as a conversion factor to calculate how much solute you need for your experiment.

$$100 \text{ mL} \times \frac{1 \text{ L}}{1000 \text{ mL}} \times \frac{375 \text{ g } CuSO_4 \cdot 5H\ 2O}{1 \text{ L solution}} = 37.5 \text{ g } CuSO_4 \cdot 5H_2O$$

Thus, you would need to measure out 37.5 g of $CuSO_4 \cdot 5H_2O$ to make 100 mL of a 1.50*M* solution.

## PRACTICE Problems

Do additional problems. Online Practice

**20.** How many grams of $CaCl_2$ would be dissolved in 1.0 L of a 0.10*M* solution of $CaCl_2$?

**21.** How many grams of $CaCl_2$ should be dissolved in 500.0 mL of water to make a 0.20*M* solution of $CaCl_2$?

**22.** What mass of NaOH is in 250 mL of a 3.0*M* NaOH solution?

**23.** **Challenge** What volume of ethanol ($C_2H_5OH$) is in 100.0 mL of 0.15*M* solution? The density of ethanol is 0.7893 g/mL.

Matt Meadows

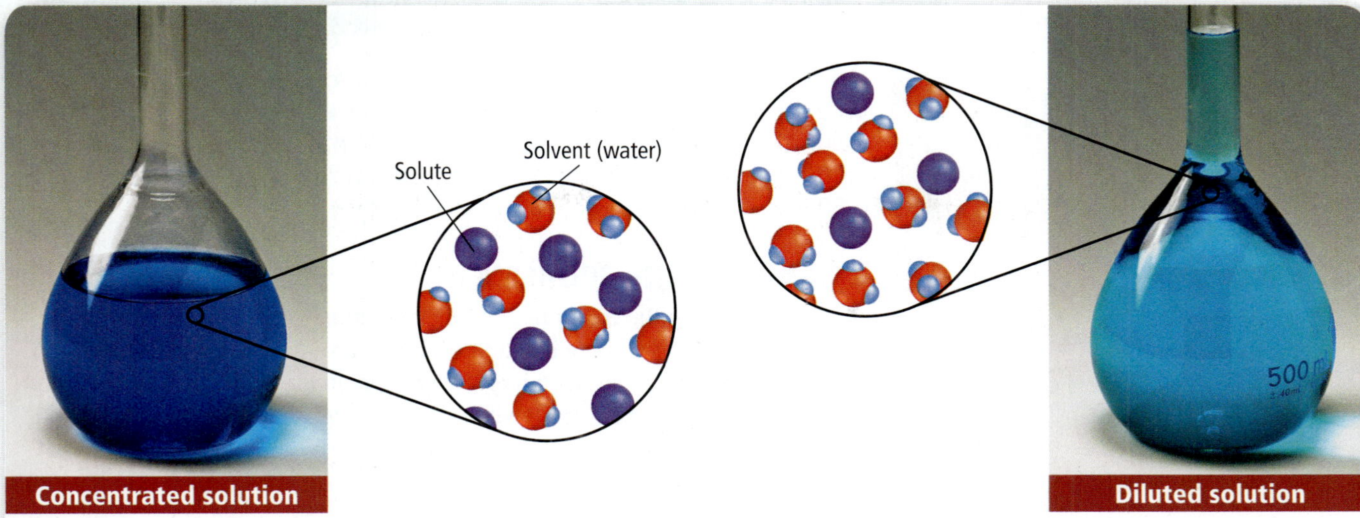


■ **Figure 7** A concentrated solution can be diluted by adding solvent. The number of moles of solute does not change when a concentrated solution is diluted.

**Diluting molar solutions** In the laboratory, you might use concentrated solutions of standard molarities, called stock solutions. For example, concentrated hydrochloric acid (HCl) is 12*M*. Recall that a concentrated solution has a large amount of solute. You can prepare a less-concentrated solution by diluting the stock solution with additional solvent. When you add solvent, you increase the number of solvent particles among which the solute particles move, as shown in **Figure 7,** thereby decreasing the solution's concentration.

How do you determine the volume of stock solution you must dilute? You can rearrange the expression of molarity to solve for moles of solute.

$$\text{molarity } (M) = \frac{\text{moles of solute}}{\text{liters of solution}}$$

$$\text{moles of solute} = \text{molarity} \times \text{liters of solution}$$

Because the total number of moles of solute does not change during dilution,

$$\text{moles of solute in the stock solution} = \text{moles of solute after dilution.}$$

Substituting moles of solute with molarity times liters of solution, the relationship can be expressed in the dilution equation.

**Dilution Equation**

$$M_1V_1 = M_2V_2$$

***M*** **represents molarity.**
***V*** **represents volume.**

For a given amount of solute, the product of the molarity and volume of the stock solution equals the product of the molarity and the volume of the dilute solution.

$M_1$ and $V_1$ represent the molarity and volume of the stock solution, and $M_2$ and $V_2$ represent the molarity and volume of the dilute solution. Before dilution, a concentrated solution contains a fairly high ratio of solute particles to solvent particles. After adding more solvent, the ratio of solute particles to solvent particles has decreased.

**VOCABULARY**
**ACADEMIC VOCABULARY**
**Concentrated**
less dilute or diffuse
*We added more water to the lemonade because it was too concentrated.*

## EXAMPLE Problem 3

**DILUTING STOCK SOLUTIONS** If you know the concentration and volume of the solution you want to prepare, you can calculate the volume of stock solution you will need. What volume, in milliliters, of 2.00*M* calcium chloride ($CaCl_2$) stock solution would you use to make 0.50 L of 0.300*M* calcium chloride solution?

### 1 ANALYZE THE PROBLEM

You are given the molarity of a stock solution of $CaCl_2$ and the volume and molarity of a dilute solution of $CaCl_2$. Use the relationship between molarities and volumes to find the volume, in liters, of the stock solution required. Then, convert the volume to milliliters.

| Known | Unknown |
|---|---|
| $M_1 = 2.00M\ CaCl_2$ | $V_1 = ?\ mL\ 2.00M\ CaCl_2$ |
| $M_2 = 0.300M$ | |
| $V_2 = 0.50\ L$ | |

### 2 SOLVE FOR THE UNKNOWN

Solve the molarity-volume relationship for the volume of the stock solution $V_1$.

$$M_1V_1 = M_2V_2$$ State the dilution equation.

$$V_1 = V_2\left(\frac{M_2}{M_1}\right)$$ Solve for $V_1$.

$$V_1 = (0.50\ L)\left(\frac{0.300M}{2.00M}\right)$$ Substitute $M_1 = 2.00M$, $M_2 = 0.300M$, and $V_2 = 0.50$ L.

$$V_1 = (0.50\ L)\left(\frac{0.300\cancel{M}}{2.00\cancel{M}}\right) = 0.075\ L$$ Multiply and divide numbers and units.

$$V_1 = (0.075\ L)\left(\frac{1000\ mL}{1\ L}\right) = 75\ mL$$ Convert to milliliters using the conversion factor 1000 mL/1 L.

To make the dilution, measure out 75 mL of the stock solution and dilute it with enough water to make the final volume 0.50 L.

### 3 EVALUATE THE ANSWER

The volume $V_1$ was calculated, and then its value was converted to milliliters. This volume should be less than the final volume of the dilute solution, and it is. Of the given information, $V_2$ had the fewest number of significant figures, with two. Thus, the volume $V_1$ should also have two significant figures, and it does.

## PRACTICE Problems

Do additional problems. Online Practice

**24.** What volume of a 3.00*M* KI stock solution would you use to make 0.300 L of a 1.25*M* KI solution?

**25.** How many milliliters of a 5.0*M* $H_2SO_4$ stock solution would you need to prepare 100.0 mL of 0.25*M* $H_2SO_4$?

**26. Challenge** If 0.50 L of 5.00*M* stock solution of HCl is diluted to make 2.0 L of solution, how much HCl, in grams, is in the solution?

**Molality** The volume of a solution changes with temperature as it expands or contracts. This change in volume alters the molarity of the solution. Masses, however, do not change with temperature. It is sometimes more useful to describe solutions in terms of how many moles of solute are dissolved in a specific mass of solvent. Such a description is called **molality**—the ratio of the number of moles of solute dissolved in 1 kg of solvent. The unit *m* is read as molal. A solution containing 1 mol of solute per kilogram of solvent is a one-molal solution.

Get help with **calculating molarity and molality.**

Personal Tutor

**Molality**

$$\text{molality } (m) = \frac{\text{moles of solute}}{\text{kg of solvent}}$$

The molality of a solution equals the moles of solute divided by kilograms of solvent.

## EXAMPLE Problem 4

**CALCULATING MOLALITY** In the lab, a student adds 4.5 g of sodium chloride (NaCl) to 100.0 g of water. Calculate the molality of the solution.

### 1 ANALYZE THE PROBLEM

You are given the mass of solute and solvent. Determine the number of moles of solute. Then, you can calculate the molality.

**Known**

mass of water ($H_2O$) = 100.0 g

mass of sodium chloride (NaCl) = 4.5 g

**Unknown**

$m$ = ? mol/kg

### 2 SOLVE FOR THE UNKNOWN

$$4.5 \text{ g NaCl} \times \frac{1 \text{ mol NaCl}}{58.44 \text{ g NaCl}} = 0.077 \text{ mol NaCl}$$

**Calculate the number of moles of solute.**

$$100.0 \text{ g } H_2O \times \frac{1 \text{ kg } H_2}{1000 \text{ g } H_2O} = 0.1000 \text{ kg } H_2O$$

**Convert the mass of $H_2O$ from grams to kilograms using the factor 1 kg/1000 g.**

Substitute the known values into the expression for molality, and solve.

$$m = \frac{\text{moles of solute}}{\text{kilograms of solvent}}$$

**Write the equation for molality.**

$$m = \frac{0.077 \text{ mol NaCl}}{0.1000 \text{ kg } H_2O} = 0.77 \text{ mol/kg}$$

**Substitute moles of solute = 0.077 mol NaCl, kilograms of solvent = 0.1000 kg $H_2O$.**

### 3 EVALUATE THE ANSWER

Because there was less than one-tenth of a mole of solute present in one-tenth of a kilogram of water, the molality should be less than one, and it is. The mass of sodium chloride was given with two significant figures; therefore, the molality is also expressed with two significant figures.

## PRACTICE Problems

Do additional problems. Online Practice

**27.** What is the molality of a solution containing 10.0 g of $Na_2SO_4$ dissolved in 1000.0 g of water?

**28. Challenge** How much ($Ba(OH)_2$), in grams, is needed to make a 1.00*m* aqueous solution?

**Hydrochloric Acid in Aqueous Solution**

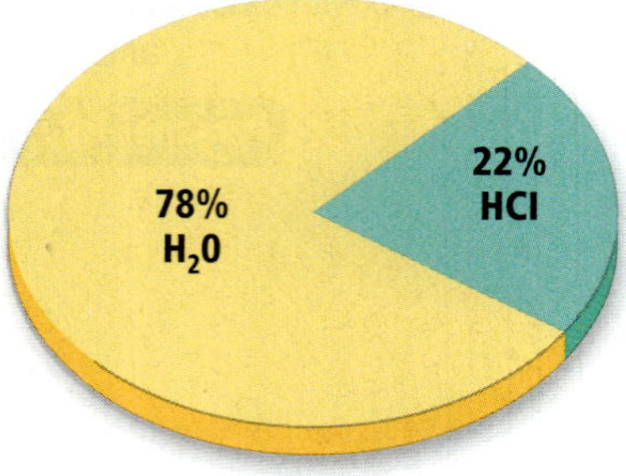


$$X_{HCl} + X_{H_2O} = 1.00$$
$$0.22 + 0.78 = 1.00$$

■ **Figure 8** The mole fraction expresses the number of moles of solute and solvent relative to the total number of moles of solution. Each mole fraction can be thought of as a percent. For example, the mole fraction of water ($X_{H_2O}$) is 0.78, which is equivalent to saying the solution contains 78% water (on a mole basis).

**Mole fraction** If you know the number of moles of solute and solvent, you can also express the concentration of a solution as a **mole fraction**—the ratio of the number of moles of solute or solvent in solution to the total number of moles of solute and solvent, as shown in **Figure 8.**

The symbol $X$ is commonly used for mole fraction, with a subscript to indicate the solvent or solute. The mole fraction for the solvent ($X_A$) and the mole fraction for the solute ($X_B$) can be expressed as follows.

### Mole Fraction

$$X_A = \frac{n_A}{n_A + n_B} \qquad X_B = \frac{n_B}{n_A + n_B}$$

**$X_A$ and $X_B$ represent the mole fractions of each substance.**
**$n_A$ and $n_B$ represent the number of moles of each substance.**

A mole fraction equals the number of moles of solute or solvent in a solution divided by the total number of moles of solute and solvent.

For example, suppose a hydrochloric acid solution contains 36 g of HCl and 64 g of $H_2O$. To convert these masses to moles, you would use the molar masses as conversion factors.

$$n_{HCl} = 36\ \cancel{g\ HCl} \times \frac{1\text{ mol HCl}}{36.5\ \cancel{g\ HCl}} = 0.99\text{ mol HCl}$$

$$n_{H_2O} = 64\ \cancel{g\ H_2O} \times \frac{1\text{ mol }H_2O}{18.0\ \cancel{g\ H_2O}} = 3.6\text{ mol }H_2O$$

The mole fractions of HCl and water can be expressed as follows.

$$X_{HCl} = \frac{n_{HCl}}{n_{HCl} + n_{H_2O}} = \frac{0.99\text{ mol HCl}}{0.99\text{ mol HCl} + 3.6\text{ mol }H_2O} = 0.22$$

$$X_{H_2O} = \frac{n_{H_2O}}{n_{HCl} + n_{H_2O}} = \frac{3.6\text{ mol }H_2O}{0.99\text{ mol HCl} + 3.6\text{ mol }H_2O} = 0.78$$

## PRACTICE Problems

**Do additional problems.** Online Practice

**29.** What is the mole fraction of NaOH in an aqueous solution that contains 22.8% NaOH by mass?

**30. Challenge** If the mole fraction of sulfuric acid ($H_2SO_4$) in an aqueous solution is 0.325, what is the percent by mass of $H_2SO_4$?

## SECTION 2 REVIEW

Section Self-Check

### Section Summary

- Concentrations can be measured qualitatively and quantitatively.
- Molarity is the number of moles of solute dissolved per liter of solution.
- Molality is the ratio of the number of moles of solute dissolved in 1 kg of solvent.
- The number of moles of solute does not change during a dilution.

**31.** MAINIDEA **Compare and contrast** five quantitative ways to describe the composition of solutions.

**32. Explain** the similarities and differences between a 1*M* solution of NaOH and a 1*m* solution of NaOH.

**33. Calculate** A can of chicken broth contains 450 mg of sodium chloride in 240.0 g of broth. What is the percent by mass of sodium chloride in the broth?

**34. Solve** How much ammonium chloride ($NH_4Cl$), in grams, is needed to produce 2.5 L of a 0.5*M* aqueous solution?

**35. Outline** the laboratory procedure for preparing a specific volume of a dilute solution from a concentrated stock solution.

SECTION 3

# Factors Affecting Solvation

**MAIN**IDEA Factors such as temperature, pressure, and polarity affect the formation of solutions.

### Essential Questions

- How do intermolecular forces affect solvation?
- What is solubility?
- Which factors affect solubility?

### Review Vocabulary

**exothermic:** a chemical reaction in which more energy is released than is required to break bonds in the initial reactants

### New Vocabulary

solvation
heat of solution
unsaturated solution
saturated solution
supersaturated solution
Henry's law

**CHEM 4 YOU** **If you have ever made microwavable soup from a dry mix, you added cold water to the dry mix and stirred. At first, only a small amount of the powdered mix dissolves in the cold water. After heating it in the microwave and stirring again, all of the powdered mix dissolves and you have soup.**

## The Solvation Process

Why are some substances soluble in each other, while others are not? To form a solution, solute particles must separate from one another and the solute and solvent particles must mix. Recall that attractive forces exist among the particles of all substances. Attractive forces exist between the pure solute particles, between the pure solvent particles, and between the solute and solvent particles. When a solid solute is placed in a solvent, the solvent particles completely surround the surface of the solid solute. If the attractive forces between the solvent and solute particles are greater than the attractive forces holding the solute particles together, the solvent particles pull the solute particles apart and surround them. These surrounded solute particles then move away from the solid solute and out into the solution.

The process of surrounding solute particles with solvent particles to form a solution is called **solvation,** as shown in **Figure 9.** Solvation in water is called hydration. "Like dissolves like" is the general rule used to determine whether solvation will occur in a specific solvent. To determine whether a solvent and solute are alike, you must examine the bonding and polarity of the particles and the intermolecular forces among particles.

■ **Figure 9** Salt begins to separate when it is dropped into water. The solute particles are pulled from the solid and surrounded by solvent particles.

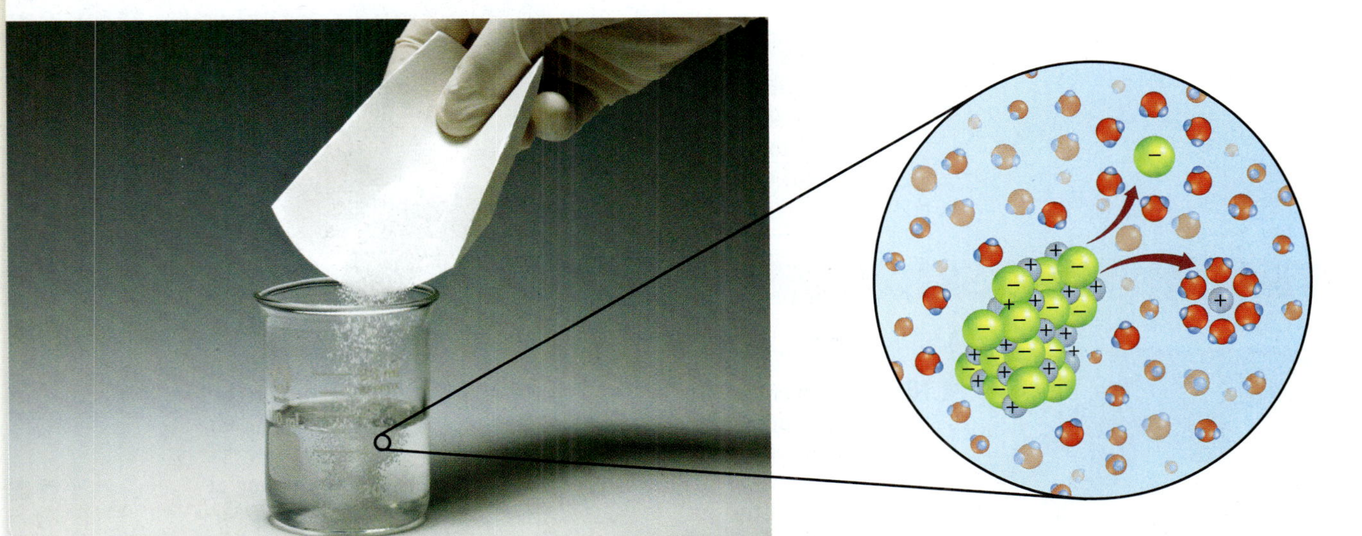

■ **Figure 10** Sodium chloride dissolves in water as the water molecules surround the sodium and chloride ions. Note how the polar water molecules orient themselves differently around the positive and negative ions.

View an **animation about the dissolution of compounds.**

**Concepts In Motion**

**Solvation Process of NaCl**

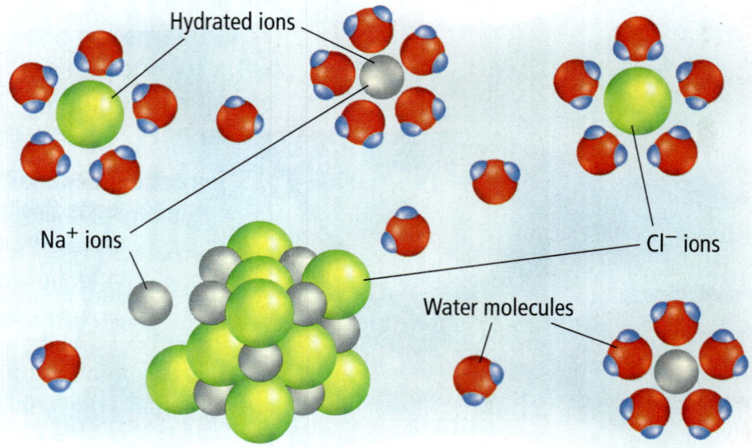


**Aqueous solutions of ionic compounds** Recall that water molecules are polar molecules and are in constant motion, as described by the kinetic-molecular theory. When a crystal of an ionic compound, such as sodium chloride (NaCl), is placed in water, the water molecules collide with the surface of the crystal. The charged ends of the water molecules attract the positive sodium ions and negative chloride ions. This attraction between the dipoles and the ions is greater than the attraction among the ions in the crystal, so the ions break away from the surface. The water molecules surround the ions, and the solvated ions move into the solution, shown in **Figure 10,** exposing more ions on the surface of the crystal. Solvation continues until the entire crystal has dissolved.

Not all ionic substances are solvated by water molecules. Gypsum is insoluble in water because the attractive forces between the ions in gypsum are so strong that they cannot be overcome by the attractive forces of the water molecules. As shown in **Figure 11,** the discoveries of specific solutions and mixtures, such as plaster made out of gypsum, have contributed to the development of many products and processes.

■ **Figure 11**

**Milestones in Solution Chemistry**

Scientists working with solutions have contributed to the development of products and processes in fields including medical technology, food preparation and preservation, and public health and safety.

1870 — 1890 — 1910

**1866** The invention of celluloid, a solution of camphor and cellulose, marks the beginning of the plastics industry.

**1883** The first successful centrifuge uses the force created by a high rate of spin to separate components of a mixture.

**1899** Newly patented technology reduces the size of fat globules dispersed in raw milk, preventing formation of a cream layer—a process called homogenization.

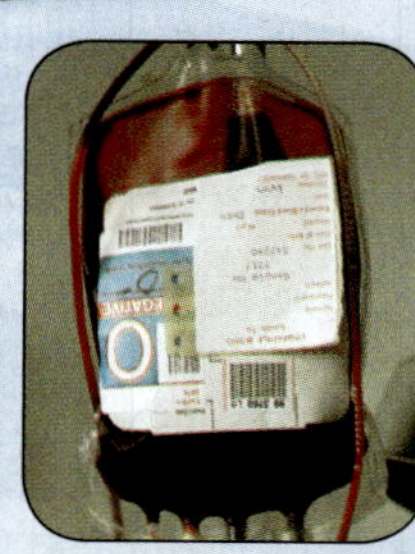

**1916** Doctors develop a glycerol solution that allows blood to be stored for up to several weeks after collection for use in transfusions.

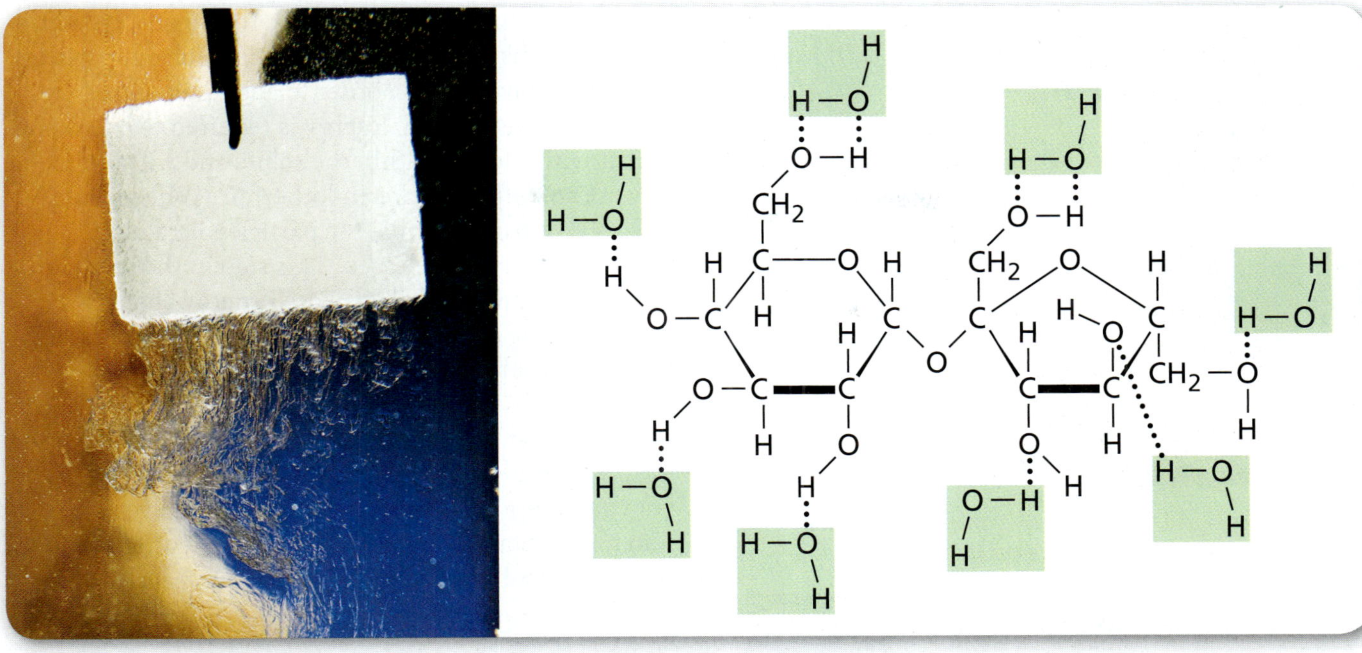

■ **Figure 12** Sucrose molecules contain eight O–H bonds and are polar. Polar water molecules form hydrogen bonds with the O–H bonds, which pulls the sucrose into solution.

**Aqueous solutions of molecular compounds** Water is also a good solvent for many molecular compounds. Table sugar is the molecular compound sucrose. As shown in **Figure 12,** sucrose molecules are polar and have several O–H bonds. As soon as the sugar crystals contact the water, water molecules collide with the outer surface of the crystal. Each O–H bond becomes a site for hydrogen bonding with water. The attractive forces among sucrose molecules are overcome by the attractive forces between polar water molecules and polar sucrose molecules. Sucrose molecules leave the crystal and become solvated by water molecules.

Oil is a substance made up primarily of carbon and hydrogen. It does not form a solution with water. There is little attraction between the polar water molecules and the nonpolar oil molecules. However, oil spills can be cleaned up with a nonpolar solvent because nonpolar solutes are more readily dissolved in nonpolar solvents.

Watch a **video about water and its solutions.**

Video

**1943** The first artificial kidney removes toxins dissolved in a patient's blood.

**1964** Stephanie Kwolek discovers a synthetic fiber, formed from liquid crystals in solution, that is stronger than steel and lighter than fiberglass.

**1980** A type of gypsum board is developed as a firewall system to separate townhome and condominium units.

**2003** Scientists develop chemical packets that remove toxic metals and pesticides and kill pathogens in drinking water. They can be distributed to survivors of natural disasters.

**2012** Initial work indicates the bacterium GFAJ-1 can use arsenic in place of phosphorus in biomolecules. However, further research shows that the bacterium, while arsenic tolerant, takes in phosphorus from solution to build these molecules.

1950 1970 1990 2010

A sugar cube in iced tea will dissolve slowly, but stirring will make the sugar cube dissolve more quickly.

Granulated sugar dissolves more quickly in iced tea than a sugar cube, and stirring will make the granulated sugar dissolve even more quickly.

Granulated sugar dissolves very quickly in hot tea.

■ **Figure 13** Agitation, surface area, and temperature affect the rate of solvation.

**Heat of solution** During the process of solvation, the solute must separate into particles. Solvent particles must also move apart in order to allow solute particles to come between them. Energy is required to overcome the attractive forces within the solute and within the solvent, so both steps are endothermic. When solute and solvent particles mix, the particles attract each other and energy is released. This step in the solvation process is exothermic. The overall energy change that occurs during the solution formation process is called the **heat of solution.**

As you observed in the Launch Lab at the beginning of this chapter, some solutions release energy as they form, whereas others absorb energy during formation. For example, after ammonium nitrate dissolves in water, its container feels cool. In contrast, after calcium chloride dissolves in water, its container feels warm.

☑ READING CHECK **Explain** why some solutions absorb energy during formation, while others release energy during formation.

## Factors That Affect Solvation

Solvation occurs only when the solute and solvent particles come in contact with each other. There are three common ways, shown in **Figure 13,** to increase the collisions between solute and solvent particles and thus increase the rate at which the solute dissolves: agitation, increasing the surface area of the solute, and increasing the temperature of the solvent.

**Agitation** Stirring or shaking—agitation of the mixture—moves dissolved solute particles away from the contact surfaces more quickly and thereby allows new collisions between solute and solvent particles to occur. Without agitation, solvated particles move away from the contact areas slowly.

**Surface area** Breaking the solute into small pieces increases its surface area. A greater surface area allows more collisions to occur. This is why a teaspoon of granulated sugar dissolves more quickly than an equal amount of sugar in cube form.

**Temperature** The rate of solvation is affected by temperature. For example, sugar dissolves more quickly in hot tea, shown in **Figure 13,** than it does in iced tea. Additionally, hotter solvents generally can dissolve more solid solute. Hot tea can hold more dissolved sugar than the iced tea. Most solids act in the same way as sugar—as temperature increases, the rate of solvation also increases. Solvation of other substances, such as gases, decreases at higher temperatures. For example, a carbonated soft drink will lose its fizz (carbon dioxide) faster at room temperature than when cold.

## Solubility

Just as solvation can be understood at the particle level, so can solubility. The solubility of a solute also depends on the nature of the solute and solvent. When a solute is added to a solvent, solvent particles collide with the solute's surface particles; solute particles begin to mix randomly among the solvent particles. At first, the solute particles are carried away from the crystal. However, as the number of solvated particles increases, the same random mixing results in increasingly frequent collisions between solvated solute particles and the remaining crystal. Some colliding solute particles rejoin the crystal, or crystallize, as illustrated in **Figure 14.** As solvation continues, the crystallization rate increases, while the solvation rate remains constant. As long as the solvation rate is greater than the crystallization rate, the net effect is continuing solvation.

Depending on the amount of solute present, the rates of solvation and crystallization might eventually equalize. No more solute appears to dissolve and a state of dynamic equilibrium exists between crystallization and solvation (as long as the temperature remains constant).

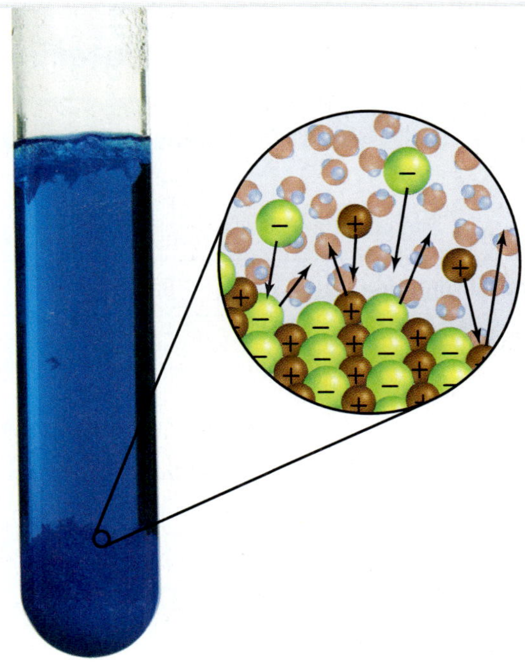

**Figure 14** In a saturated solution, the rate of solvation equals the rate of crystallization. The amount of dissolved solute does not change.

**Unsaturated solutions** An **unsaturated solution** is one that contains less dissolved solute for a given temperature and pressure than a saturated solution. In other words, more solute can be dissolved in an unsaturated solution.

**Alabama Applying Practices**

Go to the Resource tab in ConnectED to find the Applying Practices worksheet *Effect of Temperature on Solubility.*

**Saturated solutions** Although solute particles continue to dissolve and crystallize in solutions that reach equilibrium, the overall amount of dissolved solute in the solution remains constant. Such a solution, illustrated in **Figure 14,** is said to be a **saturated solution;** it contains the maximum amount of dissolved solute for a given amount of solvent at a specific temperature and pressure.

**Temperature and supersaturated solutions** Solubility is affected by raising the temperature of the solvent because the kinetic energy of its particles is increased, resulting in more-frequent collisions and collisions with greater energy than those that occur at lower temperatures. The fact that many substances are more soluble at high temperatures is demonstrated in **Figure 15.** For example, calcium chloride ($CaCl_2$) has a solubility of about 64 g $CaCl_2$ per 100 g $H_2O$ at 10°C. Increasing the temperature to approximately 27°C increases the solubility by almost 50%, to 100 g $CaCl_2$ per 100 g $H_2O$. For other substances, such as cerium sulfate, $Ce_2(SO_4)_3$, solubility initially decreases rapidly as temperature increases, but then levels off and remains constant.

**Figure 15** The solubilities of several substances as a function of temperature are shown in this graph.

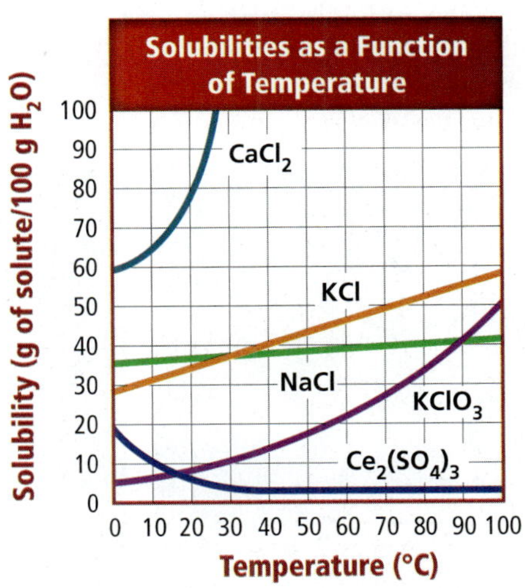


GRAPH CHECK

**Interpret** What is the solubility of NaCl at 80°C?

**Table 4** **Solubilities of Solutes in Water at Various Temperatures**

| Substance | Formula | Solubility (g/100 g $H_2O$)* | | | |
|---|---|---|---|---|---|
| | | 0°C | 20°C | 60°C | 100°C |
| Aluminum sulfate | $Al_2(SO_4)_3$ | 31.2 | 36.4 | 59.2 | 89.0 |
| Barium hydroxide | $Ba(OH)_2$ | 1.67 | 3.89 | 20.94 | – |
| Calcium hydroxide | $Ca(OH)_2$ | 0.189 | 0.173 | 0.121 | 0.076 |
| Lithium sulfate | $Li_2SO_4$ | 36.1 | 34.8 | 32.6 | – |
| Potassium chloride | KCl | 28.0 | 34.2 | 45.8 | 56.3 |
| Sodium chloride | NaCl | 35.7 | 35.9 | 37.1 | 39.2 |
| Silver nitrate | $AgNO_3$ | 122 | 216 | 440 | 733 |
| Sucrose | $C_{12}H_{22}O_{11}$ | 179.2 | 203.9 | 287.3 | 487.2 |
| Ammonia* | $NH_3$ | 1130 | 680 | 200 | – |
| Carbon dioxide* | $CO_2$ | 1.713 | 0.878 | 0.359 | – |
| Oxygen* | $O_2$ | 0.048 | 0.031 | 0.019 | – |

* L/1 L $H_2O$ of gas at standard pressure (101 kPa)

**VOCABULARY**

**WORD ORIGIN**

**Saturated**
comes from the Latin *saturatus* meaning *to fill*

The effect of temperature on solubility is also illustrated by the data in **Table 4.** Notice in **Table 4** that at 20°C, 203.9 g of sucrose ($C_{12}H_{22}O_{11}$) dissolves in 100 g of water. At 100°C, 487.2 g of sucrose dissolves in 100 g of water, a nearly 140% increase in solubility.

The fact that solubility changes with temperature and that some substances become more soluble with increasing temperature is the key to forming supersaturated solutions. A **supersaturated solution** contains more dissolved solute than a saturated solution at the same temperature. To make a supersaturated solution, a saturated solution is formed at a high temperature and then cooled slowly. The slow cooling allows the excess solute to remain dissolved in solution at the lower temperature.

**Figure 16** When a seed crystal is added to a supersaturated solution, the excess solute crystallizes out of the solution.

Stephen Frisch/McGraw-Hill Education

■ **Figure 17** Hot spring mineral deposits are an example of crystals that formed from supersaturated solutions.

Supersaturated solutions are unstable. If a tiny amount of solute, called a seed crystal, is added to a supersaturated solution, the excess solute precipitates quickly, as illustrated in **Figure 16.** Crystallization can also occur if the inside of the container is scratched or the supersaturated solution undergoes a physical shock, such as stirring or tapping the container. Using crystals of silver iodide (AgI) to seed air that is supersaturated with water vapor causes the water particles to come together and form droplets that might fall to Earth as rain. This technique is called cloud seeding. Rock candy and mineral deposits at the edges of mineral springs, such as those shown in **Figure 17,** are both formed from supersaturated solutions.

Explore **salts and solubility.**

Virtual Investigations

**Solubility of gases** In **Table 4** you can see that the gases oxygen and carbon dioxide are less soluble at higher temperatures than at lower temperatures. This is a predictable trend for all gaseous solutes in liquid solvents. Can you explain why? Recall that the kinetic energy of gas particles allows them to escape from a solution more readily at higher temperatures. Thus, as a solution's temperature increases, the solubility of a gaseous solute decreases.

**Pressure and Henry's law** Pressure affects the solubility of gaseous solutes in solutions. The solubility of a gas in any solvent increases as its external pressure (the pressure above the solution) increases. Carbonated beverages depend on this fact. Carbonated beverages contain carbon dioxide gas dissolved in an aqueous solution. In bottling or canning the beverage, carbon dioxide is dissolved in the solution at a pressure higher than atmospheric pressure. When the beverage container is opened, the pressure of the carbon dioxide gas in the space above the liquid decreases. As a result, bubbles of carbon dioxide gas form in the solution, rise to the top, and escape. Unless the container is sealed, the process will continue until the solution loses almost all of its carbon dioxide gas and goes flat. The decreased solubility of the carbon dioxide contained in the beverage after it is opened can be described by Henry's law.

**VOCABULARY**

**SCIENCE USAGE V. COMMON USAGE**

**Pressure**

***Science usage:*** the force exerted over an area

*As carbon dioxide escapes the solution, the pressure in the closed bottle increases.*

***Common usage:*** The burden of physical or mental stress

*There is a lot of pressure to do well on exams.*

**Henry's law** states that at a given temperature, the solubility ($S$) of a gas in a liquid is directly proportional to the pressure ($P$) of the gas above the liquid. When the bottle of soda is closed, as illustrated in **Figure 18,** the pressure above the solution keeps carbon dioxide from escaping the solution. You can express this relationship in the following way.

**Alabama Applying Practices**

Go to the Resource tab in ConnectED to find the Applying Practices worksheet *Investigate Solvation.*

### Henry's Law

$$\frac{S_1}{P_1} = \frac{S_2}{P_2}$$

**S represents solubility.**
**P represents pressure.**

At a given temperature, the quotient of solubility of a gas and its pressure is constant.

You will often use Henry's law to determine the solubility $S_2$ at a new pressure $P_2$, where $P_2$ is known. The basic rules of algebra can be used to solve Henry's law for any one specific variable. To solve for $S_2$, begin with the standard form of Henry's law.

$$\frac{S_1}{P_1} = \frac{S_2}{P_2}$$

Cross multiplying yields the following expression.

$$S_1P_2 = P_1S_2$$

Dividing both sides of the equation by $P_1$ yields the desired result—the equation solved for $S_2$.

$$\frac{S_1P_2}{P_1} = \frac{\cancel{P_1}S_2}{\cancel{P_1}} \qquad S_2 = \frac{S_1P_2}{P_1}$$

■ **Figure 18** Carbon dioxide ($CO_2$) is dissolved in soda. Some $CO_2$ also is found in the gas above the liquid.

**Explain** ***Why does the carbon dioxide escape from the solution when the cap is removed?***

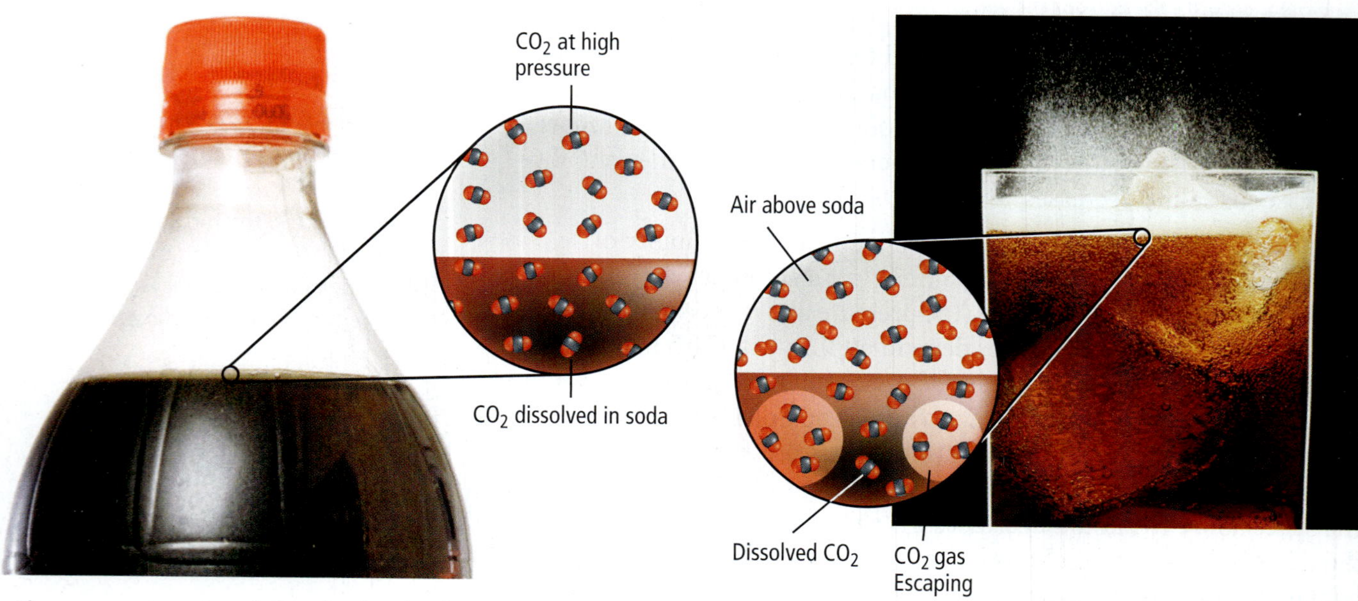


The pressure above the solution of a closed soda bottle keeps excess carbon dioxide from escaping the solution.

The pressure above the solution decreases when the cap is removed, which decreases the solubility of the carbon dioxide.

(l)Rachel Epstein/PhotoEdit; (r)©John Gillmoure/Corbis

## EXAMPLE Problem 5

EXAMPLE PROBLEM

**HENRY'S LAW** If 0.85 g of a gas at 4.0 atm of pressure dissolves in 1.0 L of water at 25°C, how much will dissolve in 1.0 L of water at 1.0 atm of pressure and the same temperature?

### 1 ANALYZE THE PROBLEM

You are given the solubility of a gas at an initial pressure. The temperature of the gas remains constant as the pressure changes. Because decreasing pressure reduces a gas's solubility, less gas should dissolve at the lower pressure.

| Known | Unknown |
|---|---|
| $S_1 = 0.85$ g/L | $S_2 = ?$ g/L |
| $P_1 = 4.0$ atm | |
| $P_2 = 1.0$ atm | |

### 2 SOLVE FOR THE UNKNOWN

$$\frac{S_1}{P_1} = \frac{S_2}{P_2}$$ **State Henry's law.**

$$S_2 = S_1\left(\frac{P_2}{P_1}\right)$$ **Solve Henry's law for $S_2$.**

$$S_2 = \left(\frac{0.85\text{ g}}{1.0\text{ L}}\right)\left(\frac{1.0\text{ atm}}{4.0\text{ atm}}\right) = 0.21\text{ g/L}$$ **Substitute $S_1 = 0.85$ g/L, $P_1 = 4.0$ atm, and $P_2 = 1.0$ atm. Multiply and divide numbers and units.**

### 3 EVALUATE THE ANSWER

The solubility decreased as expected. The pressure on the solution was reduced from 4.0 atm to 1.0 atm, so the solubility should be reduced to one-fourth its original value, which it is. The unit g/L is a solubility unit, and there are two significant figures.

## PRACTICE Problems

**Do additional problems.** Online Practice

**36.** If 0.55 g of a gas dissolves in 1.0 L of water at 20.0 kPa of pressure, how much will dissolve at 110.0 kPa of pressure?

**37.** A gas has a solubility of 0.66 g/L at 10.0 atm of pressure. What is the pressure on a 1.0-L sample that contains 1.5 g of gas?

**38. Challenge** The solubility of a gas at 7.0 atm of pressure is 0.52 g/L. How many grams of the gas would be dissolved per 1.0 L if the pressure increased 40.0 percent?

## SECTION 3 REVIEW

Section Self-Check

### Section Summary

- The process of solvation involves solute particles surrounded by solvent particles.
- Solutions can be unsaturated, saturated, or supersaturated.
- Henry's law states that at a given temperature, the solubility (*S*) of a gas in a liquid is directly proportional to the pressure (*P*) of the gas above the liquid.

**39.** MAINIDEA **Describe** factors that affect the formation of solutions.

**40. Define** *solubility.*

**41. Describe** how intermolecular forces affect solvation.

**42. Explain** on a particle basis why the vapor pressure of a solution is lower than that of the pure solvent.

**43. Summarize** If a seed crystal is added to a supersaturated solution, how would you characterize the resulting solution?

**44. Make and Use Graphs** Use the information in **Table 4** to graph the solubilities of aluminum sulfate, lithium sulfate, and potassium chloride at 0°C, 20°C, 60°C, and 100°C. Which substance's solubility is most affected by increasing temperature?

SECTION 4

# Colligative Properties of Solutions

**MAINIDEA** Colligative properties depend on the number of solute particles in a solution.

### Essential Questions

- What are colligative properties?
- What are four colligative properties of solutions?
- How are the boiling point elevation and freezing point depression of a solution determined?

### Review Vocabulary

**ion:** an atom that is electrically charged

### New Vocabulary

colligative property
vapor pressure lowering
boiling point elevation
freezing point depression
osmosis
osmotic pressure

**Alabama Applying Practices**

Go to the Resource tab in ConnectED to find the Applying Practices worksheet *Investigate Conductivity of Solutions.*

**CHEM 4 YOU** If you live in an area that experiences cold winters, you have probably noticed people spreading salt to melt icy sidewalks and roads. How does salt help make a winter's drive safer?

## Electrolytes and Colligative Properties

Solutes affect some of the physical properties of their solvents. Early researchers were puzzled to discover that the effects of a solute on a solvent depended only on how many solute particles were in the solution, not on the specific solute dissolved. Physical properties of solutions that are affected by the number of particles but not by the identity of dissolved solute particles are called **colligative properties.** The word *colligative* means *depending on the collection.* Colligative properties include vapor pressure lowering, boiling point elevation, freezing point depression, and osmotic pressure.

**Electrolytes in aqueous solution** You read previously that ionic compounds are called electrolytes because they dissociate in water to form a solution that conducts electric current, as shown in **Figure 19.** Some molecular compounds ionize in water and are also electrolytes. Electrolytes that produce many ions in a solution are called strong electrolytes; those that produce only a few ions in a solution are called weak electrolytes.

■ **Figure 19** Sodium chloride conducts electricity well because it is an electrolyte. Sucrose does not conduct electricity because it is not an electrolyte.

View an **animation about strong, weak, and nonelectrolytes.**

Concepts In Motion

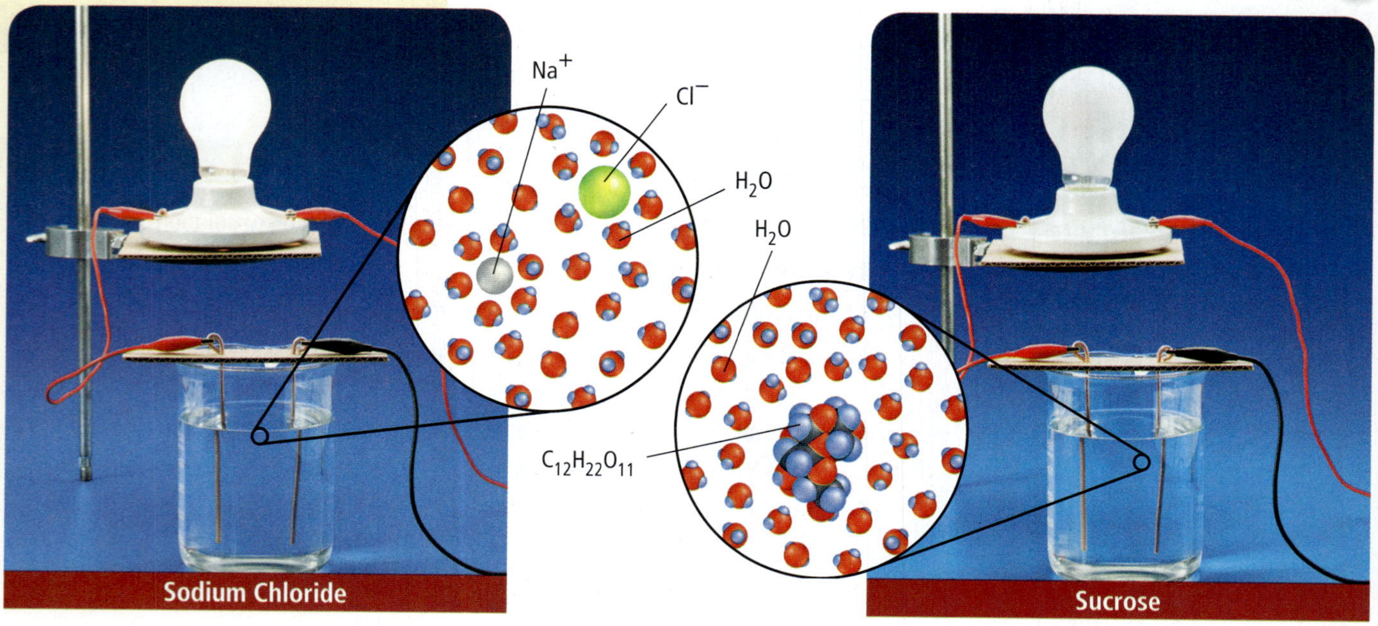

Sodium chloride is a strong electrolyte. It dissociates in solution, producing $Na^+$ and $Cl^-$ ions.

$$NaCl(s) \rightarrow Na^+(aq) + Cl^-(aq)$$

Dissolving 1 mol of NaCl in 1 kg of water would not yield a 1*m* solution of ions. Rather, there would be 2 mol of solute particles in solution—1 mol each of $Na^+$ and $Cl^-$ ions.

**Nonelectrolytes in aqueous solution** Many molecular compounds dissolve in solvents but do not ionize. Such solutions do not conduct an electric current, as shown in **Figure 19,** and the solutes are called nonelectrolytes. Sucrose is an example of a nonelectrolyte. A 1*m* sucrose solution contains only 1 mol of sucrose particles.

☑ READING CHECK **Infer** Which compound would have the greater effect on colligative properties, sodium chloride or sucrose?

## Vapor Pressure Lowering

You learned previously that vapor pressure is the pressure exerted in a closed container by particles that have escaped the liquid's surface and entered the gaseous state. In a closed container at constant temperature and pressure, the solvent particles of a pure solvent reach a state of dynamic equilibrium, escaping and reentering the liquid state at the same rate.

Experiments show that adding a nonvolatile solute (one that has little tendency to become a gas) to a solvent lowers the solvent's vapor pressure. The particles that produce vapor pressure escape the liquid phase at its surface. When a solvent is pure, as shown in **Figure 20,** its particles occupy the entire surface area. When the solvent contains solute, as also shown in **Figure 20,** a mix of solute and solvent particles occupies the surface area. With fewer solvent particles at the surface, fewer particles enter the gaseous state, and the vapor pressure is lowered. The greater the number of solute particles in a solvent, the lower the resulting vapor pressure. Thus, **vapor pressure lowering** is due to the number of solute particles in solution and is a colligative property of solutions.

You can predict the relative effect of a solute on vapor pressure based on whether the solute is an electrolyte or a nonelectrolyte. For example, 1 mol each of the solvated nonelectrolytes glucose, sucrose, and ethanol molecules has the same relative effect on the vapor pressure. However, 1 mol each of the solvated electrolytes sodium chloride (NaCl), sodium sulfate ($Na_2SO_4$), and aluminum chloride ($AlCl_3$) has an increasingly greater effect on vapor pressure because of the increasing number of ions each produces in solution.

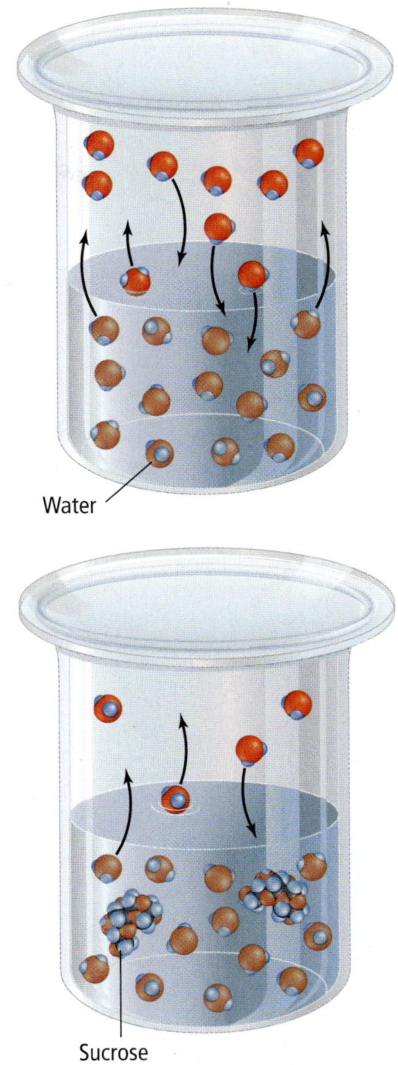


■ **Figure 20** The vapor pressure of a pure solvent is greater than the vapor pressure of a nonvolatile solution.

## Boiling Point Elevation

Because a nonvolatile solute lowers a solvent's vapor pressure, it also affects the boiling point of the solvent. Recall that liquid in a pot on a stove boils when its vapor pressure equals the atmospheric pressure. When the temperature of a solution containing a nonvolatile solute is raised to the boiling point of the pure solvent, the resulting vapor pressure is still less than the atmospheric pressure and the solution will not boil. Thus, the solution must be heated to a higher temperature to supply the additional kinetic energy needed to raise the vapor pressure to atmospheric pressure. The temperature difference between a solution's boiling point and a pure solvent's boiling point is called the **boiling point elevation.**

For nonelectrolytes, the value of the boiling point elevation, which is symbolized $\Delta T_b$, is directly proportional to the solution's molality.

### Boiling Point Elevation

$$\Delta T_b = K_b m$$

**$\Delta T_b$ represents the boiling point elevation.**
**$K_b$ represents the molal boiling elevation constant.**
**$m$ represents molality.**

The temperature difference is equal to the molal boiling point elevation constant multiplied by the solution's molality.

The molal boiling point elevation constant, $K_b$, is the difference in boiling points between a $1m$ nonvolatile, nonelectrolyte solution and a pure solvent. Boiling point elevation is expressed in units of °C/$m$ and varies for different solvents. Values of $K_b$ for several common solvents are found in **Table 5.** Note that water's $K_b$ value is 0.512°C/$m$. This means that a $1m$ aqueous solution containing a nonvolatile, nonelectrolyte solute boils at 100.512°C—a temperature just 0.512°C higher than pure water's boiling point of 100.0°C.

Explore **colligative properties.**

Virtual Investigations

Like vapor pressure lowering, boiling point elevation is a colligative property. The value of the boiling point elevation is directly proportional to the solution's solute molality; that is, the greater the number of solute particles in the solution, the greater the boiling point elevation. Because it is related to mole fraction, which involves the number of solute particles, molality is used as the concentration. Molality also uses mass of solvent rather than volume, and therefore is not affected by temperature changes. Examine **Figure 21** and notice that the curve for a solution lies below the curve for the pure solvent at any temperature.

**Table 5** Molal Boiling Point Elevation Constants ($K_b$)

| Solvent | Boiling Point (°C) | $K_b$ (°C/$m$) |
|---|---|---|
| Water | 100.0 | 0.512 |
| Benzene | 80.1 | 2.53 |
| Carbon tetrachloride | 76.7 | 5.03 |
| Ethanol | 78.5 | 1.22 |
| Chloroform | 61.7 | 3.63 |

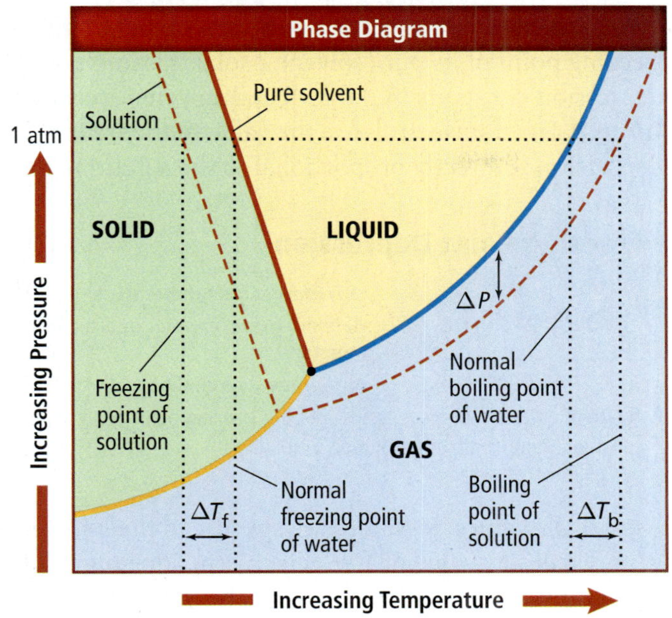


■ **Figure 21** Temperature and pressure affect solid, liquid, and gas phases of a pure solvent (solid lines) and a solution (dashed line).

**GRAPH CHECK**

**Describe** how the difference between the solid lines and dashed line corresponds to vapor pressure lowering, boiling point elevation, and freezing point depression. Use specific data from the graph to support your answer.

## Freezing Point Depression

At a solvent's freezing point temperature, the particles no longer have sufficient kinetic energy to overcome the interparticle attractive forces; the particles form into a more organized structure in the solid state. In a solution, the solute particles interfere with the attractive forces among the solvent particles. This prevents the solvent from entering the solid state at its normal freezing point.

The freezing point of a solution is always lower than that of a pure solvent. **Figure 21** shows the differences in boiling and melting points of pure water and an aqueous solution. By comparing the solid and dashed lines, you can see that the temperature range over which the aqueous solution exists as a liquid is greater than that of pure water. Two common applications of freezing point depression, shown in **Figure 22,** use salt to lower the freezing point of a water solution.

■ **Figure 22** By adding salts to the ice on a road, the freezing point of the ice is lowered, which results in the ice melting. Adding salt to ice when making ice cream lowers the freezing point of the ice, allowing the resulting water to freeze the ice cream.

**Table 6 Molal Freezing Point Depression Constants ($K_f$)**

| Solvent | Freezing Point (°C) | $K_f$ (°C/$m$) |
|---|---|---|
| Water | 0.0 | 1.86 |
| Benzene | 5.5 | 5.12 |
| Carbon tetrachloride | −23.0 | 29.8 |
| Ethanol | −114.1 | 1.99 |
| Chloroform | −63.5 | 4.68 |

A solution's **freezing point depression,** $\Delta T_f$, is the difference in temperature between its freezing point and the freezing point of its pure solvent. Molal freezing point depression constants ($K_f$) for several solvents are shown in **Table 6.** For nonelectrolytes, the value of the freezing point depression is directly proportional to the solution's molality.

**Freezing Point Depression**

$$\Delta T_f = K_f m$$

**$\Delta T_f$ represents temperature.**
**$K_f$ is the freezing point depression constant.**
**$m$ represents molality.**

The temperature difference is equal to the freezing point depression constant multiplied by the solution's molality.

As with $K_b$ values, $K_f$ values are specific to their solvents. With water's $K_f$ value of 1.86°C/$m$, a 1$m$ aqueous solution containing a nonvolatile, nonelectrolye solute freezes at −1.86°C rather than at pure water's freezing point of 0.0°C. Glycerol is a nonelectrolyte solute produced by many fish and insects to keep their blood from freezing during cold winters. Antifreeze and many de-icer solutions contain the nonelectrolyte solute ethylene glycol.

Notice that the equations for boiling point elevation and freezing point depression specify the molality of a nonelectrolyte. For electrolytes, you must make sure to use the effective molality of the solution. Example Problem 6 illustrates this point.

# MiniLAB

iLab Station

## Examine Freezing Point Depression

**How do you measure freezing point depression?**

### Procedure

1. Read and complete the lab safety form.
2. Fill two **400-mL beakers** with **crushed ice.** Add 50 mL of **cold tap water** to each beaker.
3. Measure the temperature of each beaker using a **nonmercury thermometer.**
4. Stir the contents of each beaker with a **stirring rod** until both beakers are at a constant temperature—approximately 1 min. Record the temperature.
5. Add 75 g of **rock salt (NaCl)** to one of the beakers. Continue stirring both beakers. Some of the salt will dissolve.
6. When the temperature in each beaker is constant, record the final readings.
7. To clean up, flush the contents of each beaker down the drain with excess water.

### Analysis

1. **Compare** your readings taken for the ice water and the salt water. How do you explain the observed temperature change?
2. **Explain** why salt was added to only one of the beakers.
3. **Explain** Salt is a strong electrolyte that produces two ions, $Na^+$ and $Cl^-$, when it dissociates in water. Explain why this is important to consider when calculating the colligative property of freezing point depression.
4. **Predict** whether it would be better to use coarse rock salt or fine table salt when making homemade ice cream. Explain.

## EXAMPLE Problem 6

Find help with **algebraic equations.** Math Handbook

**CHANGES IN BOILING AND FREEZING POINTS** Sodium chloride (NaCl) is often used to prevent icy roads and to freeze ice cream. What are the boiling and freezing points of a 0.029*m* aqueous solution of sodium chloride?

### 1 ANALYZE THE PROBLEM

You are given the molality of an aqueous sodium chloride solution. First, calculate $\Delta T_b$ and $\Delta T_f$ based on the number of particles in the solution. Then, to determine the elevated boiling point and the depressed freezing point, add $\Delta T_b$ to the normal boiling point and subtract $\Delta T_f$ from the normal freezing point of water.

| Known | Unknown |
|---|---|
| | **boiling point = ?°C** |
| molality of solution = 0.029*m* | **freezing point = ?°C** |
| $K_b$ = 0.512°C/m | |
| $K_f$ = 1.86°C/m | |

### 2 SOLVE FOR THE UNKNOWN

Determine the molality of the particles.

particle molality = 2 × 0.029*m* = 0.058*m*

Determine $\Delta T_b$ and $\Delta T_f$

$\Delta T_b = K_b m$

$\Delta T_f = K_f m$

**State the boiling point elevation and freezing point depression formulas.**

$\Delta T_b = (0.512°C/\cancel{m})(0.058\cancel{m}) = 0.030°C$

$\Delta T_f = (1.86°C/\cancel{m})(0.058\cancel{m}) = 0.11°C$

**Substitute $K_b$ = 0.512°C/*m*, $K_f$ = 1.86°C/*m*, and *m* = 0.058*m*.**

Determine the elevated boiling point and depressed freezing point of the solution.

**boiling point** = 100.000°C + 0.030°C
= **100.030°C**

**freezing point** = 0.00°C − 0.11°C
= **−0.11°C**

**Add $\Delta T_b$ to the normal boiling point and subtract $\Delta T_f$ from the normal freezing point.**

### 3 EVALUATE THE ANSWER

The boiling point is higher and the freezing point is lower, as expected. Because the molality of the solution has two significant figures, both $\Delta T_b$ and $\Delta T_f$ have two significant figures. Because the normal boiling point and freezing point are exact values, they do not affect the number of significant figures in the final answer.

## PRACTICE Problems

Do additional problems. Online Practice

**45.** What are the boiling point and freezing point of a 0.625*m* aqueous solution of any nonvolatile, nonelectrolyte solute?

**46.** What are the boiling point and freezing point of a 0.40*m* solution of sucrose in ethanol?

**47. Challenge** A 0.045*m* solution (consisting of a nonvolatile, nonelectrolyte solute) is experimentally found to have a freezing point depression of 0.080°C. What is the freezing point depression constant ($K_f$)? Which is most likely to be the solvent: water, ethanol, or chloroform?

## RealWorld CHEMISTRY

### Freezing Point Depression

**SALTWATER FISH** Maintaining the proper saline (salt) concentration is important to the health of saltwater fish. In the ocean, the presence of salt in arctic areas keeps the water from freezing, allowing aquatic life to be sustained.

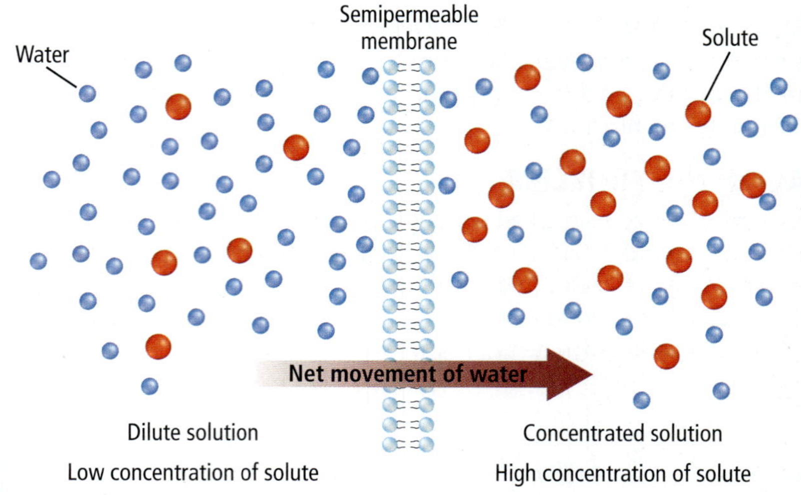


**Figure 23** Due to osmosis, solvents diffuse from a lower solute concentration to a higher solute concentration through semipermeable membranes.

View an **animation about osmosis.**

Concepts In Motion

## Osmotic Pressure

Connection to Biology Recall that diffusion is the mixing of gases or liquids resulting from their random motions. **Osmosis** is the diffusion of a solvent through a semipermeable membrane. Semipermeable membranes are barriers that allow some, but not all, particles to cross. The membranes surrounding all living cells are semipermeable membranes. Osmosis plays an important role in many biological systems, such as the uptake of nutrients by plants.

Examine a system in which a dilute solution is separated from a concentrated solution by a semipermeable membrane, illustrated in **Figure 23.** During osmosis, water molecules move in both directions across the membrane, but the solute molecules cannot cross it. Water molecules diffuse across the membrane from the dilute solution to the concentrated solution. The amount of additional pressure caused by the water molecules that moved into the concentrated solution is called the **osmotic pressure.** Osmotic pressure depends on the number of solute particles in a given volume of solution and is a colligative property of solutions.

## SECTION 4 REVIEW

Section Self-Check

### Section Summary

- Nonvolatile solutes lower the vapor pressure of a solution.
- Boiling point elevation is directly related to the solution's molality.
- A solution's freezing point depression is always lower than that of the pure solvent.
- Osmotic pressure depends on the number of solute particles in a given volume.

**48.** MAINIDEA **Explain** the nature of colligative properties.

**49. Describe** four colligative properties of solutions.

**50. Explain** why a solution has a higher boiling point than that of the pure solvent.

**51. Solve** An aqueous solution of calcium chloride ($CaCl_2$) boils at 101.3°C. How many kilograms of calcium chloride were dissolved in 1000.0 g of the solvent?

**52. Calculate** the boiling point elevation of a solution containing 50.0 g of glucose ($C_6H_{12}O_6$) dissolved in 500.0 g of water. Calculate the freezing point depression for the same solution.

**53. Investigate** A lab technician determines the boiling point elevation of an aqueous solution of a nonvolatile, nonelectrolyte to be 1.12°C. What is the solution's molality?

# Career: Environmental Chemist

## A $CO_2$ Solution

Geologic records indicate that the levels of atmospheric carbon dioxide ($CO_2$) are likely higher today than in the past 20 million years. Anthropogenic (an thruh pah JEN ihk) $CO_2$, which means $CO_2$ from human-made sources, has contributed to this high level. $CO_2$ does not remain in the atmosphere indefinitely. Oceans naturally contain $CO_2$ that comes from the atmosphere and from living organisms. Oceans have absorbed nearly 50% of anthropogenic $CO_2$. Some scientists think that over the next thousand years, as much as 90% of anthropogenic $CO_2$ will dissolve in the oceans.

**Collecting $CO_2$ data** The rate at which $CO_2$ dissolves into the oceans is influenced by many factors including temperature, concentration of $CO_2$ in the air and in the water, and the mixing of air and water due to wind and waves. A team of researchers spent years collecting and analyzing $CO_2$ data from thousands of collection points throughout the world's oceans. The data, shown in **Figure 1,** indicate that the North Atlantic has the most anthropogenic $CO_2$ per square meter of ocean surface. The combination of temperature, depth, and current make the North Atlantic an efficient absorber of anthropogenic $CO_2$.

**$CO_2$ capture and storage** One way to reduce the amount of $CO_2$ released into the atmosphere would be to capture and store the $CO_2$ produced when fossil fuels are burned. Researchers are investigating the possibility of directly injecting captured $CO_2$ into the ocean to speed up the dissolution process. This could reduce the greenhouse effect of $CO_2$ gas. However, upsetting the natural balance of dissolved $CO_2$ can have profound effects on water chemistry, which can harm or even kill marine life. For example, coral reefs throughout the world already show signs of stress due to increasing levels of dissolved $CO_2$.

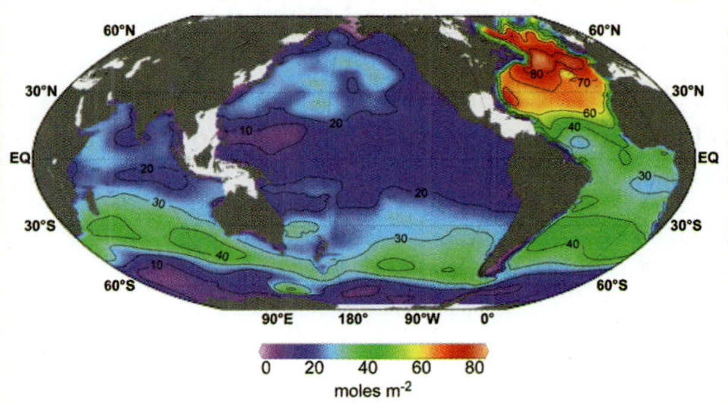


**Figure 1** The red, yellow, and green regions represent areas where high levels of anthropogenic $CO_2$ are dissolved in the water.

Data obtained from: Sabine et al. 2004. The oceanic sink for anthropogenic $CO_2$. *Science* 305: 367–371.

**Deep ocean sequestration** A proposal that might reduce atmospheric $CO_2$ and protect life in the upper ocean is to liquefy the $CO_2$ and pump it deep under water, a process known as deep ocean sequestration. It is thought that the extreme pressure at depths greater than 3000 m will cause the $CO_2$ to form a hydrate. The hydrate will dissolve into the deep ocean water, but the $CO_2$ will remain trapped for hundreds of years far from the upper ocean and atmosphere.

**Ongoing research** Scientists are working on many of the unanswered questions about deep ocean sequestration, such as the effect of $CO_2$ on deep-sea organisms. There are still many technological problems involving capturing, storing, and transporting large quantities of liquid $CO_2$. If the technological problems can be solved, the public as well as government officials will have to consider the relative dangers of releasing $CO_2$ into the air and into the ocean.

### WRITING IN Chemistry

**Brainstorm** a list of questions that must be addressed through research before deep ocean sequestration is attempted.

WebQuest

Sabine et al., 2004

# ChemLAB

iLab Station

## Investigate Factors Affecting Solubility

**Background:** **The process of making a solution involves the solvent coming in contact with the solute particles. When you add a soluble compound to water, several factors affect the rate of solution formation.**

**Question** *How do factors affect the rate of solution formation?*

### Materials

copper(II) sulfate pentahydrate
distilled water
test tubes (6)
25-mL graduated cylinder
glass stirring rod
tweezers
test tube rack
mortar and pestle
spatula
clock

### Safety Precautions

### Procedure

1. Read and complete the lab safety form.
2. Create a table to record your data.
3. Write a hypothesis that uses what you know about reaction rates to explain what you might observe during the procedure.
4. Place the 6 test tubes in the test tube rack.
5. Place one crystal of copper(II) sulfate pentahydrate in each of the first two test tubes.
6. For the remaining test tubes, use the mortar and pestle to crush the crystals. Use the spatula to scrape the crystals into the remaining test tubes.
7. Measure 15-mL of room-temperature distilled water. Pour the water into the first test tube and record the time.
8. Observe the solution in the test tube just after adding the water and after 15 min.
9. Leave the first test tube undisturbed in the rack.
10. Repeat Steps 7 and 8 for the third and fourth test tubes.
11. Use the glass stirring rod to agitate the second test tube for 1 to 2 min.
12. Leave the third test tube undisturbed.
13. Agitate the fourth test tube with the glass stirring rod for 1 to 2 min.
14. Repeat Steps 7 and 8 for the fifth test tube using cold water. Leave the fifth test tube undisturbed.
15. Repeat Steps 7 and 8 for the sixth test tube using hot water. Leave the sixth test tube undisturbed.
16. **Cleanup and Disposal** Dispose of the remaining solids and solutions as directed by your teacher. Wash and return all lab equipment to the designated locations.

### Analyze and Conclude

1. **Compare and Contrast** What effect did you observe due to the agitation of the second and fourth test tubes versus the solutions in the first and third test tubes?
2. **Observe and Infer** What factor caused the more rapid solution formation in the fourth test tube in comparison to the second test tube?
3. **Recognize Cause and Effect** Why do you think the results for the third, fifth, and sixth test tubes were different?
4. **Discuss** whether or not your data supported your hypothesis.
5. **Error Analysis** Identify a major potential source of error for this lab, and suggest an easy method to correct it.

### INQUIRY EXTENSION

**Think Critically** The observations in this lab were macroscopic in nature. Propose a submicroscopic explanation to account for these factors that affected the rate of solution formation. At the molecular level, what is occurring to speed solution formation in each case?

# CHAPTER 14 STUDY GUIDE

Vocabulary Practice

**BIGIDEA** Nearly all of the gases, liquids, and solids that make up our world are mixtures.

## SECTION 1 Types of Mixtures

**MAINIDEA** Mixtures can be either heterogeneous or homogeneous.

- The individual substances in a heterogeneous mixture remain distinct.
- Two types of heterogeneous mixtures are suspensions and colloids.
- Brownian motion is the erratic movement of colloid particles.
- Colloids exhibit the Tyndall effect.
- A solution can exist as a gas, a liquid, or a solid, depending on the solvent.
- Solutes in a solution can be gases, liquids, or solids.

**VOCABULARY**
- suspension
- colloid
- Brownian motion
- Tyndall effect
- soluble
- miscible
- insoluble
- immiscible

## SECTION 2 Solution Concentration

**MAINIDEA** Concentration can be expressed in terms of percent or in terms of moles.

- Concentrations can be measured qualitatively and quantitatively.
- Molarity is the number of moles of solute dissolved per liter of solution.

$$\textbf{molarity } (M) = \frac{\textbf{moles of solute}}{\textbf{liters of solution}}$$

- Molality is the ratio of the number of moles of solute dissolved in 1 kg of solvent.

$$\textbf{molality } (m) = \frac{\textbf{moles of solute}}{\textbf{kilograms of solvent}}$$

- The number of moles of solute does not change during a dilution.

$$M_1V_1 = M_2V_2$$

**VOCABULARY**
- concentration
- molarity
- molality
- mole fraction

## SECTION 3 Factors Affecting Solvation

**MAINIDEA** Factors such as temperature, pressure, and polarity affect the formation of solutions.

- The process of solvation involves solute particles surrounded by solvent particles.
- Solutions can be unsaturated, saturated, or supersaturated.
- Henry's law states that at a given temperature, the solubility ($S$) of a gas in a liquid is directly proportional to the pressure ($P$) of the gas above the liquid.

$$\frac{S_1}{P_1} = \frac{S_2}{P_2}$$

**VOCABULARY**
- solvation
- heat of solution
- unsaturated solution
- saturated solution
- supersaturated solution
- Henry's law

## SECTION 4 Colligative Properties of Solutions

**MAINIDEA** Colligative properties depend on the number of solute particles in a solution.

- Nonvolatile solutes lower the vapor pressure of a solution.
- Boiling point elevation is directly related to the solution's molality.

$$\Delta T_b = K_b m$$

- A solution's freezing point depression is always lower than that of the pure solvent.

$$\Delta T_f = K_f m$$

- Osmotic pressure depends on the number of solute particles in a given volume.

**VOCABULARY**
- colligative property
- vapor pressure lowering
- boiling point elevation
- freezing point depression
- osmosis
- osmotic pressure

## SECTION 1

### Mastering Concepts

**54.** Explain what is meant by the statement "not all mixtures are solutions."

**55.** What is the difference between a solute and a solvent?

**56.** What is a suspension, and how does it differ from a colloid?

**57.** How can the Tyndall effect be used to distinguish between a colloid and a solution? Why?

**58.** Name a colloid formed from a gas dispersed in a liquid.

■ **Figure 24**

**59. Salad dressing** What type of heterogenous mixture is shown in **Figure 24?** What characteristic is most useful in classifying the mixture?

**60.** What causes the Brownian motion observed in liquid colloids?

**61.** Aerosol sprays are categorized as colloids. Identify the phases of an aerosol spray.

## SECTION 2

### Mastering Concepts

**62.** What is the difference between percent by mass and percent by volume?

**63.** What is the difference between molarity and molality?

**64.** What factors must be considered when creating a dilute solution from a stock solution?

**65.** How do 0.5*M* and 2.0*M* aqueous solutions of NaCl differ?

**66.** Under what conditions might a chemist describe a solution in terms of molality? Why?

### Mastering Problems

**67.** According to lab procedure, you stir 25.0 g of $MgCl_2$ into 550 mL of water. What is the percent by mass of $MgCl_2$ in the solution?

**68.** How many grams of LiCl are in 275 g of a 15% aqueous solution of LiCl?

**69.** You need to make a large quantity of a 5% solution of HCl but have only 25 mL HCl. What volume of 5% solution can be made from this volume of HCl?

**70.** Calculate the percent by volume of a solution created by adding 75 mL of acetic acid to 725 mL of water.

**71.** Calculate the molarity of a solution that contains 15.7 g of $CaCO_3$ dissolved in enough water to make 275 mL of solution.

**72.** What is the volume of a 3.00*M* solution made with 122 g of LiF?

**73.** How many moles of BaS would be used to make $1.5 \times 10^3$ mL of a 10.0*M* solution?

**74.** How much $CaCl_2$, in grams, is needed to make 2.0 L of a 3.5*M* solution?

**75.** Stock solutions of HCl with various molarities are frequently prepared. Complete **Table 7** by calculating the volume of concentrated, or 12*M*, hydrochloric acid that should be used to make 1.0 L of HCl solution with each molarity listed.

**Table 7 HCl Solutions**

| Molarity of HCl Desired | Volume of 12*M* HCl Stock Solution Needed (mL) |
|---|---|
| 0.50 | |
| 1.0 | |
| 1.5 | |
| 2.0 | |
| 5.0 | |

**76.** How much 5.0*M* nitric acid ($HNO_3$), in milliliters, is needed to make 225 mL of 1.0*M* $HNO_3$?

**77. Experiment** In the lab, you dilute 55 mL of a 4.0*M* solution to make 250 mL of solution. Calculate the molarity of the new solution.

**78.** How many milliliters of 3.0*M* phosphoric acid ($H_3PO_4$) can be made from 95 mL of a 5.0*M* $H_3PO_4$ solution?

**79.** If you dilute 20.0 mL of a 3.5*M* solution to make 100.0 mL of solution, what is the molarity of the dilute solution?

**80.** What is the molality of a solution that contains 75.3 g of KCl dissolved in 95.0 g of water?

**81.** How many grams of $Na_2CO_3$ must be dissolved into 155 g of water to create a solution with a molality of 8.20 mol/kg?

**82.** What is the molality of a solution containing 30.0 g of naphthalene ($C_{10}H_8$) dissolved in 500.0 g of toluene?

**83.** What are the molality and mole fraction of solute in a 35.5 percent by mass aqueous solution of formic acid (HCOOH)?

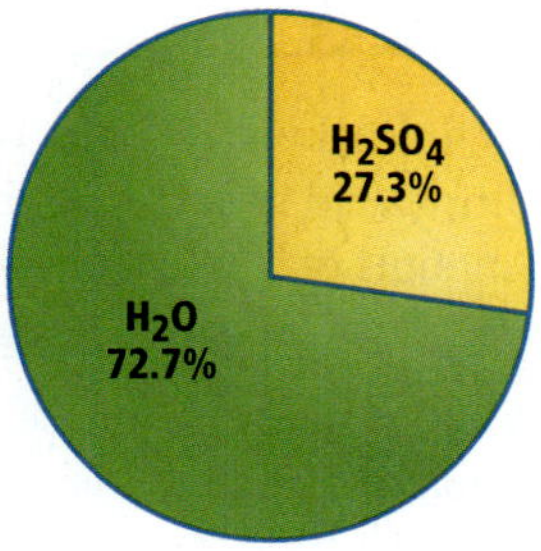


■ **Figure 25**

**84.** What is the mole fraction of $H_2SO_4$ in a solution containing the percentage of sulfuric acid and water shown in **Figure 25?**

**85.** Calculate the mole fraction of $MgCl_2$ in a solution created by dissolving 132.1 g of $MgCl_2$ in 175 mL of water.

## SECTION 3

### Mastering Concepts

**86.** Describe the process of solvation.

**87.** What are three ways to increase the rate of solvation?

**88.** Explain the difference between saturated and unsaturated solutions.

### Mastering Problems

**89.** At a pressure of 1.5 atm, the solubility of a gas is 0.54 g/L. Calculate the solubility when the pressure is doubled.

**90.** At 4.5 atm of pressure, the solubility of a gas is 9.5 g/L. How much gas, in grams, will dissolve in 1 L if the pressure is reduced by 3.5 atm?

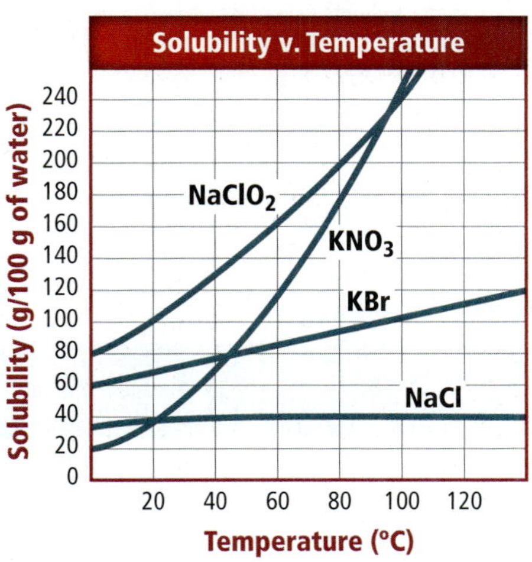


■ **Figure 26**

**91.** Using **Figure 26,** compare the solubility of potassium bromide (KBr) and potassium nitrate ($KNO_3$) at 80°C.

**92.** The solubility of a gas at 37.0 kPa is 1.80 g/L. At what pressure will the solubility reach 9.00 g/L?

**93.** Use Henry's law to complete **Table 8.**

**Table 8 Solubility and Pressure**

| Solubility (g/L) | Pressure (kPa) |
|---|---|
| 2.9 | ? |
| 3.7 | 32 |
| ? | 39 |

**94. Soft Drinks** The partial pressure of $CO_2$ inside a bottle of soft drink is 4.0 atm at 25°C. The solubility of $CO_2$ is 0.12 mol/L. When the bottle is opened, the partial pressure drops to $3.0 \times 10^{-4}$ atm. What is the solubility of $CO_2$ in the open drink? Express your answer in grams per liter.

## SECTION 4

### Mastering Concepts

**95.** Define the term *colligative property.*

**96.** Use the terms *dilute* and *concentrated* to compare the solution on both sides of a semipermeable membrane.

**97.** Identify each variable in the following formula:

$$\Delta T_b = K_b m$$

**98.** Define the term *osmotic pressure,* and explain why it is considered a colligative property.

### Mastering Problems

**99.** Calculate the freezing point of a solution of 12.1 g of naphthalene ($C_{10}H_8$) dissolved in 0.175 kg of benzene ($C_6H_6$). Refer to **Table 6** for needed data.

**100.** In the lab, you dissolve 179 g of $MgCl_2$ into 1.00 L of water. Use **Table 6** to find the freezing point of the solution.

**101. Cooking** A cook prepares a solution for boiling by adding 12.5 g of NaCl to a pot holding 0.750 L of water. At what temperature should the solution in the pot boil? Refer to **Table 5** for needed data.

**102.** The boiling point of ethanol ($C_2H_5OH$) changes from 78.5°C to 85.2°C when an amount of naphthalene ($C_{10}H_8$) is added to 1.00 kg of ethanol. How much naphthalene, in grams, is required to cause this change? Refer to **Table 5** for needed data.

**103. Ice Cream** A rock salt (NaCl), ice, and water mixture is used to cool milk and cream to make homemade ice cream. How many grams of rock salt must be added to the water to lower the freezing point by 10.0°C?

## MIXED REVIEW

**104.** Apply your knowledge of polarity and solubility to predict whether solvation is possible in each situation shown in **Table 9.** Explain your answers.

**Table 9 Is solvation possible?**

| Solute | Solvent |
|---|---|
| solid $MgCl_2$ | liquid $H_2O$ |
| liquid $NH_3$ | liquid $C_6H_6$ |
| gaseous $H_2$ | liquid $H_2O$ |
| liquid $I_2$ | liquid $Br_2$ |

**105. Household Paint** Some types of paint are colloids composed of pigment particles dispersed in oil. Based on what you know about colloids, recommend an appropriate location for storing cans of leftover household paint. Justify your recommendation.

**106.** Which solute has the greatest effect on the boiling point of 1.00 kg of water: 50.0 g of strontium chloride ($SrCl_2$) or 150.0 g of glucose ($C_6H_{12}O_6$)? Justify your answer.

**107.** Analyze solubility and temperature data in **Table 4** to determine the general trend followed by the gases in the chart. Compare this trend to the trend followed by most of the solids in the chart. Identify the solids listed that do not follow the general trend.

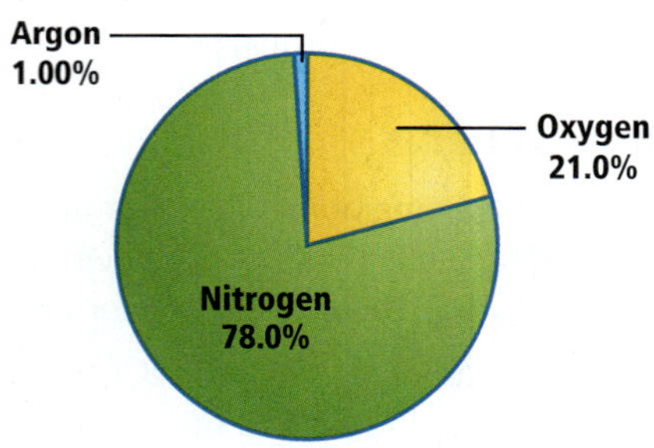


■ **Figure 27**

**108.** An air sample yields the mass percent composition shown in **Figure 27.** Calculate the mole fraction for each gas present in the sample.

**109.** If you prepared a saturated aqueous solution of potassium chloride at 25°C and then heated it to 50°C, would you describe the solution as unsaturated, saturated, or supersaturated? Explain.

**110.** How many grams of calcium nitrate ($Ca(NO_3)_2$) would you need to prepare 3.00 L of a 0.500*M* solution?

**111.** What would be the molality of the solution described in the previous problem? The density of the $Ca(NO_3)_2$ solution is 1.08 kg/L.

## THINK CRITICALLY

**112. Develop** a plan for making 1000 mL of a 5% by volume solution of sulfuric acid in water. Your plan should describe the amounts of solute and solvent necessary, as well as the steps involved in making the solution.

**113. Compare and Infer** Study the phase diagram in **Figure 21.** Compare the dotted lines surrounding $\Delta T_f$ and $\Delta T_b$, and describe the differences you observe. How might these lines be positioned differently for solutions of electrolytes and nonelectrolytes? Why?

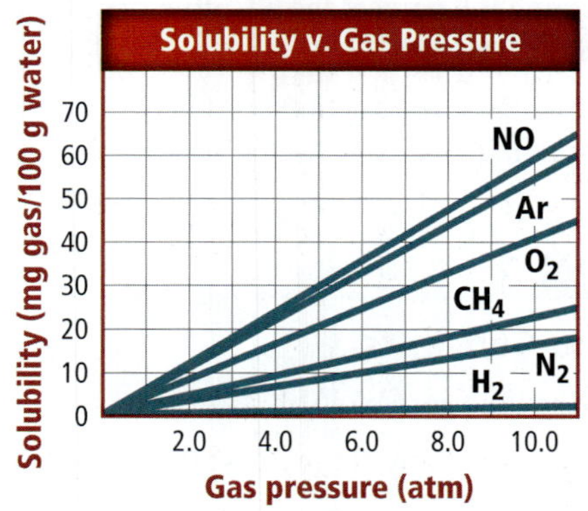


■ **Figure 28**

**114. Extrapolate** The solubility of argon in water at various pressures is shown in **Figure 28.** Extrapolate the data to 15 atm. Use Henry's law to verify the solubility determined by your extrapolation.

**115. Infer** Dehydration occurs when more fluid is lost from the body than is taken in. Scuba divers are advised to hydrate their bodies before diving. Use your knowledge of the relationship between pressure and gas solubility to explain the importance of hydration prior to a dive.

**116. Graph** **Table 10** shows solubility data that was collected in an experiment. Plot a graph of the molarity of KI versus temperature. What is the solubility of KI at 55°C?

**Table 10 Solubility of KI**

| Temperature (°C) | Grams of KI per 100.0 g Solution |
|---|---|
| 20 | 144 |
| 40 | 162 |
| 60 | 176 |
| 80 | 192 |
| 100 | 206 |

Chapter Self-Check

**117. Design an Experiment** You are given a sample of a solid solute and three aqueous solutions containing that solute. How would you determine which solution is saturated, unsaturated, and supersaturated?

**118. Compare** Which of the following solutions has the highest concentration? Rank the solutions from the greatest to the smallest boiling point elevation. Explain your answer.
**a.** 0.10 mol NaBr in 100.0 mL solution
**b.** 2.1 mol KOH in 1.00 L solution
**c.** 1.2 mol $KMnO_4$ in 3.00 L solution

## CHALLENGE PROBLEMS

**119.** Interpret the solubility data in **Table 11** using the concept of Henry's law.

**Table 11** Measurements of Solubility of a Gas

| Measurement | Solubility |
|---|---|
| 1 | 0.225 |
| 2 | 0.45 |
| 3 | 0.9 |
| 4 | 1.8 |
| 5 | 3.6 |

**120.** You have a solution containing 135.2 g of dissolved KBr in 2.3 L of water. What volume of this solution, in mL, would you use to make 1.5 L of a 0.10*M* KBr solution? What is the boiling point of this new solution?

## CUMULATIVE REVIEW

**121.** The radius of an argon atom is 94 pm. Assuming the atom is spherical, what is the volume of an argon atom in cubic nanometers? $V = 4/3\pi r^3$

**122.** Identify which molecule is polar.
**a.** $SiH_4$
**b.** $NO_2$
**c.** $SiO_2$
**d.** HBr

**123.** Name the following compounds.
**a.** NaBr
**b.** $Pb(CH_3COO)_2$
**c.** $(NH_4)_2CO_3$

**124.** A 12.0-g sample of an element contains $5.94 \times 10^{22}$ atoms. What is the unknown element?

**125.** Pure bismuth can be produced by the reaction of bismuth oxide with carbon at high temperatures.

$$2Bi_2O_3 + 3C \rightarrow 4Bi + 3CO_2$$

How many moles of $Bi_2O_3$ reacted to produce 12.6 mol of $CO_2$?

## WRITING IN ▶ Chemistry

**126. Homogenized Milk** The first homogenized milk was sold in the United States around 1919. Today, almost all milk sold in this country is homogenized in the form of a colloidal emulsion. Research the homogenization process. Write a brief article describing the process. The article should include a flowchart or diagram of the process, as well as a discussion of the reputed benefits and drawbacks associated with drinking homogenized milk.

## DBQ Document-Based Questions

**Annual Mean Dissolved Oxygen** *The data in* **Figure 29** *shows the average dissolved oxygen values, in milliliters per liter, in ocean-surface waters during a one-month period in 2001. Longitude is indicated horizontally, and latitude is indicated vertically.*

Data obtained from: National Oceanographic Data Center. 2002. *World Ocean Atlas 2001 Figures.*

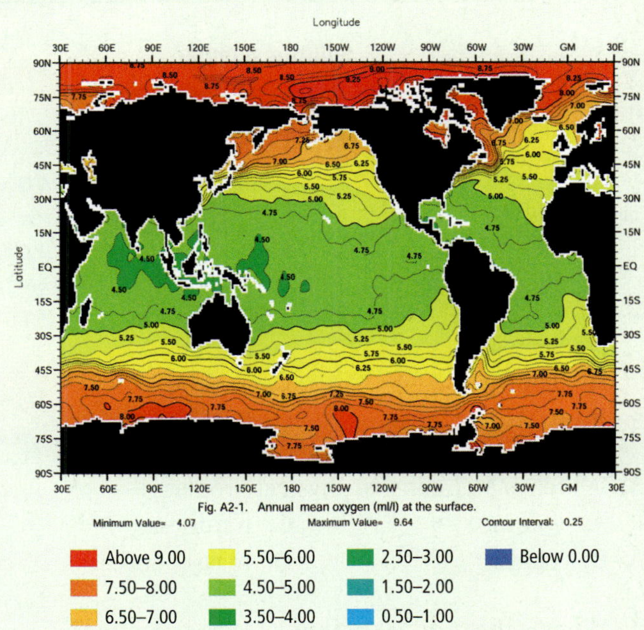


■ **Figure 29**

**127.** Are dissolved oxygen values most closely related to latitude or longitude? Why do you think this is true?

**128.** At what latitude are average dissolved oxygen values the lowest?

**129.** Describe the general trend defined by the data. Relate the trend to the relationship between gas solubility and temperature.

Courtesy NODC

## MULTIPLE CHOICE

*Use the graph below to answer Questions 1 and 2.*

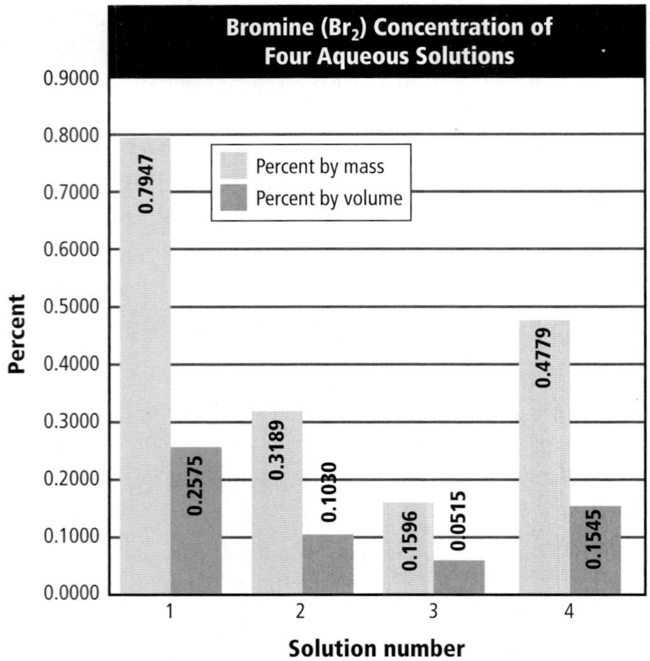


1. What is the volume of bromine ($Br_2$) in 7.000 L of Solution 1?
   A. 55.63 mL
   B. 8.808 mL
   C. 18.03 mL
   D. 27.18 mL

2. How many grams of $Br_2$ are in 55.00 g of Solution 4?
   A. 3.560 g
   B. 0.08498 g
   C. 1.151 g
   D. 0.2628 g

3. Which is an intensive physical property?
   A. volume
   B. length
   C. hardness
   D. mass

4. What is the product of this synthesis reaction?
   $Cl_2(g) + 2NO(g) \rightarrow$ ?
   A. $NCl_2$
   B. 2NOCl
   C. $N_2O_2$
   D. 2ClO

5. If 1 mol of each of the solutes listed below is dissolved in 1 L of water, which solute will have the greatest effect on the vapor pressure of its respective solution?
   A. KBr
   B. $C_6H_{12}O_6$
   C. $MgCl_2$
   D. $CaSO_4$

*Use the diagram below to answer Question 6.*

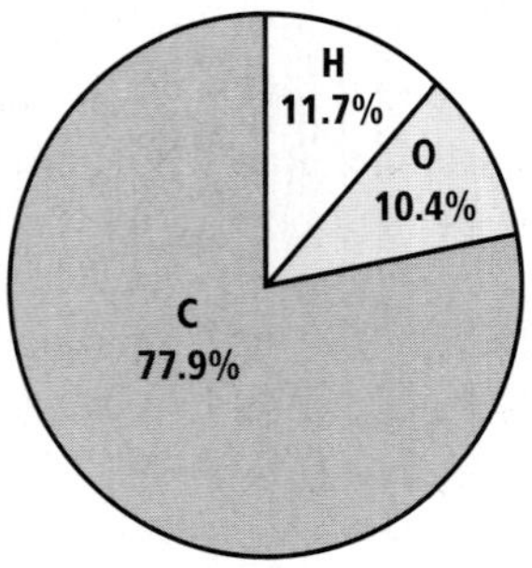


6. What is the empirical formula for this substance?
   A. $CH_2O$
   B. $C_8HO$
   C. $C_{10}H_{18}O$
   D. $C_7H_{12}O$

7. What is the correct chemical formula for the ionic compound formed by the calcium ion ($Ca^{2+}$) and the acetate ion ($C_2H_3O_2^-$)?
   A. $CaC_2H_3O_2$
   B. $CaC_4H_6O_3$
   C. $(Ca)_2C_2H_3O_2$
   D. $Ca(C_2H_3O_2)_2$

*Use the reaction below to answer Questions 8 and 9.*

$$Fe_3O_4(s) + 4H_2(g) \rightarrow 3\ Fe(s) + 4\ H_2O(l)$$

8. If 16 mol of $H_2$ are used, how many moles of Fe will be produced?
   A. 6
   B. 3
   C. 12
   D. 9

9. If 7 mol of $Fe_3O_4$ are mixed with 30 mol of $H_2$, what will be true?
   A. There will be no reactants left.
   B. 2 mol of hydrogen gas will be left over.
   C. 30 mol of water will be produced.
   D. 7 mol of Fe will be produced.

10. What is the molar mass of $Fe_3O_4$?
    A. 231.54 g/mol
    B. 71.85 g/mol
    C. 287.40 g/mol
    D. 215.56 g/mol

## SHORT ANSWER

*Use the graph below to answer Questions 11 to 13.*

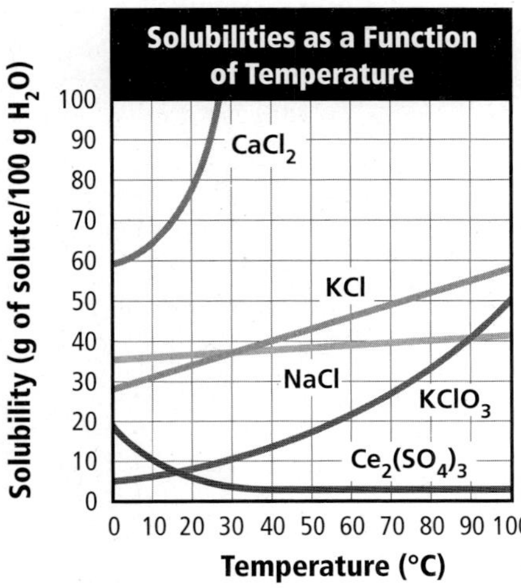


**11.** How many moles of $KClO_3$ can be dissolved in 100 g of water at 60°C?

**12.** Which can hold more solute at 20°C: a solution of NaCl or KCl? How does this compare to their solubilities at 80°C?

**13.** How many moles of $KClO_3$ would be required to make 1 L of a saturated solution of $KClO_3$ at 75°C?

## EXTENDED RESPONSE

*Use information below to answer Questions 14 and 15.*

The electron configuration for silicon is $1s^22s^22p^63s^23p^2$.

**14.** Explain how this configuration demonstrates the Aufbau principle.

**15.** Draw the orbital diagram for silicon. Explain how Hund's rule and the Pauli exclusion principle are used in constructing the orbital diagram.

## SAT SUBJECT TEST: CHEMISTRY

**16.** What volume of a 0.125*M* $NiCl_2$ solution contains 3.25 g of $NiCl_2$?
**A.** 406 mL
**B.** 32.5 mL
**C.** 38.5 mL
**D.** 26.0 mL
**E.** 201 mL

**17.** Which is NOT a colligative property?
**A.** boiling point elevation
**B.** freezing point depression
**C.** vapor pressure lowering
**D.** osmotic pressure
**E.** solubility

*Use the data table below to answer Questions 18 and 19.*

| Electronegativities of Selected Elements | | | | | | |
|---|---|---|---|---|---|---|
| H | | | | | | |
| 2.20 | | | | | | |
| Li | Be | B | C | N | O | F |
| 0.98 | 1.57 | 2.04 | 2.55 | 3.04 | 3.44 | 3.98 |
| Na | Mg | Al | Si | P | S | Cl |
| 0.93 | 1.31 | 1.61 | 1.90 | 2.19 | 2.58 | 3.16 |

**18.** What is the electronegativity difference in $Li_2O$?
**A.** 1.48
**B.** 2.46
**C.** 3.40
**D.** 4.42
**E.** 5.19

**19.** Which bond has the greatest polarity?
**A.** C–H
**B.** Si–O
**C.** Mg–Cl
**D.** Al–N
**E.** H–Cl

| NEED EXTRA HELP? | | | | | | | | | | | | | | | | | | | |
|---|---|---|---|---|---|---|---|---|---|---|---|---|---|---|---|---|---|---|---|
| If You Missed Question . . . | 1 | 2 | 3 | 4 | 5 | 6 | 7 | 8 | 9 | 10 | 11 | 12 | 13 | 14 | 15 | 16 | 17 | 18 | 19 |
| Review Section . . . | 14.2 | 14.2 | 3.1 | 9.2 | 14.4 | 10.4 | 7.3 | 11.2 | 11.3 | 10.3 | 14.3 | 14.3 | 14.3 | 5.3 | 5.3 | 14.2 | 14.4 | 8.5 | 8.5 |

CHAPTER 15

# Energy and Chemical Change

**BIGIDEA** Chemical reactions usually absorb or release energy.

## SECTIONS

Chm 12, 12a, 12b

1 Energy

2 Heat

3 Thermochemical Equations

4 Calculating Enthalpy Change

5 Reaction Spontaneity

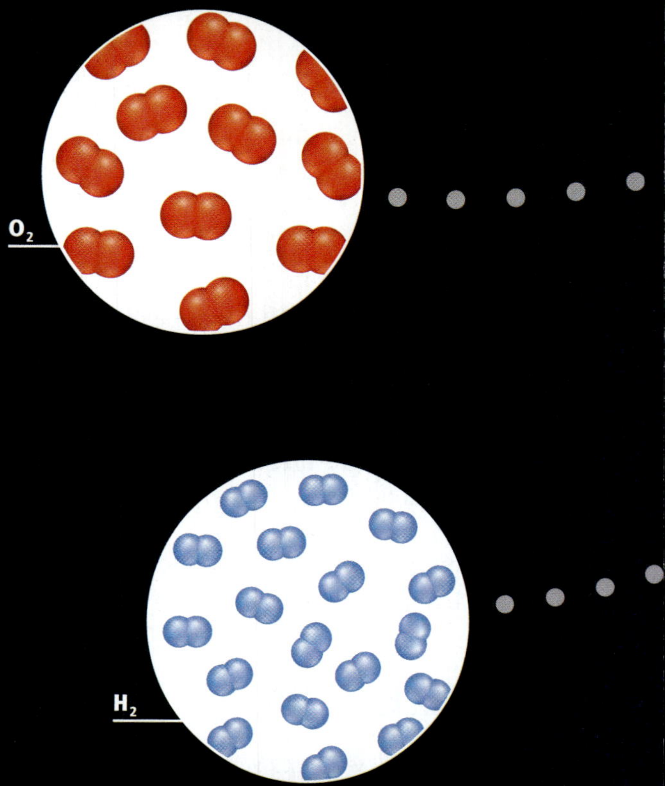


## LaunchLAB

iLab Station

### How can you make a cold pack?

Chemical cold packs are used for fast relief of pain due to injury. Some chemical cold packs contain two separate compounds that are combined in a process that absorbs heat. In this lab, you will test three chemicals to determine which would make the best chemical cold pack.

**FOLDABLES® Study Organizer**

### Gibbs Free Energy Equation

Make a concept-map book. Label it as shown. Use it to organize your study of the energy equation.

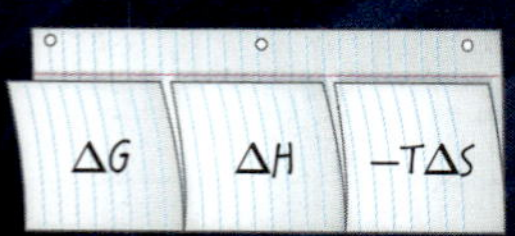


The three main engines of the space shuttle use more than 547,000 kg of liquid oxygen and approximately 92,000 kg of liquid hydrogen to lift a total mass of $2.04 \times 10^6$ kg.

Go **online!**
MHEonline.com

SECTION 1

# Energy

**MAINIDEA** Energy can change form and flow, but it is always conserved.

### Essential Questions

- What is energy?
- How do potential and kinetic energy differ?
- How can chemical potential energy be related to the heat lost or gained in chemical reactions?
- How is the amount of heat absorbed or released by a substance calculated as its temperature changes?

### Review Vocabulary

**temperature:** a measure of the average kinetic energy of the particles in a sample of matter

### New Vocabulary

energy
law of conservation of energy
chemical potential energy
heat
calorie
joule
specific heat

**CHEM 4 YOU** **Have you ever watched a roller coaster zoom up and down a track or experienced the thrill of a coaster ride? Each time a coaster climbs a steep grade or plunges down the other side, its energy changes from one form to another.**

## The Nature of Energy

You are probably familiar with the term *energy.* Perhaps you have heard someone say, "I just ran out of energy," after a strenuous game or a difficult day. Solar energy, nuclear energy, energy-efficient automobiles, and other energy-related topics are often discussed in the media.

Energy cooks the food you eat and propels the vehicles that transport you. If the day is especially hot or cold, energy from burning fuels helps maintain a comfortable temperature in your home and school. Electric energy provides light and powers devices from computers and TV sets to cellular phones, MP3 players, and calculators. Energy was involved in the manufacture and delivery of every material and device in your home. Your every movement and thought requires energy. In fact, you can think of each cell in your body as a miniature factory that runs on energy derived from the food you eat.

What is energy? **Energy** is the ability to do work or produce heat. It exists in two basic forms: potential energy and kinetic energy. Potential energy is energy due to the composition or position of an object. A macroscopic example of potential energy of position is a downhill skier poised at the starting gate for a race, as shown in **Figure 1a.** After the starting signal is given, the skier's potential energy changes to kinetic energy during the speedy trip to the finish line, as shown in **Figure 1b.** Kinetic energy is energy of motion. You can observe kinetic energy in the motion of objects and people all around you.

**Figure 1** At the top of the course, the skier in **a** has high potential energy because of her position. In **b,** the skier's potential energy changes to kinetic energy.
**Compare** ***How is the potential energy of the skier different at the starting gate and at the finish line?***

(l)Agence Zoom/Getty Images, (r)Donald Miralle/Getty Images

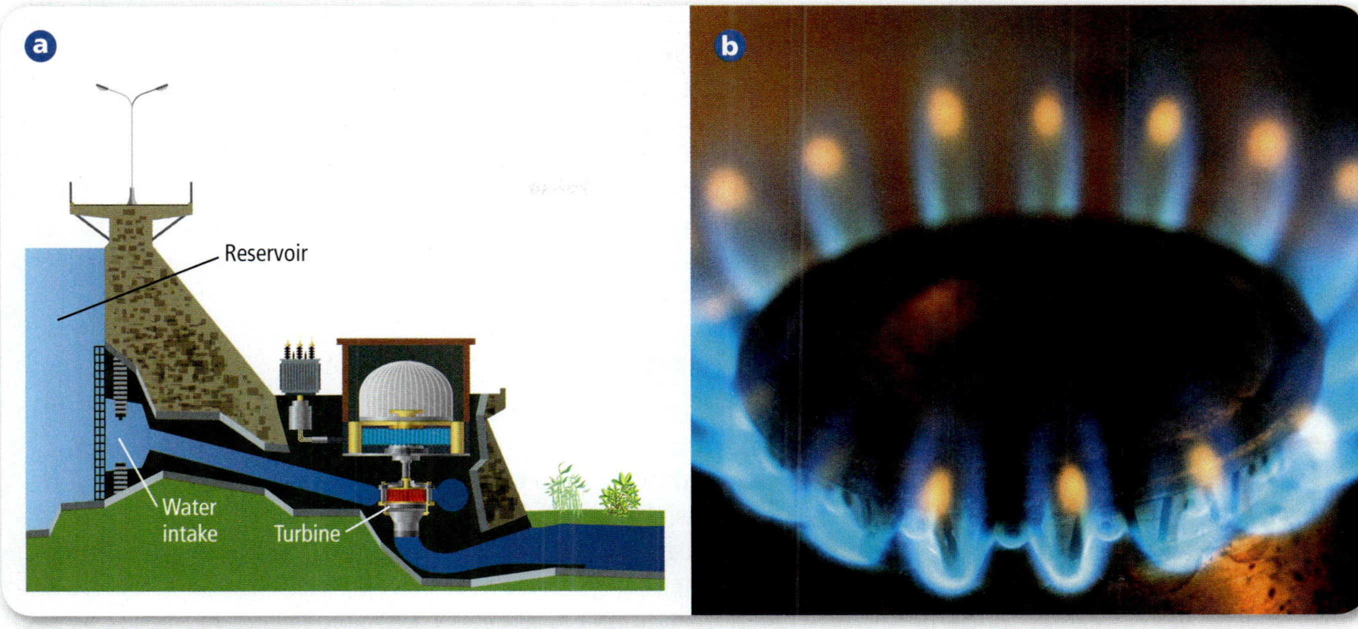


■ **Figure 2** Energy can change from one form to another but is always conserved. In **a,** the potential energy of water is converted to kinetic energy of motion as it falls through the intake from its high position in the reservoir. The rushing water spins the turbine to generate electric energy. In **b,** the potential energy stored in the bonds of propane molecules is converted to heat.

Chemical systems contain both kinetic energy and potential energy. Recall that the kinetic energy of a substance is directly related to the constant random motion of its representative particles and is proportional to temperature. As temperature increases, the motion of submicroscopic particles increases. The potential energy of a substance depends on its composition: the type of atoms in the substance, the number and type of chemical bonds joining the atoms, and the particular way the atoms are arranged.

**Law of conservation of energy** When water rushes through turbines in the hydroelectric plant shown in **Figure 2a,** some of the water's kinetic energy is converted to electric energy. Propane ($C_3H_8$) is an important fuel for cooking and heating. In **Figure 2b,** propane gas combines with oxygen to form carbon dioxide and water. Potential energy stored in the propane bonds is given off as heat. In both of these examples, energy changes from one form to another, but energy is conserved—the total amount of energy remains constant. To better understand the conservation of energy, suppose you have money in two accounts at a bank and you transfer funds from one account to the other. Although the amount of money in each account has changed, the total amount of your money in the bank remains the same. When applied to energy, this analogy embodies the law of conservation of energy. The **law of conservation of energy** states that in any chemical reaction or physical process, energy can be converted from one form to another, but it is neither created nor destroyed. This is also known as the first law of thermodynamics.

**Chemical potential energy** The energy that is stored in a substance because of its composition is called **chemical potential energy.** Chemical potential energy plays an important role in chemical reactions. For example, the chemical potential energy of propane results from the arrangement of the carbon and hydrogen atoms and the strength of the bonds that join them.

☑ **READING CHECK** **State** the law of conservation of energy in your own words.

**Table 1 Relationships Among Energy Units**

| Relationship | Conversion Factors |
|---|---|
| 1 J = 0.2390 cal | $\frac{1\ J}{0.2390\ cal}$ $\frac{0.2390\ cal}{1\ J}$ |
| 1 cal = 4.184 J | $\frac{1\ cal}{4.184\ J}$ $\frac{4.184\ J}{1\ cal}$ |
| 1 Calorie = 1 kcal | $\frac{1\ Calorie}{1000\ cal}$ $\frac{1000\ cal}{1\ Calorie}$ |

**Heat** The principal component of gasoline is octane ($C_8H_{18}$). When gasoline burns in an automobile's engine, some of octane's chemical potential energy is converted to the work of moving the pistons, which ultimately moves the wheels and propels the automobile. However, much of the chemical potential energy of octane is released as heat. The symbol $q$ is used to represent **heat,** which is energy that is in the process of flowing from a warmer object to a cooler object. When the warmer object loses energy, its temperature decreases. When the cooler object absorbs energy, its temperature rises.

## Measuring Heat

The flow of energy and the resulting change in temperature are clues to how heat is measured. In the metric system of units, the amount of energy required to raise the temperature of one gram of pure water by one degree Celsius (1°C) is defined as a **calorie** (cal). When your body breaks down sugars and fats to form carbon dioxide and water, these exothermic reactions generate heat that can be measured in Calories. Note that the nutritional Calorie is capitalized. That is because one nutritional Calorie equals 1000 calories, or one kilocalorie (kcal). Recall that the prefix *kilo-* means 1000. For example, one tablespoon of butter contains approximately 100 Calories. This means that if the butter was burned completely to produce carbon dioxide and water, 100 kcal (100,000 cal) of heat would be released.

The SI unit of of energy and of heat is the **joule** (J). One joule is the equivalent of 0.2390 calories, and one calorie equals 4.184 joules. **Table 1** summarizes the relationships between calories, nutritional Calories, joules, and kilojoules (kJ) and the conversion factors you can use to convert from one unit to another.

### EXAMPLE Problem 1

Find help with **unit conversion.** Math Handbook

EXAMPLE PROBLEM

**CONVERT ENERGY UNITS** A breakfast of cereal, orange juice, and milk might contain 230 nutritional Calories. Express this energy in joules.

**1 ANALYZE THE PROBLEM**

You are given an amount of energy in nutritional Calories. You must convert nutritional Calories to calories and then convert calories to joules.

| **Known** | **Unknown** |
|---|---|
| amount of energy = 230 Calories | **amount of energy = ? J** |

**2 SOLVE FOR THE UNKNOWN**

Convert nutritional Calories to calories.

$$230\ \cancel{Calories} \times \frac{1000\ cal}{1\ \cancel{Calorie}} = 2.3 \times 10^5\ cal$$

**Apply the relationship 1 Calorie = 1000 cal.**

Convert calories to joules.

$$2.3 \times 10^5\ \cancel{cal} \times \frac{4.184\ J}{1\ \cancel{cal}} = \mathbf{9.6 \times 10^5\ J}$$

**Apply the relationship 1 cal = 4.184 J.**

**3 EVALUATE THE ANSWER**

The minimum number of significant figures used in the conversion is two, and the answer correctly has two digits. A value of the order of $10^5$ or $10^6$ is expected because the given number of kilocalories is of the order of $10^2$ and it must be multiplied by $10^3$ to convert it to calories. Then, the calories must be multiplied by a factor of approximately 4. Therefore, the answer is reasonable.

## PRACTICE Problems

Do additional problems. Online Practice

1. A fruit-and-oatmeal bar contains 142 nutritional Calories. Convert this energy to calories.
2. An exothermic reaction releases 86.5 kJ. How many kilocalories of energy are released?
3. **Challenge** Define a new energy unit, named after yourself, with a magnitude of one-tenth of a calorie. What conversion factors relate this new unit to joules? To Calories?

## Specific Heat

You have read that one calorie, or 4.184 J, is required to raise the temperature of one gram of pure water by one degree Celsius (1°C). That quantity, 4.184 J/(g•°C), is defined as the specific heat (c) of water. The **specific heat** of any substance is the amount of heat required to raise the temperature of one gram of that substance by one degree Celsius. Because different substances have different compositions, each substance has its own specific heat.

To raise the temperature of water by one degree Celsius, 4.184 J must be absorbed by every gram of water. Much less energy is required to raise the temperature of an equal mass of concrete by one degree Celsius. You might have noticed that concrete sidewalks get hot during a sunny summer day. How hot depends on the specific heat of concrete, but other factors are also important. The specific heat of concrete is 0.84 J/(g•°C), which means that the temperature of concrete increases roughly five times more than water's temperature when equal masses of concrete and water absorb the same amount of energy. As shown in **Figure 3,** people who have been walking on hot concrete surfaces might want to cool their feet in the water of a fountain.

**Figure 3** The cooler waters of the fountain are welcome after walking on the hot concrete sidewalk. Water must absorb five times the energy of an equal mass of concrete to reach an equivalent temperature.

**Infer** ***How would the temperature change of the concrete compare to that of the water over the course of a cool night.***

(l)Stephen Chernin/Getty Images; (r)©Bob Krist/Corbis

**Table 2 Specific Heats at 298 K (25°C)**

| Substance | Specific heat J/(g•°C) |
|---|---|
| Water(l) | 4.184 |
| Ethanol(l) | 2.44 |
| Water(s) | 2.03 |
| Water(g) | 2.01 |
| Beryllium(s) | 1.825 |
| Magnesium(s) | 1.023 |
| Aluminum(s) | 0.897 |
| Concrete(s) | 0.84 |
| Granite(s) | 0.803 |
| Calcium(s) | 0.647 |
| Iron(s) | 0.449 |
| Strontium(s) | 0.301 |
| Silver(s) | 0.235 |
| Barium(s) | 0.204 |
| Lead(s) | 0.129 |
| Gold(s) | 0.129 |

**Calculating heat absorbed** Suppose that the temperature of a $5.00 \times 10^3$-g block of concrete sidewalk increased by 6.0°C. Would it be possible to calculate the amount of heat it had absorbed? Recall that the specific heat of a substance tells you the amount of heat that must be absorbed by 1 g of a substance to raise its temperature 1°C. **Table 2** shows the specific heats for some common substances. The specific heat of concrete is 0.84 J/(g•°C), so 1 g of concrete absorbs 0.84 J when its temperature increases by 1°C. To determine the heat absorbed by $5.00 \times 10^3$ g of concrete you must multiply the 0.84 J by $5.00 \times 10^3$. Then, because the concrete's temperature changed by 6.0°C, you must multiply the product of the mass and the specific heat by 6.0°C.

### Equation for Calculating Heat

$$q = c \times m \times \Delta T$$

***q*** **represents the heat absorbed or released.** ***c*** **represents the specific heat of the substance.** ***m*** **represents the mass of the sample in grams.** **$\Delta T$ is the change in temperature in °C, or $T_{final} - T_{initial}$.**

The quantity of heat absorbed or released by a substance is equal to the product of its specific heat, the mass of the substance, and the change in its temperature.

You can use this equation to calculate the heat absorbed by the concrete block.

$$q = c \times m \times \Delta T$$

$$q_{\text{concrete}} = \frac{0.84 \text{ J}}{(\cancel{\text{g}}\cdot\cancel{°\text{C}})} \times (5.00 \times 10^3 \cancel{\text{g}}) \times 6.0\cancel{°\text{C}} = 25{,}000 \text{ J or } 25 \text{ kJ}$$

The total amount of heat absorbed by the concrete block is 25,000 J or 25 kJ.

For comparison, how much heat would be absorbed by $5.00 \times 10^3$ g of the water in the fountain when its temperature is increased by 6.0°C? The calculation for $q_{\text{water}}$ is the same as it is for concrete except that you must use the specific heat of water, 4.184 J/(g•°C).

$$q_{\text{water}} = \frac{4.184 \text{ J}}{(\cancel{\text{g}}\cdot\cancel{°\text{C}})} \times (5.00 \times 10^3 \cancel{\text{g}}) \times 6.0\cancel{°\text{C}} = 1.3 \times 10^5 \text{ J or } 130 \text{ kJ}$$

If you divide the heat absorbed by the water (130 kJ) by the heat absorbed by the concrete (25 J), you will find that for the same change in temperature, the water absorbed more than five times the amount of heat absorbed by the concrete block.

**Calculating heat released** Substances can both absorb and release heat. The same equation for *q*, the quantity of heat, can be used to calculate the energy released by substances when they cool off. Suppose the $5.00 \times 10^3$-g piece of concrete reached a temperature of 74.0°C during a sunny day and cooled down to 40.0°C at night. How much heat was released? First calculate $\Delta T$.

$$\Delta T = 74.0°\text{C} - 40.0°\text{C} = 34.0°\text{C}$$

Then, use the equation for quantity of heat.

$$q = c \times m \times \Delta T$$

$$q_{\text{concrete}} = \frac{0.84 \text{ J}}{(\cancel{\text{g}}\cdot\cancel{°\text{C}})} \times (5.00 \times 10^3 \cancel{\text{g}}) \times 34.0\cancel{°\text{C}} = 140{,}000 \text{ J or } 140 \text{ kJ}$$

## EXAMPLE Problem 2

**CALCULATE SPECIFIC HEAT** In the construction of bridges and skyscrapers, gaps must be left between adjoining steel beams to allow for the expansion and contraction of the metal due to heating and cooling. The temperature of a sample of iron with a mass of 10.0 g changed from 50.4°C to 25.0°C with the release of 114 J. What is the specific heat of iron?

### 1 ANALYZE THE PROBLEM

You are given the mass of the sample, the initial and final temperatures, and the quantity of heat released. You can calculate the specific heat of iron by rearranging the equation that relates these variables to solve for *c*.

**Known**

energy released = 114 J　　$T_i$ = 50.4°C

mass of iron = 10.0 g Fe　　$T_f$ = 25.0°C

**Unknown**

**specific heat of iron, $c$ = ? J/(g•°C)**

### 2 SOLVE FOR THE UNKNOWN

Calculate $\Delta T$.

$\Delta T = 50.4°C - 25.0°C = 25.4°C$

Write the equation for calculating the quantity of heat.

$q = c \times m \times \Delta T$ — **State the equation for calculating heat.**

$\frac{c \times \cancel{m} \times \cancel{\Delta T}}{\cancel{m} \times \cancel{\Delta T}} = \frac{q}{m \times \Delta T}$ — **Solve for $c$.**

$c = \frac{q}{m \times \Delta T}$

$c = \frac{114\ J}{(10.0\ g)(25.4°C)}$ — **Substitute $q$ = 114 J, $m$ = 10.0 g, and $\Delta T$ = 25.4°C.**

$c = 0.449\ J/(g•°C)$ — **Multiply and divide numbers and units.**

### 3 EVALUATE THE ANSWER

The values used in the calculation have three significant figures, so the answer is correctly stated with three digits. The value of the denominator of the equation is approximately two times the value of the numerator, so the final result, which is approximately 0.5, is reasonable. The calculated value is the same as that recorded for iron in **Table 2.**

## RealWorld CHEMISTRY

### Specific Heat

**ABSORBING HEAT** You might have wrapped your hands around a cup of hot chocolate to stay warm at a fall football game. In much the same way, long ago, children sometimes walked to school on wintry days carrying hot, baked potatoes in their pockets. The potatoes provided warmth for cold hands, but by the time the school bell rang, the potatoes had cooled off. At lunchtime, the cold potatoes might have been placed in or on the schoolhouse stove to warm them again for eating.

Explore **phase changes.**

Virtual Investigations

## PRACTICE Problems

Do additional problems. Online Practice

4. If the temperature of 34.4 g of ethanol increases from 25.0°C to 78.8°C, how much heat has been absorbed by the ethanol? Refer to **Table 2.**
5. A 155-g sample of an unknown substance was heated from 25.0°C to 40.0°C. In the process, the substance absorbed 5696 J of energy. What is the specific heat of the substance? Identify the substance among those listed in **Table 2.**
6. **Challenge** A 4.50-g nugget of pure gold absorbed 276 J of heat. The initial temperature was 25.0°C. What was the final temperature?

■ **Figure 4** Each photoelectric cell on this panel absorbs the Sun's radiation and converts it to electricity quietly and without causing pollution.

©Helen King/Corbis

**Using the Sun's energy** Because of its high specific heat, water is sometimes used to harness the energy of the Sun. After water has been heated by solar radiation, the hot water can be circulated in homes and businesses to provide heat. Radiation from the Sun could supply all the energy needs of the world and reduce or eliminate the use of carbon dioxide-producing fuels, but several factors have delayed the development of solar technologies. For example, the Sun shines for only a part of each day. In some areas, clouds often reduce the amount of available radiation. Because of this variability, effective methods for storing energy are critical.

A more promising approach to the use of solar energy is the development of photovoltaic cells, such as those shown in **Figure 4.** These devices convert solar radiation directly to electricity. Photovoltaic cells supply power for astronauts in space, but they are not used extensively for ordinary energy needs. That is because the cost of supplying electricity by means of photovoltaic cells is high compared to the cost of burning coal or oil.

## SECTION 1 REVIEW

Section Self-Check

### Section Summary

- Energy is the capacity to do work or produce heat.
- Chemical potential energy is energy stored in the chemical bonds of a substance by virtue of the arrangement of the atoms and molecules.
- Chemical potential energy is released or absorbed as heat during chemical processes or reactions.

7. MAINIDEA **Explain** how energy changes from one form to another in an exothermic reaction. In an endothermic reaction.
8. **Distinguish** between kinetic and potential energy in the following examples: two separated magnets; an avalanche of snow; books on library shelves; a mountain stream; a stock-car race; separation of charge in a battery.
9. **Explain** how the light and heat of a burning candle are related to chemical potential energy.
10. **Calculate** the amount of heat absorbed when 5.50 g of aluminum is heated from 25.0°C to 95.0°C. The specific heat of aluminum is 0.897 J/(g·°C).
11. **Interpret Data** Equal masses of aluminum, gold, iron, and silver were left to sit in the Sun at the same time and for the same length of time. Use **Table 2** to arrange the four metals according to the increase in their temperatures from largest increase to smallest.

SECTION 2

# Heat

**MAINIDEA** The enthalpy change for a reaction is the enthalpy of the products minus the enthalpy of the reactants.

### Essential Questions

- How is a calorimeter used to measure energy that is absorbed or released?
- What do enthalpy and enthalpy change mean in terms of chemical reactions and processes?

### Review Vocabulary

**pressure:** force applied per unit area

### New Vocabulary

calorimeter
thermochemistry
system
surroundings
universe
enthalpy
enthalpy (heat) of reaction

**CHEM 4 YOU** **Think about standing under a hot shower, relaxing as your body absorbs heat from the water. When you jump into a cold pool, you might shiver as your body loses heat. In a similar way, some chemical reactions absorb heat whereas others release heat.**

## Calorimetry

Have you ever wondered how food chemists obtain the Calorie information that appears on packaged food? The packages record the results of combustion reactions carried out in calorimeters. A **calorimeter** is an insulated device used for measuring the amount of heat absorbed or released during a chemical or physical process. A known mass of water is placed in an insulated chamber to absorb the energy released from the reacting system or to provide the energy absorbed by the system. The data to be collected is the change in temperature of this mass of water. **Figure 5** shows the kind of calorimeter, called a bomb calorimeter, that is used by food chemists.

**Determining specific heat** Satisfactory results can be obtained in your calorimetry experiments using the much simpler foam-cup calorimeter. These calorimeters are open to the atmosphere, so reactions carried out in them occur at constant pressure. You can use them to determine the specific heat of an unknown metal.

Suppose you put 125 g of water into a foam-cup calorimeter and find that its initial temperature is 25.60°C. Then you heat a 50.0-g sample of the unknown metal to 115.0°C and put the metal sample into the water. Heat flows from the hot metal to the cooler water, and the temperature of the water rises. The flow of heat stops only when the temperature of the metal and the water are equal.

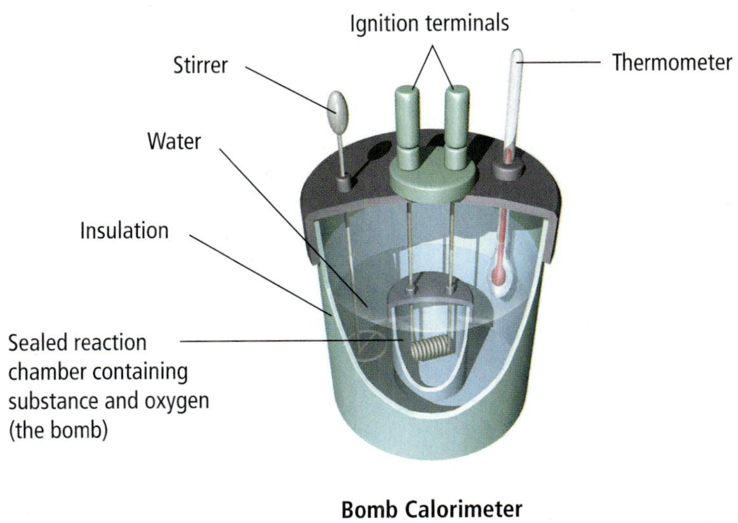


**Figure 5** A sample is positioned in a steel inner chamber called the bomb, which is filled with oxygen at high pressure. Surrounding the bomb is a measured mass of water stirred by a low-friction stirrer to ensure uniform temperature. The reaction is initiated by a spark, and the temperature is recorded until it reaches its maximum.

**Infer** ***Why is it important that the stirrer does not create friction?***

View an **animation about calorimetry.**

Concepts In Motion

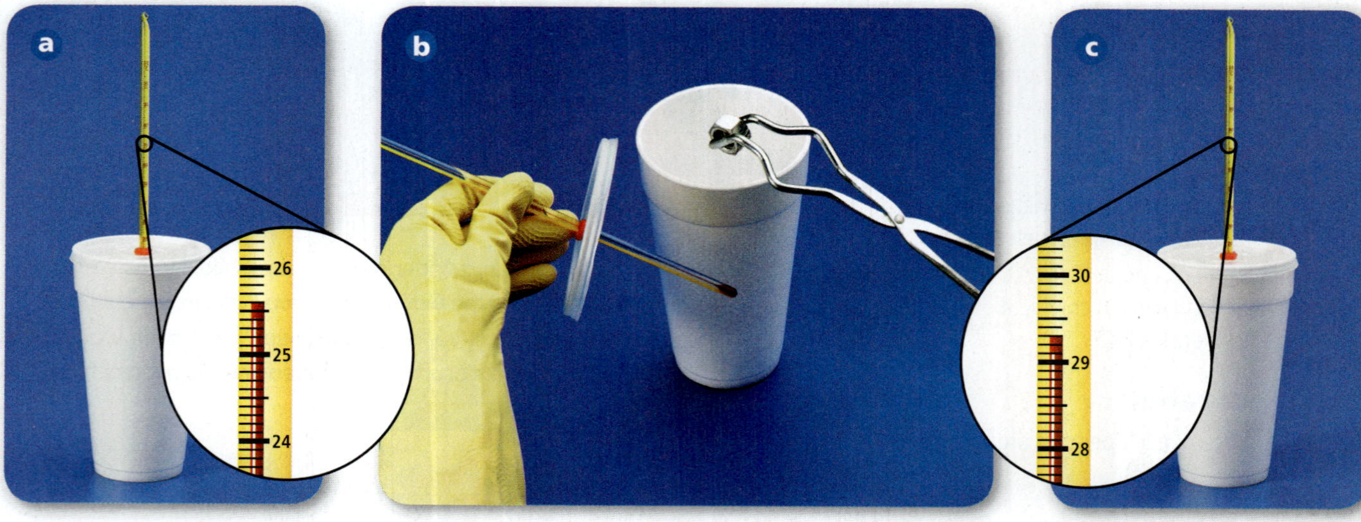


■ **Figure 6 a.** An initial temperature of 25.60°C is recorded for the 125 g of water in the calorimeter. **b.** A 50.0-g sample of an unknown metal is heated to 115.0°C and placed in the calorimeter. **c.** The metal transfers heat to the water until metal and water are at the same temperature. The final temperature is 29.30°C.

**Figure 6** shows an experimental procedure. Note that the temperature in the calorimeter becomes constant at 29.30°C, which is the final temperature attained by both the water and the metal. Assuming no heat is lost to the surroundings, the heat gained by the water is equal to the heat lost by the metal. This quantity of heat can be calculated using the equation you learned in Section 1.

$$q = c \times m \times \Delta T$$

☑ **READING CHECK Define** the four variables in the equation above.

Explore **calorimetry.**

Virtual Investigations

First, calculate the heat gained by the water. To do this, you need the specific heat of water, 4.184 J/(g•°C).

$$q_{\text{water}} = 4.184\ \text{J/(g•°C)} \times 125\ \text{g} \times (29.30°\text{C} - 25.60°\text{C})$$

$$q_{\text{water}} = 4.184\ \text{J/}(\cancel{\text{g•°C}}) \times 125\ \cancel{\text{g}} \times 3.70\cancel{°\text{C}}$$

$$q_{\text{water}} = 1940\ \text{J}$$

The heat gained by the water, 1940 J, equals the heat lost by the metal, $q_{\text{metal}}$, so you can write this equation.

$$q_{\text{metal}} = q_{\text{water}}$$

$$q_{\text{metal}} = -1940\ \text{J}$$

$$c_{\text{metal}} \times m \times \Delta T = -1940\ \text{J}$$

Now, solve the equation for the specific heat of the metal, $c_{\text{metal}}$, by dividing both sides of the equation by $m \times \Delta T$.

$$c_{\text{metal}} = \frac{-1940\ \text{J}}{m \times \Delta T}$$

The change in temperature for the metal, $\Delta T$, is the difference between the final temperature of the water and the initial temperature of the metal (29.30°C − 115.0°C = −85.7°C). Substitute the known values of $m$ and $\Delta T$ (50.0 g and −85.7°C) into the equation and solve.

$$c_{\text{metal}} = \frac{-1940\ \text{J}}{(50.0\ \text{g})(-85.7°\text{C})} = 0.453\ \text{J/(g•°C)}$$

The unknown metal has a specific heat of 0.453 J/(g•°C). **Table 2** shows that the metal could be iron.

## EXAMPLE Problem 3

Find help with **algebraic equations.** Math Handbook

**USING SPECIFIC HEAT** A piece of metal with a mass of 4.68 g absorbs 256 J of heat when its temperature increases by 182°C. What is the specific heat of the metal? Could the metal be one of the alkaline earth metals listed in **Table 2?**

### 1 ANALYZE THE PROBLEM

You are given the mass of the metal, the amount of heat it absorbs, and the temperature change. You must calculate the specific heat. Use the equation for *q*, the quantity of heat, but solve for specific heat, *c*.

**Known**

mass of metal, $m = 4.68$ g

quantity of heat absorbed, $q = 256$ J

$\Delta T = 182°C$

**Unknown**

**specific heat, $c$ = ? J/(g·°C)**

### 2 SOLVE FOR THE UNKNOWN

$q = c \times m \times \Delta T$ — State the equation for the quantity of heat, *q*.

$c = \frac{q}{m \times \Delta T}$ — Solve for *c*.

$c = \frac{256 \text{ J}}{(4.68 \text{ g})(182°\text{C})} = 0.301 \text{ J/(g·°C)}$ — Substitute $q = 256$ J, $m = 4.68$ g, and $\Delta T = 182°C$.

**Table 2** indicates that the metal could be strontium.

### 3 EVALUATE THE ANSWER

The three quantities used in the calculation have three significant figures, and the answer is correctly stated with three digits. The calculations are correct and yield the expected unit.

EXAMPLE PROBLEM

## PRACTICE Problems

Do additional problems. Online Practice

**12.** A 90.0-g sample of an unknown metal absorbed 25.6 J of heat as its temperature increased 1.18°C. What is the specific heat of the metal?

**13.** The temperature of a sample of water increases from 20.0°C to 46.6°C as it absorbs 5650 J of heat. What is the mass of the sample?

**14.** How much heat is absorbed by a $2.00 \times 10^3$ g granite boulder ($c_{granite} = 0.803$ J/(g·°C)) as its temperature changes from 10.0°C to 29.0°C?

**15. Challenge** If 335 g of water at 65.5°C loses 9750 J of heat, what is the final temperature of the water?

PRACTICE PROBLEMS

## Chemical Energy and the Universe

Virtually every chemical reaction and change of physical state either releases or absorbs heat. **Thermochemistry** is the study of heat changes that accompany chemical reactions and phase changes. The burning of fuels always produces heat. Some products have been engineered to produce heat on demand. For example, soldiers in the field use a highly exothermic reaction to heat their meals. You might have used a heat pack to warm your hands on a cold day. The energy released by a heat pack is produced by the following reaction and is shown in the equation as one of the products.

$$4Fe(s) + 3O_2(g) \rightarrow 2Fe_2O_3(s) + 1625 \text{ kJ}$$

# MiniLAB

iLab Station

## Determine Specific Heat

**How can you determine the specific heat of a metal?** You can use a coffee-cup calorimeter to determine the specific heat of a metal.

### Procedure

1. Read and complete the lab safety form.
2. Make a table to record your data.
3. Pour approximately 150 mL of **distilled water** into a **250-mL beaker.** Place the beaker on a **hot plate** set on high.
4. Use a **balance** to find the mass of a **metal cylinder.**
5. Using **crucible tongs,** carefully place the metal cylinder in the beaker on the hot plate.
6. Measure 90.0 mL of distilled water using a **graduated cylinder.**
7. Pour the water into a **polystyrene coffee cup** nested in a second **250-mL beaker.**
8. Measure and record the temperature of the water using a **nonmercury thermometer.**
9. When the water on the hot plate begins to boil, measure and record the temperature as the initial temperature of the metal.
10. Carefully add the hot metal to the cool water in the coffee cup with the crucible tongs. Do not touch the hot metal with your hands.
11. Stir, and measure the maximum temperature of the water after the metal was added.

### Analysis

1. **Calculate** the heat gained by the water. The specific heat of $H_2O$ is 4.184 J/g·°C. Because the density of water is 1.0 g/mL, use the volume of water as the mass.
2. **Calculate** the specific heat of your metal. Assume that the heat absorbed by the water equals the heat lost by the metal.
3. **Compare** this experimental value to the accepted value for your metal.
4. **Describe** major sources of error in this lab. What modifications could you make in this experiment to reduce the error?

Because you are interested in the heat given off by the chemical reaction going on inside the pack, it is convenient to think of the pack and its contents as the system. In thermochemistry, the **system** is the specific part of the universe that contains the reaction or process you wish to study. Everything in the universe other than the system is considered the **surroundings.** Therefore, the **universe** is defined as the system plus the surroundings.

$$\text{universe} = \text{system} + \text{surroundings}$$

What kind of energy transfer occurs during the exothermic heat-pack reaction? Heat produced by the reaction flows from the heat pack (the system) to your cold hands (part of the surroundings).

What happens in an endothermic reaction or process? The flow of heat is reversed. Heat flows from the surroundings to the system. When barium hydroxide and ammonium thiocyanate crystals, shown in **Figure 7,** are placed in a beaker and mixed, a highly endothermic reaction occurs. Placing the beaker on a wet board allows heat to flow from the water and board (the surroundings) into the beaker (the system). The temperature change is great enough that the beaker freezes to the board.

**Figure 7** In this endothermic reaction, the reacting mixture draws enough energy from the water and the wet board to cause the wet board to freeze to the beaker.

Matt Meadows

**Enthalpy and enthalpy changes** The total amount of energy a substance contains depends on many factors, some of which are still not completely understood. Therefore, it is impossible to know the total energy content of a substance. Fortunately, chemists are usually more interested in changes in energy during reactions than in the absolute amounts of energy contained in the reactants and products.

For many reactions, the amount of energy lost or gained can be measured conveniently in a calorimeter at constant pressure, as shown in the experiment in **Figure 6.** The foam cup is not sealed, so the pressure is constant. Many reactions take place at constant atmospheric pressure; for example, those that occur in living organisms on Earth's surface, in lakes and oceans, and those that take place in open beakers and flasks in the laboratory. The energy released or evolved from reactions carried out at constant pressure is sometimes given the symbol $q_p$. To more easily measure and study the energy changes that accompany such reactions, chemists have defined a property called enthalpy. **Enthalpy** ($H$) is the heat content of a system at constant pressure.

Although you cannot measure the actual energy or enthalpy of a substance, you can measure the change in enthalpy, which is the heat absorbed or released in a chemical reaction. The change in enthalpy for a reaction is called the **enthalpy (heat) of reaction** ($\Delta H_{rxn}$). You have already learned that a symbol preceded by the Greek letter delta ($\Delta$) means a change in the property. Thus, $\Delta H_{rxn}$ is the difference between the enthalpy of the substances that exist at the end of the reaction and the enthalpy of the substances present at the start.

$$\Delta H_{rxn} = H_{final} - H_{initial}$$

Because the reactants are present at the beginning of the reaction and the products are present at the end, $\Delta H_{rxn}$ is defined by this equation.

$$\Delta H_{rxn} = H_{products} - H_{reactants}$$

**The sign of the enthalpy of reaction** Recall the heat-pack reaction.

$$4Fe(s) + 3O_2(g) \rightarrow 2Fe_2O_3(s) + 1625 \text{ kJ}$$

According to the equation, the reactants in this exothermic reaction lose heat. Therefore, $H_{products} < H_{reactants}$. When $H_{reactants}$ is subtracted from the smaller $H_{products}$, a negative value for $\Delta H_{rxn}$ results. Enthalpy changes for exothermic reactions are always negative. The equation for the heat-pack reaction and its enthalpy change are usually written as shown.

$$4Fe(s) + 3O_2(g) \rightarrow 2Fe_2O_3(s) \quad \Delta H_{rxn} = -1625 \text{ kJ}$$

A diagram of the enthalpy change is shown in **Figure 8.**

## CAREERS IN CHEMISTRY

**Heating and Cooling Specialist**
Heating and cooling system mechanics install, maintain, and repair refrigeration and heating equipment in homes and in industry. They must understand how heat is exchanged by means of exothermic and endothermic processes. They must be able to read blueprints and use a wide range of tools, from pipe cutters to computerized diagnostic devices. Such mechanics might specialize in one aspect of this field, or become proficient in all areas.

WebQuest

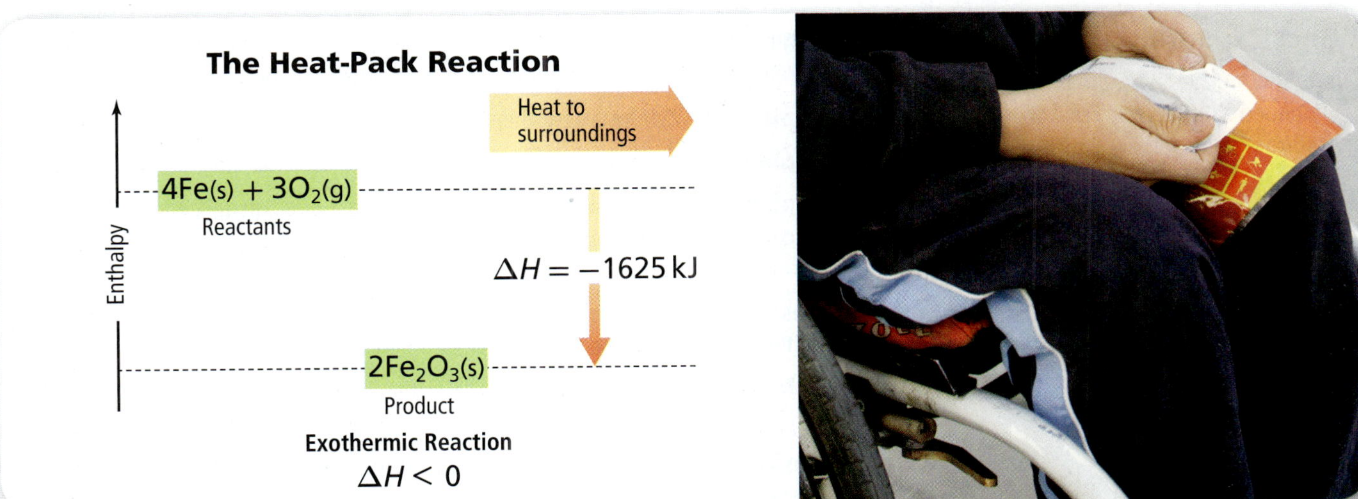


■ **Figure 8** The downward arrow shows that 1625 kJ of heat is released to the surroundings in the reaction between iron and oxygen to form $Fe_2O_3$. A heat pack utilizing this reaction of iron and oxygen provides energy for warming cold hands.

**Explain** *how the diagram shows that the reaction is exothermic.*

Tim Fuller

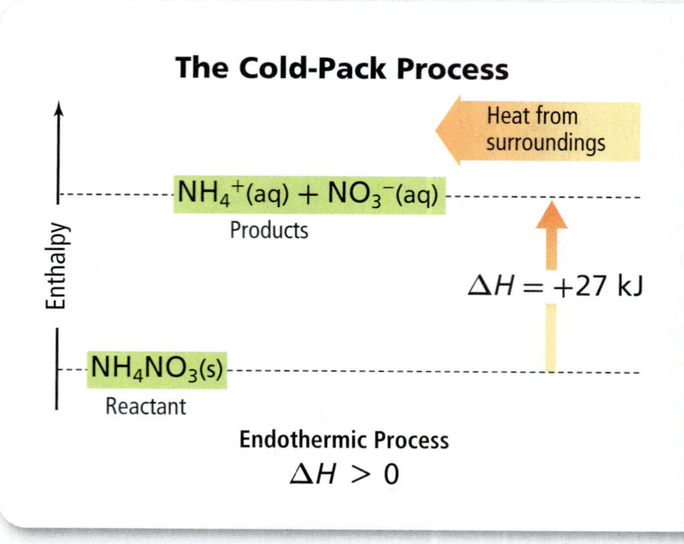


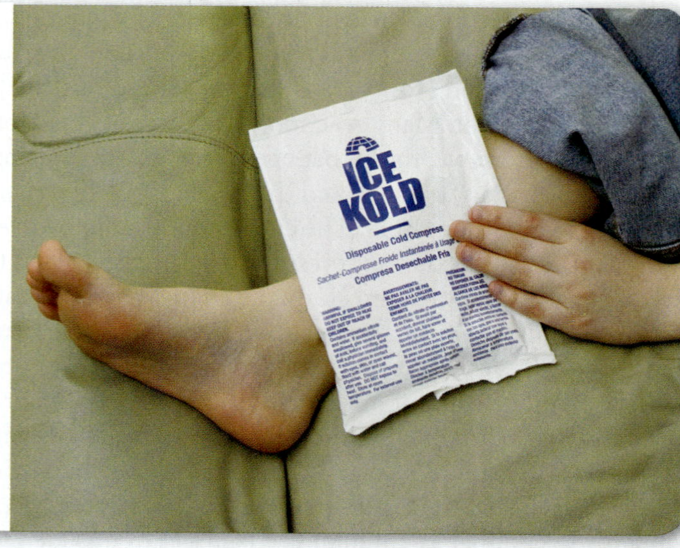


■ **Figure 9** The upward arrow shows that 27 kJ of heat is absorbed from the surroundings in the process of dissolving $NH_4NO_3$. This reaction is the basis for the cold pack. When the cold pack is placed on a person's elbow, his elbow supplies the required heat and is itself cooled.

**Determine** ***How much energy of ammonium nitrate is absorbed when a cold pack is activated?***

**Alabama Applying Practices**

Go to the Resource tab in ConnectED to find the Applying Practices worksheet *Coffee Cup Calorimetry.*

Now, recall the cold-pack process.

$$27 \text{ kJ} + NH_4NO_3(s) \rightarrow NH_4^+(aq) + NO_3^-(aq)$$

For this endothermic process, $H_{products} > H_{reactants}$. Therefore, when $H_{reactants}$ is subtracted from the larger $H_{products}$, a positive value for $\Delta H_{rxn}$ is obtained. Chemists write the equation for the cold-pack process and its enthalpy change in the following way.

$$NH_4NO_3(s) \rightarrow NH_4^+(aq) + NO_3^-(aq) \quad \Delta H_{rxn} = 27 \text{ kJ}$$

**Figure 9** shows the energy change for the cold-pack process. In this process, the enthalpy of the products is 27 kJ greater than the enthalpy of the reactant because energy is absorbed. Thus, the sign of $\Delta H_{rxn}$ for this and all endothermic reactions and processes is positive. Recall that the sign of $\Delta H_{rxn}$ for all exothermic reactions is negative.

The enthalpy change, $\Delta H$, is equal to $q_p$, the heat gained or lost in a reaction or process carried out at constant pressure. Because all reactions presented in this textbook occur at constant pressure, you can assume that $q = \Delta H_{rxn}$.

## SECTION 2 REVIEW

Section Self-Check

### Section Summary

- In thermochemistry, the universe is defined as the system plus the surroundings.
- The heat lost or gained by a system during a reaction or process carried out at constant pressure is called the change in enthalpy ($\Delta H$).
- When $\Delta H$ is positive, the reaction is endothermic. When $\Delta H$ is negative, the reaction is exothermic.

**16.** **MAIN**IDEA **Describe** how you would calculate the amount of heat absorbed or released by a substance when its temperature changes.

**17.** **Explain** why $\Delta H$ for an exothermic reaction always has a negative value.

**18.** **Explain** why a measured volume of water is an essential part of a calorimeter.

**19.** **Explain** why you need to know the specific heat of a substance in order to calculate how much heat is gained or lost by the substance as a result of a temperature change.

**20.** **Describe** what *the system* means in thermodynamics, and explain how the system is related to the surroundings and the universe.

**21.** **Calculate** the specific heat in J/(g•°C) of an unknown substance if a 2.50-g sample releases 12.0 cal as its temperature changes from 25.0°C to 20.0°C.

**22.** **Design an Experiment** Describe a procedure you could follow to determine the specific heat of a 45-g piece of metal.

Matt Meadows/Getty Images

SECTION 3

# Thermochemical Equations

**MAINIDEA** Thermochemical equations express the amount of heat released or absorbed by chemical reactions.

### Essential Questions

- How are thermochemical equations for chemical reactions and other processes written?
- How is energy lost or gained during changes of state?
- How is the heat that is absorbed or released in a chemical reaction calculated?

### Review Vocabulary

**combustion reaction:** a chemical reaction that occurs when a substance reacts with oxygen, releasing energy in the form of heat and light

### New Vocabulary

thermochemical equation
enthalpy (heat) of combustion
molar enthalpy (heat) of vaporization
molar enthalpy (heat) of fusion

**CHEM 4 YOU** Have you ever been exhausted after a hard race or other strenuous activity? If you felt as if your body had less energy than before the event, you were right. That tired feeling relates to combustion reactions that occur in the cells of your body, the same combustion you might observe in a burning campfire.

## Writing Thermochemical Equations

The change in energy is an important part of chemical reactions, so chemists include $\Delta H$ as part of many chemical equations. The heat-pack and cold-pack equations are called thermochemical equations when they are written as follows.

$$4Fe(s) + 3O_2(g) \rightarrow 2Fe_2O_3(s) \quad \Delta H = -1625 \text{ kJ}$$

$$NH_4NO_3(s) \rightarrow NH_4^+(aq) + NO_3^-(aq) \quad \Delta H = 27 \text{ kJ}$$

A **thermochemical equation** is a balanced chemical equation that includes the physical states of all reactants and products and the energy change, usually expressed as the change in enthalpy, $\Delta H$.

The highly exothermic combustion of glucose ($C_6H_{12}O_6$) occurs in the body as food is metabolized to produce energy. The thermochemical equation for the combustion of glucose is shown below.

$$C_6H_{12}O_6(s) + 6O_2(g) \rightarrow 6CO_2(g) + 6H_2O(l) \quad \Delta H_{comb} = -2808 \text{ kJ}$$

The **enthalpy (heat) of combustion** ($\Delta H_{comb}$) of a substance is the enthalpy change for the complete burning of one mole of the substance. Standard enthalpies of combustion for several substances are given in **Table 3.** Standard enthalpy changes have the symbol $\Delta H^{\circ}$. The zero superscript tells you that the enthalpy changes were determined with all reactants and products at standard conditions. Standard conditions are 1 atm pressure and 298 K (25°C) and should not be confused with standard temperature and pressure (STP).

**Table 3** Standard Enthalpies of Combustion

| Substance | Formula | $\Delta H^{\circ}_{comb}$ (kJ/mol) |
|---|---|---|
| Sucrose (table sugar) | $C_{12}H_{22}O_{11}(s)$ | −5644 |
| Octane (a component of gasoline) | $C_8H_{18}(l)$ | −5471 |
| Glucose (a simple sugar found in fruit) | $C_6H_{12}O_6(s)$ | −2808 |
| Propane (a gaseous fuel) | $C_3H_8(g)$ | −2219 |
| Methane (a gaseous fuel) | $CH_4(g)$ | −891 |

**Table 4 Standard Enthalpies of Vaporization and Fusion**

| Substance | Formula | $\Delta H^\circ_{vap}$ (kJ/mol) | $\Delta H^\circ_{fus}$ (kJ/mol) |
|---|---|---|---|
| Water | $H_2O$ | 40.7 | 6.01 |
| Ethanol | $C_2H_5OH$ | 38.6 | 4.94 |
| Methanol | $CH_3OH$ | 35.2 | 3.22 |
| Acetic acid | $CH_3COOH$ | 23.4 | 11.7 |
| Ammonia | $NH_3$ | 23.3 | 5.66 |

## Changes of State

Many processes other than chemical reactions absorb or release heat. For example, think about what happens when you step out of a hot shower. You shiver as water evaporates from your skin. That is because your skin provides the heat needed to vaporize the water.

As heat is taken from your skin to vaporize the water, you cool down. The heat required to vaporize one mole of a liquid is called its **molar enthalpy (heat) of vaporization** ($\Delta H_{vap}$). Similarly, if you want a glass of cold water, you might drop an ice cube into it. The water cools as it provides the heat to melt the ice. The heat required to melt one mole of a solid substance is called its **molar enthalpy (heat) of fusion** ($\Delta H_{fus}$). Because vaporizing a liquid and melting a solid are endothermic processes, their $\Delta H$ values are positive. Standard molar enthalpies of vaporization and fusion for five common compounds are shown in **Table 4.**

**Thermochemical equations for changes of state** The vaporization of water and the melting of ice can be described by the following equations.

$$H_2O(l) \rightarrow H_2O(g) \quad \Delta H_{vap} = 40.7 \text{ kJ}$$
$$H_2O(s) \rightarrow H_2O(l) \quad \Delta H_{fus} = 6.01 \text{ kJ}$$

The first equation indicates that 40.7 kJ of energy is absorbed when one mole of water is converted to one mole of water vapor. The second equation indicates that 6.01 kJ of energy is absorbed when one mole of ice melts to form one mole of liquid water.

What happens in the reverse processes, when water vapor condenses to liquid water or liquid water freezes to ice? The same amounts of energy are released in these exothermic processes as are absorbed in the endothermic processes of vaporization and melting. Thus, the molar enthalpy (heat) of condensation ($\Delta H_{cond}$) and the molar enthalpy of vaporization have the same numerical value but opposite signs. Similarly, the molar enthalpy (heat) of solidification ($\Delta H_{solid}$) and the molar enthalpy of fusion have the same numerical value but opposite signs.

$$\Delta H_{vap} = -\Delta H_{cond}$$
$$\Delta H_{fus} = -\Delta H_{solid}$$

These relationships are illustrated in **Figure 10.**

■ **Figure 10** The upward arrows show that the energy of the system increases as ice melts and then vaporizes. The downward arrows show that the energy of the system decreases as water vapor condenses and then solidifies.

**Phase Changes for Water**

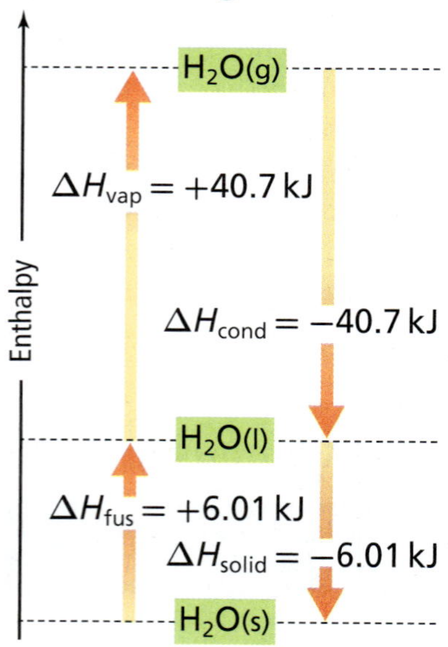


View an **animation about heat flow in endothermic and exothermic reactions.**

**Concepts In Motion**

Compare the following equations for the condensation and freezing of water with the equations on the previous page for the vaporization and melting of water.

$$H_2O(g) \rightarrow H_2O(l) \qquad \Delta H_{cond} = -40.7 \text{ kJ}$$

$$H_2O(l) \rightarrow H_2O(s) \qquad \Delta H_{solid} = -6.01 \text{ kJ}$$

Some farmers make use of the heat of fusion of water to protect fruit and vegetables from freezing. If the temperature is predicted to drop to freezing, they flood their orchards or fields with water. When the water freezes, energy ($\Delta H_{fus}$) is released and often warms the surrounding air enough to prevent frost damage. In the Problem-Solving Lab that follows, you will draw the heating curve of water and interpret it using the heats of fusion and vaporization.

**READING CHECK** **Categorize** condensation, solidification, vaporization, and fusion as exothermic or endothermic processes.

# Problem-Solving LAB

## Make and Use Graphs

**How can you derive a heating curve?** Water molecules have a strong attraction to one another because they are polar and they form hydrogen bonds. The polarity of water accounts for its high specific heat and relatively high enthalpies of fusion and vaporization.

### Analysis

Use the data in the table to plot a heating curve of temperature versus time for a 180-g sample of water as it is heated at a constant rate from −20°C to 120°C. Draw a best-fit line through the points. Note the time required for water to pass through each segment of the graph.

### Think Critically

1. **Analyze** each of the five regions of the graph, which are distinguished by an abrupt change in slope. Indicate how the absorption of heat changes the energy (kinetic and potential) of the water molecules.
2. **Calculate** the amount of heat required to pass through each region of the graph (180 g $H_2O$ = 10 mol $H_2O$, $\Delta H_{fus}$ = 6.01 kJ/mol, $\Delta H_{vap}$ = 40.7 kJ/mol, $c_{H_2O(s)}$ = 2.03 J/(g·°C), $c_{H_2O(\ell)}$ = 4.184 J/(g·°C), $c_{H_2O(g)}$ = 2.01 J/(g·°C). How does the length of time needed to pass through each region relate to the amount of heat absorbed?

**Time and Temperature Data for Water**

| Time (min) | Temperature (°C) | Time (min) | Temperature (°C) |
|---|---|---|---|
| 0.0 | −20 | 13.0 | 100 |
| 1.0 | 0 | 14.0 | 100 |
| 2.0 | 0 | 15.0 | 100 |
| 3.0 | 9 | 16.0 | 100 |
| 4.0 | 26 | 17.0 | 100 |
| 5.0 | 42 | 18.0 | 100 |
| 6.0 | 58 | 19.0 | 100 |
| 7.0 | 71 | 20.0 | 100 |
| 8.0 | 83 | 21.0 | 100 |
| 9.0 | 92 | 22.0 | 100 |
| 10.0 | 98 | 23.0 | 100 |
| 11.0 | 100 | 24.0 | 100 |
| 12.0 | 100 | 25.0 | 100 |

3. **Infer** What would the heating curve of ethanol look like? Ethanol melts at −114°C and boils at 78°C. Sketch ethanol's curve from −120°C to 90°C. What factors determine the lengths of the flat regions of the graph and the slope of the curve between the flat regions?

## EXAMPLE Problem 4

Find help with **dimensional analysis.** Math Handbook

EXAMPLE PROBLEM

**THE ENERGY RELEASED IN A REACTION** A bomb calorimeter is useful for measuring the energy released in combustion reactions. The reaction is carried out in a constant-volume bomb with a high pressure of oxygen. How much heat is evolved when 54.0 g glucose ($C_6H_{12}O_6$) is burned according to this equation?

$$C_6H_{12}O_6(s) + 6O_2(g) \rightarrow 6CO_2(g) + 6H_2O(l) \; \Delta H_{comb} = -2808 \text{ kJ}$$

### 1 ANALYZE THE PROBLEM

You are given a mass of glucose, the equation for the combustion of glucose, and $\Delta H_{comb}$. You must convert grams of glucose to moles of glucose. Because the molar mass of glucose is more than three times the mass of glucose burned, you can predict that the energy evolved will be less than one-third $\Delta H_{comb}$.

**Known**

mass of glucose = 54.0 g $C_6H_{12}O_6$

$\Delta H_{comb} = -2808$ kJ

**Unknown**

$q = ?$ kJ

### 2 SOLVE FOR THE UNKNOWN

Convert grams of $C_6H_{12}O_6$ to moles of $C_6H_{12}O_6$.

$$54.0 \cancel{\text{g } C_6H_{12}O_6} \times \frac{1 \text{ mol } C_6H_{12}O_6}{180.18 \cancel{\text{g } C_6H_{12}O_6}} = 0.300 \text{ mol } C_6H_{12}O_6$$

**Multiply by the inverse of molar mass, $\frac{1 \text{ mol}}{180.18 \text{ g}}$.**

Multiply moles of $C_6H_{12}O_6$ by the enthalpy of combustion, $\Delta H_{comb}$.

$$0.300 \cancel{\text{mol } C_6H_{12}O_6} \times \frac{2808 \text{ kJ}}{1 \cancel{\text{mol } C_6H_{12}O_6}} = \textbf{842 kJ}$$

**Multiply moles of glucose by $\frac{2808 \text{ kJ}}{1 \text{ mol } C_6H_{12}O_6}$.**

### 3 EVALUATE THE ANSWER

All values in the calculation have at least three significant figures, so the answer is correctly stated with three digits. As predicted, the released energy is less than one-third $\Delta H_{comb}$.

## PRACTICE Problems

Do additional problems. Online Practice

**23.** Calculate the heat required to melt 25.7 g of solid methanol at its melting point. Refer to **Table 4.**

**24.** How much heat evolves when 275 g of ammonia gas condenses to a liquid at its boiling point? Use **Table 4** to determine $\Delta H_{cond}$.

**25. Challenge** What mass of methane ($CH_4$) must be burned in order to liberate 12,880 kJ of heat? Refer to **Table 3**.

**Connection to Biology** When a mole of glucose is burned in a bomb calorimeter, 2808 kJ of energy is released. The same amount of energy is produced in your body when an equal mass of glucose is metabolized in the process of cellular respiration. The process takes place in every cell of your body in a series of complex steps in which glucose is broken down and carbon dioxide and water are released. These are the same products produced by the combustion of glucose in a calorimeter. The energy released is stored as chemical potential energy in the bonds of molecules of adenosine triphosphate (ATP). When energy is needed by any part of the body, molecules of ATP release their energy.

## Combustion Reactions

Combustion is the reaction of a fuel with oxygen. In biological systems, food is the fuel. **Figure 11** illustrates some of the many foods that contain glucose as well as other foods that contain carbohydrates that are readily converted to glucose in your body. You also depend on other combustion reactions to keep you warm or cool, and to transport you in vehicles. One way you might heat your home or cook your food is by burning methane gas. The combustion of one mole of methane produces 891 kJ according to this equation.

$$CH_4(g) + 2O_2(g) \rightarrow CO_2(g) + 2H_2O(l) + 891\ kJ$$

Most vehicles—cars, airplanes, boats, and trucks—run on the combustion of gasoline, which is mostly octane ($C_8H_{18}$). **Table 3** shows that the burning of one mole of octane produces 5471 kJ. The equation for the combustion of gasoline is as follows.

$$C_8H_{18}(l) + \frac{25}{2}O_2(g) \rightarrow 8CO_2(g) + 9H_2O(l) + 5471\ kJ$$

Another combustion reaction is the reaction between hydrogen and oxygen.

$$H_2(g) + \frac{1}{2}O_2(g) \rightarrow H_2O(l) + 286\ kJ$$

The combustion of hydrogen provides the energy to lift the shuttle into space, as illustrated on the opening page of this chapter.

■ **Figure 11** These foods are fuels for the body. They provide the glucose that is burned to produce 2808 kJ/mol to carry on the activities of life.

## SECTION 3 REVIEW

Section Self-Check

### Section Summary

- A thermochemical equation includes the physical states of the reactants and products and specifies the change in enthalpy.
- The molar enthalpy (heat) of vaporization, $\Delta H_{vap}$, is the amount of energy required to evaporate one mole of a liquid.
- The molar enthalpy (heat) of fusion, $\Delta H_{fus}$, is the amount of energy needed to melt one mole of a solid.

**26. MAIN IDEA Write** a complete thermochemical equation for the combustion of ethanol ($C_2H_5OH$). $\Delta H_{comb} = -1367$ kJ/mol

**27. Determine** Which of the following processes are exothermic? Endothermic?

**a.** $C_2H_5OH(l) \rightarrow C_2H_5OH(g)$

**b.** $Br_2(l) \rightarrow Br_2(s)$

**c.** $C_5H_{12}(g) + 8O_2(g) \rightarrow 5CO_2(g) + 6H_2O(l)$

**d.** $NH_3(g) \rightarrow NH_3(l)$

**e.** $NaCl(s) \rightarrow NaCl(l)$

**28. Explain** how you could calculate the heat released in freezing 0.250 mol water.

**29. Calculate** How much heat is released by the combustion of 206 g of hydrogen gas? $\Delta H_{comb} = -286$ kJ/mol

**30. Apply** The molar heat of vaporization of ammonia is 23.3 kJ/mol. What is the molar heat of condensation of ammonia?

**31. Interpret Scientific Illustrations** The reaction A → C is shown in the enthalpy diagram at right. Is the reaction exothermic or endothermic? Explain your answer.

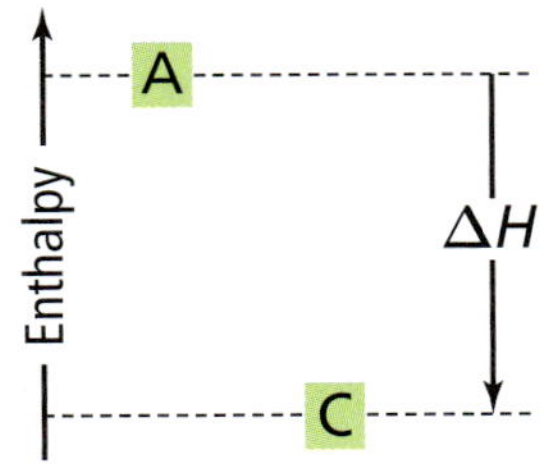

SECTION 4

# Calculating Enthalpy Change

**MAINIDEA** The enthalpy change for a reaction can be calculated using Hess's law.

### Essential Questions

- How is Hess's law applied to calculate the enthalpy change for a reaction?
- What is the basis for the table of standard enthalpies of formation?
- How is $\Delta H_{rxn}$ calculated using thermochemical equations?
- What is the enthalpy change for a reaction using standard enthalpies of formation data?

### Review Vocabulary

**allotrope:** one of two or more forms of an element with different structures and properties when they are in the same state

### New Vocabulary

Hess's law
standard enthalpy (heat) of formation

**CHEM 4 YOU** Maybe you have watched a two-act play or a two-part TV show. Each part tells some of the story, but you have to see both parts to understand the entire story. Like such a play or show, some reactions are best understood when you view them as the sum of two or more simpler reactions.

## Hess's Law

Sometimes it is impossible or impractical to measure the $\Delta H$ of a reaction by using a calorimeter. Consider the reaction in **Figure 12,** the conversion of carbon in its allotropic form, diamond, to carbon in its allotropic form, graphite.

$$C(s, diamond) \rightarrow C(s, graphite)$$

This reaction occurs so slowly that measuring the enthalpy change is impossible. Other reactions occur under conditions difficult to duplicate in a laboratory. Still others produce products other than the desired ones. For these reactions, chemists use a theoretical way to determine $\Delta H$.

Suppose you are studying the formation of sulfur trioxide in the atmosphere. You would need to determine $\Delta H$ for this reaction.

$$2S(s) + 3O_2(g) \rightarrow 2SO_3(g) \quad \Delta H = ?$$

Unfortunately, laboratory experiments to produce sulfur trioxide and determine its $\Delta H$ result in a mixture of products that is mostly sulfur dioxide ($SO_2$). In situations such as this, you can calculate $\Delta H$ by using Hess's law of heat summation. **Hess's law** states that if you can add two or more thermochemical equations to produce a final equation for a reaction, then the sum of the enthalpy changes for the individual reactions is the enthalpy change for the final reaction.

■ **Figure 12** The expression "diamonds are forever" suggests the durability of diamonds and tells you that the conversion of diamond to graphite is so slow that it would be impossible to measure its enthalpy change.

(l)©Tom Grill/Corbis; (r)Joel Arem/Science Source/Photo Researchers

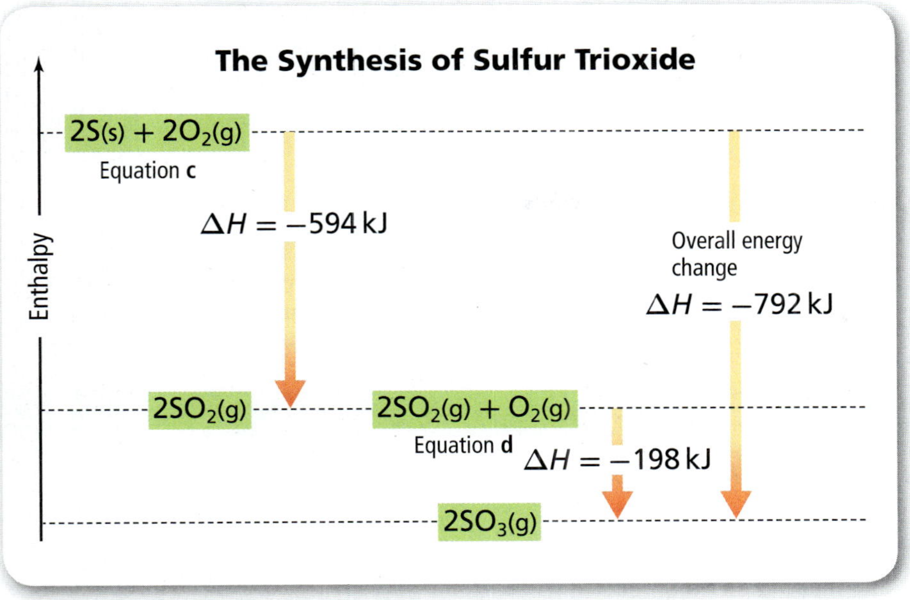


**Figure 13** The arrow on the left indicates the release of 594 kJ as S and $O_2$ react to form $SO_2$ (Equation **c**). Then, $SO_2$ and $O_2$ react to form $SO_3$ (Equation **d**) with the release of 198 kJ (middle arrow). The overall energy change (the sum of the two processes) is shown by the arrow on the right.

**Determine** *the enthalpy change for the decomposition of $SO_3$ to S and $O_2$.*

**Applying Hess's law** How can Hess's law be used to calculate the energy change for the reaction that produces $SO_3$?

$$2S(s) + 3O_2(g) \rightarrow 2SO_3(g) \quad \Delta H = ?$$

**Step 1** Chemical equations are needed that contain the substances found in the desired equation and have known enthalpy changes. The following equations contain S, $O_2$, and $SO_3$.

**a.** $S(s) + O_2(g) \rightarrow SO_2(g) \quad \Delta H = -297 \text{ kJ}$

**b.** $2SO_3(g) \rightarrow 2SO_2(g) + O_2(g) \quad \Delta H = 198 \text{ kJ}$

**Step 2** The desired equation shows two moles of sulfur reacting, so rewrite Equation **a** for two moles of sulfur by multiplying the coefficients by two. Double the enthalpy change $\Delta H$ because twice the energy will be released if two moles of sulfur react. With these changes, Equation **a** becomes the following (Equation **c**).

**c.** $2S(s) + 2O_2(g) \rightarrow 2SO_2(g) \quad \Delta H = 2(-297 \text{ kJ}) = -594 \text{ kJ}$

**Step 3** In the desired equation, sulfur trioxide is a product rather than a reactant, so reverse Equation **b.** When you reverse an equation, you must also change the sign of its $\Delta H$. Equation **b** then becomes Equation **d.**

**d.** $2SO_2(g) + O_2(g) \rightarrow 2SO_3(g) \quad \Delta H = -198 \text{ kJ}$

**Step 4** Add Equations **c** and **d** to obtain the desired reaction. Add the corresponding $\Delta H$ values. Cancel any terms that are common to both sides of the combined equation.

$$2S(s) + 2O_2(g) \rightarrow 2SO_2(g) \qquad \Delta H = -594 \text{ kJ}$$
$$2SO_2(g) + O_2(g) \rightarrow 2SO_3(g) \qquad \Delta H = -198 \text{ kJ}$$
$$2\cancel{SO_2}(g) + 2S(s) + 3O_2(g) \rightarrow 2\cancel{SO_2}(g) + 2SO_3(g) \quad \Delta H = -792 \text{ kJ}$$

The thermochemical equation for the burning of sulfur to form sulfur trioxide is as follows. **Figure 13** diagrams the energy changes.

$$2S(s) + 3O_2(g) \rightarrow 2SO_3(g) \quad \Delta H = -792 \text{ kJ}$$

Thermochemical equations are usually written and balanced for one mole of product. Often, that means that fractional coefficients must be used. For example, the thermochemical equation for the reaction between sulfur and oxygen to form one mole of sulfur trioxide is the following.

$$S(s) + \frac{3}{2}O_2(g) \rightarrow SO_3(g) \ \ \Delta H = -396 \text{ kJ}$$

**READING CHECK** **Compare** the equation above with the thermochemical equation developed on the previous page. How are they different?

## EXAMPLE Problem 5

EXAMPLE PROBLEM

**HESS'S LAW** Use thermochemical Equations **a** and **b** below to determine $\Delta H$ for the decomposition of hydrogen peroxide ($H_2O_2$), a compound that has many uses ranging from bleaching hair to powering rocket engines.

$2H_2O_2(l) \rightarrow 2H_2O(l) + O_2(g)$

**a.** $2H_2(g) + O_2(g) \rightarrow 2H_2O(l)\ \Delta H = -572$ kJ

**b.** $H_2(g) + O_2(g) \rightarrow H_2O_2(l)\ \Delta H = -188$ kJ

### 1 ANALYZE THE PROBLEM

You have been given two chemical equations and their enthalpy changes. These two equations contain all the substances found in the desired equation.

| Known | Unknown |
|---|---|
| **a.** $2H_2(g) + O_2(g) \rightarrow 2H_2O(l)\ \ \Delta H = -572$ kJ | $\Delta H = ?$ **kJ** |
| **b.** $H_2(g) + O_2(g) \rightarrow H_2O_2(l)\ \ \Delta H = -188$ kJ | |

### 2 SOLVE FOR THE UNKNOWN

$H_2O_2$ is a reactant.

$H_2O_2(aq) \rightarrow H_2(g) + O_2(g)\ \ \Delta H = 188$ kJ — **Reverse Equation b and change the sign of $\Delta H$.**

Two moles of $H_2O_2$ are needed.

**c.** $2H_2O_2(aq) \rightarrow 2H_2(g) + 2O_2(g)$ — **Multiply the reversed Equation b by two to obtain Equation c.**

$\Delta H$ for Equation **c** $= (188 \text{ kJ})(2)$ — **Multipy 188 kJ by two to obtain $\Delta H$ for Equation c.**

$= 376$ kJ

**c.** $2H_2O_2(aq) \rightarrow 2H_2(g) + 2O_2(g)\ \ \Delta H = 376$ kJ — **Write Equation c and $\Delta H$.**

Add Equations **a** and **c,** canceling any terms common to both sides of the combined equation. Add the enthalpies of Equations **a** and **c**.

**a.** $2H_2(g) + O_2(g) \rightarrow 2H_2O(l)\ \ \ \Delta H = -572$ kJ — **Write Equation a.**

**c.** $2H_2O_2(l) \rightarrow 2H_2(g) + 2O_2(g)\ \ \Delta H = 376$ kJ — **Write Equation c.**

$2H_2O_2(l) \rightarrow 2H_2O(l) + O_2(g)\ \ $ **$\Delta H = -196$ kJ** — **Add Equations a and c. Add the enthalpies.**

### 3 EVALUATE THE ANSWER

The two equations produce the desired equation. All values are accurate to the ones place, so $\Delta H$ is correctly stated.

**PRACTICE Problems** Do additional problems. Online Practice

PRACTICE PROBLEMS

**32.** Use Equations **a** and **b** to determine $\Delta H$ for the following reaction.

$2CO(g) + 2NO(g) \rightarrow 2CO_2(g) + N_2(g)$ $\Delta H = ?$

**a.** $2CO(g) + O_2(g) \rightarrow 2CO_2(g)$ $\Delta H = -566.0$ kJ

**b.** $N_2(g) + O_2(g) \rightarrow 2NO(g)$ $\Delta H = -180.6$ kJ

**33. Challenge** $\Delta H$ for the following reaction is $-1789$ kJ. Use this and Equation **a** to determine $\Delta H$ for Equation **b.**

$4Al(s) + 3MnO_2(s) \rightarrow 2Al_2O_3(s) + 3Mn(s)$ $\Delta H = -1789$ kJ

**a.** $4Al(s) + 3O_2(g) \rightarrow 2Al_2O_3(s)$ $\Delta H = -3352$ kJ

**b.** $Mn(s) + O_2(g) \rightarrow MnO_2(s)$ $\Delta H = ?$

## Standard Enthalpy (Heat) of Formation

Hess's law allows you to calculate unknown $\Delta H$ values using known reactions and their experimentally determined $\Delta H$ values. However, recording $\Delta H$ values for all known chemical reactions would be a huge and unending task. Instead, scientists record and use enthalpy changes for only one type of reaction—a reaction in which a compound is formed from its elements in their standard states. The standard state of a substance means the normal physical state of the substance at 1 atm and 298 K (25°C). For example, in their standard states, iron is a solid, mercury is a liquid, and oxygen is a diatomic gas.

The $\Delta H$ value for such a reaction is called the standard enthalpy (heat) of formation of the compound. The **standard enthalpy (heat) of formation** ($\Delta H_f^\circ$) is defined as the change in enthalpy that accompanies the formation of one mole of the compound in its standard state from its elements in their standard states. A typical standard heat of formation reaction is the formation of one mole of $SO_3$ from its elements.

$$S(s) + \frac{3}{2}O_2(g) \rightarrow SO_3(g) \quad \Delta H_f^\circ = -396 \text{ kJ}$$

The product of this equation is $SO_3$, a suffocating gas that produces acid rain when mixed with moisture in the atmosphere. The destructive results of acid precipitation are shown in **Figure 14.**

■ **Figure 14** Sulfur trioxide combines with water in the atmosphere to form sulfuric acid ($H_2SO_4$), a strong acid, which reaches Earth as acid precipitation. Acid precipitation slowly destroys trees and property.

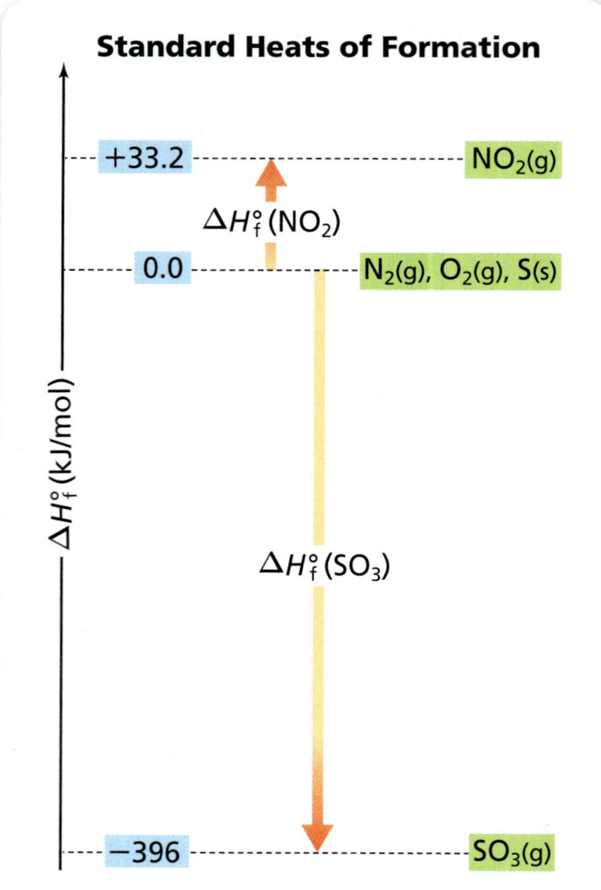


**Figure 15** $\Delta H_f^\circ$ for the elements $N_2$, $O_2$, and S is 0.0 kJ. When $N_2$ and $O_2$ react to form 1 mole of $NO_2$, 33.2 kJ is absorbed. Thus, $\Delta H_f^\circ$ for $NO_2$ is +33.2 kJ/mol. When S and $O_2$ react to form one mole of $SO_3$, 396 kJ is released. Therefore, $\Delta H_f^\circ$ for $SO_3$ is −396 kJ/mol.

**Predict** *Describe the approximate location of water on the scale. The heat of formation for the reaction* $H_2(g) + \frac{1}{2}O_2(g) \rightarrow H_2O(l)$ *is* $\Delta H_f^\circ = -286$ *kJ/mol.*

**Where do standard heats of formation come from?** When you state the height of a mountain, you do so relative to some point of reference—usually sea level. In a similar way, standard enthalpies of formation are stated based on the following assumption: Elements in their standard state have a $\Delta H_f^\circ$ of 0.0 kJ. With zero as the starting point, the experimentally determined enthalpies of formation of compounds can be placed on a scale above and below the elements in their standard states. Think of the zero of the enthalpy scale as being similar to the arbitrary assignment of 0.0°C to the freezing point of water. All substances warmer than freezing water have a temperature above zero. All substances colder than freezing water have a temperature below zero.

### Enthalpies of formation from experiments

Standard enthalpies of formation of many compounds have been measured experimentally. For example, consider the equation for the formation of nitrogen dioxide.

$$\frac{1}{2}N_2(g) + O_2(g) \rightarrow NO_2(g) \quad \Delta H_f^\circ = +33.2 \text{ kJ}$$

The elements nitrogen and oxygen are diatomic gases in their standard states, so their standard enthalpies of formation are zero. When nitrogen and oxygen gases react to form one mole of nitrogen dioxide, the experimentally determined $\Delta H$ for the reaction is +33.2 kJ. That means that 33.2 kJ of energy is absorbed in this endothermic reaction. The energy content of the product $NO_2$ is 33.2 kJ greater than the energy content of the reactants. On a scale on which $\Delta H_f^\circ$ of reactants is 0.0 kJ, $\Delta H_f^\circ$ of $NO_2$ is +33.2 kJ. **Figure 15** shows that on the scale of standard enthalpies of formation, $NO_2$ is placed 33.2 kJ above the elements from which it was formed. Sulfur trioxide ($SO_3$) is placed 396 kJ below zero on the scale because the formation of $SO_3(g)$ is an exothermic reaction. The energy content of the sulfur trioxide, $\Delta H_f^\circ$, is −396 kJ. **Table 5** lists standard enthalpies of formation for some common compounds. A more complete list is in **Table R-11**.

**Table 5 Standard Enthalpies of Formation**

| Compound | Formation Equation | $\Delta H_f^\circ$(kJ/mol) |
|---|---|---|
| $H_2S(g)$ | $H_2(g) + S(s) \rightarrow H_2S(g)$ | −21 |
| $HF(g)$ | $\frac{1}{2}H_2(g) + \frac{1}{2}F_2(g) \rightarrow HF(g)$ | −273 |
| $SO_3(g)$ | $S(s) + \frac{3}{2}O_2(g) \rightarrow SO_3(g)$ | −396 |
| $SF_6(g)$ | $S(s) + 3F_2(g) \rightarrow SF_6(g)$ | −1220 |

■ **Figure 16** Sulfur hexafluoride is used to etch minute and sometimes intricate patterns on silicon wafers in the production semiconductor devices. Semiconductors are important components of modern electronic equipment, including computers, cell phones, and MP3 players.

Jeff Maloney/Getty Images

**Using standard enthalpies of formation** Standard enthalpies of formation can be used to calculate the enthalpies of many reactions under standard conditions $\Delta H^\circ_{rxn}$ using Hess's law. Suppose you want to calculate $\Delta H^\circ_{rxn}$ for a reaction that produces sulfur hexafluoride. Sulfur hexafluoride is a stable, unreactive gas with some interesting applications, one of which is shown in **Figure 16.**

$$H_2S(g) + 4F_2(g) \rightarrow 2HF(g) + SF_6(g) \quad \Delta H^\circ_{rxn} = ?$$

**Step 1** Refer to **Table 5** to find an equation for the formation of each of the three compounds in the desired equation—HF, $SF_6$, and $H_2S$.

**a.** $\frac{1}{2}H_2(g) + \frac{1}{2}F_2(g) \rightarrow HF(g) \quad \Delta H^\circ_f = -273\ kJ$

**b.** $S(s) + 3F_2(g) \rightarrow SF_6(g) \quad \Delta H^\circ_f = -1220\ kJ$

**c.** $H_2(g) + S(s) \rightarrow H_2S(g) \quad \Delta H^\circ_f = -21\ kJ$

**Step 2** Equations **a** and **b** describe the formation of the products HF and $SF_6$ in the desired equation, so use Equations **a** and **b** in the direction in which they are written.

Equation **c** describes the formation of a product, $H_2S$, but in the desired equation, $H_2S$ is a reactant. Reverse Equation **c** and change the sign of its $\Delta H^\circ_f$.

$$H_2S(g) \rightarrow H_2(g) + S(s) \quad \Delta H^\circ_f = 21\ kJ$$

**Step 3** Two moles of HF are required. Multiply Equation **a** and its enthalpy change by two.

$$H_2(g) + F_2(g) \rightarrow 2HF(g) \quad \Delta H^\circ_f = 2(-273) = -546\ kJ$$

**Step 4** Add the three equations and their enthalpy changes. The elements $H_2$ and S cancel.

$$\begin{array}{ll} \cancel{H_2(g)} + F_2(g) \rightarrow 2HF(g) & \Delta H^\circ_f = -546\ kJ \\ \cancel{S(s)} + 3F_2(g) \rightarrow SF_6(g) & \Delta H^\circ_f = -1220\ kJ \\ H_2S(g) \rightarrow \cancel{H_2(g)} + \cancel{S(s)} & \Delta H^\circ_f = 21\ kJ \\ \hline H_2S(g) + 4F_2(g) \rightarrow 2HF(g) + SF_6(g) & \Delta H^\circ_{rxn} = -1745\ kJ \end{array}$$

**The summation equation** The stepwise procedure you have just read about shows how standard heats of formation equations combine to produce the desired equation and its $\Delta H^\circ_{rxn}$. The procedure can be summed up in the following formula.

### Summation Equation

$$\Delta H^\circ_{rxn} = \Sigma\Delta H^\circ_f(\text{products}) - \Sigma\Delta H^\circ_f(\text{reactants})$$

**$\Delta H^\circ_{rxn}$ represents the standard enthalpy of the reaction.**

**$\Sigma$ represents the sum of the terms.**

**$\Delta H^\circ_f$(products) and $\Delta H^\circ_f$(reactants) represent the standard enthalpies of formation of all the products and all the reactants.**

$\Delta H^\circ_{rxn}$ is obtained by subtracting the sum of heats of formation of the reactants from the sum of the heats of formation of the products.

You can see how this formula applies to the reaction between hydrogen sulfide and fluorine.

$$H_2S(g) + 4F_2(g) \rightarrow 2HF(g) + SF_6(g)$$

$$\Delta H^\circ_{rxn} = [(2)\Delta H^\circ_f(HF) + \Delta H^\circ_f(SF_6)] - [\Delta H^\circ_f(H_2S) + (4)\Delta H^\circ_f(F_2)]$$

$$\Delta H^\circ_{rxn} = [(2)(-273\text{ kJ}) + (-1220\text{ kJ})] - [-21\text{ kJ} + (4)(0.0\text{ kJ})]$$

$$\Delta H^\circ_{rxn} = -1745\text{ kJ}$$

## EXAMPLE Problem 6

EXAMPLE PROBLEM

**ENTHALPY CHANGE FROM STANDARD ENTHALPIES OF FORMATION** Use standard enthalpies of formation to calculate $\Delta H^\circ_{rxn}$ for the combustion of methane.

$$CH_4(g) + 2O_2(g) \rightarrow CO_2(g) + 2H_2O(l)$$

### 1 ANALYZE THE PROBLEM

You are given an equation and asked to calculate the change in enthalpy. The formula $\Delta H^\circ_{rxn} = \Sigma\Delta H^\circ_{rxn}(\text{products}) - \Sigma\Delta H^\circ_f(\text{reactants})$ can be used with data from **Table R-11**.

| **Known** | **Unknown** |
|---|---|
| $\Delta H^\circ_f(CO_2) = -394$ kJ | $\Delta H^\circ_{rxn} = ?$ kJ |
| $\Delta H^\circ_f(H_2O) = -286$ kJ | |
| $\Delta H^\circ_f(CH_4) = -75$ kJ | |
| $\Delta H^\circ_f(O_2) = 0.0$ kJ | |

### 2 SOLVE FOR THE UNKNOWN

Use the formula $\Delta H^\circ_{rxn} = \Sigma\Delta H^\circ_f(\text{products}) - \Sigma\Delta H^\circ_f(\text{reactants})$.

Expand the formula to include a term for each reactant and product. Multiply each term by the coefficient of the substance in the balanced chemical equation.

$$\Delta H^\circ_{rxn} = [\Delta H^\circ_f(CO_2) + (2)\Delta H^\circ_f(H_2O)] - [\Delta H^\circ_f(CH_4) + (2)\Delta H^\circ_f(O_2)]$$

**Substitute $CO_2$ and $H_2O$ for the products, $CH_4$ and $O_2$ for the reactants. Multiply $H_2O$ and $O_2$ by two.**

$$\Delta H^\circ_{rxn} = [(-394\text{ kJ}) + (2)(-286\text{ kJ})] - [(-75\text{ kJ}) + (2)(0.0\text{ kJ})]$$

**Substitute $\Delta H^\circ_f(CO_2) = -394$ kJ, $\Delta H^\circ_f(H_2O) = -286$ kJ, $\Delta H^\circ_f(CH_4) = -75$ kJ, and $\Delta H^\circ_f(O_2) = 0.0$ kJ into the equation.**

$$\Delta H^\circ_{rxn} = [-966\text{ kJ}] - [-75\text{ kJ}] = -966\text{ kJ} + 75\text{ kJ} = \mathbf{-891\text{ kJ}}$$

The combustion of 1 mol $CH_4$ releases 891 kJ.

## 3 EVALUATE THE ANSWER

All values are accurate to the ones place. Therefore, the answer is correct as stated. The calculated value is the same as that given in **Table 3.** You can check your answer by using the stepwise procedure shown previously.

## PRACTICE Problems

Do additional problems. Online Practice

PRACTICE PROBLEMS

**34.** Show how the sum of enthalpy of formation equations produces each of the following reactions. You do not need to look up and include $\Delta H$ values.

**a.** $2NO(g) + O_2(g) \rightarrow 2NO_2(g)$

**b.** $SO_3(g) + H_2O(l) \rightarrow H_2SO_4(l)$

**35.** Use standard enthalpies of formation from **Table R-11** to calculate $\Delta H^\circ_{rxn}$ for the following reaction.

$4NH_3(g) + 7O_2(g) \rightarrow 4NO_2(g) + 6H_2O(l)$

**36.** Determine $\Delta H^\circ_{comb}$ for butanoic acid, $C_3H_7COOH(l) + 5O_2(g) \rightarrow 4CO_2(g) + 4H_2O(l)$. Use data in **Table R-11** and the following equation.

$4C(s) + 4H_2(g) + O_2(g) \rightarrow C_3H_7COOH(l)$ $\Delta H = -534$ kJ

**37. Challenge** Two enthalpy of formation equations, **a** and **b,** combine to form the equation for the reaction of nitrogen oxide and oxygen. The product of the reaction is nitrogen dioxide: $NO(g) + \frac{1}{2}O_2(g) \rightarrow NO_2(g)$ $\Delta H^\circ_{rxn} = -58.1$ kJ

**a.** $\frac{1}{2}N_2(g) + \frac{1}{2}O_2(g) \rightarrow NO(g)$ $\Delta H^\circ_f = 91.3$ kJ

**b.** $\frac{1}{2}N_2(g) + O_2(g) \rightarrow NO_2(g)$ $\Delta H^\circ_f = ?$

What is $\Delta H^\circ_f$ for Equation **b**?

## SECTION 4 REVIEW

Section Self-Check

### Section Summary

- The enthalpy change for a reaction can be calculated by adding two or more thermochemical equations and their enthalpy changes.
- Standard enthalpies of formation of compounds are determined relative to the assigned enthalpy of formation of the elements in their standard states.

**38. MAINIDEA Explain** what is meant by Hess's law and how it is used to determine $\Delta H^\circ_{rxn}$.

**39. Explain** in words the formula that can be used to determine $\Delta H^\circ_{rxn}$ when using Hess's law.

**40. Describe** how the elements in their standard states are defined on the scale of standard enthalpies of formation.

**41. Examine** the data in **Table 5.** What conclusion can you draw about the stabilities of the compounds listed relative to the elements in their standard states? Recall that low energy is associated with stability.

**42. Calculate** Use Hess's law to determine $\Delta H$ for the reaction $NO(g) + O(g) \rightarrow NO_2(g)$ $\Delta H = ?$ given the following reactions. Show your work.

$O_2(g) \rightarrow 2O(g)$ $\Delta H = +495$ kJ

$2O_3(g) \rightarrow 3O_2(g)$ $\Delta H = -427$ kJ

$NO(g) + O_3(g) \rightarrow NO_2(g) + O_2(g)$ $\Delta H = -199$ kJ

**43. Interpret Scientific Illustrations** Use the data below to draw a diagram of standard heats of formation similar to **Figure 15** and use your diagram to determine the heat of vaporization of water at 298 K.

Liquid water: $\Delta H^\circ_f = -285.8$ kJ/mol

Gaseous water: $\Delta H^\circ_f = -241.8$ kJ/mol

SECTION 5

# Reaction Spontaneity

**MAINIDEA** Changes in enthalpy and entropy determine whether a process is spontaneous.

**Essential Questions**

- What is the difference between spontaneous and nonspontaneous processes?
- How do changes in entropy and free energy determine the spontaneity of chemical reactions and other processes?

**Review Vocabulary**

**vaporization:** the energy-requiring process by which a liquid changes to a gas or vapor

**New Vocabulary**

spontaneous process
entropy
second law of thermodynamics
free energy

**CHEM 4 YOU** How is it that some newer buildings appear to be falling apart when others that are much older seem to stand forever? It might be the level of maintenance and work put into them. Similarly, in chemistry, without a constant influx of energy, there is a natural tendency toward disorder.

## Spontaneous Processes

In **Figure 17** you can see a familiar picture of what happens to an iron object when it is left outdoors in moist air. Iron rusts slowly according to the same chemical equation that describes what happens in the heat pack you read about earlier in the chapter.

$$4Fe(s) + 3O_2(g) \rightarrow 2Fe_2O_3(s) \quad \Delta H = -1625 \text{ kJ}$$

The heat pack goes into action the moment you activate it. Similarly, unprotected iron objects rust whether you want them to or not.

Rusting is spontaneous. Any physical or chemical change that once begun, occurs with no outside intervention is a **spontaneous process.** However, for many spontaneous processes, some energy from the surroundings must be supplied to get the process started. For example, you might use a match to light a Bunsen burner in your school lab.

Suppose you reverse the direction of the equation for the rusting of iron. Recall that when you change the direction of a reaction, the sign of $\Delta H$ changes. The reaction becomes endothermic.

$$2Fe_2O_3(s) \rightarrow 4Fe(s) + 3O_2(g) \quad \Delta H = 1625 \text{ kJ}$$

Reversing the equation will not make rust decompose spontaneously into iron and oxygen under ordinary conditions. The equation represents a reaction that is not spontaneous.

**Figure 17** Left unattended, with abundant water and oxygen in the air, the iron in this boat spontaneously converts to rust ($Fe_2O_3$).

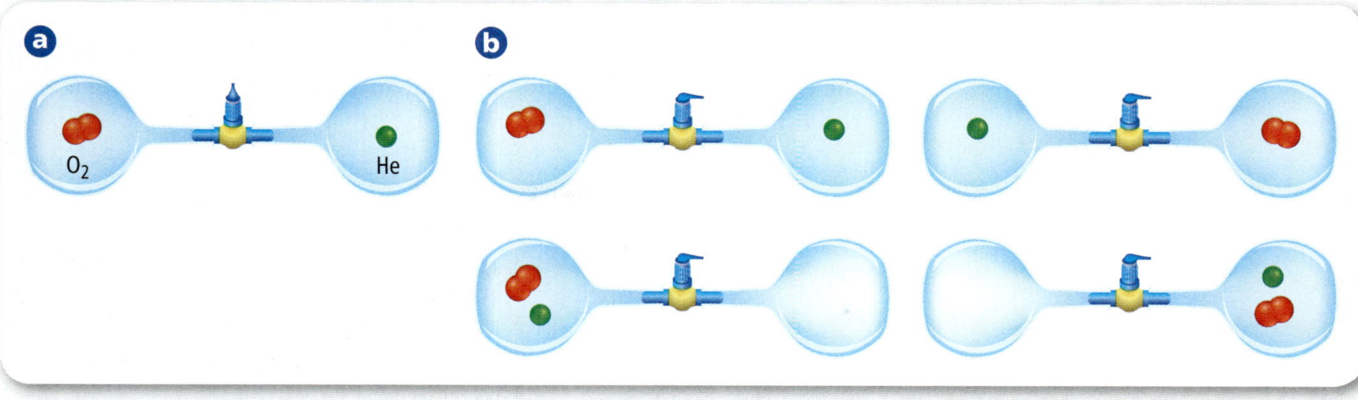


■ **Figure 18** In **a,** an oxygen molecule and a helium atom are each confined to a single bulb. When the stopcock is opened in **b,** the gas particles move freely into the double volume available. Four arrangements of the particles, which represent an increase in entropy, are possible at any given time.

The formation of rust on iron is an exothermic and spontaneous reaction. The reverse reaction is endothermic and nonspontaneous. You might conclude that all exothermic processes are spontaneous and all endothermic processes are nonspontaneous. But remember that ice melting at room temperature is a spontaneous, endothermic process. Something other than $\Delta H$ plays a role in determining whether a chemical process occurs spontaneously under a given set of conditions. That something is called entropy.

**What is entropy?** You're probably not surprised when the smell of brownies baking in the kitchen wafts to wherever you are in your home. And you know that gases tend to spread throughout Earth's atmosphere. Why do gases behave this way? When gases spread out, a system reaches a state of maximum entropy. **Entropy** ($S$) is a measure of the number of possible ways that the energy of a system can be distributed, and this is related to the freedom of the system's particles to move and number of ways they can be arranged.

Consider the two bulbs in **Figure 18.** When the stopcock is closed, one bulb contains a single molecule of oxygen. The other contains one atom of helium. When the stopcock is opened, the gas particles pass freely between the bulbs. Each gas particle can spread out into twice its original volume. The particles might be found in any of the four arrangements shown. The entropy of the system is greater with the stopcock open because the number of possible arrangements of the particles and the distribution of their energies is increased.

As the number of particles increases, the number of possible arrangements for a group of particles increases dramatically. If the two bulbs contained a total of ten particles, the number of possible arrangements would be 1024 times more than if the particles were confined to a single bulb. In general, the number of possible arrangements available to a system increases under the following conditions: when volume increases, when energy increases, when the number of particles increases, or when the particles' freedom of movement increases.

**The second law of thermodynamics** The tendency toward increased entropy is summarized in the **second law of thermodynamics,** which states that spontaneous processes always proceed in such a way that the entropy of the universe increases. Entropy is sometimes considered to be a measure of the disorder or randomness of the particles that make up a system. Particles that are more spread out are said to be more disordered, causing the system to have greater entropy than when the particles are closer together.

**Vocabulary**

**Science usage v. Common usage**

**System**

***Science usage:*** the particular reaction or process being studied

*The universe consists of the system and the surroundings.*

***Common usage:*** an organized or established procedure

*She worked out a system in which everyone would have an equal opportunity.*

Get help with **probability.**

Personal Tutor

**Predicting changes in entropy** Recall that the change in enthalpy for a reaction is equal to the enthalpy of the products minus the enthalpy of the reactants. The change in entropy ($\Delta S$) during a reaction or process is similar.

$$\Delta S_{system} = S_{products} - S_{reactants}$$

If the entropy of a system increases during a reaction or process, $S_{products} > S_{reactants}$ and $\Delta S_{system}$ is positive. Conversely, if the entropy of a system decreases during a reaction or process, $S_{products} < S_{reactants}$ and $\Delta S_{system}$ is negative.

You can sometimes predict if $\Delta S_{system}$ is positive or negative by examining the equation for a reaction or process.

1. *Entropy changes associated with changes in state can be predicted.* In solids, molecules have limited movement. In liquids, they have some freedom to move, and in gases, molecules can move freely within their container. Thus, entropy increases as a substance changes from a solid to a liquid and from a liquid to a gas. $\Delta S_{system}$ is positive as water vaporizes and methanol melts.

$$H_2O(l) \rightarrow H_2O(g) \quad \Delta S_{system} > 0$$

$$CH_3OH(s) \rightarrow CH_3OH(l) \quad \Delta S_{system} > 0$$

**VOCABULARY**

**WORD ORIGIN**

**Random**
comes from the Germanic word *rinnan,* meaning *to run;* a haphazard course

2. *The dissolving of a gas in a solvent always results in a decrease in entropy.* Gas particles have more entropy when they can move freely than when they are dissolved in a liquid or solid that limits their movements and randomness. $\Delta S_{system}$ is negative for the dissolving of oxygen in water as shown in **Figure 19.**

$$O_2(g) \rightarrow O_2(aq) \quad \Delta S_{system} < 0$$

3. *Assuming no change in physical state occurs, the entropy of a system usually increases when the number of gaseous product particles is greater than the number of gaseous reactant particles.* For the following reaction, $\Delta S_{system}$ is positive because two molecules of gas react and three molecules of gas are produced.

$$2SO_3(g) \rightarrow 2SO_2(g) + O_2(g) \quad \Delta S_{system} > 0$$

**Figure 19** In the bubbles, the nitrogen and oxygen gas molecules that make up most of the air can move more freely than when dissolved in the aquarium water.

■ **Figure 20** Sodium chloride and liquid water are pure substances, each with a degree of orderliness. When sodium chloride dissolves in water, the entropy of the system increases because sodium ions, chloride ions, and water molecules mix together to create a large number of random arrangements.

4. *With some exceptions, entropy increases when a solid or a liquid dissolves in a solvent.* The solute particles, which are together before dissolving, become dispersed throughout the solvent. The solute particles have more freedom of movement, as shown in **Figure 20** for the dissolving of sodium chloride in water. $\Delta S_{system}$ is positive.

$$NaCl(s) \rightarrow Na^{+}(aq) + Cl^{-}(aq) \quad \Delta S_{system} > 0$$

5. *The random motion of the particles of a substance increases as its temperature increases.* Recall that the kinetic energy of molecules increases with temperature. Increased kinetic energy means faster movement and more possible arrangements of particles. Therefore, the entropy of any substance increases as its temperature increases. $\Delta S_{system}$ is positive.

## PRACTICE Problems

Do additional problems. Online Practice

44. Predict the sign of $\Delta S_{system}$ for each of the following changes.
    - **a.** $ClF(g) + F_2(g) \rightarrow ClF_3(g)$
    - **b.** $NH_3(g) \rightarrow NH_3(aq)$
    - **c.** $CH_3OH(l) \rightarrow CH_3OH(aq)$
    - **d.** $C_{10}H_8(l) \rightarrow C_{10}H_8(s)$
45. **Challenge** Comment on the sign of $\Delta S_{system}$ for the following reaction. $Fe(s) + Zn^{2+}(aq) \rightarrow Fe^{2+}(aq) + Zn(s)$

### Connection to Earth Science — Earth's spontaneous processes

Volcanoes, fumaroles, hot springs, and geysers are evidence of geothermal energy in Earth's interior. Volcanoes are vents in Earth's crust from which molten rock (magma), steam, and other materials flow. When surface water moves downward through Earth's crust, it can interact with magma and/or hot rocks. Water that comes back to the surface in hot springs is heated to temperatures much higher than the surrounding air temperatures. Geysers are hot springs that spout hot water and steam into the air. Fumaroles emit steam and other gases, such as hydrogen sulfide. These geothermal processes are obviously spontaneous. Can you identify increases in entropy in these Earth processes?

■ **Figure 21** It is difficult to recognize this ancient Greek sculpture as the head of a lion. The particles of limestone that are loosened by wind and weather or dissolved by rain disperse randomly, destroying the precise representation of the image and increasing the entropy of the universe.

## Entropy, the Universe, and Free Energy

If you happen to break an egg, you know you cannot reverse the process and again make the egg whole. Similarly, an abandoned barn gradually disintegrates into a pile of decaying wood and a statue dissolves slowly in rainwater and disperses into the ground, as shown in **Figure 21.** Order turns to disorder in these processes, and the entropy of the universe increases.

What effect does entropy have on reaction spontaneity? Recall that the second law of thermodynamics states that the entropy of the universe must increase as a result of a spontaneous reaction or process. Therefore, the following is true for any spontaneous process.

$$\Delta S_{universe} > 0$$

Because the universe equals the system plus the surroundings, any change in the entropy of the universe is the sum of changes occurring in the system and surroundings.

$$\Delta S_{universe} = \Delta S_{system} + \Delta S_{surroundings}$$

In nature, $\Delta S_{universe}$ tends to be positive for reactions and processes under the following conditions.

1. *The reaction or process is exothermic, which means $\Delta H_{system}$ is negative.* The heat released by an exothermic reaction raises the temperature of the surroundings and thereby increases the entropy of the surroundings. $\Delta S_{surroundings}$ is positive.
2. *The entropy of the system increases, so $\Delta S_{system}$ is positive.*

Thus, exothermic chemical reactions accompanied by an increase in entropy are all spontaneous.

**FOLDABLES®** Incorporate information from this section into your Foldable.

**Free energy** Can you definitely determine if a reaction is spontaneous? In 1878, J. Willard Gibbs, a physicist at Yale University, defined a combined enthalpy-entropy function called Gibbs free energy that answers that question. For reactions or processes that take place at constant pressure and temperature, Gibbs free energy ($G_{system}$), commonly called **free energy,** is energy that is available to do work. Thus, free energy is useful energy. In contrast, some entropy is associated with energy that is spread out into the surroundings as, for example, random molecular motion, and cannot be recovered to do useful work. The free energy change ($\Delta G_{system}$) is the difference between the system's change in enthalpy ($\Delta H_{system}$) and the product of the kelvin temperature and the change in entropy ($T\Delta S_{system}$).

### Gibbs Free Energy Equation

$$\Delta G_{system} = \Delta H_{system} - T\Delta S_{system}$$

**$\Delta G_{system}$ represents the free energy change. $\Delta H_{system}$ represents the change in enthalpy. $T$ is temperature. $\Delta S_{system}$ represents the change in entropy.**

The free energy released or absorbed in a chemical reaction is equal to the difference between the enthalpy change and the product of the change in entropy (in joules per kelvin) and the temperature (in kelvins).

To calculate Gibbs free energy, it is usually necessary to convert units because $\Delta S$ is usually expressed in J/K, whereas $\Delta H$ is expressed in kJ.

**The sign of free energy** When a reaction or process occurs under standard conditions (298 K and 1 atm), the standard free energy change can be expressed as follows.

$$\Delta G^\circ_{system} = \Delta H^\circ_{system} - T\Delta S^\circ_{system}$$

If the sign of the free energy change ($\Delta G^\circ_{system}$) is negative, the reaction is spontaneous. If the sign of the free energy change is positive, the reaction is nonspontaneous.

Recall that free energy is energy that is available to do work. In contrast, energy related to entropy is useless because it is dispersed and cannot be harnessed to do work.

**Calculating free energy change** How do changes in enthalpy and entropy affect free energy change and spontaneity for the reaction between nitrogen and hydrogen to form ammonia?

$$N_2(g) + 3H_2(g) \rightarrow 2NH_3(g)$$

$$\Delta H^\circ_{system} = -91.8\ \text{kJ} \quad \Delta S^\circ_{system} = -197\ \text{J/K}$$

The entropy of the system decreases because 4 mol of gaseous molecules react and only 2 mol of gaseous molecules are produced. Therefore, $\Delta S^\circ_{system}$ is negative. A decrease in the entropy of the system tends to make the reaction nonspontaneous, but the reaction is exothermic ($\Delta H^\circ_{system}$ is negative), which tends to make the reaction spontaneous. To determine which of the two tendencies predominates, you must calculate $\Delta G^\circ_{system}$ for the reaction. First, convert $\Delta S^\circ_{system}$ to kilojoules.

$$\Delta S^\circ_{system} = -197\ \cancel{\text{J}}/\text{K} \times \frac{1\ \text{kJ}}{1000\ \cancel{\text{J}}} = -0.197\ \text{kJ/K}$$

Now, substitute $\Delta H^\circ_{system}$, $T$, and $\Delta S^\circ_{system}$ into the equation for $\Delta G^\circ_{system}$.

$$\Delta G^\circ_{system} = \Delta H^\circ_{system} - T\Delta S^\circ_{system}$$

$$\Delta G^\circ_{system} = -91.8\ \text{kJ} - (298\ \cancel{\text{K}})(-0.197\ \text{kJ}/\cancel{\text{K}})$$

$$\Delta G^\circ_{system} = -91.8\ \text{kJ} + 58.7\ \text{kJ} = -33.1\ \text{kJ}$$

$\Delta G^\circ_{system}$ for this reaction is negative, so the reaction is spontaneous.

The reaction between nitrogen and hydrogen demonstrates that the entropy of a system can decrease during a spontaneous process. However, it can do so only if the entropy of the surroundings increases more than the entropy of the system decreases. Thus, the entropy of the universe (system + surroundings) always increases in any spontaneous process. **Table 6** shows how reaction spontaneity depends on the signs of $\Delta H_{system}$ and $\Delta S_{system}$.

**VOCABULARY**
**ACADEMIC VOCABULARY**
**Demonstrate**
to show clearly
*People are standing by to demonstrate how the device works.*

Explore **reaction spontaneity with an interactive table.** Concepts In Motion

**Table 6 Reaction Spontaneity $\Delta G_{system} = \Delta H_{system} - T\Delta S_{system}$**

| $\Delta H_{system}$ | $\Delta S_{system}$ | $\Delta G_{system}$ | Reaction Spontaneity |
|---|---|---|---|
| negative | positive | always negative | always spontaneous |
| negative | negative | negative or positive | spontaneous at lower temperatures |
| positive | positive | negative or positive | spontaneous at higher temperatures |
| positive | negative | always positive | never spontaneous |

## EXAMPLE Problem 7

EXAMPLE PROBLEM

**DETERMINE REACTION SPONTANEITY** For a process, $\Delta H_{system}$ = 145 kJ and $\Delta S_{system}$ = 322 J/K. Is the process spontaneous at 382 K?

### 1 ANALYZE THE PROBLEM

You must calculate $\Delta G_{system}$ to determine spontaneity.

| **Known** | **Unknown** |
|---|---|
| $T$ = 382 K | sign of $\Delta G_{system}$ = ? |
| $\Delta H_{system}$ = 145 kJ | |
| $\Delta S_{system}$ = 322 J/K | |

### 2 SOLVE FOR THE UNKNOWN

Convert $\Delta S_{system}$ to kJ/K

$322\ \cancel{J}/K \times \frac{1\ kJ}{1000\ \cancel{J}} = 0.322\ kJ/K$ — **Convert $\Delta S_{system}$ to kJ/K.**

Solve the free energy equation.

$\Delta G_{system} = \Delta H_{system} - T\Delta S_{system}$ — **State the Gibbs free energy equation.**

$\Delta G_{system} = 145\ kJ - (382\ \cancel{K})(0.322\ kJ/\cancel{K})$ — **Substitute $T$ = 382 K, $\Delta H_{system}$ = 145 kJ, and $\Delta S_{system}$ = 0.322 kJ/K**

$\Delta G_{system} = 145\ kJ - 123\ kJ = 22\ kJ$ — **Multiply and subtract numbers.**

Because $\Delta G_{system}$ is positive, the reaction is nonspontaneous.

### 3 EVALUATE THE ANSWER

Because $\Delta H$ is positive and the temperature is not high enough to make the second term of the equation greater than the first, $\Delta G_{system}$ is positive. The significant figures are correct.

## PRACTICE Problems

Do additional problems. Online Practice

**46.** Determine whether each of the following reactions is spontaneous.

**a.** $\Delta H_{system}$ = −75.9 kJ, $T$ = 273 K, $\Delta S_{system}$ = 138 J/K

**b.** $\Delta H_{system}$ = −27.6 kJ, $T$ = 535 K, $\Delta S_{system}$ = −55.2 J/K

**c.** $\Delta H_{system}$ = 365 kJ, $T$ = 388 K, $\Delta S_{system}$ = −55.2 J/K

**d.** $\Delta H_{system}$ = 452 kJ, $T$ = 165 K, $\Delta S_{system}$ = 55.7 J/K

**47. Challenge** Given $\Delta H_{system}$ = −144 kJ and $\Delta S_{system}$ = −36.8 J/K for a reaction, determine the lowest temperature in kelvins at which the reaction would be spontaneous.

## SECTION 5 REVIEW

Section Self-Check

### Section Summary

- Entropy is a measure of the disorder or randomness of a system.
- Spontaneous processes always result in an increase in the entropy of the universe.
- Free energy is the energy available to do work. The sign of the free energy change indicates whether the reaction is spontaneous.

**48.** MAINIDEA **Compare and contrast** spontaneous and nonspontaneous reactions.

**49. Describe** how a system's entropy changes if the system becomes more disordered during a process.

**50. Decide** Does the entropy of a system increase or decrease when you dissolve a cube of sugar in a cup of tea? Define the system, and explain your answer.

**51. Determine** whether the system $\Delta H_{system}$ = −20.5 kJ, $T$ = 298 K, and $\Delta S_{system}$ = −35.0 J/K is spontaneous or nonspontaneous.

**52. Outline** Use the blue and red headings to outline the section. Under each heading, summarize the important ideas discussed.

HOW IT **works**

## DRIVING THE FUTURE: Flexible Fuel Vehicles

The service stations of the not-too-distant future will not only deliver various grades of gasoline, but they will also pump a fuel called E85. This fuel can be used in a flexible-fuel vehicle, or FFV. Conventional vehicles operate on 100% gasoline or on a blend of 10% ethanol and 90% gasoline. FFVs, however, operate on all these blends and E85, which is 85% ethanol. E85 has the advantage of not being highly dependent on fossil fuels.

**1 Renewable Resource** E85 is 15% gasoline and 85% ethanol by volume. Ethanol ($C_2H_5OH$) is a renewable fuel that can be produced domestically.

**2 Environmental Benefit** Compared with gasoline, burning E85 can reduce emissions of greenhouse gases such as carbon dioxide and nitrogen oxides.

**3 Combustion Requirements** The FFV engine burning E85 requires a richer mixture (more fuel, less air) than for an equal volume of gasoline. The FFV fuel injectors, therefore, must be able to inject up to 30% more fuel.

**4 Damage Prevention** The ethanol content of E85 is high enough to damage some of the material used in the construction of conventional vehicles. Therefore, the FFV fuel tank is made of stainless steel. The fuel lines are also made of stainless steel or lined with nonreactive materials.

### WRITING IN ▶ Chemistry

**Write** thermochemical equations for the complete combustion of 1 mol octane ($C_8H_{18}$), a component of gasoline, and 1 mol ethanol ($\Delta H_{comb}$ of $C_8H_{18}$ = $-5471$ kJ/mol; $\Delta H_{comb}$ of $C_2H_5OH$ = $-1367$ kJ/mol). Which releases the greater amount of energy per mole of fuel? Which releases more energy per kilogram of fuel? Discuss the significance of your findings.

WebQuest

# ChemLAB

iLab Station

## Measure Calories

**Background:** **The burning of a potato chip releases heat stored in the substances contained in the chip. Using calorimetry, you will approximate the amount of energy contained in a potato chip.**

**Question:** *How many Calories are in a potato chip?*

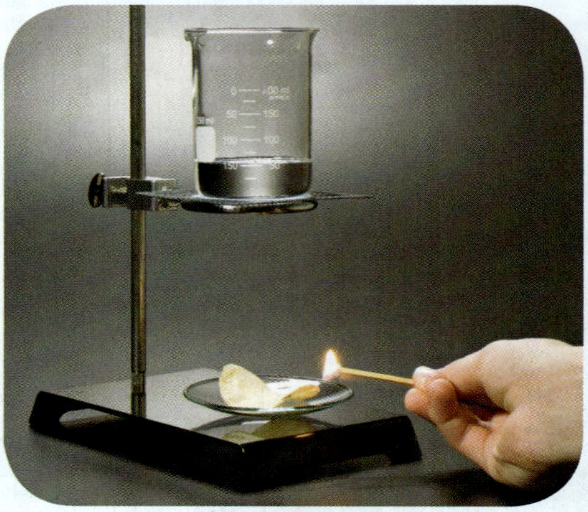

### Materials

large potato chip or other snack food
250-mL beaker
100-mL graduated cylinder
evaporating dish
nonmercury thermometer
ring stand with ring
wire gauze
matches
stirring rod
balance

### Safety Precautions

**WARNING:** ***Hot objects might not appear to be hot. Do not heat broken, chipped, or cracked glassware. Tie back long hair. Do not eat any items used in the lab.***

### Procedure

1. Read and complete the lab safety form.
2. Measure the mass of a potato chip and record it in a data table.
3. Place the potato chip in an evaporating dish on the metal base of the ring stand. Position the ring and wire gauze so that they will be 10 cm above the top of the potato chip.
4. Measure the mass of an empty 250-mL beaker and record it in your data table.
5. Using a graduated cylinder, measure 50 mL of water and pour it into the beaker. Measure the mass of the beaker and water and record it in your data table.
6. Measure and record the initial temperature of the water.
7. Place the beaker on the wire gauze on the ring stand. Use a match to ignite the bottom of the potato chip.
8. Gently stir the water in the beaker while the chip burns. Measure and record the highest temperature attained by the water.
9. **Cleanup and Disposal** Wash all lab equipment and return it to its designated place.

### Analyze and Conclude

1. **Classify** Is the reaction exothermic or endothermic? Explain how you know.
2. **Observe and Infer** Describe the reactant and products of the chemical reaction. Was the reactant (potato chip) completely consumed? What evidence supports your answer?
3. **Calculate** Determine the mass of the water and its temperature change. Use the equation $q = c \times m \times \Delta T$ to calculate how much heat, in joules, was transferred to the water by the burning of the chip.
4. **Calculate** Convert the quantity of heat from joules/chip to Calories/chip.
5. **Calculate** From the information on the chip container, determine the mass in grams of one serving. Determine how many Calories are contained in one serving. Use your data to calculate the number of Calories released by the combustion of one serving.
6. **Error Analysis** Compare your calculated Calories per serving with the value on the chip's container. Calculate the percent error.
7. **Calculate** a class average to compare to the chip container. Why would more data lead to more precise results?

### INQUIRY EXTENSION

**Predict** Do all potato chips have the same number of calories? Make a plan to test several different brands of chips.

# CHAPTER 15 STUDY GUIDE

Vocabulary Practice

**BIGIDEA** Chemical reactions usually absorb or release energy.

## SECTION 1 Energy

**MAINIDEA** Energy can change form and flow, but it is always conserved.

- Energy is the capacity to do work or produce heat.
- Chemical potential energy is energy stored in the chemical bonds of a substance by virtue of the arrangement of the atoms and molecules.
- Chemical potential energy is released or absorbed as heat during chemical processes or reactions.

$$q = c \times m \times \Delta T$$

**VOCABULARY**
- energy
- law of conservation of energy
- chemical potential energy
- heat
- calorie
- joule
- specific heat

## SECTION 2 Heat

**MAINIDEA** The enthalpy change for a reaction is the enthalpy of the products minus the enthalpy of the reactants.

- In thermochemistry, the universe is defined as the system plus the surroundings.
- The heat lost or gained by a system during a reaction or process carried out at constant pressure is called the change in enthalpy ($\Delta H$).
- When $\Delta H$ is positive, the reaction is endothermic. When $\Delta H$ is negative, the reaction is exothermic.

**VOCABULARY**
- calorimeter
- thermochemistry
- system
- surroundings
- universe
- enthalpy
- enthalpy (heat) of reaction

## SECTION 3 Thermochemical Equations

**MAINIDEA** Thermochemical equations express the amount of heat released or absorbed by chemical reactions.

- A thermochemical equation includes the physical states of the reactants and products and specifies the change in enthalpy.
- The molar enthalpy (heat) of vaporization, $\Delta H_{vap}$, is the amount of energy required to evaporate one mole of a liquid.
- The molar enthalpy (heat) of fusion, $\Delta H_{fus}$, is the amount of energy needed to melt one mole of a solid.

**VOCABULARY**
- thermochemical equation
- enthalpy (heat) of combustion
- molar enthalpy (heat) of vaporization
- molar enthalpy (heat) of fusion

## SECTION 4 Calculating Enthalpy Change

**MAINIDEA** The enthalpy change for a reaction can be calculated using Hess's law.

- The enthalpy change for a reaction can be calculated by adding two or more thermochemical equations and their enthalpy changes.
- Standard enthalpies of formation of compounds are determined relative to the assigned enthalpy of formation of the elements in their standard states.

$$\Delta H^\circ_{rxn} = \Sigma\, \Delta H^\circ_f(\text{products}) - \Sigma \Delta H^\circ_f(\text{reactants})$$

**VOCABULARY**
- Hess's law
- standard enthalpy (heat) of formation

## SECTION 5 Reaction Spontaneity

**MAINIDEA** Changes in enthalpy and entropy determine whether a process is spontaneous.

- Entropy is a measure of the disorder or randomness of a system.
- Spontaneous processes always result in an increase in the entropy of the universe.
- Free energy is the energy available to do work. The sign of the free energy change indicates whether the reaction is spontaneous.

$$\Delta G_{system} = \Delta H_{system} - T\Delta S_{system}$$

**VOCABULARY**
- spontaneous process
- entropy
- second law of thermodynamics
- free energy

# CHAPTER 15 ASSESSMENT

## SECTION 1

### Mastering Concepts

**53.** Compare and contrast temperature and heat.

**54.** How does the chemical potential energy of a system change during an endothermic reaction?

**55.** Describe a situation that illustrates potential energy changing to kinetic energy.

**56.** **Cars** How is the energy in gasoline converted and released when it burns in an automobile engine?

**57.** **Nutrition** How does the nutritional Calorie compare with the calorie? What is the relationship between the Calorie and a kilocalorie?

**58.** What quantity has the units J/(g·°C)?

■ **Figure 22**

**59.** Describe what might happen in **Figure 22** when the air above the surface of the lake is colder than the water.

**60.** Ethanol has a specific heat of 2.44 J/(g·°C). What does this mean?

**61.** Explain how the amount of energy required to raise the temperature of an object is calculated.

### Mastering Problems

**62.** **Nutrition** A food item contains 124 nutritional Calories. How many calories does the food item contain?

**63.** How many joules are absorbed in a process that absorbs 0.5720 kcal?

**64.** **Transportation** Ethanol is being used as an additive to gasoline. The combustion of 1 mol of ethanol releases 1367 kJ of energy. How many Calories are released?

**65.** To vaporize 2.00 g of ammonia, 656 calories are required. How many kilojoules are required to vaporize the same mass of ammonia?

**66.** One mole of ethanol releases 326.7 Calories of energy during combustion. How many kilojoules are released?

**67.** **Metallurgy** A 25.0-g bolt made of an alloy absorbed 250 J of heat as its temperature changed from 25.0°C to 78.0°C. What is the specific heat of the alloy?

## SECTION 2

### Mastering Concepts

**68.** Why is a foam cup used in a student calorimeter rather than a typical glass beaker?

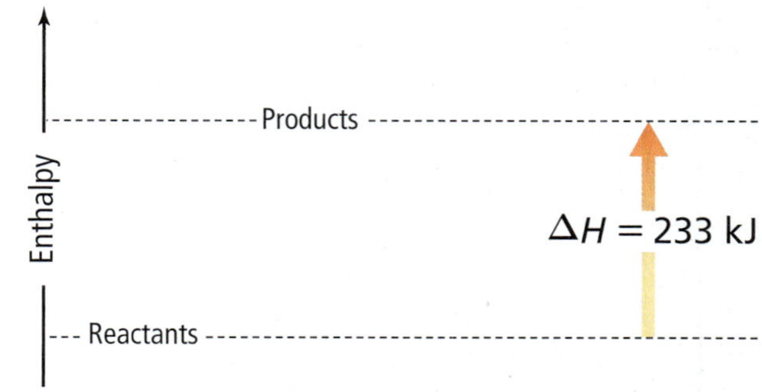


■ **Figure 23**

**69.** Is the reaction shown in **Figure 23** endothermic or exothermic? How do you know?

**70.** Give two examples of chemical systems and define the universe in terms of those examples.

**71.** Under what condition is the heat ($q$) evolved or absorbed in a chemical reaction equal to a change in enthalpy ($\Delta H$)?

**72.** The enthalpy change for a reaction, $\Delta H$, is negative. What does this indicate about the chemical potential energy of the system before and after the reaction?

**73.** What is the sign of $\Delta H$ for an exothermic reaction? An endothermic reaction?

### Mastering Problems

**74.** How many joules of heat are lost by 3580 kg of granite as it cools from 41.2°C to −12.9°C? The specific heat of granite is 0.803 J/(g·°C).

**75.** **Swimming Pool** A swimming pool measuring 20.0 m × 12.5 m is filled with water to a depth of 3.75 m. If the initial temperature is 18.4°C, how much heat must be added to the water to raise its temperature to 29.0°C? Assume that the density of water is 1.000 g/mL.

**76.** How much heat is absorbed by a 44.7-g piece of lead when its temperature increases by 65.4°C?

**77.** **Food Preparation** When 10.2 g of canola oil at 25.0°C is placed in a wok, 3.34 kJ of heat is required to heat it to a temperature of 196.4°C. What is the specific heat of the canola oil?

**78.** **Alloys** When a 58.8-g piece of hot alloy is placed in 125 g of cold water in a calorimeter, the temperature of the alloy decreases by 106.1°C, while the temperature of the water increases by 10.5°C. What is the specific heat of the alloy?

Wesley Hill/Alamy

Chapter Self-Check

## SECTION 3

### Mastering Concepts

**79.** Write the sign of $\Delta H_{system}$ for each of the following changes in physical state.
- **a.** $C_2H_5OH(s) \rightarrow C_2H_5OH(l)$
- **b.** $H_2O(g) \rightarrow H_2O(l)$
- **c.** $CH_3OH(l) \rightarrow CH_3OH(g)$
- **d.** $NH_3(l) \rightarrow NH_3(s)$

**80.** The molar enthalpy of fusion of methanol is 3.22 kJ/mol. What does this mean?

**81.** Explain how perspiration can help cool your body.

**82.** Write the thermochemical equation for the combustion of methane. Refer to **Table 3.**

### Mastering Problems

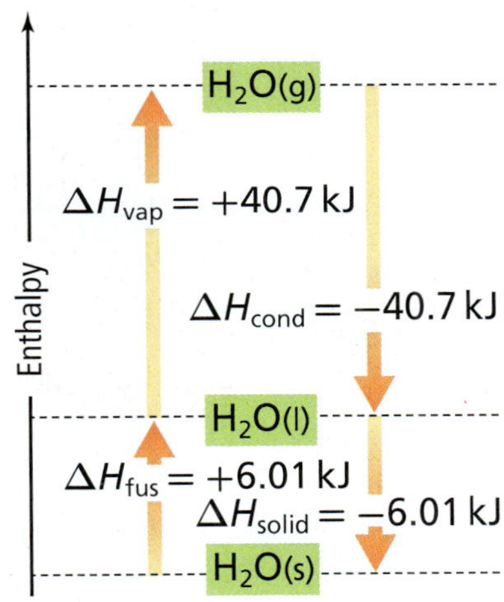


**Figure 24**

**83.** Use information from **Figure 24** to calculate how much heat is required to vaporize 4.33 mol of water at 100°C.

**84. Agriculture** Water is sprayed on oranges during a frosty night. If an average of 11.8 g of water freezes on each orange, how much heat is released?

**85. Grilling** What mass of propane ($C_3H_8$) must be burned in a barbecue grill to release 4560 kJ of heat? The $\Delta H_{comb}$ of propane is −2219 kJ/mol.

**86. Heating with Coal** How much heat is liberated when 5.00 kg of coal is burned if the coal is 96.2% carbon by mass and the other materials in the coal do not react? $\Delta H_{comb}$ of carbon is −394 kJ/mol.

**87.** How much heat is evolved when 1255 g of water condenses to a liquid at 100°C?

**88.** A sample of ammonia ($\Delta H_{solid} = -5.66$ kJ/mol) liberates 5.66 kJ of heat as it solidifies at its melting point. What is the mass of the sample?

## SECTION 4

### Mastering Concepts

**89.** For a given compound, what does the standard enthalpy of formation describe?

**90.** How does $\Delta H$ for a thermochemical equation change when the amounts of all substances are tripled and the equation is reversed?

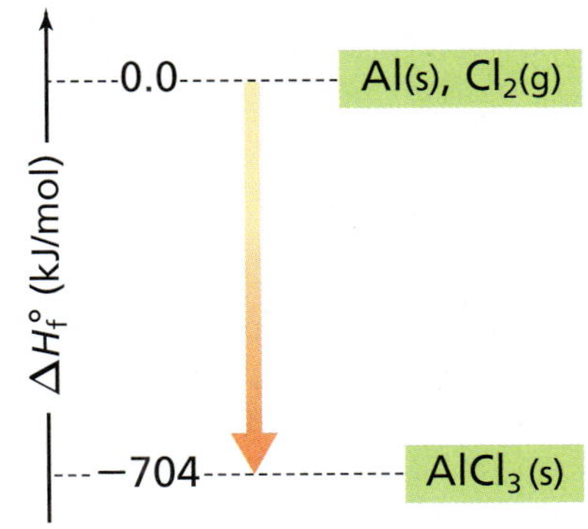


**Figure 25**

**91.** Use **Figure 25** to write the thermochemical equation for the formation of 1 mol of aluminum chloride (a solid in its standard state) from its constituent elements in their standard states.

### Mastering Problems

**92.** Use standard enthalpies of formation from **Table R-11** to calculate $\Delta H^\circ_{rxn}$ for the following reaction.

$$P_4O_6(s) + 2O_2(g) \rightarrow P_4O_{10}(s)$$

**93.** Use Hess's law and the following thermochemical equations to produce the thermochemical equation for the reaction C(s, diamond) → C(s, graphite). What is $\Delta H$ for the reaction?
- **a.** $C(s, graphite) + O_2(g) \rightarrow CO_2(g)$ $\Delta H = -394$ kJ
- **b.** $C(s, diamond) + O_2(g) \rightarrow CO_2(g)$ $\Delta H = -396$ kJ

**94.** Use Hess's law and the changes in enthalpy for the following two generic reactions to calculate $\Delta H$ for the reaction $2A + B_2C_3 \rightarrow 2B + A_2C_3$.

$$2A + \frac{3}{2}C_2 \rightarrow A_2C_3 \quad \Delta H = -1874 \text{ kJ}$$

$$2B + \frac{3}{2}C_2 \rightarrow B_2C_3 \quad \Delta H = -285 \text{ kJ}$$

## SECTION 5

### Mastering Concepts

**95.** Under what conditions is an endothermic chemical reaction in which the entropy of the system increases likely to be spontaneous?

**96.** Predict how the entropy of the system changes for the reaction $CaCO_3(s) \rightarrow CaO(s) + CO_2(g)$. Explain.

**97.** Which of these reactions would you expect to be spontaneous at relatively high temperatures? At relatively low temperatures? Explain.

**a.** $2NH_3(g) \rightarrow N_2(g) + 3H_2(g)$ $\Delta H_{system} = 92$ kJ
**b.** $2NO_2(g) \rightarrow N_2O_4(g)$ $\Delta H_{system} = -58$ kJ
**c.** $CaCO_3(g) \rightarrow CaO(s) + CO_2(g)$ $\Delta H_{system} = 178$ kJ

**98.** Explain how an exothermic reaction changes the entropy of the surroundings. Does the enthalpy change for such a reaction increase or decrease $\Delta G_{system}$? Explain.

## Mastering Problems

**99.** Calculate $\Delta G_{system}$ for each process, and state whether the process is spontaneous or nonspontaneous.

**a.** $\Delta H_{system} = 145$ kJ, $T = 293$ K, $\Delta S_{system} = 195$ J/K
**b.** $\Delta H_{system} = -232$ kJ, $T = 273$ K, $\Delta S_{system} = 138$ J/K
**c.** $\Delta H_{system} = -15.9$ kJ, $T = 373$ K, $\Delta S_{system} = -268$ J/K

**100.** Calculate the temperature at which $\Delta G_{system} = 0$ if $\Delta H_{system} = 4.88$ kJ and $\Delta S_{system} = 55.2$ J/K.

**101.** For the change $H_2O(l) \rightarrow H_2O(g)$, $\Delta G^\circ_{system}$ is 8.557 kJ and $\Delta H^\circ_{system}$ is 44.01 kJ. What is $\Delta S^\circ_{system}$ for the change?

**102.** Is the reaction to convert copper(II) sulfide to copper(II) sulfate spontaneous under standard conditions? $CuS(s) + 2O_2(g) \rightarrow CuSO_4(s)$. $\Delta H^\circ_{rxn} = -718.3$ kJ, and $\Delta S^\circ_{rxn} = -368$ J/K. Explain.

**103.** Calculate the temperature at which $\Delta G_{system} = -34.7$ kJ if $\Delta H_{system} = -28.8$ kJ and $\Delta S_{system} = 22.2$ J/K.

## MIXED REVIEW

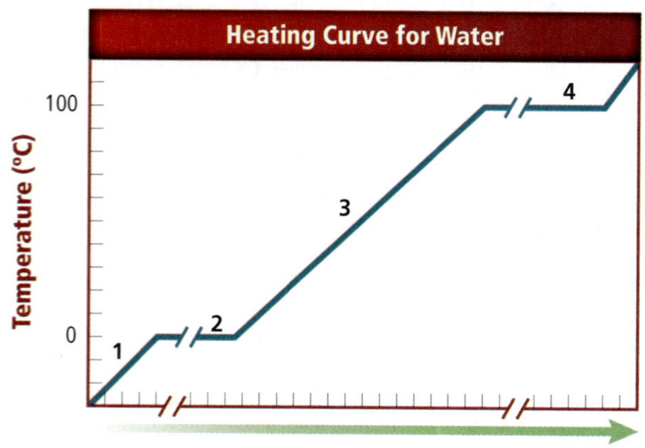


■ **Figure 26**

**104.** Heat was added consistently to a sample of water to produce the heating curve in **Figure 26.** Identify what is happening in Sections 1, 2, 3, and 4 on the curve.

**105. Bicycling** Describe the energy conversions that occur when a bicyclist coasts down a long grade, then struggles to ascend a steep grade.

**106. Hiking** Imagine that on a cold day you are planning to take a thermos of hot soup with you on a hike. Explain why you might fill the thermos with hot water first before filling it with the hot soup.

**107.** Differentiate between the enthalpy of formation of $H_2O(l)$ and $H_2O(g)$. Why is it necessary to specify the physical state of water in the following thermochemical equation $CH_4(g) + 2O_2(g) \rightarrow CO_2(g) + 2H_2O(l \text{ or } g)$ $\Delta H = ?$

## THINK CRITICALLY

■ **Figure 27**

**108. Analyze** both of the images in **Figure 27** in terms of potential energy of position, chemical potential energy, kinetic energy, and heat.

**109. Apply** Phosphorus trichloride is a starting material for the preparation of organic phosphorous compounds. Demonstrate how thermochemical equations **a** and **b** can be used to determine the enthalpy change for the reaction $PCl_3(l) + Cl_2(g) \rightarrow PCl_5(s)$.

**a.** $P_4(s) + 6Cl_2(g) \rightarrow 4PCl_3(l)$ $\Delta H = -1280$ kJ
**b.** $P_4(s) + 10Cl_2(g) \rightarrow 4PCl_5(s)$ $\Delta H = -1774$ kJ

**110. Calculate** Suppose that two pieces of iron, one with a mass exactly twice the mass of the other, are placed in an insulated calorimeter. If the original temperatures of the larger piece and the smaller piece are 90.0°C and 50.0°C, respectively, what is the temperature of the two pieces when thermal equilibrium has been established? Refer to **Table 2** for the specific heat of iron.

**111. Predict** which of the two compounds, methane gas ($CH_4$) or methanal vapor ($CH_2O$), has the greater molar enthalpy of combustion. Explain your answer. *(Hint: Write and compare the balanced chemical equations for the two combustion reactions.)*

Chapter Self-Check

## CHALLENGE PROBLEM

**112.** A sample of natural gas is analyzed and found to be 88.4% methane ($CH_4$) and 11.6% ethane ($C_2H_6$) by mass. The standard enthalpy of combustion of methane to gaseous carbon dioxide ($CO_2$) and liquid water ($H_2O$) is −891 kJ/mol. Write the equation for the combustion of gaseous ethane to carbon dioxide and water. Calculate the standard enthalpy of combustion of ethane using standard enthalpies of formation from **Table R-11.** Using that result and the standard enthalpy of combustion of methane in **Table 3,** calculate the energy released by the combustion of 1 kg of natural gas.

## CUMULATIVE REVIEW

**113.** Why is it necessary to perform repeated experiments in order to support a hypothesis?

**114.** Phosphorus has the atomic number 15 and an atomic mass of 31 amu. How many protons, neutrons, and electrons are in a neutral phosphorus atom?

**115.** What element has the electron configuration $[Ar]4s^13d^5$?

**116.** Name the following molecular compounds.

**a.** $S_2Cl_2$
**b.** $CS_2$
**c.** $SO_3$
**d.** $P_4O_{10}$

**117.** Determine the molar mass for the following compounds.

**a.** $Co(NO_3)_2 \cdot 6H_2O$
**b.** $Fe(OH)_3$

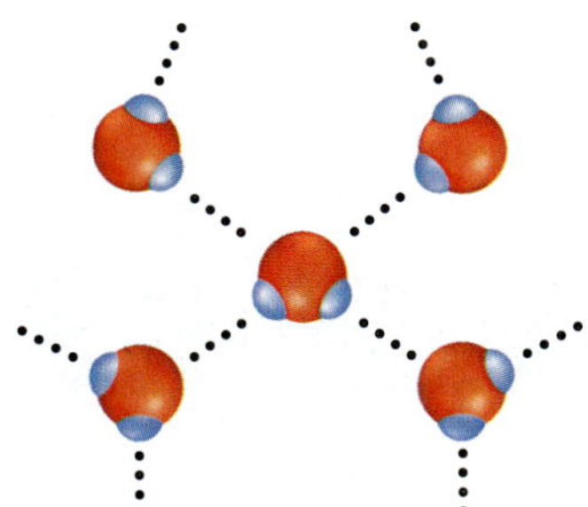

■ **Figure 28**

**118.** What kind of chemical bond is represented by the dotted lines in **Figure 28?**

**119.** A sample of oxygen gas has a volume of 20.0 $cm^3$ at −10.0°C. What volume will this sample occupy if the temperature rises to 110 °C?

**120.** What is the molarity of a solution made by dissolving 25.0 g of sodium thiocyanate (NaSCN) in enough water to make 500 mL of solution?

**121.** List three colligative properties of solutions.

## WRITING IN ▶ Chemistry

**122. Alternate Fuels** Use various sources to explain how hydrogen might be produced, transported, and used as a fuel for automobiles. Summarize the benefits and drawbacks of using hydrogen as an alternative fuel for internal combustion engines.

**123. Wind Power** Research the use of wind as a source of electrical power. Explain the possible benefits, disadvantages, and limitations of its use.

## DBQ Document-Based Questions

**Cooking Oil** *A university research group burned four cooking oils in a bomb calorimeter to determine if a relationship exists between the enthalpy of combustion and the number of double bonds in an oil molecule. Cooking oils typically contain long chains of carbon atoms linked by either single or double bonds. A chain with no double bonds is said to be saturated. Oils with one or more double bonds are unsaturated. The enthalpies of combustion of the four oils are shown in* **Table 7.** *The researchers calculated that the results deviated by only 0.6% and concluded that a link between saturation and enthalpy of combustion could not be detected by the experimental procedure used.*

Data obtained from: http: Heat of Combustion Oils. April 1998. *University of Pennsylvania.*

**Table 7 Combustion Results for Oils**

| Type of Oil | $\Delta H_{comb}$ (kJ/g) |
|---|---|
| Soy oil | 40.81 |
| Canola oil | 41.45 |
| Olive oil | 39.31 |
| Extra-virgin olive oil | 40.98 |

**124.** Which of the oils tested provided the greatest amount of energy per unit mass when burned?

**125.** According to the data, how much energy would be liberated by burning 0.554 kg of olive oil?

**126.** Assume that 12.2 g of soy oil is burned and that all the energy released is used to heat 1.600 kg of water, initially at 20.0°C. What is the final temperature of the water?

**127.** Oils can be used as fuels. How many grams of canola oil would have to be burned to provide the energy to vaporize 25.0 g of water? ($\Delta H_{vap}$ = 40.7 kJ/mol).

## MULTIPLE CHOICE

*Use the graph below to answer Questions 1 to 3.*

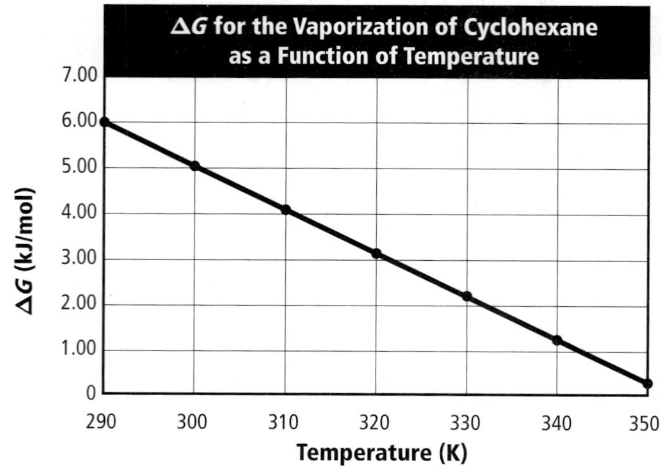


1. In the range of temperatures shown, the vaporization of cyclohexane
   **A.** does not occur at all.
   **B.** will occur spontaneously.
   **C.** is not spontaneous.
   **D.** occurs only at high temperatures.

2. What is the standard free energy of vaporization, $\Delta G^\circ_{vap}$, of cyclohexane at 300 K?
   **A.** 5.00 kJ/mol
   **B.** 4.00 kJ/mol
   **C.** 3.00 kJ/mol
   **D.** 2.00 kJ/mol

3. When $\Delta G^\circ_{vap}$ is plotted as a function of temperature, the slope of the line equals $\Delta S^\circ_{vap}$ and the *y*-intercept of the line equals $\Delta H^\circ_{vap}$.What is the approximate standard entropy of the vaporization of cyclohexane?
   **A.** −50.0 J/mol•K
   **B.** −10.0 J/mol•K
   **C.** −5.0 J/mol•K
   **D.** −100 J/mol•K

4. The metal yttrium, atomic number 39, forms
   **A.** positive ions.
   **B.** negative ions.
   **C.** both positive and negative ions.
   **D.** no ions at all.

5. Given the reaction $2Al + 3FeO \rightarrow Al_2O_3 + 3Fe$, what is the mole-to-mole ratio between iron(II) oxide and aluminum oxide?
   **A.** 2:3
   **B.** 1:1
   **C.** 3:2
   **D.** 3:1

*Use the table below to answer Question 6.*

**Electronegativity of Selected Elements**

| H | | | | | | |
|---|---|---|---|---|---|---|
| 2.20 | | | | | | |
| Li | Be | B | C | N | O | F |
| 0.98 | 1.57 | 2.04 | 2.55 | 3.04 | 3.44 | 3.98 |
| Na | Mg | Al | Si | P | S | Cl |
| 0.93 | 1.31 | 1.61 | 1.90 | 2.19 | 2.58 | 3.16 |

6. Which bond is the most electronegative?
   **A.** H–H
   **B.** H–C
   **C.** H–N
   **D.** H–O

7. Element Q has an oxidation number of +2, while Element M has an oxidation number of −3. Which is the correct formula for a compound made of elements Q and M?
   **A.** $Q_2M_3$
   **B.** $M_2Q_3$
   **C.** $Q_3M_2$
   **D.** $M_3Q_2$

8. Wavelengths of light shorter than about $4.00 \times 10^{-7}$ m are not visible to the human eye. What is the energy of a photon of ultraviolet light having a frequency of $5.45 \times 10^{16}$ $s^{-1}$? (Planck's constant is $6.626 \times 10^{-34}$J•s.)
   **A.** $3.61 \times 10^{-17}$ J
   **B.** $1.22 \times 10^{-50}$ J
   **C.** $8.23 \times 10^{49}$ J
   **D.** $3.81 \times 10^{-24}$ J

*Use the graph below to answer Question 9.*

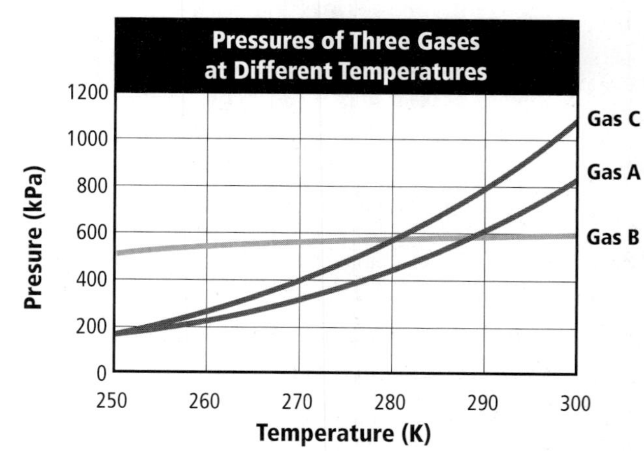


9. What is the pressure of Gas B at 295 K?
   **A.** 500 kPa
   **B.** 600 kPa
   **C.** 700 kPa
   **D.** 900 kPa

## SHORT ANSWER

*Use the figure below to answer Questions 10 to 12.*

·S:   :Cl:   :Ar:   K·   Ca·

**10.** Explain why argon is not likely to form a compound.

**11.** What is the chemical formula for calcium chloride? Explain the formation of this ionic compound using the electron-dot structures above.

**12.** Use electron-dot models to explain what charge sulfur will most likely have when it forms an ion.

## EXTENDED RESPONSE

*Use the information below to answer Questions 13 and 14.*

A sample of gas occupies a certain volume at a pressure of 1 atm. If the pressure remains constant, heating causes the gas to expand, as shown below.

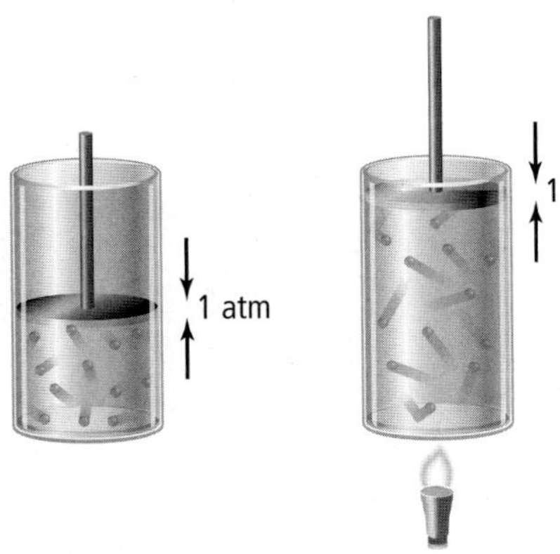


**13.** State the gas law that describes why the gas in the second canister occupies a greater volume than the gas in the first canister.

**14.** If the volume in the first container is 2.1 L at a temperature of 300.0 K, to what temperature must the second canister be heated to reach a volume of 5.4 L? Show your setup and the final answer.

## SAT SUBJECT TEST: CHEMISTRY

**15.** The specific heat of ethanol is 2.44 J/g·°C. How many kilojoules of energy are required to heat 50.0 g of ethanol from −20.0°C to 68.0°C?

**A.** 10.7 kJ
**B.** 8.30 kJ
**C.** 2.44 kJ
**D.** 1.22 kJ
**E.** 5.86 kJ

**16.** If 3.00 g of aluminum foil, placed in an oven and heated from 20.0°C to 662.0°C, absorbs 1728 J of heat, what is the specific heat of aluminum?

**A.** 0.131 J/g·°C
**B.** 0.870 J/g·°C
**C.** 0.897 J/g·°C
**D.** 2.61 J/g·°C
**E.** 0.261 J/g·°C

*Use the table below to answer Questions 17 and 18.*

**Density and Electronegativity Data for Elements**

| Element | Density (g/ml) | Electronegativity |
|---|---|---|
| Aluminum | 2.698 | 1.6 |
| Fluorine | $1.696 \times 10^{-3}$ | 4.0 |
| Sulfur | 2.070 | 2.6 |
| Copper | 8.960 | 1.9 |
| Magnesium | 1.738 | 1.3 |
| Carbon | 3.513 | 2.6 |

**17.** A sample of metal has a mass of 9.250 g and occupies a volume of 5.250 mL. Which metal is it?

**A.** aluminum
**B.** magnesium
**C.** carbon
**D.** copper
**E.** sulfur

**18.** Which pair is most likely to form an ionic bond?

**A.** carbon and sulfur
**B.** aluminum and magnesium
**C.** copper and sulfur
**D.** magnesium and fluorine
**E.** aluminum and carbon

| NEED EXTRA HELP? | | | | | | | | | | | | | | | | | | |
|---|---|---|---|---|---|---|---|---|---|---|---|---|---|---|---|---|---|---|
| **If You Missed Question . . .** | 1 | 2 | 3 | 4 | 5 | 6 | 7 | 8 | 9 | 10 | 11 | 12 | 13 | 14 | 15 | 16 | 17 | 18 |
| **Review Section . . .** | 15.5 | 15.5 | 15.5 | 8.5 | 11.1 | 8.5 | 7.3 | 5.1 | 13.1 | 5.3 | 5.3 | 5.3 | 13.1 | 13.1 | 15.2 | 15.2 | 2.1 | 8.5 |

CHAPTER 16

# Reaction Rates

**BIGIDEA** Every chemical reaction proceeds at a definite rate, but can be speeded up or slowed down by changing the conditions of the reaction.

## SECTIONS

Chm 11a, 12a

1 A Model for Reaction Rates

2 Factors Affecting Reaction Rates

3 Reaction Rate Laws

4 Instantaneous Reaction Rates and Reaction Mechanisms

## LaunchLAB

iLab Station

### How can you accelerate a reaction?

Some chemical reactions go so slowly that nothing seems to be happening. In this lab, you will study one way of speeding up a slow reaction.

**FOLDABLES® Study Organizer**

### Reaction Rates

Make a layered-look book. Label it as shown. Use it to help you organize information about factors affecting reaction rates.

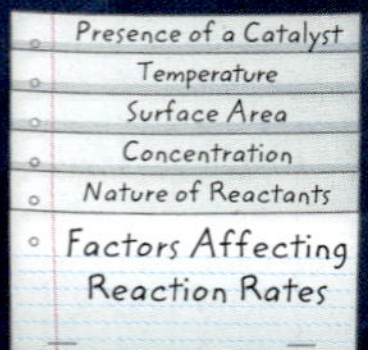

(inset)yang yu/Alamy, (bkgd)Transtock Inc/Alamy

Go **online!**
MHEonline.com

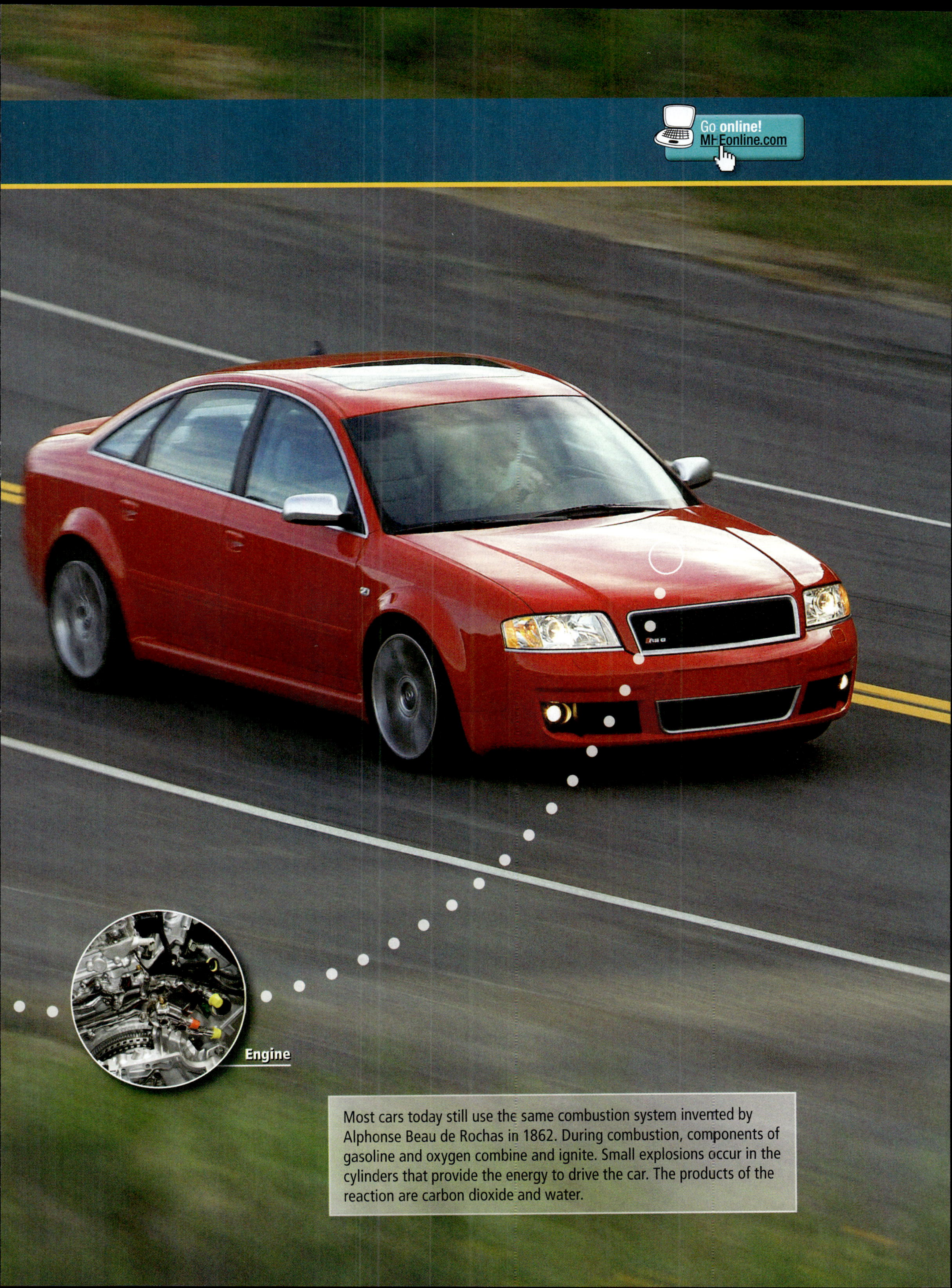


Most cars today still use the same combustion system invented by Alphonse Beau de Rochas in 1862. During combustion, components of gasoline and oxygen combine and ignite. Small explosions occur in the cylinders that provide the energy to drive the car. The products of the reaction are carbon dioxide and water.

## SECTION 1 A Model for Reaction Rates

**MAINIDEA** Collision theory is the key to understanding why some reactions are faster than others.

### Essential Questions

- How can average rates of chemical reactions be calculated from experimental data?
- How are the rates of chemical reactions related to collisions between reacting particles?

### Review Vocabulary

**energy:** the ability to do work or produce heat; it exists in two basic forms: potential energy and kinetic energy

### New Vocabulary

reaction rate
collision theory
activated complex
activation energy

**CHEM 4 YOU** **Which is faster: walking to school, or riding in a bus or car? Determining how fast a person can get to school is not all that different from calculating the rate of a chemical reaction. Either way, you are measuring change over time.**

## Expressing Reaction Rates

In the Launch Lab, you discovered that the decomposition of hydrogen peroxide can be a fast reaction, or it can be a slow one. However, *fast* and *slow* are inexact terms. Chemists, engineers, chefs, welders, concrete mixers, and others often need to be more specific. For example, a chef must know the rate at which a roast cooks to determine when it will be ready to serve. The person mixing the concrete must know the rate of mixing water, sand, gravel, and cement so that the resulting concrete can be poured at the correct consistency. Delaying pouring can result in concrete that is not strong enough for its purpose.

Think about how you express the speed or rate of a moving object. The speedometer of the speeding racer in **Figure 1** shows that the car is moving at 320 km/h. The speed of a sprinter on a track team might be expressed in meters per second (m/s). Generally, the average rate of an action or process is defined as the change in a given quantity during a specific period of time. Recall from your study of math that the Greek letter *delta* ($\Delta$) before a quantity indicates a change in the quantity. In equation form, average rate or speed is written as follows.

$$\text{average rate} = \frac{\Delta \text{quantity}}{\Delta t}$$

■ **Figure 1** The speedometer of the racer shows its speed in km/h or mph, both of which are the change in distance divided by the change in time. The sprinter's speed might be measured in m/s.

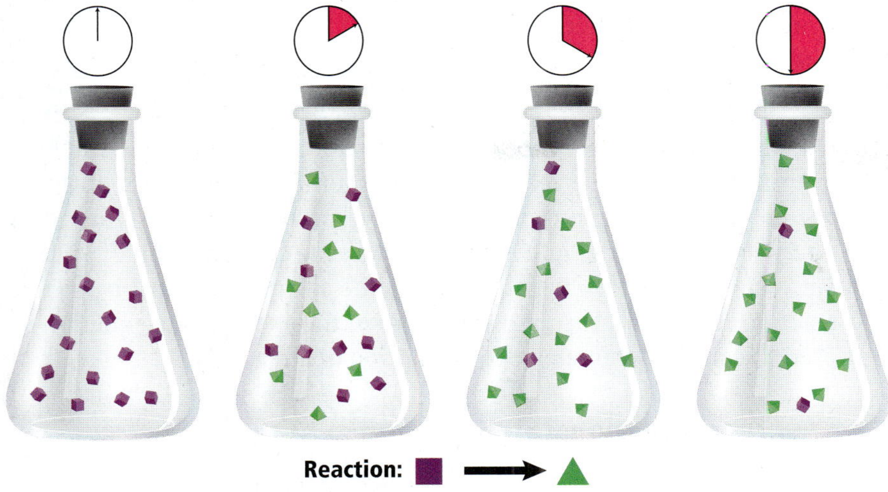


■ **Figure 2** Over time, the reactant changes to a product. The rate of a chemical reaction can be expressed as a change in the number of moles of reactant or product during an interval of time.

**Calculate** ***the rate of change for each interval.***

**Figure 2** shows how a reaction proceeds from reactant to product over time. Notice that the amount of the reactant decreases as the amount of product increases. If you know the change in a product or a reactant during a segment of time, you can calculate the average rate of the reaction. Most often, chemists are concerned with changes in the molar concentration (mol/L, $M$) of a reactant or product during a reaction. Therefore, the **reaction rate** of a chemical reaction is the change in concentration of a reactant or product per unit of time, generally expressed as mol/(L • s). Brackets around the formula for a substance denote its molar concentration. For example, $[NO_2]$ represents the molar concentration of $NO_2$.

Reaction rates are determined experimentally by measuring the concentrations of reactants and/or products as an actual chemical reaction proceeds. Reaction rates cannot be calculated from balanced equations.

Suppose you wish to express the average rate of the following reaction during the time period beginning at time $t_1$ and ending at time $t_2$.

$$CO(g) + NO_2(g) \rightarrow CO_2(g) + NO(g)$$

Calculating the rate at which the products of the reaction are produced results in a reaction rate with a positive value. The rate calculation based on the production of NO has the following form.

$$\text{Average reaction rate} = \frac{[\text{NO}] \text{ at time } t_2 - [\text{NO}] \text{ at time } t_1}{t_2 - t_1} = \frac{\Delta[\text{NO}]}{\Delta t}$$

For example, if the concentration of NO is $0.000M$ at time $t_1 = 0.00$ s and $0.010M$ two seconds after the reaction begins, the following calculation gives the average rate of the reaction expressed as moles of NO produced per liter per second.

$$\text{Average reaction rate} = \frac{0.010M - 0.000M}{2.00 \text{ s} - 0.00 \text{ s}}$$

$$= \frac{0.010M}{2.00 \text{ s}} = 0.0050 \text{ mol/(L·s)}$$

Notice how the units work out:

$$\frac{M}{\text{s}} = \frac{\text{mol}}{\text{L}} \cdot \frac{1}{\text{s}} = \frac{\text{mol}}{(\text{L·s})}$$

**VOCABULARY**

**SCIENCE USAGE V. COMMON USAGE**

**Concentration**

***Science usage:*** quantitative measure of the amount of solute in a given amount of solvent or solution

*The solution has a concentration of six moles per liter.*

***Common usage:*** the focus of attention on a single object or purpose

*The concentration of the audience was completely on the performer.*

You can also choose to state the rate of the reaction as the rate at which CO is consumed, as shown below.

$$\text{average reaction rate} = \frac{[\text{CO}] \text{ at time } t_2 - [\text{CO}] \text{ at time } t_1}{t_2 - t_1} = \frac{\Delta[\text{CO}]}{\Delta t}$$

Do you predict a positive or a negative value for this reaction rate? In this case, a negative value indicates that the concentration of CO decreases as the reaction proceeds. However, reaction rates must always be positive. When the rate is measured by the consumption of a reactant, scientists apply a negative sign to the calculation to get a positive reaction rate. Thus, the following form of the average rate equation is used to calculate the rate of consumption of a reactant.

**Average Reaction Rate Equation**

$$\text{average reaction rate} = -\frac{\Delta[\text{reactant}]}{\Delta t}$$

**$\Delta$[reactant] represents the change in concentration of a reactant.**

**$\Delta t$ represents the change in time.**

The average reaction rate for the consumption of a reactant is the negative change in the concentration of the reactant divided by the elapsed time.

## EXAMPLE Problem 1

Find help with **algebraic equations.** Math Handbook

EXAMPLE PROBLEM

**CALCULATE AVERAGE REACTION RATES** In a reaction between butyl chloride ($C_4H_9Cl$) and water, the concentration of $C_4H_9Cl$ is 0.220*M* at the beginning of the reaction. At 4.00 s, the concentration of $C_4H_9Cl$ is 0.100*M*. Calculate the average reaction rate over the given time period expressed as moles of $C_4H_9Cl$ consumed per liter per second.

### 1 ANALYZE THE PROBLEM

You are given the initial and final concentrations of the reactant $C_4H_9Cl$ and the initial and final times. You can calculate the average reaction rate of the chemical reaction using the change in concentration of butyl chloride in four seconds.

| Known | Unknown |
|---|---|
| $t_1 = 0.00$ s | Average reaction rate = ? mol/(L·s) |
| $t_2 = 4.00$ s | |
| $[C_4H_9Cl]$ at $t_1 = 0.220M$ | |
| $[C_4H_9Cl]$ at $t_2 = 0.100M$ | |

### 2 SOLVE FOR THE UNKNOWN

$$\text{Average reaction rate} = \frac{[C_4H_9Cl] \text{ at } t_2 - [C_4H_9Cl] \text{ at } t_1}{t_2 - t_1}$$

**State the average reaction rate equation.**

$$= -\frac{0.100M - 0.220M}{4.00\text{ s} - 0.00\text{ s}}$$

**Substitute $t_2 = 4.00$ s, $t_1 = 0.00$ s, $[C_4H_9Cl]$ at $t_2 = 0.100$ *M*, and $[C_4H_9Cl]$ at $t_1 = 0.220M$.**

$$= -\frac{0.100\text{ mol/L} - 0.220\text{ mol/L}}{4.00\text{ s} - 0.00\text{ s}}$$

**Substitute mol/L for *M* and perform the calculations. Note that the minus sign cancels out.**

$$\text{Average reaction rate} = \frac{0.120\text{ mol/L}}{4.00\text{ s}} = 0.0300\text{ mol/(L·s)}$$

### 3 EVALUATE THE ANSWER

The average reaction rate of 0.0300 moles $C_4H_9Cl$ consumed per liter per second is reasonable based on the starting and ending amounts. The answer is correctly expressed in three significant figures.

Get help with **reaction rate problems.** Personal Tutor

## PRACTICE Problems

Do additional problems. Online Practice

PRACTICE PROBLEMS

**Use the data in the following table to calculate the average reaction rates.**

| Experimental Data for $H_2 + Cl_2 \rightarrow 2HCl$ | | | |
|---|---|---|---|
| Time (s) | $[H_2]$ ($M$) | $[Cl_2]$ ($M$) | $[HCl]$ ($M$) |
| 0.00 | 0.030 | 0.050 | 0.000 |
| 4.00 | 0.020 | 0.040 | |

1. Calculate the average reaction rate expressed in moles $H_2$ consumed per liter per second.
2. Calculate the average reaction rate expressed in moles $Cl_2$ consumed per liter per second.
3. **Challenge** If the average reaction rate for the reaction, expressed in moles of HCl formed, is 0.0050 mol/L•s, what concentration of HCl would be present after 4.00 s?

## Collision Theory

Have you ever watched children trying to break a piñata? Each hit with a stick can result in emptying the piñata of its contents, as shown in **Figure 3.** The reactants in a chemical reaction must also collide in order to form products. **Figure 3** also represents a reaction between the molecules $A_2$ and $B_2$ to form AB. The reactant molecules must come together in a collision in order to react and produce molecules of AB. The figure is an illustration of **collision theory,** which states that atoms, ions, and molecules must collide in order to react.

**READING CHECK** **Predict** why a collision between two particles is necessary for a reaction to occur.

Look at the reaction between carbon monoxide (CO) gas and nitrogen dioxide ($NO_2$) gas at a temperature above 500 K.

$$CO(g) + NO_2(g) \rightarrow CO_2(g) + NO(g)$$

The reactant molecules collide to produce carbon dioxide ($CO_2$) gas and nitrogen monoxide (NO) gas. However, calculations of the number of molecular collisions per second yield a puzzling result: only a small fraction of collisions produce reactions.

**Figure 3** Just as a stick must hit the piñata hard enough to break it open, particles in chemical reactions must collide with a sufficient amount of energy for a reaction to occur.

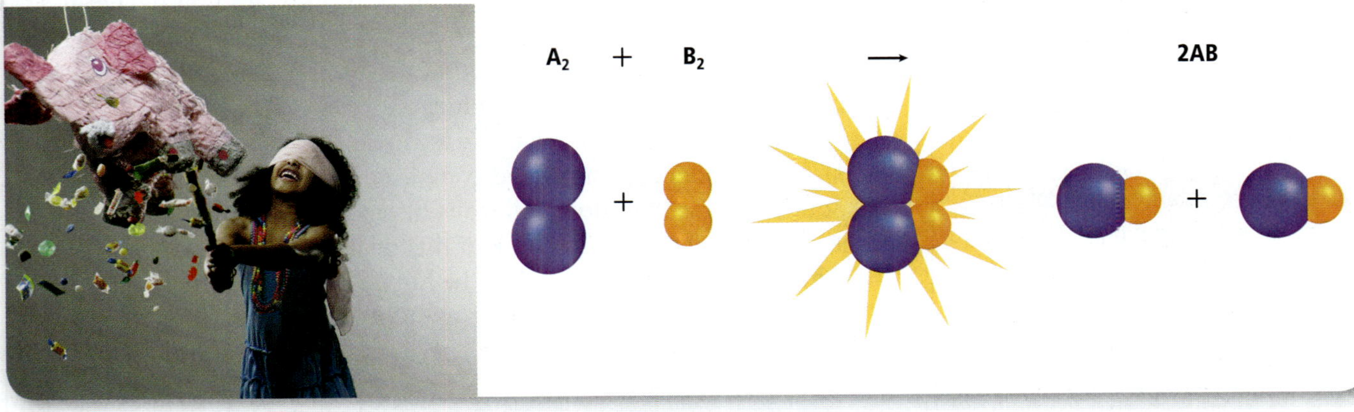


Ryan McVay/Digital Vision/ Getty Images

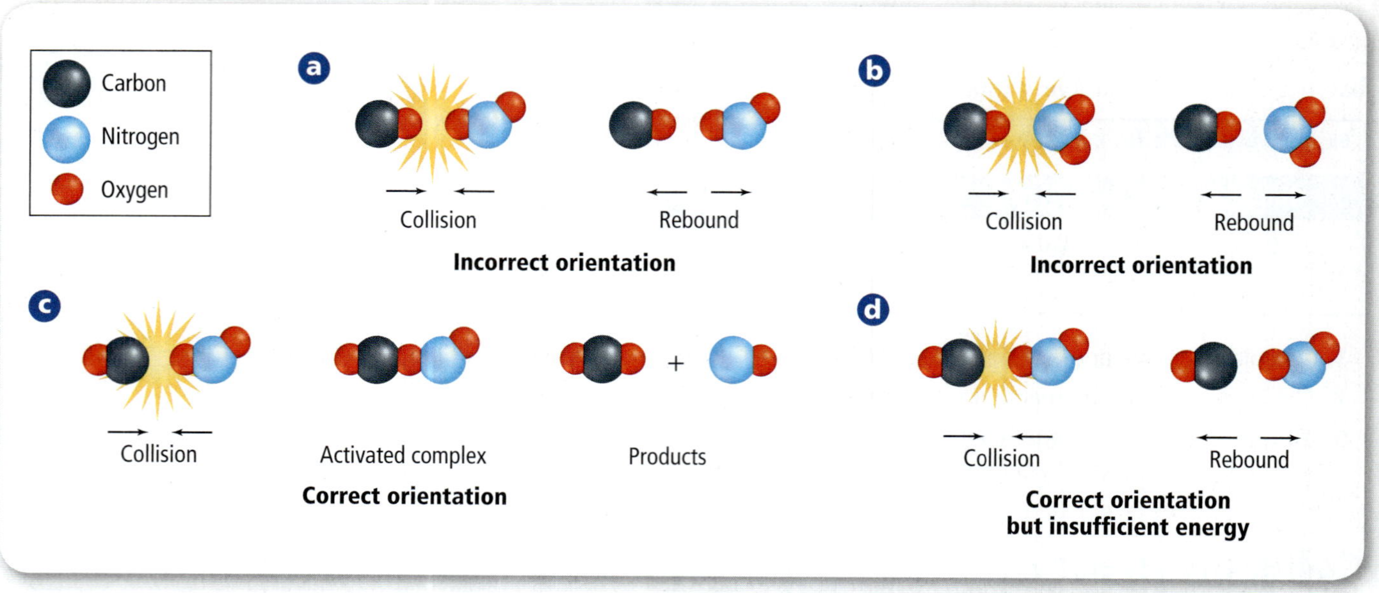


■ **Figure 4** This figure shows four different collision orientations between CO molecules and $NO_2$ molecules. The collisions in **a** and **b** do not result in a reaction because the molecules are not in position to form bonds. The molecules in **c** collide in the correct orientation, and a reaction occurs. Although the molecules in **d** are also in the correct orientation, they have insufficient energy to react.

View an **animation about molecular orientation and collision effectiveness.**

Concepts In Motion

**Collision orientation and the activated complex** Why do most collisions fail to produce products? What other factors must be considered? **Figure 4a** and **b** show one possible answer to this question. These illustrations indicate that in order for a collision to lead to a reaction, the carbon atom in a CO molecule must contact an oxygen atom in an $NO_2$ molecule at the instant of impact. This is the only way in which a temporary bond can form between the carbon atom and an oxygen atom. The collisions shown in **Figure 4a** and **b** do not lead to reactions because the molecules collide in unfavorable orientations. A carbon atom does not contact an oxygen atom at the instant of impact, so the molecules simply rebound.

When the orientation of colliding molecules is correct, as shown in **Figure 4c,** a reaction can occur. An oxygen atom is transferred from an $NO_2$ molecule to a CO molecule. When this occurs, a short-lived entity called an activated complex is formed, in this case OCONO. An **activated complex,** sometimes called a transition state, is a temporary, unstable arrangement of atoms in which old bonds are breaking and new bonds are forming. As a result, the activated complex might form products or might break apart to re-form the reactants.

**Activation energy and reaction rate** The collision depicted in **Figure 4d** does not lead to a reaction for a different reason—insufficient energy. Just as the piñata does not break open unless it is hit hard enough, no reaction occurs between the CO and $NO_2$ molecules unless they collide with sufficient energy. The minimum amount of energy that reacting particles must have to form the activated complex and lead to a reaction is called the **activation energy** ($E_a$). **Table 1** summarizes the conditions under which colliding particles can react.

A high $E_a$ means that relatively few collisions have the required energy to produce the activated complex, and the reaction rate is slow. A low $E_a$ means that more collisions have sufficient energy to react, and the reaction rate is faster. Think of this relationship in terms of a person pushing a heavy cart up a hill. If the hill is high, a substantial amount of energy is required to move the cart, and it might take a long time to get it to the top. If the hill is low, less energy is required and the task might be accomplished faster.

| **Table 1** Collision Theory Summary |
|---|
| **1.** Reacting substances (atoms, ions, or molecules) must collide. |
| **2.** Reacting substances must collide in the correct orientation. |
| **3.** Reacting substances must collide with sufficient energy to form an activated complex. |

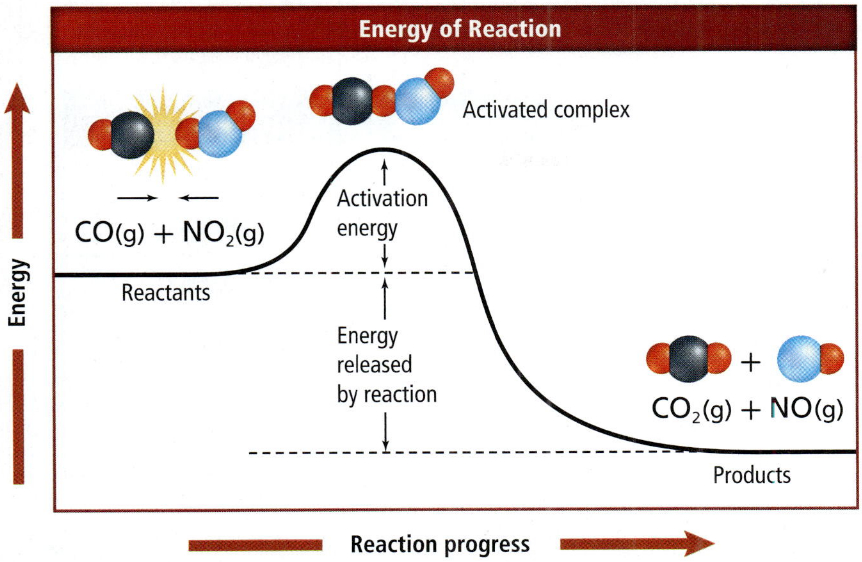


■ **Figure 5** When an exothermic reaction occurs, molecules collide with enough energy to overcome the activation energy barrier. They form an activated complex, then release energy and form products at a lower energy level.

**GRAPH CHECK**

**Explain** how you can tell from the graph that the reaction described is an exothermic reaction.

**Figure 5** shows the energy diagram for the progress of the reaction between carbon monoxide and nitrogen dioxide. Does this energy diagram look somewhat different from those you previously studied for thermochemistry? Why? This diagram shows the activation energy of the reaction. Activation energy can be thought of as a barrier the reactants must overcome in order to form the products. In this case, the CO and $NO_2$ molecules collide with enough energy to overcome the barrier, and the products formed lie at a lower energy level. Recall that reactions that lose energy are called exothermic reactions.

For many reactions, the process from reactants to products is reversible. **Figure 6** illustrates the reverse endothermic reaction between $CO_2$ and NO to re-form CO and $NO_2$. In this reaction, the reactants lie at a low energy level. They must overcome a significant activation energy to re-form CO and $NO_2$. This requires a greater input of energy than the forward reaction. If this reverse reaction is achieved, CO and $NO_2$ again lie at a higher energy level.

**Alabama Applying Practices**

Go to the Resource tab in ConnectED to find the Applying Practices worksheet *Modeling Energy in Chemical Reactions.*

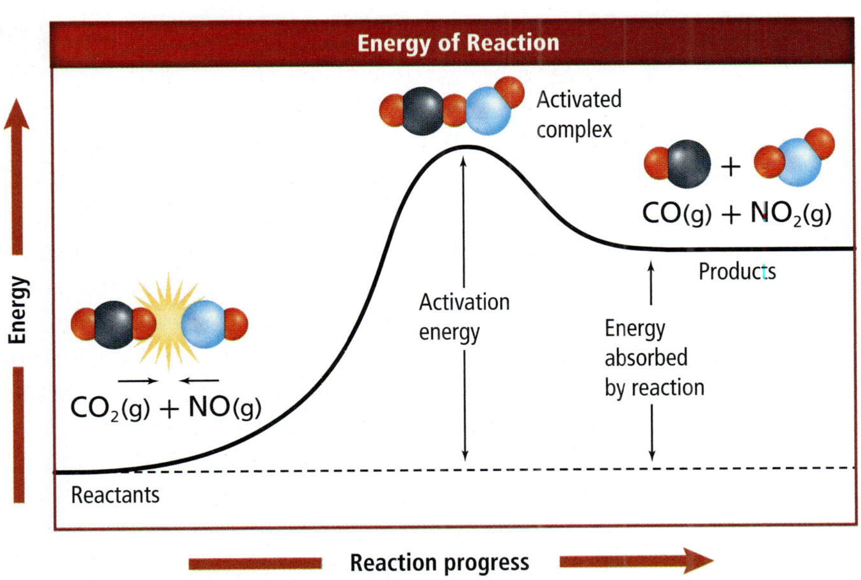


■ **Figure 6** In the reverse reaction, which is endothermic, the reactant molecules are at a lower energy than the products. To react, the reactants must absorb enough energy to overcome the activation energy barrier and form higher-energy products.

**GRAPH CHECK**

**Compare Figures 5** and **6** to determine whether the activation energy for the forward reaction is larger or smaller than the activation energy for the reverse reaction.

## Problem-Solving LAB

### Interpret Data

**How does the rate of decomposition vary over time?** The compound dinitrogen pentoxide ($N_2O_5$) decomposes in air according to the equation

$$2N_2O_5(g) \rightarrow 4NO_2(g) + O_2(g)$$

Knowing the rate of decomposition allows the concentration of dinitrogen pentoxide to be determined at any time.

#### Analysis

The table shows the results of an experiment in which the concentration of $N_2O_5$ was measured over time at normal atmospheric pressure and a temperature of 45°C.

| Time (min) | $[N_2O_5]$ (mol/L) |
|---|---|
| 0 | 0.01756 |
| 20.0 | 0.00933 |
| 40.0 | 0.00531 |
| 60.0 | 0.00295 |
| 80.0 | 0.00167 |
| 100.0 | 0.00094 |

#### Think Critically

1. **Calculate** the average reaction rate for each time interval: 0–20 min, 40–60 min, and 80–100 min. Express each rate as a positive number and in moles of $N_2O_5$ consumed per liter per minute.
2. **Express** the average reaction rate for each time interval in moles of $NO_2$ produced per liter per minute. Use the reaction equation to explain the relationship between these rates and those calculated in Question 1.
3. **Interpret** the data and your calculations in describing how the average rate of decomposition of $N_2O_5$ varies over time.
4. **Apply** collision theory to infer why the reaction rate varies as it does.

## Spontaneity and Reaction Rate

Recall that reaction spontaneity is related to change in free energy ($\Delta G$). If $\Delta G$ is negative, the reaction is spontaneous under the conditions specified. If $\Delta G$ is positive, the reaction is not spontaneous. Now consider whether spontaneity has any effect on reaction rates. Are more spontaneous reactions faster than less spontaneous ones?

**VOCABULARY**
**ACADEMIC VOCABULARY**
**Investigate**
to observe by study or close examination
*They decided to investigate how the mice were getting into the house.*

To investigate the relationship between spontaneity and reaction rate, consider the following gas-phase reaction between hydrogen and oxygen.

$$2H_2(g) + O_2(g) \rightarrow 2H_2O(g)$$

Here, $\Delta G = -458$ kJ at 298 K (25°C) and 1 atm pressure. Because $\Delta G$ is negative, the reaction is spontaneous. For the same reaction, $\Delta H = -484$ kJ, which means that the reaction is highly exothermic. You can examine the speed of this reaction by filling a tape-wrapped soda bottle with stoichiometric quantities of the two gases—two volumes hydrogen and one volume oxygen. A thermometer in the stopper allows you to monitor the temperature inside the bottle. As you watch for evidence of a reaction, the temperature remains constant for hours. Have the gases escaped, or have they failed to react?

If you remove the stopper and hold a burning splint to the mouth of the bottle, a reaction occurs explosively. Clearly, the hydrogen and oxygen gases have not escaped from the bottle. Yet, they did not react noticeably until you supplied additional energy in the form of a lighted splint.

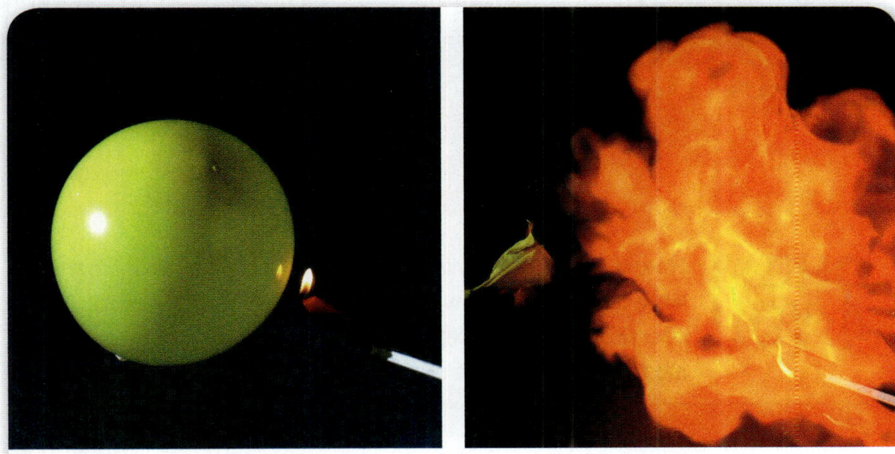

■ **Figure 7** The hydrogen and oxygen in the balloon do not react until the balloon is touched by a flame. Then, an explosive reaction occurs.
**Explain** *the role of the flame.*

**Figure 7** illustrates the reaction between hydrogen and oxygen in a similar way. The balloon is filled with a mixture of hydrogen gas and oxygen gas that appears not to react. When the lighted candle introduces additional energy, an explosive reaction occurs between the gases. Similarly, the air-fuel mixture in the cylinders of a car show little sign of reaction until a spark from a spark plug initiates a small explosion which produces energy to move the car. Logs on the forest floor combine slowly with oxygen in the air as they decompose, but they also combine with oxygen and burn rapidly in a forest fire once they are ignited.

As these examples show, reaction spontaneity in the form of $\Delta G$ implies nothing about the speed of the reaction; $\Delta G$ indicates only the natural tendency for a reaction or process to proceed. Factors other than spontaneity, however, do affect the rate of a chemical reaction. You will learn about these factors in the next section.

## SECTION 1 REVIEW

Section Self-Check

### Section Summary

- The rate of a chemical reaction is expressed as the rate at which a reactant is consumed or the rate at which a product is formed.
- Reaction rates are generally calculated and expressed in moles per liter per second (mol/(L • s)).
- In order to react, the particles in a chemical reaction must collide.
- The rate of a chemical reaction is unrelated to the spontaneity of the reaction.

4. MAINIDEA **Relate** collision theory to reaction rate.
5. **Explain** what the reaction rate indicates about a particular chemical reaction.
6. **Compare** the concentrations of the reactants and products during the course of a chemical reaction (assuming no additional reactants are added).
7. **Compare** the average reaction rate measured over an initial, short time interval to one measured over a long time interval.
8. **Describe** the relationship between activation energy and the rate of a reaction.
9. **Summarize** what happens during the brief existence of an activated complex.
10. **Apply** collision theory to explain why collisions between two reacting particles do not always result in the formation of a product.
11. **Interpret** how the speed of a chemical reaction is related to the spontaneity of the reaction.
12. **Calculate** the average rate of a reaction between hypothetical molecules A and B if the concentration of A changes from 1.00*M* to 0.50*M* in 2.00 s.

SECTION 2

# Factors Affecting Reaction Rates

**MAINIDEA** Factors such as reactivity, concentration, temperature, surface area, and catalysts affect the rate of a chemical reaction.

### Essential Questions

- What are the factors that affect the rates of chemical reactions?
- What is the role of a catalyst?

### Review Vocabulary

**concentration:** a quantitative measure of the amount of solute in a given amount of solvent or solution

### New Vocabulary

catalyst
inhibitor
heterogeneous catalyst
homogeneous catalyst

**CHEM 4 YOU** How quickly do you think a forest fire would spread if the trees were far apart or the wood were damp? Similarly, the rate of a chemical reaction is dependent on a number of factors, including the concentrations and physical properties of the reactants.

## The Nature of Reactants

Some substances react more readily than others. For example, copper and zinc are both metals and they have similar physical properties because of their relative positions on the periodic table. But they react at different rates when placed in aqueous silver nitrate solutions of equal concentration.

When a copper strip is placed in 0.05*M* silver nitrate, as shown in **Figure 8a,** the copper and silver nitrate react to form silver metal and aqueous copper(II) nitrate. When a zinc strip is placed in 0.05*M* silver nitrate, as shown in **Figure 8b,** the zinc and silver nitrate react to form silver metal and aqueous zinc nitrate. You can see that the reactions are similar. However, compare the amounts of silver formed in the two photographs, which were taken after the same number of minutes had elapsed. **Figure 8** shows that more silver formed in the reaction of zinc and silver nitrate than in the reaction of copper and silver nitrate. The reaction of zinc with silver nitrate occurs faster because zinc is higher on the activity series than copper.

Copper strip in silver nitrate

Zinc strip in silver nitrate

**Figure 8** Zinc is more reactive than copper, so it reacts with silver nitrate faster than copper does.
**Write** *the balanced equations for the reactions at right.*

The concentration of oxygen in the air surrounding the candle is about 20%.

The candle burns more rapidly because the jar contains almost 100% oxygen.

■ **Figure 9** The brighter flame in the jar containing a greater amount of oxygen indicates an increase in reaction rate. The higher oxygen concentration accounts for the faster reaction.

## Concentration

One way chemists can change the rate of a reaction is by changing the concentrations of the reactants. Remember that collision theory states that particles must collide in order to react. The more particles that are present, the more often collisions occur. Think about bumper cars at an amusement park. When more cars are in operation, the number of collisions increases. The same is true for a reaction in which Reactant A combines with Reactant B. At given concentrations of A and B, molecules of A and B collide to produce AB at a particular rate. What happens if the concentration of B is increased? Molecules of A collide with molecules of B more frequently because more molecules of B are available. More collisions ultimately increase the rate of reaction.

**FOLDABLES®** Incorporate information from this section into your Foldable.

☑ READING CHECK **Predict** what would happen to the rate of the reaction if the concentration of A was increased.

Look at the reactions shown in **Figure 9.** The wax in the candle undergoes combustion. In the first photo, the candle burns in air. How does this compare with the second photo, in which the burning candle is placed inside a jar containing nearly 100% oxygen—approximately five times the concentration of oxygen in air? According to collision theory, the higher concentration of oxygen increases the collision frequency between the wax molecules in the candle and oxygen molecules. As a result, the rate of the reaction increases, resulting in a larger, brighter flame.

## Surface Area

Now suppose you lowered a red-hot chunk of steel into a flask of oxygen gas and a red-hot bundle of steel wool into another flask of oxygen gas. What might be different? The oxygen would react with the chunk of steel much more slowly than it would with the steel wool. Using what you know about collision theory, can you explain why? You are correct if you said that, for the same mass of iron, steel wool has more surface area than the chunk of steel. The greater surface area of the steel wool allows oxygen molecules to collide with many more iron atoms per unit of time.

■ **Figure 10** The greater surface area of the steel wool means that more collisions can occur between the metal and oxygen.

For the same mass, many small particles have more total surface area than one large particle. For example, observe the reactions shown in **Figure 10.** The hot nail glows in oxygen in **Figure 10a,** but the same mass of steel wool in **Figure 10b** bursts into flames. Increasing the surface area of a reactant speeds up the rate of reaction by increasing the collision rate between reacting particles.

## Temperature

Increasing the temperature of a reaction generally increases the rate of a reaction. For example, you know that the reactions that cause foods to spoil occur faster at room temperature than when the foods are refrigerated. The graph in **Figure 11** illustrates that increasing the temperature by 10 K can approximately double the rate of a reaction. How can such a small increase in temperature have such a significant effect?

Recall that increasing the temperature of a substance increases the average kinetic energy of the particles that make up the substance. For that reason, reacting particles collide more frequently at higher temperatures than at lower temperatures. However, that fact alone does not account for the increase in reaction rate with increasing temperature.

To better understand how reaction rate varies with temperature, examine the second graph in **Figure 11.** This graph compares the numbers of particles that have sufficient energy to react at temperatures $T_1$ and $T_2$, where $T_2$ is greater than $T_1$. The dotted line indicates the activation energy ($E_a$) for the reaction. The shaded area under each curve represents the number of collisions that have energy equal to or greater than the activation energy. How do the shaded areas compare? The number of high-energy collisions at the higher temperature, $T_2$, is greater than the number at the lower temperature, $T_1$. Therefore, as the temperature increases, more collisions result in a reaction.

☑ GRAPH CHECK **Determine** the relative reaction rate at 325 K.

■ **Figure 11** Increasing the temperature of a reaction increases the frequency of collisions and therefore the rate of the reaction. Increasing the temperature also raises the kinetic energy of the particles. More of the collisions at high temperatures have enough energy to overcome the activation energy barrier and react.

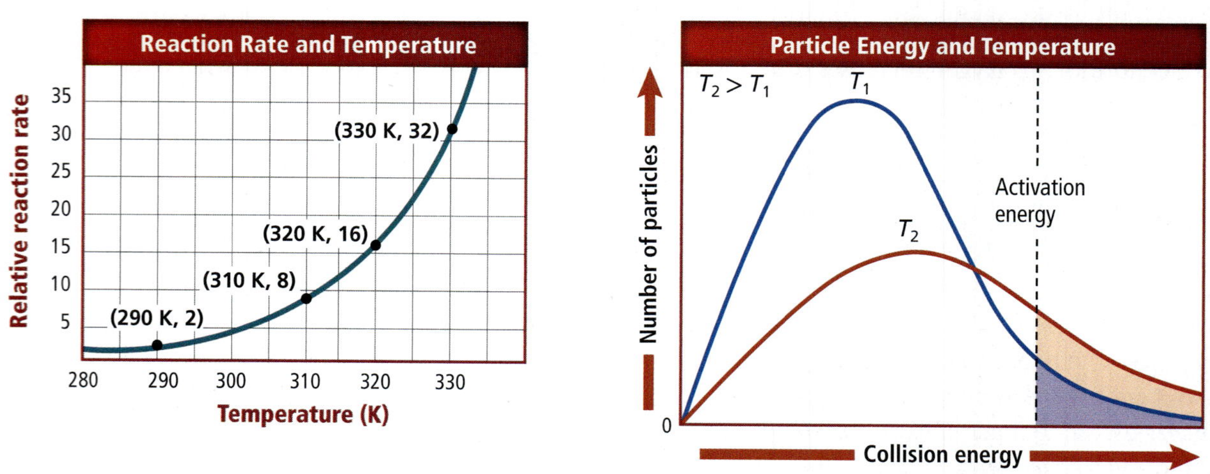

# MiniLAB

iLab Station

## Examine Reaction Rate and Temperature

**What is the effect of temperature on a common chemical reaction?**

### Procedure

1. Read and complete the lab safety form.
2. Break a single **effervescent tablet** into four equal pieces.
3. Use a **balance** to measure the mass of one piece of the tablet. Measure 50 mL of room-temperature **water** (approximately 20°C) into a **250-mL beaker.** Use a **nonmercury thermometer** to measure the temperature of the water.
4. With a **stopwatch** or a **clock with a second hand** ready, add the piece of tablet to the water. Record the amount of time elapsed between when the tablet hits the water and when all of the solid has dissolved.
5. Repeat Steps 3 and 4, this time gradually warming the 50 mL of water to about 50°C on a **hot plate.** Maintain the temperature (equilibrate) throughout the run.

### Analysis

1. **Identify** the initial mass, the final mass, and $t_1$ and $t_2$ for each trial run.
2. **Calculate** the reaction rate by finding the mass of reactant consumed per second for each run.
3. **Describe** the relationship between reaction rate and temperature for this reaction.
4. **Predict** what the reaction rate would be if the reaction were carried out at 40°C and explain the basis for your prediction. To test your prediction, repeat the reaction at 40°C using another piece of tablet.
5. **Evaluate** how well your prediction for the reaction rate at 40°C compares to the measured reaction rate.

## Catalysts and Inhibitors

The temperature and the concentration of reactants affect the rate of a reaction, but an increase in temperature is not always the best, or most practical, thing to do. For example, suppose that you want to increase the rate of the decomposition of glucose in a living cell. Increasing the temperature and/or the concentration of reactants is not an option because doing so might harm or kill the cell.

**Catalysts** Many chemical reactions in living organisms would not occur quickly enough to sustain life at normal living temperatures if it were not for the presence of enzymes. An enzyme is a type of **catalyst,** a substance that increases the rate of a chemical reaction without being consumed in the reaction. Catalysts are used extensively in manufacturing because quickly producing more of a product reduces its cost. A catalyst does not yield more product and is not included in either the reactants or the products of the reaction. Thus, catalysts are not included in chemical equations.

**Inhibitors** Another type of substance that affects reaction rates is called an inhibitor. Unlike a catalyst, which speeds up reaction rates, an **inhibitor** is a substance that slows down reaction rates, or inhibits reactions. Some inhibitors prevent a reaction from happening at all.

**How catalysts and inhibitors work** A catalyst lowers the activation energy required for a reaction to take place at a given temperature. Recall that a low activation energy means that more of the collisions between particles will have sufficient energy to overcome the activation energy barrier and bring about a reaction. By lowering the activation energy, a catalyst increases the average reaction rate.

### RealWorld CHEMISTRY

**Excluding Oxygen**

**FOOD PRESERVATION** Foods often spoil because they react with oxygen. Many methods of food preservation maintain product freshness by excluding oxygen. For example, apples stored in an atmosphere of carbon dioxide can be kept fresh long after harvest. Foods such as popcorn and crackers are often packaged in an atmosphere of an unreactive gas such as nitrogen or argon.

■ **Figure 12** The activation energy of the catalyzed reaction is lower than that of the uncatalyzed reaction. Thus the catalyzed reaction produces products at a faster rate than the uncatalyzed reaction does.

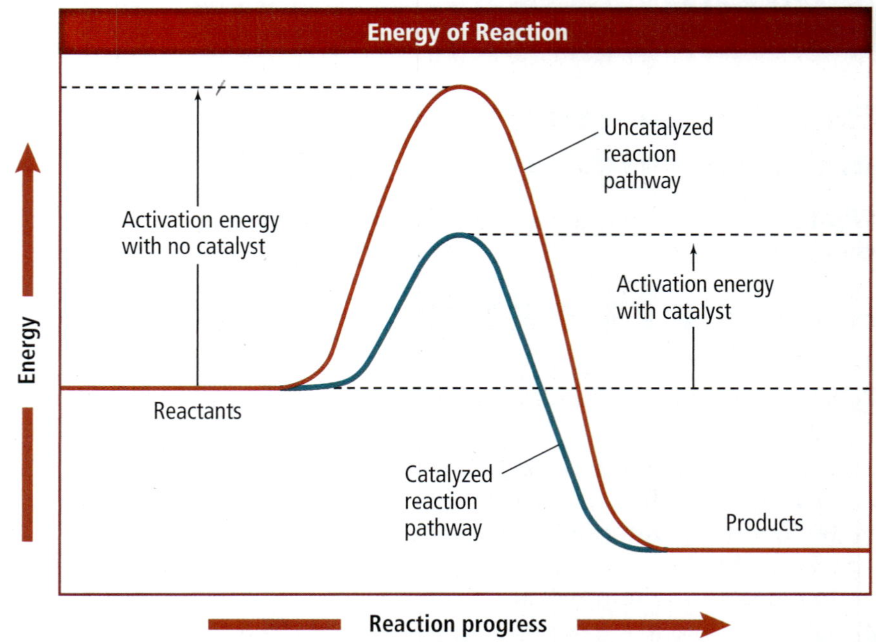


**GRAPH CHECK**

**Determine** from the graph how the use of a catalyst affects the energy released in the reaction.

**Figure 12** shows the energy diagram for an exothermic chemical reaction. The red line represents the reaction pathway with no catalyst present. The blue line represents the catalyzed reaction pathway. Note that the activation energy for the catalyzed reaction is much lower than for the uncatalyzed reaction. You can think of the reaction's activation energy as an obstacle to be cleared, as shown in **Figure 13.** In this analogy, much less energy is required for the horse and rider to clear the lower barrier than to jump the higher hurdle.

Inhibitors can act in a variety of ways. Some block lower energy pathways and thus raise the activation energy of a reaction. Others react with the catalyst and destroy it or prevent it from performing its function. In biological reactions, an inhibitor might bind the enzyme that catalyzes a reaction and prevent the reaction from occurring. In the food industry, inhibitors are called preservatives or antioxidants. Preservatives are safe to eat and give food longer shelf lives.

■ **Figure 13** A higher activation energy means that reacting particles must have more energy in order to react. The horse and rider exert little energy jumping the low barrier. Greater speed and energy are needed to clear the higher hurdle.

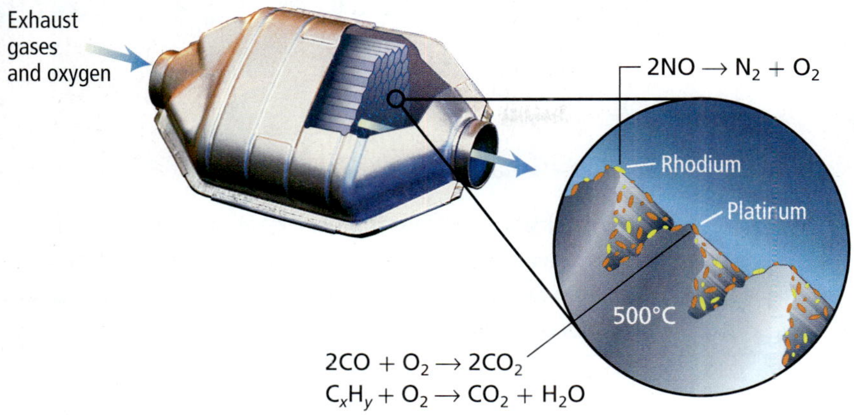


**Figure 14** The inside of a catalytic converter is coated with particles of rhodium and platinum. At 500°C, rhodium catalyzes the conversion of nitrogen oxide (NO) to nitrogen ($N_2$) and oxygen ($O_2$). Platinum catalyzes the conversion of carbon monoxide (CO) to carbon dioxide ($CO_2$) and converts any unburned gasoline, represented by $C_xH_y$, to carbon dioxide and water vapor ($H_2O$).

**Heterogeneous and homogeneous catalysts** Today's automobiles are required by law to be equipped with catalytic converters. **Figure 14** shows the reactions within a catalytic converter that convert harmful exhaust gases to acceptable substances. Nitrogen monoxide is converted to nitrogen and oxygen, carbon monoxide to carbon dioxide, and unburned gasoline to carbon dioxide and water. The most effective catalysts for this application are transition metal oxides and metals such as rhodium and platinum. Because the catalysts in a catalytic converter are solids and the reactions they catalyze are gaseous, the catalysts are called heterogeneous catalysts. A **heterogeneous catalyst** exists in a physical state different than that of the reaction it catalyzes. A catalyst that exists in the same physical state as the reaction it catalyzes is called a **homogeneous catalyst.** In the Launch Lab, you used a heterogenous catalyst (yeast) to speed up the decomposition of hydrogen peroxide. The same result can be obtained by using a potassium iodide (KI) solution. Iodide ions ($I^-$ (aq)), present in the same physical state as the hydrogen peroxide molecules, act as a homogeneous catalyst in the decomposition.

Explore **reaction rates.**

Virtual Investigations

## SECTION 2 REVIEW

Section Self-Check

### Section Summary

- Key factors that influence the rate of chemical reactions include reactivity, concentration, surface area, temperature, and catalysts.
- Raising the temperature of a reaction generally increases the rate of the reaction by increasing the collision frequency and the number of collisions that form an activated complex.
- Catalysts increase the rates of chemical reactions by lowering activation energies.

13. MAINIDEA **Explain** why magnesium metal reacts with hydrochloric acid (HCl) at a faster rate than iron does.
14. **Explain** how collision theory accounts for the effect of concentration on reaction rate.
15. **Explain** the difference between a catalyst and an inhibitor.
16. **Describe** the effect on the rate of a reaction if one of the reactants is ground to a powder rather than used as a single chunk.
17. **Infer** If increasing the temperature of a reaction by 10 K approximately doubles the reaction rate, what would be the effect of increasing the temperature by 20 K?
18. **Research** how catalysts are used in industry, in agriculture, or in the treatment of contaminated soil, waste, or water. Write a short report summarizing your findings about the role of a catalyst in one of these applications.

## SECTION 3

# Reaction Rate Laws

**MAINIDEA** The reaction rate law is an experimentally determined mathematical relationship that relates the speed of a reaction to the concentrations of the reactants.

**CHEM 4 YOU** When a bicyclist switches from first gear to second gear, the bicycle travels a greater distance with each revolution of the pedals. In the same way, when a chemist increases the concentration of a reactant, the rate of the reaction increases.

### Essential Questions

- What is the relationship between reaction rate and concentration?
- How are reaction orders determined using the method of initial rates?

### Review Vocabulary

**reactant:** the starting substance in a chemical reaction

### New Vocabulary

rate law
specific rate constant
reaction order
method of initial rates

## Writing Reaction Rate Laws

In Section 1, you learned how to calculate the average rate of a chemical reaction. The word *average* is important because most chemical reactions slow down as the reactants are consumed and fewer particles are available to collide. Chemists quantify the results of collision theory in an equation called a rate law. A **rate law** expresses the relationship between the rate of a chemical reaction and the concentration of reactants at a given temperature. For example, the reaction $A \rightarrow B$ is a one-step reaction. The rate law for this reaction is expressed as follows.

**One-Step Reaction Rate Law**

$$\text{rate} = k[A]$$

[A] represents the concentration of a reactant; $k$ is a constant.

The rate of a one-step reaction is the product of the concentration of the reactant and a constant.

The symbol $k$ is the **specific rate constant,** a numerical value that relates the reaction rate and the concentrations of reactants at a given temperature. The specific rate constant is unique for every reaction and can have a variety of units including L/(mol•s), $L^2/(mol^2 \cdot s)$, and $s^{-1}$. A rate law must be determined experimentally as illustrated in **Figure 15.**

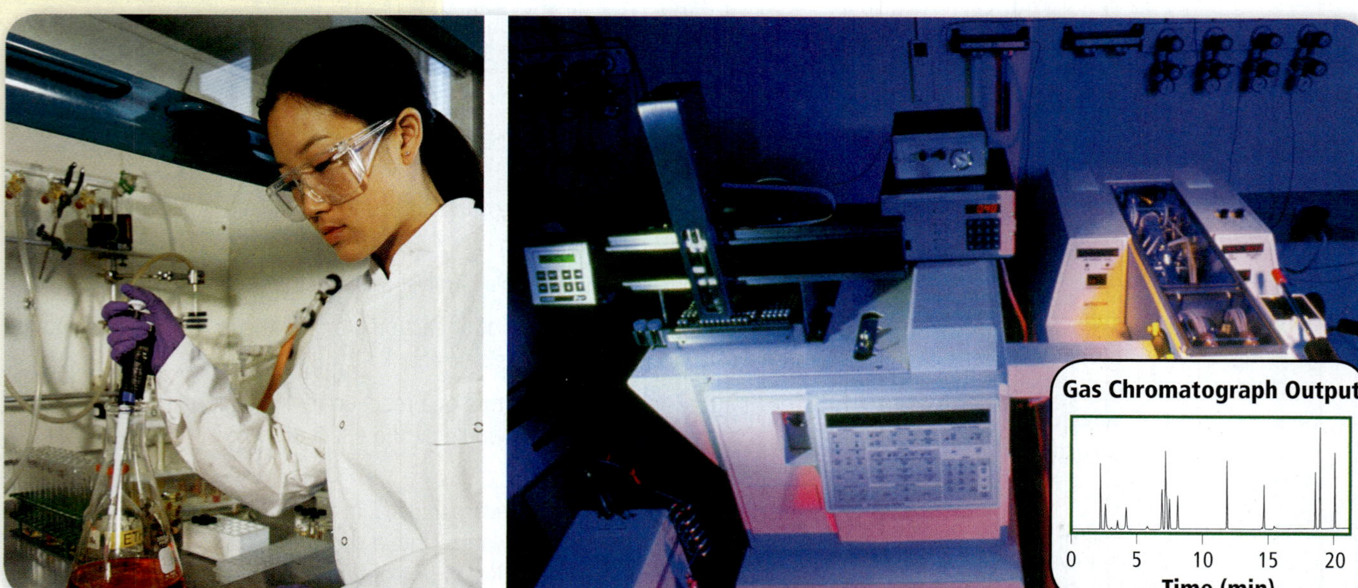


**Figure 15** Gas chromatography is one technique that can be used for reaction rate determination. To determine the rate of a reaction, samples of the reaction mixture are withdrawn at regular intervals while the reaction is proceeding. The samples are immediately injected into a gas chromatograph, which separates the components and helps identify them.

The rate law shows that the reaction rate is directly proportional to the molar concentration of A. The specific rate constant, $k$, does not change with concentration; however, $k$ does change with temperature. A large value of $k$ means that A reacts rapidly to form B.

**First-order reaction rate laws** In the expression $Rate = k[A]$, it is understood that the notation [A] means the same as $[A]^1$. For reactant A, the understood exponent 1 is called the reaction order. The **reaction order** for a reactant defines how the rate is affected by the concentration of that reactant. For example, the rate law for the decomposition of $H_2O_2$ is expressed by the following equation.

$$\text{Rate} = k[H_2O_2]$$

Because the reaction rate is directly proportional to the concentration of $H_2O_2$ raised to the first power ($[H_2O_2]^1$), the decomposition of $H_2O_2$ is said to be first order in $H_2O_2$. Because the reaction is first order in $H_2O_2$, the reaction rate changes in the same proportion that the concentration of $H_2O_2$ changes. So, if the $H_2O_2$ concentration decreases to one-half its original value, the reaction rate is also reduced by one-half.

Recall that reaction rates are determined from experimental data. Because reaction order is based on reaction rates, it follows that reaction order is also determined experimentally. Finally, because the rate constant, $k$, describes the reaction rate, it must also be determined experimentally. The graph in **Figure 16** shows how the initial reaction rate for the decomposition of $H_2O_2$ changes with the concentration of $H_2O_2$.

**READING CHECK** **Infer** If the reaction order for a reactant is first order, how will the rate of the reaction change if the concentration of the reactant is tripled?

■ **Figure 16** The graph shows a direct relationship between the concentration of $H_2O_2$ and its rate of decomposition.

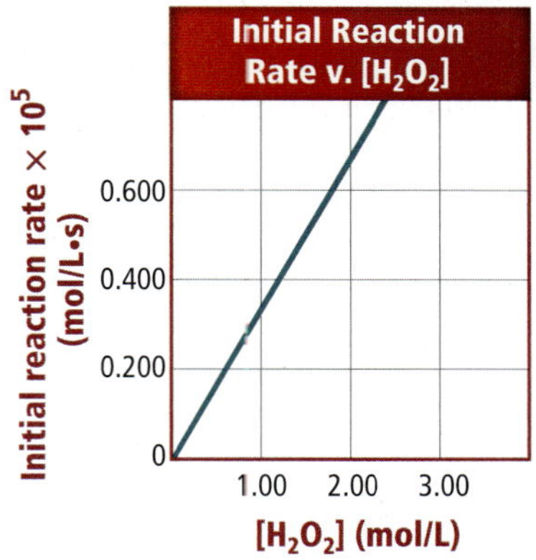


**GRAPH CHECK**

**Apply** Using the graph, determine the initial reaction rate when $[H_2O_2]$ is 1.50 mol/L.

**Other-order reaction rate laws** The overall reaction order of a chemical reaction is the sum of the orders for the individual reactants in the rate law. Many chemical reactions, particularly those that have more than one reactant, are not first-order. Consider the general form for a chemical reaction with two reactants. In this chemical equation, $a$ and $b$ are coefficients.

$$aA + bB \rightarrow \text{products}$$

The general rate law for such a reaction is described below.

**The General Rate Law**

$$\text{rate} = k[A]^m[B]^n$$

**[A] and [B] represent the concentrations of reactants A and B. The exponents $m$ and $n$ are the reaction orders.**

The rate of a reaction is equal to the product of $k$ and the concentrations of the reactants each raised to a power (order) that is determined experimentally.

Only if the reaction between A and B occurs in a single step (and with a single activated complex) does $m = a$ and $n = b$. That is unlikely, however, because single-step reactions are not common. For example, consider the multi-step reaction between nitrogen monoxide (NO) and hydrogen ($H_2$), which is described by the following equation.

$$2NO(g) + 2H_2(g) \rightarrow N_2(g) + 2H_2O(g)$$

This reaction has the following rate law.

$$\text{rate} = k[\text{NO}]^2[\text{H}_2]$$

The rate law was determined from experimental data that indicate that the rate depends on the concentration of the reactants as follows: If [NO] doubles, the rate quadruples; if $[H_2]$ doubles, the rate doubles. The reaction is described as second order in NO, first order in $H_2$, and third order overall. The overall order is the sum of the orders for the individual reactants (the sum of the exponents), which is (2 + 1), or 3.

**READING CHECK** **Explain** how you can determine the overall order of the reaction from the rate equation.

## Determining Reaction Order

**VOCABULARY**

**WORD ORIGIN**

**Initial**

adjective from Latin *initium,* meaning *of or relating to the beginning*

One common experimental method of evaluating reaction order is called the method of initial rates. The **method of initial rates** determines reaction order by comparing the initial rates of a reaction carried out with varying reactant concentrations. The initial rate measures how fast the reaction proceeds at the moment at which the reactants are mixed and the concentrations of the reactants are known. To understand how this method works, consider the general reaction $aA + bB \rightarrow$ products. Suppose that the reaction is carried out three times with varying concentrations of A and B and yields the initial reaction rates shown in **Table 2.** Recall that the general rate law for this type of reaction is as follows.

$$\text{rate} = k[\text{A}]^m[\text{B}]^n$$

To determine *m,* the exponent of [A], compare the concentrations and reaction rates in Trials 1 and 2. As you can see from the data, while the concentration of B remains constant, the concentration of A in Trial 2 is twice that of Trial 1. Note that the initial rate in Trial 2 is twice that of Trial 1. Because doubling [A] doubles the rate, the reaction must be first order in A. Since you are doubling the concentration, consider how $2^m$ affects the rate. The rate also doubles, so $2^m = 2$. Because $2^m = 2$, $m$ must equal 1. The same method is used to determine *n,* the exponent of [B], except this time Trials 2 and 3 are compared. Doubling the concentration of B causes the rate to increase by four times. Because $2^n = 4$, $n$ must equal 2. Thus the reaction is second order in B, giving the following overall rate law.

$$\text{rate} = k[\text{A}]^1[\text{B}]^2$$

The overall reaction order is third order (sum of exponents 2 + 1 = 3).

**Table 2** Experimental Initial Rates for $aA + bB \rightarrow$ products

| Trial | Initial [A](*M*) | Initial [B](*M*) | Initial Rate (mol/(L · s)) |
|---|---|---|---|
| 1 | 0.100 | 0.100 | $2.00 \times 10^{-3}$ |
| 2 | 0.200 | 0.100 | $4.00 \times 10^{-3}$ |
| 3 | 0.200 | 0.200 | $16.00 \times 10^{-3}$ |

## PRACTICE Problems

Do additional problems. Online Practice

**19.** Write the rate law for the reaction $aA \rightarrow bB$ if the reaction is third order in A. [B] is not part of the rate law.

**20.** The rate law for the reaction $2NO(g) + O_2(g) \rightarrow 2NO_2(g)$ is first order in $O_2$ and third order overall. What is the rate law for the reaction?

**21.** Given the experimental data below, use the method of initial rates to determine the rate law for the reaction $aA + bB \rightarrow$ products. *(Hint: Any number to the zero power equals one. For example, $(0.22)^0 = 1$ and $(55.6)^0 = 1$.)*

**Practice Problem 21 Experimental Data**

| Trial | Initial [A](*M*) | Initial [B](*M*) | Initial Rate (mol/(L·s)) |
|---|---|---|---|
| 1 | 0.100 | 0.100 | $2.00 \times 10^{-3}$ |
| 2 | 0.200 | 0.100 | $2.00 \times 10^{-3}$ |
| 3 | 0.200 | 0.200 | $4.00 \times 10^{-3}$ |

**22. Challenge** The rate law for the reaction $CH_3CHO(g) \rightarrow CH_4(g) + CO(g)$ is Rate $= k[CH_3CHO]^2$. Use this information to fill in the missing experimental data below.

**Practice Problem 22 Experimental Data**

| Trial | Initial $[CH_3CHO]$(*M*) | Initial Rate (mol/(L·s)) |
|---|---|---|
| 1 | $2.00 \times 10^{-3}$ | $2.70 \times 10^{-11}$ |
| 2 | $4.00 \times 10^{-3}$ | $10.8 \times 10^{-11}$ |
| 3 | $8.00 \times 10^{-3}$ | |

# SECTION 3 REVIEW

Section Self-Check

### Section Summary

- The mathematical relationship between the rate of a chemical reaction at a given temperature and the concentrations of reactants is called the rate law.
- The rate law for a chemical reaction is determined experimentally using the method of initial rates.

**23.** MAINIDEA **Explain** what the rate law for a chemical reaction tells you about the reaction.

**24. Apply** the rate-law equations to show the difference between a first-order reaction with a single reactant and a second-order reaction with a single reactant.

**25. Explain** the function of the specific rate constant in a rate-law equation.

**26. Explain** Under what circumstance is the specific rate constant ($k$) not a constant? What does the size of $k$ indicate about the rate of a reaction?

**27. Suggest** a reason why, when given the rate of a chemical reaction, it is important to know that the reaction rate is an average reaction rate.

**28. Explain** how the exponents in the rate equation for a chemical reaction relate to the coefficients in the chemical equation.

**29. Determine** the overall reaction order for a reaction between A and B for which the rate law is rate $= k[A]^2[B]^2$.

**30. Design an Experiment** Explain how you would design an experiment to determine the rate law for the general reaction $aA + bB \rightarrow$ products using the method of initial rates.

# SECTION 4

# Instantaneous Reaction Rates and Reaction Mechanisms

**MAINIDEA** The slowest step in a sequence of steps determines the rate of the overall chemical reaction.

### Essential Questions

- How are instantaneous rates of chemical reactions calculated?
- What substances and steps are involved in a reaction mechanism?
- How is the instantaneous rate of a complex reaction related to its reaction mechanism?

### Review Vocabulary

**decomposition reaction:** a chemical reaction that occurs when a single compound breaks down into two or more elements or new compounds

### New Vocabulary

instantaneous rate
complex reaction
reaction mechanism
intermediate
rate-determining step

**CHEM 4 YOU** Buying lunch in the cafeteria is a series of steps: picking up a tray and tableware, choosing food items, and paying the cashier. The first two steps might go rapidly, but a long line at the cashier will slow down the whole experience. Similarly, a reaction can go no faster than its slowest step.

## Instantaneous Reaction Rates

Chemists often need to know more than the average reaction rate. A pharmacist developing a new drug treatment might need to know the progress of a reaction at an exact instant. Consider the decomposition of hydrogen peroxide ($H_2O_2$), which is represented as follows.

$$2H_2O_2(aq) \rightarrow 2H_2O(l) + O_2(g)$$

For this reaction, the decrease in $H_2O_2$ concentration over time is shown in **Figure 17.** The curved line shows how the reaction rate decreases as the reaction proceeds. The **instantaneous rate** is the reaction rate at a specific time. It is the slope of the straight line tangent to the curve at a specific time. The expression $\Delta[H_2O_2]/\Delta t$ is one way to express the reaction rate. In other words, the rate of change in $H_2O_2$ concentration relates to one specific point (or instant) on the graph.

You can determine the instantaneous rate for a reaction in another way if you are given the reactant concentrations at a given temperature and know the experimentally determined rate law and the specific rate constant at that temperature.

■ **Figure 17** The instantaneous rate for a specific point in the reaction progress can be determined from the tangent to the curve that passes through that point.

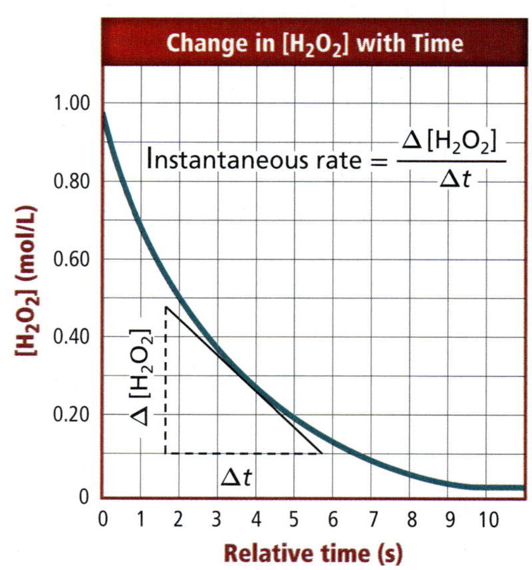


Slope of line $= \dfrac{\Delta y}{\Delta x}$

Instantaneous rate $= \dfrac{\Delta[H_2O_2]}{\Delta t}$

$\dfrac{\Delta y}{\Delta x} = \dfrac{\Delta[H_2O_2]}{\Delta t}$

**GRAPH CHECK**
**Identify** the variables that are plotted on the *y*-axis and on the *x*-axis.

Consider the decomposition of dinitrogen pentoxide ($N_2O_5$) into nitrogen dioxide ($NO_2$) and oxygen ($O_2$), which proceeds as follows.

$$2N_2O_5(g) \rightarrow 4NO_2(g) + O_2(g)$$

The experimentally determined rate law for this reaction is

$$\text{rate} = k[N_2O_5]$$

where $k = 1.0 \times 10^{-5}\ s^{-1}$. If $[N_2O_5] = 0.350M$, the instantaneous reaction rate would be calculated as

$$\text{rate} = (1.0 \times 10^{-5}\ s^{-1})(0.350\ mol/L) = 3.5 \times 10^{-6}\ mol/(L \cdot s)$$

## EXAMPLE Problem 2

Find help with **dimensional analysis.** Math Handbook

EXAMPLE PROBLEM

**CALCULATE INSTANTANEOUS REACTION RATES** The following reaction is first order in $H_2$ and second order in NO with a rate constant of $2.90 \times 10^2\ (L^2/(mol^2 \cdot s))$.

$$2NO(g) + H_2(g) \rightarrow N_2O(g) + H_2O(g)$$

Calculate the instantaneous rate when the reactant concentrations are $[NO] = 0.00200M$ and $[H_2] = 0.00400M$.

### 1 ANALYZE THE PROBLEM

The rate law can be expressed by rate $= k[NO]^2[H_2]$. Therefore, the instantaneous reaction rate can be determined by inserting reactant concentrations and the specific rate constant into the rate law equation.

**Known**

$[NO] = 0.00200M$

$[H_2] = 0.00400M$

$k = 2.90 \times 10^2\ (L^2/(mol^2 \cdot s))$

**Unknown**

**rate = ? mol/(L • s)**

### 2 SOLVE FOR THE UNKNOWN

rate $= k[NO]^2[H_2]$ — **State the rate law.**

rate $= (2.90 \times 10^2\ L^2/(mol^2 \cdot s))(0.00200\ mol/L)^2(0.00400\ mol/L)$ — **Substitute $k = 2.90 \times 10^2\ (L^2/(mol^2 \cdot s))$, $[NO] = 0.00200M$, and $[H_2] = 0.00400M$.**

rate $=$ **$4.64 \times 10^{-6}\ mol/(L \cdot s)$** — **Multiply the numbers and units.**

### 3 EVALUATE THE ANSWER

Units in the calculation cancel to give mol/(L • s), which is a common unit for reaction rates. A magnitude of approximately $10^{-6}$ mol/(L • s) fits with the quantities given and the rate law equation. The answer is correctly expressed with three significant figures.

## PRACTICE Problems

Do additional problems. Online Practice

**Use the rate law in Example Problem 2 and the concentrations given in Practice Problems 31 and 32 to calculate the instantaneous rate for the reaction between NO and $H_2$.**

**31.** $[NO] = 0.00500M$ and $[H_2] = 0.00200M$

**32.** $[NO] = 0.0100M$ and $[H_2] = 0.00125M$

**33. Challenge** Calculate [NO] for the reaction in Example Problem 2 if the rate is $9.00 \times 10^{-5}$ mol/(L • s) and $[H_2]$ is $0.00300M$.

**CAREERS IN CHEMISTRY**

**Chemical Engineer** An understanding of reaction mechanisms is vital to chemical engineers. Their jobs often include scaling up a laboratory synthesis of a substance to large-scale production in a manufacturing plant. They must design the production facility and monitor its safe and efficient operation.

WebQuest

## Reaction Mechanisms

Most chemical reactions consist of sequences of two or more simpler reactions. For example, recent evidence indicates that the reaction $2O_3 \rightarrow 3O_2$ occurs in three steps after intense ultraviolet radiation from the Sun liberates chlorine atoms from certain compounds in Earth's stratosphere. Steps 1 and 2 in this reaction might occur simultaneously or in reverse order.

1. Chlorine atoms decompose ozone according to the equation $Cl + O_3 \rightarrow O_2 + ClO$.
2. Ultraviolet radiation causes the decomposition reaction $O_3 \rightarrow O + O_2$.
3. ClO produced in the reaction in Step 1 reacts with O produced in Step 2 according to the equation $ClO + O \rightarrow Cl + O_2$.

Each of the reactions described in Steps 1 through 3 is called an elementary step. These elementary steps, illustrated in **Figure 18,** comprise the complex reaction $2O_3 \rightarrow 3O_2$. A **complex reaction** is one that consists of two or more elementary steps. A **reaction mechanism** is the complete sequence of elementary steps that makes up a complex reaction. Adding elementary Steps 1 through 3 and canceling formulas that occur in equal amounts on both sides of the reaction arrow produce the net equation for the complex reaction as shown.

| | |
|---|---|
| Elementary step: | $\cancel{Cl}(g) + O_3(g) \rightarrow \cancel{ClO}(g) + O_2(g)$ |
| Elementary step: | $O_3(g) \rightarrow \cancel{O}(g) + O_2(g)$ |
| Elementary step: | $\cancel{ClO}(g) + \cancel{O}(g) \rightarrow \cancel{Cl}(g) + O_2(g)$ |
| Complex reaction: | $2O_3 \rightarrow 3O_2$ |

Because chlorine atoms react in Step 1 and are re-formed in Step 3, chlorine is said to catalyze the decomposition of ozone. Because ClO and O are formed in Steps 1 and 2, respectively, and are consumed in the reaction in Step 3, they are called intermediates. An **intermediate** is a substance produced in one elementary step of a complex reaction and consumed in a subsequent elementary step. Like catalysts, intermediates do not appear in the net chemical equation.

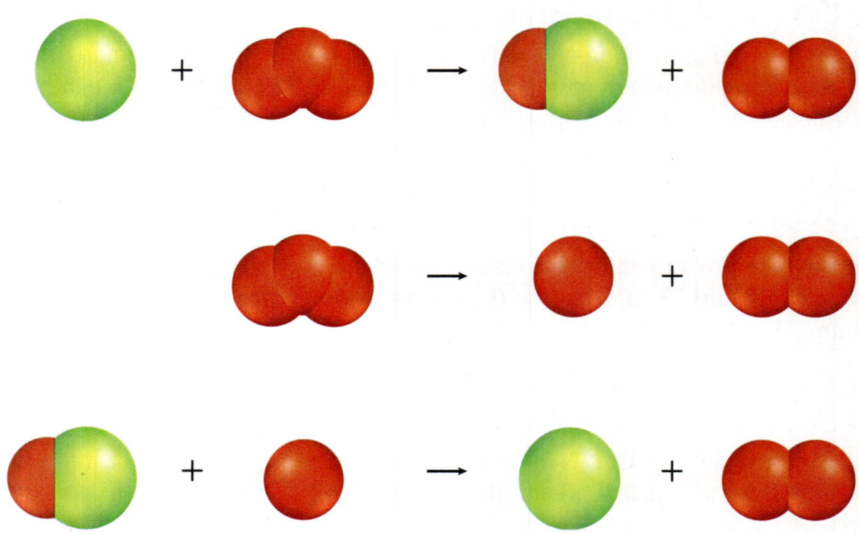

■ **Figure 18** ClO and O are intermediates in the three elementary steps of the complex reaction producing oxygen gas ($O_2$) from ozone ($O_3$).
**Infer** *What is the function of chlorine (Cl) in the complex reaction?*

### Connection to Physics Investigating reaction mechanisms

How is it possible to discover the presence of intermediates and determine their role in a chemical reaction? Learning how particles change their identities in the course of a chemical reaction means detecting evidence of bonds breaking and bonds forming. These processes take an extremely short period of time—time measured in femtoseconds. A femtosecond (fs) is one-millionth of a billionth of a second (0.000000000000001 second). Until recently, scientists could only calculate and imagine the actual atomic activity that occurs when bonds are broken and new bonds are made.

In 1999, Dr. Ahmed Zewail of the California Institute of Technology won a Nobel Prize for his achievements in the field of femtochemistry. Zewail developed an ultrafast laser device that can record pictures of chemical reactions as they happen. The laser "flashes" every 10 femtoseconds to record the movements of particles just as if they were being recorded on frames taken by a movie camera. Thus, a femtosecond recording of molecular motion could have as many as $10^{14}$ frames per second. The molecular motion corresponds to bond formation and breakage and can be related to the various possible intermediates and the products that are formed during a reaction. Zewail was able to witness an interaction between benzene ($C_6H_6$) and iodine ($I_2$) over a period of 1500 fs. A collision of iodine with benzene resulted in the breaking of the bond between the iodine atoms, after which the two atoms moved apart from one another. Technology such as this allows chemists to test their hypotheses about possible intermediates and reaction mechanisms.

**READING CHECK** **Explain** the importance of the methods of femtochemistry to the study of reaction mechanisms.

**Rate-determining step** Every complex reaction has an elementary step that is slower than all the other steps. The slowest elementary step in a complex reaction is called the **rate-determining step.** A reaction cannot go faster than its slowest elementary step. An analogy for the rate-determining step is shown in **Figure 19.**

**Figure 19** At highway toll booths, drivers must slow down and stop as tolls are paid. Although they can resume their speeds after paying the toll, the pause affects their overall rate of travel. In a similar way, the overall rate of a chemical reaction is dependent on how fast the slowest elementary step proceeds.

■ **Figure 20** The three peaks in this energy diagram correspond to activation energies to form the activated complexes. The middle peak represents the highest energy barrier to overcome; therefore, the reaction involving $N_2O_2 + H_2$ is the rate-determining step.

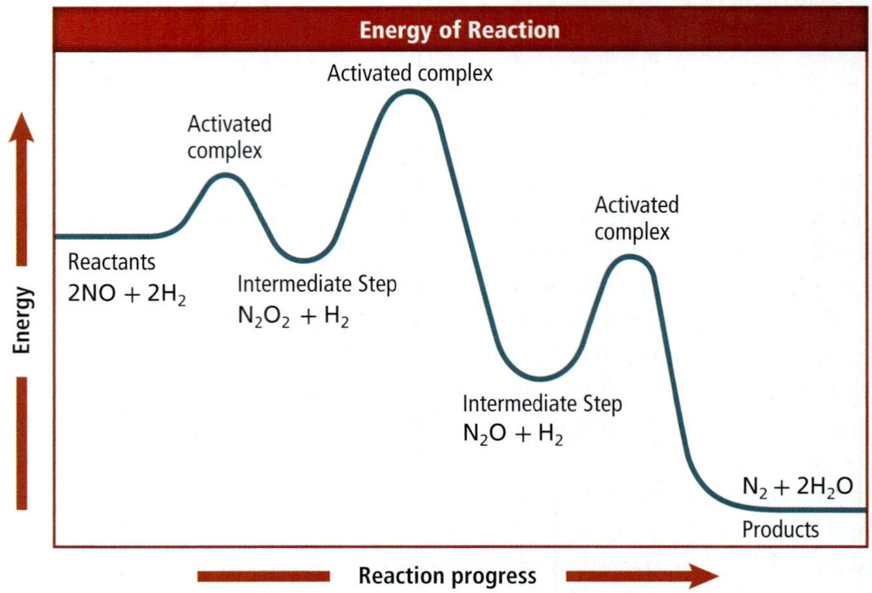


**GRAPH CHECK**

**Determine** from the graph whether the overall reaction is exothermic or endothermic.

To see how the rate-determining step affects reaction rate, consider again the gas-phase reaction from Example Problem 2. A mechanism for this reaction consists of the following elementary steps.

$$2NO \rightarrow \cancel{N_2O_2} \quad \text{(fast)}$$
$$\cancel{N_2O_2} + H_2 \rightarrow \cancel{N_2O} + H_2O \quad \text{(slow)}$$
$$\cancel{N_2O} + H_2 \rightarrow N_2 + H_2O \quad \text{(fast)}$$
$$\overline{2NO(g) + 2H_2(g) \rightarrow N_2(g) + 2H_2O(g)}$$

The first and third elementary steps occur relatively fast, so the slow middle step is the rate-determining step.

**Figure 20** shows how energy changes as this complex reaction proceeds. Each step of the reaction has its own activation energy, represented by a peak in the graph. Activation energy for Step 2 is higher than for Steps 1 and 3, which is why Step 2 is the rate-determining step.

## SECTION 4 REVIEW

Section Self-Check

### Section Summary

- The reaction mechanism of a chemical reaction must be determined experimentally.
- For a complex reaction, the rate-determining step limits the instantaneous rate of the overall reaction.

**34.** MAINIDEA **Compare and contrast** an elementary chemical reaction with a complex chemical reaction.

**35. Explain** how the rate law for a chemical reaction is used to determine the instantaneous rate of the reaction.

**36. Define** a reaction mechanism and an intermediate.

**37. Distinguish** between an intermediate and an activated complex.

**38. Relate** the size of the activation energy of an elementary step in a complex reaction to the rate of that step.

**39. Calculate** A reaction between A and B to form AB is first order in A and first order in B. The rate constant, $k$, equals 0.500 mol/(L • s). What is the rate of the reaction when $[A] = 2.00 \times 10^{-2}M$ and $[B] = 1.50 \times 10^{-2}M$?

# CHEMISTRY&health

## Reaction Rate and Body Temperature

Imagine that you're late for school and rush outside without putting on your jacket. It's a chilly day, and soon you begin to shiver. Shivering is an automatic response by your body that helps maintain your normal body temperature, which is important.

### What is normal body temperature?

Normal human body temperature is approximately 37°C, but it can vary with age, gender, time of day, and level of activity. Your temperature goes up when you engage in strenuous activities or when the temperature of the air around you is high. It can also go down when you take a cold shower or forget to wear your jacket in cold weather.

**Chemical reactions heat the body** Inside each cell of the body, food is metabolized to produce energy that is either used or stored in large molecules called adenosine triphosphate (ATP). When energy is needed, ATP splits into adenosine diphosphate (ADP) and a phosphate group ($P_i$) and energy is released.

$$\text{ATP} \rightarrow \text{ADP} + \text{P}_i + \text{Energy}$$

Reactions such as this require enzymes that regulate their rates. These enzymes are protein catalysts that are most efficient within the range of normal human body temperatures. Without the help of enzymes and a temperature near 37°C, reactions such as this one could not occur at a rate that would meet the needs of the body. Outside this temperature range, reaction rates are slower, as shown in **Figure 1.**

**Regulating body temperature** The area of the brain called the hypothalamus regulates body temperature by a complex feedback system. The system maintains a balance between the thermal energy released by chemical reactions within the body and the thermal energy exchanged between the body and the environment.

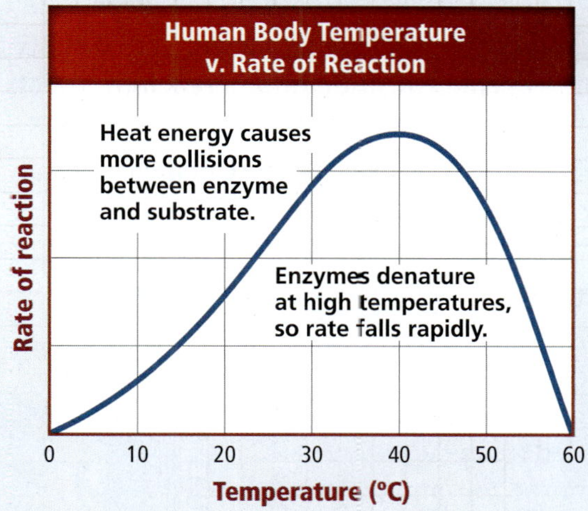


**Figure 1** Optimum temperature for humans is close to 37°C. Excessive thermal energy results in the breakdown of a protein's structure, preventing it from functioning as it should.

### Hypothermia—low body temperature

When hypothermia is detected, the hypothalamus begins actions that increase the release of thermal energy. Shivering is the rapid contractions of muscles that result from chemical reactions that release thermal energy. The body also begins actions to conserve thermal energy, including reducing blood flow to the skin.

### Hyperthermia—high body temperature

Excessive thermal energy, either from the environment or because of increased chemical reactions within the body, causes the body to respond by sweating. Blood vessels near the skin's surface dilate, and heart and lung functions increase. These actions result in an increase in the release of thermal energy to the environment. The entire system of temperature control is designed to keep reactions within the body occurring at the optimal rate.

## WRITING IN ▶ Chemistry

**Research** Write a patient-information brochure about the medical treatment of hypothermia and hyperthermia. Describe who might be at risk, any long term effects these conditions might have, and how they might be prevented.

WebQuest

# ChemLAB

iLab Station

## Observe How Concentration Affects Reaction Rate

**Background:** Collision theory describes how a change in concentration of one reactant affects the rate of a chemical reaction.

**Question:** *How does the concentration of a reactant affect the reaction rate?*

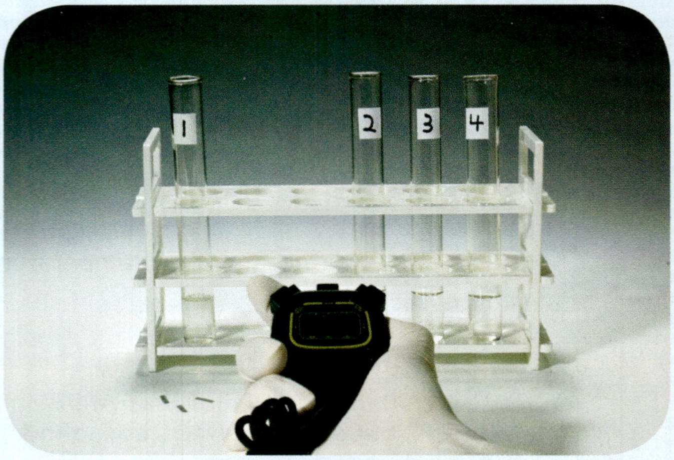


### Materials

10-mL graduated pipette
safety pipette filler
6*M* hydrochloric acid
distilled water
25-mm × 150-mm test tubes, labeled *1–4*
test-tube rack
magnesium ribbon
emery cloth or fine sandpaper
scissors
plastic ruler
tongs
watch with second hand or stopwatch
stirring rod

### Safety Precautions

WARNING: ***Never pipette any chemical by mouth. Hydrochloric acid is corrosive. Avoid contact with skin and eyes.***

### Procedure

1. Read and complete the lab safety form.
2. Use a safety pipette to draw 10 mL of 6.0*M* hydrochloric acid (HCl) into a 10-mL graduated pipette.
3. Dispense the 10 mL of 6.0*M* HCl into Test Tube 1.
4. Draw 5.0 mL of the 6.0*M* HCl from Test Tube 1 with the pipette. Dispense this acid into Test Tube 2. Use the pipette to add an additional 5.0 mL of distilled water. Mix with the stirring rod. This solution is 3.0*M* HCl.
5. Draw 5.0 mL of the 3.0*M* HCl from Test Tube 2 and dispense it into Test Tube 3. Add 5.0 mL of distilled water and stir. This solution is 1.5*M* HCl.
6. Draw 5.0 mL of the 1.5*M* HCl from Test Tube 3 and dispense it into Test Tube 4. Add 5.0 mL of distilled water and stir. This solution is 0.75*M* HCl.
7. Draw 5.0 mL of the 0.75*M* HCl from Test Tube 4. Neutralize and discard it in the sink.
8. Using tongs, place a 1-cm length of magnesium ribbon into Test Tube 1. Record in your data table the time in seconds it takes for the bubbling to stop.
9. Repeat Step 8 using the remaining three test tubes. Record the time in seconds it takes for the bubbling to stop in each test tube.
10. **Cleanup and Disposal** Place acid solutions in an acid discard container. Thoroughly wash all test tubes and lab equipment. Discard other materials as directed by your teacher. Return all lab equipment to its proper place.

### Analyze and Conclude

1. **Make a Graph** Plot the concentration of the acid on the *x*-axis and the reaction time on the *y*-axis. Draw a smooth curve through the data points.
2. **Conclude** Based on your graph, what is the relationship between the acid concentration and the reaction rate?
3. **Hypothesize** Write a hypothesis using collision theory, reaction rate, and reactant concentration to explain your results.
4. **Error Analysis** Compare your experimental results with those of other students in the laboratory. Explain the differences.

### INQUIRY EXTENSION

**Design an Experiment** Based on your observations and results, would temperature variations affect reaction rates? Plan an experiment to test your hypothesis.

# CHAPTER 16 STUDY GUIDE

Vocabulary Practice

**BIGIDEA** Every chemical reaction proceeds at a definite rate, but can be speeded up or slowed down by changing the conditions of the reaction.

## SECTION 1 A Model for Reaction Rates

**MAINIDEA** Collision theory is the key to understanding why some reactions are faster than others.

- The rate of a chemical reaction is expressed as the rate at which a reactant is consumed or the rate at which a product is formed.

$$\text{average reaction rate} = -\frac{\Delta[\text{reactant}]}{\Delta t} = \frac{\Delta[\text{product}]}{\Delta t}$$

- Reaction rates are generally calculated and expressed in moles per liter per second (mol/(L • s)).
- In order to react, the particles in a chemical reaction must collide.
- The rate of a chemical reaction is unrelated to the spontaneity of the reaction.

**VOCABULARY**
- reaction rate
- collision theory
- activated complex
- activation energy

## SECTION 2 Factors Affecting Reaction Rates

**MAINIDEA** Factors such as reactivity, concentration, temperature, surface area, and catalysts affect the rate of a chemical reaction.

- Key factors that influence the rate of chemical reactions include reactivity, concentration, surface area, temperature, and catalysts.
- Raising the temperature of a reaction generally increases the rate of the reaction by increasing the collision frequency and the number of collisions that form an activated complex.
- Catalysts increase the rates of chemical reactions by lowering activation energies.

**VOCABULARY**
- catalyst
- inhibitor
- heterogeneous catalyst
- homogeneous catalyst

## SECTION 3 Reaction Rate Laws

**MAINIDEA** The reaction rate law is an experimentally determined mathematical relationship that relates the speed of a reaction to the concentrations of the reactants.

- The mathematical relationship between the rate of a chemical reaction at a given temperature and the concentrations of reactants is called the rate law.

$$\text{rate} = k[\text{A}]$$

$$\text{rate} = k[\text{A}]^m[\text{B}]^n$$

- The rate law for a chemical reaction is determined experimentally using the method of initial rates.

**VOCABULARY**
- rate law
- specific rate constant
- reaction order
- method of initial rates

## SECTION 4 Instantaneous Reaction Rates and Reaction Mechanisms

**MAINIDEA** The slowest step in a sequence of steps determines the rate of the overall chemical reaction.

- The reaction mechanism of a chemical reaction must be determined experimentally.
- For a complex reaction, the rate-determining step limits the instantaneous rate of the overall reaction.

**VOCABULARY**
- instantaneous rate
- complex reaction
- reaction mechanism
- intermediate
- rate-determining step

# CHAPTER 16 ASSESSMENT

## SECTION 1

### Mastering Concepts

**40.** What happens to the concentrations of the reactants and products during the course of a chemical reaction?

**41.** Explain what is meant by the average rate of a reaction.

**42.** How would you express the rate of the chemical reaction $A \rightarrow B$ based on the concentration of Reactant A? How would that rate compare with the reaction rate based on the Product B?

**43.** What is the role of the activated complex in a chemical reaction?

**44.** Suppose two molecules that can react collide. Under what circumstances do the colliding molecules not react?

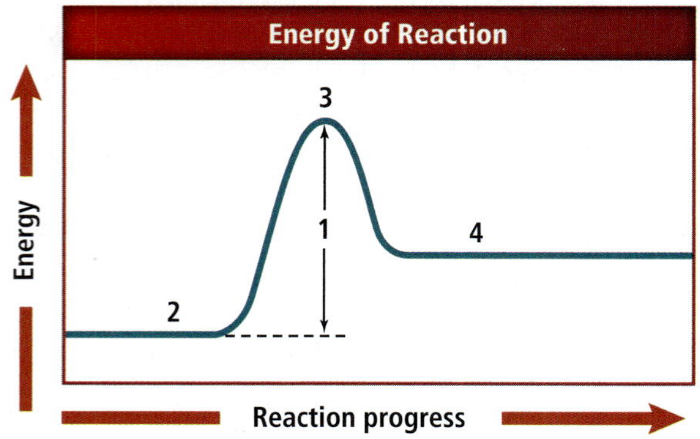


■ **Figure 21**

**45.** **Figure 21** is an energy diagram for a reaction. Match the appropriate number with the quantity it represents.
**a.** reactants
**b.** activated complex
**c.** products
**d.** activation energy

**46.** If $A \rightarrow B$ is exothermic, how does the activation energy for the forward reaction compare with the activation energy for the reverse reaction ($A \leftarrow B$)?

### Mastering Problems

**47.** In the gas-phase reaction, $I_2 + Cl_2 \rightarrow 2ICl$, $[I_2]$ changes from 0.400*M* at 0.00 min to 0.300*M* at 4.00 min. Calculate the average reaction rate in moles of $I_2$ consumed per liter per minute.

**48.** In a reaction $Mg(s) + 2HCl(aq) \rightarrow H_2(g) + MgCl_2(aq)$, 6.00 g of Mg was present at 0.00 min. After 3.00 min, 4.50 g of Mg remained. Express the average rate as mol Mg consumed/min.

**49.** If a chemical reaction occurs at the rate of $2.25 \times 10^{-2}$ moles per liter per second at 322 K, what is the rate expressed in moles per liter per minute?

## SECTION 2

### Mastering Concepts

**50.** What role does the reactivity of the reactants play in determining the rate of a chemical reaction?

**51.** In general, what is the relationship between reaction rate and reactant concentration?

**52.** Apply collision theory to explain why increasing the concentration of a reactant usually increases the reaction rate.

**53.** Explain why a crushed solid reacts with a gas more quickly than a large chunk of the same solid.

**54.** **Food Preservation** Apply collision theory to explain why foods usually spoil more slowly when refrigerated than at room temperature.

**55.** Apply collision theory to explain why powdered zinc reacts to form hydrogen gas faster than large pieces of zinc when both are placed in hydrochloric acid solution.

**56.** Hydrogen peroxide decomposes to water and oxygen gas more rapidly when manganese dioxide is added. The manganese dioxide is not consumed in the reaction. Explain the role of the manganese dioxide.

### Mastering Problems

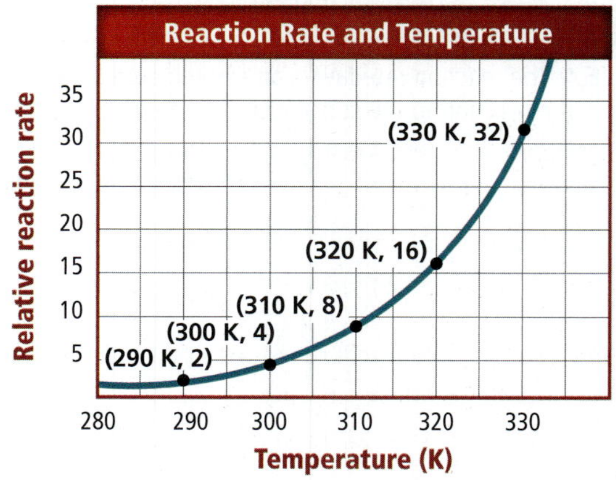


■ **Figure 22**

**57.** Examine **Figure 22,** which relates relative reaction rate and temperature. Approximately how does the reaction rate change for each increase of 10 K?

**58.** Suppose that a large volume of 3% hydrogen peroxide decomposes to produce 12 mL of oxygen gas in 100 s at 298 K. Estimate how much oxygen gas would be produced by an identical solution in 100 s at 308 K.

**59.** Using the information in Question 58, estimate how much oxygen gas would be produced in an identical solution in 100 seconds at 318 K. Estimate the time needed to produce 12 mL of oxygen gas at 288 K.

Chapter Self-Check

## SECTION 3

### Mastering Concepts

**60.** In the method of initial rates used to determine the rate law for a chemical reaction, what is the significance of the word *initial?*

**61.** Why must the rate law for a chemical reaction be based on experimental evidence rather than the balanced equation for the reaction?

**62.** Assume that the rate law for a generic chemical reaction is rate = $[A][B]^3$. What is the reaction order in A, the reaction order in B, and the overall reaction order?

**63.** Consider the generic chemical reaction: $A + B \rightarrow AB$. Based on experimental data, the reaction is second order in Reactant A. If the concentration of A is halved, and all other conditions remain unchanged, how does the reaction rate change?

### Mastering Problems

**64.** The instantaneous rate data in **Table 3** were obtained for the reaction $H_2(g) + 2NO(g) \rightarrow H_2O(g) + N_2O(g)$ at a given temperature and concentration of NO. How does the instantaneous rate of this reaction change as the initial concentration of $H_2$ is changed? Based on the data, is $[H_2]$ part of the rate law? Explain.

**Table 3 Reaction Between $H_2(g)$ and NO(g)**

| $[H_2]$ (mol/L) | Instantaneous Rate (mol/L·s) |
|---|---|
| 0.18 | $6.00 \times 10^{-3}$ |
| 0.32 | $1.07 \times 10^{-2}$ |
| 0.58 | $1.93 \times 10^{-2}$ |

**65.** Suppose that a generic chemical reaction has the rate law of rate = $[A]^2[B]^3$ and that the reaction rate under a given set of conditions is $4.5 \times 10^{-4}$ mol/(L•min). If the concentrations of both A and B are doubled and all other reaction conditions remain constant, how will the reaction rate change?

**66.** The experimental data in **Table 4** were obtained for the decomposition of azomethane ($CH_3N_2CH_3$) at a particular temperature according to the equation $CH_3N_2CH_3(g) \rightarrow C_2H_6(g) + N_2(g)$. Use the data to determine the reaction's experimental rate law.

**Table 4 Decomposition of Azomethane**

| Experiment Number | Initial $[CH_3N_2CH_3]$ | Initial Reaction Rate |
|---|---|---|
| 1 | 0.012*M* | $2.5 \times 10^{-6}$ mol/(L•s) |
| 2 | 0.024*M* | $5.0 \times 10^{-6}$ mol/(L•s) |

**67.** Use the data in **Table 4** to calculate the value of the specific rate constant, *k*.

**68.** At the same temperature, predict the reaction rate when the initial concentration of $CH_3N_2CH_3$ is 0.048*M*. Use the data in **Table 4.**

## SECTION 4

### Mastering Concepts

**69.** Distinguish between a simple and a complex reaction, an elementary step, and a reaction mechanism.

**70.** Suppose that a chemical reaction takes place in a two-step mechanism.

Step 1 (fast) $A + B \rightarrow C$

Step 2 (slow) $C + D \rightarrow E$

Which step in the reaction mechanism is the rate-determining step? Explain.

**71.** In the reaction described in Question 70, what are Steps 1 and 2 called? What is substance C called?

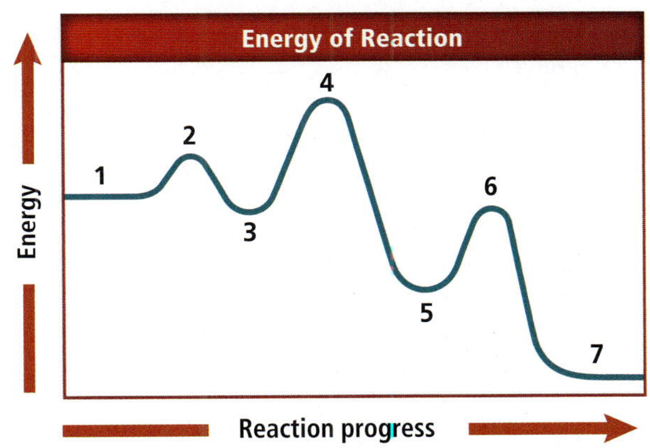


■ **Figure 23**

**72.** In **Figure 23,** identify each of the labels 1, 2, 3, 4, 5, 6, and 7 as one of the following: activated complex, intermediate, reactants, or products.

### Mastering Problems

**73.** Dinitrogen pentoxide decomposes in chloroform at a rate of $2.48 \times 10^{-4}$ mol/(L•min) at a particular temperature according to the equation $2N_2O_5 \rightarrow 4NO_2 + O_2$. The reaction is first order in $N_2O_5$. Given an initial concentration 0.400 mol/L, what is the rate constant for the reaction? What is the approximate $[N_2O_5]$ after the reaction proceeds for 1.30 h?

**74.** Radioactive decay is first order in the decaying isotope. For example, strontium-90 contained in fallout from nuclear explosions decays to yttrium-90 and a beta particle. Write the rate law for the decay of strontium-90.

## MIXED REVIEW

**75.** Evaluate the validity of this statement: You can determine the rate law for a chemical reaction by examining the mole ratio of reactants in the balanced equation. Explain your answer.

**76.** The concentration of Reactant A decreases from 0.400 mol/L at 0.00 min to 0.384 mol/L at 4.00 min. Calculate the average reaction rate during this time period. Express the rate in mol/(L • min).

**77.** The mass of a sample of magnesium is measured and the sample is placed in a container of hydrochloric acid. A chemical reaction occurs according to the equation $Mg(s) + 2HCl(aq) \rightarrow H_2(g) + MgCl_2(aq)$. Use the data in **Table 5** to calculate the volume of hydrogen gas produced at STP during the 3.00-min reaction. *(Hint: 1 mol of an ideal gas occupies 22.4 L at STP)*

**Table 5** Reaction of Magnesium and Hydrochloric Acid

| Time (min) | Mass of Magnesium (g) | Volume of Hydrogen at STP (L) |
|---|---|---|
| 0.00 | 6.00 | 0.00 |
| 3.00 | 4.50 | ? |

**78.** If the concentration of a reaction product increases from 0.0882 mol/L to 0.1446 mol/L in 12.0 minutes, what is the average reaction rate during the time interval?

**79.** A two-step mechanism has been proposed for the decomposition of nitryl chloride ($NO_2Cl$).

Step 1: $NO_2Cl(g) \rightarrow NO_2(g) + Cl(g)$

Step 2: $NO_2Cl(g) + Cl(g) \rightarrow NO_2(g) + Cl_2(g)$

What is the overall reaction? Identify any intermediates in the reaction sequence, and explain why they are called intermediates.

**80.** Compare and contrast the reaction energy diagrams for the overall decomposition of nitryl chloride by the mechanism in Problem 79 under two assumptions: A—that the first step is slower; B—that the second step is slower.

**81.** **Automobile Engine** The following reaction takes place in an automobile's engine and exhaust system.

$$NO_2(g) + CO(g) \rightarrow NO(g) + CO_2(g)$$

The reaction's rate law at a particular temperature is Rate $= 0.50\ L/(mol \cdot s)[NO_2]^2$. What is the reaction's initial, instantaneous rate when $[NO_2] = 0.0048$ mol/L?

**82.** The concentrations in a chemical reaction are expressed in moles per liter and time is expressed in seconds. If the overall rate law is third-order, what are the units for the rate and the rate constant?

## THINK CRITICALLY

**83.** **Visualize** the reaction energy diagram for a one-step, endothermic chemical reaction. Compare the heights of the activation energies for the forward and reverse reactions.

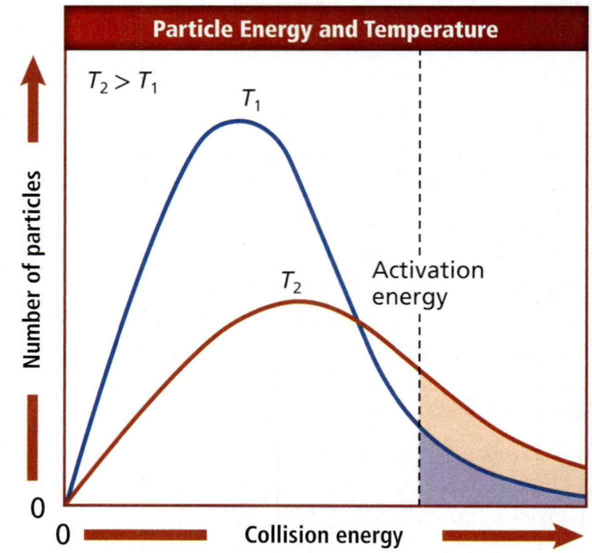


■ **Figure 24**

**84.** **Differentiate** between the shaded areas in **Figure 24** at temperatures $T_1$ and $T_2$ on the basis of the number of collisions per unit time that might occur with energy equal to or greater than the activation energy.

**85.** **Apply** the method of initial rates to determine the order of a chemical reaction with respect to Reactant X. Create a set of hypothetical experimental data that would lead you to conclude that the reaction is second order in X.

**86.** **Formulate** a rationale to explain how a complex chemical reaction might have more than one rate-determining elementary step.

**87.** **Construct** a diagram that shows all of the possible collision combinations between two molecules of Reactant A and two molecules of Reactant B. Now, increase the number of molecules of A from two to four and sketch each possible A-B collision combination. By what factor did the number of collision combinations increase? What does this imply about the reaction rate?

**88.** **Apply** collision theory to explain two reasons why increasing the temperature of a reaction by 10 K often doubles the reaction rate.

**89.** **Create** a table of concentrations, starting with 0.100*M* concentrations of all reactants, that you would propose in order to establish the rate law for the reaction $aA + bB + cD \rightarrow$ products using the method of initial rates.

## CHALLENGE PROBLEM

**90.** **Hydrocarbons** Heating cyclopropane ($C_3H_6$) converts it to propene ($CH_2{=}CHCH_3$). The rate law is first order in cyclopropane. If the rate constant at a particular temperature is $6.22 \times 10^{-4}\ s^{-1}$ and the concentration of cyclopropane is held at 0.0300 mol/L, what mass of propene is produced in 10.0 min in a volume of 2.50 L?

## CUMULATIVE REVIEW

**91.** For the following categories of elements, state the possible number(s) of electrons in their outermost orbitals in the ground state.
**a.** p-block elements
**b.** nitrogen-group elements
**c.** d-block elements
**d.** noble-gas elements
**e.** s-block elements

**92.** Classify each of the following elements as a metal, nonmetal, or metalloid.
**a.** molybdenum
**b.** bromine
**c.** arsenic
**d.** neon
**e.** cerium

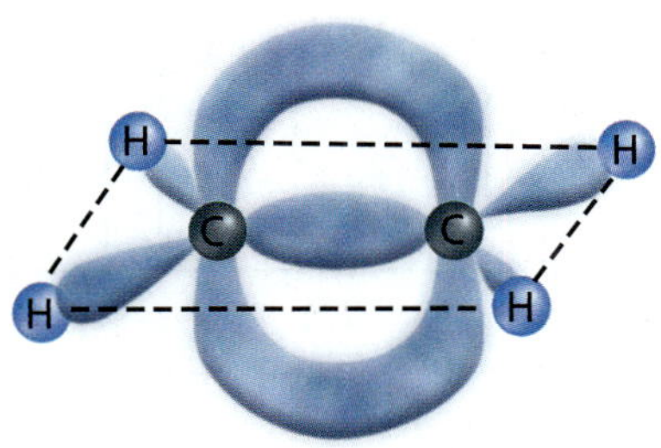


■ **Figure 25**

**93.** Using **Figure 25,** determine how many sigma and pi bonds are contained in a single ethene molecule.

**94.** Balance the following equations.
**a.** $Sn(s) + NaOH(aq) \rightarrow Na_2SnO_2(aq) + H_2(g)$
**b.** $C_8H_{18}(l) + O_2(g) \rightarrow CO_2(g) + H_2O(l)$
**c.** $Al(s) + H_2SO_4(aq) \rightarrow Al_2(SO_4)_3(aq) + H_2(g)$

**95.** What mass of iron(III) chloride is needed to prepare 1.00 L of a 0.255*M* solution?

**96.** What must you know to calculate the boiling point elevation of a solution of hexane in benzene?

**97.** $\Delta H$ for a reaction is negative. Compare the energy of the products and the reactants. Is the reaction endothermic or exothermic?

## WRITING IN ▶ Chemistry

**98.** **Pharmaceuticals** Imagine that your nation is experiencing an influenza epidemic. Fortunately, scientists have recently discovered a new catalyst that increases the rate of production of an effective flu medicine. Write a newspaper article describing how the catalyst works. Include a reaction energy diagram and an explanation detailing the importance of the discovery.

**99.** **Lawn Care** Write an advertisement that explains that Company A's fertilizer works better than Company B's fertilizer because it has smaller sized granules. Include applicable diagrams.

## DBQ Document-Based Questions

**Chemical Indicators** *Phenolphthalein is a chemical indicator used to show the presence of a base. The data in* **Table 6** *presents the decrease in phenolphthalein concentration with time when a 0.0050*M *phenolphthalein solution is added to a solution that has a concentration of hydroxide ion equal to 0.61*M.

**Table 6 Reaction Between Phenolphthalein and Excess Base**

| Concentration of Phenolphthalein (*M*) | Time (s) |
|---|---|
| 0.0050 | 0.0 |
| 0.0040 | 22.3 |
| 0.0020 | 91.6 |
| 0.0010 | 160.9 |
| 0.00050 | 230.3 |
| 0.00015 | 350.7 |

Data obtained from: Bodner Research Web. 2006. "Chemical Kinetics," *General Chemistry Help.*

**100.** What is the average rate of the reaction in the first 22.3 s expressed in moles of phenolphthalein consumed per liter per second?

**101.** What is the average rate of the reaction as the phenolphthalein concentration decreases from 0.00050*M* to 0.00015*M*?

**102.** The rate law is rate = *k*[phenolphthalein]. If the rate constant for the reaction is $1.0 \times 10^{-2}\ s^{-1}$, what is the instantaneous rate of reaction when the concentration of phenolphthalein is 0.0025*M*?

## MULTIPLE CHOICE

1. The rate of a chemical reaction is all of the following EXCEPT
   A. the speed at which a reaction takes place.
   B. the change in concentration of a reactant per unit time.
   C. the change in concentration of a product per unit time.
   D. the amount of product formed in a certain period of time.

2. How can colloids be distinguished from solutions?
   A. Dilute colloids have particles that can be seen with the naked eye.
   B. Colloid particles are much smaller than solvated particles.
   C. Colloid particles that are dispersed will settle out of the mixture in time.
   D. Colloids will scatter light beams that are shone through them.

*Use the graph below to answer Questions 3 and 4.*

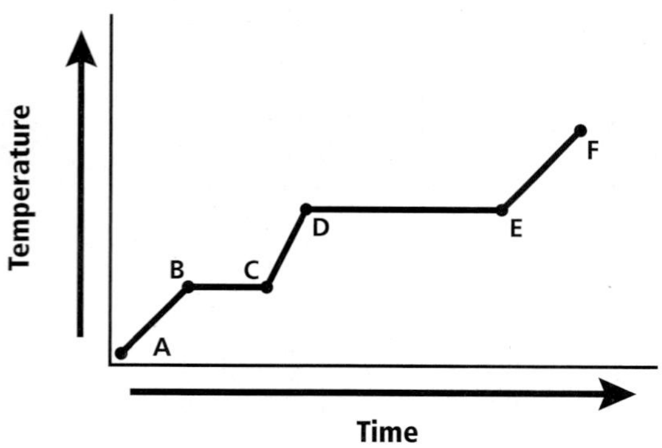


3. During which segment is this substance undergoing melting?
   A. segment AB
   B. segment BC
   C. segment CD
   D. segment DE

4. As the substance heats from point C to point D, which is true of the substance?
   A. potential energy increases, kinetic energy decreases
   B. potential energy increases, kinetic energy increases
   C. potential energy remains constant, kinetic energy increases
   D. potential energy decreases, kinetic energy remains constant

5. How much water must be added to 6.0 mL of a 0.050*M* stock solution to dilute it to 0.020*M*?
   A. 15 mL
   B. 9.0 mL
   C. 6.0 mL
   D. 2.4 mL

6. Which is NOT an acceptable unit for expressing a reaction rate?
   A. *M*/min
   B. L/s
   C. mol/(mL•h)
   D. mol/(L•min)

7. Which is the strongest type of intermolecular bond?
   A. ionic bond
   B. dipole-dipole force
   C. dispersion force
   D. hydrogen bond

*Use the diagram below to answer Questions 8 and 9.*

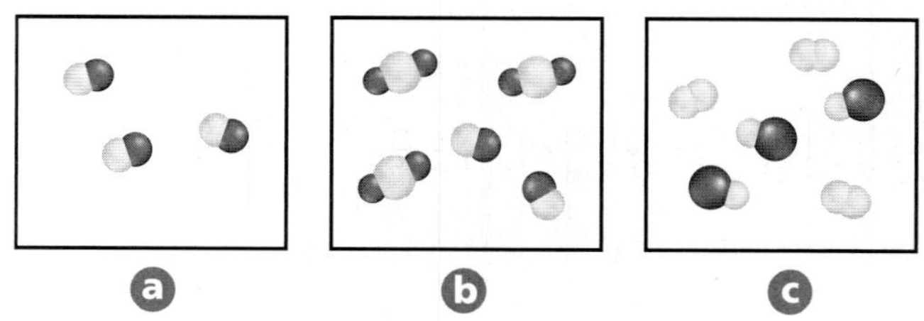


8. Which sample could contain particles of oxygen gas?
   A. a
   B. b
   C. c
   D. Both a and b

9. Which sample could contain particles of carbon dioxide?
   A. a
   B. b
   C. c
   D. Both a and b

10. How many moles are in $4.30 \times 10^2$ g of calcium phosphate ($Ca_3(PO_4)_2$)?
    A. 0.721 moles
    B. 1.39 moles
    C. 1.54 moles
    D. 3.18 moles

## SHORT ANSWER

*Use the following information to answer Question 11.*

| The complete dissociation of acid $H_3A$ takes place in three steps: | |
|---|---|
| $H_3A(aq) \rightarrow H_2A^-(aq) + H^+(aq)$ | rate $= k_1[H_3A]$<br>$k_1 = 3.2 \times 10^2\ s^{-1}$ |
| $H_2A^-(aq) \rightarrow HA^{2-}(aq) + H^+(aq)$ | rate $= k_2[H_2A^-]$<br>$k_2 = 1.5 \times 10^2\ s^{-1}$ |
| $HA^{2-}(aq) \rightarrow A^{3-}(aq) + H^+(aq)$ | rate $= k_3[HA^{2-}]$<br>$k_3 = 0.8 \times 10^2\ s^{-1}$ |
| overall reaction: $H_3A(aq) \rightarrow A^{3-}(aq) + 3H^+(aq)$ | |

**11.** When the reactant concentrations are $[H_3A] = 0.100M$, $[H_2A^-] = 0.500M$, and $[HA^{2-}] = 0.200M$, which reaction is the rate-determining step? Explain how you can tell.

**12.** The rate law for $A + B + C \rightarrow$ products is: rate $= k[A]^2[C]$. If $k = 6.92 \times 10^{-5}\ L^2/(mol^2 \cdot s)$, $[A] = 0.175M$, $[B] = 0.230M$, and $[C] = 0.315M$, what is the instantaneous reaction rate?

## EXTENDED RESPONSE

*Use the following reaction to answer Questions 13 to 15.*

Sodium nitride ($Na_3N$) breaks down to form sodium metal and nitrogen gas.

**13.** Write the balanced chemical equation for the reaction.

**14.** Classify the type of reaction. Explain your answer.

**15.** Show the steps to determine the amount of nitrogen gas that can be produced from 32.5 grams of sodium nitride.

## SAT SUBJECT TEST: CHEMISTRY

*Use the table below to answer Questions 16 to 18.*

| Reaction: $SO_2Cl_2(g) \rightarrow SO_2(g) + Cl_2(g)$ | | | |
|---|---|---|---|
| Experimental Data Collected for Reaction | | | |
| Time (min) | $[SO_2Cl_2]$ (*M*) | $[SO_2]$ (*M*) | $[Cl_2]$ (*M*) |
| 0.0 | 1.00 | 0.00 | 0.00 |
| 100.0 | 0.87 | 0.13 | 0.13 |
| 200.0 | 0.74 | ? | ? |

**16.** What is the average reaction rate for this reaction, expressed in moles $SO_2Cl_2$ consumed per liter per minute?
**A.** $1.3 \times 10^{-3}$ mol/(L • min)
**B.** $2.6 \times 10^{-1}$ mol/(L • min)
**C.** $7.4 \times 10^{-3}$ mol/(L • min)
**D.** $8.7 \times 10^{-3}$ mol/(L • min)
**E.** $2.6 \times 10^{-3}$ mol/(L • min)

**17.** On the basis of the average reaction rate, what will the concentrations of $SO_2$ and $Cl_2$ be at 200.0 min?
**A.** 0.13*M*
**B.** 0.26*M*
**C.** 0.39*M*
**D.** 0.52*M*
**E.** 0.87*M*

**18.** How long will it take for half of the original amount of $SO_2Cl_2$ to decompose at the average reaction rate?
**A.** 285 min
**B.** 335 min
**C.** 385 min
**D.** 401 min
**E.** 516 min

**19.** A sample of argon gas is compressed into a volume of 0.712 L by a piston exerting 3.92 atm of pressure. The piston is released until the pressure of the gas is 1.50 atm. What is the new volume of the gas?
**A.** 0.272 L
**B.** 3.67 L
**C.** 5.86 L
**D.** 4.19 L
**E.** 1.86 L

**NEED EXTRA HELP?**

| If You Missed Question . . . | 1 | 2 | 3 | 4 | 5 | 6 | 7 | 8 | 9 | 10 | 11 | 12 | 13 | 14 | 15 | 16 | 17 | 18 | 19 |
|---|---|---|---|---|---|---|---|---|---|---|---|---|---|---|---|---|---|---|---|
| Review Section . . . | 16.1 | 14.1 | 12.4 | 12.4 | 14.2 | 16.1 | 12.2 | 3.4 | 3.4 | 10.3 | 16.4 | 16.4 | 9.1 | 9.2 | 11.2 | 16.3 | 16.3 | 16.3 | 13.1 |

# CHAPTER 17

# Chemical Equilibrium

**BIGIDEA** Many reactions and processes reach a state of chemical equilibrium in which both reactants and products are formed at equal rates.

## SECTIONS

Chm 9

1 A State of Dynamic Balance

2 Factors Affecting Chemical Equilibrium

3 Using Equilibrium Constants

## LaunchLAB

iLab Station

### What is equal about equilibrium?

Equilibrium is a point of balance in which opposing changes cancel each other. In this lab, you will transfer different volumes of water from one container to another until equilibrium is reached.

**FOLDABLES® Study Organizer**

### Changes Affecting Equilibrium

Make a folded chart. Label it as shown. Use it to help you organize information about the factors that affect equilibrium.

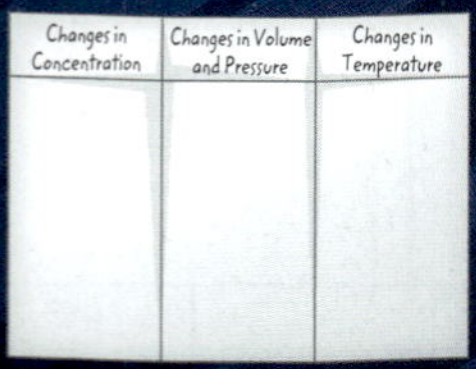

Go **online!** MHEonline.com

No other human activity causes as much air pollution as the use of motor vehicles. Over the last decade catalytic converters and changes in gasoline additives have made cars 40% cleaner.

SECTION 1

# A State of Dynamic Balance

**MAINIDEA** Chemical equilibrium is described by an equilibrium constant expression that relates the concentrations of reactants and products.

### Essential Questions

- What are the characteristics of chemical equilibrium?
- How are equilibrium expressions written for systems that are at equilibrium?
- How are equilibrium constants calculated from concentration data?

### Review Vocabulary

**free energy:** the energy that is available to do work—the difference between the change in enthalpy and the product of the entropy change and the absolute temperature

### New Vocabulary

reversible reaction
chemical equilibrium
law of chemical equilibrium
equilibrium constant
homogeneous equilibrium
heterogeneous equilibrium

**CHEM 4 YOU** **Imagine a tug-of-war between two teams. Because the rope between them is not moving, it might seem that neither team is pulling. In fact, both teams are pulling, but the forces exerted by the two teams are equal and opposite, so they are in complete balance.**

## What is equilibrium?

Often, chemical reactions reach a point of balance or equilibrium. If you performed the Launch Lab, you found that a point of balance was reached in the transfer of water from the beaker to the graduated cylinder and from the graduated cylinder to the beaker.

Consider the reaction for the formation of ammonia from nitrogen and hydrogen that you read about previously.

$$N_2(g) + 3H_2(g) \rightarrow 2NH_3(g) \quad \Delta G° = -33.1 \text{ kJ}$$

Ammonia is important in agriculture as a fertilizer and an additive to animal feed grains. In industry, it is a raw material for the manufacture of many products such as nylon, as shown in **Figure 1.**

The equation for the production of ammonia has a negative standard free energy, $\Delta G°$. Recall that a negative sign for $\Delta G°$ indicates that the reaction is spontaneous under standard conditions, defined as 298 K and 1 atm, but spontaneous reactions are not always fast. When carried out under standard conditions, this ammonia-forming reaction is much too slow. To produce ammonia at a rate that is practical, the reaction must be carried out at a much higher temperature and pressure.

**Figure 1** Ammonia reacts with both ends of a six-carbon molecule to form a diamine (1,6-diaminohexane). This is one step in the formation of the polymer nylon. Here nylon fibers, to be used in tire manufacturing, are being wound onto a spool.

Ted Horowitz/Alamy

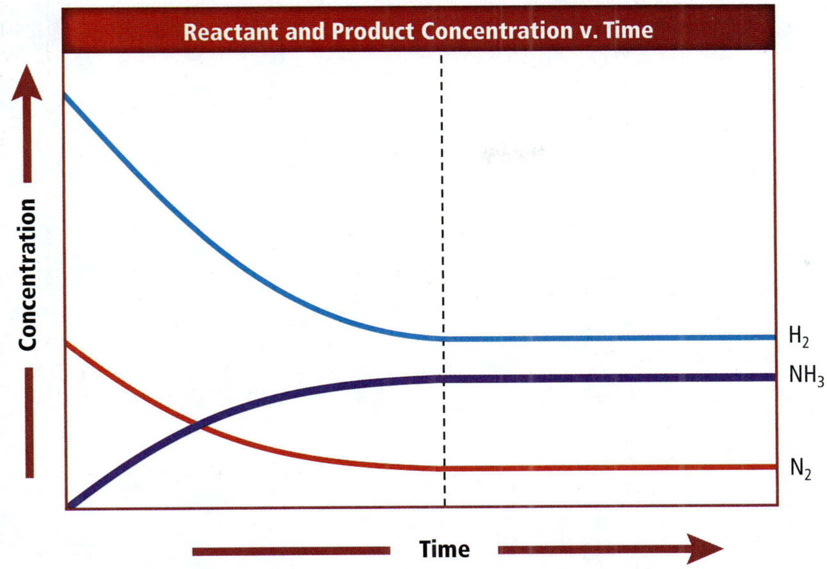


■ **Figure 2** The concentrations of the reactants ($H_2$ and $N_2$) decrease at first, while the concentration of the product ($NH_3$) increases. Then, before the reactants are used up, all concentrations become constant.

☑ GRAPH CHECK
**Explain** how the graph shows that the concentrations of the reactants and products become constant.

What happens when 1 mol of nitrogen and 3 mol of hydrogen, the number of moles shown as coefficients in the chemical equation, are placed in a closed reaction vessel at 723 K? Because the reaction is spontaneous, nitrogen and hydrogen react. **Figure 2** illustrates the progress of the reaction. Note that the concentration of the product, $NH_3$, is zero at the start and gradually increases with time. The reactants, $H_2$ and $N_2$, are consumed in the reaction, so their concentrations gradually decrease. After a period of time, however, the concentrations of $H_2$, $N_2$, and $NH_3$ no longer change. All concentrations become constant, as shown by the horizontal lines on the right side of the diagram. The concentrations of $H_2$ and $N_2$ are not zero, so not all of the reactants were converted to product, even though $\Delta G°$ for this reaction is negative.

☑ GRAPH CHECK **Describe** the slopes of the curves for the reactants and for the product on the left of the vertical dotted line. How do the slopes differ on the right of the dotted line?

**Reversible reactions and chemical equilibrium** When a reaction results in an almost complete conversion of reactants to products, chemists say that the reaction goes to completion—but most reactions do not go to completion. The reactions appear to stop because they are reversible. A **reversible reaction** is a chemical reaction that can occur in both the forward and the reverse directions.

$$\text{Forward: } N_2(g) + 3H_2(g) \rightarrow 2NH_3(g)$$
$$\text{Reverse: } N_2(g) + 3H_2(g) \leftarrow 2NH_3(g)$$

Chemists combine these two equations into a single equation that uses a double arrow to show that both reactions occur.

$$N_2(g) + 3H_2(g) \rightleftharpoons 2NH_3(g)$$

The reactants in the forward reaction are on the left of the arrows. The reactants in the reverse reaction are on the right of the arrows. In the forward reaction, hydrogen and nitrogen combine to form the product ammonia. In the reverse reaction, ammonia decomposes into the products hydrogen and nitrogen.

**VOCABULARY**
**ACADEMIC VOCABULARY**
**Convert**
to change from one form or function to another
*She converted a spare bedroom into an office.*

■ **Figure 3** The progress of a reaction to produce ammonia from hydrogen and nitrogen is shown in **a.** through **d.**

**Interpret** ***Study the diagrams to answer the following questions. In* a, *how do you know that the reaction has not yet begun? In* b, *what evidence indicates that the reverse reaction has begun? Compare* c *with* d. *How do you know that equilibrium has been reached?***

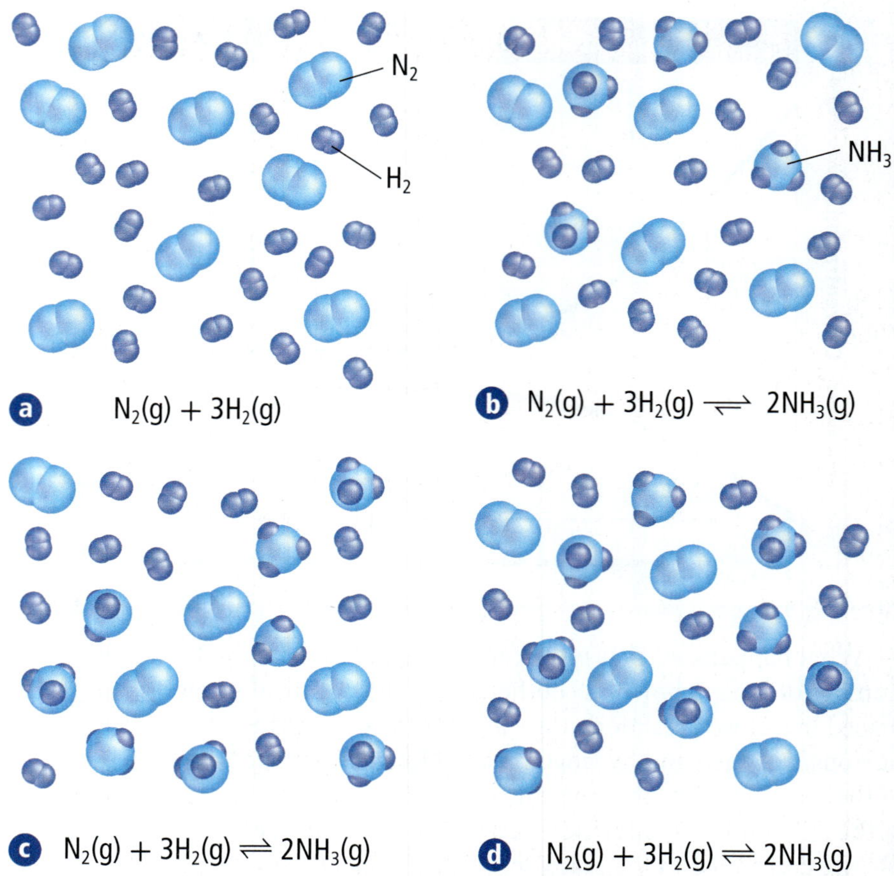


How does the reversibility of this reaction affect the production of ammonia? **Figure 3a** shows a mixture of nitrogen and hydrogen just as the reaction begins at a definite, initial rate. No ammonia is present, therefore only the forward reaction can occur.

$$N_2(g) + 3H_2(g) \rightarrow 2NH_3(g)$$

As hydrogen and nitrogen combine to form ammonia, their concentrations decrease, as shown in **Figure 3b.** You learned previously that the rate of a reaction depends on the concentration of the reactants. The decrease in the concentration of the reactants causes the rate of the forward reaction to slow. As soon as ammonia is present, the reverse reaction can occur, slowly at first, but at an increasing rate as the concentration of ammonia increases.

$$N_2(g) + 3H_2(g) \leftarrow 2NH_3(g)$$

As the reaction proceeds, the rate of the forward reaction continues to decrease and the rate of the reverse reaction continues to increase until the two rates are equal. At that point, ammonia is produced at the same rate it is decomposed, so the concentrations of $N_2$, $H_2$, and $NH_3$ remain constant, as shown in **Figures 3c** and **3d.** The system has reached a state of balance or equilibrium. The word *equilibrium* means that opposing processes are in balance. **Chemical equilibrium** is a state in which the forward and reverse reactions balance each other because they take place at equal rates.

$$\text{Rate}_{\text{forward reaction}} = \text{Rate}_{\text{reverse reaction}}$$

You can recognize that the ammonia-forming reaction reaches a state of chemical equilibrium because its chemical equation is written with a double arrow like this.

$$N_2(g) + 3H_2(g) \rightleftharpoons 2NH_3(g)$$

At equilibrium, the concentrations of reactants and products are constant, as shown in **Figures 3c** and **3d.** However, that doesn't mean that the amounts or concentrations of reactants and products are equal. That is seldom the case. In fact, it is not unusual for the equilibrium concentrations of a reactant and product to differ by a factor of one million or more.

**READING CHECK** **Explain** the meaning of a double arrow in chemical equations.

### Connection to Physics The dynamic nature of equilibrium

A push or pull on an object is a force. When you push on a door or pull on a dog's leash, you exert a force. When two or more forces are exerted on the same object in the same direction, they add together. One force subtracts from the other if the forces are in opposite directions. Thus, in a tug-of-war, when two teams pull on a rope with equal force, the resulting force has a magnitude of zero and the rope does not move. The system is said to be in equilibrium. Similarly, the people on the seesaw in **Figure 4a** represent a system in equilibrium. The equal-and-opposite forces on both ends of the seesaw are called balanced forces. If, instead, one force is greater in magnitude than the other, the combined force is greater than zero and is called an unbalanced force. An unbalanced force causes an object to accelerate, which is what has happened in **Figure 4b.**

Tim Fuller

**Figure 4** In **a,** all the forces are in perfect balance, so the position of the seesaw remains steady. In **b,** the unbalanced force on the left causes the seesaw to change its position.
**Explain** *this analogy in terms of chemical equilibrium.*

■ **Figure 5** Suppose a certain number of people are confined to the two buildings connected by this walkway and that people can walk back and forth between the buildings. The number of people in each building will remain constant only if the same number of people cross the bridge in one direction as cross in the opposite direction.

**Decide** ***whether the same people will always be in the same building. How does your answer apply to chemical equilibrium?***

Like equal forces opposing each other, equilibrium is a state of action, not inaction. For example, consider this analogy: The glassed-in walkway, shown in **Figure 5,** connects two buildings. Suppose that all entrances and exits for the buildings, except the walkway, are closed for a day. And suppose that the same number of persons cross the walkway in each direction every hour. Given these circumstances, the number of persons in each building remains constant even though people continue to cross between the two buildings. Note that the numbers of persons in the two buildings do not have to be equal. Equilibrium requires only that the number of persons crossing the walkway in one direction is equal to the number crossing in the opposite direction.

The dynamic nature of chemical equilibrium can be illustrated by placing equal masses of iodine crystals in two interconnected flasks, as shown in **Figure 6a.** The flask on the left contains iodine molecules made up entirely of the nonradioactive isotope I-127. The flask on the right contains iodine molecules made up of the radioactive isotope I-131. The radiation counters indicate the difference in the levels of radioactivity within each flask.

Each flask is a closed system. No reactant or product can enter or leave. At 298 K and 1 atm, this equilibrium is established in both flasks.

$$I_2(s) \rightleftharpoons I_2(g)$$

In the forward process, called sublimation, iodine molecules change directly from the solid phase to the gas phase. In the reverse process, gaseous iodine molecules return to the solid phase. A solid-vapor equilibrium is established in each flask.

When the stopcock in the tube connecting the two flasks is opened, as in **Figure 6b,** iodine vapor can travel back and forth between the two flasks. After a period of time, the readings on the radiation counters indicate that the flask on the left contains as many radioactive I-131 molecules as the flask on the right in both the vapor and the solid phases.

The evidence suggests that iodine molecules constantly change from the solid phase to the gas phase according to the forward process, and that gaseous iodine molecules convert back to the solid phase according to the reverse process. The constant readings on both radiation detectors indicate that equilibrium has been established in the combined volume of the two flasks.

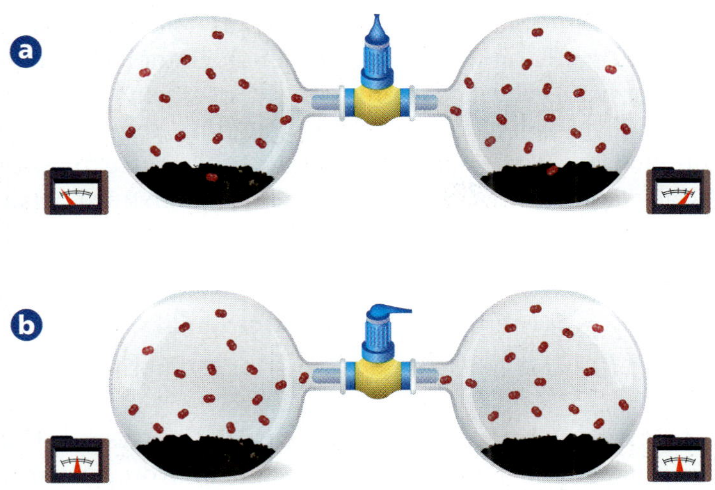

■ **Figure 6a.** Radioactive iodine molecules in the flask on the right are separated from nonradioactive iodine in the flask on the left. Note the readings on the radiation monitors. **b.** After the stopcock has been open for a time, the radiation monitors show that radioactive molecules are in both flasks. The particles must have moved back and forth between the flasks and between the solid and the gaseous phases.

## Equilibrium Expressions

Some chemical systems have little tendency to react. Others go to completion. The majority of reactions reach a state of equilibrium with some of the reactants unconsumed. If the reactants are not all consumed, then the amount of products produced is less than the amount predicted by the balanced chemical equation. According to the equation for the ammonia-producing reaction, 2 mol of ammonia should be produced when 1 mol of nitrogen and 3 mol of hydrogen react. However, because the reaction reaches a state of equilibrium, less than 2 mol of ammonia are obtained.

**Vocabulary**

**Word origin**

**Completion**
comes from the Latin verb *completus*, which means *having all necessary parts, elements, or steps.*

**The law of chemical equilibrium** In 1864, Norwegian chemists Cato Maximilian Guldberg and Peter Waage jointly proposed and developed the **law of chemical equilibrium,** which states that at a given temperature, a chemical system might reach a state in which a particular ratio of reactant and product concentrations has a constant value. The general equation for a reaction at equilibrium is as follows.

$$aA + bB \rightleftharpoons cC + dD$$

If the law of chemical equilibrium is applied to this reaction, the following ratio is obtained.

### The Equilibrium Constant Expression

$$K_{eq} = \frac{[C]^c[D]^d}{[A]^a[B]^b}$$

**[A] and [B] are the molar concentrations of the reactants. [C] and [D] are the molar concentrations of the products.**

**The exponents *a*, *b*, *c*, and *d*, are the coefficients in the balanced equation.**

The equilibrium constant expression is the ratio of the molar concentrations of the products to the molar concentrations of the reactants with each concentration raised to a power equal to its coefficient in the balanced chemical equation.

The **equilibrium constant,** $K_{eq}$, is the numerical value of the ratio of product concentrations to reactant concentrations, with each concentration raised to the power equal to its coefficient in the balanced equation. The value of $K_{eq}$ is constant only at a specified temperature.

How can you interpret the size of the equilibrium constant? Recall that a fraction with a numerator greater than its denominator has a value greater than 1. And a fraction with a numerator less than its denominator has a value less than 1. For example, compare the ratios 5/1 and 1/5. Five is a larger number than one-fifth. Because the product concentrations are in the numerator of the equilibrium expression, a numerically large $K_{eq}$ means that the equilibrium mixture contains more products than reactants. Similarly, a $K_{eq}$ less than 1 means that the equilibrium mixture contains more reactants than products.

$K_{eq} > 1$: Products are favored at equilibrium.

$K_{eq} < 1$: Reactants are favored at equilibrium.

**Expressions for homogeneous equilibria** Gaseous hydrogen iodide is produced by the equilibrium reaction of hydrogen gas with iodine. Iodine and some of its compounds have important uses in medicine, as illustrated in **Figure 7.** How would you write the equilibrium constant expression for this reaction in which hydrogen and iodine react to form hydrogen iodide?

$$H_2(g) + I_2(g) \rightleftharpoons 2HI(g)$$

This reaction is a **homogeneous equilibrium,** which means that all the reactants and products are in the same physical state. All participants are gases. First, place the product concentration in the numerator and the reactant concentrations in the denominator.

$$\frac{[HI]}{[H_2][I_2]}$$

The expression becomes equal to $K_{eq}$ when you add the coefficients from the balanced chemical equation as exponents.

$$K_{eq} = \frac{[HI]^2}{[H_2][I_2]}$$

$K_{eq}$ for this equilibrium at 731 K is 49.7. Note that 49.7 has no units. When writing equilibrium constant expressions, it is customary to omit units.

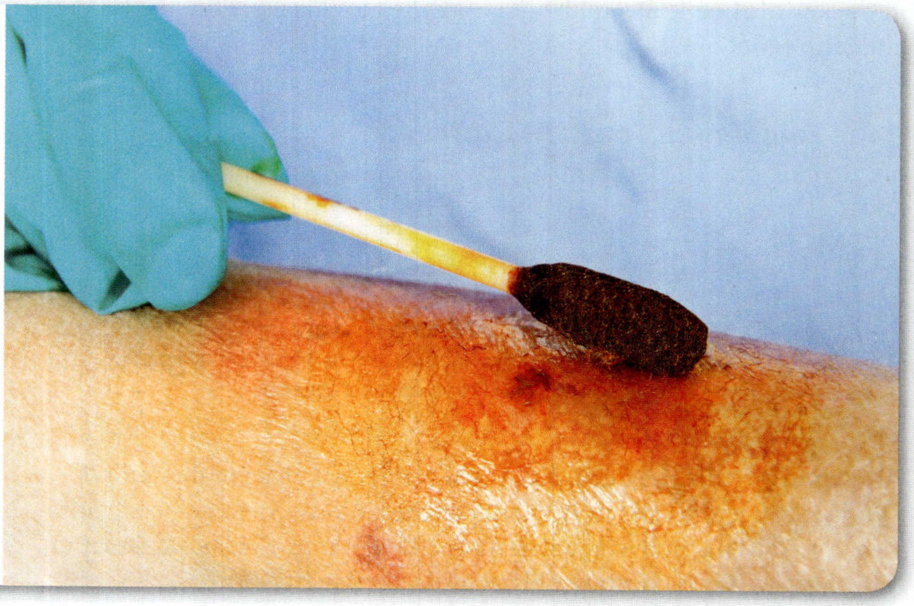



■ **Figure 7** Because of iodine's antibacterial properties, solutions of iodine and iodine compounds are used externally as antiseptics. Some iodine compounds are used internally. For example, doctors use potassium iodide (KI) in the treatment of goiter, a condition characterized by the enlargement of the thyroid gland.

## EXAMPLE Problem 1

**EQUILIBRIUM CONSTANT EXPRESSIONS FOR HOMOGENEOUS EQUILIBRIA** Millions of tons of ammonia ($NH_3$) are produced each year for use in the manufacture of products such as explosives, fertilizers, and synthetic fibers. You might have used ammonia in your home as a household cleaner, which is particularly useful for cleaning glass. Ammonia is manufactured from its elements, hydrogen and nitrogen, using the Haber process. Write the equilibrium constant expression for the following reaction.

$$N_2(g) + 3H_2(g) \rightleftharpoons 2NH_3(g)$$

### 1 ANALYZE THE PROBLEM

The equation for the reaction provides the information needed to write the equilibrium constant expression. The equilibrium is homogeneous because the reactants and product are in the same physical state.

The general form of the equilibrium constant expression is

$$K_{eq} = \frac{[C]^c}{[A]^a[B]^b}$$

**Known**

$[A] = [N_2]$, coefficient $N_2 = 1$
$[B] = [H_2]$, coefficient $H_2 = 3$
$[C] = [NH_3]$, coefficient $NH_3 = 2$

**Unknown**

$K_{eq} = ?$

### 2 SOLVE FOR THE UNKNOWN

Form a ratio of product concentration to reactant concentrations.

$$K_{eq} = \frac{[C]^c}{[A]^a[B]^b}$$ **State the general form of the equilibrium constant expression.**

$$K_{eq} = \frac{[NH_3]^c}{[N_2]^a[H_2]^b}$$ **Substitute A = $N_2$, B = $H_2$, and C = $NH_3$.**

$$K_{eq} = \frac{[NH_3]^2}{[N_2][H_2]^3}$$ **Substitute a = 1, b = 3, and c = 2.**

### 3 EVALUATE THE ANSWER

The product concentration is in the numerator and the reactant concentrations are in the denominator. Product and reactant concentrations are raised to powers equal to their coefficients.

## PRACTICE Problems

Do additional problems. Online Practice

1. Write equilibrium constant expressions for these equilibria.
   a. $N_2O_4(g) \rightleftharpoons 2NO_2(g)$
   b. $2H_2S(g) \rightleftharpoons 2H_2(g) + S_2(g)$
   c. $CO(g) + 3H_2(g) \rightleftharpoons CH_4(g) + H_2O(g)$
   d. $4NH_3(g) + 5O_2(g) \rightleftharpoons 4NO(g) + 6H_2O(g)$
   e. $CH_4(g) + 2H_2S(g) \rightleftharpoons CS_2(g) + 4H_2(g)$
2. **Challenge** Write the chemical equation that has the equilibrium constant expression $K_{eq} = \frac{[CO]^2[O_2]}{[CO_2]^2}$.

## RealWorld CHEMISTRY

### Thyroid Health

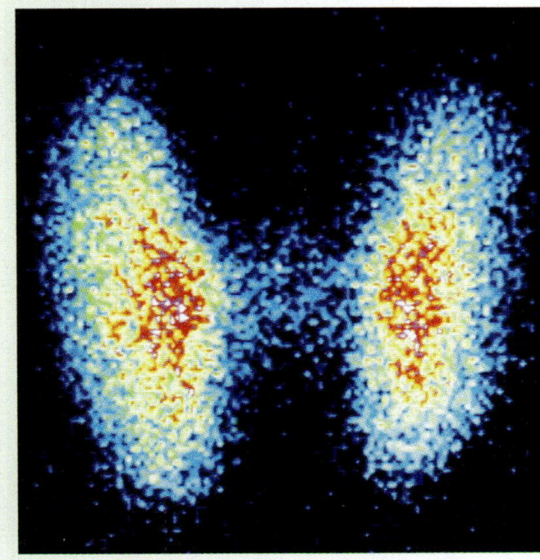

**NUCLEAR MEDICINE** Iodine-131 is a radioactive isotope that is absorbed by the thyroid gland. It is used in medicine to diagnose and treat diseases of the thyroid. When iodine-131 is administered to a patient, radiation from the isotope creates an image of the gland on film that reveals abnormalities. The image above can be used to diagnose and treat people with thyroid diseases or abnormalities, such as Graves' disease. Graves' disease is a treatable disease that is a common cause of an overactive thyroid gland.

■ **Figure 8** At equilibrium the rate of evaporation of ethanol ($C_2H_5OH$) equals the rate of condensation. This two-phase equilibrium is called a heterogeneous equilibrium. $K_{eq}$ depends only on $[C_2H_5OH(g)]$.

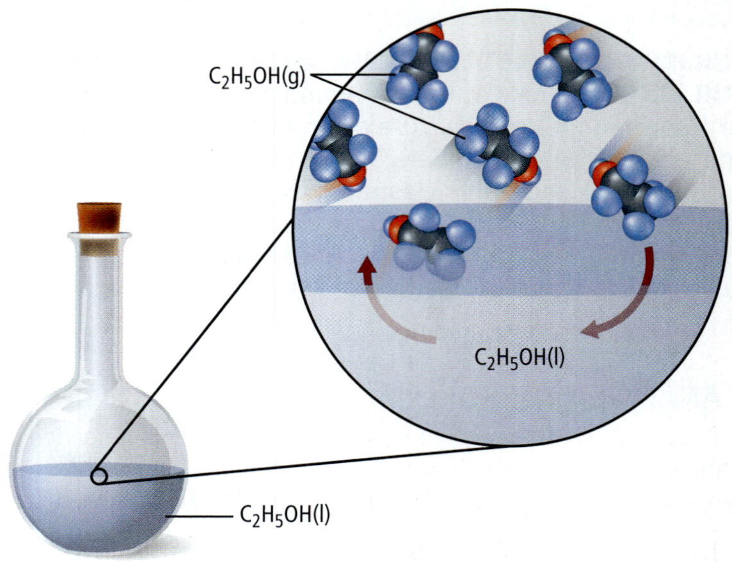


**Expressions for heterogeneous equilibria** You have learned to write $K_{eq}$ expressions for homogeneous equilibria, those in which all reactants and products are in the same physical state. When the reactants and products are present in more than one physical state, the equilibrium is called a **heterogeneous equilibrium.** When ethanol is placed in a closed flask, a liquid-vapor equilibrium is established, as illustrated in **Figure 8.**

$$C_2H_5OH(l) \rightleftharpoons C_2H_5OH(g)$$

To write the equilibrium constant expression for this process, you would form a ratio of the product to the reactant. At a given temperature, the ratio would have a constant value $K$.

$$K = \frac{[C_2H_5OH(g)]}{[C_2H_5OH(l)]}$$

Note that the concentration of liquid ethanol is in the denominator. Liquid ethanol is a pure substance, so its concentration is its density expressed in moles per liter. Recall that at any given temperature, density is constant. No matter how much or how little $C_2H_5OH$ is present, its concentration remains constant. Therefore, the term in the denominator is a constant and can be combined with $K$ in the expression for $K_{eq}$.

$$K[C_2H_5OH(l)] = [C_2H_5OH(g)] = K_{eq}$$

The equilibrium constant expression for this phase change is

$$K_{eq} = [C_2H_5OH(g)]$$

Solids are also pure substances with unchanging concentrations, so equilibria involving solids are simplified in the same way. Recall the experiment involving the sublimation of iodine crystals in **Figure 6.**

$$I_2(s) \rightleftharpoons I_2(g)$$
$$K_{eq} = [I_2(g)]$$

The equilibrium constant, $K_{eq}$, depends only on the concentration of gaseous iodine in the system.

## EXAMPLE Problem 2

**EQUILIBRIUM CONSTANT EXPRESSIONS FOR HETEROGENEOUS EQUILIBRIA** In addition to its uses in baking and as an antacid and cleaning agent, baking soda is often placed in open boxes in refrigerators to freshen the air as shown in **Figure 9.** Write the equilibrium constant expression for the decomposition of baking soda (sodium hydrogen carbonate).

$$2NaHCO_3(s) \rightleftharpoons Na_2CO_3(s) + CO_2(g) + H_2O(g)$$

### 1 ANALYZE THE PROBLEM

You are given a heterogeneous equilibrium involving gases and solids. Solids are omitted from the equilibrium constant expression.

**Known**

$[C] = [Na_2CO_3]$, coefficient $Na_2CO_3 = 1$
$[D] = [CO_2]$, coefficient $CO_2 = 1$
$[E] = [H_2O]$, coefficient $H_2O = 1$
$[A] = [NaHCO_3]$, coefficient $NaHCO_3 = 2$

**Unknown**

**equilibrium constant expression = ?**

### 2 SOLVE FOR THE UNKNOWN

Form a ratio of product concentrations to reactant concentrations.

$$K_{eq} = \frac{[C]^c[D]^d[E]^e}{[A]^a[B]^b}$$ **State the general form of the equilibrium constant expression.**

$$K_{eq} = \frac{[Na_2CO_3]^c[CO_2]^d[H_2O]^e}{[NaHCO_3]^a}$$ **Substitute $A = NaHCO_3$, $C = Na_2CO_3$, $D = CO_2$, and $E = H_2O$.**

$$K_{eq} = \frac{[Na_2CO_3]^1[CO_2]^1[H_2O]^1}{[NaHCO_3]^2}$$ **Substitute a = 2, c = 1, d = 1, and e = 1.**

$$K_{eq} = [CO_2][H_2O]$$ **Omit terms involving solid substances.**

### 3 EVALUATE THE ANSWER

The expression correctly applies the law of chemical equilibrium to the equation.

■ **Figure 9** Sodium hydrogen carbonate (baking soda) absorbs odors and freshens the air in a refrigerator. It is also a key ingredient in some toothpastes.

Get help with **equilibrium constant expressions.**

Personal Tutor

## PRACTICE Problems

Do additional problems. Online Practice

3. Write equilibrium constant expressions for these heterogeneous equilibria.
   - **a.** $C_{10}H_8(s) \rightleftharpoons C_{10}H_8(g)$
   - **b.** $H_2O(l) \rightleftharpoons H_2O(g)$
   - **c.** $CaCO_3(s) \rightleftharpoons CaO(s) + CO_2(g)$
   - **d.** $C(s) + H_2O(g) \rightleftharpoons CO(g) + H_2(g)$
   - **e.** $FeO(s) + CO(g) \rightleftharpoons Fe(s) + CO_2(g)$
4. **Challenge** Solid iron reacts with chlorine gas to form solid iron(III) chloride ($FeCl_3$). Write the balanced equation and the equilibrium constant expression for the reaction.

## CAREERS IN CHEMISTRY

**Science Writer** To convey scientific information to the nonscientific reader, a writer must have a broad knowledge of science and the ability to write clear, concise, and understandable prose. Science writers make complex subjects, such as chemical equilibrium, accessible to readers with no prior knowledge of the subject.

WebQuest

# Equilibrium Constants

For a given reaction at a given temperature, $K_{eq}$ will always be the same regardless of the initial concentrations of reactants and products. To test this statement, three experiments were carried out using the following reaction.

$$H_2(g) + I_2(g) \rightleftharpoons 2HI(g)$$

The results are summarized in **Table 1.** In Trial 1, 1.0000 mol $H_2$ and 2.0000 mol $I_2$ were placed in a 1.0000-L vessel. No HI was present at the beginning of Trial 1. In Trial 2, only HI was present at the start of the experiment. In Trial 3, each of the three substances had the same initial concentration. The reactions were carried out at 731 K.

**Equilibrium concentrations** When equilibrium was established, the concentration of each substance was determined experimentally. Note that the equilibrium concentrations are not the same in the three trials, yet when each set of equilibrium concentrations is put into the equilibrium constant expression, the value of $K_{eq}$ is the same. Each set of equilibrium concentrations represents an equilibrium position.

**The value of $K_{eq}$** Although an equilibrium system has only one value for $K_{eq}$ at a particular temperature, it has an unlimited number of equilibrium positions. Equilibrium positions depend on the initial concentrations of the reactants and products. The large value of $K_{eq}$ for the reaction $H_2(g) + I_2(g) \rightleftharpoons 2HI(g)$ means that at equilibrium the product is present in larger amount than the reactants. However, many equilibria have small $K_{eq}$ values. For the equilibrium $N_2(g) + O_2(g) \rightleftharpoons 2NO(g)$, $K_{eq}$ equals $4.6 \times 10^{-31}$ at 298 K. A $K_{eq}$ this small means that the product, NO, is practically nonexistent at equilibrium.

**Equilibrium characteristics** You might have noticed certain characteristics of all chemical reactions that reach equilibrium. First, the reaction must take place in a closed system—no reactant or product can enter or leave the system. Second, the temperature must remain constant. Third, all reactants and products are present, and they are in constant dynamic motion. This means that equilibrium is dynamic, not static.

☑ READING CHECK **Explain** why it is important that all reactants and products be present at equilibrium.

**Table 1** Experimental Data for HI Reaction Equilibrium

| | Initial Concentrations | | | Equilibrium Concentrations | | | $K_{eq}$ |
|---|---|---|---|---|---|---|---|
| Trial | $[H_2]_0$ (*M*) | $[I_2]_0$ (*M*) | $[HI]_0$ (*M*) | $[H_2]_{eq}$ (*M*) | $[I_2]_{eq}$ (*M*) | $[HI]_{eq}$ (*M*) | $\frac{[HI]^2}{[H_2][I_2]} = K_{eq}$ |
| 1 | 1.0000 | 2.0000 | 0 | 0.06587 | 1.0659 | 1.8682 | $\frac{[1.8682]^2}{[0.06587][1.0659]} = 49.70$ |
| 2 | 0 | 0 | 5.0000 | 0.5525 | 0.5525 | 3.8950 | $\frac{[3.8950]^2}{[0.5525][0.5525]} = 49.70$ |
| 3 | 1.0000 | 1.0000 | 1.0000 | 0.2485 | 0.2485 | 1.7515 | $\frac{[1.7515]^2}{[0.2485][0.2485]} = 49.70$ |

## EXAMPLE Problem 3

Find help solving algebraic equations. Math Handbook

EXAMPLE PROBLEM

**THE VALUE OF EQUILIBRIUM CONSTANTS** Calculate the value of $K_{eq}$ for the equilibrium constant expression $K_{eq} = \frac{[NH_3]^2}{[N_2][H_2]^3}$ given concentration data at one equilibrium position: $[NH_3] = 0.933$ mol/L, $[N_2] = 0.533$ mol/L, $[H_2] = 1.600$ mol/L.

### 1 ANALYZE THE PROBLEM

You have been given the equilibrium constant expression and the concentration of each reactant and product. You must calculate the equilibrium constant.

| **Known** | | **Unknown** |
|---|---|---|
| $K_{eq} = \frac{[NH_3]^2}{[N_2][H_2]^3}$ | $[N_2] = 0.533$ mol/L | $K_{eq} = ?$ |
| $[NH_3] = 0.933$ mol/L | $[H_2] = 1.600$ mol/L | |

### 2 SOLVE FOR THE UNKNOWN

$$K_{eq} = \frac{[0.933]^2}{[0.533][1.600]^3} = 0.399$$

**Substitute $[NH_3] = 0.933$ mol/L, $[N_2] = 0.533$ mol/L, and $[H_2] = 1.600$ mol/L.**

### 3 EVALUATE THE ANSWER

The answer is correctly stated with three digits. The largest concentration value is in the denominator and raised to the third power, so a value less than 1 is reasonable.

## PRACTICE Problems

Do additional problems. Online Practice

5. Calculate $K_{eq}$ for the equilibrium in Practice Problem 1a using the data $[N_2O_4] = 0.0185$ mol/L and $[NO_2] = 0.0627$ mol/L.
6. Calculate $K_{eq}$ for the equilibrium in Practice Problem 1c using the data $[CO] = 0.0613$ mol/L, $[H_2] = 0.1839$ mol/L, $[CH_4] = 0.0387$ mol/L, and $[H_2O] = 0.0387$ mol/L.
7. **Challenge** The reaction $COCl_2(g) \rightleftharpoons CO(g) + Cl_2(g)$ reaches equilibrium at 900 K. $K_{eq}$ is $8.2 \times 10^{-2}$. If the equilibrium concentrations of CO and $Cl_2$ are $0.150M$, what is the equilibrium concentration of $COCl_2$?

## SECTION 1 REVIEW

Section Self-Check

### Section Summary

- A reaction is at equilibrium when the rate of the forward reaction equals the rate of the reverse reaction.
- The equilibrium constant expression is a ratio of the molar concentrations of the products to the molar concentrations of the reactants with each concentration raised to a power equal to its coefficient in the balanced chemical equation.
- The value of the equilibrium constant expression, $K_{eq}$, is a constant for a given temperature.

8. **MAIN**IDEA **Explain** how the size of the equilibrium constant relates to the amount of product formed at equilibrium.
9. **Compare** homogeneous and heterogeneous equilibria.
10. **List** three characteristics a reaction mixture must have if it is to attain a state of chemical equilibrium.
11. **Calculate** Determine the value of $K_{eq}$ at 400 K for this equation: $PCl_5(g) \rightleftharpoons PCl_3(g) + Cl_2(g)$ if $[PCl_5] = 0.135$ mol/L, $[PCl_3] = 0.550$ mol/L, and $[Cl_2] = 0.550$ mol/L.
12. **Interpret Data** The table below shows the value of the equilibrium constant for a reaction at three different temperatures. At which temperature is the concentration of the products the greatest? Explain your answer.

| $K_{eq}$ and Temperature | | |
|---|---|---|
| 263 K | 273 K | 373 K |
| 0.0250 | 0.500 | 4.500 |

SECTION 2

# Factors Affecting Chemical Equilibrium

**MAINIDEA** When changes are made to a system at equilibrium, the system shifts to a new equilibrium position.

### Essential Questions

- Which various factors affect chemical equilibrium?
- How does Le Châtelier's principle apply to equilibrium systems?

### Review Vocabulary

**reaction rate:** the change in concentration of a reactant or product per unit time, generally calculated and expressed in moles per liter per second.

### New Vocabulary

Le Châtelier's principle

**CHEM 4 YOU** When demand for a product equals the available supply, the price remains constant. If demand exceeds supply, the price of the product increases. The price becomes constant again when supply and demand regain a state of balance. Systems at equilibrium behave in a similar way.

## Le Châtelier's Principle

Suppose the by-products of an industrial process are the gases carbon monoxide and hydrogen, and a company chemist believes these gases can be combined to produce the fuel methane ($CH_4$). When CO and $H_2$ are placed in a closed vessel at 1200 K, this exothermic reaction ($\Delta H = -206.5$ kJ) establishes equilibrium (Equilibrium Position 1).

$$\underset{0.30000M}{CO(g)} + \underset{0.10000M}{3H_2(g)} \rightleftharpoons \underset{0.05900M}{CH_4(g)} + \underset{0.02000M}{H_2O(g)} \quad \Delta H^\circ = -206.5 \text{ kJ}$$

Inserting these concentrations into the equilibrium expression gives an equilibrium constant equal to 3.933.

$$K_{eq} = \frac{[CH_4][H_2O]}{[CO][H_2]^3} = \frac{(0.05900)(0.02000)}{(0.30000)(0.10000)^3} = 3.933$$

Unfortunately, a methane concentration of 0.05900 mol/L in the equilibrium mixture is too low to be of any practical use. Could the chemist change the equilibrium position and thereby increase the amount of methane? An analogy might be the runner on a treadmill shown in **Figure 10.** If the runner increases the speed of the treadmill, she must also increase her speed to restore equilibrium.

■ **Figure 10** A runner gradually increases the speed of the treadmill. With each change, she must increase her running speed in order to restore her equilibrium at the new treadmill setting. Similarly, a chemist can change the conditions of a reaction at equilibrium in order to increase the amount of product.

In 1888, French chemist Henri-Louis Le Châtelier discovered that there are ways to control equilibria to make reactions more productive. He proposed what is now called **Le Châtelier's principle:** If a stress is applied to a system at equilibrium, the system shifts in the direction that relieves the stress. A stress is any kind of change in a system at equilibrium that upsets the equilibrium.

**FOLDABLES®**
Incorporate information from this section into your Foldable.

## Applying Le Châtelier's Principle

How could the industrial chemist apply Le Châtelier's principle to increase her yield of methane? She will need to adjust any factors that will shift the equilibrium to the product side of the reaction.

**Changes in concentration** Adjusting the concentrations of either the reactants or the products puts a stress on the equilibrium. Earlier, you read about collision theory, which states that particles must collide in order to react. The number of collisions between reacting particles depends on the concentration of the particles, so perhaps the chemist can change the equilibrium by changing concentrations.

**VOCABULARY**
**SCIENCE USAGE V. COMMON USAGE**
**Stress**
***Science usage:*** any kind of change in a system at equilibrium that upsets the equilibrium
*The stress of the addition of more reactant to the reaction mixture caused the rate of the forward reaction to increase.*
***Common usage:*** physical or mental strain or pressure
*He felt that the stress of taking on another task would be too great.*

**Adding reactants** Suppose additional carbon monoxide is injected into the reaction vessel, raising the concentration of carbon monoxide from $0.30000M$ to $1.00000M$. The higher carbon monoxide concentration immediately increases the number of effective collisions between CO and $H_2$ molecules and upsets the equilibrium. The rate of the forward reaction increases, as indicated by the longer arrow to the right.

$$CO(g) + 3H_2(g) \rightleftharpoons CH_4(g) + H_2O(g)$$

In time, the rate of the forward reaction slows down as the concentrations of CO and $H_2$ decrease. Simultaneously, the rate of the reverse reaction increases as more $CH_4$ and $H_2O$ molecules are produced. Eventually, a new equilibrium position (Position 2) is established.

$$\begin{array}{cccc} CO(g) & + \ 3H_2(g) & \rightleftharpoons \ CH_4(g) & + \ H_2O(g) \\ 0.99254M & 0.07762M & 0.06648M & 0.02746M \end{array}$$

$$K_{eq} = \frac{[CH_4][H_2O]}{[CO][H_2]^3} = \frac{(0.06648)(0.02746)}{(0.99254)(0.07762)^3} = 3.933$$

Note that although $K_{eq}$ has not changed, the new equilibrium position results in the desired effect—an increased concentration of methane. The results of this experiment are summarized in **Table 2.**

Could you have predicted this result using Le Châtelier's principle? Yes. Think of the increased concentration of CO as a stress on the equilibrium. The equilibrium system reacts to the stress by consuming CO at an increased rate. This response, called a shift to the right, forms more $CH_4$ and $H_2O$. Any increase in the concentration of a reactant results in a shift to the right and additional product.

**Table 2** At Equilibrium: $CO(g) + 3H_2(g) \rightleftharpoons CH_4(g) + H_2O(g)$

| Equilibrium position | $[CO]_{eq}$ (*M*) | $[H_2]_{eq}$ (*M*) | $[CH_4]_{eq}$ (*M*) | $[H_2O]_{eq}$ (*M*) | $K_{eq}$ |
|---|---|---|---|---|---|
| 1 | 0.30000 | 0.10000 | 0.05900 | 0.02000 | 3.933 |
| 2 | 0.99254 | 0.07762 | 0.06648 | 0.02746 | 3.933 |

■ **Figure 11** Storekeepers know that all products should be available at all times, so when stocks get low, they must be replaced.

**Explain** *this analogy in terms of Le Châtelier's principle.*

**Removing products** Suppose that rather than injecting more reactant, the chemist decides to remove a product ($H_2O$) by adding a desiccant to the reaction vessel. Recall that a desiccant is a substance that absorbs water. What does Le Châtelier's principle predict the equilibrium will do in response to a decrease in the concentration of water? The equilibrium shifts in the direction that will tend to bring the concentration of water back up. That is, the equilibrium shifts to the right and results in additional product.

Think about how supermarket shelves are kept stocked, as shown in **Figure 11.** As customers buy items from the shelves, it is someone's job to replace whatever is removed. Similarly, the equilibrium reaction restores some of the lost water by producing more water. In any equilibrium, the removal of a product results in a shift to the right and the production of more product.

**Adding products** The equilibrium position can also be shifted to the left, toward the reactants. Le Châtelier's principle predicts that if additional product is added to a reaction at equilibrium, the reaction will shift to the left. The stress is relieved by converting products to reactants. If one of the reactants is removed, a similar shift to the left will occur.

When predicting the results of a stress on an equilibrium using Le Châtelier's principle, have the equation for the reaction in view. The effects of changing concentrations are summarized in **Figure 12.**

☑ READING CHECK **Describe** how an equilibrium shifts if a reactant is removed.

**Changes in volume and pressure** Consider again the reaction for making methane from by-product gases.

$$CO(g) + 3H_2(g) \rightleftharpoons CH_4(g) + H_2O(g)$$

Can this reaction be forced to produce more methane by changing the volume of the reaction vessel? Suppose the volume can be changed using a pistonlike device similar to the one shown in **Figure 13.** If the piston is forced downward, the volume of the system decreases. Recall that Boyle's law states that decreasing the volume at constant temperature increases the pressure. The increased pressure is a stress on the reaction at equilibrium. How does the equilibrium respond to the disturbance and relieve the stress?

■ **Figure 12** The addition or removal of a reactant or product shifts the equilibrium in the direction that relieves the stress. Note the unequal arrows, which indicate the direction of the shift.

**Describe** *how the reaction would shift if you added $H_2$. If you removed $CH_4$.*

$$CO(g) + 3H_2(g) \rightleftharpoons CH_4(g) + H_2O(g)$$

**Equilibrium shifts to the right.**

1. $CO(g) + 3H_2(g) \rightleftharpoons CH_4(g) + H_2O(g)$ — $CO(g)$ Add a reactant.
2. $CO(g) + 3H_2(g) \rightleftharpoons CH_4(g) + H_2O(g)$ — Remove a product. $H_2O(g)$

**Equilibrium shifts to the left.**

3. $CO(g) + 3H_2(g) \rightleftharpoons CH_4(g) + H_2O(g)$ — $CO(g)$ Remove a reactant.
4. $CO(g) + 3H_2(g) \rightleftharpoons CH_4(g) + H_2O(g)$ — Add a product. $H_2O(g)$

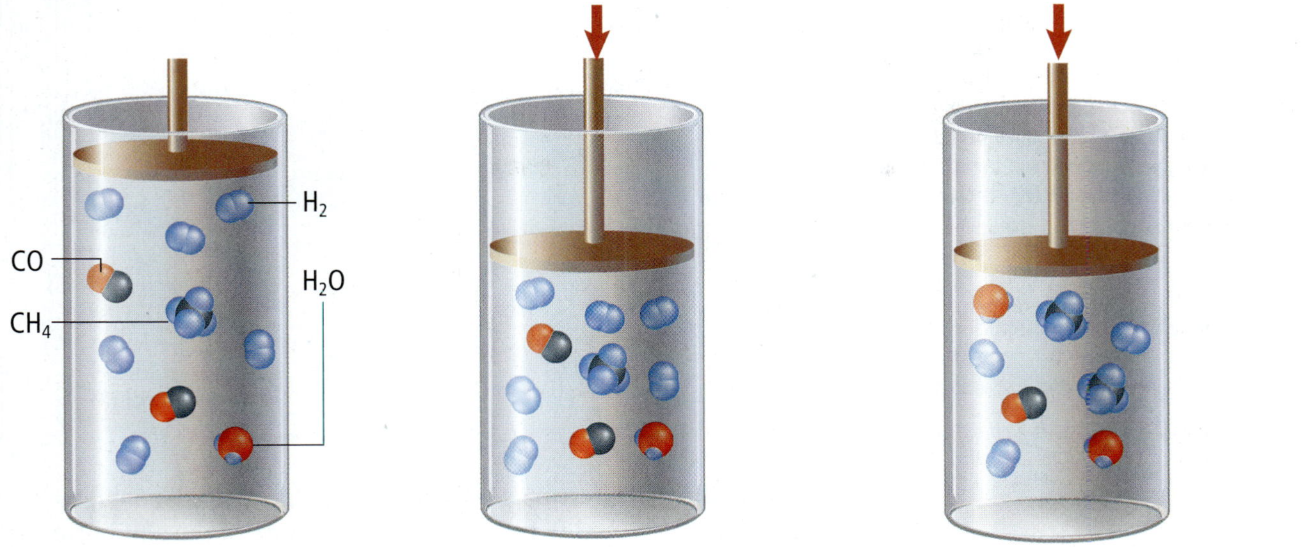


The reaction between CO and $H_2$ is at equilibrium.

Lowering the piston decreases the volume and increases the pressure.

As a result, more molecules of the products form. Their formation relieves the stress on the system.

**Figure 13** For the reaction between CO and $H_2$ at constant temperature, changing the volume of the reaction vessel changes the concentrations of gaseous reactants and products. Increasing the pressure shifts the equilibrium to the right and increases the amount of product.

**Compare** ***the numbers of product molecules on the left with the numbers on the right.***

The pressure exerted by an ideal gas depends on the number of gas particles that collide with the walls of the vessel. The more gas particles contained in the vessel, the greater the pressure will be. If the number of gas particles is increased at constant temperature, the pressure of the gas increases. If the number of gas particles is decreased, the pressure decreases. How does this relationship between numbers of gas particles and pressure apply to the reaction for making methane?

**Moles of reactant versus moles of product** Compare the number of moles of gaseous reactants in the equation to the number of moles of gaseous products. For every two moles of gaseous products, four moles of gaseous reactants are consumed, a net decrease of two moles. If you apply Le Châtelier's principle, you can see that the equilibrium can relieve the stress of increased pressure by shifting to the right. **Figure 13** shows that this shift decreases the total number of moles of gas, and thus the pressure inside the reaction vessel decreases. Although the shift to the right does not reduce the pressure to its original value, it has the desired effect—more methane is produced.

Changing the volume (and pressure) of an equilibrium system shifts the equilibrium only if the number of moles of gaseous reactants is different from the number of moles of gaseous products. If the number of moles of gas is the same on both sides of the equation, changes in volume and pressure have no effect on the equilibrium.

**Changes in temperature** A change in temperature alters both the equilibrium position and the equilibrium constant. Recall that virtually every chemical reaction is either endothermic or exothermic. The reaction for making methane has a negative $\Delta H°$, which means that the forward reaction is exothermic and the reverse reaction is endothermic.

$$CO(g) + 3H_2(g) \rightleftharpoons CH_4(g) + H_2O(g) \quad \Delta H° = -206.5 \text{ kJ}$$

In this case, you can think of heat as a product in the forward reaction and a reactant in the reverse reaction.

$$CO(g) + 3H_2(g) \rightleftharpoons CH_4(g) + H_2O(g) + \text{heat}$$

**Figure 14** When placed in a boiling-water bath (right), the equilibrium shifts in the endothermic direction, which produces more reddish-brown $NO_2$. The mixture becomes lighter in color when placed in ice (left) because the equilibrium shifts in the exothermic direction, in which more $NO_2$ is converted to colorless $N_2O_4$.

View an **animation about equilibrium shifts.**

Concepts In Motion

**Heat and equilibrium position** According to Le Châtelier's principle, if heat is added to an equilibrium system, the equilibrium shifts in the direction in which heat is used up; that is, the equilibrium shifts to the left and decreases the concentration of methane ($CH_4$). Lowering the temperature shifts the equilibrium to the right because the forward reaction liberates heat and relieves the stress. In shifting to the right, the equilibrium produces more methane.

**Temperature and $K_{eq}$** Any change in temperature results in a change in $K_{eq}$. Recall that the larger the value of $K_{eq}$, the more product is found in the equilibrium mixture. Thus, for the methane-producing reaction, $K_{eq}$ increases in value when the temperature is lowered and decreases in value when the temperature is raised.

The conversion between dinitrogen tetroxide ($N_2O_4$) and nitrogen dioxide ($NO_2$) responds to changes in temperature in an observable way. This endothermic equilibrium is described by the following equation.

$$N_2O_4(g) \rightleftharpoons 2NO_2(g) \quad \Delta H° = 57.2 \text{ kJ}$$

$N_2O_4$ is a colorless gas; $NO_2$ is a reddish-brown gas. **Figure 14** shows that the color of the equilibrium mixture in ice is much lighter than when the mixture is heated in boiling water. The removal of heat by cooling shifts the equilibrium to the left and creates more colorless $N_2O_4$. Adding heat shifts the equilibrium to the right and creates more reddish-brown $NO_2$. **Figure 15** shows the effects of heating and cooling on the reactions you have been reading about.

**Figure 15** For the exothermic reaction between CO and $H_2$, raising the temperature shifts the equilibrium to the left (Equation 1). Lowering the temperature results in a shift to the right (Equation 2). The opposite is true for the endothermic reaction involving NO and $N_2O_4$ (Equations 3 and 4).

### Exothermic Reaction

**Equilibrium shifts to the left.** (heat → Raise the temperature.)

1. $CO(g) + 3H_2(g) \rightleftharpoons CH_4(g) + H_2O(g) + \text{heat}$

**Equilibrium shifts to the right.** (heat → Lower the temperature.)

2. $CO(g) + 3H_2(g) \rightleftharpoons CH_4(g) + H_2O(g) + \text{heat}$

### Endothermic Reaction

**Equilibrium shifts to the right.** (heat → Raise the temperature.)

3. $\text{heat} + N_2O_4(g) \rightleftharpoons 2\,NO_2(g)$

**Equilibrium shifts to the left.** (heat → Lower the temperature.)

4. $\text{heat} + N_2O_4(g) \rightleftharpoons 2\,NO_2(g)$

Charles D. Winters/Science Source

# MiniLAB

iLab Station

## Observe Shifts in Equilibrium

**If a stress is placed on a reaction at equilibrium, how will the system shift to relieve the stress?**

### Procedure

1. Read and complete the lab safety form.
2. Place about 2 mL of **0.1*M* $CoCl_2$ solution** in a **test tube.** Record the color of the solution.
3. Add about 3 mL of **concentrated HCl** to the test tube. Record the color of the solution. **WARNING:** ***HCl can burn skin and clothing.***
4. Add enough **water** to the test tube to make a color change occur. Record the color.
5. Add about 2 mL of 0.1*M* $CoCl_2$ to another test tube. Add concentrated HCl a drop at a time until the solution turns purple. If the solution becomes blue, add water until it turns purple.
6. Place the test tube in an **ice bath** that has had some **table salt** sprinkled into the ice water. Record the color of the solution in the test tube.
7. Place the test tube in a **hot water bath.** Use a **nonmercury thermometer** to determine that the temperature is at least 70°C. Record the solution's color.

### Analysis

1. **Interpret** Use the equation for the reaction you just observed to explain your observations of color in Steps 2–4. The equation is as follows.

$$\underset{\text{pink}}{Co(H_2O)_6^{2+}} + 4Cl^- \rightleftharpoons \underset{\text{blue}}{CoCl_4^{2-}} + 6H_2O$$

2. **Describe** how the equilibrium shifts when energy is added or removed.
3. **Interpret** From your observations of color in Steps 6 and 7, determine whether the reaction is exothermic or endothermic.

**Catalysts and equilibrium** Changes in concentration, volume, and temperature make a difference in the amount of product formed in a reaction. Can a catalyst also affect product concentration? A catalyst speeds up a reaction, but it does so equally in both directions. Therefore, a catalyzed reaction reaches equilibrium more quickly but with no change in the amount of product formed.

**Alabama Applying Practices**

Go to the Resource tab in ConnectED to find the PBL *Food for Thought.*

# SECTION 2 REVIEW

Section Self-Check

### Section Summary

- Le Châtelier's principle describes how an equilibrium system shifts in response to a stress or a disturbance.
- When an equilibrium shifts in response to a change in concentration or volume, the equilibrium position changes but $K_{eq}$ remains constant. A change in temperature, however, alters both the equilibrium position and the value of $K_{eq}$.

13. **MAIN**IDEA **Explain** how a system at equilibrium responds to a stress and list factors that can be stresses on an equilibrium system.
14. **Explain** how decreasing the volume of the reaction vessel affects each equilibrium.
    **a.** $2SO_2(g) + O_2(g) \rightleftharpoons 2SO_3(g)$ **b.** $H_2(g) + Cl_2(g) \rightleftharpoons 2HCl(g)$
15. **Decide** whether higher or lower temperatures will produce more $CH_3CHO$ in the following equilibrium. $C_2H_2(g) + H_2O(g) \rightleftharpoons CH_3CHO(g)$ $\Delta H° = -151$ kJ
16. **Demonstrate** The table below shows the concentrations of Substances A and B in two reaction mixtures. A and B react according to the equation $2A \rightleftharpoons B$; $K_{eq} = 200$. Are the two mixtures at different equilibrium positions?

**Concentration Data in mol/L**

| Reaction | [A] | [B] |
|---|---|---|
| 1 | 0.0100 | 0.0200 |
| 2 | 0.0500 | 0.500 |

17. **Design** a concept map that shows ways in which Le Châtelier's principle can be applied to increase the products in a system at equilibrium and to increase the reactants in such a system.

## SECTION 3

# Using Equilibrium Constants

**MAINIDEA** Equilibrium constant expressions can be used to calculate concentrations and solubilities.

### Essential Questions

- How are the equilibrium concentrations of reactants and products determined?
- How is the solubility of a compound calculated from its solubility product constant?
- Why is the common ion effect important?

### Review Vocabulary

**solubility:** the maximum amount of solute that will dissolve in a given amount of solvent at a specific temperature and pressure

### New Vocabulary

solubility product constant
common ion
common ion effect

**CHEM 4 YOU** If you have ever tried to squeeze yourself into the backseat of a car already occupied by several of your friends, you know there is a limit to how many people the seat can hold. An ionic compound encounters a similar situation when being dissolved in a solution.

## Calculating Equilibrium Concentrations

How can the equilibrium constant expression be used to calculate the concentration of a product? The $K_{eq}$ for the reaction that forms $CH_4$ from $H_2$ and CO is 3.933 at 1200 K. If the concentrations of $H_2$, CO, and $H_2O$ are known, the concentration of $CH_4$ can be calculated.

$$\begin{array}{ccccccc} CO(g) & + & 3H_2(g) & \rightleftharpoons & CH_4(g) & + & H_2O(g) \\ 0.850M & & 1.333M & & ?M & & 0.286M \end{array}$$

$$K_{eq} = \frac{[CH_4][H_2O]}{[CO][H_2]^3}$$

Solve the expression for the unknown $[CH_4]$ by multiplying both sides of the equation by $[CO][H_2]^3$ and dividing both sides by $[H_2O]$.

$$[CH_4] = K_{eq} \times \frac{[CO][H_2]^3}{[H_2O]}$$

Substitute the known concentrations and the value of $K_{eq}$ (3.933).

$$[CH_4] = 3.933 \times \frac{(0.850)(1.333)^3}{(0.286)} = 27.7 \text{ mol/L}$$

The equilibrium concentration of $CH_4$ is 27.7 mol/L.

Is a yield of 27.7 mol/L sufficient to make the conversion of waste CO and $H_2$ to methane practical? That depends on the cost of methane. **Figure 16** shows a tanker transporting natural gas, which is primarily methane, to ports around the world.

**Figure 16** New port terminals are being planned to accommodate tankers, which carry increasing amounts of natural gas around the world to meet both industrial and home needs. Natural gas, which is primarily methane, is used for heating and cooking.

MsLightBox/E+/Getty Images

## EXAMPLE Problem 4

Find help with **square and cube roots.** Math Handbook

EXAMPLE PROBLEM

**CALCULATING EQUILIBRIUM CONCENTRATIONS** At a temperature of 1405 K, hydrogen sulfide, which has a foul odor resembling rotten eggs, decomposes to form hydrogen and diatomic sulfur. The equilibrium constant for the reaction is $2.27 \times 10^{-3}$.

$$2H_2S(g) \rightleftharpoons 2H_2(g) + S_2(g)$$

What is the concentration of hydrogen gas if $[S_2] = 0.0540$ mol/L and $[H_2S] = 0.184$ mol/L?

### 1 ANALYZE THE PROBLEM

You have been given $K_{eq}$ and two of the three variables in the equilibrium constant expression. The equilibrium expression can be solved for $[H_2]$. $K_{eq}$ is less than one, so more reactants than products are in the equilibrium mixture. Thus, you can predict that $[H_2]$ will be less than 0.184 mol/L, the concentration of the reactant $H_2S$.

**Known**

$K_{eq} = 2.27 \times 10^{-3}$

$[S_2] = 0.0540$ mol/L

$[H_2S] = 0.184$ mol/L

**Unknown**

$[H_2] = ?$ **mol/L**

### 2 SOLVE FOR THE UNKNOWN

$$\frac{[H_2]^2[S_2]}{[H_2S]^2} = K_{eq}$$ State the equilibrium constant expression.

Solve the equation for $[H_2]$.

$$[H_2]^2 = K_{eq} \times \frac{[H_2S]^2}{[S_2]}$$ Multiply both sides by $[H_2S]^2$. Divide both sides by $[S_2]$.

$$[H_2] = \sqrt{K_{eq} \times \frac{[H_2S]^2}{[S_2]}}$$ Take the square root of both sides.

$$[H_2] = \sqrt{(2.27 \times 10^{-3}) \times \frac{(0.184)^2}{(0.0540)}}$$ Substitute $K_{eq} = 2.27 \times 10^{-3}$, $[H_2S] = 0.184$ mol/L, and $[S_2] = 0.0540$ mol/L.

$$[H_2] = 0.0377 \text{ mol/L}$$ Multiply, divide, and take the square root.

The equilibrium concentration of $H_2$ is 0.0377 mol/L.

### 3 EVALUATE THE ANSWER

The answer is correctly stated with three significant figures. As predicted, the equilibrium concentration of $H_2$ is less than 0.184 mol/L.

## PRACTICE Problems

Do additional problems. Online Practice

PRACTICE PROBLEMS

**18.** At a certain temperature, $K_{eq} = 10.5$ for the equilibrium $CO(g) + 2H_2(g) \rightleftharpoons CH_3OH(g)$.

Calculate the following concentrations:

**a.** [CO] in an equilibrium mixture containing 0.933 mol/L $H_2$ and 1.32 mol/L $CH_3OH$

**b.** $[H_2]$ in an equilibrium mixture containing 1.09 mol/L CO and 0.325 mol/L $CH_3OH$

**c.** $[CH_3OH]$ in an equilibrium mixture containing 0.0661 mol/L $H_2$ and 3.85 mol/L CO

**19. Challenge** In a generic reaction $A + B \rightleftharpoons C + D$, 1.00 mol of A and 1.00 mol of B are allowed to react in a 1-L flask until equilibrium is established. If the equilibrium concentration of A is 0.450 mol/L, what is the equilibrium concentration of each of the other substances? What is $K_{eq}$?

■ **Figure 17** The water of the Great Salt Lake is much saltier than sea water. The high concentration of salt makes the water dense enough that most people can float in it. The Salar de Uyuni, or Uyuni Salt Flats, at right, were left behind when a similar prehistoric lake dried.

## The Solubility Product Constant

Some ionic compounds, such as sodium chloride, dissolve readily in water, and some, such as barium sulfate ($BaSO_4$) barely dissolve at all. On dissolving, all ionic compounds dissociate into ions.

$$NaCl(s) \longrightarrow Na^{+}(aq) + Cl^{-}(aq)$$

**Connection to Earth Science** Because of the high solubility of NaCl, the oceans and some lakes contain large amounts of salt. **Figure 17** shows the Great Salt Lake next to one of the Uyuni flats in Bolivia, which were left behind when a prehistoric lake dried.

Sometimes low solubility is also important. Although barium ions are toxic to humans, patients must ingest barium sulfate prior to having an X-ray of the digestive tract taken. Can patients safely ingest $BaSO_4$?

Barium sulfate dissociates in water according to this equation.

$$BaSO_4(s) \longrightarrow Ba^{2+}(aq) + SO_4{}^{2-}(aq)$$

As soon as the first product ions form, the reverse reaction begins.

$$BaSO_4(s) \longleftarrow Ba^{2+}(aq) + SO_4{}^{2-}(aq)$$

In time, equilibrium is established.

$$BaSO_4(s) \rightleftharpoons Ba^{2+}(aq) + SO_4{}^{2-}(aq)$$

For sparingly soluble compounds such as $BaSO_4$, the rates become equal when the concentrations of the aqueous ions are exceedingly small. Nevertheless, the solution at equilibrium is a saturated solution.

**Writing solubility product constant expressions** The equilibrium constant expression for the dissolving of a sparingly soluble compound is called the **solubility product constant,** $K_{sp}$. The solubility product constant expression is the product of the concentrations of the dissolved ions, each raised to the power equal to the coefficient of the ion in the chemical equation. Recall that the concentration of a pure substance is its density in moles per liter, which is constant at a given temperature. Therefore, in heterogeneous equilibria, pure solids and liquids are omitted from equilibrium expressions.

Now you can write the solubility product constant expression for the dissolving of barium sulfate ($BaSO_4$) in water. The $K_{sp}$ for the process is $1.1 \times 10^{-10}$ at 298 K.

$$K_{sp} = [Ba^{2+}][SO_4^{2-}] = 1.1 \times 10^{-10}$$

The small value of $K_{sp}$ for $BaSO_4$ indicates that products are not favored at equilibrium. The concentration of barium ions at equilibrium is only $1.0 \times 10^{-5}M$, and a patient, such as the one shown in **Figure 18,** can safely ingest a barium sulfate solution.

The solubility product constant for the antacid magnesium hydroxide ($Mg(OH)_2$) provides another example.

$$Mg(OH)_2(s) \rightleftharpoons Mg^{2+}(aq) + 2OH^-(aq)$$
$$K_{sp} = [Mg^{2+}][OH^-]^2$$

$K_{sp}$ depends only on the concentrations of the ions in the saturated solution. However, some of the undissolved solid, no matter how small the amount, must be present in the equilibrium mixture.

The solubility product constants for some ionic compounds are listed in **Table 3.** Note that they are all small numbers. Solubility product constants are measured and recorded only for sparingly soluble compounds.

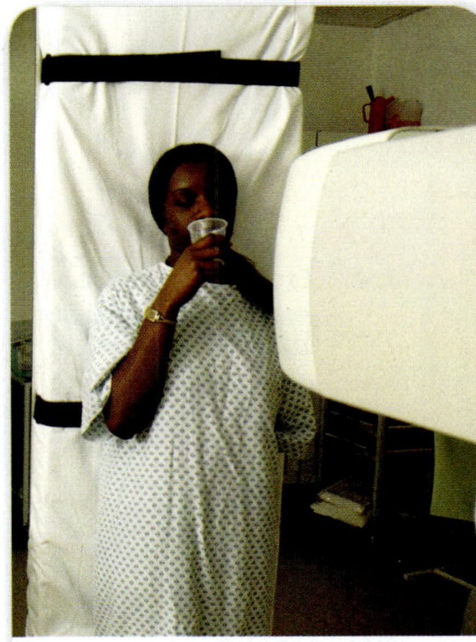

**Figure 18** Greater definition is possible in a gastrointestinal X-ray when patients drink a thick mixture containing barium sulfate. Barium sulfate is a poisonous substance, but it has such low solubility that only a minimal amount can dissolve in the patient's body.

Universal Images Group/Getty Images

**Using solubility product constants** The solubility product constants in **Table 3** have been determined through careful experiments. $K_{sp}$ values are important because they can be used to determine the solubility of a sparingly soluble compound. Recall that the solubility of a compound in water is the amount of the substance that will dissolve in a given volume of water at a given temperature.

**Table 3 Solubility Product Constants at 298 K**

| Compound | $K_{sp}$ | Compound | $K_{sp}$ | Compound | $K_{sp}$ |
|---|---|---|---|---|---|
| **Carbonates** | | **Halides** | | **Hydroxides** | |
| $BaCO_3$ | $2.6 \times 10^{-9}$ | $CaF_2$ | $3.5 \times 10^{-11}$ | $Al(OH)_3$ | $4.6 \times 10^{-33}$ |
| $CaCO_3$ | $3.4 \times 10^{-9}$ | $PbBr_2$ | $6.6 \times 10^{-6}$ | $Ca(OH)_2$ | $5.0 \times 10^{-6}$ |
| $CuCO_3$ | $2.5 \times 10^{-10}$ | $PbCl_2$ | $1.7 \times 10^{-5}$ | $Cu(OH)_2$ | $2.2 \times 10^{-20}$ |
| $PbCO_3$ | $7.4 \times 10^{-14}$ | $PbF_2$ | $3.3 \times 10^{-8}$ | $Fe(OH)_2$ | $4.9 \times 10^{-17}$ |
| $MgCO_3$ | $6.8 \times 10^{-6}$ | $PbI_2$ | $9.8 \times 10^{-9}$ | $Fe(OH)_3$ | $2.8 \times 10^{-39}$ |
| $Ag_2CO_3$ | $8.5 \times 10^{-12}$ | $AgCl$ | $1.8 \times 10^{-10}$ | $Mg(OH)_2$ | $5.6 \times 10^{-12}$ |
| $ZnCO_3$ | $1.5 \times 10^{-10}$ | $AgBr$ | $5.4 \times 10^{-13}$ | $Zn(OH)_2$ | $3 \times 10^{-17}$ |
| $Hg_2CO_3$ | $3.6 \times 10^{-17}$ | $AgI$ | $8.5 \times 10^{-17}$ | **Sulfates** | |
| **Chromates** | | **Phosphates** | | $BaSO_4$ | $1.1 \times 10^{-10}$ |
| $BaCrO_4$ | $1.2 \times 10^{-10}$ | $AlPO_4$ | $9.8 \times 10^{-21}$ | $CaSO_4$ | $4.9 \times 10^{-5}$ |
| $PbCrO_4$ | $2.3 \times 10^{-13}$ | $Ca_3(PO_4)_2$ | $2.1 \times 10^{-33}$ | $PbSO_4$ | $2.5 \times 10^{-8}$ |
| $Ag_2CrO_4$ | $1.1 \times 10^{-12}$ | $Mg_3(PO_4)_2$ | $1.0 \times 10^{-24}$ | $Ag_2SO_4$ | $1.2 \times 10^{-5}$ |

Suppose you wish to determine the solubility of silver iodide (AgI) in mol/L at 298 K. The equilibrium equation and solubility product constant expression are as follows.

$$AgI(s) \rightleftharpoons Ag^+(aq) + I^-(aq)$$
$$K_{sp} = [Ag^+][I^-] = 8.5 \times 10^{-17} \text{ at } 298 \text{ K}$$

Explore **salts and solubility.**

Virtual Investigations

It is convenient to let $s$ represent the solubility of AgI, that is, the number of moles of AgI that dissolves in one liter of solution. The equation indicates that for every mole of AgI that dissolves, an equal number of moles of $Ag^+$ ions forms in solution. Therefore, $[Ag^+]$ equals $s$. Every $Ag^+$ has an accompanying $I^-$ ion, so $[I^-]$ also equals $s$. Substituting s for $[Ag^+]$ and $[I^-]$, the $K_{sp}$ expression becomes the following.

$$[Ag^+][I^-] = (s)(s) = s^2 = 8.5 \times 10^{-17}$$
$$s = \sqrt{8.5 \times 10^{-17}} = 9.2 \times 10^{-9} \text{ mol/L}$$

The solubility of AgI is $9.2 \times 10^{-9}$ mol/L at 298 K.

## EXAMPLE Problem 5

EXAMPLE PROBLEM

**CALCULATING MOLAR SOLUBILITY** Use the $K_{sp}$ value from **Table 3** to calculate the solubility in mol/L of copper(II) carbonate ($CuCO_3$) at 298 K.

### 1 ANALYZE THE PROBLEM

You have been given the solubility product constant for $CuCO_3$. The copper and carbonate ion concentrations are in a one-to-one relationship with the molar solubility of $CuCO_3$. Use $s$ to represent the molar solubility of $CuCO_3$. Then use the solubility product constant expression to solve for the solubility. Because $K_{sp}$ is of the order of $10^{-10}$, you can predict that the solubility will be the square root of $K_{sp}$, or about $10^{-5}$.

**Known**

$K_{sp}(CuCO_3) = 2.5 \times 10^{-10}$

**Unknown**

**$s$ = ? mol/L**

### 2 SOLVE FOR THE UNKNOWN

$CuCO_3(s) \rightleftharpoons Cu^{2+}(aq) + CO_3^{2-}(aq)$ — **State the balanced chemical equation for the solubility equilibrium.**

$K_{sp} = [Cu^{2+}][CO_3^{2-}] = 2.5 \times 10^{-10}$ — **State the solubility product constant expression.**

$s = [Cu^{2+}] = [CO_3^{2-}]$ — **Relate $[Cu^{2+}]$ and $[CO_3^{2-}]$ to the solubility of $CuCO_3$, $s$.**

$(s)(s) = s^2 = 2.5 \times 10^{-10}$ — **Substitute $s$ for $[Cu^{2+}]$ and $[CO_3^{2-}]$ in the expression for $K_{sp}$.**

$s = \sqrt{2.5 \times 10^{-10}} = \mathbf{1.6 \times 10^{-5}\ mol/L}$ — **Solve for $s$, and calculate the answer.**

The molar solubility of $CuCO_3$ in water at 298 K is $1.6 \times 10^{-5}$ mol/L.

### 3 EVALUATE THE ANSWER

The $K_{sp}$ value has two significant figures, so the answer is correctly expressed with two digits. As predicted, the molar solubility of $CuCO_3$ is approximately $10^{-5}$ mol/L.

## PRACTICE Problems

Do additional problems. Online Practice

**20.** Use the data in **Table 3** to calculate the solubility in mol/L of the following ionic compounds at 298 K.

**a.** $PbCrO_4$ **b.** AgCl **c.** $CaCO_3$

**21. Challenge** The $K_{sp}$ of lead(II) carbonate ($PbCO_3$) is $7.40 \times 10^{-14}$ at 298 K. What is the solubility of lead(II) carbonate in g/L?

You have read that the solubility product constant can be used to determine the molar solubility of an ionic compound. You can apply this information as you perform the ChemLab at the end of this chapter. $K_{sp}$ can also be used to find the concentrations of the ions in a saturated solution.

## EXAMPLE Problem 6

**CALCULATING ION CONCENTRATION** Magnesium hydroxide is a white solid obtained from seawater and used in the formulation of many medications, in particular those whose function is to neutralize excess stomach acid. Determine the hydroxide ion concentration in a saturated solution of $Mg(OH)_2$ at 298 K. The $K_{sp}$ equals $5.6 \times 10^{-12}$.

### 1 ANALYZE THE PROBLEM

You have been given the $K_{sp}$ for $Mg(OH)_2$. The moles of $Mg^{2+}$ ions in solution equal the moles of $Mg(OH)_2$ that dissolved, but the moles of $OH^-$ ions in solution are two times the moles of $Mg(OH)_2$ that dissolved. You can use these relationships to write the solubility product constant expression in terms of one unknown. Because the equilibrium expression is a third-power equation, you can predict that $[OH^-]$ will be approximately the cube root of $10^{-12}$, or approximately $10^{-4}$.

| Known | Unknown |
|---|---|
| $K_{sp} = 5.6 \times 10^{-12}$ | $[OH^-] = ?$ mol/L |

### 2 SOLVE FOR THE UNKNOWN

$Mg(OH)_2(s) \rightleftharpoons Mg^{2+}(aq) + 2OH^-(aq)$ — State the equation for the solubility equilibrium.

$K_{sp} = [Mg^{2+}][OH^-]^2 = 5.6 \times 10^{-12}$ — State the $K_{sp}$ expression.

Let $x = [Mg^{2+}]$. Because there are two $OH^-$ ions for every $Mg^{2+}$ ion, $2x = [OH^-]$.

$(x)(2x)^2 = 5.6 \times 10^{-12}$ — Substitute $x = [Mg^{2+}]$ and $2x = [OH^-]$

$(x)(4)(x)^2 = 5.6 \times 10^{-12}$ — Square the terms.

$4x^3 = 5.6 \times 10^{-12}$ — Combine the terms.

$x^3 = \frac{5.6 \times 10^{-12}}{4} = 1.4 \times 10^{-12}$ — Divide.

$x = [Mg^{2+}] = \sqrt[3]{1.4 \times 10^{-12}} = 1.1 \times 10^{-4}$ mol/L — Use your calculator to determine the cube root.

Multiply $[Mg^{2+}]$ by 2 to obtain $[OH^-]$.

$[OH^-] = 2[Mg^{2+}] = 2(1.1 \times 10^{-4} \text{ mol/L}) = 2.2 \times 10^{-4}$ **mol/L**

### 3 EVALUATE THE ANSWER

The given $K_{sp}$ has two significant figures, so the answer is correctly stated with two digits. As predicted, $[OH^-]$ is about $10^{-4}$ mol/L.

## PRACTICE Problems

Do additional problems. Online Practice

**22.** Use $K_{sp}$ values from **Table 3** to calculate the following.
- **a.** $[Ag^+]$ in a solution of AgBr at equilibrium
- **b.** $[F^-]$ in a saturated solution of $CaF_2$
- **c.** $[Ag^+]$ in a solution of $Ag_2CrO_4$ at equilibrium

**23.** Calculate the solubility of $Ag_3PO_4$ ($K_{sp} = 2.6 \times 10^{-18}$).

**24. Challenge** The solubility of silver chloride (AgCl) is $1.86 \times 10^{-4}$ g/100 g of $H_2O$ at 298 K. Calculate the $K_{sp}$ for AgCl.

EXAMPLE PROBLEM

| Table 4 Ion Concentrations | |
|---|---|
| Original Solutions (mol/L) | Mixture (mol/L) |
| $[Fe^{3+}] = 0.10$ | $[Fe^{3+}] = 0.050$ |
| $[Cl^-] = 0.30$ | $[Cl^-] = 0.15$ |
| $[K^+] = 0.40$ | $[K^+] = 0.20$ |
| $[Fe(CN)_6^{4-}] = 0.10$ | $[Fe(CN)_6^{4-}] = 0.050$ |

**Predicting precipitates** Suppose equal volumes of 0.10*M* aqueous solutions of iron(III) chloride ($FeCl_3$) and potassium hexacyanoiron(II) ($K_4Fe(CN)_6$) are combined. Will a precipitate form as shown in **Figure 19?** The following double-replacement reaction might occur.

$$4FeCl_3 + 3K_4Fe(CN)_6 \rightarrow 12KCl + Fe_4(Fe(CN)_6)_3$$

You can use $K_{sp}$ to predict whether a precipitate will form when any two ionic solutions are mixed.

For the reaction above, a precipitate is likely to form only if either product, KCl or $Fe_4(Fe(CN)_6)_3$, has low solubility. You might know that KCl is a soluble compound and would be unlikely to precipitate. But $K_{sp}$ for $Fe_4(Fe(CN)_6)_3$ is a very small number, $3.3 \times 10^{-41}$, which suggests that $Fe_4(Fe(CN)_6)_3$ might precipitate if the concentrations of its ions are large enough. How large is large enough?

The following equilibrium is possible between solid $Fe_4(Fe(CN)_6)_3$—a precipitate—and its ions in solution, $Fe^{3+}$ and $Fe(CN)_6^{4-}$.

$$Fe_4(Fe(CN)_6)_3(s) \rightleftharpoons 4Fe^{3+}(aq) + 3Fe(CN)_6^{4-}(aq)$$

When the $FeCl_3$ and $Fe_4(Fe(CN)_6)_3(s)$ solutions are mixed, if the concentrations of the ions $Fe^{3+}$ and $Fe(CN)_6^{4-}$ are greater than those that can exist in a saturated solution of $Fe_4(Fe(CN)_6)_3$, the equilibrium will shift to the left and $Fe_4(Fe(CN)_6)_3(s)$ will precipitate. To predict whether a precipitate will form when the two solutions are mixed, you must first calculate the concentrations of the ions.

READING CHECK **Explain** the conditions under which you would predict that a precipitate would form.

■ **Figure 19** Because its ion-product constant ($Q_{sp}$) is greater than $K_{sp}$, you could predict that this precipitate of $Fe_4(Fe(CN)_6)_3$ would form.

View an **animation of a precipitation reaction.**

Concepts In Motion

**Calculating ion concentrations** **Table 4** shows the concentrations of the ions of reactants and products in the original solutions (0.10*M* $FeCl_3$ and 0.10*M* $K_4Fe(CN)_6$) and in the mixture immediately after equal volumes of the two solutions were mixed. Note that $[Cl^-]$ is three times as large as $[Fe^{3+}]$ because the ratio of $Cl^-$ to $Fe^{3+}$ in $FeCl_3$ is 3:1. Also note that $[K^+]$ is four times as large as $[Fe(CN)_6^{4-}]$ because the ratio of $K^+$ to $Fe(CN)_6^{4-}$ in $K_4Fe(CN)_6$ is 4:1. In addition, note that the concentration of each ion in the mixture is one-half its original concentration. This is because when equal volumes of two solutions are mixed, the same number of ions are dissolved in twice as much solution. Therefore, the concentration is reduced by one-half.

You can now use the data in the table to make a trial to see if the concentrations of $Fe^{3+}$ and $Fe(CN)_6^{4-}$ in the mixed solution exceed the value of $K_{sp}$ when substituted into the solubility product constant expression.

$$K_{sp} = [Fe^{3+}]^4[Fe(CN)_6^{4-}]^3$$

Remember that you have not determined whether the solution is saturated. When you make this substitution, it will not necessarily give the solubility product constant. Instead, it provides a number called the ion product ($Q_{sp}$). $Q_{sp}$ is a trial value that can be compared with $K_{sp}$.

$$Q_{sp} = [Fe^{3+}]^4[Fe(CN)_6^{4-}]^3 = (0.050)^4(0.050)^3 = 7.8 \times 10^{-10}$$

You can now compare $Q_{sp}$ and $K_{sp}$. This comparison can have one of three outcomes: $Q_{sp}$ can be less than $K_{sp}$, equal to $K_{sp}$, or greater than $K_{sp}$.

McGraw-Hill Education

1. If $Q_{sp} < K_{sp}$, the solution is unsaturated. No precipitate will form.
2. If $Q_{sp} = K_{sp}$, the solution is saturated, and no change will occur.
3. If $Q_{sp} > K_{sp}$, a precipitate will form, reducing the concentrations of the ions in the solution until the product of their concentrations in the $K_{sp}$ expression equals the numerical value of $K_{sp}$. Then the system is in equilibrium, and the solution is saturated.

In the case of the $Fe_4(Fe(CN)_6)_3$ equilibrium, $Q_{sp}$ ($7.8 \times 10^{-10}$) is larger than $K_{sp}$($3.3 \times 10^{-41}$) and a deeply colored blue precipitate of $Fe_4(Fe(CN)_6)_3$ forms, as shown in **Figure 19.**

## EXAMPLE Problem 7

Find help with **algebraic equations.** Math Handbook

EXAMPLE PROBLEM

**PREDICTING A PRECIPITATE** Predict whether a precipitate of $PbCl_2$ will form if 100 mL of 0.0100$M$ NaCl is added to 100 mL of 0.0200$M$ $Pb(NO_3)_2$.

### 1 ANALYZE THE PROBLEM

You have been given equal volumes of two solutions with known concentrations. The concentrations of the initial solutions allow you to calculate the concentrations of $Pb^{2+}$ and $Cl^-$ ions in the mixed solution.

| Known | Unknown |
|---|---|
| 100 mL 0.0100$M$ NaCl | $Q_{sp} > K_{sp}$? |
| 100 mL 0.0200$M$ $Pb(NO_3)_2$ | |
| $K_{sp} = 1.7 \times 10^{-5}$ | |

### 2 SOLVE FOR THE UNKNOWN

$PbCl_2(s) \rightleftharpoons Pb^{2+}(aq) + 2Cl^-(aq)$ — **State the equation for the dissolving of $PbCl_2$.**

$Q_{sp} = [Pb^{2+}][Cl^-]^2$ — **State the ion product expression, $Q_{sp}$.**

Mixing the solutions dilutes their concentrations by one-half.

$[Pb^{2+}] = \frac{0.0200M}{2} = 0.0100M$ — **Divide $[Pb^{2+}]$ by 2.**

$[Cl^-] = \frac{0.0100M}{2} = 0.00500M$ — **Divide $[Cl^-]$ by 2.**

$Q_{sp} = (0.0100)(0.00500)^2 = 2.5 \times 10^{-7}$ — **Substitute $[Pb^{2+}] = 0.0100M$ and $[Cl^-] = 0.00500M$ into $Q_{sp}$.**

$Q_{sp}\ (2.5 \times 10^{-7}) < K_{sp}\ (1.7 \times 10^{-5})$ — **Compare $Q_{sp}$ with $K_{sp}$.**

A precipitate will not form.

### 3 EVALUATE THE ANSWER

$Q_{sp}$ is less than $K_{sp}$. The $Pb^{2+}$ and $Cl^-$ ions are not present in high enough concentrations in the mixed solution to cause precipitation to occur.

## PRACTICE Problems

Do additional problems. Online Practice

**25.** Use $K_{sp}$ values from **Table 3** to predict whether a precipitate will form when equal volumes of the following solutions are mixed.

**a.** 0.10$M$ $Pb(NO_3)_2$ and 0.030$M$ NaF

**b.** 0.25$M$ $K_2SO_4$ and 0.010$M$ $AgNO_3$

**26. Challenge** Will a precipitate form when 250 mL of 0.20$M$ $MgCl_2$ is added to 750 mL of 0.0025$M$ NaOH?

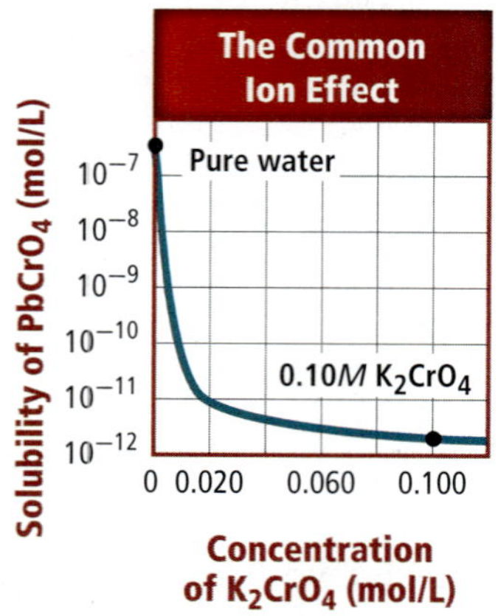


Pure water: $[Pb^{2+}] = 4.8 \times 10^{-7}$ mol/L
$[CrO_4^{2-}] = 4.8 \times 10^{-7}$ mol/L
$0.10M\ K_2CrO_4$: $[Pb^{2+}] = 2.3 \times 10^{-12}$ mol/L
$[CrO_4^{2-}] = 1.00 \times 10^{-1}$ mol/L

**Figure 20** The solubility of lead chromate becomes lower as the concentration of the potassium chromate solution in which it is dissolved increases. The change is due to the presence of $CrO_4^{2-}$ in both lead chromate and potassium chromate.

**GRAPH CHECK**

**Verify** that $K_{sp}$ does not change as the concentration of potassium chromate increases.

## The Common Ion Effect

The solubility of lead(II) chromate ($PbCrO_4$) in water is $4.8 \times 10^{-7}$ mol/L at 298 K. That means you can dissolve $4.8 \times 10^{-7}$ mol $PbCrO_4$ in 1.00 L of pure water. However, you cannot dissolve $4.8 \times 10^{-7}$ mol $PbCrO_4$ in 1.00 L of $0.10M$ aqueous potassium chromate ($K_2CrO_4$) solution at that temperature. Why is $PbCrO_4$ less soluble in an aqueous $K_2CrO_4$ solution than in pure water?

The equation for the $PbCrO_4$ solubility equilibrium and the solubility product constant expression are as follows.

$$PbCrO_4(s) \rightleftharpoons Pb^{2+}(aq) + CrO_4^{2-}(aq)$$
$$K_{sp} = [Pb^{2+}][CrO_4^{2-}] = 2.3 \times 10^{-13}$$

Recall that $K_{sp}$ is a constant at any given temperature, so if the concentration of either $Pb^{2+}$ or $CrO_4^{2-}$ increases when the system is at equilibrium, the concentration of the other ion must decrease. The product of the concentrations of the two ions must always equal $K_{sp}$. The $K_2CrO_4$ solution contains $CrO_4^{2-}$ ions before any $PbCrO_4$ dissolves. In this example, the $CrO_4^{2-}$ ion is called a common ion because it is part of both $PbCrO_4$ and $K_2CrO_4$. **Figure 20** shows the effect of the common ion, the $CrO_4^{2-}$ ion, on the solubility of $PbCrO_4$. A **common ion** is an ion that is common to two or more ionic compounds. The lowering of the solubility of a substance because of the presence of a common ion is called the **common ion effect.**

**Applying Le Châtelier's principle** A saturated solution of lead(II) chromate ($PbCrO_4$) is shown in **Figure 21a.** Note the solid-yellow $PbCrO_4$ in the bottom of the beaker. The solution and solid are in equilibrium according to the following equation.

$$PbCrO_4(s) \rightleftharpoons Pb^{2+}(aq) + CrO_4^{2-}(aq)$$

When a solution of $Pb(NO_3)_2$ is added to the saturated $PbCrO_4$ solution, more solid $PbCrO_4$ precipitates, as shown in **Figure 21b.** The $Pb^{2+}$ ion, common to both $Pb(NO_3)_2$ and $PbCrO_4$, reduces the solubility of $PbCrO_4$. Can this precipitation of $PbCrO_4$ be explained by Le Châtelier's principle? Adding $Pb^{2+}$ ion to the solubility equilibrium stresses the equilibrium. To relieve the stress, the equilibrium shifts to the left to form more solid $PbCrO_4$.

**Figure 21** Refer to **Figure 20** to see the effect of additional chromate ions on the solubility of lead(II) chromate. Adding $Pb^{2+}$ ions in the form of lead nitrate ($Pb(NO_3)_2$) also affects the solubility of lead(II) chromate. **a.** $PbCrO_4(s)$ is in equilibrium with its ions in solution. **b.** The equilibrium is stressed by the addition of $Pb(NO_3)_2$ and more $PbCrO_4$ precipitate forms.

The common ion effect also plays a role in the use of $BaSO_4$ when X-rays of the digestive system are taken. The low solubility of $BaSO_4$ helps ensure that the amount of the toxic barium ion absorbed into patient's system is small enough to be harmless. The procedure is further safeguarded by the addition of sodium sulfate ($Na_2SO_4$), a soluble ionic compound that provides a common ion, $SO_4^{2-}$.

$$BaSO_4(s) \rightleftharpoons Ba^{2+}(aq) + SO_4^{2-}(aq)$$

Le Châtelier's principle tells you that additional $SO_4^{2-}$ from the $Na_2SO_4$ shifts the equilibrium to the left to produce more solid $BaSO_4$ and reduces the number of harmful $Ba^{2+}$ ions in solution.

## PROBLEM-SOLVING STRATEGY

### Using Assumptions

In Example Problem 5, you calculated the molar solubility of $CuCO_3$ in pure water as $1.6 \times 10^{-5}$ mol/L. But suppose that $CuCO_3$ is dissolved in a solution of 0.10*M* $K_2CO_3$? A common ion is in solution. If you set up the problem the same way you did in Example Problem 5, you will need to solve a quadratic equation. Solving the quadratic equation results in the correct answer, but you can make a simple assumption that streamlines the problem-solving process.

| Concentration (*M*) | $CuCO_3$ (s) → | $Cu^{2+}$ (aq) + | $CO_3^{2-}$ (aq) |
|---|---|---|---|
| Initial | — | 0 | 0.10 |
| Change | — | $+s$ | $+s$ |
| Equilibrium | — | $s$ | $0.10 + s$ |

### Using the Quadratic Equation

1. Set up the problem

$[Cu^{2+}][CO_3^{2-}] = 2.5 \times 10^{-10}$

$(s)(0.10 + s) = 2.5 \times 10^{-10}$

2. Solve the quadratic

$0.10s + s^2 = 2.5 \times 10^{-10}$

$s^2 + 0.10s - 2.5 \times 10^{-10} = 0$

$$s = \frac{-b \pm \sqrt{b^2 - 4ac}}{2a}$$

$$= \frac{-0.10 \pm \sqrt{0.10^2 - (4)(1)(-2.5 \times 10^{-10})}}{2(1)}$$

$s = 2.5 \times 10^{-9}$ mol/L and $s = -0.10$ mol/L

### Using the Simplifying Assumption

1. Set up the problem

$[Cu^{2+}][CO_3^{2-}] = 2.5 \times 10^{-10}$

$(s)(0.10 + s) = 2.5 \times 10^{-10}$

Because $K_{sp}$ is small ($2.5 \times 10^{-10}$), assume that $s$ is negligible compared to 0.10*M*. Thus, $0.10 + s \approx 0.10$.

$(s)(0.10) = 2.5 \times 10^{-10}$

2. Solve the problem

$(s)(0.10) = 2.5 \times 10^{-10}$

$$s = \frac{2.5 \times 10^{-10}}{(0.10)} = 2.5 \times 10^{-9} \text{ mol/L}$$

The root of the quadratic that makes sense is $s = 2.5 \times 10^{-9}$ mol/L. As you can see by comparing the two answers, the assumption gave good results more quickly and easily. However, this assumption works only for sparingly soluble compounds.

### Apply the Strategy

**Calculate** the molar solubility of lead(II) fluoride in a 0.20*M* $Pb(NO_3)_2$ solution.

# Problem-Solving LAB

## Apply Scientific Explanations

**How does the fluoride ion prevent tooth decay?** During the last half century, tooth decay has decreased significantly because minute quantities of fluoride ion ($6 \times 10^{-5}M$) are being added to most public drinking-water systems, and most people are using toothpastes containing sodium fluoride or tin(II) fluoride. Use what you know about the solubility of ionic compounds and reversible reactions to explore the role of the fluoride ion in maintaining cavity-free teeth.

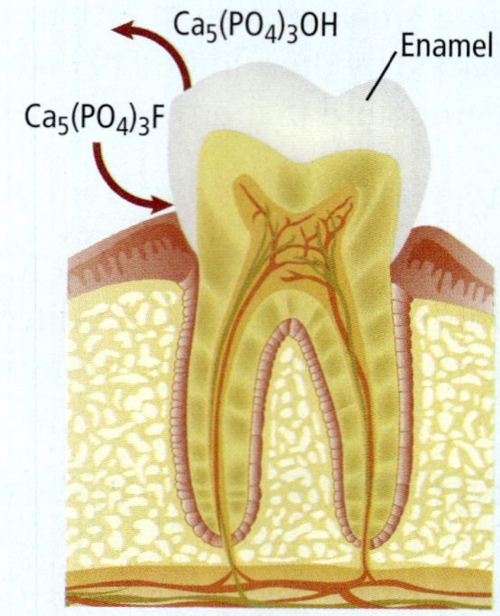


### Analysis

Enamel, the hard, protective outer layer of the tooth, is 98% hydroxyapatite ($Ca_5(PO_4)_3OH$). Although insoluble in water ($K_{sp} = 6.8 \times 10^{-37}$), demineralization, which is the dissolving of hydroxyapatite, does occur, especially when the saliva contains acids. The reverse reaction, remineralization, also occurs. Remineralization is the redepositing of tooth enamel. When hydroxyapatite is in solution with fluoride ions, a double-replacement reaction can occur. A fluoride ion replaces the hydroxide ion to form fluoroapatite ($Ca_5(PO_4)_3F$), ($K_{sp} = 1 \times 10^{-60}$). Fluoroapatite remineralizes the tooth enamel, thus partially displacing hydroxyapatite. Because fluoroapatite is less soluble than hydroxyapatite, destructive demineralization is reduced.

### Think Critically

1. **State** the equation for the dissolving of hydroxyapatite and its equilibrium constant expression. How do the conditions in the mouth differ from those of a true equilibrium?
2. **State** the equation that describes the double-replacement reaction that occurs between hydroxyapatite and sodium fluoride.
3. **Calculate** the solubility of hydroxyapatite and fluoroapatite in water. Compare the solubilities.
4. **Calculate** the ion product constant ($Q_{sp}$) for the reaction if $0.00050M$ NaF is mixed with an equal volume of $0.000015M$ $Ca_5(PO_4)_3OH$. Will a precipitate form (re-mineralization)?

## SECTION 3 REVIEW

Section Self-Check

### Section Summary

- Equilibrium concentrations and solubilities can be calculated using equilibrium constant expressions.
- $K_{sp}$ describes the equilibrium between a sparingly soluble ionic compound and its ions in solution.
- If the ion product, $Q_{sp}$, exceeds the $K_{sp}$ when two solutions are mixed, a precipitate will form.
- The presence of a common ion in a solution lowers the solubility of a dissolved substance.

27. **MAIN**IDEA **List** the information you would need in order to calculate the concentration of a product in a reaction mixture at equilibrium.
28. **Explain** how to use the solubility product constant to calculate the solubility of a sparingly soluble ionic compound.
29. **Describe** how the presence of a common ion reduces the solubility of an ionic compound.
30. **Explain** the difference between $K_{sp}$ and $Q_{sp}$. Is $Q_{sp}$ an equilibrium constant?
31. **Calculate** The $K_{sp}$ of magnesium carbonate ($MgCO_3$) is $2.6 \times 10^{-9}$. What is the solubility of $MgCO_3$ in pure water?
32. **Design an experiment** based on solubilities to demonstrate which of two ions, $Mg^{2+}$ or $Pb^{2+}$, is contained in an aqueous solution. Solubility information about ionic compounds is given in **Tables R-3** and **R-8** in the Student Resources appendix.

# CHEMISTRY&health

## Hemoglobin Rises to the Challenge

When people travel to the mountains, they often feel tired and light-headed for a time. That's because the mountain air contains fewer oxygen molecules, as shown in **Figure 1.** Over time, the fatigue lessens. The body adapts by producing more of a protein called hemoglobin.

### Hemoglobin-oxygen equilibrium

Hemoglobin (Hgb) binds with oxygen molecules that enter your bloodstream, producing oxygenated hemoglobin ($Hgb(O_2)_4$). The equilibrium of Hgb and $O_2$ is represented as follows.

$$Hgb(aq) + 4O_2(g) \rightleftharpoons Hgb(O_2)_4(aq)$$

**In the lungs** When you breathe, oxygen molecules move into your blood. The equilibrium reacts to the stress by consuming oxygen molecules at an increased rate. The equilibrium shifts to the right, increasing the blood concentration of $Hgb(O_2)_4$.

$$Hgb(aq) + 4O_2(g) \rightleftharpoons Hgb(O_2)_4(aq)$$

**In the tissues** When the $Hgb(O_2)_4$ reaches body tissues where oxygen concentrations are low, the equilibrium shifts to the left, releasing oxygen to enable the metabolic processes that produce energy.

$$Hgb(aq) + 4O_2(g) \rightleftharpoons Hgb(O_2)_4(aq)$$

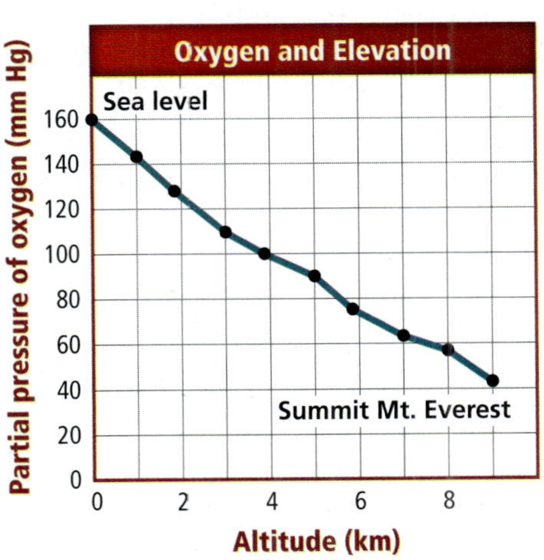


**Figure 1** On the summit, the partial pressure of $O_2$ is much lower. Each breath a person draws contains fewer $O_2$ molecules.

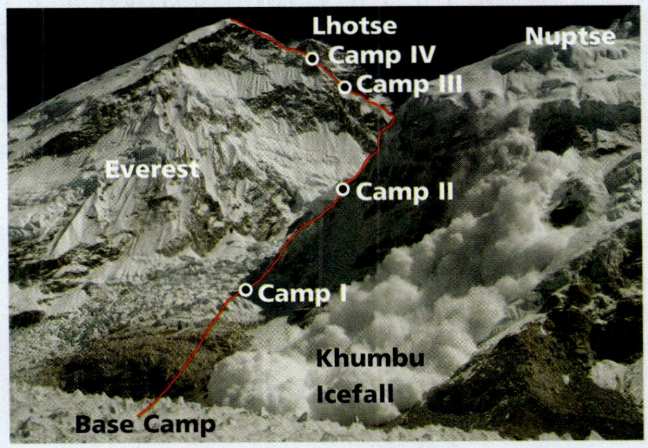


**Figure 2** On Mount Everest, a climber might ascend to Camp II, descend to Base Camp, and then ascend to Camp III over the course of several days to prepare for a summit bid.

Rafal Belzowski/Getty Images

**In the mountains** The equilibrium reacts to the stress of thin mountain air by producing oxygen at an increased rate. The shift to the left releases oxygen molecules in your lungs, leaving less oxygenated hemoglobin in your blood.

$$Hgb(aq) + 4O_2(g) \rightleftharpoons Hgb(O_2)_4(aq)$$

The lower blood concentration of oxygenated hemoglobin means that fewer oxygen molecules are released in other parts of your body. Because less energy is produced, you feel tired.

**The body adjusts** Your body responds to the lower oxygen concentration by producing more hemoglobin, part of a process known as acclimatization. More hemoglobin shifts the equilibrium position back to the right.

$$Hgb(aq) + 4O_2(g) \rightleftharpoons Hgb(O_2)_4(aq)$$

The increased concentration of $Hgb(O_2)_4(aq)$ means that more oxygen molecules can be released in your body tissues. **Figure 2** shows where climbers might adjust their bodies to high elevations before beginning their summit bid.

### WRITING IN ▶ Chemistry

**Research** the sleep disorder apnea. How would an incident of apnea affect the body's hemoglobin equilibrium?

WebQuest

# SMALL SCALE ChemLAB

iLab Station

## Compare Two Solubility Product Constants

**Background:** **By observing the formation of two precipitates in the same system, you can infer the relationship between the solubilities of the two ionic compounds and the numerical values of their solubility product constants ($K_{sp}$).**

**Question:** *How can you use Le Châtelier's principle to evaluate the relative solubilities of two precipitates?*

### Materials

$AgNO_3$ solution
NaCl solution
$Na_2S$ solution
24-well microplate
thin-stem pipettes (3)

### Safety Precautions

**WARNING:** ***Silver nitrate is highly toxic and will stain skin and clothing. Sodium sulfide is a skin irritant and should be kept away from acids.***

### Procedure

1. Read and complete the lab safety form.
2. Place 10 drops of $AgNO_3$ solution in Well A1 of a 24-well microplate. Place 10 drops of the same solution in Well A2.
3. Add 10 drops of NaCl solution to Well A1 and 10 drops to Well A2.
4. Allow the precipitates to form. Observe the wells from the top and the side and record your observations.
5. To Well A2, add 10 drops of $Na_2S$ solution.
6. Allow the precipitate to form. Record your observations of the precipitate.
7. Compare the contents of Wells A1 and A2, and record your observations.
8. **Cleanup and Disposal** Use a wash bottle to transfer the contents of the well plate into a waste beaker.

### Analyze and Conclude

1. **Analyze** Write the complete equation for the reaction that occurred when you mixed NaCl and $AgNO_3$ in Step 3. Write the net ionic equation.
2. **Analyze** Write the solubility product constant expression for the equilibrium established in Wells A1 and A2 in Step 3. $K_{sp}$ (AgCl) $= 1.8 \times 10^{-10}$.
3. **Analyze** Write the equation for the equilibrium that was established in Well A2 when you added $Na_2S$. $K_{sp}$ ($Ag_2S$) $= 8 \times 10^{-48}$.
4. **Identify** the two precipitates by color.
5. **Compare** the $K_{sp}$ values for the two precipitates. Which of the two ionic compounds is more soluble?
6. **Recognize Cause and Effect** Use Le Châtelier's principle to explain how the addition of $Na_2S$ in Step 5 affected the equilibrium established in Well A2.
7. **Calculate** the molar solubilities of the two precipitates using the $K_{sp}$ values. Which of the precipitates is more soluble?
8. **Identify** What evidence from this experiment supports your answer to Question 7? Explain.
9. **Error Analysis** Compare your observations of the well plate from the side with your observations from the top. What did you notice?
10. **Research** how industries use precipitation to remove hazardous chemicals from wastewater.

### INQUIRY EXTENSION

**Soluble v. Insoluble** The reactants that you used in this ChemLab are all soluble ionic compounds, and the precipitates are insoluble. How does soluble $Na_2S$ differ from insoluble $Ag_2S$? How does soluble NaCl differ from insoluble AgCl? Use this information, $K_{sp}$ data from **Table 3,** and other reference sources to develop general rules for solubility.

Matt Meadows

# CHAPTER 17 STUDY GUIDE

Vocabulary Practice

**BIGIDEA** Many reactions and processes reach a state of chemical equilibrium in which both reactants and products are formed at equal rates.

## SECTION 1 A State of Dynamic Balance

**MAINIDEA** Chemical equilibrium is described by an equilibrium constant expression that relates the concentrations of reactants and products.

- A reaction is at equilibrium when the rate of the forward reaction equals the rate of the reverse reaction.
- The equilibrium constant expression is a ratio of the molar concentrations of the products to the molar concentrations of the reactants with each concentration raised to a power equal to its coefficient in the balanced chemical equation.

$$K_{eq} = \frac{[C]^c[D]^d}{[A]^a[B]^b}$$

- The value of the equilibrium constant expression, $K_{eq}$, is a constant for a given temperature.

### VOCABULARY

- reversible reaction
- chemical equilibrium
- law of chemical equilibrium
- equilibrium constant
- homogeneous equilibrium
- heterogeneous equilibrium

## SECTION 2 Factors Affecting Chemical Equilibrium

**MAINIDEA** When changes are made to a system at equilibrium, the system shifts to a new equilibrium position.

- Le Châtelier's principle describes how an equilibrium system shifts in response to a stress or a disturbance.
- When an equilibrium shifts in response to a change in concentration or volume, the equilibrium position changes but $K_{eq}$ remains constant. A change in temperature, however, alters both the equilibrium position and the value of $K_{eq}$.

### VOCABULARY

- Le Châtelier's principle

## SECTION 3 Using Equilibrium Constants

**MAINIDEA** Equilibrium constant expressions can be used to calculate concentrations and solubilities.

- Equilibrium concentrations and solubilities can be calculated using equilibrium constant expressions.
- $K_{sp}$ describes the equilibrium between a sparingly soluble ionic compound and its ions in solution.
- If the ion product, $Q_{sp}$, exceeds the $K_{sp}$ when two solutions are mixed, a precipitate will form.
- The presence of a common ion in a solution lowers the solubility of a dissolved substance.

### VOCABULARY

- solubility product constant
- common ion
- common ion effect

## SECTION 1

### Mastering Concepts

**33.** Describe an equilibrium in everyday life that illustrates a state of balance between two opposing processes.

**34.** Given the fact that the concentrations of reactants and products are not changing, why is the word *dynamic* used to describe chemical equilibrium?

**35.** Explain how a person bailing out a row boat with a leak could represent a state of physical equilibrium.

**36.** Does the following equation represent a homogeneous equilibrium or a heterogeneous equilibrium? Explain. your answer.

$$H_2O(s) \rightleftharpoons H_2O(l)$$

**37.** What is an equilibrium position?

**38.** Explain how to write an equilibrium constant expression.

**39.** Why should you pay attention to the physical states of reactants and products when writing equilibrium constant expressions?

**40.** Why does a numerically large $K_{eq}$ mean that the products are favored in an equilibrium system?

**41.** What happens to $K_{eq}$ for an equilibrium system if the equation for the reaction is rewritten in the reverse?

**42.** How can an equilibrium system contain small and unchanging amounts of products yet have large amounts of reactants? What can you say about the relative size of $K_{eq}$ for such an equilibrium?

**43.** A system, which contains only molecules as reactants and products, is at equilibrium. Describe what happens to the concentrations of the reactants and products and what happens to individual reactant and product molecules.

### Mastering Problems

**44.** Write equilibrium constant expressions for these homogeneous equilibria.

**a.** $2N_2H_4(g) + 2NO_2(g) \rightleftharpoons 3N_2(g) + 4H_2O(g)$

**b.** $2NbCl_4(g) \rightleftharpoons NbCl_3(g) + NbCl_5(g)$

**45.** Write equilibrium constant expressions for these heterogeneous equilibria.

**a.** $2NaHCO_3(s) \rightleftharpoons Na_2CO_3(s) + H_2O(g) + CO_2(g)$

**b.** $C_6H_6(l) \rightleftharpoons C_6H_6(g)$

**46.** Heating limestone ($CaCO_3(s)$) forms quicklime ($CaO(s)$) and carbon dioxide gas. Write the equilibrium constant expression for this reversible reaction.

**47.** Suppose you have a cube of pure manganese metal measuring 5.25 cm on each side. You find that the mass of the cube is 1076.6 g. What is the molar concentration of manganese in the cube?

**48.** $K_{eq}$ is 3.63 for the reaction $A + 2B \rightleftharpoons C$. **Table 5** shows the concentrations of the reactants and product in two different reaction mixtures at the same temperature. Determine whether both reactions are at equilibrium.

**Table 5 Concentrations of A, B, and C**

| A (mol/L) | B (mol/L) | C (mol/L) |
|---|---|---|
| 0.500 | 0.621 | 0.700 |
| 0.250 | 0.525 | 0.250 |

**49.** When steam is passed over iron filings, solid iron(III) oxide and gaseous hydrogen are produced in a reversible reaction. Write the balanced chemical equation and the equilibrium constant expression for the reaction, which yields iron(III) oxide and hydrogen gas.

## SECTION 2

### Mastering Concepts

**50.** What is meant by a stress on a reaction at equilibrium?

**51.** How does Le Châtelier's principle describe an equilibrium's response to a stress?

**52.** Why does removing a reactant cause an equilibrium shift to the left?

**53.** When an equilibrium shifts to the right, what happens to each of the following?

**a.** the concentration of the reactants

**b.** the concentration of the products

**54. Carbonated Beverages** Use Le Châtelier's principle to explain how a shift in the equilibrium $H_2CO_3(aq) \rightleftharpoons H_2O(l) + CO_2(g)$ causes a soft drink to go flat when its container is left open.

**55.** How would each of the following changes affect the equilibrium position of the system used to produce methanol from carbon monoxide and hydrogen?

$$CO(g) + 2H_2(g) \rightleftharpoons CH_3OH(g) + \text{heat}$$

**a.** adding CO to the system

**b.** cooling the system

**c.** adding a catalyst to the system

**d.** removing $CH_3OH$ from the system

**e.** decreasing the volume of the system

**56.** Explain how a temperature increase would affect the equilibrium represented by the following equation.

$$PCl_5(g) \rightleftharpoons PCl_3(g) + Cl_2(g) + \text{heat}$$

**57.** A liquid solvent for chlorine is poured into a flask in which the following reaction is at equilibrium: $PCl_5(g) \rightleftharpoons PCl_3(g) + Cl_2(g) + \text{heat}$. How is the equilibrium affected when some of the chlorine gas dissolves?

■ **Figure 22**

**58.** **Figure 22** shows the following endothermic reaction at equilibrium at room temperature.

$$Co(H_2O)_6^{2+}(aq) + 4Cl^-(aq) \rightleftharpoons CoCl_4^{2-}(aq) + 6H_2O(l)$$

Given that $Co(H_2O)_6^{2+}(aq)$ is pink and $CoCl_4^{2-}(aq)$ is blue, what visual change would you expect to see if the flask were placed in an ice bath? Explain.

**59.** For the equilibrium described in Question 58, what visual change would you expect to see if 10 g of solid potassium chloride were added and dissolved? Explain.

**60.** Given two reactions at equilibrium:
**a.** $N_2(g) + 3H_2(g) \rightleftharpoons 2NH_3(g)$
**b.** $H_2(g) + Cl_2(g) \rightleftharpoons 2HCl(g)$,

explain why changing the volume of the systems alters the equilibrium position of **a** but has no effect on **b**.

**61.** Would you expect the numerical value of $K_{eq}$ for the following equilibrium to increase or decrease with increasing temperature? Explain your answer.

$$PCl_5(g) \rightleftharpoons PCl_3(g) + Cl_2(g) + \text{heat}$$

**62.** Explain how you would regulate the pressure to favor the products in the following equilibrium system.

$$MgCO_3(s) \rightleftharpoons MgO(s) + CO_2(g)$$

**63.** Ethylene ($C_2H_4$) reacts with hydrogen to form ethane ($C_2H_6$).

$$C_2H_4(g) + H_2(g) \rightleftharpoons C_2H_6(g) + \text{heat}$$

How would you regulate the temperature of this equilibrium in order to accomplish each of the following?
**a.** increase the yield of ethane
**b.** decrease the concentration of ethylene
**c.** increase the amount of hydrogen in the system

# SECTION 3

## Mastering Concepts

**64.** What does it mean to say that two solutions have a common ion? Give an example.

**65.** Why are compounds such as sodium chloride usually not given $K_{sp}$ values?

**66.** **X-rays** Why is barium sulfate a better choice than barium chloride for adding definition to X-rays? At 26°C, 37.5 g of $BaCl_2$ can be dissolved in 100 mL of water.

■ **Figure 23**

**67.** Explain what is happening in **Figure 23** in terms of $Q_{sp}$ and $K_{sp}$.

**68.** Explain why a common ion lowers the solubility of an ionic compound.

**69.** Describe the solution that results when two solutions are mixed and $Q_{sp}$ is found to equal $K_{sp}$. Does a precipitate form?

## Mastering Problems

**70.** Write the $K_{sp}$ expression for lead chromate ($PbCrO_4$), and calculate its solubility in mol/L. $K_{sp} = 2.3 \times 10^{-13}$.

**71.** At 350°C, $K_{eq} = 1.67 \times 10^{-2}$ for the reversible reaction $2HI(g) \rightleftharpoons H_2(g) + I_2(g)$. What is the concentration of HI at equilibrium if $[H_2]$ is $2.44 \times 10^{-3}M$ and $[I_2]$ is $7.18 \times 10^{-5}M$?

**72.** $K_{sp}$ for scandium fluoride ($ScF_3$) at 298 K is $4.2 \times 10^{-18}$. Write the chemical equation for the solubility equilibrium of scandium fluoride in water. What concentration of $Sc^{3+}$ ions is required to cause a precipitate to form if the fluoride-ion concentration is 0.076*M*?

**73.** Will a precipitate form when 62.6 mL of 0.0322*M* $CaCl_2$ and 31.3 mL of 0.0145*M* NaOH are mixed? Use data from **Table 3**. Explain your logic.

**74.** **Manufacturing** Ethyl acetate ($CH_3COOCH_2CH_3$), a solvent used in making varnishes and lacquers, can be produced by the reaction between ethanol and acetic acid. The equilibrium system is described by the equation

$$CH_3COOH + CH_3CH_2OH \rightleftharpoons CH_3COOCH_2CH_3 + H_2O.$$

Calculate $K_{eq}$ using these equilibrium concentrations: $[CH_3COOCH_2CH_3] = 2.90M$, $[CH_3COOH] = 0.316M$, $[CH_3CH_2OH] = 0.313M$, and $[H_2O] = 0.114M$.

## MIXED REVIEW

**75.** Ethyl acetate ($CH_3COOCH_2CH_3$) is produced in the equilibrium system described by the following equation.

$$CH_3COOH + CH_3CH_2OH \rightleftharpoons CH_3COOCH_2CH_3 + H_2O$$

Why does the removal of water result in the production of more ethyl acetate?

**76.** How would these equilibria be affected by decreasing the temperature?

**a.** $2O_3(g) \rightleftharpoons 3O_2(g) + \text{heat}$

**b.** $\text{heat} + H_2(g) + F_2(g) \rightleftharpoons 2HF(g)$

**77.** How would simultaneously increasing the temperature and volume of the system affect these equilibria?

**a.** $2O_3(g) \rightleftharpoons 3O_2(g) + \text{heat}$

**b.** $\text{heat} + N_2(g) + O_2(g) \rightleftharpoons 2NO(g)$

**78.** The solubility product constant for lead(II) arsenate ($Pb_3(AsO_4)_2$) is $4.0 \times 10^{-36}$ at 298 K. Calculate the molar solubility of the compound at this temperature.

**79.** Evaluate this statement: A low value for $K_{eq}$ means that both the forward and reverse reactions are occurring slowly.

**80. Food Flavoring** Benzaldehyde, known as artificial almond oil, is used in food flavorings. What is the molar concentration of benzaldehyde ($C_7H_6O$) at 298 K, when its density is 1.043 g/mL?

■ **Figure 24**

**81.** In the equilibrium system $N_2O_4(g) \rightleftharpoons 2NO_2(g)$, $N_2O_4$ is colorless and $NO_2$ is reddish-brown. Explain the different colors of the equilibrium system as shown in **Figure 24.**

**82.** Describe the process by which adding potassium hydroxide to a saturated aluminum hydroxide solution reduces the concentration of aluminum ions. Write the solubility equilibrium equation and solubility product constant expression for a saturated aqueous solution of aluminum hydroxide.

**83.** At 298 K, $K_{sp}$ for cadmium iodate ($Cd(IO_3)_2$) equals $2.3 \times 10^{-8}$. What are the molar concentrations of cadmium ions and iodate ions in a saturated solution at 298 K?

## THINK CRITICALLY

**84. Analyze** Suppose that an equilibrium system at a given temperature has a $K_{eq}$ equal to 1.000. Evaluate the possibility that such a system is made up of 50% reactants and 50% products. Explain your answer.

**85. Evaluate** Imagine that you are a chemical engineer designing a production facility for a particular process. The process will utilize a reversible reaction that reaches a state of equilibrium. Analyze the merits of a continuous-flow process or a batch process for such a reaction and determine which is preferable. As a reaction proceeds in a continuous-flow process, reactants are continuously introduced into the reaction chamber and products are continuously removed from the chamber. In a batch process, the reaction chamber is charged with reactants, the reaction is allowed to occur, and the chamber is later emptied of all materials.

**86. Interpret Data** What compound would precipitate first if a 0.500*M* sodium fluoride solution were added gradually to a solution already containing 0.500*M* concentrations of both barium ions and magnesium ions? Use the data in **Table 6.** Write the solubility equilibrium equations and solubility product constant expressions for both compounds. Explain your answer.

**Table 6 Data for Two Compounds**

| Compound | Molar Mass (g/mol) | Solubility at 25°C (g/L) |
|---|---|---|
| $BaF_2$ | 175.33 | 1.1 |
| $MgF_2$ | 62.30 | 0.13 |

**87. Apply** Smelling salts, sometimes used to revive a person who is unconscious, are made of ammonium carbonate. The equation for the endothermic decomposition of ammonium carbonate is as follows.

$$(NH_4)_2CO_3(s) \rightleftharpoons 2NH_3(g) + CO_2(g) + H_2O(g)$$

Would you expect smelling salts to work as well on a cold winter day as on a warm summer day? Explain your answer.

**88. Recognize Cause and Effect** Suppose you have 12.56 g of a mixture made up of sodium chloride and barium chloride. Explain how you could use a precipitation reaction to determine how much of each compound the mixture contains.

**89. Compare and Contrast** Which of the two solids, calcium phosphate or iron(III) phosphate, has the greater molar solubility? $K_{sp}$ ($Ca_3(PO_4)_2$) = $1.2 \times 10^{-29}$; $K_{sp}$ ($FePO_4$) = $1.0 \times 10^{-22}$. Which compound has the greater solubility, expressed in grams per liter?

Chapter Self-Check

## CHALLENGE PROBLEM

**90. Synthesis of Phosgene** Phosgene ($COCl_2$) is a toxic gas that is used in the manufacture of certain dyes, pharmaceuticals, and pesticides. Phosgene can be produced by the reaction between carbon monoxide and chlorine described by the equation $CO(g) + Cl_2(g) \rightleftharpoons COCl_2(g)$. Initially 1.0000 mol CO and 1.0000 mol $Cl_2$ are introduced into a 10.00-L reaction vessel. When equilibrium is established, both of their molar concentrations are found to be 0.0086 mol/L. What is the molar concentration of phosgene at equilibrium? What is $K_{eq}$ for the system?

## CUMULATIVE REVIEW

**91.** Explain the general trend in ionization energy as you go from left to right along Periods 1–5 of the periodic table.

**92.** How are the lengths of covalent bonds related to their strengths?

**93.** How are the chemical bonds in $H_2$, $O_2$, and $N_2$ different?

**94.** How can you tell if a chemical equation is balanced?

**95.** What mass of carbon must burn to produce 4.56 L $CO_2$ gas at STP?

$$C(s) + O_2(g) \rightarrow CO_2(g)$$

**96.** Describe a hydrogen bond. What conditions must exist for a hydrogen bond to form?

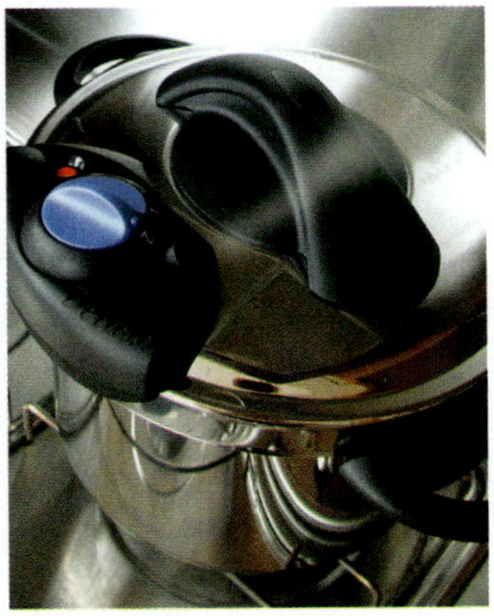

■ **Figure 25**

Marie-Louise Avery/Alamy

**97.** What gas law is exemplified in **Figure 25?** State the law.

**98.** When you reverse a thermochemical equation, why must you change the sign of $\Delta H$?

**99.** What is the sign of the free energy change, $\Delta G°_{system}$, for a spontaneous reaction?

## WRITING IN ▶ Chemistry

**100. A New Compound** Imagine that you are a scientist who has created a unique new liquid. You have named the liquid *yollane,* abbreviated *yo.* Yollane is nontoxic, inexpensive to make, and can dissolve huge volumes of gaseous carbon dioxide in the equilibrium $CO_2(g) \rightleftharpoons CO_2(yo)$, $K_{eq} = 3.4 \times 10^6$. Write a newspaper or magazine article that explains the merits of yollane in combating global warming.

**101. Kidney Stones** Research the role that solubility plays in the formation of kidney stones. Find out what compounds are found in kidney stones and their $K_{sp}$ values. Summarize your findings in a health information flyer.

**102. Hard Water** The presence of magnesium and calcium ions in water makes the water "hard." Explain in terms of solubility why the presence of these ions is often undesirable. Find out what measures can be taken to eliminate them.

## DBQ Document-Based Question

**Reducing Pollution** *Automobile exhausts contain the dangerous pollutants nitrogen monoxide (NO) and carbon monoxide (CO). An alloy catalyst offers a promising way to reduce the amounts of these gases in the atmosphere. When NO and CO are passed over this catalyst, the following equilibrium is established.*

$$2NO(g) + 2CO(g) \rightleftharpoons N_2(g) + 2CO_2(g)$$

*The equilibrium constant is found to vary with temperature as shown in* **Table 7.**

Data obtained from: Worz, et al. 2003. Cluster size-dependent mechanisms of the CO + NO reaction on small Pdn (n < or = 30) clusters on oxide surfaces. *J Am Chem Soc. 125(26): 7964–70.*

**Table 7 $K_{eq}$ v. Temperature**

| 700 K | 800 K | 900 K | 1000 K |
|---|---|---|---|
| $9.10 \times 10^{97}$ | $1.04 \times 10^{66}$ | $4.66 \times 10^{54}$ | $3.27 \times 10^{45}$ |

**103.** Write the equilibrium constant expression for this equilibrium.

**104.** Examine the relationship between $K_{eq}$ and temperature. Use Le Châtelier's principle to deduce whether the forward reaction is exothermic or endothermic.

**105.** Explain how automobile radiators plated with the alloy might help reduce the atmospheric concentrations of NO and CO.

## MULTIPLE CHOICE

1. Which describes a system that has reached chemical equilibrium?
   A. No new product is formed by the forward reaction.
   B. The reverse reaction no longer occurs in the system.
   C. The concentration of reactants in the system is equal to the concentration of products.
   D. The rate at which the forward reaction occurs equals the rate of the reverse reaction.

2. The reaction between persulfate ($S_2O_8^{2-}$) and iodide ($I^-$) ions is often studied in student laboratories because it occurs slowly enough for its rate to be measured:

   $S_2O_8^{2-}(aq) + 2I^-(aq) \rightarrow 2SO_4^{2-}(aq) + I_2(aq)$

   This reaction has been experimentally determined to be first order in $S_2O_8^{2-}$ and first order in $I^-$. Therefore, what is the overall rate law for this reaction?
   A. rate = $k[S_2O_8^{2-}]^2[I^-]$
   B. rate = $k[S_2O_8^{2-}][I^-]$
   C. rate = $k[S_2O_8^{2-}][I^-]^2$
   D. rate = $k[S_2O_8^{2-}]^2[I^-]^2$

*Use the diagrams below to answer Question 3.*

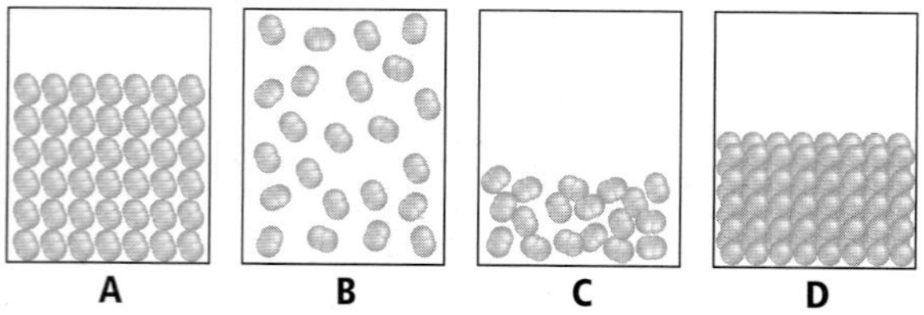


3. Which diagram shows the substance that has the weakest intermolecular forces?
   A. A  C. C
   B. B  D. D

4. Which type of intermolecular force results from a temporary imbalance in the electron density around the nucleus of an atom?
   A. ionic bonds
   B. London dispersion forces
   C. dipole-dipole forces
   D. hydrogen bonds

*Use the table below to answer Questions 5 to 7.*

| Concentration Data for the Equilibrium System $MnCO_3(s) \rightarrow Mn^{2+}(aq) + CO_3^{2-}(aq)$ at 298 K | | | | |
|---|---|---|---|---|
| Trial | $[Mn^{2+}]_0$ (*M*) | $[CO_3^{2-}]_0$ (*M*) | $[Mn^{2+}]_{eq}$ (*M*) | $[CO_3^{2-}]_{eq}$ (*M*) |
| 1 | 0.0000 | 0.00400 | $5.60 \times 10^{-9}$ | $4.00 \times 10^{-3}$ |
| 2 | 0.0100 | 0.0000 | $1.00 \times 10^{-2}$ | $2.24 \times 10^{-9}$ |
| 3 | 0.0000 | 0.0200 | $1.12 \times 10^{-9}$ | $2.00 \times 10^{-2}$ |

5. What is the $K_{sp}$ for $MnCO_3$ at 298 K?
   A. $2.24 \times 10^{-11}$  C. $1.12 \times 10^{-9}$
   B. $4.00 \times 10^{-11}$  D. $5.60 \times 10^{-9}$

6. What is the molar solubility of $MnCO_3$ at 298 K?
   A. $4.73 \times 10^{-6}M$  C. $7.48 \times 10^{-5}M$
   B. $6.32 \times 10^{-2}M$  D. $3.35 \times 10^{-5}M$

7. A 50.0-mL volume of $3.00 \times 10^{-6}M$ $K_2CO_3$ is mixed with 50.0 mL of $MnCl_2$. A precipitate of $MnCO_3$ will form only when the concentration of the $MnCl_2$ solution is greater than which of the following?
   A. $7.47 \times 10^{-6}M$  C. $2.99 \times 10^{-5}M$
   B. $1.49 \times 10^{-5}M$  D. $1.02 \times 10^{-5}M$

8. The kinetic-molecular theory describes the microscopic behavior of gases. One main point of the theory is that within a sample of gas, the frequency of collisions between individual gas particles and between the particles and the walls of their container increases if the sample is compressed. Which gas law states this relationship in mathematical terms?
   A. Gay-Lussac's law
   B. Charles's law
   C. Boyle's law
   D. Avogadro's law

9. $AB(s) + C_2(l) \rightarrow AC(g) + BC(g)$
   Which cannot be predicted about this reaction?
   A. The entropy of the system decreases.
   B. The entropy of the products is higher than that of the reactants.
   C. The change in entropy for this reaction, $\Delta S_{rxn}$, is positive.
   D. The disorder of the system increases.

## SHORT ANSWER

*Use the equation below to answer Questions 10 to 12.*

$$PCl_5 + H_2O \rightarrow HCl + H_3PO_4$$

**10.** Balance this equation, using the smallest whole-number coefficients.

**11.** Identify the mole ratio of water to phosphoric acid.

**12.** Use your balanced chemical equation to show the setup for determining the amount of hydrogen chloride produced when 25.0 g of phosphorus pentachloride is completely consumed.

## EXTENDED RESPONSE

*Use the graph below to answer Questions 13 to 15.*

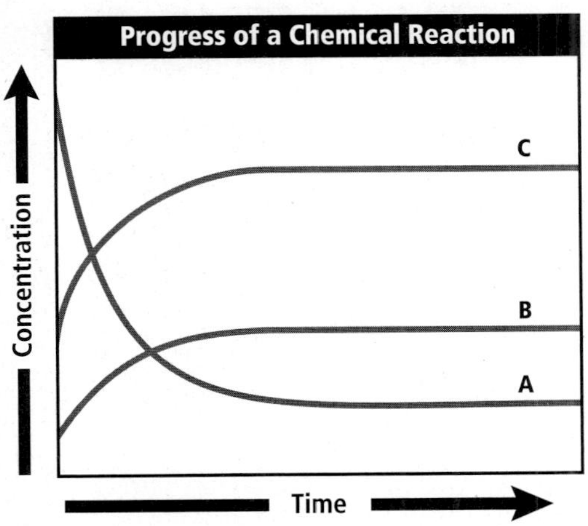


**13.** Describe the shape of the graph when equilibrium has been established.

**14.** Explain why the concentration of reactants is not zero at the end of this reaction.

**15.** Classify the type of chemical reaction that is represented in this graph. How do the data support your conclusion?

## SAT SUBJECT TEST: CHEMISTRY

**16.** The formation of perchloryl fluoride ($ClO_3F$) has an equilibrium constant of $3.42 \times 10^{-9}$ at 298 K.

$$Cl_2(g) + 3O_2(g) + F_2(g) \rightarrow 2ClO_3F(g)$$

At equilibrium, $[Cl_2] = 0.563M$, $[O_2] = 1.01M$, and $[ClO_3F] = 1.47 \times 10^{-5}M$. What is $[F_2]$?

**A.** $9.18 \times 10^{-2}M$
**B.** $3.73 \times 10^{-10}M$
**C.** $1.09 \times 10^{-1}M$
**D.** $6.32 \times 10^{-2}M$
**E.** $6.32 \times 10^{-7}M$

*Use the graph below to answer Questions 17 and 18.*

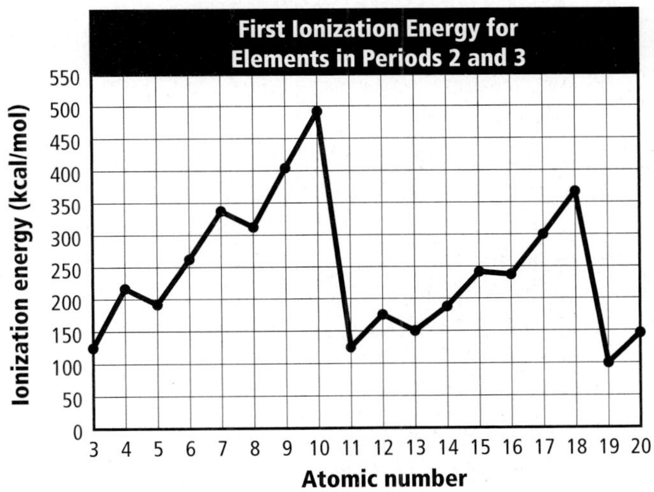


**17.** Which family of elements tends to have the lowest ionization energy in its period?
**A.** representative elements
**B.** transition elements
**C.** alkali elements
**D.** alkaline earth elements
**E.** halogens

**18.** Using the graph, what is the approximate ionization energy of the element with atomic number 7?
**A.** 300 kcal/mol
**B.** 310 kcal/mol
**C.** 325 kcal/mol
**D.** 340 kcal/mol
**E.** 390 kcal/mol

| NEED EXTRA HELP? | | | | | | | | | | | | | | | | | | |
|---|---|---|---|---|---|---|---|---|---|---|---|---|---|---|---|---|---|---|
| If You Missed Question . . . | 1 | 2 | 3 | 4 | 5 | 6 | 7 | 8 | 9 | 10 | 11 | 12 | 13 | 14 | 15 | 16 | 17 | 18 |
| Review Section . . . | 17.1 | 16.3 | 12.2 | 12.2 | 17.3 | 17.3 | 17.3 | 13.1 | 15.5 | 9.1 | 11.1 | 11.2 | 17.4 | 17.1 | 9.2 | 17.3 | 6.3 | 6.3 |

CHAPTER 18

# Acids and Bases

**BIGIDEA** Acids and bases can be defined in terms of hydrogen ions and hydroxide ions or in terms of electron pairs.

## SECTIONS

Chm 7

1 Introduction to Acids and Bases

2 Strengths of Acids and Bases

3 Hydrogen Ions and pH

4 Neutralization

## LaunchLAB

iLab Station

### What is in your cupboards?

You can learn something about the properties of products in your household by testing them with strips of paper called litmus paper. Can you separate household products into two groups? In this lab, you will determine whether a substance is an acid or a base.

**FOLDABLES® Study Organizer**

### Acids and Bases

Make a two-tab book. Label it as shown. Use it to compare the main models of acids and bases.

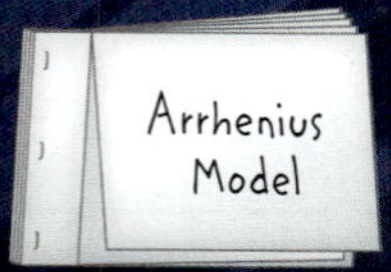

Tim Fuller; (bkgd)Gk Hart/Vikki Hart/Getty Images

Go **online!** MHEonline.com

The optimal pH for an aquarium depends on the types of organisms in it. A pH of 8.2 is generally accepted as the average pH of seawater, the habitat for the fish in the photo. On the other hand, South American cichlids, which are freshwater fish, require a pH range of 6.4 to 7.0. African cichlids, also freshwater fish, thrive in water with a pH between 8.0 and 9.2.

SECTION 1

# Introduction to Acids and Bases

**MAINIDEA** Different models help describe the behavior of acids and bases.

### Essential Questions

- What are the physical and chemical properties of acids and bases?
- How are solutions classified as acidic, basic, or neutral?
- How do the Arrhenius, Brønsted-Lowry, and Lewis models of acids and bases compare?

### Review Vocabulary

**Lewis structure:** a model that uses electron-dot structures to show how electrons are arranged in molecules

### New Vocabulary

acidic solution
basic solution
Arrhenius model
Brønsted-Lowry model
conjugate acid
conjugate base
conjugate acid-base pair
amphoteric
Lewis model

**CHEM 4 YOU** **You might not realize it, but acids and bases are two of the most common classifications of substances. You can recognize them by the tart taste of some of your favorite beverages and by the pungent odor of ammonia in some household cleaners.**

## Properties of Acids and Bases

When ants sense danger to their ant colony, they emit a substance called formic acid that alerts the entire colony. Acids dissolved in rainwater hollow out enormous limestone caverns and destroy valuable buildings and statues over time. Acids flavor many of the beverages and foods you like, and an acid in your stomach helps digest what you eat. Bases also play a role in your life. The soap you use and the antacid tablet you might take for an upset stomach are bases. Many household products, such as those you used in the Launch Lab, are acids and bases.

**Physical properties** You are probably already familiar with some of the physical properties of acids and bases. For example, you might know that acidic solutions taste sour. Carbonic and phosphoric acids give many carbonated beverages their sharp taste; citric and ascorbic acids give lemons and grapefruit their tartness; and acetic acid makes vinegar taste sour.

You might also know that basic solutions taste bitter and feel slippery. Think about how a bar of soap becomes slippery when it gets wet. You should never attempt to identify an acid or a base, or any other substance in the laboratory, by its taste or feel. **Figure 1** shows two plants growing in different soils. One grows best in acidic soil, sometimes called "sour" soil. The other thrives in basic, or alkaline, soil.

**Figure 1** Rhododendrons flourish in rich, moist soil that is moderately acidic, whereas sempervivum, commonly called hen and chicks, grows best in drier, slightly basic soil.

Acids turn blue litmus red.

Bases turn red litmus blue.

**Figure 2** The strong acid hydrochloric acid (HCl), also called muriatic acid, is used to clean bricks and concrete. The strong base sodium hydroxide (NaOH) can clear clogged drains.

**Electrical conductivity** Another physical property of acid and base solutions is the ability to conduct electricity. Pure water is a nonconductor of electricity, but the addition of an acid or base produces ions that cause the resulting solution to become a conductor.

**Chemical properties** You might have already identified acids and bases by their reaction with litmus paper. Acids can also be identified by their reactions with some metals and metal carbonates.

**Reactions with litmus** Litmus is one of the dyes commonly used to distinguish solutions of acids and bases, as shown in **Figure 2.** Aqueous solutions of acids cause blue litmus paper to turn red. Aqueous solutions of bases cause red litmus paper to turn blue.

**Reactions with metals and metal carbonates** Magnesium and zinc react with aqueous solutions of acids to produce hydrogen gas. The reaction between zinc and hydrochloric acid is described by the following equation.

$$Zn(s) + 2HCl(aq) \rightarrow ZnCl_2(aq) + H_2(g)$$

Metal carbonates and hydrogen carbonates also react with aqueous solutions of acids to produce carbon dioxide ($CO_2$) gas. When vinegar is added to baking soda, a foaming reaction occurs between acetic acid ($HC_2H_3O_2$) dissolved in the vinegar, and sodium hydrogen carbonate ($NaHCO_3$). The production of $CO_2$ gas accounts for the bubbling.

$$NaHCO_3(s) + HC_2H_3O_2(aq) \rightarrow NaC_2H_3O_2(aq) + H_2O(l) + CO_2(g)$$

Geologists identify rocks as limestone (primarily $CaCO_3$) by using a hydrochloric acid solution. If a few drops of the acid produce bubbles of carbon dioxide, the rock contains limestone.

## PRACTICE Problems

Do additional problems. Online Practice

1. Write balanced equations for the reactions between the following.
   - **a.** aluminum and sulfuric acid
   - **b.** calcium carbonate and hydrobromic acid
2. **Challenge** Write the net ionic equation for the reaction in Question 1b.

■ **Figure 3** Note how [H+] and [OH−] change simultaneously. As [H+] decreases to the right, [OH−] increases to the right.

**Identify** *the point in the diagram at which the two ion concentrations are equal.*

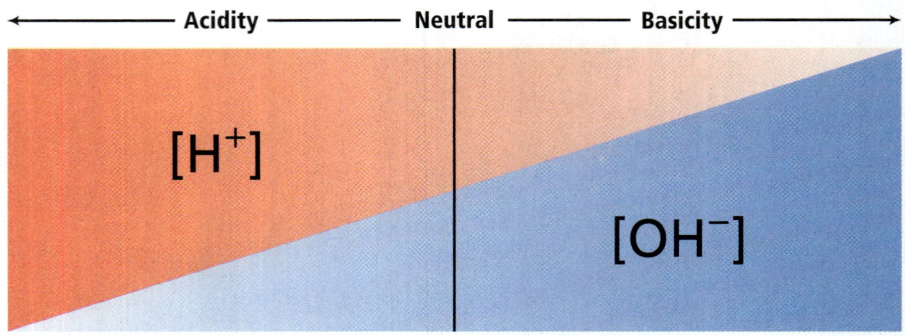


**Hydronium and hydroxide ions** All water solutions contain hydrogen ions ($H^+$) and hydroxide ions ($OH^-$). The relative amounts of the two ions determine whether an aqueous solution is acidic, basic, or neutral. Neutral solutions are neither acidic nor basic.

An **acidic solution** contains more hydrogen ions than hydroxide ions. A **basic solution** contains more hydroxide ions than hydrogen ions. A neutral solution contains equal concentrations of hydrogen ions and hydroxide ions. **Figure 3** illustrates these relationships. **Figure 4** describes how scientists developed an understanding of acids and bases.

Pure water produces equal numbers of $H^+$ ions and $OH^-$ ions in a process called self-ionization, in which water molecules react to form a hydronium ion ($H_3O^+$) and a hydroxide ion.

$$\underbrace{H_2O(l) + H_2O(l)}_{\text{Water molecules}} \rightleftharpoons \underbrace{H_3O^+(aq)}_{\text{Hydronium ion}} + \underbrace{OH^-(aq)}_{\text{Hydroxide ion}}$$

The hydronium ion is a hydrogen ion that has a water molecule attached to it by a covalent bond. The symbols $H^+$ and $H_3O^+$ can be used interchangeably, as this simplified self-ionization equation shows.

$$H_2O(l) \rightleftharpoons H^+(aq) + OH^-(aq)$$

■ **Figure 4**

## History of Acids and Bases

Current understanding of the structure and behavior of acids and bases is based on the contributions of chemists, biologists, environmental scientists, and inventors over the past 150 years.

**1865** The introduction of an antiseptic spray containing carbolic acid marks the beginning of modern antiseptic surgery.

**1869** Nucleic acids are discovered in cell nuclei. DNA and RNA are examples of nucleic acids.

1870

**1883** Svante Arrhenius proposes that acids produce hydrogen ions ($H^+$) and bases produce hydroxide ions ($OH^-$) when dissolved in water.

1890

**1909** The development of the pH scale allows scientists to define the acidity of a substance.

1910

**1923** Scientists expand and refine the definition of acids and bases, producing the definitions currently in use.

## The Arrhenius Model

If pure water itself is neutral, how does an aqueous solution become acidic or basic? The first person to answer this question was Swedish chemist Svante Arrhenius, who in 1883 proposed what is now called the Arrhenius model of acids and bases. The **Arrhenius model** states that an acid is a substance that contains hydrogen and ionizes to produce hydrogen ions in aqueous solution. A base is a substance that contains a hydroxide group and dissociates to produce a hydroxide ion in aqueous solution.

**Arrhenius acids and bases** As an example of the Arrhenius model of acids and bases, consider what happens when hydrogen chloride gas dissolves in water. HCl molecules ionize to form $H^+$ ions, which make the solution acidic.

$$HCl(g) \rightarrow H^+(aq) + Cl^-(aq)$$

When the ionic compound sodium hydroxide (NaOH) dissolves in water, it dissociates to produce $OH^-$ ions, which make the solution basic.

$$NaOH(s) \rightarrow Na^+(aq) + OH^-(aq)$$

Although the Arrhenius model is useful in explaining many acidic and basic solutions, it has some shortcomings. For example, ammonia ($NH_3$) and sodium carbonate ($Na_2CO_3$) do not contain a hydroxide group, yet both substances produce hydroxide ions in solution and are well-known bases. Sodium carbonate is the compound that causes the alkalinity of Lake Natron, Tanzania, which is shown in **Figure 5.** Clearly, a model that includes all bases is needed.

**Figure 5** Lake Natron in Africa's Great Rift Valley is a naturally basic body of water. Water, laden with dissolved sodium carbonate from surrounding volcanic rocks, drains into the lake but finds no outlet. Evaporation concentrates the mineral leaving a white crust on the surface and strongly alkaline water.

**FOLDABLES®** Incorporate information from this section into your Foldable.

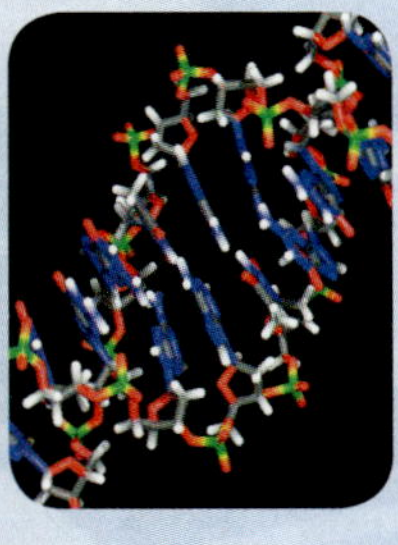

**1933–34** Scientists develop portable pH meters.

**1953** James Watson, Francis Crick, and Rosalind Franklin study the nucleic acid DNA, laying the framework for today's biotechnology industry.

**1963** Scientists discover acid rain in North America. pH measurements show polluted rain to be 100 times more acidic than unpolluted rain.

**1980s** Silicon-chip pH meters have no glass component. They are now widely used in the food, cosmetic, and pharmaceutical industries.

**2005** Scientists develop super-acids, which are more acidic than 100% sulfuric acid. Applications include producing strong plastics and high-octane gasoline.

**2010** A team of scientists at the National Institute of Standards and Technology (NIST) develops a technique to determine how a sudden change in acidity of a nanoparticle solution affects its stability. Nanoparticle production could be tailored to target tumor cells with a different acidity than that of normal cells.

1950 1970 1990 2010

## The Brønsted-Lowry Model

Danish chemist Johannes Brønsted (1879–1947) and English chemist Thomas Lowry (1843–1909) proposed a more inclusive model of acids and bases—a model that focuses on the hydrogen ion ($H^+$). In the **Brønsted-Lowry model** of acids and bases, an acid is a hydrogen-ion donor. A base is a hydrogen-ion acceptor.

**Hydrogen ion donors and acceptors** The symbols X and Y represent nonmetallic elements or negative polyatomic ions. Thus, the general formula for an acid can be represented as HX or HY. When a molecule of acid HX dissolves in water, it donates a $H^+$ ion to a water molecule. The water molecule acts as a base and accepts the $H^+$ ion.

$$HX(aq) + H_2O(l) \rightleftharpoons H_3O^+(aq) + X^-(aq)$$

Upon accepting the $H^+$ ion, the water molecule becomes an acid, $H_3O^+$. The hydronium ion ($H_3O^+$) is an acid because it has an extra $H^+$ ion that it can donate. Upon donating its $H^+$ ion, the acid HX becomes a base, $X^-$. $X^-$ is a base because it has a negative charge and can readily accept a positive hydrogen ion. Thus, an acid-base reaction in the reverse direction can occur. The acid $H_3O^+$ can react with the base $X^-$ to form water and HX, establishing the following equilibrium.

$$\underset{\text{Acid}}{HX(aq)} + \underset{\text{Base}}{H_2O(l)} \rightleftharpoons \underset{\text{Conjugate acid}}{H_3O^+(aq)} + \underset{\text{Conjugate base}}{X^-(aq)}$$

**Conjugate acids and bases** The forward reaction is the reaction of an acid and a base. The reverse reaction is also the reaction of an acid and a base. The acid and base that react in the reverse reaction are identified under the equation as a conjugate acid and a conjugate base. A **conjugate acid** is the species produced when a base accepts a hydrogen ion. The base $H_2O$ accepts a hydrogen ion from the acid HX and becomes the conjugate acid $H_3O^+$. A **conjugate base** is the species that results when an acid donates a hydrogen ion. The acid HX donates its hydrogen ion and becomes the conjugate base $X^-$. In the reaction shown above, the hydronium ion ($H_3O^+$) is the conjugate acid of the base $H_2O$. The $X^-$ ion is the conjugate base of the acid HX. Brønsted-Lowry interactions involve conjugate acid-base pairs. A **conjugate acid-base pair** consists of two substances related to each other by the donating and accepting of a single hydrogen ion.

An analogy for conjugate acid-base pairs is shown in **Figure 6.** When the father has the ball in his hand, he is an acid. He throws the ball (a hydrogen ion) to his son. Now his son is the acid because he has the ball (a hydrogen ion) to give away. The father is now a base because he is available to accept the ball (a hydrogen ion). The father is the acid and the son is the base in the forward reaction. In the reverse reaction, the son has the ball and is the conjugate acid while the father is the conjugate base.

**Figure 6** When a father throws the ball to his son, the father is like a Brønsted-Lowry acid and the son is like a base. After the son catches the ball (hydrogen ion), he becomes like a conjugate acid.

☑ READING CHECK **Explain** how the ion $HCO_3^-$ can be both an acid and a base

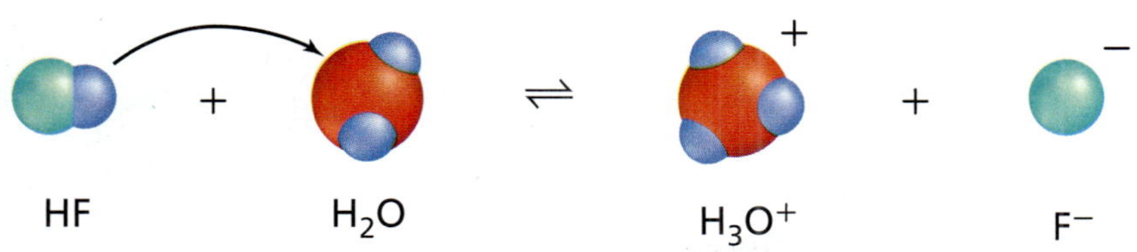


■ **Figure 7** Hydrogen fluoride donates a hydrogen ion to a water molecule, so hydrogen fluoride is an acid.
**Decide** *which species is the conjugate base of hydrogen fluoride.*

**Hydrogen fluoride—a Brønsted-Lowry acid** Consider the equation for the ionization of hydrogen fluoride (HF) in water, shown in **Figure 7.** What are the conjugate acid-base pairs? Hydrogen fluoride, the acid in the forward reaction, produces its conjugate base $F^-$, the base in the reverse reaction. Water, the base in the forward reaction, produces its conjugate acid $H_3O^+$, the acid in the reverse reaction.

$$HF(aq) + H_2O(l) \leftrightharpoons H_3O^+(aq) + F^-(aq)$$

Acid — Base — Conjugate acid — Conjugate base

Hydrogen fluoride is used to manufacture a variety of fluorine-containing compounds, such as the nonstick coating on the kitchenware shown in **Figure 8.** It is an acid according to both the Arrhenius and Brønsted-Lowry definitions.

**Ammonia—a Brønsted-Lowry base** All of the acids and bases that fit the Arrhenius definition of acids and bases also fit the Brønsted-Lowry definition. But some other substances that lack a hydroxide group and, therefore, cannot be considered bases according to the Arrhenius definition can be classified as bases according to the Brønsted-Lowry model. One example is ammonia ($NH_3$).When ammonia dissolves in water, water is a Brønsted-Lowry acid in the forward reaction. Because the $NH_3$ molecule accepts a $H^+$ ion to form the ammonium ion ($NH_4^+$), ammonia is a Brønsted-Lowry base in the forward reaction.

$$NH_3(aq) + H_2O(l) \leftrightharpoons NH_4^+(aq) + OH^-(aq)$$

Base — Acid — Conjugate acid — Conjugate base

In the reverse reaction, the ammonium ion ($NH_4^+$) gives up a $H^+$ ion to form the molecule ammonia and thus acts as a Brønsted-Lowry acid. The ammonium ion is the conjugate acid of the base ammonia. The hydroxide ion accepts a $H^+$ ion to form a water molecule and is thus a Brønsted-Lowry base. The hydroxide ion is the conjugate base of the acid water.

**Water—a Brønsted-Lowry acid and base** Recall that when HF dissolves in water, water acts a base; when $NH_3$ dissolves in water, water acts as an acid. Depending on what other substances are in the solution, water can act as either an acid or a base. Water and other substances that can act as both acids and bases are said to be **amphoteric.**

**VOCABULARY**

**WORD ORIGIN**

**Conjugate**
***con–*** prefix; from Latin, meaning *with* or *together*
***jugare,*** verb; from Latin, meaning *to join*

■ **Figure 8** To make the smooth, nonstick surface of this kitchenware, hydrogen fluoride is reacted with organic compounds called hydrocarbons to substitute fluorine atoms for hydrogen atoms.

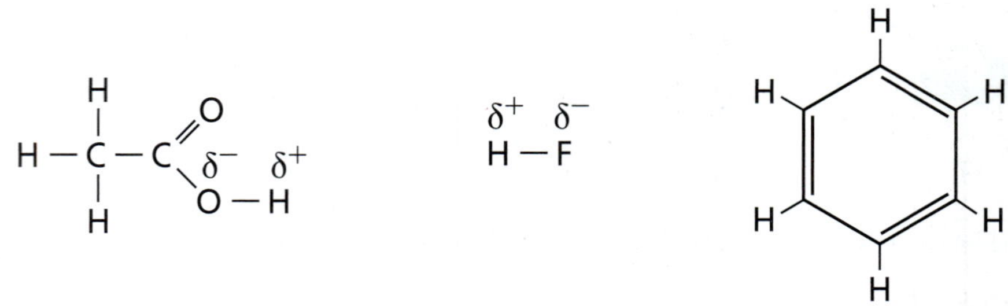

Acetic acid    Hydrogen fluoride    Benzene

■ **Figure 9** Whether a hydrogen is ionizable depends on the polarity of its bond. In acetic acid, oxygen is more electronegative than hydrogen. The bond between oxygen and hydrogen is polar, so the hydrogen atom can ionize in solution. In hydrogen fluoride, fluorine is highly electronegative, so HF is an acid in solution. In benzene, there is little electronegativity difference between the carbon and hydrogen atoms, so benzene is not an acid.

## PRACTICE Problems

Do additional problems. Online Practice

**3.** Identify the conjugate acid-base pairs in each reaction.

**a.** $NH_4^+(aq) + OH^-(aq) \rightleftharpoons NH_3(aq) + H_2O(l)$

**b.** $HBr(aq) + H_2O(l) \rightleftharpoons H_3O^+(aq) + Br^-(aq)$

**c.** $CO_3^{2-}(aq) + H_2O(l) \rightleftharpoons HCO_3^-(aq) + OH^-(aq)$

**4. Challenge** The products of an acid-base reaction are $H_3O^+$ and $SO_4^{2-}$. Write a balanced equation for the reaction and identify the conjugate acid-base pairs.

## Monoprotic and Polyprotic Acids

From the chemical formulas of HCl and HF, you know that each acid has one hydrogen ion per molecule. An acid that can donate only one hydrogen ion is called a monoprotic acid. Other monoprotic acids are perchloric acid ($HClO_4$), nitric acid ($HNO_3$), hydrobromic acid (HBr), and acetic acid ($CH_3COOH$). Because acetic acid is a monoprotic acid, its formula is often written $HC_2H_3O_2$ to emphasize the fact that only one of the four hydrogen atoms in the molecule is ionizable.

**Ionizable hydrogen atoms** The difference between acetic acid's ionizable hydrogen atom and the other three hydrogen atoms is that the ionizable atom is bonded to the element oxygen, which is more electronegative than hydrogen. The difference in electronegativity makes the bond between oxygen and hydrogen polar. The structure of acetic acid is shown in **Figure 9,** along with structures of the acid HF and the nonacid benzene ($C_6H_6$). The hydrogen atom in hydrogen fluorine is bonded to the highly electronegative fluorine atom, so the hydrogen-fluorine bond is polar and the fluorine atom is ionizable to a certain extent. However, the hydrogen atoms in benzene are each bonded to a carbon atom. Carbon atoms have about the same electronegativity as hydrogen. These bonds are nonpolar, so benzene is not an acid.

Some acids donate more than one hydrogen ion. For example, sulfuric acid ($H_2SO_4$) and carbonic acid ($H_2CO_3$) can donate two hydrogen ions. In each compound, both hydrogen atoms are attached to oxygen atoms by polar bonds. Acids that contain two ionizable hydrogen atoms per molecule are called diprotic acids. Phosphoric acid ($H_3PO_4$) and boric acid ($H_3BO_3$) contain three ionizable hydrogen atoms per molecule. Acids with three hydrogen ions to donate are called triprotic acids. The term *polyprotic acid* can be used for any acid that has more than one ionizable hydrogen atom.

**Table 1** Some Common Acids and Their Conjugate Bases

| Acid | | Congugate Base | |
|---|---|---|---|
| Name | Formula | Name | Formula |
| Hydrochloric acid | $HCl$ | Chloride ion | $Cl^-$ |
| Nitric acid | $HNO_3$ | Nitrate ion | $NO_3^-$ |
| Sulfuric acid | $H_2SO_4$ | Hydrogen sulfate ion | $HSO_4^-$ |
| Hydrogen sulfate ion | $HSO_4^-$ | Sulfate ion | $SO_4^{2-}$ |
| Hydrofluoric acid | $HF$ | Fluoride ion | $F^-$ |
| Hydrocyanic acid | $HCN$ | Cyanide | $CN^-$ |
| Acetic acid | $HC_2H_3O_2$ | Acetate ion | $C_2H_3O_2^-$ |
| Phosphoric acid | $H_3PO_4$ | Dihydrogen phosphate ion | $H_2PO_4^-$ |
| Dihydrogen phosphate ion | $H_2PO_4^-$ | Hydrogen phosphate ion | $HPO_4^{2-}$ |
| Hydrogen phosphate ion | $HPO_4^{2-}$ | Phosphate ion | $PO_4^{3-}$ |
| Carbonic acid | $H_2CO_3$ | Hydrogen carbonate ion | $HCO_3^-$ |
| Hydrogen carbonate ion | $HCO_3^-$ | Carbonate ion | $CO_3^{2-}$ |

All polyprotic acids ionize in steps. The three ionizations of phosphoric acid are described by the following equations.

$$H_3PO_4(aq) + H_2O(l) \rightleftharpoons H_3O^+(aq) + H_2PO_4^-(aq)$$

$$H_2PO_4^-(aq) + H_2O(l) \rightleftharpoons H_3O^+(aq) + HPO_4^{2-}(aq)$$

$$HPO_4^{2-}(aq) + H_2O(l) \rightleftharpoons H_3O^+(aq) + PO_4^{3-}(aq)$$

**Table 1** shows some common monoprotic and polyprotic acids.

## The Lewis Model

Notice that all substances classified as acids and bases by the Arrhenius model are classified as acids and bases by the Brønsted-Lowry model. In addition, some substances *not* classified as bases by the Arrhenius model *are* classified as bases by the Brønsted-Lowry model.

Perhaps you will not be surprised to learn that an even more general model of acids and bases was proposed by American chemist G. N. Lewis (1875–1946). Recall that Lewis developed the electron-pair theory of chemical bonding and introduced Lewis structures to keep track of the electrons in atoms and molecules. He applied his electron-pair theory of chemical bonding to acid-base reactions. Lewis proposed that an acid is an ion or molecule with a vacant atomic orbital that can accept (share) an electron pair. A base is an ion or molecule with a lone electron pair that it can donate (share). According to the **Lewis model,** a Lewis acid is an electron-pair acceptor and a Lewis base is an electron-pair donor. Note that the Lewis model includes all the substances classified as Brønsted-Lowry acids and bases and many more.

**FOLDABLES®**
Incorporate information from this section into your Foldable.

**Electron pair donors and acceptors** Consider the reaction between a hydrogen ion ($H^+$) and a fluoride ion ($F^-$) to form a hydrogen fluoride (HF) molecule. The role of the electron pair is illustrated through the following Lewis structures.

**VOCABULARY**

**ACADEMIC VOCABULARY**

**Conform**
to be similar or identical
*Their behavior conforms to the expectations of the community.*

In this reaction, the $H^+$ ion is the Lewis acid. Its vacant 1s orbital accepts an electron pair from the $F^-$ ion. The fluoride ion is the Lewis base. It donates a lone electron pair to form the hydrogen-fluorine bond in HF. Note that this reaction also conforms to the Brønsted-Lowry model of acids and bases because $H^+$ can be considered a hydrogen-ion donor and $F^-$ a hydrogen-ion acceptor.

It might surprise you to learn that the reaction of gaseous boron trifluoride ($BF_3$) with gaseous ammonia ($NH_3$) to form $BF_3NH_3$ is a Lewis acid-base reaction.

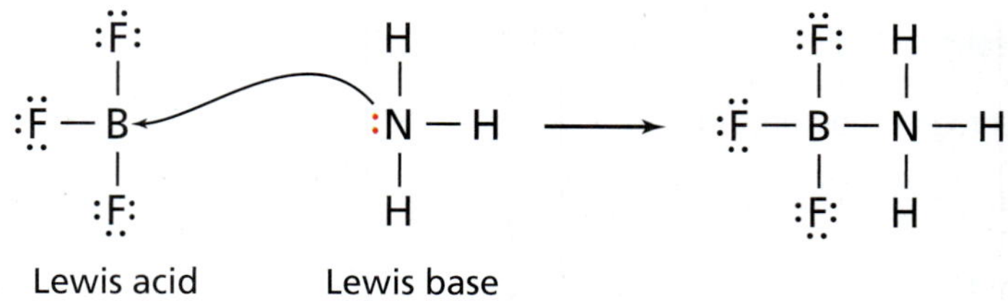


Recall that the boron atom in $BF_3$ has six valence electrons, so a vacant orbital can accept an electron pair from a Lewis base.

Another Lewis acid-base reaction occurs when gaseous sulfur trioxide ($SO_3$) is brought into contact with solid magnesium oxide (MgO).

$$SO_3(g) + MgO(s) \rightarrow MgSO_4(s)$$

The acid-base part of the reaction involves sulfur trioxide ($SO_3$) and the oxide ion ($O^{2-}$) of magnesium oxide. The product is the sulfate ion.

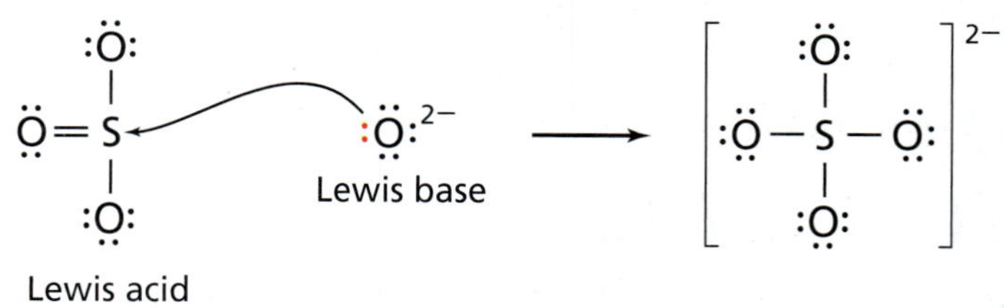


Note that the $SO_3$ molecule, a Lewis acid, accepts an electron pair from the $O^{2-}$ ion, a Lewis base. The Arrhenius, Brønsted-Lowry, and Lewis acid-base models are summarized in **Table 2.**

Explore **acids and bases with an interactive table.** Concepts In Motion

**Table 2 Three Models for Acids and Bases**

| Model | Acid Definition | Base Definition |
|---|---|---|
| Arrhenius | $H^+$ producer | $OH^-$ producer |
| Brønsted-Lowry | $H^+$ donor | $H^+$ acceptor |
| Lewis | electron-pair acceptor | electron-pair donor |

The reaction of $SO_3$ and MgO is important because it produces magnesium sulfate, a salt that forms the heptahydrate known as Epsom salt ($MgSO_4 \cdot 7H_2O$). Epsom salt has many uses, including soothing sore muscles and acting as a plant nutrient. The reaction to form magnesium sulfate also has environmental applications. When MgO is injected into the flue gases of coal-fired power plants, such as the one shown in **Figure 10,** it reacts with and removes $SO_3$. If $SO_3$ is allowed to enter the atmosphere, it can combine with water in the air to form sulfuric acid, which falls to Earth as acid precipitation.

**Connection to Earth Science** **Anhydrides** Like the $SO_3$ molecules you have been reading about, carbon dioxide gas molecules in the air also combine with water molecules in precipitation to form an acid called carbonic acid ($H_2CO_3$). When the acidic rainwater reaches the ground, some sinks into the soil and reaches limestone bedrock, where it slowly dissolves the limestone.

Over the course of thousands of years, the dissolution of limestone creates huge underground caverns. Within a cavern, groundwater might drip from the ceiling and deposit some of the dissolved limestone. Deposits shaped like icicles that form on the ceiling are called stalactites. Rounded masses rising from the floor are called stalagmites.

The formation of caverns occurs because carbon dioxide is an acid anhydride. An acid anhydride is an oxide that can combine with water to form an acid. Other oxides combine with water to form bases. For example, calcium oxide (CaO, lime) forms the base calcium hydroxide $Ca(OH)_2$. In general, oxides of metallic elements form bases; oxides of nonmetals form acids.

**Figure 10** Sulfur trioxide, a waste gas from the burning of coal, can be removed from smokestack gases by combining it with magnesium oxide in a Lewis acid-base reaction. Note that while there is a good deal of steam coming from the cooling towers, there is little visible smoke from the smokestack.



## SECTION 1 REVIEW

Section Self-Check

### Section Summary

- The concentrations of hydrogen ions and hydroxide ions determine whether an aqueous solution is acidic, basic, or neutral.
- An Arrhenius acid must contain an ionizable hydrogen atom. An Arrhenius base must contain an ionizable hydroxide group.
- A Brønsted-Lowry acid is a hydrogen ion donor. A Brønsted-Lowry base is a hydrogen ion acceptor.
- A Lewis acid accepts an electron pair. A Lewis base donates an electron pair.

5. **MAIN IDEA Explain** why many Lewis acids and bases are not classified as Arrhenius or Brønsted-Lowry acids and bases.
6. **Compare** the physical and chemical properties of acids and bases.
7. **Explain** how the concentrations of hydrogen ions and hydroxide ions determine whether a solution is acidic, basic, or neutral.
8. **Explain** why many compounds that contain one or more hydrogen atoms are not classified as Arrhenius acids.
9. **Identify** the conjugate acid-base pairs in the following equation.

$$HNO_2 + H_2O \rightleftharpoons NO_2^- + H_3O^+$$

10. **Write** the Lewis structure for phosphorus trichloride ($PCl_3$). Is $PCl_3$ a Lewis acid, a Lewis base, or neither?
11. **Interpret Scientific Illustrations** In the accompanying structural formula, identify any hydrogen atoms that are likely to be ionizable.

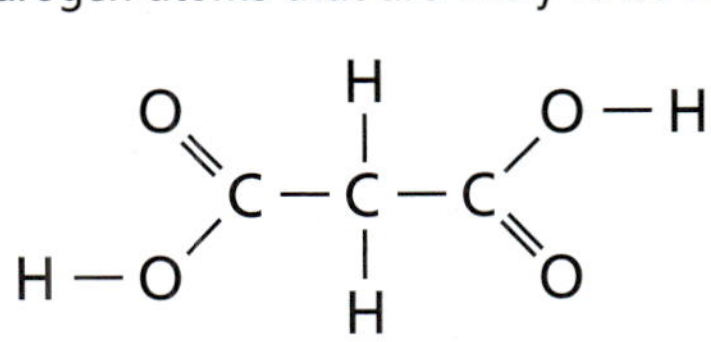

SECTION 2

# Strengths of Acids and Bases

**MAINIDEA** **In solution, strong acids and bases ionize completely, but weak acids and bases ionize only partially.**

### Essential Questions

- How is the strength of an acid or base related to its degree of ionization?
- How does the strength of a weak acid compare with the strength of its conjugate base?
- What is the relationship between the strengths of acids and bases and the values of their ionization constants?

### Review Vocabulary

**electrolyte:** an ionic compound whose aqueous solution conducts an electric current

### New Vocabulary

strong acid
weak acid
acid ionization constant
strong base
weak base
base ionization constant

**CHEM 4 YOU** **The success of a pass in a football game depends on the passer and the receiver. How ready is the passer to pass the ball? How ready is the receiver to receive the ball? Similarly, in acid and base reactions, the progress of a reaction depends on how readily the acid donates a hydrogen ion and how readily the base accepts a hydrogen ion.**

## Strengths of Acids

One of the properties of acidic and basic solutions is that they conduct electricity. What can electrical conductivity tell you about the hydrogen ions and hydroxide ions in these aqueous solutions?

Suppose you test the electrical conductivities of 0.10$M$ aqueous solutions of hydrochloric acid and acetic acid. The glow of the bulb in **Figure 11** indicates that the solution conducts electricity. However, if you compare the brightness of the bulb connected to the HCl solution in **Figure 11** with that of the bulb connected to the $HC_2H_3O_2$ solution in **Figure 12,** you should notice a difference. The 0.10$M$ HCl solution conducts electricity better than the 0.10$M$ $HC_2H_3O_2$ solution. Why is this true if the concentrations of the two acids are both 0.10$M$?

**Strong acids** The answer is that ions carry electric current through the solution and all the HCl molecules contained in the solution are ionized completely into hydronium ions and chloride ions. Acids that ionize completely are called **strong acids.** Because strong acids produce the maximum number of ions, they are good conductors of electricity.

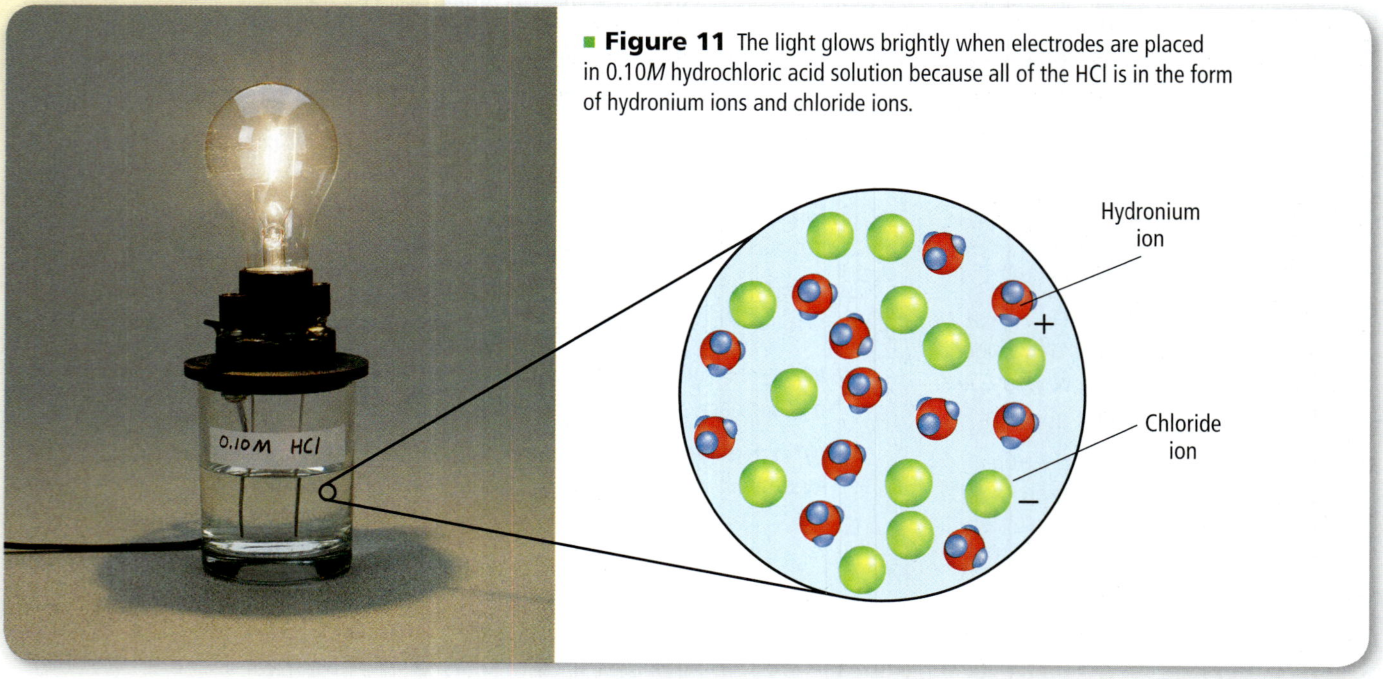


■ **Figure 11** The light glows brightly when electrodes are placed in 0.10$M$ hydrochloric acid solution because all of the HCl is in the form of hydronium ions and chloride ions.

Matt Meadows

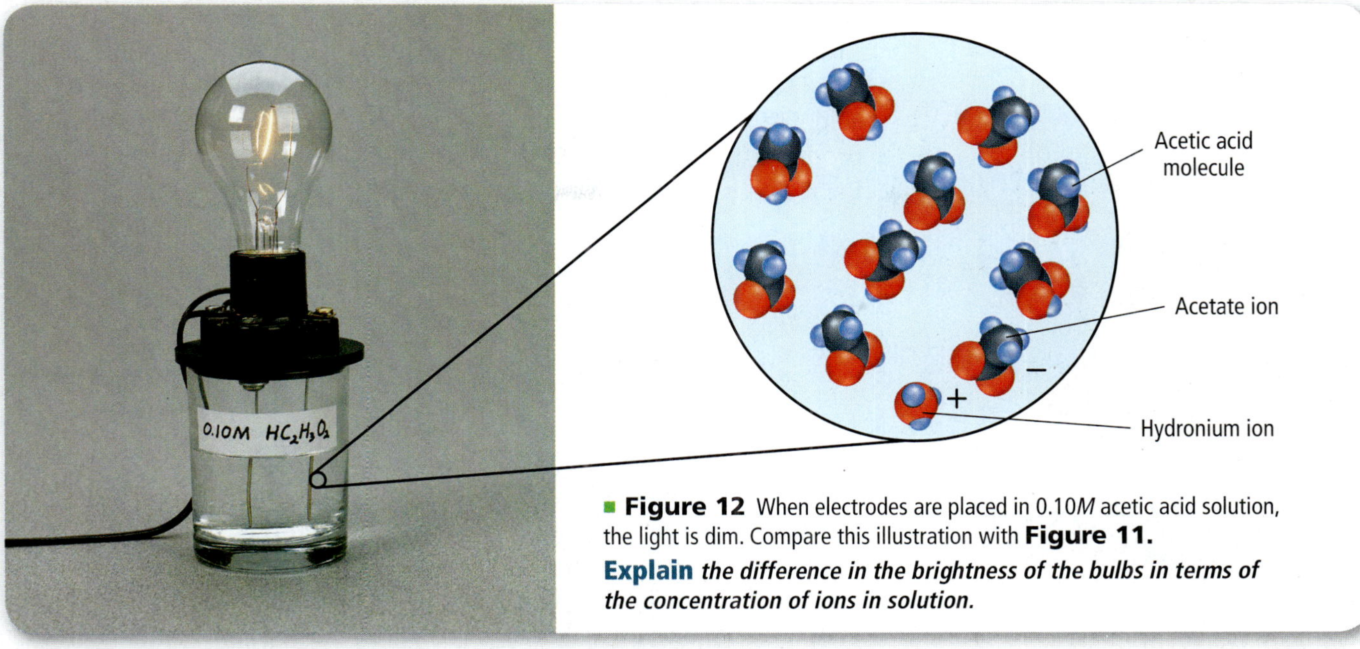


■ **Figure 12** When electrodes are placed in 0.10*M* acetic acid solution, the light is dim. Compare this illustration with **Figure 11.**
**Explain** *the difference in the brightness of the bulbs in terms of the concentration of ions in solution.*

The ionization of hydrochloric acid in water can be represented by the following equation, which has a single arrow pointing to the right. Recall that a single arrow means that a reaction goes to completion.

$$HCl(aq) + H_2O(l) \rightarrow H_3O^+(aq) + Cl^-(aq)$$

Because strong acids produce the maximum number of ions, their solutions are good conductors of electricity. The names and ionization equations for some strong acids are shown in **Table 3.**

**Weak acids** If the brightly lit bulb of the apparatus containing the HCl solution is due to the large number of ions in solution, shown in **Figure 11,** then the weakly lit bulb of the apparatus containing the $HC_2H_3O_2$ solution, shown in **Figure 12,** must mean that the acetic acid solution has fewer ions. Because the two solutions have the same molar concentrations, you can conclude that acetic acid does not ionize completely. An acid that ionizes only partially in dilute aqueous solution is a **weak acid.** Weak acids produce fewer ions and thus cannot conduct electricity as well as strong acids. **Table 3** shows ionization equations for some common weak acids.

Explore **ionization equations with an interactive table.** Concepts In Motion

**Table 3 Ionization Equations**

| Strong Acids | | Weak Acids | |
|---|---|---|---|
| **Name** | **Ionization Equation** | **Name** | **Ionization Equations** |
| Hydrochloric | $HCl \rightarrow H^+ + Cl^-$ | Hydrofluoric | $HF \rightleftharpoons H^+ + F^-$ |
| Hydroiodic | $HI \rightarrow H^+ + I^-$ | Acetic | $HC_2H_3O_2 \rightleftharpoons H^+ + C_2H_3O_2^-$ |
| Perchloric | $HClO_4 \rightarrow H^+ + ClO_4^-$ | Hydrosulfuric | $H_2S \rightleftharpoons H^+ + HS^-$ |
| Nitric | $HNO_3 \rightarrow H^+ + NO_3^-$ | Carbonic | $H_2CO_3 \rightleftharpoons H^+ + HCO_3^-$ |
| Sulfuric | $H_2SO_4 \rightarrow H^+ + HSO_4^-$ | Hypochlorous | $HClO \rightleftharpoons H^+ + ClO^-$ |

**RealWorld CHEMISTRY**

### Hydrogen Cyanide

**A DEADLY COMPOUND**
Hydrogen cyanide (HCN) is a poisonous gas found in the exhaust of vehicles, in tobacco and wood smoke, and in smoke from burning nitrogen-containing plastics. Some insects such as millipedes and burnet moths release hydrogen cyanide as a defense mechanism. A solution of hydrogen cyanide in water is called hydrocyanic acid. Fruits that have a pit, such as cherries or peaches, contain cyanohydrins, which convert to hydrocyanic acid in the digestive system if the pits are eaten. However, no hydrocyanic acid is produced in the flesh of these fruits, so the fruit can safely be eaten.

**Acid strength and the Brønsted-Lowry model** Can the Brønsted-Lowry model explain why HCl ionizes completely but $HC_2H_3O_2$ forms only a few ions? Consider the ionization of any strong acid, HX. Remember that the acid on the reactant side of the equation produces a conjugate base on the product side. Similarly, the base on the reactant side produces a conjugate acid.

$$\underset{\text{Acid}}{HX(aq)} + \underset{\text{Base}}{H_2O(l)} \rightarrow \underset{\text{Conjugate acid}}{H_3O^+(aq)} + \underset{\text{Conjugate base}}{X^-(aq)}$$

HX represents a strong acid and its conjugate base is weak. That is, HX is nearly 100% ionized because $H_2O$ is a stronger base (in the forward reaction) than is the conjugate base $X^-$ (in the reverse reaction). In other words, the ionization equilibrium lies almost completely to the right because the base $H_2O$ has a much greater attraction for the $H^+$ ion than does the base $X^-$. Think of this as the battle of the bases: Which of the two ($H_2O$ or $X^-$) has a greater attraction for the hydrogen ion? In the case of all strong acids, water is the stronger base. Notice that the equation is shown with a single arrow to the right.

How does the situation differ for any weak acid, HY?

$$\underset{\text{Acid}}{HY(aq)} + \underset{\text{Base}}{H_2O(l)} \leftrightharpoons \underset{\text{Conjugate acid}}{H_3O^+(aq)} + \underset{\text{Conjugate base}}{Y^-(aq)}$$

The ionization equilibrium for a weak acid lies far to the left because the conjugate base $Y^-$ has a greater attraction for the $H^+$ ion than does the base $H_2O$. In the battle of the bases, the conjugate base $Y^-$ (in the reverse reaction) is stronger than the base $H_2O$ (in the forward reaction) and manages to capture the $H^+$ ion. In the case of acetic acid, the conjugate base $C_3H_2O_2^-$ (in the reverse reaction) has a stronger attraction for hydrogen ions than does the base $H_2O$ (in the forward reaction).

$$HC_2H_3O_2(aq) + H_2O(l) \rightleftharpoons H_3O^+(aq) + C_2H_3O_2^-(aq)$$

Notice that the equation is shown with equilibrium arrows.

☑ READING CHECK **Summarize** the important difference between strong acids and weak acids in terms of the battle of the bases.

**Acid ionization constants** Although the Brønsted-Lowry model helps explain acid strength, the model does not provide a quantitative way to express the strength of an acid or to compare the strengths of various acids. The equilibrium constant expression provides the quantitative measure of acid strength.

As you have read, a weak acid produces an equilibrium mixture of molecules and ions in aqueous solution. Thus, the equilibrium constant, $K_{eq}$, provides a quantitative measure of the degree of ionization of the acid. Consider hydrocyanic acid (HCN), also known as prussic acid which is used in dying, engraving, and tempering steel.

The ionization equation and equilibrium constant expression for hydrocyanic acid are as follows.

$$HCN(aq) + H_2O(l) \rightleftharpoons H_3O^+(aq) + CN^-(aq)$$

$$K_{eq} = \frac{[H_3O^+][CN^-]}{[HCN][H_2O]}$$

The concentration of liquid $H_2O$ in the denominator of the expression is considered to be constant in dilute aqueous solutions, so it can be combined with $K_{eq}$ to give a new equilibrium constant, $K_a$.

$$K_{eq}[H_2O] = K_a = \frac{[H_3O^+][CN^-]}{[HCN]} = 6.2 \times 10^{-10}$$

$K_a$ is called the acid ionization constant. The **acid ionization constant** is the value of the equilibrium constant expression for the ionization of a weak acid. Like all equilibrium constants, the value of $K_a$ indicates whether reactants or products are favored at equilibrium. For weak acids, the concentrations of the ions (products) in the numerator tend to be small compared to the concentration of un-ionized molecules (reactant) in the denominator. The weakest acids have the smallest $K_a$ values because their solutions have the lowest concentrations of ions and the highest concentrations of un-ionized acid molecules. $K_a$ values and ionization equations for several weak acids are listed in **Table 4.** Note that polyprotic acids are not necessarily strong acids for any of their ionizations. Each ionization of a polyprotic acid has a $K_a$ value, and the values decrease for each successive ionization.

## CAREERS IN CHEMISTRY

**Nursery Worker** The propagation and growth of plants is the primary job of a nursery worker. This involves planting, pruning, transplanting, and selling all kinds of plant material. A nursery worker must know what nutrients are needed for optimum plant growth and what soil conditions, including acidity, foster the strongest growth for each kind of plant.

WebQuest

## PRACTICE Problems

Do additional problems. Online Practice

**12.** Write ionization equations and acid ionization constant expressions for each acid.

**a.** $HClO_2$ **b.** $HNO_2$ **c.** HIO

**13.** Write the first and second ionization equations for $H_2SeO_3$.

**14. Challenge** Given the expression $K_a = \frac{[AsO_4^{3-}][H_3O^+]}{[HAsO_4^{2-}]}$, write the balanced equation for the corresponding reaction.

**Table 4 Ionization Constants for Weak Acids**

| Acid | Ionization Equation | $K_a$ (298 K) |
|---|---|---|
| Hydrosulfuric, first ionization | $H_2S \rightleftharpoons H^+ + HS^-$ | $8.9 \times 10^{-8}$ |
| Hydrosulfuric, second ionization | $HS^- \rightleftharpoons H^+ + S^{2-}$ | $1 \times 10^{-19}$ |
| Hydrofluoric | $HF \rightleftharpoons H^+ + F^-$ | $6.3 \times 10^{-4}$ |
| Hydrocyanic | $HCN \rightleftharpoons H^+ + CN^-$ | $6.2 \times 10^{-10}$ |
| Acetic | $CH_3COOH \rightleftharpoons H^+ + CH_3COO^-$ | $1.8 \times 10^{-5}$ |
| Carbonic, first ionization | $H_2CO_3 \rightleftharpoons H^+ + HCO_3^-$ | $4.5 \times 10^{-7}$ |
| Carbonic, second ionization | $HCO_3^- \rightleftharpoons H^+ + CO_3^{2-}$ | $4.7 \times 10^{-11}$ |

# MiniLAB

iLab Station

## Compare Acid Strengths

**How can you determine the relative strengths of acid solutions?**

### Procedure

1. Read and complete the lab safety form.
2. Use a **10-mL graduated cylinder** to measure 3 mL of **glacial acetic acid.** Use a **dropping pipette** to transfer the acid into Well A1 of a **24-well microplate.**
   **WARNING:** ***Glacial acetic acid is corrosive and toxic by inhalation. Handle with caution.***
3. Lower the electrodes of a **conductivity tester** into Well A1. Record your results.
4. Rinse the graduated cylinder and pipette with water. Measure 3 mL of **6.0*M* acetic acid,** and transfer it to Well A2 of the microplate. Test and record the conductivity of the solution.
5. Repeat Step 4 with **1.0*M* acetic acid** and **0.10*M* acetic acid** using Wells A3 and A4, respectively.

### Analysis

1. **Write** the equation for the ionization of acetic acid in water and the equilibrium constant expression ($K_{eq} = 1.8 \times 10^{-5}$). What does the size of $K_{eq}$ indicate about the degree of ionization?
2. **Explain** whether the following approximate percent ionizations fit your laboratory results: glacial acetic acid, 0.1%; 6.0*M* acetic acid, 0.2%; 1.0*M* acetic acid, 0.4%; 0.1*M* acetic acid, 1.3%.
3. **State** a hypothesis that explains your observations using your answer to Question 2.
4. **Utilize** your hypothesis to draw a conclusion about the need to use large amounts of water for rinsing when acid spills on living tissue.

## Strengths of Bases

What you have read about acids can be applied to bases, except that $OH^-$ ions, rather than $H^+$ ions, are involved. For example, the conductivity of a base depends on the extent to which the base produces $OH^-$ ions in aqueous solution.

**Strong bases** A base that dissociates entirely into metal ions and hydroxide ions is known as a **strong base.** Therefore, metallic hydroxides, such as sodium hydroxide (NaOH), are strong bases.

$$NaOH(s) \rightarrow Na^+(aq) + OH^-(aq)$$

Some metallic hydroxides, such as calcium hydroxide ($Ca(OH)_2$) have low solubility and thus are poor sources of $OH^-$ ions. Note that the solubility product constant, $K_{sp}$, for calcium hydroxide ($Ca(OH)_2$) is small, indicating that few $OH^-$ ions are present in a saturated solution.

$$Ca(OH)_2(s) \rightleftharpoons Ca^{2+}(aq) + 2OH^-(aq) \quad K_{sp} = 6.5 \times 10^{-6}$$

Nevertheless, calcium hydroxide and other slightly soluble metallic hydroxides are considered strong bases because all of the compound that dissolves is completely dissociated. The dissociation equations for several strong bases are listed in **Table 5.**

**Table 5 Dissociation Equations for Strong Bases**

| Dissociation Equations for Strong Bases |
|---|
| $NaOH(s) \rightarrow Na^+(aq) + OH^-(aq)$ |
| $KOH(s) \rightarrow K^+(aq) + OH^-(aq)$ |
| $RbOH(s) \rightarrow Rb^+(aq) + OH^-(aq)$ |
| $CsOH(s) \rightarrow Cs^+(aq) + OH^-(aq)$ |
| $Ca(OH)_2(s) \rightarrow Ca^{2+}(aq) + 2OH^-(aq)$ |
| $Ba(OH)_2(s) \rightarrow Ba^{2+}(aq) + 2OH^-(aq)$ |

**Weak bases** In contrast to strong bases, a **weak base** ionizes only partially in dilute aqueous solution. For example, methylamine ($CH_3NH_2$) reacts with water to produce an equilibrium mixture of $CH_3NH_2$ molecules, $CH_3NH_3^+$ ions, and $OH^-$ ions.

$$\underset{\text{Acid}}{CH_3NH_2(aq)} + \underset{\text{Base}}{H_2O(l)} \rightarrow \underset{\text{Conjugate acid}}{CH_3NH_3^+(aq)} + \underset{\text{Conjugate base}}{OH^-(aq)}$$

**Table 6 Ionization Constants of Weak Bases**

| Base | Ionization Equation | $K_b$ (298 K) |
|---|---|---|
| Ethylamine | $C_2H_5NH_2(aq) + H_2O(l) \rightleftharpoons C_2H_5NH_3^+(aq) + OH^-(aq)$ | $5.0 \times 10^{-4}$ |
| Methylamine | $CH_3NH_2(aq) + H_2O(l) \rightleftharpoons CH_3NH_3^+(aq) + OH^-(aq)$ | $4.3 \times 10^{-4}$ |
| Ammonia | $NH_3(aq) + H_2O(l) \rightleftharpoons NH_4^+(aq) + OH^-(aq)$ | $2.5 \times 10^{-5}$ |
| Aniline | $C_6H_5NH_2(aq) + H_2O(l) \rightleftharpoons C_6H_5NH_3^+(aq) + OH^-(aq)$ | $4.3 \times 10^{-10}$ |

This equilibrium lies far to the left because the base, $CH_3NH_2$, is weak and the conjugate base, $OH^-$, is strong. The hydroxide ion has a greater attraction for a hydrogen ion than has a molecule of methylamine.

**Base ionization constants** Like weak acids, weak bases also form equilibrium mixtures of molecules and ions in aqueous solution. The equilibrium constant provides a measure of the extent of the base's ionization. The equilibrium constant for the ionization of methylamine in water is defined by the following equilibrium constant expression.

$$K_b = \frac{[CH_3NH_3^+][OH^-]}{[CH_3NH_2]}$$

The **base ionization constant,** $K_b$, is the value of the equilibrium constant expression for the ionization of a base. The smaller the value of $K_b$, the weaker the base. $K_b$ values and ionization equations for several weak bases are listed in **Table 6.**

## PRACTICE Problems

Do additional problems. Online Practice

**15.** Write ionization equations and base ionization constant expressions for the following bases.

**a.** hexylamine ($C_6H_{13}NH_2$)

**b.** propylamine ($C_3H_7NH_2$)

**c.** carbonate ion ($CO_3^{2-}$)

**d.** hydrogen sulfite ion ($HSO_3^-$)

**16. Challenge** Write an equation for a base equilibrium in which the base in the forward reaction is $PO_4^{3-}$ and the base in the reverse reaction is $OH^-$.

## SECTION 2 REVIEW

Section Self-Check

### Section Summary

- Strong acids and strong bases are completely ionized in a dilute aqueous solution. Weak acids and weak bases are partially ionized in a dilute aqueous solution.
- For weak acids and weak bases, the value of the acid or base ionization constant is a measure of the strength of the acid or base.

**17.** MAINIDEA **Describe** the contents of dilute aqueous solutions of the strong acid HI and the weak acid HCOOH.

**18. Relate** the strength of a weak acid to the strength of its conjugate base.

**19. Identify** the conjugate acid-base pairs in each equation.

**a.** $HCOOH(aq) + H_2O(l) \rightleftharpoons HCOO^-(aq) + H_3O^+(aq)$

**b.** $NH_3(aq) + H_2O(l) \rightleftharpoons NH_4^+(aq) + OH^-(aq)$

**20. Explain** what the $K_b$ for aniline ($C_6H_5NH_2$) tells you ($K_b = 4.3 \times 10^{-10}$).

**21. Interpret Data** Use the data in **Table 4** to put the seven acids in order according to increasing electrical conductivity.

SECTION 3

# Hydrogen Ions and pH

**MAIN IDEA** pH and pOH are logarithmic scales that express the concentrations of hydrogen ions and hydroxide ions in aqueous solutions.

### Essential Questions

- What are pH and pOH?
- How are pH and pOH related to the ion product constant for water?
- How are the pH and pOH of aqueous solutions calculated?

### Review Vocabulary

**Le Châtelier's principle:** states that if a stress is applied to a system at equilibrium, the system shifts in the direction that relieves the stress

### New Vocabulary

ion product constant for water
pH
pOH

**CHEM 4 YOU** Think of two children on a seesaw. When one side of a seesaw goes up, the other side goes down. Sometimes, the seesaw is balanced in the middle. The concentrations of hydrogen ions and hydroxide ions in water solutions behave in a similar way.

## Ion Product Constant for Water

Recall that pure water contains equal concentrations of $H^+$ and $OH^-$ ions produced by self-ionization. **Figure 13** shows that in self-ionization, equal numbers of hydronium and hydroxide ions are formed. The equation for the equilibrium can be simplified in the following way.

$$H_2O(l) \rightleftharpoons H^+(aq) + OH^-(aq)$$

**Writing $K_w$** The double arrow indicates that this is an equilibrium. Recall that the equilibrium constant expression is written by placing the concentrations of the products in the numerator and the concentrations of the reactants in the denominator. In this case, all terms are to the first power because all the coefficients in the balanced chemical equation are 1. The concentration of pure water is constant, so $[H_2O]$ does not appear in the denominator.

**The Ion Product of Water**

$$K_w = [H^+][OH^-]$$

**$K_w$ is the ion product constant for water. $[H^+]$ represents the concentration of the hydrogen ion. $[OH^-]$ represents the concentration of the hydroxide ion.**

In dilute aqueous solutions, the product of the concentrations of the hydrogen ion and the hydroxide ion equals $K_w$.

The expression for $K_w$ is a special equilibrium constant expression that applies only to water. The constant $K_w$ is called the ion product constant for water. The **ion product constant for water** is the value of the equilibrium constant expression for the self-ionization of water. Experiments show that in pure water at 298 K, $[H^+]$ and $[OH^-]$ are both equal to $1.0 \times 10^{-7}M$. Therefore, at 298 K, the value of $K_w$ is $1.0 \times 10^{-14}$.

$$K_w = [H^+][OH^-] = (1.0 \times 10^{-7})(1.0 \times 10^{-7})$$
$$K_w = 1.0 \times 10^{-14}$$

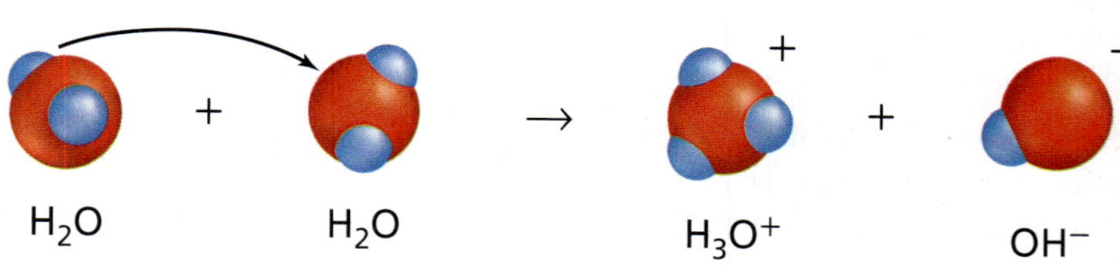


**Figure 13** In the self-ionization of water, one water molecule acts as an acid, and the other acts as a base.

**$K_w$ and Le Châtelier's Principle** The product of $[H^+]$ and $[OH^-]$ always equals $1.0 \times 10^{-14}$ at 298 K. This means that if the concentration of $H^+$ ions increases, the concentration of $OH^-$ ions must decrease. Similarly, an increase in the concentration of $OH^-$ ions causes a decrease in the concentration of $H^+$ ions. Think about these changes in terms of Le Châtelier's principle, which you read about previously. Adding extra hydrogen ions to water at equilibrium is a stress on the system. The system reacts in a way to relieve the stress. The added $H^+$ ions react with $OH^-$ ions to form more water molecules. Thus, the concentration of $OH^-$ ions decreases. Example Problem 1 shows how you can use $K_w$ to calculate the concentration of either $H^+$ or $OH^-$ if you know the concentration of the other ion.

☑ READING CHECK **Explain** why $K_w$ does not change when the concentration of hydrogen ions increases.

## EXAMPLE Problem 1

Find help with **algebraic equations.** Math Handbook

**CALCULATE $[H^+]$ AND $[OH^-]$ USING $K_W$** At 298 K, the $H^+$ ion concentration in a cup of coffee is $1.0 \times 10^{-5}M$. What is the $OH^-$ ion concentration in the coffee? Is the coffee acidic, basic, or neutral?

### 1 ANALYZE THE PROBLEM

You are given the concentration of the $H^+$ ion, and you know that $K_w$ equals $1.0 \times 10^{-14}$. You can use the ion product constant expression to solve for $[OH^-]$. Because $[H^+]$ is greater than $1.0 \times 10^{-7}$, you can predict that $[OH^-]$ will be less than $1.0 \times 10^{-7}$.

| Known | Unknown |
|---|---|
| $[H^+] = 1.0 \times 10^{-5}M$ | $[OH^-]$ = ? mol/L |
| $K_w = 1.0 \times 10^{-14}$ | |

### 2 SOLVE FOR THE UNKNOWN

Use the ion product constant expression.

$K_w = [H^+][OH^-]$ — State the ion product expression.

$[OH^-] = \dfrac{K_w}{[H^+]}$ — Solve for $[OH^-]$.

$[OH^-] = \dfrac{1.0 \times 10^{-14}}{1.0 \times 10^{-5}} = 1.0 \times 10^{-9}$ mol/L — Substitute $K_w = 1.0 \times 10^{-14}$. Substitute $[H^+] = 1.0 \times 10^{-5}M$ and solve.

Because $[H^+] > [OH^-]$, **the coffee is acidic.**

### 3 EVALUATE THE ANSWER

The answer is correctly stated with two significant figures because $[H^+]$ and $K_w$ each have two significant figures. As predicted, $[OH^-]$ is less than $1.0 \times 10^{-7}$ mol/L.

EXAMPLE PROBLEM

## PRACTICE Problems

Do additional problems. Online Practice

**22.** The concentration of either the $H^+$ ion or the $OH^-$ ion is given for four aqueous solutions at 298 K. For each solution, calculate $[H^+]$ or $[OH^-]$. State whether the solution is acidic, basic, or neutral.

**a.** $[H^+] = 1.0 \times 10^{-13}M$ **c.** $[OH^-] = 1.0 \times 10^{-3}M$

**b.** $[OH^-] = 1.0 \times 10^{-7}M$ **d.** $[H^+] = 4.0 \times 10^{-5}M$

**23. Challenge** Calculate the number of $H^+$ ions and the number of $OH^-$ ions in 300 mL of pure water at 298 K.

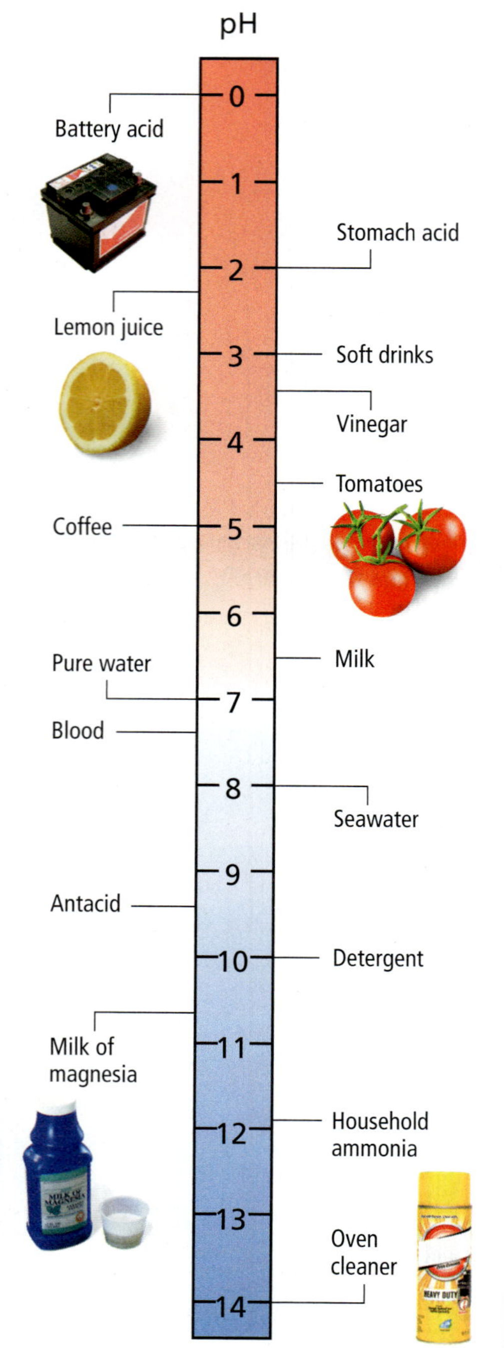


■ **Figure 14** Compare the pH values for these familiar substances.

**Determine** *whether seawater or detergent has a higher concentration of $H^+$ ions. How many times higher?*

## pH and pOH

Concentrations of $H^+$ ions are often small numbers expressed in scientific notation. Because these numbers are cumbersome, chemists adopted an easier way to express $H^+$ ion concentrations.

**What is pH?** Chemists express the concentration of hydrogen ions using a pH scale based on common logarithms. The **pH** of a solution is the negative logarithm of the hydrogen ion concentration.

**pH**

$$pH = -\log [H^+]$$

**$[H^+]$ represents the hydrogen ion concentration.**

The pH of a solution equals the negative logarithm of the hydrogen ion concentration.

At 298 K, acidic solutions have pH values below 7. Basic solutions have pH values above 7. Thus, a solution with a pH of 0.0 is strongly acidic; a solution with a pH of 14.0 is strongly basic. The logarithmic nature of the pH scale means that a change of one pH unit represents a tenfold change in ion concentration. A solution having a pH of 3.0 has ten times the hydrogen ion concentration of a solution with a pH of 4.0. The pH scale and pH values of some common substances are shown in **Figure 14.**

**What is pOH?** Sometimes it is convenient to express the basicity or alkalinity of a solution on a pOH scale that mirrors the relationship between pH and $[H^+]$. The **pOH** of a solution is the negative logarithm of the hydroxide ion concentration.

**pOH**

$$pOH = -\log [OH^-]$$

**$[OH^-]$ represents the hydroxide ion concentration.**

The pOH of a solution equals the negative logarithm of the hydroxide ion concentration.

At 298 K, a solution with a pOH less than 7.0 is basic; a solution with a pOH of 7.0 is neutral; and a solution with a pOH greater than 7.0 is acidic. As with the pH scale, a change of one pOH unit expresses a tenfold change in ion concentration.

A simple relationship between pH and pOH makes it easy to calculate either quantity if the other is known.

**How pH and pOH Are Related**

$$pH + pOH = 14.00$$

**pH represents $-\log [H^+]$.**
**pOH represents $-\log [OH^-]$.**

The sum of pH and pOH is 14.00.

**Figure 15** illustrates the relationship between pH and the $H^+$ concentration and the relationship between pOH and $OH^-$ concentration at 298 K.

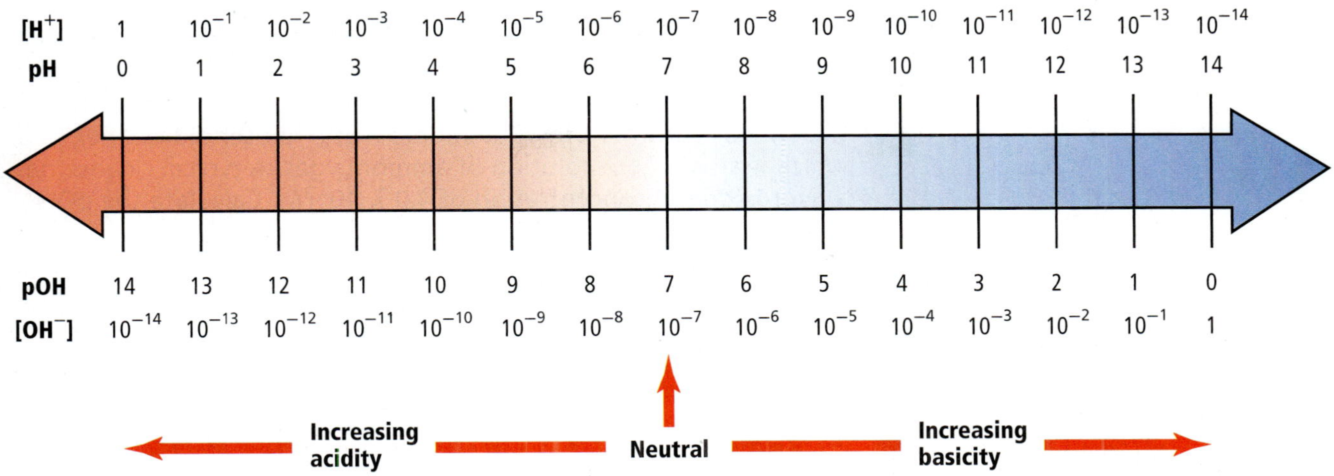


**Figure 15** Study this diagram to sharpen your understanding of pH and pOH. Note that at each vertical position, the sum of pH (above the arrow) and pOH (below the arrow) equals 14. Also note that at every position, the product of $[H^+]$ and $[OH^-]$ equals $10^{-14}$.

## EXAMPLE Problem 2

Find help with **logarithms.** Math Handbook

**CALCULATE pH FROM $[H^+]$** What is the pH of a neutral solution at 298 K?

### 1 ANALYZE THE PROBLEM

In a neutral solution at 298 K, $[H^+] = 1.0 \times 10^{-7} M$. You must find the negative log of $[H^+]$.

**Known**
$[H^+] = 1.0 \times 10^{-7}\ M$

**Unknown**
**pH = ?**

### 2 SOLVE FOR THE UNKNOWN

$\text{pH} = -\log\ [H^+]$ — State the equation for pH.

$\text{pH} = -\log\ (1.0 \times 10^{-7})$ — Substitute $[H^+] = 1.0 \times 10^{-7} M$.

The pH of the neutral solution at 298 K is **7.00**.

### 3 EVALUATE THE ANSWER

Values for pH are expressed with as many decimal places as the number of significant figures in the $H^+$ ion concentration. The pH is correctly stated with two decimal places.

## PRACTICE Problems

Do additional problems. Online Practice

**24.** Calculate the pH of solutions having the following ion concentrations at 298 K.

**a.** $[H^+] = 1.0 \times 10^{-2} M$ **b.** $[H^+] = 3.0 \times 10^{-6} M$

**25.** Calculate the pH of aqueous solutions with the following $[H^+]$ at 298 K.

**a.** $[H^+] = 0.0055 M$ **b.** $[H^+] = 0.000084 M$

**26. Challenge** Calculate the pH of a solution having $[OH^-] = 8.2 \times 10^{-6} M$.

Get help with **logarithms.** Personal Tutor

## EXAMPLE Problem 3

**CALCULATE pOH AND pH FROM [OH⁻]** In **Figure 16,** a cow is being fed straw and hay that has been treated with ammonia. The addition of ammonia to animal feed promotes protein growth in the animal. Another use of ammonia is as a household cleaner, which is an aqueous solution of ammonia gas. A typical cleaner has a hydroxide-ion concentration of $4.0 \times 10^{-3}M$. Calculate the pOH and pH of a cleaner at 298 K.

### 1 ANALYZE THE PROBLEM

You have been given the concentration of hydroxide ion and must calculate pOH and pH. First, calculate pOH using its definition. Then, calculate pH using the relationship pH + pOH = 14.00.

| Known | Unknown |
|---|---|
| $[OH^-] = 4.0 \times 10^{-3}M$ | **pOH = ?** |
| | **pH = ?** |

### 2 SOLVE FOR THE UNKNOWN

$\mathbf{pOH} = -\log [OH^-]$ — **State the equation for pOH.**

$\mathbf{pOH} = -\log (4.0 \times 10^{-3})$ — **Substitute $[OH^-] = 4.0 \times 10^{-3}M$.**

The **pOH** of the solution is **2.40.**

Use the relationship between pH and pOH to find the pH.

$\mathbf{pH} + pOH = 14.00$ — **State the equation that relates pH and pOH.**

$\mathbf{pH} = 14.00 - pOH$ — **Solve for pH.**

$\mathbf{pH} = 14.00 - 2.40 = \mathbf{11.60}$ — **Substitute pOH = 2.40.**

The **pH** of the solution is **11.60.**

### 3 EVALUATE THE ANSWER

The given concentration has two significant figures, so pH and pOH are correctly expressed with two decimal places. Because ammonia is a base, a small pOH value and a large pH value are reasonable.

**Figure 16** Farmers are able to increase the nutritional value of low-quality vegetable materials such as straw, hay, and other crop residue by immersing the materials in an atmosphere of ammonia gas for three weeks.



## PRACTICE Problems

Do additional problems. Online Practice

**27.** Calculate the pH and pOH of aqueous solutions with the following concentrations at 298 K.

**a.** $[OH^-] = 1.0 \times 10^{-6}M$

**b.** $[OH^-] = 6.5 \times 10^{-4}M$

**c.** $[H^+] = 3.6 \times 10^{-9}M$

**d.** $[H^+] = 2.5 \times 10^{-2}M$

**28.** Calculate the pH and pOH of aqueous solutions with the following concentration at 298 K.

**a.** $[OH^-] = 0.000033M$

**b.** $[H^+] = 0.0095M$

**29. Challenge** Calculate pH and pOH for an aqueous solution containing $1.0 \times 10^{-3}$ mol of HCl dissolved in 5.0 L of solution.

**Calculating ion concentrations from pH** Sometimes, you have to calculate the concentration of $H^+$ ions and $OH^-$ ions from the pH of a solution. Example Problem 4 shows how to do this.

## EXAMPLE Problem 4

Find help with **antilogarithms.** Math Handbook

EXAMPLE PROBLEM

**CALCULATE $[H^+]$ AND $[OH^-]$ FROM pH** What are $[H^+]$ and $[OH^-]$ in a healthy person's blood that has a pH of 7.40? Assume that the temperature of the blood is 298 K.

### 1 ANALYZE THE PROBLEM

You have been given the pH of a solution and must calculate $[H^+]$ and $[OH^-]$. You can obtain $[H^+]$ using the equation that defines pH. Then, subtract the pH from 14.00 to obtain the pOH and use the equation that defines pOH to get $[OH^-]$.

| **Known** | **Unknown** |
|---|---|
| pH = 7.40 | $[H^+]$ = ? mol/L |
| | $[OH^-]$ = ? mol/L |

### 2 SOLVE FOR THE UNKNOWN

Determine $[H^+]$.

| | |
|---|---|
| $pH = -\log [H^+]$ | **State the equation for pH.** |
| $-pH = \log [H^+]$ | **Multiply both sides of the equation by −1.** |
| $[H^+] = \text{antilog}(-pH)$ | **Take the antilog of each side to solve for $[H^+]$.** |
| $[H^+] = \text{antilog}(-7.40)$ | **Substitute pH = 7.40.** |
| $[H^+] = 4.0 \times 10^{-8}M$ | **A calculator shows that the antilog of −7.40 is $4.0 \times 10^{-8}$.** |

The concentration of $H^+$ ions in the blood is $4.0 \times 10^{-8}M$.

Determine $[OH^-]$.

| | |
|---|---|
| $pH + pOH = 14.00$ | **State the equation that relates pH and pOH.** |
| $pOH = 14.00 - pH$ | **Solve for pOH.** |
| $pOH = 14.00 - 7.40 = 6.60$ | **Substitute pH = 7.40.** |
| $pOH = -\log [OH^-]$ | **State the equation for pOH.** |
| $-pOH = \log [OH^-]$ | **Multiply both sides of the equation by −1.** |
| $[OH^-] = \text{antilog}(-6.60)$ | **Take the antilog of each side and substitute pOH = 6.60.** |
| $[OH^-] = 2.5 \times 10^{-7}M.$ | **A calculator shows that the antilog of −6.60 is $2.5 \times 10^{-7}$.** |

The concentration of $OH^-$ ions in the blood is $2.5 \times 10^{-7}M$.

### 3 EVALUATE THE ANSWER

The given pH has two decimal places, so the answers must have two significant figures. A $[H^+]$ less than $10^{-7}$ and a $[OH^-]$ greater than $10^{-7}$ are reasonable, given the initial pH.

## PRACTICE Problems

Do additional problems. Online Practice

**30.** Calculate $[H^+]$ and $[OH^-]$ in each of the following solutions.

**a.** Milk, pH = 6.50

**b.** Lemon juice, pH = 2.37

**c.** Milk of magnesia, pH = 10.50

**d.** Household ammonia, pH = 11.90

**31. Challenge** Calculate the $[H^+]$ and $[OH^-]$ in a sample of seawater with a pOH = 5.60.

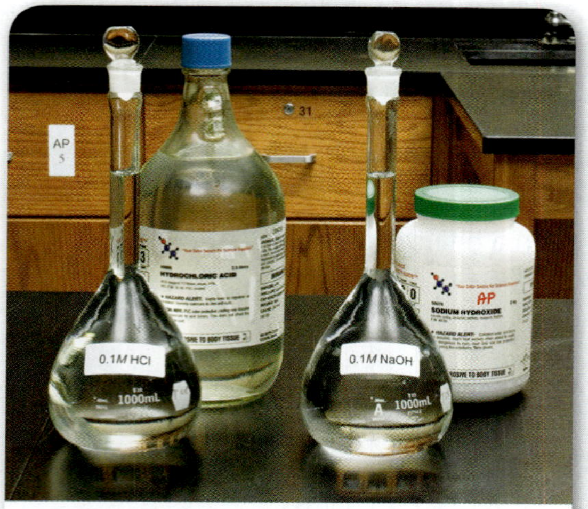


■ **Figure 17** The label on a bottle of a strong acid or a strong base tells you the concentration of hydrogen ions or hydroxide ions in the solution. That is because, in solution, strong acids and bases exist entirely as ions.

**State** *the [$H^+$] in the HCl flask and the [$OH^-$] in the NaOH flask.*

**Molarity and the pH of strong acids** Look at the flasks of acid and base solutions in **Figure 17.** The solutions have just been made up and are labeled with their molarity, which is the number of moles of molecules or formula units that were dissolved in 1 L of solution. One flask contains a strong acid (HCl), the other a strong base (NaOH). Recall that strong acids and bases are essentially 100% in the form of ions in solution. That means that the following reaction for the ionization of HCl goes to completion.

$$HCl(aq) \rightarrow H^+(aq) + Cl^-(aq)$$

Every HCl molecule produces one $H^+$ ion. The bottle labeled 0.1$M$ HCl contains 0.1 mol of $H^+$ ions per liter and 0.1 mol of $Cl^-$ ions per liter. For all strong monoprotic acids, the concentration of the acid is the concentration of $H^+$ ions. Thus, you can use the molarity of the acid to calculate pH.

**Molarity and the pH of strong bases** Similarly, the 0.1$M$ solution of the strong base NaOH in **Figure 17** is fully ionized.

$$NaOH(aq) \rightarrow Na^+(aq) + OH^-(aq)$$

One formula unit of NaOH produces one $OH^-$ ion. Thus, the concentration of the $OH^-$ ions is the same as the molarity of the solution, 0.1$M$.

Some strong bases, such as calcium hydroxide, $Ca(OH)_2$, contain two or more $OH^-$ ions in each formula unit. The concentration of $OH^-$ ion in a solution of $Ca(OH)_2$ is twice the molarity of the ionic compound. For example, the concentration of hydroxide ions in a $7.5 \times 10^{-4}M$ solution of $Ca(OH)_2$ is $7.5 \times 10^{-4}M \times 2 = 1.5 \times 10^{-3}M$.

Although strong acids and strong bases are completely ionized in dilute aqueous solutions, remember that weak acids and weak bases are only partially ionized. Therefore, you must use $K_a$ and $K_b$ values to determine the concentrations of $H^+$ and $OH^-$ ions in solutions of weak acids and bases.

☑ READING CHECK **Explain** why you cannot obtain the [$H^+$] directly from the molarity of a weak acid solution.

**Calculating $K_a$ from pH** Suppose you measured the pH of a 0.100$M$ solution of the weak acid HF and found it to be 3.20. Would you have enough information to calculate $K_a$ for HF?

$$HF(aq) \rightleftharpoons H^+(aq) + F^-(aq)$$

$$K_a = \frac{[H^+][F^-]}{[HF]}$$

From the pH, you could calculate [$H^+$]. Then, remember that for every mole per liter of $H^+$ ion there must be an equal concentration of $F^-$ ion. That means you know two of the variables in the $K_a$ expression. What about the third, [HF]? The concentration of HF at equilibrium is equal to the initial concentration of the acid (0.100$M$) minus the moles per liter of HF that dissociated, which is equal to [$H^+$].

Matt Meadows

## EXAMPLE Problem 5

Find help with **scientific notation.** Math Handbook

EXAMPLE PROBLEM

**CALCULATE $K_a$ FROM pH** Formic acid is used to process latex tapped from rubber trees into natural rubber. The pH of a 0.100*M* solution of formic acid (HCOOH) is 2.38. What is $K_a$ for HCOOH?

### 1 ANALYZE THE PROBLEM

You are given the pH of the formic acid solution, which allows you to calculate the concentration of the hydrogen ion.

$$HCOOH(aq) \rightleftharpoons H^+(aq) + HCOO^-(aq)$$

The balanced chemical equation shows that the concentration of $HCOO^-$ equals the concentration of $H^+$. The concentration of un-ionized HCOOH is the difference between the initial concentration of the acid and $[H^+]$.

**Known**
pH = 2.38
concentration of the solution = 0.100*M*

**Unknown**
$K_a$ = ?

### 2 SOLVE FOR THE UNKNOWN

Use the pH to calculate $[H^+]$.

$pH = -\log [H^+]$ — **Write the equation for pH.**

$[H^+] = \text{antilog} (-pH)$ — **Multiply both sides by −1 and take the antilog of each side.**

$[H^+] = \text{antilog} (-2.38)$ — **Substitute pH = 2.38.**

$[H^+] = 4.2 \times 10^{-3}M$ — **A calculator shows that the antilog of −2.38 is $4.2 \times 10^{-3}$.**

$[HCOO^-] = [H^+] = 4.2 \times 10^{-3}M$

[HCOOH] equals the initial concentration minus $[H^+]$.

$[HCOOH] = 0.100M - 4.2 \times 10^{-3}M = 0.096M$ — **Subtract $[H^+]$ from the initial [HCOOH].**

$$K_a = \frac{[H^+][HCOO^-]}{[HCOOH]}$$ **State the acid ionization constant expression.**

$$K_a = \frac{(4.2 \times 10^{-3})(4.2 \times 10^{-3})}{(0.096)} = 1.8 \times 10^{-4}$$ **Substitute $[H^+] = 4.2 \times 10^{-3}M$, $[HCOO^-] = 4.2 \times 10^{-3}M$, and $[HCOOH] = 0.096M$.**

The acid ionization constant for HCOOH is $1.8 \times 10^{-4}$.

### 3 EVALUATE THE ANSWER

The $K_a$ is reasonable for a weak acid. The answer is correctly reported with two significant figures.

## PRACTICE Problems

Do additional problems. Online Practice

**32.** Calculate the $K_a$ for the following acids using the given information.

**a.** 0.220*M* solution of $H_3AsO_4$, pH = 1.50 **b.** 0.0400*M* solution of $HClO_2$, pH = 1.80

**33.** Calculate the $K_a$ of the following acids using the given information.

**a.** 0.00330*M* solution of benzoic acid ($C_6H_5COOH$), pOH = 10.70

**b.** 0.100*M* solution of cyanic acid (HCNO), pOH = 11.00

**c.** 0.150*M* solution of butanoic acid ($C_3H_7COOH$), pOH = 11.18

**34. Challenge** Calculate the $K_a$ of a 0.0091*M* solution of an unknown acid (HX) having a pOH of 11.32. Use **Table 4** to identify the acid.

■ **Figure 18** The approximate pH of a solution can be obtained by wetting a piece of pH paper with the solution and comparing the color of the wet paper with a set of standard colors as shown in **a.** The portable pH meter in **b,** which is being used to measure the pH of rain water, provides a more accurate measurement in the form of a digital display of the pH.

**Measuring pH** Perhaps you have used indicator paper to measure the pH of a solution. The litmus paper you used in the Launch Lab is an example of a kind of pH paper. All pH paper is treated with one or more substances called indicators that change color depending on the concentration of hydrogen ions in a solution. While litmus paper only indicates if a substance is an acid or a base, pH papers treated with several indicators can identify a range of pH values. Phenolphthalein, which you also used in the Launch Lab, is another example of an indicator.

When a strip of pH paper is dipped into an acidic solution or a basic solution, the color of the paper changes. To determine the pH, the new color of the paper is compared with standard pH colors on a chart, as shown in **Figure 18.** The pH meter in **Figure 18** provides a more accurate measure of pH. When electrodes are placed in a solution, the meter gives a direct analog or digital readout of pH.

## SECTION 3 REVIEW

Section Self-Check

### Section Summary

- The ion product constant for water, $K_w$, equals the product of the $H^+$ ion concentration and the $OH^-$ ion concentration.
- The pH of a solution is the negative log of the hydrogen ion concentration. The pOH is the negative log of the hydroxide ion concentration. pH plus pOH equals 14.
- A neutral solution has a pH of 7.0 and a pOH of 7.0 because the concentrations of hydrogen ions and hydroxide ions are equal.

35. MAINIDEA **Explain** why the pH of an acidic solution is always a smaller number than the pOH of the same solution.
36. **Describe** how you can determine the pH of a solution if you know its pOH.
37. **Explain** the significance of $K_w$ in aqueous solutions.
38. **Explain,** using Le Châtelier's principle, what happens to the $[H^+]$ of a 0.10*M* solution of acetic acid when a drop of NaOH solution is added.
39. **List** the information needed to calculate the $K_a$ of a weak acid.
40. **Calculate** The pH of a tomato is approximately 4.50. What are $[H^+]$ and $[OH^-]$ in a tomato?
41. **Determine** the pH of a solution that contains $1.0 \times 10^{-9}$ mol of $OH^-$ ions per liter.
42. **Calculate** the pH of the following solutions.
    - **a.** 1.0*M* HI
    - **b.** 0.050*M* $HNO_3$
    - **c.** 1.0*M* KOH
    - **d.** $2.4 \times 10^{-5}M$ $Mg(OH)_2$
43. **Interpret Diagrams** Refer to **Figure 15** to answer these questions: What happens to the $[H^+]$, $[OH^-]$, pH, and pOH as a neutral solution becomes more acidic? As a neutral solution become more basic?

SECTION 4

# Neutralization

**MAINIDEA** In a neutralization reaction, an acid reacts with a base to produce a salt and water.

### Essential Questions

- What do chemical equations of neutralization reactions look like?
- How are neutralization reactions used in acid-base titrations?
- How do the properties of buffered and unbuffered solutions compare?

### Review Vocabulary

**stoichiometry:** the study of quantitative relationships between the amounts of reactants used and products formed by a chemical reaction; is based on the law of conservation of mass

### New Vocabulary

neutralization reaction
salt
titration
titrant
equivalence point
acid-base indicator
end point
salt hydrolysis
buffer
buffer capacity

**CHEM 4 YOU** When two teams in a debate present equally convincing arguments, you might find that you are neutral—favoring neither one point of view nor the other. In a similar way, a solution is neutral when the numbers of hydrogen ions and hydroxide ions are equal.

## Reactions Between Acids and Bases

If you were to experience heartburn or indigestion, you might take one of the antacids illustrated in **Figure 19** to relieve your discomfort. What kind of reaction occurs when magnesium hydroxide ($Mg(OH)_2$), the active ingredient in milk of magnesia, contacts hydrochloric acid solution ($H^+$ and $Cl^-$) produced by the stomach?

When $Mg(OH)_2$ and HCl react, a neutralization reaction occurs. A **neutralization reaction** is a reaction in which an acid and a base in an aqueous solution react to produce a salt and water. A **salt** is an ionic compound made up of a cation from a base and an anion from an acid. Neutralization is a double-replacement reaction.

**Writing neutralization equations** In the reaction between magnesium hydroxide and hydrochloric acid, magnesium replaces hydrogen in HCl and hydrogen replaces magnesium in $Mg(OH)_2$.

$$\begin{array}{ccccccc} Mg(OH)_2(aq) & + & 2HCl(aq) & \rightarrow & MgCl_2(aq) & + & 2H_2O(l) \\ \text{Base} & + & \text{Acid} & \rightarrow & \text{Salt} & + & \text{Water} \end{array}$$

Note that the cation from the base ($Mg^{2+}$) is combined with the anion from the acid ($Cl^-$) in the salt $MgCl_2$.

When writing neutralization equations, you must know whether all of the reactants and products in the solution exist as molecules or as formula units. For example, examine the formula equation and complete ionic equation for the reaction between hydrochloric acid and sodium hydroxide.

$$HCl(aq) + NaOH(aq) \rightarrow NaCl(aq) + H_2O(l)$$

**Figure 19** A dose of any of these antacids can relieve the symptoms of acid indigestion by reacting with the acidic solution in the stomach and neutralizing it.

Matt Meadows

**Figure 20** A hydronium ion transfers a hydrogen ion to a hydroxide ion. The loss of the hydrogen ion by $H_3O^+$ results in a water molecule. The gain of a hydrogen ion by $OH^-$ also results in a water molecule.

View an **animation about a neutralization reaction.**

Concepts In Motion

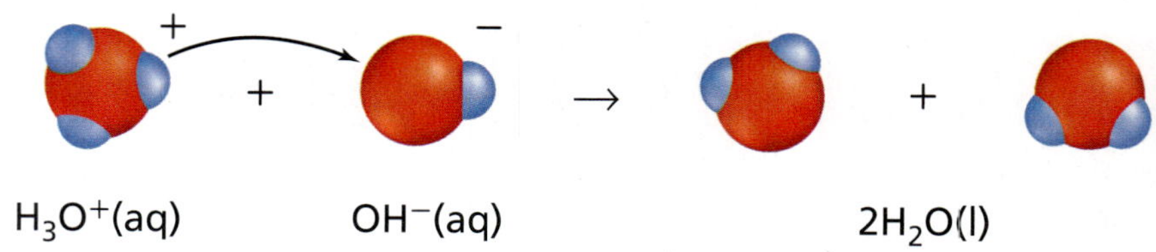


Because HCl is a strong acid, NaOH a strong base, and NaCl a soluble salt, all three compounds exist as ions in an aqueous solution.

$$H^+(aq) + Cl^-(aq) + Na^+(aq) + OH^-(aq) \rightarrow Na^+(aq) + Cl^-(aq) + H_2O(l)$$

The chloride ion and the sodium ion appear on both sides of the equation, so they are spectator ions. They can be eliminated to obtain the net ionic equation for the neutralization of a strong acid by a strong base.

$$H^+(aq) + OH^-(aq) \rightarrow H_2O(l)$$

Recall that in an aqueous solution, a $H^+$ ion exists as a $H_3O^+$ ion, so the net ionic equation for an acid-base neutralization reaction is

$$H_3O^+(aq) + OH^-(aq) \rightarrow 2H_2O(l).$$

This neutralization reaction is illustrated in **Figure 20.**

READING CHECK **Demonstrate** that the equation illustrated in **Figure 20** represents the neutralization of any strong acid by a strong base by writing the complete ionic equation and the net ionic equation for the neutralization of $HNO_3$ by KOH.

View an **animation about titration.**

Concepts In Motion

**Acid-base titration** The stoichiometry of an acid-base neutralization reaction is the same as that of any other reaction that occurs in solution. In the antacid reaction described above, 1 mol of $Mg(OH)_2$ neutralizes 2 mol of HCl.

$$Mg(OH)_2(aq) + 2HCl(aq) \rightarrow MgCl_2(aq) + 2H_2O(l)$$

In the reaction of sodium hydroxide and hydrogen chloride, 1 mol of NaOH neutralizes 1 mol of HCl.

$$NaOH(aq) + HCl(aq) \rightarrow NaCl(aq) + H_2O(l)$$

Stoichiometry provides the basis for a procedure called titration, which is used to determine the concentrations of acidic and basic solutions. **Titration** is a method for determining the concentration of a solution by reacting a known volume of that solution with a solution of known concentration. If you wish to find the concentration of an acid solution, you would titrate the acid solution with a solution of a base of known concentration. You could also titrate a base of unknown concentration with an acid of known concentration. How is an acid-base titration performed? **Figure 21** illustrates one type of setup for the titration procedure outlined on the following page. In this procedure a pH meter is used to monitor the change in the pH as the titration progresses.

**Figure 21** In the titration of an acid by a base, the pH meter measures the pH of the acid solution in the beaker as a solution of a base with a known concentration is added from the buret.

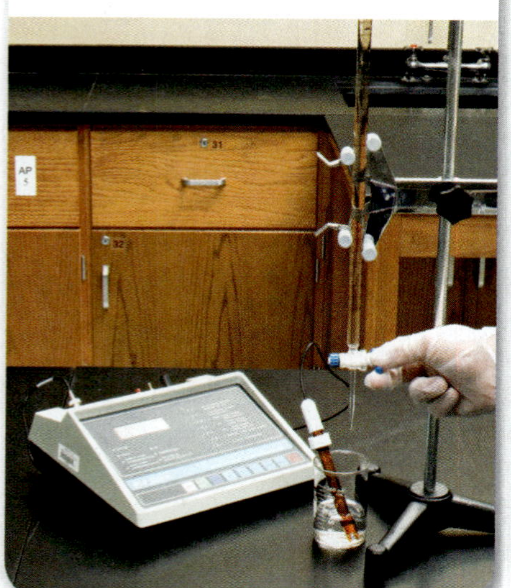

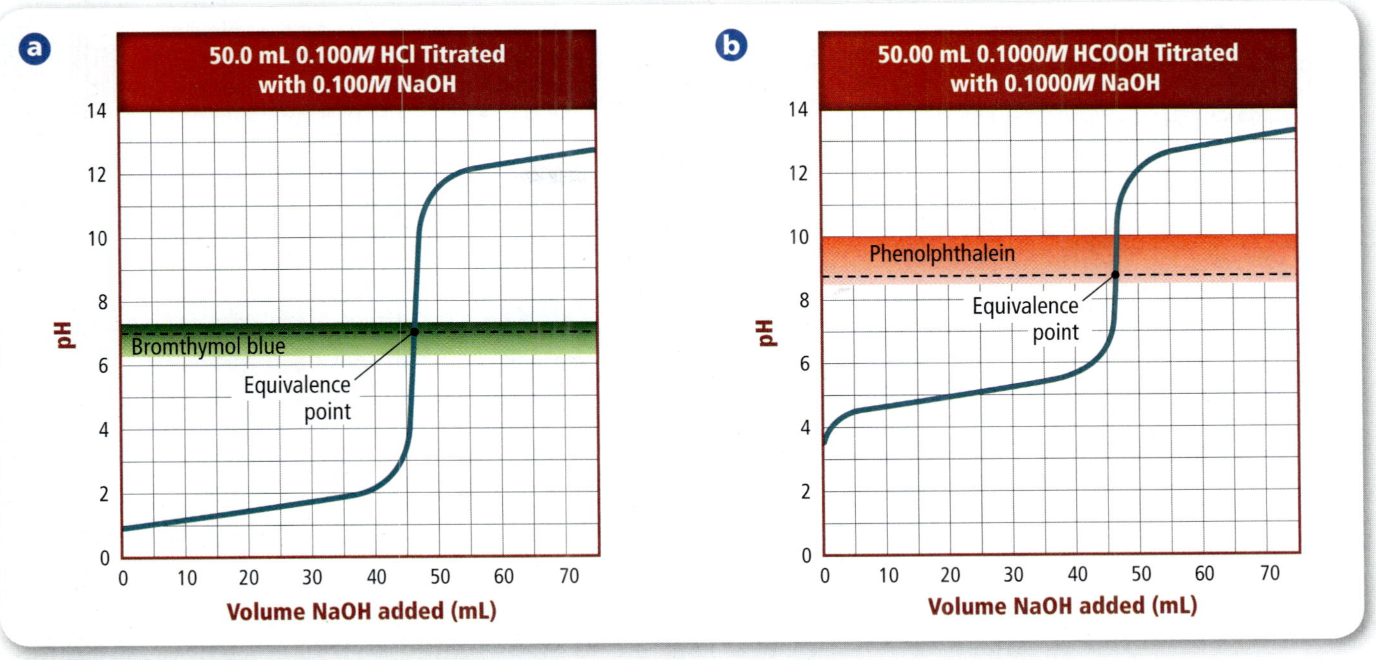


**Figure 22** In the titration of a strong acid by a strong base shown in **a,** a steep rise in the pH of the acid solution indicates that all of the $H^+$ ions from the acid have been neutralized by the $OH^-$ ions of the base. The point at which the curve flexes (at its intersection with the dashed line) is the equivalence point of the titration. Bromthymol blue is an indicator that changes color at this equivalence point. In **b,** a weak acid (HCOOH) is titrated with a strong base (NaOH). The equivalence point is not at a pH of 7. Phenolphthalein is an indicator that changes color at this equivalence point.

**Compare** ***the equivalence points in the two illustrations.***

**Titration procedure** How is an acid-base titration performed?

1. A measured volume of an acidic or basic solution of unknown concentration is placed in a beaker. The electrodes of a pH meter are immersed in this solution, and the initial pH of the solution is read and recorded.
2. A buret is filled with the titrating solution of known concentration. This is called the standard solution, or **titrant.**
3. Measured volumes of the standard solution are added slowly and mixed into the solution in the beaker. The pH is read and recorded after each addition. This process continues until the reaction reaches the **equivalence point,** which is the point at which moles of $H^+$ ion from the acid equal moles of $OH^-$ ion from the base.

**Figure 22a** shows how the pH of the solution changes during the titration of 50.0 mL of 0.100*M* HCl, a strong acid, with 0.100*M* NaOH, a strong base. The initial pH of the 0.100*M* HCl is 1.00. As NaOH is added, the acid is neutralized and the solution's pH increases gradually. However, when nearly all of the $H^+$ ions from the acid have been used up, the pH increases dramatically with the addition of an exceedingly small volume of NaOH. This abrupt increase in pH occurs at the equivalence point of the titration. Beyond the equivalence point, the addition of more NaOH again results in a gradual increase in pH.

You might think that all titrations must have an equivalence point at pH 7 because that is the point at which concentrations of hydrogen ions and hydroxide ions are equal and the solution is neutral. This is not the case, however. Some titrations have equivalence points at pH values less than 7, and some have equivalence points at pH values greater than 7. These differences occur because of reactions between the newly formed salts and water, as you will read later. **Figure 22b** shows that the equivalence point for the titration of methanoic acid (a weak acid) with sodium hydroxide (a strong base) lies between pH 8 and pH 9.

View an **animation about neutralization reactions.**

Concepts In Motion

☑ GRAPH CHECK **Identify** two ways in which the graphs in **Figure 22** are different.

■ **Figure 23** The familiar dark color of tea becomes lighter when lemon juice is added. A substance contained in tea is an indicator. Most indicators are large molecules that act as weak acids. Slight differences in bonding patterns when an indicator molecule is ionized or un-ionized account for the color changes.

**Acid-base indicators** Chemists often use a chemical dye rather than a pH meter to detect the equivalence point of an acid-base titration. Chemical dyes whose colors are affected by acidic and basic solutions are called **acid-base indicators.** Many natural substances act as indicators. If you use lemon juice in your tea, you might have noticed that the brown color of tea gets lighter when lemon juice is added, as shown in **Figure 23.** Tea contains compounds called polyphenols that have slightly ionizable hydrogen atoms and therefore are weak acids. Adding acid in the form of lemon juice to a cup of tea depresses the ionization according to Le Châtelier's principle, and the color of the un-ionized polyphenols becomes more apparent. Many of the indicators used by chemists are shown in **Figure 24.** As shown in **Figure 22,** bromthymol blue is a good choice for a titration of a strong acid with a strong base, and phenophthalein changes color at the equivalence point of a titration of a weak acid with a strong base.

■ **Figure 24** Choosing the right indicator is important. The indicator must change color at the equivalence point of the titration which is not always at pH 7.

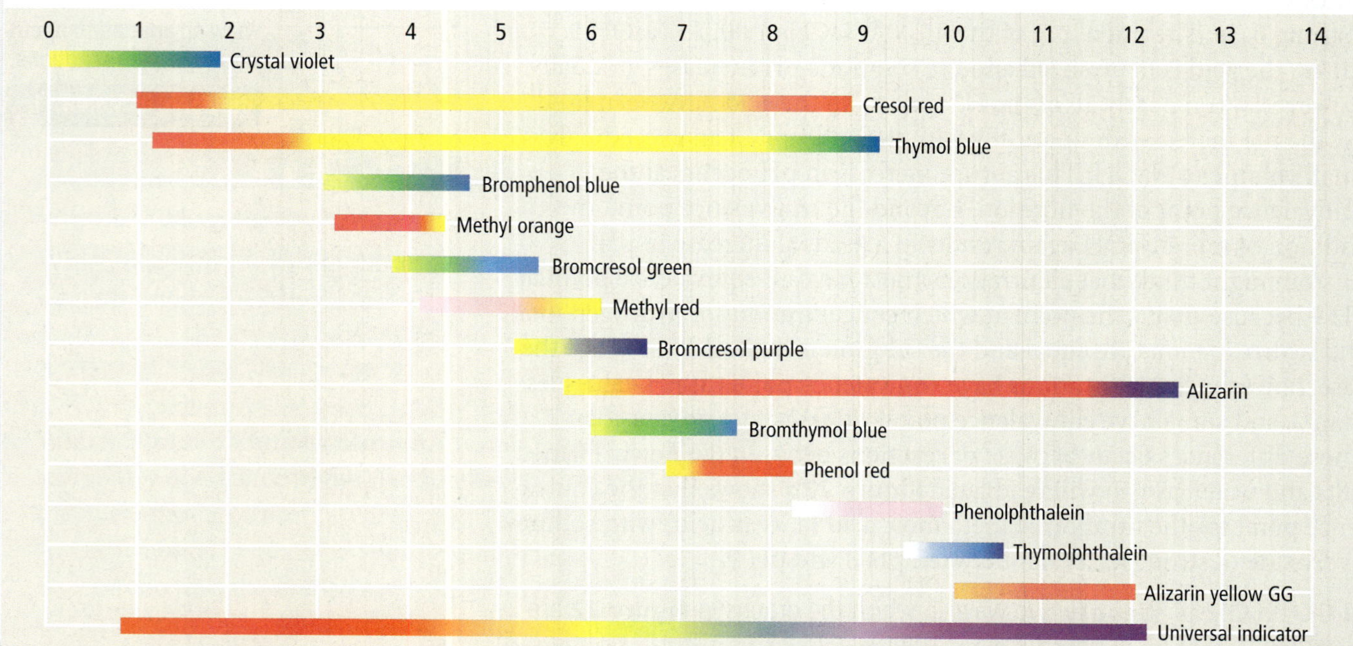


Matt Meadows

The buret contains the standard solution (0.1000*M* NaOH), and the flask contains 25.00 mL HCOOH solution along with a small amount of phenolphthalein indicator.

The standard solution is added slowly to the acid solution. The phenolphthalein indicator turns pink, but the color disappears upon mixing, until the end point is reached.

The end point of the titration is marked by a permanent, but very light, pink color. A careful reading of the buret reveals that 18.28 mL 0.1000*M* NaOH has been added.

**Figure 25** Titration is a precise procedure requiring practice. The white paper under the flask provides a background for viewing the indicator color change.

**Indicators and titration end point** Many indicators used for titration are weak acids. Each has it own particular pH or pH ranges over which it changes color. The point at which the indicator used in a titration changes color is called the **end point** of the titration. It is important to choose an indicator for a titration that will change color at the equivalence point of the titration. Remember that the role of the indicator is to indicate to you, by means of a color change, that just enough of the titrating solution has been added to neutralize the unknown solution. **Figure 25** describes the titration of an unknown solution of methanoic acid (HCOOH) with 0.1000*M* NaOH.

Watch a **video about acids and bases.**

Video

## PROBLEM-SOLVING STRATEGY

### Calculating Molarity

The balanced equation for a titration reaction is the key to calculating the unknown molarity. For example, sulfuric acid is titrated with sodium hydroxide according to this equation.

$$H_2SO_4(aq) + 2NaOH(aq) \rightarrow Na_2SO_4(aq) + 2H_2O(l)$$

1. Calculate the moles of NaOH in the standard from the titration data: molarity of the base ($M_B$) and the volume of the base ($V_B$).

   $(M_B)\ (V_B) = (\text{mol/L})(\text{L}) = \text{mol NaOH in standard}$

2. From the equation, you know that the mole ratio of NaOH to $H_2SO_4$ is 2:1. Two moles of NaOH are required to neutralize 1 mol of $H_2SO_4$.

   $\text{mol } H_2SO_4 \text{ titrated} = \text{mol NaOH in standard} \times \dfrac{1 \text{ mol } H_2SO_4}{2 \text{ mol NaOH}}$

3. $M_A$ represents the molarity of the acid and $V_A$ represents the volume of the acid in liters. $M_A = \dfrac{\text{mol } H_2SO_4 \text{ titrated}}{V_A}$

   Apply this strategy as you study Example Problem 6.

## EXAMPLE Problem 6

Find help with **dimensional analysis.** Math Handbook

EXAMPLE PROBLEM

**MOLARITY FROM TITRATION DATA** A volume of 18.28 mL of a standard solution of 0.1000*M* NaOH was required to neutralize 25.00 mL of a solution of methanoic acid (HCOOH). What is the molarity of the methanoic acid solution?

### 1 ANALYZE THE PROBLEM

You are given the molarity and volume of the NaOH solution and the volume of the methanoic acid (HCOOH) solution. The volume of base used is about four-fifths of the volume of the acid, so the molarity of the acid solution should be less than 0.1*M*.

**Known**

$V_A$ = 25.00 mL HCOOH

$V_B$ = 18.28 mL NaOH

$M_B$ = 0.1000*M*

**Unknown**

$M_A$ = ? mol/L

### 2 SOLVE FOR THE UNKNOWN

Write the balanced formula equation for the neutralization reaction.

$$HCOOH(aq) + NaOH(aq) \rightarrow HCOONa(aq) + H_2O(l)$$

1 mol NaOH neutralizes 1 mol HCOOH. — **Write the acid to base mole relationship.**

$$V_B = 18.28\ \cancel{mL} \times \frac{1\ L}{1000\ \cancel{mL}} = 0.01828\ L$$

**Convert volume of base from mL to L.**

Calculate moles of NaOH.

$$\text{Mol NaOH} = (M_B)(V_B)$$

**Apply the relationship between moles of base, molarity of base, and volume of base.**

$$\text{Mol NaOH} = (0.1000\ mol/\cancel{L})(0.01828\ \cancel{L}) = 1.828 \times 10^{-3}\ \text{mol NaOH}$$

**Substitute $M_B$ = 0.1000*M* and $V_B$ = 0.01828 L.**

Calculate moles of HCOOH.

$$1.828 \times 10^{-3}\ \cancel{\text{mol NaOH}} \times \frac{1\ \text{mol HCOOH}}{1\ \text{mol NaOH}} = 1.828 \times 10^{-3}\ \text{mol HCOOH}$$

**Apply the stoichiometric relationship.**

Calculate the molarity of HCOOH.

$$1.828 \times 10^{-3}\ \text{mol HCOOH} = (M_A)(V_A)$$

**Apply the relationship between moles of acids, molarity of acid, and volume of acid.**

$$M_A = \frac{1.828 \times 10^{-3}\ \text{mol HCOOH}}{V_A}$$

**Solve for $M_A$.**

$$V_A = 25.00\ \cancel{mL} \times \frac{1\ L}{1000\ \cancel{mL}} = 0.02500\ \text{L HCOOH}$$

**Convert volume of acid from mL to L.**

$$M_A = \frac{1.828 \times 10^{-3}\ \text{mol HCOOH}}{0.02500\ \text{L HCOOH}} = 7.312 \times 10^{-2}\ \text{mol/L}$$

**Substitute $V_A$ = 0.02500 L.**

### 3 EVALUATE THE ANSWER

The answer agrees with the prediction that the molarity of HCOOH is less than 0.1*M*, and is correctly recorded with four significant figures and the appropriate unit.

## PRACTICE Problems

Do additional problems. Online Practice

**44.** What is the molarity of a nitric acid solution if 43.33 mL of 0.1000*M* KOH solution is needed to neutralize 20.00 mL of the acid solution?

**45.** What is the concentration of a household ammonia cleaning solution if 49.90 mL of 0.5900*M* HCl is required to neutralize 25.00 mL of the solution?

**46. Challenge** How many milliliters of 0.500*M* NaOH would neutralize 25.00 mL of 0.100*M* $H_3PO_4$?

## Salt Hydrolysis

In **Figure 26,** several drops of bromthymol blue indicator solution have been added to 0.10*M* aqueous solutions of the salts ammonium chloride ($NH_4Cl$), sodium nitrate ($NaNO_3$), and potassium fluoride (KF). Sodium nitrate turns the indicator green, which means that the solution is neutral. The blue color of the KF solution means that the solution is basic, and the yellow color of the ammonium chloride solution indicates that the solution is acidic. Why are some aqueous salt solutions neutral, some basic, and some acidic? Many salts react with water in a process known as salt hydrolysis. In **salt hydrolysis,** the anions of the dissociated salt accept hydrogen ions from water or the cations of the dissociated salt donate hydrogen ions to water.

**Figure 26** The indicator bromthymol blue provides surprising results when added to three solutions of ionic salts. An $NH_4Cl$ solution is acidic, a $NaNO_3$ solution is neutral, and a KF solution is basic. The explanation has to do with the strengths of the acid and base from which each salt was formed.

**Salts that produce basic solutions** Potassium fluoride is the salt of a strong base (KOH) and a weak acid (HF). It dissociates into potassium ions and fluoride ions.

$$KF(s) \rightarrow K^+(aq) + F^-(aq)$$

The $K^+$ ions do not react with water, but the $F^-$ ion is a weak Brønsted-Lowry base. Some fluoride ions establish this equilibrium with water.

$$F^-(aq) + H_2O(l) \rightleftharpoons HF(aq) + OH^-(aq)$$

Hydrogen fluoride molecules and $OH^-$ ions are produced. The production of the $OH^-$ ions makes the solution basic.

Explore **titrations.**

Virtual Investigations

**Salts that produce acidic solutions** $NH_4Cl$ is the salt of a weak base ($NH_3$) and a strong acid (HCl). When dissolved in water, the salt dissociates into ammonium ions and chloride ions.

$$NH_4Cl(s) \rightarrow NH_4^+(aq) + Cl^-(aq)$$

The $Cl^-$ ions do not react with water, but the $NH_4^+$ ion is a weak Brønsted-Lowry acid. Ammonium ions react with water molecules to establish this equilibrium.

$$NH_4^+(aq) + H_2O(l) \rightleftharpoons NH_3(aq) + H_3O^+(aq)$$

Ammonia molecules and hydronium ions are produced. The presence of hydronium ions makes the solution acidic.

**Salts that produce neutral solutions** Sodium nitrate ($NaNO_3$) is the salt of a strong acid ($HNO_3$) and a strong base (NaOH). Little or no salt hydrolysis occurs because neither $Na^+$ nor $NO_3^-$ react with water. Therefore, a solution of sodium nitrate is neutral.

### PRACTICE Problems

Do additional problems. Online Practice

**47.** Write equations for the salt hydrolysis reactions occurring when the following salts dissolve in water. Classify each as acidic, basic, or neutral.

**a.** ammonium nitrate **c.** rubidium acetate

**b.** potassium sulfate **d.** calcium carbonate

**48. Challenge** Write the equation for the reaction that occurs in a titration of ammonium hydroxide ($NH_4OH$) with hydrogen bromide (HBr). Will the pH at the equivalence point be greater or less than 7?

■ **Figure 27** To provide a healthy environment for these jellies, the pH of the aquarium water at the Monterey Bay Aquarium must be adjusted to stay within the range of 8.1 to 8.4.

**Predict** ***what would happen if the pH were allowed to fall to 7.0.***

## Buffered Solutions

It is important for the jellies shown in **Figure 27** that the aquarium water be kept within a narrow pH range. A constant pH is also important in your body. The pH of your blood must be maintained within the range of 7.1 to 7.7. The gastric juices in your stomach must have a pH between 1.6 and 1.8 to promote digestion of certain foods. Your body maintains pH values within such narrow limits by producing buffers.

**What is a buffer?** **Buffers** are solutions that resist changes in pH when limited amounts of acid or base are added. For example, adding 0.01 mol of HCl to 1 L of pure water lowers the pH by 5.0 units, from 7.0 to 2.0. Similarly, adding 0.01 mol of NaOH to 1 L of pure water increases the pH from 7.0 to 12.0. However, if you add the same amount of either HCl or NaOH to 1 L of a buffered solution, the pH might change by no more than 0.1 unit.

**How do buffers work?** A buffer is a mixture of a weak acid and its conjugate base or a weak base and its conjugate acid. The mixture of ions and molecules in a buffer solution resists changes in pH by reacting with any hydrogen ions or hydroxide ions added to the buffered solution.

Suppose that a buffer solution contains 0.1$M$ concentrations of hydrofluoric acid (HF) and sodium fluoride (NaF). The NaF provides a 0.1$M$ concentration of $F^-$ ions. HF is the acid, and $F^-$ is its conjugate base. The following equilibrium would be established.

$$HF(aq) \rightleftharpoons H^+(aq) + F^-(aq)$$

**Adding an acid** When an acid is added to a buffered solution, the equilibrium shifts to the left. According to Le Châtelier's principle, the added $H^+$ ions from the acid are a stress on the equilibrium, which is relieved by their reaction with $F^-$ ions to form additional undissociated HF molecules.

$$HF(aq) \leftleftharpoons H^+(aq) + F^-(aq)$$

Equilibrium is established again with a larger amount of undissociated HF present. However, the pH of the solution has changed little because the shift to the left consumed most of the added $H^+$ ion.

**Adding a base** When a base is added to the hydrofluoric acid/fluoride ion buffer system, the added $OH^-$ ions react with $H^+$ ions to form $H_2O$. This decreases the concentration of $H^+$ ions, and the equilibrium shifts to the right to replace the $H^+$ ions.

$$HF(aq) \rightleftharpoons H^+(aq) + F^-(aq)$$

Although the shift to the right consumes HF molecules and produces additional $F^-$ ions, the pH remains fairly constant because the $H^+$ ion concentration has not changed appreciably.

A buffer solution's capacity to resist pH change can be exceeded by the addition of too much acid or base. The amount of acid or base a buffer solution can absorb without a significant change in pH is called the **buffer capacity** of the solution. The greater the concentrations of the buffering molecules and ions in the solution, the greater the solution's buffer capacity.

**Choosing a buffer** A buffer system is most effective when the concentrations of the conjugate acid-base pair are equal or nearly equal. Consider the $H_2PO_4^-/HPO_4^{2-}$ buffer system made by mixing equal molar amounts of $NaH_2PO_4$ and $Na_2HPO_4$.

$$H_2PO_4^- \rightleftharpoons H^+ + HPO_4^{2-}$$

What is the pH of such a buffer solution? The acid ionization constant expression for the equilibrium can provide the answer.

$$K_a = 6.2 \times 10^{-8} = \frac{[H^+][HPO_4^{2-}]}{[H_2PO^{4-}]}$$

Because the solution has been made with equal molar amounts of $Na_2HPO_4$ and $NaH_2PO_4$, $[HPO_4^{2-}]$ is equal to $[H_2PO_4^-]$. Thus, the two terms in the acid ionization expression cancel.

$$6.2 \times 10^{-8} = \frac{[H^+][\cancel{HPO_4^{2-}}]}{[\cancel{H_2PO_4^-}]} = [H^+]$$

$$pH = -\log [H^+] = -\log (6.2 \times 10^{-8}) = 7.21$$

Thus, when equimolar amounts of each of the components are present in the $HPO_4^-/H_2PO_4^{2-}$ buffer system, the system can maintain a pH close to 7.21. Note that the pH is the negative log of $K_a$. **Table 7** lists several buffer systems, with the pH at which each is effective.

**VOCABULARY**

**SCIENCE USAGE V. COMMON USAGE**

**Buffer**

***Science usage:*** a solution that resists changes in pH when limited amounts of acid or base are added

*The chemist decided to use a buffer consisting of equal molar amounts of formic acid and sodium formate.*

***Common usage:*** something that serves as a protective barrier

*For the homes along the shore, the high seawall served as a buffer against stormy seas.*

**Table 7 Buffer Systems with Equimolar Components**

| Buffer Equilibrium | Conjugate Acid-Base Pair in Buffered Solution | Buffer pH |
|---|---|---|
| $HF(aq) \rightleftharpoons H^+(aq) + F^-(aq)$ | $HF/F^-$ | 3.20 |
| $CH_3COOH(aq) \rightleftharpoons H^+(aq) + CH_3COO^-(aq)$ | $CH_3COOH/CH_3COO^-$ | 4.76 |
| $H_2CO_3(aq) \rightleftharpoons H^+(aq) + HCO_3^-(aq)$ | $H_2CO_3/HCO_3^-$ | 6.35 |
| $H_2PO_4^-(aq) \rightleftharpoons H^+(aq) + HPO_4^{2-}(aq)$ | $H_2PO_4^-/HPO_4^{2-}$ | 7.21 |
| $NH_3(aq) + H_2O(l) \rightleftharpoons NH_4^+(aq) + OH^-(aq)$ | $NH_4^+/NH_3$ | 9.4 |
| $C_2H_5NH_2(aq) + H_2O(l) \rightleftharpoons C_2H_5NH_3^+(aq) + OH^-(aq)$ | $C_2H_5NH_3^+/C_2H_5NH_2$ | 10.70 |

# Problem-Solving LAB

## Apply Scientific Explanations

**How does your blood maintain its pH?** Human blood contains three types of cells. Red blood cells deliver oxygen to every part of the body. White blood cells fight infections, and platelets aid in clotting when bleeding occurs. The critical functions of these cells are impaired if the pH of blood is not maintained within the narrow range of 7.1 to 7.7. Beyond this range, proteins in the body lose their structures and abilities to function. Fortunately, several buffers maintain the necessary acid/base balance. The carbonic acid/hydrogen carbonate ($H_2CO_3/HCO_3^-$) buffer is the most important.

$$CO_2(g) + H_2O(l) \rightleftharpoons H_2CO_3(aq) \rightleftharpoons H^+(aq) + HCO_3^-(aq)$$

As acids and bases enter the bloodstream as a result of normal activity, the blood's buffer systems shift to effectively maintain a healthful pH.

### Analysis

Depending on the body's metabolic rate and other factors, the $H_2CO_3/HCO_3^-$ equilibrium will shift according to Le Châtelier's principle. In addition, the lungs can alter the rate at which $CO_2$ is expelled from the body by breathing, and the kidneys can alter the rate of removal of $HCO_3^-$ ions.

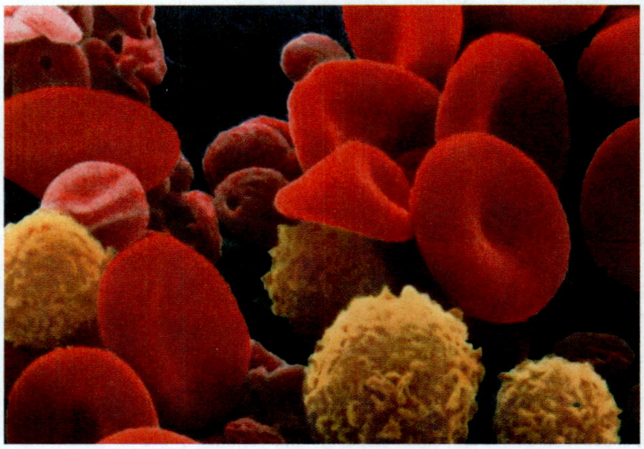

### Think Critically

1. **Determine** how many times greater the $[H^+]$ is if the blood's pH changes from pH 7.4 to 7.1.
2. **Suggest** a reason why a 20:1 ratio of $HCO_3^-$ to $CO_2$ in the blood is favorable for maintaining a healthy pH.
3. **Predict** whether, for each situation, the pH of the blood will rise or fall, and which way the $H_2CO_3/HCO_3^-$ equilibrium will shift.
   **a.** A person with a severe stomach virus vomits many times during a 24-h period.
   **b.** To combat heartburn, a person takes too much ($NaHCO_3$).

## SECTION 4 REVIEW

Section Self-Check

### Section Summary

- In a neutralization reaction, an acid and a base react to form a salt and water.
- The net ionic equation for the neutralization of a strong acid by a strong base is $H^+(aq) + OH^-(aq) \rightarrow H_2O(l)$.
- Titration is the process in which an acid-base neutralization reaction is used to determine the concentration of a solution.
- Buffered solutions contain mixtures of molecules and ions that resist changes in pH.

49. MAINIDEA **Explain** why the net ionic equation for the neutralization reaction of any strong acid with any strong base is always the same.
50. **Explain** the difference between the equivalence point and the end point of a titration.
51. **Compare** the results of two experiments: First, a small amount of base is added to an unbuffered solution with a pH of 7. Second, the same amount of base is added to a buffered solution with a pH of 7.
52. **Calculate** the molarity of a solution of hydrobromic acid (HBr) if 30.35 mL of 0.1000*M* NaOH is required to titrate 25.00 mL of the acid to the equivalence point.
53. **Interpret** What substances could be used to make a buffer solution with a pH of 9.4. How should the amounts of the substances be related. Use **Table 7.**
54. **Design an Experiment** Describe how you would design and perform a titration in which you use 0.250*M* $HNO_3$ to determine the molarity of a cesium hydroxide solution. Include the formula and net ionic equations.

**everyday**CHEMISTRY

# Acid-Base Reactions on the Rise

Do you remember how much fun it was to watch a vinegar baking soda volcano erupt? The bubbles of carbon dioxide ($CO_2$) resulted from a decomposition reaction that quickly followed the acid-base reaction between the vinegar ($HC_2H_3O_2$), an acid, and baking soda ($NaHCO_3$), a base, as shown below.

**Acid-Base Reaction**

$$HC_2H_3O_2(aq) + NaHCO_3(aq) \rightarrow NaC_2H_3O_2(aq) + H_2CO_3(aq)$$

**Decomposition**

$$H_2CO_3(aq) \rightarrow CO_2(g) + H_2O(l)$$

The release of carbon dioxide as a result of the chemical reaction between an acid and a base, as shown in **Figure 1,** is part of the reason why baked goods rise. An ingredient that causes batter to rise when baked is called a leavening agent. The two main chemical leavening agents are baking soda and baking powder.

**Baking soda** Sodium hydrogen carbonate, also called sodium bicarbonate, is the chemical name for baking soda. When used in cooking, baking soda reacts with mildly acidic liquids, and carbon dioxide bubbles form. Mildly acidic liquids include vinegar, molasses, honey, citrus juice, buttermilk, and many others.

**Figure 1** Carbon dioxide forms bubbles when baking soda, a base, is added to vinegar, an acid.

**Figure 2** Baking traps the bubbles formed during the reaction between an acid and a base, resulting in a light, airy cake.

Baking soda must be mixed with other dry ingredients and added last to a batter so that the release of carbon dioxide is uniform throughout the batter. This acid-base reaction happens quickly. If baking soda is the only leavening agent in a recipe, the batter must be baked immediately before the bubbles have a chance to escape. Baking causes the bubbles to expand, and the cake rises. As the batter firms, the bubbles are trapped, as shown in **Figure 2.**

**Baking powder** If a recipe does not include an acidic liquid, baking powder is used. Most baking powder is a mixture of baking soda and two dry acids. One of the acids reacts with the baking soda when it dissolves in the batter, and the other reacts with the baking soda when heated.

Like baking soda, baking powder is mixed with other dry ingredients and added last to a batter. However, batters made with baking powder do not have to be baked immediately.

Sometimes, batters made with mildly acidic liquids include both baking powder and baking soda. Excess acid can disrupt the action of the baking powder. The baking powder provides a reliable source of carbon dioxide, and the baking soda helps to neutralize the acid.

## WRITING IN Chemistry

**Analyze** If a recipe calls for flour, salt, sugar, bran cereal, milk, an egg, and vegetable oil, would you use baking soda or baking powder? Explain.

WebQuest

# ChemLAB

iLab Station

## Standardize a Base

**Background: Titration is a procedure by which the molarity of a base can be determined.**

**Question:** *How can you determine the molarity of a solution of a base?*

### Materials

50-mL buret
buret clamp
ring stand
sodium hydroxide pellets (NaOH)
potassium hydrogen phthalate ($KHC_8H_4O_4$)
distilled water
weighing bottle
spatula
250-mL Erlenmeyer flask
500-mL Florence flask with rubber stopper
250-mL beaker
centigram balance
wash bottle
phenolphthalein solution

### Safety Precautions

WARNING: ***Dissolving NaOH in water generates heat. Phenolphthalein is flammable. Keep away from flames.***

### Procedure

1. Read and complete the lab safety form.
2. Place about 4 g of NaOH in a 500-mL Florence flask. Add enough water to dissolve the pellets and bring the volume of the NaOH solution to about 400 mL. Stopper the flask.
3. Use the weighing bottle to mass by difference about 0.40 g of potassium hydrogen phthalate ($KHC_8H_4O_4$, molar mass = 204.32 g/mol) into a 250-mL Erlenmeyer flask. Record this mass.
4. Use a wash bottle to rinse the insides of the flask, and add about 50 mL of water. Add two drops of phenolphthalein indicator solution.
5. Rinse the buret with 10 mL of your base solution. Discard the rinse solution in a discard beaker. Attach the buret to the ring stand using the buret clamp.
6. Fill the buret with NaOH solution. The level of the liquid should be at or below the zero mark. To remove any air trapped in the tip of the buret, allow a small amount of the base to flow from the tip into the discard beaker. Read the buret to the nearest 0.02 mL, and record this initial reading.
7. Place a piece of white paper on the base of the ringstand. Swirl the flask while allowing the NaOH solution to flow slowly from the buret into the flask.

| Titration Data | |
|---|---|
| | Trial 1 |
| Mass of weighing bottle and acid | |
| Mass of weighing bottle | |
| Mass of solid acid | |
| Moles of acid | |
| Moles of base required | |
| Final reading of base buret | |
| Initial reading of base buret | |
| Volume of base used in mL | |
| Molarity of base | |

8. When the pink color begins to persist longer as the flask is swirled, add the base drop-by-drop.
9. The end point is reached when one additional drop of base turns the acid pink. The pink color should persist as the flask is swirled. Record the final volume in the buret.
10. Calculate the molarity of your base using Steps 1–4 in the Analyze and Conclude section.
11. Refill the buret. Rinse the flask with water. Repeat the titration until the calculated values of the molarity for three trials show close agreement.
12. **Cleanup and Disposal** Wash the neutralized solutions down the sink with plenty of water.

### Analyze and Conclude

1. **Interpret Data** For each titration, calculate the number of moles of acid used by dividing the mass of the sample by the molar mass of the acid.
2. **Infer** How many moles of base are required to react with the moles of acid you used?
3. **Calculate** Convert the volume of base to liters.
4. **Calculate** the molarity of the base by dividing the moles of base by the volume of base in liters.
5. **Error Analysis** Did your calculated molarities agree? Explain any irregularities.

### INQUIRY EXTENSION

**Design an Experiment** Determine the concentration of a vinegar solution without using an indicator.

# CHAPTER 18 STUDY GUIDE

Vocabulary Practice

**BIGIDEA** **Acids and bases can be defined in terms of hydrogen ions and hydroxide ions or in terms of electron pairs.**

## SECTION 1 Introduction to Acids and Bases

**MAINIDEA** Different models help describe the behavior of acids and bases.

- The concentrations of hydrogen ions and hydroxide ions determine whether an aqueous solution is acidic, basic, or neutral.
- An Arrhenius acid must contain an ionizable hydrogen atom. An Arrhenius base must contain an ionizable hydroxide group.
- A Brønsted-Lowry acid is a hydrogen ion donor. A Brønsted-Lowry base is a hydrogen ion acceptor.
- A Lewis acid accepts an electron pair. A Lewis base donates an electron pair.

**VOCABULARY**

- **acidic solution**
- **basic solution**
- **Arrhenius model**
- **Brønsted-Lowry model**
- **conjugate acid**
- **conjugate base**
- **conjugate acid-base pair**
- **amphoteric**
- **Lewis model**

## SECTION 2 Strengths of Acids and Bases

**MAINIDEA** In solution, strong acids and bases ionize completely, but weak acids and bases ionize only partially.

- Strong acids and strong bases are completely ionized in a dilute aqueous solution. Weak acids and weak bases are partially ionized in a dilute aqueous solution.
- For weak acids and weak bases, the value of the acid or base ionization constant is a measure of the strength of the acid or base.

**VOCABULARY**

- **strong acid**
- **weak acid**
- **acid ionization constant**
- **strong base**
- **weak base**
- **base ionization constant**

## SECTION 3 Hydrogen Ions and pH

**MAINIDEA** pH and pOH are logarithmic scales that express the concentrations of hydrogen ions and hydroxide ions in aqueous solutions.

- The ion product constant for water, $K_w$, equals the product of the $H^+$ ion concentration and the $OH^-$ ion concentration.

  $K_w = [H^+][OH^-]$

- The pH of a solution is the negative log of the hydrogen ion concentration. The pOH is the negative log of the hydroxide ion concentration. pH plus pOH equals 14.

  $pH = -\log [H^+]$ $\quad$ $pOH = -\log [OH^-]$

  $pH + pOH = 14.00$

- A neutral solution has a pH of 7.0 and a pOH of 7.0 because the concentrations of hydrogen ions and hydroxide ions are equal.

**VOCABULARY**

- **ion product constant for water**
- **pH**
- **pOH**

## SECTION 4 Neutralization

**MAINIDEA** In a neutralization reaction, an acid reacts with a base to produce a salt and water.

- In a neutralization reaction, an acid and a base react to form a salt and water.
- The net ionic equation for the neutralization of a strong acid by a strong base is $H^+(aq) + OH^-(aq) \rightarrow H_2O(l)$.
- Titration is the process in which an acid-base neutralization reaction is used to determine the concentration of a solution.
- Buffered solutions contain mixtures of molecules and ions that resist changes in pH.

**VOCABULARY**

- **neutralization reaction**
- **salt**
- **titration**
- **titrant**
- **equivalence point**
- **acid-base indicator**
- **end point**
- **salt hydrolysis**
- **buffer**
- **buffer capacity**

# CHAPTER 18 ASSESSMENT

## SECTION 1

### Mastering Concepts

**55.** In terms of ion concentrations, distinguish between acidic, neutral, and basic solutions.

**56.** Write a balanced chemical equation that represents the self-ionization of water.

**57.** Classify each compound as an Arrhenius acid or an Arrhenius base.

**a.** $H_2S$ **c.** $Mg(OH)_2$
**b.** RbOH **d.** $H_3PO_4$

**58. Geology** When a geologist adds a few drops of HCl to a rock, gas bubbles form. What might the geologist conclude about the nature of the gas and the rock?

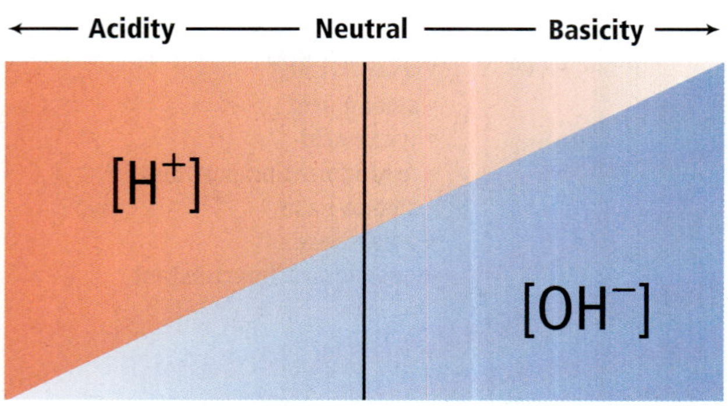


**Figure 28**

**59.** Explain the meaning of the relative sizes of the two shaded areas to the right of the dark vertical line in **Figure 28.**

**60.** Explain the difference between a monoprotic acid, a diprotic acid, and a triprotic acid. Give an example of each.

**61.** Why can $H^+$ and $H_3O^+$ be used interchangeably in chemical equations?

**62.** Use the symbols $<$, $>$, and $=$ to express the relationship between the concentrations of $H^+$ ions and $OH^-$ ions in acidic, neutral, and basic solutions.

**63.** Explain how the definition of a Lewis acid differs from the definition of a Brønsted-Lowry acid.

### Mastering Problems

**64.** Write a balanced chemical equation for each of the following.

**a.** the dissociation of solid magnesium hydroxide in water

**b.** the reaction of magnesium metal and hydrobromic acid

**c.** the ionization of propanoic acid ($CH_3CH_2COOH$) in water

**d.** the second ionization of sulfuric acid in water

## SECTION 2

### Mastering Concepts

**65.** Explain the difference between a strong acid and a weak acid.

**66.** Explain why equilibrium arrows are used in the ionization equations for some acids.

**Figure 29**

**67.** Which of the beakers shown in **Figure 29** might contain a solution of 0.1*M* hypochlorous acid? Explain your answer.

**68.** Explain how you would compare the strengths of two weak acids experimentally and using a table or a handbook.

**69.** Identify the conjugate acid-base pairs in the reaction of $H_3PO_4$ with water.

### Mastering Problems

**70. Ammonia Cleaner** Write the chemical equation and $K_b$ expression for the ionization of ammonia in water. How is it safe for a window cleaner to use a solution of ammonia, which is basic?

**71. Disinfectant** Hypochlorous acid is an industrial disinfectant. Write the chemical equation and the $K_a$ expression for the ionization of hypochlorous acid in water.

**72.** Write the chemical equation and the $K_b$ expression for the ionization of aniline in water. Aniline is a weak base with the formula $C_2H_5NH_2$.

**73.** A fictional weak base, $ZaH_2$, reacts with water to yield a solution with a $OH^-$ ion concentration of $2.68 \times 10^{-4}$ mol/L. The chemical equation for the reaction is $ZaH_2(aq) + H_2O(l) \leftrightharpoons ZaH_3^+(aq) + OH^-(aq)$. If $[ZaH_2]$ at equilibrium is 0.0997 mol/L, what is the value of $K_b$ for $ZaH_2$?

**74.** Select a strong acid, and explain how you would prepare a dilute solution of the acid. Select a weak acid, and explain how you would prepare a concentrated solution of the acid.

## SECTION 3

### Mastering Concepts

**75.** What is the relationship between the pOH and the $OH^-$ ion concentration of a solution?

**76.** Solution A has a pH of 2.0. Solution B has a pH of 5.0. Which solution is more acidic? Based on the $H^+$ ion concentrations in the two solutions, how many times more acidic?

**77.** If the concentration of $H^+$ ions in an aqueous solution decreases, what must happen to the concentration of $OH^-$ ions? Why?

**78.** Use Le Châtelier's principle to explain what happens to the equilibrium $H_2O(l) \leftrightharpoons H^+(aq) + OH^-(aq)$ when a few drops of HCl are added to pure water.

**79. Common Acids and Bases** Use the data in **Table 8** to answer the following questions.

**Table 8** pH values

| Substance | pH |
|---|---|
| Household ammonia | 11.3 |
| Lemon juice | 2.3 |
| Antacid | 9.4 |
| Blood | 7.4 |
| Soft drinks | 3.0 |

**a.** Which substance is the most basic?
**b.** Which substance is closest to neutral?
**c.** Which has a concentration of $H^+ = 4.0 \times 10^{-10}M$?
**d.** Which has a pOH of 11.0?
**e.** How many times more basic is antacid than blood?

### Mastering Problems

**80.** What is $[OH^-]$ in an aqueous solution at 298 K in which $[H^+] = 5.40 \times 10^{-3}M$?

**81.** What are the pH and pOH for the solution described in Question 80?

**82.** If 5.00 mL of 6.00$M$ HCl is added to 95.00 mL of pure water, the final volume of the solution is 100.00 mL. What is the pH of the solution?

**83.** Given two solutions, 0.10$M$ HCl and 0.10$M$ HF, which solution has the greater concentration of $H^+$ ions? Calculate pH values for the two solutions, given that $[H^+] = 7.9 \times 10^{-3}M$ in the 0.10$M$ HF.

**84. Metal Cleaner** Chromic acid is used as an industrial cleaner for metals. What is $K_a$ for the second ionization of chromic acid ($H_2CrO_4$) if a 0.040$M$ solution of sodium hydrogen chromate has a pH of 3.946?

## SECTION 4

### Mastering Concepts

**85.** What acid and base must react to produce an aqueous sodium iodide solution?

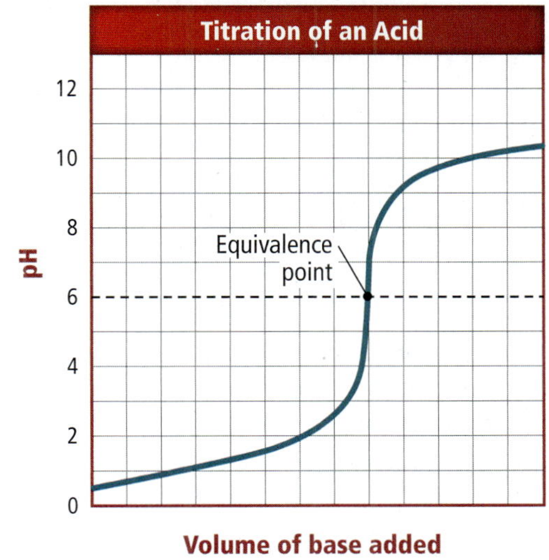


**Figure 30**

**86.** What acid-base indicators, shown in **Figure 24,** would be suitable for the neutralization reaction whose titration curve is shown in **Figure 30?** Why?

**87.** When might a pH meter be better than an indicator to determine the end point of an acid-base titration?

**88.** What happens when an acid is added to a solution containing the $HF/F^-$ buffer system?

**89.** When methyl red is added to an aqueous solution, a pink color results. When methyl orange is added to the same solution, a yellow color is produced. What is the approximate pH range of the solution? Use **Figure 24.**

**90.** Give the name and formula of the acid and the base from which each salt was formed.
**a.** NaCl **b.** $KHCO_3$ **c.** $NH_4NO_2$ **d.** CaS

### Mastering Problems

**91.** Write formula equations and net ionic equations for the hydrolysis of each salt in water.
**a.** sodium carbonate **b.** ammonium bromide

**92. Air Purifier** Lithium hydroxide is used to purify air by removing carbon dioxide. A 25.00-mL sample of lithium hydroxide solution is titrated to an end point by 15.22 mL of 0.3340$M$ hydrochloric acid solution. What is the molarity of the LiOH solution?

**93.** In an acid-base titration, 45.78 mL of a sulfuric acid solution is titrated to the end point by 74.30 mL of 0.4388$M$ sodium hydroxide solution. What is the molarity of the $H_2SO_4$ solution?

## MIXED REVIEW

**94.** Write the equation for the ionization reaction and the base ionization constant expression for ethylamine ($C_2H_5NH_2$) in water.

**95.** How many milliliters of $0.225M$ HCl would be required to titrate 6.00 g of KOH?

**96.** What is the pH of a $0.200M$ solution of hypobromous acid (HBrO)? $K_a = 2.8 \times 10^{-9}$

**97.** Which of the following are polyprotic acids? Write successive ionization equations for the polyprotic acids in water.

**a.** $H_3BO_3$
**b.** $CH_3COOH$
**c.** $HNO_3$
**d.** $H_2SeO_3$

**98.** Write balanced chemical equations for the two successive ionizations of carbonic acid in water. Identify the conjugate-base pair in each of the equations.

**99. Sugar Refining** Strontium hydroxide is used in the refining of beet sugar. Only 4.1 g of strontium hydroxide can be dissolved in 1 L of water at 273 K. Given that its solubility is so low, explain how it is possible that strontium hydroxide is considered a strong base.

**100.** What are the concentrations of $OH^-$ ions in solutions having pH values of 3.00, 6.00, 9.00, and 12.00 at 298 K? What are the pOH values for the solutions?

■ **Figure 31**

**101.** The pH probe in **Figure 31** is immersed in a $0.200M$ solution of a monoprotic acid, HA, at 303 K. What is the value of $K_a$ for the acid at 303 K?

**102.** Write the chemical equation for the reaction that would occur when a base is added to a solution containing the $H_2PO_4^-/HPO_4^{2-}$ buffer system.

**103.** An aqueous solution buffered by benzoic acid ($C_6H_5COOH$) and sodium benzoate ($C_6H_5COOHNa$) is $0.0500M$ in both compounds. Given that benzoic acid's $K_a$ equals $6.4 \times 10^{-5}$, what is the pH of the solution?

## THINK CRITICALLY

**104. Critique** the following statement: "A substance whose chemical formula contains a hydroxyl group must be considered to be a base."

**105. Analyze and Conclude** Is it possible that an a Arrhenius acid is not a Brønsted-Lowry acid? Is it possible that an acid according to the Brønsted-Lowry model is not an Arrhenius acid? Is it possible that a Lewis acid could not be classified as either an Arrhenius or a Brønsted-Lowry acid? Explain and give examples.

**106. Apply Concepts** Use the ion product constant of water at 298 K to explain why a solution with a pH of 3.0 must have a pOH of 11.0.

**107. Identify** the Lewis acids and bases in the following reactions.

**a.** $H^+ + OH^- \leftrightharpoons H_2O$
**b.** $Cl^- + BCl_3 \leftrightharpoons BCl_4^-$
**c.** $SO_3 + H_2O \leftrightharpoons H_2SO_4$

**108. Interpret Scientific Illustrations** Sketch the shape of the approximate pH v. volume curve that would result from titrating a diprotic acid with a $0.10M$ NaOH solution.

**109. Recognize Cause and Effect** Illustrate how a buffer works using the $C_2H_5NH_3^+/C_2H_5NH_2$ buffer system. Show with equations how the weak base/conjugate acid system is affected when small amounts of acid and base are added to a solution containing this buffer system.

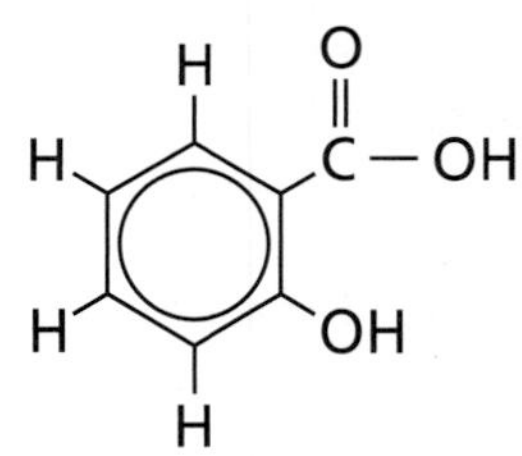

■ **Figure 32**

**110. Predict** Salicylic acid, shown in **Figure 32,** is used to manufacture acetylsalicylic acid, commonly known as aspirin. Evaluate the hydrogen atoms in the salicylic acid molecule based on your knowledge about the ionizable hydrogen in the acetic acid molecule, $CH_3COOH$. Predict which of salicylic acid's hydrogen atoms is likely to be ionizable.

**111. Apply Concepts** Like all equilibrium constants, the value of $K_w$ varies with temperature. $K_w$ equals $2.92 \times 10^{-15}$ at 10°C, $1.00 \times 10^{-14}$ at 25°C, and $2.92 \times 10^{-14}$ at 40°C. In light of this information, calculate and compare the pH values for pure water at these three temperatures. Based on your calculations, is it correct to say that the pH of pure water is always 7.0? Explain.

Chapter Self-Check

## CHALLENGE PROBLEM

**112.** You have 20.0 mL of a solution of a weak acid, HX, whose $K_a$ equals $2.14 \times 10^{-6}$. The pH of the solution is found to be 3.800. How much distilled water would you have to add to the solution to increase the pH to 4.000?

## CUMULATIVE REVIEW

**113.** What factors determine whether a molecule is polar or nonpolar?

**114.** What property of some liquids accounts for the meniscus that forms at the surface of a solution in a buret?

**115.** Which of the following physical processes are exothermic for water—freezing, boiling, condensing, subliming, evaporating?

**116.** Explain why an air pump gets hot when you pump air into a bicycle tire.

**117.** When 5.00 g of a compound was burned in a calorimeter, the temperature of 2.00 kg of water increased from 24.5°C to 40.5°C. How much heat would be released by the combustion of 1.00 mol of the compound (molar mass = 46.1 g/mol)?

**118.** What is the difference between an exothermic and an endothermic reaction?

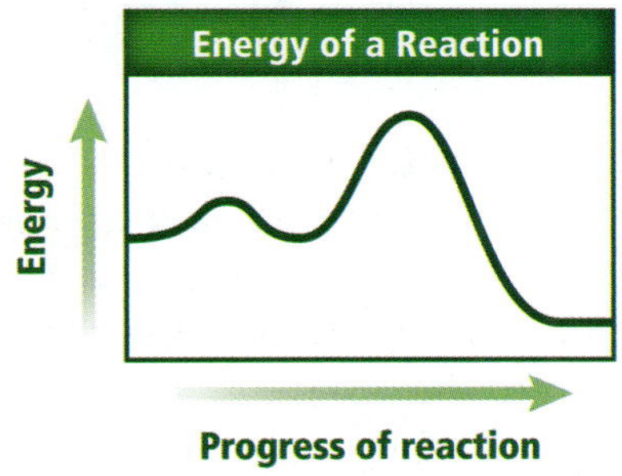


■ **Figure 33**

**119.** **Figure 33** shows how energy changes during the progress of a reaction.

**a.** Is the reaction exothermic or endothermic?

**b.** How many steps are in the reaction mechanism for the reaction?

**c.** Explain how you could use the graph to identify the rate-determining step.

**120.** Hydrogen and fluorine react to form HF according to the following equilibrium equation.

$H_2(g) + F_2(g) \leftrightharpoons 2HF \quad \Delta H = -538 \text{ kJ(g)}$

Will raising the temperature cause the amount of product to increase? Explain.

## WRITING IN ▸ Chemistry

**121.** **Acid/Base Theories** Imagine that you are the Danish chemist Johannes Brønsted. The year is 1923, and you have formulated a new theory of acids and bases. Write a letter to Swedish chemist Svante Arrhenius in which you discuss the differences between your theory and his and point out the advantages of yours.

**122.** **Amino Acids** Twenty amino acids combine to form proteins in living systems. Research the structures and $K_a$ values for five amino acids. Compare the strengths of these acids with the acids in **Table 4.**

## DBQ Document-Based Questions

**Rainwater** **Figure 34** *shows pH measurements made from a number of the monitoring sites in New York state. The pink dot represents the average of the measurements taken at all of the sites at a particular time.*

*Data obtained from: 2007 Acid Deposition Executive Summary. October, 2010. New York State Department of Environmental Conservation.*

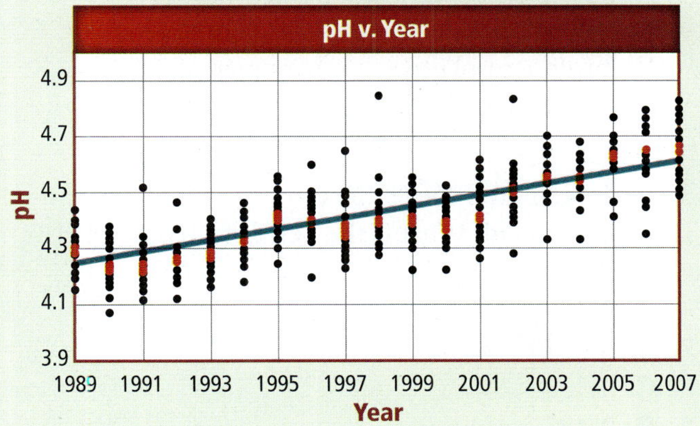


■ **Figure 34**

**123.** In general, what is the trend in the average pH for the years 1989 to 2007?

**124.** Calculate the $[H^+]$ for the lowest and the highest pH measurements recorded on the graph. How many times more acidic is the rainwater having the highest reading than the rainwater with the lowest?

**125.** What is the pH of the trend line in 2007? How much has the average pH changed between the years 1989 and 2007?

## MULTIPLE CHOICE

*Use the graph below to answer Questions 1 and 2.*

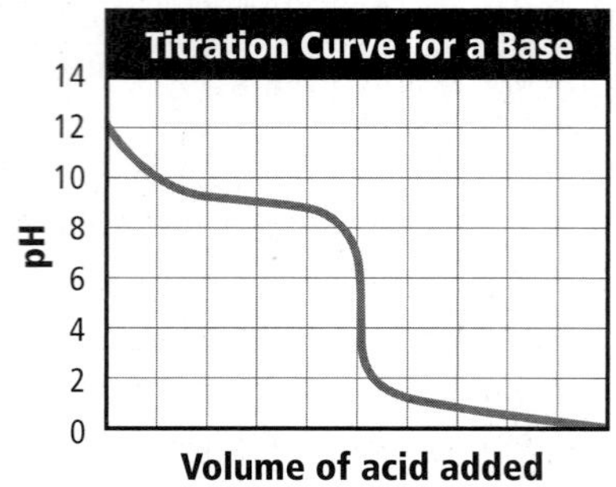


**1.** What is the pH at the equivalence point of this titration?
**A.** 10
**B.** 9
**C.** 5
**D.** 1

**2.** Which indicator would be effective for detecting the end point of this titration?
**A.** methyl orange, with a range of 3.2–4.4
**B.** phenolphthalein, with a range of 8.2–10
**C.** bromocresol green, with a range of 3.8–5.4
**D.** thymol blue, with a range of 8.0–9.6

**3.** Hydrogen bromide (HBr) is a strong, highly corrosive acid. What is the pOH of a 0.0375M HBr solution?
**A.** 12.574
**B.** 12.270
**C.** 1.733
**D.** 1.433

**4.** Cellular respiration produces about 38 mol of ATP for every mole of glucose consumed:

$$C_6H_{12}O_6 + 6O_2 \longrightarrow 6CO_2 + 6H_2O + 38ATP$$

If each mole of ATP can release 30.5 kJ of energy, how much energy can be obtained from a candy bar containing 130.0 g of glucose?
**A.** 27.4 kJ
**B.** 836 kJ
**C.** 1159 kJ
**D.** 3970 kJ

*Use the table below to answer Questions 5–7.*

**Ionization Constants and pH Data for Several Weak Organic Acids**

| Acid | pH of 1.000*M* Solution | $K_a$ |
|---|---|---|
| Formic | 1.87 | $1.78 \times 10^{-4}$ |
| Cyanoacetic | ? | $3.55 \times 10^{-3}$ |
| Propanoic | 2.43 | ? |
| Lutidinic | 1.09 | $7.08 \times 10^{-3}$ |
| Barbituric | 2.01 | $9.77 \times 10^{-5}$ |

**5.** Which acid is the strongest?
**A.** formic acid
**B.** cyanoacetic acid
**C.** lutidinic acid
**D.** barbituric acid

**6.** What is the hydronium ion concentration of the propanoic acid?
**A.** $1.4 \times 10^{-5}$
**B.** $2.43 \times 10^{0}$
**C.** $3.72 \times 10^{-3}$
**D.** $7.3 \times 10^{4}$

**7.** What is the pH of a 0.40*M* solution of cyanoacetic acid?
**A.** 2.06
**B.** 1.22
**C.** 2.45
**D.** 1.42

**8.** What does a value of $K_{eq}$ greater than 1 mean?
**A.** More reactants than products exist at equilibrium.
**B.** More products than reactants exist at equilibrium.
**C.** The rate of the forward reaction is high at equilibrium.
**D.** The rate of the reverse reaction is high at equilibrium.

**9.** Magnesium sulfate ($MgSO_4$) is often added to water-insoluble liquid products of chemical reactions to remove unwanted water. $MgSO_4$ readily absorbs water to form two different hydrates. One of them is found to contain 13.0% $H_2O$ and 87.0% $MgSO_4$. What is the name of this hydrate?
**A.** magnesium sulfate monohydrate
**B.** magnesium sulfate dihydrate
**C.** magnesium sulfate hexahydrate
**D.** magnesium sulfate heptahydrate

Online Test Practice

## SHORT ANSWER

*Use the description of an experiment below to answer Questions 10–12.*

Two 0.050-mol samples of gas at 20°C are released from the end of a long tube at the same time. One gas is xenon (Xe), and the other is sulfur dioxide ($SO_2$).

**10.** Explain which gas will have traveled farther after 5 seconds. How can you tell?

**11.** How will increasing the temperature of this experiment affect the rate of effusion of each gas?

**12.** If the pressure on the xenon at the end of the experiment is 0.092 atm, what volume will it occupy?

## EXTENDED RESPONSE

*Use the figure below to answer Question 13.*

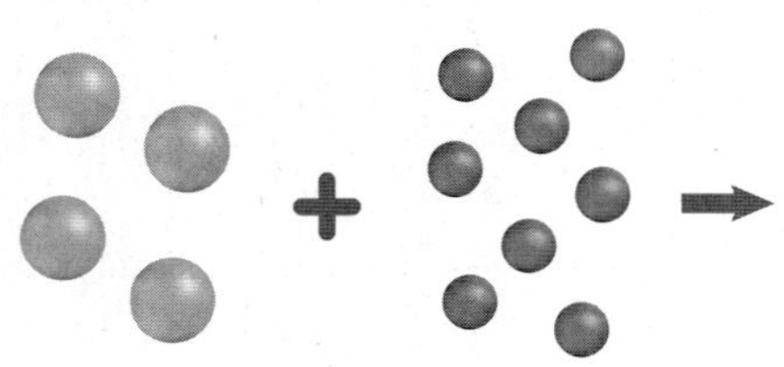


**13.** Explain how the chemical reaction shown in this figure demonstrates the law of conservation of mass.

**14.** Describe lab procedures for preparing a 0.50*M* aqueous solution of NaOH and a 0.50*m* aqueous solution of NaOH.

**15.** Explain how you could express the concentration of the 0.50*m* solution in Question 14 as a mole fraction.

## SAT SUBJECT TEST: CHEMISTRY

**16.** Water has an unusually high boiling point compared to other compounds of similar molar mass because of ____________.
**A.** hydrogen bonding
**B.** adhesive forces
**C.** covalent bonding
**D.** dispersion forces
**E.** pi bonds

*Use the graph below to answer Questions 17 and 18.*

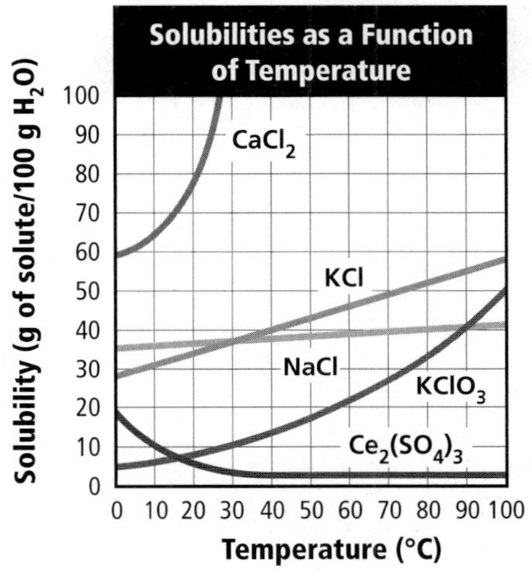


**17.** Which compound has a solubility of 38 g/100 g $H_2O$ at 50°C?
**A.** $CaCl_2$
**B.** KCl
**C.** NaCl
**D.** $KClO_3$
**E.** $Ce_2(SO_4)_3$

**18.** Which has the greatest increase in solubility as temperature increases?
**A.** $Ce_2(SO_4)_3$
**B.** $CaCl_2$
**C.** $KClO_3$
**D.** NaCl
**E.** KCl

**NEED EXTRA HELP?**

| If You Missed Question . . . | 1 | 2 | 3 | 4 | 5 | 6 | 7 | 8 | 9 | 10 | 11 | 12 | 13 | 14 | 15 | 16 | 17 | 18 |
|---|---|---|---|---|---|---|---|---|---|---|---|---|---|---|---|---|---|---|
| Review Section . . . | 18.4 | 18.4 | 18.3 | 11.2 | 18.2 | 18.2 | 18.2 | 17.3 | 10.5 | 12.1 | 12.1 | 13.2 | 11.1 | 14.2 | 14.2 | 12.2 | 14.3 | 14.3 |

# CHAPTER 19

# Redox Reactions

**BIGIDEA** Oxidation-reduction reactions—among the most-common chemical processes in both nature and industry—involve the transfer of electrons.

## SECTIONS

1 Oxidation and Reduction

2 Balancing Redox Equations

## LaunchLAB

iLab Station

### What happens when iron and copper(II) sulfate react?

Rust is the product of a reaction between iron and oxygen. Iron can also react with substances other than oxygen. In this lab, you will study a reaction between iron and copper(II) sulfate.

**FOLDABLES® Study Organizer**

### Balancing Redox Equations

Make a layered-look book. Label it as shown. Use it to help you summarize information about the different methods for balancing redox equations.

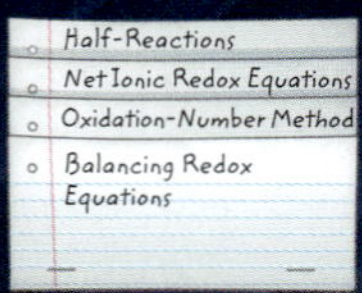


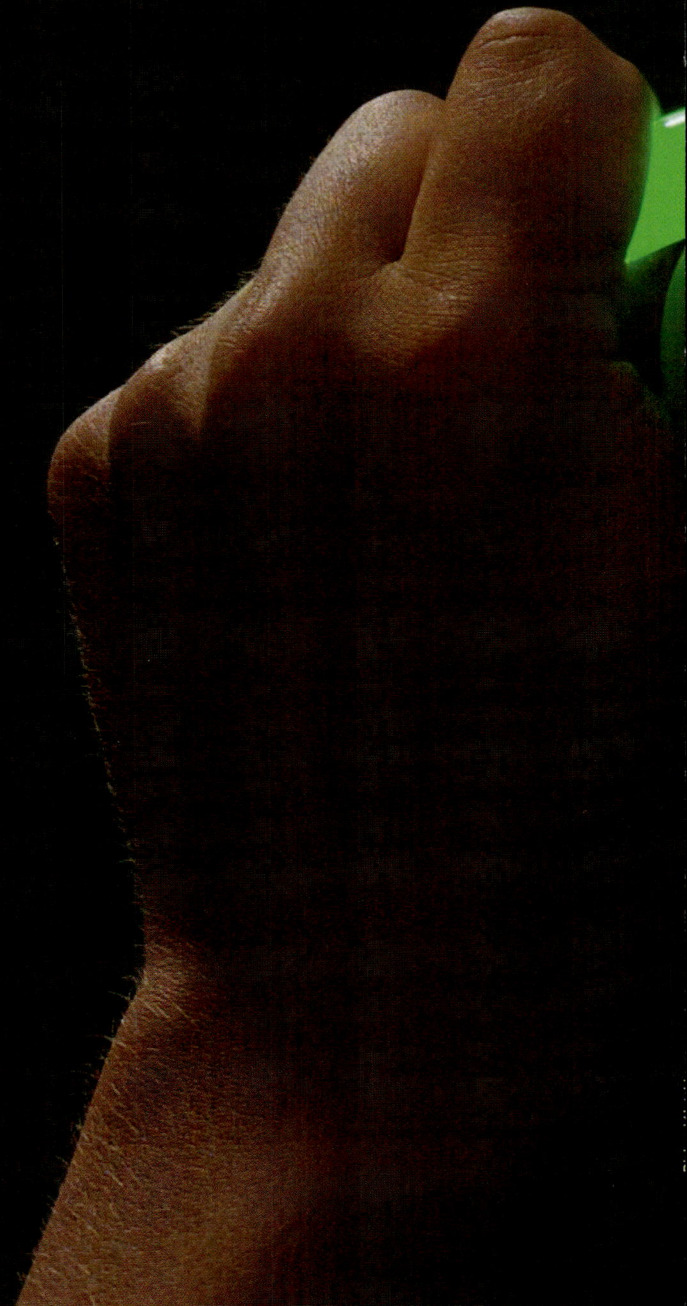

Go **online!**
MHEonline.com

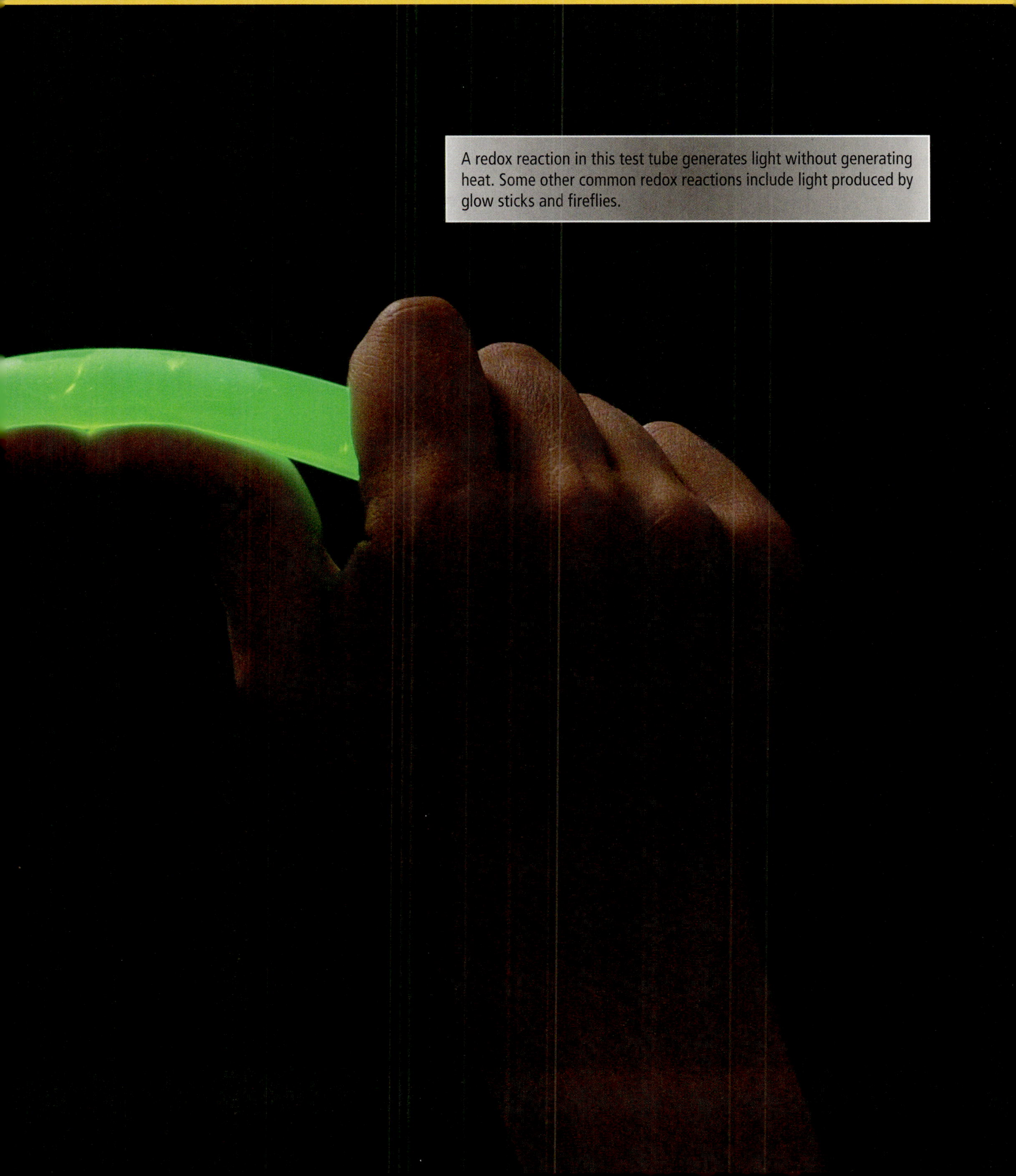

A redox reaction in this test tube generates light without generating heat. Some other common redox reactions include light produced by glow sticks and fireflies.

## SECTION 1 Oxidation and Reduction

**MAINIDEA** Oxidation and reduction are complementary—as a substance is oxidized, another substance is reduced.

### Essential Questions

- What are oxidation and reduction?
- How can oxidizing and reducing agents be identified?
- How is the oxidation number of an element in a compound determined?

### Review Vocabulary

**spectator ion:** an ion that does not participate in a reaction and is not usually shown in an ionic equation

### New Vocabulary

oxidation-reduction reaction
redox reaction
oxidation
reduction
oxidation number
oxidizing agent
reducing agent

**CHEM 4 YOU** The light produced by a glow stick is the result of a chemical reaction. When you snap the glass capsule inside the plastic case, two chemicals are mixed and electron transfer occurs. As a result of this reaction, chemical energy is converted into light energy.

## Electron Transfer and Redox Reactions

Previously, you learned that a chemical reaction can usually be classified as one of five types—synthesis, decomposition, combustion, single-replacement, or double-replacement. A defining characteristic of combustion and single-replacement reactions is that they always involve the transfer of electrons from one substance to another, as do many synthesis and decomposition reactions. For example, in the synthesis reaction in which sodium (Na) and chlorine ($Cl_2$) react to form the ionic compound sodium chloride (NaCl), an electron from each of two sodium atoms is transferred to the $Cl_2$ molecule to form two $Cl^-$ ions.

Complete chemical equation: $2Na(s) + Cl_2(g) \rightarrow 2NaCl(s)$

Net ionic equation: $2Na(s) + Cl_2(g) \rightarrow 2Na^+ + 2Cl^-$ (ions in crystal)

An example of a combustion reaction is the burning of magnesium in air, which involves the transfer of electrons.

Complete chemical equation: $2Mg(s) + O_2(g) \rightarrow 2MgO(s)$

Net ionic equation: $2Mg(s) + O_2(g) \rightarrow 2Mg^{2+} + 2O^{2-}$ (ions in crystal)

When magnesium reacts with oxygen, as illustrated in **Figure 1,** each magnesium atom transfers two electrons to each oxygen atom. The two magnesium atoms become magnesium ions ($Mg^{2+}$), and the two oxygen atoms become oxide ions ($O^{2-}$). A reaction in which electrons are transferred from one substance to another is called an **oxidation-reduction reaction,** which is also called a **redox reaction.**

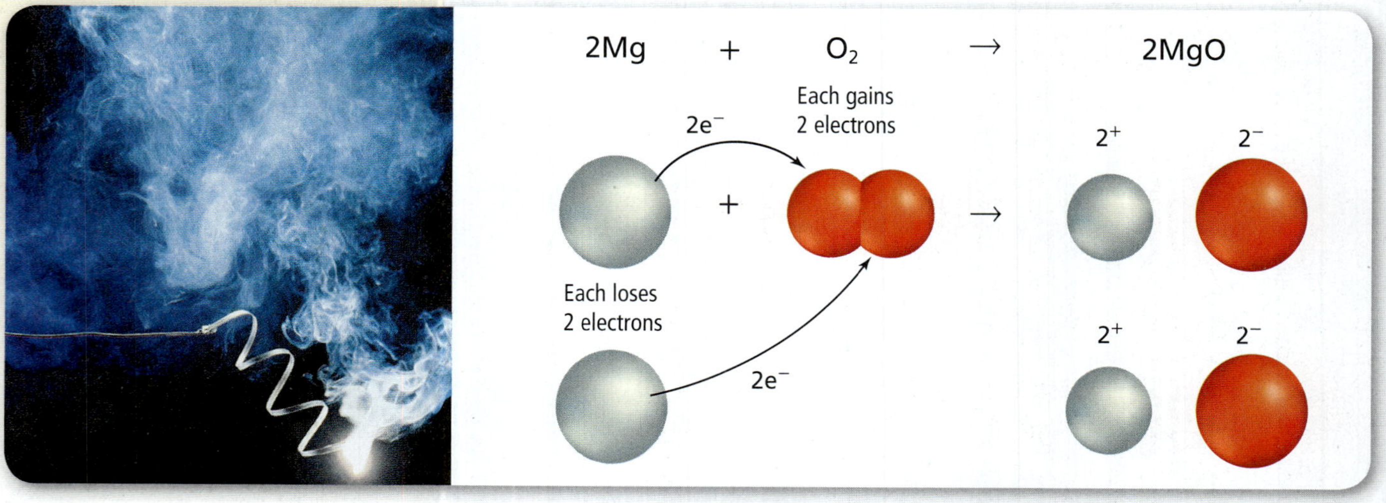


■ **Figure 1** The reaction of magnesium and oxygen involves a transfer of electrons from magnesium to oxygen. Therefore, this reaction is an oxidation-reduction reaction.
**Classify** *the reaction between magnesium and oxygen.*

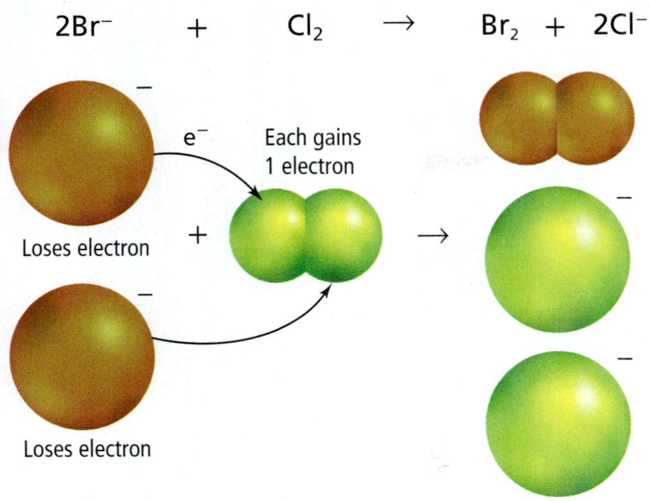


■ **Figure 2** The reaction between aqueous bromide ions and chlorine gas is a redox reaction. Here, electrons are transferred from bromide ions to chlorine.

Consider the single-replacement reaction in which chlorine in an aqueous solution reacts with bromide ions from an aqueous solution of potassium bromide, which is shown in **Figure 2.**

Complete chemical equation: $2KBr(aq) + Cl_2(aq) \longrightarrow 2KCl(aq) + Br_2(aq)$

Net ionic equation: $2Br^-(aq) + Cl_2(aq) \longrightarrow Br_2(aq) + 2Cl^-(aq)$

Note that the chlorine receives electrons from the bromide ions to become chloride ions. When the two bromide ions lose electrons, the two bromine atoms form a covalent bond with each other to produce $Br_2$ molecules.

**Oxidation and reduction** Originally, the word *oxidation* referred only to reactions in which a substance combined with oxygen. Today, **oxidation** is defined as the complete or partial loss of electrons from a reacting substance. Look again at the net ionic equation for the reaction of sodium and chlorine. Sodium is oxidized because it loses an electron.

$$\text{Oxidation: } Na \longrightarrow Na^+ + e^-$$

For oxidation to occur, the electrons lost by the substance that is oxidized must be accepted by atoms or ions of another substance. In other words, there must be an accompanying process that involves the gain of electrons. **Reduction** is the complete or partial gain of electrons by a reacting substance. Following the sodium chloride example further, the reduction reaction that accompanies the oxidation of sodium is the reduction of chlorine.

$$\text{Reduction: } Cl_2 + 2e^- \longrightarrow 2Cl^-$$

Oxidation and reduction are complementary processes: oxidation cannot occur unless reduction also occurs. It is important to recognize and distinguish between oxidation and reduction. A memory aid might help you remember the distinction. The phrase **L**oss of **E**lectrons is **O**xidation, and **G**ain of **E**lectrons is **R**eduction is shortened to **LEO GER.**

**LEO** the lion says **GER** or, for short, **LEO GER.**

View an **animation about redox reactions.**

Concepts In Motion

**VOCABULARY**

**WORD ORIGIN**

**Reduction**

comes from the Latin *re,* meaning *back,* and *ducere,* meaning *to lead*

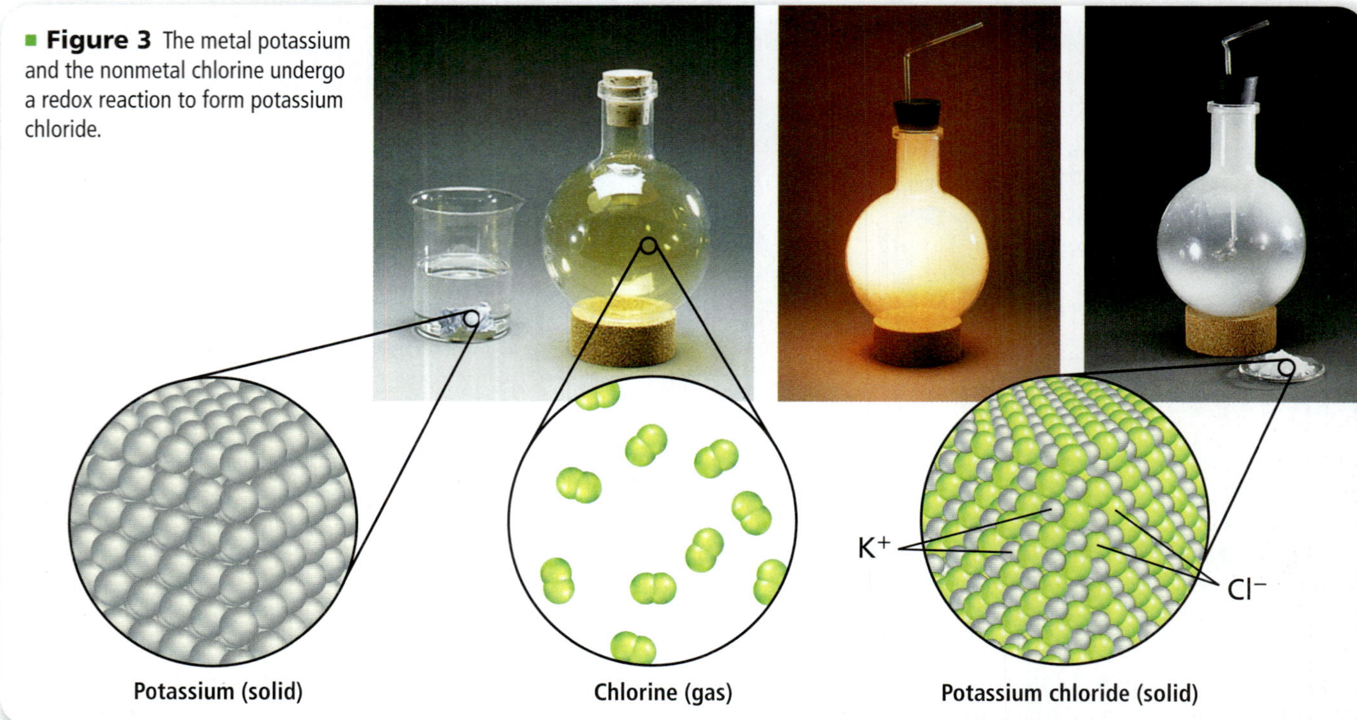


**Figure 3** The metal potassium and the nonmetal chlorine undergo a redox reaction to form potassium chloride.

**Oxidation numbers** A number assigned to an atom or ion to indicate its degree of oxidation or reduction is called its **oxidation number.** For example, the oxidation number of an element in an ionic compound is related to the number of electrons lost or gained by an atom of the element when it becomes an ion. The reaction of potassium metal with chlorine vapor, shown in **Figure 3,** is a redox reaction. The equation for the reaction is as follows.

$$\text{Complete chemical equation: } 2K(s) + Cl_2(g) \rightarrow 2KCl(s)$$
$$\text{Net ionic equation: } 2K(s) + Cl_2(g) \rightarrow 2K^+(s) + 2Cl^-(s)$$

Potassium, a group 1 element whose atoms tend to lose one electron in reactions because of its low electronegativity, is assigned an oxidation number of +1 in KCl. On the other hand, chlorine, a group 17 element whose atoms tend to gain one electron in reactions because of its high electronegativity, is assigned an oxidation number of −1 in KCl. In redox terms, you would say that potassium atoms are oxidized from 0 to the +1 state because each atom loses an electron, and chlorine atoms are reduced from 0 to the −1 state because each atom gains an electron. When an atom or ion is reduced, the numerical value of its oxidation number decreases. Conversely, when an atom or ion is oxidized, its oxidation number increases.

Oxidation numbers are tools that scientists use in written chemical equations to help them keep track of the movement of electrons in a redox reaction. Like some of the other tools you have learned about, oxidation numbers have a specific notation. Oxidation numbers are written with the positive or negative sign before the number (+3, +2), whereas ionic charge is written with the sign after the number (3+, 2+).

Oxidation number: +3 Ionic charge: 3+

☑ READING CHECK **Determine** Which element is more likely to gain electrons, potassium or chlorine?

## CAREERS IN CHEMISTRY

**Potter** A potter is an artist who makes pottery. He or she uses glazes containing metallic ions that have multiple oxidation states to achieve a variety of colors on ceramics. Glazes that contain copper ions produce a green-to-blue color when oxidized, and they produce a reddish color when reduced in a kiln.

WebQuest

## Oxidizing and Reducing Agents

The potassium-chlorine reaction in **Figure 3** can also be described by saying that "potassium is oxidized by chlorine." This description is useful because it clearly identifies both the substance that is oxidized and the substance that does the oxidizing. The substance that oxidizes another substance by accepting its electrons is called an **oxidizing agent.** This term describes the substance that is reduced. The substance that reduces another substance by losing electrons is called a **reducing agent.** A reducing agent supplies electrons to the substance being reduced (gaining electrons). The reducing agent is oxidized because it loses electrons. The reducing agent in the potassium-chlorine reaction is potassium—the substance that is oxidized.

$$\overbrace{2K(s) + Cl_2(g) \rightarrow 2KCl(s)}^{\text{oxidized}}$$
$$\underbrace{Cl_2(g) \rightarrow 2KCl(s)}_{\text{reduced}}$$

Oxidizing agent: $Cl_2$

Reducing agent: K

A common application of redox chemistry is to remove tarnish from metal objects. Other oxidizing agents and reducing agents are useful in everyday life. For example, when you add chlorine bleach to your laundry to whiten clothes, you are using an aqueous solution of sodium hypochlorite (NaClO), an oxidizing agent. It oxidizes dyes, stains, and other materials that discolor clothes. **Table 1** summarizes the different ways to describe oxidation-reduction reactions.

Explore **redox reactions with an interactive table.** Concepts In Motion

**Table 1** Summary of Redox Reactions

| | |
|---|---|
| **Process** | 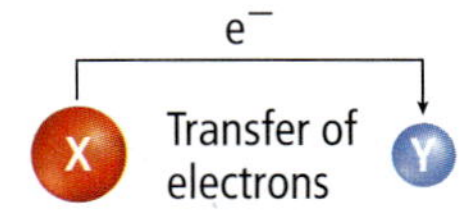  |
| **Oxidation**<br>• A reactant loses an electron.<br>• Reducing agent is oxidized.<br>• Oxidation number increases. | • X loses an electron.<br>• X is the reducing agent and becomes oxidized.<br>• The oxidation number of X increases. |
| **Reduction**<br>• Other reactant gains an electron.<br>• Oxidizing agent is reduced.<br>• Oxidation number decreases. | • Y gains an electron.<br>• Y is the oxidizing agent and becomes reduced.<br>• The oxidation number of Y decreases. |

# MiniLAB

iLab Station

### Observe a Redox Reaction

**How can tarnish be removed from silver?**

**Procedure**

1. Read and complete the lab safety form.
2. Lightly buff a piece of **aluminum foil** with **steel wool** to remove any oxide coating.
3. Wrap a **small tarnished object** in the aluminum foil, making sure that the tarnished area makes firm contact with the foil.
4. Place the wrapped object in a **400-mL beaker** and add a sufficient volume of **tap water** to cover it completely.
5. Add about 1 spoonful of **baking soda** and about 1 spoonful of **table salt** to the beaker.
6. Using **beaker tongs,** set the beaker and its contents on a **hot plate,** and heat until the water is almost boiling. Maintain the heat for approximately 15 min, until the tarnish disappears.

**Analysis**

1. **Write** the equation for the reaction of silver with hydrogen sulfide that yields silver sulfide and hydrogen.
2. **Write** the equation for the reaction of the tarnish (silver sulfide) with the aluminum foil that yields aluminum sulfide and silver.
3. **Determine** which metal, aluminum or silver, is more reactive. How do you know this from your results?
4. **Explain** why you should not use an aluminum pan to clean silver objects.

## Redox and Electronegativity

The chemistry of oxidation-reduction reactions is not limited to atoms of an element changing to ions or the reverse. Some redox reactions involve changes in molecular substances or polyatomic ions in which atoms are covalently bonded to other atoms. For example, the following equation represents the redox reaction used to manufacture ammonia ($NH_3$).

$$N_2(g) + 3H_2(g) \rightarrow 2NH_3(g)$$

This process involves neither ions nor any obvious transfer of electrons. The reactants and products are all molecular compounds. Yet, it is still a redox reaction in which nitrogen is the oxidizing agent and hydrogen is the reducing agent.

In situations such as the formation of ammonia, where two atoms share electrons, how is it possible to say that one atom lost electrons and was oxidized, while the other atom gained electrons and was reduced? To answer this, you need to know which atom attracts electrons more strongly, or, in other words, which atom is more electronegative. You might find it helpful to review the discussion of electronegativity trends on the periodic table. **Figure 4** shows that electronegativity increases left to right across a period and generally decreases down a group. Elements with low electronegativity (Groups 1 and 2) are strong reducing agents, and those with high electronegativity (Group 17 and oxygen in Group 16) are strong oxidizing agents.

Hydrogen has an electronegativity of 2.20, and nitrogen has an electronegativity of 3.04. For the purpose of studying oxidation-reduction reactions, the more-electronegative element (in this case nitrogen) is treated as if it had been reduced by gaining electrons from the other element (hydrogen). Conversely, the less-electronegative element (hydrogen) is treated as if it had been oxidized by losing electrons to the other element (nitrogen).

$$\overbrace{N_2(g) + 3\underbrace{H_2(g) \rightarrow 2NH_3(g)}_{\text{oxidized (partial loss of } e^-)}}^{\text{reduced (partial gain of } e^-)}$$

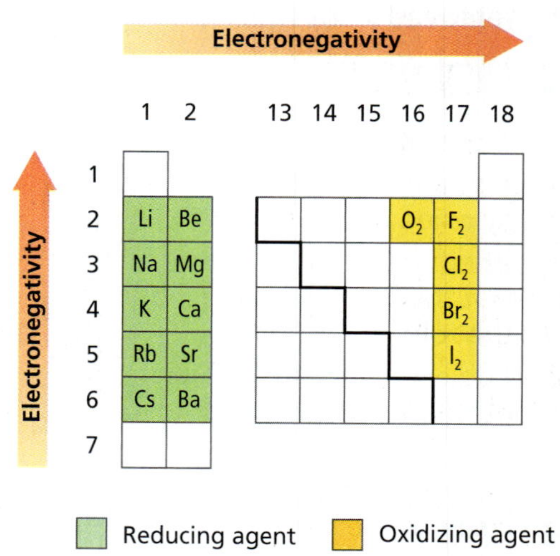


■ **Figure 4** The electronegativity of elements increases from left to right across the periodic table, and it decreases going down a group. Elements with low electronegativity are strong reducing agents, and elements with high electronegativity are strong oxidizing agents.

**Predict** ***which element would be the strongest oxidizing agent. Which is the strongest reducing agent?***

## EXAMPLE Problem 1

**Identify Oxidation-Reduction Reactions** The following equation represents the redox reaction of aluminum and iron.

$2Al + 2Fe^{3+} + 3O^{2-} \rightarrow 2Fe + 2Al^{3+} + 3O^{2-}$

Identify what is oxidized and what is reduced in this reaction. Identify the oxidizing agent and the reducing agent.

### 1 ANALYZE THE PROBLEM

You are given the reactants and products in the reaction. You must determine the electron transfers that occur. Then, you can apply the definitions of oxidizing agent and reducing agent to answer the question.

### 2 SOLVE FOR THE UNKNOWN

Identify the oxidation process and the reduction process.

$Al \rightarrow Al^{3+} + 3e^-$ (loss of $e^-$ is oxidation) — **The aluminum atom loses three electrons and becomes an aluminum ion.**

$Fe^{3+} + 3e^- \rightarrow Fe$ (gain of $e^-$ is reduction) — **The iron ion accepts the three electrons lost from aluminum and becomes an iron atom.**

Al is oxidized and is therefore the reducing agent. $Fe^{3+}$ is reduced and is therefore the oxidizing agent.

### 3 EVALUATE THE ANSWER

In this process, aluminum atoms lost electrons and were oxidized, whereas iron ions gained electrons and were reduced. The definitions of oxidation, reduction, oxidizing agent, and reducing agent apply. Note that the ionic charge of oxygen is unchanged in this reaction; therefore, oxygen is not a key factor in this problem.

## RealWorld CHEMISTRY

### Oxidation

**RUST** When moist air comes in contact with iron, the iron oxidizes. Iron oxide ($Fe_2O_3$), called rust, is common because iron combines readily with oxygen. Pure iron is uncommon in nature. Steel, a mixture that contains iron, is a commonly used form of iron. Several protective methods, such as plating, painting, and applying an enamel or plastic coating, can inhibit the production of iron oxide.

## PRACTICE Problems

Do additional problems. Online Practice

1. Identify each of the following changes as either oxidation or reduction. Recall that $e^-$ is the symbol for an electron.
   - **a.** $I_2 + 2e^- \rightarrow 2I^-$
   - **b.** $K \rightarrow K^+ + e^-$
   - **c.** $Fe^{2+} \rightarrow Fe^{3+} + e^-$
   - **d.** $Ag^+ + e^- \rightarrow Ag$
2. Identify what is oxidized and what is reduced in the following processes.
   - **a.** $2Br^- + Cl_2 \rightarrow Br_2 + 2Cl^-$
   - **b.** $2Ce + 3Cu^{2+} \rightarrow 3Cu + 2Ce^{3+}$
   - **c.** $2Zn + O_2 \rightarrow 2ZnO$
   - **d.** $2Na + 2H^+ \rightarrow 2Na^+ + H_2$
3. Identify the oxidizing agent and the reducing agent in the following equation. Explain your answer.

   $$Fe(s) + 2Ag^+(aq) \rightarrow Fe^{2+}(aq) + 2Ag(s)$$
4. **Challenge** Identify the oxidizing agent and the reducing agent in each reaction.
   - **a.** $Mg + I_2 \rightarrow MgI_2$
   - **b.** $H_2S + Cl_2 \rightarrow S + 2HCl$

■ **Figure 5** Banded iron—shown in this cross-section of rock—is a result of different oxidation states of iron, which depend on which minerals are present.

## Determining Oxidation Numbers

In order to understand all types of redox reactions, you must have a way to determine the oxidation number ($n_{element}$) of each element involved in a reaction. **Table 2** outlines the rules chemists use to make this determination easier.

Many elements other than those specified in the rules below, including most of the transition metals, metalloids, and nonmetals, can be found with different oxidation numbers in different compounds. For example, iron has different oxidation numbers, indicated by the different colors as shown in **Figure 5,** depending on which minerals are also present.

**Table 2 Rules for Determining Oxidation Numbers**

| Rule | Example | $n_{element}$ |
|---|---|---|
| **1.** The oxidation number of an atom of an uncombined element is zero. | Na, $O_2$, $Cl_2$, $H_2$ | 0 |
| **2.** The oxidation number of a monatomic ion is equal to the charge of the ion. | $Ca^{2+}$ | +2 |
| | $Br^-$ | −1 |
| **3.** The oxidation number of the more-electronegative atom in a molecule or a complex ion is the same as the charge it would have if it were an ion. | N in $NH_3$ | −3 |
| | O in NO | −2 |
| **4.** The oxidation number of the most-electronegative element, fluorine, is always −1 when it is bonded to another element. | F in LiF | −1 |
| **5.** The oxidation number of oxygen in compounds is always −2 except in peroxides, such as hydrogen peroxide ($H_2O_2$), where it is −1. When it is bonded to fluorine, the only element more electronegative than oxygen, the oxidation number of oxygen is positive. | O in $NO_2$ | −2 |
| | O in $H_2O_2$ | −1 |
| **6.** The oxidation number of hydrogen in most of its compounds is +1, except in metal hydrides; then, the oxidation number is −1. | H in NaH | −1 |
| **7.** The oxidation numbers of group 1 and 2 metals and aluminum are positive and equal to their number of valence electrons. | K | +1 |
| | Ca | +2 |
| | Al | +3 |
| **8.** The sum of the oxidation numbers in a neutral compound is zero. | $CaBr_2$ | $(+2) + 2(-1) = 0$ |
| **9.** The sum of the oxidation numbers of the atoms in a polyatomic ion is equal to the charge of the ion. | $SO_3^{2-}$ | $(+4) + 3(-2) = -2$ |

## EXAMPLE Problem 2

**Determine Oxidation Numbers** Use the rules for determining oxidation numbers to find the oxidation number of each element in potassium chlorate ($KClO_3$) and in a sulfite ion ($SO_3^{2-}$).

### 1 ANALYZE THE PROBLEM

In the rules for determining oxidation numbers, you are given the oxidation numbers of oxygen and potassium. You are also given the overall charge of the compound or ion. Using this information and applying the rules, determine the oxidation numbers of chlorine and sulfur. (Let $n_{element}$ equal the oxidation number of the element in question.)

| **Known** | **Unknown** |
|---|---|
| $KClO_3$ | $n_{Cl} = ?$ |
| $SO_3^{2-}$ | $n_S = ?$ |
| $n_O = -2$ | |
| $n_K = +1$ | |

### 2 SOLVE FOR THE UNKNOWN

Assign the known oxidation numbers to their elements, set the sum of all oxidation numbers to zero or to the ion charge, and solve for the unknown oxidation number.

$(n_K) + (n_{Cl}) + 3\,(n_O) = 0$

$(+1) + (n_{Cl}) + 3(-2) = 0$

The sum of the oxidation numbers in a neutral compound is zero. For group 1 metals, $n_{element} = +1$. Substitute $n_K = +1$, $n_O = -2$.

$1 + n_{Cl} + (-6) = 0$

$n_{Cl} = +5$

Solve for $n_{Cl}$.

$(n_S) + 3\,(n_O) = -2$

$(n_S) + 3(-2) = -2$

The sum of the oxidation numbers in a polyatomic ion equals the charge on the ion. Substitute $n_O = -2$.

$n_S + (-6) = -2$

$n_S = +4$

Solve for $n_S$.

### 3 EVALUATE THE ANSWER

The rules for determining oxidation numbers have been correctly applied. All of the oxidation numbers in each substance add up to the proper value.

## PRACTICE Problems

Do additional problems. Online Practice

5. Determine the oxidation number of the boldface element in the following formulas for compounds.
   **a.** Na**Cl**$O_4$ **b.** Al**P**$O_4$ **c.** H**N**$O_2$
6. Determine the oxidation number of the boldface element in the following formulas for ions.
   **a.** **N**$H_4^+$ **b.** **As**$O_4^{3-}$ **c.** **Cr**$O_4^{2-}$
7. Determine the oxidation number of nitrogen in each of these molecules.
   **a.** $NH_3$ **b.** KCN **c.** $N_2H_4$
8. **Challenge** Determine the net change of oxidation number of each of the elements in these redox equations.
   **a.** $C + O_2 \rightarrow CO_2$
   **b.** $Cl_2 + ZnI_2 \rightarrow ZnCl_2 + I_2$
   **c.** $CdO + CO \rightarrow Cd + CO_2$

**Table 3** Various Oxidation Numbers

| Oxidation Number | +1 | +2 | +3 | −1 | −2 |
|---|---|---|---|---|---|
| Aluminum | | | X | | |
| Barium | | X | | | |
| Bromine | | | | X | |
| Cadmium | | X | | | |
| Calcium | | X | | | |
| Cesium | X | | | | |
| Chlorine | | | | X | |
| Fluoride | | | | X | |
| Hydrogen | X | | | X | |
| Iodine | | | | X | |
| Lithium | X | | | | |
| Magnesium | | X | | | |
| Oxygen | | | | | X |
| Potassium | X | | | | |
| Sodium | X | | | | |
| Silver | X | | | | |
| Strontium | | X | | | |

## Oxidation Numbers in Redox Reactions

Having studied oxidation numbers, you should be able to relate oxidation-reduction reactions to changes in oxidation number. Refer to the equation for a reaction that you saw at the beginning of this section—the replacement of bromine in aqueous potassium bromide (KBr) by chlorine ($Cl_2$).

$$2KBr(aq) + Cl_2(aq) \rightarrow 2KCl(aq) + Br_2(aq)$$

To learn how oxidation numbers change, start by assigning numbers, using **Table 3,** to all elements in the balanced equation. Then, review the changes, as shown in the equation below.

change: −1 reduction

change: +1 oxidation

$$\overset{+1-1}{2KBr(aq)} + \overset{0}{Cl_2(aq)} \rightarrow \overset{+1-1}{2KCl(aq)} + \overset{0}{Br_2(aq)}$$

no change in oxidation number

You should notice that the oxidation number of bromine changed from −1 to 0, an increase of 1. At the same time, the oxidation number of chlorine changed from 0 to −1, a decrease of 1. Therefore, chlorine is reduced and bromine is oxidized. All redox reactions follow the same pattern. When an atom is oxidized, its oxidation number increases. When an atom is reduced, its oxidation number decreases. Note that there is no change in the oxidation number of potassium. The potassium ion takes no part in the reaction and is therefore a spectator ion.

## SECTION 1 REVIEW

Section Self-Check

### Section Summary

- Oxidation-reduction reactions involve the transfer of electrons from one atom to another.
- When an atom or ion is reduced, its oxidation number decreases. When an atom or ion is oxidized, its oxidation number increases.
- In oxidation-reduction reactions involving molecular compounds (and polyatomic ions with covalent bonds), the more-electronegative atoms are treated as if they are reduced. The less-electronegative atoms are treated as if they are oxidized.

9. **MAIN**IDEA **Explain** why oxidation and reduction must always occur together.
10. **Describe** the roles of oxidizing agents and reducing agents in a redox reaction. How is each changed in the reaction?
11. **Write** the equation for the reaction of iron metal with hydrobromic acid to form aqueous iron(III) bromide and hydrogen gas. Determine the change in oxidation number for the element that is reduced and the element that is oxidized.
12. **Determine** the oxidation number of the boldface element in these compounds.
    **a.** $HNO_3$ **c.** $Sb_2O_5$
    **b.** $Ca_3N_2$ **d.** $CuWO_4$
13. **Determine** the oxidation number of the boldface element in these ions.
    **a.** $IO_4^-$ **c.** $B_4O_7^{2-}$
    **b.** $MnO_4^-$ **d.** $NH_2^-$
14. **Make and Use Graphs** Alkali metals are strong reducing agents. Make a graph showing how the reducing abilities of the alkali metals increase or decrease as you move down the family from sodium to francium.

SECTION 2

# Balancing Redox Equations

**MAINIDEA** Redox equations are balanced when the total increase in oxidation numbers equals the total decrease in oxidation numbers of the atoms or ions involved in the reaction.

**CHEM 4 YOU** When fatty substances in foods spoil, they are referred to as rancid. Large molecules are broken down through redox reactions that result in foul-smelling products. The equation for this process is complicated but can be balanced using the same rules for simpler equations.

### Essential Questions

- How are changes in oxidation number related to the transfer of electrons?
- How can changes in oxidation numbers be used to balance redox equations?
- What are half-reactions and how can they be used to balance redox equations?

### Review Vocabulary

**net ionic equation:** an ionic equation that includes only the particles that participate in the reaction

### New Vocabulary

oxidation-number method
species
half-reaction

## The Oxidation-Number Method

Chemical equations must be balanced to show the correct quantities of reactants and products. Study the following unbalanced equation for the reaction that occurs when copper metal is placed in concentrated nitric acid. This reaction is shown in **Figure 6.** The brown gas that is produced is nitrogen dioxide ($NO_2$), from the reduction of nitrate ions ($NO_3^-$), and the blue solution is the result of the oxidation of copper (Cu) to copper(II) ions ($Cu^{2+}$).

$$Cu(s) + HNO_3(aq) \rightarrow Cu(NO_3)_2(aq) + NO_2(g) + H_2O(l)$$

Note that oxygen appears in only one reactant, $HNO_3$, but in all three products. Nitrogen appears in $HNO_3$ and in two of the products. Redox equations such as this one, in which the same element appears in several reactants and products, can be difficult to balance. As you have read, when an atom or ion loses electrons, its oxidation number increases; when an atom or ion gains electrons, its oxidation number decreases. The number of electrons transferred from atoms or ions must equal the number of electrons accepted by other atoms or ions. Therefore, the total increase in oxidation numbers (oxidation) must equal the total decrease in oxidation numbers (reduction) of the atoms or ions involved in the reaction. The balancing technique called the **oxidation-number method** is based on these principles, and is described in **Table 4.**

■ **Figure 6** Some chemical equations for redox reactions, such as the reaction between copper and nitric acid, can be difficult to balance because elements might appear more than once on each side of the equation.

Charles D. Winters/Science Source/Photo Researchers

**Table 4 The Oxidation-Number Method**

| |
|---|
| **1.** Assign oxidation numbers to all chemical elements in the equation. |
| **2.** Identify the atoms or ions that are oxidized and reduced. |
| **3.** Determine the change in oxidation number for the atoms or ions that are oxidized and for the atoms or ions that are reduced. |
| **4.** Make the change in oxidation numbers equal in magnitude by adjusting coefficients in the equation. |
| **5.** If necessary, use the conventional method to balance the remainder of the equation. |

EXAMPLE PROBLEM

## EXAMPLE Problem 3

Get help with **balancing redox equations.** Personal Tutor

**THE OXIDATION-NUMBER METHOD** Balance the following redox equation.

$$Cu + HNO_3 \rightarrow Cu(NO_3)_2 + NO_2 + H_2O$$

### 1 ANALYZE THE PROBLEM

Use the rules for determining oxidation number. The increase in oxidation number of the oxidized atoms or ions must equal the decrease in oxidation number of the reduced atoms or ions. Adjust the coefficients to balance the equation.

### 2 SOLVE FOR THE UNKNOWN

Assign oxidation numbers to all elements in the equation.

$$\overset{0}{Cu} + \overset{+1}{H}\overset{+5}{N}\overset{-2}{O_3} \rightarrow \overset{+2}{Cu}(\overset{+5}{N}\overset{-2}{O_3})_2 + \overset{+4}{N}\overset{-2}{O_2} + \overset{+1}{H_2}\overset{-2}{O}$$

**The oxidation number of copper increases from 0 to +2. The oxidation number of nitrogen decreases from +5 to +4.**

Identify which atoms or ions are oxidized, which are reduced, and which do not change.

Cu is oxidized.
N is reduced.
H does not change.
O does not change.
N does not change in the nitrate ion ($NO_3^-$).

Determine the changes in oxidation number for the atoms or ions that are oxidized and for the atoms or ions that are reduced.

Changes in oxidation number:
Oxidized: Cu +2
Reduced: N −1

**Copper loses electrons. It is oxidized. Nitrogen gains an electron. It is reduced.**

Make the changes in oxidation numbers equal in magnitude by adjusting coefficients in the equation.

$$Cu + 2HNO_3 \rightarrow Cu(NO_3)_2 + 2NO_2 + H_2O$$
$$2(-1) = -2$$

**Because the change in oxidation number for N is −1, you must add a coefficient of 2 to balance. This coefficient applies to both $HNO_3$ and $NO_2$.**

Use the conventional method to balance the remainder of the equation.

$$Cu + 2HNO_3 \rightarrow Cu(NO_3)_2 + 2NO_2 + H_2O$$

$$Cu + 4HNO_3 \rightarrow Cu(NO_3)_2 + 2NO_2 + H_2O$$

**The coefficient of $HNO_3$ must be increased from 2 to 4 to balance the four nitrogen atoms in the products.**

$$Cu(s) + 4HNO_3(aq) \rightarrow Cu(NO_3)_2(aq) + 2NO_2(g) + 2H_2O(l)$$

**Add a coefficient of 2 to $H_2O$ to balance the four hydrogen atoms on the left.**

### 3 EVALUATE THE ANSWER

The number of atoms of each element is equal on both sides of the equation. No subscripts have been changed.

## PRACTICE Problems

Do additional problems. Online Practice

**Use the oxidation-number method to balance these redox equations.**

**15.** $HCl + HNO_3 \rightarrow HOCl + NO + H_2O$

**16.** $SnCl_4 + Fe \rightarrow SnCl_2 + FeCl_3$

**17.** $NH_3(g) + NO_2(g) \rightarrow N_2(g) + H_2O(l)$

**18. Challenge** $SO_2 + Br_2 + H_2O \rightarrow HBr + H_2SO_4$

## Balancing Net Ionic Redox Equations

Sometimes, chemists prefer to express redox reactions in the simplest possible terms—as an equation showing only the oxidation and reduction processes. Refer again to the balanced equation for the oxidation of copper by nitric acid.

$$Cu(s) + 4HNO_3(aq) \rightarrow Cu(NO_3)_2(aq) + 2NO_2(g) + 2H_2O(l)$$

Note that the reaction takes place in aqueous solution, so $HNO_3$, which is a strong acid, will be ionized. Likewise, copper(II) nitrate ($Cu(NO_3)_2$) will be dissociated into ions. Therefore, the equation can also be written in ionic form.

$$Cu(s) + 4H^+(aq) + 4NO_3^-(aq) \rightarrow Cu^{2+}(aq) + 2NO_3^-(aq) + 2NO_2(g) + 2H_2O(l)$$

There are four nitrate ions among the reactants, but only two of them undergo change to form two nitrogen dioxide molecules. The other two nitrate ions are only spectator ions and can be eliminated from the equation. To simplify things when writing redox equations in ionic form, chemists usually indicate hydrogen ions by $H^+$ with the understanding that they exist in hydrated form as hydronium ions ($H_3O^+$). The equation can then be rewritten showing only the substances that undergo change.

$$Cu(s) + 4H^+(aq) + 2NO_3^-(aq) \rightarrow Cu^{2+}(aq) + 2NO_2(g) + 2H_2O(l)$$

Now look at the equation in unbalanced form.

$$Cu(s) + H^+(aq) + NO_3^-(aq) \rightarrow Cu^{2+}(aq) + NO_2(g) + H_2O(l)$$

You might also see this same reaction expressed in a way that shows only the substances that are oxidized and reduced.

$$Cu(s) + NO_3^-(aq) \rightarrow Cu^{2+}(aq) + NO_2(g) \text{ (in acidic solution)}$$

In this case, the hydrogen ion and the water molecule are eliminated because neither is oxidized nor reduced. In acidic solution, hydrogen ions ($H^+$) and water molecules are abundant and free to participate in redox reactions as either reactants or products. Some redox reactions can occur only in basic solution. When you balance equations for these reactions, you can add hydroxide ions ($OH^-$) and water molecules to either side of the equation.

# Data Analysis LAB

Based on Real Data*

## Analyze and Conclude

**How does redox lift a space shuttle?** The space shuttle gains nearly 72% of its lift from its solid rocket boosters (SRBs) during the first two minutes of launch. The two pencil-shaped SRB tanks are attached to both sides of the liquid hydrogen and oxygen fuel tank. Each SRB contains approximately 499,000 kg of propellent mixture.

### Data and Observations

**SRB Propellent Mixture**

| Component | Percent Composition |
|---|---|
| Ammonium perchlorate | 69.6 |
| Aluminum | 16 |
| Catalyst | 0.4 |
| Binder | 12.04 |
| Curing agent | 1.96 |

*Data obtained from: Dumoulin, Jim. "Solid Rocket Boosters." *NSTS Shuttle Reference Manual.* 1988

### Think Critically

1. **Balance an equation** Use the oxidation-number method to balance the chemical equation for the SRB reaction.
   $$NH_4ClO_4(s) + Al(s) \rightarrow Al_2O_3(g) + HCl(g) + N_2(g) + H_2O(g)$$
2. **State** Which elements are reduced and which are oxidized?
3. **Infer** What are the benefits of using SRBs for the first two minutes of launch?
4. **Calculate** How many moles of water vapor are produced by one SRB?

EXAMPLE PROBLEM

## EXAMPLE Problem 4

**BALANCE A NET IONIC REDOX EQUATION** Balance the following redox equation.

$ClO_4^-(aq) + Br^-(aq) \rightarrow Cl^-(aq) + Br_2(g)$ (in acidic solution)

### 1 ANALYZE THE PROBLEM

Use the rules for determining oxidation number. The increase in oxidation number of the oxidized atoms or ions must equal the decrease in oxidation number of the reduced atoms or ions. The reaction takes place under acidic conditions. Adjust the coefficients to balance the equation.

### 2 SOLVE FOR THE UNKNOWN

Assign oxidation numbers to all elements in the equation.

$\overset{+7\,-2}{ClO_4^-}(aq) + \overset{-1}{Br^-}(aq) \rightarrow \overset{-1}{Cl^-}(aq) + \overset{0}{Br_2}(g)$ (in acidic solution)

**Use the rules in Table 2.**

Identify which atoms or ions are oxidized and which are reduced.

Br is oxidized.
Cl is reduced.

**The oxidation number of bromine increases from −1 to 0. The oxidation number of chlorine decreases from +7 to −1.**

Determine the changes in oxidation number for the atoms or ions that are oxidized and reduced.

Changes in oxidation number:

Br +1
Cl −8

**Bromine loses electrons. It is oxidized.**
**Chlorine gains electrons. It is reduced.**

Make the changes in oxidation number equal in magnitude by adjusting coefficients in the equation.

$ClO_4^-(aq) + 8Br^-(aq) \rightarrow Cl^-(aq) + 4Br_2(g)$ (in acidic solution)

**Because the change in oxidation number of Br is +1, you must add the coefficient 8 to balance the equation. $4Br_2$ represents 8 Br atoms to balance the $8Br^-$ on the left side.**

Add enough hydrogen ions and water molecules to the equation to balance the oxygen atoms on both sides.

$ClO_4^-(aq) + 8Br^-(aq) + 8H^+(aq) \rightarrow Cl^-(aq) + 4Br_2(g) + 4H_2O(l)$

**Because you know the reaction takes place in acid solution, you can add $H^+$ ions on the left side of the equation.**

### 3 EVALUATE THE ANSWER

The number of atoms of each element is equal on both sides of the equation. As with any ionic equation, the net charge on the right equals the net charge on the left. No subscripts have been changed.

## PRACTICE Problems

Do additional problems. Online Practice

**Use the oxidation-number method to balance these redox equations.**

**19.** $H_2S(g) + NO_3^-(aq) \rightarrow S(s) + NO(g)$ (in acidic solution)

**20.** $Cr_2O_7^{2-}(aq) + I^-(aq) \rightarrow Cr^{3+}(aq) + I_2(s)$ (in acidic solution)

**21.** $Zn + NO_3^- \rightarrow Zn^{2+} + NO_2$ (in acidic solution)

**22. Challenge** $I^-(aq) + MnO_4^-(aq) \rightarrow I_2(s) + MnO_2(s)$ (in basic solution)

**Connection to Biology** What do many deep-sea fishes and fireflies have in common with the bacterium, *Xenorhabdus luminescens?* These and other organisms emit light. Bioluminescence is the conversion of potential energy in chemical bonds into light during a redox reaction. Depending on the species, bioluminescence is produced by different chemicals and by different means. In fireflies, shown in **Figure 7,** light results from the oxidation of the molecule luciferin.

Scientists are still unraveling the mystery of bioluminescence. Some luminescent organisms emit light constantly, whereas others emit light when they are disturbed. Deep-sea fishes and some jellyfish appear to be able to control the light they emit, and one species of mushroom is known to emit light of two different colors. Zoologists have also determined that some light-emitting organisms do not produce light themselves; they produce light by harboring bioluminescent bacteria.

■ **Figure 7** Organisms appear to use bioluminescence for different purposes. Some purposes might include attracting a mate and defense against prey. In the ocean depths, bioluminescence probably aids vision and recognition.

## Balancing Redox Equations Using Half-Reactions

In chemistry, a **species** is any kind of chemical unit involved in a process. In the equilibrium equation $NH_3 + H_2O \rightarrow NH_4^+ + OH^-$, there are four species: the two molecules $NH_3$ and $H_2O$ and the two ions $NH_4^+$ and $OH^-$. Oxidation-reduction reactions occur whenever a species that can give up electrons (reducing agent) comes in contact with another species that can accept them (oxidizing agent). For example, iron can reduce many species that are oxidizing agents, including chlorine.

$$2Fe + 3Cl_2 \rightarrow 2FeCl_3$$

In this reaction, each iron atom is oxidized by losing three electrons to become an $Fe^{3+}$ ion. At the same time, each chlorine atom in $Cl_2$ is reduced by accepting one electron to become a $Cl^-$ ion.

$$\text{Oxidation: } Fe \rightarrow Fe^{3+} + 3e^-$$
$$\text{Reduction: } Cl_2 + 2e^- \rightarrow 2Cl^-$$

Equations such as these represent half-reactions. A **half-reaction** is one of the two parts of a redox reaction—the oxidation half or the reduction half. **Table 5** shows a variety of reduction half-reactions that involve the oxidation of Fe to $Fe^{3+}$.

**VOCABULARY**
**SCIENCE USAGE V. COMMON USAGE**
**Species**
***Science usage:*** in chemistry, any kind of representative particle involved in a process
*In a synthesis reaction, two distinct species combine to form a single product.*
***Common usage:*** a class of individuals having some common characteristics or qualities; a distinct sort or kind
*Humans and chimpanzees are two different species.*

**Table 5 Redox Reactions that Oxidize Iron**

| Overall Reaction (unbalanced) | Oxidation Half-Reaction | Reduction Half-Reaction |
|---|---|---|
| $Fe + O_2 \rightarrow Fe_2O_3$ | $Fe \rightarrow Fe^{3+} + 3e^-$ | $O_2 + 4e^- \rightarrow 2O^{2-}$ |
| $Fe + F_2 \rightarrow FeF_3$ | | $F_2 + 2e^- \rightarrow 2F^-$ |
| $Fe + HBr \rightarrow H_2 + FeBr_3$ | | $2H^+ + 2e^- \rightarrow H_2$ |
| $Fe + AgNO_3 \rightarrow Ag + Fe(NO_3)_3$ | | $Ag^+ + e^- \rightarrow Ag$ |
| $Fe + CuSO_4 \rightarrow Cu + Fe_2(SO_4)_3$ | | $Cu^{2+} + 2e^- \rightarrow Cu$ |

■ **Figure 8** As a result of this redox reaction between iron and copper sulfate solution, solid copper metal is deposited on the iron. To balance the chemical equation for this reaction, you could use half-reactions.

**FOLDABLES®** Incorporate information from this section into your Foldable.

You will learn more about the importance of half-reactions when you study electrochemistry. For now, however, you can learn to use half-reactions to balance a redox equation. For example, the following unbalanced equation represents the reaction that occurs when you put an iron nail into a solution of copper(II) sulfate, as shown in **Figure 8.**

$$Fe(s) + CuSO_4(aq) \rightarrow Cu(s) + Fe_2(SO_4)_3(aq)$$

Iron atoms are oxidized as they lose electrons to the copper(II) ions. The steps for balancing redox equations by using half-reactions are shown in **Table 6.**

**VOCABULARY**
**ACADEMIC VOCABULARY**
**Method:**
a way of doing something
*Students study for an exam using different methods.*

Explore **redox titrations.**
Virtual Investigations

**Table 6 The Half-Reaction Method**

| Step |
|---|
| **1.** Write the unbalanced, net ionic equation for the reaction, omitting spectator ions. $Fe + Cu^{2+} + SO_4^{2-} \rightarrow Cu + 2Fe^{3+} + 3SO_4^{2-}$; $Fe + Cu^{2+} \rightarrow Cu + 2Fe^{3+}$ |
| **2.** Write separate, incomplete equations for the oxidation and reduction half-reactions, including oxidation numbers. $\overset{0}{Fe} \rightarrow 2\overset{+3}{Fe^{3+}}$ (oxidation)   $\overset{+2}{Cu^{2+}} \rightarrow \overset{0}{Cu}$ (reduction) |
| **3.** Balance the atoms in the half-reactions. Balance the charges in each half-reaction by adding electrons as reactants or products. $2Fe \rightarrow 2Fe^{3+} + 6e^-$   $Cu^{2+} + 2e^- \rightarrow Cu$ |
| **4.** Adjust the coefficients so that the number of electrons lost in oxidation equals the number of electrons gained in reduction. $2Fe \rightarrow 2Fe^{3+} + 6e^-$   $3Cu^{2+} + 6e^- \rightarrow 3Cu$ |
| **5.** Add the half-reactions and cancel or reduce like terms on both sides of the equation. $2Fe + 3Cu^{2+} \rightarrow 3Cu + 2Fe^{3+}$ |
| **6.** Return spectator ions, if desired. Restore state descriptions. $2Fe(s) + 3CuSO_4(aq) \rightarrow 3Cu(s) + Fe_2(SO_4)_3(aq)$ |

## EXAMPLE Problem 5

**BALANCE A REDOX EQUATION BY USING HALF-REACTIONS**
Balance the redox equation for the reaction below using half-reactions.

$$KMnO_4(aq) + SnCl_2(aq) + HCl(aq) \rightarrow MnCl_2(aq) + SnCl_4(aq) + H_2O(l) + KCl(aq)$$

### 1 ANALYZE THE PROBLEM

The reaction takes place in acidic solution. Use the rules for determining oxidation numbers and the steps for balancing by half-reactions to balance the equation for the reaction.

### 2 SOLVE FOR THE UNKNOWN

Write the unbalanced, net ionic equation for the reaction.

$MnO_4^- + Sn^{2+} \rightarrow Mn^{2+} + Sn^{4+}$

**Eliminate coefficients, spectator ions, and state symbols.**

Write incomplete equations for the oxidation and reduction half-reactions, including oxidation numbers.

$\overset{+2}{Sn^{2+}} \rightarrow \overset{+4}{Sn^{4+}}$ (oxidation)

$\overset{+7}{MnO_4^-} \rightarrow \overset{+2}{Mn^{2+}}$ (reduction)

**Use the rules in Table 2 and Table 6.**

Balance the atoms and charges in the half-reactions.

$Sn^{2+} \rightarrow Sn^{4+} + 2e^-$ (oxidation)

$5e^- + 8H^+ + MnO_4^- \rightarrow Mn^{2+} + 4H_2O$ (reduction)

**In an acid solution, $H_2O$ molecules are available in abundance and can be used to balance oxygen atoms in the half-reactions; $H^+$ ions are readily available and can be used to balance the charge.**

Adjust the coefficients so that the number of electrons lost in oxidation (2) equals the number of electrons gained in reduction (5).

$5Sn^{2+} \rightarrow 5Sn^{4+} + 10e^-$ (oxidation)

$10e^- + 16H^+ + 2MnO_4^- \rightarrow 2Mn^{2+} + 8H_2O$ (reduction)

**The least common multiple of 2 and 5 is 10. Cross-multiplying gives the balanced oxidation and reduction half-reactions.**

Add the balanced half-reactions and simplify by canceling or reducing like terms on both sides of the equation.

$5Sn^{2+} + 10e^- + 16H^+ + 2MnO_4^- \rightarrow 5Sn^{4+} + 10e^- + 2Mn^{2+} + 8H_2O$

$5Sn^{2+} + 16H^+ + 2MnO_4^- \rightarrow 5Sn^{4+} + 2Mn^{2+} + 8H_2O$

Restore state descriptions and return spectator ions ($K^+$ and $Cl^-$).

$5SnCl_2(aq) + 16HCl(aq) + 2KMnO_4(aq) \rightarrow$
$5SnCl_4(aq) + 2MnCl_2(aq) + 8H_2O(l) + 2KCl(aq)$

**Add $K^+$ ions to the two $MnO_4^-$ ions on the left and add two $K^+$ ions on the right. Add $Cl^-$ ions to the $Sn^{2+}$ and $H^+$ ions on the left and to the $Sn^{4+}$, $Mn^{2+}$, and $K^+$ ions on the right.**

### 3 EVALUATE THE ANSWER

A review of the balanced equation indicates that the number of atoms of each element is equal on both sides of the equation. No subscripts have been changed.

EXAMPLE PROBLEM

## PRACTICE Problems

Do additional problems. Online Practice

**Use the half-reaction method to balance the redox equations. Begin by writing the oxidation and reduction half-reactions. Leave the balanced equation in ionic form.**

**23.** $Cr_2O_7^{2-}(aq) + I^-(aq) \rightarrow Cr^{3+}(aq) + I_2(s)$ (in acidic solution)

**24.** $Mn^{2+}(aq) + BiO_3^-(aq) \rightarrow MnO_4^-(aq) + Bi^{2+}(aq)$ (in acidic solution)

**25. Challenge** $N_2O(g) + ClO^-(aq) \rightarrow NO_2^-(aq) + Cl^-(aq)$ (in basic solution)

## PROBLEM-SOLVING STRATEGY

### Balancing Redox Equations

Determine which species is oxidized, which species is reduced, which species is the oxidizing agent, and which species is the reducing agent.

Do the oxidized and reduced species appear more than once on either side of the equation, or does the reaction occur in an acidic or basic solution?

**No** → **Oxidation-Number Method of Balancing Redox Equations**

- Assign oxidation numbers to all of the elements.
- Adjust the coefficients in the equation so that the oxidation numbers are equal in magnitude.
- Balance the rest of the equation by the conventional method.

**Yes** → **Half-Reaction Method of Balancing Redox Equations**

- Write the unbalanced, net ionic equation for the reaction, omitting the spectator ions.
- Determine the oxidation and the reduction half-reactions.
- Balance the atoms and the charges in each half-reaction.
- Adjust the coefficients so that the number of electrons lost and the number of electrons gained is equal.
- Combine the balanced half-reactions and return spectator ions.

### Apply the Strategy

**Balance** the following equation using this flowchart.

$P_4(s) + H_2O(l) \rightarrow PH_3(g) + H_2PO_2(aq)$

## SECTION 2 REVIEW

Section Self-Check

### Section Summary

- Redox equations in which the same element appears in multiple reactants and products can be difficult to balance using the conventional method.
- The oxidation-number method is based on the number of electrons transferred from atoms or ions equaling the number of electrons accepted by other atoms or ions.
- To balance equations for reactions in an acidic solution, add hydrogen ions and water molecules.
- To balance equations for reactions in a basic solution, add hydroxide ions and water molecules.
- A half-reaction is one of the two parts of a redox reaction.

**26.** **MAIN**IDEA **Explain** how changes in oxidation number are related to the electrons transferred in a redox reaction. How are the changes related to the processes of oxidation and reduction?

**27.** **Describe** why it is important to know the conditions under which an aqueous oxidation-reduction reaction takes place in order to balance the ionic equation for the reaction.

**28.** **Explain** the steps of the oxidation-number method of balancing equations.

**29.** **State** what an oxidation half-reaction shows. What does a reduction half-reaction show?

**30.** **Write** the oxidation and reduction half-reactions for the redox equation.

$Pb(s) + Pd(NO_3)_2(aq) \rightarrow Pb(NO_3)_2(aq) + Pd(s)$

**31.** **Determine** The oxidation half-reaction of a redox reaction is $Sn^{2+} \rightarrow Sn^{4+} + 2e^-$, and the reduction half-reaction is $Au^{3+} + 3e^- \rightarrow Au$. What minimum numbers of tin(II) ions and gold(III) ions would have to react in order to have zero electrons left over?

**32.** **Apply** Balance the following equations.

**a.** $HClO_3(aq) \rightarrow ClO_2(g) + HClO_4(aq) + H_2O(l)$

**b.** $H_2SeO_3(aq) + HClO_3(aq) \rightarrow H_2SeO_4(aq) + Cl_2(g) + H_2O(l)$

**c.** $Cr_2O_7^{2-}(aq) + Fe^{2+}(aq) \rightarrow Cr^{3+}(aq) + Fe^{3+}(aq)$ (in acidic solution)

# CHEMISTRY & careers

## Career: Crime-Scene Investigator

### Blood That Glows

In Shakespeare's play *MacBeth*, Lady MacBeth washes the blood of King Duncan from her hands but can still see the bloodstains. In modern forensics, a chemical called luminol gives investigators similar visual ability.

**Blue-green whisper** Luminol oxidizes when it comes in contact with iron, as shown in **Figure 1.** In the process, the molecules release energy in the form of distinctive blue-green light. In a dark room, the faint blue glow of luminol might reveal to investigators what their eyes alone could not see—hidden traces of blood. Red blood cells consist mainly of hemoglobin—a protein that contains iron.

To use luminol, investigators mix a white powder ($C_8H_7O_3N_3$) with hydrogen peroxide ($H_2O_2$) and other chemicals. This creates a liquid that can be sprayed onto areas suspected of holding hidden blood evidence. If blood is present—even in quantities too small to detect with the eye—the luminol will glow. Forensic photographers then snap pictures with special cameras that can both capture the faint glow of the luminol and illuminate the surrounding area.

**Figure 1** The luminol oxidizes within a beaker when an iron nail is added.

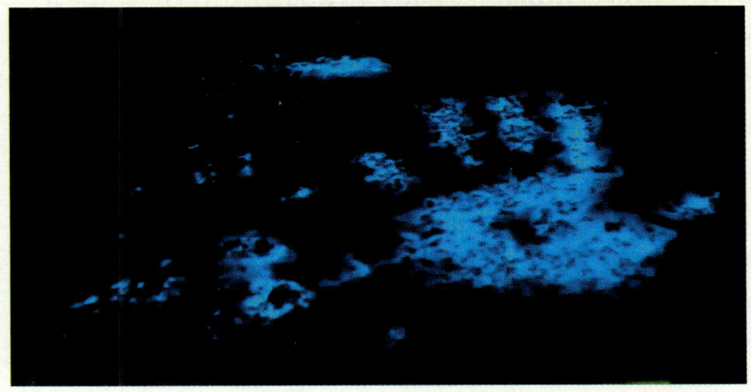

**Figure 2** A luminol impression from a murder scene can be compared to a suspect's handprint.

**Glowing evidence** Bloodstains might reveal spatter patterns, giving clues about the type of weapon used to commit a crime. Faint luminol signals on carpet might lead investigators to much larger bloodstains. Bloody handprints, such as that in **Figure 2,** might even lead investigators to the assailant.

There are other uses for luminol besides murder investigations. In a car accident, luminol might reveal whether a victim was wearing a safety belt, even after the car has been subjected to rain, cold, or direct sunlight that can greatly alter bloodstains.

**Spray of last resort** Other iron-containing substances besides blood can cause luminol to glow, although experts can usually tell the difference. More importantly, luminol might interfere with other tests. For this reason, investigators normally do not use luminol until all their other investigations are complete.

### WRITING IN ▶ Chemistry

**News Article** Research more about the use of luminol in crime scene investigations. Write a newspaper article that describes how luminol led investigators to a suspect. Describe the type of evidence that was used in the investigation.

WebQuest

SMALL SCALE **ChemLAB**

iLab Station

# Forensics: Identify the Damaging Dumper

**Background:** Something is reacting with metals found on the hulls of many boats used on a nearby creek. The investigator has determined that there are three possible culprits, each with a different source. Your job is to test the three potential pollutants and compare them with a sample from the creek. The animals that rely on the creek as their primary water source are depending on you to solve this mystery of the damaging dumper.

**Question:** *How can a series of chemical reactions be used to determine what was dumped in a water supply?*

## Materials

0.1*M* $AgNO_3$
0.1*M* HCl
0.1*M* $ZnSO_4$
unknown solution
Cu wire
Pb shot
Fe filings
Mg turnings
tongs or forceps
droppers (4)
24-well microscale reaction plate

## Safety Precautions

**WARNING:** ***Silver nitrate*** **($AgNO_3$)** ***is highly toxic and will stain skin and clothing.***

## Procedure

1. Read and complete the lab safety form.
2. Create a table to record your data.
3. Place the well plate on a sheet of white paper.
4. Place a piece of copper wire in four wells in the first row.
5. Repeat Step 4, by adding a small sample of iron filings to wells in the second row.
6. Repeat Step 4, by adding a piece of lead shot to wells in the third row
7. Repeat Step 4, by adding a piece of magnesium ribbon to wells in the fourth row.
8. Count 20 drops of the silver nitrate solution ($AgNO_3$) into each well in the first column.
9. Repeat Step 8, adding hydrochloric acid (HCl) in the second column.
10. Repeat Step 8, adding zinc sulfate ($ZnSO_4$) in the third column.

**Observations**

| | $AgNO_3$ | HCl | $ZnSO_4$ | Unknown |
|---|---|---|---|---|
| Cu | | | | |
| Fe | | | | |
| Pb | | | | |
| Mg | | | | |

11. Repeat Step 8, adding the unknown solution in the fourth column.
12. Allow the reactions to proceed for 5 min, and then describe the reactions. Write *NR* for any wells that do not have evidence of a reaction.
13. **Cleanup and Disposal** Dispose of the solids and solutions as directed by your teacher. Wash and return all lab equipment to its designated location.

## Analyze and Conclude

1. **Summarize** the results you observed in each well. How did you know a chemical reaction occurred?
2. **Model** Write a balanced equation for each of the reactions you observed. In each one, identify the species being oxidized or reduced.
3. **Conclude** Based on your data, which solution was causing damage in the creek? Justify your answer.
4. **Use Variables, Constants, and Controls** Why was it important to compare the reactions of the unknown to more than one known solution?
5. **Research** Look up the MSDS for your chemical and report on what impact this chemical would have on the ecosystem.
6. **Extend** What would you expect if a solution of lead(II) nitrate ($Pb(NO_3)_2$) was one of the solutions?
7. **Error Analysis** Compare your results with those of other students in the laboratory. Explain any differences.

## INQUIRY EXTENSION

**Design an Experiment** Hypothesize how you could remove this chemical from the creek without further damaging the ecology of the area. Design an experiment to test your hypothesis.

# CHAPTER 19 STUDY GUIDE

Vocabulary Practice

**BIGIDEA** **Oxidation-reduction reactions—among the most-common chemical processes in both nature and industry—involve the transfer of electrons.**

## SECTION 1 Oxidation and Reduction

**MAINIDEA** Oxidation and reduction are complementary—as a substance is oxidized, another substance is reduced.

- Oxidation-reduction reactions involve the complete or partial transfer of electrons from one substance to another.
- When an atom or ion is reduced, its oxidation number is lowered. When an atom or ion is oxidized, its oxidation number is raised.
- In oxidation-reduction reactions involving molecular compounds (and polyatomic ions with covalent bonds), the more-electronegative atoms are treated as if they are reduced. The less-electronegative atoms are treated as if they are oxidized.

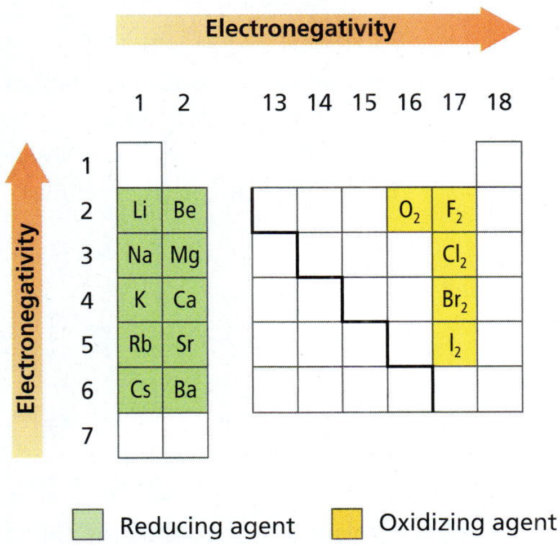


### VOCABULARY

- **oxidation-reduction reaction**
- **redox reaction**
- **oxidation**
- **reduction**
- **oxidation number**
- **oxidizing agent**
- **reducing agent**

## SECTION 2 Balancing Redox Equations

**MAINIDEA** Redox equations are balanced when the total increase in oxidation numbers equals the total decrease in oxidation numbers of the atoms or ions involved in the reaction.

- Redox equations in which the same element appears in several reactants and products can be difficult to balance using the conventional method.
- The oxidation-number method is based on the number of electrons transferred from atoms or ions equaling the number of electrons accepted by other atoms or ions.
- To balance equations for reactions in an acidic solution, add hydrogen ions and water molecules.
- To balance equations for reactions in a basic solution, add hydroxide ions and water molecules.
- A half-reaction is one of the two parts of a redox reaction.

### VOCABULARY

- **oxidation-number method**
- **species**
- **half-reaction**

# CHAPTER 19 ASSESSMENT

## SECTION 1

### Mastering Concepts

**33.** What is the main characteristic of oxidation-reduction reactions?

**34.** Explain why not all oxidation reactions involve oxygen.

**35.** In terms of electrons, what happens when an atom is oxidized? When an atom is reduced?

**36.** Define *oxidation number.*

**37. Metals** What is the oxidation number of alkaline earth metals in their compounds? Of alkali metals?

**38.** How does the oxidation number in an oxidation process relate to the number of electrons lost? How does the change in oxidation number in a reduction process relate to the number of electrons gained?

■ **Figure 9**

**39.** What probably accounts for the different forms of copper shown in **Figure 9?**

**40. Copper and air** Copper statues, such as the Statue of Liberty, begin to appear green after they have been exposed to air. In this redox process, copper metal reacts with oxygen to form solid copper oxide, which forms the green coating. Write the reaction for this redox process, and identify what is oxidized and what is reduced in the process.

### Mastering Problems

**41.** Identify the species oxidized and the species reduced in each of these redox equations.
**a.** $3Br_2 + 2Ga \rightarrow 2GaBr_3$
**b.** $2HCl + Zn \rightarrow ZnCl_2 + H_2$
**c.** $3Mg + N_2 \rightarrow Mg_3N_2$

**42.** Identify the oxidizing agent and the reducing agent in each of these redox equations.
**a.** $N_2 + 3H_2 \rightarrow 2NH_3$
**b.** $2Na + I_2 \rightarrow 2NaI$

**43.** What is the reducing agent in this balanced equation?

$$8H^+ + Sn + 6Cl^- + 4NO_3^- \rightarrow SnCl_6^{2-} + 4NO_2 + 4H_2O$$

**44.** What is the oxidation number of manganese in $KMnO_4$?

**45.** Determine the oxidation number of the boldface element in these substances and ions.
**a.** Ca**Cr**$O_4$ **c.** **N**$O_2^-$
**b.** NaH**S**$O_4$ **d.** **Br**$O_3^-$

**46.** Identify each of these half-reactions as either oxidation or reduction.
**a.** $Al \rightarrow Al^{3+} + 3e^-$
**b.** $Cu^{2+} + e^- \rightarrow Cu^+$

**47.** Which of these equations does not represent a redox reaction? Explain your answer.
**a.** $LiOH + HNO_3 \rightarrow LiNO_3 + H_2O$
**b.** $MgI_2 + Br_2 \rightarrow MgBr_2 + I_2$

**48.** Determine the oxidation number of nitrogen in each of these molecules or ions.
**a.** $NO_3^-$ **b.** $N_2O$ **c.** $NF_3$

**49.** Determine the oxidation number of each element in these compounds or ions.
**a.** $Au_2(SeO_4)_3$ (gold(III) selenate)
**b.** $Ni(CN)_2$ (nickel(II) cyanide)

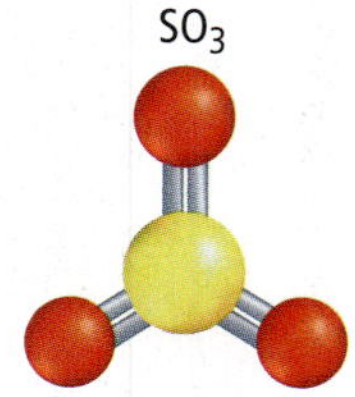


■ **Figure 10**

**50.** Explain how the sulfite ion ($SO_3^{2-}$) differs from sulfur trioxide ($SO_3$), shown in **Figure10.**

## SECTION 2

### Mastering Concepts

**51.** Compare and contrast balancing redox equations in acidic and basic solutions.

**52.** Explain why writing hydrogen ions as $H^+$ in redox reactions represents a simplification and not how they exist.

**53.** Before you attempt to balance the equation for a redox reaction, why do you need to know whether the reaction takes place in acidic or basic solution?

**54.** Explain what a spectator ion is.

**55.** Define the term *species* in terms of redox reactions.

**56.** Is the following equation balanced? Explain.

$$Fe(s) + Ag^+(aq) \rightarrow Fe^{2+}(aq) + Ag(s)$$

**57.** Does the following equation represent a reduction or an oxidation process? Explain your answer.

$$Zn^{2+} + 2e^- \rightarrow Zn$$

**58.** Describe what is happening to electrons in each half reaction of a redox process.

## Mastering Problems

**59.** Use the oxidation-number method to balance these redox equations.
**a.** $Cl_2 + NaOH \rightarrow NaCl + HOCl$
**b.** $HBrO_3 \rightarrow Br_2 + H_2O + O_2$

**60.** Balance these net ionic equations for redox reactions.
**a.** $Au^{3+}(aq) + I^-(aq) \rightarrow Au(s) + I_2(s)$
**b.** $Ce^{4+}(aq) + Sn^{2+}(aq) \rightarrow Ce^{3+}(aq) + Sn^{4+}(aq)$

**61.** Use the oxidation-number method to balance the following ionic redox equations.
**a.** $Al + I_2 \rightarrow Al^{3+} + I^-$
**b.** $MnO_2 + Br^- \rightarrow Mn^{2+} + Br_2$ (in acidic solution)

**62.** Use the oxidation-number method to balance these redox equations.
**a.** $PbS + O_2 \rightarrow PbO + SO_2$
**b.** $NaWO_3 + NaOH + O_2 \rightarrow Na_2WO_4 + H_2O$
**c.** $NH_3 + CuO \rightarrow Cu + N_2 + H_2O$
**d.** $Al_2O_3 + C + Cl_2 \rightarrow AlCl_3 + CO$

■ **Figure 11**

**63. Sapphire** The mineral corundum is comprised of aluminum oxide ($Al_2O_3$) and is dull gray. Sapphire is mostly aluminum oxide, but it contains small amounts of $Fe^{2+}$ and $Ti^{4+}$. The color of sapphire results from an electron transfer from $Fe^{2+}$ to $Ti^{4+}$. Based on **Figure 11,** write an equation that describes the reaction that occurs resulting in the mineral on the right. What are the oxidizing and reducing agents?

**64.** Write the oxidation and reduction half-reactions represented in each of these redox equations. Write the half-reactions in net ionic form if they occur in aqueous solution.
**a.** $PbO(s) + NH_3(g) \rightarrow N_2(g) + H_2O(l) + Pb(s)$
**b.** $I_2(s) + Na_2S_2O_3(aq) \rightarrow Na_2S_2O_4(aq) + NaI(aq)$
**c.** $Sn(s) + 2HCl(aq) \rightarrow SnCl_2(aq) + H_2(g)$

**65.** Write the two half-reactions that make up the following balanced redox reaction.

$$3H_2C_2O_4 + 2HAsO_2 \rightarrow 6CO_2 + 2As + 4H_2O$$

**66.** Label each half-reaction as reduction or oxidation.
**a.** $Fe^{2+}(aq) \rightarrow Fe^{3+}(aq) + e^-$
**b.** $MnO_4^- + 5e^- + 8H^+ \rightarrow Mn^{2+} + 4H_2O$
**c.** $2H^+ + 2e^- \rightarrow H_2$
**d.** $F_2 \rightarrow 2F^- + 2e^-$

■ **Figure 12**

**67. Copper** When solid copper is put into a solution of silver nitrate, as shown in **Figure 12,** silver metal appears and blue copper(II) nitrate forms. Write the corresponding, unbalanced chemical equation. Next, determine the oxidation state of each element in the equation. Write the two half-reactions, labeling each as oxidation or reduction. Finally, write a balanced equation for the reaction.

**68.** Use the oxidation-number method to balance these ionic redox equations.
**a.** $MoCl_5 + S^{2-} \rightarrow MoS_2 + Cl^- + S$
**b.** $TiCl_6^{2-} + Zn \rightarrow Ti^{3+} + Cl^- + Zn^{2+}$

**69.** Use the half-reaction method to balance these equations for redox reactions. Add water molecules and hydrogen ions or hydroxide ions as needed.
**a.** $NH_3(g) + NO_2(g) \rightarrow N_2(g) + H_2O(l)$ (in acidic solution)
**b.** $Br_2 \rightarrow Br^- + BrO_3^-$ (in basic solution)

**70.** Balance the following redox chemical equation. Rewrite the equation in full ionic form, then derive the net ionic equation and balance by the half-reaction method. Give the final answer as it is shown below but with the balancing coefficients.

$$KMnO_4(aq) + FeSO_4(aq) + H_2SO_4(aq) \rightarrow Fe_2(SO_4)_3(aq) + MnSO_4(aq) + K_2SO_4(aq) + H_2O(l)$$

**71.** Use the oxidation-number method to balance these redox equations.
**a.** $CO + I_2O_5 \rightarrow I_2 + CO_2$
**b.** $SO_2 + Br_2 + H_2O \rightarrow HBr + H_2SO_4$
**c.** $Cu + NO_3^- \rightarrow Cu^{2+} + NO$ (in acidic solution)
**d.** $Zn + NO_3^- \rightarrow Zn^{2+} + NO_2$ (in acidic solution)
**e.** $Al + OH^- + H_2O \rightarrow H_2 + AlO_2^-$

**72.** Use the half-reaction method to balance these equations. Add water molecules and hydrogen ions or hydroxide ions as needed. Write the balanced equations in net ionic form.
**a.** $Cl^-(aq) + NO_3^-(aq) \rightarrow ClO^-(aq) + NO(g)$ (in acidic solution)
**b.** $IO_3^-(aq) + Br^-(aq) \rightarrow Br_2(l) + IBr(s)$ (in acidic solution)
**c.** $I_2(s) + Na_2S_2O_3(aq) \rightarrow Na_2S_2O_4(aq) + NaI(aq)$ (in acidic solution)

## MIXED REVIEW

**73.** Determine the oxidation number of the boldface element in each of the following.
**a.** $\mathbf{O}F_2$ **b.** $\mathbf{U}O_2^{2+}$ **c.** $\mathbf{Ru}O_4$ **d.** $\mathbf{Fe}_2O_3$

**74.** Identify each of the following changes as either oxidation or reduction.
**a.** $2Cl^- \rightarrow Cl_2 + 2e^-$
**b.** $Na \rightarrow Na^+ + e^-$
**c.** $Ca^{2+} + 2e^- \rightarrow Ca$
**d.** $O_2 + 4e^- \rightarrow 2O^{2-}$

**75.** Use the rules for assigning oxidation numbers to complete **Table 7.**

**Table 7 Oxidation Number Assignment**

| Element | Oxidation Number | Rule |
|---|---|---|
| K in KBr | +1 | |
| Br in KBr | | 8 |
| Cl in $Cl_2$ | | 1 |
| K in KCl | | 7 |
| Cl in KCl | –1 | |
| Br in $Br_2$ | 0 | |

**76.** Identify the reducing agents in these equations.
**a.** $4NH_3 + 5O_2 \rightarrow 4NO + 6H_2O$
**b.** $Na_2SO_4 + 4C \rightarrow Na_2S + 4CO$
**c.** $4IrF_5 + Ir \rightarrow 5IrF_4$

**77.** Write a balanced ionic redox equation using the following pairs of redox half-reactions.
**a.** $Fe \rightarrow Fe^{2+} + 2e^-$
$Te^{2+} + 2e^- \rightarrow Te$
**b.** $IO_4^- + 2e^- \rightarrow IO_3^-$
$Al \rightarrow Al^{3+} + 3e^-$ (in acidic solution)
**c.** $I_2 + 2e^- \rightarrow 2I^-$
$N_2O \rightarrow 2NO_3^- + 8e^-$ (in acidic solution)

■ **Figure 13**

**78.** What probably accounts for the different forms of chromium shown in **Figure 13?**

**79.** Balance these ionic redox equations by any method.
**a.** $Sb^{3+} + MnO_4^- \rightarrow SbO_4^{3-} + Mn^{2+}$ (in acid solution)
**b.** $N_2O + ClO^- \rightarrow Cl^- + NO_2^-$ (in basic solution)

**80. Gemstones** Rubies are gemstones made up mainly of aluminum oxide. Their red color comes from a small amount of chromium(III) ions replacing some of the aluminum ions. Write the formula for aluminum oxide, and show the reaction in which an aluminum ion is replaced with a chromium ion. Is this a redox reaction?

**81.** Balance these ionic redox equations by any method.
**a.** $Mg + Fe^{3+} \rightarrow Mg^{2+} + Fe$
**b.** $ClO_3^- + SO_2 \rightarrow Cl^- + SO_4^{2-}$ (in acidic solution)

**82.** Balance these redox equations by any method.
**a.** $P + H_2O + HNO_3 \rightarrow H_3PO_4 + NO$
**b.** $KClO_3 + HCl \rightarrow Cl_2 + ClO_2 + H_2O + KCl$

## THINK CRITICALLY

**83. Apply** The following equations show redox reactions that are sometimes used in the laboratory to generate pure nitrogen gas and pure dinitrogen monoxide gas (nitrous oxide, $N_2O$).

$$NH_4NO_2(s) \rightarrow N_2(g) + 2H_2O(l)$$
$$NH_4NO_3(s) \rightarrow N_2O(g) + 2H_2O(l)$$

**a.** Determine the oxidation number of each element in the two equations, and then make diagrams showing the changes in oxidation numbers that occur in each reaction.
**b.** Identify the atom that is oxidized and the atom that is reduced in each of the two reactions.
**c.** Identify the oxidizing and reducing agents in each of the two reactions.
**d.** Write a sentence telling how the electron transfer taking place in these two reactions differs from that taking place in the reaction below.

$$2AgNO_3 + Zn \rightarrow Zn(NO_3)_2 + 2Ag$$

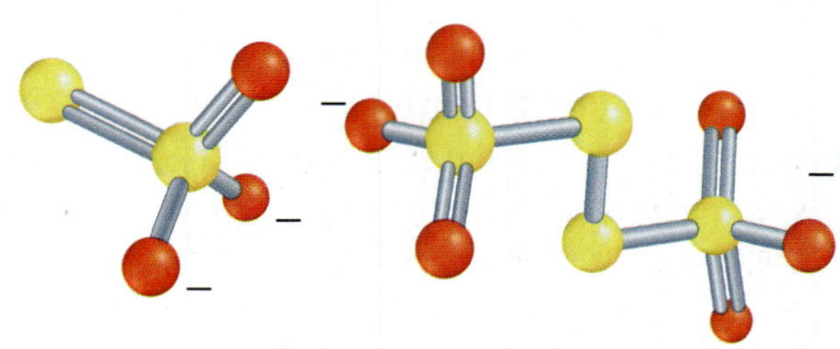

Thiosulfate ion ($S_2O_3^{2-}$) Tetrathionate ion ($S_4O_6^{2-}$)

■ **Figure 14**

**84. Analyze** Examine the net ionic equation below for the reaction that occurs when the thiosulfate ion ($S_2O_3^{2-}$) is oxidized to the tetrathionate ion ($S_4O_6^{2-}$). Balance the equation using the half-reaction method. **Figure14** will help you to determine the oxidation numbers to use.

$$S_2O_3^{2-} + I_2 \rightarrow I^- + S_4O_6^{2-} \text{ (in acidic solution)}$$

Chapter Self-Check

**85. Predict** Consider the fact that all of the following are stable compounds. What can you infer about the oxidation state of phosphorus in its compounds?

$$PH_3, PCl_3, P_2H_4, PCl_5, H_3PO_4, Na_3PO_3$$

**86. Solve** Potassium permanganate oxidizes chloride ions to chlorine gas. Balance the equation for this redox reaction taking place in acidic solution.

**87.** In the half-reaction $NO_3^- \rightarrow NH_4^+$, on which side of the equation should electrons be added? Add the correct number of electrons to the side on which they are needed, and rewrite the equation.

■ **Figure 15**

**88.** The redox reaction between dichromate ions and iodide ions in acidic solution is shown in **Figure 15.** Use the half-reaction method to balance the equation for this redox reaction.

## CHALLENGE PROBLEM

**89.** For each reaction described, write the corresponding chemical equation without putting coefficients to balance it. Next, determine the oxidation state of each element in the equation. Then, write the two half-reactions, labeling which is oxidation and which is reduction. Finally, write a balanced equation for the reaction.

   **a.** Solid mercury(II) oxide is put into a test tube and gently heated. Liquid mercury forms on the sides and in the bottom of the tube, and oxygen gas bubbles out from the test tube.

   **b.** Solid copper pieces are put into a solution of silver nitrate. Silver metal appears and blue copper(II) nitrate forms in the solution.

## CUMULATIVE REVIEW

**90.** A gaseous sample occupies 32.4 mL at −23°C and 0.75 atm. What volume will it occupy at STP?

**91.** When iron(III) chloride ($FeCl_3$) reacts in an atmosphere of pure oxygen, the following occurs:

$$4FeCl_3(s) + 3O_2(g) \rightarrow 2Fe_2O_3(s) + 6Cl_2(g)$$

If 45.0 g of $FeCl_3$ reacts and 20.5 g of iron(III) oxide is recovered, determine the percent yield.

## WRITING IN ▶ Chemistry

**92. Steel** Research the role of oxidation-reduction reactions in the manufacture of steel. Write a summary of your findings, including appropriate diagrams and equations representing the reactions.

**93. Silverware** Practice your technical writing skills by writing a procedure for cleaning tarnished silverware by a redox chemical process. Be sure to include background information describing the process as well as logical steps that would enable anyone to accomplish the task.

**94.** Copper was a useful metal even before iron, silver, and gold metals were extracted from their ores and used as tools, utensils, jewelry, and artwork. Copper was smelted to high temperatures as early as 8000 years ago. Thousands of pieces of scrap copper have been unearthed in Virginia, where in the 1600s the colonists might have traded this material for food. Compare and contrast the processing and use of copper in those older civilizations with today.

## DBQ Document-Based Questions

**Glazes** *The formation of color in ceramic glazes, such as in* **Figure 16,** *can be influenced by firing conditions. Metal ions such as copper that have more than one oxidation state can impart different colors to a glaze. In an oxidative firing, plenty of oxygen is allowed in the kiln, and copper ions present will make the glaze a green-to-blue color. Under reducing conditions, oxygen is limited and carbon dioxide is abundant. Copper ions in the glaze provide a reddish color.*

Data obtained from: Denio, Allen A. 2001. The joy of color in ceramic glazes with the help of redox chemistry. *Journal of Chemical Education.* 78 No 10.

■ **Figure16**

**95.** Write the equation for what has occurred in the pottery shown in **Figure 16.**

**96.** Based on the color of the pottery, what is the oxidation state of the copper that is reduced? Oxidized?

## MULTIPLE CHOICE

**1.** Which is NOT a reducing agent in a redox reaction?
**A.** the substance oxidized
**B.** the electron acceptor
**C.** the less-electronegative substance
**D.** the electron donor

**2.** The reaction between nickel and copper(II) chloride is shown below.

$$Ni(s) + CuCl_2(aq) \rightarrow Cu(s) + NiCl_2(aq)$$

What are the half-reactions for this redox reaction?
**A.** $Ni \rightarrow Ni^{2+} + 2e^-$, $Cl_2 \rightarrow 2Cl^- + 2e^-$
**B.** $Ni \rightarrow Ni^{2+} + e^-$, $Cu^+ + e^- \rightarrow Cu$
**C.** $Ni \rightarrow Ni^{2+} + 2e^-$, $Cu^{2+} + 2e^- \rightarrow Cu$
**D.** $Ni \rightarrow Ni^{2+} + 2e^-$, $2Cu^+ + 2e^- \rightarrow Cu$

*Use the diagram below to answer Questions 3 and 4. All four containers have a volume of 5.0 L and are at the same temperature.*

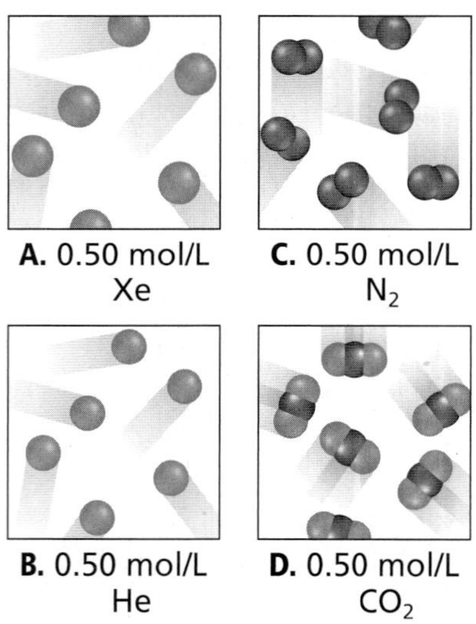


**3.** Which container contains 110 g of its gas?
**A.** A
**B.** B
**C.** C
**D.** D

**4.** If a small hole is made in each container so that the gas can escape, which container will have the fastest rate of effusion?
**A.** A
**B.** B
**C.** C
**D.** D

**5.** The following system is in equilibrium:

$$2S(s) + 5F_2(g) \leftrightarrows SF_4(g) + SF_6(g)$$

Which will cause the equilibrium to shift to the right?
**A.** increased concentration of $SF_4$
**B.** increased concentration of $SF_6$
**C.** increased pressure on the system
**D.** decreased pressure on the system

*Use the table below to answer Question 6.*

| Data for the Formation of Cobalt(II) Sulfate at 25°C | |
|---|---|
| $Co(s) + S(s) + 2O_2(g) \rightarrow CoSO_4(s)$ | |
| $\Delta H_f^\circ$ | −888.3 kJ/mol |
| $\Delta S_f^\circ$ | 118.0 J/mol·K |
| $\Delta G_f^\circ$ | ? |

**6.** What is the $\Delta G_f^\circ$ for the formation of cobalt(II) sulfate from its elements?
**A.** −853.1 kJ/mol
**B.** −885.4 kJ/mol
**C.** −891.3 kJ/mol
**D.** −923.5 kJ/mol

**7.** Which will be the result of increasing the temperature of a reaction in a system in equilibrium where the forward reaction is endothermic?
**A.** The equilibrium will shift to the left.
**B.** The equilibrium will shift to the right.
**C.** The rate of the forward reaction will be decreased.
**D.** The rate of the reverse reaction will be decreased.

**8.** The reaction between sodium iodide and chlorine is shown below.

$$2NaI(aq) + Cl_2(aq) \rightarrow 2NaCl(aq) + I_2(aq)$$

The oxidation state of sodium remains unchanged for which reason?
**A.** $Na^+$ is a spectator ion.
**B.** $Na^+$ cannot be reduced.
**C.** Na is an uncombined element.
**D.** $Na^+$ is a monatomic ion.

## SHORT ANSWER

*Use the equation below to answer Questions 9 and 10.*

The net ionic reaction between iodine and lead(IV) oxide is shown below.

$$I_2(s) + PbO_2(s) \rightarrow IO_3^-(aq) + Pb^{2+}(aq)$$

9. Identify the oxidation number in each participant in the reaction.

10. Explain how to identify which element is oxidized and which one is reduced.

## EXTENDED RESPONSE

*Use the phase diagram for $CO_2$ to answer Questions 11 to 13.*

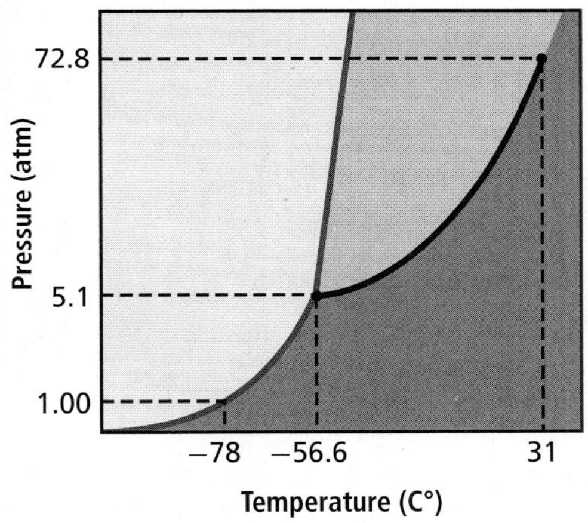


11. Explain what state or states of matter can exist at a temperature of −56.6°C and a pressure of 31.1 atm.

12. Suppose that you have a sample of $CO_2$ at 35°C and 83 atm. In what state of matter is the sample? Explain how you can predict this from the graph.

13. Is carbon dioxide denser in its liquid state or its solid state? Use the graph to explain.

## SAT SUBJECT TEST: CHEMISTRY

14. Which statement about the common ion effect is NOT true?
    - **A.** The effects of common ions on an equilibrium system can be explained by Le Châtelier's principle.
    - **B.** The decreased solubility of an ionic compound due to the presence of a common ion is called the common ion effect.
    - **C.** The addition of NaCl to a saturated solution of AgCl will produce the common ion effect.
    - **D.** The common ion effect is due to a shift in equilibrium toward the aqueous products of a system.
    - **E.** The addition of lead(II) nitrate ($Pb(NO_3)_2$) to a saturated solution of lead(II) chromate ($PbCrO_4$) will produce the common ion effect.

*Use the list below to answer Questions 15 to 18.*

Five flasks contain 500 mL of a 0.250*M* aqueous solution of the indicated chemical.

- **A.** KCl
- **B.** $CH_3OH$
- **C.** $Ba(OH)_2$
- **D.** $CH_3COOH$
- **E.** NaOH

15. Which chemical will dissociate into the greatest number of particles when in solution?

16. Which chemical has the greatest molar mass?

17. Which flask would contain 9.32 g of the labeled chemical?

18. Which flask's contents are composed of 18.7% oxygen?

| NEED EXTRA HELP? | | | | | | | | | | | | | | | | | | |
|---|---|---|---|---|---|---|---|---|---|---|---|---|---|---|---|---|---|---|
| **If You Missed Question . . .** | 1 | 2 | 3 | 4 | 5 | 6 | 7 | 8 | 9 | 10 | 11 | 12 | 13 | 14 | 15 | 16 | 17 | 18 |
| **Review Section . . .** | 19.1 | 19.2 | 13.3 | 12.1 | 17.2 | 15.5 | 17.2 | 19.1 | 19.1 | 19.1 | 12.4 | 12.4 | 12.4 | 17.2 | 14.2 | 10.2 | 10.3 | 10.4 |

## CHAPTER 20

# Electrochemistry

**BIGIDEA** Chemical energy can be converted to electric energy and electric energy to chemical energy.

## SECTIONS

1 Voltaic Cells

2 Batteries

3 Electrolysis

## LaunchLAB

iLab Station

### How can you make a battery from a lemon?

You can purchase a handy package of portable power at any convenience store—a battery. You can also light a bulb with a lemon. In this lab, you will study how these power sources are alike.

## FOLDABLES® Study Organizer

### Electrochemical Cells

Make a pocket book. Label it as shown. Use it to help you compare voltaic cells to electrolytic cells.

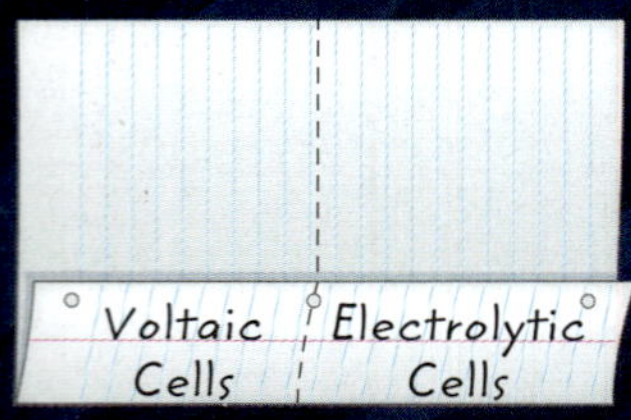

Go **online!**
MHEonline.com

A camera trap captured this image of a red fox. Camera traps are a noninvasive way to study animals by using a sensor that triggers a camera's shutter when the animal approaches. Batteries power both the camera and the sensor.

SECTION 1

# Voltaic Cells

**MAINIDEA** In voltaic cells, oxidation takes place at the anode, yielding electrons that flow to the cathode, where reduction occurs.

### Essential Questions

- How is electrical energy obtained from a redox reaction?
- What are the parts of a voltaic cell, and how does each part operate?
- How are cell potentials calculated and the spontaneity of redox reactions determined?

### Review Vocabulary

**oxidation:** the complete or partial loss of electrons from a reacting substance; increases an atom's or ion's oxidation number

**reduction:** the complete or partial gain of electrons by the atoms of a substance; decreases the atom's or ion's oxidation number

### New Vocabulary

salt bridge
electrochemical cell
voltaic cell
half-cell
anode
cathode
reduction potential
standard hydrogen electrode

**CHEM 4 YOU** What could you do with half of a dollar bill? Without the other half, you cannot spend it. Voltaic cells have two half-cells, and both are required to produce energy.

## Redox in Electrochemistry

Electrochemistry is the study of the redox processes by which chemical energy is converted to electrical energy and vice versa. Electrochemical processes are useful in industry and critically important for biological functioning.

You have learned that all redox reactions involve a transfer of electrons from the species that is oxidized to the species that is reduced. **Figure 1** and **Figure 2** illustrate the simple redox reaction in which zinc atoms are oxidized to form zinc ($Zn^{2+}$) ions. The two electrons donated from each zinc atom are accepted by a copper ($Cu^{2+}$) ion, which becomes an atom of copper metal. The following net ionic equation illustrates the electron transfer that occurs.

$$\overset{2e^-}{\longrightarrow}$$

$$Zn(s) + Cu^{2+}(aq) \rightarrow Zn^{2+}(aq) + Cu(s)$$

**Half-reactions** Two half-reactions make up this redox process:

$$Zn \rightarrow Zn^{2+} + 2e^- \text{ (oxidation half-reaction: electrons lost)}$$
$$Cu^{2+} + 2e^- \rightarrow Cu \text{ (reduction half-reaction: electrons gained)}$$

What do you think would happen if you separated the oxidation half-reaction from the reduction half-reaction? Can a redox reaction occur? Consider **Figure 1a,** in which a zinc strip is immersed in a solution of zinc sulfate and a copper strip is immersed in a solution of copper(II) sulfate.

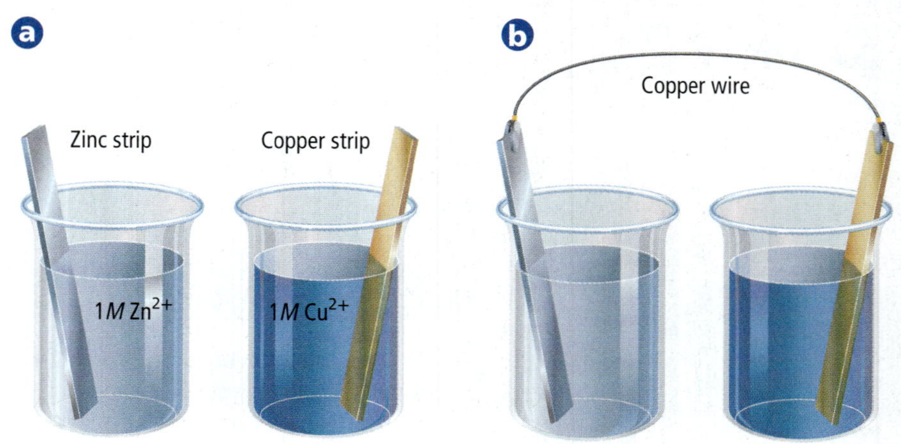


■ **Figure 1** These containers are constructed and arranged so that zinc will be oxidized on one side, while copper ions will be reduced on the other. In **a,** zinc metal is immersed in 1*M* zinc sulfate solution, and copper metal in 1*M* copper sulfate. In **b,** a wire joining the zinc and copper strips provides a pathway for the flow of electrons, but the pathway is not complete. Electron transfer is still not possible.

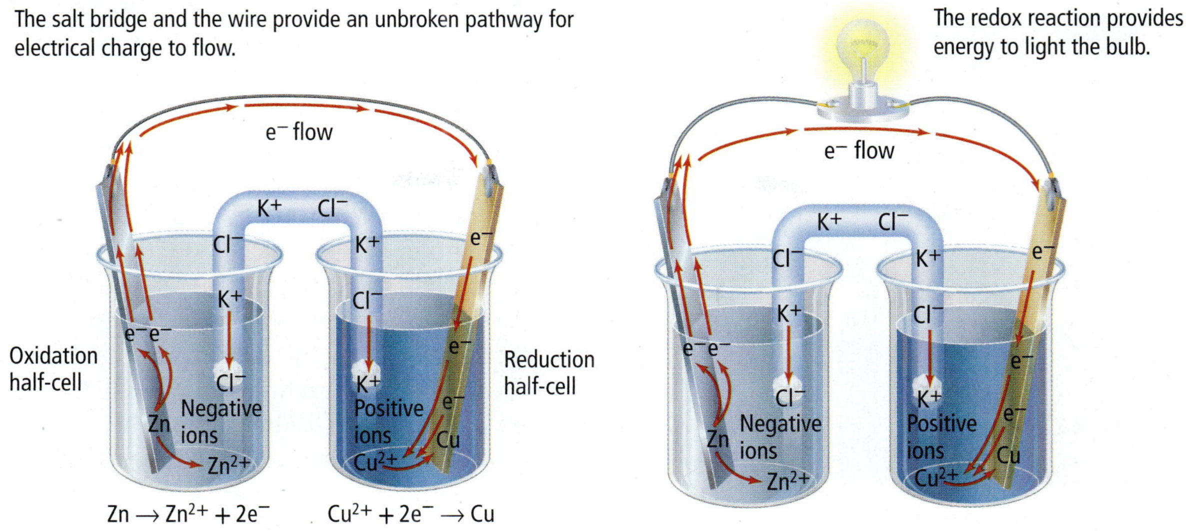


■ **Figure 2** The addition of a salt bridge completes the pathway. Negative ions move through the salt bridge to the zinc side. Positive ions move through the bridge to the copper side.

Two problems prevent a redox reaction from occurring. First, there is no way for zinc atoms to transfer electrons to copper(II) ions. This problem can be solved by connecting the zinc and copper strips with a metal wire, as shown in **Figure 1b.** The wire can serve as a pathway for electrons to flow from the zinc strip to the copper strip.

The second problem is that when the metal strips are placed in their solutions, oxidation begins at the zinc strip and reduction begins at the copper strip—but these reactions cannot continue. The reason is that as zinc is oxidized, positive zinc ions build up around the zinc electrode. As copper in the copper sulfate solution is reduced, negative sulfate ions build up around the copper electrode. The buildup of charges stops any further reaction. To solve this problem, a salt bridge must be added to the system. A **salt bridge** is a pathway to maintain solution neutrality by allowing the passage of ions from one side to another, as shown in **Figure 2.** A salt bridge consists of a tube containing a conducting solution of a soluble salt, such as KCl, held in place by an agar gel or other permeable plug. Ions can move through it, but the solutions cannot.

When the connecting metal wire and the salt bridge are in place, the spontaneous redox reaction begins. Electrons flow through the wire from the oxidation half-reaction to the reduction half-reaction, while positive and negative ions move through the salt bridge. A flow of charged particles is called an electric current. In **Figure 2,** the flow of electrons through the wire and the flow of ions through the salt bridge make up the electric current. The energy of the flowing electrons can be used to light a bulb, as shown in **Figure 2.**

**Electrochemical cells** The device shown in **Figure 2** is a type of electrochemical cell called a voltaic cell. An **electrochemical cell** is an apparatus that uses a redox reaction to produce electrical energy or uses electrical energy to cause a chemical reaction. A **voltaic cell** is a type of electrochemical cell that converts chemical energy to electrical energy by a spontaneous redox reaction. The voltaic cell, also shown in **Figure 3,** is named for Alessandro Volta (1745–1827), the Italian physicist who is credited with its invention in 1800.

View an **animation about a voltaic cell.**

Concepts In Motion

■ **Figure 3** This replica of one of Alessandro Volta's first cells consists of discs of zinc and copper arranged in alternating layers and separated by cloth or cardboard soaked in an acidic solution. Electric current increased with the number of metal discs used.

■ **Figure 4** The roller coaster at the top of the track has high potential energy relative to the track below because of the difference in height. Similarly, an electrochemical cell has potential energy to produce a current because there is a difference in the ability of the electrodes to move electrons from the anode to the cathode.

## Chemistry of Voltaic Cells

An electrochemical cell consists of two parts called **half-cells,** in which the separate oxidation and reduction reactions take place. Each half-cell contains an electrode and a solution containing ions. An electrode is an electrically conductive material, usually a metallic strip or graphite, that conducts electrons into and out of the solution in the half cell. In **Figure 2,** the beaker with the zinc electrode is where the oxidation half of the redox reaction takes place. The beaker with the copper electrode is where the reduction half of the reaction takes place. The reaction that takes place in each half-cell is called a half-cell reaction. The electrode where oxidation takes place is called the **anode.** The electrode where reduction takes place is called the **cathode.**

☑ READING CHECK **Identify** which of the beakers in **Figure 2** contains the anode.

**Voltaic cells and energy** Recall that an object's potential energy is due to its position or composition. In electrochemistry, electric potential energy is a measure of the amount of current that can be generated from a voltaic cell to do work. Electric charge can flow between two points only when a difference in electric potential energy exists between the two points. In an electrochemical cell, these two points are the two electrodes. Electrons generated at the anode, the site of oxidation, are thought to be pushed or driven toward the cathode by the electromotive force (EMF). This force is due to the difference in electric potential energy between the two electrodes and is referred to as the cell potential. A volt is a unit used to measure cell potential. The electric potential difference of a voltaic cell is an indication of the energy that is available to move electrons from the anode to the cathode.

Consider the analogy illustrated in **Figure 4.** The roller coaster is almost stationary at the top of the track. Then, it plummets from its high position because of the difference in gravitational potential energy (PE) between the top and bottom of the track. The kinetic energy (KE) attained by the roller coaster is determined by the difference in height (potential energy) between the top and bottom parts of the track. Similarly, the energy of the electrons flowing from the anode to the cathode in a voltaic cell is determined by the difference in electric potential energy between the two electrodes. In redox terms, the voltage of a cell is determined by comparing the difference in the tendency of the two electrode materials to accept electrons. The greater the difference, the greater the potential energy difference between the two electrodes and the larger the voltage of the cell will be.

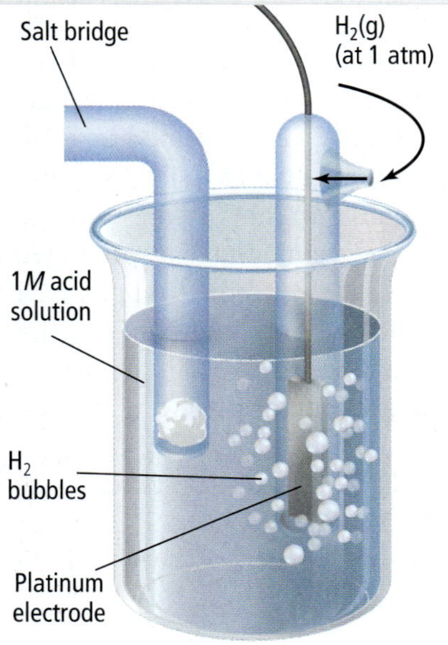


**Figure 5** A standard hydrogen electrode consists of a platinum electrode with hydrogen gas at 1 atm pressure bubbling into an acidic solution that is 1*M* in hydrogen ions. The reduction potential for this configuration is defined as 0.000 V.

The force of gravity always causes a diver to fall downward to a lower energy state, never upward to a higher energy state. When a diver steps off a diving board, his or her spontaneous motion is always downward. Similarly, in the zinc-copper cell, under standard conditions, copper(II) ions at the cathode accept electrons more readily than the zinc ions at the anode. Thus, the redox reaction occurs spontaneously only when the electrons flow from the zinc to the copper.

**Vocabulary**
**Academic Vocabulary**
**Correspond**
to be in agreement or to match
*Her directions correspond with the map.*

## Calculating Electrochemical Cell Potentials

Recall that gaining electrons is called reduction. Building on this fact, the tendency of a substance to gain electrons is its **reduction potential.** The reduction potential of an electrode cannot be determined directly because the reduction half-reaction must be coupled with an oxidation half-reaction. When two half-reactions are coupled, the voltage generated corresponds to the difference in potential between the half-reactions. The electrical potential difference between two points is expressed in volts (V).

**Foldables®**
Incorporate information from this section into your Foldable.

**The standard hydrogen electrode** Long ago, chemists decided to measure the reduction potential of all electrodes against one electrode, the standard hydrogen electrode. The **standard hydrogen electrode** consists of a small sheet of platinum immersed in a hydrochloric acid (HCl) solution that has a hydrogen-ion concentration of 1*M*. Hydrogen gas ($H_2$), at a pressure of 1 atm is bubbled in and the temperature is maintained at 25°C, as shown in **Figure 5.** The potential, also called the standard reduction potential ($E^0$), of this standard hydrogen electrode is defined as 0.000 V. This electrode can act as an oxidation half-reaction or a reduction half-reaction, depending on the half-cell to which it is connected. The two possible reactions at the hydrogen electrode are the following.

$$2H^+(aq) + 2e^- \underset{\text{Oxidation}}{\overset{\text{Reduction}}{\rightleftharpoons}} H_2(g) \qquad E^0 = 0.000\ V$$

**Half-cell potentials** Over the years, chemists have measured and recorded the standard reduction potentials of many different half-cells. **Table 1** lists some common half-cell reactions in order of increasing reduction potential. The values in the table were obtained by measuring the potential when each half-cell was connected to a standard hydrogen half-cell. All of the half-reactions in **Table 1** are written as reductions. However, in any voltaic cell, which always contains two spontaneous half-reactions, the half-reaction with the lower reduction potential will proceed in the opposite direction and will be an oxidation reaction. In other words, the half-reaction that is more positive will proceed as a reduction and the half-reaction that is more negative will proceed as an oxidation.

The electrode being measured must also be under standard conditions, that is, immersed in a 1*M* solution of its ions at 25°C and 1 atm. The superscript zero in the notation $E^0$ is a shorthand way of indicating "measured under standard conditions."

**Table 1 Standard Reduction Potentials**

| Half-Reaction | $E^0$ (V) | Half-Reaction | $E^0$ (V) |
|---|---|---|---|
| $Li^+ + e^- \rightleftarrows Li$ | −3.0401 | $Cu^{2+} + e^- \rightleftarrows Cu^+$ | +0.153 |
| $Ca^{2+} + 2e^- \rightleftarrows Ca$ | −2.868 | $Cu^{2+} + 2e^- \rightleftarrows Cu$ | +0.3419 |
| $Na^+ + e^- \rightleftarrows Na$ | −2.71 | $O_2 + 2H_2O + 4e^- \rightleftarrows 4OH^-$ | +0.401 |
| $Mg^{2+} + 2e^- \rightleftarrows Mg$ | −2.372 | $I_2 + 2e^- \rightleftarrows 2I^-$ | +0.5355 |
| $Be^{2+} + 2e^- \rightleftarrows Be$ | −1.847 | $Fe^{3+} + e^- \rightleftarrows Fe^{2+}$ | +0.771 |
| $Al^{3+} + 3e^- \rightleftarrows Al$ | −1.662 | $NO_3^- + 2H^+ + e^- \rightleftarrows NO_2 + H_2O$ | +0.775 |
| $Mn^{2+} + 2e^- \rightleftarrows Mn$ | −1.185 | $Hg_2^{2+} + 2e^- \rightleftarrows 2Hg$ | +0.7973 |
| $Cr^{2+} + 2e^- \rightleftarrows Cr$ | −0.913 | $Ag^+ + e^- \rightleftarrows Ag$ | +0.7996 |
| $2H_2O + 2e^- \rightleftarrows H_2 + 2OH^-$ | −0.8277 | $Hg^{2+} + 2e^- \rightleftarrows Hg$ | +0.851 |
| $Zn^{2+} + 2e^- \rightleftarrows Zn$ | −0.7618 | $2Hg^{2+} + 2e^- \rightleftarrows Hg_2^{2+}$ | +0.920 |
| $Cr^{3+} + 3e^- \rightleftarrows Cr$ | −0.744 | $NO_3^- + 4H^+ + 3e^- \rightleftarrows NO + 2H_2O$ | +0.957 |
| $S + 2e^- \rightleftarrows S^{2-}$ | −0.47627 | $Br_2(l) + 2e^- \rightleftarrows 2Br^-$ | +1.066 |
| $Fe^{2+} + 2e^- \rightleftarrows Fe$ | −0.447 | $Pt^{2+} + 2e^- \rightleftarrows Pt$ | +1.18 |
| $Cd^{2+} + 2e^- \rightleftarrows Cd$ | −0.4030 | $O_2 + 4H^+ + 4e^- \rightleftarrows 2H_2O$ | +1.229 |
| $PbI_2 + 2e^- \rightleftarrows Pb + 2I^-$ | −0.365 | $Cl_2 + 2e^- \rightleftarrows 2Cl^-$ | +1.35827 |
| $PbSO_4 + 2e^- \rightleftarrows Pb + SO_4^{2-}$ | −0.3588 | $Au^{3+} + 3e^- \rightleftarrows Au$ | +1.498 |
| $Co^{2+} + 2e^- \rightleftarrows Co$ | −0.28 | $MnO_4^- + 8H^+ + 5e^- \rightleftarrows Mn^{2+} + 4H_2O$ | +1.507 |
| $Ni^{2+} + 2e^- \rightleftarrows Ni$ | −0.257 | $Au^+ + e^- \rightleftarrows Au$ | +1.692 |
| $Sn^{2+} + 2e^- \rightleftarrows Sn$ | −0.1375 | $H_2O_2 + 2H^+ + 2e^- \rightleftarrows 2H_2O$ | +1.776 |
| $Pb^{2+} + 2e^- \rightleftarrows Pb$ | −0.1262 | $Co^{3+} + e^- \rightleftarrows Co^{2+}$ | +1.92 |
| $Fe^{3+} + 3e^- \rightleftarrows Fe$ | −0.037 | $S_2O_8^{2-} + 2e^- \rightleftarrows 2SO_4^{2-}$ | +2.010 |
| **$2H^+ + 2e^- \rightleftarrows H_2$** | **0.0000** | $F_2 + 2e^- \rightleftarrows 2F^-$ | +2.866 |

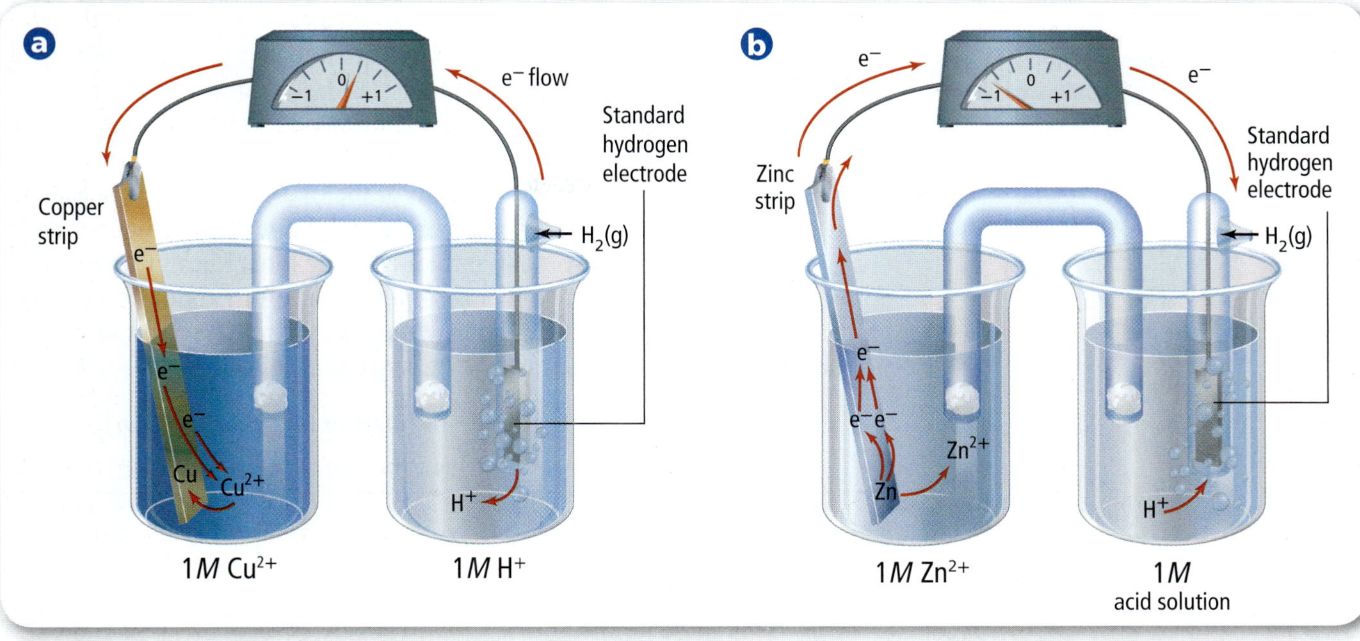


**Figure 6** **a.** When a $Cu^{2+}$ | Cu electrode is connected to the hydrogen electrode, electrons flow toward the copper strip and reduce $Cu^{2+}$ ions to Cu atoms. The voltage of this reaction is +0.342 V. **b.** When a Zn | $Zn^{2+}$ electrode is connected to the hydrogen electrode, electrons flow away from the zinc strip and zinc atoms are oxidized to $Zn^{2+}$ ions. The voltage of this reaction is −0.762 V.

**Determining electrochemical cell potentials** You can use **Table 1** to calculate the electric potential of a voltaic cell consisting of a copper electrode and a zinc electrode under standard conditions. The first step is to determine the standard reduction potential for the copper half-cell ($E^0_{Cu}$). When the copper electrode is attached to a standard hydrogen electrode, as in **Figure 6a,** electrons flow from the hydrogen electrode to the copper electrode, and copper ions are reduced to copper metal. The $E^0$, measured by a voltmeter, is +0.342 V. The positive voltage indicates that $Cu^{2+}$ ions at the copper electrode accept electrons more readily than do $H^+$ ions at the standard hydrogen electrode. Therefore, oxidation takes place at the hydrogen electrode and reduction takes place at the copper electrode. The oxidation and reduction half-cell reactions and the overall reaction are

$$H_2(g) \rightarrow 2H^+(aq) + 2e^- \text{ (oxidation half-cell reaction)}$$
$$Cu^{2+}(aq) + 2e^- \rightarrow Cu(s) \text{ (reduction half-cell reaction)}$$

---

$$H_2(g) + Cu^{2+}(aq) \rightarrow 2H^+(aq) + Cu(s) \text{ (overall redox reaction)}$$

This reaction can be written in a form called cell notation.

Reactant Product Reactant Product

$$H_2 \mid H^+ \parallel Cu^{2+} \mid Cu \qquad E^0_{Cu} = +0.342 \text{ V}$$

Oxidation half-cell    Reduction half-cell

The two participants in the oxidation reaction are written first and in the order they appear in the oxidation half-reaction—reactant | product. They are followed by a double vertical line (||) representing the wire and salt bridge connecting the half-cells. Then, the two participants in the reduction reaction are written in the same reactant | product order. Note that for positive values of $E^0$, it is customary to place a plus sign before the voltage.

Explore **electrochemical cells.**

**Virtual Investigations**

The next step is to determine the standard reduction potential for the zinc half-cell ($E^0_{Zn}$). When the zinc electrode is measured against the standard hydrogen electrode under standard conditions, as in **Figure 6b,** electrons flow from the zinc electrode to the hydrogen electrode. The $E^0$ of the zinc half-cell, measured by a voltmeter, is –0.762 V. This means that the $H^+$ ions at the hydrogen electrode accept electrons more readily than do the zinc ions. Thus, the hydrogen ions have a higher reduction potential than the zinc ions. Recall that the hydrogen electrode is assigned a zero potential, so the reduction potential of the zinc electrode must have a negative value. The two half-cell reactions and the overall reaction are written as follows.

$$Zn(s) \rightarrow Zn^{2+}(aq) + 2e^- \text{ (oxidation half-cell reaction)}$$
$$2H^+(aq) + 2e^- \rightarrow H_2(g) \text{ (reduction half-cell reaction)}$$

---

$$Zn(s) + 2H^+(aq) \rightarrow Zn^{2+}(aq) + H_2(g) \text{ (overall redox cell reaction)}$$

This reaction can be written in the following cell notation.

Reactant  Product  Reactant  Product

$$Zn \mid Zn^{2+} \parallel H^+ \mid H_2 \qquad E^0_{Zn} = -0.762 \text{ V}$$

Oxidation half-cell  Reduction half-cell

The final step in calculating electrochemical cell potential is to combine the copper and zinc half-cells as a voltaic cell. This means calculating the voltaic cell's standard potential using the following formula.

**Formula for Cell Potential**

$$E^0_{cell} = E^0_{reduction} - E^0_{oxidation}$$

**$E^0_{cell}$ represents the overall standard cell potential.**
**$E^0_{reduction}$ represents the standard half-cell potential for the reduction.**
**$E^0_{oxidation}$ represents the standard half-cell potential for the oxidation.**

The standard potential of a cell is the standard potential of the reduction half-cell minus the standard potential of the oxidation half-cell.

Because reduction occurs at the copper electrode and oxidation occurs at the zinc electrode, the $E^0$ values are substituted as follows.

$$\begin{aligned} E^0_{cell} &= E^0_{Cu^{2+}|Cu} - E^0_{Zn^{2+}|Zn} \\ &= +0.342 \text{ V} - (-0.762 \text{ V}) \\ &= +1.104 \text{ V} \end{aligned}$$

Notice that the negative sign in the formula automatically changes the sign of the oxidation half-reaction, so you do not reverse the sign of the standard reduction potentials listed in **Table 1** when they are used for the oxidation half-reaction.

The graph in **Figure 7** shows how the zinc half-cell with the lower reduction potential and the copper half-cell with the higher reduction potential are related.

■ **Figure 7** This simple graph illustrates how the overall cell potential is derived from the difference in reduction potential of two electrodes. Compare it to a number line. The potential difference between the zinc and copper electrodes is +1.104 V.

**GRAPH CHECK**
**Identify** the metal, copper or zinc, that is easier to oxidize than hydrogen.

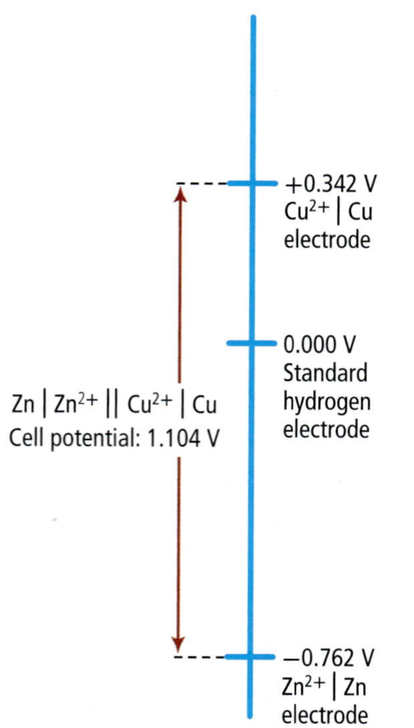

## EXAMPLE Problem 1

EXAMPLE PROBLEM

**CALCULATE A CELL POTENTIAL** The following reduction half-reactions represent the half-cells of a voltaic cell.

$$I_2(s) + 2e^- \rightarrow 2I^-(aq)$$
$$Fe^{2+}(aq) + 2e^- \rightarrow Fe(s)$$

Determine the overall cell reaction and the standard cell potential. Describe the cell using cell notation.

### 1 ANALYZE THE PROBLEM

You are given the half-cell equations and can find standard reduction potentials in **Table 1.** The half-reaction with the lower reduction potential will be an oxidation. With this information, you can write the overall cell reaction, calculate the standard cell potential, and describe the cell in cell notation.

**Known**

Standard reduction potentials for the half-cells

$E^0_{cell} = E^0_{reduction} - E^0_{oxidation}$

**Unknown**

**overall cell reaction = ?**

**$E^0_{cell}$ = ?**

**cell notation = ?**

### 2 SOLVE FOR THE UNKNOWN

Find the standard reduction potentials of each half-reaction in **Table 1.**

$I_2(s) + 2e^- \rightarrow 2I^-(aq)$ $\quad E^0_{I_2|I^-} = +0.536\ V$

$Fe^{2+}(aq) + 2e^- \rightarrow Fe(s)$ $\quad E^0_{Fe^{2+}|Fe} = -0.447\ V$

The reduction of iodine has the higher reduction potential, so this half-reaction proceeds in the forward direction as a reduction. The iron half-reaction proceeds in the reverse direction as an oxidation.

$I_2(s) + 2e^- \rightarrow 2I^-(aq)$ (reduction half-cell reaction)

$Fe(s) \rightarrow Fe^{2+}(aq) + 2e^-$ (oxidation half-cell reaction) — **Rewrite the iron half-reaction in the correct direction.**

$I_2(s) + Fe(s) \rightarrow Fe^{2+}(aq) + 2I^-(aq)$ — **Add the two equations.**

**The overall cell reaction is $I_2(s) + Fe(s) \rightarrow Fe^{2+}(aq) + 2I^-(aq)$.**

Calculate the standard cell potential.

$E^0_{cell} = E^0_{reduction} - E^0_{oxidation}$ — **State the formula for cell potential.**

$E^0_{cell} = E^0_{I_2|I^-} - E^0_{Fe^{2+}|Fe}$ — **Substitute $E^0_{I_2|I^-}$ and $E^0_{Fe^{2+}|Fe}$ in the generic equation.**

$E^0_{cell} = +0.536\ V - (-0.447\ V)$ — **Substitute $E^0_{I_2|I^-} = +0.536$ V and $E^0_{Fe^{2+}|Fe} = -0.447$ V.**

**$E^0_{cell} = +0.983\ V$**

Describe the cell using cell notation.

$Fe \mid Fe^{2+}$ — **First, write the oxidation half-reaction using cell notation: reactant then product.**

$Fe \mid Fe^{2+} \parallel I_2 \mid I^-$ — **Next, write the reduction half-reaction to the right. Separate the half-cells by a double vertical line.**

**Cell notation: $Fe \mid Fe^{2+} \parallel I_2 \mid I^-$**

### 3 EVALUATE THE ANSWER

The calculated potential is reasonable given the potentials of the half-cells. $E^0$ is reported to the correct number of significant figures.

Find help with **calculating potentials.**

Personal Tutor

## PRACTICE Problems

Do additional problems. Online Practice

**For each of these pairs of half-reactions, write the balanced equation for the overall cell reaction, and calculate the standard cell potential. Describe the reaction using cell notation. Refer to the chapter on redox reactions to review writing and balancing redox equations.**

1. $Pt^{2+}(aq) + 2e^- \rightarrow Pt(s)$ and $Sn^{2+}(aq) + 2e^- \rightarrow Sn(s)$
2. $Co^{2+}(aq) + 2e^- \rightarrow Co(s)$ and $Cr^{3+}(aq) + 3e^- \rightarrow Cr(s)$
3. $Hg^{2+}(aq) + 2e^- \rightarrow Hg(l)$ and $Cr^{2+}(aq) + 2e^- \rightarrow Cr(s)$
4. **Challenge** Write the balanced equation for the cell reaction and calculate the standard cell potential for the reaction that occurs when these half-cells are connected. Describe the reaction using cell notation.

   $NO_3^- + 4H^+ + 3e^- \rightarrow NO + 2H_2O$

   $O_2 + 2H_2O + 4e^- \rightarrow 4OH^-$

## Using Standard Reduction Potentials

The Example Problems showed you how to use the data from **Table 1** to calculate the standard potential (voltage) of voltaic cells. Another important use of standard reduction potentials is to determine if a proposed reaction under standard conditions will be spontaneous. How can standard reduction potentials indicate spontaneity? Electrons in a voltaic cell always flow from the half-cell with the lower standard reduction potential to the half-cell with the higher standard reduction potential, giving a positive cell voltage. To predict whether any proposed redox reaction will occur spontaneously, simply write the process in the form of half-reactions and look up the reduction potential of each. Use the values to calculate the potential of a voltaic cell operating with these two half-cell reactions. If the calculated potential is positive, the reaction is spontaneous. If the value is negative, the reaction is not spontaneous. However, the reverse of a nonspontaneous reaction will occur because it will have a positive cell voltage, which means that the reverse reaction is spontaneous.

☑ READING CHECK **Identify** the sign of the potential of a redox reaction that occurs spontaneously.

## PRACTICE Problems

Do additional problems. Online Practice

**Calculate the cell potential to determine if each of the following balanced redox reactions is spontaneous as written. Use Table 1 to help you determine the correct half-reactions.**

5. $Sn(s) + Cu^{2+}(aq) \rightarrow Sn^{2+}(aq) + Cu(s)$
6. $Mg(s) + Pb^{2+}(aq) \rightarrow Pb(s) + Mg^{2+}(aq)$
7. $2Mn^{2+}(aq) + 8H_2O(l) + 10Hg^{2+}(aq) \rightarrow$
   $2MnO_4-(aq) + 16H^+(aq) + 5Hg_2^{2+}(aq)$
8. $2SO_4^{2-}(aq) + Co^{2+}(aq) \rightarrow Co(s) + S_2O_8^{2-}(aq)$
9. **Challenge** Using **Table 1,** write the equation and determine the cell voltage ($E^0$) for the following cell. Is the reaction spontaneous?

   $Al \mid Al^{3+} \parallel Hg^{2+} \mid Hg_2^{2+}$

## PROBLEM-SOLVING STRATEGY

### Determining Cell Potentials

The five steps that follow summarize the procedure for calculating the potential of a voltaic cell in which a spontaneous redox reaction occurs. Suppose you must write the equation for and calculate the potential of a cell made up of these half-reactions:

$$Mn^{2+} + 2e^- \rightarrow Mn \text{ and } Fe^{3+} + 3e^- \rightarrow Fe$$

A table of reduction potentials, such as **Table 1,** is all that is required.

1. Find the two half-reactions on **Table 1.**
2. Compare the two half-cell potentials. The half-cell with the higher reduction potential is the cell in which reduction will occur. Oxidation will occur in the half-cell with the lower reduction potential.

   $Fe^{3+} + 3e^- \rightarrow Fe$ $E^0 = -0.037$ V (reduction)
   $Mn^{2+} + 2e^- \rightarrow Mn$ $E^0 = -1.185$ V (oxidation)

3. Write the equation for the reduction as it is in **Table 1.** Write the equation for the oxidation in the opposite direction.

   $Fe^{3+} + 3e^- \rightarrow Fe$ $Mn \rightarrow Mn^{2+} + 2e^-$

4. Balance the electrons in the two half-cell equations by multiplying each by a factor. Add the equations.

   Multiply by 2. $2Fe^{3+} + 6e^- \rightarrow 2Fe$
   Multiply by 3. $3Mn \rightarrow 3Mn^{2+} + 6e^-$
   Add the equations. $2Fe^{3+} + 3Mn \rightarrow 2Fe + 3Mn^{2+}$

5. Equalizing the electrons lost and gained does not affect the $E^0$ for the overall reaction because $E^0$ is an intensive property. Use the formula: $E^0_{cell} = E^0_{reduction} - E^0_{oxidation}$ to obtain the cell potential.

   $E^0_{cell} = E^0_{Fe^{3+}|Fe} - E^0_{Mn^{2+}|Mn} = -0.037\text{ V} - (-1.185\text{ V})$
   $= +1.148$ V

### Apply the Strategy

**Determine** $E^0_{cell}$ for the spontaneous redox reaction that occurs between magnesium and nickel.

## SECTION 1 REVIEW

Section Self-Check

### Section Summary

- In a voltaic cell, oxidation and reduction take place at electrodes separated from each other.
- The standard potential of a half-cell reaction is its voltage when paired with a standard hydrogen electrode under standard conditions.
- The reduction potential of a half-cell is negative if it undergoes oxidation when connected to a standard hydrogen electrode. The reduction potential of a half-cell is positive if it undergoes reduction when connected to a standard hydrogen electrode.
- The standard potential of a voltaic cell is the difference between the standard reduction potentials of the half-cell reactions.

10. MAINIDEA **Describe** the conditions under which a redox reaction causes an electric current to flow through a wire.
11. **Identify** the components of a voltaic cell. Explain the role of each component in the operation of the cell.
12. **Write** the balanced equation for the spontaneous cell reaction that occurs in a cell with these reduction half-reactions.
    a. $Ag^+(aq) + e^- \rightarrow Ag(s)$ and $Ni^{2+}(aq) + 2e^- \rightarrow Ni(s)$
    b. $Mg^{2+}(aq) + 2e^- \rightarrow Mg(s)$ and $2H^+(aq) + 2e^- \rightarrow H_2(g)$
    c. $Sn^{2+}(aq) + 2e^- \rightarrow Sn(s)$ and $Fe^{3+}(aq) + 3e^- \rightarrow Fe(s)$
    d. $PbI_2(s) + 2e^- \rightarrow Pb(s) + 2I^-(aq)$ and $Pt^{2+}(aq) + 2e^- \rightarrow Pt(s)$
13. **Determine** the standard potential for electrochemical cells in which each equation represents the overall cell reaction. Identify the reactions as spontaneous or nonspontaneous as written.
    a. $2Al^{3+}(aq) + 3Cu(s) \rightarrow 3Cu^{2+}(aq) + 2Al(s)$
    b. $Hg^{2+}(aq) + 2Cu^+(aq) \rightarrow 2Cu^{2+}(aq) + Hg(l)$
    c. $Cd(s) + 2NO_3^-(aq) + 4H^+(aq) \rightarrow Cd^{2+}(aq) + 2NO_2(g) + 2H_2O(l)$
14. **Design** a concept map for Section 1, starting with the term *electrochemical cell.* Incorporate all the new vocabulary terms in your map.

SECTION 2

# Batteries

**MAINIDEA** Batteries are voltaic cells that use spontaneous reactions to provide energy for a variety of purposes.

### Essential Questions

- What are the structure, composition, and operation of the typical carbon-zinc dry-cell battery?
- What is the difference between primary and secondary batteries and what are two examples of each type?
- What is the structure of the hydrogen-oxygen fuel cell and how does it operate?
- What is the process of corrosion of iron and what are methods to prevent corrosion?

### Review Vocabulary

**reversible reaction:** a reaction that can take place in both the forward and reverse directions

### New Vocabulary

battery
dry cell
primary battery
secondary battery
fuel cell
corrosion
galvanization

**CHEM 4 YOU** Take a moment to list some of the places where you know batteries are used. Your list might include flashlights, cars, cell phones, radios, calculators, watches, and toys, among many others. Are the batteries in these devices all the same?

## Dry Cells

Some of the spontaneous cell reactions you have been reading about provide the energy of the batteries that you use every day. A **battery** is one or more voltaic cells in a single package that generates electric current. From the time of its invention in the 1860s until recently, the most commonly used voltaic cell was the zinc-carbon dry cell, shown in **Figure 8.**

**Zinc-carbon dry cells** A **dry cell** is an electrochemical cell in which the electrolyte is a moist paste. The paste in a zinc-carbon dry cell consists of zinc chloride, manganese(IV) oxide, ammonium chloride, and a small amount of water inside a zinc case. The zinc shell is the cell's anode, where the oxidation of zinc metal occurs according to the following equation.

$$Zn(s) \rightarrow Zn^{2+}(aq) + 2e^-$$

A carbon (graphite) rod in the center of the dry cell serves as the cathode, but the reduction half-cell reaction takes place in the paste. The carbon rod in this type of dry cell is called an inactive cathode because it is made of a material that does not participate in the redox reaction. However, the inactive electrode has the important purpose of conducting electrons. The reduction half-cell reaction for this dry cell is as follows.

$$2NH_4^+(aq) + 2MnO_2(s) + 2e^- \rightarrow Mn_2O_3(s) + 2NH_3(aq) + H_2O(l)$$

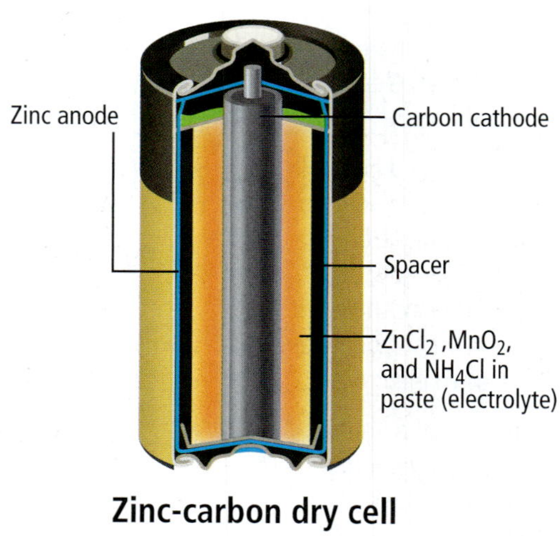


■ **Figure 8** The so-called dry cell contains a moist paste in which the cathode half-reaction takes place. In the zinc-carbon dry cell, the zinc case acts as the anode.

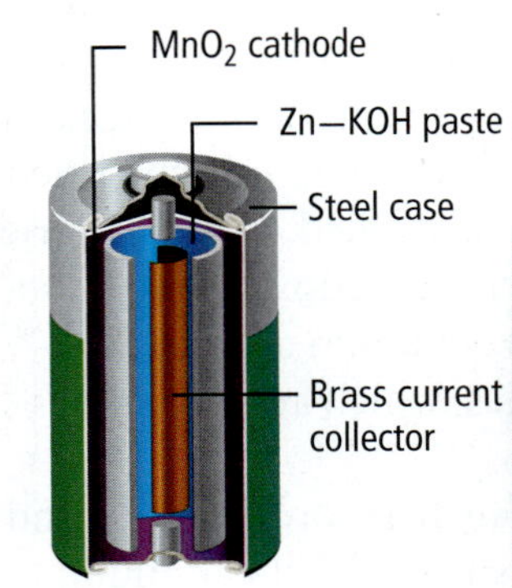


**Alkaline battery**

The alkaline battery uses powdered zinc in an alkaline paste as the anode.

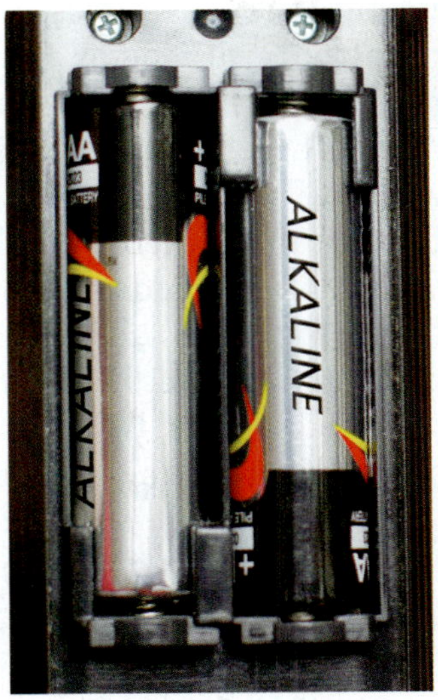


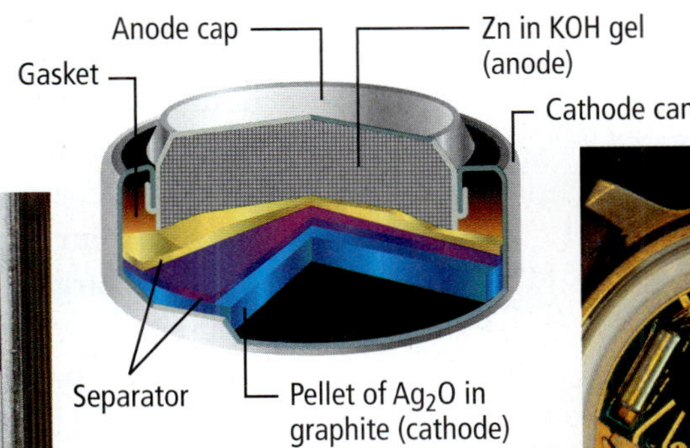


**Silver button battery**

Although it looks different, the silver battery is much like the alkaline battery. The silver battery uses silver oxide ($Ag_2O$) in the cathode instead of manganese(IV) oxide ($MnO_2$).

■ **Figure 9** Alkaline batteries are more efficient than zinc-carbon dry cells and are useful when smaller batteries are needed. Silver button batteries are even smaller, making them well suited to devices such as watches.

In the zinc-carbon dry cell, a spacer made of a porous material and damp from the liquid in the paste separates the paste from the zinc anode. The spacer acts as a salt bridge to allow the transfer of ions, much like the model voltaic cell you studied in Section 1. The zinc-carbon dry cell produces a voltage of 1.5 V until the reduction product, ammonia, comes out of its aqueous solution as a gas. At that point, the voltage drops to a level that makes the battery useless.

**Alkaline batteries** A more efficient alkaline dry cell, shown in **Figure 9,** is replacing the standard zinc-carbon dry cell in many applications. In the alkaline cell, the zinc is in a powdered form, which provides more surface area for reaction. The zinc is mixed in a paste with potassium hydroxide, a strong base, and the paste is contained in a steel case. The cathode mixture is manganese(IV) oxide, also mixed with potassium hydroxide. The anode half-cell reaction is as follows.

$$Zn(s) + 2OH^-(aq) \rightarrow ZnO(s) + H_2O(l) + 2e^-$$

The cathode half-cell reaction is as follows.

$$MnO_2(s) + 2H_2O(s) + 2e^- \rightarrow Mn(OH)_2(s) + 2OH^-(aq)$$

Alkaline batteries do not need the carbon rod cathode, so they can be made smaller and are more useful in small devices.

**Silver batteries** The silver battery shown in **Figure 9** is even smaller and is used to power devices such as hearing aids, watches, and cameras. The silver battery uses the same anode half-reaction as the alkaline battery, with the following cathode half-reaction.

$$Ag_2O(s) + H_2O(l) + 2e^- \rightarrow 2Ag(s) + 2OH^-(aq)$$

☑ READING CHECK **Identity** the half-reaction that occurs in both alkaline and silver batteries.

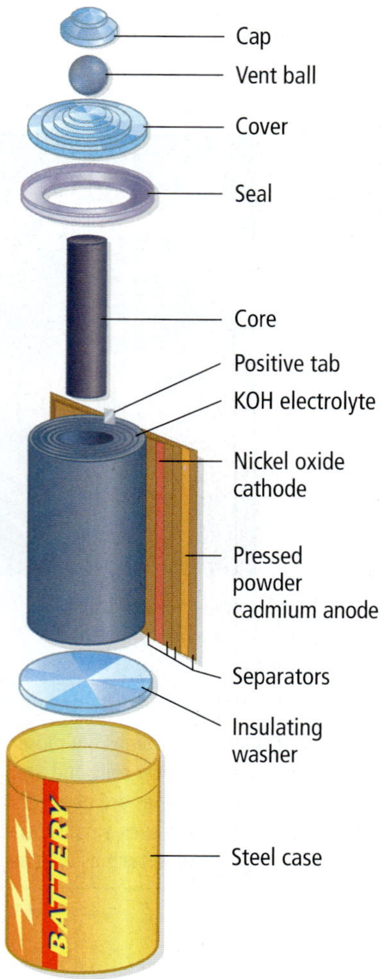


■ **Figure 10** Cordless tools and phones are often powered by rechargeable batteries, such as the NiCad battery. The battery pack is recharged by plugging it into an electric outlet, which supplies the power to drive the nonspontaneous recharge reaction.

**Primary and secondary batteries** Batteries are divided into two types, depending on their chemical processes. The zinc-carbon, alkaline-zinc, and silver cells are classified as primary batteries. **Primary batteries** produce electric energy by means of redox reactions that are not easily reversed. These cells deliver current until the reactants are gone, and then the battery must be discarded. Other batteries, called **secondary batteries,** depend on reversible redox reactions, so they are rechargeable. A car battery and the battery in a laptop computer are examples of secondary batteries, which are sometimes called storage batteries.

The storage batteries that power devices such as cordless drills and screwdrivers, shavers, and camcorders are usually nickel-cadmium rechargeable batteries, sometimes called NiCad batteries, as shown in **Figure 10.** For maximum efficiency, the anode and cathode are long, thin ribbons of material separated by a layer through which ions can pass. The ribbons are wound into a tight coil and packaged in a steel case. The anode reaction that occurs when the battery is used to generate electric current is the oxidation of cadmium in the presence of a base.

$$Cd(s) + 2OH^-(aq) \rightarrow Cd(OH)_2(s) + 2e^-$$

The cathode reaction is the reduction of nickel from the +3 to the +2 oxidation state.

$$NiO(OH)(s) + H_2O(l) + e^- \rightarrow Ni(OH)_2(s) + OH^-(aq)$$

When the battery is recharged, these reactions are reversed.

## Lead-Acid Storage Battery

Another common storage battery is the lead-acid battery used in automobiles. Most auto batteries contain six cells that generate about 2 V each for a total output of 12 V. The anode of each cell consists of two or more grids of porous lead, and the cathode consists of lead grids filled with lead(IV) oxide. This type of battery should probably be called a lead-lead(IV) oxide battery, but the term *lead-acid* is commonly used because the battery's electrolyte is a solution of sulfuric acid. The lead-acid battery is not a dry cell.

The following equation represents the oxidation half-cell reaction at the anode where lead is oxidized from the zero oxidation state to the +2 oxidation state in $PbSO_4$.

$$Pb(s) + SO_4^{2-}(aq) \rightarrow PbSO_4(s) + 2e^-$$

The reduction of lead from the +4 to the +2 oxidation state takes place at the cathode. The half-cell reaction for the cathode is

$$PbO_2(s) + 4H^+(aq) + SO_4^{2-}(aq) + 2e^- \rightarrow PbSO_4(s) + 2H_2O(l).$$

The overall reaction is

$$Pb(s) + PbO_2(s) + 4H^+(aq) + 2SO_4^{2-}(aq) \rightarrow 2PbSO_4(s) + 2H_2O(l).$$

By looking at the half-cell reactions, you can see that lead(II) sulfate ($PbSO_4$) is the reaction product in both oxidation and reduction. Also, Pb, $PbO_2$, and $PbSO_4$ are solid substances, so they stay in place where they are formed. Thus, whether the battery is discharging or charging, the reactants are available where they are needed.

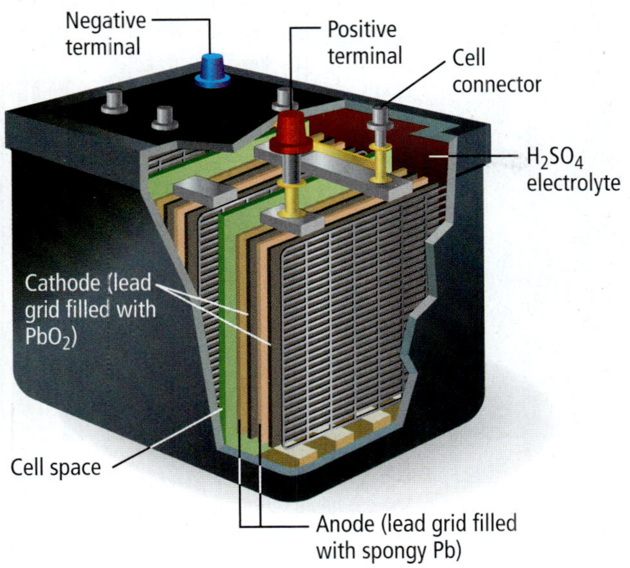


Lead-acid batteries contain lead plates and lead(IV) oxide plates. The electrolyte is a solution of sulfuric acid. When the battery is in use, the sulfuric acid is depleted and the electrolyte becomes less dense.

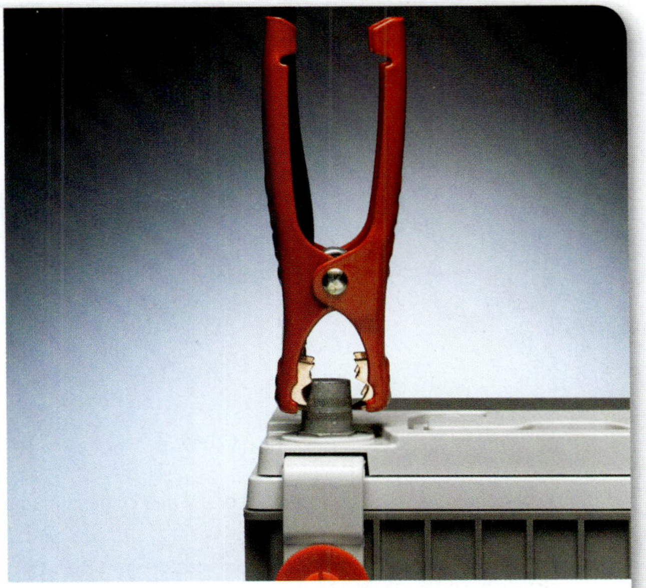

Low electrolyte levels can result in a dead battery. Jumper cables conduct current from a car with a good battery to start a car with a dead one.

■ **Figure 11** The lead-acid battery used in automobiles discharges when it starts the car and charges when the engine is running.

Sulfuric acid serves as the electrolyte in the battery, but, as the overall cell equation shows, it is depleted as the battery generates electric current. What happens when the battery is recharging? In this case, the reactions reverse, forming lead and lead(IV) oxide and releasing sulfuric acid, shown as $4H^+(aq) + 2SO_4^{2-}(aq)$ in the equation.

The lead-storage battery shown in **Figure 11** is a good choice for motor vehicles because it provides a large initial supply of energy to start the engine, has a long shelf life, and is reliable at low temperatures.

☑ READING CHECK **Identify** the species that is oxidized and the species that is reduced when the lead-acid battery is charging.

## Lithium Batteries

Although lead-acid batteries are reliable and suitable for many applications, engineers have been working to develop batteries with less mass and higher capacity to power devices from wristwatches to electric cars. For applications in which a battery is the key component and must provide a significant amount of power, such as for the operation of an electric car, lead-acid batteries are too heavy to be feasible.

The solution is to develop lightweight batteries that store a large amount of energy for their size. Engineers have focused their attention on the element lithium for two reasons: lithium is the lightest known metal and has the lowest standard reduction potential of the metallic elements, −3.04 V, as shown in **Table 1.** A battery that oxidizes lithium at the anode can generate almost 2.3 V more than a similar battery in which zinc is oxidized.

Compare the zinc and lithium oxidation half-reactions and their standard reduction potentials.

$$Zn \rightarrow Zn^{2+} + 2e^- \qquad (E^0_{Zn^{2+}|Zn} = -0.762\ V)$$
$$Li \rightarrow Li^+ + e^- \qquad (E^0_{Li^+|Li} = -3.04\ V)$$
$$E^0_{Zn^{2+}|Zn} - E^0_{Li^+|Li} = +2.28\ V$$

**VOCABULARY**
**WORD ORIGIN**
**Capacity**
***capac-, capax,*** from Latin, meaning *containing or capable of holding a great deal*

Lithium batteries often deliver either 3 V or 9 V and come in many sizes to fit different devices.

Lithium battery packs power this experimental car to a maximum speed of 113 km/h. The car has a range of over 320 km.

■ **Figure 12** The light weight, long life, and high potential of a lithium battery make it an excellent choice for a variety of purposes.

Lithium batteries can be either primary or secondary batteries, depending on which reduction reactions are coupled to the oxidation of lithium. For example, some lithium batteries use the same cathode reaction as zinc-carbon dry cells, the reduction of manganese(IV) oxide ($MnO_2$) to manganese(III) oxide ($Mn_2O_3$). These batteries produce an electric current of about 3 V compared to 1.5 V for zinc-carbon cells. Lithium batteries last much longer than other kinds of batteries. As a result, they are often used in watches, computers, and cameras to maintain time, date, memory, and personal settings—even when the device is turned off. **Figure 12** shows a range of available lithium batteries and a developing application.

**READING CHECK** **List** three advantages of lithium batteries.

**RealWorld CHEMISTRY**

**Fuel Cells**

**REDUCING POLLUTION** One of the largest sources of air pollution in many cities is vehicles. In some European cities, experimental buses powered by hydrogen fuel cells are making a difference. Exhaust from these buses contain no carbon dioxide and no oxides of nitrogen or sulfur. Pure water is their only product.

## Fuel Cells

When hydrogen burns in air, it does so explosively, with the evolution of light and heat.

$$2H_2(g) + O_2(g) \rightarrow 2H_2O(l) + \text{energy}$$

Can this reaction occur under controlled conditions inside a cell?

**Connection to Physics** A **fuel cell** is a voltaic cell in which the oxidation of a fuel is used to produce electric energy. Fuel cells differ from other batteries because they are provided with a continual supply of fuel from an external source. Many people think the fuel cell is a modern invention, but the first one was demonstrated in 1839 by William Grove (1811–1896), a British electrochemist. He called his cell a "gas battery." It was not until the 1950s, when scientists began working in earnest on the space program, that efficient, practical fuel cells were developed. If astronauts were to fly a space shuttle, supplies of water were needed to support their lives on board and a reliable source of electricity was needed to power the shuttle's many systems. Both of these primary needs were met with the development of the hydrogen fuel cell that controls the oxidation of hydrogen and provides both electricity and water. The cell produces no by-products to require disposal or storage on a space journey.

(tl)Yiap See Fat/Getty Images, (tr)Paul Rapson/Science Photo Library/Photo Researchers, (b)Alex Segre/Alamy

**How a fuel cell works** As in other voltaic cells, a fuel cell has an anode and a cathode and requires an electrolyte so that ions can migrate between electrodes. A common electrolyte in a fuel cell is an alkaline solution of potassium hydroxide. Each electrode is a hollow chamber of porous carbon walls that allows contact between the inner chamber and the electrolyte surrounding it. The following oxidation half-reaction takes place at the anode.

$$2H_2(g) + 4OH^-(aq) \rightarrow 4H_2O(l) + 4e^-$$

The reaction uses the hydroxide ions that are abundant in the alkaline electrolyte and releases electrons to the anode. Electrons from the oxidation of hydrogen flow through the external circuit to the cathode where the following reduction half-reaction takes place.

$$O_2(g) + 2H_2O(l) + 4e^- \rightarrow 4OH^-(aq)$$

The electrons reduce oxygen in the presence of water to form four hydroxide ions, which replenish the hydroxide ions used up at the anode. When the two half-reactions are combined, the equation is the same as the equation for the burning of hydrogen in oxygen.

$$2H_2(g) + O_2(g) \rightarrow 2H_2O(l)$$

Because the fuel for the cell is provided from an outside source, fuel cells never run down as batteries do. They keep producing electricity as long as fuel is available.

Some fuel cells use fuels other than hydrogen. For example, methane replaces hydrogen in some cells, but has the disadvantage of producing carbon dioxide as an exhaust gas. Fuel cells such as the one shown in **Figure 13** use a plastic sheet called a proton-exchange membrane (PEM), which eliminates the need for a liquid electrolyte.

☑ READING CHECK **Compare** fuel cells with other voltaic cells to find an important way in which they are different.

Watch a **video about hydrogen fuel cells.**

Video

CAREERS IN CHEMISTRY

**Alternative Energy Specialist**
If you like to invent new things and make them work, you might be interested in helping to develop new sources of energy for this increasingly energy-dependent world. Technologies include wind and solar power, geothermal energy, harnessing the energy of tides, utilizing the temperature gradients in bodies of water, and many more.

WebQuest

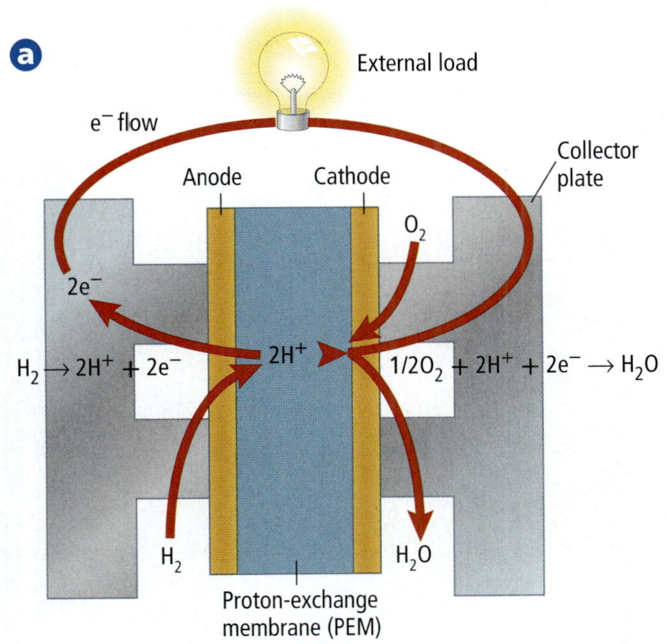

■ **Figure 13 a.** In this fuel cell, hydrogen is the fuel. The half-reactions are separated by a proton-exchange membrane so that the electrons lost in oxidation flow through an external circuit to reach the site of reduction. As electrons travel through the external circuit, they can do useful work, such as running electric motors. The by-product of this redox reaction is water. **b.** A "stack" of PEM-type cells can generate enough energy to power an electric car.

Pasquale Sorrentino/Photo Researchers

# Data Analysis LAB

Based on Real Data*

## Interpret Graphs

**How can you get electric current from microbes?** Scientists have studied the use of microbes as biofuel cells. A biofuel cell directly converts microbial metabolic energy into electric current. An electron mediator facilitates transfer of electrons to an electrode. An electron mediator is a compound that taps into the electron transport chain of cells and steals the electrons that are produced.

### Data and Observations

The graph shows the current produced in a biofuel cell with (blue line) and without (green line) the use of an electron mediator.

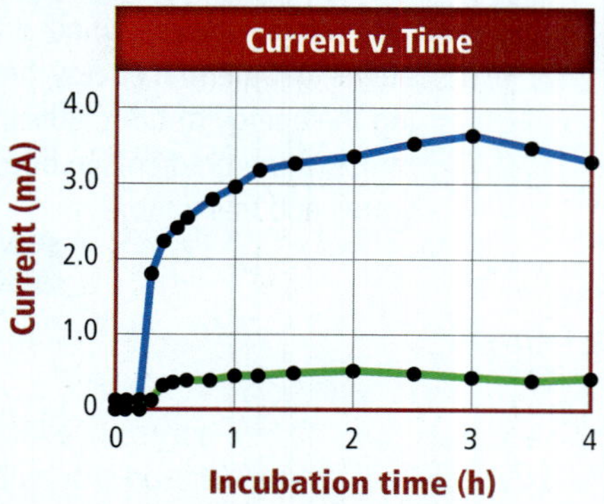


*Data obtained from: Hyun Park, Doo and J. Gregory Zeikus. April, 2000. Electricity Generation in Microbial Fuel Cells Using Neutral Red as an Electronophore. *Applied and Environmental Microbiology* 66, No. 4:1292–1297.

### Think Critically

1. **Infer** the approximate time when the electron mediator was introduced.
2. **Determine** Did the introduction of the electron mediator make a difference in the current production? Explain your answer.
3. **Analyze** What is the highest current obtained by the cell?

## Corrosion

In this chapter, you have examined the spontaneous redox reactions in voltaic cells. Spontaneous redox reactions also occur in nature. An example is the corrosion of iron, usually called rusting. **Corrosion** is the loss of metal resulting from an oxidation-reduction reaction of the metal with substances in the environment. Although rusting is usually thought of as a reaction between iron and oxygen, it is more complex. Both water and oxygen must be present for rusting to occur. For this reason, an iron object, such as the one shown in **Figure 14,** that has been left exposed to air and moisture is especially susceptible to rust. The portion that is in contact with the moist ground rusted first.

Rusting usually begins where there is a pit or a small break in the surface of the iron. This region becomes the anode of the cell as iron atoms begin to lose electrons as illustrated in **Figure 15.**

$$Fe(s) \rightarrow Fe^{2+}(aq) + 2e^-$$

The iron(II) ions become part of the water solution, while the electrons move through the iron to the cathode region. In effect, the piece of iron becomes the external circuit as well as the anode. The cathode is usually located at the edge of the water drop where water, iron, and air come in contact. Here, the electrons reduce oxygen from the air in the following half-reaction.

$$O_2(g) + 4H^+(aq) + 4e^- \rightarrow 2H_2O(l)$$

The supply of $H^+$ ions is probably furnished by carbonic acid formed when $CO_2$ from air dissolves in water.

**Figure 14** Left unattended in the presence of air and moisture, this iron barrel is slowly being oxidized to rust ($Fe_2O_3$).

i creative/Alamy

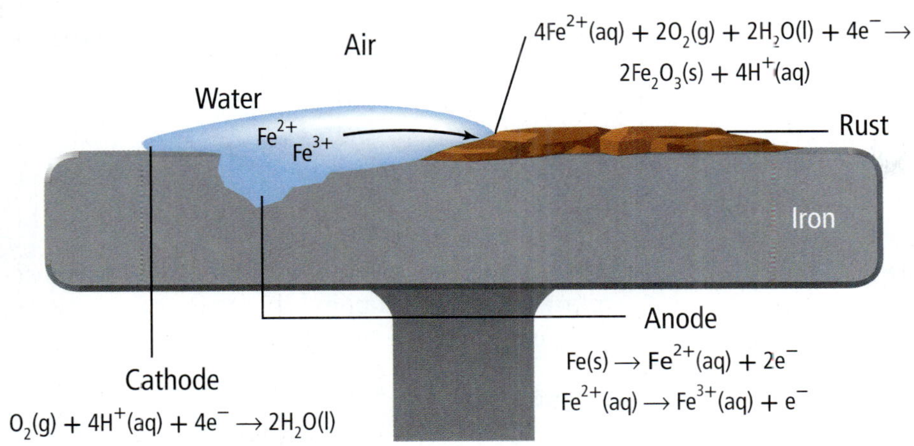


■ **Figure 15** Corrosion occurs when air, water, and iron set up a voltaic cell similar to the conditions shown at the surface of this iron I-beam. An I-beam is a large piece of iron shaped like the capital letter I and used in the construction of large buildings.

**Name** ***the two species that are oxidized at the anode.***

Next, the $Fe^{2+}$ ions in solution are oxidized to $Fe^{3+}$ ions by reacting with oxygen dissolved in the water. The $Fe^{3+}$ ions combine with oxygen to form insoluble $Fe_2O_3$, rust.

$$4Fe^{2+}(aq) + 2O_2(g) + 2H_2O(l) + 4e^- \rightarrow 2Fe_2O_3(s) + 4H^+(aq)$$

Combining the three equations yields the overall cell reaction for the corrosion of iron.

$$4Fe(s) + 3O_2(g) \rightarrow 2Fe_2O_3(s)$$

Rusting is a slow process because water droplets have few ions and therefore, are not good electrolytes. However, if the water contains abundant ions, as in seawater or in regions where roads are salted in the winter, corrosion occurs much faster because the solutions are excellent electrolytes.

**Preventing corrosion** Corrosion of cars, bridges, ships, the structures of buildings, and other metallic objects causes more than $100 billion in damage a year in the United States. For this reason, several means to minimize corrosion have been devised. One example is to apply a coat of paint to seal out both air and moisture, but because paint deteriorates over time, objects such as the bridge shown in **Figure 16** must be repainted often.

■ **Figure 16** Because corrosion can cause considerable damage, it is important to find ways to prevent rust and deterioration. Paint or another protective coating is one way to protect steel structures from corrosion.

# MiniLAB

iLab Station

## Observe Corrosion

**Which metal will corrode?**

### Procedure

1. Read and complete the lab safety form.
2. Use **sandpaper** to buff the surfaces of four **iron nails.** Wrap two nails with **magnesium ribbon** and two nails with **copper.** Wrap the metals tightly so that the nails do not slip out.
3. Place each of the nails in a separate **beaker.** Add **distilled water** to one of the beakers containing a copper-wrapped nail and one of the beakers containing a magnesium-wrapped nail. Add enough distilled water to just cover the wrapped nails. Add **salt water** to **two additional beakers.** Record your observations of the nails in each beaker.
4. Let the beakers stand overnight in the warmest place available. Examine the nails and solutions the next day, and record your observations.

### Analysis

1. **Describe** the difference between copper-wrapped nails in the distilled water and the salt water after they have been standing overnight.
2. **Describe** the difference between the magnesium-wrapped nails in the distilled water and in the salt water.
3. **Identify** the difference in corrosion between the copper- and magnesium-wrapped nails.

The steel hulls of ships are constantly in contact with salt water, so the prevention of corrosion is vital. Although the hull can be painted, another method is used to minimize corrosion. Blocks of metals, such as magnesium, aluminum, or titanium, are placed in contact with the steel hull. These blocks oxidize more easily than iron and become the anode of the corrosion cell. They are called sacrificial anodes because they are corroded, while the iron in the hull is spared. A similar technique is used to protect underground iron pipes. Magnesium bars are attached to the pipe by wires, and these bars corrode instead of the pipe, as shown in **Figure 17.**

**Figure 17** Sacrificial anodes of magnesium or other active metals are used to prevent corrosion. A magnesium rod attached to an underground iron pipe helps prevent corrosion by being oxidized itself.

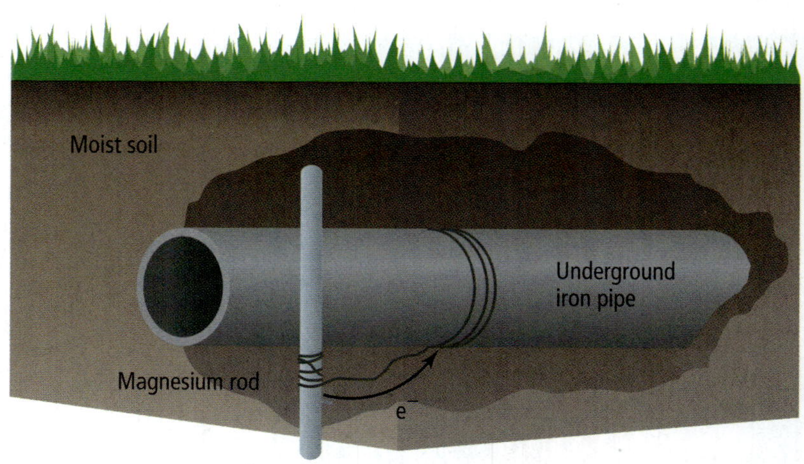


Oxidation: $Mg(s) \longrightarrow Mg^{2+}(aq) + 2e^-$

Reduction: $O_2(g) + 4H^+(aq) + 4e^- \longrightarrow 2H_2O(l)$

Geoff Butler

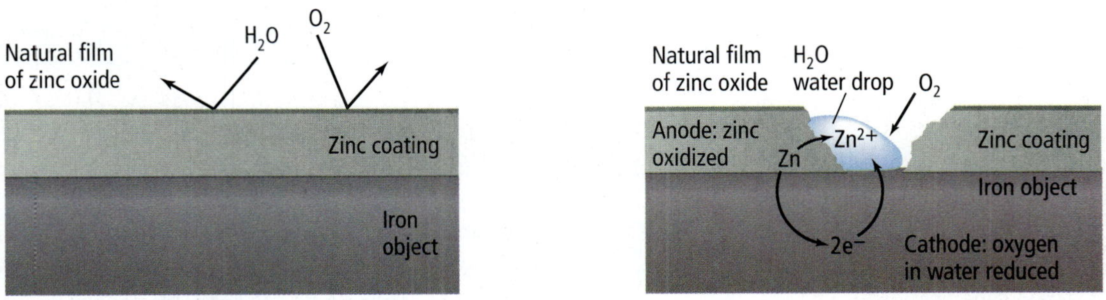


**Galvanized object with zinc coating intact**

The zinc coating seals the iron from air and water by forming a barrier of zinc oxide that repels water and oxygen.

**Galvanized object with zinc coating broken**

If the zinc coating breaks, the zinc acts as a sacrificial anode. The zinc coating, rather than the iron object, is oxidized.

**Figure 18** Galvanization helps prevent corrosion in two ways.

Another approach to preventing corrosion is to coat iron with another metal that is more resistant to corrosion. In the **galvanization** process, iron is coated with a layer of zinc by either dipping the object into molten zinc or by electroplating the zinc onto it. Although zinc is more readily oxidized than iron, it is one of the self-protecting metals, a group that also includes aluminum and chromium. When exposed to air, these metals oxidize at the surface, creating a thin metal-oxide coating that seals the metal from further oxidation.

Galvanization protects iron in two ways. As long as the zinc layer is intact, water and oxygen cannot reach the iron's surface. Inevitably, the zinc coating cracks. When this happens, zinc protects iron from rapid corrosion by becoming the anode of the voltaic cell set up when water and oxygen contact iron and zinc at the same time. **Figure 18** illustrates how these two forms of corrosion protection work.

☑ READING CHECK **Summarize** methods of preventing corrosion.

## SECTION 2 REVIEW

Section Self-Check

### Section Summary

- Primary batteries can be used only once; secondary batteries can be recharged.
- When a battery is recharged, electric energy supplied to the battery reverses the direction of the battery's spontaneous reaction.
- Fuel cells are batteries in which the substance oxidized is a fuel from an external source.
- Methods of preventing corrosion are painting, coating with another metal, or using a sacrificial anode.

15. MAINIDEA **Identify** what is reduced and what is oxidized in the zinc-carbon dry-cell battery. What features make the alkaline dry cell an improvement over the earlier type of dry-cell battery?
16. **Explain** what happens when a battery is recharged.
17. **Describe** the half-reactions that occur in a hydrogen fuel cell, and write the equation for the overall reaction.
18. **Describe** the function of a sacrificial anode. How is the function of a sacrificial anode similar to galvanization?
19. **Explain** why lithium is a good choice for the anode of a battery.
20. **Calculate** Use data from **Table 1** to calculate the cell potential of the hydrogen-oxygen fuel cell.
21. **Design an Experiment** Use your knowledge of acids to devise a method for determining whether a lead-acid battery can deliver full charge or is beginning to run down.

SECTION 3

# Electrolysis

**MAINIDEA** **In electrolysis, a power source causes nonspontaneous reactions to occur in electrochemical cells.**

### Essential Questions

- How is it possible to reverse a spontaneous redox reaction in an electrochemical cell?
- How does the molten sodium chloride electrolysis reaction compare with those in the electrolysis of brine?
- What is the importance of electrolysis in the smelting and purification of metals?

### Review Vocabulary

**redox reaction:** an oxidation-reduction reaction

### New Vocabulary

electrolysis
electrolytic cell

**CHEM 4 YOU** **When you ride a bicycle downhill, you don't have to do any work—you just coast. What is different when you ride uphill? You have to provide a lot of energy by pedaling.**

## Reversing Redox Reactions

When a battery generates electric current, electrons given up at the anode flow through an external circuit to the cathode, where they are used in a reduction reaction. A secondary battery is one that can be recharged by passing a current through it in the opposite direction. To help you understand the process, study the electrochemical cells in **Figure 19.** The beakers on the left contain zinc strips in solutions of zinc ions. The beakers on the right contain copper strips in solutions of copper ions. One electrochemical cell is supplying power to a light bulb by means of a spontaneous redox reaction. Electrons flow spontaneously from the zinc side to the copper side, creating an electric current. The reaction continues until the zinc strip is used up, and then the reaction stops. However, the cell can be regenerated if current is applied in the reverse direction using an external voltage source. The voltage source is required because the reverse reaction is nonspontaneous. If the voltage source is applied long enough, the cell will return to nearly its original strength.

The use of electrical energy to bring about a chemical reaction is called **electrolysis.** An electrochemical cell in which electrolysis occurs is called an **electrolytic cell.** For example, when a secondary battery is recharged, it is acting as an electrolytic cell.

■ **Figure 19** The zinc-copper electrochemical cell can be a voltaic cell or an electrolytic cell.
**Infer** ***In each electrochemical cell, which metal is oxidized? What is reduced?***

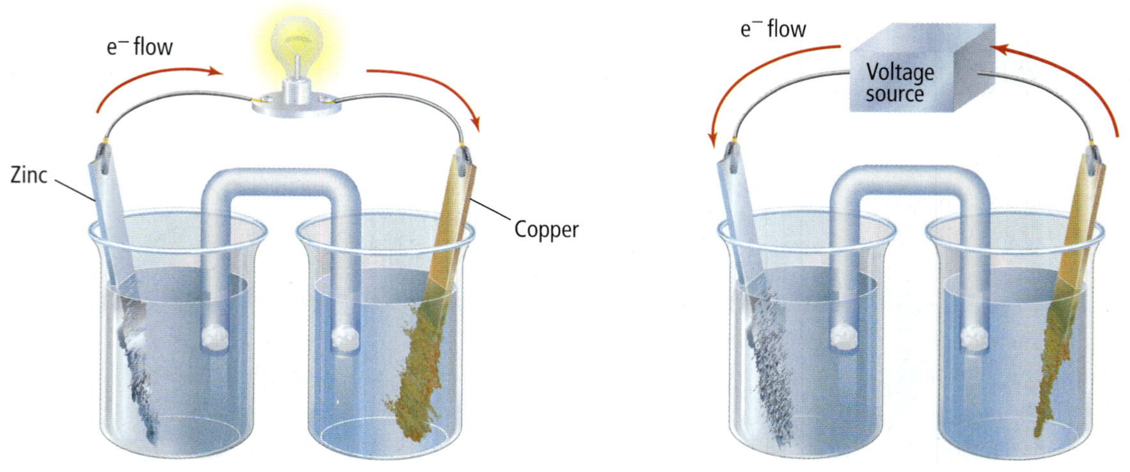


Voltaic cell

In this voltaic cell, the oxidation of zinc supplies the electrons to light the bulb and reduce copper ions. The spontaneous reaction continues until the zinc is used up.

Electrolytic cell

When an outside voltage is applied, the flow of electrons is reversed and the nonspontaneous reaction occurs, which restores the conditions of the cell.

## Applications of Electrolysis

Recall that voltaic cells convert chemical energy to electrical energy as a result of a spontaneous redox reaction. Electrolytic cells do the opposite; they use electrical energy to drive a nonspontaneous reaction. A common example is the electrolysis of water. This reaction is the opposite of consuming hydrogen and oxygen in a fuel cell.

$$2H_2O(l) \rightarrow 2H_2(g) + O_2(g)$$

The electrolysis of water is one method by which hydrogen gas can be generated for commercial use.

**Electrolysis of molten NaCl** Just as electrolysis can decompose water into its elements, it can also separate molten sodium chloride into sodium metal and chlorine gas. This process is carried out in a chamber called a Down's cell, as illustrated in **Figure 20.** The electrolyte in the cell is the molten sodium chloride itself. Remember that ionic compounds can conduct electricity only when their ions are free to move, such as when they are dissolved in water or are in the molten state.

At the anode, chloride ions are oxidized to chlorine ($Cl_2$) gas.

$$2Cl^-(l) \rightarrow Cl_2(g) + 2e^-$$

At the cathode, sodium ions are reduced to sodium metal.

$$Na^+(l) + e^- \rightarrow Na(l)$$

The net cell reaction is the following.

$$2Na^+(l) + 2Cl^-(l) \rightarrow 2Na(l) + Cl_2(g)$$

The importance of the Down's cell can best be appreciated in terms of the important roles that both sodium and chlorine play in your life. Chlorine is used throughout the world to purify water for drinking and swimming. Many cleaning products you might use, including household bleach, contain chlorine compounds. You depend on a host of other products, such as paper, plastics, insecticides, textiles, dyes, and paints, that either contain chlorine, or chlorine was used in their production.

In its pure form, sodium is used as a coolant in nuclear reactors and in sodium vapor lamps used for outdoor lighting. In its combined form in ionic compounds, you need only look on the contents list of consumer products to find a variety of sodium salts in the products you use and the foods you eat.

**READING CHECK** **Explain** why the sodium chloride must be molten in the Down's cell.

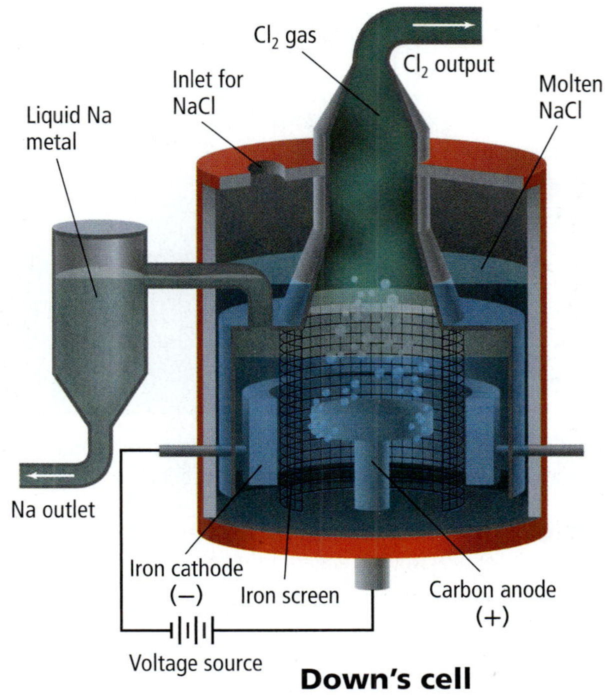


**Figure 20** In a Down's cell, electrons supplied by a generator are used to reduce sodium ions at the cathode. As electrons are removed from the anode, chloride ions are oxidized at the anode to chlorine gas.

**FOLDABLES®** Incorporate information from this section into your Foldable.

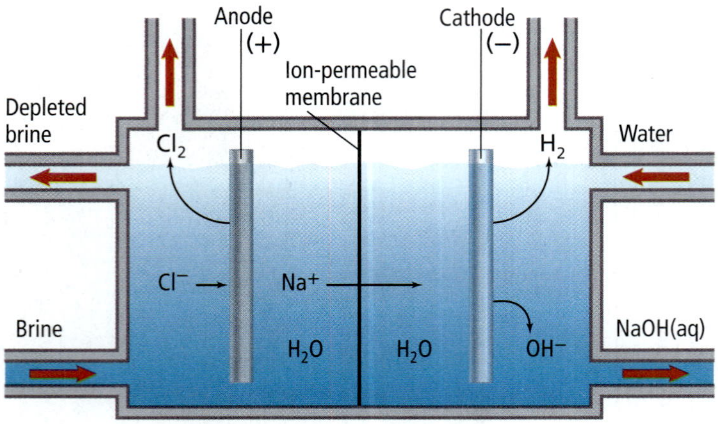


Commercial facilities use an electrolytic process to obtain hydrogen gas, chlorine gas, and sodium hydroxide from brine.

Chlorine gas is used to manufacture polyvinyl chloride products, such as these pipes for water distribution.

■ **Figure 21** In the electrolysis of brine (aqueous NaCl), sodium is not a product because water is easier to reduce.

**Electrolysis of brine** The decomposition of brine, an aqueous solution of sodium chloride, is also carried out by means of electrolysis. **Figure 21** shows a typical electrolytic cell and the products of the electrolysis. Two reactions are possible at the cathode: the reduction of sodium ions and the reduction of hydrogen in water molecules.

$$Na^+(aq) + e^- \rightarrow Na(s)$$

$$2H_2O(l) + 2e^- \rightarrow H_2(g) + 2OH^-(aq)$$

However, the reduction of sodium ($Na^+$) does not occur because water is easier to reduce, and thus is reduced preferentially.

Two reactions are also possible at the anode: the oxidation of chloride ions and the oxidation of oxygen in water molecules.

$$2Cl^-(aq) \rightarrow Cl_2(g) + 2e^-$$

$$2H_2O(l) \rightarrow O_2(g) + 4H^+(aq) + 4e^-$$

Because the desired product is chlorine ($Cl_2$), the concentration of chloride ions is kept high in order to favor this half-reaction. The overall cell reaction is as follows.

$$2H_2O(l) + 2NaCl(aq) \rightarrow H_2(g) + Cl_2(g) + 2NaOH(aq)$$

All three products are commercially important substances.

**VOCABULARY**

**SCIENCE USAGE V. COMMON USAGE**

**Reduce**

***Science usage:*** to decrease an atom's oxidation number by the addition of electrons

*Zinc reduces copper(II) ions to copper atoms by releasing two electrons.*

***Common usage:*** to diminish in size, amount, extent, or number

*The number of dancers had to be reduced because the stage was too small.*

☑ READING CHECK **Name** the species that is oxidized and the species that is reduced in the electrolysis of brine.

**Aluminum production** Until the late nineteenth century, aluminum metal was more precious than gold because no one knew how to purify it in large quantities. In 1886, 22-year-old Charles Martin Hall (1863–1914) developed a process to produce aluminum by electrolysis. He used heat from a blacksmith forge, electricity from homemade batteries, and his mother's iron skillets as electrodes. At almost the same time, one of Le Châtelier's students, Paul L. T. Héroult (1863–1914), also 22 years old, discovered the same process. Today, it is called the Hall-Héroult process and is illustrated in **Figure 22.**

In the modern version of the Hall–Héroult process, aluminum metal is obtained by electrolysis of aluminum oxide, which is refined from bauxite ore ($Al_2O_3 \cdot 2H_2O$). The aluminum oxide is dissolved at 1000°C in molten synthetic cryolite ($Na_3AlF_6$), another aluminum compound. The cell is lined with graphite, which forms the cathode for the reaction, as shown in **Figure 22.** Another set of graphite rods is immersed in the molten solution as an anode. The following half-reaction occurs at the cathode.

$$Al^{3+}(l) + 3e^- \longrightarrow Al(l)$$

The molten aluminum settles to the bottom of the cell and is drawn off periodically. Oxide ions are oxidized at the cathode in this half-reaction.

$$2O^{2-}(aq) \longrightarrow O_2(g) + 4e^-$$

Because temperatures are high, the liberated oxygen reacts with the carbon of the anode to form carbon dioxide.

$$C(s) + O_2(g) \longrightarrow CO_2(g)$$

The Hall-Héroult process uses huge amounts of electrical energy. For this reason, aluminum is often produced in plants built close to large hydroelectric power stations, where electrical energy is less expensive. The vast amount of electricity needed to produce aluminum from ore is the primary reason for recycling aluminum. Recycled aluminum has already undergone electrolysis, so the only energy required to make it usable again is the heat needed to melt it in a furnace.

**Purification of ores** Electrolysis is also used in the purification of metals such as copper. Most copper is mined in the form of the ores chalcopyrite ($CuFeS_2$), chalcocite ($Cu_2S$), and malachite ($Cu_2CO_3(OH)_2$). The sulfides are most abundant and yield copper metal when heated strongly in the presence of oxygen.

$$Cu_2S(s) + O_2(g) \longrightarrow 2Cu(l) + SO_2(g)$$

**Figure 22** The Hall-Héroult process operates at 900°C in smelters similar to this one. Note that carbon (graphite) serves as both the anode and the cathode. Recycled aluminum is often fed into the cell with the new aluminum.

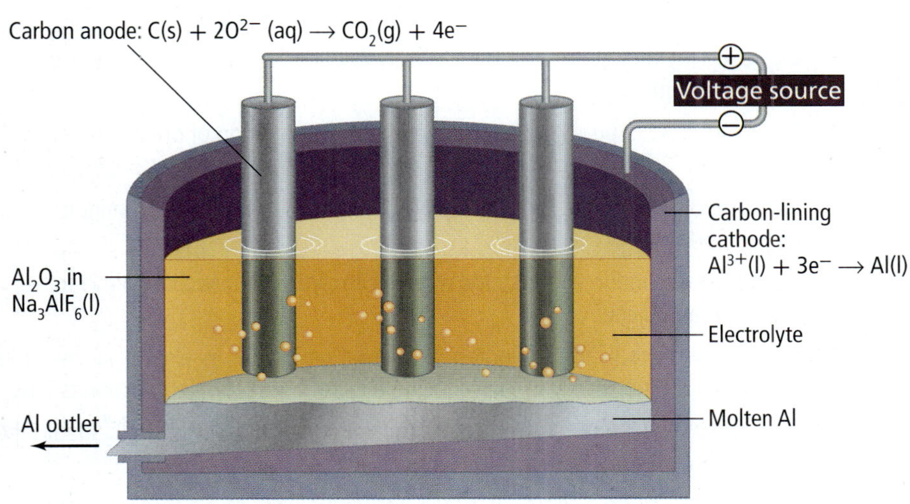

Every ton of aluminum that is recycled saves huge quantities of electrical energy that would be used to produce new aluminum from ore.

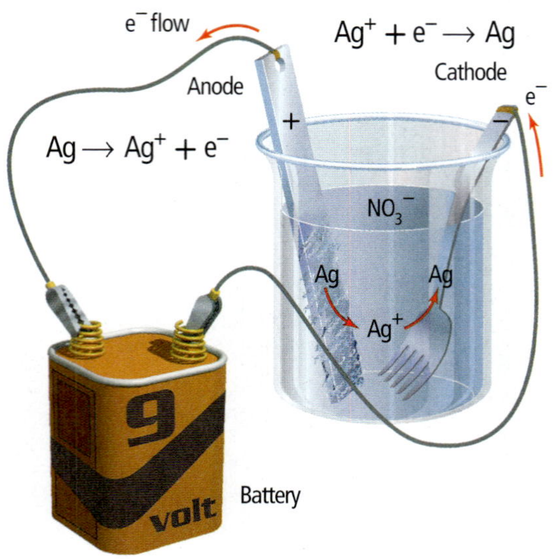


**Figure 23** Power is needed to oxidize silver at the anode and reduce silver at the cathode. In an electrolytic cell used for silver plating, the object to be plated is the cathode where silver ions in the electrolyte solution are reduced to silver metal and deposited on the object.

The copper obtained from this process contains many impurities and must be refined, so the molten copper is cast into large, thick plates. These plates are then used as an anode in an electrolytic cell containing a solution of copper(II) sulfate. The cathode of the cell is a thin sheet of pure copper. As current is passed through the cell, copper atoms in the impure anode are oxidized to copper(II) ions. The copper ions migrate through the solution to the cathode, where they are reduced to copper atoms. These atoms become part of the cathode, while impurities fall to the bottom of the cell.

**Electroplating** Objects are electroplated when a uniform coating is deposited usually as a protective or decorative layer. The substance used is typically a metal. Electroplating with a metal such as silver is accomplished with a method similar to that used to refine copper. The object to be silver-plated is the cathode of an electrolytic cell. The anode is a silver bar or sheet, as shown in **Figure 23.** At the anode, silver is oxidized to silver ions as electrons are removed by the power source. At the cathode, the silver ions are reduced to silver metal by electrons from the external power source. The silver forms a thin coating over the object being plated. Current passing through the cell must be carefully controlled in order to get a smooth, even metal coating.

Other metals are also used for electroplating. You might have costume jewelry that is electroplated with gold. Or you might admire an automobile whose steel parts such as the bumper have been made more corrosion-resistant by being electroplated first with nickel and then with chromium.

## SECTION 3 REVIEW

Section Self-Check

### Section Summary

- In an electrolytic cell, an outside source of power causes a nonspontaneous redox reaction to occur.
- The electrolysis of molten sodium chloride yields sodium metal and chlorine gas. The electrolysis of brine yields hydrogen gas, sodium hydroxide, and chlorine gas.
- Metals such as copper are purified in an electrolytic cell.
- Electrolysis is used to electroplate objects and to produce pure aluminum from its ore.

22. MAINIDEA **Define** *electrolysis* and relate the definition to the spontaneity of redox reactions.
23. **Explain** why the products of the electrolysis of brine and the electrolysis of molten sodium chloride are different.
24. **Describe** how impure copper obtained from the smelting of ore is purified by electrolysis.
25. **Explain,** by referring to the Hall-Héroult process, why recycling aluminum is very important.
26. **Describe** the anode and cathode of an electrolytic cell in which gold is to be plated on an object.
27. **Explain** why producing a kilogram of silver from its ions by electrolysis requires much less electric energy than producing a kilogram of aluminum from its ions.
28. **Calculate** Use **Table 1** to calculate the voltage of the Down's cell. Should the potential be positive or negative?
29. **Summarize** Write a short paragraph answering each of the three Essential Questions for Section 3 in your own words.

# HOW IT works

## THE PACEMAKER: Helping a Broken Heart

Your heart is made of cardiac muscle tissue that contracts and relaxes continuously. This beating results from electric impulses moving along pathways throughout your heart. A group of specialized cells in the upper wall of the heart's right atrium—upper chamber—generates electric impulses. If these cells fail to function or the electric impulse pathways are interrupted, the heart does not beat normally. A pacemaker is an electrical device that can monitor and correct an irregular heartbeat. How does it work?

**1 Leads** Insulated wires called leads carry electric signals between the heart and the pacemaker. A lead is implanted into a blood vessel and then into a chamber of the heart. Pacemakers might use one, two, or three leads, each in a different chamber.

**2 Pacemaker** Each lead is connected to the pacemaker, which contains a battery, an antenna, and computer circuits. The pacemaker is then implanted under the skin below the collarbone. Surgery is also required to replace a pacemaker when its battery is low or the circuits fail.

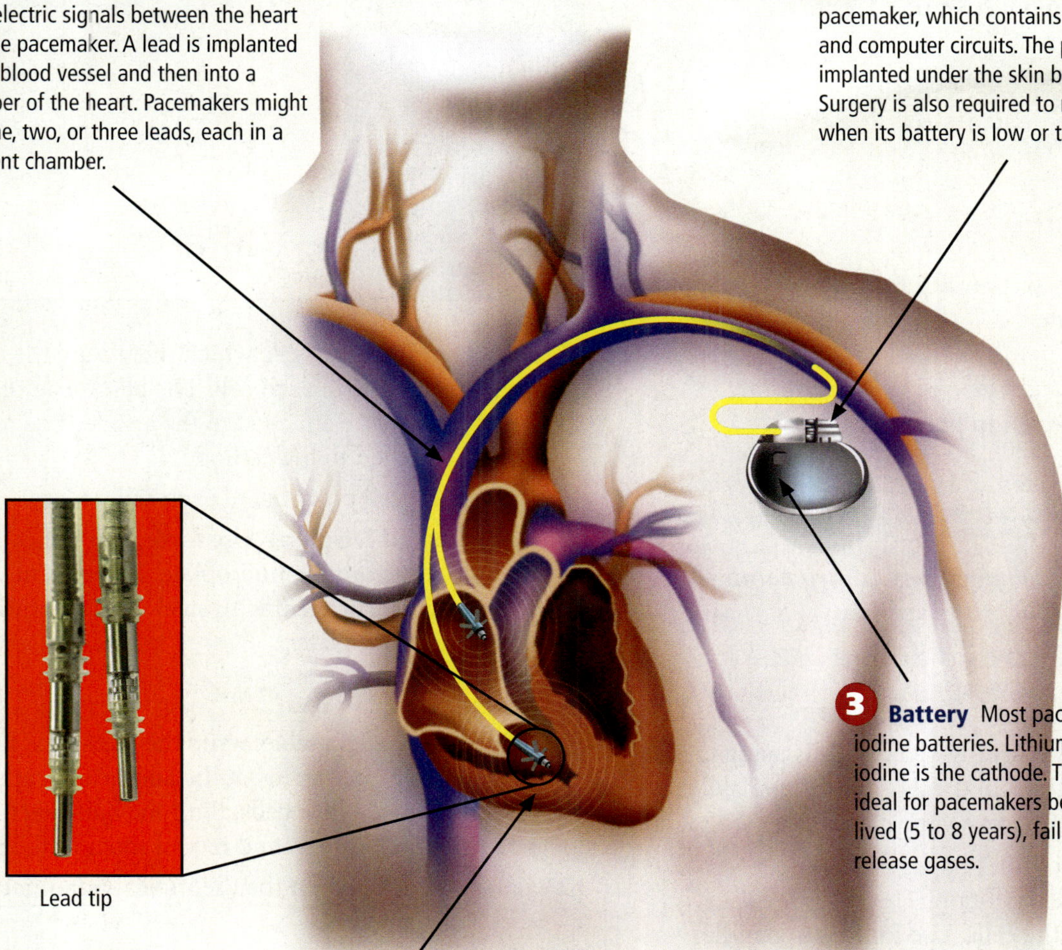

Lead tip

**3 Battery** Most pacemakers use lithium-iodine batteries. Lithium is the anode and iodine is the cathode. These batteries are ideal for pacemakers because they are long-lived (5 to 8 years), fail gradually, and do not release gases.

**4 Function** The tip of the lead attaches to heart tissue, transmits electric impulses, and senses the heartbeat. Pacemakers can stimulate a sluggish heartbeat, smooth an irregular beat, or activate the beating of an arrested heart.

### WRITING IN ▶ Chemistry

**Summarize** The pacemaker is an example of an advancement in medical technology. Write a paragraph about the ways your life or that of someone you know has changed because of medical technology and how it might change in the future.

WebQuest

SMALL SCALE ChemLAB

iLab Station

# Measure Voltaic Cell Potentials

**Background:** When two different half-cells are joined, a potential difference is produced. A voltmeter measures the potential difference of combinations of half-cells.

**Question:** *How do measured potentials of voltaic cells compare to calculated potentials?*

## Materials

metal strips of copper, aluminum, zinc, and magnesium (about 0.6 cm × 1.3 cm)
1*M* copper(II) nitrate
1*M* aluminum nitrate
1*M* zinc nitrate
1*M* magnesium nitrate
24-well microplate
Beral-type pipette (5)
voltmeter
filter paper (6 pieces 0.6 cm × 2.5 cm)
1*M* potassium nitrate
forceps
steel wool or sandpaper
table of standard reduction potentials

## Safety Precautions

WARNING: ***Steel wool might have sharp points that can pierce skin. Handle with care.***

## Procedure

1. Read and complete the lab safety form.
2. Plan how you will arrange voltaic cells using the four metal combinations in the 24-well microplate. Have your instructor approve your plan.
3. Soak the strips of filter paper in 2 mL of potassium nitrate solution. The strips are the salt bridges for the cells. Use forceps to handle the salt bridges.
4. Construct voltaic cells using the four metals and 1 mL of each of the solutions. Put the metals into the wells that contain the appropriate solution; for example, put the zinc metal in the solution with zinc nitrate. Use a different salt bridge for each voltaic cell. Connect the leads from the voltmeter to the metals. If you get a negative value for potential difference, switch the leads on the metals.

**Voltaic Cell Potential Data**

| | | | |
|---|---|---|---|
| Anode metal | | | |
| Cathode metal | | | |
| Measured cell potential (V) | | | |
| Anode half-reaction and standard potential | | | |
| Cathode half-reaction and standard potential | | | |
| Theoretical cell potential | | | |
| % Error | | | |

5. Record in the data table which metals are the anode and the cathode in each cell. The black lead of the voltmeter will be attached to the anode. The red lead will be attached to the cathode.
6. Record the cell potential of each cell.
7. **Cleanup and Disposal** Use forceps to remove the metal strips from the microplate. Rinse them with water, then clean them with steel wool or sandpaper. Rinse the microplate.

## Analyze and Conclude

1. **Apply** In the data table, write the equations for half-reactions occurring at the anode and cathode in each of the voltaic cells. Find the half-reaction potentials in **Table 1,** and record them in the table.
2. **Calculate** and record the theoretical potential for each voltaic cell.
3. **Predict** Using your data, rank the metals in order of most active to least active.
4. **Error Analysis** Calculate the percent error of the voltaic cell potential. Why is the percent error large for some voltaic cells and small for others?

### INQUIRY EXTENSION

**Design an Experiment** that would reduce the percent error discussed in Question 4.

# CHAPTER 20 STUDY GUIDE

Vocabulary Practice

**BIGIDEA** **Chemical energy can be converted to electric energy and electric energy to chemical energy.**

## SECTION 1 Voltaic Cells

**MAINIDEA** In voltaic cells, oxidation takes place at the anode, yielding electrons that flow to the cathode, where reduction occurs.

- In a voltaic cell, oxidation and reduction take place at electrodes separated from each other.
- The standard potential of a half-cell reaction is its voltage when paired with a standard hydrogen electrode under standard conditions.
- The reduction potential of a half-cell is negative if it undergoes oxidation when connected to a standard hydrogen electrode. The reduction potential of a half-cell is positive if it undergoes reduction when connected to a standard hydrogen electrode.
- The standard potential of a voltaic cell is the difference between the standard reduction potentials of the half-cell reactions.

$$E^0_{cell} = E^0_{reduction} - E^0_{oxidation}$$

**VOCABULARY**
- salt bridge
- electrochemical cell
- voltaic cell
- half-cell
- anode
- cathode
- reduction potential
- standard hydrogen electrode

## SECTION 2 Batteries

**MAINIDEA** Batteries are voltaic cells that use spontaneous reactions to provide energy for a variety of purposes.

- Primary batteries can be used only once; secondary batteries can be recharged.
- When a battery is recharged, electric energy supplied to the battery reverses the direction of the battery's spontaneous reaction.
- Fuel cells are batteries in which the substance oxidized is a fuel from an external source.
- Methods of preventing corrosion are painting, coating with another metal, or using a sacrificial anode.

**VOCABULARY**
- battery
- dry cell
- primary battery
- secondary battery
- fuel cell
- corrosion
- galvanization

## SECTION 3 Electrolysis

**MAINIDEA** In electrolysis, a power source causes nonspontaneous reactions to occur in electrochemical cells.

- In an electrolytic cell, an outside source of power causes a nonspontaneous redox reaction to occur.
- The electrolysis of molten sodium chloride yields sodium metal and chlorine gas. The electrolysis of brine yields hydrogen gas, sodium hydroxide, and chlorine gas.
- Metals such as copper are purified in an electrolytic cell.
- Electrolysis is used to electroplate objects and to produce pure aluminum from its ore.

**VOCABULARY**
- electrolysis
- electrolytic cell

# CHAPTER 20 ASSESSMENT

## SECTION 1

### Mastering Concepts

**30.** What feature of an oxidation-reduction reaction allows it to be used to generate an electric current?

**31.** Describe the process that releases electrons in a zinc-copper voltaic cell.

**32.** What is the function of a salt bridge in a voltaic cell?

**33.** What information do you need in order to determine the standard voltage of a voltaic cell?

**34.** In a voltaic cell represented by $Al|Al^{3+} || Cu^{2+}|Cu$, what is oxidized and what is reduced as the cell delivers current?

**35.** Under what conditions are standard reduction potentials measured?

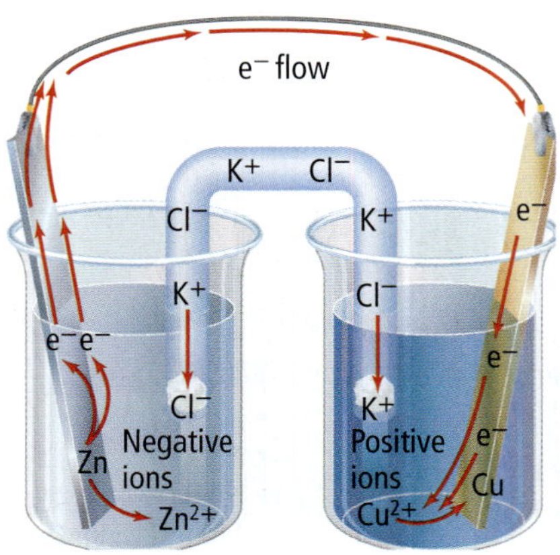


■ **Figure 24**

**36.** In **Figure 24,** identify the metal that is being oxidized. Identify the cathode.

**37.** A salt bridge is filled with $KNO_3$. Explain why it is necessary that the potassium ions move through the salt bridge to the cathode.

**38.** Recall that a reducing agent is the substance being oxidized and an oxidizing agent is the substance being reduced. Use **Table 1** to select an oxidizing agent that will convert Au to $Au^{3+}$ but will not convert $Co^{2+}$ to $Co^{3+}$.

### Mastering Problems

**39.** Using **Table 1,** write the standard cell notation for each cell in which each of the following half-cells is connected to the standard hydrogen electrode.

**a.** $Zn^{2+} | Zn$
**b.** $Hg^{2+} | Hg$
**c.** $Cu^{2+} | Cu$
**d.** $Al^{3+} | Al$

**40.** Write the balanced chemical equation for the standard cell notations listed below.

**a.** $I^- | I_2 || Fe^{3+} | Fe^{2+}$
**b.** $Sn | Sn^{2+} || Ag^+ | Ag$
**c.** $Zn | Zn^{2+} || Cd^{2+} | Cd$

**41.** Calculate the cell potentials for the following reactions.

**a.** $2Ag^+(aq) + Pb(s) \rightarrow Pb^{2+}(aq) + 2Ag(s)$
**b.** $Mn(s) + Ni^{2+}(aq) \rightarrow Mn^{2+}(aq) + Ni$
**c.** $I_2(aq) + Sn(s) \rightarrow 2I^-(aq) + Sn^{2+}(aq)$

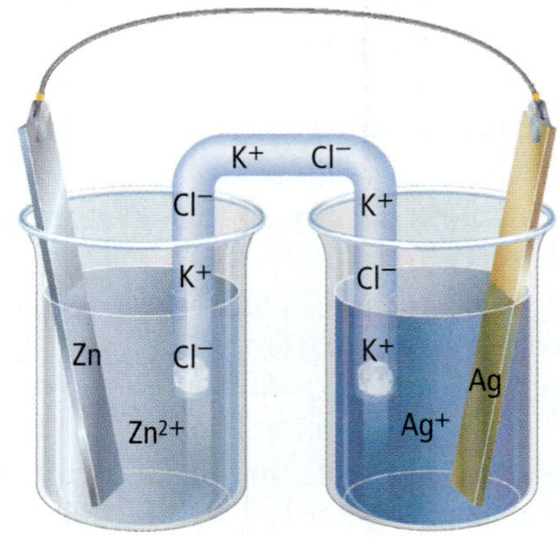


■ **Figure 25**

**42.** **Figure 25** illustrates a voltaic cell consisting of a strip of zinc in a 1.0*M* solution of zinc nitrate and a strip of silver in a 1.0*M* solution of silver nitrate. Use the diagram and **Table 1** to answer these questions.

**a.** Identify the anode.
**b.** Identify the cathode.
**c.** Where does oxidation occur?
**d.** Where does reduction occur?
**e.** In which direction is the current flowing through the connecting wire?
**f.** In which direction are positive ions flowing through the salt bridge?
**g.** What is the cell potential at 25°C and 1 atm?

## SECTION 2

### Mastering Concepts

**43.** What part of a zinc-carbon dry cell is the anode? Describe the reaction that takes place there.

**44.** How do primary and secondary batteries differ?

**45.** **Lead-Acid Battery** What substance is reduced in a lead-acid storage battery? What substance is oxidized? What substances are produced in each reaction?

**46.** **Biofuel Cell** At the cathode of a biofuel cell, $Fe^{3+}$ in potassium hexacyanoferrate(III) ($K_3[Fe(CN)_6]$) is reduced to $Fe^{2+}$ in potassium hexacyanoferrate(II) ($K_4[Fe(CN)_6]$). At the anode, reduced nicotinamide-adenine-dinucleotide (NADH) is oxidized to $NAD^+$. Use the following standard reduction potentials to determine the potential of the cell.

$NAD^+ + H^+ + 2e^- \rightarrow NADH \quad E^0 = -0.320\ V$

$[Fe(CN)_6]^{3-} + 1e^- \rightarrow [Fe(CN)_6]^{4-} \quad E^0 = +0.36\ V$

**47. Fuel Cells** List two ways in which a fuel cell differs from an ordinary battery.

**48. Galvanization** What is galvanization? How does galvanizing iron protect it from corrosion?

**49. Batteries** Explain why a lead storage battery does not produce a current when the level of $H_2SO_4$ is low.

**50. Steel Wool** is a bundle of filaments made of steel, an alloy of iron and carbon. Which would be the best way to store steel wool?
   **a.** Store it in water.
   **b.** Store it in open air.
   **c.** Store it with a desiccant.

**51. Corrosion Protection** List three ways metals can be protected from corrosion.

## Mastering Problems

**52.** Half-reactions for a lead-acid storage battery are below.

$$PbO_2(s) + SO_4^{2-}(aq) + 4H^+(aq) + 2e^- \rightarrow PbSO_4(s) + 2H_2O(l) \quad E^0 = +1.685V$$

$$PbSO_4(s) + 2e^- \rightarrow Pb(s) + SO_4^{2-}(aq) \quad E^0 = -0.3588V$$

What is the standard cell potential for one cell in a car battery?

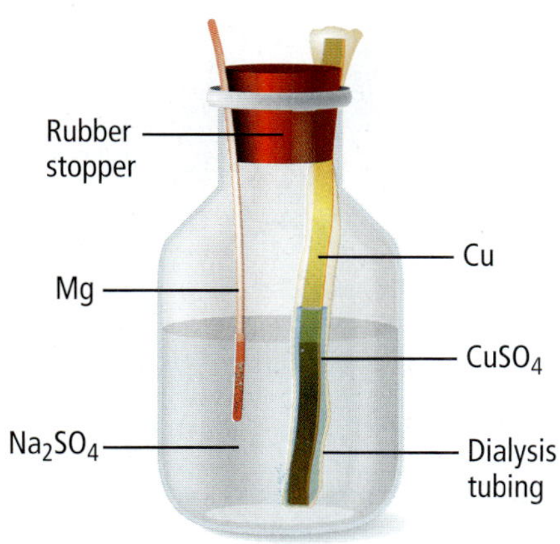


■ **Figure 26**

**53.** The setup in **Figure 26** acts as a battery.
   **a.** Determine the reaction that takes place at the copper strip.
   **b.** Determine the reaction that takes place at the magnesium wire.
   **c.** Identify the anode.
   **d.** Identify the cathode.
   **e.** Calculate the standard cell potential for this battery.

**54.** You design a battery that uses a half-cell containing Sn and $Sn^{2+}$ and another half-cell containing Cu and $Cu^{2+}$. The copper electrode is the cathode, and the tin electrode is the anode. Draw the battery and write the half-reactions that occur in each half-cell. What is the maximum voltage this battery can produce?

# SECTION 3

## Mastering Concepts

**55.** How can the spontaneous redox reaction of a voltaic cell be reversed?

**56.** Where does oxidation take place in an electrolytic cell?

**57. Down's Cell** What reaction takes place at the cathode when molten sodium chloride is electrolyzed?

**58. Industry** Explain why the electrolysis of brine is done on a large scale at many sites around the world.

**59. Recycling** Explain how recycling aluminum conserves energy.

**60.** Describe what happens at the anode and the cathode in the electrolysis of KI(aq).

## Mastering Problems

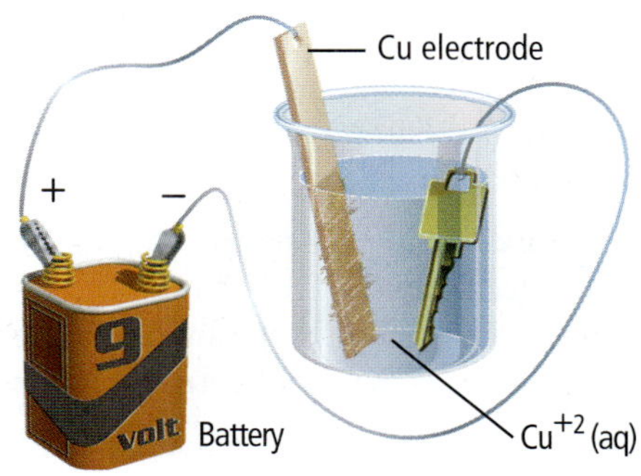


■ **Figure 27**

**61. Electroplating** **Figure 27** shows a key being electroplated with copper in an electrolytic cell. Where does oxidation occur? Explain your answer.

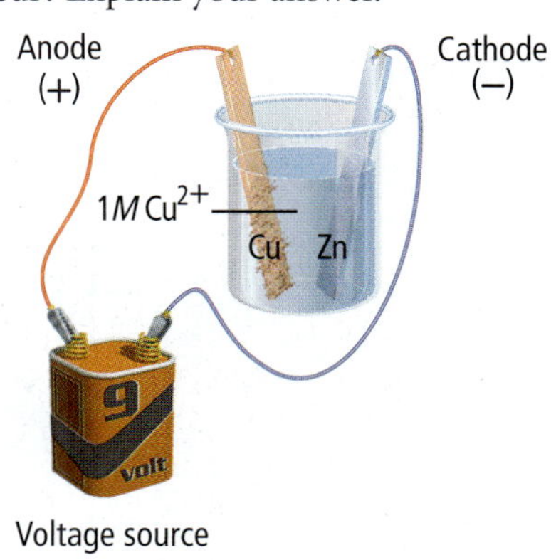


■ **Figure 28**

**62.** Answer the following questions based on **Figure 28.**
   **a.** Which electrode grows? Write the reaction that occurs at this electrode.
   **b.** Which electrode disappears? Write the reaction that occurs at this electrode.

**63.** Using **Figure 28,** explain what happens to the copper ions in solution.

## MIXED REVIEW

**64.** Why do electrons flow from one electrode to the other in a voltaic cell?

**65. Aluminum Production** What substance is electrolyzed in the industrial process to produce aluminum metal?

**66.** Write the oxidation and reduction half-reactions for a silver-chromium voltaic cell. Identify the anode, cathode, and the direction of electron flow.

**67.** Determine whether each redox reaction is spontaneous or nonspontaneous.

**a.** $Mn^{2+}(aq) + 2Br^-(aq) \rightarrow Br_2(l) + Mn(s)$
**b.** $2Fe^{2+}(aq) + Sn^{2+}(aq) \rightarrow 2Fe^{3+}(aq) + Sn(s)$
**c.** $Ni^{2+}(aq) + Mg(s) \rightarrow Mg^{2+}(aq) + Ni(s)$
**d.** $Pb^{2+}(aq) + 2Cu^+(aq) \rightarrow Pb(s) + 2Cu^{2+}(aq)$

**68.** Determine the voltage of the cell in which each half-cell is connected to a $Ag^+ \mid Ag$ half-cell.

**a.** $Be^{2+} \mid Be$
**b.** $S \mid S^{2-}$
**c.** $Au^+ \mid Au$
**d.** $I_2 \mid I^-$

**69. Corrosion** Explain why water is necessary for the corrosion of iron.

**70. Space Travel** The space shuttle uses a $H_2/O_2$ fuel cell to produce electricity.

**a.** What is the reaction at the anode? At the cathode?
**b.** What is the standard cell potential for the fuel cell?

**71. Fuel Cells** Explain how the oxidation of hydrogen in a fuel cell differs from the oxidation of hydrogen when it burns in air.

**72. Copper Refining** In the electrolytic refining of copper, what factor determines which piece of copper is the anode and which is the cathode?

**73. Storage Batteries** Lead-acid batteries and other rechargeable batteries are sometimes called storage batteries. What is being stored in these batteries?

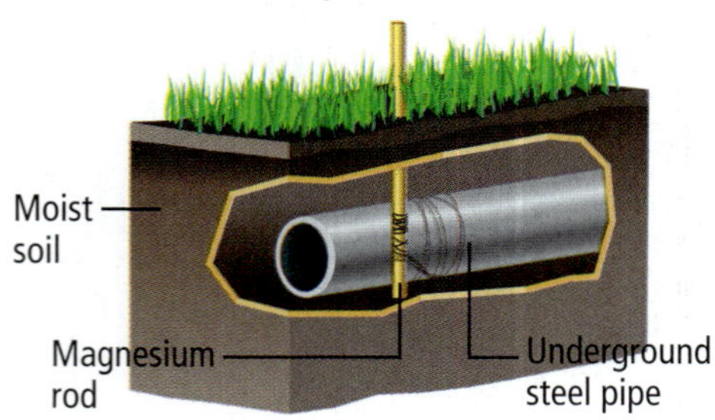


■ **Figure 29**

**74. Corrosion Prevention** **Figure 29** shows how buried steel pipes can be protected against corrosion. The steel pipe is connected to a more active metal that corrodes instead of the steel.

**a.** What is the cathode? What is the anode?
**b.** Describe how the magnesium metal protects the steel.

## THINK CRITICALLY

**75. Predict** Suppose that scientists had chosen the $Cu^{2+} \mid Cu$ half-cell as a standard instead of the $H^+ \mid H_2$ half-cell? What would the potential of the hydrogen electrode be if the copper electrode were the standard? How would the relationships among the standard reduction potentials change?

**76. Apply** Suppose that you have a voltaic cell in which one half-cell is made up of a strip of tin immersed in a solution of tin(II) ions.

**a.** How could you tell by measuring voltage whether the tin strip was acting as a cathode or an anode in the cell?
**b.** How could you tell by simple observation whether the tin strip was acting as a cathode or an anode?

**77. Hypothesize** The potential of a half-cell varies with concentration of reactants and products. For this reason, standard potentials are measured at 1*M* concentration. Maintaining a pressure of 1 atm is especially important in half-cells that involve gases as reactants or products. Suggest a reason why gas pressure is critical in these cells.

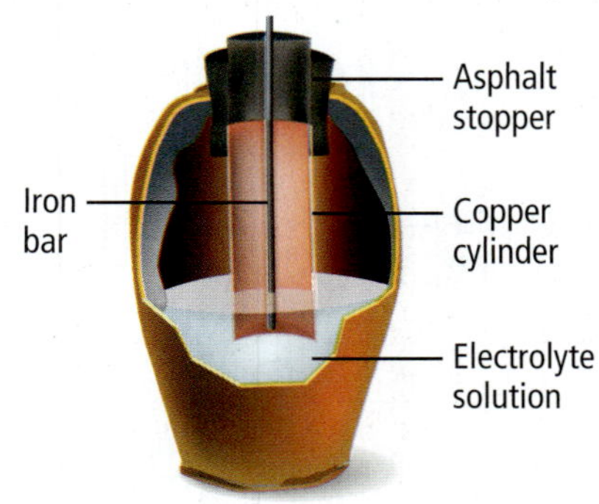


■ **Figure 30**

**78. Analyze** An earthen vessel was discovered in 1938 near Baghdad. This ancient vessel contained an iron bar surrounded by a copper cylinder, as shown in **Figure 30.** When filled with an electrolyte such as vinegar, this vessel might have acted as a battery.

**a.** Identify the cathode.
**b.** Identify the anode.
**c.** Calculate the standard cell potential of this battery.

**79. Apply** During electrolysis, an electrolytic cell releases bromine vapor and hydrogen gas. After electrolysis, the cell is found to contain a concentrated solution of potassium hydroxide. What was the composition of the cell before electrolysis began?

**80. Hypothesize** Suppose in galvanization, copper was plated on iron instead of zinc. Would copper continue to protect the iron from corrosion, as zinc does, if the copper coating became broken or cracked? Explain.

Chapter Self-Check

## Challenge Problem

**81.** A battery is assembled using tin and mercury, which have the following reduction half-reactions:
$Sn^{2+}(aq) + 2e^- \rightarrow Sn(aq)$
$Hg^{2+}(aq) + 2e^- \rightarrow Hg(l)$
  **a.** Write a balanced equation for the cell's reaction.
  **b.** What is oxidized and what is reduced? Identify the oxidizing agent and the reducing agent.
  **c.** Which reaction occurs at the anode? At the cathode?
  **d.** What is the cell potential? Use **Table 1**.
  **e.** If sodium sulfate solution is in the salt bridge, in which direction do the sulfate ions move?

## CUMULATIVE REVIEW

**82.** If the volume of a sample of chlorine gas is 8.2 L at 1.2 atm and 306 K, what volume will the gas occupy at STP?

**83.** What is meant by solvation? Explain how this process is important for the dissolving of ionic salts in water.

**84.** Explain how the molarity of a solution is different from its molality.

**85.** Define the calorie. State how the calorie is related to the Calorie and the joule.

**86.** Explain why you would find an aluminum chair to be hotter to sit on than a wooden bench after each had been in the sunlight for the same amount of time.

**87.** What does a negative sign for the free energy of a reaction tell you about the reaction? ($\Delta G_{system} = \Delta H_{system} - T\Delta S_{system}$)

**88.** According to the collision model of chemical reactions, how is it possible that two molecules can collide but not react?

**89.** List five factors that can affect the rate of a reaction.

**90.** The decomposition reaction $A_2B \rightarrow 2A + B$ proceeds to equilibrium at 499°C. Analysis of the equilibrium mixture shows $[A_2B] = 0.855$ mol/L, $[A] = 2.045$ mol/L, and $[B] = 1.026$ mol/L. What is $K_{eq}$?

**91.** What is the solubility in mol/L of silver iodide, AgI. $K_{sp}$ for AgI is $3.5 \times 10^{-17}$.

**92.** If you have a solution of a strong acid, is that the same as having a concentrated solution of the acid? Explain your answer.

**93.** What are the oxidation numbers for the elements in the ion $PO_4^{3-}$?

## WRITING IN ▶ Chemistry

**94. Sunken Ships** Study of the sunken ocean liner *Titanic* has opened the possibility that deterioration of the steel hull might be partly due to the presence of rusticle communities. Research how the biological activity of rusticle communities results in the oxidation of iron. Write an essay that describes the role of rusticle communities in the destruction of the *Titanic*.

**95. Statue of Liberty** Several years ago, the supporting structure of the Statue of Liberty became so corroded that it had to be replaced entirely. Find out what the structure was made of and why it corroded so badly. Write a report that explains the chemical processes involved and include a time line of the statue, starting in France before 1886.

## DBQ Document-Based Questions

**Electrochemical Biological Reactions** *Standard reduction potentials for some important biological reactions are given in* **Table 2.** *The strongest oxidizing agent generally available in biological systems is molecular oxygen. Consider the oxidation of reduced nicotinamide-adenine-dinucleotide (NADH) by molecular oxygen. The reaction is the following.*

$$2NADH + 2H^+ + O_2 \rightarrow 2NAD^+ + 2H_2O.$$

Data obtained from: Fromm, James Richard. 1997. "Biochemical Electrochemistry," last modified 1997, accessed September 1, 2010, http://www.3rd1000.com/chem301/chem302z.htm.

**Table 2** Aqueous Standard Reduction Potentials at 25°C and pH 7.00

| Electrode Couple | $E^0$ (V) |
|---|---|
| $2H^+ + 2e^- \rightarrow H_2(g)$ | −0.4141 |
| $NAD^+ + H^+ + 2e^- \rightarrow NADH$ | −0.320 |
| $HOOCCOCH_3$* $+ 2H^+ + 2e^- \rightarrow HOOCCHOHCH_3$** | +0.19 |
| $Fe^{3+} + e^- \rightarrow Fe^{2+}$ | +0.769 |
| $O_2(g) + 4H^+ + 4e^- \rightarrow 2H_2O$ | +0.8147 |

* $HOOCCOCH_3$ is pyruvic acid
** $HOOCCHOHCH_3$ is l-lactic acid.

**96.** Write the two half-reactions that take place in this reaction.

**97.** Calculate the cell potential of this reaction using **Table 2.**

**98.** Will $NAD^+$ oxidize $Fe^{2+}$ to $Fe^{3+}$? Explain your answer.

## MULTIPLE CHOICE

*Use the table below to answer Questions 1 to 4.*

| Selected Standard Reduction Potentials at 25°C, 1 atm, and 1*M* Ion Concentration | |
|---|---|
| Half-Reaction | $E^0$ (V) |
| $Mg^{2+} + 2e^- \longrightarrow Mg$ | −2.372 |
| $Al^{3+} + 3e^- \longrightarrow Al$ | −1.662 |
| $Pb^{2+} + 2e^- \longrightarrow Pb$ | −0.1262 |
| $Ag^+ + e^- \longrightarrow Ag$ | 0.7996 |
| $Hg^{2+} + 2e^- \longrightarrow Hg$ | 0.851 |

**1.** Which metal ion is most easily reduced?
**A.** $Mg^{2+}$
**B.** $Hg^{2+}$
**C.** $Ag^+$
**D.** $Al^{3+}$

**2.** On the basis of the standard reduction potentials shown above, which standard cell notation correctly represents its voltaic cell?
**A.** $Ag \mid Ag^+ \parallel Al^{3+} \mid Al$
**B.** $Mg \mid Mg^{2+} \parallel H^+ \mid H_2$
**C.** $H_2 \mid H^+ \parallel Pb^{2+} \mid Pb$
**D.** $Pb \mid Pb^{2+} \parallel Al^{3+} \mid Al$

**3.** A voltaic cell consists of a magnesium bar dipping into a 1*M* $Mg^{2+}$ solution and a silver bar dipping into a 1*M* $Ag^+$ solution. What is the standard potential of this cell?
**A.** 1.572 V
**B.** 3.172 V
**C.** 0.773 V
**D.** 3.971 V

**4.** Assuming standard conditions, which cell will produce a potential of 2.513 V?
**A.** $Al \mid Al^{3+} \parallel Hg^{2+} \mid Hg$
**B.** $H_2 \mid H^+ \parallel Hg^{2+} \mid Hg$
**C.** $Mg \mid Mg^{2+} \parallel Al^{3+} \mid Al$
**D.** $Pb \mid Pb^{2+} \parallel Ag^+ \mid Ag$

**5.** Which statement is NOT true of batteries?
**A.** Batteries are compact forms of voltaic cells.
**B.** Secondary batteries are storage batteries.
**C.** A battery can consist of a single cell.
**D.** The redox reaction in a rechargeable battery is not reversible.

**6.** Which is NOT a characteristic of a basic solution?
**A.** tastes bitter
**B.** conducts electricity
**C.** contains more $H^+$ ions than $OH^-$ ions
**D.** feels slippery

**7.** A carbonated soft drink has a pH of 2.5. What is the concentration of $H^+$ ions in the soft drink?
**A.** $3 \times 10^{-12}M$
**B.** $3 \times 10^{-3}M$
**C.** $4 \times 10^{-1}M$
**D.** $1 \times 10^{1}M$

**8.** Which graph correctly shows the relationship between average kinetic energy of particles and the temperature of a sample?

**A.**
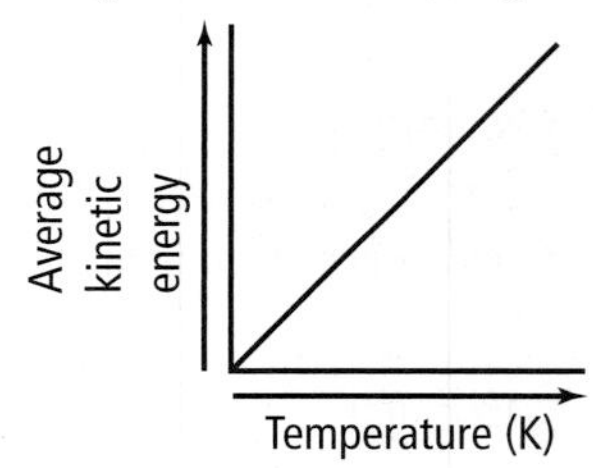


**B.**
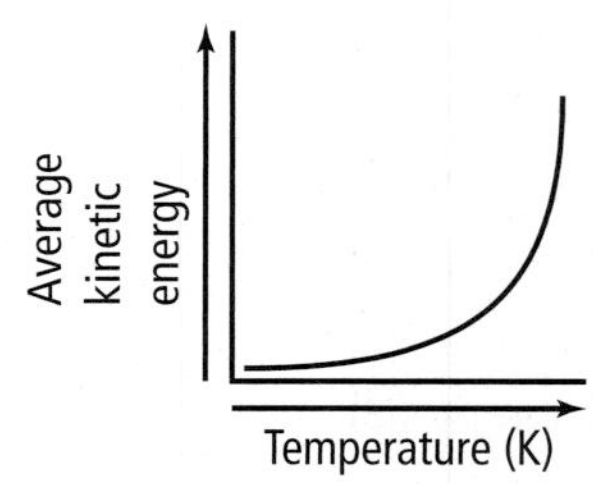


**C.**
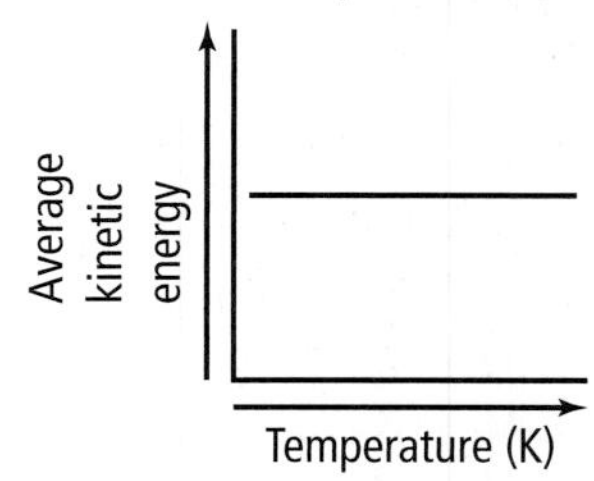


**D.**
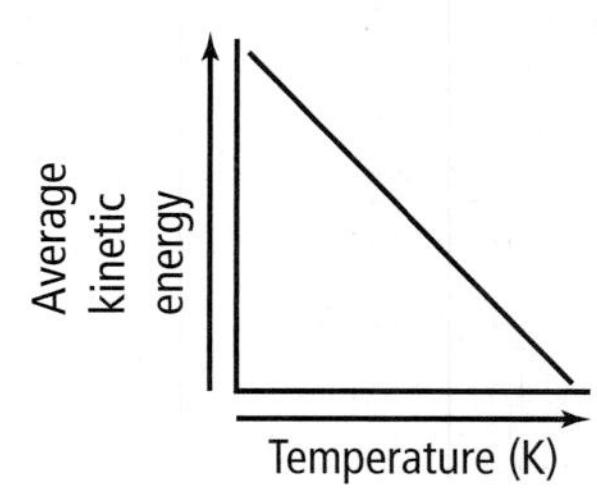

Online Test Practice

## SHORT ANSWER

*Use the description below to answer Questions 9 to 11.*

In an experimental setup, chlorine gas and nitrogen gas are in separate containers separated by a closed stopcock. One hour after the stopcock is opened, the gases have completely mixed.

**9.** Five minutes after the stopcock is opened, which gas will have traveled farther, the nitrogen or the chlorine?

**10.** Give the ratio of the speed of nitrogen gas to the speed of chlorine gas.

**11.** Evaluate this statement: After one hour, the gas particles stop moving because they have completely mixed.

## EXTENDED RESPONSE

*Use the table below to answer Question 12.*

| Standard Reduction Potentials at 25°C, 1 atm, and 1*M* Solution | |
|---|---|
| $Ag^+ + e^- \rightarrow Ag$ | 0.7996 |
| $Cr^{3+} + 3e^- \rightarrow Cr$ | −0.744 |

**12.** Based on the standard reduction potentials given above, if a silver electrode and a chromium electrode are connected in a voltaic cell, which electrode will undergo oxidation and which will undergo reduction? Explain how you can tell.

**13.** Use Le Châtelier's principle to explain why the instructions for a chemical experiment sometimes instruct the chemist to cool the reaction in an ice bath.

## SAT SUBJECT TEST: CHEMISTRY

**14.** The hydrogen sulfide produced as a by-product of petroleum refinement can be used to produce elemental sulfur: $2H_2S(g) + SO_2(g) \rightarrow 3S(l) + 2H_2O(g)$.

What is the equilibrium constant expression for this reaction?

**A.** $K_{eq} = \frac{[H_2O]}{[H_2S][SO_2]}$

**B.** $K_{eq} = \frac{[H_2S]^2[SO_2]}{[H_2O]^2}$

**C.** $K_{eq} = \frac{[H_2O]^2}{[H_2S]^2[SO_2]}$

**D.** $K_{eq} = \frac{[S]^3[H_2O]^2}{[H_2S]^2[SO_2]}$

**E.** $K_{eq} = \frac{[2H_2O]^2}{[2H_2S]^2[SO_2]}$

**15.** Which shows the correct graph of the activation energy needed for an endothermic reaction?

**A.**

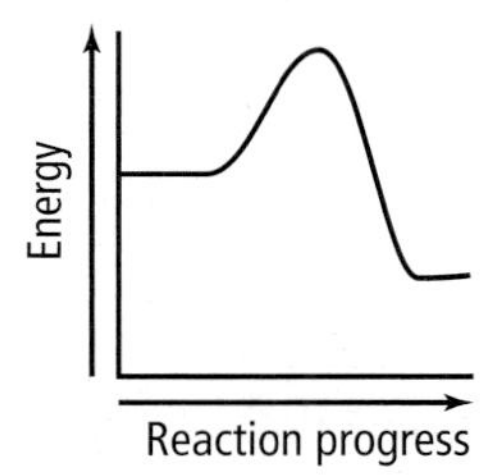


**B.**

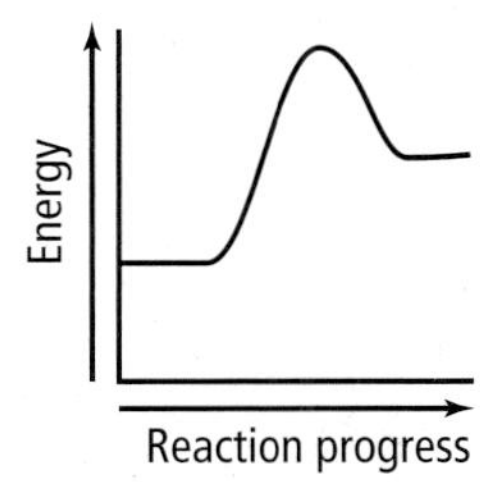


**C.**

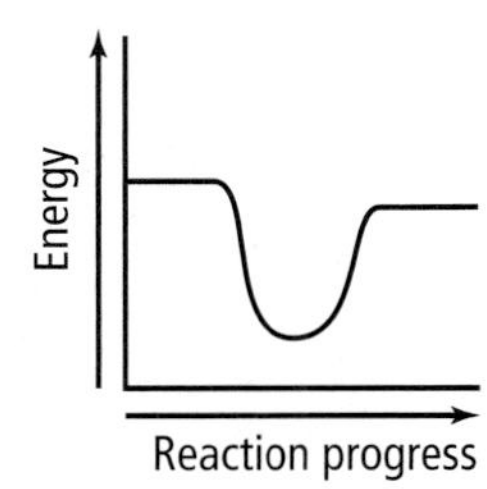


**D.**

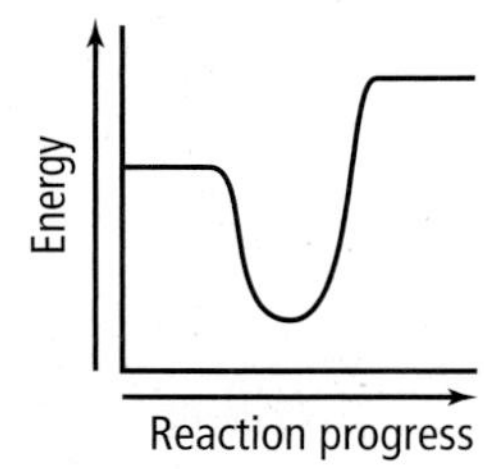


**E.**

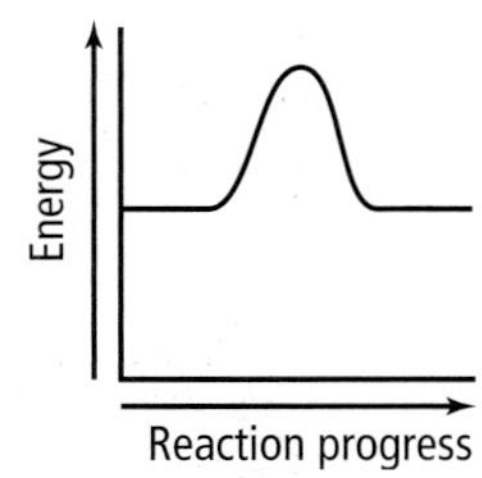


| NEED EXTRA HELP? | | | | | | | | | | | | | | | |
|---|---|---|---|---|---|---|---|---|---|---|---|---|---|---|---|
| If You Missed Question . . . | 1 | 2 | 3 | 4 | 5 | 6 | 7 | 8 | 9 | 10 | 11 | 12 | 13 | 14 | 15 |
| Review Section . . . | 20.1 | 20.1 | 20.1 | 20.1 | 20.2 | 18.1 | 18.3 | 15.1 | 12.1 | 12.1 | 12.1 | 8.5 | 17.2 | 17.3 | 15.2 |

CHAPTER 21

# Hydrocarbons

**BIGIDEA** Organic compounds called hydrocarbons differ by their types of bonds.

## SECTIONS

1. Introduction to Hydrocarbons
2. Alkanes
3. Alkenes and Alkynes
4. Hydrocarbon Isomers
5. Aromatic Hydrocarbons

## LaunchLAB

iLab Station

### How can you model simple hydrocarbons?

Hydrocarbons are made of hydrogen and carbon atoms. Recall that carbon has four valence electrons and it can form four covalent bonds. In this lab, you will build models of hydrocarbons with two, three, four, and five carbon atoms.

**FOLDABLES® Study Organizer**

### Hydrocarbon Compounds

Make a bound book. Label it as shown. Use it to help you organize information about hydrocarbon compounds.

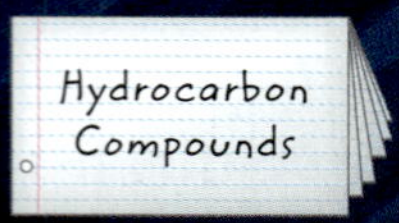


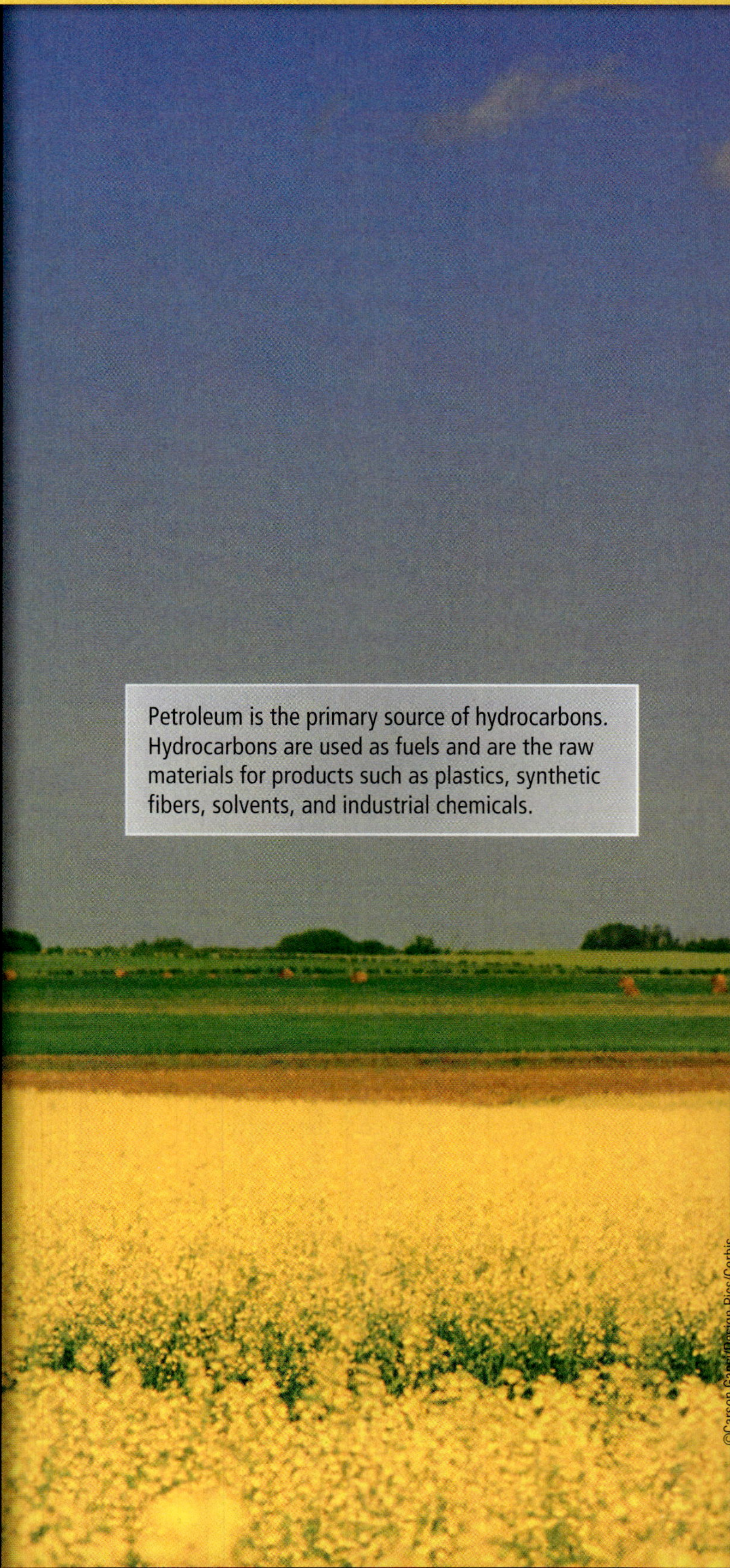

Petroleum is the primary source of hydrocarbons. Hydrocarbons are used as fuels and are the raw materials for products such as plastics, synthetic fibers, solvents, and industrial chemicals.

Go **online!**
MHEonline.com

## SECTION 1

# Introduction to Hydrocarbons

**MAINIDEA** **Hydrocarbons are carbon-containing organic compounds that provide a source of energy and raw materials.**

### Essential Questions

- What do the terms *organic compound* and *organic chemistry* mean?
- How are hydrocarbons and the models used to represent them identified?
- How are saturated and unsaturated hydrocarbons distinguished?
- Where are hydrocarbons obtained and how are they separated?

### Review Vocabulary

**microorganism:** a tiny organism, such as a bacterium or a protozoan, that cannot be seen without a microscope

### New Vocabulary

organic compound
hydrocarbon
saturated hydrocarbon
unsaturated hydrocarbon
fractional distillation
cracking

**CHEM 4 YOU** **If you have ridden in a car or a bus, you have used hydrocarbons. The gasoline and diesel fuel that are used in cars, trucks, and buses are hydrocarbons.**

## Organic Compounds

Chemists in the early nineteenth century knew that living things, such as the plants and panda shown in **Figure 1,** produce an immense variety of carbon compounds. Chemists referred to these compounds as *organic* compounds because they were produced by living organisms.

Once Dalton's atomic theory was accepted in the early nineteenth century, chemists began to understand that compounds, including those made by living organisms, consisted of arrangements of atoms bonded together in certain combinations. They were able to synthesize many new and useful substances. However, scientists were not able to synthesize organic compounds. Many scientists incorrectly concluded that they were unable to synthesize organic compounds because of vitalism. According to vitalism, organisms possessed a mysterious "vital force," enabling them to assemble carbon compounds.

**Disproving vitalism** Friedrich Wöhler (1800–1882), a German chemist, was the first scientist to realize that he had produced an organic compound, called urea, by synthesis in a laboratory. Wöhler's experiment did not immediately disprove vitalism, but it prompted a chain of similar experiments by other European chemists. Eventually, the idea that the synthesis of organic compounds required a vital force was discredited and scientists realized they could synthesize organic compounds in the laboratory.

**Figure 1** Living things contain, are made up of, and produce a variety of organic compounds.
**Identify** *two organic compounds that you have studied in a previous science course.*

**Organic chemistry** Today, the term **organic compound** is applied to all carbon-containing compounds with the primary exceptions of carbon oxides, carbides, and carbonates, which are considered inorganic. Because there are so many organic compounds, an entire branch of chemistry, called organic chemistry, is devoted to their study. Recall that carbon is an element in group 14 of the periodic table, as shown in **Figure 2.** With the electron configuration of $1s^22s^22p^2$, carbon nearly always shares its electrons and forms four covalent bonds. In organic compounds, carbon atoms are bonded to hydrogen atoms or atoms of other elements that are near carbon in the periodic table—especially nitrogen, oxygen, sulfur, phosphorus, and the halogens.

Most importantly, carbon atoms also bond to other carbon atoms and form chains from two to thousands of carbon atoms in length. Also, because carbon forms four bonds, it forms complex, branched-chain structures, ring structures, and even cagelike structures. With all of these bonding possibilities, chemists have identified millions of different organic compounds and are synthesizing more every day.

☑ READING CHECK **Explain** why carbon forms many compounds.

■ **Figure 2** Carbon is found in group 14 of the periodic table. It can bond to four other elements and form thousands of different compounds.

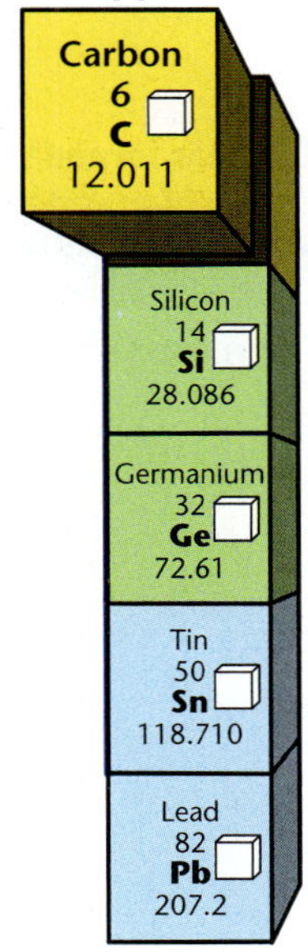


## Hydrocarbons

The simplest organic compounds are **hydrocarbons,** which contain only the elements carbon and hydrogen. How many different compounds do you think two elements can form? You might guess that only a few compounds are possible. However, thousands of hydrocarbons are known, each containing only the elements carbon and hydrogen. The simplest hydrocarbon molecule, $CH_4$, consists of a carbon atom bonded to four hydrogen atoms. This substance, called methane, is an excellent fuel and is the main component of natural gas, as shown in **Figure 3.**

☑ READING CHECK **Name** two uses of methane or natural gas in your home or community.

■ **Figure 3** Methane—a hydrocarbon found in natural gas—is the simplest hydrocarbon.

**Identify** ***In addition to hydrogen, what other elements readily bond with carbon?***

## Models of Methane

$CH_4$

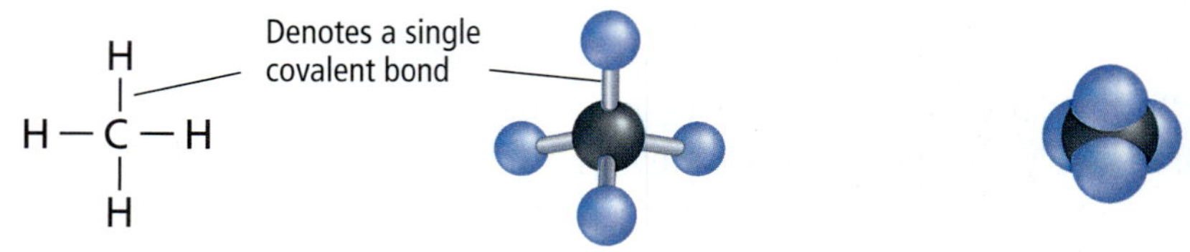


**Molecular formula** | **Structural formula** | **Ball-and-stick model** | **Space-filling model**

■ **Figure 4** Chemists use four different models to represent a methane ($CH_4$) molecule. See the Reference Tables in the Student Resources for a key to atom color conventions.

**Models and hydrocarbons** Chemists represent organic molecules in a variety of ways. **Figure 4** shows four different ways to represent a methane molecule. Covalent bonds are represented by a single straight line, which denotes two shared electrons. Most often, chemists use the type of model that best shows the information they want to highlight. As shown in **Figure 4,** molecular formulas give no information about the geometry of the molecule. A structural formula shows the general arrangement of atoms in the molecule but not the exact, three-dimensional geometry. The ball-and-stick model demonstrates the geometry of the molecule clearly, but the space-filling model gives a more realistic picture of what a molecule would look like if you could see it. Keep in mind as you look at the models that the atoms are held closely together by electron-sharing bonds.

■ **Figure 5** Carbon can bond to other carbon atoms in double and triple bonds. These Lewis structures and structural formulas show two ways to denote double and triple bonds.

One shared pair

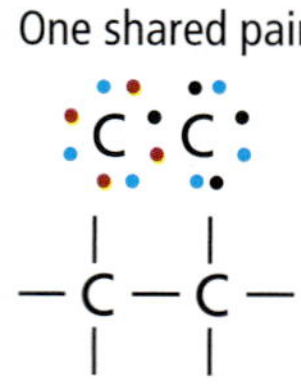

Single covalent bond

Two shared pairs

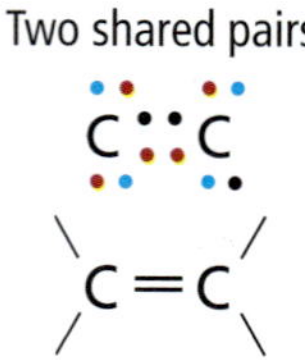

Double covalent bond

Three shared pairs

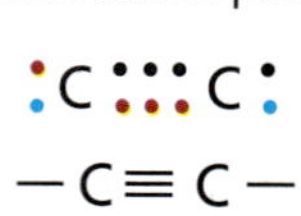

Triple covalent bond

• and • = carbon electrons
• = electron from another atom

**Multiple carbon-carbon bonds** Carbon atoms can bond to each other not only by single covalent bonds but also by double and triple covalent bonds, as shown in **Figure 5.** Recall that in a double bond, atoms share two pairs of electrons; in a triple bond, they share three pairs of electrons.

In the nineteenth century, before chemists understood bonding and the structure of organic substances, they experimented with hydrocarbons obtained from heating animal fats and plant oils. They classified these hydrocarbons according to a chemical test in which they mixed each hydrocarbon with bromine and then measured how much reacted with the hydrocarbon. Some hydrocarbons would react with a small amount of bromine, some would react with more, and some would not react with any amount of bromine. Chemists called the hydrocarbons that reacted with bromine unsaturated hydrocarbons in the same sense that an unsaturated aqueous solution can dissolve more solute. Hydrocarbons that did not react with bromine were said to be saturated.

Present-day chemists can now explain the experimental results obtained 170 years ago. Hydrocarbons that reacted with bromine had double or triple covalent bonds. Those compounds that did not react with bromine had only single covalent bonds. Today, a hydrocarbon having only single bonds is defined as a **saturated hydrocarbon.** A hydrocarbon that has at least one double or triple bond between carbon atoms is an **unsaturated hydrocarbon.** You will learn more about these different types of hydrocarbons later in this chapter.

☑ READING CHECK **Explain** the origin of the terms *saturated* and *unsaturated hydrocarbons.*

## Refining Hydrocarbons

Today, many hydrocarbons are obtained from a fossil fuel called petroleum. Petroleum formed from the remains of microorganisms that lived in Earth's oceans millions of years ago. Over time, the remains formed thick layers of mudlike deposits on the ocean floor. Heat from Earth's interior and the tremendous pressure of overlying sediments transformed this mud into oil-rich shale and natural gas. In certain kinds of geological formations, the petroleum ran out of the shale and collected in pools deep in Earth's crust. Natural gas, which formed at the same time and in the same way as petroleum, is usually found with petroleum deposits. Natural gas is composed primarily of methane, but it also contains small amounts of other hydrocarbons that have from two to five carbon atoms.

**Fractional distillation** Unlike natural gas, petroleum is a complex mixture containing more than a thousand different compounds. For this reason, raw petroleum, sometimes called crude oil, has little practical use. Petroleum is much more useful to humans when it is separated into simpler components or fractions. Separation is carried out in a process called **fractional distillation,** also called fractionation, which involves boiling the petroleum and collecting components or fractions as they condense at different temperatures. Fractional distillation is done in a fractionating tower similar to the one shown in **Figure 6.**

The temperature inside the fractionating tower is controlled so that it remains near 400°C at the bottom, where the petroleum is boiling, and gradually decreases toward the top. The condensation temperatures (boiling points) generally decrease as molecular mass decreases. Therefore, as the vapors travel up through the column, the hydrocarbons condense and are drawn off, as shown in **Figure 6.**

**VOCABULARY**

**SCIENCE USAGE V. COMMON USAGE**

**Deposit**

***Science usage:*** a natural collection of oil or ore

*There was a rich deposit of copper in the mountain.*

***Common usage:*** money placed in a bank account or the act of placing money in a bank account

*The store owner placed his deposit in the after-hours slot at the bank.*

**Figure 6** This diagram of a fractionating tower shows that fractions with lower boiling points, such as gasoline and gaseous products, are drawn off in the cooler regions near the top of the tower. Oils and greases, having much higher boiling points, stay near the bottom of the tower and are drawn off there.

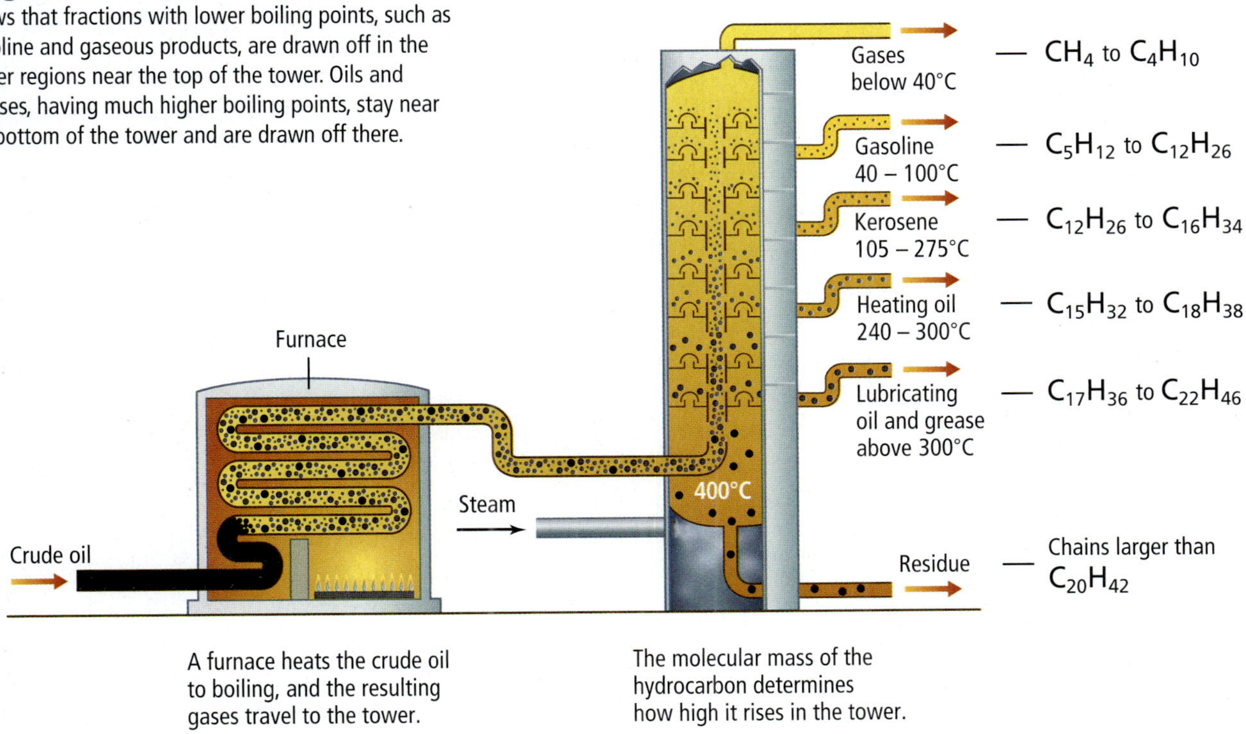

■ **Figure 7** Fractional distillation towers separate large quantities of petroleum into usable components. Thousands of products we use in our homes, for transportation, and in industry result from petroleum refining.

**Infer** ***What types of emissions must be controlled by refineries to protect the environment?***

Keith Dannemiller/Alamy

**Figure 6** also gives the names of the typical fractions separated from petroleum, along with their boiling points, hydrocarbon size ranges, and common uses. You might recognize some of the fractions because you use them every day. Unfortunately, fractional distillation towers, shown in **Figure 7,** do not yield fractions in the same proportions that they are needed. For example, distillation seldom yields the amount of gasoline desired. However, it yields more of the heavier oils than the market demands.

Many years ago, petroleum chemists and engineers developed a process to help match the supply with the demand. This process in which heavier fractions are converted to gasoline by breaking their large molecules into smaller molecules is called **cracking.** Cracking is done in the absence of oxygen and in the presence of a catalyst. In addition to breaking heavier hydrocarbons into molecules of the size range needed for gasoline, cracking also produces starting materials for the synthesis of many different products, including plastic products, films, and synthetic fabrics.

READING CHECK **Describe** the process in which large-chain hydrocarbons are broken into more-desirable smaller-chain hydrocarbons.

**Rating gasoline** None of the petroleum fractions is a pure substance. As shown in **Figure 6,** gasoline is not a pure substance, but rather a mixture of hydrocarbons. Most molecules with single covalent bonds in gasoline have 5 to 12 carbon atoms. However, the gasoline pumped into cars today is different from the gasoline used in automobiles in the early 1900s. The gasoline fraction that is distilled from petroleum is modified by adjusting its composition and adding substances to improve its performance in today's automobile engines and to reduce pollution from car exhaust.

It is critical that the gasoline-air mixture in the cylinder of an automobile engine ignite at exactly the right instant and burn evenly. If it ignites too early or too late, much energy will be wasted, fuel efficiency will drop, and the engine will wear out prematurely. Most straight-chain hydrocarbons burn unevenly and tend to ignite from heat and pressure before the piston is in the proper position and the spark plug fires. This early ignition causes a rattling or pinging noise called knocking.

CAREERS IN CHEMISTRY

**Petroleum Technician** This science technician uses instruments to measure and record physical and geological information about oil or gas wells. For example, a petroleum technician might test a geological sample to determine its petroleum content and its mineral or element composition.

WebQuest

■ **Figure 8** Octane ratings are used to give the antiknock rating of fuel. Mid-grade gasoline for cars has an octane rating of about 89. Aviation fuel has an octane rating of about 100. Racing fuel has an octane rating of about 110.

In the late 1920s, an antiknock, or octane rating, system for gasoline was established, resulting in the octane ratings posted on gasoline pumps like those shown in **Figure 8.** Mid-grade gasoline today has a rating of about 89, whereas premium gasoline has ratings of 91 or higher. Several factors determine which octane rating a car needs, including how much the piston compresses the air-fuel mixture and the altitude at which the car is driven.

**Connection to Earth Science** Since ancient times, people have found petroleum seeping from cracks in rocks. Historical records show that petroleum has been used for more than 5000 years. In the nineteenth century, as the United States entered the machine age and its population increased, the demand for petroleum products, namely kerosene for lighting and lubricants for machines, increased. In an attempt to find a reliable petroleum supply, Edwin Drake drilled the first oil well in the United States in Pennsylvania, in 1859. The oil industry flourished for a time, but when Thomas Edison introduced the electric light in 1882, investors feared that the industry was doomed. However, the invention of the automobile in the 1890s revived the industry on a massive scale.

## SECTION 1 REVIEW

Section Self-Check

### Section Summary

- Organic compounds contain carbon, which is able to form straight chains and branched chains.
- Hydrocarbons are organic substances composed of carbon and hydrogen.
- The major sources of hydrocarbons are petroleum and natural gas.
- Petroleum can be separated into components by the process of fractional distillation.

1. MAINIDEA **Identify** three applications of hydrocarbons as a source of energy and raw materials.
2. **Name** an organic compound and explain what an organic chemist studies.
3. **Identify** what each of the four molecular models highlights about a molecule.
4. **Compare and contrast** saturated and unsaturated hydrocarbons.
5. **Describe** the process of fractional distillation.
6. **Infer** Some shortening products are described as "hydrogenated vegetable oil," which are oils that reacted with hydrogen in the presence of a catalyst. Form a hypothesis to explain why hydrogen reacted with the oils.
7. **Interpret Data** Refer to **Figure 6.** What property of hydrocarbon molecules correlates to the viscosity of a particular fraction when it is cooled to room temperature?

# SECTION 2 Alkanes

**MAINIDEA** Alkanes are hydrocarbons that contain only single bonds.

### Essential Questions

- How are alkanes named when given their structures?
- How are alkane structures drawn when given their names?
- What are the properties of alkanes?

### Review Vocabulary

**IUPAC (International Union of Pure and Applied Chemistry):** an international group that aids communication between chemists by setting rules and standards in areas such as chemical nomenclature, terminology, and standardized methods

### New Vocabulary

alkane
homologous series
parent chain
substituent group
cyclic hydrocarbon
cycloalkane

**CHEM 4 YOU** Have you ever used a Bunsen burner or an outdoor gas grill? If so, you have used an alkane. Natural gas and propane are the two most common gases used in these applications, and both are alkanes.

## Straight-Chain Alkanes

Methane is the smallest member of a series of hydrocarbons known as alkanes. It is used as a fuel in homes and science labs and is a product of many biological processes. **Alkanes** are hydrocarbons that have only single bonds between atoms. Look in Section 1 to review the various models of methane. The models for ethane ($C_2H_6$), the second member of the alkane series, are shown in **Table 1.** Ethane consists of two carbon atoms bonded together with a single bond and six hydrogen atoms sharing the remaining valence electrons of the carbon atoms.

The third member of the alkane series, propane, has three carbon atoms and eight hydrogen atoms, giving it the molecular formula $C_3H_8$. The next member, butane, has four carbon atoms and the formula $C_4H_{10}$. Compare the structures of ethane, propane, and butane in **Table 1.**

Propane, also known as LP (liquefied propane) gas, is sold as a fuel for cooking and heating. Butane is used as fuel in small lighters and in some torches. It is also used in the manufacture of synthetic rubber.

**Table 1** Simple Alkanes

| Molecular Formula | Structural Formula | Ball-and-Stick Model | Space-Filling Model |
|---|---|---|---|
| Ethane ($C_2H_6$) | 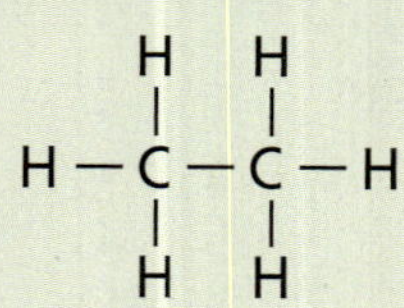 | 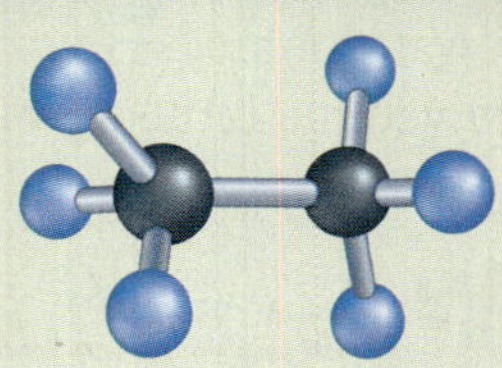 |  |
| Propane ($C_3H_8$) | 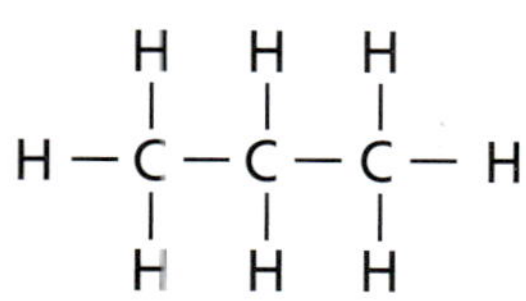 | 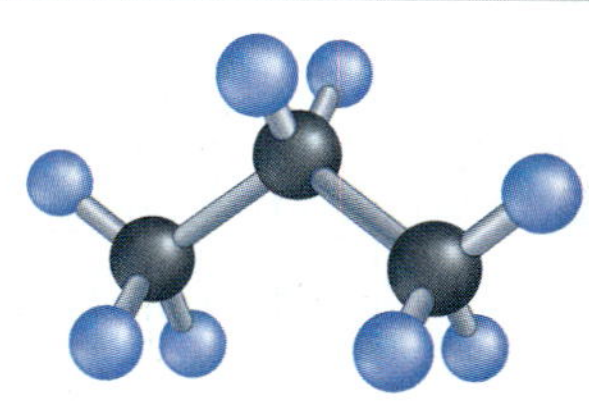 | 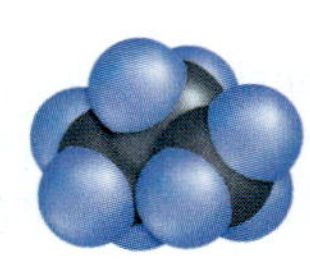 |
| Butane ($C_4H_{10}$) | 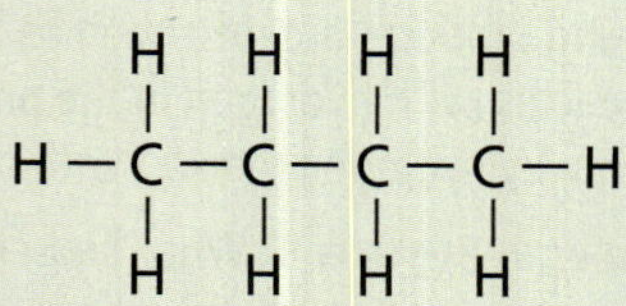 | 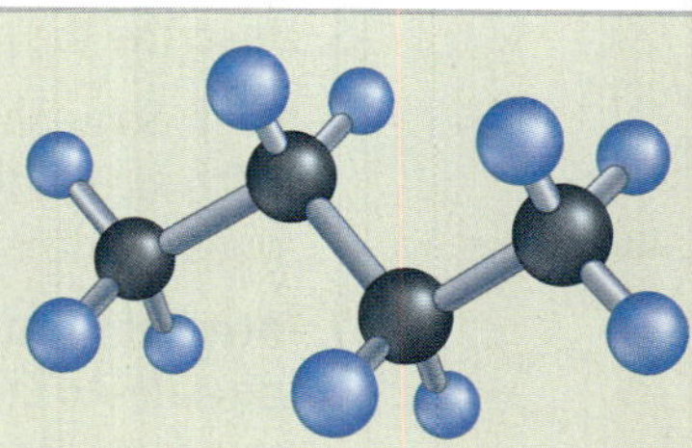 |  |

**Table 2 First Ten of the Alkane Series**

| Name | Molecular Formula | Condensed Structural Formula |
|---|---|---|
| Methane | $CH_4$ | $CH_4$ |
| Ethane | $C_2H_6$ | $CH_3CH_3$ |
| Propane | $C_3H_8$ | $CH_3CH_2CH_3$ |
| Butane | $C_4H_{10}$ | $CH_3CH_2CH_2CH_3$ |
| Pentane | $C_5H_{12}$ | $CH_3CH_2CH_2CH_2CH_3$ |
| Hexane | $C_6H_{14}$ | $CH_3CH_2CH_2CH_2CH_2CH_3$ |
| Heptane | $C_7H_{16}$ | $CH_3CH_2CH_2CH_2CH_2CH_2CH_3$ |
| Octane | $C_8H_{18}$ | $CH_3(CH_2)_6CH_3$ |
| Nonane | $C_9H_{20}$ | $CH_3(CH_2)_7CH_3$ |
| Decane | $C_{10}H_{22}$ | $CH_3(CH_2)_8CH_3$ |

**Naming straight-chain alkanes** By now, you have likely noticed that names of alkanes end in *-ane.* Also, alkanes with five or more carbons in a chain have names that use a prefix derived from the Greek or Latin word for the number of carbons in each chain. For example, *pent*ane has five carbons just as a *pent*agon has five sides, and *oct*ane has eight carbons just as an *oct*opus has eight tentacles. Because methane, ethane, propane, and butane were named before alkane structures were known, their names do not have numerical prefixes. **Table 2** shows the names and structures of the first ten alkanes. Notice the underlined prefix representing the number of carbon atoms in the molecule.

In **Table 2,** you can see that the structural formulas are written in a different way from those in **Table 1.** These formulas, called condensed structural formulas, save space by not showing how the hydrogen atoms branch off from the carbon atoms. Condensed formulas can be written in several ways. In **Table 2,** the lines between carbon atoms have been eliminated to save space.

In **Table 2,** you can see that $–CH_2–$ is a repeating unit in the chain of carbon atoms. Note, for example, that pentane has one more $–CH_2–$ unit than butane. You can further condense structural formulas by writing the $–CH_2–$ unit in parentheses followed by a subscript to show the number of units, as is done with octane, nonane, and decane.

A series of compounds that differ from one another by a repeating unit is called a **homologous series.** A homologous series has a fixed numerical relationship among the numbers of atoms. For alkanes, the relationship between the numbers of carbon and hydrogen atoms can be expressed as $C_nH_{2n+2}$, where $n$ is equal to the number of carbon atoms in the alkane. Given the number of carbon atoms in an alkane, you can write the molecular formula for any alkane. For example, heptane has seven carbon atoms, so its formula is $C_7H_{2(7)+2}$, or $C_7H_{16}$.

**VOCABULARY**
**WORD ORIGIN**
**Homologous**
comes from the Greek word *homologos* meaning *agreeing*

**READING CHECK** **Write** the molecular formula for an alkane that has 13 carbon atoms in its molecular structure.

## Branched-Chain Alkanes

The alkanes discussed so far in this chapter are called straight-chain alkanes because the carbon atoms are bonded to each other in a single line. Now look at the two structures in **Figure 9.** If you count the carbon and hydrogen atoms, you will discover that both structures have the same molecular formula, $C_4H_{10}$. Do the structures in **Figure 9** represent the same substance?

If you think that the structures represent two different substances, you are correct. The structure on the left represents butane, and the structure on the right represents a branched-chain alkane known as isobutane—a substance whose chemical and physical properties are different from those of butane. Carbon atoms can bond to one, two, three, or even four other carbon atoms. This property makes possible a variety of branched-chain alkanes.

Recall that butane is used in lighters and in torches. Isobutane is used as both an environmentally-safe refrigerant and a propellant in products such as shaving gel, as shown in **Figure 9.** In addition to these applications, both butane and isobutane are used as raw materials for many chemical processes.

☑ READING CHECK **Describe** the difference in the molecular structures of butane and isobutane.

**Alkyl groups** You have seen that both a straight-chain and a branched-chain alkane can have the same molecular formula. This fact illustrates a basic principle of organic chemistry: the order and arrangement of atoms in an organic molecule determine its identity. Therefore, the name of an organic compound must also accurately describe the molecular structure of the compound.

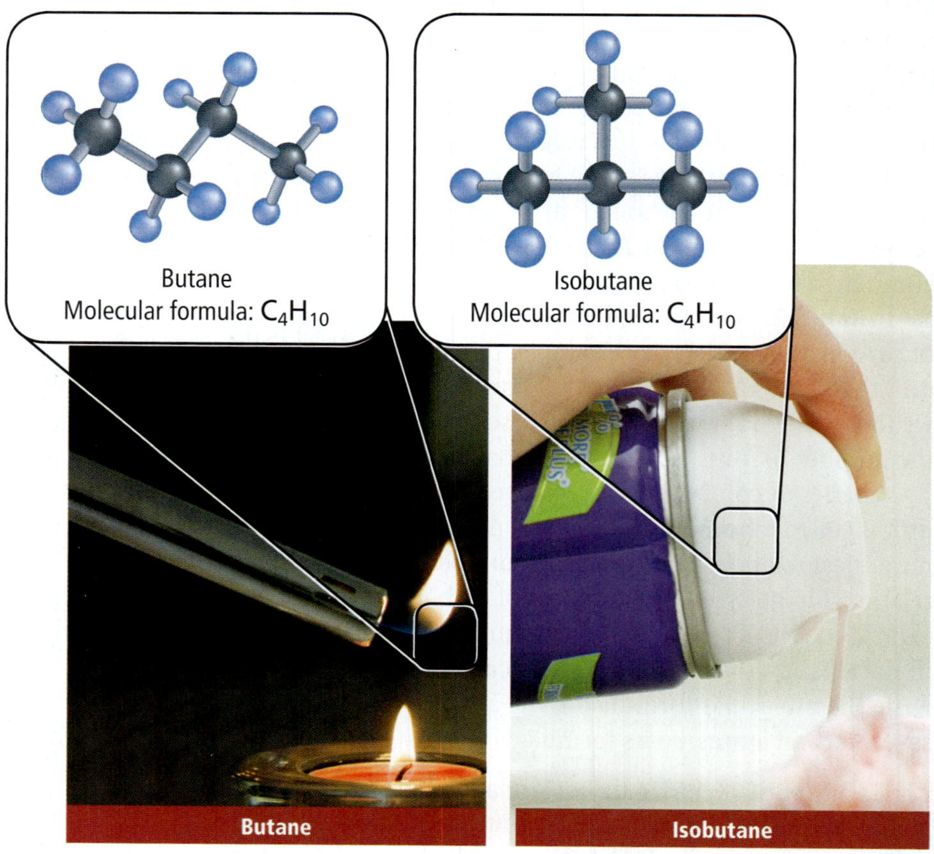


■ **Figure 9** Butane is a fuel used in lighters. Isobutane is used as a propellant in products such as shaving gel.

**Table 3** Common Alkyl Groups

| Name | Methyl | Ethyl | Propyl | Isopropyl | Butyl |
|---|---|---|---|---|---|
| Condensed structural formula | $CH_3$— | $CH_3CH_2$— | $CH_3CH_2CH_2$— | $CH_3\underset{\vert}{C}HCH_3$ | $CH_3CH_2CH_2CH_2$— |
| Structural formula | 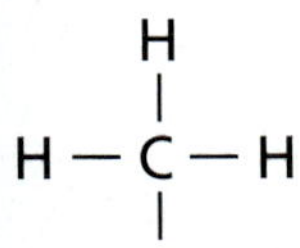 | 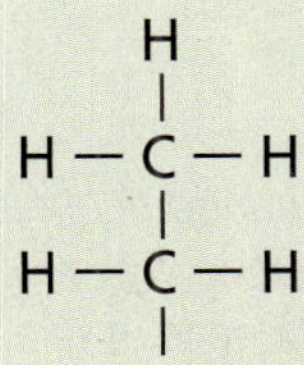 | 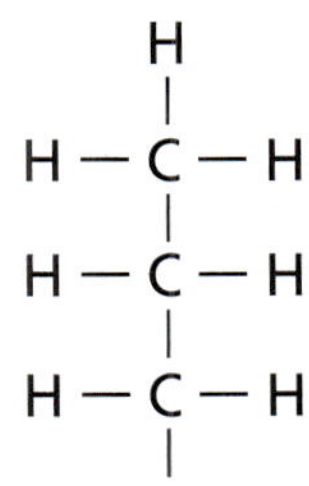 | 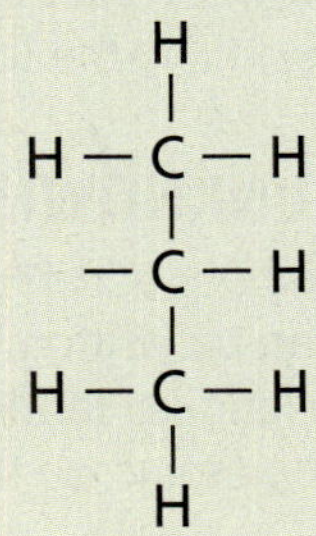 | 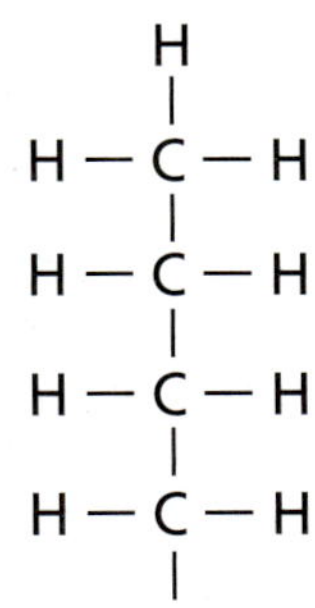 |

When naming branched-chain alkanes, the longest continuous chain of carbon atoms is called the **parent chain.** All side branches are called **substituent groups** because they appear to substitute for a hydrogen atom in the straight chain. Each alkane-based substituent group branching from the parent chain is named for the straight-chain alkane that has the same number of carbon atoms as the substituent. The ending *-ane* is replaced with the letters *-yl.* An alkane-based substituent group is called an alkyl group. Several alkyl groups are shown in **Table 3.**

**VOCABULARY**

**ACADEMIC VOCABULARY**

**Substitute**
a person or thing that takes the place of another
*A substitute teacher taught chemistry class yesterday.*

**Naming branched-chain alkanes** To name organic structures, chemists use the following systematic rules approved by the International Union of Pure and Applied Chemistry (IUPAC).

**Step 1.** *Count the number of carbon atoms in the longest continuous chain.* Use the name of the straight-chain alkane with that number of carbons as the name of the parent chain of the structure.

**Step 2.** *Number each carbon in the parent chain.* Locate the end carbon closest to a substituent group. Label that carbon *Position 1.* This step gives all the substituent groups the lowest position numbers possible.

**Step 3.** *Name each alkyl group substituent.* Place the name of the group before the name of the parent chain.

**Step 4.** *If the same alkyl group occurs more than once as a branch on the parent structure, use a prefix* (di-, tri-, tetra-, *and so on*) *before its name to indicate how many times it appears.* Then, use the number of the carbon to which each is attached to indicate its position.

**Step 5.** *When different alkyl groups are attached to the same parent structure, place their names in alphabetical order.* Do not consider the prefixes (*di-*, *tri-*, and so on) when determining alphabetical order.

**Step 6.** *Write the entire name, using hyphens to separate numbers from words and commas to separate numbers.* Do not add a space between the substituent name and the name of the parent chain.

## EXAMPLE Problem 1

EXAMPLE PROBLEM

**NAMING BRANCHED-CHAIN ALKANES**
Name the alkane shown.

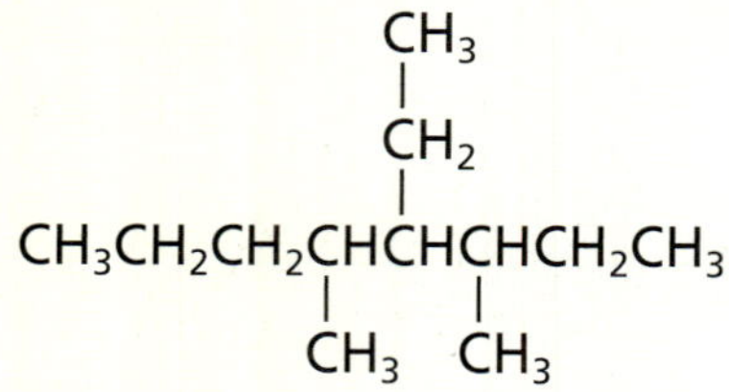

### 1 ANALYZE THE PROBLEM

You are given a structure. To determine the name of the parent chain and the names and locations of branches, follow the IUPAC rules.

### 2 SOLVE FOR THE UNKNOWN

**Step 1.** *Count the number of carbon atoms in the longest continuous chain.* Because structural formulas can be written with chains oriented in various ways, you need to be careful in finding the longest continuous carbon chain. In this case, it is easy. The longest chain has eight carbon atoms, so the parent name is *octane.*

**Step 2.** *Number each carbon in the parent chain.* Number the chain in both directions, as shown below. Numbering from the left puts the alkyl groups at Positions 4, 5, and 6. Numbering from the right puts alkyl groups at Positions 3, 4, and 5. Because 3, 4, and 5 are the lowest position numbers, they will be used in the name.

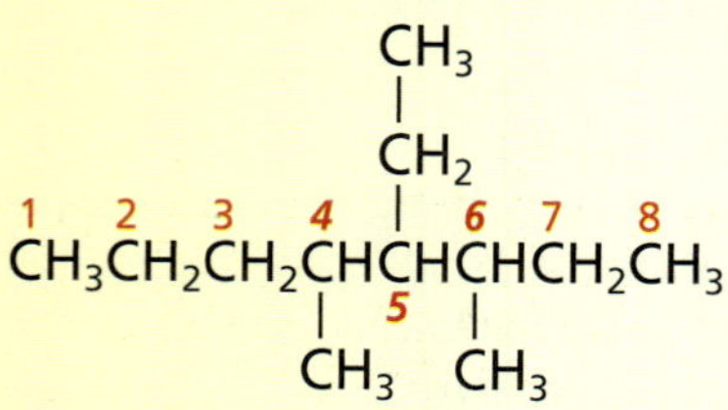

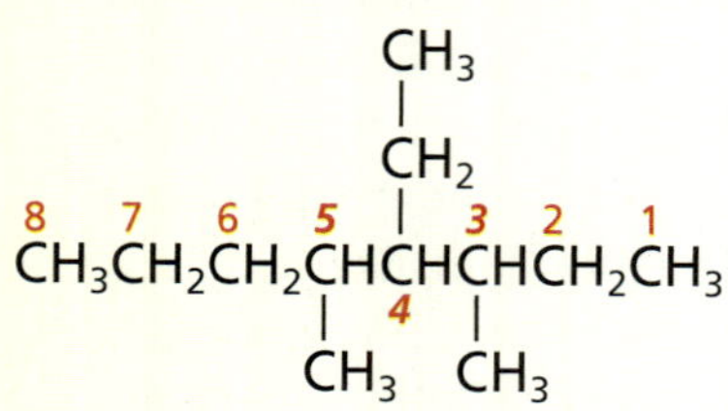

**Step 3.** *Name each alkyl group substituent.* Identify and name the alkyl groups branching from the parent chain. There are one-carbon methyl groups at Positions 3 and 5, and a two-carbon ethyl group at Position 4.

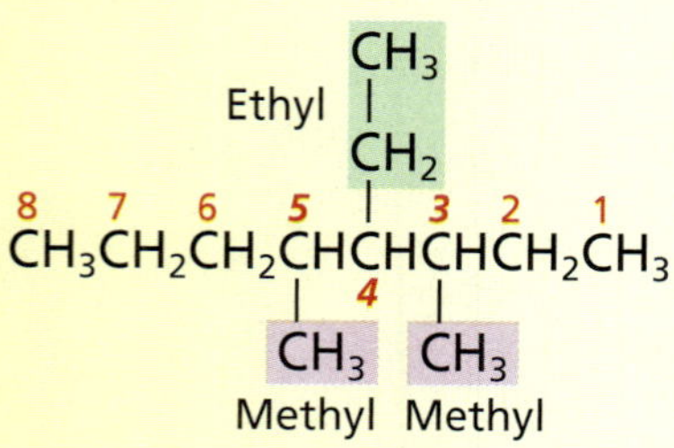


**Step 4.** *If the same alkyl group occurs more than once as a branch on the parent structure, use a prefix* (di-, tri-, tetra-, *and so on) before its name to indicate how many times it appears.* Look for and count the alkyl groups that occur more than once. Determine the prefix to use to show the number of times each group appears. In this example, the prefix *di-* will be added to the name *methyl* because two methyl groups are present. No prefix is needed for the one ethyl group. Then show the position of each group with the appropriate number.

One ethyl group: *no prefix*
Position and name: *4-ethyl*

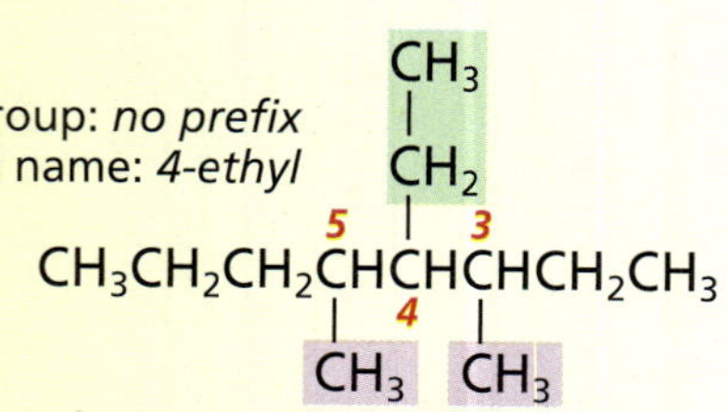

Parent chain: *octane*

Two methyl groups: use *dimethyl*
Position and name: *3,5-dimethyl*

Get help **naming hydrocarbons.**

Personal Tutor

**Step 5.** *Whenever different alkyl groups are attached to the same parent structure, place their names in alphabetical order.* Place the names of the alkyl branches in alphabetical order, ignoring the prefixes. Alphabetical order puts the name ethyl before dimethyl.

**Step 6.** *Write the entire name, using hyphens to separate numbers from words and commas to separate numbers.* Write the name of the structure, using hyphens and commas as needed. The name should be written as *4-ethyl-3,5-dimethyloctane.*

### 3 EVALUATE THE ANSWER

The longest continuous carbon chain has been found and numbered correctly. All branches have been designated with correct prefixes and alkyl group names. Alphabetical order and punctuation are correct.

## PRACTICE Problems

Do additional problems. Online Practice

**8.** Use the IUPAC rules to name the following structures.

**a.** 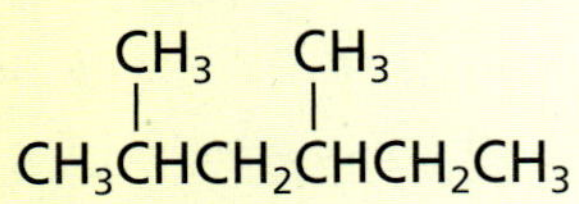

**b.** 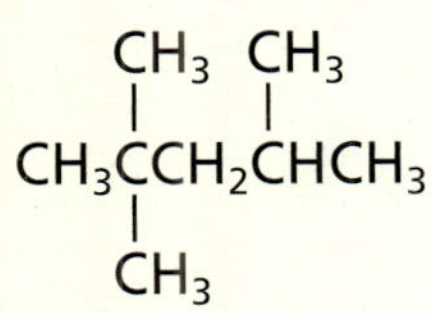

**c.** 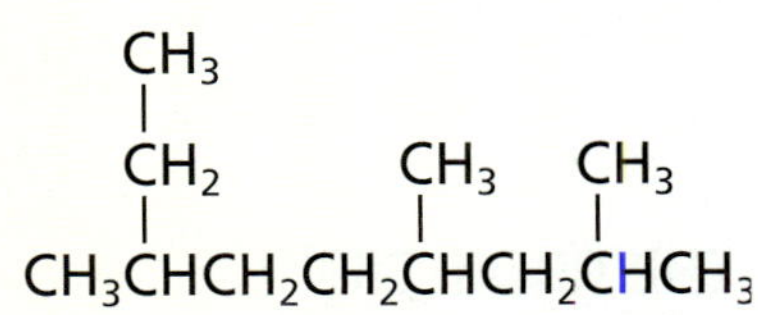

**9. Challenge** Draw the structures of the following branched-chain alkanes.

**a.** 2,3-dimethyl-5-propyldecane

**b.** 3,4,5-triethyloctane

## Cycloalkanes

One of the reasons that such a variety of organic compounds exists is that carbon atoms can form ring structures. An organic compound that contains a hydrocarbon ring is called a **cyclic hydrocarbon.** To indicate that a hydrocarbon has a ring structure, the prefix *cyclo-* is used with the hydrocarbon name. Thus, cyclic hydrocarbons that contain only single bonds are called **cycloalkanes.**

Cycloalkanes can have rings with three, four, five, six, or even more carbon atoms. The name for the six-carbon cycloalkane is *cyclohexane.* Cyclohexane, which is obtained from petroleum, is used in paint and varnish removers and for extracting essential oils to make perfume. Note that cyclohexane ($C_6H_{12}$) has two fewer hydrogen atoms than straight-chain hexane ($C_6H_{14}$) because a valence electron from each of two carbon atoms is now forming a carbon-carbon bond rather than a carbon-hydrogen bond.

**READING CHECK Evaluate** If the prefix *cyclo-* is present in the name of an alkane, what do you know about the alkane?

As shown in **Figure 10,** cyclic hydrocarbons such as cyclohexane are represented by condensed, skeletal, and line structures. Line structures show only the carbon-carbon bonds with carbon atoms understood to be at each vertex of the structure. Hydrogen atoms are assumed to occupy the remaining bonding positions unless substituents are present. Hydrogens are also not shown in skeletal structures.

**Figure 10** Cyclohexane can be represented in several ways.

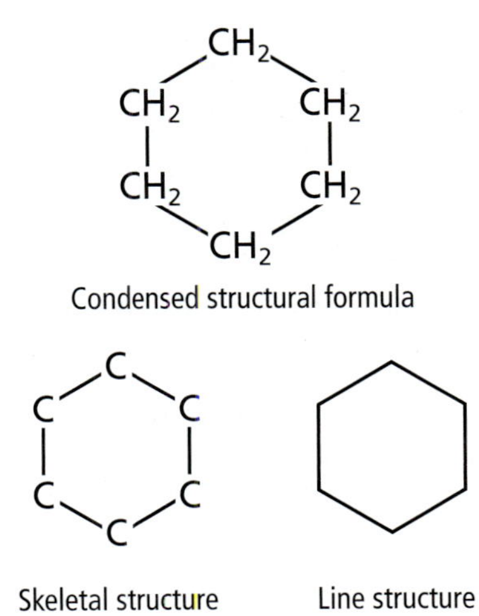

**Naming substituted cycloalkanes** Like other alkanes, cycloalkanes can have substituent groups. Substituted cycloalkanes are named by following the same IUPAC rules used for straight-chain alkanes, but with a few modifications. With cycloalkanes, there is no need to find the longest chain because the ring is always considered to be the parent chain. Because a cyclic structure has no ends, numbering is started on the carbon that is bonded to the substituent group. When there are two or more substituents, the carbons are numbered around the ring in a way that gives the lowest-possible set of numbers for the substituents. If only one group is attached to the ring, no number is necessary. The following Example Problem illustrates the naming process for cycloalkanes.

## EXAMPLE Problem 2

EXAMPLE PROBLEM

**NAMING CYCLOALKANES**
Name the cycloalkane shown.

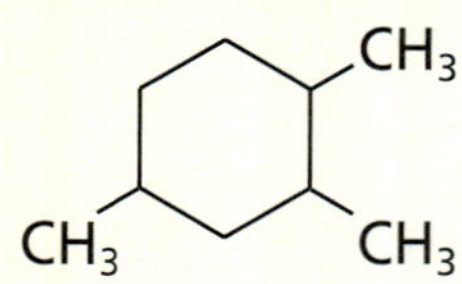


### 1 ANALYZE THE PROBLEM

You are given a structure. To determine the parent cyclic structure and the location of branches, follow the IUPAC rules.

### 2 SOLVE FOR THE UNKNOWN

**Step 1.** Count the carbons in the ring, and use the name of the parent cyclic hydrocarbon. In this case, the ring has six carbons, so the parent name is *cyclohexane.*

**Step 2.** Number the ring, starting from one of the $CH_3$– branches. Find the numbering that gives the lowest possible set of numbers for the branches. Here are two ways of numbering the ring.

A

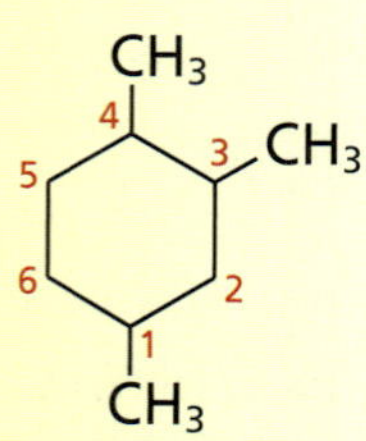


1,3,4

B

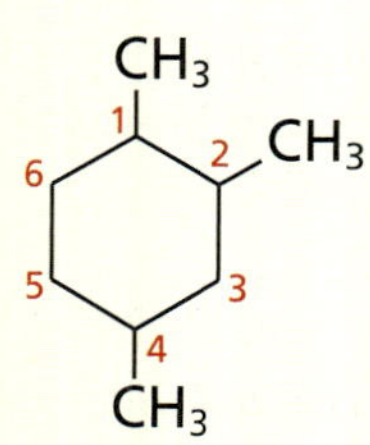


1,2,4

Numbering from the carbon atom at the bottom of the ring puts the $CH_3$– groups at Positions 1, 3, and 4 in Structure A. Numbering from the carbon at the top of the ring gives Positions 1, 2, and 4. All other numbering schemes place the $CH_3$– groups at higher position numbers. Thus, 1, 2, and 4 are the lowest possible position numbers and will be used in the name.

**Step 3.** Name the substituents. All three are the same–carbon methyl groups.

**Step 4.** Add the prefix to show the number of groups present. Three methyl groups are present, so you add the prefix *tri-* to the name *methyl* to make *trimethyl.*

**Step 5.** Alphabetical order can be ignored because only one type of group is present.

**Step 6.** Put the name together using the name of the parent cycloalkane. Use commas between separate numbers, and hyphens between numbers and words. Write the name as *1,2,4-trimethylcyclohexane.*

## 3 EVALUATE THE ANSWER

The parent-ring structure is numbered to give the branches the lowest possible set of numbers. The prefix *tri-* indicates that three methyl groups are present. No alphabetization is necessary because all branches are methyl groups.

## PRACTICE Problems

Do additional problems. Online Practice

**10.** Use IUPAC rules to name the following structures.

**a.**

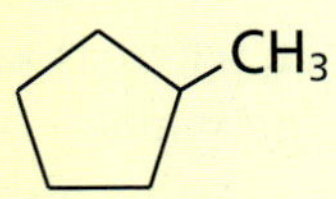


**b.**

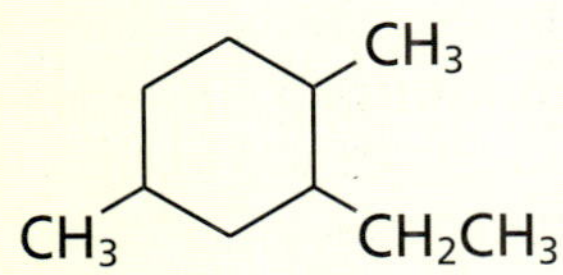


**c.**

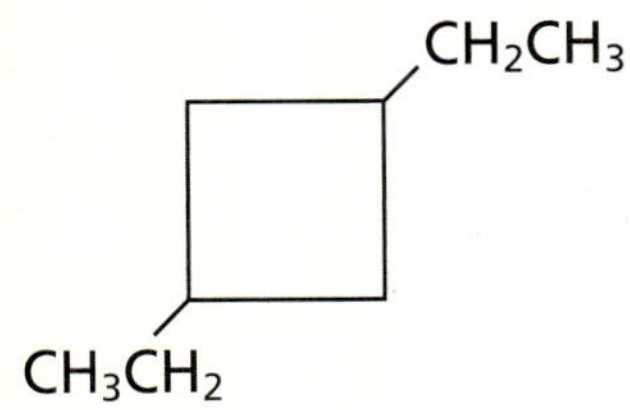


**11. Challenge** Draw the structures of the following cycloalkanes.

**a.** 1-ethyl-3-propylcyclopentane

**b.** 1,2,2,4-tetramethylcyclohexane



## Properties of Alkanes

You have learned that the structure of a molecule affects its properties. For example, the O–H bonds in a water molecule are polar, and because the H–O–H molecule has a bent geometry, the molecule itself is polar. Thus, water molecules can form hydrogen bonds with each other. As a result, the boiling and melting points of water are much higher than those of other substances having similar molecular mass and size.

What properties would you predict for alkanes? All of the bonds in alkanes are between either a carbon atom and a hydrogen atom or between two carbon atoms. A bond between two identical atoms, such as carbon, can never be polar. Also, the C–H bonds have only a small electronegativity difference and are nonpolar. Because all of the bonds in alkanes are nonpolar, alkane molecules are nonpolar, which makes them good solvents for other nonpolar substances, as shown in **Figure 11.**

**Figure 11** Many solvents—used as thinners for paints, coatings, waxes, photocopier toners, adhesives, and printer press inks—contain alkanes and cycloalkanes.

Robin Nelson/PhotoEdit

| **Table 4** Comparing Physical Properties | | |
|---|---|---|
| **Substance and formula** | **Water ($H_2O$)** | **Methane ($CH_4$)** |
| **Molecular mass** | 18 amu | 16 amu |
| **State at room temperature** | liquid | gas |
| **Boiling point** | 100°C | −162°C |
| **Melting point** | 0°C | −182°C |

**FOLDABLES®** Incorporate information from this section into your Foldable.

**Physical properties of alkanes** How do the properties of a polar and nonpolar compound compare? Refer to **Table 4,** and note that the molecular mass of methane (16 amu) is close to the molecular mass of water (18 amu). Also, water and methane molecules are similar in size. However, when you compare the melting and boiling points of methane to those of water, you can see evidence that the molecules differ in some significant way. These temperatures differ greatly because methane molecules have little intermolecular attraction compared to water molecules. This difference in attraction can be explained by the fact that methane molecules are nonpolar and do not form hydrogen bonds with each other, whereas water molecules are polar and freely form hydrogen bonds.

The difference in polarity and hydrogen bonding also explains the immiscibility of alkanes and other hydrocarbons with water. If you try to dissolve alkanes, such as lubricating oils, in water, the two liquids separate almost immediately into two phases. This separation happens because the attractive forces between alkane molecules are stronger than the attractive forces between the alkane and water molecules. Therefore, alkanes are more soluble in solvents composed of nonpolar molecules like themselves than in water, a polar solvent.

**Chemical properties of alkanes** The main chemical property of alkanes is their low reactivity. Recall that many chemical reactions occur when a reactant with a full electric charge, such as an ion, or with a partial charge, such as a polar molecule, is attracted to another reactant with the opposite charge. Molecules such as alkanes, in which atoms are connected by nonpolar bonds, have no charge. As a result, they have little attraction for ions or polar molecules. The low reactivity of alkanes can also be attributed to the relatively strong C–C and C–H bonds.

## SECTION 2 REVIEW

Section Self-Check

### Section Summary

- Alkanes contain only single bonds between carbon atoms.
- Alkanes and other organic compounds are best represented by structural formulas and can be named using systematic rules determined by the International Union of Pure and Applied Chemistry (IUPAC).
- Alkanes that contain hydrocarbon rings are called cyclic alkanes.

**12. MAIN IDEA Describe** the main structural characteristics of alkane molecules.

**13. Name** the following structures using IUPAC rules.

**a.**

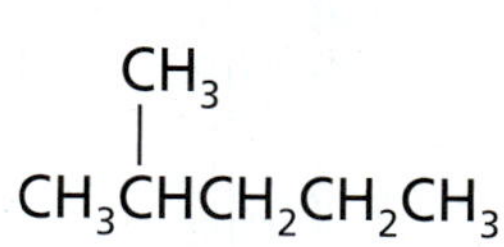


**b.**

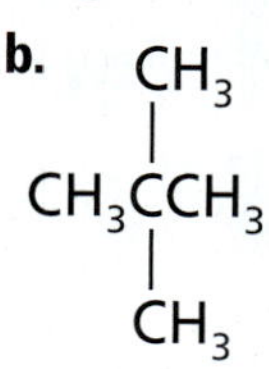


**c.**

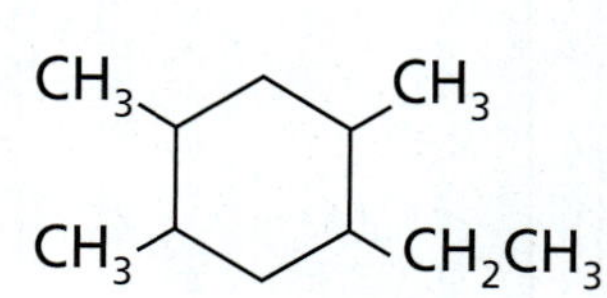


**14. Describe** the general properties of alkanes.

**15. Draw** the molecular structure for each of the following.

**a.** 3,4-diethylheptane
**b.** 4-isopropyl-3-methyldecane
**c.** 1-ethyl-4-methylcyclohexane
**d.** 1,2-dimethylcyclopropane

**16. Interpret Chemical Structures** Why is the name 3-butylpentane incorrect? Based on this name, write the structural formula for the compound. What is the correct IUPAC name for 3-butylpentane?

# SECTION 3 Alkenes and Alkynes

**Essential Questions**

- How do the properties of alkenes and alkynes compare with those of alkanes?
- How are the molecular structures of alkenes and alkynes described?
- How are alkenes and alkynes named when given their structures?
- How are alkenes and alkynes drawn when given their names?

**Review Vocabulary**

**hormone:** chemical produced in one part of an organism and transported to another part, where it causes a physiological change

**New Vocabulary**

alkene
alkyne

**MAINIDEA** Alkenes are hydrocarbons that contain at least one double bond, and alkynes are hydrocarbons that contain at least one triple bond.

**CHEM 4 YOU** Plants produce ethene as a natural ripening hormone. For efficiency in harvesting and transporting produce to market, fruits and vegetables are often picked while unripe and are exposed to ethene so they will ripen at the same time.

## Alkenes

Recall that alkanes are saturated hydrocarbons because they contain only single covalent bonds between carbon atoms, and that unsaturated hydrocarbons have at least one double or triple bond between carbon atoms. Unsaturated hydrocarbons that contain one or more double covalent bonds between carbon atoms in a chain are called **alkenes.** Because an alkene must have a double bond between carbon atoms, there is no 1-carbon alkene. The simplest alkene has two carbon atoms double bonded to each other. The remaining four electrons—two from each carbon atom—are shared with four hydrogen atoms to give the molecule ethene ($C_2H_4$).

Alkenes with only one double bond constitute a homologous series. Recall from the previous section that a homologous series has a fixed numerical relationship among the numbers of atoms. If you study the molecular formulas for the substances shown in **Table 5,** you will see that each has twice as many hydrogen atoms as carbon atoms. The general formula for the series is $C_nH_{2n}$. Each alkene has two fewer hydrogen atoms than the corresponding alkane because two electrons now form the second covalent bond and are no longer available for bonding to hydrogen atoms. What are the molecular formulas for 6-carbon and 9-carbon alkenes?

Explore **alkenes with an interactive table.** Concepts In Motion

**Table 5 Examples of Alkenes**

| Name | Ethene | Propene | 1-Butene | 2-Butene |
|---|---|---|---|---|
| Molecular formula | $C_2H_4$ | $C_3H_6$ | $C_4H_8$ | $C_4H_8$ |
| Structural formula | 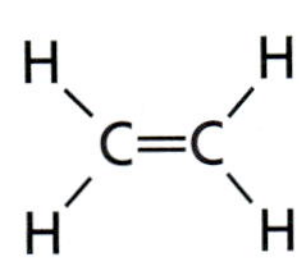 | 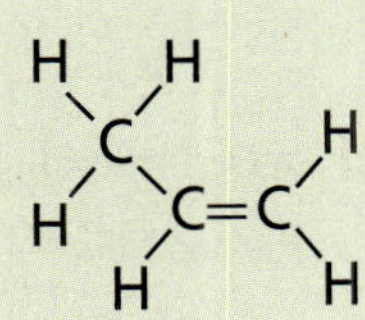 | 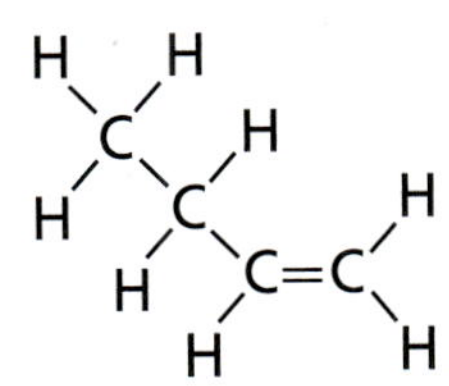 | 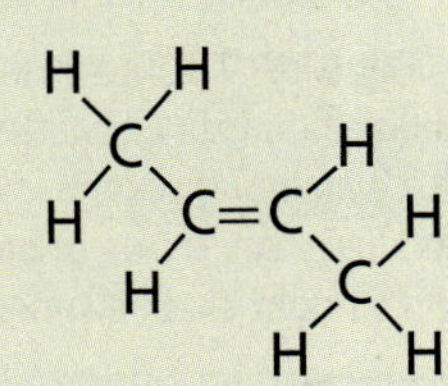 |
| Condensed structural formula | $CH_2{=}CH_2$ | $CH_3CH{=}CH_2$ | $CH_3CH_2CH{=}CH_2$ | $CH_3CH{=}CHCH_3$ |

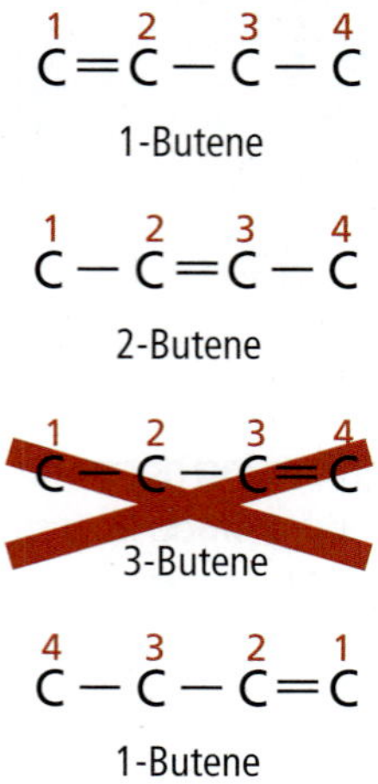


**a.** Straight-chain alkenes

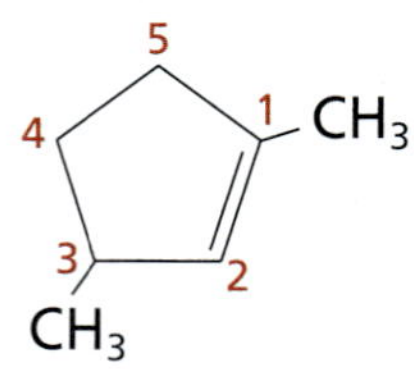

**b.** Cyclic alkenes

■ **Figure 12** When naming either branched or straight-chain alkenes, they must be numbered using IUPAC rules.

**Naming alkenes** Alkenes are named in much the same way as alkanes. Their names are formed by changing the *-ane* ending of the corresponding alkane to *-ene.* An alkane with two carbons is named eth*ane,* and an alkene with two carbons is named eth*ene.* Likewise, a three-carbon alkene is named propene. Ethene and propene have older, more common names: *ethylene* and *propylene,* respectively.

To name alkenes with four or more carbons in the chain, it is necessary to specify the location of the double bond, as shown in **Figure 12a.** This is done by numbering the carbons in the parent chain, starting at the end of the chain that will give the first carbon in the double bond the lowest number. Then, use only that number in the name.

Note that the third structure is not "3-butene" because it is identical to the first structure, 1-butene. It is important to recognize that 1-butene and 2-butene are two different substances, each with its own properties.

Cyclic alkenes are named in much the same way as cyclic alkanes; however, carbon number 1 must be one of the carbons connected by the double bond. In **Figure 12b,** note the numbering in the compound. The name of this compound is 1,3-dimethylcyclopentene.

☑ READING CHECK **Infer** why it is necessary to identify where the double bond is located in the name of an alkene.

**Naming branched-chain alkenes** When naming branched-chain alkenes, follow the IUPAC rules for naming branched-chain alkanes, but with two exceptions. First, in alkenes, the parent chain is always the longest chain that contains the double bond, whether or not it is the longest chain of carbon atoms. Second, the position of the double bond, not the branches, determines how the chain is numbered. A number specifies the location of the double bond, just as it does in straight-chain alkenes. Note that there are two 4-carbon chains in the molecule shown in **Figure 13a,** but only the one with the double bond is used as a basis for naming. This branched-chain alkene is 2-methyl-1-butene.

Some unsaturated hydrocarbons contain more than one double (or triple) bond. The number of double bonds in such molecules is shown by using a prefix (*di-*, *tri-*, *tetra-*, and so on) before the suffix *-ene.* The positions of the bonds are numbered in a way that gives the lowest set of numbers. Which numbering system would you use in the example in **Figure 13b?** Because the molecule has a seven-carbon chain, you would use the prefix *hepta-*. Because it has two double bonds, you would use the prefix *di-* before *-ene,* giving the name *heptadiene.* Adding the numbers 2 and 4 to designate the positions of the double bonds gives the name *2,4-heptadiene.*

■ **Figure 13** The positions of the double bonds in alkenes are numbered in a way that gives the lowest set of numbers. This is true of both branched and straight-chain alkenes.

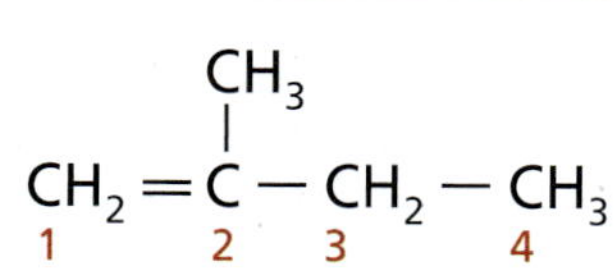

2-methyl-1-butene

**a.** Single double bond

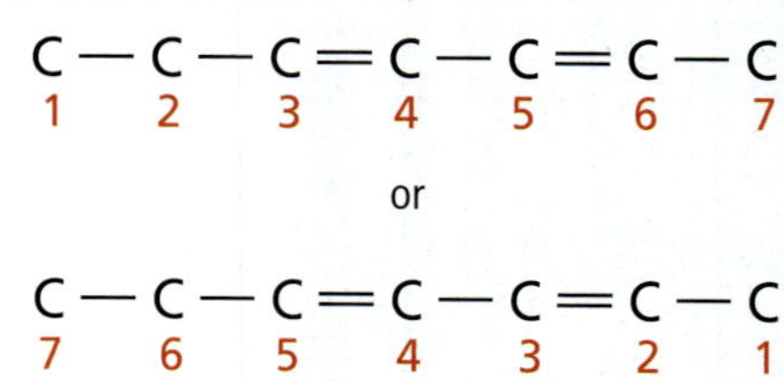


2,4-heptadiene

**b.** Two double bonds

## EXAMPLE Problem 3

**NAMING BRANCHED-CHAIN ALKENES** Name the alkene shown.

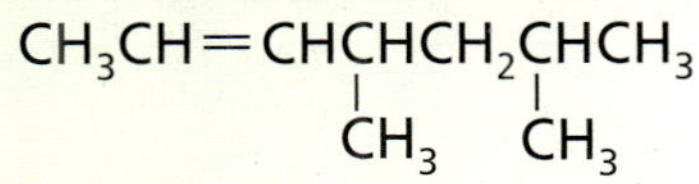

### 1 ANALYZE THE PROBLEM

You are given a branched-chain alkene that contains one double bond and two alkyl groups. Follow the IUPAC rules to name the organic compound.

### 2 SOLVE FOR THE UNKNOWN

**Step 1.** The longest continuous-carbon chain that includes the double bond contains seven carbons. The 7-carbon alkane is heptane, but the name is changed to heptene because a double bond is present.

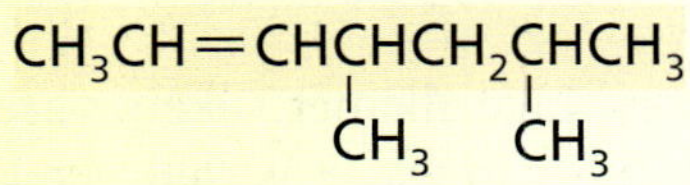

***Heptene* parent chain**

**Step 2.** Number the chain to give the lowest number to the double bond.

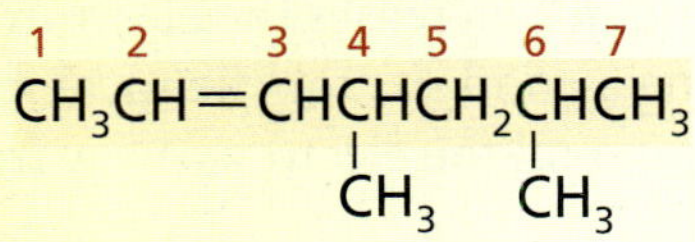

***2-Heptene* parent chain**

**Step 3.** Name each substituent.

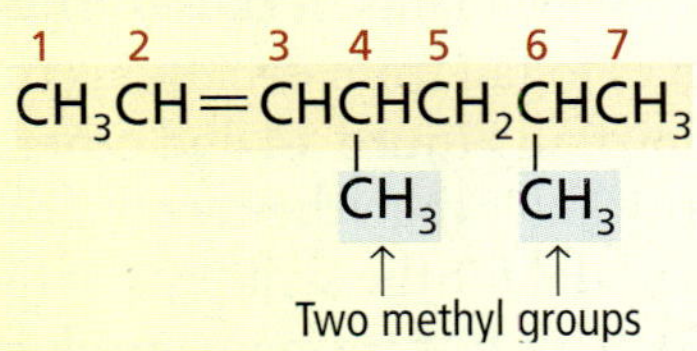


**Each substituent is a methyl group.**

**Step 4.** Determine how many of each substituent is present, and assign the correct prefix to represent that number. Then, include the position numbers to get the complete prefix.

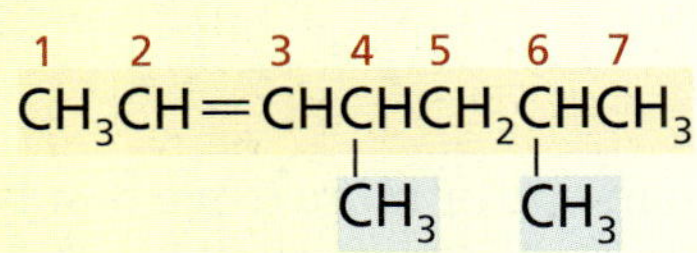

***2-Heptene* parent chain**

**Two methyl groups at Positions 4 and 6**

**Prefix is 4,6-*dimethyl***

**Step 5.** The names of substituents do not have to be alphabetized because they are the same. Apply the complete prefix to the name of the parent alkene chain. Use commas between numbers, and hyphens between numbers and words. Write the name 4,6-*dimethyl-2-heptene.*

### 3 EVALUATE THE ANSWER

The longest carbon chain includes the double bond, and the position of the double bond has the lowest possible number. Correct prefixes and alkyl group names designate the branches.

## PRACTICE Problems

Do additional problems. Online Practice

**17.** Use the IUPAC rules to name the following structures.

a. 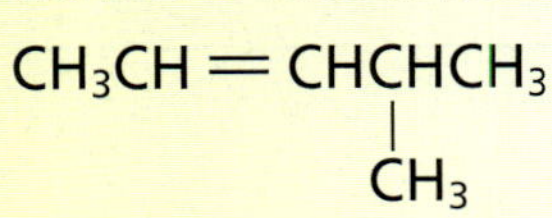

b. 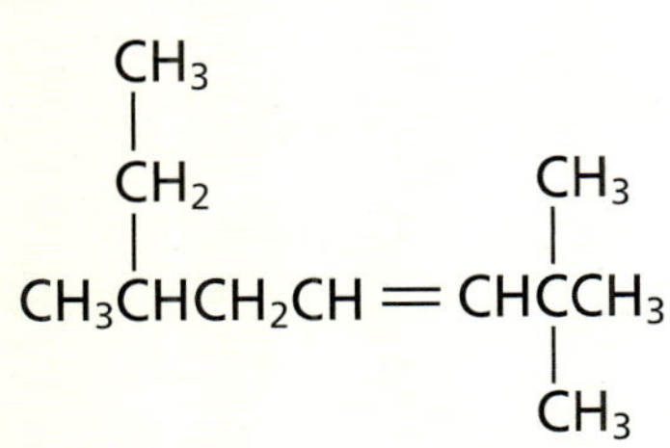

**18. Challenge** Draw the structure of 1,3-pentadiene.

■ **Figure 14** The use of ethene to ripen produce allows growers to harvest fruits and vegetables before they ripen.
**Explain** *why this is a benefit to growers.*

**Properties and uses of alkenes** Like alkanes, alkenes are nonpolar and therefore have low solubility in water as well as relatively low melting and boiling points. However, alkenes are more reactive than alkanes because the second covalent bond increases the electron density between two carbon atoms, providing a good site for chemical reactivity. Reactants that attract electrons can pull the electrons away from the double bond.

Several alkenes occur naturally in living organisms. For example, ethene is a hormone produced naturally by plants. It causes fruit to ripen and plays a part in causing leaves to fall from deciduous trees in preparation for winter. The fruits shown in **Figure 14** and other produce sold in grocery stores ripen artificially when they are exposed to ethene. Ethene is also the starting material for the synthesis of the plastic polyethylene, which is used to manufacture many products, including plastic bags, rope, and milk jugs. Other alkenes are responsible for the scents of lemons, limes, and pine trees.

## Alkynes

Unsaturated hydrocarbons that contain one or more triple bonds between carbon atoms in a chain are called **alkynes.** Triple bonds involve the sharing of three pairs of electrons. The simplest and most commonly used alkyne is ethyne ($C_2H_2$), which is widely known by its common name *acetylene.* Study the models of ethyne in **Figure 15.**

**Naming alkynes** Straight-chain alkynes and branched-chain alkynes are named in the same way as alkenes. The only difference is that the name of the parent chain ends in *-yne* rather than *-ene.* Study the examples in **Table 6.** Alkynes with one triple covalent bond form a homologous series with the general formula $C_nH_{2n-2}$.

☑ READING CHECK **Infer,** by looking at the bonds in ethyne, why it is highly reactive with oxygen.

■ **Figure 15** These three molecular models represent ethyne.

$H-C\equiv C-H$

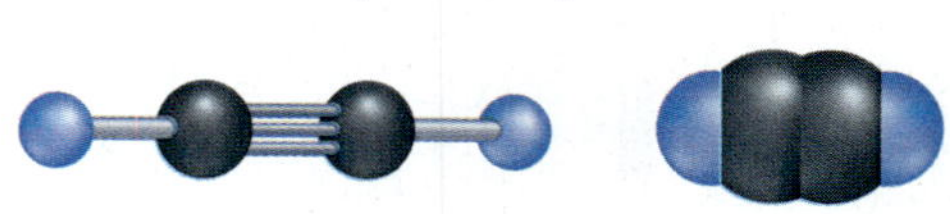

Models of ethyne (acetylene)

Explore **alkynes with an interactive table.** Concepts In Motion

**Table 6** Examples of Alkynes

| Name | Molecular Formula | Structural Formula | Condensed Structural Formula |
|---|---|---|---|
| Ethyne | $C_2H_2$ | H—C≡C—H | $CH \equiv CH$ |
| Propyne | $C_3H_4$ | 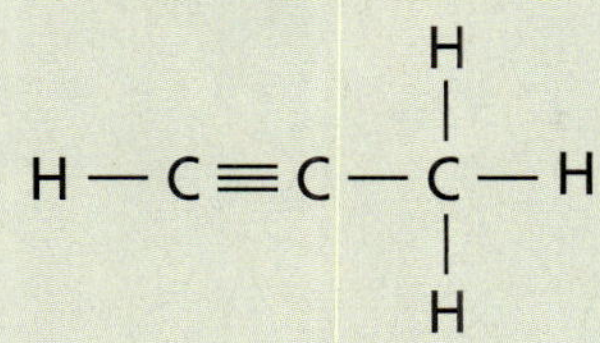 | $CH \equiv CCH_3$ |
| 1-Butyne | $C_4H_6$ | 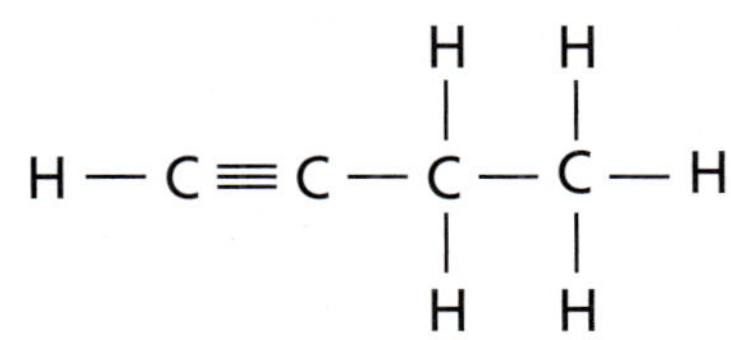 | $CH \equiv CCH_2CH_3$ |
| 2-Butyne | $C_4H_6$ | 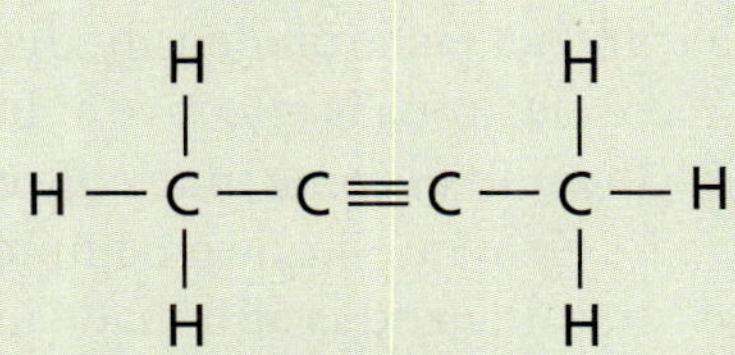 | $CH_3C \equiv CCH_3$ |

# MiniLAB

iLab Station

## Synthesize and Observe Ethyne

**Why is ethyne used in welding torches?**

### Procedure

1. Read and complete the lab safety form.
2. Use a **rubber band** to attach a **wood splint** to one end of a **ruler** that is about 40 cm long, so that about 10 cm of the splint extends beyond the ruler.
3. Place 120 mL **water** in a **150-mL beaker**, and add 5 mL **dishwashing detergent.** Mix thoroughly.
4. Use **forceps** to pick up a pea-sized lump of **calcium carbide** ($CaC_2$). Do not touch the $CaC_2$ with your fingers. **WARNING:** ***$CaC_2$ is corrosive; if $CaC_2$ dust touches your skin, wash it away immediately with a lot of water.*** Place the lump of $CaC_2$ in the beaker of detergent solution.
5. Use a **match** to light the splint while holding the ruler at the opposite end. Immediately bring the burning splint to the bubbles that have formed from the reaction in the beaker. Extinguish the splint after observing the reaction.
6. Use a **stirring rod** to dislodge a few large bubbles of ethyne. Do they float or sink in air?
7. Rinse the beaker thoroughly, then add 25 mL **distilled water** and a drop of **phenolphthalein solution.** Use forceps to place a small piece of $CaC_2$ in the solution. Observe the results.

### Analysis

1. **Infer** What can you infer about the density of ethyne compared to the density of air?
2. **Predict** The reaction of calcium carbide with water yields two products. One is ethyne gas ($C_2H_2$). What is the other product? Write a balanced chemical equation for the reaction.

■ **Figure 16** Ethyne, or acetylene, reacts with oxygen in the chemical reaction $2C_2H_2 + 5O_2 \rightarrow 4CO_2 + 2H_2O$, which produces enough heat to weld metals.

**Properties and uses of alkynes** Alkynes have physical and chemical properties similar to those of alkenes. Alkynes undergo many of the reactions alkenes undergo. However, alkynes are generally more reactive than alkenes because the triple bonds of alkynes have even greater electron density than the double bonds of alkenes. This cluster of electrons is effective at inducing dipoles in nearby molecules, causing them to become unevenly charged and thus reactive.

**FOLDABLES®** Incorporate information from this section into your Foldable.

Ethyne—known commonly as acetylene—is a by-product of oil refining and is also made in large quantities by the reaction of calcium carbide ($CaC_2$) with water. When supplied with enough oxygen, ethyne burns with an intensely hot flame that can reach temperatures as high as 3000°C. Acetylene torches are commonly used in welding, as shown in **Figure 16.** Because the triple bond makes alkynes reactive, simple alkynes like ethyne are used as starting materials in the manufacture of plastics and other organic chemicals used in industry.

## SECTION 3 REVIEW

Section Self-Check

### Section Summary

- Alkenes and alkynes are hydrocarbons that contain at least one double or triple bond, respectively.
- Alkenes and alkynes are nonpolar compounds with greater reactivity than alkanes but with other properties similar to those of alkanes.

**19.** MAINIDEA **Describe** how the molecular structures of alkenes and alkynes differ from the structure of alkanes.

**20. Identify** how the chemical properties of alkenes and alkynes differ from those of alkanes.

**21. Name** the structures shown using IUPAC rules.

**a.** $CH \equiv CCH_2$ with $CH_3$ attached to the second C

**b.** $CH_3CH_2CHCH{=}CHCH_2CH_3$ with $CH_3$ attached to the third C

**22. Draw** the molecular structure of 4-methyl-1,3-pentadiene and 2,3-dimethyl-2-butene.

**23. Infer** how the boiling and freezing points of alkynes compare with those of alkanes with the same number of carbon atoms. Explain your reasoning, then research data to see if it supports your idea.

**24. Predict** What geometric arrangement would you expect from the bonds surrounding the carbon atom in alkanes, alkenes, and alkynes? *(Hint: VSEPR theory can be used to predict the shape.)*

SECTION 4

# Hydrocarbon Isomers

**MAINIDEA** **Some hydrocarbons have the same molecular formula but have different molecular structures.**

## Essential Questions

- How can the two main categories of isomers—structural isomers and stereoisomers—be distinguished?
- How are *cis*- and *trans*- geometric isomers different?
- What is the structural variation in molecules that results in optical isomers?

### Review Vocabulary

**electromagnetic radiation:** transverse waves that carry energy through empty space

### New Vocabulary

isomer
structural isomer
stereoisomer
geometric isomer
chirality
asymmetric carbon
optical isomer
optical rotation

**CHEM 4 YOU** **Have you ever met a pair of identical twins? Identical twins have the same genetic makeup, yet they are two separate individuals with different personalities. Isomers are similar to twins—they have the same molecular formula, but different molecular structures and properties.**

## Structural Isomers

Examine the models of three alkanes in **Figure 17** to determine how they are similar and how they are different. All three have 5 carbon atoms and 12 hydrogen atoms, so they have the molecular formula $C_5H_{12}$. However, as you can see, these models represent three different arrangements of atoms and three different compounds—pentane, 2-methylbutane, and 2,2-dimethylpropane. These three compounds are isomers. **Isomers** are two or more compounds that have the same molecular formula but different molecular structures. Note that cyclopentane and pentane are not isomers because cyclopentane's molecular formula is $C_5H_{10}$.

There are two main classes of isomers. **Figure 17** shows compounds that are examples of structural isomers. **Structural isomers** have the same chemical formula, but their atoms are bonded in different arrangements. Structural isomers have different chemical and physical properties despite having the same formula. This observation supports one of the main principles of chemistry: The structure of a substance determines its properties. How does the trend in boiling points of $C_5H_{12}$ isomers relate to their molecular structures?

As the number of carbons in a hydrocarbon increases, the number of possible structural isomers increases. For example, there are nine alkanes with the molecular formula $C_7H_{16}$. There are more than 300,000 structural isomers with the formula $C_{20}H_{42}$.

■ **Figure 17** These compounds with the same molecular formula, $C_5H_{12}$, are structural isomers. Note how their boiling points differ.

View an **animation about the isomers of pentane.**

Concepts In Motion

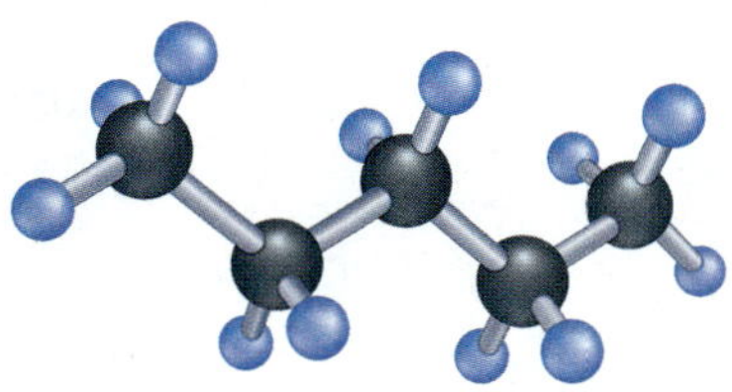

Pentane
bp = 36°C

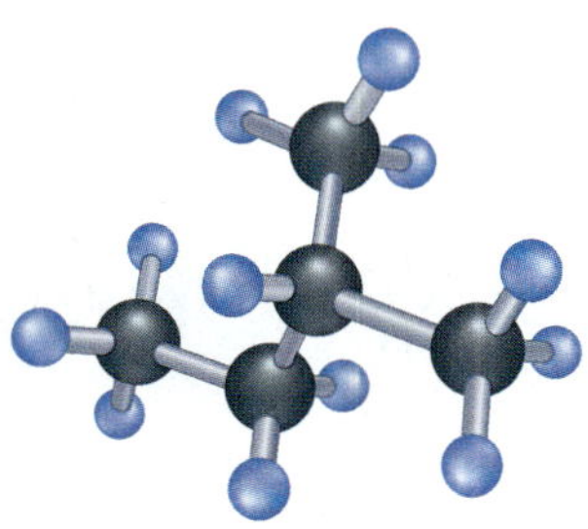

2-Methylbutane
bp = 28°C

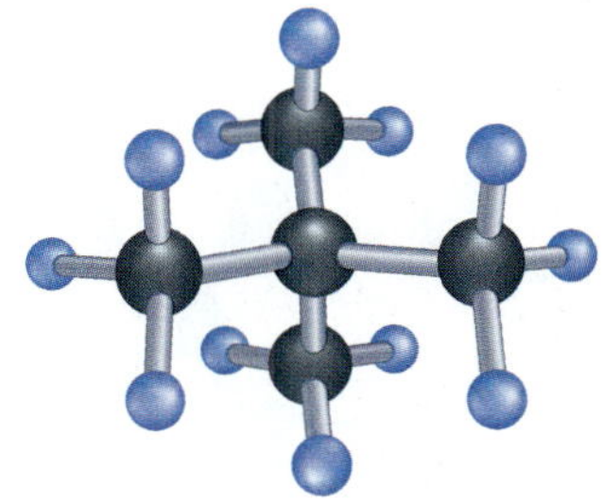

2,2-Dimethylpropane
bp = 9°C

■ **Figure 18** The single-bonded carbons in ethane are free to rotate around the bond. The double-bonded carbons in ethene resist being rotated.

**Explain** ***How do you think this difference in ability to rotate would affect atoms or groups of atoms bonded to single-bonded and double-bonded carbon atoms?***

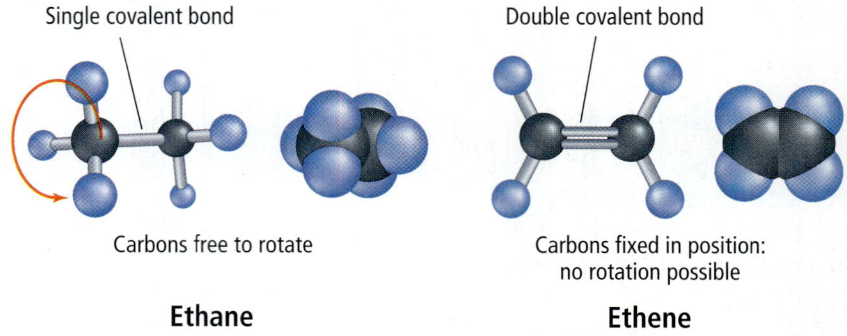


## Stereoisomers

The second class of isomers involves a more subtle difference in bonding. **Stereoisomers** are isomers in which all atoms are bonded in the same order but are arranged differently in space. There are two types of stereoisomers. One type occurs in alkenes, which contain double bonds. Two carbon atoms with a single bond between them can rotate freely in relationship to each other. However, when a second covalent bond is present, the carbons can no longer rotate; they are locked in place, as shown in **Figure 18.**

Compare the two possible structures of 2-butene shown in **Figure 19.** The arrangement in which the two methyl groups are on the same side of the molecule is indicated by the prefix *cis*-. The arrangement in which the two methyl groups are on opposite sides of the molecule is indicated by the prefix *trans*-. These terms derive from Latin: *cis* means *on the same side*, and *trans* means *across from*. Because the double-bonded carbon atoms cannot rotate, the *cis*- form cannot easily change into the *trans*- form.

Isomers resulting from different arrangements of groups around a double bond are called **geometric isomers.** Note how the difference in geometry affects the isomers' physical properties, such as melting point and boiling point. Geometric isomers differ in some chemical properties as well. If the compound is biologically active, such as a drug, the *cis*- and *trans*- isomers usually have very different effects.

☑ READING CHECK **Explain** how structural and geometric isomers differ.

■ **Figure 19** These isomers of 2-butene differ in the arrangement in space of the two methyl groups at the ends. The double-bonded carbon atoms cannot rotate with respect to each other, so the methyl groups are fixed in one of these two arrangements.

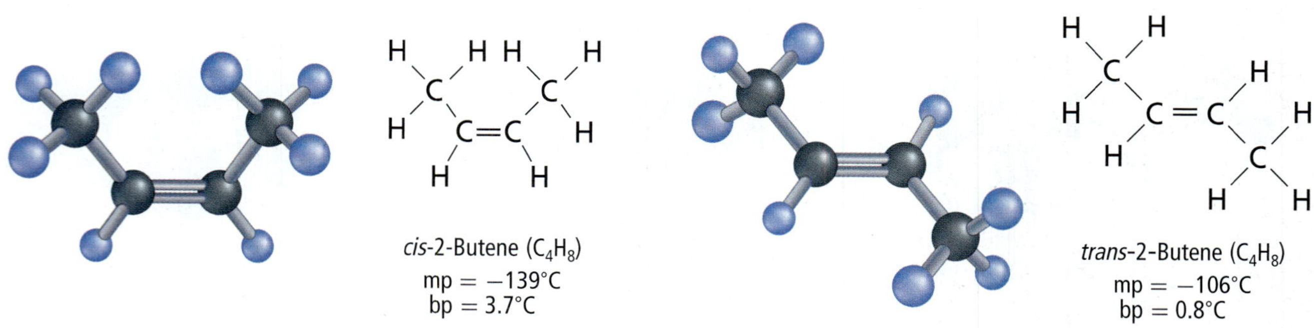

*cis*-2-Butene ($C_4H_8$)
mp = −139°C
bp = 3.7°C

*trans*-2-Butene ($C_4H_8$)
mp = −106°C
bp = 0.8°C

■ **Figure 20** The reflection of your right hand looks the same as your left hand. However, you cannot place your hands palms down with one on top of the other and have matching parts lie on top of each other.

## Chirality

**Connection to Biology** In 1848, the young French chemist Louis Pasteur (1822–1895) reported his discovery that crystals of the organic compound tartaric acid, existed in two shapes that were mirror images of each other. Because a person's hands are like mirror images, as shown in **Figure 20,** the crystals were called the right-handed and left-handed forms. The two forms of tartaric acid had the same chemical properties, melting point, density, and solubility in water, but only the left-handed form was produced by fermentation to make wine. In addition, bacteria were able to multiply only when they were fed the left-handed form as a nutrient.

The two crystalline forms of tartaric acid exist in the two arrangements as shown in **Figure 21.** Today, these two forms are called d-tartaric acid and l-tartaric acid. The letters d and l stand for the Latin prefixes *dextro-*, which means *to the right,* and *levo-*, which means *to the left.* The property in which a molecule exists in a right- and left-handed form is called **chirality.** Many of the substances found in living organisms, such as the amino acids that make up proteins, have this chirality. In general, living organisms make use of only one chiral form of a substance because only this form fits the active site of an enzyme.

### RealWorld CHEMISTRY

**Trans fats**

**ISOMERS IN THE DIET** Fats with *trans* isomers are called *trans* fats. Many pre-packaged foods are made with *trans* fats because they have a longer shelf life. Evidence suggests that *trans* fat increases the unhealthy form of cholesterol and decreases the healthy form, which increases the chance of heart disease.

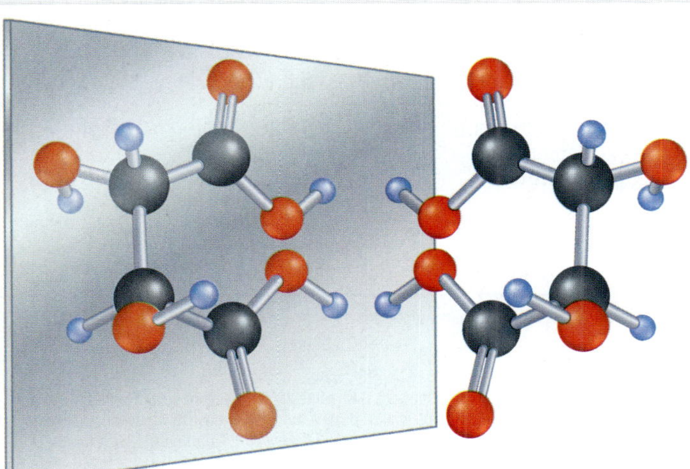

l-Tartaric acid d-Tartaric acid

■ **Figure 21** These models represent the two forms of tartaric acid that Pasteur studied. If the model of right-handed tartaric acid (d-tartaric acid) is reflected in a mirror, its image is a model of left-handed tartaric acid (l-tartaric acid).

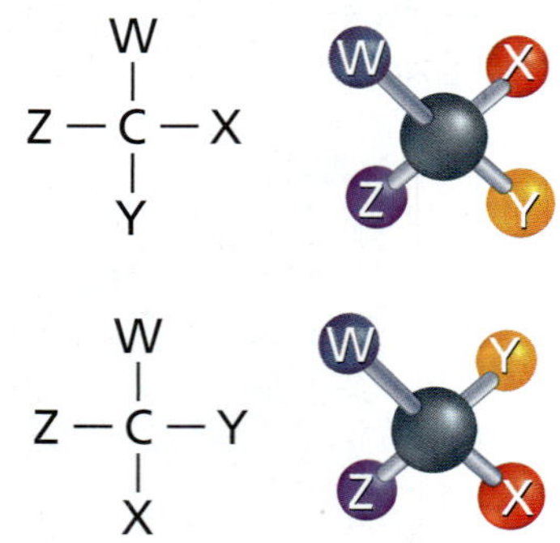


■ **Figure 22** These models represent two different molecules. Groups X and Y have switched places.

## Optical Isomers

In the 1860s, chemists realized that chirality occurs whenever a compound contains an asymmetric carbon. An **asymmetric carbon** is a carbon atom that has four different atoms or groups of atoms attached to it. The four groups can always be arranged in two different ways. Suppose that groups W, X, Y, and Z are attached to the same carbon atom in the two arrangements shown in **Figure 22.** Note that the structures differ in that groups X and Y have been exchanged. You cannot rotate the two arrangements in any way that will make them identical to each other.

Now suppose that you build models of these two structures. Is there any way you could turn one structure so that it looks the same as the other? (Whether letters appear forward or backward does not matter.) You would discover that there is no way to accomplish the task without removing X and Y from the carbon atom and switching their positions. Therefore, the molecules are different even though they look very much alike.

Isomers that result from different arrangements of four different groups around the same carbon atom represent another class of stereoisomers called optical isomers. **Optical isomers** have the same physical and chemical properties, except in chemical reactions where chirality is important, such as enzyme-catalyzed reactions in biological systems. Human cells, for example, incorporate only l-amino acids into proteins. Only the l-form of ascorbic acid is active as vitamin C. The chirality of a drug molecule can also be important. For example, only one isomer of some drugs is effective and the other isomer can be harmful.

**FOLDABLES®** Incorporate information from this section into your Foldable.

# Data Analysis LAB

Based on Real Data*

## Interpret Data

**What are the rates of oxidation of dichloroethene isomers?** *Pseudomonas butanovora* is a bacterium that uses some alkanes, alcohols, and organic acids as sources of carbon and energy. This bacteria was tested as an agent to rid groundwater of dichloroethene (DCE) contaminants. Mixtures containing various reducing agents and butane monooxygenase in *Pseudomonas butanovora* oxidized isomers of DCE.

### Data and Observations

The table shows the rate of oxidation of each compound in butane-grown *P. butanovora*.

**Rates of Oxidation**

| | Initial Rate of Oxidation (nmol $min^{-1}$ mg $protein^{-1}$) | |
|---|---|---|
| **Reducing Agent** | **1,2-cis DCE** | **1,2-trans DCE** |
| Buffer | 0.9 (1.0) | 1.6 (1.0) |
| Butyrate | 6.8 (7.6) | 2.0 (1.3) |
| Propionate | 5.9 (6.6) | 0.4 (0.3) |
| Acetate | 8.5 (9.4) | 3.8 (2.8) |
| Formate | 1.4 (1.6) | 1.2 (0.7) |
| Lactate | 11 (12.2) | 4.5 (2.8) |

Values in parentheses represent the increase (*n*-fold) above the buffer rate.

### Think Critically

1. **Compare** Which reducing agent was most useful in oxidizing each isomer?
2. **Conclude** Which isomer oxidized the slowest?

Data obtained from: Doughty, D.M. et al. 2005. Effects of dichloroethene isomers on the induction and activity of butane monooxygenase in the alkane-oxidizing Bacterium *"Pseudomonas butanovora." Applied Environmental Microbiology.* October: 6054–6059.

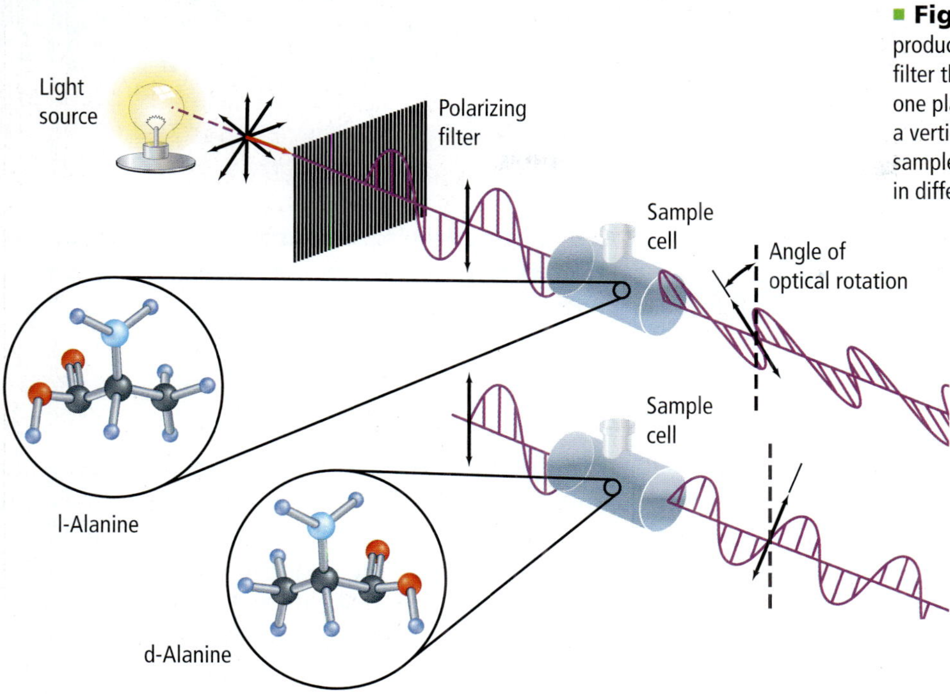


**Figure 23** Polarized light can be produced by passing ordinary light through a filter that transmits light waves that lie in only one plane. Here, the filtered light waves are in a vertical plane before they pass through the sample cells. The two isomers rotate the light in different directions.

**Optical rotation** Mirror-image isomers are called optical isomers because they affect light passing through them. Normally, the light waves in a beam from the Sun or a lightbulb move in all possible planes. However, light can be filtered or reflected in such a way that the resulting waves all lie in the same plane. This type of light is called polarized light.

When polarized light passes through a solution containing an optical isomer, the plane of polarization is rotated to the right (clockwise, when looking toward the light source) by a d-isomer or to the left (counterclockwise) by an l-isomer, producing an effect called **optical rotation.** This effect is shown in **Figure 23.**

One optical isomer that you might have used is l-menthol. This natural isomer has a strong, minty flavor, and a cooling odor and taste. The mirror-image isomer, d-menthol, does not have the same cooling effect as l-menthol.

## SECTION 4 REVIEW

Section Self-Check

### Section Summary

- Isomers are two or more compounds with the same molecular formula but different molecular structures.
- Structural isomers differ in the order in which atoms are bonded to each other.
- Stereoisomers have all atoms bonded in the same order but arranged differently in space.

**25. MAINIDEA Draw** all of the structural isomers possible for the alkane with the molecular formula $C_6H_{14}$. Show only the carbon chains.

**26. Explain** the difference between structural isomers and stereoisomers.

**27. Draw** the structures of *cis*-3-hexene and *trans*-3-hexene.

**28. Infer** why living organisms can make use of one only chiral form of a substance.

**29. Evaluate** A certain reaction yields 80% *trans*-2-pentene and 20% *cis*-2-pentene. Draw the structures of these two geometric isomers, and develop a hypothesis to explain why the isomers form in the proportions cited.

**30. Formulate Models** Starting with a single carbon atom, draw two different optical isomers by attaching the following atoms or groups to the carbon: $-H$, $-CH_3$, $-CH_2CH_3$, and $-CH_2CH_2CH_3$.

SECTION 5

# Aromatic Hydrocarbons

**MAINIDEA** Aromatic hydrocarbons are unusually stable compounds with ring structures in which electrons are shared by many atoms.

### Essential Questions

- How do the properties of aromatic and aliphatic hydrocarbons compare and contrast?
- What is a carcinogen and what are some examples?

### Review Vocabulary

**hybrid orbitals:** equivalent atomic orbitals that form during bonding by the rearrangement of valence electrons

### New Vocabulary

aromatic compound
aliphatic compound

**CHEM 4 YOU** What do bright, colorful fabrics, asphalt roofing shingles, and essential oils for perfumes have in common? They all contain aromatic hydrocarbons.

## The Structure of Benzene

Natural dyes, like those found in the fabrics in **Figure 24,** and essential oils for perfumes contain six-carbon ring structures. Compounds with these structures have been used for centuries. By the middle of the nineteenth century, chemists had a basic understanding of the structures of hydrocarbons with single, double, and triple covalent bonds. However, certain hydrocarbon ring structures remained a mystery.

The simplest example of this class of hydrocarbon is benzene, which the English physicist Michael Faraday (1791–1867) first isolated in 1825 from the gases given off when either whale oil or coal was heated. Although chemists had determined that benzene's molecular formula was $C_6H_6$, it was hard for them to determine what sort of hydrocarbon structure would give such a formula. After all, the formula of the saturated hydrocarbon with six carbon atoms, hexane, was $C_6H_{14}$. Because the benzene molecule had so few hydrogen atoms, chemists reasoned that it must be unsaturated; that is, it must have several double or triple bonds, or a combination of both. They proposed many different structures, including this one suggested in 1860.

$$CH_2=C=CH-CH=C=CH_2$$

Although this structure has a molecular formula of $C_6H_6$, such a hydrocarbon would be unstable and extremely reactive because of its many double bonds. However, benzene was fairly unreactive, and it did not react in the ways that alkenes and alkynes usually react. For that reason, chemists reasoned that structures such as the one shown above must be incorrect.

**Figure 24** Dyes used to produce brightly-colored fabrics have been used for centuries.

**Explain** *What do many natural dyes and essential oils for perfumes have in common?*

**Kekulé's dream** In 1865, the German chemist Friedrich August Kekulé (1829–1896) proposed a different kind of structure for benzene—a hexagon of carbon atoms with alternating single and double bonds. How does the molecular formula of this structure compare with that of benzene?

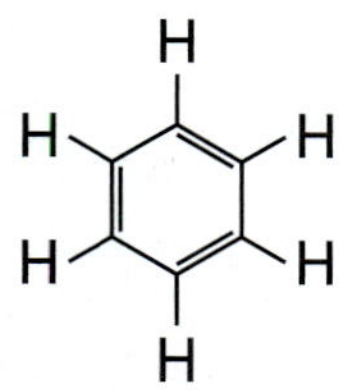

Kekulé claimed that benzene's structure came to him in a dream while he dozed in front of a fireplace in Ghent, Belgium. He said that he had dreamed of the Ouroboros, an ancient Egyptian emblem of a snake devouring its own tail, and that had made him think of a ring-shaped structure. The flat, hexagonal structure Kekulé proposed explained some of the properties of benzene, but it did not explain benzene's lack of reactivity.

**A modern model of benzene** Since the time of Kekulé's proposal, research has confirmed that benzene's molecular structure is indeed hexagonal. However, benzene's unreactivity could not be explained until the 1930s, when Linus Pauling proposed the theory of hybrid orbitals. When applied to benzene, this theory predicts that the pairs of electrons that form the second bond of each of benzene's double bonds are not localized between only two specific carbon atoms as they are in alkenes. Instead, the electron pairs are delocalized, which means they are shared among all six carbons in the ring. **Figure 25** shows that this delocalization makes the benzene molecule chemically stable because electrons shared by six carbon nuclei are harder to pull away than electrons held by only two nuclei. The six hydrogen atoms are usually not shown, but it is important to remember that they are there. In this representation, the circle in the middle of the hexagon symbolizes the cloud formed by the three pairs of electrons.

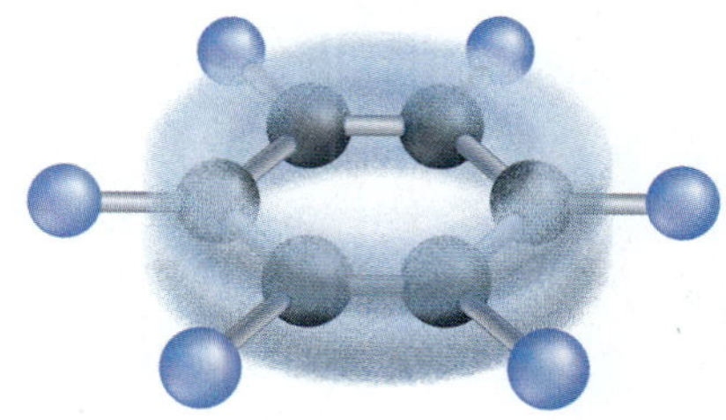

**Figure 25** Benzene's bonding electrons spread evenly in a double-donut shape around the ring instead of remaining near individual atoms.

View an **animation about bonding in benzene.**

Concepts In Motion

## Aromatic Compounds

Organic compounds that contain benzene rings as part of their structures are called **aromatic compounds.** The term *aromatic* was originally used because many of the benzene-related compounds known in the nineteenth century were found in pleasant-smelling oils that came from spices, fruits, and other plant parts. Hydrocarbons such as the alkanes, alkenes, and alkynes are called **aliphatic compounds** to distinguish them from aromatic compounds. The term *aliphatic* comes from the Greek word for *fat*, which is *aleiphatos*. Early chemists obtained aliphatic compounds by heating animal fats. What are some examples of animal fats that might contain aliphatic compounds?

**READING CHECK** **Infer** why the terms *aromatic compound* and *aliphatic compound* continue to be used by chemists today.

**VOCABULARY**
**SCIENCE USAGE V. COMMON USAGE**
**Aromatic**
***Science usage:*** an organic compound with increased chemical stability due to the delocalization of electrons
*Benzene is an aromatic compound.*

***Common usage:*** having a strong odor or smell
*The perfume was very aromatic.*

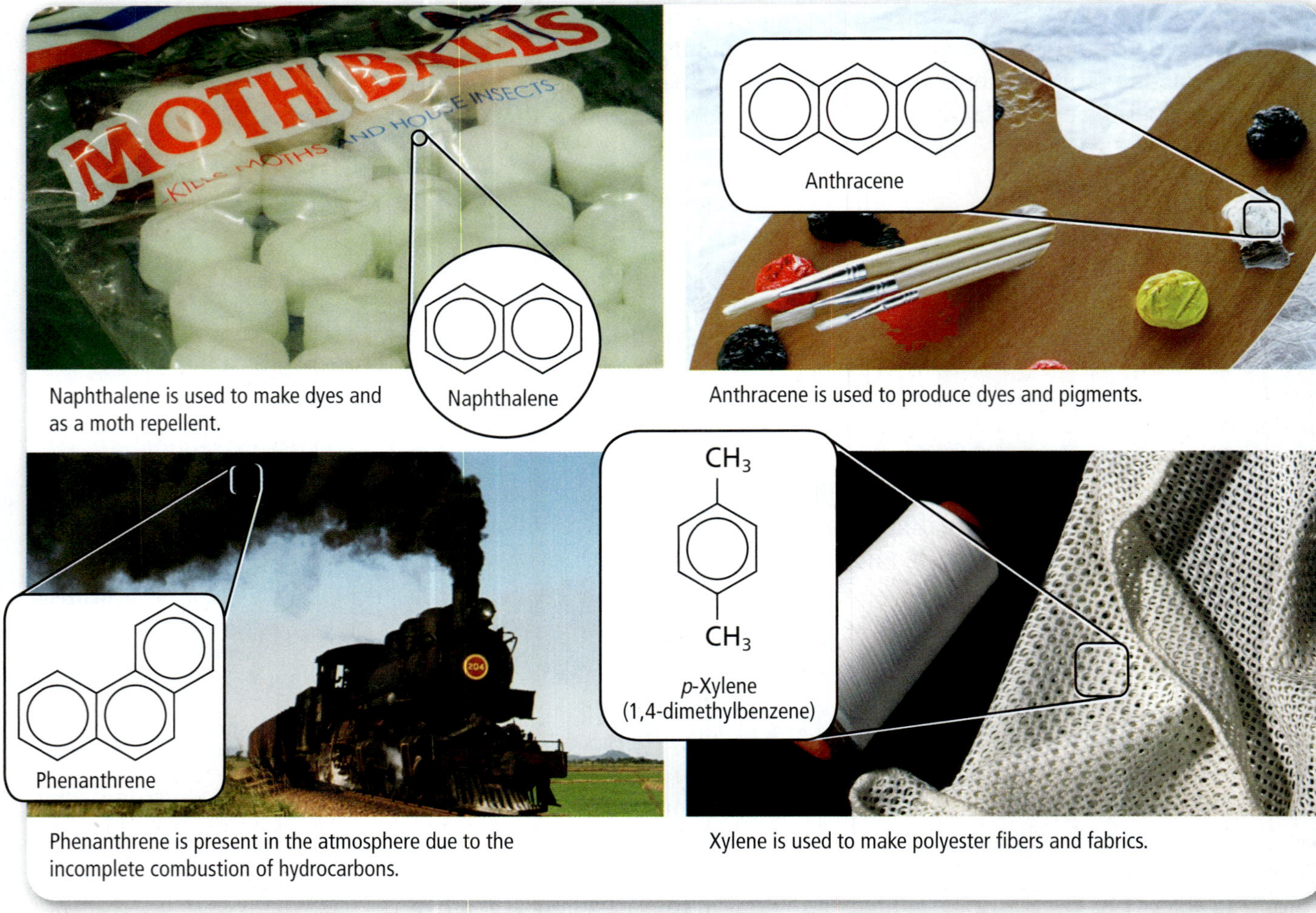


Naphthalene is used to make dyes and as a moth repellent.

Anthracene is used to produce dyes and pigments.

Phenanthrene is present in the atmosphere due to the incomplete combustion of hydrocarbons.

Xylene is used to make polyester fibers and fabrics.

■ **Figure 26** Aromatic hydrocarbons are found in the environment due to the incomplete combustion of hydrocarbons and are used to make a variety of products.

Structures of some aromatic compounds are shown in **Figure 26.** Note that naphthalene has a structure that looks like two benzene rings arranged side by side. Naphthalene is an example of a fused-ring system, in which an organic compound has two or more cyclic structures with a common side. As in benzene, electrons are shared by the carbon atoms that make up the ring systems.

**Naming substituted aromatic compounds** Like other hydrocarbons, aromatic compounds can have different groups attached to their carbon atoms. For example, methylbenzene, also known as toluene, consists of a methyl group attached to a benzene ring in place of one hydrogen atom. Whenever you see something attached to an aromatic ring system, remember that the hydrogen atom is no longer there.

Substituted benzene compounds are named in the same way as cyclic alkanes. For example, ethylbenzene has a 2-carbon ethyl group attached, and 1,4-dimethylbenzene, also known as *para*-xylene, has two methyl groups attached at positions 1 and 4.

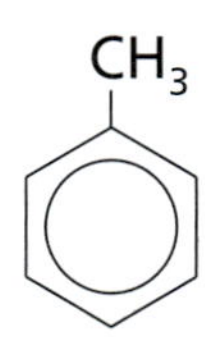


Methylbenzene (toluene)

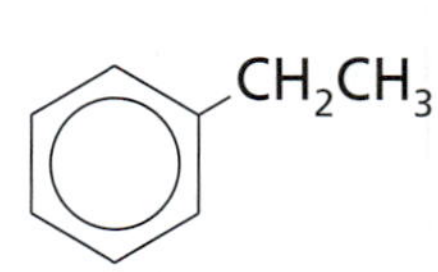


Ethylbenzene

1,4-Dimethylbenzene (*para*-xylene)

Just as with substituted cycloalkanes, substituted benzene rings are numbered in a way that gives the lowest-possible numbers for the substituents, as shown in **Figure 27.** Numbering the ring as shown gives the numbers 1, 2, and 4 for the substituent positions. Because *ethyl* comes before *methyl* in the alphabet, it is written first in the name: 2-ethyl-1,4-dimethylbenzene.

**READING CHECK** **Explain** what the circle means inside the six-membered ring structure in **Figure 27.**

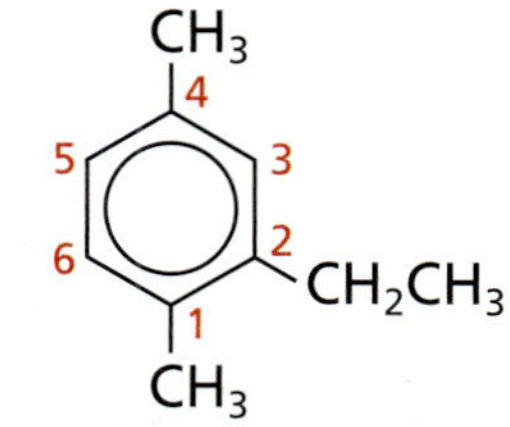


2-Ethyl-1,4-dimethylbenzene

**Figure 27** Substituted benzene rings are named in the same way as cyclic alkanes.

## EXAMPLE Problem 4

**NAMING AROMATIC COMPOUNDS** Name the aromatic compound shown.

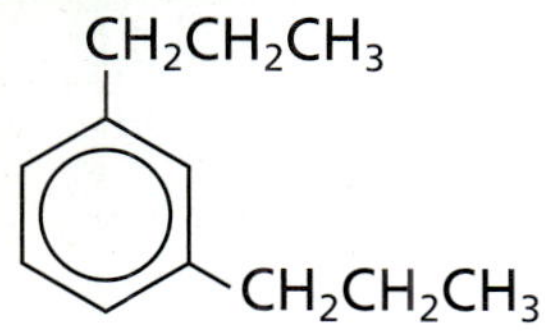


### 1 ANALYZE THE PROBLEM

You are given an aromatic compound. Follow the rules to name the aromatic compound.

### 2 SOLVE FOR THE UNKNOWN

**Step 1.** Number the carbon atoms to give the lowest numbers possible.

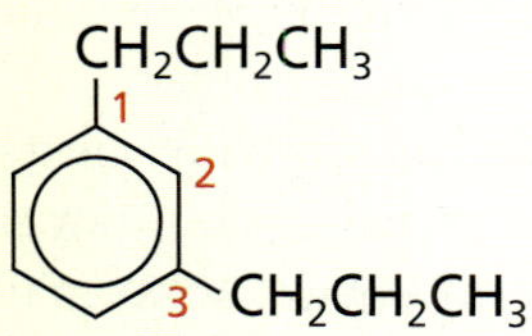


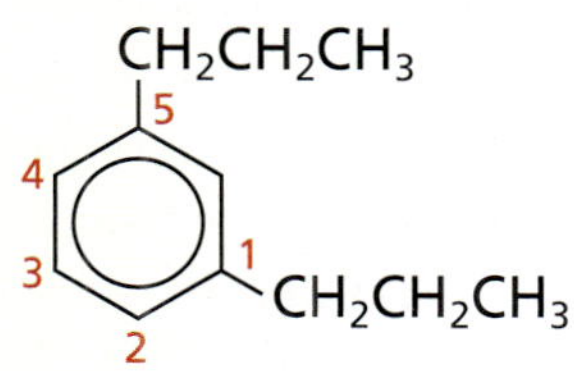


As you can see, the numbers 1 and 3 are lower than the numbers 1 and 5. So the numbers used to name the hydrocarbon should be 1 and 3.

**Step 2.** Determine the name of the substituents. If the same substituent appears more than once, add the prefix to show the number of groups present.

**Step 3.** Put the name together. Alphabetize the substituent names, and use commas between numbers and hyphens between numbers and words. Write the name as 1,3-dipropylbenzene.

### 3 EVALUATE THE ANSWER

The benzene ring is numbered to give the branches the lowest possible set of numbers. The names of the substituent groups are correctly identified.

## PRACTICE Problems

Do additional problems. Online Practice

**31.** Name the following structures.

**a.** 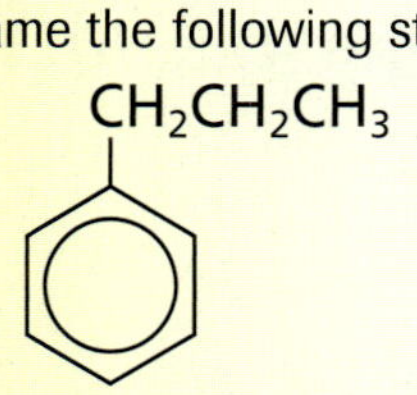


**b.** 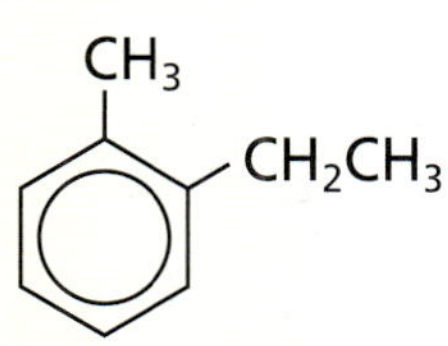


**c.** 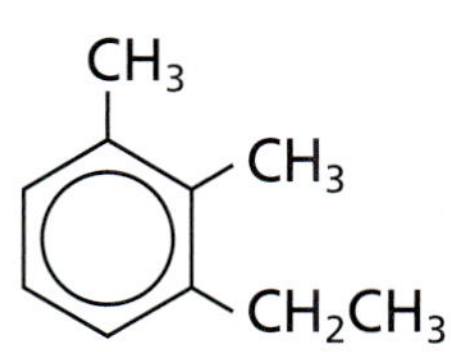


**32. Challenge** Draw the structure of 1,4-dimethylbenzene.

EXAMPLE PROBLEM

■ **Figure 28** Benzopyrene is a cancer-causing chemical that is found in soot, cigarette smoke, and car exhaust.

**Carcinogens** Many aromatic compounds, particularly benzene, toluene, and xylene, were once commonly used as industrial and laboratory solvents. However, tests have shown that the use of such compounds should be limited because they can affect the health of people who are exposed to them regularly. Health risks linked to aromatic compounds include respiratory ailments, liver problems, and damage to the nervous system. Beyond these hazards, some aromatic compounds are carcinogens, which are substances that can cause cancer.

The first known carcinogen was an aromatic substance discovered around the turn of the twentieth century in chimney soot. Chimney sweeps in Great Britain were known to have abnormally high rates of cancer. Scientists discovered that the cause of the cancer was the aromatic compound benzopyrene, shown in **Figure 28.** This compound is a by-product of the burning of complex mixtures of organic substances, such as wood and coal. Some aromatic compounds found in gasoline are also known to be carcinogenic.

**FOLDABLES®** Incorporate information from this section into your Foldable.

## SECTION 5 REVIEW

Section Self-Check

### Section Summary

- Aromatic hydrocarbons contain benzene rings as part of their molecular structures.
- The electrons in aromatic hydrocarbons are shared evenly over the entire benzene ring.

33. **MAIN IDEA Explain** benzene's structure and how it makes the molecule unusually stable.

34. **Explain** how aromatic hydrocarbons differ from aliphatic hydrocarbons.

35. **Describe** the properties of benzene that made chemists think it was not an alkene with several double bonds.

36. **Name** the following structures.

**a.**

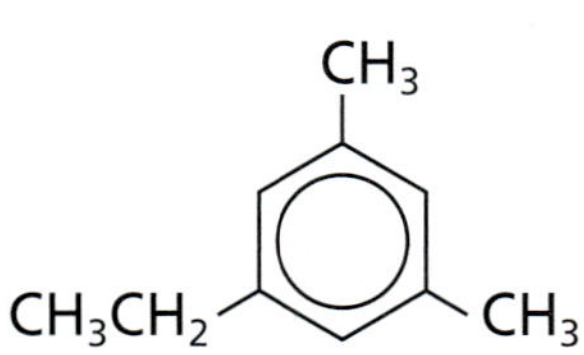


**b.**

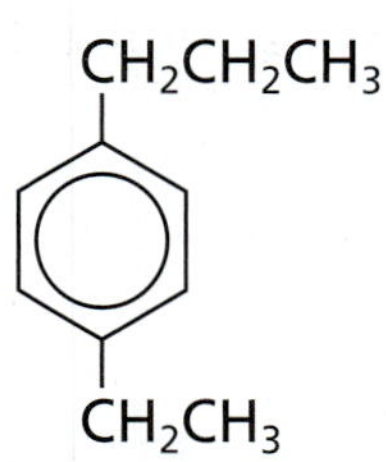


37. **Explain** why the connection between benzopyrene and cancer was significant.

# HOW IT works

Peter Titmuss/Alamy

## POOCH TO POWER: How a Methane Digester Works

Officials in San Francisco are hoping the city's pet owners will contribute their animals' wastes to a pilot project that will convert organic matter into usable energy. A methane digester converts the wastes into biogas—a mixture of methane and carbon dioxide. Burning the methane provides energy for the city.

**1 Bacteria** Animal wastes are mixed with methane-producing bacteria in the digester. These bacteria can live only under anaerobic conditions—in an oxygen-free environment. Three different anaerobic bacteria break down the wastes, first into organic acids and then into methane gas.

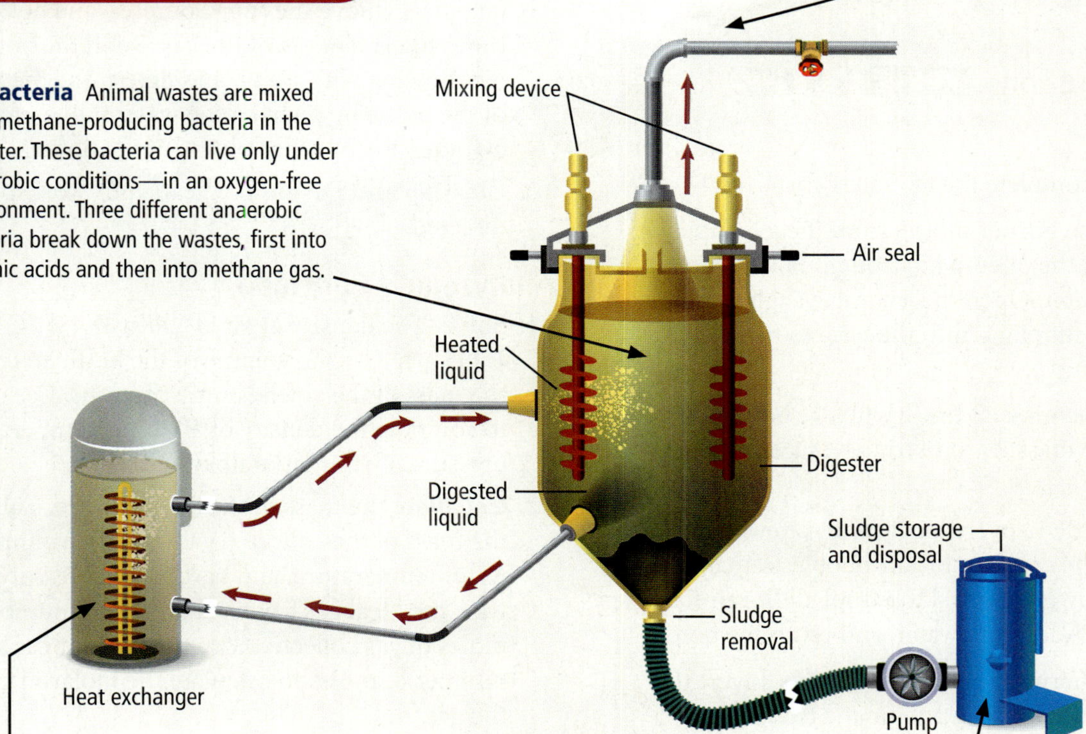


**4 Gas** Methane gas is collected, compressed, and either used immediately or stored. The methane can be used to heat homes or to generate electricity.

**3 Sludge** The bacteria cannot convert 100% of the animal wastes into methane. The remaining indigestible material, called sludge or effluent, is rich in plant nutrients and can be used as a soil conditioner.

**2 Temperature** As with any chemical reaction, temperature affects methane production. Like the bacteria in our own bodies, the bacteria in the digester are most efficient between 35°C and 37°C. An external heat exchanger, combined with insulation around the digester chamber, help to keep the temperature constant and within the optimal range.

### WRITING IN ▶ Chemistry

**Compare** Research and create a pamphlet comparing the advantages of biogas production to other forms of waste disposal for agribusinesses, such as dairies and beef, pork, and poultry producers.

WebQuest

# ChemLAB

iLab Station

## Forensics: Analyze Hydrocarbon Burner Gases

**Background:** A valve needs to be replaced in the science lab. The custodian says the gas used in the lab is propane, and the chemistry teacher says it is natural gas (methane). Use scientific methods to settle this dispute.

**Question:** *What type of alkane gas is used in the science laboratory?*

### Materials

barometer
thermometer
1-L or 2-L plastic soda bottle with cap
burner tubing
pneumatic trough
100-mL graduated cylinder
balance (0.01g)
paper towels

### Safety Precautions

### Procedure

1. Read and complete the lab safety form.
2. Connect the burner tubing from the gas supply to the inlet of the pneumatic trough. Fill the trough with tap water. Open the gas valve slightly, and let a small amount of gas into the tank to flush the air out of the tubing.
3. Measure the mass of the dry plastic bottle and cap. Record the mass, barometric pressure, and air temperature.
4. Fill the bottle to overflowing with tap water, and screw on the cap. If some air bubbles remain, tap the bottle gently on the desktop until all the air has risen to the top. Add more water, and recap the bottle.
5. Place the thermometer in the trough. Invert the capped bottle into the pneumatic trough, and remove the cap while keeping the mouth of the bottle under water. Hold the mouth of the bottle directly over the inlet opening of the trough.
6. Slowly open the gas valve, and allow gas to enter the inverted bottle until all of the water has been displaced. Close the gas valve immediately. Record the temperature of the water.
7. While the bottle is still inverted, screw on the cap. Remove the bottle from the water, and dry the outside of the bottle.
8. Measure and record the mass of the bottle containing the burner gas.

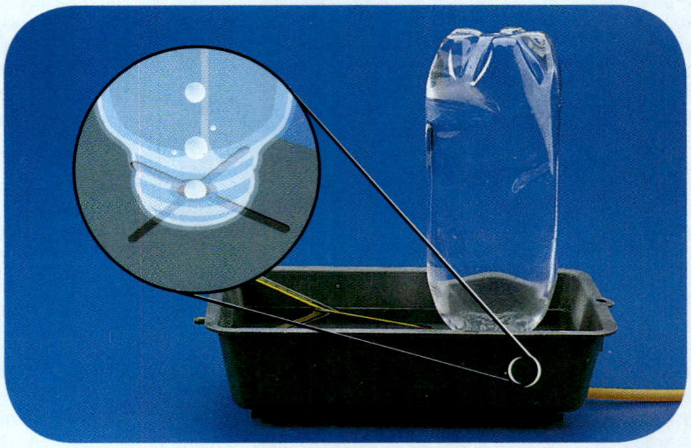

9. Place the bottle in a fume hood, turn on the exhaust fan, and remove the cap. Compress the bottle several times to expel most of the gas. Refill the bottle to overflowing with water, and determine the volume of the bottle by pouring the water into a graduated cylinder. Record the volume of the bottle.
10. **Cleanup and Disposal** Clean your workspace as directed.

### Analyze and Conclude

1. **Solve** The density of air at 1 atm and 20°C is 1.205 g/L. Use the volume of the bottle to compute the mass of the air the bottle contains. Use gas laws to compute the density of air at the temperature and pressure of your laboratory.
2. **Calculate** the mass of the empty bottle. Calculate the mass of the collected gas. Use the volume of gas, water temperature, and barometric pressure along with the ideal gas law to calculate the number of moles of gas collected. Use the mass of gas and the number of moles to calculate the molar mass of the gas.
3. **Conclude** How does your experimental molar mass compare with the molar masses of methane, ethane, and propane? Infer which gases are in the burner gas in your lab.
4. **Error Analysis** Suggest possible sources of error in the experiment.

### INQUIRY EXTENSION

**Design an Experiment** to test how one variable, such as temperature or atmospheric pressure, affects your results.

# CHAPTER 21 STUDY GUIDE

Vocabulary Practice

**BIGIDEA** Organic compounds called hydrocarbons differ by their types of bonds.

## SECTION 1 Introduction to Hydrocarbons

**MAINIDEA** Hydrocarbons are carbon-containing organic compounds that provide a source of energy and raw materials.

- Organic compounds contain the element carbon, which is able to form straight chains and branched chains.
- Hydrocarbons are organic substances composed of carbon and hydrogen.
- The major sources of hydrocarbons are petroleum and natural gas.
- Petroleum can be separated into components by the process of fractional distillation.

**VOCABULARY**

- organic compound
- hydrocarbon
- saturated hydrocarbon
- unsaturated hydrocarbon
- fractional distillation
- cracking

## SECTION 2 Alkanes

**MAINIDEA** Alkanes are hydrocarbons that contain only single bonds.

- Alkanes contain only single bonds between carbon atoms.
- Alkanes and other organic compounds are best represented by structural formulas and can be named using systematic rules determined by the International Union of Pure and Applied Chemistry (IUPAC).
- Alkanes that contain hydrocarbon rings are called cyclic alkanes.

**VOCABULARY**

- alkane
- homologous series
- parent chain
- substituent group
- cyclic hydrocarbon
- cycloalkane

## SECTION 3 Alkenes and Alkynes

**MAINIDEA** Alkenes are hydrocarbons that contain at least one double bond, and alkynes are hydrocarbons that contain at least one triple bond.

- Alkenes and alkynes are hydrocarbons that contain at least one double or triple bond, respectively.
- Alkenes and alkynes are nonpolar compounds with greater reactivity than alkanes but with other properties similar to those of alkanes.

**VOCABULARY**

- alkene
- alkyne

## SECTION 4 Hydrocarbon Isomers

**MAINIDEA** Some hydrocarbons have the same molecular formula but have different molecular structures.

- Isomers are two or more compounds with the same molecular formula but different molecular structures.
- Structural isomers differ in the order in which atoms are bonded to each other.
- Stereoisomers have all atoms bonded in the same order but arranged differently in space.

**VOCABULARY**

- isomer
- structural isomer
- stereoisomer
- geometric isomer
- chirality
- asymmetric carbon
- optical isomer
- optical rotation

## SECTION 5 Aromatic Hydrocarbons

**MAINIDEA** Aromatic hydrocarbons are unusually stable compounds with ring structures in which electrons are shared by many atoms.

- Aromatic hydrocarbons contain benzene rings as part of their molecular structures.
- The electrons in aromatic hydrocarbons are shared evenly over the entire benzene ring.

**VOCABULARY**

- aromatic compound
- aliphatic compound

# CHAPTER 21 ASSESSMENT

## SECTION 1

### Mastering Concepts

**38. Organic Chemistry** Why did Wöhler's discovery lead to the development of the field of organic chemistry?

**39.** What is the main characteristic of an organic compound?

**40.** What characteristic of carbon accounts for the large variety of organic compounds?

**41.** Name two natural sources of hydrocarbons.

**42.** Explain what physical property of petroleum compounds is used to separate them during fractional distillation.

**43.** Explain the difference between saturated hydrocarbons and unsaturated hydrocarbons.

### Mastering Problems

**44. Distillation** Rank the compounds listed in **Table 7** in the order in which they will be distilled out of a mixture. Rank the compounds in order of first to distill to last to distill.

**Table 7 Alkane Boiling Points**

| Compound | Boiling Point (°C) |
|---|---|
| hexane | 68.7 |
| methane | −161.7 |
| octane | 125.7 |
| butane | −0.5 |
| propane | −42.1 |

**45.** How many electrons are shared between two carbon atoms in each of the following carbon-carbon bonds?
**a.** single bond
**b.** double bond
**c.** triple bond

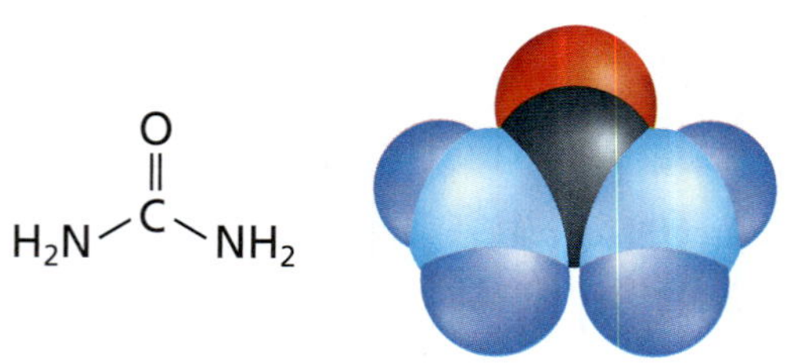

■ **Figure 29**

**46. Figure 29** shows two models of urea, a molecule that Friedrich Wöhler first synthesized in 1828.
**a.** Identify the types of models shown.
**b.** Is urea an organic or an inorganic compound? Explain your answer.

**47.** Molecules are modeled using molecular formulas, structural formulas, ball-and-stick models, and space-filling models. What are the advantages and disadvantages of each model?

## SECTION 2

### Mastering Concepts

**48.** Describe the characteristics of a homologous series of hydrocarbons.

**49. Fuels** Name three alkanes used as fuels and describe an additional application for each.

**50.** Draw the structural formula of each of the following.
**a.** ethane
**b.** hexane
**c.** propane
**d.** heptane

**51.** Write the condensed structural formulas for the alkanes in the previous question.

**52.** Write the name and draw the structure of the alkyl group that corresponds to each of the following alkanes.
**a.** methane
**b.** butane
**c.** octane

**53.** How does the structure of a cycloalkane differ from that of a straight-chain or branched-chain alkane?

**54. Freezing and Boiling Points** Use water and methane to explain how intermolecular attractions generally affect the boiling and freezing points of a substance.

### Mastering Problems

**55.** Name the compound represented by each of the following structural formulas.

**a.** $CH_3CH_2CH_2CH_2CH_3$

**b.**

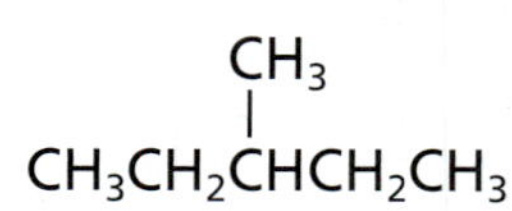

**c.**

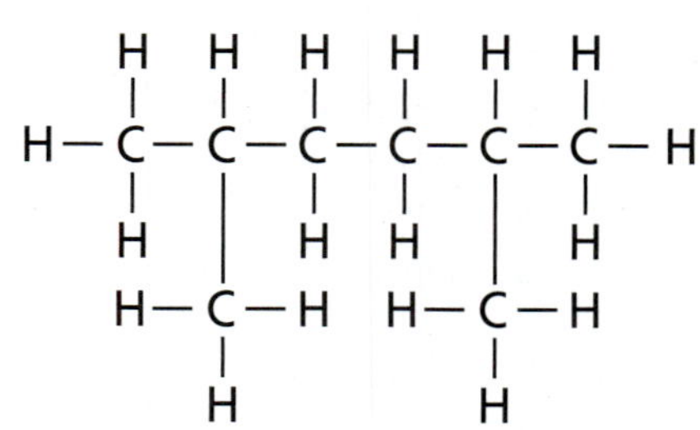

**d.**

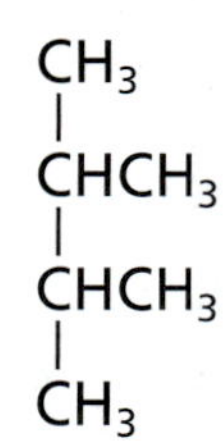

**56.** Draw full structural formulas for the following compounds.
**a.** heptane
**b.** 2-methylhexane
**c.** 2,3-dimethylpentane
**d.** 2,2-dimethylpropane

**57.** Draw condensed structural formulas for the following compounds. Use line structures for rings.
**a.** 1,2-dimethylcyclopropane
**b.** 1,1-diethyl-2-methylcyclopentane

**58.** Name the compound represented by each of the following structural formulas.

a. 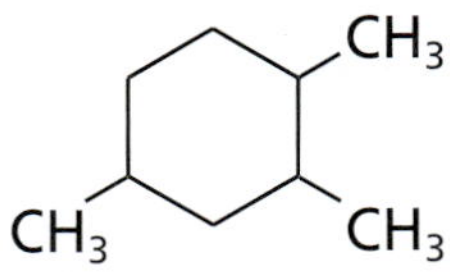

c. 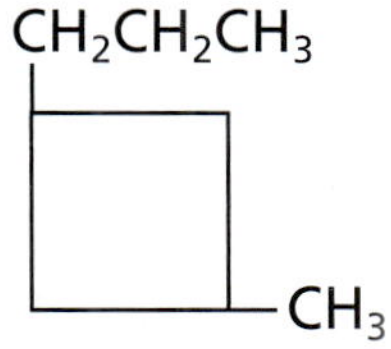

b. 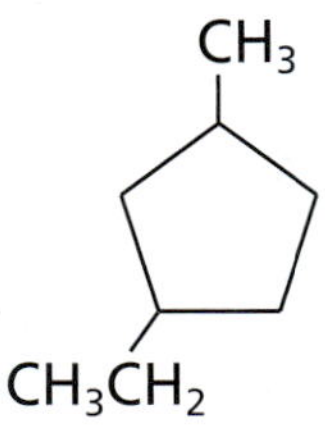

d. 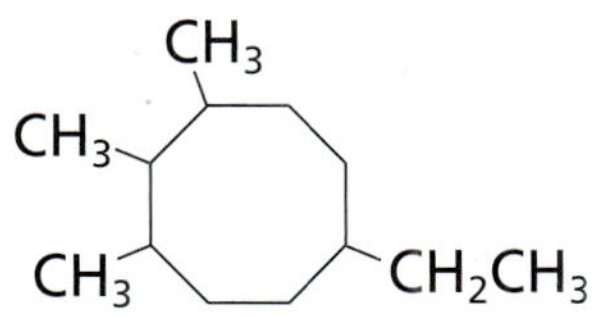

## SECTION 3

### Mastering Concepts

**59.** Explain how alkenes differ from alkanes. How do alkynes differ from both alkenes and alkanes?

**60.** The name of a hydrocarbon is based on the name of the parent chain. Explain how the determination of the parent chain when naming alkenes differs from the same determination when naming alkanes.

### Mastering Problems

**61.** Name the compound represented by each of the following condensed structural formulas.

a. 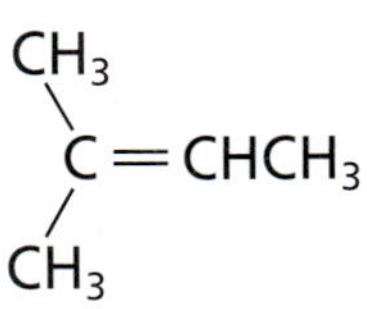

c. 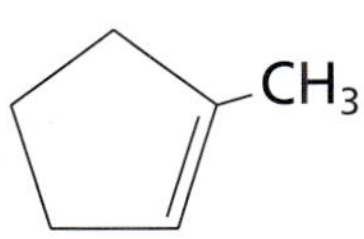

b. 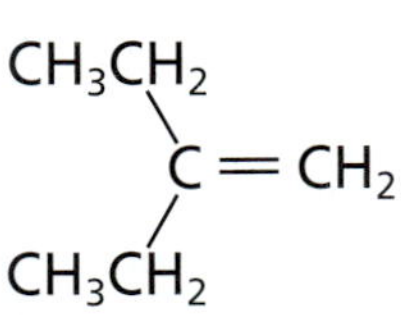

d. 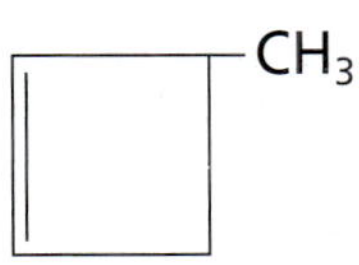

**62.** Draw condensed structural formulas for the following compounds. Use line structures for rings.

**a.** 1,4-diethylcyclohexene

**b.** 2,4-dimethyl-1-octene

**c.** 2,2-dimethyl-3-hexyne

**63.** Name the compound represented by the following condensed structural formula.

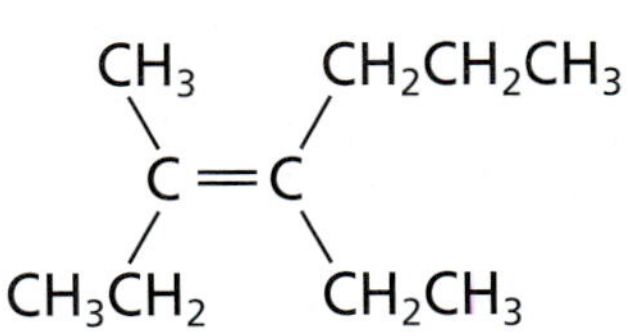

## SECTION 4

### Mastering Concepts

**64.** How are two isomers alike, and how are they different?

**65.** Describe the difference between *cis*- and *trans*- isomers in terms of geometrical arrangement.

**66.** What are the characteristics of a chiral substance?

**67. Light** How does polarized light differ from ordinary light, such as light from the Sun?

**68.** How do optical isomers affect polarized light?

### Mastering Problems

**69.** Identify the pair of structural isomers in the following group of condensed structural formulas.

a. 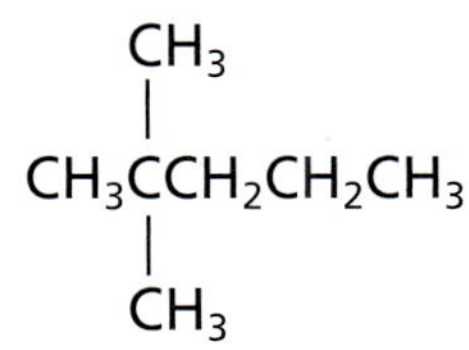

c. 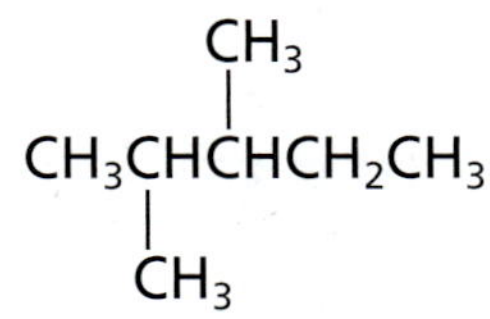

b. 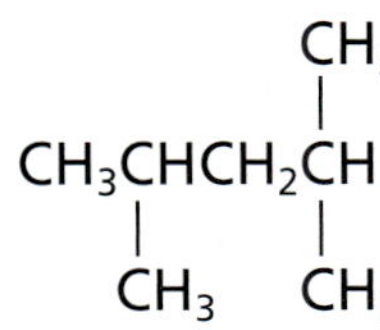

d. 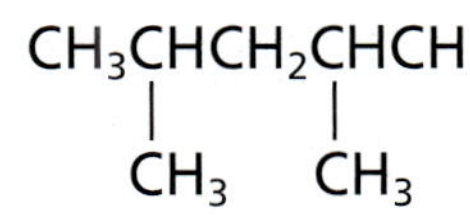

**70.** Identify the pair of geometric isomers among the following structures. Explain your selections. Explain how the third structure is related to the other two.

a. 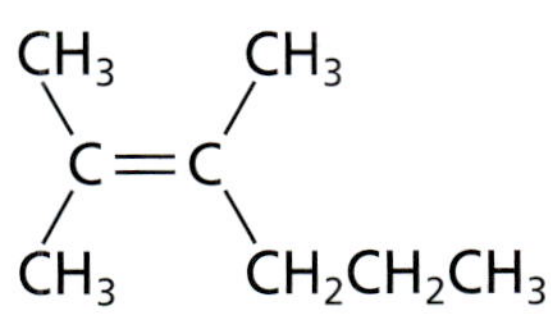

b. 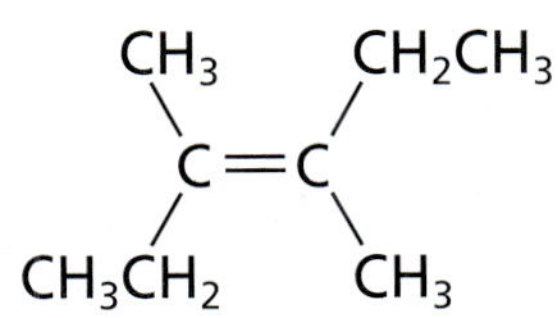

c. 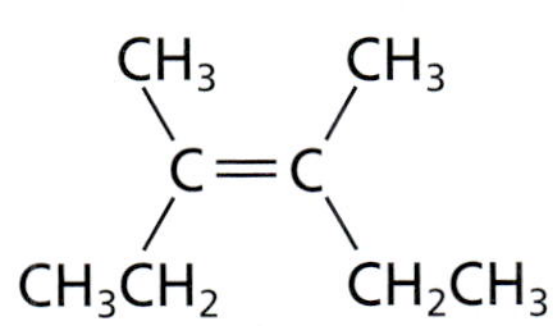

**71.** Draw condensed structural formulas for four different structural isomers with the molecular formula $C_4H_8$.

**72.** Draw and label the *cis*- and *trans*- isomers of the molecule represented by the following condensed formula.

$$CH_3CH=CHCH_2CH_3$$

**73.** Three of the following structures are exactly alike, but the fourth represents an optical isomer of the other three. Identify the optical isomer, and explain how you made your choice.

**a.**
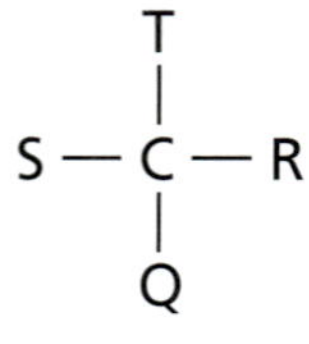

**c.**
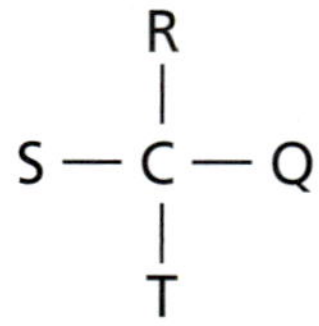

**b.**
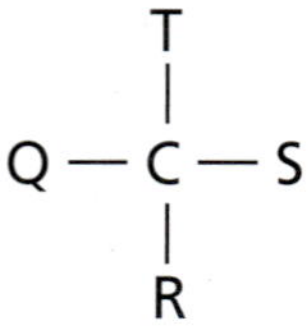

**d.**
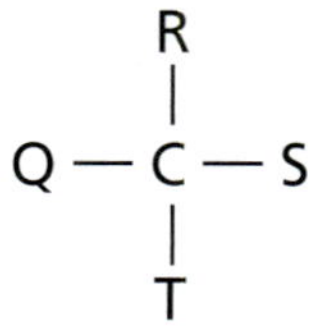

## SECTION 5

### Mastering Concepts

**74.** What structural characteristic do all aromatic hydrocarbons share?

**75.** What are carcinogens?

### Mastering Problems

**76.** Draw the structural formula of 1,2-dimethylbenzene.

**77.** Name the compound represented by each of the following structural formulas.

**a.**
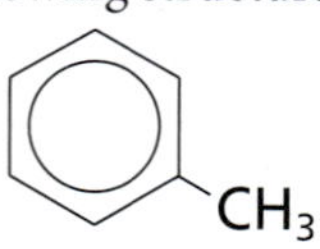

**b.**
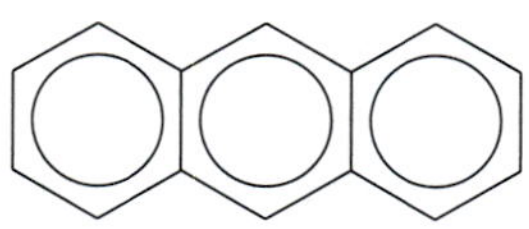

## MIXED REVIEW

**78.** Do the following structural formulas represent the same molecule? Explain your answer.

**a.**
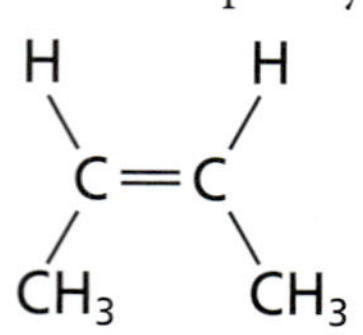

**b.**
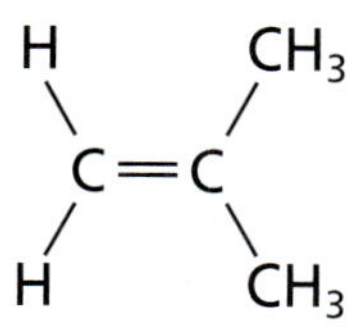

**79.** How many hydrogen atoms are in an alkane molecule with nine carbon atoms? How many are in an alkene with nine carbon atoms and one double bond?

**80.** The general formula for alkanes is $C_nH_{2n+2}$. Determine the general formula for cycloalkanes.

**81. Manufacturing** Why are unsaturated hydrocarbons more useful than saturated hydrocarbons as starting materials in chemical manufacturing?

**82.** Is cyclopentane an isomer of pentane? Explain your answer.

**83.** Determine whether each of the following structures shows the correct numbering. If the numbering is incorrect, redraw the structure with the correct numbering.

**a.**
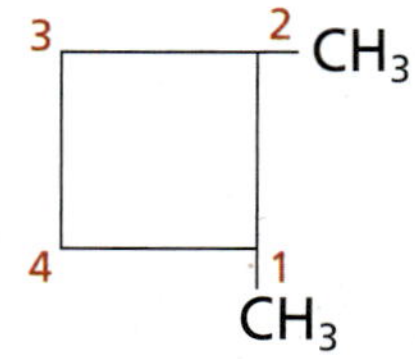

**c.**
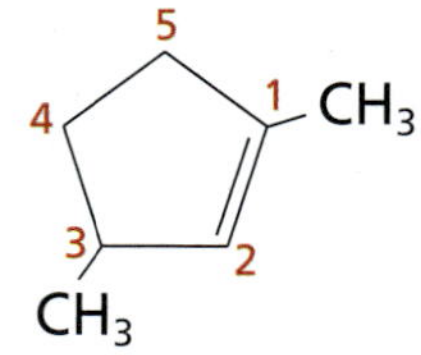

**b.**

1 2 3 4 5

$CH_3CH_2C \equiv CCH_3$

**d.**
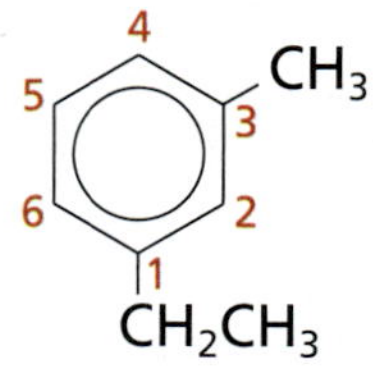

**84.** Why do chemists use structural formulas for organic compounds rather than molecular formulas, such as $C_5H_{12}$?

**85.** Which would you expect to have more similar physical properties, a pair of structural isomers or a pair of stereoisomers? Explain your reasoning.

**86.** Explain why numbers are needed in the IUPAC names of many unbranched alkenes and alkynes but not in the names of unbranched alkanes.

**87.** A compound with two double bonds is called a diene. The name of the structure shown is 1,4-pentadiene. Apply your knowledge of IUPAC nomenclature to draw the structure of 1,3-pentadiene.

$$H_2C{=}CH-CH_2-CH{=}CH_2$$

## THINK CRITICALLY

**88. Determine** which two of the following names cannot be correct, and draw the structures of the molecules.

**a.** 2-ethyl-2-butene
**b.** 1,4-dimethylcyclohexene
**c.** 1,5-dimethylbenzene

**89. Infer** The sugar glucose is sometimes called dextrose because a solution of glucose is known to be dextrorotatory. Analyze the word *dextrorotatory,* and suggest what the word means.

**90. Interpret Scientific Illustrations** Draw Kekulé's structure of benzene, and explain why it does not truly represent the actual structure.

**91. Recognize Cause and Effect** Explain why alkanes, such as hexane and cyclohexane, are effective at dissolving grease, whereas water is not.

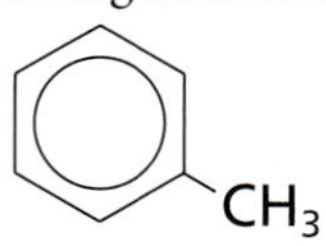

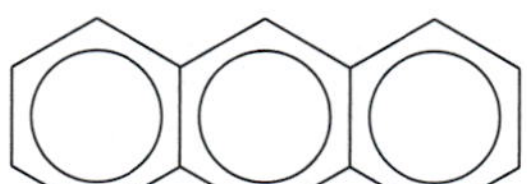

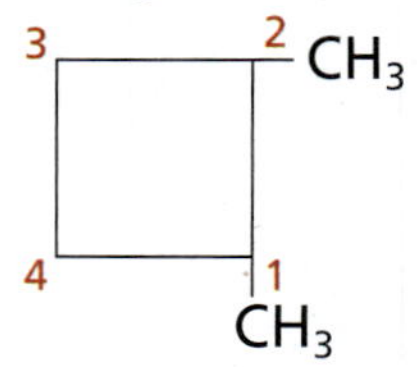

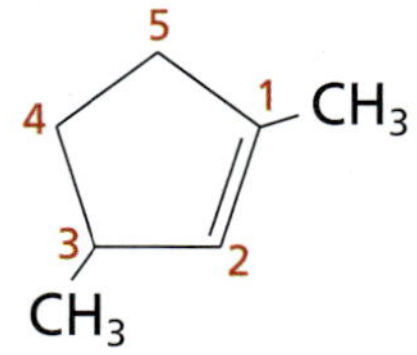

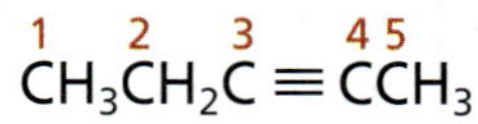

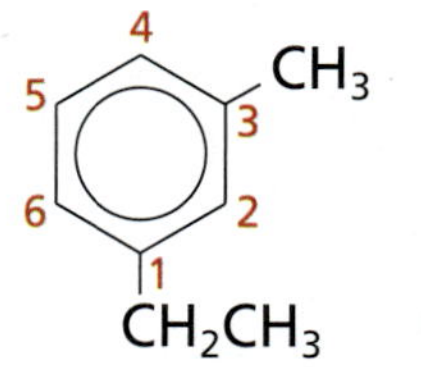

Chapter Self-Check

**92. Explain** Use **Table 8** to construct a statement explaining the relationship between numbers of carbon atoms and boiling points of the members of the alkane series shown.

**93. Graph** the information given in **Table 8.** Predict what the boiling and melting points of the 11- and 12-carbon alkanes will be. Look up the actual values and compare your predictions to the those numbers.

**Table 8** Data for Selected Alkanes

| Name | Melting Point (°C) | Boiling Point (°C) |
|---|---|---|
| $CH_4$ | −182 | −162 |
| $C_2H_6$ | −183 | −89 |
| $C_3H_8$ | −188 | −42 |
| $C_4H_{10}$ | −138 | −0.5 |
| $C_5H_{12}$ | −130 | 36 |
| $C_6H_{14}$ | −95 | 69 |
| $C_7H_{16}$ | −91 | 98 |
| $C_8H_{18}$ | −57 | 126 |
| $C_9H_{20}$ | −54 | 151 |
| $C_{10}H_{22}$ | −29 | 174 |

## CHALLENGE PROBLEM

**94. Chiral Carbons** Many organic compounds have more than one chiral carbon. For each chiral carbon in a compound, a pair of stereoisomers can exist. The total number of possible isomers for the compound is equal to $2^n$, where $n$ is the number of chiral carbons. Draw each structure, and determine how many stereoisomers are possible for each compound named below.

**a.** 3,5-dimethylnonane

**b.** 3,7-dimethyl-5-ethyldecane

## CUMULATIVE REVIEW

**95.** What element has the following ground-state electron configuration: $[Ar]4s^2 3d^6$?

**96.** What is the charge of an ion formed from the following families?

**a.** alkali metals

**b.** alkaline earth metals

**c.** halogens

**97.** Write the chemical equations for the complete combustion of ethane, ethene, and ethyne into carbon dioxide and water.

## WRITING IN ▶ Chemistry

**98. Gasoline** For many years, a principal antiknock ingredient in gasoline was the compound tetraethyllead. Research to learn about the structure of this compound, the history of its development and use, and why its use was discontinued in the United States. Find out if it is still used as a gasoline additive elsewhere in the world.

**99. Perfume** The musk used in perfumes and colognes contains many chemical compounds, including large cycloalkanes. Research and write a short report about the sources used for natural and synthetic musk compounds in these consumer products.

## DBQ Document-Based Questions

**Polycyclic Aromatic Hydrocarbons** *PAH compounds are naturally occurring, but human activities can increase the concentrations in the environment. Soil samples were collected to study PAH compounds. The core sections were dated using radionuclides to determine when each section was deposited.*

**Figure 30** *shows the concentration of polycyclic aromatic hydrocarbons (PAH) detected in Central Park in New York City.*

Data obtained from: Yan, B. et al, 2005. *Environmental Science Technology* 39 (18): 7012–7019.

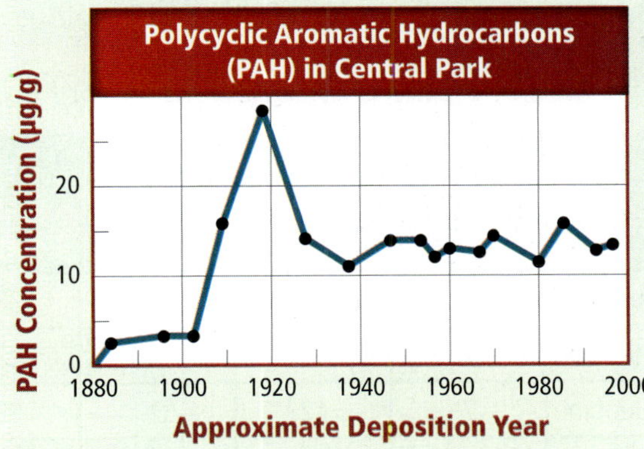


■ **Figure 30**

**100.** Compare the average PAH concentrations before 1905 and after 1925.

**101.** PAH compounds are produced in small amounts by some plants and animals, but most come from human activities, such as burning fossil fuels. Infer why the PAH levels were relatively low in the late 1800s and early 1900s.

## MULTIPLE CHOICE

1. Alanine, like most amino acids, exists in two forms:

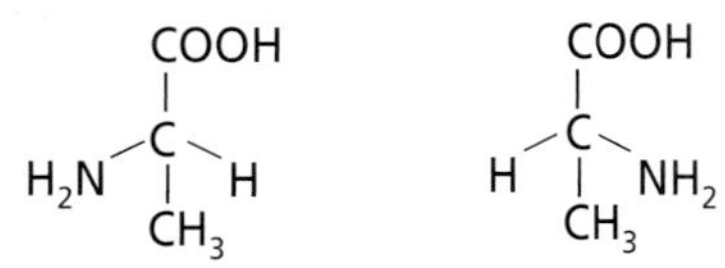

l-Alanine d-Alanine

Almost all of the amino acids found in living organisms are in the l-form. Which term best describes l-alanine and d-alanine with respect to one another?
   **A.** structural isomers
   **B.** geometric isomers
   **C.** optical isomers
   **D.** stereoisotopes

2. Which does NOT affect reaction rate?
   **A.** catalysts
   **B.** surface area of reactants
   **C.** concentration of reactants
   **D.** reactivity of products

3. What is the molality of a solution containing 0.25 g of dichlorobenzene ($C_6H_4Cl_2$) dissolved in 10.0 g of cyclohexane ($C_6H_{12}$)?
   **A.** 0.17 mol/kg
   **B.** 0.014 mol/kg
   **C.** 0.025 mol/kg
   **D.** 0.00017 mol/kg

*Use the table below to answer Questions 4 to 6.*

**Data for Various Hydrocarbons**

| Name | Number of C Atoms | Number of H Atoms | Melting Point (°C) | Boiling Point (°C) |
|---|---|---|---|---|
| Heptane | 7 | 16 | −90.6 | 98.5 |
| 1-Heptene | 7 | 14 | −119.7 | 93.6 |
| 1-Heptyne | 7 | 12 | −81 | 99.7 |
| Octane | 8 | 18 | −56.8 | 125.6 |
| 1-Octene | 8 | 16 | −101.7 | 121.2 |
| 1-Octyne | 8 | 14 | −79.3 | 126.3 |

4. Based on the information in the table, what type of hydrocarbon becomes a gas at the lowest temperature?
   **A.** alkane
   **B.** alkene
   **C.** alkyne
   **D.** aromatic

5. If *n* is the number of carbon atoms in the hydrocarbon, what is the general formula for an alkyne with one triple bond?
   **A.** $C_nH_{n+2}$
   **B.** $C_nH_{2n+2}$
   **C.** $C_nH_{2n}$
   **D.** $C_nH_{2n-2}$

6. It can be predicted from the table that nonane will have a melting point that is
   **A.** greater than that of octane.
   **B.** less than that of heptane.
   **C.** greater than that of decane.
   **D.** less than that of hexane.

7. At a pressure of 1.00 atm and a temperature of 20°C, 1.72 g $CO_2$ will dissolve in 1 L of water. How much $CO_2$ will dissolve if the pressure is raised to 1.35 atm and the temperature stays the same?
   **A.** 2.32 g/L
   **B.** 1.27 g/L
   **C.** 0.785 g/L
   **D.** 0.431 g/L

*Use the diagram below to answer Question 8.*

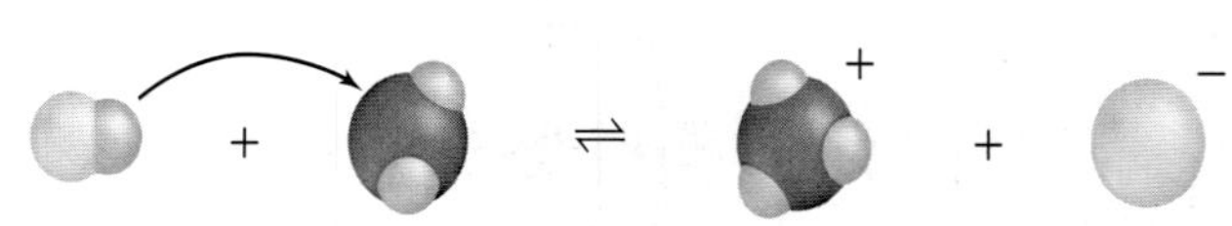

HF $H_2O$ $H_3O^+$ $F^-$

8. In the forward reaction, which substance is the Brønsted-Lowry acid?
   **A.** HF
   **B.** $H_2O$
   **C.** $H_3O^+$
   **D.** $F^-$

9. Which does NOT describe what happens as a liquid boils?
   **A.** The temperature of the system rises.
   **B.** Energy is absorbed by the system.
   **C.** The vapor pressure of the liquid is equal to atmospheric pressure.
   **D.** The liquid is entering the gas phase.

## SHORT ANSWER

*Use the diagram below to answer Questions 10 to 12.*

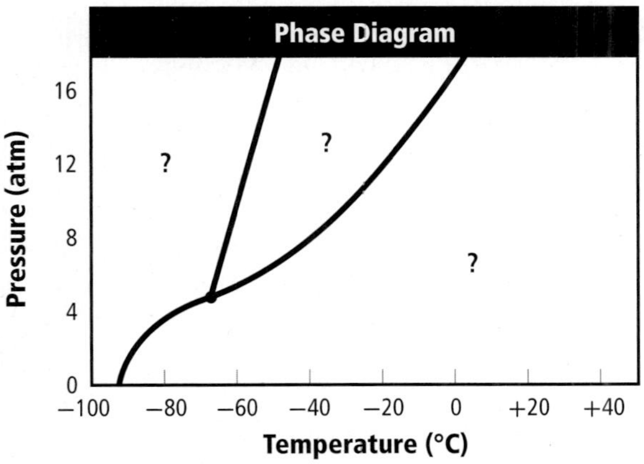


**10.** What state of matter is located at a temperature of −80°C and a pressure of 10 atm?

**11.** What are the temperature and pressure when the substance is at its triple point?

**12.** Describe the changes in molecular arrangement that occur when the pressure is increased from 8 atm to 16 atm, while the temperature is held constant at 0°C.

## EXTENDED RESPONSE

*Use the data table below to answer Questions 13 and 14.*

| Experimental Data for the Reaction between A and B | | |
|---|---|---|
| [A] Initial | [B] Initial | Initial rate (mol/L·s) |
| 0.10*M* | 0.10*M* | 7.93 |
| 0.30*M* | 0.10*M* | 23.79 |
| 0.30*M* | 0.20*M* | 95.16 |

**13.** Find the values of *m* and *n* for the rate law expression $rate = k[A]^m[B]^n$.

**14.** Determine the value of *k* for this reaction.

## SAT SUBJECT TEST: CHEMISTRY

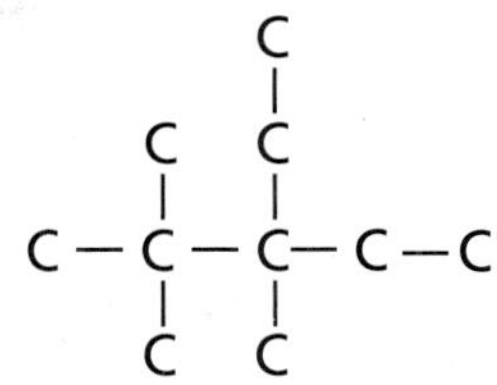

**15.** What is the name of the compound whose skeletal formula is shown above?
**A.** 2,2,3-trimethyl-3-ethylpentane
**B.** 3-ethyl-3,4,4-trimethylpentane
**C.** 2-butyl-2-ethylbutane
**D.** 3-ethyl-2,2,3-trimethylpentane
**E.** 2,2-dimethyl, 3-diethyl, 3-methylpropane

*Use the graph below to answer Questions 16 and 17.*

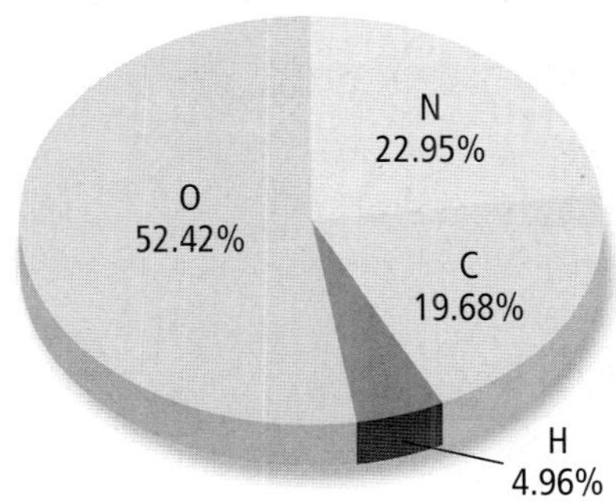


**16.** What is the formula for this compound?
**A.** $C_5H_{20}N_4O_2$
**B.** $C_8H_2N_9O_{11}$
**C.** $C_{1.6}H_5N_{1.6}O_{3.3}$
**D.** $CH_3NO_2$
**E.** $C_2H_5N_2O_5$

**17.** How many grams of nitrogen would be present in 475 g of this compound?
**A.** 33.9 g
**B.** 52.8 g
**C.** 67.9 g
**D.** 109 g
**E.** 120.0 g

| NEED EXTRA HELP? | | | | | | | | | | | | | | | | | |
|---|---|---|---|---|---|---|---|---|---|---|---|---|---|---|---|---|---|
| If You Missed Question . . . | 1 | 2 | 3 | 4 | 5 | 6 | 7 | 8 | 9 | 10 | 11 | 12 | 13 | 14 | 15 | 16 | 17 |
| Review Section . . . | 21.4 | 16.2 | 14.2 | 21.3 | 21.3 | 21.3 | 14.1 | 18.1 | 12.4 | 12.4 | 12.4 | 12.4 | 16.3 | 16.3 | 21.2 | 10.4 | 10.4 |

# CHAPTER 22

# Substituted Hydrocarbons and Their Reactions

**BIGIDEA** The substitution of different functional groups for hydrogen atoms in hydrocarbons results in a diverse group of organic compounds.

## SECTIONS


1 Alkyl Halides and Aryl Halides

2 Alcohols, Ethers, and Amines

3 Carbonyl Compounds

4 Other Reactions of Organic Compounds

5 Polymers


## LaunchLAB

iLab Station

### How do you make slime?

In addition to carbon and hydrogen, most organic substances contain other elements that give the substances unique properties. In this lab, you will determine how the properties of substances change when groups form bonds called crosslinks between the chains.

**FOLDABLES® Study Organizer**

### Functional Groups

Make a top-tab book. Label it as shown. Use it to organize information about the functional groups of organic compounds.

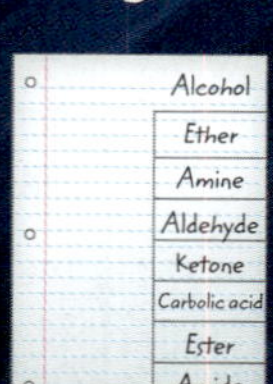


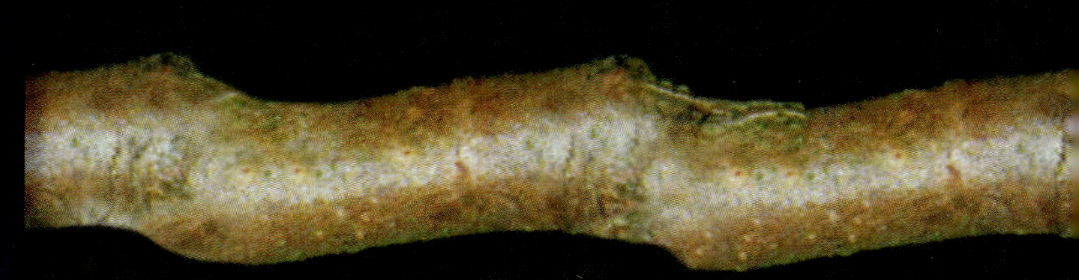

The larva of the *Cerura vinula* moth squirts formic acid when threatened. Formic acid is a substituted hydrocarbon—a type of compound that forms when one or more hydrogens in a hydrocarbon are replaced by other atoms.

Go **online!**
MHEonline.com

SECTION 1

# Alkyl Halides and Aryl Halides

**MAIN IDEA** A halogen atom can replace a hydrogen atom in some hydrocarbons.

### Essential Questions

- What are functional groups, and what are some examples?
- How do alkyl and aryl halide structures compare and contrast?
- What factors affect the boiling points of organic halides?

### Review Vocabulary

**aliphatic compound:** a nonaromatic hydrocarbon, such as an alkane, an alkene, or an alkyne

### New Vocabulary

functional group
halocarbon
alkyl halide
aryl halide
plastic
substitution reaction
halogenation

**CHEM 4 YOU** If you have ever played on a sports team, were individual players substituted during the game? For example, a player who is rested might substitute for a player who is tired. After the substitution, the characteristics of the team change.

## Functional Groups

You read previously that in hydrocarbons, carbon atoms are linked only to other carbon atoms or hydrogen atoms. But carbon atoms can also form strong covalent bonds with other elements, the most common of which are oxygen, nitrogen, fluorine, chlorine, bromine, iodine, sulfur, and phosphorus.

Atoms of these elements occur in organic substances as parts of functional groups. In an organic molecule, a **functional group** is an atom or group of atoms that always reacts in a certain way. The addition of a functional group to a hydrocarbon structure always produces a substance with physical and chemical properties that differ from those of the parent hydrocarbon. All the items—natural and synthetic—in **Figure 1** contain functional groups that give them their individual characteristics, such as smell. Organic compounds containing several important functional groups are shown in **Table 1.** The symbols R and R′ represent carbon chains or rings bonded to the functional group. An * represents a hydrogen atom, carbon chain, or carbon ring.

Keep in mind that double and triple bonds between two carbon atoms are considered functional groups even though only carbon and hydrogen atoms are involved. By learning the properties associated with a given functional group, you can predict the properties of organic compounds for which you know the structure, even if you have never studied them.

■ **Figure 1** All of these items contain at least one of the functional groups that you will study in this chapter. For example, the fruit and flowers have sweet-smelling aromas that are due to ester molecules.

Matt Meadows

Explore **functional groups with an interactive table.** Concepts In Motion

**Table 1** Organic Compounds and Their Functional Groups

| Compound Type | General Formula | Functional Group |
|---|---|---|
| Halocarbon | R—X (X = F, Cl, Br, I) | Halogen |
| Alcohol | R—OH | Hydroxyl |
| Ether | R—O—R′ | Ether |
| Amine | $R—NH_2$ | Amino |
| Aldehyde | *—C(=O)—H | Carbonyl |
| Ketone | R—C(=O)—R′ | Carbonyl |
| Carboxylic acid | *—C(=O)—OH | Carboxyl |
| Ester | *—C(=O)—O—R | Ester |
| Amide | *—C(=O)—N(H)—R | Amide |

## Organic Compounds Containing Halogens

Explore **functional groups.** Virtual Investigations

The most simple functional groups can be thought of as substituent groups attached to a hydrocarbon. Recall that a substituent group is a side branch attached to a parent chain. The elements in group 17 of the periodic table—fluorine, chlorine, bromine, and iodine—are the halogens. Any organic compound that contains a halogen substituent is called a **halocarbon.** If you replace any of the hydrogen atoms in an alkane with a halogen atom, you form an alkyl halide. An **alkyl halide** is an organic compound containing a halogen atom covalently bonded to an aliphatic carbon atom. The first four halogens—fluorine, chlorine, bromine, and iodine—are found in many organic compounds. For example, chloromethane is the alkyl halide formed when a chlorine atom replaces one of methane's four hydrogen atoms, as shown in **Figure 2.**

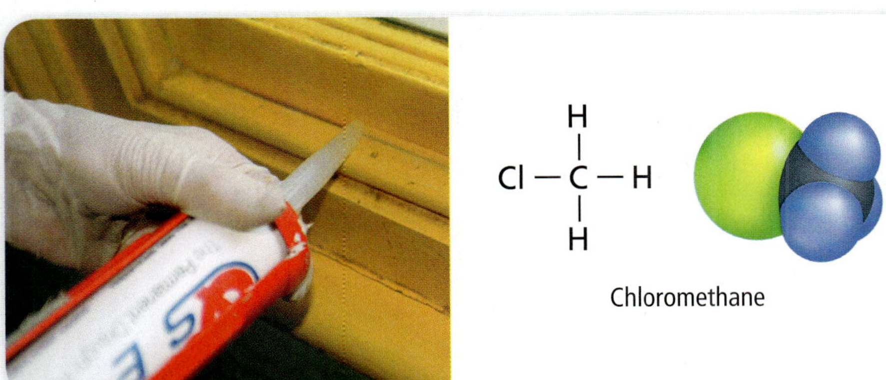


**Figure 2** Chloromethane is an alkyl halide that is used in the manufacturing process for silicone products, such as window and door sealants.

David Hoffman Photo Library/Alamy

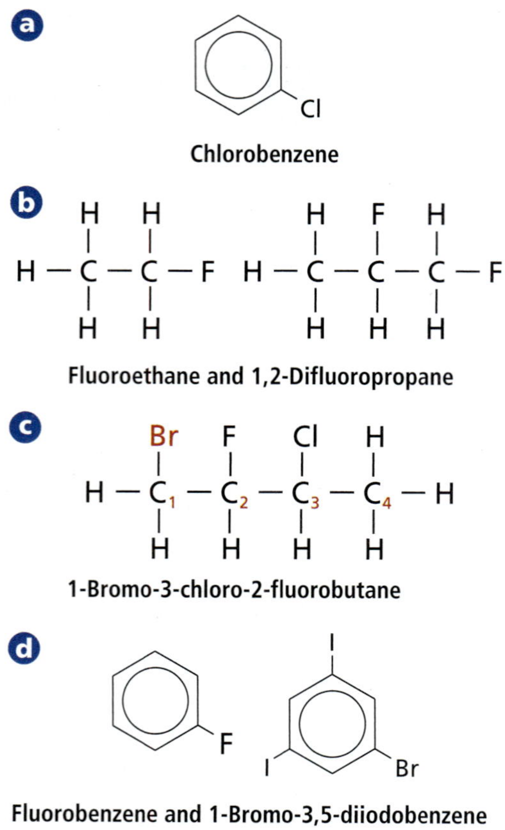


**Figure 3** Organic molecules containing functional groups are named based on their main-chain alkane structure using IUPAC conventions.

An **aryl halide** is an organic compound containing a halogen atom bonded to a benzene ring or other aromatic group. The structural formula for an aryl halide is created by first drawing the aromatic structure and then replacing its hydrogen atoms with the halogen atoms specified, as shown in **Figure 3a.**

**Connection to Earth Science** Alkyl halides are widely used as refrigerants. Until the late 1980s, alkyl halides called chlorofluorocarbons (CFCs) were widely used in refrigerators and air-conditioning systems. Recall how CFCs affect the ozone layer. CFCs have been replaced by HFCs (hydrofluorocarbons), which contain only hydrogen and fluorine atoms bonded to carbon. One of the more common HFCs is 1,1,2-trifluoroethane, also called R134a.

**Naming halocarbons** Organic molecules containing functional groups are given IUPAC names based on their main-chain alkane structures. For the alkyl halides, a prefix indicates which halogen is present. The prefixes are formed by changing the *-ine* at the end of each halogen name to *-o.* Thus, the prefix for fluorine is *fluoro-,* chlorine is *chloro-,* bromine is *bromo-,* and iodine is *iodo-,* as shown in **Figure 3b.**

If more than one kind of halogen atom is present in the same molecule, the atoms are listed alphabetically in the name. The chain also must be numbered in a way that gives the lowest position number to the substituent that comes first in the alphabet. Note how the alkyl halide in **Figure 3c** is named.

Similarly, the benzene ring in an aryl halide is numbered to give each substituent the lowest position number possible, as shown in **Figure 3d.**

**READING CHECK** **Infer** why the lowest possible position number is used to name an aryl halide instead of using a randomly chosen position number.

## PRACTICE Problems

Do additional problems. Online Practice

**Name the alkyl or aryl halide whose structure is shown.**

**1.**

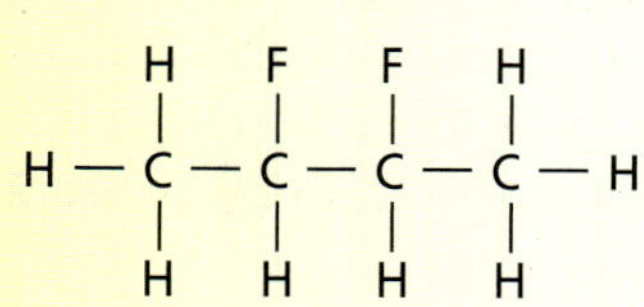

**2.**

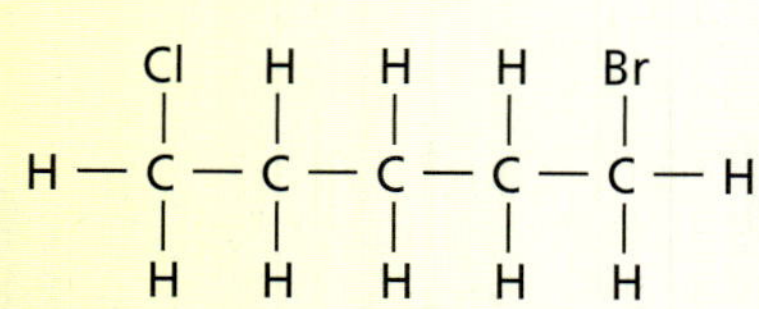

**3.**

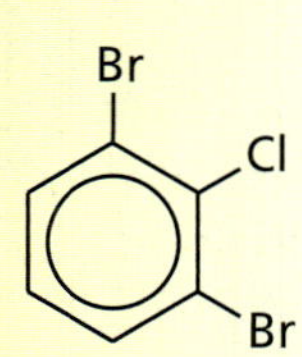

**Table 2 A Comparison of Alkyl Halides and Their Parent Alkanes**

| Structure | Name | Boiling Point (°C) | Density (g/mL) in Liquid State |
|---|---|---|---|
| $CH_4$ | methane | −162 | 0.423 at −162°C (boiling point) |
| $CH_3Cl$ | chloromethane | −24 | 0.911 at 25°C (under pressure) |
| $CH_3CH_2CH_2CH_2CH_3$ | pentane | 36 | 0.626 |
| $CH_3CH_2CH_2CH_2CH_2F$ | 1-fluoropentane | 62.8 | 0.791 |
| $CH_3CH_2CH_2CH_2CH_2Cl$ | 1-chloropentane | 108 ↓ *Increases* | 0.882 ↓ *Increases* |
| $CH_3CH_2CH_2CH_2CH_2Br$ | 1-bromopentane | 130 | 1.218 |
| $CH_3CH_2CH_2CH_2CH_2I$ | 1-iodopentane | 155 | 1.516 |

**Properties and uses of halocarbons** It is easiest to talk about properties of organic compounds containing functional groups by comparing those compounds with alkanes, whose properties you have already studied. **Table 2** lists some of the physical properties of certain alkanes and alkyl halides.

Note that each alkyl chloride has a higher boiling point and a higher density than the alkane with the same number of carbon atoms. Note also that the boiling points and densities increase as the halogen changes from fluorine to chlorine, bromine, and iodine. This trend occurs primarily because the halogens from fluorine to iodine have increasing numbers of electrons that lie farther from the halogen nucleus. These electrons shift position easily and, as a result, the halogen-substituted hydrocarbons have an increasing tendency to form temporary dipoles. Because the dipoles attract each other, the energy needed to separate the molecules also increases. Thus, the boiling points of halogen-substituted alkanes increase as the size of the halogen atom increases.

☑ READING CHECK **Explain** the relationship between the number of electrons in the halogen and the boiling point.

Organic halides are seldom found in nature, although human thyroid hormones are organic iodides. Halogen atoms bonded to carbon atoms are more reactive than the hydrogen atoms they replace. For this reason, alkyl halides are often used as starting materials in the chemical industry. Alkyl halides are also used as solvents and cleaning agents because they readily dissolve nonpolar molecules, such as greases. **Figure 4** shows an application of polytetrafluoroethene (PTFE), a plastic made from gaseous tetrafluoroethylene. A **plastic** is a polymer that can be heated and molded while relatively soft. Another plastic commonly called *vinyl* is polyvinyl chloride (PVC). It can be manufactured soft or hard, as thin sheets, or molded into objects.

☑ READING CHECK **Explain** why alkyl halides are often used in the chemical industry as starting materials instead of alkanes.

■ **Figure 4** Polytetrafluoroethene (PTFE) is made up of hundreds of units. PTFE provides a nonstick surface for many kitchen items, including bakeware.

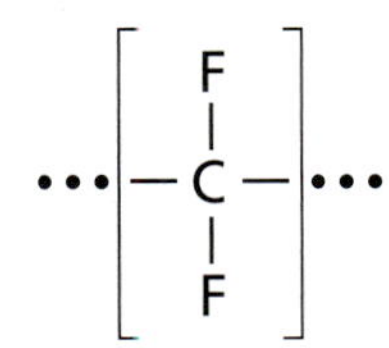


PTFE

PTFE Application

**Table 3 Substitution Reactions**

| Generic Substitution Reaction | Example of General Substitution Reaction (Halogenation) |
|---|---|
| $R{-}CH_3 + X_2 \rightarrow R{-}CH_2X + HX$<br>where X is fluorine, chlorine, or bromine | $C_2H_6 + Cl_2 \rightarrow C_2H_5Cl + HCl$<br>Ethane Chloroethane |
| **General Alkyl Halide-Alcohol Reaction**<br>$R{-}X + OH^- \rightarrow R{-}OH + X^-$<br>Alkyl halide Alcohol | **Example of an Alkyl Halide-Alcohol Reaction**<br>$CH_3CH_2Cl + OH^- \rightarrow CH_3CH_2OH + Cl^-$<br>Chloroethane Ethanol |
| **General Alkyl Halide-Ammonia Reaction**<br>$R{-}X + NH_3 \rightarrow R{-}NH_2 + HX$<br>Alkyl halide Amine | **Example of an Alkyl Halide-Ammonia Reaction**<br>$CH_3(CH_2)_6CH_2Br + NH_3 \rightarrow CH_3(CH_2)_6CH_2NH_2 + HBr$<br>1-Bromooctane 1-Octanamine |

## Substitution Reactions

From where does the immense variety of organic compounds come? Amazingly enough, the ultimate source of nearly all synthetic organic compounds is petroleum. The oil-field workers shown in **Figure 5** are drilling for petroleum, which is a fossil fuel that consists almost entirely of hydrocarbons, especially alkanes. How can alkanes be converted into compounds as different as alkyl halides, alcohols, and amines?

One way is to introduce a functional group through substitution, as shown in **Table 3.** A **substitution reaction** is one in which one atom or a group of atoms in a molecule is replaced by another atom or group of atoms. With alkanes, hydrogen atoms can be replaced by atoms of halogens, typically chlorine or bromine, in a process called **halogenation.** One example of a halogenation reaction, shown in **Table 3,** is the substitution of a chlorine atom for one of ethane's hydrogen atoms. **Figure 6** shows another halogenated hydrocarbon commonly called halothane (2-bromo-2-chloro-1,1,1-trifluoroethane), which was first used as a general anesthetic in the 1950s.

Equations for organic reactions are sometimes shown in generic form. **Table 3** shows the generic form of a substitution reaction. In this reaction, *X* can be fluorine, chlorine, or bromine, but not iodine. Iodine does not react well with alkanes.

**READING CHECK** **Draw** the molecular structure of halothane.

Keith Wood/Getty Images

**Figure 5** These oil-field workers are drilling for petroleum. A single oil well can extract more than 100 barrels per day.
**Explain** *the relationship between petroleum and synthetic organic compounds.*

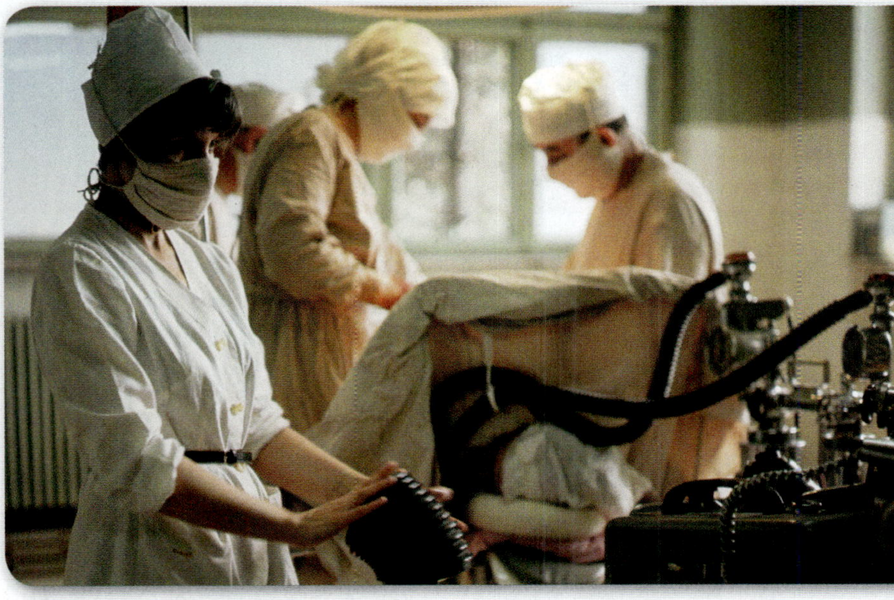

■ **Figure 6** Halothane was introduced into medicine in the 1950s as a general anesthetic for patients undergoing surgery.

**Further substitution** Once an alkane has been halogenated, the resulting alkyl halide can undergo other types of substitution reactions in which the halogen atom is replaced by another atom or group of atoms. For example, reacting an alkyl halide with a basic solution results in the replacement of the halogen atom by an –OH group, forming an alcohol. An example of an alkyl halide-alcohol reaction is shown in **Table 3.** The generic form of the alkyl halide-alcohol reaction is also shown in **Table 3.**

Reacting an alkyl halide with ammonia ($NH_3$) replaces the halogen atom with an amino group ($-NH_2$), forming an alkyl amine, also shown in **Table 3.** The alkyl amine is one of the products produced in this reaction. Some of the newly formed amines continue to react, resulting in a mixture of amines.

**FOLDABLES®** Incorporate information from this section into your Foldable.

## SECTION 1 REVIEW

Section Self-Check

### Section Summary

- The substitution of functional groups for hydrogen in hydrocarbons creates a wide variety of organic compounds.
- An alkyl halide is an organic compound that has one or more halogen atoms bonded to a carbon atom in an aliphatic compound.

4. MAINIDEA **Compare and contrast** alkyl halides and aryl halides.
5. **Draw** structures for the following molecules.
   **a.** 2-chlorobutane
   **b.** 1,3-difluorohexane
   **c.** 1,1,1-trichloroethane
   **d.** 1-bromo-4-chlorobenzene
6. **Define** *functional group* and name the group present in each of the following structures. Name the type of organic compound each substance represents.
   **a.** $CH_3CH_2CH_2OH$
   **b.** $CH_3CH_2F$
   **c.** $CH_3CH_2NH_2$
   **d.** $CH_3C(=O)—OH$
7. **Evaluate** How would you expect the boiling points of propane and 1-chloropropane to compare? Explain your answer.
8. **Interpret Scientific Illustrations** Examine the pair of substituted hydrocarbons illustrated at right, and decide whether it represents a pair of optical isomers. Explain your answer.

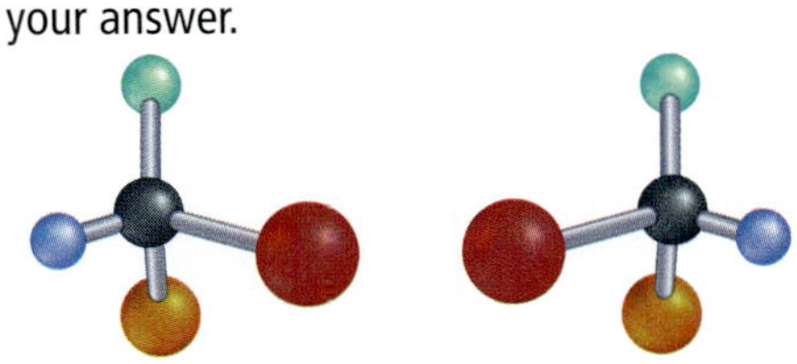

## SECTION 2

# Alcohols, Ethers, and Amines

**MAIN**IDEA **Oxygen and nitrogen are two of the most-common atoms found in organic functional groups.**

### Essential Questions

- Which functional groups define alcohols, ethers, and amines?
- How are the structures of alcohols, ethers, and amines drawn?
- What are some properties and uses of alcohols, ethers, and amines?

### Review Vocabulary

**miscible:** describes two liquids that are soluble in each other

### New Vocabulary

hydroxyl group
alcohol
denatured alcohol
ether
amine

**CHEM 4 YOU** **The last time you had a vaccination, the nurse probably disinfected your skin with an alcohol wipe before giving you the injection. Did you know that the nurse was using a substituted hydrocarbon?**

## Alcohols

Many organic compounds contain oxygen atoms bonded to carbon atoms. Because an oxygen atom has six valence electrons, it commonly forms two covalent bonds to gain a stable octet. An oxygen atom can form a double bond with a carbon atom, replacing two hydrogen atoms, or it can form one single bond with a carbon atom and another single bond with another atom, such as hydrogen. An oxygen-hydrogen group covalently bonded to a carbon atom is called a **hydroxyl group** (–OH). An organic compound in which a hydroxyl group replaces a hydrogen atom of a hydrocarbon is called an **alcohol.** As shown in **Table 4,** the general formula for an alcohol is ROH. **Table 4** also illustrates the relationship of the simplest alkane, methane, to the simplest alcohol, methanol.

Ethanol and carbon dioxide are produced by yeasts when they ferment sugars, such as those in grapes and bread dough. Ethanol is found in alcoholic beverages and medicinal products. Because it is an effective antiseptic, ethanol can be used to swab skin before an injection is given. It is also a gasoline additive and an important starting material for the synthesis of more complex organic compounds.

**Figure 7** shows a model of an ethanol molecule and a model of a water molecule. As you compare the models, notice that the covalent bonds from the oxygen in ethanol are at roughly the same angle as the bonds around the oxygen in the water molecule. Therefore, the hydroxyl groups of alcohol molecules are moderately polar, as with water, and are able to form hydrogen bonds with the hydroxyl groups of other alcohol molecules. Due to this hydrogen bonding, alcohols have much higher boiling points than hydrocarbons of similar shape and size.

**Table 4 Alcohols**

| General Formula | Simple Alcohol and Simple Hydrocarbon |
|---|---|
| ROH<br>R represents carbon chains or rings bonded to the functional group | 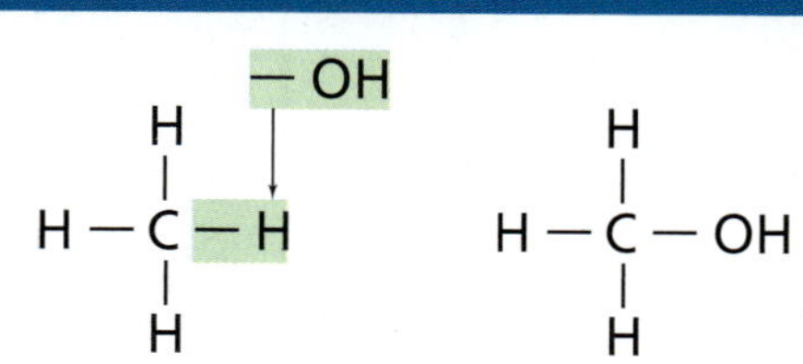<br>Methane ($CH_4$) — Alkane<br>Methanol ($CH_3OH$) — Alcohol |

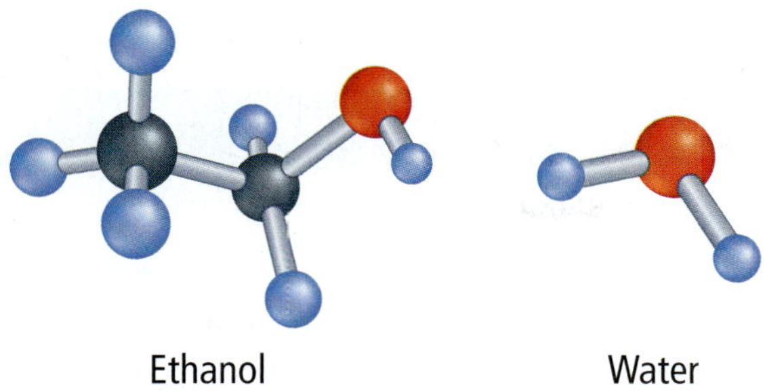


■ **Figure 7** The covalent bonds from oxygen have approximately the same bonding angle in ethanol and water.

Also, because of polarity and hydrogen bonding, ethanol is completely miscible with water. In fact, once they are mixed, it is difficult to separate water and ethanol completely. Distillation is used to remove ethanol from water, but even after that process is complete, about 5% water remains in the ethanol-water mixture.

On the shelves of drugstores, you can find bottles of ethanol labeled *denatured alcohol.* **Denatured alcohol** is ethanol to which small amounts of noxious materials, such as aviation gasoline or other organic solvents, have been added. Ethanol is denatured in order to make it unfit to drink. Because of their polar hydroxyl groups, alcohols make good solvents for other polar organic substances. For example, methanol, the smallest alcohol, is a common industrial solvent found in some paint strippers, and 2-butanol is found in some stains and varnishes.

Note that the names of alcohols are based on alkane names, like the names of alkyl halides. For example, $CH_4$ is methane and $CH_3OH$ is methanol; $CH_3CH_3$ is ethane and $CH_3CH_2OH$ is ethanol. When naming a simple alcohol based on an alkane carbon chain, the IUPAC rules call for naming the parent carbon chain or ring first and then changing the *-e* at the end of the name to *-ol* to indicate the presence of a hydroxyl group. In alcohols of three or more carbon atoms, the hydroxyl group can be at two or more positions. To indicate the position, a number is added, as shown in **Figure 8a** and **8b.**

**READING CHECK** **Explain** why *4-butanol* and *3-butanol* are not the correct names for the compounds in **Figure 8a** and **8b.**

Now look at **Figure 8c.** The compound's ring structure contains six carbons with only single bonds, so you know that the parent hydrocarbon is cyclohexane. Because an –OH group is bonded to a carbon, it is an alcohol and the name will end in *-ol.* No number is necessary because all carbons in the ring are equivalent. This compound is called cyclohexanol. It is a poisonous compound used as a solvent for certain plastics and in the manufacture of insecticides.

A carbon chain can also have more than one hydroxyl group. To name these compounds, prefixes such as *di-*, *tri-*, and *tetra-* are used before the *-ol* to indicate the number of hydroxyl groups present. The full alkane name, including *-ane,* is used before the prefix.

**Figure 8d** shows the molecule 1,2,3-propanetriol, commonly called glycerol. It is an alcohol containing more than one hydroxyl group. Glycerol is often used as an antifreeze and as an airplane deicing fluid.

**READING CHECK** **Explain** why numbers are not used to name the compound shown in **Figure 8c.**

■ **Figure 8** The names of alcohols are based on alkane names.

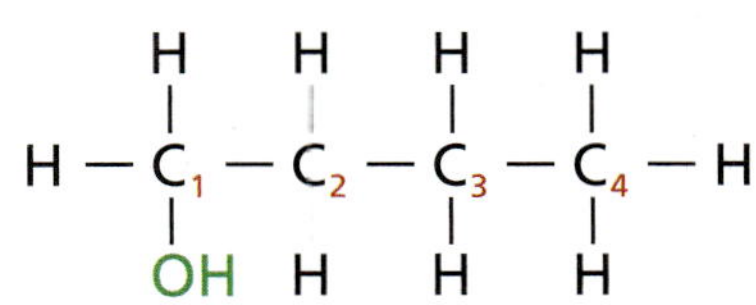

**a.** 1-Butanol

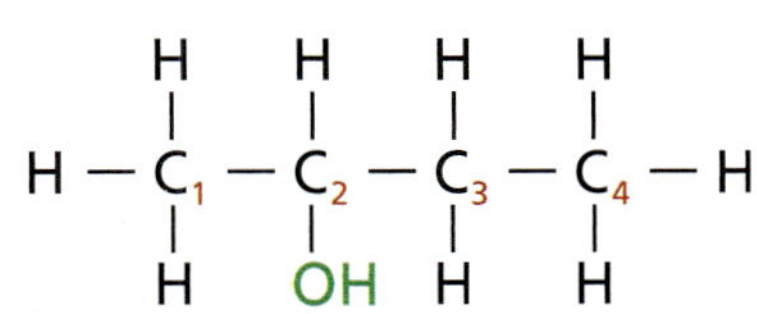

**b.** 2-Butanol

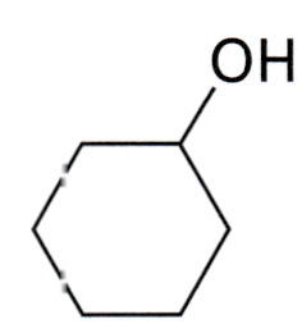

**c.** Cyclohexanol

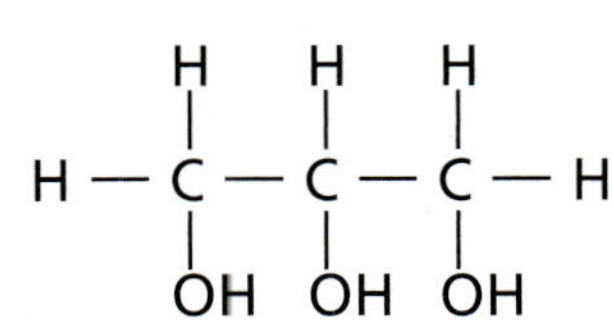

**d.** 1,2,3-Propanetriol (glycerol)

**Table 5** Ethers

| General Formula | Methanol and Methyl ether | |
|---|---|---|
| ROR′<br>where R and R′ represent carbon chains or rings bonded to functional groups | 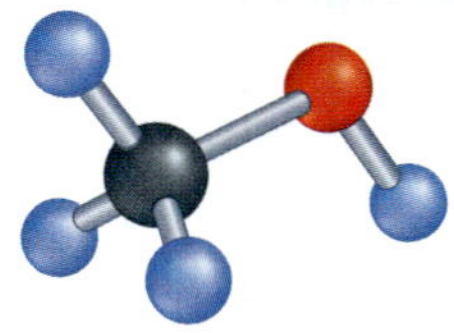<br>Methanol<br>bp = 65°C | 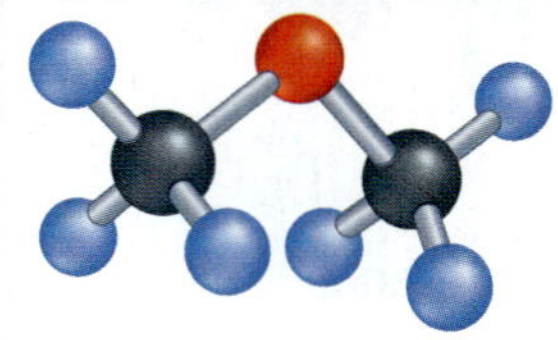<br>Methyl ether<br>bp = −25°C |

| Examples of Ethers | |
|---|---|
| 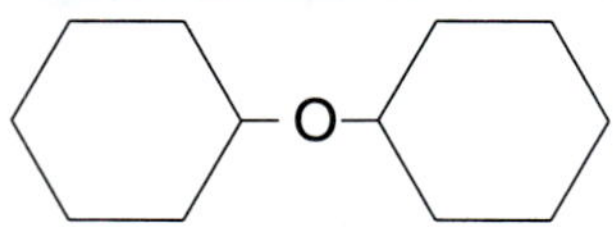<br>Cyclohexyl ether | $CH_3CH_2CH_2-O-CH_2CH_2CH_3$<br>Propyl ether |
| $CH_3CH_2-O-CH_2CH_2CH_2CH_3$<br>Butyl ethyl ether | $CH_3CH_2-O-CH_3$<br>Ethyl methyl ether |

**VOCABULARY**

**ACADEMIC VOCABULARY**

**Bond**
to connect, bind, or join
*An oxygen atom bonds to two carbon atoms in an ether.*

## Ethers

Ethers are another group of organic compounds in which oxygen is bonded to carbon. An **ether** is an organic compound containing an oxygen atom bonded to two carbon atoms. Ethers have the general formula ROR′, as shown in **Table 5.** The simplest ether is one in which oxygen is bonded to two methyl groups. Note the similarity between methanol and methyl ether shown in **Table 5.**

The term *ether* was first used in chemistry as a name for ethyl ether, the volatile, highly flammable substance that was commonly used as an anesthetic in surgery from 1842 until the twentieth century. As time passed, the term *ether* was applied to other organic substances having two hydrocarbon chains attached to the same oxygen atom.

**FOLDABLES®** Incorporate information from this section into your Foldable.

Because ethers have no hydrogen atoms bonded to the oxygen atom, their molecules cannot form hydrogen bonds with each other. Therefore, ethers are generally more volatile and have much lower boiling points than alcohols of similar size and mass. Ethers are much less soluble in water than alcohols because they have no hydrogen to donate to a hydrogen bond. However, the oxygen atom can act as a hydrogen bond receptor for the hydrogen atoms of water molecules.

☑ READING CHECK **Infer** why ethyl ether is undesirable as an anesthetic.

When naming ethers that have two identical alkyl chains bonded to oxygen, first name the alkyl group and then add the word *ether.* **Table 5** shows the structures and names of two of these symmetrical ethers, propyl ether and cyclohexyl ether. If the two alkyl groups are different, the groups are listed in alphabetical order and then followed by the word *ether.* **Table 5** contains two examples of these asymmetrical ethers, butyl ethyl ether and ethyl methyl ether.

## Amines

**Amines** contain nitrogen atoms bonded to carbon atoms in aliphatic chains or aromatic rings and have the general formula $RNH_2$, as shown in **Table 6.**

Chemists consider amines to be derivatives of ammonia ($NH_3$). Amines are considered primary, secondary, or tertiary amines depending on whether one, two, or three of the hydrogens in ammonia have been replaced by organic groups.

When naming amines, the $-NH_2$ (amino) group is indicated by the suffix *-amine.* When necessary, the position of the amino group is designated by a number, as shown in the examples in **Table 6.** When only one amino group is present, the final -e of the root hydrocarbon is dropped, as in 1-butanamine. If more than one amino group is present, the prefixes *di-*, *tri-*, *tetra-*, and so on are used to indicate the number of groups.

The amine aniline is used in the production of dyes with deep shades of color. The common name *aniline* is derived from the plant from which it was historically obtained. Cyclohexylamine and ethylamine are important in the manufacture of pesticides, plastics, pharmaceuticals, and rubber that is used to make tires.

All volatile amines have odors that humans find offensive, and amines are responsible for many of the odors characteristic of dead, decaying organisms. Two amines found in decaying human remains are putrescine and cadaverine. Specially trained dogs are used to locate human remains using these distinctive odors. Sniffer dogs are often used after catastrophic events and in forensic investigations.

**Table 6** Amines

**General Formula**

$RNH_2$

where R represents a carbon chain or ring bonded to the functional group

**Examples of Amines**

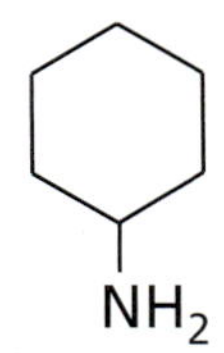

Cyclohexylamine

$CH_3CH_2$
|
$NH_2$

Ethylamine

Aniline

$NH_2$ $NH_2$
| |
$CHCH_2CH_2CH$
| |
$NH_2$ $NH_2$

1,1,4,4-Butanetetraamine

$CH_2CH_2CH_2$
| |
$NH_2$ $NH_2$

1,3-Propanediamine

## SECTION 2 REVIEW

Section Self-Check

### Section Summary

- Alcohols, ethers, and amines are formed when specific functional groups substitute for hydrogen in hydrocarbons.
- Because they readily form hydrogen bonds, alcohols have higher boiling points and higher water solubilities than other organic compounds.

**9. MAINIDEA Identify** two elements that are commonly found in functional groups.

**10. Identify** the functional group present in each of the following structures. Name the substance represented by each structure.

**a.** $NH_2$
|
$CH_3CHCH_3$

**b.**

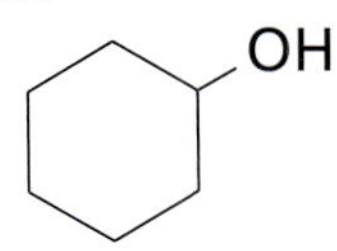

**c.** $CH_3-O-CH_2CH_2CH_3$

**11. Draw** the structure for each molecule.

**a.** 1-propanol
**b.** 1,3-cyclopentanediol
**c.** propyl ether
**d.** 1,2-propanediamine

**12. Discuss** the properties of alcohols, ethers, and amines, and give one use of each.

**13. Analyze** Based on the molecular structures below, which compound would you expect to be more soluble in water? Explain your reasoning.

$CH_3-O-CH_3$

OH
|
$CH_3CH_2$

## SECTION 3 Carbonyl Compounds

**MAINIDEA** Carbonyl compounds contain a double-bonded oxygen in the functional group.

### Essential Questions

- Which structures identify various carbonyl compounds?
- What are the properties of compounds containing the carbonyl group?

### Review Vocabulary

**electronegative:** indicates the relative ability of an element's atoms to attract electrons in a chemical bond

### New Vocabulary

carbonyl group
aldehyde
ketone
carboxylic acid
carboxyl group
ester
amide
condensation reaction

**CHEM 4 YOU** Have you ever eaten a piece of fruit-flavored candy that tasted like real fruit? Many natural fruits, such as strawberries, contain dozens of organic molecules that combine to give the distinctive aroma and flavor of fruits. The carbonyl group is found in many common types of artificial flavorings.

## Organic Compounds Containing the Carbonyl Group

The arrangement in which an oxygen atom is double-bonded to a carbon atom is called a **carbonyl group.** This group is the functional group in organic compounds known as aldehydes and ketones.

**Aldehydes** An **aldehyde** is an organic compound in which a carbonyl group located at the end of a carbon chain is bonded to a carbon atom on one side and a hydrogen atom on the other. Aldehydes have the general formula *CHO, where * represents an alkyl group or a hydrogen atom, as shown in **Table 7.**

Aldehydes are formally named by changing the final *-e* of the name of the alkane with the same number of carbon atoms to the suffix *-al.* Thus, the formal name of the compound methanal, shown in **Table 7,** is based on the one-carbon alkane methane. Because the carbonyl group in an aldehyde always occurs at the end of a carbon chain, no numbers are used in the name unless branches or additional functional groups are present. Methanal is also commonly called formaldehyde. Ethanal has the common name *acetaldehyde.* Scientists often use the common names of organic compounds because they are familiar to chemists.

**Table 7 Aldehydes**

| General Formula | Examples of Aldehydes |
|---|---|
| *CHO<br>*represents an alkyl group or a hydrogen atom | 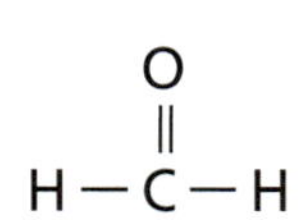 Methanal (formaldehyde)<br>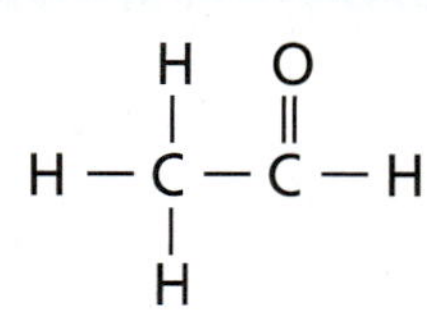 Ethanal (acetaldehyde) |
| 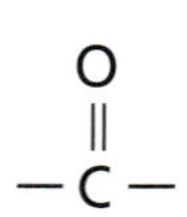 Carbonyl group | 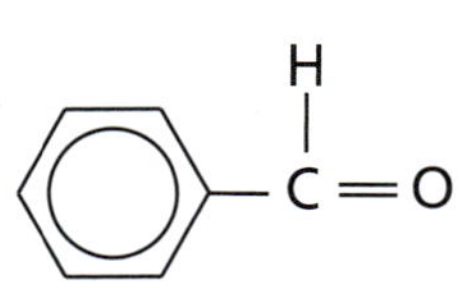 Benzaldehyde<br>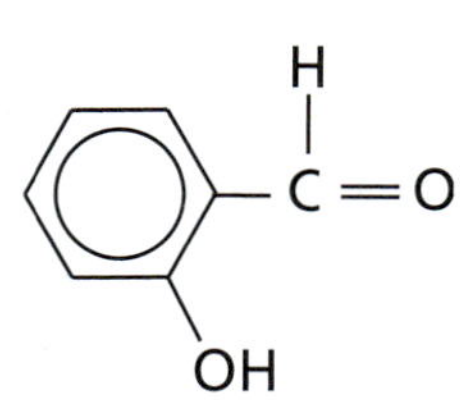 Salicylaldehyde<br>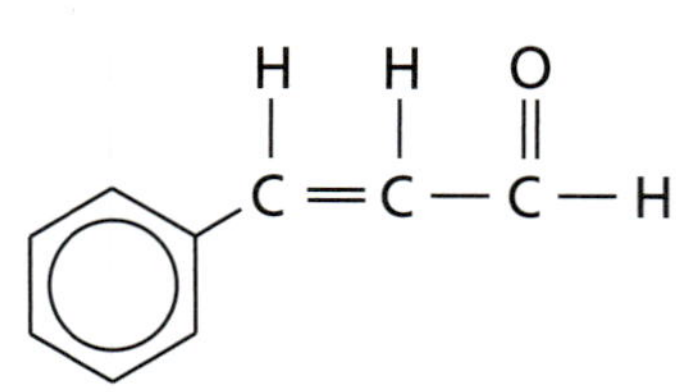 Cinnamaldehyde |

An aldehyde molecule contains a polar, reactive structure. However, like ethers, aldehyde molecules cannot form hydrogen bonds among themselves because the molecules have no hydrogen atoms bonded to an oxygen atom. Therefore, aldehydes have lower boiling points than alcohols with the same number of carbon atoms. Water molecules can form hydrogen bonds with the oxygen atom of aldehydes, so aldehydes are more soluble in water than alkanes but not as soluble as alcohols or amines.

Formaldehyde has been used for preservation for many years, as shown in **Figure 9.** Industrially, large quantities of formaldehyde are reacted with urea to manufacture a type of grease-resistant, hard plastic used to make buttons, appliance and automotive parts, and electrical outlets, as well as the glue that holds the layers of plywood together. Benzaldehyde and salicylaldehyde, shown in **Table 7,** are two components that give almonds their natural flavor. The aroma and flavor of cinnamon, a spice that comes from the bark of a tropical tree, are produced largely by cinnamaldehyde, also shown in **Table 7.**

**Figure 9** A water solution of formaldehyde was used in the past to preserve biological specimens. However, formaldehyde's use has been restricted in recent years because studies indicate it might cause cancer.

**READING CHECK** **Identify** two uses for aldehydes.

**Ketones** A carbonyl group can also be located within a carbon chain rather than at the end. A **ketone** is an organic compound in which the carbon of the carbonyl group is bonded to two other carbon atoms. Ketones have the general formula shown in **Table 8.** The carbon atoms on either side of the carbonyl group are bonded to other atoms. The simplest ketone, commonly known as acetone, has only hydrogen atoms bonded to the side carbons, as shown in **Table 8.**

Ketones are formally named by changing the *-e* at the end of the alkane name to *-one*, and including a number before the name to indicate the position of the ketone group. In the previous example, the alkane name propane is changed to propan*one*. The carbonyl group can be located only in the center, but the prefix *2-* is usually added to the name for clarity, as shown in **Table 8.**

Ketones and aldehydes share many chemical and physical properties because their structures are similar. Ketones are polar molecules and are less reactive than aldehydes. For this reason, ketones are popular solvents for other moderately polar substances, including waxes, plastics, paints, lacquers, varnishes, and glues. Like aldehydes, ketone molecules cannot form hydrogen bonds with each other but can form hydrogen bonds with water molecules. Therefore, ketones are somewhat soluble in water. Acetone is completely miscible with water.

**Table 8 Ketones**

| General Formula | Examples of Ketones | |
|---|---|---|
| 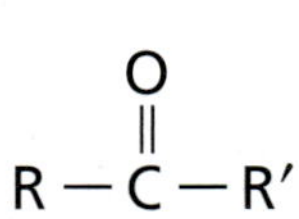 where R and R′ represent carbon chains or rings bonded to functional groups | 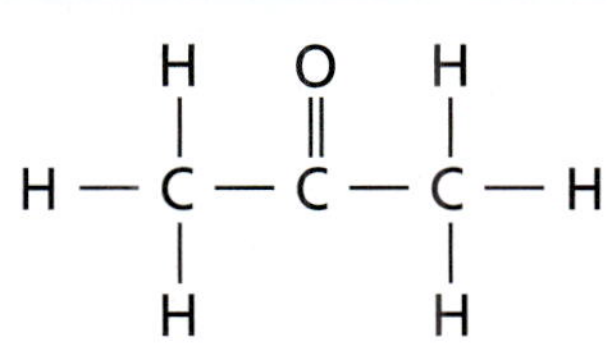 2-Propanone (acetone) | 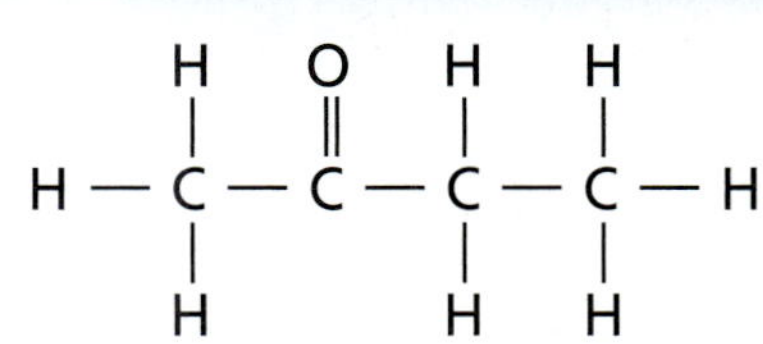 2-Butanone (methyl ethyl ketone) |

**Table 9 Carboxylic Acids**

| General Formula | Examples of Carboxylic Acids | |
|---|---|---|
| 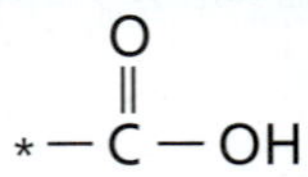 where * represents a hydrogen atom, carbon chain or ring bonded to the functional group | 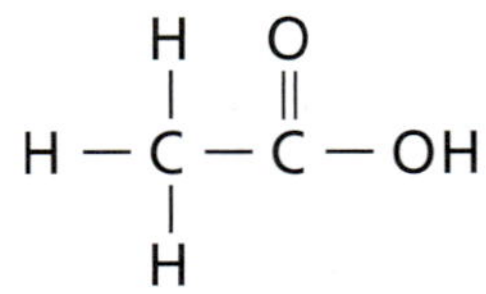 Ethanoic acid (acetic acid) | 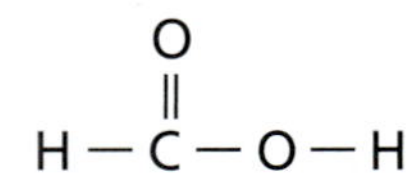 Methanoic acid (formic acid) |

## Carboxylic Acids

A **carboxylic acid** is an organic compound that has a carboxyl group. A **carboxyl group** consists of a carbonyl group bonded to a hydroxyl group. Thus, carboxylic acids have the general formula shown in **Table 9.** One diagram shown in **Table 9** is the structure of a familiar carboxylic acid—acetic acid, the acid found in vinegar. Although many carboxylic acids have common names, the formal name is formed by changing the *-ane* of the parent alkane to *-anoic acid*. Thus, the formal name of acetic acid is ethanoic acid.

A carboxyl group is usually represented in condensed form by writing –COOH. For example, ethanoic acid can be written as $CH_3COOH$. The simplest carboxylic acid consists of a carboxyl group bonded to a single hydrogen atom, HCOOH, shown in **Table 9.** Its formal name is methanoic acid, but it is more commonly known as formic acid. Some insects produce formic acid as a defense mechanism, as shown in **Figure 10.**

**READING CHECK** **Explain** how the name *ethanoic acid* is derived.

Carboxylic acids are polar and reactive. Those that dissolve in water ionize weakly to produce hydronium ions, the anion of the acid in equilibrium with water, and the unionized acid. The ionization of ethanoic acid is an example.

$$\underset{\text{Ethanoic acid (acetic acid)}}{CH_3COOH(aq)} + H_2O(l) \rightleftharpoons \underset{\text{Ethanoate ion (acetate ion)}}{CH_3COO^-(aq)} + H_3O^+(aq)$$

Carboxylic acids can ionize in water solution because the two oxygen atoms are highly electronegative and attract electrons away from the hydrogen atom in the –OH group. As a result, the hydrogen proton can transfer to another atom that has a pair of electrons not involved in bonding, such as the oxygen atom of a water molecule. Because they ionize in water, soluble carboxylic acids turn blue litmus paper red and have a sour taste.

Some important carboxylic acids, such as oxalic acid and adipic acid, have two or more carboxyl groups. An acid with two carboxyl groups is called a dicarboxylic acid. Others have additional functional groups such as hydroxyl groups, as in the lactic acid found in yogurt. Typically, these acids are more soluble in water and often more acidic than acids with only a carboxyl group.

**READING CHECK** **Evaluate** Using the information above, explain why carboxylic acids are classified as acids.

**Figure 10** Stinging ants defend themselves with a venom that contains formic acid.
**Identify** *another name for formic acid.*

**Table 10 Esters**

| General Formula | Example of an Ester |
|---|---|
| $-\overset{\displaystyle O}{\overset{\Vert}{C}}-O-R$<br>Ester group | Ethanoate group: $CH_3-\overset{\displaystyle O}{\overset{\Vert}{C}}$; Propyl group: $CH_2CH_2CH_3$<br>$CH_3-\overset{\displaystyle O}{\overset{\Vert}{C}}-O-CH_2CH_2CH_3$<br>Ester group: $-\overset{\displaystyle O}{\overset{\Vert}{C}}-O-$<br>Propyl ethanoate<br>(propyl acetate) |

## Organic Compounds Derived from Carboxylic Acids

Several classes of organic compounds have structures in which the hydrogen or the hydroxyl group of a carboxylic acid is replaced by a different atom or group of atoms. The two most common classes are esters and amides.

**Esters** An **ester** is any organic compound with a carboxyl group in which the hydrogen of the hydroxyl group has been replaced by an alkyl group, producing the arrangement shown in **Table 10.** The name of an ester is formed by writing the name of the alkyl group followed by the name of the acid with the *-ic acid* ending replaced by *-ate*, as illustrated by the example shown in **Table 10.** Note how the name *propyl* results from the structural formula. The name shown in parentheses is based on the name *acetic acid,* the common name for ethanoic acid.

Esters are polar molecules and many are volatile and sweet-smelling. Many kinds of esters are found in the natural fragrances and flavors of flowers and fruits, as shown in **Figure 11.** Natural flavors, such as apple or banana, result from mixtures of many different organic molecules, including esters, but some of these flavors can be imitated by a single ester structure. Consequently, esters are manufactured for use as flavors in many foods and beverages and as fragrances in candles, perfumes, and other scented items.

**VOCABULARY**

**SCIENCE USAGE V. COMMON USAGE**

**Class**

***Science usage:*** a group, set, or kind that share common traits

*Esters are a class of organic molecules.*

***Common usage:*** a group of students that meet at regular intervals to study the same subject

*Students meet for chemistry class during fourth period.*

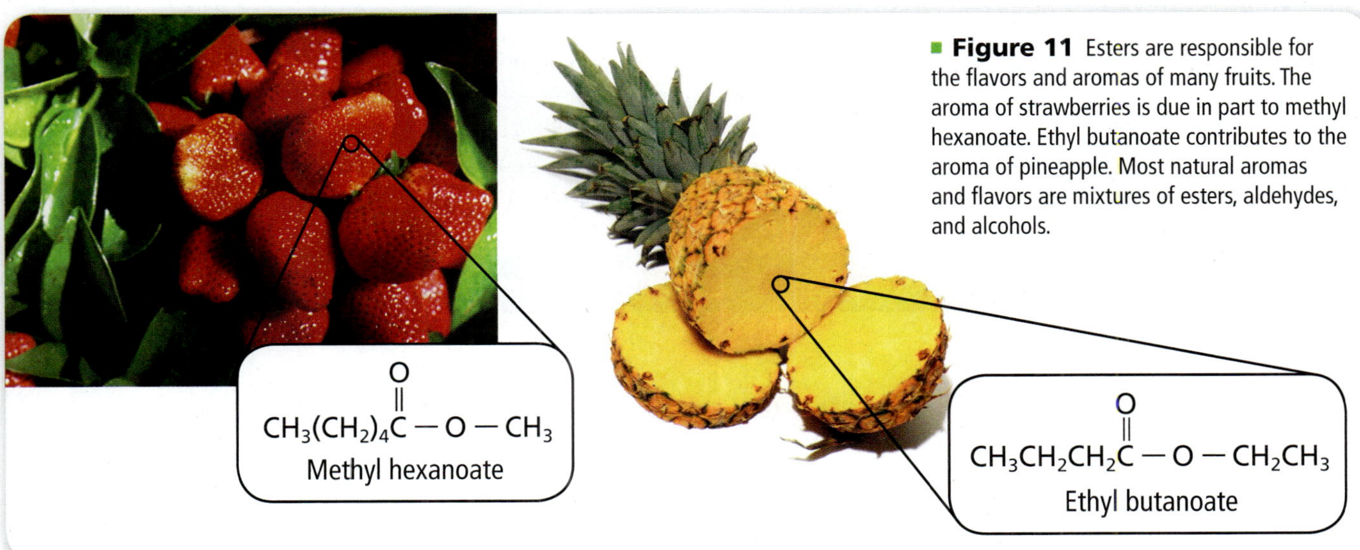


■ **Figure 11** Esters are responsible for the flavors and aromas of many fruits. The aroma of strawberries is due in part to methyl hexanoate. Ethyl butanoate contributes to the aroma of pineapple. Most natural aromas and flavors are mixtures of esters, aldehydes, and alcohols.

## MiniLAB

iLab Station

### Make an Ester

**How can you recognize an ester?**

#### Procedure

1. Read and complete the lab safety form.
2. Prepare a hot-water bath by pouring 150 mL of **tap water** into a **250-mL beaker.** Place the beaker on a **hot plate** set to medium.
3. Use a **balance** and **weighing paper** to measure 1.5 g of **salicylic acid.** Place the salicylic acid in a **small test tube** and add 3 mL of **distilled water.** Use a **10-mL graduated cylinder** to measure the water. Then add 3 mL of **methanol.** Use a **Beral pipette** to add 3 drops of **concentrated sulfuric acid** to the test tube.
**WARNING:** *Concentrated sulfuric acid can cause burns. Methanol fumes are explosive—keep away from open flame. Handle chemicals with care.*
4. When the water is hot but not boiling, place the test tube in the bath for 5 min. Use a **test-tube clamp** to remove the test tube from the bath and place in a **test-tube holder** until needed.
5. Place a **cotton ball** in a **petri dish half.** Pour the contents of the test tube onto the cotton ball. Record your observation of the odor of the product.

#### Analysis

1. **Name** The common name of the ester that you produced is *oil of wintergreen*. Name some products that you think could contain the ester.
2. **Evaluate** the advantages and disadvantages of using synthetic esters in consumer products as compared to using natural esters.

**Amides** An **amide** is an organic compound in which the –OH group of a carboxylic acid is replaced by a nitrogen atom bonded to other atoms. The general structure of an amide is shown in **Table 11.** Amides are named by writing the name of the alkane with the same number of carbon atoms, and then replacing the final *-e* with *-amide*. Thus, the amide shown in **Table 11** is called ethanamide, but it can also be named acetamide from its common name, acetic acid.

**READING CHECK** How does an amide differ from a carboxylic acid?

The amide functional group is found repeated many times in natural proteins and some synthetic materials. For example, you might have used a non-aspirin pain reliever containing acetaminophen. In the acetaminophen structure shown in **Table 11,** notice that the amide (–NH–) group connects a carbonyl group and an aromatic group.

One important amide is caramide ($NH_2CONH_2$), or urea, as it is commonly known. Urea is an end product in the metabolic breakdown of proteins in mammals. It is found in the blood, bile, milk, and perspiration of mammals. When proteins are broken down, amino groups ($NH_2$) are removed from the amino acids. The amino groups are then converted to ammonia ($NH_3$) molecules that are toxic to the body. The toxic ammonia is converted to nontoxic urea in the liver. The urea is filtered out of the blood in the kidneys and passed from the body in urine.

Because of the high nitrogen content of urea and because it is easily converted to ammonia in the soil, urea is a common commercial fertilizer. Urea is also used as a protein supplement for ruminant animals, such as cattle and sheep. These animals use urea to produce proteins in their bodies.

**READING CHECK** **Identify** an amide that is found in the human body.

**Table 11** Amides

| General Formula | Examples of Amides | |
|---|---|---|
| 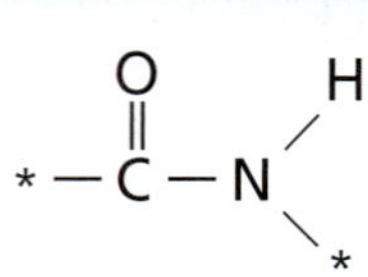 Amide group | 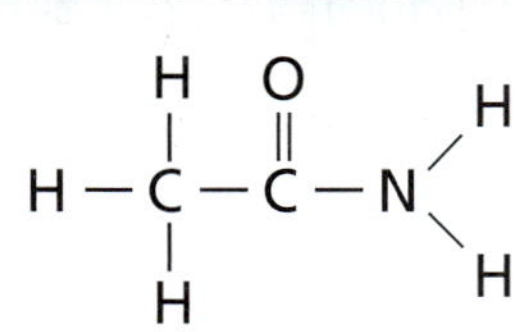 Ethanamide (acetamide) | 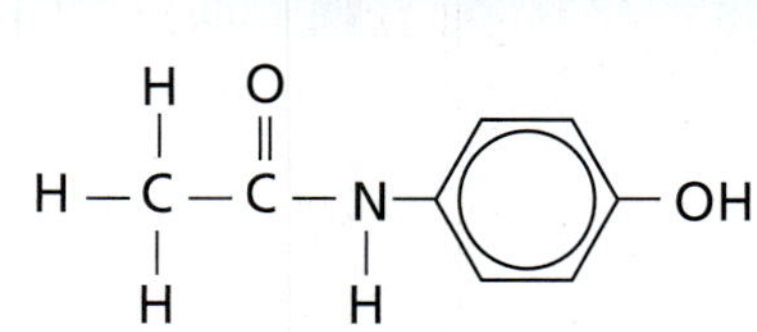 Acetaminophen |

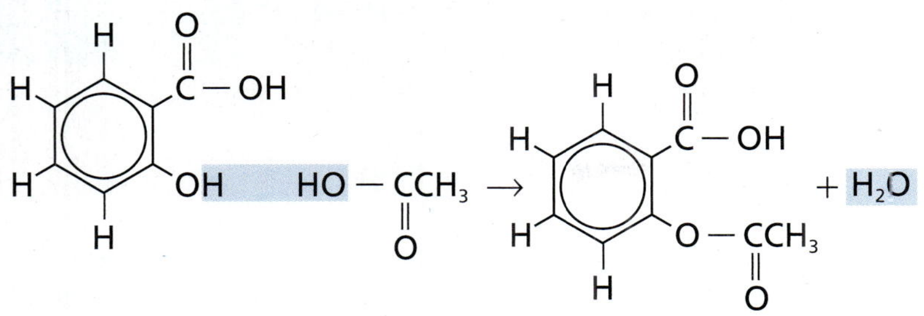


**Figure 12** To synthesize aspirin, two organic molecules are combined in a condensation reaction to form a larger molecule.

## Condensation Reactions

Many laboratory syntheses and industrial processes involve the reaction of two organic reactants to form a larger organic product, such as the aspirin shown in **Figure 12.** This type of reaction is known as a condensation reaction.

In a **condensation reaction,** two smaller organic molecules combine to form a more complex molecule, accompanied by the loss of a small molecule such as water. Typically, the molecule lost is formed from one particle from each of the reactant molecules. In essence, a condensation reaction is an elimination reaction in which a bond is formed between two atoms not previously bonded to each other.

The most common condensation reactions involve the combining of carboxylic acids with other organic molecules. A common way to synthesize an ester is by a condensation reaction between a carboxylic acid and an alcohol. Such a reaction can be represented by the following general equation.

$$RCOOH + R'OH \rightarrow RCOOR' + H_2O$$

**FOLDABLES®** Incorporate information from this section into your Foldable.

# SECTION 3 REVIEW

Section Self-Check

### Section Summary

- Carbonyl compounds are organic compounds that contain the C=O group.
- Five important classes of organic compounds containing carbonyl compounds are aldehydes, ketones, carboxylic acids, esters, and amides.

**14. MAINIDEA Classify** each of the carbonyl compounds as one of the types of organic substances you have studied in this section.

**a.** 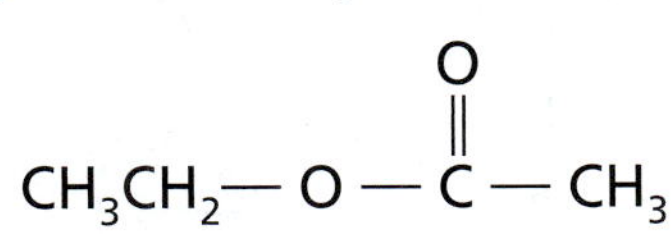

**c.** 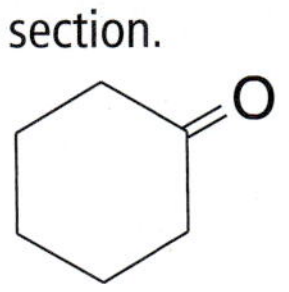

**b.** 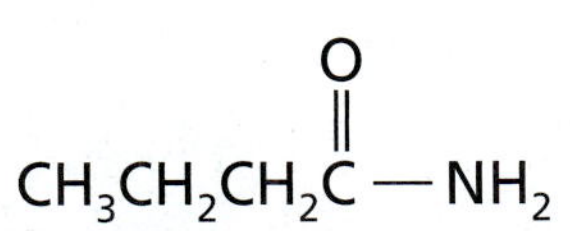

**d.** 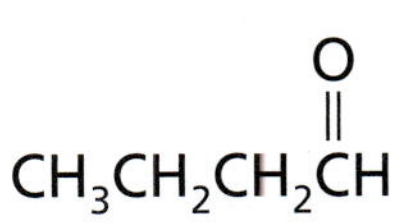

**15. Describe** the products of a condensation reaction between a carboxylic acid and an alcohol.

**16. Determine** The general formula for alkanes is $C_nH_{2n+2}$. Derive a general formula to represent an aldehyde, a ketone, and a carboxylic acid.

**17. Infer** why water-soluble organic compounds with carboxyl groups exhibit acidic properties in solutions, whereas similar compounds with aldehyde structures do not exhibit these properties.

SECTION 4

# Other Reactions of Organic Compounds

**MAINIDEA** Classifying the chemical reactions of organic compounds makes predicting products of reactions much easier.

### Essential Questions

- How are organic reactions classified?
- Why is it useful to draw structural formulas when writing equations for reactions of organic compounds?
- How can classifying a reaction help you predict the reaction's products?

### Review Vocabulary

**catalyst:** a substance that increases the rate of a chemical reaction by lowering activation energies but is not consumed in the reaction

### New Vocabulary

elimination reaction
dehydrogenation reaction
dehydration reaction
addition reaction
hydration reaction
hydrogenation reaction

**CHEM 4 YOU** As you eat lunch, the oxidation of organic compounds is probably not on your mind. However, that is exactly what is about to occur as your cells break down the food that you eat to obtain energy for your body.

## Classifying Reactions of Organic Substances

Organic chemists have discovered thousands of reactions by which organic compounds can be changed into different organic compounds. By using combinations of these reactions, chemical industries convert simple molecules from petroleum and natural gas into the large, complex organic molecules found in many useful products—including lifesaving drugs and many other consumer products as shown in **Figure 13.**

You have already read about substitution and condensation reactions. Two other important types of reactions by which organic compounds can be changed into different compounds are elimination reactions and addition reactions.

**Elimination reactions** One way to change an alkane into a chemically reactive substance is to form a second covalent bond between two carbon atoms, producing an alkene. Forming double bonds from single bonds between carbon atoms is an **elimination reaction,** a reaction in which a combination of atoms is removed from two adjacent carbon atoms, forming an additional bond between them. The atoms that are eliminated usually form stable molecules, such as $H_2O$, HCl, or $H_2$.

**READING CHECK** **Define** *elimination reaction* in your own words.

**Figure 13** Many consumer products, such as plastic containers, fibers in ropes and clothing, and oils and waxes in cosmetics, are made from petroleum and natural gas.

Cordelia Molloy/Photo Researchers

**Figure 14** Low density polyethylene (LDPE) is made from gaseous ethene under high pressure in the presence of a catalyst. LDPE is used for playground equipment because it is easy to mold into various shapes, it is easy to dye into many colors, and it is durable. The name *polyethylene* comes from *ethylene,* which is the common name for ethene.

Ethene, the starting material for the playground equipment shown in **Figure 14,** is produced by the elimination of two hydrogen atoms from ethane. A reaction that eliminates two hydrogen atoms is called a **dehydrogenation reaction.** Note that the two hydrogen atoms form a molecule of hydrogen gas.

Watch a **video about substituted hydrocarbons and their reactions.**

Video

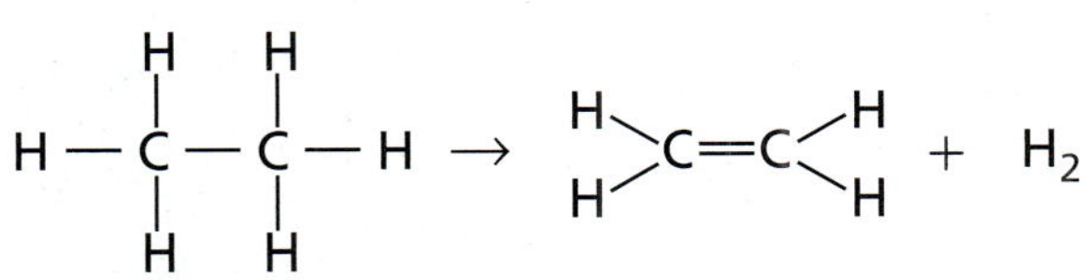


Alkyl halides can undergo elimination reactions to produce an alkene and a hydrogen halide, as shown here.

$$\underset{\text{Alkyl halide}}{R{-}CH_2{-}CH_2{-}X} \rightarrow \underset{\text{Alkene}}{R{-}CH{=}CH_2} + \underset{\text{Hydrogen halide}}{HX}$$

Get help **identifying organic reactions.**

Personal Tutor

Likewise, alcohols can also undergo elimination reactions by losing a hydrogen atom and a hydroxyl group to form water, as shown below. An elimination reaction in which the atoms removed form water is called a **dehydration reaction.** In the dehydration reaction, the alcohol is broken down into an alkene and water.

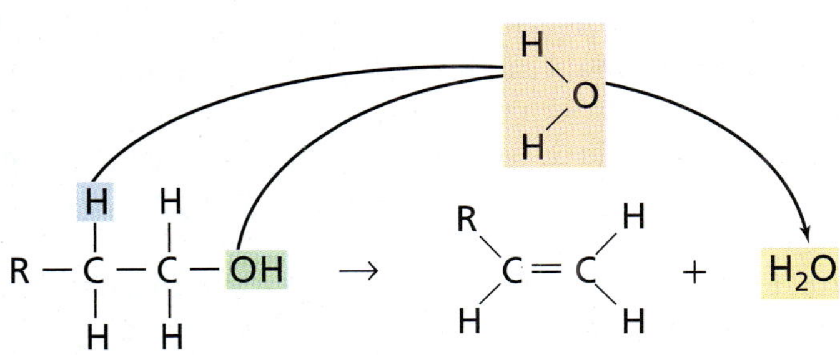


The generic form of this dehydration reaction can be written as follows.

$$R{-}CH_2{-}CH_2{-}OH \rightarrow R{-}CH{=}CH_2 + H_2O$$

Chuck Franklin/Alamy

**Table 12 Summary of Addition Reactions**

| Reactant Alkene | Addition Reactant | Product |
|---|---|---|
| 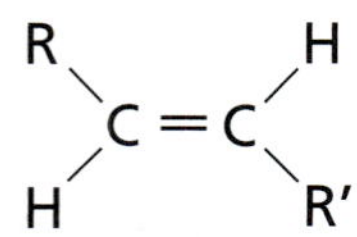 | Water (hydration)<br>H—O—H ($H_2O$) | Alcohol<br>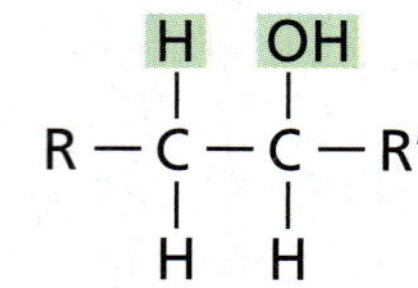 |
| | Hydrogen (hydrogenation)<br>H—H | Alkane<br>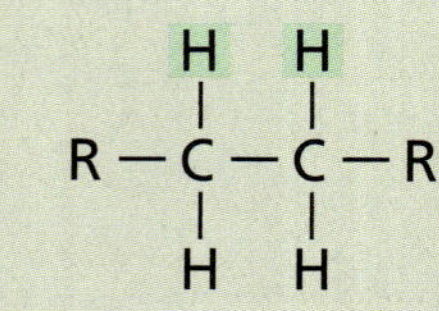 |
| | Hydrogen halide<br>H—X | Alkyl halide<br>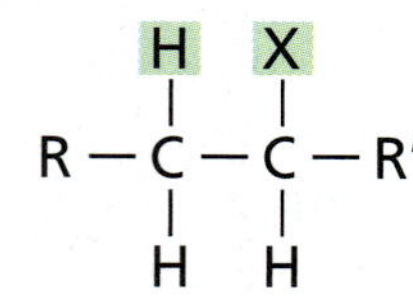 |
| | Halogen<br>X—X | Alkyl dihalide<br>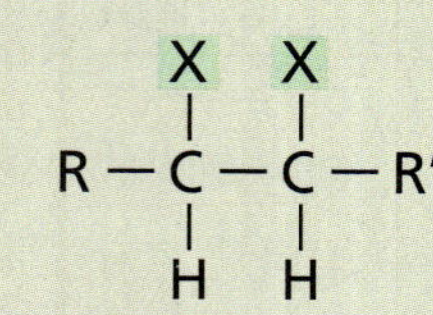 |

**Addition reactions** Another type of organic reaction appears to be an elimination reaction in reverse. An **addition reaction** results when other atoms bond to each of two atoms bonded by double or triple covalent bonds. Addition reactions typically involve double-bonded carbon atoms in alkenes or triple-bonded carbon atoms in alkynes. Addition reactions occur because double and triple bonds have a rich concentration of electrons. Therefore, molecules and ions that attract electrons tend to form bonds that use some of the electrons from the multiple bonds. The most common addition reactions are those in which $H_2O$, $H_2$, HX, or $X_2$ add to an alkene, as shown in **Table 12.**

A **hydration reaction,** also shown in **Table 12,** is an addition reaction in which a hydrogen atom and a hydroxyl group from a water molecule add to a double or triple bond. The generic equation shown in **Table 12** shows that a hydration reaction is the opposite of a dehydration reaction.

A reaction that involves the addition of hydrogen to atoms in a double or triple bond is called a **hydrogenation reaction.** One molecule of $H_2$ reacts to fully hydrogenate each double bond in a molecule. When $H_2$ adds to the double bond of an alkene, the alkene is converted to an alkane.

☑ READING CHECK **Identify** the reaction that is the reverse of a hydrogenation reaction.

■ **Figure 14** Low density polyethylene (LDPE) is made from gaseous ethene under high pressure in the presence of a catalyst. LDPE is used for playground equipment because it is easy to mold into various shapes, it is easy to dye into many colors, and it is durable. The name *polyethylene* comes from *ethylene,* which is the common name for ethene.

Ethene, the starting material for the playground equipment shown in **Figure 14,** is produced by the elimination of two hydrogen atoms from ethane. A reaction that eliminates two hydrogen atoms is called a **dehydrogenation reaction.** Note that the two hydrogen atoms form a molecule of hydrogen gas.

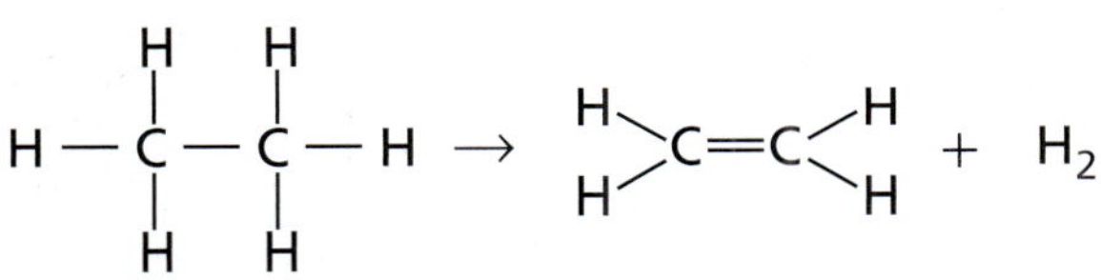

Ethane    Ethene

Watch a **video about substituted hydrocarbons and their reactions.**

Video

Alkyl halides can undergo elimination reactions to produce an alkene and a hydrogen halide, as shown here.

$$\underset{\text{Alkyl halide}}{R{-}CH_2{-}CH_2{-}X} \rightarrow \underset{\text{Alkene}}{R{-}CH{=}CH_2} + \underset{\text{Hydrogen halide}}{HX}$$

Get help **identifying organic reactions.**

Personal Tutor

Likewise, alcohols can also undergo elimination reactions by losing a hydrogen atom and a hydroxyl group to form water, as shown below. An elimination reaction in which the atoms removed form water is called a **dehydration reaction.** In the dehydration reaction, the alcohol is broken down into an alkene and water.

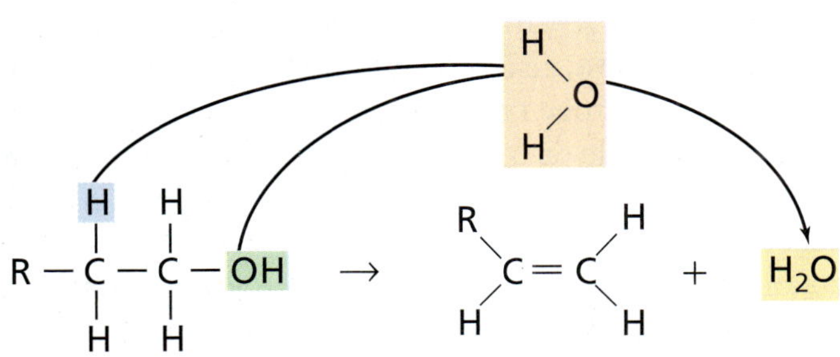

Alcohol    Alkene    Water

The generic form of this dehydration reaction can be written as follows.

$$R{-}CH_2{-}CH_2{-}OH \rightarrow R{-}CH{=}CH_2 + H_2O$$

## Table 12 Summary of Addition Reactions

| Reactant Alkene | Addition Reactant | Product |
|---|---|---|
| 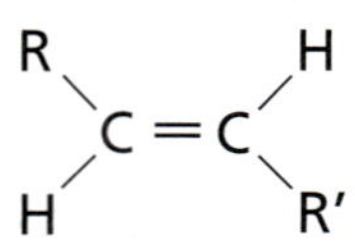 | Water (hydration)<br>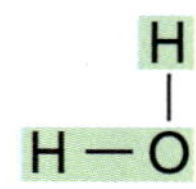 | Alcohol<br>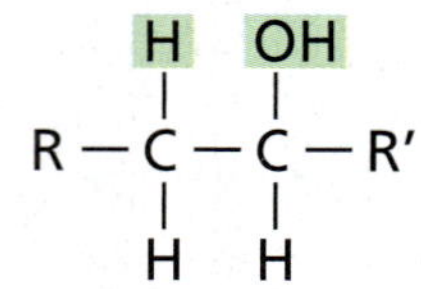 |
| | Hydrogen (hydrogenation)<br>H—H | Alkane<br>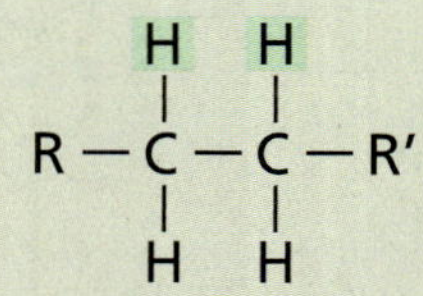 |
| | Hydrogen halide<br>H—X | Alkyl halide<br>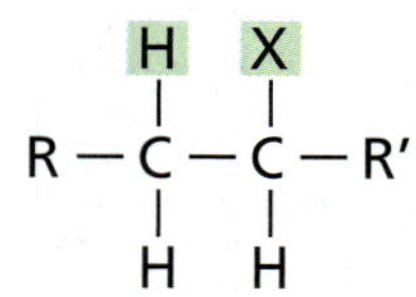 |
| | Halogen<br>X—X | Alkyl dihalide<br>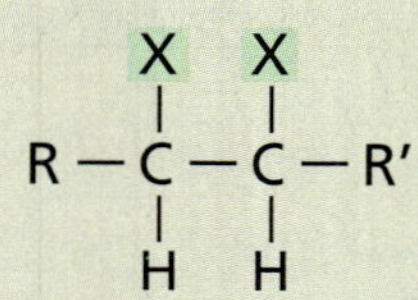 |

**Addition reactions** Another type of organic reaction appears to be an elimination reaction in reverse. An **addition reaction** results when other atoms bond to each of two atoms bonded by double or triple covalent bonds. Addition reactions typically involve double-bonded carbon atoms in alkenes or triple-bonded carbon atoms in alkynes. Addition reactions occur because double and triple bonds have a rich concentration of electrons. Therefore, molecules and ions that attract electrons tend to form bonds that use some of the electrons from the multiple bonds. The most common addition reactions are those in which $H_2O$, $H_2$, HX, or $X_2$ add to an alkene, as shown in **Table 12.**

A **hydration reaction,** also shown in **Table 12,** is an addition reaction in which a hydrogen atom and a hydroxyl group from a water molecule add to a double or triple bond. The generic equation shown in **Table 12** shows that a hydration reaction is the opposite of a dehydration reaction.

A reaction that involves the addition of hydrogen to atoms in a double or triple bond is called a **hydrogenation reaction.** One molecule of $H_2$ reacts to fully hydrogenate each double bond in a molecule. When $H_2$ adds to the double bond of an alkene, the alkene is converted to an alkane.

☑ READING CHECK **Identify** the reaction that is the reverse of a hydrogenation reaction.

Catalysts are usually needed in the hydrogenation of alkenes because the reaction's activation energy is too large without them. Catalysts such as powdered platinum or palladium provide a surface that adsorbs the reactants and makes their electrons more available to bond to other atoms.

Hydrogenation reactions are commonly used to convert the liquid unsaturated fats found in oils from plants such as soybean, corn, and peanuts into saturated fats that are solid at room temperature. These hydrogenated fats are then used to make margarine and solid shortening.

Alkynes can also be hydrogenated to produce alkenes or alkanes. One molecule of $H_2$ must be added to each triple bond in order to convert an alkyne to an alkene, as shown here.

$$\mathrm{R{-}C{\equiv}C{-}H + H_2 \rightarrow R{-}CH{=}CH_2}$$

After the first molecule of $H_2$ is added, the alkyne is converted to an alkene. A second molecule of $H_2$ follows the hydrogenation reaction.

$$\mathrm{R{-}CH{=}CH_2 + H_2 \rightarrow R{-}CH_2{-}CH_3}$$

In a similar mechanism, the addition of hydrogen halides to alkenes is an addition reaction useful to industry for the production of alkyl halides. The generic equation for this reaction is shown below.

$$\mathrm{R{-}CH{=}CH{-}R' + HX \rightarrow R{-}CHX{-}CH_2{-}R'}$$

# Data Analysis LAB

Based on Real Data*

## Interpret Data

**What are the optimal conditions to hydrogenate canola oil?** Edible vegetable oil is hydrogenated to preserve its flavor and to alter its melting properties. Because evidence suggests that *trans*-fatty acids are associated with increased risk of heart disease and cancer, the minimum amount of *trans*-fatty acids and the maximum amount of *cis*-oleic acid are desired.

Computer models were used to simulate processing conditions and to alter eight variables to optimize the output of the desirable oil. Multiple optimal operating conditions were determined. A small-scale industrial plant was used to confirm the results of the computer simulation.

### Data and Observations

The table at right shows some of the data from this investigation.

**Data for Canadian Canola Oil**

| | Computer Simulation | | Experimental | |
|---|---|---|---|---|
| Trial Run | *trans*-Fatty Acids (wt. %) | *cis*-Oleic Acid (wt. %) | *trans*-Fatty Acids (wt. %) | *cis*-Oleic Acid (wt. %) |
| 1 | 4.90 | 69.10 | 5.80 | 70.00 |
| 2 | 4.79 | 63.75 | 4.61 | 64.00 |
| 3 | 4.04 | 68.96 | 4.61 | 67.00 |
| 4 | 5.99 | 62.80 | 7.10 | 65.00 |
| 5 | 4.60 | 68.10 | 5.38 | 66.50 |

Data obtained from Izadifar, M. 2005. Application of genetic algorithm for optimization of vegetable oil hydrogenation process. *Journal of Food Engineering.* 78 (2007) 1–8.

### Think Critically

1. **Calculate** the percent yield for each of the trials shown in the table.
2. **Evaluate** Which trial(s) produced the highest yield of *cis*-oleic acid and the lowest yield of *trans*-fatty acids?
3. **Explain** why the techniques used in this investigation are useful in manufacturing processes.

## Table 13 Oxidation-Reduction Reactions

**Oxidation of an alkane to an alcohol**

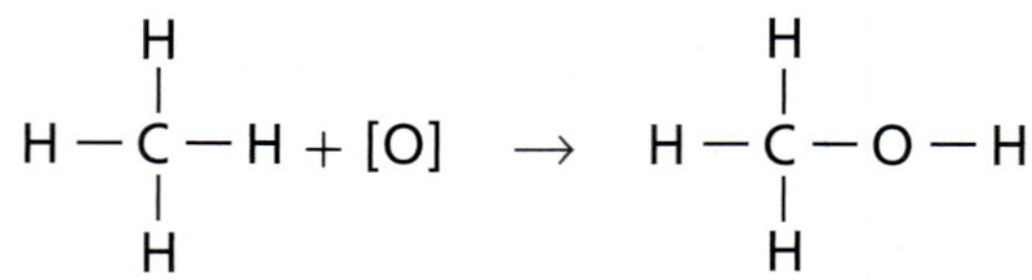

Methane — Methanol

**A sequence of oxidation reactions**

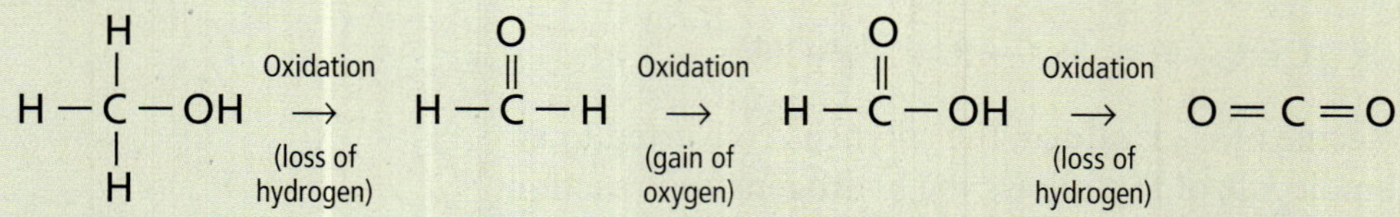

Methanol (methyl alcohol) — Methanal (formaldehyde) — Methanoic acid (formic acid) — Carbon dioxide

**Oxidation of two isomers**

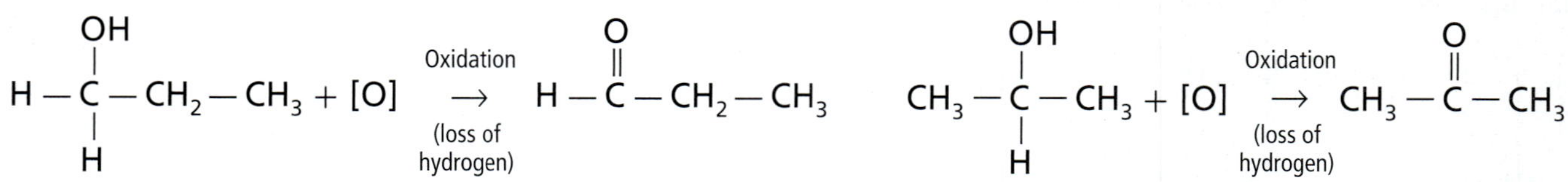

1-Propanol — Propanal — 2-Propanol — 2-Propanone

**Oxidation-reduction reactions** Many organic compounds can be converted to other compounds by oxidation and reduction reactions. For example, suppose you want to convert methane, the main constituent of natural gas, to methanol, a common industrial solvent and raw material for making formaldehyde and methyl esters. The conversion of methane to methanol can be represented by the equation shown in **Table 13,** in which [O] represents oxygen from an agent such as copper(II) oxide, potassium dichromate, or sulfuric acid.

What happens to methane in this reaction? Before answering, it might be helpful to review the definitions of oxidation and reduction. Oxidation is the loss of electrons, and a substance is oxidized when it gains oxygen or loses hydrogen. Reduction is the gain of electrons, and a substance is reduced when it loses oxygen or gains hydrogen. Thus, methane is oxidized as it gains oxygen and is converted to methanol. Of course, every redox reaction involves both an oxidation and a reduction; however, organic redox reactions are described based on the change in the organic compound.

Oxidizing the methanol shown in **Table 13** is the first step in the sequence of reactions that can be used to produce an aldehyde, which are also shown in **Table 13.** For clarity, oxidizing agents are omitted. Preparing an aldehyde by this method is not always a simple task because the oxidation might continue, forming the carboxylic acid.

☑ READING CHECK **Identify** Use **Table 13** to identify two possible products that are produced when the aldehyde is further oxidized.

However, not all alcohols can be oxidized to aldehydes and, subsequently, carboxylic acids. To understand why, compare the oxidations of 1-propanol and 2-propanol, shown in **Table 13.** Note that oxidizing 2-propanol yields a ketone, not an aldehyde. Unlike aldehydes, ketones resist further oxidation to carboxylic acids. Thus, while the propanal formed by oxidizing 1-propanol easily oxidizes to form propanoic acid, the 2-propanone formed by oxidizing 2-propanol does not react to form a carboxylic acid.

**READING CHECK** **Write** the equation using molecular structures like those in **Table 13** for the formation of propanoic acid.

How important are organic oxidations and reductions? You have seen that oxidation and reduction reactions can change one functional group into another. That ability enables chemists to use organic redox reactions, in conjunction with the substitution and addition reactions you read about earlier in the chapter, to synthesize a tremendous variety of useful products. On a personal note, all living systems—including you—depend on the energy released by oxidation reactions. Of course, some of the most dramatic oxidation-reduction reactions are combustion reactions. All organic compounds that contain carbon and hydrogen burn in excess oxygen to produce carbon dioxide and water. For example, the highly exothermic combustion of ethane is described by the following thermochemical equation.

$$2C_2H_6(g) + 7O_2(g) \rightarrow 4CO_2(g) + 6H_2O(l) \qquad \Delta H = -3120 \text{ kJ}$$

As you read previously, much of the world relies on the combustion of hydrocarbons as a primary source of energy. Our reliance on the energy from organic oxidation reactions is illustrated in **Figures 15.**

## Predicting Products of Organic Reactions

The generic equations representing the different types of organic reactions you have learned—substitution, elimination, addition, oxidation-reduction, and condensation—can be used to predict the products of other organic reactions of the same types. For example, suppose you were asked to predict the product of an elimination reaction in which 1-butanol is a reactant. You know that a common elimination reaction involving an alcohol is a dehydration reaction.

### RealWorld CHEMISTRY

**Polycyclic Aromatic Hydrocarbons (PAHs)**

**BIOLOGICAL MOLECULES** Hydrocarbons composed of multiple aromatic rings are called PAHs. They have been found in meteorites and identified in the material surrounding dying stars. Scientists simulated conditions in space and found that about 10% of the PAHs were converted to alcohols, ketones, and esters. These molecules can be used to form compounds that are important in biological systems.

**Figure 15** People around the world depend on the oxidation of hydrocarbons to get to work and to transport products.

The generic equation for the dehydration of an alcohol is as follows.

$$R—CH_2—CH_2—OH \rightarrow R—CH{=}CH_2 + H_2O$$

**FOLDABLES®**
Incorporate information from this section into your Foldable.

To determine the actual product, first draw the structure of 1-butanol. Then use the generic equation as a model to see how 1-butanol would react. The generic reaction shows that the —OH and a H— are removed from the carbon chain. Finally, draw the structure of the likely products, as shown in the following equation.

$$\underset{\text{1-Butanol}}{CH_3—CH_2—CH_2—CH_2—OH} \rightarrow \underset{\text{1-Butene}}{CH_3—CH_2—CH{=}CH_2} + H_2O$$

As another example, suppose that you wish to predict the product of the reaction between cyclopentene and hydrogen bromide. Recall that the generic equation for an addition reaction between an alkene and an alkyl halide is as follows.

$$R—CH{=}CH—R' + HX \rightarrow R—CHX—CH_2—R'$$

First, draw the structure for cyclopentene, the organic reactant, and add the formula for hydrogen bromide, the other reactant. From the generic equation, you can see that a hydrogen atom and a halide atom add across the double bond to form an alkyl halide. Finally, draw the formula for the likely product. If you are correct, you have written the following equation.

Cyclopentene Hydrogen bromide Bromocyclopentane

## SECTION 4 REVIEW

Section Self-Check

### Section Summary

- Most reactions of organic compounds can be classified into one of five categories: substitution, elimination, addition, oxidation-reduction, and condensation.
- Knowing the types of organic compounds reacting can enable you to predict the reaction products.

18. **MAINIDEA Classify** each reaction as substitution, elimination, addition, or condensation.

 a. $CH_3CH = CHCH_2CH_3 + H_2 \rightarrow CH_3CH_2—CH_2CH_2CH_3$

 b. $CH_3CH_2CH_2\underset{\displaystyle OH}{\underset{|}{C}}HCH_3 \rightarrow CH_3CH_2CH = CHCH_3 + H_2O$

19. **Identify** the type of organic reaction that would best accomplish each conversion.

 a. alkyl halide → alkene
 b. alkene → alcohol
 c. alcohol + carboxylic acid → ester
 d. alkene → alkyl dihalide

20. **Complete** each equation by writing the condensed structural formula for the product that is most likely to form.

 a. $CH_3CH = CHCH_2CH_3 + H_2 \rightarrow$

 b. $CH_3CH_2\underset{\displaystyle Cl}{\underset{|}{C}}HCH_2CH_3 + OH^- \rightarrow$

21. **Predicting Products** Explain why the hydration reaction involving 1-butene might yield two distinct products, whereas the hydration of 2-butene yields only one product.

## SECTION 5

# Polymers

**MAINIDEA** **Synthetic polymers are large organic molecules made up of repeating units that are linked together by addition or condensation reactions.**

### Essential Questions

- How does drawing a diagram help you understand the relationship between a polymer and the monomers from which it forms?
- What distinguishes addition and condensation polymerization reactions?
- How can you use molecular structures and the presence of functional groups to predict the properties of polymers?

### Review Vocabulary

**molecular mass:** the mass of one molecule of a substance

### New Vocabulary

polymer
monomer
polymerization reaction
addition polymerization
condensation polymerization
thermoplastic
thermosetting

**CHEM 4 YOU** **Think how different your life would be without plastic sandwich bags, plastic foam cups, nylon and polyester fabrics, vinyl siding on buildings, foam cushions, and a variety of other synthetic materials. These products all have at least one thing in common—they are made of polymers.**

## The Age of Polymers

The compact discs shown in **Figure 16** contain polycarbonate, which is made of extremely long molecules with groups of atoms that repeat in a regular pattern. This molecule is an example of a synthetic polymer. **Polymers** are large molecules consisting of many repeating structural units. In **Figure 16,** the letter *n* beside the structural unit of polycarbonate represents the number of structural units in the polymer chain. Because polymer *n* values vary widely, molecular masses of polymers range from less than 10,000 amu to more than 1,000,000 amu. A typical chain in nonstick coating on skillets has about 400 units, giving it a molecular mass of around 40,000 amu.

Before the development of synthetic polymers, people were limited to using natural substances such as stone, wood, metals, wool, and cotton. By the turn of the twentieth century, a few chemically treated natural polymers such as rubber and the first plastic, celluloid, had become available. Celluloid is made by treating cellulose from cotton or wood fiber with nitric acid.

The first synthetic polymer, synthesized in 1909, was a hard, brittle plastic called Bakelite. Because of its resistance to heat, it is still used today in stove-top appliances. Since 1909, hundreds of other synthetic polymers have been developed. Because of the widespread use of polymers, people might refer to this time as the Age of Polymers.

■ **Figure 16** Compact discs are made of polycarbonate and contain long chains of the structural unit shown.

Peter Smith/Alamy

■ **Figure 17** Polyethylene is a nontoxic, unbreakable polymer that is used to make toys for children.

## Reactions Used to Make Polymers

Polymers are relatively easy to manufacture. Polymers can usually be synthesized in one step in which the major reactant is a substance consisting of small, simple organic molecules called monomers. A **monomer** is a molecule from which a polymer is made.

When a polymer is made, monomers bond together one after another in a rapid series of steps. A catalyst is usually required for the reaction to take place at a reasonable pace. With some polymers, such as polyester fabric and nylon, two or more kinds of monomers bond to each other in an alternating sequence. A reaction in which monomer units are bonded together to form a polymer is called a **polymerization reaction.** The repeating group of atoms formed by the bonding of the monomers is called the structural unit of the polymer. The structural unit of a polymer made from two different monomers has the components of both monomers.

**Figure 17** shows unbreakable children's toys that are made of low-density polyethylene (LDPE), which is synthesized by polymerizing ethene under pressure. Ethene can also be made into ethylene glycol, which is the starting point for polyethylene terephthalate (PETE). PETE can be made into bottles or spun into fibers. As a fiber, it is called polyester.

**Figure 18** shows highlights of polymer development that led to the Age of Polymers. Although the first synthetic polymer was developed in 1909, the industry did not flourish until after World War II.

**READING CHECK** **Compare and contrast** a monomer and a structural unit of a polymer.

■ **Figure 18**

### The Age of Polymers

Scientists working to understand the structure and properties of organic compounds have developed products that affect people everywhere. Their contributions helped usher in the Age of Polymers.

1830 — 1860 — 1890

**1840s** Physicians begin using ether as an anesthetic during surgery.

**1865** The structure of benzene is determined. It becomes the basis for the production of aromatic compounds.

**1879** Saccharin is accidentally discovered by a chemist working with coal-tar derivatives.

**1899** Aspirin is widely distributed by physicians as a pain treatment. It quickly becomes the number-one selling drug worldwide.

**1909** The first plastic made from synthetic polymers, Bakelite, is developed.

$$n\text{HOOC}-(\text{CH}_2)_4-\text{COOH} + n\text{H}_2\text{N}-(\text{CH}_2)_6-\text{NH}_2 \rightarrow \left[\overset{\text{O}}{\overset{\|}{\text{C}}}-(\text{CH}_2)_4-\overset{\text{O}}{\overset{\|}{\text{C}}}-\text{NH}-(\text{CH}_2)_6-\text{NH}\right]_n + n\text{H}_2\text{O}$$

Adipic acid — 1,6–Diamino hexane — Nylon 6,6

**Figure 19** Nylon is a polymer consisting of thin strands that resemble silk.

**Addition polymerization** In **addition polymerization,** all of the atoms present in the monomers are retained in the polymer product. When the monomer is ethene, an addition polymerization results in the polymer polyethylene. Unsaturated bonds are broken in addition polymerization, just as they are in addition reactions. The difference is that the molecule added is a second molecule of the same substance, ethene. Note that the addition polymers in **Table 14** on the next page are similar in structure to polyethylene. That is, the molecular structure of each is equivalent to polyethylene in which other atoms or groups of atoms are attached to the chain in place of hydrogen atoms. All of these polymers are made by addition polymerization.

**Condensation polymerization** **Condensation polymerization** takes place when monomers containing at least two functional groups combine with the loss of a small by-product, usually water. Nylon and a type of bulletproof fabric are made this way. Nylon was first synthesized in 1931 and soon became popular because it is strong and can be drawn into thin strands resembling silk. Nylon 6,6 is the name of one type of nylon that is synthesized. One monomer is a chain, with the end carbon atoms being part of carboxyl groups, as shown in **Figure 19.** The other monomer is a chain having amino groups at both ends. These monomers undergo a condensation polymerization that forms amide groups linking the subunits of the polymer, as shown by the tinted box in **Figure 19.** Note that one water molecule is released for every new amide bond formed.

**1939–1945** During World War II, nylon is allocated solely for military items such as parachutes, as shown in the photo, tents, and ponchos.

**1946** Products with nonstick coating (PTFE), including bearings, bushings, gears, and cookware, become commercially available.

**1959** Spandex, an elastic fiber, is commercially produced.

**1988** The world's first polymer banknote is issued by the Reserve Bank of Australia. By 1996, all Australians use plastic money.

**2006** Researchers develop a paper-thin, radiation-resistant, liquid-crystal polymer in which electronic circuits can be imbedded, making it useful in space applications.

**2010** Scientists create a semiconducting, transparent polymer doped with fullerenes. Once the film absorbs light, it produces an electric charge. This ability could lead to applied technology, such as transparent solar cells that can be coated on windows.

1950 — 1980 — 2010

Explore **polymers** with an interactive table. **Concepts In Motion**

## Table 14 Common Polymers

| Polymer | Applications | Structural Unit |
|---|---|---|
| Polyvinyl chloride (PVC) | Plastic pipes, meat wrap, upholstery, rainwear, house siding, garden hose  | 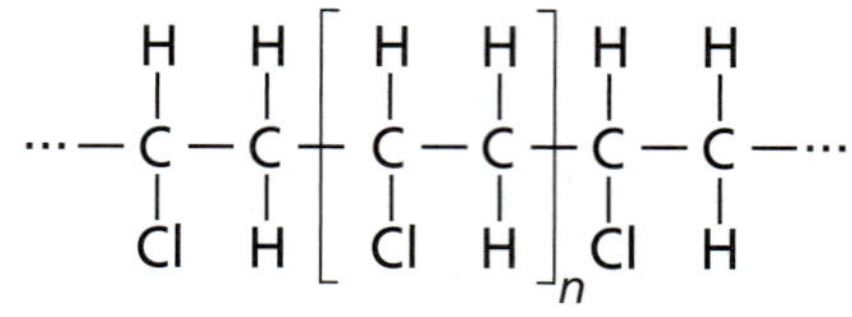 Polyvinyl chloride |
| Polyacrylonitrile | Fabrics for clothing and upholstery, carpet | 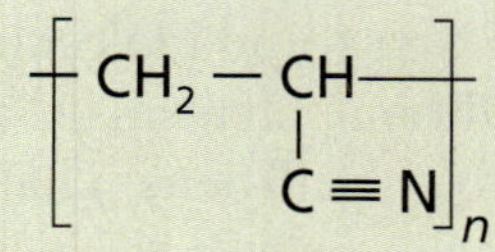 |
| Polyvinylidene chloride | Food wrap, fabrics   | 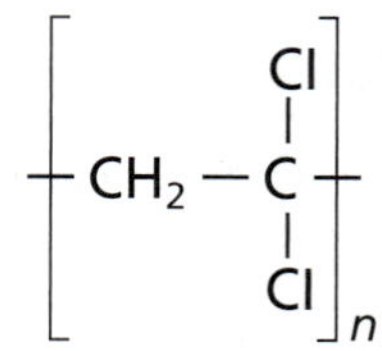 |
| Polymethyl methacrylate | "Nonbreakable" (acrylic glass) windows, inexpensive lenses, art objects  | 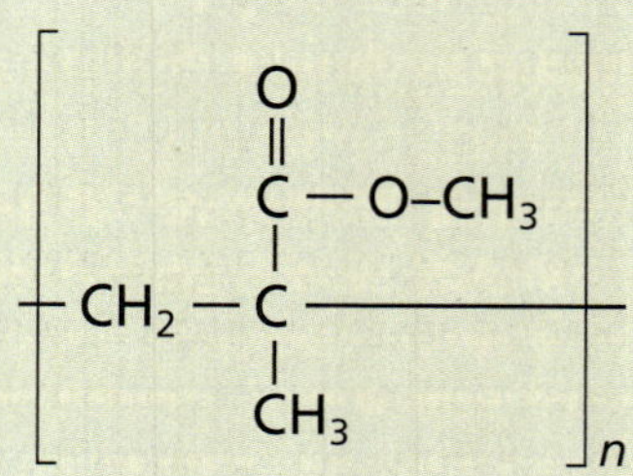 |
| Polypropylene (PP) | Beverage containers, rope, netting, kitchen appliances | 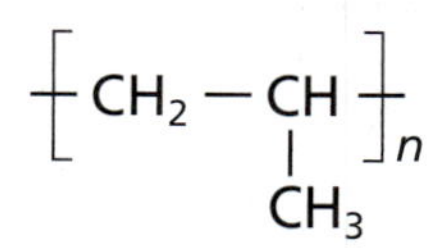 |
| Polystyrene (PS) and styrene plastic | Foam packing and insulation, plant pots, disposable food containers, model kits  | 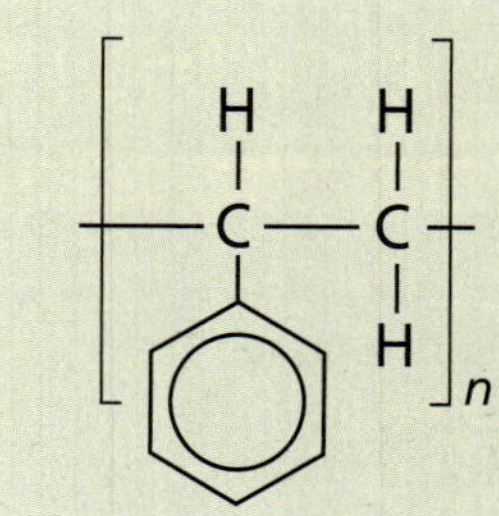 |
| Polyethylene terephthalate (PETE) | Soft-drink bottles, tire cord, clothing, recording tape, replacements for blood vessels | 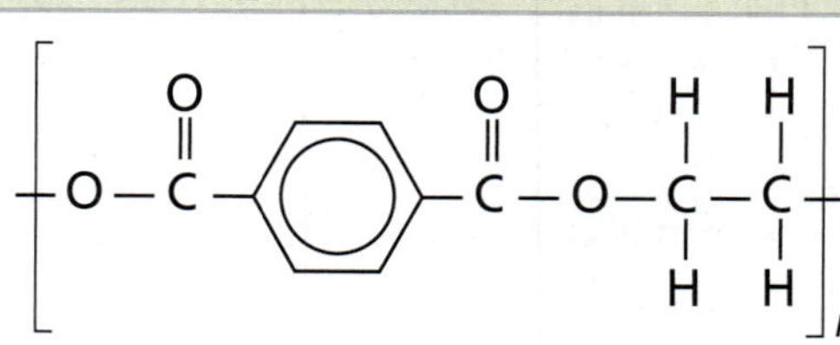 |
| Polyurethane | Foam furniture cushions, waterproof coatings, parts of shoes | 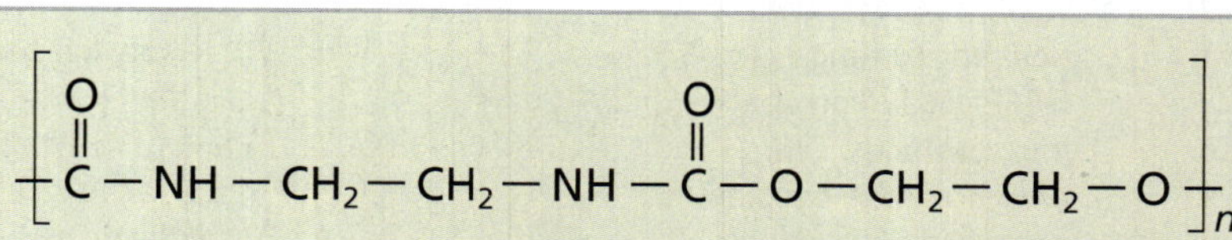 |

■ **Figure 20** Plastic lumber is made from recycled plastic, such as used soft-drink bottles, milk jugs, and other polyethylene waste.

## Properties and Recycling of Polymers

Why do we use so many different polymers today? One reason is that they are easy to synthesize. Another reason is that the starting materials used to make them are inexpensive. Still another, more important, reason is that polymers have a wide range of properties. Some polymers can be drawn into fine fibers that are softer than silk, while others are as strong as steel. Polymers do not rust like steel does, and many polymers are more durable than natural materials such as wood. Fencing and decking materials made of plastic, like those shown in **Figure 20,** do not decay and do not need to be repainted.

**Properties of polymers** Another reason why polymers are in such great demand is that it is easy to mold them into different shapes or to draw them into thin fibers. It is not easy to do this with metals and other natural materials because they must be heated either to high temperatures, do not melt at all, or are too weak to be used to form small, thin items.

As with all substances, polymers have properties that result directly from their molecular structure. For example, polyethylene is a long-chain alkane. Thus, it has a waxy feel, does not dissolve in water, is nonreactive, and is a poor electrical conductor. These properties make it ideal for use in food and beverage containers and as an insulator in electrical wire and TV cable.

Polymers fall into two different categories, based on their melting characteristics. A **thermoplastic** polymer is one that can be melted and molded repeatedly into shapes that are retained when cooled. Polyethylene and nylon are examples of thermoplastic polymers. A **thermosetting** polymer is one that can be molded when it is first prepared, but after it cools, it cannot be remelted. This property is explained by the fact that thermosetting polymers begin to form networks of bonds in many directions when they are synthesized. By the time they have cooled, thermosetting polymers have become, in essence, a single large molecule. Bakelite is an example of a thermosetting polymer. Instead of melting, Bakelite decomposes when overheated.

☑ READING CHECK **Compare and contrast** thermoplastic and thermosetting polymers.

CAREERS IN CHEMISTRY

**Polymer Chemist** Does the thought of developing new and better polymers sound inspiring and challenging to you? Polymer chemists develop new polymers and create uses or manufacturing processes for older ones.

WebQuest

**VOCABULARY**

**WORD ORIGIN**

**Thermoplastic**

*thermo-* comes from the Greek word *thermē* which means heat; *plastic* comes from the Greek word *plastikos* which means to mold or form

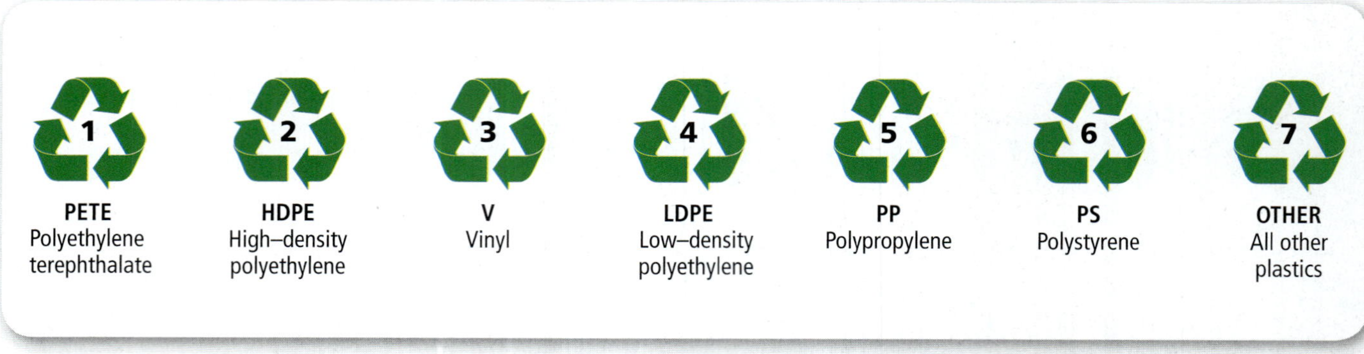


■ **Figure 21** Codes on plastic products aid in recycling because they identify the composition of the plastic.

Watch a **video about recycling.**

Video

**Recycling polymers** The starting materials for the synthesis of most polymers are derived from fossil fuels. As the supply of fossil fuels becomes depleted, recycling plastics becomes more important. Recycling and buying goods made from recycled plastics decreases the amount of fossil fuels used, which conserves fossil fuels.

Currently, about 5% of the plastics used in the United States are recycled. Plastics recycling is somewhat difficult due to the large variety of different polymers found in products. Usually, the plastics must be sorted according to polymer composition before they can be reused. Thermosetting polymers are more difficult to recycle than thermoplastic polymers because only thermoplastic materials can be melted and remolded repeatedly. The task of separating plastics can be time-consuming and expensive. That is why the plastics industry and the government have tried to improve the process by providing standardized codes that indicate the composition of each plastic product. The standardized codes for plastics are shown in **Figure 21.** These codes provide a quick way for recyclers to sort plastics.

## SECTION 5 REVIEW

Section Self-Check

### Section Summary

- Polymers are large molecules formed by combining smaller molecules called monomers.
- Polymers are synthesized through addition or condensation reactions.
- The functional groups present in polymers can be used to predict polymer properties.

**22.** MAINIDEA **Draw** the structure for the polymer that could be produced from each of the following monomers by the method stated.

**a.** Addition

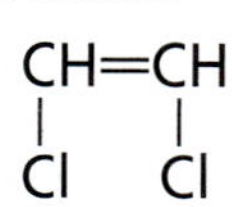

**b.** Condensation

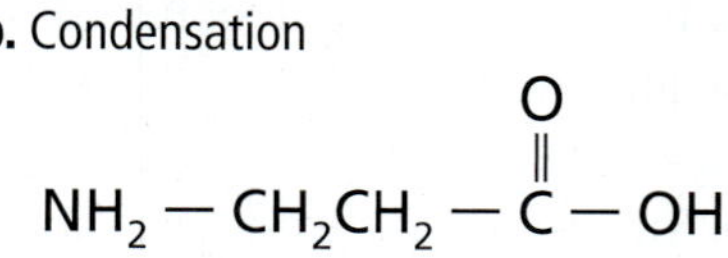

**23. Label** the following polymerization reaction as *addition* or *condensation.* Explain your answer.

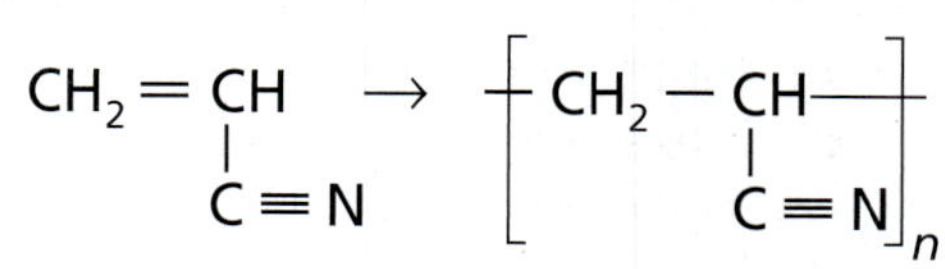

**24. Identify** Synthetic polymers often replace stone, wood, metals, wool, and cotton in many applications. Identify some advantages and disadvantages of using synthetic materials instead of natural materials.

**25. Predict** the physical properties of the polymer that is made from the following monomer. Mention solubility in water, electrical conductivity, texture, and chemical reactivity. Do you think it will be thermoplastic or thermosetting? Give reasons for your predictions.

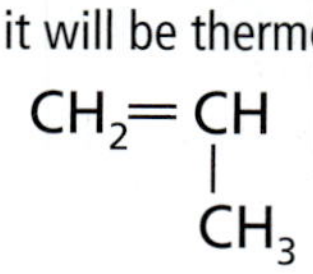

**everyday**CHEMISTRY

# Garlic: Pleasure and Pain

Did you know that the flavors of fresh and roasted garlic are very different? Fresh garlic, shown in **Figure 1,** contains substances that cause a burning sensation in your mouth. However, roasted garlic does not produce this sensation. These sensations, pleasure or pain, are results of chemical reactions.

When raw garlic is bruised, cut, or crushed, it produces a chemical called allicin, as shown in **Figure 2.** The production of allicin is a chemical defense mechanism for the garlic plant against other organisms. Allicin is an unstable compound and is converted to other compounds over time or when garlic is heated or roasted, which explains why roasted garlic does not cause the burning sensation in your mouth.

**Figure 1** Fresh garlic contains a pain-producing chemical as a defense against predators.

**Sensing temperature and pain** Temperature and pain are sensed by neurons embedded in the skin, including the skin inside your mouth. These neurons have temperature-detecting molecules on their surfaces that are called transient receptor potential (TRP) ion channels. Different TRP channels are activated by different temperature ranges. For example, when a person touches something hot, some of the TRP ion channels open and allow charged calcium ions to enter the nerve cell. This increases the charge within the nerve cell. When the charge increases enough, an electrical signal is sent to the brain, where it is interpreted as a hot sensation.

Allicin also activates neurons. Allicin apparently acts on a pair of ion channel proteins called TRPA1 and TRPV1. When the chemical allicin is present, these channels allow ions to enter the nerve cell. The additional electric charge in the nerve cell signals the brain, where the signal is interpreted by the brain as a burning sensation.

**Probing pain receptors** While it is interesting to know why tasting raw garlic is painful, the understanding of how allicin causes that pain sensation is even more interesting and useful. Researchers hope that a further understanding of how these receptors work will lead to new methods for controlling chronic pain in patients.

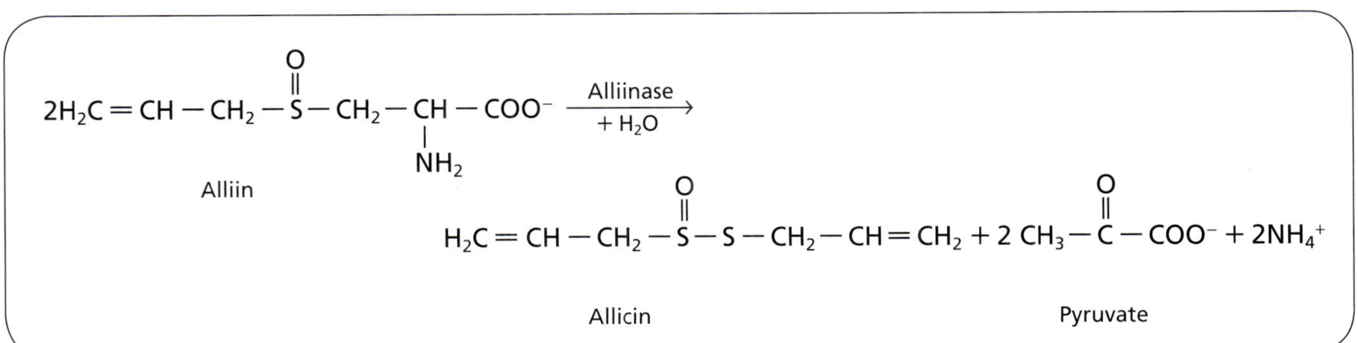


**Figure 2** When garlic is bruised or damaged, alliin and the enzyme alliinase produce allicin. When you taste fresh garlic, neurons embedded in your mouth cause an electrical signal to be sent to your brain. The brain interprets the electrical signal as a burning sensation.

**WRITING** IN ▶ **Chemistry**

**Research** and prepare a poster that shows other chemical reactions in plants.

WebQuest

SMALL SCALE **ChemLAB**

iLab Station

# Observe Properties of Alcohols

**Background:** Alcohols are organic compounds that contain the –OH functional group. How fast various alcohols evaporate indicates the strength of intermolecular forces in alcohols. The evaporation of a liquid is an endothermic process, absorbing energy from the surroundings. This means that the temperature will decrease as evaporation occurs.

**Question:** *How do intermolecular forces differ in three alcohols?*

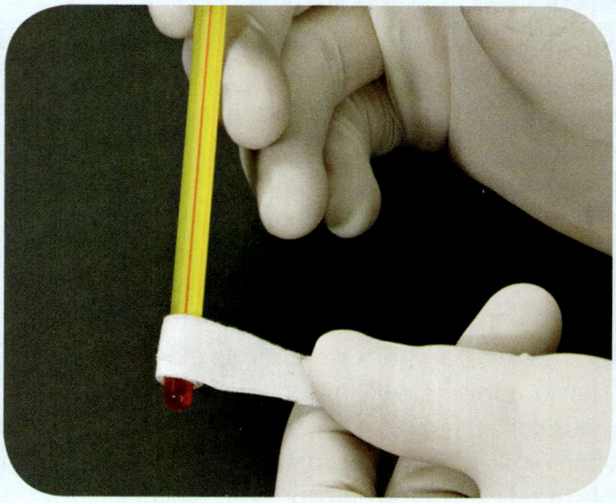

## Materials

nonmercury thermometer
stopwatch
facial tissue
cloth towel
Beral pipettes (5)
methanol
ethanol (95%)
2-propanol (99%)
wire twist tie or small rubber band
piece of cardboard for use as a fan

## Safety Precautions

WARNING: ***Alcohols are flammable. Keep liquids and vapors away from open flames and sparks.***

## Procedure

1. Read and complete the lab safety form.
2. Prepare data tables for recording data.
3. Cut five 2-cm by 6-cm strips of tissue.
4. Place a thermometer on a folded towel lying on a flat table so that the bulb of the thermometer extends over the edge of the table. Make sure the thermometer cannot roll off the table.
5. Wrap a strip of tissue around the bulb of the thermometer. Secure the tissue with a wire twist tie placed above the bulb of the thermometer.
6. Choose one person to control the stopwatch and read the temperature on the thermometer. A second person will put a small amount of the liquid to be tested into a Beral pipette.
7. When both people are ready, squeeze enough liquid onto the tissue to completely saturate it. At the same time, the other person starts the stopwatch, reads the temperature, and records it in the data table.
8. Fan the tissue-covered thermometer bulb with a piece of cardboard or other stiff paper. After 1 min, read and record the final temperature in the data table. Remove the tissue and wipe the bulb dry.
9. Repeat Steps 5 through 8 for each of the three alcohols: methanol, ethanol, and 2-propanol.
10. Obtain the classroom temperature and humidity data from your teacher.
11. **Cleanup and Disposal** Place the used tissues in the trash. Pipettes can be reused.

## Analyze and Conclude

1. **Observe and Infer** What can you conclude about the relationship between heat transfer and the differences in the temperature changes you observed?
2. **Evaluate** Molar enthalpies of vaporization (kJ/mol) for the three alcohols at 25°C are: methanol, 37.4; ethanol, 42.3; and 2-propanol, 45.4. What can you conclude about the relative strength of intermolecular forces existing in the three alcohols?
3. **Compare** Make a general statement comparing the molecular size of an alcohol in terms of the number of carbons in the carbon chain to the rate of evaporation of that alcohol.
4. **Observe and Infer** Compare your data with those of your classmates. Infer why there are differences.
5. **Error Analysis** Determine where errors might have been introduced in your procedure.

### INQUIRY EXTENSION

**Design an Experiment** Suggest a way to make this experiment more quantitative and controlled. Design an experiment using your new method.

Matt Meadows

CHAPTER 22 STUDY GUIDE

Vocabulary Practice

**BIGIDEA** The substitution of different functional groups for hydrogen atoms in hydrocarbons results in a diverse group of organic compounds.

## SECTION 1 Alkyl Halides and Aryl Halides

**MAINIDEA** A halogen atom can replace a hydrogen atom in some hydrocarbons.

- The substitution of functional groups for hydrogen in hydrocarbons creates a wide variety of organic compounds.
- An alkyl halide is an organic compound that has one or more halogen atoms bonded to a carbon atom in an aliphatic compound.

**VOCABULARY**
- functional group
- halocarbon
- alkyl halide
- aryl halide
- plastic
- substitution reaction
- halogenation

## SECTION 2 Alcohols, Ethers, and Amines

**MAINIDEA** Oxygen and nitrogen are two of the most-common atoms found in organic functional groups.

- Alcohols, ethers, and amines are formed when specific functional groups substitute for hydrogen in hydrocarbons.
- Because they readily form hydrogen bonds, alcohols have higher boiling points and higher water solubilities than other organic compounds.

**VOCABULARY**
- hydroxyl group
- alcohol
- denatured alcohol
- ether
- amine

## SECTION 3 Carbonyl Compounds

**MAINIDEA** Carbonyl compounds contain a double-bonded oxygen in the functional group.

- Carbonyl compounds are organic compounds that contain the C=O group.
- Five important classes of organic compounds containing carbonyl compounds are aldehydes, ketones, carboxylic acids, esters, and amides.

**VOCABULARY**
- carbonyl group
- aldehyde
- ketone
- carboxylic acid
- carboxyl group
- ester
- amide
- condensation reaction

## SECTION 4 Other Reactions of Organic Compounds

**MAINIDEA** Classifying the chemical reactions of organic compounds makes predicting products of reactions much easier.

- Most reactions of organic compounds can be classified into one of five categories: substitution, elimination, addition, oxidation-reduction, and condensation.
- Knowing the types of organic compounds reacting can enable you to predict the reaction products.

**VOCABULARY**
- elimination reaction
- dehydrogenation reaction
- dehydration reaction
- addition reaction
- hydration reaction
- hydrogenation reaction

## SECTION 5 Polymers

**MAINIDEA** Synthetic polymers are large organic molecules made up of repeating units linked together by addition or condensation reactions.

- Polymers are large molecules formed by combining smaller molecules called monomers.
- Polymers are synthesized through addition or condensation reactions.
- The functional groups present in polymers can be used to predict polymer properties.

**VOCABULARY**
- polymer
- monomer
- polymerization reaction
- addition polymerization
- condensation polymerization
- thermoplastic
- thermosetting

# CHAPTER 22 ASSESSMENT

## SECTION 1

### Mastering Concepts

**26.** What is a functional group?

**27.** Describe and compare the structures of alkyl halides and aryl halides.

**28.** What reactant would you use to convert methane to bromomethane?

**29.** Name the amines represented by each of the condensed formulas.
- **a.** $CH_3(CH_2)_3CH_2NH_2$
- **b.** $CH_3(CH_2)_5CH_2NH_2$
- **c.** $CH_3(CH_2)_2CH(NH_2)CH_3$
- **d.** $CH_3(CH_2)_8CH_2NH_2$

**30.** Explain why the boiling points of alkyl halides increase in order going down the column of halides in the periodic table, from fluorine through iodine.

### Mastering Problems

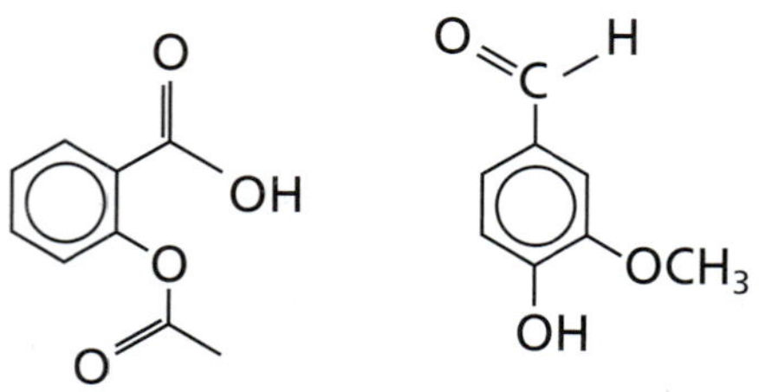

**a** Acetylsalicylic acid  **b** Vanillin

■ **Figure 22**

**31.** Name and write the general formula of each of the functional groups attached to the benzene rings shown in **Figure 22.**

**32.** Draw structures for these alkyl and aryl halides.
- **a.** chlorobenzene
- **b.** 1-bromo-4-chlorohexane
- **c.** 1,2-difluoro-3-iodocyclohexane
- **d.** 1,3-dibromobenzene
- **e.** 1,1,2,2-tetrafluoroethane

**33.** For 1-bromo-2-chloropropane:
- **a.** Draw the structure.
- **b.** Does the compound have optical isomers?
- **c.** If the compound has optical isomers, identify the chiral carbon atom.

**34.** Draw and name all of the structural isomers possible for an alkyl halide with no branches and the molecular formula $C_5H_{10}Br_2$.

**35.** Name one structural isomer created by changing the position of one or more halogen atoms in each alkyl halide.
- **a.** 2-chloropentane
- **b.** 1,1-difluropropane
- **c.** 1,3-dibromocyclopentane
- **d.** 1-bromo-2-chloroethane

## SECTION 2

### Mastering Concepts

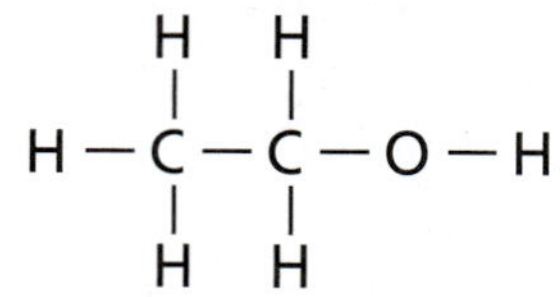

■ **Figure 23**

**36.** How is the compound shown in **Figure 23** denatured? What is the name of the compound?

**37. Practical Applications** Name one alcohol, amine, or ether that is used for each of the following purposes.
- **a.** antiseptic
- **b.** solvent in paint strippers
- **c.** antifreeze
- **d.** anesthetic
- **e.** dye production

**38.** Explain why an alcohol molecule will always have a higher solubility in water than an ether molecule having an identical molecular mass.

**39.** Explain why ethanol has a much higher boiling point than aminoethane, even though their molecular masses are nearly equal.

### Mastering Problems

**40.** Name one ether that is a structural isomer of each alcohol.
- **a.** 1-butanol
- **b.** 2-hexanol

**41.** Draw structures for the following alcohol, amine, and ether molecules.
- **a.** 1,2-butanediol
- **b.** 2-aminohexane
- **c.** isopropyl ether
- **d.** 2-methyl-1-butanol
- **e.** butyl pentyl ether
- **f.** cyclobutyl methyl ether
- **g.** 1,3-diaminobutane
- **h.** cyclopentanol

## SECTION 3

### Mastering Concepts

**42.** Draw the general structure for each of the following classes of organic compounds.
- **a.** aldehyde
- **b.** ketone
- **c.** carboxylic acid
- **d.** ester
- **e.** amide

**43. Common Uses** Name an aldehyde, ketone, carboxylic acid, ester, or amide used for each of the following purposes.
- **a.** preserving biological specimens
- **b.** solvent in fingernail polish
- **c.** acid in vinegar
- **d.** flavoring in foods and beverages

**44.** What type of reaction is used to produce aspirin from salicylic acid and acetic acid?

## Mastering Problems

**45.** Draw structures for each of the following carbonyl compounds.
- **a.** 2,2-dichloro-3-pentanone
- **b.** 4-methylpentanal
- **c.** isopropyl hexanoate
- **d.** octanamide
- **e.** 3-fluoro-2-methylbutanoic acid
- **f.** cyclopentanal
- **g.** hexyl methanoate

**46.** Name each of the following carbonyl compounds.

**a.**

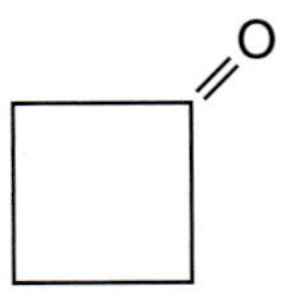

**b.** $CH_3-CH_2-CH_2-\overset{\overset{\displaystyle O}{\|}}{C}-H$

**c.** $CH_3 \leftarrow\!\!\!+ CH_2 +\!\!\!\rightarrow_4 \overset{\overset{\displaystyle O}{\|}}{C}-NH_2$

**d.** $CH_3 \leftarrow\!\!\!+ CH_2 +\!\!\!\rightarrow_4 \overset{\overset{\displaystyle O}{\|}}{C}-OH$

# SECTION 4

## Mastering Concepts

**47. Synthetic Organic Compounds** What is the starting material for making most synthetic organic compounds?

**48.** Explain the importance of classifying reactions.

**49.** List the type of organic reaction needed to perform each of the following transformations.
- **a.** alkene → alkane
- **b.** alkyl halide → alcohol
- **c.** alkyl halide → alkene
- **d.** amine + carboxylic acid → amide
- **e.** alcohol → alkyl halide
- **f.** alkene → alcohol

## Mastering Problems

**50.** Classify each of the following organic reactions as substitution, addition, oxidation-reduction elimination, or condensation.
- **a.** 2-butene + hydrogen → butane
- **b.** propane + fluorine → 2-fluoropropane + hydrogen fluoride
- **c.** 2-propanol → propene + water
- **d.** cyclobutene + water → cyclobutanol

**51.** Use structural formulas to write equations for the following reactions.
- **a.** the substitution reaction between 2-chloropropane and water yielding 2-propanol and hydrogen chloride
- **b.** the addition reaction between 3-hexene and chlorine yielding 3,4-dichlorohexane

**52.** What type of reaction converts an alcohol into each of the following types of compounds?
- **a.** ester
- **b.** alkyl halide
- **c.** alkene
- **d.** aldehyde

**53.** Use structural formulas to write the equation for the condensation reaction between ethanol and propanoic acid.

# SECTION 5

## Mastering Concepts

**54.** Explain the difference between addition polymerization and condensation polymerization.

**55.** Which type of polymer is easier to recycle, thermosetting or thermoplastic? Explain your answer.

## Mastering Problems

**56. Manufacturing Polymers** Draw the monomers that react to form each polymer.
- **a.** polyethylene
- **b.** polyvinyl chloride (PVC)
- **c.** polytetrafluoroethylene (PTFE)

**57.** Name the polymers made from the following monomers.
- **a.** $H_2C = CHCl$
- **b.** $CH_2 = CCl_2$

**58.** Choose the polymer of each pair that you expect to have the higher water solubility.

**a.**

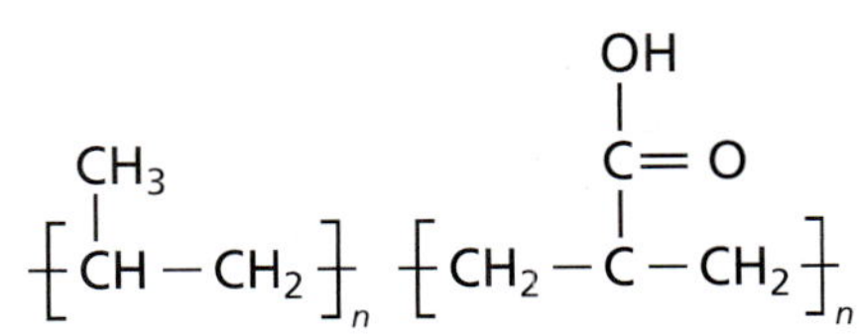

I II

**b.**

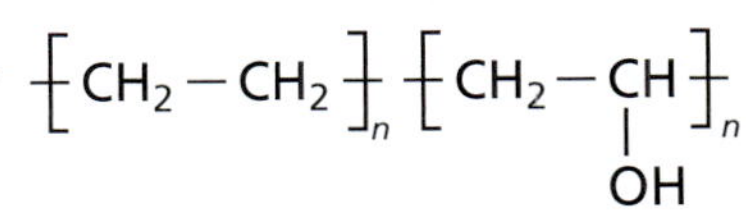

**59.** Examine the structures of the following polymers. Decide whether each is made by addition or condensation polymerization.
- **a.** nylon
- **b.** polyacrylonitrile
- **c.** polyurethane
- **d.** polypropylene

**60. Human Hormones** Which halogen is found in hormones made by a normal human thyroid gland?

## MIXED REVIEW

**61.** Describe the properties of carboxylic acids.

**62.** Draw structures of the following compounds.
- **a.** butanone
- **b.** propanal
- **c.** hexanoic acid
- **d.** heptanamide

**63.** Name the type of organic compound formed by each of the following reactions.
- **a.** elimination from an alcohol
- **b.** addition of hydrogen chloride to an alkene
- **c.** addition of water to an alkene
- **d.** substitution of a hydroxyl group for a halogen atom

**64.** List two uses for each of the following polymers.
- **a.** polypropylene
- **b.** polyurethane
- **c.** polytetrafluoroethylene
- **d.** polyvinyl chloride

**65.** Draw structures of and supply names for the organic compounds produced by reacting ethene with each of the following substances.
- **a.** water
- **b.** hydrogen
- **c.** hydrogen chloride
- **d.** fluorine

**66. Environmentally-Safe Propellants** Hydrofluoroalkanes (HFAs) are replacing chlorofluorocarbons in hand-held asthma inhalers, because of CFC damage to the ozone layer. Draw the structures of the HFAs listed below.
- **a.** 1,1,1,2,3,3,3-heptafluoropropane
- **b.** 1,1,1,2-tetrafluoroethane

## THINK CRITICALLY

**67. Interpret Scientific Illustrations** List all the functional groups present in each of the following complex organic molecules.

**a.**

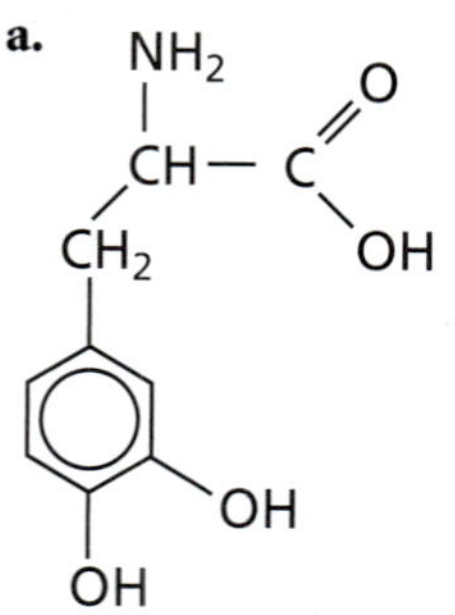


Levadopa

**b.**

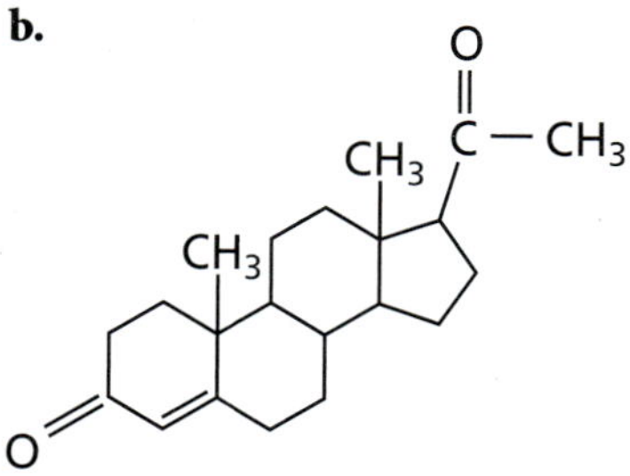


Progesterone

**68. Evaluate** Ethanoic acid (acetic acid) is very soluble in water. However, naturally occurring long-chain carboxylic acids, such as palmitic acid ($CH_3(CH_2)_{14}COOH$), are insoluble in water. Explain.

**69. Communicate** Write structural formulas for all structural isomers of molecules having the following formulas. Name each isomer.
- **a.** $C_3H_8O$
- **b.** $C_2H_4Cl_2$

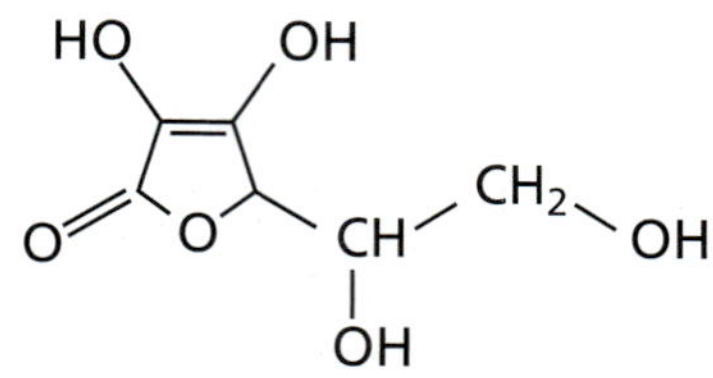


Vitamin C

**■ Figure 24**

**70. Interpret Scientific Illustrations** Human cells require vitamin C to properly synthesize materials that make up connective tissue such as that found in ligaments. List the functional groups present in the vitamin C molecule shown in **Figure 24.**

**71. Identify** Draw the structure of an example of an organic molecule that has four carbons and falls into each of the compound types listed.
- **a.** ester
- **b.** aldehyde
- **c.** ether
- **d.** alcohol

**72. Predict** A monohalogenation reaction describes a substitution reaction in which a single hydrogen atom is replaced by a halogen. A dihalogenation reaction is a reaction in which two hydrogen atoms are replaced by two halogen atoms.
- **a.** Draw the structures of all the possible monohalogenation products that can form when pentane reacts with $Cl_2$.
- **b.** Draw the structures of all the possible dihalogenation products that can form when pentane reacts with $Cl_2$.

**Table 15 Alcohol Solubility in Water (mol/100 g $H_2O$)**

| Name | Alcohol | Solubility |
|---|---|---|
| Methanol | $CH_3OH$ | infinite |
| Ethanol | $C_2H_5OH$ | infinite |
| Propanol | $C_3H_7OH$ | infinite |
| Butanol | $C_4H_9OH$ | 0.11 |
| Pentanol | $C_5H_{11}OH$ | 0.030 |
| Hexanol | $C_6H_{13}OH$ | 0.0058 |
| Heptanol | $C_7H_{15}OH$ | 0.0008 |

**73. Evaluate** Examine **Table 15** comparing some alcohols and their solubility in water. Use the table to answer the following questions.
- **a.** What type of bond forms between the –OH group of alcohols and water?
- **b.** State a relationship between water solubility and alcohol size from the data in the table.
- **c.** Provide an explanation for the relationship you stated in Part b.

**74. Recognize** Most useful organic molecules are made from raw materials using several steps. This is called a multistep synthesis pathway. Label the types of reaction or process taking place in each step of the multistep synthesis pathway below.

petroleum ⟶ ethane ⟶ chloroethane ⟶ ethene ⟶ ethanol ⟶ ethanoic (acetic) acid

## CHALLENGE PROBLEM

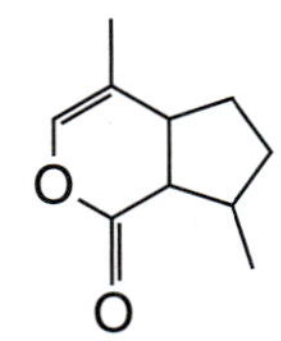

■ **Figure 25**

**75. Animal Pheromones** Catnip contains an organic chemical known as *nepetalactone,* shown in **Figure 25,** that is thought to mimic feline sex pheromones. Cats will rub in it, roll over it, paw at it, chew it, lick it, leap about, then purr loudly, growl, and meow for several minutes before losing interest. It takes up to two hours for the cat to "reset" and then have the same response to the catnip.

**a.** What type of organic compound is nepetalactone?

**b.** Draw the structural formula for nepetalactone on a sheet of paper and then draw in all the missing hydrogen atoms. Remember that carbon atoms must have four bonds to be stable.

**c.** Write the molecular formula for nepetalactone.

## CUMULATIVE REVIEW

**76.** Explain why the concentration of ozone over Antarctica decreases at about the same time every year.

**77.** Why do the following characteristics apply to transition metals?

**a.** Ions vary in charge.

**b.** Many of their solids are colored.

**c.** Many are hard solids.

**78.** Determine the number of atoms in each of the following.

**a.** 56.1 g Al **b.** 2 moles C

**79.** What is a rate-determining step?

**80.** According to Le Châtelier's principle, how would increasing the volume of the reaction vessel affect the equilibrium $2SO_2(g) + O_2(g) \leftrightarrows 2SO_3(g)$?

**81.** Compare and contrast saturated and unsaturated hydrocarbons.

## WRITING IN ▸ Chemistry

**82. Historical Perspective** Write a short story describing how your life would differ if you lived in the 1800s, before the development of synthetic polymers.

## DBQ Document-Based Questions

**Pharmaceutical Propellants** *Many inhaled medications used to treat asthma contained chlorofluorocarbons (CFCs). However, the Montreal Protocol called for a ban of CFCs as a propellant in pharmaceutical products. Two hydrofluoroalkanes (HFAs) appear to be effective in delivering asthma medications to the lungs. However, the medication dosage had to be cut in half with the new HFA propellents.*

**Figure 26** *shows the concentration after one dose of the drug beclomethasone in the blood of volunteers using a CFC or an HFA propellant in the inhaler.*

Data obtained from: Anderson, P.J. 2006. *Chest: The Cardiopulmonary and Critical Care Journal.* 120:89–93

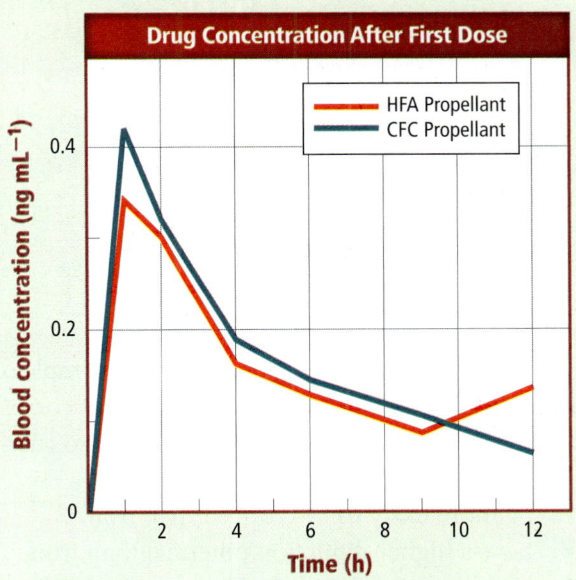


■ **Figure 26**

**83.** After one dose of the drug beclomethasone was given, which propellant resulted in the highest concentration of medication in the blood, HFA or CFC?

**84.** When does the drug reach its peak concentration?

**85.** Only one-half the amount of medication is needed with the HFA propellant when compared to the CFC propellant to achieve a similar blood-concentration level. Infer the advantages of using a lower dose of medication to get similar results.

## MULTIPLE CHOICE

1. What are the products of this reaction?

   $CH_3CH_2CH_2Br + NH_3 \rightarrow ?$
   A. $CH_3CH_2CH_2NH_2Br$ and $H_2$
   B. $CH_3CH_2CH_2NH_3$ and $Br_2$
   C. $CH_3CH_2CH_2NH_2$ and $HBr$
   D. $CH_3CH_2CH_3$ and $NH_2Br$

2. What kind of reaction is this?

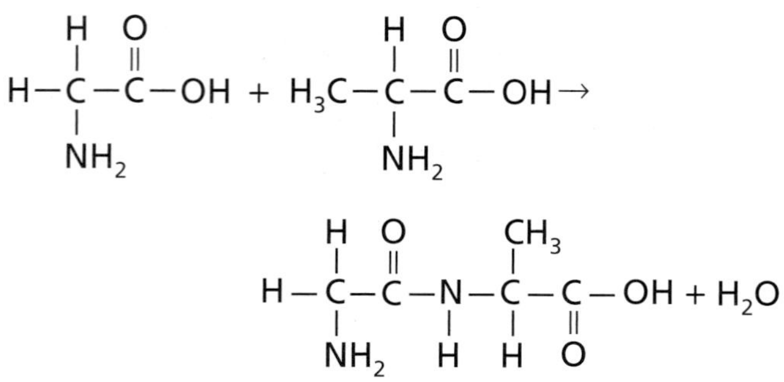

   A. substitution
   B. condensation
   C. addition
   D. elimination

3. What are the oxidation numbers of the elements in $CuSO_4$?
   A. Cu = +2, S = +6, O = −2
   B. Cu = +3, S = +5, O = −2
   C. Cu = +2, S = +2, O = −1
   D. Cu = +2, S = 0, O = −2

4. The corrosion, or rusting, of iron is an example of a naturally occurring voltaic cell. To prevent corrosion, sacrificial anodes are sometimes attached to rust-susceptible iron. Sacrificial anodes must
   A. be more likely to be reduced than iron.
   B. have a higher reduction potential than iron.
   C. be more porous and abraded than iron.
   D. lose electrons more easily than iron.

5. What type of compound does this molecule represent?

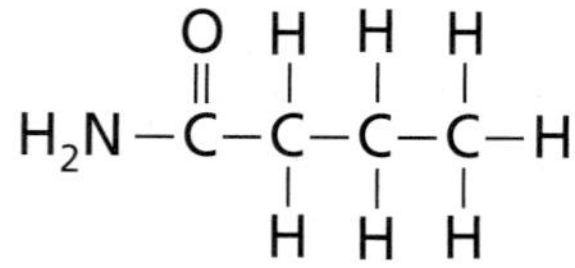

   A. amine
   B. amide
   C. ester
   D. ether

6. Diprotic succinic acid ($H_2C_4H_4O_4$) is an important part of the process that converts glucose to energy in the human body. What is the $K_a$ expression for the second ionization of succinic acid?
   A. $K_a = [H_3O^+][HC_4H_4O_4^-] / [H_2C_4H_4O_4]$
   B. $K_a = [H_3O^+][C_4H_4O_4^{2-}] / [HC_4H_4O_4^-]$
   C. $K_a = [H_2C_4H_4O_4] / [H_3O^+][HC_4H_4O_4^-]$
   D. $K_a = [H_2C_4H_4O_4] / [H_3O^+][C_4H_4O_4^{2-}]$

*Use the figure below to answer Question 7.*

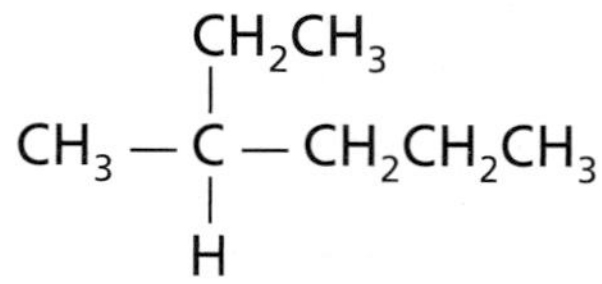

7. Which is the correct name for this compound?
   A. 3-methylhexane
   B. 2-ethylpentane
   C. 2-propylbutane
   D. 1-ethyl-1-methylbutane

8. A strip of metal X is immersed in a 1*M* solution of $X^+$ ions. When this half-cell is connected to a standard hydrogen electrode, a voltmeter reads a positive reduction potential. Which is true of the X electrode?
   A. It accepts electrons more readily than $H^+$ ions.
   B. It is undergoing oxidation.
   C. It is adding positive $X^+$ ions to its solution.
   D. It acts as the anode in the cell.

9. What is the mass of one formula unit of barium hexafluorosilicate ($BaSiF_6$)?
   A. $4.64 \times 10^{-22}$ g
   B. $1.68 \times 10^{26}$ g
   C. $2.16 \times 10^{21}$ g
   D. $6.02 \times 10^{-23}$ g

10. Which type of compound accepts $H^+$ ions?
    A. an Arrhenius acid
    B. an Arrhenius base
    C. a Brønsted-Lowry acid
    D. a Brønsted-Lowry base

11. Which substituted hydrocarbon has the general formula R–OH?
    A. alcohol
    B. amine
    C. ketone
    D. carboxylic acid

Online Test Practice

## SHORT ANSWER

*Use the figure below to answer Questions 12 and 13.*

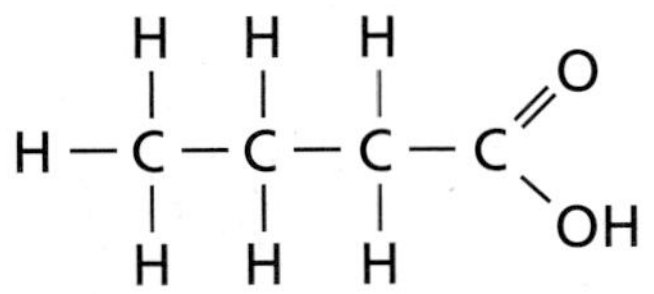

**12.** What is the functional group present in this compound?

**13.** Give the name for this compound.

## EXTENDED RESPONSE

*Use the graph below to answer Question 14.*

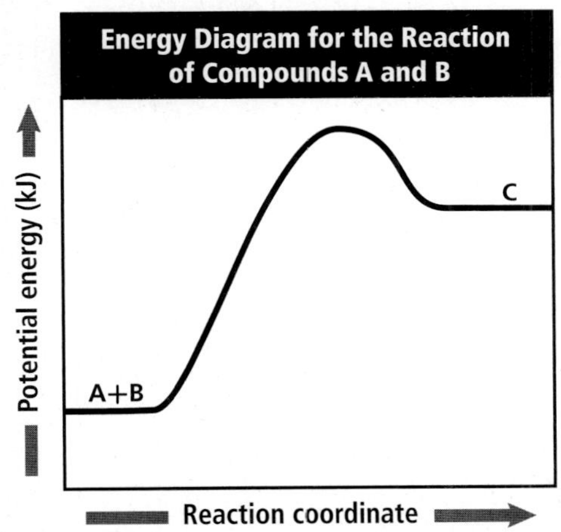


**14.** Discuss the reaction that results in the shape of the energy graph shown.

*Use the figure below to answer Question 15.*

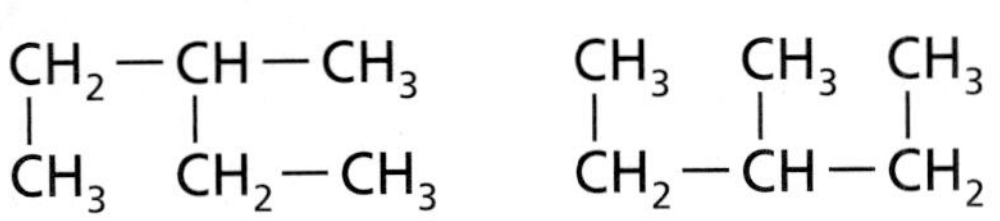

**15.** The two structures above both have the molecular formula $C_6H_{14}$. Are they isomers of one another? Explain how you can tell.

## SAT SUBJECT TEST: CHEMISTRY

**16.** To electroplate an iron fork with silver,

**A.** the silver electrode must have more mass than the fork.
**B.** the iron fork must act as the anode in the cell.
**C.** electric current must be applied to the iron fork.
**D.** iron ions must be present in the cell solution.
**E.** the electric current must be pulsed.

**17.** Which type of reaction is shown below?

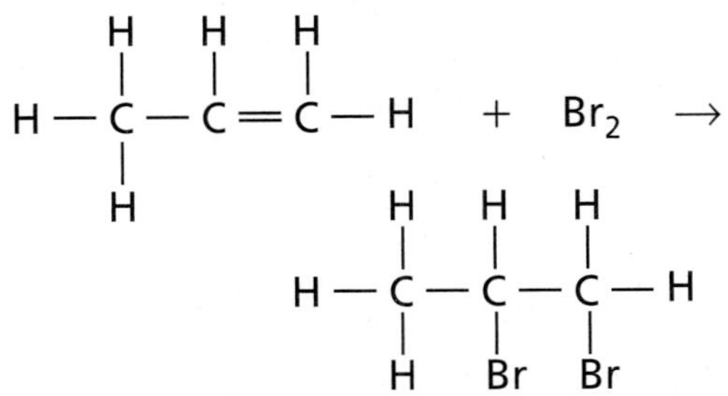

**A.** condensation
**B.** dehydration
**C.** polymerization
**D.** halogenation
**E.** hydration

*Use the table below to answer Question 18.*

| Experimental Data for A + B → C | | | |
|---|---|---|---|
| Time | [A]*M* | [B]*M* | [C]*M* |
| 0.00 sec | 0.35 | 0.50 | 0.00 |
| 5.00 sec | 0.15 | 0.30 | 0.40 |

**18.** Which is the rate of this reaction in terms of product produced in mol/(L•s)?

**A.** 0.40 mol/(L•s)
**B.** 0.85 mol/(L•s)
**C.** 0.08 mol/(L•s)
**D.** 0.17 mol/(L•s)
**E.** 0.93 mol/(L•s)

| NEED EXTRA HELP? | | | | | | | | | | | | | | | | | | |
|---|---|---|---|---|---|---|---|---|---|---|---|---|---|---|---|---|---|---|
| If You Missed Question . . . | 1 | 2 | 3 | 4 | 5 | 6 | 7 | 8 | 9 | 10 | 11 | 12 | 13 | 14 | 15 | 16 | 17 | 18 |
| Review Section . . . | 22.4 | 22.4 | 19.1 | 20.1 | 22.2 | 18.2 | 21.2 | 20.1 | 10.3 | 18.1 | 22.2 | 22.1 | 22.3 | 16.1 | 21.4 | 20.1 | 22.4 | 16.3 |

CHAPTER 23

# The Chemistry of Life

**BIGIDEA** Biological molecules—proteins, carbohydrates, lipids, and nucleic acids—interact to carry out activities necessary to living cells.

## SECTIONS

1 Proteins

2 Carbohydrates

3 Lipids

4 Nucleic Acids

5 Metabolism

## LaunchLAB

iLab Station

### How do you test for simple sugars?

Many different food sources supply the energy that your body uses constantly. This energy is stored in the bonds of molecules called simple sugars. In this lab, you will test several mixtures to determine if a simple sugar is present.

FOLDABLES® Study Organizer

### Biological Molecules

Make a concept-map book. Label it as shown. Use it to help you organize information about biological molecules.

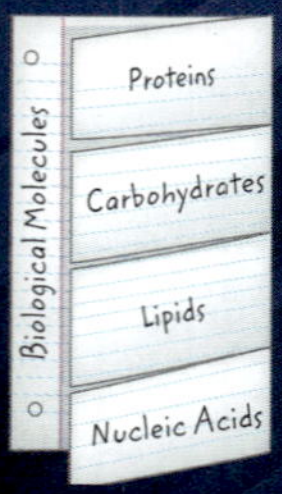

(l)Eye of Science/Science Photo Library/Photo Researchers, (t,r)Steve Gschmeissner/Photo Researchers, (bkgd)Alistair Berg/Digital Vision/Getty Images

Go **online!**
MHEonline.com

Fats provide more than twice the energy per gram as carbohydrates and proteins. Special lipids, called phospholipids, make up the cellular membranes of living cells.

SECTION 1

# Proteins

**MAINIDEA** Proteins perform essential functions, including structural support, transport of materials, muscle contractions, and regulation of chemical reactions.

### Essential Questions

- How can the structures of amino acids and proteins be described?
- What are the roles of proteins in cells?

### Review Vocabulary

**polymer:** large molecules composed of many repeating units called monomers

### New Vocabulary

protein
amino acid
peptide bond
peptide
denaturation
enzyme
substrate
active site

**CHEM 4 YOU** Some cleaning products, such as contact lens cleaning solution, contain enzymes. Did you ever wonder what an enzyme was?

## Protein Structure

Enzymes form a class of proteins. **Proteins** are organic polymers made of amino acids linked together in a specific order. Proteins are not just large, randomly arranged chains of amino acids. To function properly, each protein must be folded into a specific three-dimensional structure. All living organisms, including the mountain goat and the plants shown in **Figure 1,** are composed of proteins. In this section, you will read about how proteins are made from their amino-acid building blocks and how different types of proteins function.

**Amino acids** As you read previously, many different functional groups are found in organic compounds. **Amino acids,** as their name implies, are organic molecules that have both an amino group and an acidic carboxyl group. The general structure of an amino acid is shown below.

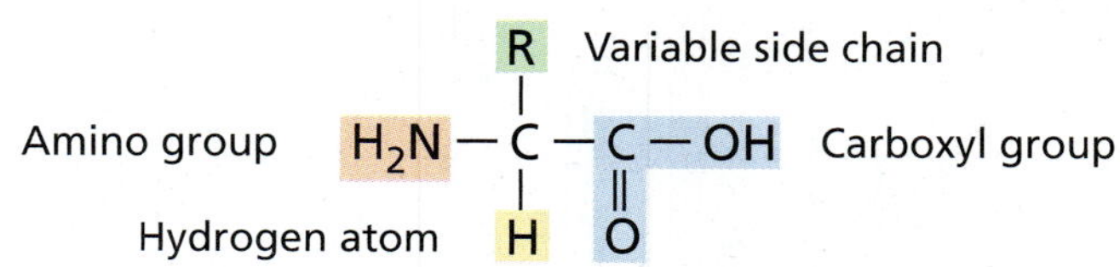


Each amino acid has a central carbon atom around which four groups are arranged: an amino group ($—NH_2$), a carboxyl group ($—COOH$), a hydrogen atom, and a variable side chain, R. The side chains range from a single hydrogen atom to a complex double-ring structure.

**Figure 1** All living organisms contain proteins. A goat's hair, hooves, and muscles are made up of structural proteins, as are the roots and leaves of plants.

Explore **amino acids** with an **interactive table.** Concepts In Motion

**Table 1** Amino Acid Examples

| | | | |
|---|---|---|---|
| 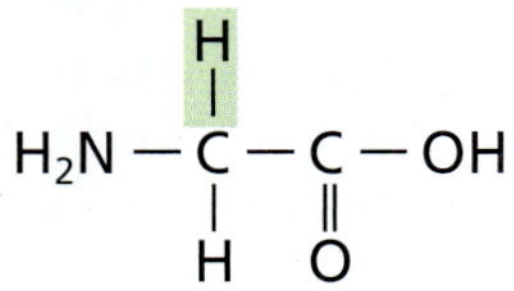<br>Glycine | 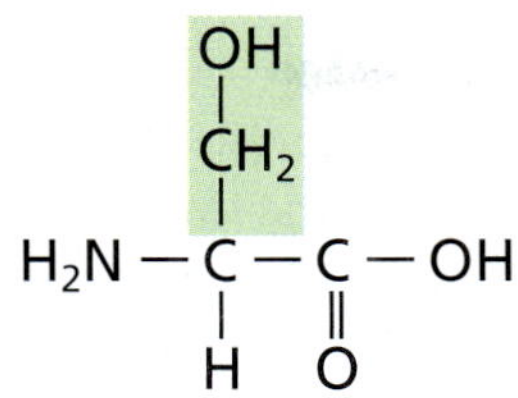<br>Serine | 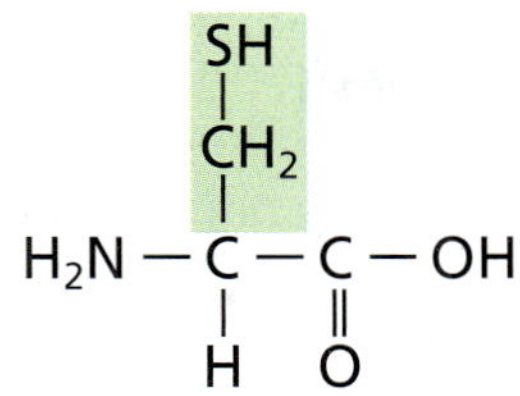<br>Cysteine | 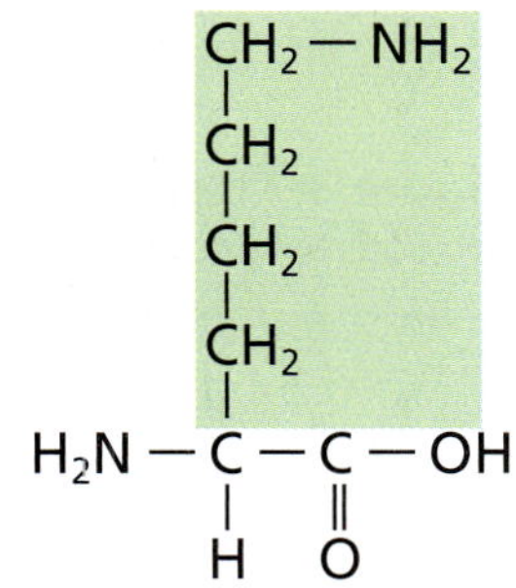<br>Lysine |
| 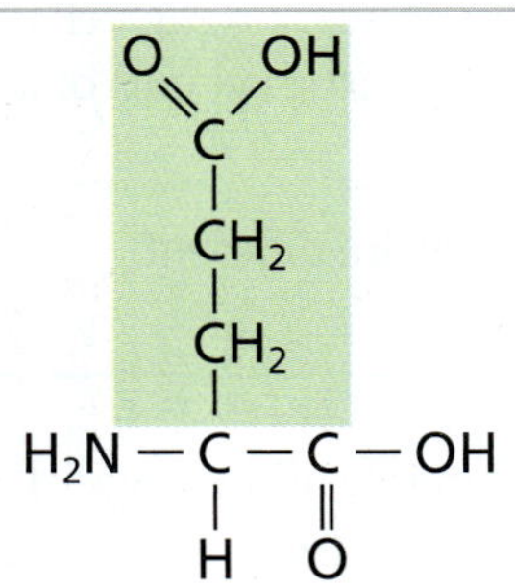<br>Glutamic acid | 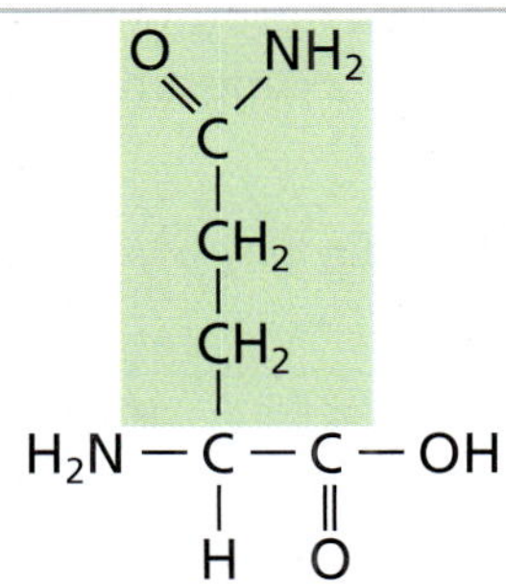<br>Glutamine | 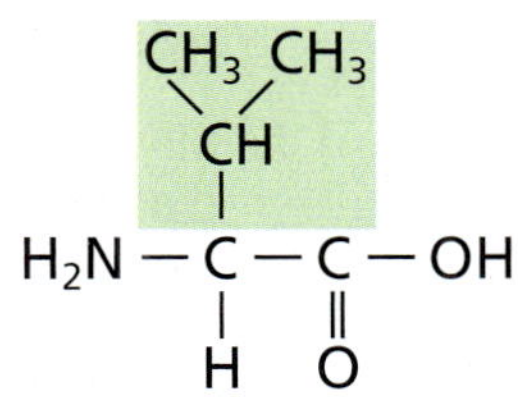<br>Valine | 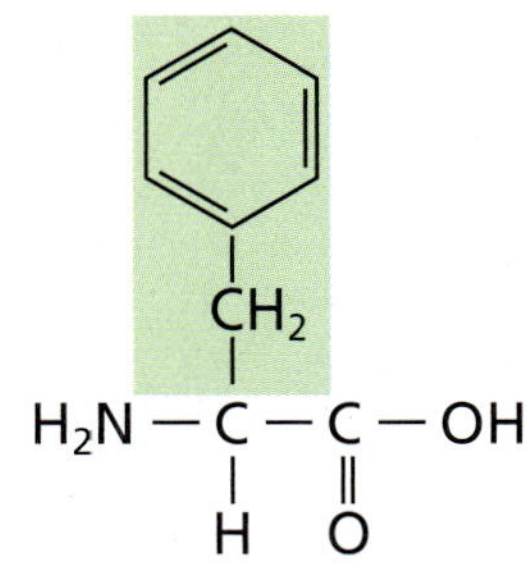<br>Phenylalanine |

Examine the different side chains of the amino acids shown in **Table 1.** Identify the nonpolar alkanes, polar hydroxyl groups, acidic and basic groups such as carboxyl and amino groups, aromatic rings, and sulfur-containing groups. This wide range of side chains gives the different amino acids a large variety of chemical and physical properties and is an important reason why proteins can perform so many different functions.

**The peptide bond** The amino and carboxyl groups provide convenient bonding sites for linking amino acids together. Because an amino acid is both an amine and a carboxylic acid, two amino acids can combine to form an amide, releasing water in the process. This reaction is a condensation reaction. As **Figure 2** shows, the carboxyl group of one amino acid reacts with the amino group of another amino acid to form an amide functional group.

☑ READING CHECK **Explain** how an amide functional group forms.

■ **Figure 2** The amino group of one amino acid bonds to the carboxyl group of another amino acid to form a dipeptide and water. The organic functional group formed is an amide linkage called a peptide bond.

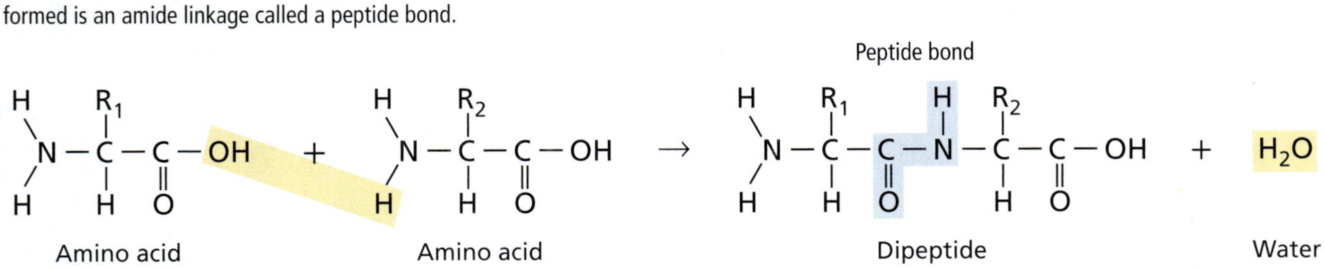

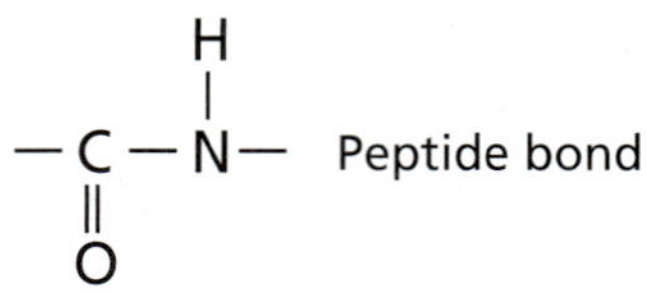


■ **Figure 3** A peptide bond joins two amino acids to form a dipeptide.

The amide bond that joins two amino acids, shown in **Figure 3,** is referred to as a **peptide bond.** A chain of two or more amino acids linked together by peptide bonds is called a **peptide.** A molecule that consists of two amino acids bound together by a peptide bond is called a dipeptide. **Figure 4a** shows the structure of a dipeptide that is formed from the amino acids glycine (Gly) and phenylalanine (Phe). **Figure 4b** shows a different dipeptide, also formed by linking together glycine and phenylalanine. Is Gly-Phe the same compound as Phe-Gly? No, they're different. Examine these two dipeptides to see that the order in which amino acids are linked in a dipeptide is important.

Each end of the two-amino-acid unit in a dipeptide still has a free group—one end has a free amino group and the other end has a free carboxyl group. Each of those groups can be linked to the opposite end of yet another amino acid, forming more peptide bonds. Living cells always build peptides by adding amino acids to the carboxyl end of a growing chain.

☑ READING CHECK **Explain** why Gly-Phe and Phe-Gly are different dipeptides.

**Polypeptides** As peptide chains increase in length, other ways of referring to them become necessary. A chain of ten or more amino acids joined by peptide bonds is referred to as a polypeptide. When a chain reaches a length of about 50 amino acids, it is called a protein.

Because there are only 20 different amino acids that form proteins, it might seem reasonable to think that only a limited number of different protein structures are possible. However, a protein can have as few as 50 or more than a 1000 amino acids, arranged in any possible sequence. To calculate the number of possible sequences these amino acids can have, consider that each position on the chain can have any of 20 possible amino acids. For a peptide that contains $n$ amino acids, there are $20^n$ possible sequences of the amino acids. So a dipeptide, with only two amino acids, can have $20^2$, or 400, different possible amino acid sequences. Even the smallest protein, containing only 50 amino acids, has $20^{50}$, or more than $1 \times 10^{65}$, possible arrangements of amino acids! It is estimated that human cells make between 80,000 and 100,000 different proteins. You can see that this is only a small fraction of the total number of proteins possible.

☑ READING CHECK **Calculate** the possible number of sequences for a peptide chain comprised of four amino acids.

■ **Figure 4** Glycine (Gly) and phenylalanine (Phe) can combine in two configurations.
**Describe** *the difference between the configuration of the peptide bonds in these two dipeptides.*

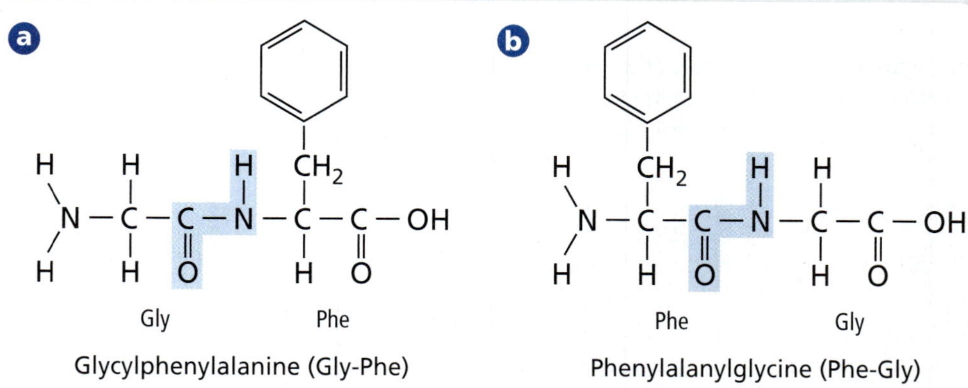

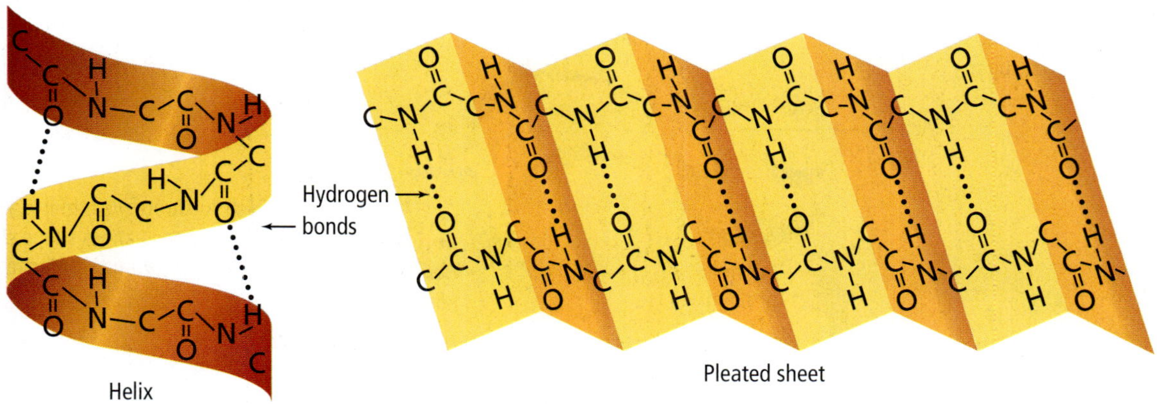


■ **Figure 5** The folding of polypeptide chains into both helices and sheets involves amino acids in the chain held in position by hydrogen bonds. Other interactions among the various side chains are not shown here but play an important role in determining the three-dimensional shape of a polypeptide.

**Three-dimensional protein structure** Long chains of amino acids start to fold into unique three-dimensional shapes before they are fully synthesized. The three-dimensional shape is determined by the interactions among the amino acids. Some areas of a polypeptide might twirl into helices, which are similar to the coils on a telephone cord. Other areas might bend back and forth repeatedly into a pleated sheet structure, like the folds of an accordion. A polypeptide chain might also fold back on itself and change direction. A given protein might have several helices, sheets, and turns, or none at all. **Figure 5** shows the folding patterns of a typical helix and a sheet. The overall three-dimensional shape of many proteins is globular—shaped like an irregular sphere. Other proteins have a long, fibrous shape. The shape is important to the function of the protein. If the shape of the protein changes, it might not be able to carry out its function in the cell.

**Denaturation** Changes in temperature, ionic strength, pH, and other factors result in the unfolding and uncoiling of a protein. **Denaturation** is the process in which a protein's natural three-dimensional structure is disrupted. Cooking often denatures the proteins in foods. When an egg is hard-boiled, the protein-rich egg white solidifies due to the denaturation of its protein. Because proteins function properly only when folded, denatured proteins are generally inactive.

## The Many Functions of Proteins

Proteins play many roles in living cells. They are involved in speeding up chemical reactions, transport of substances, regulation of cellular processes, structural support of cells, communication within cells and among cells, cellular motion, and even serving as an energy source when other sources are scarce.

**Speeding up reactions** In most organisms, the largest number of proteins function as enzymes, catalyzing the many reactions that occur in living cells. An **enzyme** is a biological catalyst. You read previously that a catalyst speeds up a chemical reaction without being consumed in the reaction. A catalyst usually lowers the activation energy of a reaction by stabilizing the transition state.

### RealWorld CHEMISTRY

**Enzymes**

**PAPAIN** An example of an enzyme you might have used is papain, found in papayas, pineapples, and other plant sources. This enzyme catalyzes a reaction that breaks down protein molecules into free amino acids. Papain is the active ingredient in many meat tenderizers. When you sprinkle the dried form of papain onto moist meat, the papain forms a solution that breaks down the tough protein fibers in the meat, making the meat more tender.

Watch a **video about surfactants and enzymes.**

Video

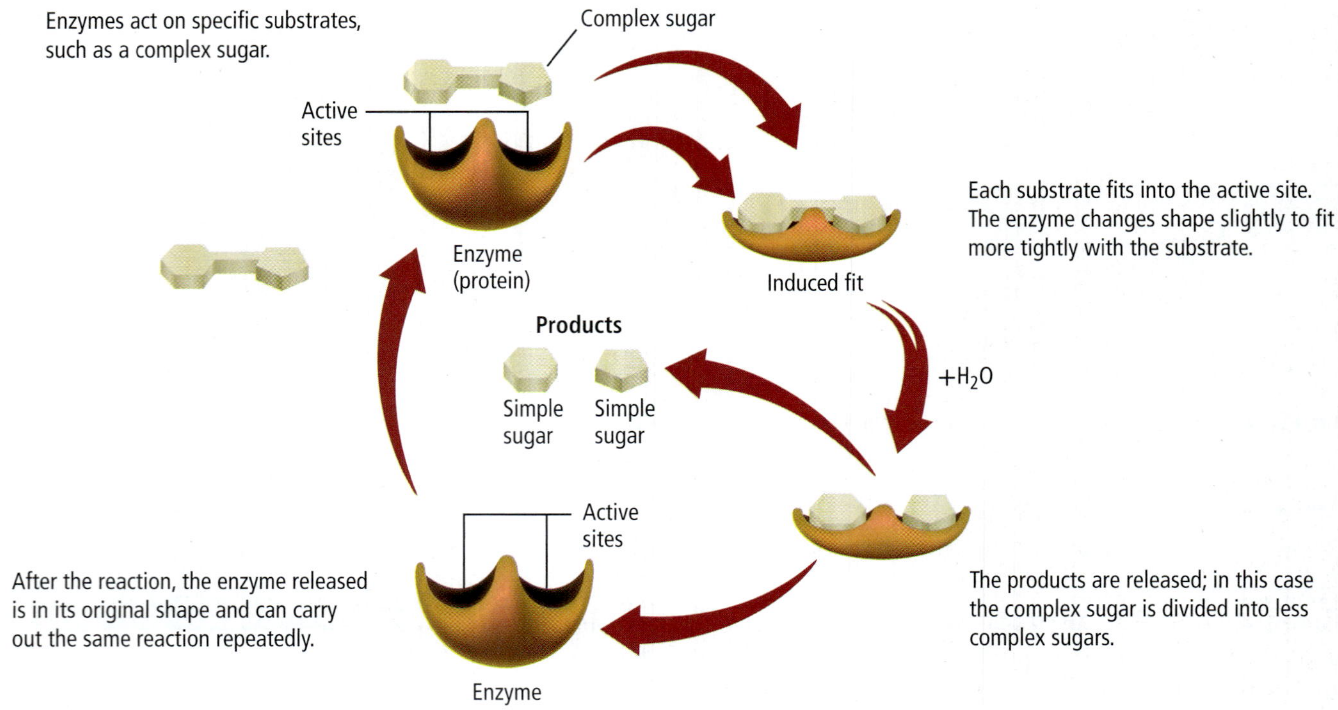


■ **Figure 6** Enzymes lower the activation energy needed for a reaction to occur. Enzymes change the speed at which chemical reactions occur without being altered themselves in the reaction.

How do enzymes function? The term **substrate** refers to a reactant in an enzyme-catalyzed reaction, as shown in **Figure 6.** Substrates bind to specific sites on enzyme molecules, usually pockets or crevices. The spot to which the substrates bind is called the **active site** of the enzyme. After the substrates bind to the active site, the active site changes shape slightly to fit more tightly around the substrates. This recognition process is called induced fit. The shapes of the substrates must fit the shape of the active site, in the same way that puzzle pieces or a lock and key fit together. A molecule that is only slightly different in shape from an enzyme's normal substrate will not bind as well to the active site and might not undergo the catalyzed reaction.

The structure that forms when substrates are bound to an enzyme is called an enzyme-substrate complex. The large size of enzyme molecules allows them to form multiple bonds with their substrates, and the large variety of amino acid side chains in the enzyme allows a number of different intermolecular forces to form. These intermolecular forces lower the activation energy needed for the reaction in which bonds are broken and the substrates are converted to product.

☑ READING CHECK **Describe** in your own words how an enzyme works.

■ **Figure 7** Hemoglobin is a globular protein with four polypeptide chains, each containing an iron group (called a heme) to which oxygen binds.

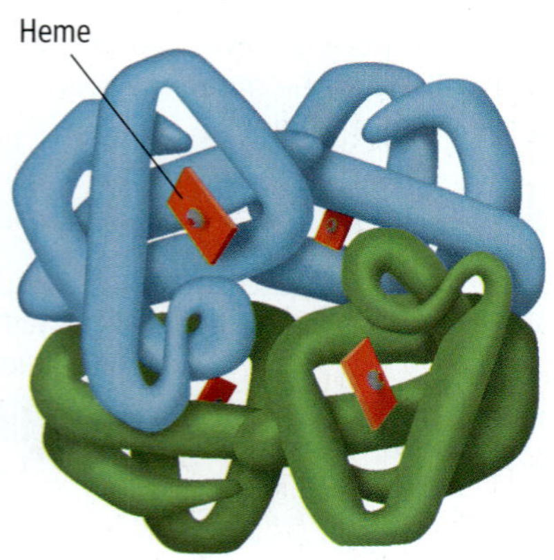


**Transport proteins** Some proteins are involved in transporting smaller particles throughout the body. **Figure 7** shows the protein hemoglobin, which carries oxygen in the blood from the lungs to the rest of the body. Other proteins combine with biological molecules called lipids to transport them from one part of the body to another through the bloodstream. You will learn about lipids later in this chapter.

■ **Figure 8** Human hair is made up of a fibrous structural protein called keratin.

**Structural support** The sole function of certain proteins is to form structures vital to organisms. These molecules are known as structural proteins. The most abundant structural protein in most animals is collagen, which is part of skin, ligaments, tendons, and bones. Other structural proteins make up feathers, fur, wool, hooves, fingernails, cocoons, and hair, as shown in **Figure 8.**

**Communication** Hormones are messenger molecules that carry signals from one part of the body to another. Some hormones are proteins. Insulin, a familiar example, is a small (51 amino acids) protein hormone made by pancreas cells. When insulin is released into the bloodstream, it signals body cells that blood sugar is abundant and should be stored. A lack of insulin often results in diabetes, a disease that results when there is too much sugar in the bloodstream.

Because modern technology has made possible the laboratory synthesis of proteins, some protein hormones are being synthetically produced for use as medicines. Insulin, thyroid hormones, and growth hormones are some examples. Both natural and synthetic proteins are used in a variety of products—from meat tenderizer to cleaning solutions to health and beauty aids.

**FOLDABLES®**
Incorporate information from this section into your Foldable.

## SECTION 1 REVIEW

Section Self-Check

### Section Summary

- Proteins are biological polymers made of amino acids that are linked by peptide bonds.
- Protein chains fold into intricate three-dimensional structures.
- Proteins have many functions in the human body, including functions within cells, functions between cells, and functions of structural support.

1. **MAINIDEA** **Describe** three proteins and identify their functions.
2. **Compare** the structures of amino acids, dipeptides, polypeptides, and proteins. Which has the largest molecular mass? The smallest?
3. **Draw** the structure of the dipeptide Gly-Ser, circling the peptide bond.
4. **Evaluate** How do the properties of proteins make them such useful catalysts? How do they differ from other catalysts you have studied?
5. **Explain** how a change in temperature might affect a protein's function.
6. **Categorize** Identify an amino acid from **Table 1** that can be classified into each of the categories in the following pairs.
   **a.** nonpolar side chain v. polar side chain
   **b.** aromatic v. aliphatic
   **c.** acidic v. basic

SECTION 2

# Carbohydrates

**MAIN IDEA** Carbohydrates provide energy and structural material for living things.

### Essential Questions

- How can the structures of monosaccharides, disaccharides, and polysaccharides be described?
- What are the functions of carbohydrates in living things?

### Review Vocabulary

**stereoisomers:** a class of isomers whose atoms are bonded in the same order but are arranged differently in space

### New Vocabulary

carbohydrate
monosaccharide
disaccharide
polysaccharide

**CHEM 4 YOU** A lot of media attention has been focused on carbohydrates. Some recommend low-carb diets as a way of controlling weight. However, carbohydrates are an important energy source for the body.

## Kinds of Carbohydrates

Analyzing the term *carbohydrate* offers a hint about the structure of this group of molecules. Early observations that these compounds have the general chemical formula $C_n(H_2O)_n$ and appear to be hydrates of carbon led to their being called carbohydrates. Although scientists now know that there are no full water molecules attached to carbohydrates, the name has stayed.

The main function of carbohydrates in living organisms is as a source of energy, both immediate and stored. Foods rich in carbohydrates include pasta, milk, fruit, bread, and potatoes. **Carbohydrates** are compounds that contain multiple hydroxyl groups (—OH) as well as a carbonyl functional group (C=O). These molecules range in size from single monomers to polymers made of hundreds or even thousands of monomer units.

**Monosaccharides** The simplest carbohydrates, often called simple sugars, are **monosaccharides.** The most common monosaccharides have either five or six carbon atoms. Examples of monosaccharides are shown in **Figure 9.** Notice that they have a carbonyl group on one carbon and hydroxyl groups on most of the other carbons. The presence of a carbonyl group makes these compounds either aldehydes or ketones, depending on the location of the carbonyl group. Multiple polar groups make monosaccharides water-soluble and give them high melting points.

**Figure 9** Glucose, galactose, and fructose are monosaccharides. In aqueous solutions, they exist in an equilibrium between their open-chain and cyclic forms.

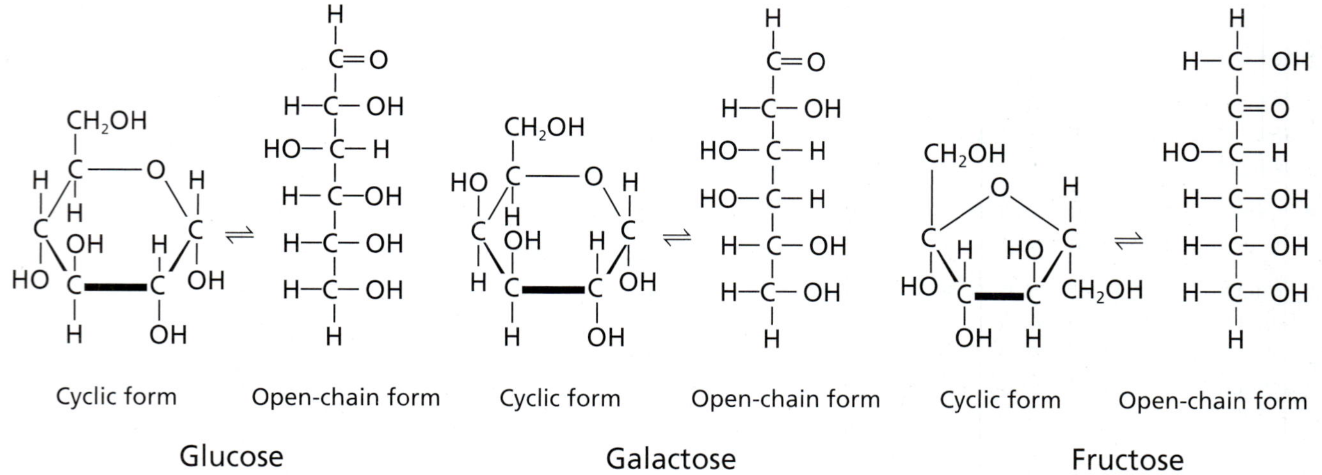

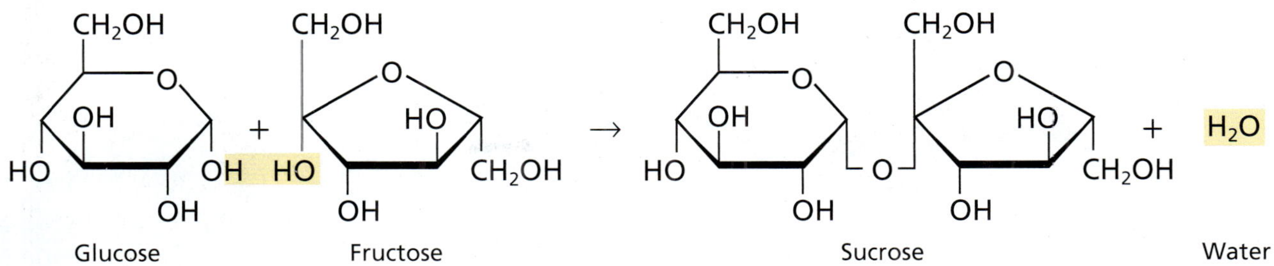


■ **Figure 10** When glucose and fructose bond, the disaccharide sucrose forms. Note that water is also a product of this condensation reaction. Remember that each ring structure is made of carbon atoms, which are not shown for simplicity.

Glucose is a six-carbon sugar that has an aldehyde structure. Glucose is present in high concentration in blood because it serves as the major source of immediate energy for the body. For this reason, glucose is often called blood sugar. Closely related to glucose is galactose, which differs only in how a hydrogen and a hydroxyl group are oriented in space around one of the six carbon atoms. Recall that this relationship makes glucose and galactose stereoisomers. Fructose, also known as fruit sugar because it is the major carbohydrate in most fruits, is a six-carbon monosaccharide that has a ketone structure. Fructose is a structural isomer of glucose.

When monosaccharides are in aqueous solution, they exist in both open-chain and cyclic structures, but they rapidly interconvert forms. The cyclic structures are more stable and are the predominant form of monosaccharides at equilibrium. Note in **Figure 9** that the carbonyl groups are present only in the open-chain structures. In the cyclic structures, they are converted to hydroxyl groups.

**VOCABULARY**

**WORD ORIGIN**

**Polysaccharide**
comes from the Greek word *polys,* which means *many* and the ancient Sanskrit word *śarkarā,* which means *sugar*

**Disaccharides** Like amino acids, monosaccharides can be linked together by a condensation reaction in which water is released. When two monosaccharides bond together, a **disaccharide** is formed, as shown in **Figure 10.** The new bond formed is an ether functional group (C–O–C).

One common disaccharide is sucrose, also known as table sugar because sucrose is used mainly as a sweetener. Sucrose is formed by the linking of glucose and fructose. Another common disaccharide is lactose, the most important carbohydrate in milk. It is often called milk sugar. Lactose is formed when glucose and galactose bond.

**Polysaccharides** *Complex carbohydrate* is a term used in some nutrition books and journal articles. Another name for a complex carbohydrate is **polysaccharide,** which is a polymer of simple sugars that contains 12 or more monomers, or subunits. The same type of bond that joins two monosaccharides in a disaccharide also links the monomers in a polysaccharide. Glycogen, shown in **Figure 11,** is a polysaccharide. It is composed of glucose subunits. It stores energy and is found mostly in the liver and muscles of humans and other animals. It is also found in some species of microorganisms including bacteria and fungi.

■ **Figure 11** The glycogen found in the muscle and liver of animals is a polysaccharide made of glucose.

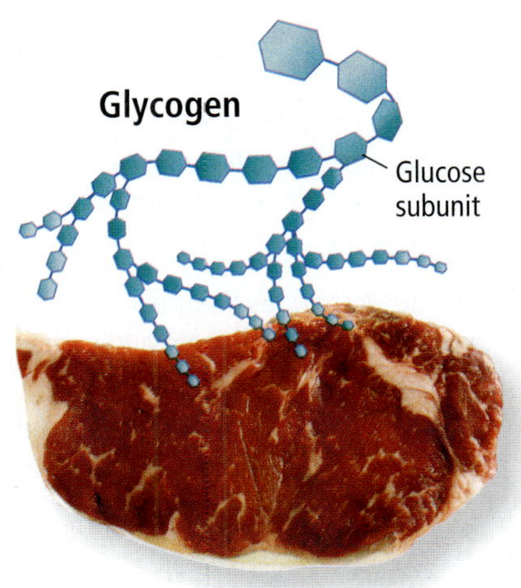


☑ READING CHECK **Explain** the differences among a monosaccharide, a disaccharide, and a polysaccharide.

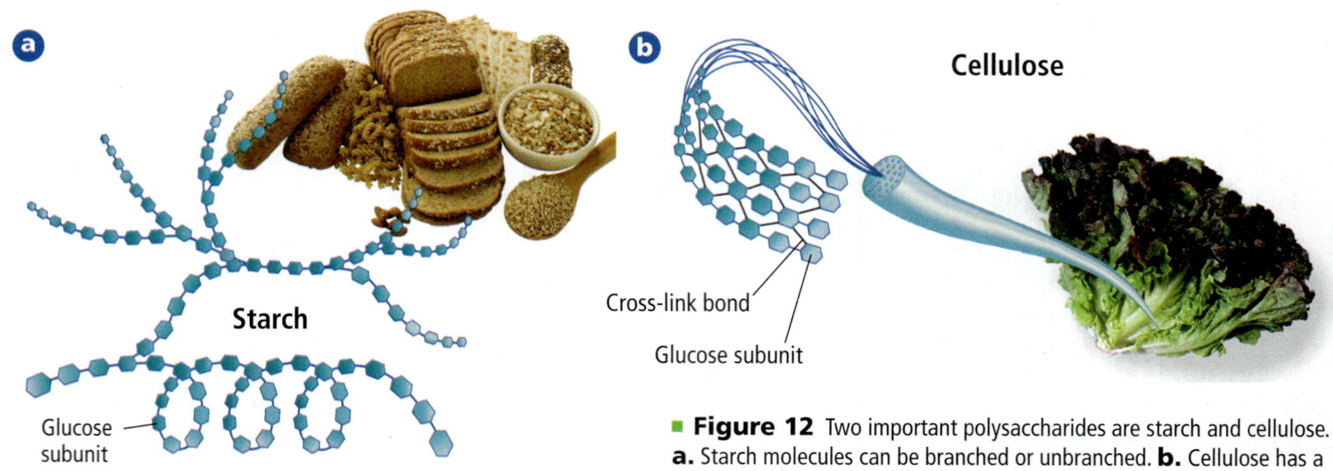


■ **Figure 12** Two important polysaccharides are starch and cellulose. **a.** Starch molecules can be branched or unbranched. **b.** Cellulose has a linear, unbranched structure that resembles a chain-link fence.

Two other important polysaccharides are starch and cellulose, shown in **Figure 12.** Starch and cellulose are also composed solely of glucose subunits. However, that is the only similarity among the three polysaccharides, as all three have different properties and functions. Plants make both starch and cellulose. Starch is a soft, water-insoluble molecule used to store energy, whereas cellulose is a water-insoluble polymer that forms rigid plant-cell walls, such as those found in wood.

**FOLDABLES®** Incorporate information from this section into your Foldable.

Glycogen, starch, and cellulose are composed of glucose subunits, but they have different properties. The bonds that link the subunits together are oriented differently in space. Because of this difference in bond shape, humans can digest glycogen and starch but not cellulose. Digestive enzymes cannot fit cellulose into their active sites. The cellulose in the fruits, vegetables, and grains that we eat is called *dietary fiber* because it passes through the digestive system largely unchanged.

## SECTION 2 REVIEW

Section Self-Check

### Section Summary

- Carbohydrates are compounds that contain multiple hydroxyl groups (–OH) and a carbonyl functional group (C=O).
- Carbohydrates range in size from single monomers to polymers composed of hundreds or thousands of monomers.
- Monosaccharides in aqueous solution exist in both open-chain and cyclic structures.

7. **MAINIDEA Explain** the functions of carbohydrates in living things.
8. **Describe** the structures of monosaccharides, disaccharides, and polysaccharides. Which has the largest molecular mass? The smallest?
9. **Compare and contrast** the structures of starch and cellulose. How do the structural differences affect our ability to digest these two polysaccharides?
10. **Calculate** If a carbohydrate has $2^n$ possible isomers, where $n$ is equal to the number of chiral carbon atoms in the structure, calculate the number of possible isomers for the following monosaccharides: galactose, glucose, and fructose.
11. **Interpret Scientific Illustrations** Copy the illustration of sucrose on a separate sheet of paper, and circle the ether functional group that bonds the monomer sugars together.

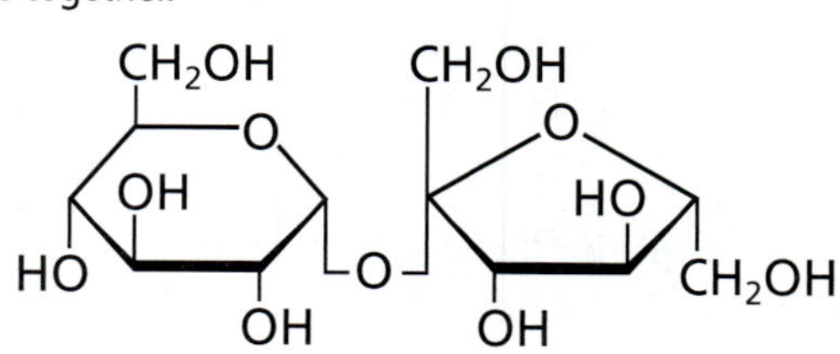

SECTION 3

# Lipids

**MAINIDEA** Lipids make cell membranes, store energy, and regulate cellular processes.

### Essential Questions

- How can the structures of fatty acids, triglycerides, phospholipids, and steroids be described?
- What are the functions of lipids in living organisms?
- What are some reactions that fatty acids undergo?
- How are the structure and function of cell membranes related?

### Review Vocabulary

**nonpolar:** without separate positive and negative areas or dipoles

### New Vocabulary

lipid
fatty acid
triglyceride
saponification
phospholipid
wax
steroid

**CHEM 4 YOU** The wax used to polish cars, the fat that drips out of hamburgers, and the vitamin D that fortifies the milk people drink—what do these things have in common? They are all lipids.

## What is a lipid?

A **lipid** is a large, nonpolar biological molecule. Because lipids are nonpolar, they are insoluble in water. Lipids have two major functions in living organisms. They store energy efficiently, and they make up most of the structure of cell membranes. Unlike proteins and carbohydrates, lipids are not polymers with repeated monomer subunits.

**Fatty acids** Although lipids are not polymers, many lipids have a major building block in common. This building block is the **fatty acid,** a long-chain carboxylic acid. Most naturally occurring fatty acids contain between 12 and 24 carbon atoms. Their structure can be represented by the following formula.

$$CH_3(CH_2)_nCOOH$$

Most fatty acids have an even number of carbon atoms, which is a result of being constructed two carbons at a time in enzymatic reactions.

Fatty acids can be grouped into two main categories, depending on the presence or absence of double bonds between carbon atoms. Fatty acids that contain no double bonds are referred to as saturated. Those that have one or more double bonds are called unsaturated. The structures of two common fatty acids are shown in **Figure 13.**

**READING CHECK** **Explain** why oleic acid is described as *unsaturated.*

**Figure 13** Two fatty acids, which are found in many foods, including butter, are the 18-carbon unsaturated oleic acid and the 18-carbon saturated stearic acid.
**Explain** *how the structure of the molecule is affected by the presence of a double bond.*

**Oleic acid**

O=C(HO)CH$_2$CH$_2$CH$_2$CH$_2$CH$_2$CH$_2$CH$_2$CH=CHCH$_2$CH$_2$CH$_2$CH$_2$CH$_2$CH$_2$CH$_2$CH$_3$

**Stearic acid**

O=C(HO)CH$_2$CH$_2$CH$_2$CH$_2$CH$_2$CH$_2$CH$_2$CH$_2$CH$_2$CH$_2$CH$_2$CH$_2$CH$_2$CH$_2$CH$_2$CH$_2$CH$_3$

D. Hurst/Alamy

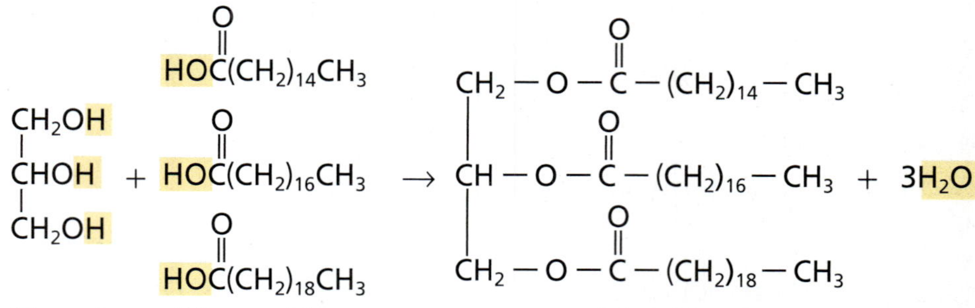


■ **Figure 14** Ester bonds in a triglyceride are formed when the hydroxyl groups of glycerol combine with the carboxyl groups of the fatty acids.

An unsaturated fatty acid can become saturated if it reacts with hydrogen. As you read previously, hydrogenation is an addition reaction in which hydrogen gas reacts with carbon atoms that are linked by multiple bonds. Each unsaturated carbon atom can pick up one hydrogen atom to become saturated. For example, oleic acid, shown in **Figure 13,** can be hydrogenated to form stearic acid.

The double bonds in naturally occurring fatty acids are almost all in the *cis* geometric isomer form. Recall that the *cis* isomer has identical groups oriented on the same side of the molecule around a double bond. Because of the *cis* orientation, unsaturated fatty acids have a kink, or bend, in their structure that prevents them from packing together. They do not form as many intermolecular attractions as saturated fatty acid molecules. As a result, unsaturated fatty acids have lower melting points.

**Triglycerides** Although fatty acids are abundant in living organisms, they are rarely found alone. They are most often found bonded to glycerol, a molecule with three carbons, each containing a hydroxyl group. When three fatty acids are bonded to a glycerol backbone through ester bonds, a **triglyceride** is formed. The formation of a triglyceride is shown in **Figure 14.** Triglycerides can be either solids or liquids at room temperature, as shown in **Figure 15.** If liquid, they are usually called oils. If solid at room temperature, they are called fats.

☑ **READING CHECK** **Describe** the difference between fatty acids and triglycerides.

**VOCABULARY**

**SCIENCE USAGE V. COMMON USAGE**

**Saturate**

***Science usage:*** to add something to the point that no more can be absorbed, dissolved, or retained

*The salt water solution was saturated with salt.*

***Common usage:*** to furnish a market with a product or products to its full purchasing capacity

*The shops in the coastal town are saturated with sea shell craft items.*

■ **Figure 15** Most mixtures of triglycerides from plant sources are liquids because the triglycerides contain unsaturated fatty acids. Animal fats contain a larger proportion of saturated fatty acids. They are usually solids at room temperature.

Maximilian Stock Ltd./Photolibrary/Getty Images

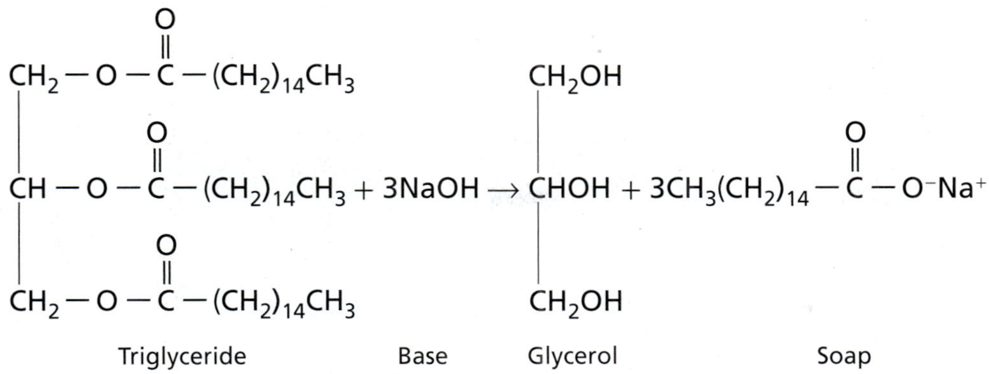


■ **Figure 16** Soap forms by the reaction of a triglyceride and a strong base.

Fatty acids are stored in the fat cells of your body as triglycerides. When energy is abundant, fat cells store the excess energy in the fatty acids of triglycerides. When energy is scarce, the cells break down the triglycerides, releasing the energy used to form them.

Although enzymes break down triglycerides in living cells, the reaction can be duplicated outside of cells by using a strong base, such as sodium hydroxide. This reaction—the hydrolysis of a triglyceride using an aqueous solution of a strong base to form carboxylate salts and glycerol—is **saponification,** as shown in **Figure 16.** Saponification is used to make soaps, which are usually the sodium salts of fatty acids. A soap molecule has both a polar end and a nonpolar end. Soaps are used with water to clean nonpolar dirt and oil because the nonpolar dirt and oil bond to the nonpolar end of the soap molecules, and the polar end of the soap molecules is soluble in water. Thus, the dirt-laden soap molecules can be rinsed away with the water.

# MiniLAB

iLab Station

## Observe a Saponification Reaction

**How is soap made?** The reaction between a triglyceride and a strong base is called saponification. A sample chemical reaction is shown in **Figure 16.**

### Procedure

1. Read and complete the lab safety form.
2. Place a **250-mL beaker** on a **hot plate.** Add 25 g **solid vegetable shortening** to the beaker. Turn on the hot plate to a medium setting.
3. As the vegetable shortening melts, use a **25-mL graduated cylinder** to slowly add 12 mL **ethanol** and then 5 mL **6.0*M* NaOH** to the beaker.

   **WARNING:** *Ethanol is flammable. NaOH causes skin burns. Wear gloves.*
4. Heat the mixture for about 15 min. Use a **stirring rod** to occasionally stir the mixture. Do not allow it to boil.
5. When the mixture begins to thicken, use **tongs** to remove the beaker from the heat. Allow the beaker to cool for 5 min, then place it in a **cold water bath** in a **600-mL beaker.**
6. Add 25 mL **saturated NaCl solution** to the mixture in the beaker. The soap is not very soluble and will appear as small clumps.
7. Collect the solid soap clumps by filtering them through a **cheesecloth-lined funnel.**
8. Using gloved hands, press the soap into an **evaporating dish.** Remove your gloves and wash your hands.

### Analysis

1. **Explain** What type of bonds present in the triglycerides are broken during the saponification reaction?
2. **Identify** the type of salt formed in this chemical reaction.
3. **Determine** which is the polar end and which is the nonpolar end of the soap molecule.

■ **Figure 17** A phospholipid has a polar head and two nonpolar tails. The membranes of living cells are formed by a double layer of lipids, called a bilayer. The polar heads are on the outer and inner perimeter of the membrane and the tails are on the inside of the bilayer.

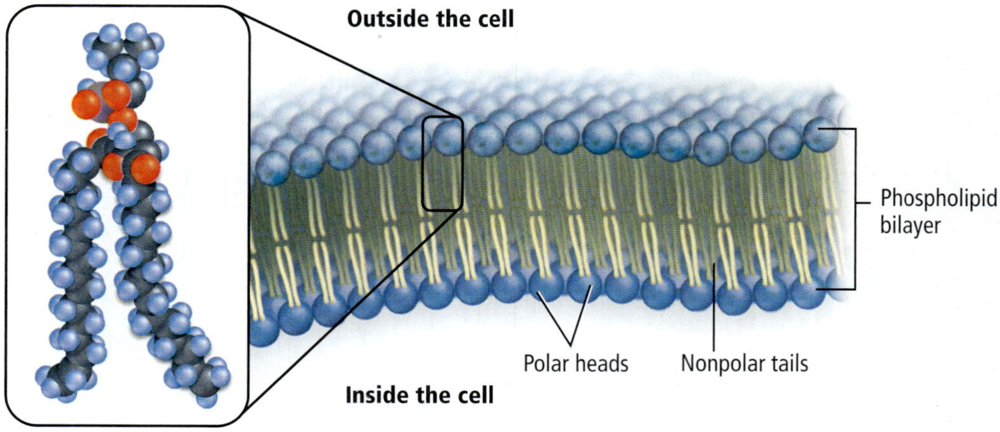


**Phospholipids** Another important type of triglyceride, a phospholipid, is found in greatest abundance in cellular membranes. A **phospholipid** is a triglyceride in which one of the fatty acids is replaced by a polar phosphate group. As shown in **Figure 17,** the polar part of the molecule forms a head and the nonpolar fatty acids look like tails. A typical cell membrane has two layers of phospholipids, which are arranged with their nonpolar tails pointing inward and their polar heads pointing outward. This arrangement is called a lipid bilayer. Because the lipid bilayer structure acts as a barrier, the cell is able to regulate the materials that enter and leave through the membrane.

**Connection to Biology** The venom of poisonous snakes contains a class of enzymes known as phospholipases. These enzymes catalyze the breakdown of phospholipids. The venom of the eastern diamond-back rattlesnake contains a phospholipase that hydrolyzes the ester bond at the middle carbon of phospholipids. If the larger of the two breakdown products of this reaction gets into the bloodstream, it dissolves the membranes of red blood cells, causing them to rupture. Because the venom destroys the blood cells, it is referred to as a hemotoxic venom. (The prefix *hemo-* indicates blood.) A bite from the eastern diamond-back can lead to death if it is not treated immediately.

■ **Figure 18** Plants produce a wax that coats their leaves. The wax protects the leaves from drying out.

**Waxes** Another type of lipid, wax, also contains fatty acids. A **wax** is a lipid that is formed by combining a fatty acid with a long-chain alcohol. The general structure of these soft, solid fats with low melting points is shown below, with *x* and *y* representing variable numbers of $CH_2$ groups.

$$CH_3(CH_2)_x - \overset{\overset{\displaystyle O}{\|}}{C} - O - (CH_2)_yCH_3$$

Both plants and animals make waxes. Plant leaves are often coated with wax, which prevents water loss. Notice in **Figure 18** how raindrops bead up on the leaves of a plant, indicating the presence of the waxy layer. The honeycombs that bees make are also made of a wax, commonly called beeswax. Combining the 16-carbon fatty acid palmitic acid and a 30-carbon alcohol chain makes a common form of beeswax. Candles are sometimes made of beeswax because it tends to burn slowly and evenly.

■ **Figure 19** This Giant Marine toad uses a steroid toxin called bufotoxin as a defense mechanism. The toxin is fatal to some animals, including dogs and cats.

Chris Johnson/Alamy

**Steroids** Not all lipids contain fatty acid chains. **Steroids** are lipids that have multiple cyclic rings in their structures. All steroids are built from the basic four-ring steroid structure shown below.

**FOLDABLES®** Incorporate information from this section into your Foldable.

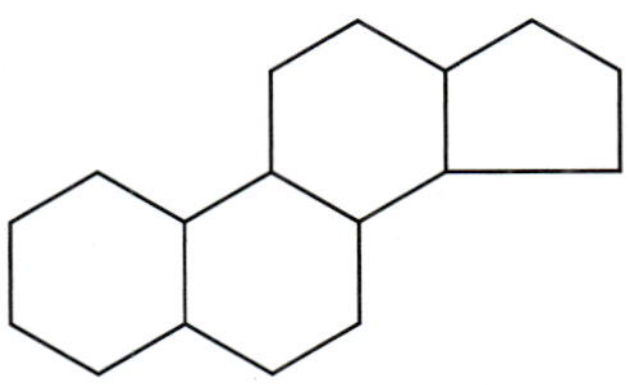

Some hormones, such as many sex hormones, are steroids that function to regulate metabolic processes. Cholesterol, another steroid, is an important structural component of cell membranes. Vitamin D also contains the four-ring steroid structure and plays a role in the formation of bones. The Giant Marine toad, *Bufo marinus,* shown in **Figure 19** uses a steroid called bufotoxin as a defense mechanism. The toad secretes the toxin from warts on its back and from glands just behind the eye. The toxin is only an irritant for humans, but in small animals the toxin causes drooling, loss of coordination, convulsions, and death.

## SECTION 3 REVIEW

Section Self-Check

### Section Summary

- Fatty acids are long-chain carboxylic acids that usually have between 12 and 24 carbon atoms.
- Saturated fatty acids have no double bonds; unsaturated fatty acids have one or more double bonds.
- Fatty acids can be linked to glycerol backbones to form triglycerides.
- Steroids are lipids that have multiple-ring structures.

**12. MAIN IDEA Describe** the function of lipids.

**13. Describe** the structures of fatty acids, triglycerides, phospholipids, and steroids.

**14. List** an important function of each of these types of lipids.

**a.** triglycerides
**b.** phospholipids
**c.** waxes
**d.** steroids

**15. Identify** two reactions that fatty acids undergo.

**16. Describe** the structure and function of cell membranes.

**17. Compare and contrast** the structures of a steroid, a phospholipid, and a wax.

**18. Write** the equation for the complete hydrogenation of the polyunsaturated fatty acid linoleic acid, $CH_3(CH_2)_4CH{=}CHCH_2CH{=}CH(CH_2)_7COOH$.

**19. Interpret Scientific Illustrations** Draw the general structure of a phospholipid. Label the polar and nonpolar portions of the structure.

SECTION 4

# Nucleic Acids

**MAIN**IDEA Nucleic acids store and transmit genetic information.

### Essential Questions

- What are the structural components of nucleic acids?
- How is the function of DNA related to its structure?
- What are the structure and function of RNA?

### Review Vocabulary

**genetic information:** an inherited sequence of RNA or DNA that causes traits or characteristics to pass from one generation to the next

### New Vocabulary

nucleic acid
nucleotide

**CHEM 4 YOU** DNA testing is becoming more routine in medicine, forensic science, genealogy, and identification of victims in disasters. Modern techniques have made it possible to get a useful DNA sample from surprising sources, such as a strand of hair or dried saliva on a postage stamp.

## Structure of Nucleic Acids

Nucleic acids comprise a fourth class of biological molecules. They are the information-storage molecules of the cell. This group of molecules got its name from the cellular location in which the molecules are primarily found—the nucleus. It is from this control center of cells that nucleic acids carry out their major functions. A **nucleic acid** is a nitrogen-containing biological polymer that is involved in the storage and transmission of genetic information. The monomer that makes up a nucleic acid is called a **nucleotide.** Each nucleotide has three parts: an inorganic phosphate group, a five-carbon monosaccharide sugar, and a nitrogen-containing structure called a nitrogenous base. Examine each part of **Figure 20a.** Although the phosphate group is the same in all nucleotides, the sugar and the nitrogen base vary.

In a nucleic acid, the sugar of one nucleotide is bonded to the phosphate of another nucleotide, as shown in **Figure 20b.** Thus, the nucleotides are strung together in a chain, or strand, containing alternating sugar and phosphate groups. Each sugar is also bonded to a nitrogen base that sticks out from the chain. The nitrogen bases on adjoining nucleotide units are stacked one above the other in a slightly askew position, much like the steps in a staircase. This orientation is shown in **Figure 20b.** Intermolecular forces hold each nitrogen base close to the nitrogen bases above and below it.

**Figure 20** Nucleotides are the monomers from which nucleic acid polymers are formed.

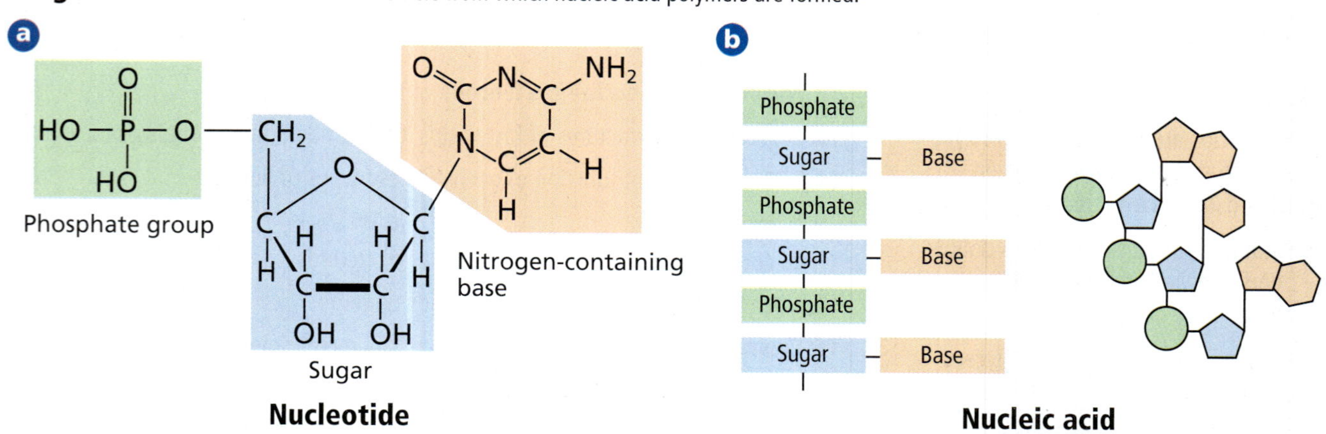


Each nucleotide contains a nitrogen-containing base, a five-carbon sugar, and a phosphate group.

Nucleic acids are linear chains of alternating sugars and phosphates. Attached to every sugar is a nitrogen base. Because the nucleotides are offset, the chains resemble steps in a staircase.

## DNA: The Double Helix

You might have heard of DNA (deoxyribonucleic acid), one of the two types of nucleic acids found in living cells. DNA contains the master plans for building all the proteins in an organism's body.

**The structure of DNA** DNA consists of two long chains of nucleotides wound together to form a spiral structure, as shown in **Figure 21.** Each nucleotide in DNA contains a phosphate group, the five-carbon sugar deoxyribose, and a nitrogenous base. The alternating sugar and phosphate groups in each chain make up the outside, or backbone, of the spiral structure. The nitrogen bases are on the inside of the structure. Because the spiral structure is composed of two chains, it is known as a double helix.

DNA contains four different nitrogenous bases: adenine (A), thymine (T), cytosine (C), and guanine (G). As **Figure 22** shows, both adenine and guanine contain a double ring. Thymine and cytosine are single-ring structures. Looking again at **Figure 22,** notice that each nitrogen base on one strand of the helix is oriented next to a nitrogen base on the opposite strand, in the same way that the teeth of a zipper are oriented. The side-by-side base pairs are close enough so that hydrogen bonds form between them. Because each nitrogen base has a unique arrangement of organic functional groups that can form hydrogen bonds, the nitrogen bases always pair in a specific way so that the optimum number of hydrogen bonds form. As **Figure 22** shows, guanine always binds to cytosine, and adenine always binds to thymine. The G–C and A–T pairs are called complementary base pairs.

☑ READING CHECK **Describe** what forms the teeth of the DNA zipper.

Because of complementary base pairing, the amount of adenine in a molecule of DNA always equals the amount of thymine, and the amount of cytosine always equals the amount of guanine. In 1953, James Watson and Francis Crick used this observation to make one of the greatest scientific discoveries of the twentieth century when they determined the double-helix structure of DNA. They accomplished this feat without performing many laboratory experiments themselves. Instead, they analyzed and synthesized the work of numerous scientists who had carefully carried out studies on DNA.

View an **animation about the structure of DNA.**

Concepts In Motion

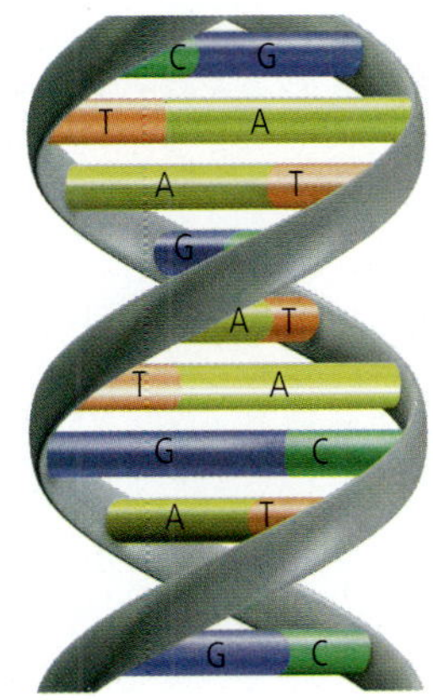


■ **Figure 21** The structure of DNA is a double helix that resembles a twisted zipper. The two sugar-phosphate backbones form the outsides of the zipper.

■ **Figure 22** In DNA, base pairing exists between a double-ringed base and a single-ringed base. Adenine and thymine always pair, forming two hydrogen bonds between them. Guanine and cytosine always form three hydrogen bonds when they pair.

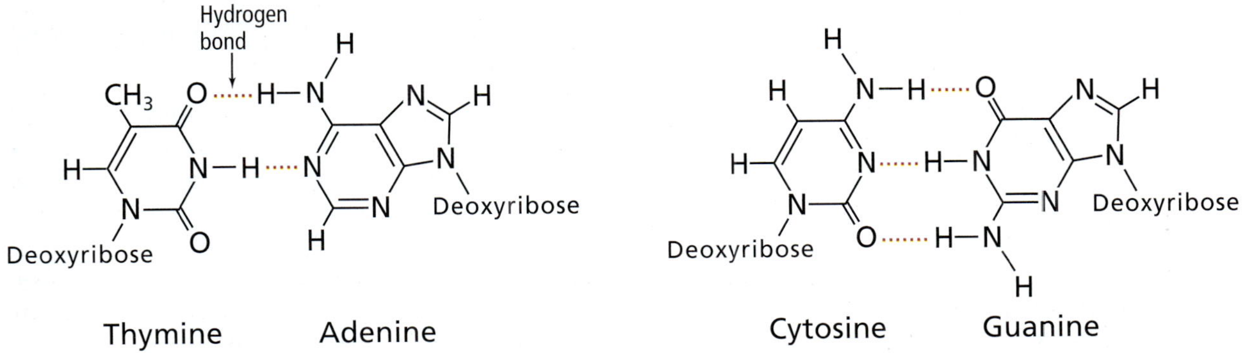

**The function of DNA** Watson and Crick used their model to predict how DNA's chemical structure enables it to function. DNA stores the genetic information of a cell in the cell's nucleus. Before the cell divides, the DNA is copied so that the new generation of cells gets the same genetic information. Having determined that the two chains of the DNA helix are complementary, Watson and Crick realized that complementary base pairing provides a mechanism by which the genetic material of a cell is copied.

The four nitrogenous bases of DNA serve as the letters of the alphabet in the information-storage language of living cells. The specific sequence of these letters represents an organism's master instructions, just as the sequence of letters in the words of this sentence convey special meaning. The sequence of bases is different in every species of organism, allowing for an enormous diversity of life-forms—all from a language that uses only four letters. It is estimated that the DNA in a human cell has about three billion complementary base pairs, arranged in a sequence unique to humans.

# Problem-Solving LAB

## Formulate a Model

**How does DNA replicate?** DNA replicates before a cell divides so that each of the two newly formed cells has a complete set of genetic instructions. When DNA begins to replicate, the two nucleotide strands start to unzip. An enzyme breaks the hydrogen bonds between the nitrogenous bases, and the strands separate. Other enzymes deliver free nucleotides from the surrounding medium to the exposed nitrogenous bases, adenine hydrogen-bonding with thymine, and cytosine bonding with guanine. Thus, each strand builds a complementary strand by base-pairing with free nucleotides. This process is shown in the top diagram at the right. When the free nucleotides have been hydrogen-bonded into place, their sugars and phosphates bond covalently to those on adjacent nucleotides to form the new backbone. Each strand of the original DNA molecule is now bonded to a new strand.

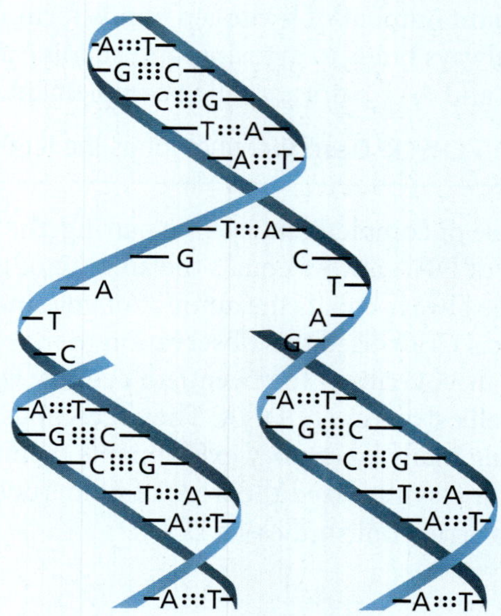


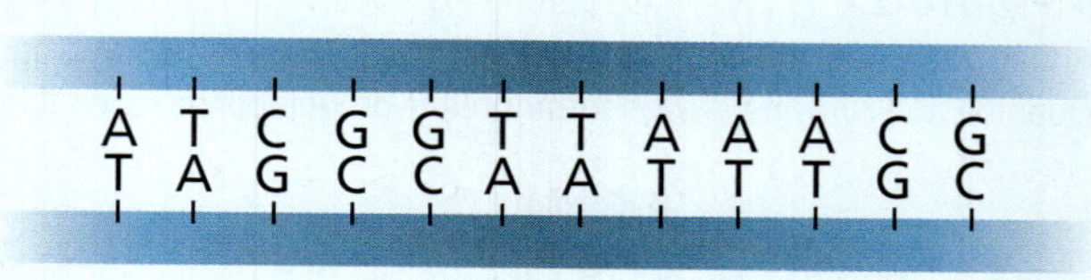


### Analysis

The bottom diagram shows a small segment of a DNA molecule. Copy the base sequence onto a clean sheet of paper, being careful not to make copying errors. Show the steps of replication to produce two segments of the DNA.

### Think Critically

1. **Describe** how the base sequence of a newly synthesized strand compares with the original strand to which it is bonded.
2. **Explain** If the original DNA segment is colored red and the free nucleotides are colored blue, what pattern of colors will the newly replicated DNA segments have? Will all new segments have the same color pattern?
3. **Explain** how an organism might be affected if an error occurs during replication of its DNA. Are the effects permanent? Explain.

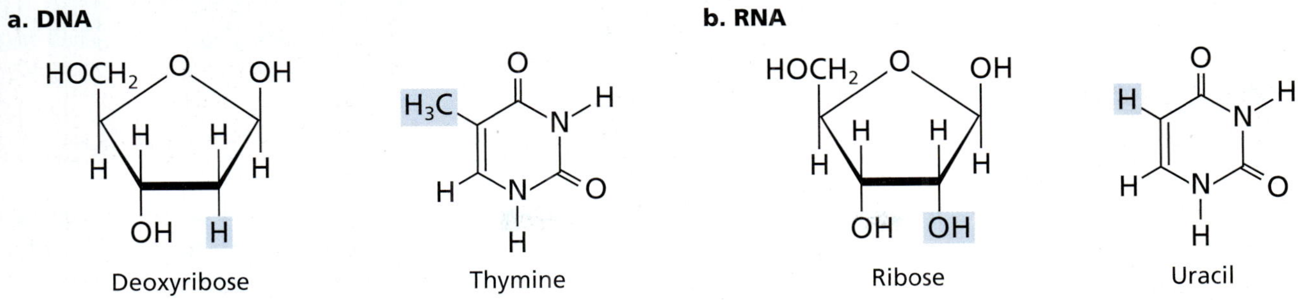


■ **Figure 23** DNA and RNA differ in their components. The two structures on the left are found in DNA. The two structures on the right are found in RNA.
**Identify** *two differences in the structures of RNA and DNA.*

## RNA

RNA (ribonucleic acid) is also a nucleic acid. Its general structure differs from that of DNA in three important ways, as shown in **Figure 23.** First, as you have read, DNA contains the nitrogen bases adenine, cytosine, guanine, and thymine. RNA contains adenine, cytosine, guanine, and uracil. Thymine is never found in RNA. Second, RNA contains the sugar ribose. DNA contains the sugar deoxyribose, which has a hydrogen atom in place of a hydroxyl group at one position.

The third difference between DNA and RNA is a result of these structural differences. DNA is normally arranged in a double helix in which hydrogen bonding links the two chains together through their bases. RNA is usually single-stranded, with no such hydrogen bonds forming among the bases.

Whereas DNA functions to store genetic information, RNA allows cells to use the information found in DNA. You have read that the genetic information of a cell is contained in the sequence of nitrogen bases in the DNA molecule. Cells use this base sequence to make RNA with a corresponding sequence. The RNA is then used to make proteins, each with an amino-acid sequence that is determined by the order of nitrogen bases in RNA. The sequences of bases are referred to as the genetic code. Because proteins are the molecular tools that carry out most activities in a cell, the DNA double helix is ultimately responsible for controlling the thousands of chemical reactions that take place in cells.

**FOLDABLES®**
Incorporate information from this section into your Foldable.

## SECTION 4 REVIEW

Section Self-Check

### Section Summary

- Nucleic acids are polymers of nucleotides, which consist of a nitrogen base, a phosphate group, and a sugar.
- DNA and RNA are the information-storage molecules of a cell.
- DNA is double stranded, and RNA is single stranded.

**20.** MAINIDEA **Explain** the primary function of RNA and DNA.

**21. Identify** the specific structural components of both RNA and DNA.

**22. Relate** the function of DNA to its structure.

**23. Relate** the function of RNA to its structure.

**24. Analyze** the structure of nucleic acids to determine what structural feature makes them acidic.

**25. Predict** what might happen if the DNA that coded for a protein contained the wrong base sequence.

SECTION 5

# Metabolism

**MAINIDEA** Metabolism involves many thousands of reactions in living cells.

**Essential Questions**

- How do anabolism and catabolism compare?
- What is the role of ATP in metabolism?
- How can the processes of photosynthesis, cellular respiration, and fermentation be compared and contrasted?

**Review Vocabulary**

**redox process:** a chemical reaction in which electrons are transferred from one atom to another

**New Vocabulary**

metabolism
catabolism
anabolism
ATP
photosynthesis
cellular respiration
fermentation

**CHEM 4 YOU** You have studied the four major kinds of biological molecules and learned that they are all present in the food you eat. What happens to these molecules after they enter your body?

## Anabolism and Catabolism

Many thousands of chemical reactions take place in the cells of a living organism. The set of chemical reactions that occur within an organism is its **metabolism.** Why are so many reactions involved in metabolism? Living organisms must accomplish two major functions in order to survive. They have to extract energy from nutrients in forms that they can use immediately as well as store for future use. In addition, they have to use nutrients to make building blocks for synthesizing all of the molecules needed to perform their life functions. These processes are summarized in **Figure 24.**

The term **catabolism** refers to the metabolic reactions that break down complex biological molecules such as proteins, polysaccharides, triglycerides, and nucleic acids for the purposes of forming smaller building blocks and extracting energy. After you eat a meal of spaghetti and meatballs, your body immediately begins to break down the starch polymer in the pasta into glucose. The glucose is then broken down into smaller molecules in a series of energy-releasing catabolic reactions. Meanwhile, the protein polymers in the meatballs are catabolized into amino acids.

The term **anabolism** refers to the metabolic reactions that use energy and small building blocks to synthesize the complex molecules needed by an organism. After your body has extracted the energy from the starch in the pasta, it uses that energy and the amino-acid building blocks produced from the meat proteins to synthesize the specific proteins that allow your muscles to contract, catalyze metabolic reactions, and perform many other functions in your body.

**Figure 24** A large number of different metabolic reactions take place in living cells. Some involve breaking down nutrients to extract energy; these are catabolic processes. Others involve using energy to build large biological molecules; these reactions are anabolic processes.

**Describe** *Choose one food that you ate recently, and describe how it was metabolized.*

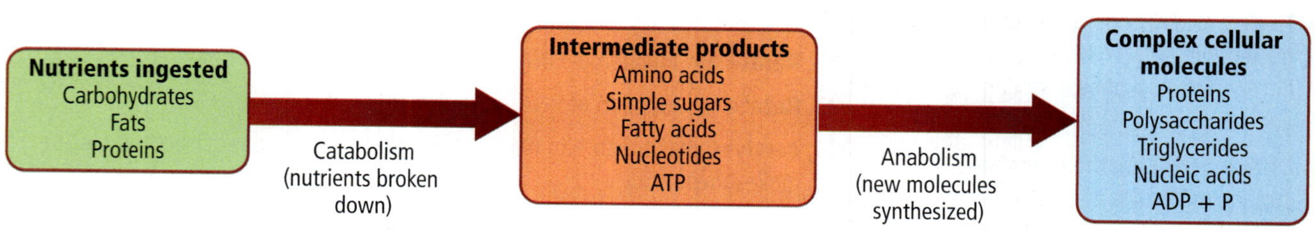

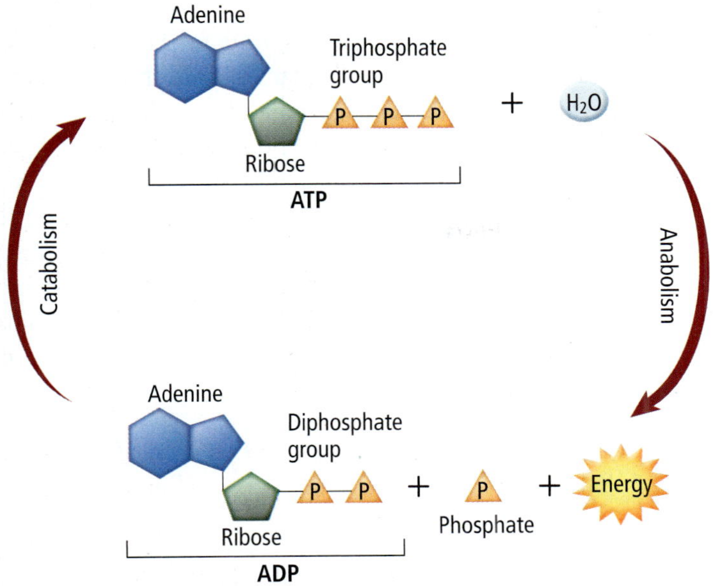


■ **Figure 25** The breakdown of ATP provides energy for cellular processes in living organisms.
**Explain** *where the energy is stored in ATP.*

**Figure 24** shows the relationship between catabolism and anabolism. The nutrients listed on the left side of the diagram are broken down into intermediate products. These intermediate products are used as building blocks for the products listed on the right side of the diagram. Another way of conceptualizing this process is to view the nutrients ingested as the raw materials for the complex cellular molecules formed in a living organism.

**VOCABULARY**
**ACADEMIC VOCABULARY**
**Conceptualize**
visualizing or conceiving an abstract idea in the mind
*The atomic cloud model is hard to conceptualize.*

☑ READING CHECK **Explain** how the terms *metabolism, catabolism,* and *anabolism* are related.

**ATP** Catabolism and anabolism are linked by common building blocks that catabolic reactions produce and anabolic reactions use. A common form of potential chemical energy also links the two processes, as shown in **Figure 25.** **ATP** (adenosine triphosphate) is a nucleotide that functions as the universal energy-storage molecule in living cells. During catabolic reactions, cells harness the chemical energy of foods and store it in the bonds of ATP. When these bonds are broken, the chemical energy is released and used by cells to drive anabolic reactions that might not otherwise occur. Most cellular reactions have an efficiency of only about 40% at best; the remaining 60% of the energy in food is lost as heat, which your body uses to keep warm.

During catabolic reactions, cells produce ATP by adding an inorganic phosphate group to the nucleotide adenosine diphosphate (ADP) in an endothermic reaction. One mole of ATP stores approximately 30.5 kJ of energy under normal cellular conditions. During anabolism, the reverse reaction occurs. ATP is broken down to form ADP and inorganic phosphate in an exothermic reaction. Approximately 30.5 kJ of energy is released from each mole of ATP.

☑ READING CHECK **Describe** what occurs when ATP becomes ADP.

■ **Figure 26** Grass and other green plants use certain wavelengths of sunlight as an energy source. Other living organisms, such as cows, obtain energy by eating plants or eating other organisms that eat plants.

## Photosynthesis

What is the source of energy that fuels metabolism? For most living things, including the grass shown in **Figure 26,** certain wavelengths of sunlight provide this energy. Some bacteria and the cells of all plants and algae are able to capture light energy and convert some of it into chemical energy. Animals cannot capture light energy, so they get energy by eating plants or by eating other animals that eat plants. The process that converts energy from sunlight to chemical energy in the bonds of carbohydrates is called **photosynthesis.** During the complex process of photosynthesis, carbon dioxide and water yield a carbohydrate (glucose) and oxygen gas. The following net reaction takes place during photosynthesis.

$$\underset{\text{Carbon dioxide}}{6CO_2} + \underset{\text{Water}}{6H_2O} + \text{light energy} \rightarrow \underset{\text{Glucose}}{C_6H_{12}O_6} + \underset{\text{Oxygen}}{6O_2}$$

Photosynthesis results in the reduction of the carbon atoms in carbon dioxide as glucose is formed. During this redox process, oxygen atoms in water are oxidized to oxygen gas.

## Cellular Respiration

Most organisms need oxygen to live. Oxygen that is produced during photosynthesis is used by living things during **cellular respiration,** the process in which glucose is broken down to form carbon dioxide, water, and large amounts of energy. Cellular respiration is the major energy-producing process in living organisms. **Figure 27** shows one use of energy in the body. This energy is stored in the bonds of ATP. Cellular respiration is a redox process; the carbon atoms in glucose are oxidized while oxygen atoms in oxygen gas are reduced to the oxygen in water. The net reaction that takes place during cellular respiration is as follows.

$$\underset{\text{Glucose}}{C_6H_{12}O_6} + \underset{\text{Oxygen}}{6O_2} \rightarrow \underset{\text{Carbon dioxide}}{6CO_2} + \underset{\text{Water}}{6H_2O} + \text{energy}$$

Get help with **photosynthesis and respiration.**

Personal Tutor

■ **Figure 27** Swimmers need large amounts of energy when they compete in a race. This energy is stored in the bonds of ATP in their cells.

■ **Figure 28** Carbon dioxide formed during fermentation, leaves holes in the bread. These holes give bread a light, less-dense texture.

## Fermentation

During cellular respiration, glucose is completely oxidized, and oxygen gas is required to act as the oxidizing agent. Cells can extract energy from glucose in the absence of oxygen, but not nearly as efficiently. Without oxygen, only a fraction of the chemical energy of glucose can be released. Whereas cellular respiration produces 38 mol of ATP for every 1 mol of glucose catabolized in the presence of oxygen, only 2 mol of ATP are produced per mole of glucose that is catabolized in the absence of oxygen. This provides enough energy for oxygen-deprived cells so that they do not die. The process by which glucose is broken down in the absence of oxygen is known as **fermentation.** There are two common kinds of fermentation. In one, ethanol and carbon dioxide are produced. In the other, lactic acid is produced.

**Alcoholic fermentation** Yeast and some bacteria can ferment glucose to produce the alcohol ethanol.

$$\underset{\text{Glucose}}{C_6H_{12}O_6} \rightarrow \underset{\text{Ethanol}}{2CH_3CH_2OH} + \underset{\text{Carbon dioxide}}{2CO_2} + \text{energy}$$

This reaction, called alcoholic fermentation, is important in producing some foods, as shown in **Figure 28.** Alcoholic fermentation is needed to make bread dough rise, form tofu from soybeans, and produce the ethanol in alcoholic beverages. Another use of the ethanol is as an additive to gasoline, as shown in **Figure 29.**

### CAREERS IN CHEMISTRY

**Baker** Using a variety of chemical processes to create tasty and often beautiful creations is the job of a baker. Not only do breads and doughnuts undergo fermentation processes, but cakes and other pastries are often acid-base reactions.

WebQuest

■ **Figure 29** Ethanol is often added to gasoline and used as a fuel in some cars and trucks. Ethanol is made from grain.
**Explain** *how the use of ethanol can reduce the dependence on fossil fuels.*

(t)Henn Photography/Getty Images, (b)Alex Farnsworth/The Image Works

■ **Figure 30** During strenuous activity, oxygen can be depleted in cells. Then, energy is produced without oxygen and lactic acid is produced. Soreness in muscles a day or two after the activity is a sign of lactic acid formation.

©Wally McNamee/Corbis

**Lactic acid fermentation** Have you ever experienced muscle fatigue while running a race, like the person shown in **Figure 30?** During strenuous activity, muscle cells often use oxygen faster than it can be supplied by the blood. When the supply of oxygen is depleted, cellular respiration stops. Although animal cells cannot undergo alcoholic fermentation, they can produce lactic acid and a small amount of energy from glucose through lactic acid fermentation.

$$\underset{\text{Glucose}}{C_6H_{12}O_6} \rightarrow \underset{\text{Lactic acid}}{2CH_3CH(OH)COOH} + \text{energy}$$

The lactic acid that is produced is moved from the muscles through the blood to the liver. There, it is converted back into glucose that can be used in catabolic processes to yield more energy once oxygen becomes available. However, if lactic acid builds up in muscle cells at a faster rate than the blood can remove it, muscle fatigue results. An immediate burning sensation and soreness a few days later is an indication that lactic acid was produced in the muscles during exercise.

## SECTION 5 REVIEW

Section Self-Check

### Section Summary

- Living organisms undergo catabolism and anabolism.
- Photosynthesis directly or indirectly provides almost all living things with energy.
- The net equation for cellular respiration is the reverse of the net equation for photosynthesis.

26. **MAINIDEA Explain** why metabolism is important to living cells.
27. **Compare and contrast** the processes of anabolism and catabolism.
28. **Explain** the role of ATP in the metabolism of living organisms.
29. **Compare and contrast** the processes of photosynthesis, cellular respiration, and fermentation.
30. **Determine** whether each process is anabolic or catabolic.
    **a.** photosynthesis
    **b.** cellular respiration
    **c.** fermentation
31. **Evaluate** Why is it necessary to use sealed casks when making wine?
32. **Calculate** How many moles of ATP would a yeast cell produce if 6 mol of glucose were oxidized completely in the presence of oxygen? How many moles of ATP would the yeast cell produce from 6 mol of glucose if the cell were deprived of oxygen?

# CHEMISTRY & careers

## Career: Molecular Paleontologist

### Acid Test Reveals Surprise

"No right-thinking paleontologist would do what Mary did. We don't go to all this effort to dig this stuff out of the ground and then destroy it in acid." So says a colleague about Mary Schweitzer, the scientist who used the techniques of molecular biology to discover soft tissue where none should be—in the thighbone of a 68-million-year-old *Tyrannosaurus rex.*

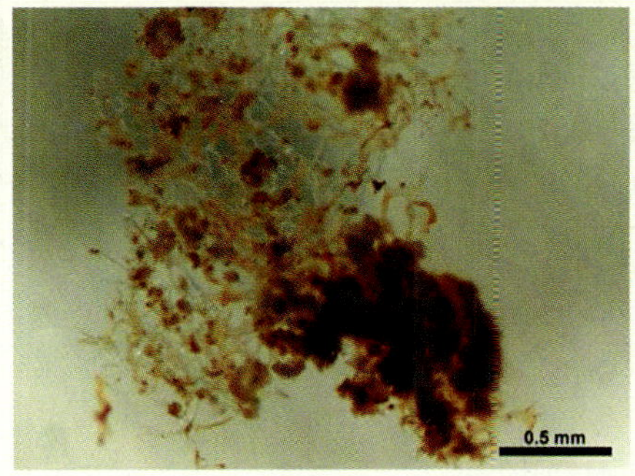


**Figure 2** Scientists also found blood vessels and individual cells in the soft tissue of the *T. rex.*

**Mother Bob** When the fossilized *T. rex,* nicknamed Bob, was recovered by paleontologists in 2003 from a remote section of Montana, the bones were encased in plaster for protection during transport. However, the bones and plaster weighed more than the helicopter could lift. So the paleontologists were forced to break the intact thighbone to move the dinosaur out of the remote area. Schweitzer took small fragments from the broken thighbone for further study.

The first surprise came quickly. "Bob" was a female, and she had been producing eggs at the time of her death. The bone Schweitzer studied is called medullary bone. Previously, this bone tissue was known only in birds, as shown in **Figure 1.** Ovulating hens produce medullary bone, then later use the calcium stored in the bone to make eggshells. After egg production, the bone disappears. **Figure 1** shows the medullary bone found in the *T. rex* thighbone.

**Figure 1** The hen bone and *T. rex* bone both have a hard outer bone called cortical bone (CB) and softer medullary bone (MB).

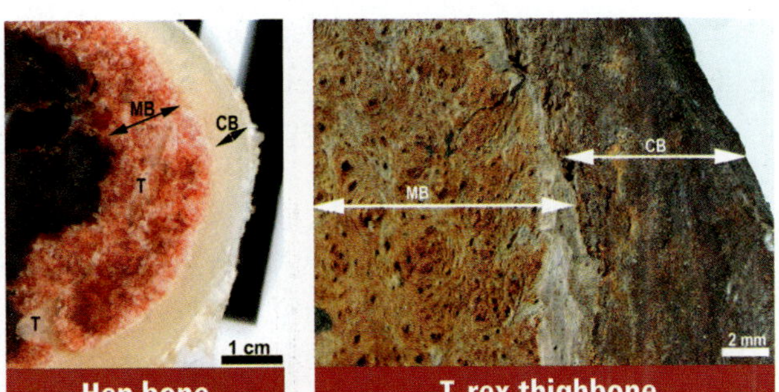




**The acid test** To study the medullary bone more closely, Schweitzer dissolved fragments of the bone in dilute acid to remove calcium phosphate—a technique normally used to examine fresh tissue. Because a fossilized bone has usually mineralized, it was assumed that the bone would completely dissolve in dilute acid. Yet this step yielded astonishing results—within the bone was soft tissue. Under the microscope, the tissue showed what looked like preserved blood vessels and even individual cells, as shown in **Figure 2.** But how could soft tissue have survived 68 million years in the ground?

**More work** Schweitzer has since subjected other bones to the same acid test, and found similar soft tissue and fine structures. No one knows yet just what these fine structures are showing, but, says a colleague, "there may be a lot of things out there that we've missed because of our assumption of how preservation works." Clearly, more research is needed.

### WRITING IN ▶ Chemistry

**Persuasive Writing** It is unlikely that dinosaur DNA will be found in these soft tissues. Even so, the discovery brings up the question: Should extinct animals be cloned from recovered DNA? Write a persuasive essay expressing your opinion.

WebQuest

# ChemLAB

iLab Station

## Observe Temperature and Enzyme Action

**Background:** **Enzymes are natural catalysts used by living things to speed reactions. These proteins have specialized structures that enable them to interact with specific substances.**

**Question:** *How does temperature affect the action of enzymes?*

### Materials

red-skin potato pulp
hydrogen peroxide (3% $H_2O_2$)
water
250-mL beaker (4)
test tubes (4)
test-tube rack
test-tube clamp
25-mL graduated cylinder
thermometer
ice
ruler
clock
hot plate
raw fresh liver pulp

### Safety Precautions

### Procedure

1. Read and complete the lab safety form.
2. Write a hypothesis that identifies the temperature at which the enzymes are the most active.
3. Copy the data table on a separate sheet of paper.
4. Place the four test tubes in the test-tube rack.
5. Measure and place 2.0 mL of red-skin potato pulp into each test tube.
6. Using the hot plate and ice, prepare water baths in the beakers at four different temperatures: ice water, room-temperature water, body-temperature water, and gently boiling water at or near 100°C.
7. Place one test tube in each water bath using a test-tube clamp.
8. Measure and record the temperature of each water bath.
9. After 5 min in the water baths, measure and place 5.0 mL of 3% $H_2O_2$ in each test tube.
10. Allow the reaction to proceed for 5 min.
11. Measure the height of the foam produced in each test tube.
12. Dispose of the contents of the test tubes as directed by your teacher and wash the test tubes.
13. Repeat Steps 4–12 using 2.0 mL of beef liver pulp instead of potato pulp.

**Data Table**

| Water Bath | Temperature (°C) | Height of Foam (cm) |
|---|---|---|
| **Potato** | | |
| Ice water | | |
| Room-temperature water | | |
| Body-temperature water | | |
| Boiling water (near 100°C) | | |
| **Liver** | | |
| Ice water | | |
| Room-temperature water | | |
| Body-temperature water | | |
| Boiling water (near 100°C) | | |

14. **Cleanup and Disposal** Dispose of the remaining solutions as directed by your teacher. Wash and return all lab equipment to its designated location.

### Analyze and Conclude

1. **Make and Use Graphs** Make a line graph with temperature on the *x*-axis and height of foam on the *y*-axis. Use a different color for the potato and liver data points and lines.
2. **Summarize** How does temperature affect the action of enzymes? Infer why the maximum reaction occurred at the temperature in which it did for the potato and liver.
3. **Recognize Cause and Effect** Which water bath produced the least amount of foam for each material? Propose explanations for why this happened.
4. **Compare and Contrast** Did the experimental data support your hypothesis in Step 2? Explain.
5. **Model** Write a balanced reaction for the decomposition of hydrogen peroxide for each reaction. How are the reactions similar? Infer why they are similar.
6. **Error Analysis** Identify potential sources of errors for this investigation and suggest methods to correct them.

### INQUIRY EXTENSION

**Design an Experiment** Would a change in pH affect the results? Design an experiment to find out.

# CHAPTER 23 STUDY GUIDE

Vocabulary Practice

**BIG**IDEA **Biological molecules—proteins, carbohydrates, lipids, and nucleic acids—interact to carry out activities necessary to living cells.**

## SECTION 1 Proteins

**MAIN**IDEA Proteins perform essential functions, including regulation of chemical reactions, structural support, transport of materials, and muscle contractions.

- Proteins are biological polymers made of amino acids that are linked by peptide bonds.
- Protein chains fold into intricate three-dimensional structures.
- Proteins have many functions in the human body, including functions within cells, functions between cells, and functions of structural support.

**VOCABULARY**
- protein
- amino acid
- peptide bond
- peptide
- denaturation
- enzyme
- substrate
- active site

## SECTION 2 Carbohydrates

**MAIN**IDEA Carbohydrates provide energy and structural material for living things.

- Carbohydrates are compounds that contain multiple hydroxyl groups (–OH) and a carbonyl functional group (C=O).
- Carbohydrates range in size from single monomers to polymers composed of hundreds or thousands of monomers.
- Monosaccharides in aqueous solution exist in both open-chain and cyclic structures.

**VOCABULARY**
- carbohydrate
- monosaccharide
- disaccharide
- polysaccharide

## SECTION 3 Lipids

**MAIN**IDEA Lipids make cell membranes, store energy, and regulate cellular processes.

- Fatty acids are long-chain carboxylic acids that usually have between 12 and 24 carbon atoms.
- Saturated fatty acids have no double bonds; unsaturated fatty acids have one or more double bonds.
- Fatty acids can be linked to glycerol backbones to form triglycerides.
- Steroids are lipids that have multiple-ring structures.

**VOCABULARY**
- lipid
- fatty acid
- triglyceride
- saponification
- phospholipid
- wax
- steroid

## SECTION 4 Nucleic Acids

**MAIN**IDEA Nucleic acids store and transmit genetic information.

- Nucleic acids are polymers of nucleotides, which consist of a nitrogen base, a phosphate group, and a sugar.
- DNA and RNA are the information-storage molecules of a cell.
- DNA is double stranded, and RNA is single stranded.

**VOCABULARY**
- nucleic acid
- nucleotide

## SECTION 5 Metabolism

**MAIN**IDEA Metabolism involves many thousands of reactions in living cells.

- Living organisms undergo catabolism and anabolism.
- Photosynthesis directly or indirectly provides almost all living things with energy.
- The net equation for cellular respiration is the reverse of the net equation for photosynthesis.

**VOCABULARY**
- metabolism
- catabolism
- anabolism
- ATP
- photosynthesis
- cellular respiration
- fermentation

# CHAPTER 23 ASSESSMENT

## SECTION 1

### Mastering Concepts

**33.** What should you call a chain of eight amino acids? A chain of 200 amino acids?

**34.** Name the two types of functional groups that react together to form a peptide bond, and name the functional group in the peptide bond itself.

**35.** Using the symbols below to represent four amino acids, draw peptide structures for four-member chains that link them together in different orders.

Amino acid 1: ■ Amino acid 3: ◆

Amino acid 2: ▲ Amino acid 4: ●

**36. Human Anatomy** Name five parts of the body that contain structural proteins.

**37.** List four major functions of proteins, and give one example of a protein that carries out each function.

**38.** Describe two common shapes found in the three-dimensional folding of proteins.

**39.** Name the organic functional groups in the side chains of the following amino acids.
**a.** glutamine **c.** glutamic acid
**b.** serine **d.** lysine

**40.** Explain how the active site of an enzyme functions.

**41.** Name an example of an amino acid that has an aromatic ring in its side chain.

**42.** Name two amino acids with nonpolar side chains and two amino acids with polar side chains.

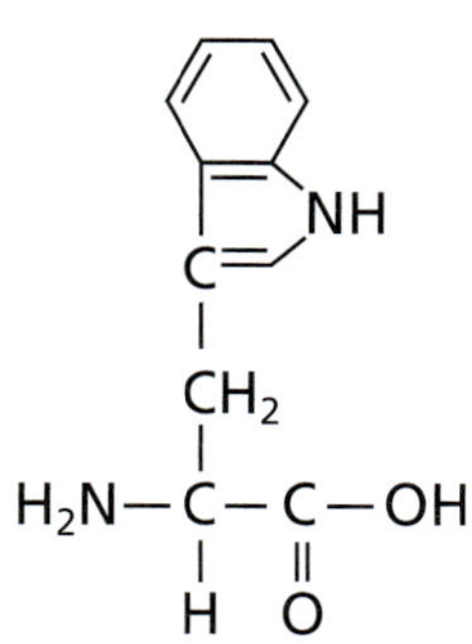


■ **Figure 31**

**43.** The structure shown in **Figure 31** is tryptophan. Describe some of the properties you would expect tryptophan to have, based on its structure. In what class of large molecules is tryptophan a member? Explain.

**44.** Is the dipeptide lysine-valine the same compound as the dipeptide valine-lysine? Explain.

**45. Enzymes** How do enzymes lower the activation energy for a reaction?

**46. Cellular Chemistry** Most proteins with a globular shape are oriented so that they have mostly amino acids with nonpolar side chains located on the inside and amino acids with polar side chains located on the outer surface. Does this make sense in terms of the nature of the cellular environment? Explain.

### Mastering Problems

**47.** How many different ways can you arrange three different amino acids in a peptide? Four amino acids? Five amino acids?

**48.** How many peptide bonds are present in a peptide that has five amino acids?

**49. Proteins** The average molecular weight of an amino acid residue in a polypeptide is 110 amu. What is the approximate molecular weight of the following proteins?
**a.** Insulin (51 amino acids)
**b.** Myosin (1750 amino acids)

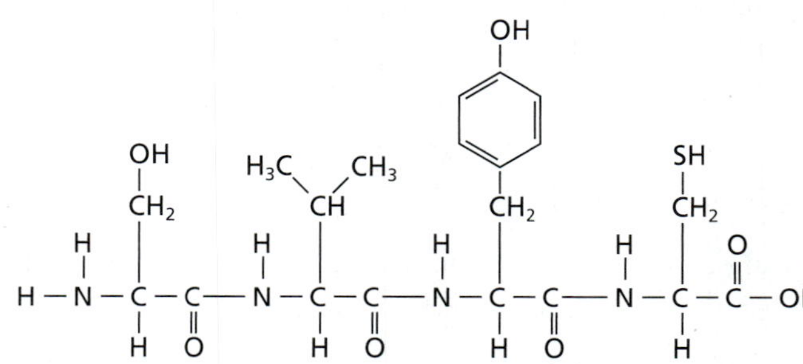


■ **Figure 32**

**50.** Determine how many amino acids and peptide bonds are in the peptide shown in **Figure 32.**

**51.** The average molar mass of an amino acid is 110 g/mol. Calculate the approximate number of amino acids in a protein that has a molar mass of 36,500 g/mol.

## SECTION 2

### Mastering Concepts

**52. Carbohydrates** Classify the following carbohydrates as monosaccharides, disaccharides, or polysaccharides.
**a.** starch **d.** ribose **g.** fructose
**b.** glucose **e.** cellulose **h.** lactose
**c.** sucrose **f.** glycogen

**53.** Name two isomers of glucose.

**54.** What kind of bond is formed when two monosaccharides combine to form a disaccharide?

**55. Sugars** Give a scientific term for each of the following.
**a.** blood sugar **c.** table sugar
**b.** fruit sugar **d.** milk sugar

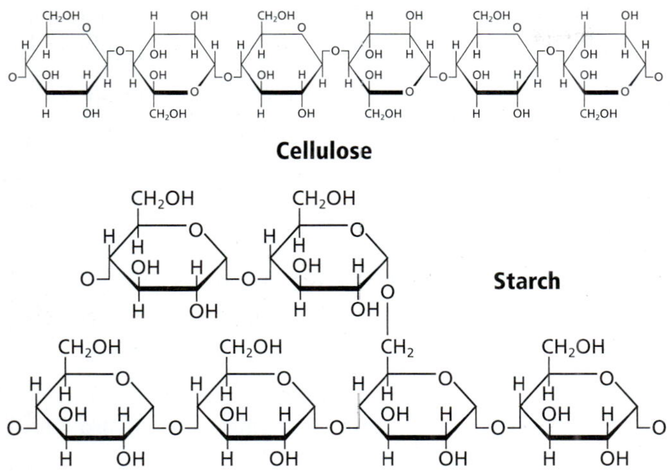


■ **Figure 33**

**56. Cellulose and Starch** The molecular structures of cellulose and starch are shown in **Figure 33.** Compare and contrast their molecular structures.

**57. Chemistry in Plants** Compare and contrast the functions of starch and cellulose in plants. Explain why their molecular structures are important to their functions.

**58.** Infer how the different bonding arrangements in cellulose and starch give them such different properties.

**59.** The disaccharide maltose is formed from two glucose monomers. Draw its structure.

**60.** The hydrolysis of cellulose, glycogen, and starch produces only one monosaccharide. Why is this so? What monosaccharide is produced?

**61. Digestion** Disaccharides and polysaccharides cannot be broken down in the absence of water. Why do you think this is so? Include an equation in your answer.

**62.** Draw the structure of the open-chain form of fructose. Circle all chiral carbons, and then calculate the number of stereoisomers with the same formula as fructose.

**63. Sugars** Compare and contrast the molecular formula, molecular weight, and functional groups found in glucose and fructose.

**64. Historical Perspective** Carbohydrates are not hydrates of carbon as the name suggests. Explain how this misconception occurred.

## Mastering Problems

**65. Complex Carbohydrates** Stachyose is a tetrasaccharide that contains two D-galactose units, one D-glucose unit, and one D-fructose unit. Each sugar unit has a molecular weight of 180 g/mol before it is linked together in this tetrasaccharide, and one water molecule is released for each two sugar units that come together. What is the molecular weight of stachyose?

# SECTION 3

## Mastering Concepts

**66.** Compare and contrast the structures of a triglyceride and a phospholipid.

**67.** Predict whether a triglyceride from beef fat or a triglyceride from olive oil will have a higher melting point. Explain your reasoning.

**68. Soaps and Detergents** Explain how the structure of soaps makes them effective cleaning agents.

**69.** Draw a portion of a lipid bilayer membrane, labeling the polar and nonpolar parts of the membrane.

**70.** Where and in what form are fatty acids stored in the human body?

**71.** What type of lipid does not contain fatty acid chains? Why are these molecules classified as lipids?

**72. Soap** Draw the structure of the soap sodium palmitate (palmitate is the conjugate base of the 16-carbon saturated fatty acid, palmitic acid). Label its polar and nonpolar ends.

**73.** Determine whether each structure is a fatty acid, triglyceride, phospholipid, steroid, or wax. Explain your reasoning.

**a.**

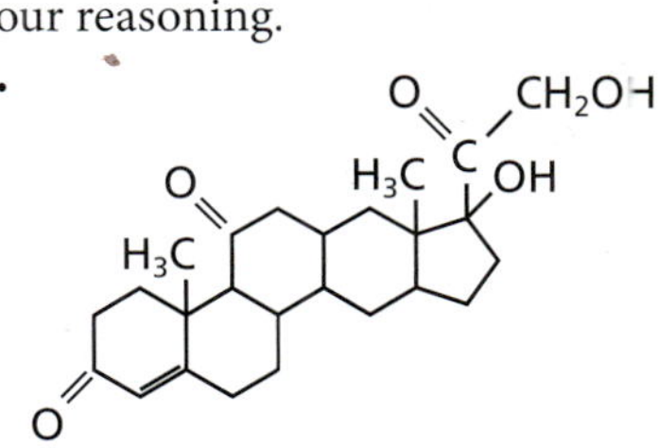

**b.**

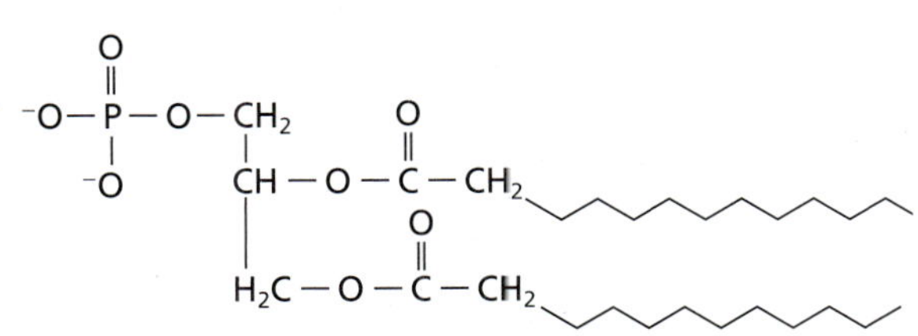

## Mastering Problems

**74.** The fatty acid palmitic acid has a density of 0.853 g/mL at 62°C. What will be the mass of a 0.886-L sample of palmitic acid at that temperature?

**75. Polyunsaturated Fats** How many moles of hydrogen gas are required for complete hydrogenation of 1 mol of linolenic acid, whose structure is shown below? Write a balanced equation for the hydrogenation reaction.

$$CH_3CH_2CH = (CHCH_2CH)_2 = CH(CH_2)_7COOH$$

## SECTION 4

### Mastering Concepts

**76.** What three structures make up a nucleotide?

**77.** Name two nucleic acids found in organisms.

**78.** Explain the roles of DNA and RNA in the production of proteins.

**79.** Where in living cells is DNA found?

**80.** Describe the types of bonds and attractions that link the monomers together in a DNA molecule.

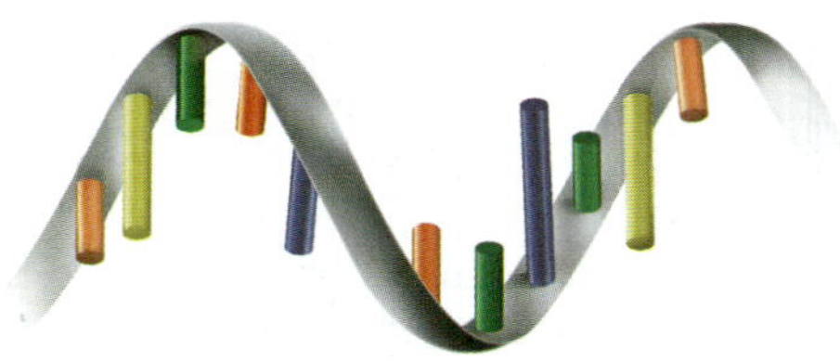


■ **Figure 34**

**81.** Classify the nucleic acid structure shown in **Figure 34** as DNA or RNA and explain your reasoning.

**82.** In the double-helical structure of DNA, the base guanine is always bonded to cytosine, and adenine is always bonded to thymine. What do you expect to be the relative proportional amounts of A, T, C, and G in a given length of DNA?

**83. DNA Replication** One strand in a DNA molecule has the following base sequence. What is the base sequence of the other strand in the DNA molecule?

C-C-G-T-G-G-A-C-A-T-T-A

### Mastering Problems

**84.** A triplet code is a sequence of three bases in RNA that codes for one amino acid in a peptide chain or protein. How many RNA bases are required to code for a protein that contains 577 amino acids?

**85. DNA Comparisons** A cell of the bacterium *Escherichia coli* has about $4.2 \times 10^6$ base pairs of DNA, whereas each human cell has about $3 \times 10^9$ base pairs of DNA. What percentage of the size of the human genome does the *E. coli* DNA represent?

## SECTION 5

### Mastering Problems

**86. Energy** Calculate and compare the total energy in kJ that is converted to ATP during the processes of cellular respiration and fermentation.

**87.** How many grams of glucose can be oxidized completely by 2.0 L of $O_2$ gas at STP during cellular respiration?

**88. Life Processes** Compare the net reactions for photosynthesis and cellular respiration with respect to reactants, products, and energy.

## MIXED REVIEW

**89.** Draw the carbonyl functional groups present in glucose and fructose. How are the groups similar? How are the groups different?

**90.** List the names of the monomers that make up proteins, complex carbohydrates, and nucleic acids.

**91.** Describe the functions of proteins, carbohydrates, lipids, and nucleic acids in living cells.

**92.** Write balanced equations for photosynthesis, cellular respiration, and the hydrolysis of lactose.

**93.** Write a balanced equation for the synthesis of sucrose from glucose and fructose.

## THINK CRITICALLY

**94. Make and Use Graphs** A number of saturated fatty acids and values for some of their physical properties are listed in **Table 2.**

**a.** Make a graph plotting melting point versus number of carbon atoms.
**b.** Graph density versus the number of carbon atoms.
**c.** Draw conclusions about the relationships between the number of carbon atoms in a saturated fatty acid and its density and melting point values.
**d.** Predict the approximate melting point of a saturated fatty acid that has 24 carbon atoms.

**Table 2** Physical Properties of Saturated Fatty Acids

| Name | Number of Carbon Atoms | Melting Point (°C) | Density (g/mL) (values at 60–80°C) |
|---|---|---|---|
| Palmitic acid | 16 | 63 | 0.853 |
| Myristic acid | 14 | 58 | 0.862 |
| Arachidic acid | 20 | 77 | 0.824 |
| Caprylic acid | 8 | 16 | 0.910 |
| Docosanoic acid | 22 | 80 | 0.822 |
| Stearic acid | 18 | 70 | 0.847 |
| Lauric acid | 12 | 44 | 0.868 |

Chapter Self-Check

**95. Calculate** Approximately 38 mol of ATP are formed when glucose is completely oxidized during cellular respiration. If the heat of combustion for 1 mol of glucose is $2.82 \times 10^3$ kJ/mol and each mole of ATP stores 30.5 kJ of energy, what is the efficiency of cellular respiration in terms of the percentage of available energy that is stored in the chemical bonds of ATP?

**96. Recognize Cause and Effect** Some diets suggest severely restricting the intake of lipids. Why is it not a good idea to eliminate all lipids from the diet?

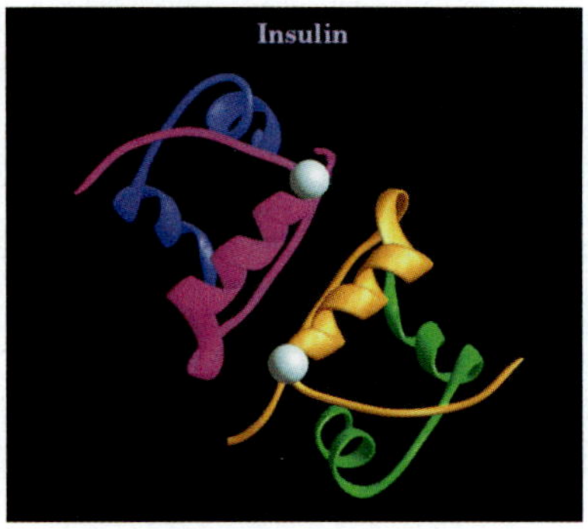


■ **Figure 35**

**97. Analyze** Insulin is a protein that functions as an important hormone in the human body. People who are diabetic often do not produce enough insulin, and must inject themselves with an insulin solution to maintain their health. Use **Figure 35** to infer how a person should care for a bottle of insulin.

**98. Calculate** If a double-stranded section of DNA has adenine as 20% of its bases, what percent of the other three bases are present in the DNA strand?

## CHALLENGE PROBLEM

**99.** Calculate how many moles of ATP a human body can produce from the sugar in a bushel of medium-sized Red Delicious apples. Use the Internet to find the information you need to solve this problem.

## CUMULATIVE REVIEW

**100. a.** Write the balanced equation for the synthesis of ethanol from ethene and water.

**b.** If 448 L of ethene gas reacts with excess water at STP, how many grams of ethanol will be produced?

**101.** Identify whether each of the reactants in these reactions is acting as an acid or a base.

**a.** $HBr + H_2O \rightarrow H_3O^+ + Br^-$

**b.** $NH_3 + HCOOH \rightarrow NH_4^+ + HCOO^-$

**c.** $HCO_3^- + H_2O \rightarrow CO_3^{2-} + H_3O^+$

**102.** What is a voltaic cell?



## WRITING IN ▶ Chemistry

**103. Cholesterol** Use the library or the Internet to research cholesterol. Write a newspaper article about cholesterol that is written for a teenage audience. Make sure the following questions are answered in the article. Where is this molecule used in your body? What is its function? Why is too much dietary cholesterol considered to be bad for you? Is genetics a factor in high cholesterol?

## DBQ Document-Based Questions

**Fatty Acids** *Omega-3 and omega-6 fatty acids are fatty acids that get their names from their structures. They contain a double bond either three or six carbon atoms from the end of the fatty acid chain. These fatty acids have a beneficial effect on health because they lower bad cholesterol levels and raise good cholesterol levels in the blood. Levels of omega-3 and omega-6 fatty acids were studied in salmon from three different sources as well as in the feed used in salmon farming.*

**Figure 36** *shows the percent of omega-3 and omega-6 fatty acids compared to the total amount of lipids in the samples.*

Data obtained from: Hamilton, M.C. et al. 2005. *Environmental Science Technology* 39: 8622–8629.

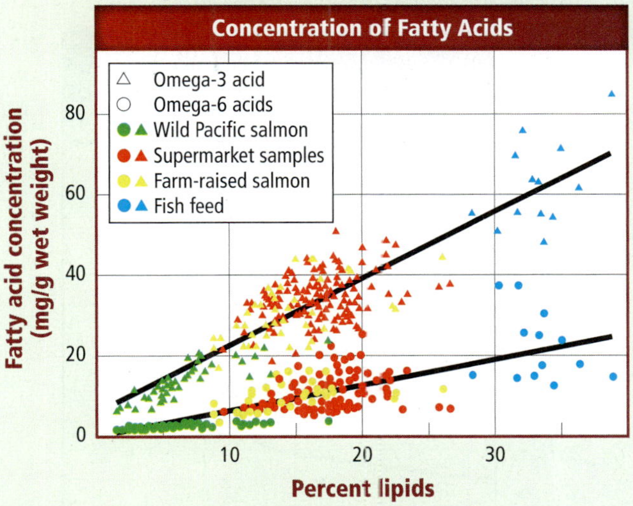


■ **Figure 36**

**104.** Which type of fish had the most omega fatty acids?

**105.** Based on this study, which type of salmon would you recommend to people who want to maximize the amounts of omega-3 and omega-6 fatty acids in their diets?

**106.** Infer from the graph why the farm-raised and supermarket salmon contains more omega-3 and omega-6 fatty acids than wild salmon.

## MULTIPLE CHOICE

**1.** Which is NOT true of carbohydrates?
**A.** Monosaccharides in aqueous solutions convert continuously between an open-chain structure and a cyclic structure.
**B.** The monosaccharides in starch are linked together by the same kind of bond that links the monosaccharides in lactose.
**C.** All carbohydrates have the general chemical formula $C_n(H_2O)_n$.
**D.** Cellulose, made only by plants, is easily digestible by humans.

**2.** Which is NOT a difference between RNA and DNA?
**A.** DNA contains the sugar deoxyribose, while RNA contains the sugar ribose.
**B.** RNA contains the nitrogen base uracil, while DNA does not.
**C.** RNA is usually single-stranded, while DNA is usually double-stranded.
**D.** DNA contains the nitrogen base adenine, while RNA does not.

*Use the graph below to answer Question 3.*

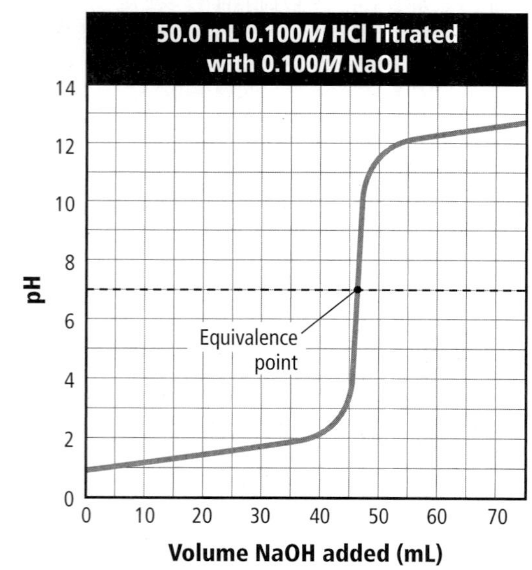


**3.** Which chemical indicator would be most effective in identifying the equivalence point of this titration?
**A.** methyl orange, pH range of 3.2–4.4
**B.** phenolphthalein, pH range of 8.2–10
**C.** thymol blue, pH range of 8.0–9.6
**D.** bromothymol blue, pH range of 6.0–7.6

**4.** How much NaOH, in grams, is required to completely neutralize 50.0 mL of 0.100*M* HCl?
**A.** 0.200 g
**B.** 5.00 g
**C.** 0.125 g
**D.** 200 g

*Use the table below to answer Questions 5 to 7.*

**Nucleotide Data for Samples of Double-Stranded DNA**

| Sample | Content of Each Nucleotide | A | G | C | T |
|---|---|---|---|---|---|
| I | number | 165 | ? | 231 | ? |
| | percent | 20.8 | ? | 29.2 | ? |
| II | number | ? | 402 | ? | ? |
| | percent | ? | 32.5 | ? | ? |
| III | number | ? | ? | 194 | 234 |
| | percent | ? | ? | 22.7 | 27.3 |
| IV | number | 266 | 203 | ? | ? |
| | percent | 28.4 | 21.6 | ? | ? |

**5.** What is the % T of Sample IV?
**A.** 28.4%
**B.** 78.4%
**C.** 71.6%
**D.** 21.6%

**6.** Every nitrogen base found in a DNA molecule is part of a nucleotide of that molecule. The A nucleotide, C nucleotide, G nucleotide, and T nucleotide have molar masses of 347.22 g/mol, 323.20 g/mol, 363.23 g/mol, and 338.21 g/mol respectively. What is the mass of 1 mol of Sample I?
**A.** $2.79 \times 10^5$ g
**B.** $2.7001 \times 10^5$ g
**C.** $2.6390 \times 10^5$ g
**D.** $2.72 \times 10^5$ g

**7.** How many molecules of adenine are in one molecule of Sample II?
**A.** 402
**B.** 434
**C.** 216
**D.** 175

**8.** Which is not a structural isomer of $CH_2{=}CHCH_2CH{=}CHCH_3$?
**A.** $CH_2{=}CHCH_2CH_2CH{=}CH_2$
**B.** $CH_3CH{=}CHCH_2CH{=}CH_2$
**C.** $CH_3CH{=}CHCH{=}CHCH_3$
**D.** $CH_2{=}C{=}CHCH_2CH_2CH_3$

## SHORT ANSWER

**9.** The sequence of bases in RNA determines the sequence of amino acids in a protein. Three bases code for a single amino acid; for example, CAG is the code for glutamine. How many amino acids are coded for in a strand of RNA $2.73 \times 10^4$ bases long?

*Use the diagram below to answer Question 10.*

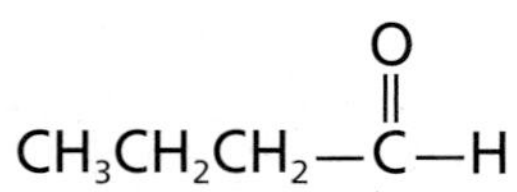

**10.** Which type of functional group is in this compound?

*Use the diagram below to answer Question 11.*

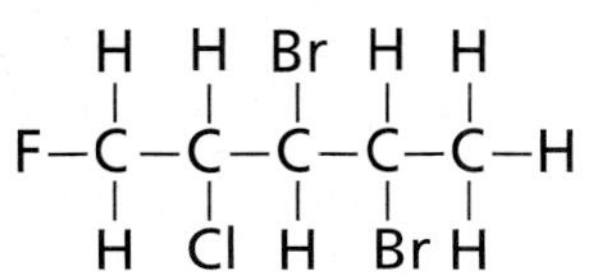

**11.** Give the IUPAC name for this organic compound.

**12.** What is the condensed structural formula of heptane?

## EXTENDED RESPONSE

*Use the diagram below to answer Question 13.*

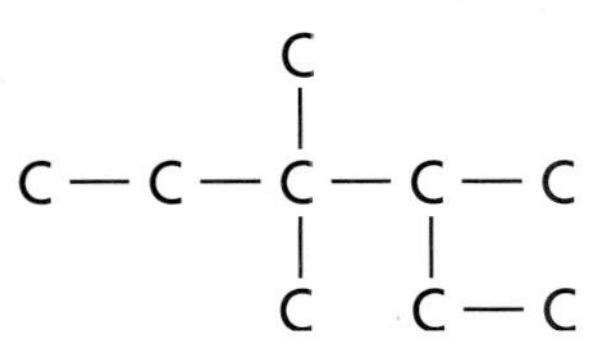

**13.** A student records the name of the alkane represented by this carbon skeleton as 2-ethyl 3,3-dimethyl pentane. Evaluate whether this is the correct name for the compound.

**14.** Compare and contrast aliphatic and aromatic compounds.

## SAT SUBJECT TEST: CHEMISTRY

*Use the table below to answer Questions 15–17.*

**Data for Elements in the Redox Reaction**
**$Zn + HNO_3 \rightarrow Zn(NO_3)_2 + NO_2 + H_2O$**

| Element | Oxidation Number | Complex Ion of which Element is a Part |
|---|---|---|
| Zn | 0 | none |
| Zn in $Zn(NO_3)_2$ | +2 | none |
| H in $HNO_3$ | +1 | none |
| H in $H_2O$ | ? | none |
| N in $HNO_3$ | ? | $NO_3^-$ |
| N in $NO_2$ | +4 | none |
| N in $Zn(NO_3)_2$ | ? | $NO_3^-$ |
| O in $HNO_3$ | −2 | $NO_3^-$ |
| O in $NO_2$ | ? | none |
| O in $Zn(NO_3)_2$ | ? | $NO_3^-$ |
| O in $H_2O$ | −2 | none |

**15.** Which element forms a monatomic ion that is a spectator in the redox reaction?
**A.** Zn
**B.** O
**C.** N
**D.** H
**E.** $O_2$

**16.** What is the oxidation number of N in $Zn(NO_3)_2$?
**A.** +1
**B.** +2
**C.** +3
**D.** +5
**E.** +6

**17.** What is the element that is oxidized in this reaction?
**A.** Zn
**B.** O
**C.** N
**D.** H
**E.** $O_3$

**NEED EXTRA HELP?**

| If You Missed Question . . . | 1 | 2 | 3 | 4 | 5 | 6 | 7 | 8 | 9 | 10 | 11 | 12 | 13 | 14 | 15 | 16 | 17 |
|---|---|---|---|---|---|---|---|---|---|---|---|---|---|---|---|---|---|
| Review Section . . . | 23.2 | 23.4 | 19.4 | 19.4 | 23.4 | 23.4 | 23.4 | 21.4 | 23.4 | 22.1 | 23.1 | 21.2 | 21.2 | 21.5 | 9.3 | 19.1 | 19.1 |

CHAPTER 24

# Nuclear Chemistry

**BIGIDEA** Nuclear chemistry has a vast range of applications, from the production of electricity to the diagnosis and treatment of diseases.

## SECTIONS


1 Nuclear Radiation

2 Radioactive Decay

3 Nuclear Reactions

4 Applications and Effects of Nuclear Reactions


## LaunchLAB

iLab Station

### How do chain reactions occur?

When the products of one nuclear reaction cause additional nuclear reactions to occur, the resulting chain reaction can release large amounts of energy in a short period of time. In this lab, you will explore chain reactions by modeling them with dominoes.

**FOLDABLES® Study Organizer**

### Types of Radiation

Make a layered-look book. Label it as shown. Use it to help you organize information about the different types of radiation.

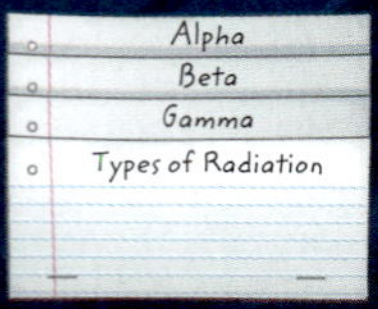

There's a good chance that you or someone you know has been helped by nuclear chemistry. From the magnetic properties of protons that enable MRIs, to the radiation used in X-rays, PET scans, and hundreds of other tests and treatments, research in nuclear chemistry has resulted in some of the most powerful tools available to modern medicine.

Go online!
MHEonline.com
PET scan—brain
MRI—spine
X-Ray—knee

SECTION 1

# Nuclear Radiation

**MAINIDEA** **Under certain conditions, some nuclei can emit alpha, beta, or gamma radiation.**

### Essential Questions

- How was radioactivity discovered and studied?
- What are the key properties of alpha, beta, and gamma radiations?

### Review Vocabulary

**nucleus:** the extremely small, positively charged, dense center of an atom that contains positively charged protons, neutral neutrons, and is surrounded by empty space through which one or more negatively charged electrons move

### New Vocabulary

radioisotope
X-ray
penetrating power

**CHEM 4 YOU** **If you wake up while it is still dark, the glowing numbers on your clock let you know what time it is. Many clocks use a type of radiation to make the numbers glow. The word *radiation* might cause you to think about nuclear power plants or dangerous, highly radioactive substances. However, less dangerous forms of radiation are often used in everyday objects, such as clocks.**

## The Discovery of Radioactivity

You have studied various forms of chemical reactions. Atoms can gain, lose, or share valence electrons, but the identity of the atoms does not change. Nuclear reactions, which you will study in this chapter, are different. Nuclear chemistry is concerned with the structure of atomic nuclei and the changes they undergo. Whereas chemical reactions involve only small energy changes, nuclear reactions involve much larger energy changes. **Table 1** offers a comparison of chemical reactions and nuclear reactions.

In 1895, German physicist Wilhelm Roentgen (1845–1923) found that invisible rays were emitted when electrons bombarded the surface of certain materials. These invisible rays caused photographic plates to darken, and Roentgen named these high-energy emissions *X-rays.* At that time, French physicist Henri Becquerel (1852–1908) was studying minerals that emit light after being exposed to sunlight, a phenomenon called phosphorescence. Building on Roentgen's work, Becquerel wanted to determine whether phosphorescent minerals also emitted X-rays.

**Table 1** Comparison of Chemical and Nuclear Reactions

| Chemical Reactions | | Nuclear Reactions | |
|---|---|---|---|
|  | • Occur when bonds are broken and formed<br>• Involve only valence electrons<br>• Associated with small energy changes<br>• Atoms keep the same identity although they might gain, lose, or share electrons, and form new substances<br>• Temperature, pressure, concentration, and catalysts affect reaction rates |  | • Occur when nuclei combine, split, and emit radiation<br>• Can involve protons, neutrons, and electrons<br>• Associated with large energy changes<br>• Atoms of one element are often converted into atoms of another element<br>• Temperature, pressure, and catalysts do not normally affect reaction rates |

Becquerel discovered by chance that phosphorescent uranium salts produced spontaneous emissions that darkened photographic plates. He observed this phenomenon even when the uranium salts were not exposed to light. Chemist Marie Curie (1867–1934) and her husband Pierre Curie (1859–1906) took Becquerel's mineral sample, called pitchblende, and isolated the components emitting the rays. They concluded that the darkening of the photographic plates was due to rays emitted from the uranium atoms present in the mineral sample. Marie Curie named the process by which materials give off such rays *radioactivity;* the rays and particles emitted by a radioactive source are called radiation. **Figure 1** shows the darkening of photographic film that is exposed to radiation emitted by radium salts.

The work of Marie and Pierre Curie was extremely important in establishing the origin of radioactivity and developing the field of nuclear chemistry. In 1898, the Curies identified two new elements, polonium and radium, on the basis of their radioactivity. Henri Becquerel and the Curies shared the 1903 Nobel Prize in Physics for their work. Marie Curie also received the 1911 Nobel Prize in Chemistry for her work with polonium and radium.

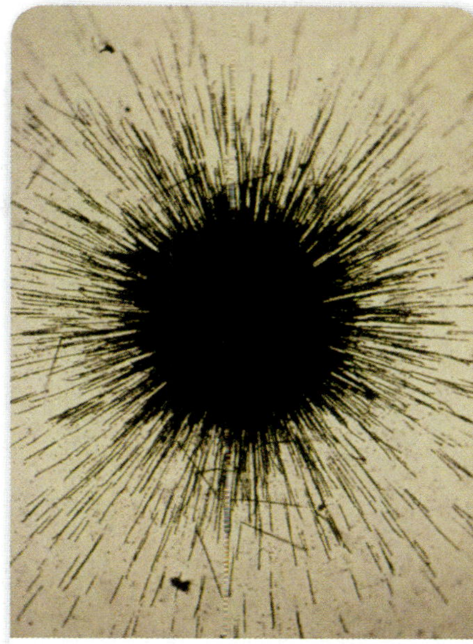

**Figure 1** Radium salts are placed on a special emulsion on a photographic plate. After the plate is developed, the emulsion shows the dark tracks left by radiation emitted by the radium salts.

READING CHECK **Explain** what Marie and Pierre Curie concluded about the darkening of the photographic plates.

## Types of Radiation

After reading about the discovery of radioactivity, you might wonder what types of radiation are emitted by radioactive nuclei or which nuclei are radioactive.

Recall that isotopes are atoms of the same element that have different numbers of neutrons. Isotopes of atoms with unstable nuclei are called **radioisotopes.** These unstable nuclei emit radiation to attain more stable atomic configurations in a process called radioactive decay. During radioactive decay, unstable atoms lose energy by emitting radiation. The three most common types of radiation are alpha ($\alpha$), beta ($\beta$), and gamma ($\gamma$). **Table 2** summarizes some of their important properties. Later in this chapter, you will learn about other types of radiation that can be emitted in a nuclear reaction.

**Table 2** Properties of Alpha, Beta, and Gamma Radiation

| Property | Alpha Radiation | Beta Radiation | Gamma Radiation |
|---|---|---|---|
| Symbol | $\alpha$ | $\beta$ | $\gamma$ |
| Composition | alpha particles | beta particles | high-energy electromagnetic radiation |
| Description of radiation | helium nuclei, $^{4}_{2}He$ | electrons | photons |
| Charge | 2+ | 1− | 0 |
| Mass | $6.64 \times 10^{-27}$ kg | $9.11 \times 10^{-31}$ kg | 0 |
| Approximate Energy | 5 MeV | 0.05 to 1 MeV | 1 MeV |
| Relative penetrating power | blocked by paper | blocked by metal foil | not completely blocked by lead or concrete |

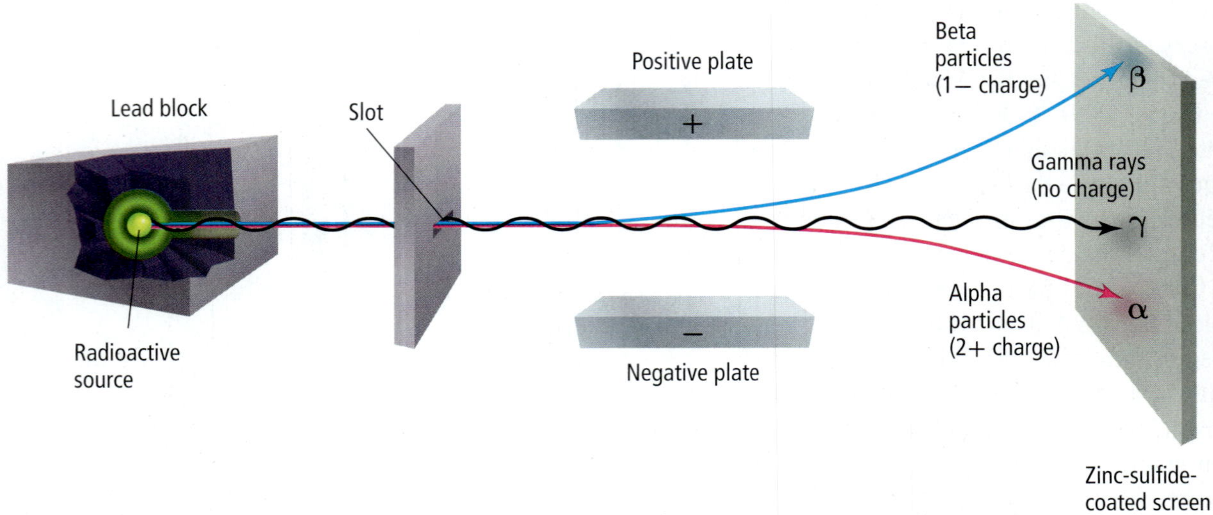


■ **Figure 2** The effect of an electric field depends on the charge and mass of the radiation. Positively charged alpha particles deflect toward the negative plate. Negatively charged beta particles deflect toward the positive plate. The lighter beta particles undergo the larger deflection. Gamma rays have no charge and are not affected by an electric field.

Ernest Rutherford (1871–1937), who performed the famous gold foil experiment that helped define modern atomic structure, identified alpha, beta, and gamma radiation when studying the effects of an electric field on the emissions from a radioactive source. As you can see in **Figure 2,** gamma rays carry no charge and are not affected by the electric field. Alpha particles carry a 2+ charge and are deflected toward the negatively charged plate. Beta particles carry a 1− charge and are deflected toward the positively charged plate. Because beta particles are less massive than alpha particles, they undergo a larger deflection.

☑ READING CHECK **Explain** how Rutherford determined whether each of the three types of radiation had a positive or negative charge or was unchanged.

FOLDABLES® Incorporate information from this section into your Foldable.

**Alpha particles** An alpha particle (α) has the same composition as a helium nucleus—two protons and two neutrons—and is therefore given the symbol $^{4}_{2}He$. The charge of an alpha particle is 2+ due to the presence of the two protons. Alpha radiation consists of a stream of alpha particles. Because of their mass and charge, alpha particles are relatively slow-moving compared with other types of radiation. Thus, alpha particles are not very penetrating—a single sheet of paper stops alpha particles. As you can see in **Figure 3,** radium-226, an atom whose nucleus contains 88 protons and 138 neutrons, undergoes alpha decay by emitting an alpha particle.

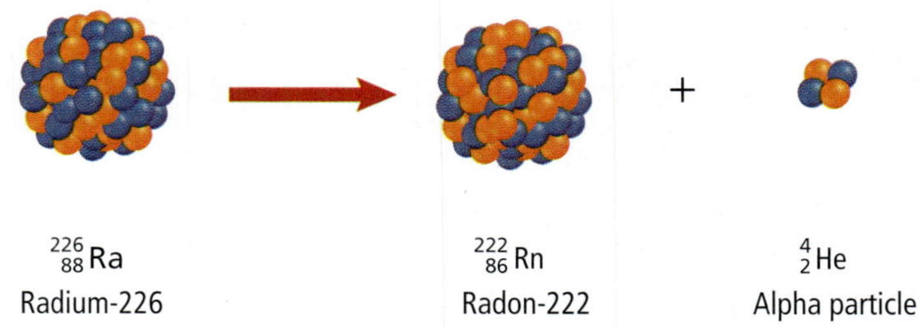


■ **Figure 3** A radium-226 nucleus undergoes alpha decay to form radon-222 and an alpha particle.
**Evaluate** ***What is the number of protons and neutrons in radium-226 and radon-222?***

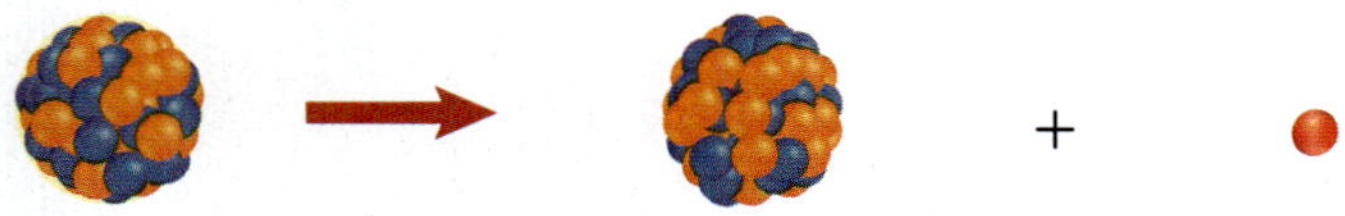

$^{131}_{53}I$ Iodine-131 → $^{131}_{54}Xe$ Xenon-131 + β Beta particle

■ **Figure 4** An iod ne-131 nucleus undergoes beta decay to form xenon-131 and a beta particle.
**Explain** ***How does beta decay affect the mass number of the decaying nucleus?***

In examining **Figure 3,** note that the reaction is balanced. That is, the sum of the mass numbers (superscripts) and the sum of the atomic numbers (subscripts) on each side of the arrow are equal. Also note that when a radioactive nucleus emits an alpha particle, the product nucleus has an atomic number that is lower by 2 and a mass number that is lower by 4.

**Beta particles** A beta particle is a very fast-moving electron that is emitted when a neutron in an unstable nucleus converts into a proton. Beta particles are represented by the symbol β or $e^-$. They have a 1− charge. Their mass is so small compared with the mass of nuclei involved in nuclear reactions that it can be approximated to zero. Beta radiation consists of a stream of fast-moving electrons. An example of the beta decay process is the decay of iodine-131 into xenon-131 by beta-particle emission, as shown in **Figure 4.** Note that the mass number of the product nucleus is the same as that of the original nucleus (they are both 131), but its atomic number has increased by 1 (54 instead of 53). This change in atomic number occurs because a neutron is converted into a proton, as shown by the following equation.

$$n \rightarrow p + \beta$$

As you might recall, the number of protons in an atom determines its identity. Thus, the formation of an additional proton results in the transformation from iodine-131 to xenon-131. Also, note that the electric charge in the equation above is conserved. The neutron is neutral. The proton has a 1+ charge and the beta particle has a 1− charge. Because beta particles are both lightweight and fast-moving, they have greater penetrating power than alpha particles. A thin sheet of metal foil is required to stop beta particles.

**VOCABULARY**
**WORD ORIGIN**
**Radiation**
comes from the Latin word *radiare* which means *to radiate*

**Gamma rays** Gamma rays are photons, which are high-energy (short wavelength) electromagnetic radiation. They are denoted by the symbol γ. Because photons have no mass and no charge, the emission of gamma rays does not change the atomic number or mass number of a nucleus. Gamma rays almost always accompany alpha and beta radiation, as they account for most of the energy loss that occurs as a nucleus decays. For example, gamma rays accompany the alpha-decay reaction of uranium-238.

$$^{238}_{92}U \rightarrow ^{234}_{90}Th + ^{4}_{2}He + 2\gamma$$

The 2 in front of the γ symbol indicates that two gamma rays of different frequencies are emitted. Because gamma rays have no effect on mass number or atomic number, it is customary to omit them from nuclear equations.

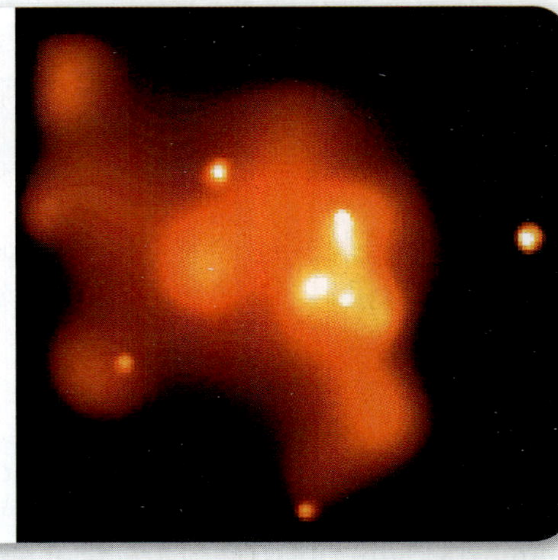

■ **Figure 5** The *Chandra Observatory,* launched in July 1999, photographed X-rays emitted from a cool gas cloud surrounding the black hole at the center of a neighboring galaxy.

As you have learned, the discovery of X-rays helped set the stage for the discovery of radioactivity. **X-rays,** like gamma rays, are a form of high-energy electromagnetic radiation. However, X-rays are not produced by radioactive sources and their energy is lower than that of gamma rays. They are emitted when inner electrons are knocked out and electrons from higher energy levels drop down to fill the vacancy. **Figure 5** shows an X-ray image taken in space. It allows astronomers to observe objects not visible in optical images. The presence of X-rays indicates phenomena such as exploding stars or black holes. Hospitals and dentists have machines that produce X-rays when a beam of electrons strikes a metal target. The familiar X-ray images are produced as the beam of X-rays passes easily through soft tissue but is partly blocked by hard tissue, such as bone.

READING CHECK **Compare and contrast** X-rays and gamma rays.

**Penetrating power** The ability of radiation to pass through matter is called **penetrating power.** Alpha particles have a low penetrating power because they move slowly due to their large mass, and their 2+ charge causes them to lose energy quickly through interactions with other particles. The penetrating power of beta particles is higher because they are smaller and faster than alpha particles. However, they can still interact with particles and can be stopped by thin shielding. Gamma rays are highly penetrating. Because they have no charge and no mass, the probability of matter stopping them is low.

## SECTION 1 REVIEW

Section Self-Check

### Section Summary

- Wilhelm Roentgen discovered X-rays in 1895.
- Henri Becquerel, Marie Curie, and Pierre Curie pioneered the fields of radioactivity and nuclear chemistry.
- Radioisotopes emit radiation to attain more stable atomic configurations.

1. **MAIN**IDEA **List** the different types of radiation and their charges.
2. **Compare** the subatomic particles involved in nuclear and chemical reactions.
3. **Explain** how you know whether the reaction is chemical or nuclear when an atom undergoes a reaction and attains a more-stable form.
4. **Calculate** **Table 2** gives approximate energy values in units of MeV. Convert each value into joules using the following conversion factor: $1 \text{MeV} = 1.6 \times 10^{-13} \text{J}$.
5. **Summarize** Make a time line that summarizes the major events that led to the understanding of alpha, beta, and gamma radiation.

SECTION 2

# Radioactive Decay

**MAINIDEA** Unstable nuclei can break apart spontaneously, changing the identity of atoms.

### Essential Questions

- Why are certain nuclei radioactive?
- How are nuclear equations balanced?
- How can you use radioactive decay rates to analyze samples of radioisotopes?

### Review Vocabulary

**radioactivity:** the process by which some substances spontaneously emit radiation

### New Vocabulary

transmutation
nucleon
strong nuclear force
band of stability
positron emission
positron
electron capture
radioactive decay series
half-life
radiochemical dating

**CHEM 4 YOU** To make sure that containers have the correct amount of fluid, some manufacturing processes use radioactivity. Detectors measure the number of particles produced by radioactive decay after they pass through the containers. For instance, a half-full bottle of juice would allow too much radiation to pass through and would not pass inspection.

## Nuclear Stability

Except for the emission of gamma radiation, radioactive decay involves the conversion of an element into another element. Such a reaction, in which an atom's atomic number is altered, is called **transmutation.** Whether an atom spontaneously decays and what type of radiation it emits depends on its neutron-to-proton ratio.

An atom's nucleus contains positively charged protons and neutral neutrons. Protons and neutrons are referred to as **nucleons.** Despite the strong electrostatic repulsion forces among protons, all nucleons remain bound in the dense nucleus because of the strong nuclear force. The **strong nuclear force** acts on subatomic particles that are extremely close together and overcomes the electrostatic repulsion among protons.

The fact that the strong nuclear force acts on both protons and neutrons is important. Two protons repel each other, but because neutrons are neutral, a neutron that is adjacent to a positively charged proton creates no repulsive electrostatic force. Yet these two adjacent particles are held together by the strong nuclear force. Likewise, two adjacent neutrons create no electrostatic force, but they, too, are held together by the strong nuclear force. Thus, the presence of neutrons adds an attractive force within the nucleus, as illustrated in **Figure 6.** The number of neutrons in a nucleus is important because nuclear stability is related to the balance between electrostatic and strong nuclear forces.

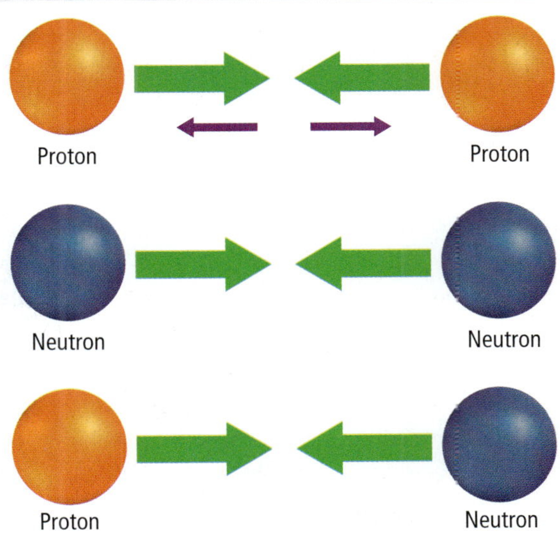


**Figure 6** The electrostatic force, represented by the purple arrows, acts between two charged particles. It is repulsive between two protons. The strong nuclear force, represented by the green arrows, acts between any two or more nucleons and is always attractive.

**Infer** *What is the effect of the electrostatic force between two neutrons? Between a proton and an electron?*

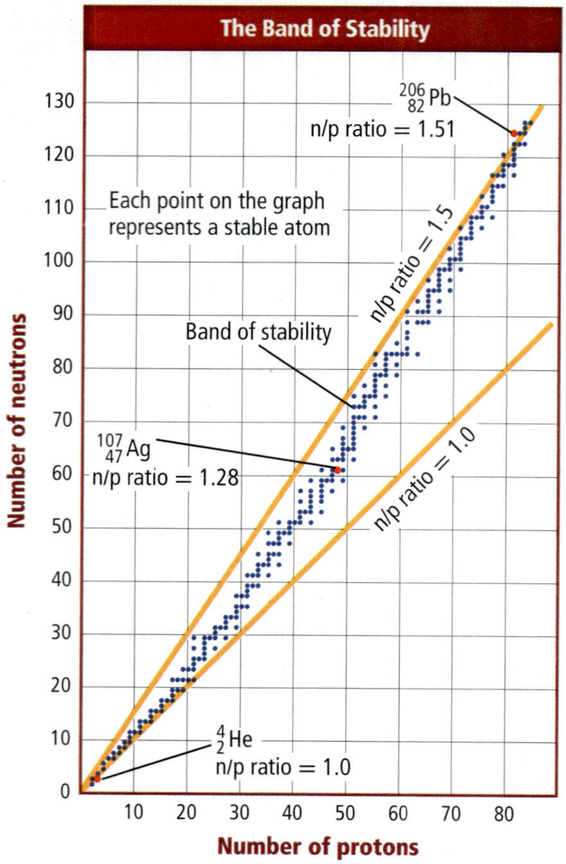


■ **Figure 7** The band of stability is the region where all stable nuclei fall when plotting the number of neutrons versus the number of protons. As the atomic number increases, the neutron-to-proton ratio (n/p) increases from 1:1 to 1.5:1.

☑ **GRAPH CHECK**

**Find** the number of protons above which the neutron-to-proton ratio starts to differ from 1:1.

**Neutron-to-proton ratio** To a certain degree, the stability of a nucleus can be correlated to its neutron-to-proton (n/p) ratio. For atoms with low atomic numbers (<20), the most stable nuclei are those with neutron-to-proton ratios of 1:1. For example, helium ($^{4}_{2}$He) has two neutrons and two protons, and a neutron-to-proton ratio of 1:1. As atomic number increases, more and more neutrons are needed to produce a strong nuclear force that is sufficient to balance the electrostatic repulsion force between protons. Therefore, the neutron-to-proton ratio for stable atoms gradually increases, reaching a maximum of approximately 1.5:1 for the largest atoms. An example of this is lead ($^{206}_{82}$Pb). With 124 neutrons and 82 protons, lead has a neutron-to-proton ratio of 1.51:1.

☑ READING CHECK **Explain** why the neutron-to-proton ratio of stable nuclei increases as the atomic number increases.

**The band of stability** Examine the plot of the number of neutrons versus the number of protons for all known stable nuclei shown in **Figure 7.** Notice that the slope of the plot indicates that the number of neutrons required for a nucleus to be stable increases as the number of protons increases. This correlates with the increase in the neutron-to-proton ratio of stable nuclei with increasing atomic number. The area on the graph within which all stable nuclei are found is known as the **band of stability.** As shown in **Figure 7,** $^{4}_{2}$He and $^{206}_{82}$Pb are both positioned within the band of stability although they have a different neutron-to-proton ratio. All nuclei outside the band of stability—either above or below—are radioactive and undergo decay in order to gain stability. After decay, the new atom is positioned more closely to, if not within, the band of stability. The band of stability ends at lead-208; all elements with atomic numbers greater than 82 are radioactive.

☑ READING CHECK **Define** the band of stability and relate it to the value of the neutron-to-proton ratio.

## Types of Radioactive Decay

The type of radioactive decay a particular radioisotope undergoes depends to a large degree on the underlying causes for its instability. Atoms lying above the band of stability generally have too many neutrons to be stable, whereas atoms lying below the band of stability tend to have too many protons to be stable. Depending on the relative number of neutrons and protons, atoms can undergo different types of decay—beta decay, alpha decay, positron emission, or electron capture—to gain stability.

**Beta decay** A radioisotope that lies above the band of stability is unstable because it has too many neutrons relative to its number of protons. For example, unstable $^{14}_{6}C$ has a neutron-to-proton ratio of 1.33:1, whereas stable elements of similar mass, such as $^{12}_{6}C$ and $^{14}_{7}N$, have neutron-to-proton ratios of approximately 1:1. It is not surprising, then, that $^{14}_{6}C$ undergoes beta decay, as this type of decay decreases the number of neutrons in the nucleus.

$$^{14}_{6}C \rightarrow {}^{14}_{7}N + \beta$$

**Figure 8a** shows the beta decay of carbon-14 into nitrogen-14. Note that the atomic number of the product nucleus, $^{14}_{7}N$, has increased by one. The nitrogen-14 atom now has a stable neutron-to-proton ratio of 1:1. Thus, beta emission has the effect of increasing the stability of a neutron-rich atom by increasing its atomic number, that is by lowering its neutron-to-proton ratio. The resulting atom is closer to, if not within, the band of stability.

READING CHECK **Explain** why radioisotopes above the band of stability are unstable.

**Alpha decay** All nuclei with more than 82 protons are radioactive and decay spontaneously. Both the number of neutrons and the number of protons must be reduced in order to make these radioisotopes stable. These very heavy nuclei often decay by emitting alpha particles. For example, polonium-210 spontaneously decays into lead-206 by emitting an alpha particle.

$$^{210}_{84}Po \rightarrow {}^{206}_{82}Pb + {}^{4}_{2}He$$

**Figure 8b** shows the alpha decay of polonium-210 into lead-206. The atomic number of $^{210}_{84}Po$ decreases by 2 and the mass number decreases by 4 as the nucleus decays into $^{206}_{82}Pb$.

READING CHECK **Calculate** how the neutron-to-proton ratio changes when polonium-210 decays into lead-206.

**Figure 8** Depending on where nuclei lie on the band of stability, they can emit a beta particle or an alpha particle.
**Compare and contrast** *beta decay and alpha decay in terms of the atomic number of the nuclei involved in the reaction.*

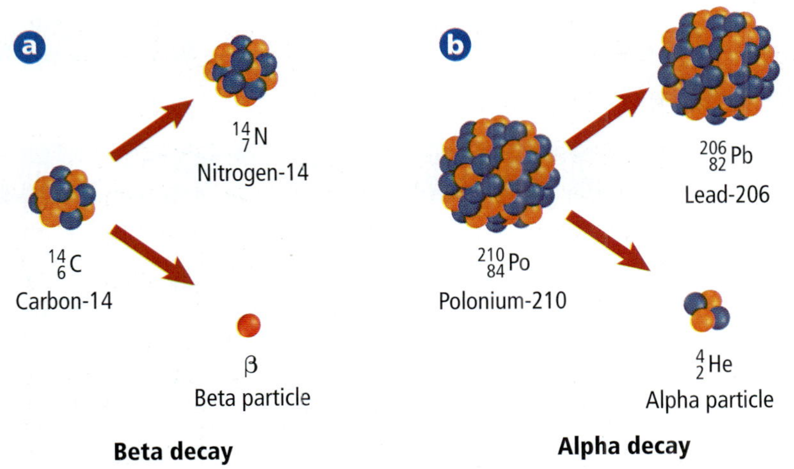


### Vocabulary
**Science usage v. Common usage**

**Unstable**

***Science usage:*** spontaneously radioactive
*Unstable atoms decay to reach a more stable state.*

***Common usage:*** not firm or fixed in one place
*The chair is unstable because one of its legs is shorter than the other.*

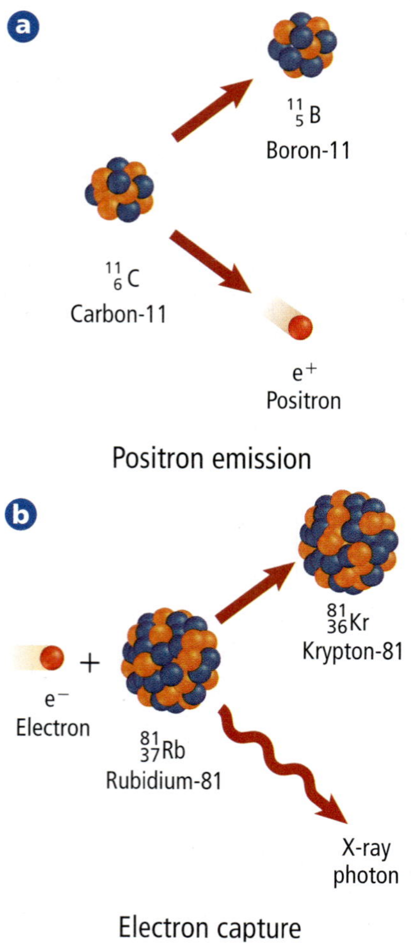


■ **Figure 9** When a nucleus undergoes positron emission or captures an electron, the number of protons decreases by one.

**Compare and contrast** *how the number of protons and neutrons change during positron emission and electron capture.*

**Positron emission and electron capture** For nuclei with low neutron-to-proton ratios, two common radioactive decay processes occur: positron emission and electron capture. These two processes tend to increase the neutron-to-proton ratio of the neutron-poor atom, bringing the atom closer to, if not within, the band of stability.

**Positron emission** is a radioactive decay process that involves the emission of a positron from a nucleus. A **positron** is a particle with the same mass as an electron but opposite charge; thus, it is represented by the symbol $\beta^+$ or $e^+$. During positron emission, a proton in the nucleus is converted into a neutron and a positron, and then the positron is emitted.

$$p \rightarrow n + e^+$$

**Figure 9** shows the positron emission of a carbon-11 nucleus. Carbon-11 lies below the band of stability and has a low neutron-to-proton ratio of approximately 0.8:1. Carbon-11 undergoes positron emission to form boron-11. Positron emission decreases the number of protons from six to five, and increases the number of neutrons from five to six. The resulting atom, $^{11}_{5}B$, has a neutron-to-proton ratio of 1.2:1, which is within the band of stability.

Electron capture is the other common radioactive-decay process that decreases the number of protons in unstable nuclei lying below the band of stability. **Electron capture** occurs when the nucleus of an atom draws in a surrounding electron, usually one from the lowest energy level. This captured electron combines with a proton to form a neutron.

$$p + e^- \rightarrow n$$

The atomic number of the nucleus decreases by 1 as a consequence of electron capture. The formation of the neutron also results in an X-ray photon being emitted. These two characteristics of electron capture are shown in the electron capture of rubidium-81 in **Figure 9.** The balanced nuclear equation for the reaction is shown below.

$$e^- + {}^{81}_{37}Rb \rightarrow {}^{81}_{36}Kr + \text{X-ray photon}$$

The five types of radioactive decay you have read about in this chapter are summarized in **Table 3.**

☑ READING CHECK **List** the decay processes that result in an increased neutron-to-proton ratio and a decreased neutron-to-proton ratio.

Explore **radioactive decay with an interactive table.** Concepts In Motion

**Table 3** Summary of Radioactive Decay Processes

| Type of Radioactive Decay | Particle Emitted | Change in Mass Number | Change in Atomic Number |
|---|---|---|---|
| Alpha decay | $^{4}_{2}He$ | decreases by 4 | decreases by 2 |
| Beta decay | $\beta$ or $e^-$ | no change | increases by 1 |
| Positron emission | $\beta^+$ or $e^+$ | no change | decreases by 1 |
| Electron capture | X-ray photon | no change | decreases by 1 |
| Gamma emission | $\gamma$ | no change | no change |

## Writing and Balancing Nuclear Equations

The radioactive decay processes you have just read about are all examples of nuclear reactions. Nuclear reactions are expressed by balanced nuclear equations just as chemical reactions are expressed by balanced chemical equations. However, in balanced chemical equations, numbers and types of atoms are conserved; in balanced nuclear equations, mass numbers and charges are conserved.

### EXAMPLE Problem 1

Find help with **algebraic equations.** Math Handbook

EXAMPLE PROBLEM

**BALANCING A NUCLEAR EQUATION** NASA uses the alpha decay of plutonium-238 $\left({}^{238}_{94}\text{Pu}\right)$ as a heat source on spacecraft. Write a balanced equation for this decay.

#### 1 ANALYZE THE PROBLEM

You are given that a plutonium atom undergoes alpha decay and forms an unknown product. Plutonium-238 is the initial reactant, while the alpha particle is one of the products of the reaction. The reaction is summarized below.

$$^{238}_{94}\text{Pu} \rightarrow {}^{A}_{Z}X + {}^{4}_{2}\text{He}$$

You must determine the unknown product of the reaction, *X*.

| **Known** | **Unknown** |
|---|---|
| reactant: plutonium-238 $({}^{238}_{94}\text{Pu})$ | mass number of the product $A = ?$ |
| decay type: alpha particle emission $({}^{4}_{2}\text{He})$ | atomic number of the product $Z = ?$ |
| | reaction product $X = ?$ |

#### 2 SOLVE FOR THE UNKNOWN

$238 = A + 4$ — **Apply the conservation of mass number.**

$A = 238 - 4 = 234$ — **Solve for A.**

Thus, the mass number of *X* is **234.**

$94 = Z + 2$ — **Apply the conservation of charges.**

$Z = 94 - 2 = 92$ — **Solve for Z.**

Thus, the atomic number of *X* is **92.**

The periodic table identifies the element as uranium (U).

$^{238}_{94}\text{Pu} \rightarrow {}^{234}_{92}\text{U} + {}^{4}_{2}\text{He}$ — **Write the balanced nuclear equation.**

#### 3 EVALUATE THE ANSWER

The correct formula for an alpha particle is used. The sums of the superscripts and subscripts on each side of the equation are equal. Therefore, the charge and the mass number are conserved. The nuclear equation is balanced.

### PRACTICE Problems

Do additional problems. Online Practice

6. Write a balanced nuclear equation for the reaction in which oxygen-15 undergoes positron emission.
7. Thorium-229 is used to increase the lifetime of fluorescent bulbs. What type of decay occurs when thorium-229 decays to form radium-225?
8. **Challenge** The figure at right shows one way that bismuth-212 can decay, producing isotopes **A** and **B.**
   a. Write a balanced nuclear equation for this decay.
   b. Identify the isotopes **A** and **B** that are produced.

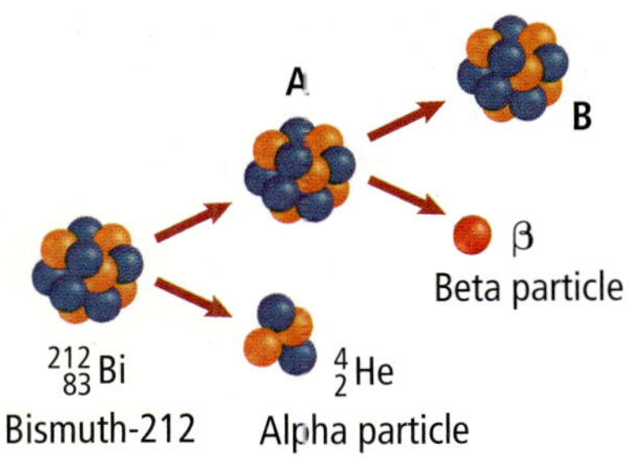

**Figure 10** Uranium-238 undergoes 14 different radioactive decay steps before forming stable lead-206.

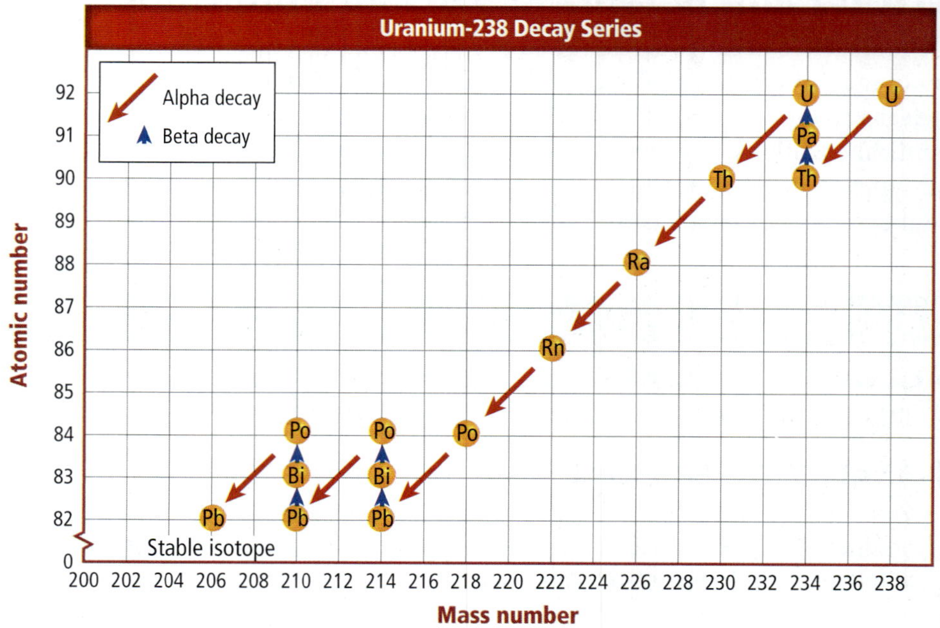


## Radioactive Series

A **radioactive decay series** is a series of nuclear reactions that begins with an unstable nucleus and results in the formation of a stable nucleus. As **Figure 10** shows, uranium-238 first decays to thorium-234, which in turn decays to protactinium-234. Decay reactions continue until a stable nucleus, lead-206, is formed.

GRAPH CHECK **List** each step in the decay of uranium-238. Include the type of decay and the resulting product.

## Radioactive Decay Rates

You might wonder how there could be any naturally occurring radioisotopes found on Earth. After all, if radioisotopes undergo continuous radioactive decay, won't they eventually disappear? Furthermore, radioisotopes have been decaying for about 4.6 billion years—the span of Earth's existence. Yet, naturally occurring radioisotopes are not uncommon on Earth. Some radioisotopes, such as carbon-14, are continuously formed in the upper atmosphere of Earth. Others are formed in the universe, during stellar nucleosynthesis for instance. Radioisotopes can also be synthesized in laboratories. The differing decay rates of isotopes also contribute to their presence on Earth.

Radioactive decay rates are measured in half-lives. A **half-life** is the time required for one-half of a radioisotope's nuclei to decay into its products. For example, the half-life of the radioisotope strontium-90 is 29 years. If you had 10.0 g of strontium-90 today, 29 years from now you would have 5.0 g left. **Table 4** shows how this decay continues through four half-lives of strontium-90. **Figure 11** presents the data from the table in terms of the percent of strontium-90 remaining after each half-life. The decay continues until a negligible amount of strontium-90 remains.

Explore **half-life.**

Virtual Investigations

READING CHECK **Define** the term *half-life.*

**Table 4 The Decay of Strontium-90**

| Number of Half-Lives | Elapsed Time | Amount of Strontium-90 Present |
|---|---|---|
| 0 | 0 y | 10.0 g |
| 1 | 29 y | $10.0\text{ g} \times \left(\frac{1}{2}\right) = 5.00\text{ g}$ |
| 2 | 58 y | $10.0\text{ g} \times \left(\frac{1}{2}\right)\left(\frac{1}{2}\right) = 2.50\text{ g}$ |
| 3 | 87 y | $10.0\text{ g} \times \left(\frac{1}{2}\right)\left(\frac{1}{2}\right)\left(\frac{1}{2}\right) = 1.25\text{ g}$ |
| 4 | 116 y | $10.0\text{ g} \times \left(\frac{1}{2}\right)\left(\frac{1}{2}\right)\left(\frac{1}{2}\right)\left(\frac{1}{2}\right) = 0.625\text{ g}$ |

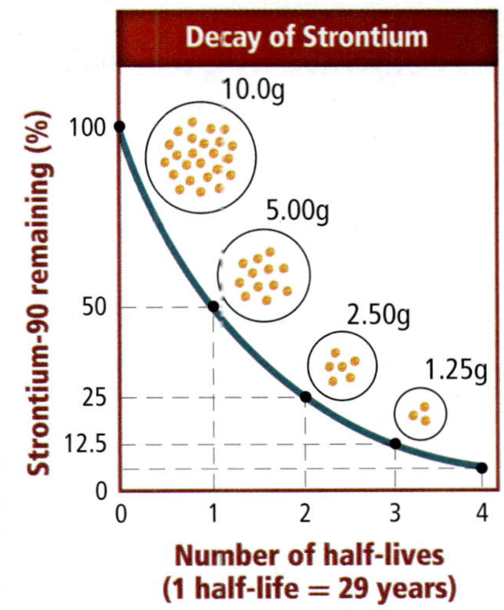


**Figure 11** The graph shows how the amount of strontium in a sample changes as a function of the number of half-lives.

Get help with **exponential graphing.**

Personal Tutor

The data in **Table 4** can be summarized in a simple equation representing the decay of any radioactive element.

**Remaining Amount of Radioactive Element**

$$N = N_0\left(\frac{1}{2}\right)^n$$

**N is the remaining amount.**
**$N_0$ is the inital amount.**
***n* is the number of half-lives that have passed.**

The amount remaining is equal to the initial amount times one-half raised the number of half-lives that have passed.

The exponent $n$ can also be replaced with the equivalent quantity $t/T$, where $t$ is the elapsed time and $T$ is the duration of the half-life. Note that $t$ and $T$ must have the same units of time.

$$N = N_0\left(\frac{1}{2}\right)^{t/T}$$

This type of expression is known as an exponential decay function. **Figure 11** shows the graph of a typical exponential decay function—in this case, the decay curve for strontium-90.

**GRAPH CHECK** **Infer** how much strontium remains after 1.5 half-lives.

Each radioisotope has its own characteristic half-life. Half-lives for several radioisotopes are given in **Table 5.** Notice the large range of values for half-lives, from millionths of a second to billions of years!

**Table 5 Half-Lives of Several Radioisotopes**

| Radioisotope | Symbol | Half-Life |
|---|---|---|
| Polonium-214 | $^{214}_{84}Po$ | 163.7 $\mu$s |
| Cobalt-60 | $^{60}_{27}Co$ | 5.272 y |
| Radon-222 | $^{222}_{86}Ra$ | 3.8 d |
| Phosphorus-32 | $^{32}_{15}P$ | 14.28 d |
| Carbon-14 | $^{14}_{6}C$ | 5730 y |
| Uranium-238 | $^{238}_{92}U$ | $4.46 \times 10^9$ y |

## EXAMPLE Problem 2

Find help with **fractions.** Math Handbook

EXAMPLE PROBLEM

**CALCULATING THE AMOUNT OF REMAINING ISOTOPE** Krypton-85 is used in indicator lights of appliances. The half-life of krypton-85 is 11 y. How much of a 2.000-mg sample remains after 33 y?

### 1 ANALYZE THE PROBLEM

You are given a known mass of a radioisotope with a known half-life. You must first determine the number of half-lives that passed during the 33-year period. Then, use the exponential decay equation to calculate the amount of the sample remaining.

**Known**

Initial amount = 2.000 mg

Elapsed time ($t$) = 33 y

Half-life ($T$) = 11 y

**Unknown**

**Amount remaining = ? mg**

### 2 SOLVE FOR THE UNKNOWN

$$\text{Number of half-lives } (n) = \frac{\text{elapsed time}(t)}{\text{half-life}(T)}$$

**Determine the number of half-lives passed during the 33 y.**

$$n = \frac{33\text{ y}}{11\text{ y}} = 3.0\text{ half-lives}$$

**Substitute t = 33 y and T = 11 y.**

$$\textbf{Amount remaining} = (\text{initial amount})\left(\frac{1}{2}\right)^n$$

**Write the exponential decay equation.**

$$\textbf{Amount remaining} = (2.000\text{ mg})\left(\frac{1}{2}\right)^{3.0}$$

**Substitute initial amount = 2.000 mg and $n$ = 3.**

$$\textbf{Amount remaining} = (2.000\text{ mg})\left(\frac{1}{8}\right) = \textbf{0.2500 mg}$$

### 3 EVALUATE THE ANSWER

Three half-lives are equivalent to $\left(\frac{1}{2}\right)\left(\frac{1}{2}\right)\left(\frac{1}{2}\right)$, or $\left(\frac{1}{8}\right)$. The answer (0.25 mg) is equal to $\left(\frac{1}{8}\right)$ of the initial amount. The answer has two significant figures because the number of years has two significant figures. $n$ does not affect the number of significant figures.

## PRACTICE Problems

Do additional problems. Online Practice

PRACTICE PROBLEMS

9. Bandages can be sterilized by exposure to gamma radiation from cobalt-60, which has a half-life of 5.27 y. How much of a 10.0-mg sample of cobalt-60 is left after one half-life? Two half-lives? Three half-lives?

10. If the passing of five half-lives leaves 25.0 mg of a strontium-90 sample, how much was present in the beginning?

11. **Challenge** The table shows the amounts of radioisotopes in three different samples. To the nearest gram, how much will be in Sample B and Sample C when Sample A has 16.2 g remaining?

| Sample | Radioisotope | Half-life | Amount (g) |
|---|---|---|---|
| A | cobalt-60 | 5.27 y | 64.8 |
| B | tritium | 12.32 y | 58.4 |
| C | strontium-90 | 28.79 y | 37.6 |

MiniLAB

iLab Station

## Model Radioactive Decay

**How do radioactive isotopes decay?**

### Procedure

1. Read and complete the lab safety form.
2. Place **100 pennies** in a **plastic cup.**
3. Place your hand over the top of the cup and shake the cup several times.
4. Pour the pennies into a **shoebox.** Remove all the pennies that land heads-up. These pennies represent atoms of the radioisotope that have undergone radioactive decay.
5. Prepare a data table to record the number of remaining pennies (tails-up pennies).
6. Count the number of pennies that remain, and record this number in your data table.
7. Place all of the tails-up pennies back in the plastic cup.
8. Repeat Steps 3 through 7 as many times as needed until no pennies remain.

### Analysis

1. **Construct** a graph of *Trial Number* v. *Number of Pennies Remaining* from your data table. Draw a curve through the plotted points.
2. **Calculate** how many trials it took for 50%, 75%, and 90% of the sample to decay.
3. **Evaluate** the half-life of the radioisotope if the time between each trial is 1 min.
4. **Determine** how the results would change if you used 100 dice instead of pennies. In this case, you would assume that any dice that lands with the six side facing up represents a decayed atom and is removed.

**Radiochemical dating** Chemical reaction rates are greatly affected by changes in temperature, pressure, and concentration, and by the presence of a catalyst. In contrast, nuclear reaction rates remain constant regardless of such changes. In fact, the half-life of any particular radioisotope is constant. Because of this, radioisotopes can be used to determine the age of an object. The process of determining the age of an object by measuring the amount of a certain radioisotope remaining in that object is called **radiochemical dating.**

**Connection to Biology** A type of radiochemical dating known as carbon dating is used to measure the age of artifacts that were once part of a living organism. Carbon dating makes use of the radioactive decay of carbon-14, which is formed by cosmic rays in the upper atmosphere at a fairly constant rate. These carbon-14 atoms become evenly spread throughout Earth's biosphere, where they mix with stable carbon-12 and carbon-13 atoms. Plants use carbon dioxide from the environment, which contains all carbon isotopes, to build more complex molecules through the process of photosynthesis. When animals eat plants, the carbon-14 atoms that were part of the plant become part of the animal. Because organisms are constantly taking in carbon compounds, they contain the same ratio of carbon-14 to carbon-12 and carbon-13 found in the atmosphere. However, after they die, organisms no longer ingest new carbon compounds, and the carbon-14 they already contain continues to decay. The carbon-14 undergoes beta decay to form nitrogen-14.

$$^{14}_{6}C \rightarrow ^{14}_{7}N + \beta$$

Carbon-14 has a half-life of 5730 years. Because the amount of stable carbon in the dead organism remains constant while the carbon-14 continues to decay, the ratio of unstable carbon-14 to stable carbon-12 and carbon-13 decreases.

■ **Figure 12** Using the radiocarbon dating method on organic materials, such as ash and charcoal found at the Great Pyramid of Giza, scientists estimate the pyramid to be more than 4000 years old.

By measuring this ratio and comparing it to the nearly constant ratio present in the atmosphere, the age of an object can be estimated. For example, if an object's C-14 to (C-12 + C-13) ratio is one-quarter of the ratio measured in the atmosphere, the object is approximately two half-lives, or 11,460 years old. Carbon-14 dating is limited to accurately dating objects up to approximately 45,000 years of age. This method was used to date the Great Pyramid of Giza, shown in **Figure 12.**

**Connection to Earth Science** The decay process of a different radioisotope, uranium-238 to lead-206, is commonly used to date objects such as rocks. Because the half-life of uranium-238 is $4.5 \times 10^9$ years, it can be used to estimate the age of objects that are too old to be dated using carbon-14. By radiochemical dating of meteorites, the age of the solar system has been estimated at $4.6 \times 10^9$ years.

## SECTION 2 REVIEW

Section Self-Check

### Section Summary

- The conversion of an atom of one element to an atom of another by radioactive decay processes is called transmutation.
- Atomic number and mass number are conserved in nuclear reactions.
- A half-life is the time required for half of the atoms in a radioactive sample to decay.
- Radiochemical dating is a technique for determining the age of an object by measuring the amount of certain radioisotopes remaining in the object.

**12.** MAINIDEA **Describe** what happens to unstable nuclei.

**13. Explain** how you can predict whether or not an isotope is likely to be stable if you know its number of neutrons and protons.

**14. Describe** the forces acting on the particles within a nucleus and explain why neutrons are the glue holding the nucleus together.

**15. Predict** the nuclear equation for the alpha decay of radium-226 used on the tips of older lightning rods.

**16. Calculate** how much of a 10.0-g sample of americium-241 remains after four half-lives. Americium-241 is a radioisotope commonly used in smoke detectors and has a half-life of 430 y.

**17. Calculate** After 2.00 y, 1.986 g of a radioisotope remains from a sample that had an original mass of 2.000 g.

**a.** Calculate the half-life.

**b.** How much of the radioisotope remains after 10.00 y?

**18. Graph** A sample of polonium-214 originally has a mass of 1.0 g. Express the mass remaining as a percent of the original sample after a period of one, two, and three half-lives. Graph the percent remaining versus the number of half-lives. Approximately how much time has elapsed when 20% of the original sample remains?

SECTION 3

# Nuclear Reactions

**MAIN IDEA** Fission, the splitting of nuclei, and fusion, the combining of nuclei, release tremendous amounts of energy.

### Essential Questions

- How are mass and energy related?
- How do nuclear fission and nuclear fusion compare and contrast?
- What is the process by which nuclear reactors generate electricity?

### Review Vocabulary

**mass number:** the number after an element's name, representing the sum of its protons and neutrons

### New Vocabulary

induced transmutation
transuranium element
mass defect
nuclear fission
critical mass
breeder reactor
nuclear fusion
thermonuclear reaction

**CHEM 4 YOU** On a hot summer day, you step outside and feel the intense heat of the Sun. Nuclear reactions within the Sun release enough energy to warm Earth and other planets in the solar system for billions of years. It is no surprise, then, that scientists are trying to use this same type of nuclear reaction to produce electricity.

## Induced Transmutation

All nuclear reactions, or transmutations, that have been described thus far are examples of radioactive decay, where one element is converted into another element by the spontaneous emission of radiation. However, transmutations can also be forced, or induced, by bombarding a stable nucleus with a neutron or with high-energy alpha, beta, or gamma radiation. In 1919, Ernest Rutherford performed the first laboratory conversion of one element into another element. By bombarding nitrogen-14 with high-speed alpha particles, oxygen-17 and hydrogen-1 were formed. This transmutation reaction is illustrated in **Figure 13** and the reaction is shown below.

$$^{14}_{7}N + ^{4}_{2}He \rightarrow ^{17}_{8}O + ^{1}_{1}H$$

As Rutherford demonstrated, nuclear reactions can be induced, in other words, produced artificially. The process, which involves striking nuclei with high-velocity particles, is called **induced transmutation.** In the case of charged particles, such as the alpha particles used by Rutherford, the incident particles must be moving at extremely high speeds to overcome the electrostatic repulsion between themselves and the target nucleus. Because of this, scientists have developed methods to accelerate charged particles to extreme speeds by using very strong electrostatic fields and magnetic fields. Particle accelerators are machines built to produce the high-speed particles needed to induce transmutation. Since Rutherford's first experiments involving induced transmutation, scientists have used the technique to synthesize hundreds of new isotopes in the laboratory.

**Figure 13** When an alpha particle bombards a nitrogen-14 atom, an atom of oxygen-17 and an atom of hydrogen-1 are produced.

$^{4}_{2}He$ Bombarding alpha particle | $^{14}_{7}N$ Target nitrogen atom | $^{17}_{8}O$ Oxygen atom | $^{1}_{1}H$ Hydrogen atom

**Transuranium elements** The elements immediately following uranium in the periodic table—elements with atomic numbers 93 and greater—are known as the **transuranium elements.** All transuranium elements have been produced in the laboratory by induced transmutation and are radioactive. Many transuranium elements have been named in honor of their discoverers or the laboratories at which they were created. Scientists continue their ongoing efforts to synthesize new transuranium elements and study their properties.

## EXAMPLE Problem 3

EXAMPLE PROBLEM

**INDUCED TRANSMUTATION REACTION EQUATIONS** Write a balanced nuclear equation for the induced transmutation of oxygen-16 into nitrogen-13 by proton bombardment. An alpha particle is emitted from the nitrogen atom in the reaction.

### 1 ANALYZE THE PROBLEM

You are given all of the particles involved in an induced transmutation reaction. Because the proton bombards the oxygen atom, they are reactants and must appear on the reactant side of the reaction arrow.

**Known**

reactants: oxygen-16 and a proton

products: nitrogen-13 and an $\alpha$-particle

**Unknown**

**nuclear equation for the reactant = ?**

### 2 SOLVE FOR THE UNKNOWN

Nuclear formula for oxygen-16: ${}^{16}_{8}O$ — **Use the periodic table to obtain the atomic number of oxygen.**

Nuclear formula for nitrogen-13: ${}^{13}_{7}N$ — **Use the periodic table to obtain the atomic number of nitrogen.**

Nuclear formula for proton: p

Nuclear formula for alpha particle: ${}^{4}_{2}He$

$${}^{16}_{8}O + p \rightarrow {}^{13}_{7}N + {}^{4}_{2}He$$

**Write the balanced nuclear equation.**

### 3 EVALUATE THE ANSWER

A proton has a charge of 1+ and a mass number of 1. Therefore, both charge and mass number are conserved. The formula for each participant in the reaction is also correct. The nuclear equation is written correctly.

## PRACTICE Problems

PRACTICE PROBLEMS

**19.** Write the balanced nuclear equation for the induced transmutation of aluminum-27 into sodium-24 by neutron bombardment. An alpha particle is released in the reaction.

**20.** Write the balanced nuclear equation for the alpha-particle bombardment of ${}^{239}_{94}Pu$. One of the reaction products is a neutron.

**21. Challenge** Archeologists sometimes use a procedure called neutron activation analysis to identify elements in artifacts. The figure at right shows one type of reaction that can occur when an artifact is bombarded with neutrons. If the product of the process is cadmium-110, what was the target and unstable isotope? Write balanced nuclear equations for the process to support your answer.

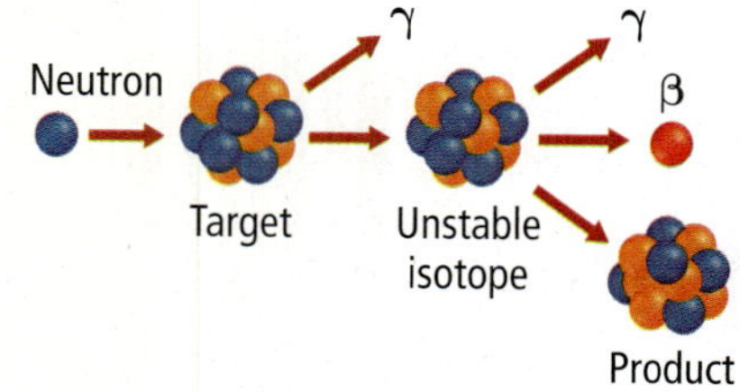

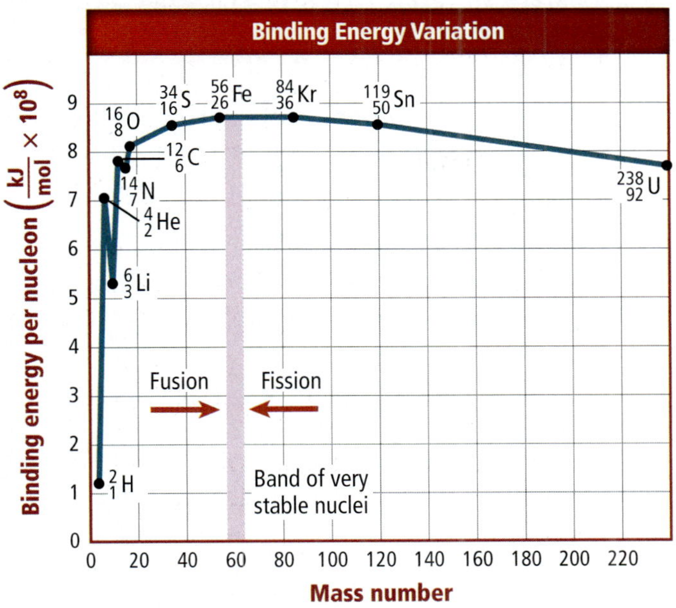


■ **Figure 14** The binding energy per nucleon is a function of the mass number. Light nuclei gain stability by undergoing nuclear fusion. Heavy nuclei gain stability by undergoing nuclear fission.

**GRAPH CHECK**

**Describe** how the binding energy varies as a function of the mass number.

## Nuclear Reactions and Energy

In your study of chemical reactions, you read that mass is conserved. For most practical situations this is true—but, it is not accurate.

**Einstein's equation** Albert Einstein's equation relates mass and energy. It states that any reaction produces or consumes energy due to a loss or gain in mass. Energy and mass are equivalent. Note that because $c^2$ is large, a small change in mass results in a large change in energy.

### Energy Equivalent of Mass

$$\Delta E = \Delta mc^2$$

**$\Delta E$ is the change in energy, in Joules. $\Delta m$ is the change in mass, in kg. $c$ is the speed of light.**

The change in energy is equal to the change in mass times the square of the speed of light.

**Mass defect and binding energy** Scientists have determined that the mass of the nucleus is always less than the sum of the masses of the individual protons and neutrons that comprise it. This difference in mass between a nucleus and its component nucleons is called the **mass defect.**

When nucleons combine together to form an atom, the energy corresponding to the mass defect is released. Conversely, energy is needed to break apart a nucleus into its nucleons. The nuclear binding energy can be defined as the amount of energy needed to break one mole of nuclei into individual nucleons. The larger the binding energy per nucleon, the more strongly the nucleons are held together, and the more stable the nucleus is. Less-stable atoms have lower binding energies per nucleon. In other words, it is harder to break apart a nucleus with a high binding energy than a nucleus with a low binding energy.

**Figure 14** shows the average binding energy per nucleon versus the mass number. Note that the binding energy per nucleon reaches a maximum around a mass number of 60. Elements with a mass number near 60 are the most stable.

## PROBLEM-SOLVING STRATEGIES

### Calculating Mass Defect

You can calculate the mass defect of an isotope if you know the mass of the isotope and the number and masses of its components. Applying the equation $\Delta E = \Delta mc^2$, you can then derive the equivalent binding energy.

Mass defect $= m_{nucleus} - [N_p m_p + N_n m_n]$

where $m_{nucleus}$ is the mass of the nucleus, $m_p$ is the mass of a proton, $m_n$ is the mass of a neutron, $N_p$ is the number of protons, and $N_n$ is the number of neutrons.

If you start with the mass of the atom, you have to take into account the mass of the electrons. To do so, the mass of a hydrogen atom, which is composed of a proton and an electron, is used instead of the mass of a proton. The equation is then:

Mass defect $= m_{isotope} - [N_p m_H - N_n m_n]$

Use the following values for the calculations: $m_H = 1.007825$ amu and $m_n = 1.008665$ amu. The accepted value for $c$ is $3.00 \times 10^8$ m/s.

To calculate the energy in Joules, you can convert the masses into kilograms using 1 amu $= 1.660540 \times 10^{-27}$ kg.

### Apply the Strategy

**Calculate** the mass defect and binding energy of lithium-7. The mass of lithium-7 is 7.016003 amu.

In typical chemical reactions, the energy produced or consumed is so small that the accompanying changes in mass are negligible. In contrast, the mass changes and associated energy changes in nuclear reactions are significant. For example, the energy released from the nuclear reaction of 1 kg of uranium is equivalent to the energy released during the chemical combustion of about four billion kilograms of coal.

## Nuclear Fission

Binding energies in **Figure 14** indicate that heavy nuclei tend to be unstable. To gain stability, they can fragment into several smaller nuclei. Because atoms with mass numbers around 60 are the most stable, heavy atoms (those with mass numbers greater than 60) tend to fragment into smaller atoms in order to increase their stability. The splitting of a nucleus into fragments is known as **nuclear fission.** The fission of a nucleus is accompanied by a very large release of energy.

Nuclear power plants use nuclear fission to generate power. The first nuclear fission reaction discovered involved uranium-235. As you can see in **Figure 15,** when a neutron strikes a uranium-235 nucleus, it undergoes fission. Barium-141 and krypton-92 are just two of the many possible products of this fission reaction. In fact, scientists have identified more than 200 different product isotopes from the fission of a uranium-235 nucleus.

**VOCABULARY**

**ACADEMIC VOCABULARY**

**Generate**
to bring into existence, to originate by a physical or chemical process
*Fire generates a lot of heat.*

☑ READING CHECK **Explain** why heavy atoms undergo nuclear fission.

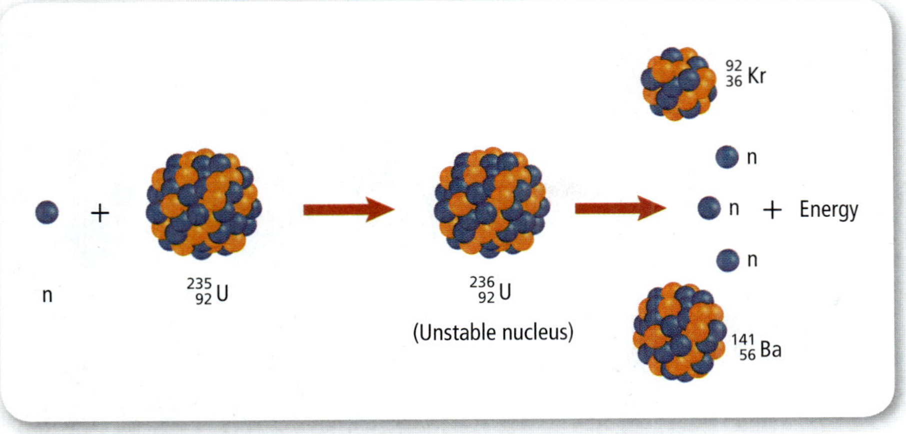


**Figure 15** When bombarded with a neutron, uranium-235 forms unstable uranium-236, which then splits into two lighter nuclei and additional neutrons. The fission of uranium-235 is accompanied by a large release of energy.

**Chain reactions** Each fission of uranium-235 releases additional neutrons, as shown in **Figure 15.** If one fission reaction produces two neutrons, these two neutrons can cause two additional fissions. If those two fissions release four neutrons, those four neutrons could then produce four more fissions, and so on, as shown in **Figure 16.** This self-sustaining process in which one reaction initiates the next is called a chain reaction. As you might imagine, the number of fissions and the amount of energy released can increase rapidly. The explosion from an atomic bomb is an example of an uncontrolled chain reaction.

**Figure 16** When uranium nuclei undergo fission, they release neutrons, which trigger more fission reactions. The ongoing reactions are characteristics of a nuclear chain reaction.

View an **animation about chain reactions.**

Concepts In Motion

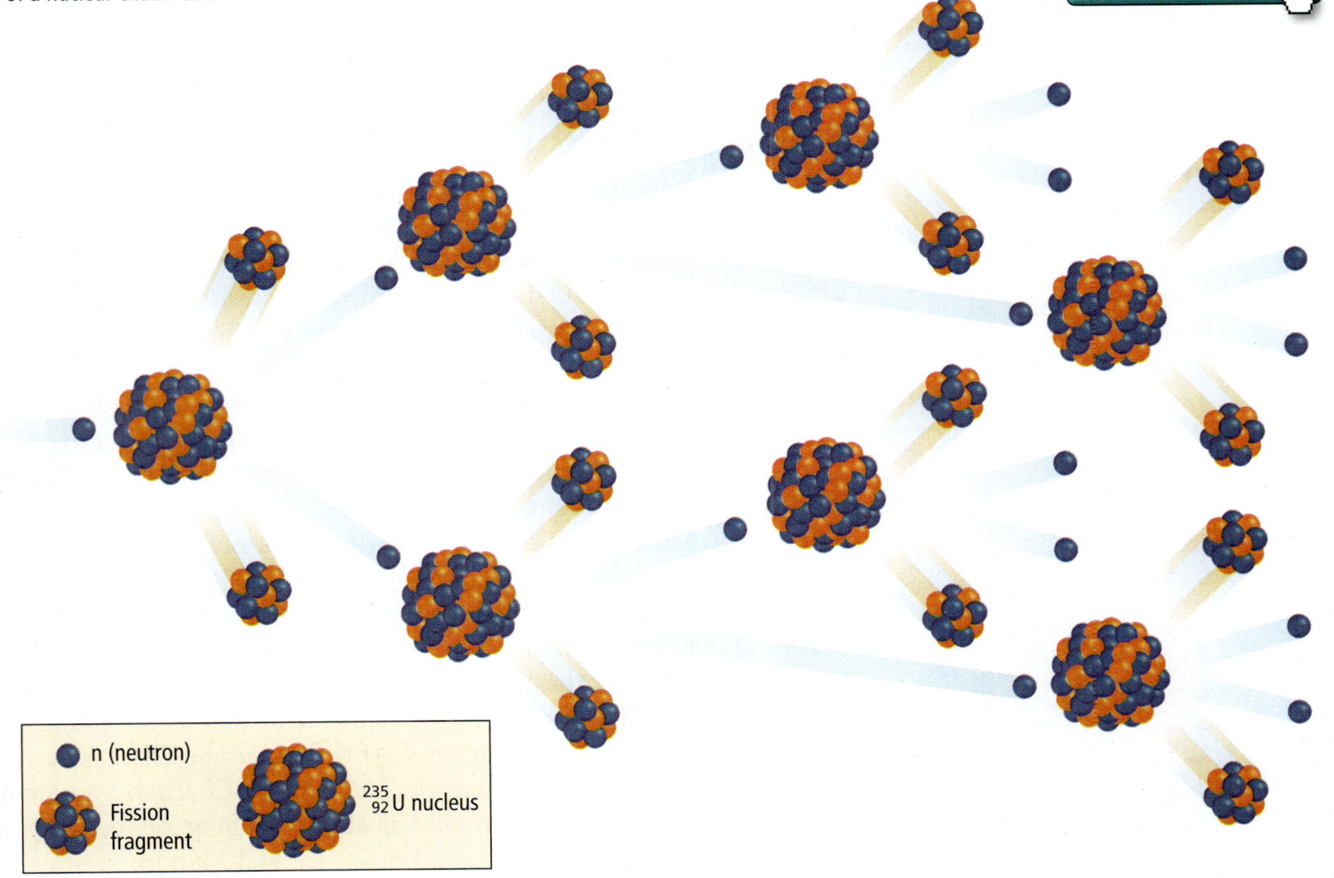

■ **Figure 17** Whether a nuclear reaction can be sustained depends on the amount of matter present. In a subcritical mass, the chain reaction does not start because neutrons escape before causing enough fission to sustain the chain reaction. In a supercritical mass, neutrons cause more and more fissions and the chain reaction accelerates.

View an **animation about critical mass.**

Concepts In Motion

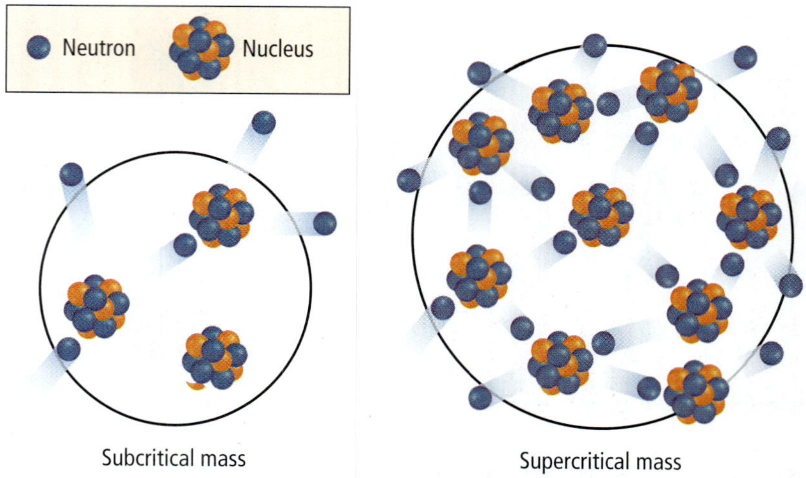


A sample of fissionable material must have sufficient mass in order for a chain reaction to occur. If it does not, neutrons escape from the sample before they can start the chain reaction by striking other nuclei. A sample that is not massive enough to sustain a chain reaction is said to have subcritical mass. A sample that is massive enough to sustain a chain reaction has **critical mass.** When a critical mass is present, the neutrons released in one fission cause other fissions to occur. If much more mass than the critical mass is present, the chain reaction rapidly escalates. This can lead to a violent nuclear explosion. A sample of fissionable material with a mass greater than the critical mass is said to have supercritical mass. **Figure 17** shows the effect of mass on the initiation and progression of a fission reaction.

☑ READING CHECK **Compare** subcritical mass and critical mass.

■ **Figure 18** The main parts of a nuclear power plant are the reactor under the dome and the cooling tower.

vario images GmbH & Co.KG/Alamy

## Nuclear Reactors

Nuclear fission produces the energy generated by nuclear reactors. This energy is primarily used to generate electricity at nuclear power plants, such as the one shown in **Figure 18**. A common fuel is fissionable uranium (IV) oxide ($UO_2$) encased in corrosion-resistant rods. U-238 is the most abundant isotope (99%) of uranium. U-235, which makes up 0.7% of the natural uranium, has the rare property of being able to undergo induced fission; U-235 atoms undergo fission when hit by a neutron. The fuel used in nuclear power plants is enriched to contain 3% uranium-235, the amount required to sustain a chain reaction, and is called enriched uranium. Additional rods, often made of cadmium or boron, control the fission process inside the reactor by absorbing neutrons released during the reaction.

Keeping the chain reaction going while preventing it from racing out of control requires precise monitoring and continual adjusting of the control rods. Much of the concern about nuclear power plants focuses on the risk of losing control of the nuclear reactor, possibly resulting in the accidental release of harmful levels of radiation. The Three Mile Island accident in the United States in 1979 and the Chernobyl accident in Ukraine in 1986 provide examples of why controlling the reactor is critical. **Figure 19** shows the city of Pripyat, located 3 km from Chernobyl. The city was completely abandoned after the accident.

■ **Figure 19** The city of Pripyat was deserted after the accident at the Chernobyl power plant.

The fission within a nuclear reactor is started by a neutron-emitting source and is stopped by positioning the control rods to absorb all of the neutrons produced in the reaction. The reactor core contains a reflector that acts to reflect neutrons back into the core, where they will react with the fuel elements, also called fuel rods. A coolant, usually water, circulates through the reactor core, to carry off the heat generated by the nuclear fission reactions. The hot coolant heats water that is used to power steam-driven turbines, which produce electric power.

Nuclear power plants and fossil-fuel burning power plants are similar; heat from a reaction—nuclear fission or chemical combustion of coal—is used to generate steam. The steam then drives turbines that produce electricity, as shown in the nuclear power plant illustrated in **Figure 20.** The other major components of a nuclear power plant are also illustrated in **Figure 20.**

View an **animation about nuclear power plants.**

Concepts In Motion

■ **Figure 20** A nuclear reactor produces heat that drives the formation of steam. The energy from the steam spins a turbine which produces electricity. The steam is eventually cooled and recycled. The water used to cool the steam enters the cooling tower where steam is released to the atmosphere.

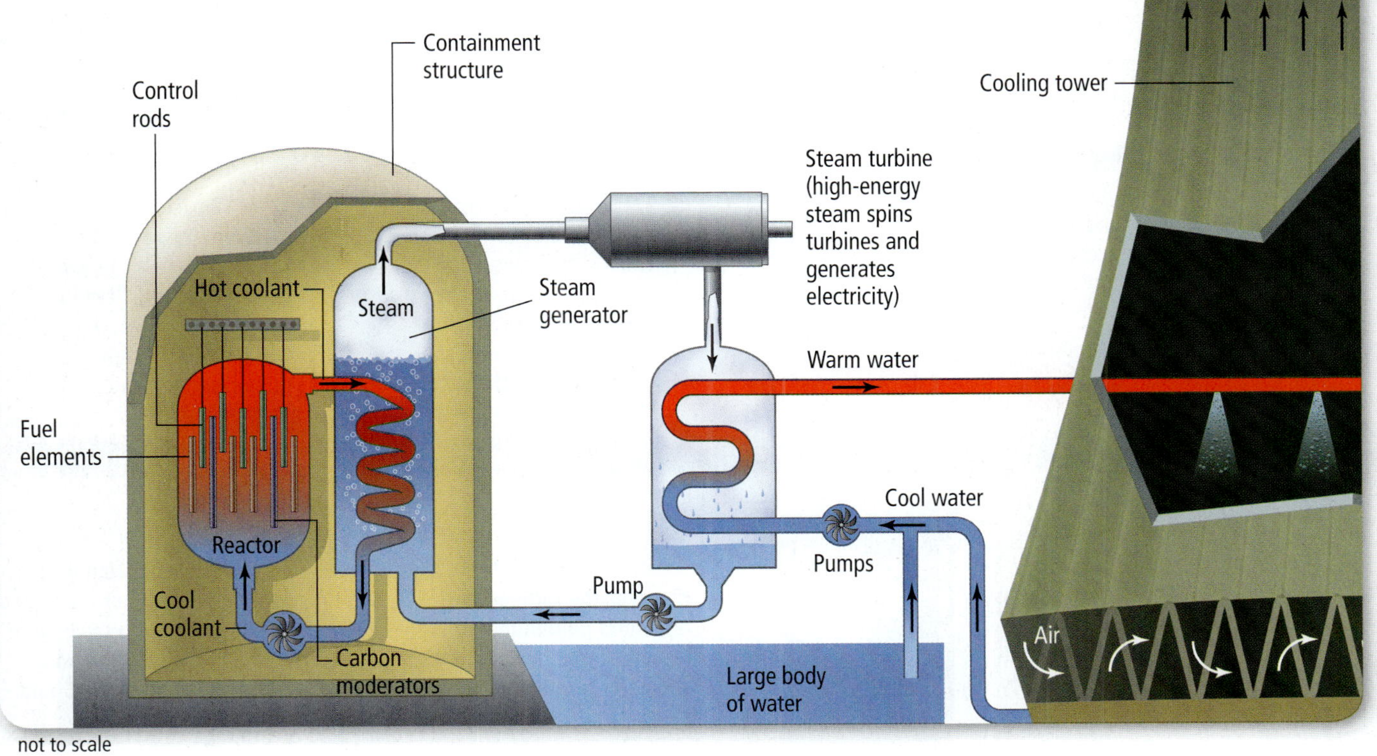


not to scale

■ **Figure 21** The interior of a reactor is filled with water. A crane is used to extract and replace fuel rods.

Because of the hazardous radioactive fuels and fission products present at nuclear power plants, a dense concrete structure is usually built to enclose the reactor. The main purpose of the containment structure is to shield personnel and nearby residents from harmful radiation.

As the reactor operates, the fuel rods are gradually depleted and products from the fission reactions accumulate. Because of this, the reactor must be serviced periodically. Spent fuel rods are extracted from the reactor, as shown in **Figure 21,** and can be reprocessed and repackaged to make new fuel rods. Some fission products, however, are extremely radioactive and cannot be used again. These products must be stored as nuclear waste.

Risks of accidents, such as the ones mentioned in **Figure 22,** have to be taken into account when operating nuclear power plants. However, the storage of highly radioactive nuclear waste is still one of the major issues surrounding the debate over the use of nuclear power. Approximately 20 half-lives are required for the radioactivity of nuclear waste materials to reach levels acceptable for biological exposure. For some types of nuclear fuels, the wastes remain substantially radioactive for thousands of years. A considerable amount of scientific research is devoted to the disposal of radioactive wastes. Highly radioactive materials from the reactor core are first treated with advanced technologies that ensure the materials will not deteriorate over a very long period of time. Treated wastes are then stored in sealed containers that are buried deep underground.

Another issue is the limited supply of the uranium-235 used in the fuel rods. One option is to build reactors that produce new quantities of fissionable fuels. Reactors able to produce more fuel than they use are called **breeder reactors.** Although the design of breeder reactors poses many difficult technical problems, they are currently in operation in several countries.

☑ READING CHECK **Infer** how the storage of nuclear wastes affects the environment.

■ **Figure 22**

## The Nuclear Age

The discovery of X-rays in 1895 initiated a series of breakthroughs in understanding atomic nuclei. Today, nuclear chemistry applications involving medicine, weaponry, and energy affect the lives of people worldwide.

1890 — 1900 — 1920 — 1940

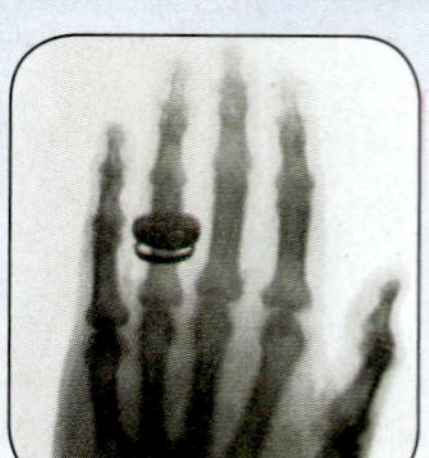

**1895** The first X-ray photographs fuel intense interest among the scientific community.

**1898** Marie and Pierre Curie discover the radioactive elements polonium and radium. Their work establishes the early framework for the study of nuclear chemistry.

**1919** The first artificially induced nuclear reaction causes the transmutation of nitrogen into an isotope of oxygen by bombarding nitrogen gas with alpha particles.

**1934** Enrico Fermi's experiments result in the world's first nuclear fission reaction. Fermi's subsequent research will pioneer nuclear power generation.

**1941–45** Manhattan Project scientists develop uranium and plutonium bombs, which were dropped on Hiroshima and Nagasaki, Japan, in 1945 and ended World War II.

## Nuclear Fusion

Recall from the binding energy diagram in **Figure 14** that a mass number of about 60 has the most stable atomic configuration. Thus, it is possible to bind together two or more light (mass number less than 60) and less-stable nuclei to form a single more-stable nucleus. The combining of atomic nuclei is called **nuclear fusion.** Nuclear fusion reactions, which are responsible for the production of the heaviest elements, are capable of releasing very large amounts of energy. You already have some everyday knowledge of this fact—the Sun is powered by a series of fusion reactions as hydrogen atoms fuse to form helium atoms.

$$4\,^{1}_{1}\text{H} \rightarrow 2\beta + {}^{4}_{2}\text{He} + \text{energy}$$

Scientists have spent several decades researching nuclear fusion. It is a promising source of energy and has several advantages compared to nuclear fission. Lightweight isotopes used to fuel the reactions, such as hydrogen, are abundant. Fusion reaction products are not generally radioactive. Nuclear fusion produces large amounts of energy. Fusion reactions produce more energy per unit of mass of fuel than fission reactions. This could solve the problem of the increasing needs for electricity in the world's societies.

Unfortunately, there are major problems that must be overcome on a commercially viable scale. One such problem is that fusion requires extremely high energies to initiate and sustain a reaction. The required energy, which is achieved only at extremely high temperatures, is needed to overcome the electrostatic repulsion between the nuclei in the reaction. Because of the energy requirements, fusion reactions are also known as **thermonuclear reactions.** A temperature of 5,000,000 K is required to fuse hydrogen atoms. This temperature—and even higher temperatures—have been achieved using an atomic explosion to initiate the fusion process, but this approach is not practical for controlled electric power generation.

### RealWorld CHEMISTRY

### Nuclear Fusion

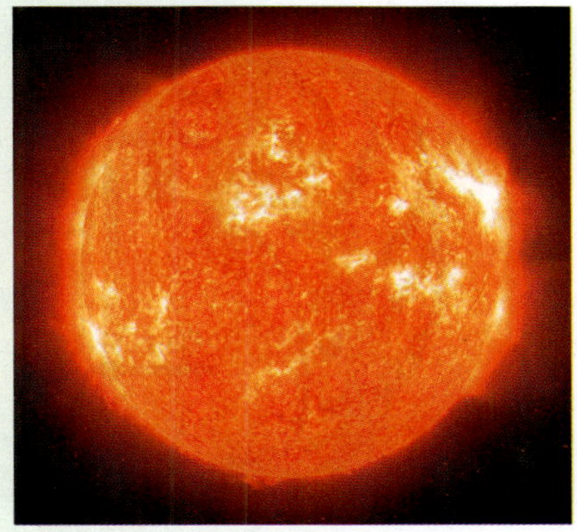

**SOLAR FUSION** Nuclear fusion reactions are responsible for the glow and heat from stars such as the Sun. The temperature of the Sun's core is about 15,000,000 K. It is so hot and dense that hydrogen nuclei fuse to produce helium. After billions of years, the Sun's hydrogen will be mostly depleted. Its temperature will rise to about 100,000,000 K, and the fusion process will then change helium into carbon.

**1949** Radiocarbon dating allows scientists to determine the age of artifacts made from plant-based materials as old as 45,000 years.

**2006** The *Cassini* spacecraft explores the Saturn system. *Cassini* is powered by technology that converts heat from the radioactive decay of plutonium into electricity.

1960 — 1980 — 2000

**1960s** Scientists research using high-energy radiation to treat cancer. Clinical trials bring dramatic improvement in the treatment and cure of malignant tumors.

**1979, 1986** Nuclear power plant accidents at Three Mile Island, Pennsylvania, and Chernobyl, Ukraine, focus world attention on the dangers associated with nuclear power.

**2010** Scientists using NASA's Fermi Gamma-ray Space Telescope discover two massive bubbles, one above and one below the core of the Milky Way. They are thought to be the result of activity from either a black hole or star formation.

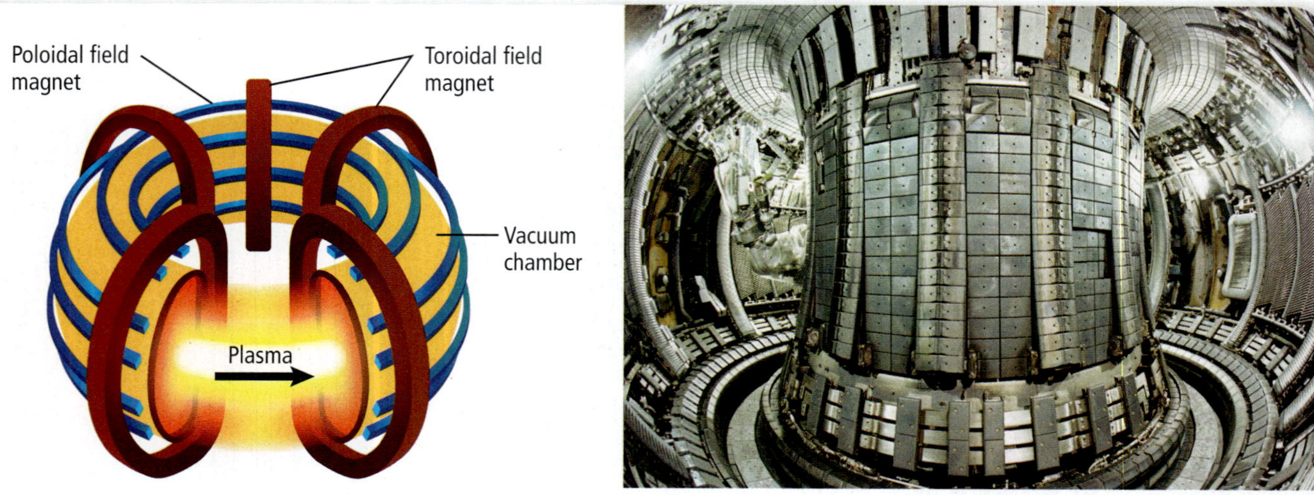


■ **Figure 23** A tokamak reactor, a ring-shaped reactor, uses strong magnetic fields to contain the intensely hot fusion reaction and keep it from direct contact with the reactor interior walls. The poloidal magnets follow the shape of the reactor and the toroidal magnets wrap around the reactor.

Obviously, many problems must be resolved before fusion becomes a practical energy source. Another significant problem is confinement of the reaction. There are currently no materials capable of withstanding the tremendous temperatures that are required by a fusion reaction. Much of the current research centers around an apparatus called a tokamak reactor. The name *tokamak* comes from Russian and means *toroidal chamber with an axial magnetic field.* A tokamak reactor, shown in **Figure 23,** is a donut-shaped device that uses strong magnetic fields to contain the fusion reaction. While significant progress has been made in the field of fusion, temperatures high enough for continuous fusion have not yet been sustained for long periods of time.

## SECTION 3 REVIEW

Section Self-Check

### Section Summary

- Induced transmutation is the bombardment of nuclei with particles in order to create new elements.
- In a chain reaction, one reaction induces others to occur. A sufficient mass of fissionable material is necessary to initiate the chain reaction.
- Fission and fusion reactions release large amounts of energy.

**22.** MAINIDEA **Compare and contrast** nuclear fission and nuclear fusion reactions. Describe the particles that are involved in each type of reaction and the changes they undergo.

**23. Describe** the process that occurs during a nuclear chain reaction and explain how to monitor a chain reaction in a nuclear reactor.

**24. Explain** how nuclear fission can be used to generate electric power.

**25. Formulate** an argument supporting or opposing nuclear power as your state's primary power source. Assume the primary source of power currently is the burning of fossil fuels.

**26. Calculate** What is the energy change ($\Delta E$) associated with a change in mass ($\Delta m$) of 1.00 mg?

**27. Interpret Graphs** Use the graph in **Figure 14** to answer the following questions.

**a.** Why is the isotope $^{56}_{26}Fe$ highest on the curve?

**b.** Are more stable isotopes located higher or lower on the curve?

**c.** Compare the stability of Li-6 and He-4.

SECTION 4

# Applications and Effects of Nuclear Reactions

**MAINIDEA** Nuclear reactions have many useful applications, but they also have harmful biological effects.

### Essential Questions

- What are several methods used to detect and measure radiation?
- How is radiation used in the treatment of disease?
- What are some of the damaging effects of radiation on biological systems?

### Review Vocabulary

**isotope:** an atom of the same element with the same number of protons but different number of neutrons

### New Vocabulary

ionizing radiation
radiotracer

**CHEM 4 YOU** Almost everyone gets cuts or scrapes from time to time. Usually, the first thing you do is clean the injury and cover it with a bandage to keep out germs. One of the many uses of radiation is to sterilize medical bandages.

## Detecting Radioactivity

You read earlier that Becquerel discovered radioactivity because of the effect of radiation on photographic plates. Since this discovery, several other methods have been devised to detect radiation. People who work near radioactive sources, for example, might be required to wear a thermoluminescent dosimeter (TLD) badge, which contains a tiny crystal. Radiation excites electrons within the crystal. To determine the radiation dose, the crystal is heated, and the electrons return to their ground states, emitting light. Radioactivity readers detect this light as a measure of the radiation dose to which a worker has been exposed. Monitoring the radiation dose received by people who work near radioactive sources is important to ensure their safety.

Radiation energetic enough to ionize matter with which it collides is called **ionizing radiation.** The Geiger counter is an ionizing radiation detection device. As shown in **Figure 24,** a Geiger counter consists of a metal tube filled with a gas. In the center of the tube is a wire that is connected to a power supply. When ionizing radiation penetrates the end of the tube, the gas inside the tube absorbs the radiation and forms ions and free electrons. The free electrons are attracted to the wire, causing an electric current. A meter built into the Geiger counter measures the current flow through the ionized gas. This current measurement is used to determine the amount of ionizing radiation present.

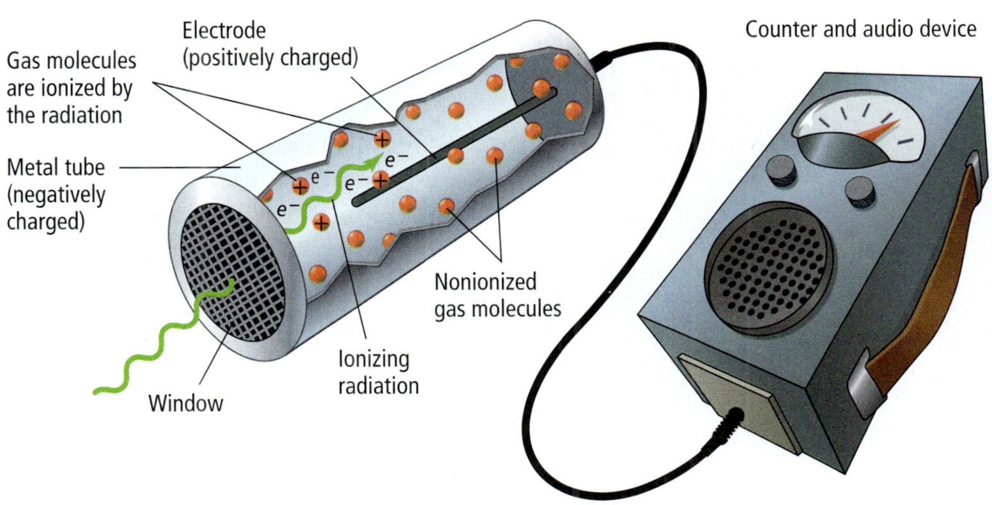


**Figure 24** A Geiger counter is used to detect and measure radiation levels. Ionizing radiation produces an electric current in the counter. The current is displayed on a scaled meter, whereas a speaker produces audible sounds.

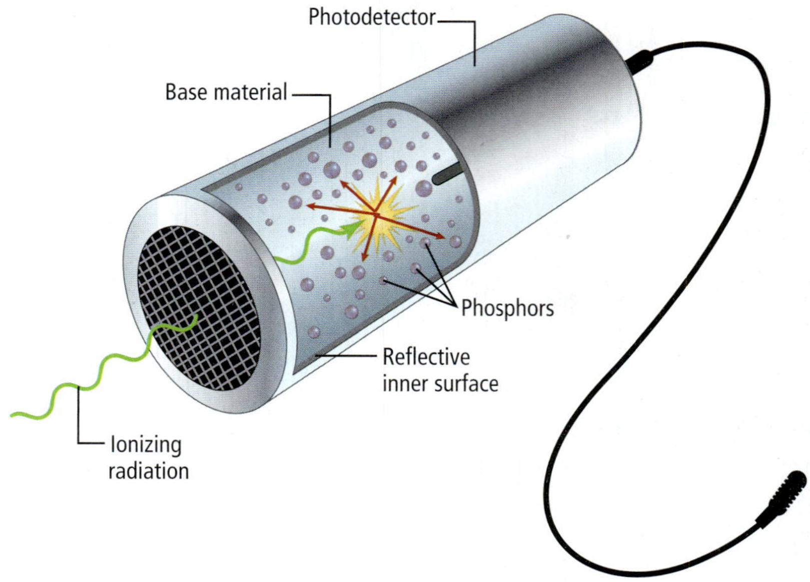


■ **Figure 25** Scintillation counters are used to detect the presence of ionizing radiation. An ionizing radiation excites the electrons in the phosphors. As the electrons return to their ground states, they emit photons, which are then detected by the photodetector.

Another detection device is a scintillation counter. Scintillations are brief flashes of light produced when ionizing radiation excites the electrons in certain types of atoms or molecules called phosphors. A scintillation counter contains a base material—often a plastic, a crystal, or a liquid—containing phosphors, as shown in **Figure 25.** Ionizing radiation that strikes the scintillation counter can transfer energy either directly to the phosphors or to the base material, which then transfers the energy to the phosphors. This energy excites electrons in the phosphors. As these electrons return to their ground states, they release energy in the form of light. This light is transmitted through the base material to a photodetector that convert the light to an electrical signal. The number and brightness of the scintillations give a measure of the amount of ionizing radiation.

☑ READING CHECK **Summarize** how a scintillation detector works.

## Uses of Radiation

With proper safety procedures, radiation can be useful in many scientific experiments and industrial applications. For instance, neutron activation analysis is used to detect trace amounts of elements present in a sample. Computer-chip manufacturers use this technique to analyze the composition of highly purified silicon wafers. In the process, the sample is bombarded with a beam of neutrons from a radioactive source, causing some of the atoms in the sample to become radioactive. The type and amount of radiation emitted by the sample is used to determine the types and quantities of elements present. Neutron activation analysis is a highly sensitive measurement technique capable of detecting quantities of less than $1 \times 10^{-9}$ atoms in a sample. Beta emission is another application of radiation. It is used to measure paper thickness, as shown in **Figure 26.**

■ **Figure 26** Gauges such as the one pictured use beta emission from krypton, promethium, or strontium. The radioactive source is placed on one side of the paper, and a detector is on the other side. Most beta particles are absorbed by the paper, but the percentage that are able to travel through to the detector indicates the thickness of the paper.

Martin Bond/Science Photo Library/Photo Researchers

**Using radioisotopes** Radioisotopes can also be used to follow the course of an element through a chemical reaction. For example, $CO_2$ gas containing radioactive carbon-14 isotopes has been used to study glucose formation in photosynthesis.

$$6CO_2 + 6H_2O \xrightarrow{\text{sunlight}} C_6H_{12}O_6 + 6O_2$$

Because the $CO_2$ containing carbon-14 is used to trace the progress of carbon through the reaction, it is referred to as a radiotracer. A **radiotracer** is a radioisotope that emits non-ionizing radiation and is used to signal the presence of an element or specific substance. The fact that all of an element's isotopes have the same chemical properties makes the use of radioisotopes possible. Thus, replacing a stable atom of an element in a reaction with one of its isotopes does not alter the reaction. Radiotracers are important in a number of areas of chemical research, particularly in analyzing the reaction mechanisms of complex, multistep reactions.

Radiotracers also have important uses in medicine. Iodine-131, for example, is commonly used to detect diseases associated with the thyroid gland. If a problem is suspected, the patient will drink a solution containing a small amount of iodine-131. After the iodine is absorbed, the amount of iodine taken up by the thyroid is measured and used to monitor the functioning of the thyroid gland.

READING CHECK **Define** *radiotracer.*

**Treating cancer** Radiation can pose serious health problems for humans because it can damage or destroy healthy cells. However, radiation can also destroy unhealthy cells, such as cancer cells. All cancers are characterized by the rapid growth of abnormal cells. This growth can produce masses of abnormal tissue, called malignant tumors. Radiation therapy is used to treat cancer by destroying the cancer cells. In fact, cancer cells are more susceptible to destruction by radiation than healthy ones. **Figure 27** shows a brain scan with a malignant tumor. After radiation treatment, the baseline returns to normal. Unfortunately, in the process of destroying unhealthy cells, radiation also destroys some healthy cells. Despite this major drawback, radiation therapy has become one of the most effective treatment options in the fight against cancer.

## CAREERS IN CHEMISTRY

**Radiation Therapist** Under the supervision of a physician, a radiation therapist administers radiation treatment to patients. Radiation therapists work closely with patients and must be compassionate and supportive. Training programs prepare radiation therapists to use particle accelerators and other forms of technology. Knowledge of radiation hazards is an important part of this job.

WebQuest

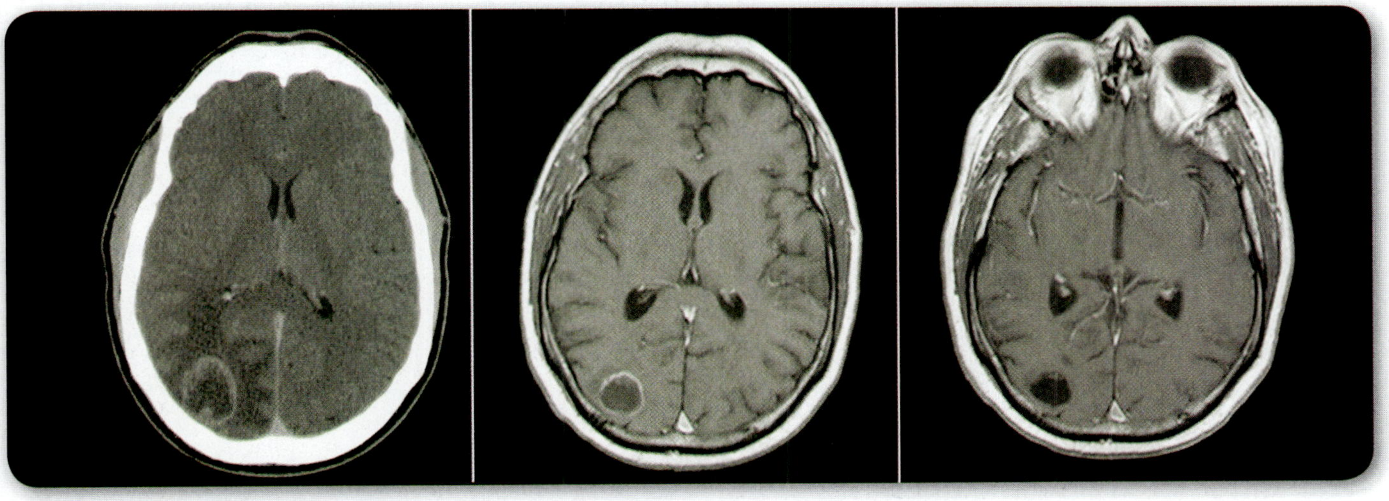

**Figure 27** Radiation can be used to treat cancer. MRI images taken before treatment and after 4 and 10 months of treatment show the decrease in the size of the tumor.

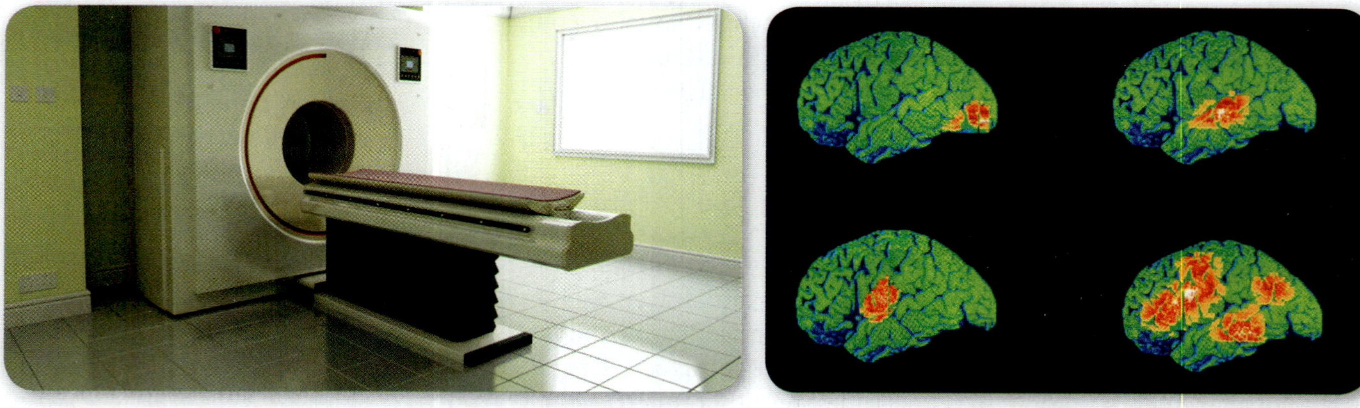

■ **Figure 28** Gamma rays emitted by the radiotracers absorbed by the patients are measured with this detector. The image on the right shows different areas of the brain emitting gamma rays. These images might help doctors locate a tumor or observe a brain function.

**Using positron emission** Another radiation-based medical diagnostic tool is called positron emission transaxial tomography (PET). In this procedure, a radiotracer that decays by positron emission is injected into the patient's bloodstream. Positrons emitted by the radiotracer cause gamma-ray emissions that are then detected by an array of sensors surrounding the patient, as shown in **Figure 28.** PET scans can be used to diagnose diseases or study the parts of the brain that are activated under given circumstances, also shown in **Figure 28.**

## Biological Effects of Radiation

Although radiation has a number of medical and scientific applications, it can be very harmful. The damage produced from ionizing radiation absorbed by the body depends on several factors, such as the type of radiation, its energy, the type of tissue absorbing the radiation, the penetrating power, and the distance from the source. **Figure 29** shows an example of such damage.

**Connection to Biology** High-energy ionizing radiation is dangerous because it can fragment and ionize molecules within biological tissue. A free radical is an atom or molecule that contains one or more unpaired electrons and is one example of the highly reactive products of ionizing radiation. In a biological system, free radicals can affect a large number of other molecules and ultimately disrupt the operation of normal cells. Ionizing radiation damage to living systems can be classified as either somatic or genetic. Somatic damage affects only nonreproductive body tissue. It includes burns and cancer caused by damage to the cell's growth mechanism. Genetic damage can affect offspring by damaging reproductive tissue. Such damage is difficult to study because it might not become apparent for several generations.

■ **Figure 29** Radiation can disrupt cell processes and damage skin.

**Infer** *Is the lesion pictured here somatic or genetic?*

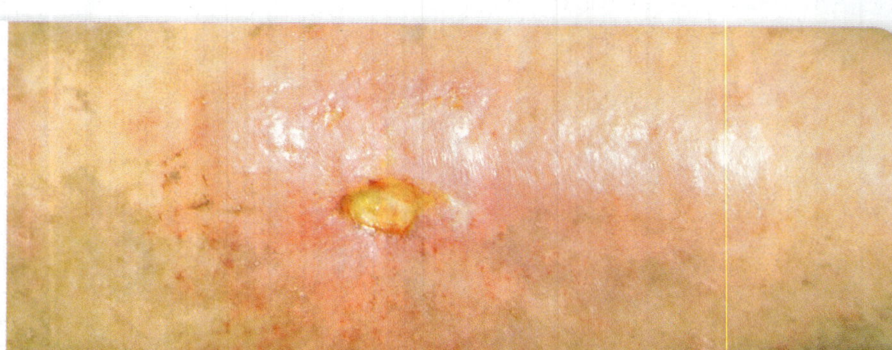

**Dose of radiation** A dose of radiation refers to the amount of radiation a body absorbs from a radioactive source. Two units, the rad and the rem, are commonly used to measure doses. The rad, which stands for radiation-absorbed dose, is a measure of the amount of radiation that results in the absorption of 0.01 J of energy per kilogram of tissue. The dose in rads, however, does not account for the energy of the radiation, the type of living tissue absorbing the radiation, or the time of the exposure. To account for these factors, the dose in rads is multiplied by a numerical factor that is related to the radiation's effect on the tissue involved. The result of this multiplication is a unit called the rem. The rem, which stands for roentgen equivalent for man, is named after Wilhelm Roentgen, who discovered X-rays in 1895. **Table 6** summarizes the short-term effects of radiation on humans, depending on the dose.

A variety of sources constantly bombard your body with radiation. Your exposure to these sources results in an average annual radiation exposure of 100–300 millirems of high-energy radiation or 0.1–0.3 rems. **Table 7** shows your annual exposure to common radiation sources.

**Intensity and distance** The intensity of radiation depends on the distance from the source as shown by the equation below. The farther away the source, the lower the intensity. The intensity of radiation is measured in amount of radiation per unit of time and/or surface, such as mrem/s•m$^2$.

**Radiation Intensity and Distance**

$$I_1 d_1^2 = I_2 d_2^2$$

$d_1$ and $d_2$ are two distances from the source. $I_1$ is the intensity at $d_1$, and $I_2$ is the intensity at $d_2$.

The intensity of a radiation at a distance $d_1$ from the source multiplied by the square of the distance equals the intensity of the radiation at a distance $d_2$ multiplied by the square of the distance.

**Table 7 Average Annual Radiation Exposure**

| Source | Average Exposure (mrem/y) |
|---|---|
| Cosmic radiation | 20–50 |
| Radiation from the ground | 25–175 |
| Radiation from buildings | 10–160 |
| Radiation from air | 20–260 |
| Human body (internal) | ~20 |
| Medical and dental X-rays | 50–75 |
| Nuclear weapon testing | <1 |
| Air travel | 5 |
| Total average | 100–300 |

**Table 6 Effects of Short-term Radiation Exposure**

| Dose (rem) | Effects on Humans |
|---|---|
| 0–25 | no detectable effects |
| 25–50 | temporary decrease in white-blood-cell population |
| 100–200 | nausea, substantial decrease in white-blood-cell population |
| 500 | 50% chance of death within 30 days of exposure |

Watch a **video about radioisotopes.**

Video

# Problem-Solving LAB

## Interpret Graphs

**How does distance affect radiation exposure?** When one of the reactors at the Chernobyl nuclear power plant exploded, the immediate vicinity of the power plant was highly contaminated and declared a dead zone. The radiation spread over thousands of kilometers. However, the intensity of the radiation decreased with the distance from the reactor.

### Analysis

The graph to the right shows the intensity of a radioactive source versus the distance from the source. Note how the intensity of the radiation varies with the distance from the source. The unit of radiation intensity is millirems per second per square meter. This is the amount of radiation striking a square meter of area each second.

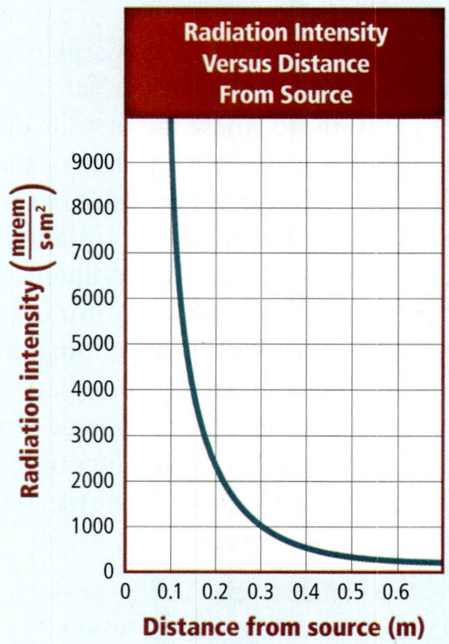


### Think Critically

1. **Evaluate** How does the radiation exposure change as the distance doubles from 0.1 m to 0.2 m? How does it change as the distance quadruples from 0.1 m to 0.4 m?
2. **Formulate** in words the mathematical relationship described in your answer to Question 1.
3. **Interpret Graphs** Determine the distance from the source at which the radiation decreased to 0.69 mrem/s•m$^2$. This intensity is the maximum radiation exposure intensity considered safe. (*Hint: Use the equation* $I_1/I_2 = d_2^2/d_1^2$.)

## SECTION 4 REVIEW

Section Self-Check

### Section Summary

- Different types of counters are used to detect and measure radiation.
- Radiotracers are used to diagnose disease and to analyze chemical reactions.
- Short-term and long-term radiation exposure can cause damage to living cells.

28. **MAIN**IDEA **Explain** one way in which nuclear chemistry is used to diagnose or treat disease.
29. **Describe** several methods used to detect and measure radiation.
30. **Compare and contrast** somatic and genetic biological damage.
31. **Explain** why it is safe to use radioisotopes to diagnose medical problems.
32. **Calculate** A lab worker receives an average radiation dose of 21 mrem each month. Her allowed dose is 5,000 mrem/y. On average, what fraction of her yearly dose does she receive?
33. **Interpret Data** Look at the data in **Table 7.** Suppose someone is exposed to the maximum values listed for average annual radiation from the ground, from buildings, and from the air. What fraction would the person receive of the minimum short-term dose (25 rem) that causes a temporary decrease in white blood cell population?

# CHEMISTRY&careers

## Career: Archaeologist

### Neutron Activation Analysis

In the Andes mountain range, more than 500 years ago, a young girl was sacrificed to appease the gods. As was the custom of the ancient Incas, pottery and other artifacts were buried with her. Neutron activation analysis performed on pottery such as the vessels in **Figure 1** allowed archaeologists to determine the origin of the soil from which the pottery was made.

**Figure 1** Neutron activation analysis lets scientists compare soil and pottery to determine where the pottery was made.

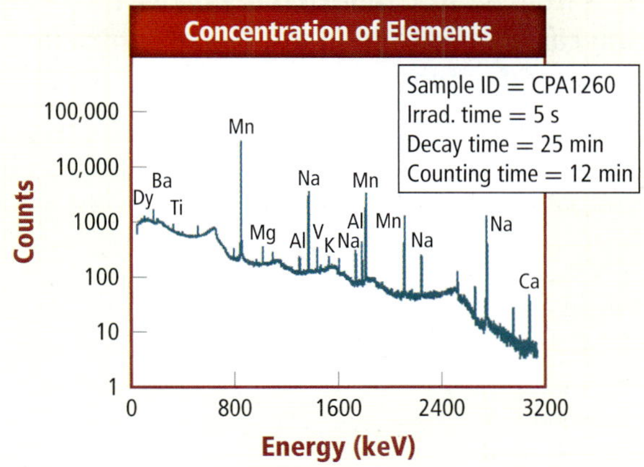


**Figure 2** A gamma-ray spectrum indicates the concentration of different elements in a sample.

**Detecting elements** Neutron activation analysis is a method of detecting elements in a material. A small sample of the material is first exposed to a strong neutron source. Neutron bombardment produces radioisotopes in about three-fourths of the elements. When the radio-isotopes decay, they emit gamma rays with energies that are characteristic of the element.

A gamma detector is used to measure the sample's radiation output. Gamma rays of different energies produce peaks at different places on graphs, such as the one in **Figure 2.** Each peak corresponds to a specific element. Some elements have more than one peak because they emit gamma rays of different energies. The height of the peak, or the area under the peak, indicates the concentration of the element in the sample.

This method can be used to search for just one element or many elements in a sample. The process can detect extremely low concentrations of elements, as low as parts per billion.

**Advantages** Most forms of chemical analysis require vaporization, dissolution, or alteration of the analyzed sample in some way. Neutron activation analysis is a nondestructive process that can be used to study liquid, solid, or gaseous samples. Sensitive items, such as forensic evidence, meteorites, or artifacts, can be analyzed without harm.

**Uses** Analyzing the composition of artifacts such as pottery allowed scientists to establish the origin of the clay used to make the objects that were buried with the young, sacrificed girl. Astonishingly, the clay did not come from local soil but from the Incan capital and other religious centers. Representatives from the Incan Empire traveled to remote places, bringing pottery and other artifacts with them, to perform rituals.

### WRITING IN ▶ Chemistry

**Analyze** Look at the graph in **Figure 2.** Write an explanation about how a technician could use the graph to determine the elements present in the irradiated sample. Is the height of the peaks important? Which element is found in the greatest concentration in the sample? What are the approximate energies of gamma rays emitted by this element?

WebQuest

# ChemLAB

iLab Station

## Investigate Radiation Dosage

**Background:** **Radiation is a term that causes fearful responses in people. However, not all radiation is dangerous. We are surrounded by radiation from space and from natural radioactivity on Earth. Radiation can also be used in a safe and controlled way for medical purposes.**

**Question:** *What methods are effective in minimizing exposure to radiation?*

### Materials

alpha source
beta source
gamma source
Geiger counter
piece of cardboard
piece of plastic
meterstick
clock

### Safety Precautions

WARNING: ***Radioactive sources can be harmful. Wash hands and arms thoroughly before handling objects which go to the mouth, nose, or eyes. Do not eat or drink in laboratories where radioactive sources are used. Do not handle radioactive sources if you have a break in the skin below the wrist. Do not use—and immediately report to your teacher—any sealed disc containing a radioactive source which is damaged.***

### Procedure

1. Read and complete the lab safety form.
2. Using what you know about types of radiation, write a hypothesis about how pieces of cardboard and plastic will affect the radiation dose.
3. Create a table to record your data.
4. Place the meterstick on the lab station with the Geiger counter at the zero-end.
5. Place the alpha source at the 10-cm mark, and record the highest reading on the Geiger counter.
6. Repeat the measurement with the source at 20 cm and 30 cm.
7. Repeat Steps 5 and 6 with the beta source and gamma source.
8. Place the alpha source on the 10-cm mark, and place a heavy piece of cardboard between the source and the Geiger counter.
9. Measure and record the highest reading.
10. Place the source on the 30-cm mark, and place the piece of cardboard on the 10-cm mark. Measure and record the radiation.
11. Place the piece of cardboard on the 20-cm mark, and repeat the measurement.
12. Place the piece of plastic between the source, and counter and record the highest reading.
13. Repeat Steps 8–12 with the beta source and the gamma source.
14. **Cleanup and Disposal** Return all lab equipment and radiation sources to the designated location. Remember to wash your hands with soap and water after completing the lab.

### Analyze and Conclude

1. **Summarize** How does distance affect the amount of radiation from a source?
2. **Compare and Contrast** Does the experimental data support your hypothesis?
3. **Explain** Based on the data, explain why you were required to wear goggles and a lab apron in this lab.
4. **Recognize Cause and Effect** Which radiation source was least affected by the cardboard and plastic shields? Explain why this source is different from the other two sources.
5. **Infer** Did the position of the piece of cardboard influence the results? Explain why or why not.
6. **Observe and Infer** What can you say about the penetrating power of X-rays based on the fact that you have to wear a lead shield at the dentist to protect your body from the radiation?

### INQUIRY EXTENSION

**Research** Find references that list and quantify the exposure to radiation that we receive in everyday life. Calculate your average annual exposure, and describe methods that could reduce this dosage.

# CHAPTER 24 STUDY GUIDE

Vocabulary Practice

**BIGIDEA** Nuclear chemistry has a vast range of applications, from the production of electricity to the diagnosis and treatment of diseases.

## SECTION 1 Nuclear Radiation

**MAINIDEA** Under certain conditions, some nuclei can emit alpha, beta, or gamma radiation.

- Wilhelm Roentgen discovered X-rays in 1895.
- Henri Becquerel, Marie Curie, and Pierre Curie pioneered the fields of radioactivity and nuclear chemistry.
- Radioisotopes emit radiation to attain more-stable atomic configurations.

**VOCABULARY**
- radioisotope
- X-ray
- penetrating power

## SECTION 2 Radioactive Decay

**MAINIDEA** Unstable nuclei can break apart spontaneously, changing the identity of atoms.

- The conversion of an atom of one element to an atom of another by radioactive decay processes is called transmutation.
- Atomic number and mass number are conserved in nuclear reactions.
- A half-life is the time required for half of the atoms in a radioactive sample to decay.

$$N = N_0\left(\frac{1}{2}\right)^n \text{ or } N = N_0\left(\frac{1}{2}\right)^{t/T}$$

- Radiochemical dating is a technique for determining the age of an object by measuring the amount of certain radioisotopes remaining in the object.

**VOCABULARY**
- transmutation
- nucleon
- strong nuclear force
- band of stability
- positron emission
- positron
- electron capture
- radioactive decay series
- half-life
- radiochemical dating

## SECTION 3 Nuclear Reactions

**MAINIDEA** Fission, the splitting of nuclei, and fusion, the combining of nuclei, release tremendous amounts of energy.

- Induced transmutation is the bombardment of nuclei with particles in order to create new elements.
- In a chain reaction, one reaction induces others to occur. A sufficient mass of fissionable material is necessary to initiate the chain reaction.
- Fission and fusion reactions release large amounts of energy.

$$E = mc^2$$

**VOCABULARY**
- induced transmutation
- transuranium element
- mass defect
- nuclear fission
- critical mass
- breeder reactor
- nuclear fusion
- thermonuclear reaction

## SECTION 4 Applications and Effects of Nuclear Reactions

**MAINIDEA** Nuclear reactions have many useful applications, but they also have harmful biological effects.

- Different types of counters are used to detect and measure radiation.
- Radiotracers are used to diagnose disease and to analyze chemical reactions.
- Short-term and long-term radiation exposure can cause damage to living cells.

$$I_1 d_1^2 = I_2 d_2^2$$

**VOCABULARY**
- ionizing radiation
- radiotracer

# CHAPTER 24 ASSESSMENT

## SECTION 1

### Mastering Concepts

**34.** Compare and contrast chemical reactions and nuclear reactions in terms of energy changes and the particles involved.

**35.** Match each numbered choice on the right with the correct radiation type on the left.

| | |
|---|---|
| **a.** alpha | **1.** high-speed electrons |
| **b.** beta | **2.** 2+ charge, blocked easily |
| **c.** gamma | **3.** no charge, electromagnetic radiation |

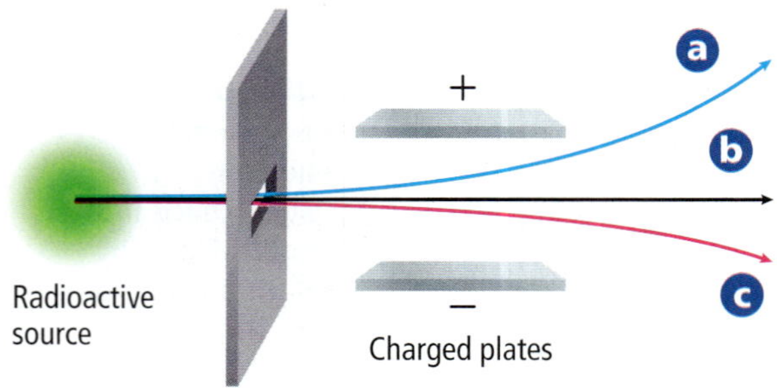


■ **Figure 30**

**36.** **Figure 30** shows alpha particles, beta particles, and gamma rays passing through a screen and between two charged plates. What can you infer about the identity of **a, b,** and **c?** Explain your answer.

**37.** What is the difference between X-rays and gamma rays?

### Mastering Problems

**38. Dental crown** Uranium-234 is used to make dental crowns appear brighter. The alpha decay of uranium-234 produces what isotope?

**39. Detecting Material Flaws** Flaws in welded metal parts of airplanes can be identified by placing the isotope iridium-192 on one side of the weld and photographic film on the other side to detect gamma rays that pass through. How does the gamma ray emission affect the atomic number and mass number of the iridium?

**40. Colored Glass** Thorium-230 can be used to provide coloring in glass objects. One method of producing thorium-230 is through the radioactive decay of actinium-230. Is this an example of alpha decay or beta decay? How do you know?

**41. Plastic Bags** Thin sheets of plastic are used to make items such as grocery bags. The sheets move under a source of promethium-147, emitting beta particles. The radiation intensity, measured under the plastic sheets, is used to monitor the thickness of the plastic. During this process, promethium changes into which element?

## SECTION 2

### Mastering Concepts

**42.** What is the strong nuclear force? On which particles does it act?

**43.** Explain the difference between positron emission and electron capture.

**44.** Categorize each type of radioactive decay.
- **a.** Mass number and atomic number are unchanged.
- **b.** Mass number remains the same and atomic number decreases.

**45.** What is the significance of the band of stability?

**46.** What is a radioactive decay series? When does it end?

**47. Radioisotopes** What are the factors that determine the amount of a given radioisotope in nature?

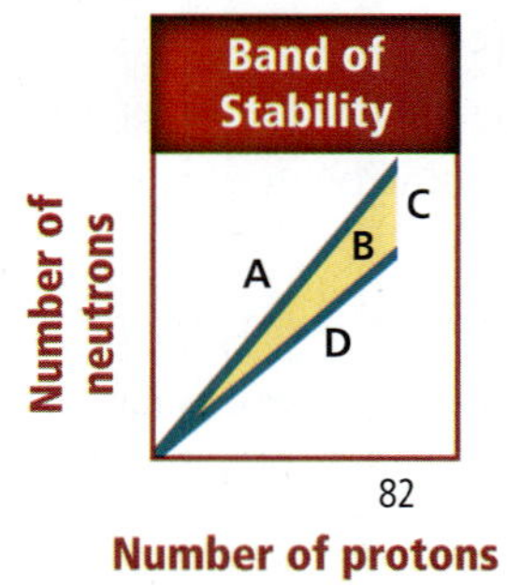


■ **Figure 31**

**48.** In which region(s) in **Figure 31** are you likely to find
- **a.** stable nuclei?
- **b.** nuclei that undergo alpha decay?
- **c.** nuclei that undergo beta decay?
- **d.** nuclei that undergo positron emission?

**49. Carbon-14 Dating** Carbon-14 dating makes use of a specific ratio of two different radioisotopes. Define the ratio used in carbon-14 dating. Why is this ratio constant in living organisms?

### Mastering Problems

**50.** Calculate the neutron-to-proton ratio for each atom.

| | |
|---|---|
| **a.** tin-134 | **c.** carbon-12 |
| **b.** silver-107 | **d.** carbon-14 |

**51.** Complete the following equations.

**a.** ${}^{214}_{83}Bi \rightarrow {}^{4}_{2}He + ?$ **b.** ${}^{239}_{93}Np \rightarrow {}^{239}_{94}Pu + ?$

**52.** Write a balanced nuclear equation for the alpha decay of americium-241.

**53.** Write a balanced nuclear equation for the beta decay of cesium-137.

**54. Bone Formation** The electron capture of strontium-85 can be used by physicians to study bone formation. Write a balanced nuclear equation for this reaction.

55. **Nuclear Safety** The half-life of tritium ($^{3}_{1}H$) is 12.3 y. If 48.0 mg of tritium is released from a nuclear power plant during the course of a mishap, what mass of the nuclide will remain after 49.2 y? After 98.4 y?

56. **Static Charge** Static charge can interfere with the production of plastic products by attracting dust and dirt. To reduce it, manufacturers expose the area to polonium-210, which has a half-life of 138 days. How much of a 25.0-g sample will remain after one year (365 days)?

57. The half-life of polonium-218 is 3.0 min. If you start with 20.0 g, how long will it be before only 1.0 g remains?

58. An unknown radioisotope exhibits 8540 decays per second. After 350.0 min, the number of decays has decreased to 1250 per second. What is the half-life?

## SECTION 3

### Mastering Concepts

59. Define *transmutation.* Are all nuclear reactions also transmutation reactions? Explain.

60. Relate binding energy per nucleon to mass number.

61. Referring to **Figure 7,** would you expect $^{39}_{20}Ca$ to be radioactive? Explain.

62. What is a chain reaction? Give an example of a nuclear chain reaction.

63. Explain the purpose of control rods in a nuclear reactor.

64. Why is the fuel of a nuclear reactor enriched?

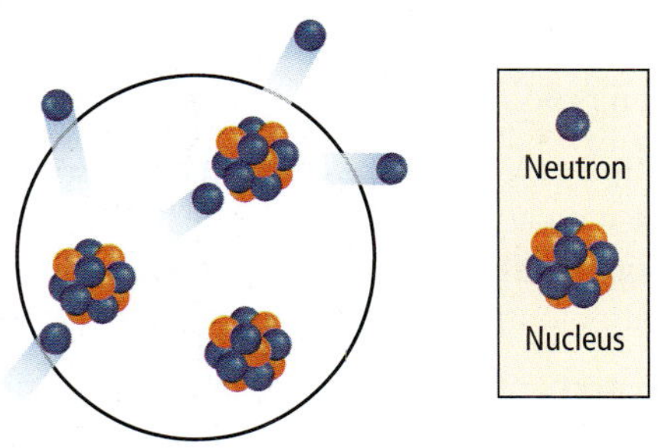


■ **Figure 32**

65. Describe what is meant by the terms *critical mass, subcritical mass,* and *supercritical mass.* Which is shown in **Figure 32?** How can you tell?

66. Explain how it is possible that fission (the splitting of nuclei) and fusion (the combining of nuclei) both release tremendous amounts of energy.

67. Describe the current limitations of fusion as a power source.

68. Why does nuclear fusion require so much heat? How is heat contained within a tokamak reactor?

### Mastering Problems

69. **Smoke Detectors** Americium-241, a radioisotope used in smoke detectors, is produced by bombarding plutonium-238 with neutrons to produce plutonium-240, which is bombarded with neutrons to produce plutonium-241. The plutonium-241 decays to americium-241. Write the balanced nuclear equations for each reaction.

70. **Exit Signs** Exit signs are coated with a paint containing phosphors. These phosphors are activated by the radioisotope tritium ($^{3}_{1}H$), produced by bombarding lithium-6 with neutrons to produce lithium-7. The lithium-7 then undergoes alpha decay to produce the tritium. Write balanced nuclear equations for both steps.

71. **Control Rods** Bombarding uranium-235 with neutrons produces samarium-149, which is used in nuclear reactor control rods. What other element is produced?

72. **The Sun** $^{1}_{1}H + ^{2}_{1}H \rightarrow ^{3}_{2}He + \gamma$ is one of the fusion reactions in the Sun. The mass of $^{1}_{1}H$ is 1.007825 amu, the mass of $^{2}_{1}H$ is 2.014102 amu, and the mass of $^{3}_{2}He$ is 3.016029 amu.
   **a.** What is the mass defect of $^{3}_{2}He$?
   **b.** What energy is released by the process?

## SECTION 4

### Mastering Concepts

73. What property of isotopes allows radiotracers to be useful in studying chemical reactions?

74. Which unit of radiation dose, rem or rad, is most useful for describing the effect of radiation on living tissue?

75. **PET Scans** In PET scans, the radiotracer emits positrons, which travel a few millimeters before interacting with electrons. How can the original radiotracer be detected?

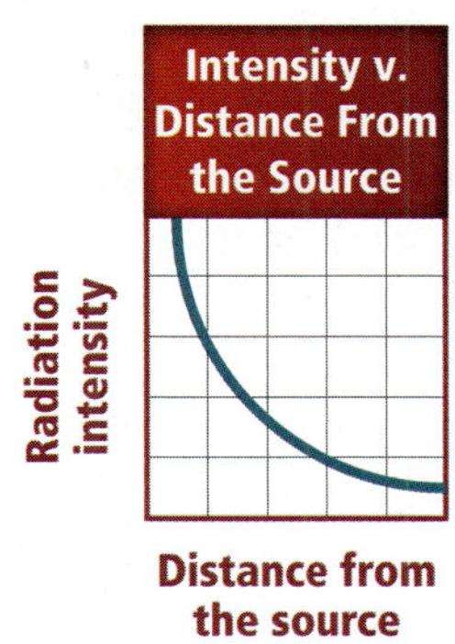


■ **Figure 33**

76. **Figure 33** shows a simplified graph of radiation intensity versus distance from the source. Explain this graph and what it implies about a method of reducing the effects of radiation exposure.

## Mastering Problems

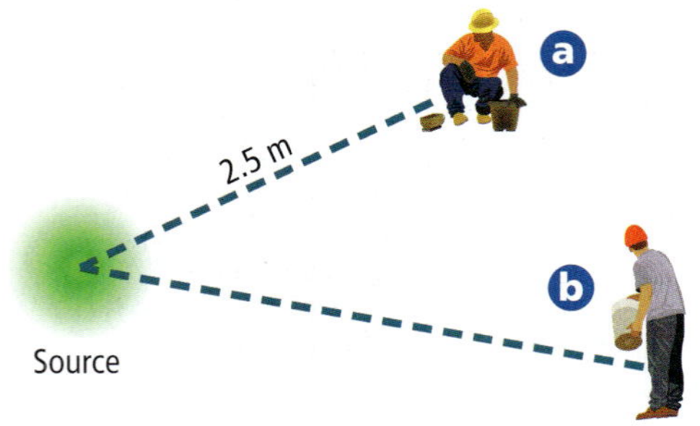


■ **Figure 34**

**77.** **Figure 34** shows the position of two workers near a radioactive gamma source. The worker at Position A is 2.5 m from the source and receives an exposure of 0.98 mrem/s•m$^2$. The worker at Position B receives an exposure of 0.50 mrem/s•m$^2$. What is the distance of the worker at Position B from the source?

**78.** A worker stands near a machine that uses a cobalt-60 gamma source to sterilize medical equipment. The worker's dose 2.0 m from the source is 0.85 mrem/s•m$^2$. What is the worker's dose at a distance of 3.5 m?

**79.** **Safe Exposure** The intensity of a radioactive source is 1.15 mrem/s•m$^2$ at a distance of 0.50 m. What is the minimum distance a person could be from the source to have a maximum exposure of 0.65 mrem/s•m$^2$?

## MIXED REVIEW

**80.** Technetium-104 has a half-life of 18.0 min. How much of a 165.0 g sample remains after 90.0 minutes have passed?

**81.** A bromine-80 nucleus can decay by gamma emission, positron emission, or electron capture. What is the product nucleus in each case?

**82.** The half-life of plutonium-239 is 24,000 y. How much nuclear waste generated today will remain in 1000 years?

**83.** **Red Blood Cells** A medical researcher is using a chromium-51 source to study red blood cells. The gamma-emission intensity at a distance of 1.0 m is 0.75 mrem/s•m$^2$. At what distance would the intensity drop to 0.15 mrem/s•m$^2$?

**84.** The binding energy per nucleon reaches a maximum around what mass number? Explain how this number is related to the fission and fusion processes.

**85.** You have an alpha source, a beta source, and a gamma source. Design a plan to use a Geiger counter, paper, and foil to determine the identity of each source.

**86.** What is the half-life of radon-222 if a sample initially contains 150 mg and only 18.7 mg after 11.4 days?

**87.** **Sheet Metal** A company plans to monitor the thickness of sheet metal during production. What would you recommend the company do to determine a safe distance for workers from the gamma source?

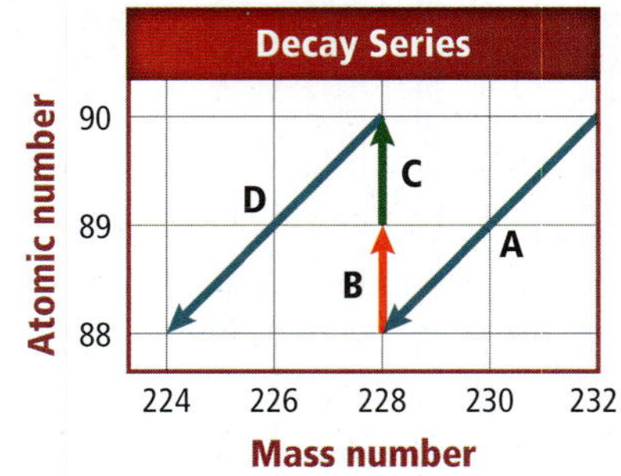


■ **Figure 35**

**88.** **Figure 35** shows part of the decay series of a radioisotope. For each segment on the graph, tell whether alpha decay or beta decay occurs, and identify the change in atomic number and mass number.

## THINK CRITICALLY

**89.** **Make and Use Graphs** Thorium-231 decays to lead-207 by emitting the following particles in successive steps: $\beta$, $\alpha$, $\alpha$, $\beta$, $\alpha$, $\alpha$, $\alpha$, $\beta$, $\beta$, $\alpha$. Plot each step of the decay series on a graph of mass number versus atomic number. Label each plotted point with the symbol of the radioisotope.

**90.** **Apply** Chemical treatment is often used to destroy harmful chemicals. For example, bases neutralize acids. Why can't chemical treatment be applied to destroy the fission products produced in a nuclear reactor?

**91.** **Compare** A biological concern about working around some radioactive materials is the radioactive dust a person might inhale. Compare the effect of alpha radiation outside the body and inside the body.

**92.** **Interpret** Small radioactive sources are often used for laboratory experiments. The radioactive substance is enclosed in a metal container with a small window. A gamma source might be covered with a stainless steel window. What would you expect the window of an alpha source to be like? Why?

**93.** **Analyze** Some radioisotopes used for medical imaging have half-lives as short as several hours. Why is a short half-life beneficial? Why is it a problem?

**94.** **Infer** The production of electricity at nuclear fission reactor facilities is controversial. Think about the benefits and dangers of this technology. Explain your opinion about whether nuclear reactors should be used.

Chapter Self-Check

## CHALLENGE PROBLEM

**95.** Use the information in **Table 8** to calculate the mass defect and binding energy of deuterium ($^{2}_{1}H$), a hydrogen isotope involved in fusion reactions in the Sun.

**Table 8 Mass of Particles**

| Particle | Mass (amu) |
|---|---|
| Hydrogen | 1.007941 |
| Deuterium | 2.014102 |
| Neutron | 1.008665 |

**a.** Find the mass of the nucleons.
**b.** Find the mass defect by subtracting the mass of the nucleons from the mass of the deuterium.
**c.** Find the binding energy using the conversion 1 amu = 931.49 MeV.

## CUMULATIVE REVIEW

**96.** Identify each property as chemical or physical.
**a.** The element mercury has a high density.
**b.** Solid carbon dioxide sublimes at room temperature.
**c.** Zinc oxidizes when exposed to air.
**d.** Sucrose is a white crystalline solid.

**97.** Why does the second period of the periodic table contain eight elements?

**98.** Draw each molecule and show the locations of hydrogen bonds between the molecules.
**a.** two water molecules
**b.** two ammonia molecules
**c.** one water molecule and one ammonia molecule

**99.** What process takes place in each situation?
**a.** a solid air-freshener cube getting smaller and smaller
**b.** dewdrops forming on leaves in the morning
**c.** steam rising from a hot spring
**d.** a crust of ice forming on top of a pond

**100.** If the volume of a sample of chlorine gas is 4.5 L at 0.65 atm and 321 K, what volume will the gas occupy at STP?

**101.** The temperature of 756 g of water in a calorimeter increases from 23.2°C to 37.6°C. How much heat was given off by the reaction in the calorimeter?

**102.** Explain what a buffer is and why buffers are found in body fluids.

**103.** Explain how the structure of benzene can be used to explain its unusually high stability compared to other unsaturated cyclic hydrocarbons.

## WRITING IN ▶ Chemistry

**104. Marie Curie and Irene Curie Joliot** Research and report on the lives of Marie Curie and her daughter, Irene Curie Joliot. What kind of scientific training did each receive? What was it like to be a female chemist in their time? What discoveries did each make?

**105. Nuclear Waste** Evaluate environmental issues associated with nuclear wastes. Research the Yucca Mountain nuclear waste disposal plan, the Hanford nuclear site, or a local nuclear facility. Prepare a poster or multimedia presentation on your findings.

**106. Radioactive Sources** Students in your school might not realize how beneficial radioactive sources can be. Create a poster showing some common, beneficial uses of radioactive sources. Be sure to point out safeguards that are taken to ensure the sources are safe.

## DBQ Document-Based Questions

**Half-Lives** *The National Institute of Standards and Technology (NIST) maintains a database of radionuclide half-lives. In 1992, researchers at NIST measured the half-lives shown in* **Table 9.**

Data obtained from: Unterweger, M.P., Hoppes, D.D., and Schima, F.J. 1992. New and revised half-life measurements results, *Nucl. Instrum. Meth. Phys. Res.* A312:349-352.

**Table 9 Half-Lives**

| Radionuclide | Half-life |
|---|---|
| Fluorine-18 | 1.82951 h |
| Molybdenum-99 | 65.9239 h |
| Samarium-153 | 46.2853 h |

**107.** Fluorine-18 is used in medical imaging. If a lab has a sample containing 15 g of fluorine-18, how much fluorine-18 will remain in the sample after 8.0 h?

**108.** Technetium-99 can be used for diagnostic tests of the heart and lungs. Because of technetium-99's very short half-life, medical facilities produce it from molybdenum-99. If the facility has a 25-g sample of molybdenum-99, how much will it have one week (168 h) later?

**109.** Samarium-153 is used in the production of a drug to treat pain from bone tumors. Radiation released by the samarium hinders the tumor growth, thereby reducing pain. How much of a 1.0 g sample of samarium-153 is left after 4 days (96 h)?

## MULTIPLE CHOICE

**1.** Geologists use the decay of potassium-40 in volcanic rocks to determine their ages. Potassium-40 has a half-life of $1.26 \times 10^9$ years, so it can be used to date very old rocks. If a sample of rock $3.15 \times 10^8$ years old contains $2.73 \times 10^{-7}$ g of potassium-40 today, how much potassium-40 was originally present in the rock?

**A.** $1.71 \times 10^{-8}$ g
**B.** $2.30 \times 10^{-7}$ g
**C.** $3.25 \times 10^{-7}$ g
**D.** $4.37 \times 10^{-6}$ g

**2.** In the early 1930s, van de Graaf generators were used to generate neutrons by bombarding stable beryllium atoms with deuterons ($^{2}_{1}H$), the nuclei of deuterium atoms. A neutron is released in the reaction. Which is the balanced nuclear equation describing this induced transmutation?

**A.** $^{9}_{4}Be + ^{2}_{1}H \rightarrow ^{10}_{5}B + n$

**B.** $^{6}_{4}Be + ^{2}_{1}H \rightarrow ^{8}_{5}B + n$

**C.** $^{9}_{4}Be \rightarrow ^{10}_{5}B + ^{2}_{1}H + n$

**D.** $^{9}_{4}Be + ^{2}_{1}H \rightarrow ^{11}_{5} + n$

*Use the figure below to answer question 3.*

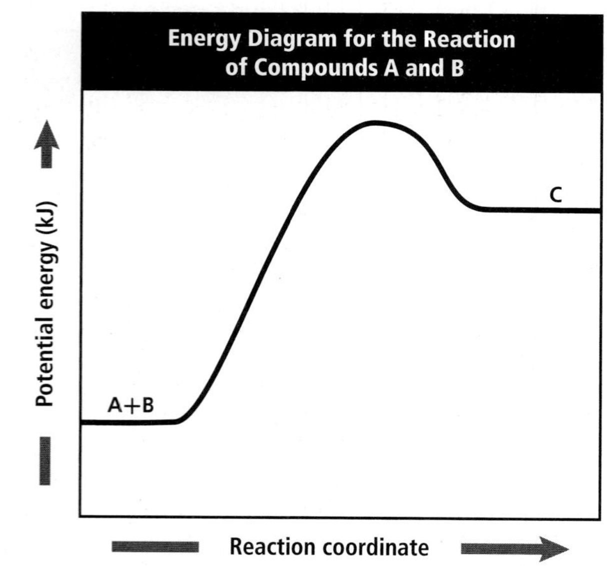


**3.** Which is NOT a correct description of this reaction?

**A.** This is a synthesis reaction.
**B.** This reaction releases energy.
**C.** This reaction is endothermic.
**D.** This reaction will occur spontaneously.

**4.** Which statement is NOT true of alpha particles?

**A.** They carry a charge of 2+.
**B.** They are represented by the symbol $^{4}_{2}He$.
**C.** They are more penetrating than β particles.
**D.** They have the same composition as helium nuclei.

*Use the graph below to answer questions 5 and 6.*

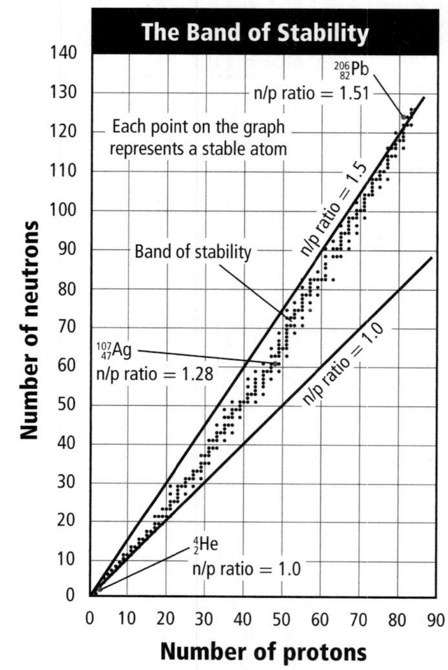


**5.** Why will calcium-35 undergo positron emission?

**A.** It lies above the line of stability.
**B.** It lies below the line of stability.
**C.** It has a high neutron-to-proton ratio.
**D.** It has an overabundance of neutrons.

**6.** Based on its position relative to the band of stability, which process will $^{70}_{30}Zn$ undergo?

**A.** beta decay
**B.** electron capture
**C.** nuclear fusion
**D.** positron emission nuclear fusion

**7.** A solution of 0.600*M* HCl is used to titrate 15.00 mL of KOH solution. The end point of the titration is reached after the addition of 27.13 mL of HCl. What is the concentration of the KOH solution?

**A.** 9.00*M*
**B.** 1.09*M*
**C.** 0.332*M*
**D.** 0.0163*M*

Online Test Practice

## SHORT ANSWER

*Use the figure below to answer questions 8 to 10.*

8. Identify the anode and cathode of this apparatus.

9. Write the oxidation half-reaction.

10. Explain the function of the salt bridge in this apparatus.

11. Predict the products of this reaction.
$Al(NO_3)_3(aq) + CaSO_4(aq) \rightarrow$

## EXTENDED RESPONSE

*Use the figure below to answer Questions 12 and 13.*

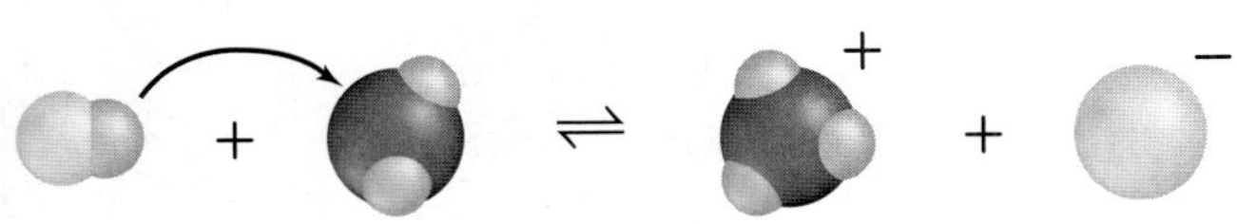


12. Identify the acid and the base for the forward reaction. Explain how you can tell.

13. Explain how you can identify the conjugate acid and conjugate base for the forward reaction. What are they?

## SAT SUBJECT TEST: CHEMISTRY

*Use the figure below to answer Questions 14 and 15.*

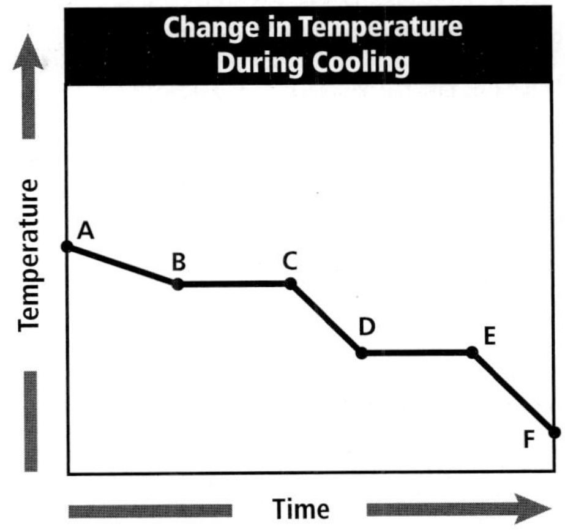


14. During which segments are particles changing states of matter?
**A.** AB, CD, EF
**B.** AB, EF
**C.** BC, CD, DE
**D.** BC, EF
**E.** BC, DE

15. During which segments are particles losing kinetic energy?
**A.** BC, DE
**B.** AB, DE
**C.** AB, CD, EF
**D.** BC, DE, EF
**E.** AB, CD, DE

16. In the first steps of its radioactive decay series, thorium-232 decays to radium-228, which then decays to actinium-228. What are the balanced nuclear equations describing these first two decay steps?
**A.** ${}^{232}_{90}Th \rightarrow {}^{228}_{88}Ra + e^-$, ${}^{228}_{88}Ra \rightarrow {}^{228}_{89}Ac + e^+$
**B.** ${}^{232}_{90}Th \rightarrow {}^{228}_{88}Ra + {}^{4}_{2}He$, ${}^{228}_{88}Ra \rightarrow {}^{228}_{89}Ac + e^-$
**C.** ${}^{232}_{90}Th \rightarrow {}^{228}_{88}Ra + e^+$, ${}^{228}_{88}Ra \rightarrow {}^{228}_{89}Ac + e^-$
**D.** ${}^{232}_{90}Th \rightarrow {}^{228}_{88}Ra + {}^{4}_{2}He$, ${}^{228}_{88}Ra + e^- \rightarrow {}^{228}_{89}Ac$
**E.** ${}^{232}_{90}Th + e^- \rightarrow {}^{228}_{88}Ra$, ${}^{228}_{88}Ra \rightarrow {}^{228}_{89}Ac + e^-$

**NEED EXTRA HELP?**

| If You Missed Question . . . | 1 | 2 | 3 | 4 | 5 | 6 | 7 | 8 | 9 | 10 | 11 | 12 | 13 | 14 | 15 | 16 |
|---|---|---|---|---|---|---|---|---|---|---|---|---|---|---|---|---|
| Review Section . . . | 24.2 | 24.3 | 15.5 | 24.1 | 24.2 | 24.2 | 18.4 | 20.1 | 20.1 | 20.1 | 9.2 | 18.1 | 18.1 | 12.4 | 12.4 | 24.2 |

# STUDENT RESOURCES

## Elements Handbook 901


Hydrogen . . . . . 904
Group 1: Alkali Metals . . . . . 906
Group 2: Alkaline Earth Metals . . . . . 910
Groups 3–12: Transition Elements . . . . . 916
Group 13: Boron Group . . . . . 922
Group 14: Carbon Group . . . . . 926
Group 15: Nitrogen Group . . . . . 932
Group 16: Oxygen Group . . . . . 936
Group 17: Halogen Group . . . . . 940
Group 18: Noble Gases . . . . . 944


## Math Handbook 946


Scientific Notation . . . . . 946
Operations with Scientific Notation . . . . . 948
Square and Cube Roots . . . . . 949
Significant Figures . . . . . 949
Solving Algebraic Equations . . . . . 954
Dimensional Analysis . . . . . 956
Unit Conversion . . . . . 957
Drawing Line Graphs . . . . . 959
Using Line Graphs . . . . . 961
Ratios, Fractions, and Percents . . . . . 964
Operations Involving Fractions . . . . . 965
Logarithms and Antilogarithms . . . . . 966


## Reference Tables 968


**R-1** Color Key . . . . . 968
**R-2** Symbols and Abbreviations . . . . . 968
**R-3** Solubility Product Constants . . . . . 969
**R-4** Physical Constants . . . . . 969
**R-5** Names and Charges of Polyatomic Ions . . . . . 970
**R-6** Ionization Constants . . . . . 970
**R-7** Properties of Elements . . . . . 971
**R-8** Solubility Guidelines . . . . . 974
**R-9** Specific Heat Values . . . . . 975
**R-10** Molal Freezing Point Depression and Boiling Point Elevation Constants . . . . . 975
**R-11** Heat of Formation Values . . . . . 975
**R-12** Safety Symbols . . . . . 976


## Supplemental Practice Problems 977

## Selected Solutions 993

## Glossary • Glosario 1006

## Index 1032

# ELEMENTS HANDBOOK

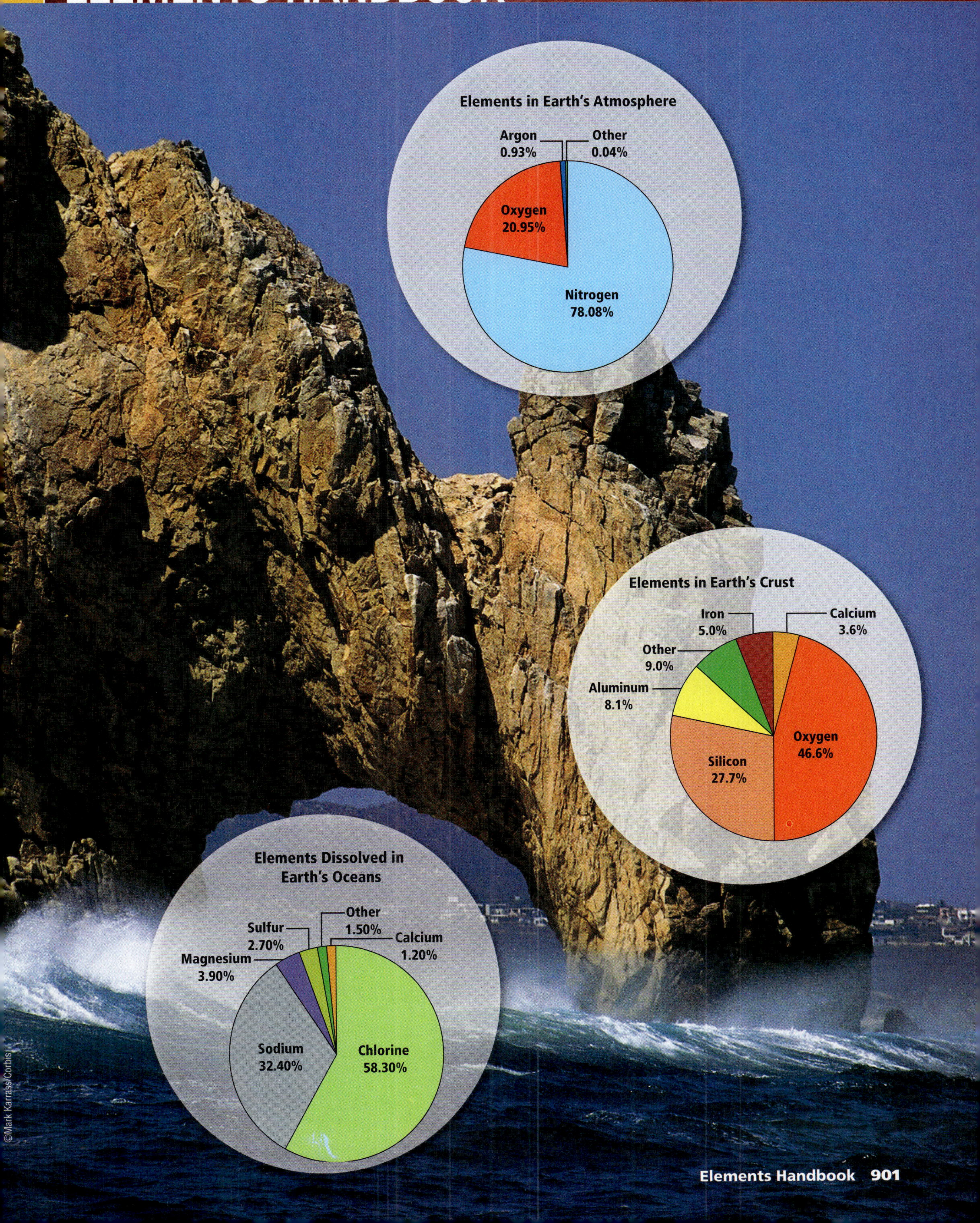

Elements Handbook

# Table of Contents

**How This Handbook Is Organized** *The Elements Handbook is divided into 10 sections: hydrogen and groups 1, 2, 3–12, 13, 14, 15, 16, 17, and 18. You will discover physical and atomic properties, common reactions, analytical tests, and real-world applications of the elements in each section. Questions at the end of each section will assess your understanding of the elements.*


Hydrogen .......... 904
Group 1: Alkali Metals .......... 906
Group 2: Alkaline Earth Metals .......... 910
Groups 3–12: Transition Elements .......... 916
Group 13: Boron Group .......... 922
Group 14: Carbon Group .......... 926
Group 15: Nitrogen Group .......... 932
Group 16: Oxygen Group .......... 936
Group 17: Halogen Group .......... 940
Group 18: Noble Gases .......... 944


# How to Use Element Boxes

*Each element box on the periodic table contains useful information. In the Elements Handbook, each element box has an element name, symbol, atomic number, and electron configuration. At the beginning of each section, each element box also identifies the state of matter at 25°C and 1 atm. A typical box from the handbook is shown below.*

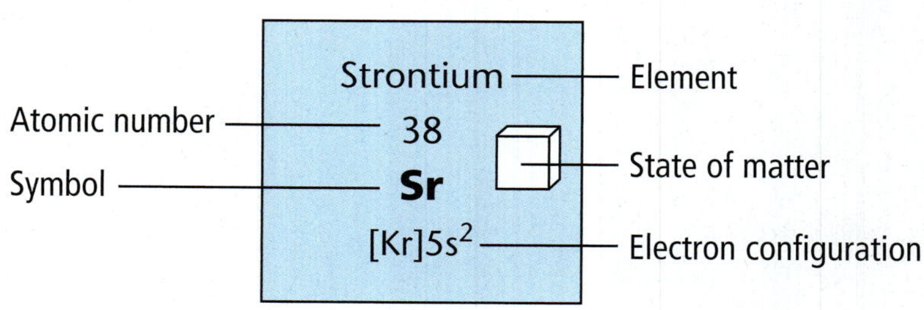


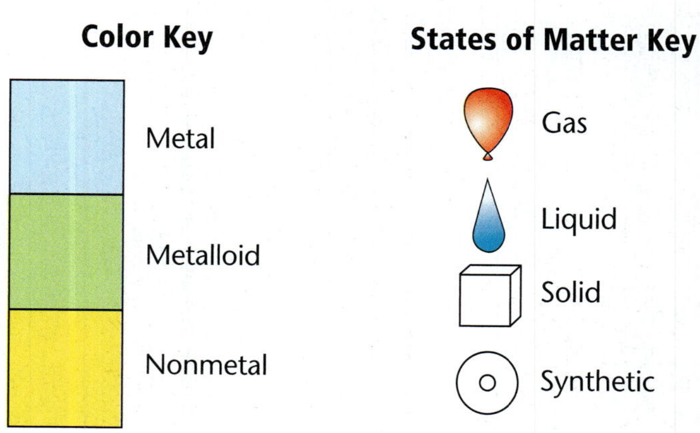


View **animations about the elements.**

Concepts In Motion

*When you read the Elements Handbook, you need to read for information. Here are some tools that the Elements Handbook has to help you find that information.*

See how a group fits in the **Periodic Table.**

Discover the **Physical Properties** and **Atomic Properties** of the elements in a group.

Summarize **Common Reactions** for the elements within a group.

Identify elements by **Analytical Tests.**

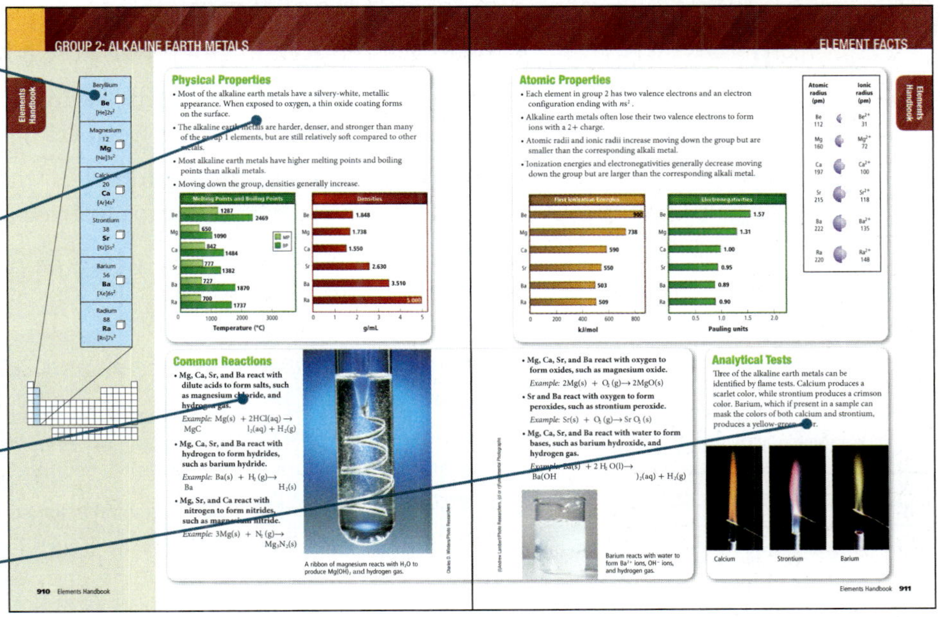


Source: Elements Handbook, p. 910–911

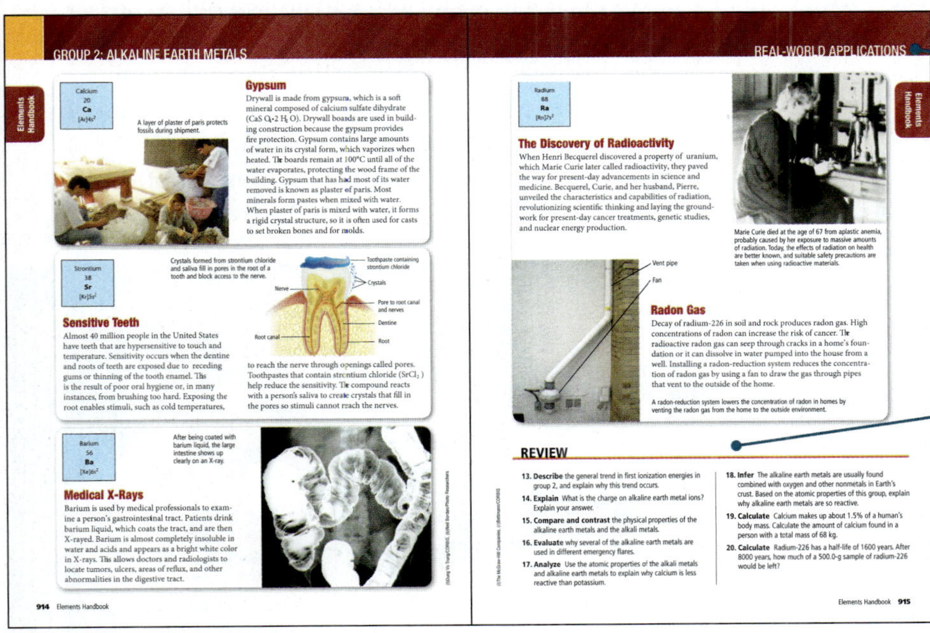


Source: Elements Handbook, p. 914–915

Learn how elements are used every day in **Real-World Applications.**

Test your knowledge of the elements by answering **Assessment** questions.

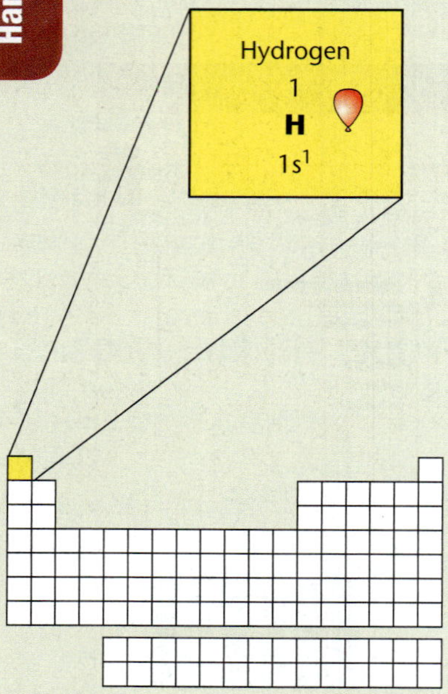


## Physical and Atomic Properties

- At constant temperature and pressure, hydrogen gas ($H_2$) has the lowest density of any gas.
- At very high pressures, such as the interior of planet Jupiter, hydrogen might exist as a solid metal.
- Hydrogen is placed in group 1 because it has one valence electron.
- Hydrogen shares some properties with the group 1 metals. It can lose an electron to form a hydrogen ion ($H^+$).
- Hydrogen also shares some properties with the group 17 nonmetals. It can gain an electron to form a hydride ion ($H^-$).
- There are three common hydrogen isotopes. Protium, the most common isotope, has one proton, one electron, and no neutrons. Deuterium, also called heavy hydrogen, has one proton, one neutron, and one electron. Tritium, which is radioactive, has one proton, two neutrons, and one electron.

| Physical and Atomic Properties of Hydrogen | |
|---|---|
| Melting point | −259°C |
| Boiling point | −253°C |
| Density | $8.24 \times 10^{-5}$ g/mL |
| Atomic radius | 78 pm |
| First ionization energy | 1312 kJ/mol |
| Electronegativity | 2.20 Pauling units |

## Common Reactions

- **When ignited, hydrogen reacts with oxygen to form water.**

  *Example:* $2H_2(g) + O_2(g) \rightarrow 2H_2O(l)$
- **Hydrogen reacts with sulfur to form hydrogen sulfide.**

  *Example:* $H_2(g) + S(g) \rightarrow H_2S(g)$
- **Hydrogen reacts with nitrogen at high temperatures and pressures to form ammonia.**

  *Example:* $3H_2(g) + N_2(g) \xrightarrow{\text{catalyst}} 2NH_3(g)$

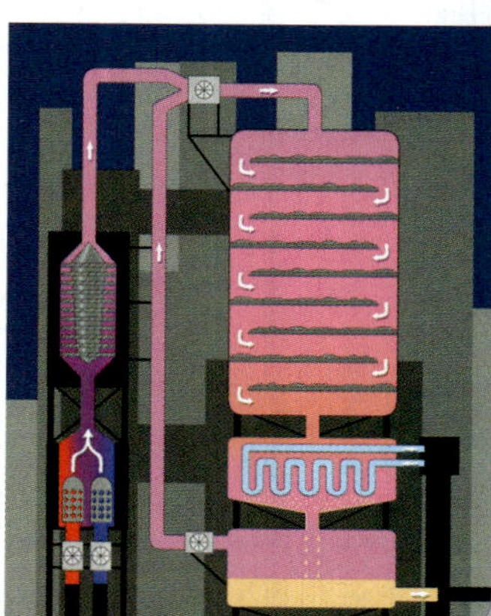

Hydrogen gas in the red tube and nitrogen gas in the blue tube are mixed, then compressed under high pressure and temperature to form liquid ammonia in the orange tube at bottom right.

## Analytical Tests

pH is a measure of the hydrogen ion ($H^+$) concentration of aqueous solutions. When the hydrogen ion concentration is expressed in moles per liter, pH is the negative logarithm of the hydrogen ion concentration, $-\log[H^+]$. For example, if the hydrogen ion concentration is $1 \times 10^{-2}$ mol/L, the pH is 2.

Some common household items are bases or acids, depending on their $H^+$ concentrations. The greater the $H^+$ concentration, the lower the pH.

| Hydrogen |
| --- |
| 1 |
| **H** |
| $1s^1$ |

## Identifying Hydrogen in Stars

Spectroscopy is the study of the spectral lines present in an electromagnetic spectrum. The colored lines in an emission spectrum represent the emission of energy. How do scientists know that more than 90% of the atoms in the universe are hydrogen atoms? By analyzing the emission spectra of light from stars or galaxies, astronomers can identify hydrogen. The spectrum of hydrogen consists of four distinct color lines at different wavelengths. The lines are produced when electrons in a gas move to different energy levels in an atom by absorbing and then emitting energy. Each element can be identified by characteristic patterns of spectral lines.

The colorful cloud that makes up this nebula is composed of hydrogen gas.

## Hydrogen Fuel Cells

Hydrogen fuel cells produce electricity by combining hydrogen ($H_2$) and oxygen ($O_2$) without burning. Water and heat are the only by-products of this process. Current demonstration projects that use hydrogen fuel cells as their energy sources include laptop computers, cars, buses, classrooms, and musical instruments. In the future, it might be possible to use a pen-sized container filled with hydrogen gas to power a laptop computer. Or, you might drive a fuel cell car to a filling station and fill a high-pressure gas cylinder with hydrogen gas.

Hydrogen fuel cells provide the energy to power this electric guitar.

## REVIEW

1. **Compare and contrast** hydrogen isotopes.
2. **Write** the balanced equation for the reaction between hydrogen gas and oxygen gas in a fuel cell.
3. **Explain** what happens when hydrogen reacts with a nonmetal element.
4. **Evaluate** at least one advantage and one possible disadvantage of hydrogen fuel cells compared to conventional petroleum engines.
5. **Infer** Hydrogen can gain one electron to reach a stable electron configuration. Why isn't hydrogen placed with the group 17 elements that share this behavior?
6. **Apply** A solution's hydrogen ion concentration is $3.2 \times 10^{-4}$ mol/L. Determine if this solution is an acid or a base. What is the pH of this solution?

(t)European Southern Observatory/Photo Researchers, (b)Melanie Stetson Freeman/The Christian Science Monitor/Getty Images

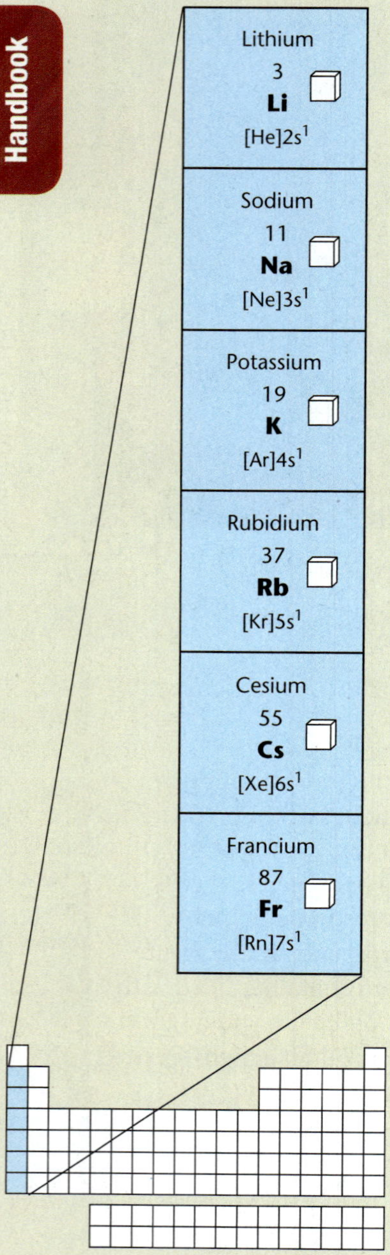


## Physical Properties

- Pure alkali metals have a silvery, metallic appearance.
- Solid alkali metals are soft enough to cut with a knife.
- Most of the alkali metals have low densities compared to the solid form of elements from other groups. Lithium, sodium, and potassium metals are less dense than water.
- Compared to other metals, such as silver or gold, alkali metals have low melting points.

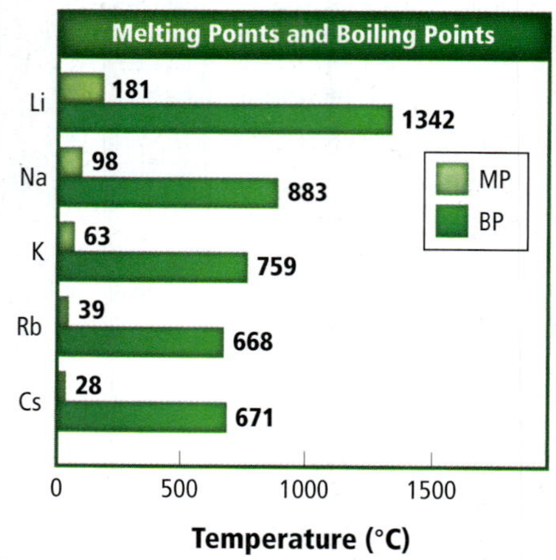


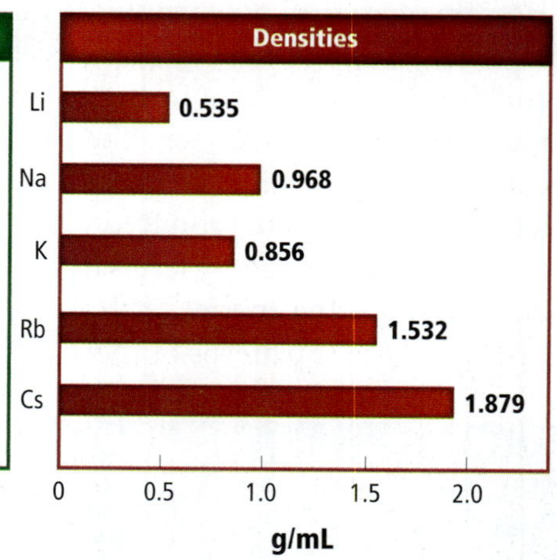


## Common Reactions

- **Li, Na, K, Rb, and Cs react vigorously with halogens to form salts, such as lithium chloride.**

  *Example:* $2Li(s) + Cl_2(g) \rightarrow 2LiCl(s)$

- **Li, Na, K, Rb, and Cs react with oxygen to form oxides ($X_2O$), peroxides ($X_2O_2$), or superoxides ($XO_2$). For example, sodium reacts with oxygen to form sodium peroxide.**

  *Example:* $2Na(s) + O_2(g) \rightarrow Na_2O_2(s)$

- **Li, Na, K, Rb, and Cs react vigorously with water to form metal hydroxides, such as potassium hydroxide, and hydrogen gas.**

  *Example:* $2K(s) + 2H_2O(l) \rightarrow 2KOH(aq) + H_2(g)$

Potassium reacts violently with water, producing enough heat to ignite the hydrogen gas produced.

Stephen Frisch/McGraw-Hill Education

## Atomic Properties

- Each element in group 1 has one valence electron and an electron configuration ending with $ns^1$.
- Group 1 elements lose their valence electrons to form ions with a 1+ charge.
- Going down the elements in group 1, the atomic radii and ionic radii increase.
- Electronegativity decreases going down the elements in group 1.
- The alkali metals are so reactive that they are not found in nature as free metals.
- All the alkali metals have at least one radioactive isotope.
- Because francium is rare and decays rapidly, its properties are not well known.

| Atomic radius (pm) | Ionic radius (pm) |
|---|---|
| Li 152 | $Li^{1+}$ 76 |
| Na 186 | $Na^{1+}$ 102 |
| K 227 | $K^{1+}$ 138 |
| Rb 248 | $Rb^{1+}$ 152 |
| Cs 265 | $Cs^{1+}$ 167 |
| Fr 270 | $Fr^{1+}$ 180 |

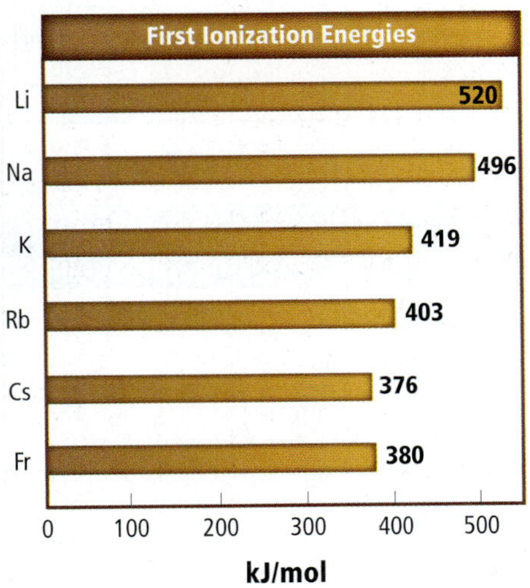


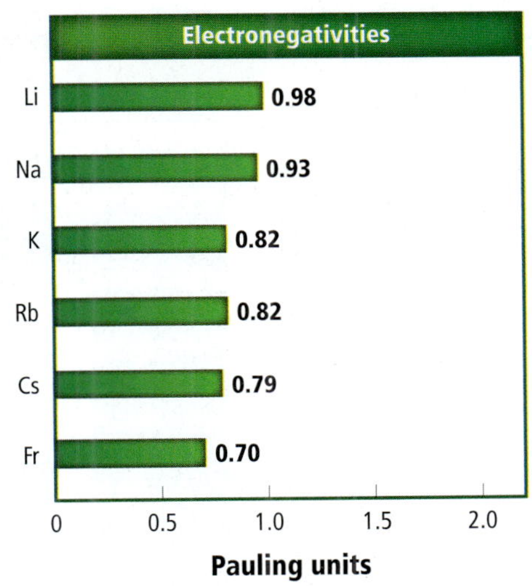


## Analytical Tests

Alkali metals can be qualitatively identified by flame tests. Lithium produces a red flame. Sodium produces an orange flame. Potassium, rubidium, and cesium produce violet flames.

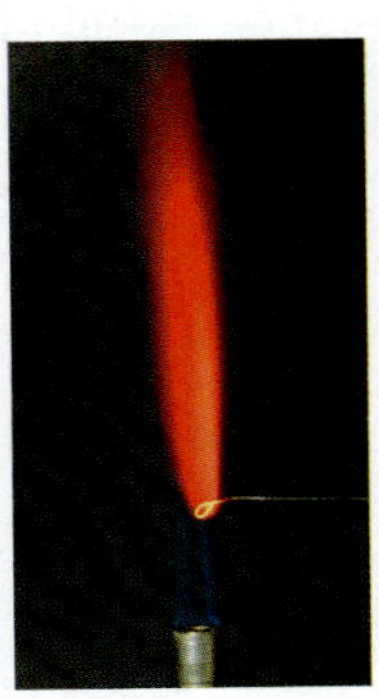
Lithium

Sodium

Potassium

Rubidium

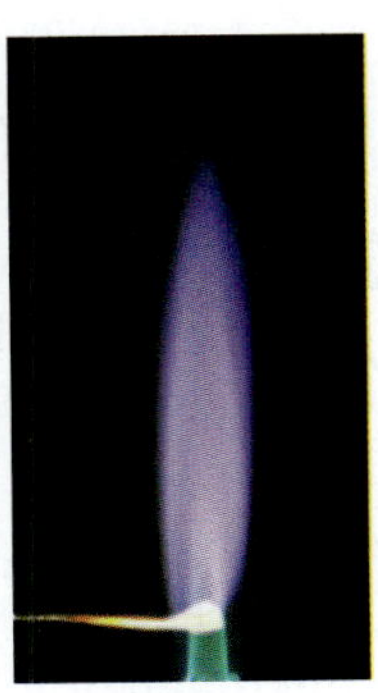
Cesium

Lithium
3
**Li**
[He]$2s^1$

The Mars rovers, Spirit and Opportunity, use solar energy to recharge lithium-ion batteries.

## Environmentally Friendly Batteries

Someday, electric cars might be powered by lightweight lithium-ion batteries. Lithium batteries have several advantages compared to lead-acid batteries. Unlike lead-acid batteries, lithium batteries do not contain toxic metals or corrosive acids, making them safer for the environment. Lithium's light weight is also an advantage for electric vehicles. However, lithium batteries do have some disadvantages. Researchers are trying to find ways to make lithium batteries that recharge more rapidly. Cost is also a drawback. Lithium batteries are currently used for small applications such as laptop computers, but they will need to be less expensive before they can be routinely used in larger, more energy-demanding applications such as electric or hybrid vehicles.

Sodium
11
**Na**
[Ne]$3s^1$

## Dietary Salt

In 2006, the American Medical Association recommended that the amount of sodium in processed and restaurant foods be reduced by one-half over the next decade. Sodium is essential for humans, but too much might contribute to high blood pressure and heart failure. Current guidelines advise consuming less than 2300 mg of sodium per day, which is about one teaspoon. However, Americans typically consume 4000 to 6000 mg of sodium per day. Foods that contain more than 480 mg of sodium per serving are considered high-sodium foods. To be labeled as low sodium, foods must contain 140 mg or less per serving. The table lists some common foods that are either high or low in sodium.

**Sodium Content of Some Common Foods**

| | Food | Sodium Content (mg) per Serving |
|---|---|---|
| High sodium | fast-food submarine sandwich with cold cuts | 1310 |
| | canned chicken noodle soup | 1106 |
| | fast-food biscuit with egg and sausage | 1080 |
| | cottage cheese | 851 |
| | dill pickle | 833 |
| | fast-food cheeseburger | 740 |
| | canned corn | 571 |
| | beef hotdog | 513 |
| | fried fish fillet | 484 |
| Low sodium | wheat bread | 133 |
| | low-fat fruit yogurt | 132 |
| | fast-food salad with cheese and egg, no dressing | 119 |
| | pound cake | 111 |
| | oatmeal cookie | 96 |
| | raw carrots | 76 |
| | canned peaches | 16 |
| | frozen corn | 2 |

Sodium
11
**Na**
[Ne]3s[1]

Potassium
19
**K**
[Ar]4s[1]

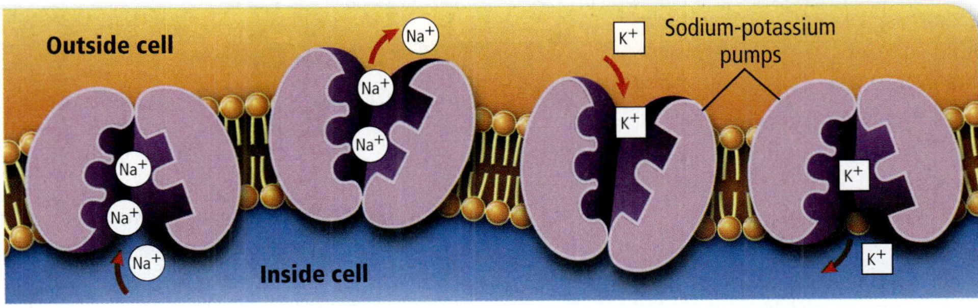


The sodium-potassium pump brings two $K^+$ ions into a cell for every three $Na^+$ ions it moves out of a cell.

## The Sodium-Potassium Pump

Humans and other vertebrates need to maintain a negative potential charge inside their cells in order to survive. This process requires sodium ions, potassium ions, and a membrane-bound enzyme called sodium/potassium ATPase. Sodium/potassium ATPase uses energy from the hydrolysis of ATP to pump sodium ions out of cells and pump potassium ions into cells. Because of the action of this pump, the sodium ion concentration is low inside cells and high outside cells. The potassium ion concentration is high inside cells and low outside cells. In fact, potassium ions are the most common ions inside living cells. For every three sodium ions pumped out of a cell, sodium/potassium ATPase pumps two potassium ions into the cell. The net result is a negative charge inside the cell and concentration gradients across the cell membrane for both potassium and sodium ions.

Cesium
55
**Cs**
[Xe]6s[1]

The cesium fountain atomic clock at NIST is accurate to about 1 second over a period of 100 million years.

## Cesium Atomic Clocks

One of the most accurate clocks in the world is located at the United States National Institute of Standards and Technology (NIST) in Boulder, Colorado. This cesium fountain atomic clock provides the official time for the United States. The clock is based on the natural resonance frequency of the cesium atom (9,192,631,770 Hz), which defines the second.

## REVIEW

7. **Describe** the trend in density of the alkali metals as atomic number increases.
8. **Compare** lithium-ion batteries and lead-acid batteries.
9. **Write** a balanced equation for the reaction between lithium and water.
10. **Predict** the reactivity of lithium metal with water.
11. **Analyze** Are lithium's atomic and ionic radii closer to sodium or closer to magnesium, which is in group 2? Explain.
12. **Organize** Make a table to summarize the data for physical and atomic properties of the group 1 elements based on the trend of increasing atomic number.

Geoffrey Wheeler

# GROUP 2: ALKALINE EARTH METALS



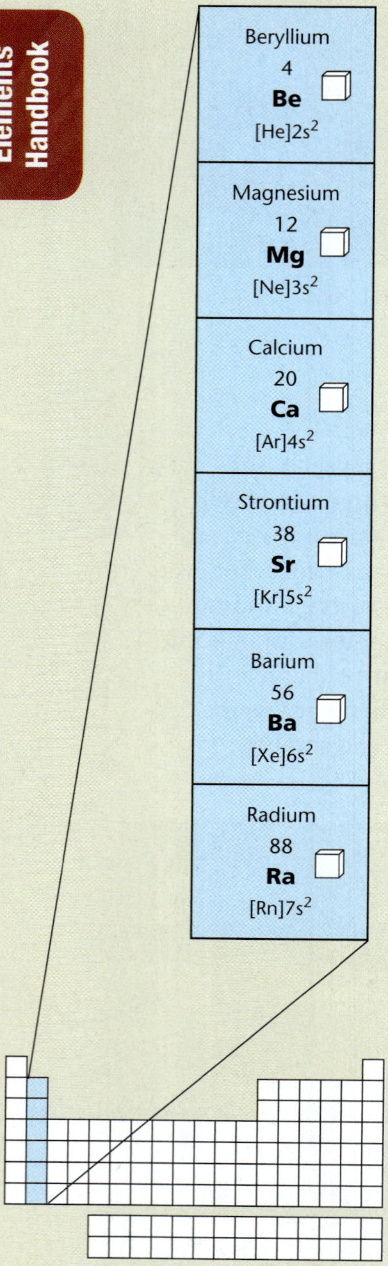


## Physical Properties

- Most of the alkaline earth metals have a silvery-white, metallic appearance. When exposed to oxygen, a thin oxide coating forms on the surface.
- The alkaline earth metals are harder, denser, and stronger than many of the group 1 elements, but are still relatively soft compared to other metals.
- Most alkaline earth metals have higher melting points and boiling points than alkali metals.
- Moving down the group, densities generally increase.

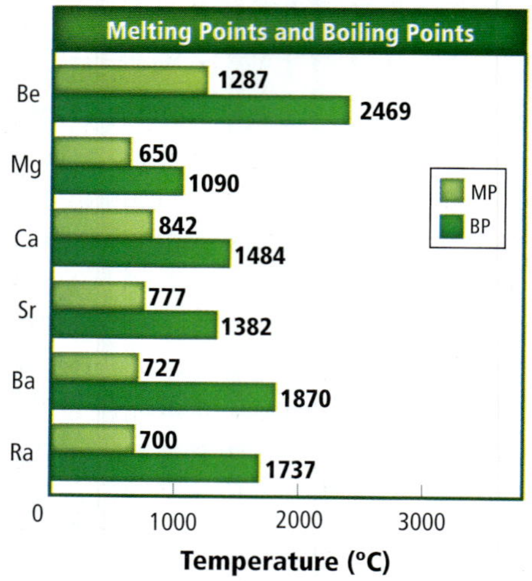


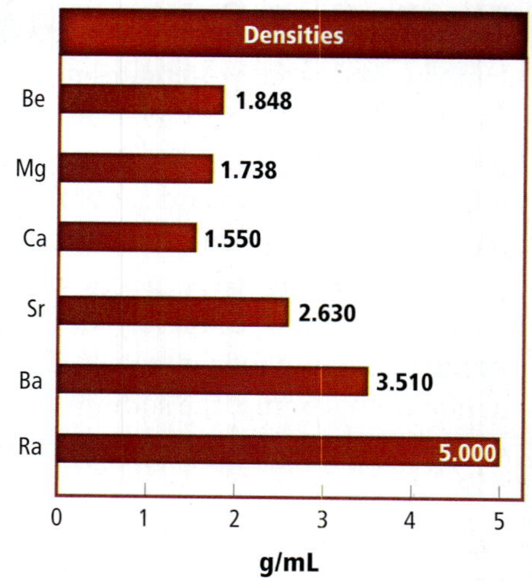


## Common Reactions

- **Mg, Ca, Sr, and Ba react with dilute acids to form salts, such as magnesium chloride, and hydrogen gas.**

  *Example:* $Mg(s) + 2HCl(aq) \rightarrow MgCl_2(aq) + H_2(g)$

- **Mg, Ca, Sr, and Ba react with hydrogen to form hydrides, such as barium hydride.**

  *Example:* $Ba(s) + H_2(g) \rightarrow BaH_2(s)$

- **Mg, Sr, and Ca react with nitrogen to form nitrides, such as magnesium nitride.**

  *Example:* $3Mg(s) + N_2(g) \rightarrow Mg_3N_2(s)$

A ribbon of magnesium reacts with $H_2O$ to produce $Mg(OH)_2$ and hydrogen gas.

Charles D. Winters/Photo Researchers

## Atomic Properties

- Each element in group 2 has two valence electrons and an electron configuration ending with $ns^2$.
- Alkaline earth metals often lose their two valence electrons to form ions with a 2+ charge.
- Atomic radii and ionic radii increase moving down the group but are smaller than the corresponding alkali metal.
- Ionization energies and electronegativities generally decrease moving down the group but are larger than the corresponding alkali metal.

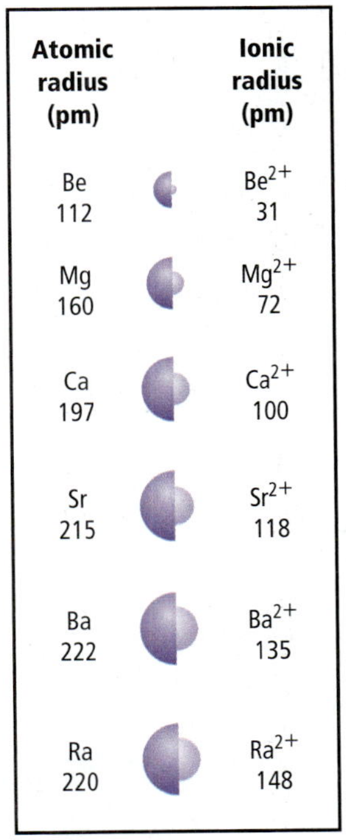


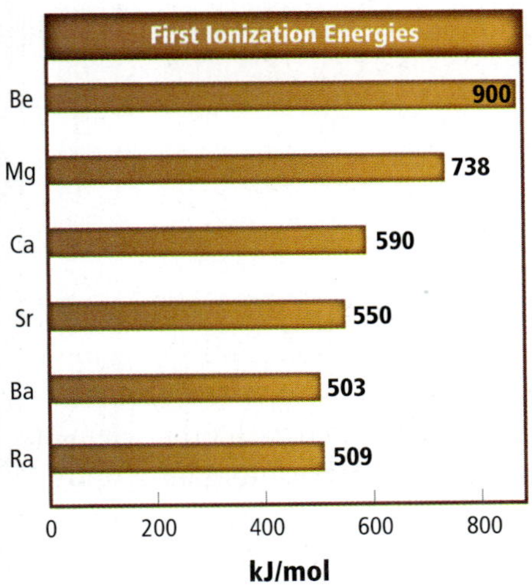


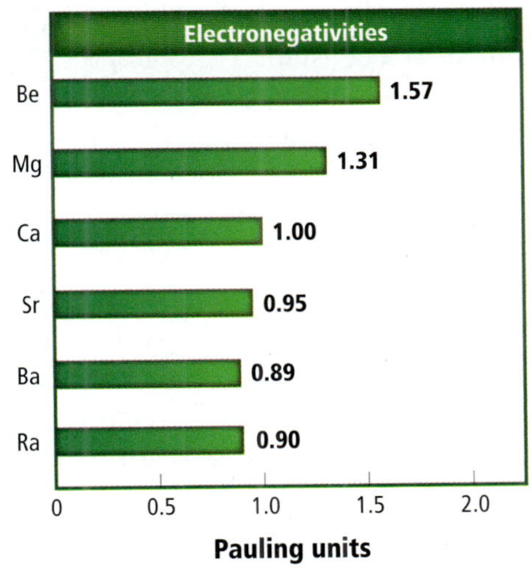


- **Mg, Ca, Sr, and Ba react with oxygen to form oxides, such as magnesium oxide.**

  *Example:* $2Mg(s) + O_2(g) \rightarrow 2MgO(s)$
- **Sr and Ba react with oxygen to form peroxides, such as strontium peroxide.**

  *Example:* $Sr(s) + O_2(g) \rightarrow SrO_2(s)$
- **Mg, Ca, Sr, and Ba react with water to form bases, such as barium hydroxide, and hydrogen gas.**

  *Example:* $Ba(s) + 2H_2O(l) \rightarrow Ba(OH)_2(aq) + H_2(g)$

Barium reacts with water to form $Ba^{2+}$ ions, $OH^-$ ions, and hydrogen gas.

## Analytical Tests

Three of the alkaline earth metals can be identified by flame tests. Calcium produces a scarlet color, while strontium produces a crimson color. Barium, which if present in a sample can mask the colors of both calcium and strontium, produces a yellow-green color.

Calcium

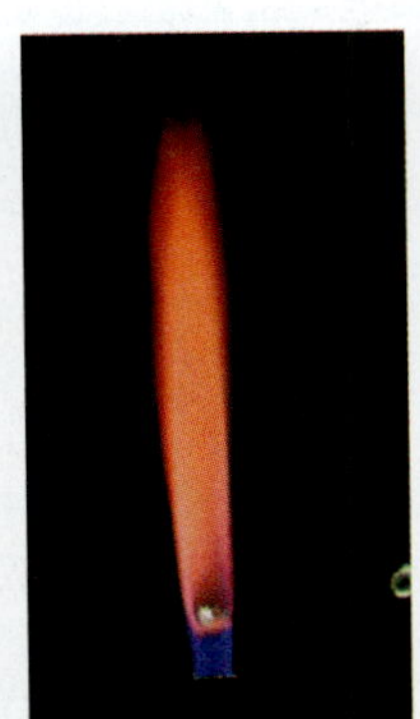

Strontium

Barium

Beryllium
4
**Be**
$[He]2s^2$

## Space Telescopes

Beryllium and beryllium alloys have properties that make them useful for applications in space: they are hard, they are lighter than aluminum, and they are stable over a wide temperature range. The *Hubble Space Telescope's* reaction plate is made of lightweight beryllium. The reaction plate carries heaters that keep the main mirror at a constant temperature. Beryllium is also being used in the *Hubble's* replacement—the *James Webb Space Telescope (JWST)*.

The JWST's large mirror is composed of 18 hexagonal beryllium plates.

◀ Emerald beryl

## Precious Gems

Emerald ($Be_3Al_2Si_6O_{18}$), one of the world's most valuable gemstones, belongs to a family of gemstones known as beryls. Pure beryls are clear, colorless crystals. Beryls tinted with other elements form gems such as aquamarine, morganite, and emerald. Trace amounts of chromium or vanadium give emeralds their unique green color.

Magnesium
12
**Mg**
$[Ne]3s^2$

**Amount of Magnesium Removed by Crops from One Hectare of Soil**

| Crop | Magnesium Removed from Soil (kg) |
|---|---|
| Alfalfa | 44 |
| Corn | 58 |
| Cotton | 25 |
| Oranges | 25 |
| Peanuts | 27 |
| Rice | 15 |
| Soybeans | 27 |
| Tomatoes | 40 |
| Wheat | 20 |

## Chlorophyll and Crop Yields

In the early 1900s, German chemist Richard Willstätter discovered that a molecule of chlorophyll has a magnesium ion at its center. Chlorophyll, the green pigment in plants, is responsible for photosynthetic processes, which convert sunlight to chemical energy. It is this chemical energy that supports life on Earth. Notice in the table that an average yield of common crops removes large amounts of magnesium from just one hectare of soil. Once the importance of magnesium was revealed, soils deficient in magnesium were fertilized, greatly increasing crop yields. Willstätter's work won him the Nobel Prize in Chemistry in 1915.

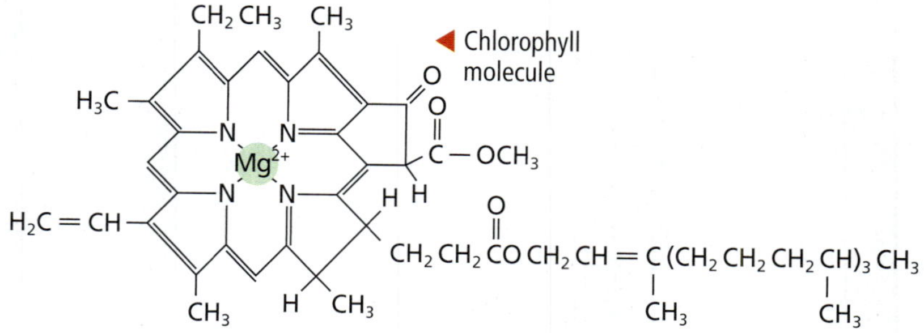

◀ Chlorophyll molecule

| Magnesium | Calcium | Strontium | Barium |
|---|---|---|---|
| 12 | 20 | 38 | 56 |
| **Mg** | **Ca** | **Sr** | **Ba** |
| [Ne]$3s^2$ | [Ar]$4s^2$ | [Kr]$5s^2$ | [Xe]$6s^2$ |

## Fireworks

The four main components of fireworks are a container, a fuse, a bursting charge, and stars. Stars contain the chemical compounds needed to produce light of brilliant colors. Many of these compounds contain alkaline earth metals, such as barium chloride ($BaCl_2$), strontium carbonate ($SrCO_3$), and calcium chloride ($CaCl_2$). The table identifies which metals are needed to make the colors seen during a fireworks display.

**Metals Used in Fireworks**

| Color | Metal |
|---|---|
| Red | strontium, lithium |
| Orange | calcium |
| Gold | iron (with carbon) |
| Yellow | sodium |
| White | white-hot magnesium or aluminum, barium |
| Green | barium |
| Blue | copper |
| Purple | mixture of strontium (red) and copper (blue) |
| Silver | aluminum, titanium, or magnesium powder or flakes |

## New Engineering Alloys

Magnesium alloys are used when strong, but lightweight, materials are needed, such as in backpack frames and aircraft. These alloys also enable automotive engineers to design lighter, more fuel-efficient cars. A new magnesium alloy, introduced in the engine cradle of some 2006 automotive models, replaces traditional aluminum. This alloy reduces the engine cradle's mass by approximately one-third, creating a vehicle that is both agile and controllable. Considered a breakthrough in engineering technology, the new alloy is currently being evaluated for use in other applications.

The magnesium-alloy engine cradle is lighter than the aluminum model, yet it can still withstand the high temperatures produced by the car's engine.

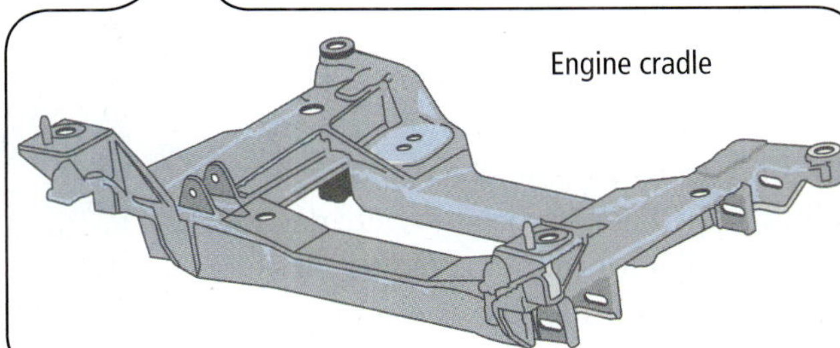

| Calcium |
|---|
| 20 |
| **Ca** |
| $[Ar]4s^2$ |

A layer of plaster of paris protects fossils during shipment.

## Gypsum

Drywall is made from gypsum, which is a soft mineral composed of calcium sulfate dihydrate ($CaSO_4 \cdot 2H_2O$). Drywall boards are used in building construction because the gypsum provides fire protection. Gypsum contains large amounts of water in its crystal form, which vaporizes when heated. The boards remain at 100°C until all of the water evaporates, protecting the wood frame of the building. Gypsum that has had most of its water removed is known as plaster of paris. Most minerals form pastes when mixed with water. When plaster of paris is mixed with water, it forms a rigid crystal structure, so it is often used for casts to set broken bones and for molds.

| Strontium |
|---|
| 38 |
| **Sr** |
| $[Kr]5s^2$ |

Crystals formed from strontium chloride and saliva fill in pores in the root of a tooth and block access to the nerve.

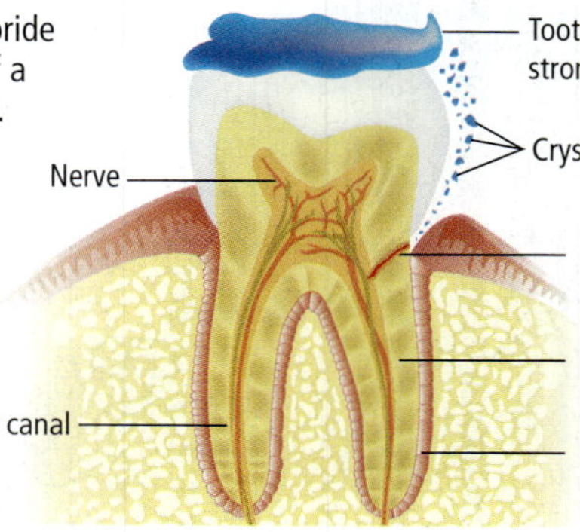


## Sensitive Teeth

Almost 40 million people in the United States have teeth that are hypersensitive to touch and temperature. Sensitivity occurs when the dentine and roots of teeth are exposed due to receding gums or thinning of the tooth enamel. This is the result of poor oral hygiene or, in many instances, from brushing too hard. Exposing the root enables stimuli, such as cold temperatures, to reach the nerve through openings called pores. Toothpastes that contain strontium chloride ($SrCl_2$) help reduce the sensitivity. The compound reacts with a person's saliva to create crystals that fill in the pores so stimuli cannot reach the nerves.

| Barium |
|---|
| 56 |
| **Ba** |
| $[Xe]6s^2$ |

After being coated with barium liquid, the large intestine shows up clearly on an X-ray.

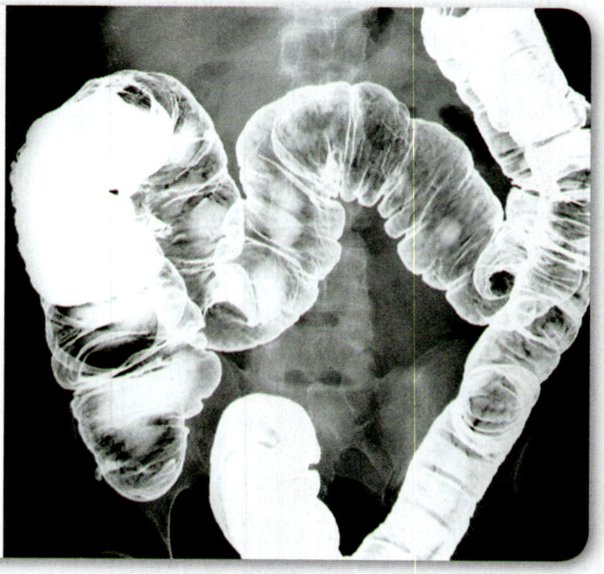

## Medical X-Rays

Barium is used by medical professionals to examine a person's gastrointestinal tract. Patients drink barium liquid, which coats the tract, and are then X-rayed. Barium is almost completely insoluble in water and acids and appears as a bright white color in X-rays. This allows doctors and radiologists to locate tumors, ulcers, areas of reflux, and other abnormalities in the digestive tract.

| Radium |
| --- |
| 88 |
| **Ra** |
| $[Rn]7s^2$ |

## The Discovery of Radioactivity

When Henri Becquerel discovered a property of uranium, which Marie Curie later called radioactivity, they paved the way for present-day advancements in science and medicine. Becquerel, Curie, and her husband, Pierre, unveiled the characteristics and capabilities of radiation, revolutionizing scientific thinking and laying the groundwork for present-day cancer treatments, genetic studies, and nuclear energy production.

Marie Curie died at the age of 67 from aplastic anemia, probably caused by her exposure to massive amounts of radiation. Today, the effects of radiation on health are better known, and suitable safety precautions are taken when using radioactive materials.

## Radon Gas

Decay of radium-226 in soil and rock produces radon gas. High concentrations of radon can increase the risk of cancer. The radioactive radon gas can seep through cracks in a home's foundation or it can dissolve in water pumped into the house from a well. Installing a radon-reduction system reduces the concentration of radon gas by using a fan to draw the gas through pipes that vent to the outside of the home.

A radon-reduction system lowers the concentration of radon in homes by venting the radon gas from the home to the outside environment.

## REVIEW

**13. Describe** the general trend in first ionization energies in group 2, and explain why this trend occurs.

**14. Explain** What is the charge on alkaline earth metal ions? Explain your answer.

**15. Compare and contrast** the physical properties of the alkaline earth metals and the alkali metals.

**16. Evaluate** why several of the alkaline earth metals are used in different emergency flares.

**17. Analyze** Use the atomic properties of the alkali metals and alkaline earth metals to explain why calcium is less reactive than potassium.

**18. Infer** The alkaline earth metals are usually found combined with oxygen and other nonmetals in Earth's crust. Based on the atomic properties of this group, explain why alkaline earth metals are so reactive.

**19. Calculate** Calcium makes up about 1.5% of a human's body mass. Calculate the amount of calcium found in a person with a total mass of 68 kg.

**20. Calculate** Radium-226 has a half-life of 1600 years. After 8000 years, how much of a 500.0-g sample of radium-226 would be left?

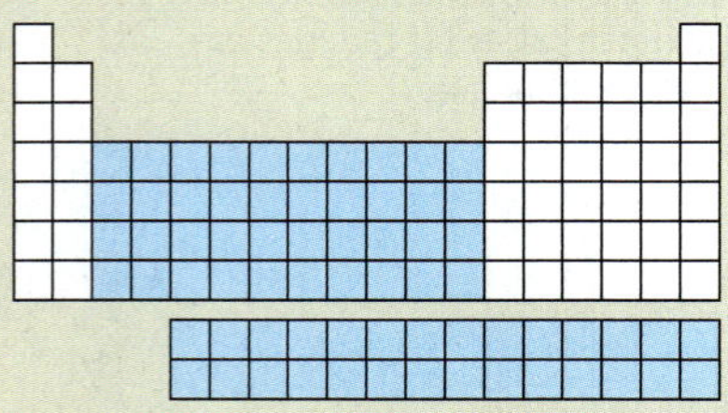

## Physical Properties

- The main transition elements include four series of d-block elements with atomic numbers between 21–30, 39–48, 72–80, and 104–112. The inner transition elements include the f-block (rare earth) elements in the lanthanide series (atomic numbers 57–71) and actinide series (atomic numbers 89–103.) All are metals.
- Like many other metals, transition elements are generally good conductors of electricity and heat. They are ductile, which means they can be pulled into wires. Transition metals are also malleable, which means they can be hammered into thin sheets.
- In general, the transition elements have high densities, high melting points, and low vapor pressure. All transition elements are solids at room temperature except for mercury, which is a liquid.
- Strength and availability make transition elements, such as iron, good structural materials.
- Many transition elements reflect visible light at specific wavelengths making some compounds appear brightly colored.
- Transition elements are often paramagnetic, which means they are attracted to an applied magnetic field. Three transition elements—iron, cobalt, and nickel—are ferromagnetic. These elements can form their own magnetic fields.

When exposed to a magnet, iron filings become magnetic and are attracted to the magnet and to each other.

## Common Reactions

- **Most transition elements can form stable complex ions and coordinate covalent compounds.** A complex ion is an ion in which a central metal ion is surrounded by weakly bound molecules or ions called ligands.

  *Example:* Prussian blue, an intense blue pigment used in paints, is a coordinate compound made of iron(III) and an iron(II) cyanide complex: $Fe_4[Fe(CN)_6]_3$.
- **Some transition elements form solutions called alloys.**

  *Examples:*
  - Brass is an alloy of copper and zinc.
  - Bronze is an alloy of copper and tin.
- **Transition elements and their compounds are often useful as catalysts.**

  *Example:* Nickel is used as a catalyst in converting unsaturated fats to saturated fats.
- **Transition elements can react with oxygen to form oxides.**

  *Example:* In the presence of water, iron reacts with oxygen to form rust. The overall reaction is: $4Fe + 3O_2 \rightarrow 2Fe_2O_3$.
- **Some transition elements are important in biochemical reactions.**

  *Example:* In the protein hemoglobin, iron binds to $O_2$ to transport oxygen from the lungs to the rest of the body.

## Atomic Properties

- The more unpaired electrons in the d sublevel, the greater the hardness and the higher the melting and boiling points. Unpaired d and f electrons produce paramagnetism.
- For transition elements, there is little variation in atomic size, electronegativity, and ionization energy across a period due to shielding. The d electrons shield the outer electrons from an increase in nuclear charge as atomic number increases. Although the number of protons in the nucleus increases across a period, the outer electrons in each atom experience a similar nuclear charge.
- Transition elements can lose s electrons and form ions with a 1+ and/or 2+ charge. Because the s and d electrons are close in energy, many transition elements can also form ions with a charge of 3+ or higher.
- Compounds containing transition elements with partially filled d sublevels absorb specific wavelengths of visible light making compounds appear brightly colored.

**Oxidation Numbers of the First Row of Transition Elements**

| | | | | | | | | |
|---|---|---|---|---|---|---|---|---|
| Sc | 0 | | | +3 | | | | |
| Ti | 0 | | +2 | +3 | +4 | | | |
| V | 0 | +1 | +2 | +3 | +4 | +5 | | |
| Cr | 0 | +1 | +2 | +3 | +4 | +5 | +6 | |
| Mn | 0 | +1 | +2 | +3 | +4 | +5 | +6 | +7 |
| Fe | 0 | +1 | +2 | +3 | +4 | | +6 | |
| Co | 0 | +1 | +2 | +3 | +4 | +5 | | |
| Ni | 0 | +1 | +2 | +3 | +4 | | | |
| Cu | 0 | +1 | +2 | +3 | | | | |
| Zn | 0 | | +2 | | | | | |

## Analytical Tests

Notice the colorful compounds of transition metals in the photo. These compounds absorb different wavelengths of light. The color you see depends on the wavelengths of light that are reflected—the wavelengths least absorbed. Visible spectroscopy uses light absorption at specific wavelengths to measure the concentration of colored compounds in solution. The amount of light absorbed at specific wavelengths relates to the concentration of compound in the sample. Because many transition element compounds are colored, this technique can be used in transition element analysis.

The compounds of transition metals appear colored because of the partially filled d sublevels. The electrons in these sublevels can absorb visible light of specific wavelengths. Compounds with empty or filled d sublevels do not produce brilliant colors.

Titanium
22
**Ti**
$[Ar]3d^24s^2$

## Lighter but Stronger than Steel

The curved surfaces of the Guggenheim Museum in Bilbao, Spain, are covered with 32,000 $m^2$ of 0.4 mm-thick titanium panels. Titanium's reflective properties give the building a warm look that is ever changing. Titanium is also three times stronger than steel, more resistant to weathering, and weighs less than steel.

The titanium panels that cover the outside of the Guggenheim Museum in Bilbao, Spain, were chosen for the metal's physical properties.

| Chromium 24 **Cr** $[Ar]3d^54s^1$ | Manganese 25 **Mn** $[Ar]3d^54s^2$ | Cobalt 27 **Co** $[Ar]3d^74s^2$ | Tungsten 74 **W** $[Xe]4f^{14}5d^46s^2$ | Platinum 78 **Pt** $[Xe]4f^{14}5d^96s^1$ |
|---|---|---|---|---|

## Strategic and Critical Materials

Transition metals, such as chromium, manganese, cobalt, and platinum, play a vital role in the economy of many countries because they have a wide variety of uses. As the uses of transition metals increase, so does the demand for these valuable materials. Ores that contain transition metals are located throughout the world.

The United States now imports more than 60 materials that are classified as "strategic and critical" because industry and the military are dependent on these materials.

Colin Walton/Alamy

| Iron<br>26<br>**Fe**<br>$[Ar]3d^64s^2$ | Nickel<br>28<br>**Ni**<br>$[Ar]3d^84s^2$ |
|---|---|

## Earth's Iron Core

Earth's core is a solid iron sphere about the size of the Moon. Surrounding the inner core, there is an outer liquid core that contains a nickel-iron alloy. Scientists think the iron core formed when multiple collisions during Earth's early history resulted in enough heat to melt metals. In the molten state, the densest materials, including iron and nickel, settled to the center and became Earth's core. The less-dense materials remained at the surface. As Earth cooled, the outer layers solidified, creating Earth's mantle and crust.

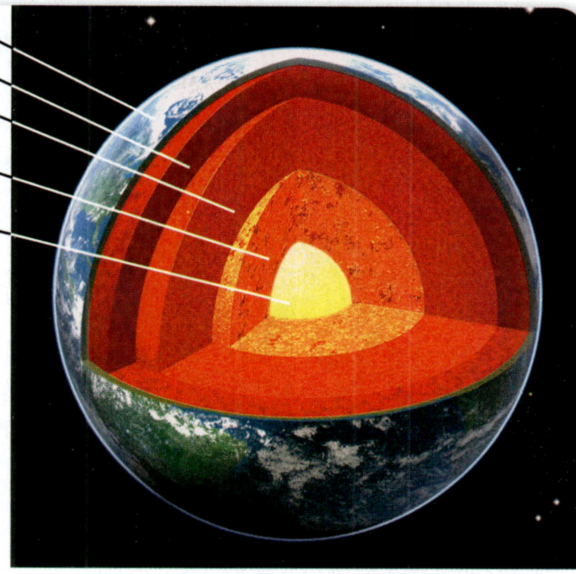


Earth's crust and mantle insulate the hot iron core.

| Copper<br>29<br>**Cu**<br>$[Ar]3d^{10}4s^1$ |
|---|

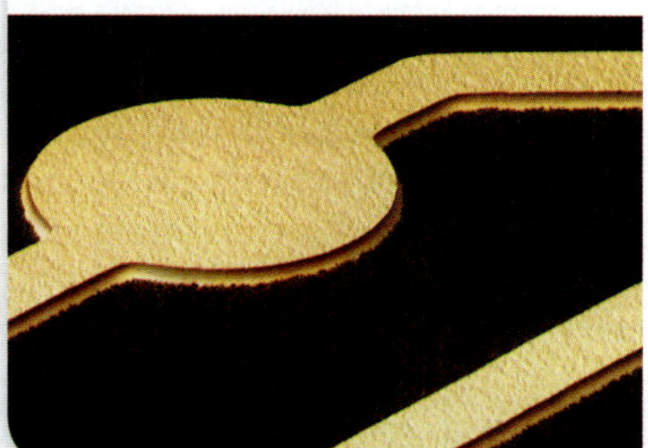

## Copper Microchips

For many years, aluminum was used to make computer microchips. Although copper is a better electrical conductor than aluminum, it was not until the late 1990s that the technology existed to use copper in microchips. Combined with the extremely small size of copper wires, this allows copper microchips to be smaller and to operate 25 to 30 times faster than other kinds of microchips. To make wires this small, the copper must be between 99.999% and 99.9999% pure. Copper microchips are part of circuit boards that are used in handheld games, computers, and other electronic devices.

To create a copper microchip, first a layer of tantalum coats a silicon substrate. Then, copper is deposited using a vacuum process.

| Titanium<br>22<br>**Ti**<br>$[Ar]3d^24s^2$ | Chromium<br>24<br>**Cr**<br>$[Ar]3d^54s^1$ | Iron<br>26<br>**Fe**<br>$[Ar]3d^64s^2$ | Cobalt<br>27<br>**Co**<br>$[Ar]3d^74s^2$ | Copper<br>29<br>**Cu**<br>$[Ar]3d^{10}4s^1$ |
|---|---|---|---|---|

## Paint Pigments

Paints are a mixture of particles of pigment in a liquid base. Once the liquid evaporates, the pigment particles coat a painted surface. Transition elements and their compounds are often used as paint pigments. Iron oxides are used as red, yellow, and brown pigments. Chromium, copper, and cobalt compounds produce green and blue pigments. Titanium dioxide is often used for white paint.

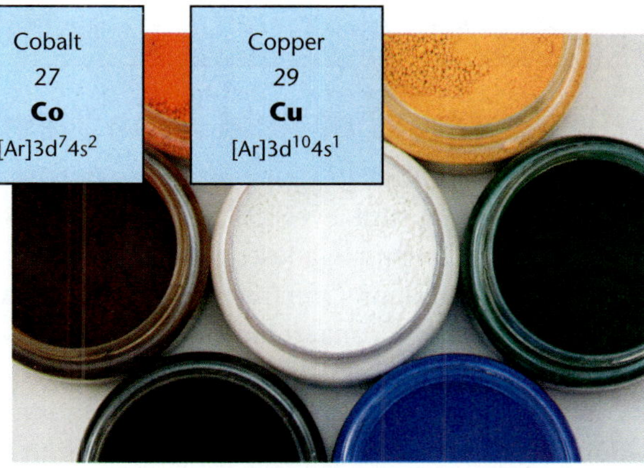

Artists can create their own paints by mixing dry pigments in a liquid base such as oil, latex, or even egg yolk.

(t)Roger Harris/Photo Researchers, (c)Eye of Science/Science Source, (b)Kaicoba/Alamy

| Gold |
|---|
| 79 |
| **Au** |
| [Xe]$4f^{14}5d^{10}6s^1$ |

## Gilding

Covering an ordinary object with gold foil or gold leaf can make the object look like it is made of solid gold. The process, which is called gilding, has been used for more than 5000 years. To create gold foil, gold is hammered until it is very thin. The thinnest sheets are called gold leaf. They can be as thin as 0.1 mm thick. It takes skill and a special gilder's brush to handle sheets this thin, but the results can be spectacular.

Egyptian King Tutankhamun's coffin was made of wood covered with gold foil. It has lasted more than 3000 years.

| Cadmium | Gold |
|---|---|
| 48 | 79 |
| **Cd** | **Au** |
| [Kr]$4d^{10}5s^2$ | [Xe]$4f^{14}5d^{10}6s^1$ |

## Touch Sensors for Robot Fingers

Imagine a surgeon using a robot for microsurgery. In the future, it might be possible for the surgeon to feel what is happening as the robot makes a microsuture. Future robots might use thin, film sensors to mimic the human sense of touch. These sensors are built on a glass base from alternating layers of nanoparticles of gold and cadmium sulfide separated by layers of plastic. The entire sensor is only 100 nm thick and works by transmitting an electroluminescent signal and electric current when regions of the sensor are touched.

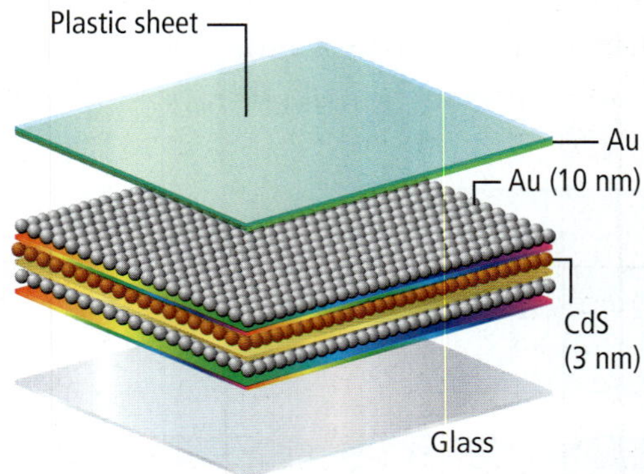


This touch sensor is made from nanoparticles of gold and cadmium sulfide.

| Manganese | Iron | Copper | Zinc | Silver | Cadmium |
|---|---|---|---|---|---|
| 25 | 26 | 29 | 30 | 47 | 48 |
| **Mn** | **Fe** | **Cu** | **Zn** | **Ag** | **Cd** |
| [Ar]$3d^5 4s^2$ | [Ar]$3d^6 4s^2$ | [Ar]$3d^{10}4s^1$ | [Ar]$3d^{10}4s^2$ | [Kr]$4d^{10}5s^1$ | [Kr]$4d^{10}5s^2$ |

## Biotreatment of Acid Mine Wastes

Mining operations can generate acidic wastewater that contain harmful levels of dissolved transition metals, including manganese, iron, copper, zinc, silver, and cadmium. One treatment method uses naturally occurring anaerobic bacteria to remove all of the oxygen. Then sulfate-reducing bacteria convert sulfuric acid in the mine waste to sulfide. Sulfide reacts with metals in the wastewater to form metal sulfide precipitates, which can be recovered and processed for commercial use.

Untreated acid mine drainage can contaminate streams with harmful concentrations of transition metals. The red-orange color of the water comes from iron compounds.

| Gadolinium |
| --- |
| 64 |
| **Gd** |
| $[Xe]4f^75d^16s^2$ |

## Magnetic Resonance Imaging

Gadolinium contrast agents are compounds that enhance differences between normal tissue and abnormal tissue, such as tumors, in magnetic resonance imaging (MRI) scans. The gadolinium compounds are injected directly into the bloodstream prior to an MRI scan. Tumors accumulate more of the gadolinium compounds than normal tissue. Gadolinium enhances MRI images because it is paramagnetic. Magnetic resonance imaging uses a strong magnetic field and radio waves to stimulate water molecules to an excited state. The MRI image is formed as water molecules relax back to their normal state. Gadolinium speeds up the relaxation rate, which improves the contrast between normal and abnormal tissue.

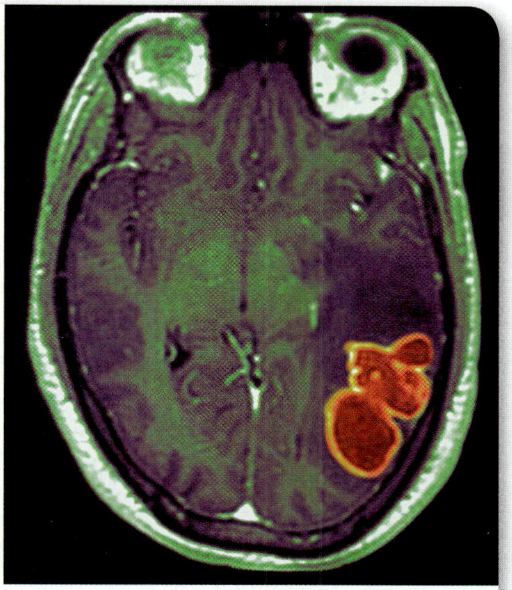

This gadolinium-enhanced MRI scan shows an abcess (red patch) that causes seizures in the patient.

| Thorium | Lawrencium |
| --- | --- |
| 90 | 103 |
| **Th** | **Lr** |
| $[Rn]6d^27s^2$ | $[Rn]5f^{14}6d^17s^2$ |

## Reorganizing the Periodic Table

The actinides are a row of radioactive elements from thorium to lawrencium. They were not always separated into their own row in the periodic table. Originally, the actinides were located within the d-block following actinium. In 1944, Glenn Seaborg proposed a reorganization of the periodic chart to reflect what he knew about the chemistry of the actinide elements. He placed the actinide series elements in their own row directly below the lanthanide series. Seaborg had played a major role in the discovery of plutonium in 1941. His reorganization of the periodic table made it possible for him and his coworkers to predict the properties of possible new elements and facilitated the synthesis of nine additional transuranium elements.

Seaborg won the Nobel Prize in Chemistry in 1951 for his work. Element 106, seaborgium, was named in his honor.



## REVIEW

**21. Compare** the electron configurations of the main transition elements and the inner transition elements.

**22. Explain** how some transition metals can form ions with more than one charge.

**23. Identify** countries that export only one "strategic and critical" transition metal to the United States.

**24. Predict** Which elements would you expect to have properties most closely related to gold?

**25. Calculate** A particular copper-chip manufacturing process specifies that the copper must be 99.999% to 99.9999% pure. Calculate the maximum limit for impurities in the copper in parts per million (ppm).

**26. Hypothesize** Silver is the best conductor of electricity. Hypothesize why silver is not used for electric wires if it is such a good conductor of electricity.

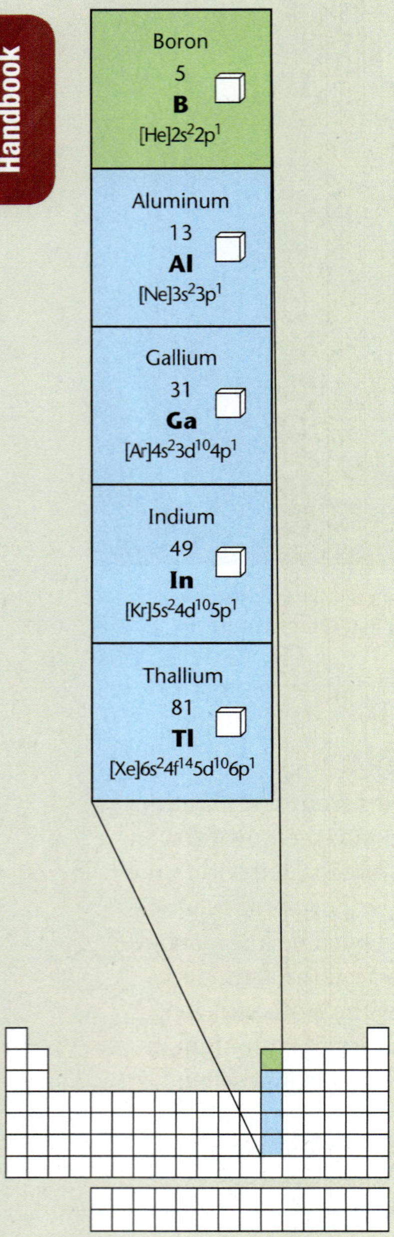


## Physical Properties

- Most of the elements in group 13 are metals that have a silvery-white appearance. The exception is boron, which is pure black. Thallium is initially silvery, but oxidizes quickly.
- Boron is a metalloid. The remaining group 13 elements are metals.
- Elements in this group are relatively lightweight and soft, except for boron. Boron is extremely hard—almost as hard as diamond.
- The group 13 elements are solids at room temperature. Gallium melts slightly above room temperature.
- They have higher boiling points than the alkaline earth metals and lower boiling and melting points than the carbon group elements.

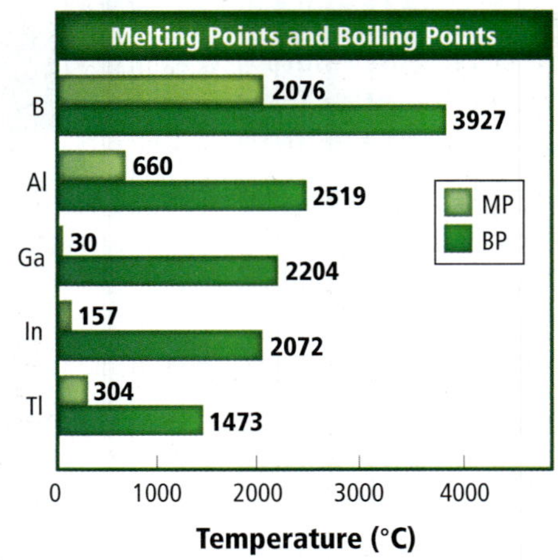


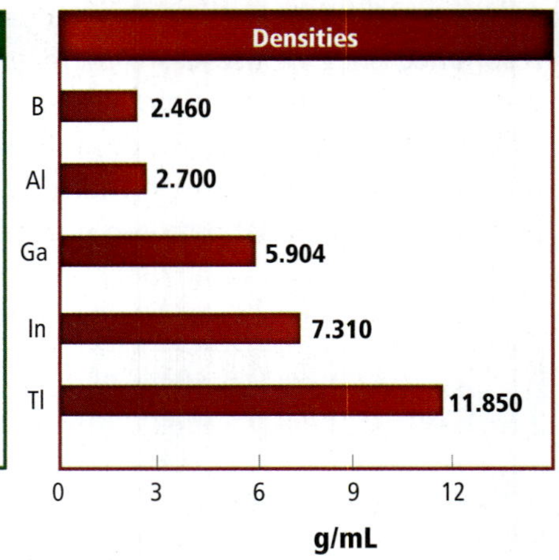


## Common Reactions

- **B, Al, Ga, In, and Tl react with oxygen when heated to form metal(III) oxides, such as aluminum(III) oxide.**

  *Example:* $4Al(s) + 3O_2(g) \rightarrow 2Al_2O_3(s)$
- **B and Al react with nitrogen to form nitrides, such as boron nitride.**

  *Example:* $2B(s) + N_2(g) \rightarrow 2BN(s)$
- **B, Al, Ga, and In react with halogens to form metal(III) halides, such as gallium(III) fluoride.**

  *Example:* $2Ga(s) + 3F_2(g) \rightarrow 2GaF_3(g)$
- **Tl reacts with halogens to form metal(I) halides, such as thallium(I) fluoride.**

  *Example:* $2Tl(s) + F_2(g) \rightarrow 2TlF(s)$
- **Aluminum metal is produced when aluminum(III) oxide reacts with carbon in a redox reaction.**

  *Example:* $2Al_2O_3(s) + 3C(s) \rightarrow 4Al(s) + 3CO_2(g)$
- **Tl reacts with water to form thallium hydroxide and hydrogen gas.**

  *Example:* $2Tl(s) + 2H_2O(l) \rightarrow 2TlOH(aq) + H_2(g)$

## Atomic Properties

- Each element in group 13 has three valence electrons and an electron configuration ending with $ns^2np^1$.
- Except for boron and thallium, the group 13 elements lose their three valence electrons to form ions with a 3+ charge. Some of the elements (Ga, In, and Tl) also have the ability to lose just one of their valence electrons to form ions with a 1+ charge.
- Boron participates only in covalent bonding.
- Atomic radii and ionic radii generally increase going down the group and are similar in size to the group 14 elements.
- Except for boron, the group 13 elements have similar first ionization energies.

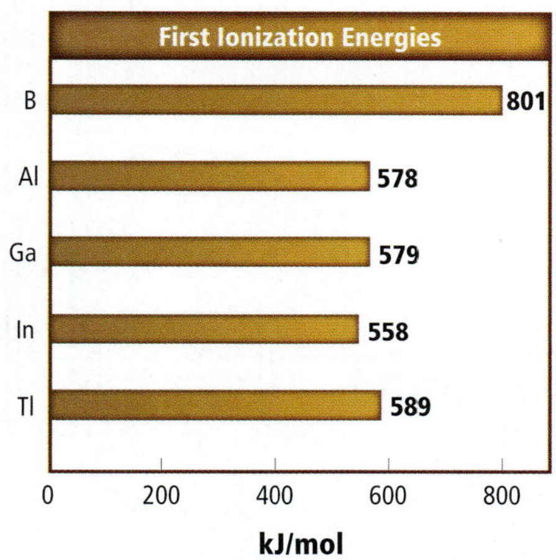


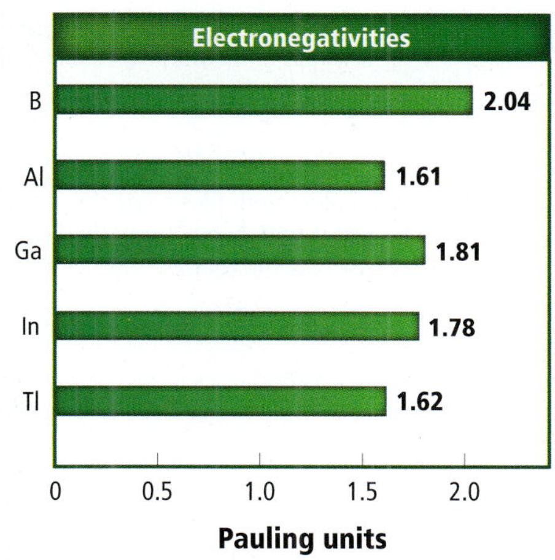


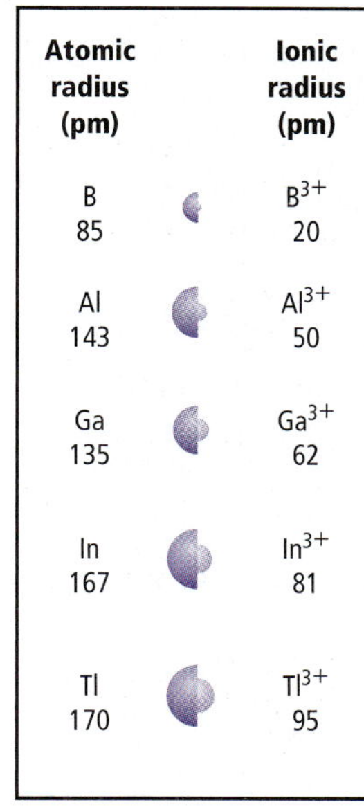


## Analytical Tests

With the exception of aluminum, which is one of the most abundant elements in Earth's crust, most of the boron group elements are rare. None of the elements are found free in nature. Three can be identified by flame tests, as shown in the table. Boron produces a bright green color, while indium produces an indigo blue color. Thallium produces a green color. More precise identification methods involve advanced spectral and imaging techniques.

**Flame Test Results**

| Element | Color of Flame |
|---|---|
| Boron | initial bright green flash |
| Indium | indigo blue |
| Thallium | green |

Indium was named after its distinct indigo blue spectral line.

Boron
5
**B**
$[He]2s^2 2p^1$

## Detergent

Sodium perborate ($NaBO_3 \cdot H_2O$ or $NaBO_3 \cdot 4H_2O$) is one of the key ingredients in powdered laundry detergent. The hydrate, formed by combining borax pentahydrate ($Na_2B_4O_7 \cdot 5H_2O$) with hydrogen peroxide and sodium hydroxide, releases oxygen during the laundering process to help make clothes whiter and brighter. Sodium perborate is the chemical of choice because it remains stable over long periods of time, helps maintain wash water pH, and increases the solubility of detergent ingredients.

Many powder laundry detergents contain boron compounds that help make clothes cleaner.

Aluminum
13
**Al**
$[Ne]3s^2 3p^1$

A thin aluminum film coats the depressions embedding information in a compact disc and makes the surface of a CD shiny.

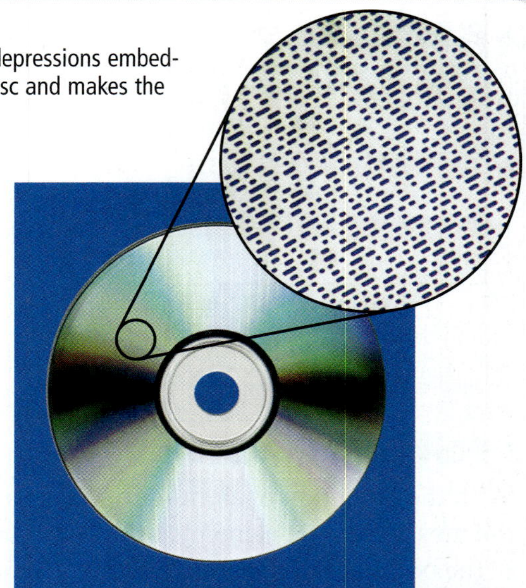

## CDs and DVDs

Have you ever wondered what your CDs and DVDs are made of? The inside is made of plastic, about 1 mm thick. A machine embeds digital information, such as sound recordings, into the plastic as a series of bumps and then coats the plastic with aluminum. That is what makes CDs and DVDs so shiny. A thin layer of acrylic protects the aluminum. The shiny surface allows the laser from the CD or DVD player to read the information reflected off the disc's surface.

Gallium
31
**Ga**
$[Ar]4s^2 3d^{10} 4p^1$

## HD DVDs

Videos in high-definition (HD) have higher quality sound and pictures than regular DVDs. However, HD technology requires more information than can be stored on regular DVDs. A red laser is used to read and write data on a regular DVD. Blue lasers made from gallium nitride (GaN) are used to read and write data on HD DVDs. Blue light has a shorter wavelength than red light, so a blue laser can read more densely packed information, allowing more information to be stored in the same amount of space.

HD DVDs store up to 50 gigabytes (GB) of information, compared to 4.7 GB on a regular DVD.

| Indium |
| --- |
| 49 |
| **In** |
| $[Kr]5s^24d^{10}5p^1$ |

## Flat-Screen Televisions

Known as ITO in the electronics industry, indium-tin oxide has proven to be the cornerstone of liquid crystal display (LCD) technology. During production, a thin layer of indium-tin oxide (a mixture of $In_2O_3$ and $SnO_2$) is used to coat the glass contained within an LCD flat-screen panel. This allows the glass to be both conductive and transparent. About half of the world's indium is used to make LCDs.

Indium-tin oxide is one of the main components in LCD flat-panel televisions.

| Thallium |
| --- |
| 81 |
| **Tl** |
| $[Xe]6s^24f^{14}5d^{10}6p^1$ |

## Cardiac Scans

Thallium-201 is a radioisotope used by medical professionals to determine the health of a person's heart. During a thallium-201 scan, also called a heart stress test, a patient performs physical activity and is injected with thallium-201 one to two minutes before stopping the activity. The isotope emits gamma rays that are recorded by a detector to display a two-dimensional image of the heart and its blood supply. If gamma rays are not detected in certain areas in and around the heart, the areas are considered "cold." This means that the blood supply has been impeded or blocked, a condition that often leads to heart attack or stroke.

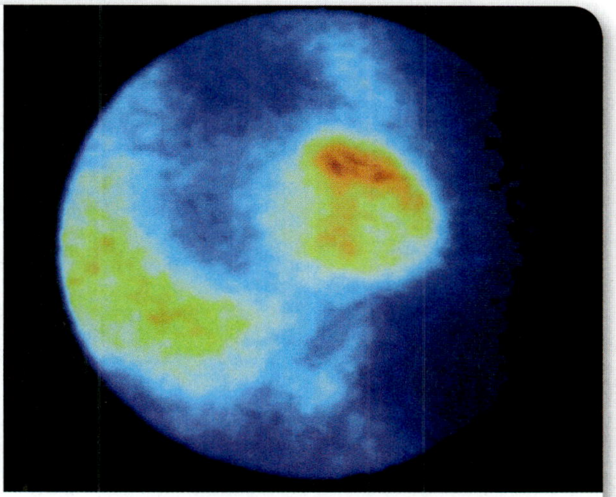

The blue areas in this thallium-201 scan of a heart are areas with low blood supply.

## REVIEW

27. **Describe** how the properties of boron are different from the other group 13 elements.
28. **Identify** what an unknown element would be if it produced a green flash of color at the beginning of a flame test.
29. **Describe** any trends in the first ionization energies of the group 13 elements.
30. **Explain** why HD DVDs can store more information than regular DVDs.
31. **Summarize** how "cold" areas in thallium-201 scans could correspond to artery blockages.
32. **Calculate** It is estimated that 123,000 aluminum cans are recycled each minute. Assume that each can has a mass of 14 g. Determine how much aluminum (kg) is recycled during the month of September.

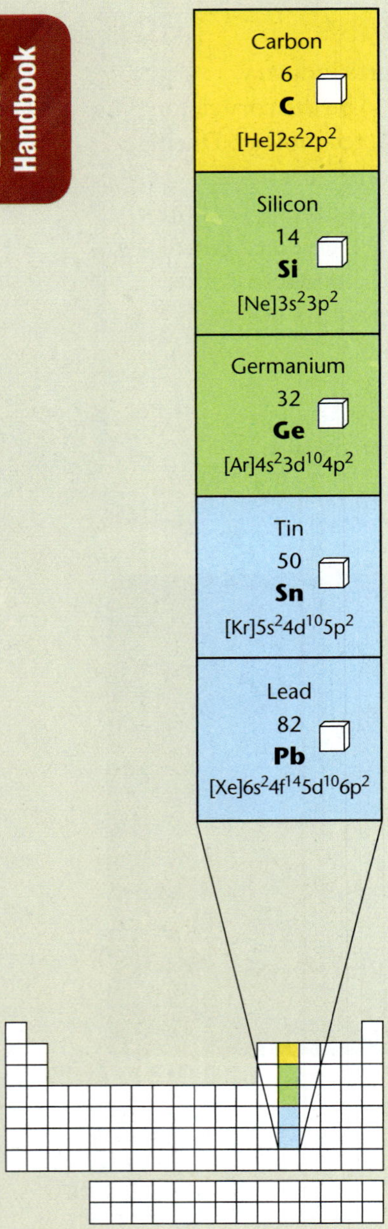


## Physical Properties

- Elements in the carbon group increase in metallic character going down the group. Carbon is a nonmetal. Silicon and germanium are metalloids. Tin and lead are metals.
- Carbon can be a black powder; a soft, slippery gray solid; a hard, transparent solid; or an orange-red solid.
- Silicon can be a brown powder or a shiny-gray solid.
- Germanium is a shiny, gray-white solid that breaks easily.
- Tin also occurs in two forms. One form is a silvery-white solid, while the other is a shiny-gray solid. Both forms are ductile and malleable.
- Lead is a shiny-gray solid. It is soft, malleable, and ductile.
- Moving down the group, melting and boiling points decrease and densities increase.

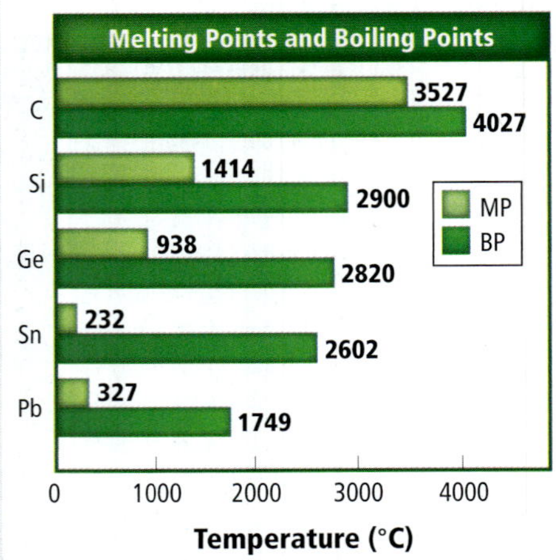


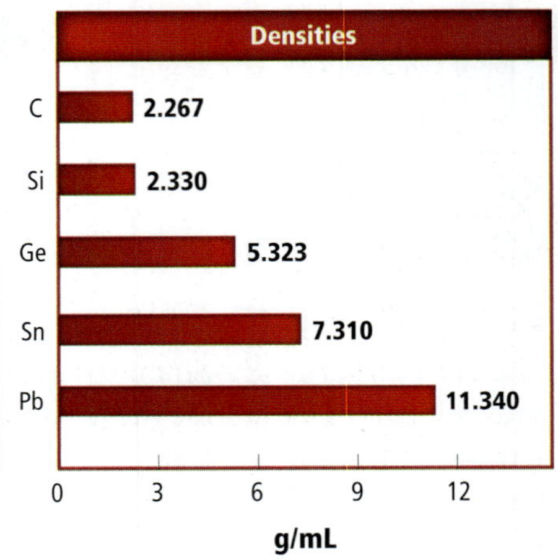


## Common Reactions

At room temperature, carbon group elements are generally unreactive. Reactions do occur under elevated temperature conditions.

- **C, Si, Ge, and Sn react with oxygen to form oxides, such as carbon dioxide.**

  *Example:* $C(s) + O_2(g) \rightarrow CO_2(g)$

- **C, Si, Ge, and Sn react with halogens to form halides, such as silicon chloride.**

  *Example:* $Si(s) + 2Cl_2(l) \rightarrow SiCl_4(g)$

- **Sn and Pb react with bases to form hydroxo ions and hydrogen gas.**

  *Example:*
  $Sn(s) + KOH(aq) + 2H_2O(l) \rightarrow K^+(aq) + Sn(OH)_3^-(aq) + H_2(g)$

Silicon chloride ($SiCl_4$) reacts with water to form silicon dioxide and hydrochloric acid, which turns litmus paper pink.

## Atomic Properties

- Each element in group 14 has four valence electrons and an electron configuration ending with $ns^2np^2$.
- Carbon group elements participate in covalent bonding with an oxidation number of +4. Carbon, germanium, tin, and lead can also have an oxidation number of +2. Carbon and silicon have an oxidation number of −4 in some compounds.
- Carbon, silicon, and tin occur as allotropes.
- Atomic and ionic radii increase moving down the group and are similar to their corresponding group 13 elements.
- Except for carbon, the group 14 elements have similar ionization energies and no distinct pattern of electronegativities.

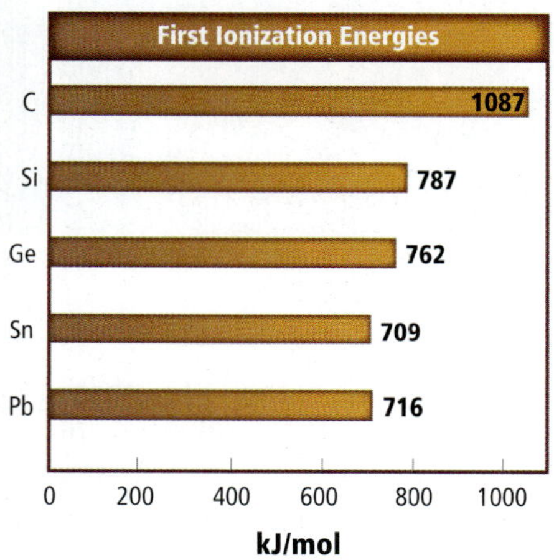


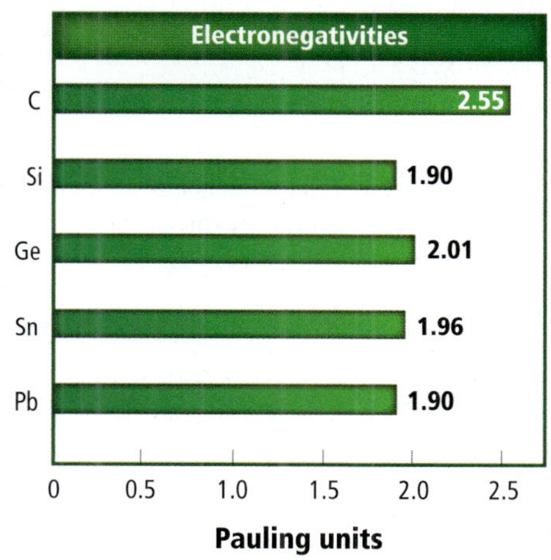


| Atomic radius (pm) | Ionic radius (pm) |
|---|---|
| C 77 | $C^{4+}$ 15 |
| Si 118 | $Si^{4+}$ 41 |
| Ge 122 | $Ge^{4+}$ 53 |
| Sn 140 | $Sn^{4+}$ 71 |
| Pb 146 | $Pb^{4+}$ 84 |

- **C reacts with water to form carbon monoxide and hydrogen gas.**

  *Example:* $C(s) + H_2O(g) \rightarrow CO(g) + H_2(g)$
- **Si reacts with water to form silicon dioxide and hydrogen gas.**

  *Example:* $Si(s) + 2H_2O(l) \rightarrow SiO_2(s) + 2H_2(g)$
- **Sn and Pb react with acids to form hydrogen gas.**

  *Example:* $Pb(s) + 2HBr(aq) \rightarrow PbBr_2(aq) + H_2(g)$
- **C reacts with hydrogen to form hydrocarbons, such as propane.**

  *Example:* $3C(s) + 4H_2(g) \rightarrow C_3H_8(g)$

## Analytical Tests

Because the group 14 elements bond covalently, they do not lend themselves to identification through flame tests. The exception is lead, which produces a light-blue color. The carbon group elements can be identified through analysis of their physical properties (melting point, boiling point, density), emission spectra, or reactions with other chemicals. For example, tin and lead form precipitates when added to specific solutions.

If lead nitrate is added to potassium iodide, a yellow precipitate of lead iodide forms.

Carbon
6
**C**
$[He]2s^22p^2$

## Graphite Golf Shafts

Some golf shafts are created by fusing sheets of graphite together with a binding material. The use of graphite instead of traditional steel allows greater versatility in club design and construction. Graphite sheets can be layered to vary the weight and stiffness of the club, which for many golfers translates into greater shot distance and overall performance. Graphite also offers greater durability than steel for golfers with powerful swings.

Graphite can be easily formed into sheets due to its atomic structure.

## Diamond Cutting

The way a diamond is cut is one of the "4 Cs" that gemologists use to determine a diamond's value. If diamond is the hardest mineral on Earth, then how is it possible to cut a diamond? Diamond cutters use other diamonds and lasers to create facets that reflect and refract light. The more precisely the cuts are made, the greater the gem's brilliance. If a diamond cut is too shallow or too deep, light escapes from the diamond without traveling back to the eye, resulting in a lackluster appearance.

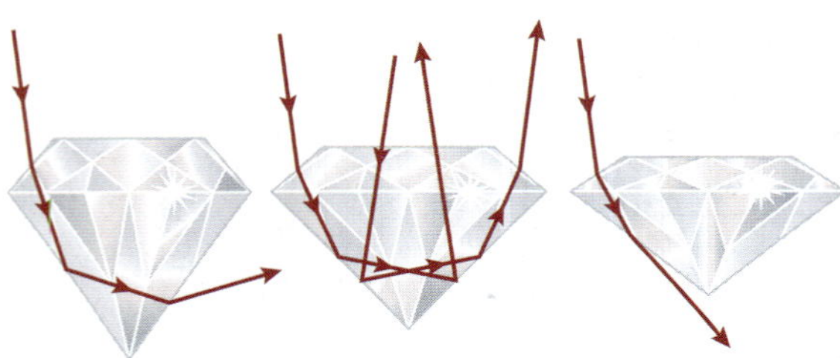


The way a diamond is cut determines how well light is reflected and refracted within the gemstone.

## Nanotubes

Fullernes form a group of carbon allotropes. There are spherical fullerenes nicknamed buckyballs and cylindrical fullerenes known as buckytubes or nanotubes. Fullerenes have yet to display all of their capabilities to scientists. One of the most promising areas of fullerene research involves the creation of nanotubes. Nanotubes are sheets of carbon that are rolled up into cylinders. These cylinders are strong—due to the hexagonal structure of the carbon atoms—and have unique conducting properties. Fullerene nano-technology on the horizon includes the development of faster computer chips, smaller electronic components, and more advanced space-exploration vehicles.

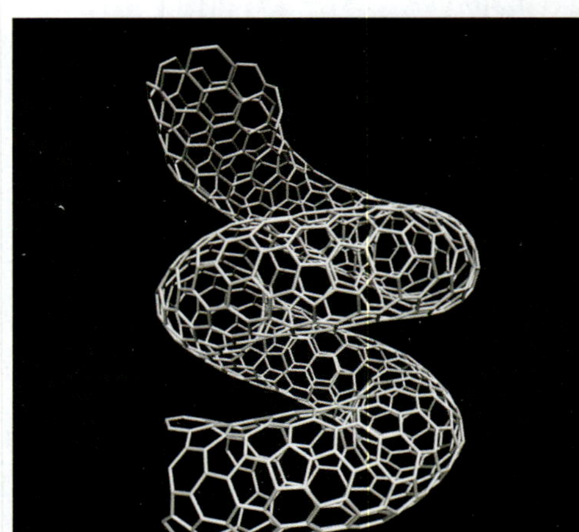

The hexagonal structure of carbon atoms gives extraordinary strength to carbon nanotubes.

| Silicon |
|---|
| 14 |
| **Si** |
| $[Ne]3s^23p^2$ |

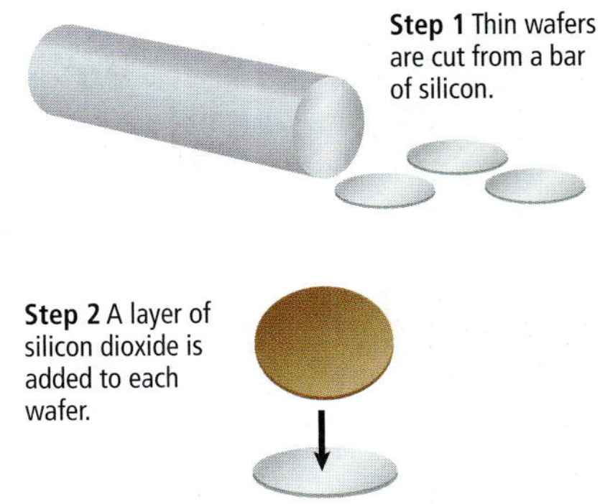


More than 250 steps are needed to create one computer chip.

## Computer Chips

Computer chips are everywhere. From pet-identification systems to laptop computers—any device that can be programmed contains a computer chip. Silicon's abundance and ability as a semiconductor make it an ideal material for the production of computer chips. The first step in making a computer chip involves cutting pure silicon into wafer-like pieces. Silicon dioxide ($SiO_2$) is then cultivated on each wafer. Layers upon layers of silicon dioxide and other chemicals are used to create chips for specific functions.

## Glass

Almost 40% of the sand produced in the United States is used for glass production. Glass is created by first melting silicon dioxide ($SiO_2$) obtained from sand with sodium carbonate and then supercooling the mixture. This results in a solid whose structure resembles a liquid and whose physical properties make it ideal for glassmaking. For manufacturing purposes, sand that yields at least 95% $SiO_2$ with no impurities is required for making glass products, such as exterior panels on buildings, automotive windshields, and commercial beverage containers. Manufacturers of high precision optical instruments, such as telescopes and microscopes, require sand that contains more than 99.5% $SiO_2$.

Sand dunes in Michigan provide millions of metric tons of sand each year.

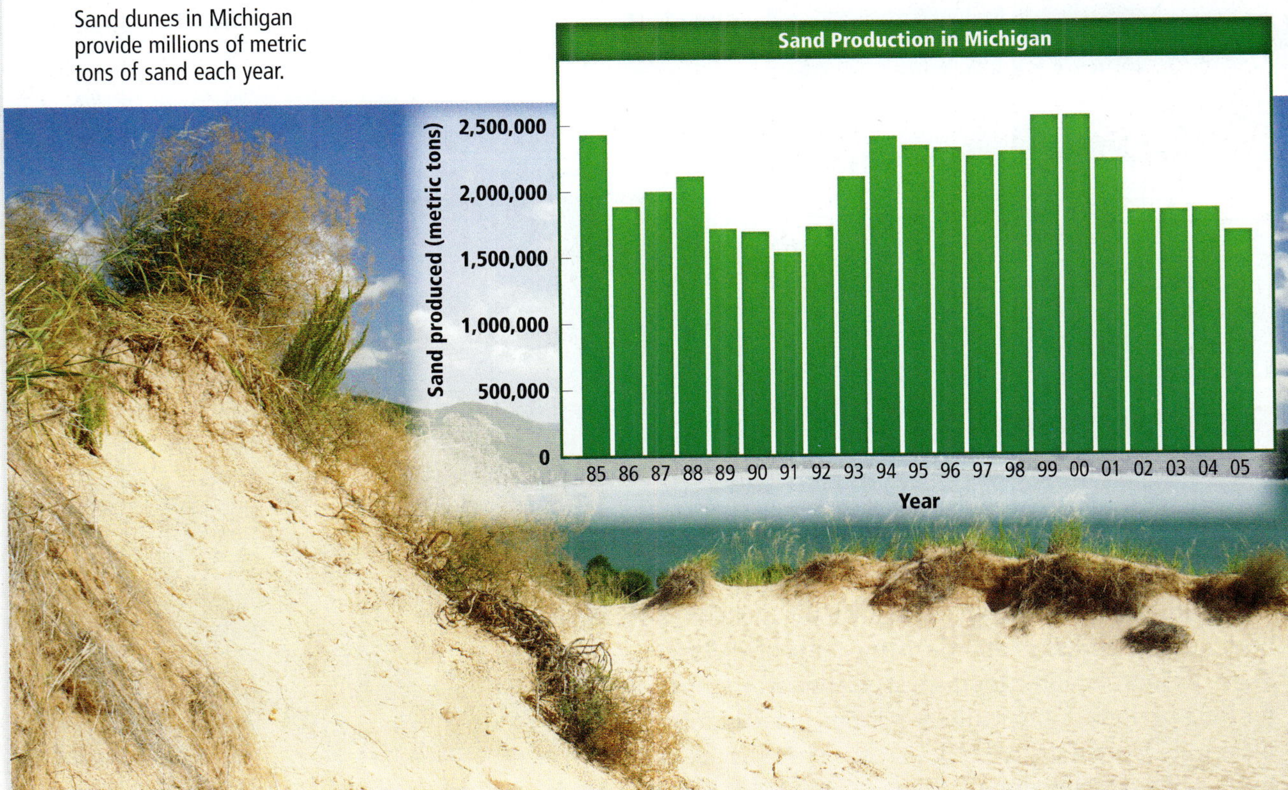


gabes1976/iStock/Getty Images

| Germanium |
| --- |
| 32 |
| **Ge** |
| $[Ar]4s^23d^{10}4p^2$ |

## Night Vision

Lenses that contain germanium are found in an array of night vision equipment including goggles, binoculars, and cameras. Unlike ordinary glass lenses, germanium-containing lenses are transparent to infrared radiation. Infrared radiation is emitted by objects that radiate heat. Infrared radiation is part of the electromagnetic spectrum, a region distinct from the visible spectrum, so special equipment is needed to detect it. Night vision is used for military and security applications, to monitor wildlife, to navigate roads, and to locate objects that have been hidden by criminals.

A night scope is used to observe a military transport plane that takes off with lights out in a conflict zone.

## Fiber Optic Cables

Fiber optic cables are responsible for the transmission of information both across the street and across the globe. These cables are made of extremely pure glass that allows light signals to travel the span of the cable without losing a significant amount of energy. Each fiber optic cable consists of three main parts: a core, cladding, and a buffer coating. The core is made by exposing gaseous germanium tetrachloride ($GeCl_4$) to oxygen, resulting in germanium dioxide ($GeO_2$). The germanium dioxide helps the light signal move effectively along the cable.

Germanium is added to the core of a fiber optic cable to improve the efficiency of the light signal.

| Tin |
| --- |
| 50 |
| **Sn** |
| $[Kr]5s^24d^{10}5p^2$ |

## Food Packaging

A quick trip to the grocery store reveals that many different foods are stored in cans. Soft drinks, fruits, vegetables, and even meats can be stored in cans. Cans are made from sheets of steel that are coated on both sides with pure tin. Known as tinplate, the metal is both durable and resistant to rusting and corrosion. These properties allow foods to stay fresh on the shelf for long periods of time, and to be transported long distances. More than 200 million cans are used per day in the United States alone.

More than 2500 different products are packaged in cans.

Lead
82
**Pb**
$[Xe]6s^2 4f^{14} 5d^{10} 6p^2$

## Leaded or Unleaded?

In the early 1900s, the automotive industry needed to solve a problem that people complained about when they drove their cars—knocking in the engine. At the time, little was known about the chemistry of fuels and fuel additives. Researchers spent seven years searching for a gasoline additive that effectively reduced knocking before discovering tetraethyl lead ($Pb(C_2H_5)_4$). Further research revealed the health and environmental risks posed by lead, leading to the development of unleaded fuels that reduce knocking.

Unleaded fuels reduce knocking in car engines and do not have the health and environmental concerns posed by leaded fuels.

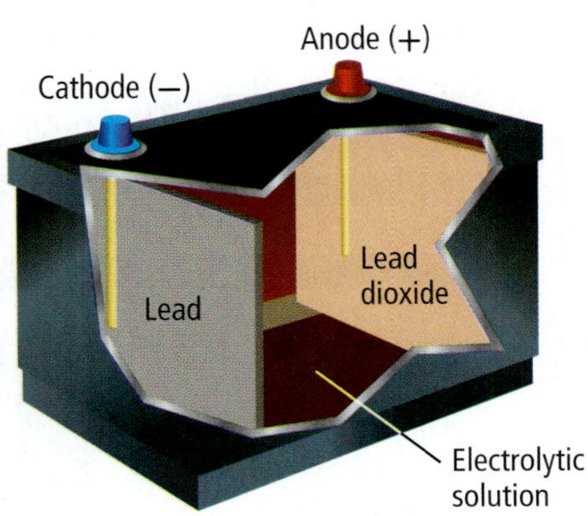


Eighty-five percent of the lead used in the United States goes into making lead-acid batteries.

## Batteries

A car battery is composed of three main parts: one electrode made of lead, one electrode made of lead dioxide ($PbO_2$), and an electrolytic solution made with sulfuric acid ($H_2SO_4$). That is why car batteries are also called lead-acid batteries. The battery's energy comes from the chemical reactions occurring between the electrodes and the electrolyte. During the chemical reaction, electrons are produced that accumulate on the lead electrode. When a wire connects the electrodes, electrons flow freely from the lead electrode to the lead-dioxide electrode, and the battery discharges. Applying a current reverses the reaction, recharging the battery.

## REVIEW

**33. Write** the electron configuration of tin.

**34. Summarize** the physical properties of the elements in group 14.

**35. Compare and contrast** the atomic properties of the group 13 and group 14 elements.

**36. Predict** what product or products will be formed if bromine gas reacts with solid carbon under elevated temperature conditions.

**37. Consider** why graphite is the most suitable carbon allotrope for golf clubs.

**38. Calculate** Pure diamond has a density of 3.52 $g/cm^3$, while graphite has a density of 2.20 $g/cm^3$. Recall that density = mass/volume. Samples of diamond and graphite each displace 4.60 mL of water. What is the mass of each sample?

# GROUP 15: NITROGEN GROUP

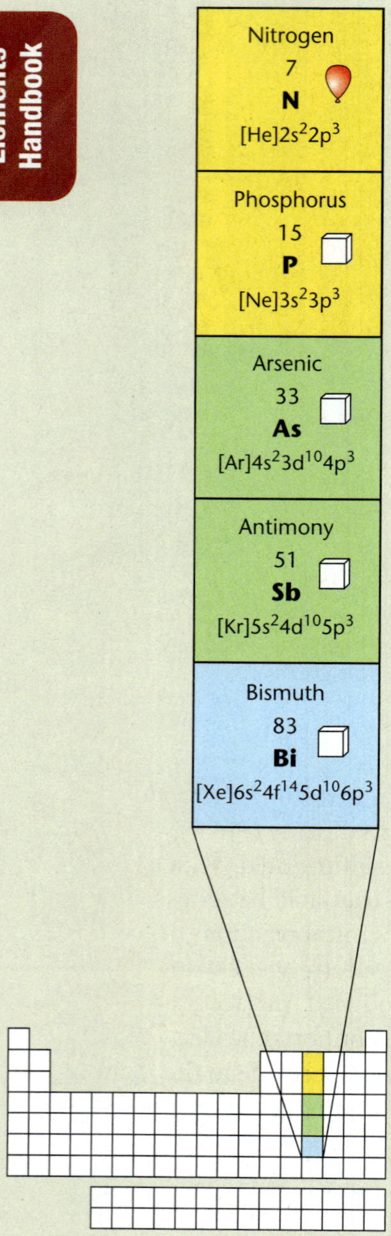


## Physical Properties

- Like the elements in group 14, the group 15 elements increase in metallic character going down the group. Nitrogen and phosphorus are nonmetals. Arsenic and antimony are metalloids. Bismuth is a metal.
- Also like group 14, the nitrogen group elements vary in appearance.
- Nitrogen is a colorless, odorless gas ($N_2$).
- Phosphorus exists in three allotropic forms, which are all solids. The forms are white, red, and black in color.
- Arsenic is a shiny, gray solid that is brittle. Under certain conditions, it can become a dull, yellow solid. Arsenic sublimates when heated.
- Antimony is a shiny, silver-gray solid that is very brittle.
- Bismuth is a shiny, gray solid that has a pink cast to it. It is one of the least conductive metals on the periodic table and is also brittle.
- Boiling points and densities of the group 15 elements generally increase going down the group.

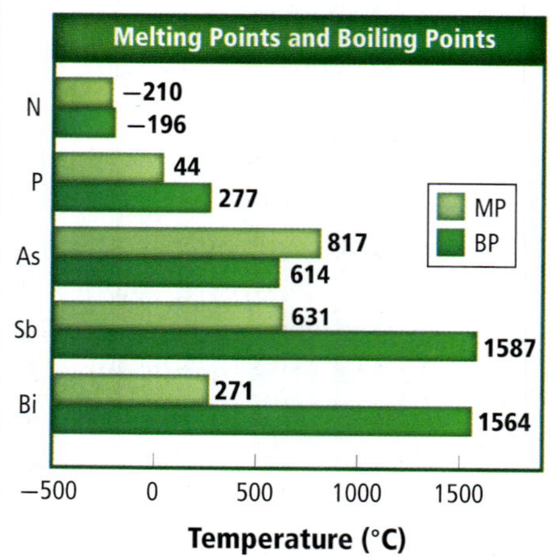


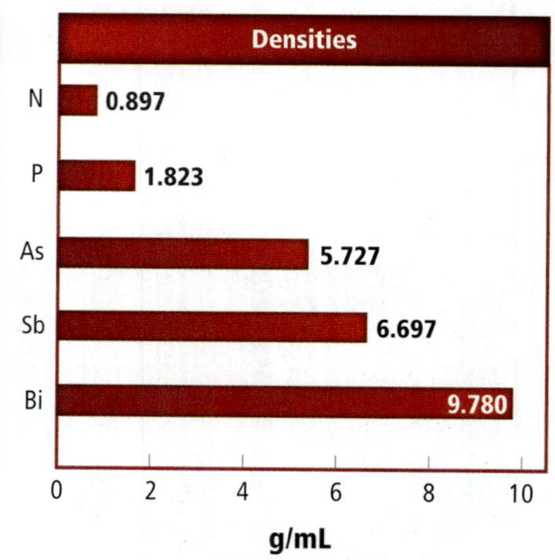


## Common Reactions

- **As high temperatures are increased, nitrogen reacts with oxygen to form nitric oxide.**

  *Example:* $N_2(g) + O_2(g) \rightarrow 2NO(g)$
- **At high temperature and pressure, nitrogen reacts with hydrogen to form ammonia.**

  *Example:* $N_2(g) + 3H_2(g) \rightarrow 2NH_3(g)$
- **P reacts with an excess of oxygen to form phosphorus(V) oxide.**

  *Example:* $P_4(s) + 5O_2(g) \rightarrow P_4O_{10}(s)$
- **P, As, Sb, and Bi react with oxygen to form element(III) oxides.**

  *Example:* $P_4(s) + 3O_2(g) \rightarrow P_4O_6(s)$
- **P, As, Sb, and Bi react with halogens to form trihalides.**

  *Example:* $2Sb(s) + 3Cl_2(g) \rightarrow 2SbCl_3(s)$

## Atomic Properties

- Each element in group 15 has five valence electrons and an electron configuration ending with $ns^2p^3$.
- Nitrogen is diamagnetic, meaning it is repelled by magnetic fields. This indicates that all of nitrogen's electrons are paired.
- Nitrogen can have oxidation numbers ranging from −3 to +5.
- Phosphorus, arsenic, and antimony can have oxidation numbers of −3, +3, and +5.
- Bismuth commonly has an oxidation number of +3.
- Going down the group, first ionization energies and electronegativities decrease and atomic radii increase.

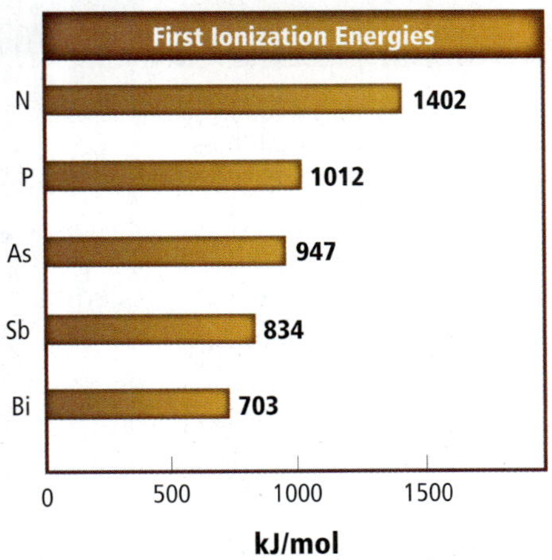


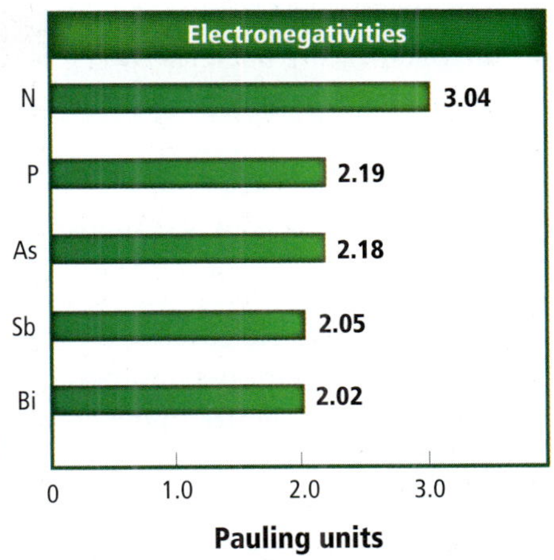


| Atomic radius (pm) | Ionic radius (pm) |
|---|---|
| N 75 | $N^{3-}$ 146 |
| P 110 | $P^{3-}$ 212 |
| As 120 | $As^{3-}$ 222 |
| Sb 140 | $Sb^{5+}$ 62 |
| Bi 150 | $Bi^{5+}$ 74 |

## Analytical Tests

Because group 15 elements bond covalently and most are nonmetallic in nature, they do not lend themselves to identification through flame tests. The exceptions are antimony and bismuth. Antimony produces a faint green or blue color when placed in a flame, while bismuth produces a light purple-blue color.

The nitrogen group elements can be identified through analysis of their physical properties (melting point, boiling point, density), emission spectra, or reactions with other chemicals. For example, bismuth ions precipitate when added to tin(II) hydroxide and sodium hydroxide. Another example is the test for ammonium compounds. These compounds, which contain nitrogen, can be identified by their distinct smell when added to sodium hydroxide and by the color change observed when red litmus paper is placed at the opening of the test tube.

The ammonia vapor produced by mixing ammonium compounds ($NH_4^+$) with sodium hydroxide changes red litmus paper to blue.

| Nitrogen |
|---|
| 7 |
| **N** |
| $[He]2s^2 2p^3$ |

## Nitrogen-Fixing Bacteria

Although nitrogen makes up about 78% of Earth's atmosphere, it occurs in a form that plants cannot use. Some bacteria in the soil convert nitrogen gas ($N_2$) from the air into a usable form by breaking the molecule's triple bond. This creates a form of nitrogen that plants uptake into their root systems. Plants need nitrogen to build cellular components, to participate in photosynthesis, and to transfer energy effectively. Commercial fertilizers mimic the action of nitrogen-fixing bacteria by providing nitrogen and other nutrients in forms that are easily incorporated into the plant system.

Nitrogen-fixing bacteria are found in protective nodules along plant roots.

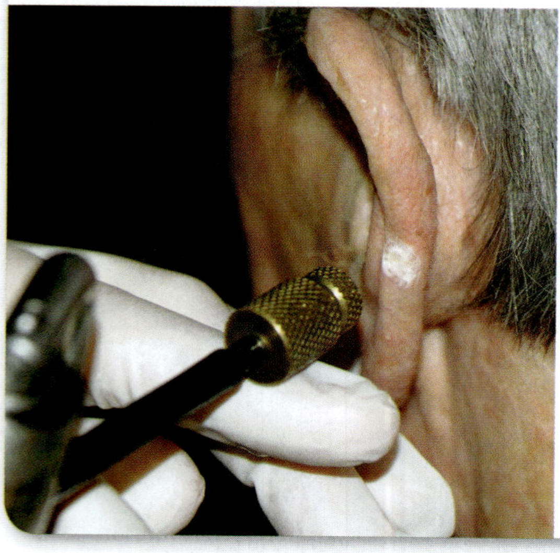

## Liquid Nitrogen Cryotherapy

Cryotherapy, also called cryosurgery, is a medical procedure used to remove a variety of skin lesions, including carcinomas, warts, and other tissue abnormalities. The procedure involves dabbing liquid nitrogen onto the affected area to freeze and kill the cells. This is then repeated over time until all of the affected tissue is gone. Research has shown that patients who undergo cryotherapy treatment for certain types of lesions experience a lower recurrence rate than patients who receive radiation or surgical removal.

Doctors use liquid nitrogen as one of the treatment options to remove certain types of skin cancer. More than 1.3 million new cases of skin cancer are recorded each year in the United States.

| Phosphorus |
|---|
| 15 |
| **P** |
| $[Ne]3s^2 3p^3$ |

## Safety Matches

Safety matches consist of two main parts: the tip and the textured strip on the side of the box. The tip contains potassium chlorate, and the textured strip contains red phosphorus. When these two chemicals come in contact, a chemical reaction occurs, and fire is produced. In safety matches, the chemicals needed for reaction are separate from each other. In strike-anywhere matches, both chemicals are contained in the matchstick so that ignition can occur using almost any surface.

The strike of a match initiates a chemical reaction that produces a flame.

Antimony
51
**Sb**
$[Kr]5s^2 4d^{10} 5p^3$

## Flame Retardants

Antimony trioxide ($Sb_2O_3$) is used along with brominated or chlorinated compounds in the making of flame retardants that protect plastics, paints, and some textile products. Antimony trioxide increases the effectiveness of the halogen compounds in preventing the spread of a fire. Research shows that approximately 5000 deaths in the United States are caused by fire each year. The use of flame retardants improves escape time, releases less toxic gases and heat, and decreases fire damage.

Antimony trioxide fire retardants coat electrical wires and components found in a variety of everyday appliances.

Bismuth
83
**Bi**
$[Xe]6s^2 4f^{14} 5d^{10} 6p^3$

## Soothing Upset Stomachs

Originally named *Mixture Cholera Infantum,* the popular pink medicine now used for upset stomachs was created to combat cholera. This mixture, whose active ingredient was bismuth subsalicylate ($C_7H_5BiO_4$), proved effective in treating the nausea and vomiting associated with infant cholera. However, it could not cure the disease itself. Nonetheless, the product became a wide success. As science advanced and doctors realized that cholera was contracted from bacteria (which could be treated with antibiotics), bismuth subsalicylate found its way into medical treatments for a variety of other stomach problems, including heartburn, indigestion, and ulcers.

Bismuth subsalicylate ($C_7H_5BiO_4$) is the active ingredient in some medicines used to treat stomach problems.



## REVIEW

**39. Identify** which elements in the nitrogen group are metals, nonmetals, or metalloids.

**40. Infer** why nitrogen does not react with other elements under normal temperature conditions.

**41. Explain** why a compound of antimony is used in flame retardants that protect plastic products.

**42. Describe** how fertilizers mimic the action of nitrogen-fixing bacteria.

**43. Write** a balanced chemical equation for the reaction between potassium chlorate ($KClO_3$) and red phosphorus ($P_4$). The reaction produces potassium chloride (KCl) and phosphorus pentoxide ($P_4O_{10}$).

**44. Predict** what product will be formed when bismuth is combined with chlorine.

**45. Calculate** A 35-kg bag of fertilizer contains 5.25 kg of nitrogen. What percentage of the fertilizer is nitrogen?

# GROUP 16: OXYGEN GROUP



Oxygen
8
**O**
$[He]2s^22p^4$

Sulfur
16
**S**
$[Ne]3s^23p^4$

Selenium
34
**Se**
$[Ar]4s^23d^{10}4p^4$

Tellurium
52
**Te**
$[Kr]5s^24d^{10}5p^4$

Polonium
84
**Po**
$[Xe]6s^24f^{14}5d^{10}6p^4$

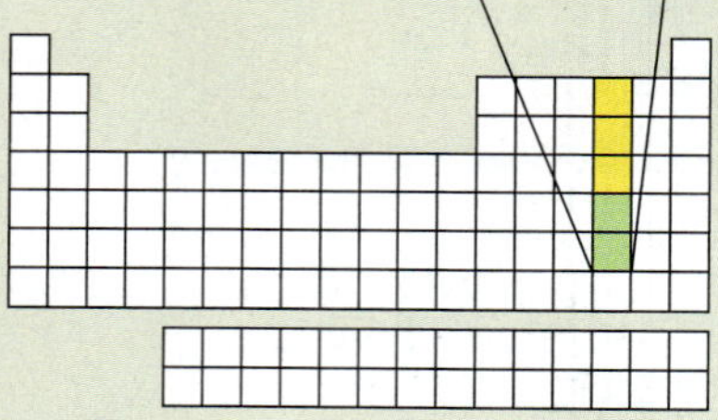

## Physical Properties

- At room temperature, oxygen is a clear, odorless gas, while the other group 16 elements are solids.
- Some of the group 16 elements have several common allotropic forms. Oxygen can exist as either $O_2$ or $O_3$ (ozone). Sulfur has many allotropes. Selenium has three common allotropes: amorphous gray, red crystalline, and red/black powder.
- Oxygen, sulfur, and selenium are nonmetals. Tellurium and polonium are metalloids.
- $O_2$ is paramagnetic, which means that a strong magnet will attract oxygen molecules.
- Except for polonium, boiling points and melting points of the group 16 elements increase with increasing atomic number. Density increases with increasing atomic number for all group 16 elements.

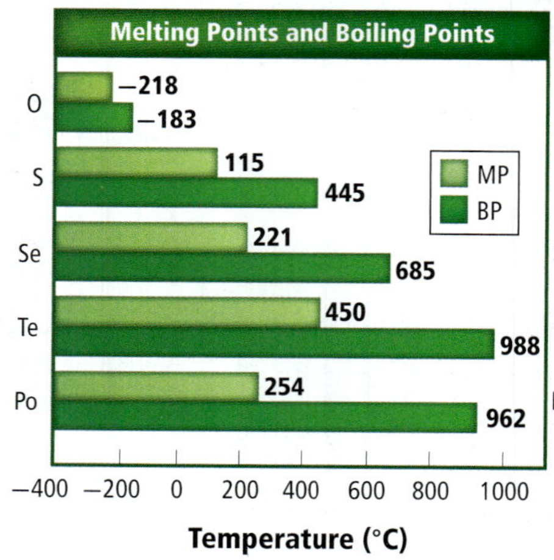


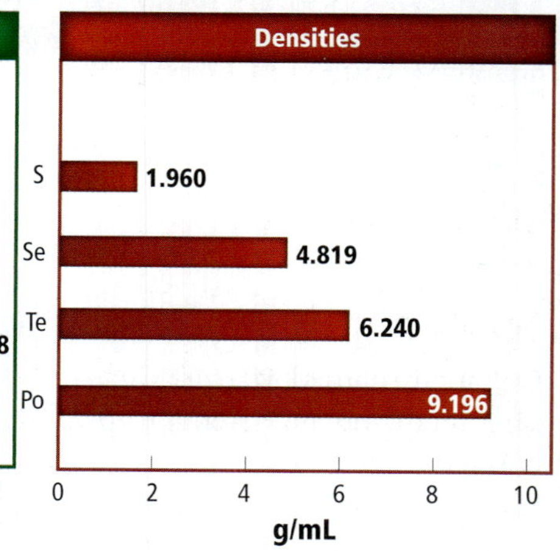


## Common Reactions

- **S, Se, Te, and Po react with oxygen to form oxides, such as selenium oxide.**

  *Example:* $Se(s) + O_2(g) \rightarrow SeO_2(s)$

- **Oxygen also reacts with hydrogen and most of the elements in groups 1, 2, 13, 14, 15, and 17 to form oxides, such as silicon dioxide and magnesium oxide.**

  *Examples:* $Si + O_2 \rightarrow SiO_2$
  $2Mg + O_2 \rightarrow 2MgO$

- **O, S, Se, Te, and Po react with halogens to form halides, such as sulfur(VI) fluoride.**

  *Example:* $S(s) + 3F_2(g) \rightarrow SF_6(l)$

### Oxides of Main Group Elements

| | |
|---|---|
| H | $H_2O$, $H_2O_2$ |
| 1 | $Li_2O$, $Na_2O$, $K_2O$, $Rb_2O$, $Cs_2O$, $Fr_2O$ |
| 2 | BeO, MgO, CaO, SrO, BaO, RaO |
| 13 | $B_2O_3$, $Al_2O_3$, $Ga_2O_3$, $In_2O_3$, $In_2O$, $Ti_2O$ |
| 14 | $CO_2$, $SiO_2$, $GeO_2$, $SnO_2$, SnO, $PbO_2$, PbO |
| 15 | $N_2O_5$, $N_2O_3$, $N_2O$, NO, $NO_2$, $P_4O_{10}$, $P_4O_6$, $As_2O_5$, $As_4O_6$, $Sb_2O_5$, $Sb_4O_6$, $Bi_2O_3$ |
| 17 | $Cl_2O_7$, $Cl_2O$, $Br_2O$, $I_2O_5$ |

## Atomic Properties

- Each element in group 16 has six valence electrons and an electron configuration ending with $ns^2np^4$.
- Group 16 elements can have many different oxidation numbers. For example, oxygen can have oxidation numbers of −2 and −1, and sulfur can have oxidation numbers of +6, +4, and −2.
- Going down the elements in group 16, the atomic radii and ionic radii increase.
- Electronegativity and first ionization energy decrease going down the elements in group 16.
- Polonium has 27 known isotopes. All are radioactive.

| Atomic radius (pm) | Ionic radius (pm) |
|---|---|
| O 73 | $O^{2-}$ 140 |
| S 103 | $S^{2-}$ 184 |
| Se 119 | $Se^{2-}$ 198 |
| Te 142 | $Te^{2-}$ 221 |
| Po 168 | |

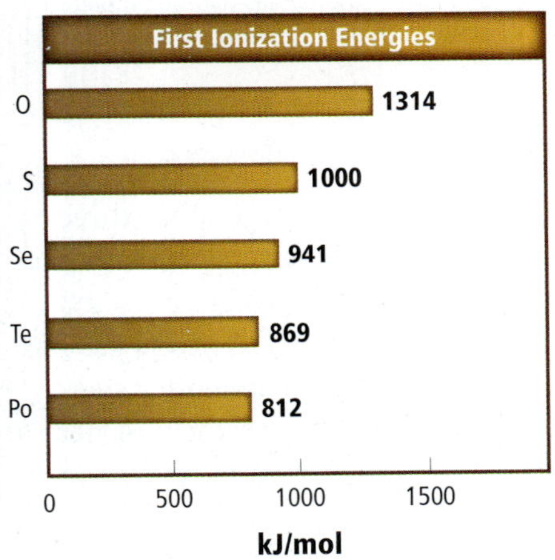


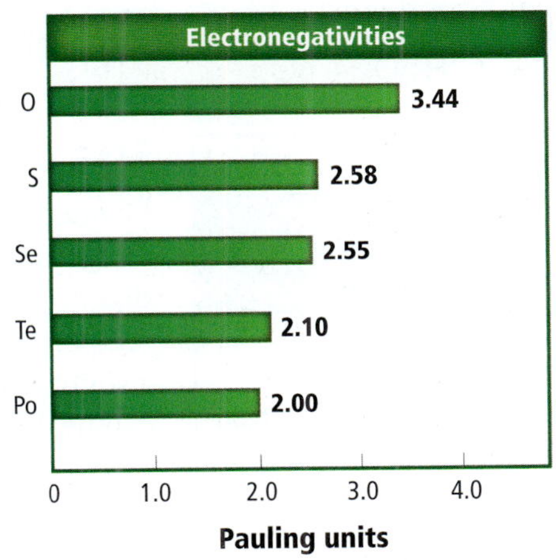


- **Group 16 elements are involved in many important industrial reactions, such as the formation of sulfuric acid.**

  *Example:* Sulfuric-acid production is a three-step process.

  1) $S(s) + O_2(g) \rightarrow SO_2(g)$

  2) $2SO_2(g) + O_2(g) \rightarrow 2SO_3(g)$

  3) $SO_3(g) + H_2O(l) \rightarrow H_2SO_4(l)$

## Analytical Tests

Oxygen can be measured in many different ways and in many different environments. For example, dissolved-oxygen meters measure oxygen in water samples. Dissolved-oxygen meters use an electrochemical reaction that reduces oxygen molecules to hydroxide ions. The meter measures the electric current produced during this reaction. The higher the oxygen concentration, the larger the current.

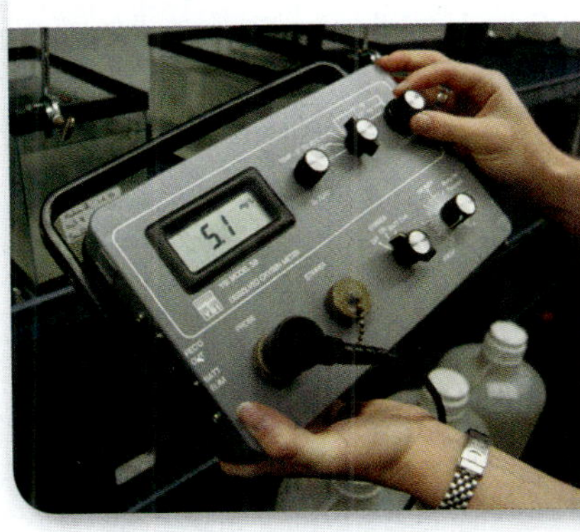

Dissolved-oxygen tests help scientists determine biological activity in water samples.

Oxygen
8
**O**
$[He]2s^22p^4$

Photosynthesis captures energy from sunlight and provides hydrogen ions to synthesize glucose from carbon dioxide.

## Photosynthesis Produces $O_2$ from $H_2O$

Earth's atmosphere is 21% oxygen by volume. Most of the oxygen in the atmosphere comes from photosynthesis. Photosynthetic organisms, including plants and cyanobacteria, use energy from sunlight to oxidize water. The result is hydrogen ions ($H^+$) and oxygen ($O_2$). The reactions involved in this part of photosynthesis are called light reactions because they depend on light energy to proceed. During the dark reactions of photosynthesis, the hydrogen ions derived during the light reactions are combined with carbon dioxide ($CO_2$) to form glucose ($C_6H_{12}O_6$). The overall reaction for photosynthesis follows:

$$6H_2O + 6CO_2 \rightarrow C_6H_{12}O_6 + 6O_2$$

**Air Quality Index for Ozone**

| Index Values | Levels of Health Concern | Cautionary Statements |
|---|---|---|
| 0–50 | good | none |
| 51–100 | moderate | Unusually sensitive people should consider reducing prolonged or heavy exertion outdoors. |
| 101–150 | unhealthy for sensitive groups | Active children and adults, and people with lung disease, such as asthma, should reduce prolonged or heavy exertion outdoors. |
| 151–200 | unhealthy | Active children and adults, and people with lung disease should avoid prolonged or heavy exertion outdoors. Everyone else should reduce prolonged or heavy exertion outdoors. |
| 201–300 | very unhealthy | Active children and adults, and people with lung disease, such as asthma, should avoid all outdoor exertion. Everyone else should avoid prolonged or heavy exertion outdoors. |
| 301–500 | hazardous | Everyone should avoid all physical activity outdoors. |

Data obtained from: Patient Exposure and the Air Quality Index. *U.S. E.P.A.* March 2006

## The Dual Nature of Ozone

Ozone ($O_3$), an allotrope of oxygen, has three oxygen atoms per molecule instead of two. Like diatomic oxygen ($O_2$), ozone is a gas at room temperature. However, unlike $O_2$, ozone gas has a slight blue color and a distinctive odor that can be detected during a thunderstorm or near a high-voltage electric motor. Ozone is also more reactive than diatomic oxygen. At ground level, ozone can be a serious potential health hazard, irritating eyes and lungs. High ground-level ozone concentrations are a particular threat on hot sunny days. The table illustrates how ozone affects air quality and health. On the other hand, stratospheric ozone protects Earth from harmful UV radiation by absorbing UV rays from sunlight.

Many cities issue air-quality alerts when ground-level ozone levels are high.

| Sulfur |
|---|
| 16 |
| **S** |
| $[Ne]3s^23p^4$ |

## An Economic Indicator

Sulfuric acid is one of the world's most important industrial raw materials. In the United States, more sulfuric acid is produced than any other industrial chemical. Most sulfuric acid is used in the production of phosphate fertilizers. Sulfuric acid is also important in extracting metals from ore, oil refining, waste treatment, chemical synthesis, and as a component in lead-acid batteries. Sulfuric acid is so important that economists use its production as a measure of a nation's industrial development.

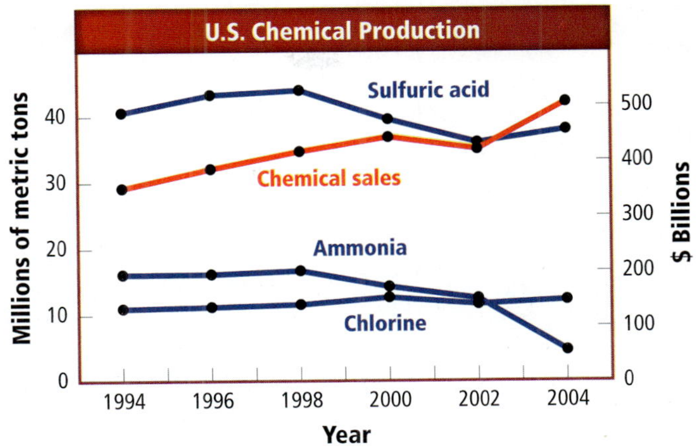


Data obtained from: *Chemical & Engineering News* 83 (2005) and 84 (2006).

Sulfuric acid production in the United States is used to track chemical economic trends.

| Selenium |
|---|
| 34 |
| **Se** |
| $[Ar]4s^23d^{10}4p^4$ |

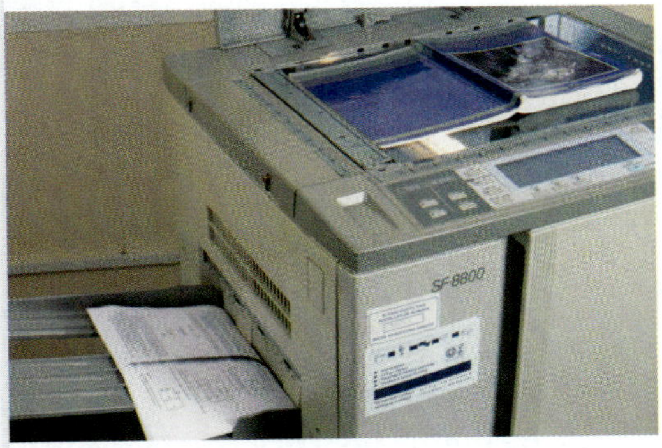


Gray selenium is a key component in many photocopiers.

## Photocopies

Gray selenium is a photoconductor, which means it conducts electricity more efficiently in the presence of light than in the dark. Some photocopiers use this property to copy images. In a photocopier, a bright light shines on the original. Mirrors reflect the dark and light areas onto a drum coated with a thin layer of selenium. Because selenium is a photoconductor, the light areas conduct electricity, while the dark areas do not. As current flows through the drum, the light areas develop a negative charge and the dark areas develop a positive charge. Negatively charged toner particles are attracted to the positively charged dark areas to create a copy of the original image. Some of this same technology has been applied in developing new high-resolution digital detectors that use selenium as a photoconductor.



## REVIEW

46. **Identify** the molecule that is the source of oxygen atoms for $O_2$ production during photosynthesis.

47. **Explain** why high ozone concentrations are harmful at ground level but beneficial in the upper atmosphere.

48. **Calculate** Approximately 90% of the sulfur used in the United States is used to make sulfuric acid. In 2004, 38.0 million metric tons of sulfuric acid were produced. How much sulfur did the United States use in 2004?

49. **Apply** Coal and petroleum products are sometimes contaminated with sulfur. When coal or petroleum containing sulfur is burned, sulfur dioxide ($SO_2$) can be released into the atmosphere. Use the information about the reactions involved in industrial sulfuric acid production to infer how atmospheric sulfur dioxide contributes to acid precipitation.

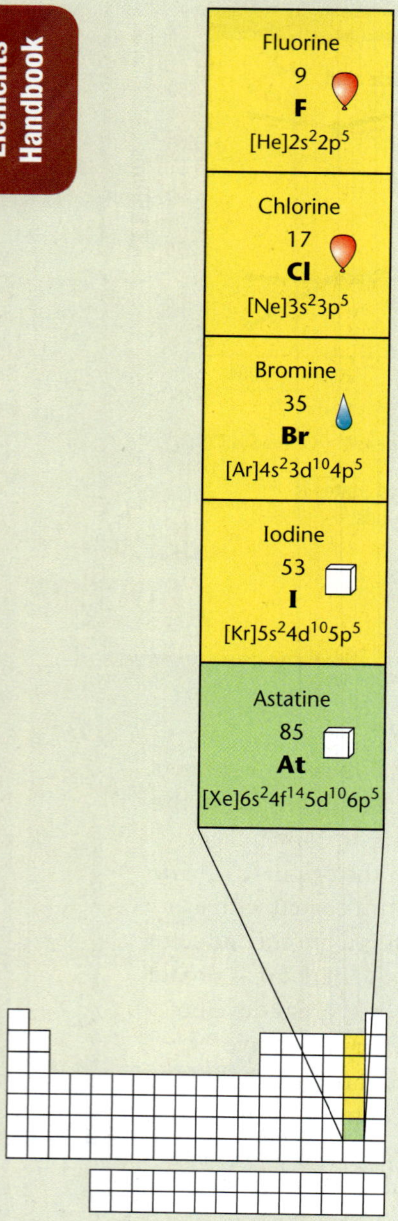


## Physical Properties

- Fluorine and chlorine are gases at room temperature. Along with mercury, bromine is one of only two elements that are liquid at room temperature. Iodine is a solid that easily sublimes at room temperature.
- Fluorine gas is pale yellow. Chlorine gas is yellow-green. Bromine is a red-brown liquid. Iodine is a blue-black solid.
- Both boiling points and melting points of the group 17 elements increase with increasing atomic number.

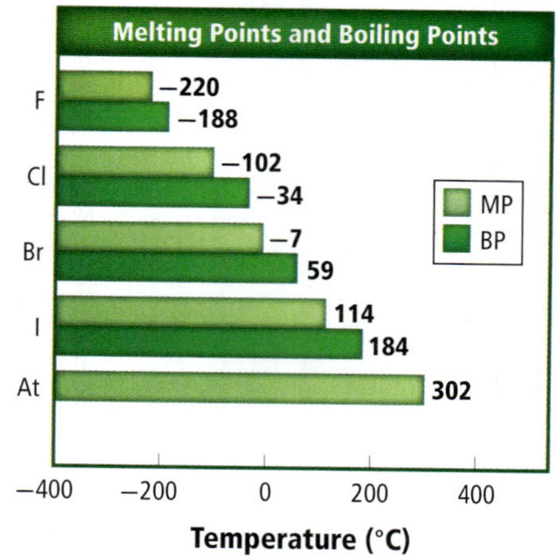

Iodine crystals are a blue-black color. They produce a violet vapor when they sublime at room temperature.

## Common Reactions

- **The halogens react with alkali metals and alkaline earth metals to form salts, such as potassium bromide and calcium chloride.**

  *Examples:* $2K(s) + Br_2(g) \rightarrow 2KBr(s)$ and $Ca(s) + Cl_2(g) \rightarrow CaCl_2(s)$
- **The halogens can form acids, such as hydrochloric acid, by hydrolysis in water.**

  *Example:* $Cl_2(g) + H_2O(l) \rightarrow HClO(aq) + HCl(aq)$
- **Several important plastic polymers, including nonstick coatings and polyvinyl chloride, contain group 17 elements.**

  *Example:* Polyvinyl chloride (vinyl) is made by a three-step process.

  1) Ethene reacts with chlorine to form dichloroethane.
     $C_2H_4(g) + Cl_2(g) \rightarrow C_2H_4Cl_2(l)$
  2) At high temperature and pressure, dichloroethane is converted to vinyl chloride and HCl gas.
     $C_2H_4Cl_2(l) \rightarrow C_2H_3Cl(l) + HCl(g)$
  3) Vinyl chloride polymerizes to form polyvinyl chloride.
     $2n(C_2H_3Cl)(l) \rightarrow (\text{—}CH_2\text{–}CHCl\text{–}CH_2\text{–}CHCl\text{—})_n(l)$
- **Fluorine is the most active of all the elements and reacts with every element except helium, neon, and argon.**

  *Example:* $2Al(s) + 3F_2(g) \rightarrow 2AlF_3(s)$

## Atomic Properties

- Each element in group 17 has seven valence electrons and an electron configuration ending with $ns^2np^5$.
- Electronegativities and first ionization energies decrease going down the elements in group 17.
- Fluorine is the most electronegative element on the periodic table. Therefore, it has the greatest tendency to attract electrons.
- Astatine is a radioactive element with no known uses.
- The atomic radii and ionic radii of the group 17 elements increase going down the group.

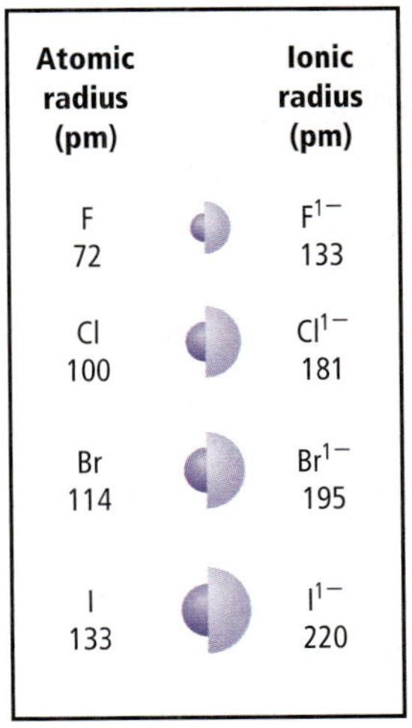


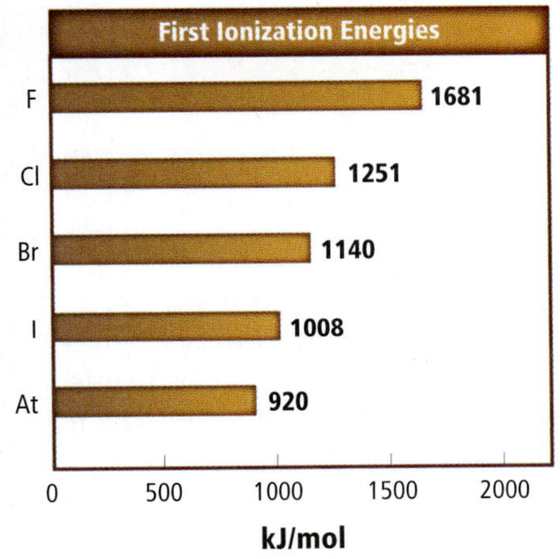


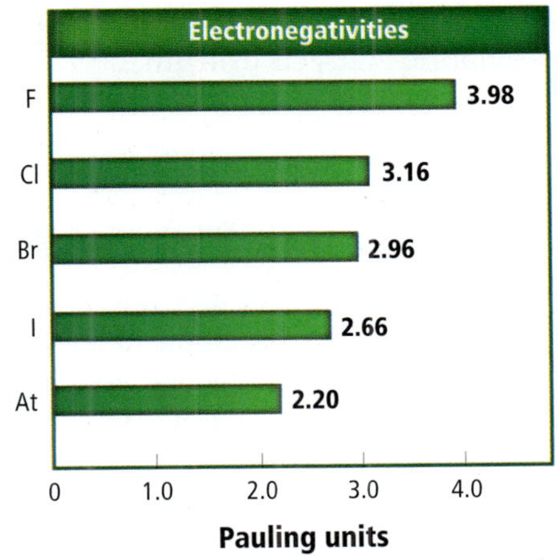


## Analytical Tests

Three of the halogens can be identified through precipitation reactions. Chlorine, bromine, and iodine react with silver nitrate, forming distinctive precipitates. Silver chloride is a white precipitate, silver bromide is a cream-colored precipitate, and silver iodide is a yellow precipitate.

Chlorine, bromine, and iodine can also be identified when they dissolve in cyclohexane. As shown in the photo, when these halogens are dissolved in cyclohexane, the solution turns yellow for chlorine, orange for bromine, and violet for iodine.

The halogens are only slightly soluble in water (bottom layer). However, in cyclohexane (top layer), chlorine (yellow), bromine (orange), and iodine (violet) readily dissolve.

Fluorine
9
**F**
$[He]2s^2 2p^5$

Many brands of toothpaste contain either stannous fluoride or sodium fluoride, which, like fluoridated water, strengthen teeth and provide protection from cavities.

## Fluoridation

Fluorine compounds added to toothpaste and public drinking-water supplies have greatly reduced the incidence of cavities. Fluoride protects teeth in two ways. As teeth form, fluoride from food and drink is incorporated into the enamel layer. The fluoride makes the enamel stronger and more resistant to decay. Once teeth are present in the mouth, fluoride in saliva bonds to teeth and strengthens the surface enamel. This surface fluoride attracts calcium, which helps to fill in areas where decay has begun.

Chlorine
17
**Cl**
$[Ne]3s^2 3p^5$

## How Chlorine Bleach Is Made

Chlorine compounds are widely used as bleaching agents by the textile and paper industries. Some chlorine compounds can bleach materials by oxidizing colored molecules. Chlorine compounds are also used as disinfectants. Household bleach is a 5.25% solution of sodium hypochlorite (NaOCl) in water. Chlorine bleach is prepared commercially by passing an electric current through a solution of sodium chloride in water. As the sodium chloride breaks down, sodium hydroxide collects at the cathode and chlorine gas is generated at the anode. Sodium hydroxide and chlorine can then be combined to form sodium hypochlorite.

Household chlorine bleach is made by reacting chlorine gas or liquid chlorine with sodium hydroxide to form sodium hypochlorite.

Bromine
35
**Br**
$[Ar]4s^2 3d^{10} 4p^5$

Iodine
53
**I**
$[Kr]5s^2 4d^{10} 5p^5$

Halogen lamps use bromine or other halogen molecules to capture tungsten vapor and return tungsten atoms to the filament.

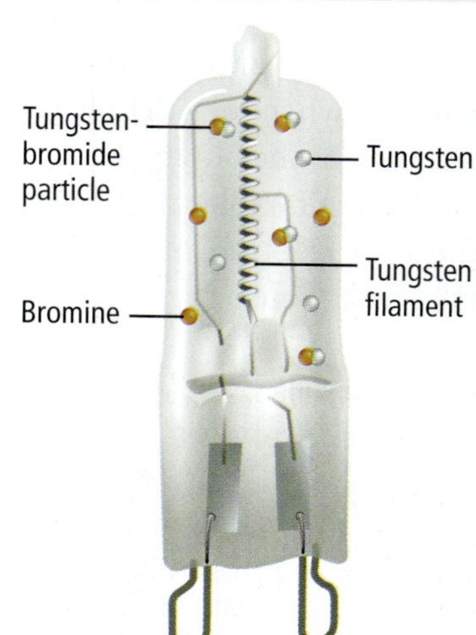


## Halogen Lightbulbs

Halogen lightbulbs include a halogen gas, such as iodine or bromine. Compared to standard lightbulbs, halogen bulbs are brighter and last longer and can be more energy efficient. During the operation of a normal lightbulb, some of the tungsten in the filament evaporates and is deposited on the inside surface of the bulb. In a halogen lamp, the evaporated tungsten reacts with the halogen gas and is redeposited back on the filament. This extends the life of the filament.

Isolated Products(John Boud)/Alamy

Iodine
53
I
$[Kr]5s^24d^{10}5p^5$

## Combating Iodine Deficiency with Salt

The thyroid gland is the only part of the body that absorbs iodine. Thyroid cells use iodine to produce thyroid hormones, which regulate metabolism. Low levels of iodine in the diet can lead to thyroid-hormone deficiencies and goiters, which are enlarged thyroid glands. In serious cases, low levels of thyroid hormones can cause birth defects and brain damage. In the United States, potassium iodide is added to most table salt to protect against dietary iodine deficiency. Even small amounts of added iodine can prevent iodine-deficiency disorders. However, there are parts of the world in which iodine deficiency is still prevalent.

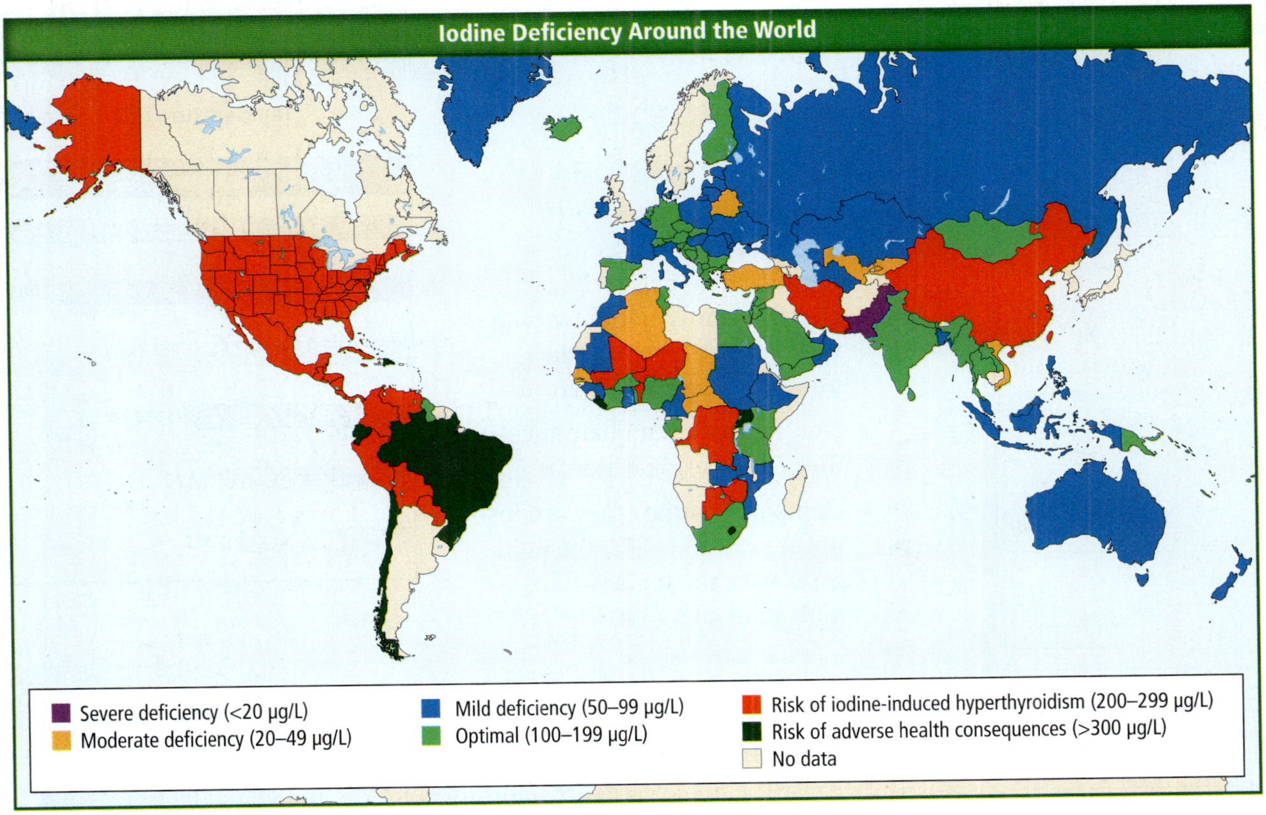


A significant percentage of the world's population was at risk for iodine deficiency in 2004. In 2005, the World Health Organization launched a program to eliminate iodine deficiency worldwide.

## REVIEW

50. **Compare** the risks for iodine deficiency in Europe, Africa, and the United States.

51. **Explain** why fluorine is the most reactive of all the elements.

52. **Evaluate** Why does a tungsten filament last longer in a halogen lightbulb than in a normal lightbulb?

53. **Calculate** Household bleach is typically a 5.25% solution of sodium hypochlorite in water. How many grams of sodium hypochlorite would there be in 300 mL of bleach?

54. **Hypothesize** In 1962, Neil Bartlett synthesized the first noble gas compound using $PtF_6$. Hypothesize why Bartlett used a fluorine compound for this synthesis.

Helium
2
**He**
$1s^2$

Neon
10
**Ne**
$[He]2s^22p^6$

Argon
18
**Ar**
$[Ne]3s^23p^6$

Krypton
36
**Kr**
$[Ar]4s^23d^{10}4p^6$

Xenon
54
**Xe**
$[Kr]5s^24d^{10}5p^6$

Radon
86
**Rn**
$[Xe]6s^24f^{14}5d^{10}6p^6$

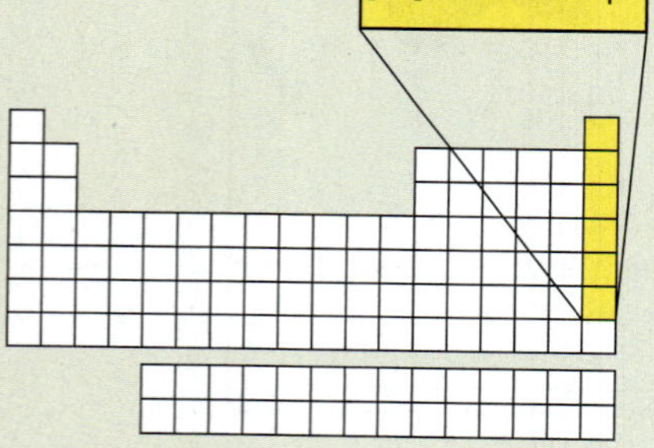

## Physical Properties

- The group 18 elements are colorless, odorless gases.
- They are all nonmetals.
- Their melting points and boiling points increase going down the group, but are much lower than those of the other groups in the periodic table.

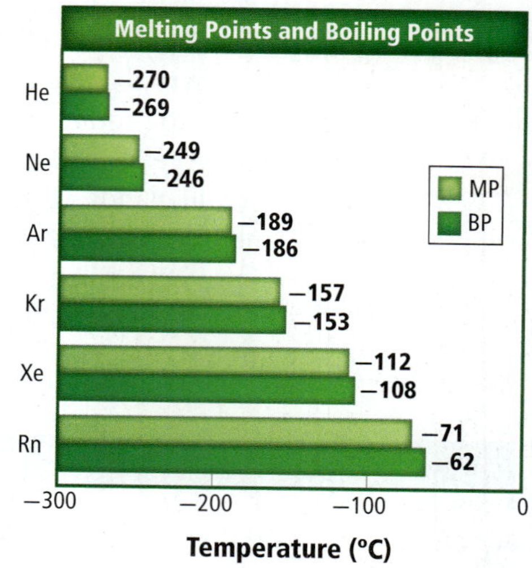


## Atomic Properties

- Each element in group 18 has eight valence electrons, producing an octet with an electron configuration ending with $ns^2np^6$, except for helium, which has two electrons.
- Noble gases are monatomic—they exist as single atoms.
- Compared to the other groups in the periodic table, the noble gases have the highest first ionization energies.

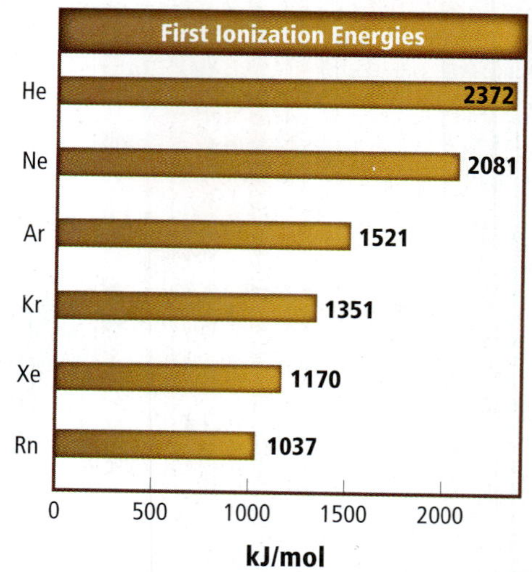


## Analytical Tests

Because the noble gases are odorless, colorless and generally unreactive, many of the common analytical tests used for identifying elements are not useful. However, the noble gases do emit light of certain colors when exposed to an electric current and have characteristic emission line spectra.

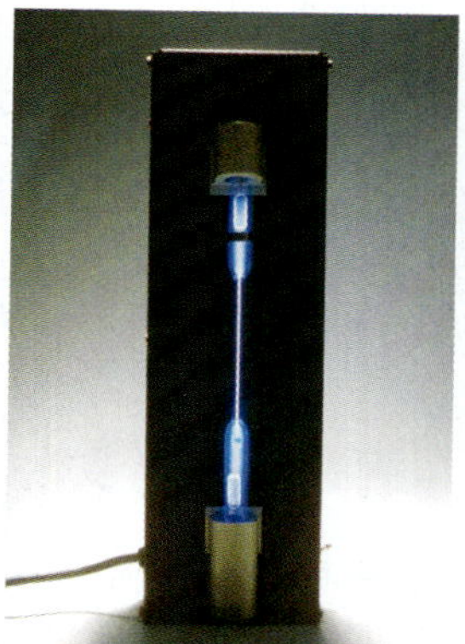

When an electric current passes through xenon, it exhibits a characteristic color (blue) and line spectrum.

## Common Reactions

Although the noble gases are also known as inert gases, a few compounds can be formed if conditions are favorable. Generally, however, noble gases are nonreactive.

(l)Charles D. Winters/Photo Researchers, (r)Ted Kinsman/Science Photo Library/Photo Researchers

| Helium |
|---|
| 2 |
| **He** |
| $1s^2$ |

## The Sun

Only 150 million km away (considered close in astronomical terms), the Sun provides the energy needed to support life on Earth. The Sun makes its energy through the fusion of hydrogen to make helium. Scientists have determined that the core of the Sun is composed of approximately 50% helium, leaving enough hydrogen for the Sun to burn for another 5 billion years.

The Sun's energy comes from a nuclear reaction that produces helium.

| Neon | Argon | Krypton | Xenon |
|---|---|---|---|
| 10 | 18 | 36 | 54 |
| **Ne** | **Ar** | **Kr** | **Xe** |
| $[He]2s^22p^6$ | $[Ne]3s^23p^6$ | $[Ar]4s^23d^{10}4p^6$ | $[Kr]5s^24d^{10}5p^6$ |

The noble gases are found in many different light sources.

## Lighting

Neon, argon, krypton, and xenon are all used in different lighting applications. Neon signs are found in many businesses to advertise products or display the name of the business. Although true neon signs glow with a red-orange color, the term *neon sign* has also come to represent the collection of gas tubes that contain gases that display other colors. Argon is found in everyday lightbulbs such as those in lamps. Because argon is inert, it provides an ideal atmosphere for the filament. Krypton and xenon bulbs produce whiter, sharper light and last longer than traditional argon bulbs. These bulbs are commonly found in chandeliers, flashlights, and luxury car headlights.



## REVIEW

**55. Describe** three physical properties of the noble gases.

**56. Write** the reaction for the production of xenon tetroxide.

**57. Analyze** why the noble gases have the highest first ionization energies compared to the rest of the elements on the periodic table.

**58. Hypothesize** why argon is used in everyday lighting even though krypton and xenon produce whiter light and last longer.

**59. Calculate** If the Sun is 150 million km away and light travels at $3.00 \times 10^5$ km/s, how long does it take for sunlight to reach Earth?

# MATH HANDBOOK

*Mathematics is a language used in science to express and solve problems. Calculations you perform during your study of chemistry require arithmetic operations, such as addition, subtraction, multiplication, and division. Use this handbook to review basic math skills and to reinforce some math skills presented in more depth in the chapters.*

## Scientific Notation

Scientists must use extremely small and extremely large numbers to describe the objects in **Figure 1.** The mass of the proton at the center of a hydrogen atom is 0.000000000000000000000000001673 kg. HIV, the virus that causes AIDS, is about 0.00000011 m. The temperature at the center of the Sun reaches 15,000,000 K. Such small and large numbers are difficult to read and hard to work with in calculations. Scientists have adopted a method of writing exponential numbers called scientific notation. It is easier than writing numerous zeros when numbers are very large or very small. It is also easier to compare the relative size of numbers when they are written in scientific notation.

A number written in scientific notation has two parts.

$$N \times 10^n$$

The first part ($N$) is a number in which only one digit is placed to the left of the decimal point and all remaining digits are placed to the right of the decimal point. The second part is an exponent of ten ($10^n$) by which the decimal portion is multiplied. For example, the number $2.53 \times 10^6$ is written in scientific notation.

$$2.53 \times 10^6$$

Number between one and ten (2.53) — Exponent of ten ($10^6$)

The decimal portion is 2.53 and the exponent is $10^6$.

Positive exponents are used to express large numbers, and negative exponents are used to express small numbers.

**Figure 1** Scientific notation provides a convenient way to express data with extremely large or small numbers. Scientists can express the mass of a proton, the length of HIV, and the temperature of the Sun in scientific notation.

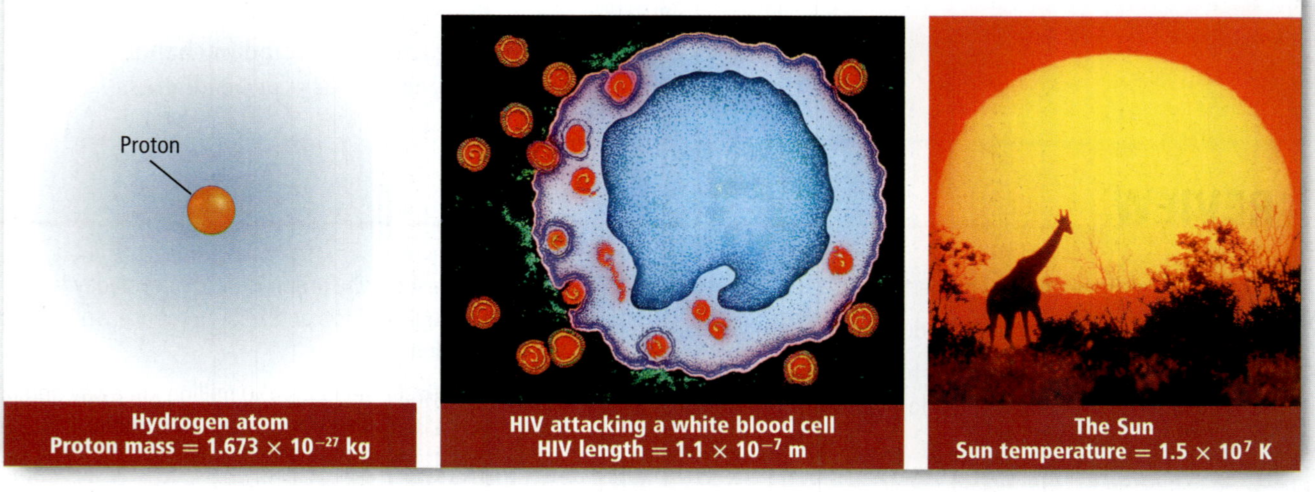

## Positive exponents

When scientists discuss the physical properties of the Moon, shown in **Figure 2,** the numbers are enormously large. A positive exponent of 10, ($n$) tells how many times a number must be multiplied by 10 to give the long form of the number.

$$\begin{aligned} &2.53 \times 10^6 \\ &= 2.53 \times 10 \times 10 \times 10 \times 10 \times 10 \times 10 \\ &= 2{,}530{,}000 \end{aligned}$$

You can also think of the positive exponent of 10 as the number of places you move the decimal to the left until only one nonzero digit is to the left of the decimal point.

2,530,000. — **The decimal point moves six places to the left.**

**Figure 2** The mass of the Moon is $7.349 \times 10^{22}$ kg.

To convert the number 567.98 to scientific notation, first write the number as an exponential number by multiplying by $10^0$.

$$567.98 \times 10^0$$

(Remember that multiplying any number by $10^0$ is the same as multiplying the number by 1.) Move the decimal point to the left until there is only one digit to the left of the decimal. At the same time, increase the exponent by the same number as the number of places the decimal is moved.

$567.98 \times 10^{0+2}$ — **The decimal point moves two places to the left.**

Thus, 567.98 written in scientific notation is $5.6798 \times 10^2$.

## Negative exponents

Measurements can also have negative exponents, such as shown by the X-rays in **Figure 3.** Negative exponents are used for numbers that are very small. A negative exponent of 10 tells how many times a number must be divided by 10 to give the long form of the number.

$$6.43 \times 10^{-4} = \frac{6.43}{10 \times 10 \times 10 \times 10} = 0.000643$$

A negative exponent of 10 is the number of places you move the decimal to the right until it is just past the first nonzero digit.

When converting a number that requires the decimal to be moved to the right, the exponent is decreased by the appropriate number. For example, the expression of 0.0098 in scientific notation is as follows:

$0.0098 \times 10^0$

$0.0098 \times 10^{0-3}$ — **The decimal point moves three places to the right.**

$9.8 \times 10^{-3}$

Thus, 0.0098 written in scientific notation is $9.8 \times 10^{-3}$.

**Figure 3** Because of their short wavelengths ($10^{-8}$ m to $10^{-13}$ m), X-rays can pass through some objects.

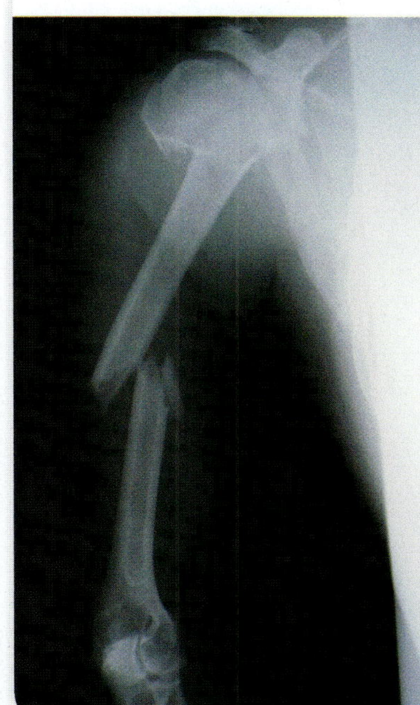

## Operations with Scientific Notation

The arithmetic operations performed with ordinary numbers can be done with numbers written in scientific notation. However, the exponential portion of the numbers must also be considered.

### 1. Addition and subtraction

Before numbers in scientific notation can be added or subtracted, the exponents must be equal. Remember that the decimal is moved to the left to increase the exponent and to the right to decrease the exponent.

$$\begin{aligned}(3.4 \times 10^2) + (4.57 \times 10^3) &= (0.34 \times 10^3) + (4.57 \times 10^3) \\ &= (0.34 + 4.57) \times 10^3 \\ &= 4.91 \times 10^3\end{aligned}$$

$$\begin{aligned}(7.52 \times 10^{-4}) - (9.7 \times 10^{-5}) &= (7.52 \times 10^{-4}) - (0.97 \times 10^{-4}) \\ &= (7.52 - 0.97) \times 10^{-4} \\ &= 6.55 \times 10^{-4}\end{aligned}$$

### 2. Multiplication

When numbers in scientific notation are multiplied, only the decimal portion is multiplied. The exponents are added.

$$\begin{aligned}(2.00 \times 10^3)(4.00 \times 10^4) &= (2.00)(4.00) \times 10^{3+4} \\ &= 8.00 \times 10^7\end{aligned}$$

### 3. Division

When numbers in scientific notation are divided, only the decimal portion is divided, while the exponents are subtracted as follows:

$$\begin{aligned}\frac{9.60 \times 10^7}{1.60 \times 10^4} &= \frac{9.60}{1.60} \times 10^{7-4} \\ &= 6.00 \times 10^3\end{aligned}$$

**PRACTICE Problems**

**1.** Express the following numbers in scientific notation.

**a.** 5800

**b.** 453,000

**c.** 0.0005877

**d.** 0.0036

**2.** Perform the following operations.

**a.** $(5.0 \times 10^6) + (3.0 \times 10^7)$

**b.** $(1.8 \times 10^9) + (2.0 \times 10^8)$

**c.** $(3.89 \times 10^{12}) - (1.9 \times 10^{11})$

**d.** $(6.0 \times 10^{-8}) - (4.0 \times 10^{-9})$

**3.** Perform the following operations.

**a.** $(6.0 \times 10^{-4}) \times (4.0 \times 10^{-6})$

**b.** $(4.5 \times 10^9) \times (6.0 \times 10^{-10})$

**c.** $\frac{4.5 \times 10^{-8}}{1.5 \times 10^{-4}}$

**d.** $\frac{9.6 \times 10^8}{1.6 \times 10^{-6}}$

**e.** $\frac{(2.5 \times 10^6)(7.2 \times 10^4)}{1.8 \times 10^{-5}}$

**f.** $\frac{(6.2 \times 10^{12})(6.0 \times 10^{-7})}{1.2 \times 10^6}$

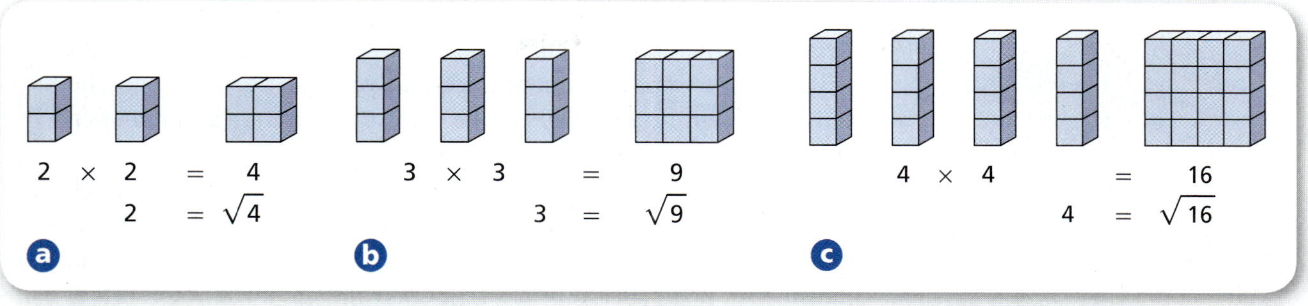




## Square and Cube Roots

A square root is one of two identical factors of a number. As shown in **Figure 4a,** the number 4 is the product of two identical factors—2. Thus, the square root of 4 is 2. The symbol $\sqrt{\ }$, called a radical sign, is used to indicate a square root. Most scientific calculators have a square root key labeled $\sqrt{\ }$.

$$\sqrt{4} = \sqrt{2 \times 2} = 2$$

This equation is read "the square root of 4 equals 2." What is the square root of 9, shown in **Figure 4b?**

There can be more than two identical factors of a number. You know that $2 \times 4 = 8$. Are there any other factors of the number 8? It is the product of $2 \times 2 \times 2$. A cube root is one of three identical factors of a number. Thus, what is the cube root of 8? It is 2. A cube root is also indicated by a radical.

$$\sqrt[3]{8} = \sqrt[3]{2 \times 2 \times 2} = 2$$

Check your calculator handbook for more information on finding roots.

■ **Figure 4 a.** The number 4 can be expressed as two groups of 2. The identical factors are 2. **b.** The number 9 can be expressed as three groups of 3. Thus, 3 is the square root of 9. **c.** 4 is the square root of 16.

**Determine** ***the cube root of 16 using your calculator.***

## Significant Figures

Accuracy reflects how close the measurements you make in the laboratory come to the real value. Precision describes the degree of exactness of your measurements. Which ruler in **Figure 5** would give you the most precise length? The top ruler, with the millimeter markings, would allow your measurements to come closer to the actual length of the pencil. The measurement would be more precise.

■ **Figure 5** The estimated digit must be read between the millimeter markings on the top ruler.

**Evaluate** ***Why is the bottom ruler less precise?***

■ **Figure 6** If you determine that the length of this pencil is 27.65 cm, that measurement has four significant figures.

Measuring tools are never perfect, nor are the people doing the measuring. Therefore, whenever you measure a physical quantity, there will always be some amount of uncertainty in the measurement. The number of significant figures in the measurement indicates the uncertainty of the measuring tool.

The number of significant figures in a measured quantity is all of the certain digits plus the first uncertain digit. For example, the pencil in **Figure 6** has a length that is between 27.6 and 27.7 cm. You can read the ruler to the nearest millimeter (27.6 cm), but after that you must estimate the next digit in the measurement. If you estimate that the next digit is 5, you would report the measured length of the pencil as 27.65 cm. Your measurement has four significant figures. The first three are certain, and the last is uncertain. The ruler used to measure the pencil has precision to the nearest tenth of a millimeter.

## How many significant figures?

When a measurement is provided, the following series of rules will help you to determine how many significant figures there are in that measurement.

1. *All nonzero figures are significant.*
2. *When a zero falls after the decimal point and after a significant figure, that zero is significant.*
3. *When a zero falls between significant figures, the zero is also significant.*
4. *When a zero is used merely to indicate the position of the decimal, it is not significant.*
5. *All counting numbers and exact numbers are treated as if they have an infinite number of significant figures.*

Examine each of the following measurements. Use the rules above to check that all of them have three significant figures.

| | |
|---|---|
| 245 K | **Rule 1** |
| 18.0 L | **Rule 2** |
| 308 km | **Rule 3** |
| 0.00623 g | **Rule 4** |
| 186,000 m | **Rule 4** |

Suppose you must do a calculation using the measurement 200 L. You cannot be certain which zero was estimated. To indicate the significance of digits, especially zeros, write measurements in scientific notation. In scientific notation, all digits in the decimal portion are significant. Which measurement is most precise?

200 L has one significant figure.
$2 \times 10^2$ L has one significant figure.
$2.0 \times 10^2$ L has two significant figures.
$2.00 \times 10^2$ L has three significant figures.

The greater the number of digits in a measurement expressed in scientific notation, the more precise the measurement is. In this example, $2.00 \times 10^2$ L is the most precise data.

## EXAMPLE Problem 1

**SIGNIFICANT FIGURES** How many significant figures are in the measurement 0.00302 g? 60 min? 5.620 m? $9.80 \times 10^2$ m/s$^2$?

### 1 ANALYZE THE PROBLEM

To determine the number of significant digits in a series of numbers, review the rules for significant figures.

### 2 SOLVE FOR THE UNKNOWN

0.00302 g

Not significant (Rule 4) — Significant (Rules 1 and 3)

The measurement 0.00302 g has three significant figures.

60 min
Unlimited significant figures
(Rule 5)

5.620 m
Significant
(Rules 1 and 2)

The measurement 5.620 m has four significant figures.

$9.80 \times 10^2$ m/s$^2$
Significant
(Rules 1 and 2)

### 3 EVALUATE THE ANSWER

The measurements 0.00302 g and $9.80 \times 10^2$ m/s$^2$ have three significant figures. The measurement 60 min has unlimited significant figures. The measurement 5.620 m has four significant figures.

## PRACTICE Problems

**4.** Determine the number of significant figures in each measurement:

**a.** 35 g
b. 3.57 m
c. 3.507 km
d. 0.035 kg
e. 0.246 L
f. 0.004 m$^3$
g. 24.068 kPa
h. 268 K
i. 20.04080 g
j. 20 dozen
k. 730,000 kg
l. 6.751 g
**m.** 0.157 kg
**n.** 28.0 mL
**o.** 2500 m
**p.** 0.070 mol
**q.** 30.07 nm
**r.** 0.106 cm
**s.** 0.0076 g
**t.** 0.0230 cm$^3$
**u.** 26.509 cm
**v.** 54.52 cm$^3$
**w.** $2.40 \times 10^6$ kg
**x.** $4.07 \times 10^{16}$ m

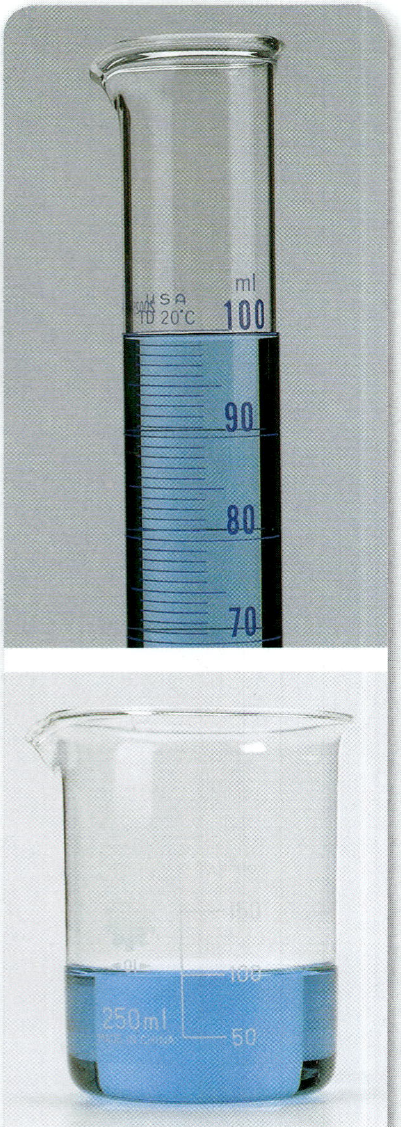


■ **Figure 7** Compare the markings on the graduated cylinder at the top with the markings on the beaker at the bottom.

**Analyze** ***Which piece of glassware will yield more precise measurements?***

## Rounding

Arithmetic operations that involve measurements are done the same way as operations involving any other numbers. However, the results must correctly indicate the uncertainty in the calculated quantities. Perform all of the calculations, and then round the result to the least number of significant figures in any of the measurements used in the calculations. To round a number, use the following rules.

1. *When the left most digit to be dropped is less than 5, that digit and any digits that follow are dropped. Then, the last digit in the rounded number remains unchanged.* For example, when rounding the number 8.7645 to three significant figures, the left most digit to be dropped is 4. Therefore, the rounded number is 8.76.
2. *When the left most digit to be dropped is greater than 5, that digit and any digits that follow are dropped, and the last digit in the rounded number is increased by one.* For example, when rounding the number 8.7676 to three significant figures, the left most digit to be dropped is 7. Therefore, the rounded number is 8.77.
3. *When the left most digit to be dropped is 5 followed by a nonzero number, that digit and any digits that follow are dropped. The last digit in the rounded number increases by one.* For example, 8.7519 rounded to two significant figures equals 8.8.
4. *If the digit to the right of the last significant figure is equal to 5 and is not followed by a nonzero digit, look at the last significant figure. If it is odd, increase it by one; if even, do not round up.* For example, 92.350 rounded to three significant figures equals 92.4, and 92.25 equals 92.2.

## Calculations with significant figures

Look at the glassware in **Figure 7.** Would you expect to measure a more precise volume with the beaker or the graduated cylinder? When you perform any calculation using measured quantities such as volume or mass, it is important to remember that the result can never be more precise than the least-precise measurement. That is, your answer cannot have more significant figures than the least precise measurement. Note that it is important to perform all calculations before dropping any insignificant digits.

The following rules determine how to use significant figures in calculations that involve measurements.

1. *To add or subtract measurements, first perform the mathematical operation, then round off the result to the least-precise value.* There should be the same number of digits to the right of the decimal as the measurement with the least number of decimal digits.
2. *To multiply or divide measurements, first perform the calculation, then round the answer to the same number of significant figures as the measurement with the least number of significant figures.* The answer should contain no more significant figures than the fewest number of significant figures in any of the measurements in the calculation.

## EXAMPLE Problem 2

**CALCULATING WITH SIGNIFICANT FIGURES** Air contains oxygen ($O_2$), nitrogen ($N_2$), carbon dioxide ($CO_2$), and trace amounts of other gases. Use the known pressures in **Table 1** to calculate the partial pressure of oxygen.

### 1 ANALYZE THE PROBLEM

The data in **Table 1** contains the gas pressure for nitrogen gas, carbon dioxide gas, and trace gases. To add or subtract measurements, first perform the operation, then round off the result to correspond to the least-precise value involved.

### 2 SOLVE FOR THE UNKNOWN

$$P_{O_2} = P_{total} - (P_{N_2} + P_{CO_2} + P_{trace})$$
$$P_{O_2} = 101.3 \text{ kPa} - (79.10 \text{ kPa} + 0.040 \text{ kPa} + 0.94 \text{ kPa})$$
$$P_{O_2} = 101.3 \text{ kPa} - 80.080 \text{ kPa}$$
$$P_{O_2} = 21.220 \text{ kPa}$$

The total pressure ($P_{total}$) was measured to the tenths place. It is the least precise measurement. Therefore, the result should be rounded to the nearest tenth of a kilopascal. The pressure of oxygen is $P_{O_2} = 21.2$ kPa.

### 3 EVALUATE THE ANSWER

By adding the gas pressure of all the gases, including oxygen, the total gas pressure is 101.3 kPa.

**Table 1 Pressures of Gases in Air**

| | Pressure (kPa) |
|---|---|
| Nitrogen gas | 79.10 |
| Carbon dioxide gas | 0.040 |
| Trace gases | 0.94 |
| Total gases | 101.3 |

Math Handbook

## PRACTICE Problems

5. Round off the following measurements to the number of significant figures indicated in parentheses.
   - **a.** 2.7518 g (3)
   - **b.** 8.6439 m (2)
   - **c.** 13.841 g (2)
   - **d.** 186.499 m (5)
   - **e.** 634,892.34 (4)
   - **f.** 355,500 g (2)
6. Perform the following operations.
   - **a.** (2.475 m) + (3.5 m) + (4.65 m)
   - **b.** (3.45 m) + (3.658 m) + (47 m)
   - **c.** $(5.36 \times 10^{-4} \text{ g}) - (6.381 \times 10^{-5} \text{ g})$
   - **d.** $(6.46 \times 10^{12} \text{ m}) - (6.32 \times 10^{11} \text{ m})$
   - **e.** $(6.6 \times 10^{12} \text{ m}) \times (5.34 \times 10^{18} \text{ m})$
   - **f.** $\dfrac{5.634 \times 10^{11} \text{ m}}{3.0 \times 10^{12} \text{ m}}$
   - **g.** $\dfrac{(4.765 \times 10^{11} \text{ m})(5.3 \times 10^{-4} \text{ m})}{7.0 \times 10^{-5} \text{ m}}$

## Solving Algebraic Equations

When you are given a problem to solve, it often can be written as an algebraic equation. You can use letters to represent measurements or unspecified numbers in the problem. The laws of chemistry are often written in the form of algebraic equations. For example, the ideal gas law relates pressure, volume, moles, and temperature of the gases. The ideal gas law is written as follows.

$$PV = n\mathrm{R}T$$

The variables are pressure ($P$), volume ($V$), number of moles ($n$), and temperature ($T$). R is a constant. This is a typical algebraic equation that can be manipulated to solve for any of the individual variables.

When you solve algebraic equations, any operation that you perform on one side of the equal sign must be performed on the other side of the equation. Suppose you are asked to use the ideal gas law to find the pressure of a gas ($P$). To solve for, or isolate, $P$ requires you to divide the left-hand side of the equation by $V$. This operation must be performed on the right-hand side of the equation as well, as shown in the second equation below.

$$PV = n\mathrm{R}T$$

$$\frac{PV}{V} = \frac{n\mathrm{R}T}{V}$$

The $V$s on the left-hand side of the equation cancel each other out.

$$\frac{PV}{V} = \frac{n\mathrm{R}T}{V}$$

$$P \times \frac{\cancel{V}}{\cancel{V}} = \frac{n\mathrm{R}T}{V}$$

$$P = \frac{n\mathrm{R}T}{V}$$

The ideal gas law equation is now written in terms of pressure. That is, $P$ has been isolated.

### Order of operations

When isolating a variable in an equation, it is important to remember that arithmetic operations have an order of operations, as shown in **Figure 8,** that must be followed. Operations in parentheses (or brackets) take precedence over multiplication and division, which in turn take precedence over addition and subtraction. For example, in the following equation

$$a + b \times c$$

variable $b$ must be multiplied first by variable $c$. Then, the resulting product is added to variable $a$. If the equation is written

$$(a + b) \times c$$

the operation in parentheses or brackets must be done first. In the equation above, variable $a$ is added to variable $b$ before the sum is multiplied by variable $c$.

**Figure 8** When faced with an equation that contains more than one operation, use this flowchart to determine the order in which to perform your calculations.

**Order of Operations**

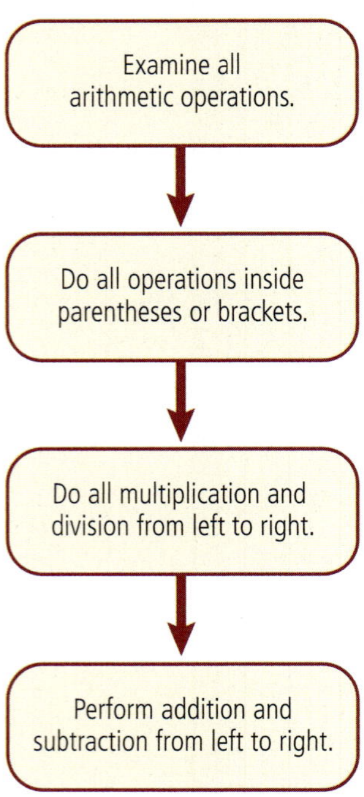

To see the difference order of operations makes, try replacing $a$ with 2, $b$ with 3, and $c$ with 4.

$$a + (b \times c) = 2 + (3 \times 4) = 14$$

$$(a + b) \times c = (2 + 3) \times 4 = 20$$

To solve algebraic equations, you also must remember the distributive property. To remove parentheses to solve a problem, any number outside the parentheses is distributed across the parentheses as follows.

$$6(x + 2y) = 6(x) + 6(2y) = 6x + 12y$$



## EXAMPLE Problem 3

**ORDER OF OPERATIONS** The temperature on a cold day was 25°F. What was the temperature on the Celsius scale?

### 1 ANALYZE THE PROBLEM

The temperature in Celsius can be calculated by using the equation for converting from the Celsius temperature to Fahrenheit temperature. The Celsius temperature is the unknown variable. The known variable is 25°F.

### 2 SOLVE FOR THE UNKNOWN

Determine the equation for calculating the temperature in Celsius.

$$°F = \frac{9}{5}°C + 32$$

$$°F - 32 = \frac{9}{5}°C + 32 - 32$$

**Rearrange the equation to isolate °C. Begin by subtracting 32 from both sides.**

$$°F - 32 = \frac{9}{5}°C$$

$$5 \times (°F - 32) = 5 \times \frac{9}{5}°C$$

**Then, multiply both sides by 5.**

$$5 \times (°F - 32) = 9°C$$

$$\frac{5 \times (°F - 32)}{9} = \frac{9°C}{9}$$

**Finally, divide both sides by 9.**

$$°C = \frac{5}{9}(°F - 32)$$

$$= \frac{5}{9}(25 - 32)$$

**Substitute the known Fahrenheit temperature.**

$$= -3.9°C$$

The Celsius temperature is −3.9°C.

### 3 EVALUATE THE ANSWER

To determine if the answer is correct, place the answer, −3.9°C, into the original equation. If the Fahrenheit temperature is 25°, the calculation was done correctly.

## PRACTICE Problems

**Isolate the indicated variable in each equation.**

**7.** $PV = nRT$ for R

8. $3 = 4(x + y)$ for $y$

9. $z = x(4 + 2y)$ for $y$

**10.** $\frac{2}{x} = 3 + y$ for $x$

**11.** $\frac{2x + 1}{3} = 6$ for $x$

## Dimensional Analysis

The dimensions of a measurement refer to the type of units attached to a quantity. For example, length is a dimensional quantity that can be measured in meters, centimeters, and kilometers. Dimensional analysis is the process of solving algebraic equations for units as well as numbers. It is a way of checking to ensure that you have used the correct equation, and that you have correctly applied the rules of algebra when solving the equation. It can also help you to choose and set up the correct equation, as shown on the next page, when you learn how to do unit conversions. It is good practice to make dimensional analysis a habit by always stating the units as well as the numerical values whenever substituting values into an equation.

■ **Figure 9** Aluminum is a metal that is useful from the kitchen to the sculpture garden.

ABN Stock Images/Alamy

## EXAMPLE Problem 4

**DIMENSIONAL ANALYSIS** The sculpture in **Figure 9** is made from aluminum. The density ($D$) of aluminum is 2700 kg/m$^3$. Determine the mass ($m$) of a piece of aluminum of volume ($V$) 0.20 m$^3$.

### 1 ANALYZE THE PROBLEM

The facts of the problem are density (2700 kg/m$^3$), volume (0.20 m$^3$), and the density equation, $D = m/V$.

### 2 SOLVE FOR THE UNKNOWN

Determine the equation for mass by rearranging the density equation.

The equation for density is

$$D = \frac{m}{V}$$

$$DV = \frac{mV}{V}$$

$$DV = \frac{\cancel{V}}{\cancel{V}} \times m$$

$$m = DV$$

**Multiply both sides of the equation by *V*, and isolate *m*.**

$$m = (2700\ \text{kg}/\cancel{\text{m}^3})(0.20\ \cancel{\text{m}^3}) = 540\ \text{kg}$$

**Substitute the known values for *D* and *V*.**

### 3 EVALUATE THE ANSWER

Notice that the unit m$^3$ cancels out, leaving mass in kg, a unit of mass.

## PRACTICE Problems

**Determine whether the following equations are dimensionally correct. Explain.**

**12.** $v = s \times t$ where $v = 24$ m/s, $s = 12$ m, and $t = 2$ s.

**13.** $R = \frac{nT}{PV}$ where R is in L•atm/mol•K, $n$ is in mol, $T$ is in K, $P$ is in atm, and $V$ is in L.

**14.** $t = \frac{v}{s}$ where $t$ is in seconds, $v$ is in m/s, and $s$ is in m.

**15.** $s = \frac{at^2}{2}$ where $s$ is in m, $a$ is in m/s$^2$, and $t$ is in s.

Math Handbook

## Unit Conversion

Recall that the universal unit system used by scientists is called Le Système Internationale d'Unités, or SI. It is a metric system based on seven base units—meter, second, kilogram, kelvin, mole, ampere, and candela—from which all other units are derived. The size of a unit in the metric system is indicated by a prefix related to the difference between that unit and the base unit. For example, the base unit for length in the metric system is the meter. One-tenth of a meter is a decimeter, where the prefix *deci-* means *one-tenth*. One thousand meters is a kilometer, where the prefix *kilo-* means *one thousand*.

You can use the information in **Table 2** to express a measured quantity in different units. For example, how is 65 m expressed in centimeters? **Table 2** indicates one centimeter and one-hundredth meter are equivalent, that is, 1 cm = $10^{-2}$ m. This information can be used to form a conversion factor. A conversion factor is a ratio equal to one that relates two units. You can make the following conversion factors from the relationship between meters and centimeters. Be sure when you set up a conversion factor that the measurement in the numerator (the top of the ratio) is equivalent to the measurement in the denominator (the bottom of the ratio).

$$1 = \frac{1\text{ cm}}{10^{-2}\text{ m}} \text{ and } 1 = \frac{10^{-2}\text{ m}}{1\text{ cm}}$$

**Table 2** Common SI Prefixes

| Prefix | Symbol | Exponential Notation | Prefix | Symbol | Exponential Notation |
|---|---|---|---|---|---|
| Peta | P | $10^{15}$ | Deci | d | $10^{-1}$ |
| Tera | T | $10^{12}$ | Centi | c | $10^{-2}$ |
| Giga | G | $10^{9}$ | Milli | m | $10^{-3}$ |
| Mega | M | $10^{6}$ | Micro | μ | $10^{-6}$ |
| Kilo | k | $10^{3}$ | Nano | n | $10^{-9}$ |
| Hecto | h | $10^{2}$ | Pico | p | $10^{-12}$ |
| Deka | da | $10^{1}$ | Femto | f | $10^{-15}$ |

Recall that the value of a quantity does not change when it is multiplied by 1. To convert 65 m to centimeters, multiply 65 m by the conversion factor for centimeters.

$$65\ \cancel{m} \times \frac{1\ cm}{10^{-2}\ \cancel{m}}$$
$$= 65 \times 10^{2}\ cm$$
$$= 6.5 \times 10^{3}\ cm$$

Note the conversion factor is set up so that the unit meters cancels and the answer is in centimeters as required. When setting up a unit conversion, use dimensional analysis to check that the units cancel to give an answer in the desired units. Always check your answer to be certain the units make sense.

You make unit conversions every day when you determine how many quarters are needed to make a dollar or how many feet are in a yard. One unit that is often used in calculations in chemistry is the mole. Equivalent relationships among moles, grams, and the number of representative particles (atoms, molecules, formula units, or ions) must be used. For example, 1 mol of a substance contains $6.02 \times 10^{23}$ representative particles. Try the next Example Problem to see how this information can be used in a conversion factor to determine the number of atoms in a sample of manganese.

■ **Figure 10** The mass of one mole of manganese equals 54.94 g.
**Determine** *How many significant figures are in this measurement?*

## EXAMPLE Problem 5

**UNIT CONVERSIONS** One mole of manganese (Mn), shown in **Figure 10,** has a mass of 54.94 g. How many atoms are in 2.0 mol of manganese?

### 1 ANALYZE THE PROBLEM

You are given the mass of 1 mol of manganese. In order to convert to the number of atoms, you must set up a conversion factor relating the number of moles and the number of atoms.

### 2 SOLVE FOR THE UNKNOWN

The conversion factors for moles and atoms are shown below.

$$\frac{1\ mol}{6.02 \times 10^{23}\ atoms} \text{ and } \frac{6.02 \times 10^{23}\ atoms}{1\ mol}$$

Choose the conversion factor that cancels units of moles and gives an answer in number of atoms.

$$2.0\ \cancel{mol} \times \frac{6.02 \times 10^{23}\ atoms}{1\ \cancel{mol}} = 12.04 \times 10^{23}\ atoms$$
$$= 1.2 \times 10^{24}\ atoms$$

### 3 EVALUATE THE ANSWER

The answer is expressed in the desired units (number of atoms). It is expressed in two significant figures because the number of moles (2.0) has two significant figures.

## PRACTICE Problems

**16.** Convert the following measurements as indicated.

**a.** 4 m = ____cm

**b.** 50.0 cm = ____m

**c.** 15 cm = ____mm

**d.** 567 mg = ____g

**e.** 324 mL = ____L

**f.** 28 L = ____mL

**g.** $4.6 \times 10^3$ m = ____mm

**h.** $8.3 \times 10^4$ g = ____kg

**i.** $2.7 \times 10^2$ L = ____mL

**j.** $7.3 \times 10^5$ mL = ____L

**k.** $8.4 \times 10^{10}$ m = ____km

**l.** $3.8 \times 10^4$ $m^2$ = ____$mm^2$

**m.** $6.9 \times 10^{12}$ $cm^2$ = ____$m^2$

**n.** $6.3 \times 10^{21}$ $mm^3$ = ____$cm^3$

**o.** $9.4 \times 10^{12}$ $cm^3$ = ____$m^3$

**p.** $5.7 \times 10^{20}$ $cm^3$ = ____$km^3$

# Drawing Line Graphs

Scientists, such as the one shown in **Figure 11,** as well as you and your classmates, use graphing to analyze data gathered in experiments. Graphs provide a way to visualize data in order to determine the mathematical relationship between the variables in your experiment. Line graphs are used most often.

**Figure 11** also shows a line graph. Line graphs are drawn by plotting variables along two axes. Plot the independent variable on the *x*-axis (horizontal axis), also called the abscissa. The independent variable is the quantity controlled by the person doing the experiment. Plot the dependent variable on the *y*-axis (vertical axis), also called the ordinate. The dependent variable is the variable that depends on the independent variable. Label the axes with the variables being plotted and the units attached to those variables.

**Figure 11** Once experimental data have been collected, they must be analyzed to determine the relationships between the measured variables.

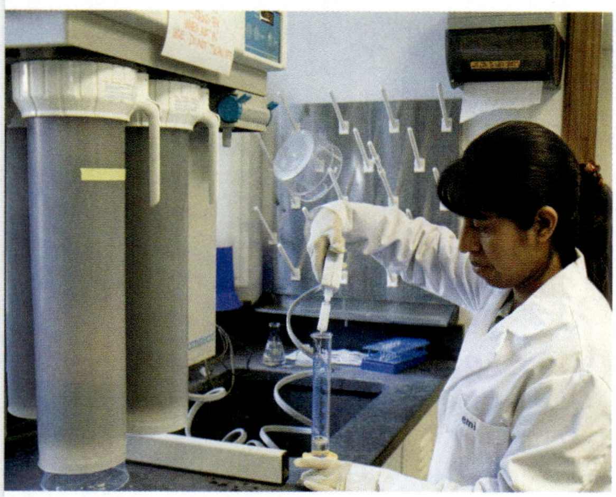

This research scientist might use graphs to analyze the data she collects on ultrapure water.

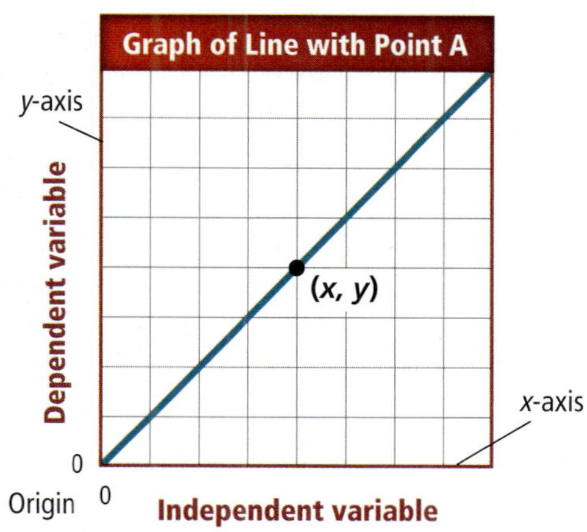


Any graph of your data should include labeled *x*- and *y*-axes, a suitable scale, and a title.

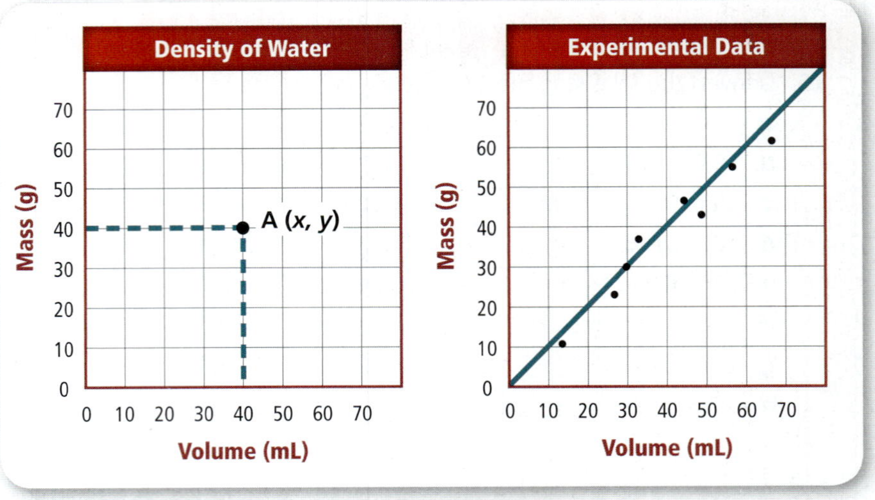


■ **Figure 12** To plot a point on a graph, place a dot at the location for each ordered pair (*x*, *y*) determined by your data. In the *Density of Water* graph, the dot marks the ordered pair (40 mL, 40 g). Generally, the line or curve that you draw will not include all of your experimental data points, as shown in the *Experimental Data* graph.

## Determining a scale

An important part of graphing is the selection of a scale. Scales should be easy to plot and easy to read. First, examine the data to determine the highest and lowest values. Assign each division on the axis (the square on the graph paper) with an equal value so that all data can be plotted along the axis. Scales divided into multiples of 1, 2, 5, or 10, or decimal values, are often the most convenient. It is not necessary to start at zero, nor is it necessary to plot both variables to the same scale. Scales must, however, be labeled clearly with the appropriate numbers and units.

## Plotting data

The values of the independent and dependent variables form ordered pairs of numbers, called the *x*-coordinate and the *y*-coordinate (*x*, *y*), that correspond to points on the graph. The first number in an ordered pair always corresponds to the *x*-axis; the second number always corresponds to the *y*-axis. The ordered pair (0,0) is always the origin. Sometimes, the points are named by using a letter. In **Figure 12,** Point A on the *Density of Water* graph corresponds to Point (*x*, *y*).

After the scales are chosen, plot the data. To graph or plot an ordered pair means to place a dot at the point that corresponds to the values in the ordered pair. The *x*-coordinate indicates how many units to move right (if the number is positive) or left (if the number is negative). The *y*-coordinate indicates how many units to move up or down. Which direction is positive on the *y*-axis? Negative? Locate each pair of *x*- and *y*-coordinates by placing a dot, as shown in **Figure 12** in the *Density of Water* graph. Sometimes, a pair of rulers, one extending from the *x*-axis and the other from the *y*-axis, can ensure that data are plotted correctly.

## Drawing a curve

Once the data is plotted, a straight line or a curve is drawn. It is not necessary to make it go through every point plotted, or even any of the points, as shown in the *Experimental Data* graph in **Figure 12.** Graphing data is an averaging process. If the points do not fall along a line, the best-fit line or most-probable smooth curve through the points is drawn. Note that curves do not always go through the origin (0, 0).

## Naming a graph

Last but not least, give each graph a title that describes what is being graphed. The title should be placed at the top of the page, or in a box on a clear area of the graph. It should not cross the data curve.

# Using Line Graphs

Once the data from an experiment has been collected and plotted, the graph must be interpreted. Much can be learned about the relationship between the independent and dependent variables by examining the shape and slope of the curve. Four common types of curves are shown in **Figure 13.** Each type of curve corresponds to a mathematical relationship between the independent and dependent variables.

## Direct and inverse relationships

In your study of chemistry, the most common curves are the linear, representing the direct relationship ($y \propto x$), and the inverse, representing the inverse relationship ($y \propto 1/x$), where $x$ represents the independent variable and $y$ represents the dependent variable. In a direct relationship, $y$ increases in value as $x$ increases in value, or $y$ decreases when $x$ decreases. In an inverse relationship, $y$ decreases in value as $x$ increases.

An example of a typical direct relationship is the increase in volume of a gas with increasing temperature. When the gases inside a hot-air balloon are heated, the balloon gets larger. As the balloon cools, its size decreases. However, a plot of the decrease in pressure as the volume of a gas increases yields a typical inverse curve.

You might also encounter exponential and root curves in your study of chemistry. See **Figure 13.** An exponential curve describes a relationship in which one variable is expressed by an exponent. A root curve describes a relationship in which one variable is expressed by a root.

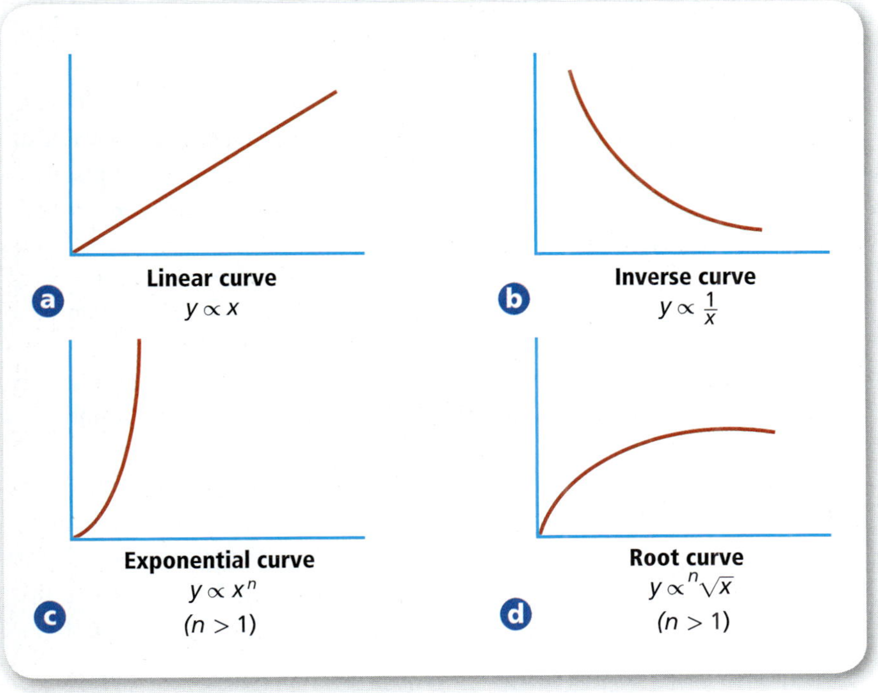


**Figure 13** The shape of the curve formed by a plot of experimental data indicates how the variables are related.

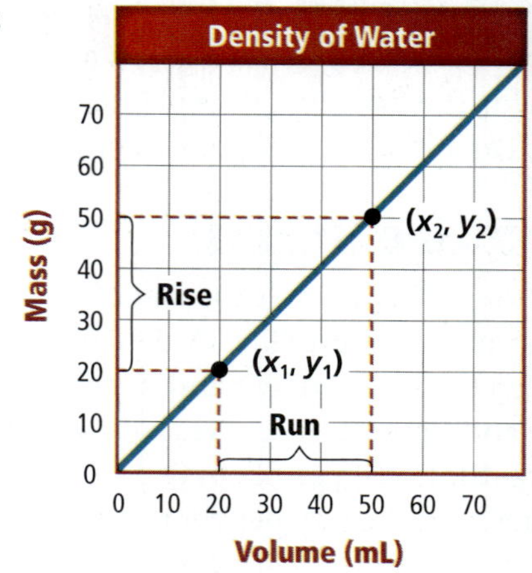


■ **Figure 14** A steep slope indicates that the dependent variable changes rapidly with a change in the independent variable.

**Infer** *What would an almost flat line indicate?*

## The linear graph

The linear graph is useful in analyzing data because a linear relationship can be translated easily into equation form using the equation for a straight line.

$$y = mx + b$$

In the equation, $y$ stands for the dependent variable, $m$ is the slope of the line, $x$ stands for the independent variable, and $b$ is the $y$-intercept, the point where the curve crosses the $y$-axis.

The slope of a linear graph is the steepness of the line. Slope is defined as the ratio of the vertical change (the rise) to the horizontal change (the run) as you move from one point to the next along the line. Use the graph in **Figure 14** to calculate slope. Choose any two points on the line, $(x_1, y_1)$ and $(x_2, y_2)$. The two points need not be actual data points, but both must fall somewhere on the straight line. After selecting two points, calculate slope, $m$, using the following equation.

$$m = \frac{\text{rise}}{\text{run}} = \frac{\Delta y}{\Delta x} = \frac{y_2 - y_1}{x_2 - x_1}, \text{ where } x_1 \neq x_2$$

The symbol $\Delta$ stands for change, $x_1$ and $y_1$ are the coordinates or values of the first point, and $x_2$ and $y_2$ are the coordinates of the second point.

Choose any two points along the graph of mass v. volume in **Figure 15,** and calculate its slope.

$$m = \frac{135 \text{ g} - 54 \text{ g}}{50.0 \text{ cm}^3 - 20.0 \text{ cm}^3} = 2.7 \text{ g/cm}^3$$

Note that the units for the slope are the units for density. Plotting a graph of mass versus volume is one way of determining the density of a substance.

Apply the general equation for a straight line to the graph in **Figure 15.**

$$y = mx + b$$
$$mass = (2.7 \text{ g/cm}^3)(volume) + 0$$
$$mass = (2.7 \text{ g/cm}^3)(volume)$$

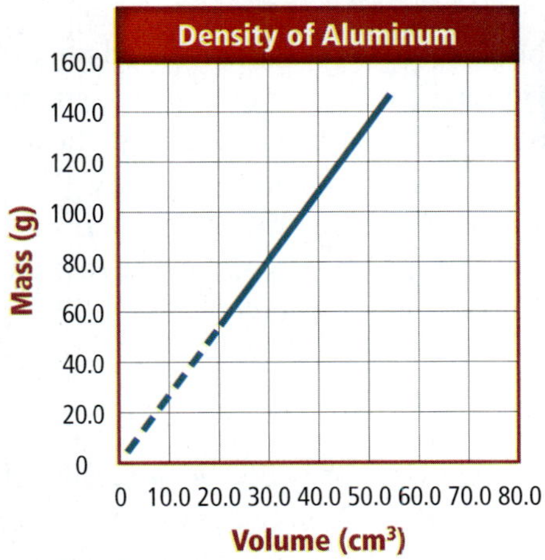


| Data | |
|---|---|
| Volume ($cm^3$) | Mass (g) |
| 20.0 | 54.0 |
| 30.0 | 81.0 |
| 50.0 | 135.0 |

**Figure 15** Interpolation and extrapolation will help you determine the values of points you did not plot.

Once the data from the graph in **Figure 15** has been placed in the general equation for a straight line, this equation verifies the direct relationship between mass and volume. For any increase in volume, the mass also increases.

## Interpolation and extrapolation

Graphs also serve functions other than determining the relationship between variables. They permit interpolation, the prediction of values of the independent and dependent variables. For example, you can see in the table in **Figure 15** that the mass of 40.0 $cm^3$ of aluminum was not measured. However, you can interpolate from the graph that the mass would be 108 g.

Graphs also permit extrapolation, which is the determination of points beyond the measured points. To extrapolate, draw a broken line to extend the curve to the desired point. In **Figure 15,** you can determine that the mass at 10.0 $cm^3$ equals 27 g. One caution regarding extrapolation—some straight-line curves do not remain straight indefinitely. So, extrapolation should only be done where there is a reasonable likelihood that the curve does not change.

### PRACTICE Problems

**17.** Plot the data in each table. Explain whether the graphs represent direct or inverse relationships.

**Table 3 Gas Pressure and Volume**

| Pressure (mm Hg) | Volume (mL) |
|---|---|
| 3040 | 5.0 |
| 1520 | 10.0 |
| 1013 | 15.0 |
| 760 | 20.0 |

**Table 4 Gas Pressure and Temperature**

| Pressure (mm Hg) | Temperature (K) |
|---|---|
| 3040 | 1092 |
| 1520 | 546 |
| 1013 | 410 |
| 760 | 273 |

Math Handbook

■ **Figure 16** The mass of one lime would be one-twelfth the mass of one dozen limes.

## Ratios, Fractions, and Percents

When you analyze data, you may be asked to compare measured quantities. Or, you may be asked to determine the relative amounts of elements in a compound. Suppose, for example, you are asked to compare the molar masses of the diatomic gases, hydrogen ($H_2$) and oxygen ($O_2$). The molar mass of hydrogen gas equals 2.00 g/mol; the molar mass of oxygen equals 32.00 g/mol. The relationship between molar masses can be expressed in three ways: a ratio, a fraction, or a percent.

### Ratios

You make comparisons by using ratios in your daily life. For example, if the mass of a dozen limes is shown in **Figure 16,** how does it compare to the mass of one lime? The mass of one dozen limes is 12 times larger than the mass of one lime. In chemistry, the chemical formula for a compound compares the elements that make up that compound, as shown in **Figure 17.** A ratio is a comparison of two numbers by division. One way it can be expressed is with a colon (:). The comparison between the molar masses of oxygen and hydrogen can be expressed as follows.

$$\text{molar mass } (H_2)\text{:molar mass } (O_2)$$
$$2.00 \text{ g/mol:}32.00 \text{ g/mol}$$
$$2.00\text{:}32.00$$
$$1\text{:}16$$

Notice that the ratio 1:16 is the smallest integer (whole number) ratio. It is obtained by dividing both numbers in the ratio by the smaller number, and then rounding the larger number to remove the digits after the decimal. The ratio of the molar masses is 1 to 16. In other words, the ratio indicates that the molar mass of diatomic hydrogen gas is 16 times smaller than the molar mass of diatomic oxygen gas.

■ **Figure 17** In a crystal of table salt (sodium chloride), each sodium ion is surrounded by chloride ions, yet the ratio of sodium ions to chloride ions is 1:1. The formula for sodium chloride is NaCl.

### Fractions

Ratios are often expressed as fractions in simplest form. A fraction is a quotient of two numbers. To express the comparison of the molar masses as a fraction, place the molar mass of hydrogen over the molar mass of oxygen as follows.

$$\frac{\text{molar mass } H_2}{\text{molar mass } O_2}$$
$$= \frac{2.0 \cancel{\text{g/mol}}}{32.00 \cancel{\text{g/mol}}}$$
$$= \frac{2.00}{32.00}$$
$$= \frac{1}{16}$$

In this case, the simplified fraction is calculated by dividing both the numerator (top of the fraction) and the denominator (bottom of the fraction) by 2.00. This fraction yields the same information as the ratio. That is, diatomic hydrogen gas has one-sixteenth the mass of diatomic oxygen gas.

## Percents

A percent is a ratio that compares a number to 100. The symbol for percent is %. You also are used to working with percents in your daily life. The number of correct answers on an exam can be expressed as a percent. If you answered 90 out of 100 questions correctly, you would receive a grade of 90%. Signs like the one in **Figure 18** indicate a reduction in price. If the item's regular price is \$100, how many dollars would you save? Sixty percent means 60 of every 100, so you would save \$60. How much would you save if the sign said 75% off?

The comparison between molar mass of hydrogen gas and the molar mass of oxygen gas described on the previous page can also be expressed as a percent by taking the fraction, converting it to decimal form, and multiplying by 100 as follows.

$$\frac{\text{molar mass } H_2}{\text{molar mass } O_2} \times 100 = \frac{2.00 \cancel{\text{g/mol}}}{32.00 \cancel{\text{g/mol}}} \times 100 = 0.0625 \times 100 = 6.25\%$$

Diatomic hydrogen gas has 6.25% of the mass of diatomic oxygen gas.

■ **Figure 18** Stores often use percentages when advertising sales.
**Analyze** ***Would the savings be large at this sale? How would you determine the sale price?***

# Operations Involving Fractions

Fractions are subject to the same type of operations as other numbers. Remember that the number on the top of a fraction is the numerator and the number on the bottom is the denominator. **Figure 19** shows an example of a fraction.

### 1. Addition and subtraction

Before two fractions can be added or subtracted, they must have a common denominator. Common denominators are found by finding the least common multiple of the two denominators. Finding the least common multiple is often as easy as multiplying the two denominators together. For example, the least common multiple of the denominators of the fractions $\frac{1}{2}$ and $\frac{1}{3}$ is $2 \times 3$ or 6.

$$\frac{1}{2} + \frac{1}{3} = \left(\frac{3}{3} \times \frac{1}{2}\right) + \left(\frac{2}{2} \times \frac{1}{3}\right) = \frac{3}{6} + \frac{2}{6} = \frac{5}{6}$$

Sometimes, one of the denominators will divide into the other, which makes the larger of the two denominators the least common multiple. For example, the fractions $\frac{1}{2}$ and $\frac{1}{6}$ have 6 as the least common multiple denominator.

$$\frac{1}{2} + \frac{1}{6} = \left(\frac{3}{3} \times \frac{1}{2}\right) + \frac{1}{6} = \frac{3}{6} + \frac{1}{6} = \frac{4}{6}$$

In other situations, both denominators will divide into a number that is not the product of the two. For example, the fractions $\frac{1}{4}$ and $\frac{1}{6}$ have the number 12 as their least common multiple denominator, rather than 24, the product of the two denominators.

The least common denominator can be deduced as follows:

$$\frac{1}{6} + \frac{1}{4} = \left(\frac{4}{4} \times \frac{1}{6}\right) + \left(\frac{6}{6} \times \frac{1}{4}\right) = \frac{4}{24} + \frac{6}{24} = \frac{2}{12} + \frac{3}{12} = \frac{5}{12}$$

Because both fractions can be simplified by dividing numerator and denominator by 2, the least common multiple must be 12.

■ **Figure 19** When two numbers are divided, the one on top is the numerator and the one on the bottom is the denominator. The result is called the quotient. When you perform calculations with fractions, the quotient can be expressed as a fraction or a decimal.

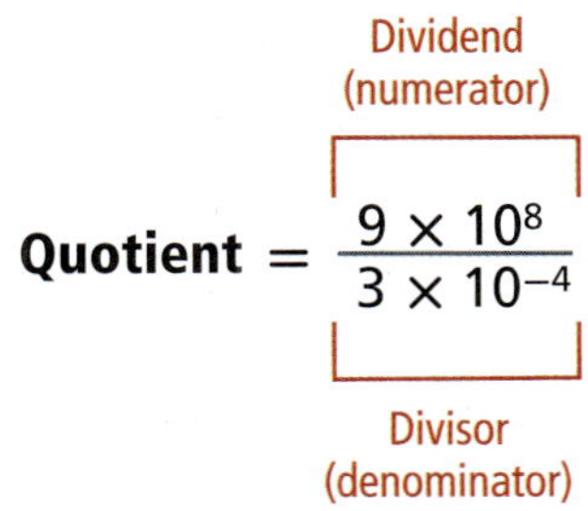


Math Handbook

### 2. Multiplication and division

When multiplying fractions, the numerators and denominators are multiplied together as follows:

$$\frac{1}{2} \times \frac{2}{3} = \frac{1 \times 2}{2 \times 3} = \frac{2}{6} = \frac{1}{3}$$

Note the final answer is simplified by dividing the numerator and denominator by 2.

When dividing fractions, the divisor is inverted and multiplied by the dividend as follows:

$$\frac{2}{3} \div \frac{1}{2} = \frac{2}{3} \times \frac{2}{1} = \frac{2 \times 2}{3 \times 1} = \frac{4}{3}$$

**PRACTICE Problems**

**18.** Perform the indicated operation:

**a.** $\frac{2}{3} + \frac{3}{4}$

**b.** $\frac{4}{5} + \frac{3}{10}$

**c.** $\frac{1}{4} - \frac{1}{6}$

**d.** $\frac{7}{8} - \frac{5}{6}$

**e.** $\frac{1}{3} \times \frac{3}{4}$

**f.** $\frac{3}{5} \times \frac{2}{7}$

**g.** $\frac{5}{8} \div \frac{1}{4}$

**h.** $\frac{4}{9} \div \frac{3}{8}$

## Logarithms and Antilogarithms

When you perform calculations, such as the pH of the products in **Figure 20,** you might need to use the log or antilog function on your calculator. A logarithm (log) is the power or exponent to which a number, called a base, must be raised in order to obtain a given positive number.

This textbook uses common logarithms based on a base of 10. Therefore, the common log of any number is the power to which 10 is raised to equal that number. Examine **Table 5** to compare logs and exponents. Note the log of each number is the power of 10 for the exponent of that number. For example, the common log of 100 is 2, and the common log of 0.01 is −2.

**Table 5** Comparison Between Exponents and Logs

| Exponent | Logarithm |
|---|---|
| $10^0 = 1$ | $\log 1 = 0$ |
| $10^1 = 10$ | $\log 10 = 1$ |
| $10^2 = 100$ | $\log 100 = 2$ |
| $10^{-1} = 0.1$ | $\log 0.1 = -1$ |
| $10^{-2} = 0.01$ | $\log 0.01 = -2$ |

$$\log 10^2 = 2$$
$$\log 10^{-2} = -2$$

A common log can be written in the following general form.

$$\text{If } 10^n = y, \text{ then } \log y = n.$$

In each example in **Table 5,** the log can be determined by inspection. How do you express the common log of $5.34 \times 10^5$? Because logarithms are exponents, they have the same properties as exponents, as shown in **Table 6** on the next page.

$$\log 5.34 \times 10^5 = \log 5.34 + \log 10^5$$

**Table 6 Properties of Exponents**

| Exponential Notation | Logarithm |
|---|---|
| $10^A \times 10^B = 10^{A+B}$ | $\log (A \times B) = \log A + \log B$ |
| $10^A \div 10^B = 10^{A-B}$ | $\log (A \div B) = \log A - \log B$ |
| $A^B$ | $(\log A) \times B$ |



## Significant figures and logarithms

Most scientific calculators have a button labeled *log* and, in most cases, you enter the number and push the log button to display the log of the number. Note that there is the same number of digits after the decimal in the log as there are significant figures in the original number entered.

$$\log 5.34 \times 10^5 = \log 5.34 + \log 10^5 = 0.728 + 5 = 5.728$$

## Antilogarithms

Suppose the pH of the aqueous ammonia in **Figure 20** is 9.54 and you are asked to find the concentration of the hydrogen ions in that solution. By definition, $\text{pH} = -\log [\text{H}^+]$. Compare this to the general equation for the common log.

| | |
|---|---|
| Equation for pH: | $\text{pH} = -\log [\text{H}^+]$ |
| General equation: | $y = \log 10^n$ |

To solve the equation for $[\text{H}^+]$, you must follow the reverse process and calculate the antilogarithm (antilog) of $-9.54$ to find $[\text{H}^+]$.

Antilogs are the reverse of logs. To find the antilog, use a scientific calculator to input the value of the log. Then, use the inverse function and press the log button. The number of digits after the decimal in the log equals the number of significant figures in the antilog. An antilog can be written in the following general form.

$$\text{If } n = \text{antilog } y, \text{ then } y = 10^n.$$

$$\text{Thus, } [\text{H}^+] = \text{antilog}(-9.54) = 10^{-9.54} = 10^{(0.46 - 10)}$$
$$= 10^{0.46} \times 10^{-10}$$
$$= 2.9 \times 10^{-10} M$$

Check the instruction manual for your calculator. The exact procedure to calculate logs and antilogs might vary.

**PRACTICE Problems**

**19.** Find the log of each of the following numbers.

**a.** 367 **b.** 4078 **c.** $X^n$

**20.** Find the antilog of each of the following logs.

**a.** 4.663 **b.** 2.367 **c.** 0.371 **d.** −1.588

**Figure 20** Ammonia is a base. That means its hydrogen ion concentration is less than $10^{-7}M$.

# REFERENCE TABLES



**Table R-1 Color Key**

| | | |
|---|---|---|
| Carbon | Bromine | Sodium/ Other metals |
| Hydrogen | Iodine | Gold |
| Oxygen | Sulfur | Copper |
| Nitrogen | Phosphorus | Electron |
| Chlorine | Silicon | Proton |
| Fluorine | Helium | Neutron |

**Table R-2 Symbols and Abbreviations**

| Symbol | | Meaning |
|---|---|---|
| $\alpha$ | = | rays from radioactive materials, helium nuclei |
| $\beta$ | = | rays from radioactive materials, electrons |
| $\gamma$ | = | rays from radioactive materials, high-energy quanta |
| $\Delta$ | = | change in |
| $\lambda$ | = | wavelength |
| $\nu$ | = | frequency |
| A | = | ampere *(electric current)* |
| amu | = | atomic mass unit |
| Bq | = | becquerel *(nuclear disintegration)* |
| °C | = | Celsius degree *(temperature)* |
| C | = | coulomb *(quantity of electricity)* |
| *c* | = | speed of light |
| cd | = | candela *(luminous intensity)* |
| c | = | specific heat |
| *D* | = | density |
| *E* | = | energy, electromotive force |
| *F* | = | force |
| *G* | = | free energy |
| g | = | gram *(mass)* |
| Gy | = | gray *(radiation)* |
| *H* | = | enthalpy |
| Hz | = | hertz *(frequency)* |
| *h* | = | Planck's constant |
| h | = | hour *(time)* |
| J | = | joule *(energy)* |
| K | = | kelvin *(temperature)* |
| $K_a$ | = | ionization constant *(acid)* |
| $K_b$ | = | ionization constant *(base)* |
| $K_{eq}$ | = | equilibrium constant |
| $K_{sp}$ | = | solubility product constant |
| kg | = | kilogram *(mass)* |
| *M* | = | molarity |
| *m* | = | mass, molality |
| m | = | meter *(length)* |
| mol | = | mole *(amount)* |
| min | = | minute *(time)* |
| N | = | newton *(force)* |
| $N_A$ | = | Avogadro's number |
| *n* | = | number of moles |
| *P* | = | pressure, power |
| Pa | = | pascal *(pressure)* |
| *q* | = | heat |
| $Q_{sp}$ | = | ion product |
| *R* | = | ideal gas constant |
| *S* | = | entropy |
| s | = | second *(time)* |
| Sv | = | sievert *(absorbed radiation)* |
| *T* | = | temperature |
| *V* | = | volume |
| V | = | volt *(electric potential)* |
| *v* | = | velocity |
| W | = | watt *(power)* |
| *w* | = | work |
| *X* | = | mole fraction |

**Table R-3** Solubility Product Constants at 298 K

| Compound | $K_{sp}$ | Compound | $K_{sp}$ | Compound | $K_{sp}$ |
|---|---|---|---|---|---|
| **Carbonates** | | **Halides** | | **Hydroxides** | |
| $BaCO_3$ | $2.6 \times 10^{-9}$ | $CaF_2$ | $3.5 \times 10^{-11}$ | $Al(OH)_3$ | $4.6 \times 10^{-33}$ |
| $CaCO_3$ | $3.4 \times 10^{-9}$ | $PbBr_2$ | $6.6 \times 10^{-6}$ | $Ca(OH)_2$ | $5.0 \times 10^{-6}$ |
| $CuCO_3$ | $2.5 \times 10^{-10}$ | $PbCl_2$ | $1.7 \times 10^{-5}$ | $Cu(OH)_2$ | $2.2 \times 10^{-20}$ |
| $PbCO_3$ | $7.4 \times 10^{-14}$ | $PbF_2$ | $3.3 \times 10^{-8}$ | $Fe(OH)_2$ | $4.9 \times 10^{-17}$ |
| $MgCO_3$ | $6.8 \times 10^{-6}$ | $PbI_2$ | $9.8 \times 10^{-9}$ | $Fe(OH)_3$ | $2.8 \times 10^{-39}$ |
| $Ag_2CO_3$ | $8.5 \times 10^{-12}$ | AgCl | $1.8 \times 10^{-10}$ | $Mg(OH)_2$ | $5.6 \times 10^{-12}$ |
| $ZnCO_3$ | $1.5 \times 10^{-10}$ | AgBr | $5.4 \times 10^{-13}$ | $Zn(OH)_2$ | $3 \times 10^{-17}$ |
| $Hg_2CO_3$ | $3.6 \times 10^{-17}$ | AgI | $8.5 \times 10^{-17}$ | **Sulfates** | |
| **Chromates** | | **Phosphates** | | $BaSO_4$ | $1.1 \times 10^{-10}$ |
| $BaCrO_4$ | $1.2 \times 10^{-10}$ | $AlPO_4$ | $9.8 \times 10^{-21}$ | $CaSO_4$ | $4.9 \times 10^{-5}$ |
| $PbCrO_4$ | $2.3 \times 10^{-13}$ | $Ca_3(PO_4)_2$ | $2.1 \times 10^{-33}$ | $PbSO_4$ | $2.5 \times 10^{-8}$ |
| $Ag_2CrO_4$ | $1.1 \times 10^{-12}$ | $Mg_3(PO_4)_2$ | $1.0 \times 10^{-24}$ | $Ag_2SO_4$ | $1.2 \times 10^{-5}$ |
| **Iodates** | | $FePO_4 \cdot 2H_2O$ | $9.91 \times 10^{-16}$ | **Arsenates** | |
| $Cd(IO_3)_2$ | $2.3 \times 10^{-8}$ | $Ni_3(PO_4)_2$ | $4.7 \times 10^{-32}$ | $Pb_3(AsO_4)_2$ | $4.0 \times 10^{-36}$ |

**Table R-4** Physical Constants

| Quantity | Symbol | Value |
|---|---|---|
| Atomic mass unit | amu | $1.6605 \times 10^{-27}$ kg |
| Avogadro's number | $N_A$ | $6.022 \times 10^{23}$ particles/mole |
| Ideal gas constant | $R$ | 8.31 L•kPa/mol•K<br>0.0821 L•atm/mol•K<br>62.4 mm Hg•L/mol•K<br>62.4 torr•L/mol•K |
| Mass of an electron | $m_e$ | $9.109 \times 10^{-31}$ kg<br>$5.485799 \times 10^{-4}$ amu |
| Mass of a neutron | $m_n$ | $1.6749 \times 10^{-27}$ kg<br>1.008665 amu |
| Mass of a proton | $m_p$ | $1.6726 \times 10^{-27}$ kg<br>1.007276 amu |
| Molar volume of ideal gas at STP | $V$ | 22.414 L/mol |
| Normal boiling point of water | $T_b$ | 373.15 K<br>100.0°C |
| Normal freezing point of water | $T_f$ | 273.15 K<br>0.00°C |
| Planck's constant | $h$ | $6.6260693 \times 10^{-34}$ J•s |
| Speed of light in a vacuum | $c$ | $2.997925 \times 10^{8}$ m/s |

**Table R-5 Names and Charges of Polyatomic Ions**

| 1− | 2− | 3− | 4− | 1+ | 2+ |
|---|---|---|---|---|---|
| Acetate, $CH_3COO^-$ | Carbonate, $CO_3^{2-}$ | Arsenate, $AsO_4^{3-}$ | Hexacyanoferrate(II), $Fe(CN)_6^{4-}$ | Ammonium, $NH_4^+$ | Mercury(I), $Hg_2^{2+}$ |
| Amide, $NH_2^-$ | Chromate, $CrO_4^{2-}$ | Arsenite, $AsO_3^{3-}$ | Orthosilicate, $SiO_4^{4-}$ | Neptunyl(V), $NpO_2^+$ | Neptunyl(VI), $NpO_2^{2+}$ |
| Astatate, $AtO_3^-$ | Dichromate, $Cr_2O_7^{2-}$ | Borate, $BO_3^{3-}$ | Diphosphate, $P_2O_7^{4-}$ | Plutonyl(V), $PuO_2^+$ | Plutonyl(VI), $PuO_2^{2+}$ |
| Azide, $N_3^-$ | Hexachloroplatinate, $PtCl_6^{2-}$ | Citrate, $C_6H_5O_7^{3-}$ | | Uranyl(V), $UO_2^+$ | Uranyl(VI), $UO_2^{2+}$ |
| Benzoate, $C_6H_5COO^-$ | Hexafluorosilicate, $SiF_6^{2-}$ | Hexacyanoferrate(III), $Fe(CN)_6^{3-}$ | | Vanadyl(V), $VO_2^+$ | Vanadyl(IV), $VO^{2+}$ |
| Bismuthate, $BiO_3^-$ | Molybdate, $MoO_4^{2-}$ | Phosphate, $PO_4^{3-}$ | | | |
| Bromate, $BrO_3^-$ | Oxalate, $C_2O_4^{2-}$ | Phosphite, $PO_3^{3-}$ | | | |
| Chlorate, $ClO_3^-$ | Peroxide, $O_2^{2-}$ | | | | |
| Chlorite, $ClO_2^-$ | Peroxydisulfate, $S_2O_8^{2-}$ | | | | |
| Cyanide, $CN^-$ | Ruthenate, $RuO_4^{2-}$ | | | | |
| Formate, $HCOO^-$ | Selenate, $SeO_4^{2-}$ | | | | |
| Hydroxide, $OH^-$ | Selenite, $SeO_3^{2-}$ | | | | |
| Hypobromite, $BrO^-$ | Silicate, $SiO_3^{2-}$ | | | | |
| Hypochlorite, $ClO^-$ | Sulfate, $SO_4^{2-}$ | | | | |
| Hypophosphite, $H_2PO_2^-$ | Sulfite, $SO_3^{2-}$ | | | | |
| Iodate, $IO_3^-$ | Tartrate, $C_4H_4O_6^{2-}$ | | | | |
| Nitrate, $NO_3^-$ | Tellurate, $TeO_4^{2-}$ | | | | |
| Nitrite, $NO_2^-$ | Tellurite, $TeO_3^{2-}$ | | | | |
| Perbromate, $BrO_4^-$ | Tetraborate, $B_4O_7^{2-}$ | | | | |
| Perchlorate, $ClO_4^-$ | Thiosulfate, $S_2O_3^{2-}$ | | | | |
| Periodate, $IO_4^-$ | Tungstate, $WO_4^{2-}$ | | | | |
| Permanganate, $MnO_4^-$ | | | | | |
| Perrhenate, $ReO_4^-$ | | | | | |
| Thiocyanate, $SCN^-$ | | | | | |
| Vanadate, $VO_3^-$ | | | | | |

**Table R-6 Ionization Constants**

| Substance | Ionization Constant | Substance | Ionization Constant | Substance | Ionization Constant |
|---|---|---|---|---|---|
| $HCOOH$ | $1.77 \times 10^{-4}$ | $HBO_3^{2-}$ | $1.58 \times 10^{-14}$ | $HS^-$ | $1.00 \times 10^{-19}$ |
| $CH_3COOH$ | $1.75 \times 10^{-5}$ | $H_2CO_3$ | $4.5 \times 10^{-7}$ | $HSO_4^-$ | $1.02 \times 10^{-2}$ |
| $CH_2ClCOOH$ | $1.36 \times 10^{-3}$ | $HCO_3^-$ | $4.68 \times 10^{-11}$ | $H_2SO_3$ | $1.29 \times 10^{-2}$ |
| $CHCl_2COOH$ | $4.47 \times 10^{-2}$ | $HCN$ | $6.17 \times 10^{-10}$ | $HSO_3^-$ | $6.17 \times 10^{-8}$ |
| $CCl_3COOH$ | $3.02 \times 10^{-1}$ | $HF$ | $6.3 \times 10^{-4}$ | $HSeO_4^-$ | $2.19 \times 10^{-2}$ |
| $HOOCCOOH$ | $5.36 \times 10^{-2}$ | $HNO_2$ | $5.62 \times 10^{-4}$ | $H_2SeO_3$ | $2.29 \times 10^{-3}$ |
| $HOOCCOO^-$ | $1.55 \times 10^{-4}$ | $H_3PO_4$ | $7.08 \times 10^{-3}$ | $HSeO_3^-$ | $4.79 \times 10^{-9}$ |
| $CH_3CH_2COOH$ | $1.34 \times 10^{-5}$ | $H_2PO_4^-$ | $6.31 \times 10^{-8}$ | $HBrO$ | $2.51 \times 10^{-9}$ |
| $C_6H_5COOH$ | $6.25 \times 10^{-5}$ | $HPO_4^{2-}$ | $4.17 \times 10^{-13}$ | $HClO$ | $2.9 \times 10^{-8}$ |
| $H_3AsO_4$ | $6.03 \times 10^{-3}$ | $H_3PO_3$ | $5.01 \times 10^{-2}$ | $HIO$ | $3.16 \times 10^{-11}$ |
| $H_2AsO_4^-$ | $1.05 \times 10^{-7}$ | $H_2PO_3^-$ | $2.00 \times 10^{-7}$ | $NH_3$ | $5.62 \times 10^{-10}$ |
| $H_3BO_3$ | $5.75 \times 10^{-10}$ | $H_3PO_2$ | $5.89 \times 10^{-2}$ | $H_2NNH_2$ | $7.94 \times 10^{-9}$ |
| $H_2BO_3^-$ | $1.82 \times 10^{-13}$ | $H_2S$ | $9.1 \times 10^{-8}$ | $H_2NOH$ | $1.15 \times 10^{-6}$ |

## Table R-7 Properties of Elements

| Element | Symbol | Atomic Number | Atomic Mass* (amu) | Melting Point (°C) | Boiling Point (°C) | Density ($g/cm^3$) (gases measured at STP) | Atomic Radius (pm) | First Ionization Energy (kJ/mol) | Standard Reduction Potential (V) (for elements from or to oxidation state indicated) | Enthalpy of Fusion | Specific Heat | Enthalpy of Vaporization | Abundance in Earth's Crust | Major Oxidation States |
|---|---|---|---|---|---|---|---|---|---|---|---|---|---|---|
| Actinium | Ac | 89 | [227] | 1050 | 3300 | 10.07 | --- | 499 | (3+)−2.13 | 14 | 0.120 | 400 | --- | 3+ |
| Aluminum | Al | 13 | 26.981539 | 660.32 | 2519 | 2.7 | 143 | 577.5 | (3+)−1.68 | 10.789 | 0.897 | 294 | 8.2 | 3+ |
| Americium | Am | 95 | [243] | 1176 | 2607 | 13.67 | --- | 578 | (3+)−2.07 | 14.39 | 0.110 | --- | --- | 2+, 3+, 4+ |
| Antimony | Sb | 51 | 121.760 | 630.6 | 1587 | 6.697 | 140 | 834 | (3+)+0.15 | 19.79 | 0.207 | 68 | $2 \times 10^{-5}$ | 3+, 5+ |
| Argon | Ar | 18 | 39.948 | −189.3 | −185.8 | 0.001784 | 98 | 1521 | --- | 1.18 | 0.520 | 6.43 | $1.5 \times 10^{-4}$ | --- |
| Arsenic | As | 33 | 74.92160 | 817 | 614 | 5.727 | 120 | 947 | (3+)+0.24 | 24.44 | 0.329 | 32.4 | $2.1 \times 10^{-4}$ | 3+, 5+ |
| Astatine | At | 85 | [210] | 302 | --- | --- | 140 | 920 | (1−)+0.2 | 6 | --- | 40 | --- | 1−, 5+ |
| Barium | Ba | 56 | 137.327 | 727 | 1870 | 3.51 | 222 | 502.9 | (2+)−2.92 | 7.12 | 0.204 | 140 | 0.034 | 2+ |
| Berkelium | Bk | 97 | [247] | 986 | --- | 14.78 | --- | 601 | (3+)−2.01 | --- | --- | --- | --- | 3+, 4+ |
| Beryllium | Be | 4 | 9.012182 | 1287 | 2469 | 1.848 | 112 | 899.5 | (2+)−1.97 | 7.895 | 1.825 | 297 | $2 \times 10^{-4}$ | 2+ |
| Bismuth | Bi | 83 | 208.98040 | 271.3 | 1564 | 9.78 | 150 | 703 | (3+)+0.317 | 11.145 | 0.122 | 151 | $3 \times 10^{-7}$ | 3+, 5+ |
| Bohrium | Bh | 107 | [264] | --- | --- | --- | --- | --- | --- | --- | --- | --- | --- | --- |
| Boron | B | 5 | 10.811 | 2076 | 3927 | 2.46 | 85 | 800.6 | (3+)−0.89 | 50.2 | 1.026 | 480 | $9 \times 10^{-4}$ | 3+ |
| Bromine | Br | 35 | 79.904 | −7.3 | 59 | 3.119 | 114 | 1139.9 | (1−)+1.065 | 10.57 | 0.474 | 29.96 | $3 \times 10^{-4}$ | 1−, 1+, 3+, 5+ |
| Cadmium | Cd | 48 | 112.411 | 321.07 | 767 | 8.65 | 151 | 867.8 | (2+)−0.4025 | 6.21 | 0.232 | 99.87 | $1.5 \times 10^{-5}$ | 2+ |
| Calcium | Ca | 20 | 40.078 | 842 | 1484 | 1.55 | 197 | 589.8 | (2+)−2.84 | 8.54 | 0.647 | 155 | 5.00 | 2+ |
| Californium | Cf | 98 | [251] | 900 | --- | 15.1 | --- | 608 | (3+)−1.93 | --- | --- | --- | --- | 3+, 4+ |
| Carbon | C | 6 | 12.0107 | 3527 | 4027 | 2.267 | 77 | 1086.5 | (4−)+0.132 | 117 | 0.709 | 715 | 0.018 | 4−, 2+, 4+ |
| Cerium | Ce | 58 | 140.116 | 795 | 3360 | 6.689 | --- | 534.4 | (3+)−2.34 | 5.46 | 0.192 | 350 | 0.006 | 3+, 4+ |
| Cesium | Cs | 55 | 132.905451 | 28.4 | 671 | 1.879 | 265 | 375.7 | (1+)−2.923 | 2.09 | 0.242 | 65 | $1.9 \times 10^{-4}$ | 1+ |
| Chlorine | Cl | 17 | 35.453 | −101.5 | −34 | 0.003 | 100 | 1251.2 | (1−)+1.358 | 6.40 | 0.479 | 20.41 | 0.017 | 1−, 1+, 3+, 5+ |
| Chromium | Cr | 24 | 51.9961 | 1907 | 2671 | 7.14 | 128 | 652.9 | (3+)−0.74 | 21.0 | 0.449 | 339 | 0.014 | 2+, 3+, 6+ |
| Cobalt | Co | 27 | 58.9332 | 1495 | 2927 | 8.9 | 125 | 760.4 | (2+)−0.28 | 16.06 | 0.421 | 375 | 0.003 | 2+, 3+ |
| Copernicium | Cn | 112 | [285] | --- | --- | --- | --- | --- | --- | --- | --- | --- | --- | --- |
| Copper | Cu | 29 | 63.546 | 1084.62 | 2570 | 8.92 | 128 | 745.5 | (2+)+0.34 | 12.93 | 0.385 | 300 | 0.0068 | 1+, 2+ |
| Curium | Cm | 96 | [247] | 1340 | 3110 | 13.51 | --- | 581 | (3+)−2.06 | --- | --- | --- | --- | 3+, 4+ |
| Darmstadtium | Ds | 110 | [281] | --- | --- | --- | --- | --- | --- | --- | --- | --- | --- | --- |
| Dubnium | Db | 105 | [262] | --- | --- | --- | --- | --- | --- | --- | --- | --- | --- | --- |
| Dysprosium | Dy | 66 | 162.5 | 1407 | 2567 | 8.551 | --- | 573 | (3+)−2.29 | 11.06 | 0.173 | 280 | $6 \times 10^{-4}$ | 2+, 3+ |
| Einsteinium | Es | 99 | [252] | 860 | --- | --- | --- | 619 | (3+)−2 | --- | --- | --- | --- | 3+ |
| Erbium | Er | 68 | 167.259 | 1497 | 2868 | 9.066 | --- | 589.3 | (3+)−2.32 | 19.9 | 0.168 | 285 | $3 \times 10^{-4}$ | 3+ |
| Europium | Eu | 63 | 151.964 | 826 | 1527 | 5.244 | --- | 547.1 | (3+)−1.99 | 9.21 | 0.182 | 175 | $1.8 \times 10^{-4}$ | 2+, 3+ |
| Fermium | Fm | 100 | [257] | 1527 | --- | --- | --- | 627 | (3+)−1.96 | --- | --- | --- | --- | 2+, 3+ |
| Flerovium | Fl | 114 | [289] | --- | --- | --- | --- | --- | --- | --- | --- | --- | --- | --- |
| Fluorine | F | 9 | 18.9984032 | −219.62 | −188.12 | 0.001696 | 71 | 1681 | (1−)+2.87 | 0.51 | 0.824 | 6.62 | 0.054 | 1− |
| Francium | Fr | 87 | [223] | --- | --- | --- | 270 | 380 | (1+)−2.92 | 2 | --- | 65 | --- | 1+ |
| Gadolinium | Gd | 64 | 157.25 | 1312 | 3250 | 7.901 | --- | 593.4 | (3+)−2.28 | 10.0 | 0.236 | 305 | $5.2 \times 10^{-4}$ | 3+ |
| Gallium | Ga | 31 | 69.723 | 29.76 | 2204 | 5.904 | 135 | 578.8 | (3+)−0.53 | 5.576 | 0.373 | 254 | 0.0019 | 1+, 3+ |
| Germanium | Ge | 32 | 72.64 | 938.3 | 2820 | 5.323 | 122 | 762 | (4+)+0.124 | 36.94 | 0.320 | 334 | $1.4 \times 10^{-4}$ | 2+, 4+ |
| Gold | Au | 79 | 196.966569 | 1064 | 2856 | 19.3 | 144 | 890.1 | (3+)+1.52 | 12.72 | 0.129 | 324 | $3 \times 10^{-7}$ | 1+, 3+ |

*[ ] indicates mass of longest-lived isotope

**Table R-7** Properties of Elements (continued)

| Element | Symbol | Atomic Number | Atomic Mass* (amu) | Melting Point (°C) | Boiling Point (°C) | Density (g/cm³) (gases measured at STP) | Atomic Radius (pm) | First Ionization Energy (kJ/mol) | Standard Reduction Potential (V) (for elements from or to oxidation state indicated) | Enthalpy of Fusion | Specific Heat | Enthalpy of Vaporization | Abundance in Earth's Crust | Major Oxidation States |
|---|---|---|---|---|---|---|---|---|---|---|---|---|---|---|
| Hafnium | Hf | 72 | 178.49 | 2233 | 4603 | 13.31 | 159 | 658.5 | (4+)−1.70 | 27.2 | 0.144 | 630 | $3 \times 10^{-4}$ | 4+ |
| Hassium | Hs | 108 | [277] | −272.2 | --- | 0.0001785 | --- | 2372.3 | --- | --- | --- | 0.083 | $5.5 \times 10^{-4}$ | --- |
| Helium | He | 2 | 4.002602 | −269.7 (2536 kPa) | −268.93 | 0.00017847 | 31 | 2372 | --- | 0.021 | 5.193 | 0.08 | --- | --- |
| Holmium | Ho | 67 | 164.93032 | 1461 | 2720 | 8.795 | --- | 581 | (3+)−2.33 | 17.0 | 0.165 | 265 | $1.2 \times 10^{-4}$ | 3+ |
| Hydrogen | H | l | 1.00794 | −259.14 | −252.87 | 0.0000899 | 37 | 1312 | (1+)0.000 | 0.12 | 14.304 | 0.90 | 0.15 | 1−, 1+ |
| Indium | In | 49 | 114.818 | 156.6 | 2072 | 7.31 | 167 | 558.3 | (3+)−0.3382 | 3.281 | 0.233 | 230 | $1.6 \times 10^{-5}$ | 1+, 3+ |
| Iodine | I | 53 | 126.90447 | 113.7 | 184.3 | 4.94 | 133 | 1008.4 | (1−)+0.535 | 15.52 | 0.214 | 41.57 | $4.9 \times 10^{-5}$ | 1−, 1+, 5+, 7+ |
| Iridium | Ir | 77 | 192.217 | 2466 | 4428 | 22.65 | 136 | 880 | (4+)+0.926 | 41.12 | 0.131 | 560 | $4 \times 10^{-7}$ | 3+, 4+, 5+ |
| Iron | Fe | 26 | 55.845 | 1538 | 2861 | 7.874 | 126 | 762.5 | (3+)−0.04 | 13.81 | 0.449 | 347 | 6.3 | 2+, 3+ |
| Krypton | Kr | 36 | 83.798 | −157.36 | −153.22 | 0.0037493 | 112 | 1350.8 | --- | 1.64 | 0.248 | 9.08 | $1.5 \times 10^{-7}$ | --- |
| Lanthanum | La | 57 | 138.9055 | 920 | 3470 | 6.146 | 187 | 538.1 | (3+)−2.38 | 6.20 | 0.195 | 400 | 0.0034 | 3+ |
| Lawrencium | Lr | 103 | [262] | 1627 | --- | --- | --- | --- | (3+)−2 | --- | --- | --- | --- | 3+ |
| Lead | Pb | 82 | 207.2 | 327.46 | 1749 | 11.34 | 146 | 715.6 | (2+)−0.1251 | 4.782 | 0.130 | 179.5 | 0.001 | 2+, 4+ |
| Lithium | Li | 3 | 6.941 | 180.54 | 1342 | 0.535 | 152 | 520.2 | (1+)−3.040 | 3.00 | 3.582 | 147 | 0.0017 | 1+ |
| Livermorium | Lv | 116 | [291] | --- | --- | --- | --- | --- | --- | --- | --- | --- | --- | --- |
| Lutetium | Lu | 71 | 174.967 | 1652 | 3402 | 9.841 | 160 | 523.5 | (3+)−2.3 | 22 | 0.154 | 415 | $5.6 \times 10^{-5}$ | 3+ |
| Magnesium | Mg | 12 | 24.305 | 650 | 1090 | 1.738 | 160 | 737.7 | (2+)−2.356 | 8.48 | 1.023 | 128 | 2.9 | 2+ |
| Manganese | Mn | 25 | 54.938045 | 1246 | 2061 | 7.47 | 127 | 717.3 | (2+)−1.18 | 12.91 | 0.479 | 220 | 0.11 | 2+, 3+, 4+, 6+, 7+ |
| Meitnerium | Mt | 109 | [268] | --- | --- | --- | --- | --- | --- | --- | --- | --- | --- | --- |
| Mendelevium | Md | 101 | [258] | 827 | --- | --- | --- | 635 | (3+)−1.7 | --- | --- | --- | --- | 2+, 3+ |
| Mercury | Hg | 80 | 200.59 | −38.83 | 356.73 | 13.6 | 151 | 1007.1 | (2+)+0.8535 | 2.29 | 0.140 | 59.11 | $6.7 \times 10^{-6}$ | 1+, 2+ |
| Molybdenum | Mo | 42 | 95.94 | 2623 | 4639 | 10.28 | 139 | 684.3 | (6+)+0.114 | 37.48 | 0.251 | 600 | $1.1 \times 10^{-4}$ | 4+, 5+, 6+ |
| Neodymium | Nd | 60 | 144.24 | 1024 | 3100 | 6.8 | --- | 533.1 | (3+)−2.32 | 7.14 | 0.190 | 285 | 0.0033 | 2+,3+ |
| Neon | Ne | 10 | 20.1797 | −248.59 | −246.08 | 0.0008999 | 71 | 2080.7 | --- | 0.328 | 1.030 | 1.71 | --- | --- |
| Neptunium | Np | 93 | [237] | 637 | 4000 | 20.45 | --- | 604.5 | (4+)−1.30 | 3.20 | 0.120 | 335 | --- | 2+, 3+, 4+, 5+, 6+ |
| Nickel | Ni | 28 | 58.6934 | 1455 | 2913 | 8.908 | 124 | 737.1 | (2+)−0.257 | 17.04 | 0.444 | 378 | 0.009 | 2+, 3+, 4+ |
| Niobium | Nb | 41 | 92.90638 | 2477 | 4744 | 8.57 | 146 | 652.1 | (5+)−0.65 | 30 | 0.265 | 690 | 0.0017 | 4+, 5+ |
| Nitrogen | N | 7 | 14.0067 | −210.1 | −195.79 | 0.0012506 | 75 | 1402.3 | (2−)−0.23 | 0.71 | 1.040 | 5.57 | 0.002 | 3−, 2−, 1−, 1+, 2+, 3+, 4+, 5+ |
| Nobelium | No | 102 | [259] | 827 | --- | --- | --- | 642 | (2+)−2.5 | --- | --- | --- | --- | 2+, 3+ |
| Osmium | Os | 76 | 190.23 | 3033 | 5012 | 22.61 | 135 | 840 | (4+)+0.687 | 57.85 | 0.130 | 630 | $1.8 \times 10^{-7}$ | 4+, 6+, 8+ |
| Oxygen | 0 | 8 | 15.9994 | −218.79 | −182.9 | 0.001308 | 73 | 1313.9 | (2−)+1.23 | 0.44 | 0.918 | 6.82 | 46.0 | 2−, 1− |
| Palladium | Pd | 46 | 106.42 | 1554.9 | 2963 | 12.023 | 137 | 804.4 | (2+)+0.915 | 16.74 | 0.246 | 380 | $6.3 \times 10^{-7}$ | 2+, 4+ |
| Phosphorus | P | 15 | 30.973462 | 44.2 | 277 | 1.823 | 110 | 1011.8 | (3−)−0.063 | 0.66 | 0.769 | 12.4 | 0.10 | 3−, 3+, 5+ |
| Platinum | Pt | 78 | 195.078 | 1768.3 | 3825 | 21.09 | 138 | 870 | (4+)+1.15 | 22.17 | 0.133 | 490 | $3.7 \times 10^{-7}$ | 2+, 4+ |
| Plutonium | Pu | 94 | [244] | 639.4 | 3230 | 19.816 | --- | 584.7 | (4+)−1.25 | 2.82 | 0.130 | 325 | --- | 3+, 4+, 5+, 6+ |
| Polonium | Po | 84 | [209] | 254 | 962 | 9.196 | 168 | 812.1 | (4+)+0.73 | 13 | --- | 100 | --- | 2−, 2+, 4+, 6+ |

*[ ] indicates mass of longest-lived isotope

## Table R-7 Properties of Elements (continued)

| | | | | | | | | | | | | | | |
|---|---|---|---|---|---|---|---|---|---|---|---|---|---|---|
| Potassium | K | 19 | 39.0983 | 63.38 | 759 | 0.856 | 227 | 418.8 | (1+)−2.925 | 2.33 | 0.757 | 76.9 | 1.50 | 1+ |
| Praseodymium | Pr | 59 | 140.90765 | 935 | 3290 | 6.64 | --- | 527 | (3+)−2.35 | 6.89 | 0.193 | 330 | $8.7 \times 10^{-4}$ | 3+, 4+ |
| Promethium | Pm | 61 | [145] | 1100 | 3000 | 7.264 | --- | 540 | (3+)−2.29 | 7.7 | --- | 290 | --- | 3+ |
| Protactinium | Pa | 91 | 231.03588 | 1568 | --- | 15.37 | --- | 568 | (5+)−1.19 | 12.34 | --- | 470 | trace | 3+, 4+, 5+ |
| Radium | Ra | 88 | [226] | 700 | 1737 | 5 | 220 | 509.3 | (2+)−2.916 | 8 | 0.095 | 125 | trace | 2+ |
| Radon | Rn | 86 | [222] | −71 | −61.7 | 0.00973 | 140 | 1037 | --- | 3 | 0.094 | 17 | --- | 3+ |
| Rhenium | Re | 75 | 186.207 | 3186 | 5596 | 21.02 | 137 | 760 | (7+)+0.415 | 60.43 | 0.137 | 705 | $2.6 \times 10^{-7}$ | 3+, 4+, 6+, 7+ |
| Rhodium | Rh | 45 | 102.9055 | 1964 | 3695 | 12.45 | 134 | 719.7 | (3+)+0.76 | 26.59 | 0.243 | 495 | $7 \times 10^{-8}$ | 3+, 4+, 5+ |
| Roentgenium | Rg | 111 | [272] | --- | --- | --- | --- | --- | --- | --- | --- | --- | --- | --- |
| Rubidium | Rb | 37 | 85.4678 | 39.31 | 688 | 1.532 | 248 | 403 | (1+)−2.924 | 2.19 | 0.363 | 72 | 0.006 | 1+ |
| Ruthenium | Ru | 44 | 101.07 | 2334 | 4150 | 12.37 | 134 | 710.2 | (4+)+0.68 | 38.59 | 0.238 | 580 | $1 \times 10^{-7}$ | 2+, 3+, 4+, 5+ |
| Rutherfordium | Rf | 104 | [261] | --- | --- | --- | --- | --- | --- | --- | --- | --- | --- | --- |
| Samarium | Sm | 62 | 150.36 | 1072 | 1803 | 7.353 | --- | 544.5 | (3+)−2.3 | 8.62 | 0.197 | 175 | $6 \times 10^{-4}$ | 2+, 3+ |
| Scandium | Sc | 21 | 44.95591 | 1541 | 2830 | 2.985 | 162 | 633.1 | (3+)−2.03 | 14.1 | 0.568 | 318 | 0.0026 | 3+ |
| Seaborgium | Sg | 106 | [266] | --- | --- | --- | --- | --- | --- | --- | --- | --- | --- | --- |
| Selenium | Se | 34 | 78.96 | 221 | 685 | 4.819 | 119 | 941 | (1−)−0.11 | 6.69 | 0.321 | 95.48 | $5 \times 10^{-6}$ | 2−, 2+, 4+, 6+ |
| Silicon | Si | 14 | 28.0588 | 1414 | 2900 | 2.33 | 118 | 786.5 | (4−)−0.143 | 50.21 | 0.712 | 359 | 27.0 | 2+, 4+ |
| Silver | Ag | 47 | 107.8682 | 961.78 | 2162 | 10.49 | 144 | 731 | (1+)+0.7991 | 11.28 | 0.235 | 255 | $8 \times 10^{-6}$ | 1+ |
| Sodium | Na | 11 | 22.989769 | 97.72 | 883 | 0.968 | 186 | 495.8 | (1+)−2.713 | 2.60 | 1.228 | 97.7 | 2.3 | 1+ |
| Strontium | Sr | 38 | 87.62 | 777 | 1382 | 2.63 | 215 | 549.5 | (2+)−2.89 | 7.43 | 0.306 | 137 | 0.036 | 2+ |
| Sulfur | S | 16 | 32.065 | 115.2 | 444.7 | 1.96 | 103 | 999.6 | (2−)−0.14 | 1.72 | 0.708 | 45 | 0.042 | 2−, 4+, 6+ |
| Tantalum | Ta | 73 | 180.9479 | 3017 | 5458 | 16.65 | 146 | 761 | (5+)−0.81 | 36.57 | 0.140 | 735 | $1.7 \times 10^{-4}$ | 4+, 5+ |
| Technetium | Tc | 43 | [98] | 2157 | 4265 | 11.5 | 136 | 702 | (6+)+0.83 | 33.29 | 0.240 | 550 | --- | 2+, 4+, 6+, 7+ |
| Tellurium | Te | 52 | 127.60 | 449.51 | 988 | 6.24 | 142 | 869.3 | (2−)−1.14 | 17.49 | 0.202 | 114.1 | $1 \times 10^{-7}$ | 2−, 2+, 4+, 6+ |
| Terbium | Tb | 65 | 158.92534 | 1356 | 3230 | 8.219 | --- | 565.8 | (3+)−2.31 | 10.15 | 0.182 | 295 | $1 \times 10^{-4}$ | 3+, 4+ |
| Thallium | Tl | 81 | 204.3822 | 304 | 1473 | 11.85 | 170 | 589.4 | (1+)−0.3363 | 4.14 | 0.129 | 165 | $5.3 \times 10^{-5}$ | 1+, 3+ |
| Thorium | Th | 90 | 232.0381 | 1842 | 4820 | 11.72 | --- | 587 | (4+)−1.83 | 13.81 | 0.118 | 530 | $6 \times 10^{-4}$ | 4+ |
| Thulium | Tm | 69 | 168.93421 | 1545 | 1950 | 9.321 | --- | 596.7 | (3+)−2.32 | 16.84 | 0.160 | 250 | $5 \times 10^{-5}$ | --- |
| Tin | Sn | 50 | 118.710 | 231.93 | 2602 | 7.31 | 140 | 708.6 | (4+)+0.15 | 7.173 | 0.227 | 290 | $2.2 \times 10^{-4}$ | 2+, 4+ |
| Titanium | Ti | 22 | 47.867 | 1668 | 3287 | 4.507 | 147 | 658.8 | (4+)−0.86 | 14.15 | 0.523 | 425 | 0.66 | 2+, 3+, 4+ |
| Tungsten | W | 74 | 183.84 | 3422 | 5555 | 19.25 | 139 | 770 | (6+)−0.09 | 52.31 | 0.132 | 800 | $1.1 \times 10^{-4}$ | 4+, 5+, 6+ |
| Ununoctium | Uuo | 118 | [294] | --- | --- | --- | --- | --- | --- | --- | --- | --- | --- | --- |
| Ununpentium | Uup | 115 | [288] | --- | --- | --- | --- | --- | --- | --- | --- | --- | --- | --- |
| Ununseptium | Uus | 117 | [294] | --- | --- | --- | --- | --- | --- | --- | --- | --- | --- | --- |
| Ununtrium | Uut | 113 | [284] | --- | --- | --- | --- | --- | --- | --- | --- | --- | --- | --- |
| Uranium | U | 92 | 238.02891 | 1132.2 | 3927 | 19.05 | --- | 597.6 | (4+)−1.38 | 9.14 | 0.116 | 420 | $1.8 \times 10^{-4}$ | 3+, 4+, 5+, 6+ |
| Vanadium | V | 23 | 50.9415 | 1910 | 3407 | 6.11 | 134 | 650.9 | (5+)−0.236 | 21.5 | 0.489 | 453 | 0.019 | 2+, 3+, 4+, 5+ |
| Xenon | Xe | 54 | 131.293 | −111.7 | −108 | 0.0058971 | 131 | 1170.4 | (6+)+2.12 | 2.27 | 0.158 | 12.57 | trace | --- |
| Ytterbium | Yb | 70 | 173.04 | 824 | 1196 | 6.57 | --- | 603.4 | (3+)−2.22 | 7.66 | 0.155 | 160 | $2.8 \times 10^{-4}$ | 2+, 3+ |
| Yttrium | Y | 39 | 88.90585 | 1526 | 3336 | 4.472 | 180 | 600 | (3+)−2.37 | 11.4 | 0.298 | 380 | 0.0029 | 3+ |
| Zinc | Zn | 30 | 65.409 | 419.53 | 907 | 7.14 | 134 | 906.4 | (2+)−0.7926 | 7.068 | 0.388 | 119 | 0.0079 | 2+ |
| Zirconium | Zr | 40 | 91.224 | 1855 | 4409 | 6.511 | 160 | 640.1 | (4+)−1.55 | 21.00 | 0.278 | 580 | 0.013 | 4+ |

*[ ] indicates mass of longest-lived isotope

Reference Tables

**Table R-8** Solubility Guidelines

A substance is considered soluble if more than three grams of the substance dissolves in 100 mL of water. The more common rules are listed below.

1. All common salts of the group 1 elements and ammonium ions are soluble.
2. All common acetates and nitrates are soluble.
3. All binary compounds of group 17 elements (other than F) with metals are soluble except those of silver, mercury(I), and lead.
4. All sulfates are soluble except those of barium, strontium, lead, calcium, silver, and mercury(I).
5. Except for those in Rule 1, carbonates, hydroxides, oxides, sulfides, and phoshates are insoluble.

**Solubility of Compounds in Water**

| | Acetate | Bromide | Carbonate | Chlorate | Chloride | Chromate | Hydroxide | Iodide | Nitrate | Oxide | Perchlorate | Phosphate | Sulfate | Sufide |
|---|---|---|---|---|---|---|---|---|---|---|---|---|---|---|
| Aluminum | S | S | — | S | S | — | I | S | S | I | S | I | S | D |
| Ammonium | S | S | S | S | S | S | S | S | S | — | S | S | S | S |
| Barium | S | S | P | S | S | I | S | S | S | S | S | I | I | D |
| Calcium | S | S | P | S | S | S | S | S | S | P | S | P | I | P |
| Copper(II) | S | S | — | S | S | — | I | — | S | I | S | I | S | I |
| Hydrogen | S | S | — | S | S | — | — | S | S | S | S | S | S | S |
| Iron(II) | — | S | P | S | S | — | I | S | S | I | S | I | S | I |
| Iron(III) | — | S | — | S | S | I | I | S | S | I | S | P | P | D |
| Lead(II) | S | I | — | S | I | I | P | P | S | P | S | I | I | I |
| Lithium | S | S | S | S | S | ? | S | S | S | S | S | P | S | S |
| Magnesium | S | S | P | S | S | S | I | S | S | I | S | P | S | D |
| Manganese(II) | S | S | P | S | S | — | I | S | S | I | S | P | S | I |
| Mercury(I) | P | I | I | S | I | P | — | I | S | I | S | I | I | I |
| Mercury(II) | S | S | — | S | S | P | I | P | S | P | S | I | D | I |
| Potassium | S | S | S | S | S | S | S | S | S | S | S | S | S | S |
| Silver | P | I | I | S | I | P | — | I | S | P | S | I | I | I |
| Sodium | S | S | S | S | S | S | S | S | S | D | S | S | S | S |
| Strontium | S | S | P | S | S | P | S | S | S | S | S | I | I | S |
| Tin(II) | D | S | — | S | S | O | | S | D | I | S | I | S | I |
| Tin(IV) | S | S | — | — | S | S | I | D | — | I | S | — | S | I |
| Zinc | S | S | P | S | S | P | P | S | S | P | S | I | S | I |

S – soluble P – partially soluble I – insoluble D – decomposes

Reference Tables

## Table R-9 Specific Heat Values (J/g·K)

| Substance | c | Substance | c | Substance | c |
|---|---|---|---|---|---|
| $AlF_3$ | 0.8948 | $Fe_3C$ | 0.5898 | $NaVO_3$ | 1.540 |
| $BaTiO_3$ | 0.79418 | $FeWO_4$ | 0.37735 | $Ni(CO)_4$ | 1.198 |
| BeO | 1.020 | HI | 0.22795 | $PbI_2$ | 0.1678 |
| $CaC_2$ | 0.9785 | $K_2CO_3$ | 0.82797 | $SF_6$ | 0.6660 |
| $CaSO_4$ | 0.7320 | $MgCO_3$ | 0.8957 | SiC | 0.6699 |
| $CCl_4$ | 0.85651 | $Mg(OH)_2$ | 1.321 | $SiO_2$ | 0.7395 |
| $CH_3OH$ | 2.55 | $MgSO_4$ | 0.8015 | $SrCl_2$ | 0.4769 |
| $CH_2OHCH_2OH$ | 2.413 | MnS | 0.5742 | $Tb_2O_3$ | 0.3168 |
| $CH_3CH_2OH$ | 2.4194 | $Na_2CO_3$ | 1.0595 | $TiCl_4$ | 0.76535 |
| CdO | 0.3382 | NaF | 1.116 | $Y_2O_3$ | 0.45397 |
| $CuSO_4{\cdot}5H_2O$ | 1.12 | | | | |

## Table R-10 Molal Freezing Point Depression and Boiling Point Elevation Constants

| Substance | $K_{fp}$ (°C/*m*) | Freezing Point (°C) | $K_{bp}$ (°C/*m*) | Boiling Point (°C) |
|---|---|---|---|---|
| Acetic acid | 3.90 | 16.66 | 3.22 | 117.90 |
| Benzene | 5.12 | 5.533 | 2.53 | 80.100 |
| Camphor | 37.7 | 178.75 | 5.611 | 207.42 |
| Cyclohexane | 20.0 | 6.54 | 2.75 | 80.725 |
| Cyclohexanol | 39.3 | 25.15 | --- | --- |
| Nitrobenzene | 6.852 | 5.76 | 5.24 | 210.8 |
| Phenol | 7.40 | 40.90 | 3.60 | 181.839 |
| Water | 1.86 | 0.000 | 0.512 | 100.000 |

## Table R-11 Heat of Formation Values

$\Delta H_f^\circ$ (kJ/mol) (concentration of aqueous solutions is 1*M*)

| Substance | $\Delta H_f^\circ$ | Substance | $\Delta H_f^\circ$ | Substance | $\Delta H_f^\circ$ | Substance | $\Delta H_f^\circ$ |
|---|---|---|---|---|---|---|---|
| Ag(s) | 0 | CsCl(s) | −443.0 | $H_3PO_4$(aq) | −1271.7 | NaBr(s) | −361.1 |
| AgCl(s) | −127.0 | $Cs_2SO_4$(s) | −1443.0 | $H_2S$(g) | −20.6 | NaCl(s) | −411.2 |
| AgCN(s) | 146.0 | CuI(s) | −67.8 | $H_2SO_3$(aq) | −608.8 | $NaHCO_3$(s) | −950.8 |
| $Al_2O_3$ | −1675.7 | CuS(s) | −53.1 | $H_2SO_4$(aq) | −814.0 | $NaNO_3$(s) | −467.9 |
| $BaCl_2$(aq) | −855.0 | $Cu_2S$(s) | −79.5 | $HgCl_2$(s) | −224.3 | NaOH(s) | −425.8 |
| $BaSO_4$ | −1473.2 | $CuSO_4$(s) | −771.4 | $Hg_2Cl_2$(s) | −265.4 | $Na_2CO_3$(s) | −1130.7 |
| BeO(s) | −609.4 | $F_2$(g) | 0 | $Hg_2SO_4$(s) | −743.1 | $Na_2S$(s) | −364.8 |
| $BiCl_3$(s) | −379.1 | $FeCl_3$(s) | −399.49 | $I_2$(s) | 0 | $Na_2SO_4$(s) | −1387.1 |
| $Bi_2S_3$(s) | −143.1 | FeO(s) | −272.0 | K(s) | 0 | $NH_4Cl$(s) | −314.4 |
| $Br_2$ | 0 | FeS(s) | −100.0 | KBr(s) | −393.8 | $O_2$(g) | 0 |
| $CCl_4$(l) | −128.2 | $Fe_2O_3$(s) | −824.2 | $KMnO_4$(s) | −837.2 | $P_4O_6$(s) | −1640.1 |
| $CH_4$(g) | −74.6 | $Fe_3O_4$(s) | −1118.4 | KOH | −424.6 | $P_4O_{10}$(s) | −2984.0 |
| $C_2H_2$(g) | 227.4 | H(g) | 218.0 | LiBr(s) | −351.2 | $PbBr_2$(s) | −278.7 |
| $C_2H_4$(g) | 52.4 | $H_2$(g) | 0 | LiOH(s) | −487.5 | $PbCl_2$(s) | −359.4 |
| $C_2H_6$(g) | −84.0 | HBr(g) | −36.3 | Mn(s) | 0 | $SF_6$(g) | −1220.5 |
| CO(g) | −110.5 | HCl(g) | −92.3 | $MnCl_2$(aq) | −555.0 | $SO_2$(g) | −296.8 |
| $CO_2$(g) | −393.5 | HCl(aq) | −167.159 | $Mn(NO_3)_2$(aq) | −635.5 | $SO_3$(g) | −395.7 |
| $CS_2$(l) | 89.0 | HCN(aq) | 108.9 | $MnO_2$(s) | −520.0 | SrO(s) | −592.0 |
| Ca(s) | 0 | HCHO | −108.6 | MnS(s) | −214.2 | $TiO_2$(s) | −944.0 |
| $CaCO_3$(s) | −1206.9 | HCOOH | −425.0 | $N_2$(g) | 0 | TlI(s) | −123.8 |
| CaO(s) | −634.9 | HF(g) | −273.3 | $NH_3$(g) | −45.9 | $UCl_4$(s) | −1019.2 |
| $Ca(OH)_2$(s) | −985.2 | HI(g) | 26.5 | $NH_4Br$(s) | −270.8 | $UCl_6$(s) | −1092.0 |
| $Cl_2$(g) | 0 | $H_2O$(l) | −285.8 | NO(g) | 91.3 | Zn(s) | 0 |
| $Co_3O_4$(s) | −891.0 | $H_2O$(g) | −241.8 | $NO_2$(g) | 33.2 | $ZnCl_2$(aq) | −415.1 |
| CoO(s) | −237.9 | $H_2O_2$(l) | −187.8 | $N_2O$(g) | 81.6 | ZnO(s) | −350.5 |
| $Cr_2O_3$(s) | −1139.7 | $H_3PO_2$(l) | −595.4 | Na(s) | 0 | $ZnSO_4$(s) | −982.8 |

# Safety Symbols

Safety symbols in the following table are used in the lab activities to indicate possible hazards. Learn the meaning of each symbol. **It is recommended that you wear safety goggles and apron at all times in the lab. This might be required in your school district.**

| Safety Symbols | Hazard | Examples | Precaution | Remedy |
|---|---|---|---|---|
| **Disposal** | Special disposal procedures need to be followed. | certain chemicals, living organisms | Do not dispose of these materials in the sink or trash can. | Dispose of wastes as directed by your teacher. |
| **Biological** | Organisms or other biological materials that might be harmful to humans | bacteria, fungi, blood, unpreserved tissues, plant materials | Avoid skin contact with these materials. Wear mask or gloves. | Notify your teacher if you suspect contact with material. Wash hands thoroughly. |
| **Extreme Temperature** | Objects that can burn skin by being too cold or too hot | boiling liquids, hot plates, dry ice, liquid nitrogen | Use proper protection when handling. | Go to your teacher for first aid. |
| **Sharp Object** | Use of tools or glassware that can easily puncture or slice skin | razor blades, pins, scalpels, pointed tools, dissecting probes, broken glass | Practice common-sense behavior and follow guidelines for use of the tool. | Go to your teacher for first aid. |
| **Fume** | Possible danger to respiratory tract from fumes | ammonia, acetone, nail polish remover, heated sulfur, moth balls | Be sure there is good ventilation. Never smell fumes directly. Wear a mask. | Leave foul area and notify your teacher immediately. |
| **Electrical** | Possible danger from electrical shock or burn | improper grounding, liquid spills, short circuits, exposed wires | Double-check setup with teacher. Check condition of wires and apparatus. Use GFI-protected outlets. | Do not attempt to fix electrical problems. Notify your teacher immediately. |
| **Irritant** | Substances that can irritate the skin or mucous membranes of the respiratory tract | pollen, moth balls, steel wool, fiberglass, potassium permanganate | Wear dust mask and gloves. Practice extra care when handling these materials. | Go to your teacher for first aid. |
| **Chemical** | Chemicals that can react with and destroy tissue and other materials | bleaches such as hydrogen peroxide; acids such as sulfuric acid, hydrochloric acid; bases such as ammonia, sodium hydroxide | Wear goggles, gloves, and an apron. | Immediately flush the affected area with water and notify your teacher. |
| **Toxic** | Substance may be poisonous if touched, inhaled, or swallowed. | mercury, many metal compounds, iodine, poinsettia plant parts | Follow your teacher's instructions. | Always wash hands thoroughly after use. Go to your teacher for first aid. |
| **Flammable** | Flammable chemicals may be ignited by open flame, spark, or exposed heat. | alcohol, kerosene, potassium permanganate | Avoid open flames and heat when using flammable chemicals. | Notify your teacher immediately. Use fire safety equipment if applicable. |
| **Open Flame** | Open flame in use, may cause fire. | hair, clothing, paper, synthetic materials | Tie back hair and loose clothing. Follow teacher's instruction on lighting and extinguishing flames. | Notify your teacher immediately. Use fire safety equipment if applicable. |

**Eye Safety**
Proper eye protection should be worn at all times by anyone performing or observing science activities.

**Clothing Protection**
This symbol appears when substances could stain or burn clothing.

**Animal Safety**
This symbol appears when safety of animals and students must be ensured.

**Radioactivity**
This symbol appears when radioactive materials are used.

**Handwashing**
After the lab, wash hands with soap and water before removing goggles.

# SUPPLEMENTAL PRACTICE PROBLEMS

## CHAPTER 2

**Section 1**

1. The density of a substance is 4.8 g/mL. What is the volume of a sample that is 19.2 g?

2. A 2.00-mL sample of Substance A has a density of 18.4 g/mL, and a 5.00-mL sample of Substance B has a density of 35.5 g/mL. Do you have an equal mass of Substances A and B?

**Section 2**

3. Express the following quantities in scientific notation.
   - **a.** 5,453,000 m
   - **b.** 300.8 kg
   - **c.** 0.00536 ng
   - **d.** 0.0120325 km
   - **e.** 34,800 s
   - **f.** 332,080,000 cm
   - **g.** 0.0002383 ms
   - **h.** 0.3048 mL

4. Solve the following problems. Express your answers in scientific notation.
   - **a.** $3 \times 10^{2}$ m $+ 5 \times 10^{2}$ m
   - **b.** $8 \times 10^{-5}$ m $+ 4 \times 10^{-5}$ m
   - **c.** $6.0 \times 10^{5}$ m $+ 2.38 \times 10^{6}$ m
   - **d.** $2.3 \times 10^{-3}$ L $+ 5.78 \times 10^{-2}$ L
   - **e.** $2.56 \times 10^{2}$ g $- 1.48 \times 10^{2}$ g
   - **f.** $5.34 \times 10^{-3}$ L $- 3.98 \times 10^{-3}$ L
   - **g.** $7.623 \times 10^{5}$ nm $- 8.32 \times 10^{4}$ nm
   - **h.** $9.052 \times 10^{-2}$ s $- 3.61 \times 10^{-3}$ s

5. Solve the following problems. Express your answers in scientific notation.
   - **a.** $(8 \times 10^{3}$ m$) \times (1 \times 10^{5}$ m$)$
   - **b.** $(4 \times 10^{2}$ m$) \times (2 \times 10^{4}$ m$)$
   - **c.** $(5 \times 10^{-3}$ m$) \times (3 \times 10^{4}$ m$)$
   - **d.** $(3 \times 10^{-4}$ m$) \times (3 \times 10^{-2}$ m$)$
   - **e.** $(8 \times 10^{4}$ g$) \div (4 \times 10^{3}$ mL$)$
   - **f.** $(6 \times 10^{-3}$ g$) \div (2 \times 10^{-1}$ mL$)$
   - **g.** $(1.8 \times 10^{-2}$ g$) \div (9 \times 10^{-5}$ mL$)$
   - **h.** $(4 \times 10^{-4}$ g$) \div (1 \times 10^{3}$ mL$)$

6. Perform the following conversions.
   - **a.** 96 kg to g
   - **b.** 155 mg to g
   - **c.** 15 cg to kg
   - **d.** 584 μs to s
   - **e.** 188 dL to L
   - **f.** 3600 m to km
   - **g.** 24 g to pg
   - **h.** 85 cm to nm

7. How many minutes are there in 5 days?

8. A car is traveling at 118 km/h. What is its speed in Mm/h?

**Section 3**

9. Three measurements of 34.5 m, 38.4 m, and 35.3 m are taken. If the accepted value of the measurement is 36.7 m, what is the percent error for each measurement?

10. Three measurements of 12.3 mL, 12.5 mL, and 13.1 mL are taken. The accepted value for each measurement is 12.8 mL. Calculate the percent error for each measurement.

Supplemental Practice Problems

**11.** Determine the number of significant figures in each measurement.

**a.** 340,438 g
**b.** 87,000 ms
**c.** 4080 kg
**d.** 961,083,110 m
**e.** 1.040 s
**f.** 0.0483 m
**g.** 0.2080 mL
**h.** 0.0000481 g

**12.** Write the following in three significant figures.

**a.** 0.0030850 km
**b.** 3.0823 g
**c.** 5808 mL
**d.** 34.654 mg

**13.** Write the answers in scientific notation.

**a.** 0.005832 g
**b.** 386,808 ns
**c.** 0.0005800 km
**d.** 2086 L

**14.** Use rounding rules when you complete the following.

**a.** 34.3 m + 35.8 m + 33.7 m
**b.** 0.056 kg + 0.0783 kg + 0.0323 kg
**c.** 309.1 mL + 158.02 mL + 238.1 mL
**d.** 1.03 mg + 2.58 mg + 4.385 mg
**e.** 8.376 km − 6.153 km
**f.** 34.24 s − 12.4 s
**g.** 804.9 dm − 342.0 dm
**h.** $6.38 \times 10^2$ m − $1.57 \times 10^2$ m

**15.** Complete the following calculations. Round off the answers to the correct number of significant figures.

**a.** 34.3 cm × 12 cm
**b.** 0.054 mm × 0.3804 mm
**c.** 45.1 km × 13.4 km
**d.** 45.5 g ÷ 15.5 mL
**e.** 35.43 g ÷ 24.84 mL
**f.** 0.0482 g ÷ 0.003146 mL

## CHAPTER 3

**Section 2**

**1.** A 3.5-kg iron shovel is left outside through the winter. The shovel, now orange with rust, is rediscovered in the spring. Its mass is 3.7 kg. How much oxygen combined with the iron?

**2.** When 5.0 g of tin reacts with hydrochloric acid, the mass of the products, tin chloride and hydrogen, totals 8.1 g. How many grams of hydrochloric acid were used?

**Section 4**

**3.** A compound is analyzed and found to be 50.0% sulfur and 50.0% oxygen. If the total amount of the sulfur oxide compound is 12.5 g, how many grams of sulfur are there?

**4.** Two unknown compounds are analyzed. Compound I contains 5.63 g of tin and 3.37 g of chlorine, while Compound II contains 2.5 g of tin and 2.98 g of chlorine. Are the compounds the same?

## CHAPTER 4

**Section 3**

**1.** How many protons and electrons are in each of the following atoms?

**a.** gallium
**b.** silicon
**c.** cesium
**d.** calcium
**e.** molybdenum
**f.** titanium

2. What is the atomic number of each of the following elements?
   - **a.** an atom that contains 37 electrons
   - **b.** an atom that contains 72 protons
   - **c.** an atom that contains 1 electron
   - **d.** an atom that contains 85 protons

3. Use the periodic table to write the name and the symbol for each element identified in Question 2.

4. An isotope of copper contains 29 electrons, 29 protons, and 36 neutrons. What is the mass number of this isotope?

5. An isotope of uranium contains 92 electrons and 144 neutrons. What is the mass number of this isotope?

6. Use the periodic table to write the symbols for each of the following elements. Then, determine the number of electrons, protons, and neutrons each contains.
   - **a.** yttrium-88
   - **b.** arsenic-75
   - **c.** xenon-129
   - **d.** bromine-79
   - **e.** gold-197
   - **f.** helium-4

7. An element has two naturally occurring isotopes: $^{14}X$ and $^{15}X$. $^{14}X$ has a mass of 14.00307 amu and a relative abundance of 99.63%. $^{15}X$ has a mass of 15.00011 amu and a relative abundance of 0.37%. Identify the unknown element.

8. Silver has two naturally occurring isotopes. Ag-107 has an abundance of 51.82% and a mass of 106.9 amu. Ag-109 has a relative abundance of 48.18% and a mass of 108.9 amu. Calculate the atomic mass of silver.

## CHAPTER 5

**Section 1**

1. What is the frequency of an electromagnetic wave that has a wavelength of $4.55 \times 10^{-3}$ m? $1.00 \times 10^{-12}$ m?

2. Calculate the wavelength of an electromagnetic wave with a frequency of $8.68 \times 10^{16}$ Hz; $5.0 \times 10^{14}$ Hz; and $1.00 \times 10^{6}$ Hz.

3. What is the energy of a quantum of visible light having a frequency of $5.45 \times 10^{14}\ s^{-1}$?

4. An X-ray has a frequency of $1.28 \times 10^{18}\ s^{-1}$. What is the energy of a quantum of the X-ray?

**Section 3**

5. Write the ground-state electron configuration for the following.
   - **a.** nickel
   - **b.** cesium
   - **c.** boron
   - **d.** krypton

6. What element has the following ground-state electron configuration $[He]2s^2$? $[Xe]6s^24f^{14}5d^{10}6p^1$?

7. Which element in period 4 has four electrons in its electron-dot structure?

8. Which element in period 2 has six electrons in its electron-dot structure?

9. Draw the electron-dot structure for each element in Question 5.

Supplemental Practice Problems

## CHAPTER 6

**Section 2**

**1.** Identify the group, period, and block of an atom with the following electron configurations.

**a.** $[He]2s^2 2p^1$ **b.** $[Kr]5s^2 4d^5$ **c.** $[Xe]6s^2 5f^{14} 6d^5$

**2.** Write the electron configuration for the element fitting each of the following descriptions.

**a.** a noble gas in the first period
**b.** a group 4 element in the fifth period
**c.** a group 14 element in the sixth period
**d.** a group 1 element in the seventh period

**Section 3**

**3.** Using the periodic table, rank each group of elements in order of increasing size.

**a.** calcium, magnesium, and strontium
**b.** oxygen, lithium, and fluorine
**c.** fluorine, cesium, and calcium
**d.** selenium, chlorine, and tellurium
**e.** iodine, krypton, and beryllium

## CHAPTER 7

**Section 2**

**1.** Explain the formation of an ionic compound from zinc and chlorine.

**2.** Explain the formation of an ionic compound from barium and nitrogen.

**Section 3**

**3.** Write the chemical formula of an ionic compound composed of the following pairs of ions.

**a.** calcium and arsenide
**b.** iron(III) and chloride
**c.** magnesium and sulfide
**d.** barium and iodide
**e.** gallium and phosphide

**4.** Determine the formula for ionic compounds composed of the following ions.

**a.** copper(II) and acetate **c.** calcium and hydroxide
**b.** ammonium and phosphate **d.** gold(III) and cyanide

**5.** Name the following compounds.

**a.** $Co(OH)_2$ **c.** $Na_3PO_4$ **e.** $SrI_2$
**b.** $Ca(ClO_3)_2$ **d.** $K_2Cr_2O_7$ **f.** $HgF_2$

## CHAPTER 8

**Section 1**

**1.** Draw the Lewis structure for each of the following molecules.

**a.** $CCl_2H_2$ **b.** $HF$ **c.** $PCl_3$ **d.** $CH_4$

**Section 2**

**2.** Name the following binary compounds.

**a.** $S_4N_2$ **c.** $SF_6$ **e.** $SiO_2$
**b.** $Cl_2O$ **d.** $NO$ **f.** $IF_7$

**3.** Name the following acids: $H_3PO_4$, $HBr$, $HNO_3$.

**Section 3**

4. Draw the Lewis structure for each of the following.

   **a.** CO  **c.** $N_2O$  **e.** $SiO_2$

   **a.** $CH_2O$  **d.** $Cl_2O$  **f.** $AlBr_3$

5. Draw the Lewis resonance structure for $CO_3^{2-}$.

6. Draw the Lewis resonance structure for $CH_3CO_2^-$.

7. Draw the Lewis structure for NO and $IF_4^-$.

**Section 4**

8. Determine the molecular geometry, bond angles, and hybrid of each molecule in Question 4.

**Section 5**

9. Determine whether each of the following molecules is polar or nonpolar.

   **a.** $CH_2O$  **b.** $BF_3$  **c.** $SiH_4$  **d.** $H_2S$

## CHAPTER 9

**Section 1**

**Write skeleton equations for the following reactions.**

1. Solid barium and oxygen gas react to produce solid barium oxide.

2. Solid iron and aqueous hydrogen sulfate react to produce aqueous iron(III) sulfate and gaseous hydrogen.

**Write balanced chemical equations for the following reactions.**

3. Liquid bromine reacts with solid phosphorus ($P_4$) to produce solid diphosphorus pentabromide.

4. Aqueous lead(II) nitrate reacts with aqueous potassium iodide to produce solid lead(II) iodide and aqueous potassium nitrate.

5. Solid carbon reacts with gaseous fluorine to produce gaseous carbon tetrafluoride.

6. Aqueous carbonic acid reacts to produce liquid water and gaseous carbon dioxide.

7. Gaseous hydrogen chloride reacts with gaseous ammonia to produce solid ammonium chloride.

8. Solid copper(II) sulfide reacts with aqueous nitric acid to produce aqueous copper(II) sulfate, liquid water, and nitrogen dioxide gas.

**Section 2**

**Classify each of the following reactions into as many types as possible.**

9. $2Mo(s) + 3O_2(g) \rightarrow 2MoO_3(s)$

10. $N_2H_4(l) + 3O_2(g) \rightarrow 2NO_2(g) + 2H_2O(l)$

**Write balanced chemical equations for the following decomposition reactions.**

11. Aqueous hydrogen chlorite decomposes to produce water and gaseous chlorine(III) oxide.

12. Calcium carbonate(s) decomposes to produce calcium oxide(s) and carbon dioxide(g).

**Use the activity series to predict whether each of the following single-replacement reactions will occur.**

13. $Al(s) + FeCl_3(aq) \rightarrow AlCl_3(aq) + Fe(s)$

14. $Br_2(l) + 2LiI(aq) \rightarrow 2LiBr(aq) + I_2(aq)$

15. $Cu(s) + MgSO_4(aq) \rightarrow Mg(s) + CuSO_4(aq)$

**Write chemical equations for the following chemical reactions.**

16. Bismuth(III) nitrate(aq) reacts with sodium sulfide(aq), yielding bismuth(III) sulfide(s) plus sodium nitrate(aq).

17. Magnesium chloride(aq) reacts with potassium carbonate(aq), yielding magnesium carbonate(s) plus potassium chloride(aq).

**Section 3**

**Write net ionic equations for the following reactions.**

18. Aqueous solutions of barium chloride and sodium fluoride are mixed to form a precipitate of barium fluoride.

19. Aqueous solutions of copper(I) nitrate and potassium sulfide are mixed to form insoluble copper(I) sulfide.

20. Hydrobromic acid reacts with aqueous lithium hydroxide.

21. Perchloric acid reacts with aqueous rubidium hydroxide.

22. Nitric acid reacts with aqueous sodium carbonate.

23. Hydrochloric acid reacts with aqueous lithium cyanide.



## CHAPTER 10

**Section 1**

1. Determine the number of atoms in 3.75 mol of Fe.

2. Calculate the number of formula units in 12.5 mol of $CaCO_3$.

3. How many moles of $CaCl_2$ contain $1.26 \times 10^{24}$ formula units of $CaCl_2$?

4. How many moles of Ag contain $4.59 \times 10^{25}$ atoms of Ag?

**Section 2**

5. Determine the mass in grams of 0.0458 mol of sulfur.

6. Calculate the mass in grams of $2.56 \times 10^{-3}$ mol of iron.

7. Determine the mass in grams of 125 mol of neon.

8. How many moles of titanium are contained in 71.4 g?

9. How many moles of lead are equivalent to $9.51 \times 10^{3}$ g of Pb?

10. Determine the number of moles of arsenic in 1.90 g of As.

11. Determine the number of atoms in $4.56 \times 10^{-2}$ g of sodium.

12. How many atoms of gallium are in $2.85 \times 10^{3}$ g of gallium?

13. Determine the mass in grams of $5.65 \times 10^{24}$ atoms of Se.

14. What is the mass in grams of $3.75 \times 10^{21}$ atoms of Li?

**Section 3**

15. How many moles of each element are in 0.0250 mol of $K_2CrO_4$?

16. How many moles of ammonium ions are in 4.50 mol of $(NH_4)_2CO_3$?

17. Determine the molar mass of silver nitrate.

18. Calculate the molar mass of acetic acid ($CH_3COOH$).

19. Determine the mass of 8.57 mol of sodium dichromate ($Na_2Cr_2O_7$).

20. Calculate the mass of 42.5 mol of potassium cyanide.

21. Determine the number of moles present in 456 g of $Cu(NO_3)_2$.

22. Calculate the number of moles in 5.67 g of potassium hydroxide.

23. Calculate the number of each atom in 40.0 g of methanol ($CH_3OH$).

24. What mass of sodium hydroxide contains $4.58 \times 10^{23}$ formula units?

**Section 4**

25. What is the percent by mass of each element in sucrose ($C_{12}H_{22}O_{11}$)?

26. Which compound has a greater percent by mass of chromium, $K_2CrO_4$ or $K_2Cr_2O_7$?

27. Analysis of a compound indicates the percent composition 42.07% Na, 18.89% P, and 39.04% O. Determine its empirical formula.

28. A colorless liquid was found to contain 39.12% C, 8.76% H, and 52.12% O. Determine the empirical formula of the substance.

29. Analysis of a compound used in cosmetics reveals the compound contains 26.76% C, 2.21% H, 71.17% O and has a molar mass of 90.04 g/mol. Determine the molecular formula for this substance.

30. Eucalyptus leaves are the food source for panda bears. Eucalyptol is an oil found in these leaves. Analysis of eucalyptol indicates it has a molar mass of 154 g/mol and contains 77.87% C, 11.76% H, and 10.37% O. Determine the molecular formula of eucalyptol.

31. Beryl is a hard mineral that occurs in a variety of colors. A 50.0-g sample of beryl contains 2.52 g Be, 5.01 g Al, 15.68 g Si, and 26.79 g O. Determine its empirical formula.

32. Analysis of a 15.0-g sample of a compound used to leach gold from low-grade ores is 7.03 g Na, 3.68 g C, and 4.29 g N. Determine the empirical formula for this substance.

**Section 5**

33. Analysis of a hydrate of iron(III) chloride revealed that in a 10.00-g sample of the hydrate, 6.00 g is anhydrous iron(III) chloride and 4.00 g is water. Determine the formula and name of the hydrate.

34. When 25.00 g of a hydrate of nickel(II) chloride was heated, 11.37 g of water was released. Determine the name and formula of the hydrate.

## CHAPTER 11

**Section 1**

**Interpret the following balanced chemical equations in terms of particles, moles, and mass.**

1. $Mg + 2HCl \rightarrow MgCl_2 + H_2$

2. $2Al + 3CuSO_4 \rightarrow Al_2(SO_4)_3 + 3Cu$

3. $Cu(NO_3)_2 + 2KOH \rightarrow Cu(OH)_2 + 2KNO_3$

4. Write and balance the equation for the decomposition of aluminum carbonate. Determine the possible mole ratios.

5. Write and balance the equation for the formation of magnesium hydroxide and hydrogen from magnesium and water. Determine the possible mole ratios.

**Section 2**

6. Some antacid tablets contain aluminum hydroxide. The aluminum hydroxide reacts with stomach acid according to the equation: $Al(OH)_3 + 3HCl \rightarrow AlCl_3 + 3H_2O$. Determine the moles of acid neutralized if a tablet contains 0.200 mol of $Al(OH)_3$.

7. Chromium reacts with oxygen according to the equation: $4Cr + 3O_2 \rightarrow 2Cr_2O_3$. Determine the moles of chromium(III) oxide produced when 4.58 mol of chromium is allowed to react.

8. Space vehicles use solid lithium hydroxide to remove exhaled carbon dioxide according to the equation: $2LiOH + CO_2 \rightarrow Li_2CO_3 + H_2O$. Determine the mass of carbon dioxide removed if the space vehicle carries 42.0 mol of LiOH.

9. Some of the sulfur dioxide released into the atmosphere is converted to sulfuric acid according to the equation: $2SO_2 + 2H_2O + O_2 \rightarrow 2H_2SO_4$. Determine the mass of sulfuric acid formed from 3.20 mol of sulfur dioxide.



10. How many grams of carbon dioxide are produced when 2.50 g of sodium hydrogen carbonate reacts with excess citric acid according to the equation: $3NaHCO_3 + H_3C_6H_5O_7 \rightarrow Na_3C_6H_5O_7 + 3CO_2 + 3H_2O$?

11. Aspirin ($C_9H_8O_4$) is produced when salicylic acid ($C_7H_6O_3$) reacts with acetic anhydride ($C_4H_6O_3$) according to the equation: $C_7H_6O_3 + C_4H_6O_3 \rightarrow C_9H_8O_4 + HC_2H_3O_2$. Determine the mass of aspirin produced when 150.0 g of salicylic acid reacts with an excess of acetic anhydride.

**Section 3**

12. Chlorine reacts with benzene to produce chlorobenzene and hydrogen chloride, $Cl_2 + C_6H_6 \rightarrow C_6H_5Cl + HCl$. Determine the limiting reactant if 45.0 g of benzene reacts with 45.0 g of chlorine, the mass of the excess reactant after the reaction is complete, and the mass of chlorobenzene produced.

13. Nickel reacts with hydrochloric acid to produce nickel(II) chloride and hydrogen according to the equation: $Ni + 2HCl \rightarrow NiCl_2 + H_2$. If 5.00 g of Ni and 2.50 g of HCl react, determine the limiting reactant, the mass of the excess reactant after the reaction is complete, and the mass of nickel(II) chloride produced.

**Section 4**

14. Tin(IV) iodide is prepared by reacting tin with iodine. Write the balanced chemical equation for the reaction. Determine the theoretical yield if a 5.00-g sample of tin reacts in an excess of iodine. Determine the percent yield if 25.0 g of $SnI_4$ was recovered.

15. Gold is extracted from gold-bearing rock by adding sodium cyanide in the presence of oxygen and water, according to the reaction: $4Au(s) + 8NaCN(aq) + O_2(g) + 2H_2O(l) \rightarrow 4NaAu(CN)_2(aq) + NaOH(aq)$. Determine the theoretical yield of $NaAu(CN)_2$ if 1000.0 g of gold-bearing rock is used, which contains 3.00% gold by mass. Determine the percent yield of $NaAu(CN)_2$ if 38.790 g of $NaAu(CN)_2$ is recovered.

## CHAPTER 12

**Section 1**

1. Calculate the ratio of effusion rates for methane ($CH_4$) and nitrogen.
2. Calculate the molar mass of butane. Butane's rate of diffusion is 3.8 times slower than that of helium.
3. What is the total pressure in a canister that contains oxygen gas at a partial pressure of 804 mm Hg, nitrogen at a partial pressure of 220 mm Hg, and hydrogen at a partial pressure of 445 mm Hg?
4. Calculate the partial pressure of neon in a flask that has a total pressure of 1.87 atm. The flask contains krypton at a partial pressure of 0.77 atm and helium at a partial pressure of 0.62 atm.

## CHAPTER 13

**Section 1**

1. The pressure of air in a 2.25-L container is 1.20 atm. What is the new pressure if the sample is transferred to a 6.50-L container? Temperature is constant.
2. The volume of a sample of hydrogen gas at 0.997 atm is 5.00 L. What will be the new volume if the pressure is decreased to 0.977 atm? Temperature is constant.
3. A gas at 55.0°C occupies a volume of 3.60 L. What volume will it occupy at 30.0°C? Pressure is constant.
4. The volume of a gas is 0.668 L at 66.8°C. At what Celsius temperature will the gas have a volume of 0.942 L, assuming pressure remains constant?
5. The pressure in a bicycle tire is 1.34 atm at 33.0°C. At what temperature will the pressure inside the tire be 1.60 atm? Volume is constant.
6. If a sample of oxygen gas has a pressure of 810 torr at 298 K, what will be its pressure if its temperature is raised to 330 K?
7. Air in a tightly sealed bottle has a pressure of 0.978 atm at 25.5°C. What will be its pressure if the temperature is raised to 46.0°C?
8. Hydrogen gas at a temperature of 22.0°C that is confined in a 5.00-L cylinder exerts a pressure of 4.20 atm. If the gas is released into a 10.0-L reaction vessel at a temperature of 33.6°C, what will be the pressure inside the reaction vessel?
9. A sample of neon gas at a pressure of 1.08 atm fills a flask with a volume of 250 mL at a temperature of 24.0°C. If the gas is transferred to another flask at 37.2°C and a pressure of 2.25 atm, what is the volume of the new flask?

**Section 2**

10. What volume of beaker contains exactly $2.23 \times 10^{-2}$ mol of nitrogen gas at STP?
11. How many moles of air are in a 6.06-L tire at STP?
12. How many moles of oxygen are in a 5.5-L canister at STP?
13. What mass of helium is in a 2.00-L balloon at STP?
14. What volume will 2.3 kg of nitrogen gas occupy at STP?

15. Calculate the number of moles of gas that occupy a 3.45-L container at a pressure of 150 kPa and a temperature of 45.6°C.

16. What is the pressure in torr that a 0.44-g sample of carbon dioxide gas will exert at a temperature of 46.2°C when it occupies a volume of 5.00 L?

17. What is the molar mass of a gas that has a density of 1.02 g/L at 0.990 atm pressure and 37°C?

18. Calculate the grams of oxygen gas present in a 2.50-L sample kept at 1.66 atm pressure and a temperature of 10.0°C.

**Section 3**

19. What volume of oxygen gas is needed to completely combust 0.202 L of butane gas ($C_4H_{10}$)?

20. Determine the volume of methane gas ($CH_4$) needed to react completely with 0.660 L of $O_2$ gas to form methanol ($CH_3OH$).

21. Calculate the mass of hydrogen peroxide needed to obtain 0.460 L of oxygen gas at STP. $2H_2O_2(aq) \rightarrow 2H_2O(l) + O_2(g)$

22. When potassium chlorate is heated in the presence of a catalyst such as manganese dioxide, it decomposes to form solid potassium chloride and oxygen gas: $2KClO_3(s) \rightarrow 2KCl(s) + 3O_2(g)$. How many liters of oxygen will be produced at STP if 1.25 kg of potassium chlorate decomposes completely?



## CHAPTER 14

**Section 2**

1. What is the percent by mass of a sample of ocean water that is found to contain 1.36 g of magnesium ions per 1000 g?

2. What is the percent by mass of iced tea containing 0.75 g of aspartame in 250 g of water?

3. A bottle of hydrogen peroxide is labeled 3%. If you pour out 50 mL of hydrogen peroxide solution, what volume is actually hydrogen peroxide?

4. If 50 mL of pure acetone is mixed with 450 mL of water, what is the percent by volume of acetone?

5. Calculate the molarity of 1270 g of $K_3PO_4$ in 4.0 L aqueous solution.

6. What is the molarity of 90.0 g of $NH_4Cl$ in 2.25 L aqueous solution?

7. Which is more concentrated, 25 g of NaCl dissolved in 500 mL of water or a 10% solution of NaCl (percent by mass)?

8. Calculate the mass of NaOH required to prepare a 0.343*M* solution dissolved in 2500 mL of water.

9. Calculate the volume required to dissolve 11.2 g of $CuSO_4$ to prepare a 0.140*M* solution.

10. How would you prepare 500 mL of a solution that has a new concentration of 4.5*M* if the stock solution is 11.6*M*?

11. Caustic soda is 19.1*M* NaOH and is diluted for household use. What is the household concentration if 10 mL of the concentrated solution is diluted to 400 mL?

12. What is the molality of a solution containing 63.0 g of $HNO_3$ in 0.500 kg of water?

13. What is the molality of an acetic acid solution containing 0.500 mol of $HC_2H_3O_2$ in 0.800 kg of water?

14. What mass of ethanol ($C_2H_5OH$) will be required to prepare a 2.00$m$ solution in 8.00 kg of water?

15. A mixture of gases contains 0.215 mol $N_2$, 0.345 mol $O_2$, 0.023 mol $CO_2$, and 0.014 mol $SO_2$. What is the mole fraction of $N_2$?

16. A necklace contains 4.85 g of gold, 1.25 g of silver, and 2.40 g of copper. What is the mole fraction of each metal?

**Section 3**

17. Calculate the mass of gas dissolved at 150.0 kPa, if 0.35 g of the gas dissolves in 2.0 L of water at 30.0 kPa.

18. At which depth, 10 m or 40 m, will a scuba diver have more nitrogen dissolved in the bloodstream?

**Section 4**

19. Calculate the freezing point and boiling point of a solution containing 6.42 g of sucrose ($C_{12}H_{22}O_{11}$) in 100.0 g of water.

20. Calculate the freezing point and boiling point of a solution containing 23.7 g of copper(II) sulfate in 250.0 g of water.

21. Calculate the freezing point and boiling point of a solution containing 0.15 mol of the molecular compound naphthalene in 175 g of benzene ($C_6H_6$).

Supplemental Practice Problems

## CHAPTER 15

**Section 1**

1. What is the equivalent in joules of 126 Calories?

2. Convert 455 kilojoules to kilocalories.

3. How much heat is required to warm 122 g of water by 23.0°C?

4. The temperature of 55.6 g of a material decreases by 14.8°C when it loses 3080 J of heat. What is its specific heat?

5. What is the specific heat of a metal if the temperature of a 12.5-g sample increases from 19.5°C to 33.6°C when it absorbs 37.7 J of heat?

**Section 2**

6. A 75.0-g sample of a metal is placed in boiling water until its temperature is 100.0°C. A calorimeter contains 100.00 g of water at a temperature of 24.4°C. The metal sample is removed from the boiling water and immediately placed in water in the calorimeter. The final temperature of the metal and water in the calorimeter is 34.9°C. Assuming that the calorimeter provides perfect insulation, what is the specific heat of the metal?

**Section 3**

7. Use **Table 4** to determine how much heat is released when 1.00 mol of gaseous methanol condenses to a liquid.

8. Use **Table 4** to determine how much heat must be supplied to melt 4.60 g of ethanol.

**Section 4**

**9.** Calculate $\Delta H_{rxn}$ for the reaction $2C(s) + 2H_2(g) \rightarrow C_2H_4(g)$, given the following thermochemical equations:

$2CO_2(g) + 2H_2O(l) \rightarrow C_2H_4(g) + 3O_2(g)$ $\Delta H = 1411$ kJ

$C(s) + O_2(g) \rightarrow CO_2(g)$ $\Delta H = -393.5$ kJ

$2H_2(g) + O_2(g) \rightarrow 2H_2O(l)$ $\Delta H = -572$ kJ

**10.** Calculate $\Delta H_{rxn}$ for the reaction $HCl(g) + NH_3(g) \rightarrow NH_4Cl(s)$, given the following thermochemical equations:

$H_2(g) + Cl_2(g) \rightarrow 2HCl(g)$ $\Delta H = -184$ kJ

$N_2(g) + 3H_2(g) \rightarrow 2NH_3(g)$ $\Delta H = -92$ kJ

$N_2(g) + 4H_2(g) + Cl_2(g) \rightarrow 2NH_4Cl(s)$ $\Delta H = -628$ kJ

**Use standard enthalpies of formation from Table 5 and Table R-11 to calculate $\Delta H°_{rxn}$ for each of the following reactions.**

**11.** $2HF(g) \rightarrow H_2(g) + F_2(g)$

**12.** $2H_2S(g) + 3O_2(g) \rightarrow 2H_2O(l) + 2SO_2(g)$

**Section 5**

**Predict the sign of $\Delta S_{system}$ for each reaction or process.**

**13.** $FeS(s) \rightarrow Fe^{2+}(aq) + S^{2-}(aq)$

**14.** $SO_2(g) + H_2O(l) \rightarrow H_2SO_3(aq)$

**Determine if each of the following processes or reactions is spontaneous or nonspontaneous.**

**15.** $\Delta H_{system} = 15.6$ kJ, $T = 415$ K, $\Delta S_{system} = 45$ J/K

**16.** $\Delta H_{system} = 35.6$ kJ, $T = 415$ K, $\Delta S_{system} = 45$ J/K

## CHAPTER 16

**Section 1**

**1.** In the reaction $A \rightarrow 2B$, suppose that [A] changes from 1.20 mol/L at time = 0 to 0.60 mol/L at time = 3.00 min and that [B] = 0.00 mol/L at time = 0.

**a.** What is the average rate at which A is consumed in mol/(L•min)?

**b.** What is the average rate at which B is produced in mol/(L•min)?

**Section 3**

**2.** What are the overall reaction orders in Practice Problems 19 to 22 in this section on reaction rate laws?

**3.** If halving [A] in the reaction $A \rightarrow B$ causes the initial rate to decrease to one-fourth its original value, what is the probable rate law for the reaction?

**4.** Use the data below and the method of initial rates to determine the rate law for the reaction $2NO(g) + O_2(g) \rightarrow 2NO_2(g)$.

**Formation of $NO_2$ Data**

| Trial | Initial [NO] (*M*) | Initial [$O_2$] (*M*) | Initial Rate (mol/(L•s)) |
|---|---|---|---|
| 1 | 0.030 | 0.020 | 0.0041 |
| 2 | 0.060 | 0.020 | 0.0164 |
| 3 | 0.060 | 0.040 | 0.0328 |

Supplemental Practice Problems

**Section 4**

5. When 1 mol of cyclobutane ($C_4H_8$) decomposes to 2 mol of ethylene ($C_2H_4$) at 1273 K the rate law for the reaction is Rate = (87 $s^{-1}$) [$C_4H_8$]. What is the instantaneous rate of this reaction when
   a. [$C_4H_8$] = 0.0100 mol/L?
   b. [$C_4H_8$] = 0.200 mol/L?

# CHAPTER 17

**Section 1**

**Write equilibrium constant expressions for the following equilibria.**

1. $N_2(g) + O_2(g) \rightleftharpoons 2NO(g)$
2. $3O_2(g) \rightleftharpoons 2O_3(g)$
3. $P_4(g) + 6H_2(g) \rightleftharpoons 4PH_3(g)$
4. $CCl_4(g) + HF(g) \rightleftharpoons CFCl_3(g) + HCl(g)$
5. $4NH_3(g) + 5O_2(g) \rightleftharpoons 4NO(g) + 6H_2O(g)$

**Write equilibrium constant expressions for the following equilibria.**

6. $NH_4Cl(s) \rightleftharpoons NH_3(g) + HCl(g)$
7. $SO_3(g) + H_2O(l) \rightleftharpoons H_2SO_4(l)$
8. $2Na_2O_2(s) + 2CO_2(g) \rightleftharpoons 2Na_2CO_3(s) + O_2(g)$

**Calculate $K_{eq}$ for the following equilibria.**

9. $H_2(g) + I_2(g) \rightleftharpoons 2HI(g)$
   $[H_2] = 0.0109$, $[I_2] = 0.00290$, $[HI] = 0.0460$
10. $I_2(s) \rightleftharpoons I_2(g)$
    $[I_2(g)] = 0.0665$

**Section 3**

11. At a certain temperature, $K_{eq} = 0.0211$ for the equilibrium $PCl_5(g) \rightleftharpoons PCl_3(g) + Cl_2(g)$.
    a. What is [$Cl_2$] in an equilibrium mixture containing 0.865 mol/L $PCl_5$ and 0.135 mol/L $PCl_3$?
    b. What is [$PCl_5$] in an equilibrium mixture containing 0.100 mol/L $PCl_3$ and 0.200 mol/L $Cl_2$?
12. Use the $K_{sp}$ value for zinc carbonate given in **Table 3** to calculate its molar solubility at 298 K.
13. Use the $K_{sp}$ value for iron(II) hydroxide given in **Table 3** to calculate its molar solubility at 298 K.
14. Use the $K_{sp}$ value for silver carbonate given in **Table 3** to calculate [$Ag^+$] in a saturated solution at 298 K.
15. Use the $K_{sp}$ value for calcium phosphate given in **Table 3** to calculate [$Ca^{2+}$] in a saturated solution at 298 K.
16. Does a precipitate form when equal volumes of 0.0040*M* $MgCl_2$ and 0.0020*M* $K_2CO_3$ are mixed? If so, identify the precipitate.
17. Does a precipitate form when equal volumes of $1.2 \times 10^{-4}M$ $AlCl_3$ and $2.0 \times 10^{-3}M$ NaOH are mixed? If so, identify the precipitate.

## CHAPTER 18

**Section 1**

**1.** Write the balanced formula equation for the reaction between zinc and nitric acid.

**2.** Write the balanced formula equation for the reaction between magnesium carbonate and sulfuric acid.

**3.** Identify the base in the reaction
$H_2O(l) + CH_3NH_2(aq) \rightarrow OH^-(aq) + CH_3NH_3^+(aq)$.

**4.** Identify the conjugate acid-base pair described in the reactions in Practice Problem 1.

**5.** Write the steps in the complete ionization of hydrosulfuric acid.

**6.** Write the steps in the complete ionization of carbonic acid.

**Section 2**

**7.** Write the acid ionization equation and ionization constant expression for formic acid (HCOOH).

**8.** Write the acid ionization equation and ionization constant expression for the hydrogen carbonate ion ($HCO_3^-$).

**9.** Write the base ionization constant expression for ammonia.

**10.** Write the base ionization expression for aniline ($C_6H_5NH_2$).



**Section 3**

**11.** Is a solution in which $[H^+] = 1.0 \times 10^{-5}M$ acidic, basic, or neutral?

**12.** Is a solution in which $[OH^-] = 1.0 \times 10^{-11}M$ acidic, basic, or neutral?

**13.** What is the pH of a solution in which $[H^+] = 4.5 \times 10^{-4}M$?

**14.** Calculate the pH and pOH of a solution in which $[OH^-] = 8.8 \times 10^{-3}M$.

**15.** Calculate the pH and pOH of a solution in which $[H^+] = 2.7 \times 10^{-6}M$.

**16.** What is $[H^+]$ in a solution having a pH of 2.92?

**17.** What is $[OH^-]$ in a solution having a pH of 13.56?

**18.** What is the pH of a 0.00067*M* $H_2SO_4$ solution?

**19.** What is the pH of a 0.000034*M* NaOH solution?

**20.** The pH of a 0.200*M* HBrO solution is 4.67. What is the acid's $K_a$?

**21.** The pH of a 0.030*M* $C_2H_5COOH$ solution is 3.20. What is the acid's $K_a$?

**Section 4**

**22.** Write the formula equation for the reaction between hydroiodic acid and beryllium hydroxide.

**23.** Write the formula equation for the reaction between perchloric acid and lithium hydroxide.

**24.** In a titration, 15.73 mL of 0.2346*M* HI solution neutralizes 20.00 mL of a LiOH solution. What is the molarity of the LiOH?

**25.** What is the molarity of a caustic soda (NaOH) solution if 35.00 mL of solution is neutralized by 68.30 mL of 1.250*M* HCl?

26. Write the chemical equation for the hydrolysis reaction that occurs when sodium hydrogen carbonate is dissolved in water. Is the resulting solution acidic, basic, or neutral?

27. Write the chemical equation for any hydrolysis reaction that occurs when cesium chloride is dissolved in water. Is the resulting solution acidic, basic, or neutral?

# CHAPTER 19

**Section 1**

**Identify the following information for each problem. What element is oxidized? Reduced? What is the oxidizing agent? Reducing agent?**

1. $2P + 3Cl_2 \rightarrow 2PCl_3$

2. $C + H_2O \rightarrow CO + H_2$

3. $2ClO_3^- + 3AsO_2^- \rightarrow 3AsO_4^{3-} + 2Cl^-$

4. Determine the oxidation number for each element in the following compounds.
   a. $Na_2SeO_3$
   b. $HAuCl_4$
   c. $H_3BO_3$

5. Determine the oxidation number for the following compounds or ions.
   a. $P_4O_8$
   b. $Na_2O_2$ *(Hint: This is like $H_2O_2$.)*
   c. $AsO_4^{-3}$

**Section 2**

6. How many electrons will be lost or gained in each of the following half-reactions? Identify whether each is an oxidation or reduction.
   a. $Cr \rightarrow Cr^{3+}$
   b. $O_2 \rightarrow O^{2-}$
   c. $Fe^{+2} \rightarrow Fe^{3+}$

7. Balance the following reaction by the oxidation number method: $MnO_4^- + CH_3OH \rightarrow MnO_2 + HCHO$ (acidic). *(Hint: Assign the oxidation of hydrogen and oxygen as usual, and solve for the oxidation number of carbon.)*

8. Balance the following reaction by the oxidation number method: $Zn + HNO_3 \rightarrow ZnO + NO_2 + NH_3$

9. Use the oxidation number method to balance these net ionic equations.
   a. $SeO_3^{2-} + I^- \rightarrow Se + I_2$ (acidic solution)
   b. $NiO_2 + S_2O_3^{2-} \rightarrow Ni(OH)_2 + SO_3^{2-}$ (acidic solution)

**Use the half-reaction method to balance the following redox equations.**

10. $Zn(s) + HCl(aq) \rightarrow ZnCl_2(aq) \rightarrow H_2(g)$

11. $MnO_4^-(aq) + H_2SO_3(aq) \rightarrow Mn^{2+}(aq) + HSO_4^-(aq) + H_2O(l)$ (acidic solution)

12. $NO_2(aq) + OH^-(aq) \rightarrow NO_2^-(aq) + NO_3^-(aq) + H_2O(l)$ (basic solution)

13. $HS^-(aq) + IO_3^-(aq) \rightarrow I^-(aq) + S(s) + H_2O(l)$ (acidic solution)

Supplemental Practice Problems

## CHAPTER 20

**Section 1**

1. Calculate the cell potential for each of the following.
   a. $Co^{2+}(aq) + Al(s) \rightarrow Co(s) + Al^{3+}(aq)$
   b. $Hg^{2+}(aq) + Cu(s) \rightarrow Cu^{2+}(aq) + Hg(s)$
   c. $Zn(s) + Br_2(l) \rightarrow 2Br^-(aq) + Zn^{2+}(aq)$

2. Calculate the cell potential to determine whether the reaction will occur spontaneously or not spontaneously. For each reaction that is not spontaneous, correct the reactants or products so that a reaction would occur spontaneously.
   a. $Ni^{2+}(aq) + Al(s) \rightarrow Ni(s) + Al^{3+}(aq)$
   b. $Ag^+(aq) + H_2(g) \rightarrow Ag(s) + H^+(aq)$
   c. $Fe^{2+}(aq) + Cu(s) \rightarrow Fe(s) + Cu^{2+}(aq)$

## CHAPTER 21

**Section 2**

1. Draw the structure of the following branched alkanes.
   a. 2,2,4-trimethylheptane
   b. 4-isopropyl-2-methylnonane

2. Draw the structure of each of the following cycloalkanes.
   a. 1-ethyl-2-methylcyclobutane
   b. 1,3-dibutylcyclohexane



**Section 3**

3. Draw the structure of each of the following alkenes.
   a. 1,4-hexadiene
   b. 2,3-dimethyl-2-butene
   c. 4-propyl-1-octene
   d. 2,3-diethylcyclohexene

## CHAPTER 22

**Section 1**

1. Draw the structures of the following alkyl halides.
   a. chloroethane
   b. chloromethane
   c. 1-fluoropentane
   d. 1,3-dibromocyclohexane
   e. 1,2-dibromo-3-chloropropane

## CHAPTER 24

**Section 2**

1. Write balanced equations for each of the following decay processes.
   a. alpha emission of $^{244}_{96}Cm$
   b. positron emission of $^{70}_{33}As$
   c. beta emission of $^{210}_{83}Bi$
   d. electron capture by $^{116}_{51}Sb$

2. $^{47}_{20}Ca \rightarrow \beta + ?$

3. $^{240}_{95}Am + ? \rightarrow ^{243}_{97}Bk + n$

4. How much time has passed if 1/8 of an original sample of radon-222 is left? Use **Table 5** for half-life information.

5. If a basement air sample contains 3.64 μg of radon-222, how much radon will remain after 19 days?

6. Cobalt-60, with a half-life of 5 years, is used in cancer radiation treatments. If a hospital purchases 30.0 g, how much would be left after 15 years?

# SELECTED SOLUTIONS

## Solutions to Selected Practice Problems

## CHAPTER 1

No practice problems

## CHAPTER 2

**1.** No; the density of aluminum is 2.7 g/cm$^3$; the density of the cube is $\frac{20\text{ g}}{5\text{ cm}^3} = 4\text{ g/cm}^3$.

**3.** volume $= \frac{\text{mass}}{\text{density}} = \frac{147\text{ g}}{7.00\text{ g/mL}} = 21.0\text{ mL}$
volume $= 20.0\text{ mL} + 21.0\text{ mL} = 41.0\text{ mL}$

**11.** **a.** $7 \times 10^2$ **b.** $3.8 \times 10^4$ **c.** $4.5 \times 10^6$ **d.** $6.85 \times 10^{11}$ **e.** $5.4 \times 10^{-3}$ **f.** $6.87 \times 10^{-6}$ **g.** $7.6 \times 10^{-8}$ **h.** $8 \times 10^{-10}$

**13.** **a.** $7 \times 10^{-5}$ **b.** $3 \times 10^8$ **c.** $2 \times 10^2$ **d.** $5 \times 10^{-12}$

**15.** **a.** $(4 \times 1) \times 10^{2+8} = 4 \times 10^{10}$
**b.** $(2 \times 3) \times 10^{-4+2} = 6 \times 10^{-2}$
**c.** $(6 \div 2) \times 10^{2-1} = 3 \times 10^1$
**d.** $(8 \div 4) \times 10^{4-1} = 2 \times 10^3$

**17.** **a.** $\frac{16\text{ g salt}}{100\text{ g solution}}; \frac{100\text{ g solution}}{16\text{ g salt}}$
**b.** $\frac{1.25\text{ g}}{1\text{ mL}}; \frac{1\text{ mL}}{1.25\text{ g}}$
**c.** $\frac{25\text{ m}}{1\text{ s}}; \frac{1\text{s}}{25\text{ m}}$

**19.** **a.** $360\text{ s} \times \frac{1000\text{ ms}}{1\text{ s}} = 360{,}000\text{ ms}$
**b.** $4800\text{ g} \times \frac{1\text{ kg}}{1000\text{ g}} = 4.8\text{ kg}$
**c.** $5600\text{ dm} \times \frac{1\text{ m}}{10\text{ dm}} = 560\text{ m}$
**d.** $72\text{ g} \times \frac{1000\text{ mg}}{1\text{ g}} = 72{,}000\text{ mg}$
**e.** $2.45 \times 10^2\text{ ms} \times \frac{1\text{ s}}{1000\text{ ms}} = 0.245\text{ s}$
**f.** $5\ \mu\text{m} \times \frac{1\text{ mm}}{1000\ \mu\text{m}} \times \frac{1\text{ m}}{1000\text{ mm}} \times \frac{1\text{ km}}{1000\text{ m}}$
$= 5 \times 10^{-9}\text{ km}$
**g.** $6.800 \times 10^3\text{ cm} \times \frac{1\text{ m}}{100\text{ cm}} \times \frac{1\text{ km}}{1000\text{ m}}$
$= 6.800 \times 10^{-2}\text{ km}$
**h.** $2.5 \times 10^1\text{ kg} \times \frac{1\text{ Mg}}{1000\text{ kg}} = 0.025\text{ Mg}$

**21.** $\frac{65\text{ mi}}{1\text{ h}} \times \frac{1\text{ km}}{0.62\text{ mi}} = 1.0 \times 10^2\text{ km/h}$

**23.** mass = (volume)(density) = (185 mL)(1.02 g/mL)
mass = 189 g vinegar
$(189\text{ g vinegar})\left(\frac{5.00\text{ g acetic acid}}{100\text{ g vinegar}}\right) = 9.45\text{ g acetic acid}$

**33.** $\frac{0.11}{1.59} \times 100 = 6.92\%$
$\frac{0.10}{1.59} \times 100 = 6.29\%$
$\frac{0.12}{1.59} \times 100 = 7.55\%$

Note: The answers are reported in three significant figures because student error is the difference between the actual value (1.59 g/cm$^3$) and the measured value.

**35.** **a.** 4 **b.** 7 **c.** 5 **d.** 3

**37.** two significant figures: $1.0 \times 10^1$, $1.0 \times 10^2$, $1.0 \times 10^3$
three significant figures: $1.00 \times 10^1$, $1.00 \times 10^2$, $1.00 \times 10^3$
four significant figures: $1.000 \times 10^1$, $1.000 \times 10^2$, $1.000 \times 10^3$

**39.** **a.** $5.482 \times 10^{-4}$ g **b.** $1.368 \times 10^5$ kg **c.** $3.087 \times 10^8$ mm **d.** 2.014 mL

**41.** **a.** $4.32 \times 10^3\text{ cm} - 1.6 \times 10^6\text{ mm}$
$= 4.32 \times 10^3\text{ cm} - 16 \times 10^6\text{ cm}$
$= 4.32 \times 10^3\text{ cm} - 16{,}000 \times 10^3\text{ cm}$
$= -15{,}995.68 \times 10^3\text{ cm} = -16.0 \times 10^6\text{ cm}$
**b.** $2.12 \times 10^7\text{ mm} + 1.8 \times 10^3\text{ cm}$
$= 2.12 \times 10^7\text{ mm} + 1.8 \times 10^4\text{ mm}$
$= 2120 \times 10^4\text{ mm} + 1.8 \times 10^4\text{ mm}$
$= 2121.8 \times 10^4\text{ mm} = 2.12 \times 10^7\text{ mm}$

**43.** **a.** 2.0 m/s **b.** 3.00 m/s **c.** 2.00 m/s **d.** 2.9 m/s

## CHAPTER 3

**5.** amount of bromine that reacted = 100.0 g − 8.5 g = 91.5 g; amount of compound formed = 100.0 g + 10.3 g − 8.5 g = 101.8 g

**7.** $\text{mass}_{\text{reactants}} = \text{mass}_{\text{products}}$
$\text{mass}_{\text{sodium}} + \text{mass}_{\text{chlorine}} = \text{mass}_{\text{sodium chloride}}$
$\text{mass}_{\text{sodium}} = 15.6\text{ g}$
$\text{mass}_{\text{sodium chloride}} = 39.7\text{ g}$
Substituting and solving for $\text{mass}_{\text{chlorine}}$ yields
$15.6\text{ g} + \text{mass}_{\text{chlorine}} = 39.7\text{ g}$
$\text{mass}_{\text{chlorine}} = 39.7\text{ g} - 15.6\text{ g} = 24.1\text{ g}$ used in the reaction.

Because the sodium reacts with excess chlorine, all of the sodium is used in the reaction; that is, 15.6 g of sodium are used in the reaction.

**9.** 156.3 g − 106.5 g = 49.8 g
Yes. Mass of reactants equals mass of products.

**19.** percent by $\text{mass}_{\text{hydrogen}} = \frac{\text{mass}_{\text{hydrogen}}}{\text{mass}_{\text{compound}}} \times 100$
percent by $\text{mass}_{\text{hydrogen}} = \frac{12.4\text{ g}}{78.0\text{ g}} \times 100 = 15.9\%$

**21.** $mass_{XY} = 3.50\ g + 10.5\ g = 14.0\ g$

$$\text{percent by mass}_X = \frac{mass_X}{mass_{XY}} \times 100$$

$$\text{percent by mass}_X = \frac{3.50\cancel{g}}{14.0\cancel{g}} \times 100 = 25.0\%$$

$$\text{percent by mass}_Y = \frac{mass_Y}{mass_{XY}} \times 100$$

$$\text{percent by mass}_Y = \frac{10.5\cancel{g}}{14.0\cancel{g}} \times 100 = 75.0\%$$

**23.** No, you cannot be sure. Having the same mass percentage of a single element does not guarantee that the composition of each compound is the same.

## CHAPTER 4

**13.** dysprosium

**15.** Yes. 9

**17.** 25 protons, 25 electrons, 30 neutrons, manganese

**19.** N-14 is more abundant because the atomic mass is closer to 14 than 15.

## CHAPTER 5

**1.** $c = \lambda\nu$

$\nu = c / \lambda$

$$\nu = \frac{3.00 \times 10^8\ m/s}{4.90 \times 10^{-7}\ m} = 6.12 \times 10^{14}\ Hz$$

**3.** $3.00 \times 10^8$ m/s

**5. a.** $E_{photon} = \lambda\nu = (6.626 \times 10^{-34}\ J\cdot\cancel{s})(6.32 \times 10^{20}\ \cancel{s^{-1}}) = 4.19 \times 10^{-13}\ J$

**b.** $E_{photon} = \lambda\nu = (6.626 \times 10^{-34}\ J\cdot\cancel{s})(9.50 \times 10^{13}\ \cancel{s^{-1}}) = 6.29 \times 10^{-20}\ J$

**c.** $E_{photon} = \lambda\nu = (6.626 \times 10^{-34}\ J\cdot\cancel{s})(1.05 \times 10^{16}\ \cancel{s^{-1}}) = 6.96 \times 10^{-18}\ J$

**7.** $E_{photon} = hc / \lambda$

$$E_{photon} = \frac{(6.626 \times 10^{-34}\ J\cdot\cancel{s})(3.00 \times 10^8\ \cancel{m}/\cancel{s})}{1.25 \times 10^{-1}\ \cancel{m}}$$

$= 1.59 \times 10^{-24}\ J$

**21. a.** bromine (35 electrons): $[Ar]4s^2 3d^{10} 4p^5$

**b.** strontium (38 electrons): $[Kr]5s^2$

**c.** antimony (51 electrons): $[Kr]5s^2 4d^{10} 5p^3$

**d.** rhenium (75 electrons): $[Xe]6s^2 4f^{14} 5d^5$

**e.** terbium (65 electrons): $[Xe]6s^2 4f^9$

**f.** titanium (22 electrons): $[Ar]4s^2 3d^2$

**23.** Sulfur (15 electrons) has the electron configuration $[Ne]3s^2 3p^4$. Therefore, 6 electrons are in orbitals related to the third energy level of the sulfur atom.

**25.** $[Xe]6s^2$; barium

**27.** aluminum; 3 electrons

## CHAPTER 6

**9. a.** Sc, Y, La, Ac

**b.** N, P, As, Sb, Bi

**c.** Ne, Ar, Kr, Xe, Rn

**17.** B. The atomic radius increases when going down a group so helium is the smallest and radon is the biggest.

**19. a.** the element in period 2, group 1

**b.** the element in period 5, group 2

**c.** the element in period 6, group 15

**d.** the element in period 4, group 18

## CHAPTER 7

**7.** Three Na atoms each lose 1 $e^-$, forming 1+ ions. One N atom gains 3 $e^-$, forming a 3− ion. The ions attract, forming $Na_3N$.

$$3\ \cancel{\text{Na ions}}\left(\frac{1+}{\cancel{\text{Na ion}}}\right) + 1\ \cancel{\text{N ion}}\left(\frac{3-}{\cancel{\text{N ion}}}\right)$$

$= 3(1+) + 1(3-) = 0$

The overall charge on one formula unit of $Na_3N$ is zero.

**9.** One Sr atom loses 2 $e^-$, forming a 2+ ion. Two F atoms each gain 1 $e^-$, forming 1− ions. The ions attract, forming $SrF_2$.

$$1\ \cancel{\text{Sr ion}}\left(\frac{2+}{\cancel{\text{Sr ion}}}\right) + 2\ \cancel{\text{F ions}}\left(\frac{1-}{\cancel{\text{F ion}}}\right)$$

$= 1(2+) + 2(1-) = 0$

The overall charge on one formula unit of $SrF_2$ is zero.

**11.** Three group 1 atoms lose 1 $e^-$, forming 1+ ions. One group 15 atom gains 3 $e^-$, forming a 3− ion. The ions attract, forming $X_3Y$, where X represents a group 1 atom and Y represents a group 15 atom.

**19.** KI

**21.** $AlBr_3$

**23.** The general formula is $XY_2$, where X represents the group 2 element and Y represents the group 17 element.

**25.** $Ca(ClO_3)_2$

**27.** $MgCO_3$; answers will vary

**29.** calcium chloride

**31.** copper(II) nitrate

**33.** ammonium perchlorate

## CHAPTER 8

**1.**

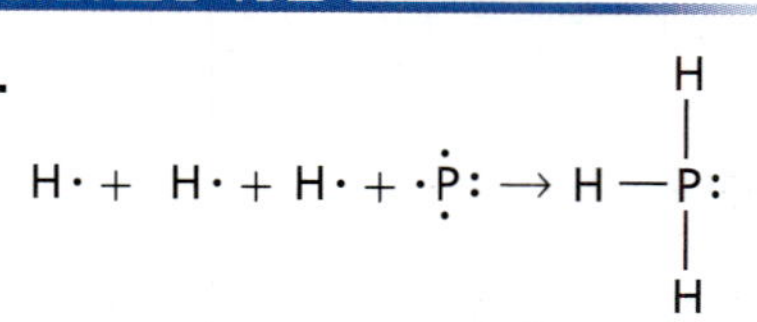

**3.**

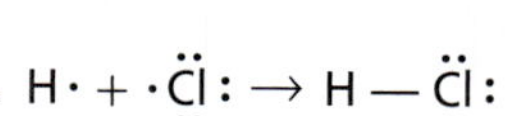

**5.**

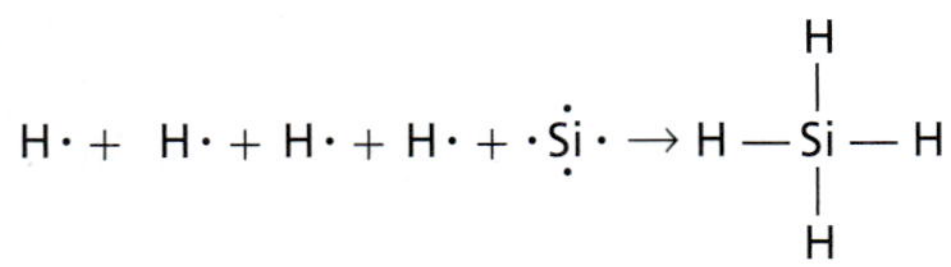

**15.** sulfur dioxide

**17.** carbon tetrachloride

**19.** hydroiodic acid

**21.** chlorous acid

**23.** hydrosulfuric acid

**25.** AgCl

**27.** $ClF_3$

**29.** $S_2F_{10}$

**37.**

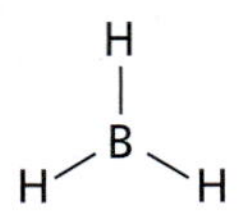

**39.**

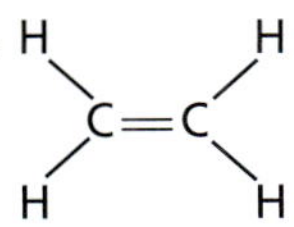

**41.**

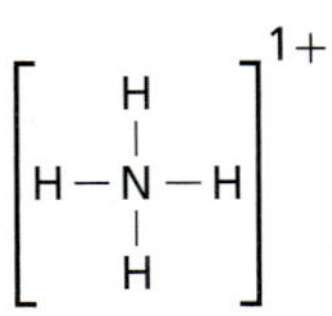

**43.**

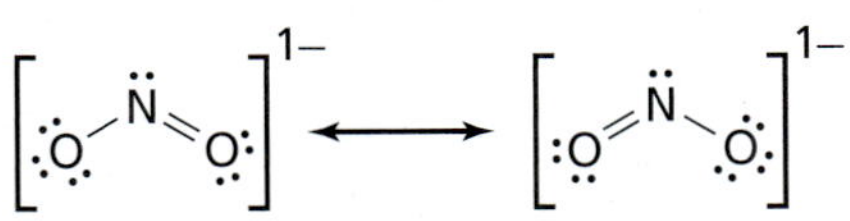

**45.**

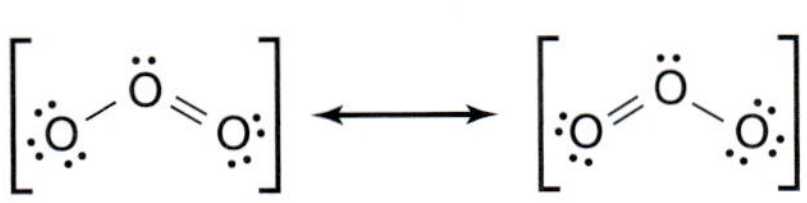

**47.**

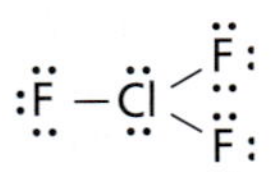

**49.**

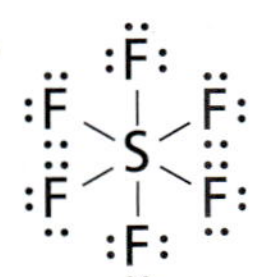

**57.** bent, 104.5°, $sp^3$

**59.** tetrahedral, 109°, $sp^3$

## CHAPTER 9

**1.** $H_2(g) + Br_2(g) \rightarrow HBr(g)$

**3.** potassium chlorate(s) → potassium chloride(s) and oxygen(g)
$KClO_3(s) \rightarrow KCl(s) + O_2(g)$

**5.** $CS_2(l) + 3O_2(g) \rightarrow CO_2(g) + 2SO_2(g)$

**15.** $H_2O(l) + N_2O_5(g) \rightarrow 2HNO_3(aq)$; synthesis

**17.** $H_2SO_4(aq) + 2NaOH(aq) \rightarrow Na_2SO_4(aq) + 2H_2O(l)$

**19.** $Ni(OH)_2(s) \rightarrow NiO(s) + H_2O(l)$

**21.** Yes. K is above Zn in the metal activity series.
$2K(s) + ZnCl_2(aq) \rightarrow Zn(s) + 2KCl(aq)$

**23.** No. Fe is below Na in the metal activity series.

**25.** $LiI(aq) + AgNO_3(aq) \rightarrow AgI(s) + LiNO_3(aq)$

**27.** $Na_2C_2O_4(aq) + Pb(NO_3)_2(aq) \rightarrow PbC_2O_4(s) + 2NaNO_3(aq)$

**35.** chemical equation: $KI(aq) + AgNO_3(aq) \rightarrow KNO_3(aq) + AgI(s)$

complete ionic equation:
$K^+(aq) + I^-(aq) + Ag^+(aq) + NO_3^-(aq) \rightarrow K^+(aq) + NO_3^-(aq) + AgI(s)$

net ionic equation: $I^-(aq) + Ag^+(aq) \rightarrow AgI(s)$

**37.** chemical equation: $AlCl_3(aq) + 3NaOH(aq) \rightarrow Al(OH)_3(s) + 3NaCl(aq)$

complete ionic equation:
$Al^{3+}(aq) + 3Cl^-(aq) + 3Na^+(aq) + 3OH^2(aq) \rightarrow Al(OH)_3(s) + 3Na^+(aq) + 3Cl^-(aq)$

net ionic equation: $Al^{3+}(aq) + 3OH^-(aq) \rightarrow Al(OH)_3(s)$

**39.** chemical equation: $5Na_2CO_3(aq) + 2MnCl_5(aq) \rightarrow 10NaCl(aq) + Mn_2(CO_3)_5(s)$

complete ionic equation:
$10Na^+(aq) + 5CO_3^{2-}(aq) + 2Mn^{5+}(aq) + 10Cl^-(aq) \rightarrow 10Na^+(aq) + 10Cl^-(aq) + Mn_2(CO_3)_5(s)$

net ionic equation: $5CO_3^{2-}(aq) + 2Mn^{5+}(aq) \rightarrow Mn_2(CO_3)_5(s)$

**41.** chemical equation: $2HCl(aq) + Ca(OH)_2(aq) \rightarrow 2H_2O(l) + CaCl_2(aq)$

complete ionic equation:
$2H^+(aq) + 2Cl^-(aq) + Ca^{2+}(aq) + 2OH^-(aq) \rightarrow 2H_2O(l) + Ca^{2+}(aq) + 2Cl^-(aq)$

net ionic equation: $H^+(aq) + OH^-(aq) \rightarrow H_2O(l)$

**43.** chemical equation: $H_2S(aq) + 1\ Ca(OH)_2(aq) \rightarrow 2H_2O(l) + CaS(aq)$

complete ionic equation:
$2H^+(aq) + S^{2-}(aq) + Ca^{2+}(aq) + 2OH^-(aq) \rightarrow 2H_2O(l) + Ca^{2+}(aq) + S^{2-}(aq)$

net ionic equation: $H^+(aq) + OH^-(aq) \rightarrow H_2O(l)$

**45.** chemical equation: $2HClO_4(aq) + K_2CO_3(aq) \rightarrow H_2O(l) + CO_2(g) + 2KClO_4(aq)$

complete ionic equation:
$2H^+(aq) + 2ClO_4^-(aq) + 2K^+(aq) + CO_3^{2-}(aq) \rightarrow H_2O(l) + CO_2(g) + 2K^+(aq) + 2ClO_4^-(aq)$

net ionic equation: $2H^+(aq) + CO_3^{2-}(aq) \rightarrow H_2O(l) + CO_2(g)$

**47.** chemical equation: $2HBr(aq) + (NH_4)_2CO_3(aq) \rightarrow H_2O(l) + CO_2(g) + 2NH_4Br(aq)$

complete ionic equation:

$$2H^+(aq) + 2Br^-(aq) + 2NH_4^+(aq) + CO_3^{2-}(aq) \rightarrow H_2O(l) + CO_2(g) + 2NH_4^+(aq) + 2Br^-(aq)$$

net ionic equation: $2H^+(aq) + CO_3^{2-}(aq) \rightarrow H_2O(l) + CO_2(g)$

**49.** chemical equation: $2KI(aq) + Pb(NO_3)_2(aq) \rightarrow 2KNO_3(aq) + PbI_2(s)$

complete ionic equation:

$$2K^+(aq) + 2I^-(aq) + Pb^{2+}(aq) + 2NO_3^-(aq) \rightarrow 2K^+(aq) + 2NO_3^-(aq) + PbI_2(s)$$

net ionic equation: $Pb^{2+}(aq) + 2I^-(aq) \rightarrow PbI_2(s)$

## CHAPTER 10

**1.** $2.50\ \text{mol Zn} \times \dfrac{6.02 \times 10^{23}\ \text{atoms}}{1\ \text{mol}}$

$= 1.51 \times 10^{24}$ atoms of Zn

**3.** $3.25\ \text{mol } AgNO_3 \times \dfrac{6.02 \times 10^{23}\ \text{formula units}}{1\ \text{mol}}$

$= 1.96 \times 10^{24}$ formula units of $AgNO_3$

**5. a.** $5.75 \times 10^{24}\ \text{atoms Al} \times \dfrac{1\ \text{mol}}{6.02 \times 10^{23}\ \text{atoms}}$

$= 9.55$ mol Al

**b.** $2.50 \times 10^{20}\ \text{atoms Fe} \times \dfrac{1\ \text{mol}}{6.02 \times 10^{23}\ \text{atoms}}$

$= 4.15 \times 10^{-4}$ mol Fe

**15. a.** $3.57\ \text{mol Al} \times \dfrac{26.98\ \text{g Al}}{1\ \text{mol Al}} = 96.3\ \text{g Al}$

**b.** $42.6\ \text{mol Si} \times \dfrac{28.09\ \text{g Si}}{1\ \text{mol Si}} = 1.20 \times 10^3\ \text{g Si}$

**17. a.** $25.5\ \text{g Ag} \times \dfrac{1\ \text{mol Ag}}{107.9\ \text{g Ag}} = 0.236\ \text{mol Ag}$

**b.** $300.0\ \text{g S} \times \dfrac{1\ \text{mol S}}{32.07\ \text{g S}} = 9.355\ \text{mol S}$

**19. a.** $55.2\ \text{g Li} \times \dfrac{1\ \text{mol Li}}{6.94\ \text{g Li}} \times \dfrac{6.02 \times 10^{23}\ \text{atoms}}{1\ \text{mol}}$

$= 4.79 \times 10^{24}$ atoms Li

**b.** $0.230\ \text{g Pb} \times \dfrac{1\ \text{mol Pb}}{207.2\ \text{g Pb}} \times \dfrac{6.02 \times 10^{23}\ \text{atoms}}{1\ \text{mol}}$

$= 6.68 \times 10^{20}$ atoms Pb

**c.** $11.5\ \text{g Hg} \times \dfrac{1\ \text{mol Hg}}{200.6\ \text{g Hg}} \times \dfrac{6.02 \times 10^{23}\ \text{atoms}}{1\ \text{mol}}$

$= 3.45 \times 10^{22}$ atoms Hg

**21. a.** $4.56 \times 10^3\ \text{g Si} \times \dfrac{1\ \text{mol Si}}{28.09\ \text{g Si}} \times \dfrac{6.02 \times 10^{23}\ \text{atoms}}{1\ \text{mol}}$

$= 9.77 \times 10^{25}$ atoms Si

**b.** $0.120\ \text{kg Ti} \times \dfrac{1000\ \text{g Ti}}{1\ \text{kg Ti}} \times \dfrac{1\ \text{mol Ti}}{47.87\ \text{g Ti}}$

$\times \dfrac{6.02 \times 10^{23}\ \text{atoms}}{1\ \text{mol}} = 1.51 \times 10^{24}$ atoms Ti

**29.** $2.50\ \text{mol } ZnCl_2 \times \dfrac{2\ \text{mol } Cl^-}{1\ \text{mol } ZnCl_2} = 5.00\ \text{mol } Cl^-$

**31.** $3.00\ \text{mol } Fe_2(SO_4)_3 \times \dfrac{3\ \text{mol } SO_4^{2-}}{1\ \text{mol } Fe_2(SO_4)_3} = 9.00\ \text{mol } SO_4^{2-}$

**33.** $1.15 \times 10^1\ \text{mol } H_2O \times \dfrac{2\ \text{mol H}}{1\ \text{mol } H_2O} = 23.0\ \text{mol H}$

$= 2.30 \times 10^1$ mol H

**35. a.**

$2\ \text{mol C} \times \dfrac{12.01\ \text{g C}}{1\ \text{mol C}} = 24.02\ \text{g C}$

$6\ \text{mol H} \times \dfrac{1.008\ \text{g H}}{1\ \text{mol H}} = 6.048\ \text{g H}$

$1\ \text{mol O} \times \dfrac{16.00\ \text{g O}}{1\ \text{mol O}} = \underline{16.00\ \text{g O}}$

molar mass $C_2H_5OH$ = 46.07 g/mol

**b.**

$1\ \text{mol H} \times \dfrac{1.008\ \text{g H}}{1\ \text{mol H}} = 1.008\ \text{g H}$

$1\ \text{mol C} \times \dfrac{12.01\ \text{g C}}{1\ \text{mol C}} = 12.01\ \text{g C}$

$1\ \text{mol N} \times \dfrac{14.01\ \text{g N}}{1\ \text{mol N}} = \underline{14.01\ \text{g N}}$

molar mass HCN = 27.03 g/mol

**c.**

$1\ \text{mol C} \times \dfrac{12.01\ \text{g C}}{1\ \text{mol C}} = 12.01\ \text{g C}$

$4\ \text{mol Cl} \times \dfrac{35.45\ \text{g Cl}}{1\ \text{mol Cl}} = \underline{141.80\ \text{g Cl}}$

molar mass $CCl_4$ = 153.81 g/mol

**37.** Step 1: Find the molar mass of $H_2SO_4$.

$2\ \text{mol H} \times \dfrac{1.008\ \text{g H}}{1\ \text{mol H}} = 2.016\ \text{g H}$

$1\ \text{mol S} \times \dfrac{32.07\ \text{g S}}{1\ \text{mol S}} = 32.07\ \text{g S}$

$4\ \text{mol O} \times \dfrac{16.00\ \text{g O}}{1\ \text{mol O}} = \underline{64.00\ \text{g O}}$

molar mass $H_2SO_4$ = 98.09 g/mol

Step 2: Make mole → mass conversion.

$3.25\ \text{mol } H_2SO_4 \times \dfrac{98.09\ \text{g } H_2SO_4}{1\ \text{mol } H_2SO_4} = 319\ \text{g } H_2SO_4$

**39.** Potassium permanganate has a formula of $KMnO_4$.

Step 1: Find the molar mass of $KMnO_4$.

$1\ \text{mol K} \times \dfrac{39.10\ \text{g K}}{1\ \text{mol K}} = 39.10\ \text{g K}$

$1\ \text{mol Mn} \times \dfrac{54.94\ \text{g Mn}}{1\ \text{mol Mn}} = 54.94\ \text{g Mn}$

$4\ \text{mol O} \times \dfrac{16.00\ \text{g O}}{1\ \text{mol O}} = \underline{64.00\ \text{g O}}$

molar mass $KMnO_4$ = 158.04 g/mol

Step 2: Make mole → mass conversion.

$2.55\ \text{mol } KMnO_4 \times \dfrac{158.04\ \text{g } KMnO_4}{1\ \text{mol } KMnO_4} = 403\ \text{g } KMnO_4$

**41. a.** ionic compound

Step 1: Find the molar mass of $Fe_2O_3$.

$$2 \text{ mol Fe} \times \frac{55.85 \text{ g Fe}}{1 \text{ mol Fe}} = 111.70 \text{ g Fe}$$

$$3 \text{ mol O} \times \frac{16.00 \text{ g O}}{1 \text{ mol O}} = \underline{48.00 \text{ g O}}$$

$$\text{molar mass } Fe_2O_3 = 159.70 \text{ g/mol}$$

Step 2: Make mass → mole conversion.

$$2500 \text{ g } Fe_2O_3 \times \frac{1 \text{ mol } Fe_2O_3}{159.70 \text{ g } Fe_2O_3} = 1.57 \times 10^1 \text{ mol } Fe_2O_3$$

**b.** ionic compound

Step 1: Find the molar mass of $PbCl_4$.

$$1 \text{ mol Pb} \times \frac{207.2 \text{ g Pb}}{1 \text{ mol Pb}} = 207.2 \text{ g Pb}$$

$$4 \text{ mol Cl} \times \frac{35.45 \text{ g Cl}}{1 \text{ mol Cl}} = \underline{141.80 \text{ g CI}}$$

$$\text{molar mass } PbCl_4 = 349.0 \text{ g/mol}$$

Step 2: Make mass → mole conversion.

$$25.4 \text{ mg } PbCl_4 \times \frac{1 \text{ g } PbCl_4}{1000 \text{ mg } PbCl_4} \times \frac{1 \text{ mol } PbCI_4}{349.0 \text{ g } PbCl_4} =$$

$7.28 \times 10^{-5}$ mol $PbCl_4$

**43. a.** Step 1: Find the molar mass of $Na_2SO_3$

$$2 \text{ mol Na} \times \frac{22.99 \text{ g Na}}{1 \text{ mol Na}} = 45.98 \text{ g Na}$$

$$1 \text{ mol S} \times \frac{32.07 \text{ g S}}{1 \text{ mol S}} = 32.07 \text{ g S}$$

$$3 \text{ mol O} \times \frac{16.00 \text{ g O}}{1 \text{ mol O}} = \underline{48.00 \text{ g O}}$$

$$\text{molar mass } Na_2SO_3 = 126.05 \text{ g/mol}$$

Step 2: Make mass → mole conversion.

$$2.25 \text{ g } Na_2SO_3 \times \frac{1 \text{ mol } Na_2SO_3}{126.05 \text{ g } Na_2SO_3}$$

$= 0.0179$ mol $Na_2SO_3$

Step 3: Make mole → formula unit conversion.

$$0.0179 \text{ mol } Na_2SO_3 \times \frac{6.02 \times 10^{23} \text{ formula units}}{1 \text{ mol } Na_2SO_3}$$

$= 1.08 \times 10^{22}$ formula units $Na_2SO_3$

Step 4: Determine the number of $Na^+$ ions.

$1.08 \times 10^{22}$ formula units $Na_2SO_3$ ×

$$\frac{2 \text{ } Na^+ \text{ ions}}{1 \text{ formula unit } Na_2SO_3} = 2.16 \times 10^{22} \text{ } Na^+ \text{ ions}$$

**b.** $1.08 \times 10^{22}$ formula units $Na_2SO_3$ ×

$$\frac{1 \text{ } SO_3^{2-} \text{ ion}}{1 \text{ formula unit } Na_2SO_3} = 1.08 \times 10^{22} \text{ } SO_3^{2-} \text{ ions}$$

**c.** $$\frac{126.05 \text{ g } Na_2SO_3}{1 \text{ mol } Na_2SO_3} \times \frac{1 \text{ mol } Na_2SO_3}{6.02 \times 10^{23} \text{ formula unit } Na_2SO_3}$$

$= 2.09 \times 10^{-22}$ g $Na_2SO_3$/formula unit

**45.** Step 1: Find the number of moles of NaCl.

$4.59 \times 10^{24}$ formula units NaCl ×

$$\frac{1 \text{ mol NaCl}}{6.02 \times 10^{23} \text{ formula unit NaCl}}$$

$= 7.62$ mol NaCl

Step 2: Find the molar mass of NaCl.

$$1 \text{ mol Na} \times \frac{22.99 \text{ g Na}}{1 \text{ mol Na}} = 22.99 \text{ g Na}$$

$$1 \text{ mol Cl} \times \frac{35.45 \text{ g Cl}}{1 \text{ mol Cl}} = \underline{35.45 \text{ g CI}}$$

$$\text{molar mass NaCl} = 58.44 \text{ g/mol}$$

Step 3: Make mole → mass conversion.

$$7.62 \text{ mol NaCl} \times \frac{58.44 \text{ g NaCl}}{1 \text{ mol NaCl}} = 445 \text{ g NaCl}$$

**55.** Steps 1 and 2: Assume 1 mole; calculate molar mass of $H_2SO_3$.

$$2 \text{ mol H} \times \frac{1.008 \text{ g H}}{1 \text{ mol H}} = 2.016 \text{ g H}$$

$$1 \text{ mol S} \times \frac{32.06 \text{ g S}}{1 \text{ mol S}} = 32.06 \text{ g S}$$

$$3 \text{ mol O} \times \frac{16.00 \text{ g O}}{1 \text{ mol O}} = \underline{48.00 \text{ g O}}$$

$$\text{molar mass } H_2SO_3 = 82.08 \text{ g/mol}$$

Step 3: Determine percent by mass of S.

$$\text{percent S} = \frac{32.06 \text{ g S}}{82.08 \text{ g } H_2SO_3} \times 100 = 39.06\% \text{ S}$$

Repeat steps 1 and 2 for $H_2S_2O_8$. Assume 1 mole; calculate molar mass of $H_2S_2O_8$.

$$2 \text{ mol H} \times \frac{1.008 \text{ g H}}{1 \text{ mol H}} = 2.016 \text{ g H}$$

$$2 \text{ mol S} \times \frac{32.06 \text{ g S}}{1 \text{ mol S}} = 64.12 \text{ g S}$$

$$8 \text{ mol O} \times \frac{16.00 \text{ g O}}{1 \text{ mol O}} = \underline{128.00 \text{ g O}}$$

$$\text{molar mass } H_2S_2O_8 = 194.14 \text{ g/mol}$$

Step 3: Determine percent by mass of S.

$$\text{percent S} = \frac{64.12 \text{ g S}}{194.14 \text{ g } H_2S_2O_8} \times 100 = 33.02\% \text{ S}$$

$H_2SO_3$ has a larger percent by mass of S.

**57. a.** sodium, sulfur, and oxygen; $Na_2SO_4$

**b.** ionic

**c.** Steps 1 and 2: Assume 1 mole; calculate molar mass of $Na_2SO_4$.

$$2 \text{ mol Na} \times \frac{22.99 \text{ g Na}}{1 \text{ mol Na}} = 45.98 \text{ g Na}$$

$$1 \text{ mol S} \times \frac{32.07 \text{ g S}}{1 \text{ mol S}} = 32.07 \text{ g S}$$

$$4 \text{ mol O} \times \frac{16.00 \text{ g O}}{1 \text{ mol O}} = \underline{64.00 \text{ g O}}$$

$$\text{molar mass } Na_2SO_4 = 142.05 \text{ g/mol}$$

Step 3: Determine percent by mass of each element.

$$\text{percent Na} = \frac{45.98 \text{ g Na}}{142.05 \text{ g } Na_2SO_4} \times 100 = 32.37\% \text{ Na}$$

$$\text{percent S} = \frac{32.07 \text{ g S}}{142.05 \text{ g } Na_2SO_4} \times 100 = 22.58\% \text{ S}$$

$$\text{percent O} = \frac{64.00 \text{ g O}}{142.05 \text{ g } Na_2SO_4} \times 100 = 45.05\% \text{ O}$$

**59.** Step 1: Assume 100 g sample; calculate moles of each element.

$$35.98 \text{ g Al} \times \frac{1 \text{ mol Al}}{26.98 \text{ g Al}} = 1.333 \text{ mol Al}$$

$$64.02 \text{ g S} \times \frac{1 \text{ mol S}}{32.07 \text{ g S}} = 1.996 \text{ mol S}$$

Step 2: Calculate mole ratios.

$$\frac{1.333 \text{ mol Al}}{1.333 \text{ mol Al}} = \frac{1.000 \text{ mol Al}}{1.000 \text{ mol Al}} = \frac{1 \text{ mol Al}}{1 \text{ mol Al}}$$

$$\frac{1.996 \text{ mol S}}{1.333 \text{ mol Al}} = \frac{1.497 \text{ mol S}}{1.000 \text{ mol Al}} = \frac{1.5 \text{ mol S}}{1 \text{ mol Al}}$$

The simplest ratio is 1 mol Al: 1.5 mol S.

Step 3: Convert decimal fraction to whole number.

In this case, multiply by 2 because $1.5 \times 2 = 3$. Therefore, the empirical formula is $Al_2S_3$.

**61.** Step 1: Assume 100 g sample; calculate moles of each element.

$$60.00 \text{ g C} \times \frac{1 \text{ mol C}}{12.01 \text{ g C}} = 5.000 \text{ mol C}$$

$$4.44 \text{ g H} \times \frac{1 \text{ mol H}}{1.008 \text{ g H}} = 4.40 \text{ mol H}$$

$$35.56 \text{ g O} \times \frac{1 \text{ mol O}}{16.00 \text{ g O}} = 2.22 \text{ mol O}$$

Step 2: Calculate mole ratios.

$$\frac{5.00 \text{ mol C}}{2.22 \text{ mol O}} = \frac{2.25 \text{ mol C}}{1.00 \text{ mol O}} = \frac{2.25 \text{ mol C}}{1 \text{ mol O}}$$

$$\frac{4.40 \text{ mol H}}{2.22 \text{ mol O}} = \frac{1.98 \text{ mol H}}{1.00 \text{ mol O}} = \frac{2 \text{ mol H}}{1 \text{ mol O}}$$

$$\frac{2.22 \text{ mol O}}{2.22 \text{ mol O}} = \frac{1.00 \text{ mol O}}{1.00 \text{ mol O}} = \frac{1 \text{ mol O}}{1 \text{ mol O}}$$

The simplest ratio is 2.25 mol C: 2 mol H: 1 mol O.

Step 3: Convert decimal fraction to whole number.

In this case, multiply by 4 because $2.25 \times 4 = 9$. Therefore, the empirical formula is $C_9H_8O_4$.

**63.** Step 1: Assume 100 g sample; calculate moles of each element.

$$46.68 \text{ g N} \times \frac{1 \text{ mol N}}{14.01 \text{ g N}} = 3.332 \text{ mol N}$$

$$53.32 \text{ g O} \times \frac{1 \text{ mol O}}{16.00 \text{ g O}} = 3.332 \text{ mol O}$$

Step 2: Calculate mole ratios.

$$\frac{3.332 \text{ mol N}}{3.332 \text{ mol N}} = \frac{1.000 \text{ mol N}}{1.000 \text{ mol N}} = \frac{1 \text{ mol N}}{1 \text{ mol N}}$$

$$\frac{3.332 \text{ mol O}}{3.332 \text{ mol N}} = \frac{1.000 \text{ mol O}}{1.000 \text{ mol N}} = \frac{1 \text{ mol O}}{1 \text{ mol N}}$$

The simplest ratio is 1 mol N: 1 mol O.

The empirical formula is NO.

Step 3: Calculate the molar mass of the empirical formula.

$$1 \text{ mol N} \times \frac{14.01 \text{ g N}}{1 \text{ mol N}} = 14.01 \text{ g N}$$

$$1 \text{ mol O} \times \frac{16.00 \text{ g O}}{1 \text{ mol O}} = \underline{16.00 \text{ g O}}$$

$$\text{molar mass NO} = 30.01 \text{ g/mol}$$

Step 4: Determine whole number multiplier.

$$\frac{60.01 \text{ g/mol}}{30.01 \text{ g/mol}} = 2.000$$

The molecular formula is $N_2O_2$.

**65.** Step 1: Assume 100 g sample; calculate moles of each element.

$$65.45 \text{ g C} \times \frac{1 \text{ mol C}}{12.01 \text{ g C}} = 5.450 \text{ mol C}$$

$$5.45 \text{ g H} \times \frac{1 \text{ mol H}}{1.008 \text{ g H}} = 5.41 \text{ mol H}$$

$$29.09 \text{ g O} \times \frac{1 \text{ mol O}}{16.00 \text{ g O}} = 1.818 \text{ mol O}$$

Step 2: Calculate mole ratios.

$$\frac{5.450 \text{ mol C}}{1.818 \text{ mol O}} = \frac{3.000 \text{ mol C}}{1.000 \text{ mol O}} = \frac{3 \text{ mol C}}{1 \text{ mol O}}$$

$$\frac{5.41 \text{ mol H}}{1.818 \text{ mol O}} = \frac{2.98 \text{ mol H}}{1.00 \text{ mol O}} = \frac{3 \text{ mol H}}{1 \text{ mol O}}$$

$$\frac{1.818 \text{ mol O}}{1.818 \text{ mol O}} = \frac{1.000 \text{ mol O}}{1.000 \text{ mol O}} = \frac{1 \text{ mol O}}{1 \text{ mol O}}$$

The simplest ratio is 3 mol C: 3 mol H: 1 mol O.

Therefore, the empirical formula is $C_3H_3O$.

Step 3: Calculate the molar mass of the empirical formula.

$$3 \text{ mol C} \times \frac{12.01 \text{ g C}}{1 \text{ mol C}} = 36.03 \text{ g C}$$

$$3 \text{ mol H} \times \frac{1.008 \text{ g H}}{1 \text{ mol H}} = 3.024 \text{ g H}$$

$$1 \text{ mol O} \times \frac{16.00 \text{ g O}}{1 \text{ mol O}} = \underline{16.00 \text{ g O}}$$

$$\text{molar mass } C_3H_3O = 55.05 \text{ g/mol}$$

Step 4: Determine whole number multiplier.

$$\frac{110.00 \text{ g/mol}}{55.05 \text{ g/mol}} = 1.998, \text{ or } 2$$

The molecular formula is $C_6H_6O_2$.

**75.** Step 1: Calculate the mass of $CoCl_2$ remaining.

$$0.0712 \text{ mol } CoCl_2 \times \frac{129.83 \text{ g } CoCl_2}{1 \text{ mol } CoCl_2} = 9.24 \text{ g } CoCl_2$$

Step 2: Calculate the mass of water driven off.

mass of hydrated compound − mass of anhydrous compound remaining

$$= 11.75 \text{ g } CoCl_2{\cdot}xH_2O - 9.24 \text{ g } CoCl_2 = 2.51 \text{ g } H_2O$$

Step 3: Calculate moles of each component.

$$9.24\ \text{g CoCl}_2 \times \frac{1\ \text{mol CoCl}_2}{129.83\ \text{g CoCl}_2}$$

$$= 0.0712\ \text{mol CoCl}_2$$

$$2.51\ \text{g H}_2\text{O} \times \frac{1\ \text{mol H}_2\text{O}}{18.02\ \text{g H}_2\text{O}} = 0.139\ \text{mol H}_2\text{O}$$

Step 4: Calculate mole ratios.

$$\frac{0.139\ \text{mol H}_2\text{O}}{0.0712\ \text{mol CoCl}_2} = \frac{1.95\ \text{mol H}_2\text{O}}{1.00\ \text{mol CoCl}_2} = \frac{2\ \text{mol H}_2\text{O}}{1\ \text{mol CoCl}_2}$$

The formula of the hydrate is $CoCl_2 \cdot 2H_2O$. Its name is cobalt(II) chloride dehydrate.

# CHAPTER 11

**1. a.** 1 molecule $N_2$ + 3 molecules $H_2$ → 2 molecules $NH_3$

1 mole $N_2$ + 3 moles $H_2$ → 2 moles $NH_3$

28.014 g $N_2$ + 6.048 g $H_2$ → 34.062 g $NH_3$

**b.** 1 molecule HCl + 1 formula unit KOH → 1 formula unit KCl + 1 molecule $H_2O$

1 mole HCl + 1 mole KOH → 1 mole KCl + 1 mole $H_2O$

36.461 g HCl + 56.105 g KOH → 74.551 g KCl + 18.015 g $H_2O$

**c.** 2 atoms Mg + 1 molecule $O_2$ → 2 formula units MgO

2 moles Mg + 1 mole $O_2$ → 2 moles MgO

48.610 g Mg + 31.998 g $O_2$ → 80.608 g MgO

**3. a.** $\frac{4\ \text{mol Al}}{3\ \text{mol O}_2}\ \frac{3\ \text{mol O}_2}{2\ \text{mol Al}_2\text{O}_3}\ \frac{2\ \text{mol Al}_2\text{O}_3}{4\ \text{mol Al}}$

$\frac{3\ \text{mol O}_2}{4\ \text{mol Al}}\ \frac{2\ \text{mol Al}_2\text{O}_3}{3\ \text{mol O}_2}\ \frac{4\ \text{mol Al}}{2\ \text{mol Al}_2\text{O}_3}$

**b.** $\frac{3\ \text{mol Fe}}{4\ \text{mol H}_2\text{O}}\ \frac{3\ \text{mol Fe}}{4\ \text{mol H}_2}\ \frac{3\ \text{mol Fe}}{1\ \text{mol Fe}_3\text{O}_4}$

$\frac{4\ \text{mol H}_2\text{O}}{3\ \text{mol Fe}}\ \frac{4\ \text{mol H}_2}{3\ \text{mol Fe}}\ \frac{1\ \text{mol Fe}_3\text{O}_4}{3\ \text{mol Fe}}$

$\frac{1\ \text{mol Fe}_3\text{O}_4}{4\ \text{mol H}_2}\ \frac{1\ \text{mol Fe}_3\text{O}_4}{4\ \text{mol H}_2\text{O}}\ \frac{4\ \text{mol H}_2\text{O}}{4\ \text{mol H}_2}$

$\frac{4\ \text{mol H}_2}{1\ \text{mol Fe}_3\text{O}_4}\ \frac{4\ \text{mol H}_2\text{O}}{1\ \text{mol Fe}_3\text{O}_4}\ \frac{4\ \text{mol H}_2}{4\ \text{mol H}_2\text{O}}$

**c.** $\frac{2\ \text{mol HgO}}{2\ \text{mol Hg}}\ \frac{1\ \text{mol O}_2}{2\ \text{mol Hg}}\ \frac{1\ \text{mol O}_2}{2\ \text{mol HgO}}$

$\frac{2\ \text{mol Hg}}{2\ \text{mol HgO}}\ \frac{2\ \text{mol Hg}}{1\ \text{mol O}_2}\ \frac{2\ \text{mol HgO}}{1\ \text{mol O}_2}$

**11. a.** $2CH_4(g) + S_8(s) \rightarrow 2CS_2(l) + 4H_2S(g)$

**b.** $1.50\ \text{mol S}_8 \times \frac{2\ \text{mol CS}_2}{1\ \text{mol S}_8} = 3.00\ \text{mol CS}_2$

**c.** $1.50\ \text{mol S}_8 \times \frac{4\ \text{mol H}_2\text{S}}{1\ \text{mol S}_8} = 6.00\ \text{mol H}_2\text{S}$

**13.** Step 1: Balance the chemical equation.

$2NaCl(s) \rightarrow 2Na(s) + Cl_2(g)$

Step 2: Make mole → mole conversion.

$$2.50\ \text{mol NaCl} \times \frac{1\ \text{mol Cl}_2}{2\ \text{mol NaCl}} = 1.25\ \text{mol Cl}_2$$

Step 3: Make mole → mass conversion.

$$1.25\ \text{mol Cl}_2 \times \frac{70.9\ \text{g Cl}_2}{1\ \text{mol Cl}_2} = 88.6\ \text{g Cl}_2$$

**15.** $2NaN_3(s) \rightarrow 2Na(s) + 3N_2(g)$

Step 1: Make mass → mole conversion.

$$100.0\ \text{g NaN}_3 \times \frac{1\ \text{mol NaN}_3}{65.02\ \text{g NaN}_3} = 1.538\ \text{mol NaN}_3$$

Step 2: Make mole → mole conversion.

$$1.538\ \text{mol NaN}_3 \times \frac{3\ \text{mol N}_2}{2\ \text{mol NaN}_3} = 2.307\ \text{mol N}_2$$

Step 3: Make mole → mass conversion.

$$2.307\ \text{mol N}_2 \times \frac{28.02\ \text{g N}_2}{1\ \text{mol N}_2} = 64.64\ \text{g N}_2$$

**23.** Step 1: Make mass → mole conversion.

$$100.0\ \text{g Na} \times \frac{1\ \text{mol Na}}{22.99\ \text{g Na}} = 4.350\ \text{mol Na}$$

$$100.0\ \text{g Fe}_2\text{O}_3 \times \frac{1\ \text{mol Fe}_2\text{O}_3}{159.7\ \text{g Fe}_2\text{O}_3} = 0.6261\ \text{mol Fe}_2\text{O}_3$$

Step 2: Make mole ratio comparison.

$$\frac{0.6261\ \text{mol Fe}_2\text{O}_3}{4.350\ \text{mol Na}} \quad \text{compared to} \quad \frac{1\ \text{mol Fe}_2\text{O}_3}{6\ \text{mol Na}}$$

0.1439 compared to 0.1667

**a.** The actual ratio is less than the needed ratio, so iron(III) oxide is the limiting reactant.

**b.** Sodium is the excess reactant.

**c.** Step 1: Make mole → mole conversion.

$$0.6261\ \text{mol Fe}_2\text{O}_3 \times \frac{2\ \text{mol Fe}}{1\ \text{mol Fe}_2\text{O}_3} = 1.252\ \text{mol Fe}$$

Step 2: Make mole → mass conversion.

$$1.252\ \text{mol Fe} \times \frac{55.85\ \text{g Fe}}{1\ \text{mol Fe}} = 69.92\ \text{g Fe}$$

**d.** Step 1: Make mole → mole conversion.

$$0.6261\ \text{mol Fe}_2\text{O}_3 \times \frac{6\ \text{mol Na}}{1\ \text{mol Fe}_2\text{O}_3}$$

= 3.757 mol Na needed

Step 2: Make mole → mass conversion.

$$3.757\ \text{mol Na} \times \frac{22.9\ \text{g Na}}{1\ \text{mol Na}} = 86.37\ \text{g Na needed}$$

100.0 g Na given − 86.37 g Na needed

= 13.6 g Na in excess

**29. a.** Step 1: Write the balanced chemical equation.

$Zn(s) + I_2(s) \rightarrow ZnI_2(s)$

Step 2: Make mole → mole conversion.

$$1.912\ \text{mol Zn} \times \frac{1\ \text{mol ZnI}_2}{1\ \text{mol Zn}} = 1.912\ \text{mol ZnI}_2$$

Step 3: Make mole → mass conversion.

$$1.912 \text{ mol } ZnI_2 \times \frac{319.2 \text{ g } ZnI_2}{1 \text{ mol } ZnI_2} = 610.3 \text{ g } ZnI_2$$

610.3 g of $ZnI_2$ is the theoretical yield.

**b.** $\% \text{ yield} = \dfrac{515.6 \text{ g } ZnI_2}{610.3 \text{ g } ZnI_2} \times 100$

$= 84.48\%$ yield of $ZnI_2$

## CHAPTER 12

**1.** $\dfrac{\text{Rate}_{\text{nitrogen}}}{\text{Rate}_{\text{neon}}} = \sqrt{\dfrac{20.2 \text{ g/mol}}{28.0 \text{ g/mol}}} = 0.849$

**3.** Rearrange Graham's law to solve for $\text{Rate}_A$.

$$\text{Rate}_A = \text{Rate}_B \times \sqrt{\frac{\text{molar mass}_B}{\text{molar mass}_A}}$$

$\text{Rate}_B = 3.6 \text{ mol/min}$

$\dfrac{\text{molar mass}_B}{\text{molar mass}_A} = \dfrac{1}{2}$

$\text{Rate}_A = 3.6 \text{ mol/min} \times \sqrt{\dfrac{1}{2}}$

$= 2.5 \text{ mol/min}$

**5.** $P_{\text{total}} = 5.00 \text{ kPa} + 4.56 \text{ kPa} + 3.02 \text{ kPa} + 1.20 \text{ kPa}$
$= 13.78 \text{ kPa}$

**7.** $N_2 = 760 \text{ mm Hg} \times 0.78 = 590 \text{ mm Hg}$;
$O_2 = 760 \text{ mm Hg} \times 0.21 = 160 \text{ mm Hg}$;
$Ar = 760 \text{ mm Hg} \times 0.01 = 8 \text{ mm Hg}$

## CHAPTER 13

**1.** $V_2 = \dfrac{V_1P_1}{P_2} = \dfrac{(300.0 \text{ mL})(99.0 \text{ kPa})}{188 \text{ kPa}} = 158 \text{ mL}$

**3.** $P_2 = 1.08 \text{ atm} + (1.08 \text{ atm} \times 0.25) = 1.35 \text{ atm}$

$V_2 = \dfrac{V_1P_1}{P_2} = \dfrac{(145.7 \text{ mL})(1.08 \text{ atm})}{1.35 \text{ atm}} = 117 \text{ mL}$

**5.** $T_1 = 89°C + 273 = 362 \text{ K}$

$T_2 = \dfrac{T_1V_2}{V_1} = \dfrac{(362 \text{ K})(1.12 \text{ L})}{0.67 \text{ L}} = 605 \text{ K}$

$605 - 273 = 332°C = 330°C$

**7.** $V_2 = 0.67 \text{ L} - (0.67 \text{ L} \times 0.45) = 0.37 \text{ L}$

$T_2 = \dfrac{T_1V_2}{V_1} = \dfrac{(350 \text{ K})(0.37 \text{ L})}{0.67 \text{ L}} = 190 \text{ K}$

**9.** $T_2 = 36.5°C + 273 = 309.5 \text{ K}$

$T_1 = \dfrac{T_2P_1}{P_2} = \dfrac{(309.5 \text{ K})(1.12 \text{ atm})}{2.56 \text{ atm}} = 135 \text{ K}$

$135 \text{ K} - 273 = -138°C$

**11.** $T_1 = 22.0°C + 273 = 295 \text{ K}$

$T_2 = 100.0°C + 273 = 373 \text{ K}$

$V_1 = \dfrac{V_2T_1P_2}{T_2P_1} = \dfrac{(0.224 \text{ mL})(295 \text{ K})(1.23 \text{ atm})}{(373 \text{ K})(1.02 \text{ atm})} = 0.214 \text{ mL}$

**13.** $T_1 = 0.00°C + 273 = 273 \text{ K}$

$T_2 = 30.0°C + 273 = 303 \text{ K}$

$\dfrac{V_2}{V_1} = \dfrac{P_1T_2}{P_2T_1} = \dfrac{(1.00 \text{ atm})(303 \text{ K})}{(1.20 \text{ atm})(273 \text{ K})} = 0.92$

This is a ratio, so there are no units. The final volume is less than the original volume, so the piston will move down.

**21.** $1.0 \text{ L} \times \dfrac{1 \text{ mol}}{22.4 \text{ L}} = 0.045 \text{ mol}$

$0.045 \text{ mol} \times \dfrac{44.0 \text{ g}}{1 \text{ mol}} = 2.0 \text{ g}$

**23.** $0.416 \text{ g} \times \dfrac{1 \text{ mol}}{83.80 \text{ g}} = 0.00496 \text{ mol}$

$0.00496 \text{ mol} \times \dfrac{22.4 \text{ L}}{1 \text{ mol}} = 0.111 \text{ L}$

**25.** $0.860 \text{ g} - 0.205 \text{ g} = 0.655 \text{ g He remaining}$

Set up the problem as a ratio.

$\dfrac{V}{0.655 \text{ g}} = \dfrac{19.2 \text{ L}}{0.860 \text{ g}}$

Solve for $V$.

$V = \dfrac{(19.2 \text{ L})(0.655 \text{ g})}{0.860 \text{ g}} = 14.6 \text{ L}$

**27.** $V = \dfrac{nRT}{P} = \dfrac{(0.323 \text{ mol})\left(0.0821 \dfrac{\text{L·atm}}{\text{mol·K}}\right)(265 \text{ K})}{0.900 \text{ atm}} = 7.81 \text{ L}$

**29.** $n = \dfrac{PV}{RT} = \dfrac{(3.81 \text{ atm})(0.44 \text{ L})}{\left(0.0821 \dfrac{\text{L·atm}}{\text{mol·K}}\right)(298 \text{ K})} = 6.9 \times 10^{-3} \text{ mol}$

**39.** $2H_2(g) + O_2(g) \rightarrow 2H_2O(g)$

$5.00 \text{ L } O_2 \times \dfrac{2 \text{ volumes } H_2}{1 \text{ volume } O_2} = 10.0 \text{ L } H_2$

**41.** $N_2 + O_2 = N_2O$

$2N_2 + O_2 = 2N_2O$

$34 \text{ L } N_2O \times \dfrac{1 \text{ volume } O_2}{2 \text{ volumes } N_2O} = 17 \text{ L } O_2$

**43.** $2.38 \text{ kg} \times \dfrac{1000 \text{ g}}{1 \text{ kg}} \times \dfrac{1 \text{ mol } CaCO_3}{100.09 \text{ g}} \times \dfrac{1 \text{ mol } CO_2}{1 \text{ mol } CaCO_3}$

$\times \dfrac{22.4 \text{ L}}{1 \text{ mol}} = 533 \text{ L } CO_2$

**45.** Molecular mass of sodium bicarbonate = 83.9 g/mol

$28 \text{ g } NaHCO_3 \times \dfrac{1 \text{ mol } NaHCO_3}{83.9 \text{ g}} = 0.33 \text{ mol } NaHCO_3$

For each mole of sodium bicarbonate, one mole of $CO_2$ is produced, so 0.33 mol $NaHCO_3$ will produce 0.33 mol $CO_2$.

For an ideal gas, molar volume is 22.4 L at 273 K and 1 atm.

$T = 20°C + 273 = 293 \text{ K}$

$0.33 \text{ mol } CO_2 \times \dfrac{22.4 \text{ L}}{1 \text{ mol}} \times \dfrac{293 \text{ K}}{273 \text{ K}} = 7.9 \text{ L of } CO_2$

# CHAPTER 14

**9.** 600.0 mL $H_2O \times 1.0$ g/mL = 600.0 g $H_2O$

$$\frac{20.0 \text{ g NaHCO}_3}{600.0 \text{ g H}_2\text{O} + 20.0 \text{ g NaHCO}_3} \times 100 = 3\%$$

**11.** 1500.0 g − 54.3 g = 1445.7 g solvent

**13.** $$\frac{35 \text{ mL}}{155 \text{ mL} + 35 \text{ mL}} \times 100 = 18\%$$

**15.** $$15\% = \frac{18 \text{ mL}}{x \text{ mL solution}} \times 100 = 120 \text{ mL}$$

**17.** $$\text{mol KBr} = 1.55 \text{ g} \times \frac{1 \text{ mol}}{119.0 \text{ g}} = 0.0130 \text{ mol KBr}$$

$$\text{molarity} = \frac{\text{mol KBr}}{1.60 \text{ L solution}} = \frac{0.0130 \text{ mol}}{1.60 \text{ L}}$$

$= 8.13 \times 10^{-3}M$

**19.** $$0.25M = \frac{x \text{ mol Ca(OH)}_2}{1.5 \text{ L solution}}$$

$x = 0.38$ mol $Ca(OH)_2$

$$0.38 \text{ mol Ca(OH)}_2 \times \frac{74.08 \text{ g}}{1 \text{ mol}}$$

$= 28$ g $Ca(OH)_2$

**21.** $$\text{mol CaCl}_2 = 500.0 \text{ mL} \times \frac{1 \text{ L}}{1000 \text{ mL}} \times 0.20M$$

$$= 500.0 \text{ mL} \times \frac{1 \text{ L}}{1000 \text{ mL}} \times \frac{0.20 \text{ mol}}{1 \text{ L}} = 0.10 \text{ mol}$$

$$\text{mass CaCl}_2 = 0.10 \text{ mol CaCl}_2 \times \frac{110.98 \text{ g}}{1 \text{ mol}}$$

=11 g

**23.** $$100 \text{ mL} \times \frac{1 \text{ L}}{1000 \text{ mL}} \times \frac{0.15 \text{ mol ethanol}}{1 \text{ L solution}} \times \frac{46 \text{ g ethanol}}{1 \text{ mol ethanol}}$$

$$\times \frac{1 \text{ mL ethanol}}{0.7893 \text{ g ethanol}} = 0.87 \text{ mL ethanol}$$

**25.** $(5.0M)V_1 = (0.25M)(100.0 \text{ mL})$

$$V_1 = \frac{(0.25M)(100.0 \text{ mL})}{5.0M} = 5.0 \text{ mL}$$

**27.** $$\text{mol Na}_2\text{SO}_4 = 10.0 \text{ g Na}_2\text{SO}_4 \times \frac{1 \text{ mol}}{142.04 \text{ g Na}_2\text{SO}_4}$$

$= 0.0704$ mol $Na_2SO_4$

$$\text{molality} = \frac{0.0704 \text{ mol Na}_2\text{SO}_4}{1.0000 \text{ kg H}_2\text{O}} = 0.0704m$$

**29.** $$22.8\% = \frac{\text{mass NaOH}}{\text{mass NaOH + mass H}_2\text{O}} \times 100$$

Assume 100.0 g sample.

Then, mass NaOH = 22.8 g

mass $H_2O$ = 100.0 g − (mass NaOH) = 77.2 g

$$\text{mol NaOH} = 22.8 \text{ g} \times \frac{1 \text{ mol}}{40.00 \text{ g}} = 0.570 \text{ mol NaOH}$$

$$\text{mol H}_2\text{O} = 77.2 \text{ g} \times \frac{1 \text{ mol}}{18.02 \text{ g}} = 4.28 \text{ mol H}_2\text{O}$$

$$\text{mol fraction NaOH} = \frac{\text{mol NaOH}}{\text{mol NaOH + mol H}_2\text{O}}$$

$$= \frac{0.570 \text{ mol NaOH}}{0.570 \text{ mol NaOH} + 4.28 \text{ mol H}_2\text{O}} = \frac{0.570}{4.85}$$

$= 0.118$

The mole fraction of NaOH is 0.118.

**37.** $$S_2 = \frac{1.5 \text{ g}}{1.0 \text{ L}} = 1.5 \text{ g/L}$$

$$P_2 = P_1 \times \frac{S_2}{S_1} = 10.0 \text{ atm} \times \frac{1.5 \text{ g/L}}{0.66 \text{ g/L}} = 23 \text{ atm}$$

**45.** $\Delta T_b = 0.512°\text{C}/m \times 0.625m = 0.320°\text{C}$

$T_b = 100°\text{C} + 0.320°\text{C} = 100.320°\text{C}$

$\Delta T_f = 1.86°\text{C}/m \times 0.625m = 1.16°\text{C}$

$T_f = 0.0°\text{C} - 1.16°\text{C} = -1.16°\text{C}$

**47.** $$K_f = \frac{\Delta T_f}{m}$$

$$= \frac{0.080°\text{C}}{0.045\ m}$$

$= 1.8°\text{C}/m$

It is most likely water because the calculated value is closest to 1.86°C/*m*.

# CHAPTER 15

**1.** 142 Calories = 142 kcal

$$142 \text{ kcal} \times \frac{1000 \text{ cal}}{1 \text{ kcal}} = 142{,}000 \text{ cal}$$

**3.** Unit X = 0.1 cal

1 cal = 4.184 J

X = (0.1 cal)(4.184 J/cal) = 0.4184 J

1 cal = 0.001 Calorie

X = (0.1 cal)(1 Cal/1000 cal) = 0.0001 Calorie

**5.** $q = c \times m \times \Delta T$

5696 J = $c \times 155$ g $\times$ 15.0°C

$c = 2.45$ J/(g·°C)

The specific heat is very close to the value for ethanol.

**13.** $q = c \times m \times \Delta T$

5650 J = 4.184 J/(g·°C) × $m$ × 26.6°C

$m = 50.8$ g

**15.** $q = c \times m \times \Delta T$

9750 J = 4.184 J/(g·°C) × 335 g × $\Delta T$

$\Delta T = 6.96°\text{C}$

Because the water lost heat, let $\Delta T = -6.96°\text{C}$.

$\Delta T = -6.96°\text{C} = T_f - 65.5°\text{C}$

$T_f = 58.5°\text{C}$

**23.** $$25.7 \text{ g CH}_3\text{OH} \times \frac{1 \text{ mol CH}_3\text{OH}}{32.04 \text{ g CH}_3\text{OH}} \times \frac{3.22 \text{ kJ}}{1 \text{ mol CH}_3\text{OH}}$$

= 2.58 kJ

**25.** $$12{,}880 \text{ kJ} = m \times \frac{1 \text{ mol CH}_4}{16.04 \text{ g CH}_4} \times \frac{891 \text{ kJ}}{1 \text{ mol CH}_4}$$

$$m = 12{,}880 \text{ kJ} \times \frac{16.04 \text{ g CH}_4}{1 \text{ mol CH}_4} \times \frac{1 \text{ mol CH}_4}{891 \text{ kJ}}$$

$m = 232$ g $CH_4$

Selected Solutions

**33. a.** $4Al(s) + 3O_2(g) \rightarrow 2Al_2O_3(s) \quad \Delta H = -3352\ kJ$

**b.** $\Delta H$ for Equation **b** $= -x$ kJ

Add Equation **a** to Equation **b** reversed and tripled.

$4Al(s) + 3O_2(g) \rightarrow 2Al_2O_3(s) \quad \Delta H = -3352\ kJ$

$3MnO_2(s) \rightarrow 3Mn(s) + 3O_2(g) \quad \Delta H = 3x\ kJ$

$4Al(s) + 3MnO_2(s) \rightarrow 2Al_2O_3(s) + 3Mn(s)$

$-1789\ kJ = 3x\ kJ + (-3352\ kJ)$

$3x\ kJ = -1789\ kJ + 3352\ kJ = +1563\ kJ$

$x = \dfrac{1563\ kJ}{3} = +521\ kJ$

Because the direction of Equation **b** was changed, $\Delta H$ for Equation **b** $= -521$ kJ.

**35.** $\Delta H^\circ_{rxn} = [4(33.18\ kJ) + 6(-285.83\ kJ)] - 4(-46.11)\ kJ = -1398\ kJ$

**37.** Reverse Equation **a** and change the sign of $\Delta H^\circ_f$ to obtain Equation **c**.

Add equation **b**.

**c.** $NO(g) \rightarrow \frac{1}{2}N_2(g) + \frac{1}{2}O_2(g)\ \Delta H^\circ_f = -91.3\ kJ$

**b.** $\frac{1}{2}N_2(g) + O_2(g) \rightarrow NO_2(g)\ \Delta H^\circ_f = ?$

Add the equations.

$NO(g) + \frac{1}{2}O_2(g) \rightarrow NO_2(g)$

$\Delta H^\circ_{rxn} = -58.1\ kJ = \Delta H^\circ_f(\mathbf{c}) + \Delta H^\circ_f(\mathbf{b})$

$-58.1\ kJ = -91.3\ kJ + \Delta H^\circ_f(\mathbf{b})$

$\Delta H^\circ_f(\mathbf{b}) = -58.1\ kJ + 91.3\ kJ = 33.2\ kJ$

**45.** The states of the two reactants are the same on both sides of the equation, so it is impossible from the equation alone to predict the sign of $\Delta S_{system}$.

**47.** For a spontaneous reaction, $\Delta G_{system} < 0$.

$\Delta H_{system} - T\Delta S_{system} < 0$

$T > \dfrac{\Delta H_{system}}{\Delta S_{system}}$

$T > \dfrac{-144\ KJ}{(36.8\ J/K)(1\ KJ/1000\ J)}$

$T > 3910\ K$

At any temperature above 3910 K, the reaction is spontaneous.

## CHAPTER 16

**1.** $H_2$ is consumed. Average reaction rate expression should be negative.

Average reaction rate =

$-\dfrac{[H_2]\text{ at time } t_2 - [H_2]\text{ at time } t_1}{t_2 - t_1} = -\dfrac{\Delta[H_2]}{\Delta t}$

Average reaction rate $= -\dfrac{0.020M - 0.030M}{4.00\ s - 0.00\ s}$

$= -\dfrac{-0.010M}{4.00\ s} = 0.0025\ mol/(L{\cdot}s)$

**3.** HCl is formed so the average rate expression should be positive.

Average reaction rate =

$\dfrac{[HCl]\text{ at time } t_2 - [HCl]\text{ at time } t_1}{t_2 - t_1} = 0.0050\ mol/(L{\cdot}s)$

$[HCl]_{\text{at time } t_2} =$

$(0.0050\ mol/(L{\cdot}s))(t_2 - t_1) + [HCl]_{\text{at time } t_1}$

$= (0.0050\ mol/L{\cdot}s)(4.00\ s - 0.00\ s) + 0.00\ s$

$= 0.020M$

**19.** Rate $= k[A]^3$

**21.** Examining trials 1 and 2, doubling [A] has no effect on the rate; therefore, the reaction is zero order in A. Examining trials 2 and 3, doubling [B] doubles the rate; therefore, the reaction is first order in B.

Rate $= k[A]^0[B] = k[B]$

**31.** $[NO] = 0.00500M$

$[H_2] = 0.00200M$

$k = 2.90 \times 10^2\ L^2/(mol^2{\cdot}s)$

Rate $= k\ [NO]^2[H_2]$

$= [2.90 \times 10^2 L^2/(mol^2{\cdot}s)](0.00500\ mol/L)^2(0.00200\ mol/L)$

$= [2.90 \times 10^2\ \cancel{L^2}/(\cancel{mol^2}{\cdot}s)](0.00500\ \cancel{mol}/\cancel{L})^2$

$(0.00200\ mol/L)$

$= 1.45 \times 10^{-5}\ mol/(L{\cdot}s)$

**33.** Rate $= k\ [NO]^2[H_2]$

$[NO] = \sqrt{\dfrac{\text{Rate}}{k[H_2]}}$

$= \sqrt{\dfrac{9.00 \times 10^{-5}\ mol/(L{\cdot}s)}{(2.90 \times 10^2\ L^2/mol^2{\cdot}s)(0.00300mol/L)}}$

$= 1.02 \times 10^{-2}M$

## CHAPTER 17

**1. a.** $K_{eq} = \dfrac{[NO_2]^2}{[N_2O_4]}$

**b.** $K_{eq} = \dfrac{[H_2]^2[S_2]}{[H_2S]^2}$

**c.** $K_{eq} = \dfrac{[CH_4][H_2O]}{[CO][H_2]^3}$

**d.** $K_{eq} = \dfrac{[NO]^4[H_2O]^6}{[NH_3]^4[O_2]^5}$

**e.** $K_{eq} = \dfrac{[CS_2][H_2]^4}{[CH_4][H_2S]^2}$

**3. a.** $K_{eq} = [C_{10}H_8(g)]$

**b.** $K_{eq} = [H_2O(g)]$

**c.** $K_{eq} = [CO_2(g)]$

**d.** $K_{eq} = \dfrac{[CO(g)][H_2(g)]}{[H_2O(g)]}$

**e.** $K_{eq} = \dfrac{[CO_2(g)]}{[CO(g)]}$

**5.** $K_{eq} = \dfrac{[NO_2]^2}{[N_2O_4]} = \dfrac{0.0627^2}{0.0185} = 0.213$

**7.** $\dfrac{[CO][Cl_2]}{[COCl_2]} = 8.2 \times 10^{-2}$

Selected Solutions

$$\frac{(0.150)(0.150)}{[COCl_2]} = 8.2 \times 10^{-2}$$

$$[COCl_2] = \frac{(0.150)(0.150)}{8.2 \times 10^{-2}} = 0.27M$$

**19.** According to the stoichiometry of the equation, the concentration of B is 0.450$M$; C and D are $1.00 - 0.450 = 0.550M$.

$$K_{eq} = \frac{(0.550)(0.550)}{(0.450)(0.450)} = 1.49$$

**21.** $K_{sp} = [Pb^{2+}][CO_3^{2-}] = 7.40 \times 10^{-14}$

$(s)(s) = 7.40 \times 10^{-14}$

$s = \sqrt{7.40 \times 10^{-14}} = 2.72 \times 10^{-7}M$

$s = 2.72 \times 10^{-7}$ ~~mol~~/L × 267.2 g/~~mol~~ $= 7.27 \times 10^{-5}$ g/L

**23.** $K_{sp} = [Ag^+]^3[PO_4^{3-}] = 2.6 \times 10^{-18}$

$[PO_4^{3-}] = s$, $[Ag^+] = 3s$

$(3s)^3(s) = (27s^3)(s) = 27s^4 = 2.6 \times 10^{-18}$

$$s = \sqrt[4]{\frac{2.6 \times 10^{-18}}{27}} = 1.8 \times 10^{-5} \text{ mol/L}$$

**25. a.** $PbF_2(s) \rightleftharpoons Pb^{2+}(aq) + 2F^-(aq)$

$Q_{sp} = [Pb^{2+}][F^-]^2 = (0.050M)(0.015M)^2 = 1.12 \times 10^{-5}$

$K_{sp} = 3.3 \times 10^{-8}$

$Q_{sp} > K_{sp}$, so a precipitate of $PbF_2$ will form.

**b.** $Ag_2SO_4(s) \rightleftharpoons 2Ag^+(aq) + SO_4^{2-}(aq)$

$Q_{sp} = [Ag^+]^2[SO_4^{2-}] = (0.0050M)^2(0.125M) = 3.1 \times 10^{-6}$

$K_{sp} = 1.2 \times 10^{-5}$

$Q_{sp} < K_{sp}$, so a precipitate will not form.

## CHAPTER 18

**1. a.** $2Al(s) + 3H_2SO_4(aq) \rightarrow Al_2(SO_4)_3(aq) + 3H_2(g)$

**b.** $CaCO_3(s) + 2HBr(aq) \rightarrow CaBr_2(aq) + H_2O(l) + CO_2(g)$

**3.**

| Acid | Conjugate base | Base | Conjugate acid |
|---|---|---|---|
| **a.** $NH_4^+$ | $NH_3$ | $OH^-$ | $H_2O$ |
| **b.** HBr | $Br^-$ | $H_2O$ | $H_3O^+$ |
| **c.** $H_2O$ | $OH^-$ | $CO_3^{2-}$ | $HCO_3^-$ |

**13.** $H_2SeO_3(aq) + H_2O(l) \rightleftharpoons HSeO_3^-(aq) + H_3O^+(aq)$

$HSeO_3^-(aq) + H_2O(l) \rightleftharpoons SeO_3^{2-}(aq) + H_3O^+(aq)$

**15. a.** $C_6H_{13}NH_2(aq) + H_2O(l) \rightleftharpoons C_6H_{13}NH_3^-(aq) + OH^-(aq)$

$$K_b = \frac{[C_6H_{13}NH_3^+][OH^-]}{[C_6H_{13}NH_2]}$$

**b.** $C_3H_7NH_2(aq) + H_2O(l) \rightleftharpoons C_3H_7NH_3^-(aq) + OH^-(aq)$

$$K_b = \frac{[C_3H_7NH_3^+][OH^-]}{[C_3H_7NH_2]}$$

**c.** $CO_3^{2-}(aq) + H_2O(l) \rightleftharpoons HCO_3^-(aq) + OH^-(aq)$

$$K_b = \frac{[HCO_3^-][OH^-]}{[CO_3^{2-}]}$$

**d.** $HSO_3^-(aq) + H_2O(l) \rightleftharpoons H_2SO_3(aq) + OH^-(aq)$

$$K_b = \frac{[H_2SO_3^-][OH^-]}{[HSO_3^-]}$$

**23.** At 298 K, $[H^+] = [OH^-] = 1.0 \times 10^{-7}M$

$$\text{mol } H^- = \frac{1.0 \times 10^{-7} \text{ mol}}{1 \cancel{L}} \times \frac{1 \cancel{L}}{1000 \cancel{mL}} \times 300 \cancel{mL} = 3.0 \times 10^{-8} \text{ mol}$$

$$3.0 \times 10^{-8} \cancel{\text{mol } H^+ \text{ ions}} \times \frac{6.02 \times 10^{23} \text{ } H^+ \text{ ions}}{1 \cancel{\text{mol } H^+ \text{ ions}}} = 1.8 \times 10^{16} \text{ } H^+ \text{ ions}$$

Number of $H^+$ = number of $OH^- = 1.8 \times 10^{16}$ ions

**25. a.** $[H^+] = 0.0055M$

$pH = -\log [H^+]$

$pH = -\log 0.0055$

$pH = 2.26$

**b.** $[H^+] = 0.000084M$

$pH = -\log [H^+]$

$pH = -\log 0.000084$

$pH = 4.08$

**27. a.** $[OH^-] = 1.0 \times 10^{-6}M$

$pOH = -\log [OH^-]$

$pOH = -\log(1.0 \times 10^{-6})$

$pOH = 6.00$

$pH = 14.00 - pOH = 14.00 - 6.00 = 8.00$

**b.** $[OH^-] = 6.5 \times 10^{-4}M$

$pOH = -\log [OH^-]$

$pOH = -\log(6.5 \times 10^{-4})$

$pOH = 3.19$

$pH = 14.00 - pOH = 14.00 - 3.19 = 10.81$

**c.** $[H^-] = 3.6 \times 10^{-9}M$

$pH = -\log [H^+]$

$pH = -\log(3.6 \times 10^{-9})$

$pH = 8.44$

$pOH = 14.00 - pH = 14.00 - 8.44 = 5.56$

**d.** $[H^+] = 2.5 \times 10^{-2}M$

$pH = -\log(-2.5 \times 10^{-2})$

$pH = 1.60$

$pOH = 14.00 - pH = 14.00 - 1.60 = 12.40$

**29.** $$[HCl] = [H^+] = \frac{1.0 \times 10^{-3} \text{ mol}}{5.0 \text{ L}} = 0.00020M = 2.0 \times 10^{-4}M$$

$pH = -\log(2.0 \times 10^{-4}) = -(-3.70) = 3.70$

$pOH = 14.00 - 3.70 = 10.30$

**31.** $[OH^-] = \text{antilog}(-pOH)$

$[OH^-] = \text{antilog}(-5.60) = 2.5 \times 10^{-6}M$

$pH = 14.00 - 5.60 = 8.40$

$[H^+] = \text{antilog}(-8.40) = 4.0 \times 10^{-9}M$

**33. a.** $pH = 14.00 - pOH$

$pH = 14.00 - 10.70 = 3.30$

$[H^+] = \text{antilog}(-pH)$

$[H^+] = \text{antilog}(-3.30) = 5.0 \times 10^{-4}M$

$[C_6H_5COO^-] = [H^+] = 5.0 \times 10^{-4}M$

$[C_6H_5COOH] = 0.00330M - 5.0 \times 10^{-4}M =$ $0.0028M$

$$K_a = \frac{[H^+][C_6H_5COO^-]}{[C_6H_5COOH]} = \frac{(5.0 \times 10^{-4})(5.0 \times 10^{-4})}{(22.8 \times 10^{-3})}$$

$K_a = 8.9 \times 10^{-5}$

**b.** $pH = 14.00 - pOH$

$pH = 14.00 - 11.00 = 3.00$

$[H^+] = \text{antilog}(-pH)$

$[H^+] = \text{antilog}(-3.00) = 1.0 \times 10^{-3}M$

$[CNO^-] = [H^+] = 1.0 \times 10^{-3}M$

$[HCNO] = 0.100 - 1.0 \times 10^{-3}M = 0.099M$

$$K_a = \frac{[H^+][CNO^-]}{[HCNO]} = \frac{(1.0 \times 10^{-3})(1.0 \times 10^{-3})}{(0.099)}$$

$K_a = 1.0 \times 10^{-5}$

**c.** $pH = 14.00 - pOH$

$pH = 14.00 - 11.18 = 2.82$

$[H^+] = \text{antilog}(-pH)$

$[H^+] = \text{antilog}(-2.82) = 1.5 \times 10^{-3}M$

$[C_3H_7COO^-] = [H^+] = 1.5 \times 10^{-3}M$

$[C_3H_7COOH] = 0.150M - 1.5 \times 10^{-3}M = 0.149M$

$$K_a = \frac{[H^+][C_3H_7COO^-]}{[C_3H_7COOH]} = \frac{(1.5 \times 10^{-3})(1.5 \times 10^{-3})}{(0.149)}$$

$K_a = 1.5 \times 10^{-5}$

**45.** $$49.90\ \cancel{\text{mL HCl}} \times \frac{1\ \cancel{\text{L}}}{1000\ \cancel{\text{mL}}} \times \frac{0.5900\ \text{mol HCl}}{1\ \cancel{\text{L HCl}}} = 2.944 \times 10^{-2}\ \text{mol HCl}$$

$$2.944 \times 10^{-2}\ \cancel{\text{mol HCl}} \times \frac{1\ \text{mol NH}_3}{1\ \cancel{\text{mol HCl}}} = 2.944 \times 10^{-2}\ \text{mol NH}_3$$

$$M_{NH_3} = \frac{2.944 \times 10^{-2}\ \text{mol NH}_3}{0.02500\ \text{L NH}_3} = 1.178M$$

**47. a.** $NH_4^+(aq) + H_2O(l) \rightleftharpoons NH_3(aq) + H_3O^+(aq)$

The solution is acidic.

**b.** $SO_4^{2-}(aq) + H_2O(l) \rightleftharpoons HSO_4^-(aq) + OH^-(aq)$

The solution is neutral.

**c.** $CH_3COO^-(aq) + H_2O(l) \rightleftharpoons CH_3COOH(aq) + OH^-(aq)$

The solution is basic.

**d.** $CO_3^{2-}(aq) + H_2O(l) \rightleftharpoons HCO_3^-(aq) + OH^-(aq)$

The solution is basic.

## CHAPTER 19

**1. a.** reduction **b.** oxidation **c.** oxidation **d.** reduction

**3.** $Ag^+$ is the oxidizing agent, Fe is the reducing agent; $Ag^+$ is reduced, Fe is oxidized

**5. a.** +7 **b.** +5 **c.** +3

**7. a.** −3 **b.** −3 **c.** −2

**15.** Oxidation numbers: $\overset{+1\,-1}{HCl} + \overset{+1\,+5\,-2}{HNO_3} \rightarrow \overset{+1\,-2\,+1}{HOCl} + \overset{+2\,-2}{NO} + \overset{+1\,-2}{H_2O}$; Cl: 3(+2); N: 2(−3)

$3HCl + 2HNO_3 \rightarrow 3HOCl + 2NO + H_2O$

**17.** Oxidation numbers: $\overset{-3\,+1}{NH_3(g)} + \overset{+4\,-2}{NO_2(g)} \rightarrow \overset{0}{N_2(g)} + \overset{+1\,-2}{H_2O(l)}$; 4(+3)(2); 3(−4)(2)

$8NH_3(g) + 6NO_2(g) \rightarrow 7N_2(g) + 12H_2O(l)$

**19.** Oxidation numbers: $\overset{+1\,-2}{H_2S(g)} + \overset{+5\,-2}{NO_3^-(aq)} \rightarrow \overset{0}{S(s)} + \overset{+2\,-2}{NO(g)}$; 3(+2); 2(−3)

$2H^+(aq) + 3H_2S(g) + 2NO_3^-(aq) \rightarrow 3S(s) + 2NO(g) + 4H_2O(l)$

**21.** Oxidation numbers: $\overset{0}{Zn} + \overset{+5\,-2}{2NO_3^-} + 4H^+ \rightarrow \overset{+2}{Zn^{2+}} + \overset{+4\,-2}{2NO_2} + 2H_2O$; +2; (−1)

$Zn + 2NO_3^- + 4H^+ \rightarrow Zn^{2+} + 2NO_2 + 2H_2O$

**23.** $2I^-(aq) \rightarrow I_2(s) + 2e^-$ (oxidation)

$14H^+(aq) + 6e^- + Cr_2O_7^{2-}(aq) \rightarrow 2Cr^{3+}(aq) + 7H_2O(l)$ (reduction)

Multiply oxidation half-reaction by 3 and add to reduction half-reaction.

$14H^+(aq) + 6e^- + CrO_7^{2-}(aq) + 6I^-(aq) \rightarrow 3I_2(s) + 2Cr^{3+}(aq) + 7H_2O(l) + 6e^-$

$14H^+(aq) + CrO_7^{2-}(aq) + 6I^-(aq) \rightarrow 3I_2(s) + 2Cr^{3+}(aq) + 7H_2O(l)$

**25.** $6OH^-(aq) + N_2O(g) \rightarrow 2NO_2^-(aq) + 4e^- + 3H_2O(l)$ (oxidation)

$ClO^-(aq) + 2e^- + H_2O(l) \rightarrow Cl^-(aq) + 2OH^-(aq)$ (reduction)

Multiply reduction half-reaction by 2 and add to oxidation half-reaction.

$$6OH^-(aq) + N_2O(g) + 2ClO^-(aq) + 4e^- + 2H_2O(l) \longrightarrow 2NO_2^-(aq) + 4e^- + 3H_2O(l) + 2Cl^-(aq) + 4OH^-(aq)$$

$$N_2O(g) + 2ClO^-(aq) + 2OH^-(aq) \longrightarrow 2NO_2^-(aq) + 2Cl^-(aq) + H_2O(l)$$

## CHAPTER 20

**1.** $Pt^{2+}(aq) + Sn(s) \longrightarrow Pt(s) + Sn^{2+}(aq)$

$E^\circ_{cell} = +1.18\ V - (-0.1375\ V)$
$E^\circ_{cell} = +1.32\ V$
$Sn|Sn^{2+}||Pt^{2+}|Pt$

**3.** $Hg^{2+}(aq) + Cr(s) \longrightarrow Hg(l) + Cr^{2+}(aq)$
$E^\circ_{cell} = +0.851\ V - (-0.913\ V)$
$E^\circ_{cell} = +1.764\ V$
$Cr|Cr^{2+}||Hg^{2+}|Hg$

**5.** $E^\circ_{cell} = +0.3419\ V - (-0.1375\ V)$
$E^\circ_{cell} = +0.4794\ V$
$E^\circ_{cell} > 0$; spontaneous

**7.** $E^\circ_{cell} = 0.920\ V - (+1.507\ V)$
$E^\circ_{cell} = -0.587\ V$
$E^\circ_{cell} < 0$; not spontaneous

**9.** $Al|Al^{3+}||Hg^{2+}|Hg_2{}^{2+}$
$2Al(s) + 6Hg^{2+}(aq) \longrightarrow 2Al^{3+}(aq) + 3Hg_2{}^{2+}(aq)$
$E^\circ_{cell} = 0.920\ V - (-1.662\ V) = +2.582\ V$
The reaction is spontaneous.

## CHAPTER 21

**1. a.**

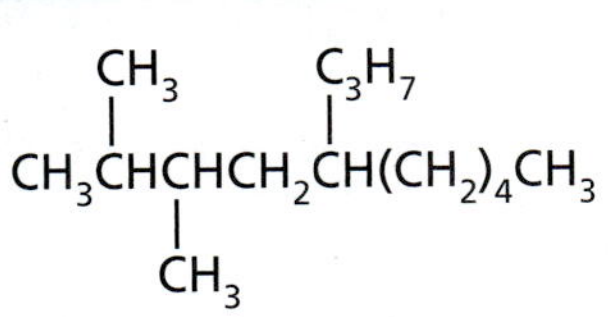

**b.**

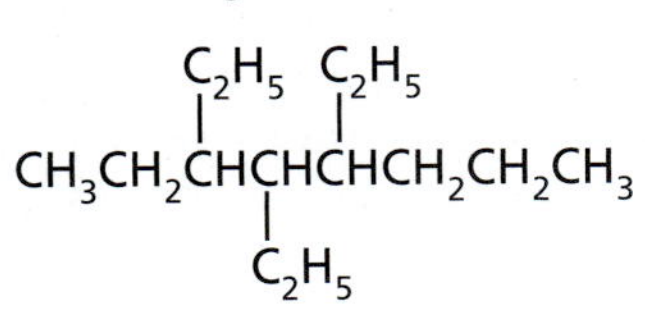

**11. a.**

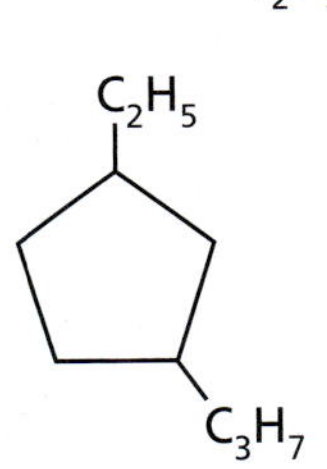

**b.**

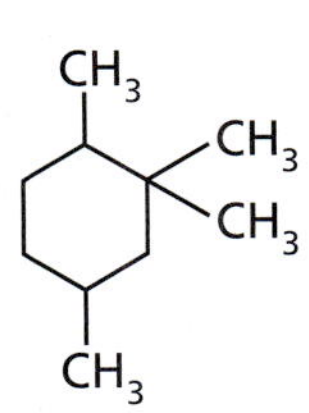

**17. a.** 4-methyl-2-pentene **b.** 2,2,6-trimethyl-3-octene

**31. a.** propylbenzene
**b.** 1-ethyl-2-methylbenzene
**c.** 1-ethyl-2,3-dimethylbenzene

## CHAPTER 22

**1.** 2,3-difluorobutane

**3.** 1,3-dibromo-2-chlorobenzene

## CHAPTER 23

No practice problems

## CHAPTER 24

**7.** ${}^{229}_{90}Th \longrightarrow {}^{4}_{2}He + {}^{225}_{88}Ra$

Alpha decay

**9.** For one half-life, amount remaining = (initial amount) $\left(\frac{1}{2}\right)^n = (10.0\ mg)\left(\frac{1}{2}\right)^1 = 5.00$ mg.

For two half-lives, amount remaining = (initial amount)$\left(\frac{1}{2}\right)^n = (10.0\ mg)\left(\frac{1}{2}\right)^2 = 2.50$ mg.

For three half-lives, amount remaining = (initial amount)$\left(\frac{1}{2}\right)^n = (10.0\ mg)\left(\frac{1}{2}\right)^3 = 1.25$ mg.

**11.** For Sample A: $16.2\ g = 64.8\ g\left(\frac{1}{2}\right)^n$; $n = 2$

$t = 2 \times 5.7$ years $= 10.54$ years

For Sample B, amount remaining = (initial amount)$\left(\frac{1}{2}\right)^{\frac{t}{T}} = (58.4\ g)\left(\frac{1}{2}\right)^{\frac{10.54y}{12.32y}} \approx 32.3$ g

For Sample C, amount remaining = (initial amount)$\left(\frac{1}{2}\right)^{\frac{t}{T}} = (37.6\ g)\left(\frac{1}{2}\right)^{\frac{10.54y}{28.79y}} \approx 29.2$ g

**19.** ${}^{27}_{13}Al + {}^{1}_{0}n \longrightarrow {}^{24}_{11}Na + {}^{4}_{2}He$

**21.** Let T = target and I = unstable isotope. Then,
${}^{1}_{0}n + T = I$ and $I = {}^{0}_{-1}\beta + {}^{110}_{48}Cd$
Balancing the second equation gives:
${}^{110}_{47}Ag = {}^{0}_{-1}\beta + {}^{110}_{48}Cd$
The first equation must then be: ${}^{1}_{0}n + T = {}^{110}_{47}Ag$
Balancing this equation gives: ${}^{1}_{0}n + {}^{109}_{47}Ag = {}^{110}_{47}Ag$
The target, then, was silver-109, and the unstable isotope was silver-110.

# GLOSSARY • GLOSARIO

Multilingual eGlossary

The multilingual science glossary includes Arabic, Bengali, Chinese, English, Haitian, Creole, Hmong, Korean, Portuguese, Russian, Tagalog, Urdu, and Vietnamese.

**Como usar el glosario en espanol:**

1. Busca el termino en ingles que desees encontrar.
2. El termino en espanol, junto con la definicion, se encuentran en la columna de la derecha.

## Pronunciation Key

Use the following key to help you sound out words in the glossary.

| | | | |
|---|---|---|---|
| **a** | **back** (BAK) | **ew** | **food** (FEWD) |
| **ay** | **day** (DAY) | **yoo** | **pure** (PYOOR) |
| **ah** | **father** (FAH thur) | **yew** | **few** (FYEW) |
| **ow** | **flower** (FLOW ur) | **uh** | comm**a** (CAHM uh) |
| **ar** | **car** (CAR) | **u** (+con) | r**u**b (RUB) |
| **e** | l**e**ss (LES) | **sh** | **sh**elf (SHELF) |
| **ee** | l**ea**f (LEEF) | **ch** | na**t**ure (NAY chur) |
| **ih** | tr**i**p (TRIHP) | **g** | **g**ift (GIHFT) |
| **i** (i+con+e) | **i**dea, l**i**fe (i DEE uh, life) | **j** | **g**em (JEM) |
| **oh** | g**o** (GOH) | **ing** | s**ing** (SING) |
| **aw** | s**o**ft (SAWFT) | **zh** | vi**si**on (VIHZH un) |
| **or** | **or**bit (OR but) | **k** | ca**k**e (KAYK) |
| **oy** | c**oi**n (COYN) | **s** | **s**eed, **c**ent (SEED, SENT) |
| **oo** | f**oo**t (FOOT) | **z** | **z**one, rai**s**e (ZOHN, RAYZ) |

## ENGLISH — A — ESPAÑOL

**absolute zero** **(p. 445)** Zero on the Kelvin scale, which represents the lowest possible theoretical temperature; atoms are all in the lowest possible energy state.

**cero absoluto** **(pág. 445)** Equivale a cero grados en la escala de Kelvin y representa la temperatura teórica más fría posible; a esta temperatura todos los átomos se encuentran en el menor estado energético posible.

**accuracy** **(p. 47)** Refers to how close a measured value is to an accepted value.

**exactitud** **(pág. 47)** Se refiere a la cercanía entre un valor medido y el valor aceptado.

**acid-base indicator** **(p. 662)** A chemical dye whose color is affected by acidic and basic solutions.

**indicador ácido-base** **(pág. 662)** tinción química cuyo color cambia al entrar en contacto con soluciones ácidas y básicas.

**acidic solution** **(p. 636)** Contains more hydrogen ions than hydroxide ions.

**solución ácida** **(pág. 636)** Solución que contiene más iones hidrógeno que iones hidróxido.

**acid ionization constant** **(p. 647)** The value of the equilibrium constant expression for the ionization of a weak acid.

**constante ácida de ionización** **(pág. 647)** Valor de la expresión de la constante de equilibrio para la ionización de un ácido débil.

**actinide series** **(p. 180)** In the periodic table, the f-block elements from period 7 that follow the element actinium.

**serie de actínidos** **(pág. 180)** Elementos del bloque F del período 7 de la tabla periódica que aparecen después del elemento actinio.

**activated complex** **(p. 564)** A short-lived, unstable arrangement of atoms that can break apart and re-form the reactants or can form products; also sometimes referred to as the transition state.

**complejo activado** **(pág. 564)** Complejo efímero e inestable de átomos que se puede romper para volver a formar los reactivos o para formar los productos; a veces también se le llama estado de transición.

**activation energy** **(p. 564)** The minimum amount of energy required by reacting particles in order to form the activated complex and lead to a reaction.

**energía de activación** **(pág. 564)** La cantidad mínima de energía que requieren las partículas de una reacción para formar el complejo activado y producir la reacción.

**active site** **(p. 830)** The pocket or crevice to which a substrate binds in an enzyme-catalyzed reaction.

**sitio activo** **(pág. 830)** Saliente o hendidura a la que se enlaza un sustrato durante una reacción catalizada por enzimas.

**actual yield (p. 385)** The amount of product produced when a chemical reaction is carried out.

**addition polymerization (p. 811)** Occurs when all the atoms present in the monomers are retained in the polymer product.

**addition reaction (p. 804)** A reaction that occurs when other atoms bond to each of two atoms bonded by double or triple covalent bonds.

**alcohol (p. 792)** An organic compound in which a hydroxyl group replaces a hydrogen atom of a hydrocarbon.

**aldehyde (p. 796)** An organic compound containing the structure in which a carbonyl group at the end of a carbon chain is bonded to a carbon atom on one side and a hydrogen atom on the other side.

**aliphatic compounds (a luh FA tihk • KAHM pownd) (p. 771)** Nonaromatic hydrocarbons, such as the alkanes, alkenes, and alkynes.

**alkali metals (p. 177)** Group 1 elements, except for hydrogen, they are reactive and usually exist as compounds with other elements.

**alkaline earth metals (p. 177)** Group 2 elements in the modern periodic table and are highly reactive.

**alkane (p. 750)** Hydrocarbon that contains only single bonds between atoms.

**alkene (p. 759)** An unsaturated hydrocarbon, such as ethene ($C_2H_4$), with one or more double covalent bonds between carbon atoms in a chain.

**alkyl halide (p. 787)** An organic compound containing a halogen atom covalently bonded to an aliphatic carbon atom.

**alkyne (p. 762)** An unsaturated hydrocarbon, such as ethyne ($C_2H_2$), with one or more triple bonds between carbon atoms in a chain.

**allotrope (p. 423)** One of two or more forms of an element with different structures and properties when they are in the same state—solid, liquid, or gas.

**alloy (p. 227)** A mixture of elements that has metallic properties; most commonly forms when the elements are either similar in size (substitutional alloy) or the atoms of one element are much smaller than the atoms of the other (interstitial alloy).

**alpha particle (p. 123)** A particle with two protons and two neutrons, with a 2+ charge; is equivalent to a helium-4 nucleus, can be represented by $\alpha$; and is emitted during radioactive decay.

**alpha radiation (p. 123)** Radiation that is made up of alpha particles; is deflected toward a negatively charged plate when radiation from a radioactive source is directed between two electrically charged plates.

**amide (AM ide) (p. 800)** An organic compound in which the $-H$ group of a carboxylic acid is replaced by a nitrogen atom bonded to other atoms.

**amines (A meen) (p. 795)** Organic compounds that contain nitrogen atoms bonded to carbon atoms in aliphatic chains or aromatic rings and have the general formula $RNH_2$.

**amino acid (p. 826)** An organic molecule that has both an amino group ($-NH_2$) and a carboxyl group ($-COOH$).

**rendimiento real (pág. 385)** Cantidad de producto que se obtiene al realizar una reacción química.

**polimerización de adición (pág. 811)** Ocurre cuando todos los átomos presentes en los monómeros forman parte del producto polimérico.

**reacción de adición (pág. 804)** Reacción que ocurre cuando dos átomos unidos entre sí por enlaces covalentes dobles o triples se unen con otros átomos.

**alcohol (pág. 792)** Compuesto orgánico en el que un grupo hidroxilo reemplaza a un átomo de hidrógeno de un hidrocarburo.

**aldehído (pág. 796)** Compuesto orgánico que contiene una estructura en la que un grupo carbonilo, situado al final de una cadena de carbonos, se une a un átomo de carbono por un lado y a un átomo de hidrógeno por el lado opuesto.

**compuestos alifáticos (pág. 771)** Hidrocarburos no aromáticos como los alcanos, los alquenos y los alquinos.

**metales alcalinos (pág. 177)** Incluyen los elementos del grupo 1, a excepción del hidrógeno. Son reactivos y generalmente existen como compuestos con otros elementos.

**metales alcalinotérreos (pág. 177)** Elementos altamente reactivos del grupo 2 de la tabla periódica moderna.

**alcano (pág. 750)** Hidrocarburo que sólo contiene enlaces sencillos entre sus átomos.

**alqueno (pág. 759)** Hidrocarburo no saturado, como el eteno ($C_2H_4$), que tiene uno o más enlaces covalentes dobles entre los átomos de carbono en una cadena.

**haluro de alquilo (pág. 787)** Compuesto orgánico que contiene un átomo de halógeno enlazado covalentemente a un átomo de carbono alifático.

**alquino (pág. 762)** Hidrocarburo no saturado, como el acetileno ($C_2H_2$), que tiene uno o más enlaces triples entre los átomos de carbono en una cadena.

**alótropos (pág. 423)** Formas de un elemento que tienen estructura y propiedades distintas cuando están en el mismo estado: sólido, líquido o gaseoso.

**aleación (pág. 227)** Mezcla de elementos que posee propiedades metálicas; en general se forman cuando los elementos tienen un tamaño similar (aleación de sustitución) o cuando los átomos de un elemento son mucho más pequeños que los átomos del otro (aleación intersticial).

**partícula alfa (pág. 123)** Partícula con dos protones y dos neutrones que tiene una carga 2+; equivale a un núcleo de helio 4, se puede representar como $\alpha$ y es emitida durante la desintegración radiactiva.

**radiación alfa (pág. 123)** Radiación compuesta de partículas alfa; si la radiación proveniente de una fuente radiactiva es dirigida hacia dos placas cargadas eléctricamente, este tipo de radiación se desvía hacia la placa con carga negativa.

**amida (pág. 800)** Compuesto orgánico en el que el grupo $-H$ de un ácido carboxílico es sustituido por un átomo de nitrógeno unido a otros átomos.

**aminas (pág. 795)** Compuestos orgánicos que contienen átomos de nitrógeno unidos a átomos de carbono en cadenas alifáticas o anillos aromáticos; su fórmula general es $RNH_2$.

**amino ácido (pág. 826)** Molécula orgánica que posee un grupo amino ($-NH_2$) y un grupo carboxilo ($-COOH$).

**amorphous solid (p. 424)** A solid in which particles are not arranged in a regular, repeating pattern that often is formed when molten material cools too quickly to form crystals.

**amphoteric (AM foh TAR ihk) (p. 639)** Describes water and other substances that can act as both acids and bases.

**amplitude (p. 137)** The height of a wave from the origin to a crest, or from the origin to a trough.

**anabolism (ah NAB oh lih zum) (p. 844)** Refers to the metabolic reactions through which cells use energy and small building blocks to build large, complex molecules needed to carry out cell functions and for cell structures.

**anion (AN i ahn) (p. 209)** An ion that has a negative charge.

**anode (p. 710)** In an electrochemical cell, the electrode where oxidation takes place.

**applied research (p. 17)** A type of scientific investigation that is undertaken to solve a specific problem.

**aqueous solution (p. 299)** A solution in which the solvent is water.

**aromatic compounds (p. 771)** Organic compounds that contain one or more benzene rings as part of their molecular structure.

**Arrhenius model (ah REE nee us • MAH dul) (p. 637)** A model of acids and bases; states that an acid is a substance that contains hydrogen and ionizes to produce hydrogen ions in aqueous solution and a base is a substance that contains a hydroxide group and dissociates to produce a hydroxide ion in aqueous solution.

**aryl halide (p. 788)** An organic compound that contains a halogen atom bonded to a benzene ring or another aromatic group

**asymmetric carbon (p. 768)** A carbon atom that has four different atoms or groups of atoms attached to it; occurs in chiral compounds.

**atmosphere (p. 407)** The unit that is often used to report air pressure.

**atom (p. 106)** The smallest particle of an element that retains all the properties of that element; is electrically neutral, spherically shaped, and composed of electrons, protons, and neutrons.

**atomic emission spectrum (p. 144)** A set of frequencies of electromagnetic waves given off by atoms of an element; consists of a series of fine lines of individual colors.

**atomic mass (p. 119)** The weighted average mass of the isotopes of that element.

**atomic mass unit (amu) (p. 119)** One-twelfth the mass of a carbon-12 atom.

**atomic number (p. 115)** The number of protons in an atom.

**atomic orbital (p. 152)** A three-dimensional region around the nucleus of an atom that describes an electron's probable location.

**ATP (p. 845)** Adenosine triphosphate—a nucleotide that functions as the universal energy-storage molecule in living cells.

**sólido amorfo (pág. 424)** Sólido cuyas partículas no están ordenadas de modo que formen un patrón regular repetitivo; a menudo se forma cuando el material fundido se enfría demasiado rápido como para formar cristales.

**anfotérico (pág. 639)** Término que describe al agua y otras sustancias que pueden actuar como ácidos y bases.

**amplitud (pág. 137)** Altura de una onda desde el origen hasta una cresta o desde el origen hasta un valle.

**anabolismo (pág. 844)** Reacciones metabólicas en las que las células usan energía y pequeñas unidades básicas para formar las moléculas grandes y complejas que requieren para realizar sus funciones celulares y para construir sus estructuras.

**anión (pág. 209)** Ion con carga negativa.

**ánodo (pág. 710)** Electrodo donde sucede la oxidación en una celda electroquímica.

**investigación aplicada (pág. 17)** Tipo de investigación científica que se realiza para resolver un problema concreto.

**solución acuosa (pág. 299)** Solución en la que el agua funciona como disolvente.

**compuestos aromáticos (pág. 771)** Compuestos orgánicos que contienen uno o más anillos de benceno como parte de su estructura molecular.

**modelo de Arrhenius (pág. 637)** Modelo de ácidos y bases; establece que un ácido es una sustancia que contiene hidrógeno y se ioniza para producir iones hidrógeno en solución acuosa, y que una base es una sustancia que contiene un grupo hidróxido y se disocia para producir un ion hidróxido en solución acuosa.

**haluro de arilo (pág. 788)** Compuesto orgánico que contiene un átomo de halógeno unido a un anillo de benceno u otro grupo aromático.

**carbono asimétrico (pág. 768)** Átomo de carbono que está unido a cuatro átomos o grupos de átomos diferentes; se hallan en compuestos quirales.

**atmósfera (pág. 407)** Unidad que a menudo se usa para reportar la presión atmosférica.

**átomo (pág. 106)** La partícula más pequeña de un elemento que retiene todas las propiedades de ese elemento; es eléctricamente neutro, de forma esférica y está compuesto de electrones, protones y neutrones.

**espectro de emisión atómica (pág. 144)** Conjunto de frecuencias de ondas electromagnéticas que emiten los átomos de un elemento; consta de una serie de líneas finas de distintos colores.

**masa atómica (pág. 119)** La masa promedio ponderada de los isótopos de un elemento.

**unidad de masa atómica (uma) (pág. 119)** La doceava parte de la masa de un átomo de carbono 12.

**número atómico (pág. 115)** El número de protones en un átomo.

**orbital atómico (pág. 152)** Región tridimensional alrededor del núcleo de un átomo que describe la ubicación probable de un electrón.

**ATP (pág. 845)** Trifosfato de adenosina; nucleótido que sirve como la molécula universal de almacenamiento de energía en las células vivas.

**aufbau principle** **(p. 156)** States that each electron occupies the lowest energy orbital available.

**Avogadro's number** **(p. 321)** The number $6.0221367 \times 10^{23}$, which is the number of representative particles in a mole, and can be rounded to three significant digits $6.02 \times 10^{23}$.

**Avogadro's principle** **(p. 452)** States that equal volumes of gases at the same temperature and pressure contain equal numbers of particles.

**principio de aufbau** **(pág. 156)** Establece que cada electrón ocupa el orbital de energía más bajo disponible.

**número de Avogadro** **(pág. 321)** Equivale al número $6.0221367 \times 10^{23}$; es el número de partículas representativas en un mol; se puede redondear a tres dígitos significativos: $6.02 \times 10^{23}$.

**principio de Avogadro** **(pág. 452)** Establece que los volúmenes iguales de gases, a la misma temperatura y presión, contienen igual número de partículas.

## B

**band of stability** **(p. 866)** The region on a graph within which all stable nuclei are found when plotting the number of neutrons versus the number of protons.

**barometer** **(p. 407)** An instrument that is used to measure atmospheric pressure.

**base ionization constant** **(p. 649)** The value of the equilibrium constant expression for the ionization of a base.

**base unit** **(p. 33)** A defined unit in a system of measurement that is based on an object or event in the physical world and is independent of other units.

**basic solution** **(p. 636)** Contains more hydroxide ions than hydrogen ions.

**battery** **(p. 718)** One or more electrochemical cells in a single package that generates electrical current.

**beta particle** **(p. 123)** A high-speed electron with a 1− charge that is emitted during radioactive decay.

**beta radiation** **(p. 123)** Radiation that is made up of beta particles; is deflected toward a positively charged plate when radiation from a radioactive source is directed between two electrically charged plates.

**boiling point** **(p. 427)** The temperature at which a liquid's vapor pressure is equal to the external or atmospheric pressure.

**boiling-point elevation** **(p. 500)** The temperature difference between a solution's boiling point and a pure solvent's boiling point.

**Boyle's law** **(p. 442)** States that the volume of a fixed amount of gas held at a constant temperature varies inversely with the pressure.

**breeder reactor** **(p. 882)** A nuclear reactor that is able to produce more fuel than it uses.

**Brønsted-Lowry model** **(p. 638)** A model of acids and bases in which an acid is a hydrogen-ion donor and a base is a hydrogen-ion acceptor.

**Brownian motion** **(p. 477)** The erratic, random, movements of colloid particles that results from collisions of particles of the dispersion medium with the dispersed particles.

**buffer** **(p. 666)** A solution that resists changes in pH when limited amounts of acid or base are added.

**buffer capacity** **(p. 667)** The amount of acid or base a buffer solution can absorb without a significant change in pH.

**banda de estabilidad** **(pág. 866)** Región de una gráfica en la que se hallan todos los núcleos estables cuando se grafica el número de neutrones contra el número de protones.

**barómetro** **(pág. 407)** Instrumento que se utiliza para medir la presión atmosférica.

**constante de ionización básica** **(pág. 649)** El valor de la expresión de la constante de equilibrio para la ionización de una base.

**unidad básica** **(pág. 33)** Unidad definida en un sistema de medidas; está basada en un objeto o evento del mundo físico y es independiente de otras unidades.

**solución básica** **(pág. 636)** Solución que contiene más iones hidróxido que iones hidrógeno.

**batería** **(pág. 718)** Una o más celdas electroquímicas contenidas en una sola unidad que genera corriente eléctrica.

**partícula beta** **(pág. 123)** Electrón de alta velocidad con una carga 1− que es emitido durante la desintegración radiactiva.

**radiación beta** **(pág. 123)** Radiación compuesta de partículas beta; si la radiación proveniente de una fuente radiactiva es dirigida hacia dos placas cargadas eléctricamente, este tipo de radiación se desvía hacia la placa con carga positiva.

**punto de ebullición** **(pág. 427)** Temperatura a la cual la presión de vapor de un líquido es igual a la presión externa o atmosférica.

**elevación del punto de ebullición** **(pág. 500)** Diferencia de temperatura entre el punto de ebullición de una solución y el punto de ebullición de un disolvente puro.

**ley de Boyle** **(pág. 442)** Establece que el volumen de una cantidad dada de gas a temperatura constante varía inversamente según la presión.

**reactor generador** **(pág. 882)** Reactor nuclear capaz de producir más combustible del que utiliza.

**modelo de Brønsted-Lowry** **(pág. 638)** Modelo de ácidos y bases en el que un ácido es un donante de iones hidrógeno y una base es un receptor de iones hidrógeno.

**movimiento browniano** **(pág. 477)** Movimientos erráticos, aleatorios de las partículas coloidales, producidos por el choque entre las partículas del medio de dispersión con las partículas dispersas.

**amortiguador** **(pág. 666)** Solución que resiste los cambios de pH cuando se agregan cantidades moderadas del ácido o la base.

**capacidad amortiguadora** **(pág. 667)** Cantidad de ácido o base que una solución amortiguadora puede absorber sin sufrir un cambio significativo en el pH.

## C

**calorie (p. 518)** The amount of heat required to raise the temperature of one gram of pure water by one degree Celsius.

**calorimeter (p. 523)** An insulated device that is used to measure the amount of heat released or absorbed during a physical or chemical process.

**carbohydrates (p. 832)** Compounds that contain multiple hydroxyl groups, plus an aldehyde or a ketone functional group, and function in living things to provide immediate and stored energy.

**carbonyl group (p. 796)** Arrangement in which an oxygen atom is double-bonded to a carbon atom.

**carboxyl group (p. 798)** Consists of a carbonyl group bonded to a hydroxyl group.

**carboxylic acid (p. 798)** An organic compound that contains a carboxyl group and is polar and reactive.

**catabolism (kuh TAB oh lih zum) (p. 844)** Refers to metabolic reactions that break down complex biological molecules for the purpose of forming smaller building blocks and extracting energy.

**catalyst (p. 571)** A substance that increases the rate of a chemical reaction by lowering activation energies but is not itself consumed in the reaction.

**cathode (p. 710)** In an electrochemical cell, the electrode where reduction takes place.

**cathode ray (p. 108)** Radiation that originates from the cathode and travels to the anode of a cathode-ray tube.

**cation (KAT i ahn) (p. 207)** An ion that has a positive charge.

**cellular respiration (p. 846)** The process in which glucose is broken down in the presence of oxygen gas to produce carbon dioxide, water, and energy.

**Charles's law (p. 445)** States that the volume of a given mass of gas is directly proportional to its kelvin temperature at constant pressure.

**chemical bond (p. 206)** The force that holds two atoms together; may form by the attraction of a positive ion for a negative ion or by sharing electrons.

**chemical change (p. 77)** A process involving one or more substances changing into new substances; also called a chemical reaction.

**chemical equation (p. 285)** A statement using chemical formulas to describe the identities and relative amounts of the reactants and products involved in the chemical reaction.

**chemical equilibrium (p. 596)** The state in which forward and reverse reactions balance each other because they occur at equal rates.

**chemical potential energy (p. 517)** The energy stored in a substance because of its composition; most is released or absorbed as heat during chemical reactions or processes.

**chemical property (p. 74)** The ability or inability of a substance to combine with or change into one or more new substances.

**caloría (pág. 518)** Cantidad de calor que se requiere para elevar un grado centígrado la temperatura de un gramo de agua pura.

**calorímetro (pág. 523)** Dispositivo aislado que sirve para medir la cantidad de calor liberada o absorbida durante un proceso físico o químico.

**carbohidratos (pág. 832)** Compuestos que contienen múltiples grupos hidroxilo, además de un grupo funcional aldehído o cetona, cuya función en los seres vivos es proporcionar energía inmediata o almacenada.

**grupo carbonilo (pág. 796)** Grupo formado por un átomo de oxígeno unido por un enlace doble a un átomo de carbono.

**grupo carboxilo (pág. 798)** Consiste en un grupo carbonilo unido a un grupo hidroxilo.

**ácido carboxílico (pág. 798)** Compuesto orgánico que contiene un grupo carboxilo; es polar y reactivo.

**catabolismo (pág. 844)** Reacciones metabólicas en las que se desdoblan moléculas biológicas complejas para obtener unidades básicas más pequeñas y energía.

**catalizador (pág. 571)** Sustancia que aumenta la velocidad de una reacción química al reducir su energía de activación; el catalizador no es consumido durante la reacción.

**cátodo (pág. 710)** Electrodo donde sucede la reducción en una celda electroquímica.

**rayo catódico (pág. 108)** Radiación que se origina en el cátodo y viaja hacia el ánodo de un tubo de rayos catódicos.

**catión (pág. 207)** Ion con carga positiva.

**respiración celular (pág. 846)** Proceso en el cual la glucosa es desdoblada en presencia del gas oxígeno para producir dióxido de carbono, agua y energía.

**Ley de Charles (pág. 445)** Establece que el volumen de una masa dada de gas es directamente proporcional a su temperatura Kelvin a presión constante.

**enlace químico (pág. 206)** La fuerza que mantiene a dos átomos unidos; puede formarse por la atracción de un ion positivo por un ion negativo compartiendo electrones.

**cambio químico (pág. 77)** Proceso que involucra una o más sustancias que se transforman en sustancias nuevas; también se conoce como reacción química.

**ecuación química (pág. 285)** Expresión que utiliza fórmulas químicas para describir las identidades y cantidades relativas de los reactivos y productos presentes en una reacción química.

**equilibrio químico (pág. 596)** Estado en el que se equilibran mutuamente las reacciones en sentido directo e inverso de una reacción química debido a que suceden a tasas iguales.

**energía potencial química (pág. 517)** La energía almacenada en una sustancia debido a su composición; la mayoría es liberada o absorbida como calor durante reacciones o procesos químicos.

**propiedad química (pág. 74)** La capacidad de una sustancia de combinarse con una o más sustancias nuevas o de transformarse en una o más sustancias nuevas.

**chemical reaction (p. 282)** The process by which the atoms of one or more substances are rearranged to form different substances; occurrence can be indicated by changes in temperature, color, odor, and physical state.

**chemistry (p. 4)** The study of matter and the changes that it undergoes.

**chirality (p. 767)** A property of a compound to exist in both left (l—) and right (d—) forms; occurs whenever a compound contains an asymmetric carbon.

**chromatography (p. 83)** A technique that is used to separate the components of a mixture based on the tendency of each component to travel or be drawn across the surface of a fixed substrate.

**coefficient (p. 285)** In a chemical equation, the number written in front of a reactant or product; in a balanced equation describes the lowest whole-number ratio of the amounts of all reactants and products.

**colligative property (kol LIHG uh tihv • PRAH pur tee) (p. 498)** A physical property of a solution that depends on the number, but not the identity, of the dissolved solute particles.

**collision theory (p. 563)** States that atoms, ions, and molecules must collide in order to react.

**colloids (p. 477)** A heterogeneous mixture of intermediate-sized particles (between atomic-size of solution particles and the size of suspension particles).

**combined gas law (p. 449)** A single law combining Boyle's, Charles's, and Gay-Lussac's laws that states the relationship among pressure, volume, and temperature of a fixed amount of gas.

**combustion reaction (p. 290)** A chemical reaction that occurs when a substance reacts with oxygen, releasing energy in the form of heat and light.

**common ion (p. 620)** An ion that is common to two or more ionic compounds.

**common ion effect (p. 620)** The lowering of the solubility of a substance by the presence of a common ion.

**complete ionic equation (p. 301)** An ionic equation that shows all the particles in a solution as they realistically exist.

**complex reaction (p. 580)** A chemical reaction that consists of two or more elementary steps.

**compound (p. 85)** A chemical combination of two or more different elements; can be broken down into simpler substances by chemical means and has properties different from those of its component elements.

**concentration (p. 480)** A measure of how much solute is dissolved in a specific amount of solvent or solution.

**conclusion (p. 15)** A judgment based on the information obtained.

**condensation (p. 428)** The energy-releasing process by which a gas or vapor becomes a liquid.

**condensation polymerization (p. 811)** Occurs when monomers containing at least two functional groups combine with the loss of a small by-product, usually water.

**reacción química (pág. 282)** Proceso por el cual los átomos de una o más sustancias se reordenan para formar sustancias diferentes; su pueden identificar cuando suceden cambios en temperatura, color, olor o estado físico.

**química (pág. 4)** El estudio de la materia y los cambios que ésta experimenta.

**quiralidad (pág. 767)** Propiedad de un compuesto para existir en forma levógira (i—) o dextrógira (d—); ocurre cuando un compuesto contiene un carbono asimétrico.

**cromatografía (pág. 83)** Técnica que sirve para separar los componentes de una mezcla según la tendencia de cada componente a desplazarse o ser atraído a lo largo de la superficie de otro material.

**coeficiente (pág. 285)** Número que precede a un reactivo o un producto en una ecuación química; en una ecuación equilibrada, indica la razón más pequeña expresada en números enteros de las cantidades de reactivos y productos en dicha reacción.

**propiedad coligativa (pág. 498)** Propiedad física de una solución que depende del número, pero no de la identidad, de las partículas de soluto disueltas.

**teoría de colisión (pág. 563)** Establece que los átomos, iones y moléculas deben chocar para reaccionar.

**coloides (pág. 477)** Mezcla heterogénea de partículas de tamaño intermedio (entre el tamaño atómico de partículas en solución y el de partículas en suspensión).

**ley combinada de los gases (pág. 449)** Ley que combina las leyes de Boyle, Charles y de Gay-Lussac; indica la relación entre la presión, el volumen y la temperatura de una cantidad constante de gas.

**reacción de combustión (pág. 290)** Reacción química que ocurre al reaccionar una sustancia con el oxígeno, liberando energía en forma de calor y luz.

**ion común (pág. 620)** Ion común a dos o más compuestos iónicos.

**efecto del ion común (pág. 620)** Disminución de la solubilidad de una sustancia debida a la presencia de un ion común.

**ecuación iónica total (pág. 301)** Ecuación iónica que muestra cómo existen realmente todas las partículas en una solución.

**reacción compleja (pág. 580)** Reacción química que consiste en dos o más pasos elementales.

**compuesto (pág. 85)** Combinación química de dos o más elementos diferentes; puede ser separado en sustancias más sencillas por medios químicos y exhibe propiedades que difieren de los elementos que lo componen.

**concentración (pág. 480)** Medida de la cantidad de soluto que se disuelve en una cantidad dada de disolvente o solución.

**conclusión (pág. 15)** Juicio basado en la información obtenida.

**condensación (pág. 428)** El proceso de liberación de energía mediante el cual un gas o vapor se convierte en líquido.

**polimerización por condensación (pág. 811)** Ocurre cuando monómeros que contienen al menos dos grupos funcionales se combinan y pierden un producto secundario pequeño, generalmente agua.

**condensation reaction (p. 801)** Occurs when two smaller organic molecules combine to form a more complex molecule, accompanied by the loss of a small molecule such as water.

**conjugate acid (p. 638)** The species produced when a base accepts a hydrogen ion from an acid.

**conjugate acid-base pair (p. 638)** Consists of two substances related to each other by the donating and accepting of a single hydrogen ion.

**conjugate base (p. 638)** The species produced when an acid donates a hydrogen ion to a base.

**control (p. 14)** In an experiment, the standard that is used for comparison.

**conversion factor (p. 44)** A ratio of equivalent values used to express the same quantity in different units; is always equal to 1 and changes the units of a quantity without changing its value.

**coordinate covalent bond (p. 259)** Forms when one atom donates a pair of electrons to be shared with an atom or ion that needs two electrons to become stable.

**corrosion (p. 724)** The loss of metal that results from an oxidation-reduction reaction of the metal with substances in the environment.

**covalent bond (p. 241)** A chemical bond that results from the sharing of valence electrons.

**cracking (p. 748)** The process by which heavier fractions of petroleum are converted to gasoline by breaking their large molecules into smaller molecules.

**critical mass (p. 880)** The minimum mass of a sample of fissionable material necessary to sustain a nuclear chain reaction.

**crystal lattice (p. 214)** A three-dimensional geometric arrangement of particles in which each positive ion is surrounded by negative ions and each negative ion is surrounded by positive ions; vary in shape due to sizes and relative numbers of the ions bonded.

**crystalline solid (p. 420)** A solid whose atoms, ions, or molecules are arranged in an orderly, geometric, three-dimensional structure.

**crystallization (p. 83)** A separation technique that produces pure solid particles of a substance from a solution that contains the dissolved substance.

**cyclic hydrocarbon (p. 755)** An organic compound that contains a hydrocarbon ring.

**cycloalkane (p. 755)** Cyclic hydrocarbons that contain single bonds only and can have rings with three, four, five, six, or more carbon atoms.

**reacción de condensación (pág. 801)** Ocurre cuando dos moléculas orgánicas pequeñas se combinan para formar una molécula más compleja; esta reacción es acompañada de la pérdida de una molécula pequeña como el agua.

**ácido conjugado (pág. 638)** Especie que se produce cuando una base acepta un ion hidrógeno de un ácido.

**par ácido-base conjugado (pág. 638)** Consiste en dos sustancias que se relacionan entre sí mediante la donación y aceptación de un solo ion hidrógeno.

**base conjugada (pág. 638)** Especie que se produce cuando un ácido dona un ion hidrógeno a una base.

**control (pág. 14)** Estándar de comparación en un experimento.

**factor de conversión (pág. 44)** Razón de valores equivalentes que sirve para expresar una misma cantidad en unidades diferentes; siempre es igual a 1 y cambia las unidades de una cantidad sin cambiar su valor.

**enlace covalente coordinado (pág. 259)** Se forma cuando un átomo dona un par de electrones para compartirlos con un átomo o un ion que requieren dos electrones para adquirir estabilidad.

**corrosión (pág. 724)** Pérdida de metal producida por una reacción de óxido-reducción del metal con sustancias en el ambiente.

**enlace covalente (pág. 241)** Enlace químico que se produce al compartir electrones de valencia.

**cracking (pág. 748)** Proceso por el cual las fracciones más pesadas de petróleo son convertidas en gasolina al romper las moléculas grandes en moléculas más pequeñas.

**masa crítica (pág. 880)** La masa mínima de una muestra de material fisionable que se necesita para sostener una reacción nuclear en cadena.

**red cristalina (pág. 214)** Ordenamiento geométrico tridimensional de partículas en el que cada ion positivo queda rodeado de iones negativos y cada ion negativo queda rodeado de iones positivos; su forma varía según el tamaño y número de iones enlazados.

**sólido cristalino (pág. 420)** Sólido cuyos átomos, iones o moléculas forman una estructura tridimensional, ordenada y geométrica.

**cristalización (pág. 83)** Técnica de separación que produce partículas sólidas puras de una sustancia a partir de una solución que contiene dicha sustancia en solución.

**hidrocarburo cíclico (pág. 755)** Compuesto orgánico que contiene un anillo de hidrocarburos.

**cicloalcano (pág. 755)** Hidrocarburos cíclicos que sólo contienen enlaces simples; pueden formar anillos con tres, cuatro, cinco, seis o más átomos de carbono.

## D

**Dalton's atomic theory (p. 104)** States that matter is composed of extremely small particles called atoms; atoms are invisible and indestructable; atoms of a given element are identical in size, mass, and chemical properties; atoms of a specific element are different from those of another element; different atoms combine in simple whole-number ratios to form compounds; in a chemical reaction, atoms are separated, combined, or rearranged.

**teoría atómica de Dalton (pág. 104)** Establece que la materia se compone de partículas extremadamente pequeñas denominadas átomos; los átomos son invisibles e indestructibles; los átomos de un elemento dado son idénticos en tamaño, masa y propiedades químicas; los átomos de un elemento específico difieren de los de otros elementos; átomos diferentes se combinan en razones simples de números enteros para formar compuestos; los átomos se separan, se combinan o se reordenan durante una reacción química.

**Dalton's law of partial pressures (p. 408)** States that the total pressure of a mixture of gases is equal to the sum of the pressures of all the gases in the mixture.

**de Broglie equation (p. 150)** Predicts that all moving particles have wave characteristics and relates each particle's wavelength to its frequency, its mass, and Planck's constant.

**decomposition reaction (p. 292)** A chemical reaction that occurs when a single compound breaks down into two or more elements or new compounds.

**dehydration reaction (p. 803)** An elimination reaction in which the atoms removed form water.

**dehydrogenation reaction (p. 803)** A reaction that eliminates two hydrogen atoms, which form a hydrogen molecule of gas.

**delocalized electrons (p. 225)** The electrons involved in metallic bonding that are free to move easily from one atom to the next throughout the metal and are not attached to a particular atom.

**denaturation (p. 829)** The process in which a protein's natural, intricate three-dimensional structure is disrupted.

**denatured alcohol (p. 793)** Ethanol to which noxious substances have been added in order to make it unfit to drink.

**density (p. 36)** The amount of mass per unit volume; a physical property.

**dependent variable (p. 14)** In an experiment, the variable whose value depends on the independent variable.

**deposition (p. 429)** The energy-releasing process by which a substance changes from a gas or vapor to a solid without first becoming a liquid.

**derived unit (p. 35)** A unit defined by a combination of base units.

**diffusion (p. 404)** The movement of one material through another from an area of higher concentration to an area of lower concentration.

**dimensional analysis (p. 44)** A systematic approach to problem solving that uses conversion factors to move from one unit to another.

**dipole-dipole forces (p. 412)** The attractions between oppositely charged regions of polar molecules.

**disaccharide (p. 833)** Forms when two monosaccharides bond together.

**dispersion forces (p. 412)** The weak forces resulting from temporary shifts in the density of electrons in electron clouds.

**distillation (p. 82)** A technique that can be used to physically separate most homogeneous mixtures based on the differences in the boiling points of the substances.

**double-replacement reaction (p. 296)** A chemical reaction that involves the exchange of ions between two compounds and produces either a precipitate, a gas, or water.

**dry cell (p. 718)** An electrochemical cell that contains a moist electrolytic paste inside a zinc shell.

**ley de Dalton de las presiones parciales (pág. 408)** Establece que la presión total de una mezcla de gases es igual a la suma de las presiones de todos los gases en la mezcla.

**ecuación de deBroglie (pág. 150)** Predice que todas las partículas móviles tienen características ondulatorias y relaciona la longitud de onda de cada partícula con su frecuencia, su masa y la constante de Planck.

**reacción de descomposición (pág. 292)** Reacción química que ocurre cuando un solo compuesto se divide en dos o más elementos o nuevos compuestos.

**reacción de deshidratación (pág. 803)** Una reacción de eliminación en la que los átomos que se pierden forman agua.

**reacción de deshidrogenación (pág. 803)** Reacción orgánica en la que se pierden dos átomos de hidrógeno, los cuales se unen y forman una molécula de hidrógeno.

**electrones deslocalizados (pág. 225)** Los electrones que forman un enlace metálico; estos electrones pasan fácilmente de un átomo a otro a través del metal y no están unidos a ningún átomo en particular.

**desnaturalización (pág. 829)** Proceso que afecta la estructura tridimensional, compleja y natural de una proteína.

**alcohol desnaturalizado (pág. 793)** Etanol al cual se añaden sustancias nocivas para evitar que se pueda beber.

**densidad (pág. 36)** La cantidad de masa por unidad de volumen; una propiedad física.

**variable dependiente (pág. 14)** Es la variable de un experimento cuyo valor depende de la variable independiente.

**depositación (pág. 429)** Proceso de liberación de energía por el cual una sustancia cambia de gas o vapor a sólido sin antes convertirse en un líquido.

**unidad derivada (pág. 35)** Unidad definida por una combinación de unidades básicas.

**difusión (pág. 404)** El movimiento de un material a través de otro en dirección al área de menor concentración.

**análisis dimensional (pág. 44)** Un enfoque sistemático para resolver un problema en el que se usan factores de conversión para pasar de una unidad a otra.

**fuerzas dipolo-dipolo (pág. 412)** La atracción entre regiones con cargas opuestas de moléculas polares.

**disacárido (pág. 833)** Se forma a partir de la unión de dos monosacáridos.

**fuerzas de dispersión (pág. 412)** Fuerzas débiles causadas por los cambios temporales en la densidad de electrones en las nubes electrónicas.

**destilación (pág. 82)** Técnica que se usa para separar físicamente la mayoría de las mezclas homogéneas según las diferencias en los puntos de ebullición de las sustancias.

**reacción de sustitución doble (pág. 296)** Reacción química en la que dos compuestos intercambian iones positivos, produciendo un precipitado, un gas o agua.

**pila seca (pág. 718)** Celda electroquímica que contiene una pasta electrolítica húmeda dentro de un armazón de zinc.

## E

**elastic collision (p. 403)** Collision in which no kinetic energy is lost; kinetic energy can be transferred between the colliding particles, but the total kinetic energy of the two particles remains the same.

**choque elástico (pág. 403)** Colisión en que no se pierde energía cinética; la energía cinética es transferida entre las partículas en choque, pero la energía cinética total de las dos partículas permanece igual.

**electrochemical cell** **(p. 709)** An apparatus that uses a redox reaction to produce electrical energy or uses electrical energy to cause a chemical reaction.

**electrolysis** **(p. 728)** The process that uses electrical energy to bring about a chemical reaction.

**electrolyte** **(p. 215)** An ionic compound whose aqueous solution conducts an electric current.

**electrolytic cell** **(p. 728)** An electrochemical cell in which electrolysis occurs.

**electromagnetic radiation** **(p. 137)** A form of energy exhibiting wavelike behavior as it travels through space; can be described by wavelength, frequency, amplitude, and speed.

**electromagnetic spectrum** **(p. 139)** Includes all forms of electromagnetic radiation; the types of radiation differ in their frequencies and wavelengths.

**electron** **(p. 108)** A negatively charged, fast-moving particle with an extremely small mass that is found in all forms of matter and moves through the empty space surrounding an atom's nucleus.

**electron capture** **(p. 868)** A radioactive decay process that occurs when an atom's nucleus draws in a surrounding electron, which combines with a proton to form a neutron, resulting in an X-ray photon being emitted.

**electron configuration** **(p. 156)** The arrangement of electrons in an atom, which is prescribed by three rules—the aufbau principle, the Pauli exclusion principle, and Hund's rule.

**electron-dot structure** **(p. 161)** Consists of an element's symbol, representing the atomic nucleus and inner-level electrons, that is surrounded by dots, representing the atom's valence electrons.

**electron sea model** **(p. 225)** Proposes that all metal atoms in a metallic solid contribute their valence electrons to form a "sea" of electrons, and can explain properties of metallic solids such as malleability, conduction, and ductility.

**electronegativity** **(p. 194)** Indicates the relative ability of an element's atoms to attract electrons in a chemical bond.

**element** **(p. 84)** A pure substance that cannot be broken down into simpler substances by physical or chemical means.

**elimination reaction** **(p. 802)** A reaction of organic compounds that occurs when a combination of atoms is removed from two adjacent carbon atoms forming an additional bond between the atoms.

**empirical formula** **(p. 344)** A formula that shows the smallest whole-number mole ratio of the elements of a compound, and might or might not be the same as the actual molecular formula.

**endothermic** **(p. 247)** A chemical reaction or process in which a greater amount of energy is required to break the existing bonds in the reactants than is released when the new bonds form in the product molecules.

**end point** **(p. 663)** The point at which the indicator that is used in a titration changes color.

**celda electroquímica** **(pág. 709)** Aparato que usa una reacción redox para producir energía eléctrica o que utiliza energía eléctrica para causar una reacción química.

**electrólisis** **(pág. 728)** Proceso que emplea energía eléctrica para producir una reacción química.

**electrolito** **(pág. 215)** Compuesto iónico cuya solución acuosa conduce una corriente eléctrica.

**celda electrolítica** **(pág. 728)** Celda electroquímica en donde ocurre la electrólisis.

**radiación electromagnética** **(pág. 137)** Forma de energía que exhibe un comportamiento ondulatorio al viajar por el espacio; se puede describir por su longitud de onda, su frecuencia, su amplitud y su rapidez.

**espectro electromagnético** **(pág. 139)** Incluye toda forma de radiación electromagnética; los distintos tipos de radiación difirien en sus frecuencias y sus longitudes de onda.

**electrón** **(pág. 108)** Partícula móvil rápida, de carga negativa y con una masa extremadamente pequeña que se encuentra en todas las formas de materia y que se mueve a través del espacio vacío que rodea el núcleo de un átomo.

**captura electrónica** **(pág. 868)** Proceso de desintegración radiactiva que ocurre cuando el núcleo de un átomo atrae un electrón circundante, que luego se combina con un protón para formar un neutrón, provocando la emisión de un fotón de rayos X.

**configuración electrónica** **(pág. 156)** El ordenamiento de los electrones en un átomo; está determinado por tres reglas: el principio de aufbau, el principio de exclusión de Pauli y la regla de Hund.

**estructura de puntos de electrones** **(pág. 161)** Consiste en el símbolo del elemento, que representa al núcleo atómico y los electrones de los niveles internos, rodeado por puntos que representan los electrones de valencia del átomo.

**modelo del mar de electrones** **(pág. 225)** Propone que todos los átomos de metal en un sólido metálico contribuyen con sus electrones de valencia para formar un "mar" de electrones.

**electronegatividad** **(pág. 194)** Indica la capacidad relativa de los átomos de un elemento para atraer electrones en un enlace químico.

**elemento** **(pág. 84)** Sustancia pura que no puede separarse en sustancias más sencillas por medios físicos ni químicos.

**reacción de eliminación** **(pág. 802)** Reacción de compuestos orgánicos que ocurre cuando se pierden un conjunto de átomos en dos átomos adyacentes de carbono, al formarse un enlace entre dichos átomos de carbono.

**fórmula empírica** **(pág. 344)** Fórmula que muestra la proporción molar más pequeña expresada en números enteros de los elementos de un compuesto; puede ser distinta de la fórmula molecular real.

**endotérmica** **(pág. 247)** Reacción o proceso químico que requiere una mayor cantidad de energía para romper los enlaces existentes en los reactivos, que la que se se libera al formarse los enlaces nuevos en las moléculas del producto.

**punto final** **(pág. 663)** Punto en el que el indicador que se utiliza en una titulación cambia de color.

**energy (p. 516)** The capacity to do work or produce heat; exists as potential energy, which is stored in an object due to its composition or position, and kinetic energy, which is the energy of motion.

**energy sublevels (p. 153)** The energy levels contained within a principal energy level.

**enthalpy (p. 527)** The heat content of a system at constant pressure.

**enthalpy (heat) of combustion (p. 529)** The enthalpy change for the complete burning of one mole of a given substance.

**enthalpy (heat) of reaction (p. 527)** The change in enthalpy for a reaction—the difference between the enthalpy of the substances that exist at the end of the reaction and the enthalpy of the substances present at the start .

**entropy (p. 543)** A measure of the number of possible ways that the energy of a system can be distributed; related to the freedom of the system's particles to move and the number of ways they can be arranged.

**enzyme (p. 829)** A biological catalyst.

**equilibrium constant (p. 599)** $K_{eq}$ is the numerical value that describes the ratio of product concentrations to reactant concentrations, with each raised to the power corresponding to its coefficient in the balanced equation.

**equivalence point (p. 661)** The point at which the moles of $H^+$ ions from the acid equals moles of $OH^-$ ions from the base.

**error (p. 48)** The difference between an experimental value and an accepted value

**ester (p. 799)** An organic compound with a carboxyl group in which the hydrogen of the hydroxyl group is replaced by an alkyl group; may be volatile and sweet-smelling and is polar.

**ether (p. 794)** An organic compound that contains an oxygen atom bonded to two carbon atoms.

**evaporation (p. 426)** The process in which vaporization occurs only at the surface of a liquid.

**excess reactant (p. 379)** A reactant that remains after a chemical reaction stops.

**exothermic (p. 247)** A chemical reaction or process in which more energy is released than is required to break bonds in the initial reactants.

**experiment (p. 14)** A set of controlled observations that test a hypothesis.

**extensive property (p. 73)** A physical property, such as mass, length, and volume, that is dependent upon the amount of substance present.

**energía (pág. 516)** Capacidad de realizar trabajo o producir calor; existe como energía potencial (almacenada en un objeto debido a su composición o posición) o como energía cinética (energía del movimiento).

**subniveles de energía (pág. 153)** Los niveles de energía dentro de un nivel principal de energía.

**entalpía (pág. 527)** El contenido de calor en un sistema a presión constante.

**entalpía (calor) de combustión (pág. 529)** El cambio de entalpía causado por la combustión completa de un mol de una sustancia dada.

**entalpía (calor) de reacción (pág. 527)** El cambio en la entalpía que ocurre en una reacción; es decir, la diferencia entre la entalpía de las sustancias que existen al final de la reacción y la entalpía de las sustancias presentes al comienzo de la misma.

**entropía (pág. 543)** Una medida de las formas posibles en que se puede distribuir la energía de un sistema; está relacionada con la libertad de movimiento de las partículas del sistema y el número de maneras en que éstas se pueden ordenar.

**enzima (pág. 829)** Catalizador biológico.

**constante de equilibrio (pág. 599)** $K_{eq}$ es el valor numérico que describe la razón de las concentraciones de los productos con respecto a las concentraciones de los reactivos, cada una de ellas elevada a la potencia correspondiente a su coeficiente en la ecuación equilibrada.

**punto de equivalencia (pág. 661)** Punto en el cual los moles de iones $H^+$ del ácido equivalen a los moles de iones $OH^-$ de la base.

**error (pág. 48)** La diferencia entre el valor experimental y el valor aceptado.

**éster (pág. 799)** Compuesto orgánico con un grupo carboxilo en el que el hidrógeno del grupo de hidroxilo es reemplazado por un grupo alquilo; es polar y puede ser volátil y de olor dulce.

**éter (pág. 794)** Compuesto orgánico que contiene un átomo de oxígeno unido a dos átomos de carbono.

**evaporación (pág. 426)** Proceso en el cual la vaporización ocurre sólo en la superficie de un líquido.

**reactivo en exceso (pág. 379)** Reactivo que sobra luego de finalizar una reacción química.

**exotérmica (pág. 247)** Reacción o proceso químico en el que se libera más energía que la requerida para romper los enlaces en los reactivos iniciales.

**experimento (pág. 14)** Conjunto de observaciones controladas que se realizan para probar una hipótesis.

**propiedad extensiva (pág. 73)** Propiedades físicas, como la masa, la longitud y el volumen, que dependen de la cantidad de sustancia presente.

## F

**fatty acid (p. 835)** A long-chain carboxylic acid that usually has between 12 and 24 carbon atoms and can be saturated (no double bonds), or unsaturated (one or more double bonds).

**ácido graso (pág. 835)** Ácido carboxílico de cadena larga que tiene generalmente entre 12 y 24 átomos de carbono; puede ser saturado (sin enlaces dobles) o insaturado o no saturado (con uno o más enlaces dobles).

**fermentation (p. 847)** The process in which glucose is broken down in the absence of oxygen, producing either ethanol, carbon dioxide, and energy (alcoholic fermentation) or lactic acid and energy (lactic acid fermentation).

**filtration (p. 82)** A technique that uses a porous barrier to separate a solid from a liquid.

**formula unit (p. 218)** The simplest ratio of ions represented in an ionic compound.

**fractional distillation (p. 747)** The process by which petroleum can be separated into simpler components, called fractions, as they condense at different temperatures.

**free energy (p. 546)** The energy available to do work—the difference between the change in enthalpy and the product of the entropy change and the kelvin temperature.

**freezing point (p. 428)** The temperature at which a liquid is converted into a crystalline solid.

**freezing-point depression (p. 502)** The difference in temperature between a solution's freezing point and the freezing point of its pure solvent.

**frequency (p. 137)** The number of waves that pass a given point per second.

**fuel cell (p. 722)** A voltaic cell in which the oxidation of a fuel, such as hydrogen gas, is used to produce electric energy.

**functional group (p. 786)** An atom or group of atoms that always reacts in a certain way in an organic molecule.

**fermentación (pág. 847)** Proceso en el cual la glucosa es desdoblada en ausencia de oxígeno produciendo etanol, dióxido de carbono y energía (fermentación alcohólica) o ácido láctico y energía (fermentación del ácido láctico).

**filtración (pág. 82)** Técnica que utiliza una barrera porosa para separar un sólido de un líquido.

**fórmula unitaria (pág. 218)** La razón más simple de iones representados en un compuesto iónico.

**destilación fraccionaria (pág. 747)** Proceso mediante el cual se separa el petróleo en componentes más simples llamados fracciones, las cuales se condensan a temperaturas diferentes.

**energía libre (pág. 546)** Energía disponible para hacer trabajo: la diferencia entre el cambio en la entalpía y el producto del cambio de entropía por la temperatura kelvin.

**punto de congelación (pág. 428)** La temperatura a la cual un líquido se convierte en un sólido cristalino.

**depresión del punto de congelación (pág. 502)** Diferencia de temperatura entre el punto de congelación de una solución y el punto de congelación de su disolvente puro.

**frecuencia (pág. 137)** Número de ondas que pasan por un punto dado en un segundo.

**celda de combustible (pág. 722)** Celda voltaica en la cual la oxidación de un combustible, como el gas hidrógeno, se utiliza para producir energía eléctrica.

**grupo funcional (pág. 786)** Átomo o grupo de átomos que siempre reaccionan de cierta manera en una molécula orgánica.

## G

**galvanization (p. 727)** The process in which an iron object is dipped into molten zinc or electroplated with zinc to make the iron more resistant to corrosion.

**gamma rays (p. 124)** High-energy radiation that has no electrical charge and no mass, is not deflected by electric or magnetic fields, usually accompanies alpha and beta radiation, and accounts for most of the energy lost during radioactive decay.

**gas (p. 72)** A form of matter that flows to conform to the shape of its container, fills the container's entire volume, and is easily compressed.

**Gay-Lussac's law (p. 447)** States that the pressure of a fixed mass of gas varies directly with the kelvin temperature when the volume remains constant.

**geometric isomers (p. 766)** A category of stereoisomers that results from different arrangements of groups around a double bond.

**Graham's law of effusion (p. 404)** States that the rate of effusion for a gas is inversely proportional to the square root of its molar mass.

**graph (p. 55)** A visual display of data.

**ground state (p. 146)** The lowest allowable energy state of an atom.

**group (p. 177)** A vertical column of elements in the periodic table arranged in order of increasing atomic number; also called a family.

**galvanizado (pág. 727)** Proceso en el cual un objeto de hierro en sumergido o galvanizado en zinc para aumentar la resistencia del hierro a la corrosión.

**rayos gamma (pág. 124)** Radiación de alta energía sin carga eléctrica ni masa; no es desviada por campos eléctricos ni magnéticos; acompaña generalmente a la radiación alfa y beta; representa la mayor parte de la energía perdida durante la desintegración radiactiva.

**gas (pág. 72)** Forma de la materia que fluye para adaptarse a la forma de su contenedor, llena el volumen entero del recipiente y se comprime fácilmente.

**ley de Gay-Lussac (pág. 447)** Establece que la presión de una masa dada de gas varía directamente con la temperatura en grados Kelvin cuando el volumen permanece constante.

**isómeros geométricos (pág. 766)** Categoría de estereoisómeros originada por los diversos ordenamientos posibles de grupos alrededor de un enlace doble.

**ley de efusión de Graham (pág. 404)** Establece que la tasa de efusión de un gas es inversamente proporcional a la raíz cuadrada de su masa molar.

**gráfica (pág. 55)** Representación visual de datos.

**estado base (pág. 146)** Estado de energía más bajo posible de un átomo.

**grupo (pág. 177)** Columna vertical de los elementos en la tabla periódica ordenados en sentido creciente según su número atómico; llamado también familia.

# H

**half-cells (p. 710)** The two parts of an electrochemical cell in which the separate oxidation and reduction reactions occur.

**half-life (p. 870)** The time required for one-half of a radioisotope's nuclei to decay into its products.

**half-reaction (p. 693)** One of two parts of a redox reaction—the oxidation half, which shows the number of electrons lost when a species is oxidized, or the reduction half, which shows the number of electrons gained when a species is reduced.

**halocarbon (p. 787)** Any organic compound containing a halogen substituent.

**halogen (p. 180)** A highly reactive group 17 element.

**halogenation (p. 790)** A process by which hydrogen atoms are replaced by halogen atoms.

**heat (p. 518)** A form of energy that flows from a warmer object to a cooler object.

**heat of solution (p. 492)** The overall energy change that occurs during the solution formation process.

**Heisenberg uncertainty principle (p. 151)** States that it is not possible to know precisely both the velocity and the position of a particle at the same time.

**Henry's law (p. 496)** States that at a given temperature, the solubility of a gas in a liquid is directly proportional to the pressure of the gas above the liquid.

**Hess's law (p. 534)** States that if two or more thermochemical equations can be added to produce a final equation for a reaction, then the sum of the enthalpy changes for the individual reactions is the enthalpy change for the final reaction.

**heterogeneous catalyst (p. 573)** A catalyst that exists in a different physical state than the reaction it catalyzes.

**heterogeneous equilibrium (p. 602)** A state of equilibrium that occurs when the reactants and products of a reaction are present in more than one physical state.

**heterogeneous mixture (p. 81)** One that does not have a uniform composition and in which the individual substances remain distinct.

**homogeneous catalyst (p. 573)** A catalyst that exists in the same physical state as the reaction it catalyzes.

**homogeneous equilibrium (p. 600)** A state of equilibrium that occurs when all the reactants and products of a reaction are in the same physical state.

**homogeneous mixture (p. 81)** One that has a uniform composition throughout and always has a single phase; also called a solution.

**homologous series (p. 751)** Describes a series of compounds that differ from one another by a repeating unit.

**Hund's rule (p. 157)** States that single electrons with the same spin must occupy each equal-energy orbital before additional electrons with opposite spins can occupy the same orbitals.

**semiceldas (pág. 710)** Las dos partes de una celda electroquímica en las que ocurren las reacciones separadas de oxidación y reducción.

**vida media (pág. 870)** Tiempo requerido para que la mitad de los núcleos de un radioisótopo se desintegren en sus productos.

**semirreacción (pág. 693)** Una de dos partes de una reacción redox: la correspondiente a la oxidación muestra el número de electrones que se pierden al oxidarse una especie y la correspondiente a la reducción muestra el número de electrones que se ganan al reducirse una especie.

**halocarbono (pág. 787)** Cualquier compuesto orgánico que contiene un sustituyente halógeno.

**halógeno (pág. 180)** Elemento sumamente reactivo del grupo 17.

**halogenación (pág. 790)** Proceso mediante el cual se reemplazan átomos de hidrógeno por átomos de halógeno.

**calor (pág. 518)** Forma de energía que fluye hacia cuerpos más fríos.

**calor de solución (pág. 492)** El cambio global de energía que ocurre durante el proceso de formación de una solución.

**principio de incertidumbre de Heisenberg (pág. 151)** Establece que no es posible saber con precisión y al mismo tiempo la velocidad y la posición de una partícula.

**ley de Henry (pág. 496)** Establece que a una temperatura dada, la solubilidad de un gas en un líquido es directamente proporcional a la presión del gas sobre el líquido.

**ley de Hess (pág. 534)** Establece que si para producir la ecuación final para una reacción se pueden sumar dos o más ecuaciones termoquímicas, entonces la suma de los cambios de entalpía para las reacciones individuales equivale al cambio de entalpía de la reacción final.

**catalizador heterogéneo (pág. 573)** Catalizador que existe en un estado físico diferente al de la reacción que cataliza.

**equilibrio heterogéneo (pág. 602)** Estado de equilibrio que ocurre cuando los reactivos y los productos de una reacción están presentes en más de un estado físico.

**mezcla heterogénea (pág. 81)** Aquella que no tiene una composición uniforme y en la que las sustancias individuales permanecen separadas.

**catalizador homogéneo (pág. 573)** Catalizador que existe en el mismo estado físico de la reacción que cataliza.

**equilibrio homogéneo (pág. 600)** Estado de equilibrio que ocurre cuando todos los reactivos y productos de una reacción están en el mismo estado físico.

**mezcla homogénea (pág. 81)** Aquella que tiene una composición uniforme y siempre tiene una sola fase; también llamada solución.

**serie homóloga (pág. 751)** Describe una serie de compuestos que difieren entre sí por una unidad repetitiva.

**regla de Hund (pág. 157)** Establece que los electrones individuales con igual rotación deben ocupar cada uno orbitales distintos con la misma energía, antes de que electrones adicionales con rotación opuesta puedan ocupar los mismos orbitales.

**hybridization (p. 262)** A process in which atomic orbitals are mixed to form new, identical hybrid orbitals.

**hibridación (pág. 262)** Proceso mediante el cual se mezclan los orbitales atómicos para formar orbitales híbridos nuevos e idénticos.

**hydrate (p. 351)** A compound that has a specific number of water molecules bound to its atoms.

**hidrato (pág. 351)** Compuesto que tiene un número específico de moléculas de agua unidas a sus átomos.

**hydration reaction (p. 804)** An addition reaction in which a hydrogen atom and a hydroxyl group from a water molecule add to a double or triple bond.

**reacción de hidratación (pág. 804)** Reacción de adición en la que se añaden el átomo de hidrógeno y el grupo hidroxilo de una molécula de agua a un enlace doble o triple.

**hydrocarbon (p. 745)** Simplest organic compound composed only of the elements carbon and hydrogen.

**hidrocarburo (pág. 745)** El compuesto orgánico más simple; está formado sólo por los elementos carbono e hidrógeno.

**hydrogenation reaction (p. 804)** An addition reaction in which hydrogen is added to atoms in a double or triple bond; usually requires a catalyst.

**reacción de hidrogenación (pág. 804)** Reacción de adición en la que se agrega hidrógeno a los átomos que forman un enlace doble o triple; requiere generalmente de un catalizador.

**hydrogen bond (p. 413)** A strong dipole-dipole attraction between molecules that contain a hydrogen atom bonded to a small, highly electronegative atom.

**enlace de hidrógeno (pág. 413)** Fuerte atracción dipolo-dipolo entre moléculas que contienen un átomo de hidrógeno unido a un átomo pequeño, sumamente electronegativo.

**hydroxyl group (p. 792)** An oxygen-hydrogen group covalently bonded to a carbon atom.

**grupo hidroxilo (pág. 792)** Un grupo hidrógeno-oxígeno unido covalentemente a un átomo de carbono.

**hypothesis (p. 13)** A tentative, testable statement or prediction about what has been observed.

**hipótesis (pág. 13)** Enunciado tentativo y comprobable o predicción acerca de lo que ha sido observado.

## I

**ideal gas constant (R) (p. 454)** An experimentally determined constant whose value in the ideal gas equation depends on the units that are used for pressure.

**constante de los gases ideales (R) (pág. 454)** Constante determinada experimentalmente cuyo valor en la ecuación de los gases ideales depende de las unidades en las que se expresa la presión.

**ideal gas law (p. 454)** Describes the physical behavior of an ideal gas in terms of pressure, volume, temperature, and number of moles of gas.

**ley de los gases ideales (pág. 454)** Describe el comportamiento físico de un gas ideal en términos de la presión, el volumen, la temperatura y el número de moles del gas.

**immiscible (ih MIHS ih bul) (p. 479)** Describes two liquids that can be mixed together but separate shortly after you cease mixing them.

**inmiscible (pág. 479)** Describe dos líquidos que se pueden mezclar entre sí, pero que se separan poco después de que se cesa de mezclarlos.

**independent variable (p. 14)** In an experiment, the variable that the experimenter plans to change.

**variable independiente (pág. 14)** La variable de un experimento que el experimentador piensa cambiar.

**induced transmutation (p. 875)** The process in which nuclei are bombarded with high-velocity charged particles in order to create new elements.

**transmutación inducida (pág. 875)** Proceso en cual se bombardean núcleos con partículas cargadas de alta velocidad para crear elementos nuevos.

**inhibitor (p. 571)** A substance that slows down the reaction rate of a chemical reaction or prevents a reaction from happening.

**inhibidor (pág. 571)** Sustancia que reduce la tasa de reacción de una reacción química o evita que ésta suceda.

**inner transition metal (p. 180)** A type of group B element that is contained in the f-block of the periodic table and is characterized by a filled outermost orbital, and filled or partially filled 4f and 5f orbitals.

**metal de transición interna (pág. 180)** Tipo de elemento del grupo B contenido dentro del bloque F de la tabla periódica; se caracteriza por tener el orbital más externo lleno y los orbitales 4f y 5f parcialmente llenos.

**insoluble (p. 479)** Describes a substance that cannot be dissolved in a given solvent.

**insoluble (pág. 479)** Describe una sustancia que no se puede disolver en un disolvente dado.

**instantaneous rate (p. 578)** The rate of decomposition at a specific time, calculated from the rate law, the specific rate constant, and the concentrations of all the reactants.

**velocidad instantánea (pág. 578)** La tasa de descomposición en un tiempo dado, se calcula a partir de la ley de velocidad de la reacción, la constante de velocidad de la reacción y las concentraciones de los reactivos.

**intensive property (p. 73)** A physical property that remains the same no matter how much of a substance is present.

**propiedad intensiva (pág. 73)** Propiedad física que permanece igual sea cual sea la cantidad de sustancia presente.

**intermediate (p. 580)** A substance produced in one elementary step of a complex reaction and consumed in a subsequent elementary step.

**intermediario (pág. 580)** Sustancia producida en un paso elemental de una reacción compleja y que es consumida en un paso elemental subsecuente.

**ion** **(p. 189)** An atom or bonded group of atoms with a positive or negative charge.

**ionic bond** **(p. 210)** The electrostatic force that holds oppositely charged particles together in an ionic compound.

**ionic compounds** **(p. 210)** Compounds that contain ionic bonds.

**ionization energy** **(p. 191)** The energy required to remove an electron from a gaseous atom; generally increases in moving from left-to-right across a period and decreases in moving down a group.

**ionizing radiation** **(p. 885)** Radiation that is energetic enough to ionize matter it collides with.

**ion product constant for water** **(p. 650)** The value of the equilibrium constant expression for the self-ionization of water.

**isomers** **(p. 765)** Two or more compounds that have the same molecular formula but have different molecular structures.

**isotopes** **(p. 117)** Atoms of the same element with different numbers of neutrons.

**ion** **(pág. 189)** Átomo o grupo de átomos unidos que tienen carga positiva o negativa.

**enlace iónico** **(pág. 210)** Fuerza electrostática que mantiene unidas las partículas con carga opuesta en un compuesto iónico.

**compuestos iónicos** **(pág. 210)** Compuestos que contienen enlaces iónicos.

**energía de ionización** **(pág. 191)** Energía que se requiere para separar un electrón de un átomo en estado gaseoso; generalmente aumenta al moverse de izquierda a derecha a lo largo de un período de la tabla periódica y disminuye al moverse hacia abajo a lo largo de un grupo.

**radiación ionizante** **(pág. 885)** Radiación que posee suficiente energía como para ionizar la materia con la que choca.

**constante del producto iónico del agua** **(pág. 650)** Valor de la expresión de la constante de equilibrio de la ionización del agua.

**isómeros** **(pág. 765)** Dos o más compuestos que tienen la misma fórmula molecular pero poseen estructuras moleculares diferentes.

**isótopos** **(pág. 117)** Átomos del mismo elemento con diferente número de neutrones.

## J

**joule** **(p. 518)** The SI unit of heat and energy.

**julio** **(pág. 518)** La unidad SI de medida del calor y la energía.

## K

**kelvin** **(p. 35)** The SI base unit of temperature.

**ketone** **(p. 797)** An organic compound in which the carbon of the carbonyl group is bonded to two other carbon atoms.

**kilogram** **(p. 34)** The SI base unit for mass.

**kinetic-molecular theory** **(p. 402)** Describes the behavior of gases in terms of particles in motion; makes several assumptions about size, motion, and energy of gas particles.

**kelvin** **(pág. 35)** Unidad básica de temperatura del SI.

**cetona** **(pág. 797)** Compuesto orgánico en el que el carbono del grupo carbonilo está unido a otros dos átomos de carbono.

**kilogramo** **(pág. 34)** Unidad básica de masa del SI.

**teoría cinético-molecular** **(pág. 402)** Explica el comportamiento de los gases en términos de partículas en movimiento; hace varias suposiciones acerca del tamaño, movimiento y energía de las partículas de gas.

## L

**lanthanide series** **(p. 180)** In the periodic table, the f-block elements from period 6 that follow the element lanthanum.

**lattice energy** **(p. 216)** The energy required to separate one mole of the ions of an ionic compound, which is directly related to the size of the ions bonded and is also affected by the charge of the ions.

**law of chemical equilibrium** **(p. 599)** States that at a given temperature, a chemical system may reach a state in which a particular ratio of reactant and product concentrations has a constant value.

**law of conservation of energy** **(p. 517)** States that in any chemical reaction or physical process, energy may change from one form to another, but it is neither created nor destroyed.

**law of conservation of mass** **(p. 77)** States that mass is neither created nor destroyed during a chemical reaction but is conserved.

**serie de los lantánidos** **(pág. 180)** Los elementos del bloque F del período 6 de la tabla periódica que siguen al elemento lantano.

**energía reticular** **(pág. 216)** Energía que se requiere para separar un mol de los iones de un compuesto iónico; está directamente relacionada con el tamaño de los iones enlazados y es afectada también por la carga de los iones.

**ley del equilibrio químico** **(pág. 599)** Establece que a una temperatura dada, un sistema químico puede alcanzar un estado en el que la razón particular de las concentraciones del reactivo y el producto tiene un valor constante.

**ley de conservación de la energía** **(pág. 517)** Establece que en toda reacción química y en todo proceso físico la energía puede cambiar de una forma a otra, pero no puede ser creada ni destruida.

**ley de conservación de la masa** **(pág. 77)** Establece que durante una reacción química la masa no se crea ni se destruye, sino que se conserva.

**law of definite proportions** **(p. 87)** States that, regardless of the amount, a compound is always composed of the same elements in the same proportion by mass.

**law of multiple proportions** **(p. 89)** States that when different compounds are formed by the combination of the same elements, different masses of one element combine with the same fixed mass of the other element in a ratio of small whole numbers.

**Le Châtelier's principle** **(luh SHAHT uh lee yays • PRIHN sih puhl) (p. 607)** States that if a stress is applied to a system at equilibrium, the system shifts in the direction that relieves the stress.

**Lewis model** **(p. 641)** A Lewis acid is an electron-pair acceptor and a Lewis base is an electron-pair donor.

**Lewis structure** **(p. 242)** A model that uses electron-dot structures to show how electrons are arranged in molecules. Pairs of dots or lines represent bonding pairs.

**limiting reactant** **(p. 379)** A reactant that is totally consumed during a chemical reaction, limits the extent of the reaction, and determines the amount of product.

**lipids** **(p. 835)** Large, nonpolar biological molecules that vary in structure, store energy in living organisms, and make up most of the structure of cell membranes.

**liquid** **(p. 71)** A form of matter that flows, has constant volume, and takes the shape of its container.

**liter** **(p. 35)** The metric unit for volume equal to one cubic decimeter.

**ley de las proporciones definidas** **(pág. 87)** Establece que, independientemente de la cantidad, un compuesto siempre se compone de los mismos elementos en la misma proporción por masa.

**ley de las proporciones múltiples** **(pág. 89)** Establece que cuando la combinación de los mismos elementos forma compuestos diferentes, una masa dada de uno de los elementos se combina con masas diferentes del otro elemento de acuerdo con una razón que se expresa en números enteros pequeños.

**Principio de Le Châtelier** **(pág. 607)** Establece que si se aplica una perturbación a un sistema en equilibrio, el sistema cambia en la dirección que reduce la perturbación.

**modelo de Lewis** **(pág. 641)** Un ácido es un receptor de pares de electrones y una base es un donante de pares de electrones.

**estructura de Lewis** **(pág. 242)** Modelo que utiliza diagramas de puntos de electrones para mostrar la disposición de los electrones en las moléculas. Los pares de puntos o líneas representan pares de electrones enlazados.

**reactivo limitante** **(pág. 379)** Reactivo que se consume completamente durante una reacción química, limita la duración de la reacción y determina la cantidad del producto.

**lípidos** **(pág. 835)** Moléculas biológicas no polares de gran tamaño que varían en estructura, almacenan energía en los seres vivos y conforman la mayor parte de la estructura de las membranas celulares.

**líquido** **(pág. 71)** Forma de materia que fluye, tiene volumen constante y toma la forma de su envase.

**litro** **(pág. 35)** Unidad de volumen del sistema métrico; equivale a un decímetro cúbico.

## M



**mass** **(p. 9)** A measure that reflects the amount of matter.

**mass defect** **(p. 877)** The difference in mass between a nucleus and its component nucleons.

**mass number** **(p. 117)** The number after an element's name, representing the sum of its protons and neutrons.

**matter** **(p. 4)** Anything that has mass and takes up space.

**melting point** **(p. 426)** For a crystalline solid, the temperature at which the forces holding a crystal lattice together are broken and it becomes a liquid.

**metabolism** **(p. 844)** The sum of the many chemical reactions that occur in living cells.

**metal** **(p. 177)** An element that is solid at room temperature, a good conductor of heat and electricity, and generally is shiny; most metals are ductile and malleable.

**metallic bond** **(p. 225)** The attraction of a metallic cation for delocalized electrons.

**metalloid** **(p. 181)** An element that has physical and chemical properties of both metals and nonmetals.

**meter** **(p. 33)** The SI base unit for length.

**masa** **(pág. 9)** Medida que refleja la cantidad de materia.

**defecto másico** **(pág. 877)** La diferencia de masa entre un núcleo y los nucleones que lo componen.

**número de masa** **(pág. 117)** El número que va después del nombre de un elemento; representa la suma de sus protones y neutrones.

**materia** **(pág. 4)** Cualquier cosa que tiene masa y ocupa espacio.

**punto de fusión** **(pág. 426)** Para un sólido cristalino, es la temperatura a la que se rompen las fuerzas que mantienen unida la red cristalina y el sólido se convierte en líquido.

**metabolismo** **(pág. 844)** El conjunto de las numerosas reacciones químicas que ocurren en las células vivas.

**metal** **(pág. 177)** Elemento sólido a temperatura ambiente, es buen conductor de calor y electricidad y generalmente es brillante; la mayoría de los metales son dúctiles y maleables.

**enlace metálico** **(pág. 225)** Atracción de un catión metálico por los electrones deslocalizados.

**metaloide** **(pág. 181)** Elementos que tienen las propiedades físicas y químicas de metales y de no metales.

**metro** **(pág. 33)** Unidad básica de longitud del SI.

**method of initial rates (p. 576)** Determines the reaction order by comparing the initial rates of a reaction carried out with varying reactant concentrations.

**miscible (p. 479)** Describes two liquids that are soluble in each other.

**mixture (p. 80)** A physical blend of two or more pure substances in any proportion in which each substance retains its individual properties; can be separated by physical means.

**model (p. 10)** A visual, verbal, and/or mathematical explanation of data collected from many experiments.

**molality (p. 487)** The ratio of the number of moles of solute dissolved in one kilogram of solvent; also known as molal concentration.

**molar enthalpy (heat) of fusion (p. 530)** The amount of heat required to melt one mole of a solid substance.

**molar enthalpy (heat) of vaporization (p. 530)** The amount of heat required to vaporize one mole of a liquid.

**molarity (p. 482)** The number of moles of solute dissolved per liter of solution; also known as molar concentration.

**molar mass (p. 326)** The mass in grams of one mole of any pure substance.

**molar volume (p. 452)** For a gas, the volume that one mole occupies at 0.00°C and 1.00 atm pressure.

**mole (p. 321)** The SI base unit used to measure the amount of a substance, abbreviated mol; the number of carbon atoms in exactly 12 g of pure carbon; one mole is the amount of a pure substance that contains $6.02 \times 10^{23}$ representative particles.

**molecular formula (p. 346)** A formula that specifies the actual number of atoms of each element in one molecule of a substance.

**molecule (p. 241)** Forms when two or more atoms covalently bond and is lower in potential energy than its constituent atoms.

**mole fraction (p. 488)** The ratio of the number of moles of solute in solution to the total number of moles of solute and solvent.

**mole ratio (p. 371)** In a balanced equation, the ratio between the numbers of moles of any two substances.

**monatomic ion (p. 218)** An ion formed from only one atom.

**monomer (p. 810)** A molecule from which a polymer is made.

**monosaccharides (p. 832)** The simplest carbohydrates, also called simple sugars.

**método de las velocidades iniciales (pág. 576)** Determina el orden de la reacción al comparar las velocidades iniciales de una reacción realizada con diversas concentraciones de reactivo.

**miscible (pág. 479)** Describe dos líquidos que son solubles entre sí.

**mezcla (pág. 80)** Combinación física de dos o más sustancias puras en cualquier proporción en la que cada sustancia retiene sus propiedades individuales; las sustancias se pueden separar por medios físicos.

**modelo (pág. 10)** Explicación matemática, verbal o visual de datos recolectados en muchos experimentos.

**molalidad (pág. 487)** La razón del número de moles de soluto disueltos en un kilogramo de disolvente; también se conoce como concentración molal.

**entalpía (calor) molar de fusión (pág. 530)** Cantidad requerida de calor para fundir un mol de una sustancia sólida.

**entalpía (calor) molar de vaporización (pág. 530)** Cantidad requerida de calor para vaporizar un mol de un líquido.

**molaridad (pág. 482)** Número de moles de soluto disueltos por litro de solución; también se conoce como concentración molar.

**masa molar (pág. 326)** Masa en gramos de un mol de cualquier sustancia pura.

**volumen molar (pág. 452)** Para un gas, es el volumen que ocupa un mol a 0.00°C y una presión de 1.00 atm.

**mol (pág. 321)** Unidad básica del SI para medir la cantidad de una sustancia, se abrevia mol; el número de átomos de carbono en 12 g exactos de carbono puro; un mol es la cantidad de sustancia pura que contiene $6.02 \times 10^{23}$ partículas representativas.

**fórmula molecular (pág. 346)** Fórmula que especifica el número real de átomos de cada elemento en una molécula de la sustancia.

**molécula (pág. 241)** Se forma cuando dos o más átomos se unen covalentemente y posee menor energía potencial que los átomos que la conforman.

**fracción molar (pág. 488)** La razón del número de moles de soluto en solución al número total de moles de soluto y disolvente.

**razón molar (pág. 371)** En una ecuación equilibrada, se refiere a la razón entre el número de moles de dos sustancias cualesquiera.

**ion monoatómico (pág. 218)** Ion formado de un sólo átomo.

**monómero (pág. 810)** Molécula a partir de la cual se forma un polímero.

**monosacáridos (pág. 832)** Los carbohidratos más simples; se llaman también azúcares simples.

## N

**net ionic equation (p. 301)** An ionic equation that includes only the particles that participate in the reaction.

**neutralization reaction (p. 659)** A reaction in which an acid and a base react in aqueous solution to produce a salt and water.

**ecuación iónica neta (pág. 301)** Ecuación iónica que incluye sólo las partículas que participan en la reacción.

**reacción de neutralización (pág. 659)** Reacción en la que un ácido y una base reaccionan en una solución acuosa para producir sal y agua.

**neutron (p. 113)** A neutral, subatomic particle in an atom's nucleus that has a mass nearly equal to that of a proton.

**noble gas (p. 180)** An extremely unreactive group 18 element.

**nonmetals (p. 180)** Elements that are generally gases or dull, brittle solids that are poor conductors of heat and electricity.

**nuclear equation (p. 123)** A type of equation that shows the atomic number and mass number of the particles involved.

**nuclear fission (p. 878)** The splitting of a nucleus into smaller, more stable fragments, accompanied by a large release of energy.

**nuclear fusion (p. 883)** The process of binding smaller atomic nuclei into a single, larger, and more stable nucleus.

**nuclear reaction (p. 122)** A reaction that involves a change in the nucleus of an atom.

**nucleic acid (p. 840)** A nitrogen-containing biological polymer that is involved in the storage and transmission of genetic information.

**nucleons (p. 865)** The positively charged protons and neutral neutrons contained in an atom's nucleus.

**nucleotide (p. 840)** The monomer that makes up a nucleic acid; consists of a nitrogen base, an inorganic phosphate group, and a five-carbon monosaccharide sugar.

**nucleus (p. 112)** The extremely small, positively charged, dense center of an atom that contains positively charged protons and neutral neutrons.

**neutrón (pág. 113)** Partícula subatómica neutral en el núcleo de un átomo que tiene una masa casi igual a la de un protón.

**gas noble (pág. 180)** Elemento extremadamente no reactivo del grupo 18.

**no metales (pág. 180)** Elementos que generalmente son gases o sólidos quebradizos, sin brillo y malos conductores de calor y electricidad.

**ecuación nuclear (pág. 123)** Tipo de ecuación que muestra el número atómico y el número de masa de las partículas involucradas.

**fisión nuclear (pág. 878)** Ruptura de un núcleo en fragmentos más pequeños y más estables; se acompaña de una gran liberación de energía.

**fusión nuclear (pág. 883)** Proceso de unión de núcleos atómicos pequeños en un solo núcleo más grande y más estable.

**reacción nuclear (pág. 122)** Reacción que implica un cambio en el núcleo de un átomo.

**ácido nucleico (pág. 840)** Polímero biológico que contiene nitrógeno y que participa en el almacenamiento y transmisión de información genética.

**nucleones (pág. 865)** Los protones de carga positiva y los neutrones sin carga que contiene el núcleo de un átomo.

**nucleótido (pág. 840)** Monómeros que forman los ácidos nucleicos; consisten de una base nitrogenada, un grupo fosfato inorgánico y un azúcar monosacárido de cinco carbonos.

**núcleo (pág. 112)** El diminuto y denso centro con carga positiva de un átomo; contiene protones con su carga positiva y neutrones sin carga.

## O



**octet rule (p. 193)** States that atoms lose, gain, or share electrons in order to acquire the stable electron configuration of a noble gas.

**optical isomers (p. 768)** Result from different arrangements of four different groups around the same carbon atom and have the same physical and chemical properties except in chemical reactions where chirality is important.

**optical rotation (p. 769)** An effect that occurs when polarized light passes through a solution containing an optical isomer and the plane of polarization is rotated to the right by a d-isomer or to the left by an l-isomer.

**organic compounds (p. 745)** All compounds that contain carbon with the primary exceptions of carbon oxides, carbides, and carbonates, all of which are considered inorganic.

**osmosis (p. 504)** The diffusion of solvent particles across a semipermeable membrane from an area of higher solvent concentration to an area of lower solvent concentration.

**osmotic pressure (p. 504)** The pressure caused when water molecules move into or out of a solution.

**regla del octeto (pág. 193)** Establece que los átomos pierden, ganan o comparten electrones para adquirir la configuración electrónica estable de un gas noble.

**isómeros ópticos (pág. 768)** Son resultado de los distintos ordenamientos que adquieren los cuatro grupos diferentes que rodean a un mismo átomo de carbono; todos poseen las mismas propiedades químicas y físicas, excepto en las reacciones químicas donde la quiralidad es importante.

**rotación óptica (pág. 769)** Efecto que ocurre cuando la luz polarizada atraviesa una solución que contiene un isómero óptico y el plano de polarización rota a la derecha en los isómeros dextrógiros (–d) y a la izquierda en los isómeros levógiros (–l).

**compuestos orgánicos (pág. 745)** Todo compuesto que contiene carbono; las excepciones más importantes son los óxidos de carbono, los carburos y los carbonatos, todos los cuales se consideran inorgánicos.

**osmosis (pág. 504)** Difusión de partículas de disolvente a través de una membrana semipermeable hacia el área donde la concentración del disolvente es menor.

**presión osmótica (pág. 504)** La presión que causan las moléculas de agua al entrar o salir de una solución.

**oxidation (p. 681)** The complete or partial loss of electrons from a reacting substance; increases an atom's oxidation number.

**oxidation number (p. 219)** The positive or negative charge of a monatomic ion.

**oxidation-number method (p. 689)** The technique that can be used to balance more difficult redox reactions, based on the fact that the number of electrons or ions transferred from atoms must equal the number of electrons or ions accepted by other atoms.

**oxidation-reduction reaction (p. 680)** Any chemical reaction in which electrons are transferred from one substance to another; also called a redox reaction.

**oxidizing agent (p. 683)** The substance that oxidizes another substance by accepting its electrons.

**oxyacid (p. 250)** Any acid that contains hydrogen and an oxyanion.

**oxyanion (ahk see AN i ahn) (p. 222)** A polyatomic ion composed of an element, usually a nonmetal, bonded to one or more oxygen atoms.

**oxidación (pág. 681)** Pérdida de electrones de los átomos de una sustancia; aumenta el número de oxidación de un átomo.

**número de oxidación (pág. 219)** La carga positiva o negativa de un ion monoatómico.

**método del número de oxidación (pág. 689)** Técnica que sirve para equilibrar las reacciones redox más difíciles; se basa en el hecho de que el número de electrones transferidos por los átomos debe ser igual al número de electrones aceptados por otros átomos.

**reacción de oxidación-reducción (pág. 680)** Toda reacción química en la que sucede transferencia de electrones de una sustancia a otro; también se llama reacción redox.

**agente oxidante (pág. 683)** Sustancia que oxida otra sustancia al aceptar sus electrones.

**oxiácido (pág. 250)** Todo ácido que contiene hidrógeno y un oxianión.

**oxianión (pág. 222)** Ion poliatómico compuesto de un elemento, generalmente un no metal, unido a uno o a más átomos de oxígeno.

## P

**parent chain (p. 753)** The longest continuous chain of carbon atoms in a branched-chain alkane, alkene, or alkyne.

**pascal (p. 407)** The SI unit of pressure; one pascal (Pa) is equal to a force of one newton per square meter.

**Pauli exclusion principle (p. 157)** States that a maximum of two electrons can occupy a single atomic orbital but only if the electrons have opposite spins.

**penetrating power (p. 864)** The ability of radiation to pass through matter.

**peptide (p. 828)** A chain of two or more amino acids linked by peptide bonds.

**peptide bond (p. 828)** The amide bond that joins two amino acids.

**percent by mass (p. 87)** A percentage determined by the ratio of the mass of each element to the total mass of the compound.

**percent composition (p. 342)** The percent by mass of each element in a compound.

**percent error (p. 48)** The ratio of an error to an accepted value.

**percent yield (p. 386)** The ratio of actual yield (from an experiment) to theoretical yield (from stoichiometric calculations) expressed as a percent.

**period (p. 177)** A horizontal row of elements in the modern periodic table.

**periodic law (p. 176)** States that when the elements are arranged by increasing atomic number, there is a periodic repetition of their properties.

**periodic table (p. 85)** A chart that organizes all known elements into a grid of horizontal rows (periods) and vertical columns (groups or families) arranged by increasing atomic number.

**cadena principal (pág. 753)** La cadena continua más larga de átomos de carbono en un alcano, un alqueno o un alquino ramificados.

**pascal (pág. 407)** La unidad SI de presión; un pascal (Pa) es igual a una fuerza de un newton por metro cuadrado.

**principio de exclusión de Pauli (pág. 157)** Establece que cada orbital atómico sólo puede ser ocupado por un máximo de dos electrones, pero sólo si los electrones tienen giros opuestos.

**poder de penetración (pág. 864)** La capacidad de la radiación de atravesar la materia.

**péptido (pág. 828)** Cadena de dos o más aminoácidos unidos por enlaces peptídicos.

**enlace peptídico (pág. 828)** Enlace amida que une dos aminoácidos.

**porcentaje en masa (pág. 87)** Porcentaje determinado por la razón de la masa de cada elemento respecto a la masa total del compuesto.

**composición porcentual (pág. 342)** Porcentaje en masa de cada elemento en un compuesto.

**porcentaje de error (pág. 48)** La razón del error al valor aceptado.

**porcentaje de rendimiento (pág. 386)** Razón del rendimiento real (de un experimento) al rendimiento teórico (de cálculos estequiométricos) expresada como porcentaje.

**período (pág. 177)** Fila horizontal de elementos en la tabla periódica moderna.

**ley periódica (pág. 176)** Establece que al ordenar los elementos por número atómico en sentido ascendente, existe una repetición periódica de sus propiedades.

**tabla periódica (pág. 85)** Tabla en la que se organizan todos los elementos conocidos en una cuadrícula de filas horizontales (períodos) y columnas verticales (grupos o familias), ordenados según su número atómico en sentido ascendente.

**pH (p. 652)** The negative logarithm of the hydrogen ion concentration of a solution; acidic solutions have pH values between 0 and 7, basic solutions have values between 7 and 14, and a solution with a pH of 7.0 is neutral.

**phase change (p. 76)** A transition of matter from one state to another.

**phase diagram (p. 429)** A graph of pressure versus temperature that shows which phase a substance exists in under different conditions of temperature and pressure.

**phospholipid (p. 838)** A triglyceride in which one of the fatty acids is replaced by a polar phosphate group.

**photoelectric effect (p. 142)** A phenomenon in which photoelectrons are emitted from a metal's surface when light of a certain frequency shines on the surface.

**photon (p. 143)** A particle of electromagnetic radiation with no mass that carries a quantum of energy.

**photosynthesis (p. 846)** The complex process that converts energy from sunlight to chemical energy in the bonds of carbohydrates.

**physical change (p. 76)** A type of change that alters the physical properties of a substance but does not change its composition.

**physical property (p. 73)** A characteristic of matter that can be observed or measured without changing the sample's composition—for example, density, color, taste, hardness, and melting point.

**pi bond (p. 245)** A bond that is formed when parallel orbitals overlap to share electrons.

**Planck's constant (*h*) (p. 142)** $6.626 \times 10^{-34}$ J•s, where J is the symbol for the joule.

**plastic (p. 789)** A polymer that can be heated and molded while relatively soft.

**pOH (p. 652)** The negative logarithm of the hydroxide ion concentration of a solution; a solution with a pOH above 7.0 is acidic, a solution with a pOH below 7.0 is basic, and a solution with a pOH of 7.0 is neutral.

**polar covalent bond (p. 266)** A type of bond that forms when electrons are not shared equally.

**polyatomic ion (p. 221)** An ion made up of two or more atoms bonded together that acts as a single unit with a net charge.

**polymerization reaction (p. 810)** A reaction in which monomer units are bonded together to form a polymer.

**polymers (p. 809)** Large molecules formed by combining many repeating structural units (monomers); are synthesized through addition or condensation reactions.

**polysaccharide (p. 833)** A complex carbohydrate, which is a polymer of simple sugars that contains 12 or more monomer units.

**positron (p. 868)** A particle that has the same mass as an electron but an opposite charge.

**positron emission (p. 868)** A radioactive decay process in which a proton in the nucleus is converted into a neutron and a positron, and then the positron is emitted from the nucleus.

**pH (pág. 652)** El logaritmo negativo de la concentración de iones hidrógeno de una solución; las soluciones ácidas poseen valores de pH entre 0 y 7, las soluciones básicas tienen valores entre 7 y 14 y una solución con un pH de 7.0 es neutra.

**cambio de fase (pág. 76)** La transición de la materia de un estado a otro.

**diagrama de fase (pág. 429)** Gráfica de presión contra temperatura que muestra la fase en la que se encuentra una sustancia bajo distintas condiciones de temperatura y presión.

**fosfolípido (pág. 838)** Triglicérido en el que uno de los ácidos grasos es sustituido por un grupo fosfato polar.

**efecto fotoeléctrico (pág. 142)** Fenómeno en el cual la superficie de un metal emiten fotoelectrones cuando una luz de cierta frecuencia ilumina su superficie.

**fotón (pág. 143)** Partícula de radiación electromagnética sin masa que transporta un cuanto de energía.

**fotosíntesis (pág. 846)** Proceso complejo que convierte la energía de la luz solar en la energía química de los enlaces en carbohidratos.

**cambio físico (pág. 76)** Tipo de cambio que altera las propiedades físicas de una sustancia pero no cambia su composición.

**propiedad física (pág. 73)** Característica de la materia que se puede observar o medir sin cambiar la composición de una muestra de la materia; por ejemplo, la densidad, el color, el sabor, la dureza y el punto de fusión.

**enlace pi (pág. 245)** Enlace que se forma cuando orbitales paralelos se superponen para compartir electrones.

**constante de Planck (*h*) (pág. 142)** $6.626 \times 10^{-34}$ J•s, donde J es el símbolo de julios.

**plástico (pág. 789)** Polímero que se puede calentar y moldear mientras esté relativamente suave.

**pOH (pág. 652)** El logaritmo negativo de la concentración de iones hidróxido de una solución; una solución con un pOH mayor que 7.0 es ácida, una solución con un pOH menor que 7.0 es básica y una solución con un pOH de 7.0 es neutra.

**enlace covalente polar (pág. 266)** Tipo de enlace que se forma cuando los electrones no se comparten de manera equitativa.

**ion poliatómico (pág. 221)** Ion compuesto de dos o más átomos unidos entre sí que actúan como una unidad con carga neta.

**reacción de polimerización (pág. 810)** Reacción en la cual los monómeros se unen para formar un polímero.

**polímeros (pág. 809)** Moléculas grandes formadas por la unión de muchas unidades estructurales repetidas (monómeros); se sintetizan a través de reacciones de adición o de condensación.

**polisacárido (pág. 833)** Carbohidrato complejo; es un polímero de azúcares simples que contiene 12 ó más monómeros.

**positrón (pág. 868)** Partícula que tiene la misma masa que un electrón pero carga opuesta.

**emisión de positrones (pág. 868)** Proceso de desintegración radiactiva en el que un protón del núcleo se convierte en un neutrón y un positrón y luego el positrón es emitido del núcleo.

**precipitate (p. 296)** A solid produced during a chemical reaction in a solution.

**precision (p. 47)** Refers to how close a series of measurements are to one another; precise measurements show little variation over a series of trials but might not be accurate.

**pressure (p. 406)** Force applied per unit area.

**primary battery (p. 720)** A type of battery that produces electric energy by redox reactions that are not easily reversed, delivers current until the reactants are gone, and then is discarded.

**principal energy levels (p. 153)** The major energy levels of an atom.

**principal quantum number (*n*) (p. 153)** Assigned by the quantum mechanical model to indicate the relative sizes and energies of atomic orbitals.

**product (p. 283)** A substance formed during a chemical reaction.

**protein (p. 826)** An organic polymer made up of animo acids linked together by peptide bonds that can function as an enzyme, transport important chemical substances, or provide structure in organisms.

**proton (p. 113)** A subatomic particle in an atom's nucleus that has a positive charge of 1+.

**pure research (p. 17)** A type of scientific investigation that seeks to gain knowledge for the sake of knowledge itself.

**precipitado (pág. 296)** Sólido que se produce durante una reacción química en una solución.

**precisión (pág. 47)** Se refiere a la cercanía de una serie de medidas entre sí; las medidas precisas muestran poca variación durante una serie de pruebas, incluso si no son exactas.

**presión (pág. 406)** Fuerza aplicada por unidad de área.

**batería primaria (pág. 720)** Tipo de batería que produce energía eléctrica por reacciones redox que no son fácilmente reversibles, produce corriente hasta que se agotan los reactivos y luego se desecha.

**niveles energéticos principales (pág. 153)** Los niveles energéticos más importantes de un átomo.

**número cuántico principal (*n*) (pág. 153)** Asignado por el modelo mecánico cuántico para indicar el tamaño y la energía relativas de los orbitales atómicos.

**producto (pág. 283)** Sustancia que se forma durante una reacción química.

**proteína (pág. 826)** Polímero orgánico compuesto de aminoácidos unidos por enlaces peptídicos; puede funcionar como enzima, transportar sustancias químicas importantes o ser parte de la estructura en los organismos.

**protón (pág. 113)** Partícula subatómica en el núcleo de un átomo con carga positiva 1+.

**investigación pura (pág. 17)** Tipo de investigación científica que busca obtener conocimiento sin otro interés que satisfacer el interés científico.

## Q

**qualitative data (p. 13)** Information describing color, odor, shape, or some other physical characteristic.

**quantitative data (p. 13)** Numerical information describing how much, how little, how big, how tall, or how fast.

**quantum (p. 141)** The minimum amount of energy that can be gained or lost by an atom.

**quantum mechanical model of the atom (p. 152)** An atomic model in which electrons are treated as waves; also called the wave mechanical model of the atom.

**quantum number (p. 147)** The number assigned to each orbit of an electron.

**datos cualitativos (pág. 13)** Información que describe el color, el olor, la forma o alguna otra característica física.

**datos cuantitativos (pág. 13)** Información numérica que describe cantidad, tamaño o rapidez.

**cuanto (pág. 141)** La cantidad mínima de energía que puede ganar o perder un átomo.

**modelo mecánico cuántico del átomo (pág. 152)** Modelo atómico en el cual los electrones se estudian como si fueran ondas; también se denomina modelo mecánico ondulatorio del átomo.

**número cuántico (pág. 147)** Número que se asigna a cada órbita de un electrón.

## R

**radiation (p. 122)** The rays and particles—alpha and beta particles and gamma rays—that are emitted by radioactive materials.

**radioactive decay (p. 122)** A spontaneous process in which unstable nuclei lose energy by emitting radiation.

**radioactive decay series (p. 870)** A series of nuclear reactions that starts with an unstable nucleus and results in the formation of a stable nucleus.

**radioactivity (p. 122)** The process in which some substances spontaneously emit radiation.

**radiochemical dating (p. 873)** The process that is used to determine the age of an object by measuring the amount of a certain radioisotope remaining in that object.

**radiación (pág. 122)** Los rayos y partículas que emiten los materiales radiactivos (partículas alfa y beta y rayos gamma).

**desintegración radiactiva (pág. 122)** Proceso espontáneo en el que los núcleos inestables pierden energía al emitir radiación.

**serie de desintegración radiactiva (pág. 870)** Serie de reacciones nucleares que empieza con un núcleo inestable y produce la formación de un núcleo estable.

**radiactividad (pág. 122)** Proceso en el que algunas sustancias emiten radiación espontáneamente.

**datación radioquímica (pág. 873)** Proceso que sirve para determinar la edad de un objeto al medir la cantidad restante de cierto radioisótopo en dicho objeto.

**radioisotopes (p. 861)** Isotopes of atoms that have unstable nuclei and emit radiation to attain more stable atomic configurations.

**radiotracer (p. 887)** An isotope that emits non-ionizing radiation and is used to signal the presence of an element or specific substance; can be used to analyze complex chemical reactions mechanisms and to diagnose disease.

**rate-determining step (p. 581)** The slowest elementary step in a complex reaction; limits the instantaneous rate of the overall reaction.

**rate law (p. 574)** The mathematical relationship between the rate of a chemical reaction at a given temperature and the concentrations of reactants.

**reactant (p. 283)** The starting substance in a chemical reaction.

**reaction mechanism (p. 580)** The complete sequence of elementary steps that make up a complex reaction.

**reaction order (p. 575)** For a reactant, describes how the rate is affected by the concentration of that reactant.

**reaction rate (p. 561)** The change in concentration of a reactant or product per unit time, generally calculated and expressed in moles per liter per second.

**redox reaction (p. 680)** An oxidation-reduction reaction.

**reducing agent (p. 683)** The substance that reduces another substance by losing electrons.

**reduction (p. 681)** The complete or partial gain of electrons by a reacting substance; decreases an atom's oxidation number.

**reduction potential (p. 711)** The tendency of a substance to gain electrons.

**representative elements (p. 177)** Elements from groups 1, 2, and 13–18 in the modern periodic table, possessing a wide range of chemical and physical properties.

**resonance (p. 258)** Condition that occurs when more than one valid Lewis structure exists for the same molecule.

**reversible reaction (p. 595)** A reaction that can take place in both the forward and reverse directions; leads to an equilibrium state where the forward and reverse reactions occur at equal rates and the concentrations of reactants and products remain constant.

**radioisótopos (pág. 861)** Isótopos de átomos que poseen núcleos inestables y emiten radiación para obtener una configuración atómica más estable.

**radiolocalizador (pág. 887)** Isótopo que emite radiación no ionizante y se utiliza para señalar la presencia de un elemento o sustancia específica; se usan para analizar los mecanismos de reacciones químicas complejas y para diagnosticar enfermedades.

**paso determinante de la velocidad de reacción (pág. 581)** El paso elemental más lento en una reacción compleja; limita la velocidad instantánea de la reacción general.

**ley de velocidad de la reacción (pág. 574)** Relación matemática entre la velocidad de una reacción química a una temperatura dada y las concentraciones de los reactivos.

**reactivo (pág. 283)** Sustancia inicial en una reacción química.

**mecanismo de reacción (pág. 580)** Sucesión completa de pasos elementales que componen una reacción compleja.

**orden de la reacción (pág. 575)** Describe cómo la concentración de un reactivo afecta la velocidad de la reacción para dicho reactivo.

**tasa de reacción (pág. 561)** Cambio en la concentración de un reactivo o producto por unidad de tiempo, generalmente se calcula y expresa en moles por litro por segundo.

**reacción redox (pág. 680)** Una reacción de oxidorreducción.

**agente reductor (pág. 683)** Sustancia que reduce otra sustancia al perder electrones.

**reducción (pág. 681)** Ganancia de electrones por los átomos de una sustancia; reduce el número de oxidación de los átomos.

**potencial de reducción (pág. 711)** Tendencia de una sustancia a ganar electrones.

**elementos representativos (pág. 177)** Elementos de los grupos 1, 2 y 13 a 18 de la tabla periódica moderna; poseen una gran variedad de propiedades químicas y físicas.

**resonancia (pág. 258)** Condición que ocurre cuando existe más de una estructura válida de Lewis para una misma molécula.

**reacción reversible (pág. 595)** Reacción que puede ocurrir en direcciones normal e inversa; produce un estado de equilibrio donde las reacciones en sentido normal e inverso ocurren a tasas iguales, ocasionando que la concentración de reactivos y productos permanezcan constantes.

## S

**salt (p. 659)** An ionic compound made up of a cation from a base and an anion from an acid.

**salt bridge (p. 709)** A pathway constructed to maintain solution neutrality by allowing positive and negative ions to move from one solution to another.

**salt hydrolysis (p. 665)** The process in which anions of the dissociated salt accept hydrogen ions from water, or the cations of the dissociated salt donate hydrogen ions to water.

**sal (pág. 659)** Compuesto iónico formado por un catión proveniente de una base y un anión proveniente de un ácido.

**puente salino (pág. 709)** Medio que permite el movimiento de iones positivos y negativos de una solución a otra.

**hidrólisis de sales (pág. 665)** Proceso en el que los aniones de una sal disociada aceptan iones hidrógeno del agua o en el que los cationes de la sal disociada donan iones hidrógeno al agua.

**saponification (suh pahn ih fih KAY shuhn) (p. 837)** The hydrolysis of the ester bonds of a triglyceride using an aqueous solution of a strong base to form carboxylate salts and glycerol.

**saturated hydrocarbon (p. 746)** A hydrocarbon that contains only single bonds.

**saturated solution (p. 493)** Contains the maximum amount of dissolved solute for a given amount of solvent at a specific temperature and pressure.

**scientific law (p. 16)** Describes a relationship in nature that is supported by many experiments.

**scientific methods (p. 12)** A systematic approach used in scientific study; an organized process used by scientists to do research and to verify the work of others.

**scientific notation (p. 40)** Expresses any number as a number between 1 and 10 (known as a coefficient) multiplied by 10 raised to a power (known as an exponent).

**second (p. 33)** The SI base unit for time.

**second law of thermodynamics (p. 543)** The spontaneous processes always proceed in such a way that the entropy of the universe increases.

**secondary battery (p. 720)** A rechargeable battery that depends on reversible redox reactions.

**sigma bond (p. 244)** A single covalent bond that is formed when an electron pair is shared by the direct overlap of bonding orbitals.

**significant figures (p. 50)** The number of all known digits reported in measurements plus one estimated digit.

**single-replacement reaction (p. 293)** A chemical reaction that occurs when the atoms of one element replace the atoms of another element in a compound.

**solid (p. 71)** A form of matter that has its own definite shape and volume, is incompressible, and expands only slightly when heated.

**solubility (p. 614)** The maximum amount of solute that will dissolve in a given amount of solvent at a specific temperature and pressure.

**solubility product constant (p. 614)** $K_{sp}$, which is an equilibrium constant for the dissolving of a sparingly soluble ionic compound in water.

**soluble (p. 479)** Describes a substance that can be dissolved in a given solvent.

**solute (p. 299)** One or more substances dissolved in a solution.

**solution (p. 81)** A uniform mixture that can contain solids, liquids, or gases; also called a homogeneous mixture.

**solvation (p. 489)** The process of surrounding solute particles with solvent particles to form a solution; occurs only where and when the solute and solvent particles come in contact with each other.

**solvent (p. 299)** The substance that dissolves a solute to form a solution; the most plentiful substance in the solution.

**species (p. 693)** Any kind of chemical unit involved in a process.

**saponificación (pág. 837)** La hidrólisis de los enlaces éster de un triglicérido, usando una solución acuosa de una base fuerte, para formar sales de carboxilato y glicerol.

**hidrocarburo saturado (pág. 746)** Hidrocarburo que sólo contiene enlaces sencillos.

**solución saturada (pág. 493)** Solución que contiene la cantidad máxima de soluto disuelto para una cantidad dada de disolvente a una temperatura y presión específicas.

**ley científica (pág. 16)** Describe una relación natural demostrada en muchos experimentos.

**métodos científicos (pág. 12)** Enfoque sistemático que se usa en los estudios científicos; proceso organizado que siguen los científicos para realizar sus investigaciones y verificar el trabajo realizado por otros científicos.

**notación científica (pág. 40)** Expresa cualquier número como un número entre 1 y 10 (conocido como coeficiente) multiplicado por 10 elevado a alguna potencia (conocida como exponente).

**segundo (pág. 33)** Unidad básica de tiempo del SI.

**segunda ley de la termodinámica (pág. 543)** Los procesos espontáneos siempre proceden de una forma que aumenta la entropía del universo.

**batería secundaria (pág. 720)** Batería recargable que depende de reacciones redox reversibles.

**enlace sigma (pág. 244)** Enlace covalente simple que se forma cuando se comparte un par de electrones mediante la superposición directa de los orbitales del enlace.

**cifras significativas (pág. 50)** El número de dígitos conocidos que se reportan en medidas, más un dígito estimado.

**reacción de sustitución simple (pág. 293)** Reacción química que ocurre cuando los átomos de un elemento reemplazan a los átomos de otro elemento en un compuesto.

**sólido (pág. 71)** Forma de la materia que tiene su propia forma y volumen, es incompresible y sólo se expande levemente cuando se calienta.

**solubilidad (pág. 614)** Cantidad máxima de soluto que se disolverá en una cantidad dada de disolvente a una temperatura y presión específicas.

**constante de producto de solubilidad (pág. 614)** Se representa como $K_{sp}$; es la constante de equilibrio para la disolución de un compuesto iónico moderadamente soluble en agua.

**soluble (pág. 479)** Describe una sustancia que se puede disolver en un disolvente dado.

**soluto (pág. 299)** Una o más sustancias disueltas en una solución.

**solución (pág. 81)** Mezcla uniforme que puede contener sólidos, líquidos o gases; llamada también mezcla homogénea.

**solvatación (pág. 489)** Proceso de rodear las partículas de soluto con partículas del disolvente para formar una solución; ocurre sólo en los lugares y en el momento en que las partículas de soluto y disolvente entran en contacto.

**disolvente (pág. 299)** Sustancia que disuelve un soluto para formar una solución; la sustancia más abundante en la solución.

**especie (pág. 693)** Cualquier clase de unidad química que participa en un proceso.

**specific heat (p. 519)** The amount of heat required to raise the temperature of one gram of a given substance by one degree Celsius.

**specific rate constant (p. 574)** A numerical value that relates reaction rate and concentration of reactant at a specific temperature.

**spectator ion (p. 301)** Ion that does not participate in a reaction.

**spontaneous process (p. 542)** A physical or chemical change that occurs without outside intervention and may require energy to be supplied to begin the process.

**standard enthalpy (heat) of formation (p. 537)** The change in enthalpy that accompanies the formation of one mole of a compound in its standard state from its constituent elements in their standard states.

**standard hydrogen electrode (p. 711)** The standard electrode against which the reduction potential of all electrodes can be measured.

**states of matter (p. 71)** The physical forms in which all matter naturally exists on Earth—most commonly as a solid, a liquid, or a gas.

**stereoisomers (p. 766)** A class of isomers whose atoms are bonded in the same order but are arranged differently in space.

**steroids (p. 839)** Lipids that have multiple cyclic rings in their structures.

**stoichiometry (p. 368)** The study of quantitative relationships between the amounts of reactants used and products formed by a chemical reaction; is based on the law of conservation of mass.

**strong acid (p. 644)** An acid that ionizes completely in aqueous solution.

**strong base (p. 648)** A base that dissociates entirely into metal ions and hydroxide ions in aqueous solution.

**strong nuclear force (p. 865)** A force that acts on subatomic particles that are extremely close together.

**structural formula (p. 253)** A molecular model that uses symbols and bonds to show relative positions of atoms; can be predicted for many molecules by drawing the Lewis structure.

**structural isomers (p. 765)** A class of isomers whose atoms are bonded in different orders with the result that they have different chemical and physical properties despite having the same formula.

**sublimation (p. 83)** The energy-requiring process by which a solid changes directly to a gas without first becoming a liquid.

**substance (p. 5)** Matter that has a definite composition; also known as a chemical.

**substituent groups (p. 753)** The side branches that extend from the parent chain; they appear to substitute for a hydrogen atom in the straight chain.

**substitution reaction (p. 790)** A reaction of organic compounds in which one atom or group of atoms in a molecule is replaced by another atom or group of atoms.

**calor específico (pág. 519)** Cantidad de calor requerida para elevar la temperatura de un gramo de una sustancia dada en un grado centígrado (Celsius).

**constante de velocidad de la reacción (pág. 574)** Valor numérico que relaciona la velocidad de la reacción y la concentración de reactivos a una temperatura específica.

**ion espectador (pág. 301)** Ion que no participa en una reacción.

**proceso espontáneo (pág. 542)** Cambio físico o químico que ocurre sin intervención externa; la iniciación del proceso puede requerir un suministro de energía.

**entalpía (calor) estándar de formación (pág. 537)** Cambio en la entalpía que acompaña la formación de un mol de un compuesto en su estado normal, a partir de sus elementos constituyentes en su estado normal.

**electrodo normal de hidrógeno (pág. 711)** Electrodo estándar que sirve de referencia para medir el potencial de reducción de todos los electrodos.

**estados de la materia (pág. 71)** Las formas físicas en las que la materia existe naturalmente en la Tierra, más comúnmente como sólido, líquido o gas.

**estereoisómeros (pág. 766)** Clase de isómeros cuyos átomos se unen en el mismo orden, pero con distinta disposición espacial.

**esteroides (pág. 839)** Lípidos con múltiples anillos en sus estructuras.

**estequiometría (pág. 368)** El estudio de las relaciones cuantitativas entre las cantidades de reactivos utilizados y los productos formados durante una reacción química; se basa en la ley de la conservación de la masa.

**ácido fuerte (pág. 644)** Ácido que se ioniza completamente en solución acuosa.

**base fuerte (pág. 648)** Base que se disocia enteramente en iones metálicos e iones hidróxido en solución acuosa.

**fuerza nuclear fuerte (pág. 865)** Fuerza que actúa sólo en las partículas subatómicas que se encuentran extremadamente cercanas.

**fórmula estructural (pág. 253)** Modelo molecular que usa símbolos y enlaces para mostrar las posiciones relativas de los átomos; esta fórmula se puede predecir para muchas moléculas al trazar su estructura de Lewis.

**isómeros estructurales (pág. 765)** Clase de isómeros cuyos átomos están unidos en distinto orden, por lo que tienen propiedades químicas y físicas diferentes a pesar de tener la misma fórmula.

**sublimación (pág. 83)** Proceso que requiere de energía en el que un sólido se convierte directamente en gas, sin convertirse primero en un líquido.

**sustancia (pág. 5)** Materia con una composición definida; también se conoce como sustancia química.

**grupos sustituyentes (pág. 753)** Las ramas laterales que se extienden desde la cadena principal y parecen sustituir un átomo de hidrógeno de la cadena recta.

**reacción de sustitución (pág. 790)** Reacción de compuestos orgánicos en la cual un átomo o un grupo de átomos en una molécula son sustituidos por otro átomo o grupo de átomos.

**substrate (p. 830)** A reactant in an enzyme-catalyzed reaction that binds to specific sites on enzyme molecules.

**supersaturated solution (p. 494)** Contains more dissolved solute than a saturated solution at the same temperature.

**surface tension (p. 418)** The energy required to increase the surface area of a liquid by a given amount; results from an uneven distribution of attractive forces.

**surfactant (p. 419)** A compound, such as soap, that lowers the surface tension of water by disrupting hydrogen bonds between water molecules; also called a surface active agent.

**surroundings (p. 526)** In thermochemistry, includes everything in the universe except the system.

**suspension (p. 476)** A type of heterogeneous mixture whose particles settle out over time and can be separated from the mixture by filtration.

**synthesis reaction (p. 289)** A chemical reaction in which two or more substances react to yield a single product.

**system (p. 526)** In thermochemistry, the specific part of the universe containing the reaction or process being studied.

**sustrato (pág. 830)** Reactivo en una reacción catalizada por enzimas que se enlaza a sitios específicos en las moléculas de la enzima.

**solución sobresaturada (pág. 494)** Aquella que contiene más soluto disuelto que una solución saturada a la misma temperatura.

**tensión superficial (pág. 418)** Energía requerida para aumentar el área superficial de un líquido en una cantidad dada; es producida por una distribución desigual de las fuerzas de atracción.

**surfactante (pág. 419)** Compuesto, como el jabón, que reduce la tensión superficial del agua al romper los enlaces de hidrógeno entre las moléculas de agua; llamado también agente tensioactivo.

**alrededores (pág. 526)** En termoquímica, incluye todo el universo a excepción del sistema.

**suspensión (pág. 476)** Tipo de mezcla heterogénea cuyas partículas se asientan con el tiempo y pueden separarse de la mezcla por filtración.

**reacción de síntesis (pág. 289)** Reacción química en la que dos o más sustancias reaccionan para generar un solo producto.

**sistema (pág. 526)** En termoquímica, se refiere a la parte específica del universo que contiene la reacción o el proceso en estudio.

## T

**technology (p. 9)** The practical use of scientific information.

**temperature (p. 403)** A measure of the average kinetic energy of the particles in a sample of matter.

**theoretical yield (p. 385)** In a chemical reaction, the maximum amount of product that can be produced from a given amount of reactant.

**theory (p. 16)** An explanation supported by many experiments; is still subject to new experimental data, can be modified, and is considered valid if it can be used to make predictions that are proven true.

**thermochemical equation (p. 529)** A balanced chemical equation that includes the physical states of all the reactants and the energy change, usually expressed as the change in enthalpy.

**thermochemistry (p. 525)** The study of heat changes that accompany chemical reactions and phase changes.

**thermonuclear reaction (p. 883)** A nuclear fusion reaction.

**thermoplastic (p. 813)** A type of polymer that can be melted and molded repeatedly into shapes that are retained when it is cooled.

**thermosetting (p. 813)** A type of polymer that can be molded when it is first prepared but when cool cannot be remelted.

**titrant (p. 661)** A solution of known concentration used to titrate a solution of unknown concentration; also called the standard solution.

**titration (p. 660)** The process in which an acid-base neutralization reaction is used to determine the concentration of a solution of unknown concentration.

**tecnología (pág. 9)** Uso práctico de la información científica.

**temperatura (pág. 403)** Medida de la energía cinética promedio de las partículas en una muestra de materia.

**rendimiento teórico (pág. 385)** La cantidad máxima de producto que se puede producir a partir de una cantidad dada de reactivo, durante una reacción química.

**teoría (pág. 16)** Explicación respaldada por muchos experimentos; está sujeta a los resultados obtenidos en nuevos experimentos, se puede modificar y se considera válida si permite hacer predicciones verdaderas.

**ecuación termoquímica (pág. 529)** Ecuación química equilibrada que incluye el estado físico de todos los reactivos y el cambio de energía, este último usualmente expresado como el cambio en entalpía.

**termoquímica (pág. 525)** El estudio de los cambios de calor que acompañan a las reacciones químicas y a los cambios de fase.

**reacción termonuclear (pág. 883)** Reacción de fusión nuclear.

**termoplástico (pág. 813)** Tipo de polímero que se puede fundir y moldear repetidas veces en formas que el plástico mantiene al enfriarse.

**fraguado (pág. 813)** Tipo de polímero que se puede moldear la primera vez que es producido, pero que no puede fundirse de nuevo una vez que se ha enfriado.

**solución tituladora (pág. 661)** Solución de concentración conocida que se usa para titular una solución de concentración desconocida; también conocida como solución estándar.

**titulación (pág. 660)** Proceso en el que se usa una reacción de neutralización ácido-base para determinar la concentración de una solución de concentración desconocida.

**transition elements (p. 177)** Elements in groups 3–12 of the modern periodic table and are further divided into transition metals and inner transition metals.

**transition metal (p. 180)** An element in groups 3–12 that is contained in the d-block of the periodic table and, with some exceptions, is characterized by a filled outermost s orbital of energy level *n*, and filled or partially filled d orbitals of energy level $n - 1$.

**transition state (p. 564)** Term used to describe an activated complex because the activated complex is as likely to form reactants as it is to form products.

**transmutation (p. 865)** The conversion of an atom of one element to an atom of another element.

**transuranium element (p. 876)** An element with an atomic number of 93 or greater in the periodic table.

**triglyceride (p. 836)** Forms when three fatty acids are bonded to a glycerol backbone through ester bonds; can be either solid or liquid at room temperature.

**triple point (p. 429)** The point on a phase diagram representing the temperature and pressure at which the three phases of a substance (solid, liquid, and gas) can coexist.

**Tyndall effect (TIHN duhl • ee FEKT) (p. 478)** The scattering of light by colloidal particles.

**elementos de transición (pág. 177)** Elementos de los grupos 3 al 12 de la tabla periódica moderna; se subdividen en metales de transición y metales de transición interna.

**metal de transición (pág. 180)** Elementos de los grupos 3 al 12 del bloque d de la tabla periódica; con algunas excepciones, se caracterizan por tener lleno el orbital externo s del nivel de energía *n* y por tener orbitales d llenos o parcialmente llenos en el nivel de energía $n - 1$.

**estado de transición (pág. 564)** Término que se usa para describir un complejo activado por su probabilidad de formar tanto reactivos como productos.

**transmutación (pág. 865)** Conversión de un átomo de un elemento a un átomo de otro elemento.

**elemento transuránico (pág. 876)** Elementos de la tabla periódica con un número atómico igual o mayor que 93.

**triglicérido (pág. 836)** Se forma cuando tres ácidos grasos se enlazan a un cadena principal de glicerol por enlaces éster; puede ser sólido o líquido a temperatura ambiente.

**punto triple (pág. 429)** El punto en un diagrama de fase que representa la temperatura y la presión en la que coexisten las tres fases de una sustancia (sólido, líquido y gas).

**efecto Tyndall (pág. 478)** Dispersión de la luz causada por las partículas coloidales.

## U

**unit cell (p. 421)** The smallest arrangement of atoms in a crystal lattice that has the same symmetry as the whole crystal; a small representative part of a larger whole.

**universe (p. 526)** In thermochemistry, is the system plus the surroundings.

**unsaturated hydrocarbon (p. 746)** A hydrocarbon that contains at least one double or triple bond between carbon atoms.

**unsaturated solution (p. 493)** Contains less dissolved solute for a given temperature and pressure than a saturated solution; has further capacity to hold more solute.

**celda unitaria (pág. 421)** El conjunto más pequeño de átomos en una red cristalina que posee la simetría de todo el cristal; pequeña parte representativa de un entero mayor.

**universo (pág. 526)** En termoquímica, se refiere el sistema más los alrededores.

**hidrocarburo no saturado (pág. 746)** Hidrocarburo que contiene por lo menos un enlace doble o triple entre sus átomos de carbono.

**solución no saturada (pág. 493)** Aquella que contiene menos soluto disuelto a una temperatura y presión dadas que una solución saturada; puede contener cantidades adicionales del soluto.



## V

**valence electrons (p. 161)** The electrons in an atom's outermost orbitals; determine the chemical properties of an element.

**vapor (p. 72)** Gaseous state of a substance that is a liquid or a solid at room temperature.

**vaporization (p. 426)** The energy-requiring process by which a liquid changes to a gas or vapor.

**vapor pressure (p. 427)** The pressure exerted by a vapor over a liquid.

**vapor pressure lowering (p. 499)** The lowering of vapor pressure of a solvent by the addition of a nonvolatile solute to the solvent.

**viscosity (p. 417)** A measure of the resistance of a liquid to flow, which is affected by the size and shape of particles, and generally increases as the temperature decreases and as intermolecular forces increase.

**electrones de valencia (pág. 161)** Los electrones en el orbital más externo de un átomo; determinan las propiedades químicas de un elemento.

**vapor (pág. 72)** Estado gaseoso de una sustancia que es líquida o sólida a temperatura ambiente.

**vaporización (pág. 426)** Proceso que requiere energía en el que un líquido se convierte en gas o vapor.

**presión de vapor (pág. 427)** Presión que ejerce un vapor sobre un líquido.

**disminución de la presión de vapor (pág. 499)** Reducción de la presión de vapor de un disolvente por la adición de un soluto no volátil al disolvente.

**viscosidad (pág. 417)** Medida de la resistencia de un líquido a fluir; es afectada por el tamaño y la forma de las partículas y en general aumenta cuando disminuye temperatura y cuando aumentan las fuerzas intermoleculares.

**voltaic cell (p. 709)** A type of electrochemical cell that converts chemical energy into electrical energy by a spontaneous redox reaction.

**VSEPR model (p. 261)** Valence Shell Electron Pair Repulsion model, which is based on an arrangement that minimizes the repulsion of shared and unshared pairs of electrons around the central atom.

**pila voltaica (pág. 709)** Tipo de celda electroquímica que convierte la energía química en energía eléctrica mediante una reacción redox espontánea.

**modelo RPCEV (pág. 261)** Modelo de Repulsión de los Pares Electrónicos de la Capa de Valencia; se basa en un ordenamiento que minimiza la repulsión de los pares de electrones compartidos y no compartidos alrededor del átomo central.

## W

**wavelength (p. 137)** The shortest distance between equivalent points on a continuous wave; is usually expressed in meters, centimeters, or nanometers.

**wax (p. 838)** A type of lipid that is formed by combining a fatty acid with a long-chain alcohol; is made by both plants and animals.

**weak acid (p. 645)** An acid that ionizes only partially in dilute aqueous solution.

**weak base (p. 648)** A base that ionizes only partially in dilute aqueous solution to form the conjugate acid of the base and hydroxide ion.

**weight (p. 9)** A measure of an amount of matter and also the effect of Earth's gravitational pull on that matter.

**longitud de onda (pág. 137)** La distancia más corta entre puntos equivalentes en una onda continua; se expresa generalmente en metros, centímetros o nanómetros.

**cera (pág. 838)** Tipo de lípido que se forma al combinarse un ácido graso con un alcohol de cadena larga; son elaborados por plantas y animales.

**ácido débil (pág. 645)** Ácido que se ioniza parcialmente en una solución acuosa diluida.

**base débil (pág. 648)** Base que se ioniza parcialmente en una solución acuosa diluida para formar el ácido conjugado de la base y el ion hidróxido.

**peso (pág. 9)** Medida de la cantidad de materia y también del efecto de la fuerza gravitatoria de la Tierra sobre esa materia.

## X

**X-ray (p. 864)** A form of high-energy, penetrating electromagnetic radiation emitted from some materials that are in an excited electron state.

**rayos X (pág. 864)** Forma de radiación electromagnética penetrante de alta energía que emiten algunos materiales que se encuentran en un estado electrónico excitado.

# INDEX



*Italic numbers* = *illustration/photo* **Bold numbers = vocabulary term**
*act.* = activity *prob.* = problem

## A


Absolute zero, **445**
Absorption spectrum, 145, 164 *act.*
Accelerants, 91
Accuracy, **47**–48
Acetaldehyde, 796
Acetaminophen, 800
Acetic acid, 634, 798, 800
Acetone, 432 *act.*, 797
Acetylene. *See* Ethyne
Acid anhydrides, 643
Acid-base chemistry, 633 *act.*, 634–668; acid-base titration, 660–663, 664 *prob.*, 670 *act.*; acids, strength of, 644–647, 648 *act.*; Arrhenius model, 637, 642 *table*; bases, strength of, 648–649; Brønsted-Lowry model, 638–640, 642 *table*; buffers, 666–667, 668 *act.*; chemical properties of acids and bases, 635; hydronium and hydroxide ions, 636; ion-product of water and, 650 *prob.*, 650–651; Lewis model, 641–643, 642 *table*; litmus paper and, 633 *act.*, 635, 658; milestones in understanding, *636–637*; molarity and pH, 656; monoprotic and polyprotic acids, 640–641, 641 *table*; neutralization reactions, 659–660; pH and, 633 *act.*, 652, 653, 653 *prob.*, 654 *prob.*; physical properties of acids and bases, 634–635; pOH and, 652, 653; salt hydrolysis, 665
Acid-base indicators, 658, **663**, 664
Acid-base titration. *See* Titration
Acid hydrolysis, 665
Acidic solutions, **636**
Acid ionization constant ($K_a$), **647**, 647 *table*, 970 *table*; calculate from pH, 656, 657 *prob.*
Acid mine waste, biotreatment of, 920
Acidosis, 666
Acid rain, *637*
Acids. *See also* Acid-base chemistry; acid ionization constant ($K_a$), 647, 647 *table*, 656, 657 *prob.*; anhydrides, 643; Arrhenius, 637; Brønsted-Lowry, 638–639, 646; chemical properties, 635; conjugate, 638; electrical conductivity, 635; in household items, 633 *act.*; ionization equations, 645, 645 *table*; molarity and pH of strong, 656; monoprotic, 640, 640 *table*; naming, 250–251, 252; pH of. *See* pH; physical properties, 634–635; polyprotic, 640–641, 641 *table*; strength of, 644–647, 648 *act.*; strong, 644; titration of. *See* Titration; weak, 645
Actinide series, **180**, *185*, 921
Activated complex, **564**
Activation energy ($E_a$), **564**–566, 571–572
Active site, **830**
Activities. *See* CHEMLABs; Data Analysis Labs; Launch Labs; MiniLabs; Problem-Solving Labs
Activity series, 293–294, 310 *act.*
Actual yield, **385**
Addition: scientific notation and, 42, 948; significant figures and, 53, 53 *prob.*, 952, 953 *prob.*
Addition polymerization, **811**
Addition reactions, 804 *table*, **804**–805
Adenine (A), 841
Adenosine diphosphate (ADP), 845
Adenosine triphosphate (ATP), 532, **845**
Adhesion, 419
Adipic acid, 798
ADP (adenosine diphosphate), 845
Age of Polymers. *See* Polymers
Agitation, 492
AIDS, 389
Air masses, density of and weather, 37
Air pressure, 406; deep sea diving and, 408 *act.*; measurement of, 406–407; units of, 407
Alcoholic fermentation, 847
Alcohols, **792**–793; denatured, 793; elimination reactions, 803; evaporation of, 432 *act.*, 816 *act.*; functional groups, 787 *table*; layering of in graduated cylinder, 31 *act.*; naming, 793; properties, 792–793, 816 *act.*
Aldehydes, 787 *table*, 796 *table*, **796**–797
Algal blooms, 250
Algebraic equations, 954–955, 955 *prob.*
Aliphatic compounds, **771**. *See also* Alkanes; Alkenes; Alkynes
Alkali metals (Group 1A), **177**, 906–909
Alkaline batteries, 719
Alkaline earth metals (Group 2A), **177**, 910–915
Alkanes, **750**–758; alkyl halides and, 789; branched-chain, 752–753, 754–755 *prob.*; burner gas analysis, 776 *act.*; chemical properties, 758; condensed structural formulas, 751; cycloalkanes, 755–756, 756–757 *prob.*; hydrogenation reactions, 805; naming, 751, 752–753, 754–755 *prob.*; nonpolarity of, 757, 758; physical properties, 758; solubility, 758; straight-chain, 750–751
Alkenes, **759**; addition reactions involving, 804; naming, 760, 761 *prob.*; properties, 762; stereoisomers, 766; uses, 762
Alkyl groups, 752, 753 *table*
Alkyl halides, **787**; dehydrogenation reactions, 803; naming, 788; parent alkanes v., 789 *table*; substitution reactions, 791
Alkynes, **763**–764; ethyne, synthesize and observe, 762 *act.*; examples, 763 *table*; hydrogenation reactions, 805; naming, 763; properties, 764; uses, 764
Allotropes, 423, 928
Alloys, 81, **227**–228; commercially important, 228 *table*; interstitial, 228; magnesium, 913; substitutional, 228; transition metal, 916
Alnico, 228 *table*
Alpha decay, 867, 868 *table*
Alpha particles, **123**, 861 *table*, 862, 864, 888 *table*
Alpha radiation, **123**, 124 *table*, 861, 861 *table*, 862, 868 *table*
Alternative energy specialist, 729
Aluminum, 159 *table*, 226 *table*, 730–731, 922, 923, 924
Aluminum oxide, 212
Amide functional group, 787 *table*
Amides, 787 *table*, **800**, 800 *table*
Amines, 787 *table*, **795**, 795 *table*
Amino acids, **826**–827, 827 *table*
Amino functional group, 787 *table*, 826
Ammonia: as Brønsted-Lowry base, 639; evaporation of, 432 *act.*; Lewis structure, *243*, 255 *prob.*; polarity of, 268; production of, *290*, 462, 594, 596, 597; sigma bonds in, 244, *245*
Ammoniated cattle feed, 601
Ammonium, 221 *table*
Amorphous solids, **424**
Amphoteric, **639**
Amplitude, **137**
Anabolism, **844**–845
Analytical balance, 77, 79
Analytical chemistry, 11 *table*, 79, 341
Anhydrides, 643
Anhydrous calcium chloride, 354


Index

Aniline, 795
Anions, **209**
Anodes, 107, **710**
Antacids, 659
Antarctica, ozone hole over, 7, 20–21
Anthracene, 772
Antilogarithms, 966–967
Antimony, 932, 933, 935
Applied research, **17**
Aqueous solutions, **299**–308. *See also* Solutions; electrolytes in and colligative properties, 498–499; ionic compounds in, 300; ionic equations and, 301, 302 *prob.*; molecular compounds in, 299; nonelectrolytes in and colligative properties, 499; overall equations for reactions in, 307; reactions producing water in, 303, 304 *prob.*; reactions that form gases, 281 act., 304–305, 306 *prob.*; reactions that form precipitates in, 300, 301 *act.*, 302 *prob.*; solvation of ionic compounds in, 490; solvation of molecular compounds in, 491
Aragonite, *214*
Archaeologist, 891
Argon, 159 *table*, 185 *table*, 944, 945
Aristotle, 103, 103 *table*, 416
Aromatic compounds, **771**–774; benzene, 770–771; carcinogenic, 774; fused-ring systems, 772; naming, 772–773, 773 *prob.*
Arrhenius model of acid-base chemistry, **637**, 642 *table*
Arrhenius, Svante, *636*, 637
Arsenic, 932, 933
Arson investigator, 91
Art restorer, 23
Aryl halides, **788**
Aspirin, *810*
Astatine, 940, 941
Asymmetric carbon, **768**
Atmosphere (atm), **407**, 407 *table*
Atmosphere, Earth's: cycling of carbon dioxide in, 505; elements in, 901; layers of, 5; ozone layer and, 5–8
Atomic bomb, 879
Atomic distances, 113 *act.*
Atomic emission spectrum, **144**–145, 164 *act.*
Atomic Force Microscope, *291*
Atomic mass, **119**–120, 121 *prob.*, 126 *act.*
Atomic mass unit (amu), **119**, 325, 969 *table*
Atomic nucleus, **112**; discovery of, 112; nuclear model of mass and, 326 *act.*
Atomic number, **115**, 116 *prob.*, 118 *prob.*
Atomic orbitals, **152**, 154, 262
Atomic radii, trends in, 187, 188, 189 *prob.*
Atomic solids, 422, 422 *table*
Atomic structure: Bohr model of, 146–148, 150 *act.*; Dalton's model of, 104 *table*, 104–105; Democritus' early idea of, 103; Greek philosophers' views of, 102–103, 103 *table*; milestones in understanding, *110–111*; nuclear atomic model, 112–114, 136; plum pudding model, 110; quantum mechanical model, 149–152; try to determine, 135 *act.*
Atomic weapons, *111*
Atoms, 10, **106**–107; atom-to-mass conversions, 331 *prob.*; determining structure of. *See* Atomic structure; excited state, 146, 147; ground state, 146; mass-to-atom conversions, 329–330, 330 *prob.*; size of, 106, 112; stability of, 240; subatomic particles, 113–114, 114 *table*; viewing, 107
ATP (adenosine triphosphate), **845**
Aufbau diagram, 156–157, 157 *table*, 160
Aufbau principle, **156**, 157 *table*
Automobile air safety bags, 292
Average rate of reaction, 560–562, 562 *prob.*
Avogadro's number, **321**, 326 *act.*, 969 *table*
Avogadro's principle, **452**

## B

Bacteria, nitrogen-fixing, 934
Bakelite, 809, *810*, 813
Baker, 847
Baking, acid-base chemistry and, 669
Baking powder, 669
Baking soda, 378 *act.*, 669
Balanced chemical equations: conservation of mass and, 285, 288; deriving, 285–286, 286 *table*, 287 *prob.*, 288, *288*. *See also* Stoichiometry; mole ratios and, 371–372; particle and mole relationships in, 368–369; relationships derived from, 369 *table*
Balanced forces, 597
Ball-and-stick molecular models, *253*, 746
Balmer (visible) series, *147*, 148, 150 *act.*
Band of stability, **866**
Bar graphs, 56
Barite, *214*
Barium, 226 *table*, 910–911, 913, 914
Barium carbonate, 302, 302 *prob.*
Barium chloride, 913
Barium sulfate, 621
Barometers, **407**, *416*
Base hydrolysis, 665
Base ionization constant ($K_b$), 649, 649 *table*, 970 *table*
Bases. *See also* Acid-base chemistry; antacids, 659; Arrhenius, 637; base ionization constant ($K_b$), 649, 649 *table*; Brønsted-Lowry, 638–639; chemical properties, 635; conjugate, 638; in household items, 633 *act.*; molarity and pH of strong, 656; physical properties, 634–635; strength of, 648–649; strong, 649; titration of. *See* Titration; weak, 649
Base units, **33**, 35–37
Basic solutions, **636**
Batteries, **717**, 718–723; dry cells, 718–720; fuel cells, 722–723; lead-acid, 720–721, 930; lemon battery, 707 *act.*; lithium, 721–722
Becquerel, Henri, 860–861, 885
Beetles, bioluminescent, 309
Bent molecular shape, 263 *table*
Benzaldehyde, 796 *table*, 797
Benzene, 770–771; carcinogenic nature of, 774; naming of substituted, 772–773
Benzopyrene, 774
Bernoulli, Daniel, *416*
Beryl, *214*
Beryllium, 158 *table*, 161 *table*, 910–911, 912
Beryls, 912
Best-fit line, 56–57
Beta decay, 867, 868 *table*
Beta particles, **123**, 861 *table*, 863, 864, 888 *table*
Beta radiation, **123**, 124 *table*, 861, 861 *table*, 862, 863, 868 *table*
Binary acids, 250, 252
Binary ionic compounds, 210, 219
Binary molecular compounds, 248–250, 249 *prob.*, 252
Binding energy, 877, 878
Biochemist, 308
Biochemistry, 11 *table*
Biofuel cells, 724 *act.*
Biofuels, 774 *act.*, 775
Biogas, 775
Biological metabolism. *See* Metabolism
Bioluminescence, 309, 693
Biomolecules: carbohydrates, 825 *act.*, 832–834; lipids, 835–839; nucleic acids, 840–843; proteins, 826–831
Bioremediation, 920
Bismuth, 932, 933, 935
Bismuth subsalicylate, 935

Blocks, periodic table, 183–185. *See also* Specific blocks
Blood, pH of, 666, 668 *act.*
Bloodstains, detecting, 697
Body temperature, reaction rate and, 583
Bohr atomic model, 146–148, 150 *act.*
Bohr, Niels, *110*, 146
Boiling, **427**
Boiling point, 77, **427**; of alkanes, 758; of covalent compounds, 270; of halocarbons, 789; as physical property, 73
Boiling point elevation, **500**–501, 503 *prob.*
Boltzmann, Ludwig, 402
Bond angles, 261
Bond character, 266
Bond dissociation energies, 247
Bonding pairs, 242
Bonds. *See* Chemical bonds
Book preservation, 661
Borates, 214
Boron, 158 *table*, 161 *table*, 184, 922, 923, 924
Boron group (Group 13), 922–925
Bose-Einstein condensate, *417*
Bose, Satyendra Nath, *417*
Boyle, Robert, 442
Boyle's law, **442**–443, 443 *prob.*, 444 *act.*, 451 *table*
Branched-chain alkanes, 752–753; alkyl groups, 752; naming, 752–753, 754–755 *prob.*, 760, 761 *prob.*
Brass, 228 *table*
Breathing, Boyle's law and, 444 *act.*
Breeder reactors, **882**
Brine, electrolysis of, 730
Bromate, 221 *table*
Bromine, 120, 180, 940, 941, 942
Brønsted, Johannes, 638
Brønsted-Lowry model, **638**–640, 642 *table*, 646
Bronze, 228 *table*
Brownian motion, **477**
Brown, Robert, 477
Buckminsterfullerene, 928
Buckyballs, 928
Buffer capacity, **667**
Buffers, **666**–667, 668 *act.*
Buffer systems, 666–667, 668 *act.*
Bufotoxin, 839
Burner gas analysis, 776 *act.*
Butane, 750, 751, 751 *table*
1-Butene, 759 *table*
2-Butene, 759 *table*
Butyl group, 753 *table*

## C

Cadaverine, 795
Cadmium, 920
Calcium, 177, 195, 910–911, 913, 914
Calcium chloride, 913
Calcium hydroxide, 287
Calibration technician, 56
Calorie (cal), **518**
Calorimeter, **523**–524, 525 *prob.*, 532 *prob.*
Calx of mercury, 79
Cancer, 163, 887
Canola oil, hydrogenation of, 805 *act.*
Capillary action, 419
Caramide, 800
Carbohydrates, **832**–834; disaccharides, 833; functions of, 832; monosaccharides, 832–833; polysaccharides, 833–834; test for simple sugars, 825 *act.*
Carbolic acid, *636*
Carbon. *See also* Organic compounds; abundance of, 84; analytical tests for, 926–927; atomic properties, 158 *table*, 161 *table*, 926–927; common reactions involving, 926–927; in human body, 195; organic compounds and, 745; physical properties, 926; uses of, 928
Carbonated beverages, 495
Carbonates, 214
Carbon dating, 873–874, *883*
Carbon dioxide, 256 *prob.*, 430, 505
Carbon group (Group 4A), 926–931, 932–935
Carbonic acid, 634
Carbon tetrachloride, 20, 267–268
Carbonyl compounds, 796–801; aldehydes, 796–797; carboxylic acids, 798; ketones, 797
Carbonyl group, 787 *table*, **796**
Carboxyl group, 787 *table*, **798**, 798 *table*, 826
Carboxylic acids, **798**, 798 *table*; condensation reactions, 801; functional groups, 787 *table*; naming, 798; organic compounds derived from, 799–800, 800 *act.*; properties, 798
Carcinogens, 774
Cardiac scans, 925
Careers. *See* Careers in Chemistry; In the Field
Careers in Chemistry: alternative energy specialist, 729; baker, 847; biochemist, 308; calibration technician, 56; chemical engineer, 580; chemistry teacher, 123; environmental chemist, 7; flavor chemist, 267; food scientist, 219; heating and cooling specialist, 527; materials scientist, 81; medicinal chemist, 342; metallurgist, 423; meteorologist, 447; nursery worker, 646; petroleum technician, 748; pharmacist, 381; pharmacy technician, 483; polymer chemist, 813; potter, 682; radiation therapist, 887; research chemist, 185; science writer, 604; spectroscopist, 139
Cast iron, 228 *table*
Catabolism, **844**–845
Catalysts, **571**–573. *See also* Enzymes; chemical equilibrium and, 611; hydrogenation reactions and, 805; temperature and, 850 *act.*
Catalytic converters, 573
Cathode rays, **108**
Cathode-ray tubes, 107–108
Cathodes, 107, **710**
Cations, **207**–208
Cattle feed, 601
Cave formation, 643
CDs, 924
Cell membrane, 838
Cell notation, 713
Cell potential: applications of, 716; calculate, 713–714, 715 *prob.*, 717; measure, 734 *act.*
Cellular respiration, **846**
Celluloid, *490*
Cellulose, 834
Celsius scale, 34
Centrifuge, *490*
CERN, *111*
Cesium, 194, 906, 907, 909
Cesium clock, 909
CFCs. *See* Chlorofluorocarbons (CFCs)
Chadwick, James, *110*, 113
Chain reactions, 859 *act.*, 879, 880
Chance, scientific discoveries and, 18
Charles, Jacques, 444
Charles's law, 441 *act.*, 444–**445**, 446 *prob.*, 451 *table*
Chelation therapy, 229
Chemical bonds, **206**; character of, 266; covalent. *See* Covalent bonds; electron affinity and, 265; ionic. *See* Ionic bonds; melting point and, 242 *act.*; metallic. *See* Metallic bonds; valence electrons and, 207
Chemical changes, 69 *act.*, **77**, 92 *act.*, 281 *act. See also* Chemical reactions
Chemical engineer, 580
Chemical equations, **285**. *See also* Ionic equations; Nuclear equations; Redox equations; Stoichiometry; Thermochemical equations;

balancing, 285–286, 286 *table*, 287 *prob.*, 288; coefficients in, 369; interpretation, 370 *prob.*; mole ratios and, 371–372; products, 283; reactants, 283; relationships derived from, 369; symbols used in, 283, 283 *table*
Chemical equilibrium, **596**; addition of products and, 608; addition of reactants and, 607; catalysts and, 611; changes affecting, 593 *act.*; characteristics of, 604; common ion effect and, 620–621; concentration and, 607; determine point of, 593 *act.*; dynamic nature of, 597–598; equilibrium constant ($K_{eq}$), 599–600, 604, 605 *prob.*; equilibrium expressions, 600, 601 *prob.*, 602, 603 *prob.*; hemoglobin-oxygen equilibrium in body, 623; law of, 599–600; Le Châtelier's principle and, 606–611; moles of reactant v. moles of product and, 609; removal of products and, 608; reversible reactions and, 595–596; temperature and, 609–610, 611 *act.*; volume and pressure and, 608–609
Chemical formulas, 85; for binary ionic compounds, 219, 220 *prob.*; empirical. *See* Empirical formula; for hydrates, 351 *table*, 352, 353 *prob.*, 356 *act.*; for ionic compounds, 218–219, 220 *prob.*, 221, 222 *prob.*; molecular. *See* Molecular formulas; mole relationship to, 333–334, 334–335 *prob.*; for monatomic ions, 218–219; name of molecular compound from, 251; percent composition from, 342, 343 *prob.*; for polyatomic ionic compounds, 221, 222 *prob.*; structural. *See* Structural formulas
Chemical potential energy, **517**
Chemical properties, **74**
Chemical reaction rates. *See* Reaction rates
Chemical reactions, 77, **282**–288; actual yield from, 385; addition, 804–805; in aqueous solutions, 299–301, 302 *prob.*, 303–305, 306 *prob.*, 307–308; classification of, 291 *prob.*; combustion, 290–291, 532 *prob.*, 533; condensation, 801; conservation of mass and, 77, 78 *prob.*, 79, 285, 288; decomposition, 292, 292 *prob.*; dehydration, 803; dehydrogenation, 803; elimination, 802; endothermic, 216, 247; equations for, 283 *table*, 283–285; evidence of, 69 *act.*, 77, 282–283, 367 *act.*; excess reactants in, 379, 384; exothermic, 216, 247; heat from. *See* Thermochemistry; limiting reactants, 379–381, 382–383 *prob.*; milestones in understanding, *290–291*; neutralization, 659–660; nuclear reactions v., 860 *table*; organic. *See* Organic reactions; oxidation reduction reactions, 806–807; percent yield from, 386, 386 *prob.*, 388; products of, identify, 92 *act.*; products of, predict, 298, 298 *table*, 807–808; rates of. *See* Reaction rates; redox. *See* Redox reactions; replacement, 293–294, 295 *prob.*, 296–297; spontaneity of, 542–545, 546–547, 548 *prob.*, 566–567; stoichiometry in. *See* Stoichiometry; substitution, 790–791; synthesis, 289; theoretical yield from, 385
Chemical symbols, 84
Chemistry, **4**, 11; benefits of studying, 22; branches of, 11, 11 *table*; symbols and abbreviations used in, 968 *table*
Chemistry & Health: elements of the body, 195; evolution and HIV, 389; hemoglobin-oxygen equilibrium, 623; hyperbaric oxygen therapy, 465; laser scissors, 163; PA-457 anti-HIV drug, 389; rate of reaction and body temperature, 583; toxicology, 59
Chemistry teacher, 123
CHEMLABs, *228*. *See also* Data Analysis Labs; Launch Labs; MiniLabs; absorption and emission spectra, 164 *act.*; alcohols, properties of, 816 *act.*; atomic mass of unknown element, 126 *act.*; burner gas analysis, 776 *act.*; calorimetry, 550 *act.*; density, dating coins by, 60 *act.*; descriptive chemistry, 196 *act.*; enzyme action and temperature, 850 *act.*; evaporation, compare rates of, 432 *act.*; gas, identify an unknown, 466 *act.*; hydrate, determine formula for, 356 *act.*; hydrocarbon burner gas analysis, 776 *act.*; ionic compounds, formation of, 230 *act.*; metals, reactivity of, 310 *act.*; molar solubility, calculate and compare, 624 *act.*; molecular shape, 272 *act.*; mole ratios, determine, 390 *act.*; products of chemical reaction, identify, 92 *act.*; reaction rate, affect of concentration on, 584 *act.*; redox and the damaging dumper, 698 *act.*; solubility rate, factors affecting, 506 *act.*; vapor pressure and popcorn popping, 466 *act.*; voltaic cell potentials, measure, 734 *act.*; water analysis, 24 *act.*
Chernobyl, 880, *883*, 889 *act.*
Chewing gum, percent composition, 342 *act.*
Chimney soot, 774
Chirality, **767**, 768
Chlorate, 221 *table*
Chlorine, 89–90, 119–120, 159 *table*, 180, 940, 941, 942
Chlorine bleach, 942
Chlorite, 221 *table*
Chlorofluorocarbons (CFCs), 7–8, 17, 20, *291*, 788
Chloromethane, 787
Chlorophyll, 912
Chocolate, 431
Chromatograms, polarity and, 269 *act.*
Chromatography, 82 *act.*, **83**, 269 *act.*
Chrome, 328
Chromium, 160, 328, 918, 919
Cinnameldehyde, 796 *table*, 797
Circle graphs, 55
*cis-* isomers, 766
Clay, 476
Clay roofing tiles, 302
Clouds, 428
Cloud seeding, 495
Cobalt, 918, 919
Coefficients, **285**; balancing equations and, 285; scientific notation and, 40–41
Cohesion, 419
Cold-packs, 515 *act.*, 528
Collagen, 831
Colligative properties, **498**–504; boiling point elevation, 500–501; electrolytes and, 498–499; freezing point depression, 501–502, 502 *act.*, 503 *prob.*; osmotic pressure, 504; vapor pressure lowering, 499–500
Collision theory, **563**–564, 564 *table*
Colloids, **477**, 477 *table*, 478
Color: change in as evidence of chemical reaction, 283; as physical property, 73
Combined gas law, **449**, 450 *prob.*, 451 *table*, 454
Combustion engines, *290*
Combustion reactions, **290**–291, 532 *prob.*, 533
Common ion, **620**
Common ion effect, **620**–621
Complementary base pairs, 841, 842
Complete ionic equations, **301**, 302 *prob.*, 304 *prob.*
Complex carbohydrates. *See* Polysaccharides
Complex reactions, **580**
Compounds, **85**–87; compare melting points of, 242 *act.*; formulas for. *See* Formulas; ionic. *See* Ionic compounds; law of definite proportions and, 87–88; law of multiple proportions and, 89–90; mass-to-mole

conversions, 337, 337 *prob.*; molar mass of, 335, 335 *prob.*; mole-to-mass conversions, 336, 336 *prob.*; percent composition and. *See* Percent composition; properties of, 86; separating components of, 86; stability of, 240
Computer chips, 181, 929
Concentration, 475 *act.*, **480**–488. *See* Solution concentration; calculate from equilibrium constant expression, 612, 613 *prob.*; chemical equilibrium and, 607; qualitative descriptions of, 480; ratios of. *See* Concentration ratios; reaction rate and, 569, 574–576, 584 *act.*
Concentration ratios: molality, 480 *table*, 487, 487 *prob.*; molarity, 480 *table*, 482, 483 *prob.*; mole fraction, 480 *table*, 488; percent by mass, 480 *table*, 481, 481 *prob.*; percent by volume, 480 *table*, 482
Conclusions, **15**
Condensation, *76*, **428**
Condensation polymerization, **811**
Condensation reactions, **801**
Condensed structural formulas, 751
Conductivity: among types of elements 177–181; as physical property, 73; explanation of, 226; of ionic compounds in solution, 215, 498–499
Conjugate acid-base pair, **638**
Conjugate acids, **638**, 641 *table*
Conjugate bases, **638**, 641 *table*
Conservation of energy. *See* Law of conservation of energy
Conservation of mass. *See* Law of conservation of mass
Constant, 14
Controls, **14**
Conversion factors, **44**–46, 46 *prob.*, 319 *act.*
Coordinate covalent bonds, **259**
Copper: acid mine waste, 920; electron configuration, 160; in fireworks, 913; flame test for, 92 *act.*; law of multiple proportions and, 89–90; melting and boiling point, 226 *table*; in microchip wiring, 919; as paint pigment, 919; properties of, 74 *table*; purification of, 731–732
Core, iron in Earth's, 919
Corn oil, 31 *act.*
Corrosion, **724**–727, 726 *act.*
Counting units, 320
Covalent bonds, **241**–247; bond angle, 261, 263 *table*; coordinate, 259; double, 245; electron affinity and, 265; electronegativity and, 266; energy in, 247; formation of, 241; hybridization and, 262; length of, 246; nonpolar, 266; pi bonds and, 245; polar, 266, 267–268; sigma bonds and, 244, *245*; single, 242–244; strength of, 246–247; super ball properties, 239 *act.*; triple, 245
Covalent compounds: boiling points of, 270; formulas from names of, 251; intermolecular forces in, 269–270; Lewis structures for, 253–260, 255 *prob.*, 256 *prob.*, 257 *prob.*, 258 *prob.*, 260 *prob.*; melting points of, 242 *act.*, 270; naming, 248–251, 249 *prob.*, *252*; polarity of and chromatograms, 269 *act.*; properties of, 270; shape of (VSEPR model), 261–262, 263 *table*, 264 *prob.*
Covalent gases, 270
Covalent molecular solids, 270
Covalent network solids, 270, 422, 422 *table*, 423
Cracking, **748**
*CRC Handbook of Chemistry and Physics*, 75, 77
Crick, Francis, *637*, 841–842
Crime-scene investigator, 697
Critical mass, **880**
Critical point, 429
Crookes, Sir William, 108
Crude oil. *See* Petroleum
Crust, Earth's, 901
Cryosurgery, 934
Cryotherapy, 934
Crystal lattices, **214**, 270, 420–421, 422 *act.*
Crystalline solids, **420**–421, 422 *table*; categories, 422 *table*, 422–423; crystal unit cells, 421, 422 *act.*
Crystallization, **83**
Cube root, 949
Cubic unit cells, 421 *table*
Curie, Marie, 861, *882*, 915
Curie, Pierre, 861, *882*
Cyanide, 221 *table*
Cyclic hydrocarbons, **755**
Cycloalkanes, **755**–756, 756–757 *prob.*
Cyclohexane, 755
Cyclohexanol, 793
Cyclohyexylamine, 795
Cysteine, 827 *table*
Cytosine (C), 841

## D

Dalton, John, *417*
Dalton's atomic theory, 104 *table*, **104**–105, 109
Dalton's law of partial pressures, **408**, 409 *prob.*, 410
Data, 13
Data Analysis Labs. *See also* CHEMLABs; Launch Labs; MiniLabs; Problem-Solving Labs; antimicrobial properties of polymers, 216 *act.*; atomic distances in highly ordered pyrolytic graphite (HOPG), 113 *act.*; biofuel cells, 724 *act.*; gas pressure and deep sea diving, 408 *act.*; hydrogenation of canola oil, 805 *act.*; microbes, electric current from, 724 *act.*; oxidation rate of dichloroethene isomers, 768 *act.*; oxygen in moon rocks, 387 *act.*; ozone levels in Antarctica, 21 *act.*; polarity and chromatograms, 269 *act.*; redox reactions and space shuttle launch, 691 *act.*; turbidity and Tyndall effect, 478 *act.*
d-block elements, 185, 916
de Broglie equation, **150**
de Broglie, Louis, 149
Decane, 751 *table*
Decomposition reactions, **292**, 292 *prob.*, 566 *act.*
Deep sea diving, gas pressure and, 408 *act.*
Dehydration reactions, **803**
Delocalized electrons, **225**
Democritus, 103, 103 *table*, 416
Denaturation, **829**
Denatured alcohol, **793**
Density, **36**–37; calculate, 37; date coins by, 60 *act.*; of gases, 403, 456, 457 *act.*; identification of unknowns by, 37, 38 *prob.*, 39 *act.*; of liquids, 31 *act.*, 415; as physical property, 73; of solids, 420; units of, 36
Dental amalgams, 228 *table*
Deoxyribonucleic acid. *See* DNA (deoxyribonucleic acid)
Deoxyribose sugar, 841
Dependent variables, **14**, 56
Deposition, **429**
Derived units, **35**–36, 44
Desalination, 730
Descriptive chemistry, 196 *act.*
Dessicants, 354
Detergents, 13 *act.*, 419, 924
Deuterium, 904
Diamonds, 423, 928
Diatomic molecules, 241
Dichloroethene, 768 *act.*
Dietary salt, 908
Diffusion, **404**, 405
Dilute solutions, 485, 486 *prob.*
Dimensional analysis, **44**–46, 46 *prob.*, 956, 956 *prob.*
Dinitrogen pentoxide, 565 *act.*
Dipeptides, 828

Dipole-dipole forces, 269, 411, **412**–413
Direct relationships, 961
Disaccharides, **833**
Dispersion forces, 269, 411, **412**
Dispersion medium, 477 *table*
Dissociation energy, 247
Dissociation equations, strong bases, 648, 648 *table*
Distillation, **82**
Distilled water: electrical conductivity of, 205 *act.*; evaporation of, 432 *act.*
Diving, gas pressure and, 408 *act.*
Division operations, 54
DNA (deoxyribonucleic acid), 841–842, 842 *act.*, 843
Dobson, G. M. B., 6
Dobson units (DU), 6
d orbitals, 154
Dose of radiation, 889–890
Dose-response curve, 59
Double covalent bonds, 245, 246
Double helix, DNA, 841
Double-replacement reactions, **296**–297, 297 *prob.*, 297 *table*
Down's cells, 729
Drake, Edwin, 749
Dry cells, **718**–720; alkaline batteries, 719; primary batteries, 720; secondary batteries, 720; silver batteries, 719; zinc-carbon, 718–719
Dry ice, 428
Drywall, 914
Ductility, 226
DVDs, 924

## E

Earth: atmosphere of, 5, 901; elements in core of, 919; elements in crust of, 84, 901; elements in oceans of, 901; entropy and geologic changes on, 545
Effusion, 404–405, 405 *prob.*
Egyptian cubits, 46 *prob.*
Einstein, Albert, 143, *417*, 877
Elastic collisions, **403**
Electrical conductivity: of acids and bases, 635; of ionic compounds, 214–215; of metals, 180, 226; of strong acids, 645; of various compounds, 205 *act.*; of weak acids, 645, 648 *table*
Electric charge, observe, 101 *act.*
Electrochemical cell potentials, 711–717, 734 *act.*; calculate, 713–714, 715 *prob.*, 717; cell notation, 713; half-cell potentials, 712, 712 *table*; of standard hydrogen electrode, 711
Electrochemical cells, 707 *act.*, **709**, 709–711; alkaline batteries, 719; chemistry of, 710–711; dry cells, 718–720; electrochemical cell potentials, 711–714, 715 *prob.*, 716–717; electrolysis and, 728–732; half-cells, 710; lead-acid batteries, 720–721; lithium batteries, 721–722; primary and secondary batteries, 720; silver batteries, 719
Electrochemistry: batteries, 717, 718–723; biofuel cells, 724 *act.*; corrosion, 724–727; electrochemical cell potentials, 711–714, 715 *prob.*, 716–717; electrochemical cells, 707 *act.*, 709; electrolysis, 728–732; lemon battery, 707 *act.*; redox reactions in, 708–709; voltaic cell chemistry, 710–711
Electrolysis, 86, **728**–732; aluminum production, 730–731; desalination by, 730; electroplating and, 732; of molten NaCl, 729; ore purification and, 731–732
Electrolytes, **215**; colligative properties of aqueous solutions and, 498–499; strong, 498; weak, 498
Electrolytic cells, **728**; aluminum production and, 730–731; electrolysis of brine and, 730; electrolysis of molten NaCl and, 729; electroplating and, 732; purification of ores and, 731–732
Electromagnetic radiation, **137**–139, 140 *prob.*, 861 *table*, 863–864
Electromagnetic (EM) spectrum, 138–**139**
Electromagnetic wave relationship, 137, 150
Electromotive force (emf), 710
Electron affinity, 265
Electron capture, **868**, 868 *table*
Electron configuration notation, 158–159; first period elements, 158 *table*; second period elements, 158 *table*; third period elements, 159 *table*
Electron configurations, **156**–162; aufbau principle and, 156–157, 157 *table*; electron configuration notation, 158–159; electron-dot structures, 161, 162 *prob.*; exceptions to predicted, 160; ground state, 156; Hund's rule and, 157; Noble-gas notation, 159; orbital diagrams of, 158; Pauli exclusion principle and, 157; periodic table trends, 182–185, 186 *prob.*; valence electrons, 161
Electron-dot structures, **161**, 161 *table*, 162 *prob.*, 207. *See also* Lewis structures
Electronegativity, **194**, 265; bond character and, 266, 266 *table*; bond polarity and, 266, 267; periodic table trends, 194, 265; redox and, 684
Electronegativity scale, 194, *212*, 265
Electron mediator, 724 *act.*
Electrons, **108**; charge of, 108–109; discovery of, 107–109; energy levels and, 146–148; location of around nucleus, 152; mass of, 108–109, 119, 969 *table*; photoelectric effect and, 142; properties of, 114 *table*; quantum mechanical model of atom and, 150–152; valence, 161
Electron sea model, **225**
Electroplating, 732
Electrostatic force, 865
Elements, 10, **84**–85, 87; abundance of various, 84; in atmosphere, 901; atomic number of, 115, 116 *prob.*, 118 *prob.*; chemical symbols for, 84; color key, 968 *table*; in Earth's atmosphere, 901; in Earth's core, 919; in Earth's crust, 84, 901; in Earth's oceans, 901; emission spectra of, 164 *act.*; in the human body, 195; isotopes, 117; law of definite proportions, 87–88; law of multiple proportions, 89–90; periodic table of. *See* Periodic table; physical states of, 84; properties of, 180 *act.*, 196 *act.*, 971–974 *table*; representative, 177, 196 *act.*
Elimination reactions, **802**
Emeralds, 912
Emission spectra, 164 *act.*
Empirical formulas, **344**; from mass data, 349–350 *prob.*; from percent composition, 344, 345 *prob.*, *347*
Endothermic reactions, 216, **247**, 528, 528 *table*
End point (titration), **663**
Energy, **516**–522; bond dissociation, 247; change during solution formation, 475 *act.*, 492; changes of state and, 530–530, 531 *act.*, 532 *prob.*; chemical cold pack and, 515 *act.*; chemical potential, 517; flow of as heat, 518. *See also* Heat; kinetic, 402, 403, 516–517, 710; lattice, 216–217; law of conservation of, 517; potential, 516–517; quantized, 141–143, 146; solar, 522; units of, 518, 518 *prob.*, 518 *table*; uses of, 516; voltaic cells and, 710–711
Energy levels, 153
Energy sublevels, **153**–154
English units, 32
Enthalpy (*H*), **527**; calculate changes in (Hess's law), 534–536, 536 *prob.*; calorimetry measurement of, 550 *act.*; changes of state and, 530–533,

531 *act.*, 532 *prob.*; thermochemical equations and, 529
Enthalpy (heat) of combustion ($\Delta H_{comb}$), **529**, 529 *table*
Enthalpy (heat) of reaction ($\Delta H_{rxn}$), **527**–528
Entropy (S), **543**; Earth's geologic processes and, 545; predict changes in, 544–545; reaction spontaneity and, 546–547, 548 *prob.*; second law of thermodynamics and, 543
Environmental chemist, 7
Environmental chemistry, 11 *table*
Enzymes, 826, **829**–830. *See also* Catalysts; Proteins; affect on reaction rate, 571; chirality and, 767, 768; temperature and, 850 *act.*
Enzyme-substrate complex, 830
Equations: algebraic, 954–955, 955 *prob.*; atomic number, 115; average rate of reaction, 562; boiling point elevation, 500; Boyle's law, 443; cell potential, 714; Charles's law, 445; chemical. *See* Chemical equations; Dalton's law of partial pressures, 409; density, 37; dilution, 485; Einstein's ($E=mc^2$), 877; electromagnetic wave relationship, 137, 150; energy of a photon, 143; energy of a quantum, 142; error, 48; Gay-Lussac's law, 447; general rate law, 575; Gibbs free energy, 515 *act.*, 546; Graham's law of effusion, 404; Henry's law, 496; ideal gas law, 454; induced transmutation, 876 *prob.*; ionic, 301; ion-product of water, 650; law of conservation of mass, 77; mass number, 117; molality, 487; molarity, 482; mole fraction, 488; neutralization, 659–660; nuclear, 123, 869, 869 *prob.*; overall, 307; percent by mass, 87, 481; percent by mass from the chemical formula, 342; percent by volume, 482; percent error, 48; percent yield, 386; pH, 652; pH and pOH, relationship between, 652; pOH, 652; quantum, energy of, 142; radiation, intensity and distance of, 890; radioactive element, remaining amount of, 871; rate law, 574; skeleton, 284; slope of a line, 57, 962; specific heat, 520; summation, 540; symbols used in, 283 *table*; thermochemical, 529–533; word, 284
Equilibrium. *See* Chemical equilibrium; Solubility equilibrium
Equilibrium concentrations, calculate, 612, 613 *prob.*
Equilibrium constant ($K_{eq}$), **599**–600, 604, 605 *prob.*
Equilibrium constant expressions, 599–600; calculate concentrations from, 612, 613 *prob.*; for heterogeneous equilibrium, 602, 603 *prob.*; for homogeneous equilibrium, 600, 601 *prob.*; Le Châtelier's principle and, 606–611; solubility product constant expressions. *See* Solubility product constant expressions
Equivalence point, **661**
Error, **48**
Essential elements, 383
Essential oils, 770
Esterification, 806 *table*
Esters, 787 *table*, **799**, 799 *table*, 800 *act.*
Ethanal, 796
Ethanamide, 800
Ethane, 750, 751 *table*, 793
Ethanol, 432 *act.*, 792–793, 816 *act.*
Ethene, 759 *table*, 762, 803
Ether functional group, 787 *table*
Ethers, 787 *table*, **794**, 794 *table*
Ethylamine, 795
Ethyl group, 753 *table*
Ethyne (acetylene), 762 *act.*, 763, *763*, 764
Evaporation, **426**–427, 432 *act.*, 816 *act.*
Everyday Chemistry: baking soda and baking powder and cooking, 669; chocolate, manufacture of, 431; garlic and pain receptors, 815; history in a glass of water, 355; killer fashion, 229
Example Problems: algebraic equations, 955 *prob.*; alkanes, naming, 754–755 *prob.*; alkenes, naming, 761 *prob.*; aromatic compounds, naming, 773 *prob.*; atomic mass, 121 *prob.*; atomic number, 116 *prob.*, 118 *prob.*; atomic radii trends, 189 *prob.*; atom-to-mass conversions, 330 *prob.*; average rate of reaction, 562 *prob.*; balancing equations, 287 *prob.*; boiling point elevation, 503 *prob.*; Boyle's law, 443 *prob.*; branched-chain alkanes, naming, 754–755 *prob.*; cell potential, calculate, 715 *prob.*; Charles's law (gas temperature and volume relationship), 446 *prob.*; chemical equations, interpret, 370 *prob.*; combined gas law, 450 *prob.*; combustion reactions, energy released by, 532 *prob.*; concentration from equilibrium constant expression, 613 *prob.*; conservation of mass, 78 *prob.*; conversion factors, 46 *prob.*; cycloalkanes, naming, 756–757 *prob.*; density and volume to find mass, 38 *prob.*; dimensional analysis, 956 *prob.*; electron configuration and the periodic *table*, 186 *prob.*; electron-dot structure, 162 *prob.*; empirical formula from mass data, 349–350 prob.; empirical formula from percent composition, 345 *prob.*; energy of a photon, 143 *prob.*; energy units, convert, 518 *prob.*; equilibrium constant expression for heterogeneous equilibrium, 603 *prob.*; equilibrium constant expression for homogeneous equilibrium, 601 *prob.*; equilibrium constants, value of, 605 *prob.*; formula for polyatomic compound, 222 *prob.*; formulas for ionic compound, 220 *prob.*; freezing point depression, 503 *prob.*; gas stoichiometry, 461 *prob.*; Gay-Lussac's law, 448 *prob.*; Graham's law of effusion, 405 *prob.*; half-reaction method, 695 *prob.*; heat absorbed, calculate, 521 *prob.*; hydrates, determine formula for, 353 *prob.*; ideal gas law, 455 *prob.*; induced transmutation equations, 876 *prob.*; instantaneous reaction rates, 579 *prob.*; ionic equations and precipitation reactions, 302 *prob.*; ionic equations for aqueous solutions forming gases, 306 *prob.*; ionic equations for aqueous solutions forming water, 304 *prob.*; ion product constant, 651 *prob.*; ion product constant $Q_{sp}$, 619 *prob.*; Lewis structure for covalent compound with multiple bonds, 256 *prob.*; Lewis structure for covalent compound with single bond, 255 *prob.*; Lewis structures, 244 *prob.*; limiting reactant, determine, 382–383 *prob.*; mass number, 118 *prob.*; mass-to-atom conversions, 330 *prob.*; mass-to-mass stoichiometric conversion, 377 *prob.*; mass-to-mole conversions, 329 *prob.*; mass-to-mole conversions for compounds, 337 *prob.*; mass to moles to particles conversions, 338–339 *prob.*; molality, 487 *prob.*; molarity, 483 *prob.*; molarity from titration data, 664 *prob.*; molar solubility, 616 *prob.*; molar volume, 453 *prob.*; molecular formula from percent composition, 348–349 *prob.*; molecular shape, 264 *prob.*; mole relationship from a chemical formula, 334 *prob.*; mole-to-mass conversion, 328 *prob.*; mole-to-mass conversions for compounds, 336 *prob.*; mole-to-mass stoichiometric conversion, 376 *prob.*; mole-to-mole stoichiometric conversion, 375 *prob.*; net ionic redox equation, balance, 692; nuclear equations, balancing, 869 *prob.*; oxidation number, determine, 687 *prob.*; oxidation-number method, 690 *prob.*; particles, convert to moles, 324 *prob.*; percent

Index

by mass, 481; percent error, 49 *prob.*; percent yield, 386 *prob.*; pH, calculate, 653 *prob.*, 654 *prob.*; pOH, calculate, 654 *prob.*; radioactive element, remaining amount of, 872 *prob.*; reaction spontaneity, 548 *prob.*; redox reactions, identify, 685 *prob.*; scientific notation, 41 *prob.*, 43 *prob.*; significant figures, 51 *prob.*, 53 *prob.*, 54 *prob.*; significant figures and, 951 *prob.*, 953 *prob.*; single-replacement reactions, 295 *prob.*; standard enthalpy (heat) of formation, 540 *prob.*; unit conversion, 958 *prob.*; wavelength of EM wave, 140 *prob.*
Excess reactants, **379**, 384
Exothermic reactions, 216, **247**; activation energy and, 565; enthalpy and, 527, 528 *table*
Expanded octets, 259
Experimental data, percent composition from, 341–342, 342 *act.*
Experiments, **14**. *See also* CHEMLABs; MiniLabs; Problem-Solving Labs; laboratory safety and, 18, 19 *table*
Exponents, 40–41
Extensive properties, **73**
Extrapolation, 57, 963

## F

Fahrenheit scale, 34
Families, periodic *table*. *See* Groups
Faraday, Michael, 770
Fasteners, arrange, 173 *act.*
Fats. *See* Lipids
Fatty acids, 767, **835**–836, 837
f-Block elements, 185, 916
Femtochemistry, 581
Fermentation, **847**–848; alcoholic, 847; lactic acid, 848
Fermi, Enrico, *882*
Fermionic condensate, *417*
Ferromagnetism, 916
Fertilizers, 250, 388, 462
Fiber-optic cable, 930
Filtration, **82**
Fire extinguishers, ideal gas law and, 456, 457 *act.*
Firefly, bioluminescence, 309
Fireworks, 913
First period elements: electron configuration notation, 158 *table*; orbital diagrams, 158 *table*
Fission, *111*
Flame retardant fabric, 935
Flame tests, 92 *act.*, 144 *act.*, 907, 923
Flat-screen televisions, 925
Flavor chemist, 267
Fleming, Alexander, 18
Flexible-fuel vehicles (FFV), 549
Fluidity, 416
Fluids, 416
Fluoridation, 622 *act.*, 942
Fluoride, 180
Fluorine: analytical tests for, 941; atomic properties, 941; common reactions involving, 940; electron configuration and orbital diagram, 158 *table*; electron-dot structure, 161 *table*; electronegativity of, 194, 265; isotopes, 120; physical properties, 940
Fluoroapatite, 622 *act.*
Fog, 428
Foldables: acid-base chemistry, 633 *act.*; atoms, 101 *act.*; biomolecules, 825 act.; bond character, 239 *act.*; chemical reactions, 281 *act.*; concentration of solutions, 475 *act.*; electrochemical cells, 707 *act.*; electron configuration, 135 *act.*; equilibrium, changes affecting, 593 *act.*; functional groups, 785 act.; gas laws, 441 *act.*; Gibbs free energy equation, 515 *act.*; hydrocarbon compounds, 743 *act.*; hydrocarbons, 743 *act.*; ionic compounds, 205 *act.*; mole conversion factors, 319 *act.*; periodic trends, 173 *act.*; properties and changes, 69 *act.*; reaction rates, 559 *act.*; redox equations, balance, 679 *act.*; scientific method, 3 *act.*; states of matter, 401 *act.*; stoichiometric calculations, 367 *act.*; types of graphs, 31 *act.*; types of radiation, 859 *act.*
Food: from fermentation, 847; homogenization, *490*; measure calories in, 550 *act.*; preservation of, 571; test for simple sugars in, 825 *act.*
Food scientist, 219
f orbitals, 154
Forces: balanced, 597; dipole-dipole, 269, 411, 412–413; dispersion, 269, 411, 412; intermolecular, 411–414
Forensic accelerant detection, 91
Forensics CHEMLABs: density, dating coins by, 60 *act.*; hydrocarbon burner gases, identify, 776 *act.*; identify the damaging dumper, 698 *act.*; water source, determine, 24 *act.*
Forensics, luminol and, 697
Formaldehyde, 796, 797
Formic acid, 634
Formulas. *See* Chemical formulas; Structural formulas
Formula unit, **218**
Fossil fuels: natural gas, 416, 745, 747; petroleum, 747–748
Fractional distillation, **747**–748
Fractionation, 747–748
Fractions, 964, 965–966
Francium, 84, 180 *act.*, 194, 265, 906, 907
Franklin, Rosalind, *637*
Free energy ($G_{system}$), **546**–547; calculate, 547, 548 *prob.*; sign of, 547, 547 *table*
Freezing, 428
Freezing point, **428**
Freezing point depression, 501–**502**, 502 *act.*, 503 *prob.*
Frequency, **137**
Fructose, *832*, 833
Fuel cells, **722**–723, 905
Fuel rods, nuclear reactor, 880–882
Functional groups, 785 *act.*, **786**, 787 *table*; amide group, 800; carbonyl group, 796; carboxyl group, 798; hydroxyl group, 792
Fused-ring systems, 772
Fusion, molar enthalpy (heat) of ($\Delta H_{fus}$), **530**
Fusion nuclear reactions, 883–884
Fusion (phase change), 425–426, *See also* Melting

## G

Gadolinium, 921
Galactose, *832*, 833
Gallium, 922, 923, 924
Galvanization, **727**
Gamma radiation, 124, 861, 861 *table*, 862, 863, 868 *table*
Gamma rays, **124**, 863, 864
Garlic, 815
Gases, **72**, 402–410; compression and expansion of, 72 *act.*, 404; Dalton's law of partial pressures and, 408, 409 *prob.*, 410; density of, 403; diffusion and effusion of, 404–405; formation of in aqueous solutions, 281 *act.*, 304–305, 306 *prob.*; gas laws. *See* Gas laws; identify an unknown, 466 *act.*; kinetic-molecular theory and, 402–403; molar volume of, 452, 453 *prob.*; pressure and volume relationship (Boyle's law), 442–443, 443 *prob.*, 444 *act.*; real v. ideal, 457–459; solubility of, 495–496, 497 *prob.*; temperature and volume relationship, 441 *act.*
Gas grills, 375, 461
Gas laws, 442–451; Boyle's law (pressure and volume relationship), 442–443, 443 *prob.*, 444 *act.*; Charles's law (temperature and volume), 441 *act.*, 444–445, 446 *prob.*; combined gas law,

449, 450 *prob.*, 454; Gay-Lussac's law (temperature and pressure relationship), 447, 448 *prob.*; ideal gas law, 454, 455 *prob.*, 456; summary of, 451 *table*; temperature scales and, 451
Gasoline octane rating system, 748–749
Gas particles, 403; kinetic energy of, 403; motion of, 403; size of, 403
Gas pressure, 406–410; air pressure and, 406–407; Boyle's law (pressure and volume relationship), 442–443, 443 *prob.*, 444 *act.*; Charles's law (temperature and volume), 441 *act.*, 444–445, 446 *prob.*; combined gas law, 449, 450 *prob.*, 454; Dalton's law of partial pressures and, 408, 409 *prob.*, 410; deep sea diving and, 408 *act.*; Gay-Lussac's law (temperature and pressure relationship), 447, 448 *prob.*; ideal gas law, 454, 455 *prob.*, 456
Gas stoichiometry, 460–464; industrial applications of, 464; volume-mass problems, 462, 462–463 *prob.*; volume-volume problems, 460–461, 461 *prob.*
Gay-Lussac's law, **447**, 448 *prob.*, 451 *table*
Geckos, grip of, 271
Geiger counters, 885
Gemstones, 912
Geometric isomers, **766**
Germanium, 181, 926–927, 930
Germanium tetrachloride, 930
GFP (green fluorescent protein), 309
Gibbs free energy ($G_{system}$), 515 *act.*, 546–547, 548 *prob.*
Gibbs, J. Willard, 546
Glass, 929
Glucose, 532, 532 *prob.*, 832, 833
Glutamic acid, 827 *table*
Glutamine, 827 *table*
Glycerol, 31 *act.*, 793
Glycine, 827 *table*, 828
Glycogen, 834. *See also* Polysaccharides
Goiter, 943
Gold, 228 *table*, 920
Gold foil experiment, Rutherford's, *110*, 111–112, 113, 862
Gold leaf, 920
Graduated cylinder, layers of liquids in, 31 *act.*
Graham's law of effusion, **404**–405, 405 *prob.*
Graham, Thomas, 404
Grams (g), 34
Graphite, 423
Graphite golf shafts, 928
Graphs, **55**–58; bar, 56; circle, 55; interpreting, 57–58; line, 56–57, 959–963
Gravimetric analysis, 341
Gravitation, law of universal, 16
Great Smog (London), *291*
Greek philosophers, ideas on structure of matter, 102–103, 103 *table*
Green fluorescent protein (GFP), 309
Ground state, **146**
Ground-state electron configuration, 143 *prob.*
Ground-state electron configurations, 156–160; aufbau principle and, 156–157, 157 *table*; electron configuration notation, 158–159; exceptions to predicted, 160; Hund's rule and, 157; noble-gas notation, 159; orbital diagrams of, 158; Pauli exclusion principle and, 157; problem-solving strategy, 160
Group 1 elements (Alkali metals), 182 *table*, 182–184, 192, 207 *table*, 208, 208 *table*, 906, 906–909; (representative elements), 177
Group 2 elements (Alkaline earth metals), 182, 183, 184, 207 *table*, 208, 208 *table*, 218 *table*, 219 *table*, 910–915
Group 13 elements (Boron group), 184, 207 *table*, 208, 208 *table*, 219 *table*, 922–925
Group 14 elements (Carbon group), 184, 207 *table*, 219 *table*, 243, 926–931
Group 15 elements (Nitrogen group), 184, 207 *table*, 209, 209 *table*, 218 *table*, 243, 932–935
Group 16 elements (Oxygen group), 184, 207 *table*, 209 *table*, 218 *table*, 243, 936–939
Group 17 elements (Halogens), 184, 207 *table*, 209, 209 *table*, 218 *table*, 243, 940–943
Group 18 elements (Noble gases), 180, 184, 185 *table*, 192, 207 *table*, 944–945
Groups (families), periodic table, **177**; atomic radii trends, 188, 189 *prob.*; electron configuration and position on periodic table, 183; ionic radii trends, 191
Grove, William, 722
Guanine (G), 841
Gypsum, 490, *491*, 914

## H

Haber-Bosch process, *290*
Hahn, Otto, *111*
Half-cells, **710**
Half-life, **870**–871, 871 *table*
Half-reaction method, 693–694, 694 *table*, 695 *prob.*
Half-reactions, **693**
Halides, 214
Hall, Charles Martin, 730
Hall-Héroult process, 730–731
Halocarbons, 787 *table*, **787**–789; alkyl halides, 787; aryl halides, 788; functional group, 787, 787 *table*; naming, 788; properties of, 789; substitution reactions forming, 790; uses of, 789
Halogenated hydrocarbons, 940
Halogenation, **790**
Halogen functional group, 787 *table*, 787–788
Halogen light bulbs, 942
Halogens, **180**
Halogens (Group 17 elements), 184, 207 *table*, 209, 209 *table*, 218 *table*, 243, 940–943
Halogens, 940–943; analytical tests for, 941; applications of, 942–943; atomic properties, 941; common reactions involving, 940; covalent bonding in, 243; physical properties of, 940; predict reactivity of, 294 *act.*; single-replacement reactions involving, 294, 294 *act.*
Halothane, **790**, *791*
Hardness, as physical property, 73
Hard water, 24 *act.*
HD DVDs, 924
Heart stress test, 925
Heat (*q*), **518**. *See also* Thermochemistry; absorption of by chemical reactions. *See* Endothermic reactions; calorimetry and, 523–524, 525 *prob.*, 550 *act.*; release of by chemical reactions. *See* Exothermic reactions; specific heat, 519–520, 521 *prob.*, 522, 526 *act.*; thermochemical systems and, 523–524; units of, 518, 518 *prob.*
Heating and cooling specialist, 527
Heating curves, 531 *act.*
Heat of combustion ($\Delta H_{comb}$), **529**, 529 *table*
Heat of reaction ($\Delta H_{rxn}$), **527**–528
Heat of solution, 475 *act.*, **492**
Heat-pack reaction, 527, 542
Heat-treated steel, 227 *act.*
Heavy hydrogen (deuterium), 904
Heisenberg uncertainty principle, **151**
Helium, 158 *table*, 159, 183, 185 *table*, 192, 944, 945
Hemoglobin, 623, 830
Henry's law, 495–**496**, 497 *prob.*
Heptane, 751, 751 *table*
Héroult, Paul L. T., 730
Hertz (Hz), 137
Hess's law, **534**–536, 536 *prob.*

Heterogeneous catalysts, **573**
Heterogeneous equilibrium, **602**, 603 *prob.*
Heterogeneous mixtures, **81**, *87*, 476–478; colloids, 477, 477 *table*; separating components of, 82–83; suspensions, 476
Hexagonal unit cells, 421 *table*, 422 *act.*
Hexane, 751 *table*
HFCs (hydrofluorocarbons), 788
Hill, Julian, 18
HIV, 389
Homogeneous catalysts, **573**
Homogeneous equilibrium, **600**, 601 *prob.*
Homogeneous mixtures, **81**, 82–83, *87*, 478–479
Homogenization, *490*
Homologous series, **751**
Hope Diamond, 40
HOPG, atomic distances in, 113 *act.*
Hormones, 831, 839
Household items, acidity of, 633 *act.*
How It Works: bioluminescence, 309; flexible-fuel vehicles (FFV), 549; gecko grip, 271; mass spectrometer, 125; methane digester, 775; pacemaker, 733
*Hubble Space Telescope*, 912
Human body, elements in, 84, 195
Human immunodeficiency virus (HIV), 389
Hund's rule, **157**
Hybridization, **262**
Hybrid orbitals, 262
Hydrates, **351**–354; formulas for, 351 *table*, 352, 353 *prob.*, 356 *act.*; naming, 351; uses for, 354
Hydration (solvation in water), 489
Hydration reactions, 804, 804 *table*
Hydrocarbons, 291, **745**–749. *See also specific types*; alkanes. *see* Alkanes; alkenes. See Alkenes; alkynes, 763–764; aromatic. *See* Aromatic compounds; burner gas analysis, 776 *act.*; chirality of, 767; Foldable, 743 *act.*; halogenated, 940; isomers of, 765–766, 768–769; models of, 743 *act.*, 746; refinement of petroleum, 747–748; saturated, 746; substituted. *See* Substituted hydrocarbons; unsaturated, 746
Hydrofluorocarbons (HFCs), 788
Hydrogen, 904–905; abundance of, 84; atomic properties, 153–155, 158 *table*, 904; Bohr model of, 146–148, 147 *table*; emission spectrum, 144, *145*, 147–148, 150 *act.*; in human body, 195; isotopes of, 904; physical properties, 904; single-replacement reactions involving, 293; in stars, 905
Hydrogenated fats, 805
Hydrogenation, 767, 836
Hydrogenation reactions, 804 *table*, **804**–805, 805 *act.*
Hydrogen bonds, 411, **413**–414
Hydrogen carbonate, 221 *table*
Hydrogen cyanide, 647
Hydrogen fluoride, *244*, 244 *prob.*, 639
Hydrogen fuel cells, 905
Hydrogen peroxide, 89
Hydrometers, 37
Hydronium ions, 636, 652; calculate concentration of from pH, 655 *prob.*; calculate concentrations from, 651, 651 *prob.*; calculate pH from concentration of, 653 *prob.*, 654 *prob.*
Hydroxide ions, 221 *table*, 636, 652; calculate concentration of from pH, 655 *prob.*; calculate concentrations from, 651, 651 *prob.*; calculate pOH from concentration of, 654 *prob.*
Hydroxyl group, 787 *table*, **792**, 816 *act.*
Hyperbaric oxygen therapy, 465
Hyperthermia, 583
Hypochlorite, 221 *table*
Hypothermia, 583
Hypotheses, **13**

## I

Ice, 420, 425–426
Ideal gas constant (*R*), **454**, 969 *table*
Ideal gases, real versus, 457–459
Ideal gas law, **454**, 455 *prob.*, 456; density and, 456; derive other laws from, 458; exceptions to, 458–459; fire extinguishers and, 457 *act.*; molar mass and, 456
Immiscible, **479**
Independent variables, **14**, 56
Indicators, acid-base, 658, 663, 654
Indium, 922, 923, 925
Indium-tin oxide, 925
Induced fit, 830
Induced transmutation, **875**, 882; equations representing, 876 *prob.*; transuranium elements, 876
Industrial chemistry, 11 *table*, 341, 464
Infrared (Paschen) series, *147*, 148, 150 *act.*
Inhibitors, **571**
Initial rates, method of, **576**, 577 *prob.*
Inner transition metals, **180**, 185, 916, 917
Inorganic chemistry, 11 *table*
Insoluble, **479**
Instantaneous reaction rates, **578**–579, 579 *prob.*
Insulin, 831
Intensive properties, **73**, 77
Intermediates, **580**
Intermolecular forces, 411–414; covalent compounds and, 269–270; dipole-dipole, 411, 412–413; dispersion, 411, 412; evaporation and, 432 *act.*; grip of a gecko and, 271; hydrogen bonds, 411, 413–414
International Union of Pure and Applied Chemistry (IUPAC), naming conventions. *See* Naming conventions
Interpolation, 57, 963
Interstitial alloys, 228
In the Field: archaeologist, 891; arson investigator, 91; art restorer, 23; crime-scene investigator, 697; environmental chemist, 505; molecular paleontologist, 849
Intramolecular forces, comparison of, 411 *table*
Inverse relationships, 961
Iodate, 221 *table*
Iodine, 86, 940, 941, 943
Iodine-131, 887
Iodine deficiency, 943
Ion concentration: from $K_{sp}$, 617 *prob.*, 618–619; from pH, 655, 655 *prob.*
Ionic bonds, **210**; electronegativity and, 266; energy in, 216–217, 217 *table*
Ionic compounds, **210**–215; in aqueous solutions, 300; binary, 210; formation of, 211–212, 212 *prob.*, 216, 230 *act.*; formulas for, 218–219, 220 *prob.*, 221, 221 *prob.*, 222 *prob.*; lattice energies of, 216–217, 217 *table*; melting point of, 242 *act.*; milestones in understanding, *212–213*; naming, 222, 223–224; oxidation number of, 219; physical properties, 212, 214–215, 230 *act.*; physical structure, 212–214; polyatomic. *See* Polyatomic ions; solvation of aqueous solutions of, 490; study organizer, 205 *act.*
Ionic crystals, 215
Ionic equations, 301, 302 *prob.*, 304 *prob.*; complete, 301; for reactions forming gases, 304–305, 306 *prob.*; for reactions forming water, 303, 304 *prob.*; net, 301
Ionic liquids, 229
Ionic radii, periodic table trends, 189–191, 189–191
Ionic solids, 422, 422 *table*, 423
Ionization constants. *See* Acid ionization constant; base ionization constant

Ionization energy, **191**–194; chemical bonds and, 207; periodic table trends, 193
Ionizing radiation, **885**, 886; biological effects of, 888–890; medical uses of, 886–887
Ion product constant ($Q_{sp}$), 618–619, 619 *prob.*
Ion product constant for water, **650**–651, 651 *prob.*
Ions, **189**; anion formation, 209; cation formation, 207; formula for monatomic, 218–219; ionic radii periodic table trends, 189–191; metal, 208; monatomic. *See* Monatomic ions; naming, 222–223; oxidation number of, 219; polyatomic, 221, 222 *prob.*; pseudo-noble gas configuration, 208; stability of, 240; transition metal, 208
Iron: in acid mine waste, 920; Earth's core and, 919; as paint pigment, 919; redox reactions oxidizing, 693 *table*; rust formation, 74, 77, 679 *act.*
Iron oxide. *See* Rusting
Isobutane, 752
Isomers, **765**; *cis*-, 766; geometric, 766; optical, 768–769; stereoisomers, 766; structural, 765; *trans*-, 766; trans-fatty acid, 767
Isopropyl alcohol, 432 *act.*
Isopropyl group, 753 *table*
Isotopes, **117**, 118 *prob.*. *See also* Radioactivity; abundance of, 117, 120; atomic mass and, 117, 118 *act.*, 119–120, 121 *prob.*, 126 *act.*; mass of, 117; modeling, 120 *act.*; notation for, 117; radioactive. *See* Radioisotopes
IUPAC naming conventions. *See* Naming conventions

## J

*James Webb Space Telescope* (JWST), 912
Jin, Deborah S., *417*
Joule (J), 142, **518**

## K

Kekule, Friedrich August, 771
Kelvin (K), **35**, 451
Kelvin scale, 35, 451
Ketones, 787 *table*, **797**, 797 *table*
Kilns, 461
Kilocalorie (kcal), 518
Kilogram (kg), **34**
Kilometer (km), 33
Kinetic energy (KE), 516–517; kinetic-molecular theory and, 402, 403, 517; voltaic cells and, 710
Kinetic-molecular theory, **402**–403; assumptions of, 403; compression and expansion of gases and, 404; density of gases and, 403; diffusion and effusion of gases and, 404–405; liquids and, 415
Knocking, 748
Krypton, 185 *table*, 944, 945
Kwolek, Stephanie, *491*

## L

Lab activities. *See* CHEMLABs; Data Analysis Labs; Launch Labs; MiniLabs; Problem-Solving Labs
Laboratory safety, 18, 19 *table*
Lactic acid fermentation, 848
Lactose, 833
Lanthanide series, **180**, 185, 916
Large Hadron Collider, *111*
Laser scissors, 163
Lattice energy, **216**–217, 217 *table*
Launch Labs: arrange items, 172 *act.*; atomic structure, 134 *act.*; chemical change, evidence of, 280 *act.*; chemical change, observe, 68 *act.*; chemical cold pack, 514 *act.*; chemical reaction, observe, 366 *act.*; covalent bonding (super ball properties), 238 *act.*; electrical conductivity of solutions, 204 *act.*; electric charge, observe, 100 *act.*; equilibrium point, 592 *act.*; hydrocarbons, model, 742 *act.*; lemon battery, 706 *act.*; liquids, layering of (density), 30; liquids, properties of, 400 *act.*; mole conversion factors, 318 *act.*; nuclear chain reactions, 858 *act.*; reaction rates, speeding, 558 *act.*; rust formation, 678 *act.*; slime, make, 784 *act.*; solution formation, energy change and, 474 *act.*; sugars, test for simple, 824 *act.*; temperature and gas volume (Charles's Law), 440 *act.*; viscosity of liquids, 400 *act.*; Where is it? (conservation of matter), 2 *act.*
Lavoisier, Antoine, 79, 174, 174 *table*, *184*, *290*
Law, **16**
Law of chemical equilibrium, **599**–600
Law of conservation of energy, **517**
Law of conservation of mass, **77**, 78 *prob.*, 79; balancing equations and, 285, 288; Dalton's experimental evidence of, 105; molar mass and, 335; stoichiometry and, 368
Law of definite proportions, **87**–88
Law of multiple proportions, **89**–90
Law of octaves, 175
Law of universal gravitation, 16
Lawrencium, 921
LCD panels, 925
Lead, 229, 926–927, 930; poisoning, 229
Lead-acid storage batteries, 720–721, 930
Lead shot, 228 *table*
Le Châtelier, Henri-Louis, 607
Le Châtelier's principle, **607**; chemical equilibrium and, 606–611; common ion effect and, 620–621; ion-product of water and, 650, 650 *prob.*; molar solubility and, 624 *act.*
Lecithin, 431
Lemon battery, 707 *act.*
Length, 33, 33 *table*
LEO GER, **681**
Lewis, G. N., 161, *212*, 641
Lewis model, **641**–643, 642 *table*
Lewis structures, **242**, 244 *prob.*, 253–260. *See also* Electron-dot structures; covalent compound with multiple bond, 256 *prob.*; covalent compound with single bond, 255 *prob.*; modeling, 272 *act.*; octet rule exceptions and, 258–259, 260 *prob.*; polyatomic ions, 256, 257 *prob.*; resonance and, 258
Light: continuous spectrum of, 138; dual nature of, 143; electromagnetic spectrum, 138–139; particle nature of, 141–143; speed of ($c$), 137; visible spectrum of, 139; wave nature of, 137–139, 140 *prob.*, 143
"Like dissolves like", 489
Limestone, 635, 643
Limiting reactants, **379**–381; calculating product with, 380–381, 382–383 *prob.*; determining, 380
Linear molecular shape, *261*, 263 *table*
Line graphs, 56–57, 58, 959–963
Line, slope of, 57, 962
Line spectra. *See* Emission spectra
Lipid bilayer, 838
Lipids, 13 *act.*, 830, **835**–839; fatty acids, 835–836, 837; phospholipids, 838; saponification of, 837, 837 *act.*; steroids, 839; triglycerides, 836–837; waxes, 838
Liquids, **71**, 415–419; adhesion and cohesion of, 419; attractive forces in, 417; capillary action, 419; compression of, 415; density of, 31 *act.*, 415; evaporation of, 426–427, 432 *act.*; fluidity of, 416; properties of, compare, 401 *act.*; shape and size of particles in, 417; surface tension, 418–419; viscosity of, 401 *act.*, 417, 418

Liter (L), **35**
Lithium, 136, 158 *table*, 161 *table*, 177, 226 *table*, 906, 907, 913
Lithium batteries, 721–722, 908
Litmus paper, 633 *act.*, 635, 658
Logarithms, 966–967
London forces. *See* Dispersion forces
London, Fritz, 412
Lowry, Thomas, 638
LP (liquefied propane) gas, 750
Luciferin, 309
Luminol, 697
Lunar missions, oxygen in moon rocks, 387 *act.*
Lyman (ultraviolet) series, *147*, 148, 150 *act.*
Lysine, 827 *table*

## M

Magnesium, 159 *table*, 177, 910–911, 912, 913
Magnesium oxide, 210, 217 *table*
Magnetic resonance imaging, 921
Malleability, 226
Manganese, 918, 920
Manhattan Project, *882*
Manometers, 407
Mass, **9**–10; determine from density and volume, 38 *prob.*; identify an unknown by, 50 *act.*; law of conservation of, 77, 78 *prob.*, 79, 105; mass-to-atom conversions, 329–330, 330 *prob.*; mass-to-mole conversions, 329 *prob.*; mass-to-mole conversions for compounds, 337, 337 *prob.*; mass-to-moles-to-particles conversions, 338, 338–339 *prob.*; molar. *See* Molar mass; mole-to-mass conversions, 328 *prob.*; SI base unit for, 33 *table*, 34; volume-mass gas stoichiometry, 462, 462–463 *prob.*; weight v., 9–10
Mass defect, **877**
Mass number, **117**, 118 *prob.*
Mass spectrometry, 125, 327
Mass-to-mass stoichiometric conversions, 374, 377, 377 *prob.*
Material Safety Data Sheets (MSDS), 59
Materials scientist. *See* Careers in Chemistry; In the Field
Math Handbook, 946–967; algebraic equations, 954–955, 955 *prob.*; antilogarithms, 967; dimensional analysis, 956 *prob.*; fractions, 964, 965–966; line graphs, 959–963; logarithms, 966–967; percents, 965; ratios, 964; scientific notation, 946–948; significant figures, 949–950, 951 *prob.*; square and cube roots, 949; unit conversion, 957–958, 958 *prob.*
Matter: categories of, *87*; characteristics of, 9–10; chemical changes in, 69 *act.*, 77; chemical properties of, 74; Greek philosophers' theories of, 102–103; mixtures of. *See* Mixtures; physical changes in, 76–77; physical properties of, 73; properties of, observe, 74–75; pure substances. *See* Pure substances; states of. *See* States of matter; study of chemistry and, 4
Maxwell, James, 402
Measurement, 295; accuracy of, 47–48; precision of, 47–48; significant figures and, 50–51; units of, 32–37
Medicinal chemist, 342
Meitner, Lise, *111*
Melting, 425–426, 530
Melting point, 77, **426**
Melting points: of alkanes, 758; bond type and, 242 *act.*; of covalent compounds, 270; of metals, 226, 226 *table*; as physical property, 73
Mendeleev, Dmitri, 85, 175, 176 *table*, *184*
Mercury, 73 *table*, 226
Mercury(II) oxide, 79
Metabolism, **844**–848; anabolism, 844–845; ATP and, 845; catabolism, 844–845; cellular respiration, 846; fermentation, 847–848; photosynthesis, 846
Metal alloys, 227–228
Metal carbonates, 635
Metal ions: formation of, 208; monatomic, 218, 219, 219 *table*
Metallic bonds, **225**
Metallic hydroxids, 648
Metallic solids, 422, 422 *table*, 423
Metalloids, **181**, 196 *act.*
Metallurgist, 423
Metals, **177**. *See also* Alkali metals; Alkaline earth metals; Inner transition metals; Transition metals; acid-base reactions and, 635; activities of, 310 *act.*; boiling points, 226, 226 *table*; bonding in, 225; ductility of, 177, 226; durability of, 226; electrical conductivity of, 177, 226; fireworks and, 913; hardness and strength of, 226; malleability of, 177, 226; melting points, 226, 226 *table*; periodic table position, 177; properties of, 177, 196 *act.*, 226, 226 *table*; purification of by electrolysis, 731–732; reactivity of, 293–294, 310 *act.*; single-replacement reactions involving, 293–294; specific heat of, 526 *act.*; thermal conductivity of, 226
Meteorologist, 447
Meter (m), **33**, 33 *table*
Methanal, 796
Methane, *243*, 244, *245*, 291, 745, 747, 750, 751, 751 *table*
Methane digester, 775
Methanol, 793, 816 *act.*
Method of initial rates, **576**, 577 *prob.*
Methylbenzene, 772
Methyl chloroform, 20
Methyl group, 753 *table*
Methyl red, *662*
Meyer, Lothar, 175, 176 *table*, *184*
Microbes, electric current from, 724 *act.*
Microchips, 919
Microwaves, 137, 140 *prob.*
Midgley, Thomas Jr., 7
Milligrams (mg), 34
Millikan, Robert, 109
Milliliters (ml), 33 *table*, 36
Millimeter (mm), 33, 33 *table*
Mineralogists, 214
Minerals, 383; classification of, 214; crystal lattice structure, 214
Mineral supplements, 220
MiniLabs. *See also* CHEMLABs; Data Analysis Labs; Problem-Solving Labs; acid strengths, compare, 648 *act.*; bond type and melting point, 242 *act.*; chemical equilibrium, stress and, 611 *act.*; corrosion, 726 *act.*; crystal unit cells, model, 422 *act.*; density of unknown objects, 39 *act.*; esters, recognize, 800 *act.*; ethyne, synthesize and observe, 762 *act.*; flame test, 144 *act.*; freezing point depression, 502 *act.*; halogens, predict reactivity of, 294 *act.*; heat-treated steel, properties of, 227 *act.*; isotopes, model, 120 *act.*; molar volume and mass (fire extinguisher), 457 *act.*; observation skills, develop, 13 *act.*; paper chromatography, 82 *act.*; percent composition of chewing gum, 342 *act.*; periodic trends, model, 193 *act.*; precipitate-forming reaction, observe, 301 *act.*; radioactive decay, model, 873 *act.*; reaction rate and temperature, 571 *act.*; saponification (soap making), 837 *act.*; specific heat, 526 *act.*; stoichiometry of baking soda decomposition, 378 *act.*; tarnish removal (redox reaction), 683 *act.*
Miscible, **479**
Mixtures, **80**–83, *87*; heterogeneous, 81, 476–478; homogeneous, 81, 478–479; separate components of, 80, 82 *act.*, 82–83

Index

Mobile phase, chromatography, 83
Model, **10**, 15
Molal boiling point elevation constant ($K_b$), 500, 500 *table*, 976 *table*
Molal freezing point elevation constant ($K_f$), 502, 502 *table*, 976 *table*
Molality ($m$), 480 *table*, **487**, 487 *prob.*
Molar calculations, history in a glass of water and, 355
Molar enthalpy (heat) of condensation, 530
Molar enthalpy (heat) of fusion, **530**
Molar enthalpy (heat) of vaporization, **530**, 531 *act.*
Molarity ($M$), 480 *table*, **482**, 483 *prob.*; from titration, 663, 664 *prob.*, 670 *act.*
Molar mass, **326**–332; atom-to-mass conversions, 331 *prob.*; of compounds, 335, 335 *prob.*; effusion rate and, 404, 405 *prob.*; ideal gas law and, 456; mass-to-atom conversions, 329–330, 330 *prob.*; mass-to-mole conversions, 329 *prob.*; mole-to-mass conversions, 327–328, 328 *prob.*; nuclear model of mass and, 326 *act.*
Molar solubility, 615–617, 616 *prob.*, 621, 624 *act.*
Molar solutions, preparation of, 484, 485, 486 *prob.*
Molar volume, **452**, 453 *prob.*, 969 *table*
Mole (mol), **321**–324; chemical formulas and, 333–334, 334–335 *prob.*; conversion factors, 319 *act.*; convert particles to, 323, 323 *prob.*, 324 *prob.*; convert to particles, 322; as counting unit, 319 *act.*, 320; mass-to-mole conversions, 329 *prob.*; mass-to-mole conversions for compounds, 337, 337 *prob.*; mass to moles to particles conversions, 338, 338–339 *prob.*; molar mass and, 326–332; mole-to-mass conversions, 327–328, 327–328, 328 *prob.*; mole-to-mass conversions for compounds, 336, 336 *prob.*
Molecular compounds: in aqueous solutions, 299; formation of, 241; formulas from names of, 251; Lewis structures for, 253–260, 255 *prob.*, 256 *prob.*, 257 *prob.*, 258 *prob.*, 260 *prob.*; naming, 248–251, 249 *prob.*, *252*; shape of (VSEPR model), 261–262, 263 table, 264 *prob.*, 272 *act.*; solvation of aqueous solutions of, 491
Molecular formulas, *253*, **346**–347; of organic compounds, 746; from percent composition, 346–347, 348–349 *prob.*
Molecular manufacturing, 107
Molecular paleontologist, 849
Molecular shape, 261–262, 263 *table*, 264 *prob.*, 267–268
Molecular solids, 422, 422 *table*
Molecules, **241**; diatomic, 241; shape of, 261–262, 263 *table*, 264 *prob.*, 267–268
Mole fraction, 480 *table*, **488**, 488 *prob.*
Mole ratios, **371**–372, 390 *act.*
Mole-to-mass stoichiometric conversions, 374, 376, 376 *prob.*
Mole-to-mole stoichiometric conversions, 373–374, 375 *prob.*
Monatomic ions, **218**; formulas for, 218–219; oxidation number of, 219
Monoclinic unit cells, 421 *table*, 422 *act.*
Monomers, **810**
Monoprotic acids, 640, 641 *table*
Monosaccharides, 825 *act.*, **832**–833
Montreal Protocol, 20
Moon rocks, oxygen in, 387 *act.*
Moseley, Henry, 115, 176, 176 *table*, *184*
Mothballs, **428**
Motor oil, viscosity of, 417, 418
Multidrug therapy, 389
Multiple covalent bonds, 245–246
Multiplication, 54, 54 *prob.*

## N

Naming conventions: acids, 250–251, 250–251, *252*; alcohols, 793; aldehydes, 796; alkenes, 760, 761 *prob.*; alkynes, 764; amides, 800; amines, 795; aromatic compounds, 772–773, 773 *prob.*; binary molecular compounds, 248–250, 249 *prob.*, *252*; branched-chain alkanes, 752–753, 754–755 *prob.*; carboxylic acids, 798; cycloalkanes, 756, 756–757 *prob.*; esters, 799; halocarbons, 788; hydrates, 351; ionic compounds, 223–224; ions, 222–223; ketones, 797; oxyanions, 222 *table*, 222–223; straight-chain alkanes, 751
Nanoparticles, 216 *act.*
Nanotechnology, 107
Nanotubes, 928
Naphthalene, 772
National Oceanic and Atmospheric Administration (NOAA), 20, 21 *act.*
Natural gas, 416, 745, 747
Negatively charged ions. *See* Anions
Neon, 143, 158 *table*, 161 *table*, 185 *table*, 944, 945
Net ionic equations, **301**, 302 *prob.*, 304 *prob.*
Net ionic redox equations, balancing, 691, 692 *prob.*
Network solids, 270
Neutralization equations, 659–660
Neutralization reactions, **659**–660
Neutral solutions, 636
Neutron activation analysis, 886, 891
Neutrons, **113**, 114 *table*, 119, 969 *table*
Neutron-to-proton ratio, nuclear stability and, 865, 866
Newlands, John, 175, 176 *table*
Newton, Sir Isaac, 16
NiCad batteries, 720
Nickel, 919
Night-vision lenses, 930
Nitrate, 221 *table*
Nitrite, 221 *table*
Nitrogen, 158 *table*, 161 *table*, 195, 932, 933, 934
Nitrogen cryotherapy, 934
Nitrogen-fixation, 462, 934
Nitrogenous bases, 841, 843
Noble gases (Group 18), **180**, 183, 184, 185 *table*, 207, 944–945
Noble-gas notation, 159
Nonane, 751 *table*
Nonmetals, **180**; ions of, 209; periodic *table* position, 177; properties of, 196 *act.*
Nonpolar covalent bonds, 266
Nonpolar molecules, 267–268, 269
Nuclear atomic model, 112–113, 136
Nuclear chain reactions. *See* Chain reactions
Nuclear equations, **123**, 869, 869 *prob.*
Nuclear fission, **878**–880; chain reactions and, 879–880; nuclear reactors and, 880–882
Nuclear fusion, 883–884
Nuclear power plants, 878, 880–882
Nuclear reactions, **122**; balanced equations representing, 863, 869, 869 *prob.*; chain reactions, 859 *act.*, 879–880; chemical reactions vs., 860 *table*; induced transmutation, 875–876, 876 *prob.*; mass defect and binding energy, 877–878; milestones in understanding, 882–883; nuclear fission, 878–880; nuclear fusion, 883–884; radioactive decay series, 870; thermonuclear reactions, **883**
Nuclear reactors, 878, 880–882
Nuclear stability, 124, 865–866
Nuclear waste, storage of, 882
Nucleic acids, *636*, **840**–843; DNA, 841–842, 842 *act.*; RNA, 843
Nucleons, **865**
Nucleotides, **840**
Nucleus (atomic), **112**; discovery of, 112; nuclear model of mass and, 326 *act.*; size of, 112

Nutritional calories, 518
Nylon, 18, 594, 811

## O

Observation, 13, 13 *act.*
Oceans: elements in, 901; sequestration of carbon dioxide in, 505
Octahedral molecular shape, *261*, 263 *table*
Octane, 751, 751 *table*
Octane rating system, 748–749
Octet rule, **193**, 240; exceptions to, 258–259, 260 *prob.*
Odor, 73, 283
Oil drop experiment, Milikan's, 109
Oil of wintergreen, 800 *act.*
Oleic acid, 835
Optical isomers, **768**–769
Optical rotation, **769**
Orbital diagrams, 158, 158 *table*, 159 *table*
Orbitals, **152**, 154, 262
Order of operations, algebraic, 954–955, 955 *prob.*
Ores, 731–732
Organic chemistry, 11 *table*, 745
Organic compounds, 744–**745**. *See also* Hydrocarbons; carbon-carbon bonds in, 746; models of, 746; reactions forming. *See* Organic reactions
Organic reactions: addition reactions, 804–805; condensation reactions, 801; dehydration reactions, 803; dehydrogenation reaction, 803; elimination reactions, 802; oxidation reduction reactions, 806–807; products of, predict, 807–808; substitution reactions, 790–791
Organosilicon oxide, 239 *act.*
Orthorhombic unit cell, 421 *table*, 422 *act.*
Osmosis, **504**
Osmotic pressure, **504**
Overall equations, 307
Oxalic acid, 798
Oxidation, **681**
Oxidation number, **219**, 682; determine, 686, 686 *table*, 687 *prob.*; monatomic ion formulas and, 219; in redox reactions, 688; of various elements, 688 *table*
Oxidation-number method, **689**, 689 *table*, 690 *prob.*
Oxidation reduction reactions, **680**. *See also* Redox reactions
Oxidizing agent, **683**
Oxyacids, **250**–251, 252
Oxyanions, **222**, 223
Oxygen: abundance of, 84; analytical tests for, 937; atomic properties, 937; common reactions involving, 936–937; diatomic, 241; electron configuration and orbital diagram, 158 *table*; electron-dot structure, 161 *table*; in human body, 195, 623; photosynthesis and, 846, 912, 938; physical properties, 73 *table*, 936
Oxygen group (group 16), 184, 207 *table*, 209 *table*, 218 *table*, 243, 936–939
Ozone, 5, 6, 21 *act.*, 938
Ozone depletion, 20–21
Ozone hole, 7, 20–21, 21 *act.*
Ozone layer, 5–8, 938; chlorofluorocarbons (CFCs) and, 7–8, 20; formation of ozone in, 6; thinning of, 7, 20, 21 *act.*

## P

PA-457 anti-HIV drug, 389
Pacemakers, 733
Pain receptors, temperature and, 815
Painting restoration, 23
Paint pigments, 919
Paleontologist, 849
Papain, 829
Paper chromatography, 82 *act.*, 83, 269 *act.*
Paraffin, 270
Paramagnetism, 916, 917
Parent chain, **752**
Partial pressure, Dalton's law of, 408, 409 *prob.*, 410
Particle accelerators, 875
Particle model of light, 141–143
Particles: convert moles to, 322, 323 *prob.*; convert to moles, 323, 324 *prob.*; counting, 320–321; mass-to-moles-to-particles conversions, 338, 338–339 *prob.*; representative, 321
Pascal (Pa), **407**
Paschen (infrared) series, *147*, 148, 150 *act.*
Pasteur, Louis, 767
Pauli exclusion principle, **157**
Pauling, Linus, 194, 771
Paulings, 194
Pauli, Wolfgang, 157
p-Block elements, 184
Penetrating power, **864**; of alpha particles, 862; of beta particles, 863; of X rays, 864
Penicillin, 18
Pennies: dating by density, 60 *act.*; model isotopes with, 120 *act.*
Pentane, 751, 751 *table*
Peptide bond, 827–**828**
Peptides, **828**
Percent by mass concentration ratio, **87**–88, 480 *table*
Percent by volume concentration ratio, 480 *table*, 482, 482 *prob.*
Percent composition, 341–**342**; from chemical formula, 342, 343 *prob.*; empirical formula from, 344, 345 *prob.*; from experimental data, 341–342, 342 *act.*; molecular formula from, 346–347, 348–349 *prob.*
Percent error, 48–49, 49 *prob.*
Percents, 965; as conversion factors, 44
Percent yield, **386**, 386 *prob.*, 388
Perchlorate, 221 *table*
Perfumes, 770
Periodic law, **176**
Periodic table of the elements, **85**, 173 *act.*, 174–177, *178–179*, 180–181; atomic radii trends, 187–188, 189 *prob.*; blocks on, 183–185; boxes on, 177; electron configuration of elements and, 182–185, 186 *prob.*; electronegativity trends, 194, 265; groups (families), 177; history of development of, 174–177, 176 *table*, *184–185*; ionic radii trends, 189–191; ionization energy trends, 193; model periodic trends, 193 *act.*; model trends, 173 *act.*; nonmetals, 180; periods (rows), 177, 182; predict element properties from, 180 *act.*
Periods, periodic table, 85, **177**; atomic radii trends, 188, 189 *prob.*; electron configuration, 182 *table*; ionic radii trends, 190; ionization energies, 192 *table*; valence electrons and, 182
Permaganate, 221 *table*
Perspiration, 426
Petroleum, 747–749, 790
Petroleum technician, 748
PET scans, 888
Pewter, 228 *table*
pH, **652**, *653*; acid ionization constant ($K_a$) from, 656, 657 *prob.*; of familiar substances, *652*; of household items, 633 *act.*; ion concentration from, 655, 655 *prob.*; from ion concentrations, 653 *prob.*, 654 *prob.*; measurement of, 633 *act.*, 635, 658
Pharmacist, 381
Pharmacy technician, 483
Phase changes, **76**–77, 425–430; boiling, 427; condensation, 428; deposition,

429; evaporation, 426–427, 432 *act.*; freezing, 428; melting, 425–426; phase diagrams and, 429–430; six possible transitions, *425*; sublimation, 428; thermochemical equations for, 530–531, 531 *act.*; vaporization, 426–427
Phase diagrams, **429**–430
Phenanthrene, *772*
Phenolthphalein, 658, *662*
Phenylalanine, 827 *table*, 828
pH meters, *637*, 658
Phosphate ion structure, 257 *prob.*
Phosphates, 250
Phospholipases, 838
Phospholipids, **838**
Phosphoric acid, 634
Phosphors, 180, 886
Phosphorus, 159 *table*, 932, 933, 934
Phosphorus trihydride, 264 *prob.*
Photocopies, 939
Photoelectric effect, **142**–143
Photoelectrons. *See* Electrons
Photons, **143**, 143 *prob.*
Photosynthesis, **846**, 912, 938
Photovoltaic cells, 142, 522
pH paper, 633 *act.*, 635, 658
pH scale, *636*
Physical changes, **76**–77
Physical chemistry, 11 *table*
Physical constants, 969 *table*
Physical properties, **73**; of common substances, 73 *table*; extensive, 73; intensive, 73, 77; mineral identification by, 73; observe, 74–75
Pi bond, **245**–246
Pie charts, 55
Planck, Max, 141–142
Planck's constant, 142, 969 *table*
Plants: hydrogen cyanide in, 647; nitrogen-fixation, 462, 934; photosynthesis, 846, 912, 938; waxes, 838
Plasma, 71, *417*
Plastics, **789**, 802, 810–811, 814
Plastic viscosity, 431
Platinum, 918
Plum pudding model, 110
pOH, **652**, *653*, 654 *prob.*
Polar covalent bonds, **266**, 267–268
Polarized light, **769**
Polar molecules, 267–268; chromatograms and, 269 *act.*; ideal gas law and, 459; shape of, 267–268; solubility of, 268
Polonium, *882*, 936, 937
Polyacrylonitrile, 812 *table*
Polyatomic ions, **221**, 970 *table*; common, 221 *table*; formulas for, 221, 222 *prob.*; Lewis structures, 256, 257 *prob.*; naming, 222–223
Polycarbonate, 809
Polycyclic aromatic hydrocarbons (PAHs), 807
Polyethylene, 762, 810, 811
Polyethylene terephthalate (PET), 810, 812 *table*
Polymer chemist, 813
Polymer chemistry, 11 *table*
Polymerization reactions, **810**–811
Polymers, **809**–814; antimicrobial properties of, 216 *act.*; common, 812 *table*; milestones in understanding, *810–811*; properties of, 813; reactions forming, 810–811; recycling of, 814; synthetic, 809
Polymethyl methacrylate, 812 *table*
Polypeptides, 828
Polyphenols, 662
Polypropylene, 812 *table*
Polyprotic acids, 640–641, 641 *table*
Polysaccharides, **833**–834
Polyurethane, 812 *table*
Polyvinyl chloride (PVC), 812 *table*
Polyvinylidene chloride, 812 *table*
Popcorn, 466 *act.*
p orbitals, 154
Positive ions. *See* Cations
Positron, **868**
Positron emission, **868**, 868 *table*, 888
Positron emission transaxial tomography (PET), 888
Potassium, 86, 117, 136, 906, 907
Potential energy, 516–517
Potter, 682
Pottery kilns, 461
Practice Problems: acid-metal reactions, 635 *prob.*; acids, naming, 251 *prob.*; aromatic compounds, naming, 773 *prob.*; atomic mass, 121 *prob.*; atomic number, 116 *prob.*, 118 *prob.*; atomic radii trends, 189 *prob.*; atoms-to-mass conversions, 331 *prob.*; average reaction rates, 563 *prob.*; balanced chemical equations, interpret, 371 *prob.*; binary molecular compounds, naming, 249 *prob.*; Boyle's law (pressure and volume relationship), 443 *prob.*; branched-chain alkanes, naming, 755 *prob.*; branched-chain alkenes, naming, 761 *prob.*; calorimetry data, 525 *prob.*; Charles's law, 446 *prob.*; chemical equations, write, 287 *prob.*; chemical reactions, classify, 291 *prob.*; combined gas law, 450 *prob.*; conjugate acid-base pairs, 640 *prob.*; cycloalkanes, naming, 757 *prob.*; decomposition reactions, 292 *prob.*; dilute stock solutions, 486 *prob.*; double-replacement reactions, 297 *prob.*; electron configuration and the periodic table, 186 *prob.*; electron-dot structures, 162 *prob.*; empirical formula from mass data, 350 *prob.*; empirical formula from percent composition, 346 *prob.*; energy released by reaction, 532 *prob.*; energy units, convert, 519 *prob.*; equilibrium concentrations, 613 *prob.*; equilibrium constant expressions, 601 *prob.*, 603 *prob.*; equilibrium constants, value of, 605 *prob.*; expanded octets, 260 *prob.*; formulas from names of molecular compounds, 251 *prob.*; freezing and boiling point depressions, 503 *prob.*; gas-forming reactions, 306 *prob.*; Gay-Lussac's law, 448 *prob.*; Graham's law of effusion, 405 *prob.*; ground-state electron configuration, 160 *prob.*; half-cell potentials, 716 *prob.*; half-reaction method, 695 *prob.*; halocarbons, naming, 788 *prob.*; Henry's law, 497 *prob.*; Hess's law, 537 *prob.*; hydrate, determine formula for, 353 *prob.*; ideal gas law, 455 *prob.*; induced transmutation, 876 *prob.*; instantaneous reaction rates, 579 *prob.*; ion concentrations, 617 *prob.*; ion concentrations from pH, 655 *prob.*; ionic compound formation, 212 *prob.*; ionic compounds, formulas for, 221 *prob.*, 222 *prob.*; ionic compounds, naming, 223 *prob.*; ionization constant of water, 651 *prob.*; ionization equations and base ionization constants, 649 *prob.*; isotopes, amount of remaining, 872 *prob.*; law of conservation of mass, 78 *prob.*; law of definite proportions, 88 *prob.*; Lewis structures, 244 *prob.*, 255 *prob.*, 256 *prob.*, 257 *prob.*, 258 *prob.*, 260 *prob.*; limiting reactant, determine, 383 *prob.*; mass number, 118 *prob.*; mass-to-mass stoichiometry, 377 *prob.*; mass-to-mole conversions, 329 *prob.*; mass-to-mole conversions for compounds, 337 *prob.*; mass-to-moles-to-particles conversions, 339 *prob.*; molality, 487 *prob.*; molarity, 483 *prob.*; molarity from titration data, 664 *prob.*; molar mass and, 335 *prob.*; molar solubility, 616 *prob.*; molar solutions, 484 *prob.*; molar volume, 453 *prob.*; molecular shape, 264 *prob.*; mole fraction, 488 *prob.*; mole ratios, 372 *prob.*; mole relationships from a chemical formula, 335 *prob.*; moles, convert to particles,

Index

323 *prob.*; mole-to-mass conversions, 328 *prob.*; mole-to-mass conversions for compounds, 336 *prob.*; mole-to-mass stoichiometry, 376 *prob.*; mole-to-mole stoichiometry, 375 *prob.*; nuclear equations, balancing, 869 *prob.*; oxidation number, 687 *prob.*; oxidation-number method, 690 *prob.*, 692 *prob.*; oxidation-reduction reactions, 685 *prob.*; partial pressure of a gas, 409 *prob.*; particles, convert to moles, 324 *prob.*; percent by mass, 481 *prob.*; percent by volume, 482 *prob.*; percent composition, 344 *prob.*; percent yield, 387 *prob.*; pH, acid dissociation constant from, 657 *prob.*; pH from $[H^+]$, 653 *prob.*; photon, energy of, 143 *prob.*; pOH and pH from $[OH^-]$, 654 *prob.*; precipitate-forming reactions, 302 *prob.*; precipitates, predicting, 619 *prob.*; rate laws, 577 *prob.*; reaction spontaneity, 545 *prob.*, 548 *prob.*; resonance structures, 258 *prob.*; salt hydrolysis, 665 *prob.*; single-replacement reactions, 295 *prob.*; skeleton equations, 284 *prob.*; specific heat, 521 *prob.*; standard enthalpies of formation, 541 *prob.*; volume-mass gas stoichiometry, 463 *prob.*; volume-volume problems, 461 *prob.*; water-forming reactions, 304 *prob.*; wavelength, 140 *prob.*
Precipitates, **296**; determine with $K_{sp}$, 618, 619 *prob.*; reactions in aqueous solutions forming, 300, 301 *act.*, 302 *prob.*
Precipitation, 428
Precision, **47**–48, 50
Pressure, **406**; chemical equilibrium and, 608–609; combined gas law and, 449, 450 *prob.*; extreme and ideal gas law, 458, 466 *act.*; gas temperature and (Gay-Lussac's law), 447, 448 *prob.*; gas volume and (Boyle's law), 442–443, 443 *prob.*, 444 *act.*; partial pressure of a gas, 408, 409 *prob.*, 410; popcorn popping and, 466 *act.*; solubility of gases and (Henry's law), 495–496, 497 *prob.*; units of, 407, 407 *table*
Primary batteries, **720**
Principal energy levels, **153**, 154
Principal quantum numbers ($n$), **153**
Problem-Solving Labs: Bohr model of the atom, 150 *act.*; Boyle's law and breathing, 444 *act.*; decomposition rate, variation in, 566 *act.*; DNA replication, 842 *act.*; elements, predict properties of by periodic table position, 180 *act.*; fluoride ions and prevention of tooth decay, 622 *act.*; francium, predict properties of, 180 *act.*; gas, release of compressed, 72 *act.*; identify an unknown by mass and volume, 50 *act.*; molar enthalpy (heat) of vaporization, 531 *act.*; molar mass, Avogadro's number, and atomic nucleus, 326 *act.*; pH of blood, 668 *act.*; radiation exposure, distance and, 890 *act.*; rate of decomposition of dinitrogen pentoxide, 566 *act.*
Problem-Solving Strategies: ground-state electron configuration, 160; halogens, predict reactivity of, 294 *act.*; ideal gas law, derive other laws from, 458; ionic compound naming flowchart, 224; Lewis structures, 254; mass defect and binding energy, 878; molarity from titration, 663; molar solubility, streamlining calculation of, 621; potential of voltaic cell, 717; redox equations, balance, 696; rounding numbers, 52; significant figures, recognizing, 51; stoichiometry, 374
Products, 77, **283**; addition of and chemical equilibrium, 608; calculating when reactant is limiting, 380–381, 382–383 *prob.*; identifying, 92 *act.*; predicting, 298, 298 *table*; removal of and chemical equilibrium, 608
Propane, 750, 751, 751 *table*; chemical equation for, 370 *prob.*; gas grills and, 375
Propanol, 816 *act.*
Propene, 759 *table*
Propyl group, 753 *table*
Proteins, **826**–831; amino acid building blocks, 826–827; denaturation of, 829; enzymes, 826, 829–830; peptide bonds in, 827–828; polypeptides, 828; protein hormones, 831; structural proteins, 831; three-dimensional structure, 829; transport proteins, 830
Protium, 904
Protons, **113**, 114 *table*, 119, 969 *table*
Prussian blue, 916
Pseudo-noble gas configurations, 208
PTFE (nonstick coating), *811*
Pure covalent bond, 266
Pure research, **17**
Pure substances, 70, *87*. *See also* Substances; compounds. *See* Compounds; elements. *See* Elements; mixtures of. *See* Mixtures; physical properties of, 73
Putrescine, 795

## Q

Qualitative data, **13**
Quantitative data, **13**
Quantized energy, 141–143, 146
Quantum, **141**–142
Quantum mechanical model of atom, 149–**152**
Quantum number ($n$), **147**
Quarks, *111*, 114

## R

Rad, 889
Radiation, **122**; alpha, 123, 124 *table*, 861, 861 *table*, 862, 888 *table*; average annual exposure to, 890 *table*; beta, 123, 124 *table*, 861, 861 *table*, 862, 863, 888 *table*; biological effects of, 888–890, 889 *table*; detection of, 885–886; discovery of, 860–861; distance and, 889 *act.*, 890; dose of, 889–890; gamma, 124, 861, 861 *table*, 862, 863, 888 *table*; intensity of and distance, 889 *act.*, 890; ionizing, 885; medical uses of, 886–887; neutron activation analysis, 891; scientific uses of, 886; types of, 123–124, 859 *act.*, 861 *table*, 861–864
Radiation-detection tools, 885–886
Radiation therapist, 887
Radiation therapy, 887
Radioactive decay, **122**, 861; model, 873 *act.*; nuclear stability and, 865–866; radiochemical dating and, 873–874; rate of, 870–871, 872 *prob.*, 873–874; transmutation, 865; types of, 866–868, 868 *table*
Radioactive decay series, **870**
Radioactivity, **122**. *See also* Radiation; detection of, 885–886; discovery of, 860–861, 915
Radiocarbon dating. *See* Carbon dating
Radiochemical dating, **873**–874
Radioisotopes, **861**; half-life of, 870–871, 871 *table*; medical uses of, 887–888; radioactive decay of. *See* Radioactive decay; radiochemical dating and, 873–874
Radiotracers, **887**
Radium, *882*, 910–911, 915
Radium-226, 862
Radon, 944
Radon gas, 915
Rainbows, 138
Rare Earth elements. *See* f-Block elements
Rate constant ($k$), 574

Rate-determining steps, **581**–582
Rate laws, **574**–576
Rates, reaction. *See* Reaction rates
Ratios, 964
Reactants, 77, **283**; addition of and chemical equilibrium, 607; calculate product when limited, 380–381, 382–383 *prob.*
Reaction mechanisms, **580**–582; complex reactions, 580; intermediates, 580; rate-determining steps, 581–582
Reaction order, **575**–577; determination of, 576, 577 *prob.*; first-order reaction rate laws and, 575; other-order reaction rate laws and, 575–576
Reaction rate laws. *See* Rate laws
Reaction rates, **561**–567; activation energy and, 564–566; average rate of, 560–562, 562 *prob.*; catalysts and, 571–573; collision theory and, 563, 564; concentration and, 569, 584 *act.*; decomposition of dinitrogen pentoxide, 565 *act.*; factors affecting, 559 *act.*; inhibitors and, 571; instantaneous, 578–579, 579 *prob.*; rate-determining steps, 581–582; rate laws, 574–576; reactivity of reactants and, 566–567; speeding, 559 *act.*; spontaneity and, 542–545, 566–567; surface area and, 569–570; temperature and, 570, 571 *act.*
Reaction spontaneity ($\Delta G$), 542–545; Earth's geologic processes and, 545; entropy and, 544–545, 545 *prob.*; free energy and, 548 *prob.*; Gibbs free energy and, 546–547; reaction rate and, 566–567
Real-World Chemistry: algal blooms and phosphates, 250; ammoniated cattle feed, 601; book preservation and, 661; cathode ray, 108; chrome and chromium, 328; clay roofing tiles, 302; enzymes (papain), 829; food preservation, 571; fuel cells, 722; gas grills, 375, 461; Gay-Lussac's law and pressure cookers, 448; hydrogen cyanide, 647; iron oxidation, 685; kilns, 461; liquid density measurement, 37; mineral identification, 73; mineral supplements, 220; perspiration, 426; photoelectric effect, 142; polycyclic aromatic hydrocarbons (PAHs), 807; reef aquariums, 287; saltwater fish and freezing point depression, 503; scuba diving and helium, 192; solar energy, 142; solar fusion, 883; specific heat, 521; sunscreen, protection from UV radiation, 5; trans-fatty acids, 767; zinc-plating, 295
Reaumur scale, 451
Recycling, 814
Redox equations, balancing, 679 *act.*, 689–696; half-reaction method, 693–693, 695 *prob.*; net ionic redox equations, 691, 692 *prob.*; oxidation-number method, 689, 689 *table*, 690 *prob.*; problem-solving flow-chart, 696
Redox reactions, **680**–688, 806–807; bioluminescence, 693; in electrochemistry, 707 *act.*, 708–709, 711; electronegativity and, 684; electron transfer and, 680–682; forensics and, 697, 698 *act.*; identify, 685 *prob.*; oxidation, 681; oxidation number, 219, 682, 686, 686 *table*, 687 *prob.*, 688; oxidizing agents, 683; reducing agents, 683; reduction, 681; reversal of (electrolysis), 728; rust formation, 679 *act.*; space shuttle launch and, 691 *act.*; summary of, 683 *table*; tarnish removal, 683 *act.*
Reduction, **681**
Reduction agent, **683**
Reduction potential, **711**
Reef aquariums, 287
Refrigerators, CFCs and, 7–8
Rem, 889
Replacement reactions, 293–294, 296–297; double-replacement, 296–297; single-replacement, 293–294, 295 *prob.*
Representative elements, **177**, 184, 196 *act.*
Representative particles, 321; convert moles to, 322; convert to moles, 323, 323 *prob.*, 324 *prob.*; mass to moles to particles conversions, 338, 338–339 *prob.*
Research: applied, 17; pure, 17
Research chemist, 185
Resonance, **258**
Reversible reactions, **595**
Rhombohedral unit cells, 421 *table*, 422 *act.*
RNA (ribonucleic acid), 843
Roentgen, Wilhelm, 860, 889
Rubber, 762
Rubidium, 906, 907
Rusting, 74, 77, 724–727; observe, 726 *act.*; prevent, 685, 725–727; redox reactions in, 679 *act.*, 724–725; as spontaneous process, 542–543
Rutherford, Ernest, *110*, 111–112, 112–113, 862, 875
Rutherfordium, *185*

## S

Saccharin, *810*
Sacrificial anodes, 726
Safety, lab, 18, 19 *table*
Safety matches, 934
Salicylaldehyde, 796 *table*, 797
Salt bridges, **709**
Salt hydrolysis, **665**
Saltwater fish, 503
Saponification, 837, 837 *act.*
Saturated fats, 805
Saturated fatty acids, 835–836
Saturated hydrocarbons, **746**
Saturated solutions, **493**
s-Block elements, 184
Scandium, 185
Scanning tunneling microscope (STM), 107, *213*
Schrodinger wave equation, 152
Science writer, 604
Scientific investigations. *See also* CHEMLABs; Data Analysis Labs; MiniLabs; Problem-Solving Labs; accidental discoveries and, 18; applied research, 17; pure research, 17; safety and, 18; scientific method and, 12–16
Scientific law, **16**
Scientific methods, **12**–16; conclusion, 15; experiments, 14–15; hypothesis, 13; observation, 13, 13 *act.*; scientific law and, 16; theory and, 16
Scientific notation, **40**–43, 946–948; addition and subtraction and, 41 *prob.*, 42, 948; multiplication and division and, 43, 43 *prob.*, 948
Scintillation counter, 886
Scuba diving, helium and, 192
Seaborg, Glenn, 921
Second (s), **33**
Secondary batteries, **720**
Second ionization energy, 192
Second law of thermodynamics, **543**, 546
Second period elements, 158 *table*, 161 *table*
Seed crystal, 495
Selenium, 936, 937, 939
Semimetals. *See* Metalloids
Sensitive teeth, 914
Serine, 827 *table*
Sex hormones, 839
Shape-memory alloys, *213*
Ships, corrosion of hulls of, 725–726
Side chains, amino acid, 827
Sigma bonds, **244**, *245*
Significant figures, **50**–51, 51 *prob.*,

949–950, 951 *prob.*; adding and subtracting, 53, 53 *prob.*, 952, 953 *prob.*; atomic mass values and, 328; multiplication and division and, 54, 54 *prob.*, 952; rounding numbers and, 52, 952
Silicates, 214
Silicon, 84, 159 *table*, 181, 926–927, 929
Silicon computer chips, 929
Silicon dioxide, 929
Silver, 226 *table*, 920
Silver batteries, 719
Silver nitrate flame test, 92 *act.*
Simple sugars. *See* Monosaccharides
Single covalent bonds, 242–244
Single-replacement reactions, **293**–294, 295 *prob.*; metal replaces hydrogen, 293; metal replaces metal, 293–294, 310 *act.*; nonmetal replaces nonmetal, 294, 294 *act.*
SI units, 32–37, 958 *table*
Skeleton equations, 284
Slime, 785 *act.*
Slope, line, 57, 962
Soap, 419, 634, 837 *act.*
Sodium, 136, 159, 159 *table*, 177, 906, 907, 908, 913
Sodium bicarbonate, 308
Sodium carbonate, 378 *act.*
Sodium chloride, 70, 73 *table*, 85, 205 *act.*, 210, 211 *table*, 213, 729
Sodium hypochlorite, 683
Sodium perborate, 924
Sodium/potassium ATPase, 909
Sodium-potassium pump, 909
Soft water, 24 *act.*
Solar energy, 142, 354, 522
Solar fusion, 883
Solidification, 76. *See also* Freezing
Solids, **71**, 420–424; amorphous, 424; crystalline, 420–423, 422 *act.*, 422 *table*; density of, 39 *act.*, 420; molecular, 422
Solubility, 479, 493–497; factors affecting, 492–494, 506 *act.*; of gases, 495–496, 497 *prob.*; guidelines for, 974 *table*; of polar molecules, 268; saturated solutions and, 493; supersaturated solutions and, 494–495; temperature and, 493–494, 494 *table*; unsaturated solutions and, 493
Solubility product constant ($K_{sp}$), **614**–619, 969 *table*; compare, 624 *act.*; ion concentrations from, 617, 617 *prob.*, 618–619; ion product constant ($Q_{sp}$) and, 618–619, 619 *prob.*; molar solubility from, 615–617, 616 *prob.*; predicting precipitates, 618
Solubility product constant expressions, 614–619; ion concentrations from, 617, 618–619, 619 *prob.*; molar solubility from, 616 *prob.*, 616–617; predicting precipitates, 618, 619 *prob.*; writing, 614–615
Soluble, **479**
Solutes, **299**
Solution concentration. *See* Concentration
Solution formation. *See* Solvation
Solutions, **81**, 478–479; acidic. *See* Acidic solutions; aqueous. *See* Aqueous solutions; basic. *See* Basic solutions; boiling point elevation, 500–501, 503 *prob.*; concentration, 475 *act.*, 480–488; dilution of, 485, 486 *prob.*; electrolytes and colligative properties, 498–499; formation (solvation), 489–492; freezing point depression, 501–502, 502 *act.*, 503 *prob.*; heat of solution, 475 *act.*, 492; milestones in understanding, *490–491*; molar. *See* Molar solutions; neutral, 636; osmotic pressure and, 504; saturated, 493; solubility and. *See* Solubility; supersaturated, 494–495; types of, 81 *table*, 479 *table*; unsaturated, 493; vapor pressure lowering and, 499–500
Solution systems, 81, 81 *table*
Solvation, **489**–492; aqueous solutions of ionic compounds, 490; aqueous solutions of molecular compounds, 491; factors affecting, 492–494, 506 *act.*; heat of solution, 475 *act.*, 492; "like dissolves like", 489
Solvents, **299**
s orbitals, 154
Space-filling molecular model, *253*, 746
Space shuttle, 691 *act.*, 722
Space telescopes, 912
Spandex, *811*
Species, **693**
Specific heat, **519**–520, 522, 976 *table*; calorimetry and, 523–524, 525 *prob.*, 526 *act.*; heat absorbed, calculate, 520, 521 *prob.*; heat released, calculate, 520; solar energy and, 522; of various substances, 520 *table*
Specific rate constant ($k$), **574**
Spectator ions, **301**
Spectroscopist, 139
Speed of light ($c$), 137, 969 *table*
Spontaneous processes, **542**. *See also* Reaction spontaneity ($\Delta G$)
Spontaneity, reaction rate and. *See* Reaction spontaneity ($\Delta G$)
Square root, 949
Stainless steel, 228 *table*
Standard enthalpy (heat) of formation, **537**–541, 538 *table*, 540 *prob.*
Standard hydrogen electrode, **711**
Standardized Test Practice, 28–29, 66–67, 98–99, 132–133, 170–171, 202–203, 236–237, 278–279, 316–317, 364–365, 398–399, 438–439, 472–473, 512–513, 556–557, 590–591, 630–631, 676–677, 704–705, 740–741, 782–783, 822–823, 856–857, 898–899
Standard reduction potentials, 712; applications of, 716; calculate, 713–714, 715 *prob.*; determine, 712, 712 *table*; measure, 734 *act.*
Standard temperature and pressure (STP), 452
Starch, 834
States of matter, **71**–72; gases, 72, 72 *act.*, 402–410; liquids, 71, 401 *act.*, 415–419; milestones in understanding, *416–417*; phase changes, 76–77, 425–430; solids, 71, 420–424; summarize information on, 401 *act.*
Stationary phase, chromatography, 83
Stearic acid, 835
Steel, 227, 227 *act.*
Stereoisomers, **766**. *See also* Optical isomers
Sterling silver, 228 *table*
Steroids, **839**
Steroid toxins, 839
Stock solutions, dilution of, 485, 486 *prob.*
Stoichiometry, **368**–378; actual yield and, 385; baking soda decomposition, 378 *act.*; interpret chemical equations, 370 *prob.*; mass-to-mass conversions, 377, 377 *prob.*; mole ratios and, 371–372, 390 *act.*; mole-to-mass conversions, 376, 376 *prob.*; mole-to-mole conversions, 373–374, 375 *prob.*; particle and mole relationships and, 368–369; percent yield and, 386, 386 *prob.*, 388; problem-solving flow chart, 374; product, calculate when reactant is limiting, 380–381, 382–383 *prob.*; reactions involving gases. *See* Gas stoichiometry; theoretical yield and, 385; titration and. *See* Titration
Storage batteries, 720
Straight-chain alkanes, 750–751
Stratosphere, 5
Straussman, Fritz, *111*
Strong acids, **644**, 656
Strong bases, **648**, 656
Strong electrolytes, 498
Strong nuclear force, **865**
Strontium, 186 *prob.*, 910–911, 913, 914

Strontium-90, 870, 871 *table*
Strontium carbonate, 913
Strontium chloride, 914
Structural formulas, **253**, 253, 746, 751
Structural isomers, **765**
Structural proteins, 831
Subatomic particles, 114 *table*, 119 *table*
Sublimation, **83**, 428
Suboctets, 259
Substances, **5**, 70
Substituent groups, **752**
Substituted cycloalkanes, naming, 756, 756–757 *prob.*
Substituted hydrocarbons: alcohols, 792–793; aldehydes, 796–797; amides, 800; amines, 795; carboxylic acids, 798; chemical reactions involving. *See* Organic reactions; crosslinks (make slime), 785 *act.*; esters, 799, 800 *act.*; ethers, 794; functional groups, 785 *act.*, 786, 787 *table*; halocarbons, 787–791; ketones, 797
Substitutional alloys, 228
Substitution reactions, **790**–791
Substrates, **830**
Subtraction: scientific notation and, 42; significant figures and, 53
Sucrose, 73 *table*, 88, 205 *act.*, 833
Sulfur, 159 *table*, 195, 936–937, 939
Sulfuric acid, manufacture of, 388, 939
Sunburn, 5
Sunlight, continuous spectrum of, 138
Sunscreen, 5
Sun, solar fusion in, 883
Superacids, *637*
Super ball, properties of, 239 *act.*
Supercritical mass, 880
Supersaturated solutions, **494**–495
Surface area: reaction rate and, 569–570; solvation and, 492
Surface tension, **418**–419
Surfactants, **419**
Surroundings (thermochemical), **526**
Suspensions, **476**
Synthesis reactions, **289**
System (thermochemical), **526**
Systeme International d'Unites. *See* SI units

## T

Table salt. *See* Sodium chloride
Tap water, hard and soft, 24 *act.*
Tarnish removal, 683, 683 *act.*
Tartaric acid, 767
Taste, 262
Taste buds, 262
Television, 108
Tellurium, 936, 937
Temperature, **403**; change in as evidence of chemical reaction, 282; chemical equilibrium and, 609–610, 611 *act.*; combined gas law and, 449, 450 *prob.*; enzyme action and, 850 *act.*; evaporation rate and, 432 *act.*; extreme and ideal gas law, 458; gas pressure and (Gay-Lussac's law), 447, 448 *prob.*; gas volume and (Charles's Law), 441 *act.*, 444–445, 446 *prob.*; pain receptors and, 815; reaction rate and, 570, 571 *act.*, 583; solubility and, 493–494, 494 *table*; viscosity and, 418
Temperature inversion, 428
Temperature scales, 34–35; convert between, 34, 35; gas laws and, 451
Tetraethyl lead, 930
Tetragonal unit cell, 421 *table*, 422 *act.*
Tetrahedral molecular shape, *261*, 263 *table*
Thallium, 922, 923, 925
Theoretical chemistry, 11 *table*
Theoretical yield, **385**
Theory, **16**
Thermal conductivity, 226
Thermochemical equations, **529**–533; for changes of state, 530–531, 531 *act.*; Hess's law, 534–536, 536 *prob.*; standard enthalpy (heat) of formation, 537–541, 540 *prob.*; writing, 529
Thermochemical universe, **526**, 546
Thermochemistry, **523**–528; combustion reactions, 532 *prob.*, 533; enthalpy and enthalpy changes, 526–528; enthalpy (heat) of reaction, 527–528; Hess's law, 534–536, 536 *prob.*; molar enthalpy (heat) of fusion, 530–531; molar enthalpy (heat) of vaporization, 530; phase changes and, 530–531; surroundings, 526; systems, 526; thermochemical equations, 529–533
Thermocouples, 34
Thermodynamics, second law of, 543
Thermoluminescent dosimeter (TLD), 885
Thermonuclear reactions, **883**
Thermoplastic polymers, **813**
Thermosetting polymers, **813**
Third ionization energy, 192
Third period elements, 159 *table*
Thixotropic substances, 476
Thomson, J. J., 108–109, 110, *212*
Thomson, William (Lord Kelvin), 35
Thorium, 921
Three Mile Island, 880, *883*
Thymine (T), 841
Time, 33
Tin, 226 *table*, 926–927, 930
Tinplate, 930
Titanium, 180, *181*, *228*, 918, 919
Titrant, **661**
Titration, **660**–663; acid-base indicators and, 662, 663; end point of, 663; molarity from, 663, 664 *prob.*, 670 *act.*; steps in, 661
Tokamak reactor, 884
Tolerances, 49
Toluene, 774
Tools, zinc plating of, 295
Tooth decay, fluoride and, 622 *act.*
Torricelli, Evangelista, 406
Touch sensors, 920
Toxicologist, 59
Toxicology, 59
Trace elements, 195
Transactinide elements, *185*
Trans-fatty acids, 767
*trans-* isomers, 766
Transition elements, **177**, 916–921; analytical tests for, 917; applications of, 918–921; atomic properties, 917; common reactions involving, 916; inner transition metals, 180; locations of strategic, 918; physical properties of, 916; transition metals, 180
Transition metal ions, 208, 219, 219 *table*
Transition metals, **180**, 185
Transition state, **564**
Transmutation, **865**, 875
Transport proteins, 830
Transuranium elements, **876**
Triclinic unit cells, 421 *table*
Triglycerides, **836**–837, 837 *act.*; phospholipids, 838; saponification of, 838, 838 *act.*
Trigonal bipyramidal molecular shape, 263 *table*
Trigonal planar molecular shape, *261*, 263 *table*
Trigonal pyramid molecular shape, *261*, 263 *table*
Triple covalent bonds, 245, 246
Triple point, **429**
Tritium, 904
Troposphere, 5
Tungsten, 226, 918
Turbidity, 478 *act.*
Tyndall effect, **478**, 478 *act.*

## U

Ultraviolet radiation: overexposure to, damage from, 5; ozone layer and, 5, 6
Ultraviolet (Lyman) series, *147*, 150 *act.*
Unbalanced forces, 597
Unit cell, **421**, 421 *table*, 422 *act.*
Units, 32–37; base SI, 33–35; converting between, 957–958, 958 *prob.*; derived SI, 35–37; English, 32
Universe (thermochemical), **526**, 546
Unsaturated fatty acids, 835–836
Unsaturated hydrocarbons, **746**
Unsaturated solutions, **493**
Ununquadium, *185*
Uranium-235, 878–879, 880
Uranium-238, 863, 880
Urea, 800
UV-B radiation, 5

## V

Valence electrons, **161**; chemical bonds and, 207; periodic *table* trends, 182–185, 186 *prob.*
Valence Shell Electron Pair Repulsion (VSEPR) theory. *See* VSEPR model
Valine, 827 *table*
van der Waals forces, 269–270, 271
Vapor, **72**
Vaporization, **426**–427; molar enthalpy (heat) of vaporization, 530, 531 *act.* *See also* Boiling, Evaporation
Vapor pressure, **427**
Vapor pressure lowering, **499**–500
Variables, 14; controlling, 14–15; dependent, 14, 56; independent, 14
Venom, 838
Vinegar-baking soda volcano, 669
Viscosity, 401 *act.*, **417**, 418
Visible (Balmer) series, *147*, 148, 150 *act.*
Visible spectroscopy, 917
Visible spectrum, 138–139
Vitalism, 744
Vitamins, 383
Vocabulary margin features: alloy, 227; anhydrous, 352; aromatic, 771; atom, 103; attain, 243; aufbau, 157; bond, 794; buffer, 667; capacity, 721; cis-, 766; class, 799; combustion, 290; completion, 599; complex, 845; compound, 300; concentrated, 485; concentration, 561; concept, 113; conceptualize, 845; conduct, 215; conductor, 180; conform, 642; conjugate, 639; convert, 595; correspond, 711; demonstrate, 547; deposit, 747; derive, 372; disaccharide, 833; element, 85; eliminate, 751; environment, 75; evolve, 5; force, 419; formula, 284; gases, 403; generate, 878; homologous, 751; indicators, 663; initial, 576; investigate, 566; meter, 33; method, 694; mixture, 81; mole, 321, 456; monosaccharide, 833; neutral, 113; orient, 412; overlap, 244; ozone, 5; percent, 48; period, 159; periodic, 176; phenomenon, 141; polysaccharide, 833; potential, 714; pressure, 495; product, 381; radiation, 863; random, 544; ratio, 333, 462; recover, 21; reduce, 730; reduction, 681; resonance, 258; saturated, 494; species, 693; specific, 119; stoichiometry, 369; stress, 607; structure, 184; sum, 42; system, 543; *trans*-, 766; transfer, 219; trigonal planar, 262; unstable, 867; weight, 10
Volt, 710
Volta, Alessandro, 709
Voltaic cell potentials. *See* Electrochemical cell potentials
Voltaic cells, **709**–711; chemistry of, 710–711; electrochemical cell potentials, 711–714, 715 *prob.*, 716–717, 734 *act.*; half-cells, 710
Voltaic pile, *709*
Volume: chemical equilibrium and, 608–609; combined gas law and, 449, 450 *prob.*; determine mass of object from, 38 *prob.*; gas pressure and (Boyle's law), 442–443, 443 *prob.*, 444 *act.*; gas stoichiometry and, 460–461, 461 *prob.*, 462, 462–463 *prob.*; gas temperature and (Charles' Law), 441 *act.*, 444–445, 446 *prob.*; identify an unknown by, 50 *act.*; SI units for, 35–36
Volumetric analysis, 341
VSEPR model, **261**–262, 263 *table*, 264 *prob.*, 272 *act.*

## W

Warfarin, 59
Water: adhesion and cohesion of, 419; amphoteric nature of, 639; boiling of, 427, 969 *table*; capillary action, 419; changes of state and, 76, 425–428; chemical properties, 75; condensation of, 428; covalent bonds in, 240, 243; density of solid, 420; electrical conductivity of, 205 *act.*; electrolysis of, 86; evaporation of, 426–427, 432 *act.*; formation of in aqueous solutions, 303, 304 *prob.*; freezing, 428, 969 *table*; hard v. soft, 24 *act.*; history in a glass of, 355; hydration reactions forming, 804; hydrogen bonds in, 413–414; ion product constant for ($K_w$), 650–651, 651 *prob.*; law of multiple proportions and, 89; layering of in graduated cylinder, 31 *act.*; Lewis structure, *243*; melting of, 425–426; phase diagram, 429, *430*; physical properties, 73 *table*, 75; polarity of, 267–268; as pure substance, 70; sigma bonds in, 244, *245*; solutions of. *See* Aqueous solutions; surface tension of, 419; thermochemistry, 530–531, 531 *act.*; turbidity and Tyndall effect, 478 *act.*; vaporization of, 426
Watson, James, *637*, 841–842
Wavelength, **137**, 140 *prob.*
Wave mechanical model of the atom. *See* Quantum mechanical model of atom
Wave model of light, 137–139; atomic emission spectrum and, 144–145; dual nature of light and, 143
Waves, 137–138; amplitude of, 137; electromagnetic wave relationship, 137; frequency of, 137; wavelength of, 137, 140 *prob.*
Waxes, **838**
Weak acids, **645**, 648 *table*
Weak bases, **649**
Weak electrolytes, 498
Weather balloons, *449*
Weather patterns, density of air masses and, 37
Weight, **9**–10
Willstater, Richard, 912
Wohler, Friedrich, 744
Word equations, 284

## X

Xenon, 944, 945
X-ray crystallography, *212*
X rays, 137, **864**, 914
Xylene, *772*, 774

## Z

Zewail, Ahmed, 581
Zinc, 208, 920
Zinc-carbon dry cells, **718**–719
Zinc plating, 295

# PERIODIC TABLE OF THE ELEMENTS

Element — Hydrogen
Atomic number — 1
Symbol — H
Atomic mass — 1.008
State of matter

Gas
Liquid
Solid
Synthetic

| | 1 | 2 | 3 | 4 | 5 | 6 | 7 | 8 | 9 |
|---|---|---|---|---|---|---|---|---|---|
| 1 | Hydrogen<br>1<br>**H**<br>1.008 | | | | | | | | |
| 2 | Lithium<br>3<br>**Li**<br>6.941 | Beryllium<br>4<br>**Be**<br>9.012 | | | | | | | |
| 3 | Sodium<br>11<br>**Na**<br>22.990 | Magnesium<br>12<br>**Mg**<br>24.305 | | | | | | | |
| 4 | Potassium<br>19<br>**K**<br>39.098 | Calcium<br>20<br>**Ca**<br>40.078 | Scandium<br>21<br>**Sc**<br>44.956 | Titanium<br>22<br>**Ti**<br>47.867 | Vanadium<br>23<br>**V**<br>50.942 | Chromium<br>24<br>**Cr**<br>51.996 | Manganese<br>25<br>**Mn**<br>54.938 | Iron<br>26<br>**Fe**<br>55.847 | Cobalt<br>27<br>**Co**<br>58.933 |
| 5 | Rubidium<br>37<br>**Rb**<br>85.468 | Strontium<br>38<br>**Sr**<br>87.62 | Yttrium<br>39<br>**Y**<br>88.906 | Zirconium<br>40<br>**Zr**<br>91.224 | Niobium<br>41<br>**Nb**<br>92.906 | Molybdenum<br>42<br>**Mo**<br>95.94 | Technetium<br>43<br>**Tc**<br>(98) | Ruthenium<br>44<br>**Ru**<br>101.07 | Rhodium<br>45<br>**Rh**<br>102.906 |
| 6 | Cesium<br>55<br>**Cs**<br>132.905 | Barium<br>56<br>**Ba**<br>137.327 | Lanthanum<br>57<br>**La**<br>138.905 | Hafnium<br>72<br>**Hf**<br>178.49 | Tantalum<br>73<br>**Ta**<br>180.948 | Tungsten<br>74<br>**W**<br>183.84 | Rhenium<br>75<br>**Re**<br>186.207 | Osmium<br>76<br>**Os**<br>190.23 | Iridium<br>77<br>**Ir**<br>192.217 |
| 7 | Francium<br>87<br>**Fr**<br>(223) | Radium<br>88<br>**Ra**<br>(226) | Actinium<br>89<br>**Ac**<br>(227) | Rutherfordium<br>104<br>**Rf**<br>(261) | Dubnium<br>105<br>**Db**<br>(262) | Seaborgium<br>106<br>**Sg**<br>(266) | Bohrium<br>107<br>**Bh**<br>(264) | Hassium<br>108<br>**Hs**<br>(277) | Meitnerium<br>109<br>**Mt**<br>(268) |

The number in parentheses is the mass number of the longest lived isotope for that element.

| | | | | | | |
|---|---|---|---|---|---|---|
| **Lanthanide series** | Cerium<br>58<br>**Ce**<br>140.115 | Praseodymium<br>59<br>**Pr**<br>140.908 | Neodymium<br>60<br>**Nd**<br>144.242 | Promethium<br>61<br>**Pm**<br>(145) | Samarium<br>62<br>**Sm**<br>150.36 | Europium<br>63<br>**Eu**<br>151.965 |
| **Actinide series** | Thorium<br>90<br>**Th**<br>232.038 | Protactinium<br>91<br>**Pa**<br>231.036 | Uranium<br>92<br>**U**<br>238.029 | Neptunium<br>93<br>**Np**<br>(237) | Plutonium<br>94<br>**Pu**<br>(244) | Americium<br>95<br>**Am**<br>(243) |

Metal

Metalloid

Nonmetal

Recently observed

| 10 | 11 | 12 | 13 | 14 | 15 | 16 | 17 | 18 |
|---|---|---|---|---|---|---|---|---|
| | | | | | | | | Helium<br>2<br>**He**<br>4.003 |
| | | | Boron<br>5<br>**B**<br>10.811 | Carbon<br>6<br>**C**<br>12.011 | Nitrogen<br>7<br>**N**<br>14.007 | Oxygen<br>8<br>**O**<br>15.999 | Fluorine<br>9<br>**F**<br>18.998 | Neon<br>10<br>**Ne**<br>20.180 |
| | | | Aluminum<br>13<br>**Al**<br>26.982 | Silicon<br>14<br>**Si**<br>28.086 | Phosphorus<br>15<br>**P**<br>30.974 | Sulfur<br>16<br>**S**<br>32.066 | Chlorine<br>17<br>**Cl**<br>35.453 | Argon<br>18<br>**Ar**<br>39.948 |
| Nickel<br>28<br>**Ni**<br>58.693 | Copper<br>29<br>**Cu**<br>63.546 | Zinc<br>30<br>**Zn**<br>65.39 | Gallium<br>31<br>**Ga**<br>69.723 | Germanium<br>32<br>**Ge**<br>72.61 | Arsenic<br>33<br>**As**<br>74.922 | Selenium<br>34<br>**Se**<br>78.96 | Bromine<br>35<br>**Br**<br>79.904 | Krypton<br>36<br>**Kr**<br>83.80 |
| Palladium<br>46<br>**Pd**<br>106.42 | Silver<br>47<br>**Ag**<br>107.868 | Cadmium<br>48<br>**Cd**<br>112.411 | Indium<br>49<br>**In**<br>114.82 | Tin<br>50<br>**Sn**<br>118.710 | Antimony<br>51<br>**Sb**<br>121.757 | Tellurium<br>52<br>**Te**<br>127.60 | Iodine<br>53<br>**I**<br>126.904 | Xenon<br>54<br>**Xe**<br>131.290 |
| Platinum<br>78<br>**Pt**<br>195.08 | Gold<br>79<br>**Au**<br>196.967 | Mercury<br>80<br>**Hg**<br>200.59 | Thallium<br>81<br>**Tl**<br>204.383 | Lead<br>82<br>**Pb**<br>207.2 | Bismuth<br>83<br>**Bi**<br>208.980 | Polonium<br>84<br>**Po**<br>208.982 | Astatine<br>85<br>**At**<br>209.987 | Radon<br>86<br>**Rn**<br>222.018 |
| Darmstadtium<br>110<br>**Ds**<br>(281) | Roentgenium<br>111<br>**Rg**<br>(272) | Copernicium<br>112<br>**Cn**<br>(285) | Ununtrium<br>* 113<br>**Uut**<br>(284) | Flerovium<br>114<br>**Fl**<br>(289) | Ununpentium<br>* 115<br>**Uup**<br>(288) | Livermorium<br>116<br>**Lv**<br>(291) | Ununseptium<br>* 117<br>**Uus**<br>(294) | Ununoctium<br>* 118<br>**Uuo**<br>(294) |

* The names and symbols for elements 113, 115, 117, and 118 are temporary. Final names will be approved by IUPAC (International Union of Pure and Applied Chemistry).

| | | | | | | | |
|---|---|---|---|---|---|---|---|
| Gadolinium<br>64<br>**Gd**<br>157.25 | Terbium<br>65<br>**Tb**<br>158.925 | Dysprosium<br>66<br>**Dy**<br>162.50 | Holmium<br>67<br>**Ho**<br>164.930 | Erbium<br>68<br>**Er**<br>167.259 | Thulium<br>69<br>**Tm**<br>168.934 | Ytterbium<br>70<br>**Yb**<br>173.04 | Lutetium<br>71<br>**Lu**<br>174.967 |
| Curium<br>96<br>**Cm**<br>(247) | Berkelium<br>97<br>**Bk**<br>(247) | Californium<br>98<br>**Cf**<br>(251) | Einsteinium<br>99<br>**Es**<br>(252) | Fermium<br>100<br>**Fm**<br>(257) | Mendelevium<br>101<br>**Md**<br>(258) | Nobelium<br>102<br>**No**<br>(259) | Lawrencium<br>103<br>**Lr**<br>(262) |

- Metal
- Metalloid
- Nonmetal
- Recently observed

| 10 | 11 | 12 | 13 | 14 | 15 | 16 | 17 | 18 |
|---|---|---|---|---|---|---|---|---|
| | | | | | | | | Helium 2 **He** 4.003 |
| | | | Boron 5 **B** 10.811 | Carbon 6 **C** 12.011 | Nitrogen 7 **N** 14.007 | Oxygen 8 **O** 15.999 | Fluorine 9 **F** 18.998 | Neon 10 **Ne** 20.180 |
| | | | Aluminum 13 **Al** 26.982 | Silicon 14 **Si** 28.086 | Phosphorus 15 **P** 30.974 | Sulfur 16 **S** 32.066 | Chlorine 17 **Cl** 35.453 | Argon 18 **Ar** 39.948 |
| Nickel 28 **Ni** 58.693 | Copper 29 **Cu** 63.546 | Zinc 30 **Zn** 65.39 | Gallium 31 **Ga** 69.723 | Germanium 32 **Ge** 72.61 | Arsenic 33 **As** 74.922 | Selenium 34 **Se** 78.96 | Bromine 35 **Br** 79.904 | Krypton 36 **Kr** 83.80 |
| Palladium 46 **Pd** 106.42 | Silver 47 **Ag** 107.868 | Cadmium 48 **Cd** 112.411 | Indium 49 **In** 114.82 | Tin 50 **Sn** 118.710 | Antimony 51 **Sb** 121.757 | Tellurium 52 **Te** 127.60 | Iodine 53 **I** 126.904 | Xenon 54 **Xe** 131.290 |
| Platinum 78 **Pt** 195.08 | Gold 79 **Au** 196.967 | Mercury 80 **Hg** 200.59 | Thallium 81 **Tl** 204.383 | Lead 82 **Pb** 207.2 | Bismuth 83 **Bi** 208.980 | Polonium 84 **Po** 208.982 | Astatine 85 **At** 209.987 | Radon 86 **Rn** 222.018 |
| Darmstadtium 110 **Ds** (281) | Roentgenium 111 **Rg** (272) | Copernicium 112 **Cn** (285) | Ununtrium * 113 **Uut** (284) | Flerovium 114 **Fl** (289) | Ununpentium * 115 **Uup** (288) | Livermorium 116 **Lv** (291) | Ununseptium * 117 **Uus** (294) | Ununoctium * 118 **Uuo** (294) |

* The names and symbols for elements 113, 115, 117, and 118 are temporary. Final names will be approved by IUPAC (International Union of Pure and Applied Chemistry).

| | | | | | | | |
|---|---|---|---|---|---|---|---|
| Gadolinium 64 **Gd** 157.25 | Terbium 65 **Tb** 158.925 | Dysprosium 66 **Dy** 162.50 | Holmium 67 **Ho** 164.930 | Erbium 68 **Er** 167.259 | Thulium 69 **Tm** 168.934 | Ytterbium 70 **Yb** 173.04 | Lutetium 71 **Lu** 174.967 |
| Curium 96 **Cm** (247) | Berkelium 97 **Bk** (247) | Californium 98 **Cf** (251) | Einsteinium 99 **Es** (252) | Fermium 100 **Fm** (257) | Mendelevium 101 **Md** (258) | Nobelium 102 **No** (259) | Lawrencium 103 **Lr** (262) |